Developmental Mathematics
Basic Mathematics and Algebra

Second Edition

Margaret L. Lial
American River College

John Hornsby
University of New Orleans

Terry McGinnis

Stanley A. Salzman
American River College

Diana L. Hestwood
Minneapolis Community and Technical College

Addison-Wesley
OCN 245025227
Boston • New York • San Francisco
London • Toronto • Sydney • Tokyo • Singapore • Madrid
Mexico City • Munich • Paris • Cape Town • Hong Kong • Montreal

Editorial Director	Christine Hoag
Editor in Chief	Maureen O'Connor
Executive Project Manager	Kari Heen
Project Editor	Courtney Slade
Editorial Assistant	Mary Gallagher
Senior Managing Editor	Karen Wernholm
Senior Production Supervisor	Kathleen A. Manley
Senior Designer	Barbara T. Atkinson
Photo Researcher	Beth Anderson
Supplements Production	Marianne Groth and Kayla Smith-Tarbox
Media Producers	Ceci Fleming, Lin Mahoney, and Jean Choe
Software Development	Eric Gregg, MathXL; Mary Durnwald, TestGen
Senior Marketing Manager	Michelle Renda
Marketing Assistant	Alicia Frankel
Senior Author Support/Technology Specialist	Joe Vetere
Senior Prepress Supervisor	Caroline Fell
Senior Media Buyer	Ginny Michaud
Rights and Permissions Advisor	Shannon Barbe
Senior Manufacturing Buyer	Carol Melville
Composition/Production Coordination	Nesbitt Graphics, Inc.
Cover Image	Autumn Canopy Copyright © Lorraine Cota Manley

Library of Congress Cataloging-in-Publication Data

Developmental mathematics: basic mathematics and algebra / Margaret L. Lial . . . [et al.].—2nd ed.

    p. cm.

ISBN-13: 978-0-321-59920-9 (student edition)

ISBN-10: 0-321-59920-9 (student edition)

1. Arithmetic—Textbooks. 2. Algebra—Textbooks. 3. Problem sovling—Textbooks. I. Lial, Margaret L.

QA107.2.D48 2009

513—dc22   2008038404

1 2 3 4 5 6 7 8 9 10—CRK—12 11 10 09

Addison-Wesley
is an imprint of

www.pearsonhighered.com

ISBN 10: 0-321-59920-9
ISBN 13: 978-0-321-59920-9

CONTENTS

The second edition of *Developmental Mathematics: Basic Mathematics and Algebra* exemplifies our commitment to provide the best possible text and supplements package to help instructors teach and students succeed. To that end, we have addressed the diverse needs of today's students through an attractive design, up-to-date applications and graphs, helpful features, careful explanation of concepts, and an expansive package of supplements and study aids.

The text is designed for students who need to review basic mathematics before moving on to introductory algebra topics. This book begins with basic mathematics concepts, and then introduces geometry, statistics, and introductory algebra topics.

Students will benefit from the text's student-oriented approach. Of particular interest to students and instructors will be the **NEW** pointers in examples, Study Skills activities, Math in the Media feature, and Solutions section.

This text is part of a series that also includes the following books:

- *Essential Mathematics*, Third Edition, by Lial and Salzman

- *Basic College Mathematics*, Eighth Edition, by Lial, Salzman, and Hestwood

- *Prealgebra*, Fourth Edition, by Lial and Hestwood

- *Introductory Algebra,* Ninth Edition, by Lial, Hornsby, and McGinnis

- *Intermediate Algebra*, Ninth Edition, by Lial, Hornsby, and McGinnis

- *Introductory and Intermediate Algebra*, Fourth Edition, by Lial, Hornsby, and McGinnis

- *Prealgebra and Introductory Algebra*, Third Edition, by Lial, Hestwood, Hornsby, and McGinnis

Hallmark Features

We are pleased to offer the following features, each of which is designed to increase ease-of-use by students and actively engage them in learning mathematics.

▶ *Chapter Openers* New and updated chapter openers feature real-world applications of mathematics that are relevant to students and tied to specific material within the chapters. Examples of topics include Americans' personal savings rate, the Olympics, and student credit card debt. (See pp. 615, 697, and 771—Chapters 9, 10, and 11.)

▶ *Real-Life Applications* We are always on the lookout for interesting data to use in real-life applications. As a result, we have included many new or updated examples and exercises throughout the text that focus on real-life applications of mathematics. Students are often asked to find data in a table, chart, graph, or advertisement. (See pp. 280, 298, 377, and 800.) These applied problems provide an up-to-date flavor that will appeal to and motivate students.

▶ *Figures, Photos, and NEW Hand-Drawn Graphs* Today's students are more visually oriented than ever. Thus, we have made a concerted effort to include mathematical figures, diagrams, tables, and graphs, including the new "hand-drawn" style of graphs, whenever possible. (See pp. 325, 376, and 787.) Many of the graphs use a style similar to that seen by students in today's print and electronic media. Even more photos have been incorporated to enhance applications in examples and exercises. (See pp. 376 and 397.)

▶ *Emphasis on Problem Solving* A six-step method for solving applied problems with basic mathematics is introduced at the end of Chapter 1 (see p. 89), and this method is modified for solving applied problems with algebra in Chapter 10 (see p. 723). These steps are emphasized in boldface type and repeated in examples and exercises to reinforce the problem-solving process for students. (See pp. 424, 729, and 965.) Problem-Solving Hint boxes provide students with helpful problem-solving tips and strategies. (See pp. 624 and 723.)

▶ *Learning Objectives* Each section begins with clearly stated, numbered objectives, and the included material is directly keyed to these objectives so that students know exactly what is covered in each section. (See pp. 368 and 941.)

▶ *Examples* The new edition of the text features a multitude of step-by-step, worked-out examples that include pedagogical color, helpful side comments, and **NEW** pointers. We give increased attention to checking example solutions—more checks, designated using a special **Check** tag, are included than in past editions. (See pp. 713 and 1027.)

▶ *Margin Problems* Margin problems, with answers immediately available at the bottom of the page, are found in every section of the text. (See pp. 392 and 1078.) This popular feature allows students to immediately practice the material covered in the examples in preparation for the exercise sets. We have added more margin problems in this edition.

▶ *Cautions and Notes* One of the most popular features of the previous edition, **CAUTION** and **Note** boxes warn students about common errors and emphasize important ideas throughout the exposition. (See pp. 319, 322, and 1043.) The text design makes them easy to spot: Cautions are highlighted in bright yellow and Notes are highlighted in purple.

▶ *Calculator Tips* Optional Calculator Tips, marked with calculator icons, offer basic information and instruction for students using calculators in the course. (See p. 425.)

▶ *Ample and Varied Exercise Sets* One of the most commonly mentioned strengths of this text is its exercise sets. The text contains a wealth of exercises to provide students with opportunities to practice, apply, connect, and extend the algebraic concepts and skills they are learning. Numerous illustrations, tables, graphs, and photos have been added to the exercise sets to help students visualize the problems they are solving. Problem types include writing, estimation, and calculator exercises as well as applications and multiple-choice, matching, true/false, and fill-in-the-blank problems. In the *Annotated Instructor's Edition* of the text, writing exercises are marked with ✍ icons so that teachers may assign these problems at their discretion. (See p. 428.) Exercises suitable for calculator work are marked in both the student and teacher editions with ▦ icons. (See pp. 429 and 1152.) Students can watch an instructor work through the solutions for exercises marked with the ⊙ DVD icon on the Videos on DVD or in MyMathLab.

▶ *Relating Concepts Exercises* These sets of exercises help students tie together topics and develop problem-solving skills as they compare and contrast ideas, identify and describe patterns, and extend concepts to new situations. (See pp. 390 and 878.) These exercises make great collaborative activities for pairs or small groups of students.

▶ *Test Your Word Power* To help students understand and master mathematical vocabulary, this feature can be found in each chapter summary. Key terms from the chapter are presented along with four possible definitions in a multiple-choice format. Answers and examples illustrating each term are provided. (See p. 446.)

▶ *Ample Opportunity for Review* Each chapter concludes with a Chapter Summary that features Key Terms with definitions and helpful graphics, New Symbols, Test Your Word Power, and a Quick Review of each section's content with additional examples. A comprehensive set of Chapter Review Exercises, keyed to individual sections, is included, as are Mixed Review Exercises and a Chapter Test. Chapters 8 and 17 each conclude with a set of Cummulative Review Exercises. (See pp. 611–614 and 1227–1231.) Students can watch an instructor work through the full solutions for all Chapter Test exercises on the Chapter Test Prep Video CD that accompanies each new copy of the text.

What's New in This Edition?

Throughout this edition of the text, we are pleased to offer the following new student-oriented features:

NEW *Math in the Media* These new one-page activities provide a relevant application of mathematics as it is found in various media forms, such as newspapers, movies, and TV. Designed to help teachers answer the often-asked question, "Why do I need to learn this?", these activities are well-suited for individual or collaborative work, as well as class discussions. We hope both students and instructors will enjoy them. (See pp. 338 and 682.)

NEW *Study Skills* Poor study skills are a major reason why students do not succeed in mathematics. These new two-page activities provide helpful information, tips, and strategies on a variety of essential study skills, including *Using Your Textbook, Taking Lecture Notes, Tips for Taking Math Tests*, and *Managing Your Time*. While most of the activities are concentrated in the early chapters of the text, each has been designed independently and can be used at most any point in your course with individuals or small groups of students, or as a source of material for in-class discussions. (See pp. 315–316.)

NEW *Solutions to Selected Exercises* Exercise numbers enclosed in a blue square, such as **11.** , indicate that a step-by-step, worked-out solution for the problem is included at the back of the text. These solutions are given for selected exercises that extend the skills and concepts presented in the section examples—actually providing students with a pool of examples for exercises that include some kind of twist or are a bit more difficult. (See pp. S-1 through S-28.)

NEW *Pointers* Pointers from the authors have been added to examples and provide students with important on-the-spot reminders and warnings about common pitfalls. (See pp. 327 and 956.)

NEW *Chapter Test Prep Video CD* The Chapter Test Prep Video CD provides students with the opportunity to watch instructors work through step-by-step solutions to all the Chapter Test exercises from the textbook. The Chapter Test Prep Video CD is included with each new student text.

A primary focus of this revision of the text was to polish and enhance individual presentations of topics and exercise sets, based on user and reviewer feedback, and we have worked hard to do this throughout the book. Some of the specific content changes you may notice include the following:

- The exercise sets received special attention. There are new and updated exercises, including problems that check conceptual understanding, focus on skill development, and provide review.

- Real-world data in the examples and exercises have been updated.

- There is an increased emphasis on the difference between expressions and equations, including a new Caution at the beginning of Section 10.1. Throughout the text, we have reformatted many example solutions to use a "drop down" layout in order to further emphasize for students the difference between simplifying expressions and solving equations.

- There is an increased emphasis on checking solutions and answers, as indicated by the new **Check** tag in the exposition and examples.

- The presentation on solving linear equations in Sections 10.1–10.3 includes five new examples, with new margin problems and corresponding exercises.

- Interval notation is used to graph the solution sets of linear inequalities in Section 10.6. We now introduce solving three-part linear inequalities in this section as well.

- When a new type of graph is introduced (Sections 11.2, 11.5, and 17.4), a new "hand-drawn" graph style is used to simulate what a student might actually sketch on graph paper.

- Presentations of the following topics have also been enhanced and expanded:

  Applications from geometry (Sections 10.4 and 10.5)
  Slopes of lines (Section 11.3)
  Slope-intercept form of the equation of a line (Section 11.4)
  Solving systems of equations with decimal coefficients (Section 15.2)
  Scientific notation (Section 12.8)
  Solving quadratic equations by factoring (Section 13.6)
  Solving work applications (Section 14.7)
  Solving problems involving formulas with radicals (Section 16.6)

What Supplements Are Available?

For a comprehensive list of the supplements and study aids that accompany *Developmental Mathematics: Basic Mathematics and Algebra*, Second Edition, see pages xiii and xiv.

Acknowledgments

The comments, criticisms, and suggestions of users, nonusers, instructors, and students have positively shaped this textbook over the years, and we are most grateful for the many responses we have received. We especially wish to thank the following reviewers whose valuable contributions have helped to refine this and the previous edition of this text.

Mary Kay Abbey, *Montgomery College*
Carla Ainsworth, *Salt Lake Community College*
George Alexander, *University of Wisconsin College*
Randall Allbritton, *Daytona Beach Community College*
Sonya Armstrong, *West Virginia State College*
Jannette Avery, *Monroe Community College*
Linda Beattie, *Western New Mexico University*
Linda Beller, *Brevard Community College*
Solveig R. Bender, *William Rainey Harper College*
Carla J. Bissell, *University of Nebraska at Omaha*
Jean Bolyard, *Fairmont State University*
Vernon Bridges, *Durham Technical Community College*
Barbara Brown, *Anoka-Ramsey Community College*
Kim Brown, *Tarrant County College—Northeast Campus*
Hien Bui, *Hillsborough Community College*
Tim C. Caldwell, *Meridian Community College*
Russell Campbell, *Fairmont State University*
Ernie Chavez, *Gateway Community College*
John Close, *Salt Lake Community College*
Terry Joe Collins, *Hinds Community College*
Dawn Cox, *Cochise College*
Jane Cuellar, *Taft College*
Martha Daniels, *Central Oregon Community College*
Ky Davis, *Muskingum Area Technical College*
Julie Dewan, *Mowhawk Valley Community College*
Bill Dunn, *Las Positas College*
Lucy Edwards, *Las Positas College*
Rob Farinelli, *Community College of Allegheny—Boyce Campus*
Scott Fallstrom, *Shoreline Community College*
Matthew Flacche, *Camden Community College*
Donna Foster, *Piedmont Technical College*
Randy Gallaher, *Lewis and Clark Community College*
Lourdes Gonzalez, *Miami-Dade Community College*
Mark Gollwitzer, *Greenville Technical College*

J. Lloyd Harris, *Gulf Coast Community College*
Terry Haynes, *Eastern Oklahoma State College*
Edith Hays, *Texas Woman's University*
Anthony Hearn, *Community College of Philadelphia*
Karen Heavin, *Morehead State University*
Lance Hemlow, *Raritan Valley Community College*
Elizabeth Heston, *Monroe Community College*
Joe Howe, *St. Charles County Community College*
Matthew Hudock, *St. Phillip's College*
Sharon Jackson, *Brookhaven College*
Rose Kaniper, *Burlington County College*
Rosemary Karr, *Collin County Community College*
Harriet Kiser, *Floyd College*
Jeffrey Kroll, *Brazosport College*
Babara Krueger, *Cochise College*
Valerie Lazzara, *Palm Beach Community College*
Christine Heinecke Lehmann, *Purdue University–North Central*
Douglas Lewis, *Yakima Valley Community College*
Sandy Lofstock, *California Lutheran University*
Lou Ann Mahaney, *Tarrant County College–Northeast Campus*
Valerie H. Maley, *Cape Fear Community College*
Susan McClory, *San Jose State University*
Gary McCracken, *Shelton State Community College*
Judy Mee, *Oklahoma City Community College*
Pam Miller, *Phoenix College*
Wayne Miller, *Lee College*
Jeffrey Mills, *Ohio State University*
Elizabeth Morrison, *Valencia Community College—West Campus*
Linda J. Murphy, *Northern Essex Community College*
Celia Nippert, *Western Oklahoma State College*
Elizabeth Olgilvie, *Horry-Georgetown Technical College*
Ted Panitz, *Cape Code Community College*
Claire Peacock, *Chattanooga State Technical Community College*
Kathy Peay, *Sampson Community College*
Faith Peters, *Miami-Dade Community College*
Thea Philliou, *College of Santa Fe*
Larry Potanski, *Pueblo Community College*
Manoj Raghunandanan, *Temple University*
Serban Raianu, *California State University—Dominguez Hills*
Janice Rech, *University of Nebraska at Omaha*
Janalyn Richards, *Idaho State University*
Jane Roads, *Moberly Area Community College*
Diann Robinson, *Ivy Tech State College–Lafayette*
Richard D. Rupp, *Del Mar College*
Rachael Schettenhelm, *Southern Connecticut State University*
Ellen Sawyer, *College of DuPage*
Lois Schuppig, *College of Mount St. Joseph*
Mary Lee Seitz, *Erie Community College—City Campus*
Julia Simms, *Southern Illinois University—Edwardsville*
Sounny Slitine, *Palo Alto College*
Dwight Smith, *Prestonburg Community College*
Lee Ann Spahr, *Durham Technical Community College*
Julia Speights, *Shelton State Community College*
Theresa Stalder, *University of Illinois—Chicago*
Carol Stewart, *Fairmont State University*
Kathryn Taylor, *Santa Ana College*
Sharon Testone, *Onondaga Community College*
Shae Thompson, *Montana State University*

Mike Tieleman-Ward, *Anoka Technical College*
Mark Tom, *College of the Sequoias*
Sven Trenholm, *North County Community College*
Bettie A. Truitt, *Black Hawk College*
Cora S. West, *Florida Community College at Jacksonville*
Cheryl Wilcox, *Diablo Valley College*
Johanna Windmueller, *Seminole Community College*
Jackie Wing, *Angelina College*
Gabriel Yimesghen, *Community College of Philadelphia*
Kevin Yokoyama, *College of the Redwoods*
Karl Zilm, *Lewis and Clark Community College*

Over the years, we have come to rely on an extensive team of experienced professionals. Our sincere thanks go to these dedicated individuals at Addison-Wesley, who worked long and hard to make this revision a success: Greg Tobin, Maureen O'Connor, Michelle Renda, Kari Heen, Courtney Slade, Kathy Manley, Barbara Atkinson, Beth Anderson, Lin Mahoney, Ceci Fleming, Alicia Frankel, and Mary Gallagher.

Abby Tanenbaum did an outstanding job helping us with manuscript preparation and prepared the index. We are truly grateful for her contributions to so many of our books over the years. Janette Krauss, Bonnie Boehme, and Nesbitt Graphics, Inc. provided excellent production work on the challenging format of these books. We thank Jeff Cole, who continues to provide accurate, helpful solutions manuals and Barb Brown, who helped us update the real-data applications. Janis Cimperman, De Cook, Shannon d'Hemecourt, Perian Herring, Paul Lorczak, Ann Ostberg, Sarah Sponholz, and Sharon Testone did a wonderful and timely job accuracy checking.

Special thanks go to Linda Russell, who wrote the Study Skills activities that appear in the text. In her roles as both an instructor and a specialist in reading and study skills at Minneapolis Community and Technical College, she created and revised these activities based on her work with hundreds of developmental-level students.

As an author team, we are committed to providing the best possible text and supplements package to help students succeed and instructors teach. As we continue to work toward this goal, we would welcome any comments or suggestions you might have via e-mail to *math@pearson.com*.

Margaret L. Lial
John Hornsby
Terry McGinnis
Stan Salzman
Diana L. Hestwood

Student Supplements

Student's Solutions Manual
- By Jeffery A. Cole, *Anoka-Ramsey Community College*
- Provides detailed solutions to the odd-numbered, section-level exercises and to all margin, Relating Concepts, Summary, Chapter Review, Chapter Test, and Cumulative Review Exercises
  ISBNs: 0-321-59970-5, 978-0-321-59970-4

Videos on DVD
- Feature an engaging team of lecturers
- Include a complete set of lectures for each section of the text on DVD for student use at home or on campus
- Ideal for distance learning or supplemental instruction
- Watch an instructor work through the complete solution for all exercises marked with a DVD icon ◉
- Include optional English and Spanish subtitles
  ISBNs: 0-321-59961-6, 978-0-321-59961-2

Worksheets for Classroom or Lab Practice
- Provide extra practice exercises for every section of the text with ample space for students to show their work
- List the learning objectives and key vocabulary terms for every text section, along with vocabulary practice problems
  ISBNs: 0-321-59973-X, 978-0-321-59973-5

InterAct Math Tutorial Website   *www.interactmath.com*
- Offers online practice and tutorial help
- Retry an exercise with new values each time for unlimited practice and mastery
- Every exercise is accompanied by an interactive guided solution that gives helpful feedback when an incorrect answer is entered
- Allows students to view steps of a worked-out sample problem similar to those in the text

Chapter Test Prep Video CD
- Watch instructors work through step-by-step solutions to all the Chapter Test exercises from the textbook
- Included with each new student text
- Available with optional English subtitles

Instructor Supplements

Annotated Instructor's Edition
- Provides answers to all text exercises in color next to the corresponding problem
- Includes icons to identify writing 🖉 and calculator 🧮 exercises
  ISBNs: 0-321-59965-9, 978-0-321-59965-0

Instructor's Solutions Manual
- By Jeffery A. Cole, *Anoka-Ramsey Community College*
- Provides complete answers to all the exercises in the text
  ISBNs: 0-321-59969-1, 978-0-321-59969-8

Additional Teaching Resources
Includes resources to help both new and adjunct faculty with course preparation and classroom management by offering helpful teaching tips correlated to the sections of the text
Available for download at *www.pearsonhighered.com*

Instructor's Resource Manual with Tests
- By James Ball, *Indiana State University*
- Contains a test bank with two diagnostic pretests, six free-response and two multiple-choice test forms per chapter, and two final exams
- Also contains a mini-lecture for each section of the text with objectives, key examples, and teaching tips
- Includes a correlation guide from the first to the second edition and phonetic spellings for all key terms in the text
  ISBNs: 0-321-59960-8, 978-0-321-59960-5

PowerPoint® Lecture Slides
- Present key concepts and definitions from the text
- Available for download at *www.pearsonhighered.com*

TestGen®   *www.pearsonhighered.com/testgen*
- Enables instructors to build, edit, print, and administer tests using a computerized bank of questions developed to cover all text objectives
- Allows instructors to create multiple but equivalent versions of the same question or test with the click of a button
- Allows instructors to modify test bank questions or add new questions
- Tests can be printed or administered online

Pearson Math Adjunct Support Center
http://www.pearsontutorservices.com/math-adjunct.html
Staffed by qualified instructors with more than 50 years of combined experience at both the community college and university levels. Assistance provided for faculty in the following areas:
- Suggested syllabus consultation
- Tips on using materials packed with your book
- Book-specific content assistance
- Teaching suggestions, including advice on classroom strategies

Available for Students and Instructors

MyMathLab®   MyMathLab is a series of text-specific, easily customizable online courses for Pearson Education's textbooks in mathematics and statistics. Powered by Course-Compass™ (our online teaching and learning environment) and MathXL® (our online homework, tutorial, and assessment system), MyMathLab provides the tools needed to deliver all or a portion of a course online, whether students are in a lab setting or working from home. MyMathLab provides a rich and flexible set of course materials, featuring free-response exercises that are algorithmically generated for unlimited practice and mastery. Students can also use online tools, such as video lectures, animations, and a multimedia textbook, to independently improve their understanding and performance. Instructors can use MyMathLab's homework and test managers to select and assign online exercises correlated directly to the textbook, and they can also create and assign their own online exercises and import TestGen tests for added flexibility. MyMathLab's online gradebook—designed specifically for mathematics and statistics—automatically tracks students' homework and test results and gives the instructor control over how to calculate final grades. Instructors can also add offline (paper-and-pencil) grades to the gradebook. MyMathLab also includes access to the **Pearson Tutor Center** (*www.pearsontutorservices.com*). The Tutor Center is staffed by qualified mathematics instructors who provide textbook-specific tutoring for students via toll-free phone, fax, email, and interactive Web sessions. MyMathLab is available to qualified adopters. For more information, visit our Web site at *www.mymathlab.com,* or contact your sales representative.

MathXL®   MathXL is a powerful online homework, tutorial, and assessment system that accompanies Pearson Education's textbooks in mathematics or statistics. With MathXL, instructors can create, edit, and assign online homework and tests using algorithmically generated exercises correlated at the objective level to the textbook. They can also create and assign their own online exercises and import TestGen tests for added flexibility. All student work is tracked in MathXL's online gradebook. Students can take chapter tests in MathXL and receive personalized study plans based on their test results. The study plan diagnoses weaknesses and links students directly to tutorial exercises for the objectives they need to study and retest. Students can also access supplemental animations and video clips directly from selected exercises. MathXL is available to qualified adopters. For more information, visit our Web site at *www.mathxl.com,* or contact your sales representative.

MathXL® Tutorials on CD   This interactive tutorial CD-ROM provides algorithmically generated practice exercises that are correlated at the objective level to the exercises in the textbook. Every practice exercise is accompanied by an example and a guided solution designed to involve students in the solution process. Selected exercises may also include a video clip to help students visualize concepts. The software provides helpful feedback for incorrect answers and can generate printed summaries of students' progress.
ISBNs:
0-321-59966-7
978-0-321-59966-7

Study Skills

Your brain knows how to learn, just as your lungs know how to breathe; however, there are important things you can do to maximize your brain's ability to do its work. This short introduction will help you choose effective strategies for learning mathematics. This is a simplified explanation of a complex process.

Your brain's outer layer is called the **neocortex,** which is where higher level thinking, language, reasoning, and purposeful behavior occur. The neocortex has about 100 billion (100,000,000,000) brain cells called **neurons.**

▶ As you learn something new, threadlike branches grow out of each neuron. These branches are called **dendrites.**

▶ When the dendrite from one neuron grows close enough to the dendrite from another neuron, a connection is made. There is a small gap at the connection point called a **synapse.** One dendrite sends an electrical signal across the gap to another dendrite.

▶ *Learning = growth and connecting of dendrites.*

▶ When you practice a skill just once or twice, the connections between neurons are very weak. If you do not practice the skill again, the dendrites at the connection points wither and die back. You have forgotten the new skill!

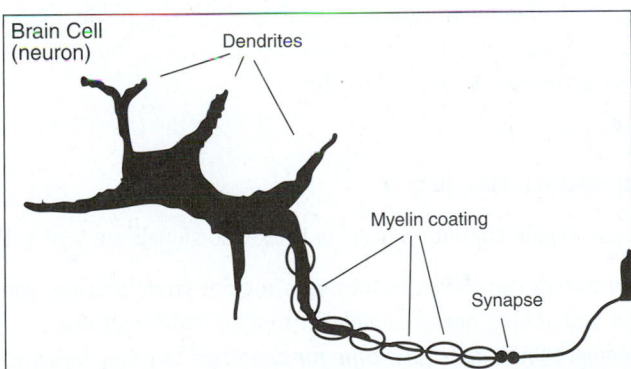

A neuron with several dendrites: one dendrite has developed a myelin coating through repeated practice.

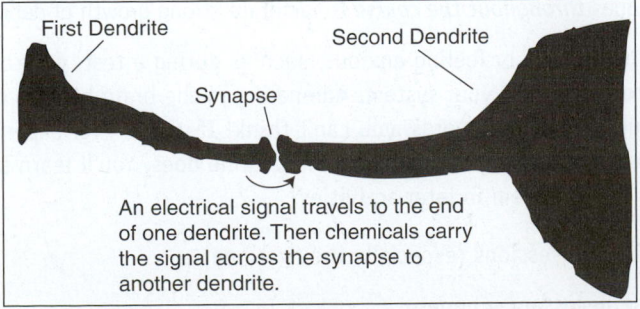

A close up view of the connection (synapse) between two dendrites.

▶ If you practice a new skill many times, the dendrites for that skill become coated with a fatty protein called **myelin.** Each time one dendrite sends a signal to another dendrite, the myelin coating becomes thicker and smoother, allowing the signals to move faster and with less interference. Thinking can now occur more quickly and easily, and *you will remember the skill for a long time* because the dendrite connections are very strong.

Become An Effective Student

▶ You grow dendrites specifically for the thing you are studying. If you practice dividing fractions, you will grow specialized dendrites just for dividing fractions. If you *watch other people* solve fraction problems, *you will grow dendrites for watching, not for solving.* So, be sure you are actively learning and practicing.

▶ If you practice something the *wrong* way, you will develop strong dendrite connections for doing it the wrong way! So, as you study, check frequently that you are getting correct answers.

▶ As you study a new topic that is related to things you already know, you will grow new dendrites, but your brain will also send signals throughout the network of dendrites for the related topics. In this way, you build a complex **neural network** that allows you to apply concepts, see differences and similarities between ideas, and understand relationships between concepts.

In the first few chapters of this textbook you will find "brain friendly" activities that are designed to help you grow and develop your own reliable neural networks for mathematics. Since you must grow your own dendrites (no one can grow them for you), these activities show you how to

▶ develop new dendrites,

▶ strengthen existing ones, and

▶ encourage the myelin coating to become thicker so signals are sent with less effort.

When you incorporate the activities into your regular study routine, you will discover that you understand better, remember longer, and forget less.

Also remember that *it does take time for dendrites to grow.* Trying to cram in several new concepts and skills at the last minute is not possible. Your dendrites simply can't grow that quickly. You can't expect to develop huge muscles by lifting weights for just one evening before a body building competition! In the same way, practice the study techniques *throughout the course* to facilitate strong growth of dendrites.

When Anxiety Strikes

If you are under stress or feeling anxious, such as during a test, your body secretes **adrenaline** into your system. Adrenaline in the brain blocks connections between neurons. In other words, you can't think! If you've ever experienced "blanking out" on a test, you know what adrenaline does. You'll learn several solutions to that problem in later activities.

Start Your Course Right!

▶ Attend all class sessions (especially the first one).

▶ Gather the necessary supplies.

▶ Carefully read the syllabus for the course, and ask questions if you don't understand.

1

Whole Numbers

In 1965, seventeen-year-old Fred DeLuca had just graduated from high school. DeLuca and a family friend, Dr. Peter Buck, opened the first SUBWAY sandwich shop with a total investment of $1000. From the beginning, the freshest ingredients were used, with DeLuca driving many miles in his Volkswagen bug to purchase vegetables directly from the produce mart. Today, with over 27,000 stores in 83 countries, SUBWAY is the largest submarine sandwich chain in the world. The company operates more stores in the United States, Canada, and Australia than McDonald's. In this chapter, we discuss whole numbers, which are used daily in our lives. (See Exercises 97–100 in **Section 1.3,** Exercise 1 and 31 in **Section 1.10,** and Exercise 123 in the **Chapter 1 Review Exercises.**)

1

1.1 ▶▶▶ Reading and Writing Whole Numbers

Knowing how to read and write numbers is an important step in learning mathematics.

OBJECTIVE **1** **Identify whole numbers.** The **decimal system** of writing numbers uses the ten digits

$$0, 1, 2, 3, 4, 5, 6, 7, 8, 9$$

to write any number. For example, these digits can be used to write the **whole numbers:**

$$0, 1, 2, 3, 4, 5, 6, 7, 8, 9, 10, 11, 12, 13 \ldots$$

The three dots indicate that the list goes on forever.

OBJECTIVE **2** **Give the place value of a digit.** Each digit in a whole number has a **place value,** depending on its position in the whole number. The following place value chart shows the names of the different places used most often and has the whole number 153,524 entered.

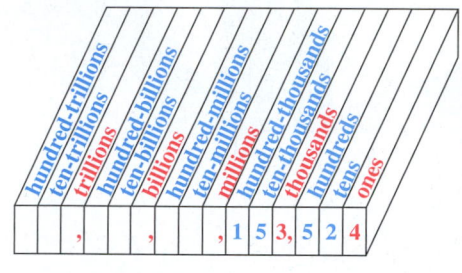 

Starbucks sells 153,524 pounds of coffee each day. Each of the 5s in 153,524 represents a different amount because of its position, or **place value** within the number. The **place value** of the 5 on the left is 5 ten-thousands (50,000). The **place value** of the 5 on the right is 5 hundreds (500).

EXAMPLE 1 Identifying Place Values

Identify the place value of 8 in each whole number.

> Each "8" has a different value.

(a) 28   **(b)** 85   **(c)** 869

    8 ones   8 tens   8 hundreds

Notice that the value of 8 in each number is different, depending on its location (place) in the number.

◀ Work Problem **1** at the Side.

EXAMPLE 2 Identifying Place Values

Identify the place value of each digit in the number 725,283.

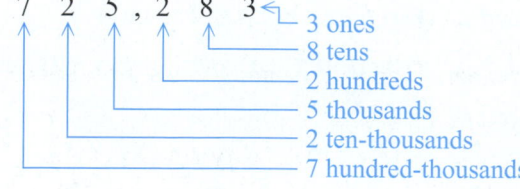

---

1 Identify the place value of the 4 in each whole number.

(a) 341

(b) 714

(c) 479

Notice the comma between the hundreds and thousands position in the number 725,283 in Example 2.

Work Problem **2** *at the Side.* ▶

Using Commas

Commas are used to separate each group of three digits, starting from the right. This makes numbers easier to read. (An exception: Commas are frequently omitted in four-digit numbers such as 9748 or 1329.) Each three-digit group is called a **period**. Some instructors prefer to just call them **groups.**

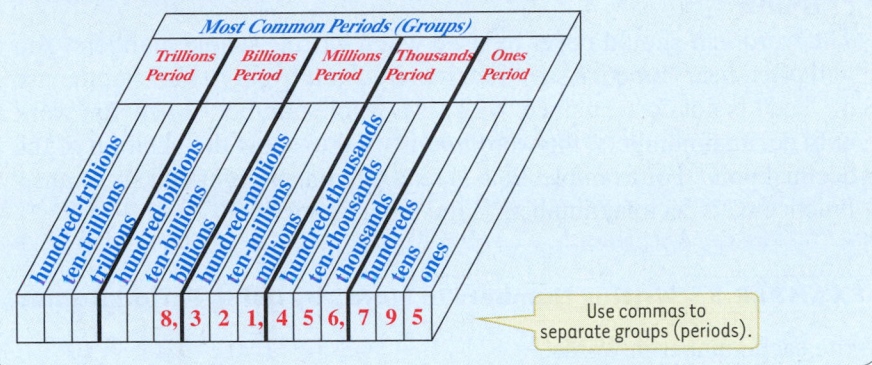

Use commas to separate groups (periods).

EXAMPLE 3 **Knowing the Period or Group Names**

Write the digits in each period of 8,321,456,795.

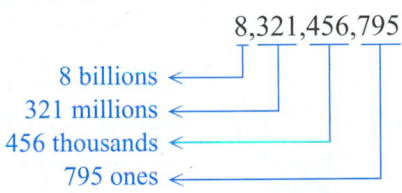

8,321,456,795

8 billions
321 millions
456 thousands
795 ones

Work Problem **3** *at the Side.* ▶

Use the following rule to read a number with more than three digits.

Writing Numbers in Words

Start at the left when writing a number in words or saying it aloud. Write or say the digit names in each period (group), followed by the name of the period, except for the period name "ones," which is *not* used.

OBJECTIVE 3 Write a number in words or digits. The following examples show how to write names for whole numbers.

EXAMPLE 4 **Writing Numbers in Words**

Write each number in words.

(a) 57

This number means 5 tens and 7 ones, or 50 ones and 7 ones. Write the number as

fifty-seven.

—— **Continued on Next Page**

---

2 Identify the place value of each digit.

(a) 14,218

(b) 460,329

3 In the number 3,251,609,328 identify the digits in each period (group).

(a) billions period

(b) millions period

(c) thousands period

(d) ones period

ANSWERS

2. **(a)** 1 : ten-thousands
    4 : thousands
    2 : hundreds
    1 : tens
    8 : ones
  (b) 4 : hundred-thousands
    6 : ten-thousands
    0 : thousands
    3 : hundreds
    2 : tens
    9 : ones
3. **(a)** 3 **(b)** 251 **(c)** 609 **(d)** 328

4 Write each number in words.

(a) 18

(b) 36

(c) 418

(d) 902

5 Write each number in words.

(a) 3104

(b) 95,372

(c) 100,075,002

(d) 11,022,040,000

6 Rewrite each number using digits.

(a) one thousand, four hundred thirty-seven

(b) nine hundred seventy-one thousand, six

(c) eighty-two million, three hundred twenty-five

ANSWERS

4. (a) eighteen
   (b) thirty-six
   (c) four hundred eighteen
   (d) nine hundred two
5. (a) three thousand, one hundred four
   (b) ninety-five thousand, three hundred seventy-two
   (c) one hundred million, seventy-five thousand, two
   (d) eleven billion, twenty-two million, forty thousand
6. (a) 1437   (b) 971,006   (c) 82,000,325

(b) 94

> ninety-four

(c) 874

> *Remember: Start at the left to read a number.*

> eight hundred seventy-four

(d) 601

> six hundred one

◀ Work Problem **4** at the Side.

CAUTION

The word *and* should never be used when writing whole numbers. You will often hear someone say "five hundred *and* twenty-two," but the use of "and" is not correct since "522" is a whole number. When you work with decimal numbers, the word *and* is used to show the position of the decimal point. For example, 98.6 is read as "ninety-eight *and* six tenths." Practice with decimal numbers is the topic of **Section 4.1.**

EXAMPLE 5 **Writing Numbers in Words by Using Period Names**

Write each number in words.

(a) 725,283

> seven hundred twenty-five **thousand,** two hundred eighty-three
>
> Number in period | Name of period | Number in period (not necessary to write "ones")

(b) 7252

> *Careful: Do not use "and" when reading a whole number.*
>
> seven **thousand,** two hundred fifty-two
>
> Name of period | No period name needed

(c) 111,356,075

> one hundred eleven **million,** three hundred fifty-six **thousand,** seventy-five

(d) 17,000,017,000

> seventeen **billion,** seventeen **thousand**

◀ Work Problem **5** at the Side.

EXAMPLE 6 **Writing Numbers in Digits**

Rewrite each number using digits.

(a) six **thousand,** twenty-two

> 6022   *With 4 digits or fewer, no comma is needed.*

(b) two hundred fifty-six **thousand,** six hundred twelve

> 256,612

(c) nine **million,** five hundred fifty-nine

> 9,000,559
>
> Zeros indicate there are no thousands.

◀ Work Problem **6** at the Side.

🖩 **Calculator Tip**   Does your calculator show a comma between each group of three digits? Probably not, but try entering a long number such as 34,629,075. Notice that there is no key with a comma on it, so you do not enter commas. A few calculators may show the position of the commas *above* the digits, like this

34'629'075

Most of the time you will have to write in the commas where needed.

OBJECTIVE **4** **Read a table.**   A common way of showing number values is by using a **table.** Tables organize and display facts so that they are more easily understood. The following table shows some past facts and future predictions for the United States.

These estimated numbers give us a glimpse of what we can expect in the 21st century.

NUMBERS FOR THE 21ST CENTURY

Year	1990	2010	2020*
U.S. population	261 million	307 million	338 million
Births	16 million	14 million	13 million
Household income	$42,936	$49,564	$53,375
Average salary	$21,129	$25,743	$28,050

*Estimated figures
Source: Family Circle magazine; U.S. Census Bureau.

If you read from left to right along the row labeled "U.S. population," you find that the population in 1990 was 261 million, then the population in 2010 was 307 million, and the estimated population for 2020 is 338 million.

EXAMPLE 7 **Reading a Table**

Use the table to find each number, and write the number in words.

(a) The estimated household income in the year 2020
Read from left to right along the row labeled "Household income" until you reach the 2020 column and find $53,375.

Fifty-three thousand, three hundred seventy-five dollars

(b) The average salary in 1990.
Read from left to right along the row labeled "Average salary." In the 1990 column you find $21,129.

Remember: use hyphens when necessary.

Twenty-one thousand, one hundred twenty-nine dollars

Work Problem **7** *at the Side.* ▶

Note
Notice in Example 7 that hyphens are used when writing numbers in words. A hyphen is used when writing the numbers 21 through 99 (twenty-one through ninety-nine), except for numbers ending in zero (20, 30, 40, ... 90).

7 Use the table to find each number, and write the number in digits when given in words, or write the number in words when given in digits.

(a) The number of births in 2010

(b) The estimated number of births in 2020

(c) Household income in 1990

(d) The estimated average salary in 2020

Math in the Media

THE TOLL OF WEDDING BELLS

The Wedding Report recently released statistics showing the average wedding costs in the United States in 2008. In 2004, the cost of the average wedding was $24,168. This cost has increased as the average number of wedding guests has grown to over 200. The graph gives most of the costs involved in a wedding. Use this information to answer the questions that follow.

Numbers in the News

'Til Debt Do You Part

With over 200 guests, the cost of an average wedding has continued to grow. Most of the money is spent on the following:

Transportation	$426
Wedding Stationery	$881
Music	$991
Favors and Gifts	$1166
Flowers	$2048
Wedding Jewelry	$2148
Ceremony	$2627
Wedding Attire	$2710
Photography and Video	$3846
Wedding Reception	$14,037

Source: The Wedding Report, *Wedding Statistics and Research for the Wedding Industry.*

1. What is the total of the costs shown in the graph?

2. How much more expensive was a wedding in 2008 compared with 2004?

3. The groom pays for the photography and video, the flowers, the wedding jewelry, the clergy ($500), and the groom's formal wear ($95). What is the total amount spent by the groom?

4. If you budgeted $65 per person for the wedding reception and you invited 225 guests to a wedding in 2004, how much would you have spent compared to the 2008 wedding reception costs?

5. If you budget $6000 for the wedding reception and the cost per person is $37, how many guests can you invite and how much of your budgeted amount will be left over?

6. If you budget $1000 for the wedding reception and the cost per person is $15, how many guests can you invite and how much of your budgeted amount will be left over?

7. What kind of an arithmetic problem did you work to get the answers to Problems 5 and 6? What is the mathematical term for the "left over" budget?

1.1 ▶▶▶ **Exercises**

FOR
EXTRA
HELP

*Write the digit for the given **place value** in each whole number. See Examples 1 and 2.*

1. 3065
thousands
tens

2. 4681
thousands
ones

3. 18,015
ten-thousands
hundreds

4. 86,332
ten-thousands
ones

5. 7,628,592,183
millions
thousands

6. 1,700,225,016
billions
millions

*Write the digits for the given **period** (group) in each whole number. See Example 3.*

7. 3,561,435
millions
thousands
ones

8. 28,785,203
millions
thousands
ones

9. 60,000,502,109
billions
millions
thousands
ones

10. 100,258,100,006
billions
millions
thousands
ones

11. Do you think the fact that humans have four fingers and a thumb on each hand explains why we use a number system based on ten digits? Explain.

12. The decimal system uses ten digits. Fingers and toes are often referred to as digits. In your opinion, is there a relationship here? Explain.

Write each number in words. See Examples 4 and 5.

13. 23,115

14. 37,886

15. 346,009

16. 218,033

17. 25,756,665

18. 999,993,000

Write each number using digits. See Example 6.

19. sixty-three thousand, one hundred sixty-three

20. ninety-five thousand, one hundred eleven

21. ten million, two hundred twenty-three

22. one hundred million, two hundred

Write the numbers from each sentence using digits. See Example 6.

23. There are three million, two hundred thousand parachute jumps in the United States each year. (*Source:* History Channel.)

24. The United States Postal Service set a record of two hundred eighty million, four hundred eighty-nine thousand postmarked pieces of mail on a single day. (*Source:* U.S. Postal Service.)

25. The number of cans of Pepsi Cola sold each day is fifty million, fifty-one thousand, five hundred seven. (*Source:* Andy Rooney, *60 Minutes*.)

26. The number of motorcycle owners in the world is twenty-three million, five hundred thirty-five thousand. (*Source:* Motorcycle Industry Council.)

27. There are fifty-four million, seven hundred fifty thousand Hot Wheels sold each year. (*Source:* Andy Rooney, *60 Minutes*.)

28. Burger King sells two million, four hundred thousand hamburgers each day, (*Source:* Andy Rooney, *60 Minutes*.)

29. Rewrite eight hundred trillion, six hundred twenty-one million, twenty thousand, two hundred fifteen by using digits.

30. Rewrite 70,306,735,002,102 in words

The table at the right shows various ways people get to work. Use the table to answer Exercises 31–34. See Example 7.

31. Which method of transportation is least used? Write the number in words.

32. Which method of transportation is most used? Write the number in words.

33. Find the number of people who walk to work or work at home, and write it in words.

34. Find the number of people who carpool, and write it in words.

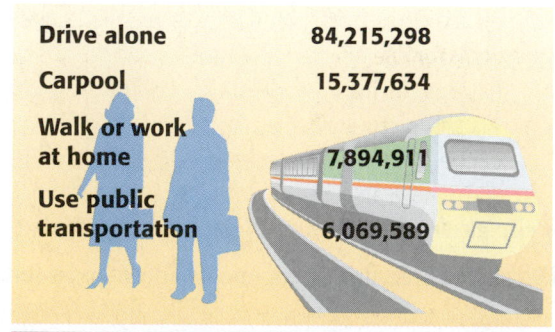

GETTING TO WORK

How workers 16 and over get to work:

Drive alone	84,215,298
Carpool	15,377,634
Walk or work at home	7,894,911
Use public transportation	6,069,589

Source: U.S. Census Bureau.

Study Skills

USING YOUR TEXTBOOK

> Be sure to read *Your Brain Can Learn Mathematics* before this activity. You'll find out how your brain learns and remembers.

Your textbook can be very helpful. Find out what it has to offer. First, let's look at some general features that will help in all chapters.

Look at page iii in the very front for the Table of Contents. You should look at the Preface, which begins on page vii. On page xiii is a list of Supplementary Resources for students. If you are interested in any of these, ask your instructor if they are available.

Each chapter is divided into sections, and each section has a number, such as 1.3 or 3.5. Your instructor will use these numbers to assign readings and homework.

Chapter 3 → **3.5** ← **Section 5 within Chapter 3**

There are four features to pay special attention to as you work in your book.

- **Objectives.** Each section lists the objectives in the upper corner of the first page. The objectives are listed again as each one is introduced. An objective tells you *what you will be able to do after you complete the section*. An excellent way to check your learning is to go back to the list of objectives when you are finished with a section and ask yourself if you can do them all.
- **Margin Exercises.** The exercises in the shaded margins of the pages in your textbook give you immediate practice. **This is a perfect way to get your dendrites growing right away!** The answers are given at the bottom, so you can check yourself easily.
- **Cautions, Pointers, Notes, and Calculator Tips.**
  - **Caution!** A bright yellow box is a comment about a common error that students make or a common trouble spot you will want to avoid.
  - **Pointers** are little "clouds" next to worked examples. They point to specific places where common mistakes are made and give on-the-spot reminders.
  - Look for the specially marked purple **Note** boxes. They contain hints, explanations, or interesting side comments about a topic.
  - A small picture of a red calculator 🔢 appears several places. In the main part of the chapter, the icon means that there is a **Calculator Tip,** which helps you learn more about using your calculator. A calculator in an Exercise section is a recommendation to use your calculator to work that exercise.

Go back to the Table of Contents again. What is listed at the end of each chapter?

- **Chapter Summary.** Turn to page 97 to find the Summary for Chapter 1. It lists the chapter's **Key Terms** (arranged in the order that they appear in the chapter) and **New Symbols** and/or **New Formulas.** Then, **Test Your Word Power** checks your understanding of the math vocabulary. Next is a **Quick Review section.** It lists each topic in the chapter and shows a worked example, with tips.

OBJECTIVES

1. Explain the meaning of text features such as section numbering, objectives, margin exercises.
2. Locate the Index, Answers, and Solutions sections.

Table of Contents

Section Numbering

Chapter Features

End of Chapter Features

9

End of Chapter Features
(continued)

How will you make good use of the features at the end of each chapter?

_____

_____

Answers

Flag the Answers section with a sticky note or other device, so that you can turn to it quickly.

Solutions

Index

Why Are These Features Brain Friendly?

The textbook authors included text features that make it easier for you to understand the mathematics. **Your brain naturally seeks organization and predictability.** When you pay attention to the regular features of your textbook, you are allowing your brain to get familiar with all of the helpful tips, suggestions, and explanations that your book has to offer. You will make the best possible use of your textbook.

▶ **Review Exercises** Use these exercises as a way to check your understanding of all the concepts in the chapter. You can practice every type of problem. If you get stuck, the red numbers in brackets tell you which section of the chapter to go back to for more explanations. Make sure you do the **Mixed Review Exercises** to practice for tests.

▶ **Chapter Test.** Plan to take the test as a practice exam. That way you can be sure you really know how to work all types of problems without looking back at the chapter.

▶ **Cumulative Review (after Chapters 8 and 17)** These exercises help you maintain the skills you've learned in previous chapters. Working on previous skills throughout the course will be a big help on the final exam.

How do you find out if you've worked the exercises correctly? Your textbook provides many of the answers. Throughout each chapter you should work the sample problems in the **margins.** The answers for those are at the **bottom of each page** in the margin area.

For homework, you can find the answers to all of the **odd-numbered section exercises** in the **Answers to Selected Exercises** section near the end of your textbook. Also, *all* of the answers are given for the Chapter Review Exercises and Chapter Tests. Check your textbook now, and find the page on which the Answers section begins.

The Solutions section near the end of the book shows how to solve some of the harder odd-numbered exercises step by step. Look for the problem numbers with a square of blue shading around them. These are the ones that have a solution.

The Index is the last thing in your book. It lists all the topics, vocabulary, and concepts in alphabetical order. For example, look up the words below. Go to the page or pages listed and find each word. Write down the page that introduces or defines each one. There may be several subheadings listed under the main word, or several page numbers listed. Usually, the *first* place that a word appears in the textbook is where it is introduced and defined. So, the earliest page number is a good place to start.

Commutative property of multiplication is defined on page _____.
Factors of numbers are defined on page _____.
Rounding of mixed numbers is explained on page _____.

List a page number from Chapter 1 for each of these features:

A *Caution* appears on page_____.

A *Pointer* appears on page_____.

A *Note* appears on page_____.

A *Calculator Tip* appears on page_____.

1.2 ▶▶▶ Adding Whole Numbers

There are four soccer balls at the left and two at the right. In all, there are six soccer balls.

The process of finding the total is called **addition.** Here 4 and 2 were added to get 6. Addition is written with a + sign, so that

$$4 + 2 = 6.$$

OBJECTIVE 1 Add two single-digit numbers.   In addition, the numbers being added are called **addends,** and the resulting answer is called the **sum** or **total.**

$$
\begin{array}{r}
4 \leftarrow \text{Addend} \\
+\ 2 \leftarrow \text{Addend} \\
\hline
6 \leftarrow \text{Sum (total)}
\end{array}
$$

Addition problems can also be written horizontally, as follows.

$$
\underset{\text{Addend}}{4} \ + \ \underset{\text{Addend}}{2} \ = \ \underset{\text{Sum}}{6}
$$

> **Commutative Property of Addition**
> By the **commutative property of addition,** changing the order of the addends in an addition problem does not change the sum.

For example, the sum of $4 + 2$ is the same as the sum of $2 + 4$. This allows the addition of the same numbers in a different order.

EXAMPLE 1   Adding Two Single-Digit Numbers

Add, and then change the order of numbers to write another addition problem.

(a) $5 + 3 = 8$ and $3 + 5 = 8$

(b) $7 + 8 = 15$ and $8 + 7 = 15$ ⟨ Changing the order in addition does not change the sum.

(c) $8 + 3 = 11$ and $3 + 8 = 11$

(d) $8 + 8 = 16$

Work Problem **1** *at the Side.* ▶

> **Associative Property of Addition**
> By the **associative property of addition,** changing the grouping of the addends in an addition problem does not change the sum.

For example, the sum of $3 + 5 + 6$ may be found as follows.

$$(3 + 5) + 6 = 8 + 6 = 14 \qquad \text{Parentheses tell us to add } 3 + 5 \text{ first.}$$

Another way to add the same numbers is   ⟨ Changing the grouping of addends does not change the sum.

$$3 + (5 + 6) = 3 + 11 = 14. \qquad \text{Parentheses tell us to add } 5 + 6 \text{ first.}$$

Either grouping gives a sum of 14 because of the associative property of addition.

OBJECTIVES

1 Add two single-digit numbers.

2 Add more than two numbers.

3 Add when regrouping (carrying) is not required.

4 Add with regrouping (carrying).

5 Use addition to solve application problems.

6 Check the answer in addition.

1 Add, and then change the order of numbers to write another addition problem.

(a) $2 + 6$

(b) $9 + 5$

(c) $4 + 7$

(d) $6 + 9$

ANSWERS

1. **(a)** $8; 6 + 2 = 8$ **(b)** $14; 5 + 9 = 14$
   (c) $11; 7 + 4 = 11$ **(d)** $15; 9 + 6 = 15$

2 Add each column of numbers.

(a)   3
8
5
4
+ 6

(b)   5
6
3
2
+ 4

(c)   9
6
8
7
+ 3

(d)   3
8
6
4
+ 8

OBJECTIVE 2 Add more than two numbers. To add several numbers, first write them in a column. Add the first number to the second. Add this sum to the third number. Continue until all the numbers are used.

EXAMPLE 2   **Adding More Than Two Numbers**

Add 2, 5, 6, 1, and 4.

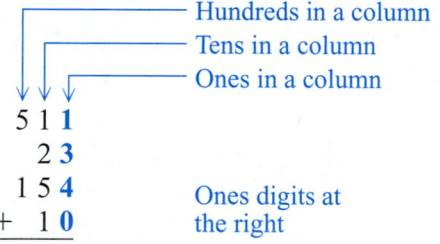

◀ *Work Problem* **2** *at the Side.*

Note

By the commutative and associative properties of addition, numbers may also be added starting at the bottom of a column. Adding from the top or adding from the bottom will give the same answer.

OBJECTIVE 3 Add when regrouping (carrying) is not required. If numbers have two or more digits, you must arrange the numbers in columns so that the ones digits are in the same column, tens are in the same column, hundreds are in the same column, and so on. Next, you add column by column starting at the right.

EXAMPLE 3   **Adding without Regrouping**

Add $511 + 23 + 154 + 10$.

First line up the numbers in columns, with the ones column at the right.

```
                      Hundreds in a column
                      Tens in a column
                      Ones in a column
          ↓ ↓ ↓
        5 1 1
          2 3
        1 5 4          Ones digits at
      +   1 0          the right
```

Now start at the right and add the ones digits. Add the tens digits next, and finally, the hundreds digits.

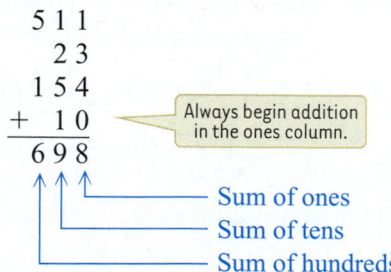

The sum of the four numbers is 698.

Work Problem ③ *at the Side.* ▶

OBJECTIVE 4 Add with regrouping (carrying). If the sum of the digits in any column is more than 9, use **regrouping** (sometimes called **carrying**).

EXAMPLE 4 Adding with Regrouping

Add 47 and 29.

Add ones.

$$
\begin{array}{r}
47 \\
+\ 29 \\
\hline
\end{array}
$$

↑ ———— 7 ones and 9 ones = 16 ones

Regroup 16 ones as 1 ten and 6 ones. Write 6 ones in the ones column and write 1 ten in the tens column.

1 ←——————————— Write 1 ten in the tens column.
$$
\begin{array}{r}
47 \\
+\ 29 \\
\hline
6 \\
\end{array}
$$
7 ones and 9 ones = 16 ones ——— Write 6 ones in the ones column.

Add the digits in the tens column, including the regrouped 1.

1
$$
\begin{array}{r}
47 \\
+\ 29 \\
\hline
76 \\
\end{array}
$$
↑ ——— 1 ten + 4 tens + 2 tens = 7 tens

Work Problem ④ *at the Side.* ▶

EXAMPLE 5 Adding with Regrouping

Add 324 + 7855 + 23 + 7 + 86.

Step 1 Add the digits in the ones column.

2 ←
$$
\begin{array}{r}
324 \\
7855 \\
23 \\
7 \\
+\ 86 \\
\hline
5 \\
\end{array}
$$
Write 2 tens in the tens column.

Sum of the ones column is 25 ones.

Write 5 ones in the ones column.

Notice that 25 ones are regrouped as 2 tens and 5 ones.

Step 2 Add the digits in the tens column, including the regrouped 2.

1 2
$$
\begin{array}{r}
324 \\
7855 \\
23 \\
7 \\
+\ 86 \\
\hline
95 \\
\end{array}
$$
Write 1 hundred in the hundreds column.

Sum of the tens column is 19 tens.

Notice that 19 tens are regrouped as 1 hundred and 9 tens.

Write 9 in the tens column.

Continued on Next Page

---

③ Add.

(a)
$$
\begin{array}{r}
26 \\
+\ 73 \\
\hline
\end{array}
$$

(b)
$$
\begin{array}{r}
534 \\
+\ 265 \\
\hline
\end{array}
$$

(c)
$$
\begin{array}{r}
42{,}305 \\
+\ 11{,}563 \\
\hline
\end{array}
$$

④ Add with regrouping

(a)
$$
\begin{array}{r}
66 \\
+\ 27 \\
\hline
\end{array}
$$

(b)
$$
\begin{array}{r}
58 \\
+\ 33 \\
\hline
\end{array}
$$

(c)
$$
\begin{array}{r}
56 \\
+\ 37 \\
\hline
\end{array}
$$

(d)
$$
\begin{array}{r}
34 \\
+\ 49 \\
\hline
\end{array}
$$

ANSWERS

3. (a) 99 (b) 799 (c) 53,868
4. (a) 93 (b) 91 (c) 93 (d) 83

⑤ Add by regrouping as necessary.

(a) 42
     651
     396
   + 87

(b)   162
     4271
      372
   + 8976

(c) 57
      4
    392
    804
     51
   + 27

(d) 7821
      435
       72
      305
   + 1693

⑥ Add by regrouping mentally.

(a)   816
      363
       17
        2
        5
   + 7654

(b) 3305
      650
      708
       29
       40
        6
   +   3

(c) 15,829
        765
         78
         15
          9
          7
   + 13,179

Step 3   Add the hundreds column, including the regrouped 1.

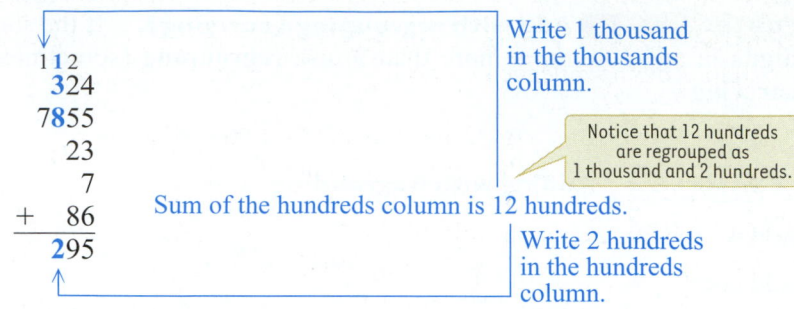

     1 1 2
      324
     7855
       23
        7
    + 86
      2 95

Write 1 thousand in the thousands column.

Notice that 12 hundreds are regrouped as 1 thousand and 2 hundreds.

Sum of the hundreds column is 12 hundreds.

Write 2 hundreds in the hundreds column.

Step 4   Add the thousands column, including the regrouped 1.

     1 1 2
      324
     7855
       23
        7
    + 86
     8295

Sum of the thousands column is 8.

Finally, $324 + 7855 + 23 + 7 + 86 = 8295$.

◀ Work Problem **⑤** at the Side.

Note

For additional speed, try to regroup mentally. Do not write the regrouped number, but just remember it as you move to the top of the next column. Try this method. If it works for you, use it.

◀ Work Problem **⑥** at the Side.

OBJECTIVE ⑤ Use addition to solve application problems.   In **Section 1.10** we will describe how to solve application problems in more detail. The next two examples are application problems that require adding.

EXAMPLE 6   **Applying Addition Skills**

On this map of the Walt Disney World area in Florida, the distance in miles from one location to another is written alongside of the road. Find the shortest route from Altamonte Springs to Clear Lake.

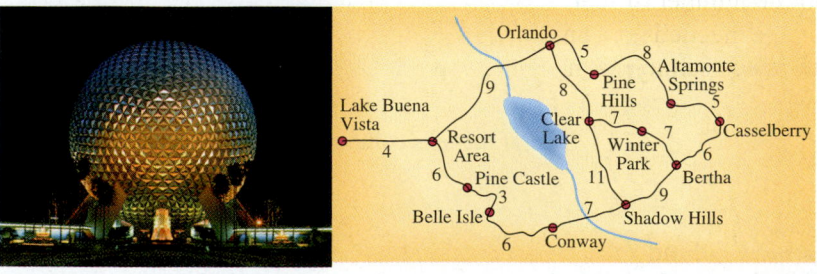

Continued on Next Page

Approach   Add the mileage along various routes to determine the distances from Altamonte Springs to Clear Lake. Then select the shortest route.

Solution   One way from Altamonte Springs to Clear Lake is through Orlando. Add the mileage numbers along this route.

8	Altamonte Springs to Pine Hills
5	Pine Hills to Orlando
+ 8	Orlando to Clear Lake
21 →	miles from Altamonte Springs to Clear Lake, going through Orlando

> Remember: Shortest distance is the fewest total miles.

Another way is through Casselberry, Bertha, and Winter Park. Add the mileage numbers along this route.

5	Altamonte Springs to Casselberry
6	Casselberry to Bertha
7	Bertha to Winter Park
+ 7	Winter Park to Clear Lake
25 →	miles from Altamonte Springs to Clear Lake through Bertha and Winter Park

The shortest route from Altamonte Springs to Clear Lake is 21 miles through Orlando.

Work Problem **7** *at the Side.* ▶

EXAMPLE 7   **Finding a Total Distance**

Use the map in Example 6 to find the total distance from Shadow Hills to Casselberry to Orlando and back to Shadow Hills.

Approach   Add the mileage from Shadow Hills to Casselberry to Orlando and back to Shadow Hills to find the total distance.

Solution   Use the numbers from the map.

9	Shadow Hills to Bertha
6	Bertha to Casselberry
5	Casselberry to Altamonte Springs
8	Altamonte Springs to Pine Hills
5	Pine Hills to Orlando
8	Orlando to Clear Lake
+ 11	Clear Lake to Shadow Hills
52 →	miles from Shadow Hills to Casselberry to Orlando and back to Shadow Hills

Work Problem **8** *at the Side.* ▶

EXAMPLE 8   **Finding a Perimeter**

Find the number of floating pipe needed to contain the farm-raised salmon in the habitat shown.

The short way to write *feet* is *ft*.

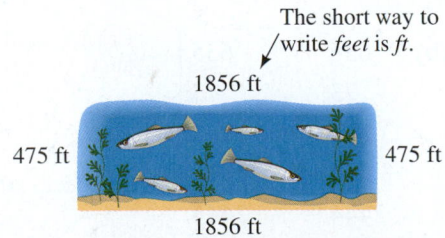

1856 ft

475 ft   475 ft

1856 ft

Approach   Find the **perimeter,** or total distance around the habitat, by adding the lengths of all the sides.

Continued on Next Page

---

7 Use the map in Example 6 to find the shortest route from Lake Buena Vista to Conway.

8 The road is closed between Orlando and Clear Lake, so this route cannot be used. Use the map in Example 6 to find the next shortest route from Orlando to Clear Lake.

ANSWERS

7. 19 miles
8.

5	Orlando to Pine Hills
8	Pine Hills to Altamonte Springs
5	Altamonte Springs to Casselberry
6	Casselberry to Bertha
7	Bertha to Winter Park
+7	Winter Park to Clear Lake
38 miles	

⑨ Solve the problem. Find the number of feet of fencing needed to enclose the solar electricity generating project shown.

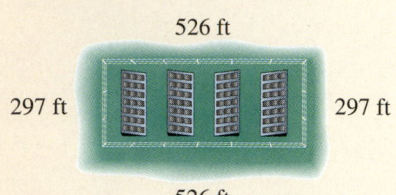

526 ft

297 ft   297 ft

526 ft

⑩ Check the following additions. If an answer is incorrect, find the correct answer.

(a)   63
          4
          9
       + 28
        104

(b)   927
        395
         64
      + 251
       1637

(c)   79
        218
          7
      + 639
        953

(d)   21,892
      11,746
    + 43,925
      79,563

Solution Use the lengths shown.

          1856
           475
          1856
        + 475
          4662 ft

The amount of floating pipe needed is 4662 ft, which is the perimeter of (distance around) the habitat.

◀ *Work Problem* ⑨ *at the Side.*

OBJECTIVE **6** **Check the answer in addition.** Checking the answer is an important part of problem solving. A common method for checking addition is to re-add from bottom to top. This is an application of the commutative and associative properties of addition.

EXAMPLE 9 **Checking Addition**

Check the following addition.

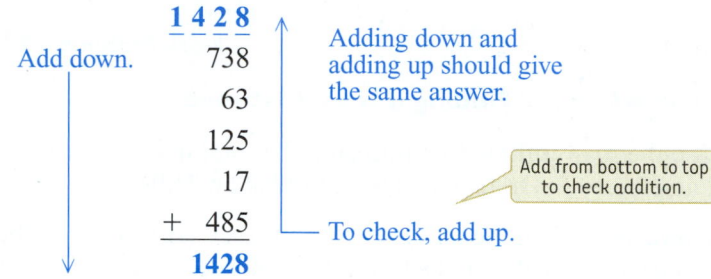

                    1428
    Add down.   738   Adding down and
                     63   adding up should give
                    125   the same answer.
                     17
                  + 485 ── To check, add up.   Add from bottom to top
                   1428   to check addition.

Here the answers agree, so the sum is probably correct.

EXAMPLE 10 **Checking Addition**

Check the following additions. Are they correct?

(a)   785   **1033**   Correct, because both
        63   785   answers are the same.
     + 185   63
       1033   + 185 ── To check, add up.
               1033

(b)   635   **2454**   Error, because the
        73   635   answers are different.
       831   73
     + 915   831
       2444   + 915 ── To check, add up.   Avoid wrong answers by
               2444   checking your work.

Re-add to find that the correct sum is 2454.

◀ *Work Problem* ⑩ *at the Side.*

1.2 ▶▶▶ **Exercises**

FOR EXTRA HELP  Math XL PRACTICE WATCH DOWNLOAD READ REVIEW

Add. See Examples 1–3.

1. 43
+ 54

2. 18
+ 11

3. 56
+ 33

4. 83
+ 15

5. 317
+ 572

6. 574
+ 325

7. 318
151
+ 420

8. 135
253
+ 410

9. 6310
252
+ 1223

10. 121
5705
+ 3163

11. 932 + 44 + 613

12. 517 + 131 + 250

13. 1251 + 4311 + 2114

14. 3241 + 1513 + 2014

15. 12,142 + 43,201 + 23,103

16. 41,124 + 12,302 + 23,500

17. 3213 + 5715

18. 6344 + 1655

19. 38,204 + 21,020

20. 63,251 + 36,305

Add, regrouping as necessary. See Examples 4 and 5.

21. 87
+ 63

22. 19
+ 92

23. 86
+ 69

24. 37
+ 85

25. 47
+ 74

26. 97
+ 79

27. 67
+ 78

28. 96
+ 47

29. 73
+ 29

30. 68
+ 37

31. 746
+ 905

32. 621
+ 359

33. 306
+ 848

34. 798
+ 206

35. 278
+ 135

36. 172
+ 156

37. 928
+ 843

38. 686
+ 726

39. 526
+ 884

40. 116
+ 897

41. 3574
    + 2817

42. 6871
    + 7528

43. 7896
    + 3728

44. 9382
    + 7586

45. 9625
    + 7986

46. 5718
    5623
    + 7436

47. 9056
    78
    6089
    + 731

48. 4022
    709
    8621
    + 37

49. 18
    708
    9286
    + 636

50. 1708
    321
    61
    + 8926

51. 422
    6074
    435
    + 8663

52. 6505
    173
    7044
    + 168

53. 321
    9603
    8
    21
    + 1604

54. 7631
    5983
    7
    36
    + 505

55. 2109
    63
    16
    3
    + 9887

56. 322
    6508
    93
    745
    18
    + 2005

57. 553
    97
    2772
    437
    63
    + 328

58. 3187
    810
    527
    76
    2665
    + 317

59. 413
    85
    9919
    602
    31
    + 1218

60. 576
    7934
    60
    781
    5968
    + 371

Check each addition. If an answer is incorrect, find the correct answer. See Examples 9 and 10.

61. ____
    832
    468
    + 791
    2091

62. ____
    326
    852
    + 679
    1857

63. ____
    179
    214
    + 376
    759

64. ____
    17
    296
    713
    + 94
    1220

65. ____
    4713
    28
    615
    + 64
    5420

66. ____
    3 628
    72
    564
    + 7 319
    11,583

67. ____
    678
    7 952
    56
    718
    + 2 173
    11,377

68. ____
    516
    8 760
    24
    189
    + 1 723
    11,212

69. ____
    4 714
    27
    77
    8 878
    + 636
    14,332

70. ____
    6 715
    283
    9 617
    13
    + 81
    16,719

71. Explain the commutative property of addition in your own words. How is this used when checking an addition problem?

72. Explain the associative property of addition. How can this be used when adding columns of numbers?

For Exercises 73–76, use the map to find the shortest route between each pair of cities. See Examples 6 and 7.

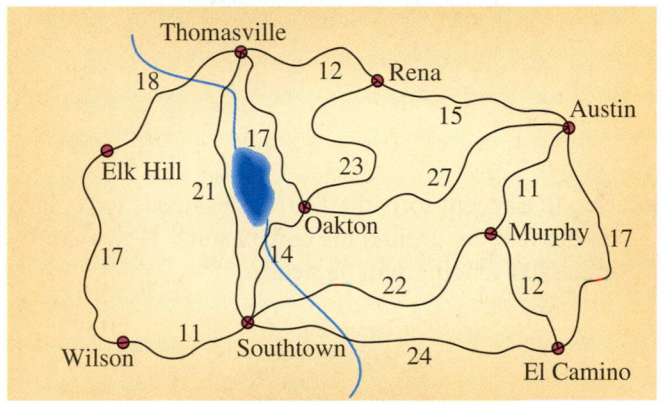

73. Southtown and Rena

74. Elk Hill and Oakton

75. Thomasville and Murphy

76. Murphy and Thomasville

Solve each application problem.

77. The Twin Lakes Food Bank raised $3482 at a flea market and $12,860 at their annual auction. Find the total amount raised at these two events.

78. A clothing store ordered 75 tops and 52 pairs of shorts. How many items were ordered?

79. There are 413 women and 286 men on the sales staff. How many people are on the sales staff?

80. One department in an office building has 283 employees while another department has 218 employees. How many employees are in the two departments?

81. This semester there are 13,786 students enrolled in on-campus day classes, 3497 students enrolled in night classes, and 2874 student's enrolled in on-line classes. Find the total number of students enrolled.

82. Robert and Crystal Hernandez have a car loan balance of $10,329 and a balance owed on their credit card of $2685. After receiving a home loan of $169,760, find the total amount of their loans.

Solve each problem involving perimeter. See Example 8.

83. Find the total distance around a lot that has been developed as a go-cart track.

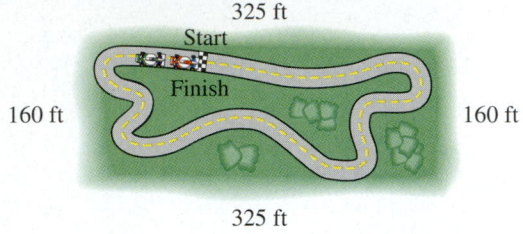

84. Because of heavy snowfall this winter, Maria needs to put new rain gutters around her entire roof. How many feet of gutters will she need?

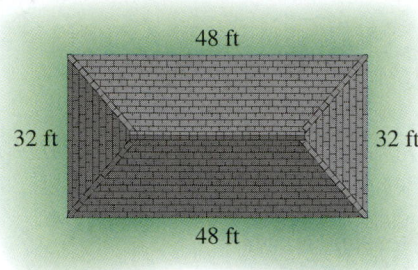

85. Martin plans to frame his back patio with redwood lumber. How many feet of lumber will he need?

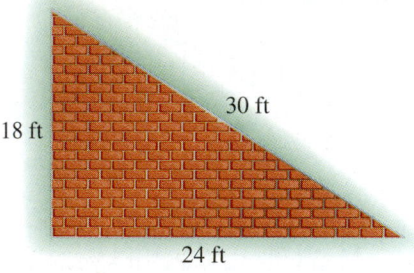

86. Due to a recent tornado, Carl Jones needs to replace all the fencing around his cow pasture. How many meters of fencing will he need?

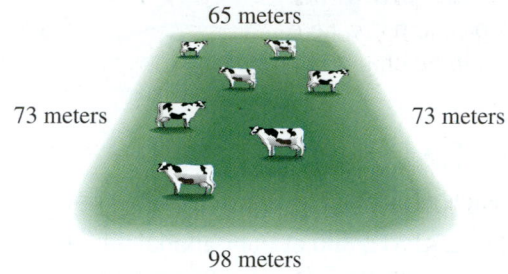

Relating Concepts (Exercises 87–94)   For Individual or Group Work

*Recall the place values of digits discussed in Section 1.1 and **work Exercises 87–94 in order.***

87. Write the largest four-digit number possible using the digits 4, 1, 9, and 2. Use each digit once.

88. Using the digits 4, 1, 9, and 2, write the smallest four-digit number possible. Use each digit once.

89. Write the largest five-digit number possible using the digits 6, 2, and 7. Use each digit at least once.

90. Using the digits 6, 2, and 7, write the smallest five-digit number possible. Use each digit at least once.

91. Write the largest seven-digit number possible using the digits 4, 3, and 9. Use each digit at least twice.

92. Using the digits 4, 3, and 9, write the smallest seven-digit number possible. Use each digit at least twice.

93. Explain your rule or procedure for writing the largest number in Exercise 91.

94. Explain your rule or procedure for writing the smallest number in Exercise 92.

1.3 ▷▷▷ Subtracting Whole Numbers

Suppose you have $9, and you spend $2 for parking. You then have $7 left. There are two different ways of looking at these numbers.

> *As an addition problem:*
>
> $$\$2 \quad + \quad \$7 \quad = \quad \$9$$
>
> Amount   Amount   Original
> spent   left   amount
>
> *As a subtraction problem:*
>
> $$\$9 \quad - \quad \$2 \quad = \quad \$7$$
>
> Original   Subtraction   Amount   Amount
> amount   symbol   spent   left

OBJECTIVES

1. Change addition problems to subtraction and subtraction problems to addition.

2. Identify the minuend, subtrahend, and difference.

3. Subtract when no regrouping (borrowing) is needed.

4. Check subtraction answers by adding.

5. Subtract with regrouping (borrowing).

6. Solve application problems with subtraction.

OBJECTIVE 1 Change addition problems to subtraction and subtraction problems to addition. As shown in the box above, an addition problem can be changed to a subtraction problem and a subtraction problem can be changed to an addition problem.

EXAMPLE 1 Changing Addition Problems to Subtraction

Change each addition problem to a subtraction problem.

(a) $4 + 1 = 5$

Two subtraction problems are possible:

$$5 - 1 = 4 \quad \text{or} \quad 5 - 4 = 1$$

These figures show each subtraction problem.

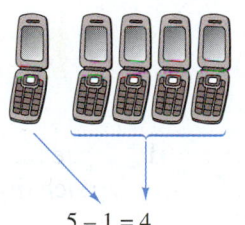 

$5 - 1 = 4$ $5 - 4 = 1$

(b) $8 + 7 = 15$

$$15 - 7 = 8 \quad \text{or} \quad 15 - 8 = 7$$

Work Problem **1** *at the Side.* ▶

EXAMPLE 2 Changing Subtraction Problems to Addition

Change each subtraction problem to an addition problem.

(a) $8 - 3 = 5$

$$8 = 3 + 5$$

It is also correct to write $8 = 5 + 3$.

—— **Continued on Next Page**

1 Write two subtraction problems for each addition problem.

(a) $7 + 2 = 9$

(b) $7 + 4 = 11$

(c) $15 + 22 = 37$

(d) $23 + 55 = 78$

ANSWERS

1. **(a)** $9 - 2 = 7$ or $9 - 7 = 2$
   (b) $11 - 4 = 7$ or $11 - 7 = 4$
   (c) $37 - 22 = 15$ or $37 - 15 = 22$
   (d) $78 - 55 = 23$ or $78 - 23 = 55$

2 Write an addition problem for each subtraction problem.

(a) $7 - 5 = 2$

(b) $9 - 4 = 5$

(c) $21 - 15 = 6$

(d) $58 - 42 = 16$

3 Subtract.

(a)
$$\begin{array}{r} 74 \\ - 43 \\ \hline \end{array}$$

(b)
$$\begin{array}{r} 68 \\ - 24 \\ \hline \end{array}$$

(c)
$$\begin{array}{r} 429 \\ - 318 \\ \hline \end{array}$$

(d)
$$\begin{array}{r} 3927 \\ - 2614 \\ \hline \end{array}$$

(e)
$$\begin{array}{r} 5464 \\ - 324 \\ \hline \end{array}$$

ANSWERS

2. (a) $7 = 5 + 2$ or $7 = 2 + 5$
(b) $9 = 4 + 5$ or $9 = 5 + 4$
(c) $21 = 15 + 6$ or $21 = 6 + 15$
(d) $58 = 42 + 16$ or $58 = 16 + 42$
3. (a) 31 (b) 44 (c) 111
(d) 1313 (e) 5140

(b) $18 - 13 = 5$

$18 = 13 + 5$ or $18 = 5 + 13$

(c) $29 - 13 = 16$

$29 = 13 + 16$ or $29 = 16 + 13$

◄ *Work Problem* **2** *at the Side.*

OBJECTIVE **2** **Identify the minuend, subtrahend, and difference.**
In subtraction, as in addition, the numbers in a problem have names. For example, in the problem $8 - 5 = 3$, the number 8 is the **minuend,** 5 is the **subtrahend,** and 3 is the **difference** or answer.

$$\underset{\underset{\text{Minuend}}{\uparrow}}{8} \quad - \quad \underset{\underset{\text{Subtrahend}}{\uparrow}}{5} \quad = \quad 3 \leftarrow \text{Difference}$$

The answer in subtraction is the difference.

$$\begin{array}{r} 8 \leftarrow \text{Minuend} \\ - 5 \leftarrow \text{Subtrahend} \\ \hline 3 \leftarrow \text{Difference} \end{array}$$

OBJECTIVE **3** **Subtract when no regrouping (borrowing) is needed.** Subtract two numbers by lining up the numbers in columns so the digits in the ones place are in the same column. Next, subtract by columns, starting at the right with the ones column.

EXAMPLE 3 **Subtracting Two Numbers**

Subtract.

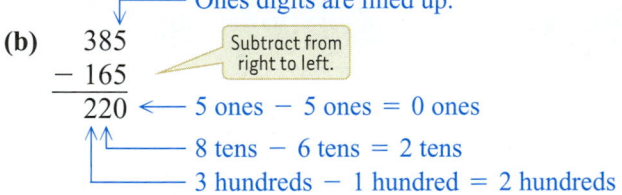

(a)
Ones digits are lined up in the same column.
$$\begin{array}{r} 5\mathbf{3} \\ - 2\mathbf{1} \\ \hline 3\mathbf{2} \end{array}$$
3 ones − 1 one = 2 ones
5 tens − 2 tens = 3 tens

(b)
Ones digits are lined up.
$$\begin{array}{r} 385 \\ - 165 \\ \hline 220 \end{array}$$
Subtract from right to left.
← 5 ones − 5 ones = 0 ones
8 tens − 6 tens = 2 tens
3 hundreds − 1 hundred = 2 hundreds

(c)
$$\begin{array}{r} 9437 \\ - 210 \\ \hline 9227 \end{array}$$
← 7 ones − 0 ones = 7 ones
3 tens − 1 ten = 2 tens
4 hundreds − 2 hundreds = 2 hundreds
9 thousands − 0 thousands = 9 thousands

◄ *Work Problem* **3** *at the Side.*

OBJECTIVE **4** **Check subtraction answers by adding.** Use addition to check your answer to a subtraction problem. For example, check $8 - 3 = 5$ by *adding* 3 and 5.

$$3 + 5 = 8, \quad \text{so} \quad 8 - 3 = 5 \quad \text{is correct.}$$

EXAMPLE 4 **Checking Subtraction by Using Addition**

Use addition to check each answer. If the answer is incorrect, find the correct answer.

(a)
$$\begin{array}{r} 89 \\ -\,47 \\ \hline 42 \end{array}$$

Rewrite as an addition problem, as shown in Example 2.

Subtraction problem
$$\left. \begin{array}{r} 89 \\ -\,47 \\ \hline 42 \\ \hline 89 \end{array} \right\}$$
Addition problem
$$\begin{array}{r} 47 \\ +\,42 \\ \hline 89 \end{array}$$

Because $47 + 42 = 89$, the subtraction was done correctly.

> Avoid errors by checking answers.

(b) $72 - 41 = 21$

Rewrite as an addition problem.

$$72 = 41 + 21$$

But, $41 + 21 = 62$, **not** 72, so the subtraction was done **incorrectly.** Rework the original subtraction to get the correct answer, 31. Then, $41 + 31 = 72$.

(c)
$$\begin{array}{r} 374 \\ -\,141 \\ \hline 233 \end{array} \xleftarrow{\text{Match}}$$
$$141 + 233 = 374$$

The answer checks.

———— *Work Problem* **4** *at the Side.* ▶

OBJECTIVE **5** **Subtract with regrouping (borrowing).** When a digit in the minuend is less than the one directly below it, **regrouping** is necessary (also called **borrowing**).

EXAMPLE 5 **Subtracting with Regrouping**

Subtract 19 from 57.

Write the problem vertically.

$$\begin{array}{r} 57 \\ -\,19 \end{array}$$

In the ones column, 7 is **less** than 9, so in order to subtract, we must regroup 1 ten as 10 ones.

5 tens − 1 ten = 4 tens → 4 17 ← 1 ten = 10 ones, and
10 ones + 7 ones = 17 ones
$$\begin{array}{r} \cancel{5}\;\cancel{7} \\ -\,1\;\;9 \end{array}$$

Now subtract 9 ones from 17 ones in the ones column. Then subtract 1 ten from 4 tens in the tens column.

$$\begin{array}{r} 4\;\;17 \\ \cancel{5}\;\cancel{7} \\ -\,1\;\;9 \\ \hline 3\;\;8 \end{array} \quad \text{Difference}$$

Finally, $57 - 19 = 38$. Check by adding 19 and 38; you should get 57.

———— *Work Problem* **5** *at the Side.* ▶

4 Use addition to determine whether each answer is correct. If incorrect, what should it be?

(a)
$$\begin{array}{r} 76 \\ -\,45 \\ \hline 31 \end{array}$$

(b)
$$\begin{array}{r} 53 \\ -\,22 \\ \hline 21 \end{array}$$

(c)
$$\begin{array}{r} 374 \\ -\,251 \\ \hline 113 \end{array}$$

(d)
$$\begin{array}{r} 7531 \\ -\,4301 \\ \hline 3230 \end{array}$$

5 Subtract.

(a)
$$\begin{array}{r} 58 \\ -\,19 \end{array}$$

(b)
$$\begin{array}{r} 86 \\ -\,38 \end{array}$$

(c)
$$\begin{array}{r} 41 \\ -\,27 \end{array}$$

(d)
$$\begin{array}{r} 863 \\ -\,47 \end{array}$$

(e)
$$\begin{array}{r} 762 \\ -\,157 \end{array}$$

ANSWERS

4. (a) correct **(b)** incorrect; should be 31
(c) incorrect; should be 123 **(d)** correct
5. (a) 39 **(b)** 48 **(c)** 14 **(d)** 816
(e) 605

6 Subtract.

(a) 927
− 43

(b) 675
− 86

(c) 477
− 389

(d) 1417
− 988

(e) 8739
− 3892

EXAMPLE 6 **Subtracting with Regrouping**

Subtract by regrouping when necessary.

(a) 7856
− 137

Regroup 1 ten as 10 ones. ⟶   ⟵ 10 ones + 6 ones = 16 ones

$$
\begin{array}{r}
\overset{4\ \ 16}{7\ 8\ \cancel{5}\ \cancel{6}} \\
-\ \ \ 1\ 3\ 7 \\
\hline
7\ 7\ 1\ 9
\end{array}
$$
Difference

(b) 635
− 546

Regroup 1 ten as 10 ones. ⟶ ⟵ 10 ones + 5 ones = 15 ones

$$
\begin{array}{r}
\overset{2\ \ 15}{6\ \cancel{3}\ \cancel{5}} \\
-\ 5\ 4\ 6 \\
\hline
9
\end{array}
$$
Need to regroup further because 2 is less than 4 in the tens column.

Regroup 1 hundred as 10 tens ⟶ ⟵ 10 tens + 2 tens = 12 tens

$$
\begin{array}{r}
\overset{5\ \ 12\ \ 15}{\cancel{6}\ \cancel{3}\ \cancel{5}} \\
-\ 5\ 4\ 6 \\
\hline
8\ 9
\end{array}
$$
Difference

(c) 647
− 489

$$
\begin{array}{r}
\overset{3\ \ 17}{6\ \cancel{4}\ 7} \\
-\ 4\ 8\ 9 \\
\hline
8
\end{array}
$$
Need to regroup further because 3 is less than 8 in the tens column.

$$
\begin{array}{r}
\overset{5\ \ 13\ \ 17}{\cancel{6}\ \cancel{4}\ \cancel{7}} \\
-\ 4\ 8\ 9 \\
\hline
1\ 5\ 8
\end{array}
$$
Difference

◀ *Work Problem* **6** *at the Side.*

Sometimes a minuend has zeros in some of the positions. In such cases, regrouping may be a little more complicated than what we have shown so far.

EXAMPLE 7 **Regrouping with Zeros**

Subtract.

4607
− 3168

There are no tens that can be regrouped into ones. So you must first regroup 1 hundred as 10 tens.

Regroup 1 hundred as 10 tens. ⟶ ⟵ Write 10 tens.

$$
\begin{array}{r}
\overset{5\ \ 10}{4\ \cancel{6}\ \cancel{0}\ 7} \\
-\ 3\ 1\ 6\ 8
\end{array}
$$

Now we may regroup from the tens position.

Regroup 1 ten as 10 ones; 10 tens − 1 ten = 9 tens.

$$
\begin{array}{r}
\overset{\ \ \ \ \ \ 9}{\overset{5\ \ 10\ \ 17}{4\ \cancel{6}\ \cancel{0}\ \cancel{7}}} \\
-\ 3\ 1\ 6\ 8 \\
\hline
9
\end{array}
$$

10 ones + 7 ones = 17 ones

Continued on Next Page

Complete the problem.

$$
\begin{array}{r}
\overset{\scriptstyle 9}{} \\
5\ \cancel{10}\ 17 \\
4\ \cancel{6}\ \cancel{0}\ \cancel{7} \\
-\ 3\ 1\ 6\ 8 \\
\hline
1\ 4\ 3\ 9
\end{array}
\qquad \text{Difference}
$$

Check by adding 1439 and 3168; you should get 4607.

Work Problem **7** *at the Side.* ▶

EXAMPLE 8   **Regrouping with Zeros**

Subtract.

(a)
$$
\begin{array}{r}
708 \\
-\ 149 \\
\end{array}
$$

Write 10 tens. ⟶ Regroup 1 ten as 10 ones.

Regroup 1 hundred as ⟶ 6 $\overset{9}{\cancel{10}}$ 18 ← 10 ones + 8 ones = 18 ones
10 tens.

$$
\begin{array}{r}
\cancel{7}\ \cancel{0}\ \cancel{8} \\
-\ 1\ 4\ 9 \\
\hline
5\ 5\ 9
\end{array}
$$

(b)
$$
\begin{array}{r}
380 \\
-\ 276 \\
\end{array}
$$

Regroup 1 ten as 10 ones. ⟶   ⟵ Write 10 ones.

$$
\begin{array}{r}
7\ 10 \\
3\ \cancel{8}\ \cancel{0} \\
-\ 2\ 7\ 6 \\
\hline
1\ 0\ 4
\end{array}
$$

(c)
$$
\begin{array}{r}
9000 \\
-\ 6999 \\
\end{array}
$$

$$
\begin{array}{r}
8\ \overset{9}{\cancel{10}}\ \overset{9}{\cancel{10}}\ 10 \\
\cancel{9}\ \cancel{0}\ \cancel{0}\ \cancel{0} \\
-\ 6\ 9\ 9\ 9 \\
\hline
2\ 0\ 0\ 1
\end{array}
$$

> Be extra careful when zeros are involved.

Work Problem **8** *at the Side.* ▶

As we have seen, an answer to a subtraction problem can be checked by adding.

EXAMPLE 9   **Checking Subtraction by Using Addition**

Use addition to check each answer.

Check

(a)
$$
\begin{array}{r}
\mathbf{613} \\
-\ 275 \\
\hline
338
\end{array}
\qquad
\begin{array}{r}
275 \\
+\ 338 \\
\hline
\mathbf{613}
\end{array}
$$

Match   Correct

Continued on Next Page

---

9 Use addition to check each answer. If the answer is incorrect, find the correct answer.

(a) 357
    − 168
    ───
     189

(b) 570
    − 328
    ───
     252

(c) 14,726
    − 8 839
    ─────
     5 887

(b)   1915
     − 1635   Match
     ─────
      280

Check
 1635
+ 280
─────
 1915   Correct

(c)   15,803
     − 7 325   No Match
     ──────
      8 578

Check
 7 325
+ 8 578
──────
 15,903   Error

> It's always a good idea to check your work.

Rework the original problem to get the correct answer, 8478. Then, 7325 + 8478 **does** give 15,803.

◀ *Work Problem* **9** *at the Side.*

OBJECTIVE 6 **Solve application problems with subtraction.** As shown in the next example, subtraction can be used to solve an application problem.

EXAMPLE 10 **Applying Subtraction Skills**

Use the table to find how much more, on average, a person with an Associate of Arts degree earns each year than a high school graduate.

10 Use the table from Example 10 to find, on average,

(a) how much more a person with a Bachelor's degree earns each year than a person with an Associate of Arts degree.

(b) how much more a person with an Associate of Arts degree earns each year than a person who is not a high school graduate.

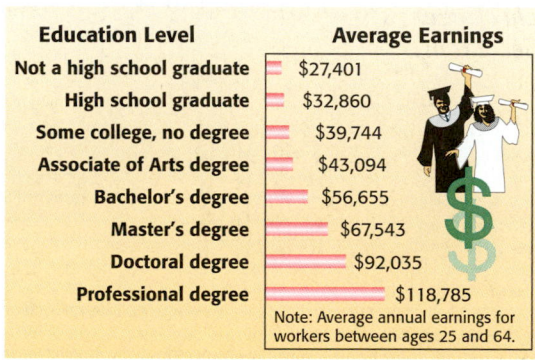

EDUCATION PAYS

The more education adults get, the higher their annual earnings.

Education Level	Average Earnings
Not a high school graduate	$27,401
High school graduate	$32,860
Some college, no degree	$39,744
Associate of Arts degree	$43,094
Bachelor's degree	$56,655
Master's degree	$67,543
Doctoral degree	$92,035
Professional degree	$118,785

Note: Average annual earnings for workers between ages 25 and 64.

Source: U.S. Census Bureau.

Approach   The average earnings for a person with an Associate of Arts degree is $43,094 each year and the average for a high school graduate is $32,860. Find how much more a college graduate earns by subtracting $32,860 from $43,094.

Solution $43,094 ← Associate of Arts degree
              − $32,860 ← High school graduate
              ────────
               $10,234 ← More earnings

> Education pays.

On average, a person with an Associate of Arts degree earns $10,234 more each year than a high school graduate.

◀ *Work Problem* **10** *at the Side.*

1.3 ▶▶▶ **Exercises**

FOR EXTRA HELP

 PRACTICE   WATCH   DOWNLOAD   READ   REVIEW

Work each subtraction problem. Use addition to check each answer. See Examples 3 and 4.

1. 48 − 32	**2.** 17 − 13	**3.** 86 − 53	**4.** 78 − 35	**5.** 77 − 60
6. 87 − 63	**7.** 335 − 122	**8.** 602 − 301	**9.** 552 − 451	**10.** 888 − 215
11. 7352 − 241	**12.** 4420 − 310	**13.** 5546 − 2134	**14.** 1875 − 1362	**15.** 6259 − 4148
16. 9654 − 4323	**17.** 24,392 − 11,232	**18.** 57,921 − 34,801	**19.** 46,253 − 5 143	**20.** 75,904 − 3 702

Use addition to check each subtraction problem. If an answer is not correct, find the correct answer. See Example 4.

21. 54 − 42 ⎯⎯ 12	**22.** 87 − 43 ⎯⎯ 44	**23.** 89 − 27 ⎯⎯ 63	**24.** 47 − 35 ⎯⎯ 13	**25.** 382 − 261 ⎯⎯ 131
26. 754 − 342 ⎯⎯ 412	**27.** 4683 − 3542 ⎯⎯ 1141	**28.** 5217 − 4105 ⎯⎯ 1132	**29.** 8643 − 1421 ⎯⎯ 7212	**30.** 9428 − 3124 ⎯⎯ 6324

Subtract, regrouping when necessary. See Examples 5–8.

31. 75 − 37	**32.** 86 − 28	**33.** 94 − 49	**34.** 68 − 39	**35.** 57 − 38
36. 47 − 29	**37.** 828 − 547	**38.** 916 − 618	**39.** 771 − 252	**40.** 973 − 788

41. 7538
       − 479

42. 5863
       − 1295

43. 9988
       − 2399

44. 3576
       − 1658

45. 38,335
       − 29,476

46. 82,731
       − 14,826

47. 40
       − 37

48. 80
       − 73

49. 60
       − 37

50. 70
       − 27

51. 308
       − 289

52. 600
       − 599

53. 4041
       − 1208

54. 4602
       − 2063

55. 9305
       − 1530

56. 7120
       − 6033

57. 1580
       − 1077

58. 3068
       − 2105

59. 2006
       − 1850

60. 8203
       − 5365

61. 8240
       − 6056

62. 7050
       − 6045

63. 8503
       − 2816

64. 16,004
       − 5 087

65. 80,705
       − 61,667

66. 81,000
       − 55,456

67. 66,000
       − 34,444

68. 77,000
       − 65,308

69. 20,080
       − 13,496

70. 80,056
       − 23,869

Use addition to check each subtraction problem. If an answer is incorrect, find the correct answer. See Example 9.

71. 9428
       − 4509
       ─────
       4919

72. 1671
       − 1325
       ─────
       1346

73. 2548
       − 2278
       ─────
       270

74. 5274
       − 1130
       ─────
       4144

75. 93,758
       − 52,869
       ──────
       40,889

76. 82,357
       − 14,396
       ──────
       68,961

77. 36,778
       − 17,405
       ──────
       19,373

78. 34,821
       − 17,735
       ──────
       17,735

79. An addition problem can be changed to a subtraction problem and a subtraction problem can be changed to an addition problem. Give two examples of each to demonstrate this.

80. Can you use the commutative and the associative properties in subtraction? Explain.

Solve each application problem. See Example 10.

81. A man burns 187 calories during 60 minutes of sitting at a computer while a woman burns 140 calories at the same activity. How many fewer calories does a woman burn than a man? (*Source:* www.cookinglight.com)

82. A woman burns 302 calories during 60 minutes of walking, while a man burns 403 calories doing the same activity. How many more calories does a man burn than a woman? (*Source:* www.cookinglight.com)

83. Toronto's skyline is dominated by the CN Tower, which rises 1821 ft. The Sears Tower in Chicago is 1454 ft high. Find the difference in height between the two structures. (*Source:* Trizec Properties; *World Almanac*.)

1821 ft
d
1454 ft

CN Tower Sears Tower

84. The fastest animal in the world, the peregrine falcon, dives at 217 miles per hour (mph). A Boeing 747 cruises at 580 mph. How much faster is the plane?

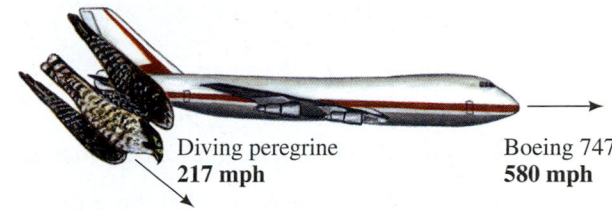

Diving peregrine
217 mph

Boeing 747
580 mph

85. A cruise ship has 1815 passengers. When in port at Grand Cayman, 1348 passengers go ashore for the day while the others remain on the ship. How many passengers remain on the ship?

86. In a recent three-month period there were 81,465 Ford Explorers and 70,449 Jeep Grand Cherokees sold. Which vehicle had greater sales? By how much? (*Source:* J. D. Power and Associates.)

87. Six years ago there were 6970 bridge and lock-tender jobs across the United States. Today there are 3700 that remain. How many of these jobs have been eliminated? (*Source:* Bureau of Labor Statistics.)

88. In 1964, its first year on the market, the Ford Mustang sold for $2500. In 2009, the Ford Mustang sold for $33,065. Find the increase in price. (*Source:* eBay.)

89. Patriot Flag Company manufactured 14,608 U.S. flags and sold 5069. How many flags remain unsold?

90. Eye exams have been given to 14,679 children in the school district. If there are 23,156 students in the school district, how many have not received eye exams?

91. The Jordanos now pay rent of $650 per month. If they buy a house, their housing expense will be $913 per month. How much more will they pay per month if they buy a house?

92. A retired couple who used to receive a Social Security payment of $1479 per month now receives $1568 per month. Find the amount of the monthly increase.

93. On Monday, 11,594 people visited Arcade Amusement Park, and 12,352 people visited the park on Tuesday. Which day had more people visit the park? How many more?

94. In the year 2020 it is predicted that we will need 2,820,000 nurses in the United States, while only 1,810,000 nurses will be available. Find the shortage in the number of nurses. (*Source:* American Hospital Association.)

Solve each application problem. Add or subtract as necessary.

95. A survey of large hotels found that the average salary for a general manager of a deluxe spa and tennis resort is one hundred one thousand, five hundred dollars per year, while spa and tennis directors earn $44,000. How much more does a general manager earn than a spa and tennis director?

96. This year there were 555,800 knee surgeries performed in the United States. The number of knee surgeries performed six years ago was 328,900. How many more of these surgeries were performed this year than six years ago? (*Source:* Agency for Healthcare Research and Quality.)

SUBWAY promotes healthy food choices by offering eight sandwiches that are low in fat. The nutritional information, printed on every SUBWAY napkin, appears below and includes information to answer Exercises 97–100. (Source: SUBWAY.)

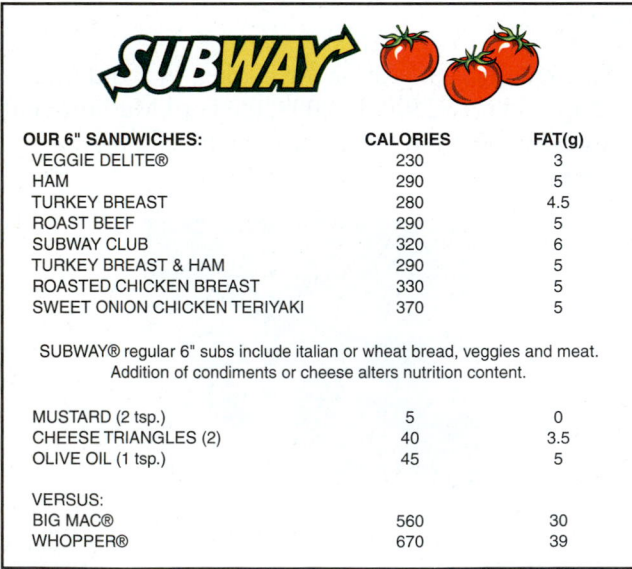

OUR 6" SANDWICHES:	CALORIES	FAT(g)
VEGGIE DELITE®	230	3
HAM	290	5
TURKEY BREAST	280	4.5
ROAST BEEF	290	5
SUBWAY CLUB	320	6
TURKEY BREAST & HAM	290	5
ROASTED CHICKEN BREAST	330	5
SWEET ONION CHICKEN TERIYAKI	370	5

SUBWAY® regular 6" subs include italian or wheat bread, veggies and meat. Addition of condiments or cheese alters nutrition content.

MUSTARD (2 tsp.)	5	0
CHEESE TRIANGLES (2)	40	3.5
OLIVE OIL (1 tsp.)	45	5
VERSUS:		
BIG MAC®	560	30
WHOPPER®	670	39

97. How many fewer calories and grams of fat are in a 6-inch Veggie Delite sandwich than a Big Mac?

98. How many fewer calories and grams of fat are in a 6-inch Turkey Breast and Ham sandwich than a Whopper?

99. Find the total number of calories and grams of fat in a Roasted Chicken Breast sandwich with mustard and olive oil.

100. A customer ate two sandwiches, one with the least calories and one with the most calories. Find the total number of calories and grams of fat in the two sandwiches.

1.4 ▶▶▶ Multiplying Whole Numbers

Suppose we want to know the total number of exercise bicycles available at the gym. The bicycles are arranged in four columns with three stations in each column. Adding the number 3 a total of 4 times gives 12.

$$3 + 3 + 3 + 3 = 12$$

This result can also be shown with a figure.

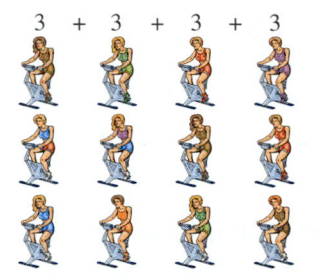

3 + 3 + 3 + 3

3 bicycles in each column

4 columns

OBJECTIVE 1 Identify the parts of a multiplication problem.
Multiplication is a shortcut for repeated addition. In the exercise bicycle example, instead of *adding* $3 + 3 + 3 + 3$ to get 12, we can *multiply* 3 by 4 to get 12. The numbers being multiplied are called **factors.** The answer is called the **product.** For example, the product of 3 and 4 can be written with the symbol $\times$, a raised dot, or parentheses, as follows.

$$\begin{array}{r} 3 \leftarrow \text{Factor (also called } multiplicand) \\ \times\ 4 \leftarrow \text{Factor (also called } multiplier) \\ \hline 12 \leftarrow \text{Product (answer)} \end{array}$$

$$3 \times 4 = 12 \quad or \quad 3 \cdot 4 = 12 \quad or \quad (3)(4) = 12 \quad or \quad 3(4) = 12$$

Work Problem **1** at the Side. ▶

Commutative Property of Multiplication   Multiply numbers in any order.
By the **commutative property of multiplication,** the product (answer) remains the same when the order of the factors is changed. For example,

$$3 \times 5 = 15 \quad and \quad 5 \times 3 = 15.$$

CAUTION
Recall that addition also has a commutative property. For example, $4 + 2$ gives the same sum as $2 + 4$. Subtraction, however, is **not** commutative.

EXAMPLE 1   Multiplying Two Numbers

Multiply.

(a) $3 \times 4 = 12$

(b) $6 \cdot 0 = 0$ (The product of any number and 0 is 0; if you give no money to each of 6 relatives, you give no money.)

(c) $4(8) = 32$

— **Continued on Next Page**

Continued on Next Page

OBJECTIVES

1 Identify the parts of a multiplication problem.

2 Do chain multiplication.

3 Multiply by single-digit numbers.

4 Use multiplication shortcuts for numbers ending in zeros.

5 Multiply by numbers having more than one digit.

6 Solve application problems with multiplication.

1 Identify the factors and the product in each multiplication problem.

(a) $8 \times 5 = 40$

(b) $6(4) = 24$

(c) $7 \cdot 6 = 42$

(d) $(3)(9) = 27$

ANSWERS
1. (a) factors: 8, 5; product: 40
   (b) factors: 6, 4; product: 24
   (c) factors: 7, 6; product: 42
   (d) factors: 3, 9; product: 27

2 Multiply.

(a) 7×4

(b) 0×9

(c) $8(5)$

(d) $6 \cdot 5$

(e) $(3)(8)$

3 Multiply.

(a) $3 \times 2 \times 5$

(b) $4 \cdot 7 \cdot 1$

(c) $(8)(3)(0)$

You may want to review this multiplication table.

Multiplication Table

×	1	2	3	4	5	6	7	8	9
1	1	2	3	4	5	6	7	8	9
2	2	4	6	8	10	12	14	16	18
3	3	6	9	12	15	18	21	24	27
4	4	8	12	16	20	24	28	32	36
5	5	10	15	20	25	30	35	40	45
6	6	12	18	24	30	36	42	48	54
7	7	14	21	28	35	42	49	56	63
8	8	16	24	32	40	48	56	64	72
9	9	18	27	36	45	54	63	72	81

Know the "Times Table."

◀ *Work Problem* **2** *at the Side.*

OBJECTIVE 2 Do chain multiplication. Some multiplications involve more than two factors.

Associative Property of Multiplication

By the **associative property of multiplication,** grouping the factors differently does not change the product.

EXAMPLE 2 **Multiplying Three Numbers**

Multiply $2 \times 3 \times 5$.

$$(2 \times 3) \times 5 \quad \text{Parentheses show what to do first.}$$
$$6 \quad \times 5 = 30$$

Also,

$$2 \times (3 \times 5)$$
$$2 \times \quad 15 = 30$$

Either grouping results in the same product.

◀ *Work Problem* **3** *at the Side.*

Calculator Tip The calculator approach to Example 2 uses chain calculations.

$$2 \; \boxed{\times} \; 3 \; \boxed{\times} \; 5 \; \boxed{=} \; \mathbf{30}$$

A problem with more than two factors, such as the one in Example 2, is called a **chain multiplication** problem.

ANSWERS

2. (a) 28 **(b)** 0 **(c)** 40 **(d)** 30 **(e)** 24
3. (a) 30 **(b)** 28 **(c)** 0

OBJECTIVE 3 Multiply by single-digit numbers. Regrouping may be needed in multiplication problems with larger factors.

EXAMPLE 3 Multiplying with Regrouping

Multiply.

(a)   53
        × 4

Start by multiplying in the ones column.

     1 ←
    53 ───────┐   Write 1 ten in the tens column.
  × 4   4 × 3 = **12** ones
     2 ←───────┘ Write 2 ones in the ones column.

Next, multiply 4 times 5 tens.

     1
    5 3
  × 4   4 × 5 tens = **20** tens
     2

Add the 1 ten that was written at the top of the tens column.

     1 ─ ─ ─ ─ ┐
    5 3   ¦
  × 4   ↓
    212   20 tens + 1 ten = **21** tens
     ↑_____¦

(b)   724
       × 5

Work as shown.

     12
    724   ┌─────────────┐
  × 5   │ Use regrouping here. │
   3620 ←─ 5 × 4 = **20** ones; write 0 ones; write **2** tens in the
    ↑↑   tens column.
    ││└── 5 × 2 = **10** tens; add the 2 regrouped tens to get 12 tens;
    │   write 2 tens; write 1 hundred in the hundreds column.
    └──── 5 × 7 hundreds = **35** hundreds; add the 1 regrouped
              hundred to get 36 hundreds.

Work Problem **4** *at the Side.* ▶

OBJECTIVE 4 Use multiplication shortcuts for numbers ending in zeros. The product of two whole number factors is also called a **multiple** of either factor. For example, since 4 • 2 = 8, the whole number 8 is a multiple of both 4 and 2. *Multiples of 10 are very useful when multiplying.* A **multiple of 10** is a whole number that ends in 0, such as 10, 20, or 30; 100, 200, or 300; 1000, 2000, or 3000; and so on. There is a short way to multiply by these multiples of 10. Look at the following examples.

$$26 \times 1 = 26$$
$$26 \times 10 = 26\textbf{0}$$
$$26 \times 100 = 26\textbf{00}$$
$$26 \times 1000 = 26\textbf{,000}$$

Do you see a pattern? These examples suggest the rule that follows.

4 Multiply.

(a)   53
        × 5

(b)   79
        × 0

(c)   758
        × 8

(d)   2831
        × 7

(e)   4714
        × 8

5 Multiply.

(a) 63×10

(b) 305×100

(c) 714×1000

6 Multiply.

(a) 16×50

(b) 73×400

(c) $\quad 180$
$\underline{\times \ \ 30}$

(d) $\quad 4200$
$\underline{\times \ \ \ 80}$

(e) $\quad 800$
$\underline{\times \ 600}$

> **Multiplying by Multiples of 10**
> To multiply a whole number by 10, 100, or 1000, attach one, two, or three zeros, respectively, to the right of the whole number.

EXAMPLE 4 Using Multiples of 10 to Multiply

Multiply.

(a) $59 \times 10 = 590$
      └── Attach 0.

(b) $74 \times 100 = 7400$
      └── Attach 00.

(c) $803 \times 1000 = 803,000$ ← Attach 000.

◀ *Work Problem* **5** *at the Side.*

You can also find the product of other multiples of 10 by attaching zeros.

EXAMPLE 5 Using Multiples of 10 to Multiply

Multiply.

(a) 75×3000
Multiply 75 by 3, and then attach three zeros.

$$75 \times 3000 = 225,000$$

$\quad 75$
$\underline{\times \ \ 3}$
$\quad 225$ ── └── Attach 000.

Use useful shortcuts.

(b) 150×70
Multiply 15 by 7, and then attach two zeros.

$$150 \times 70 = 10,500 \leftarrow \text{Attach 00.}$$

$\quad 15$
$\underline{\times \ \ 7}$
$\quad 105$

◀ *Work Problem* **6** *at the Side.*

OBJECTIVE 5 **Multiply by numbers having more than one digit.**
The next example shows multiplication when both factors have more than one digit.

EXAMPLE 6 Multiplying with More Than One Digit

Multiply 46 and 23.

First multiply 46 by 3.

$\quad \overset{1}{46}$ Regrouping is needed here.
$\underline{\times \ \ 3}$
$\quad 138$ ← $46 \times 3 = 138$

Continued on Next Page

Now multiply 46 by 20.

$$
\begin{array}{r}
\overset{1}{4}6 \\
\times\ 20 \\
\hline
920 \leftarrow 46 \times 20 = 920
\end{array}
$$

Add the results.

$$
\begin{array}{r}
46 \\
\times\ 23 \\
\hline
138 \leftarrow 46 \times 3 \\
+\ 920 \leftarrow 46 \times 20 \\
\hline
1058
\end{array}
$$

⎣— Add.

Both 138 and 920 are called **partial products.** To save time, the 0 in 920 is usually not written.

$$
\begin{array}{r}
46 \\
\times\ 23 \\
\hline
138 \\
92 \\
\hline
1058
\end{array}
$$

{ 0 not written. Be very careful to place the 2 in the tens column.

Work Problem **7** *at the Side.* ▶

EXAMPLE 7 **Using Partial Products**

Multiply.

(a)
$$
\begin{array}{r}
233 \\
\times\ 132 \\
\hline
466 \\
699 \\
233 \\
\hline
30{,}756 \leftarrow \text{Product}
\end{array}
$$

(Tens lined up)
(Hundreds lined up)

Be certain to align numbers in columns.

(b)
$$
\begin{array}{r}
538 \\
\times\ 46
\end{array}
$$

First multiply by 6.

$$
\begin{array}{r}
\overset{24}{5}38 \\
\times\ 46 \\
\hline
3228
\end{array}
$$
← Regrouping is needed here.

Now multiply by 4, being careful to line up the tens.

$$
\begin{array}{r}
\overset{13}{\overset{24}{5}}38 \\
\times\ 46 \\
\hline
3228 \\
2152 \\
\hline
24{,}748
\end{array}
$$

⎤— Finally, add the partial products.

Work Problem **8** *at the Side.* ▶

---

7 Complete each multiplication.

(a)
$$
\begin{array}{r}
35 \\
\times\ 54 \\
\hline
140 \\
175
\end{array}
$$

(b)
$$
\begin{array}{r}
76 \\
\times\ 49 \\
\hline
684 \\
304
\end{array}
$$

8 Multiply.

(a)
$$
\begin{array}{r}
52 \\
\times\ 16
\end{array}
$$

(b)
$$
\begin{array}{r}
81 \\
\times\ 49
\end{array}
$$

(c)
$$
\begin{array}{r}
75 \\
\times\ 63
\end{array}
$$

(d)
$$
\begin{array}{r}
234 \\
\times\ 73
\end{array}
$$

(e)
$$
\begin{array}{r}
835 \\
\times\ 189
\end{array}
$$

ANSWERS

7. **(a)** 1890 **(b)** 3724
8. **(a)** 832 **(b)** 3969 **(c)** 4725
  (d) 17,082 **(e)** 157,815

9 Multiply.

(a)   28
     × 60

(b)   728
    × 50

(c)   562
    × 109

(d)   3526
    × 6002

10 Find the total cost of the following items.

(a) 289 redwood planters at $12 per planter

(b) 58 compound miter saws priced at $129 each

(c) 12 delivery vans at $28,300 per van

When 0 appears in the multiplier, be sure to move the partial products to the left to account for the position held by the 0.

EXAMPLE 8   **Multiplying with Zeros**

Multiply.

(a)   1 3 7
     ×   3 0 6
        8 2 2
        0 0 0   (Tens lined up)
      4 1 1   (Hundreds lined up)
     4 1,9 2 2

(b)
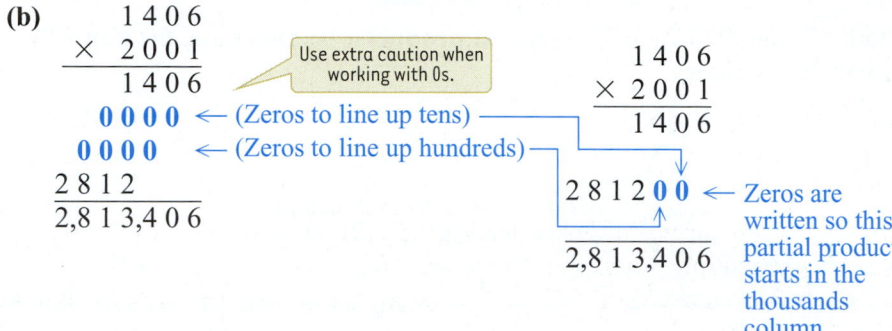

        1 4 0 6
     × 2 0 0 1
        1 4 0 6   Use extra caution when working with 0s.
        0 0 0 0 ← (Zeros to line up tens)
        0 0 0 0 ← (Zeros to line up hundreds)
      2 8 1 2
      2,8 1 3,4 0 6

        1 4 0 6
     × 2 0 0 1
        1 4 0 6
      2 8 1 2 0 0 ← Zeros are written so this partial product starts in the thousands column.
      2,8 1 3,4 0 6

> **Note**
>
> In Example 8(b) in the alternative method on the right, zeros were inserted so that thousands were placed in the thousands column. This is a commonly used shortcut.

◀ **Work Problem** **9** **at the Side.**

OBJECTIVE **6** **Solve application problems with multiplication.** The next example shows how multiplication can be used to solve an application problem.

EXAMPLE 9   **Applying Multiplication Skills**

Find the total cost of 53 portable DVD players priced at $78 each.

Approach   To find the cost of all the DVD players, multiply the number of players (53) by the cost of one player ($78).

Solution   Multiply 53 by 78.

            53
         × 78
           424
          371
         $4134

The total cost of the portable DVD players is $4134.

⊞ **Calculator Tip**   If you are using a calculator for Example 9, you will do this calculation.

    53 ⊗ 78 ⊜ 4134

◀ **Work Problem** **10** **at the Side.**

1.4 ▶▶▶ Exercises

Work each chain multiplication. See Example 2.

1. $2 \times 6 \times 2$

2. $3 \times 5 \times 3$

3. $8 \times 6 \times 1$

4. $2 \times 4 \times 5$

5. $7 \cdot 8 \cdot 0$

6. $9 \cdot 0 \cdot 5$

7. $4 \cdot 1 \cdot 6$

8. $1 \cdot 5 \cdot 7$

9. $(4)(5)(2)$

10. $(4)(1)(9)$

11. $(3)(0)(7)$

12. $(0)(9)(4)$

13. Explain in your own words the commutative property of multiplication. How do the commutative properties of addition and multiplication compare to each other?

14. Explain in your own words the associative property of multiplication. How do the associative properties of addition and multiplication compare to each other?

Multiply. See Example 3.

15. $\begin{array}{r} 35 \\ \times\ 6 \\ \hline \end{array}$

16. $\begin{array}{r} 53 \\ \times\ 7 \\ \hline \end{array}$

17. $\begin{array}{r} 34 \\ \times\ 7 \\ \hline \end{array}$

18. $\begin{array}{r} 76 \\ \times\ 5 \\ \hline \end{array}$

19. $\begin{array}{r} 642 \\ \times\ \ \ 5 \\ \hline \end{array}$

20. $\begin{array}{r} 472 \\ \times\ \ \ 4 \\ \hline \end{array}$

21. $\begin{array}{r} 624 \\ \times\ \ \ 3 \\ \hline \end{array}$

22. $\begin{array}{r} 852 \\ \times\ \ \ 7 \\ \hline \end{array}$

23. $\begin{array}{r} 2153 \\ \times\ \ \ \ 4 \\ \hline \end{array}$

24. $\begin{array}{r} 1137 \\ \times\ \ \ \ 3 \\ \hline \end{array}$

25. $\begin{array}{r} 2521 \\ \times\ \ \ \ 4 \\ \hline \end{array}$

26. $\begin{array}{r} 2544 \\ \times\ \ \ \ 3 \\ \hline \end{array}$

27. $\begin{array}{r} 2561 \\ \times\ \ \ \ 8 \\ \hline \end{array}$

28. $\begin{array}{r} 7326 \\ \times\ \ \ \ 5 \\ \hline \end{array}$

29. $\begin{array}{r} 36{,}921 \\ \times\ \ \ \ \ \ \ 7 \\ \hline \end{array}$

30. $\begin{array}{r} 28{,}116 \\ \times\ \ \ \ \ \ \ 4 \\ \hline \end{array}$

Multiply. See Examples 4 and 5.

31. $\begin{array}{r} 40 \\ \times\ 7 \\ \hline \end{array}$

32. $\begin{array}{r} 20 \\ \times\ 7 \\ \hline \end{array}$

33. $\begin{array}{r} 80 \\ \times\ 6 \\ \hline \end{array}$

34. $\begin{array}{r} 70 \\ \times\ 5 \\ \hline \end{array}$

35. $\begin{array}{r} 740 \\ \times\ \ \ 3 \\ \hline \end{array}$

36. $\begin{array}{r} 400 \\ \times\ \ \ 8 \\ \hline \end{array}$

37. $\begin{array}{r} 600 \\ \times\ \ \ 6 \\ \hline \end{array}$

38. $\begin{array}{r} 860 \\ \times\ \ \ 7 \\ \hline \end{array}$

39. $\begin{array}{r} 125 \\ \times\ \ 30 \\ \hline \end{array}$

40. $\begin{array}{r} 246 \\ \times\ \ 50 \\ \hline \end{array}$

41. 1635
     × 40

42. 7311
     × 50

☗ **43.** 900
     × 300

44. 400
     × 700

45. 43,000
     × 2 000

46. 11,000
     × 9 000

47. 970 · 50

48. 730 · 40

49. 800 · 900

50. 850 · 700

51. 9700 · 200

52. 10,050 · 300

Multiply. See Examples 6–8.

53. 28
     × 17

54. 16
     × 34

☗ **55.** 75
     × 32

56. 82
     × 32

57. 83
     × 45

58. (75)(21)

59. (58)(41)

60. (82)(67)

61. (67)(92)

62. (26)(33)

63. (28)(564)

64. (58)(312)

65. (619)(35)

66. (681)(47)

67. (55)(286)

68. 286
     × 574

69. 735
     × 112

70. 621
     × 415

71. 538
     × 342

72. 3228
     × 751

73. 9352
     × 264

74. 528
     × 106

☗ **75.** 215
     × 307

76. 218
     × 106

77. 428
     × 201

78. 3706
     × 208

79. 6310
     × 3078

80. 3533
     × 5001

81. 2195
     × 1038

82. 1502
     × 2009

83. A classmate of yours is not clear on how to use a shortcut to multiply a whole number by 10, by 100, or by 1000. Write a short note explaining how this can be done.

84. Show two ways to multiply when a 0 is in the multiplier. Use the problem 291×307 to show this.

Solve each application problem. See Example 9.

85. Carepanian Company, a health care supplier, purchased 300 cartons of Thera Bond Gym Balls. If there are 10 balls in each carton, find the total number of balls purchased.

86. A medical supply house has 30 bottles of vitamin C tablets, with each bottle containing 500 tablets. Find the total number of vitamin C tablets in the supply house.

87. Annie's Restaurant buys 15 cartons of eggs. If each carton contains 36 eggs, find the number of eggs purchased.

88. A hummingbird's wings beat about 65 times per second. How many times do the hummingbird's wings beat in 30 seconds?

89. The average amount of water used per person each day in the United States is 66 gallons. How much water does the average person use in one year? (1 year = 365 days). (*Source:* Okfam.)

90. Squid are being hauled out of the Santa Barbara Channel by the ton. They are then processed, renamed calamari, and exported. Last night 27 fishing boats each hauled out 40 tons of squid. What was the total catch for the night? (*Source: Santa Barbara News Press.*)

Find the total cost of the following items. See Examples 7–9.

91. 75 first-aid kits
at $8 per kit

92. 38 gardeners at
$64 per day

93. 65 rebuilt alternators at
$24 per alternator

94. 62 wheelchair cushions
at $44 per cushion

95. 206 desktop computers at
$548 per computer

96. 520 printers at
$219 per printer

Multiply.

97. $21 \cdot 43 \cdot 56$

98. $(600)(8)(75)(40)$

Use addition, subtraction, or multiplication to solve each application problem.

99. In a forest-planting project, 450 trees are planted on each acre. Find the number of trees needed to plant 85 acres.

100. The largest living land mammal is the African elephant, and the largest mammal of all time is the blue whale. An African elephant weighs 15,225 pounds and a blue whale weighs 28 times that amount. Find the weight of the blue whale.

101. New York City has a population of 8,214,426, the largest in the country. Boston, in twenty-second place, has a population of 590,763. How many more people live in New York City than in Boston? (*Source:* Analysis of Census Bureau Estimates.)

102. Los Angeles, the second largest city in the country, has a population of 3,849,378. Dallas, at ninth largest, has a population of 1,232,940. Find the difference in the population of these two cities. (*Source:* Analysis of Census Bureau Estimates.)

103. A medical center purchased six laptop computers at $880 each, six printers at $235 each, and five fax machines at $140 each. Find the total cost of this equipment.

104. A motorcycle club traveled 640 miles on Saturday, 438 miles on Sunday, and 535 miles on Monday. Find the total number of miles traveled.

Relating Concepts (Exercises 105–114) For Individual or Group Work

Work Exercises 105–114 in order.

105. Add.

(a) $189 + 263$

(b) $263 + 189$

106. Your answers to Exercise 105(a) and (b) should be the same. This shows that the order of numbers in an addition problem does not change the sum. This is known as the _____ property of addition.

107. Add. Recall that parentheses show you what to do first.

(a) $(65 + 81) + 135$

(b) $65 + (81 + 135)$

108. Since the answers to Exercise 107(a) and (b) are the same, we see that grouping the numbers differently when adding does not change the sum. This is known as the _____ property of addition.

109. Multiply.

(a) 220×72

(b) 72×220

110. Since the answers to Exercise 109(a) and (b) are the same, we see that the product remains the same when the order of the factors is changed. This is known as the _____ property of multiplication.

111. Multiply. Recall that parentheses tell you what to do first.

(a) $(26 \times 18) \times 14$

(b) $26(18 \times 14)$

112. Since the answers to Exercise 111(a) and (b) are the same, we see that grouping the numbers differently when multiplying does not change the product. This is known as the _____ property of multiplication.

113. Do the commutative and associative properties apply to subtraction? Explain your answer using several examples.

114. Do you think that the commutative and associative properties will apply to division? Explain your answer using several examples.

1.5 ▶▶▶ Dividing Whole Numbers

Suppose the cost of lunch at a SUBWAY is $18 and is to be divided equally by three friends. Each person would pay $6, as shown here.

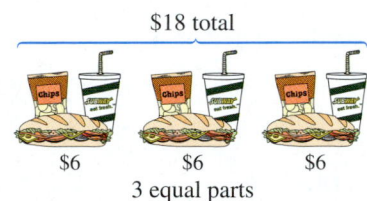

$18 total

$6 $6 $6
3 equal parts

OBJECTIVE 1 Write division problems in three ways.   Just as $3 \cdot 6$, 3×6, and $(3)(6)$ are different ways of indicating the multiplication of 3 and 6, there are several ways to write 18 divided by 3.

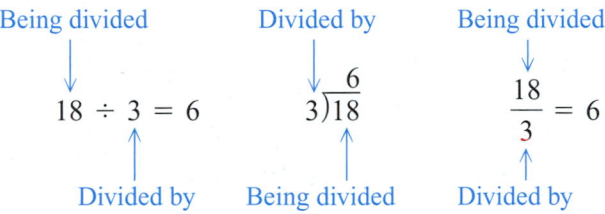

Being divided   Divided by   Being divided

$$18 \div 3 = 6 \qquad 3\overline{)18}^{\,6} \qquad \frac{18}{3} = 6$$

Divided by   Being divided   Divided by

We will use all three division symbols, $\div$, $\overline{)}$, and —. In courses such as algebra, a slash symbol, /, or a fraction bar, —, is most often used.

EXAMPLE 1 Using Division Symbols

Write each division problem using two other symbols.

(a) $18 \div 6 = 3$
This division can also be written as shown below.

$$6\overline{)18}^{\,3} \quad \text{or} \quad \frac{18}{6} = 3$$

Remember the three division symbols.

(b) $\dfrac{15}{5} = 3$

$$15 \div 5 = 3 \quad \text{or} \quad 5\overline{)15}^{\,3}$$

(c) $5\overline{)20}^{\,4}$

$$20 \div 5 = 4 \quad \text{or} \quad \frac{20}{5} = 4$$

Work Problem **1** *at the Side.* ▶

OBJECTIVE 2 Identify the parts of a division problem.   In division, the number being divided is the **dividend,** the number divided by is the **divisor,** and the answer is the **quotient.**

$$\textbf{dividend} \div \textbf{divisor} = \textbf{quotient}$$

$$\textbf{divisor}\overline{)\textbf{dividend}}^{\,\textbf{quotient}} \qquad \frac{\textbf{dividend}}{\textbf{divisor}} = \textbf{quotient}$$

OBJECTIVES

1. Write division problems in three ways.

2. Identify the parts of a division problem.

3. Divide 0 by a number.

4. Recognize that a number cannot be divided by 0.

5. Divide a number by itself.

6. Divide a number by 1.

7. Use short division.

8. Use multiplication to check the answer to a division problem.

9. Use tests for divisibility.

---

1 Write each division problem using two other symbols.

(a) $24 \div 6 = 4$

(b) $9\overline{)36}^{\,4}$

(c) $48 \div 6 = 8$

(d) $\dfrac{42}{6} = 7$

ANSWERS

1. **(a)** $6\overline{)24}^{\,4}$ and $\dfrac{24}{6} = 4$

   (b) $36 \div 9 = 4$ and $\dfrac{36}{9} = 4$

   (c) $6\overline{)48}^{\,8}$ and $\dfrac{48}{6} = 8$

   (d) $6\overline{)42}^{\,7}$ and $42 \div 6 = 7$

2 Identify the dividend, divisor, and quotient.

(a) $15 \div 3 = 5$

(b) $18 \div 6 = 3$

(c) $\dfrac{28}{7} = 4$

(d) $9\overline{)27}$ ³

EXAMPLE 2 Identifying the Parts of a Division Problem

Identify the dividend, divisor, and quotient.

(a) $35 \div 7 = 5$

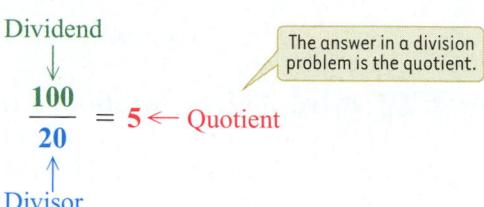

$35 \div 7 = 5 \leftarrow$ Quotient
Dividend   Divisor

(b) $\dfrac{100}{20} = 5$

Dividend
$$\dfrac{100}{20} = 5 \leftarrow \text{Quotient}$$
Divisor

The answer in a division problem is the quotient.

(c) $8\overline{)72}$ ⁹

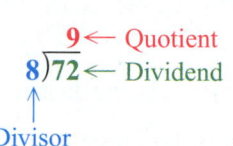

$8\overline{)72}$
$9 \leftarrow$ Quotient
$72 \leftarrow$ Dividend
Divisor

◀ *Work Problem* **2** *at the Side.*

---

3 Divide.

(a) $0 \div 5$

(b) $\dfrac{0}{9}$

(c) $\dfrac{0}{24}$

(d) $37\overline{)0}$

OBJECTIVE 3 Divide 0 by a number. If no money, or $0, is divided equally among five people, each person gets $0. The general rule for dividing 0 follows.

> **Dividing 0 by a Number**
> The number **0** divided by any nonzero number is **0**.

EXAMPLE 3 Dividing 0 by a Number

Divide.

(a) $0 \div 12 = 0$

(b) $0 \div 1728 = 0$

(c) $\dfrac{0}{375} = 0$ Zero divided by any nonzero number is zero.

(d) $129\overline{)0}$ ⁰

◀ *Work Problem* **3** *at the Side.*

Just as a subtraction such as $8 - 3 = 5$ can be written as the addition $8 = 3 + 5$, any division can be written as a multiplication. For example, $12 \div 3 = 4$ can be written as

$$3 \times 4 = 12 \quad \text{or} \quad 4 \times 3 = 12.$$

EXAMPLE 4   **Changing Division Problems to Multiplication**

Change each division problem to a multiplication problem.

(a) $\dfrac{20}{4} = 5$ becomes $4 \cdot 5 = 20$ or $5 \cdot 4 = 20$

(b) $8\overline{)48}$ (quotient 6)   becomes $8 \cdot 6 = 48$ or $6 \cdot 8 = 48$

(c) $72 \div 9 = 8$ becomes $9 \cdot 8 = 72$ or $8 \cdot 9 = 72$

Work Problem **4** *at the Side.* ▶

OBJECTIVE **4** **Recognize that a number cannot be divided by 0.**
Division of any number by 0 cannot be done. To see why, try to find

$$9 \div 0 = ?$$

As we have just seen, any division problem can be converted to a multiplication problem so that

divisor • quotient = dividend.

If you convert the preceding problem to its multiplication counterpart, it reads as follows.

$$0 \cdot ? = 9$$

You already know that 0 times any number must always be 0. Try any number you like to replace the "**?**" and you'll aways get 0 instead of 9. Therefore, the division problem $9 \div 0$ cannot be done. Mathematicians say it is *undefined* and have agreed never to divide by 0. However, $0 \div 9$ *can* be done. Check by rewriting it as a multiplication problem.

$$0 \div 9 = 0 \quad \text{because} \quad 9 \cdot 0 = 0 \text{ is true.}$$

Dividing a Number by 0

Since dividing any number by 0 cannot be done, we say that division by **0** is *undefined.* It is impossible to compute an answer.

EXAMPLE 5   **Dividing Numbers by 0**

All the following are undefined.

(a) $\dfrac{6}{0}$ is undefined.

(b) $0\overline{)8}$ is undefined.

(c) $18 \div 0$ is undefined. — You **cannot** divide a number by zero.

(d) $\dfrac{3}{0}$ is undefined.

---

4 Write each division problem as a multiplication problem.

(a) $5\overline{)15}$ (quotient 3)

(b) $\dfrac{32}{4} = 8$

(c) $45 \div 9 = 5$

---

5 Divide. If the division is not possible, write "undefined."

(a) $\dfrac{4}{0}$

(b) $\dfrac{0}{4}$

(c) $0\overline{)36}$

(d) $36\overline{)0}$

(e) $100 \div 0$

(f) $0 \div 100$

6 Divide.

(a) $8 \div 8$

(b) $15\overline{)15}$

(c) $\dfrac{37}{37}$

ANSWERS

5. **(a)** undefined
   (b) 0
   (c) undefined
   (d) 0
   (e) undefined
   (f) 0
6. all 1

Division Involving 0

$$0 \div \text{nonzero number} = 0 \quad \text{and} \quad \frac{0}{\text{nonzero number}} = 0$$

but

$$\text{nonzero number} \div 0 \quad \text{and} \quad \frac{\text{nonzero number}}{0} \text{ are } \textbf{undefined.}$$

CAUTION
When 0 is the divisor in a problem, you write "undefined" as the answer. Never divide by 0.

◀ *Work Problem* **5** *at the Side.*

⊞ **Calculator Tip** Try these two problems on your calculator. Jot down your answers.

$$9 \; ⊕ \; 0 \; ⊜ \; \underline{\hspace{1.5cm}} \qquad 0 \; ⊕ \; 9 \; ⊜ \; \underline{\hspace{1.5cm}}$$

When you try to divide by 0, the calculator cannot do it, so it shows the word "Error" or the letter "E" (for error) in the display. But, when you divide 0 by 9 the calculator displays 0, which is the correct answer.

OBJECTIVE 5 Divide a number by itself. What happens when a number is divided by itself? For example, what is $4 \div 4$ or $97 \div 97$?

Dividing a Number by Itself
Any *nonzero* number divided by itself is **1**.

EXAMPLE 6 **Dividing a Nonzero Number by Itself**

Divide.

(a) $16 \div 16 = 1$

(b) $32\overline{)32}$ with quotient 1

(c) $\dfrac{57}{57} = 1$ ◁ A nonzero number divided by itself is 1.

◀ *Work Problem* **6** *at the Side.*

OBJECTIVE 6 Divide a number by 1. What happens when a number is divided by 1? For example, what is $5 \div 1$ or $86 \div 1$?

Dividing a Number by 1
Any number divided by 1 is itself.

EXAMPLE 7 Dividing Numbers by 1

Divide.

(a) $5 \div 1 = 5$

(b) $1)\overline{26}$ with 26 on top

(c) $\dfrac{41}{1} = 41$ ◁ A number divided by one is itself.

Work Problem **7** *at the Side.* ▶

OBJECTIVE 7 Use short division. **Short division** is a method of dividing a number by a one-digit divisor.

EXAMPLE 8 Using Short Division

Divide: $3)\overline{96}$.

First, divide 9 by 3.

$$\begin{array}{r} 3 \\ 3)\overline{96} \end{array} \leftarrow \frac{9}{3} = 3$$

Next, divide 6 by 3.

$$\begin{array}{r} 32 \\ 3)\overline{96} \end{array} \leftarrow \frac{6}{3} = 2$$

Work Problem **8** *at the Side.* ▶

When two numbers do not divide exactly, the leftover portion is called the **remainder.** The remainder must always be less than the divisor.

EXAMPLE 9 Using Short Division with a Remainder

Divide 147 by 4.
Write the problem.

$$4)\overline{147}$$

Because 1 cannot be divided by 4, divide 14 by 4. Notice that the 3 is placed over the 4 in 14.

$$\begin{array}{r} 3 \\ 4)\overline{14^27} \end{array} \qquad \frac{14}{4} = 3 \text{ with 2 left over}$$

Next, divide 27 by 4. The final number left over is the remainder. Use **R** to indicate the remainder, and write the remainder to the side.

$$\begin{array}{r} 3\ 6\ \mathbf{R3} \\ 4)\overline{14^27} \end{array} \qquad \frac{27}{4} = 6 \text{ with 3 left over}$$

Work Problem **9** *at the Side.* ▶

7 Divide.

(a) $9 \div 1$

(b) $1)\overline{18}$

(c) $\dfrac{43}{1}$

8 Divide.

(a) $2)\overline{24}$

(b) $3)\overline{93}$

(c) $4)\overline{88}$

(d) $2)\overline{624}$

9 Divide.

(a) $2)\overline{125}$

(b) $3)\overline{215}$

(c) $4)\overline{538}$

(d) $\dfrac{819}{5}$

ANSWERS

7. (a) 9 **(b)** 18 **(c)** 43
8. (a) 12 **(b)** 31 **(c)** 22 **(d)** 312
9. (a) 62 **R1** **(b)** 71 **R2** **(c)** 134 **R2**
   (d) 163 **R4**

10 Divide.

(a) $4\overline{)523}$

(b) $\dfrac{515}{7}$

(c) $3\overline{)1885}$

(d) $6\overline{)1415}$

EXAMPLE 10 **Dividing with a Remainder**

Divide 1809 by 7.

Divide 18 by 7.

$$7\overline{)18^409}\qquad \dfrac{18}{7}=2\text{ with 4 left over}$$

Divide 40 by 7.

$$7\overline{)18^40^59}\qquad \dfrac{40}{7}=5\text{ with 5 left over}$$

Divide 59 by 7.

$$7\overline{)18^40^59}\ \mathbf{R}3\qquad \dfrac{59}{7}=8\text{ with 3 left over}$$

> The remainder must be less than the divisor.

◀ Work Problem 10 at the Side.

> **Note**
> Short division takes practice but is useful in many situations.

OBJECTIVE 8 **Use multiplication to check the answer to a division problem.** **Check** the answer to a division problem as follows.

> **Checking Division**
>
> $$(\text{divisor} \times \text{quotient}) + \text{remainder} = \text{dividend}$$
>
> Parentheses tell you what to do first: Multiply the divisor by the quotient, then add the remainder.

EXAMPLE 11 **Checking Division by Using Multiplication**

Check each answer.

(a) $5\overline{)458}$ quotient 91 **R**3

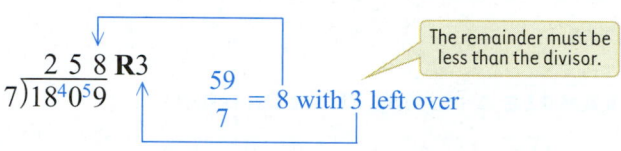

$(\text{divisor} \times \text{quotient}) + \text{remainder} = \text{dividend}$

$$(5 \times 91) + 3$$
$$455 + 3 = 458$$

> Be careful! Always add the remainder when checking division.

Matches original dividend, so the division was done correctly.

— **Continued on Next Page**

(b) $6\overline{)1437}$ $\quad$ 239 **R4**

(divisor $\times$ quotient) + remainder = dividend

$$(6 \quad \times \quad 239) \quad + \quad 4$$

$$1434 \quad\quad + \quad 4 \quad = \quad \textbf{1438}$$

Does not match original dividend.

The answer does **not** check. Rework the original problem to get the correct answer, 239 **R3**. Then, $(6 \times 239) + 3$ **does** give 1437.

> **CAUTION**
> A common error when checking division is to forget to add the remainder. Be sure to add any remainder when checking a division problem.

Work Problem *at the Side.* ▶

OBJECTIVE 9 Use tests for divisibility. It is often important to know whether a number is *divisible* by another number. You will find this useful in Chapter 2 when writing fractions in lowest terms.

> **Divisibility**
> One whole number is **divisible** by another if the remainder is 0.

Use the following tests to decide whether one number is divisible by another number.

> **Tests for Divisibility**
>
> **A number is divisible by**
>
> 2 $\quad$ if it ends in 0, 2, 4, 6, or 8. These are the even numbers.
>
> 3 $\quad$ if the sum of its digits is divisible by 3.
>
> 4 $\quad$ if the last two digits make a number that is divisible by 4.
>
> 5 $\quad$ if it ends in 0 or 5.
>
> 6 $\quad$ if it is divisible by both 2 and 3.
>
> > The most often used rules are for 2, 5, 10, and occasionally 3.
>
> 7 $\quad$ has no simple test.
>
> 8 $\quad$ if the last three digits make a number that is divisible by 8.
>
> 9 $\quad$ if the sum of its digits is divisible by 9.
>
> 10 $\quad$ if it ends in 0.

The most commonly used tests are those for 2, 3, 5, and 10.

---

11 Use multiplication to check each division. If an answer is incorrect, give the correct answer.

(a) $2\overline{)65}$ $\quad$ 32 **R**1

(b) $7\overline{)586}$ $\quad$ 83 **R**4

(c) $3\overline{)1223}$ $\quad$ 407 **R**2

(d) $5\overline{)2383}$ $\quad$ 476 **R**3

12 Which numbers are divisible by 2?

(a) 258

(b) 307

(c) 4216

(d) 73,000

Divisibility by 2

A number is divisible by **2** if the number ends in 0, 2, 4, 6, or 8. All even numbers are divisible by 2.

EXAMPLE 12 **Testing for Divisibility by 2**

Are the following numbers divisible by 2?

(a) 986

↑
└── Ends in 6

> All even numbers are divisible by 2.

Because the number ends in 6, which is an even number, the number 986 is divisible by 2.

(b) 3255 is not divisible by 2.

↑
└── Ends in 5, and not in 0, 2, 4, 6, or 8

◀ *Work Problem* **12** *at the Side.*

Divisibility by 3

A number is divisible by **3** if the sum of its digits is divisible by **3.**

13 Which numbers are divisible by 3?

(a) 743

(b) 5325

(c) 374,214

(d) 205,633

EXAMPLE 13 **Testing for Divisibility by 3**

Are the following numbers divisible by 3?

(a) 4251
Add the digits.

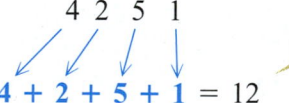

4 2 5 1

$4 + 2 + 5 + 1 = 12$

> If the sum of the digits is divisible by 3, the number is divisible by 3.

Because 12 is divisible by 3, the number 4251 is also divisible by 3.

(b) 29,806
Add the digits.

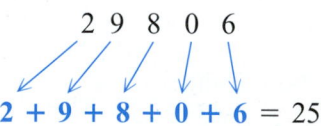

2 9 8 0 6

$2 + 9 + 8 + 0 + 6 = 25$

Because 25 is *not* divisible by 3, the number 29,806 is *not* divisible by 3.

CAUTION
Be careful when testing for divisibility by adding the digits. This method works only for the numbers 3 and 9.

◀ *Work Problem* **13** *at the Side.*

Divisibility by 5 and by 10

A number is divisible by **5** if it ends in 0 or 5.

A number is divisible by **10** if it ends in 0.

EXAMPLE 14 **Testing for Divisibility by 5**

Are the following numbers divisible by 5?

(a) 12,900 ends in 0 and is divisible by 5.

> If the number ends in 0 or 5, it's divisible by 5.

(b) 4325 ends in 5 and is divisible by 5.

(c) 392 ends in 2 and is not divisible by 5.

Work Problem 14 *at the Side.* ▶

EXAMPLE 15 **Testing for Divisibility by 10**

Are the following numbers divisible by 10?

> If the number ends in 0, it's divisible by 10.

(a) 700 and 9140 both end in 0 and are divisible by 10.

(b) 355 and 18,743 do not end in 0 and are not divisible by 10.

Work Problem 15 *at the Side.* ▶

---

14 Which numbers are divisible by 5?

(a) 180

(b) 635

(c) 8364

(d) 206,105

15 Which numbers are divisible by 10?

(a) 270

(b) 495

(c) 5030

(d) 14,380

ANSWERS

14. all but (c)

15. all but (b)

Math in the Media

STUDY TIME

As a college student, you need to plan a schedule to accommodate many responsibilities. These will include: preparing for and attending class, studying for exams, traveling to and from college, part-time work, family responsibilities, and personal time. Learning how to better manage your time is a skill presented by Stephen R. Covey at his Web site, stephencovey.com, and in his book *First Things First*. Dr. Covey says, "We're constantly making choices about the way we spend our time, from the major seasons to the individual moments of our lives. We're also living with the consequences of those choices." One of the first things you must do is calculate the amount of time dedicated to each of your obligations.

As an example, suppose that you are a full-time student enrolled in 12 credit hours of class: 4 credits of biology, 4 credits of computer science, 3 credits of mathematics, and 1 credit of physical education. Biology and computer science each have 3 hours of lecture and a 3-hour lab each week. Your biology and mathematics instructors recommend that you spend an additional 2 hours per week of study time for each hour of lecture time. Your physical education class is only 1 credit, but you are in class 3 hours each week.

Activity	Hours per Week
Class time	
Lab time	
Study time for mathematics and biology	
Travel time to and from college	5
Part-time work (including travel time)	25
Sleep (8 hours per day)	
Meals (3 hours per day)	
Hygiene (showers, dressing, etc.)	7
Other (housecleaning, laundry, etc.)	14

1. How many total hours are in one week?

2. Fill in the table entries for the number of hours in a week spent on class time, lab time, study time for biology and mathematics, sleep, and meals.

3. How many hours per week are spent on college-related activities?

4. How much more time is required for personal time (sleeping, eating, hygiene) than for college-related activities?

5. Based on the table data, how many hours per week are spent on the activities listed?

6. How many hours per week are available for other activities, such as dating, shopping, family responsibilities, and so on?

Write each division problem using two other symbols. See Example 1.

1. $24 \div 4 = 6$

2. $36 \div 3 = 12$

3. $\dfrac{45}{9} = 5$

4. $\dfrac{56}{8} = 7$

5. $2\overline{)16}^{\,8}$

6. $8\overline{)48}^{\,6}$

Divide. If the division is not possible, write "undefined." See Examples 3–7.

7. $9 \div 9$

8. $36 \div 9$

9. $\dfrac{14}{2}$

10. $\dfrac{10}{0}$

11. $22 \div 0$

12. $6 \div 6$

13. $\dfrac{24}{1}$

14. $\dfrac{12}{1}$

15. $15\overline{)0}$

16. $\dfrac{0}{12}$

17. $0\overline{)43}$

18. $\dfrac{8}{0}$

19. $\dfrac{15}{1}$

20. $\dfrac{6}{0}$

21. $\dfrac{8}{1}$

22. $\dfrac{0}{5}$

Divide by using short division. Use multiplication to check each answer. See Examples 8–10.

23. $3\overline{)75}$

24. $5\overline{)85}$

25. $7\overline{)126}$

26. $6\overline{)168}$

27. $4\overline{)1216}$

28. $5\overline{)2305}$

29. $4\overline{)2509}$

30. $8\overline{)1335}$

31. $6\overline{)9137}$

32. $9\overline{)8371}$

33. $6\overline{)1854}$

34. $8\overline{)856}$

35. 12,020 ÷ 4   **36.** 8012 ÷ 4   **37.** 30,036 ÷ 6   **38.** 32,008 ÷ 8

39. 2434 ÷ 3   **40.** 5993 ÷ 7   **41.** 12,947 ÷ 5   **42.** 33,285 ÷ 9

43. 29,298 ÷ 4   **44.** 17,937 ÷ 6   **45.** 12,630 ÷ 4   **46.** 46,560 ÷ 7

47. 21,040 ÷ 8   **48.** $\dfrac{8199}{9}$ **49.** $\dfrac{74,751}{6}$ **50.** $\dfrac{72,543}{5}$

51. $\dfrac{71,776}{7}$ **52.** $\dfrac{77,621}{3}$ **53.** $\dfrac{128,645}{7}$ **54.** $\dfrac{172,255}{4}$

Use multiplication to check each answer. If an answer is incorrect, find the correct answer.
See Example 11.

55. $5)\overline{1877}$ 375 **R2**   **56.** $3)\overline{1282}$ 427 **R1**   **57.** $3)\overline{5725}$ 1908 **R2**   **58.** $5)\overline{2158}$ 432 **R3**

59. $7)\overline{4692}$ 650 **R2**   **60.** $9)\overline{5974}$ 663 **R5**   **61.** $6)\overline{21,409}$ 3 568 **R2**   **62.** $6)\overline{3192}$ 532

63. $8)\overline{16,019}$ 2 002 **R3**   **64.** $8)\overline{33,664}$ 4 208   **65.** $6)\overline{69,140}$ 11,523 **R2**   **66.** $3)\overline{82,598}$ 27,532 **R1**

67. $9)\overline{86,655}$ 9 628 **R7**   **68.** $7)\overline{50,809}$ 7 258 **R4**   **69.** $8)\overline{222,576}$ 27,822   **70.** $4)\overline{311,216}$ 77,804

71. Explain in your own words how to check a division problem using multiplication. Be sure to include what must be done if the quotient includes a remainder.

72. Describe the three divisibility rules that you feel might be most useful to you and tell why.

Solve each application problem.

73. The Carnival Cruise Line has 2624 linen napkins. If it takes eight napkins to set each table, find the number of tables that can be set. (*Source: USA Today.*)

74. A school district will distribute 1620 new science books equally among 12 schools. How many books will each school receive?

75. In one 8-hour day Dreyer's Edy's can produce 76,800 ice cream drumsticks. How many are produced each hour? (*Source:* History Channel, *Modern Marvels: Snack Food Tech.*)

76. Tootsie Roll Industries produces 415,000,000 Tootsie Rolls in a 5-day week. Find the number of Tootsie Rolls produced each day. (*Source:* History Channel, *Modern Marvels: Snack Food Tech.*)

77. Lottery winnings of $436,500 are divided equally among nine Starbucks employees. Find the amount received by each employee.

78. How many 5-pound bags of organic whole wheat flour can be filled from a 17,175-pound bin of flour?

79. If 8 gallons of fertilizer are needed for each acre of land, find the number of acres that can be fertilized with 1080 gallons of fertilizer.

80. A roofing contractor has purchased 1134 squares of roofing material. If each cabin needs 9 squares of material, find the number of cabins that can be roofed. (1 square measures 10 ft by 10 ft.)

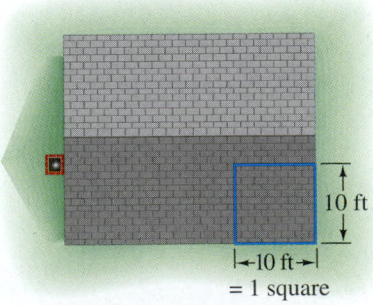

10 ft

|←10 ft→|
= 1 square

81. A class-action lawsuit settlement of $6,825,000 will be divided evenly among six injured people. Find the amount received by each person.

82. A 12,000-square foot condominium at the edge of Central Park in Manhattan sold for a record $45,000,000. If the buyer pays for the condominium in eight equal payments, find the amount of each payment. (*Source: USA Today.*)

83. The state of Maryland has the highest annual median household income of $65,148. How much income is this each month? (*Source:* U.S. Census Bureau.)

84. A professional basketball player has signed a 4-year contract for $21,937,500. How much is this each year?

Put a ✓ mark in the blank if the number at the left is divisible by the number at the top.
Put an X in the blank if the number is not divisible by the number at the top.
See Examples 12–15.

	2	3	5	10			2	3	5	10
85. 60	___	___	___	___	**86.** 35		___	___	___	___
87. 92	___	___	___	___	**88.** 96		___	___	___	___
89. 445	___	___	___	___	**90.** 897		___	___	___	___
91. 903	___	___	___	___	**92.** 500		___	___	___	___
93. 5166	___	___	___	___	**94.** 8302		___	___	___	___
95. 21,763	___	___	___	___	**96.** 32,472		___	___	___	___

1.6 ▶▶▶ Long Division

If the total cost of 42 Olympus digital cameras is $3066, we can find the cost of each camera using **long division.** Long division is used to divide by a number with more than one digit.

OBJECTIVE 1 Do long division.   In long division, estimate the various numbers by using a **trial divisor,** which is used to get a **trial quotient.**

OBJECTIVES

1 Do long division.

2 Divide numbers ending in 0 by numbers ending in 0.

3 Use multiplication to check division answers.

EXAMPLE 1 **Using a Trial Divisor and a Trial Quotient**

Divide. $42\overline{)3066}$

Because 42 is closer to 40 than to 50, use the first digit of the divisor as a trial divisor.

Try to divide the first digit of the dividend by 4. Since 3 cannot be divided by 4, use the first *two* digits, 30.

$$\frac{30}{4} = 7 \text{ with remainder 2}$$

$$7 \leftarrow \text{Trial quotient}$$

$$42\overline{)3066}$$

7 goes over the 6, because $\dfrac{306}{42}$ is about 7.

Multiply 7 and 42 to get 294; next, subtract 294 from 306.

$$
\begin{array}{r}
7 \\
42\overline{)3066} \\
294 \leftarrow 7 \times 42 \\
\hline
12 \leftarrow 306 - 294
\end{array}
$$

Bring down the 6 at the right.

$$
\begin{array}{r}
7 \\
42\overline{)3066} \\
294\downarrow \\
\hline
126 \leftarrow 6 \text{ brought down}
\end{array}
$$

Use the trial divisor, 4.

First two digits of 126 → $\dfrac{12}{4} = 3$

$$
\begin{array}{r}
73 \\
42\overline{)3066} \\
294 \\
\hline
126 \\
126 \leftarrow 3 \times 42 = 126 \\
\hline
0
\end{array}
$$

The cost of each camera is $73.
Check the answer by multiplying 42 and 73. The product should be 3066.

1 Divide.

(a) $28\overline{)2296}$

(b) $16\overline{)1024}$

(c) $61\overline{)8784}$

(d) $\dfrac{2697}{93}$

2 Divide.

(a) $24\overline{)1344}$

(b) $72\overline{)4472}$

(c) $65\overline{)5416}$

(d) $89\overline{)6649}$

CAUTION
The *first digit* of the quotient in long division must be placed in the proper position over the dividend.

◀ *Work Problem* **1** *at the Side.*

EXAMPLE 2   **Dividing to Find a Trial Quotient**

Divide. $58\overline{)2730}$

Use 6 as a trial divisor, since 58 is closer to 60 than to 50.

First two digits of dividend → $\dfrac{27}{6} = 4$ with 3 left over

4 ← Trial quotient

$58\overline{)2730}$
$\underline{232}$ ← $4 \times 58 = 232$
41 ← $273 - 232 = 41$ (smaller than 58, the divisor)

Bring down the 0.

$$\begin{array}{r} 4 \\ 58\overline{)2730} \\ \underline{232}\downarrow \\ 410 \end{array}$$ ← 0 brought down

First two digits of 410 → $\dfrac{41}{6} = 6$ with 5 left over

46 ← Trial quotient

$58\overline{)2730}$
$\underline{232}$
410
$\underline{348}$ ← $6 \times 58 = 348$
62 ← Greater than 58

Do not leave a remainder that is **greater** than the divisor.

The remainder, 62, is greater than the divisor, 58, so 7 should be used instead of 6.

$$\begin{array}{r} 4\mathbf{7}\ \mathbf{R4} \\ 58\overline{)2730} \\ \underline{232} \\ 410 \\ \underline{406} \\ 4 \end{array}$$
← $7 \times 58 = 406$
← $410 - 406$

Now the remainder, 4, is *less* than the divisor, 58.

◀ *Work Problem* **2** *at the Side.*

ANSWERS

1. (a) 82   (b) 64   (c) 144   (d) 29
2. (a) 56   (b) 62 **R8**
   (c) 83 **R21**   (d) 74 **R63**

Sometimes it is necessary to write a 0 in the quotient.

(**EXAMPLE 3**) **Writing Zeros in the Quotient**

Divide: $34\overline{)7068}$

Start as in Examples 1 and 2.

$$
\begin{array}{r}
2 \\
34\overline{)7068} \\
68 \\
\hline
2
\end{array}
$$

← 2 × 34 = 68
← 70 − 68 = 2

Bring down the 6.

$$
\begin{array}{r}
2 \\
34\overline{)7068} \\
68\downarrow \\
\hline
26
\end{array}
$$

← 6 brought down

Since 26 cannot be divided by 34, write a 0 in the quotient as a placeholder.

$$
\begin{array}{r}
\mathbf{20} \\
34\overline{)7068} \\
68 \\
\hline
26
\end{array}
$$

← 0 in quotient

> Use a zero to hold a place in the quotient.

Bring down the final digit, the 8.

$$
\begin{array}{r}
20 \\
34\overline{)7068} \\
68\downarrow \\
\hline
268
\end{array}
$$

← 8 brought down

Complete the problem.

$$
\begin{array}{r}
207\ \mathbf{R}30 \\
34\overline{)7068} \\
68 \\
\hline
268 \\
238 \\
\hline
30
\end{array}
$$

The quotient is 207 **R**30.

> **CAUTION**
> There *must be a digit* in the quotient (answer) above every digit in the dividend once the answer has begun. Notice in Example 3 that a **0** was used to assure a digit in the quotient above every digit in the dividend.

Work Problem (**3**) *at the Side.* ▶

OBJECTIVE **2** **Divide numbers ending in 0 by numbers ending in 0.** When the divisor and dividend both contain zeros at the far right, recall that these numbers are multiples of 10. As with multiplication, there is a short way to divide these multiples of 10. Look at the following examples.

$$26,000 \div 1 = 26,000$$
$$26,000 \div 10 = 2600$$
$$26,000 \div 100 = 260$$
$$26,000 \div 1000 = 26$$

Do you see a pattern? These examples suggest the following rule.

---

3 Divide.

(a) $17\overline{)1823}$

(b) $23\overline{)4791}$

(c) $39\overline{)15,933}$

(d) $78\overline{)23,462}$

4 Divide.

(a) $70 \div 10$

(b) $2600 \div 100$

(c) $505,000 \div 1000$

Dividing a Whole Number by 10, 100, or 1000

Divide a whole number by 10, 100, or 1000 by dropping the appropriate number of zeros from the whole number.

EXAMPLE 4 Dividing by Multiples of 10

Divide.

(a) $60 \div 1\mathbf{0} = 6$ — One 0 in divisor — 0 dropped

(b) $3500 \div 1\mathbf{00} = 35$ — Two zeros in divisor — 00 dropped

(c) $915,000 \div 1\mathbf{000} = 915$ — Three zeros in divisor — 000 dropped

◀ Work Problem **4** at the Side.

Now we'll find the quotients for other multiples of 10 by dropping zeros.

EXAMPLE 5 Dividing by Multiples of 10

Divide:

5 Divide.

(a) $50\overline{)6250}$

(b) $130\overline{)131,040}$

(c) $3400\overline{)190,400}$

(a) $4\mathbf{0}\overline{)11,00\mathbf{0}}$ Drop one zero from the divisor and the dividend.

$$
\begin{array}{r}
275 \\
4\overline{)1100} \\
\underline{8} \\
30 \\
\underline{28} \\
20 \\
\underline{20} \\
0
\end{array}
$$

Since $1100 \div 4$ is 275, then $11,000 \div 40$ is also 275.

(b) $35\mathbf{00}\overline{)31,5\mathbf{00}}$ Drop two zeros from the divisor and the dividend.

$$
\begin{array}{r}
9 \\
35\overline{)315} \\
\underline{315} \\
0
\end{array}
$$

Since $315 \div 35$ is 9, then $31,500 \div 3500$ is also 9.

Note

Dropping zeros when dividing by multiples of 10 **does not** change the quotient (answer).

◀ Work Problem **5** at the Side.

OBJECTIVE **3** **Use multiplication to check division answers.**
Answers in long division can be checked just as answers in short division were checked.

EXAMPLE 6 **Checking Division by Using Multiplication**

Check each answer.

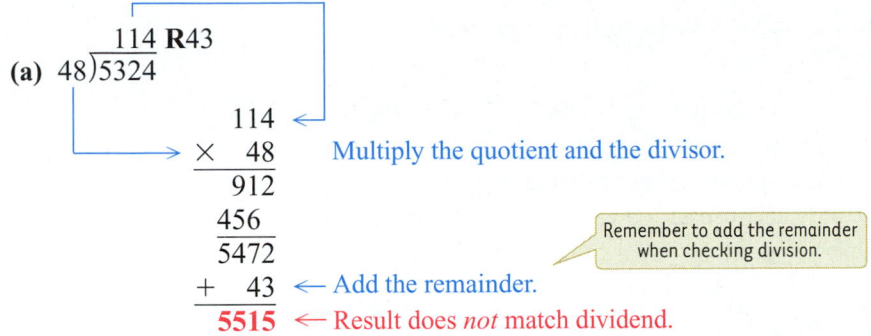

(a)

$$48\overline{)5324}^{\,114\,\text{R}43}$$

$$
\begin{array}{r}
114 \\
\times\ \ 48 \\
\hline
912 \\
456 \\
\hline
5472 \\
+\ \ \ 43 \\
\hline
\mathbf{5515}
\end{array}
$$

Multiply the quotient and the divisor.

Remember to add the remainder when checking division.

← Add the remainder.

← Result does *not* match dividend.

The answer does **not** check. Rework the original problem to get 110 **R**44. Then (110 × 48) + 44 **does** give 5324.

(b)

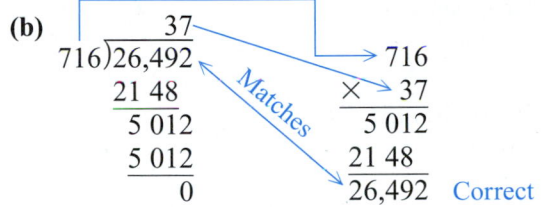

$$
\begin{array}{r}
37 \\
716\overline{)26{,}492} \\
21\ 48 \\
\hline
5\ 012 \\
5\ 012 \\
\hline
0
\end{array}
\qquad
\begin{array}{r}
716 \\
\times\ \ \ 37 \\
\hline
5\ 012 \\
21\ 48 \\
\hline
26{,}492
\end{array}
$$

Matches

Correct

🖩 **Calculator Tip** To check the answer to Example 6(a), don't forget to add the remainder.

48 ⊗ 110 ⊕ 44 ⊜ **5324**

Add the remainder.

CAUTION
When checking a division problem, first multiply the quotient and the divisor. Then be sure to **add any remainder** before checking it against the original dividend.

Work Problem **6** *at the Side.* ▶

6 Decide whether each answer is correct. If the answer is incorrect, find the correct answer.

(a)
$$
\begin{array}{r}
38 \\
16\overline{)608} \\
48 \\
\hline
128 \\
128 \\
\hline
0
\end{array}
$$

(b)
$$
\begin{array}{r}
42\ \text{R}178 \\
426\overline{)19{,}170} \\
17\ 040 \\
\hline
1\ 130 \\
952 \\
\hline
178
\end{array}
$$

(c)
$$
\begin{array}{r}
57\ \text{R}18 \\
514\overline{)29{,}316} \\
25\ 700 \\
\hline
3\ 616 \\
3\ 598 \\
\hline
18
\end{array}
$$

Math in the Media

LIGHTS ARE STILL ON AT THE DRIVE-IN MOVIES

It's just beginning to get dark, you're with family or friends in your car, you have plenty of pizza, popcorn, and other snacks, and the movie is about to start. You are at a drive-in movie theater.

A recent article in *USA Today* included some interesting facts about drive-ins. The first drive-in movie theater was opened in 1933 by the owner of an auto parts company and gas station to attract customers at night. It consisted of a sheet hanging between two trees. The popularity of drive-ins peaked in 1960, declined in the years that followed, but has remained fairly stable in recent years. The biggest reason for the decline of drive-ins has been the rising value of land and not a decline in popularity.

Use the drive-in movie theater facts below to answer the questions that follow.

Drive-in Movie Facts	States with the Most Open Drive-ins in 2008
1933: First drive-in movie	Pennsylvania: 35
1941: 12 open drive-ins	Ohio: 33
1958: 4063 open drive-ins	New York: 32
1960: 5000 open drive-ins	Indiana: 21
2008: 407 open drive-ins	California: 19
Typical screen size: 35 feet by 70 feet	Texas: 18
Typical popcorn machine cost: $12,500	Tennessee: 17
	Michigan: 14
	Illinois: 13
	Missouri: 13

(*Source: USA Today*)

1. What was the increase in the number of drive-in movies from 1941 to 1958?

2. Find the increase in the number of drive-in movies from 1958 to 1960.

3. How many less drive-ins were there in 2008 than in the peak year?

4. Find **(a)** the total number of drive-ins in the ten states with the most open drive-ins and **(b)** the number of open drive-ins in the remaining 40 states.

5. Find the cost of ten typical popcorn machines.

6. What is the area of the typical drive-in movie screen. (*Hint:* Area = width × height. The answer is to be expressed in square feet.)

7. Using your answer to Question 4(a), find the average number of open drive-ins in the ten states with the most open drive-ins. (The answer includes a remainder.)

Decide where the first digit in the quotient would be located. Then without finishing the division, you can tell which of the three choices is the correct answer. Circle your choice. See Examples 1 and 2.

1. $50\overline{)2650}$
   5   53   530

2. $14\overline{)476}$
   3   34   304

3. $18\overline{)4500}$
   2   25   250

4. $35\overline{)5600}$
   16   160   1600

5. $86\overline{)10,327}$
   12   120 **R7**   1200

6. $46\overline{)24,026}$
   5   52   522 **R14**

7. $26\overline{)28,735}$
   11   110   1105 **R5**

8. $12\overline{)116,953}$
   974 **R2**   9746 **R1**   97,460

9. $21\overline{)149,826}$
   71   713   7134 **R12**

10. $64\overline{)208,138}$
   325 **R2**   3252 **R10**   32,521

11. $523\overline{)470,800}$
   9 **R100**   90 **R100**   900 **R100**

12. $230\overline{)253,230}$
   11   110   1101

Divide by using long division. Use multiplication to check each answer. See Examples 1–3, 5, and 6.

13. $18\overline{)1319}$

14. $58\overline{)3654}$

15. $23\overline{)10,963}$

16. $83\overline{)39,692}$

17. $26\overline{)62,583}$

18. $28\overline{)84,249}$

19. $74\overline{)84,819}$

20. $238\overline{)186,948}$

21. $153\overline{)509,725}$

22. $308\overline{)26,796}$

23. $420\overline{)357,000}$

24. $900\overline{)153,000}$

Use multiplication to check each answer. If an answer is incorrect, find the correct answer.
See Example 6.

25. $\underset{35)\overline{3549}}{\overset{101\ \textbf{R}4}{}}$

26. $\underset{64)\overline{2712}}{\overset{42\ \textbf{R}26}{}}$

27. $\underset{28)\overline{18,424}}{\overset{658\ \textbf{R}9}{}}$

28. $\underset{145)\overline{34,776}}{\overset{239\ \textbf{R}121}{}}$

29. $\underset{614)\overline{38,068}}{\overset{62\ \textbf{R}3}{}}$

30. $\underset{557)\overline{97,286}}{\overset{174\ \textbf{R}368}{}}$

31. Describe in your own words a shortcut you can use to divide multiples of 10 by 10, by 100, or by 1000. Write an example problem and solve it.

32. Suppose you have a division problem with a remainder in the answer. Explain how to check your answer by writing an example problem that has a remainder.

Solve each application problem by using addition, subtraction, multiplication, or division as needed. See Examples 3–5.

33. Scientists using high-tech instruments have traced the travels of a tiger shark from Australia to South Africa—a total of 4950 miles in 99 days. On average, how far did the shark travel each day? (*Source:* Discovery Channel, *Shark Week.*)

34. A new bridge from Owensboro, Kentucky, to Rockport, Indiana, is 2200 feet long and cost $55,998,800. Find the construction cost per foot. (*Source:* Federal Highway Administration.)

35. Don Gracey, the Mountain Timesmith, has serviced and repaired 636 clocks this year. He has worked on 272 wall clocks and 308 table clocks. The rest were standing floor clocks. Find the number of floor clocks he worked on this year.

36. There are 24,000,000 business enterprises in the United States. If 7000 of these are larger businesses (over 500 employees), find the number of businesses that are small to mid-size. (*Source:* U.S. Census Bureau.)

37. To complete her college education, Judy Martinez received education loans of $34,080 including interest. Find her monthly payment if the loan is to be paid off in 96 months (8 years).

38. A consultant charged $13,050 for evaluating a school's compliance with the Americans with Disabilities Act. If the consultant worked 225 hours, find the rate charged per hour.

39. Each minute there is one diamond ring sold on eBay's U.S. site. Find the number of diamond rings sold in 30 days. (*Source: Time Style and Design.*)

40. A retired milkman in Indianapolis has eaten a Twinkie every day for the last 60 years. How many Twinkies has he eaten over this time period? *Hint:* Use a 365-day year, ignoring leap years. (*Source:* History Channel, *Modern Marvels: Snack Food Tech.*)

41. The average U.S. household of 2.5 people spent $2028 eating away from home last year. Find the average weekly household cost of eating away from home. *Hint:* 1 year equals 52 weeks. (*Source:* U.S. Bureau of Labor Statistics consumer expenditure surveys.)

42. Former professional basketball player Junior Bridgeman now owns 120 Wendy's restaurants with 4080 employees. Find the average number of employees at each restaurant. (*Source:* National Basketball Retired Players Association.)

Relating Concepts (Exercises 43—50) For Individual or Group Work

Knowing and using the rules of divisibility is necessary in problem solving.
Work Exercises 43–50 in order.

43. If you have $0 and you divide this amount among three people, how much will each receive?

44. When 0 is divided by any nonzero number, the result is _____.

45. Divide.
$8 \div 0$

46. We say that division by 0 is *undefined* because it is _____ to compute the answer. Give an
(possible/impossible)
example involving cookies that will support your answer.

47. Divide.
(a) $14 \div 1$
(b) $1\overline{)17}$
(c) $\dfrac{38}{1}$

48. Any number divided by 1 is the number itself. Is this also true when multiplying by 1? Give three examples that support your answer.

49. Divide.
(a) $32{,}000 \div 10$
(b) $32{,}000 \div 100$
(c) $32{,}000 \div 1000$

50. Write a rule that explains the shortcut for doing divisions like the ones in Exercise 49.

Math in the Media

POST OFFICE FACTS

The United States Postal Service posts a Web page on the Internet that gives a list of facts about their service. Some of those facts are listed in the table below.

Resource/Service	Number
1. Mail collection boxes	326,000
2. Post offices	36,895
3. Delivery points	146 million
4. Pieces of First Class mail delivered each year	213 billion
5. Processing plants sorting and shipping the mail	331
6. Pounds of mail carried on commercial airline flights annually	2.7 billion
7. Number of new deliveries each day	3500
8. Miles driven to move the mail annually	1.2 billion
9. Vehicles to pick up, transport, and deliver the mail	216,450
10. Customers each day who transact business at a post office	9 million

Source: United States Postal Service, www.usps.com

1. Write the number of delivery points entirely in digits. (Do not use the word "million" in your answer.)

2. Write the number of miles driven annually to move the mail entirely in digits.

3. Round the number of post offices to the nearest thousand.

4. Use your rounded number from Question 3 to compute the average number of customers who transact business each day per post office. Round the answer to the nearest person. (*Hint:* To find the average, divide the total number of customers per day by the number of post offices).

5. Find the total number of new deliveries each year. Write the result entirely in words. (*Hint:* Assume that new deliveries are added 365 days per year.)

6. Mail is sorted and shipped from processing plants to the post offices. Which of the following is a rough estimate of the number of post offices served by each processing plant: 10, 100, or 1000? Explain your choice.

UNITED STATES POSTAL SERVICE ®

1.7 ▶▶▶ Rounding Whole Numbers

One way to get a quick check on an answer is to *round* the numbers in the problem. **Rounding** a number means finding a number that is close to the original number, but easier to work with.

For example, the county planning commissioner might be discussing the need for more affordable housing. To demonstrate this, she probably would not need to say that the county is in need of 8235 more affordable housing units— she probably could say that the county needs 8200 or even 8000 housing units.

OBJECTIVE 1 Locate the place to which a number is to be rounded. The first step in rounding a number is to locate the *place to which the number is to be rounded.*

EXAMPLE 1 Finding the Place to Which a Number Is to Be Rounded

Locate and draw a line under the place to which each number is to be rounded.

(a) Round 83 to the nearest ten. Is 83 closer to 80 or to 90?

83 is closer to 80.

83 is closer to 80 than to 90.

Tens place

(b) Round 54,702 to the nearest thousand. Is it closer to 54,000 or to 55,000?

54,702 is closer to 55,000.

Thousands place

(c) Round 2,806,124 to the nearest hundred-thousand. Is it closer to 2,800,000 or to 2,900,000?

2,806,124 is closer to 2,800,000.

Hundred-thousands place

Work Problem **1** *at the Side.* ▶

OBJECTIVE 2 Round numbers. Use the following rules for rounding whole numbers.

> **Rounding Whole Numbers**
>
> *Step 1* Locate the *place* to which the number is to be rounded. Draw a line under that place.
>
> *Step 2(a)* Look only at the next digit to the right of the one you underlined. If it is *5 or more, increase* the underlined digit by 1.
>
> *Step 2(b)* If the next digit to the right is *4 or less, do not change* the digit in the underlined place.
>
> *Step 3* *Change* all digits to the right of the underlined place to zeros.

EXAMPLE 2 Using Rounding Rules for 4 or Less

Round 349 to the nearest hundred.

Step 1 Locate the place to which the number is being rounded. Draw a line under that place.

349

Hundreds place

Continued on Next Page

OBJECTIVES

1 **Locate the place to which a number is to be rounded.**

2 **Round numbers.**

3 **Round numbers to estimate an answer.**

4 **Use front end rounding to estimate an answer.**

1 Locate and draw a line under the place to which each number is to be rounded. Then answer the question.

(a) 373 (nearest ten)

Is it closer to 370 or to 380?

_____

(b) 1482 (nearest thousand)

Is it closer to 1000 or to 2000?

_____

(c) 89,512 (nearest hundred)

Is it closer to 89,500 or to 89,600?

_____

(d) 546,325 (nearest ten-thousand)

Is it closer to 540,000 or to 550,000?

_____

ANSWERS

1. (a) 373 is closer to 370.
(b) 1482 is closer to 1000.
(c) 89,512 is closer to 89,500.
(d) 546,325 is closer to 550,000.

2 Round to the nearest ten.

(a) 62

(b) 94

(c) 134

(d) 7543

3 Round to the nearest thousand.

(a) 3683

(b) 6502

(c) 84,621

(d) 55,960

Step 2 Because the next digit to the right of the underlined place is 4, which is 4 or less, do *not* change the digit in the underlined place.

Next digit is 4 or less.

349

4 or less, do not change underlined digit.

3 remains 3.

Step 3 Change all digits to the right of the underlined place to zeros.

349 rounded to the nearest hundred is 300.

In other words, 349 is closer to 300 than to 400.

◀ Work Problem **2** at the Side.

EXAMPLE 3 Using Rounding Rules for 5 or More

Round 36,833 to the nearest thousand.

Step 1 Find the place to which the number is to be rounded and draw a line under that place.

36,833

Thousands

Step 2 Because the next digit to the right of the underlined place is 8, which is 5 or more, add 1 to the underlined place.

Next digit is 5 or more.

36,833

5 or more, add 1 to underlined digit.

Change 6 to 7.

Step 3 Change all digits to the right of the underlined place to zeros.

Change to 0.

36,833 rounded to the nearest thousand is 37,000.

Change 6 to 7.

In other words, 36,833 is closer to 37,000 than to 36,000.

◀ Work Problem **3** at the Side.

EXAMPLE 4 Using Rounding Rules

(a) Round 2382 to the nearest ten.

Step 1 2382

Tens place

Step 2 The next digit to the right is 2, which is 4 or less.

Next digit is 4 or less.

2382

Leave 8 as 8.

Step 3 2382   Change to 0.

2382 rounded to the nearest ten is 2380.
In other words, 2382 is closer to 2380 than to 2390.

Continued on Next Page

(b) Round 13,961 to the nearest hundred.

Step 1   13,961
— Hundreds place

Step 2   The next digit to the right is 6.

— Next digit is 5 or more.

13,961

— Change 9 to 10; write 0 and regroup 1 into thousands place.
— 3 + regrouped 1 = 4

— Change to 0

Step 3   14,061

13,961 rounded to the nearest hundred is 14,000.
In other words, 13,961 is closer to 14,000 than to 13,900.

> **Note**
>
> In Step 2 of Example 4(b), notice that the first three digits increased from 139 to 140 when we added 1 to the hundreds place.
>
> (13,9)61 rounded to (14,0)00

Work Problem **4** *at the Side.* ▶

EXAMPLE 5   **Rounding Large Numbers**

(a) Round 37,892 to the nearest ten-thousand.

Step 1   37,892

— Ten-thousands place.

> Remember to *underline* the place to which you are rounding.

Step 2   The next digit to the right is 7.

— Next digit is 5 or more.

37892

— Change 3 to 4.

— Change 0.

Step 3   47,892

37,892 rounded to the nearest ten-thousand is 40,000.

(b) Round 528,498,675 to the nearest million.

Step 1   528,498,675

— Millions place

— Next digit is 4 or less.

Step 2   528,498,675

— Leave 8 as 8

— Change to 0.

> Remember to change *everything* to the right of the place you have rounded to 0.

Step 3   528,498,675

528,498,675 rounded to the nearest million is 528,000,000.

Work Problem **5** *at the Side.* ▶

4 Round each number as indicated.

(a) 3458 to the nearest ten

(b) 6448 to the nearest hundred

(c) 73,077 to the nearest hundred

(d) 85,972 to the nearest hundred

5 Round each number as indicated.

(a) 14,598 to the nearest ten-thousand

(b) 724,518,715 to the nearest million

ANSWERS

4. (a) 3460   **(b)** 6400
   (c) 73,100   **(d)** 86,000
5. (a) 10,000   **(b)** 725,000,000

6 Round each number to the nearest ten and to the nearest hundred.

(a) 458

(b) 549

(c) 9308

Sometimes a number must be rounded to different places.

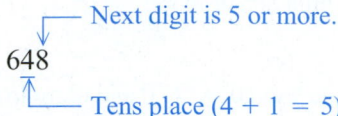 **EXAMPLE 6**   **Rounding to Different Places**

Round 648 **(a)** to the nearest ten and **(b)** to the nearest hundred.

(a) to the nearest ten

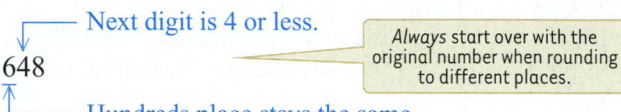

648 rounded to the nearest ten is 650.

(b) to the nearest hundred

648 rounded to the nearest hundred is 600.

Notice that if 648 is rounded to the nearest ten (650), and then 650 is rounded to the nearest hundred, the result is 700. If, however, 648 is rounded directly to the nearest hundred, the result is 600 (not 700).

◀ *Work Problem* **6** *at the Side.*

> **CAUTION**
> Before rounding to a different place, always go back to the *original,* unrounded number.

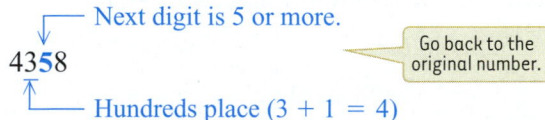 **EXAMPLE 7**   **Applying Rounding Rules**

Round each number to the nearest ten, nearest hundred, and nearest thousand.

(a) 4358
First round 4358 to the nearest ten.

4358 rounded to the nearest ten in 4360.
Now go back to 4358, the *original* number, before rounding to the nearest hundred.

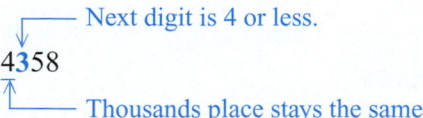

4358 rounded to the nearest hundred is 4400.
Again, go back to the *original* number before rounding to the nearest thousand.

4358 rounded to the nearest thousand is 4000.

Continued on Next Page

(b) 680,914
First, round to the nearest ten.

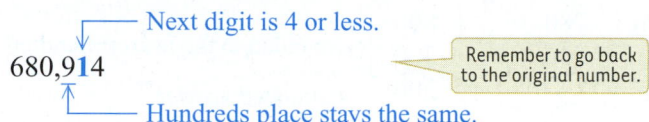

Next digit is 4 or less.

680,91**4**

Tens place stays the same.

680,914 rounded to the nearest ten is 680,910.
    Go back to 680,914, the *original* number, to round to the nearest hundred.

Next digit is 4 or less.

680,**9**14

                                                            Remember to go back
                                                            to the original number.

Hundreds place stays the same.

680,914 rounded to the nearest hundred is 680,900.
    Go back to the *original* number to round to the nearest thousand.

Next digit is 5 or more.

680,**9**14

Thousands place $(0 + 1 = 1)$

680,914 rounded to the nearest thousand is 681,000.

Work Problem **7** *at the Side.* ▶

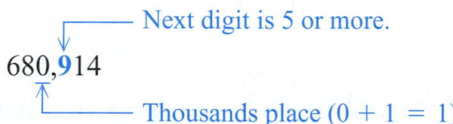

OBJECTIVE 3 Round numbers to estimate an answer. Numbers may be rounded to **estimate** an answer. An estimated answer is one that is close to the exact answer and may be used as a check when the exact answer is found. The "≈" sign is often used to show that an answer has been rounded or estimated and is almost equal to the exact answer; ≈ means "approximately equal to."

EXAMPLE 8 **Using Rounding to Estimate an Answer**

Estimate each answer by rounding to the nearest ten.

(a)
$$
\begin{array}{r}
76 \longrightarrow 80 \\
53 \longrightarrow 50 \\
38 \longrightarrow 40 \\
+91 \longrightarrow +90 \\
\hline
260
\end{array}
$$
} Rounded to the nearest ten

260 Estimated answer

(b)
$$
\begin{array}{r}
27 \\
-14 \\
\hline
\end{array}
\quad
\begin{array}{r}
30 \\
-10 \\
\hline
20
\end{array}
$$
} Rounded to the nearest ten

20 Estimated answer

(c)
$$
\begin{array}{r}
16 \\
\times 21 \\
\hline
\end{array}
\quad
\begin{array}{r}
20 \\
\times 20 \\
\hline
400
\end{array}
$$
} Rounded to the nearest ten

400 Estimated answer

Work Problem **8** *at the Side.* ▶

---

7 Round each number to the nearest ten, nearest hundred, and nearest thousand.

  (a) 4078

  (b) 46,364

  (c) 268,328

8 Estimate the answers by rounding each number to the nearest ten.

  (a)
$$
\begin{array}{r}
16 \\
74 \\
58 \\
+31 \\
\hline
\end{array}
$$

  (b)
$$
\begin{array}{r}
53 \\
-19 \\
\hline
\end{array}
$$

  (c)
$$
\begin{array}{r}
46 \\
\times 74 \\
\hline
\end{array}
$$

ANSWERS

7. **(a)** 4080; 4100; 4000
   (b) 46,360; 46,400; 46,000
   (c) 268,330; 268,300; 268,000
8. **(a)** $20 + 70 + 60 + 30 = 180$
   (b) $50 - 20 = 30$
   (c) $70 \times 50 = 3500$

9 Estimate the answers by rounding each number to the nearest hundred.

(a) 358
743
822
+ 978

(b) 842
− 475

(c) 723
× 478

10 Use front end rounding to estimate each answer.

(a) 36
3852
749
+ 5474

(b) 2583
− 765

(c) 648
× 67

ANSWERS

9. **(a)** $400 + 700 + 800 + 1000 = 2900$
(b) $800 − 500 = 300$
(c) $500 \times 700 = 350,000$
10. **(a)** $40 + 4000 + 700 + 5000 = 9740$
(b) $3000 − 800 = 2200$
(c) $70 \times 600 = 42,000$

EXAMPLE 9 **Using Rounding to Estimate an Answer**

Estimate each answer by rounding to the nearest hundred.

(a) 252 ⟶ 300
749 ⟶ 700
576 ⟶ 600   } Rounded to the nearest hundred
+ 819 ⟶ + 800
                  2400   Estimated answer

The hundreds position is 3 places to the left.

(b) 780   800
− 536   − 500   } Rounded to the nearest hundred
                 300   Estimated answer

(c) 664   700
× 834   × 800   } Rounded to the nearest hundred
            560,000   Estimated answer

◀ *Work Problem* **9** *at the Side.*

OBJECTIVE 4 Use front end rounding to estimate an answer.
A convenient way to estimate an answer is to use *front end rounding*. With **front end rounding,** we round to the highest possible place so that all the digits become 0 except the first one. For example, suppose you want to buy a big flat-screen television for $2449, a home theater system for $1759, and a reclining chair for $525. Using front end rounding, you can estimate the total cost of these purchases.

Television	$2449 →	2000
Home theater system	$1759 →	2000
Reclining chair	$525 →	+ 500
		$4500 ← Estimated total cost

EXAMPLE 10 **Using Front End Rounding to Estimate an Answer**

Estimate each answer using front end rounding.

(a) 3825   4000
72   70   } All digits changed to 0 except first digit, which is rounded
565   600
+ 2389   + 2000
             6670   Estimated answer

(b) 6712   7000
− 825   − 800   } First digit rounded and all others changed to 0
             6200   Estimated answer

Notice: Front end rounding leaves *only* one nonzero digit.

(c) 725   700
× 86   × 90
           63,000   Estimated answer

Note
When using front end rounding, all the digits become 0 except the highest-place digit (the first digit).

◀ *Work Problem* **10** *at the Side.*

1.7 ▸▸▸ **Exercises**

FOR EXTRA HELP

MyMathLab

Math XL
PRACTICE

WATCH

DOWNLOAD

READ

REVIEW

Round each number as indicated. See Examples 1–5.

1. 624 to the nearest ten

2. 509 to the nearest ten

3. 855 to the nearest ten

4. 946 to the nearest ten

5. 6771 to the nearest hundred

6. 5847 to the nearest hundred

7. 86,813 to the nearest hundred

8. 17,211 to the nearest hundred

9. 28,472 to the nearest hundred

10. 18,273 to the nearest hundred

11. 5996 to the nearest hundred

12. 4452 to the nearest hundred

13. 15,758 to the nearest thousand

14. 28,465 to the nearest thousand

15. 78,499 to the nearest thousand

16. 14,314 to the nearest thousand

17. 7,760,058,721 to the nearest billion

18. 44,706,892 to the nearest ten-million

19. 12,987 to the nearest ten-thousand

20. 6599 to the nearest ten-thousand

21. 595,008 to the nearest ten-thousand

22. 725,182 to the nearest ten-thousand

23. 4,860,220 to the nearest million

24. 13,713,409 to the nearest million

Round each number to the nearest ten, nearest hundred, and nearest thousand. See Examples 6 and 7.

	Ten	Hundred	Thousand			Ten	Hundred	Thousand
25. 4476	_____	_____	_____		**26.** 6483	_____	_____	_____
27. 3374	_____	_____	_____		**28.** 7632	_____	_____	_____
29. 6048	_____	_____	_____		**30.** 7065	_____	_____	_____

	Ten	Hundred	Thousand
31. 5343	_____	_____	_____
33. 19,539	_____	_____	_____
35. 26,292	_____	_____	_____
37. 93,706	_____	_____	_____

	Ten	Hundred	Thousand
32. 7456	_____	_____	_____
34. 59,806	_____	_____	_____
36. 78,519	_____	_____	_____
38. 84,639	_____	_____	_____

39. Write in your own words the three steps that you would use to round a number when the digit to the right of the place to which you are rounding is 5 or more.

40. Write in your own words the three steps that you would use to round a number when the digit to the right of the place to which you are rounding is 4 or less.

Estimate the answer by rounding each number to the nearest ten. Then find the exact answer. See Example 8.

41. *Estimate:*   *Exact:*

Rounds to

$$
\begin{array}{r}
\leftarrow \quad 25 \\
\leftarrow \quad 63 \\
\leftarrow \quad 47 \\
+ \underline{} \quad \leftarrow \quad + \underline{84} \\
\end{array}
$$

42. *Estimate:*   *Exact:*

$$
\begin{array}{r}
56 \\
24 \\
85 \\
+ \underline{} \quad + \underline{71} \\
\end{array}
$$

43. *Estimate:*   *Exact:*

$$
\begin{array}{r}
78 \\
- \underline{} \quad - \underline{43} \\
\end{array}
$$

44. *Estimate:*   *Exact:*

$$
\begin{array}{r}
57 \\
- \underline{} \quad - \underline{24} \\
\end{array}
$$

45. *Estimate:*   *Exact:*

$$
\begin{array}{r}
67 \\
\times \underline{} \quad \times \underline{34} \\
\end{array}
$$

46. *Estimate:*   *Exact:*

$$
\begin{array}{r}
53 \\
\times \underline{} \quad \times \underline{75} \\
\end{array}
$$

Estimate the answer by rounding each number to the nearest hundred. Then find the exact answer. See Example 9.

47. *Estimate:*   *Exact:*

Rounds to

$$
\begin{array}{r}
\leftarrow \quad 863 \\
\leftarrow \quad 735 \\
\leftarrow \quad 438 \\
+ \underline{} \quad \leftarrow \quad + \underline{792} \\
\end{array}
$$

48. *Estimate:*   *Exact:*

$$
\begin{array}{r}
623 \\
362 \\
189 \\
+ \underline{} \quad + \underline{736} \\
\end{array}
$$

49. Estimate: Exact:

                              883
         ___   − 448
                             _____

50. Estimate: Exact:

                              614
         ___   − 276
                             _____

51. Estimate: Exact:

                              752
         × ___   × 375
                             _____

52. Estimate: Exact:

                              845
         × ___   × 396
                             _____

Estimate each answer using front end rounding. Then find the exact answer. See Example 10.

53. Estimate: Exact:

                    Rounds to
         ←_____   8215
         ←_____   56
         ←_____   729
     + ___ ←_____ + 3605
                             _____

54. Estimate: Exact:

                             2685
                              73
                             592
     + ___   + 7183
                             _____

55. Estimate: Exact:

                              687
         − ___   − 529
                             _____

56. Estimate: Exact:

                              543
         − ___   − 174
                             _____

57. Estimate: Exact:

                              939
         × ___   × 29
                             _____

58. Estimate: Exact:

                              864
         × ___   × 74
                             _____

59. The number 3492 rounded to the nearest hundred is 3500, and 3500 rounded to the nearest thousand is 4000. But when 3492 is rounded directly to the nearest thousand it becomes 3000. Why is this true? Explain.

60. The use of rounding is helpful when estimating the answer to a problem. Why is this true? Give an example using either addition, subtraction, multiplication, or division to show how this works.

61. In 1900, the population of the United States was 76 million. Today it's 303 million. Round each of these numbers to the nearest ten-million. (*Source: Reiman Publications and U.S. Census Bureau.*)

62. In 1900, the average workweek in the United States was 59 hours. Today it's 38 hours. Round each of these numbers to the nearest ten. (*Source: Reiman Publications.*)

63. There are 348,900 streets named Elm Street in the United States. Round this number to the nearest thousand and nearest ten-thousand. (*Source: Expo Design Center.*)

64. The most expensive item ever sold at a Costco Wholesale warehouse was a $235,000 diamond ring. Round this number to the nearest ten-thousand and the nearest hundred-thousand. (*Source: Costco Stores magazine.*)

65. In Chicago, the Sears Tower tenants recycled 1,667,300 pounds of paper last year. Round this number to the nearest ten-thousand, nearest hundred-thousand, and nearest million. (*Source:* Trizec Properties.)

66. The population of India is 1,129,866,154 people. Round the number to the nearest ten-thousand, nearest hundred-thousand, and nearest million. (*Source:* U.S. Census Bureau.)

67. American pharmaceutical companies spent $25,765,475,000 last year to develop new products. Round this amount to the nearest hundred-thousand, nearest hundred-million, and nearest billion. (*Source:* American Demographics.)

68. In one year the U.S. Federal Food Assistance Program paid out $18,915,762,568 in food stamps. Round this amount to the nearest hundred-thousand, nearest hundred-million, and nearest ten-billion. (*Source:* U.S. Department of Agriculture.)

Relating Concepts (Exercises 69–75) For Individual or Group Work

To see how both rounding and front end rounding are used in solving problems, **work Exercises 69–75 in order.**

69. A number rounded to the nearest thousand is 72,000. What is the *smallest* whole number this could have been before rounding?

70. A number rounded to the nearest thousand is 72,000. What is the *largest* whole number this could have been before rounding?

71. When front end rounding is used, a whole number rounds to 8000. What is the *smallest* possible original number?

72. When front end rounding is used, a whole number rounds to 8000. What is the *largest* possible original number?

The graph below shows the number of personal injuries in the United States each year for people participating in common activities.

WATCH YOUR STEP

Accidents while participating in common activities

2,788,000

534,883

3925 11,243 15,974 17,916

Brushing Reading Stapling Bowling Bicycling Driving
Teeth   Paper

Source: AARP Magazine.

73. Round the number of accidents occurring in each activity to the nearest ten.

74. Use front end rounding to round the number of accidents in each activity.

75. (a) What is one advantage of using front end rounding instead of rounding to the nearest ten?

(b) What is one disadvantage?

1.8 ▶▶▶   Exponents, Roots, and Order of Operations

OBJECTIVE 1 Identify an exponent and a base. The product $3 \cdot 3$ can be written as 3^2 (read as "3 squared"). The small raised number 2, called an **exponent,** says to use 2 factors of 3. The number 3 is called the **base.** Writing 3^2 as 9 is called *simplifying the expression.*

EXAMPLE 1   Simplifying Expressions

Identify the exponent and the base, and then simplify each expression.

(a) 4^3

The small raised number is the exponent.

$\text{Base} \rightarrow 4^3 \leftarrow \text{Exponent}$ $4^3 = 4 \times 4 \times 4 = 64$

(b) $2^5 = 2 \times 2 \times 2 \times 2 \times 2 = 32$
The base is 2 and the exponent is 5.

Work Problem **1** *at the Side.* ▶

OBJECTIVE 2 Find the square root of a number. Because $3^2 = 9$, the number 3 is called the **square root** of 9. The square root of a number is one of two identical factors of that number. Square roots of numbers are written with the symbol $\sqrt{}$.

$$\sqrt{9} = 3$$

Square Root

$$\sqrt{\text{number} \cdot \text{number}} = \sqrt{\text{number}^2} = \text{number}$$

For example: $\sqrt{36} = \sqrt{6 \cdot 6} = \sqrt{6^2} = 6$ is the square root of 36.

To find the square root of 64 ask, "What number can be multiplied by itself (that is, *squared*) to give 64?" The answer is 8, so

$$\sqrt{64} = \sqrt{8 \cdot 8} = \sqrt{8^2} = 8.$$

A **perfect square** is a number that is the square of a *whole number.* The first few perfect squares are listed here.

Perfect Squares Table

$0 = 0^2$	$16 = 4^2$	$64 = 8^2$	$144 = 12^2$
$1 = 1^2$	$25 = 5^2$	$81 = 9^2$	$169 = 13^2$
$4 = 2^2$	$36 = 6^2$	$100 = 10^2$	$196 = 14^2$
$9 = 3^2$	$49 = 7^2$	$121 = 11^2$	$225 = 15^2$

EXAMPLE 2   Using Perfect Squares

Find each square root.

(a) $\sqrt{16}$ Because $4^2 = 16$, $\sqrt{16} = 4$.   **(b)** $\sqrt{49} = 7$
(c) $\sqrt{0} = 0$ **(d)** $\sqrt{169} = 13$

Work Problem **2** *at the Side.* ▶

OBJECTIVE 3 Use the order of operations. Frequently problems may have parentheses, exponents, and square roots, and may involve more than one operation. Work these problems by following the **order of operations.**

OBJECTIVES

1 Identify an exponent and a base.

2 Find the square root of a number.

3 Use the order of operations.

---

1 Identify the exponent and the base, and then simplify each expression.

(a) 4^2

(b) 5^3

(c) 3^4

(d) 2^6

2 Find each square root.

(a) $\sqrt{4}$

(b) $\sqrt{25}$

(c) $\sqrt{36}$

(d) $\sqrt{225}$

(e) $\sqrt{1}$

ANSWERS

1. (a) 2; 4; 16   **(b)** 3; 5; 125
    (c) 4; 3; 81   **(d)** 6; 2; 64
2. (a) 2   **(b)** 5   **(c)** 6   **(d)** 15   **(e)** 1

3 Simplify each expression.

(a) $4 + 5 + 2^2$

(b) $3^2 + 2^3$

(c) $4 \cdot 6 \div 12 - 2$

(d) $60 \div \sqrt{36} \div 2$

(e) $8 + 6(14 \div 2)$

Order of Operations

1. Do all operations inside *parentheses* or *other grouping symbols.*
2. Simplify any expressions with *exponents* and find any *square roots.*
3. *Multiply* or *divide,* proceeding from left to right.
4. *Add* or *subtract,* proceeding from left to right.

EXAMPLE 3 Understanding the Order of Operations

Use the order of operations to simplify each expression.

(a) $8^2 + 5 + 2$

$$8^2 + 5 + 2$$
$$8 \cdot 8 + 5 + 2 \qquad \text{Evaluate exponent first; } 8^2 \text{ is } 8 \cdot 8.$$
$$64 + 5 + 2 \qquad \text{Add from left to right.}$$
$$69 + 2 = 71$$

(b) $35 \div 5 \cdot 6 \qquad$ Divide first (start at left).
$$7 \cdot 6 = 42 \qquad \text{Multiply.}$$

(c) $9 + (20 - 4) \cdot 3 \qquad$ Work inside parentheses first.
$$9 + 16 \cdot 3 \qquad \text{Multiply.}$$
$$9 + 48 = 57 \qquad \text{Add last.}$$

(d) $12 \cdot \sqrt{16} - 8(4) \qquad$ Find the square root first.
$$12 \cdot 4 - 8(4) \qquad \text{Multiply from left to right.}$$
$$48 - 32 = 16 \qquad \text{Subtract last.}$$

◀ *Work Problem* **3** *at the Side.*

---

4 Simplify each expression.

(a) $12 - 6 + 4^2$

(b) $2^3 + 3^2 - (5 \cdot 3)$

(c) $2 \cdot \sqrt{64} - 5 \cdot 3$

(d) $20 \div 2 + (7 - 5)$

(e) $15 \cdot \sqrt{9} - 8 \cdot \sqrt{4}$

EXAMPLE 4 Using the Order of Operations

Use the order of operations to simplify each expression.

(a) $15 - 4 + 2 \qquad$ Subtract first (start at left).
$$11 + 2 = 13 \qquad \text{Add.}$$

(b) $8 + (7 - 3) \div 2 \qquad$ Work inside parentheses first.
$$8 + 4 \div 2 \qquad \text{Divide.}$$
$$8 + 2 = 10 \qquad \text{Add last.}$$

> Add or subtract last.

(c) $4^2 \cdot 2^2 + (7 + 3) \cdot 2 \qquad$ Work inside parentheses first.
$$4^2 \cdot 2^2 + 10 \cdot 2 \qquad \text{Evaluate exponents.}$$
$$16 \cdot 4 + 10 \cdot 2 \qquad \text{Multiply from left to right.}$$
$$64 + 20 = 84 \qquad \text{Add last.}$$

(d) $4 \cdot \sqrt{25} - 7 \cdot 2 + \dfrac{0}{5} \qquad$ Find the square root first.

> Remember: Zero divided by any nonzero number is zero.

$$4 \cdot 5 - 7 \cdot 2 + \dfrac{0}{5} \qquad \text{Multiply or divide from left to right.}$$
$$20 - 14 + 0 = 6 \qquad \text{Add or subtract last.}$$

Note

Getting a correct answer depends on following the order of operations.

◀ *Work Problem* **4** *at the Side.*

ANSWERS

3. (a) 13 **(b)** 17 **(c)** 0 **(d)** 5 **(e)** 50
4. (a) 22 **(b)** 2 **(c)** 1 **(d)** 12 **(e)** 29

1.8 ▶▶▶ Exercises

Identify the exponent and the base, and then simplify each expression. See Example 1.

1. 3^2

2. 2^3

3. 5^2

4. 4^2

5. 8^2

6. 10^3

7. 15^2

8. 11^3

Use the Perfect Squares Table on page 75 to find each square root. See Example 2.

9. $\sqrt{16}$

10. $\sqrt{25}$

11. $\sqrt{64}$

12. $\sqrt{36}$

13. $\sqrt{100}$

14. $\sqrt{49}$

15. $\sqrt{144}$

16. $\sqrt{225}$

Fill in each blank. See Example 2.

17. $6^2 = $ _____ so $\sqrt{} = 6$

18. $9^2 = $ _____ so $\sqrt{} = 9$

19. $20^2 = $ _____ so $\sqrt{} = 20$

20. $30^2 = $ _____ so $\sqrt{} = 30$

21. $35^2 = $ _____ so $\sqrt{} = 35$

22. $38^2 = $ _____ so $\sqrt{} = 38$

23. $25^2 = $ _____ so $\sqrt{} = 25$

24. $50^2 = $ _____ so $\sqrt{} = 50$

25. $100^2 = $ _____ so $\sqrt{} = 100$

26. $60^2 = $ _____ so $\sqrt{} = 60$

27. Describe in your own words a perfect square. Of the two numbers 25 and 50, identify which is a perfect square and explain why.

28. Use the following list of words and phrases to write the four steps in the order of operations.

add	square root
exponents	subtract
multiply	divide

parentheses or other grouping symbols

Simplify each expression by using the order of operations. See Examples 3 and 4.

29. $3^2 + 8 - 5$

30. $5^2 + 5 - 6$

31. $3 \cdot 7 - 6$

32. $5 \cdot 7 - 7$

33. $8 \cdot 5 \div 10$

34. $6 \cdot 8 \div 8$

35. $25 \div 5(8 - 4)$

36. $36 \div 18(7 - 3)$

37. $5 \cdot 3^2 + \dfrac{0}{8}$

38. $8 \cdot 3^2 - \dfrac{10}{2}$

39. $4 \cdot 1 + 8(9 - 2) + 3$

40. $3 \cdot 2 + 7(3 + 1) + 5$

41. $2^2 \cdot 3^3 + (20 - 15) \cdot 2$

42. $4^2 \cdot 5^2 + (20 - 9) \cdot 3$

43. $5\sqrt{36} - 2(4)$

44. $2 \cdot \sqrt{100} - 3(4)$

45. $8(2) + 3 \cdot 7 - 7 =$

46. $10(3) + 6 \cdot 5 - 20$

47. $2^3 \cdot 3^2 + 3(14 - 4)$

48. $3^2 \cdot 4^2 + 2(15 - 6)$

49. $7 + 8 \div 4 + \dfrac{0}{7}$

50. $6 + 8 \div 2 + \dfrac{0}{8}$

51. $3^2 + 6^2 + (30 - 21) \cdot 2$

52. $4^2 + 5^2 + (25 - 9) \cdot 3$

53. $7 \cdot \sqrt{81} - 5 \cdot 6$

54. $6 \cdot \sqrt{64} - 6 \cdot 5$

55. $8 \cdot 2 + 5(3 \cdot 4) - 6$

56. $5 \cdot 2 + 3(5 + 3) - 6$

57. $4 \cdot \sqrt{49} - 7(5 - 2)$

58. $3 \cdot \sqrt{25} - 6(3 - 1)$

59. $7(4 - 2) + \sqrt{9}$

60. $5(4 - 3) + \sqrt{9}$

61. $7^2 + 3^2 - 8 + 5$

62. $3^2 - 2^2 + 3 - 2$

63. $5^2 \cdot 2^2 + (8 - 4) \cdot 2$

64. $5^2 \cdot 3^2 + (30 - 20) \cdot 2$

65. $5 + 9 \div 3 + 6 \cdot 3$

66. $8 + 3 \div 3 + 6 \cdot 3$

67. $8 \cdot \sqrt{49} - 6(9 - 4)$

68. $8 \cdot \sqrt{49} - 6(5 + 3)$

69. $5^2 - 4^2 + 3 \cdot 6$

70. $3^2 + 6^2 - 5 \cdot 8$

71. $8 + 8 \div 8 + 6 + \dfrac{5}{5}$

72. $3 + 14 \div 2 + 7 + \dfrac{8}{8}$

73. $6 \cdot \sqrt{25} - 7(2)$

74. $8 \cdot \sqrt{36} - 4(6)$

75. $9 \cdot \sqrt{16} - 3 \cdot \sqrt{25}$

76. $6 \cdot \sqrt{81} - 3 \cdot \sqrt{49}$

77. $7 \div 1 \cdot 8 \cdot 2 \div (21 - 5)$

78. $12 \div 4 \cdot 5 \cdot 4 \div (15 - 13)$

79. $15 \div 3 \cdot 2 \cdot 6 \div (14 - 11)$

80. $9 \div 1 \cdot 4 \cdot 2 \div (11 - 5)$

81. $6 \cdot \sqrt{25} - 4 \cdot \sqrt{16}$

82. $10 \cdot \sqrt{49} - 4 \cdot \sqrt{64}$

83. $5 \div 1 \cdot 10 \cdot 4 \div (17 - 9)$

84. $15 \div 3 \cdot 8 \cdot 9 \div (12 - 8)$

85. $8 \cdot 9 \div \sqrt{36} - 4 \div 2 + (14 - 8)$

86. $3 - 2 + 5 \cdot 4 \cdot \sqrt{144} \div \sqrt{36}$

87. $2 + 1 - 2 \cdot \sqrt{1} + 4 \cdot \sqrt{81} - 7 \cdot 2$

88. $6 - 4 + 2 \cdot 9 - 3 \cdot \sqrt{225} \div \sqrt{25}$

89. $5 \cdot \sqrt{36} \cdot \sqrt{100} \div 4 \cdot \sqrt{9} + 8$

90. $9 \cdot \sqrt{36} \cdot \sqrt{81} \div 2 + 6 - 3 - 5$

Study Skills

▶▶▶ TAKING LECTURE NOTES

Study the set of sample math notes in this section, and read the comments about them. Then try to incorporate the techniques into your own math note taking in class.

OBJECTIVES

1 Apply note taking strategies, such as writing problems as well as explanations.

2 Use appropriate abbreviations in notes.

▶ The **date and title** of the day's lecture topic are always at the top of every page. **Always begin a new day with a new page.**

▶ Note the **definitions** of base and exponent are written in parentheses—don't trust your memory!

▶ **Skipping lines** makes the notes easier to read.

▶ See how the **direction word** (*simplify*) is emphasized and explained.

▶ A **star marks an important concept.** This is a warning to avoid future mistakes. **Note the underlining,** too, which highlights the importance.

▶ Notice the two columns, which allow for the example and its explanation to be close together. **Whenever you know you'll be given a series of steps to follow, try the two-column method.**

▶ Note the **brackets and arrows,** which clearly show how the problem is set up to be simplified.

January 2   *Exponents*

Exponents used to show repeated multiplication.

$$3 \cdot 3 \cdot 3 \cdot 3 \text{ can be written } 3^4 \quad \begin{array}{l}\text{exponent} \\ \text{(how many times} \\ \text{it's multiplied)}\end{array}$$

base (the number being multiplied)

Read 3^2 as 3 to the 2nd power or 3 squared

3^3 as 3 to the 3rd power or 3 cubed

3^4 as 3 to the 4th power

etc.

Simplifying an expression with exponents

→ *actually do the repeated multiplication*

2^3 means $2 \cdot 2 \cdot 2$ and $2 \cdot 2 \cdot 2 = 8$

★ *Careful!* 5^2 means $5 \cdot 5$ <u>NOT</u> $5 \cdot 2$

so $5^2 = 5 \cdot 5 = 25$ BUT $5^2 \neq 10$

Example	*Explanation*
Simplify $2^4 \cdot 3^2$	Exponents mean <u>multiplication</u>.
$2 \cdot 2 \cdot 2 \cdot 2 \cdot 3 \cdot 3$	Use 2 as a factor 4 times.
	Use 3 as a factor 2 times.
16 · 9	$2 \cdot 2 \cdot 2 \cdot 2$ is 16
	$3 \cdot 3$ is 9 16 · 9 is 144
144	Simplified result is 144 (no exponents left)

81

Why Are These Notes Brain Friendly?

The notes are **easy to look at,** and you know that the brain responds to things that are visually pleasing. Other techniques that are visually memorable are the use of spacing (the two columns), stars, underlining, and circling. All of these methods **allow your brain to take note of important concepts and steps.**

The notes are also **systematic,** which means that they use certain techniques regularly. This way, your brain easily recognizes the topic of the day, the signals that show an important point, and the steps to follow for procedures. When you develop a system that you always use in your notes, your notes are easy to understand later when you are reviewing for a test.

Find one or two people in your math class to work with. Compare each other's lecture notes over a period of a week or so. Ask yourself the following questions as you examine the notes.

1. What are you doing in your notes to show the **main points** or larger concepts? (Such as underlining, boxing, using stars, capital letters, etc.)

2. In what ways do you **set off the explanations** for worked problems, examples, or smaller ideas (subpoints)? (Such as indenting, using arrows, circling or boxing)

3. What does **your instructor do** to show that he or she is moving from one idea to the next? (Such as saying "Next" or "Any questions," "Now," or erasing the board, etc.)

4. **How do you mark** that in your notes? (Such as skipping lines, using dashes or numbers, etc.)

5. What **explanations (in words) do you give yourself** in your notes, so when those new dendrites you grew in lecture are fading, you can read your notes and still remember the new concepts later when you try to do your homework?

6. What **did you learn** by examining your classmates' notes?

   - _____
   - _____
   - _____

7. What **will you try** in your own note taking? List **four** techniques that you will use next time you take notes in math class.

   - _____
   - _____
   - _____
   - _____

1.9 ▶▶▶ Reading Pictographs, Bar Graphs, and Line Graphs

We have all heard the saying "A picture is worth a thousand words," and there may be some truth in this. Today, so much information and data are being presented in the form of pictographs, circle graphs, bar graphs, and line graphs that it is important to be able to read and understand these tools.

OBJECTIVE ➊ Read and understand a pictograph. A **pictograph** is a graph that uses pictures or symbols. It displays information that can be compared easily. However, since a symbol is used to represent a certain quantity, it can be difficult to determine the amount represented by a fraction of a symbol.

The Elvis Presley U.S. postage stamp retains the title as the most popular stamp of all time. The pictograph below compares the number of U.S. postage stamps sold in the five top releases. In this pictograph, it is difficult to determine what fractional amount is represented by the partial stamps.

ELVIS IS STILL KING

Five of the most popular U.S. postage stamps of all time.

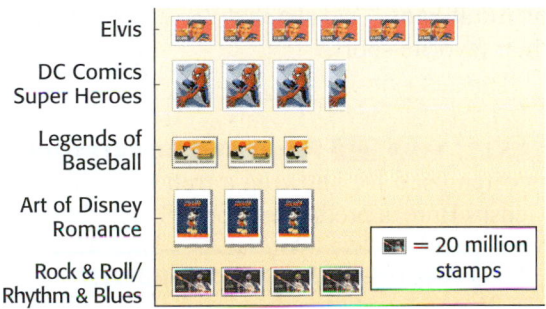

Source: United States Postal Service.

EXAMPLE 1 **Using a Pictograph**

Use the pictograph to answer each question.

> Each stamp symbol represents 20 million stamps.

(a) Which of the U.S. postage stamps shown has the lowest number sold?

The row representing Legends of Baseball has the fewest symbols. This means that the lowest number of U.S. postage stamps sold was Legends of Baseball.

(b) Approximately how many more Elvis stamps were sold than the Art of Disney Romance stamps?

The row representing the Elvis stamps has three more symbols than that for the Art of Disney stamps. This means that 3 • 20 million or 60,000,000 more Elvis stamps were sold than the Art of Disney Romance stamps.

Work Problem ➊ *at the Side.* ▶

OBJECTIVE ➋ Read and understand a bar graph. Bar graphs are useful for showing comparisons. For example, the following bar graph shows how many people out of every 100 fans surveyed chose each sport as their favorite.

OBJECTIVES

➊ Read and understand a pictograph.

➋ Read and understand a bar graph.

➌ Read and understand a line graph.

➊ Use the pictograph to answer each question.

(a) Which of the U.S. stamps had the second greatest number of sales?

(b) Approximately how many more Rock and Roll/Rhythm and Blues stamps were sold than Art of Disney Romance stamps?

ANSWERS

2. **(a)** Rock and Roll/Rhythm & Blues
   (b) about 20 million (or 20,000,000) more stamps

2 Use the bar graph to find the approximate number of fans who picked each sport as their favorite.

(a) College football

(b) Pro baseball

(c) Pro basketball

(d) College basketball

(e) Golf

3 Use the line graph to find the predicted population of the United States for each year.

(a) 2050

(b) 2075

(c) 2100

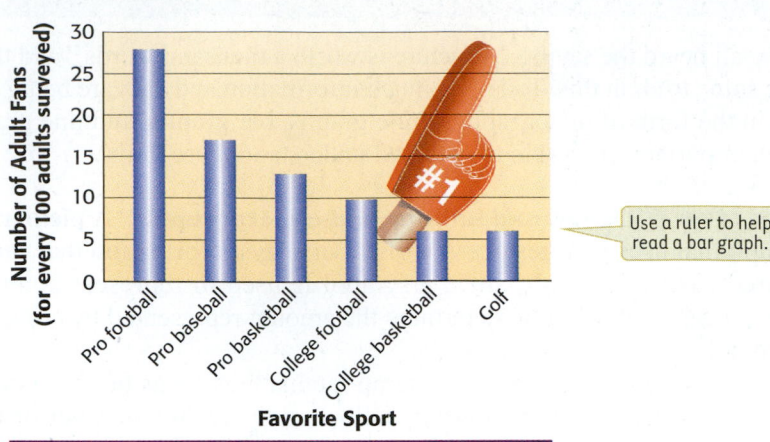

FAN APPEAL

Source: The Harris Poll.

EXAMPLE 2 Using a Bar Graph

Use the bar graph to find the number of fans who picked pro football as their favorite sport.

Use a ruler or straightedge to line up the top of the bar labeled "Pro football," with the numbers on the left edge of the graph, labeled "Number of Adult Fans." We see that 28 out of 100 adult fans picked pro football as their favorite sport.

◀ *Work Problem* **2** *at the Side.*

OBJECTIVE 3 Read and understand a line graph. A **line graph** is often used for showing a trend. The following line graph shows the U.S. Census Bureau predictions for U.S. population growth to the year 2100.

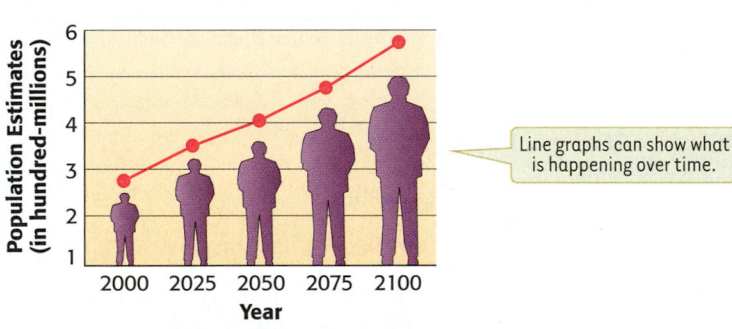

ANOTHER CENTURY OF GROWTH

Source: U.S. Census Bureau.

EXAMPLE 3 Using a Line Graph

Use the line graph to answer each question.

(a) What trend or pattern is shown in the graph?
    The population will continue to increase.

(b) What is the estimated population for 2025?
    Use a ruler or straightedge to line up the dot above the year labeled 2025 on the horizontal line with the numbers along the left edge of the graph. Notice that the label on the left side says "in hundred-millions." Since the 2025 dot is halfway between 3 and 4, the population in 2025 is halfway between 3 • 100,000,000 and 4 • 100,000,000 or 300,000,000 and 400,000,000. That means that the predicted population in 2025 is about 350,000,000 people.

 ◀ *Work Problem* **3** *at the Side.*

FOR
EXTRA
HELP
MyMathLab   Math XL PRACTICE   WATCH   DOWNLOAD   READ   REVIEW

The following pictograph shows the number of retail stores for the seven companies with the greatest number of outlets. Use the pictograph to answer Exercises 1–6. See Example 1.

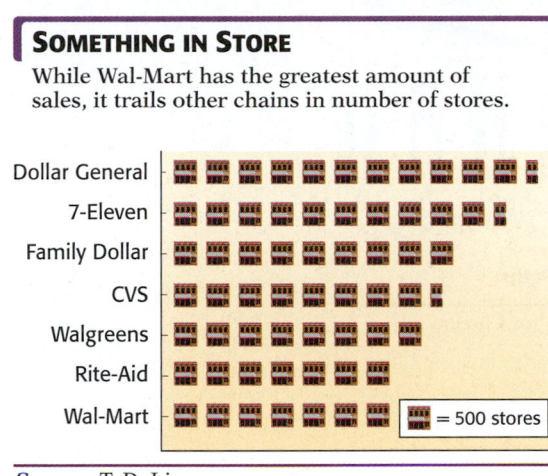

SOMETHING IN STORE
While Wal-Mart has the greatest amount of sales, it trails other chains in number of stores.

Dollar General
7-Eleven
Family Dollar
CVS
Walgreens
Rite-Aid
Wal-Mart

▦ = 500 stores

Source: T. D. Linx.

1. Find the number of Family Dollar retail stores.

2. Approximately how many retail stores does 7-Eleven have?

3. Which company has the greatest number of retail stores? How many is that?

4. Which companies have the least number of retail stores? How many does each one have?

5. How many fewer stores does Family Dollar have than Dollar General?

6. How many more retail stores does Walgreens have than Wal-Mart?

The following bar graph shows the results of a survey that was taken of 100 working adults to determine how they chose their careers. Use the bar graph to answer Exercises 7–12. See Example 2.

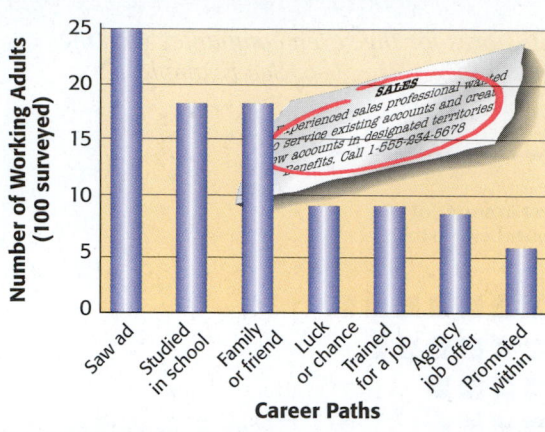

HOW DID YOU CHOOSE YOUR CAREER?

Source: Market Facts/TeleNation for Career
Education Corporation.

7. How many people found their careers as a result of training for a job?

8. How many people found their careers because they studied for the career in school?

9. (a) Which career path was taken by the greatest number of people?

(b) How many people used this path?

10. (a) Which career path was taken by the least number of people?

(b) How many people used this path?

11. How many more people found their careers as a result of "Studied in school" than "Luck or chance"?

12. Find the total number of people who found their careers as a result of either "Studied in school" or "Trained for a job."

Foxworthy Reforestation Company collected tree planting data and prepared the following line graph. Remembering that the tree planting data is shown in thousands of trees, use the line graph to answer Exercises 13–18. See Example 3.

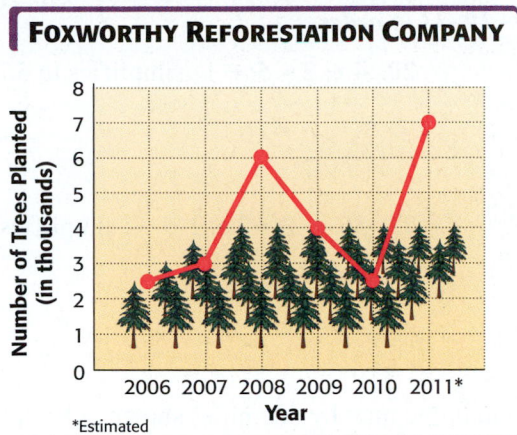

13. Which year had the greatest number of trees planted? How many trees were planted?

14. Which two years had the least number of trees planted? How many trees were planted in each of those years?

15. Find the increase in the number of trees planted from 2010 to 2011.

16. Find the decrease in the number of trees planted from 2009 to 2010.

17. Give three possible explanations for the decrease in trees planted in 2009.

18. Give three possible explanations for the increase in trees planted in 2011.

Relating Concepts (Exercises 19—23) For Individual or Group Work

Getting a correct answer in mathematics always depends on following the order of operations. Insert grouping symbols (parentheses) so that each given expression will result in the given number when simplified. **Work Exercises 19—23 in order.**

19. $7 - 2 \cdot 3 - 6$; simplifies to 9

20. $4 + 2 \cdot 5 + 1$; simplifies to 36

21. $36 \div 3 \cdot 3 \cdot 4$; simplifies to 16

22. $56 \div 2 \cdot 2 \cdot 2 + \dfrac{0}{6}$; simplifies to 7

23. The Good Shepherd Ranch owns one section of land (one mile by one mile) shown below as Parcel 1. Parcels 2 and 3 are leased from neighbors.

 (a) Use the order of operations to write an expression for the distance around the combined parcels.

 (b) How many feet of barbed wire are needed for a three-strand barbed wire fence around the parcels?

 (c) How many miles of barbed wire is this? (*Hint:* 1 mile $=$ 5280 feet.)

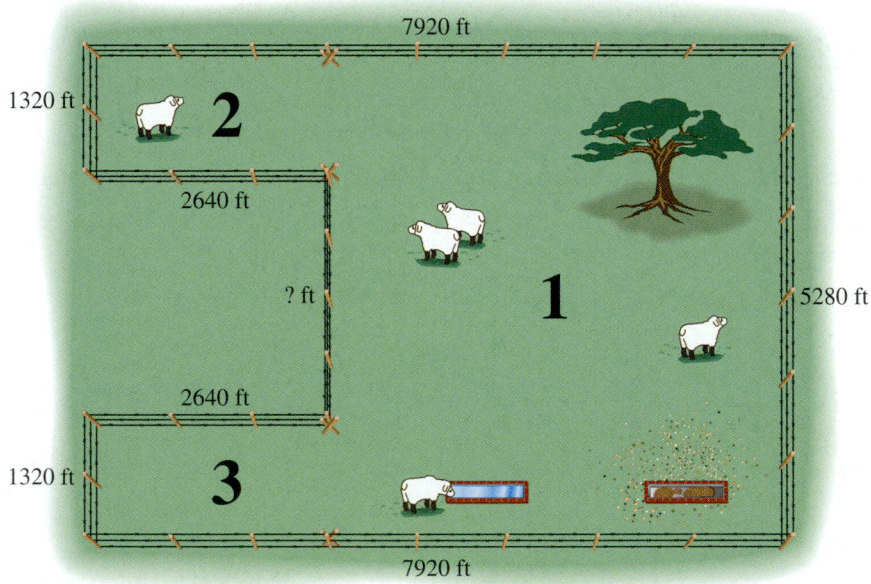

1.10 ▶▶▶ Solving Application Problems

Most problems involving applications of mathematics are written out in sentence form. You need to read the problem carefully to decide how to solve it.

OBJECTIVE 1 Find indicator words in application problems.
As you read an application problem, look for **indicator words** that help you determine whether to use addition, subtraction, multiplication, or division. Some of these indicator words are shown here.

Addition	Subtraction	Multiplication	Division	Equals
plus	less	product	divided by	is
more	subtract	double	divided into	the same as
more than	subtracted from	triple	quotient	equals
added to	difference	times	goes into	equal to
increased by	less than	of	divide	yields
sum	fewer	twice	divided equally	results in
total	decreased by	twice as much	per	are
sum of	loss of			
increase of	minus			
gain of	take away			

CAUTION
The word *and* does not always indicate addition, so it does not appear as an indicator word in the preceding table. Notice how the "and" shows the location of several different operation signs below.

The sum of 6 *and* 2 is 6 + 2.
The difference of 6 *and* 2 is 6 − 2.
The product of 6 *and* 2 is 6 · 2.
The quotient of 6 *and* 2 is 6 ÷ 2.

OBJECTIVE 2 Solve application problems. Solve application problems by using the following six steps.

Solving an Application Problem

Step 1 **Read** the problem carefully and be certain you *understand* what the problem is asking. It may be necessary to read the problem several times.

Read and **understand** a problem before you begin.

Step 2 Before doing any calculations, **work out a plan** and try to visualize the problem. Draw a sketch if possible. Know which facts are given and which must be found. Use *indicator words* to help decide on the *plan* (whether you will need to add, subtract, multiply, or divide).

Step 3 **Estimate** a *reasonable answer* by using rounding.

Always **estimate** the final answer.

Step 4 **Solve** the problem by using the facts given and your plan.

Step 5 **State the answer.**

Step 6 **Check** your work. If the answer does not seem reasonable, begin again by reading the problem.

Check your answer to see if it is *reasonable*.

1 Pick the most reasonable answer for each problem.

(a) A grocery clerk's hourly wage: $1.40; $14; $140

(b) The total length of five sport-utility vehicles: 8 ft; 18 ft; 80 ft; 800 ft

(c) The cost of heart bypass surgery: $1000; $100,000; $10,000,000

2 Solve each problem.

(a) On a recent geology field trip, 84 fossils were collected. If the fossils are divided equally among John, Sean, Jenn, and Kara, how many fossils will each receive?

(b) This week there are 408 children attending a winter sports camp. If 12 children are assigned to each camp counselor, how many counselors are needed?

> **CAUTION**
> Be certain that you know what the problem is asking before you try to solve it.

OBJECTIVE **3** **Estimate an answer.** The six problem-solving steps give a systematic approach for solving word problems. Each of the steps is important, but special emphasis should be placed on Step 3, estimating a *reasonable answer.* Many times an "answer" just does not fit the problem.

What is a reasonable answer? Read the problem and try to determine the approximate size of the answer. Should the answer be part of a dollar, a few dollars, hundreds, thousands, or even millions of dollars? For example, if a problem asks for the cost of a man's shirt, would an answer of $20 be reasonable? $2000? $2? $200?

> **CAUTION**
> Always estimate the answer, then look at your final result to be sure it fits your estimate and is reasonable. This step will give greater success in problem solving.

◀ *Work Problem* **1** *at the Side.*

EXAMPLE 1 **Applying Division**

A community group has raised $8260 for charity. Equal amounts are given to the Food Bank, Children's Center, Boy Scouts of America, and the Women's Shelter. How much did each group receive?

Step 1 **Read.** A reading of the problem shows that the four charities divided $8260 equally.

Step 2 **Work out a plan.** The indicator words, *divided equally,* show that the amount each received can be found by dividing $8260 by 4.

Step 3 **Estimate.** Round $8260 to $8000. Then $8000 ÷ 4 = $2000, so a reasonable answer would be a little greater than $2000 each.

Step 4 **Solve.** Find the actual answer by dividing $8260 by 4.

$$\begin{array}{r} 2065 \\ 4\overline{)8260} \end{array}$$

Step 5 **State the answer.** Each charity received $2065.

Step 6 **Check.** The exact answer of $2065 is reasonable, as $2065 is close to the estimated answer of $2000. Is the answer $2065 correct? Check by multiplying.

$2065 ← Amount received by each charity

× 4 ← Number of charities

$8260 ← Total raised; matches number given in problem

> Remember: Check your work.

◀ *Work Problem* **2** *at the Side.*

EXAMPLE 2 **Applying Addition**

One week, Andrea Abriani, operations manager, decided to total the stroller production at Safe T First Strollers. The daily production figures were 7642 strollers on Monday, 8150 strollers on Tuesday, 7916 strollers on Wednesday, 8419 strollers on Thursday, and 7704 strollers on Friday. Find the total production for the week.

Step 1 **Read.** In this problem, the production for each day is given and the total production for the week must be found.

Step 2 **Work out a plan.** Add the daily production figures to arrive at the weekly total.

Step 3 **Estimate.** Because the production was about 8000 strollers per day for a week of five days, a reasonable estimate would be $5 \cdot 8000 = 40{,}000$ strollers.

Step 4 **Solve.** Find the exact answer by adding the production numbers for the 5 days.

$$
\begin{array}{r}
\mathbf{39{,}831} \leftarrow \text{Check by} \\
7642 \quad \text{adding up.} \\
8150 \\
7916 \\
8419 \\
+\ 7704 \\
\hline
39{,}831 \leftarrow \text{Number of strollers for the week}
\end{array}
$$

Add from bottom to top to check addition.

Step 5 **State the answer.** Abriani's total production figure for the week was 39,831 strollers.

Step 6 **Check.** The exact answer of 39,831 strollers is close to the estimate of 40,000 strollers, so it is reasonable. Add up the columns to check the exact answer.

> 🖩 **Calculator Tip**   The calculator solution to Example 2 uses chain calculations.
>
> 7642 ⊕ 8150 ⊕ 7916 ⊕ 8419 ⊕ 7704 ⊜ **39,831**

Work Problem **3** *at the Side.* ▶

EXAMPLE 3 **Determining Whether Subtraction Is Necessary**

The number of miles driven this year is 3028 fewer than the number driven last year. The miles driven last year was 16,735. Find the number of miles driven this year.

Step 1 **Read.** In this problem, the miles driven decreased from last year to this year. The miles driven last year and the decrease in miles driven are given. This year's miles driven must be found.

Step 2 **Work out a plan.** The indicator word, *fewer,* shows that subtraction must be used to find the number of miles driven this year.

Step 3 **Estimate.** Because the driving last year was about 17,000 miles, and the decrease in driving is about 3000 miles, a reasonable estimate would be $17{,}000 - 3000 = 14{,}000$ miles.

Continued on Next Page

---

3 Solve each problem.

(a) During the semester, Cindy received the following points on examinations and quizzes: 92, 81, 83, 98, 15, 14, 15, and 12. Find her total points for the semester.

(b) Stephanie Dixon works at the telephone order desk of a catalog sales company. One week she had the following number of customer contacts: Monday, 78; Tuesday, 64; Wednesday, 118; Thursday, 102; and Friday, 196. How many customer contacts did she have that week?

4 Solve each problem.

(a) A home has a living area of 1450 square feet, while an apartment has 980 square feet. Find the difference between the number of square feet in the two living areas.

Step 4 **Solve.** Find the exact answer by subtracting 3078 from 16,735.

$$\begin{array}{r} 16,735 \\ -3078 \\ \hline 13,707 \end{array}$$

Step 5 **State the answer.** The driving this year is 13,707 miles.

Step 6 **Check.** The exact answer of 13,707 is reasonable, as it is close to the estimate of 14,000. Check by adding.

$$\begin{array}{r} 13,707 \\ +3028 \\ \hline 16,735 \end{array}$$
 ← miles driven this year
 ← decrease in miles driven
 ← miles driven last year; matches number given in problem

◄ *Work Problem* **4** *at the Side.*

(b) The Antique Military Vehicle Collectors (AMVC) had $14,863 in their club treasury bank account. After writing a check for $1180 to rent a display hall, find the amount remaining in the club account.

EXAMPLE 4 **Solving a Two-Step Problem**

In May, a landlord received $720 from each of eight tenants. After paying $2180 in expenses, how much rent money did the landlord have left?

Step 1 **Read.** The problem asks for the amount of rent remaining after expenses have been paid.

Step 2 **Work out a plan.** The wording *from each of eight tenants* indicates that the eight rents must be totaled. Since the rents are all the same, use multiplication to find the total rent received. Then, subtract expenses.

Step 3 **Estimate.** The amount of rent is about $700, making the total rent received about $700 • 8 = $5600. The expenses are about $2000. A reasonable estimate of the amount remaining is $5600 − $2000 = $3600.

5 Solve each problem.

(a) Brenda is paid $685 for each kitchen remodeling job that she sells. If she sold 6 remodeling jobs and had $320 in sales expense deducted, how much did she make?

Step 4 **Solve.** Find the exact amount by first multiplying $720 by 8 (the number of tenants).

$$\begin{array}{r} \$720 \\ \times8 \\ \hline \$5760 \end{array}$$

Then subtract the $2180 in expenses from $5760.

$$\begin{array}{r} \$5760 \\ -\$2180 \\ \hline \$3580 \end{array}$$
The exact answer is close to the estimate and reasonable.

Step 5 **State the answer.** The amount remaining is $3580.

(b) An Internet book company had sales of 12,628 books with a profit of $6 for each book sold. If 863 books were returned, how much profit remains?

Step 6 **Check.** The exact answer of $3580 is reasonable, since it is close to the estimated answer of $3600. Check by adding the expenses to the amount remaining and then dividing by 8.

$$\$3580 + \$2180 = \$5760$$

Always check your work.

$$\begin{array}{r} \$720 \\ 8\overline{)5760} \end{array}$$
Matches the rent amount given in the problem

◄ *Work Problem* **5** *at the Side.*

1.10 ▶▶▶ Exercises

Solve each application problem. First use front end rounding to estimate the answer. Then find the exact answer. See Examples 1–4.

1. Last week, SUBWAY sold 602 Veggie Delite sandwiches, 935 ham sandwiches, 1328 turkey breast sandwiches, 757 roast beef sandwiches, and 1586 SUBWAY club sandwiches. Find the total number of sandwiches sold.

Estimate:

Exact:

2. During a recent week, Radio Flyer, Inc. manufactured 32,815 Model #18 wagons, 4875 steel miniwagons, 1975 wood 40-inch wagons, 15,308 scooters, and 9815 new-design plastic wagons. Find the total number of units manufactured.

Estimate:

Exact:

The graph shows the number of contaminated meat recalls by the U.S. Department of Agriculture over a five-year period. Use the graph for Exercises 3–4.

3. How many more meat recalls were there in 2003 than in 2007?

Estimate:

Exact:

4. How many fewer meat recalls were there in 2006 than in 2008?

Estimate:

Exact:

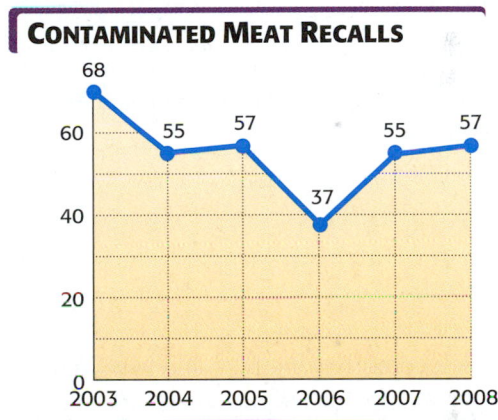

CONTAMINATED MEAT RECALLS

Source: U.S. Department of Agriculture.

5. A packing machine can package 236 first-aid kits each hour. At this rate, find the number of first-aid kits packaged in 24 hours.

Estimate:

Exact:

6. If 450 admission tickets to a classic car show are sold each day, how many tickets are sold in a 12-day period?

Estimate:

Exact:

7. Clarence Hanks, coordinator of Toys for Tots, has collected 2628 toys. If his group can give the same number of toys to each of 657 children, how many toys will each child receive?

Estimate:

Exact:

8. If profits of $680,000 are divided evenly among a firm's 1000 employees, how much money will each employee receive?

Estimate:

Exact:

9. The number of boaters and campers at the lake was 8392 on Friday. If this was 4218 more than the number of people at the lake on Wednesday, how many were there on Wednesday?

Estimate:

Exact:

10. The community has raised $52,882 for the homeless shelter. If the total amount needed for the shelter is $75,650, find the additional amount needed.

Estimate:

Exact:

11. Turn down the thermostat in the winter and you can save money and energy. In the upper Midwest, setting back the thermostat from 68° to 55° at night can save $34 per month on fuel. Find the amount of money saved in five months.

Estimate:

Exact:

12. The cost of tuition and fees at a community college is $785 per quarter. If Gale Klein has five quarters remaining, find the total amount that she will need for tuition and fees.

Estimate:

Exact:

The table shows the average annual earnings of those in careers that do not require a four year degree. Refer to the table to answer Exercises 13–16. First use front end rounding to estimate the answer. Then find the exact answer.

13. How much more does a dental hygienist earn than a flight attendent?

Estimate:

Exact:

14. How much more does an air traffic controller earn than a court reporter?

Estimate:

Exact:

No Degree? Apply Here?

Four years of college may be your best ticket to a high-paying career, but these solid jobs don't require an undergraduate degree:

Profession	Median Annual Earnings		
Air traffic controller	$87,930	Locomotive engineer	46,540
Nuclear power reactor operator	60,180	Telecom equipment installer/repairer*	46,390
Dental hygienist*	54,700	Funeral director*	42,010
Elevator installer/ repairer	51,630	Aircraft mechanic*	41,990
Real estate broker	51,380	Brick mason	41,590
Commercial pilot (non-airline)	47,410	Police officer	40,970
Electrical power line installer/repairer	47,210	Electrician	40,770
		Flight attendant	40,600
		Court reporter*	40,410
		Real estate appraiser*	38,950

*Requires associate's degree or vocational diploma

Source: U.S. Department of Labor.

15. Mr. White is a locomotive engineer and Mrs. White is a court reporter. Mr. Easterly is an aircraft mechanic and Mrs. Easterly is a real estate broker.

(a) Which couple has higher earnings?

Estimate:

Exact:

(b) Find the difference in the earnings.

Estimate:

Exact:

16. Mr. Means is a funeral director and Mrs. Means is an electrician. Mr. Strong is a police officer and Mrs. Strong is a commercial pilot.

(a) Which couple has higher earnings?

Estimate:

Exact:

(b) Find the difference in the earnings.

Estimate:

Exact:

17. Ronda Biondi decides to establish a monthly budget. She will spend $695 for rent, $340 for food, $435 for child care, $240 for transportation, $180 for other expenses, and she will put the remainder in savings. If her monthly take-home pay is $2240, find her monthly savings.

Estimate:

Exact:

18. Robert Heisner had $2874 in his checking account. He wrote checks for $308 for auto repairs, $580 for child support, and $778 for an insurance payment. Find the amount remaining in his account.

Estimate:

Exact:

19. There are 43,560 square feet in one acre. How many square feet are there in 138 acres?

Estimate:

Exact:

20. The number of gallons of water polluted each day in an industrial area is 209,670. How many gallons of water are polluted each year? (Use a 365-day year.)

Estimate:

Exact:

The Internet was used to find the following minivan optional features and the price of each feature. Use this information to answer Exercises 21–24.

Safety and Security Options		Convenience and Comfort Options	
Option	Cost	Option	Cost
Integrated child bench seats	$400	Power sliding door	$400
Supplemental side air bags	$1395	8-way power seat	$395
Security alarm	$170	Roof rack	$250
Hands-free communication	$360	Power door locks	$315
Power adjustable pedal	$195	Keyless entry	$150
Full-size spare tire	$160	AM/FM stereo and CD	$350

Source: www.edmunds.com

21. Find the total cost of all Safety and Security Options listed.

Estimate:

Exact:

22. Find the total cost of all Convenience and Comfort Options listed.

Estimate:

Exact:

23. A new-car dealer offers an option value package that includes integrated child bench seats, security alarm, keyless entry, and a power sliding door at a cost of $980. If Jill buys the value package instead of paying for each option separately, how much will she save?

Estimate:

Exact:

24. A new-car dealer offers an option package that includes integrated child bench seats, security alarm, full-size spare tire, power door locks, 8-way power seat, and a roof rack for a total of $1550. How much will Samuel save if he buys the option package instead of paying for each option separately?

Estimate:

Exact:

25. The Enabling Supply House purchased 6 wheelchairs at $1256 each and 15 speech compression recorder-players at $895 each. Find the total cost.

Estimate:

Exact:

26. A college bookstore buys 17 desktop computers at $506 each and 13 printers at $482 each. Find the total cost.

Estimate:

Exact:

27. Being able to identify indicator words is helpful in determining how to solve an application problem. Write three indicator words for each of these operations: add, subtract, multiply, and divide. Write two indicator words that mean equals.

28. Identify and explain the six steps used to solve an application problem. You may refer to the text if you need help, but use your own words.

29. Write in your own words why it is important to estimate a reasonable answer. Give three examples of what might be a reasonable answer to a math problem from your daily activities.

30. First estimate by rounding to thousands, then find the exact answer to the following problem.

$$7438 + 6493 + 2380$$

Do the two answers vary by more than 1000? Why? Will estimated answers always vary from exact answers?

Estimate:

Exact:

Solve each application problem. See Examples 1–4.

31. Steve Edwards, manager, decided to total his sales at SUBWAY. The daily sales figures were $2358 on Monday, $3056 on Tuesday, $2515 on Wednesday, $1875 on Thurdsay, $3978 on Friday, $3219 on Saturday, and $3008 on Sunday. Find his total sales for the week.

32. The numbers of visitors at a war veteran's memorial during one week are 5318; 2865; 4786; 1998; 3899; 2343; and 7221. Find the total attendance for the week.

33. A car weighs 2425 pounds. If its 582-pound engine is removed and replaced with a 634-pound engine, what will the car weigh?

34. Barbara has $2324 in her preschool operating account. She spends $734 from this account, and then the class parents raise $568 in a rummage sale. Find the balance in the account after she deposits the money from the rummage sale.

35. In a recent survey of Reno/Lake Tahoe hotels, the cost per night at Harrah's in Reno was $45, while the cost at Harrah's in Lake Tahoe was $99 per night. Find the amount saved on a 7-night stay at Harrah's in Reno instead of staying at Harrah's in Lake Tahoe. (*Source:* Harrah's Casinos and Hotels.)

36. The most expensive hotel room in a recent study was the Ritz-Carlton at $645 per night, while the least expensive was Motel 6 at $74 per night. Find the amount saved in a 4-night stay at Motel 6 instead of staying at the Ritz-Carlton. (*Source:* Ritz-Carlton/Motel 6.)

37. A youth soccer association raised $7588 through fund-raising projects. After expenses of $838 were paid, the balance of the money was divided evenly among the 18 teams. How much did each team receive?

38. Feather Farms Egg Ranch collected 3545 eggs in the morning and 2575 eggs in the afternoon. If the eggs are packed in flats containing 30 eggs each, find the number of flats needed for packing.

39. A theater owner wants to provide enough seating for 1250 people. The main floor has 30 rows of 25 seats in each row. If the balcony has 25 rows, how many seats must be in each balcony row to satisfy the owner's seating requirements?

40. Jennie makes 24 grapevine wreaths per week to sell to gift shops. She works 40 weeks a year and packages six wreaths per box. If she ships equal quantities to each of five shops, find the number of boxes each store will receive.

Chapter 1 ▷▷▷ Summary

▶ **Key Terms**

1.1	**whole numbers**	The whole numbers are 0, 1, 2, 3, 4, 5, 6, 7, 8, and so on.
	place value	The place value of each digit in a whole number is determined by its position in the whole number.
	table	A table is a display of facts in rows and columns.
1.2	**addition**	The process of finding the total is addition.
	addends	The numbers being added in an addition problem are addends.
	sum (total)	The answer in an addition problem is called the sum.
	commutative property of addition	The commutative property of addition states that the order of numbers in an addition problem can be changed without changing the sum.
	associative property of addition	The associative property of addition states that grouping the addition of numbers differently does not change the sum.
	regrouping	The process of regrouping is used in an addition problem when the sum of the digits in a column is greater than 9.
	perimeter	The perimeter is the distance around the outside edges of a figure.
1.3	**minuend**	The number from which another number (the subtrahend) is being subtracted is the minuend.
	subtrahend	The subtrahend is the number being subtracted in a subtraction problem.
	difference	The answer in a subtraction problem is called the difference.
	regrouping	The process of regrouping is used in subtraction if a digit is less than the one directly below it.
1.4	**factors**	The numbers being multiplied are called factors. For example, in $3 \times 4 = 12$, both 3 and 4 are factors.
	product	The answer in a multiplication problem is called the product.
	commutative property of multiplication	The commutative property of multiplication states that changing the order of the factors in a multiplication problem does not change the product.
	associative property of multiplication	The associative property of multiplication states that grouping the numbers differently does not change the product.
	chain multiplication problem	A multiplication problem having more than two factors is a chain multiplication problem.
	multiple	The product of two whole number factors is a multiple of those numbers.
1.5	**dividend**	The number being divided by another number in a division problem is the dividend.
	divisor	The divisor is the number doing the dividing in a division problem.
	quotient	The answer in a division problem is called the quotient.
	short division	A method of dividing a number by a one-digit divisor is short division.
	remainder	The remainder is the number left over when two numbers do not divide exactly.
1.6	**long division**	The process of long division is used to divide by a number with more than one digit.
1.7	**rounding**	Rounding is used to find a number that is close to the original number, but easier to work with. Use the $\approx$ sign, which means "approximately equal to."
	estimate	An estimated answer is one that is close to the exact answer.
	front end rounding	Rounding to the highest possible place so that all the digits become zeros except the first one is front end rounding.
1.8	**square root**	The square root of a whole number is the number that can be multiplied by itself to produce the given number.
	perfect square	A number that is the square of a whole number is a perfect square.
	order of operations	For problems or expressions with more than one operation, the order of operations tells what to do first, second, and so on to get the correct answer.
1.9	**pictograph**	A graph that uses pictures or symbols to show data is a pictograph.
	bar graph	A graph that uses bars of various heights to show quantity is a bar graph.
	line graph	A graph that uses dots connected by lines to show trends is a line graph.
1.10	**indicator words**	Words in a problem that indicate the necessary operations—addition, subtraction, multiplication, or division—are indicator words.

▶ New Symbols

$\approx$ This sign is used to show that an answer has been estimated. It means "is approximately equal to."

$\sqrt{}$ The symbol for square root.

5^2 The small raised 2 is an exponent; it tells how many times to use 5 (the base) as a factor in multiplication.

▶ Test Your Word Power

See how well you have learned the vocabulary in this chapter. Answers follow the Quick Review.

1. When using **addends** you are performing
   - A. division
   - B. subtraction
   - C. addition
   - D. multiplication.

2. The subtrahend is the
   - A. number being multiplied
   - B. number being subtracted
   - C. number being added
   - D. answer in division.

3. A **factor** is
   - A. the answer in an addition problem
   - B. one of two or more numbers being added
   - C. one of two or more numbers being multiplied
   - D. one of two or more numbers being divided.

4. The **divisor** is
   - A. the number being rounded
   - B. the number being multiplied
   - C. always the largest number
   - D. the number doing the dividing.

5. We use **rounding** to
   - A. avoid solving a problem
   - B. purely guess at the answer
   - C. help estimate a reasonable answer
   - D. find the remainder.

6. A **perfect square** is
   - A. the square of a whole number
   - B. the same as square root
   - C. similar to a perfect triangle

▶ Quick Review

Concepts	Examples

1.1 Reading and Writing Whole Numbers

Do not use the word *and* when writing a whole number. Commas help divide the periods or groups for ones, thousands, millions, and billions. A comma is not needed when a number has four digits or fewer.

795 is written *seven hundred ninety-five*.

9,768,002 is written *nine million, seven hundred sixty-eight thousand, two*

1.2 Adding Whole Numbers

Add from top to bottom, starting with the ones column and working left. To check, add from bottom to top.

(Add up to check.)

$$\begin{array}{r} \mathbf{1\,1\,4\,0} \\ 6\,8\,7 \\ 2\,6 \\ 9 \\ +\ 4\,1\,8 \\ \hline 1\,1\,4\,0 \end{array}$$

Addends

Sum

1.2 Commutative Property of Addition

Changing the order of the addends in an addition problem does not change the sum.

$$2 + 4 = 6$$
$$4 + 2 = 6$$

By the commutative property, the sum is the same.

1.2 Associative Property of Addition

Grouping the addends differently when adding does not change the sum.

$$(2 + 3) + 4 = 5 + 4 = 9$$
$$2 + (3 + 4) = 2 + 7 = 9$$

By the associative property, the sum is the same.

1.3 Subtracting Whole Numbers

Subtract the subtrahend from the minuend to get the difference, using regrouping when necessary. To check, add the difference to the subtrahend to get the minuend.

Problem

$$\begin{array}{r} \overset{6\ 12\ 18}{4\,7\,3\,8} \leftarrow \text{Minuend} \\ -\quad 6\,4\,9 \quad \text{Subtrahend} \\ \hline 4\,0\,8\,9 \quad \text{Difference} \end{array}$$

Check

$$\begin{array}{r} 4\,0\,8\,9 \\ +\quad 6\,4\,9 \\ \hline 4\,7\,3\,8 \end{array}$$

Concepts	Examples

1.4 Multiplying Whole Numbers

Use $\times$, $\cdot$ (a raised dot), or parentheses to indicate multiplication.

The numbers being multiplied are called *factors*. The multiplicand is being multiplied by the multiplier, giving the product. When the multiplier has more than one digit, partial products must be used and added to find the product.

$$3 \times 4 \quad \text{or} \quad 3 \cdot 4 \quad \text{or} \quad (3)(4) \quad \text{or} \quad 3(4)$$

$$
\begin{array}{r}
78 \\
\times \quad 24 \\
\hline
312 \\
156 \\
\hline
1872
\end{array}
$$

78 — Multiplicand ⎱ Factors
24 — Multiplier ⎰
312 — Partial product
156 — Partial product (move one position left)
1872 — Product

1.4 Commutative Property of Multiplication

The product in a multiplication problem remains the same when the order of the factors is changed.

$$3 \times 4 = 12$$
$$4 \times 3 = 12$$

By the commutative property, the product is the same.

1.4 Associative Property of Multiplication

Grouping the factors differently when multiplying does not change the product.

$$(2 \times 3) \times 4 = 6 \times 4 = 24$$
$$2 \times (3 \times 4) = 2 \times 12 = 24$$

By the associative property, the product is the same.

1.5 Dividing Whole Numbers

$\div$ and $\overline{)}$ mean divide.

Also a —, as in $\frac{25}{5}$, means to divide the top number (dividend) by the bottom number (divisor).

$$
\begin{array}{r}
22 \quad \leftarrow \text{Quotient} \\
\text{Divisor} \rightarrow 4\overline{)88} \quad \leftarrow \text{Dividend} \\
88 \\
\hline
0
\end{array}
$$

Dividend

$$\frac{88}{4} = 22 \quad \leftarrow \text{Quotient}$$

$$88 \div 4 = 22$$

Dividend — Quotient

Divisor

1.7 Rounding Whole Numbers

Rules for Rounding:

Step 1 Locate the place to be rounded, and draw a line under it.

Step 2 If the next digit to the right is 5 or more, increase the underlined digit by 1. If the next digit is 4 or less, do not change the underlined digit.

Step 3 Change all digits to the right of the underlined place to zeros.

Round 726 to the nearest ten.

Next digit is 5 or more.

726

Tens place increases by 1 $(2 + 1 = 3)$.

726 rounds to 730.

Round 1,498,586 to the nearest million.

Next digit is 4 or less.

1,498,586

Millions place does not change.

1,498,586 rounds to 1,000,000.

1.7 Front End Rounding

Front end rounding is rounding to the highest possible place so that all the digits become 0 except the first digit.

Round each number using front end rounding.

76 rounds to 80.

348 rounds to 300.

6512 rounds to 7000.

23,751 rounds to 20,000.

652,179 rounds to 700,000.

Concepts	Examples

1.8 Order of Operations

Problems may have several operations. Work these problems using the order of operations.

1. Do all operations inside parentheses or other grouping symbols.
2. Simplify any expressions with exponents and find any square roots $\left(\sqrt{} \right)$.
3. Multiply or divide proceeding from left to right.
4. Add or subtract proceeding from left to right.

Simplify, using the order of operations.

$$7 \cdot \sqrt{9} - 4 \cdot 5 \qquad \text{Find the square root.}$$
$$7 \cdot 3 - 4 \cdot 5 \qquad \text{Multiply from left to right.}$$
$$21 - 20 = 1 \qquad \text{Subtract.}$$

1.9 Reading Pictographs, Bar Graphs, and Line Graphs

A *pictograph* uses pictures or symbols to show data.

When reading a pictograph, be certain that you determine the quantity represented by each picture or symbol.

A *bar graph* uses bars of various heights to show quantity.

When reading a bar graph, use a straightedge to line up the top of the bar with the numbers along the left edge of the graph.

A *line graph* uses dots connected by lines to show trends.

When reading a line graph, use a straightedge to line up the dot with the numbers along the left edge of the graph.

1.10 Application Problems

Steps for Solving an Application Problem

Step 1 **Read** the problem carefully, perhaps several times.

Step 2 **Work out a plan** before starting. Draw a sketch if possible.

Step 3 **Estimate** a reasonable answer.

Step 4 **Solve** the problem.

Step 5 **State the answer.**

Step 6 **Check** your work. If the answer is not reasonable, start over.

Manuel earns $118 on Sunday, $87 on Monday, and $63 on Tuesday. Find his total earnings for the 3 days.

Step 1 The earnings for each day are given, and the total for the 3 days must be found.

Step 2 Add the daily earnings to find the total.

Step 3 Since the earnings were about $100 + $90 + $60 = $250, a reasonable estimate would be approximately $250.

Step 4 **$268**   Check by adding up
$$\begin{array}{r} \$118 \\ 87 \\ +\ \ 63 \\ \hline \$268 \end{array}$$ Total earnings

Step 5 Manuel's total earnings are $268.

Step 6 The exact answer is reasonable, because it is close to the estimate of $250.

ANSWERS TO TEST YOUR WORD POWER

1. C; *Example:* In $2 + 3 = 5$, the 2 and the 3 are addends.
2. B; *Example:* In $5 - 4 = 1$, the 4 is the subtrahend.
3. C; *Example:* In $3 \times 5 = 15$, the numbers 3 and 5 are factors.
4. D; *Example:* In $8 \div 4 = 2$, $\frac{8}{4} = 2$, and $4\overline{)8}$, the 4 is the divisor.
5. C; *Example:* We can use rounding to estimate our answer and then determine whether the exact answer is reasonable.
6. A; *Example:* 25 is a perfect square because $5^2 = 25$ and 5 is a whole number.

Chapter 1 ▶▶▶ Review Exercises

If you need help with any of these Review Exercises, look in the section indicated in brackets.

[1.1] *Write the digits for the given period or group in each number.*

1. 6573

thousands

ones

2. 36,215

thousands

ones

3. 105,724

thousands

ones

4. 1,768,710,618

billions

millions

thousands

ones

Rewrite each number in words.

5. 728

6. 15,310

7. 319,215

8. 62,500,005

Rewrite each number in digits.

9. ten thousand, eight

10. two hundred million, four hundred fifty-five

[1.2] *Add.*

11.
```
   72
 + 38
```

12.
```
   54
 + 67
```

13.
```
   807
  4606
 +  51
```

14.
```
   8215
      9
 + 7433
```

15.
```
   2130
    453
   8107
 +  296
```

16.
```
   5684
    218
   2960
 +  983
```

17.
```
    5 732
   11,069
       37
    1 595
 + 22,169
```

18.
```
    3 451
   12,286
       43
    1 291
 + 32,784
```

[1.3] *Subtract.*

19. $\begin{array}{r} 64 \\ -28 \\ \hline \end{array}$	**20.** $\begin{array}{r} 46 \\ -19 \\ \hline \end{array}$	**21.** $\begin{array}{r} 375 \\ -186 \\ \hline \end{array}$	**22.** $\begin{array}{r} 573 \\ -389 \\ \hline \end{array}$

23. $\begin{array}{r} 7416 \\ -567 \\ \hline \end{array}$	**24.** $\begin{array}{r} 5210 \\ -883 \\ \hline \end{array}$	**25.** $\begin{array}{r} 2210 \\ -1986 \\ \hline \end{array}$	**26.** $\begin{array}{r} 99{,}704 \\ -73{,}838 \\ \hline \end{array}$

[1.4] *Multiply.*

27. $\begin{array}{r} 7 \\ \times 7 \\ \hline \end{array}$	**28.** $\begin{array}{r} 8 \\ \times 0 \\ \hline \end{array}$	**29.** $8\,(4)$	**30.** $8\,(8)$

31. $(5)\,(9)$	**32.** $(6)\,(7)$	**33.** $7 \cdot 8$	**34.** $9 \cdot 9$

Work each chain multiplication.

35. $5 \times 4 \times 2$	**36.** $9 \times 1 \times 5$	**37.** $4 \times 4 \times 3$	**38.** $2 \times 2 \times 2$

39. $(6)\,(0)\,(8)$	**40.** $(7)\,(1)\,(6)$	**41.** $6 \cdot 1 \cdot 8$	**42.** $7 \cdot 7 \cdot 0$

Multiply.

43. $\begin{array}{r} 28 \\ \times 3 \\ \hline \end{array}$	**44.** $\begin{array}{r} 46 \\ \times 8 \\ \hline \end{array}$	**45.** $\begin{array}{r} 58 \\ \times 9 \\ \hline \end{array}$	**46.** $\begin{array}{r} 98 \\ \times 1 \\ \hline \end{array}$

47. $\begin{array}{r} 625 \\ \times 8 \\ \hline \end{array}$	**48.** $\begin{array}{r} 374 \\ \times 8 \\ \hline \end{array}$	**49.** $\begin{array}{r} 1349 \\ \times 4 \\ \hline \end{array}$	**50.** $\begin{array}{r} 9163 \\ \times 5 \\ \hline \end{array}$

51. $\begin{array}{r} 7456 \\ \times 2 \\ \hline \end{array}$	**52.** $\begin{array}{r} 2880 \\ \times 7 \\ \hline \end{array}$	**53.** $\begin{array}{r} 93{,}105 \\ \times 5 \\ \hline \end{array}$	**54.** $\begin{array}{r} 21{,}873 \\ \times 8 \\ \hline \end{array}$

55. 35
× 25

56. 74
× 32

57. 98
× 12

58. 68
× 75

59. 472
× 33

60. 392
× 77

61. 4051
× 219

62. 1527
× 328

Find each total cost.

63. 30 scientific calculators at $12 per calculator

64. 76 subscribers at $14 per subscription

65. 318 drill bit sets at $64 per set

66. 114 earplugs at $6 per earplug

Multiply by using the shortcut for multiples of 10.

67. 280
× 50

68. 340
× 70

69. 517
× 400

70. 637
× 500

71. 16,000
× 8 000

72. 43,000
× 2 100

[1.5] *Divide. If the division is not possible, write "undefined."*

73. $20 \div 4$

74. $35 \div 5$

75. $42 \div 7$

76. $18 \div 9$

77. $\dfrac{54}{9}$

78. $\dfrac{36}{9}$

79. $\dfrac{49}{7}$

80. $\dfrac{0}{6}$

81. $\dfrac{148}{0}$

82. $\dfrac{0}{23}$

83. $\dfrac{64}{8}$

84. $\dfrac{81}{9}$

[1.5–1.6] *Divide.*

85. $4\overline{)328}$

86. $3\overline{)294}$

87. $6\overline{)26,532}$

88. $76\overline{)26,752}$

89. $2704 \div 18$

90. $15,525 \div 125$

[1.7] *Round as indicated.*

91. 817 to the nearest ten

92. 15,208 to the nearest hundred

93. 20,643 to the nearest thousand

94. 67,485 to the nearest ten-thousand

Round each number to the nearest ten, nearest hundred, and nearest thousand.
Remember to round from the original number.

	Ten	**Hundred**	**Thousand**
95. 3487	_____	_____	_____
96. 20,065	_____	_____	_____
97. 98,201	_____	_____	_____
98. 352,118	_____	_____	_____

[1.8] *Find each square root by using the Perfect Squares Table on page 75.*

99. $\sqrt{16}$ **100.** $\sqrt{49}$ **101.** $\sqrt{144}$ **102.** $\sqrt{196}$

Identify the exponent and the base, and then simplify each expression.

103. 7^3 **104.** 3^6 **105.** 5^3 **106.** 4^5

Simplify each expression by using the order of operations.

107. $7^2 - 15$ **108.** $6^2 - 10$ **109.** $2 \cdot 3^2 \div 2$

110. $9 \div 1 \cdot 2 \cdot 2 \div (11 - 2)$ **111.** $\sqrt{9} + 2\,(3)$ **112.** $6 \cdot \sqrt{16} - 6 \cdot \sqrt{9}$

[1.9] *The bar graph shows the number of parents out of 100 surveyed who nag their children about performing certain household chores.*

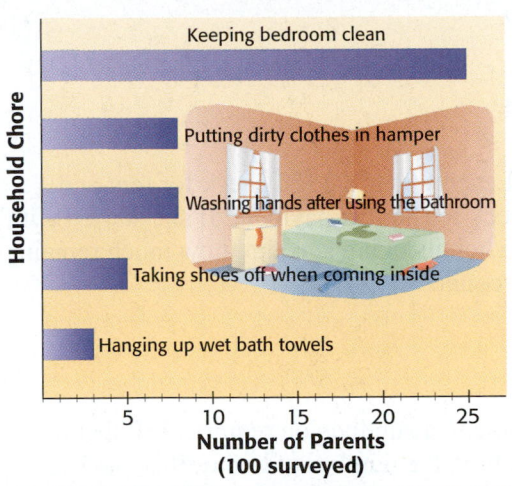

CLEAN UP YOUR ROOM

Keeping bedroom clean

Putting dirty clothes in hamper

Washing hands after using the bathroom

Taking shoes off when coming inside

Hanging up wet bath towels

Household Chore (vertical axis)

5   10   15   20   25

**Number of Parents
(100 surveyed)**

Source: Opinion Research Corporation for the Soap and Detergent Association.

113. How many parents nagged their children about washing hands after using the bathroom?

114. Find the number of parents who nagged their children about taking shoes off when coming inside.

115. Which household chore was nagged about by the greatest number of parents? How many were there?

116. Which household chore was nagged about by the least number of parents? How many did this?

[1.10] *Solve each application problem. First use front end rounding to estimate the answer. Then find the exact answer.*

117. Bank of America processes 40 million checks each day. Find the number of checks processed by the bank in a year. Use a 365-day year. (*Source:* Bank of America.)

Estimate:

Exact:

118. A pulley on an evaporative cooler turns 1400 revolutions per minute. How many revolutions will the pulley turn in 60 minutes?

Estimate:

Exact:

119. There are 144 plastic forks in a box. Find the number of plastic forks in 15 boxes.

Estimate:

Exact:

120. A drum contains 6000 brackets. How many brackets are in 30 drums?

Estimate:

Exact:

121. It takes 2000 hours of work to build one home. How many hours of work are needed to build 12 homes?

Estimate:

Exact:

122. A Japanese bullet train travels 80 miles in 1 hour. Find the number of miles traveled in 5 hours.

Estimate:

Exact:

123. Find the total cost if SUBWAY buys 32 baking ovens at $1538 each and 28 warming ovens at $887 each.

Estimate:

Exact:

124. A newspaper carrier has 62 customers who take the paper daily and 21 customers who take the paper on weekends only. A daily customer pays $16 per month and a weekend-only customer pays $7 per month. Find the total monthly collections.

Estimate:

Exact:

125. This holiday season, the average amount consumers plan to spend on holiday shopping for others is $620. If they plan to spend $107 on holiday shopping for themselves, how much more do they plan to spend on others than themselves? (*Source:* National Retail Federation.)

Estimate:

Exact:

126. A stamping machine produces 986 license plates each hour. How long will it take to produce 32,538 license plates?

Estimate:

Exact:

127. A food canner uses 1 pound of pork for every 175 cans of pork and beans. How many pounds of pork are needed for 8750 cans?

Estimate:

Exact:

128. Rachel Leach writes a $520 check for rent and a $385 check for her car payment. If she started with $1924 in her checking account, how much remains in her account.

Estimate:

Exact:

129. Nitrogen sulfate is used in farming to enrich nitrogen-poor soil. If 625 pounds of nitrogen sulfate are spread per acre, how many acres can be spread with 32,500 pounds of nitrogen sulfate?

Estimate:

Exact:

130. Each home in a subdivision requires 180 feet of fencing. Find the number of homes that can be fenced with 5760 feet of fencing material.

Estimate:

Exact:

> ▶▶▶ **Mixed Review Exercises***

Perform the indicated operations.

131. 4 (83)

132. 7 (64)

133.
```
  309
-  56
```

134.
```
  835
- 247
```

135.
```
  662
+ 379
```

136.
```
  789
+ 872
```

137.
```
 38,140
- 6 078
```

138.
```
 29,156
- 4 209
```

139. 21 ÷ 7

140. $\frac{42}{6}$

141.
```
  7 218
      3
     18
  1 791
 82,623
+  1 982
```

142.
```
  3 812
      5
     22
  1 836
 75,134
+  2 369
```

143. $\frac{9}{0}$

144. $\frac{7}{1}$

145. 27,600 ÷ 4

---

* The order of exercises in this final group does not correspond to the order in which topics occur in the chapter. This random ordering should help you prepare for the chapter test in yet another way.

146. $18{,}480 \div 8$ **147.** $\begin{array}{r} 8430 \\ \times\ 128 \\ \hline \end{array}$ **148.** $\begin{array}{r} 21{,}702 \\ \times\ 6 \\ \hline \end{array}$ ▦ **149.** $34\overline{)3672}$ ▦ **150.** $68\overline{)14{,}076}$

151. Rewrite 376,853 in words.

152. Rewrite 408,610 in words.

153. Round 8749 to the nearest hundred.

154. Round 400,503 to the nearest thousand.

Find each square root.

155. $\sqrt{64}$

156. $\sqrt{81}$

Find each total cost.

157. 308 pairs of knee guards at $18 per pair

158. 84 dishwashers at $370 per dishwasher

159. 208 baseball hats at $11 per hat

160. 607 boxes of avocados at $26 per box

Solve each application problem.

161. There are 52 playing cards in a deck. How many cards are there in nine decks?

162. Your college bookstore receives textbooks packed 20 books per carton. How many textbooks are received in a delivery of 180 cartons?

163. Push-type gasoline-powered lawn mowers cost $100 less than self-propelled mowers that you walk behind. If a self-propelled mower costs $380, find the cost of a push-type mower.

164. The Country Day School wants to raise $218,450 to construct and equip a computer lab. If $103,815 has already been raised, how much more is needed?

American River Raft Rentals lists the following daily raft rental fees. Notice that there is an additional $2 launch fee payable to the park system for each raft rented. Use this information to solve Exercises 165 and 166.

AMERICAN RIVER RAFT RENTALS

Size	Rental Fee	Launch Fee
4-person	$28	$2
6-person	$38	$2
10-person	$70	$2
12-person	$75	$2
16-person	$85	$2

Source: American River Raft Rentals.

165. On a recent Tuesday the following rafts were rented: 6 4-person; 15 6-person; 10 10-person; 3 12-person; and 2 16-person. Find the total receipts, including the $2 per-raft launch fee.

166. On the 4th of July the following rafts were rented: 38 4-person; 73 6-person; 58 10-person; 34 12-person; and 18 16-person. Find the total receipts, including the $2 per-raft launch fee.

Chicago plans to build the world's tallest structure, the Chicago Spire. The pictogram below shows the height of the tallest buildings in the United States. Use this information to solve Exercises 167–170.

167. How much taller is the Chicago Spire (planned) than the Trump Tower in Chicago?

168. How much taller is the Freedom Tower in New York than the San Francisco Towers (planned)?

169. (a) What is the combined height of all the buildings shown? Include the planned buildings.

(b) Is the combined height of these six buildings greater or less than a mile? How much greater or less than a mile? (*Hint*: 1 mile = 5280 ft)

170. The height of the San Francisco Towers (planned) is equivalent to the length of how many football fields? (*Hint:* A football field is 100 yards long and 1 yard = 3 feet.)

Building Higher and Higher

Developers in San Francisco have submitted a proposal that includes two 1,200-foot towers, which would be the tallest buildings west of Chicago. How some planned buildings compare with the nation's tallest (in feet):

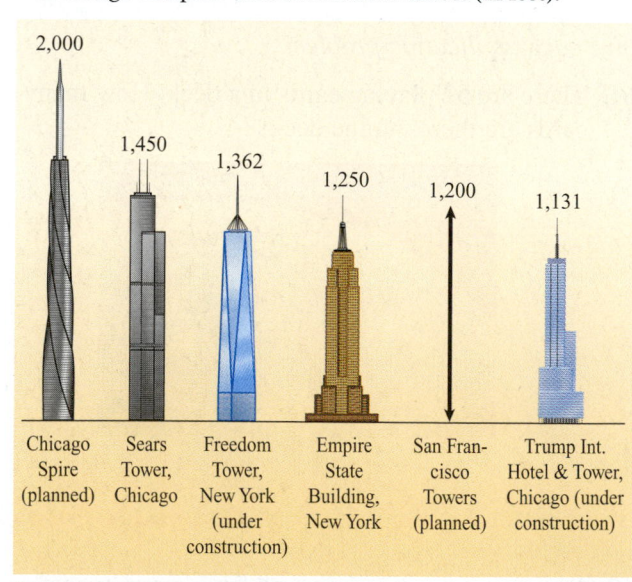

Source: USA Today research.

Write each number in words.

1. 9205

2. 25,065

3. Use digits to write four hundred twenty-six thousand, five.

Add.

4.
$$\begin{array}{r} 853 \\ 66 \\ 4022 \\ + \ 3589 \\ \hline \end{array}$$

5.
$$\begin{array}{r} 17{,}063 \\ 7 \\ 12 \\ 1\ 505 \\ 93{,}710 \\ + \ \ \ \ 333 \\ \hline \end{array}$$

Subtract.

6.
$$\begin{array}{r} 9009 \\ - \ 7964 \\ \hline \end{array}$$

7.
$$\begin{array}{r} 9075 \\ - \ 2869 \\ \hline \end{array}$$

Multiply

8. $7 \times 6 \times 4$

9. $57 \cdot 3000$

10. $85\,(19)$

11.
$$\begin{array}{r} 7381 \\ \times \ 603 \\ \hline \end{array}$$

Divide. If the division is not possible, write "undefined."

12. $16\overline{)112{,}752}$

13. $\dfrac{835}{0}$

14. $19{,}241 \div 42$

15. $280\overline{)44{,}800}$

Round as indicated.

16. 6347 to the nearest ten

17. 76,502 to the nearest thousand

1. _____

2. _____

3. _____

4. _____

5. _____

6. _____

7. _____

8. _____

9. _____

10. _____

11. _____

12. _____

13. _____

14. _____

15. _____

16. _____

17. _____

18. _____

19. _____

20. *Estimate:* _____
    Exact: _____

21. *Estimate:* _____
    Exact: _____

22. *Estimate:* _____
    Exact: _____

23. *Estimate:* _____
    Exact: _____

24. _____

25. _____

Simplify each expression.

18. $5^2 + 8\,(2)$

19. $7 \cdot \sqrt{64} - 14 \cdot 2$

Solve each application problem. First use front end rounding to estimate the answer. Then find the exact answer.

20. Amy collects the following monthly rents from the tenants in her fourplex: $485, $500, $515, and $425. After she pays expenses of $785, how much does she have left?

21. The major producer of ethanol made from corn is the United States. If 374 gallons of ethanol can be produced from the corn grown on one acre of land, how many acres are needed to produce 86,394 gallons. (*Source:* Earth Policy Institute.)

22. Barb Weaks paid $528 for tires, $195 for brakes, and $235 for a timing belt. If this money was withdrawn from her checking account, which had a balance of $1906, find her new balance.

23. The technicians at a chicken ranch identify the sex of 48 baby chicks each minute for 4 hours in the morning and 36 baby chicks each minute for 3 hours in the afternoon. Find the total number of baby chicks identified. (*Source:* Discovery Channel, *Dirty Jobs.*)

24. Explain in your own words the rules for rounding numbers. Give an example of rounding a number to the nearest ten-thousand.

25. List the six steps for solving application problems.

2

Multiplying and Dividing Fractions

Most recipes include ingredients that use common fractions and mixed numbers in their measurements. The recipe shown below uses Jelly Belly jelly beans and will make $2\frac{1}{2}$ dozen (30) cookies. But suppose you wanted to make 3 dozen, 10 dozen, or even $3\frac{1}{2}$ dozen cookies? You would have to multiply or divide each of the ingredients to arrive at the proper amounts needed. In this chapter we discuss multiplication and division of fractions, which you need to know when cooking or baking.

(See Exercise 39 in **Section 2.7** and Exercises 25–28 in **Section 2.8**.)

NEST COOKIES

$1\frac{1}{2}$ cups all purpose
flour

1 tsp. baking powder

$\frac{1}{2}$ tsp. salt

$\frac{1}{2}$ cup shortening

$\frac{1}{2}$ cup sugar

1 egg white

1 tsp. vanilla

2 Tbs. milk

2 cups shredded
coconut

5 oz. Jelly Belly
jelly beans
assorted flavors

Heat oven to 375° F. Sift together flour, baking powder, and salt and set aside. In large bowl beat shortening, sugar, egg white, and vanilla until well blended. Add flour mixture and milk until blended. Stir in coconut.

Roll dough into a ball, divide in half. Roll 15 one-inch balls from each half and place on ungreased baking sheet. Make thumb print depression in center of each ball to form nest. Bake 6 minutes. Remove from oven and place 4 Jelly Belly beans in center of each cookie. Return to oven and bake 5 more minutes. Transfer cookies to wire rack to cool. Makes 30 cookies.

2.1 ▶▶▶ Basics of Fractions

OBJECTIVES

1 Use a fraction to show which part of a whole is shaded.

2 Identify the numerator and denominator.

3 Identify proper and improper fractions.

In **Chapter 1** we discussed whole numbers. Many times, however, we find that parts of whole numbers are considered. One way to write parts of a whole is with **fractions.** Another way is with decimals, which is discussed in **Chapter 4**.

OBJECTIVE 1 **Use a fraction to show which part of a whole is shaded.** The number $\frac{1}{8}$ is a fraction that represents 1 of 8 equal parts. Read $\frac{1}{8}$ as "one eighth."

A fraction represents part of a whole.

EXAMPLE 1 **Identifying Fractions**

Use fractions to represent the shaded portions and the unshaded portions of each figure.

(a) The figure on the left has 6 equal parts. The 1 shaded part is represented by the fraction $\frac{1}{6}$. The *un*shaded part is $\frac{5}{6}$.

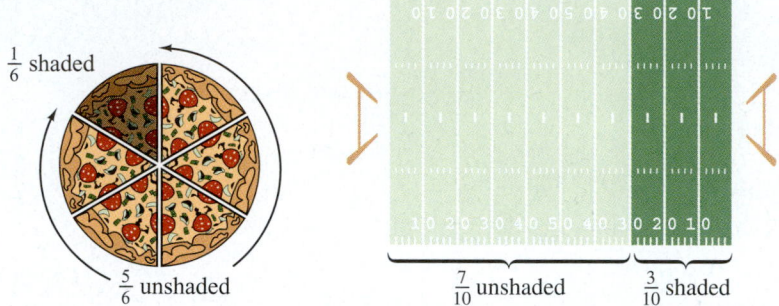

$\frac{1}{6}$ shaded

$\frac{5}{6}$ unshaded

$\frac{7}{10}$ unshaded $\frac{3}{10}$ shaded

(b) The 3 shaded parts of the 10-part figure on the right are represented by the fraction $\frac{3}{10}$. The *un*shaded part is $\frac{7}{10}$.

◀ Work Problem **1** at the Side.

Fractions can be used to represent more than one whole object.

EXAMPLE 2 **Representing Fractions Greater Than 1**

Use a fraction to represent the shaded part of each figure.

(a)
$\frac{1}{4}$

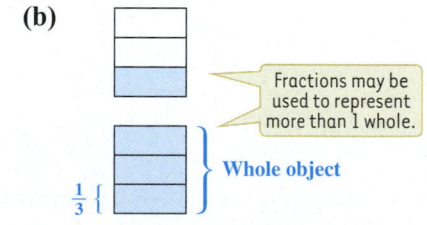

Whole object

(b)

Fractions may be used to represent more than 1 whole.

Whole object

$\frac{1}{3}$

An area equal to 5 of the $\frac{1}{4}$ parts is shaded. Write this as $\frac{5}{4}$.

An area equal to 4 of the $\frac{1}{3}$ parts is shaded, so $\frac{4}{3}$ is shaded.

◀ Work Problem **2** at the Side.

OBJECTIVE 2 **Identify the numerator and denominator.** In the fraction $\frac{2}{3}$, the number 2 is the **numerator** and 3 is the **denominator**. The bar between the numerator and the denominator is the *fraction bar*.

Fraction bar → $\dfrac{2 \leftarrow \text{Numerator}}{3 \leftarrow \text{Denominator}}$

---

1 Write fractions for the shaded portions and the unshaded portions of each figure.

(a)
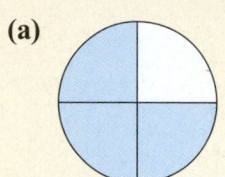

(b)
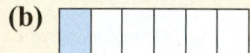

(c)
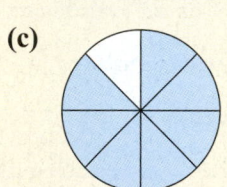

2 Write fractions for the shaded portions of each figure.

(a)
$\frac{1}{7}$
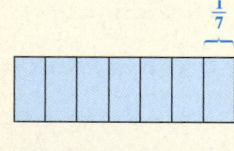
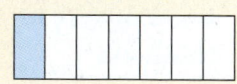

(b)
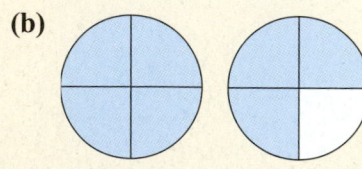

ANSWERS

1. **(a)** $\frac{3}{4}$; $\frac{1}{4}$ **(b)** $\frac{1}{6}$; $\frac{5}{6}$ **(c)** $\frac{7}{8}$; $\frac{1}{8}$

2. **(a)** $\frac{8}{7}$ **(b)** $\frac{7}{4}$

Numerators and Denominators

The **denominator** of a fraction shows the number of equivalent parts in the whole, and the **numerator** shows how many parts are being considered.

Note

Remember that a bar, —, is one of the division symbols, and that division by 0 is undefined. A fraction with a denominator of 0 is also undefined.

EXAMPLE 3   **Identifying Numerators and Denominators**

Identify the numerator and denominator in each fraction.

(a) $\dfrac{3}{4}$ **(b)** $\dfrac{8}{5}$

$$\dfrac{3}{4} \begin{matrix} \leftarrow \text{Numerator} \\ \leftarrow \text{Denominator} \end{matrix} \qquad \dfrac{8}{5} \begin{matrix} \leftarrow \text{Numerator} \\ \leftarrow \text{Denominator} \end{matrix}$$

Work Problem ③ *at the Side.* ▶

OBJECTIVE ③ **Identify proper and improper fractions.**   Fractions are sometimes called *proper* or *improper* fractions.

Proper and Improper Fractions

If the numerator of a fraction is *less* than the denominator, the fraction is a **proper fraction.** A proper fraction is less than 1 whole.

If the numerator is *greater than or equal to* the denominator, the fraction is an **improper fraction.** An improper fraction is greater than or equal to 1 whole.

Proper Fractions	Improper Fractions
$\dfrac{5}{8}\quad\dfrac{3}{5}\quad\dfrac{23}{24}$	$\dfrac{6}{5}\quad\dfrac{10}{10}\quad\dfrac{115}{112}$

EXAMPLE 4   **Classifying Types of Fractions**

(a) Identify all proper fractions in this list.

$$\dfrac{3}{4}\quad\dfrac{5}{9}\quad\dfrac{17}{5}\quad\dfrac{9}{7}\quad\dfrac{3}{3}\quad\dfrac{12}{25}\quad\dfrac{1}{9}\quad\dfrac{5}{3}$$

Proper fractions have a numerator that is smaller than the denominator. The proper fractions are shown below.

$$\dfrac{3}{4} \leftarrow 3 \text{ is smaller than 4.} \qquad \dfrac{5}{9}\quad\dfrac{12}{25}\quad\dfrac{1}{9} \quad \boxed{\text{A proper fraction is less than 1.}}$$

(b) Identify all improper fractions in the list in part (a).

Improper fractions have a numerator that is equal to or greater than the denominator. The improper fractions are shown below.

$$\dfrac{17}{5} \leftarrow 17 \text{ is greater than 5.} \qquad \dfrac{9}{7}\quad\dfrac{3}{3}\quad\dfrac{5}{3} \quad \boxed{\begin{array}{l}\text{An improper fraction is equal to or greater than 1.}\end{array}}$$

Work Problem ④ *at the Side.* ▶

---

③ Identify the numerator and the denominator. Draw a picture with shaded parts to show each fraction. Your drawings may vary, but they should have the correct number of shaded parts.

(a) $\dfrac{2}{3}$

(b) $\dfrac{1}{4}$

(c) $\dfrac{8}{5}$

(d) $\dfrac{5}{2}$

④ From the following group of fractions:

$$\dfrac{2}{3}\quad\dfrac{4}{3}\quad\dfrac{3}{4}\quad\dfrac{8}{8}\quad\dfrac{3}{1}\quad\dfrac{1}{3}$$

(a) list all proper fractions;

(b) list all improper fractions.

ANSWERS

3. (a) N: 2; D: 3

(b) N: 1; D: 4 ▭▭▭▭

(c) N: 8; D: 5

(d) N: 5; D: 2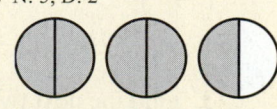

4. (a) $\dfrac{2}{3}, \dfrac{3}{4}, \dfrac{1}{3}$ **(b)** $\dfrac{4}{3}, \dfrac{8}{8}, \dfrac{3}{1}$

Math in the Media

QUILT PATTERNS

People who make quilts often base their designs on a block that is cut into a grid of 4, 9, 16, 25, or 49 squares. The quilter chooses various colors for the pieces. Each quilt design shown was selected from Antique Quilt Designs at their Web site http://earlywomanmasters.net/quilts/index.html.

1. Identify the makeup of the block as 4, 9, 16, etc. Each color is what fractional part of the block?

Aunt Eliza's Star

Block size: _____

Tan: _____ Purple: _____

Green: _____

Barbara Fritchie Star

Block size: _____

Blue: _____ White: _____

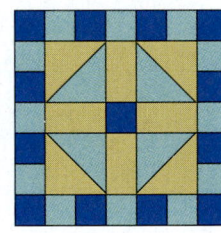

Cobwebs

Block size: _____

Mauve: _____ Purple: _____

Yellow: _____

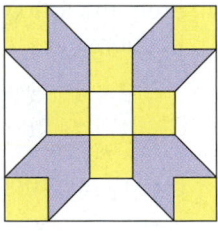

Farmer's Daughter

Block size: _____

Lavender: _____ White: _____

Yellow: _____

Tall Ships

Block size: _____

Blue: _____ Green: _____

Light gray: _____

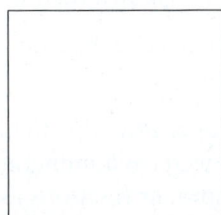

Lincoln's Platform

Block size: _____

Blue: _____ Aqua: _____

Gold: _____

2. Use the blocks to design and color your own quilt patterns. Tell the fractional part of the block that is represented by each color.

3. Find the next two numbers in this pattern: 4, 9, 16, 25, _____, _____.

4. Explain how the pattern works:

Write fractions to represent the shaded and unshaded portions of each figure. See Examples 1 and 2.

1.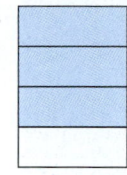

2.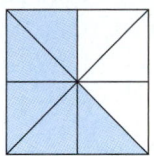

3.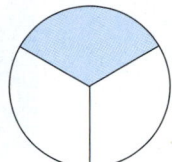

4.

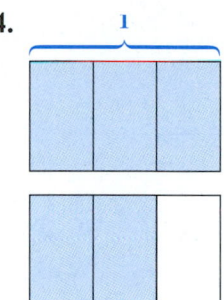

5.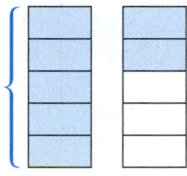

6.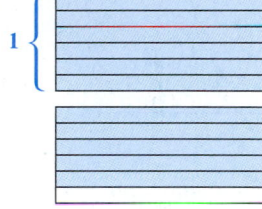

7. What fraction of these 6 bills has a life span of 2 years or greater? What fraction has a life of 4 years or less? What fraction has a life of 9 years?

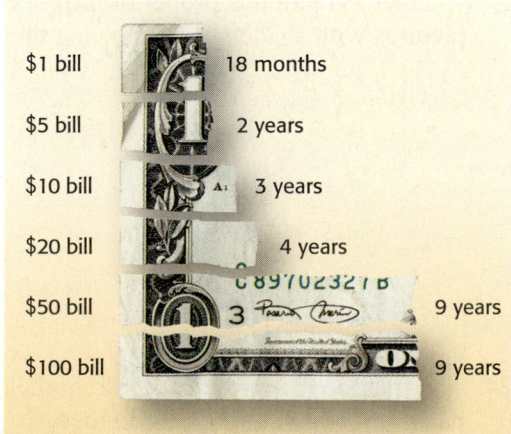

A BILL'S LIFE

A $1 bill lasts about 18 months as compared with the average lifespan of other denominations:

$1 bill	18 months
$5 bill	2 years
$10 bill	3 years
$20 bill	4 years
$50 bill	9 years
$100 bill	9 years

Source: Federal Reserve System; Bureau of Engraving and Printing.

8. What fraction of the 5 pets shown are dogs? What fraction are cats? What fraction of the pets are tan?

9. In an American Sign Language (ASL) class of 25 students, 8 are hearing impaired. What fraction of the students are hearing impaired?

10. Of 98 bicycles in a bike rack, 67 are mountain bikes. What fraction of the bicycles are not mountain bikes?

11. There are 520 rooms in a hotel. If 217 of the rooms are reserved for nonsmokers, what fraction of the rooms are for smokers?

12. Of the 46 employees at the college bookstore, 15 are full-time while the rest are part-time student help. What fraction of the bookstore employees work part-time?

Identify the numerator and denominator. See Example 3.

	Numerator	Denominator		Numerator	Denominator
13. $\frac{4}{5}$	_____	_____	**14.** $\frac{5}{6}$	_____	_____
15. $\frac{9}{8}$	_____	_____	**16.** $\frac{7}{5}$	_____	_____

List the proper and improper fractions in each group. See Example 4.

		Proper	Improper
17.	$\frac{8}{5}$ $\frac{1}{3}$ $\frac{5}{8}$ $\frac{6}{6}$ $\frac{12}{2}$ $\frac{7}{16}$	_____	_____
18.	$\frac{1}{3}$ $\frac{3}{8}$ $\frac{16}{12}$ $\frac{10}{8}$ $\frac{6}{6}$ $\frac{3}{4}$	_____	_____
19.	$\frac{3}{4}$ $\frac{3}{2}$ $\frac{5}{5}$ $\frac{9}{11}$ $\frac{7}{15}$ $\frac{19}{18}$	_____	_____
20.	$\frac{12}{12}$ $\frac{15}{11}$ $\frac{13}{12}$ $\frac{11}{8}$ $\frac{17}{17}$ $\frac{19}{12}$	_____	_____

21. Write a fraction of your own choice. Label the parts of the fraction and write a sentence describing what each part represents. Draw a picture with shaded parts showing your fraction.

22. Give one example of a proper fraction and one example of an improper fraction. What determines whether a fraction is proper or improper? Draw pictures with shaded parts showing these fractions.

Fill in the blanks to complete each sentence.

23. The fraction $\frac{3}{8}$ represents _____ of the _____ equal parts into which a whole is divided.

24. The fraction $\frac{7}{16}$ represents _____ of the _____ equal parts into which a whole is divided.

25. The fraction $\frac{5}{24}$ represents _____ of the _____ equal parts into which a whole is divided.

26. The fraction $\frac{24}{32}$ represents _____ of the _____ equal parts into which a whole is divided.

Study Skills

▶▶▶ HOMEWORK: HOW, WHY, AND WHEN

It is best for your brain if you keep up with the reading and homework in your math class. Remember that the more times you work with the information, the more dendrites you grow! So, give yourself every opportunity to read, work problems, and review your mathematics.

You have two choices for reading your math textbook. Read the short descriptions below and decide which will be best for you.

Maddy learns best by listening to her teacher explain things. She "gets it" when she sees the instructor work problems on the board. She likes to ask questions in class and put the information in her notes. She has learned that it helps if she has *previewed* the section before the lecture, so she knows generally what to expect in class. *But after the class instruction*, when Maddy gets home, she finds that she can understand the math textbook easily. She remembers what her teacher said, and she can double-check her notes if she gets confused. So, Maddy does her **careful** reading of the section in her text **after** hearing the classroom lecture on the topic.

De'Lore, on the other hand, feels he learns well by reading on his own. He prefers to read the section and try working the example problems before coming to class. That way, he already knows what the teacher is going to talk about. Then, he can follow the teacher's examples more easily. It is also easier for him to take notes in class. De'Lore likes to have his questions answered right away, which he can do if he has already read the chapter section. So, De'Lore **carefully** reads the section in his text **before** he hears the classroom lecture on the topic.

Notice that there is **no one right way** to work with your textbook. You always must figure out what works best for you. Note also that both Maddy and De'Lore work with one section at a time. **The key is that you read the textbook regularly!** The rest of this activity will give you some ideas of how to make the most of your reading.

Try the following steps as you **read** your math textbook.

▶ Read slowly. Read only one section—or even part of a section—at a time.

▶ Do the sample problems in the margins **as you go.** Check them right away. The answers are at the bottom of the page.

▶ If your mind wanders, work problems on separate paper and write explanations in your own words.

▶ Make study cards as you read each section. Pay special attention to the yellow and blue boxes in the book. Make cards for new vocabulary, rules, procedures, formulas, and sample problems.

▶ **NOW**, you are ready to do your homework assignment!

OBJECTIVES

1 Select an appropriate strategy for homework.

2 Use textbook features effectively.

Preview before Class; Read Carefully after Class

Read Carefully before Class

Why Are These Reading Techniques Brain Friendly?

The steps at the left encourage you to be **actively working with the material** in your text. Your brain grows dendrites when it is doing something.

These methods require you to **try several different techniques,** not just the same thing over and over. Your brain loves variety!

Also, the techniques allow you to **take small breaks** in your learning. Those rest periods are crucial for good dendrite growth.

Now Try This ▶▶▶

Which steps for reading this book will be most helpful for you?

1. _____

2. _____

3. _____

Homework

Why Are These Homework Suggestions Brain Friendly?

Your brain will grow dendrites as you study the worked examples in the text and **try doing them yourself** on separate paper. So, when you see similar problems in the homework, you will already have dendrites to work from.

Giving yourself a practice test by trying to remember the steps (without looking at your card) is an excellent way to reinforce what you are learning.

Correcting errors right away is how you learn and reinforce the correct procedures. It is hard to unlearn a mistake, so always check to see that you are on the right track.

Teachers assign homework so you can grow your own dendrites (learn the material) and then coat the dendrites with myelin through practice (remember the material). Really! In learning, you get good at what you practice. So, completing homework every day will strengthen your neural network and prepare you for exams.

If you have read each section in your textbook according to the steps above, you will probably encounter few difficulties with the exercises in the homework. Here are some additional suggestions that will help you succeed with the homework.

▶ If you **have trouble with a problem,** find a similar worked example in the section. Pay attention to *every line* of the worked example to see how to get from step to step. Work it yourself too, on separate paper; don't just look at it.

▶ If it is **hard to remember the steps** to follow for certain procedures, write the steps on a separate card. Then write a short explanation of each step. Keep the card nearby while you do the exercises, but try *not* to look at it.

▶ If you **aren't sure you are working the assigned exercises correctly,** choose two or three odd-numbered problems that are a similar type and work them. Then check the answers in the Answers section of your book and see if you are doing them correctly. If you aren't, go back to the section in the text and review the examples and find out how to correct your errors. Finally, when you are sure you understand, try the assigned problems again.

▶ **Make sure you do some homework every day,** even if the math class does not meet each day!

Now Try This ▶▶▶

What are your biggest homework concerns?

List your two main concerns and a **brain friendly solution** for each one.

1. Concern: _____

   Solution: _____

2. Concern: _____

   Solution: _____

2.2 ▶▶▶ Mixed Numbers

Suppose you had three whole trays of muffins and half of another tray. You would state this as a whole number and a fraction.

OBJECTIVE 1 Identify mixed numbers. When a whole number and a fraction are written together, the result is a **mixed number.** For example, the mixed number

$$3\frac{1}{2} \qquad \text{represents} \qquad 3 + \frac{1}{2}$$

or 3 wholes and $\frac{1}{2}$ of a whole. Read $3\frac{1}{2}$ as "three and one-half." As this figure shows, the mixed number $3\frac{1}{2}$ is equal to the improper fraction $\frac{7}{2}$.

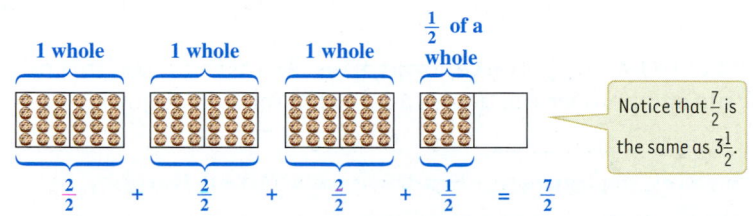

| 1 whole | 1 whole | 1 whole | $\frac{1}{2}$ of a whole |

Notice that $\frac{7}{2}$ is the same as $3\frac{1}{2}$.

$$\frac{2}{2} + \frac{2}{2} + \frac{2}{2} + \frac{1}{2} = \frac{7}{2}$$

*Work Problem **1** at the Side.* ▶

OBJECTIVE 2 Write mixed numbers as improper fractions. Use the following steps to write $3\frac{1}{2}$ as an improper fraction without drawing a figure.

Step 1 Multiply 3 and 2.

$$3\frac{1}{2} \qquad 3 \cdot 2 = 6$$

Step 2 Add 1 to the product.

$$3\frac{1}{2} \qquad 6 + 1 = 7$$

Step 3 Use 7, from Step 2, as the numerator and 2 as the denominator.

$$3\frac{1}{2} = \frac{7}{2}$$

Same denominator

In summary, use the following steps to *write a mixed number as an improper fraction.*

Writing a Mixed Number as an Improper Fraction

Step 1 *Multiply* the denominator of the fraction and the whole number.

Step 2 *Add* to this product the numerator of the fraction.

Step 3 Write the result of Step 2 as the *numerator* and the original denominator as the *denominator.*

---

1 **(a)** Use these diagrams to write $1\frac{2}{3}$ as an improper fraction.

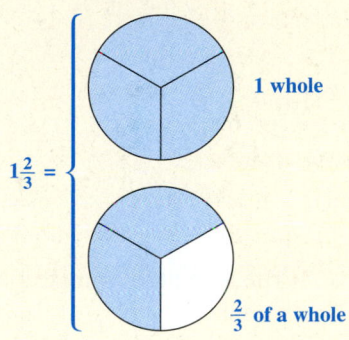

$$1\frac{2}{3} =$$

1 whole

$\frac{2}{3}$ of a whole

(b) Use these diagrams to write $2\frac{1}{4}$ as an improper fraction.

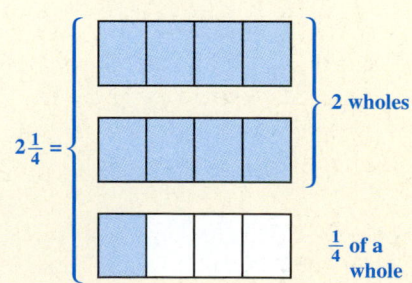

$$2\frac{1}{4} =$$

2 wholes

$\frac{1}{4}$ of a whole

ANSWERS

1. **(a)** $\frac{5}{3}$ **(b)** $\frac{9}{4}$

2 Write as improper fractions.

(a) $6\frac{1}{2}$

(b) $7\frac{3}{4}$

(c) $4\frac{7}{8}$

(d) $8\frac{5}{6}$

EXAMPLE 1 **Writing a Mixed Number as an Improper Fraction**

Write $7\frac{2}{3}$ as an improper fraction (numerator greater than denominator).

Step 1 $7\frac{2}{3}$ $7 \cdot 3 = 21$ Multiply 7 and 3.

Step 2 $7\frac{2}{3}$ $21 + 2 = 23$ Add 2. The numerator is 23.

Step 3 $7\frac{2}{3} = \frac{23}{3}$ Use the same denominator.

> Always use the same denominator.

◀ *Work Problem* **2** *at the Side.*

OBJECTIVE **3** **Write improper fractions as mixed numbers.**
Write an improper fraction as a mixed number as follows.

> **Writing an Improper Fraction as a Mixed Number**
>
> Write an **improper fraction** as a mixed number by dividing the numerator by the denominator. The quotient is the whole number (of the mixed number), the remainder is the numerator of the fraction part, and the denominator remains unchanged.

---

3 Write as whole or mixed numbers.

(a) $\frac{6}{5}$

(b) $\frac{9}{4}$

(c) $\frac{35}{5}$

(d) $\frac{78}{7}$

EXAMPLE 2 **Writing Improper Fractions as Mixed Numbers**

Write each improper fraction as a mixed number.

> Divide numerator by denominator.

(a) $\frac{17}{5}$ Divide 17 by 5.   ⟶ $5)\overline{17}$ ← **Whole number part** (quotient **3**)
$\underline{15}$
2 ← Remainder

The quotient **3** is the whole number part of the mixed number. The remainder **2** is the numerator of the fraction, and the denominator stays as **5**.

$$\frac{17}{5} = 3\frac{2}{5}$$ ← Remainder
Same denominator

We can check this by using a diagram in which $\frac{17}{5}$ is shaded.

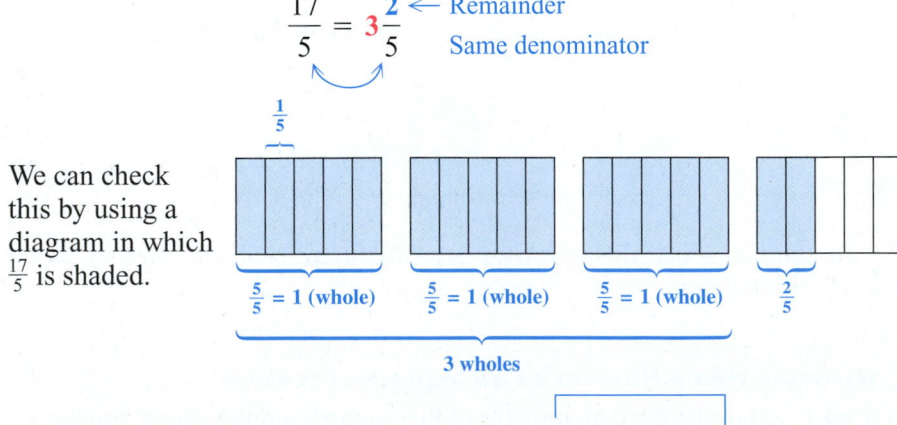

$\frac{5}{5} = 1$ (whole) $\frac{5}{5} = 1$ (whole) $\frac{5}{5} = 1$ (whole) $\frac{2}{5}$

3 wholes

(b) $\frac{24}{4}$ Divide 24 by 4.   ⟶ $4)\overline{24}$ (quotient 6) so $\frac{24}{4} = 6$
$\underline{24}$
0 ← No remainder

◀ *Work Problem* **3** *at the Side.*

ANSWERS

2. (a) $\frac{13}{2}$ (b) $\frac{31}{4}$ (c) $\frac{39}{8}$ (d) $\frac{53}{6}$

3. (a) $1\frac{1}{5}$ (b) $2\frac{1}{4}$ (c) 7 (d) $11\frac{1}{7}$

2.2 ▶▶▶ Exercises

Write each mixed number as an improper fraction. See Example 1.

1. $1\frac{1}{4}$

2. $2\frac{1}{2}$

3. $4\frac{3}{5}$

4. $5\frac{2}{3}$

5. $6\frac{1}{2}$

6. $5\frac{3}{5}$

7. $8\frac{1}{4}$

8. $8\frac{1}{2}$

9. $1\frac{7}{11}$

10. $4\frac{3}{7}$

11. $6\frac{1}{3}$

12. $8\frac{2}{3}$

13. $10\frac{1}{8}$

14. $12\frac{2}{3}$

15. $10\frac{3}{4}$

16. $5\frac{4}{5}$

17. $3\frac{3}{8}$

18. $2\frac{8}{9}$

19. $8\frac{3}{5}$

20. $3\frac{4}{7}$

21. $4\frac{10}{11}$

22. $11\frac{5}{8}$

23. $32\frac{3}{4}$

24. $15\frac{3}{10}$

25. $18\dfrac{5}{12}$

26. $19\dfrac{8}{11}$

27. $17\dfrac{14}{15}$

28. $9\dfrac{5}{16}$

29. $7\dfrac{19}{24}$

30. $9\dfrac{7}{12}$

Write each improper fraction as a whole or mixed number. See Example 2.

31. $\dfrac{4}{3}$

32. $\dfrac{11}{9}$

◑ 33. $\dfrac{9}{4}$

34. $\dfrac{7}{2}$

35. $\dfrac{48}{6}$

36. $\dfrac{64}{8}$

37. $\dfrac{38}{5}$

38. $\dfrac{33}{7}$

39. $\dfrac{39}{8}$

40. $\dfrac{40}{9}$

41. $\dfrac{27}{3}$

42. $\dfrac{78}{78}$

43. $\dfrac{63}{4}$

44. $\dfrac{19}{5}$

45. $\dfrac{47}{9}$

46. $\dfrac{65}{9}$

47. $\dfrac{65}{8}$

48. $\dfrac{37}{6}$

49. $\dfrac{84}{5}$

50. $\dfrac{92}{3}$

51. $\dfrac{112}{4}$

52. $\dfrac{117}{9}$

53. $\dfrac{183}{7}$

54. $\dfrac{212}{11}$

55. Your classmate asks you how to change a mixed number to an improper fraction. Write a couple of sentences and give an example to show her how this is done.

56. Explain in a sentence or two how to change an improper fraction to a mixed number. Give an example to show how this is done.

Write each mixed number as an improper fraction.

57. $250\dfrac{1}{2}$

58. $185\dfrac{3}{4}$

59. $333\dfrac{1}{3}$

60. $138\dfrac{4}{5}$

61. $522\dfrac{3}{8}$

62. $622\dfrac{1}{4}$

Write each improper fraction as a whole or mixed number.

63. $\dfrac{617}{4}$

64. $\dfrac{760}{8}$

65. $\dfrac{2565}{15}$

66. $\dfrac{2915}{16}$

67. $\dfrac{3917}{32}$

68. $\dfrac{5632}{64}$

Relating Concepts (Exercises 69–74) For Individual or Group Work

Knowing the basics of fractions is necessary in problem solving. **Work Exercises 69–74 in order.**

69. Which of these fractions are proper fractions?

$$\frac{2}{3} \quad \frac{4}{5} \quad \frac{8}{5} \quad \frac{3}{4} \quad \frac{6}{6} \quad \frac{7}{10}$$

70. (a) The proper fractions in Exercise 69 are the ones where the _____ is less than the _____.

(b) Draw a picture with shaded parts to show each proper fraction in Exercise 69.

(c) The proper fractions in Exercise 69 are all _____ than 1.
(less/greater)

71. Which of these fractions are improper fractions?

$$\frac{5}{5} \quad \frac{3}{4} \quad \frac{10}{3} \quad \frac{2}{3} \quad \frac{5}{6} \quad \frac{6}{5}$$

72. (a) The improper fractions in Exercise 71 are the ones where the _____ is equal to or greater than the _____.

(b) Draw a picture with shaded parts to show each improper fraction in Exercise 71.

(c) The improper fractions in Exercise 71 are all equal to or _____ than 1.
(less/greater)

73. Identify which of these fractions can be written as whole or mixed numbers, and then write them as whole or mixed numbers.

$$\frac{5}{3} \quad \frac{7}{8} \quad \frac{7}{7} \quad \frac{11}{6} \quad \frac{4}{5} \quad \frac{15}{16}$$

74. (a) The fractions that can be written as whole or mixed numbers in Exercise 73 are _____ fractions, and their value is
(proper/improper)
always _____ 1.
(less than/greater than or equal to)

(b) Draw a picture with shaded parts to show each whole or mixed number in Exercise 73.

(c) Explain how to write an improper fraction as a whole or mixed number.

2.3 ▷▷▷ Factors

OBJECTIVES

1 Find factors of a number.

2 Identify prime numbers.

3 Find prime factorizations.

OBJECTIVE 1 Find factors of a number. You will recall that numbers multiplied to give a product are called **factors**. Because 2 • 5 = 10, both 2 and 5 are factors of 10. The numbers 1 and 10 are also factors of 10, because

$$1 \cdot 10 = 10$$

The various tests for divisibility show that 1, 2, 5, and 10 are the only whole number factors of 10. The products 2 • 5 and 1 • 10 are called **factorizations** of 10.

> **Note**
>
> The tests to decide whether one number is divisible by another number were shown in **Section 1.5.** You might want to review these. The tests that you will want to remember are those for 2, 3, 5, and 10.

EXAMPLE 1 Using Factors

Find all possible two-number factorizations of each number.

(a) 12

$$1 \cdot 12 = 12 \qquad 2 \cdot 6 = 12 \qquad 3 \cdot 4 = 12$$

The factors of 12 are 1, 2, 3, 4, 6, and 12.

(b) 60

$$
\begin{array}{ll}
1 \cdot 60 = 60 & 2 \cdot 30 = 60 \\
3 \cdot 20 = 60 & 4 \cdot 15 = 60 \\
5 \cdot 12 = 60 & 6 \cdot 10 = 60
\end{array}
$$

> The factors of a number all divide evenly into that number.

The factors of 60 are 1, 2, 3, 4, 5, 6, 10, 12, 15, 20, 30, and 60.

Work Problem **1** *at the Side.* ▶

> **Composite Numbers**
>
> A number with a factor other than itself or 1 is called a **composite number.**

EXAMPLE 2 Identifying Composite Numbers

Which of the following numbers are composite?

(a) 6

Because 6 has factors of **2** and **3**, numbers other than 6 or 1, the number 6 is composite.

(b) 11

The number 11 has only two factors, 11 and 1. It is *not* composite.

(c) 25

A factor of 25 is **5**, so 25 is composite.

Work Problem **2** *at the Side.* ▶

OBJECTIVE 2 Identify prime numbers. Whole numbers that are not composite are called **prime numbers,** except 0 and 1, which are neither prime nor composite.

> **Prime Numbers**
>
> A **prime number** is a whole number that has exactly *two different* factors, *itself* and *1.*

1 Find all the whole number factors of each number.

(a) 18

(b) 16

(c) 36

(d) 80

2 Which of these numbers are composite?

2, 4, 5, 6, 8, 10, 11, 13, 19, 21, 27, 28, 33, 36, 42

ANSWERS

1. (a) 1, 2, 3, 6, 9, 18 **(b)** 1, 2, 4, 8, 16
   (c) 1, 2, 3, 4, 6, 9, 12, 18, 36
   (d) 1, 2, 4, 5, 8, 10, 16, 20, 40, 80
2. 4, 6, 8, 10, 21, 27, 28, 33, 36, 42

3 Which of the following are prime?

4, 7, 9, 13, 17, 19, 29, 33

4 Find the prime factorization of each number.

(a) 8

(b) 28

(c) 18

(d) 40

The number 3 is a prime number, since it can be divided evenly only by itself and 1. The number 8 is not a prime number (it is composite), since 8 can be divided evenly by 2 and 4, as well as by itself and 1.

> **CAUTION**
> A prime number has **only two** different factors, itself and 1. The number 1 is not a prime number because it does not have *two different* factors; the only factor of 1 is 1.

EXAMPLE 3 **Finding Prime Numbers**

Which of the following numbers are prime?

2   5   11   15   27

Only the number itself and 1 will divide evenly into a prime number.

The number 15 can be divided by 3 and 5, so it is not prime. Also, because 27 can be divided by 3 and 9, then 27 is not prime. The other numbers in the list, 2, 5, and 11, are divisible only by themselves and 1, so they are prime.

◀ *Work Problem* **3** *at the Side.*

OBJECTIVE 3 Find prime factorizations. For reference, here are the prime numbers less than 50.

2	3	5	7	11
13	17	19	23	29
31	37	41	43	47

These are the prime numbers less than 50.

> **CAUTION**
> All prime numbers are odd numbers except the number 2. Be careful, though, because *all odd numbers are not prime numbers*. For example, 9, 15, and 21 are odd numbers but are *not* prime numbers.

The **prime factorization** of a number can be especially useful when we are adding or subtracting fractions and need to find a common denominator or write a fraction in lowest terms.

> **Prime Factorization**
> A **prime factorization** of a number is a factorization in which every factor is a *prime number.*

EXAMPLE 4 **Determining the Prime Factorization**

Find the prime factorization of 12.
  Try to divide 12 by the first prime, 2.

$$12 \div 2 = 6,$$

└— First prime

so

$$12 = 2 \cdot 6$$

Try to divide 6 by the prime, 2.

$$6 \div 2 = 3,$$

so

$$12 = 2 \cdot \underline{2 \cdot 3}$$

└— Factorization of 6

Because all factors are prime, the prime factorization of 12 is

$$2 \cdot 2 \cdot 3$$

All these factors are prime.

◀ *Work Problem* **4** *at the Side.*

EXAMPLE 5   **Factoring by Using the Division Method**

Find the prime factorization of 48.

$2\overline{)48}$ Divide 48 by 2 (first prime).

$2\overline{)24}$ Divide 24 by 2.

> The divisors are all prime factors.

$2\overline{)12}$ Divide 12 by 2.

$2\overline{)6}$ Divide 6 by 2.

$3\overline{)3}$ Divide 3 by 3 (second prime).

1   Continue to divide until the quotient is 1.

Because all factors (divisors) are prime, the prime factorization of 48 is

$$\mathbf{2 \cdot 2 \cdot 2 \cdot 2 \cdot 3}$$

In Chapter 1, we wrote $2 \cdot 2 \cdot 2 \cdot 2$ as 2^4, so the prime factorization of 48 can be written, using exponents, as

$$48 = \mathbf{2 \cdot 2 \cdot 2 \cdot 2} \cdot 3 = \mathbf{2^4} \cdot 3$$

Work Problem **5** *at the Side.* ▶

Note

When using the division method of factoring, the last quotient found is 1. The "1" is never used as a prime factor because 1 is neither prime nor composite. Besides, 1 times any number is the number itself.

EXAMPLE 6   **Using Exponents with Prime Factorization**

Find the prime factorization of 225.

$3\overline{)225}$ 225 is not divisible by 2; use 3.

$3\overline{)75}$ Divide 75 by 3.

> All the divisors are prime factors.

$5\overline{)25}$ 25 is not divisible by 3; use 5.

$5\overline{)5}$ Divide by 5.

> 1 is **not** a prime factor.

1   Continue to divide until the quotient is 1.

Write the prime factorization.

$$255 = \mathbf{3 \cdot 3 \cdot 5 \cdot 5}$$

Or, using exponents,

$$255 = \mathbf{3^2 \cdot 5^2}$$

Work Problem **6** *at the Side.* ▶

---

5 Find the prime factorization of each number. Write the factorization with exponents.

(a) 36

(b) 54

(c) 60

(d) 81

6 Write the prime factorization of each number using exponents.

(a) 48

(b) 44

(c) 90

(d) 120

(e) 180

ANSWERS

5. **(a)** $2^2 \cdot 3^2$ **(b)** $2 \cdot 3^3$ **(c)** $2^2 \cdot 3 \cdot 5$
   (d) 3^4
6. **(a)** $2^4 \cdot 3$ **(b)** $2^2 \cdot 11$ **(c)** $2 \cdot 3^2 \cdot 5$
   (d) $2^3 \cdot 3 \cdot 5$ **(e)** $2^2 \cdot 3^2 \cdot 5$

Another method of factoring is a *factor tree*.

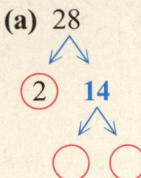

7 Complete each factor tree and give the prime factorization.

(a) 28

(b) 35

(c) 78

7. **(a)**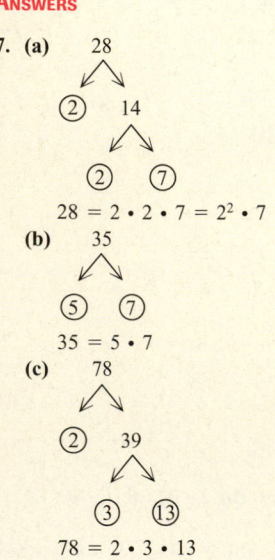
$28 = 2 \cdot 2 \cdot 7 = 2^2 \cdot 7$
(b) 35
$35 = 5 \cdot 7$
(c) 78
$78 = 2 \cdot 3 \cdot 13$

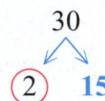**EXAMPLE 7** **Factoring by Using a Factor Tree**

Find the prime factorization of each number using a factor tree.

(a) 30

Try to divide by the first prime, 2. Write the factors under the 30. Circle the 2, since it is a prime.

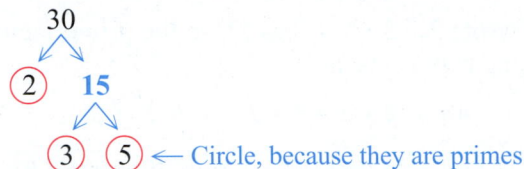

Since 15 cannot be divided evenly by 2, try the next prime, 3.

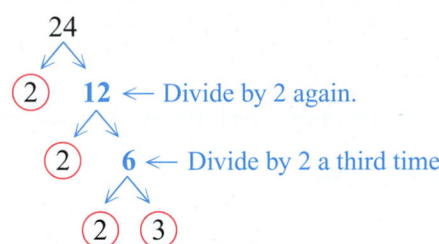

← Circle, because they are primes.

No uncircled factors remain, so the prime factorization (the circled factors) has been found.

$$30 = 2 \cdot 3 \cdot 5$$

(b) 24

Divide by 2.

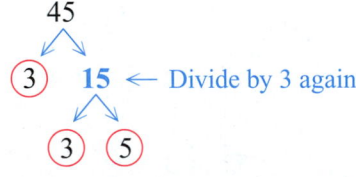

← Divide by 2 again.

← Divide by 2 a third time.

$24 = 2 \cdot 2 \cdot 2 \cdot 3$ or, using exponents, $24 = 2^3 \cdot 3$

(c) 45

Because 45 cannot be divided by 2, try 3.

45

← Divide by 3 again.

$45 = 3 \cdot 3 \cdot 5$ or, using exponents, $45 = 3^2 \cdot 5$

Note

The diagrams used in Example 7 look like tree branches, and that is why this method is referred to as using a factor tree.

◀ *Work Problem* **7** *at the Side.*

2.3 ▶▶▶ Exercises

Find all the factors of each number. See Example 1.

1. 8

2. 12

3. 15

4. 28

5. 48

6. 30

7. 36

8. 20

9. 40

10. 60

11. 64

12. 84

Decide whether each number is prime *or* composite. *See Examples 2 and 3.*

13. 6	**14.** 9	**15.** 5	**16.** 7
17. 16	**18.** 10	**19.** 13	**20.** 65
21. 19	**22.** 17	**23.** 25	**24.** 26
25. 48	**26.** 47	**27.** 45	**28.** 53

Find the prime factorization of each number. Write answers with exponents when repeated factors appear. See Examples 4–7.

29. 8

30. 6

31. 20

32. 40

33. 36

34. 18

35. 25

36. 56

37. 68

38. 70

39. 72

40. 64

41. 44

42. 104

43. 100

44. 112

45. 125

46. 135

47. 180

48. 300

49. 320

50. 480

51. 360

52. 400

53. Give a definition in your own words of both a composite number and a prime number. Give three examples of each. Which whole numbers are neither prime nor composite?

54. With the exception of the number 2, all prime numbers are odd numbers. Nevertheless, all odd numbers are not prime numbers. Explain why these statements are true.

55. Explain the difference between finding all possible factors of 24 and finding the prime factorization of 24.

56. Use the division method to find the prime factorization of 36. Can you divide by 3s before you divide by 2s? Does the order of division change the answers?

Find the prime factorization of each number. Write answers using exponents.

57. 350

58. 640

59. 960

60. 1000

61. 1560

62. 2000

63. 1260

64. 2200

Relating Concepts (Exercises 65–70) For Individual or Group Work

An understanding of factors and factorization will be needed to solve fraction problems.
Work Exercises 65–70 in order.

65. A prime number is a whole number that has exactly two different factors, itself and 1. List all prime numbers less than 50.

66. Explain what it is about the numbers in Exercise 65 that makes them prime.

67. The number 2 is an even number and a prime number. Can any other even numbers be prime numbers? Explain.

68. Can a multiple of a prime number be prime (for example, 6, 9, 12, and 15 are multiples of 3)? Explain.

69. Find the prime factorization of 2100. Do not use exponents in your answer.

70. Write the answer to Exercise 69 using exponents for repeated factors.

2.4 ▶▶▶ Writing a Fraction in Lowest Terms

When working problems involving fractions, we must often compare two fractions to determine whether they represent the same portion of a whole. Look at the two cases of soda.

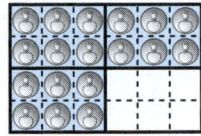

$\frac{3}{4}$ **full** $\frac{18}{24}$ **full**

The cases of soda show areas that are $\frac{3}{4}$ full and $\frac{18}{24}$ full. Because the full areas are equivalent (the same portion), the fractions $\frac{3}{4}$ and $\frac{18}{24}$ are **equivalent fractions.** They each represent the same portion of the whole.

$$\frac{3}{4} = \frac{18}{24}$$

Because the numbers 18 and 24 both have 6 as a factor, 6 is called a **common factor** of the numbers. Other common factors of 18 and 24 are 1, 2, and 3.

Work Problem **1** *at the Side.* ▶

OBJECTIVE 1 Tell whether a fraction is written in lowest terms.
The fraction $\frac{3}{4}$ is written in *lowest terms* because the numerator and denominator have no common factor other than 1. However, the fraction $\frac{18}{24}$ is *not* in lowest terms because its numerator and denominator have common factors of 6, 3, 2, and 1.

> **Writing a Fraction in Lowest Terms**
>
> A fraction is written in **lowest terms** when the numerator and denominator have no common factor other than 1.

EXAMPLE 1 Understanding Lowest Terms

Are the following fractions in lowest terms?

(a) $\frac{3}{8}$ — No common factors other than 1.

The numerator and denominator have no common factor other than 1, so the fraction is in lowest terms.

(b) $\frac{21}{36}$ — 3 is a common factor of 21 and 36.

The numerator and denominator have a common factor of 3, so the fraction is **not in lowest terms.**

Work Problem **2** *at the Side.* ▶

OBJECTIVE 2 Write a fraction in lowest terms using common factors. There are two common methods for writing a fraction in lowest terms. These methods are shown in the next examples. The first method works best when the numerator and denominator are small numbers.

---

1 Decide whether the number in blue is a common factor of the other two numbers.

(a) 6, 12; **2**

(b) 32, 64; **8**

(c) 32, 56; **16**

(d) 75, 81; **1**

2 Are the following fractions in lowest terms?

(a) $\frac{4}{5}$

(b) $\frac{6}{18}$

(c) $\frac{9}{15}$

(d) $\frac{17}{46}$

ANSWERS

1. **(a)** yes **(b)** yes **(c)** no **(d)** yes
2. **(a)** yes **(b)** no **(c)** no **(d)** yes

3 Write in lowest terms.

(a) $\dfrac{8}{16}$

(b) $\dfrac{9}{12}$

(c) $\dfrac{28}{42}$

(d) $\dfrac{30}{80}$

(e) $\dfrac{16}{40}$

ANSWERS

3. (a) $\dfrac{1}{2}$ (b) $\dfrac{3}{4}$ (c) $\dfrac{2}{3}$ (d) $\dfrac{3}{8}$ (e) $\dfrac{2}{5}$

EXAMPLE 2 **Writing Fractions in Lowest Terms**

Write each fraction in lowest terms.

(a) $\dfrac{18}{24}$

The greatest common factor of 18 and 24 is 6. Divide both numerator and denominator by **6**.

$$\frac{18}{24} = \frac{18 \div 6}{24 \div 6} = \frac{3}{4}$$

Divide by the greatest common factor, 6.

(b) $\dfrac{30}{50} = \dfrac{30 \div 10}{50 \div 10} = \dfrac{3}{5}$ Divide both numerator and denominator by 10.

(c) $\dfrac{24}{42} = \dfrac{24 \div 6}{42 \div 6} = \dfrac{4}{7}$ Divide both numerator and denominator by 6.

(d) $\dfrac{60}{72}$

Suppose we thought that 4 was the greatest common factor of 60 and 72. Dividing by 4 would give

$$\frac{60}{72} = \frac{60 \div 4}{72 \div 4} = \frac{15}{18} \quad \leftarrow \text{Not in lowest terms.}$$

But $\frac{15}{18}$ is not in lowest terms, because 15 and 18 have a common factor of 3. So we divide by 3.

$$\frac{15}{18} = \frac{15 \div 3}{18 \div 3} = \frac{5}{6}$$

Continue dividing until there is no common factor other than 1.

The fraction $\frac{60}{72}$ could have been written in lowest terms in one step by dividing by 12, the greatest common factor of 60 and 72.

$$\frac{60}{72} = \frac{60 \div 12}{72 \div 12} = \frac{5}{6} \quad \leftarrow \text{Same answer as above.}$$

Continue dividing until the fraction is in lowest terms.

> **Note**
>
> Dividing the numerator and denominator by the same number results in an equivalent fraction.

In Example 2, we wrote fractions in lowest terms by dividing by a common factor. This method is summarized in the following steps.

> **The Method of Dividing by a Common Factor**
>
> *Step 1* Find the greatest number that will divide evenly into both the numerator and denominator. This number is a *common factor.*
>
> *Step 2* *Divide* both numerator and denominator by the common factor.
>
> *Step 3* *Check* to see whether the new fraction has any common factors (besides 1). If it does, repeat Steps 2 and 3. If the only common factor is 1, the fraction is in lowest terms.

◀ *Work Problem* **3** *at the Side.*

OBJECTIVE **3** **Write a fraction in lowest terms using prime factors.**   The method of writing a fraction in lowest terms by division works well for fractions with small numerators and denominators. For larger numbers, when common factors are not obvious, use the method of *prime factors,* which is shown in the next example.

EXAMPLE 3   **Using Prime Factors**

Write each fraction in lowest terms.

(a) $\dfrac{24}{42}$

Write the prime factorization of both numerator and denominator. See **Section 2.3** for help.

$$\frac{24}{42} = \frac{2 \cdot 2 \cdot 2 \cdot 3}{2 \cdot 3 \cdot 7}$$

Just as with the method used in Example 2, divide both numerator and denominator by any common factors. Write a **1** by each factor that has been divided.

2 ÷ 2 is 1   3 ÷ 3 is 1

$$\frac{24}{42} = \frac{\overset{1}{\cancel{2}} \cdot 2 \cdot 2 \cdot \overset{1}{\cancel{3}}}{\underset{1}{\cancel{2}} \cdot \underset{1}{\cancel{3}} \cdot 7}$$

Multiply the remaining factors in both numerator and denominator.

$$\frac{24}{42} = \frac{\mathbf{1 \cdot 2 \cdot 2 \cdot 1}}{\mathbf{1 \cdot 1 \cdot 7}} = \frac{4}{7} \leftarrow \text{Lowest terms}$$

Finally, $\frac{24}{42}$ written in lowest terms is $\frac{4}{7}$.

(b) $\dfrac{162}{54}$

Write the prime factorization of both numerator and denominator.

$$\frac{162}{54} = \frac{2 \cdot 3 \cdot 3 \cdot 3 \cdot 3}{2 \cdot 3 \cdot 3 \cdot 3}$$

Now divide by the common factors. ***Do not forget to write the 1s.***

$$\frac{162}{54} = \frac{\overset{1}{\cancel{2}} \cdot \overset{1}{\cancel{3}} \cdot \overset{1}{\cancel{3}} \cdot \overset{1}{\cancel{3}} \cdot 3}{\underset{1}{\cancel{2}} \cdot \underset{1}{\cancel{3}} \cdot \underset{1}{\cancel{3}} \cdot \underset{1}{\cancel{3}}}$$

Remember to write in the 1s when dividing by a common factor.

$$= \frac{\mathbf{1 \cdot 1 \cdot 1 \cdot 1 \cdot 3}}{\mathbf{1 \cdot 1 \cdot 1 \cdot 1}} = \frac{3}{1} = 3$$

(c) $\dfrac{18}{90}$

$$\frac{18}{90} = \frac{\overset{1}{\cancel{2}} \cdot \overset{1}{\cancel{3}} \cdot \overset{1}{\cancel{3}}}{\underset{1}{\cancel{2}} \cdot \underset{1}{\cancel{3}} \cdot \underset{1}{\cancel{3}} \cdot 5} = \frac{\mathbf{1 \cdot 1 \cdot 1}}{\mathbf{1 \cdot 1 \cdot 1 \cdot 5}} = \frac{1}{5}$$

CAUTION

In Example 3(c) above, all factors of the numerator were divided. But $1 \cdot 1 \cdot 1$ is still 1, so the final answer is $\frac{1}{5}$ (***not*** 5).

4 Use the method of prime factors to write each fraction in lowest terms.

(a) $\dfrac{12}{36}$

(b) $\dfrac{32}{56}$

(c) $\dfrac{74}{111}$

(d) $\dfrac{124}{340}$

5 Is each pair of fractions equivalent?

(a) $\dfrac{24}{48}$ and $\dfrac{36}{72}$

(b) $\dfrac{45}{60}$ and $\dfrac{50}{75}$

(c) $\dfrac{20}{4}$ and $\dfrac{110}{22}$

(d) $\dfrac{120}{220}$ and $\dfrac{180}{320}$

In Example 3, we wrote fractions in lowest terms using prime factors. This method is summarized as follows.

> **Step 1**   Write the ***prime factorization*** of both numerator and denominator.
>
> **Step 2**   Use slashes to show you are ***dividing*** both numerator and denominator by common factors.
>
> **Step 3**   ***Multiply*** the remaining factors in the numerator and denominator.

◀ *Work Problem* **4** *at the Side.*

OBJECTIVE **4** **Determine whether two fractions are equivalent.** The next example shows how to decide whether two fractions are equivalent.

EXAMPLE 4 **Determining Whether Two Fractions Are Equivalent**

Determine whether each pair of fractions is equivalent. In other words, do both fractions represent the same part of a whole?

(a) $\dfrac{16}{48}$ and $\dfrac{24}{72}$

Use the method of prime factors to write each fraction in lowest terms.

$$\frac{16}{48} = \frac{\cancel{2} \cdot \cancel{2} \cdot \cancel{2} \cdot \cancel{2}}{\cancel{2} \cdot \cancel{2} \cdot \cancel{2} \cdot \cancel{2} \cdot 3} = \frac{1 \cdot 1 \cdot 1 \cdot 1}{1 \cdot 1 \cdot 1 \cdot 1 \cdot 3} = \frac{1}{3}$$

Equivalent $\left(\dfrac{1}{3} = \dfrac{1}{3} \right)$

$$\frac{24}{72} = \frac{\cancel{2} \cdot \cancel{2} \cdot \cancel{2} \cdot \cancel{3}}{\cancel{2} \cdot \cancel{2} \cdot \cancel{2} \cdot \cancel{3} \cdot 3} = \frac{1 \cdot 1 \cdot 1 \cdot 1}{1 \cdot 1 \cdot 1 \cdot 1 \cdot 3} = \frac{1}{3}$$

(b) $\dfrac{32}{52}$ and $\dfrac{64}{112}$

$$\frac{32}{52} = \frac{\cancel{2} \cdot \cancel{2} \cdot 2 \cdot 2 \cdot 2}{\cancel{2} \cdot \cancel{2} \cdot 13} = \frac{2 \cdot 2 \cdot 2}{1 \cdot 1 \cdot 13} = \frac{8}{13}$$

Not equivalent $\left(\dfrac{8}{13} \ne \dfrac{4}{7} \right)$

$$\frac{64}{112} = \frac{\cancel{2} \cdot \cancel{2} \cdot \cancel{2} \cdot \cancel{2} \cdot 2 \cdot 2}{\cancel{2} \cdot \cancel{2} \cdot \cancel{2} \cdot \cancel{2} \cdot 7} = \frac{1 \cdot 1 \cdot 1 \cdot 1 \cdot 2 \cdot 2}{1 \cdot 1 \cdot 1 \cdot 1 \cdot 7} = \frac{4}{7}$$

(c) $\dfrac{75}{15}$ and $\dfrac{60}{12}$

$$\frac{75}{15} = \frac{\cancel{3} \cdot \cancel{5} \cdot 5}{\cancel{3} \cdot \cancel{5}} = \frac{1 \cdot 1 \cdot 5}{1 \cdot 1} = 5$$

Equivalent $(5 = 5)$

$$\frac{60}{12} = \frac{\cancel{2} \cdot \cancel{2} \cdot \cancel{3} \cdot 5}{\cancel{2} \cdot \cancel{2} \cdot \cancel{3}} = \frac{1 \cdot 1 \cdot 1 \cdot 5}{1 \cdot 1 \cdot 1} = 5$$

◀ *Work Problem* **5** *at the Side.*

Put a ✓ mark in the blank if the number at the left is divisible by the number at the top. Put an ✗ in the blank if the number is not divisible by the number at the top. (For help, see **Section 1.5.**)

	2	3	5	10			2	3	5	10
1. 60	——	——	——	——		**2.** 90	——	——	——	——
3. 48	——	——	——	——		**4.** 36	——	——	——	——
5. 160	——	——	——	——		**6.** 175	——	——	——	——
7. 138	——	——	——	——		**8.** 150	——	——	——	——

Write each fraction in lowest terms. See Example 2.

9. $\dfrac{6}{8}$ **10.** $\dfrac{6}{12}$ **11.** $\dfrac{3}{12}$ **12.** $\dfrac{4}{12}$

13. $\dfrac{15}{25}$ **14.** $\dfrac{32}{48}$ **15.** $\dfrac{36}{42}$ **16.** $\dfrac{22}{33}$

17. $\dfrac{56}{64}$ **18.** $\dfrac{21}{35}$ **19.** $\dfrac{180}{210}$ **20.** $\dfrac{72}{80}$

21. $\dfrac{72}{126}$ **22.** $\dfrac{73}{146}$ **23.** $\dfrac{12}{600}$ **24.** $\dfrac{8}{400}$

25. $\dfrac{96}{132}$ **26.** $\dfrac{165}{180}$ **27.** $\dfrac{60}{108}$ **28.** $\dfrac{112}{128}$

Write the numerator and denominator of each fraction as a product of prime factors and divide by the common factors. Then write the fraction in lowest terms. See Example 3.

29. $\dfrac{18}{24}$ **30.** $\dfrac{16}{64}$ **31.** $\dfrac{35}{40}$

32. $\dfrac{20}{32}$ **33.** $\dfrac{90}{180}$ **34.** $\dfrac{36}{48}$

35. $\dfrac{36}{12}$

36. $\dfrac{192}{48}$

37. $\dfrac{72}{225}$

38. $\dfrac{65}{234}$

Write each fraction in lowest terms. Then state whether the fractions are equivalent or not equivalent. See Example 4.

39. $\dfrac{3}{6}$ and $\dfrac{18}{36}$

40. $\dfrac{3}{8}$ and $\dfrac{27}{72}$

41. $\dfrac{10}{24}$ and $\dfrac{12}{30}$

42. $\dfrac{15}{35}$ and $\dfrac{18}{40}$

43. $\dfrac{15}{24}$ and $\dfrac{35}{52}$

44. $\dfrac{21}{33}$ and $\dfrac{9}{12}$

45. $\dfrac{14}{16}$ and $\dfrac{35}{40}$

46. $\dfrac{27}{90}$ and $\dfrac{24}{80}$

47. $\dfrac{48}{6}$ and $\dfrac{72}{8}$

48. $\dfrac{33}{11}$ and $\dfrac{72}{24}$

49. $\dfrac{25}{30}$ and $\dfrac{65}{78}$

50. $\dfrac{24}{72}$ and $\dfrac{30}{90}$

51. What does it mean when a fraction is expressed in lowest terms? Give three examples.

52. Explain what equivalent fractions are, and give an example of a pair of equivalent fractions. Show that they are equivalent.

Write each fraction in lowest terms.

53. $\dfrac{160}{256}$

54. $\dfrac{363}{528}$

55. $\dfrac{238}{119}$

56. $\dfrac{570}{95}$

Study Skills

USING STUDY CARDS

You may have used "flash cards" in other classes before. In math, study cards can be helpful, too. However, they are different because the main things to remember in math are *not* necessarily terms and definitions; they are *sets of steps to follow* to solve problems (and how to know which set of steps to follow) and *concepts about how math works* (principles). So, the cards will look different but will be just as useful.

In this two-part activity, you will find four types of study cards to use in math. Look carefully at what kinds of information to put on them and where to put it. Then use them the way you would any flash card:

▶ to quickly review when you have a few minutes,

▶ to do daily reviews,

▶ to review before a test.

Remember, the most helpful thing about study cards is making them. While you are making them, you have to do the kind of thinking that is most brain friendly, which improves your neural network of dendrites. After each card description you will find an assignment to try. It is marked **NOW TRY THIS.**

For **new vocabulary cards,** put the word (spelled correctly) and the page number where it is found on the front of the card. On the back, write:

▶ the definition (in your own words if possible),

▶ an example, an exception (if there are any),

▶ any related words, and

▶ a sample problem (if appropriate).

New Vocabulary Cards

Front of Card

| Prime Numbers | p. 122 |

Back of Card

Definition: Whole numbers that can only be divided by themselves and 1.

★Must divide evenly, NO remainders!

Ex: 2, 3, 5, 7, 11, 13, 17 are the first few primes.

→ NOT 0 or 1
→ Used in factoring
→ Related word: composite number

Why Are Study Cards Brain Friendly?

• Making cards is **active.**

• Cards are **visually** appealing.

• **Repetition** is good for your brain.

For details see Using Study Cards Revisited following **Section 2.8.**

Now Try This ▶▶▶

List four new vocabulary words/concepts you need to learn right now. Make a card for each one.

_____   _____   _____   _____

Procedure ("Steps") Cards

For **procedure cards,** write the name of the procedure at the top on the front of the card. Then write each step *in words*. If you need to know abbreviations for some words, include them along with the whole words written out. On the back, put an example of the procedure, showing each step you need to take. You can review by looking at the front and practicing a new worked example, or by looking at the back and remembering what the procedure is called and what the steps are.

Writing a fraction in lowest terms using prime factors

Use this method with larger denominators.

Step 1: Write prime factorization of numerator and denominator.

Step 2: Divide out all common factors.

Step 3: Multiply remaining factors.

Front of Card

Example: Write this fraction in lowest terms.

$$\frac{64}{112} = \frac{2 \cdot 2 \cdot 2 \cdot 2 \cdot 2 \cdot 2}{2 \cdot 2 \cdot 2 \cdot 2 \cdot 7}$$

Prime factors of 64.

Prime factors of 112.

Divide out common factors of 2

$$= \frac{\overset{1}{2} \cdot \overset{1}{2} \cdot \overset{1}{2} \cdot \overset{1}{2} \cdot 2 \cdot 2}{\underset{1}{2} \cdot \underset{1}{2} \cdot \underset{1}{2} \cdot \underset{1}{2} \cdot 7} = \frac{1 \cdot 1 \cdot 1 \cdot 1 \cdot 2 \cdot 2}{1 \cdot 1 \cdot 1 \cdot 1 \cdot 7} = \frac{4}{7}$$

multiply remaining factors

lowest terms

Back of Card

Now Try This ▶▶▶

What procedure are you learning right now? Make a "steps" card for it.

Procedure: _____

2.5 ▶▶▶ Multiplying Fractions

OBJECTIVE 1 Multiply fractions. Suppose that you give $\frac{1}{2}$ of your Energy Bar to your kickboxing partner Jennifer. Then Jennifer gives $\frac{1}{2}$ of her share to Tony. How much of the Energy Bar does Tony get to eat?

Start with a sketch showing the Energy Bar cut in half (2 equal pieces).

1

$\frac{1}{2}$ to Jennifer

OBJECTIVES

1 Multiply fractions.

2 Use a multiplication shortcut.

3 Multiply a fraction and a whole number.

4 Find the area of a rectangle.

Next, take $\frac{1}{2}$ of the shaded area. (Here we are dividing $\frac{1}{2}$ into 2 equal parts and shading one darker than the other.)

$\frac{1}{2}$ $\frac{1}{2}$

$\frac{1}{2}$ of $\frac{1}{2}$ to Tony

The sketch shows that Tony gets $\frac{1}{4}$ of the Energy Bar.

Tony gets $\frac{1}{2}$ of $\frac{1}{2}$ of the Energy Bar. When used between two fractions, the word **of** tells us to multiply.

$$\frac{1}{2} \text{ of } \frac{1}{2} \quad \text{means} \quad \frac{1}{2} \cdot \frac{1}{2}$$

Tony's share of the Energy Bar is

$$\frac{1}{2} \cdot \frac{1}{2} = \frac{1}{4}$$

Work Problem **1** *at the Side.* ▶

The rule for multiplying fractions follows.

1 Use these figures to find $\frac{1}{4}$ of $\frac{1}{2}$.

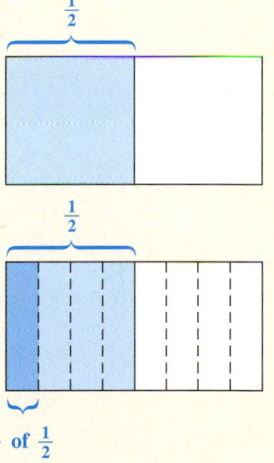

$\frac{1}{2}$

$\frac{1}{2}$

$\frac{1}{4}$ of $\frac{1}{2}$

> **Multiplying Fractions**
> Multiply two fractions by multiplying the numerators and multiplying the denominators.

2 Multiply. Write answers in lowest terms.

(a) $\dfrac{1}{2} \cdot \dfrac{3}{4}$

(b) $\dfrac{3}{5} \cdot \dfrac{1}{3}$

(c) $\dfrac{5}{6} \cdot \dfrac{1}{2} \cdot \dfrac{1}{8}$

(d) $\dfrac{1}{2} \cdot \dfrac{3}{4} \cdot \dfrac{3}{8}$

Use this rule to find the product of $\frac{2}{3}$ and $\frac{1}{3}$ (multiply $\frac{2}{3}$ by $\frac{1}{3}$).

$$\dfrac{2}{3} \cdot \dfrac{1}{3} = \dfrac{2 \cdot 1}{3 \cdot 3} \quad \leftarrow \text{Multiply numerators.}$$
$$\phantom{\dfrac{2}{3} \cdot \dfrac{1}{3} = \dfrac{2 \cdot 1}{3 \cdot 3}} \quad \leftarrow \text{Multiply denominators.}$$

Finish multiplying.

$$\dfrac{2}{3} \cdot \dfrac{1}{3} = \dfrac{2 \cdot 1}{3 \cdot 3} = \dfrac{2}{9} \quad \begin{matrix}\leftarrow 2 \cdot 1 = 2 \\ \leftarrow 3 \cdot 3 = 9\end{matrix}$$

Check that the final result is in lowest terms. $\frac{2}{9}$ is in lowest terms because 2 and 9 have no common factor other than 1.

EXAMPLE 1 **Multiplying Fractions**

Multiply. Write answers in lowest terms.

(a) $\dfrac{5}{8} \cdot \dfrac{3}{4}$

Multiply the numerators and multiply the denominators.

$$\dfrac{5}{8} \cdot \dfrac{3}{4} = \dfrac{5 \cdot 3}{8 \cdot 4} = \dfrac{15}{32} \quad \longleftarrow \boxed{\text{Already in lowest terms}}$$

Notice that 15 and 32 have no common factors other than 1, so the answer is in lowest terms.

(b) $\dfrac{4}{7} \cdot \dfrac{2}{5}$

$$\dfrac{4}{7} \cdot \dfrac{2}{5} = \dfrac{4 \cdot 2}{7 \cdot 5} = \dfrac{8}{35} \quad \leftarrow \text{Lowest terms}$$

(c) $\dfrac{5}{8} \cdot \dfrac{3}{4} \cdot \dfrac{1}{2}$

$$\dfrac{5}{8} \cdot \dfrac{3}{4} \cdot \dfrac{1}{2} = \dfrac{5 \cdot 3 \cdot 1}{8 \cdot 4 \cdot 2} = \dfrac{15}{64} \quad \leftarrow \text{Lowest terms}$$

◀ *Work Problem* **2** *at the Side.*

OBJECTIVE **2** **Use a multiplication shortcut.** A multiplication **shortcut** that can be used with fractions is shown in Example 2.

EXAMPLE 2 **Using the Multiplication Shortcut**

Multiply $\frac{5}{6}$ and $\frac{9}{10}$. Write the answer in lowest terms.

$$\dfrac{5}{6} \cdot \dfrac{9}{10} = \dfrac{5 \cdot 9}{6 \cdot 10} = \dfrac{45}{60} \quad \leftarrow \text{Not in lowest terms}$$

The numerator and denominator have a common factor other than 1, so write the prime factorization of each number.

$$\dfrac{5}{6} \cdot \dfrac{9}{10} = \dfrac{5 \cdot 9}{6 \cdot 10} = \dfrac{5 \cdot 3 \cdot 3}{2 \cdot 3 \cdot 2 \cdot 5} \quad \longleftarrow \boxed{\begin{matrix}\text{Write the prime} \\ \text{factorization of} \\ \text{each number.}\end{matrix}}$$

Continued on Next Page

ANSWERS

2. (a) $\dfrac{3}{8}$ (b) $\dfrac{1}{5}$ (c) $\dfrac{5}{96}$ (d) $\dfrac{9}{64}$

Next, divide by the common factors of 5 and 3.

$$\frac{5}{6} \cdot \frac{9}{10} = \frac{5 \cdot 9}{6 \cdot 10} = \frac{\overset{1}{\cancel{5}} \cdot \overset{1}{\cancel{3}} \cdot 3}{2 \cdot \underset{1}{\cancel{3}} \cdot 2 \cdot \underset{1}{\cancel{5}}}$$

Finally, multiply the remaining factors in the numerator and in the denominator.

$$\frac{5}{6} \cdot \frac{9}{10} = \frac{\mathbf{1 \cdot 1} \cdot 3}{2 \cdot \mathbf{1} \cdot 2 \cdot \mathbf{1}} = \frac{3}{4} \quad \leftarrow \text{Lowest terms}$$

As a shortcut, instead of writing the prime factorization of each number, find the product of $\frac{5}{6}$ and $\frac{9}{10}$ as follows.

First, divide by 5, a common factor of both 5 and 10.

$$\frac{\overset{1}{\cancel{5}}}{6} \cdot \frac{9}{\underset{2}{\cancel{10}}}$$

Next, divide by 3, a common factor of both 6 and 9.

$$\frac{\overset{1}{\cancel{5}}}{\underset{2}{\cancel{6}}} \cdot \frac{\overset{3}{\cancel{9}}}{\underset{2}{\cancel{10}}}$$

Finally, multiply numerators and multiply denominators.

$$\frac{1 \cdot 3}{2 \cdot 2} = \frac{3}{4}$$

CAUTION

When using the multiplication shortcut, you are dividing a numerator and a denominator by a common factor. Be certain that you divide a numerator and a denominator ***by the same number.*** If you do all possible divisions, your answer will be in lowest terms.

EXAMPLE 3 **Using the Multiplication Shortcut**

Use the multiplication shortcut to find each product. Write the answers in lowest terms and as mixed numbers where possible.

(a) $\dfrac{6}{11} \cdot \dfrac{7}{8}$

Divide both 6 and 8 by their common factor of 2. Notice that 7 and 11 have no common factor. Then multiply.

$$\frac{\overset{3}{\cancel{6}}}{11} \cdot \frac{7}{\underset{4}{\cancel{8}}} = \frac{3 \cdot 7}{11 \cdot 4} = \frac{21}{44} \quad \leftarrow \text{Lowest terms}$$

(b) $\dfrac{7}{10} \cdot \dfrac{20}{21}$

Divide a numerator and a denominator by the same number.

Divide 7 and 21 by 7, and divide 10 and 20 by 10.

$$\frac{\overset{1}{\cancel{7}}}{\underset{1}{\cancel{10}}} \cdot \frac{\overset{2}{\cancel{20}}}{\underset{3}{\cancel{21}}} = \frac{1 \cdot 2}{1 \cdot 3} = \frac{2}{3} \quad \leftarrow \text{Lowest terms}$$

Continued on Next Page

3 Use the multiplication shortcut to find each product.

(a) $\dfrac{3}{4} \cdot \dfrac{2}{3}$

(b) $\dfrac{6}{11} \cdot \dfrac{33}{21}$

(c) $\dfrac{20}{4} \cdot \dfrac{3}{40} \cdot \dfrac{1}{3}$

(d) $\dfrac{18}{17} \cdot \dfrac{1}{36} \cdot \dfrac{2}{3}$

(c) $\dfrac{35}{12} \cdot \dfrac{32}{25}$

$$\dfrac{\overset{7}{\cancel{35}}}{\underset{3}{\cancel{12}}} \cdot \dfrac{\overset{8}{\cancel{32}}}{\underset{5}{\cancel{25}}} = \dfrac{7 \cdot 8}{3 \cdot 5} = \dfrac{56}{15} \quad \text{or} \quad 3\dfrac{11}{15} \leftarrow \text{Mixed number}$$

(d) $\dfrac{2}{3} \cdot \dfrac{8}{15} \cdot \dfrac{3}{4}$

$$\dfrac{\overset{1}{\cancel{2}}}{\underset{1}{\cancel{3}}} \cdot \dfrac{\overset{4}{\cancel{8}}}{15} \cdot \dfrac{\overset{1}{\cancel{3}}}{\underset{\underset{1}{2}}{\cancel{4}}} = \dfrac{1 \cdot 4 \cdot 1}{1 \cdot 15 \cdot 1} = \dfrac{4}{15} \leftarrow \text{Lowest terms}$$

This shortcut is especially helpful when the fractions involve large numbers.

> **Note**
>
> There is no specific order that must be used when dividing numerators and denominators, as long as both the numerator and the denominator are divided by the *same* number.

◀ *Work Problem* **3** *at the Side.*

OBJECTIVE 3 Multiply a fraction and a whole number. The rule for multiplying a fraction and a whole number follows.

> **Multiplying a Whole Number and a Fraction**
>
> Multiply a whole number and a fraction by writing the whole number as a fraction with a denominator of 1.

For example, write the whole numbers 8, 10, and 25 as follows.

$$8 = \dfrac{8}{1} \qquad 10 = \dfrac{10}{1} \qquad 25 = \dfrac{25}{1} \longleftarrow \boxed{\text{Write the whole number over 1.}}$$

EXAMPLE 4 Multiplying by Whole Numbers

Multiply. Write answers in lowest terms and as whole numbers where possible.

(a) $8 \cdot \dfrac{3}{4}$

Write 8 as $\dfrac{8}{1}$ and multiply.

$$8 \cdot \dfrac{3}{4} = \dfrac{\overset{2}{\cancel{8}}}{1} \cdot \dfrac{3}{\underset{1}{\cancel{4}}} = \dfrac{2 \cdot 3}{1 \cdot 1} = \dfrac{6}{1} = 6 \longleftarrow \boxed{\dfrac{6}{1} \text{ is the same as } 6 \div 1, \text{ which equals 6.}}$$

Continued on Next Page

ANSWERS

3. (a) $\dfrac{\overset{1}{\cancel{3}}}{\underset{2}{\cancel{4}}} \cdot \dfrac{\overset{1}{\cancel{2}}}{\underset{1}{\cancel{3}}} = \dfrac{1}{2}$ (b) $\dfrac{\overset{2}{\cancel{6}}}{\underset{1}{\cancel{11}}} \cdot \dfrac{\overset{3}{\cancel{33}}}{\underset{7}{\cancel{21}}} = \dfrac{6}{7}$

(c) $\dfrac{\overset{1}{\cancel{20}}}{4} \cdot \dfrac{\overset{1}{\cancel{3}}}{\underset{2}{\cancel{40}}} \cdot \dfrac{1}{\underset{1}{\cancel{3}}} = \dfrac{1}{8}$

(d) $\dfrac{\overset{1}{\cancel{18}}}{17} \cdot \dfrac{1}{\underset{\underset{1}{2}}{\cancel{36}}} \cdot \dfrac{\overset{1}{\cancel{2}}}{3} = \dfrac{1}{51}$

(b) $15 \cdot \dfrac{5}{6}$

$$15 \cdot \frac{5}{6} = \frac{\overset{5}{\cancel{15}}}{1} \cdot \frac{5}{\underset{2}{\cancel{6}}} = \frac{5 \cdot 5}{1 \cdot 2} = \frac{25}{2} = 12\frac{1}{2}$$

Work Problem **4** *at the Side.* ▶

OBJECTIVE **4** **Find the area of a rectangle.** To find the area of a rectangle (the amount of surface inside the rectangle), use the following formula.

> **Area of a Rectangle**
> The area of a rectangle is equal to the length multiplied by the width.
> $$\textbf{Area} = \textbf{length} \cdot \textbf{width}$$

For example, the rectangle shown here has an area of 12 square feet (ft²).

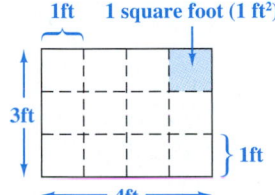

Area = length • width
Area = 4 ft • 3 ft
Area = 12 ft²

Area is the amount of surface.

Other units for measuring area are square inches (in.²), square yards (yd²), and square miles (mi²). (See **Section 8.3** for more information on area.)

EXAMPLE 5 **Applying Fraction Skills**

Find the area of each rectangle.

(a) Find the area of each shower tile.

Area = length • width

$$\text{Area} = \frac{11}{12} \cdot \frac{3}{4}$$

$$= \frac{11}{\underset{4}{\cancel{12}}} \cdot \frac{\overset{1}{\cancel{3}}}{4} \quad \text{Divide numerator and denominator by 3.}$$

$$= \frac{11}{16} \text{ ft}^2 \quad \boxed{\text{This is less than 1 ft}^2.}$$

Continued on Next Page

---

4 Multiply. Write answers in lowest terms and as whole numbers or mixed numbers where possible.

(a) $8 \cdot \dfrac{1}{8}$

(b) $\dfrac{3}{4} \cdot 5 \cdot \dfrac{5}{3}$

(c) $\dfrac{3}{5} \cdot 40$

(d) $\dfrac{3}{25} \cdot \dfrac{5}{11} \cdot 99$

ANSWERS

4. **(a)** 1 **(b)** $6\dfrac{1}{4}$ **(c)** 24 **(d)** $5\dfrac{2}{5}$

5 Find the area of each rectangle.

(a)

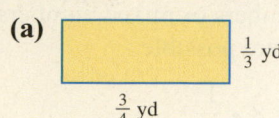

$\frac{1}{3}$ yd

$\frac{3}{4}$ yd

(b) a community college campus that is $\frac{3}{8}$ mile by $\frac{1}{3}$ mile

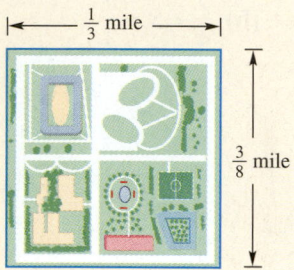

$\frac{1}{3}$ mile

$\frac{3}{8}$ mile

(c) a parcel of subdivided land that is $\frac{9}{7}$ mile by $\frac{7}{12}$ mile

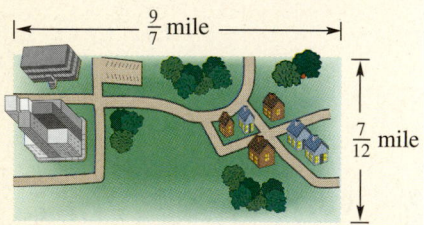

$\frac{9}{7}$ mile

$\frac{7}{12}$ mile

(b) Find the area of this polished diamond plate SUV running board.

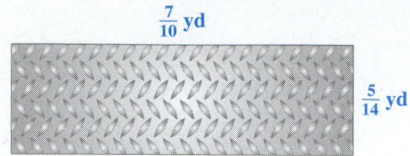

$\frac{7}{10}$ yd

$\frac{5}{14}$ yd

Multiply the length and width.

$$\text{Area} = \frac{7}{10} \cdot \frac{5}{14}$$

$$= \frac{\overset{1}{\cancel{7}}}{\underset{2}{\cancel{10}}} \cdot \frac{\overset{1}{\cancel{5}}}{\underset{2}{\cancel{14}}} \qquad \begin{array}{l}\text{Divide 7 and 14 by 7.}\\ \text{Divide 10 and 5 by 5.}\end{array}$$

$$= \frac{1}{4} \text{ square yard (yd}^2)$$

◀ *Work Problem* **5** *at the Side.*

ANSWERS

5. (a) $\frac{1}{4}$ yd^2

(b) $\frac{1}{8}$ mi^2

(c) $\frac{3}{4}$ mi^2

2.5 ▶▶▶ Exercises

Multiply. Write answers in lowest terms. See Examples 1–3.

1. $\dfrac{1}{3} \cdot \dfrac{3}{4}$

2. $\dfrac{2}{5} \cdot \dfrac{3}{4}$

3. $\dfrac{2}{7} \cdot \dfrac{1}{5}$

4. $\dfrac{2}{3} \cdot \dfrac{1}{2}$

 5. $\dfrac{8}{5} \cdot \dfrac{15}{32}$

6. $\dfrac{5}{9} \cdot \dfrac{4}{3}$

7. $\dfrac{2}{3} \cdot \dfrac{7}{12} \cdot \dfrac{9}{14}$

8. $\dfrac{7}{8} \cdot \dfrac{16}{21} \cdot \dfrac{1}{2}$

9. $\dfrac{3}{4} \cdot \dfrac{5}{6} \cdot \dfrac{2}{3}$

10. $\dfrac{2}{5} \cdot \dfrac{3}{8} \cdot \dfrac{2}{3}$

11. $\dfrac{9}{22} \cdot \dfrac{11}{16}$

12. $\dfrac{5}{12} \cdot \dfrac{7}{10}$

13. $\dfrac{5}{8} \cdot \dfrac{16}{25}$

14. $\dfrac{6}{11} \cdot \dfrac{22}{15}$

15. $\dfrac{14}{25} \cdot \dfrac{65}{48} \cdot \dfrac{15}{28}$

16. $\dfrac{35}{64} \cdot \dfrac{32}{15} \cdot \dfrac{27}{72}$

17. $\dfrac{16}{25} \cdot \dfrac{35}{32} \cdot \dfrac{15}{64}$

18. $\dfrac{39}{42} \cdot \dfrac{7}{13} \cdot \dfrac{7}{24}$

Multiply. Write answers in lowest terms and as whole or mixed numbers where possible.
See Example 4.

 19. $5 \cdot \dfrac{4}{5}$

20. $20 \cdot \dfrac{3}{4}$

21. $\dfrac{5}{8} \cdot 64$

22. $\dfrac{5}{6} \cdot 24$

 23. $36 \cdot \dfrac{2}{3}$

24. $30 \cdot \dfrac{3}{10}$

25. $36 \cdot \dfrac{5}{8} \cdot \dfrac{9}{15}$

26. $35 \cdot \dfrac{3}{5} \cdot \dfrac{1}{2}$

27. $100 \cdot \dfrac{21}{50} \cdot \dfrac{3}{4}$

28. $400 \cdot \dfrac{7}{8}$

29. $\dfrac{2}{5} \cdot 200$

30. $\dfrac{6}{7} \cdot 245$

31. $142 \cdot \dfrac{2}{3}$

32. $\dfrac{12}{25} \cdot 430$

33. $\dfrac{28}{21} \cdot 640 \cdot \dfrac{15}{32}$

34. $\dfrac{21}{13} \cdot 520 \cdot \dfrac{7}{20}$

35. $\dfrac{54}{38} \cdot 684 \cdot \dfrac{5}{6}$

36. $\dfrac{76}{43} \cdot 473 \cdot \dfrac{5}{19}$

Find the area of each rectangle. See Example 5.

37.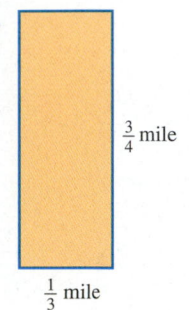

$\frac{3}{4}$ mile

$\frac{1}{3}$ mile

38.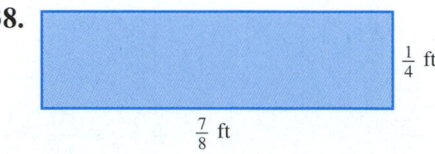

$\frac{1}{4}$ ft

$\frac{7}{8}$ ft

39. 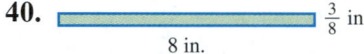 $\frac{3}{4}$ meter

12 meters

40. $\frac{3}{8}$ in.

8 in.

41. $\frac{3}{14}$ in.

$\frac{7}{5}$ in.

42.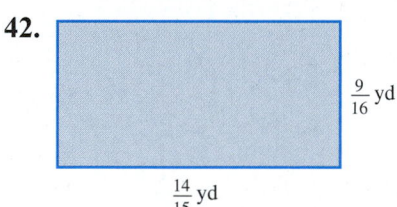

$\frac{9}{16}$ yd

$\frac{14}{15}$ yd

43. Write in your own words the rule for multiplying fractions. Make up an example problem to show how this works.

44. A useful shortcut when multiplying fractions is to divide a numerator and a denominator by the same number. Describe how this works and give an example.

Find the area of each rectangle in these application problems. Write answers in lowest terms and as whole or mixed numbers where possible. See Example 5.

45. Find the area of a heating-duct grill having a length of 2 yd and a width of $\frac{3}{4}$ yd.

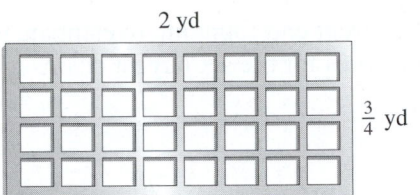

2 yd

$\frac{3}{4}$ yd

46. Find the area of the top of an office desk having a length of 2 yd and a width of $\frac{7}{8}$ yd.

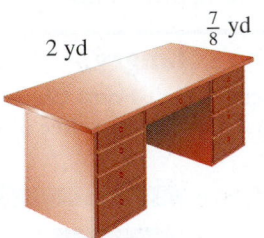

2 yd

$\frac{7}{8}$ yd

47. A wildfire is contained in an area measuring $\frac{7}{8}$ mile by 4 miles. Find the total area of the containment.

48. A motorcycle racecourse is $\frac{2}{3}$ mile wide by 4 miles long. Find the area of the racecourse.

49. The Sunny Side Soccer Park is $\frac{1}{4}$ mile long and $\frac{3}{16}$ mile wide, while the Creek Side Soccer Park is $\frac{3}{8}$ mile long and $\frac{1}{8}$ mile wide. Which park has the larger area?

50. The Rocking Horse Ranch is $\frac{3}{4}$ mile long and $\frac{2}{3}$ mile wide. The Silver Spur Ranch is $\frac{5}{8}$ mile long and $\frac{4}{5}$ mile wide. Which ranch has the larger area?

Relating Concepts (Exercises 51–56) For Individual or Group Work

Front end rounding can be used to estimate an answer when multiplying fractions.
Work Exercises 51–56 *in order*. *Round exact answers to the nearest whole number.*

The bar graph shows the six largest fast-food chains by the number of stores.

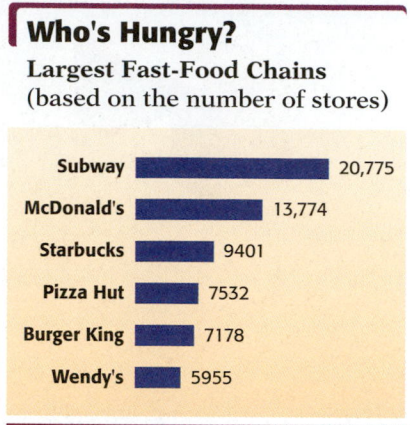

Who's Hungry?

Largest Fast-Food Chains
(based on the number of stores)

Subway	20,775
McDonald's	13,774
Starbucks	9401
Pizza Hut	7532
Burger King	7178
Wendy's	5955

Source: Technomic.

51. Use front end rounding to estimate the total number of stores for the six largest chains.

52. Find the exact total number of stores for the six largest chains.

53. If $\frac{2}{3}$ of the Starbucks stores are in medium to large population areas, use front end rounding to estimate, and then find the exact number of stores in these population areas.

Estimate:

Exact:

54. If $\frac{2}{5}$ of the Pizza Hut stores are in shopping centers and food courts, use front end rounding to estimate, and then find the exact number of stores in these locations.

Estimate:

Exact:

Compare the estimated answers and the exact answers in Exercises 53 and 54. How can you get an estimated answer that is closer to the exact answer? Try rounding the number of stores to some multiple of the denominator in the fraction that has two *nonzero digits.*

55. Refer to Exercise 53. Round the number of Starbucks locations to some multiple of the denominator in $\frac{2}{3}$ that has *two* nonzero digits. Now estimate the answer, showing your work.

56. Refer to Exercise 54. Round the number of Pizza Hut stores to some multiple of the denominator in $\frac{2}{5}$ that has *two* nonzero digits. Now estimate the answer, showing your work.

2.6 ▶▶▶ Applications of Multiplication

OBJECTIVE 1 Solve fraction application problems using multiplication. Many application problems are solved by multiplying fractions. Use the following indicator words for multiplication.

product
double
triple
times
of (when "of" follows a fraction)
twice
twice as much

> Always look for indicator words.

OBJECTIVE

1 Solve fraction application problems using multiplication.

Look for these indicator words in the following examples.

EXAMPLE 1 Applying Indicator Words

Lois Stevens gives $\frac{1}{10}$ of her income to her church. One month she earned $2980. How much did she give to the church that month?

Step 1 **Read** the problem. The problem asks us to find the amount of money given to the church.

Step 2 **Work out a plan.** The indicator word is *of:* Stevens gave $\frac{1}{10}$ *of* her income. When it follows a fraction, the word *of* indicates multiplication, so find the amount given to the church by multiplying $\frac{1}{10}$ and $2980.

Step 3 **Estimate** a reasonable answer. Round the income of $2980 to $3000. Then divide $3000 by 10 to find $\frac{1}{10}$ of the income (one of 10 equal parts). Our estimate is $3000 ÷ 10 = $300. (Recall the shortcut for dividing by 10; drop one 0 from the dividend.)

Step 4 **Solve** the problem.

$$\text{amount} = \frac{1}{\overset{}{\underset{1}{10}}} \cdot \frac{\overset{298}{2980}}{1} = \frac{298}{1} = 298$$

Step 5 **State the answer.** Stevens gave $298 to her church that month.

Step 6 **Check.** The exact answer, $298, is close to our estimate of $300.

Work Problem **1** *at the Side.* ▶

EXAMPLE 2 Solving a Fraction Application Problem

Of the 39 students in Carol Dixon's high school biology class, $\frac{2}{3}$ plan to go to college. How many plan to go to college?

Step 1 **Read** the problem. The problem asks us to find the number of students who plan to go to college.

Step 2 **Work out a plan.** Reword the problem to read

$\frac{2}{3}$ **of** the students plan to go to college.

Indicator word for multiplication
when it follows a fraction

Continued on Next Page

1 Solve each problem.

(a) Eric and Sabrina Means are saving $\frac{3}{8}$ of their income for the down payment on their first home. If they have a combined annual income of $81,576, how much can they save in a year?

(b) A retiring firefighter will receive $\frac{5}{8}$ of her highest annual salary as retirement income. If her highest annual salary is $62,504, how much will she receive as retirement income?

2 At Sid's Pharmacy, $\frac{5}{16}$ of the prescriptions are paid by a third party (insurance company). If 3696 prescriptions are filled, find the number paid by a third party.

Use the six problem-solving steps.

Step 3 **Estimate** a reasonable answer. Round the number of students in the class from 39 to 40. Then, $\frac{1}{2}$ of 40 is 20. Since $\frac{2}{3}$ is more than $\frac{1}{2}$, our estimate is that "more than 20 students" plan to go to college.

Step 4 **Solve** the problem. Find the number who plan to go to college by multiplying $\frac{2}{3}$ and 39.

$$\text{number who plan to go} = \frac{2}{3} \cdot 39$$

$$= \frac{2}{\underset{1}{\cancel{3}}} \cdot \frac{\overset{13}{\cancel{39}}}{1} = \frac{26}{1} = 26$$

Step 5 **State the answer.** 26 students plan to go to college.

Step 6 **Check.** The exact answer, 26, fits our estimate of "more than 20."

◀ Work Problem **2** at the Side.

3 At our college, $\frac{1}{3}$ of the students speak a foreign language. Of those speaking a foreign language, $\frac{3}{4}$ speak Spanish. What fraction of the students speak Spanish?

Use the six problem-solving steps.

EXAMPLE 3 Finding a Fraction of a Fraction

In her will, a woman divides her estate into 6 equal parts. Five of the 6 parts are given to relatives. Of the sixth part, $\frac{1}{3}$ goes to the Salvation Army. What fraction of her total estate goes to the Salvation Army?

Step 1 **Read** the problem. The problem asks for the fraction of an estate that goes to the Salvation Army.

Step 2 **Work out a plan.** Reword the problem to read

the Salvation Army gets $\frac{1}{3}$ **of** $\frac{1}{6}$.

Indicator word for multiplication
when it follows a fraction

Step 3 **Estimate** a reasonable answer. If the estate is divided into 6 equal parts and each of these parts was divided into 3 equal parts, we would have $6 \cdot 3 = 18$ equal parts. Our estimate is $\frac{1}{18}$.

Step 4 **Solve** the problem. The Salvation Army gets $\frac{1}{3}$ **of** $\frac{1}{6}$.

Indicator word

To find the fraction that the Salvation Army is to receive, multiply $\frac{1}{3}$ and $\frac{1}{6}$.

$$\text{fraction to Salvation Army} = \frac{1}{3} \cdot \frac{1}{6}$$

$$= \frac{1}{18}$$

Step 5 **State the answer.** The Salvation Army gets $\frac{1}{18}$ of the total estate.

Step 6 **Check.** The exact answer, $\frac{1}{18}$, matches our estimate.

Remember to check your work.

◀ Work Problem **3** at the Side.

ANSWERS

2. 1155 prescriptions

3. $\frac{1}{4}$ speak Spanish

EXAMPLE 4   **Using Fractions with a Circle Graph**

The circle graph, or pie chart, shows where children 8 to 17 years of age make food purchases when away from home. If 2500 children were in the survey, find the number of children who buy food in the school cafeteria.

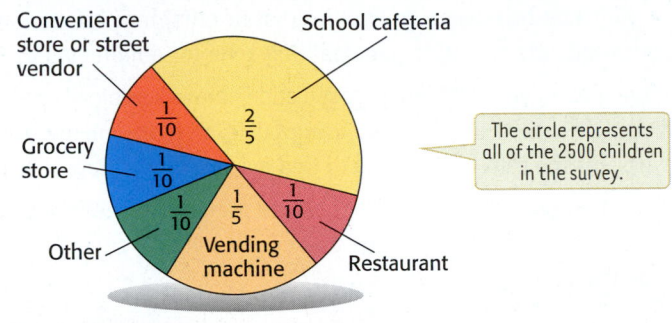

FINDING FOOD

Children ages 8 to 17 made food purchases at the following locations:

The circle represents all of the 2500 children in the survey.

Source: Pursuant Inc. for American Dietetic Association Foundation.

Step 1 **Read** the problem. The problem asks for the number of children who buy food in the school cafeteria.

Step 2 **Work out a plan.** Reword the problem to read

$\dfrac{2}{5}$ **of** 2500 children buy food in the school cafeteria.

Indicator word for multiplication when it follows a fraction

Step 3 **Estimate** a reasonable answer. $\frac{1}{2}$ of 2500 people is 1250 people. $\frac{2}{5}$ is less than $\frac{1}{2}$, so our estimate is "less than 1250 people."

Step 4 **Solve** the problem. Find the number who buy food in the school cafeteria by multiplying $\frac{2}{5}$ and 2500.

$$\text{number in school cafeteria} = \frac{2}{5} \cdot 2500$$

$$= \frac{2}{5} \cdot \frac{2500}{1}$$

$$= \frac{2}{\overset{1}{\cancel{5}}} \cdot \frac{\overset{500}{\cancel{2500}}}{1} \quad \text{Divide both numerator and denominator by 5.}$$

$$= \frac{1000}{1} = 1000$$

Step 5 **State the answer.** 1000 children buy food in the school cafeteria.

Step 6 **Check.** The exact answer, 1000 children, fits our estimate of "less than 1250 children."

Work Problem ④ *at the Side.* ▶

④ Solve each problem using the six problem-solving steps. Use the circle graph in Example 4.

(a) What fraction of the children buy food from vending machines?

(b) What number of children buy food from vending machines?

(c) What fraction of the children buy food from a convenience store or street vendor?

(d) What number of children buy food from a convenience store or street vendor?

ANSWERS

4. (a) $\dfrac{1}{5}$ (b) 500 children (c) $\dfrac{1}{10}$

(d) 250 children

Math in the Media

The Kaiser Permanente *Healthwise Handbook* says that "no one can prescribe the perfect fitness plan for you." However, your personal fitness plan should include aerobic conditioning that strengthens your heart and lungs. Activities that provide aerobic conditioning include brisk walking, running, stair climbing, biking, swimming, aerobic dance, or anything else that raises your heart rate to within your training zone for a minimum of 12 minutes.

If you train at the higher end of the training zone, you will burn glycogen and improve aerobic fitness. Training for longer periods at the lower end of the training zone results in your body using fat reserves for energy. Wickipedia, The Free Encyclopedia, at www.wickipedia.org, explains the calculations used to find maximum heart rates and training zones.

Example: The training zone (TZ) is based on your heart rate (HR) for one minute. To see if you are in the training zone, measure your heart rate for 15 seconds. Compare it to the 15-second training zone. Find the exact answer, and then round to the nearest whole number.

Instruction	Calculation	Example (age 22)
Calculate maximum heart rate (MHR).	$220 - $ your age	$220 - 22 = 198$
Calculate lower limit of training zone (TZ).	$\frac{3}{5} \times$ (MHR)	$\frac{3}{5} \times (198) = \frac{594}{5} = 118\frac{4}{5}$
Calculate upper limit of training zone (TZ).	$\frac{4}{5} \times$ (MHR)	$\frac{4}{5} \times (198) = \frac{792}{5} = 158\frac{2}{5}$
Calculate the exact 15-second training zone. Round the results to the nearest whole number.	$\left(\frac{1}{4} \times \text{lower TZ},\right.$ $\left.\frac{1}{4} \times \text{Upper TZ}\right)$	$\frac{1}{4} \times \frac{594}{5} = 29\frac{7}{10}; \frac{1}{4} \times \frac{792}{5} = 39\frac{3}{5}$ HR between $29\frac{7}{10}$ and $39\frac{3}{5}$ HR between 30 and 40

1. Suppose you work in a physical fitness center and decide to design a poster to remind the clients of the training zone for their age. Compute the exact 15-second training zone for people of each of the ages in the table. Write fractions in lowest terms. Then round the answers to the nearest whole number.

2. Explain why the lower and upper training zones (TZ) are multiplied by $\frac{1}{4}$.

Age	MHR	Lower Limit of TZ	Upper Limit of TZ	15-Second TZ (exact)	15-Second TZ (rounded)
18					
25					
30					
40					
50					
60					

2.6 ▶▶▶ Exercises

FOR EXTRA HELP   **MyMathLab**   Math XL PRACTICE   WATCH   DOWNLOAD   READ   REVIEW

Solve each application problem. Look for indicator words. See Examples 1–4.

1. A file cabinet top is $\frac{3}{4}$ yd by $\frac{2}{3}$ yd. Find its area.

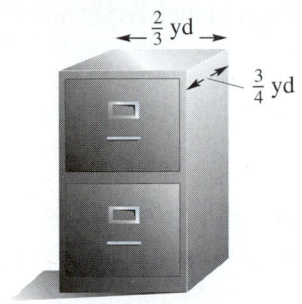

2. The rectangular floor of Darby's house measures $\frac{14}{15}$ yd by $\frac{3}{4}$ yd. Find its area.

3. A cookie sheet is $\frac{4}{3}$ ft by $\frac{2}{3}$ ft. Find its area.

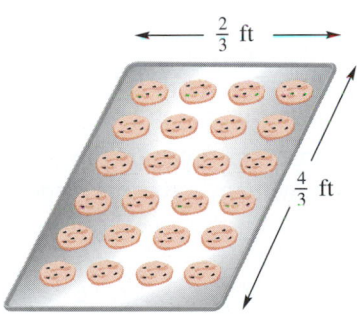

4. Each day there are 16 million people who shop at flea markets. If $\frac{2}{5}$ of these people purchase produce at the flea market, how many purchase produce? (*Source:* National Flea Market Association.)

5. Pete is helping Colin make a rectangular mahogany lamp table for Carolyn's birthday. Find the area of the top of the table if it is $\frac{4}{5}$ yd long by $\frac{3}{8}$ yd wide.

6. A convenience store sells 1650 items, of which $\frac{12}{25}$ are classified as junk food. How many of the items are junk food?

7. Dan Crump had expenses of $6848 during one semester of college. His part-time job provided $\frac{3}{8}$ of the amount he needed. How much did he earn on his job?

8. Erin Hernandez produces $5680 in profits for her employer. If her personal earnings are $\frac{2}{5}$ of these profits, find the amount of her earnings.

9. The city with the most expensive daily parking fee is New York City (Midtown) at $45. The daily parking fee in Boston (third most expensive) is $\frac{4}{5}$ as much as New York City. Find the daily parking fee in Boston. (*Source:* Colliers International.)

10. The daily parking fee in New York City (Downtown) is $36. In San Francisco (fifth most expensive), the daily parking fee is $\frac{3}{4}$ the cost of New York City (Downtown). How much is the daily parking fee in San Francisco? (*Source:* Colliers International.)

11. At the Garlic Festival Fun Run, $\frac{7}{12}$ of the runners are women. If there are 1560 runners, how many are women?

12. A hotel has 408 rooms. Of these rooms, $\frac{9}{17}$ are for nonsmokers. How many rooms are for nonsmokers?

Almost $\frac{1}{3}$ of all television owners own a TiVo or some other DVR recording device. The circle graph below shows how the owners of this device have changed their television viewing time. Use this information to work Exercises 13–18.

TV with TiVo

Do you watch more or less TV now that you have this device? 1000 people surveyed.

Much more $\frac{1}{10}$

Much less $\frac{1}{10}$

Don't know $\frac{1}{20}$

Somewhat more $\frac{1}{2}$

Somewhat less $\frac{1}{4}$

Source: Greenfield Online Omnibus

13. Which response was given by the least number of people? How many people gave this response?

14. Which response was given by the greatest number of people? How many gave this response?

15. Find the total number of people who said they watch less television.

16. Find the total number of people who said they watch more television.

17. Without actually adding the fractions given for all the groups, explain why their sum has to be 1.

18. Refer to Exercise 17. Suppose you added all the fractions for the groups and did not get 1 as an answer. List some possible explanations.

*The table shows the earnings for the Owens family last year and the circle graph shows
how they spent their earnings. Use this information to answer Exercises 19–24.*

Month	Earnings	Month	Earnings
January	$4575	July	$5540
February	$4312	August	$3732
March	$4988	September	$4170
April	$4530	October	$5512
May	$4320	November	$4965
June	$4898	December	$6458

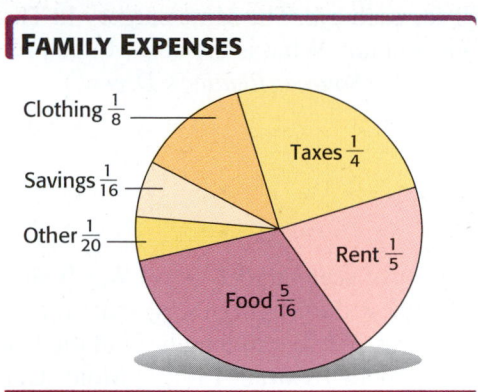

FAMILY EXPENSES

Clothing $\frac{1}{8}$

Savings $\frac{1}{16}$

Other $\frac{1}{20}$

Taxes $\frac{1}{4}$

Rent $\frac{1}{5}$

Food $\frac{5}{16}$

19. Find the Owens family's total income for the year.

20. How much of their annual earnings went to taxes?

21. Find the amount of their rent for the year.

22. How much did they spend for food during the year?

23. Find their annual savings.

24. How much of their annual income was spent on clothing?

25. Here is how one student solved a multiplication problem. Find the error and solve the problem correctly.

$$\frac{9}{10} \times \frac{20}{21} = \frac{\overset{3}{\cancel{9}}}{\underset{1}{\cancel{10}}} \times \frac{\overset{2}{\cancel{20}}}{\underset{3}{\cancel{21}}} = \frac{6}{3} = 2$$

26. When two whole numbers are multiplied, the product is always larger than the numbers being multiplied. When two proper fractions are multiplied, the product is always smaller than the numbers being multiplied. Are these statements true? Why or why not?

Solve each application problem.

27. The cost of Lasik eye surgery in the United States is $2000 for each eye. The same surgery in Thailand is $\frac{3}{8}$ of this amount. What is the cost of this procedure in Thailand? (*Source: Reader's Digest.*)

28. A knee replacement in the United States costs $36,300, while in Mexico the same procedure costs $\frac{3}{50}$ of this amount. Find the cost of a knee replacement in Mexico. (*Source: Reader's Digest.*)

29. A collector of scale model World War II ships wants to know the length of a $\frac{1}{128}$ scale model of a ship that was 256 feet in length. Find the length of the scale model. (*Source:* Lilliput Motor Company, LTD.)

30. Howard Martin, manager of Bayside Fishing, is adding $\frac{3}{32}$ of a quart of 2-cycle oil to each gallon of gasoline. How many quarts of oil will he need to add to 760 gallons of gasoline?

31. LaDonna Washington is running for city council. She needs to get $\frac{2}{3}$ of her votes from senior citizens and 27,000 votes in all to win. How many votes does she need from voters other than the senior citizens?

32. The start-up cost of a Subs and Sandwich Shop is $32,000. If the bank will loan you $\frac{9}{16}$ of the start up and you must pay the balance, how much more will you need to open a shop?

33. A will states that $\frac{7}{8}$ of an estate is to be divided among relatives. Of the remaining estate, $\frac{1}{4}$ goes to the American Cancer Society. What fraction of the estate goes to the American Cancer Society?

34. A couple has $\frac{2}{5}$ of their total investments in real estate. Of the remaining investments, $\frac{1}{3}$ is invested in bonds. What fraction of the total investments is in bonds?

2.7 ▶▶▶ Dividing Fractions

OBJECTIVE **1** **Find the reciprocal of a fraction.** To divide fractions, we need to know how to find the **reciprocal** of a fraction.

> ### Reciprocal of a Fraction
> Two numbers are reciprocals of each other if their product is 1. To find the reciprocal of a fraction, interchange the numerator and denominator.

For example, the reciprocal of $\frac{3}{4}$ is $\frac{4}{3}$.

$$\text{Fraction} \quad \frac{3}{4} \diagdown \diagup \frac{4}{3} \quad \text{Reciprocal}$$

> ### Note
> Notice that you invert, or "flip," a fraction to find its reciprocal.

EXAMPLE 1 **Finding Reciprocals**

Find the reciprocal of each fraction.

> Flip a fraction to find the reciprocal.

(a) The reciprocal of $\frac{1}{4}$ is $\frac{4}{1}$ because $\frac{1}{4} \cdot \frac{4}{1} = \frac{4}{4} = 1$

(b) The reciprocal of $\frac{2}{3}$ is $\frac{3}{2}$ because $\frac{2}{3} \cdot \frac{3}{2} = \frac{6}{6} = 1$

(c) The reciprocal of $\frac{3}{5}$ is $\frac{5}{3}$ because $\frac{3}{5} \cdot \frac{5}{3} = \frac{15}{15} = 1$

(d) The reciprocal of 8 is $\frac{1}{8}$ because $\frac{8}{1} \cdot \frac{1}{8} = \frac{8}{8} = 1$ Think of 8 as $\frac{8}{1}$.

Work Problem **1** *at the Side.* ▶

> ### Note
> Every number has a reciprocal except 0. The number 0 has no reciprocal because there is no number that can be multiplied by 0 to get 1.
>
> $$0 \cdot (\text{reciprocal}) = 1$$
>
> ↑
>
> There is no number to use here that will give an answer of 1. When you multiply by 0, you always get 0.

1 Find the reciprocal of each fraction.

(a) $\frac{3}{4}$

(b) $\frac{3}{8}$

(c) $\frac{9}{4}$

(d) 16

ANSWERS

1. **(a)** $\frac{4}{3}$ **(b)** $\frac{8}{3}$ **(c)** $\frac{4}{9}$ **(d)** $\frac{1}{16}$

In **Chapter 1,** we saw that the division problem $12 \div 3$ asks how many 3s are in 12. In the same way, the division problem $\frac{2}{3} \div \frac{1}{6}$ asks how many $\frac{1}{6}$s are in $\frac{2}{3}$. The figure illustrates $\frac{2}{3} \div \frac{1}{6}$.

$\frac{2}{3}$

$\frac{1}{6}$

Ask: How many $\frac{1}{6}$s are in $\frac{2}{3}$?

The figure shows that there are 4 of the $\frac{1}{6}$ pieces in $\frac{2}{3}$, or

$$\frac{2}{3} \div \frac{1}{6} = 4$$

OBJECTIVE ② **Divide fractions.** We will use reciprocals to divide fractions.

> **Dividing Fractions**
>
> To divide two fractions, multiply the first fraction by the reciprocal of the second fraction.

EXAMPLE 2 **Dividing One Fraction by Another**

Divide. Write answers in lowest terms and as mixed numbers where possible.

(a) $\dfrac{7}{8} \div \dfrac{15}{16}$

The reciprocal of $\dfrac{15}{16}$ is $\dfrac{16}{15}$

Reciprocals

$$\frac{7}{8} \div \frac{15}{16} = \frac{7}{8} \cdot \frac{16}{15}$$

Use $\frac{16}{15}$, the reciprocal of $\frac{15}{16}$, and multiply.

Change division to multiplication.

$$= \frac{7}{\overset{}{\underset{1}{8}}} \cdot \frac{\overset{2}{16}}{15} \qquad \text{Divide the numerator and denominator by 8.}$$

$$= \frac{7 \cdot 2}{1 \cdot 15} \qquad \text{Multiply.}$$

$$= \frac{14}{15} \quad \leftarrow \text{Lowest terms}$$

Continued on Next Page

(b) $\dfrac{\frac{4}{5}}{\frac{3}{10}}$

$$\dfrac{\frac{4}{5}}{\frac{3}{10}} = \frac{4}{5} \div \frac{3}{10} \qquad \text{\textcolor{blue}{Rewrite by using the} } \div \text{\textcolor{blue}{ symbol for division.}}$$

$$= \frac{4}{\cancel{5}_{1}} \cdot \frac{\cancel{10}^{2}}{3} \qquad \text{\textcolor{blue}{The reciprocal of } \frac{3}{10} \text{ is } \frac{10}{3}. \text{ Change ``} \div \text{'' to ``} \cdot \text{''}.}$$
$$\text{\textcolor{blue}{Divide the numerator and denominator by 5.}}$$

$$= \frac{4 \cdot 2}{1 \cdot 3} \qquad \text{\textcolor{blue}{Multiply.}}$$

$$= \frac{8}{3} = 2\frac{2}{3} \qquad \text{\textcolor{blue}{Rewrite the answer as a mixed number.}}$$

> **CAUTION**
> Be certain that the divisor fraction is changed to its reciprocal *before* you divide numerators and denominators by common factors.

Work Problem **2** *at the Side.* ▶

EXAMPLE 3 **Dividing with a Whole Number**

Divide. Write all answers in lowest terms and as whole or mixed numbers where possible.

(a) $5 \div \dfrac{1}{4}$

Write 5 as $\frac{5}{1}$. Next, use the reciprocal of $\frac{1}{4}$, which is $\frac{4}{1}$.

$$5 \div \frac{1}{4} = \frac{5}{1} \cdot \frac{4}{1} \qquad \text{\textcolor{blue}{Reciprocal of } \frac{1}{4} \text{ is } \frac{4}{1}.}$$

$$\underset{\text{Reciprocals}}{}$$

$$= \frac{5 \cdot 4}{1 \cdot 1} \qquad \text{\textcolor{blue}{Multiply.}}$$

$$= \frac{20}{1} = 20 \qquad \text{\textcolor{blue}{Rewrite the answer as a whole number.}}$$

Continued on Next Page

---

2 Divide. Write answers in lowest terms and as mixed numbers where possible.

(a) $\dfrac{1}{4} \div \dfrac{2}{3}$

(b) $\dfrac{3}{8} \div \dfrac{5}{8}$

(c) $\dfrac{\frac{2}{3}}{\frac{4}{5}}$

(d) $\dfrac{\frac{5}{6}}{\frac{7}{12}}$

ANSWERS

2. (a) $\dfrac{3}{8}$ (b) $\dfrac{3}{5}$ (c) $\dfrac{5}{6}$ (d) $1\dfrac{3}{7}$

3 Divide. Write answers in lowest terms and as whole or mixed numbers where possible.

(a) $10 \div \dfrac{1}{2}$

(b) $6 \div \dfrac{6}{7}$

(c) $\dfrac{4}{5} \div 6$

(d) $\dfrac{3}{8} \div 4$

4 Solve each problem using the six problem-solving steps.

(a) How many $\frac{3}{4}$-quart leafblower fuel tanks can be filled from 15 quarts of fuel?

(b) Find the number of $\frac{4}{5}$-quart bottles that can be filled from a 120-quart cask.

(b) $\dfrac{2}{3} \div 6$

Write 6 as $\frac{6}{1}$. The reciprocal of $\frac{6}{1}$ is $\frac{1}{6}$.

$$\frac{2}{3} \div \frac{6}{1} = \frac{2}{3} \cdot \frac{1}{6}$$

Reciprocals

> Careful! Divide out common factors **after** changing to the reciprocal and to multiplication.

$$= \frac{\overset{1}{2}}{3} \cdot \frac{1}{\underset{3}{6}}$$

Divide the numerator and denominator by 2, then multiply.

$$= \frac{1 \cdot 1}{3 \cdot 3} = \frac{1}{9}$$

Lowest terms

◀ Work Problem **3** at the Side.

OBJECTIVE 3 Solve application problems in which fractions are divided. Many application problems require division of fractions. Recall that typical indicator words for division are *goes into, per, divide, divided by, divided equally,* and *divided into.*

EXAMPLE 4 Applying Fraction Skills

Goldie, the manager of the Burnside Deli, must fill a 10-gallon kosher dill pickle crock with salt brine. She has only a $\frac{2}{3}$-gallon container to use. How many times must she fill the $\frac{2}{3}$-gallon container and empty it into the 10-gallon crock?

Step 1 **Read** the problem. We need to find the number of times Goldie needs to use a $\frac{2}{3}$-gallon container in order to fill a 10-gallon crock.

Step 2 **Work out a plan.** We can solve the problem by finding the number of times 10 can be divided by $\frac{2}{3}$.

Step 3 **Estimate** a reasonable answer. Round $\frac{2}{3}$ gallon to 1 gallon. In order to fill the 10-gallon container, she would have to use the 1-gallon container 10 times, so our estimate is 10.

Step 4 **Solve** the problem.

Reciprocals

$$10 \div \frac{2}{3} = \frac{10}{1} \cdot \frac{3}{2}$$

The reciprocal of $\frac{2}{3}$ is $\frac{3}{2}$. Change "÷" to " • ."

$$= \frac{\overset{5}{10}}{1} \cdot \frac{3}{\underset{1}{2}}$$

Divide the numerator and denominator by 2, and then multiply.

$$= \frac{15}{1} = 15$$

Step 5 **State the answer.** Goldie must fill the container 15 times.

Step 6 **Check.** The exact answer, 15 times, is reasonably close to our estimate of 10 times.

◀ Work Problem **4** at the Side.

ANSWERS

3. (a) 20 (b) 7 (c) $\dfrac{2}{15}$ (d) $\dfrac{3}{32}$

4. (a) 20 tanks (b) 150 bottles

EXAMPLE 5 **Applying Fraction Skills**

At the Happi-Time Day Care Center, $\frac{6}{7}$ of the total budget goes to classroom operation. If there are 18 classrooms and each one receives the same amount, what fraction of the operating amount does each classroom receive?

Step 1 **Read** the problem. Since $\frac{6}{7}$ of the total budget must be split into 18 parts, we must find the fraction of the classroom operating amount received by each classroom.

Step 2 **Work out a plan.** We must divide the fraction of the total budget going to classroom operation ($\frac{6}{7}$) by the number of classrooms (18).

Step 3 **Estimate** a reasonable answer. Round $\frac{6}{7}$ to 1. If all of the operating expenses (1 whole) were divided between 18 classrooms, each classroom would receive $\frac{1}{18}$ of the operating expenses, our estimate.

Step 4 **Solve** the problem. We solve by dividing $\frac{6}{7}$ by 18.

$$\frac{6}{7} \div 18 = \frac{6}{7} \div \frac{18}{1} \quad \boxed{\text{Write 18 as } \frac{18}{1}}$$

$$= \frac{\overset{1}{\cancel{6}}}{7} \cdot \frac{1}{\underset{3}{\cancel{18}}} \quad \text{The reciprocal of } \frac{18}{1} \text{ is } \frac{1}{18}. \text{ Change "}\div\text{" to "} \cdot \text{."}$$
$$\text{Divide the numerator and denominator by 6.}$$

$$= \frac{1}{21} \quad \text{Multiply.}$$

Step 5 **State the answer.** Each classroom receives $\frac{1}{21}$ of the total budget.

Step 6 **Check.** The exact answer, $\frac{1}{21}$, is close to our estimate of $\frac{1}{18}$.

Work Problem **5** *at the Side.* ▶

---

5 Solve each problem using the six problem-solving steps.

(a) The top 12 employees at Mayfield Manufacturing will divide $\frac{3}{4}$ of the annual bonus money. What fraction of the bonus money will each employee receive?

(b) A winning lottery ticket was purchased by 8 employees of United States Marketing and Promotions (USMP). They will donate $\frac{1}{5}$ of the total winnings to pay the medical expenses of a fellow employee, and then divide the remaining winnings evenly. What fraction of the prize money will each receive?

ANSWERS

5. **(a)** $\frac{1}{16}$ of the bonus money

   (b) $\frac{1}{10}$ of the lottery prize money

Math in the Media

Mathematics teachers attending conferences in New Orleans, Louisiana, and San Jose, California, found the following information about hotel rates on Internet Web sites. To find the hotel room rates in New Orleans, they went to www.hotel-rate.com/us/louisiana/neworleans/, while the hotel room rates in San Jose were found at www.sanjose.com.

Hotel New Orleans	Single	Double	Triple	Quad	Suites
Marriott	$229	$239	$249	$259	$806
Sheraton	$165	$174	$193	$223	$748
San Jose					
Crowne Plaza	$209	$209	$219	$219	—
Hilton Towers	$179	$179	$199	$219	—
Sainte Claire	$149	$149	$169	$189	$329

1. The double rate is for two people sharing a room. What fractional part does each person pay?

2. **(a)** Multiply the double rate at the Hilton Towers in San Jose by $\frac{1}{2}$. What is the result?

   (b) Divide the double rate at the Hilton Towers in San Jose by 2. What is the result?

   (c) Explain what happened. How much money would one person owe if he or she shared a double room at the Hilton Towers in San Jose?

3. The triple rate is for three people sharing a room. What fractional part does each person pay?

4. How much money would one person owe if he or she shared a triple room at the Sainte Claire in San Jose? How much money would each person save if he or she could book a triple room at the Sainte Claire instead of the Crowne Plaza? Find your answer using two different methods, based on your observations in Problem 2.

5. The quad rate is for four people sharing a room. What fractional part does each person pay?

6. How much money would one person owe if he or she shared a quad room at the Sheraton in New Orleans? Find your answer using two different methods, based on your observations in Problem 2.

7. How many people would have to share a suite at the Sheraton in New Orleans for the cost per person to be less than sharing a quad room at the same hotel? Round the answer to the nearest whole number. (*Hint:* Estimate the cost per person for a quad room first. Then estimate the number of people needed to share the cost of the suite. Check your work using actual values.)

DO NOT DISTURB

NO MOLESTE

PRIÈRE DE NE PAS DÉRANGER

BITTE NICHT STÖREN

Find the reciprocal of each number. See Example 1.

1. $\dfrac{3}{8}$ **2.** $\dfrac{2}{5}$ **3.** $\dfrac{5}{6}$ **4.** $\dfrac{12}{7}$

5. $\dfrac{8}{5}$ **6.** $\dfrac{13}{20}$ **7.** 4 **8.** 10

Divide. Write answers in lowest terms and as whole or mixed numbers where possible. See Examples 2 and 3.

9. $\dfrac{1}{2} \div \dfrac{3}{4}$ **10.** $\dfrac{5}{8} \div \dfrac{7}{8}$ **11.** $\dfrac{7}{8} \div \dfrac{1}{3}$ **12.** $\dfrac{7}{8} \div \dfrac{3}{4}$

13. $\dfrac{3}{4} \div \dfrac{5}{3}$ **14.** $\dfrac{4}{5} \div \dfrac{9}{4}$ **15.** $\dfrac{7}{9} \div \dfrac{7}{36}$ **16.** $\dfrac{5}{8} \div \dfrac{5}{16}$

17. $\dfrac{15}{32} \div \dfrac{5}{64}$ **18.** $\dfrac{7}{12} \div \dfrac{14}{15}$ **19.** $\dfrac{\frac{13}{20}}{\frac{4}{5}}$ **20.** $\dfrac{\frac{9}{10}}{\frac{3}{5}}$

21. $\dfrac{\frac{5}{6}}{\frac{25}{24}}$ **22.** $\dfrac{\frac{28}{15}}{\frac{21}{5}}$ **23.** $12 \div \dfrac{2}{3}$ **24.** $7 \div \dfrac{1}{4}$

25. $\dfrac{\frac{18}{1}}{\frac{3}{4}}$ **26.** $\dfrac{\frac{12}{1}}{\frac{3}{4}}$ **27.** $\dfrac{\frac{4}{7}}{8}$ **28.** $\dfrac{\frac{7}{10}}{3}$

Solve each application problem by using division. See Examples 4 and 5.

29. Veterinarian Jasmine Cato has $\frac{8}{9}$ quart of medication. She wishes to prescribe this medication for 4 cats in her pet hospital. If she divides the medication evenly, how much will each cat receive?

30. Harold Pishke, barber, has 15 quarts of conditioning shampoo. If he wants to put this shampoo into $\frac{3}{8}$-quart containers, how many containers can be filled?

31. Some college roommates want to make pancakes for their neighbors. They need 5 cups of flour, but have only a $\frac{1}{3}$-cup measuring cup. How many times will they need to fill their measuring cup?

32. A cross-country bike rider completes $\frac{5}{6}$ of her trip in 15 days. What fraction of her total trip does she complete each day?

33. How many $\frac{1}{8}$-ounce eye drop dispensers can be filled with 11 ounces of eye drops?

34. It is estimated that each guest at a party will eat $\frac{5}{16}$ pound of peanuts. How many guests may be served with 10 pounds of peanuts?

35. Pam Trizlia had a small pickup truck that could carry $\frac{2}{3}$ cord of firewood. Find the number of trips needed to deliver 40 cords of wood.

36. Manuel Servin has a 200-yard roll of weather stripping material. Find the number of pieces of weather stripping $\frac{5}{8}$ yard in length that may be cut from the roll.

37. Your classmate is confused on how to divide by a fraction. Write a short note telling him how this should be done.

38. If you multiply positive proper fractions, the product is smaller than the fractions multiplied. When you divide by a proper fraction, is the quotient smaller than the numbers in the problem? Prove your answer with examples.

Solve each application problem using multiplication or division.

39. The recipe for a Jelly Belly Express loafcake calls for $\frac{3}{4}$ pound of Jelly Belly jelly beans in assorted colors. If you want to make 16 cakes, how many pounds of Jelly Belly jelly beans will you need?

40. In a recent study, it was found that one month after leaving the hospital only $\frac{7}{8}$ of the 1520 heart attack patients were still taking the life-saving drugs prescribed for them. How many of these patients were still taking their drugs? (*Source:* Dr. Michael Ho, Denver Veterans Medical Center.)

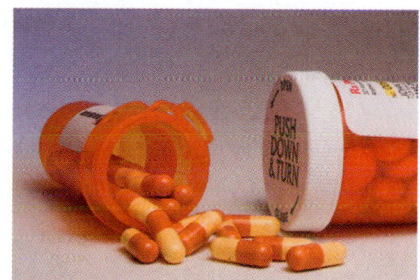

41. Broadly Plumbing finds that $\frac{3}{4}$ can of pipe joint compound is needed to plumb each new home. How many homes can be plumbed with 156 cans of compound?

42. A mechanic uses an average of $\frac{2}{3}$ gallon of gear lube to service each tractor differential. Find the number of tractors that can be serviced with 28 gallons of gear lube.

43. In recordings of 186 patient visits, doctors failed to mention a new drug's side effects or how long to take the drug in $\frac{2}{3}$ of the visits. In how many visits did doctors fail to discuss these issues with patients? (*Source:* Dr. Deijieng Tam, UCLA.)

44. Laura has been working on a job that will require 81 hours to complete. If she has completed $\frac{7}{9}$ of the job, how many hours has she worked?

45. A dish towel manufacturer requires $\frac{3}{8}$ yard of cotton fabric for each towel. Find the number of dish towels that can be made from 912 yards of fabric.

46. Each patient will receive $\frac{7}{10}$ vial of medication. How many patients can be treated with 3150 vials of medication?

Relating Concepts (Exercises 47—52) For Individual or Group Work

Many application problems are solved using multiplication and division of fractions.
Work Exercises 47—52 in order.

47. Perhaps the most common indicator word for multiplication is the word *of*. Circle the words in the list below that are also indicator words for multiplication.

more than   per

double   twice

times   product

less than   difference

equals   twice as much

48. Circle the words in the list below that are indicator words for division.

fewer   sum of

goes into   divide

per   quotient

equals   double

loss of   divided by

49. To divide two fractions, multiply the first fraction by the _____ of the second fraction.

50. Find the reciprocals for each number.

$$\frac{3}{4} \quad \frac{7}{8} \quad 5 \quad \frac{12}{19}$$

The size of a U.S.A. first-class Forever postage stamp is shown here. Use this to answer Exercises 51 and 52.

$\longleftarrow \frac{15}{16}$ in. $\longrightarrow$

$\frac{15}{16}$ in.

51.(a) Explain how to find the perimeter of any regular 3-, 4-, 5-, or 6-sided figure using multiplication.

(b) Find the perimeter of the stamp using multiplication.

52. Find the area of the postage stamp. Explain how to find the area of any rectangle.

| 2.8 ▶▶▶ | Multiplying and Dividing Mixed Numbers |

In **Section 2.2** we worked with mixed numbers—a whole number and a fraction written together. Many of the fraction problems you encounter in everyday life involve mixed numbers.

OBJECTIVE 1 Estimate the answer and multiply mixed numbers. When multiplying mixed numbers, it is a good idea to estimate the answer first. Then multiply the mixed numbers by using the following steps.

> **Multiplying Mixed Numbers**
> **Step 1** *Change* each mixed number to an improper fraction.
> **Step 2** *Multiply* as fractions.
> **Step 3** *Simplify* the answer, which means to write it in *lowest terms,* and change it to a mixed number or whole number where possible.

To estimate the answer, round each mixed number to the nearest whole number. If the numerator is *half* of the denominator or *more,* round up the whole number part. If the numerator is *less* than half the denominator, leave the whole number as it is.

$$1\frac{5}{8} \leftarrow \text{5 is more than 4.} \atop \leftarrow \text{Half of 8 is 4.} \quad 1\frac{5}{8} \text{ rounds up to 2}$$

Round mixed numbers to the nearest whole number when estimating.

$$3\frac{2}{5} \leftarrow \text{2 is less than } 2\frac{1}{2}. \atop \leftarrow \text{Half of 5 is } 2\frac{1}{2}. \quad 3\frac{2}{5} \text{ rounds to 3}$$

Work Problem **1** *at the Side.* ▶

EXAMPLE 1 Multiplying Mixed Numbers

First estimate the answer. Then multiply to get an exact answer. Simplify your answers.

(a) $2\frac{1}{2} \cdot 3\frac{1}{5}$

Estimate the answer by rounding the mixed numbers.

$$2\frac{1}{2} \text{ rounds to 3} \quad \text{and} \quad 3\frac{1}{5} \text{ rounds to 3}$$
$$3 \cdot 3 = 9 \quad \text{Estimated answer}$$

To find the exact answer, change each mixed number to an improper fraction.

Step 1 $\quad 2\frac{1}{2} = \frac{5}{2} \quad \text{and} \quad 3\frac{1}{5} = \frac{16}{5}$

Continued on Next Page

OBJECTIVES
1 Estimate the answer and multiply mixed numbers.
2 Estimate the answer and divide mixed numbers.
3 Solve application problems with mixed numbers.

1 Round each mixed number to the nearest whole number.

(a) $4\frac{2}{3}$

(b) $3\frac{2}{5}$

(c) $5\frac{3}{4}$

(d) $4\frac{7}{12}$

(e) $1\frac{1}{2}$

(f) $8\frac{4}{9}$

ANSWERS
1. (a) 5 (b) 3 (c) 6 (d) 5 (e) 2 (f) 8

2 First estimate the answer. Then multiply to find the exact answer. Simplify your answers.

(a) $3\dfrac{1}{2}$ • $6\dfrac{1}{3}$

$\downarrow$ ⠀⠀ $\downarrow$

_____ • _____

= _____ estimate

(b) $4\dfrac{2}{3}$ • $2\dfrac{3}{4}$

$\downarrow$ ⠀⠀ $\downarrow$

_____ • _____

= _____ estimate

(c) $3\dfrac{3}{5}$ • $4\dfrac{4}{9}$

$\downarrow$ ⠀⠀ $\downarrow$

_____ • _____

= _____ estimate

(d) $5\dfrac{1}{4}$ • $3\dfrac{3}{5}$

$\downarrow$ ⠀⠀ $\downarrow$

_____ • _____

= _____ estimate

ANSWERS

2. **(a)** *Estimate:* $4 \cdot 6 = 24$; *Exact:* $22\dfrac{1}{6}$

(b) *Estimate:* $5 \cdot 3 = 15$; *Exact:* $12\dfrac{5}{6}$

(c) *Estimate:* $4 \cdot 4 = 16$; *Exact:* 16

(d) *Estimate:* $5 \cdot 4 = 20$; *Exact:* $18\dfrac{9}{10}$

Next, multiply.

Step 1 ⠀⠀⠀ *Step 2* ⠀⠀⠀ *Step 3*

$$2\dfrac{1}{2} \cdot 3\dfrac{1}{5} = \dfrac{5}{2} \cdot \dfrac{16}{5} = \dfrac{\cancel{5}^{1}}{\cancel{2}_{1}} \cdot \dfrac{\cancel{16}^{8}}{\cancel{5}_{1}} = \dfrac{1 \cdot 8}{1 \cdot 1} = \dfrac{8}{1} = 8$$

> Remember: Change to improper fractions, then divide out common factors, and finally multiply.

The estimated answer is 9 and the exact answer is 8. The exact answer is reasonable.

(b) $3\dfrac{5}{8} \cdot 4\dfrac{4}{5}$

$3\dfrac{5}{8}$ rounds to 4⠀⠀ and⠀⠀ $4\dfrac{4}{5}$ rounds to 5

$\downarrow$⠀$\downarrow$

$4 \cdot 5 = 20$⠀⠀ Estimated answer

Now find the exact answer.

Step 1 ⠀⠀⠀⠀⠀⠀ *Step 2*

$$3\dfrac{5}{8} \cdot 4\dfrac{4}{5} = \dfrac{29}{8} \cdot \dfrac{24}{5} = \dfrac{29}{\cancel{8}_{1}} \cdot \dfrac{\cancel{24}^{3}}{5} = \dfrac{29 \cdot 3}{1 \cdot 5} = \dfrac{87}{5}$$

Step 3

$$\dfrac{87}{5} = 17\dfrac{2}{5}$$

> Simplify this answer by writing it as a mixed number.

The estimate was 20, so the exact answer of $17\dfrac{2}{5}$ is reasonable.

(c) $1\dfrac{3}{5} \cdot 3\dfrac{1}{3}$

$1\dfrac{3}{5}$ rounds to 2⠀⠀ and⠀⠀ $3\dfrac{1}{3}$ rounds to 3

$\downarrow$⠀$\downarrow$

$2 \cdot 3 = 6$⠀⠀ Estimated answer

The exact answer is shown below.

$$1\dfrac{3}{5} \cdot 3\dfrac{1}{3} = \dfrac{8}{\cancel{5}_{1}} \cdot \dfrac{\cancel{10}^{2}}{3} = \dfrac{8 \cdot 2}{1 \cdot 3} = \dfrac{16}{3} = 5\dfrac{1}{3}$$

The estimate was 6, so the exact answer of $5\dfrac{1}{3}$ is reasonable.

◀ *Work Problem* **2** *at the Side.*

OBJECTIVE **2** **Estimate the answer and divide mixed numbers.**
Just as you did when multiplying mixed numbers, it is also a good idea to estimate the answer when dividing mixed numbers. To divide mixed numbers, use the following steps.

Dividing Mixed Numbers

Step 1   *Change* each mixed number to an improper fraction.

Step 2   Use the *reciprocal* of the second fraction (divisor).

Step 3   *Multiply.*

Step 4   *Simplify* the answer, which means to write it in *lowest terms,* and change it to a mixed number or whole number where possible.

Note

Recall that the reciprocal of a fraction is found by interchanging the numerator and the denominator.

EXAMPLE 2   **Dividing Mixed Numbers**

First estimate the answer. Then divide to find the exact answer. Simplify your answers.

(a) $2\dfrac{2}{5} \div 1\dfrac{1}{2}$

First estimate the answer by rounding each mixed number to the nearest whole number.

$$2\dfrac{2}{5} \quad \div \quad 1\dfrac{1}{2}$$

$$\downarrow \quad \text{Rounded} \quad \downarrow$$

$$2 \quad \div \quad 2 = 1 \qquad \text{Estimated answer}$$

To find the exact answer, first change each mixed number to an improper fraction.

$$\overset{\textit{Step 1}}{2\dfrac{2}{5} \div 1\dfrac{1}{2} = \dfrac{12}{5} \div \dfrac{3}{2}}$$

Next, use the reciprocal of the second fraction and multiply.

$$\overset{\textit{Step 2} \quad \textit{Step 3} \quad \textit{Step 4}}{\dfrac{12}{5} \div \dfrac{3}{2} = \dfrac{\overset{4}{\cancel{12}}}{5} \cdot \dfrac{2}{\underset{1}{\cancel{3}}} = \dfrac{4 \cdot 2}{5 \cdot 1} = \dfrac{8}{5} = 1\dfrac{3}{5}} \qquad \text{Exact answer simplified}$$

Reciprocals

Remember: Use the reciprocal of the *second* fraction.

The estimate was 1, so the exact answer of $1\frac{3}{5}$ is reasonable.

Continued on Next Page

3 First estimate the answer. Then divide to find the exact answer. Simplify all answers.

(a) $3\dfrac{1}{8}$ ÷ $6\dfrac{1}{4}$

↓ ↓

_____ ÷ _____

= _____ estimate

(b) $10\dfrac{1}{3}$ ÷ $2\dfrac{1}{2}$

↓ ↓

_____ ÷ _____

= _____ estimate

(c) 8 ÷ $5\dfrac{1}{3}$

↓ ↓

_____ ÷ _____

= _____ estimate

(d) $13\dfrac{1}{2}$ ÷ 18

↓ ↓

_____ ÷ _____

= _____ estimate

ANSWERS

3. (a) *Estimate:* $3 \div 6 = \dfrac{1}{2}$; *Exact:* $\dfrac{1}{2}$

(b) *Estimate:* $10 \div 3 = 3\dfrac{1}{3}$; *Exact:* $4\dfrac{2}{15}$

(c) *Estimate:* $8 \div 5 = 1\dfrac{3}{5}$; *Exact:* $1\dfrac{1}{2}$

(d) *Estimate:* $14 \div 18 = \dfrac{7}{9}$; *Exact:* $\dfrac{3}{4}$

(b) $8 \div 3\dfrac{3}{5}$

$$8 \quad \div \quad 3\dfrac{3}{5}$$

↓ Rounded ↓

$$8 \quad \div \quad 4 = 2 \qquad \text{Estimate}$$

Now find the exact answer.

Reciprocals

$$8 \div 3\dfrac{3}{5} = \dfrac{8}{1} \div \dfrac{18}{5} = \dfrac{\overset{4}{\cancel{8}}}{1} \cdot \dfrac{5}{\underset{9}{\cancel{18}}} = \dfrac{20}{9} = 2\dfrac{2}{9}$$

Write 8 as $\dfrac{8}{1}$.

> Divide out common factors *only after* you have changed to the reciprocal and are multiplying.

The estimate was 2, so the exact answer of $2\dfrac{2}{9}$ is reasonable.

(c) $4\dfrac{3}{8} \div 5$

$$4\dfrac{3}{8} \quad \div \quad 5$$

↓ Rounded ↓ Reciprocals

$$4 \quad \div \quad 5 = \dfrac{4}{1} \div \dfrac{5}{1} = \dfrac{4}{1} \cdot \dfrac{1}{5} = \dfrac{4}{5} \qquad \text{Estimate}$$

The exact answer is shown below.

Reciprocals

$$4\dfrac{3}{8} \div 5 = \dfrac{35}{8} \div \dfrac{5}{1} = \dfrac{35}{8} \cdot \dfrac{1}{\underset{1}{\cancel{5}}} = \dfrac{7}{8}$$

Write 5 as $\dfrac{5}{1}$.

The estimate was $\dfrac{4}{5}$, so the exact answer of $\dfrac{7}{8}$ is reasonable.

◀ *Work Problem* **3** *at the Side.*

OBJECTIVE 3 Solve application problems with mixed numbers.
The next two examples show how to solve application problems involving mixed numbers.

EXAMPLE 3 **Applying Multiplication Skills**

The local Habitat for Humanity chapter is looking for 11 contractors who will each donate $3\dfrac{1}{4}$ days of labor to a community building project. How many days of labor will be donated in all?

Step 1 **Read** the problem. The problem asks for the total days of labor donated by the 11 contractors.

Step 2 **Work out a plan.** Multiply the number of contractors (11) and the amount of labor that each donates ($3\dfrac{1}{4}$ days).

—— **Continued on Next Page**

Step 3 **Estimate** a reasonable answer. Round $3\frac{1}{4}$ days to 3 days. Multiply 3 days by 11 contractors ($3 \cdot 11$) to get an estimate of 33 days.

Step 4 **Solve** the problem. Find the exact answer.

$$11 \cdot 3\frac{1}{4} = 11 \cdot \frac{13}{4}$$

$$= \frac{11}{1} \cdot \frac{13}{4} = \frac{143}{4} = 35\frac{3}{4}$$

> Always check to see if the answer is close to the estimate.

Step 5 **State the answer.** The community building project will receive $35\frac{3}{4}$ days of donated labor.

Step 6 **Check.** The exact answer, $35\frac{3}{4}$ days, is close to our estimate of 33 days.

Work Problem **4** *at the Side.* ▶

EXAMPLE 4 **Applying Division Skills**

A dome tent for backpacking requires $7\frac{1}{4}$ yards of nylon material. How many tents can be made from $65\frac{1}{4}$ yards of material?

Step 1 **Read** the problem. The problem asks how many tents can be made from $65\frac{1}{4}$ yards of material.

Step 2 **Work out a plan.** Divide the number of yards of cloth ($65\frac{1}{4}$ yd) by the number of yards needed for one tent ($7\frac{1}{4}$ yd).

Step 3 **Estimate** a reasonable answer.

$$65\frac{1}{4} \quad \div \quad 7\frac{1}{4}$$

$$\downarrow \quad \text{Rounded} \quad \downarrow$$

$$65 \quad \div \quad 7 \approx 9 \text{ tents} \qquad \text{Estimate}$$

Step 4 **Solve** the problem.

$$65\frac{1}{4} \div 7\frac{1}{4} = \frac{261}{4} \div \frac{29}{4}$$

$$= \frac{\overset{9}{\cancel{261}}}{\underset{1}{\cancel{4}}} \cdot \frac{\overset{1}{\cancel{4}}}{\underset{1}{\cancel{29}}} = \frac{9}{1} = 9 \qquad \text{Matches estimate}$$

Step 5 **State the answer.** 9 tents can be made from $65\frac{1}{4}$ yards of cloth.

Step 6 **Check.** The exact answer, 9, matches our estimate.

Work Problem **5** *at the Side.* ▶

Note

When rounding mixed numbers to estimate the answer to a problem, the estimated answer usually varies somewhat from the exact answer. However, the importance of the estimated answer is that it will show you whether your exact answer is reasonable or not.

---

4 Use the six problem-solving steps. Simplify all answers.

(a) If one automobile requires $2\frac{5}{8}$ quarts of paint, find the number of quarts needed to paint 15 cars.

(b) Clare earns $\$9\frac{1}{4}$ per hour. How much would she earn in $6\frac{1}{2}$ hours? Write the answer as a mixed number.

5 Use the six problem-solving steps. Simplify all answers.

(a) The manufacture of one outboard engine propeller requires $4\frac{3}{4}$ pounds of brass. How many propellers can be manufactured from 57 pounds of brass?

(b) Jack Armstrong Trucking uses $21\frac{3}{4}$ quarts of motor oil for each oil change on his diesel engine truck. Find the number of oil changes that can be made with 609 quarts of oil.

ANSWERS

4. **(a)** *Estimate:* $3 \cdot 15 = 45$; *Exact:* $39\frac{3}{8}$ quarts

   (b) *Estimate:* $9 \cdot 7 = 63$; *Exact:* $\$60\frac{1}{8}$

5. **(a)** *Estimate:* $57 \div 5 \approx 11$;
   Exact: 12 propellers

   (b) *Estimate:* $600 \div 22 \approx 27$;
   Exact: 28 oil changes

Math in the Media

RECIPES

Alaska Burgers

1 pound 93% lean ground beef
$\frac{1}{2}$ medium Spanish onion, minced or processed
4 shakes Worcestershire sauce
$\frac{1}{4}$ teaspoon allspice (1 good pinch)
$\frac{1}{2}$ teaspoon ground cumin (two good pinches)
Cracked black pepper
$\frac{1}{3}$ pound brick of smoked cheddar cheese, cut into $\frac{1}{2}$ inch slices

4 fresh, crusty onion rolls
Thick-sliced tomato and lettuce to top

Mix beef, onion, Worcestershire, allspice, cumin, and black pepper in a bowl. Separate a quarter of the mixture. Take a slice of the smoked cheese and place it in the middle of the mixture. Form the pattie shape around the cheese filling. Patties should be no more than $\frac{3}{4}$ inch thick. Repeat with rest of mixture to have a total of 4 patties.

Heat a nonstick griddle or frying pan to medium hot. Cook burgers 5 to 6 minutes on each side. Meat should be cooked through and cheese melted. Check each burger with an instant-read thermometer for an internal temp of 170°F for well done if undercooking concerns you. Or cut into one and check the color of the meat.

Salt burgers after preparation to your taste. (Salting beef before cooking draws out juices and flavor.) Top with tomato slices and lettuce. Serves 4.

Rachael Ray is a Food Network television host, bestselling cookbook author, and the editor of her own lifestyle magazine. Rachael Ray's recipes can be found on her Web site, www.rachaelray.com, on her television show, or in one of her many cookbooks. The recipe at the side is from her book Rachael Ray: 30-Minute Meals.

1. Following the recipe, **(a)** what is the weight of one $\frac{1}{2}$-inch slice of cheese, and **(b)** what is the thickness of a $\frac{1}{3}$-pound brick of smoked cheddar cheese?

2. According to the recipe, **(a)** how many teaspoons of allspice are in 8 good pinches, and **(b)** how many teaspoons of ground cumin are in 7 good pinches?

3. Suppose you are preparing Alaska Burgers for 18 guests. By what factor will you change the ingredient amounts?

4. You know that of the 15 guests at your next party, 5 large eaters will eat $1\frac{1}{2}$ burgers each, 5 children will eat $\frac{1}{2}$ burger each, and the rest of the guests will each eat 1 burger. **(a)** How many burgers will you need, and **(b)** by what factor will you change the ingredient amounts?

5. Fill in the blanks with the ingredient amounts needed to make 9 servings of Alaska Burgers

Lean ground beef	_____
Spanish onions	_____
Worcestershire sauce	_____
Allspice	_____
Ground cumin	_____
Cheddar cheese	_____

6. If you have $5\frac{3}{4}$ pounds of beef, how many servings of Alaska Burgers can be prepared?

2.8 ▶▶▶ Exercises

First estimate the answer. Then multiply to find the exact answer. Simplify all answers.
See Example 1.

1. *Exact:*

$$4\frac{1}{2} \cdot 1\frac{3}{4}$$

Estimate:

____ • ____ = ____

2. *Exact:*

$$2\frac{1}{2} \cdot 2\frac{1}{4}$$

Estimate:

____ • ____ = ____

3. *Exact:*

$$1\frac{2}{3} \cdot 2\frac{7}{10}$$

Estimate:

____ • ____ = ____

4. *Exact:*

$$4\frac{1}{2} \cdot 2\frac{1}{4}$$

Estimate:

____ • ____ = ____

5. *Exact:*

$$3\frac{1}{9} \cdot 1\frac{2}{7}$$

Estimate:

____ • ____ = ____

6. *Exact:*

$$6\frac{1}{4} \cdot 3\frac{1}{5}$$

Estimate:

____ • ____ = ____

7. *Exact:*

$$8 \cdot 6\frac{1}{4}$$

Estimate:

____ • ____ = ____

8. *Exact:*

$$6 \cdot 2\frac{1}{3}$$

Estimate:

____ • ____ = ____

9. *Exact:*

$$4\frac{1}{2} \cdot 2\frac{1}{5} \cdot 5$$

Estimate:

____ • ____ • ____ = ____

10. *Exact:*

$$5\frac{1}{2} \cdot 1\frac{1}{3} \cdot 2\frac{1}{4}$$

Estimate:

____ • ____ • ____ = ____

11. *Exact:*

$$3 \cdot 1\frac{1}{2} \cdot 2\frac{2}{3}$$

Estimate:

____ • ____ • ____ = ____

12. *Exact:*

$$\frac{2}{3} \cdot 3\frac{2}{3} \cdot \frac{6}{11}$$

Estimate:

____ • ____ • ____ = ____

First estimate the answer. Then divide to find the exact answer. Simplify all answers.
See Example 2.

13. *Exact:*

$$1\frac{1}{4} \div 3\frac{3}{4}$$

Estimate:

____ ÷ ____ = ____

14. *Exact:*

$$1\frac{1}{8} \div 2\frac{1}{4}$$

Estimate:

____ ÷ ____ = ____

15. *Exact:*

$$2\frac{1}{2} \div 3$$

Estimate:

____ ÷ ____ = ____

16. *Exact:*

$$2\frac{3}{4} \div 2$$

Estimate:

____ ÷ ____ = ____

17. *Exact:*

$$9 \div 2\frac{1}{2}$$

Estimate:

____ ÷ ____ = ____

18. *Exact:*

$$5 \div 1\frac{7}{8}$$

Estimate:

____ ÷ ____ = ____

19. *Exact:*

$$\frac{5}{8} \div 1\frac{1}{2}$$

Estimate:

____ ÷ ____ = ____

20. *Exact:*

$$\frac{3}{4} \div 2\frac{1}{2}$$

Estimate:

____ ÷ ____ = ____

21. *Exact:*

$$1\frac{7}{8} \div 6\frac{1}{4}$$

Estimate:

____ ÷ ____ = ____

22. *Exact:*

$$8\frac{2}{5} \div 3\frac{1}{2}$$

Estimate:

____ ÷ ____ = ____

23. *Exact:*

$$5\frac{2}{3} \div 6$$

Estimate:

____ ÷ ____ = ____

24. *Exact:*

$$5\frac{3}{4} \div 2$$

Estimate:

____ ÷ ____ = ____

For Exercises 25–42, first estimate the answer. Then solve each application problem by using the six problem-solving steps. Simplify all answers. See Examples 3 and 4. Use the recipe for Carrot Cake Cupcakes to work Exercises 25–28.

CARROT CAKE CUPCAKES

12 paper bake cups
$1\frac{3}{4}$ cups flour
1 cup packed brown sugar
1 tsp. baking powder
1 tsp. baking soda
1 tsp. ground cinnamon
$\frac{1}{2}$ tsp. salt
1 cup shredded carrots
$\frac{3}{4}$ cup applesauce

$\frac{1}{3}$ cup vegetable oil
1 large egg
$\frac{1}{2}$ tsp. vanilla extract
1 container (16 oz.) ready-to-
   spread cream cheese frosting
Shredded coconut,
   tinted green
3 oz. Jelly Belly jelly beans,
   Orange Sherbet flavor

Preheat oven to 350° F. Place 12 bake cups in a muffin pan; set aside. In a large bowl, using a wire whisk, stir together flour, brown sugar, baking powder, baking soda, cinnamon, and salt. In a medium bowl, combine carrots, applesauce, oil, egg, and vanilla until blended. Add carrot mixture to flour mixture, stir well. Spoon batter into bake cups, filling $\frac{2}{3}$ full. Bake until toothpick inserted in center comes out clean,

20–25 minutes. Cool cupcakes in pan 10 minutes; remove to wire rack and cool completely. Frost with cream cheese frosting. To make carrot design on cupcakes, place gourmet Jelly Belly jelly beans in a carrot shape on each cupcake, top with green coconut for carrot top. Makes 12 cupcakes.

25. If 30 cupcakes are baked ($2\frac{1}{2}$ times the recipe), find the amount of each ingredient.

(a) Applesauce

Estimate:

Exact:

(b) Salt

Estimate:

Exact:

(c) Flour

Estimate:

Exact:

26. If 18 cupcakes are baked ($1\frac{1}{2}$ times the recipe), find the amount of each ingredient.

(a) Flour

Estimate:

Exact:

(b) Applesauce

Estimate:

Exact:

(c) Vegetable oil

Estimate:

Exact:

27. How much of each ingredient is needed if you bake one-half of the recipe?

(a) Vanilla extract

Estimate:

Exact:

(b) Applesauce

Estimate:

Exact:

(c) Flour

Estimate:

Exact:

28. How much of each ingredient is needed if you bake one-third of the recipe?

(a) Flour

Estimate:

Exact:

(b) Salt

Estimate:

Exact:

(c) Applesauce

Estimate:

Exact:

29. A new condominium conversion project requires $11\frac{3}{4}$ gallons of paint for each unit. How many units can be painted with 1316 gallons of paint?

Estimate:

Exact:

30. According to an old English system of time units, a moment is one and one-half minutes. How many moments are there in an 8-hour work day? (8 hours = 480 minutes). (*Source:* hightechscience.org/funfacts.htm)

Estimate:

Exact:

31. A manufacturer of floor jacks is ordering steel tubing to make the handles for the jack shown below. How much steel tubing is needed to make 135 of these jacks? The symbol for inches is ″, as in 5″ means 5 inches. (*Source:* Harbor Freight Tools.)

Estimate:

Exact:

32. A wheelbarrow manufacturer uses handles made of hardwood. Find the amount of wood that is necessary to make 182 handles. The longest dimension shown in the advertisement below is the handle length. (*Source:* Harbor Freight Tools.)

Estimate:

Exact:

2-TON COMPACT FLOOR JACK

LOT NO. 36119

4000 LB. CAPACITY

- 19 1/2″ handle
- Lifts 5″ to 15 1/4″
- 21″ x 9 1/2″ x 6″
- Compact size & lightweight for portability— perfect for the trunk

6.0 CUBIC FT. WHEEL BARROW

LOT NO. 46852

- Steel construction with hardwood handles
- 14″ tubeless pneumatic tire
- Fully rolled edge for added tray strength
- Overall dimensions: 61 1/2″ L x 27″ W x 24.9″ H

33. Write the three steps for multiplying mixed numbers. Use your own words.

34. Refer to Exercise 33. In your own words, write the additional step that must be added to the rule for multiplying mixed numbers to make it the rule for dividing mixed numbers.

35. A tire manufacturer uses $20\frac{3}{4}$ pounds of rubber to make a tire. Find the number of tires that can be manufactured with 51,460 pounds of rubber.

Estimate:

Exact:

36. The manager of the flooring department at The Home Depot determines that each apartment unit requires $62\frac{1}{2}$ square yards of carpet. Find the number of apartment units that can be carpeted with 6750 square yards of carpet.

Estimate:

Exact:

37. Mother Nature, manufacturer of bird feeders, cuts spacers from a tube that is $9\frac{3}{4}$ inches long. How many spacers can be cut from the tube if each spacer must be $\frac{3}{4}$ inch thick?

Estimate:

Exact:

38. A building contractor must move 12 tons of sand. If his truck can carry $\frac{3}{4}$ ton of sand, how many trips must he make to move the sand?

Estimate:

Exact:

Use the information on bottle jacks in the advertisement to answer Exercises 39–40. The " symbol is for inches.

39. A mechanic needs a hydraulic lift that will raise a car 4 times as high as the standard jack shown.

   (a) How high must it lift?

   (b) Will a mechanic 6 feet tall be able to fit under the vehicle without bending down? *Hint:* 1 ft = 12 in.

   (a) *Estimate:*   **(b)**

      Exact:

40. A race car driver needs a jack that will raise a vehicle only $\frac{1}{3}$ as high as the low-profile jack pictured.

   (a) How high must it lift?

   (b) Will a 6-inch part fit under the car?

   (a) *Estimate:*   **(b)**

      Exact:

41. A grape grower uses $6\frac{3}{4}$ gallons of a chemical for each acre of grapes. If she has $25\frac{1}{2}$ acres of grapes, how many gallons of the chemical are needed?

Estimate:

Exact:

42. A flooring contractor needs $24\frac{2}{7}$ boxes of tile to cover a kitchen floor. If there are 24 homes in a subdivision, how many boxes of tile are needed to cover all of the floors?

Estimate:

Exact:

Study Skills

This is the second part of the Study Cards activity. As you get further into a chapter, you can choose particular problems that will serve as a good test review. Here are two more types of study cards that will help you.

When you are doing your homework and find yourself saying, "This is really hard," or "I'm having trouble with this," make a **tough problem** study card! On the front, write out the procedure to work the type of problem *in words*. If there are special notes (like what *not* to do), include them. On the back, work at least one example; make sure you label what you are doing.

OBJECTIVES

1 Create study cards for difficult problems.

2 Create study cards of quiz problems.

Tough Problems Card

Front of Card

> Warning: Division is NOT commutative. The order in which you write the numbers DOES matter.
>
> Example: $1\frac{1}{3} \div 4 = \frac{1\cancel{4}}{3} \cdot \frac{1}{4} = \frac{1}{3}$ ← Very different
>
> But $4 \div 1\frac{1}{3} = \frac{1\cancel{4}}{1} \cdot \frac{3}{\cancel{4}_1} = \frac{3}{1} = 3$ ← answers!
>
> In an application problem, do NOT assume the numbers are given in the correct order. Use estimation to check that the answer is reasonable!

Back of Card

> Maite painted 4 windows using $1\frac{1}{3}$ cans of paint. How much paint did she use on each window?
>
> Try $4 \div 1\frac{1}{3}$
>
>   ↓   ↓ (round)
>
> $4 \div 1 = 4$ cans on each window ← Estimate
>
> Not reasonable — she only used $1\frac{1}{3}$ cans in all!
>
> Need to find: paint on each window
>
>   ↓   ↓   ↓
>
> $1\frac{1}{3}$ cans $\div$ $4 = \frac{1\cancel{4}}{3} \cdot \frac{1}{\cancel{4}_1} = \frac{1}{3}$ can
>
>   Reasonable!

Choose three types of difficult problems, and work them out on study cards. *Be sure to put the words for solving the problem on one side and the worked problem on the other side.*

◀◀◀ **Now Try This**

Practice Quiz Cards

Make up a few **quiz cards** for each type of problem you learn, and use them to prepare for a test. Choose two or three problems from the different sections of the chapter. Be sure you don't just choose the easiest problems! Put the problem **with the direction words** (like *solve, simplify, estimate*) on the front, and work the problem on the back. If you like, put the page number from the text there, too. When you review, you work the problem on a separate paper, and check it by looking at the back.

Solve this application problem.

Tiffany's monthly income is $1275. She spends $\frac{2}{5}$ of her income on rent and utilities. How much does she pay for rent and utilities?

Front of Card

Proper fraction followed by "of" indicates multiplication.

She spends $\frac{2}{5}$ of her income

$$\frac{2}{5} \cdot 1275 = \frac{2}{\overset{1}{\cancel{5}}} \cdot \frac{\overset{255}{\cancel{1275}}}{1} = \frac{510}{1} = 510$$

Tiffany spends $510 on rent and utilities.

Back of Card

Why Are Study Cards Brain Friendly?

First, making the study cards is an **active technique** that really gets your dendrites growing. You have to make decisions about what is most important and how to put it on the card. This kind of thinking is more in depth than just memorizing, and as a result, you will understand the concepts better and remember them longer.

Second, the cards are **visually appealing** (if you write neatly and try some color). Your brain responds to pleasant visual images, and again, you will remember longer and may even be able to "picture in your mind" how your cards look. This will help you during tests.

Third, because study cards are small and portable, you can review them easily whenever you have a few minutes. Even while you're waiting for a bus or have a few minutes between classes you can take out your cards and read them to yourself. Your **brain really benefits from repetition;** each time you review your cards your dendrites are growing thicker and stronger. After a while, the information will become automatic and you will remember it for a long time.

Chapter 2 ▶▶▶ Summary

▶ Key Terms

2.1	**numerator**	The number above the fraction bar in a fraction is called the numerator. It shows how many of the equivalent parts are being considered.
	denominator	The number below the fraction bar in a fraction is called the denominator. It shows the number of equal parts in a whole.
	proper fraction	In a proper fraction, the numerator is smaller than the denominator. The fraction is less than 1.
	improper fraction	In an improper fraction, the numerator is greater than or equal to the denominator. The fraction is equal to or greater than 1.
2.2	**mixed number**	A mixed number includes a fraction and a whole number written together.
2.3	**factors**	Numbers that are multiplied to give a product are factors.
	composite number	A composite number has at least one factor other than itself and 1.
	prime number	A prime number is a whole number other than 0 and 1 that has exactly two factors, itself and 1.
	factorizations	The numbers that can be multiplied to give a specific number (product) are factorizations of that number.
	prime factorization	In a prime factorization, every factor is a prime number.
2.4	**equivalent fractions**	Two fractions are equivalent when they represent the same portion of a whole.
	common factor	A common factor is a number that can be divided evenly into two or more whole numbers.
	lowest terms	A fraction is written in lowest terms when its numerator and denominator have no common factor other than 1.
2.5	**multiplication shortcut**	When multiplying or dividing fractions, the process of dividing a numerator and denominator by a common factor can be used as a shortcut.
2.6	**reciprocal**	Two numbers are reciprocals of each other if their product is 1. To find the reciprocal of a fraction, interchange the numerator and the denominator.

▶ New Formula

Area of a rectangle: Area = length $\cdot$ width

▶ Test Your Word Power

See how well you have learned the vocabulary in this chapter. Answers follow the Quick Review.

1. A **numerator** is
   A. a number greater than 5
   B. the number above the fraction bar in a fraction
   C. any number
   D. the number below the fraction bar in a fraction.

2. A **proper fraction**
   A. has a value less than 1
   B. has a whole number and a fraction
   C. has a value greater than 1
   D. is equal to 1.

3. A **mixed number** is
   A. equal to 1
   B. less than 1
   C. a whole number and a fraction written together
   D. a number multiplied by another number.

4. A **factor** is
   A. one of two or more numbers that are added to get another number
   B. the answer in division
   C. one of two or more numbers that are multiplied to get another number
   D. the answer in multiplication.

5. A whole number greater than 1 is **prime** if
   A. it cannot be factored
   B. it has just one factor
   C. it has only itself and 1 as factors
   D. it has more than two different factors.

6. A **common factor** can
   A. only be divided by itself and 1
   B. be divided evenly into two or more whole numbers
   C. never be divided by 2
   D. only be divided by the numbers 5 and 10.

7. A fraction is in **lowest terms** when
   A. it cannot be divided
   B. it is a common fraction
   C. its numerator and denominator have no common factor other than 1
   D. it has a value less than 1.

8. To find the **reciprocal** of a fraction.
   A. multiply it by itself
   B. interchange the numerator and the denominator
   C. change it to an improper fraction
   D. change it to lowest terms.

▶ Quick Review

Concepts	Examples

2.1 Types of Fractions

Proper
Numerator smaller than denominator; a value less than 1

$$\frac{2}{3} \quad \frac{3}{4} \quad \frac{15}{16} \quad \frac{1}{8} \qquad \text{Proper fractions}$$

Improper
Numerator equal to or greater than denominator; a value equal to or greater than 1

$$\frac{17}{8} \quad \frac{19}{12} \quad \frac{11}{2} \quad \frac{5}{3} \quad \frac{7}{7} \qquad \text{Improper fractions}$$

2.2 Converting Fractions

Mixed to Improper
Multiply denominator by whole number, add numerator, and place over denominator.

$$7\frac{2}{3} = \frac{23}{3} \quad \leftarrow 3 \times 7 + 2$$

Same denominator

Improper to Mixed
Divide numerator by denominator and place remainder over denominator.

$$\frac{17}{5} = 3\frac{2}{5}$$

Same denominator

$$5)\overline{17} \quad \begin{array}{c} 3\frac{2}{5} \end{array} \quad \leftarrow \text{Divide numerator by denominator.}$$
$$\underline{15}$$
$$2$$

Concepts	Examples

2.3 Prime Numbers

Determine whether a whole number is evenly divisible only by itself and 1. (By definition, 0 and 1 are not prime.)

The prime numbers less than 100 are 2, 3, 5, 7, 11, 13, 17, 19, 23, 29, 31, 37, 41, 43, 47, 53, 59, 61, 67, 71, 73, 79, 83, 89, and 97.

2.3 Finding the Prime Factorization of a Number

Divide each factor by a prime number using a diagram that forms the shape of tree branches.

Find the prime factorization of 30. Use a factor tree.

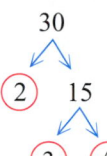

Prime factors are circled.

$30 = 2 \cdot 3 \cdot 5$

2.4 Writing Fractions in Lowest Terms

Divide the numerator and denominator by the greatest common factor.

Write $\frac{30}{42}$ in lowest terms.

$$\frac{30}{42} = \frac{30 \div 6}{42 \div 6} = \frac{5}{7}$$

2.5 Multiplying Fractions

1. Multiply the numerators and multiply the denominators.
2. Write answers in lowest terms if the multiplication shortcut was not used.

Multiply.

$$\frac{6}{11} \cdot \frac{7}{8} = \frac{\overset{3}{\cancel{6}}}{11} \cdot \frac{7}{\underset{4}{\cancel{8}}} = \frac{3 \cdot 7}{11 \cdot 4} = \frac{21}{44}$$

2.7 Finding the Reciprocal

To find the reciprocal of a fraction, interchange the numerator and denominator.

Find the reciprocal of each fraction.

$\frac{3}{4}$ The reciprocal of $\frac{3}{4}$ is $\frac{4}{3}$.

$\frac{8}{5}$ The reciprocal of $\frac{8}{5}$ is $\frac{5}{8}$.

9 The reciprocal of 9 is $\frac{1}{9}$.

2.7 Dividing Fractions

Use the reciprocal of the second fraction (divisor) and multiply as fractions.

Divide.

$$\frac{25}{36} \div \frac{15}{18} = \frac{\overset{5}{\cancel{25}}}{\underset{2}{\cancel{36}}} \cdot \frac{\overset{1}{\cancel{18}}}{\underset{3}{\cancel{15}}} = \frac{5 \cdot 1}{2 \cdot 3} = \frac{5}{6}$$

Reciprocals

Concepts	Examples

(2.8) Multiplying Mixed Numbers

First estimate the answer. Then follow these steps.
Step 1 *Change* each mixed number to an improper fraction.
Step 2 *Multiply.*
Step 3 *Simplify* the answer, which means to write it in *lowest terms,* and change it to a mixed number or whole number where possible.

First estimate the answer. Then multiply to get the exact answer.

Estimate: *Exact:*

$$1\frac{3}{5} \quad \cdot \quad 3\frac{1}{3}$$

Rounded

$$2 \quad \cdot \quad 3 = \mathbf{6}$$

$$1\frac{3}{5} \cdot 3\frac{1}{3} = \frac{8}{\overset{}{\underset{1}{5}}} \cdot \frac{\overset{2}{10}}{3}$$

$$= \frac{8 \cdot 2}{1 \cdot 3}$$

$$= \frac{16}{3} = \mathbf{5\frac{1}{3}}$$

Close to estimate

(2.8) Dividing Mixed Numbers

First estimate the answer. Then follow these steps.
Step 1 *Change* each mixed number to an improper fraction.
Step 2 Use the *reciprocal* of the second fraction (divisor).
Step 3 *Multiply.*
Step 4 *Simplify* the answer, which means to write it in *lowest terms,* and change it to a mixed number or whole number where possible.

First estimate the answer. Then divide to get the exact answer.

Estimate: *Exact:*

$$3\frac{5}{9} \quad \div \quad 2\frac{2}{5}$$

Rounded

$$4 \quad \div \quad 2 = \mathbf{2}$$

$$3\frac{5}{9} \div 2\frac{2}{5} = \frac{32}{9} \div \frac{\mathbf{12}}{\mathbf{5}}$$

$$= \frac{\overset{8}{32}}{9} \cdot \frac{\mathbf{5}}{\underset{3}{\mathbf{12}}} = \frac{40}{27}$$

$$= \mathbf{1\frac{13}{27}}$$

Close to estimate

ANSWERS TO TEST YOUR WORD POWER

1. B; *Example:* In $\frac{3}{8}$, the numerator is 3.

2. A; *Example:* $\frac{1}{2}, \frac{3}{4}$, and $\frac{7}{8}$ are all proper fractions with a value less than 1.

3. C; *Example:* $2\frac{3}{8}$ and $5\frac{3}{4}$ are mixed numbers.

4. C; *Example:* Since $3 \cdot 5 = 15$, the numbers 3 and 5 are factors of 15.

5. C; *Example:* 3, 5, and 11 are prime numbers; 4, 8, and 12 are composite numbers.

6. B; *Example:* 3 is a common factor of both 6 and 9 because it can be evenly divided into each of them.

7. C; *Example:* $\frac{3}{8}, \frac{4}{5}$, and $\frac{5}{6}$ are in lowest terms but $\frac{6}{8}, \frac{3}{6}$, and $\frac{2}{4}$ are not.

8. B; *Example:* The reciprocal of $\frac{3}{8}$ is $\frac{8}{3}$, the reciprocal of $\frac{25}{4}$ is $\frac{4}{25}$, and the reciprocal of 6 or $\frac{6}{1}$ is $\frac{1}{6}$.

Chapter 2 ▶▶▶ Review Exercises

[2.1] *Write the fraction that represents each shaded portion.*

1.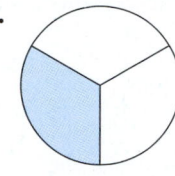

2.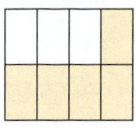

3.

List the proper and improper fractions in each group.

	Proper	**Improper**

4. $\dfrac{1}{8}$ $\dfrac{4}{3}$ $\dfrac{5}{5}$ $\dfrac{3}{4}$ $\dfrac{2}{3}$ _____   _____

5. $\dfrac{6}{5}$ $\dfrac{15}{16}$ $\dfrac{16}{13}$ $\dfrac{1}{8}$ $\dfrac{5}{3}$ _____   _____

[2.2] *Write each mixed number as an improper fraction. Write each improper fraction as a mixed number.*

6. $4\dfrac{3}{4}$

7. $9\dfrac{5}{6}$

8. $\dfrac{27}{8}$

9. $\dfrac{63}{5}$

[2.3] *Find all factors of each number.*

10. 6

11. 24

12. 55

13. 90

Write the prime factorization of each number by using exponents.

14. 27

15. 150

16. 420

Simplify each expression.

17. 5^2

18. $6^2 \cdot 2^3$

19. $8^2 \cdot 3^3$

20. $4^3 \cdot 2^5$

[2.4] *The fineness (purity) of gold is regulated by law and is the same in all parts of the world. Exercises 21–24 show the fineness of 24-kt, 18-kt, 14-kt, and 10-kt gold by comparing the parts of gold to the parts of alloy (metals other than gold). Write a fraction in lowest terms to show the portion that is gold. (Source: Costco Wholesale.)*

21. 24 kt. = (24 parts gold, 0 parts alloy)

22. 18 kt. = (18 parts gold, 6 parts alloy)

23. 14 kt. = (14 parts gold, 10 parts alloy)

24. 10 kt. = (10 parts gold, 14 parts alloy)

Write the numerator and denominator of each fraction as a product of prime factors. Then, write the fraction in lowest terms.

25. $\dfrac{25}{60}$

26. $\dfrac{384}{96}$

Decide whether each pair of fractions is equivalent or not equivalent, using the method of prime factors.

27. $\dfrac{3}{4}$ and $\dfrac{48}{64}$

28. $\dfrac{5}{8}$ and $\dfrac{70}{120}$

29. $\dfrac{2}{3}$ and $\dfrac{360}{540}$

[2.5–2.8] *Multiply. Write answers in lowest terms, and as mixed numbers or whole numbers where possible.*

30. $\dfrac{4}{5} \cdot \dfrac{3}{4}$

31. $\dfrac{3}{10} \cdot \dfrac{5}{8}$

32. $\dfrac{70}{175} \cdot \dfrac{5}{14}$

33. $\dfrac{44}{63} \cdot \dfrac{3}{11}$

34. $\dfrac{5}{16} \cdot 48$

35. $\dfrac{5}{8} \cdot 1000$

Divide. Write answers in lowest terms, and as mixed numbers or whole numbers where possible.

36. $\dfrac{2}{3} \div \dfrac{1}{2}$

37. $\dfrac{5}{6} \div \dfrac{1}{2}$

38. $\dfrac{\frac{15}{18}}{\frac{10}{30}}$

39. $\dfrac{\frac{3}{4}}{\frac{3}{8}}$

40. $7 \div \dfrac{7}{8}$

41. $18 \div \dfrac{3}{4}$

42. $\dfrac{5}{8} \div 3$

43. $\dfrac{2}{3} \div 5$

44. $\dfrac{\frac{12}{13}}{3}$

Find the area of each rectangle.

45.

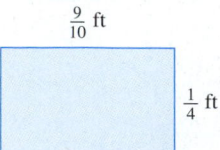

$\frac{9}{10}$ ft

$\frac{1}{4}$ ft

46.

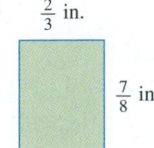

$\frac{2}{3}$ in.

$\frac{7}{8}$ in.

47. Ceramic tile is being installed on the floor of a meeting hall that is 108 ft long and $72\frac{3}{4}$ ft wide. Find the area.

48. Find the area of a display shelf that is a rectangle measuring 6 ft long and $\frac{11}{12}$ ft wide.

First estimate the answer. Then multiply or divide to find the exact answer. Simplify all answers.

49. *Exact:*

$$5\frac{1}{2} \cdot 1\frac{1}{4}$$

Estimate:

____ • ____ = ____

50. *Exact:*

$$2\frac{1}{4} \cdot 7\frac{1}{8} \cdot 1\frac{1}{3}$$

Estimate:

____ • ____ • ____ = ____

51. *Exact:*

$$15\frac{1}{2} \div 3$$

Estimate:

____ ÷ ____ = ____

52. *Exact:*

$$4\frac{3}{4} \div 6\frac{1}{3}$$

Estimate:

____ ÷ ____ = ____

Solve each application problem by using the six problem-solving steps.

53. Blue Diamond Almonds has 320 tons of almonds. How many $\frac{5}{8}$ ton bins will be needed to store the almonds?

54. An estate is divided so that each of 5 children receives equal shares of $\frac{2}{3}$ of the estate. What fraction of the total estate will each receive?

55. How many window-blind pull cords can be made from $157\frac{1}{2}$ yards of cord if $4\frac{3}{8}$ yards of cord are needed for each blind? First estimate, and then find the exact answer.

Estimate:

Exact:

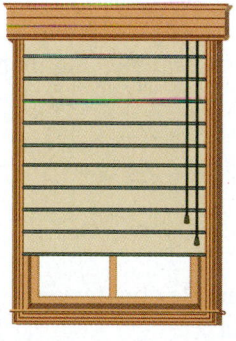

56. A gallon of water weighs $8\frac{1}{3}$ pounds. Find the weight of the water in two 50-gallon aquariums. First estimate, and then find the exact answer.

Estimate:

Exact:

57. Ebony Wilson purchased 100 pounds of rice at the food co-op. After selling $\frac{1}{4}$ of this to her neighbor, she gives $\frac{2}{3}$ of the remaining rice to her parents. How many pounds of rice does she have left?

58. Sheila Spinney, a recent college graduate, receives a salary of $2976 each month. She pays $\frac{3}{8}$ of this amount in taxes, social security, and a retirement plan. Of the remainder, $\frac{9}{10}$ goes for basic living expenses. How much money remains?

59. The Citrus Heights Park District will divide $\frac{3}{4}$ of its budget among 8 local parks. What fraction of the total budget will each park receive?

60. In a morning of deep-sea fishing, 5 fishermen catch $\frac{4}{5}$ ton of salmon. If they divide the fish evenly, how much will each receive?

> ▶▶▶ **Mixed Review Exercises**

Multiply or divide as indicated. Simplify all answers.

61. $\dfrac{1}{2} \cdot \dfrac{3}{4}$

62. $\dfrac{2}{3} \cdot \dfrac{3}{5}$

63. $12\dfrac{1}{2} \cdot 2\dfrac{1}{2}$

64. $12\dfrac{1}{2} \cdot 2\dfrac{1}{4}$

65. $\dfrac{\frac{4}{5}}{8}$

66. $\dfrac{\frac{5}{8}}{4}$

67. $\dfrac{15}{31} \cdot 62$

68. $3\dfrac{1}{4} \div 1\dfrac{1}{4}$

Write each mixed number as an improper fraction. Write each improper fraction as a mixed number.

69. $\dfrac{8}{5}$

70. $\dfrac{153}{4}$

71. $5\dfrac{2}{3}$

72. $38\dfrac{3}{8}$

Write the numerator and denominator of each fraction as a product of prime factors; then write the fraction in lowest terms.

73. $\dfrac{8}{12}$

74. $\dfrac{108}{210}$

Write each fraction in lowest terms.

75. $\dfrac{75}{90}$

76. $\dfrac{48}{72}$

77. $\dfrac{44}{110}$

78. $\dfrac{87}{261}$

Solve each application problem.

79. The directions on a can of fabric glue say to apply $3\dfrac{3}{5}$ ounces of glue to each square yard of fabric. How many ounces are needed for $43\dfrac{3}{4}$ square yards? First estimate, and then find the exact answer.

Estimate:

Exact:

80. Valley Farms purchased some diesel fuel additive. The instructions say to use $7\dfrac{1}{4}$ quarts of additive for each tank of fuel. How many quarts are needed for $25\dfrac{1}{2}$ tanks? First estimate, and then find the exact answer.

Estimate:

Exact:

81. The U.S.A. Breast Cancer stamp is $1\dfrac{3}{4}$ in. by $\dfrac{7}{8}$ in. Find its area.

82. A section of marble countertop is $\dfrac{1}{2}$ yard by $\dfrac{5}{8}$ yard. What is its area?

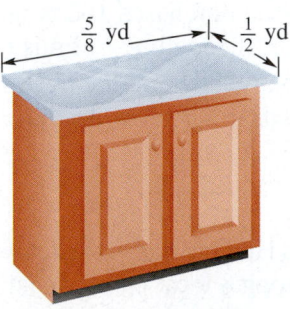

Study Skills

▶▶▶ REVIEWING A CHAPTER

This activity is really about **preparing for tests.** Some of the suggestions are ideas that you will learn to use a little later in the term, but get started trying them out now. Often, the first chapters in your math textbook will be review, so it is good to practice some of the study techniques on material that is not too challenging.

Use these **chapter reviewing techniques.**

▶ **Make a study card for each vocabulary word and concept.** Include a definition, an example, a sketch, and a page reference. Include the symbol or formula if there is one. See the *Using Study Cards* activity for a quick look at some sample study cards.

▶ **Go back to the section** to find more explanations or information about any new vocabulary, formulas, or symbols.

▶ **Use the Chapter Summary** to practice each type of problem. Do not expect the Summary to substitute for reading and working through the whole chapter! First, take the "Test Your Word Power" quiz to check your understanding of new vocabulary. The answers are at the end of the Quick Review. Then read the Quick Review. **Pay special attention to the red headings.** Check the explanations for the solutions to problems given. Try to think about how all the topics in the **whole chapter** are related.

▶ **Study your lecture notes** to see what your instructor has emphasized in class. Then review that material in your text.

▶ **Do the Review Exercises.**
   ✓ Check your answers **after** you're done with each **section of exercises.**
   ✓ If you get stuck on a problem, **first** check the Chapter Summary. If that doesn't clear up your confusion, then check the section and your lecture notes.
   ✓ Pay attention to **direction words** for the problems, such as *simplify, round, solve,* and *estimate.*
   ✓ Make **study cards for especially difficult problems.**

▶ **Do the Mixed Review exercises.** This is a good check to see if you can still do the problems when they are in mixed-up order. **Check your answers carefully** in the Answers section in the back of your book. Are your answers **exact** and **complete?** Make sure you are **labeling** answers correctly, using the right **units.** For example, does your answer need to include *$, cm², ft,* and so on?

OBJECTIVES

1 **Use the Chapter Summary to practice every type of problem.**

2 **Create study cards for vocabulary.**

3 **Practice by doing review and mixed review exercises.**

4 **Take the Chapter Test as a practice test.**

Chapter Reviewing Techniques

Why Are These Review Activities Brain Friendly?

You have already become familiar with the features of your textbook. This activity requires you to make good use of them. Your **brain needs repetition** to strengthen dendrites and the connections between them. By following the steps outlined here, you will be reinforcing the concepts, procedures, and skills you need to use for tests (and for the next chapters).

The combination of techniques provides repetition in different ways. That **promotes good branching of dendrites** instead of just relying on one branch, or route, to connect to the other dendrites. A thorough review of each chapter will **solidify your dendrite connections.** It will help you be sure that you understand the concepts **completely and accurately.** Also, taking the Chapter Test will **simulate the testing situation,** which gives you practice in test taking conditions.

Now Try This ▶▶▶

▶ **Take the Chapter Test as if it is a real test.** If your instructor has skipped sections in the chapter, figure out which problems to skip on the test before you start.

✓ **Time yourself** just as you would for a real test.

✓ **Use a calculator or notes** just as you would be permitted to (or not) on a real test.

✓ **Take the test in one sitting,** just like a real test is given in one sitting.

✓ **Show all your work.** Practice showing your work just the way your instructor has asked you to show it.

✓ **Practice neatness.** Can someone else follow your steps?

✓ **Check your answers** in the back of the book.

Notice that reviewing a chapter will take some time. Remember that it takes time for dendrites to grow! You cannot grow a good network of dendrites by rushing through a review in one night. But if you use the suggestions over a few days or evenings, you will notice that you understand the material more thoroughly and remember it longer.

Follow the reviewing techniques listed above for your next test. For each technique, write a comment *about how it worked for you in the spaces below.*

1. **Make a study card for each vocabulary word and concept.**

2. **Go back to the section** to find more explanations or information.

3. **Take the Test Your Word Power quiz and use the Quick Review** to review each concept in the chapter.

4. **Study your lecture notes** to see what your instructor has emphasized in class.

5. **Do the Review Exercises,** following the specific suggestions on the previous page.

6. **Do the Mixed Review exercises.**

7. **Take the Chapter Test** as if it is a real test.

Write a fraction to represent each shaded portion.

1.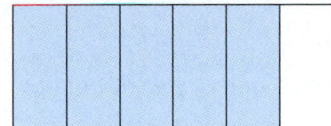

2.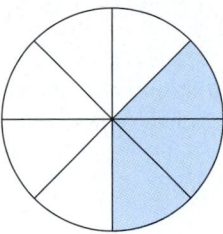

3. Identify all the proper fractions in this list:

   $\dfrac{2}{3}$ $\dfrac{4}{4}$ $\dfrac{6}{7}$ $\dfrac{5}{2}$ $\dfrac{1}{4}$ $\dfrac{5}{8}$ $\dfrac{30}{18}$

4. Write $3\dfrac{3}{8}$ as an improper fraction.

5. Write $\dfrac{123}{4}$ as a mixed number.

6. Find all factors of 18.

Find the prime factorization of each number. Write the answers using exponents.

7. 45   8. 144   9. 500

Write each fraction in lowest terms.

10. $\dfrac{36}{48}$ 11. $\dfrac{60}{72}$

12. The method of prime factors is used to write a fraction in lowest terms. Briefly explain how this is done. Use the fraction $\frac{56}{84}$ to show how this works.

1. _____

2. _____

3. _____

4. _____

5. _____

6. _____

7. _____

8. _____

9. _____

10. _____

11. _____

12. _____

13. _____

14. _____

15. _____

16. _____

17. _____

18. _____

19. _____

20. _____

21. *Estimate:* _____

   Exact: _____

22. *Estimate:* _____

   Exact: _____

23. *Estimate:* _____

   Exact: _____

24. *Estimate:* _____

   Exact: _____

25. *Estimate:* _____

   Exact: _____

13. Explain how to multiply fractions. What additional step must be taken when dividing fractions?

Multiply or divide. Write answers in lowest terms, and as mixed numbers or whole numbers where possible.

14. $\dfrac{3}{4} \cdot \dfrac{4}{9}$

15. $54 \cdot \dfrac{2}{3}$

16. A rectangular barbecue grill is $\dfrac{15}{16}$ yard by $\dfrac{4}{9}$ yard. Find the area of the grill.

17. The Sierra College Conservation Club planted 8760 Douglass Fir seedlings. If $\dfrac{3}{8}$ of these seedlings are not expected to survive, find the number of seedlings that do survive.

18. $\dfrac{3}{4} \div \dfrac{5}{6}$

19. $\dfrac{\dfrac{7}{4}}{9}$

20. To complete a custom-designed cabinet, oak trim pieces must be cut exactly $2\dfrac{1}{4}$ inches long so that they can be used as dividers in a spice rack. Find the number of pieces that can be cut from a piece of oak that is 54 inches in length.

First estimate the answer. Then find the exact answer. Simplify all answers.

21. $4\dfrac{1}{8} \cdot 3\dfrac{1}{2}$

22. $1\dfrac{5}{6} \cdot 4\dfrac{1}{3}$

23. $9\dfrac{3}{5} \div 2\dfrac{1}{4}$

24. $\dfrac{8\dfrac{1}{2}}{1\dfrac{2}{3}}$

25. A new vaccine is synthesized at the rate of $2\dfrac{1}{2}$ grams per day. How many grams can be synthesized in $12\dfrac{1}{4}$ days?

Study Skills

Many things besides studying can improve your test scores. You may not realize that eating the right foods, and getting enough exercise and sleep, can also improve your scores. Your brain (and therefore your ability to think) is affected by the condition of your whole body. So, part of your preparation for tests includes keeping yourself in good physical shape as well as spending time on the actual course material. Try these suggestions and see the difference.

OBJECTIVES

1 Restate the importance of sleep and good nutrition as it affects learning.

2 Explain the effect of anxiety and stress on learning.

Performance Health Tips

Performance Health Tips to Improve Your Test Score	Explanation
Get *seven to eight hours of sleep* the night before the exam. (It's helpful to get that much sleep *every* night.)	*Fatigue and exhaustion* reduce efficiency. They also cause poor memory and recall. If you didn't sleep much the night before a test, 20 minutes of relaxation or meditation can help.
Eat a *small, high-energy meal* about two hours before the test. Start the meal with a small amount of protein such as fish, chicken, or nonfat yogurt. Include carbohydrates if you like, but no high-fat foods.	Just 3 to 4 ounces of protein increases the amount of a chemical in the brain called tyrosine, which *improves your alertness, accuracy, and motivation.* High-fat foods dull your mind and slow down your brain.
Drink plenty of water. Don't wait until you feel thirsty; your body is already dehydrated by the time you feel it.	Research suggests that staying well hydrated improves the electrochemical communications in your brain.
Give your brain the time it needs to grow dendrites!	*Cramming doesn't work;* your brain cannot grow dendrites that quickly. *Studying every day* is the way to give your brain the time it needs.

Anxiety Prevention Tips

To Prevent Anxiety	Explanation
Practice slow, deep breathing for five minutes each day. Then do a minute or two of deep breathing right before the test. Also, if you feel your anxiety building during the test, stop for a minute, close your eyes, and do some deep breathing.	When *test anxiety* hits, you breathe more quickly and shallowly, which causes hyperventilation. Symptoms may be confusion, inability to concentrate, shaking, dizziness, and more. Slow, deep breathing will *calm you and prevent panic.*

Anxiety Prevention Tips
(*continued*)

To Prevent Anxiety	Explanation
Do 15 to 20 minutes of **moderate exercise** (like walking) shortly before the test. Daily exercise is even better!	**Exercise reduces stress** and will help prevent "blanking out" on a test. Exercise also increases your alertness, clear thinking, and energy.
To help you sleep the night before the test, or any time you need to calm down, **eat high carbohydrate foods** such as popcorn, bread, rice, crackers, muffins, bagels, pasta, corn, baked potatoes (not fries or chips), and cereals.	Carbohydrates increase the level of a chemical in the brain called serotonin, which has a *calming effect on the mind*. It reduces feelings of tension and stress and improves your ability to concentrate. You only need to eat a small amount, like half a bagel, to get this effect.
Before the test, **go easy on caffeinated beverages** such as coffee, tea, and soft drinks. Do not eat candy bars or other sugary snacks.	Extra caffeine can **make you jittery,** "hyper," and shaky for the test. It can increase the tendency to panic. Too much sugar causes negative emotional reactions in some people.

Now Try This ▶▶▶

What will you do to improve your next test score? List the three or four tips you think will help you the most.

1. _____

2. _____

3. _____

4. _____

What changes will you have to make in order to try the tips you chose?

_____

_____

_____

_____

See *Tips for Taking Math Tests* **and** *Preparing for Your Final Exam* **for more ideas about managing anxiety.** (Check the Table of Contents to find their locations.)

3

Adding and Subtracting Fractions

When Bryan Berg was a small boy, his grandfather taught him how to build a simple house of cards. Berg has set 8 Guinness world records for the tallest freestanding card structures—no glue, no tape, no hidden supports, just the force of gravity and a very steady hand. While a high school student, he set his first record of 14 feet, 6 inches tall. That house of cards required 4 days, 208 decks of cards, and a scaffold to build. Berg is a graduate of the Harvard Graduate School of Design and makes his living "stacking cards." At the 2007 State of Texas Fair, he set a new world record of 25 feet, $9\frac{7}{16}$ inches. As an architect and when *Stacking the Deck* (the title of his book), Berg will be adding and subtracting fractions. (See Exercises 49 and 66 in **Section 3.4.**) (*Source: Reader's Digest.*)

193

3.1 ▶▶▶ Adding and Subtracting Like Fractions

OBJECTIVES

1 Define like and unlike fractions.

2 Add like fractions.

3 Subtract like fractions.

In **Chapter 2** we looked at the basics of fractions and then practiced with multiplication and division of fractions and mixed numbers. In this chapter we will work with addition and subtraction of fractions and mixed numbers.

OBJECTIVE 1 **Define like and unlike fractions.**   Fractions with the same denominators are **like fractions.** Fractions with different denominators are **unlike fractions.**

EXAMPLE 1   **Identifying Like and Unlike Fractions**

(a) $\frac{3}{4}, \frac{1}{4}, \frac{5}{4}, \frac{6}{4}$, and $\frac{4}{4}$ are **like** fractions.

↑ ↑ ↑ ↑   ↑ ─── All denominators are the same.

(b) $\frac{7}{12}$ and $\frac{12}{7}$ are **unlike** fractions.

↑   ↑ ─── Denominators are different.

Note

Like fractions have the *same* denominator.

◀ *Work Problem* 1 *at the Side.*

1 Next to each pair of fractions write *like* or *unlike*.

(a) $\frac{2}{5}$ $\frac{3}{5}$ _____

(b) $\frac{2}{3}$ $\frac{3}{4}$ _____

(c) $\frac{7}{12}$ $\frac{11}{12}$ _____

(d) $\frac{3}{8}$ $\frac{3}{16}$ _____

OBJECTIVE 2 **Add like fractions.**   The figures below show you how to add the fractions $\frac{2}{7}$ and $\frac{4}{7}$.

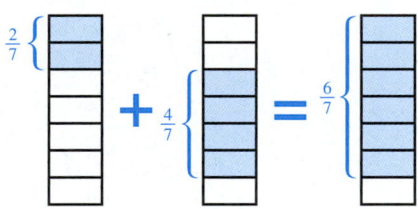

As the figures show,

$$\frac{2}{7} + \frac{4}{7} = \frac{6}{7}$$

Add like fractions as follows.

Adding Like Fractions

Step 1   Add the numerators to find the numerator of the sum.

Step 2   Write the denominator of the like fractions as the denominator of the sum.

Step 3   Write the sum in lowest terms.

ANSWERS

1. **(a)** like   **(b)** unlike   **(c)** like   **(d)** unlike

EXAMPLE 2 **Adding Like Fractions**

Add and write the sum in lowest terms.

(a) $\dfrac{1}{5} + \dfrac{2}{5}$

Add numerators.

$$\dfrac{1}{5} + \dfrac{2}{5} = \dfrac{\overbrace{1+2}}{5} = \dfrac{3}{5} \leftarrow \text{Same denominator}$$

(b) $\dfrac{1}{12} + \dfrac{7}{12} + \dfrac{1}{12}$ | Fractions are ready to be added if they are *like* fractions.

Add numerators.

Step 1 $\dfrac{\overbrace{1+7+1}}{12}$

Step 2 $= \dfrac{9}{12} \begin{array}{l}\leftarrow \text{Sum of numerators} \\ \leftarrow \text{Same denominator}\end{array}$

Step 3 $= \dfrac{9 \div \mathbf{3}}{12 \div \mathbf{3}} = \dfrac{3}{4}$ In lowest terms

---

CAUTION
Fractions may be added **only** if they have like denominators.

Work Problem **2** *at the Side.* ▶

OBJECTIVE **3** **Subtract like fractions.**   The figures below show $\frac{7}{8}$ broken into $\frac{4}{8}$ and $\frac{3}{8}$.

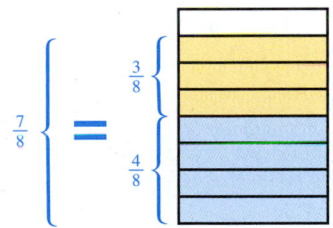

Subtracting $\frac{3}{8}$ from $\frac{7}{8}$ gives the answer $\frac{4}{8}$, or

$$\dfrac{7}{8} - \dfrac{3}{8} = \dfrac{4}{8}$$

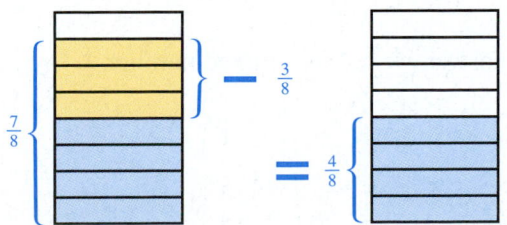

2 Add and write the sums in lowest terms.

(a) $\dfrac{3}{8} + \dfrac{1}{8}$

(b) $\begin{array}{r} \dfrac{2}{9} \\[4pt] + \dfrac{5}{9} \\ \hline \end{array}$

(c) $\dfrac{3}{16} + \dfrac{1}{16}$

(d) $\dfrac{3}{10} + \dfrac{1}{10} + \dfrac{4}{10}$

3 Find the difference and simplify.

(a) $\dfrac{5}{6} - \dfrac{1}{6}$

(b) $\begin{array}{r} \dfrac{16}{10} \\[2mm] -\,\dfrac{7}{10} \\ \hline \end{array}$

(c) $\dfrac{15}{3} - \dfrac{5}{3}$

(d) $\begin{array}{r} \dfrac{25}{32} \\[2mm] -\,\dfrac{6}{32} \\ \hline \end{array}$

Write $\frac{4}{8}$ in lowest terms.

$$\frac{7}{8} - \frac{3}{8} = \frac{4 \div 4}{8 \div 4} = \frac{1}{2}$$

The steps for subtracting like fractions are very similar to those for adding like fractions.

Subtracting Like Fractions

Step 1   Subtract the numerators to find the numerator of the difference.

Step 2   Write the denominator of the like fractions as the denominator of the difference.

Step 3   Write the answer in lowest terms.

EXAMPLE 3   Subtracting Like Fractions

Find the difference and simplify the answer.

(a) $\dfrac{15}{16} - \dfrac{3}{16}$

Subtract numerators.

Step 1 $\dfrac{15}{16} - \dfrac{3}{16} = \dfrac{\overbrace{15 - 3}}{16}$

Step 2 $= \dfrac{12}{16}$ ← Difference of numerators
              ← Same denominator

Step 3 $= \dfrac{12 \div 4}{16 \div 4} = \dfrac{3}{4}$,   In lowest terms

(b) $\dfrac{13}{4} - \dfrac{6}{4}$

> Fractions are ready to be subtracted if they are *like* fractions.

Subtract numerators.

$$\dfrac{13}{4} - \dfrac{6}{4} = \dfrac{\overbrace{13 - 6}}{4}$$ ← Difference of numerators
← Same denominator

$$= \dfrac{7}{4}$$

To simplify the answer, write $\frac{7}{4}$ as a mixed number.

$$\frac{7}{4} = 1\frac{3}{4}$$

CAUTION

Fractions may be subtracted *only* if they have like denominators.

◀ *Work Problem* **3** *at the Side.*

ANSWERS

3. (a) $\dfrac{2}{3}$ (b) $\dfrac{9}{10}$ (c) $3\dfrac{1}{3}$ (d) $\dfrac{19}{32}$

3.1 ▶▶▶ Exercises

Find the sum and simplify the answer. See Example 2.

1. $\dfrac{3}{8} + \dfrac{2}{8}$

2. $\dfrac{1}{5} + \dfrac{3}{5}$

3. $\dfrac{2}{6} + \dfrac{3}{6}$

4. $\dfrac{9}{11} + \dfrac{1}{11}$

5. $\dfrac{1}{4} + \dfrac{1}{4}$

6. $\dfrac{1}{14} + \dfrac{1}{14}$

7. $\begin{aligned} &\dfrac{9}{10}\\ +&\dfrac{3}{10}\end{aligned}$

8. $\begin{aligned} &\dfrac{13}{12}\\ +&\dfrac{5}{12}\end{aligned}$

9. $\begin{aligned} &\dfrac{2}{9}\\ +&\dfrac{1}{9}\end{aligned}$

10. $\dfrac{7}{12} + \dfrac{3}{12}$

11. $\dfrac{6}{20} + \dfrac{4}{20} + \dfrac{3}{20}$

12. $\dfrac{1}{7} + \dfrac{2}{7} + \dfrac{3}{7}$

13. $\dfrac{4}{15} + \dfrac{2}{15} + \dfrac{5}{15}$

14. $\dfrac{5}{11} + \dfrac{1}{11} + \dfrac{4}{11}$

15. $\dfrac{3}{8} + \dfrac{7}{8} + \dfrac{2}{8}$

16. $\dfrac{4}{9} + \dfrac{1}{9} + \dfrac{7}{9}$

17. $\dfrac{2}{54} + \dfrac{8}{54} + \dfrac{12}{54}$

18. $\dfrac{7}{64} + \dfrac{15}{64} + \dfrac{20}{64}$

Find the difference and simplify the answer. See Example 3.

19. $\dfrac{7}{8} - \dfrac{4}{8}$

20. $\dfrac{2}{3} - \dfrac{1}{3}$

21. $\dfrac{10}{11} - \dfrac{4}{11}$

22. $\dfrac{4}{5} - \dfrac{3}{5}$

23. $\dfrac{9}{10} - \dfrac{3}{10}$

24. $\dfrac{7}{14} - \dfrac{3}{14}$

25. $\begin{aligned} &\dfrac{31}{21}\\ -&\dfrac{7}{21}\end{aligned}$

26. $\begin{aligned} &\dfrac{43}{24}\\ -&\dfrac{13}{24}\end{aligned}$

27. $\begin{aligned} &\dfrac{27}{40}\\ -&\dfrac{19}{40}\end{aligned}$

28. $\begin{aligned} &\dfrac{38}{55}\\ -&\dfrac{16}{55}\end{aligned}$

29. $\dfrac{47}{36} - \dfrac{5}{36}$

30. $\dfrac{76}{45} - \dfrac{21}{45}$

31. $\dfrac{73}{60} - \dfrac{7}{60}$

32. $\dfrac{181}{100} - \dfrac{31}{100}$

33. In your own words, write an explanation of how to add like fractions. Consider using three steps in your explanation.

34. Describe in your own words the difference between *like* fractions and *unlike* fractions. Give three examples of each type.

Solve each application problem. Write answers in lowest terms.

35. The Fair Oaks Save the Bluffs Committee raised $\frac{2}{9}$ of their target goal last year and another $\frac{5}{9}$ of the goal this year. What fraction of their goal has been raised?

36. After an initial payment to a winner at an arcade fundraiser, the organization still owed the winner $\frac{7}{10}$ of her total winnings. If the organization pays the winner another $\frac{3}{10}$ of the winnings, what fraction is still owed?

37. Robert Hernandez has saved $\frac{3}{7}$ of the amount needed for his children's college education fund. In the next five years he plans to save another $\frac{2}{7}$ of the total amount needed. What fraction of the amount needed will he have saved?

38. Phil Fravesi, general contractor, completes $\frac{3}{12}$ of a hobby and toy train addition to his home in April. In May, he completes another $\frac{5}{12}$ of the project. What portion of the project has he completed?

39. An organic farmer purchased $\frac{9}{10}$ acre of land one year and $\frac{3}{10}$ acre the next year. She then planted carrots on $\frac{7}{10}$ acre of the land and squash on the remainder. How much land is planted in squash?

40. A forester planted $\frac{5}{12}$ acre in seedlings in the morning and $\frac{11}{12}$ acre in the afternoon. If $\frac{7}{12}$ acre of seedlings were destroyed by frost, how many acres of seedlings remained?

Study Skills

▶▶▶ MANAGING YOUR TIME

Many college students find themselves juggling a difficult schedule and multiple responsibilities. Perhaps you are going to school, working part time, and managing family demands. Here are some tips to help you develop good time management skills and habits.

▶ **Read the syllabus for each class.** Check on class policies, such as attendance, late homework, and make-up tests. Find out how you are graded. Keep the syllabus in your notebook.

▶ **Make a semester or quarter calendar.** Put test dates and major due dates for *all* your classes on the same calendar. That way you will see which weeks are the really busy ones. Try using a different color pen for each class. Your brain responds well to the use of color. A semester calendar is on the next page.

▶ **Make a weekly schedule.** After you fill in your classes and other regular responsibilities (such as work, picking up kids from school, etc.), block off some study periods during the day that you can guarantee you will use for studying. Aim for 2 hours of study for each 1 hour you are in class.

▶ **Make "To Do" lists.** Then use them by crossing off the tasks as you complete them. You might even number them in the order they need to be done (most important ones first).

▶ **Break big assignments into smaller chunks.** They won't seem so big that way. Make deadlines for each small part so you stay on schedule.

▶ **Give yourself small breaks in your studying.** Do not try to study for hours at a time! Your brain needs rest between periods of learning. Try to give yourself a 10 minute break each hour or so. You will learn more and remember it longer.

▶ **If you get off schedule, just try to get back on schedule tomorrow.** We all slip from time to time. All is not lost! Make a new "To Do" list and start doing the most important things first.

▶ **Get help when you need it.** Talk with your instructor during office hours. Also, most colleges have some kind of Learning Center, tutoring center, or counseling office. If you feel lost and overwhelmed, ask for help. Someone can help you decide what to do first and what to spend your time on right away.

Which two or three of the suggestions above will you try this week? How do you think they will help you?

1. _____

2. _____

3. _____

OBJECTIVES

1 Create a semester schedule.

2 Create a "to do" list.

Why Are These Techniques Brain Friendly?

Your brain appreciates some order. It enjoys a little routine, for example, choosing the same study time and place each day. You will find that you quickly settle in to your reading or homework.

Also, your brain **functions better when you are calm.** Too much rushing around at the last minute to get your homework and studying done sends hostile chemicals to your brain and makes it more difficult for you to learn and remember. So, a little planning can really pay off.

Building rest into your schedule is good for your brain. Remember, it takes time for dendrites to grow.

We've suggested using color on your calendars. This too, is brain friendly. Remember, your brain **likes pleasant colors and visual material** that are nice to look at. Messy and hard to read calendars will not be helpful, and you probably won't look at them often.

SEMESTER CALENDAR

WEEK	MON	TUES	WED	THUR	FRI	SAT	SUN
1							
2							
3							
4							
5							
6							
7							
8							
9							
10							
11							
12							
13							
14							
15							
16							

3.2 ▶▶▶ Least Common Multiples

Only *like* fractions can be added or subtracted. Because of this, we must rewrite *unlike* fractions as *like* fractions before we can add or subtract them.

OBJECTIVE 1 Find the least common multiple. We can rewrite unlike fractions as like fractions by finding the *least common multiple* of the denominators.

> **Least Common Multiple (LCM)**
> The **least common multiple (LCM)** of two whole numbers is the smallest whole number divisible by both of those numbers.

EXAMPLE 1 Finding the Least Common Multiple

Find the least common multiple of 6 and 9.
 First, find the multiples of 6.

$$\underbrace{6 \cdot 1}_{6}, \quad \underbrace{6 \cdot 2}_{12}, \quad \underbrace{6 \cdot 3}_{18}, \quad \underbrace{6 \cdot 4}_{24}, \quad \underbrace{6 \cdot 5}_{30}, \quad \underbrace{6 \cdot 6}_{36}, \quad \underbrace{6 \cdot 7}_{42}, \quad \underbrace{6 \cdot 8}_{48}, \ldots$$

(The three dots at the end of the list show that the list continues in the same pattern without stopping.) Now, find the multiples of 9.

$$\underbrace{9 \cdot 1}_{9}, \quad \underbrace{9 \cdot 2}_{18}, \quad \underbrace{9 \cdot 3}_{27}, \quad \underbrace{9 \cdot 4}_{36}, \quad \underbrace{9 \cdot 5}_{45}, \quad \underbrace{9 \cdot 6}_{54}, \quad \underbrace{9 \cdot 7}_{63}, \quad \underbrace{9 \cdot 8}_{72}, \ldots$$

The smallest number found in *both* lists is 18, so 18 is the **least common multiple** of 6 and 9; the number 18 is the smallest whole number divisible by both 6 and 9.

Multiples of 6: 6, 12, **18**, 24, 30, 36, 42, 48, . . .

Multiples of 9: 9, **18**, 27, 36, 45, 54, 63, 72, . . .

> The *smallest* number in *both* lists is the LCM.

18 is the smallest number found in both lists. **18** is the least common multiple (LCM) of 6 and 9.

Work Problem **1** *at the Side.* ▶

OBJECTIVE 2 Find the least common multiple using multiples of the largest number. There are several ways to find the least common multiple. If the numbers are small, the least common multiple can often be found by inspection. Can you think of a number that can be divided evenly by both 3 and 4? The number 12 will work; it is the least common multiple of the numbers 3 and 4. A method that works well to find the least common multiple is to write multiples of the larger number.
 In this case, 4 is larger than 3, so write the multiples of 4.

$$4, 8, 12, 16, 20, \ldots$$

Now, check each multiple of 4 to see if it is divisible by 3.

4 is *not* divisible by 3
8 is *not* divisible by 3
12 *is* divisible by 3

The first multiple of 4 that is divisible by 3 is 12, so 12 is the least common multiple of 3 and 4.

OBJECTIVES

1. Find the least common multiple.

2. Find the least common multiple using multiples of the largest number.

3. Find the least common multiple using prime factorization.

4. Find the least common multiple using an alternative method.

5. Write a fraction with an indicated denominator.

---

1 (a) List the multiples of 5.

5, ____, ____, ____,

____, ____, ____,

____, . . .

(b) List the multiples of 8.

8, ____, ____, ____,

____, ____, ____, . . .

(c) Find the least common multiple of 5 and 8.

ANSWERS
1. (a) 10, 15, 20, 25, 30, 35, 40, . . .
   (b) 16, 24, 32, 40, 48, 56, . . .
   (c) 40

2 Use multiples of the larger number to find the least common multiple in each set of numbers.

(a) 2 and 5

(b) 3 and 9

(c) 6 and 8

(d) 4 and 7

3 Use prime factorization to find the LCM for each pair of numbers.

(a) 15 and 18

(b) 12 and 20

EXAMPLE 2   **Finding the Least Common Multiple**

Use multiples of the larger number to find the least common multiple of 6 and 9.

We start by writing the first few multiples of 9.

Multiples of 9

$$9, 18, 27, 36, 45, 54, \ldots$$

Now, we check each multiple of 9 to see if it is divisible by 6. The first multiple of 9 that is divisible by 6 is 18.

$$9, \mathbf{18}, 27, 36, 45, 54, \ldots$$

First multiple divisible by 6, because $18 \div 6 = 3$

The least common multiple of the numbers 6 and 9 is 18.

◀ *Work Problem* **2** *at the Side.*

OBJECTIVE **3** **Find the least common multiple using prime factorization.**   Example 2 shows how to find the least common multiple of two numbers by making a list of the multiples of the *larger* number. Although this method works well if both numbers are fairly small, it is usually easier to find the least common multiple for larger numbers by using *prime factorization,* as shown in the next example.

EXAMPLE 3   **Applying Prime Factorization Knowledge**

Use prime factorization to find the least common multiple of 9 and 12.

We start by finding the prime factorization of each number.

Factors of 9

$$9 = 3 \bullet 3$$
$$12 = 2 \bullet 2 \bullet 3 \qquad LCM = 3 \bullet 3 \bullet 2 \bullet 2 = 36$$

Factors of 12

Check to see that 36 is divisible by 9 (yes) and by 12 (yes). The smallest whole number divisible by both 9 and 12 is 36.

> **CAUTION**
> Notice that we did **not** have to repeat the factors that 9 and 12 have in common. In this case, the **3** in 2 • 2 • **3** = 12 was **not** used because 3 is already included in 3 • 3 = 9.

◀ *Work Problem* **3** *at the Side.*

ANSWERS

2. (a) 10 **(b)** 9 **(c)** 24 **(d)** 28

3. (a)
$$15 = 3 \bullet 5 \qquad LCM = 2 \bullet 3 \bullet 3 \bullet 5 = 90$$
$$18 = 2 \bullet 3 \bullet 3$$

(b)
$$12 = 2 \bullet 2 \bullet 3 \qquad LCM = 2 \bullet 2 \bullet 3 \bullet 5 = 60$$
$$20 = 2 \bullet 2 \bullet 5$$

EXAMPLE 4 **Using Prime Factorization**

Find the least common multiple of 12, 18, and 20.

Find the prime factorization of each number. Then use the prime factors to build the LCM.

12

$12 = 2 \cdot 2 \cdot 3$
$18 = 2 \cdot 3 \cdot 3$ LCM $= 2 \cdot 2 \cdot 3 \cdot 3 \cdot 5 = 180$
$20 = 2 \cdot 2 \cdot 5$

18

20

Check to see that 180 is divisible by 12 (yes) and by 18 (yes) and by 20 (yes). This smallest whole number divisible by 12, 18, and 20 is 180.

> **Note**
> The LCM did *not* repeat the factors that 12, 18, and 20 have in common.

Work Problem **4** *at the Side.* ▶

EXAMPLE 5 **Finding the Least Common Multiple**

Find the least common multiple for each set of numbers.

(a) 5, 6, 35

Find the prime factorization for each number.

5

$5 = 5$
$6 = 2 \cdot 3$ LCM $= 2 \cdot 3 \cdot 5 \cdot 7 = 210$
$35 = 5 \cdot 7$

6   35

The least common multiple of 5, 6, and 35 is 210.

(b) 10, 20, 24

Find the prime factorization for each number.

10
20

$10 = 2 \cdot 5$
$20 = 2 \cdot 2 \cdot 5$ LCM $= 2 \cdot 2 \cdot 2 \cdot 3 \cdot 5 = 120$
$24 = 2 \cdot 2 \cdot 2 \cdot 3$

24

The least common multiple of 10, 20, and 24 is 120.

Work Problem **5** *at the Side.* ▶

OBJECTIVE **4** **Find the least common multiple using an alternative method.** Some people like the following *alternative method* for finding the least common multiple for larger numbers. Try both methods, and *use the one you prefer.* As a review, a list of the first few prime numbers follows.

$$2, 3, 5, 7, 11, 13, 17$$

4 Find the least common multiple of the denominators in each set of fractions.

(a) $\dfrac{3}{8}$ and $\dfrac{6}{5}$

(b) $\dfrac{5}{6}$ and $\dfrac{1}{14}$

(c) $\dfrac{4}{9}, \dfrac{5}{18},$ and $\dfrac{7}{24}$

5 Find the least common multiple for each set of numbers.

(a) 4, 8, 9

(b) 3, 6, 8

(c) 9, 36, 48

(d) 15, 20, 30, 40

ANSWERS

4. (a)
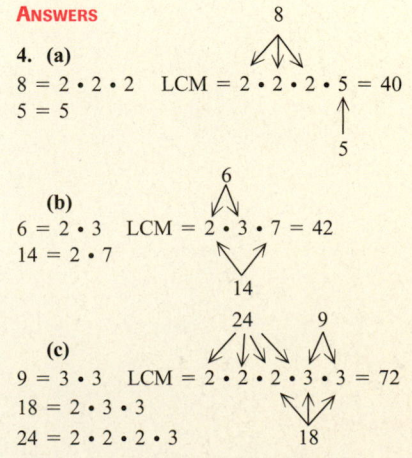
8
$8 = 2 \cdot 2 \cdot 2$ LCM $= 2 \cdot 2 \cdot 2 \cdot 5 = 40$
$5 = 5$
5

(b)
6
$6 = 2 \cdot 3$ LCM $= 2 \cdot 3 \cdot 7 = 42$
$14 = 2 \cdot 7$
14

(c)
24   9
$9 = 3 \cdot 3$ LCM $= 2 \cdot 2 \cdot 2 \cdot 3 \cdot 3 = 72$
$18 = 2 \cdot 3 \cdot 3$
$24 = 2 \cdot 2 \cdot 2 \cdot 3$ 18

5. (a) 72 **(b)** 24 **(c)** 144 **(d)** 120

6 In the following problems, the divisions have already been worked out. Multiply the prime numbers on the left to find the least common multiple.

(a)
$$
\begin{array}{c|cc}
2 & 6 & \cancel{15} \\
3 & 3 & 15 \\
5 & \cancel{1} & 5 \\
\hline
 & 1 & 1 \\
\end{array}
$$

(b)
$$
\begin{array}{c|cc}
2 & 20 & 36 \\
2 & 10 & 18 \\
3 & \cancel{5} & 9 \\
3 & \cancel{5} & 3 \\
5 & 5 & \cancel{1} \\
\hline
 & 1 & 1 \\
\end{array}
$$

EXAMPLE 6 **Alternative Method for Finding the Least Common Multiple**

Find the least common multiple for each set of numbers.

(a) 14 and 21

Start by trying to divide 14 and 21 by the first prime number in the list of prime numbers: 2, 3, 5, 7, 11, 13, and 17. Use the following shortcut. Divide by 2, the first prime.

$$
\begin{array}{c|cc}
\mathbf{2} & 14 & \cancel{21} \\
\hline
 & 7 & 21 \\
\end{array}
$$

Because 21 cannot be divided evenly by 2, cross out 21 and bring it down. Divide by 3, the second prime.

$$
\begin{array}{c|cc}
\mathbf{2} & 14 & \cancel{21} \\
\mathbf{3} & \cancel{7} & 21 \\
\hline
 & 7 & 7 \\
\end{array}
$$

Since 7 cannot be divided evenly by the third prime, 5, skip 5 and divide by the next prime, 7.
Divide by 7, the fourth prime.

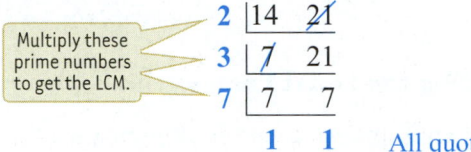

Multiply these prime numbers to get the LCM.

$$
\begin{array}{c|cc}
\mathbf{2} & 14 & \cancel{21} \\
\mathbf{3} & \cancel{7} & 21 \\
\mathbf{7} & 7 & 7 \\
\hline
 & 1 & 1 \\
\end{array}
$$

All quotients are 1.

When all quotients are 1, multiply the prime numbers on the left side.

$$\text{least common multiple} = \mathbf{2 \cdot 3 \cdot 7 = 42}$$

The least common multiple of 14 and 21 is 42.

(b) 6, 15, 18

Divide by 2.

$$
\begin{array}{c|ccc}
\mathbf{2} & 6 & \cancel{15} & 18 \\
\hline
 & 3 & 15 & 9 \\
\end{array}
$$

Cross out 15 and bring it down.

Divide by 3.

$$
\begin{array}{c|ccc}
\mathbf{2} & 6 & \cancel{15} & 18 \\
\mathbf{3} & 3 & 15 & 9 \\
\hline
 & 1 & 5 & 3 \\
\end{array}
$$

Divide by 3 again, since the remaining 3 can be divided.

$$
\begin{array}{c|ccc}
\mathbf{2} & 6 & \cancel{15} & 18 \\
\mathbf{3} & 3 & 15 & 9 \\
\mathbf{3} & \cancel{1} & \cancel{5} & 3 \\
\hline
 & 1 & 5 & 1 \\
\end{array}
$$

Finally, divide by 5.

$$
\begin{array}{c|ccc}
\mathbf{2} & 6 & \cancel{15} & 18 \\
\mathbf{3} & 3 & 15 & 9 \\
\mathbf{3} & \cancel{1} & \cancel{5} & 3 \\
\mathbf{5} & \cancel{1} & 5 & \cancel{1} \\
\hline
 & 1 & 1 & 1 \\
\end{array}
$$

All quotients are 1.

Multiply the prime numbers on the left side.

$$\mathbf{2 \cdot 3 \cdot 3 \cdot 5} = 90 \leftarrow \text{Least common multiple}$$

◀ *Work Problem* **6** *at the Side.*

Work Problem **7** *at the Side.* ▶

OBJECTIVE **5** **Write a fraction with an indicated denominator.**
Before we can add or subtract *unlike* fractions, we must find the least common multiple, which is then used as the denominator of the fractions.

EXAMPLE 7 **Writing a Fraction with an Indicated Denominator**

Write the fraction $\frac{2}{3}$ with a denominator of 15
 Find a numerator, so that these fractions are equivalent.

$$\frac{2}{3} = \frac{?}{15}$$

To find the new numerator, first divide **15** by **3.**

$$\frac{2}{3} = \frac{?}{15} \qquad 15 \div 3 = 5$$

Multiply both numerator and denominator of the fraction $\frac{2}{3}$ by 5

$$\frac{2}{3} = \frac{2 \cdot 5}{3 \cdot 5} = \frac{10}{15}$$

 This process is just the opposite of writing a fraction in lowest terms. Check the answer by writing $\frac{10}{15}$ in lowest terms; you should get $\frac{2}{3}$ again.

EXAMPLE 8 **Writing Fractions with a New Denominator**

Rewrite each fraction with the indicated denominator.

(a) $\frac{3}{8} = \frac{?}{48}$

 Divide 48 by 8, getting 6. Now multiply both the numerator and the denominator of $\frac{3}{8}$ by 6.

$$\frac{3}{8} = \frac{3 \cdot 6}{8 \cdot 6} = \frac{18}{48}$$ Multiply numerator and denominator by 6.

Recall that multiplying a number by 1 does *not* change the number, and $\frac{6}{6} = 1$.

That is, $\frac{3}{8} = \frac{18}{48}$. As a check, write $\frac{18}{48}$ in lowest terms; you should get $\frac{3}{8}$ again.

(b) $\frac{5}{6} = \frac{?}{42}$

 Divide 42 by 6, getting 7. Next, multiply both the numerator and the denominator of $\frac{5}{6}$ by 7.

Notice that $\frac{7}{7} = 1$.

$$\frac{5}{6} = \frac{5 \cdot 7}{6 \cdot 7} = \frac{35}{42}$$ Multiply numerator and denominator by 7.

This shows that $\frac{5}{6} = \frac{35}{42}$. As a check, write $\frac{35}{42}$ in lowest terms. Did you get $\frac{5}{6}$ again?

Continued on Next Page

7 Find the least common multiple of each set of numbers. Use whichever method you prefer.

(a) 3, 6, 10

(b) 15 and 40

(c) 9, 24

(d) 8, 21, 24

ANSWERS

7. (a) 30 **(b)** 120 **(c)** 72 **(d)** 168

8 Rewrite each fraction with the indicated denominator.

(a) $\dfrac{1}{4} = \dfrac{?}{16}$

Note

In Example 7, on the previous page, the fraction $\frac{2}{3}$ was multiplied by $\frac{5}{5}$. In Example 8, the fraction $\frac{3}{8}$ was multiplied by $\frac{6}{6}$ and the fraction $\frac{5}{6}$ was multiplied by $\frac{7}{7}$. The fractions, $\frac{5}{5}$, $\frac{6}{6}$, and $\frac{7}{7}$ are all equal to 1.

$$\frac{5}{5} = 1 \qquad \frac{6}{6} = 1 \qquad \frac{7}{7} = 1$$

Recall that any number multiplied by 1 is the number itself.

◀ *Work Problem* **8** *at the Side.*

(b) $\dfrac{2}{3} = \dfrac{?}{15}$

(c) $\dfrac{7}{16} = \dfrac{?}{32}$

(d) $\dfrac{6}{11} = \dfrac{?}{33}$

ANSWERS

8. (a) $\dfrac{4}{16}$ (b) $\dfrac{10}{15}$ (c) $\dfrac{14}{32}$ (d) $\dfrac{18}{33}$

3.2 ▶▶▶ **Exercises**

FOR EXTRA HELP

 PRACTICE   WATCH   DOWNLOAD   READ   REVIEW

Use multiples of the larger number to find the least common multiple in each set of numbers. See Examples 1 and 2.

1. 3 and 6

2. 2 and 4

3. 3 and 5

4. 3 and 7

5. 4 and 9

6. 4 and 10

7. 2 and 7

8. 6 and 8

9. 6 and 10

10. 12 and 16

11. 20 and 50

12. 25 and 75

Find the least common multiple of each set of numbers. Use any method. See Examples 2–6.

13. 4, 10

14. 8, 10

15. 12, 20

16. 9 and 15

17. 6, 9, 12

18. 20, 24, 30

19. 4, 6, 8, 10

20. 8, 9, 12, 18

21. 12, 15, 18, 20

22. 6, 8, 9, 27, 36

23. 8, 10, 12, 16, 36

24. 5, 6, 8, 25, 30

Rewrite each fraction with a denominator of 24. See Examples 7 and 8.

25. $\dfrac{2}{3} =$

26. $\dfrac{3}{8} =$

27. $\dfrac{3}{4} =$

28. $\dfrac{5}{12} =$

29. $\dfrac{5}{6} =$

30. $\dfrac{7}{8} =$

Rewrite each fraction with the indicated denominator.

31. $\dfrac{1}{2} = \dfrac{}{6}$

32. $\dfrac{2}{3} = \dfrac{}{9}$

33. $\dfrac{3}{4} = \dfrac{}{16}$

34. $\dfrac{7}{10} = \dfrac{}{30}$

35. $\dfrac{7}{8} = \dfrac{}{32}$

36. $\dfrac{5}{12} = \dfrac{}{48}$

37. $\dfrac{3}{16} = \dfrac{}{64}$

38. $\dfrac{7}{8} = \dfrac{}{96}$

39. $\dfrac{8}{5} = \dfrac{}{20}$

40. $\dfrac{5}{8} = \dfrac{}{40}$

41. $\dfrac{9}{7} = \dfrac{}{56}$

42. $\dfrac{3}{2} = \dfrac{}{64}$

43. $\dfrac{7}{4} = \dfrac{}{48}$

44. $\dfrac{7}{6} = \dfrac{}{120}$

45. $\dfrac{8}{11} = \dfrac{}{132}$

46. $\dfrac{4}{15} = \dfrac{}{165}$

47. $\dfrac{3}{16} = \dfrac{}{144}$

48. $\dfrac{7}{16} = \dfrac{}{112}$

49. There are several methods for finding the least common multiple (LCM). Do you prefer the method using multiples of the largest number or the method using prime factorizations? Why? Would you ever use the other method?

50. Explain in your own words how to write a fraction with an indicated denominator. As part of your explanation, show how to change $\frac{3}{4}$ to a fraction having 12 as a denominator.

Find the least common multiple of the denominators of each pair of fractions.

51. $\dfrac{25}{400}, \dfrac{38}{1800}$

52. $\dfrac{53}{600}, \dfrac{115}{4000}$

53. $\dfrac{109}{1512}, \dfrac{23}{392}$

54. $\dfrac{61}{810}, \dfrac{37}{1170}$

Relating Concepts (Exercises 55—62) For Individual or Group Work

Most people think that addition and subtraction of fractions is more difficult than multiplication and division of fractions. This is probably because a common denominator must be used. **Work Exercises 55–62 in order.**

55. Fractions with the same denominators are _____ fractions and fractions with different denominators are _____ fractions.

56. To subtract like fractions, first subtract the _____ to find the numerator of the difference. Write the denominator of the like fractions as the _____ of the difference. Finally, write the answer in _____ terms.

57. The _____ common multiple (LCM) of two numbers is the _____ whole number
(smallest/largest)
divisible by both those numbers.

58. The following shows the common multiples for both 8 and 10. What is the least common multiple for these two numbers?

Multiples of 8:   8, 16, 24, 32, 40, 48, 56, 64, 72, 80, 88, . . .

Multiples of 10:   10, 20, 30, 40, 50, 60, 70, 80, 90, . . .

Find the least common multiple for each set of numbers.

59. 5, 7, 14, 10

60. 25, 18, 30, 5

61. Explain why the least common multiple for 8, 3, 5, 4, and 10 is not 240. Find the least common multiple.

62. Demonstrate that the least common multiple of 55 and 1760 is 1760.

3.3 ►►► Adding and Subtracting Unlike Fractions

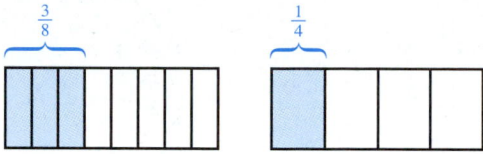

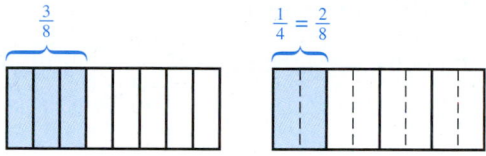

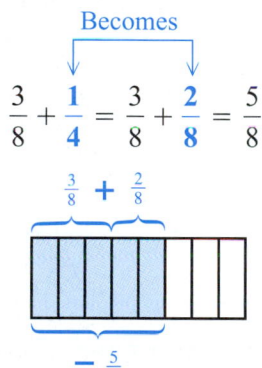

OBJECTIVE 1 Add unlike fractions. In this section, we add and subtract unlike fractions. To add unlike fractions, we must first change them to like fractions (fractions with the same denominator). For example, the figures below show $\frac{3}{8}$ and $\frac{1}{4}$.

OBJECTIVES

1 Add unlike fractions.

2 Add unlike fractions vertically.

3 Subtract unlike fractions.

4 Subtract unlike fractions vertically.

These fractions can be added by changing them to like fractions. Make like fractions by changing $\frac{1}{4}$ to the equivalent fraction $\frac{2}{8}$.

Now you can add the fractions.

Becomes

$$\frac{3}{8} + \frac{1}{4} = \frac{3}{8} + \frac{2}{8} = \frac{5}{8}$$

$$= \frac{5}{8}$$

Use the following steps to add or subtract unlike fractions.

Adding or Subtracting Unlike Fractions

Step 1 Rewrite the *unlike fractions* as *like fractions* with the least common multiple as their new denominator. This new denominator is called the **least common denominator (LCD).**

Step 2 Add or subtract as with like fractions.

Step 3 Simplify the answer by writing it in lowest terms and as a whole or mixed number where possible.

EXAMPLE 1 **Adding Unlike Fractions**

Add $\frac{2}{3}$ and $\frac{1}{9}$.

The least common multiple of 3 and 9 is 9, so first write the fractions as like fractions with a denominator of 9. This is the *least common denominator (LCD)* of 3 and 9.

Continued on Next Page

1 Add.

(a) $\dfrac{1}{2} + \dfrac{3}{8}$

(b) $\dfrac{3}{4} + \dfrac{1}{8}$

(c) $\dfrac{3}{5} + \dfrac{3}{10}$

(d) $\dfrac{1}{12} + \dfrac{5}{6}$

2 Add. Simplify all answers.

(a) $\dfrac{3}{10} + \dfrac{1}{5}$

(b) $\dfrac{5}{8} + \dfrac{1}{3}$

(c) $\dfrac{1}{10} + \dfrac{1}{3} + \dfrac{1}{6}$

Step 1 $\dfrac{2}{3} = \dfrac{?}{9}$

Divide 9 by 3, getting 3. Next, multiply numerator and denominator by 3.

$$\dfrac{2}{3} = \dfrac{2 \cdot 3}{3 \cdot 3} = \dfrac{6}{9}$$

Now, add the like fractions $\dfrac{6}{9}$ and $\dfrac{1}{9}$.

Becomes

> Both fractions must have the *same* denominator **before** you add them.

Step 2 $\dfrac{2}{3} + \dfrac{1}{9} = \dfrac{6}{9} + \dfrac{1}{9} = \dfrac{6+1}{9} = \dfrac{7}{9}$

Step 3   Step 3 is not needed because $\dfrac{7}{9}$ is already in lowest terms.

◀ *Work Problem* **1** *at the Side.*

EXAMPLE 2 **Adding Fractions**

Add each pair of fractions using the three steps. Simplify all answers.

(a) $\dfrac{1}{3} + \dfrac{1}{6}$

The least common multiple of 3 and 6 is 6. Rewrite both fractions as fractions with a least common denominator of 6.

Rewritten as like fractions

Step 1 $\dfrac{1}{3} + \dfrac{1}{6} = \dfrac{2}{6} + \dfrac{1}{6}$

> 6 is the LCD (least common denominator).

Add numerators.

Step 2 $\dfrac{2}{6} + \dfrac{1}{6} = \dfrac{2+1}{6} = \dfrac{3}{6}$

> Only *like* fractions can be added.

Step 3 $\dfrac{3}{6} = \dfrac{1}{2}$ ← In lowest terms

(b) $\dfrac{6}{15} + \dfrac{3}{10}$

The least common multiple of 15 and 10 is 30, so rewrite both fractions with a least common denominator of 30.

Rewritten as like fractions

Step 1 $\dfrac{6}{15} + \dfrac{3}{10} = \dfrac{12}{30} + \dfrac{9}{30}$

> The least common multiple of the denominators is the LCD.

Add numerators.

Step 2 $\dfrac{12}{30} + \dfrac{9}{30} = \dfrac{12+9}{30} = \dfrac{21}{30}$

Step 3 $\dfrac{21}{30} = \dfrac{7}{10}$ ← In lowest terms

◀ *Work Problem* **2** *at the Side.*

ANSWERS

1. (a) $\dfrac{7}{8}$ (b) $\dfrac{7}{8}$ (c) $\dfrac{9}{10}$ (d) $\dfrac{11}{12}$

2. (a) $\dfrac{1}{2}$ (b) $\dfrac{23}{24}$ (c) $\dfrac{3}{5}$

OBJECTIVE **2** **Add unlike fractions vertically.** Fractions can also be added vertically (one fraction written below the other).

EXAMPLE 3 Vertical Addition of Fractions

Add the following fractions vertically.

(a)
$$\frac{3}{8} = \frac{3 \cdot 3}{8 \cdot 3} = \frac{9}{24} \quad\longleftarrow$$

$$+\frac{7}{12} = \frac{7 \cdot 2}{12 \cdot 2} = \frac{14}{24} \quad\longleftarrow$$

24 is the LCD.

Rewritten as like fractions

$$\frac{23}{24} \quad\longleftarrow \text{ Add the numerators.}$$
$$\quad\longleftarrow \text{ Denominator is 24, the LCD.}$$

(b)
$$\frac{2}{9} = \frac{2 \cdot 4}{9 \cdot 4} = \frac{8}{36} \quad\longleftarrow$$

$$+\frac{1}{4} = \frac{1 \cdot 9}{4 \cdot 9} = \frac{9}{36} \quad\longleftarrow$$

Rewritten as like fractions

$$\frac{17}{36} \quad\longleftarrow \text{ Add the numerators.}$$
$$\quad\longleftarrow \text{ Denominator is 36, the LCD.}$$

Work Problem **3** *at the Side.* ▶

OBJECTIVE **3** **Subtract unlike fractions.** The next example shows subtraction of unlike fractions.

EXAMPLE 4 Subtracting Unlike Fractions

Subtract. Simplify all answers.

As with addition, rewrite unlike fractions with a least common denominator.

(a) $\frac{3}{4} - \frac{3}{8}$

Rewritten as like fractions

Step 1
$$\frac{3}{4} - \frac{3}{8} = \frac{6}{8} - \frac{3}{8}$$

The LCD is 8.

Subtract numerators.

Step 2
$$\frac{6}{8} - \frac{3}{8} = \frac{6 - 3}{8} = \frac{3}{8}$$

Step 3 Not needed because $\frac{3}{8}$ is in lowest terms.

Continued on Next Page

3 Add the following fractions vertically.

(a)
$$\frac{5}{8}$$
$$+\frac{1}{12}$$

(b)
$$\frac{7}{16}$$
$$+\frac{1}{4}$$

4 Subtract. Simplify all answers.

(a) $\dfrac{5}{8} - \dfrac{1}{4}$

(b) $\dfrac{4}{5} - \dfrac{3}{4}$

(b) $\dfrac{3}{4} - \dfrac{7}{12}$

Rewritten as like fractions

Step 1 $\dfrac{3}{4} - \dfrac{7}{12} = \dfrac{9}{12} - \dfrac{7}{12}$

Subtract numerators.

Step 2 $\dfrac{9}{12} - \dfrac{7}{12} = \dfrac{9-7}{12} = \dfrac{2}{12}$ ← Subtract the numerators.
$\dfrac{2}{12}$ ← Denominator is 12, the LCD.

Step 3 $\dfrac{2}{12} = \dfrac{1}{6}$ ← Lowest terms

> Always write the final answer in lowest terms.

◀ *Work Problem* **4** *at the Side.*

OBJECTIVE **4** Subtract unlike fractions vertically.

EXAMPLE 5 **Vertical Subtraction of Fractions**

Subtract the following fractions vertically.

(a)
$$\dfrac{4}{5} = \dfrac{4 \cdot 8}{5 \cdot 8} = \dfrac{32}{40}$$
$$-\dfrac{3}{8} = \dfrac{3 \cdot 5}{8 \cdot 5} = \dfrac{15}{40}$$

Rewritten as like fractions

$\dfrac{17}{40}$ ← Subtract numerators.
← Denominator is 40, the LCD.

(b)
$$\dfrac{3}{7} = \dfrac{3 \cdot 12}{7 \cdot 12} = \dfrac{36}{84}$$
$$-\dfrac{5}{12} = \dfrac{5 \cdot 7}{12 \cdot 7} = \dfrac{35}{84}$$

Rewritten as like fractions

$\dfrac{1}{84}$ ← Subtract numerators.
← Denominator is 84, the LCD.

◀ *Work Problem* **5** *at the Side.*

5 Subtract vertically. Simplify all answers.

(a) $\dfrac{7}{8}$
$-\dfrac{2}{3}$

(b) $\dfrac{5}{6}$
$-\dfrac{1}{12}$

ANSWERS

4. (a) $\dfrac{3}{8}$ (b) $\dfrac{1}{20}$

5. (a) $\dfrac{5}{24}$ (b) $\dfrac{3}{4}$

3.3 ▶▶▶ Exercises

Add the following fractions. Simplify all answers. See Examples 1–3.

1. $\dfrac{3}{4} + \dfrac{1}{8}$

2. $\dfrac{1}{6} + \dfrac{2}{3}$

3. $\dfrac{2}{3} + \dfrac{2}{9}$

4. $\dfrac{3}{7} + \dfrac{1}{14}$

5. $\dfrac{9}{20} + \dfrac{3}{10}$

6. $\dfrac{5}{8} + \dfrac{1}{4}$

7. $\dfrac{3}{5} + \dfrac{3}{8}$

8. $\dfrac{5}{7} + \dfrac{3}{14}$

9. $\dfrac{2}{9} + \dfrac{5}{12}$

10. $\dfrac{5}{8} + \dfrac{1}{12}$

11. $\dfrac{1}{3} + \dfrac{3}{5}$

12. $\dfrac{2}{5} + \dfrac{3}{7}$

13. $\dfrac{1}{4} + \dfrac{2}{9} + \dfrac{1}{3}$

14. $\dfrac{3}{7} + \dfrac{2}{5} + \dfrac{1}{10}$

15. $\dfrac{3}{10} + \dfrac{2}{5} + \dfrac{3}{20}$

16. $\dfrac{1}{3} + \dfrac{3}{8} + \dfrac{1}{4}$

17. $\dfrac{4}{15} + \dfrac{1}{6} + \dfrac{1}{3}$

18. $\dfrac{5}{12} + \dfrac{2}{9} + \dfrac{1}{6}$

19. $\begin{array}{r} \dfrac{1}{4} \\ + \dfrac{1}{8} \\ \hline \end{array}$

20. $\begin{array}{r} \dfrac{7}{12} \\ + \dfrac{1}{8} \\ \hline \end{array}$

21. $\begin{array}{r} \dfrac{5}{12} \\ + \dfrac{1}{16} \\ \hline \end{array}$

22. $\begin{array}{r} \dfrac{3}{7} \\ + \dfrac{1}{3} \\ \hline \end{array}$

Subtract the following fractions. Simplify all answers. See Example 4.

23. $\dfrac{5}{6} - \dfrac{1}{3}$

24. $\dfrac{3}{4} - \dfrac{5}{8}$

25. $\dfrac{2}{3} - \dfrac{1}{6}$

26. $\dfrac{3}{4} - \dfrac{5}{8}$

⊙ 27. $\dfrac{2}{3} - \dfrac{1}{5}$

28. $\dfrac{5}{6} - \dfrac{7}{9}$

29. $\dfrac{5}{12} - \dfrac{1}{4}$

30. $\dfrac{5}{7} - \dfrac{1}{3}$

31. $\dfrac{8}{9} - \dfrac{7}{15}$

32. $\begin{array}{r} \dfrac{4}{5} \\ -\dfrac{1}{3} \\ \hline \end{array}$

33. $\begin{array}{r} \dfrac{7}{8} \\ -\dfrac{4}{5} \\ \hline \end{array}$

34. $\begin{array}{r} \dfrac{5}{8} \\ -\dfrac{1}{3} \\ \hline \end{array}$

35. $\begin{array}{r} \dfrac{5}{12} \\ -\dfrac{1}{16} \\ \hline \end{array}$

36. $\begin{array}{r} \dfrac{7}{12} \\ -\dfrac{1}{3} \\ \hline \end{array}$

Solve each application problem.

Use the newspaper advertisement for this 4-piece chisel set to answer Exercises 37–38. (*Source:* Harbor Freight Tools.)

37. Find the difference in the cutting-edge width of the two chisels with the widest blades. The symbol ″ is for inches.

38. Find the difference in the cutting-edge width of the two chisels with the narrowest blades. The symbol ″ is for inches.

39. A sports and entertainment center has $\frac{4}{5}$ of its total area devoted to seating of fans and guests. If $\frac{3}{8}$ of the seating area is used for general admission seating and the rest for reserved seating, find the fraction of the total area used for reserved seating.

40. Della Daniel wants to open a day care center and has saved $\frac{2}{5}$ of the amount needed for start-up costs. If she saves another $\frac{1}{8}$ of the amount needed and then $\frac{1}{6}$ more, find the total portion of the start-up costs she has saved.

41. When installing cabinets for The Home Depot, Sarah Bryn must be certain that the proper type and size of mounting screw is used. Find the total length of the screw shown.

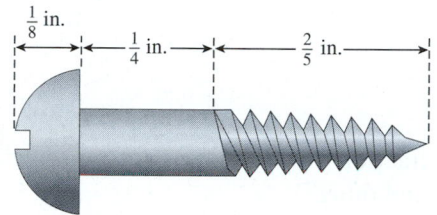

42. When installing a computer chassis, Bonnie Bottorff must be certain that the proper type and size of bolt is used. Find the total length of the bolt shown.

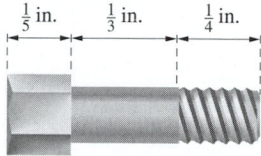

43. Bill Newton is a general contractor and begins a job with $\frac{3}{4}$ of a tank of fuel in his back-hoe. If he uses $\frac{1}{3}$ of the tank in the morning and $\frac{3}{8}$ of the tank in the afternoon, what fraction of the tank of fuel remains?

44. Adrian Ortega drives a tanker for the British Petroleum Company. He leaves the refinery with his tanker filled to $\frac{7}{8}$ of capacity. If he delivers $\frac{1}{4}$ of the tanker's capacity at the first stop and $\frac{1}{3}$ of the tanker's capacity at the second stop, find the fraction of the tanker's capacity remaining.

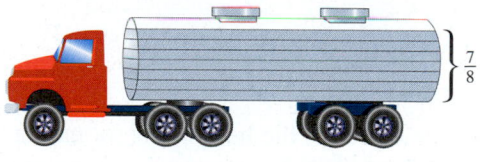

45. Step 1 in adding or subtracting unlike fractions is to rewrite the fractions so they have the least common multiple as a denominator. Explain in your own words why this is necessary.

46. Briefly list the three steps used for addition and subtraction of unlike fractions.

Refer to the circle graph to answer Exercises 47–50.

THE DAY OF THE STUDENT
(One day = 24 hours)

Work and Travel $\frac{1}{3}$

Class $\frac{1}{6}$

Study $\frac{1}{8}$

Sleep $\frac{7}{24}$

Other $\frac{1}{12}$

47. What fraction of the day was spent in class and study?

48. What fraction of the day was spent in work and travel and other?

49. In which activity was the greatest amount of time spent? How many hours did this activity take? What fraction of the day was spent on this activity and class time?

50. In which activity was the least amount of time spent? How many hours did this activity take? What fraction of the day was spent on this activity and studying?

51. Find the diameter of the hole in the mounting bracket shown. (The diameter is the distance across the center of the hole.)

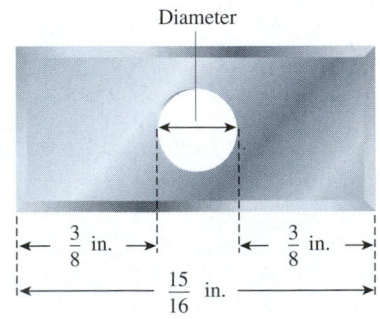

Diameter

$\frac{3}{8}$ in. $\frac{3}{8}$ in.

$\frac{15}{16}$ in.

52. Chakotay is fitting a turquoise stone into a bear claw pendant. Find the diameter of the hole in the pendant. (The diameter is the distance across the center of the hole.)

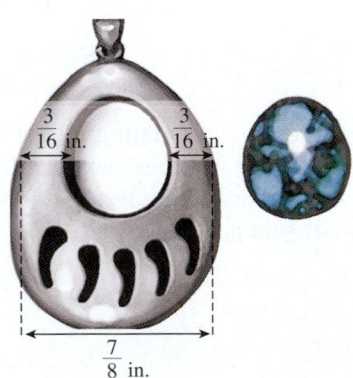

$\frac{3}{16}$ in. $\frac{3}{16}$ in.

$\frac{7}{8}$ in.

Study Skills

Mind mapping is a visual way to show information that you have learned. It is an excellent way to review. Mapping is flexible and can be personalized, which is helpful for your memory. Your brain likes to see things that are **pleasing** to look at, **colorful,** and that **show connections** between ideas. Take advantage of that by creating maps that

▶ are easy to read,

▶ use color in a systematic way, and

▶ clearly show you how different concepts are related (using arrows or dotted lines, for example).

Here are some general directions for making a map. After you read them, work on completing the map that has been started for you on the next page. It is from Chapters 2 and 3: Fractions.

▶ To begin a mind map, write the concept in the center of a piece of paper and either circle it or draw a box around it.

▶ Make a line out from the center concept, and draw a box large enough to write the definition of the concept.

▶ Think of the other aspects (subpoints) of the concept that you have learned, such as procedures to follow or formulas. Make a separate line and box connecting each subpoint to the center.

▶ From each of the new boxes, add the information you've learned. You can continue making new lines and boxes or circles, or you can list items below the new information.

▶ Use color to highlight the major points. For example, everything related to one subpoint might be the same color. That way you can easily see related ideas.

▶ You may also use arrows, underlining, or small drawings to help yourself remember.

OBJECTIVES

1 **Create mind maps for appropriate concepts.**

2 **Visually show how concepts relate to each other using arrows or lines.**

Directions for Making a Mind Map

Why Is Mapping Brain Friendly?

Remember that your brain grows dendrites when you are **actively thinking** about and working with information. Making a map requires you to think hard about **how to place the information, how to show connections** between parts of the map, and **how color will be useful.** It also takes a lot of thinking to fill in all related details and **show how those details connect to the larger concept.** All that thinking will let your brain grow a complex, many-branched neural network of interconnected dendrites. It is time well spent.

Try This Fractions Mind Map Using Sections 2.5, 2.7, and 3.3

On a separate paper, make a map that summarizes Computations with Fractions. Follow the directions and use the starter map below.

▶ The longest rectangles are *instructions* for all four operations. (The first one starts "Rewrite all numbers as fractions..." and the second one is at the bottom of the map.)

▶ Notice the wavy dividing lines that separate the map into two sides.

▶ Your job is to complete the map by writing the steps used in multiplying and dividing fractions (from **Sections 2.5 and 2.7**) and the steps used in adding and subtracting fractions (from **Section 3.3**).

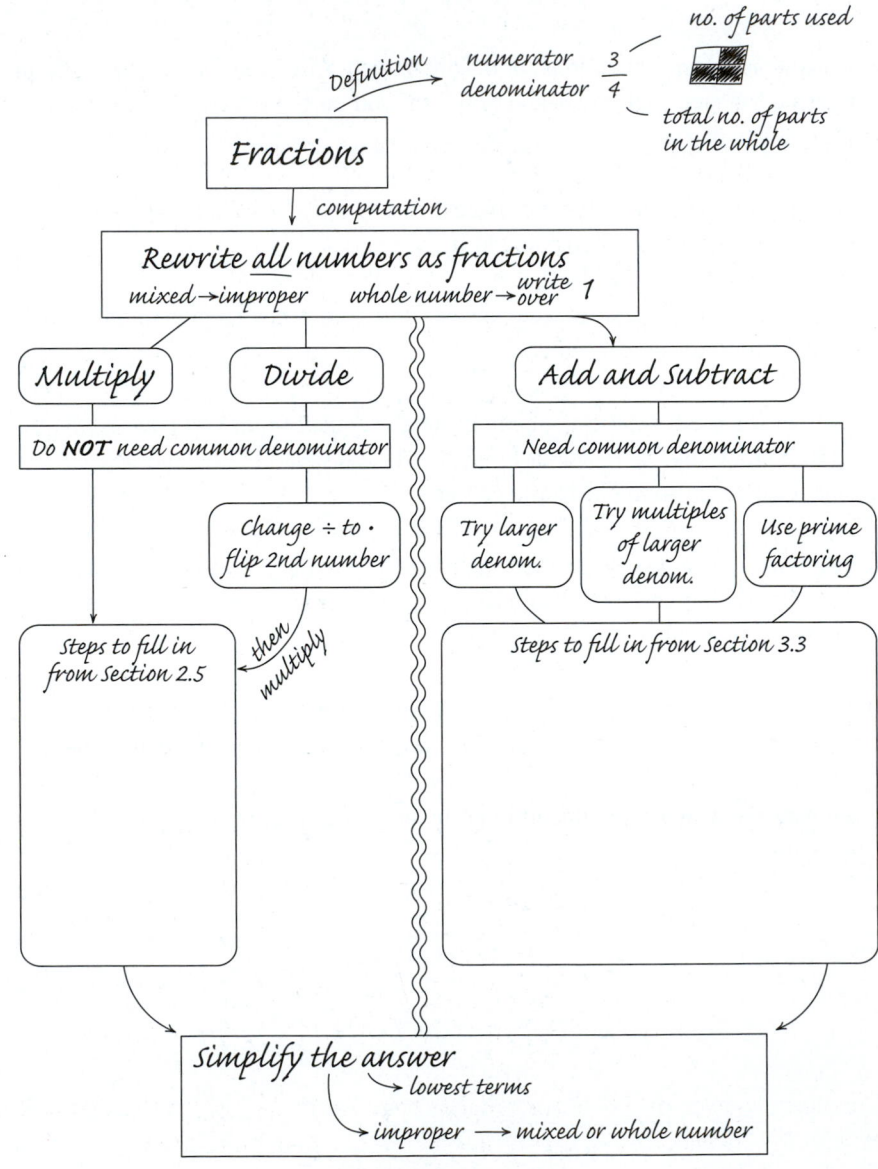

3.4 ▶▶▶ Adding and Subtracting Mixed Numbers

Recall that a mixed number is the sum of a whole number and a fraction. For example,

$$3\frac{2}{5} \quad \text{means} \quad 3 + \frac{2}{5}$$

OBJECTIVE 1 Estimate an answer, then add or subtract mixed numbers. Add or subtract mixed numbers by adding or subtracting the fraction parts and then the whole number parts. It is a good idea to estimate the answer first, as we did when multiplying and dividing mixed numbers in **Section 2.8.**

Work Problem **1** *at the Side.* ▶

---

1 Estimate an answer, then add or subtract mixed numbers.

2 Estimate an answer, then subtract mixed numbers by regrouping.

3 Add or subtract mixed numbers using an alternate method.

EXAMPLE 1 **Adding and Subtracting Mixed Numbers**

First estimate the answer. Then add or subtract to find the exact answer.

(a) $16\frac{1}{8} + 5\frac{5}{8}$

Estimate: *Exact:*

$$16 \xleftarrow{\text{Rounds to}} \left\{ 16\frac{1}{8} \right.$$

$$+ \ 6 \xleftarrow{\text{Rounds to}} \left\{ + \ 5\frac{5}{8} \right.$$

First, estimate the answer. 22 $21\frac{6}{8} = 21\frac{3}{4} \leftarrow$ Lowest terms

Sum of ——┘ └—— Sum of fractions
whole numbers

In lowest terms $\frac{6}{8}$ is $\frac{3}{4}$, so the exact answer of $21\frac{3}{4}$ is in lowest terms. The exact answer is *reasonable* because it is close to the estimate of 22.

(b) $8\frac{5}{8} - 3\frac{1}{12}$

Estimate: *Exact:*

$$9 \xleftarrow{\text{Rounds to}} \left\{ 8\frac{5}{8} = \ 8\frac{15}{24} \right. \leftarrow$$

$$- 3 \xleftarrow{\text{Rounds to}} \left\{ -3\frac{1}{12} = -3\frac{2}{24} \right. \leftarrow$$

24 is the least common denominator.

$$6 \qquad\qquad 5\frac{13}{24}$$

Subtract ——┘ └—— Subtract fractions.
whole numbers. ——┘

The exact answer of $5\frac{13}{24}$ is *reasonable* because it is close to the estimated answer of 6. Check by adding $5\frac{13}{24}$ and $3\frac{1}{12}$; the sum should be $8\frac{5}{8}$.

Continued on Next Page

---

1 As a review of mixed numbers, write each mixed number as an improper fraction and each improper fraction as a mixed number.

(a) $\frac{9}{2}$

(b) $\frac{8}{3}$

(c) $4\frac{3}{4}$

(d) $3\frac{7}{8}$

ANSWERS

1. (a) $4\frac{1}{2}$ (b) $2\frac{2}{3}$ (c) $\frac{19}{4}$ (d) $\frac{31}{8}$

2 First estimate, and then add or subtract to find the exact answer.

(a) *Estimate:*   *Exact:*

$$7 \xleftarrow{\text{Rounds to}} \begin{cases} 6\frac{7}{8} = 6\frac{7}{8} \end{cases}$$

$$+\,2 \xleftarrow{\text{Rounds to}} \begin{cases} +\,2\frac{1}{4} = 2\frac{2}{8} \end{cases}$$

(b) $25\frac{3}{5} + 12\frac{3}{10}$

Estimate:

_____ + _____ = _____

Exact: _____

(c) *Estimate:*   *Exact:*

$$\xleftarrow{\text{Rounds to}} \begin{cases} 4\frac{7}{9} \end{cases}$$

$$-\quad\quad \xleftarrow{\text{Rounds to}} \begin{cases} -\,2\frac{2}{3} \end{cases}$$

3 First estimate, and then add to find the exact answer.

(a) *Estimate:*   *Exact:*

$$\xleftarrow{\text{Rounds to}} \begin{cases} 9\frac{3}{4} \end{cases}$$

$$+\quad\quad \xleftarrow{\text{Rounds to}} \begin{cases} +\,7\frac{1}{2} \end{cases}$$

(b) *Estimate:*   *Exact:*

$$\xleftarrow{\text{Rounds to}} \begin{cases} 15\frac{4}{5} \end{cases}$$

$$+\quad\quad \xleftarrow{\text{Rounds to}} \begin{cases} +\,12\frac{2}{3} \end{cases}$$

ANSWERS

2. **(a)** $7 + 2 = 9;\ 9\frac{1}{8}$

(b) $26 + 12 = 38;\ 37\frac{9}{10}$

(c) $5 - 3 = 2;\ 2\frac{1}{9}$

3. **(a)** $10 + 8 = 18;\ 17\frac{1}{4}$

(b) $16 + 13 = 29;\ 28\frac{7}{15}$

Note

When estimating, if the numerator is *half* of the denominator or *more,* round up the whole number part. If the numerator is *less* than *half* the denominator, leave the whole number part as it is.

◀ Work Problem **2** at the Side.

When you add the fraction parts of mixed numbers, the sum may be greater than 1. If this happens, simplify the fraction and regroup in the whole number column. (You cannot leave a whole number along with an improper fraction as the answer.)

EXAMPLE 2   **Simplify and Regroup When Adding Mixed Numbers**

First estimate, and then add $9\frac{5}{8} + 13\frac{7}{8}$.

Estimate: *Exact:*

$$10 \xleftarrow{\text{Rounds to}} \begin{cases} 9\frac{5}{8} \end{cases}$$

$$+\,14 \xleftarrow{\text{Rounds to}} \begin{cases} +\,13\frac{7}{8} \end{cases}$$

$$24 \qquad\qquad 22\frac{12}{8}$$

First, add the fractions, then the whole numbers.

Sum of whole numbers   Sum of fractions

The improper fraction $\frac{12}{8}$ can be written in lowest terms as $\frac{3}{2}$. Because $\frac{3}{2} = 1\frac{1}{2}$, the simplified sum is

Becomes

$$22\frac{12}{8} = 22 + \frac{12}{8} = 22 + 1\frac{1}{2} = 23\frac{1}{2}.$$

The estimate was 24, so the exact answer of $23\frac{1}{2}$ is reasonable.

Note

When adding mixed numbers, first add the fraction parts, then add the whole number parts. Then combine the two answers, simplifying the fraction part when necessary.

◀ Work Problem **3** at the Side.

OBJECTIVE 2 Estimate an answer, then subtract mixed numbers by regrouping. When subtracting mixed numbers, **regrouping** is necessary when the fraction part of the first number is less than the fraction part of the second number.

EXAMPLE 3   **Regroup When Subtracting Mixed Numbers**

First estimate, and then subtract to find the exact answer.

(a) $7 - 2\frac{5}{6}$

Continued on Next Page

Estimate: *Exact:*

7   ←Rounds to $\{$ 7   There is no fraction here from which to subtract $\frac{5}{6}$.

$-\;3$ ←Rounds to $\{$ $-\,2\frac{5}{6}$

$\overline{\quad4\quad}$

It is **not** possible to subtract $\frac{5}{6}$ without regrouping the whole number **7** first.

Regroup 7 as 6 + 1

$$7 = \overbrace{6 + \mathbf{1}}$$

$1 = \frac{6}{6}$

$$= 6 + \frac{\mathbf{6}}{\mathbf{6}}$$

$$= 6\frac{6}{6}$$

Now you can subtract.

$$7 = \quad 6\frac{6}{6}$$ 7 was written as $6\frac{6}{6}$.

$$-\,2\frac{5}{6} = \quad -\,2\frac{5}{6}$$

$$\overline{\qquad\qquad 4\frac{1}{6}}$$

The estimate was 4, so the exact answer of $4\frac{1}{6}$ is reasonable.

(b) $8\frac{1}{3} - 4\frac{3}{5}$

Estimate: *Exact:*

8   ←Rounds to $\{$ $8\frac{1}{3} = \quad 8\frac{5}{15}$ ←

$-\;5$ ←Rounds to $\{$ $-\,4\frac{3}{5} = -\,4\frac{9}{15}$ ←   15 is the least common denominator

$\overline{\quad3\quad}$ $\overline{\qquad\qquad\qquad}$

It is **not** possible to subtract $\frac{9}{15}$ from $\frac{5}{15}$, so regroup the whole number **8.**

Regroup 8 as 7 + 1

$$8\frac{5}{15} = 8 + \frac{5}{15} = \overbrace{7 + \mathbf{1}} + \frac{5}{15}$$

$1 = \frac{15}{15}$

$$= 7 + \frac{\mathbf{15}}{\mathbf{15}} + \frac{5}{15}$$

$$= 7 + \frac{\mathbf{20}}{\mathbf{15}} \leftarrow \frac{15}{15} + \frac{5}{15}$$

$$= 7\frac{20}{15}$$

Continued on Next Page

4 First estimate and then subtract to find the exact answer.

(a) *Estimate:*   *Exact:*

$$\underleftarrow{\text{Rounds to}} \left\{ 7\frac{1}{3} \right.$$

$$- \underleftarrow{\text{Rounds to}} \left\{ -4\frac{5}{6} \right.$$

(b) *Estimate:*   *Exact:*

$$\longleftarrow \left\{ 4\frac{5}{8} \right.$$

$$- \longleftarrow \left\{ -2\frac{15}{16} \right.$$

(c) *Estimate:*   *Exact:*

$$\longleftarrow \left\{ 15 \right.$$

$$- \longleftarrow \left\{ -6\frac{4}{9} \right.$$

5 Add or subtract by changing mixed numbers to improper fractions. Simplify answers.

(a)
$$3\frac{3}{8}$$
$$+\ 2\frac{1}{2}$$

(b)
$$6\frac{3}{4}$$
$$-\ 4\frac{2}{3}$$

ANSWERS

4. **(a)** $7 - 5 = 2;\ 2\frac{1}{2}$

   (b) $5 - 3 = 2;\ 1\frac{11}{16}$

   (c) $15 - 6 = 9;\ 8\frac{5}{9}$

5. **(a)** $\frac{47}{8} = 5\frac{7}{8}$

   (b) $\frac{25}{12} = 2\frac{1}{12}$

Now you can subtract.

$$8\frac{1}{3} = 8\frac{5}{15} = 7\frac{20}{15}$$
$$-\ 4\frac{3}{5} = 4\frac{9}{15} = 4\frac{9}{15}$$
$$\overline{3\frac{11}{15}}$$

The exact answer is $3\frac{11}{15}$ (lowest terms), which is reasonable because it is close to the estimate of 3.

◀ *Work Problem* **4** *at the Side.*

OBJECTIVE **3** **Add or subtract mixed numbers using an alternate method.** An alternate method for adding or subtracting mixed numbers is to first change the mixed numbers to improper fractions. Then rewrite the unlike fractions as like fractions. Finally, add or subtract the numerators and write the answer in lowest terms.

EXAMPLE 4 **Adding or Subtracting Mixed Numbers**

Add or subtract.

> Rewrite $2\frac{3}{8}$ as $\frac{19}{8}$ and $3\frac{3}{4}$ as $\frac{15}{4}$.

(a)
$$2\frac{3}{8} = \frac{19}{8} = \frac{19}{8}$$
$$+\ 3\frac{3}{4} = \frac{15}{4} = \frac{30}{8}$$

8 is the least common denominator.

$$\frac{49}{8} = 6\frac{1}{8} \quad \text{Answer as mixed number}$$

↑ Change to improper fractions.

(b)
$$4\frac{2}{3} = \frac{14}{3} = \frac{70}{15}$$
$$-\ 2\frac{1}{5} = \frac{11}{5} = \frac{33}{15}$$

15 is the least common denominator.

$$\frac{37}{15} = 2\frac{7}{15}$$

> Simplify the answer by writing it as a mixed number.

↑ Improper fractions

◀ *Work Problem* **5** *at the Side.*

Note

The advantage of this alternate method of adding or subtracting mixed numbers is that it eliminates the need to regroup. It is also the most useful method for working with algebraic fractions. (See **Section 14.4.**) However, if the mixed numbers are large, then the numerators of the improper fractions may become so large that they are difficult to work with. In such cases, you may want to keep the numbers as mixed numbers.

3.4 ▶▶▶ **Exercises**

First estimate the answer. Then add to find the exact answer. Write answers as mixed numbers. See Examples 1 and 2.

1. *Estimate:*   *Exact:*

Rounds to → $\left\{ 5\dfrac{1}{2} \right.$

$+$ ___   Rounds to → $\left\{ + 3\dfrac{1}{3} \right.$

___

2. *Estimate:*   *Exact:*

$6\dfrac{3}{5}$

$+$ ___ $+ 7\dfrac{1}{10}$

___

3. *Estimate:*   *Exact:*

$7\dfrac{1}{3}$

$+$ ___ $+ 4\dfrac{1}{6}$

___

4. *Estimate:*   *Exact:*

$10\dfrac{1}{4}$

$+$ ___ $+ 5\dfrac{5}{8}$

___

5. *Estimate:*   *Exact:*

$\dfrac{5}{8}$

$+$ ___ $+ 3\dfrac{7}{12}$

___

6. *Estimate:*   *Exact:*

$12\dfrac{4}{5}$

$+$ ___ $+ \dfrac{7}{10}$

___

7. *Estimate:*   *Exact:*

$24\dfrac{5}{6}$

$+$ ___ $+ 18\dfrac{5}{6}$

___

8. *Estimate:*   *Exact:*

$14\dfrac{6}{7}$

$+$ ___ $+ 15\dfrac{1}{2}$

___

9. *Estimate:*   *Exact:*

$33\dfrac{3}{5}$

$+$ ___ $+ 18\dfrac{1}{2}$

___

10. *Estimate:*   *Exact:*

$18\dfrac{5}{8}$

$+$ ___ $+ 6\dfrac{2}{3}$

___

11. *Estimate:*   *Exact:*

$22\dfrac{3}{4}$

$+$ ___ $+ 15\dfrac{3}{7}$

___

12. *Estimate:*   *Exact:*

$7\dfrac{1}{4}$

$+$ ___ $+ 25\dfrac{7}{8}$

___

13. *Estimate:*   *Exact:*

$$12\frac{8}{15}$$

$$18\frac{3}{5}$$

$+$ _____ $+\ 14\frac{7}{10}$

_____   _____

14. *Estimate:*   *Exact:*

$$14\frac{9}{10}$$

$$8\frac{1}{4}$$

$+$ _____ $+\ 13\frac{3}{5}$

_____   _____

First estimate the answer. Then subtract to find the exact answer. Simplify all answers.
See Examples 1 and 3.

15. *Estimate:*   *Exact:*

$$14\frac{7}{8}$$

$-$ _____ $-\ 12\frac{1}{4}$

_____   _____

16. *Estimate:*   *Exact:*

$$14\frac{3}{4}$$

$-$ _____ $-\ 11\frac{3}{8}$

_____   _____

17. *Estimate:*   *Exact:*

$$12\frac{2}{3}$$

$-$ _____ $-\ 1\frac{1}{5}$

_____   _____

18. *Estimate:*   *Exact:*

$$11\frac{9}{20}$$

$-$ _____ $-\ 4\frac{3}{5}$

_____   _____

19. *Estimate:*   *Exact:*

$$28\frac{3}{10}$$

$-$ _____ $-\ 6\frac{1}{15}$

_____   _____

20. *Estimate:*   *Exact:*

$$15\frac{7}{20}$$

$-$ _____ $-\ 6\frac{1}{8}$

_____   _____

21. *Estimate:*   *Exact:*

$$17$$

$-$ _____ $-\ 6\frac{5}{8}$

_____   _____

22. *Estimate:*   *Exact:*

$$22$$

$-$ _____ $-\ 4\frac{5}{8}$

_____   _____

23. *Estimate:*   *Exact:*

$$18\frac{3}{4}$$

$$-\ 5\frac{4}{5}$$

$-$ _____   _____

24. *Estimate:*   *Exact:*

$$14\frac{5}{8}$$

$$-\underline{\quad\quad} \quad\quad -\ 3\frac{2}{3}$$

◐ 25. *Estimate:*   *Exact:*

$$19\frac{2}{3}$$

$$-\underline{\quad\quad} \quad\quad -\ 11\frac{3}{4}$$

26. *Estimate:*   *Exact:*

$$20\frac{3}{5}$$

$$-\underline{\quad\quad} \quad\quad -\ 12\frac{7}{15}$$

Add or subtract by changing mixed numbers to improper fractions. Write answers as mixed numbers when possible. See Example 4.

27. $7\frac{5}{8}$

$+\ 1\frac{1}{2}$

28. $8\frac{3}{4}$

$+\ 1\frac{5}{8}$

◐ 29. $4\frac{2}{3}$

$+\ 6\frac{5}{6}$

30. $7\frac{5}{12}$

$+\ 6\frac{2}{3}$

31. $2\frac{2}{3}$

$+\ 1\frac{1}{6}$

32. $4\frac{1}{2}$

$+\ 2\frac{3}{4}$

33. $3\frac{1}{4}$

$+\ 3\frac{2}{3}$

34. $2\frac{4}{5}$

$+\ 5\frac{1}{3}$

35. $1\frac{3}{8}$

$+\ 6\frac{3}{4}$

36. $1\frac{5}{12}$

$+\ 1\frac{7}{8}$

37. $3\frac{1}{2}$

$-\ 2\frac{2}{3}$

38. $4\frac{1}{4}$

$-\ 3\frac{7}{12}$

39. $8\dfrac{3}{4}$
$-\,5\dfrac{7}{8}$

40. $12\dfrac{2}{3}$
$-\,7\dfrac{11}{12}$

41. $7\dfrac{1}{4}$
$-\,4\dfrac{2}{3}$

42. $4\dfrac{1}{10}$
$-\,3\dfrac{7}{8}$

43. $9\dfrac{1}{5}$
$-\,3\dfrac{3}{4}$

44. $10\dfrac{2}{7}$
$-\,5\dfrac{5}{14}$

45. $6\dfrac{3}{7}$
$-\,2\dfrac{2}{3}$

46. $8\dfrac{2}{15}$
$-\,6\dfrac{1}{2}$

47. In your own words, explain the steps you would take to add two large mixed numbers.

48. When subtracting mixed numbers, explain when you need to regroup. Explain how to regroup using your own example.

First estimate the answer. Then solve each application problem.

49. At the beginning of this chapter you read about Bryan Berg, who builds houses of cards. While in high school, Bryan built a house of cards $14\frac{1}{2}$ ft tall, setting his first world record. Today his current world record is $25\frac{3}{4}$ ft tall. How much taller is his current world record than his first world record?
(*Source: Guinness World Records.*)

Estimate:

Exact:

50. The heaviest marine mammal in the world is the blue whale at $143\frac{3}{10}$ tons. The sixth heaviest is the humpback whale at $29\frac{1}{5}$ tons. Find the difference in their weights. (*Source: Top 10 of Everything.*)

Estimate:

Exact:

Use the newspaper advertisement for this 6-piece jumbo wrench set to answer Exercises 51–54. The symbol " is for inches. (*Source:* Harbor Freight Tools.)

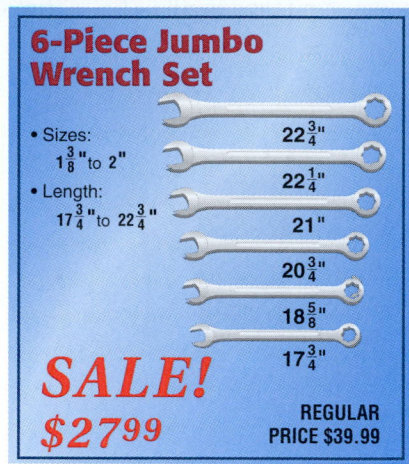

6-Piece Jumbo Wrench Set

- Sizes: $1\frac{3}{8}$" to 2"
- Length: $17\frac{3}{4}$" to $22\frac{3}{4}$"

$22\frac{3}{4}$"

$22\frac{1}{4}$"

21"

$20\frac{3}{4}$"

$18\frac{5}{8}$"

$17\frac{3}{4}$"

SALE!
$27⁹⁹

REGULAR PRICE $39.99

51. How much longer is the longest wrench than the second-to-shortest wrench?

Estimate:

Exact:

52. Find the difference in length between the shortest wrench and the second-to-longest wrench.

Estimate:

Exact:

53. What is the total length of the three longest wrenches?

Estimate:

Exact:

54. What is the total length of the three shortest wrenches?

Estimate:

Exact:

Storehouse 34 PC. GEAR HOSE CLAMP ASSORTMENT

LOT NO. 1420

Includes: two 2-3/4", two 2-1/2", four 2-1/4", four 2", six 1-3/4", six 1-1/2", six 1-1/4", four 9/16" through 1-1/16"

SALE!
$5⁹⁷

SAVE 40%

REGULAR PRICE $9.99

A mechanic buys the 34-piece hose clamp assortment shown at the left. Use the advertisement to answer Exercises 55–56. The symbol " is for inches. (*Source:* Harbor Freight Tools.)

55. Find the difference in size between the largest hose clamp and the smallest hose clamp.

Estimate:

Exact:

56. Find the difference in size between the second to largest hose clamp and the smallest hose clamp.

Estimate:

Exact:

57. The four sides of Andre Herrebout's vegetable garden are $15\frac{1}{2}$ feet, $18\frac{3}{4}$ feet, $24\frac{1}{4}$ feet, and $30\frac{1}{2}$ feet. How many feet of fencing are needed to go around the garden?

Estimate:

Exact:

58. On a recent vacation to Canada, Erin Gavin drove for $7\frac{3}{4}$ hours on the first day, $5\frac{1}{4}$ hours on the second day, $6\frac{1}{2}$ hours on the third day, and 9 hours on the fourth day. How many hours did she drive altogether?

Estimate:

Exact:

59. A craftsperson must attach a lead strip around all four sides of a stained glass window before it is installed. Find the length of lead stripping needed.

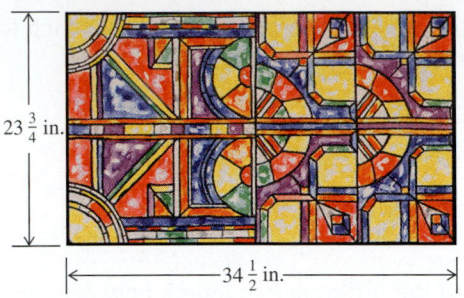

$23\frac{3}{4}$ in.

$34\frac{1}{2}$ in.

Estimate:

Exact:

60. To complete a custom order, Zak Morten of Home Depot must find the number of inches of brass trim needed to go around the four sides of the lamp base plate shown. Find the length of brass trim needed.

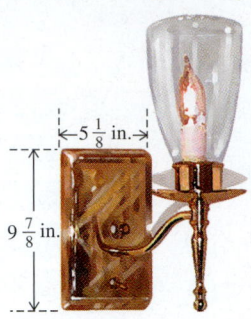

$5\frac{1}{8}$ in.

$9\frac{7}{8}$ in.

Estimate:

Exact:

61. A museum humidifier contains 100 gallons of water. The system uses $10\frac{1}{4}$ gallons of water on Monday, $13\frac{1}{2}$ gallons on Tuesday, $8\frac{7}{8}$ gallons on Wednesday, $18\frac{3}{4}$ gallons on Thursday, $12\frac{3}{8}$ gallons on Friday, $9\frac{1}{2}$ gallons on Saturday, and $14\frac{1}{8}$ gallons on Sunday. Find the total number of gallons of water remaining.

Estimate:

Exact:

62. Marv Levinson bought 15 yards of Italian silk fabric. He made two shirts, needing a total of $3\frac{3}{4}$ yards of the material, a suit for his wife with $4\frac{1}{8}$ yards, and a jacket with $3\frac{7}{8}$ yards. Find the number of yards of material remaining.

Estimate:

Exact:

63. The exercise yard at the correction center has four sides and is surrounded by $527\frac{1}{24}$ ft of security fencing. If three sides of the yard measure $107\frac{2}{3}$ ft, $150\frac{3}{4}$ ft, and $138\frac{5}{8}$ ft, find the length of the fourth side.

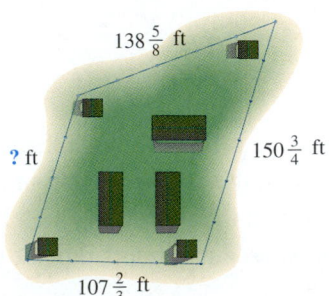

$138\frac{5}{8}$ ft

? ft

$150\frac{3}{4}$ ft

$107\frac{2}{3}$ ft

Estimate:
Exact:

64. Three sides of a parking lot are $108\frac{1}{4}$ ft, $162\frac{3}{8}$ ft, and $143\frac{1}{2}$ ft. If the distance around the lot is $518\frac{3}{4}$ ft, find the length of the fourth side.

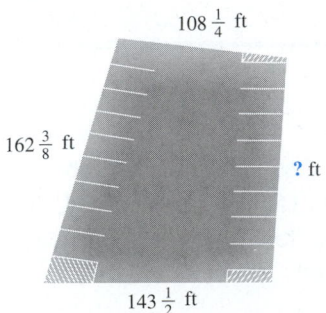

$108\frac{1}{4}$ ft

$162\frac{3}{8}$ ft

? ft

$143\frac{1}{2}$ ft

Estimate:

Exact:

65. A freight car is loaded with Morton Salt products consisting of $58\frac{1}{2}$ tons of coarse rock salt, $23\frac{5}{8}$ tons of medium rock salt, $16\frac{5}{6}$ tons of table salt, and $29\frac{1}{4}$ tons of animal salt lick blocks. If the weight of the freight car is $58\frac{1}{3}$ tons, find the weight of the loaded freight car.

Estimate:

Exact:

66. Bryan Berg, from the opening page of this chapter, built houses of cards reaching heights of $14\frac{1}{2}$ ft, $19\frac{3}{8}$ ft, $23\frac{5}{12}$ ft, and $25\frac{3}{4}$ ft (his current world record). Find the total height of these four houses of cards. (*Source: Reader's Digest.*)

Estimate:

Exact:

Find the unknown length, labeled with a question mark, in each figure.

67.

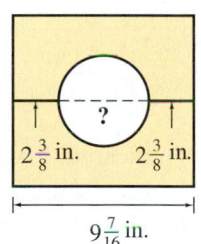

$2\frac{3}{8}$ in. ? $2\frac{3}{8}$ in.

$9\frac{7}{16}$ in.

68.

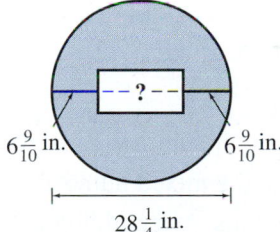

?

$6\frac{9}{10}$ in. $6\frac{9}{10}$ in.

$28\frac{1}{4}$ in.

69.

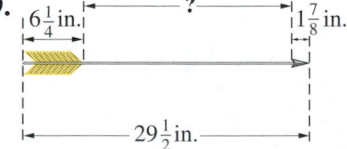

$6\frac{1}{4}$ in.   ? $1\frac{7}{8}$ in.

$29\frac{1}{2}$ in.

70.

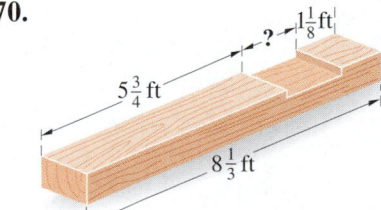

$1\frac{1}{8}$ ft

?

$5\frac{3}{4}$ ft

$8\frac{1}{3}$ ft

Relating Concepts (Exercises 71–76) For Individual or Group Work

Most fraction problems include fractions with different denominators.
Work Exercises 71–76 in order.

71. To add or subtract fractions, we must first rewrite them as like fractions. Rewrite each fraction with the indicated denominator.

(a) $\dfrac{5}{9} = \dfrac{}{54}$ (b) $\dfrac{7}{12} = \dfrac{}{48}$

(c) $\dfrac{5}{8} = \dfrac{}{40}$ (d) $\dfrac{11}{5} = \dfrac{}{120}$

72. When rewriting unlike fractions as like fractions with the least common multiple as a denominator, the new denominator is called the _____ _____ _____, or LCD.

73. Add or subtract as indicated. Write answers in lowest terms.

(a) $\dfrac{5}{8} + \dfrac{1}{3}$ (b) $\dfrac{19}{20} - \dfrac{5}{12}$

(c)
$$\begin{aligned}&\dfrac{7}{12}\\&\dfrac{3}{16}\\+&\dfrac{3}{24}\\\hline\end{aligned}$$

(d)
$$\begin{aligned}&\dfrac{6}{7}\\-&\dfrac{2}{3}\\\hline\end{aligned}$$

74. A common method for adding or subtracting mixed numbers is to add or subtract the _____ _____ and then add or subtract the whole number parts

75. Another method for adding or subtracting mixed numbers is to first change the mixed numbers to _____ fractions. After adding or subtracting, write the answer in lowest terms and as a mixed number when possible. This method is difficult to use if the mixed numbers are _____.
(large/small)

76. Add or subtract these fractions as indicated. First use the method where you add or subtract fraction parts and then whole number parts. Then use the method where you change each mixed number to an improper fraction before adding or subtracting. Do you get the same answer using both methods? Which method do you prefer?

(a)
$$\begin{aligned}&4\tfrac{5}{8}\\+&3\tfrac{3}{4}\\\hline\end{aligned}$$

(b)
$$\begin{aligned}&12\tfrac{2}{5}\\-&8\tfrac{7}{8}\\\hline\end{aligned}$$

3.5 ▶▶▶ Order Relations and the Order of Operations

There are times when we want to compare the size of two numbers. For example, we might want to know which is the greater amount, the larger size, or the longer distance.

Fractions, like whole numbers, can be graphed on a number line. For fractions, divide the space between whole numbers into equal parts.

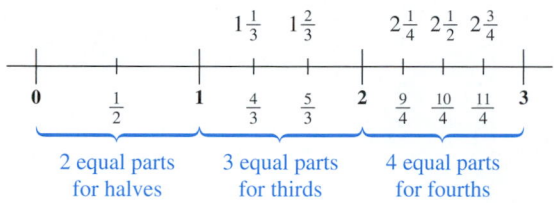

2 equal parts for halves   3 equal parts for thirds   4 equal parts for fourths

OBJECTIVES

1. **Identify the greater of two fractions.**
2. **Use exponents with fractions.**
3. **Use the order of operations with fractions.**

OBJECTIVE **1** **Identify the greater of two fractions.** To compare the size of two numbers, place the two numbers on a number line and use the following rule.

> **Comparing the Size of Two Numbers**
> The number farther to the *left* on the number line is always *less*, and the number farther to the *right* on the number line is always *greater*.

For example, on the number line above, $\frac{1}{2}$ is to the *left* of $\frac{4}{3}$, so $\frac{1}{2}$ is *less than* $\frac{4}{3}$.

Work Problem **1** *at the Side.* ▶

Write *order relations* using the symbols shown below.

> **Symbols Used to Show Order Relations**
> $<$ is less than $>$ is greater than

EXAMPLE 1 **Using Less-Than and Greater-Than Symbols**

Rewrite the following using $<$ and $>$ symbols.

(a) $\frac{1}{2}$ is less than $\frac{4}{3}$

$\frac{1}{2}$ **is less than** $\frac{4}{3}$ is written as $\frac{1}{2} < \frac{4}{3}$ ◀ The number farther to the *left* on the number line is *less*.

(b) $\frac{9}{4}$ is greater than 1

$\frac{9}{4}$ **is greater than** 1 is written as $\frac{9}{4} > 1$ ◀ The number farther to the *right* on the number line is *greater*.

(c) $\frac{5}{3}$ is less than $\frac{11}{4}$

$\frac{5}{3}$ **is less than** $\frac{11}{4}$ is written as $\frac{5}{3} < \frac{11}{4}$

— **Continued on Next Page**

1 Locate each fraction on the number line.

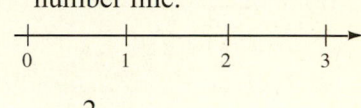

(a) $\frac{2}{3}$

(b) $1\frac{1}{2}$

(c) $2\frac{3}{4}$

ANSWER

1.

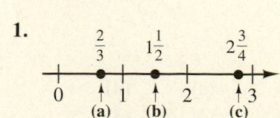

2 Use the number line on the previous page to help you write < or > in each blank to make a true statement.

(a) 1 _____ $\dfrac{5}{4}$

(b) $\dfrac{8}{3}$ _____ $\dfrac{3}{2}$

(c) 0 _____ 1

(d) $\dfrac{17}{8}$ _____ $\dfrac{8}{4}$

3 Write < or > in each blank to make a true statement.

(a) $\dfrac{7}{8}$ _____ $\dfrac{3}{4}$

(b) $\dfrac{13}{8}$ _____ $\dfrac{15}{9}$

(c) $\dfrac{9}{4}$ _____ $\dfrac{7}{3}$

(d) $\dfrac{9}{10}$ _____ $\dfrac{14}{15}$

ANSWERS

2. (a) < (b) > (c) < (d) >
3. (a) > (b) < (c) < (d) <

> **Note**
> A number line is a very useful tool when working with order relations.

◀ Work Problem **2** at the Side.

The fraction $\frac{7}{8}$ represents 7 of 8 equivalent parts, while $\frac{3}{8}$ means 3 of 8 equivalent parts. Because $\frac{7}{8}$ represents more of the equivalent parts, $\frac{7}{8}$ is greater than $\frac{3}{8}$.

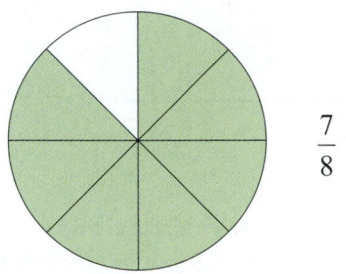 $\dfrac{7}{8} > \dfrac{3}{8}$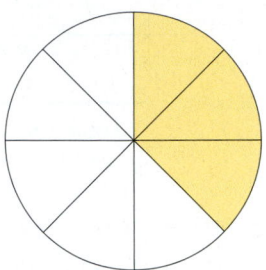

To identify the greater fraction, use the following steps.

> **Identifying the Greater Fraction**
>
> **Step 1** Write the fractions as like fractions (same denominators).
>
> **Step 2** Compare the numerators. The fraction with the greater numerator is the greater fraction.

EXAMPLE 2 **Identifying the Greater Fraction**

Determine which fraction in each pair is greater.

(a) $\dfrac{7}{8}, \dfrac{9}{10}$

First, write the fractions as like fractions. The least common multiple for 8 and 10 is 40.

$$\dfrac{7}{8} = \dfrac{7 \cdot \mathbf{5}}{8 \cdot \mathbf{5}} = \dfrac{35}{40} \quad \text{and} \quad \dfrac{9}{10} = \dfrac{9 \cdot \mathbf{4}}{10 \cdot \mathbf{4}} = \dfrac{36}{40}$$

> Rewrite both fractions with 40 as the denominator.

Look at the numerators. Because 36 is greater than 35, $\frac{36}{40}$ is greater than $\frac{35}{40}$. Then, because $\frac{36}{40}$ is equivalent to $\frac{9}{10}$,

$$\dfrac{9}{10} > \dfrac{7}{8} \quad \text{or} \quad \dfrac{7}{8} < \dfrac{9}{10}$$

The greater fraction is $\frac{9}{10}$.

(b) $\dfrac{8}{5}, \dfrac{23}{15}$

The least common multiple of 5 and 15 is 15.

$$\dfrac{8}{5} = \dfrac{8 \cdot \mathbf{3}}{5 \cdot \mathbf{3}} = \dfrac{24}{15} \quad \text{and} \quad \dfrac{23}{15} = \dfrac{23}{15}$$

This shows that $\frac{8}{5}$ is greater than $\frac{23}{15}$, or

$$\dfrac{8}{5} > \dfrac{23}{15}$$

◀ Work Problem **3** at the Side.

OBJECTIVE 2 Use exponents with fractions. Exponents were used in **Section 1.8** to write repeated multiplication. For example,

$$3^2 = \underbrace{3 \cdot 3}_{\substack{\text{Two} \\ \text{factors of 3}}} = 9 \quad \text{and} \quad 5^3 = \underbrace{5 \cdot 5 \cdot 5}_{\substack{\text{Three} \\ \text{factors of 5}}} = 125$$

The next example shows exponents used with fractions.

EXAMPLE 3 Using Exponents with Fractions

Simplify.

(a) $\left(\dfrac{1}{2}\right)^3$

$$\left(\dfrac{1}{2}\right)^3 = \overbrace{\dfrac{1}{2} \cdot \dfrac{1}{2} \cdot \dfrac{1}{2}}^{\text{Three factors of } \frac{1}{2}} = \dfrac{1}{8}$$

(b) $\left(\dfrac{5}{8}\right)^2$

$$\left(\dfrac{5}{8}\right)^2 = \overbrace{\dfrac{5}{8} \cdot \dfrac{5}{8}}^{\text{Two factors of } \frac{5}{8}} = \dfrac{25}{64}$$

(c) $\left(\dfrac{3}{4}\right)^2 \cdot \left(\dfrac{2}{3}\right)^3$

$$\left(\dfrac{3}{4}\right)^2 \cdot \left(\dfrac{2}{3}\right)^3 = \left(\dfrac{3}{4} \cdot \dfrac{3}{4}\right) \cdot \left(\dfrac{2}{3} \cdot \dfrac{2}{3} \cdot \dfrac{2}{3}\right)$$

$$= \dfrac{1 \cdot 1 \cdot 1 \cdot 1 \cdot 1}{4 \cdot 4 \cdot 3 \cdot 3 \cdot 3} \quad \text{Divide out all the common factors}$$

$$= \dfrac{1}{6} \quad \boxed{\text{The fraction is in lowest terms after all common factors are divided out.}}$$

Work Problem **4** *at the Side.* ▶

OBJECTIVE 3 Use the order of operations with fractions. Recall the *order of operations* from **Section 1.8.**

Order of Operations
1. Do all operations inside *parentheses or other grouping symbols.*
2. Simplify any expressions with *exponents* and find any *square roots.*
3. *Multiply* or *divide,* proceeding from left to right.
4. *Add* or *subtract,* proceeding from left to right.

4 Simplify.

(a) $\left(\dfrac{1}{2}\right)^4$

(b) $\left(\dfrac{3}{4}\right)^2$

(c) $\left(\dfrac{1}{2}\right)^3 \cdot \left(\dfrac{2}{3}\right)^2$

(d) $\left(\dfrac{1}{5}\right)^2 \cdot \left(\dfrac{5}{3}\right)^2$

The next example shows how to apply the order of operations with fractions.

5 Simplify by using the order of operations.

(a) $\dfrac{5}{9} - \dfrac{3}{4}\left(\dfrac{2}{3}\right)$

EXAMPLE 4 **Using the Order of Operations with Fractions**

Simplify by using the order of operations.

(a) $\dfrac{1}{3} + \dfrac{1}{2}\left(\dfrac{4}{5}\right)$

Multiply $\frac{1}{2}\left(\frac{4}{5}\right)$ first because multiplication and division are done *before* adding.

Do **not** add $\frac{1}{3} + \frac{1}{2}$ as the first step.

$$\frac{1}{3} + \frac{1}{\overset{}{\underset{1}{2}}}\left(\frac{\overset{2}{4}}{5}\right) = \frac{1}{3} + \frac{2}{5}$$

Next, add. The least common denominator of 3 and 5 is 15.

$$\frac{1}{3} + \frac{2}{5} = \frac{5}{15} + \frac{6}{15} = \frac{11}{15}$$

(b) $\dfrac{3}{4}\left(\dfrac{2}{3} \cdot \dfrac{3}{5}\right)$

(b) $\dfrac{3}{8}\left(\dfrac{1}{2} + \dfrac{1}{3}\right)$

$$\frac{3}{8}\left(\frac{1}{2} + \frac{1}{3}\right) = \frac{3}{8}\underbrace{\left(\frac{3}{6} + \frac{2}{6}\right)}$$

Work inside parentheses first.

$$= \frac{3}{8}\left(\frac{5}{6}\right)$$

$$= \frac{\overset{1}{\cancel{3}}}{8}\left(\frac{5}{\underset{2}{\cancel{6}}}\right)$$ Divide numerator and denominator by 3.

(c) $\dfrac{7}{8}\left(\dfrac{2}{3}\right) - \left(\dfrac{1}{2}\right)^2$

$$= \frac{5}{16}$$ Multiply.

(c) $\left(\dfrac{2}{3}\right)^2 - \dfrac{4}{5}\left(\dfrac{1}{2}\right)$

Simplify the expression with the exponent. $\frac{2}{3} \cdot \frac{2}{3}$ is $\frac{4}{9}$.

No work inside parentheses, so operations with exponents are next.

$$\left(\frac{2}{3}\right)^2 - \frac{4}{5}\left(\frac{1}{2}\right) = \frac{4}{9} - \frac{4}{5}\left(\frac{1}{2}\right)$$

(d) $\dfrac{\left(\dfrac{5}{6}\right)^2}{\dfrac{4}{3}}$

$$= \frac{4}{9} - \frac{\overset{2}{\cancel{4}}}{5}\left(\frac{1}{\underset{1}{\cancel{2}}}\right)$$ Now, multiply.

$$= \frac{4}{9} - \frac{2}{5}$$

$$= \frac{20}{45} - \frac{18}{45}$$ Subtract last. (Least common denominator is 45.)

$$= \frac{2}{45}$$

◄ *Work Problem* **5** *at the Side.*

3.5 ▶▶▶ Exercises

Locate each fraction in Exercises 1–12 on the following number line. See Margin Problem 1.

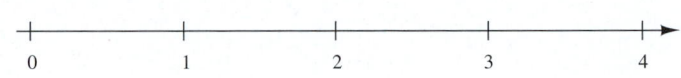

1. $\dfrac{1}{2}$ 2. $\dfrac{1}{4}$ 3. $\dfrac{3}{2}$ 4. $\dfrac{5}{4}$ 5. $\dfrac{7}{3}$ 6. $\dfrac{11}{4}$

7. $2\dfrac{1}{6}$ 8. $3\dfrac{4}{5}$ 9. $\dfrac{7}{2}$ 10. $\dfrac{7}{8}$ 11. $3\dfrac{1}{4}$ 12. $1\dfrac{7}{8}$

Write < or > to make a true statement. See Examples 1 and 2.

13. $\dfrac{1}{2}$ —— $\dfrac{3}{8}$ 14. $\dfrac{5}{8}$ —— $\dfrac{3}{4}$ 15. $\dfrac{5}{6}$ —— $\dfrac{11}{12}$ 16. $\dfrac{13}{18}$ —— $\dfrac{5}{6}$

17. $\dfrac{5}{12}$ —— $\dfrac{3}{8}$ 18. $\dfrac{7}{15}$ —— $\dfrac{9}{20}$ 19. $\dfrac{7}{12}$ —— $\dfrac{11}{18}$ 20. $\dfrac{17}{24}$ —— $\dfrac{5}{6}$

21. $\dfrac{11}{18}$ —— $\dfrac{5}{9}$ 22. $\dfrac{13}{15}$ —— $\dfrac{8}{9}$ 23. $\dfrac{37}{50}$ —— $\dfrac{13}{20}$ 24. $\dfrac{7}{12}$ —— $\dfrac{11}{20}$

Simplify. See Example 3.

25. $\left(\dfrac{1}{3}\right)^2$ 26. $\left(\dfrac{2}{3}\right)^2$ 27. $\left(\dfrac{5}{8}\right)^2$

28. $\left(\dfrac{7}{8}\right)^2$ 29. $\left(\dfrac{3}{4}\right)^2$ 30. $\left(\dfrac{3}{5}\right)^3$

31. $\left(\dfrac{4}{5}\right)^3$ 32. $\left(\dfrac{4}{7}\right)^3$ 33. $\left(\dfrac{3}{2}\right)^4$

34. $\left(\dfrac{4}{3}\right)^4$ 35. $\left(\dfrac{3}{4}\right)^4$ 36. $\left(\dfrac{2}{3}\right)^5$

37. Describe in your own words what a number line is, and draw a picture of one. Be sure to include how it works and how it can be used.

38. You have used the order of operations with whole numbers and again with fractions. List from memory the steps in the order of operations.

Use the order of operations to simplify each expression. See Example 4.

39. $2^4 - 4\,(3)$

40. $3^2 + 4\,(1)$

41. $3 \cdot 2^2 - \dfrac{6}{3}$

42. $5 \cdot 2^3 - \dfrac{6}{2}$

43. $\left(\dfrac{1}{2}\right)^2 \cdot 4$

44. $\left(\dfrac{1}{4}\right)^2 \cdot 4$

45. $\left(\dfrac{3}{4}\right)^2 \cdot \left(\dfrac{1}{3}\right)$

46. $\left(\dfrac{2}{3}\right)^3 \cdot \left(\dfrac{1}{2}\right)$

47. $\left(\dfrac{4}{5}\right)^2 \cdot \left(\dfrac{5}{6}\right)^2$

48. $\left(\dfrac{5}{8}\right)^2 \cdot \left(\dfrac{4}{25}\right)^2$

49. $6\left(\dfrac{2}{3}\right)^2 \left(\dfrac{1}{2}\right)^3$

50. $9\left(\dfrac{1}{3}\right)^3 \left(\dfrac{4}{3}\right)^2$

51. $\dfrac{3}{5}\left(\dfrac{1}{3}\right) + \dfrac{2}{5}\left(\dfrac{3}{4}\right)$

52. $\dfrac{1}{4}\left(\dfrac{3}{4}\right) + \dfrac{3}{8}\left(\dfrac{4}{3}\right)$

53. $\dfrac{1}{2} + \left(\dfrac{1}{2}\right)^2 - \dfrac{3}{8}$

54. $\dfrac{2}{3} + \left(\dfrac{1}{3}\right)^2 - \dfrac{5}{9}$

55. $\left(\dfrac{1}{3} + \dfrac{1}{6}\right) \cdot \dfrac{1}{2}$

56. $\left(\dfrac{3}{5} - \dfrac{3}{20}\right) \cdot \dfrac{4}{3}$

57. $\dfrac{9}{8} \div \left(\dfrac{2}{3} + \dfrac{1}{12}\right)$

58. $\dfrac{6}{5} \div \left(\dfrac{3}{5} - \dfrac{3}{10}\right)$

59. $\left(\dfrac{7}{8} - \dfrac{3}{4}\right) \div \dfrac{3}{2}$

60. $\left(\dfrac{4}{5} - \dfrac{3}{10}\right) \div \dfrac{4}{5}$

61. $\dfrac{3}{8}\left(\dfrac{1}{4} + \dfrac{1}{2}\right) \cdot \dfrac{32}{3}$

62. $\dfrac{1}{3}\left(\dfrac{4}{5} - \dfrac{3}{10}\right) \cdot \dfrac{4}{2}$

63. $\left(\dfrac{3}{4}\right)^2 - \left(\dfrac{1}{2} - \dfrac{1}{6}\right) \div \dfrac{4}{3}$

64. $\left(\dfrac{2}{3}\right)^2 - \left(\dfrac{5}{8} - \dfrac{1}{2}\right) \div \dfrac{3}{2}$

65. $\left(\dfrac{7}{8} - \dfrac{1}{4}\right) - \dfrac{2}{3}\left(\dfrac{3}{4}\right)^2$

66. $\left(\dfrac{5}{6} - \dfrac{7}{12}\right) - \dfrac{3}{4}\left(\dfrac{1}{3}\right)^2$

67. $\left(\dfrac{3}{4}\right)^2\left(\dfrac{2}{3} - \dfrac{5}{9}\right) - \dfrac{1}{4}\left(\dfrac{1}{8}\right)$

68. $\left(\dfrac{2}{3}\right)^2\left(\dfrac{1}{2} - \dfrac{1}{8}\right) - \dfrac{2}{3}\left(\dfrac{1}{8}\right)$

Solve each application problem.

69. The population of Las Vegas, Nevada has had an increase of $\frac{4}{25}$ since the turn of the century. During this same period, the population in Atlanta, Georgia has had an increase of $\frac{5}{30}$. Which city has had a higher rate of population growth? (*Source:* Census Bureau estimates.)

70. Two of the cities with the most expensive average home prices are Newport Beach, California ($1\frac{5}{8}$ million) and Santa Barbara, California ($1\frac{21}{32}$ million). Which city has the highest average home price? (*Source:* Coldwell Banker Home Price Comparison index.)

Relating Concepts (Exercises 71–80) For Individual or Group Work

You often need to use order relations and the order of operations when solving problems.
Work Exercises 71–80 in order.

71. When comparing the size of two numbers, the symbol ―― means **is less than** and the symbol ―― means **is greater than.**

72. (a) To identify the greater of two or more fractions, we must first write the fractions as _____ fractions and then compare the _____. The fraction with the greater _____ is the greater fraction.

(b) Write four pairs of fractions, all with different denominators. Write the symbol for **less than** or for **greater than** between each pair.

73. Fill in the blanks to complete the order of operations.

1. Do all operations inside _____ or other grouping symbols.

2. Simplify any expressions with _____ and find any _____ roots.

3. _____ or _____ proceeding from left to right.

4. _____ or _____ proceeding from left to right.

74. Use the order of operations to simplify the following.

$$\left(\frac{2}{3}\right)^2 - \left(\frac{4}{5} - \frac{3}{10}\right) \div \frac{5}{4}$$

Simplify, then place the results on the number line.

75. $\left(\dfrac{2}{3}\right)^2$

76. $\left(\dfrac{3}{2}\right)^2$

77. $\left(\dfrac{3}{5}\right)^3$

78. $\left(\dfrac{5}{4}\right)^2$

79. $4 + 2 - 2^2$

80. $\left(\dfrac{5}{8}\right)^2 + \left(\dfrac{7}{8} - \dfrac{1}{4}\right) \div \dfrac{1}{4}$

Chapter 3 ▷▷▷ Summary

▶ Key Terms

3.1	**like fractions**	Fractions with the same denominator are called *like fractions*.
	unlike fractions	Fractions with different denominators are called *unlike fractions*.
3.2	**least common multiple**	Given two or more whole numbers, the least common multiple is the smallest whole number that is divisible by all the numbers.
	LCM	The abbreviation for *least common multiple* is LCM.
3.3	**least common denominator**	When unlike fractions are rewritten as like fractions having the least common multiple as the denominator, the new denominator is the least common denominator.
	LCD	The abbreviation for *least common denominator* is LCD.
3.4	**regrouping when adding fractions**	Regrouping is used in the addition of mixed numbers when the sum of the fractions is greater than 1.
	regrouping when subtracting fractions	Regrouping is used in the subtraction of mixed numbers when the fraction part of the minuend is less than the fraction part of the subtrahend.

▶ New Symbols

$<$ is less than ($2 < 5$) $>$ is greater than ($4 > 2$)

▶ Test Your Word Power

See how well you have learned the vocabulary in this chapter. Answers follow the Quick Review.

1. **Like fractions** are
   A. fractions that are equivalent
   B. fractions that are not equivalent
   C. fractions that have the same numerator
   D. fractions that have the same denominator.

2. Two or more fractions are **unlike fractions** if
   A. they are not equivalent
   B. they have different numerators
   C. they have different denominators
   D. they are improper fractions.

3. The abbreviation **LCM** stands for
   A. the largest common multiple
   B. the longest common multiple
   C. the most likely common multiplier
   D. the least common multiple.

4. The **least common multiple** is
   A. the smallest whole number that is divisible by each of two or more numbers
   B. the smallest numerator
   C. the smallest denominator
   D. the smallest whole number that is not divisible by a group of numbers.

5. The abbreviation **LCD** stands for
   A. the largest common denominator
   B. the least common denominator
   C. the least common divisor
   D. the most likely common denominator.

6. The **least common denominator** is
   A. needed when multiplying fractions
   B. needed when dividing fractions
   C. the least common multiple of the denominators in a fraction problem
   D. any denominator that is common to a group of fractions.

▶ Quick Review

Concepts	Examples

3.1 Adding Like Fractions

Add numerators and keep the same denominator. Write the result in lowest terms.

$$\frac{3}{4} + \frac{1}{4} + \frac{5}{4} = \frac{3+1+5}{4} = \frac{9}{4} = 2\frac{1}{4}$$

3.1 Subtracting Like Fractions

Subtract numerators and keep the same denominator. Write the result in lowest terms.

$$\frac{7}{8} - \frac{5}{8} = \frac{7-5}{8} = \frac{2}{8} = \frac{2 \div 2}{8 \div 2} = \frac{1}{4}$$

3.2 Finding the Least Common Multiple (LCM)

Method of using multiples of the larger number: List the first few multiples of the larger number. Check each one until you find the multiple that is divisible by the smaller number.

$$\frac{1}{3} + \frac{1}{4} \qquad 4, 8, 12, 16, \ldots \longleftarrow \text{Multiples of 4}$$

First multiple divisible by 3 ($12 \div 3 = 4$)

The least common multiple (LCM) of 3 and 4 is 12.

3.2 Finding the Least Common Multiple (LCM)

Method of prime numbers: First find the prime factorization of each number. Then use the prime factors to build the least common multiple.

Factors of 9

$$9 = 3 \cdot 3$$
$$\text{LCM} = 3 \cdot 3 \cdot 5 = 45$$
$$15 = 3 \cdot 5$$

Factors of 15

The least common multiple (LCM) of 9 and 15 is 45.

3.3 Adding Unlike Fractions

Step 1 Find the least common multiple (LCM).
Step 2 Rewrite the fractions with the least common multiple as the denominator.
Step 3 Add the numerators, placing the sum over the common denominator, and simplify the answer.

$$\frac{1}{3} + \frac{1}{4} + \frac{1}{10} \qquad \text{LCM} = 60$$

$$\frac{1}{3} = \frac{20}{60} \quad \frac{1}{4} = \frac{15}{60} \quad \frac{1}{10} = \frac{6}{60}$$

$$\frac{20}{60} + \frac{15}{60} + \frac{6}{60} = \frac{41}{60} \qquad \text{Lowest terms}$$

3.3 Subtracting Unlike Fractions

Step 1 Find the least common multiple (LCM).
Step 2 Rewrite the fractions with the least common multiple as the denominator.
Step 3 Subtract the numerators, place the difference over the common denominator, and simplify the answer.

$$\frac{5}{8} - \frac{1}{3} \qquad \text{LCM} = 24$$

$$\frac{5}{8} = \frac{15}{24} \quad \frac{1}{3} = \frac{8}{24}$$

$$\frac{15}{24} - \frac{8}{24} = \frac{7}{24} \qquad \text{Lowest Terms}$$

Concepts	Examples

[3.4] Adding Mixed Numbers

Round the numbers and estimate the answer. Then find the exact answer using these steps.

Step 1 Add the fractions using a common denominator.

Step 2 Add the whole numbers.

Step 3 Combine the sums of the whole numbers and the fractions, simplifying the fraction part when necessary.

Compare the exact answer to the estimate to see if it is reasonable.

Estimate: *Exact:*

$$10 \xleftarrow{\text{Rounds to}} \begin{cases} 9\frac{2}{3} = 9\frac{8}{12} \end{cases}$$

$$+\,7 \xleftarrow{\text{Rounds to}} \begin{cases} +\,6\frac{3}{4} = 6\frac{9}{12} \end{cases}$$

$$17 \qquad\qquad 15\frac{17}{12} = 16\frac{5}{12}$$

The exact answer of $16\frac{5}{12}$ is reasonable because it is close to the estimate of 17.

[3.4] Subtracting Mixed Numbers

Round the numbers and estimate the answer. Then find the exact answer using these steps.

Step 1 Subtract the fractions, regrouping if necessary.

Step 2 Subtract the whole numbers.

Step 3 Combine the differences of the whole numbers and the fractions, simplifying the fraction part when necessary.

Compare the exact answer to the estimate to see if it is reasonable.

Estimate: *Exact:*

$$9 \xleftarrow{\text{Rounds to}} \begin{cases} 8\frac{5}{8} = 8\frac{15}{24} = 7\frac{39}{24} \end{cases}$$

$$-\,4 \xleftarrow{\text{Rounds to}} \begin{cases} -\,3\frac{11}{12} = 3\frac{22}{24} = 3\frac{22}{24} \end{cases}$$

$$5 \qquad\qquad 4\frac{17}{24}$$

The exact answer of $4\frac{17}{24}$ is reasonable because it is close to the estimate of 5.

[3.4] Adding or Subtracting Mixed Numbers Using an Alternate Method

Step 1 Change the mixed numbers to improper fractions.

Step 2 Rewrite the unlike fractions as like fractions.

Step 3 Add or subtract the numerators and simplify the answer.

Add.

$$2\frac{2}{3} = \frac{8}{3} = \frac{64}{24}$$

$$+\,1\frac{3}{8} = \frac{11}{8} = +\,\frac{33}{24}$$

24 is the least common denominator.

$$\frac{97}{24} = 4\frac{1}{24} \quad \text{Answer as mixed number}$$

Improper fractions

Subtract.

$$8\frac{2}{3} = \frac{26}{3} = \frac{104}{12}$$

$$-\,5\frac{3}{4} = \frac{23}{4} = -\,\frac{69}{12}$$

12 is the least common denominator.

$$\frac{35}{12} = 2\frac{11}{12} \quad \text{Answer as mixed number}$$

Improper fractions

Concepts	Examples

3.5 Identifying the Greater of Two Fractions

With unlike fractions, change to like fractions first. The fraction with the greater numerator is the greater fraction. Use these symbols:

$<$ is less than

$>$ is greater than

Identify the greater fraction.

$$\frac{7}{8}, \frac{9}{10}$$

$$\frac{7}{8} = \frac{7 \cdot 5}{8 \cdot 5} = \frac{35}{40}$$

$$\frac{9}{10} = \frac{9 \cdot 4}{10 \cdot 4} = \frac{36}{40}$$

$\frac{35}{40}$ is smaller than $\frac{36}{40}$, so $\frac{7}{8} < \frac{9}{10}$ or $\frac{9}{10} > \frac{7}{8}$.
$\frac{9}{10}$ is greater.

3.5 Using the Order of Operations with Fractions

Follow the order of operations.

1. Do all operations inside parentheses or other grouping symbols.
2. Simplify any expressions with exponents and find any square roots.
3. Multiply or divide, proceeding from left to right.
4. Add or subtract, proceeding from left to right.

Simplify by using the order of operations.

$$\frac{1}{2}\left(\frac{2}{3}\right) - \left(\frac{1}{4}\right)^2 \qquad \text{Simplify fraction with exponent.}$$

$$= \frac{1}{\overset{1}{\cancel{2}}}\left(\frac{\overset{1}{\cancel{2}}}{3}\right) - \frac{1}{16} \qquad \text{Next, multiply.}$$

$$= \quad \frac{1}{3} \quad - \frac{1}{16}$$

$$= \quad \frac{16}{48} \quad - \frac{3}{48} \qquad \begin{array}{l}\text{Change to common}\\ \text{denominator and subtract.}\end{array}$$

$$= \quad \frac{13}{48}$$

ANSWERS TO TEST YOUR WORD POWER

1. D; *Example:* Because the fractions $\frac{3}{8}$ and $\frac{10}{8}$ both have 8 as a denominator, they are like fractions.

2. C; *Example:* The fractions $\frac{2}{3}$ and $\frac{3}{4}$ are unlike fractions because they have different denominators.

3. D; *Example:* LCM is the abbreviation for least common multiple.

4. A; *Example:* The least common multiple of 4 and 5 is 20 because 20 is the smallest number into which 4 and 5 will divide evenly.

5. B; *Example:* LCD is the abbreviation for least common denominator.

6. C; *Example:* The least common denominator of the fractions $\frac{2}{3}$ and $\frac{1}{2}$ is 6 because 6 is the least common multiple of 3 and 2.

When written using the least common denominator, $\frac{2}{3}$ and $\frac{1}{2}$ become $\frac{4}{6}$ and $\frac{3}{6}$, respectively.

Chapter 3 ▶▶▶ Review Exercises

[3.1] *Add or subtract. Write answers in lowest terms.*

1. $\dfrac{5}{7} + \dfrac{1}{7}$

2. $\dfrac{4}{9} + \dfrac{3}{9}$

3. $\dfrac{1}{8} + \dfrac{3}{8} + \dfrac{2}{8}$

4. $\dfrac{5}{16} - \dfrac{3}{16}$

5. $\dfrac{5}{10} + \dfrac{3}{10}$

6. $\dfrac{5}{12} - \dfrac{3}{12}$

7. $\dfrac{36}{62} - \dfrac{10}{62}$

8. $\dfrac{68}{75} - \dfrac{43}{75}$

Solve each application problem. Write answers in lowest terms.

9. Jaime Villagranna earns $\frac{7}{12}$ of his income installing kitchen cabinets for the Home Depot and $\frac{4}{12}$ of his income by operating his own cabinet business. What fraction of his total income comes from the two activities?

10. The Koats for Kids committee members completed $\frac{5}{8}$ of their Web-page design in the morning and $\frac{3}{8}$ in the afternoon. How much less did they complete in the afternoon than in the morning?

[3.2] *Find the least common multiple of each set of numbers.*

11. 5, 2

12. 3, 4

13. 10, 12, 20

14. 3, 8, 4

15. 6, 8, 5, 15

16. 15, 9, 20

Rewrite each fraction using the indicated denominator.

17. $\dfrac{2}{3} = \dfrac{}{12}$

18. $\dfrac{3}{8} = \dfrac{}{56}$

19. $\dfrac{2}{5} = \dfrac{}{25}$

20. $\dfrac{5}{9} = \dfrac{}{81}$

21. $\dfrac{4}{5} = \dfrac{}{40}$

22. $\dfrac{5}{16} = \dfrac{}{64}$

[3.1–3.3] *Add or subtract. Write answers in lowest terms.*

23. $\dfrac{1}{2} + \dfrac{1}{3}$

24. $\dfrac{1}{5} + \dfrac{3}{10} + \dfrac{3}{8}$

25. $\begin{array}{r} \dfrac{5}{12} \\ + \dfrac{5}{24} \\ \hline \end{array}$

26. $\dfrac{2}{3} - \dfrac{1}{4}$

27. $\begin{array}{r} \dfrac{7}{8} \\ - \dfrac{1}{3} \\ \hline \end{array}$

28. $\begin{array}{r} \dfrac{11}{12} \\ - \dfrac{4}{9} \\ \hline \end{array}$

Solve each application problem.

29. The San Juan School District operates an after school program for students. This year $\frac{2}{5}$ of the students played after school sports, $\frac{1}{6}$ participated in arts and crafts, and $\frac{1}{3}$ spent their time in tutoring and study hall. What fraction of the total students participated in these activities?

30. Lynn Couch Catering serves food and beverages when and where you like it. She finds that $\frac{1}{3}$ of her business is for company events, $\frac{3}{8}$ for wedding parties, and $\frac{1}{4}$ for club and membership events. What portion of her business comes from these three categories?

[3.4] *First estimate the answer. Then add or subtract to find the exact answer. Write exact answers as mixed numbers.*

31. *Estimate:*   *Exact:*

<u>←Rounds to</u> $\left\{ 18\dfrac{5}{8} \right.$

$+$ <u>←Rounds to</u> $\left\{ +\ 13\dfrac{3}{4} \right.$

_____   _____

32. *Estimate:*   *Exact:*

$22\dfrac{2}{3}$

$+$ _____ $+\ 15\dfrac{4}{9}$

_____   _____

33. *Estimate:*   *Exact:*

$12\dfrac{3}{5}$

$8\dfrac{5}{8}$

$+$ _____ $+\ 10\dfrac{5}{16}$

_____   _____

34. *Estimate:*   *Exact:*

$31\dfrac{3}{4}$

$-$ _____ $-\ 14\dfrac{2}{3}$

_____   _____

35. *Estimate:*   *Exact:*

34

$-$ _____ $-\ 15\dfrac{2}{3}$

_____   _____

36. *Estimate:*   *Exact:*

$215\dfrac{7}{16}$

$-$ _____ $-\ 136$

_____   _____

Add or subtract by changing mixed numbers to improper fractions. Simplify all answers.

37. $5\dfrac{2}{5}$

$+\ 3\dfrac{7}{10}$

_____

38. $4\dfrac{3}{4}$

$+\ 5\dfrac{2}{3}$

_____

39. 5

$-\ 1\dfrac{3}{4}$

_____

40. $6\dfrac{1}{2}$

$-\ 4\dfrac{5}{6}$

_____

41. $8\dfrac{1}{3}$

$-\ 2\dfrac{5}{6}$

_____

42. $5\dfrac{5}{12}$

$-\ 2\dfrac{5}{8}$

_____

First estimate the answer and then find the exact answer for each application problem.

43. Two long-distance runners began an $18\frac{3}{4}$ mile run. They ran uphill $5\frac{5}{8}$ miles, downhill $7\frac{1}{3}$ miles, and the rest of the course was level. Find the distance of the level portion of the course.

Estimate:

Exact:

44. The Boys and Girls Clubs of America collected $28\frac{2}{3}$ tons of newspapers on Saturday and $24\frac{3}{4}$ tons on Sunday. Find the total weight of the newspapers collected.

Estimate:

Exact:

45. In a recent bass-fishing derby, Darrel Holmes caught three largemouth bass weighing $8\frac{7}{8}$ pounds, $9\frac{1}{2}$ pounds, and $6\frac{3}{4}$ pounds. Find their total weight.

Estimate:

Exact:

46. A developer wants to build a shopping center. She bought two parcels of land, one, $1\frac{11}{16}$ acres, and the other, $2\frac{3}{4}$ acres. If she needs a total of $8\frac{1}{2}$ acres for the center, how much additional land does she need to buy?

Estimate:

Exact:

[3.5] *Locate each fraction in Exercises 47–50 on the number line.*

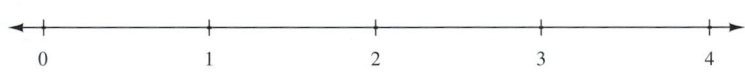

47. $\frac{3}{8}$ **48.** $\frac{7}{4}$ **49.** $\frac{8}{3}$ **50.** $3\frac{1}{5}$

Write < or > in each blank to make a true statement.

51. $\frac{2}{3}$ ____ $\frac{3}{4}$ **52.** $\frac{3}{4}$ ____ $\frac{7}{8}$ **53.** $\frac{1}{2}$ ____ $\frac{7}{15}$ **54.** $\frac{7}{10}$ ____ $\frac{8}{15}$

55. $\frac{9}{16}$ ____ $\frac{5}{8}$ **56.** $\frac{7}{20}$ ____ $\frac{8}{25}$ **57.** $\frac{19}{36}$ ____ $\frac{29}{54}$ **58.** $\frac{19}{132}$ ____ $\frac{7}{55}$

Simplify each expression.

59. $\left(\frac{1}{2}\right)^2$ **60.** $\left(\frac{2}{3}\right)^2$ **61.** $\left(\frac{3}{10}\right)^3$ **62.** $\left(\frac{3}{8}\right)^4$

Simplify by using the order of operations.

63. $8\left(\dfrac{1}{4}\right)^2$

64. $12\left(\dfrac{3}{4}\right)^2$

65. $\left(\dfrac{2}{3}\right)^2 \cdot \left(\dfrac{3}{8}\right)^2$

66. $\dfrac{7}{8} \div \left(\dfrac{1}{8} + \dfrac{3}{4}\right)$

67. $\left(\dfrac{1}{2}\right)^2 \cdot \left(\dfrac{1}{4} + \dfrac{1}{2}\right)$

68. $\left(\dfrac{1}{4}\right)^3 + \left(\dfrac{5}{8} + \dfrac{3}{4}\right)$

▶▶▶ Mixed Review Exercises

Simplify. Use the order of operations as necessary.

69. $\dfrac{7}{8} - \dfrac{1}{8}$

70. $\dfrac{7}{10} - \dfrac{3}{10}$

71. $\dfrac{29}{32} - \dfrac{5}{16}$

72. $\dfrac{1}{4} + \dfrac{1}{8} + \dfrac{5}{16}$

73. $\begin{array}{r} 6\frac{2}{3} \\ -\ 4\frac{1}{2} \\ \hline \end{array}$

74. $\begin{array}{r} 9\frac{1}{2} \\ +\ 16\frac{3}{4} \\ \hline \end{array}$

75. $\begin{array}{r} 7 \\ -\ 1\frac{5}{8} \\ \hline \end{array}$

76. $\begin{array}{r} 2\frac{3}{5} \\ 8\frac{5}{8} \\ +\ \frac{5}{16} \\ \hline \end{array}$

77. $\begin{array}{r} 32\frac{5}{12} \\ -17 \\ \hline \end{array}$

78. $\dfrac{7}{22} + \dfrac{3}{22} + \dfrac{3}{11}$

79. $\left(\dfrac{1}{4}\right)^2 \cdot \left(\dfrac{2}{5}\right)^3$

80. $\dfrac{3}{8} \div \left(\dfrac{1}{2} + \dfrac{1}{4}\right)$

81. $\left(\dfrac{2}{3}\right)^2 \cdot \left(\dfrac{1}{3} + \dfrac{1}{6}\right)$

82. $\left(\dfrac{2}{3}\right)^3 + \left(\dfrac{2}{3} - \dfrac{5}{9}\right)$

Write < or > in each blank to make a true statement.

83. $\dfrac{2}{3}$ _____ $\dfrac{7}{12}$

84. $\dfrac{8}{9}$ _____ $\dfrac{15}{8}$

85. $\dfrac{17}{30}$ _____ $\dfrac{36}{60}$

86. $\dfrac{5}{8}$ _____ $\dfrac{17}{30}$

Find the least common multiple of each set of numbers.

87. 12, 18

88. 6, 8, 10, 12

89. 9, 14, 21

Rewrite each fraction using the indicated denominator.

90. $\dfrac{2}{3} = \dfrac{}{27}$

91. $\dfrac{9}{12} = \dfrac{}{144}$

92. $\dfrac{4}{5} = \dfrac{}{75}$

First estimate the answer and then find the exact answer for each application problem.

93. A cement contractor needs $13\frac{1}{2}$ ft of wire mesh for a concrete walkway and $22\frac{3}{8}$ ft of wire mesh for a driveway. If the contractor starts with a roll of wire that is $92\frac{3}{4}$ ft long, find the number of feet remaining after the two jobs have been completed.

Estimate:

Exact:

94. A baker had four 50-pound bags of sugar. If she used $68\frac{1}{2}$ pounds of sugar to bake cakes, $76\frac{5}{8}$ pounds for baking pies, and $33\frac{1}{4}$ pounds for baking cookies, how many pounds of sugar remain?

Estimate:

Exact:

Chapter 3 ▶▶▶ Test

Use the Chapter Test Prep Video CD to see fully worked-out solutions to any of the exercises you want to review

Add or subtract. Write answers in lowest terms.

1. $\dfrac{5}{8} + \dfrac{1}{8}$

2. $\dfrac{1}{16} + \dfrac{7}{16}$

3. $\dfrac{7}{10} - \dfrac{3}{10}$

4. $\dfrac{7}{12} - \dfrac{5}{12}$

Find the least common multiple of each set of numbers.

5. 2, 3, 4

6. 6, 3, 5, 15

7. 6, 9, 27, 36

Add or subtract. Write answers in lowest terms.

8. $\dfrac{3}{8} + \dfrac{1}{4}$

9. $\dfrac{2}{9} + \dfrac{5}{12}$

10. $\dfrac{7}{8} - \dfrac{2}{3}$

11. $\dfrac{2}{5} - \dfrac{3}{8}$

First estimate the answer. Then add or subtract to find the exact answer; simplify exact answers.

12. $7\dfrac{2}{3} + 4\dfrac{5}{6}$

13. $16\dfrac{2}{5} - 11\dfrac{2}{3}$

14. $18\dfrac{3}{4} + 9\dfrac{2}{5} + 12\dfrac{1}{3}$

15. $24 - 18\dfrac{3}{8}$

1. _____

2. _____

3. _____

4. _____

5. _____

6. _____

7. _____

8. _____

9. _____

10. _____

11. _____

12. Estimate: _____

    Exact: _____

13. Estimate: _____

    Exact: _____

14. Estimate: _____

    Exact: _____

15. Estimate: _____

    Exact: _____

16. _____

16. Most students say that "addition and subtraction of fractions is more difficult than multiplication and division of fractions." Why do you think they say this? Do you agree with these students?

17. _____

17. Devise and explain a method of estimating an answer to addition and subtraction problems involving mixed numbers. Might your estimated answer vary from the exact answer? If it did, what would the estimation accomplish?

First estimate the answer and then find the exact answer for each application problem.

18. *Estimate:* _____

   Exact: _____

18. In one week, a kennel owner used $10\frac{3}{8}$ pounds of puppy chow, $84\frac{1}{2}$ pounds of dry kibble, $36\frac{5}{6}$ pounds of high-protein mature dog mix, and $8\frac{1}{3}$ pounds of fresh ground meat products. Find the total number of pounds used.

19. *Estimate:* _____

   Exact: _____

19. A painting contractor arrived at a 6-unit apartment complex with $147\frac{1}{2}$ gallons of exterior paint. If his crew sprayed $68\frac{1}{2}$ gallons on the wood siding, rolled $37\frac{3}{8}$ gallons on the masonry exterior, and brushed $5\frac{3}{4}$ gallons on the trim, find the number of gallons of paint remaining.

Write $<$ or $>$ between each pair of fractions to make a true statement.

20. _____

20. $\dfrac{3}{4}$ ——— $\dfrac{17}{24}$

21. $\dfrac{19}{24}$ ——— $\dfrac{17}{36}$

21. _____

Simplify. Use the order of operations as needed.

22. _____

22. $\left(\dfrac{1}{3}\right)^{3} \cdot 54$

23. $\left(\dfrac{3}{4}\right)^{2} - \left(\dfrac{7}{8} \cdot \dfrac{1}{3}\right)$

23. _____

24. _____

25. _____

24. $4\left(\dfrac{7}{8} - \dfrac{7}{16}\right)$

25. $\dfrac{5}{6} + \dfrac{4}{3}\left(\dfrac{3}{8}\right)$

Study Skills

CRITICAL —

TIPS FOR TAKING MATH TESTS

OBJECTIVES

1. Apply suggestions to tests and quizzes.
2. Develop a set of "best practices" to apply while testing.

Improving Your Test Score

To Improve Your Test Score	Comments
Come prepared with a pencil, eraser, and calculator, if allowed. If you are easily distracted, sit in the corner farthest from the door.	*Working in pencil lets you erase,* keeping your work neat and readable.
Scan the entire test, note the point value of different problems, and plan your time accordingly. Allow at least five minutes to check your work at the end of the testing time.	If you have 50 minutes to do 20 problems, $50 \div 20 = 2.5$ minutes per problem. *Spend less time on easy ones,* more time on problems with higher point values.
Read directions carefully, and circle any significant words. When you finish a problem, read the directions again to make sure you did what was asked.	*Pay attention to announcements* written on the board or made by your instructor. Ask if you don't understand. You don't want to get problems wrong because you misread the directions!
Show your work. Most math teachers give partial credit if some of the steps in your work are correct, even if the final answer is wrong. *Write neatly.* If you like to scribble when first working or checking a problem, do it on scratch paper.	*If your teacher can't read your writing, you won't get credit for it.* If you need more space to work, ask if you can use extra pieces of paper that you hand in with your test paper.
Check that the *answer to an application problem is reasonable* and makes sense. Read the problem again to make sure you've answered the question.	*Use common sense.* Can the father really be seven years old? Would a month's rent be $32,140? Label your answer: $, years, inches, etc.
To check for careless errors, you need to *rework the problem again, without looking at your previous work.* Cover up your work with a piece of scratch paper, and pretend you are doing the problem for the first time. Then compare the two answers.	If you just "look over" your work, your mind can make the same mistake again without noticing it. Reworking the problem from the beginning *forces you to rethink it.* If possible, use a different method to solve the problem the second time.

Reducing Anxiety

To Reduce Anxiety	Comments
Do not try to review up until the last minute before the test. Instead, go for a walk, do some deep breathing, and arrive just in time for the test. Ignore other students.	Listening to anxious classmates before the test *may cause you to panic.* Moderate exercise and deep breathing will calm your mind.

Reducing Anxiety
(continued)

Why Are These Suggestions Brain Friendly?

Several suggestions address anxiety. **Reducing anxiety allows your brain to make the connections between dendrites;** in other words, you can think clearly.

Remember that **your brain continues to work on a difficult problem** even if you skip it and go on to the next one. Your subconscious mind will come through for you if you are open to the idea!

Some of the suggestions ask you to **use your common sense.** Follow the directions, show your work, write neatly, and pay attention to whether your answers really make sense.

To Reduce Anxiety	Comments
Do a "knowledge dump" as soon as you get the test. Write important notes to yourself in a corner of the test paper: formulas, or common errors you want to watch out for.	Writing down tips and things that you've memorized *lets you relax;* you won't have to worry about forgetting those things and can refer to them as needed.
Do the easy problems first in order to build confidence. If you feel your anxiety starting to build, *immediately* stop for a minute, close your eyes, and take several slow, deep breaths.	Greater confidence helps you *get the easier problems correct.* Anxiety causes shallow breathing, which leads to confusion and reduced concentration. Deep breathing calms you.
As you work on more difficult problems, *notice your "inner voice."* You may have negative thoughts such as, "I can't do it," or "who cares about this test anyway." In your mind, yell, "STOP" and take several deep, slow breaths. Or, replace the negative thought with a positive one.	Here are *examples of positive statements.* Try writing one of them on the top of your test paper. • I know I can do it. • I can do this one step at a time. • I've studied hard, and I'll do the best I can.
If you still can't solve a difficult problem when you come back to it the second time, *make a guess and do not change it.* In this situation, your first guess is your best bet. Do not change an answer just because you're a little unsure. *Change it only if you find an obvious mistake.*	If you are thinking about changing an answer, be sure you have a good reason for changing it. If you cannot find a specific error, leave your first answer alone. *When the tests are returned, check to see if changing answers helped or hurt you.*
Read the harder problems twice. Write down *anything* that might help solve the problem: a formula, a picture, etc. If you still can't get it, circle the problem and *come back to it later.* Do *not* erase any of the things you wrote down.	If you know even a *little* bit about the problem, write it down. The *answer may come to you* as you work on it, or you may get partial credit. Don't spend too long on any one problem. Your subconscious mind will work on the tough problem while you go on with the test.
Ignore students who finish early. Use the entire test time. *You do not get extra credit for finishing early.* Use the extra time to rework problems and correct careless errors.	Students who leave early are often the ones who didn't study or who are too anxious to continue working. If they bother you, *sit as far from the door as possible.*

4

Decimals

Over 41 million Americans go fishing at least once a year, making it America's sixth most popular recreational activity. (*Source:* National Sporting Goods Association.) Record-size fish caught include a 67.5-pound muskie in Wisconsin, a 58-pound channel catfish in South Carolina, and a 97.25-pound chinook salmon in Alaska.

In **Section 4.3**, Exercises 47–50, this father and daughter will use decimal numbers when paying for new fishing equipment. But will decimals help them catch their limit? (See **Section 4.1**, Exercises 59–62, and **Section 4.6**, Exercises 65–68.)

4.1 ▶▶▶ Reading and Writing Decimals

Fractions are used to represent parts of a whole. In this chapter, **decimals** are used as another way to show parts of a whole. For example, our money system is based on decimals. One dollar is divided into 100 equivalent parts. One cent ($0.01) is one of the parts, and a dime ($0.10) is 10 of the parts. Metric measurement (see **Appendix A**) is also based on decimals.

OBJECTIVE 1 Write parts of a whole using decimals. Decimals are used when a whole is divided into 10 equivalent parts, or into 100 or 1000 or 10,000 equivalent parts. In other words, decimals are fractions with denominators that are a power of 10. For example, the square below is cut into 10 equivalent parts. Written as a fraction, each part is $\frac{1}{10}$ of the whole. Written as a decimal, each part is 0.1. Both $\frac{1}{10}$ and 0.1 are read as "*one tenth.*"

$\frac{1}{10}$ ◀ ▶ 0.1

One-tenth of the square is shaded.

The dot in 0.1 is called the **decimal point.**

0.1

Decimal point ⟶↑

1 There are 10 dimes in one dollar. Each dime is $\frac{1}{10}$ of a dollar. Write the yellow shaded portion of each dollar as a fraction, as a decimal, and in words.

(a)

The square at the right has **7** of its 10 parts shaded.

Written as a *fraction,* $\frac{7}{10}$ of the square is shaded.

Written as a *decimal,* 0.7 of the square is shaded.

Both $\frac{7}{10}$ and 0.7 are read as "*seven tenths.*"

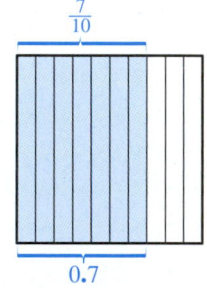

$\frac{7}{10}$

0.7

Seven-tenths of the square is shaded.

(b)

◀ **Work Problem 1 at the Side.**

The square below is cut into 100 equivalent parts. Written as a *fraction,* each part is $\frac{1}{100}$ of the whole.

Written as a decimal, each part is **0.01** of the whole. Both $\frac{1}{100}$ and 0.01 are read as "one hundredth."

(c)

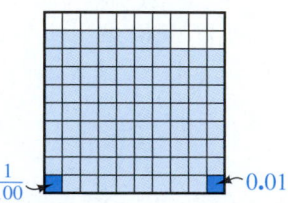

$\frac{1}{100}$ ◀ ▶ 0.01

The square above has 87 parts shaded.

Written as a fraction, $\frac{87}{100}$ of the total area is shaded.

Written as a decimal, **0.87** of the total area is shaded.

Both $\frac{87}{100}$ and 0.87 are read as "*eighty-seven hundredths.*"

Work Problem **2** at the Side. ▶

Example 1 below shows several numbers written as fractions, as decimals, and in words.

EXAMPLE 1   **Using the Decimal Forms of Fractions**

Fraction	Decimal	Read As
(a) $\frac{4}{10}$	0.4	four tenths
(b) $\frac{9}{100}$	0.09	nine hundredths
(c) $\frac{71}{100}$	0.71	seventy-one hundredths
(d) $\frac{8}{1000}$	0.008	eight thousandths
(e) $\frac{45}{1000}$	0.045	forty-five thousandths
(f) $\frac{832}{1000}$	0.832	eight hundred thirty-two thousandths

Work Problem **3** at the Side. ▶

OBJECTIVE 2 **Identify the place value of a digit.**   The decimal point separates the *whole number part* from the *fractional part* in a decimal number. In the chart below, you see that the **place value** names for fractional parts are similar to those on the whole number side, but end in "***ths***."

Decimal Place Value Chart

hundred-thousands	ten-thousands	thousands	hundreds	tens	ones		tenths	hundredths	thousandths	ten-thousandths	hundred-thousandths
100,000	10,000	1000	100	10	1	.	$\frac{1}{10}$	$\frac{1}{100}$	$\frac{1}{1000}$	$\frac{1}{10,000}$	$\frac{1}{100,000}$

Whole number part ← → Decimal point (Read "and")   Fractional part →

Note

Notice that the **ones** place is at the center of the place value chart. There is no "oneths" place.

Also notice that each place is 10 times the value of the place to its right.

Finally, be sure to write a hyphen (dash) in ten-**thousandths** and hundred-**thousandths**.

---

2 Write the portion of each square that is shaded as a fraction, as a decimal, and in words.

(a)

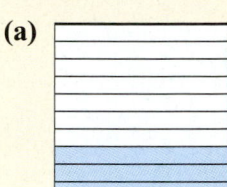

(b)

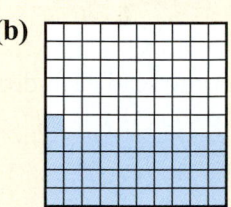

3 Write each decimal as a fraction.

(a) 0.7

(b) 0.2

(c) 0.03

(d) 0.69

(e) 0.047

(f) 0.351

4 Identify the place value of each digit.

(a) 971.54

(b) 0.4

(c) 5.60

(d) 0.0835

5 Tell how to read each decimal in words.

(a) 0.6

(b) 0.46

(c) 0.05

(d) 0.409

(e) 0.0003

(f) 0.2703

(g) 0.088

ANSWERS

4. (a)
hundreds	tens	ones		tenths	hundredths
9	7	1	.	5	4

(b)
ones		tenths
0	.	4

(c)
ones		tenths	hundredths
5	.	6	0

(d)
ones		tenths	hundredths	thousandths	ten-thousandths
0	.	0	8	3	5

4. (a) six tenths
(b) forty-six hundredths
(c) five hundredths
(d) four hundred nine thousandths
(e) three ten-thousandths
(f) two thousand seven hundred three ten-thousandths
(g) eighty-eight thousandths

CAUTION

If a number does *not* have a decimal point, it is a *whole number*. A whole number has no fractional part. If you want to show the decimal point in a whole number, it is just to the **right** of the digit in the ones place. Here are two examples.

$$8 = 8. \qquad\qquad 306 = 306.$$

Decimal point   Decimal point

EXAMPLE 2 **Identifying the Place Value of a Digit**

Identify the place value of each digit.

(a) 178.36   (b) 0.00935

Notice in Example 2(b) that we do *not* use commas on the right side of the decimal point.

◀ Work Problem **4** at the Side.

OBJECTIVE 3 Read and write decimals in words.   A decimal is read according to its form as a fraction.

We read 0.9 as "nine tenths" because 0.9 is the same as $\frac{9}{10}$. Notice that 0.9 ends in the tenths place.

We read 0.02 as "two hundredths" because 0.02 is the same as $\frac{2}{100}$. Notice that 0.02 ends in the hundredths place.

EXAMPLE 3 **Reading Decimal Numbers**

Tell how to read each decimal in words.

(a) 0.3

Because $0.3 = \frac{3}{10}$, read the decimal as three ten**ths**.

(b) 0.49   Read it as: forty-nine hundred**ths**.

(c) 0.08   Read it as: eight hundred**ths**.

> Think: $0.08 = \frac{8}{100}$ so write *hundredths*.

(d) 0.918   Read it as: nine hundred eighteen thousand**ths**.

(e) 0.0106   Read it as: one hundred six ten-thousand**ths**.

> Think: $0.0106 = \frac{106}{10,000}$

◀ Work Problem **5** at the Side.

Reading Decimal Numbers

Step 1   Read any whole number part to the *left* of the decimal point as you normally would.

Step 2   Read the decimal point as "*and.*"

Step 3   Read the part of the number to the *right* of the decimal point as if it were an ordinary whole number.

Step 4   Finish with the place value name of the rightmost digit; these names all end in "*ths.*"

Note

If there is *no whole number part,* you will use only Steps 3 and 4.

EXAMPLE 4   **Reading Decimals**

Read each decimal.

(a)

9 is in tenths place.

16.9

sixteen **and** nine **tenths**

16.9 is read "sixteen and nine tenths."

Remember to say or write "and" *only* when you see a decimal point.

(b)

5 is in hundredths place.

482.35

four hundred eighty-two **and** thirty-five **hundredths**

482.35 is read "four hundred eighty-two and thirty-five hundredths."

3 is in thousandths place.

(c) 0.063 is "sixty-three **thousandths**." (No whole number part.)

(d) 11.1085 is "eleven **and** one thousand eighty-five **ten-thousandths**."

CAUTION

Use "and" *only* when reading a decimal point. A common mistake is to read the whole number 405 as "four hundred *and* five." But there is *no decimal point* shown in 405, so it is read "four hundred five."

Work Problem **6** *at the Side.* ▶

OBJECTIVE **4** **Write decimals as fractions or mixed numbers.**
Knowing how to read decimals will help you when writing decimals as fractions.

Writing Decimals as Fractions or Mixed Numbers

Step 1   The digits to the right of the decimal point are the numerator of the fraction.

Step 2   The denominator is 10 for tenths, 100 for hundredths, 1000 for thousandths, 10,000 for ten-thousandths, and so on.

Step 3   If the decimal has a whole number part, it will be written as a mixed number with the same whole number part.

---

6 Tell how to read each decimal in words.

(a) 3.8

(b) 15.001

(c) 0.0073

(d) 64.309

7 Write each decimal as a fraction or mixed number.

(a) 0.7

(b) 12.21

(c) 0.101

(d) 0.007

(e) 1.3717

8 Write each decimal as a fraction or mixed number in lowest terms.

(a) 0.5

(b) 12.6

(c) 0.85

(d) 3.05

(e) 0.225

(f) 420.0802

EXAMPLE 5 Writing Decimals as Fractions or Mixed Numbers

Write each decimal as a fraction or mixed number.

(a) 0.19

The digits to the right of the decimal point, 19, are the numerator of the fraction. The denominator is 100 for hundredths because the rightmost digit is in the hundredths place.

$$0.19 = \frac{19}{100} \leftarrow 100 \text{ for hundredths}$$

Hundredths place ⟍

(b) 0.863

$$0.863 = \frac{863}{1000} \leftarrow 1000 \text{ for thousandths}$$

Thousandths place ⟍

(c) 4.0099

The whole number part stays the same.

$$4.0099 = 4\frac{99}{10{,}000} \leftarrow 10{,}000 \text{ for ten-thousandths}$$

Ten-thousandths place ⟍

◀ Work Problem **7** at the Side.

CAUTION

After you write a decimal as a fraction or a mixed number, make sure the fraction is in lowest terms.

EXAMPLE 6 Writing Decimals as Fractions or Mixed Numbers in Lowest Terms

Write each decimal as a fraction or mixed number in lowest terms.

(a) $0.4 = \frac{4}{10} \leftarrow 10 \text{ for tenths}$ Write $\frac{4}{10}$ in lowest terms.

$$\frac{4}{10} = \frac{4 \div 2}{10 \div 2} = \frac{2}{5} \leftarrow \text{Lowest terms}$$

(b) $0.75 = \frac{75}{100} = \frac{75 \div 25}{100 \div 25} = \frac{3}{4} \leftarrow \text{Lowest terms}$

The whole number part stays the same.

(c) $18.105 = 18\frac{105}{1000} = 18\frac{105 \div 5}{1000 \div 5} = 18\frac{21}{200} \leftarrow \text{Lowest terms}$

(d) $42.8085 = 42\frac{8085}{10{,}000} = 42\frac{8085 \div 5}{10{,}000 \div 5} = 42\frac{1617}{2000} \leftarrow \text{Lowest terms}$

◀ Work Problem **8** at the Side.

Calculator Tip In this book we will write a 0 in the ones place for decimal fractions. We write **0**.45 instead of just .45, to emphasize that there is no whole number. Many calculators show these zeros also. Try entering ⊙ ④ ⑤ ; the display probably shows 0.45 even though you did not press 0. For comparison, enter the whole number 45 by pressing ④ ⑤ ⊕ and notice where the decimal point is shown in the display. (It automatically appears to the *right* of the 5.)

ANSWERS

7. (a) $\frac{7}{10}$ (b) $12\frac{21}{100}$ (c) $\frac{101}{1000}$
(d) $\frac{7}{1000}$ (e) $1\frac{3717}{10{,}000}$

8. (a) $\frac{1}{2}$ (b) $12\frac{3}{5}$ (c) $\frac{17}{20}$ (d) $3\frac{1}{20}$
(e) $\frac{9}{40}$ (f) $420\frac{401}{5000}$

Identify the digit that has the given place value. See Example 2.

1. 70.489
tens
ones
tenths

2. 135.296
ones
tenths
tens

3. 0.2518
hundredths
thousandths
ten-thousandths

4. 0.9347
hundredths
thousandths
ten-thousandths

5. 93.01472
thousandths
ten-thousandths
tenths

6. 0.51968
tenths
ten-thousandths
hundredths

7. 314.658
tens
tenths
hundreds

8. 51.325
tens
tenths
hundredths

9. 149.0832
hundreds
hundredths
ones

10. 3458.712
hundreds
hundredths
tenths

11. 6285.7125
thousands
thousandths
hundredths

12. 5417.6832
thousands
thousandths
ones

Write the decimal number that has the specified place values. See Example 2.

13. 0 ones, 5 hundredths, 1 ten, 4 hundreds, 2 tenths

14. 7 tens, 9 tenths, 3 ones, 6 hundredths, 8 hundreds

15. 3 thousandths, 4 hundredths, 6 ones, 2 ten-thousandths, 5 tenths

16. 8 ten-thousandths, 4 hundredths, 0 ones, 2 tenths, 6 thousandths

17. 4 hundredths, 4 hundreds, 0 tens, 0 tenths, 5 thousandths, 5 thousands, 6 ones

18. 7 tens, 7 tenths, 6 thousands, 6 thousandths, 3 hundreds, 3 hundredths, 2 ones

Write each decimal as a fraction or mixed number in lowest terms. See Examples 5 and 6.

19. 0.7   **20.** 0.1   ☉ **21.** 13.4   **22.** 9.8   **23.** 0.25

24. 0.55   **25.** 0.66   **26.** 0.33   **27.** 10.17   **28.** 31.99

29. 0.06   **30.** 0.08   ☉ **31.** 0.205   **32.** 0.805

33. 5.002   **34.** 4.008   **35.** 0.686   **36.** 0.492

Tell how to read each decimal in words. See Examples 3 and 4.

37. 0.5   **38.** 0.2

39. 0.78   **40.** 0.55
☉

41. 0.105   **42.** 0.609

43. 12.04   **44.** 86.09

45. 1.075   **46.** 4.025

Write each decimal in numbers. See Examples 3 and 4.

47. six and seven tenths

48. eight and twelve hundredths

49. thirty-two hundredths

50. one hundred eleven thousandths

51. four hundred twenty and eight thousandths

52. two hundred and twenty-four thousandths

53. seven hundred three ten-thousandths

54. eight hundred and six hundredths

55. seventy-five and thirty thousandths

56. sixty and fifty hundredths

57. Anne read the number 4302 as "four thousand three hundred and two." Explain what is wrong with the way Anne read the number.

58. Jerry read the number 9.0106 as "nine and one hundred and six ten-thousandths." Explain the error he made.

The father on the first page of this chapter needs to select the correct fishing line for his daughter's reel. Fishing line is sold according to how many pounds of "pull" the line can withstand before breaking. Use the table to answer Exercises 59–62. Write all fractions in lowest terms. (Note: The diameter of the fishing line is its thickness.)

RELATING FISHING LINE DIAMETER TO TEST STRENGTH

Test Strength (pounds)	Average Diameter (inches)
4	0.008
8	0.010
12	0.013
14	0.014
17	0.015
20	0.016

Source: Berkley Outdoor Technologies Group.

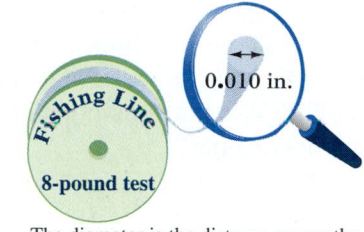

The diameter is the distance across the end of the line (or its thickness).

59. Write the diameter of 8-pound test line in words and as a fraction.

60. Write the diameter of 17-pound test line in words and as a fraction.

61. What is the test strength of the line with a diameter of $\frac{13}{1000}$ inch?

62. What is the test strength of the line with a diameter of sixteen thousandths inch?

Suppose your job is to take phone orders for precision parts. Use the table, and in Exercises 63–66, write the correct part number that matches what you hear the customer say over the phone. In Exercises 67–68, write the words you would say to the customer.

Part Number	Size in Centimeters
3-A	0.06
3-B	0.26
3-C	0.6
3-D	0.86
4-A	1.006
4-B	1.026
4-C	1.06
4-D	1.6
4-E	1.602

63. "Please send the six-tenths centimeter bolt."

Part number _____.

64. "The part missing from our order was the one and six hundredths size."

Part number _____.

65. "The size we need is one and six thousandths centimeters."

Part number _____.

66. "Do you still stock the twenty-six hundredths centimeter bolt?"

Part number _____.

67. "What size is part number 4-E?" Write your answer in words.

68. "What size is part number 4-B?" Write your answer in words.

Relating Concepts (Exercises 69–76) For Individual or Group Work

Use your knowledge of place value to **work Exercises 69–76 in order.**

69. Look back at the decimal place value chart on page 267. What do you think would be the names of the next four places to the *right* of hundred-thousandths? What information did you use to come up with these names?

70. A common mistake is to think that the first place to the right of the decimal point is "oneths" and the second place is "tenths." Why might someone make that mistake? How would you explain why there is no "oneths" place?

71. Use your answer to Exercise 69 to write 0.72436955 in words.

72. Use your answer to Exercise 69 to write 0.000678554 in words.

73. Write 8006.500001 in words.

74. Write 20,060.000505 in words.

75. Write this decimal in numbers.

three hundred two thousand forty ten-millionths

76. Write this decimal in numbers.

nine billion, eight hundred seventy-six million, five hundred forty-three thousand, two hundred ten and one hundred million two hundred thousand three hundred billionths

4.2 ▶▶▶   Rounding Decimals

Section 1.7 showed how to round whole numbers. For example, 89 rounded to the nearest ten is 90, and 8512 rounded to the nearest hundred is 8500.

OBJECTIVE 1 Learn the rules for rounding decimals.   It is also important to be able to **round** decimals. For example, a store is selling 2 candy mints for $0.75 but you want only one mint. The price of each mint is $0.75 ÷ 2, which is $0.375, but you cannot pay part of a cent. Is $0.375 closer to $0.37 or to $0.38? Actually, it's exactly halfway between. When this happens in everyday situations, the rule is to round *up*. The store will charge you $0.38 for the mint.

Rounding Decimals

Step 1 Find the place to which the rounding is being done. Draw a "cut off" line *after* that place to show that you are cutting off and dropping the rest of the digits.

Step 2 Look *only* at the *first* digit you are cutting off.

Step 3(a) If this digit is *4 or less,* the part of the number you are keeping *stays the same.*

Step 3(b) If this digit is *5 or more,* you must *round up* the part of the number you are keeping.

Step 4 You can use the ≈ symbol or the ≐ symbol to indicate that the rounded number is now an approximation (close, but not exact). Both symbols mean "is approximately equal to." In this book we will use the ≈ symbol.

CAUTION
Do *not* move the decimal point when rounding.

OBJECTIVE 2 Round decimals to any given place.   The following examples show you how to round decimals.

EXAMPLE 1   **Rounding a Decimal Number**

Round 14.39652 to the nearest thousandth.

Step 1 Draw a "cut-off" line after the thousandths place.

$$1\ 4\ .\ 3\ 9\ 6\ \not{|}\ 5\ 2$$

Thousandths ⟶

You are cutting off the 5 and 2. They will be dropped.

Step 2 Look *only* at the *first* digit you are cutting off. Ignore the other digits you are cutting off.

$$1\ 4\ .\ 3\ 9\ 6\ \not{|}\ \mathbf{5}\ 2$$

Look *only* at the 5. Ignore the 2.

Continued on Next Page

1 Round to the nearest thousandth.

(a) 0.33492

(b) 8.00851

(c) 265.42038

(d) 10.70180

Step 3 If the first digit you are cutting off is *5 or more,* round up the part of the number you are keeping.

$$14.396 \:\:\:\: 5 \: 2$$
$$+0.001$$
$$\overline{14.397}$$

> First digit cut is *5 or more,* so round up by adding 1 thousandth to the part you are keeping.

> Think: Rounding to *thousandths* means *three* decimal places.

So, 14.39652 rounded to the nearest thousandth is 14.397
We can write 14.39652 ≈ 14.397

> **CAUTION**
> When rounding whole numbers in **Section 1.7,** you kept all the digits but changed some to zeros. With decimals, you cut off and *drop the extra digits.* In Example 1 above, 14.39652 rounds to 14.397, *not* 14.39700.

◄ *Work Problem* **1** *at the Side.*

In Example 1, the rounded number 14.397 had *three decimal places.* **Decimal places** are the number of digits to the *right* of the decimal point. The first decimal place is tenths, the second is hundredths, the third is thousandths, and so on.

─ **EXAMPLE 2** **Rounding Decimals to Different Places**

Round to the place indicated.

(a) 5.3496 to the nearest tenth

> *Tenths is one decimal place.*

Step 1 Draw a cut-off line after the tenths place.

$$5 . 3 \:\:\:\: 4 \: 9 \: 6$$
Tenths

> You are cutting off the 4, 9, and 6. They will be dropped.

Step 2
$$5 . 3 \:\:\:\: \mathbf{4} \: 9 \: 6$$

> Look *only* at the 4.
> Ignore these digits.

Step 3
$$\underline{5 . 3} \:\:\:\: \mathbf{4} \: 9 \: 6$$
$$5 . 3 \leftarrow \text{Stays the same}$$

> First digit cut is *4 or less,* so the part you are keeping stays the same.

5.3496 rounded to the nearest tenth is 5.3 (*one* decimal place for *tenths*).
We can write 5.3496 ≈ 5.3
Notice: 5.3496 does *not* round to 5.3000, which would be ten-thousandths.

(b) 0.69738 to the nearest hundredth

Step 1
$$0 . 6 \: 9 \mid 7 \: 3 \: 8$$
Hundredths

> Draw a cut-off line after the hundredths place.

Step 2
$$0 . 6 \: 9 \mid \mathbf{7} \: 3 \: 8$$

> Look *only* at the 7.

─ **Continued on Next Page**

Step 3 $0.69\underset{\underline{}}{7}38$

First digit cut is *5 or more,* so round up by adding 1 hundredth to the part you are keeping.

$$
\begin{array}{r}
\overset{1}{} \\
0.69 \\
+\,0.01 \\
\hline
0.70
\end{array}
$$

← Keep this part.
← To round up, add 1 hundredth.
← 9 + 1 is 10; write 0 and regroup 1 to the tenths place.

0.69738 rounded to the nearest hundredth is 0.70. Hundredths is *two* decimal places so you *must* write the 0 in the hundredths place.
We can write $0.69738 \approx 0.70$

Think: Rounding to *hundredths* means *two* decimal places.

CAUTION

If a *rounded* number has a 0 in the rightmost place, you *must* keep the 0. As shown above, 0.69738 rounded to the nearest hundredth is 0.7**0**. Do ***not*** write 0.7, which is rounded to tenths instead of hundredths.

(c) 0.01806 to the nearest thousandth

First digit cut is *4 or less,* so the part you are keeping stays the same.

$$0.018\,|\,06$$

$0.018 \leftarrow$ Stays the same

0.01806 rounded to the nearest thousandth is 0.018 (*three* decimal places for *thousandths*). We can write $0.01806 \approx 0.018$

(d) 57.976 to the nearest tenth

First digit cut is *5 or more,* so round up by adding 1 tenth to the part you are keeping.

$$
\begin{array}{r}
57.9\,|\,76 \\
57.9 \\
+\quad 0.1 \\
\hline
58.0
\end{array}
$$

Be sure to write the 0 in the tenths place.

← 9 + 1 is 10; write the 0 and regroup the 1 to the ones place.

57.976 rounded to the nearest tenth is 58.0. We can write $57.976 \approx 58.0$
You *must* write the 0 in the tenths place to show that the number was rounded to the nearest tenth.

CAUTION

Check that your rounded answer shows *exactly* the number of decimal places asked for in the problem. Be sure your answer shows *one* decimal place if you rounded to *tenths,* *two* decimal places for *hundredths,* *three* decimal places for *thousandths,* and so on.

Work Problem **2** *at the Side.* ▶

OBJECTIVE 3 Round money amounts to the nearest cent or nearest dollar. In many everyday situations, such as shopping in a store, money amounts are rounded to the nearest cent. There are 100 cents in a dollar.

$$\text{Each cent is } \frac{1}{100} \text{ of a dollar.}$$

Another way to write $\frac{1}{100}$ is 0.01. So rounding to the *nearest cent* is the same as rounding to the *nearest hundredth of a dollar.*

---

2 Round to the place indicated.

(a) 0.8988 to the nearest hundredth

(b) 5.8903 to the nearest hundredth

(c) 11.0299 to the nearest thousandth

(d) 0.545 to the nearest tenth

ANSWERS

2. (a) 0.90 **(b)** 5.89 **(c)** 11.030 **(d)** 0.5

3 Round each money amount to the nearest cent.

(a) $14.595

(b) $578.0663

(c) $0.849

(d) $0.0548

EXAMPLE 3 **Rounding to the Nearest Cent**

How much will you pay in each shopping situation? Round each money amount to the nearest cent.

(a) $2.4238 (Is it closer to $2.42 or to $2.43?)

> First digit cut is *4 or less,* so the part you are keeping stays the same.

$2.42 | **38**

$2.42 ⟵ You pay.

> Rounding to the *nearest cent* is rounding to *hundredths.*

You pay $2.42 because $2.4238 is closer to $2.42 than to $2.43.

(b) $0.695 (Is it closer to $0.69 or to $0.70?)

> *5 or more;* round up

$0.69 | **5**

$0.69
+ $0.01 ⟵ To round up, add 1 hundredth (1 cent).
———
$0.70 ⟵ You pay.

◀ Work Problem **3** at the Side.

Note

Some stores round *all* money amounts up to the next higher cent, even if the next digit is *4 or less.* In Example 3(a) above, some stores would round $2.4238 *up* to $2.43, even though it is closer to $2.42.

It is also common to round money amounts to the nearest dollar. For example, you can do that on your federal and state income tax returns to make the calculations easier.

EXAMPLE 4 **Rounding to the Nearest Dollar**

Round to the nearest dollar.

(a) $48.69 (Is it closer to $48 or to $49?)

> First digit cut is *5 or more,* so round up by adding $1.

$48. | **69**

$48
+ 1
———
$49

> Write $49 *not* $49.00

$48.69 is closer to $49 than to $48.
So $48.69 rounded to the nearest dollar is $49

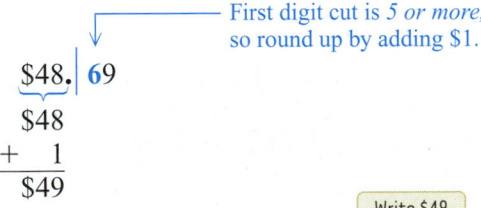

CAUTION

$48.69 rounded to the nearest dollar is $49. Write the answer as $49 to show that the rounding is to the *nearest dollar.* Writing $49.00 would show rounding to the *nearest cent.*

—— **Continued on Next Page**

(b) $594.36 (Is it closer to $594 or $595?)

First digit cut is *4 or less,* so the part you keep stays the same.

$594. | 36

$594

Careful!
Write $594,
not $594.00

$594.36 rounded to the nearest dollar is $594

(c) $399.88 (Is it closer to $399 or to $400?)

5 or more, so round up by adding $1.

$399. | 88

$399
+   1
─────
$400

You are rounding to the nearest whole number, so do *not* show any decimal places.

$399.88 rounded to the nearest dollar is $400

(d) $2689.50 (Is it closer to $2689 or $2690?)

5 or more, so round up by adding $1.

$2689. | 50

$2689
+   1
──────
$2690

$2689.50 rounded to the nearest dollar is $2690

> **Note**
> When rounding $2689.50 to the nearest dollar, above, notice that it is exactly halfway between $2689 and $2690. When this happens in everyday situations, the rule is to round *up.* (Scientists working with technical data may use a more complicated rule when rounding numbers that are exactly in the middle.)

(e) $0.61 (Is it closer to $0 or to $1?)

5 or more, so round up.

$0. | 61

Write the rounded amount as $1, *not* $1.00

$0.61 rounded to the nearest dollar is $1

🖩 **Calculator Tip**   Accountants and other people who work with money amounts often set their calculators to automatically round to two decimal places (nearest cent) or to round to zero decimal places (nearest dollar). Your calculator may have this feature.

Work Problem **4** *at the Side.* ▶

---

4 Round to the nearest dollar.

(a) $29.10

(b) $136.49

(c) $990.91

(d) $5949.88

(e) $49.60

(f) $0.55

(g) $1.08

Math in the Media

LAWN FERTILIZER

Gotta Be Green

A lot's being written about personal responsibility these days, and the idea seems to be ending up on the front lawn—literally! Each spring, homeowners across the country gear up to green up their lawns, and the increased use of fertilizer has a lot of environmentalists concerned about the potential effects of chemical runoff into nearby rivers and streams.

Every year, according to a study conducted by the University of Minnesota's Department of Agriculture, each household in the Minneapolis/St. Paul metro area uses an average of 36 pounds of lawn fertilizer. That adds up to 25,529,295 pounds, or 12,765 tons. Add to that another 193,000 pounds of weed killer and you're looking at the total picture for keeping it green in the Twin Cities.

Source: Minneapolis Star Tribune.

1. According to the article,

   (a) How many pounds of lawn fertilizer are used each year in the *entire metro area?*

   (b) Do a division on your calculator to find the number of *households* in the metro area.

   (c) Why does it make sense to round your answer to part (b)? How would you round it?

2. There are 2000 pounds in one ton.

   (a) Find the number of tons equivalent to 25,529,295 pounds of fertilizer.

   (b) Does your answer match the figure given in the article? If not, what did the author of the article do to get 12,765 tons?

   (c) Is the author's figure accurate? Why or why not?

3. (a) When the average amount of lawn fertilizer per household was calculated, the answer was probably not *exactly* 36 pounds. List six different values that are *less than* 36 that would round to 36. List two values with one decimal place; two values with two decimal places; and two values with three decimal places.

   (b) List six different values that are *greater than* 36 that would round to 36. List two values each with one, two, and three decimal places.

4.2 ▶▶▶ Exercises

Round each number to the place indicated. See Examples 1 and 2.

1. 16.8974 to the nearest tenth

2. 193.845 to the nearest hundredth

3. 0.95647 to the nearest thousandth

4. 96.81584 to the nearest ten-thousandth

5. 0.799 to the nearest hundredth

6. 0.952 to the nearest tenth

7. 3.66062 to the nearest thousandth

8. 1.5074 to the nearest hundredth

9. 793.988 to the nearest tenth

10. 476.1196 to the nearest thousandth

11. 0.09804 to the nearest ten-thousandth

12. 176.004 to the nearest tenth

13. 48.512 to the nearest one

14. 3.385 to the nearest one

15. 9.0906 to the nearest hundredth

16. 30.1290 to the nearest thousandth

17. 82.000151 to the nearest ten-thousandth

18. 0.400594 to the nearest ten-thousandth

Nardos is grocery shopping. The store will round the amount she pays for each item to the nearest cent. Write the rounded amounts. See Example 3.

19. Soup is three cans for $2.45, so one can is $0.81666. Nardos pays _____.

20. Orange juice is two cartons for $3.89, so one carton is $1.945. Nardos pays _____.

21. Facial tissue is four boxes for $4.89, so one box is $1.2225. Nardos pays _____.

22. Muffin mix is three packages for $1.75, so one package is $0.58333. Nardos pays _____.

23. Candy bars are six for $2.99, so one bar is $0.4983. Nardos pays _____.

24. Spaghetti is four boxes for $4.39, so one box is $1.0975. Nardos pays _____.

As she gets ready to do her income tax return, Ms. Chen rounds each amount to the nearest dollar. Write the rounded amounts. See Example 4.

25. Income from job, $48,649.60

26. Income from interest on bank account, $69.58

27. Union dues, $310.08

28. Federal withholding, $6064.49

29. Donations to charity, $848.91

30. Medical expenses, $609.38

Round each money amount as indicated.

31. $499.98 to the nearest dollar.

32. $9899.59 to the nearest dollar

33. $0.996 to the nearest cent

34. $0.09929 to the nearest cent.

35. $999.73 to the nearest dollar.

36. $9999.80 to the nearest dollar.

The table lists speed records for various types of transportation. Use the table to answer Exercises 37–40.

Record	Speed (miles per hour)
Land speed record (specially built car)	763.04
Motorcycle speed record (specially adapted motorcycle)	322.16
Fastest roller coaster	106.9
Fastest military jet	2193.167
Boeing 737-300 airplane (regular passenger service)	495
Indianapolis 500 auto race (fastest average winning speed)	185.981
Daytona 500 auto race (fastest average winning speed)	177.602

Sources: Guinness World Records and The World Almanac.

37. Round these speed records to the nearest whole number.

   (a) Motorcycle

   (b) Roller coaster

38. Round these speed records to the nearest hundredth.

   (a) Daytona 500 average winning speed

   (b) Indianapolis 500 average winning speed

39. Round these speed records to the nearest tenth.

   (a) Indianapolis 500 average winning speed

   (b) Land speed record

40. Round these speed records to the nearest hundred.

   (a) military jet

   (b) Boeing 737-300 airplane

Relating Concepts (Exercises 41—44) For Individual or Group Work

Use your knowledge about rounding money amounts to **work Exercises 41–44 in order.**

41. Explain what happens when you round $0.499 to the nearest dollar. Why does this happen?

42. Look again at Exercise 41. How else could you round $0.499 that would be more helpful? What kind of guideline does this suggest about rounding to the nearest dollar?

43. Explain what happens when you round $0.0015 to the nearest cent. Why does this happen?

44. Suppose you want to know which of these amounts is less, so you round them both to the nearest cent.

      $0.5968 $0.6014

Explain what happens. Describe what you could do instead of rounding to the nearest cent.

4.3 ▶▶▶ Adding and Subtracting Decimals

OBJECTIVE 1 Add decimals. When adding or subtracting *whole* numbers **(Sections 1.2** and **1.3),** you lined up the numbers in columns so that you were adding ones to ones, tens to tens, and so on. A similar idea applies to adding or subtracting *decimal* numbers. With decimals, you line up the decimal points to make sure you are adding tenths to tenths, hundredths to hundredths, and so on.

OBJECTIVES

1 Add decimals.

2 Subtract decimals.

3 Estimate the answer when adding or subtracting decimals.

> **Adding and Subtracting Decimals**
>
> **Step 1** Write the numbers in columns with the decimal points lined up.
>
> **Step 2** If necessary, write in zeros so both numbers have the same number of decimal places. Then add or subtract as if they were whole numbers.
>
> **Step 3** Line up the decimal point in the answer directly below the decimal points in the problem.

1 Find each sum.

EXAMPLE 1 Adding Decimal Numbers

Find each sum.

(a) 16.92 and 48.34

Step 1 Write the numbers in columns with the decimal points lined up.

```
  tens ones . tenths hundredths
   1   6  .   9     2
+  4   8  .   3     4
```
Decimal points are lined up.

Step 2 Add as if these were whole numbers.

```
   11
  16.92
+ 48.34
```
Step 3
```
  65.26
```
Decimal point in answer is lined up under decimal points in problem.

(b) 5.897 + 4.632 + 12.174
Write the numbers vertically with decimal points lined up. Then add.

```
  11 21
   5.897
   4.632
+ 12.174
  22.703
```
Decimal points are lined up.

(a) 2.86 + 7.09

(b) 13.761 + 8.325

(c) 0.319 + 56.007 + 8.252

(d) 39.4 + 0.4 + 177.2

─────── *Work Problem* **1** *at the Side.* ▶

In Example 1(a) above, both numbers had *two* decimal places (two digits to the right of the decimal point). In Example 1(b), all the numbers had *three decimal places* (three digits to the right of the decimal point). That made it easy to add tenths to tenths, hundredths to hundredths, and so on.

2 Find each sum.

(a) $6.54 + 9.8$

If the number of decimal places does *not* match, you can write in zeros as placeholders to make them match. This is shown in Example 2.

EXAMPLE 2 **Writing Zeros as Placeholders before Adding**

Find each sum.

(a) $7.3 + 0.85$

There are two decimal places in 0.85 (tenths and hundredths), so write a 0 in the hundredths place in 7.3 so that it has two decimal places also.

$$\begin{array}{r} 7.3\mathbf{0} \leftarrow \text{One 0 is} \\ +\ 0.85 \quad \text{written in.} \\ \hline 8.15 \end{array}$$

7.30 is equivalent to 7.3 because

$7\dfrac{30}{100}$ in lowest terms is $7\dfrac{3}{10}$

(b) $0.831 + 222.2 + 10$

(b) $6.42 + 9 + 2.576$

Write in zeros so that all the addends have three decimal places. Notice how the whole number 9 is written with the decimal point at the *far right* side. (If you put the decimal point on the *left* side of the 9, you would turn it into the decimal fraction 0.9.)

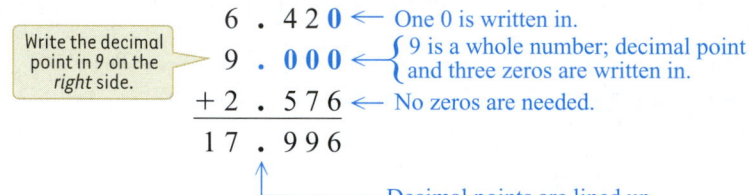

Write the decimal point in 9 on the *right* side.

$$\begin{array}{r} 6 \ . \ 4\ 2\ \mathbf{0} \leftarrow \text{One 0 is written in.} \\ 9 \ . \ \mathbf{0\ 0\ 0} \leftarrow \left\{\begin{array}{l}\text{9 is a whole number; decimal point}\\ \text{and three zeros are written in.}\end{array}\right. \\ +\ 2 \ . \ 5\ 7\ 6 \leftarrow \text{No zeros are needed.} \\ \hline 1\ 7 \ . \ 9\ 9\ 6 \end{array}$$

Decimal points are lined up.

(c) $8.64 + 39.115 + 3.0076$

Note

Writing zeros to the right of a *decimal* number does *not* change the value of the number, as shown in Example 2(a) above.

◀ *Work Problem* **2** *at the Side.*

OBJECTIVE **2** **Subtract decimals.** Subtraction of decimals is done in much the same way as addition of decimals. You can check the answers to subtraction problems using addition, as you did with whole numbers (see **Section 1.3**).

(d) $5 + 429.823 + 0.76$

EXAMPLE 3 **Subtracting Decimal Numbers**

Find each difference. Check your answers using addition.

(a) 15.82 from 28.93

Step 1

$$\begin{array}{r} 28\ .\ 93 \\ -\ 15\ .\ 82 \end{array}$$

Line up decimal points. Then you will be subtracting hundredths from hundredths and tenths from tenths.

Continued on Next Page

Step 2

$$
\begin{array}{r}
28\,.\,93 \\
-\ 15\,.\,82 \\
\hline
13\,.\,11
\end{array}
$$

Both numbers have two decimal places; no need to write in zeros.

13.11 ← Subtract as if they were whole numbers.

↑ Decimal point in answer is lined up.

Step 3

Check the answer by adding 13.11 and 15.82. If the subtraction is done correctly, the sum will be 28.93.

(b) 146.35 minus 58.98
Regrouping is needed here.

$$
\begin{array}{r}
\overset{0\ 13\ 15\quad\ 12\ 15}{1\,4\,6\,.\,3\,5} \\
-\ \ 5\,8\,.\,9\,8 \\
\hline
8\,7\,.\,3\,7
\end{array}
$$

Line up decimal points.

Check the answer by adding 87.37 and 58.98. If you did the subtraction correctly, the sum will be 146.35. (If it *isn't*, you need to rework the problem.)

Work Problem ③ *at the Side.* ▶

EXAMPLE 4 **Writing Zeros as Placeholders before Subtracting**

Find each difference.

(a) 16.5 from 28.362
Use the same steps as in Example 3 above. Remember to write in zeros so both numbers have three decimal places.

Line up decimal points.

16.500 is equivalent to 16.5

$$
\begin{array}{r}
28.362 \\
-\ 16.5\mathbf{00} \\
\hline
11.862
\end{array}
$$

← Write two zeros.
← Subtract as usual.

Check the answer by adding.

$$
\begin{array}{r}
16.500 \\
+\ 11.862 \\
\hline
28.362
\end{array}
$$

← Matches minuend in original problem.

(b) 59.7 − 38.914

$$
\begin{array}{r}
59.7\mathbf{00} \\
-\ 38.914 \\
\hline
20.786
\end{array}
$$

← Write two zeros.

← Subtract as usual.

(c) 12 less 5.83

12.00 is equivalent to 12

$$
\begin{array}{r}
12.\mathbf{00} \\
-\ 5.83 \\
\hline
6.17
\end{array}
$$

← Write a decimal point and two zeros.

← Subtract as usual.

Work Problem ④ *at the Side.* ▶

OBJECTIVE **3** **Estimate the answer when adding or subtracting decimals.** A common error in working decimal problems by hand is to misplace the decimal point in the answer. Or, when using a calculator, you may accidentally press the wrong key. **Estimating** the answer will help you avoid these mistakes. Start by using *front end rounding* on each number (as you did in **Section 1.7**). Here are several examples. Notice that in the rounded numbers, only the leftmost digit is something other than 0.

3.25	rounds to	3		6.812	rounds to	7
532.6	rounds to	500		26.397	rounds to	30
7094.2	rounds to	7000		351.24	rounds to	400

---

③ Find each difference. Check your answers using addition.

(a) 22.7 from 72.9

(b) 6.425 from 11.813

(c) 20.15 − 19.67

④ Find each difference. Check your answers using addition.

(a) 18.651 from 25.3

(b) 5.816 − 4.98

(c) 40 less 3.66

(d) 1 − 0.325

ANSWERS

3. **(a)** 50.2; 50.2 + 22.7 = 72.9
   (b) 5.388; 5.388 + 6.425 = 11.813
   (c) 0.48; 0.48 + 19.67 = 20.15
4. **(a)** 6.649; 6.649 + 18.651 = 25.3
   (b) 0.836; 0.836 + 4.98 = 5.816
   (c) 36.34; 36.34 + 3.66 = 40
   (d) 0.675; 0.675 + 0.325 = 1

⑤ First, use front end rounding and estimate each answer. Then add or subtract to find the exact answer.

(a) $2.83 + 5.009 + 76.1$

Estimate:

Exact:

(b) 11.365 from 58

Estimate:

Exact:

(c) $398.81 + 47.658 + 4158.7$

Estimate:

Exact:

(d) Find the difference between 12.837 meters and 46.091 meters.

Estimate:

Exact:

(e) $19.28 plus $1.53

Estimate:

Exact:

EXAMPLE 5 **Estimating Decimal Answers**

Use front end rounding to round each number. Then add or subtract the rounded numbers to get an estimated answer. Finally, find the exact answer.

(a) Find the sum of 194.2 and 6.825.

Estimate: *Exact:*

$$\begin{array}{r} \textbf{200} \xleftarrow{\text{Rounds to}} \textbf{194.200} \\ +\ \ \ \ \textbf{7} \xleftarrow{\text{Rounds to}} +\ \ \textbf{6.825} \\ \hline 207 \qquad\qquad 201.025 \end{array}$$

The estimate goes out to the hundreds place (three places to the *left* of the decimal point), and so does the exact answer. Therefore, the decimal point is probably in the correct place in the exact answer.

(b) $69.42 + $13.78

Estimate: *Exact:*

$$\begin{array}{r} \$70 \xleftarrow{\text{Rounds to}} \$69.42 \\ +\ \ 10 \xleftarrow{\text{Rounds to}} +\ \ 13.78 \\ \hline \$80 \qquad\qquad \$83.20 \end{array}$$
← Exact answer is close to estimate, so it is probably correct.

(c) Find the difference between 0.92 ft and 8 ft.

Use subtraction to find the difference between two numbers. The larger number, 8, is written on top.

Estimate: *Exact:*

$$\begin{array}{r} 8 \xleftarrow{\text{Rounds to}} 8.\textbf{00} \\ -\ 1 \xleftarrow{\text{Rounds to}} -\ 0.92 \\ \hline 7 \qquad\qquad 7.08 \text{ ft} \end{array}$$
8.**00** ← Write a decimal point and two zeros.
7.08 ft ← Exact answer is close to estimate.

(d) Subtract 1.8614 from 7.3.

Estimate: *Exact:*

$$\begin{array}{r} 7 \xleftarrow{\text{Rounds to}} 7.\textbf{3000} \\ -\ 2 \xleftarrow{\text{Rounds to}} -\ 1.8614 \\ \hline 5 \qquad\qquad 5.4386 \end{array}$$
7.3**000** ← Write three zeros.
5.4386 ← Exact answer is close to estimate.

◀ **Work Problem ⑤ at the Side.**

> 🖩 **Calculator Tip** If you are *adding* numbers, you can enter them in any order on your calculator. Try these; jot down the answers.
>
>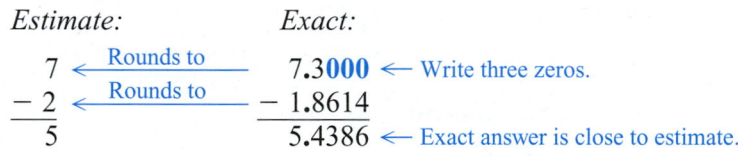
> 9.82 ⊕ 1.86 ⊜ _____   1.86 ⊕ 9.82 ⊜ _____
>
> The answers are the same because addition is *commutative*. (See **Section 1.2.**) But subtraction is *not* commutative. It *does* matter which number you enter first. Try these:
>
> 9.82 ⊖ 1.86 ⊜ _____   1.86 ⊖ 9.82 ⊜ _____
>
> The second answer has a negative sign (−) next to it. A negative number is *less* than 0. If it was in your checkbook, you'd be "in the hole" by $7.96. (See **Chapter 9** for more about negative numbers.)

Find each sum or difference. See Examples 1–4.

1. $5.69 + 11.79$

2. $372.1 - 33.7$

3. $24.008 - 0.995$

4. $0.7759 + 9.8883$

5. $8.263 - 0.5$

6. $47.658 - 20.9$

7. $76.5 + 0.506$

8. $1.87 + 9.749$

9. $21 - 0.896$

10. $9 - 1.183$

11. Subtract 0.291 from 0.4

12. Subtract 0.088 from 0.35

13. $39.76005 + 182 + 4.799 + 98.31 + 5.9999$

14. $489.76 + 0.9993 + 38 + 8.55087 + 80.697$

This drawing of a human skeleton shows the average length of the longest bones, in inches. Use the drawing to answer Exercises 15–18. (Source: Top Ten of Everything.)

15. (a) What is the combined length of the humerus and radius bones?

   (b) What is the difference in the lengths of these two bones?

16. (a) What is the total length of the femur and tibia bones?

   (b) How much longer is the femur than the tibia?

17. (a) Find the sum of the lengths of the humerus, ulna, femur, and tibia.

   (b) How much shorter is the 8th rib than the 7th rib?

18. (a) What is the difference in the lengths of the two bones in the lower arm?

   (b) What is the difference in the lengths of the two bones in the lower leg?

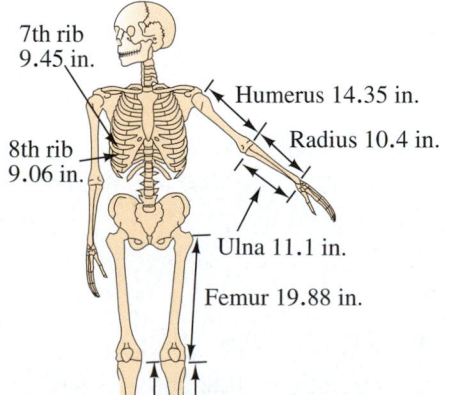

7th rib 9.45 in.
Humerus 14.35 in.
Radius 10.4 in.
8th rib 9.06 in.
Ulna 11.1 in.
Femur 19.88 in.
Tibia 16.94 in.
Fibula 15.94 in.

19. Explain and correct
the error that a student
made when he added
$0.72 + 6 + 39.5$ this way:

$$
\begin{array}{r}
0.72 \\
6 \\
+\ 39.50 \\
\hline
40.28
\end{array}
$$

20. Explain the difference between saying "subtract 2.9 from 8" and saying "2.9 minus 8."

Use front end rounding to round each number. Then add or subtract the rounded numbers to get an estimated answer. Finally, find the exact answer. See Example 5.

21. *Estimate:*   *Exact:*

$$
\begin{array}{r}
\$19.74 \\
-\underline{} \qquad -\ \ 6.58 \\
\end{array}
$$

22. *Estimate:*   *Exact:*

$$
\begin{array}{r}
\$27.96 \\
-\underline{} \qquad -\ \ 8.39 \\
\end{array}
$$

23. *Estimate:*   *Exact:*

$$
\begin{array}{r}
392.7 \\
0.865 \\
+\underline{} \qquad +\ 21.08 \\
\end{array}
$$

24. *Estimate:*   *Exact:*

$$
\begin{array}{r}
38.55 \\
7.716 \\
+\underline{} \qquad +\ \ 0.6 \\
\end{array}
$$

25. What is 8.6 less 3.751?

   Estimate: *Exact:*

26. What is 31.7 less 4.271?

   Estimate: *Exact:*

27. *Estimate:*   *Exact:*

$$
\begin{array}{r}
62.8173 \\
539.99 \\
+\underline{} \qquad +\ \ 5.629 \\
\end{array}
$$

28. *Estimate:*   *Exact:*

$$
\begin{array}{r}
332.607 \\
12.5 \\
+\underline{} \qquad +\ 823.3949 \\
\end{array}
$$

Use your estimation skills to pick the most reasonable answer for each example.
*Do **not** solve the problems. Circle your choice.*

29. $12 - 11.725$

   2.75   0.275   27.5

30. $20 - 1.37$

   0.1863   1.863   18.63

31. $6.5 + 0.007$

   6.507   0.6507   65.07

32. $9.67 + 0.09$

   0.976   9.76   0.00976

33. $456.71 - 454.9$

   18.1   181   1.81

34. $803.25 - 0.6$

   802.65   0.80265   8.0265

35. $6004.003 + 52.7172$

   60.567202   605.67202   6056.7202

36. $128.35 + 97.0093$

   2253.593   225.3593   0.2253593

First use front end rounding to round each number and estimate the answer. Then find the
exact answer. Use the information in the table below for Exercises 37–40.

INTERNET USERS IN SELECTED COUNTRIES

Country	Number of Users
United States	197.8 million
China	119.5 million
Japan	86.3 million
India	50.6 million
South Korea	34 million
Canada	21.9 million
Mexico	16.9 million
World total	**1081.1 million**

Source: Computer Industry Almanac.

37. How many fewer Internet users are there in Canada
compared to South Korea?

Estimate:

Exact:

38. How many more users are there in China compared
to India?

Estimate:

Exact:

39. How many Internet users are there in all the
countries listed in the table?

Estimate:

Exact:

40. Using the answer from Exercise 39, calculate the
number of worldwide Internet users in countries
other than the ones in the table.

Estimate:

Exact:

41. The tallest known land mammal is a prehistoric
ancestor of the rhino measuring 6.4 meters. Find the
combined heights of these NBA basketball stars:
Allen Iverson at 1.83 meters, Shaquille O'Neal at
2.16 meters, and Kevin Garnett at 2.11 meters. Is
their combined height greater or less than the
prehistoric rhino? By how much?
(*Source:* www.NBA.com/players)

6.4 meters

Estimate:

Exact:

42. At a bakery, Sue Chee bought $7.42 worth of
muffins and $10.09 worth of croissants for a staff
party and a $0.69 cookie for herself. How much
change did she receive from two $10 bills?

Estimate:

Exact:

43. Namiko is comparing two boxes of chicken nuggets.
One box weighs 9.85 ounces and the other weighs
10.5 ounces. What is the difference in the weight of
the two boxes?

Estimate:

Exact:

44. Sammy works in a veterinarian's office. He weighed
two young kittens. One was 3.9 ounces and the
other was 4.05 ounces. What was the difference in
the weight of the two kittens?

Estimate:

Exact:

Find the perimeter of (distance around) each figure by adding the lengths of the sides.

45.

19.75 in.

6.3 in.   6.3 in.

19.75 in.

Estimate:

Exact:

46.

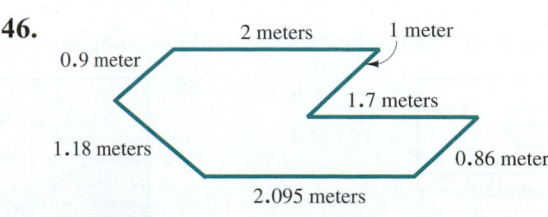

2 meters   1 meter

0.9 meter

1.7 meters

1.18 meters

0.86 meter

2.095 meters

Estimate:

Exact:

The father and daughter on the first page of this chapter are buying fishing equipment. They brought along the store's sale insert from the Sunday paper. Use the information below on sale prices to answer Exercises 47–50. When estimating, round to the nearest whole number.

★ **Fishing Opener Sale** ★
Catch your limit of savings!

Bobbers 3 for 87¢

8-pound test fishing line
regular $4.84
invisible $7.47   No-See Line
fluorescent $5.14

Environmentally safe
tin split shot
$2.07

Tackle boxes
Two trays $7.96
Three trays $9.96

Leaded split shot
94¢

Spinning reels: $9.88, $12.54, $18.84, $24.96
Spinning rods: $9.97, $18.97, $22.96, $28.94

Source: Wal-Mart.

47. What is the difference in price between the fluorescent and regular fishing line?

Estimate:

Exact:

48. How much more does the least expensive spinning rod cost than the least expensive spinning reel?

Estimate:

Exact:

49. Find the total cost of the second highest priced spinning reel, two packages of tin split shot, and a three-tray tackle box. Sales tax for all the items was $2.31.

Estimate:

Exact:

50. The father bought three bobbers on sale. He also bought some SPF15 sunscreen for $7.53 and a flotation vest for $44.96. Sales tax was $3.74. How much did he spend in all?

Estimate:

Exact:

Olivia Sanchez kept track of her expenses for one month. Use her list to answer Exercises 51–56.

MONTHLY EXPENSES

Rent	$994
Car payment	$190.78
Car repairs, gas	$205
Cable TV	$39.95
Internet access	$19.95
Electricity	$40.80
Cell phone	$57.32
Groceries	$186.81
Entertainment	$97.75
Clothing, laundry	$107

51. What were Olivia's total expenses for the month?

52. How much did Olivia pay for cell phone, cable TV, and Internet access?

53. What was the difference in the amounts spent for groceries and for the car payment?

54. Compare the amount Olivia spent on entertainment to the amount spent on car repairs and gas. What is the difference?

55. How much more did Olivia spend on rent than on all her car expenses?

56. How much less did Olivia spend on clothing and laundry than on all her car expenses?

Find the length of the dashed line in each rectangle or circle.

57.

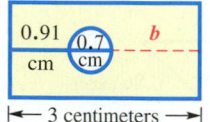

0.91 cm   0.7 cm   *b*
|← 3 centimeters →|

58.

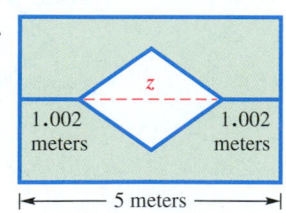

z
1.002 meters   1.002 meters
|← 5 meters →|

59.

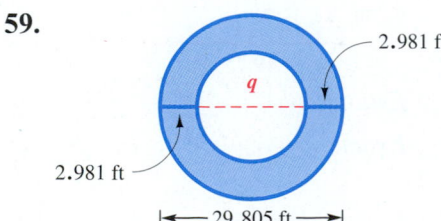

2.981 ft
q
2.981 ft
|← 29.805 ft →|

4.4 ►►► Multiplying Decimals

OBJECTIVE 1 Multiply decimals. The decimals 0.3 and 0.07 can be multiplied by writing them as fractions.

$$0.\underset{\uparrow}{3} \quad \times \quad 0.0\underset{\vee}{7} \quad = \frac{3}{10} \times \frac{7}{100} = \frac{3 \times 7}{10 \times 100} = \frac{21}{1000} = 0.0\underset{\uparrow\uparrow\uparrow}{21}$$

1 decimal place + 2 decimal places → 3 decimal places

Can you see a way to multiply decimals without writing them as fractions? Try these steps. Remember that each number in a multiplication problem is called a *factor,* and the answer is called the *product.*

Multiplying Decimals

Step 1 Multiply the numbers (the factors) as if they were whole numbers.

Step 2 Find the *total* number of decimal places in *both* factors.

Step 3 Write the decimal point in the product (the answer) so it has the same number of decimal places as the total from Step 2. You may need to write in extra zeros on the left side of the product to get the correct number of decimal places.

Note

When multiplying decimals, you do **not** need to line up decimal points. (You **do** need to line up decimal points when adding or subtracting.)

EXAMPLE 1 Multiplying Decimal Numbers

Find the product of 8.34 and 4.2

Step 1 Multiply the numbers as if they were whole numbers.

```
      8.3 4     ⎫ You do not have to
    ×   4.2     ⎬ line up decimal points
    ───────     ⎭ when multiplying.
      1 6 6 8
    3 3 3 6
    ─────────
    3 5 0 2 8
```

Step 2 Count the total number of decimal places in both factors.

```
      8.3 4 ← 2 decimal places
    ×   4.2 ← 1 decimal place
    ─────────
      1 6 6 8    3 total decimal places
    3 3 3 6
    ─────────
    3 5 0 2 8
```

Step 3 Count over 3 places in the product and write the decimal point. Count from *right to left.*

```
      8.3 4 ← 2 decimal places
    ×   4.2 ← 1 decimal place
    ─────────
      1 6 6 8    3 total decimal places
    3 3 3 6
    ─────────
    3 5.0 2 8 ← 3 decimal places in product
```

Count over 3 places from right to left to position the decimal point.

Work Problem **1** *at the Side.* ►

OBJECTIVES

1 Multiply decimals.

2 Estimate the answer when multiplying decimals.

1 Find each product.

(a)   2.6
     × 0.4

(b)   45.2
     × 0.25

(c)   0.104 ← 3 decimal places
     ×   7 ← 0 decimal places
              ← 3 decimal places
                in the product

(d)   3.18
     × 2.23

(e)   611
     × 3.7

ANSWERS

1. (a) 1.04   (b) 11.300   (c) 0.728
(d) 7.0914   (e) 2260.7

2 Find each product.

(a) 0.04×0.09

(b) $(0.2)(0.008)$

(c) $(0.003)^2$ *Hint:* Recall that the 2 is an exponent. See **Section 1.8.**

(d) $(0.0081)(0.003)$

(e) $(0.11)(0.0005)$

3 First use front end rounding and estimate the answer. Then find the exact answer.

(a) $(11.62)(4.01)$

(b) $(5.986)(33)$

(c) $8.31(4.2)$

(d) 58.6×17.4

EXAMPLE 2   **Writing Zeros as Placeholders in the Product**

Find the product: $(0.042)(0.03)$
  Start by multiplying, then count decimal places.

$$
\begin{array}{r}
0.0\,4\,2 \leftarrow \text{3 decimal places} \\
\times \quad 0.0\,3 \leftarrow \text{2 decimal places} \\
\hline
1\,2\,6 \leftarrow \text{5 decimal places needed in product}
\end{array}
$$

After multiplying, the answer has only three decimal places, but five are needed, so write two zeros on the *left* side of the answer.

$$
\begin{array}{r}
0.0\,4\,2 \\
\times \quad 0.0\,3 \\
\hline
\mathbf{0\,0}\,1\,2\,6
\end{array}
\qquad
\begin{array}{r}
0.0\,4\,2 \leftarrow \text{3 decimal places} \\
\times \quad 0.0\,3 \leftarrow \text{2 decimal places} \\
\hline
.0\,0\,1\,2\,6 \leftarrow \text{5 decimal places}
\end{array}
$$

Write two zeros on *left* side of answer.   Now count over 5 places and write in the decimal point.

The final product is 0.00126, which has five decimal places.

◀ Work Problem **2** at the Side.

OBJECTIVE **2** **Estimate the answer when multiplying decimals.**
If you are doing multiplication problems by hand, estimating the answer helps you check that the decimal point is in the right place. When you are using a calculator, estimating helps you catch an error like pressing the ÷ key instead of the × key.

EXAMPLE 3   **Estimating before Multiplying**

First estimate the answer to $(76.34)(12.5)$ using front end rounding. Then find the exact answer.

Estimate: *Exact:*

$$
\begin{array}{r}
80 \quad \overset{\text{Rounds to}}{\longleftarrow} \\
\times 10 \quad \overset{\text{Rounds to}}{\longleftarrow} \\
\hline
800
\end{array}
\qquad
\begin{array}{r}
7\,6.3\,4 \leftarrow \text{2 decimal places} \\
\times \quad 1\,2.5 \leftarrow \text{1 decimal place} \\
\hline
3\,8\,1\,7\,0 \quad \text{3 decimal places needed in product} \\
1\,5\,2\,6\,8 \quad \\
7\,6\,3\,4 \quad\quad \\
\hline
9\,5\,4.2\,5\,0 \quad
\end{array}
$$

Both the estimate and the exact answer go out to the hundreds place, so the decimal point in 954.250 is probably in the correct place.

◀ Work Problem **3** at the Side.

▦ **Calculator Tip**   When working with money amounts, you may need to write a 0 in your answer. For example, try multiplying $\$3.54 \times 5$ on your calculator. Write down the result.

$$3.54 \; \times \; 5 \; = \; \underline{\hspace{2cm}}$$

Notice that the result is 17.7, which is *not* the way to write a money amount. You have to write the 0 in the hundredths place: $\$17.7\mathbf{0}$ is correct. The calculator does not show the "extra" 0 because:

$$17.70 \text{ or } 17\frac{70}{100} \quad \text{simplifies to} \quad 17\frac{7}{10} \text{ or } 17.7$$

So keep an eye on your calculator—it doesn't know when you're working with money amounts.

ANSWERS

2. (a) 0.0036 (b) 0.0016 (c) 0.000009
   (d) 0.0000243 (e) 0.000055
3. (a) $(10)(4) = 40$; 46.5962
   (b) $(6)(30) = 180$; 197.538
   (c) $8(4) = 32$; 34.902
   (d) $60 \times 20 = 1200$; 1019.64

4.4 ▶▶▶   **Exercises**

FOR EXTRA HELP       PRACTICE   WATCH   DOWNLOAD   READ   REVIEW

Find each product. See Example 1.

 1.   0.042
       × 3.2

2.   0.571
       × 2.9

3.   21.5
       × 7.4

4.   85.4
       × 3.5

5. (0.666) (23.4)

6. (0.799) (0.896)

7. $51.88
       ×   665

8. $736.75
       ×   118

Use the fact that 72 × 6 = 432 *to solve Exercises 9–16 by simply counting decimal places and writing the decimal point in the correct location. See Examples 1 and 2.*

9. 72 × 0.6 =   4 3 2

10. 7.2 × 6 =   4 3 2

11. (7.2) (0.06) =   4 3 2

12. (0.72) (0.6) =   4 3 2

13. 0.72 (0.06) =   4 3 2

14. 72 (0.0006) =   4 3 2

15. 0.0072 (0.6) =   4 3 2

16. 0.072 (0.006) =   4 3 2

Find each product. See Example 2.

 **17.** (0.006) (0.0052)

18. (0.0052) (0.009)

19. $(0.005)^2$

20. $(0.03)^2$

Relating Concepts (Exercises 21–22) For Individual or Group Work

Look for patterns in the multiplications as you **work Exercises 21 and 22 in order.**

21. Do these multiplications:

(5.96) (10) = _____   (3.2) (10) = _____

(0.476) (10) = _____   (80.35) (10) = _____

(722.6) (10) = _____   (0.9) (10) = _____

What pattern do you see? Write a "rule" for multiplying by 10. What do you think the rule is for multiplying by 100? by 1000? Write the rules and try them out on the numbers above.

22. Do these multiplications:

(59.6) (0.1) = _____   (3.2) (0.1) = _____

(0.476) (0.1) = _____   (80.35) (0.1) = _____

(65) (0.1) = _____   (523) (0.1) = _____

What pattern do you see? Write a "rule" for multiplying by 0.1. What do you think the rule is for multiplying by 0.01? by 0.001? Write the rules and try them out on the numbers above.

First use front end rounding to round each number and estimate the answer. Then find the exact answer. See Example 3.

23. *Estimate:*   *Exact:*

      ←Rounds to 39.6
  × ___ ←Rounds to × 4.8

24. *Estimate:*   *Exact:*

          18.7
  × ___   × 2.3

25. *Estimate:*   *Exact:*

          37.1
  × ___   × 42

26. *Estimate:*   *Exact:*

          5.08
  × ___   × 71

27. *Estimate:*   *Exact:*

          6.53
  × ___   × 4.6

28. *Estimate:*   *Exact:*

          7.51
  × ___   × 8.2

29. *Estimate:*   *Exact:*

          2.809
  × ___   × 6.85

30. *Estimate:*   *Exact:*

          73.52
  × ___   × 22.34

Even with most of the problem missing, you can tell whether or not these answers are reasonable. Circle *reasonable or unreasonable. If the answer is unreasonable, move the decimal point, or insert a decimal point, to make the answer reasonable.*

31. How much was his car payment? $28.90

  reasonable

  unreasonable, should be _____

32. How many hours did she work today? 25 hours

  reasonable

  unreasonable, should be _____

33. How tall is her son? 60.5 in.

  reasonable

  unreasonable, should be _____

34. How much does he pay for rent now? $6.92

  reasonable

  unreasonable, should be _____

35. What is the price of one gallon of milk? $419

  reasonable

  unreasonable, should be _____

36. How long is the living room? 16.8 feet

  reasonable

  unreasonable, should be _____

37. How much did Mrs. Brown's baby weigh? 0.095 pound

  reasonable

  unreasonable, should be _____

38. What was the sale price of the jacket? $1.49

  reasonable

  unreasonable, should be _____

Solve each application problem. Round money answers to the nearest cent when necessary.

39. LaTasha worked 50.5 hours over the last two weeks. She earns $18.73 per hour. How much did she make?

40. Michael's time card shows 42.2 hours at $10.03 per hour. What are his earnings?

41. Sid needs 0.6 meter of canvas material to make a carry-all bag that fits on his wheelchair. If canvas is $4.09 per meter, how much will Sid spend? (*Note:* $4.09 *per* meter means $4.09 for *one* meter.)

42. How much will Mrs. Nguyen pay for 3.5 yards of lace trim that costs $0.87 per yard?

43. Michelle filled the tank of her pickup truck with regular unleaded gas. Use the information shown on the pump to find how much she paid for gas.

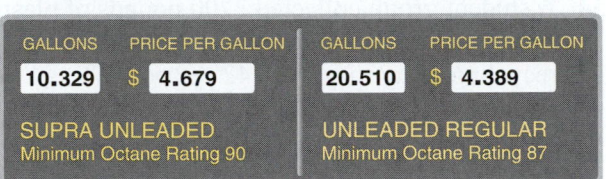

GALLONS	PRICE PER GALLON	GALLONS	PRICE PER GALLON
10.329	$ 4.679	20.510	$ 4.389
SUPRA UNLEADED		UNLEADED REGULAR	
Minimum Octane Rating 90		Minimum Octane Rating 87	

Source: Holiday.

44. Ground beef and chicken legs are on sale. Juma bought 1.7 pounds of legs. Use the information in the ad to find the amount she paid.

BIG ONE FOODS   Sale

Ground Beef $2.09 per pound   For juicy burgers

Chicken Legs $0.98 per pound

PRICES GOOD THROUGH SUNDAY!

45. Ms. Rolack is a real estate broker who helps people sell their homes. Her fee is 0.07 times the price of the home. What was her fee for selling a $289,500 home?

46. Manny Ramirez of the Boston Red Sox had a batting average of 0.296 in the 2007 season. He went to bat 483 times. How many hits did he make? (*Hint:* Multiply the number of times at bat by his batting average.) Round to the nearest whole number. (*Source: World Almanac.*)

Paper money in the United States has not always been the same size. Shown below are the measurements of bills printed before 1929 and the measurements from 1929 on. Use this information to answer Exercises 47–50. Recall that perimeter is the total distance around the edges of a figure (see **Section 1.2**). *The area of a rectangle is found by multiplying the length by the width (see* **Section 2.5**). (*Source:* www.moneyfactory.com)

Before 1929
3.125 in.
7.4218 in.

From 1929 on
2.61 in.
6.14 in.

47. (a) Find the area of each bill, rounded to the nearest tenth.

(b) What is the difference in the rounded areas?

48. (a) Find the perimeter of each bill, to the nearest hundredth.

(b) How much less is the perimeter of today's bills than the bills printed before 1929?

49. The thickness of one piece of today's paper money is 0.0043 inch.

(a) If you had a pile of 100 bills, how high would the pile be?

(b) How high would a pile of 1000 bills be?

50. (a) Use your answers from Exercise 49 to find the number of bills in a pile that is 43 inches high.

(b) How much money would you have if the pile is all $20 bills?

51. Judy Lewis pays $38.96 per month for basic cable TV. The one-time installation fee was $49. How much will she pay for cable over two years? How much would she pay in two years for the deluxe cable package that costs $89.95 per month?

52. Chuck's car payment is $420.27 per month for four years. He also made a down payment of $5000 at the time he bought the car. How much will he pay altogether?

53. Paper for the copy machine at the library costs $0.015 per sheet. How much will the library pay for 5100 sheets?

54. A student group collected 2200 pounds of plastic as a fund-raiser. How much will they make if the recycling center pays $0.142 per pound?

55. Barry bought 16.5 meters of rope at $0.47 per meter and three meters of wire at $1.05 per meter. How much change did he get from three $5 bills?

56. Susan bought a 42-inch plasma HDTV that cost $1999.99. She paid $68.83 per month for 36 months. How much could she have saved by paying for the HDTV when she bought it?

Use the information from the Look Smart mail-order catalog to answer Exercises 57–60.

43–2A 43–2B
43–3A 43–3B

Knit Shirt Ordering Information			
43-2A	Short sleeved, solid colors	$14.75 each	
43-2B	Short sleeved, stripes	$16.75 each	
43-3A	Long sleeved, solid colors	$18.95 each	
43-3B	Long sleeved, stripes	$21.95 each	
XXL size, add $2 per shirt.			
Monogram, $4.95 each. Gift box, $5 each.			

Total Price of All Items (excluding monograms and gift boxes)	Shipping, Packing, and Handling
$0–25.00	$3.50
$25.01–75.00	$5.95
$75.01–125.00	$7.95
$125.01 +	$9.95
Shipping to each additional address, add $4.25.	

57. Find the total cost of ordering four long-sleeved, solid-color shirts and two short-sleeved, striped shirts, all in the XXL size, and all shipped to your home.

58. What is the total cost of eight long-sleeved shirts, size L, five in solid colors and three striped? Include the cost of shipping the solid shirts to your home and the striped shirts to your brother's home.

59. (a) What is the total cost, including shipping, of sending three short-sleeved, solid-color shirts, size M, with monograms, in a gift box to your aunt for her birthday?

(b) How much did the monograms, gift box, and shipping add to the cost of your gift?

60. (a) Suppose you order one of each type of shirt for yourself, adding a monogram to each of the solid-color shirts. At the same time, you order three long-sleeved striped shirts, in the XXL size, shipped to your dad in a gift box. Find the total cost of your order.

(b) What is the difference in total cost (excluding shipping) between the shirts for yourself and the gift for your dad?

4.5 ▶▶▶ Dividing Decimals

There are two kinds of decimal division problems: those in which a decimal is divided by a whole number, and those in which a number is divided by a decimal. First recall the parts of a division problem from **Section 1.5.**

$$
\begin{array}{r}
8 \leftarrow \text{Quotient} \\
\text{Divisor} \rightarrow 4\overline{)33} \leftarrow \text{Dividend} \\
\underline{32} \\
1 \leftarrow \text{Remainder}
\end{array}
$$

OBJECTIVE 1 Divide a decimal by a whole number. When the divisor is a whole number, use these steps.

> **Dividing Decimals by Whole Numbers**
>
> **Step 1** Write the decimal point in the quotient (answer) directly above the decimal point in the dividend.
>
> **Step 2** Divide as if both numbers were whole numbers.

> **EXAMPLE 1** Dividing Decimals by Whole Numbers

Find each quotient. Check the quotients by multiplying.

(a) 21.93 by 3

Dividend ⌣ └ Divisor

Rewrite the division problem. $3\overline{)21.93}$

Step 1 Write the decimal point in the quotient directly above the decimal point in the dividend.

Decimal points lined up

$3\overline{)21\ .\ 93}$

Step 2 Divide as if the numbers were whole numbers.

$$
\begin{array}{r}
7.31 \\
3\overline{)21.93}
\end{array}
$$

Check by multiplying the quotient times the divisor.

Matches, so 7.31 is correct.

Check:
$$
\begin{array}{r}
7.31 \\
\times\ \ \ 3 \\
\hline
21.93
\end{array}
$$

The quotient (answer) is 7.31.

(b) $9\overline{)470.7}$

Divisor ┘ └ Dividend

Write the decimal point in the quotient above the decimal point in the dividend. Then divide as if the numbers were whole numbers.

Decimal points lined up

$$
\begin{array}{r}
52.3 \\
9\overline{)470.7} \\
\underline{45} \\
20 \\
\underline{18} \\
27 \\
\underline{27} \\
0
\end{array}
$$

Check:
$$
\begin{array}{r}
52.3 \\
\times\ \ \ 9 \\
\hline
470.7
\end{array}
$$

Matches

Multiply the quotient by the divisor. The result should match the dividend.

The quotient is 52.3.

Work Problem **1** at the Side. ▶

1 Find each quotient. Check the quotients by multiplying.

(a) $4\overline{)93.6}$

(b) $6\overline{)6.804}$

(c) $11\overline{)278.3}$

(d) $0.51835 \div 5$

(e) $213.45 \div 15$

ANSWERS

1. **(a)** 23.4; (23.4)(4) = 93.6
   (b) 1.134; (1.134)(6) = 6.804
   (c) 25.3; (25.3)(11) = 278.3
   (d) 0.10367; (0.10367)(5) = 0.51835
   (e) 14.23; (14.23)(15) = 213.45

2 Divide. Check each quotient by multiplying.

(a) $5\overline{)6.4}$

(b) $30.87 \div 14$

(c) $\dfrac{259.5}{30}$

(d) $0.3 \div 8$

EXAMPLE 2   **Writing Extra Zeros to Complete a Division**

Divide 1.5 by 8. Check the quotient by multiplying.

Keep dividing until the remainder is 0, or until the digits in the quotient begin to repeat in a pattern. In Example 1(b), you ended up with a remainder of 0. But sometimes you run out of digits in the dividend before that happens. If so, write extra zeros on the right side of the dividend so you can continue dividing.

$$
\begin{array}{r}
0.1 \\
8\overline{)1.5} \\
\underline{8} \\
7 \\
\end{array}
$$
$\leftarrow$ All digits have been used.
$\leftarrow$ Remainder is not yet 0.

Write a 0 after the 5 in the dividend so you can continue dividing. Keep writing more zeros in the dividend, if needed. Recall that writing zeros to the *right* of a decimal number does ***not*** change its value.

Check:

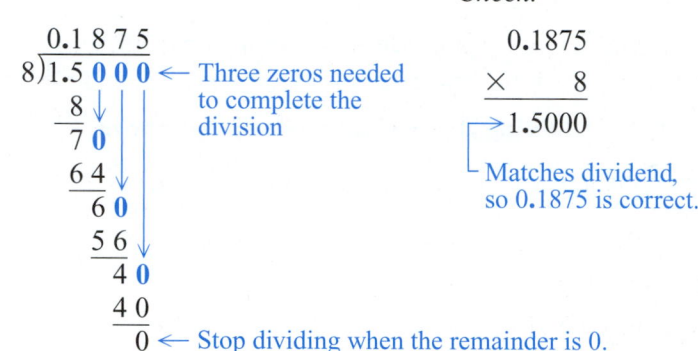

$\leftarrow$ Three zeros needed to complete the division

$\leftarrow$ Stop dividing when the remainder is 0.

Matches dividend, so 0.1875 is correct.

> **CAUTION**
>
> When dividing decimals, notice that the dividend might *not* be the larger number. In Example 2 above, the dividend is 1.5, which is *smaller* than the divisor 8.

◀ *Work Problem* **2** *at the Side.*

▦ **Calculator Tip**   When *multiplying* numbers, you can enter them in any order because multiplication is commutative (see **Section 1.4**). But division is *not* commutative. It *does* matter which number you enter first. Try Example 2 both ways; jot down your answers.

$$1.5 \;÷\; 8 \;=\; \underline{\hspace{1cm}} \qquad 8 \;÷\; 1.5 \;=\; \underline{\hspace{1cm}}$$

Notice that the first answer, 0.1875, matches the result from Example 2. But the second answer is much different: 5.333333333. Be careful to enter the dividend first.

The next example shows a quotient (answer) that must be rounded because you will never get a remainder of 0.

EXAMPLE 3 **Rounding a Decimal Quotient**

Divide 4.7 by 3. Round the quotient to the nearest thousandth. Write extra zeros in the dividend so you can continue dividing.

```
      1.5 6 6 6
   3)4.7 0 0 0  ← Three zeros added so far
     3
     ─
     1 7
     1 5
     ───
       2 0
       1 8
       ───
         2 0
         1 8
         ───
           2 0
           1 8
           ───
             2  ← Remainder is still not 0.
```

Notice that the digit 6 in the answer is repeating. It will continue to do so. The remainder will *never be 0*. There are two ways to show that the answer is a **repeating decimal** that goes on forever. You can write three dots after the answer, or you can write a bar above the digits that repeat (in this case, the 6).

$$1.5\underbrace{666}\ldots \quad \text{or} \quad 1.5\overset{-}{6} \quad \overset{\leftarrow \text{ Bar above}}{\text{repeating digit}}$$

Three dots

When repeating decimals occur, round the quotient according to the directions in the problem. In this example, to round to thousandths, divide out one *more* place, to ten-thousandths.

$$4.7 \div 3 = 1.5666\ldots \quad \text{rounds to} \quad 1.567$$

> Nearest thousandth is *three* decimal places.

Check the answer by multiplying 1.567 by 3. Because 1.567 is a rounded answer, the check will not give exactly 4.7, but it should be very close.

$$(1.567)(3) = 4.701 \leftarrow \text{Does not equal exactly 4.7} \\ \text{because 1.567 was rounded}$$

CAUTION
When checking quotients that you've rounded, the check will *not* match the dividend exactly, but it should be very close.

Work Problem **3** *at the Side.* ▶

OBJECTIVE 2 Divide a number by a decimal. To divide by a *decimal* divisor, first change the divisor to a whole number. Then divide as before. To see how this is done, write the problem in fraction form. Here is an example.

$$1.2)\overline{6.36} \quad \text{can be written} \quad \frac{6.36}{1.2}$$

In **Section 3.2** you learned that multiplying the numerator and denominator by the same number gives an equivalent fraction. We want the divisor (1.2) to be a whole number. Multiplying by 10 will accomplish that.

$$\underset{\substack{\text{Decimal} \\ \text{divisor}}}{\frac{6.36}{1.2}} = \frac{(6.36)(10)}{(1.2)(10)} = \underset{\substack{\text{Whole number} \\ \text{divisor}}}{\frac{63.6}{12}}$$

3 Divide. Round quotients to the nearest thousandth. If it is a repeating decimal, also write the answer using a bar. Check your quotients by multiplying.

(a) $13)\overline{267.01}$

(b) $6)\overline{20.5}$

(c) $\dfrac{10.22}{9}$

(d) $16.15 \div 3$

(e) $116.3 \div 11$

ANSWERS

3. (a) 20.539 (rounded); no repeating digits visible on calculator; (20.539)(13) = 267.007
   (b) 3.417 (rounded); 3.41$\overline{6}$; (3.417)(6) = 20.502
   (c) 1.136 (rounded); 1.13$\overline{5}$; (1.136)(9) = 10.224
   (d) 5.383 (rounded); 5.38$\overline{3}$; (5.383)(3) = 16.149
   (e) 10.573 (rounded); 10.5$\overline{72}$; (10.573)(11) = 116.303

The short way to multiply by 10 is to move the decimal point *one place* to the *right* in both the divisor and the dividend.

$$1.2\,)\overline{6.3\,6} \quad \text{is equivalent to} \quad 12\,)\overline{63.6}$$

4 Divide. If the quotient does not come out even, round to the nearest hundredth.

> **Note**
> Moving the decimal points the **same** number of places in **both** the divisor and dividend will **not** change the answer.

(a) $0.2\,)\overline{1.04}$

Dividing by Decimals

Step 1 Count the number of decimal places in the divisor and move the decimal point that many places to the *right*. (This changes the divisor to a whole number.)

Step 2 Move the decimal point in the dividend the *same* number of places to the *right*. (Write in extra zeros if needed.)

Step 3 Write the decimal point in the quotient directly above the decimal point in the dividend. Then divide as usual.

(b) $0.06\,)\overline{1.8072}$

(c) $0.005\,)\overline{32}$

EXAMPLE 4 **Dividing by Decimals**

(a) $0.003\,)\overline{27.69}$

Move the decimal point in the divisor *three* places to the *right* so 0.003 becomes the whole number 3. To move the decimal point in the dividend the same number of places, write in an extra 0.

(d) $8.1 \div 0.025$

$$0.003\,)\overline{27.690}$$

Move decimal points in divisor and dividend. Then line up the decimal point in the quotient.

Moving decimal point three places to the right is the same as multiplying by 1000. ⟶

$$\begin{array}{r} 9230. \\ 3\,)\overline{27690.} \end{array}$$

Divide as usual.

(e) $\dfrac{7}{1.3}$

(b) Divide 5 by 4.2. Round to the nearest hundredth.

Move the decimal point in the divisor one place to the right so 4.2 becomes the whole number 42. The decimal point in the dividend starts on the right side of 5 and is also moved one place to the right.

(f) $5.3091 \div 6.2$

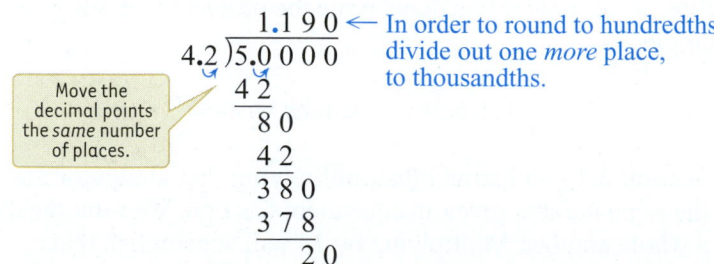

← In order to round to hundredths, divide out one *more* place, to thousandths.

Move the decimal points the *same* number of places.

$$\begin{array}{r} 1.1\,9\,0 \\ 4.2\,)\overline{5.0\,0\,0\,0} \\ \underline{4\,2} \\ 8\,0 \\ \underline{4\,2} \\ 3\,8\,0 \\ \underline{3\,7\,8} \\ 2\,0 \end{array}$$

Round the quotient. It is 1.19 (rounded to the nearest hundredth).

◀ *Work Problem* **4** *at the Side.*

OBJECTIVE 3 **Estimate the answer when dividing decimals.**
Estimating answers helps you catch errors. Compare the estimate to your exact answer. If they are very different, work the problem again.

EXAMPLE 5 Estimating before Dividing

First use front end rounding to round each number and estimate the answer. Then divide to find the exact answer.

$$580.44 \div 2.8$$

Here is how one student solved this problem. She rounded 580.44 to 600 and rounded 2.8 to 3 to estimate the answer.

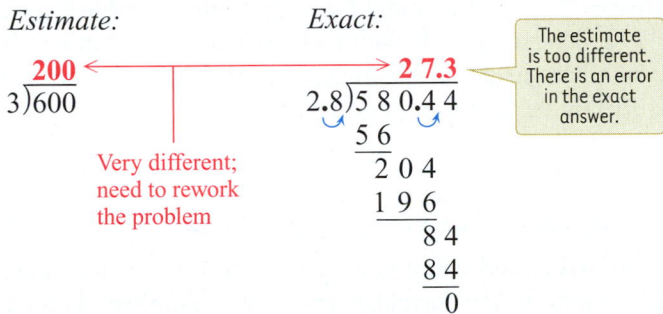

Estimate:

200
3)600

Very different; need to rework the problem

Exact:

2 7.3
2.8)5 8 0.4 4
　5 6
　2 0 4
　1 9 6
　　8 4
　　8 4
　　　0

The estimate is too different. There is an error in the exact answer.

Notice that the estimate, which is in the hundreds, is very different from the exact answer, which is only in the tens. This tells the student that she needs to rework the problem. Can you find the error?
(The exact answer should be 207.3, which fits with the estimate of 200.)

Work Problem **5** *at the Side.* ▶

OBJECTIVE 4 **Use the order of operations with decimals.** Use the order of operations when a decimal problem involves more than one operation, as you did with whole numbers in **Section 1.8.**

Order of Operations
1. Do all operations inside *parentheses* or *other grouping symbols.*
2. Simplify any expressions with *exponents* and find any *square roots.*
3. *Multiply* or *divide,* proceeding from left to right.
4. *Add* or *subtract,* proceeding from left to right.

EXAMPLE 6 Using the Order of Operations

Use the order of operations to simplify each expression.

(a) $2.5 + \mathbf{6.3^2} + 9.62$　Apply the exponent: $(6.3)(6.3)$ is 39.69

$\mathbf{2.5 + 39.69} + 9.62$　Add from left to right.

$42.19 + 9.62$

51.81

(b) $1.82 + \mathbf{(6.7 - 5.2)}(5.8)$　Work inside parentheses.

$1.82 + \mathbf{(1.5)(5.8)}$　Multiply next.

$1.82 + 8.7$　Add last.

10.52

Continued on Next Page

---

5 Decide whether each answer is reasonable by using front end rounding to estimate the answer. If the exact answer is *not* reasonable, find and correct the error.

(a) $42.75 \div 3.8 = 1.125$

Estimate:

(b) $807.1 \div 1.76 = 458.580$
to nearest thousandth

Estimate:

(c) $48.63 \div 52 = 93.519$
to nearest thousandth

Estimate:

(d) $9.0584 \div 2.68 = 0.338$

Estimate:

ANSWERS
5. **(a)** Estimate is $40 \div 4 = 10$; answer is not reasonable; should be 11.25
(b) Estimate is $800 \div 2 = 400$; answer is reasonable.
(c) Estimate is $50 \div 50 = 1$; answer is not reasonable, should be 0.935 (rounded)
(d) Estimate is $9 \div 3 = 3$; answer is not reasonable, should be 3.38

6 Use the order of operations to simplify each expression.

(a) $4.6 - 0.79 + 1.5^2$

(c)

$\underbrace{3.7^2} - 1.8 \div 5(1.5)$	Apply the exponent.
$13.69 - \underline{1.8 \div 5}(1.5)$	Multiply and divide from left to right, so first divide 1.8 by 5
$13.69 - \underline{0.36(1.5)}$	Then multiply 0.36 by 1.5
$\underbrace{13.69 - \quad 0.54}$	Subtract last.
$\underbrace{13.15}$	

◀ *Work Problem* **6** *at the Side.*

(b) $3.64 \div 1.3 \times 3.6$

🔢 **Calculator Tip** Most scientific calculators that have parentheses keys ⓧ ⓧ can handle calculations like those in Example 6 if you just enter the numbers in the order given. For example, the keystrokes for Example 6(b) on the previous page are:

Parentheses

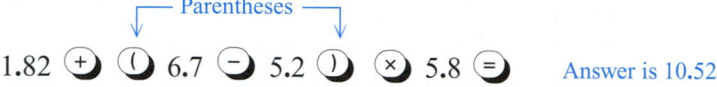

Answer is 10.52

Standard, four-function calculators generally do not have parentheses keys and will *not* give the correct answer if you simply enter the numbers in the order given.

Check the instruction manual that came with your calculator for information on "order of calculations" to see if your model has the rules for order of operations built into it. For a quick check, try entering this problem.

$$2 \; ⊕ \; 2 \; ⊗ \; 2 \; ⊜$$

If the result is 6, the calculator follows the order of operations. If the result is 8, it does *not* have the rules built into it. To see why this test works, do the calculations by hand.

(c) $0.08 + 0.6(3 - 2.99)$

Follow the order of operations.	Work from left to right.
$2 + \underbrace{2 \times 2}$ Multiply before adding.	$\underbrace{2 + 2} \times 2$
$\underbrace{2 + \quad 4}$	$\underbrace{4 \quad \times 2}$
6 ⟵——— **Correct**	✗ 8 ⟵———**Incorrect**

(d) $10.85 - 2.3(5.2) \div 3.2$

4.5 ▶▶▶ Exercises

Find each quotient. See Examples 1 and 4.

 1. $7\overline{)27.3}$

2. $8\overline{)50.4}$

 3. $\dfrac{4.23}{9}$

4. $\dfrac{1.62}{6}$

5. $0.05\overline{)20.01}$

6. $0.08\overline{)16.04}$

7. $1.5\overline{)54}$

8. $2.4\overline{)132}$

Use the fact that $108 \div 18 = 6$ to work Exercises 9–12 simply by moving decimal points.

9. $0.108 \div 1.8$

10. $10.8 \div 18$

11. $0.018\overline{)108}$

12. $0.18\overline{)1.08}$

Divide. Round quotients to the nearest hundredth if necessary. See Examples 3 and 4.

13. $4.6\overline{)116.38}$

14. $2.6\overline{)4.992}$

15. $\dfrac{3.1}{0.006}$

16. $\dfrac{1.7}{0.09}$

Divide. Round quotients to the nearest thousandth.

17. $240.8 \div 9$

18. $76.43 \div 7$

19. $0.034\overline{)342.81}$

20. $0.043\overline{)1748.4}$

Relating Concepts (Exercises 21–22) For Individual or Group Work

*Look for patterns as you **work Exercises 21 and 22 in order.***

21. Do these division problems:

$3.77 \div 10 = $ _____ $9.1 \div 10 = $ _____

$0.886 \div 10 = $ _____ $30.19 \div 10 = $ _____

$406.5 \div 10 = $ _____ $6625.7 \div 10 = $ _____

(a) What pattern do you see? Write a "rule" for dividing by 10. What do you think the rule is for dividing by 100? by 1000? Write the rules and try them out on the numbers above.

(b) Compare your rules to the ones you wrote in **Section 4.4,** Exercise 21. How are they different?

22. Do these division problems:

$40.2 \div 0.1 = $ _____ $7.1 \div 0.1 = $ _____

$0.339 \div 0.1 = $ _____ $15.77 \div 0.1 = $ _____

$46 \div 0.1 = $ _____ $873 \div 0.1 = $ _____

(a) What pattern do you see? Write a "rule" for dividing by 0.1. What do you think the rule is for dividing by 0.01? by 0.001? Write the rules and try them out on the numbers above.

(b) Compare your rules to the ones you wrote in **Section 4.4,** Exercise 22. How are they different?

Decide whether each answer is reasonable or unreasonable by rounding the numbers and estimating the answer. If the exact answer is not reasonable, find the correct answer. See Example 5.

23. $37.8 \div 8 = 47.25$

Estimate:

24. $345.6 \div 3 = 11.52$

Estimate:

25. $54.6 \div 48.1 = 1.135$

Estimate:

26. $2428.8 \div 4.8 = 56$

Estimate:

27. $307.02 \div 5.1 = 6.2$

Estimate:

28. $395.415 \div 5.05 = 78.3$

Estimate:

29. $9.3 \div 1.25 = 0.744$

Estimate:

30. $78 \div 14.2 = 5493$

Estimate:

Solve each application problem. Round money answers to the nearest cent, if necessary.

31. Rob discovered that his daughter's favorite brand of tights are on sale. He decided to buy one pair as a surprise for her. How much did he pay?

32. The bookstore has a special price on notepads. How much did Randall pay for one notepad?

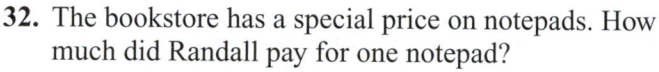

33. It will take 21 equal monthly payments for Aimee to pay off her credit card balance of $1408.68. How much is she paying each month?

34. Marcella Anderson bought 2.6 meters of microfiber woven suede fabric for $33.77. How much did she pay per meter?

35. Adrian Webb bought 619 bricks to build a barbecue pit, paying $185.70. Find the cost per brick. (*Hint:* Cost *per* brick means the cost for *one* brick.)

36. Lupe Wilson is a newspaper distributor. Last week she paid the newspaper $130.51 for 842 copies. Find the cost per copy.

37. Darren Jackson earned $476.80 for 40 hours of work. Find his earnings per hour.

38. At a CD manufacturing company, 400 CDs cost $289. Find the cost per CD.

39. It took 16.35 gallons of gas to fill the gas tank of Kim's car. She had driven 346.2 miles since her last fill-up. How many miles per gallon did her car get? Round to the nearest tenth.

40. Mr. Rodriquez pays $53.19 each month to Household Finance. How many months will it take him to pay off $1436.13?

Use the table of women's longest long jumps (through the year 2007) to answer Exercises 41–46. To find an average, add up the values you are interested in and then divide the sum by the number of values. Round your answers to the nearest hundredth.

Athlete	Country	Year	Length (meters)
Galina Christyakova	USSR	1988	7.52
Jackie Joyner-Kersee	U.S.	1994	7.49
Heike Drechsler	Germany	1992	7.48
Jackie Joyner-Kersee	U.S.	1987	7.45
Jackie Joyner-Kersee	U.S.	1988	7.40
Jackie Joyner-Kersee	U.S.	1991	7.32
Jackie Joyner-Kersee	U.S.	1996	7.20
Chioma Ajunwa	Nigeria	1996	7.12
Fiona May	Italy	2000	7.09
Tatyana Lebedeva	Russia	2004	7.07

Source: CNNSI.com

41. Find the average length of the long jumps made by Jackie Joyner-Kersee.

42. Find the average length of all the long jumps listed in the table.

43. How much longer was the fifth-longest jump than the sixth-longest jump?

44. If the first-place athlete made five jumps of the same length, what would be the total distance jumped?

45. What was the total length jumped by the top three athletes in the table?

46. How much less was the last-place jump than the next-to-last-place jump?

Use the order of operations to simplify each expression. See Example 6.

47. $7.2 - 5.2 + 3.5^2$

48. $6.2 + 4.3^2 - 9.72$

49. $38.6 + 11.6(13.4 - 10.4)$

50. $2.25 - 1.06(4.85 - 3.95)$

51. $8.68 - 4.6(10.4) \div 6.4$

52. $25.1 + 11.4 \div 7.5(3.75)$

53. $33 - 3.2(0.68 + 9) - 1.3^2$

54. $0.6 + (1.89 + 0.11) \div 0.004(0.5)$

Solve each application problem.

55. Soup is on sale at six cans for $3.25, or you may purchase individual cans for $0.57. How much will you save per can if you buy six cans? Round to the nearest cent.

56. Nadia's diet says she can eat 3.5 ounces of chicken nuggets. The package weighs 10.5 ounces and contains 15 nuggets. How many nuggets can Nadia eat?

57. The U.S. Treasury prints about 38,000,000 pieces of paper money each day. The printing presses run 24 hours a day. How many pieces of money are printed, to the nearest whole number:

(a) each hour?

(b) each minute?

(c) each second?

38,000,000 pieces of paper money are printed each day.

58. Mach 1 is the speed of sound. Dividing a vehicle's speed by the speed of sound gives its speed on the Mach scale. In 1997, a specially built car with two 110,000-horsepower engines broke the world land speed record by traveling 763.035 miles per hour. The speed of sound that day was 748.11 miles per hour. What was the car's Mach speed, to the nearest hundredth? (*Source:* Associated Press.)

(*Source:* www.moneyfactory.com)

General Mills will give a school 10¢ for each box top logo from its cereals and other products. A school can earn up to $10,000 per year. Use this information to answer Exercises 59–62. Round your answers to the nearest whole number when necessary. (Source: General Mills.)

59. How many box tops would a school need to collect in one year to earn the maximum amount?

60. (Complete Exercise 59 first.) If a school has 550 children, how many box tops would each child need to collect in one year to reach the maximum?

61. How many box tops would need to be collected during each of the 38 weeks in the school year to reach the maximum amount?

62. How many box tops would each of the 550 children need to collect during each of the 38 weeks of school to reach the maximum amount?

4.6 ▶▶▶ Writing Fractions as Decimals

Writing fractions as equivalent decimals can help you do calculations more easily or compare the size of two numbers.

OBJECTIVE 1 Write fractions as equivalent decimals. Recall that a fraction is one way to show division (see **Section 1.5**). For example, $\frac{3}{4}$ means $3 \div 4$. If you are doing the division by hand, write it as $4\overline{)3}$. When you do the division, the result is 0.75, the decimal equivalent of $\frac{3}{4}$.

> **Writing a Fraction as an Equivalent Decimal**
>
> **Step 1** Divide the numerator of the fraction by the denominator.
>
> **Step 2** If necessary, round the answer to the place indicated.

Work Problem **1** *at the Side.* ▶

EXAMPLE 1 Writing Fractions or Mixed Numbers as Decimals

(a) Write $\frac{1}{8}$ as a decimal.

$\frac{1}{8}$ means $1 \div 8$. Write it as $8\overline{)1}$. The decimal point in the dividend is on the *right* side of the 1. Write extra zeros in the dividend so you can continue dividing until the remainder is 0.

$$\frac{1}{8} \;\longrightarrow\; 1 \div 8 \;\longrightarrow\; 8\overline{)1} \;\longrightarrow\;
\begin{array}{r}
0.125 \\
8\overline{)1.000} \\
\underline{8} \\
20 \\
\underline{16} \\
40 \\
\underline{40} \\
0
\end{array}$$

Decimal points lined up ← Three extra zeros needed ← Remainder is 0.

Therefore, $\frac{1}{8} = 0.125$.

To check this, write 0.125 as a fraction, then change it to lowest terms.

$$0.125 = \frac{125}{1000} \qquad \text{In lowest terms} \qquad \frac{125 \div \mathbf{125}}{1000 \div \mathbf{125}} = \frac{1}{8} \;\leftarrow\; \begin{array}{l}\text{Original}\\ \text{fraction}\end{array}$$

> 🖩 **Calculator Tip** When using your calculator to write fractions as decimals, enter the numbers from the top down. Remember that the order in which you enter the numbers *does* matter in division. Example 1(a) above works like this:
>
> $\frac{1}{8}$ ↓ Top down   Enter 1 ÷ 8 =   Answer is 0.125
>
> What happens if you enter 8 ÷ 1 = ? Do you see why that cannot possibly be correct? (Answer: $8 \div 1 = 8$. A proper fraction like $\frac{1}{8}$ *cannot* be equivalent to a whole number.)

Continued on Next Page

OBJECTIVES

1 Write fractions as equivalent decimals.

2 Compare the size of fractions and decimals.

---

1 Rewrite each fraction so you could do the division by hand. Do *not* complete the division.

(a) $\frac{1}{9}$ is written $9\overline{)}$

(b) $\frac{2}{3}$ is written $\overline{)}$

(c) $\frac{5}{4}$ is written $\overline{)}$

(d) $\frac{3}{10}$ is written $\overline{)}$

(e) $\frac{21}{16}$ is written $\overline{)}$

(f) $\frac{1}{50}$ is written $\overline{)}$

ANSWERS

1. **(a)** $9\overline{)1}$ **(b)** $3\overline{)2}$ **(c)** $4\overline{)5}$
   (d) $10\overline{)3}$ **(e)** $16\overline{)21}$ **(f)** $50\overline{)1}$

2 Write each fraction or mixed number as a decimal.

(a) $\frac{1}{4}$

(b) $2\frac{1}{2}$

(c) $\frac{5}{8}$

(d) $4\frac{3}{5}$

(e) $\frac{7}{8}$

(b) Write $2\frac{3}{4}$ as a decimal.

One method is to divide 3 by 4 to get 0.75 for the fraction part. Then add the whole number part to 0.75.

$$\frac{3}{4} \longrightarrow \begin{array}{r} 0.75 \\ 4\overline{)3.00} \\ \underline{2\,8} \\ 20 \\ \underline{20} \\ 0 \end{array} \quad \text{Fraction part} \longrightarrow \begin{array}{r} 2.00 \leftarrow \text{Whole number part} \\ +\,0.75 \\ \hline 2.75 \end{array}$$

So, $2\frac{3}{4} = 2.75$ *Check:* $2.75 = 2\frac{75}{100} = 2\frac{3}{4} \leftarrow$ Lowest terms

Whole number parts match.

A second method is to first write $2\frac{3}{4}$ as an improper fraction and then divide numerator by denominator.

$$2\frac{3}{4} = \frac{11}{4}$$

$$\frac{11}{4} \longrightarrow 11 \div 4 \longrightarrow 4\overline{)11} \longrightarrow \begin{array}{r} 2.75 \\ 4\overline{)11.00} \\ \underline{8} \\ 3\,0 \\ \underline{2\,8} \\ 20 \\ \underline{20} \\ 0 \end{array} \leftarrow \text{Two extra zeros needed}$$

Whole number parts match.

So, $2\frac{3}{4} = 2.75$

$\frac{3}{4}$ is equivalent to $\frac{75}{100}$ or 0.75

◀ *Work Problem* **2** *at the Side.*

EXAMPLE 2   **Writing a Fraction as a Decimal with Rounding**

Write $\frac{2}{3}$ as a decimal and round to the nearest thousandth.

$\frac{2}{3}$ means $2 \div 3$. To round to thousandths, divide out one *more* place, to ten-thousandths.

$$\frac{2}{3} \longrightarrow 2 \div 3 \longrightarrow 3\overline{)2} \longrightarrow \begin{array}{r} 0.6666 \\ 3\overline{)2.0000} \\ \underline{1\,8} \\ 20 \\ \underline{18} \\ 20 \\ \underline{18} \\ 20 \\ \underline{18} \\ 2 \end{array} \leftarrow \begin{array}{l}\text{Four zeros needed} \\ \text{for ten-thousandths}\end{array}$$

Be careful to divide in the correct order!

Written as a repeating decimal, $\frac{2}{3} = 0.\overline{6} \leftarrow$ Bar above repeating digit

Rounded to the nearest thousandth, $\frac{2}{3} \approx 0.667$

Continued on Next Page

▦ Calculator Tip   Try Example 2 on your calculator. Enter 2 ÷ 3. Which answer do you get?

| 0.666666667 | or | 0.6666666 |

Many scientific calculators will show a 7 as the last digit. Because the sixes keep on repeating forever, the calculator automatically rounds in the last decimal place it has room to show. If you have a 10-digit display space, the calculator is rounding as shown below.

0.6666666666 (11 digits)   rounds to   0.66666666**7**

Next digit is 5 or more, so 6 rounds to 7.

Other calculators, especially standard, four-function ones, may *not* round. They just cut off, or *truncate,* the extra digits. Such a calculator would show 0.6666666 in the display.

    Would this difference in calculators show up when changing $\frac{1}{3}$ to a decimal? Why not? (Answer: The repeating digit is 3, which is *4 or less,* so it stays as 3 whether it's rounded or not.)

Work Problem **3** *at the Side.* ▶

OBJECTIVE 2 Compare the size of fractions and decimals.
You can use a number line to compare fractions and decimals. For example, the number line below shows the space between 0 and 1. The locations of some commonly used fractions are marked, along with their decimal equivalents.

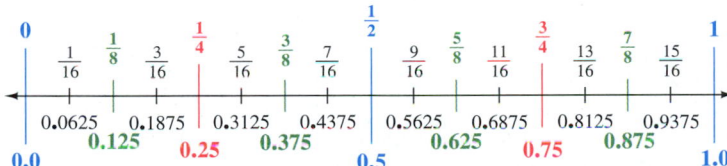

    The next number line shows the locations of some commonly used fractions between 0 and 1 that are equivalent to *repeating* decimals. The decimal equivalents use a bar above repeating digits.

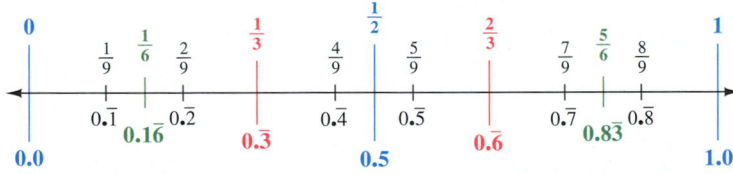

EXAMPLE 3   Using a Number Line to Compare Numbers

Use the number lines above to decide whether to write >, <, or = in the blank between each pair of numbers.

(a) 0.6875 _____ 0.625
    You learned in **Section 3.5** that the number farther to the right on the number line is the greater number. Look at the first number line above. Because 0.6875 is to the *right* of 0.625, use the > symbol.

       0.6875 **is greater than** 0.625   0.6875 > 0.625

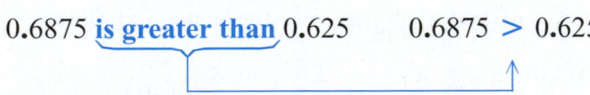

    Continued on Next Page

---

3 Write as decimals. Round to the nearest thousandth.

(a) $\frac{1}{3}$

(b) $2\frac{7}{9}$

(c) $\frac{10}{11}$

(d) $\frac{3}{7}$

(e) $3\frac{5}{6}$

ANSWERS

1. All answers are rounded.
    (a) 0.333 **(b)** 2.778 **(c)** 0.909
    (d) 0.429 **(e)** 3.833

4 Use the number lines on the previous page to help you decide whether to write $<$, $>$, or $=$ in each blank.

(a) 0.4375 _____ 0.5

(b) 0.75 _____ 0.6875

(c) 0.625 _____ 0.0625

(d) $\dfrac{2}{8}$ _____ 0.375

(e) $0.8\overline{3}$ _____ $\dfrac{5}{6}$

(f) $\dfrac{1}{2}$ _____ $0.\overline{5}$

(g) $0.\overline{1}$ _____ $0.1\overline{6}$

(h) $\dfrac{8}{9}$ _____ $0.\overline{8}$

(i) $0.\overline{7}$ _____ $\dfrac{4}{6}$

(j) $\dfrac{1}{4}$ _____ 0.25

5 Arrange each group in order from least to greatest.

(a) 0.7, 0.703, 0.7029

(b) 6.39, 6.309, 6.401, 6.4

(c) 1.085, $1\dfrac{3}{4}$, 0.9

(d) $\dfrac{1}{4}, \dfrac{2}{5}, \dfrac{3}{7}$, 0.428

ANSWERS

4. (a) $<$ (b) $>$ (c) $>$ (d) $<$ (e) $=$
   (f) $<$ (g) $<$ (h) $=$ (i) $>$ (j) $=$
5. (a) 0.7, 0.7029, 0.703
   (b) 6.309, 6.39, 6.4, 6.401
   (c) 0.9, 1.085, $1\dfrac{3}{4}$
   (d) $\dfrac{1}{4}, \dfrac{2}{5}$, 0.428, $\dfrac{3}{7}$

(b) $\dfrac{3}{4}$ _____ 0.75

On the first number line, $\dfrac{3}{4}$ and 0.75 are at the same point on the number line. They are equivalent.

$$\tfrac{3}{4} = 0.75$$

(c) 0.5 _____ $0.\overline{5}$

On the second number line, 0.5 is to the *left* of $0.\overline{5}$ (which is actually 0.555 . . .) so use the $<$ symbol.

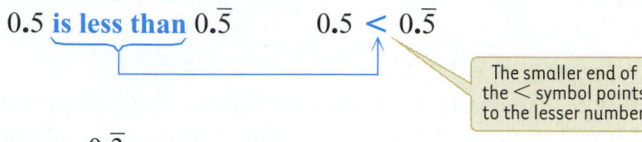

0.5 **is less than** $0.\overline{5}$ $0.5 < 0.\overline{5}$

The smaller end of the $<$ symbol points to the lesser number.

(d) $\dfrac{2}{6}$ _____ $0.\overline{3}$

Write $\dfrac{2}{6}$ in lowest terms as $\dfrac{1}{3}$.
On the second number line you can see that $\dfrac{1}{3} = 0.\overline{3}$.

◀ *Work Problem* **4** *at the Side.*

Fractions can also be compared by first writing each one as a decimal. The decimals can then be compared by writing each one with the same number of decimal places.

EXAMPLE 4 **Arranging Numbers in Order**

Write each group of numbers in order, from least to greatest.

(a) 0.49   0.487   0.4903

It is easier to compare decimals if they are all tenths, or all hundredths, and so on. Because 0.4903 has four decimal places (ten-thousandths), write zeros to the right of 0.49 and 0.487 so they also have four decimal places. Writing zeros to the right of a decimal number does *not* change its value (see **Section 4.3**). Then find the least and greatest number of ten-thousandths.

$0.49 = 0.4900 = $ **4900** ten-thousandths $\leftarrow$ 4900 is in the middle.

$0.487 = 0.4870 = $ **4870** ten-thousandths $\leftarrow$ 4870 is the least.

$0.4903 = $ **4903** ten-thousandths $\leftarrow$ 4903 is the greatest.

From least to greatest, the correct order is shown below.

0.487   0.49   0.4903

(b) $2\dfrac{5}{8}$ 2.63   2.6

Write $2\dfrac{5}{8}$ as $\dfrac{21}{8}$ and divide $8\overline{)21}$ to get the decimal form, 2.625. Then, because 2.625 has three decimal places, write zeros so all the numbers have three decimal places.

$2\dfrac{5}{8} = 2.625 = 2$ and **625** thousandths $\leftarrow$ 625 is in the middle.

$2.63 = 2.630 = 2$ and **630** thousandths $\leftarrow$ 630 is the greatest.

$2.6 = 2.600 = 2$ and **600** thousandths $\leftarrow$ 600 is the least.

From least to greatest, the correct order is shown below.

2.6 $2\dfrac{5}{8}$ 2.63

◀ *Work Problem* **5** *at the Side.*

4.6 ▶▶▶ Exercises

Write each fraction or mixed number as a decimal. Round to the nearest thousandth if necessary. See Examples 1 and 2.

1. $\dfrac{1}{2}$

2. $\dfrac{1}{4}$

3. $\dfrac{3}{4}$

4. $\dfrac{1}{10}$

5. $\dfrac{3}{10}$

6. $\dfrac{7}{10}$

7. $\dfrac{9}{10}$

8. $\dfrac{4}{5}$

9. $\dfrac{3}{5}$

10. $\dfrac{2}{5}$

11. $\dfrac{7}{8}$

12. $\dfrac{3}{8}$

13. $2\dfrac{1}{4}$

14. $1\dfrac{1}{2}$

15. $14\dfrac{7}{10}$

16. $23\dfrac{3}{5}$

17. $3\dfrac{5}{8}$

18. $2\dfrac{7}{8}$

19. $6\dfrac{1}{3}$

20. $5\dfrac{2}{3}$

21. $\dfrac{5}{6}$

22. $\dfrac{1}{6}$

23. $1\dfrac{8}{9}$

24. $5\dfrac{4}{7}$

Relating Concepts (Exercises 25–28)   For Individual or Group Work

*Use your knowledge of fractions and decimals to **work Exercises 25–28 in order.***

25. (a) Explain how you can tell that Keith made an error *just by looking at his final answer*. Here is his work.

$$\frac{5}{9} = 5\overline{)9.0}^{\,1.8} \quad \text{so} \quad \frac{5}{9} = 1.8$$

(b) Show the correct way to change $\frac{5}{9}$ to a decimal. Explain why your answer makes sense.

26. (a) How can you prove to Sandra that $2\frac{7}{20}$ is *not* equivalent to 2.035? Here is her work.

$$2\frac{7}{20} = 20\overline{)7.00}^{\,0.35} \quad \text{so} \quad 2\frac{7}{20} = 2.035$$

(b) What is the correct answer? Show how to prove that it is correct.

27. Ving knows that $\frac{3}{8} = 0.375$. How can he write $1\frac{3}{8}$ as a decimal *without* having to do a division? How can he write $3\frac{3}{8}$ as a decimal? $295\frac{3}{8}$? Explain your answer.

28. Iris has found a shortcut for writing mixed numbers as decimals.

$$2\frac{7}{10} = 2.7 \qquad 1\frac{13}{100} = 1.13$$

Does her shortcut work for all mixed numbers? Explain when it works and why it works.

Find each decimal or fraction equivalent. Write fractions in lowest terms.

Fraction	Decimal	Fraction	Decimal
29. _____	0.4	**30.** _____	0.75
31. _____	0.625	**32.** _____	0.111
33. _____	0.35	**34.** _____	0.9
35. $\frac{7}{20}$	_____	**36.** $\frac{1}{40}$	_____
37. _____	0.04	**38.** _____	0.52
39. _____	0.15	**40.** _____	0.85
41. $\frac{1}{5}$	_____	**42.** $\frac{1}{8}$	_____
43. _____	0.09	**44.** _____	0.02

Solve each application problem.

45. The average length of a newborn baby is 20.8 inches. Charlene's baby is 20.08 inches long. Is her baby longer or shorter than the average? By how much?

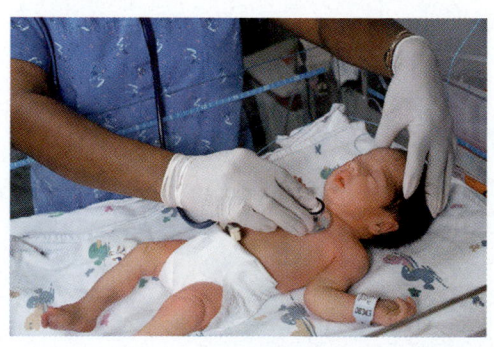

46. The patient in room 830 is supposed to get 8.3 milligrams of medicine. She was actually given 8.03 milligrams. Did she get too much or too little medicine? What was the difference?

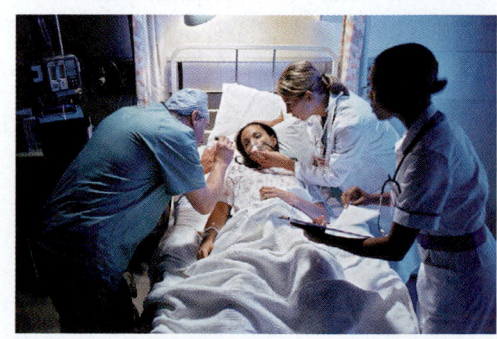

47. The label on the bottle of vitamins says that each capsule contains 0.5 gram of calcium. When checked, each capsule had 0.505 gram of calcium. Was there too much or too little calcium? What was the difference?

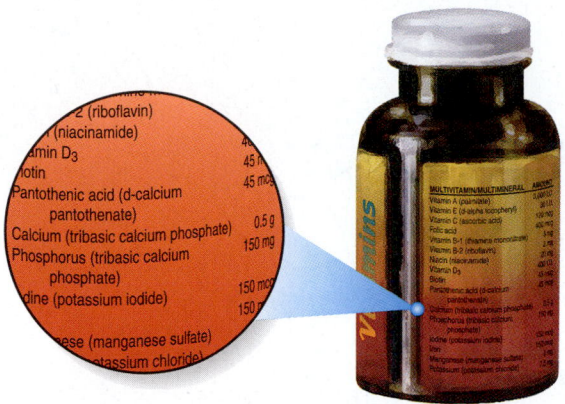

48. The glass mirror of the Hubble telescope had to be repaired in space because it would not focus properly. The problem was that the mirror's outer edge had a thickness of 0.6248 centimeter when it was supposed to be 0.625 centimeter. Was the edge too thick or too thin? By how much? (*Source:* NASA.)

49. Precision Medical Parts makes an artificial heart valve that must measure between 0.998 centimeter and 1.002 centimeters. Circle the lengths that are acceptable:

1.01 cm   0.9991 cm   1.0007 cm   0.99 cm

50. The white rats in a medical experiment must start out weighing between 2.95 ounces and 3.05 ounces. Circle the weights that can be used:

3.0 ounces   2.995 ounces   3.055 ounces

3.005 ounces

51. Ginny Brown hoped her crops would get $3\frac{3}{4}$ inches of rain this month. The newspaper said the area received 3.8 inches of rain. Was that more or less than Ginny had hoped for? By how much?

52. The rats in the experiment in Exercise 50 gained $\frac{3}{8}$ ounce. They were expected to gain 0.3 ounce. Was their actual gain more or less than expected? By how much?

Arrange each group of numbers in order, from least to greatest. See Example 4.

53. 0.54, 0.5455, 0.5399

54. 0.76, 0.7, 0.7006

55. 5.8, 5.79, 5.0079, 5.804

56. 12.99, 12.5, 13.0001, 12.77

57. 0.628, 0.62812, 0.609, 0.6009

58. 0.27, 0.281, 0.296, 0.3

59. 5.8751, 4.876, 2.8902, 3.88

60. 0.98, 0.89, 0.904, 0.9

61. 0.043, 0.051, 0.006, $\dfrac{1}{20}$

62. 0.629, $\dfrac{5}{8}$, 0.65, $\dfrac{7}{10}$

63. $\dfrac{3}{8}$, $\dfrac{2}{5}$, 0.37, 0.4001

64. 0.1501, 0.25, $\dfrac{1}{10}$, $\dfrac{1}{5}$

Four boxes of fishing line are in the sale bin. The thicker the line, the stronger it is.
The diameter of the fishing line is its thickness. Use the information on the boxes to answer
Exercises 65–68.

65. Which color box has the strongest line?

66. Which color box has the line with the least strength?

67. Which color box has the line that is $\dfrac{1}{125}$ inch in diameter?

68. What is the difference in line diameter between the blue and purple boxes?

Some rulers for technical occupations show each inch divided into tenths. Use this scale
drawing for Exercises 69–74. Change the measurements on the drawing to decimals and
round them to the nearest tenth of an inch.

69. Length **(a)** is _____

70. Length **(b)** is _____

71. Length **(c)** is _____

72. Length **(d)** is _____

73. Length **(e)** is _____

74. Length **(f)** is _____

(a) $1\dfrac{7}{16}$ in.

(d) $\dfrac{1}{2}$ in.

$1\dfrac{1}{8}$ in. (b)

(e) $\dfrac{3}{8}$ in.

$\dfrac{11}{16}$ in.
(f)

$\dfrac{1}{4}$ in.
(c)

Chapter 4 ▷▷▷ Summary

▶ Key Terms

4.1	**decimals**	Decimals, like fractions, are used to show parts of a whole.
	decimal point	A decimal point is the dot that is used to separate the whole number part from the fractional part of a decimal number.
	place value	A place value is assigned to each place to the right or left of the decimal point. Whole numbers such as ones and tens, are to the *left* of the decimal point. Fractional parts, such as tenths and hundredths, are to the *right* of the decimal point.
4.2	**rounding**	Rounding is "cutting off" a number after a certain place, such as rounding to the nearest hundredth. The rounded number is less accurate than the original number. You can use the symbol "≈" to mean "is approximately equal to."
	decimal places	Decimal places are the number of digits to the *right* of the decimal point. For example, 6.37 has two decimal places, and 4.706 has three decimal places.
4.3	**estimating**	Estimating is the process of rounding the numbers in a problem and getting an approximate answer. This helps you check that the decimal point is in the correct place in the exact answer.
4.5	**repeating decimal**	A repeating decimal (like the 6 in 0.1666 . . .) is a decimal number with one or more digits that repeat forever; it never ends. Use three dots to indicate that it is a repeating decimal. Or, write the number with a bar above the repeating digits, as in 0.1$\overline{6}$. (Use the dots or the bar, but not both.)

▶ New Symbols

3.8$\overline{6}$ ⟵ Bar above repeating digit(s) in a decimal number   3.866 . . .   Three dots indicate a repeating decimal

▶ Test Your Word Power

See how well you have learned the vocabulary in this chapter. Answers follow the Quick Review.

1. **Decimal numbers** are like fractions in that they both
   A. must be written in lowest terms
   B. need common denominators
   C. have decimal points
   D. represent parts of a whole.

2. **Decimal places** refer to
   A. the digits from 0 to 9
   B. digits to the left of the decimal point
   C. digits to the right of the decimal point
   D. the number of zeros in a decimal number.

3. When a decimal number is **rounded,** it
   A. always ends in 0
   B. has the same number of decimal places as the original number
   C. is less accurate than the original number
   D. is less than one whole.

4. The **decimal point**
   A. separates the whole number part from the fractional part
   B. is always moved when finding a quotient
   C. separates tenths from hundredths
   D. is at the far left side of a whole number.

5. The number 0.$\overline{3}$ is an example of
   A. an estimate
   B. a repeating decimal
   C. a rounded number
   D. a truncated number.

6. The **place value** names on the right side of the decimal point are
   A. ones, tens, hundreds, and so on
   B. ones, tenths, hundredths, and so on
   C. zero, one, two, three, four, and so on
   D. tenths, hundredths, thousandths, and so on.

▶ Quick Review

Concepts	Examples

4.1 Reading and Writing Decimals

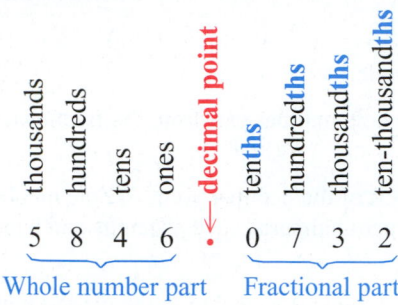

Whole number part   Fractional part

Write each decimal in words.

> 8 is in hundredths place.

15.38

fifteen **and** thirty-eight **hundredths**

> 3 is in ten-thousandths place.

0.0103

one hundred three **ten-thousandths**

4.1 Writing Decimals as Fractions

The digits to the right of the decimal point are the numerator. The place value of the rightmost digit determines the denominator.

Always write the fractions in lowest terms.

Write 0.45 as a fraction in lowest terms.

The numerator is 45. The rightmost digit, 5, is in the hundredths place, so the denominator is 100. Then write the fraction in lowest terms.

$$\frac{45}{100} = \frac{45 \div 5}{100 \div 5} = \frac{9}{20} \leftarrow \text{Lowest terms}$$

4.2 Rounding Decimals

Find the place to which you are rounding. Draw a cut-off line to the right of that place; the rest of the digits will be dropped. Look *only* at the first digit being cut. If it is *4 or less*, the part you are keeping stays the same. If it is *5 or more*, the part you are keeping rounds up. Do not move the decimal point when rounding. Write "≈" to mean "is approximately equal to."

Round 0.17952 to the nearest thousandth.

┌─ First digit cut is *5 or more,*
│   so round up.

0.179│52

0.179 ← Keep this part.
+ 0.001 ← To round up, add 1 thousandth.
0.180

0.17952 rounds to 0.180. Write $0.17952 \approx 0.180$

4.3 Adding and Subtracting Decimals

Use front end rounding to round each number and estimate the answer.

To find the exact answer, line up the decimal points. If needed, write in zeros as placeholders. Add or subtract as if they were whole numbers. Line up the decimal point in the answer directly below the decimal points in the problem.

Add $5.68 + 785.3 + 12 + 2.007$.

Estimate: *Exact:*

6 ←	5.68**0**
800 ←	785.3**00**
10 ←	12.**000**
+ 2 ←	+ 2.007
818	804.987

Use zeros as placeholders so that all numbers have three decimal places.
Line up decimal points.

The estimate and exact answer are both in the hundreds, so the decimal point is probably in the correct place.

4.4 Multiplying Decimals

Step 1 Multiply as you would for whole numbers.
Step 2 Count the total number of decimal places in both factors.
Step 3 Write the decimal point in the product so it has the same number of decimal places as the total from Step 2. You may need to write extra zeros on the left side of the product to get enough decimal places.

Multiply 0.169×0.21.

0.169	← 3 decimal places
× 0.21	← 2 decimal places
169	5 total decimal places
338	
.03549	← 5 decimal places in product

Write a 0 in the product so you can count over 5 decimal places. The final answer is 0.03549.

Concepts	Examples

4.5 Dividing by Decimals

Step 1 Change the divisor to a whole number by moving the decimal point to the right.

Step 2 Move the decimal point in the dividend the same number of places to the right.

Step 3 Write the decimal point in the quotient directly above the decimal point in the dividend.

Step 4 Divide as with whole numbers.

Divide 52.8 by 0.75.

$$
\begin{array}{r}
7\,0.4 \\
0.75\overline{)5\,2.8\,0\,0} \\
\underline{5\,2\,5} \\
3\,0\,0 \\
\underline{3\,0\,0} \\
0
\end{array}
$$

Move the decimal point two places to the right in the divisor and dividend. Write zeros in the dividend so you can move the decimal point and continue dividing until the remainder is 0.

To check your answer, multiply 70.4 times 0.75. If the result matches the dividend (52.8), you solved the problem correctly.

4.6 Writing Fractions as Decimals

Divide the numerator by the denominator. If necessary, round to the place indicated.

Write $\frac{1}{8}$ as a decimal.

$\frac{1}{8}$ means $1 \div 8$. Write it as $8\overline{)1}$.

The decimal point is on the right side of 1.

$$
\begin{array}{r}
0.125 \\
8\overline{)1.000} \\
\underline{8} \\
20 \\
\underline{16} \\
40 \\
\underline{40} \\
0
\end{array}
$$

Write a decimal point and three zeros so you can continue dividing.

Therefore, $\frac{1}{8}$ is equivalent to 0.125

4.6 Comparing the Size of Fractions and Decimals

Step 1 Write any fractions as decimals.

Step 2 Write zeros so that all the numbers being compared have the same number of decimal places.

Step 3 Use < to mean "is less than," > to mean "is greater than," or list the numbers from least to greatest.

Arrange in order from least to greatest.

$$0.505 \qquad \frac{1}{2} \qquad 0.55$$

$0.505 = 505$ thousandths ← 505 is in the middle.

$\frac{1}{2} = 0.5 = 0.500 = 500$ thousandths ← 500 is least.

$0.55 = 0.550 = 550$ thousandths ← 550 is greatest.

$$(\text{least})\ \frac{1}{2} \qquad 0.505 \qquad 0.55\ (\text{greatest})$$

ANSWERS TO TEST YOUR WORD POWER

1. D; *Example:* For 0.7, the whole is cut into ten parts, and you are interested in 7 of the parts.

2. C; *Examples:* The number 6.87 has two decimal places; 0.309 has three decimal places.

3. C; *Example:* When 0.815 is rounded to 0.8 it is accurate only to the nearest tenth, while the original number was accurate to the nearest thousandth.

4. A; *Example:* In 5.42, the whole number part is 5 ones, and the decimal part is 42 hundredths.

5. B; *Example:* The bar above the 3 in $0.\overline{3}$ indicates that the 3 repeats forever.

6. D; *Example:* In 6.219, the 2 is in the tenths place, the 1 is in the hundredths place, and the 9 is in the thousandths place.

Math in the Media

DOLLAR-COST AVERAGING

Making money in the stock market can be difficult. You want to buy shares of stock when the price is low and sell them when the price is high. Predicting the right time to buy is tricky. One strategy is dollar-cost averaging. You invest the same amount at regular intervals, like the first of each month. Over time you will usually buy more shares at lower prices, though there is no guarantee of making a profit.

The table below shows Microsoft's closing share price on the first day of each month in 2007. See what happens if you invest $100 each month rather than buying a lot of shares at one time.

Microsoft Corp. (MSFT)	Amount Invested	Price Per Share	Number of Shares Bought
January	$100	30.86	($100 ÷ 30.86) ≈ 3.2404 shares
February	$100	28.17	
March	$100	27.87	
April	$100	29.94	
May	$100	30.69	
June	$100	29.47	
July	$100	28.99	
August	$100	28.73	
September	$100	29.46	
October	$100	36.81	
November	$100	34.74	
December	$100	33.60	
Total investment			

Source: http://finance.yahoo.com

Use a calculator to help answer these questions.

1. Calculate the total amount invested and enter the value in the table.

2. Calculate the number of shares bought each month. Round the number of shares to the nearest ten-thousandth. The calculation for January is shown.

3. Calculate the average market price per share. (*Hint:* Add the monthly prices per share and divide by 12.)

4. Calculate the average price based on *dollar-cost averaging*. (*Hint:* Divide the total amount invested by the total number of shares.)

5. Rank the following scenarios in order of which was the best investment (most profit or least loss). Show the value of each investment in December 2007 as a basis for your answer.

   (a) $1200 invested in Microsoft in January 2007

   (b) $1200 invested in Microsoft using *dollar-cost averaging*

   (c) $1200 invested in Microsoft in October 2007

Chapter 4 ▸▸▸ Review Exercises

[4.1] *Name the digit that has the given place value.*

1. 243.059
tenths
hundredths

2. 0.6817
ones
tenths

3. $5824.39
hundreds
hundredths

4. 896.503
tenths
tens

5. 20.73861
tenths
ten-thousandths

Write each decimal as a fraction or mixed number in lowest terms.

6. 0.5

7. 0.75

8. 4.05

9. 0.875

10. 0.027

11. 27.8

Write each decimal in words.

12. 0.8

13. 400.29

14. 12.007

15. 0.0306

Write each decimal in numbers.

16. eight and three tenths

17. two hundred five thousandths

18. seventy and sixty-six ten-thousandths

19. thirty hundredths

[4.2] *Round each decimal to the place indicated.*

20. 275.635 to the nearest tenth

21. 72.789 to the nearest hundredth

22. 0.1604 to the nearest thousandth

23. 0.0905 to the nearest thousandth

24. 0.98 to the nearest tenth

Round each money amount to the nearest cent.

25. $15.8333

26. $0.698

27. $17,625.7906

Round each income or expense item to the nearest dollar.

28. Income from pancake breakfast was $350.48.

29. Members paid $129.50 in dues.

30. Refreshments cost $99.61.

31. Bank charges were $29.37.

[4.3] *First use front end rounding to round each number and estimate the answer. Then find the exact answer.*

32. *Estimate:*   *Exact:*

$$
\begin{array}{r}
5.81 \\
423.96 \\
+ \quad 15.09 \\
\hline
\end{array}
$$

$+$ _____

🖩 **33.** *Estimate:*   *Exact:*

$$
\begin{array}{r}
75.6 \\
1.29 \\
122.045 \\
0.88 \\
+ \quad 33.7 \\
\hline
\end{array}
$$

$+$ _____

34. *Estimate:*   *Exact:*

$$
\begin{array}{r}
308.5 \\
- \quad 17.8 \\
\hline
\end{array}
$$

$-$ _____

35. *Estimate:*   *Exact:*

$$
\begin{array}{r}
9.2 \\
- \quad 7.9316 \\
\hline
\end{array}
$$

$-$ _____

36. American's favorite household pet is a cat. There are about 90.5 million pet cats, 73.9 million pet dogs, and 16.6 million pet birds. How many more pet cats are there than pet birds? (*Source:* American Pet Products Manufacturers Association.)

Estimate:

Exact:

37. Today, Jasmin had $306 in her checking account. She wrote a check to the day care center for $215.53 and a check for $44.67 at the grocery store. What is the new balance in her account?

Estimate:

Exact:

38. Joey spent $1.59 for toothpaste, $5.33 for a gift, and $18.94 for a toaster. He gave the clerk three $10 bills. How much change did he get?

Estimate:

Exact:

39. Roseanne is training for a wheelchair race. She raced 2.3 kilometers on Monday, 4 kilometers on Wednesday, and 5.25 kilometers on Friday. How far did she race altogether?

Estimate:

Exact:

[4.4] *First use front end rounding to round each number and estimate the answer. Then find the exact answer.*

40. *Estimate:*   *Exact:*

$$
\begin{array}{r}
6.138 \\
\times \quad 3.7 \\
\hline
\end{array}
$$

$\times$ _____

41. *Estimate:*   *Exact:*

$$
\begin{array}{r}
42.9 \\
\times \quad 3.3 \\
\hline
\end{array}
$$

$\times$ _____

Find each product.

42. $(5.6)(0.002)$

43. $0.071(0.005)$

[4.5] *Decide whether each answer is reasonable by rounding the numbers and estimating the answer. If the exact answer is not reasonable, find and correct the error.*

44. $706.2 \div 12 = 58.85$

Estimate:

45. $26.6 \div 2.8 = 0.95$

Estimate:

Divide. Round each quotient to the nearest thousandth if necessary.

46. $3\overline{)43.4}$

47. $\dfrac{72}{0.06}$

48. $0.00048 \div 0.0012$

[4.4–4.5] *Solve each application problem.*

49. Adrienne worked 46.5 hours this week. Her hourly wage is $14.24 for the first 40 hours and 1.5 times that rate over 40 hours. Find her total earnings to the nearest dollar.

50. A book of 12 tickets costs $35.89 at the State Fair midway. What is the cost per ticket, to the nearest cent?

51. Stock in MathTronic sells for $3.75 per share. Kenneth is thinking of investing $500. How many whole shares could he buy?

52. Grapes are on sale at $0.99 per pound. How much will Ms. Lee pay for 3.5 pounds of grapes, to the nearest cent?

Simplify each expression.

53. $3.5^2 + 8.7(1.95)$

54. $11 - 3.06 \div (3.95 - 0.35)$

[4.6] *Write each fraction or mixed number as a decimal. Round to the nearest thousandth when necessary.*

55. $3\dfrac{4}{5}$

56. $\dfrac{16}{25}$

57. $1\dfrac{7}{8}$

58. $\dfrac{1}{9}$

Arrange each group of numbers in order from least to greatest.

59. $3.68, 3.806, 3.6008$

60. $0.215, 0.22, 0.209, 0.2102$

61. $0.17, \dfrac{3}{20}, \dfrac{1}{8}, 0.159$

> ▶▶▶ **Mixed Review Exercises**

Add, subtract, multiply, or divide as indicated.

62. $89.19 + 0.075 + 310.6 + 5$

63. 72.8×3.5

64. $1648.3 \div 0.46$ Round to the nearest thousandth.

65. $30 - 0.9102$

66. $4.38(0.007)$

67. $0.005\overline{)0.047}$

68. $72.105 + 8.2 + 95.37$

69. $81.36 \div 9$

70. $(5.6 - 1.22) + 4.8 (3.15)$

71. 0.455×18

72. $(1.6)(0.58)$

73. $0.218\overline{)7.63}$

74. $21.059 - 20.8$

75. $18.3 - 3^2 \div 0.5$

Use the information in the ad to answer Exercises 76–80. Round money answers to the nearest cent. (Disregard any sales tax.)

76. How much would one pair of men's socks cost?

77. How much more would one pair of men's socks cost than one pair of children's socks?

78. How much would Fernando pay for a dozen pair of men's socks?

79. How much would Akiko pay for five pairs of teen jeans and four pairs of women's jeans?

80. What is the difference between the cheapest sale price for athletic shoes and the highest regular price?

To decrease your risk of clogged arteries, it is recommended that you get at least 2 milligrams of vitamin B-6 each day. Use the information in the table to answer Exercises 81–82.

SOURCES OF VITAMIN B-6
(AMOUNTS IN MILLIGRAMS)

1 cup orange juice	0.11
1 banana	0.66
1 baked potato with skin	0.7
$\frac{1}{2}$ cup strawberries	0.45
3 ounces skinless chicken	0.5
3 ounces water-packed tuna	0.3
$\frac{1}{2}$ cup chickpeas	0.57

Source: Harvard Health Letter.

81. (a) Which food item has the highest amount of vitamin B-6?

(b) Which food item has the lowest amount?

(c) What is the difference in the amount of vitamin B-6 between the food items with the highest and lowest amounts?

82. (a) Suppose you ate a banana, 3 ounces of skinless chicken, and 1 cup of strawberries. How many milligrams of vitamin B-6 would you get?

(b) Did you get more or less than the recommended daily amount? By how much?

Chapter 4 ▶▶▶ **Test**  Use the Chapter Test Prep Video CD to see fully worked-out solutions to any of the exercises you want to review.

Write each decimal as a fraction or mixed number in lowest terms.

1. 18.4

2. 0.075

Write each decimal in words.

3. 60.007

4. 0.0208

Round each decimal to the place indicated.

5. 725.6089 to the nearest tenth

6. 0.62951 to the nearest thousandth

7. $1.4945 to the nearest cent

8. $7859.51 to the nearest dollar

First use front end rounding to round each number and estimate the answer. Then find the exact answer.

9. 7.6 + 82.0128 + 39.59

10. 79.1 − 3.602

11. 5.79 (1.2)

12. 20.04 ÷ 4.8

Find the exact answer.

13. 53.1 + 4.631 + 782 + 0.031

14. 670 − 0.996

15. (0.0069) (0.007)

16. 0.15)‾72‾

1. _____

2. _____

3. _____

4. _____

5. _____

6. _____

7. _____

8. _____

9. *Estimate:* _____

   Exact: _____

10. *Estimate:* _____

    Exact: _____

11. *Estimate:* _____

    Exact: _____

12. *Estimate:* _____

    Exact: _____

13. _____

14. _____

15. _____

16. _____

17. _____

17. Write $2\frac{5}{8}$ as a decimal. Round to the nearest thousandth, if necessary.

Arrange in order from least to greatest.

18. _____

18. 0.44, 0.451, $\frac{9}{20}$, 0.4506

Simplify this expression.

19. _____

19. $6.3^2 - 5.9 + 3.4\,(0.5)$

Solve each application problem.

20. _____

20. Jennifer had $71.15 in her checking account. Yesterday her account earned $0.95 interest for the month, and she deposited a paycheck for $390.77. The bank charged her $16 for new checks. What is the new balance in her account?

21. _____

21. Three types of ducks that are hunted in the United States are gadwalls, wigeons, and pintails. The estimated populations of these ducks are 2.5 million gadwalls, 2.551 million wigeons, and 2.56 million pintails. List the ducks in order from the greatest number to the least. (_Source:_ U.S. Fish and Wildlife Service.)

22. _____

22. Mr. Yamamoto bought 1.85 pounds of cheese at $2.89 per pound. What was the total amount he paid for the cheese, to the nearest cent?

23. _____

23. Loren's baby had a temperature of 102.7 degrees. Later in the day it was 99.9 degrees. How much had the baby's temperature dropped?

24. _____

24. Pat bought 6.5 ft of decorative gold chain to hang a light over her dining table. She paid $11.64. What was the cost per foot, to the nearest cent?

25. _____

25. Write your own application problem using decimals. Make it different from Problems 20–24. Then show how to solve your problem.

Study Skills

After taking a test, many students heave a big sigh of relief and try to forget it ever happened. Don't fall into this trap! An exam is a learning opportunity. It gives you clues about *what your instructor thinks is important,* what *concepts and skills are valued* in mathematics, and *if you are on the right track.*

Jot down problems that caused you trouble. Find out how to solve them by checking your textbook, notes, or asking your instructor or tutor (if available). You might see those same problems again on a final exam.

Find out what you got wrong and why you had points deducted. Write down the problem so you can learn how to do it correctly. Sometimes you only have a short time in class to review your test. *If you need more time,* ask your instructor if you can look at the test in his or her office.

Here is a list of typical reasons for making errors on math tests.

1. You read the directions wrong.
2. You read the question wrong or skipped over something.
3. You made a computation error (maybe even an easy one).
4. Your answer is not accurate.
5. Your answer is not complete.
6. You labeled your answer wrong. For example, you labeled it "feet" and it should have been "feet^2."
7. You didn't show your work.
8. *You didn't understand the concept.
9. *You were unable to go from words (in a word problem) to setting up the problem.
10. *You were unable to apply a procedure to a new situation.
11. You were so anxious that you made errors even when you knew the material.

The first seven errors are **test-taking errors.** They are easy to correct if you decide to carefully read test questions and directions, proofread or rework your problems, show all your work, and double check units and labels every time.

The three starred errors (*) are **test preparation errors.** Remember that to grow a complex neural network, you need to practice the kinds of problems that you will see on the tests. So, for example, if application problems are difficult for you, you must *do more application problems!* If you have practiced the study skills techniques, however, you are less likely to make these kinds of errors on tests because you will have a deeper understanding of course concepts and you will be able to remember them better.

The last error isn't really an error. **Anxiety** can play a big part in your test results. Go back to the *Preparing for Tests* activity and read the suggestions about exercise and deep breathing. Recall from the *Your Brain Can Learn Mathematics* activity that when you are anxious, your body produces adrenaline. The presence of *adrenaline in the brain blocks connections* between dendrites. If you can *reduce*

OBJECTIVES

1 Determine the reason for errors.

2 Develop a plan to avoid test taking errors.

3 Review material to correct misunderstandings.

Immediately After the Test

After the Test Is Returned

Find Out Why You Made the Errors You Made

the *adrenaline* in your system, you will be able to *think more clearly* during your test. Just five minutes of brisk walking right before your test can help do that. Also, *practicing a relaxation technique while you do your homework* will make it more likely that you can benefit from using the technique during a test. *Deep breathing* is helpful because it gets *oxygen into your brain*. When you are anxious you tend to breathe more shallowly, which can make you feel confused and easily distracted.

Make a Plan for the Next Test

Make a plan for your next test based on your results from this test. You might review the Chapter Summary and work the problems in the Chapter Review Exercises or the Chapter Test. Ask your instructor or a tutor (if available) for more help if you are confused about any of the problems.

Now Try This ▶▶▶

Below is a record sheet to track your progress in test taking. Use it to find out if you make particular kinds of errors. Then you can work specifically on correcting them. Just check in the box when you made one of the errors. If you take more than five tests, make your own grid on separate paper.

Test Taking Errors

Test #	Read directions wrong	Read question wrong	Computation error	Not exact or accurate	Not complete	Labeled wrong	Didn't show work
1							
2							
3							
4							
5							

Test Preparation Errors

Test #	Didn't understand concept	Didn't set up problem correctly	Couldn't apply concept to new situation
1			
2			
3			
4			
5			

Anxiety

Test #	Felt anxious *before* the exam	Felt anxious *during* the exam	Blanked out on questions	Got questions wrong that I knew how to do
1				
2				
3				
4				
5				

What will you do to avoid Test Taking Errors?

_____

_____

_____

What will you do to avoid Test Preparation Errors?

_____

_____

_____

What will you do to reduce anxiety?

_____

_____

_____

Ratio and Proportion

Three out of every four Americans have cellular phones. Worldwide, about 1 out of 3 people use a cell phone! (*Source:* International Telecommunication Union.)

You can use unit rates to find the best deal on cell phone service plans. But most plans do not cover long-distance calls to certain U.S. locations or to other countries. Then you can use unit rates to find the cheapest long-distance calling card. (See **Section 5.2,** Exercises 29–32 for calling cards and Exercises 37–40 for cell phone plans.)

5.1 ▶▶▶ Ratios

A **ratio** compares two quantities. You can compare two numbers, such as 8 and 4, or two measurements that have the *same* type of units, such as 3 *days* and 12 *days*. (*Rates* compare measurements with different types of units and are covered in the next section.)

Ratios can help you see important relationships. For example, if the ratio of your monthly expenses to your monthly income is 10 to 9, then you are spending $10 for every $9 you earn and going deeper into debt.

OBJECTIVE 1 Write ratios as fractions.   A ratio can be written in three ways.

> **Writing a Ratio**
> The ratio of $7 **to** $3 can be written:
>
> $$7 \text{ to } 3 \qquad \text{or} \quad 7{:}3 \quad \text{or} \quad \frac{7}{3} \leftarrow \text{Fraction bar indicates "to."}$$
>
> "**:**" indicates "**to**"

Writing a ratio as a fraction is the most common method, and the one we will use here. All three ways are read, "the ratio of 7 to 3." The word **to** separates the quantities being compared.

> **Writing a Ratio as a Fraction**
> Order is important when writing a ratio. The quantity mentioned **first** is the **numerator**. The quantity mentioned **second** is the **denominator**. For example:
>
> $$\text{The ratio of } 5 \text{ to } 12 \text{ is written } \frac{5}{12}$$

EXAMPLE 1   **Writing Ratios**

Ancestors of the Pueblo Indians built multistory apartment towns in New Mexico about 1100 years ago. A room might measure 14 ft long, 11 ft wide, and 15 ft high.

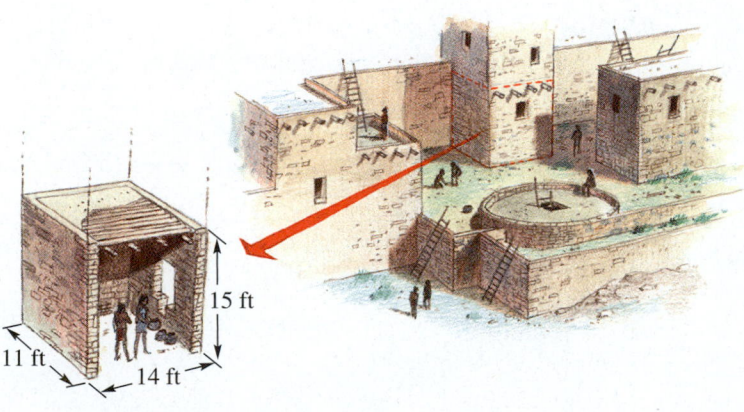

15 ft

11 ft   14 ft

Continued on Next Page

Write each ratio as a fraction, using the room measurements.

(a) Ratio of length to width

The ratio of **length to width** is $\dfrac{14 \ \cancel{ft}}{11 \ \cancel{ft}} = \dfrac{14}{11}$

> *Do not rewrite the ratio as $1\frac{3}{11}$.*

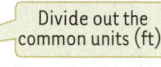

Numerator (mentioned first)   Denominator (mentioned second)

You can divide out common *units* just like you divided out common *factors* when writing fractions in lowest terms. (See **Section 2.4.**) However, do *not* rewrite the fraction as a mixed number. Keep it as the ratio of 14 to 11.

(b) Ratio of width to height

> *Divide out the common units (ft).*

The ratio of width **to** height is $\dfrac{11 \ \cancel{ft}}{15 \ \cancel{ft}} = \dfrac{11}{15}$

> **CAUTION**
> Remember, the order of the numbers is important in a ratio. Look for the words "ratio of *a* to *b*." Write the ratio as $\dfrac{a}{b}$, *not* $\dfrac{b}{a}$. The quantity mentioned first is the numerator.

Work Problem 1 at the Side. ▶

Any ratio can be written as a fraction. Therefore, you can write a ratio in *lowest terms,* just as you do with any fraction.

EXAMPLE 2   **Writing Ratios in Lowest Terms**

Write each ratio in lowest terms.

(a) 60 days to 20 days

The ratio is $\frac{60}{20}$. Write this ratio in lowest terms by dividing the numerator and denominator by 20.

$$\frac{60}{20} = \frac{60 \div 20}{20 \div 20} = \frac{3}{1} \quad \left\{ \begin{array}{l}\text{Ratio in} \\ \text{lowest terms}\end{array}\right.$$

So, the ratio of 60 days to 20 days is 3 to 1 or, written as a fraction, $\frac{3}{1}$.

> **CAUTION**
> In the fractions chapters you would have rewritten $\frac{3}{1}$ as 3. But a *ratio* compares *two* quantities, so you need to keep both parts of the ratio and write it as $\frac{3}{1}$.

(b) 50 ounces of medicine to 120 ounces of medicine

The ratio is $\frac{50}{120}$. Divide the numerator and denominator by 10.

$$\frac{50}{120} = \frac{50 \div 10}{120 \div 10} = \frac{5}{12} \quad \left\{ \begin{array}{l}\text{Ratio in} \\ \text{lowest terms}\end{array}\right.$$

So, the ratio of 50 ounces to 120 ounces is $\frac{5}{12}$.

Continued on Next Page

---

1 Shane spent \$14 on meat, \$5 on milk, and \$7 on fresh fruit. Write each ratio as a fraction.

(a) The ratio of amount spent on fruit to amount spent on milk.

(b) The ratio of amount spent on milk to amount spent on meat.

(c) The ratio of amount spent on meat to amount spent on milk.

ANSWERS

1. (a) $\dfrac{7}{5}$ (b) $\dfrac{5}{14}$ (c) $\dfrac{14}{5}$

2 Write each ratio as a fraction in lowest terms.

(a) 9 hours to 12 hours

(b) 100 meters to 50 meters

(c) Write the ratio of width to length for this rectangle.

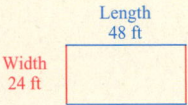

Length
48 ft

Width
24 ft

3 Write each ratio as a ratio of whole numbers in lowest terms.

(a) The price of Tamar's favorite brand of lipstick increased from $5.50 to $7.00. Find the ratio of the increase in price to the original price.

(b) Last week, Lance worked 4.5 hours each day. This week he cut back to 3 hours each day. Find the ratio of the decrease in hours to the original number of hours.

ANSWERS

2. **(a)** $\frac{3}{4}$ **(b)** $\frac{2}{1}$ **(c)** $\frac{1}{2}$

3. **(a)** $\frac{1.50 \times 100}{5.50 \times 100} = \frac{150 \div 50}{550 \div 50} = \frac{3}{11}$

(b) $\frac{1.5 \times 10}{4.5 \times 10} = \frac{15 \div 15}{45 \div 15} = \frac{1}{3}$

(c) 15 people in a large van to 6 people in a small van

The ratio is $\dfrac{15}{6} = \dfrac{15 \div 3}{6 \div 3} = \dfrac{5}{2}$ ← { Ratio in lowest terms

> **Note**
> Although $\frac{5}{2} = 2\frac{1}{2}$, ratios are *not* written as mixed numbers. Nevertheless, in Example 2(c) above, the ratio $\frac{5}{2}$ does mean the large van holds $2\frac{1}{2}$ times as many people as the small van.

◀ Work Problem **2** at the Side.

OBJECTIVE 2 Solve ratio problems involving decimals or mixed numbers. Sometimes a ratio compares two decimal numbers or two fractions. It is easier to understand if we rewrite the ratio as a ratio of two whole numbers.

EXAMPLE 3 Using Decimal Numbers in a Ratio

The price of a Sunday newspaper increased from $1.50 to $1.75. Find the ratio of the increase in price **to** the original price.

The words increase in price are mentioned first, so the increase will be the numerator. How much did the price go up? Use subtraction.

new price − original price = increase

$1.75 − $1.50 = $0.25

The words the original price are mentioned second, so the original price of $1.50 is the denominator.

The ratio of increase in price **to** original price is shown below.

$$\frac{0.25}{1.50}\quad\begin{array}{l}\leftarrow \text{increase in price}\\ \leftarrow \text{original price}\end{array}$$

Now rewrite the ratio as a ratio of whole numbers. Recall that if you multiply both the numerator and denominator of a fraction by the same number, you get an equivalent fraction. The decimals in this example are hundredths, so multiply by 100 to get whole numbers. (If the decimals are tenths, multiply by 10. If thousandths, multiply by 1000.) Then write the ratio in lowest terms.

$$\frac{0.25}{1.50} = \frac{0.25 \times 100}{1.50 \times 100} = \frac{25}{150} = \frac{25 \div 25}{150 \div 25} = \frac{1}{6}\quad \left\{ \begin{array}{l}\text{Ratio in}\\ \text{lowest terms}\end{array}\right.$$

Ratio as two whole numbers

◀ Work Problem **3** at the Side.

EXAMPLE 4 Using Mixed Numbers in Ratios

Write each ratio as a comparison of whole numbers in lowest terms.

(a) 2 days to $2\frac{1}{4}$ days

Write the ratio as follows. Divide out the common units.

$$\frac{2 \text{ days}}{2\frac{1}{4} \text{ days}} = \frac{2}{2\frac{1}{4}}$$

Continued on Next Page

Next, write 2 as $\frac{2}{1}$ and $2\frac{1}{4}$ as the improper fraction $\frac{9}{4}$.

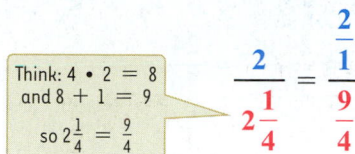

$$\frac{2}{2\frac{1}{4}} = \frac{\frac{2}{1}}{\frac{9}{4}}$$

Think: $4 \cdot 2 = 8$
and $8 + 1 = 9$
so $2\frac{1}{4} = \frac{9}{4}$

Now rewrite the problem in horizontal format, using the "÷" symbol for division. Finally, multiply by the reciprocal of the divisor, as you did in **Section 2.7.**

$$\frac{\frac{2}{1}}{\frac{9}{4}} = \frac{2}{1} \div \frac{9}{4} = \frac{2}{1} \cdot \frac{4}{9} = \frac{8}{9}$$

Reciprocals

The ratio, in lowest terms, is $\frac{8}{9}$.

(b) $3\frac{1}{4}$ to $1\frac{1}{2}$

Write the ratio as $\dfrac{3\frac{1}{4}}{1\frac{1}{2}}$. Then write $3\frac{1}{4}$ and $1\frac{1}{2}$ as improper fractions.

$$3\frac{1}{4} = \frac{13}{4} \quad \text{and} \quad 1\frac{1}{2} = \frac{3}{2}$$

The ratio is shown below.

$$\frac{3\frac{1}{4}}{1\frac{1}{2}} = \frac{\frac{13}{4}}{\frac{3}{2}}$$

Rewrite as a division problem in horizontal format, using the "÷" symbol. Then multiply by the reciprocal of the divisor.

$$\frac{13}{4} \div \frac{3}{2} = \frac{13}{\underset{2}{4}} \cdot \frac{\overset{1}{2}}{3} = \frac{13}{6} \quad \left\{ \begin{array}{l} \text{Ratio in} \\ \text{lowest terms} \end{array} \right.$$

Work Problem (4) *at the Side.* ▶

OBJECTIVE **3** **Solve ratio problems after converting units.**
When a ratio compares measurements, both measurements must be in the *same* units. For example, *feet* must be compared to *feet, hours* to *hours, pints* to *pints,* and *inches* to *inches.*

EXAMPLE 5 **Ratio Applications Using Measurement**

(a) Write the ratio of the length of the shorter board on the left to the length of the longer board on the right. Compare in inches.

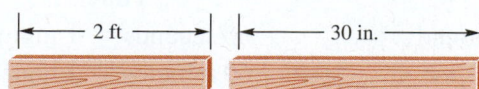

|← 2 ft →| |← 30 in. →|

First, express 2 ft in inches. Because 1 ft has 12 in., 2 ft is

$$2 \cdot \textbf{12 in.} = 24 \text{ in.}$$

Continued on Next Page

---

4 Write each ratio as a ratio of whole numbers in lowest terms.

(a) $3\frac{1}{2}$ to 4

(b) $5\frac{5}{8}$ pounds to $3\frac{3}{4}$ pounds

(c) $3\frac{1}{2}$ in. to $\frac{7}{8}$ in.

5 Write each ratio as a fraction in lowest terms. (*Hint:* Recall that it is usually easier to write the ratio using the smaller measurement unit.)

(a) 9 in. to 6 ft

(b) 2 days to 8 hours

(c) 7 yd to 14 ft

(d) 3 quarts to 3 gallons

(e) 25 minutes to 2 hours

(f) 4 pounds to 12 ounces

On the previous page, the length of the board on the left is 24 in., so the ratio of the lengths is

$$\frac{2 \text{ ft}}{30 \text{ in.}} = \frac{24 \text{ in.}}{30 \text{ in.}} = \frac{24}{30}$$

> Once the units match, you can divide them out.

Write the ratio in lowest terms.

$$\frac{24}{30} = \frac{24 \div 6}{30 \div 6} = \frac{4}{5} \quad \left\{ \begin{array}{l} \text{Ratio in} \\ \text{lowest terms} \end{array} \right.$$

The shorter board is $\frac{4}{5}$ the length of the longer board.

> **Note**
>
> Notice in the example above that we wrote the ratio using the smaller unit (inches are smaller than feet). Using the smaller unit will help you avoid working with fractions. If we wrote the ratio using feet, then
>
> $$30 \text{ in.} = 2\frac{1}{2} \text{ ft}$$
>
> So the ratio in feet is shown below.
>
> $$\frac{2 \text{ ft}}{2\frac{1}{2} \text{ ft}} = \frac{2}{1} \div \frac{5}{2} = \frac{2}{1} \cdot \frac{2}{5} = \frac{4}{5} \quad \leftarrow \text{Same result}$$
>
> The ratio is the same, but it takes more steps to get the answer. Using the smaller unit is usually easier.

(b) Write the ratio of 28 days to 3 weeks.

Since it is easier to write the ratio using the smaller measurement unit, compare in *days* because days are shorter than weeks.

First express 3 weeks in days. Because 1 week has 7 days, 3 weeks is

$$3 \cdot 7 \text{ days} = 21 \text{ days}.$$

So the ratio in days is shown below.

$$\frac{28 \text{ days}}{3 \text{ weeks}} = \frac{28 \text{ days}}{21 \text{ days}} = \frac{28}{21} = \frac{28 \div 7}{21 \div 7} = \frac{4}{3} \quad \left\{ \begin{array}{l} \text{Ratio in} \\ \text{lowest terms} \end{array} \right.$$

The following table will help you set up ratios that compare measurements. You will work with these measurements again in **Appendix A.**

Measurement Comparisons

Length	Capacity (Volume)
12 inches = 1 foot	2 cups = 1 pint
3 feet = 1 yard	2 pints = 1 quart
5280 feet = 1 mile	4 quarts = 1 gallon
Weight	**Time**
16 ounces = 1 pound	60 seconds = 1 minute
2000 pounds = 1 ton	60 minutes = 1 hour
	24 hours = 1 day
	7 days = 1 week

◀ *Work Problem* **5** *at the Side.*

5.1 ▶▶▶ **Exercises**

FOR EXTRA HELP
   PRACTICE   WATCH   DOWNLOAD   READ   REVIEW

Write each ratio as a fraction in lowest terms. See Examples 1 and 2.

1. 8 days to 9 days

2. $11 to $15

 3. $100 to $50

4. 35¢ to 7¢

5. 30 minutes to 90 minutes

6. 9 pounds to 36 pounds

7. 80 miles to 50 miles

8. 300 people to 450 people

9. 6 hours to 16 hours

10. 45 books to 35 books

Write each ratio as a ratio of whole numbers in lowest terms. See Examples 3 and 4.

11. $4.50 to $3.50

12. $0.08 to $0.06

13. 15 to $2\frac{1}{2}$

14. 5 to $1\frac{1}{4}$

 15. $1\frac{1}{4}$ to $1\frac{1}{2}$

16. $2\frac{1}{3}$ to $2\frac{2}{3}$

Write each ratio as a fraction in lowest terms. For help, use the table of measurement relationships on the previous page. See Example 5.

17. 4 ft to 30 in.

18. 8 ft to 4 yd

19. 5 minutes to 1 hour

20. 8 quarts to 5 pints

21. 15 hours to 2 days

22. 3 pounds to 6 ounces

23. 5 gallons to 5 quarts

24. 3 cups to 3 pints

The table shows the number of greeting cards that Americans buy for various occasions. Use the information to answer Exercises 25–30. Write each ratio as a fraction in lowest terms.

Holiday/Event	Cards Sold
Valentine's Day	900 million
Mother's Day	150 million
Father's Day	95 million
Graduation	60 million
Thanksgiving	30 million
Halloween	25 million

Source: Hallmark Cards.

25. Find the ratio of Thanksgiving cards to graduation cards.

26. Find the ratio of Halloween cards to Mother's Day cards.

27. Find the ratio of Valentine's Day cards to Halloween cards.

28. Find the ratio of Mother's Day cards to Father's Day cards.

29. Explain how you might use the information in the table if you owned a shop selling gifts and greeting cards.

30. Why is the ratio of Valentine's Day cards to graduation cards $\frac{15}{1}$? Give two possible reasons.

The bar graph shows worldwide sales of the most popular songs of all time. Use the graph to complete Exercises 31–34. Write each ratio as a fraction in lowest terms.

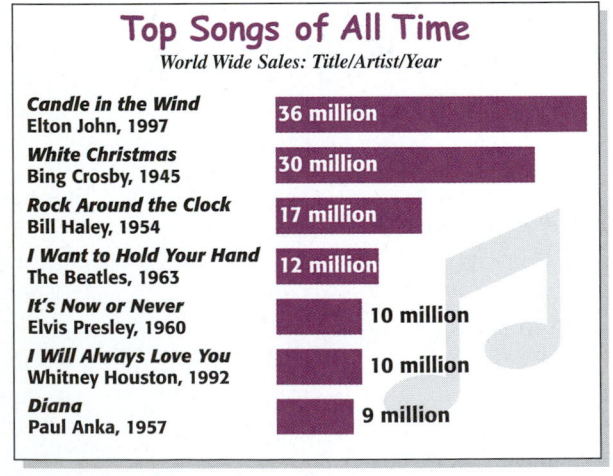

Source: The Music Information Database.

31. Write a ratio that compares the top-selling song to the second best seller, and a ratio that compares the top-selling song to the third best seller.

32. Write two ratios that compare sales of Elvis Presley's song with sales of the songs just ahead and just behind it in the graph.

33. Sales of which two songs give a ratio of $\frac{3}{1}$? There may be more than one correct answer.

34. Sales of which two songs give a ratio of $\frac{5}{6}$? There may be more than one correct answer.

Use the circle graph of one person's monthly budget to complete Exercises 35–36.
Write each ratio as a fraction in lowest terms.

35. Find the ratio of
  (a) rent to utilities

  (b) rent to food

  (c) rent and utilities to total budget.

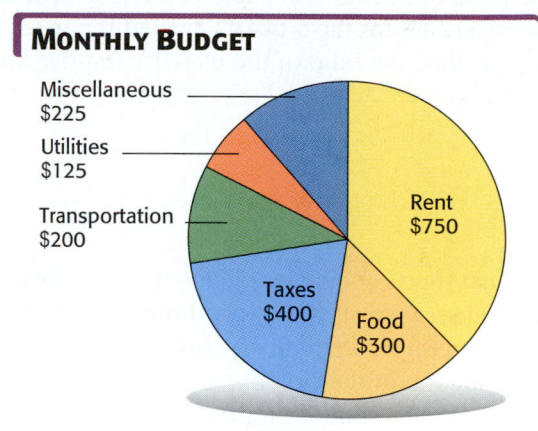

MONTHLY BUDGET

Miscellaneous $225
Utilities $125
Transportation $200
Rent $750
Taxes $400
Food $300

36. Find the ratio of

  (a) taxes to rent

  (b) food to transportation

  (c) taxes and food to rent and utilities.

For each figure, find the ratio of the length of the longest side to the length of the shortest
side. Write each ratio as a fraction in lowest terms.

37.

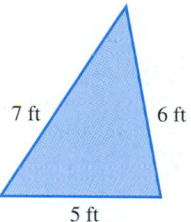

7 ft   6 ft
5 ft

38.

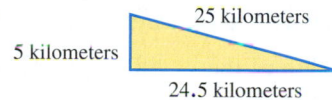

25 kilometers
5 kilometers
24.5 kilometers

39.

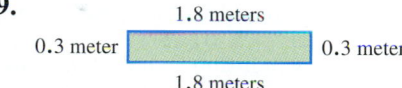

1.8 meters
0.3 meter   0.3 meter
1.8 meters

40.

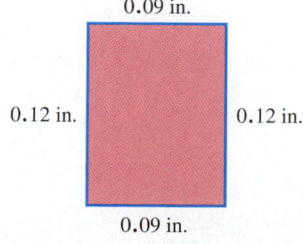

0.09 in.
0.12 in.   0.12 in.
0.09 in.

41.

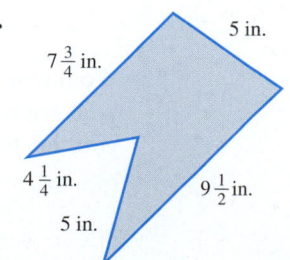

5 in.
$7\frac{3}{4}$ in.
$4\frac{1}{4}$ in. $9\frac{1}{2}$ in.
5 in.

42.

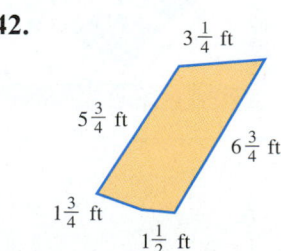

$3\frac{1}{4}$ ft
$5\frac{3}{4}$ ft
$6\frac{3}{4}$ ft
$1\frac{3}{4}$ ft
$1\frac{1}{2}$ ft

Write each ratio as a fraction in lowest terms.

43. The price of automobile engine oil has gone from $10 to $12.50 for the 5 quarts needed for an oil change. Find the ratio of the increase in price to the original price.

44. The price that a pharmacy pays for an antibiotic decreased from $8.80 to $5.60 for 10 tablets. Find the ratio of the decrease in price to the original price.

45. The first time a movie was made in Minnesota, the cast and crew spent $59\frac{1}{2}$ days filming winter scenes. The next year, another movie was filmed in $8\frac{3}{4}$ weeks. Find the ratio of the first movie's filming time to the second movie's time. Compare in weeks.

46. The percheron, a large draft horse, measures about $5\frac{3}{4}$ ft at the shoulder. The prehistoric ancestor of the horse measured only $15\frac{3}{4}$ in. at the shoulder. Find the ratio of the percheron's height to its prehistoric ancestor's height. Compare in inches. (*Source: Eyewitness Books: Horse.*)

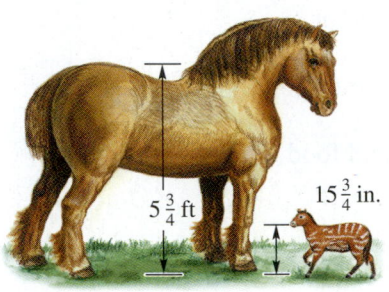

Relating Concepts (Exercises 47–50)   For Individual or Group Work

*Use your knowledge of ratios to **work Exercises 47–50 in order.***

47. In this painting, what is the ratio of the length of the longest side to the length of the shortest side? What other measurements could the painting have and still maintain the same ratio?

48. The ratio of my son's age to my daughter's age is 4 to 5. One possibility is that my son is 4 years old and my daughter is 5 years old. Find six other possibilities that fit the 4 to 5 ratio.

49. Amelia said that the ratio of her age to her mother's age is 5 to 3. Is this possible? Explain your answer.

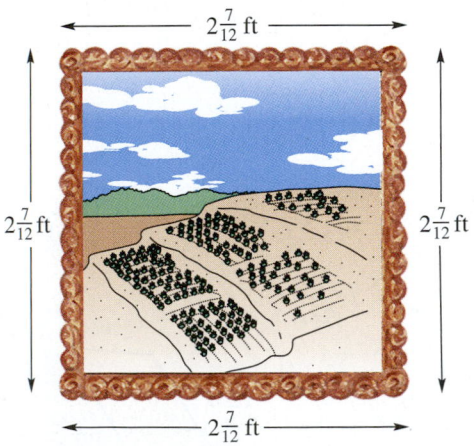

50. Would you prefer that the ratio of your income to your friend's income be 1 to 3 or 3 to 1? Explain your answers.

5.2 ▶▶▶ Rates

A *ratio* compares two measurements with the same type of units, such as 9 feet **to** 12 feet (both length measurements). But many of the comparisons we make use measurements with different types of units, such as the following.

160 dollars **for** 8 hours   (money to time)

450 miles **on** 15 gallons   (distance to capacity)

This type of comparison is called a **rate**.

OBJECTIVE 1 **Write rates as fractions.** Suppose you hiked 18 miles **in** 4 hours. The *rate* at which you hiked can be written as a fraction in lowest terms.

$$\frac{18 \text{ miles}}{4 \text{ hours}} = \frac{18 \text{ miles} \div 2}{4 \text{ hours} \div 2} = \frac{9 \text{ miles}}{2 \text{ hours}} \left.\begin{cases} \\ \end{cases}\right\} \begin{array}{l} \text{Rate in} \\ \text{lowest terms} \end{array}$$

In a rate, you often find these words separating the quantities you are comparing.

in   for   on   per   from

> **CAUTION**
> When writing a rate, always include the units, such as miles, hours, dollars, and so on. Because the units in a rate are different, the units do *not* divide out.

EXAMPLE 1 **Writing Rates in Lowest Terms**

Write each rate as a fraction in lowest terms.

(a) 5 gallons of chemical **for** $60.

$$\frac{5 \text{ gallons} \div 5}{60 \text{ dollars} \div 5} = \frac{1 \text{ gallon}}{12 \text{ dollars}} \quad \begin{array}{l}\text{Write the units:} \\ \text{gallons and dollars.}\end{array}$$

(b) $1500 wages **in** 10 weeks

$$\frac{1500 \text{ dollars} \div 10}{10 \text{ weeks} \div 10} = \frac{150 \text{ dollars}}{1 \text{ week}}$$

Be sure to write the units in a *rate*: dollars, miles, gallons, and so on.

(c) 2225 miles **on** 75 gallons of gas

$$\frac{2225 \text{ miles} \div 25}{75 \text{ gallons} \div 25} = \frac{89 \text{ miles}}{3 \text{ gallons}}$$

Work Problem **1** *at the Side.* ▶

OBJECTIVE 2 **Find unit rates.** When the *denominator* of a rate is 1, it is called a **unit rate**. We use unit rates frequently. For example, you earn $12.75 for *1 hour* of work. This unit rate is written:

$12.75 **per** hour   or $12.75/hour.

Or, you drive 28 miles on *1 gallon* of gas. This unit rate is written

28 miles **per** gallon   or   28 miles/gallon.

Use **per** or a slash mark (**/**) when writing unit rates.

OBJECTIVES

1 **Write rates as fractions.**

2 **Find unit rates.**

3 **Find the best buy based on cost per unit.**

1 Write each rate as a fraction in lowest terms.

(a) $6 for 30 packages

(b) 500 miles in 10 hours

(c) 4 teachers for 90 students

(d) 1270 bushels from 30 acres

ANSWERS

1. **(a)** $\frac{1 \text{ dollar}}{5 \text{ packages}}$ **(b)** $\frac{50 \text{ miles}}{1 \text{ hour}}$

(c) $\frac{2 \text{ teachers}}{45 \text{ students}}$ **(d)** $\frac{127 \text{ bushels}}{3 \text{ acres}}$

2 Find each unit rate.

(a) $4.35 for 3 pounds of cheese

EXAMPLE 2   **Finding Unit Rates**

Find each unit rate.

(a)   337.5 miles on 13.5 gallons of gas
Write the rate as a fraction.

$$\frac{337.5 \text{ miles}}{13.5 \text{ gallons}} \leftarrow \text{The fraction bar indicates division.}$$

Divide 337.5 by 13.5 to find the unit rate.

$$13.5\overline{)337.5} \quad \begin{array}{r} 2\,5. \end{array}$$

$$\frac{337.5 \text{ miles} \div \textbf{13.5}}{13.5 \text{ gallons} \div \textbf{13.5}} = \frac{25 \text{ miles}}{1 \text{ gallon}}$$

The unit rate is 25 miles **per** gallon, or 25 miles/gallon.

(b) 304 miles on 9.5 gallons of gas

(b)   549 miles in 18 hours.

$$\frac{549 \text{ miles}}{18 \text{ hours}} \qquad \text{Divide: } 18\overline{)549.0} \quad \begin{array}{r} 30.5 \end{array}$$

The unit rate is 30.5 miles/hour

(c) $810 in 6 days

$$\frac{810 \text{ dollars}}{6 \text{ days}} \qquad \text{Divide: } 6\overline{)810} \quad \begin{array}{r} 135 \end{array}$$

Use *per* or a slash mark to write unit rates.

The unit rate is $135/day.

◀ *Work Problem* **2** *at the Side.*

(c) $850 in 5 days

OBJECTIVE 3 **Find the best buy based on cost per unit.**   When shopping for groceries, household supplies, and health and beauty items, you will find many different brands and package sizes. You can save money by finding the lowest *cost per unit.*

Cost per Unit

Cost per unit is a rate that tells how much you pay for *one* item or *one* unit. Examples are $3.25 per gallon, $47 per shirt, and $2.98 per pound.

(d) 24-pound turkey for 15 people

EXAMPLE 3   **Determining the Best Buy**

The local store charges the following prices for pancake syrup. Find the best buy.

Continued on Next Page

The best buy is the container with the *lowest* cost per unit. All the containers are measured in *ounces* (oz), so you first need to find the *cost per ounce* for each one. Divide the price of the container by the number of ounces in it. Round to the nearest thousandth, if necessary.

Let the *order* of the *words* help you set up the rate.

cost   ⟶   **$1.28**
per (means divide)⟶ ─────
ounce   ⟶   **12 ounces**

Size	Cost per Unit (rounded)
12 ounces	$\dfrac{\$1.28}{12 \text{ ounces}} \approx \$0.107$ per ounce (highest)
24 ounces	$\dfrac{\$1.81}{24 \text{ ounces}} \approx \$0.075$ per ounce (lowest)
36 ounces	$\dfrac{\$2.73}{36 \text{ ounces}} \approx \$0.076$ per ounce

The lowest cost per ounce is $0.075, so the 24-ounce container is the best buy.

> **Note**
> Earlier we rounded money amounts to the nearest hundredth (nearest cent). But when comparing unit costs, rounding to the nearest thousandth will help you see the difference between very similar unit costs. Notice that the 24-ounce and 36-ounce syrup containers above would both have rounded to $0.08 per ounce if we had rounded to hundredths.

Work Problem **3** *at the Side.* ▶

🖩 **Calculator Tip**   When using a calculator to find unit prices, remember that division is *not* commutative. In Example 3 you wanted to find cost per ounce. Let the *order* of the *words* help you enter the numbers in the correct order.

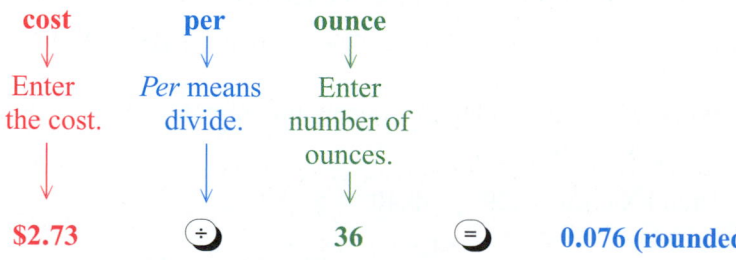

cost	**per**	**ounce**
↓	↓	↓
Enter the cost.	*Per* means divide.	Enter number of ounces.
↓	↓	↓
$2.73	÷	**36** = **0.076 (rounded)**

If you entered 36 ÷ 2.73 = , you'd get the number of *ounces* per *dollar*. How could you use that information to find the best buy? (*Answer:* The best buy would be to get the greatest number of ounces per dollar.)

Finding the best buy is sometimes a complicated process. Things that affect the cost per unit can include "cents off" coupons and differences in how much use you'll get out of each unit.

---

3 Find the best buy (lowest cost per unit) for each purchase.

  (a) 2 quarts for $3.25
      3 quarts for $4.95
      4 quarts for $6.48

  (b) 6 cans of cola for $1.99
      12 cans of cola for $3.49
      24 cans of cola for $7

ANSWERS

3. **(a)** 4 quarts, at $1.62 per quart
   (b) 12 cans, at $0.291 per can (rounded)

4 Solve each problem.

(a) Some batteries claim to last longer than others. If you believe these claims, which brand is the best buy?

Four-pack of AA-size batteries for $2.79

One AA-size battery for $1.19; lasts twice as long

(b) Which tube of toothpaste is the best buy? You have a coupon for 85¢ off Brand C and a coupon for 20¢ off Brand D.

Brand C is $3.89 for 6 ounces.

Brand D is $1.59 for 2.5 ounces.

EXAMPLE 4 **Solving Best Buy Applications**

Solve each application problem.

(a) There are many brands of liquid laundry detergent. If you feel they all do a good job of cleaning your clothes, you can base your purchase on cost per unit. But some brands are "concentrated" so you can use less detergent for each load of clothes. Which of the choices shown below is the best buy?

To find Sudzy's unit cost, divide $3.99 by 64 ounces, not 50 ounces. You're getting as many clothes washed as if you bought 64 ounces. Similarly, to find White-O's unit cost, divide $9.89 by 256 ounces (twice 128 ounces, or 2 • 128 ounces = 256 ounces).

Sudzy $\dfrac{\$3.99}{64 \text{ ounces}}$ ≈ $0.062 per ounce

White-O $\dfrac{\$9.89}{256 \text{ ounces}}$ ≈ $0.039 per ounce

> The best buy is the *lower* cost per ounce.

White-O has the lower cost per ounce and is the better buy. (However, if you try it and it really doesn't get out all the stains, Sudzy may be worth the extra cost.)

(b) "Cents-off" coupons also affect the best buy. Suppose you are looking at these choices for "extra-strength" pain reliever. Both brands have the same amount of pain reliever in each tablet.

Brand X is $2.29 for 50 tablets.

Brand Y is $10.75 for 200 tablets.

You have a 40¢ coupon for Brand X and a 75¢ coupon for Brand Y. Which choice is the best buy?

To find the best buy, first subtract the coupon amounts, then divide to find the lower cost per ounce.

Brand X costs $2.29 − $0.40 = $1.89

$\dfrac{\$1.89}{50 \text{ tablets}}$ ≈ $0.038 per tablet

Brand Y costs $10.75 − $0.75 = $10.00

> Look for the *lower* cost per tablet.

$\dfrac{\$10.00}{200 \text{ tablets}}$ = $0.05 per tablet

Brand X has the lower cost per tablet and is the better buy.

◀ *Work Problem* **4** *at the Side.*

5.2 ▶▶▶ Exercises

Write each rate as a fraction in lowest terms. See Example 1.

1. 10 cups for 6 people

2. $12 for 30 pens

3. 15 feet in 35 seconds

4. 100 miles in 30 hours

5. 72 miles on 4 gallons

6. 132 miles on 8 gallons

Find each unit rate. See Example 2.

7. $60 in 5 hours

8. $2500 in 20 days

9. 7.5 pounds for 6 people

10. 44 bushels from 8 trees

11. $413.20 for 4 days

12. $74.25 for 9 hours

Earl kept the following record of the gas he bought for his car. For each entry, find the number of miles he traveled and the unit rate. Round your answers to the nearest tenth.

	Date	Odometer at Start	Odometer at End	Miles Traveled	Gallons Purchased	Miles per Gallon
13.	2/4	27,432.3	27,758.2		15.5	
14.	2/9	27,758.2	28,058.1		13.4	
15.	2/16	28,058.1	28,396.7		16.2	
16.	2/20	28,396.7	28,704.5		13.3	

Source: Author's car records.

Find the best buy (based on the cost per unit) for each item. See Example 3.
(*Sources:* Cub Foods, Target, Rainbow Foods.)

17. Black pepper

18. Shampoo

19. Cereal
12 ounces for $2.49
14 ounces for $2.89
18 ounces for $3.96

20. Soup (same size cans)
2 cans for $2.18
3 cans for $3.57
5 cans for $5.29

21. Chunky peanut butter
12 ounces for $1.29
18 ounces for $1.79
28 ounces for $3.39
40 ounces for $4.39

22. Baked beans
8 ounces for $0.59
16 ounces for $0.99
21 ounces for $1.29
28 ounces for $1.89

23. Suppose you are choosing between two brands of chicken noodle soup. Brand A is $0.88 per can and Brand B is $0.98 per can. The cans are the same size but Brand B has more chunks of chicken in it. Which soup is the better buy? Explain your choice.

24. A small bag of potatoes costs $0.19 per pound. A large bag costs $0.15 per pound. But there are only two people in your family, so half the large bag would probably rot before you used it up. Which bag is the better buy? Explain.

Solve each application problem. See Examples 2–4.

25. Makesha lost 10.5 pounds in six weeks. What was her rate of loss in pounds per week?

26. Enrique's taco recipe uses three pounds of meat to feed 10 people. Give the rate in pounds per person.

27. Russ works 7 hours to earn $85.82. What is his pay rate per hour?

28. Find the cost of 1 gallon of Hawaiian Punch beverage if 18 gallons for a graduation party cost $55.62.

The table lists information about three long-distance calling cards. The connection fee is charged each time you make a call, no matter how long the call lasts. Use the table to answer Exercises 29–32. Round answers to the nearest thousandth when necessary.

LONG-DISTANCE CALLING CARDS (U.S.)

Card Name	Cost per Minute	Connection Fee
Radiant Penny	$0.01	$0.39
IDT Special	0.022	0.14
Access America	0.047	0.00

Source: www.1callcard.com

29. **(a)** Find the *actual* total cost, including the connection charge, for a five-minute call using each card.
    (b) Find the cost per minute for this call using each card and select the best buy.

30. **(a)** Find the *actual* total cost, including the connection charge, for a 30-minute call using each card.
    (b) Find the cost per minute for this call using each card and select the best buy.

31. Find the *actual* total cost per minute for a 15-minute call and a 25-minute call using each card. Then select the best buy for each call.

32. All the cards round calls up to the next full minute.
    (a) Suppose you call the wrong number. How much would you pay for this 40-*second* call on each card?
    (b) How much would you save on this call by using Access America instead of Radiant Penny?

33. If you believe the claims that some batteries last longer, which is the better buy?

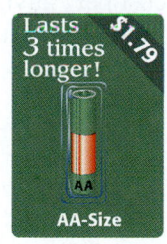

34. Which is the better buy, assuming these laundry detergents both clean equally well?

35. Three brands of cornflakes are available. Brand G is priced at $2.39 for 10 ounces. Brand K is $3.99 for 20.3 ounces and Brand P is $3.39 for 16.5 ounces. You have a coupon for 50¢ off Brand P and a coupon for 60¢ off Brand G. Which cereal is the best buy based on cost per unit?

36. Two brands of facial tissue are available. Brand K is specially priced at three boxes of 175 tissues each for $5. Brand S is priced at $1.29 per box of 125 tissues. You have a coupon for 20¢ off one box of Brand S and a coupon for 45¢ off one box of Brand K. How can you get the best buy on one box of tissue?

Relating Concepts (Exercises 37–40) For Individual or Group Work

On the first page of this chapter, we said that unit rates can help you get the best deal on cell phone service. Use the information in the table to **work Exercises 37–40 in order.**

CELL PHONE SERVICE PLANS

Company	Anytime Minutes	One-time Activation Fee	Monthly Charge	Termination Fee
Verizon	400	$35	$59.99	$175
T-Mobile	600	$35	$39.99	$200
Nextel	500	$35	$45.99	$200
Sprint	500	$36	$55	$150

Source: Advertisements appearing in *Minneapolis Star Tribune.*

Notes:

1. All companies require that you sign a contract for one year of service and charge a one-time termination fee if you quit early.
2. Unused minutes cannot be carried over to the next month.

37. How much will each company's activation fee cost you on a *monthly* basis during the one-year contract?

38. All the plans allow unlimited calls on nights and weekends, so you would be using the "anytime minutes" on weekdays. Figure out the average number of weekdays per month. Then, for each plan, how many minutes could you use per weekday? Round to the nearest whole minute.

39. Find the actual average cost per "anytime minute" during the one-year contract for each company, including the activation fee. Assume you use all the minutes and no more. Decide how to round your answers so you can find the best buy.

40. Suppose that after two months you canceled your service because you found that you only used 100 "anytime minutes" per month. Under those conditions, find the actual cost per "anytime minute" for each company, to the nearest cent.

5.3 ▶▶▶ Proportions

OBJECTIVE **1** **Write proportions.** A **proportion** states that two ratios (or rates) are equivalent. For example,

$$\frac{\$20}{4 \text{ hours}} = \frac{\$40}{8 \text{ hours}}$$

is a proportion that says the rate $\frac{\$20}{4 \text{ hours}}$ is equivalent to the rate $\frac{\$40}{8 \text{ hours}}$.
As the amount of money doubles, the number of hours also doubles. This proportion is read:

20 dollars **is to** 4 hours **as** 40 dollars **is to** 8 hours.

OBJECTIVES

1 **Write proportions.**

2 **Determine whether proportions are true or false.**

3 **Find cross products.**

EXAMPLE 1 **Writing Proportions**

Write each proportion.

(a) 6 ft is to 11 ft **as** 18 ft is to 33 ft.

$$\frac{6 \text{ ft}}{11 \text{ ft}} = \frac{18 \text{ ft}}{33 \text{ ft}} \quad \text{so} \quad \frac{6}{11} = \frac{18}{33} \qquad \text{The common units (ft) divide out and are not written.}$$

(b) \$9 is to 6 liters **as** \$3 is to 2 liters.

$$\frac{\$9}{6 \text{ liters}} = \frac{\$3}{2 \text{ liters}} \qquad \text{The units do \textit{not} match so you must write them in the proportion.}$$

Work Problem **1** *at the Side.* ▶

OBJECTIVE **2** **Determine whether proportions are true or false.**
There are two ways to see whether a proportion is true. One way is to *write both of the ratios in lowest terms.*

EXAMPLE 2 **Writing Both Ratios in Lowest Terms**

Determine whether each proportion is true or false by writing both ratios in lowest terms.

(a) $\dfrac{5}{9} = \dfrac{18}{27}$

Write each ratio in lowest terms.

$$\frac{5}{9} \; \leftarrow \; \begin{matrix}\text{Already in}\\\text{lowest terms}\end{matrix} \qquad \frac{18 \div 9}{27 \div 9} = \frac{2}{3} \; \leftarrow \; \begin{matrix}\text{Lowest}\\\text{terms}\end{matrix}$$

Because $\frac{5}{9}$ is *not* equivalent to $\frac{2}{3}$, the proportion is *false.*

(b) $\dfrac{16}{12} = \dfrac{28}{21}$

Write each ratio in lowest terms.

$$\frac{16 \div 4}{12 \div 4} = \frac{4}{3} \quad \text{and} \quad \frac{28 \div 7}{21 \div 7} = \frac{4}{3}$$

Both ratios are equivalent to $\frac{4}{3}$, so the proportion is *true.*

Work Problem **2** *at the Side.* ▶

---

1 Write each proportion.

(a) \$7 is to 3 cans as \$28 is to 12 cans

(b) 9 meters is to 16 meters as 18 meters is to 32 meters

(c) 5 is to 7 as 35 is to 49

(d) 10 is to 30 as 60 is to 180

2 Determine whether each proportion is true or false by writing both ratios in lowest terms.

(a) $\dfrac{6}{12} = \dfrac{15}{30}$

(b) $\dfrac{20}{24} = \dfrac{3}{4}$

(c) $\dfrac{25}{40} = \dfrac{30}{48}$

(d) $\dfrac{35}{45} = \dfrac{12}{18}$

ANSWERS

1. **(a)** $\dfrac{\$7}{3 \text{ cans}} = \dfrac{\$28}{12 \text{ cans}}$ **(b)** $\dfrac{9}{16} = \dfrac{18}{32}$
   (c) $\dfrac{5}{7} = \dfrac{35}{49}$ **(d)** $\dfrac{10}{30} = \dfrac{60}{180}$
2. **(a)** $\dfrac{1}{2} = \dfrac{1}{2}$; true **(b)** $\dfrac{5}{6} \neq \dfrac{3}{4}$; false
   (c) $\dfrac{5}{8} = \dfrac{5}{8}$; true **(d)** $\dfrac{7}{9} \neq \dfrac{2}{3}$; false

OBJECTIVE 3 **Find cross products.** Another way to test whether the ratios in a proportion are equivalent is to compare *cross products*.

Using Cross Products to Determine Whether a Proportion Is True

To see whether a proportion is true, first multiply along one diagonal, then multiply along the other diagonal, as shown here.

$$5 \cdot 4 = \mathbf{20}$$

$$\frac{2}{5} = \frac{4}{10}$$

Cross products are equal.

$$2 \cdot 10 = \mathbf{20}$$

In this case the **cross products** are both 20. When cross products are *equal,* the proportion is *true.* If the cross products are *unequal,* the proportion is *false.*

Note

The cross products test is based on rewriting both fractions with a common denominator of $5 \cdot 10$, or 50.

$$\frac{2 \cdot \mathbf{10}}{5 \cdot \mathbf{10}} = \frac{20}{50} \quad \text{and} \quad \frac{4 \cdot \mathbf{5}}{10 \cdot \mathbf{5}} = \frac{20}{50}$$

We see that $\frac{2}{5}$ and $\frac{4}{10}$ are equivalent because both can be rewritten as $\frac{20}{50}$. The cross product test takes a shortcut by comparing only the two numerators ($20 = 20$).

EXAMPLE 3 **Using Cross Products**

Use cross products to see whether each proportion is true or false.

(a) $\dfrac{3}{5} = \dfrac{12}{20}$

Multiply along one diagonal, then multiply along the other diagonal.

$$5 \cdot 12 = \mathbf{60}$$

$$\frac{3}{5} = \frac{12}{20}$$

Equal cross products; proportion is *true.*

$$3 \cdot 20 = \mathbf{60}$$

The cross products are *equal,* so the proportion is *true.*

CAUTION

Use cross products *only* when working with *proportions.* Do **not** use cross products when multiplying fractions, adding fractions, or writing fractions in lowest terms.

Continued on Next Page

(b) $\dfrac{2\frac{1}{3}}{3\frac{1}{3}} = \dfrac{9}{16}$

Find the cross products.

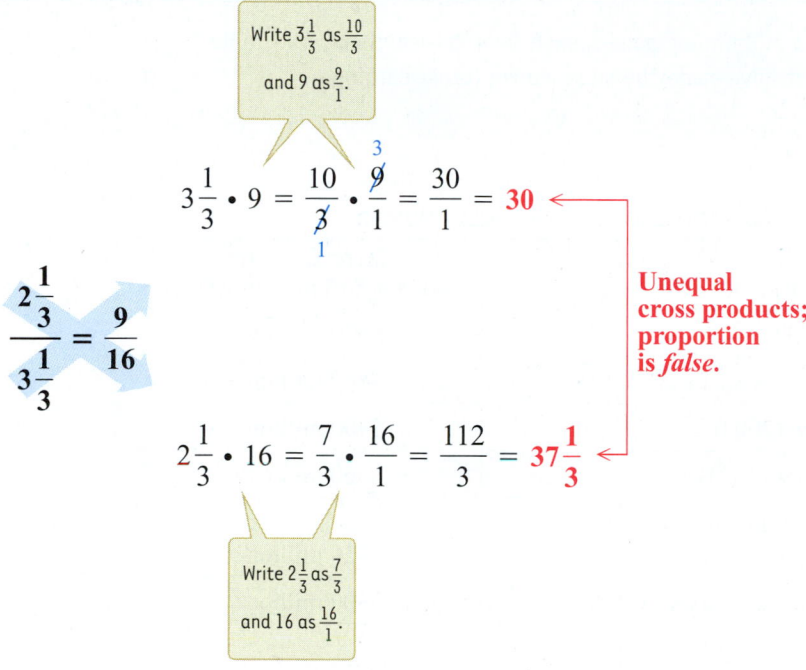

Write $3\frac{1}{3}$ as $\frac{10}{3}$ and 9 as $\frac{9}{1}$.

$$3\frac{1}{3} \cdot 9 = \frac{10}{3} \cdot \frac{9}{1} = \frac{30}{1} = 30$$

Unequal cross products; proportion is *false*.

$$\dfrac{2\frac{1}{3}}{3\frac{1}{3}} = \dfrac{9}{16}$$

$$2\frac{1}{3} \cdot 16 = \frac{7}{3} \cdot \frac{16}{1} = \frac{112}{3} = 37\frac{1}{3}$$

Write $2\frac{1}{3}$ as $\frac{7}{3}$ and 16 as $\frac{16}{1}$.

The cross products are *unequal,* so the proportion is *false.*

> **Note**
> The numbers in a proportion do *not* have to be whole numbers. They can be fractions, mixed numbers, decimal numbers, and so on.

Work Problem **3** *at the Side.* ▶

3 Find the cross products to see whether each proportion is true or false.

(a) $\dfrac{5}{9} = \dfrac{10}{18}$

(b) $\dfrac{32}{15} = \dfrac{16}{8}$

(c) $\dfrac{10}{17} = \dfrac{20}{34}$

(d) $\dfrac{2.4}{6} = \dfrac{5}{12}$

(6)(5) =

(2.4)(12) =

(e) $\dfrac{3}{4.25} = \dfrac{24}{34}$

(f) $\dfrac{1\frac{1}{6}}{2\frac{1}{3}} = \dfrac{4}{8}$

Math in the Media

MOVIES MAKE BIG BUCKS

Americans love movies and spend millions of dollars each year to see them. The table below lists some of the movies that have made the most money (gross earnings).

SELECTED ALL-TIME TOP GROSSING AMERICAN MOVIES

Title (Year Released)	Gross Earnings (nearest $10 million dollars)
Titanic (1997)	$600 million
Star Wars: Episode IV (1977)	$460 million
Spider-Man (2002)	$400 million
Jurassic Park (1993)	$360 million
Harry Potter and the Sorcerer's Stone (2001)	$320 million
Pirates of the Caribbean (2003)	$300 million

Source: The World Almanac.

Write all ratios as fractions in lowest terms.

1. **(a)** Write a ratio to compare *Titanic*'s earnings to the earnings from *Pirates of the Caribbean*.

   (b) Write a ratio to compare the earnings from *Pirates of the Caribbean* to *Titanic*'s earnings.

   (c) Look at the two ratios you wrote. What is special about them? What is the mathematical word that describes the relationship between these two ratios?

2. **(a)** Write a ratio to compare *Jurassic Park's* earnings to those of the movie just below it.

   (b) One way to use the ratio you wrote in part (a) is to say, "For every $9 earned by *Jurassic Park,* the *Harry Potter* movie earned $8." Now use the ratios you wrote in Problem 1 above to talk about those movies in the same way.

   (c) Find a pair of movies where one of them earned $4 to every $3 earned by the other.

   (d) Find a pair of movies where one of them earned $4 to every $5 earned by the other.

Write each proportion. See Example 1.

1. $9 is to 12 cans as $18 is to 24 cans.

2. 28 people is to 7 cars as 16 people is to 4 cars.

3. 200 adults is to 450 children as 4 adults is to 9 children.

4. 150 trees is to 1 acre as 1500 trees is to 10 acres.

🌐 **5.** 120 ft is to 150 ft as 8 ft is to 10 ft.

6. $6 is to $9 as $10 is to $15.

Determine whether each proportion is true or false by writing the ratios in lowest terms. Show the simplified ratios and then write true *or* false. *See Example 2.*

7. $\dfrac{6}{10} = \dfrac{3}{5}$

8. $\dfrac{1}{4} = \dfrac{9}{36}$

9. $\dfrac{5}{8} = \dfrac{25}{40}$

10. $\dfrac{2}{3} = \dfrac{20}{27}$

🌐 **11.** $\dfrac{150}{200} = \dfrac{200}{300}$

12. $\dfrac{100}{120} = \dfrac{75}{100}$

13. $\dfrac{42}{15} = \dfrac{28}{10}$

14. $\dfrac{18}{16} = \dfrac{36}{32}$

15. $\dfrac{32}{18} = \dfrac{48}{27}$

16. $\dfrac{15}{48} = \dfrac{10}{24}$

17. $\dfrac{7}{6} = \dfrac{54}{48}$

18. $\dfrac{28}{21} = \dfrac{44}{33}$

Use cross products to determine whether each proportion is true or false. Show the cross products and then circle true *or* false. *See Example 3.*

19. $\dfrac{2}{9} = \dfrac{6}{27}$

True   False

20. $\dfrac{20}{25} = \dfrac{4}{5}$

True   False

21. $\dfrac{20}{28} = \dfrac{12}{16}$

True   False

22. $\dfrac{16}{40} = \dfrac{22}{55}$

True   False

23. $\dfrac{110}{18} = \dfrac{160}{27}$

True   False

24. $\dfrac{600}{420} = \dfrac{20}{14}$

True   False

25. $\dfrac{3.5}{4} = \dfrac{7}{8}$

True   False

26. $\dfrac{36}{23} = \dfrac{9}{5.75}$

True   False

27. $\dfrac{18}{16} = \dfrac{2.8}{2.5}$

True   False

28. $\dfrac{0.26}{0.39} = \dfrac{1.3}{1.9}$

True   False

29. $\dfrac{6}{3\frac{2}{3}} = \dfrac{18}{11}$

True   False

30. $\dfrac{16}{13} = \dfrac{2}{1\frac{5}{8}}$

True   False

31. $\dfrac{2\frac{5}{8}}{3\frac{1}{4}} = \dfrac{21}{26}$

True   False

32. $\dfrac{28}{17} = \dfrac{9\frac{1}{3}}{5\frac{2}{3}}$

True   False

33. $\dfrac{\frac{2}{3}}{2} = \dfrac{2.7}{8}$

True   False

34. $\dfrac{3.75}{1\frac{1}{4}} = \dfrac{7.5}{2\frac{1}{2}}$

True   False

35. $\dfrac{2\frac{3}{10}}{8.05} = \dfrac{\frac{1}{4}}{0.9}$

True   False

36. $\dfrac{3}{\frac{5}{6}} = \dfrac{1.5}{\frac{7}{12}}$

True   False

37. Suppose Joe Mauer of the Minnesota Twins had 17 hits in 50 times at bat and Freddy Sanchez of the Pittsburgh Pirates was at bat 450 times and got 153 hits. Paul is trying to convince Jamie that the two men hit equally well. Show how you could use a proportion and cross products to see whether Paul is correct.

38. Jay worked 3.5 hours and packed 91 cartons. Craig packed 126 cartons in 5.25 hours. To see whether the men worked equally fast, Barry set up this proportion:

$$\frac{3.5}{91} = \frac{126}{5.25}$$

Explain what is wrong with Barry's proportion and write a correct one. Is the correct proportion true or false?

5.4 ▶▶▶ Solving Proportions

OBJECTIVES

1 Find the unknown number in a proportion.

2 Find the unknown number in a proportion with mixed numbers or decimals.

OBJECTIVE 1 Find the unknown number in a proportion. Four numbers are used in a proportion. If any three of these numbers are known, the fourth can be found. For example, find the unknown number that will make this proportion true.

$$\frac{3}{5} = \frac{x}{40}$$

The *x* represents the unknown number. Start by finding the cross products.

$$\frac{3}{5} \underset{3 \cdot 40}{\overset{5 \cdot x}{\bowtie}} \frac{x}{40} \quad \biggr] \text{Cross products}$$

To make the proportion true, the cross products must be equal.

$$5 \cdot x = 3 \cdot 40$$
$$5 \cdot x = 120$$

The equal sign says that $5 \cdot x$ and 120 are equivalent. If $5 \cdot x$ and 120 are *both* divided by 5, the results will still be equivalent.

$$\frac{5 \cdot x}{5} = \frac{120}{5} \quad \leftarrow \text{Divide both sides by 5.}$$

On the left side, divide out the common factor of 5; slashes indicate the divisions. $\dfrac{\overset{1}{\cancel{5}} \cdot x}{\underset{1}{\cancel{5}}} = 24$ On the right side, divide 120 by 5 to get 24.

Multiplying by 1 does *not* change a number, so in the numerator on the left side, $1 \cdot x$ is the same as *x*.

$$\frac{x}{1} = 24$$

Dividing by 1 does *not* change a number, so on the left side, $\frac{x}{1}$ is the same as *x*.

$$x = 24$$

The unknown number in the proportion is 24. The complete proportion is shown below.

$$\frac{3}{5} = \frac{24}{40} \quad \leftarrow x \text{ is 24.}$$

Check by finding the cross products. If they are equal, you solved the problem correctly. If they are unequal, rework the problem.

$$\frac{3}{5} \underset{3 \cdot 40 = \textbf{120}}{\overset{5 \cdot 24 = \textbf{120}}{\bowtie}} \frac{24}{40} \quad \biggr] \text{Equal; proportion is true.}$$

The cross products are equal, so the solution, $x = 24$, is correct.

> **CAUTION**
> The solution is 24, which is the unknown number in the proportion. 120 is *not* the solution; it is the cross product you get when *checking* the solution.

Solve a proportion for an unknown number by using the following steps.

Finding an Unknown Number in a Proportion

Step 1 Find the cross products.

Step 2 Show that the cross products are equivalent.

Step 3 Divide both products by the number multiplied by x (the number next to x).

Step 4 Check by writing the solution in the *original* proportion and finding the cross products.

EXAMPLE 1 **Solving Proportions for Unknown Numbers**

Find the unknown number in each proportion. Round answers to the nearest hundredth when necessary.

(a) $\dfrac{16}{x} = \dfrac{32}{20}$

Recall that ratios can be rewritten in lowest terms. If desired, you can do that *before* finding the cross products. In this example, write $\frac{32}{20}$ in lowest terms as $\frac{8}{5}$, which gives the proportion $\dfrac{16}{x} = \dfrac{8}{5}$.

Step 1
$$\dfrac{16}{x} = \dfrac{8}{5}$$
$x \cdot 8$
$16 \cdot 5$ ← Find the cross products.

Step 2
$x \cdot 8 = \underbrace{16 \cdot 5}$ ← Show that cross products are equivalent.
$x \cdot 8 = 80$

Step 3
$$\dfrac{x \cdot \overset{1}{\cancel{8}}}{\cancel{8}} = \dfrac{80}{8}$$ ← Divide both sides by 8.
$x = 10$ ← Find x. (No rounding necessary)

The unknown number in the proportion is 10.

Step 4 Write the solution in the *original* proportion and check by finding cross products.

$10 \cdot 32 = \textbf{320}$

x is 10 → $\dfrac{16}{10} = \dfrac{32}{20}$ Equal: proportion is true.

$16 \cdot 20 = \textbf{320}$

> The solution is 10, **not** 320.

The cross products are equal, so 10 is the correct solution.

Note

It is not necessary to write the ratios in lowest terms before solving. However, if you do, you will work with smaller numbers.

Continued on Next Page

(b) $\dfrac{7}{12} = \dfrac{15}{x}$

Step 1

$12 \cdot 15 = 180$ ⟵

$\dfrac{7}{12} \bcancel{=} \dfrac{15}{x}$ Find the cross products.

$7 \cdot x$ ⟵

Step 2 $7 \cdot x = 180$ ⟵ Show that cross products are equal.

Step 3 $\dfrac{\overset{1}{\cancel{7}} \cdot x}{\underset{1}{\cancel{7}}} = \dfrac{180}{7}$ ⟵ Divide both sides by 7.

$x \approx 25.71$ ⟵ Rounded to nearest hundredth

When the division does not come out even, check for directions on how to round your answer. Divide out one more place, then round.

$$\begin{array}{r} 25.714 \\ 7\overline{)180.000} \end{array}$$ ⟵ Divide out to thousandths so you can round to hundredths.

The unknown number in the proportion is 25.71 (rounded).

Step 4 Write the solution in the original proportion and check by finding the cross products.

$12 \cdot 15 = \mathbf{180}$ ⟵

$\dfrac{7}{12} \bcancel{=} \dfrac{15}{25.71}$ Very close, but *not* equal due to rounding the solution.

$7 \cdot 25.71 = \mathbf{179.97}$ ⟵

The cross products are slightly different because you rounded the value of x. However, they are close enough to see that the problem was done correctly and that 25.71 is the approximate solution.

Work Problem (**1**) *at the Side.* ▶

OBJECTIVE **2** **Find the unknown number in a proportion with mixed numbers or decimals.** The next example shows how to work with mixed numbers or decimals in a proportion.

EXAMPLE 2 **Solving Proportions with Mixed Numbers and Decimals**

Find the unknown number in each proportion.

(a) $\dfrac{2\frac{1}{5}}{6} = \dfrac{x}{10}$ $\dfrac{2\frac{1}{5}}{6} \bcancel{=} \dfrac{x}{10}$

$6 \cdot x$ ⟵

$2\frac{1}{5} \cdot 10$ ⟵   Find the cross products.

Find $2\frac{1}{5} \cdot 10$.

$$2\frac{1}{5} \cdot 10 = \frac{11}{5} \cdot \frac{10}{1} = \frac{11}{\cancel{5}} \cdot \frac{\overset{2}{\cancel{10}}}{1} = \frac{22}{1} = 22$$

Changed to improper fraction

Continued on Next Page

---

1 Find the unknown numbers. Round to hundredths when necessary. Check your solutions by finding the cross products.

(a) $\dfrac{1}{2} = \dfrac{x}{12}$

(b) $\dfrac{6}{10} = \dfrac{15}{x}$

(c) $\dfrac{28}{x} = \dfrac{21}{9}$

(d) $\dfrac{x}{8} = \dfrac{3}{5}$

(e) $\dfrac{14}{11} = \dfrac{x}{3}$

ANSWERS

1. **(a)** $x = 6$ **(b)** $x = 25$
   (c) $x = 12$ **(d)** $x = 4.8$
   (e) $x \approx 3.82$ (rounded to nearest hundredth)

2 Find the unknown numbers. Round to hundredths on the decimal problems, if necessary. Check your solutions by finding the cross products.

(a) $\dfrac{3\frac{1}{4}}{2} = \dfrac{x}{8}$

(b) $\dfrac{x}{3} = \dfrac{1\frac{2}{3}}{5}$

(c) $\dfrac{0.06}{x} = \dfrac{0.3}{0.4}$

(d) $\dfrac{2.2}{5} = \dfrac{13}{x}$

(e) $\dfrac{x}{6} = \dfrac{0.5}{1.2}$

(f) $\dfrac{0}{2} = \dfrac{x}{7.092}$

Show that the cross products are equivalent.

$$6 \cdot x = 22$$

Divide both sides by 6.

$$\dfrac{\cancel{6}^{1} \cdot x}{\cancel{6}_{1}} = \dfrac{22}{6}$$

Write the solution as a mixed number in lowest terms.

$$x = \dfrac{22 \div 2}{6 \div 2} = \dfrac{11}{3} = 3\dfrac{2}{3}$$

The unknown number is $3\frac{2}{3}$.

Write the solution in the proportion and check by finding the cross products.

$$\dfrac{2\frac{1}{5}}{6} \bowtie \dfrac{3\frac{2}{3}}{10}$$

$$6 \cdot 3\dfrac{2}{3} = \dfrac{\cancel{6}^{2}}{1} \cdot \dfrac{11}{\cancel{3}_{1}} = \dfrac{22}{1} = \mathbf{22} \longleftarrow$$

$$2\dfrac{1}{5} \cdot 10 = \dfrac{11}{\cancel{5}_{1}} \cdot \dfrac{\cancel{10}^{2}}{1} = \dfrac{22}{1} = \mathbf{22} \longleftarrow$$

Equal

> The solution is $3\frac{2}{3}$, **not** 22.

The cross products are equal, so $3\frac{2}{3}$ is the correct solution.

(b) $\dfrac{1.5}{0.6} = \dfrac{2}{x}$

Show that cross products are equivalent.

$$(1.5)(x) = \underbrace{(0.6)(2)}$$
$$(1.5)(x) = \quad 1.2$$

Divide both sides by 1.5

$$\dfrac{\cancel{(1.5)}^{1}(x)}{\cancel{1.5}_{1}} = \dfrac{1.2}{1.5}$$

$$x = \dfrac{1.2}{1.5}$$

Complete the division.

$$x = 0.8 \qquad 1.5)\overline{1.20} \;\; .8$$

So the unknown number is 0.8. Write the solution in the original proportion and check it by finding the cross products.

$$(0.6)(2) = \mathbf{1.2} \longleftarrow$$

$$\dfrac{1.5}{0.6} \bowtie \dfrac{2}{0.8}$$

Equal

$$(1.5)(0.8) = \mathbf{1.2} \longleftarrow$$

> The solution is 0.8, **not** 1.2.

The cross products are equal, so 0.8 is the correct solution.

◀ *Work Problem* **2** *at the Side.*

Find the unknown number in each proportion. Round your answers to hundredths, if necessary. Check your answers by finding the cross products. See Examples 1 and 2.

1. $\dfrac{1}{3} = \dfrac{x}{12}$

2. $\dfrac{x}{6} = \dfrac{15}{18}$

3. $\dfrac{15}{10} = \dfrac{3}{x}$

4. $\dfrac{5}{x} = \dfrac{20}{8}$

5. $\dfrac{x}{11} = \dfrac{32}{4}$

6. $\dfrac{12}{9} = \dfrac{8}{x}$

7. $\dfrac{42}{x} = \dfrac{18}{39}$

8. $\dfrac{49}{x} = \dfrac{14}{18}$

9. $\dfrac{x}{25} = \dfrac{4}{20}$

10. $\dfrac{6}{x} = \dfrac{4}{8}$

11. $\dfrac{8}{x} = \dfrac{24}{30}$

12. $\dfrac{32}{5} = \dfrac{x}{10}$

13. $\dfrac{99}{55} = \dfrac{44}{x}$

14. $\dfrac{x}{12} = \dfrac{101}{147}$

15. $\dfrac{0.7}{9.8} = \dfrac{3.6}{x}$

16. $\dfrac{x}{3.6} = \dfrac{4.5}{6}$

17. $\dfrac{250}{24.8} = \dfrac{x}{1.75}$

18. $\dfrac{4.75}{17} = \dfrac{43}{x}$

Find the unknown number in each proportion. Write your answers as whole or mixed numbers when possible. See Example 2.

19. $\dfrac{15}{1\frac{2}{3}} = \dfrac{9}{x}$

20. $\dfrac{x}{\frac{3}{10}} = \dfrac{2\frac{2}{9}}{1}$

21. $\dfrac{2\frac{1}{3}}{1\frac{1}{2}} = \dfrac{x}{2\frac{1}{4}}$

22. $\dfrac{1\frac{5}{6}}{x} = \dfrac{\frac{3}{14}}{\frac{6}{7}}$

Solve each proportion two different ways. First change all the numbers to decimal form and solve. Then change all the numbers to fraction form and solve; write your answers in lowest terms.

23. $\dfrac{\frac{1}{2}}{x} = \dfrac{2}{0.8}$

24. $\dfrac{\frac{3}{20}}{0.1} = \dfrac{0.03}{x}$

25. $\dfrac{x}{\frac{3}{50}} = \dfrac{0.15}{1\frac{4}{5}}$

26. $\dfrac{8\frac{4}{5}}{1\frac{1}{10}} = \dfrac{x}{0.4}$

Relating Concepts (Exercises 27–28) For Individual or Group Work

*Work Exercises 27–28 in order. First prove that the proportions are **not** true. Then create four true proportions for each exercise by changing one number at a time.*

27. $\dfrac{10}{4} = \dfrac{5}{3}$

28. $\dfrac{6}{8} = \dfrac{24}{30}$

5.5 ▶▶▶ Solving Application Problems with Proportions

OBJECTIVE 1 Use proportions to solve application problems.
Proportions can be used to solve a wide variety of problems. Watch for problems in which you are given a ratio or rate and then are asked to find part of a corresponding ratio or rate. Remember that a ratio or rate compares two quantities and often includes one of the following indicator words.

in	**for**	**on**	**per**	**from**	**to**

Use the six problem-solving steps you learned in **Section 1.10.**

Step 1 **Read** the problem.	*Step 4* **Solve** the problem.
Step 2 **Work out a plan.**	*Step 5* **State the answer.**
Step 3 **Estimate** a reasonable answer.	*Step 6* **Check** your work.

OBJECTIVE

1 Use proportions to solve application problems.

EXAMPLE 1 Solving a Proportion Application

Mike's car can travel 163 **miles** **on** 6.4 **gallons** of gas. How far can it travel on a full tank of 14 **gallons** of gas? Round to the nearest whole mile.

Step 1 **Read** the problem. The problem asks for the number of miles the car can travel on 14 gallons of gas.

Step 2 **Work out a plan.** Decide what is being compared. This example compares **miles** to **gallons**. Write a proportion using the two rates. Be sure that *both* rates compare miles to gallons in the same order. In other words, miles is in both numerators and gallons is in both denominators. Use a letter to represent the unknown number.

Matching units

This rate compares **miles** to **gallons**. $\quad \dfrac{163 \text{ \textbf{miles}}}{6.4 \text{ \textbf{gallons}}} = \dfrac{x \text{ \textbf{miles}}}{14 \text{ \textbf{gallons}}} \quad$ This rate compares **miles** to **gallons**.

Matching units

Step 3 **Estimate** a reasonable answer. To estimate the answer, notice that 14 gallons is a little more than *twice as much* as 6.4 gallons, so the car should travel a little more than *twice as far*. So use 2 • 163 miles = 326 miles as the estimate.

Step 4 **Solve** the problem. Ignore the units while solving for x.

$$\frac{163 \text{ miles}}{6.4 \text{ gallons}} = \frac{x \text{ miles}}{14 \text{ gallons}}$$

$$(6.4)(x) = \underbrace{(163)(14)} \qquad \text{Show that cross products are equivalent.}$$

$$(6.4)(x) = 2282$$

$$\frac{(6.4)(x)}{6.4} = \frac{2282}{6.4} \qquad \text{Divide both sides by 6.4.}$$

$$x = 356.5625 \qquad \text{Round to 357.} \quad \text{— Check the problem for rounding directions; this one asks for nearest whole number.}$$

Continued on Next Page

1 Set up and solve a proportion for each problem.

(a) If 2 pounds of fertilizer will cover 50 square feet of garden, how many pounds are needed for 225 square feet?

(b) A U.S. map has a scale of 1 inch to 75 miles. Lake Superior is 4.75 inches long on the map. What is the lake's actual length to the nearest whole mile?

(c) Cough syrup is to be given at the rate of 30 milliliters for each 100 pounds of body weight. How much should be given to a 34-pound child? Round to the nearest whole milliliter.

ANSWERS

1. **(a)** $\dfrac{2 \text{ pounds}}{50 \text{ square feet}} = \dfrac{x \text{ pounds}}{225 \text{ square feet}}$

   $x = 9$ pounds

   (b) $\dfrac{1 \text{ inch}}{75 \text{ miles}} = \dfrac{4.75 \text{ inches}}{x \text{ miles}}$

   $x = 356.25$ miles, rounds to 356 miles

   (c) $\dfrac{30 \text{ milliliters}}{100 \text{ pounds}} = \dfrac{x \text{ milliliters}}{34 \text{ pounds}}$

   $x \approx 10$ milliliters (rounded)

Step 5 **State the answer.** Rounded to the nearest mile, the car can travel about 357 miles on a full tank of gas.

Step 6 **Check** your work. The answer, 357 miles, is a little more than the estimate of 326 miles, so it is reasonable.

> **CAUTION**
> When setting up a proportion do *not* mix up the units in the rates.
>
> $$\underbrace{\frac{163 \text{ miles}}{6.4 \text{ gallons}}}_{\substack{\text{compares \textbf{miles}}\\\text{to \textbf{gallons}}}} \neq \underbrace{\frac{14 \text{ gallons}}{x \text{ miles}}}_{\substack{\text{compares \textbf{gallons}}\\\text{to \textbf{miles}}}}$$
>
> These rates do *not* compare things in the same order and *cannot* be set up as a proportion.

◀ Work Problem **1** at the Side.

EXAMPLE 2 **Solving a Proportion Application**

A newspaper report says that 7 out of 10 people surveyed watch the news on TV. At that rate, how many of the 3200 people in town would you expect to watch the news?

Step 1 **Read** the problem. The problem asks how many of the 3200 people in town would be expected to watch TV news.

Step 2 **Work out a plan.** You are comparing people who watch the news to people surveyed. Set up a proportion using the two rates described in the example. Be sure that both rates make the same comparison. "People who watch the news" is mentioned first, so it should be in the numerator of *both* rates.

People who watch news → $\dfrac{7}{10} = \dfrac{x}{3200}$ ← People who watch news

Total group → (people surveyed) ⠀⠀⠀⠀ ← Total group (people in town)

Step 3 **Estimate** a reasonable answer. To estimate the answer, notice that 7 out of 10 people is more than half the people, but less than all the people. Half of 3200 people is $3200 \div 2 = 1600$, so our estimate is between 1600 and 3200 people.

Step 4 **Solve** the problem. Solve for the unknown number in the proportion.

$$\frac{7}{10} = \frac{x}{3200}$$

$10 \cdot x = 7 \cdot 3200$ Show that cross products are equivalent.

$10 \cdot x = 22{,}400$

$\dfrac{\cancel{10} \cdot x}{\cancel{10}} = \dfrac{22{,}400}{10}$ Divide both sides by 10.

$x = 2240$ No rounding is needed here.

Continued on Next Page

Step 5 **State the answer.** You would expect 2240 people in town to watch the news on TV.

Step 6 **Check** your work. The answer, 2240 people, is between 1600 and 3200, as called for in the estimate.

> **CAUTION**
>
> Always check that your answer is reasonable. If it is not, look at the way your proportion is set up. Be sure you have matching units in the numerators and matching units in the denominators.
>
> For example, suppose you had set up the last proportion ***incorrectly*** as shown here.
>
> $$\frac{7}{10} = \frac{3200}{x} \quad \leftarrow \textbf{\textcolor{red}{Incorrect setup}}$$
>
> $$7 \cdot x = 10 \cdot 3200$$
>
> $$\frac{\overset{1}{\cancel{7}} \cdot x}{\underset{1}{\cancel{7}}} = \frac{32{,}000}{7}$$
>
> $$x \approx 4571 \text{ people} \quad \leftarrow \textbf{\textcolor{red}{Unreasonable answer}}$$
>
> This answer is ***unreasonable*** because there are only 3200 people in the town; it is ***not*** possible for 4571 people to watch the news.

Work Problem **2** *at the Side.* ▶

2 Solve each problem to find a reasonable answer. Then flip one side of your proportion to see what answer you get with an *incorrect* setup. Explain why the second answer is *unreasonable.*

(a) A survey showed that 2 out of 3 people would like to lose weight. At this rate, how many people in a group of 150 want to lose weight?

(b) In one state, 3 out of 5 college students receive financial aid. At this rate, how many of the 4500 students at Central Community College receive financial aid?

(c) An advertisement says that 9 out of 10 dentists recommend sugarless gum. If the ad is true, how many of the 60 dentists in our city would recommend sugarless gum?

ANSWERS

2. **(a)** 100 people (reasonable); incorrect setup gives 225 people (only 150 people in the group).
   (b) 2700 students (reasonable); incorrect setup gives 7500 students (only 4500 students at the college).
   (c) 54 dentists (reasonable); incorrect setup gives about 67 dentists (only 60 dentists in the city).

Math in the Media

FEEDING HUMMINGBIRDS

After getting a hummingbird feeder, the next step is to fill it! You have two choices at this point: you can either buy one of the commercial mixtures or you can make your own solution. See the recipe at the right.

The concentration of the sugar is important. The 1 to 4 ratio of sugar to water is recommended because it approximates the ratio of sugar to water found in the nectar of many hummingbird flowers.

Boiling the solution helps retard fermentation. Sugar-and-water solutions are subject to rapid spoiling, especially in hot weather.

Source: The Hummingbird Book.

> **Recipe for Homemade Mixture:**
> 1 part sugar (not honey)
> 4 parts water
> Boil for 1 to 2 minutes. Cool.
> Store extra in refrigerator.

A recipe can be used to make as much of a mixture as you need as long as the ingredients are kept proportional. Use the recipe for a homemade mixture of sugar water for hummingbird feeders to answer these problems.

1. What is the ratio of sugar to water in the recipe?

   What is the ratio of water to sugar in the recipe?

2. Complete each table.

Sugar	Water
1 cup	4 cups
	5 cups
	6 cups
	7 cups
2 cups	8 cups

Sugar	Water
1 cup	4 cups
	3 cups
	2 cups
	1 cup

3. How much water would you need if you used
   (a) 3 cups of sugar?
   (b) 4 cups of sugar?
   (c) $\frac{1}{3}$ cup of sugar?

4. As you change the amounts of water and sugar, should you change the length of time that you boil the mixture? Explain your answer.

5.5 ▶▶▶ **Exercises**

FOR EXTRA HELP

MyMathLab

Math XL
PRACTICE

WATCH

DOWNLOAD

READ

REVIEW

Set up and solve a proportion for each application problem. See Example 1.

1. Caroline can sketch four cartoon strips in five hours. How long will it take her to sketch 18 strips?

2. The Cosmic Toads recorded eight songs on their first CD in 26 hours. How long will it take them to record 14 songs for their second CD?

3. Sixty newspapers cost $27. Find the cost of 16 newspapers.

4. Twenty-two guitar lessons cost $528. Find the cost of 12 lessons.

5. If three pounds of fescue grass seed cover about 350 square feet of ground, how many pounds are needed for 4900 square feet?

6. Anna earns $1242.08 in 14 days. How much does she earn in 260 days?

7. Tom makes $672.80 in 5 days. How much does he make in 3 days?

8. If 5 ounces of a medicine must be mixed with 8 ounces of water, how many ounces of medicine would be mixed with 20 ounces of water?

9. The bag of rice noodles below makes 7 servings. At that rate, how many ounces of noodles do you need for 12 servings, to the nearest ounce?

10. This can of sweet potatoes is enough for 4 servings. How many ounces are needed for 9 servings, to the nearest ounce?

11. Three quarts of a latex enamel paint will cover about 270 square feet of wall surface. How many quarts will you need to cover 350 square feet of wall surface in your kitchen and 100 square feet of wall surface in your bathroom?

12. One gallon of clear gloss wood finish covers about 550 square feet of surface. If you need to apply three coats of finish to 400 square feet of surface, how many gallons do you need, to the nearest tenth?

Use the floor plan shown to complete Exercises 13–16. On the plan, one inch represents four feet.

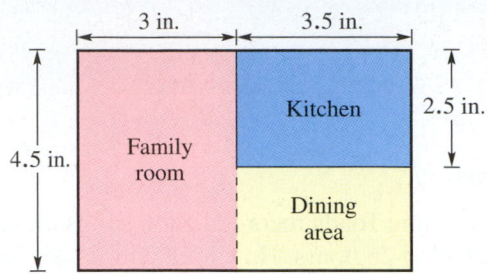

13. What is the actual length and width of the kitchen?

14. What is the actual length and width of the family room?

15. What is the actual length and width of the dining area?

16. What is the actual length and width of the entire floor plan?

The table below lists recommended amounts of food to order for 25 party guests. Use the table to answer Exercises 17 and 18. (Source: CubFoods.)

FOOD FOR 25 GUESTS

Item	Amount
Fried chicken	40 pieces
Lasagna	14 pounds
Deli meats	4.5 pounds
Sliced cheese	$2\frac{1}{3}$ pounds
Bakery buns	3 dozen
Potato salad	6 pounds

17. How much of each food item should Nathan and Amanda order for a graduation party with 60 guests?

18. Taisha is having 20 neighbors over for a Fourth of July picnic. How much food should she buy?

Set up a proportion to solve each problem. Check to see whether your answer is reasonable. Then flip one side of your proportion to see what answer you get with an incorrect setup. Explain why the second answer is unreasonable. See Example 2.

19. About 7 out of 10 people entering a community college need to take a refresher math course. If there are 2950 entering students, how many will probably need refresher math? (*Source:* Minneapolis Community and Technical College.)

20. In a survey, only 3 out of 100 people like their eggs poached. At that rate, how many of the 60 customers who ordered eggs at Soon-Won's restaurant this morning asked to have them poached? Round to the nearest whole person.

21. About 1 out of 3 people choose vanilla as their favorite ice cream flavor. If 250 people attend an ice cream social, how many would you expect to choose vanilla? Round to the nearest whole person.

22. In a test of 200 sewing machines, only one had a defect. At that rate, how many of the 5600 machines shipped from the factory have defects?

23. About 98 out of 100 U.S. households have at least one TV set. There were 113,100,000 U.S. households in 2005. How many households have one or more TVs? (*Source:* Nielsen Media Research.)

24. In a survey, 3 out of 100 dog owners wash their pets by having the dogs go into the shower with them. If the survey is accurate, how many of the 31,200,000 dog owners in the United States use this method? (*Source:* Teledyne Water Pik, American Veterinary Medical Association.)

Set up and solve a proportion for each problem.

25. The stock market report says that 5 stocks went up for every 6 stocks that went down. If 750 stocks went down yesterday, how many went up?

26. The human body contains 90 pounds of water for every 100 pounds of body weight. How many pounds of water are in a child who weighs 80 pounds?

27. The ratio of the length of an airplane wing to its width is 8 to 1. If the length of a wing is 32.5 meters, how wide must it be? Round to the nearest hundredth.

28. The Rosebud School District wants a student-to-teacher ratio of 19 to 1. How many teachers are needed for 1850 students? Round to the nearest whole number.

29. The number of calories you burn is proportional to your weight. A 150-pound person burns 222 calories during 30 minutes of tennis. How many calories would a 210-pound person burn, to the nearest whole number? (*Source: Wellness Encyclopedia.*)

30. (Complete Exercise 29 first.) A 150-pound person burns 189 calories during 45 minutes of grocery shopping. How many calories would a 115-pound person burn, to the nearest whole number? (*Source: Wellness Encyclopedia.*)

31. At 3 P.M., Coretta's shadow is 1.05 meters long. Her height is 1.68 meters. At the same time, a tree's shadow is 6.58 meters long. How tall is the tree? Round to the nearest hundredth.

32. Refer to Exercise 31. Later in the day, Coretta's shadow was 2.95 meters long. How long a shadow did the tree have at that time? Round to the nearest hundredth.

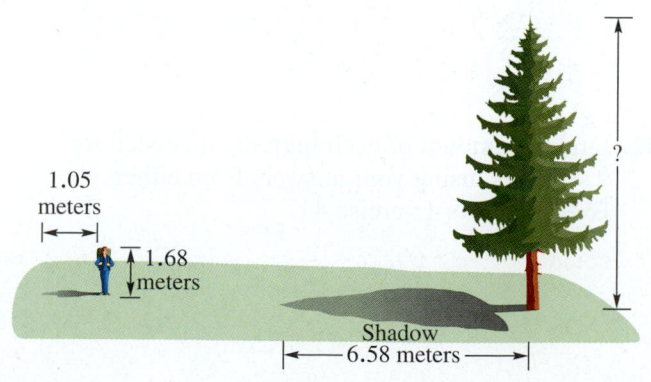

33. Can you set up a proportion to solve this problem? Explain why or why not. Jim is 25 years old and weighs 180 pounds. How much will he weigh when he is 50 years old?

34. Write your own application problem that can be solved by setting up a proportion. Also show the proportion and the steps needed to solve your problem.

35. A survey of college students shows that 4 out of 5 drink coffee. Of the students who drink coffee, 1 out of 8 adds cream to it. How many of the 50,500 students at Ohio State University would be expected to use cream in their coffee?

36. About 9 out of 10 adults think it is a good idea to exercise regularly. But of the ones who think it is a good idea, only 1 in 6 actually exercises at least three times a week. At this rate, how many of the 300 employees in our company exercise regularly?

37. The nutrition information on a bran cereal box says that a $\frac{1}{3}$-cup serving provides 80 calories and 8 grams of dietary fiber. At that rate, how many calories and grams of fiber are in a $\frac{1}{2}$-cup serving? (*Source:* Kraft Foods, Inc.)

38. A $\frac{2}{3}$-cup serving of penne pasta has 210 calories and 2 grams of dietary fiber. How many calories and grams of fiber would be in a 1-cup serving? (*Source:* Borden Foods.)

Relating Concepts (Exercises 39—42) For Individual or Group Work

A box of instant mashed potatoes has the list of ingredients shown in the table. Use this information to **work Exercises 39–42 in order.**

Ingredient	For 12 Servings
Water	$3\frac{1}{2}$ cups
Margarine	6 Tbsp
Milk	$1\frac{1}{2}$ cups
Potato flakes	4 cups

Source: General Mills.

39. Find the amount of each ingredient needed for 6 servings. Show *two* different methods for finding the amounts. One method should use proportions.

40. Find the amount of each ingredient needed for 18 servings. Show *two* different methods for finding the amounts, one using proportions and one using your answers from Exercise 39.

41. Find the amount of each ingredient needed for 3 servings, using your answers from either Exercise 39 or Exercise 40.

42. Find the amount of each ingredient needed for 9 servings, using your answers from either Exercise 40 or Exercise 41.

Chapter 5 ▷▷▷ Summary

▶ Key Terms

5.1	**ratio**	A ratio compares two quantities having the same type of units. For example, the ratio of 6 apples to 11 apples is written in fraction form as $\frac{6}{11}$. The common units (apples) divide out.
5.2	**rate**	A rate compares two measurements with different types of units. Examples are 96 dollars for 8 hours, or 450 miles on 18 gallons.
	unit rate	A unit rate has 1 in the denominator.
	cost per unit	Cost per unit is a rate that tells how much you pay for one item or one unit. The lowest cost per unit is the best buy.
5.3	**proportion**	A proportion states that two ratios or rates are equivalent.
	cross products	Multiply along one diagonal and then multiply along the other diagonal to find the cross products of a proportion. If the cross products are equal, the proportion is true.

▶ Test Your Word Power

See how well you have learned the vocabulary in this chapter. Answers follow the Quick Review.

1. **A ratio**
   A. can be written only as a fraction
   B. compares two quantities that have the same type of units
   C. compares two quantities that have different types of units
   D. is the reciprocal of a rate.

2. **A rate**
   A. can be written only as a decimal
   B. compares two quantities that have the same type of units
   C. compares two quantities that have different types of units
   D. is the reciprocal of a ratio.

3. **A unit rate**
   A. has a numerator of 1
   B. has a denominator of 1
   C. is found by cross multiplying
   D. is usually written in fraction form.

4. **Cost per unit** is
   A. the best buy
   B. a ratio written in lowest terms
   C. found by comparing cross products
   D. the price of one item or one unit.

5. **A proportion**
   A. shows that two ratios or rates are equivalent
   B. contains only whole numbers or decimals
   C. always has one unknown number
   D. states that two improper fractions are equivalent.

6. **Cross products** are
   A. used only with ratios, not with rates
   B. equal when a proportion is false.
   C. used to find the best buy
   D. equal when a proportion is true.

▶ Quick Review

Concepts	Examples

5.1 **Writing a Ratio**

A ratio compares two quantities that have the same type of units. A ratio is usually written as a fraction with the number that is mentioned first in the numerator. The common units divide out and are not written in the answer. Check that the fraction is in lowest terms.

Write this ratio as a fraction in lowest terms.

60 ounces of medicine **to** 160 ounces of medicine

$$\frac{60 \text{ ounces}}{160 \text{ ounces}} = \frac{60 \div 20}{160 \div 20} = \frac{3}{8} \quad \left\{ \begin{array}{l} \text{Ratio in lowest} \\ \text{terms} \end{array} \right.$$

Divide out common units.

5.1 Using Mixed Numbers in a Ratio

If a ratio has mixed numbers, change the mixed numbers to improper fractions. Rewrite the problem in horizontal format using the "÷" symbol for division. Finally, multiply by the reciprocal of the divisor.

Write as a ratio of whole numbers in lowest terms.

$$2\frac{1}{2} \text{ to } 3\frac{3}{4}$$

$$\frac{2\frac{1}{2}}{3\frac{3}{4}} = \frac{\frac{5}{2}}{\frac{15}{4}}$$

Reciprocal

$$= \frac{5}{2} \div \frac{15}{4} = \frac{5}{2} \cdot \frac{4}{15}$$

$$= \frac{\overset{1}{\cancel{5}}}{\underset{1}{\cancel{2}}} \cdot \frac{\overset{2}{\cancel{4}}}{\underset{3}{\cancel{15}}} = \frac{2}{3} \quad \leftarrow \text{Ratio in lowest terms}$$

5.1 Using Measurements in Ratios

When a ratio compares measurements, both measurements must be in the *same* units. It is usually easier to compare the measurements using the smaller unit, for example, inches instead of feet.

Write as a ratio in lowest terms.

8 in. to 6 ft

Compare using the smaller unit, inches. Because 1 ft has 12 in., 6 ft is

$$6 \cdot \textbf{12 in.} = 72 \text{ in.}$$

The ratio is shown below.

$$\frac{8 \text{ in.}}{72 \text{ in.}} = \frac{8 \div 8}{72 \div 8} = \frac{1}{9}$$

Divide out common units.

5.2 Writing Rates

A rate compares two measurements with different types of units. The units do *not* divide out, so you must write them as part of the rate.

Write the rate as a fraction in lowest terms.

475 miles in 10 hours

$$\frac{475 \text{ miles} \div 5}{10 \text{ hours} \div 5} = \frac{95 \text{ miles}}{2 \text{ hours}} \quad \begin{array}{l}\leftarrow \text{Must write units:} \\ \leftarrow \text{miles and hours}\end{array}$$

5.2 Finding a Unit Rate

A unit rate has 1 in the denominator. To find the unit rate, divide the numerator by the denominator. Write unit rates using the word **per** or a / mark.

Write as a unit rate: $1278 in 9 days.

$$\frac{\$1278}{9 \text{ days}} \quad \leftarrow \text{The fraction bar indicates division.}$$

$$9)\overline{1278}^{\,142} \quad \text{so} \quad \frac{\$1278 \div 9}{9 \text{ days} \div 9} = \frac{\$142}{1 \text{ day}}$$

Write the answer as $142 **per** day or $142/day.

Concepts	Examples

5.2 Finding the Best Buy

The best buy is the item with the lowest cost per unit. Divide the price by the number of units. Round to thousandths, if necessary. Then compare to find the lowest cost per unit.

Find the best buy on cheese. You have a coupon for 50¢ off on 2 pounds or 75¢ off on 3 pounds.

$$2 \text{ pounds for } \$2.75$$

$$3 \text{ pounds for } \$4.15$$

Find the cost per unit (cost per pound) after subtracting the coupon.

$$2 \text{ pounds cost } \$2.75 - \$0.50 = \$2.25$$

$$\frac{\$2.25}{2} = \$1.125 \text{ per pound}$$

$$3 \text{ pounds cost } \$4.15 - \$0.75 = \$3.40$$

$$\frac{\$3.40}{3} \approx \$1.133 \text{ per pound}$$

The lower cost per pound is $1.125, so 2 pounds of cheese is the best buy.

5.3 Writing Proportions

A proportion states that two ratios or rates are equivalent. The proportion "5 is to 6 as 25 is to 30" is written as shown below.

$$\frac{5}{6} = \frac{25}{30}$$

To see whether a proportion is true or false, multiply along one diagonal, then multiply along the other diagonal. If the two cross products are equal, the proportion is true. If the two cross products are unequal, the proportion is false.

Write as a proportion: 8 is to 40 as 32 is to 160

$$\frac{8}{40} = \frac{32}{160}$$

Is this proportion true or false?

$$\frac{6}{8\frac{1}{2}} = \frac{24}{34}$$

Find the cross products.

$$8\frac{1}{2} \cdot 24 = \frac{17}{2} \cdot \frac{\overset{12}{\cancel{24}}}{1} = \mathbf{204}$$

$$\frac{6}{8\frac{1}{2}} = \frac{24}{34}$$

$$6 \cdot 34 = \mathbf{204} \longleftarrow \text{Equal}$$

The cross products are equal, so the proportion is true.

5.4 Solving Proportions

Solve for an unknown number in a proportion by using the steps shown on the next page.

Find the unknown number.

$$\frac{12}{x} = \frac{6}{8}$$

Write $\frac{6}{8}$ in lowest terms as $\frac{3}{4}$

$$\frac{12}{x} = \frac{3}{4}$$

(continued)

Concepts	Examples

5.4 Solving Proportions (*continued*)

Step 1 Find the cross products. (If desired, you can rewrite the ratios in lowest terms before finding the cross products.)

Step 1
$$\frac{12}{x} = \frac{3}{4}$$
$x \cdot 3$
$12 \cdot 4$
Find cross products

Step 2 Show that the cross products are equivalent.

Step 2
$$x \cdot 3 = \underbrace{12 \cdot 4}$$ Show that cross products are equivalent.
$$x \cdot 3 = 48$$

Step 3 Divide both products by the number multiplied by x (the number next to x).

Step 3
$$\frac{x \cdot \overset{1}{\cancel{3}}}{\underset{1}{\cancel{3}}} = \frac{48}{3}$$ Divide both sides by 3.
$$x = 16$$

Step 4 Check by writing the solution in the original proportion and finding the cross products.

Step 4
x is 16. →
$$\frac{12}{16} = \frac{6}{8}$$
$16 \cdot 6 = \textbf{\color{red}96}$
$12 \cdot 8 = \textbf{\color{red}96}$
Equal

The cross products are equal, so 16 is the correct solution (**not** 96).

5.5 Solving Application Problems with Proportions

Decide what is being compared. Set up and solve a proportion using the two rates described in the problem. Be sure that *both* rates compare things in the *same order*. Use a letter, like x, to represent the unknown number.

If 3 pounds of grass seed cover 450 square feet of lawn, how much seed is needed for 1500 square feet of lawn?

Use the six problem-solving steps.

Step 1 **Read** the problem carefully.

Step 1 The problem asks for the pounds of grass seed needed for 1500 square feet of lawn.

Step 2 **Work out a plan.**

Step 2 Pounds of seed is compared to square feet of lawn. Set up and solve a proportion using the two given rates. Be sure that pounds of seed is in both numerators and square feet of lawn is in both denominators.

Step 3 **Estimate** a reasonable answer.

Step 3 Because 1500 square feet is about three times as much lawn as 450 square feet, about three times as much seed is needed. So, 3 • 3 pounds = 9 pounds as our estimate.

Step 4 **Solve** the problem.

Step 4 With the proportion set up correctly, solve for the unknown number.

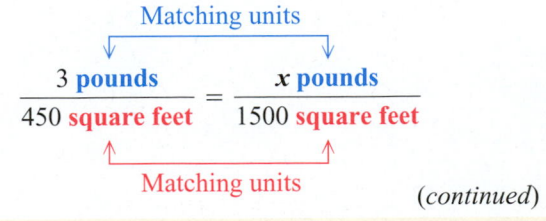

(*continued*)

Concepts

5.5 Solving Application Problems with Proportions (*continued*)

Step 5   **State the answer.**

Step 6   **Check** your work.

Examples

Both sides compare pounds to square feet. Ignore the units while finding the cross products and solving for x.

$$450 \cdot x = \underbrace{3 \cdot 1500}$$ Show that cross products are equivalent.

$$450 \cdot x = 4500$$

$$\frac{\overset{1}{\cancel{450}} \cdot x}{\underset{1}{\cancel{450}}} = \frac{4500}{450}$$ Divide both sides by 450.

$$x = 10$$

Step 5   10 pounds of grass seed are needed.

Step 6   The exact answer, 10 pounds of seed, is close to our estimate of 9 pounds, so it is reasonable.

ANSWERS TO TEST YOUR WORD POWER

1. B; *Example:*   The ratio of 3 miles to 4 miles is $\frac{3}{4}$; the common units (miles) divide out.

2. C; *Example:* $4.50 for 3 pounds is a rate comparing dollars to pounds.

3. B; *Example:* $\frac{\$1.79}{1 \text{ pound}}$ is a unit rate. We write it as $1.79 per pound or $1.79/pound.

4. D; *Example:* $3.95 per gallon tells the price of one gallon (one unit).

5. A; *Example:*   The proportion $\frac{5}{6} = \frac{25}{30}$ says that $\frac{5}{6}$ is equivalent to $\frac{25}{30}$.

6. D; *Example:*   The cross products for $\frac{5}{6} = \frac{25}{30}$ are $6 \cdot 25 = 150$ and $5 \cdot 30 = 150$.

Math in the Media

CURRENCY EXCHANGE

When you travel between countries, you will exchange U.S. dollars for the local currency. The exchange rate between currencies changes daily, and you can easily find the updated rates using the Internet or any major newspaper. The table shown below has been extracted from the Oanda Web page, www.oanda.com. It shows how much of each country's currency was equivalent to 1 U.S. dollar on April 16, 2008.

NORTH AMERICA/CARIBBEAN CURRENCY RATES (APRIL 16, 2008)

Currency	Symbol	Value
Canadian dollar	CAD	1.0198
Cayman Islands dollar	KYD	0.833
Jamaican dollar	JMD	74.75
Mexican peso	MXN	10.5
United States dollar	USD	1.00

From the table, $1.00 U.S. was equivalent to 10.5 Mexican pesos. You can set up a proportion to convert dollars to pesos. For example, suppose you want to determine the number of pesos that is equivalent to $50.00.

$$\frac{\$1}{10.5 \text{ pesos}} = \frac{\$50}{x \text{ pesos}} \quad \text{or} \quad \frac{1}{10.5} = \frac{50}{x}$$

$$(1)(x) = (10.5)(50)$$

$$x = 525.0 \text{ pesos}$$

So $50 buys 525 pesos.

1. Based on the currency exchange rates for April 16, 2008, find the amount of each local currency that is equivalent to $50 U.S. and find the number of U.S. dollars that is equivalent to 200 units of each local currency. Round your answers to the nearest hundredth.

   (a) $50 = _____ Canadian dollars, and 200 Canadian dollars = _____ U.S. dollars.

   (b) $50 = _____ Cayman Islands dollars, and 200 Cayman Islands dollars = _____ U.S. dollars.

   (c) $50 = _____ Jamaican dollars, and 200 Jamaican dollars = _____ U.S. dollars.

2. Set up a proportion to find the number of U.S. dollars that was equivalent to 1 Mexican Peso. Round your answer to the nearest cent. 1 Mexican peso was equivalent to $_____ (U.S.).

3. From Problem 2, you should recognize the conversion rate based on 1 Mexican peso as the expression $\frac{1}{10.5}$. What is the mathematical word that describes the relationship between the conversion rates 10.5 and $\frac{1}{10.5}$?

Chapter 5 ▶▶▶ Review Exercises

[5.1] *Write each ratio as a fraction in lowest terms. Change to the same units when necessary, using the table of measurement comparisons in* **Section 5.1.** *Use the information in the graph to answer Exercises 1–3.*

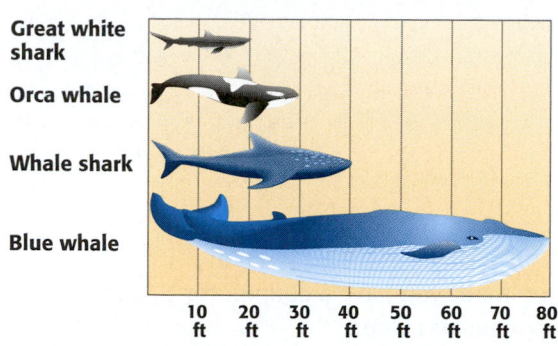

AVERAGE LENGTH OF SHARKS AND WHALES

Great white shark

Orca whale

Whale shark

Blue whale

10 ft 20 ft 30 ft 40 ft 50 ft 60 ft 70 ft 80 ft

Source: Grolier Multimedia Encyclopedia.

1. Ratio of orca whale's length to whale shark's length

2. Ratio of blue whale's length to great white shark's length

3. Which two animals' lengths give a ratio of $\frac{1}{2}$? There are several answers.

4. $2.50 to $1.25

5. $0.30 to $0.45

6. $1\frac{2}{3}$ cups to $\frac{2}{3}$ cup

7. $2\frac{3}{4}$ miles to $16\frac{1}{2}$ miles

8. 5 hours to 100 minutes

9. 9 in. to 2 ft

10. 1 ton to 1500 pounds

11. 8 hours to 3 days

12. Jake sold $350 worth of his kachina figures. Ramona sold $500 worth of her pottery. What is the ratio of Ramona's sales to Jake's sales?

13. Ms. Wei's new car gets 35 miles per gallon. Her old car got 25 miles per gallon. Find the ratio of the new car's mileage to the old car's mileage.

14. This fall, 6000 students are taking math courses and 7200 students are taking English courses. Find the ratio of math students to English students.

[5.2] *Write each rate as a fraction in lowest terms.*

15. $88 for 8 dozen

16. 96 children in 40 families

17. When entering data into his computer, Patrick can type four pages in 20 minutes. Give his rate in pages per minute and minutes per page.

18. Elena made $24 in three hours. Give her earnings in dollars per hour and hours per dollar.

Find the best buy.

19. Minced onion
   8 ounces for $4.98
   3 ounces for $2.49
   2 ounces for $1.89

20. Dog food; you have a coupon for $1 off on 8 pounds or more.
   35.2 pounds for $36.96
   17.6 pounds for $18.69
   3.5 pounds for $4.25

[5.3] *Use either the method of writing in lowest terms or the method of finding cross products to decide whether each proportion is true or false. Show your work and then write* true *or* false.

21. $\dfrac{6}{10} = \dfrac{9}{15}$

22. $\dfrac{6}{48} = \dfrac{9}{36}$

23. $\dfrac{47}{10} = \dfrac{98}{20}$

24. $\dfrac{64}{36} = \dfrac{96}{54}$

25. $\dfrac{1.5}{2.4} = \dfrac{2}{3.2}$

26. $\dfrac{3\frac{1}{2}}{2\frac{1}{3}} = \dfrac{6}{4}$

[5.4] *Find the unknown number in each proportion. Round answers to the nearest hundredth, if necessary.*

27. $\dfrac{4}{42} = \dfrac{150}{x}$

28. $\dfrac{16}{x} = \dfrac{12}{15}$

29. $\dfrac{100}{14} = \dfrac{x}{56}$

30. $\dfrac{5}{8} = \dfrac{x}{20}$

31. $\dfrac{x}{24} = \dfrac{11}{18}$

32. $\dfrac{7}{x} = \dfrac{18}{21}$

33. $\dfrac{x}{3.6} = \dfrac{9.8}{0.7}$

34. $\dfrac{13.5}{1.7} = \dfrac{4.5}{x}$

35. $\dfrac{0.82}{1.89} = \dfrac{x}{5.7}$

[5.5] *Set up and solve a proportion for each application problem.*

36. The ratio of cats to dogs at the animal shelter is 3 to 5. If there are 45 dogs, how many cats are there?

37. Danielle had 8 hits in 28 times at bat during last week's games. If she continues to hit at the same rate, how many hits will she gets in 161 times at bat?

38. If 3.5 pounds of ground beef cost $9.77, what will 5.6 pounds cost? Round to the nearest cent.

39. About 4 out of 10 students are expected to vote in campus elections. There are 8247 students. How many are expected to vote? Round to the nearest whole number.

40. The scale on Brian's model railroad is 1 in. to 16 ft. One of the scale model boxcars is 4.25 in. long. What is the length of a real boxcar in feet?

41. Marvette makes necklaces to sell at a local gift shop. She made 2 dozen necklaces in $16\frac{1}{2}$ hours. How long will it take her to make 40 necklaces?

42. A 180-pound person burns 284 calories playing basketball for 25 minutes. At this rate, how many calories would the person burn in 45 minutes, to the nearest whole number? (*Source: Wellness Encyclopedia.*)

43. In the hospital pharmacy, Michiko sees that a medicine is to be given at the rate of 3.5 milligrams for every 50 pounds of body weight. How much medicine should be given to a patient who weighs 210 pounds?

▶▶▶ Mixed Review Exercises

Find the unknown number in each proportion. Round answers to the nearest hundredth, if necessary.

44. $\dfrac{x}{45} = \dfrac{70}{30}$

45. $\dfrac{x}{52} = \dfrac{0}{20}$

46. $\dfrac{64}{10} = \dfrac{x}{20}$

47. $\dfrac{15}{x} = \dfrac{65}{100}$

48. $\dfrac{7.8}{3.9} = \dfrac{13}{x}$

49. $\dfrac{34.1}{x} = \dfrac{0.77}{2.65}$

Find cross products to decide whether each proportion is true or false. Show the cross products and then circle true *or* false.

50. $\dfrac{55}{18} = \dfrac{80}{27}$

True   False

51. $\dfrac{5.6}{0.6} = \dfrac{18}{1.94}$

True   False

52. $\dfrac{\frac{1}{5}}{2} = \dfrac{1\frac{1}{6}}{11\frac{2}{3}}$

True   False

Write each ratio as a fraction in lowest terms. Change to the same units when necessary.

53. 4 dollars to 10 quarters

54. $4\frac{1}{8}$ in. to 10 in.

55. 10 yd to 8 ft

56. $3.60 to $0.90

57. 12 eggs to 15 eggs

58. 37 meters to 7 meters

59. 3 pints to 4 quarts

60. 15 minutes to 3 hours

61. $4\frac{1}{2}$ miles to $1\frac{3}{10}$ miles

62. Nearly 7 out of 8 fans buy something to drink at rock concerts. How many of the 28,500 fans at today's concert would be expected to buy a beverage? Round to the nearest hundred fans.

63. Emily spent $150 on car repairs and $400 on car insurance. What is the ratio of the amount spent on insurance to the amount spent on repairs?

64. Antonio is choosing among three packages of plastic wrap. Is the best buy 25 ft for $0.78; 75 ft for $1.99; or 100 ft for $2.59? He has a coupon for 50¢ off either of the larger two packages.

65. On this scale drawing of a backyard patio, 0.5 in. represents 6 ft. If the patio measures 1.75 in. long and 1.25 in. wide on the drawing, what will be the actual length and width of the patio when it is built?

0.5 in. = 6 ft

66. A lawn mower uses 0.8 gallon of gas every 3 hours. How long can the mower run on 2 gallons of gas?

67. An antibiotic is to be given at the rate of $1\frac{1}{2}$ teaspoons for every 24 pounds of body weight. How much should be given to an infant who weighs 8 pounds?

68. Charles made 251 points during 169 minutes of playing time last year. At that same rate, how many points would you expect him to make if he plays 14 minutes in tonight's game? Round to the nearest whole number.

69. Refer to Exercise 67. Explain each step you took in solving the problem. Be sure to tell how you decided which way to set up the proportion and how you checked your answer.

70. A vitamin supplement for cats is to be given at the rate of 1000 milligrams for a 5-pound cat. (*Source:* St. Jon Pet Care Products.)

(a) How much should be given to a 7-pound cat?

(b) How much should be given to an 8-ounce kitten?

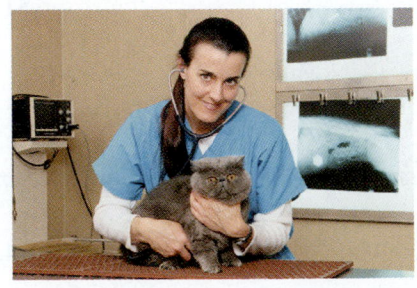

Chapter 5 ▶▶▶ **Test**  Use the Chapter Test Prep Video CD to see fully worked-out solutions to any of the exercises you want to review.

Write each rate or ratio as a fraction in lowest terms. Change to the same units when necessary.

1. 16 fish to 20 fish

2. 300 miles on 15 gallons

3. $15 for 75 minutes

4. 3 hours to 40 minutes

5. The little theater at our college has 320 seats. The auditorium has 1200 seats. Find the ratio of auditorium seats to theater seats.

6. Use the information in the table about Quiznos chicken with bacon sub sandwich to find the best buy.

Size	Length of Sub	Price
Small	5 inches	$4.39
Regular	8 inches	$5.79
Large	11 inches	$8

7. Find the best buy on spaghetti sauce. You have a coupon for 75¢ off Brand X and a coupon for 50¢ off Brand Y.

   26 ounces of Brand X for $3.89

   16 ounces of Brand Y for $1.89

   14 ounces of Brand Z for $1.29

8. Suppose the ratio of your income last year to your income this year is 3 to 2. Explain what this means. Give an example of the dollars earned last year and this year that fits the 3 to 2 ratio.

Determine whether each proportion is true or false. Show your work and then write true *or* false.

9. $\dfrac{6}{14} = \dfrac{18}{45}$

10. $\dfrac{8.4}{2.8} = \dfrac{2.1}{0.7}$

1. _____

2. _____

3. _____

4. _____

5. _____

6. _____

7. _____

8. _____

9. _____

10. _____

Find the unknown number in each proportion. In Problems 11–13, round the answers to the nearest hundredth, if necessary.

11. _____

11. $\dfrac{5}{9} = \dfrac{x}{45}$

12. _____

12. $\dfrac{3}{1} = \dfrac{8}{x}$

13. _____

13. $\dfrac{x}{20} = \dfrac{6.5}{0.4}$

14. _____

14. $\dfrac{2\frac{1}{3}}{x} = \dfrac{\frac{8}{9}}{4}$

15. _____

Set up and solve a proportion for each application problem.

15. Pedro entered 18 orders into his computer in thirty minutes at his job. At that rate, how many orders could he enter in forty minutes?

16. _____

16. Just 0.8 ounce of wildflower seeds is enough for 50 square feet of ground. What weight of seeds is needed for a garden with 225 square feet? (*Source:* White Swan Ltd.)

17. _____

17. About 2 out of every 15 people are left-handed. How many of the 650 students in our school would you expect to be left-handed? Round to the nearest whole number.

18. _____

18. A student set up the proportion for Problem 17 this way and arrived at an answer of 4875.

$15 \cdot 650 = 9750$

$\dfrac{2}{15} = \dfrac{650}{x}$ *Check:* $\dfrac{2}{15} \times \dfrac{650}{4875}$

$2 \cdot 4875 = 9750$

Because the cross products are equal, the student said the answer is correct. Is the student right? Explain why or why not.

19. _____

19. A medication is given at the rate of 8.2 grams for every 50 pounds of body weight. How much should be given to a 145-pound person? Round to the nearest tenth.

20. _____

20. On a scale model, 1 in. represents 8 ft. If a building in the model is 7.5 in. tall, what is the actual height of the building in feet?

Percent

P ercents are widely used in our everyday lives. For example, interest rates on savings and investments, automobile loans, home loans, and other installment loans are almost always given as percents. Stores often advertise sale prices as being a certain percent off the regular price. Sales tax, commission rates, and current government figures about inflation, recession, and unemployment are also reported as percents. An everyday example of the importance of understanding percent is knowing how to calculate the sales tax on items that you purchase. This allows you to know the true cost of anything you buy. (See Examples 1 and 2, Exercises 27–30, 47, and 48 in **Section 6.6**.)

6

6.1 ▶▶▶ Basics of Percent

OBJECTIVES

1 Learn the meaning of percent.

2 Write percents as decimals.

3 Write decimals as percents.

4 Understand 100%, 200%, and 300%.

5 Use 50%, 10%, and 1%.

Notice that the figure below has one hundred squares of equal size. Eleven of the squares are shaded. The shaded portion is $\frac{11}{100}$, or 0.11, of the total figure.

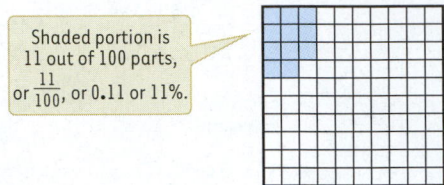

Shaded portion is 11 out of 100 parts, or $\frac{11}{100}$, or 0.11 or 11%.

The shaded portion is also 11% of the total, or "eleven parts out of 100 parts." Read **11%** as "eleven percent."

OBJECTIVE **1** **Learn the meaning of percent.** As we just saw, a percent is a ratio with a denominator of 100.

> **The Meaning of Percent**
>
> **Percent** means *per one hundred*. The "%" sign is used to show the number of parts out of one hundred parts.

1 Write as percents.

(a) In a group of 100 people, 46 are homeowners. What percent are homeowners?

EXAMPLE 1 **Understanding Percent**

(a) If *43 out of 100* students are men, then *43 per 100* or $\frac{43}{100}$ or **43%** of the students are men.

(b) If a person pays a tax of $7 on every $100 of purchases, then the tax rate is $7 per $100. The ratio is $\frac{7}{100}$ and the percent of tax is **7%**.

◀ *Work Problem* **1** *at the Side.*

(b) The sales tax is $6 per $100. What percent is this?

OBJECTIVE **2** **Write percents as decimals.** If 8% means 8 parts out of 100 parts or $\frac{8}{100}$, then $p\%$ means p parts out of 100 parts or $\frac{p}{100}$. Because $\frac{p}{100}$ is another way to write the division $p \div 100$, we have

$$p\% = \frac{p}{100} = p \div 100$$

> **Writing a Percent as a Decimal**
>
> $$p\% = \frac{p}{100} \qquad \text{or} \qquad p\% = p \div 100$$
>
> As a fraction $\qquad\qquad$ As a decimal

(c) Out of 100 students, 68 are working full- or part-time. What percent are working?

EXAMPLE 2 **Writing Percents as Decimals**

Write each percent as a decimal.

(a) 47%

$$p\% = p \div 100$$
$$47\% = 47 \div 100 = 0.47 \leftarrow \text{Decimal form}$$

0.47 is $\frac{47}{100}$ which is equivalent to 47%

ANSWERS

1. **(a)** 46% **(b)** 6% **(c)** 68%

Continued on Next Page

(b) 76% 76% = 76 ÷ 100 = 0.76 ← Decimal form

(c) 28.2% 28.2% = 28.2 ÷ 100 = 0.282 ← Decimal form

(d) 100% 100% = 100 ÷ 100 = 1.00 ← Decimal form

CAUTION

In Example 2(d) above, notice that 100% is 1.00, or 1, which is a whole number. Whenever you have a percent that is *100% or greater,* the equivalent decimal number will be *1 or greater than 1.*

Work Problem **2** *at the Side.* ▶

The answers in Example 2 above suggest these steps for writing a percent as a decimal.

Writing a Percent as a Decimal

Step 1 Drop the percent sign.

Step 2 Divide by 100.

Note

Recall from **Section 4.6** that a quick way to divide a number by 100 is to move the decimal point **two places to the left.**

EXAMPLE 3 **Writing Percents as Decimals by Moving the Decimal Point**

Write each percent as a decimal by moving the decimal point two places to the left.

(a) 17%

17% = 17.% Decimal point starts at far right side.

0.17 ← Percent sign is dropped. (Step 1)

Decimal point is moved two places to the left. (Step 2)

17% = 0.17

(b) 160%

1.60 is equivalent to 1.6 because $1\frac{60}{100}$ simplifies to $1\frac{6}{10}$

160% = 160.% = 1.60 or 1.6   Decimal point starts at far right side.

(c) 4.9%

.049   0 is attached so the decimal point can be moved two places to the left.

4.9% = 0.049

Continued on Next Page

---

2 Write each percent as a decimal.

(a) 68%

(b) 34%

(c) 58.5%

(d) 175%

(e) 200%

3 Write each percent as a decimal.

(a) 96%

(d) 0.6%

$$0.6\% = 0.\underset{\smile}{006}$$ Two zeros are attached so the decimal point can be moved two places to the left.

> **CAUTION**
> Look at Example 3(d) above, where 0.6% is less than 1%. Because 0.6% is $\frac{6}{10}$ of 1%, it is *less than 1%*. Any fraction of a percent is *less than 1%*.

◀ *Work Problem* **3** *at the Side.*

(b) 6%

OBJECTIVE 3   **Write decimals as percents.**   You can write any decimal as a percent. For example, the decimal 0.78 is the same as the fraction

$$\frac{78}{100}$$

This fraction means 78 out of 100 parts, or 78%. The following steps give the same result.

> **Writing a Decimal as a Percent**
> *Step 1*   Multiply by 100.
> *Step 2*   Attach a percent sign.

(c) 24.8%

> **Note**
> A quick way to divide or multiply a number by 100 is to move the decimal point two places to the left or two places to the right, respectively.
>
> Multiply decimal by 100; move decimal point two places to the *right*.
>
> **Decimal**   **Percent**
>
> Divide percent by 100; move decimal point two places to the *left*.

(d) 0.9%

EXAMPLE 4   **Writing Decimals as Percents by Moving the Decimal Point**

Write each decimal as a percent by moving the decimal point two places to the right.

(a) 0.21

$$0.\underset{\curvearrowright}{21}$$

Decimal point is moved two places to the right. (Step 1)

$0.21 = 21\%$ ← Percent sign is attached. (Step 2)   Remember to attach the percent (%) sign.

Decimal point is not written with whole number percents.

— **Continued on Next Page**

(b) $0.529 = 52.9\%$

(c) $1.92 = 192\%$

(d) 2.5

 $2.5\underset{\curvearrowright}{0}$ 0 is attached so the decimal point can be moved two places to the right.

 $2.5 = 250\,\%$ Attach % sign.

When necessary, attach zeros so you can move the decimal point.

(e) 3

 $3. = 3.\underset{\curvearrowright}{00}$ Two zeros are attached so the decimal point can be moved two places to the right.

 so $3 = 300\,\%$ Attach % sign.

> **CAUTION**
> Look at Examples 4(c), 4(d), and 4(e) above, where 1.92, 2.5, and 3 are greater than 1. Because the number 1 is equivalent to 100%, all numbers greater than 1 will be *greater than 100%*.

Work Problem **4** *at the Side.* ▶

OBJECTIVE 4 Understand 100%, 200%, and 300%. When working with percents, it is helpful to have several reference points. 100%, 200%, and 300% are three such helpful reference points.

 100% means 100 parts out of 100 parts. That's **all** of the parts. If 100% of the 18 people attending last week's meeting attended this week's meeting, then 18 people (**all** of them) attended this week.

 If attendance at the meeting this week is 200% of last week's attendance of 18 people, then this week's attendance is 36 people, or *two* times as many people ($2 \cdot 18 = 36$). Likewise, if attendance is 300% of last week's attendance, then *three* times as many people, or 54 people, attended ($3 \cdot 18 = 54$).

EXAMPLE 5 Finding 100%, 200%, and 300% of a Number

Fill in the blanks.

Notice that 100% of something is all of it (the whole thing).

(a) 100% of 82 people is _____ .
  100% is **all** of the people. So, 100% of 82 people is <u>82 people</u> .

(b) 200% of $63 is _____ .
  200% is twice (2 times) as much money.
  So, 200% of $63 is $2 \cdot \$63 = \underline{\$126}$.

(c) 300% of 32 employees is _____ .
  300% is 3 times as many employees.
  So, 300% of 32 employees is $3 \cdot 32 = \underline{96\ employees}$.

Work Problem **5** *at the Side.* ▶

---

4 Write each decimal as a percent.

 (a) 0.74   **(b)** 0.15

 (c) 0.09   **(d)** 0.617

 (e) 0.834   **(f)** 5.34

 (g) 2.8   **(h)** 4

5 Fill in the blanks.

 (a) 100% of $7.80 is.

 _____ .

 (b) 100% of 1850 workers is.

 _____ .

 (c) 200% of 24 photographs is _____ .

 (d) 300% of 8 miles is

 _____ .

ANSWERS

4. **(a)** 74% **(b)** 15% **(c)** 9% **(d)** 61.7%
  (e) 83.4% **(f)** 534% **(g)** 280%
  (h) 400%
5. **(a)** $7.80 **(b)** 1850 workers
  (c) 48 photographs **(d)** 24 miles

6 Fill in the blanks.

(a) 50% of 200 patients is
_____.

(b) 50% of 64 e-mails is
_____.

(c) 10% of 3850 elm trees
is _____.

(d) 10% of 7 pounds
is _____.

(e) 1% of 240 ft is
_____.

(f) 1% of $3000
is _____.

OBJECTIVE 5 Use 50%, 10%, and 1%. 50% means 50 parts out of 100 parts, which is *half* of the parts ($\frac{50}{100} = \frac{1}{2}$). So, 50% of $18 is $9 (*half* of the money).

When using 10%, we have 10 parts out of 100 parts, which is $\frac{1}{10}$ of the parts ($\frac{10}{100} = \frac{1}{10}$). To find 10% or $\frac{1}{10}$ of a number, we move the decimal point **one** place to the left. 10% of $285 is $28.50 (because $28 5. = $28.50).

To find 1% of a number ($\frac{1}{100}$), we move the decimal point **two** places to the left. 1% of $198 is $1.98 (because $19 8. = $1.98).

EXAMPLE 6 Finding 50%, 10%, and 1% of a Number

Fill in the blanks.

> Think: 50% of something is $\frac{50}{100}$ or $\frac{1}{2}$ of it.

(a) 50% of 24 hours is _____.
50% is half of the hours. So, 50% of 24 hours is 12 hours.

(b) 10% of 280 pages is _____.
10% is $\frac{1}{10}$ of the pages. Move the decimal point *one* place to the left. So, 10% of 280. pages is 28 pages .

(c) 1% of $540 is _____.
1% is $\frac{1}{100}$ of the money. Move the decimal point *two* places to the left. So, 1% of $5 40. is $5.40.

◀ *Work Problem* **6** *at the Side.*

6.1 ▶▶▶ Exercises

Write each percent as a decimal. See Examples 2 and 3.

1. 12%   **2.** 57%   **3.** 70%   **4.** 40%

5. 25%   **6.** 35%   **7.** 140%   **8.** 250%

9. 5.5%   **10.** 6.7%   **11.** 100%   **12.** 600%

13. 0.5%   **14.** 0.25%   **15.** 0.35%   **16.** 0.75%

Write each decimal as a percent. See Example 4.

17. 0.6   **18.** 0.9   **19.** 0.58   **20.** 0.25

21. 0.01   **22.** 0.07   **23.** 0.125   **24.** 0.875

25. 0.375   **26.** 0.625   **27.** 2   **28.** 5

29. 3.7   **30.** 2.2   **31.** 0.0312   **32.** 0.0625

33. 4.162   **34.** 8.715   **35.** 0.0028   **36.** 0.0064

37. Fractions, decimals, and percents are all used to describe a part of something. The use of percents is much more common than fractions and decimals. Why do you suppose this is true?

38. List five uses of percent that are or will be part of your life. Consider the activities of working, shopping, saving, and planning for the future.

Write each percent as a decimal and each decimal as a percent. See Examples 2–4.

39. When asked, 38% of those 50 years of age or older don't think they need a flu shot.

40. At College of DuPage, 82% of the students work part-time.

41. Lack of parking spaces bothers 47% of holiday shoppers.

42. There was a 43.2% voter turnout at the election.

43. The property tax rate in Alpine County is 0.035.

44. A church building fund has 0.89 of the money needed.

45. The number of people successfully completing CPR training this session is 2 times that of the last session.

46. The number of newspaper subscribers was 4 times as great as last quarter.

47. Only 0.005 of the total population has this genetic defect.

48. The return rate of defective keyboards is 0.0075 of total output.

49. The patient's blood pressure was 153.6% of normal.

50. Success with the diet was 248.7% greater than anticipated.

Fill in the blanks. Remember that 100% is all of something, 200% is two times as many, and 300% is three times as many. See Example 5.

51. There are 20 children in the preschool class. 100% of the children are served breakfast and lunch. How many children are served both meals?

_____

52. When 500 adults were asked, "Do you think your taxes are too high," 100% said yes. How many said yes?

_____

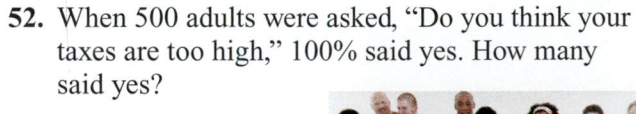

53. Last year we had 210 employees. This year we have 200% of that number. How many employees do we have this year? _____

54. This week's expenses are 200% of last week's $380. This week's expenses are _____.

55. Last week 90 chairs were used for the meeting. This week we need 300% of that number of chairs. We'll need _____.

56. Wayman's new hybrid car gets 300% of the 12 miles per gallon that his old car got. His new car gets

_____.

Fill in the blanks. Remember that 50% is half of something, 10% is found by moving the decimal point one place to the left, and 1% is found by moving the decimal point two places to the left. See Example 6.

57. Jacob owes $755 for tuition. Financial aid will pay 50% of the cost. Financial aid will pay _____.

58. Elizabeth gets 3200 "off-peak" minutes of calling time on her cell phone plan. Last month she used 50% of her allowed minutes. The number of minutes she used was _____.

59. Only 10% of 8200 commuters are carpooling to work. How many commuters carpool? _____

60. Sarah expects that 10% of the 240 dozen plants in her greenhouse will not be sold. The expected number of unsold plants is _____.

61. The naturalist said that 1% of the 2600 plants in the park are poisonous. How many plants are poisonous? _____

62. Of the 4800 accidents, only 1% were caused by mechanical failure. How many accidents were caused by mechanical failure? _____.

63. (a) Describe a shortcut method of finding 100% of a number.

(b) Show an example using your shortcut.

64. (a) Describe a shortcut method of finding 50% of a number.

(b) Show an example using your shortcut.

65. (a) Describe a shortcut method of finding 200% of a number.

(b) Show an example using your shortcut.

66. (a) Describe a shortcut method of finding 300% of a number.

(b) Show an example using your shortcut.

67. (a) Describe a shortcut method of finding 10% of a number.

(b) Show an example using your shortcut.

68. (a) Describe a shortcut method of finding 1% of a number.

(b) Show an example using your shortcut.

More than 7.4 million households will dress up their pets (dogs and cats) in Halloween costumes this year. The bar graph shows the ranking of the top pet costumes and the percent of pet owners selecting each costume. Use this graph to answer Exercises 69–72. Write each answer as a percent and as a decimal. (Source: BIGresearch survey of 8877 adult pet owners.)

69. What portion of the pet owners selected the devil costume for their pet?

70. What portion of the pet owners selected the pirate costume for their pet?

71. (a) What was the third-most-popular costume?

(b) Write the portion of the pet owners who selected this costume.

72. (a) What was the second-most-popular costume?

(b) Write the portion of the pet owners who selected this costume.

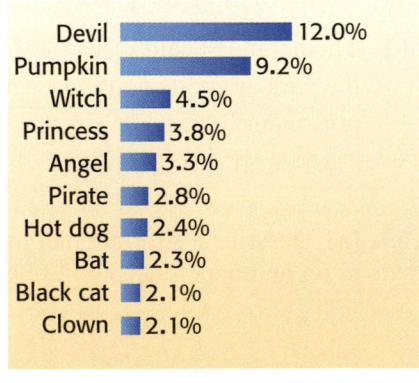

Pets get dressed up

This year, 7.4 million households plan to put their furry friends into a Halloween costume. Top outfits:

Costume	Percent
Devil	12.0%
Pumpkin	9.2%
Witch	4.5%
Princess	3.8%
Angel	3.3%
Pirate	2.8%
Hot dog	2.4%
Bat	2.3%
Black cat	2.1%
Clown	2.1%

Source: BIGresearch survey of 8,877 adults, Sept. 4–11.

There are more than 200,000 women serving in the U.S. military. The circle graph shows the percent of these women serving in each branch of the armed forces. Use this graph to answer Exercises 73–76. Write each answer as a percent and as a decimal.

73. What portion of the women in the U.S. military are in the Air Force?

74. What portion of the women in the U.S. military are in the Navy?

75. (a) Which branch of the U.S. military has the lowest portion of women?

(b) What portion is this?

76. (a) Which branch of the U.S. military has the highest portion of women?

(b) What portion is this?

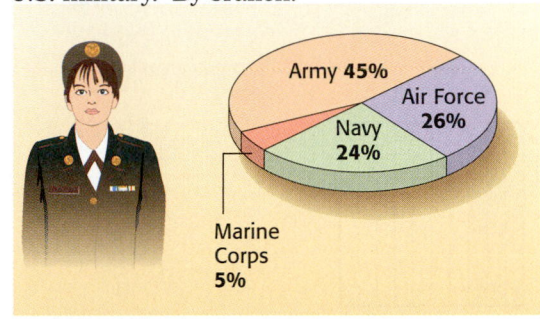

ENLISTED WOMEN
Over 200,000 women serve in the U.S. military. By branch:

Source: U.S. Department of Defense.

In the United States 12.7% of the population is 65 years of age or older. The bar graph shows the countries with the highest percent of people 65 years of age or older. Use this graph to answer Exercises 77–80. Write each answer as a percent and as a decimal.

77. What portion of the population of Spain is 65 or older?

78. What portion of the population of Greece is 65 or older?

79. (a) Which two countries in the graph have the lowest portion of the population 65 or older?

(b) What portion is this?

80. (a) Which country has the highest portion of the population 65 or older?

(b) What portion is this?

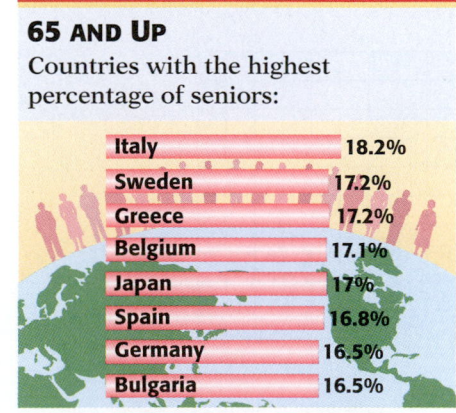

65 AND UP
Countries with the highest percentage of seniors:

Country	Percentage
Italy	18.2%
Sweden	17.2%
Greece	17.2%
Belgium	17.1%
Japan	17%
Spain	16.8%
Germany	16.5%
Bulgaria	16.5%

Source: U.S. Bureau of the Census International Programs Center.

Write a percent for both the shaded and unshaded parts of each figure.

81.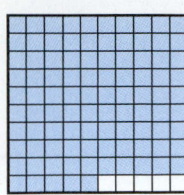

82.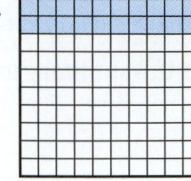

83.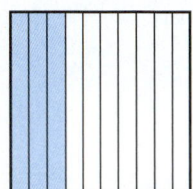

84.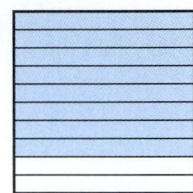

85.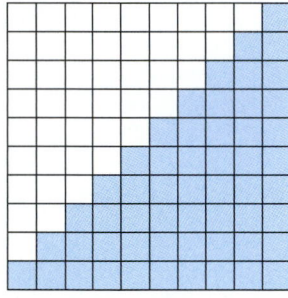

86.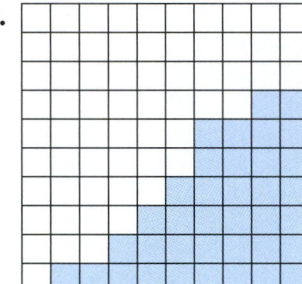

87.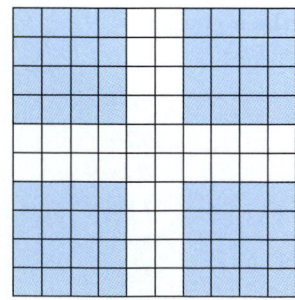

88.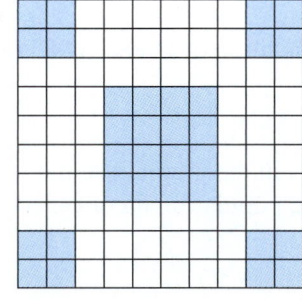

6.2 ▶▶▶ Percents and Fractions

OBJECTIVE 1 Write percents as fractions. Percents can be written as fractions by using what we learned in the previous section.

OBJECTIVES

1 Write percents as fractions.

2 Write fractions as percents.

3 Use the table of percent equivalents.

> **Writing a Percent as a Fraction**
> $$p\% = \frac{p}{100}, \quad \text{as a fraction.}$$

EXAMPLE 1 Writing Percents as Fractions

Write each percent as a fraction or mixed number in lowest terms.

(a) 25%

As we saw in the last section, 25% can be written as a decimal.

$$25\% = 25 \div 100 = 0.25 \qquad \text{Percent sign dropped}$$

Because 0.25 means 25 hundredths,

$$0.25 = \frac{25}{100} = \frac{25 \div 25}{100 \div 25} = \frac{1}{4} \qquad \text{Lowest terms}$$

It is not necessary, however, to write 25% as a decimal first. Just write

$$25\% = \frac{25}{100} \qquad \text{25 per 100}$$
$$= \frac{1}{4} \longleftarrow \text{Lowest terms}$$

(b) 76%

The percent becomes the numerator.

Write 76% as $\frac{76}{100}$

The *denominator* is always 100 because percent means *parts per 100.*

Write $\frac{76}{100}$ in lowest terms.

To write a fraction in lowest terms, divide numerator and denominator by the same number.

$$\frac{76 \div 4}{100 \div 4} = \frac{19}{25} \longleftarrow \text{Lowest terms}$$

(c) 150%

$$150\% = \frac{150}{100} = \frac{150 \div 50}{100 \div 50} = \frac{3}{2} = 1\frac{1}{2} \longleftarrow \text{Mixed number}$$

> **Note**
> Remember that percent means *per 100.*

—————— Work Problem ① at the Side. ▶

1 Write each percent as a fraction or mixed number in lowest terms.

(a) 50%

(b) 75%

(c) 48%

(d) 23%

(e) 125%

(f) 250%

ANSWERS

1. **(a)** $\frac{1}{2}$ **(b)** $\frac{3}{4}$ **(c)** $\frac{12}{25}$ **(d)** $\frac{23}{100}$
  (e) $1\frac{1}{4}$ **(f)** $2\frac{1}{2}$

The next example shows how to write decimal and fraction percents as fractions.

2 Write each percent as a fraction in lowest terms.

(a) 37.5%

(b) 62.5%

(c) 4.5%

(d) $66\frac{2}{3}\%$

(e) $10\frac{1}{3}\%$

(f) $87\frac{1}{2}\%$

EXAMPLE 2 Writing Decimal or Fraction Percents as Fractions

Write each percent as a fraction in lowest terms.

(a) 15.5%

Write 15.5 over 100.

$$15.5\% = \frac{15.5}{100}$$

To get a whole number in the numerator, multiply the numerator and denominator by 10. (Recall that multiplying by $\frac{10}{10}$ is the same as multiplying by 1.)

$$\frac{15.5}{100} = \frac{15.5\,(10)}{100\,(10)} = \frac{155}{1000}$$

> Move the decimal point 1 place to the *right* to multiply by 10.

Now write the fraction in lowest terms.

$$\frac{155 \div 5}{1000 \div 5} = \frac{31}{200}$$

(b) $33\frac{1}{3}\%$

Write $33\frac{1}{3}$ over 100.

$$33\frac{1}{3}\% = \frac{33\frac{1}{3}}{100}$$

When we have a mixed number in the numerator, we must write the mixed number as an improper fraction.

$$\frac{33\frac{1}{3}}{100} = \frac{\frac{100}{3}}{100}$$

> Write $33\frac{1}{3}$ as $\frac{100}{3}$

Next, rewrite the division problem in a horizontal form. Finally, multiply by the reciprocal of the divisor.

Reciprocals

$$\frac{\frac{100}{3}}{100} = \frac{100}{3} \div 100 = \frac{100}{3} \div \frac{100}{1} = \frac{\overset{1}{\cancel{100}}}{3} \cdot \frac{1}{\underset{1}{\cancel{100}}} = \frac{1}{3}$$

> **Note**
>
> In Example 2(a) at the top of the page we could have changed 15.5% to $15\frac{1}{2}\%$ and then written it as the improper fraction $\frac{31}{2}$ over 100. But it is usually easier *not* to change decimals to fractions but to leave decimal percents as they are.

◄ Work Problem **2** at the Side.

ANSWERS

2. (a) $\frac{3}{8}$ (b) $\frac{5}{8}$ (c) $\frac{9}{200}$ (d) $\frac{2}{3}$

(e) $\frac{31}{300}$ (f) $\frac{7}{8}$

OBJECTIVE **2** **Write fractions as percents.**   We will use the formula from the beginning of this section to write fractions as percents.

$$p\% = \frac{p}{100}$$

EXAMPLE 3   **Writing Fractions as Percents**

Write each fraction as a percent. Round to the nearest tenth if necessary.

(a) $\frac{3}{5}$

Write $\frac{3}{5}$ as a percent by solving for p in the proportion below.

$$\frac{3}{5} = \frac{p}{100}$$

Find cross products and show that they are equivalent.

$$5 \cdot p = 3 \cdot 100$$
$$5 \cdot p = 300$$

$$\frac{\overset{1}{\cancel{5}} \cdot p}{\underset{1}{\cancel{5}}} = \frac{300}{5}$$ ⟵ Divide both sides by 5.

$$p = 60$$

This result means that $\frac{3}{5} = \frac{60}{100}$ or 60%.

> **Note**
> Solving proportions can be reviewed in **Section 5.4.**

(b) $\frac{7}{8}$

Write a proportion.

$$\frac{7}{8} = \frac{p}{100}$$

$$8 \cdot p = 7 \cdot 100$$ Show that cross products are equivalent.

$$8 \cdot p = 700$$

$$\frac{\overset{1}{\cancel{8}} \cdot p}{\underset{1}{\cancel{8}}} = \frac{700}{8}$$ Divide both sides by 8.

$$p = 87.5$$

So, $\frac{7}{8} = 87.5\%$.

> **Note**
> If you think of $\frac{700}{8}$ as an improper fraction, changing it to a mixed number gives an answer of $87\frac{1}{2}$. So $\frac{7}{8} = 87.5\%$ or $87\frac{1}{2}\%$.

Continued on Next Page

3 Write as percents. Round to the nearest tenth if necessary.

(a) $\frac{1}{4}$

(b) $\frac{3}{10}$

(c) $\frac{6}{25}$

(d) $\frac{5}{8}$

(e) $\frac{1}{6}$

(f) $\frac{2}{9}$

(c) $\frac{5}{6}$

Start with a proportion.

$$\frac{5}{6} = \frac{p}{100}$$

$6 \cdot p = 5 \cdot 100$ Show that cross products are equivalent.

$6 \cdot p = 500$

$$\frac{\overset{1}{\cancel{6}} \cdot p}{\underset{1}{\cancel{6}}} = \frac{500}{6}$$ Divide both sides by 6.

$p = 83.\overline{3}$ A bar over the 3 indicates that the decimal keeps repeating 83.3333 forever.

$p \approx 83.3$ Round to the nearest tenth.

So, $\frac{5}{6} = 83.\overline{3}\% \approx 83.3\%$ (rounded)   The " $\approx$ " symbol shows that 83.3% is rounded.

Note

You can change $\frac{500}{6}$ to a mixed number to get an exact answer of $83\frac{1}{3}\%$.

◀ Work Problem **3** at the Side.

OBJECTIVE 3 Use the table of percent equivalents. The table on the next two pages shows common and not so common fractions and mixed numbers and their decimal and percent equivalents. The more you work with them, the more familiar they will become.

EXAMPLE 4 Using the Table of Percent Equivalents

Read the following from the table.

(a) $\frac{1}{12}$ as a percent

Find $\frac{1}{12}$ in the "fraction" column. The equivalent percent is 8.3% (rounded) or $8\frac{1}{3}\%$ (exact).

(b) 0.375 as a fraction

Look in the "decimal" column for 0.375. The equivalent fraction is $\frac{3}{8}$.

(c) $\frac{13}{16}$ as a percent

Find $\frac{13}{16}$ in the "fraction" column. The equivalent percent is 81.25% or $81\frac{1}{4}\%$.

 **Calculator Tip**   Example 4 (c) above can be solved on a calculator as shown below.

$$13 \div 16 = 0.8125 \times 100 = \mathbf{81.25}$$

Multiplying by 100 changes the decimal to a percent.
Or, if your calculator has a percent key, follow these steps.

$$13 \div 16 \,\%\, \mathbf{81.25}$$

↑—— **Press % key instead of = key.**

On scientific calculators, you may need to press the **2nd** key to access the % function.

ANSWERS

3. (a) 25% (b) 30% (c) 24% (d) 62.5%

(e) 16.7% (rounded); $16\frac{2}{3}\%$ (exact)

(f) 22.2% (rounded); $22\frac{2}{9}\%$ (exact)

Note

When a fraction like $\frac{1}{12}$ is changed to a decimal, it is a *repeating decimal* that goes on forever, $0.083333\ldots$. In the table below these decimals are rounded to the nearest thousandth. When the decimal is then changed to a percent, it will be to the nearest tenth of a percent. Decimals that do not repeat are usually not rounded.

Work Problem **4** *at the Side.* ▶

Percent, Decimal, and Fraction Equivalents

Percent (rounded to tenths when necessary)	Decimal	Fraction
1%	0.01	$\frac{1}{100}$
2%	0.02	$\frac{1}{50}$
4%	0.04	$\frac{1}{25}$
5%	0.05	$\frac{1}{20}$
6.25% or $6\frac{1}{4}$%	0.0625	$\frac{1}{16}$
8.3% (rounded) or $8\frac{1}{3}$% (exact)	$0.08\overline{3}$ rounds to 0.083	$\frac{1}{12}$
10%	0.1	$\frac{1}{10}$
12.5% or $12\frac{1}{2}$%	0.125	$\frac{1}{8}$
16.7% (rounded) or $16\frac{2}{3}$% (exact)	$0.1\overline{6}$ rounds to 0.167	$\frac{1}{6}$
18.75% or $18\frac{3}{4}$%	0.1875	$\frac{3}{16}$
20%	0.2	$\frac{1}{5}$
25%	0.25	$\frac{1}{4}$
30%	0.3	$\frac{3}{10}$
31.25% or $31\frac{1}{4}$%	0.3125	$\frac{5}{16}$
33.3% (rounded) or $33\frac{1}{3}$% (exact)	$0.\overline{3}$ rounds to 0.333	$\frac{1}{3}$
37.5% or $37\frac{1}{2}$%	0.375	$\frac{3}{8}$
40%	0.4	$\frac{2}{5}$

(continued)

4 Read the following fractions, mixed numbers, decimals, and percents from the table on this or the next page. If you already know the answer or can solve for the answer quickly, don't use the table.

(a) $\frac{3}{4}$ as a percent

(b) 10% as a fraction

(c) $0.\overline{6}$ as a fraction

(d) $37\frac{1}{2}$% as a fraction

(e) $\frac{7}{8}$ as a percent

(f) $\frac{1}{2}$ as a percent

(g) $33\frac{1}{3}$% as a fraction

(h) $1\frac{3}{4}$ as a percent

ANSWERS

4. **(a)** 75% **(b)** $\frac{1}{10}$ **(c)** $\frac{2}{3}$ **(d)** $\frac{3}{8}$
(e) 87.5% **(f)** 50% **(g)** $\frac{1}{3}$ **(h)** 175%

Percent, Decimal, and Fraction Equivalents (continued)

Percent	Decimal	Fraction
43.75% or $43\frac{3}{4}$%	0.4375	$\frac{7}{16}$
50%	0.5	$\frac{1}{2}$
56.25% or $56\frac{1}{4}$%	0.5625	$\frac{9}{16}$
60%	0.6	$\frac{3}{5}$
62.5% or $62\frac{1}{2}$%	0.625	$\frac{5}{8}$
66.7% (rounded) or $66\frac{2}{3}$% (exact)	$0.\overline{6}$ rounds to 0.667	$\frac{2}{3}$
68.75% or $68\frac{3}{4}$%	0.6875	$\frac{11}{16}$
70%	0.7	$\frac{7}{10}$
75%	0.75	$\frac{3}{4}$
80%	0.8	$\frac{4}{5}$
81.25% or $81\frac{1}{4}$%	0.8125	$\frac{13}{16}$
83.3% (rounded) or $83\frac{1}{3}$% (exact)	$0.8\overline{3}$ rounds to 0.833	$\frac{5}{6}$
87.5% or $87\frac{1}{2}$%	0.875	$\frac{7}{8}$
90%	0.9	$\frac{9}{10}$
93.75% or $93\frac{3}{4}$%	0.9375	$\frac{15}{16}$
100%	1.0	1
110%	1.1	$1\frac{1}{10}$
125%	1.25	$1\frac{1}{4}$
133.3% (rounded) or $133\frac{1}{3}$% (exact)	$1.\overline{3}$ rounds to 1.333	$1\frac{1}{3}$
150%	1.5	$1\frac{1}{2}$
166.7% (rounded) or $166\frac{2}{3}$% (exact)	$1.\overline{6}$ rounds to 1.667	$1\frac{2}{3}$
175%	1.75	$1\frac{3}{4}$
200%	2.0	2

6.2 ▶▶▶ Exercises

Write each percent as a fraction or mixed number in lowest terms. See Examples 1 and 2.

1. 25%

2. 30%

3. 75%

4. 80%

5. 85%

6. 45%

7. 62.5%

8. 87.5%

9. 6.25%

10. 43.75%

11. $16\frac{2}{3}\%$

12. $66\frac{2}{3}\%$

13. $6\frac{2}{3}\%$

14. $46\frac{2}{3}\%$

15. 0.5%

16. 0.8%

17. 180%

18. 140%

19. 375%

20. 225%

Write each fraction as a percent. Round percents to the nearest tenth if necessary. See Example 3.

21. $\frac{1}{2}$

22. $\frac{4}{10}$

23. $\frac{4}{5}$

24. $\frac{3}{10}$

25. $\frac{7}{10}$

26. $\frac{3}{4}$

27. $\frac{37}{100}$

28. $\frac{63}{100}$

29. $\frac{5}{8}$

30. $\frac{1}{8}$

31. $\frac{7}{8}$

32. $\frac{3}{8}$

33. $\frac{12}{25}$

34. $\frac{15}{25}$

35. $\frac{23}{50}$

36. $\frac{18}{50}$

37. $\dfrac{7}{20}$ **38.** $\dfrac{9}{20}$ **39.** $\dfrac{5}{6}$ **40.** $\dfrac{1}{6}$

41. $\dfrac{5}{9}$ **42.** $\dfrac{7}{9}$ **43.** $\dfrac{1}{7}$ **44.** $\dfrac{5}{7}$

Complete the chart. Round decimals to the nearest thousandth and percents to the nearest tenth if necessary. See Examples 3 and 4.

Fraction	Decimal	Percent
45. _____	0.5	_____
46. $\dfrac{1}{4}$	_____	_____
47. _____	_____	87.5%
48. $\dfrac{3}{4}$	_____	_____
49. _____	0.8	_____
50. _____	_____	60%
51. $\dfrac{1}{6}$	_____	_____
52. $\dfrac{1}{3}$	_____	_____
53. _____	0.7	_____

Fraction	Decimal	Percent
54. _____	_____	37.5%
55. _____	_____	12.5%
56. _____	0.625	_____
57. $\dfrac{2}{3}$	_____	_____
58. $\dfrac{5}{6}$	_____	_____
59. $\dfrac{3}{50}$	_____	_____
60. $\dfrac{3}{10}$	_____	_____
61. $\dfrac{8}{100}$	_____	_____
62. _____	_____	100%
63. $\dfrac{1}{200}$	_____	_____
64. $\dfrac{1}{400}$	_____	_____

Fraction	Decimal	Percent
65. _____	2.5	_____
66. _____	1.7	_____
67. $3\frac{1}{4}$	_____	_____
68. $2\frac{4}{5}$	_____	_____

69. Select a decimal percent and write it as a fraction. Select a different fraction and write it as a percent. Write an explanation of each step of your work.

70. Prepare a table showing fraction, decimal, and percent equivalents for five fractions and mixed numbers of your choice.

In the following application problems, write the answer as a fraction in lowest terms, as a decimal, and as a percent.

71. Many pet owners say they have used the Internet to find pet information. Of 500 people who used the Internet for this purpose, 90 said they used it when buying a pet. What portion used the Internet when buying a pet? (*Source:* American Animal Hospital Association.)

72. About $\frac{1}{3}$ of all books purchased last year were for children. Of these children's books, 27 of every 100 purchased included a coloring activity. What portion of the children's books included a coloring activity? (*Source:* Consumer Research Study on Book Publishing, the American Booksellers, and the Book Industry Study Group.)

73. Only 13 out of every 100 adults consume the recommended 1000 milligrams of calcium daily. What portion consumes the recommended daily amount? (*Source:* Market Facts for Milk Mustache.)

74. In a survey on how people learn to parent, 360 parents out of 800 said they were most influenced by relatives, friends, and spouses. What portion learned to parent this way? (*Source:* Bama Research.)

75. In a recent survey, "Attitudes in the American Workplace," 750 workers were asked if they would hire their own boss if they were in charge. A total of 150 workers said no, they would not. What portion of the workers said no? (*Source:* Marlin's 13th annual Attitudes in the Workplace Survey.)

76. When 1500 adults were asked what was the most important factor to consider when relocating after retirement, 675 said that climate was most important. What portion consider climate to be most important? (*Source:* Longevity Alliance Retirement and Relocation Survey.)

77. An insurance office has 80 employees. If 64 of the employees have cell phones, what portion of the employees do *not* have cell phones?

78. A zoo has 125 types of animals, including 25 that are endangered. What portion is *not* endangered?

79. An antibiotic is used to treat 380 people. If 342 people do not have side effects from the antibiotic, find the portion that do have side effects.

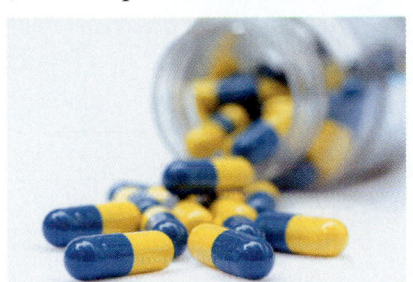

80. A medical group includes 340 doctors. While electronic recordkeeping was used most often by the younger doctors, overall, only 85 of the doctors used it. What portion of the doctors did not use it? (*Source:* National Center for Health Statistics.)

The circle graph shows how much money the 10,800 students at a college expect to earn this summer. Use this graph to answer Exercises 81–84, giving each answer as a fraction, as a decimal, and as a percent.

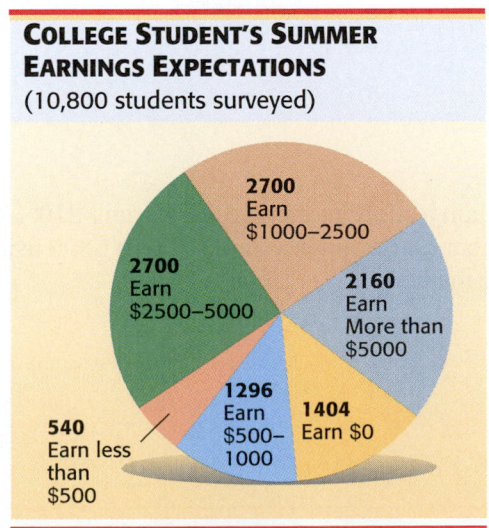

COLLEGE STUDENT'S SUMMER EARNINGS EXPECTATIONS
(10,800 students surveyed)

2700 Earn $1000–2500
2700 Earn $2500–5000
2160 Earn More than $5000
1296 Earn $500–1000
1404 Earn $0
540 Earn less than $500

Source: U.S. Collegeclub.com

81. What portion of the students expect to earn $2500 to $5000?

82. What portion of the students expect to earn $500 to $1000?

83. Find the portion who expect to earn less than $500.

84. Find the portion who expect to earn more than $5000.

*Work Exercises 85–94 **in order** to review the basics of percent.*

85. 100% of a number means all of the parts or
_____ parts out of _____ parts.
200% means two times as many parts and 300%
means three times as many parts.

86. Fill in the blanks.

    (a) 100% of 765 workers is _____.

    (b) 200% of 48 letters is _____.

    (c) 300% of 7 videos is _____.

87. 50% of a number is _____ parts out of

_____ parts, which is _____ of the parts.

88. 10% of a number is _____ parts out of _____
parts and can be found quickly by moving the
decimal point _____ place to the _____.

89. 1% of a number is _____ part out of
_____ parts and can be found quickly by moving
the decimal point _____ places to the _____.

90. Fill in the blanks.

    (a) 50% of 1050 homes is _____.

    (b) 10% of 370 printers is _____.

    (c) 1% of $8 is _____.

In Exercises 91–94, use the shortcut methods for finding 1%, 10%, 50%, 100%, 200%, *and* 300%.

91. Devise a shortcut method for finding 15% of a
number. Use your method to find 15% of $160.

92. Devise a shortcut method for finding 150% of a
number. Use your method to find 150% of $160.

93. Explain a shortcut method for finding 90% of a
number. Show how to find 90% of $450 using your
method.

94. Explain a shortcut method for finding 210% of a
number. Show how to find 210% of $800 using
your method.

6.3 ▸▸▸ Using the Percent Proportion and Identifying the Components in a Percent Problem

There are two ways to solve percent problems. One method uses proportions and is discussed in this and the next section. The other method uses the percent equation and is explained in **Section 6.5.**

OBJECTIVE 1 Learn the percent proportion. We have seen that a statement of two equivalent ratios is called a proportion.

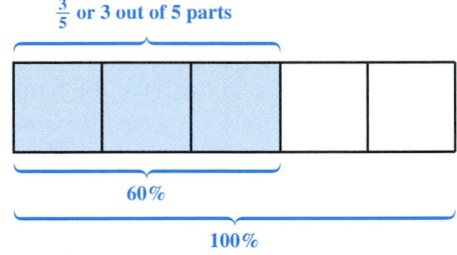

$\frac{3}{5}$ or 3 out of 5 parts

60%

100%

For example, the fraction $\frac{3}{5}$ is the same as the ratio 3 to 5, and 60% is the same as the ratio 60 to 100. As the figure above shows, these two ratios are equivalent and make a proportion.

Work Problem **1** *at the Side.* ▶

The percent proportion can be used to solve percent problems.

Percent Proportion

Part is to *whole* as *percent* is to *100.*

$$\frac{\textbf{part}}{\textbf{whole}} = \frac{\textbf{percent}}{\textbf{100}} \quad \leftarrow \text{Always 100 because percent means per 100}$$

In the figure at the top of the page, the **whole** is 5 (the entire quantity), the **part** is 3 (the part of the whole), and the **percent** is 60. Write the percent proportion as follows.

$$\textbf{part} \rightarrow \frac{3}{5} = \frac{60}{100} \leftarrow \textbf{percent}$$
$$\textbf{whole} \rightarrow \qquad \leftarrow \textbf{100}$$

Remember: Percent means *per 100.*

OBJECTIVE 2 Solve for an unknown value in a percent proportion. As shown in **Section 5.4,** if any three of the four values in a proportion are known, the fourth can be found by solving the proportion.

EXAMPLE 1 Using the Percent Proportion

Use the percent proportion and solve for the unknown value. Let x represent the unknown value.

(a) part = 12, percent = 25; find the whole.

$$\frac{\text{part}}{\text{whole}} = \frac{\text{percent}}{100} \qquad \text{Percent proportion}$$

Percent

$$\text{Part} \rightarrow \frac{12}{x} = \frac{25}{100} \quad \text{or} \quad \frac{12}{x} = \frac{1}{4} \qquad \frac{25}{100} \text{ is } \frac{1}{4} \text{ in lowest terms.}$$
Whole (unknown) $\rightarrow$

Continued on Next Page

---

1 As a review of proportions, use the method of comparing cross products to decide whether each proportion is *true* or *false.* Show the cross products.

(a) $\dfrac{1}{2} = \dfrac{25}{50}$

(b) $\dfrac{3}{4} = \dfrac{150}{200}$

(c) $\dfrac{7}{8} = \dfrac{180}{200}$

(d) $\dfrac{32}{53} = \dfrac{160}{265}$

(e) $\dfrac{112}{41} = \dfrac{332}{123}$

2 Use the percent proportion
$$\left(\frac{\text{part}}{\text{whole}} = \frac{\text{percent}}{100}\right) \text{ and}$$
solve for the unknown value.

(a) part = 12, percent = 16

(b) part = 30, whole = 120

(c) whole = 210, percent = 20

(d) whole = 4000, percent = 32

(e) part = 74, whole = 185

First find the cross products.

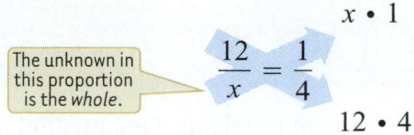

Show that the cross products are equivalent.

$$x \cdot 1 = 12 \cdot 4$$
$$x = 48$$

The whole is 48.

> **CAUTION**
> You **cannot** divide out a common factor from the numerator of one ratio and the denominator of the other ratio in **a proportion.** This can be done *only* when you are multiplying fractions.

(b) part = 30, whole = 50; find the percent.
Use the percent proportion.

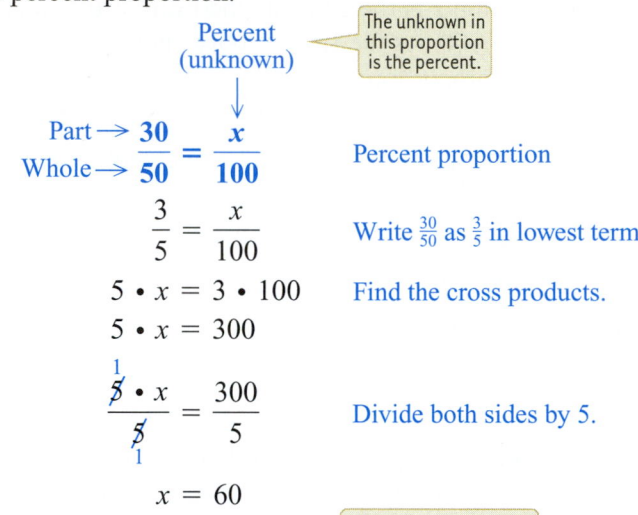

Part → $\dfrac{30}{50} = \dfrac{x}{100}$ Percent proportion
Whole →

$$\frac{3}{5} = \frac{x}{100} \qquad \text{Write } \tfrac{30}{50} \text{ as } \tfrac{3}{5} \text{ in lowest terms.}$$

$$5 \cdot x = 3 \cdot 100 \qquad \text{Find the cross products.}$$
$$5 \cdot x = 300$$

$$\frac{\overset{1}{\cancel{5}} \cdot x}{\underset{1}{\cancel{5}}} = \frac{300}{5} \qquad \text{Divide both sides by 5.}$$

$$x = 60$$

The percent is 60, written as 60%.   Remember to write the % sign when solving for an unknown percent.

(c) whole = 150, percent = 18; find the part.

Percent

Part (unknown) → $\dfrac{x}{150} = \dfrac{18}{100}$ or $\dfrac{x}{150} = \dfrac{9}{50}$ Write $\tfrac{18}{100}$ as $\tfrac{9}{50}$ in lowest terms.
Whole →

$$x \cdot 50 = 150 \cdot 9 \qquad \text{Find the cross products.}$$
$$x \cdot 50 = 1350$$

$$\frac{x \cdot \overset{1}{\cancel{50}}}{\underset{1}{\cancel{50}}} = \frac{1350}{50} \qquad \text{Divide both sides by 50.}$$

$$x = 27$$

The part is 27.

ANSWERS

2. **(a)** whole = 75
(b) percent = 25
   (so, the percent is 25%)
(c) part = 42
(d) part = 1280
(e) percent = 40
   (so, the percent is 40%)

◀ **Work Problem** **2** *at the Side.*

As a help in solving percent problems, keep in mind this basic idea.

> **Percent Problems**
>
> All percent problems involve a comparison between a part of something and the whole.

Solving these problems requires identifying the three components of a percent proportion: part, whole, and percent.

OBJECTIVE 3 Identify the percent.   Look for the percent first. It is the easiest to identify.

> **Percent**
>
> The **percent** is the ratio of a part to a whole, with 100 as the denominator. In a problem, the percent appears with the word **percent** or with the symbol "**%**" after it.

EXAMPLE 2   **Finding the Percent in Percent Problems**

Find the percent in the following.

(a) 32% of the 900 men were too large for the imported car.

Percent ⟵ Look for the percent sign or the word percent.

The percent is 32. The number 32 appears with the symbol %.

(b) $150 is 25 percent of what number?

Percent

The percent is 25 because 25 appears with the word *percent*.

(c) What percent of 7000 pounds is 3500 pounds?

Percent (unknown)

The word *percent* has no number with it, so the percent is the unknown part of the problem.

Work Problem **3** *at the Side.* ▶

OBJECTIVE 4 Identify the whole.   Next, look for the whole.

> **Whole**
>
> The **whole** is the entire quantity. In a percent problem, the whole often appears after the word **of**.

---

3 Identify the percent.

(a) Of the 900 blood glucose tests, 25% will be completed by Adrian.

(b) Of the 620 preschool students, 65% will be served breakfast and lunch.

(c) Find the amount of sales tax by multiplying $590 and $6\frac{1}{2}$ percent.

(d) 8500 tons of recyclables is 42% of what number of tons?

(e) What percent of the 380 guests will return this year?

Identify the whole.

(a) Of the 900 blood glucose tests, 25% will be completed by Adrian.

(b) Of the 620 preschool students, 65% will be served breakfast and lunch.

(c) Find the amount of sales tax by multiplying sales of $590 and $6\frac{1}{2}$ percent.

(d) 8500 tons of recyclables is 42% of what number of tons?

(e) What percent of the 380 guests will return this year?

 Identify the part, then set up the percent proportion.

(a) Of the 900 blood glucose tests, 25% or 225 will be completed by Adrian.

(b) Of the 620 preschool students, 65% or 403 will be served breakfast and lunch.

(c) Find the sales tax by multiplying $590 and $6\frac{1}{2}$ percent.

(d) 8500 tons of recyclables is 42% of what number of tons?

(e) 80% of the 380 guests will return this year.

ANSWERS

4. (a) 900 (b) 620 (c) 590
   (d) what number (an unknown) (e) 380

5. (a) 225; $\text{Part} \to \dfrac{225}{900} = \dfrac{25}{100} \begin{array}{l}\leftarrow \text{Percent} \\ \leftarrow \text{Always 100}\end{array}$ $\text{Whole} \to$

   (b) 403; $\text{Part} \to \dfrac{403}{620} = \dfrac{65}{100} \begin{array}{l}\leftarrow \text{Percent} \\ \leftarrow \text{Always 100}\end{array}$ $\text{Whole} \to$

   (c) Unknown;
   $\begin{array}{l}\text{Part} \to \\ \text{Whole} \to\end{array} \dfrac{\text{unknown}}{590} = \dfrac{6\frac{1}{2}}{100} \begin{array}{l}\leftarrow \text{Percent} \\ \leftarrow \text{Always 100}\end{array}$

   (d) 8500;
   $\begin{array}{l}\text{Part} \to \\ \text{Whole} \to\end{array} \dfrac{8500}{\text{unknown}} = \dfrac{42}{100} \begin{array}{l}\leftarrow \text{Percent} \\ \leftarrow \text{Always 100}\end{array}$

   (e) unknown;
   $\begin{array}{l}\text{Part} \to \\ \text{Whole} \to\end{array} \dfrac{\text{unknown}}{380} = \dfrac{80}{100} \begin{array}{l}\leftarrow \text{Percent} \\ \leftarrow \text{Always 100}\end{array}$

EXAMPLE 3 **Finding the Whole in Percent Problems**

Identify the whole in the following.

(a) 32% **of** the 900 men were too large for the imported car.

    ↓ Whole

The whole is 900. The number 900 appears after the word *of.*

(b) $150 is 25 percent **of** what number?

    ↓ Whole   The whole is the unknown part of the problem.

(c) What percent **of** 7000 pounds is 3500 pounds?

    ↓ Whole

◄ *Work Problem* **4** *at the Side.*

OBJECTIVE **5** **Identify the part.**   Finally, look for the part.

Part
The **part** is the portion being compared with the whole.

Note
If you have trouble identifying the part, find the percent and whole first. The remaining number is the part.

EXAMPLE 4 **Finding the Part in Percent Problems**

Identify the part. Then set up the percent proportion. (Do **not** solve the proportions.)

(a) 54% **of** 700 students is 378 students.
    First find the percent and the whole.

54% **of** 700 students is 378 students.

Percent; with % sign   Whole; follows "of"

Find the percent and then the whole. The remaining number, 378, is the part.

The remaining number, 378, is the part.

54% **of** 700 students is 378 students. $\begin{array}{l}\text{Part} \to \\ \text{Whole} \to\end{array} \dfrac{378}{700} = \dfrac{54}{100} \begin{array}{l}\leftarrow \text{Percent} \\ \leftarrow \text{Always 100}\end{array}$

Percent   Whole   Part

(b) $150 is 25% **of** what number? $\begin{array}{l}\text{Part} \to \\ \text{Whole} \to\end{array} \dfrac{150}{\text{unknown}} = \dfrac{25}{100} \begin{array}{l}\leftarrow \text{Percent} \\ \leftarrow \text{Always 100}\end{array}$

    Percent   Whole (unknown)

$150 is the remaining number, so the part is $150.

(c) 85% **of** 7000 is what number? $\begin{array}{l}\text{Part} \to \\ \text{Whole} \to\end{array} \dfrac{\text{unknown}}{7000} = \dfrac{85}{100} \begin{array}{l}\leftarrow \text{Percent} \\ \leftarrow \text{Always 100}\end{array}$

    Percent   Whole   Part (unknown)

◄ *Work Problem* **5** *at the Side.*

Find the unknown value in the percent proportion $\dfrac{part}{whole} = \dfrac{percent}{100}$. *Round to the nearest tenth if necessary. If the answer is a percent, be sure to include a percent sign (%). See Example 1.*

1. part = 5, percent = 10

2. part = 20, percent = 25

3. part = 30, percent = 20

4. part = 25, percent = 25

5. part = 28, percent = 40

6. part = 11, percent = 5

 7. part = 15, whole = 60

8. part = 105, whole = 35

9. part = 36, whole = 24

10. part = 1.5, whole = 4.5

11. part = 9.25, whole = 27.75

12. part = 12.8, whole = 9.6

13. whole = 52, percent = 50

14. whole = 160, percent = 35

15. whole = 72, percent = 30

16. whole = 115, percent = 38

17. whole = 94.4, part = 25

18. whole = 89.6, part = 50

Solve each problem. If the answer is a percent, be sure to include a percent sign (%).
See Examples 2–4.

19. Find the whole if the part is 46 and the percent is 40.

20. The percent is 45 and the whole is 160. Find the part.

21. The whole is 5000 and the part is 20. Find the percent.

22. Suppose the part is 15 and the whole is 2500. Find the percent.

23. Find the percent if the whole is 4300 and the part is $107\frac{1}{2}$.

24. What is the part, if the percent is $12\frac{3}{4}$ and the whole is 5600?

25. The whole is 6480 and the part is 19.44. Find the percent.

26. Suppose the part is 281.25 and the percent is $1\frac{1}{4}$. Find the whole.

*In Exercises 27–42, set up the percent proportion, and write "unknown" for any value that is not given. Recall that the percent proportion is $\dfrac{part}{whole} = \dfrac{percent}{100}$. Do **not** try to solve for the unknowns. See Examples 2–4.*

27. 10% of how many bicycles is 60 bicycles?

28. 58% of how many preschoolers is 203 preschoolers?

29. 75% of $800 is $600.

30. 93% of $1500 is $1395.

31. What is 25% of $970?

32. What is 61% of 830 homes?

33. 12 injections is 20% of what number of injections?

34. 92 servings is 26% of what number of servings?

35. 34 trophies is 50% of 68 trophies.

36. 410 pallets is $33\dfrac{1}{3}$% of 1230 pallets.

37. What percent of $296 is $177?

38. What percent of $121 is $30?

39. 54.34 is 3.25% of what number?

40. 16.74 is 11.9% of what number?

41. 0.68% of $487 is what amount?

42. What amount is 6.21% of $704.35?

43. Identify the three components in a percent problem. In your own words, write a sentence telling how you will identify each of these three components.

44. Write one short sentence using numbers and words. The sentence should include a percent, a whole, and a part. Identify each of these three components.

*Set up the percent proportion for each application problem. Do **not** try to solve for any unknowns.*

45. Of the 1802 television sets sold at a Best Buy, 585 were high-definition sets. What percent of the televisions were high-definition sets?

46. Ivory Soap is $99\frac{44}{100}$% pure. If a bar of Ivory Soap weighs 4 ounces, how many ounces are pure? (*Source:* Procter & Gamble.)

47. Of the 142 people attending a movie theater, 86 bought buttered popcorn. What percent bought buttered popcorn?

48. On her first check from the Pizza Hut Restaurant, 15% is withheld from Maria's total earnings of $225. What amount is withheld?

49. Of the customers buying a salad at McDonald's, 23% prefer Newman's Own Light Salad Dressing. If the total number of salad customers is 610, find the number who prefer Newman's Own Light Dressing.

50. Of the total candy bars contained in a vending machine, 240 bars have been sold. If 25% of the bars have been sold, find the total number of candy bars that were in the machine.

51. There are 680 computer chips in a secured storage area designed to hold 2000 computer chips. What percent of the storage area is filled?

52. There have been 36 cups of coffee served from a banquet-sized coffee pot. If this is 30% of the capacity of the pot, find the capacity of the pot.

53. In a recent survey of 480 adults, 55% said that they would prefer to have their wedding at a religious site. How many said they would prefer the religious site? (*Source:* National Family Opinion Research.)

54. Sue Ann needs 64 credits to graduate. If she has completed 48 of the credits needed, what percent of the credits has she already completed?

55. In a poll of 822 people, 49.5% said that they get their news from television. Find the number of people who said they get their news from television. (*Source: Brills Content.*)

56. The sales tax on a new car is $1575. If the sales tax rate is 7%, find the price of the car before the sales tax is added.

57. A medical clinic found that 16.8% of the patients were late for their appointments. The number of patients who were late was 504. Find the total number of patients.

58. The state troopers tested 924 cars for safety. There were 231 cars that failed the safety test for one or more reasons. Find the percent of cars that failed the test.

59. Jerry Azzaro has listed 680 antique toys on eBay. If 45% of these were antique toy trains, find the number that were toy trains.

60. In a recent survey, 141 people said that they save money for up to two years to pay for a vacation that is at least a week long. If the total number of people in the survey was 1008 people, what percent of them save for up to two years to pay for their vacation? (*Source:* Capital One Direct Banking.)

6.4 ▶▶▶ Using Proportions to Solve Percent Problems

This is the percent proportion that you learned about in the previous section.

$$\frac{\text{part}}{\text{whole}} = \frac{\text{percent}}{100}$$

Recall that if one of the values is unknown, you can find it by solving the percent proportion.

OBJECTIVE 1 Use the percent proportion to find the part. The first example shows how to use the percent proportion to find the part.

EXAMPLE 1 **Finding the Part with the Percent Proportion**

Find 15% of $160.

Here the percent is 15 and the whole is 160. (Recall that the whole often comes after the word *of.*) Now find the part. Let x represent the unknown part.

$$\frac{\text{part}}{\text{whole}} = \frac{\text{percent}}{100} \quad \text{so} \quad \frac{x}{160} = \frac{15}{100} \quad \text{or} \quad \frac{x}{160} = \frac{3}{20}$$

Write $\frac{15}{100}$ as $\frac{3}{20}$ in lowest terms.

Find the cross products in the proportion and show that they are equivalent.

$$x \cdot 20 = 160 \cdot 3 \qquad \text{Cross products}$$

$$x \cdot 20 = 480$$

$$\frac{x \cdot \overset{1}{\cancel{20}}}{\underset{1}{\cancel{20}}} = \frac{480}{20} \qquad \text{Divide both sides by 20.}$$

$$x = 24 \qquad \text{The unknown part is 24.}$$

15% of $160 is **$24**.

───────── *Work Problem* ① *at the Side.* ▶

Just as with some of the fraction application problems in **Section 2.6,** the word *of* may be an indicator word meaning *multiply.* Here is an example.

$$\textbf{15\% of } 160$$
$$\downarrow$$
$$15\% \cdot 160$$

In this type of example, there is another way to find the part.

┌───┐
Finding the Part Using Multiplication

To find the part:

Step 1 Identify the percent. Write the percent as a decimal.

Step 2 Multiply this decimal by the whole.
└───┘

OBJECTIVES

1 Use the percent proportion to find the part.

2 Find the whole using the percent proportion.

3 Find the percent using the percent proportion.

1 Use the percent proportion to find the part.

(a) 8% of 400 patients

(b) 15% of $3220

(c) 7% of 2700 miles

(d) 48% of 1580 kilowatts

ANSWERS

1. (a) 32 patients **(b)** $483
   (c) 189 miles **(d)** 758.4 kilowatts

2 Use multiplication to find the part.

(a) 55% of 10,000 injections

(b) 16% of 120 miles

(c) 135% of 60 dosages

(d) 0.5% of $238

EXAMPLE 2 Finding the Part Using Multiplication

Use multiplication to find the part.

(a) Find **42% of** 830 yards.

Step 1 Here, the percent is 42. Write 42% as the decimal 0.42.

Step 2 Multiply 0.42 and the whole, which is 830.

$$\text{part} = (0.42)(830)$$
$$= 348.6 \text{ yd}$$

When the percent and whole are given, the part must be found.

It is a good idea to estimate the answer, to make sure no mistakes were made with decimal points. Round 42% to 40% or 0.4, and round 830 as 800. Next, 40% of 800 is

$$(0.4)(800) = 320 \leftarrow \text{Estimate}$$

so the exact answer of 348.6 is reasonable.

(b) Find **25% of** 1680 cars.
Identify the percent as 25. Write 25% in decimal form as 0.25. Now, multiply 0.25 and 1680.

$$\text{part} = (0.25)(1680) = 420 \text{ cars} \quad \text{Multiply.}$$

You can also use a shortcut to find the answer. Since 25% means 25 parts out of 100 parts, this is the same as $\frac{1}{4}$ of the whole ($\frac{25}{100} = \frac{1}{4}$). Do you see a shortcut here? You can find $\frac{1}{4}$ of a number by dividing the number by 4. So, this shortcut gives us the exact answer, $1680 \div 4 = 420$.

(c) Find **140% of** 60 miles.
In this problem, the percent is 140. Write 140% as the decimal 1.40. Next, multiply 1.40 and 60.

$$\text{part} = (1.40)(60) = 84 \text{ miles} \quad \text{Multiply.}$$

You can estimate the answer by realizing that 140% is close to 150% (which is $1\frac{1}{2}$) and $1\frac{1}{2}$ times 60 is 90. So, 84 miles is a reasonable answer.

(d) Find **0.4% of** 50 kilometers.

$$\text{part} = (0.004)(50) = 0.2 \text{ kilometer} \quad \text{Multiply.}$$

Write 0.4% as a decimal.

Estimate the answer by realizing that 0.4% is less than 1%.

$$1\% \text{ of } 50 \text{ kilometers} = .50. = 0.5 \text{ kilometer}$$

So our exact answer should be *less than* 0.5 kilometer, and 0.2 kilometer fits this requirement.

◀ *Work Problem* **2** *at the Side.*

EXAMPLE 3 Solving for the Part in an Application Problem

Raley's Markets has 850 employees. Of these employees, 28% are students. How many of the employees are students? Use the six problem-solving steps.

Continued on Next Page

Step 1 **Read** the problem. The problem asks us to find the number of employees who are students.

Step 2 **Work out a plan.** Look for the word *of* as an indicator word for multiplication.

28% of the employees are students.

⬑ —— Indicator word

   The total number of employees is 850, so the whole is 850. The percent is 28. To find the number of students, find the part.

Step 3 **Estimate** a reasonable answer. You can estimate the answer by rounding 28% to 25% and 850 to 900. Remember that 25% is 25 parts out of 100, which is equivalent to $\frac{1}{4}$. So divide 900 by 4.

$900 \div 4 = 225$ students ← Estimate

Step 4 **Solve** the problem.

part $= (0.28)(850) = 238$ Multiply.

⬑ —— Write 28% as a decimal.

> Notice that the decimal point was moved two places to the *left*.

Step 5 **State the answer.** Raley's Markets has 238 student employees.

Step 6 **Check.** The exact answer, 238 students, is close to our estimate of 225 students.

Work Problem ③ *at the Side.* ▶

🖩 **Calculator Tip**   If you are using a calculator, you could solve Example 3 above like this.

0.28 ⊗ 850 ⊜ 238

Or, you can use this alternate approach on calculators with a % key.

850 ⊗ 28 ⊛ ⊜ 238

OBJECTIVE 2 Find the whole using the percent proportion.   The next example shows how to use the percent proportion to find the whole.

> **Note**
> Remember, the *whole* is the entire quantity.

EXAMPLE 4   **Finding the Whole with the Percent Proportion**

(a) 8 iPods is 4% of what number of iPods?
   Here the percent is 4, the whole is unknown, and the part is 8. Use the percent proportion to find the whole. Let x represent the unknown whole.

$$\frac{\text{part}}{\text{whole}} = \frac{\text{percent}}{100} \quad \text{so} \quad \frac{8}{x} = \frac{4}{100} \quad \text{or} \quad \frac{8}{x} = \frac{1}{25}$$

Write $\frac{4}{100}$ as $\frac{1}{25}$ in lowest terms.

$$x \cdot 1 = 8 \cdot 25 \quad \text{Cross products}$$

$$x = 200$$

8 iPods is 4% of **200 iPods**.

Continued on Next Page

---

③ Use the six problem-solving steps to solve each problem.

(a) One day on Jacob's mail route there were 2920 pieces of mail. If 45% of those were advertising pieces, find the number of advertising pieces.

(b) There are 9750 students at the college. If 12% of them wear glasses or contact lenses, how many students wear glasses or contact lenses?

ANSWERS

3. **(a)** 1314 advertising pieces
   (b) 1170 wear glasses or contact lenses

4 Use the percent proportion to find the unknown whole.

(a) 750 Super Lotto Tickets is 25% of what number of tickets?

(b) 28 antiques is 35% of what number of antiques?

(c) 387 customers is 36% of what number of customers?

🔢 **(d)** 292.5 miles is 37.5% of what number of miles?

(b) 135 tourists is 15% of what number of tourists?
The percent is 15 and the part is 135.

$$\text{Part} \rightarrow \frac{135}{x} = \frac{15}{100} \leftarrow \text{Percent} \qquad \text{Whole (unknown)} \rightarrow \qquad \leftarrow \text{Always 100}$$

If the part and percent are given, the **whole** *must be found.*

$$\frac{135}{x} = \frac{3}{20} \qquad \text{Write } \tfrac{15}{100} \text{ as } \tfrac{3}{20} \text{ in lowest terms.}$$

$$x \cdot 3 = 135 \cdot 20 \qquad \text{Cross products}$$

$$x \cdot 3 = 2700$$

$$\frac{x \cdot \overset{1}{\cancel{3}}}{\underset{1}{\cancel{3}}} = \frac{2700}{3} \qquad \text{Divide both sides by 3.}$$

$$x = 900$$

135 tourists is 15% of **900 tourists**.

◀ *Work Problem* **4** *at the Side.*

EXAMPLE 5   **Applying the Percent Proportion**

At Newark Salt Works, 78 employees are absent because of illness. If this is 5% of the total number of employees, how many employees does the company have? Use the six problem solving steps.

Step 1   **Read** the problem. The problem asks for the total number of employees.

Step 2   **Work out a plan.** From the information in the problem, the percent is 5 and the part of the total number of employees is 78. The total number of employees or entire quantity, which is the whole, is the unknown.

Step 3   **Estimate** a reasonable answer. Round the number of employees from 78 to 80. Then, 5% is equivalent to the fraction $\tfrac{1}{20}$, and 80 is $\tfrac{1}{20}$ of the total number of employees.

$$80 \cdot 20 = 1600 \text{ employees} \leftarrow \text{Estimate}$$

Step 4   **Solve** the problem. Use the percent proportion to find the whole (the total number of employees).

$$\text{Part} \rightarrow \frac{78}{x} = \frac{5}{100} \leftarrow \text{Percent} \qquad \text{Whole (unknown)} \rightarrow \qquad \leftarrow \text{Always 100}$$

$$\frac{78}{x} = \frac{1}{20} \qquad \text{Write } \tfrac{5}{100} \text{ as } \tfrac{1}{20} \text{ in lowest terms.}$$

$$x \cdot 1 = 78 \cdot 20 \qquad \text{Cross products}$$

$$x = 1560$$

Step 5   **State the answer.** The company has **1560 employees**.

Step 6   **Check.** The exact answer, 1560 employees, is close to our estimate of 1600 employees.

Continued on Next Page

Note

To estimate the answer to Example 5 on the previous page, the 5% was changed to its fraction equivalent, $\frac{1}{20}$. Because 80 (rounded) is $\frac{1}{20}$ of the total employees, 80 was multiplied by 20 to get 1600, the estimated answer.

Work Problem *at the Side.* ▶

OBJECTIVE **3** **Find the percent using the percent proportion.** If the part and the whole are known, the percent proportion can be used to find the percent.

EXAMPLE 6 **Using the Percent Proportion to Find the Percent**

(a) 13 coupons is what percent of 52 coupons?
The whole is 52 (follows *of*) and the part is 13. Next, find the percent.

$$\frac{\text{part}}{\text{whole}} = \frac{\text{percent}}{100}$$

$$\text{Part} \rightarrow \frac{13}{52} = \frac{x}{100} \quad \begin{matrix} \leftarrow \text{Percent (unknown)} \\ \leftarrow \text{Always 100} \end{matrix}$$

Write $\frac{13}{52}$ as $\frac{1}{4}$ in lowest terms. $\frac{1}{4} = \frac{x}{100}$

Find the cross products.

$$4 \cdot x = 1 \cdot 100 \qquad \text{Cross products}$$

$$\frac{\overset{1}{\cancel{4}} \cdot x}{\underset{1}{\cancel{4}}} = \frac{100}{4} \qquad \text{Divide both sides by 4.}$$

$$x = 25$$

13 coupons is **25%** of 52 coupons.

(b) What percent of $500 is $100?
The whole is 500 (follows *of*) and the part is 100.

$$\frac{100}{500} = \frac{x}{100} \quad \longleftarrow \text{Percent (unknown)}$$

$$\frac{1}{5} = \frac{x}{100} \qquad \begin{matrix} \text{Write } \frac{100}{500} \text{ as } \frac{1}{5} \text{ in} \\ \text{lowest terms.} \end{matrix}$$

$$5 \cdot x = 1 \cdot 100 \qquad \text{Cross products}$$

$$5 \cdot x = 100$$

$$\frac{\overset{1}{\cancel{5}} \cdot x}{\underset{1}{\cancel{5}}} = \frac{100}{5} \qquad \text{Divide both sides by 5.}$$

$$x = 20$$

20% of $500 is $100.
Remember to write the % symbol in the answer.

Continued on Next Page

---

5 Use the six problem-solving steps and the percent proportion to solve each problem.

(a) A freeze resulted in a loss of 52% of an avocado crop. If the loss was 182 tons, find the total number of tons in the crop.

(b) A factory batch of cake mix contains 900 pounds of sugar, which is 18%, by weight, of the entire batch. What is the total weight of the batch?

ANSWERS

5. **(a)** 350 tons **(b)** 5000 pounds

6 Use the percent proportion to solve each problem.

(a) $21 is what percent of $105?

(b) What percent of 320 Internet companies is 48 Internet companies?

(c) What percent of 2280 court trials is 1026 trials?

(d) 432 snowboarders is what percent of 108 snowboarders?

7 Solve each problem.

(a) The bid price on an auction item is $289 while the minimum acceptable price is $425. The bid price is what percent of the minimum?

(b) A laboratory technician completes 80 tests in one day. If 52 of these tests were completed in the morning, what percent of the tests were completed in the morning?

> **CAUTION**
> When finding the percent, be sure to label your answer with the percent symbol (%).

◀ Work Problem **6** at the Side.

EXAMPLE 7 Applying the Percent Proportion

A roof is expected to last 20 years before needing replacement. If the roof is now 15 years old, what percent of the roof's life has been used?

Step 1 **Read** the problem. The problem asks for the percent of the roof's life that is already used.

Step 2 **Work out a plan.** The expected life of the roof is the entire quantity or *whole,* which is 20. The *part* of the roof's life that is already used is 15. Use the percent proportion to find the percent of the roof's life used.

Step 3 **Estimate** a reasonable answer. Since the roof is 15 years old, it is $\frac{15}{20}$ or $\frac{3}{4}$ used. Remember that $\frac{3}{4}$ is equivalent to 75%, so our estimate is 75%.

Step 4 **Solve** the problem. Let x represent the unknown percent.

$$\text{Part} \rightarrow \frac{15}{20} = \frac{x}{100} \quad \text{or} \quad \frac{3}{4} = \frac{x}{100} \qquad \text{Write } \frac{15}{20} \text{ as } \frac{3}{4} \text{ in lowest terms.}$$
$$\text{Whole} \rightarrow$$

$$4 \cdot x = 3 \cdot 100 \qquad \text{Cross products}$$
$$4 \cdot x = 300$$

$$\frac{\overset{1}{\cancel{4}} \cdot x}{\underset{1}{\cancel{4}}} = \frac{300}{4} \qquad \text{Divide both sides by 4.}$$

$$x = 75$$

Step 5 **State the answer. 75%** of the roof's life has been used.

Step 6 **Check.** The exact answer, 75%, matches our estimate of 75%.

◀ Work Problem **7** at the Side.

EXAMPLE 8 Applying the Percent Proportion

Rainfall this year was 33 inches, while normal rainfall is only 30 inches. What percent of normal rainfall is this year's rainfall?

Step 1 **Read** the problem. The problem asks us to find what percent this year's rainfall is of normal rainfall.

Step 2 **Work out a plan.** The normal rainfall is the *whole,* which is 30. This year's rainfall is *all of normal rainfall and more,* or 33 (part = 33). You need to find the percent that this year's rainfall is of normal rainfall.

Continued on Next Page

ANSWERS

6. (a) 20% **(b)** 15% **(c)** 45% **(d)** 400%
7. (a) 68% **(b)** 65%

Step 3 **Estimate** a reasonable answer. The increase in rainfall is 3 inches and the whole is 30 inches. The increase is $\frac{3}{30}$ or $\frac{1}{10}$ which is 10%. The whole is 100%, so 100% + 10% = 110%, our estimate.

Step 4 **Solve** the problem. Let x represent the unknown percent.

$$\frac{33}{30} = \frac{x}{100} \quad \text{or} \quad \frac{11}{10} = \frac{x}{100} \qquad \text{Write } \tfrac{33}{30} \text{ as } \tfrac{11}{10} \text{ in lowest terms.}$$

$$10 \cdot x = 11 \cdot 100 \qquad \text{Cross products}$$
$$10 \cdot x = 1100$$

$$\frac{\overset{1}{\cancel{10}} \cdot x}{\underset{1}{\cancel{10}}} = \frac{1100}{10} \qquad \text{Divide both sides by 10.}$$

$$x = 110$$

> If the part is *greater than* the whole, the percent is greater than 100.

Step 5 **State the answer.** This year's rainfall is **110%** of normal rainfall.

Step 6 **Check.** The exact answer, 110%, matches our estimate of 110%.

Work Problem **8** *at the Side.* ▶

8 Solve each problem.

(a) A new Toyota Prius Hybrid gets 32 miles per gallon on the highway and 48 miles per gallon around town. What percent of the highway mileage does the car get around town?

(b) The service department set a goal of 360 service calls this week. If they made 432 service calls, find the percent of their goal that they completed.

Math in the Media

EDUCATIONAL TAX INCENTIVES

The government sponsors tax incentive programs to make education more affordable. To qualify for the programs, you have to have an adjusted gross income below a certain level (most recently $47,000). You can find specific information at the Internal Revenue Service Web site.

- The Hope Scholarship offers 100% of the first $1200 spent for certain expenses, such as tuition and books, during the first year of college, plus 50% of the next $1200 incurred during the second year of college. The scholarship money is payable as a tax refund. The student cannot have completed the first two years of post-secondary education and must meet certain educational goals and workload criteria.

Suppose you are paying your own educational costs, and your adjusted gross income meets the guidelines to qualify for the Hope Scholarship. Your goals are to earn an Associate of Arts degree from a community college and then transfer to a state university to complete a Bachelor's degree. Tuition costs for resident students at North Harris Montgomery Community College District (NHMCCD) in Texas are used as an example of educational expenses.

Residents of NHMCCD pay a $12 registration fee for each semester enrolled plus tuition of $36 per semester hour. Assume that you must study a total of 15 semester hours in developmental work in mathematics, reading, and writing, and an additional 60 semester hours to complete an Associate of Arts degree. You decide to limit your course load to 15 credit hours each semester. Assume that one course is 3 semester hours, and you will have to purchase books at an approximate cost of $90 per course.

1. How many semesters and how many courses will it take you to finish the requirements for an Associate of Arts degree?

2. What is the total cost to complete the Associate of Arts degree for **(a)** books and **(b)** tuition and fees?

3. Calculate the total cost for tuition, fees, and books during the first two years (four semesters). What is the maximum tax incentive payable under the Hope Scholarship during **(a)** the first year and **(b)** the second year?

6.4 ▶▶▶ Exercises

FOR EXTRA HELP

MyMathLab

Math XL
PRACTICE

WATCH

DOWNLOAD

READ

REVIEW

Find the part using the multiplication shortcut. See Example 2.

1. 35% of 120 test tubes

2. 20% of 1800 rentals

🌐 **3.** 45% of 4080 military personnel

4. 12% of 3650 Web sites

5. 4% of 120 ft

6. 9% of $150

7. 150% of 210 files

8. 130% of 60 trees

9. 52.5% of 1560 trucks

10. 38.2% of 4250 loads

11. 2% of $164

12. 6% of $434

13. 225% of 680 tables

14. 110% of 150 apartments

15. 17.5% of 1040 cell phones

16. 46.1% of 843 kilograms

17. 0.9% of $2400

18. 0.3% of $1400

Find the whole using the percent proportion. See Example 4.

🌐 **19.** 80 e-mails is 25% of what number of e-mails?

20. 32 medical exams is 5% of what number of medical exams?

21. 30% of what number of hay bales is 48 hay bales?

22. 55% of what number of experiments is 209 experiments?

23. 495 successful students is 90% of what number of students?

24. 84 letters is 28% of what number of letters?

25. 462 mountain bikes is 140% of what number of mountain bikes?

26. 1496 graduates is 110% of what number of graduates?

27. $12\frac{1}{2}$% of what number is 350?

$\left(\textit{Hint:}\ \text{Write}\ 12\frac{1}{2}\%\ \text{as } 12.5\%.\right)$

28. $5\frac{1}{2}$% of what number is 176?

$\left(\textit{Hint:}\ \text{Write}\ 5\frac{1}{2}\%\ \text{as } 5.5\%.\right)$

Find the percent using the percent proportion. Round your answers to the nearest tenth if necessary. See Example 6.

29. 18 bean burritos is what percent of 36 bean burritos?

30. 62 hospital rooms is what percent of 248 hospital rooms?

31. 390 SUVs is what percent of 750 SUVs?

32. 650 liters is what percent of 1000 liters?

33. 32 patients is what percent of 400 patients?

34. 7 bridges is what percent of 350 bridges?

35. 54 CDs is what percent of 3600 CDs?

36. 60 cartons is what percent of 2400 cartons?

37. What percent of $344 is $64?

38. What percent of $398 is $14?

39. What percent of 250 tires is 23 tires?

40. What percent of 105 employees is 54 employees?

41. A student turned in the following answers on a test. You can tell that two of the answers are incorrect without even working the problems. Find the incorrect answers and explain how you identified them (without actually solving the problems).

50% of $84 is $42 .

150% of $30 is $20 .

25% of $16 is $32 .

100% of $217 is $217 .

42. Write a percent problem on any topic you choose. Be sure to include only two of the three components so that you can solve for the third component. Identify each component of the problem and then solve it.

Solve each application problem. Round percent answers to the nearest tenth if necessary. See Examples 3, 5, 7, and 8.

43. Aimee Toit, who works part-time, earns $240 per week and has 22% of this amount withheld for taxes, Social Security, and Medicare. Find the amount withheld.

44. An estimated 29.5% of automobile crashes are caused by driver distractions such as mobile communications devices. If there are 16,450 automobile crashes in a study, what number would be caused by driver distractions? Round to the nearest whole number. (*Source:* National Conference of State Legislatures.)

45. The guided-missile destroyer USS *Sullivans* has a 335-person crew, of which 13% are female. Find the number of female crew members. Round to the nearest whole number. (*Source:* U.S. Navy.)

46. In a survey on where to hold their weddings, 45% of 480 adults preferred a nonreligious site. How many adults said they would prefer a religious site? (*Source:* National Family Opinion Research.)

47. This year, there are 550 scholarship applicants. If 40% of the applicants will receive a scholarship, find the number of students who will receive a scholarship.

48. A U.S. Food and Drug Administration (FDA) biologist found that canned tuna is "relatively clean." Extraneous matter was found in 5% of the 1600 cans of tuna tested. How many cans of tuna contained extraneous matter?

49. There are 1,094,751 active lawyers living in the United States. If 71.4% of these lawyers are male, find **(a)** the percent of the lawyers who are female and **(b)** the number of lawyers who are female. Round to the nearest whole number. (*Source:* American Bar Association.)

50. According to the National Association of Realtors, 56% of first-time buyers make down payments of less than 5% of the purchase price. If 2.6 million first-time buyers bought homes, find **(a)** the percent who had down payments of 5% or more and **(b)** the number of buyers having down payments of less than 5% (*Source:* NAR Research.)

The bar graph below shows the percent of children 6–11 years of age who are neglecting dental hygiene. Use the information to answer Exercises 51–54.

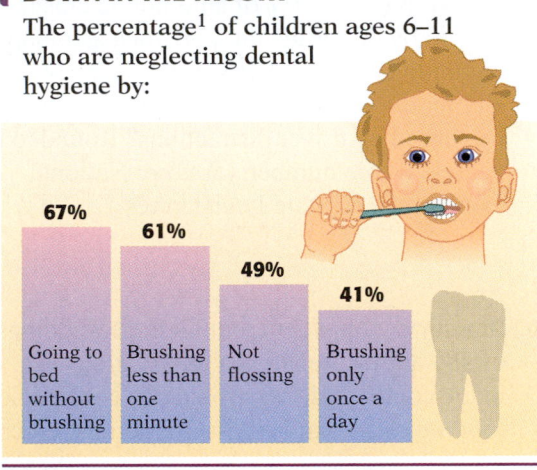

DOWN IN THE MOUTH

The percentage[1] of children ages 6–11 who are neglecting dental hygiene by:

- 67% — Going to bed without brushing
- 61% — Brushing less than one minute
- 49% — Not flossing
- 41% — Brushing only once a day

[1]Respondents allowed to choose multiple answers.
Source: Services for Crest.

51. What percent of the children brush their teeth before going to bed?

52. What percent of the children floss their teeth?

53. If 3400 children answered the questions for this survey, how many of the children brush less than one minute?

54. How many of the 3400 children in the survey brush only once a day?

55. A recent study examined 48,000 military jobs, such as Army attack helicopter pilot or Navy gunner's mate. It was found that only 960 of these jobs are filled by women. What percent of these jobs are filled by women? (*Source:* Rand's National Defense Research Institute.)

56. There are more than 55,000 words in *Webster's Dictionary,* but most educated people can identify only 20,000 of these words. What percent of the words in the dictionary can these people identify?

57. Ebony Durrant has 7.5% of her earnings deposited into her retirement plan. If $240 per month is deposited in the plan, find her monthly and yearly earnings.

58. About 61% of the 43,000,000 people who receive Social Security benefits are paid with a direct deposit to their bank. How many of the people receiving benefits are paid with a direct deposit?

The circle graph shows the percent of various ice cream brands purchased by the 1582 Americans in a recent survey. Use this information to answer Exercises 59–62.

GET THE SCOOP

Last year, Americans spent a total of $4.5 billion on ice cream. These are the brands they purchased.

22.3% Private label

37.7% Other

14.7% Breyers

10.6% Dreyers/ Edy's Grand

4.5% Ben & Jerry's

4.8% Häagen-Dazs

5.4% Blue Bell

Source: Information Resources Inc.; NPD Group.

59. Of the specific brands purchased, (not "Other" or "Private Label,") which brand was purchased most often?

60. What percent of ice cream purchases were "Private Label" or "Other" brands?

61. Find the number of people in the survey who said they purchase Häagen-Dazs. Round to the nearest whole number.

62. How many more people said they purchase Blue Bell brand than Ben & Jerry's brand? Round to the nearest whole number.

The circle graph shows the sales at fast-food hamburger chains as a percent of total fast-food hamburger sales. Use this information to answer Exercises 63–66.

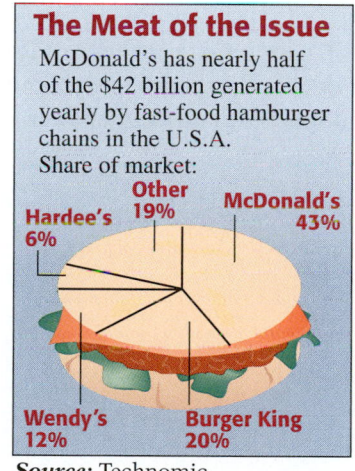

The Meat of the Issue

McDonald's has nearly half of the $42 billion generated yearly by fast-food hamburger chains in the U.S.A. Share of market:

Other 19% **McDonald's** 43%

Hardee's 6%

Wendy's 12% **Burger King** 20%

Source: Technomic.

63. Which of the hamburger chains had the lowest sales?

64. What percent of the total hamburger sales were made by the two companies having the lowest sales?

65. Find the total annual sales for McDonald's.

66. Find the total annual sales for Burger King.

67. A collection agency, specializing in collecting past-due child support, charges $25 as an application fee plus 20% of the amount collected. What is the total charge for collecting $3100 in past-due child support?

68. The income earned on an investment is 8.5% of the amount invested. If the income is $12,750, find the amount of the investment.

69. Marketing Intelligence Service says that there were 15,401 new products introduced last year. If 86% of the products introduced last year failed to reach their business objectives, find the number of products that were successful. (Round to the nearest whole number.)

70. A family of four with a monthly income of $2900 spends 90% of its earnings and saves the balance. Find **(a)** the monthly savings and **(b)** the annual savings of this family.

Relating Concepts (Exercises 71–77) For Individual or Group Work

Knowing and using the percent proportion is useful when solving percent problems.
Work Exercises 71–77 in order.

71. In the percent proportion, part is to _____ as percent is to _____ .

72. All percent problems involve a comparison between a part of something and the _____ .

Use this Ramen Noodles label of nutrition facts to answer Exercises 73–77. Read the label very carefully. Round to the nearest tenth of a percent.

Ramen Noodles

Nutrition Facts	Amount/serving	%DV*	Amount/serving	%DV*
Serving Size 1/2 Pkg. (15 oz/42.5 g)	Total Fat 8g	12%	Total Carbohydrates 27g	9%
	Saturated Fat 4g	20%	Dietary Fiber Less Than 1g	3%
Servings Per Package 2	Cholesterol 0mg	0%	Sugars Less Than 1g	
Calories 190	Sodium 670mg	28%	Protein 4g	
Calories from Fat 70				
*Percent Daily Values (DV) are based on a 2,000 calorie diet.	Vitamin A 0% • Vitamin C 0% • Calcium 2% • Iron 4%			
	Calories Per Gram Fat 9 • Carbohydrates 4 • Protein 4			

Source: Ramen Noodles package.

73. How many calories per serving are from total carbohydrates?

74. The package label shows that the 27 grams (g) of carbohydrates in one serving are 9% of the recommended Daily Value (DV). What is the recommended daily value of carbohydrates?

75. Find the number of grams of total fat that are needed to meet the recommended Daily Value (DV) of fat. Round to the nearest gram.

76. Will a person eating two packages of Ramen Noodles in one day exceed their recommended daily value of sodium? Explain your answer.

77. How many packages of Ramen Noodles must be eaten in a day to meet the recommended daily value of fiber? Would this be possible? Would this result in good nutrition? (Round to the nearest whole package.)

6.5 ▶▶▶ Using the Percent Equation

In the last section you used a proportion to solve percent problems. In this section we show another way to solve these problems by using the **percent equation.** The percent equation is just a rearrangement of the percent proportion.

OBJECTIVES

1 Use the percent equation to find the part.

2 Find the whole using the percent equation.

3 Find the percent using the percent equation.

> **Percent Equation**
>
> $$\text{part} = \text{percent} \cdot \text{whole}$$
>
> *Be sure to write the percent as a decimal before using the equation.*

When using the percent proportion, we did *not* have to write the percent as a decimal because 100 was used in the denominator of the proportion. However, because there is no 100 in the percent *equation, we must* first write the percent as a decimal by dividing by 100.

Some of the examples solved earlier will be reworked by using the percent equation. If you want to, you can look back at **Section 6.4** to see how some of these same problems were solved using proportions. This will give you a comparison of the two methods.

OBJECTIVE 1 Use the percent equation to find the part. The first example shows how to find the part.

EXAMPLE 1 Finding the Part

(a) Find 15% of $160.
Write 15% as the decimal 0.15. The whole, which comes after the word *of,* is 160. Next, use the percent equation. Let x represent the unknown part.

$$\text{part} = \text{percent} \cdot \text{whole}$$
$$x = (0.15)(160)$$

Multiply 0.15 and 160.

> 15% *must* be written as the decimal 0.15

$$x = 24$$

15% of $160 is **$24.**

(b) Find 110% of 80 cases.
Write 110% as the decimal 1.10. The whole is 80. Let x represent the unknown part.

$$\text{part} = \text{percent} \cdot \text{whole}$$
$$x = (1.10)(80)$$
$$x = 88$$

110% of 80 cases is **88 cases.**

(c) Find 0.4% of 250 patients.
Write 0.4% as the decimal 0.004. The whole is 250. Let x represent the unknown part.

$$\text{part} = \text{percent} \cdot \text{whole}$$
$$x = (0.004)(250)$$

> Write 0.4% as a decimal.
> 0.4% = 0.004

$$x = 1$$

0.4% of 250 patients is **1 patient.**

Continued on Next Page

1 Use the percent equation to find the part.

(a) 15% of 880 policyholders

(b) 23% of 840 gallons

(c) 120% of $220

(d) 135% of $1080

(e) 0.5% of 1200 fruit cups

(f) 0.25% of 1600 lab tests

To estimate the answer, think of 0.4% as approximately 0.5% or $\frac{1}{2}$ of 1%. Because 1% is $\frac{1}{100}$, 1% of 250 is

$$250 \div 100 = 2.5$$

Since 1% of 250 is 2.5 then 0.5% of 250 is 1.25 (because $2.5 \div 2 = 1.25$). So, the exact answer of 1 patient is reasonable.

◀ Work Problem **1** at the Side.

CAUTION

When using the percent equation, the percent must always be *changed to a decimal* before multiplying.

OBJECTIVE 2 Find the whole using the percent equation.   The next example shows how to use the percent equation to find the whole.

Note

When the word *of* follows a percent, it is an indicator word for *multiply*.

EXAMPLE 2 Solving for the Whole

(a) 8 tables is 4% of what number of tables?

The part is 8 and the percent is 4% or the decimal 0.04. The whole is unknown.

8   is   **4% of** what number?
                    ↑
                    └── Indicator word

Now, use the percent equation.

part = percent • whole

$$8 = (0.04)(x) \qquad \text{Let } x \text{ represent the unknown whole.}$$

$$\frac{8}{0.04} = \frac{\overset{1}{\cancel{(0.04)}}(x)}{\underset{1}{\cancel{0.04}}} \qquad \text{Divide both sides by 0.04.}$$

$$200 = x \longleftarrow \text{Whole}$$

8 tables is 4% of **200 tables**.

(b) 135 tourists is 15% of what number of tourists?

Write 15% as 0.15. The part is 135. Use the percent equation to find the whole.

part = percent • whole

$$135 = (0.15)(x) \qquad \text{Let } x \text{ represent the unknown whole.}$$

$$\frac{135}{0.15} = \frac{\overset{1}{\cancel{(0.15)}}(x)}{\underset{1}{\cancel{0.15}}} \qquad \text{Divide both sides by 0.15.}$$

$$900 = x \longleftarrow \text{Whole}$$

135 tourists is 15% of **900 tourists.**

ANSWERS

1. (a) 132 policyholders **(b)** 193.2 gallons
   (c) $264 **(d)** $1458 **(e)** 6 fruit cups
   (f) 4 lab tests

Continued on Next Page

(c) $8\frac{1}{2}\%$ of what number is 102?

Write $8\frac{1}{2}\%$ as 8.5%, or the decimal 0.085. The part is 102. Use the percent equation.

> Write **8.5%** as a decimal
> $8.5\% = 0.085$

$$\textbf{part} = \textbf{percent} \cdot \textbf{whole}$$

$$102 = (0.085)(x) \qquad \text{Let } x \text{ represent the unknown whole.}$$

$$\frac{102}{0.085} = \frac{(0.085)(x)}{0.085} \qquad \text{Divide both sides by 0.085.}$$

$$1200 = x \longleftarrow \text{Whole}$$

102 is $8\frac{1}{2}\%$ of **1200**.

Estimate the answer. Notice that $8\frac{1}{2}\%$ is close to 10%. If 102 is 10% of a number, then the number is 10 times 102, or 1020. So the exact answer, 1200, is reasonable.

CAUTION

In Example 2(c) above, $8\frac{1}{2}\%$ was first written as 8.5%, which is the decimal form of $8\frac{1}{2}\%$ $(8\frac{1}{2} = 8.5)$. **The percent sign still remained** in 8.5%. Then 8.5% was changed to the decimal 0.085 before solving the equation.

Work Problem **2** *at the Side.* ▶

OBJECTIVE 3 Find the percent using the percent equation.
The final example shows how to use the percent equation to find the percent.

EXAMPLE 3 **Finding the Percent**

(a) 13 auto mechanics is what percent of 52 auto mechanics?

Because 52 follows *of,* the whole is 52. The part is 13, and the percent is unknown. Use the percent equation.

$$\textbf{part} = \textbf{percent} \cdot \textbf{whole}$$

$$13 = x \cdot 52 \qquad \text{Let } x \text{ represent the unknown percent.}$$

$$\frac{13}{52} = \frac{x \cdot 52}{52} \qquad \text{Divide both sides by 52.}$$

$$0.25 = x$$

$$0.25 \text{ is } 25\%$$

> You *must* write the decimal answer as a percent. $0.25 = 25\%$

13 auto mechanics is **25%** of 52 auto mechanics.

The equation can also be set up using *of* as an indicator word for multiplication and *is* as an indicator word for "is equal to."

13 is what percent of 52?

$$13 = \qquad x \qquad \cdot 52$$

$$13 = x \cdot 52 \qquad \text{Same equation as above}$$

Continued on Next Page

---

2 Find the whole using the percent equation.

(a) 18 supervisors is 45% of what number of supervisors?

(b) 67.5 containers is 27% of what number of containers?

(c) 666 inoculations is 45% of what number of inoculations?

(d) $5\frac{1}{2}\%$ of what number of policies is 66 policies?

ANSWERS

2. **(a)** 40 supervisors **(b)** 250 containers
  (c) 1480 inoculations **(d)** 1200 policies

3 Find the percent using the percent equation.

(a) What percent of 35 monitors is 7 monitors?

(b) 34 post office boxes is what percent of 85 post office boxes?

(c) What percent of 920 invitations is 1288 invitations?

(d) 9 world-class runners is what percent of 1125 runners?

(b) What percent of $500 is $100?

The whole is 500 and the part is 100. Let x represent the unknown percent.

$$\textbf{part = percent} \bullet \textbf{whole}$$

$$100 = x \bullet 500 \qquad \text{Let } x \text{ represent the unknown percent.}$$

$$\frac{100}{500} = \frac{x \bullet \overset{1}{\cancel{500}}}{\underset{1}{\cancel{500}}} \qquad \text{Divide both sides by 500.}$$

$$0.20 = x$$

0.20 is 20%

> Write the decimal as a percent.
> 0.20 = 20%

20% of $500 is $100.

(c) What percent of $300 is $390?

The whole is 300 and the part is 390. Let x represent the unknown percent.

$$\textbf{part = percent} \bullet \textbf{whole}$$

$$390 = x \bullet 300 \qquad \text{Let } x \text{ represent the unknown percent.}$$

$$\frac{390}{300} = \frac{x \bullet \overset{1}{\cancel{300}}}{\underset{1}{\cancel{300}}} \qquad \text{Divide both sides by 300.}$$

$$1.3 = x$$

1.3 is 130%

> Write the decimal as a percent
> 1.3 = 1.30 = 130%

130% of $300 is $390.

(d) 6 ladders is what percent of 1200 ladders?

Since 1200 follows *of,* the whole is 1200. The part is 6.

$$\textbf{part = percent} \bullet \textbf{whole}$$

$$6 = x \bullet 1200 \qquad \text{Let } x \text{ represent the unknown percent.}$$

$$\frac{6}{1200} = \frac{x \bullet \overset{1}{\cancel{1200}}}{\underset{1}{\cancel{1200}}} \qquad \text{Divide both sides by 1200.}$$

$$0.005 = x$$

0.005 is 0.5%

> Write the decimal as a percent
> 0.005 = 0.5%

6 ladders is **0.5%** of 1200 ladders.

You can estimate the answer because 1% of 1200 ladders is found by moving the decimal point two places to the left in 1200, resulting in 12. Since 6 ladders is half of 12 ladders, our answer should be $\frac{1}{2}$ of 1% or 0.5%. Our exact answer matches the estimate.

CAUTION

When you use the percent equation to solve for an unknown percent, the answer will always be in decimal form. Notice that in Example 3(a), (b), (c), and (d) above, **the decimal answer had to be changed to a percent** by multiplying by 100 and attaching the percent sign. The answers became: (a) 0.25 = 25%; (b) 0.20 = 20%; (c) 1.3 = 130%; and (d) 0.005 = 0.5%.

ANSWERS

3. (a) 20% (b) 40% (c) 140% (d) 0.8%

◀ *Work Problem* **3** *at the Side.*

6.5   ►►►►   Exercises

Find the part using the percent equation. See Example 1.

1. 25% of 1080 blood donors

2. 19% of 700 MP3 players

3. 45% of 3000 bath towels

4. 75% of 360 dosages

5. 32% of 260 quarts

6. 44% of 430 liters

7. 140% of 2500 air bags

8. 145% of 580 hamburgers

9. 12.4% of 8300 meters

10. 26.4% of 4700 miles

11. 0.8% of $520

12. 0.3% of $480

Find the whole using the percent equation. See Example 2.

13. 24 patients is 15% of what number of patients?

14. 32 classrooms is 20% of what number of classrooms?

15. 40% of what number of salads is 130 salads?

16. 75% of what number of wrenches is 675 wrenches?

17. 476 circuits is 70% of what number of circuits?

18. 270 lab tests is 45% of what number of lab tests?

19. $12\frac{1}{2}$% of what number of people is 135 people?

20. $18\frac{1}{2}$% of what number of circuit breakers is 370 circuit breakers?

21. $1\frac{1}{4}$% of what number of gallons is 3.75 gallons?

22. $2\frac{1}{4}$% of what number of files is 9 files?

Find the percent using the percent equation. See Example 3.

23. 70 shipments is what percent of 140 shipments?

24. 180 telemarketers is what percent of 450 telemarketers?

25. 114 tuxedos is what percent of 150 tuxedos?

26. 75 PDAs is what percent of 125 PDAs?

27. What percent of $264 is $330?

28. What percent of $480 is $696?

29. What percent of 160 liters is 2.4 liters?

30. What percent of 600 meters is 7.5 meters?

31. 170 cartons is what percent of 68 cartons?

32. 612 orders is what percent of 425 orders?

33. When using the percent equation, the percent must always be changed to a decimal before doing any calculations. Show and explain how to change a fraction percent to a decimal. Use $2\frac{1}{2}\%$ in your explanation.

34. Suppose a problem on your homework assignment was, "Find $\frac{1}{2}\%$ of $1300." Your classmates got answers of $0.65, $6.50, $65, and $650. Which answer is correct? How and why are they getting all of these answers? Explain.

Solve each application problem.

35. A study of office workers found that 27% would like more storage space. If there are 14 million office workers, how many want more storage space? (*Source:* Steelcase Workplace Index.)

36. Most shampoos contain 75% to 90% water. If a 16-ounce bottle of shampoo contains 78% water, find the number of ounces of water in the bottle. Round to the nearest tenth of an ounce.

37. The household lubricant WD-40 is used in 79% of U.S. homes. If there are 115.8 million U.S. homes, find the number of homes in which WD-40 is used. Round to the nearest tenth of a million.

38. Three Spam Mobiles travel throughout the country promoting Spam and Spam Lite. By using these kitchens on wheels, the annual goal is to give 1.5 million taste samples to the public. If 58.6% of the goal has been met, find the number of samples that have been given. (*Source:* Hormel Foods Corporation.)

39. In the United States, 98% of all households have a refrigerator. Out of 18,000 households, **(a)** how many are expected to have a refrigerator and **(b)** how many are expected to not have a refrigerator? (*Source:* American Housing Authority.)

40. In the United States, 56% of the households have a dishwasher. Out of 214,500 households **(a)** how many are expected to have a dishwasher and **(b)** how many are expected to not have a dishwasher? (*Source:* American Housing Authority.)

41. In the United States, 2 of the 50 states (Alaska and Louisiana) do not have any drive-in movies. The remaining states do have drive-in movies. (*Source:* USA Today)

    (a) What percent of the states do not have drive-in movies?

    (b) What percent have drive-in movies?

42. Among the 50 companies receiving the greatest number of U.S. patents last year, 18 were Japanese companies. (*Source: Wall Street Journal.*)

    (a) What percent of the companies were Japanese companies?

    (b) What percent of the companies were not Japanese companies?

43. In a survey of 1250 Americans, 461 rated their health as excellent. What percent of these Americans rate their health as excellent? Round to the nearest tenth of a percent. (*Source:* National Health Interview Survey.)

44. General Nutrition Center now has 3200 stores and plans to add 450 more stores. Find the percent of additional stores that they have planned.

The graph shows the average time spent preparing weekday dinners. Assume that 5400 people were surveyed to gather this data. Use this graph to answer Exercises 45–48.

DINNER TIME
Average time required to prepare a weekday dinner.

16–30 minutes
39%

31–45 minutes
35%

6%
15 minutes or less

4%
61+ minutes

0.5%
No response

0.5%
Other

15%
46–60 minutes

Source: The NPD Group.

45. Find the number of people who said they spend "16–30 minutes" preparing weekday dinners.

46. What total number of people answered "No response" and "Other"?

47. How many people said they spend over 30 minutes preparing weekday dinners?

48. Find the number of people who said they spend 30 minutes or less preparing weekday dinners.

49. Chemical Banking Corporation made $338 million worth of mortgage loans to minorities last year. If this represented 18.6% of all their mortgages, find the total value of all mortgages that they made last year. Round to the nearest tenth of a million.

50. Rachel Williams has 8.5% of her monthly earnings deposited into the credit union. If this amounts to $131.75 per month, find her annual earnings.

51. The Chevy Camaro was introduced in 1967. Sales that year were 220,917 Camaros, which was 46.2% of the number of Ford Mustangs sold in the same year. Find the number of Mustangs sold in 1967. Round to the nearest whole number.

52. Chris Goodwin is a waiter and has sales of $822.25 on Saturday. If this is 28.6% of his sales for the week, find his weekly sales.

53. J & K Mustang has increased the sale of auto parts by $32\frac{1}{2}$% over last year. If the sale of parts last year amounted to $385,200, find the volume of sales this year.

54. An ad for steel-belted radial tires promises 15% better mileage. If mileage had been 25.6 miles per gallon in the past, what mileage could be expected after these tires are installed? Round to the nearest tenth of a mile.

55. A Polaris Vac-Sweep is priced at $524 with an allowed trade-in of $125 on an old unit. If sales tax of $7\frac{3}{4}$% is charged on the price of the new Polaris unit before the trade-in, find the total cost to the customer after receiving the trade-in. (*Hint:* Trade-in is subtracted last.)

56. General Motors car sales in China were 36.7% greater than last year's sales of 629,778 cars. Find this year's sales. Round to the nearest whole number. (*Source:* General Motors Corporation.)

6.6 ▶▶▶ Solving Application Problems with Percent

Percent has many applications in our daily lives. This section discusses percent as it applies to sales tax, commissions, discounts, and the percent of change (increase and decrease).

OBJECTIVE 1 Find sales tax. States, countries, and cities often collect taxes on sales to customers. The **sales tax** is a percent of the total sale. The following formula for finding sales tax is based on the percent equation.

OBJECTIVES

1 Find sales tax.

2 Find commissions.

3 Find the discount and sale price.

4 Find the percent of change.

> **Sales Tax Formula**
>
> **part** = **percent** • **whole**
> ↓ ↓ ↓
> amount of sales tax = rate of tax • cost of item

EXAMPLE 1 Solving for Sales Tax

Office Max sells a flat panel computer monitor for $299. If the sales tax rate is 5%, how much tax is paid? What is the total cost of the flat panel computer monitor? Use the six problem-solving steps.

Step 1 **Read** the problem. The problem asks for the total cost of the flat panel display including the sales tax.

Step 2 **Work out a plan.** Use the sales tax formula to find the amount of sales tax. Write the tax rate (5%) as a decimal (0.05). The cost of the item is $299. Use the letter *a* to represent the unknown *amount* of tax. Add the sales tax to the cost of the item.

Step 3 **Estimate** a reasonable answer. Round $299 to $300. Recall that 5% is equivalent to $\frac{1}{20}$, so divide $300 by 20 to estimate the tax.

$$\$300 \div 20 = \$15 \text{ tax}$$

The total estimated cost is $300 + $15 = $315. ← Estimate

Step 4 **Solve** the problem.

part = **percent** • **whole**
↓ ↓ ↓
amount of sales tax = **rate of tax** • **cost of item**

$$a = (5\%)(\$299)$$
$$a = (0.05)(\$299)$$
$$a = \mathbf{\$14.95} \quad \text{Sales tax}$$

The tax paid on the flat panel computer monitor is $14.95. The customer would pay a total cost of $299 + **$14.95** = $313.95.

Step 5 **State the answer.** The total cost of the flat panel computer monitor is $313.95.

Step 6 **Check.** The exact answer, $313.95, is close to our estimate of $315.

Work Problem **1** *at the Side.* ▶

---

1 Suppose the sales tax rate in your state is 6%. Find the amount of the tax and the total you would pay for each item.

(a) $29 Little League bat

(b) $89 Guitar Hero: Legends of Rock bundle

(c) $1287 leather chair and ottoman

(d) $19,300 pickup truck

2 Find the rate of sales tax.

(a) The tax on a $320 patio set is $25.60.

EXAMPLE 2 Finding the Sales Tax Rate

The sales tax on a $14,800 Honda Civic is $962. Find the rate of the sales tax.

Step 1 **Read** the problem. This problem asks us to find the sales tax rate.

Step 2 **Work out a plan.** Use the sales tax formula.

$$\text{sales tax} = \text{rate of tax} \cdot \text{cost of item}$$

Solve for the rate of tax, which is the percent. The cost of the Honda Civic (the whole) is $14,800, and the amount of sales tax (the part) is $962. Use r to represent the unknown *rate* of tax (the percent).

Step 3 **Estimate** a reasonable answer. Round $14,800 to $15,000 and round $962 to $1000. The sales tax is $\frac{1000}{15,000}$ or $\frac{1}{15}$ of the cost of the car. So divide 1 by 15 to estimate the percent (rate) of sales tax.

$$\frac{1}{15} = 0.06\overline{6} \approx 7\% \leftarrow \text{Rounded estimate}$$

(b) The tax on a $24 sweatshirt is $1.56.

Step 4 **Solve** the problem.

$$\text{sales tax} = \text{rate of tax} \cdot \text{cost of item}$$
$$\$962 = r \cdot \$14,800$$

$$\frac{962}{14,800} = \frac{r \cdot \overset{1}{\cancel{14,800}}}{\underset{1}{\cancel{14,800}}} \quad \text{Divide both sides by 14,800.}$$

$$0.065 = r$$
$$0.065 \text{ is } 6.5\%$$

> Write the decimal as a percent.
> $0.065 = 6.5\%$.

Step 5 **State the answer.** The sales tax rate is 6.5% or $6\frac{1}{2}$%.

Step 6 **Check.** The exact answer, $6\frac{1}{2}$%, is close to our estimate of 7%.

◀ Work Problem **2** at the Side.

(c) The tax on a $22,620 Ford Escape is $904.80.

Note

You can use the sales tax formula to find the amount of sales tax, the cost of an item, or the rate of sales tax (the percent).

OBJECTIVE 2 **Find commissions.** Many salespeople are paid by *commission* rather than an hourly wage. If you are paid by **commission,** you are paid a certain percent of your total sales dollars. The formula below for finding the commission is based on the percent equation.

Commission Formula

$$\text{part} = \text{percent} \cdot \text{whole}$$

amount of commission = rate of commission · amount of sales

ANSWERS

2. (a) 8% (b) 6.5% or $6\frac{1}{2}$% (c) 4%

EXAMPLE 3 **Determining the Amount of Commission**

Scott Samuels had pharmaceutical sales of $42,500 last month. If his commission rate is 9%, find the amount of his commission.

Step 1 **Read** the problem. The problem asks for the amount of commission that Samuels earned.

Step 2 **Work out a plan.** Use the commission formula. Write the rate of commission (9%) as a decimal (0.09). The amount of Samuels' sales ($42,500) is the whole. Use c to represent the unknown *amount* of commission.

Step 3 **Estimate** a reasonable answer. Round the commission rate of 9% to 10%. Round the amount of sales from $42,500 to $40,000. Since 10% is equivalent to $\frac{1}{10}$, divide $40,000 by 10 to estimate the amount of commission.

$$\$40,000 \div 10 = \$4000 \leftarrow \text{Estimate}$$

Step 4 **Solve** the problem.

amount of commission = rate of commission • amount of sales

$$c = (9\%)(\$42,500)$$
$$c = (0.09)(\$42,500)$$
$$c = \$3825 \quad \text{Amount of commission}$$

Step 5 **State the answer.** Samuels earned a commission of $3825 for selling the pharmaceuticals.

Step 6 **Check.** The exact answer, $3825, is close to our estimate of $4000.

Work Problem **3** *at the Side.* ▶

EXAMPLE 4 **Finding the Rate of Commission**

Chris Knudson earned a commission of $510 for selling $17,000 worth of shipping supplies. Find the rate of commission.

Step 1 **Read** the problem. In this problem, we must find the rate (percent) of commission.

Step 2 **Work out a plan.** You could use the commission formula. Another approach is to use the percent proportion. The *whole* is $17,000, the *part* is $510, and the *percent* is unknown. (The rate of commission is the percent.)

Step 3 **Estimate** a reasonable answer. Round the commission, $510, to $500, and round $17,000 to $20,000. The commission in fraction form is $\frac{\$500}{\$20,000}$, which simplifies to $\frac{1}{40}$. Changing $\frac{1}{40}$ to a percent gives $2\frac{1}{2}\%$ (rounded), as our estimate.

Step 4 **Solve** the problem.

$$\frac{\text{part}}{\text{whole}} = \frac{x}{100} \leftarrow \text{Percent (unknown)}$$

> Think: The rate of the commission is the percent.

$$\frac{510}{17,000} = \frac{x}{100}$$
$$17,000 \cdot x = 510 \cdot 100 \quad \text{Cross products.}$$

$$\frac{\overset{1}{\cancel{17,000}} \cdot x}{\underset{1}{\cancel{17,000}}} = \frac{51,000}{17,000} \quad \text{Divide both sides by 17,000.}$$

$$x = 3$$

Continued on Next Page

3 Find the amount of commission.

(a) Jill Buteo sells dental equipment at a commission rate of 12% and has sales for the month of $28,750.

(b) Last month Janis Cimperman sold a home for $220,500 for a client and earned a commission of 6%.

4 Find the rate of commission.

(a) A commission of $450 is earned on one sale of computer products worth $22,500.

(b) Jamal Story earns $2898 for selling office furniture worth $32,200.

5 Find the amount of the discount and the sale price.

(a) An Easy-Boy leather recliner originally priced at $950 is offered at a 42% discount.

(b) Wal-Mart has women's sweater sets on sale at 35% off. One sweater set was originally priced at $30.

Step 5 **State the answer.** The rate of commission is 3%.

Step 6 **Check.** The exact answer of 3% is close to our estimate of $2\frac{1}{2}$%.

◀ *Work Problem* **4** *at the Side.*

OBJECTIVE 3 Find the discount and sale price. Most of us prefer buying things when they are on sale. A store will reduce prices, or **discount,** to attract additional customers. Use the following formula to find the discount and the sale price.

> **Discount Formula and Sale Price Formula**
> amount of discount = rate (or percent) of discount • original price
> sale price = original price − amount of discount

EXAMPLE 5 Finding a Sale Price

Whitings Oak Furniture Store has a home theater cabinet with an original price of $840 on sale at 15% off. Find the sale price of the cabinet.

Step 1 **Read** the problem. This problem asks for the price of a home theater cabinet after a discount of 15%.

Step 2 **Work out a plan.** The problem is solved in two steps. First, find the amount of the discount, that is, the amount that will be "taken off" (subtracted), by multiplying the original price ($840) by the rate of the discount (15%). The second step is to subtract the amount of discount from the original price. This gives you the sale price, which is what you will actually pay for the home theater cabinet.

Step 3 **Estimate** a reasonable answer. Round the original price from $840 to $800, and the rate of discount from 15% to 20%. Since 20% is equivalent to $\frac{1}{5}$, the estimated discount is $800 ÷ 5 = $160, so the estimated sale price is $800 − $160 = $640.

Step 4 **Solve** the problem. First find the exact amount of the discount.

amount of discount = rate of discount • original price

$a = (0.15)($840)$ Write 15% as a decimal.

$a = 126 Amount of discount

Now find the sale price of the home theater cabinet by subtracting the amount of the discount ($126) from the original price.

sale price = original price − amount of discount

$= $840 − 126

$= 714 Sale price

Step 5 **State the answer.** The sale price of the home theater cabinet is $714.

Step 6 **Check.** The exact answer, $714, is close to our estimate of $640.

◀ *Work Problem* **5** *at the Side.*

ANSWERS

4. **(a)** 2% **(b)** 9%
5. **(a)** $399; $551 **(b)** $10.50; $19.50

▦ **Calculator Tip** In Example 5 on the previous page, you can use a calculator to find the amount of discount and subtract the discount from the original price, all in one step.

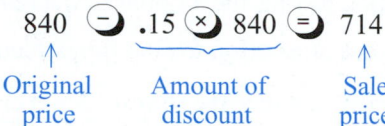

$$840 \ominus .15 \otimes 840 = 714$$

   ↑   ↑   ↑
  Original Amount of Sale
   price   discount   price

A scientific calculator observes the order of operations, so it will automatically do the multiplication before the subtraction.

OBJECTIVE 4 Find the percent of change. We are often interested in looking at increases or decreases in sales, production, population, and many other items. This type of problem involves finding the *percent of change*. Use the following steps to find the **percent of increase.**

> **Finding the Percent of Increase**
>
> **Step 1** Use subtraction to find the amount of increase.
>
> **Step 2** Use the percent proportion to find the percent of increase.
>
> $$\frac{\textbf{amount of increase (part)}}{\textbf{original value (whole)}} = \frac{\textbf{percent}}{\textbf{100}}$$

EXAMPLE 6 **Finding the Percent of Increase**

Attendance at county parks climbed from 18,300 last month to 56,730 this month. Find the percent of increase.

Step 1 **Read** the problem. The problem asks for the percent of increase.

Step 2 **Work out a plan.** Subtract the attendance last month (18,300) from the attendance this month (56,730) to find the amount of increase in attendance. Next, use the percent proportion. The whole is 18,300 (last month's original attendance), the part is 38,430 (amount of increase in attendance), and the percent is unknown.

Step 3 **Estimate** a reasonable answer. Round 18,300 to 20,000 and 56,730 to 60,000. The amount of increase is $60,000 - 20,000 = 40,000$. Since 40,000 (the increase) is *twice* as large as the original amount, the estimated percent of increase is 200%.

Step 4 **Solve** the problem.

$$56,730 - 18,300 = 38,430$$ ◀ Subtract to find the **amount of increase** in attendance

Amount of increase → $$\frac{38,430}{18,300} = \frac{x}{100}$$ Percent proportion

Use the *original* value of 18,300 (**not** 56,730).

Solve this proportion to find that $x = 210$.

Step 5 **State the answer.** The percent of increase is 210%.

Step 6 **Check.** The exact answer, 210%, is close to our estimate of 200%.

Work Problem **6** *at the Side.* ▶

---

6 Find the percent of increase.

(a) A manufacturer of snowboards increased production from 14,100 units last year to 19,035 this year.

(b) The number of flu cases rose from 496 cases last week to 620 this week.

7 Find the percent of decrease.

(a) The number of service calls fell from 380 last month to 285 this month.

Use the following steps to find the **percent of decrease.**

> **Finding the Percent of Decrease**
>
> **Step 1** Use subtraction to find the amount of decrease.
>
> **Step 2** Use the percent proportion to find the percent of decrease.
>
> $$\frac{\text{amount of decrease (part)}}{\text{original value (whole)}} = \frac{\text{percent}}{100}$$

EXAMPLE 7 Finding the Percent of Decrease

The number of production employees this week fell to 1406 people from 1480 people last week. Find the percent of decrease.

Step 1 **Read** the problem. The problem asks for the percent of decrease.

Step 2 **Work out a plan.** Subtract the number of employees this week (1406) from the number of employees last week (1480) to find the amount of decrease. Then, use the percent proportion. The whole is 1480 (last week's *original* number of employees), the part is 74 (amount of decrease in employees), and the percent is unknown.

Step 3 **Estimate** a reasonable answer. Estimate the answer by rounding 1406 to 1400 and 1480 to 1500. The decrease is $1500 - 1400 = 100$. Since 100 is $\frac{1}{15}$ of 1500, our estimate is $1 \div 15 \approx 0.07$ or 7%.

Step 4 **Solve** the problem.

(b) The number of workers applying for unemployment fell from 4850 last month to 3977 this month.

$$1480 - 1406 = 74 \quad \text{Subtract to find the \textbf{amount of decrease} in number of employees.}$$

Amount of decrease → $\dfrac{74}{1480} = \dfrac{x}{100}$ Percent proportion

Use the original value of 1480 (**not** 1406).

Solve this proportion to find that $x = 5$.

Step 5 **State the answer.** The percent of decrease is 5%.

Step 6 **Check.** The exact answer, 5%, is close to our estimate of 7%.

> **CAUTION**
> When solving for percent of increase or decrease, the *whole is always the original value* or *value before the change occurred.* The part is the change in values, that is, how much something went up or went down.

◀ *Work Problem* **7** *at the Side.*

6.6 ▷▷▷ Exercises

Find the amount of sales tax or the tax rate and the total cost (amount of sale + amount of tax = total cost). Round money answers to the nearest cent if necessary. See Examples 1 and 2.

	Amount of Sale	Tax Rate	Amount of Tax	Total Cost
1.	$6	4%	_____	_____
2.	$45	5%	_____	_____
3.	$425	_____	$12.75	_____
4.	$322	_____	$19.32	_____
5.	$284	_____	$14.20	_____
6.	$84	_____	$5.88	_____
7.	$12,229	$5\frac{1}{2}$%	_____	_____
8.	$11,789	$7\frac{1}{2}$%	_____	_____

Find the commission earned or the rate of commission. Round money answers to the nearest cent if necessary. See Examples 3 and 4.

	Sales	Rate of Commission	Commission
9.	$280	8%	_____
10.	$660	10%	_____
11.	$3000	_____	$600
12.	$7800	_____	$1170
13.	$6183.50	3%	_____
14.	$4416.70	7%	_____
15.	$73,500	9%	_____
16.	$55,800	6%	_____

Find the amount or rate of discount and the sale price after the discount. Round money answers to the nearest cent if necessary. See Example 5.

Original Price	Rate of Discount	Amount of Discount	Sale Price
17. $199.99	10%	_____	_____
18. $29.95	15%	_____	_____
19. $180	_____	$54	_____
20. $38	_____	$9.50	_____
21. $17.50	25%	_____	_____
22. $76	60%	_____	_____
23. $58.40	15%	_____	_____
24. $99.80	30%	_____	_____

25. You are trying to decide between Company A paying a 10% commission and Company B paying an 8% commission. For which company would you prefer to work? What considerations other than commission rate would be important to you?

26. Give four examples of where you might use the percent of increase or the percent of decrease in your own personal activities. Think in terms of work, school, home, hobbies, and sports.

Solve each application problem. Round money answers to the nearest cent and rates to the nearest tenth of a percent if necessary. See Examples 1–7.

Country Store has a unique selection of merchandise that it sells by mail and over the Internet. Use the shipping and insurance delivery chart at the right and a sales tax rate of 5% to solve Exercises 27–30. There is no sales tax on shipping and insurance. (*Source:* Country Store Catalog.)

SHIPPING AND INSURANCE DELIVERY CHART

Up to $15.00	add $4.99
$15.01 to $25.00	add $6.99
$25.01 to $35.00	add $7.99
$35.01 to $50.00	add $8.99
$50.01 to $70.00	add $10.99
$70.01 to $99.99	add $12.99
$100.00 or more	add $14.99

27. Find the total cost of six Small Fry Handi-Pan electric skillets priced at $29.99 each.

28. A customer ordered five sets of flour-sack towels priced at $12.99 per set. What is the total cost?

29. Find the total cost of three pop-up hampers at $9.99 each and four nonstick mini doughnut pans at $10.99 each.

30. What is the total cost of five coach lamp bird feeders at $19.99 each and six garden weather centers at $14.99 each?

31. An Anderson wood-frame French door is priced at $1980 with a sales tax of $99. Find the rate of sales tax.

32. Textbooks for three classes cost $245 plus tax of $17.15. Find the sales tax rate.

33. Today there are 635,000 women motorcyclists in the United States, up from 467,400 just eight years ago. Find the percent of increase in the number of women motorcyclists. (*Source:* Motorcycle Industry Council.)

34. Americans are eating more fish. This year the average American will eat 15.5 pounds compared to only 12.5 pounds per year a decade ago. Find the percent of increase. (*Source: Consumer Reports.*)

35. The number of industrial accidents this month fell to 989 accidents from 1276 accidents last month. Find the percent of decrease.

36. The average number of hours worked in manufacturing jobs last week fell from 41.1 to 40.9. Find the percent of decrease.

37. A "60% off sale" begins today at Wanda's Women's Wear. What is the sale price of women's wool coats normally priced at $335?

38. What is the sale price of a $769 Kenmore washer/dryer set with a discount of 25%?

A weekly sales report for the top four sales people at Active Sports is shown below. Use this information to answer Exercises 39–42.

Employee	Sales	Rate of Commission	Commission
Strong, A.	$18,960	3%	_____
Ferns, K.	$21,460	3%	_____
Keyes, B.	$17,680	_____	$707.20
Vargas, K.	$23,104	_____	$1152.20

39. Find the commission for Strong.

40. Find the commission for Ferns.

41. What is the rate of commission for Keyes?

42. What is the rate of commission for Vargas?

43. A Sony Micro Hi Fi Component System was priced at $390 and is on sale at 22% off. Find the discount and the sale price.

44. A Honda Pilot is offered at 12% off the manufacturer's suggested retail price. Find the discount and the sale price of this SUV, originally priced at $32,500.

45. The price per share of Toys Я Us stock fell from $35.50 to $33.50. Find the percent of decrease in price.

46. In the past five years, the cost of generating electricity from the sun has been brought down from 24 cents per kilowatt hour to 8 cents (less than the newest nuclear power plants.) Find the percent of decrease.

47. College students are offered a 6% discount on a dictionary that sells for $18.50. If the sales tax is 6%, find the cost of the dictionary including the sales tax.

48. A fax machine priced at $398 is marked down 7% to promote the new model. If the sales tax is also 7%, find the cost of the fax machine including sales tax.

49. A real estate agent sells a condominium for $129,605. A sales commission of 6% is charged. The agent gets 55% of this commission. How much money does the agent get?

50. The local real estate agents' association collects a fee of 2% on all money received by its members. The members charge 6% of the selling price of a property as their fee. How much does the association get, if its members sell property worth a total of $8,680,000?

51. What is the total price of a ski boat with an original price of $15,321, if it is sold at a 15% discount? The sales tax rate is $7\frac{3}{4}$%.

52. A commercial security alarm system originally priced at $10,800 is discounted 22%. Find the total price of the system if the sales tax rate is $7\frac{1}{4}$%.

Relating Concepts (Exercises 53–58) For Individual or Group Work

Knowing how to use the percent equation is important when solving application problems involving sales tax. **Work Exercises 53–58 in order.**

53. The percent equation is

part = _____ • _____

54. The formula used to find sales tax is an application of the percent equation. The sales tax formula is

sales tax = _____ • _____.

*In the United States there are certain items on which an excise tax is charged in addition to a sales tax. A table of federal excise taxes is shown here. Use this table to answer Exercises 55–58. (Excise tax is calculated on the amount of the sale **before** sales tax is added.) Round answers to the nearest cent.*

FEDERAL EXCISE TAXES

Product or Service	Rate	Product or Service	Rate
Telephone service	3%	Tires (by weight)	
Teletypewriter service	3%	Under 40 pounds	No tax
Air transportation	7.5%	40–69 pounds	15¢/pound over 40 pounds
International air travel	$13.40/person	70–89 pounds	$4.50 plus 30¢/pound over 70 pounds
Air freight	6.25%	90 pounds and more	$10.50 plus 50¢/pound over 90 pounds
		Truck and trailer, chassis and bodies	12%
Fishing rods	10%	Inland waterways fuel	24.4¢/gallon
Bows and arrows	12.4%	Ship passenger tax	$3/passenger
Gasoline	18.4¢ gallon	Vaccines	75¢/dose
Diesel fuel	24.4¢ gallon		
Aviation fuel	21.9¢ gallon		

Source: Publication 510, I.R.S., Excise Taxes.

55. Some archery equipment (bows and arrows) is priced at $123. Use the federal excise tax table and a sales tax rate of $6\frac{1}{2}$% to find the cost of the equipment, including both taxes. (Round to the nearest cent.)

56. Refer to Exercise 55. Calculate the two taxes separately and then add them together. Now, add the two tax rates together and then find the tax. Are your answers the same? Why or why not? (*Hint:* Recall the commutative and associative properties of multiplication.)

57. The price of an international airline ticket is $1248. Use the federal excise tax table and a sales tax rate of $7\frac{3}{4}$% to find the total cost of one ticket. (Sales tax is not charged on the $13.40 federal excise tax.)

58. Refer to Exercise 57. Can the federal excise tax be added to the sales tax rate to find the total tax? Why or why not?

6.7 ▷▷▷ Simple Interest

When we open a savings account, we are actually lending money to the bank or credit union. It will in turn lend this money to individuals and businesses. These people then become borrowers. The bank or credit union pays a fee to the savings account holders and charges a higher fee to its borrowers. These fees are called *interest.*

Interest is a fee paid or a charge made for lending or borrowing money. The amount of money borrowed is called the **principal.** The charge for interest is often given as a percent, called the interest rate or **rate of interest.** The rate of interest is assumed to be *per year,* unless stated otherwise. Time is always expressed in years or fractions of a year.

OBJECTIVE 1 Find the simple interest on a loan. In most cases, interest on a loan is computed on the *original principal* and is called **simple interest.** We use the following **interest formula** to find simple interest.

> **Formula for Simple Interest**
>
> $$\text{Interest} = \text{principal} \cdot \text{rate} \cdot \text{time}$$
>
> The formula is usually written using letters.
>
> $$I = p \cdot r \cdot t$$

> **Note**
>
> Simple interest is used for most short-term business loans, most real estate loans, and many automobile and consumer loans.

EXAMPLE 1   Finding Interest for a Year

Find the interest on $5000 at 6% for 1 year.

The amount borrowed, or principal (p), is $5000. The interest rate (r) is 6%, which is 0.06 as a decimal, and the time of the loan (t) is 1 year. Use the formula.

$$I = \quad p \cdot r \cdot t$$
$$I = (5000)(0.06)(1) \quad \text{←} \quad \boxed{\text{Notice that "1" is used as the time for 1 year.}}$$
$$I = 300$$

The interest is $300.

Work Problem **1** *at the Side.* ▶

EXAMPLE 2   Finding Interest for More Than a Year

Find the interest on $4200 at 8% for three and a half years.

The principal (p) is $4200. The rate ($r$) is 8%, or 0.08 as a decimal, and the time (t) is $3\frac{1}{2}$ or 3.5 years. Use the formula.

$$I = \quad p \cdot r \cdot t$$
$$I = (4200)(0.08)(3.5) \quad \text{←} \quad \boxed{\text{3.5 years is equivalent to } 3\frac{1}{2} \text{ years because 3.5 is } 3\frac{5}{10} \text{ which simplifies to } 3\frac{1}{2}.}$$
$$I = 1176$$

The interest is $1176.

Work Problem **2** *at the Side.* ▶

OBJECTIVES

1 Find the simple interest on a loan.

2 Find the total amount due on a loan.

1 Find the interest.

(a) $1000 at 5% for 1 year

(b) $3650 at 2% for 1 year

2 Find the interest.

(a) $820 at 6% for $3\frac{1}{2}$ years

(b) $4850 at 8% for $2\frac{1}{2}$ years

(c) $16,800 at 3% for $2\frac{3}{4}$ years

ANSWERS

1. **(a)** $50 **(b)** $73
2. **(a)** $172.20 **(b)** $970 **(c)** $1386

③ Find the interest.

  (a) $1800 at 3% for 4 months

  (b) $28,000 at $9\frac{1}{2}$% for 3 months

④ Find the total amount due on each loan.

  (a) $3800 at $6\frac{1}{2}$% for 6 months

  (b) $12,400 at 5% for 5 years

  (c) $2400 at 11% for $2\frac{3}{4}$ years

Interest rates are given **per year.** For loan periods of less than one year, be careful to express time as a fraction of a year.

If time is given in months, for example, use a denominator of 12, because there are 12 months in a year. A loan for 9 months would be for $\frac{9}{12}$ of a year.

EXAMPLE 3   **Finding Interest for Less Than 1 Year**

Find the interest on $840 at $8\frac{1}{2}$% for 9 months.

The principal is $840. The rate is $8\frac{1}{2}$% or 0.085, and the time is $\frac{9}{12}$ of a year. Use the formula $I = p \cdot r \cdot t$.

$$I = \underbrace{(840)(0.085)}\left(\frac{9}{12}\right)$$

       9 months = $\frac{9}{12}$ of a year

$$= 71.4 \left(\frac{3}{4}\right)$$

       Write $\frac{9}{12}$ in lowest terms as $\frac{3}{4}$.

$$= \frac{(71.4)(3)}{4}$$

$$= \frac{214.2}{4} = 53.55$$

The interest is $53.55.

> ▦ **Calculator Tip**   The calculator solution to Example 3 above uses chain calculations.
>
> 840 ⊗ .085 ⊗ 9 ⊘ 12 ⊜ **53.55**

◀ Work Problem **③** at the Side.

OBJECTIVE **2** **Find the total amount due on a loan.**   When a loan is repaid, the interest is added to the original principal to find the total amount due.

> **Formula for Amount Due**
>
> amount due = principal + interest

EXAMPLE 4   **Calculating the Total Amount Due**

A loan of $3240 was made at 12% for 3 months. Find the total amount due.

First find the interest. Then add the principal and the interest to find the total amount due.

$$I = (3240)(0.12)\left(\frac{3}{12}\right)$$

  3 months = $\frac{3}{12}$ of a year.

$$I = \$97.20$$

The interest is $97.20.

      Remember that the *total amount due* is the amount of the loan plus the interest.

amount due = principal + interest

     = $3240 + $97.20 = $3337.20

The total amount due is $3337.20

◀ Work Problem **④** at the Side.

6.7 ▶▶▶ Exercises

Find the interest. See Examples 1 and 2.

	Principal	Rate	Time in Years	Interest
1.	$100	6%	1	_____
2.	$200	3%	1	_____
3.	$700	5%	3	_____
4.	$900	2%	4	_____
5.	$240	4%	3	_____
6.	$190	3%	2	_____
7.	$2300	$8\frac{1}{2}\%$	$2\frac{1}{2}$	_____
8.	$4700	$5\frac{1}{2}\%$	$1\frac{1}{2}$	_____
9.	$10,800	$7\frac{1}{2}\%$	$2\frac{3}{4}$	_____
10.	$12,400	$6\frac{1}{2}\%$	$3\frac{3}{4}$	_____

Find the interest. Round to the nearest cent if necessary. See Example 3.

	Principal	Rate	Time in Months	Interest
11.	$400	5%	6	_____
12.	$600	2%	5	_____
13.	$820	6%	12	_____

Principal	Rate	Time in Months	Interest
14. $780	8%	24	_____
15. $940	3%	18	_____
16. $178	4%	12	_____
17. $1225	$5\frac{1}{2}\%$	3	_____
18. $2660	$7\frac{1}{2}\%$	3	_____
19. $15,300	$7\frac{1}{4}\%$	7	_____
20. $13,700	$3\frac{3}{4}\%$	11	_____

Find the total amount due on the following loans. Round to the nearest cent if necessary. See Example 4.

Principal	Rate	Time	Total Amount Due
21. $200	5%	1 year	_____
22. $400	2%	6 months	_____
23. $740	6%	9 months	_____
24. $1180	3%	2 years	_____

	Principal	Rate	Time	Total Amount Due
25.	$1800	9%	18 months	_____
26.	$9000	6%	7 months	_____
27.	$3250	10%	6 months	_____
28.	$7600	5%	1 year	_____
29.	$16,850	$7\frac{1}{2}\%$	9 months	_____
30.	$19,450	$5\frac{1}{2}\%$	6 months	_____

31. The amount of interest paid on savings accounts and charged on loans can vary from one institution to another. However, when the amount of interest is calculated, three factors are used in the calculation. Name these three factors and describe them in your own words.

32. Interest rates are usually given as a rate per year (annual rate). Explain what must be done when time is given in months. Write your own problem where time is given in months and then show how to solve it.

Solve each application problem. Round to the nearest cent if necessary.

33. Reann Chang deposits $825 at 5% for 1 year. How much interest will she earn?

34. The Jidobu family invests $18,000 at 9% for 6 months. What amount of interest will the family earn?

35. CITI Bank loans $150,000 to Estelle Class to expand the size of her jewelry store. If the loan is for 30 months at 7%, how much interest will the bank earn?

36. Esther Albert, a professional dancer, deposits $68,000 of her earnings at 5% for 5 years. How much interest will she earn?

37. A student borrows $1200 at 8% for 5 months to pay for books and tuition. Find the total amount due.

38. Sarah Brynski borrows $2750 from her dad for a used car. The loan will be paid back with 8% interest at the end of 9 months. Find the total amount due.

39. Nicholas Thomas deposits $14,800 in his school credit union account for 10 months. If the credit union pays $2\frac{1}{4}$% interest, find the amount of interest he will earn.

40. Sid and Shirley Kordell, owners of the Nut House, borrow $54,000 to update their store. If the loan is for 42 months at $7\frac{1}{4}$%, find the amount of interest they will owe.

41. An investment fund pays $7\frac{1}{4}$% interest. If Beverly Habecker deposits $8800 in her account for $\frac{1}{4}$ year, find the amount of interest she will earn.

42. Pat Carper owes $1900 in taxes. She is charged a penalty of $12\frac{1}{4}$% annual interest and pays the taxes and penalty after 6 months. Find the total amount she must pay.

43. A gift shop owner invests his profits of $11,500 at $8\frac{3}{4}$% interest for $\frac{3}{4}$ year. Find the total amount in his account at the end of this time.

44. A pawn shop owner lends $35,400 to another business for $\frac{1}{2}$ year at an interest rate of 14.9%. How much interest will be earned on the loan?

45. The owners of Clear Lake Marina bought six canoes to use as rentals at a cost of $550 per canoe. If they borrowed 60% of the total cost for 9 months at $12\frac{1}{2}$% interest, find the amount due.

46. The owners of Baily and Daughters Excavating purchased four earth movers at a cost of $485,000 each. If they borrowed 80% of the total purchase price for $2\frac{1}{2}$ years at $10\frac{1}{2}$% interest, find the total amount due.

6.8 ▶▶▶ Compound Interest

The interest we studied in **Section 6.7** was *simple interest* (interest only on the original principal). A common type of interest used with savings accounts and most investments is **compound interest** or interest paid on past interest as well as on the principal.

OBJECTIVE 1 Understand compound interest.   Suppose that you make a single deposit of $1000 in a savings account that earns 5% per year. What will happen to your savings over 3 years? At the end of the first year, 1 year's interest on the original deposit is found. Use the simple interest formula.

> Interest = principal • rate • time

Year 1 ($1000)(0.05)(1) = **$50**

Add the interest to the $1000 to find the amount in your account at the end of the first year. $1000 + **$50** = **$1050**

> In year 1, interest is calculated on principal.

The interest for the second year is found on $1050; that is, the interest is **compounded.**

Year 2 (**$1050**)(0.05)(1) = **$52.50**

Add this interest to the $1050 to find the amount in your account at the end of the second year. **$1050** + **$52.50** = **$1102.50**. The interest for the third year is found on $1102.50.

> In year 2 and thereafter, interest is calculated on the principal and all past interest.

Year 3 (**$1102.50**)(0.05)(1) ≈ **$55.13**

Add this interest to the **$1102.50.** So, **$1102.50** + **$55.13** = **$1157.63**

At the end of 3 years, you will have **$1157.63** in your savings account. The $1157.63 that you have in your account is called the **compound amount.**

If you had earned only *simple* interest for 3 years, your interest would be as follows.

$$I = (\$1000)(0.05)(3)$$
$$= \$150 \leftarrow \text{Simple interest}$$

At the end of 3 years, you would have $1000 + $150 = $1150 in your account. Compounding the interest increased your earnings by $7.63 because $1157.63 − $1150 = $7.63.

With *compound* interest, the interest earned during the second year is greater than that earned during the first year, and the interest earned during the third year is greater than that earned during the second year. This happens because the interest earned each year is *added* to the principal, and the new total is used to find the amount of interest in the next year.

> **Compound Interest**
> Interest paid on principal plus past interest is called **compound interest.**

OBJECTIVES
1 Understand compound interest.
2 Understand compound amount.
3 Find the compound amount.
4 Use a compound interest table.
5 Find the compound amount and the amount of compound interest.

1 Find the compound amount given the following deposits. Round to the nearest cent if necessary.

(a) $500 at 4% for 2 years

(b) $2000 at 7% for 3 years

OBJECTIVE 2 Understand compound amount. Find the compound amount as shown below.

EXAMPLE 1 Finding the Compound Amount

Nancy Wegener deposits $3400 in an account that pays 6% interest compounded annually for 4 years. Find the compound amount. Round to the nearest cent when necessary.

Year	Interest	Compound Amount
1	($3400)(0.06)(1) = **$204** $3400 + **$204** =	**$3604**
2	($3604)(0.06)(1) = **$216.24** $3604 + **$216.24** =	**$3820.24**
3	($3820.24)(0.06)(1) ≈ **$229.21** $3820.24 + **$229.21** =	**$4049.45**
4	($4049.45)(0.06)(1) ≈ **$242.97** $4049.45 + **$242.97** =	**$4292.42**

The compound amount is $4292.42.

◀ *Work Problem* **1** *at the Side.*

OBJECTIVE 3 Find the compound amount. A more efficient way of finding the compound amount is to add the interest rate to 100% and then multiply by the original deposit. Notice that in Example 1 above, at the end of the first year, you will have $3400 (100% of the original deposit) plus 6% (of the original deposit) or 106% (because 100% + 6% = 106%).

2 Find the compound amount by multiplying the original deposit by 100% plus the compound interest rate. Round to the nearest cent if necessary.

(a) $1800 at 2% for 3 years

$(1800)(1.02)(1.02)(1.02) =$

_____

(b) $900 at 3% for 2 years

EXAMPLE 2 Finding the Compound Amount

Find the compound amount in Example 1 using multiplication.

Year 1 Year 2 Year 3 Year 4
$(\$3400)(1.06)(1.06)(1.06)(1.06) \approx \$4292.42$

Original deposit   100% + 6% = 106% = 1.06   Compound amount

This method works well with a calculator.

Our answer, $4292.42, is the same as in Example 1 above.

◀ *Work Problem* **2** *at the Side.*

Note

By adding the compound interest rate to 100%, we can then multiply by the original deposit. This will give us the compound amount at the end of each compound interest period.

(c) $2500 at 5% for 4 years

Calculator Tip If you use a calculator for Example 2 above, you can use the y^x key (exponent key).

3400 ⊗ 1.06 y^x 4 ⊜ **4292.42 (rounded)**

The 4 following the y^x key represents the number of compound interest periods.

OBJECTIVE 4 Use a compound interest table. The calculation of compound interest can be quite tedious. For this reason, compound interest tables have been developed.

Suppose you deposit $1 in a savings account today that earns 4% compounded annually and you allow the deposit to remain for 3 years. The diagram below shows the compound amount at the end of each of the 3 years.

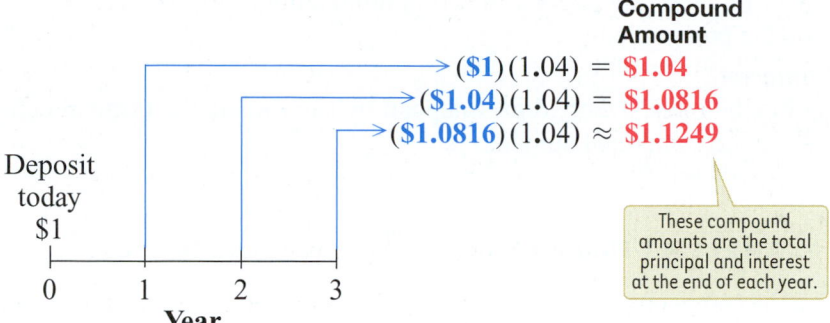

Compound
Amount
($1)(1.04) = **$1.04**
($1.04)(1.04) = **$1.0816**
($1.0816)(1.04) ≈ **$1.1249**

These compound amounts are the total principal and interest at the end of each year.

Deposit today $1

| 0 | 1 | 2 | 3 |
Year

Using the compound amounts for $1, a table can be formed. Look at the table below and find the column headed 4%. The first three numbers for years 1, 2, and 3 are the same as those we have calculated for $1 at 4% for 3 years. This table, giving the compound amounts on a $1 deposit for given lengths of time and interest rates, can be used for finding the compound amount on any amount of deposit.

COMPOUND INTEREST

Time Periods	3.00%	3.50%	4.00%	4.50%	5.00%	5.50%	6.00%	8.00%	Time Periods
1	1.0300	1.0350	1.0400	1.0450	1.0500	1.0550	1.0600	1.0800	1
2	1.0609	1.0712	1.0816	1.0920	1.1025	1.1130	1.1236	1.1664	2
3	1.0927	1.1087	1.1249	1.1412	1.1576	1.1742	1.1910	1.2597	3
4	1.1255	1.1475	1.1699	1.1925	1.2155	1.2388	1.2625	1.3605	4
5	1.1593	1.1877	1.2167	1.2462	1.2763	1.3070	1.3382	1.4693	5
6	1.1941	**1.2293**	1.2653	1.3023	1.3401	1.3788	1.4185	1.5869	6
7	1.2299	1.2723	1.3159	1.3609	1.4071	1.4547	1.5036	1.7138	7
8	1.2668	1.3168	1.3686	1.4221	1.4775	1.5347	1.5938	1.8509	8
9	1.3048	1.3629	1.4233	1.4861	1.5513	1.6191	1.6895	1.9990	9
10	1.3439	1.4106	1.4802	1.5530	**1.6289**	1.7081	1.7908	2.1589	10
11	1.3842	1.4600	1.5395	1.6229	1.7103	1.8021	1.8983	2.3316	11
12	1.4258	1.5111	1.6010	1.6959	1.7959	1.9012	2.0122	2.5182	12

EXAMPLE 3 Using a Compound Interest Table

Find each compound amount using the compound interest table. Round answers to the nearest cent.

(a) $1 is deposited at a 5% interest rate for 10 years.

Look down the column headed 5%, and across to row 10 (because 10 years = 10 time periods). At the intersection of the column and row, read the compound amount, **1.6289**, which can be rounded to $1.63.

(b) $1 is deposited at $3\frac{1}{2}$% for 6 years.

The intersection of the $3\frac{1}{2}$% (3.50%) column and row 6 shows **1.2293** as the compound amount. Round this to $1.23.

Work Problem **3** *at the Side.* ▶

---

3 Find the compound amount using the compound interest table. Round to the nearest cent.

(a) $1 at 3% for 6 years

(b) $1 at 6% for 8 years

(c) $1 at $4\frac{1}{2}$% for 12 years

ANSWERS

3. (a) $1.19 (rounded) **(b)** $1.59 (rounded)
(c) $1.70 (rounded)

4 Use the compound interest table to find the compound amount and the interest.

(a) $4000 at 3% for 10 years

(b) $12,600 at $3\frac{1}{2}$% for 8 years

(c) $32,700 at 6% for 12 years

OBJECTIVE **5** **Find the compound amount and the amount of compound interest.** Find the compound amount and interest as follows.

Finding the Compound Amount and the Interest

Compound Amount
Multiply the principal by the compound amount for $1 (from the table on the previous page).

Interest
Find the interest earned on a deposit by subtracting the original deposit from the compound amount.

EXAMPLE 4 **Finding Compound Amount and Interest**

Use the compound interest table to find the compound amount and the interest.

(a) $1000 at $5\frac{1}{2}$% interest for 12 years.

Look in the table on the previous page for $5\frac{1}{2}$% (5.50%) and 12 periods to find the number **1.9012** but do *not* round it. Multiply this number and the principal of $1000.

> Never round the numbers found in the table.

$$(\$1000)(1.9012) = \mathbf{\$1901.20}$$

The account will contain $1901.20 after 12 years.

Find the interest by subtracting the original deposit from the compound amount.

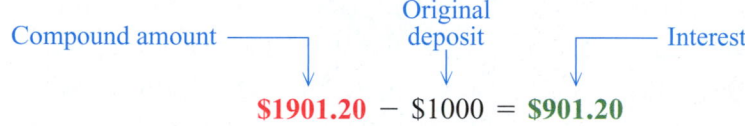

Compound amount ——   Original deposit   —— Interest

$$\mathbf{\$1901.20} - \$1000 = \mathbf{\$901.20}$$

(b) $6400 at 8% for 7 years

Look in the table for 8% and 7 periods to find **1.7138**. Multiply.

$$(\$6400)(1.7138) = \mathbf{\$10,968.32} \quad \text{Compound amount}$$

Subtract the original deposit from the compound amount.

$$\mathbf{\$10,968.32} - \$6400 = \mathbf{\$4568.32} \quad \text{Interest}$$

> Remember:
> Compound amount −
> Principal = Interest.

A total of $4568.32 in interest was earned.

◀ *Work Problem* **4** *at the Side.*

6.8 ▷▷▷ Exercises

Find the compound amount given the following deposits. Calculate the interest each year, then add it to the previous year's amount. See Example 1.

1. $500 at 4% for 2 years

2. $1500 at 5% for 3 years

3. $1800 at 3% for 3 years

4. $2000 at 8% for 3 years

5. $3500 at 7% for 4 years

6. $5500 at 6% for 4 years

Find each compound amount by multiplying the original deposit by 100% plus the compound rate given in the following. See Example 2. Round answers to the nearest cent if necessary.

7. $1000 at 5% for 2 years

8. $500 at 4% for 3 years

9. $1400 at 6% for 5 years

10. $2500 at 3% for 4 years

11. $1180 at 7% for 8 years

12. $12,800 at 6% for 7 years

13. $10,940 at 8% for 6 years

14. $15,710 at 10% for 8 years

Use the table on page 439 to find the compound amount and the interest. Interest is compounded annually. Round answers to the nearest cent if necessary. See Examples 3 and 4.

15. $1000 at 4% for 5 years

16. $10,000 at 3% for 4 years

17. $8000 at 6% for 10 years

18. $7800 at 5% for 8 years

19. $8428.17 at $4\frac{1}{2}$% for 6 years

20. $10,472.88 at $5\frac{1}{2}$% for 12 years

21. Write a definition for compound interest. Describe in your own words what compound interest means to you.

22. What is the difference between the compound amount and compound interest?

Use the table on page 439 to solve each application problem. Round answers to the nearest cent if necessary. See Examples 3 and 4.

23. Jane Chavez deposited $8450 in an account that pays 6% interest compounded annually. Find the amount she will have (compound amount) at the end of 8 years.

24. Yen Lee borrowed $32,800 from her family to open a small restaurant named Shanghai Winds. Her plan is to repay the loan at the end of 4 years at $3\frac{1}{2}$% interest compounded annually. Find the total amount that she must repay.

25. Al Granard lends $76,000 to the owner of Rick's Limousine Service. He will be repaid at the end of 9 years at 6% interest compounded annually. Find **(a)** the total amount that he should be repaid and **(b)** the amount of interest earned.

26. Vincent and Shannon Zagorin have $48,500 in profit from the sale of their first home. They have invested all of it at 8% interest compounded annually for 6 years. Find **(a)** the total amount they will have at the end of 6 years and **(b)** the amount of interest earned.

27. Jennifer Barrister deposits $30,000 at 6% interest compounded annually. Two years after she makes the first deposit, she deposits another $40,000, also at 6% compounded annually.

  (a) What total amount will she have five years after her first deposit?

  (b) What amount of interest will she have earned?

28. Kara Ivee invests $25,000 at 8% interest compounded annually. Three years after she makes the first deposit, she deposits another $25,000, also at 8% compounded annually.

  (a) What total amount will she have five years after her first deposit?

  (b) What amount of interest will she have earned?

Relating Concepts (Exercises 29–34) For Individual or Group Work

Knowing how to solve interest problems is important to businesspeople and consumers alike. **Work Exercises 29–34 in order.**

29. Simple interest calculation is used for most short-term business loans, most real estate loans, and many automobile and consumer loans. The formula for simple interest is

Interest = _____ • _____ • _____

or I = _____ • _____ • _____ .

30. When a loan is repaid, the interest is added to the original principal. The formula for amount due is

amount due = _____ + _____ .

31. Compound interest is paid on most savings accounts and many other types of investments. Compound interest is interest calculated on _____ plus past _____ .

32. The compound amount is the total amount in an account at the end of a period of time. Compound amount is the original _____ + compound _____ .

Use the table on page 439 to answer Exercises 33 and 34.

33. Donna Gonsalves has two choices. She can invest $4350 at 6% simple interest for 6 years, or she can invest the same amount at 6% interest compounded annually for 6 years.

  (a) Find the difference in the amount of interest earned in these two accounts.

  (b) If the length of time is doubled from 6 to 12 years, will the difference in the interest earned also double?

  (c) Use your own example to determine that what you found in part (b) is true with a different interest rate and length of time.

34. One account is opened with $20,000 at 8% simple interest for 12 years. Another account is opened with $16,000 at 8% interest compounded annually for 12 years.

  (a) At the end of the 12 years, which account has the higher balance and by how much?

  (b) What does this tell you about compound interest? Why?

Chapter 6 ▶▶▶ Summary

▶ Key Terms

6.1	**percent**	Percent means "per one hundred." A percent is a ratio with a denominator of 100.
6.3	**percent proportion**	The proportion $\dfrac{\text{part}}{\text{whole}} = \dfrac{\text{percent}}{100}$ is used to solve percent problems.
	whole	The whole in a percent problem is the entire quantity, the total, or the base.
	part	The part in a percent problem is the portion being compared with the whole.
6.5	**percent equation**	The percent equation is: part = percent • whole. It is another way to solve percent problems.
6.6	**sales tax**	Sales tax is a percent of the total sales charged as a tax.
	commission	Commission is a percent of the dollar value of total sales paid to a salesperson.
	discount	Discount is often expressed as a percent of the original price; it is then deducted from the original price, resulting in the sale price.
	percent of increase or decrease	Percent of increase or decrease is the amount of change (increase or decrease) expressed as a percent of the original amount.
6.7	**interest**	Interest is a fee paid or a charge made for lending or borrowing money
	interest formula	The interest formula is used to calculate interest. It is: Interest = principal • rate • time or $I = p \cdot r \cdot t$.
	simple interest	Interest that is computed only on the original principal is simple interest.
	principal	Principal is the amount of money on which interest is earned.
	rate of interest	Often referred to as "rate," it is the charge for interest and is given as a percent.
6.8	**compound interest**	Compound interest is interest paid on both the past interest and on the principal.
	compound amount	Compound amount is the total amount in an account including compound interest and the original principal.
	compounding	Interest that is compounded once each year is compounded annually.

▶ New Symbols

%   percent (per one hundred)

▶ New Formulas

To write percents as decimals: $p\% = p \div 100$ **To write percents as fractions:** $p\% = \dfrac{p}{100}$

Percent proportion: $\dfrac{\text{part}}{\text{whole}} = \dfrac{\text{percent}}{100}$ **Percent equation:** part = percent • whole

Amount of sales tax: amount of sales tax = rate of tax • cost of item

Amount of commission: amount of commission = rate of commission • amount of sales

Amount of discount: amount of discount = rate of discount • original price

Sale price: sale price = original price − amount of discount

Percent of increase: $\dfrac{\text{amount of increase}}{\text{original value}} = \dfrac{\text{percent}}{100}$

Percent of decrease: $\dfrac{\text{amount of decrease}}{\text{original value}} = \dfrac{\text{percent}}{100}$

Interest: Interest = principal • rate • time   or $I = p \cdot r \cdot t$

Amount due: amount due = principal + interest

▶ Test Your Word Power

See how well you have learned the vocabulary in this chapter. Answers follow the Quick Review.

1. To write a **percent as a decimal,** you drop the percent sign
   A. after finding the decimal point
   B. and drop the decimal point
   C. and move the decimal point two places to the right
   D. after moving the decimal point two places to the left.

2. To write a **decimal as a percent,** you add the percent sign
   A. after finding the decimal point
   B. after removing the decimal point
   C. after moving the decimal point two places to the right
   D. after moving the decimal point two places to the left.

3. **Percent** means
   A. the same as interest
   B. per every ten
   C. per one thousand
   D. per one hundred.

4. When you use
   $$\frac{\text{part}}{\text{whole}} = \frac{\text{percent}}{100} \text{ to solve}$$
   percent problems, you are using the
   A. simple interest formula
   B. percent proportion
   C. percent equation
   D. percent sales tax formula.

5. The **percent equation** is
   A. part = percent • whole
   B. $I = p \cdot r \cdot t$
   C. $p\% = \dfrac{p}{100}$
   D. amount due = principal + interest.

6. In a **percent of increase** problem, the increase is a percent of
   A. the largest amount
   B. the original amount
   C. the new or most recent amount
   D. all the amounts.

7. In the formula $I = p \cdot r \cdot t,$ the p stands for
   A. proportion
   B. product
   C. principal
   D. percent.

8. The term **rate** in an interest problem represents the
   A. whole
   B. percent
   C. part
   D. amount of interest.

▶ Quick Review

Concepts	Examples

(6.1) Basics of Percent

Writing a Percent as a Decimal
To write a percent as a decimal, drop the percent sign and move the decimal point two places to the left.

50%   (.50%) = 0.50 or just 0.5

3%   (.03%) = 0.03

12.5%   (12.5%) = 0.125

Writing a Decimal as a Percent
To write a decimal as a percent, move the decimal point two places to the right and attach a % sign.

0.75   (0.75) = 75%

0.875   (0.875) = 87.5%

3.6   (3.60) = 360%

(6.2) Writing a Fraction as a Percent

Use a proportion and solve for p to change a fraction to percent.

$$\frac{2}{5} = \frac{p}{100} \qquad \text{Proportion}$$

$$5 \cdot p = 2 \cdot 100 \qquad \text{Cross products}$$

$$5 \cdot p = 200$$

$$\frac{\overset{1}{\cancel{5}} \cdot p}{\underset{1}{\cancel{5}}} = \frac{200}{5} \qquad \text{Divide both sides by 5.}$$

$$p = 40$$

$$\frac{2}{5} = 40\% \qquad \text{Attach \% sign.}$$

Concepts	Examples

6.3 Learning the Percent Proportion

Part is to whole as percent is to 100.

$$\frac{\textbf{part}}{\textbf{whole}} = \frac{\textbf{percent}}{\textbf{100}} \;\leftarrow \text{Always 100}$$

Use the percent proportion to solve for the unknown value.
part = 30, whole = 50; find the percent.

$$\begin{array}{l} \text{Part} \rightarrow \\ \text{Whole} \rightarrow \end{array} \frac{30}{50} = \frac{\overset{\text{Percent (unknown)}}{x}}{100} \;\leftarrow \text{Always 100}$$

$$\frac{3}{5} = \frac{x}{100} \qquad \text{Write } \tfrac{30}{50} \text{ as } \tfrac{3}{5} \text{ in lowest terms.}$$

$$5 \cdot x = 3 \cdot 100 \qquad \text{Cross products}$$
$$5 \cdot x = 300$$

$$\frac{\overset{1}{\cancel{5}} \cdot x}{\underset{1}{\cancel{5}}} = \frac{300}{5} \qquad \text{Divide both sides by 5.}$$

$$x = 60$$

The percent is 60, which is written as 60**%**.

6.3 Identifying Percent, Whole, and Part in a Percent Problem

The percent appears with the word **percent** or with the symbol **%**.

The whole often appears after the word **of**. The whole is the entire quantity or total.

The part is the portion of the total. If the percent and the whole are found first, the remaining number is the part.

Find the percent, whole, and part in the following.

10% **of** the 500 pies is how many pies?
Percent · Whole · · Part (unknown)

20 cats is 5% **of** what number of cats?
Part · Percent · Whole (unknown)

What percent **of** $220 is $33?
Percent (unknown) · Whole · Part

6.4 Applying the Percent Proportion

Read the problem and identify the percent, whole, and part. Use the percent proportion to solve for the unknown quantity.

A liquid mixture in a tank contains 35% distilled water. If 28 gallons of distilled water are in the tank when it is full, find the capacity of the tank.

$$\text{percent} = 35 \quad \text{and} \quad \text{part} = 28$$

Use the percent proportion to find the whole.

$$\begin{array}{l} \text{Whole} \\ \text{(unknown)} \end{array} \rightarrow \frac{\text{part}}{x} = \frac{\text{percent}}{100}$$

$$\frac{28}{x} = \frac{35}{100}$$

$$\frac{28}{x} = \frac{7}{20} \qquad \text{Write } \tfrac{35}{100} \text{ as } \tfrac{7}{20} \text{ in lowest terms.}$$

$$x \cdot 7 = 560 \qquad \text{Cross products}$$

$$\frac{x \cdot \overset{1}{\cancel{7}}}{\underset{1}{\cancel{7}}} = \frac{560}{7} \qquad \text{Divide both sides by 7.}$$

$$x = 80$$

The capacity of the tank is 80 gallons.

Concepts	Examples

6.5 Using the Percent Equation

The percent equation is part = percent • whole. Identify the percent, whole, and part and solve for the unknown quantity. Always write the percent as a decimal before using the equation.

Solve each problem.

(a) Find 20% of 220 applicants.

$$\textbf{part (unknown)} = \text{percent} \cdot \text{whole}$$
$$x = (0.2)(220)$$
$$x = 44$$

20% of 220 applicants is 44 applicants.

(b) 8 balls is 4% of what number of balls?

$$\text{part} = \text{percent} \cdot \textbf{whole (unknown)}$$
$$8 = (0.04)(x)$$
$$\frac{8}{0.04} = \frac{\overset{1}{\cancel{(0.04)}}(x)}{\underset{1}{\cancel{0.04}}}$$
$$x = 200$$

8 balls is 4% of 200 balls.

(c) $13 is what percent of $52?

$$\text{part} = \textbf{percent (unknown)} \cdot \text{whole}$$
$$13 = x \cdot 52$$
$$\frac{13}{52} = \frac{x \cdot \overset{1}{\cancel{52}}}{\underset{1}{\cancel{52}}}$$
$$x = 0.25 = 25\%$$

$13 is 25% of $52.

6.6 Solving Application Problems with Proportions

To solve for **sales tax,** use this formula.

amount of sales tax = rate of tax • cost of item

The price of a 37-inch plasma HD television is $699, and the sales tax is 5%. Find the sales tax.

$$\textbf{amount of sales tax} = (5\%)(\$699)$$
$$= (0.05)(\$699) = \$34.95$$

To find **commissions,** use this formula.

**amount of commission =
rate of commission • amount of sales**

The sales are $92,000 with a commission rate of 3%. Find the commission.

$$\textbf{amount of commission} = (3\%)(\$92,000)$$
$$= (0.03)(\$92,000)$$
$$= \$2760$$

To find the **discount** and the **sale price,** use these formulas.

amount of discount = rate of discount • original price

sale price = original price − amount of discount

A gas oven originally priced at $480 is offered at a 25% discount. Find the amount of the discount and the sale price.

$$\textbf{discount} = (0.25)(\$480) = \textbf{\$120}$$
$$\textbf{sale price} = \$480 - \textbf{\$120} = \textbf{\$360}$$

Concepts	Examples

6.6 Solving Application Problems with Proportions (continued)

To find the **percent of change,** subtract to find the amount of change (increase or decrease), which is the part. The whole is the original value or value before the change.

The number of parking violations rose from 1980 violations to 2277. Find the percent of increase.

$$2277 - 1980 = 297 \quad \text{Increase}$$

$$\text{Original value} \rightarrow \frac{297}{1980} = \frac{\textbf{percent}}{100}$$

Solve the proportion to find that the percent = 15, so the percent of increase is 15%.

6.7 Finding Simple Interest

Use the formula $\quad I = p \cdot r \cdot t$

$$\textbf{Interest} = \textbf{principal} \cdot \textbf{rate} \cdot \textbf{time}$$

Time (t) is in years. When the time is given in months, use a fraction with 12 in the denominator because there are 12 months in a year.

$2800 is deposited at 8% for 3 months. Find the amount of interest.

$$I = p \cdot r \cdot t$$

$$= \underbrace{(2800)(0.08)}\left(\frac{3}{12}\right)$$

$$= \quad (224) \quad \left(\frac{1}{4}\right) = \frac{(224)(1)}{4} = \$56$$

6.8 Finding Compound Amount and Compound Interest

There are three methods for finding the compound amount.

Find the compound amount and interest if $1500 is deposited at 5% interest for 3 years.

1. Calculate the interest for each compound interest period, then add it back to the principal.

1.

	Interest	Compound Amount
Year 1	($1500)(0.05)(1) = **$75**	
	$1500 + **$75** = **$1575**	
Year 2	($1575)(0.05)(1) = **$78.75**	
	$1575 + **$78.75** = **$1653.75**	
Year 3	($1653.75)(0.05) ≈ **$82.69**	
	$1653.75 + **$82.69** = **$1736.44**	

2. Multiply the original deposit by 100% plus the compound interest rate.

2. ($1500)$\underbrace{(1.05)(1.05)(1.05)}$ ≈ $1736.44

Original deposit   Compound amount

$$100\% + 5\% = 105\% = 1.05$$

3. Use the table on page 439 to find the interest on $1. Then, multiply the table value by the principal.

The compound interest is found with this formula.

$$\begin{matrix} \textbf{compound} \\ \textbf{interest} \end{matrix} = \begin{matrix} \textbf{compound} \\ \textbf{amount} \end{matrix} - \begin{matrix} \textbf{original} \\ \textbf{deposit} \end{matrix}$$

3. Locate 5% across the top of the table and 3 periods at the left. The table value is **1.1576**

$$\text{compound amount} = (\$1500)\,(\textbf{1.1576}) = \textbf{\$1736.40}^*$$

$$\text{interest} = \$1736.40 - \$1500 = \$236.40$$

---

* The difference in the compound amount results from rounding in the table.

ANSWERS TO TEST YOUR WORD POWER

1. D; *Example:* 50% written as a decimal is 0 50. or 0.5.

2. C; *Example:* 0.25 written as a percent is 0.25 or 25%.
3. D; *Example:* 8% means 8 per 100.

4. B; *Example:* Part = 4, and whole = 25. To find the percent, $\frac{4}{25} = \frac{x}{100}$; $x = 16$ or 16%.

5. A; *Example:* Percent = 25, and whole = 300. To find the part, $(0.25)(300) = 75$.
6. B; *Example:* Original value = $200, and amount of increase = $40. To find the percent of increase,

$$\frac{40}{200} = \frac{x}{100}; x = 0.20 \text{ or } 20\% \text{ increase.}$$

7. C; *Example:* Principal (p) = $800, rate ($r$) = 5%, and time ($t$) = 1 year. Then, $I = p \cdot r \cdot t$ so $I = (\$800)(0.05)(1) = \$40$.

8. B; *Example:* Principal (p) = $1650, rate ($r$) = 4%, and time ($t$) = $\frac{1}{2}$ year. To find the interest (I),

$$I = (\$1650)(0.04)\left(\frac{1}{2}\right) = \$33.$$

Chapter 6 ▷▷▷ Review Exercises

[6.1] *Write each percent as a decimal and each decimal as a percent.*

1. 35%   **2.** 150%   **3.** 99.44%   **4.** 0.085%

5. 3.15   **6.** 0.02   **7.** 0.875   **8.** 0.002

[6.2] *Write each percent as a fraction or mixed number in lowest terms and each fraction as a percent.*

9. 15%   **10.** 37.5%   **11.** 175%   **12.** 0.25%

13. $\frac{3}{4}$ **14.** $\frac{5}{8}$ **15.** $3\frac{1}{4}$ **16.** $\frac{1}{200}$

Complete this chart.

Fraction	Decimal	Percent
$\frac{1}{8}$	**17.** ___	**18.** ___
19. ___	0.25	**20.** ___
21. ___	**22.** ___	180%

[6.3] *Find the unknown value in the percent proportion* $\frac{part}{whole} = \frac{percent}{100}$.

23. part = 25, percent = 10

24. whole = 480, percent = 5

Identify each component and then set up each problem using the percent proportion, $\frac{part}{whole} = \frac{percent}{100}$. *Do **not** try to solve for the unknown value.*

25. 35% of 820 mailboxes is 287 mailboxes.

26. 73 DVDs is what percent of 90 DVDs

27. Find 14% of 160 Magellan Road Mates.

28. 418 curtains is 16% of what number of curtains?

29. A golfer lost three of his eight golf balls. What percent were lost?

30. Only 88% of the door keys cut will operate properly. If there are 1280 keys cut, find the number of keys that will operate properly.

[6.4] *Find the part using the percent proportion or the multiplication shortcut.*

31. 18% of 950 programs

32. 60% of 1450 reference books

33. 0.6% of 5200 acres

34. 0.2% of 1400 kilograms

Find the whole using the percent proportion.

35. 105 crates is 14% of what number of crates?

36. 348 test tubes is 15% of what number of test tubes?

37. 677.6 miles is 140% of what number of miles?

38. 2.5% of what number of cases is 425 cases?

Find the percent using the percent proportion. Round percent answers to the nearest tenth if necessary.

39. 649 tulip bulbs is what percent of 1180 tulip bulbs?

40. What percent of 1620 dinner rolls is 85 dinner rolls?

41. What percent of 380 pairs of socks is 36 pairs?

42. What percent of 650 soup cans is 200 soup cans?

[6.1–6.4] *Solve each application problem. Round percent answers to the nearest tenth if necessary.*

43. The average cost of 30 seconds of advertising during the Super Bowl three years ago was $2.3 million. If the increase in cost over the last three years has been 17.4%, find the average cost of 30 seconds of advertising during the Super Bowl this year. Round the cost to the nearest tenth of a million. (*Source:* NFL Research; CBS Advertising Sales; www.docsports.com)

44. Scientists tell us that there are 9600 species of birds and that 1000 of these species are in danger of extinction. What percent of the bird species are in danger of extinction?

[6.5] *Use the percent equation to answer each question.*

45. 32% of $454 is what amount?

46. 155% of 120 trucks is how many trucks?

47. 0.128 ounce is what percent of 32 ounces?

48. 304.5 meters is what percent of 174 meters?

49. 33.6 miles is 28% of what number of miles?

50. $92 is 16% of what number?

[6.6] *Find the amount of sales tax or the tax rate and the total cost. Round to the nearest cent if necessary.*

Amount of Sale	Tax Rate	Amount of Tax	Total Cost
51. $630	5%	_____	_____
52. $780	_____	$58.50	_____

Find the commission earned or the rate of commission.

Sales	Rate of Commission	Commission
53. $3450	8%	_____
54. $65,300	_____	$3265

Find the amount or rate of discount and the sale price. Round to the nearest cent if necessary.

Original Price	Rate of Discount	Amount of Discount	Sale Price
55. $112.50	30%	_____	_____
56. $252	_____	$63	_____

[6.7] *Find the simple interest due on each loan.*

Principal	Rate	Time in Years	Interest
57. $200	4%	1	_____
58. $1080	5%	$1\frac{1}{4}$	_____

▦ *Find the simple interest paid on each investment.*

Principal	Rate	Time in Months	Interest
59. $400	7%	3	_____
60. $1560	$6\frac{1}{2}\%$	18	_____

▦ *Find the total amount due on each simple interest loan.*

Principal	Rate	Time	Total Amount Due
61. $750	$5\frac{1}{2}\%$	2 years	_____
62. $1530	6%	9 months	_____

[6.8] ▦ *Find the compound amount and compound interest. Interest is compounded annually. You may use the table on page 439. Round answers to the nearest cent if necessary.*

Principal	Rate	Time in Years	Compound Amount	Compound Interest
63. $4000	3%	10	_____	_____
64. $1870	4%	4	_____	_____
65. $3600	8%	3	_____	_____
66. $12,500	$5\frac{1}{2}\%$	5	_____	_____

▶▶▶ Mixed Review Exercises

Find the unknown value in the percent proportion $\dfrac{part}{whole} = \dfrac{percent}{100}$.

67. whole = 80, percent = 15

68. part = 738, percent = 45

Use the percent proportion or percent equation to answer each question.

69. 12% of 194 meters is how many meters?

70. 327 cars is what percent of 218 cars?

71. 0.6% of $85 is what amount?

72. 99 employees is 5% of what number of employees?

73. 76 chickens is what percent of 190 chickens?

74. 214.484 liters is 43% of what number of liters?

Write each percent as a decimal and each decimal as a percent.

75. 55%   **76.** 300%   **77.** 5   **78.** 4.71

79. 8.6%   **80.** 0.621   **81.** 0.375%   **82.** 0.0006

Write each percent as a fraction in lowest terms and each fraction as a percent.

83. $\dfrac{3}{4}$ **84.** 42%   **85.** 87.5%   **86.** $\dfrac{3}{8}$

87. $32\dfrac{1}{2}\%$ **88.** $\dfrac{3}{5}$ **89.** 0.25%   **90.** $3\dfrac{3}{4}$

Solve each application problem. Round percent answers to the nearest tenth and money answers to the nearest cent if necessary.

91. Richard Zanotti deposits $20,500 in his credit union savings accounts. If he earns $6\dfrac{1}{2}\%$ simple interest for 30 months, how much interest will be earned?

92. Jim Havey borrows $14,750 at 8% simple interest for 18 months to buy a small Lionel train collection. Find the total amount due.

93. A Hotpoint refrigerator has a capacity of 11.5 cubic feet in the refrigerator and 5.5 cubic feet in the freezer. What percent of the total capacity is the capacity of the freezer?

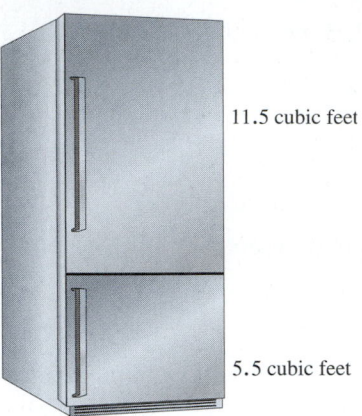

11.5 cubic feet

5.5 cubic feet

94. Tommy Downs invests the $12,500 he inherited from his aunt at 6% interest compounded annually for 4 years.

  (a) Find the compound amount at the end of 4 years. Do not use the table.

  (b) Find the amount of interest that he earned.

95. Tom Dugally, a real estate agent, sold two properties, one for $125,000 and the other for $290,000. He receives a commission of $1\frac{1}{2}$% of total sales. Find the commission that he earned.

96. Vending machines on campus must include healthy food choices such as fruits, fruit juices, and healthy snacks. Sales of healthy foods in the vending machines increased from 4320 items last month to 5107 items this month. Find the percent of increase.

97. A Sears Kenmore washer/dryer set priced at $958 is marked down 18%. If the sales tax is 8%, find the cost of the washer/dryer set including the sales tax.

98. In a recent insurance company study of boaters who had lost items overboard, 88 boaters or 8% said that they lost their cell phones. Find the total number of boaters in the survey. (*Source:* Progressive Groups of Insurance Companies.)

99. Jack and Jill Ahearn begin to budget 25% for rent, 20% for food, 7% for education, 5% for clothing, 10% for transportation, 12% for travel and recreation, 4% for miscellaneous, and the remainder for savings. Jack takes home $2850 per month, and Jill takes home $42,300 per year. How much money will the couple save in a year?

100. The mileage on a hybrid car dropped from 42.8 miles per gallon in the city to 28.5 miles per gallon on the highway. Find the percent of decrease.

Chapter 6 ▷▷▷▷ **Test**  Use the Chapter Test Prep Video CD to see fully worked-out solutions to any of the exercises you want to review.

Write each percent as a decimal and each decimal as a percent.

1. 65%

2. 0.8

3. 1.75

4. 0.875

5. 300%

6. 2%

Write each percent as a fraction in lowest terms.

7. 12.5%

8. 0.25%

Write each fraction or mixed number as a percent.

9. $\dfrac{3}{5}$

10. $\dfrac{5}{8}$

11. $2\dfrac{1}{2}$

Solve each problem.

12. 32 sacks is 4% of what number of sacks?

13. $680 is what percent of $3400?

14. There are still 100,000 households in the United States that do not have electricity. If this is 0.08% of the homes, find the total number of households. (*Source: Time* magazine.)

15. The price of a diamond engagement ring is $3240 plus sales tax of $6\frac{1}{2}$%. Find the total cost of the engagement ring including sales tax.

16. An insurance company pays its salespeople on commission. If a commission of $628 is earned on insurance sales of $7850, find the rate of commission.

1. _____

2. _____

3. _____

4. _____

5. _____

6. _____

7. _____

8. _____

9. _____

10. _____

11. _____

12. _____

13. _____

14. _____

15. _____

16. _____

17. _____

18. _____

19. _____

20. _____

21. _____

22. _____

23. _____

24. _____

25. (a) _____

(b) _____

17. Attendance at the homecoming game decreased from 5760 fans last year to 4320 fans this year. Find the percent of decrease.

18. A problem includes last year's salary, this year's salary, and asks for the percent of increase. Explain how you would identify the part, the whole, and the percent in the problem. Show the percent proportion that you would use.

19. Write the formula used to find interest. Explain the difference in what to do if the time is expressed in months or in years. Write a problem that involves finding interest for 9 months and another problem that involves finding interest for $2\frac{1}{2}$ years. Use your own numbers for the principal and the rate. Show how to solve your problems.

Find the amount of discount and the sale price.

Original Price	Rate of Discount
20. $96	12%
21. $280	32.5%

Find the simple interest on each loan.

Principal	Rate	Time
22. $4200	6%	$1\frac{1}{2}$ years
23. $6400	9%	4 months

24. A parent borrows $5300 to help her son start college. The loan is for 9 months at 9% interest. Find the total amount due on the loan.

25. The River City School PTA Emergency Fund deposited $4000 at 6% interest compounded annually. Two years after the first deposit, they deposit another $5000, also at 6% interest compounded annually. Use the compound interest table on page 439.

(a) What total amount will they have 4 years after their first deposit? Round to the nearest dollar.

(b) What amount of interest will they have earned?

Study Skills

Your math final exam is likely to be a **comprehensive exam.** This means that it will cover material from the **entire term.** The end of the term will be less stressful if you **make a plan** for how you will prepare for each of your exams.

First, figure out the **score you need to earn on the final exam** to get the course grade you want. Check your course syllabus for grading policies, or ask your instructor if you are not sure of them. This allows you to set a goal for yourself.

> How many points do you need to earn on your mathematics final exam to get the grade you want? _____
>
> _____

Second, create a **final exam week plan for your work and personal life.** If you need to make an adjustment in your work schedule, do it in advance, so you aren't scrambling at the last minute. If you have family members to care for, you might want to enlist some help from others so you can spend extra time studying. Try to plan in advance so you don't create additional stress for yourself. You will have to set some priorities, and studying has to be at the top of the list! Although life doesn't stop for finals, some things can be ignored for a short time. You don't want to "burn out" during final exam week; **get enough sleep and healthy food so you can perform your best.**

> What adjustments in your personal life do you need to make for final exam week? _____
>
> _____

Third, use the following suggestions to guide your studying and reviewing.

- ▶ **Know exactly which chapters and sections will be on the final exam.**
- ▶ **Divide up the chapters,** and decide how much you will review each day.
- ▶ Begin your reviewing **several days** before the exam.
- ▶ **Use returned quizzes and tests** to review earlier material (if you have them).
- ▶ **Practice all types of problems,** but emphasize the types that are most difficult for you. Use the **Cumulative Reviews** that are at the end of each chapter in your textbook.
- ▶ **Rewrite your notes or make mind maps** to create summaries.
- ▶ **Make study cards for all types of problems.** Be sure to use the same **direction words** (such as *simplify, solve, estimate*) that your exam will use. Carry the cards with you and review them whenever you have a few spare minutes.

OBJECTIVES

1 Create a final exam week plan.

2 Break studying into chunks and study over several days.

3 Practice all types of problems.

Create a Plan

Study and Review

Managing Stress

Of course, a week of final exams produces stress. **Students who develop skills for reducing and managing stress do better on their final exams and are less likely to "bomb" an exam.** You already know the damaging effect of adrenaline on your ability to think clearly. But several days (or weeks) of elevated stress is also harmful to your brain and your body. You will feel better if you make a conscious effort to reduce your stress level. Even if it takes you away from studying for a little while each day, the time will be well spent.

Reducing Physical Stress

Examples of ways to reduce **physical stress** are listed below. Can you add any of your own ideas to the list?

> *Which techniques will you try?*
>
> _____
>
> _____
>
> _____

▶ *Laugh until your eyes water.* Watching your favorite funny movie, exchanging a joke with a friend, or viewing a comedy bit on the Internet are all ways to generate a healthy laugh. Laughing raises the level of *calming* chemicals (endorphins) in your brain.

▶ *Exercise for 20 to 30 minutes.* If you normally exercise regularly, do NOT stop during final exam week! Exercising helps relax muscles, diffuses adrenaline, and raises the level of endorphins in your body. If you don't exercise much, get some gentle exercise, such as a daily walk, to help you relax.

▶ *Practice deep breathing.* Several minutes of deep, smooth breathing will calm you. Close your eyes too.

▶ *Visualize a relaxing scene.* Choose something that you find peaceful and picture it. Imagine what it feels like and sounds like. Try to put yourself in the picture.

▶ If you feel stress in your muscles, such as your shoulders or back, *slowly squeeze the muscles as much as you can, and then release them.* Sometimes we don't realize we are clenching our teeth or holding tension in our shoulders until we consciously work with them. Try to notice what it feels like when they are relaxed and loose. Squeezing and then releasing muscles is also something you can do during an exam if you feel yourself tightening up.

Reducing Mental Stress

Mental stress reduction is also a powerful tool both before and during an exam. In addition to these suggestions, do you have any of your own techniques?

> *Which techniques will you try?*
>
> _____
>
> _____
>
> _____

▶ *Talk positively to yourself.* Tell yourself you will get through it.

▶ *Reward yourself.* Give yourself small breaks, a little treat—something that makes you happy—every day of final exam week.

▶ *Make a list of things to do* and feel the sense of accomplishment when you cross each item off.

▶ When you take time to relax or exercise, *make sure you are relaxing your mind too.* Use your mind for something *completely* different from the kind of thinking you do when you study. Plan your garden, play your favorite music, walk your dog, read a good book.

▶ *Visualize.* Picture yourself completing exams and projects successfully. Picture yourself taking the test calmly and confidently.

Geometry

An important part of managing our parks and forests is taking an inventory of the trees. The circumference of each tree is measured at chest height. Then, using the circle formulas in **Section 7.5,** the diameter is calculated. This information helps analyze growth patterns and tree age. (See Exercise 31 in **Section 7.5.**)

7

7.1 ▶▶▶ Lines and Angles

Geometry starts with the idea of a point. A **point** can be described as a location in space. It has no length or width. A point is represented by a dot and is named by writing a capital letter next to the dot.

Point P

OBJECTIVE 1 Identify and name lines, line segments, and rays.
A **line** is a straight row of points that goes on forever in both directions. A line is drawn by using arrowheads to show that it never ends. The line is named by using the letters of any two points on the line.

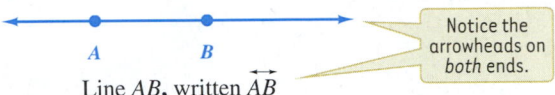

Line AB, written $\overleftrightarrow{AB}$

Notice the arrowheads on *both* ends.

A piece of a line that has two endpoints is called a **line segment.** A line segment is named for its endpoints. The segment with endpoints P and Q is shown below. It can be named $\overline{PQ}$ or $\overline{QP}$.

Line segment PQ, written $\overline{PQ}$

There are *no* arrowheads.

A **ray** is a part of a line that has only one endpoint and goes on forever in one direction. A ray is named by using the endpoint and some other point on the ray. The endpoint is always mentioned first.

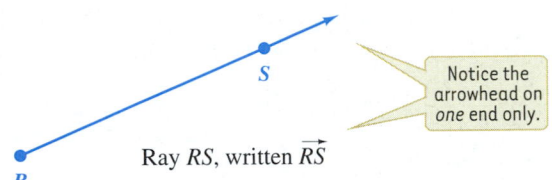

Ray RS, written $\overrightarrow{RS}$

Notice the arrowhead on *one* end only.

1 Identify each figure as a line, line segment, or ray, and name it.

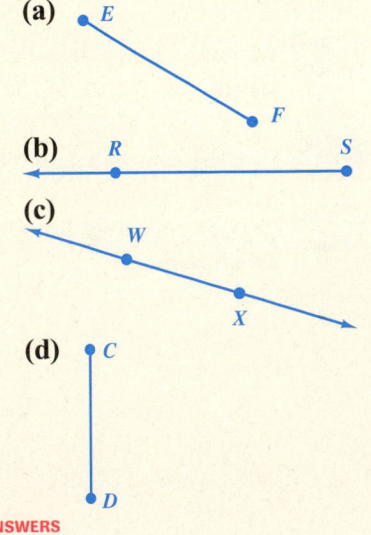

(a) E, F

(b) R, S

(c) W, X

(d) C, D

EXAMPLE 1 **Identifying and Naming Lines, Rays, and Line Segments**

Identify each figure below as a line, line segment, or ray and name it using the appropriate symbol.

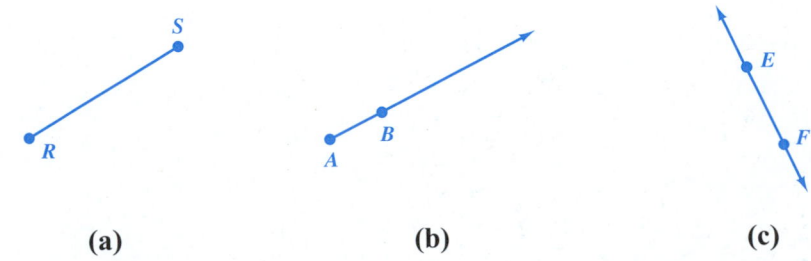

(a)   (b)   (c)

Figure **(a)** has two endpoints, so it is a *line segment* named $\overline{RS}$ or $\overline{SR}$.

Figure **(b)** starts at point A and goes on forever in one direction, so it is a *ray* named $\overrightarrow{AB}$.

Figure **(c)** goes on forever in both directions, so it is a *line* named $\overleftrightarrow{EF}$ or $\overleftrightarrow{FE}$.

◀ *Work Problem* **1** *at the Side.*

OBJECTIVE **2** **Identify parallel and intersecting lines.** A *plane* is an infinitely large, flat surface. A floor or a wall is part of a plane. Lines that are in the *same plane,* but that never intersect (never cross), are called **parallel lines,** while lines that cross are called **intersecting lines.** (Think of an intersection, where two streets cross each other.)

EXAMPLE 2 **Identifying Parallel and Intersecting Lines**

Label each pair of lines as appearing to be parallel or as intersecting.

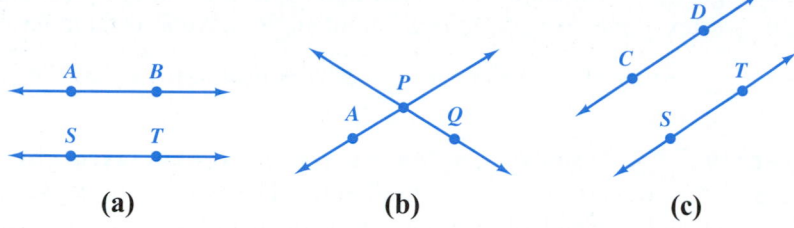

| (a) | (b) | (c) |

The lines in Figures **(a)** and **(c)** do not intersect; they appear to be *parallel lines*.

The lines in Figure **(b)** cross at *P*, so they are *intersecting lines*.

> **CAUTION**
> Appearances may be deceiving! Do not assume that lines are parallel unless it is stated that they are parallel.

Work Problem **2** *at the Side.* ▶

OBJECTIVE **3** **Identify and name angles.** An **angle** is made up of two rays that start at a common endpoint. This common endpoint is called the *vertex*.

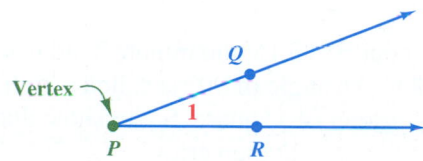

$\overrightarrow{PQ}$ and $\overrightarrow{PR}$ are called the *sides* of the angle. The angle can be named in four different ways, as shown below.

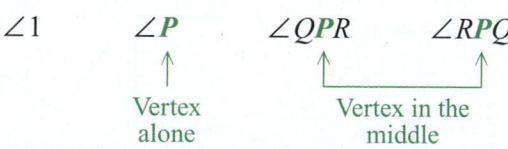

Naming an Angle

To name an angle, write the vertex alone or write the vertex in the middle of two other points, one from each side. If two or more angles have the *same vertex,* as in Example 3 on the next page, do *not* use the vertex alone to name an angle.

---

2 Label each pair of lines as appearing to be parallel or as intersecting.

(a)

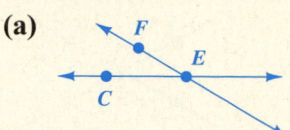

(b)

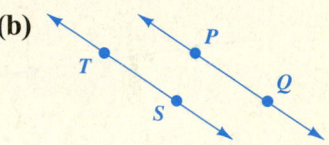

(c)

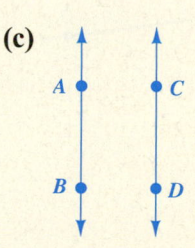

3 (a) Name the highlighted angle in three different ways.

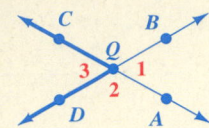

(b) Darken the rays that make up $\angle ZTW$.

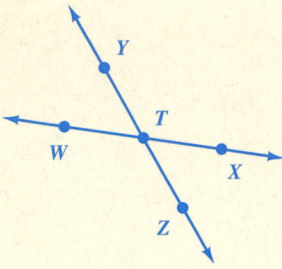

(c) Name this angle in four different ways.

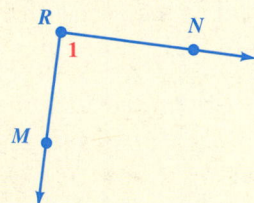

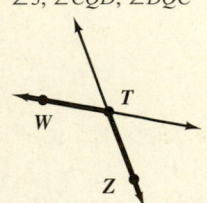

EXAMPLE 3 **Identifying and Naming an Angle**

Name the highlighted angle in three different ways.

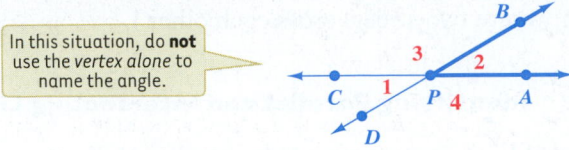

In this situation, do **not** use the *vertex alone* to name the angle.

The angle can be named $\angle BPA$, $\angle APB$, or $\angle 2$. It *cannot* be named $\angle P$, using the vertex alone, because four different angles have P as their vertex.

◄ *Work Problem* **3** *at the Side.*

OBJECTIVE 4 Classify angles as right, acute, straight, or obtuse. Angles can be measured in **degrees.** The symbol for degrees is a small, raised circle °. Think of the minute hand on a clock as a ray of an angle. Suppose it is at 12:00. During one hour of time, the minute hand moves around in a complete circle. It moves 360 *degrees,* or 360°. In half an hour, at 12:30, the minute hand has moved halfway around the circle, or 180°. An angle of 180° is called a **straight angle.** When two rays go in opposite directions and form a straight line, then the rays form a straight angle.

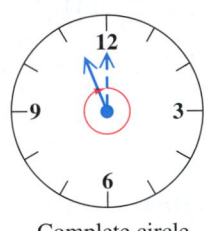

Complete circle
360°

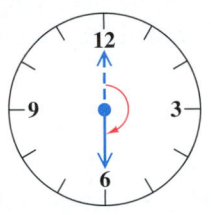

Straight angle
(half a circle)
180°

In a quarter of an hour, at 12:15, the minute hand has moved $\frac{1}{4}$ of the way around the circle, or 90°. An angle of 90° is called a **right angle.** The rays of a right angle form one corner of a square. So, to show that an angle is a **right angle**, we draw a **small square** at the vertex.

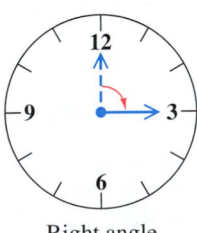

Right angle
($\frac{1}{4}$ of a circle)
90°

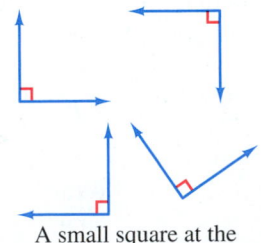

A small square at the
vertex identifies right angles.

An angle that measures 1° is shown below. You can see that an angle of 1° is very small.

1° angle

Some other terms used to describe angles are shown below.

Acute angles measure less than 90°.

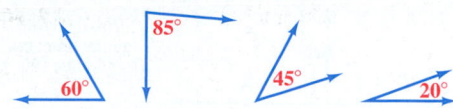

Examples of acute angles

Obtuse angles measure more than 90° but less than 180°.

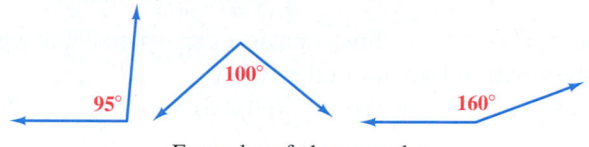

Examples of obtuse angles

Section 8.1 shows you how to use a tool called a *protractor* to measure the number of degrees in an angle.

Classifying Angles

Acute angles measure less than 90°.

Right angles measure *exactly* 90°.

Obtuse angles measure more than 90° but less than 180°.

Straight angles measure *exactly* 180°.

Note

Angles can also be measured in radians, which you will learn about in a later math course.

EXAMPLE 4   **Classifying an Angle**

Label each angle as acute, right, obtuse, or straight.

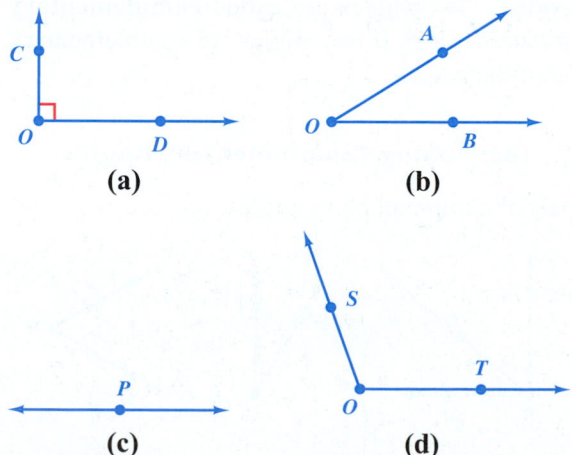

(a)

(b)

(c)

(d)

Figure **(a)** shows a *right angle* (exactly 90° and identified by a small square at the vertex).

Figure **(b)** shows an *acute angle* (less than 90°).

Figure **(c)** shows a *straight angle* (exactly 180°).

Figure **(d)** shows an *obtuse angle* (more than 90° but less than 180°).

Work Problem **4** *at the Side.* ▶

---

4 Label each angle as acute, right, obtuse, or straight. State the number of degrees in the right angle and in the straight angle.

(a)

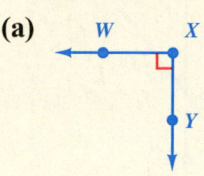

(b)

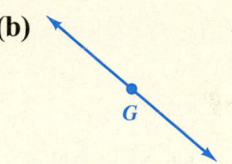

(c)

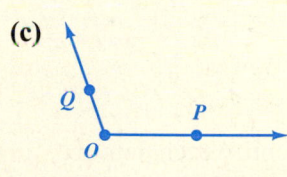

(d)

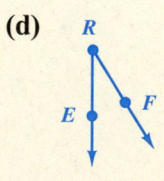

ANSWERS

4. (a) right; 90° **(b)** straight; 180°
  (c) obtuse **(d)** acute

5 Which pair of lines is perpendicular? How can you describe the other pair of lines?

(a)

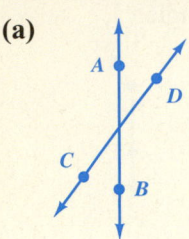

(b)

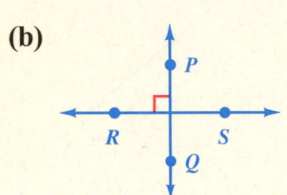

6 Identify each pair of complementary angles.

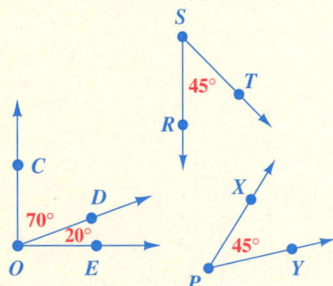

OBJECTIVE **5** **Identify perpendicular lines.**   Two lines are called **perpendicular lines** if they intersect to form a right angle.

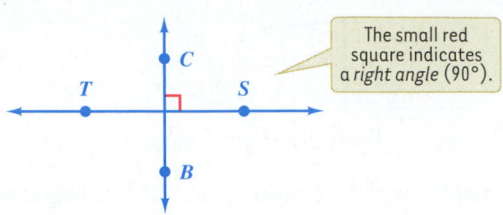

> The small red square indicates a *right angle* (90°).

$\overleftrightarrow{CB}$ and $\overleftrightarrow{ST}$ are **perpendicular** lines because they intersect at right angles, as indicated by the small red square in the figure.

Perpendicular lines can be written in the following way: $\overleftrightarrow{CB} \perp \overleftrightarrow{ST}$.

EXAMPLE 5   **Identifying Perpendicular Lines**

Which pairs of lines are perpendicular?

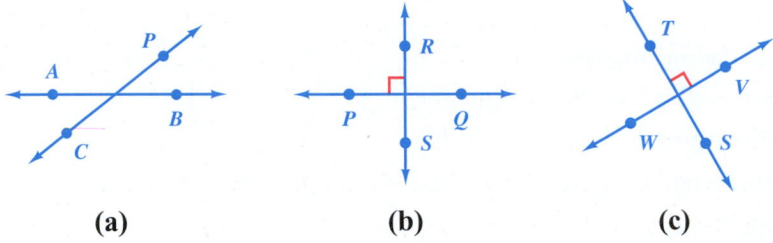

| (a) | (b) | (c) |

The lines in Figures **(b)** and **(c)** are *perpendicular* to each other, because they intersect at right angles.

The lines in Figure **(a)** are *intersecting lines,* but they are *not* perpendicular because they do *not* form a right angle.

◀ *Work Problem* **5** *at the Side.*

OBJECTIVE **6** **Identify complementary angles and supplementary angles and find the measure of a complement or supplement of a given angle.**   Two angles are called **complementary angles** if the sum of their measures is 90°. If two angles are complementary, each angle is the *complement* of the other.

EXAMPLE 6   **Identifying Complementary Angles**

Identify each pair of complementary angles.

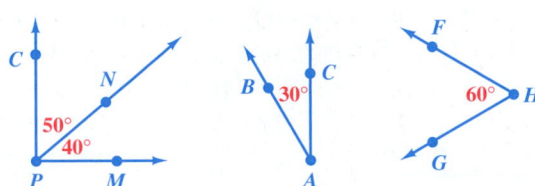

$\angle MPN$ (40°) and $\angle NPC$ (50°) are complementary angles because

$$40° + 50° = \mathbf{90°}$$

$\angle CAB$ (30°) and $\angle FHG$ (60°) are complementary angles because

$$30° + 60° = \mathbf{90°}$$

◀ *Work Problem* **6** *at the Side.*

EXAMPLE 7 **Finding the Complement of Angles**

Find the complement of each angle.

(a) 30°
Find the complement of 30° by subtracting. $90° - 30° = 60°$ ← Complement

(b) 75°
Find the complement of 75° by subtracting. $90° - 75° = 15°$ ← Complement

———————————————— *Work Problem* **7** *at the Side.* ▶

Two angles are called **supplementary angles** if the sum of their measures is 180°. If two angles are supplementary, each angle is the *supplement* of the other.

EXAMPLE 8 **Identifying Supplementary Angles**

Identify each pair of supplementary angles.

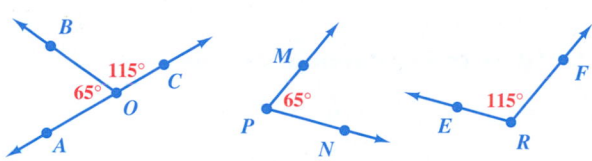

$\angle BOA$ and $\angle BOC$, because $65° + 115° = 180°$

$\angle BOA$ and $\angle ERF$, because $65° + 115° = 180°$

$\angle BOC$ and $\angle MPN$, because $115° + 65° = 180°$

$\angle MPN$ and $\angle ERF$, because $65° + 115° = 180°$

———————————————— *Work Problem* **8** *at the Side.* ▶

EXAMPLE 9 **Finding the Supplement of Angles**

Find the supplement of each angle.

(a) 70°
Find the supplement of 70° by subtracting. $180° - 70° = 110°$ ← Supplement

(b) 140°
Find the supplement of 140° by subtracting. $180° - 140° = 40°$ ← Supplement

———————————————— *Work Problem* **9** *at the Side.* ▶

OBJECTIVE 7 Identify congruent angles and vertical angles and use this knowledge to find the measures of angles. Two angles are called **congruent angles** if they measure the same number of degrees. If two angles are congruent, this is written as $\angle A \cong \angle B$ and read as, "angle A **is congruent to** angle B." Here is an example.

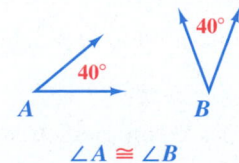

$$\angle A \cong \angle B$$

Example of congruent angles

---

7 Find the complement of each angle.

(a) 35°

(b) 80°

8 Identify each pair of supplementary angles. (*Hint:* There are four pairs.)

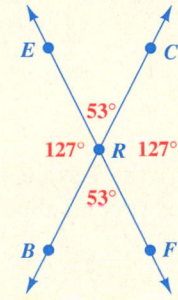

9 Find the supplement of each angle.

(a) 175°

(b) 30°

ANSWERS
7. **(a)** 55° **(b)** 10°
8. $\angle CRF$ and $\angle BRF$; $\angle CRE$ and $\angle ERB$; $\angle BRF$ and $\angle BRE$; $\angle CRE$ and $\angle CRF$
9. **(a)** 5° **(b)** 150°

10 Identify the angles that are congruent.

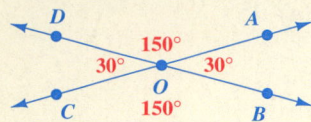

EXAMPLE 10 Identifying Congruent Angles

Identify the angles that are congruent.

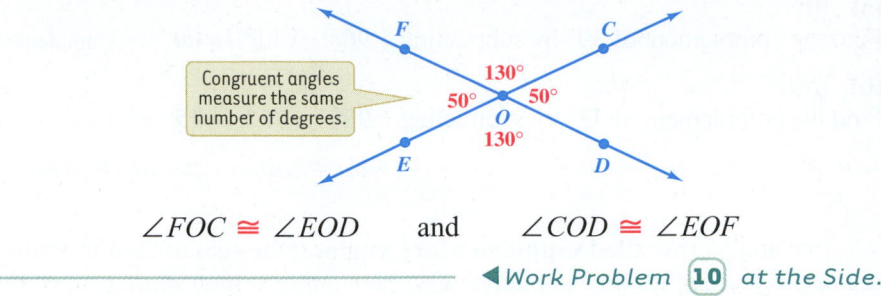

Congruent angles measure the same number of degrees.

$$\angle FOC \cong \angle EOD \qquad \text{and} \qquad \angle COD \cong \angle EOF$$

◄ *Work Problem* **10** *at the Side.*

Angles that share a common side and a common vertex are called *adjacent* angles, such as ∠*FOC* and ∠*COD* in Example 10 above. Angles that do *not* share a common side are called *nonadjacent* angles. Two nonadjacent angles formed by two intersecting lines are called **vertical angles.**

EXAMPLE 11 Identifying Vertical Angles

Identify the vertical angles in this figure.

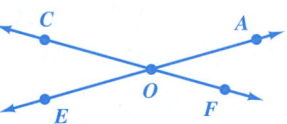

∠*AOF* and ∠*COE* are vertical angles because they do *not* share a common side and they are formed by two intersecting lines ($\overleftrightarrow{CF}$ and $\overleftrightarrow{EA}$).

∠*COA* and ∠*EOF* are also vertical angles.

11 Identify the vertical angles. What is special about vertical angles?

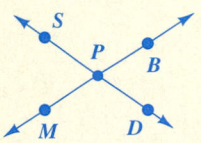

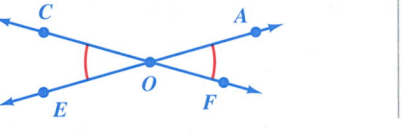

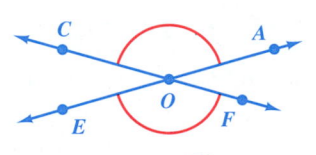

◄ *Work Problem* **11** *at the Side.*

Look back at Example 10 at the top of the page. Notice that the two *congruent* angles that measure 130° are also *vertical* angles. Also, the two congruent angles that measure 50° are vertical angles. This illustrates the following property.

> **Vertical Angles Are Congruent**
>
> If two angles are *vertical* angles, they are *congruent;* that is, they measure the same number of degrees.

EXAMPLE 12 **Finding the Measures of Vertical Angles**

In the figure below, find the measure of each unlabeled angle.

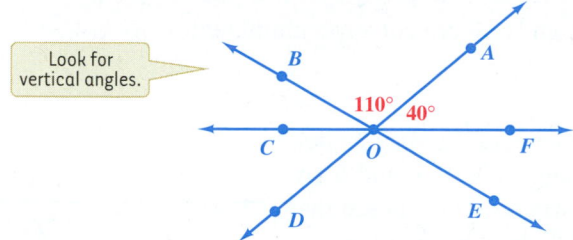

(a) ∠COD
∠COD and ∠AOF are vertical angles, so they are congruent. This means they measure the same number of degrees.

The measure of ∠AOF is 40°   so   the measure of ∠COD is **40°** also.

(b) ∠DOE
∠DOE and ∠BOA are vertical angles, so they are congruent.

The measure of ∠BOA is 110°   so   the measure of ∠DOE is **110°** also.

(c) ∠COB
Look at ∠COB, ∠BOA, and ∠AOF. Notice that $\overrightarrow{OC}$ and $\overrightarrow{OF}$ go in opposite directions. Therefore, ∠COF is a straight angle and measures 180°. To find the measure of ∠COB, subtract the sum of the other two angles from 180°.

$$180° - (110° + 40°) = 180° - (150°) = 30°$$

The measure of ∠COB is **30°**.

(d) ∠EOF
∠EOF and ∠COB are vertical angles, so they are congruent. We know from part (c) above that the measure of ∠COB is 30° so the measure of ∠EOF is **30°** also.

Work Problem **12** *at the Side.* ▶

OBJECTIVE **8** **Identify corresponding angles and alternate interior angles and use this knowledge to find the measures of angles.**   We can also find congruent angles (angles with the same measure) when two *parallel lines* are crossed by a third line, called a *transversal*. When a transversal crosses two *parallel* lines, eight angles are formed, as shown below. There are special names for certain pairs of angles.

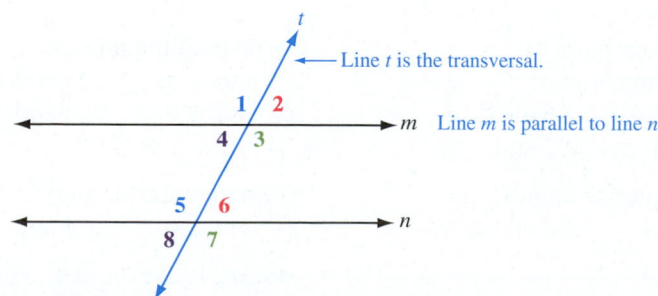

∠1 and ∠5 are called **corresponding angles.** Notice that they are both on the same side of the transversal (line *t*) and in the same relative position. *Corresponding angles are congruent,* so ∠1 and ∠5 measure the same number of degrees. There are four pairs of corresponding angles.

12 In the figure below, find the measure of each unlabeled angle. Write the angle measures on the figure.

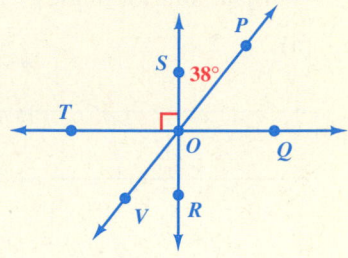

(a) ∠TOS

(b) ∠QOR

(c) ∠VOR

(d) ∠POQ

(e) ∠TOV

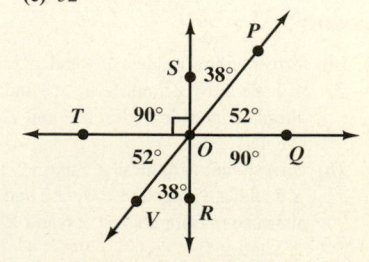

13 In each figure below, line m is parallel to line n. Identify all pairs of corresponding angles and all pairs of alternate interior angles.

(a)

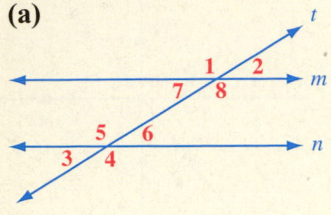

(b)

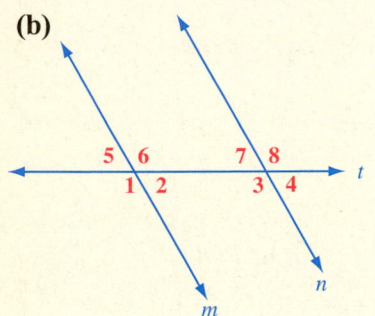

∠1 and ∠5 are corresponding angles, so ∠1 ≅ ∠5.

∠2 and ∠6 are corresponding angles, so ∠2 ≅ ∠6.

∠3 and ∠7 are corresponding angles, so ∠3 ≅ ∠7.

∠4 and ∠8 are corresponding angles, so ∠4 ≅ ∠8.

When a transversal crosses two parallel lines, angles 3, 4, 5, and 6 are called *interior angles*. You can see that they are "inside" the *parallel* lines.

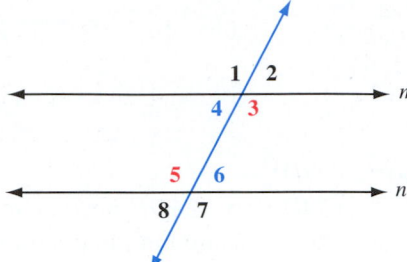

∠3 and ∠5 are alternate interior angles.

∠4 and ∠6 are alternate interior angles.

When two lines are *parallel,* then **alternate interior angles** *are congruent* (they have the same measure). Notice that alternate interior angles are on opposite sides of the transversal.

$$∠3 ≅ ∠5 \quad \text{and} \quad ∠4 ≅ ∠6$$

> **Angles Formed by Parallel Lines and a Transversal**
>
> When two parallel lines are crossed by a transversal:
> 1. Corresponding angles are congruent, and
> 2. Alternate interior angles are congruent.

EXAMPLE 13   **Identifying Corresponding Angles and Alternate Interior Angles**

In each figure, line m is parallel to line n. Identify all pairs of corresponding angles and all pairs of alternate interior angles.

(a)

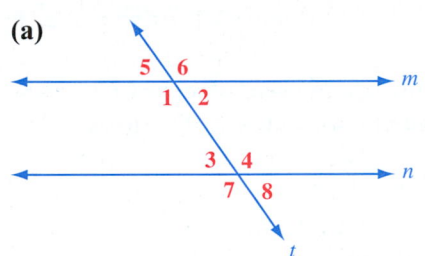

There are four pairs of corresponding angles:

∠5 and ∠3   ∠6 and ∠4

∠1 and ∠7   ∠2 and ∠8

Alternate interior angles:

∠1 and ∠4   ∠2 and ∠3

(b)

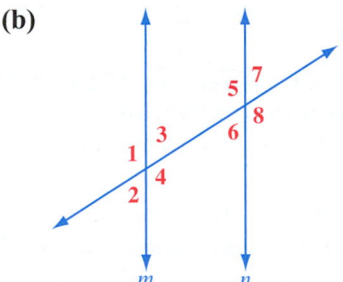

Corresponding angles:

∠1 and ∠5   ∠3 and ∠7

∠2 and ∠6   ∠4 and ∠8

Alternate interior angles:

∠3 and ∠6   ∠4 and ∠5

◀ *Work Problem* **13** *at the Side.*

Recall that two angles are supplementary angles if the sum of their measures is 180°. Also remember that two rays that form a 180° angle form a straight line. Now you can combine your knowledge about supplementary angles with the information on parallel lines.

EXAMPLE 14 **Working with Parallel Lines**

In the figure at the right, line *m* is parallel to line *n* and the measure of ∠4 is 70°. Find the measures of the other angles.

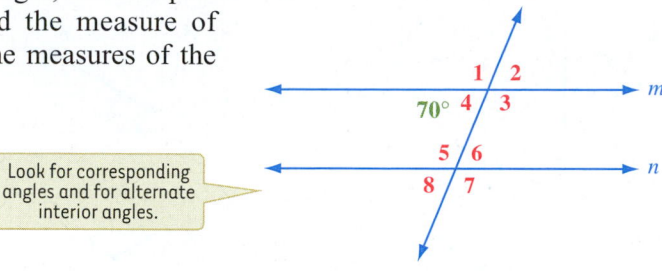

Look for corresponding angles and for alternate interior angles.

As you find the measure of each angle, write it on the figure.

∠4 ≅ ∠8 (corresponding angles), so the measure of ∠8 is also 70°.
∠4 ≅ ∠6 (alternate interior angles), so the measure of ∠6 is also 70°.
∠6 ≅ ∠2 (corresponding angles), so the measure of ∠2 is also 70°.

Notice that the exterior sides of ∠4 and ∠3 form a straight line, that is, a straight angle of 180°. Therefore, ∠4 and ∠3 are supplementary angles and the sum of their measures is 180°. If ∠4 is 70° then ∠3 must be 110° because 180° − 70° = 110°. So the measure of ∠3 is 110°.

∠3 ≅ ∠7 (corresponding angles), so the measure of ∠7 is also 110°.
∠3 ≅ ∠5 (alternate interior angles), so the measure of ∠5 is also 110°.
∠5 ≅ ∠1 (corresponding angles), so the measure of ∠1 is also 110°.

With the measures of all the angles labeled, you can double-check that each pair of angles that forms a straight angle also adds up to 180°.

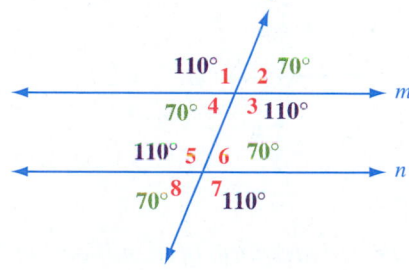

Work Problem **14** *at the Side.* ▶

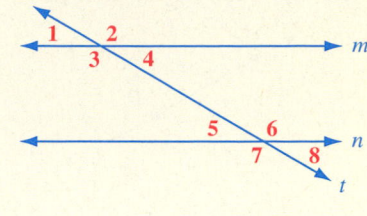

14 In each figure below, line *m* is parallel to line *n*.

(a) The measure of ∠6 is 150°. Find the measures of the other angles.

(b) The measure of ∠1 is 45°. Find the measures of the other angles.

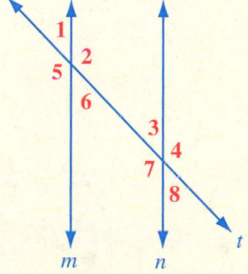

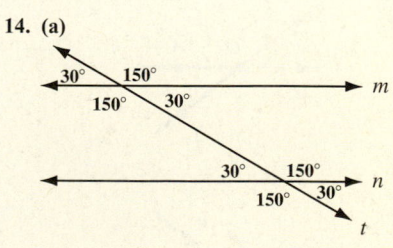
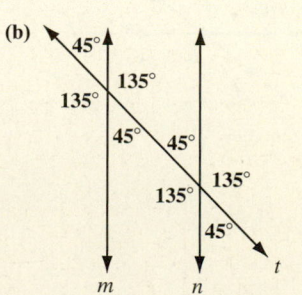

7.1 ▶▶▶ Exercises

FOR
EXTRA
HELP

MyMathLab

Math XL
PRACTICE

WATCH

DOWNLOAD

READ

REVIEW

Identify each figure as a line, line segment, *or* ray *and name it using the appropriate symbol. See Example 1.*

1.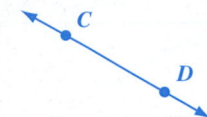

2.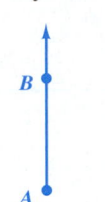

3.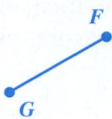

4.

5.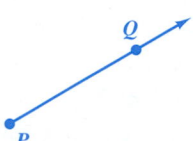

6.

Label each pair of lines as appearing to be parallel, *as* perpendicular, *or as* intersecting. *See Examples 2 and 5.*

7.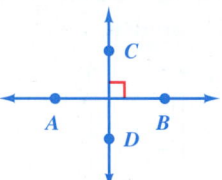

8.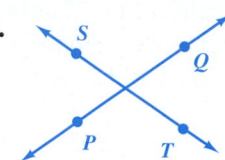

9.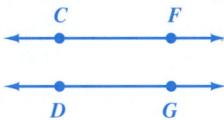

10.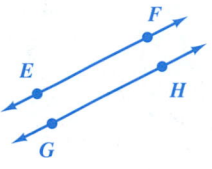

11.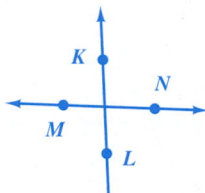

12.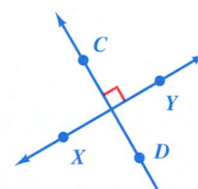

Name each highlighted angle by using the three-letter form of identification. See Example 3.

13.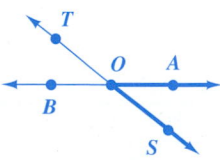

14.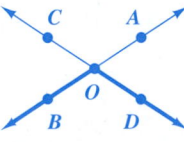

15.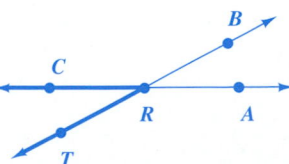

16.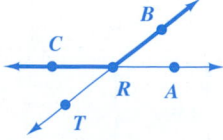

17.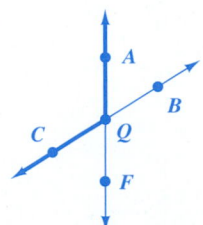

18.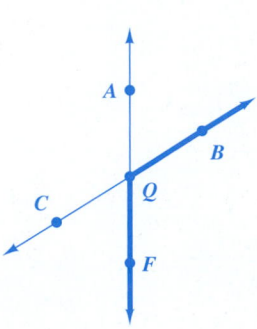

Label each angle as acute, right, obtuse, *or* straight. *For right angles and straight angles, indicate the number of degrees in the angle. See Example 4.*

19.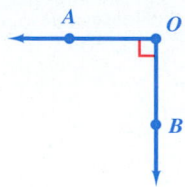

20.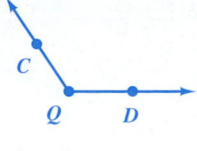

21.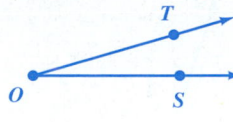

22.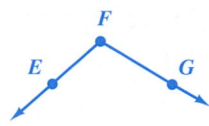

23.

24.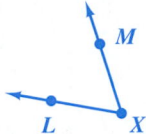

Identify each pair of complementary angles. See Example 6.

25.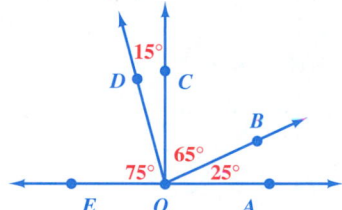

26.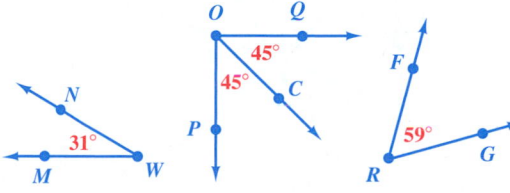

Identify each pair of supplementary angles. See Example 8.

27.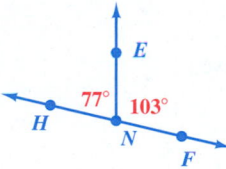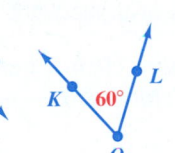

28.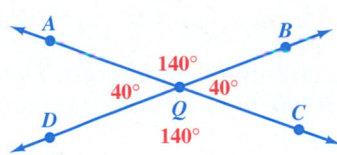

Find the complement of each angle. See Example 7.

29. 40°   **30.** 35°   **31.** 86°   **32.** 59°

Find the supplement of each angle. See Example 9.

33. 130°   **34.** 75°   **35.** 90°   **36.** 5°

In Exercises 37 and 38, identify the angles that are congruent. See Examples 10 and 11.

37.

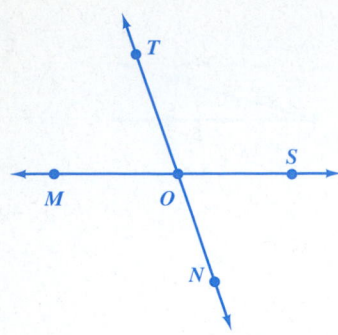

38.

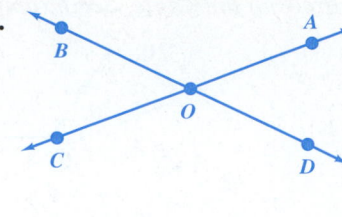

In Exercises 39 and 40, find the measure of each of the angles. See Example 12.

39. In the figure below, ∠AOH measures 37° and ∠COE measures 63°.

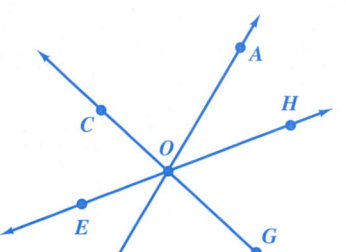

40. In the figure below, ∠POU measures 105° and ∠UOT measures 40°.

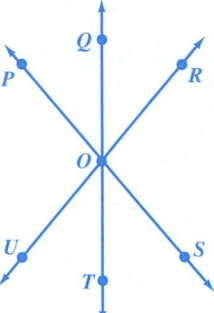

Relating Concepts (Exercises 41–46) For Individual or Group Work

Use the figure to **work Exercises 41–46 in order.** *Decide whether each statement is* true *or* false. *If it is true, explain why. If it is false, rewrite it to make a true statement.*

41. ∠UST is 90°.

42. $\overleftrightarrow{SQ}$ and $\overleftrightarrow{PQ}$ are perpendicular.

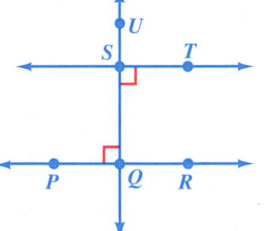

43. The measure of ∠USQ is less than the measure of ∠PQR.

44. $\overleftrightarrow{ST}$ and $\overleftrightarrow{PR}$ are intersecting.

45. $\overleftrightarrow{QU}$ and $\overleftrightarrow{TS}$ are parallel.

46. ∠UST and ∠UQR measure the same number of degrees.

In each figure, line m is parallel to line n. Identify all pairs of corresponding angles and all pairs of alternate interior angles. See Example 13.

47.

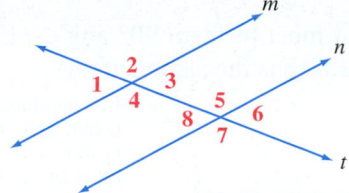

48.

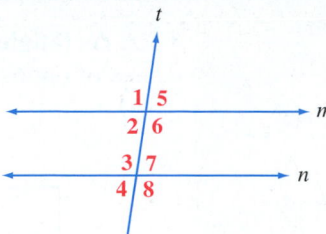

In each figure, line m is parallel to line n. Find the measure of each angle. See Example 14.

49. ∠8 measures 130°.

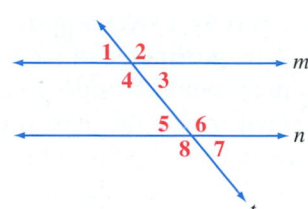

50. ∠2 measures 80°.

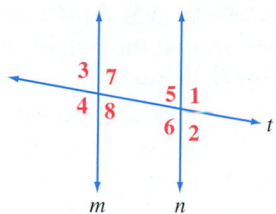

51. ∠6 measures 47°.

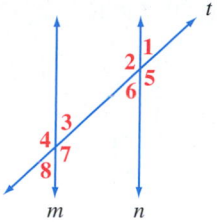

52. ∠2 measures 108°.

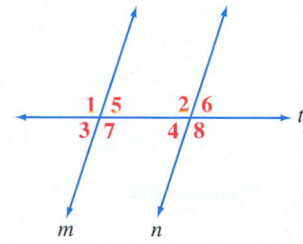

53. ∠6 measures 114°.

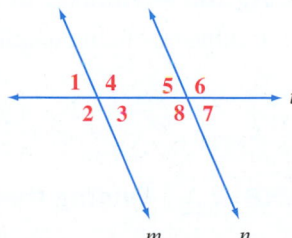

54. ∠3 measures 59°.

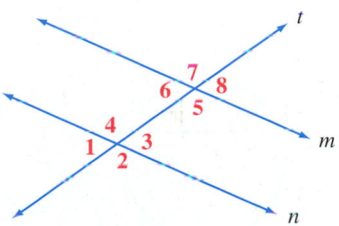

In each figure, $\overrightarrow{BA}$ is parallel to $\overrightarrow{CD}$. Find the measure of each numbered angle.

55.

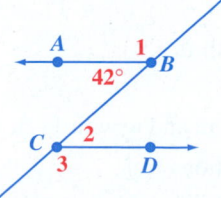

56.

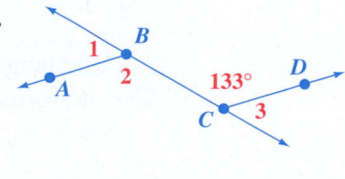

7.2 ▶▶▶ Rectangles and Squares

◀ Work Problem 1 at the Side.

OBJECTIVES

1. **Find the perimeter and area of a rectangle.**
2. **Find the perimeter and area of a square.**
3. **Find the perimeter and area of a composite figure.**

A **rectangle** is a figure with four sides that meet to form 90° angles. Each set of opposite sides is *parallel* and *congruent* (has the same length).

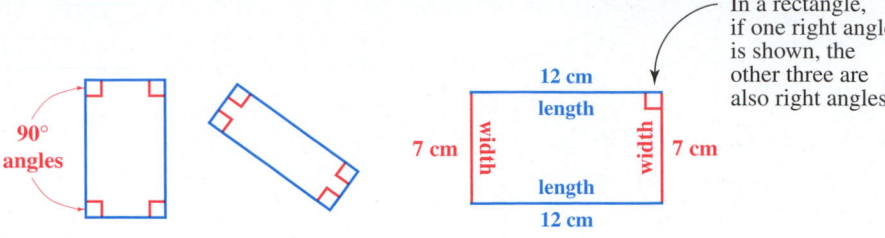

In a rectangle, if one right angle is shown, the other three are also right angles.

Each longer side of a rectangle is called the length (l) and each shorter side is called the width (w).

OBJECTIVE 1 Find the perimeter and area of a rectangle. The distance around the outside edges of a figure is the **perimeter** of the figure. Think of how much fence you would need to put around the sides of a garden plot, or how far you would walk if you walked around the outside edges of your living room. In either case you would add up the lengths of the sides. Look at the rectangle above that has the lengths of the sides labeled. To find its perimeter, you add the lengths of the sides.

$$\text{Perimeter} = \textbf{12 cm} + \textbf{12 cm} + \textbf{7 cm} + \textbf{7 cm} = 38 \text{ cm}$$

Because the two long sides are both 12 cm, and the two short sides are both 7 cm, you can also use this formula.

> **Finding the Perimeter of a Rectangle**
>
> Perimeter of a rectangle = length + length + width + width
> $$P = \quad (2 \cdot \text{length}) \quad + \quad (2 \cdot \text{width})$$
> $$P = 2 \cdot l + 2 \cdot w$$

EXAMPLE 1 Finding the Perimeter of a Rectangle

Find the perimeter of each rectangle.

(a)

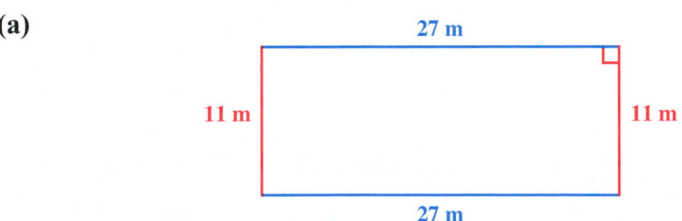

The length of this rectangle is **27 m** and the width is **11 m**.
Use the formula $P = 2 \cdot l + 2 \cdot w$

$P = 2 \cdot \quad l \quad + 2 \cdot \quad w$ — Replace l with 27 m and w with 11 m.

$P = 2 \cdot \textbf{27 m} + 2 \cdot \textbf{11 m}$ — Do the multiplications first.

$P = \quad 54 \text{ m} \quad + \quad 22 \text{ m}$ — Add last.

$P = 76 \text{ m}$

The perimeter of the rectangle (the distance you would walk around the outside edges of the rectangle) is 76 m.

Continued on Next Page

---

1 Identify all the rectangles.

(a)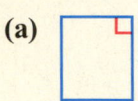

(b)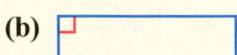

(c)

(d)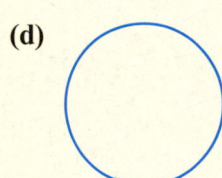

(e)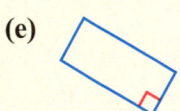

(f)

(g)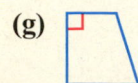

ANSWERS

1. **(a), (b)**, and **(e)** are rectangles; **(c), (d), (f)**, and **(g)** are not.

As a check, you can add up the lengths of the four sides.

$$P = \mathbf{27\ m} + \mathbf{27\ m} + \mathbf{11\ m} + \mathbf{11\ m}$$

$$P = 76\ m \leftarrow \text{Same result as using the formula}$$

(b) A rectangle 8.9 ft by 12.3 ft
You can use the formula, as shown below.

$$P = 2 \bullet \quad l \quad + 2 \bullet \quad w$$

$$P = \underbrace{2 \bullet \mathbf{12.3\ ft}} + \underbrace{2 \bullet \mathbf{8.9\ ft}}$$

$$P = \quad 24.6\ ft \quad + \quad 17.8\ ft$$

$$P = 42.4\ ft$$

> Be sure to write **ft** in the answer.

Or, you can add up the lengths of the four sides.

$$P = \mathbf{12.3\ ft} + \mathbf{12.3\ ft} + \mathbf{8.9\ ft} + \mathbf{8.9\ ft}$$

$$P = 42.4\ ft \leftarrow \text{Same result as using the formula}$$

Either method will give you the correct result.

Work Problem **2** *at the Side.* ▶

The *perimeter* of a rectangle is the distance around the *outside edges*. The **area** of a rectangle is the amount of surface *inside* the rectangle. We measure area by seeing how many squares of a certain size are needed to cover the surface inside the rectangle. Think of covering the floor of a rectangular living room with carpet. Carpet is measured in square yards, that is, square pieces that measure 1 yard along each side. Here is a drawing of a living room floor.

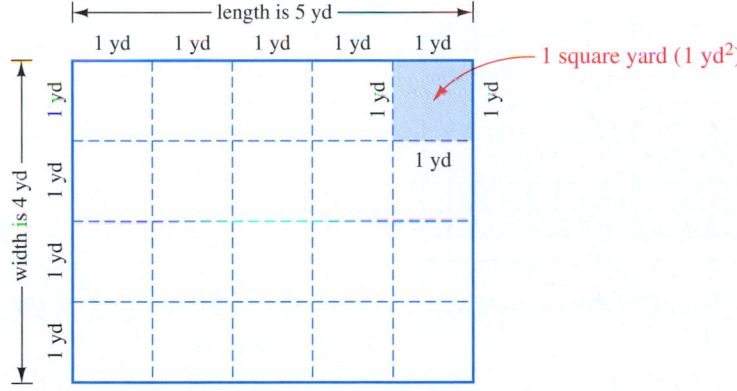

You can see from the drawing that it takes 20 squares to cover the floor. We say that the area of the floor is 20 *square yards*. A shorter way to write square yards is yd^2.

20 **square yards**   can be written as   20 **yd^2**

To find the number of squares, you can count them, or you can multiply the number of squares in the length (5) times the number of squares in the width (4) to get 20. The formula is given below.

Finding the Area of a Rectangle

$$\text{Area of a rectangle} = \text{length} \bullet \text{width}$$

$$A = l \bullet w$$

Remember to use *square units* when measuring area.

---

2 Find the perimeter of each rectangle by using the formula or by adding the lengths of the sides.

(a)
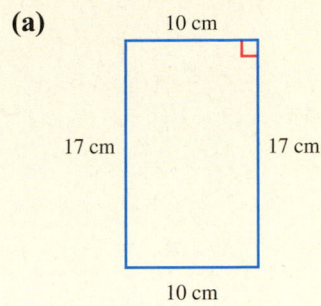

(b)
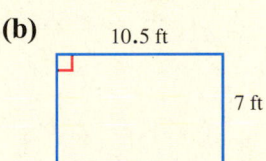

(c) 6 m wide and 11 m long

(d) 0.9 km by 2.8 km

Squares of many sizes can be used to measure area. For smaller areas, you might use the ones shown below.

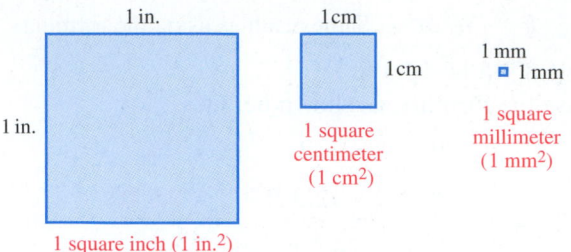

Actual-size drawings

Other sizes of squares that are often used to measure area are listed here, but they are too large to draw on this page.

1 square meter (1 m^2)	1 square foot (1 ft^2)
1 square kilometer (1 km^2)	1 square yard (1 yd^2)
	1 square mile (1 mi^2)

CAUTION

The raised 2 in 4^2 means that you multiply $4 \cdot 4$ to get 16. The raised 2 in cm^2 or yd^2 is a short way to write the word *square*. When you see 5 cm^2, say "five square centimeters." Do *not* multiply $5 \cdot 5$. The exponent applies to cm, *not* to the number.

EXAMPLE 2 **Finding the Area of a Rectangle**

Find the area of each rectangle.

(a)

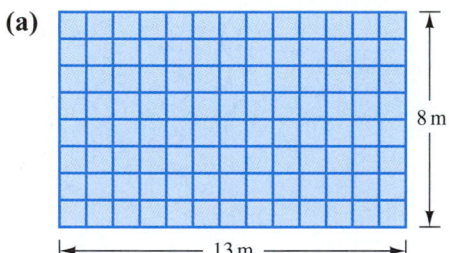

8 m

13 m

The length of this rectangle is 13 m and the width is 8 m. Use the formula $A = l \cdot w$.

$A = \quad l \quad \cdot \quad w$ Replace l with 13 m and w with 8 m.

$A = \mathbf{13\ m} \cdot \mathbf{8\ m}$ Multiply.

$A = 104$ square meters

> Write **m^2** in the answer.

"Square meters" can be written as m^2, so the area is 104 m^2.

(b) A rectangle measuring 7 cm by 21 cm

First make a sketch of the rectangle. The length is 21 cm (the longer measurement) and the width is 7 cm. Then use the formula for the area of a rectangle, $A = l \cdot w$.

7 cm

21 cm

$A = 21$ cm $\cdot$ 7 cm

$A = 147\ \mathbf{cm^2}$ ← Square units for area

The area of the rectangle is 147 cm^2.

Continued on Next Page

Work Problem *at the Side.* ▶

OBJECTIVE 2 Find the perimeter and area of a square.
A **square** is a rectangle with all sides the same length. Two squares are shown below. Notice the 90° angles.

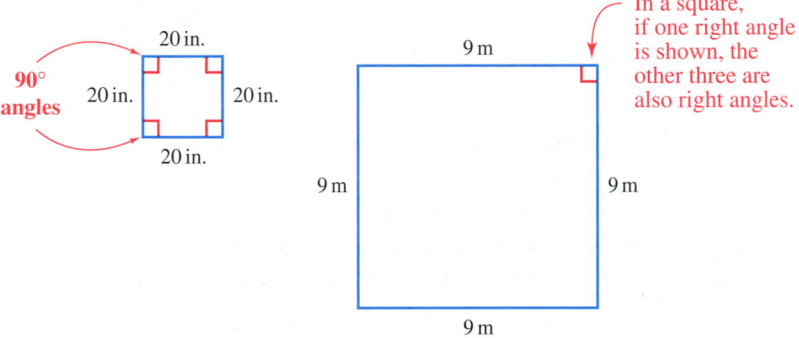

90° angles

20 in.   20 in.   20 in.   20 in.

9 m   9 m   9 m   9 m

In a square, if one right angle is shown, the other three are also right angles.

To find the *perimeter* of (distance around) the square on the right, you could add 9 m + 9 m + 9 m + 9 m to get 36 m. A shorter way is to multiply the length of one side times 4, because all four sides are the same length.

Finding the Perimeter of a Square

Perimeter of a square = side + side + side + side

or, $P = 4 \cdot$ side

$P = 4 \cdot s$

As with a rectangle, you can multiply length times width to find the *area* of (surface inside) a square. Because the length and the width are the same in a square, the formula is written as shown below.

Finding the Area of a Square

Area of a square = side $\cdot$ side

$A = s \cdot s$

$A = s^2$

Remember to use **square units** when measuring area.

EXAMPLE 3 Finding the Perimeter and Area of a Square

(a) Find the perimeter of the square shown above where each side measures 9 m.

Use the formula.   Or add up the four sides.

$P = 4 \cdot s$ $P = 9\,m + 9\,m + 9\,m + 9\,m$

$P = 4 \cdot 9\,m$ $P = 36\,m$

$P = 36\,m$

Be careful to write **m** in the answer.

Same answer

Continued on Next Page

---

3 Find the area of each rectangle.

(a)
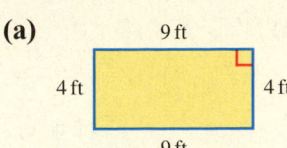
9 ft
4 ft   4 ft
9 ft

(b) A rectangle is 6 m long and 0.5 m wide. (First make a sketch of the rectangle and label the lengths of the sides.)

(c) A rectangular patio measures 3.5 yd by 2.5 yd. (First make a sketch of the patio and label the lengths of the sides.)

ANSWERS

3. **(a)** $A = 36$ ft^2
   (b) $A = 3$ m^2
   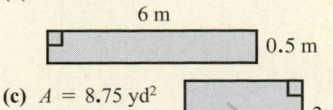
   6 m
   0.5 m
   (c) $A = 8.75$ yd^2
   2.5 yd
   3.5 yd

4 Find the perimeter and area of each square.

(a)

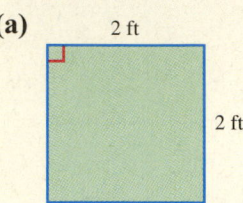

2 ft

2 ft

(b) 10.5 cm on each side (Make a sketch of the square.)

(c) 2.1 mi on a side (Make a sketch of the square.)

(b) Find the area of the same square where each side measures 9 m.

$$A = s^2$$
$$A = s \cdot s$$
$$A = \mathbf{9\ m} \cdot \mathbf{9\ m}$$
$$A = 81\ \mathbf{m^2} \leftarrow \text{Square units for area}$$

> **CAUTION**
> Be careful! s^2 means $s \cdot s$. It does ***not*** mean $2 \cdot s$. In Example 3(b) above, s is 9 m, so s^2 is 9 m $\cdot$ 9 m $= 81\ \text{m}^2$. It is ***not*** $2 \cdot 9$ m $= 18$ m.

◀ *Work Problem* **4** *at the Side.*

OBJECTIVE 3 Find the perimeter and area of a composite figure. As with any other shape, you can find the perimeter of (distance around) an irregular shape by adding up the lengths of the sides. To find the area (surface inside the shape), try to break it up into pieces that are squares or rectangles. Find the area of each piece and then add them together.

> **CAUTION**
> **Perimeter** is the ***distance around the outside edges*** of a flat shape. It is always measured in *linear units* such as cm, m, yd, and so on.
>
> **Area** is the amount of ***surface inside*** a flat shape. It is always measured in *square units* such as cm², m², yd², and so on.

EXAMPLE 4 **Finding the Perimeter and Area of a Composite Figure**

The floor of a room has the shape shown below.

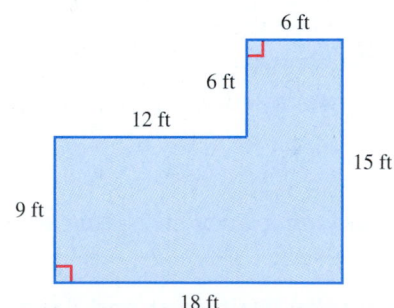

6 ft

6 ft

12 ft

15 ft

9 ft

18 ft

(a) Suppose you want to put a new wallpaper border along the top of all the walls. How much material do you need?

Find the perimeter of the room by adding up the lengths of the sides.

$$P = 9\ \text{ft} + 12\ \text{ft} + 6\ \text{ft} + 6\ \text{ft} + 15\ \text{ft} + 18\ \text{ft}$$
$$P = 66\ \text{ft}$$

You need 66 ft of wallpaper border.

— **Continued on Next Page**

ANSWERS

4. **(a)** $P = 8$ ft; $A = 4$ ft²
   (b) $P = 42$ cm;
   $A = 110.25$ cm²

   10.5 cm

   10.5 cm

   (c) $P = 8.4$ mi;
   $A = 4.41$ mi²

   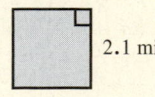

   2.1 mi

   2.1 mi

(b) The carpet you like costs $20.50 per square yard. How much will it cost to carpet the room?

First change the measurements from feet to yards, because the carpet is sold in square yards. There are 3 ft in 1 yd, so multiply by the unit fraction that allows you to divide out feet. Let's start with 9 ft.

$$\frac{\overset{3}{\cancel{9\ \text{ft}}}}{1} \cdot \frac{1\ \text{yd}}{\underset{1}{\cancel{3\ \text{ft}}}} = 3\ \text{yd}$$

       Divide out ft.
       Divide 9 and 3 by 3.

Use the same unit fraction to change the other measurements from feet to yards.

$$\frac{\overset{4}{\cancel{12\ \text{ft}}}}{1} \cdot \frac{1\ \text{yd}}{\underset{1}{\cancel{3\ \text{ft}}}} = 4\ \text{yd} \qquad \frac{\overset{2}{\cancel{6\ \text{ft}}}}{1} \cdot \frac{1\ \text{yd}}{\underset{1}{\cancel{3\ \text{ft}}}} = 2\ \text{yd}$$

$$\frac{\overset{5}{\cancel{15\ \text{ft}}}}{1} \cdot \frac{1\ \text{yd}}{\underset{1}{\cancel{3\ \text{ft}}}} = 5\ \text{yd} \qquad \frac{\overset{6}{\cancel{18\ \text{ft}}}}{1} \cdot \frac{1\ \text{yd}}{\underset{1}{\cancel{3\ \text{ft}}}} = 6\ \text{yd}$$

Next, break up the room into two pieces. Use just the measurements for the length and width of each piece.

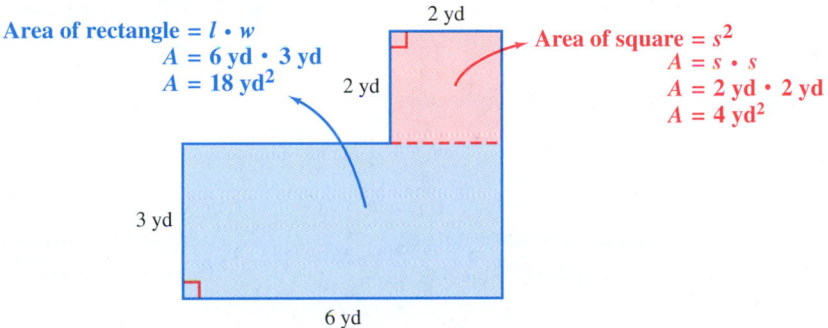

Area of rectangle = $l \cdot w$
$A = 6\ \text{yd} \cdot 3\ \text{yd}$
$A = 18\ \text{yd}^2$

Area of square = s^2
$A = s \cdot s$
$A = 2\ \text{yd} \cdot 2\ \text{yd}$
$A = 4\ \text{yd}^2$

Total area = **$18\ \text{yd}^2$** + **$4\ \text{yd}^2$** = $22\ \text{yd}^2$

Multiply to find the cost of the carpet.

$$\frac{22\ \cancel{\text{yd}^2}}{1} \cdot \frac{\$20.50}{1\ \cancel{\text{yd}^2}} = \$451.00$$

It will cost $451.00 to carpet the room.

You could have cut the room into two rectangles as shown below. The total area is the same

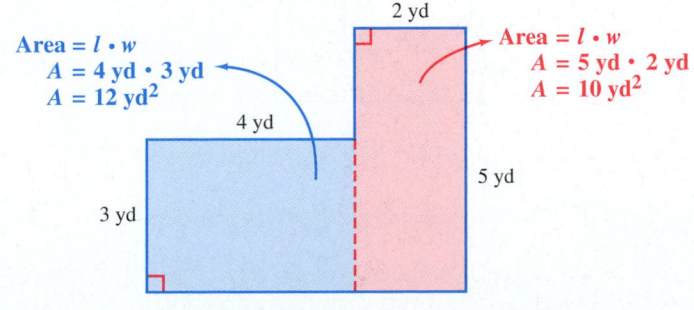

Area = $l \cdot w$
$A = 4\ \text{yd} \cdot 3\ \text{yd}$
$A = 12\ \text{yd}^2$

Area = $l \cdot w$
$A = 5\ \text{yd} \cdot 2\ \text{yd}$
$A = 10\ \text{yd}^2$

Total area = **$12\ \text{yd}^2$** + **$10\ \text{yd}^2$** = $22\ \text{yd}^2$ ← Same answer as above

Work Problem **5** *at the Side.* ▶

---

5 Carpet costs $19.95 per square yard. Find the cost of carpeting each room. Round your answers to the nearest cent if necessary.

(a)
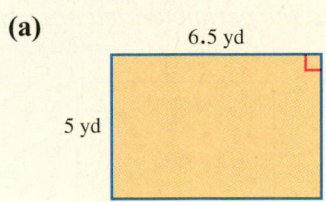

(b)
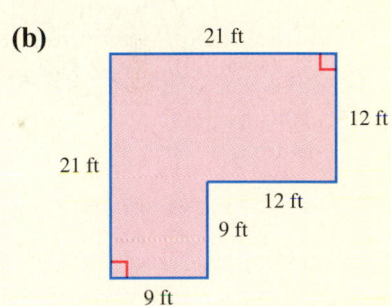

(c) A rectangular classroom is 24 ft long and 18 ft wide. (Make a sketch of the classroom.)

ANSWERS

5. **(a)** 32.5 yd² costs $648.38 (rounded)
   (b) 37 yd² costs $738.15
   (c) 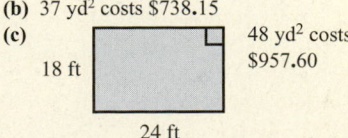 48 yd² costs $957.60

7.2 ▶▶▶ Exercises

Find the perimeter and area of each rectangle or square. See Examples 1–3.

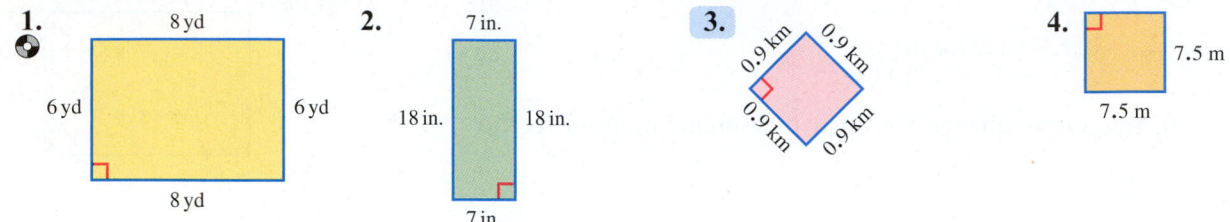

1.
8 yd
6 yd · 6 yd
8 yd

2.
7 in.
18 in. · 18 in.
7 in.

3.
0.9 km · 0.9 km · 0.9 km · 0.9 km

4.
7.5 m
7.5 m

Draw a sketch of each square or rectangle and label the lengths of the sides. Then find the perimeter and the area. (Sketches may vary, show your sketches to your instructor.)

5. 10 ft by 10 ft

6. 8 cm by 17 cm

7. 14 m by 0.5 m

8. 2.35 km by 8.4 km

9. A storage building that is 76.1 ft by 22 ft

10. A science lab measuring 12 m by 12 m

11. A square nature preserve 3 mi wide

12. A square of cardboard 20.3 cm on a side

Find the perimeter and area of each figure. See Example 4.

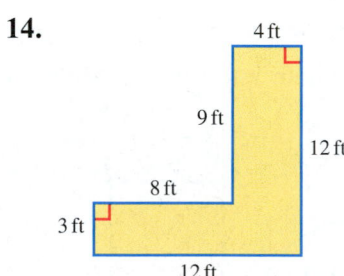

13.
7 m
3 m
5 m
12 m
9 m
2 m

14.
4 ft
9 ft
12 ft
8 ft
3 ft
12 ft

15.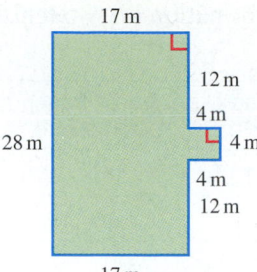

17 m
12 m
4 m
28 m
4 m
4 m
12 m
17 m

16.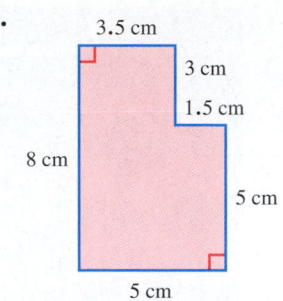

3.5 cm
3 cm
1.5 cm
8 cm
5 cm
5 cm

First find the length of the unlabeled side in each figure. Then find the perimeter and area of each figure.

17.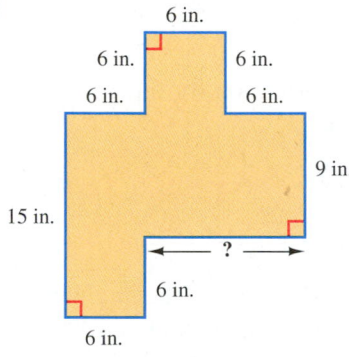

6 in.
6 in.   6 in.
6 in.   6 in.
9 in.
15 in.
?
6 in.
6 in.

18.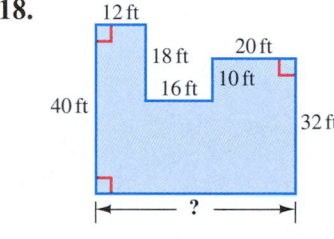

12 ft
18 ft   20 ft
16 ft   10 ft
40 ft
32 ft
?

Solve each application problem. In Exercises 19–24, draw a sketch for each problem and label it with the appropriate measurements. (Sketches may vary; show your sketches to your instructor.)

19. Gymnastic floor exercises are performed on a square mat that is 12 meters on a side. Find the perimeter and area of a mat. (*Source:* www.nist.gov)

20. A regulation volleyball court is 18 meters by 9 meters. Find the perimeter and area of a regulation court. (*Source:* www. nist. gov)

21. The Wang's family room measures 20 ft by 25 ft. They are covering the floor with square tiles that measure 1 ft on a side and cost $0.92 each. How much will they spend on tile?

22. A page in this book measures 27.5 cm from top to bottom and 21 cm from side to side. Find the perimeter and the area of the page.

23. Tyra's kitchen is 4.4 m wide and 5.1 m long. She is pasting a decorative border strip that costs $4.99 per meter around the top edge of all the walls. How much will she spend?

24. Mr. and Mrs. Gomez are buying carpet for their square-shaped bedroom that is 5 yd wide. The carpet is $23 per square yard and padding and installation is another $6 per square yard. How much will they spend in all?

25. Regulation soccer fields can measure 50 to 100 yards wide and 100 to 130 yards long, depending upon the age and skill level of the players. Find the area of the smallest soccer field and the area of the largest soccer field. What is the difference in the playing room between the smallest and largest fields?

26. The table below shows information on two tents for camping.

Tents	Coleman Family Dome	Eddie Bauer Dome Tent
Dimensions	13 ft × 13 ft	12 ft × 12 ft
Sleeps	8 campers	6 campers
Sale price	$127	$99

Source: target.com

(a) For the Coleman tent, find the perimeter, area, and number of square feet of floor space for each camper. Round to the nearest whole number.

(b) Find the same information for the Eddie Bauer tent.

27. A regulation football field is rectangular, 100 yd long (excluding end zones), and has an area of 5300 yd^2. Find the width of the field. (*Source:* National Football League.)

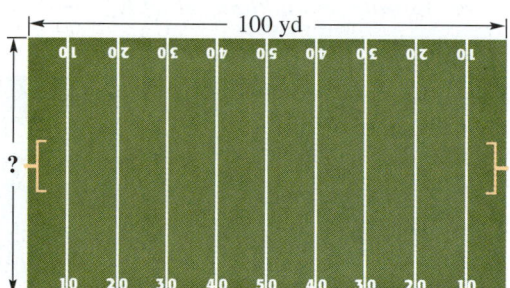

28. There are 14,790 ft^2 of ice in the rectangular playing area for a major league hockey game (excluding the area behind the goal lines). If the playing area is 85 ft wide, how long is it? (*Source:* National Hockey League.)

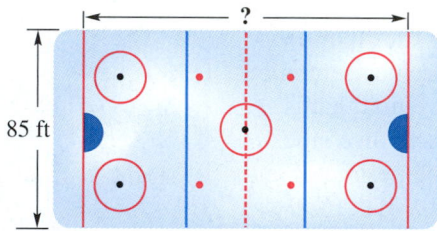

29. A rectangular lot is 124 ft by 172 ft. County rules require that nothing be built on land within 12 ft of any edge of the lot. First, add labels to the sketch of the lot, showing the land that cannot be build on. Then find the area of the land that cannot be built on.

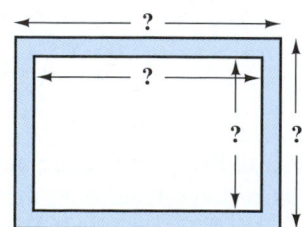

30. Find the cost of fencing needed for this rectangular field. Fencing along the country roads costs $4.25 per foot. Fencing for the other two sides costs $2.75 per foot.

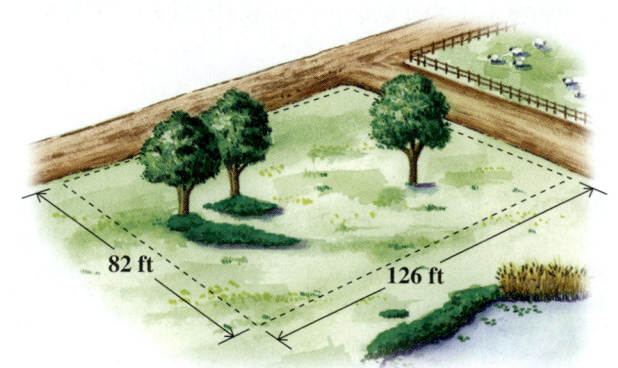

Relating Concepts (Exercises 31—36) For Individual or Group Work

Use your knowledge of perimeter and area to **work Exercises 31–36 in order.**

31. Suppose you have 12 ft of fencing to make a square or rectangular garden plot. Draw sketches of *all* the possible plots that use exactly 12 ft of fencing and label the lengths of the sides. Use only *whole number* lengths. (*Hint:* There are three possibilities.)

32. (a) Find the area of each plot in Exercise 31.

(b) Which plot has the greatest area?

33. Repeat Exercise 31 using 16 ft of fencing. Be sure to draw *all* possible plots that have whole number lengths for the sides.

34. (a) Find the area of each plot in Exercise 33.

(b) Compare your results to those from Exercise 32. What do you notice about the plots with the greatest area?

35. (a) Draw a sketch of a rectangular plot 3 ft by 2 ft. Find the perimeter and area.

(b) Suppose you *double* the length of the plot and *double* the width. Draw a sketch of the enlarged plot and find the perimeter and area.

(c) The *perimeter* of the enlarged plot is how many times greater than the perimeter of the original plot? The *area* of the enlarged plot is how many times greater than the original area?

36. (a) Refer to part (a) of Exercise 35. Suppose you *triple* the length and width of the original plot. Draw a sketch of the enlarged plot and find the perimeter and area.

(b) How many times greater is the *perimeter* of the enlarged plot? How many times greater is the *area* of the enlarged plot?

(c) Suppose you make the length and width *four times greater* in the enlarged plot. What would you predict will happen to the perimeter and area, compared to the original plot?

7.3 ▶▶▶ Parallelograms and Trapezoids

A **parallelogram** is a four-sided figure with opposite sides parallel, such as the ones shown below. Notice that opposite sides have the same length.

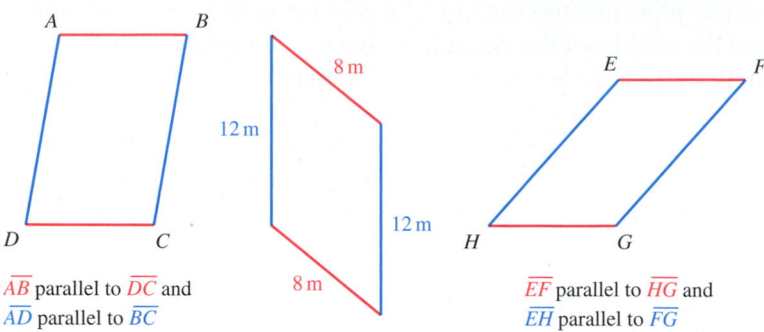

$\overline{AB}$ parallel to $\overline{DC}$ and $\overline{AD}$ parallel to $\overline{BC}$

$\overline{EF}$ parallel to $\overline{HG}$ and $\overline{EH}$ parallel to $\overline{FG}$

1 Find the perimeter of each parallelogram

(a)

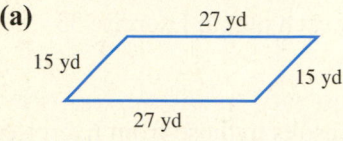

OBJECTIVE 1 Find the perimeter and area of a parallelogram.
Perimeter is the distance around a flat shape, so the easiest way to find the perimeter of a parallelogram is to add the lengths of the four sides.

EXAMPLE 1 Finding the Perimeter of a Parallelogram

Find the perimeter of the middle parallelogram above.

> You must write **m** in the answer.

$$P = \mathbf{12\ m} + \mathbf{12\ m} + \mathbf{8\ m} + \mathbf{8\ m} = 40\ m$$

◀ *Work Problem* **1** *at the Side.*

To find the area of a parallelogram, first draw a dashed line inside the figure as shown here.

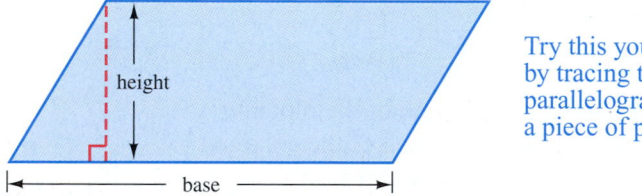

Try this yourself by tracing this parallelogram onto a piece of paper.

The length of the dashed line is the *height* of the parallelogram. It forms a *right angle* with the base. The height is the shortest distance between the base and the opposite side.

Now cut off the triangle created on the left side of the parallelogram above and move it to the right side, as shown below.

(b)

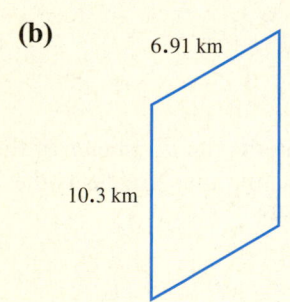

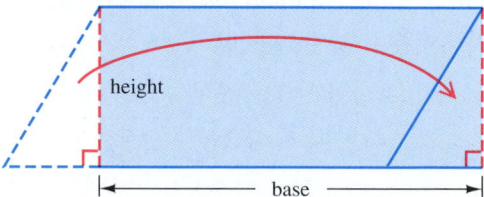

The parallelogram has been made into a rectangle. You can see that the area of the parallelogram and the rectangle are the same.

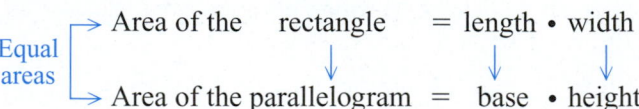

Finding the Area of a Parallelogram

Area of a parallelogram = base • height

$$A = b \bullet h$$

Remember to use *square units* when measuring area.

EXAMPLE 2 **Finding the Area of Parallelograms**

Find the area of each parallelogram.

(a)

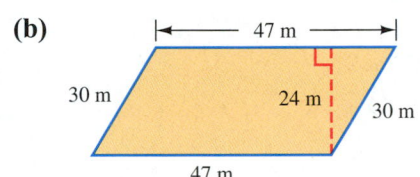

The base is 24 cm and the height is 19 cm. Use the formula $A = b \bullet h$.

$A = \quad b \quad \bullet \quad h$

$A = 24 \text{ cm} \bullet 19 \text{ cm}$

$A = 456 \text{ cm}^2$

Careful! Write **cm²** in the answer.

(b)

$A = 47 \text{ m} \bullet 24 \text{ m}$

$A = 1128 \text{ m}^2$

Write **m²** in the answer.

Notice that you do *not* use the 30 m sides when finding the area. But you would use them when finding the *perimeter* of the parallelogram.

Work Problem **2** *at the Side.* ▶

OBJECTIVE **2** **Find the perimeter and area of a trapezoid.**
A **trapezoid** is a four-sided figure with exactly one pair of parallel sides, such as the figures shown below. Unlike parallelograms, opposite sides of a trapezoid might *not* have the same length.

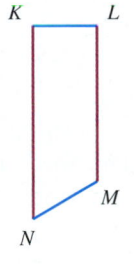

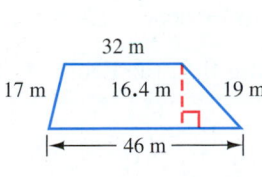

$\overline{KN}$ is parallel to $\overline{LM}$.

The 32 m and 46 m sides are parallel.

$\overline{PQ}$ is parallel to $\overline{SR}$.

EXAMPLE 3 **Finding the Perimeter of a Trapezoid**

Find the perimeter of the middle trapezoid above.
You can find the perimeter of any flat shape by adding the lengths of the sides.

The height of 16.4 m is **not** one of the sides.

$$P = 17 \text{ m} + 32 \text{ m} + 19 \text{ m} + 46 \text{ m}$$

$$P = 114 \text{ m}$$

Notice that the height (16.4 m) is *not* part of the perimeter, because the height is *not* one of the *outside edges* of the shape.

Work Problem **3** *at the Side.* ▶

2 Find the area of each parallelogram.

(a)

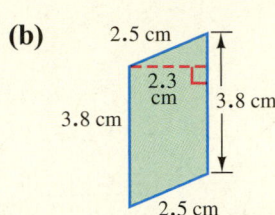

(b)

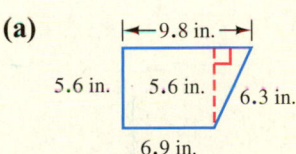

(c) A parallelogram with base $12\frac{1}{2}$ mi and height $4\frac{3}{4}$ mi (*Hint:* Write $12\frac{1}{2}$ as 12.5 and $4\frac{3}{4}$ as 4.75.)

3 Find the perimeter of each trapezoid.

(a)

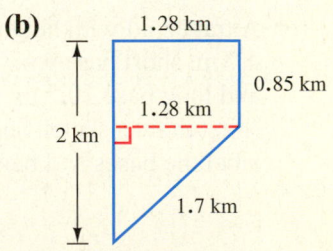

(b)

(c) A trapezoid with sides 39.7 cm, 29.2 cm, 74.9 cm, and 16.4 cm

ANSWERS

2. (a) $A = 2100 \text{ ft}^2$ (b) $A = 8.74 \text{ cm}^2$
   (c) $A = 59\frac{3}{8} \text{ mi}^2$ or 59.375 mi^2

3. (a) $P = 28.6$ in. (b) $P = 5.83$ km
   (c) $P = 160.2$ cm

4 Find the area of each trapezoid.

(a)

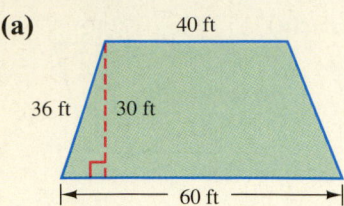

(b)

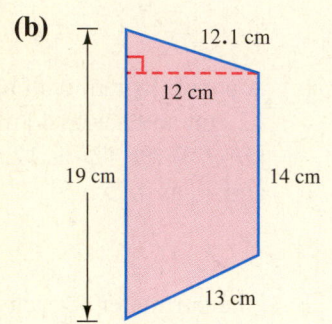

(c) A trapezoid with height 4.7 m, short base 9 m, and long base 10.5 m (First draw a sketch and label the bases and height.)

Use this formula to find the *area* of a trapezoid.

Finding the Area of a Trapezoid

$$\text{Area} = \frac{1}{2} \cdot \text{height} \cdot (\text{short base} + \text{long base})$$

$$A = \frac{1}{2} \cdot h \cdot (b + B)$$

or $\quad A = 0.5 \cdot h \cdot (b + B)$

Remember to use **square units** when measuring area.

EXAMPLE 4 **Finding the Area of a Trapezoid**

Find the area of this trapezoid. The short base and long base are the *parallel* sides.

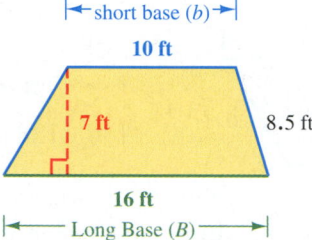

short base (*b*)

10 ft

7 ft   8.5 ft

16 ft

Long Base (*B*)

The height (**h**) is **7 ft**, the short base (**b**) is **10 ft**, and the long base (**B**) is **16 ft**. You do *not* need the lengths of the other two sides to find the area.

$$A = \frac{1}{2} \cdot h \cdot (b + B)$$

$$A = \frac{1}{2} \cdot 7 \text{ ft} \cdot (10 \text{ ft} + 16 \text{ ft}) \qquad \text{Work inside parentheses first.}$$

$$A = \frac{1}{\overset{}{2}} \cdot 7 \text{ ft} \cdot (\overset{13}{2\!\!\!6} \text{ ft})$$

Be sure to write **ft²** in the answer.

$$A = 91 \text{ ft}^2 \qquad \text{Square units for area}$$

You can also find the area by using 0.5, the decimal equivalent for $\frac{1}{2}$, in the formula.

$$A = 0.5 \cdot h \cdot (b + B)$$
$$A = 0.5 \cdot 7 \cdot (10 + 16)$$
$$A = 0.5 \cdot 7 \cdot \quad 26$$
$$A = 91 \text{ ft}^2 \qquad \text{Same answer as above}$$

▦ Calculator Tip Use the parentheses keys on your scientific calculator to work Example 4 above.

0.5 ⊗ 7 ⊗ ⦅ 10 ⊕ 16 ⦆ ⊜ **91**

What happens if you do *not* use the parentheses keys? What order of operations will the calculator follow then? (Answer: The calculator will multiply 0.5 times 7 times 10, and then add 16, giving an *incorrect* answer of 51.)

◀ *Work Problem* **4** *at the Side.*

EXAMPLE 5 **Finding the Area of a Composite Figure**

Find the area of this figure.

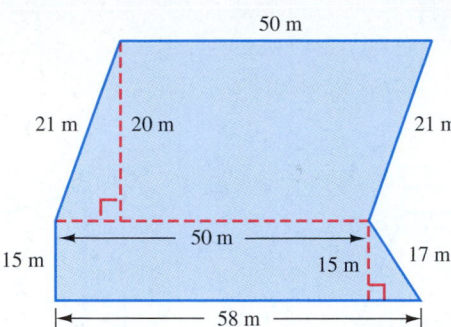

Break the figure into two pieces, a parallelogram and a trapezoid. Find the area of each piece, and then add the areas.

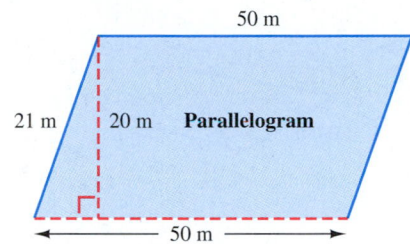

Area of parallelogram
$A = b \quad \bullet \quad h$
$A = 50 \text{ cm} \bullet 20 \text{ cm}$
$A = 1000 \text{ m}^2$

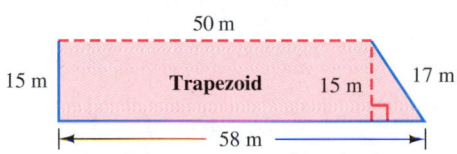

Area of trapezoid
$A = \dfrac{1}{2} \bullet h \bullet (b + B)$
$A = 0.5 \bullet 15 \text{ m} \bullet (50 \text{ m} + 58 \text{ m})$
$A = 810 \text{ m}^2$

Total area = **1000 m² + 810 m² = 1810 m²**  Write **m²** in the answer.

The area of the figure is 1810 m².

Work Problem **5** *at the Side.* ▶

EXAMPLE 6 **Applying Knowledge of Area**

Suppose the figure in Example 5 above represents the floor plan of a hotel lobby. What is the cost of labor to install tile on the floor if the labor charge is $35.11 per square meter?

From Example 5, the floor area is 1810 m². To find the labor cost, multiply the number of square meters times the cost of labor per square meter.

$$\text{cost} = \frac{1810 \ \cancel{m^2}}{1} \bullet \frac{\$35.11}{1 \ \cancel{m^2}}$$

$$\text{cost} = \$63,549.10$$

The cost of the labor is $63,549.10.

Work Problem **6** *at the Side.* ▶

---

5 Find the area of each floor.

(a)

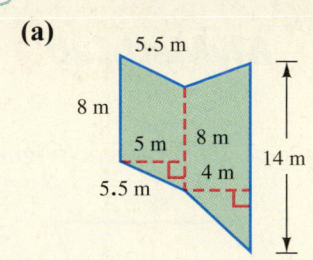

(b)

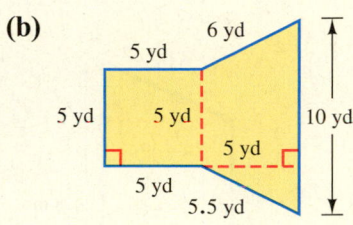

6 Find the cost of carpeting the floors in Problem 5 above. The cost of carpet is as follows:

(a) Floor (a), $18.50 per square meter.

(b) Floor (b), $28 per square yard.

ANSWERS

5. **(a)** $A = 40 \text{ m}^2 + 44 \text{ m}^2 = 84 \text{ m}^2$
   (b) $A = 25 \text{ yd}^2 + 37.5 \text{ yd}^2 = 62.5 \text{ yd}^2$
6. **(a)** $1554   **(b)** $1750

Find the perimeter of each figure. See Examples 1 and 3.

1.
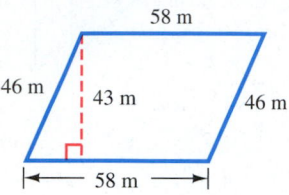
58 m, 46 m, 43 m, 46 m, 58 m

2.
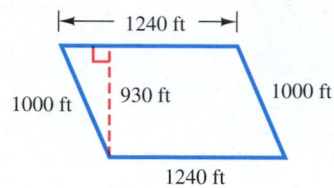
1240 ft, 1000 ft, 930 ft, 1000 ft, 1240 ft

3.
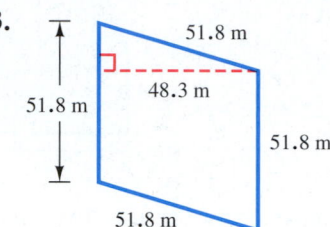
51.8 m, 48.3 m, 51.8 m, 51.8 m, 51.8 m

4.
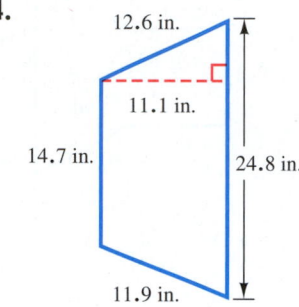
12.6 in., 11.1 in., 14.7 in., 24.8 in., 11.9 in.

5.
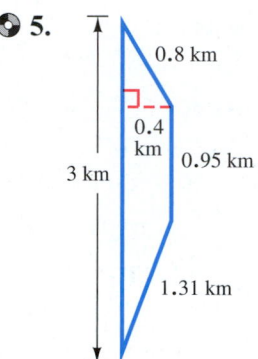
0.8 km, 0.4 km, 0.95 km, 3 km, 1.31 km

6.
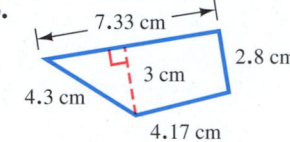
7.33 cm, 2.8 cm, 3 cm, 4.3 cm, 4.17 cm

Find the area of each figure. See Examples 2 and 4.

7.
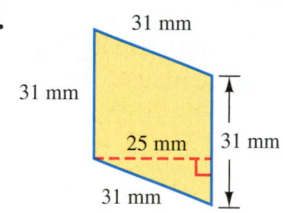
31 mm, 31 mm, 25 mm, 31 mm, 31 mm

8.
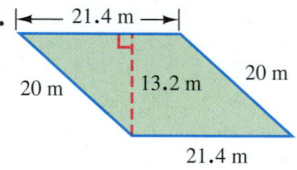
21.4 m, 20 m, 13.2 m, 20 m, 21.4 m

9.
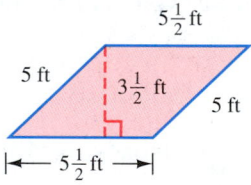
$5\frac{1}{2}$ ft, 5 ft, $3\frac{1}{2}$ ft, 5 ft, $5\frac{1}{2}$ ft

10.
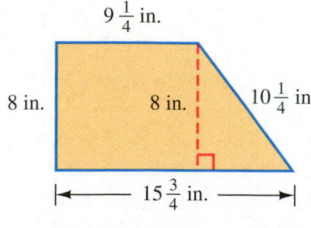
$9\frac{1}{4}$ in., 8 in., 8 in., $10\frac{1}{4}$ in., $15\frac{3}{4}$ in.

11.
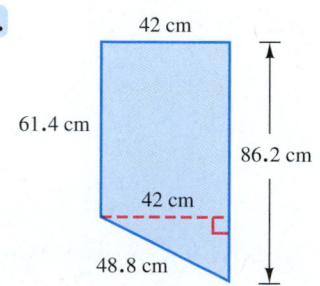
42 cm, 61.4 cm, 86.2 cm, 42 cm, 48.8 cm

12.
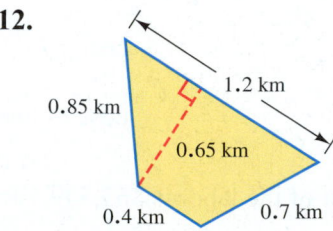
1.2 km, 0.85 km, 0.65 km, 0.4 km, 0.7 km

Solve each application problem. First label the bases and heights on the sketches. See Example 6.

13. The backyard of a new home is shaped like a trapezoid with a height of 45 ft and bases of 80 ft and 110 ft. What is the cost of putting sod on the yard if the landscaper charges $0.33 per square foot for sod?

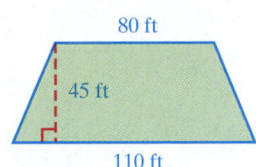

14. A swimming pool is in the shape of a parallelogram with a height of 9.6 m and base of 12.4 m. Find the labor cost to make a custom solar cover for the pool at a cost of $4.92 per square meter.

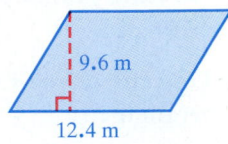

15. A piece of fabric for a quilt design is in the shape of a parallelogram. The base is 5 in. and the height is 3.5 in. What is the total area of the 25 parallelogram pieces needed for the quilt?

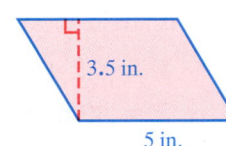

16. An accountant is paying $832 per month to rent an office in an old building. Her office is shaped like a trapezoid, with bases of 32 ft and 20 ft and a height of 20 ft. How much rent is she paying per square foot?

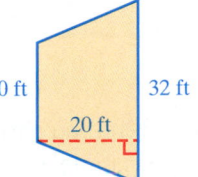

*Find **two** errors in each student's solution below. Write a sentence explaining each error. Then show how to work the problem correctly.*

17.

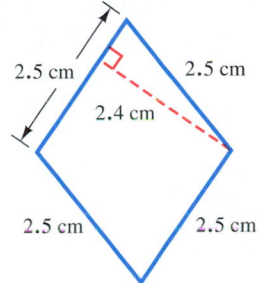

$P = 2.5 \text{ cm} + 2.4 \text{ cm} + 2.5 \text{ cm} + 2.5 \text{ cm} + 2.5 \text{ cm}$
$P = 12.4 \text{ cm}^2$

18.

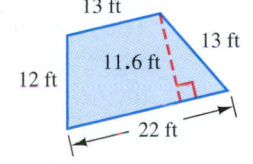

$A = (0.5)(11.6 \text{ ft}) \cdot (12 \text{ ft} + 13 \text{ ft})$
$A = 145 \text{ ft}$

Find the area of each figure. See Example 5.

19.

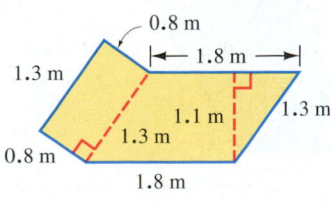

20.

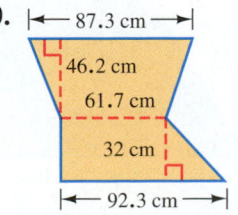

21.

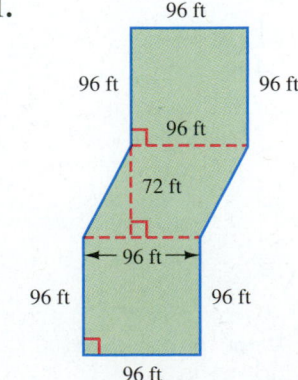

7.4 ▶▶▶ Triangles

OBJECTIVES

1 Find the perimeter of a triangle.

2 Find the area of a triangle.

3 Given the measures of two angles in a triangle, find the measure of the third angle.

A **triangle** is a figure with exactly three sides. Some examples are shown below.

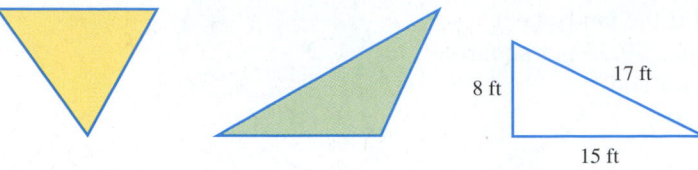

OBJECTIVE 1 Find the perimeter of a triangle. To find the perimeter of a triangle (the distance around the edges), add the lengths of the three sides.

EXAMPLE 1 Finding the Perimeter of a Triangle

Find the perimeter of the triangle above on the right.

$$P = 8 \text{ ft} + 15 \text{ ft} + 17 \text{ ft}$$
$$P = 40 \text{ ft}$$

Careful! Write **ft** in the answer.

◀ *Work Problem* **1** *at the Side.*

As with parallelograms, you can find the *height* of a triangle by measuring the distance from one vertex of the triangle to the opposite side (the base). The height line must be *perpendicular* to the base; that is, it must form a right angle with the base. Sometimes you have to extend the base in order to draw the height perpendicular to it, as shown below in the figure on the right.

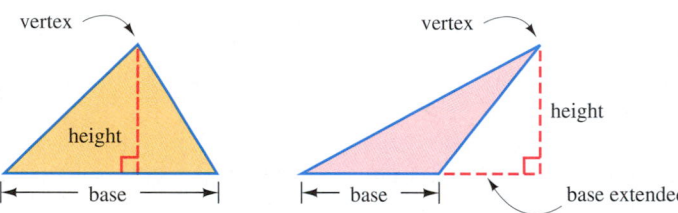

If you cut out two identical triangles and turn one upside down, you can fit them together to form a parallelogram, as shown below.

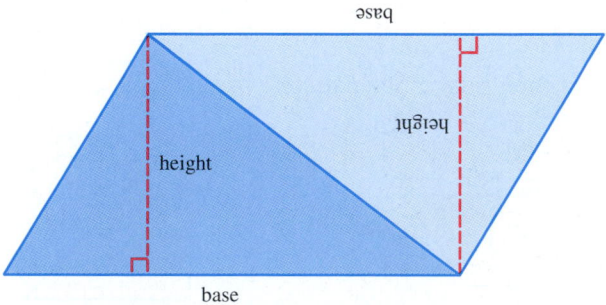

Recall from **Section 7.3** that the area of the parallelogram is *base* times *height*. Because each triangle is *half* of the parallelogram, the area of one triangle is

$$\frac{1}{2} \text{ of base times height}.$$

1 Find the perimeter of each triangle.

(a)
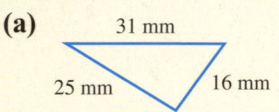

(b)
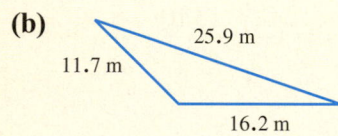

(c) A triangle with sides of $6\frac{1}{2}$ yd, $9\frac{3}{4}$ yd, and $11\frac{1}{4}$ yd

ANSWERS

1. (a) $P = 72$ mm **(b)** $P = 53.8$ m

(c) $P = 27\frac{1}{2}$ yd or 27.5 yd

OBJECTIVE **2** **Find the area of a triangle.** Use the following formula to find the *area* of a triangle.

> **Finding the Area of a Triangle**
>
> $$\text{Area of a triangle} = \frac{1}{2} \cdot \text{base} \cdot \text{height}$$
>
> $$A = \frac{1}{2} \cdot b \cdot h$$
>
> or $A = 0.5 \cdot b \cdot h$
>
> Remember to use *square units* when measuring area.

EXAMPLE 2 **Finding the Area of Triangles**

Find the area of each triangle.

(a)

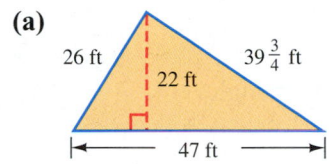

The base is 47 ft and the height is 22 ft. You do *not* need the 26 ft or $39\frac{3}{4}$ ft sides to find the area.

$$A = \frac{1}{2} \cdot \quad b \quad \cdot \quad h$$

$$A = \frac{1}{2} \cdot 47 \text{ ft} \cdot \overset{11}{\cancel{22}} \text{ ft} \qquad \begin{array}{l}\text{Divide out} \\ \text{common} \\ \text{factor of 2.}\end{array}$$

> This is *area*, so write **ft²** in the answer.

$$A = 517 \text{ ft}^2 \qquad \text{Square units for area}$$

(b)

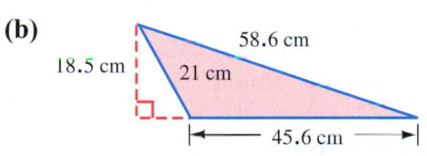

$$A = 0.5 \cdot 45.6 \text{ cm} \cdot 18.5 \text{ cm}$$

$$A = 421.8 \text{ cm}^2$$

The base must be extended to draw the height. However, still use 45.6 cm for b in the formula. Because the measurements are decimal numbers, it is easier to use 0.5 (the decimal equivalent of $\frac{1}{2}$) in the formula.

(c)

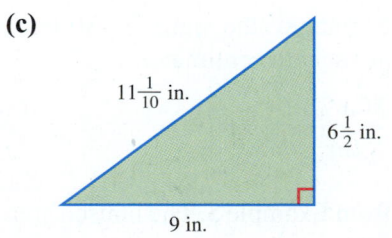

Because two sides of the triangle are perpendicular to each other, use those sides as the base and the height. (Recall that the height must be perpendicular to the base.) You can use fractions or use the equivalent decimal numbers.

Using fractions $A = \dfrac{1}{2} \cdot 9 \text{ in.} \cdot 6\dfrac{1}{2} \text{ in.} = \dfrac{1}{2} \cdot \dfrac{9 \text{ in.}}{1} \cdot \dfrac{13 \text{ in.}}{2} = \dfrac{117}{4} \text{ in.}^2 = 29\dfrac{1}{4} \text{ in.}^2$

Using decimals $A = 0.5 \cdot 9 \text{ in.} \cdot 6.5 \text{ in.} = 29.25 \text{ in.}^2 \leftarrow$ Equivalent

Work Problem **2** *at the Side.* ▶

---

2 Find the area of each triangle.

(a)

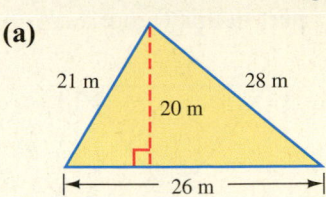

(b)

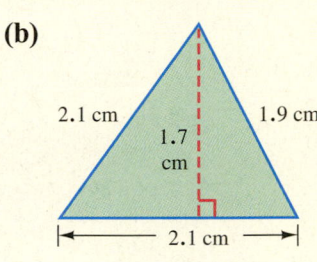

(c)

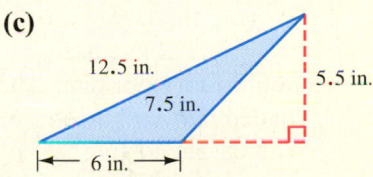

(d)

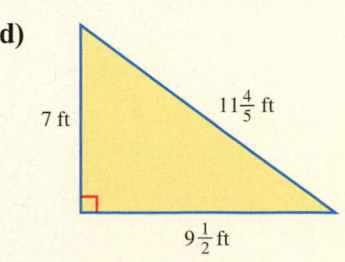

ANSWERS

2. **(a)** $A = 260 \text{ m}^2$ **(b)** $A = 1.785 \text{ cm}^2$
(c) $A = 16.5 \text{ in.}^2$
(d) $A = 33.25 \text{ ft}^2$ or $33\frac{1}{4} \text{ ft}^2$

3 Find the area of the shaded part in this figure.

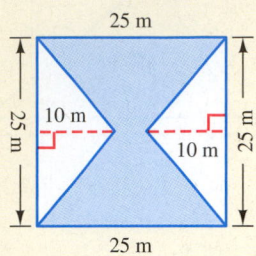

25 m
25 m
10 m
10 m
25 m
25 m

EXAMPLE 3 Using the Concept of Area

Find the area of the shaded part in this figure.

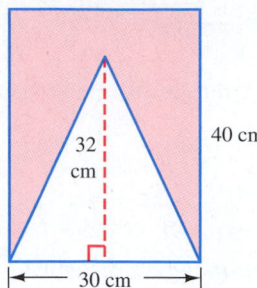

32 cm
40 cm
30 cm

The *entire* figure is a rectangle. Find the area of the rectangle.

$$A = l \cdot w$$
$$A = 30 \text{ cm} \cdot 40 \text{ cm}$$
$$A = 1200 \text{ cm}^2$$

The *un*shaded part is a triangle. Find the area of the triangle.

$$A = \frac{1}{\overset{}{2}} \cdot \overset{15}{\cancel{30}} \text{ cm} \cdot 32 \text{ cm}$$
$$A = 480 \text{ cm}^2$$

Subtract to find the area of the shaded part.

$$A = \overset{\text{Entire area}}{\overbrace{1200 \text{ cm}^2}} - \overset{\text{Unshaded part}}{\overbrace{480 \text{ cm}^2}} = \overset{\text{Shaded part}}{\overbrace{720 \text{ cm}^2}}$$

This is *area*, so write **cm²** in the answer.

The area of the shaded part of the figure is 720 cm².

◀ Work Problem **3** at the Side.

4 Suppose the figure in Problem 3 above is an auditorium floor plan. The shaded part will be covered with carpet costing $27 per square meter. The rest will be covered with vinyl floor covering costing $18 per square meter. What is the total cost of covering the floor?

EXAMPLE 4 Applying the Concept of Area

The Department of Transportation cuts triangular signs out of rectangular pieces of metal using the measurements shown above in Example 3. If the metal costs $0.02 per square centimeter, how much does the metal cost for the sign? What is the cost of the metal that is *not* used?

From Example 3 above, the area of the triangle (the sign) is 480 cm². Multiply the area of the sign times the cost per square centimeter.

$$\text{cost of sign} = \frac{480 \text{ cm}^2}{1} \cdot \frac{\$0.02}{1 \text{ cm}^2} = \$9.60$$

The metal that is *not* used is the *shaded* part from Example 3. The unused area is 720 cm².

$$\text{cost of unused mental} = \frac{720 \text{ cm}^2}{1} \cdot \frac{\$0.02}{1 \text{ cm}^2} = \$14.40$$

The cost of the metal for the triangular sign is $9.60. The unused metal costs $14.40.

◀ Work Problem **4** at the Side.

ANSWERS
3. $A = 625 \text{ m}^2 - 125 \text{ m}^2 - 125 \text{ m}^2 = 375 \text{ m}^2$
4. $\$10{,}125 + \$2250 + \$2250 = \$14{,}625$

OBJECTIVE 3 **Given the measures of two angles in a triangle, find the measure of the third angle.** The *tri* in *triangle* means *three*. So the name tells you that a triangle has three angles. The sum of the measures of the three angles in any triangle is *always* 180°. You can see it by drawing a triangle, cutting off the three angles, and rearranging them to make a straight angle (180°).

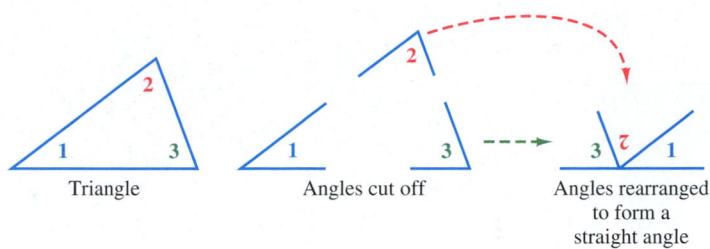

Triangle   Angles cut off   Angles rearranged
                                      to form a
                                      straight angle

> **Finding the Unknown Angle Measurement in a Triangle**
>
> *Step 1*   Add the number of degrees in the measures of the two given angles.
>
> *Step 2*   Subtract the sum from 180°.

EXAMPLE 5   **Finding an Angle Measurement in Triangles**

Find the number of degrees in the indicated angle.

(a) Angle *R*

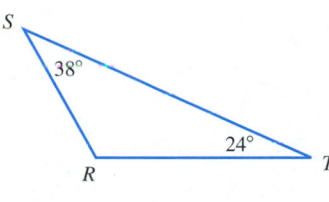

Step 1   Add the two angle measurements you are given.

$$38° + 24° = 62°$$

Step 2   Subtract the sum from 180°.

$$\mathbf{180°} - 62° = 118°$$

∠*R* measures 118°.

(b) Angle *F*

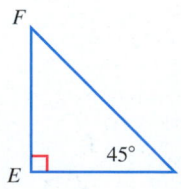

∠*E* is a right angle, so it measures 90°.

Step 1 $90° + 45° = 135°$

Step 2 $\mathbf{180°} - 135° = 45°$

∠*F* measures 45°.

> Write a small, raised circle for "degrees."

Work Problem **5** *at the Side.* ▶

5 Find the number of degrees in the third angle of each triangle.

(a)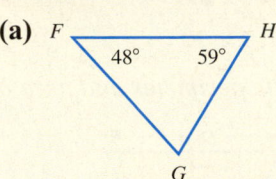

(b)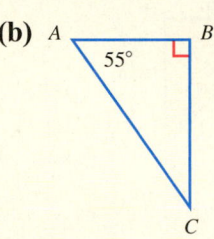

(c)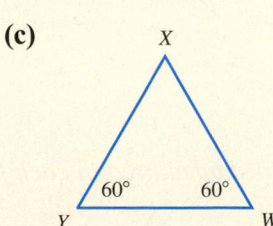

7.4 ▶▶▶ Exercises

Find the perimeter and area of each triangle. See Examples 1 and 2.

1.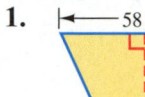
58 m
66 m
72 m   72 m

2.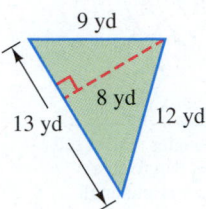
9 yd
8 yd
13 yd   12 yd

◐ 3.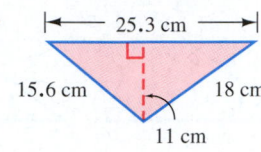
25.3 cm
15.6 cm   18 cm
11 cm

4.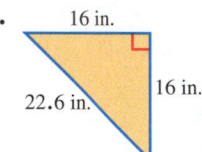
16 in.
16 in.
22.6 in.

5.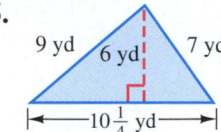
9 yd   6 yd   7 yd
$10\frac{1}{4}$ yd

6.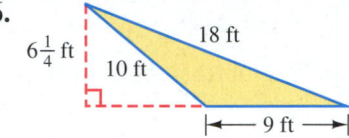
18 ft
$6\frac{1}{4}$ ft   10 ft
9 ft

7.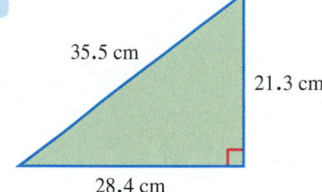
35.5 cm
21.3 cm
28.4 cm

8.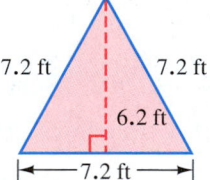
7.2 ft   7.2 ft
6.2 ft
7.2 ft

Find the shaded area in each figure. See Example 3.

9.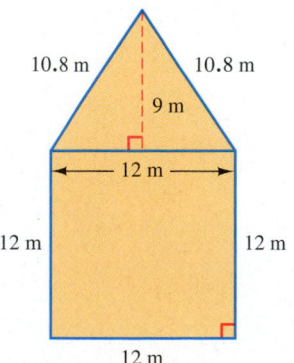
10.8 m   10.8 m
9 m
12 m
12 m   12 m
12 m

10.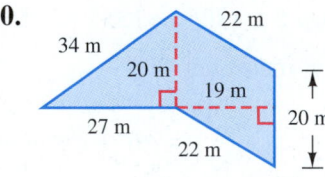
22 m
34 m
20 m
19 m
27 m   20 m
22 m

11.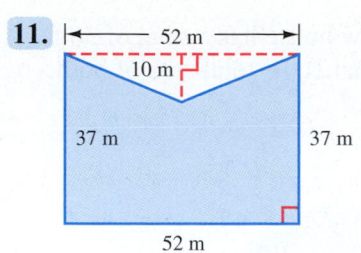

12.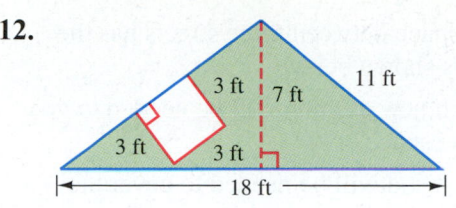

Find the number of degrees in the third angle of each triangle. See Example 5.

13.

14.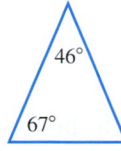

🌐 **15.**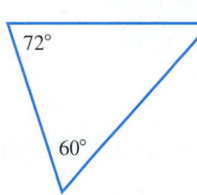

16.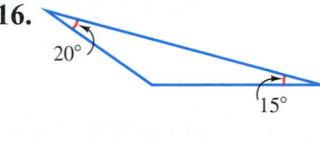

17. Can a triangle have two right angles? Explain your answer.

18. In your own words, explain where the $\frac{1}{2}$ comes from in the formula for area of a triangle. Draw a sketch to illustrate your explanation.

Solve each application problem. See Example 4.

19. A triangular tent flap measures $3\frac{1}{2}$ ft along the base and has a height of $4\frac{1}{2}$ ft. How much canvas is needed to make the flap?

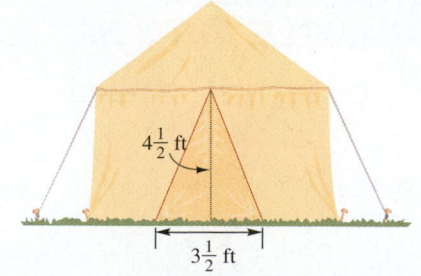

20. A wooden sign in the shape of a right triangle has perpendicular sides measuring 1.5 m and 1.2 m. How much surface area does the sign have?

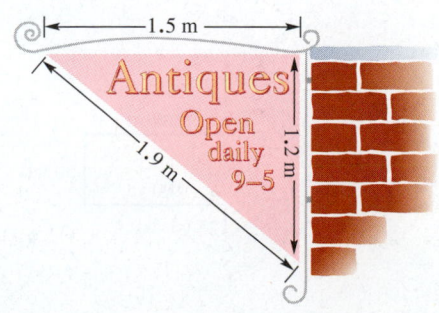

21. A triangular space between three streets has the measurements shown below.

   (a) How much new curbing will be needed to go around the space?

   (b) How much sod will be needed to cover the space?

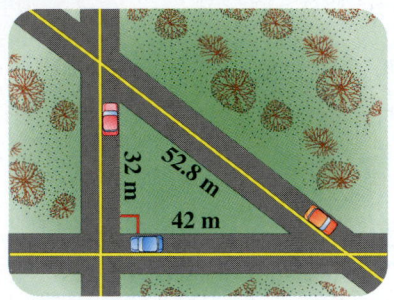

22. Each gable end of a new house has a span of 36 ft and a rise of 9.5 ft. What is the total area of both gable ends of the house?

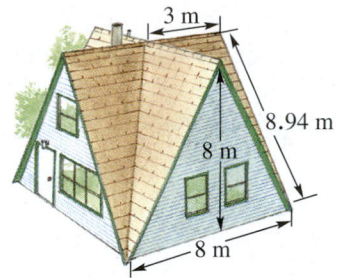

23. All sides of the house below are congruent and all roof sections are congruent.

   (a) Find the area of one side of the house.

   (b) Find the area of one roof section.

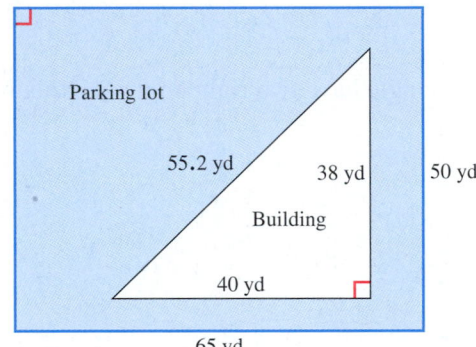

24. The sketch shows the plan for an office building. The shaded area will be a parking lot. What is the cost of building the parking lot if the contractor charges $35 per square yard for materials and labor?

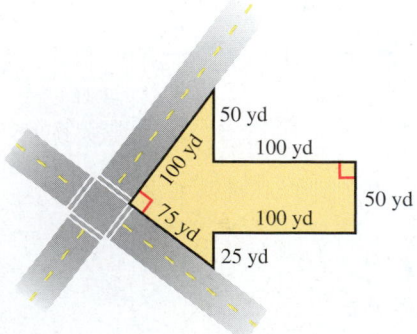

25. A city lot with an unusual shape is shown below.

   (a) How much frontage (distance along streets) does the lot have?

   (b) What is the area of the lot?

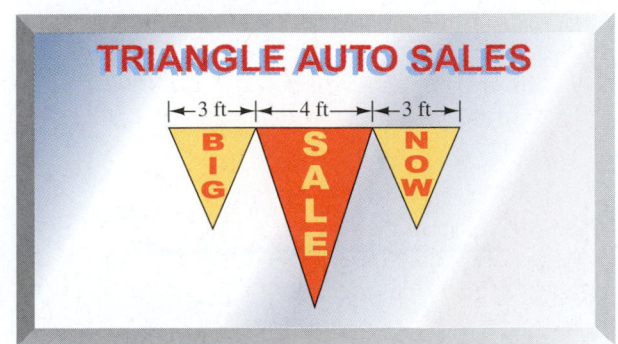

26. A car dealership wants three nylon pennants to hang in its front window. The height of the two smaller pennants is 3.5 ft, and the larger pennant has a height of 6.5 ft. How much nylon fabric is needed for all the pennants?

Math in the Media

SCREEN DISPLAYS

The world's largest plasma screen panel display is now used in airports, sports stadiums, and resorts in the United States, Europe, Japan, and China. The table below shows some data available on-line about the plasma screen. *Source:* www.eurosell.it

WORLD'S LARGEST PLASMA SCREEN PANEL DISPLAY

Measurements	Width 2269 mm Height 1277 mm
Weight	Approximately 215 kg
Service life	60,000 hours

Use the data in the table to answer Questions 1–5, rounding to the nearest tenth if necessary.

1. The screen is how many meters wide and how many meters high?

2. Using your answers from Question 1, convert the width and height to feet, using the fact that $1 \text{ m} \approx 3.28 \text{ ft}$.

3. Now find the perimeter and area of the screen using your results from Question 2.

4. Convert the weight to pounds, using the fact that $1 \text{ kg} \approx 2.20 \text{ lb}$.

5. (a) If the screen is never turned off, how long is the service life in days?

   (b) How long is the service life in years? (Use 365 days per year.)

COMPUTER MONITOR DISPLAYS

Type of Monitor	Width of screen	Height of screen	Viewing area
Dell PC monitor	$12\frac{3}{4}$ in.	$9\frac{1}{2}$ in.	
ViewSonic LCD monitor	16.1 in.	10.0 in.	

Source: Companies' Web sites.

6. The table above shows the measurements of several typical, rectangular, computer monitor screens. Find the viewing area of each screen, to the nearest whole number, and write it in the table. For the Dell PC monitor, first change the measurements to decimal numbers before finding the viewing area.

7. Draw a sketch of the Dell monitor. Label the lengths of the sides. Label the right angles. Which sides are parallel?

7.5 ▶▶▶ Circles

OBJECTIVES

1 Find the radius and diameter of a circle.

2 Find the circumference of a circle.

3 Find the area of a circle.

4 Become familiar with Latin and Greek prefixes used in math terminology.

OBJECTIVE **1** **Find the radius and diameter of a circle.** Suppose you start with one dot on a piece of paper. Then you draw many dots that are each 2 cm away from the first dot. If you draw enough dots (points) you'll end up with a *circle*. Each point on the circle is exactly 2 cm away from the *center* of the circle. The 2 cm distance is called the *radius, r*, of the circle. The distance across the circle (passing through the center) is called the *diameter, d*, of the circle.

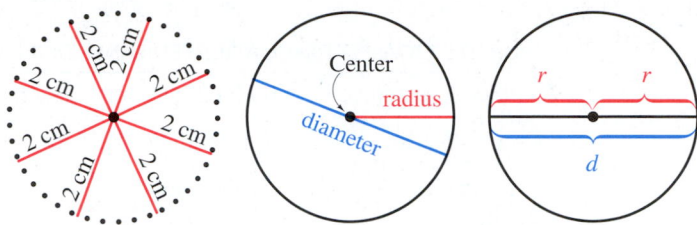

Circle, Radius, and Diameter

A **circle** is a two-dimensional (flat) figure with all points the same distance from a fixed center point.

The **radius** (*r*) is the distance from the center of the circle to any point on the circle.

The **diameter** (*d*) is the distance across the circle passing through the center.

Using the circle above on the right as a model, you can see some relationships between the radius and diameter.

Finding the Diameter and Radius of a Circle

$$\text{diameter} = 2 \cdot \text{radius}$$
$$d = 2 \cdot r$$

$$\text{and} \quad \text{radius} = \frac{\text{diameter}}{2} \quad \text{or} \quad r = \frac{d}{2}$$

EXAMPLE 1 **Finding the Diameter and Radius of a Circle**

Find the unknown length of the diameter or radius in each circle.

(a)

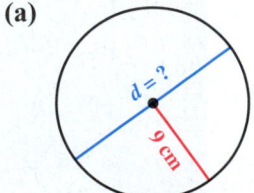

Because the radius is 9 cm, the diameter is twice as long.

$$d = 2 \cdot \ r$$
$$d = 2 \cdot 9 \text{ cm}$$
$$d = 18 \text{ cm}$$

Continued on Next Page

(b)

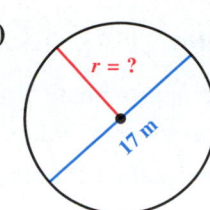

The radius is half the diameter.

$$r = \frac{d}{2}$$

$$r = \frac{17 \text{ m}}{2}$$

$$r = 8.5 \text{ m} \quad \text{or} \quad 8\frac{1}{2} \text{ m}$$

> $8.5 = 8\frac{1}{2}$ because $8\frac{5}{10}$ simplifies to $8\frac{1}{2}$.

Work Problem (1) *at the Side.* ▶

OBJECTIVE 2 Find the circumference of a circle. The perimeter of a circle is called its **circumference.** Circumference is the distance around the edge of a circle.

The diameter of the can in the drawing is about 10.6 cm, and the circumference of the can is about 33.3 cm. Dividing the circumference of the circle by the diameter gives an interesting result.

$$\frac{\text{circumference}}{\text{diameter}} = \frac{33.3}{10.6} \approx 3.14 \qquad \text{Rounded to the nearest hundredth}$$

Dividing the circumference of *any* circle by its diameter *always* gives an answer close to 3.14. This means that going around the edge of any circle is a little more than 3 times as far as going straight across the circle.

This ratio of circumference to diameter is called π (the Greek letter **pi**, pronounced PIE). There is no decimal that is exactly equal to π, but here is the *approximate* value.

$$\pi \approx 3.14159265359$$

Rounding the Value of *Pi* (π)

We usually round π to 3.14. Therefore, calculations involving π will give approximate answers and should be written using the $\approx$ symbol.

Use the following formulas to find the *circumference* of a circle.

Finding the Circumference (Distance around a Circle)

Circumference = π • diameter

$$C = \pi \bullet d$$

or, because $d = 2 \bullet r$, then $C = \pi \bullet 2 \bullet r$ usually written $C = 2 \bullet \pi \bullet r$

Remember to use linear units such as ft, yd, m, and cm when measuring circumference (**not** square units).

---

(1) Find the unknown length of the diameter or radius in each circle.

(a)

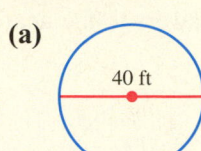

(b)

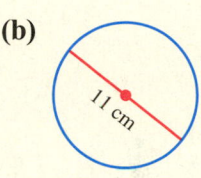

(c)

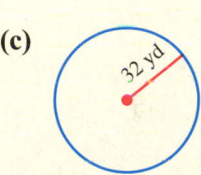

(d)

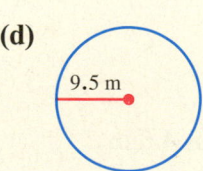

2 Find the circumference of each circle. Use 3.14 as the approximate value for π. Round answers to the nearest tenth.

(a)

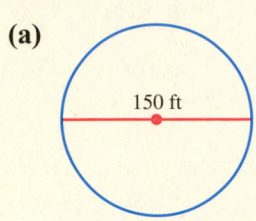

150 ft

(b)

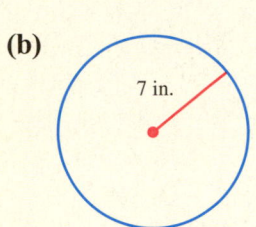

7 in.

(c) diameter 0.9 km

(d) radius 4.6 m

EXAMPLE 2 Finding the Circumference of Circles

Find the circumference of each circle. Use 3.14 as the approximate value for π. Round answers to the nearest tenth.

(a)

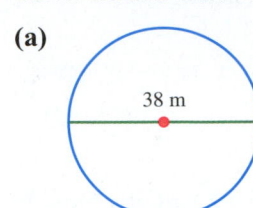

38 m

The *diameter* is 38 m, so use the formula with *d* in it.

$$C = \pi \cdot d$$
$$C \approx 3.14 \cdot 38 \text{ m}$$

Be sure to write **m** in the answer.

$$C \approx 119.3 \text{ m} \quad \text{Rounded}$$

(b)

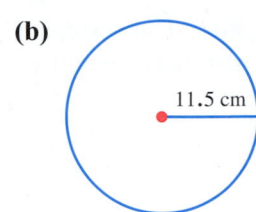

11.5 cm

In this example, the *radius* is labeled, so it is easier to use the formula with *r* in it.

$$C = 2 \cdot \pi \cdot r$$
$$C \approx 2 \cdot 3.14 \cdot 11.5 \text{ cm}$$

Write **cm** in the answer.

$$C \approx 72.2 \text{ cm} \quad \text{Rounded}$$

Calculator Tip Many *scientific* calculators have a π key. Try pressing it. With a 10-digit display, you'll see the value of π to the nearest billionth.

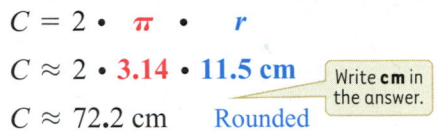

3.141592654

But this is still an approximate value, although it is more precise than rounding π to 3.14. Try finding the circumference in Example 2(a) above using the π key.

π $\times$ 38 $=$ Answer is 119.3805208; rounds to 119.4

When you used 3.14 as the approximate value of π, the result rounded to 119.3, so the answers are slightly different. In this book, we will use 3.14 instead of the π key. Our measurements of radius and diameter are given as whole numbers or with tenths, so it is acceptable to round π to hundredths. And, some students may be using standard calculators without a π key or doing the calculations by hand.

◀ Work Problem **2** at the Side.

OBJECTIVE 3 Find the area of a circle. To find the formula for the area of a circle, start by cutting two circles into many pie-shaped pieces.

Circumference (distance around) is $2 \cdot \pi \cdot r$.

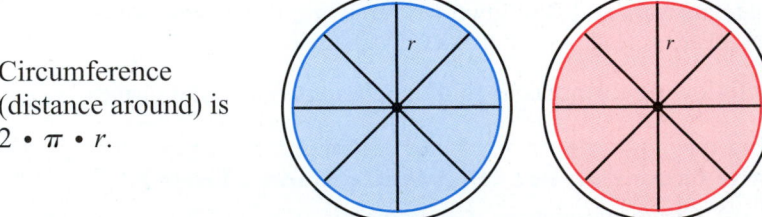

Unfold the circles, much as you might "unfold" a peeled orange, and put them together as shown here.

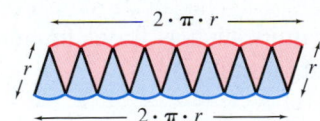

(continued)

ANSWERS

2. **(a)** $C \approx 471$ ft   **(b)** $C \approx 44.0$ in.
   (c) $C \approx 2.8$ km   **(d)** $C \approx 28.9$ m

The figure is approximately a parallelogram with height r (the radius of the original circle) and base $2 \cdot \pi \cdot r$ (the circumference of the original circle). The area of the "parallelogram" is base times height.

$$\text{Area} = \overbrace{b \quad \cdot \underbrace{h}}$$
$$\text{Area} = 2 \cdot \pi \cdot r \cdot r$$
$$\text{Area} = 2 \cdot \pi \cdot r^2 \leftarrow \text{Recall that } r \cdot r \text{ is } r^2$$

Because the "parallelogram" was formed from *two* circles, the area of *one* circle is half as much.

$$\frac{1}{\cancel{2}} \cdot \cancel{2} \cdot \pi \cdot r^2 = 1 \cdot \pi \cdot r^2 \quad \text{or simply} \quad \pi \cdot r^2$$

> **Finding the Area of a Circle**
> $$\text{Area of a circle} = \pi \cdot \text{radius} \cdot \text{radius}$$
> $$A = \pi \cdot r^2$$
> Remember to use **square units** when measuring area.

EXAMPLE 3   **Finding the Area of Circles**

Find the area of each circle. Use 3.14 for π. Round your answers to the nearest tenth.

(a) A circle with a radius of 8.2 cm
Use the formula $A = \pi \cdot r^2$, which means $\pi \cdot r \cdot r$.

$$A = \pi \cdot r \cdot r$$
$$A \approx 3.14 \cdot 8.2 \text{ cm} \cdot 8.2 \text{ cm}$$
This is *area*, so write **cm²** in the answer.
$$A \approx 211.1 \text{ cm}^2 \quad \text{Rounded}$$

(b)

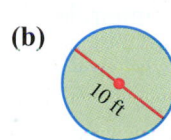

To use the area formula, you need to know the *radius* (r). In this circle, the *diameter* is 10 ft. First find the radius.

$$r = \frac{d}{2}$$
You *cannot* use the *diameter* measurement in the area formula, so find the *radius*.

$$r = \frac{10 \text{ ft}}{2} = 5 \text{ ft}$$

Now find the area.

$$A \approx 3.14 \cdot 5 \text{ ft} \cdot 5 \text{ ft}$$
$$A \approx 78.5 \text{ ft}^2$$
Write square units (**ft²**) in your answer.

CAUTION
When finding *circumference,* you can start with either the radius or the diameter. When finding *area,* you must use the *radius.* If you are given the diameter, divide it by 2 to find the radius. Then find the area.

Work Problem **3** *at the Side.* ▶

---

3 Find the area of each circle. Use 3.14 for π. Round your answers to the nearest tenth.

(a)

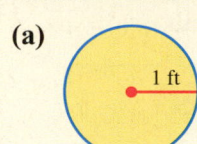
1 ft

(b)

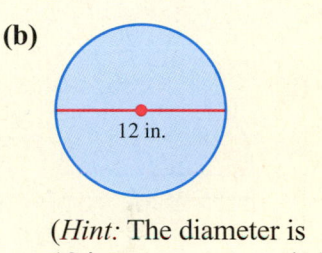
12 in.

(*Hint:* The diameter is 12 in. so $r = \underline{\quad}$ in.)

(c)

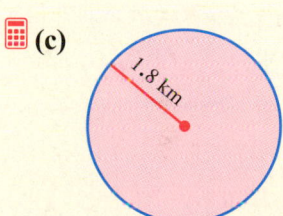
1.8 km

(d)

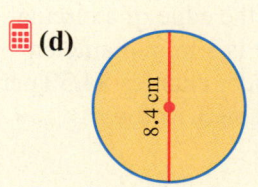
8.4 cm

ANSWERS

3. **(a)** $A \approx 3.1 \text{ ft}^2$
   (b) $r = 6$ in.; $A \approx 113.0 \text{ in.}^2$
   (c) $A \approx 10.2 \text{ km}^2$ **(d)** $A \approx 55.4 \text{ cm}^2$

4 Find the area of each semicircle. Use 3.14 for π. Round your answers to the nearest tenth.

(a)

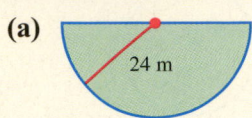

24 m

(b)

|←——35.4 ft——→|

(c)

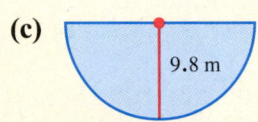

9.8 m

---

5 Find the cost of binding around the edge of a circular rug that is 3 m in diameter. The binder charges $4.50 per meter. Use 3.14 for π.

ANSWERS

4. (a) $A \approx 904.3 \text{ m}^2$ **(b)** $A \approx 491.9 \text{ ft}^2$
  (c) $A \approx 150.8 \text{ m}^2$
5. $42.39

⌨ **Calculator Tip**   You can use your calculator to find the area of the circle in Example 3(a) on the previous page. The first method works on both scientific and standard calculators:

$$3.14 \; \textcircled{\times} \; 8.2 \; \textcircled{\times} \; 8.2 \; \textcircled{=} \qquad \text{Answer is } 211.1336$$

You round the answer to 211.1 (nearest tenth).

On a *scientific* calculator you can also use the $\boxed{x^2}$ key, which automatically squares the number you enter (that is, multiplies the number times itself):

$$3.14 \; \textcircled{\times} \; 8.2 \; \boxed{x^2} \; \underbrace{67.24} \; \textcircled{=} \qquad \text{Answer is } 211.1336$$

Appears automatically;
8.2×8.2 is 67.24

In the next example, we find the area of a *semicircle,* which is half the area of a circle.

EXAMPLE 4 **Finding the Area of a Semicircle**

Find the area of the semicircle. Use 3.14 for π. Round your answer to the nearest tenth.

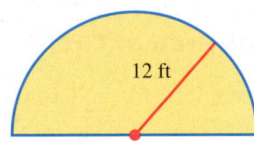

12 ft

First, find the area of a whole circle with a radius of 12 ft.

$$A = \pi \cdot r \cdot r$$
$$A \approx 3.14 \cdot 12 \text{ ft} \cdot 12 \text{ ft}$$
$$A \approx 452.16 \text{ ft}^2 \leftarrow \text{Do not round yet.}$$

Divide the area of the whole circle by 2 to find the area of the semicircle.

$$\frac{452.16 \text{ ft}^2}{2} = 226.08 \text{ ft}^2$$

The *last* step is rounding 226.08 to the nearest tenth.   [Write **ft²** in the answer.]

Area of semicircle $\approx 226.1 \text{ ft}^2$ Rounded

◀ *Work Problem* **4** *at the Side.*

EXAMPLE 5 **Applying the Concept of Circumference**

A circular rug is 8 ft in diameter. The cost of fringe for the edge is $2.25 per foot. What will it cost to add fringe to the rug? Use 3.14 for π.

$$\text{Circumference} = \pi \cdot d$$
$$C \approx 3.14 \cdot 8 \text{ ft}$$
$$C \approx 25.12 \text{ ft}$$

$$\text{cost} = \text{cost per foot} \cdot \text{circumference}$$
$$\text{cost} = \frac{\$2.25}{1 \text{ ft}} \cdot \frac{25.12 \text{ ft}}{1}$$
$$\text{cost} = \$56.52$$

The cost of adding fringe to the rug is $56.25.

◀ *Work Problem* **5** *at the Side.*

EXAMPLE 6 | **Applying the Concept of Area**

Find the cost of covering the rug in Example 5 (on the previous page) with a plastic cover. The material for the cover costs $1.50 per square foot. Use 3.14 for π.

First find the radius. $\qquad r = \dfrac{d}{2} = \dfrac{8 \text{ ft}}{2} = 4 \text{ ft}$

Then find the area. $\qquad A = \pi \cdot r^2$

$\qquad\qquad\qquad\qquad A \approx 3.14 \cdot 4 \text{ ft} \cdot 4 \text{ ft}$

$\qquad\qquad\qquad\qquad A \approx 50.24 \text{ ft}^2$

$$\text{cost} = \frac{\$1.50}{1 \text{ ft}^2} \cdot \frac{50.24 \text{ ft}^2}{1} = \$75.36$$

The cost of the plastic cover is $75.36.

Work Problem 6 *at the Side.* ▶

OBJECTIVE 4 **Become familiar with Latin and Greek prefixes used in math terminology.** Many English words are built from Latin or Greek root words and prefixes. Knowing the meaning of the more common ones can help you figure out the meaning of terms in many subject areas, including mathematics.

EXAMPLE 7 | **Using Prefixes to Understand Math Terms**

(a) Listed below are some Latin and Greek root words and prefixes with their meanings in parentheses. You've already seen math terms in this textbook that use these prefixes. List at least one math term and one nonmathematical word that use each prefix or root word.

cent- (100): *cent*imeter; *cent*ury

circum- (around): *circum*ference; *circum*vent

de- (down): *de*nominator; *de*cline

dec- (10): *dec*imal; *Dec*ember (originally the 10th month in the old calendar)

There are many answers. These are some of the possibilities.

(b) Suppose you have trouble remembering which part of a fraction is the denominator. How could your knowledge of prefixes help in this situation?

The *de-* prefix in *de*nominator means "down" so the denominator is the number *down* below the fraction bar.

Work Problem 7 *at the Side.* ▶

Note

Here are some additional prefixes and root words and their meanings that you will see in the rest of this chapter, later chapters in this book, and in other math classes. An example of a math term and a nonmathematical word are shown for each one.

equ- (equal): *equ*ation; *equi*nox

hemi- (half): *hemi*sphere; *hemi*trope

lateral (side): quadri*lateral*; bi*lateral*

re- (back or again): *re*ciprocal; *re*duce

---

6 | Find the cost of covering the underside of the rug in margin problem 5 with a nonslip rubber backing. The rubber backing costs $3.89 per square meter.

7 | **(a)** Here are some more prefixes you have seen in this textbook. List at least one math term and one nonmathematical word that use each prefix.

dia- (through):

fract- (break):

par- (beside):

per- (divide):

peri- (around);

rad- (ray):

rect- (right):

sub- (below):

(b) How could you use your knowledge of prefixes to remember the difference between perimeter and area?

ANSWERS

6. $27.48
7. **(a)** Some possibilities are:
   *dia*meter; *dia*gonal
   *fract*ion; *fract*ure
   *par*allel; *par*amedic
   *per*cent; *per* capita
   *peri*meter; *peri*scope
   *rad*ius; *rad*iate
   *rect*angle; *rect*ify
   *sub*tract; *sub*marine
   (b) *Peri-* in perimeter means "around," so perimeter is the distance *around* the edges of a shape.

7.5 ▶▶▶ Exercises

Find the unknown length in each circle. See Example 1.

1.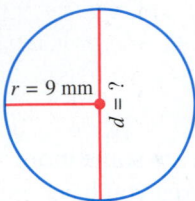

$r = 9$ mm $d = ?$

2.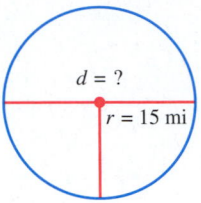

$d = ?$ $r = 15$ mi

3.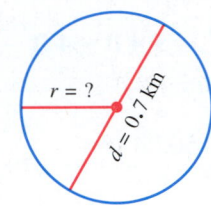

$r = ?$ $d = 0.7$ km

4.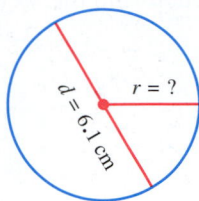

$d = 6.1$ cm $r = ?$

Find the circumference and area of each circle. Use 3.14 as the approximate value for π. Round your answers to the nearest tenth. See Examples 2 and 3.

5.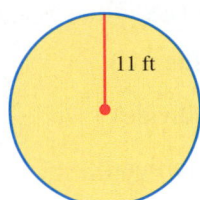

11 ft

6.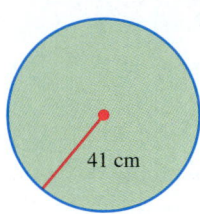

41 cm

7.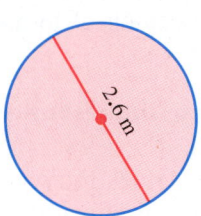

2.6 m

8.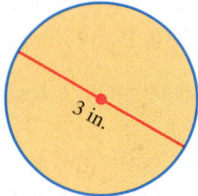

3 in.

Find the circumference and area of circles having the following diameters. Use 3.14 as the approximate value of π. Round your answers to the nearest tenth. See Examples 2 and 3.

9. $d = 15$ cm

10. $d = 39$ ft

11. $d = 7\frac{1}{2}$ ft

12. $d = 4\frac{1}{2}$ yd

13. $d = 8.65$ km

14. $d = 19.5$ mm

Find each shaded area. Note that Exercises 15–18 all contain semicircles. Use 3.14 as the approximate value of π. Round your answers to the nearest tenth if necessary. See Example 4.

15.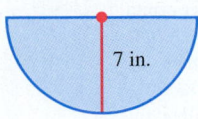

7 in.

16.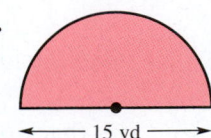

15 yd

17.

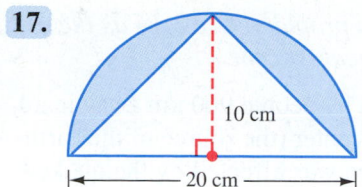

18.

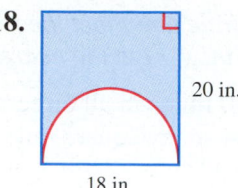

19. How would you explain π to a friend who is not in your math class? Write an explanation. Then make up a test question that requires the use of π, and show how to solve it.

20. Explain how circumference and perimeter are alike. How are they different? Make up two problems, one involving perimeter, the other circumference. Show how to solve your problems.

Solve each application problem. Use 3.14 as the approximate value of π. Round your answers to the nearest tenth. See Examples 5 and 6.

21. An irrigation system moves around a center point to water a circular area for crops. If the irrigation system is 50 yd long, how large is the watered area?

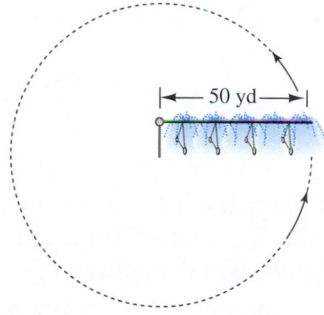

22. If you swing a ball held at the end of a string 20 cm long, how far will the ball travel on each turn?

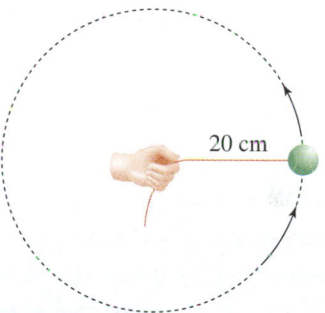

23. A Michelin Cross Terrain SUV tire has an overall diameter of 29.10 inches. How far will a point on the tire tread move in one complete turn? (*Source:* Michelin.) *Bonus question:* How many revolutions does the tire make per mile? Round to the nearest whole number.

24. In August 2005, Hurricane Katrina slammed into coastal Mississippi and flooded New Orleans. The diameter of the circular storm was 210 miles. In July 2005, Hurricane Dennis, with a diameter of only 80 miles, hit the Florida Panhandle. What was the area of each storm, to the nearest hundred square miles? (*Source:* Associated Press.)

For Exercises 25–30, first draw a circle and label the radius or diameter. Then solve the problem. Use 3.14 as the approximate value of π and round answers to the nearest tenth, money answers to the nearest cent.

25. A radio station can be heard 150 miles in all directions during evening hours. How many square miles are in the station's broadcast area?

26. An earthquake was felt by people 900 km away in all directions from the epicenter (the source of the earthquake). How much area was affected by the quake?

27. The diameter of Diana Hestwood's wristwatch is 1 in. and the radius of the clock face on her kitchen wall is 3 in. Find the circumference and the area of the watch face and the clock face.

28. The diameter of the largest known ball of twine is 12 ft 9 in. The sign posted near the ball says it has a circumference of 40 ft. Is the sign correct? *Hint:* First change 9 in. to feet and add it to 12 ft. (*Source: Guinness World Records.*)

29. Blaine Fenstad wants to buy a pair of two-way radios. Some models have a range of 2 miles under ideal conditions. More expensive models have a range of 5 miles. What is the difference in the area covered by the 2-mile and 5-mile models? (*Source:* Best Buy.)

30. The National Audubon Society holds an end-of-year bird count. Volunteers count all the birds they see in a circular area during a 24-hour period. Each circle has a diameter of 15 miles. About 1700 circular areas are counted across the United States each December. What is the total area covered by the count?

31. On the first page of this chapter, you read about a forester measuring the circumferences of trees.

   (a) If the circumference of one tree is 144 cm, what is the diameter?

   (b) Explain how you solved part (a).

32. In Atlanta, Interstate 285 circles the city and is known as the "perimeter." If the circumference of the circle made by the highway is 62.8 miles, find:

   (a) the diameter of the circle.

   (b) the area inside the circle.

   (*Source: Greater Atlanta Newcomer's Guide.*)

33. Find the cost of sod, at $0.49 per square foot, for this playing field that has a semicircle on each end.

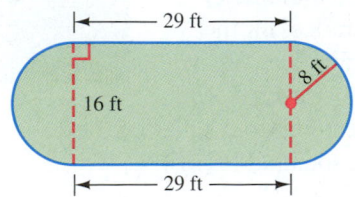

34. Find the area of this skating rink.

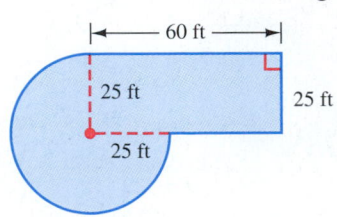

Use the information about prefixes in Example 7 to answer Exercises 35 and 36.

35. Explain how you could use the information about prefixes to remember the difference between radius, diameter, and circumference.

36. Explain how you could use the information about prefixes to avoid confusion between parallel and perpendicular lines.

Relating Concepts (Exercises 37–42) For Individual or Group Work

Use the table below to **work Exercises 37–42 in order.**

Find the best buy for each type of pizza. The best buy is the lowest cost per square inch of pizza. All the pizzas are circular in shape, and the measurement given on the menu board is the diameter of the pizza in inches. Use 3.14 *as the approximate value of* π*. Round the area to the nearest tenth. Round cost per square inch to the nearest thousandth.*

Pizza Menu	Small 7½″	Medium 13″	Large 16″
Cheese only	$2.80	$ 6.50	$ 9.30
"The Works"	$3.70	$ 8.95	$14.30
Deep-dish combo	$4.35	$10.95	$15.65

37. Find the area of a small pizza.

38. Find the area of a medium pizza.

39. Find the area of a large pizza.

40. What is the cost per square inch for each size of cheese pizza? Which size is the best buy?

41. What is the cost per square inch for each size of "The Works" pizza? Which size is the best buy?

42. You have a coupon for 95¢ off any small pizza. What is the cost per square inch for each size of deep-dish combo pizza? Which size is the best buy?

7.6 ▶▶▶ Volume and Surface Area

OBJECTIVE 1 Find the volume of a rectangular solid. A shoe box and a cereal box are examples of three-dimensional (or solid) figures. The three dimensions are length, width, and height. (A rectangle or square is a two-dimensional figure. The two dimensions are length and width.)

If you want to know how much a shoe box will hold, you find its *volume*. We measure volume by seeing how many cubes of a certain size will fill the space inside the box. Three sizes of *cubic units* are shown below. Notice that all the edges of a cube have the same length and all the sides meet at right angles.

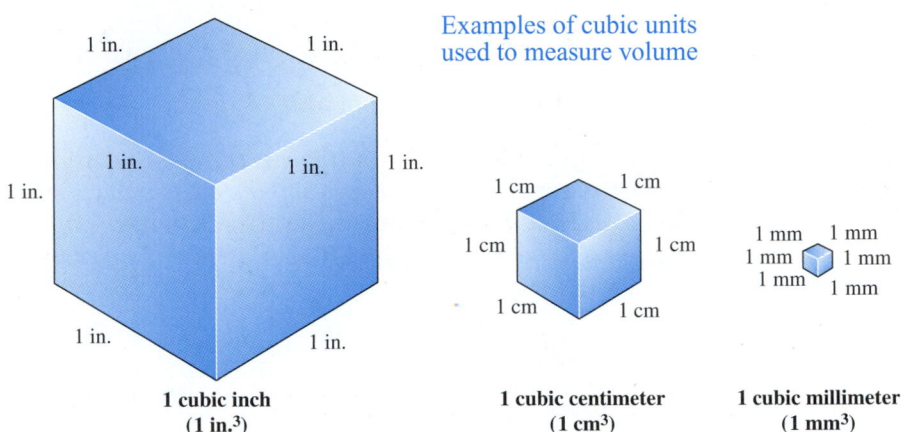

Examples of cubic units used to measure volume

1 cubic inch (1 in.³)   **1 cubic centimeter (1 cm³)**   **1 cubic millimeter (1 mm³)**

Some other sizes of cubes that are used to measure volume are 1 cubic foot (1 ft³), 1 cubic yard (1 yd³), and 1 cubic meter (1 m³).

> **CAUTION**
> The raised 3 in 4^3 means that you multiply $4 \cdot 4 \cdot 4$ to get 64. The raised 3 in cm³ or ft³ is a short way to write the word *cubic*. When you see 5 cm³, say "five *cubic* centimeters." Do *not* multiply $5 \cdot 5 \cdot 5$. The exponent applies to cm, *not* to the number.

> **Volume**
> **Volume** is a measure of the space inside a solid shape. The volume of a solid is how many cubic units it takes to fill the solid.

Use the formula below to find the *volume* of *rectangular solids* (box-like shapes).

> **Finding the Volume of Rectangular Solids**
> Volume of a rectangular solid = length · width · height
> $$V = l \cdot w \cdot h$$
> Remember to use **cubic units** when measuring volume.

| EXAMPLE 1 | **Finding the Volume of Rectangular Solids** |

Find the volume of each box.

(a)

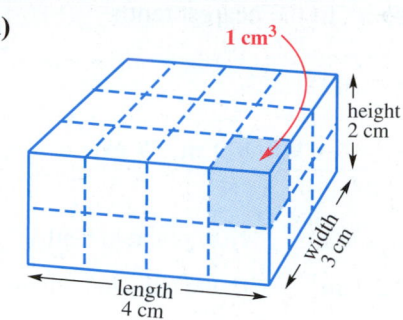

Each cube that fits in the box is 1 cubic centimeter (1 cm³). To find the volume, you can count the number of cubes.

Bottom layer has 12 cubes. } total of 24 cubes (24 cm³)

Top layer has 12 cubes.

Or, you can use the formula for rectangular solids.

$$V = l \cdot w \cdot h$$

$$V = \textbf{4 cm} \cdot \textbf{3 cm} \cdot \textbf{2 cm}$$

$$V = 24 \text{ cm}^3 \qquad \text{Cubic units for volume}$$

(b)

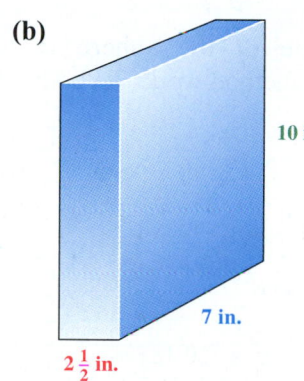

Use the formula $V = l \cdot w \cdot h$.

$$V = \textbf{7 in.} \cdot \textbf{2}\frac{1}{2} \textbf{ in.} \cdot \textbf{10 in.}$$

$$V = \frac{7 \text{ in.}}{1} \cdot \frac{\overset{5}{\cancel{5} \text{ in.}}}{\underset{1}{\cancel{2}}} \cdot \frac{\overset{5}{\cancel{10} \text{ in.}}}{1} = 175 \text{ in.}^3$$

If you like, use 2.5 instead of $2\frac{1}{2}$; they are equivalent.

Write **in.³** for volume.

$$V = 7 \text{ in.} \cdot \textbf{2.5 in.} \cdot 10 \text{ in.} = 175 \text{ in.}^3$$

Work Problem ① *at the Side.* ▶

| OBJECTIVE 2 | **Find the volume of a sphere.** A *sphere* is shown below. Examples of spheres include baseballs, oranges, and Earth. (They aren't perfect spheres, but they're close.)

As with circles, the *radius* of a sphere is the distance from the center to the edge of the sphere. Use the following formula to find the *volume* of a *sphere*.

Finding the Volume of a Sphere

$$\text{Volume of a sphere} = \frac{4}{3} \cdot \pi \cdot r \cdot r \cdot r$$

$$V = \frac{4}{3} \cdot \pi \cdot r^3 \quad \text{or} \quad \frac{4 \cdot \pi \cdot r^3}{3}$$

Remember to use *cubic units* when measuring volume.

① Find the volume of each box. Round your answers to the nearest tenth if necessary.

(a)

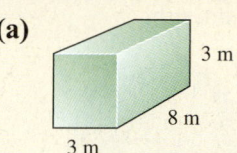

▦ **(b)**

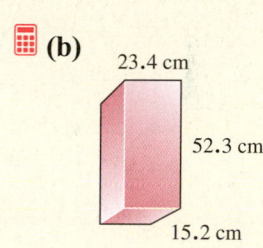

▦ **(c)** Length $6\frac{1}{4}$ ft, width $3\frac{1}{2}$ ft, height 2 ft

ANSWERS

1. **(a)** $V = 72 \text{ m}^3$ **(b)** $V \approx 18{,}602.1 \text{ cm}^3$
   (c) $V = 43\frac{3}{4} \text{ ft}^3$ or $V \approx 43.8 \text{ ft}^3$

2 Find the volume of each sphere. Use 3.14 for π. Round your answers to the nearest tenth.

(a)

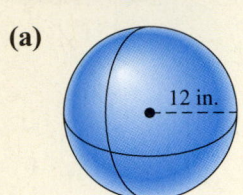

12 in.

(b)

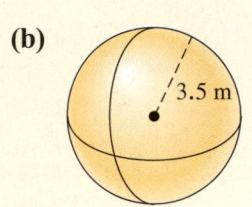

3.5 m

(c) Sphere with a radius of 2.7 cm

EXAMPLE 2 **Finding the Volume of Spheres**

Find the volume of each sphere with the help of a calculator. Use 3.14 as the approximate value of π. Round your answers to the nearest tenth.

(a)

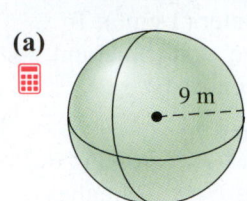

9 m

$$V = \frac{4}{3} \cdot \boldsymbol{\pi} \cdot \boldsymbol{r^3}$$

$$V \approx \frac{4 \cdot \mathbf{3.14} \cdot \mathbf{9\ m} \cdot \mathbf{9\ m} \cdot \mathbf{9\ m}}{3}$$

$V \approx 3052.08$ m^3 Now round to tenths.

$V \approx 3052.1$ **m^3**   Cubic units for volume

(b)

4.2 ft

$$V \approx \frac{4 \cdot \mathbf{3.14} \cdot \mathbf{4.2\ ft} \cdot \mathbf{4.2\ ft} \cdot \mathbf{4.2\ ft}}{3}$$

$V \approx 310.18176$ ft^3 Now round to tenths.

$V \approx 310.2$ **ft^3**   Cubic units for volume

Calculator Tip   You can find the volume of the sphere in Example 2(b) above on your calculator. The first method works on both scientific and standard calculators:

4 $\times$ 3.14 $\times$ 4.2 $\times$ 4.2 $\times$ 4.2 $\div$ 3 $=$ Answer is 310.18176

Round the answer to 310.2 ft^3.

On a *scientific* calculator you can use the y^x key to calculate r^3 (to multiply the radius times itself three times).

4 $\times$ 3.14 $\times$ 4.2 y^x 3 $\div$ 3 $=$ Answer is 310.18176
$\underbrace{\qquad\qquad\qquad}_{r^3}$

Recall that we are using 3.14 as the approximate value for π instead of using the π key.

You can also use the y^x key with other exponents. For example:

To find 2^5, press 2 y^x 5 $=$ Answer is 32

To find 6^4, press 6 y^x 4 $=$ Answer is 1296

◀ *Work Problem* **2** *at the Side.*

Half a sphere is called a *hemisphere.* The volume of a hemisphere is *half* the volume of a sphere. Use the following formula to find the *volume* of a hemisphere.

Finding the Volume of a Hemisphere

$$\text{Volume of a hemisphere} = \frac{1}{\underset{1}{2}} \cdot \frac{\overset{2}{\cancel{4}}}{3} \cdot \pi \cdot r^3$$

$$V = \frac{2}{3} \cdot \pi \cdot r^3 \quad \text{or} \quad \frac{2 \cdot \pi \cdot r^3}{3}$$

Remember to use ***cubic units*** when measuring volume.

EXAMPLE 3 **Finding the Volume of a Hemisphere**

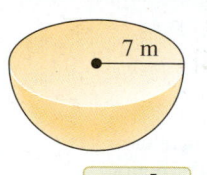 Find the volume of the hemisphere with the help of a calculator. Use 3.14 for π. Round your answer to the nearest tenth.

7 m

$$V = \frac{2 \cdot \pi \cdot r^3}{3}$$

$$V \approx \frac{2 \cdot 3.14 \cdot 7 \text{ m} \cdot 7 \text{ m} \cdot 7 \text{ m}}{3}$$

Write **m³** for volume.

$$V \approx 718.0 \text{ m}^3 \qquad \text{Rounded to nearest tenth}$$

Work Problem **3** *at the Side.* ▶

OBJECTIVE **3** **Find the volume of a cylinder.** Several *cylinders* are shown below.

The height must be perpendicular to the circular top and bottom of the cylinder.

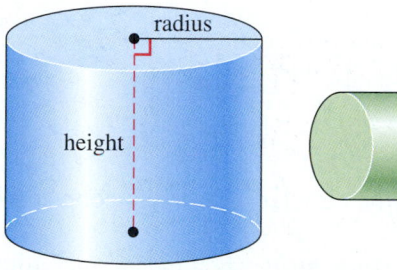

radius

height

These are called *right circular cylinders* because the top and bottom are circles, and the side makes a right angle with the top and bottom. Examples of cylinders are a soup can, a home water heater, and a piece of pipe.

Use the formula below to find the *volume* of a *cylinder*. Notice that the first part of the formula, $\pi \cdot r^2$, is the area of the circular base.

Finding the Volume of a Cylinder

$$\text{Volume of a cylinder} = \pi \cdot r \cdot r \cdot h$$
$$V = \pi \cdot r^2 \cdot h$$

Remember to use **cubic units** when measuring volume.

EXAMPLE 4 **Finding the Volume of Cylinders**

Find the volume of each cylinder. Use 3.14 as the approximate value of π. Round your answers to the nearest tenth if necessary.

(a)

20 m

9 m

The diameter is 20 m so the radius is $\frac{20 \text{ m}}{2} = 10$ m. The height is 9 m. Use the formula to find the volume.

$$V = \pi \cdot r^2 \cdot h$$
$$V \approx 3.14 \cdot 10 \text{ m} \cdot 10 \text{ m} \cdot 9 \text{ m}$$
$$V \approx 2826 \text{ m}^3 \qquad \text{Cubic units for volume}$$

(b)

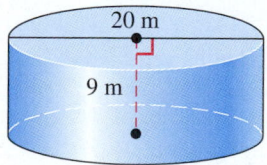

6.2 cm

38.4 cm

$$V \approx 3.14 \cdot 6.2 \text{ cm} \cdot 6.2 \text{ cm} \cdot 38.4 \text{ cm}$$
$$V \approx 4634.94144 \qquad \text{Now round to tenths.}$$
$$V \approx 4634.9 \text{ cm}^3 \qquad \text{Cubic units for volume}$$

Work Problem **4** *at the Side.* ▶

---

3 Find the volume of each hemisphere. Use 3.14 for π. Round your answers to the nearest tenth.

(a)

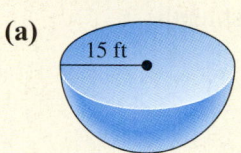

15 ft

(b)

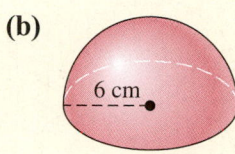

6 cm

---

4 Find the volume of each cylinder. Use 3.14 for π. Round your answers to the nearest tenth.

(a)

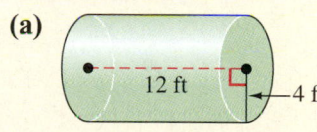

12 ft   4 ft

(b)

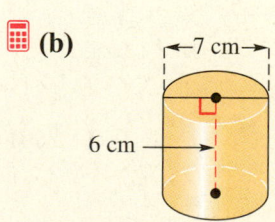

7 cm

6 cm

---

5 Find the volume of a cone with base radius 2 ft and height 11 ft. Use 3.14 for π. Round your answer to the nearest tenth.

OBJECTIVE 4 **Find the volume of a cone and a pyramid.** A cone and a pyramid are shown below. Notice that the height line is perpendicular to the base in both solids.

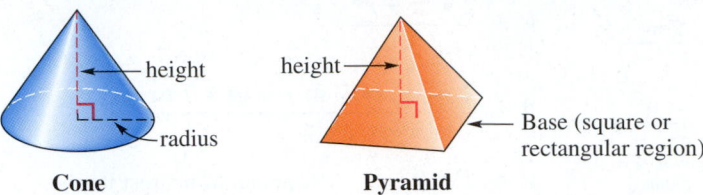

Cone

Pyramid

Use the formula below to find the *volume* of a cone.

Finding the Volume of a Cone

$$\text{Volume of a cone} = \frac{1}{3} \cdot B \cdot h$$

$$\text{or} \quad V = \frac{B \cdot h}{3}$$

where B is the area of the circular base of the cone and h is the height of the cone.

Remember to use **cubic units** when measuring volume.

EXAMPLE 5 **Finding the Volume of a Cone**

Find the volume of the cone. Use 3.14 for π. Round your answer to the nearest tenth.

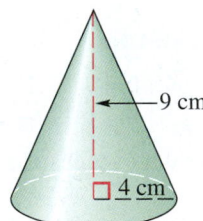

First find the value of B in the formula, which is the *area of the circular base*. Recall that the formula for the area of a circle is πr^2.

$$B = \pi \cdot r \cdot r$$

$$B \approx 3.14 \cdot 4 \text{ cm} \cdot 4 \text{ cm}$$

$$\boldsymbol{B \approx 50.24 \text{ cm}^2} \leftarrow \text{Do not round to tenths yet.}$$

Now find the volume. The height is 9 cm.

$$V = \frac{\boldsymbol{B \cdot h}}{3}$$

$$V \approx \frac{\boldsymbol{50.24 \text{ cm}^2 \cdot 9 \text{ cm}}}{3}$$

$$V \approx 150.72 \text{ cm}^3 \qquad \text{Now round to tenths.}$$

Write **cm³** in the answer.

$$V \approx 150.7 \text{ cm}^3 \qquad \text{Cubic units for volume}$$

◀ *Work Problem* **5** *at the Side.*

ANSWER

5. $V \approx 46.1 \text{ ft}^3$

Use the same formula to find the *volume* of a *pyramid* as you did to find the *volume* of a *cone*.

> **Finding the Volume of a Pyramid**
>
> $$\text{Volume of a pyramid} = \frac{1}{3} \cdot B \cdot h$$
>
> $$\text{or} \quad V = \frac{B \cdot h}{3}$$
>
> where B is the area of the square or rectangular base of the pyramid and h is the height of the pyramid.
>   Remember to use **cubic units** when measuring volume.

> **Note**
>
> In this book, we will work only with pyramids that have a base with four sides (square or rectangle). In later math courses you may work with pyramids that have a base with three sides (triangle), five sides (pentagon), six sides (hexagon), and so on.

EXAMPLE 6   **Finding the Volume of a Pyramid**

Find the volume of this pyramid with a rectangular base. Round your answer to the nearest tenth.

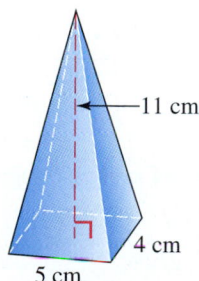

—11 cm

4 cm

5 cm

   First find the value of B in the formula, which is the *area of the rectangular base.* Recall that the area of a rectangle is found by multiplying length times width.

$$B = 5 \text{ cm} \cdot 4 \text{ cm}$$

$$B = 20 \text{ cm}^2 \quad \boxed{\text{Find the } \textit{area} \text{ of the base first.}}$$

Now find the volume.

$$V = \frac{B \cdot h}{3}$$

$$V = \frac{20 \text{ cm}^2 \cdot 11 \text{ cm}}{3} \quad \boxed{\text{Now find the } \textit{volume} \text{ of the pyramid.}}$$

$$V \approx 73.3 \text{ cm}^3 \qquad \text{Rounded to nearest tenth}$$

Work Problem **6** *at the Side.* ▶

---

6 Find the volume of a pyramid with a square base 10 m by 10 m. The height of the pyramid is 8 m. Round your answer to the nearest tenth.

ANSWER

6. $V \approx 266.7 \text{ m}^3$

8 Find the volume and surface area of each rectangular solid.

(a)

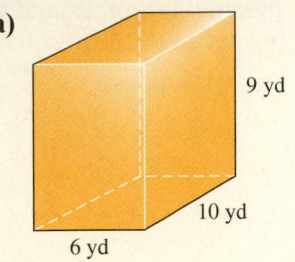

9 yd

10 yd

6 yd

(b)

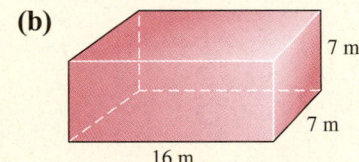

7 m

7 m

16 m

OBJECTIVE 5 Find the surface area of a rectangular solid.
Earlier in this section, you learned how to find the *volume* of a rectangular solid. For example, the volume of the cereal box shown below is $V = lwh = (7 \text{ in.})(2 \text{ in.})(10 \text{ in.}) = 140 \text{ in.}^3$ But if your company makes cereal boxes, you also need to know how much cardboard is needed for each box. You need to find the *surface area* of the box.

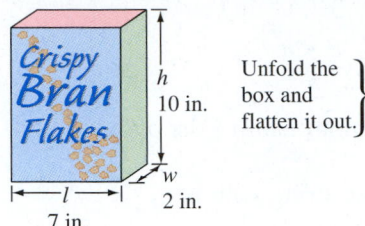

Unfold the box and flatten it out.

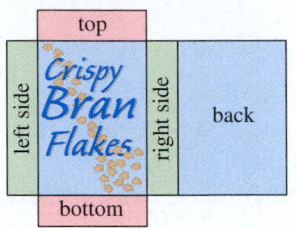

The unfolded box is made up of six rectangles: front, back, top, bottom, left side, right side.

Surface area is the area on the surface of a three-dimensional object (a solid). For a rectangular solid like the cereal box, the surface area is the sum of the areas of the six rectangular sides. Notice that the top and bottom have the same area, the front and back have the same area, and the left and right sides have the same area.

$$SA = l \cdot w + l \cdot w + l \cdot h + l \cdot h + w \cdot h + w \cdot h$$
$$SA = 2lw + 2lh + 2wh$$

Finding the Surface Area of a Rectangular Solid

Surface Area $= (2 \cdot l \cdot w) + (2 \cdot l \cdot h) + (2 \cdot w \cdot h)$

$$SA = 2lw + 2lh + 2wh$$

Remember that area is measured in square units, so use **square units** when measuring *surface* area.

EXAMPLE 8 Finding the Volume and Surface Area of a Rectangular Solid

Find the volume and surface area of this shipping carton.

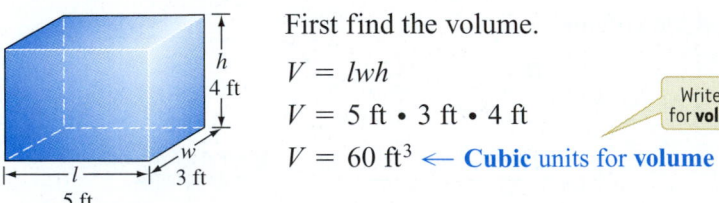

h
4 ft

w
3 ft

l
5 ft

First find the volume.

$V = lwh$

$V = 5 \text{ ft} \cdot 3 \text{ ft} \cdot 4 \text{ ft}$ *Write ft³ for volume.*

$V = 60 \text{ ft}^3 \leftarrow$ **Cubic** units for **volume**

Next find the surface area.

$SA = 2lw + 2lh + 2wh$
$SA = (2 \cdot 5 \text{ ft} \cdot 3 \text{ ft}) + (2 \cdot 5 \text{ ft} \cdot 4 \text{ ft}) + (2 \cdot 3 \text{ ft} \cdot 4 \text{ ft})$
$SA = 30 \text{ ft}^2 + 40 \text{ ft}^2 + 24 \text{ ft}^2$ *Write ft² for area.*
$SA = 94 \text{ ft}^2 \leftarrow$ **Square** units for **area**

◀ *Work Problem* **8** *at the Side.*

ANSWERS

8. (a) $V = 540 \text{ yd}^3$ (*cubic* yd for *volume*)
  $SA = 408 \text{ yd}^2$ (*square* yd for *area*)
(b) $V = 784 \text{ m}^3$
  $SA = 546 \text{ m}^2$

OBJECTIVE 6 **Find the surface area of a cylinder.** You can use the same idea of "unfolding" a shape to find the surface area of a cylinder, such as the soup can shown below. Finding the surface area will tell you how much aluminum you need to make the can.

The unfolded soup can is made up of a rectangular side, a circular top, and a circular bottom.

Remember that the formula for the area of a circle is πr^2.

$$Surface\ Area = \text{Side} + \text{Top} + \text{Bottom}$$

Circumference of can ($2\pi r$)

$$SA = 2\pi r \cdot h + \pi r^2 + \pi r^2$$
$$SA = 2\pi rh + 2\pi r^2$$

Finding the Surface Area of a Right Circular Cylinder

$$\text{Surface Area} = (2 \cdot \pi \cdot r \cdot h) + (2 \cdot \pi \cdot r \cdot r)$$
$$SA = 2\pi rh + 2\pi r^2$$

Remember that area is measured in square units, so use **square units** when measuring *surface* area.

EXAMPLE 9 **Finding the Volume and Surface Area of a Right Circular Cylinder**

Find the volume and surface area of this water tank. Use 3.14 as the approximate value for π. Round your answers to the nearest tenth when necessary.

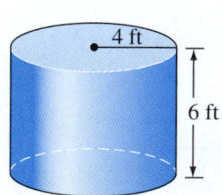

First find the volume.

$$V = \pi r^2 h$$
$$V \approx 3.14 \cdot 4\ ft \cdot 4\ ft \cdot 6\ ft$$
$$V \approx 301.44\ ft^3 \leftarrow \text{Now round to tenths.}$$
$$V \approx 301.4\ ft^3 \leftarrow \textbf{Cubic units for volume}$$

ft³ for volume

Now find the surface area.

$$SA = 2\pi rh + 2\pi r^2$$
$$SA \approx (2 \cdot 3.14 \cdot 4\ ft \cdot 6\ ft) + (2 \cdot 3.14 \cdot 4\ ft \cdot 4\ ft)$$
$$SA \approx 150.72\ ft^2 + 100.48\ ft^2$$
$$SA \approx 251.2\ ft^2 \leftarrow \textbf{Square units for area}$$

ft² for area

Work Problem 9 at the Side. ▶

---

9 Find the volume and surface area of each cylinder. Use 3.14 for π. Round your answers to the nearest tenth.

(a)

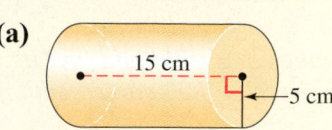

15 cm

5 cm

(b)

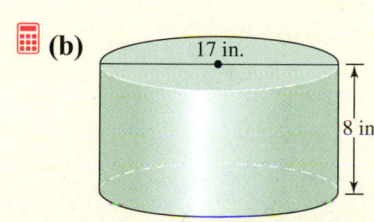

17 in.

8 in.

(*Hint:* You are given the *diameter* of the cylinder. Start by finding the *radius*.)

ANSWERS

9. **(a)** $V \approx 1177.5\ cm^3$ (cubic units for volume)
   $SA \approx 628\ cm^2$ (square units for area)
   (b) $V \approx 1814.9\ in.^3$
   $SA \approx 880.8\ in.^2$

7.6 ▶▶▶ Exercises

FOR EXTRA HELP

   PRACTICE   WATCH   DOWNLOAD   READ   REVIEW

Name each solid and find its volume. Use 3.14 as the approximate value of π. Round your answers to the nearest tenth if necessary. See Examples 1–6.

1.

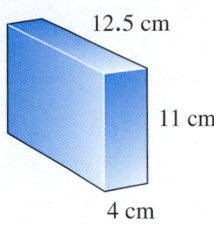

12.5 cm
11 cm
4 cm

2.

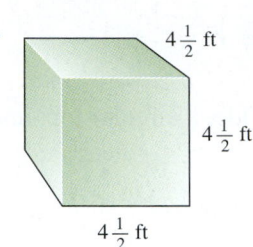

$4\frac{1}{2}$ ft
$4\frac{1}{2}$ ft
$4\frac{1}{2}$ ft

3.

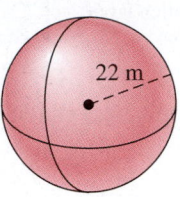

22 m

4.

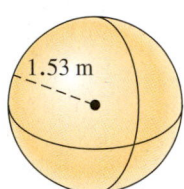

1.53 m

5.

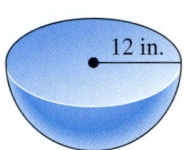

12 in.

6.

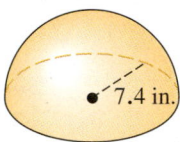

7.4 in.

7.

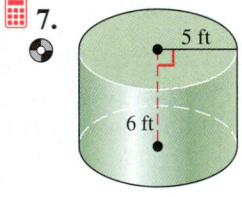

5 ft
6 ft

8.

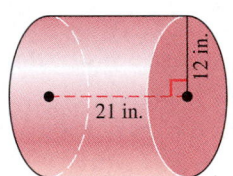

12 in.
21 in.

9.

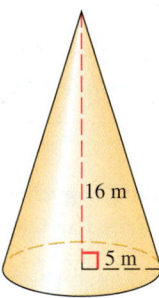

16 m
5 m

10.

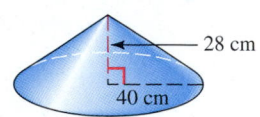

28 cm
40 cm

11.

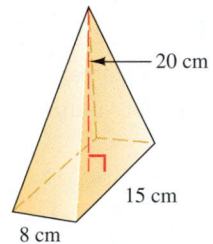

20 cm
15 cm
8 cm

12.

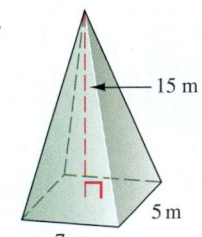

15 m
5 m
7 m

13. Find the volume.

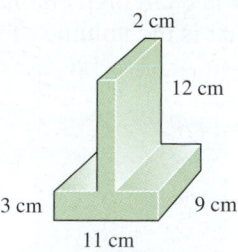

2 cm
12 cm
3 cm   9 cm
11 cm

14. Find the volume. (*Hint:* Notice the square hole that goes through the center of the shape.)

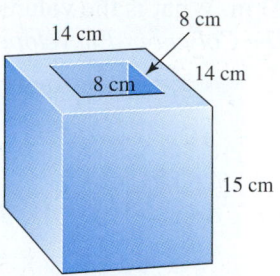

14 cm   8 cm
8 cm   14 cm
15 cm

Find the volume and surface area of each cylinder or rectangular solid. Use 3.14 as the approximate value of π. Round your answers to the nearest tenth when necessary. See Examples 7–9.

15.

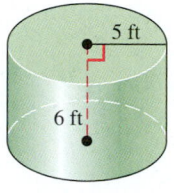

5 ft
6 ft

16.

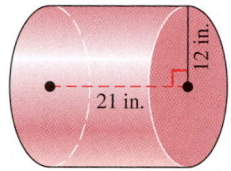

21 in.   12 in.

17.

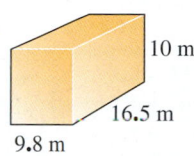

10 m
16.5 m
9.8 m

18.

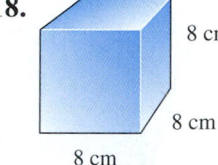

8 cm
8 cm
8 cm

19.

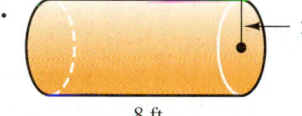

18 in.   3 in.

20.

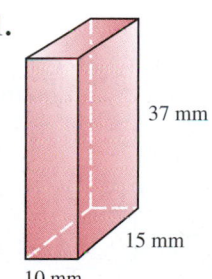

2 ft
8 ft

21.

37 mm
15 mm
10 mm

22.

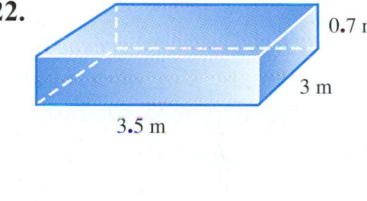

0.7 m
3 m
3.5 m

Solve each application problem. Use 3.14 as the approximate value of π. Round your final answers to the nearest tenth if necessary.

23. A pencil box measures 3 in. by 8 in. by $\frac{3}{4}$ in. high. Find the volume of the box. (*Source:* Faber Castell.)

24. A train is being loaded with shipping crates. Each one is 6 m long, 3.4 m wide, and 2 m high. How much space will each crate take?

25. An oil candle globe made of hand-blown glass has a diameter of 16.8 cm. What is the volume of the globe?

26. A metal sphere used as part of a fountain has a diameter of $6\frac{1}{2}$ ft. Find its volume.

27. One of the ancient stone pyramids in Egypt has a square base that measures 145 m on each side. The height is 93 m. What is the volume of the pyramid? (*Source: The Columbia Encyclopedia.*)

28. A cylindrical woven basket made by a Northwest Coast tribe is 8 cm high and has a diameter of 11 cm. What is the volume of the basket?

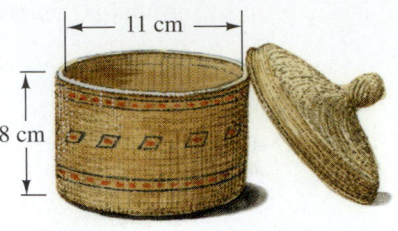

29. A city sewer pipe has a diameter of 5 ft and a length of 200 ft. Find the volume of the pipe.

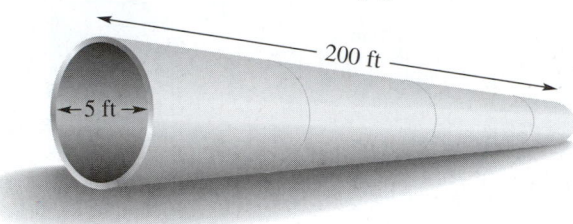

30. An ice cream cone has a diameter of 2 in. and a height of 4 in. Find its volume.

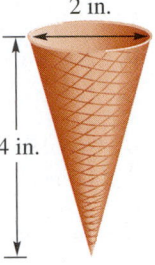

31. A box for graham crackers measures 5.5 in. by 2.8 in. by 8 in. high. Find the amount of cardboard needed to make the box.

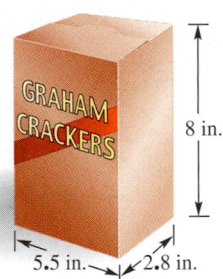

32. A soda can is 12.5 cm tall and the round top has a diameter of 6.7 cm. How much aluminum is needed to make the can? (Assume the can has a flat top and bottom.)

33. Explain the *two* errors made by a student in finding the volume of a cylinder with a diameter of 7 cm and a height of 5 cm. Find the correct answer.

$$V \approx 3.14 \cdot 7 \cdot 7 \cdot 5$$

$$V \approx 769.3 \text{ cm}^2$$

34. **(a)** Look again at Exercise 18 on the previous page. The figure is a *cube*. What is special about the measurements of a cube?

(b) Find a shortcut you can use to calculate the surface area of a cube.

7.7 ▶▶▶ Pythagorean Theorem

In **Section 7.2** you used this formula for the area of a square, $A = s^2$. The blue square below has an area of 25 cm^2 because 5 cm • 5 cm = 25 cm^2.

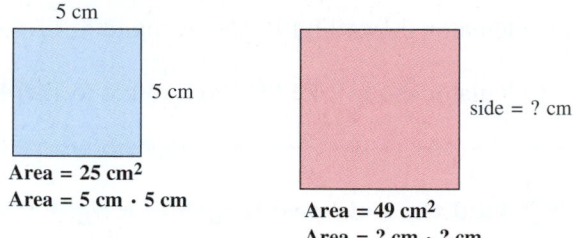

5 cm

5 cm

Area = 25 cm^2
Area = 5 cm • 5 cm

side = ? cm

Area = 49 cm^2
Area = ? cm • ? cm

The red square above has an area of 49 cm^2. To find the length of a side, ask yourself, "What number can be multiplied by itself to give 49?" Because 7 • 7 = 49, the length of each side is 7 cm.

Also, because 7 • 7 = 49, we say that 7 is the *square root* of 49, or $\sqrt{49} = 7$. Also, $\sqrt{81} = 9$, because 9 • 9 = 81. (See **Section 1.8** for further review.)

Work Problem **1** *at the Side.* ▶

A number that has a whole number as its square root is called a *perfect square*. For example, 9 is a perfect square because $\sqrt{9} = 3$, and 3 is a whole number.

The first few perfect squares are listed below.

> **The First Twelve Perfect Squares**
>
> | | | | |
> |---|---|---|---|
> | $\sqrt{1} = 1$ | $\sqrt{16} = 4$ | $\sqrt{49} = 7$ | $\sqrt{100} = 10$ |
> | $\sqrt{4} = 2$ | $\sqrt{25} = 5$ | $\sqrt{64} = 8$ | $\sqrt{121} = 11$ |
> | $\sqrt{9} = 3$ | $\sqrt{36} = 6$ | $\sqrt{81} = 9$ | $\sqrt{144} = 12$ |

OBJECTIVE **1** **Find square roots using the square root key on a calculator.** If a number is *not* a perfect square, then you can find its *approximate* square root by using a calculator with a square root key.

▦ **Calculator Tip** To find a square root, use the $\boxed{\sqrt{}}$ key on a standard calculator or the $\boxed{\sqrt{x}}$ key on a scientific calculator. In either case, you do *not* need to use the $\boxed{=}$ key. Try these. Jot down your answers.

To find $\sqrt{16}$ press: 16 $\boxed{\sqrt{x}}$ Answer is 4

To find $\sqrt{7}$ press: 7 $\boxed{\sqrt{x}}$ Answer is 2.645751311

For $\sqrt{7}$, your calculator shows 2.645751311, which is an *approximate* answer. (Some calculators show more or fewer digits.) We will be rounding to the nearest thousandth, so $\sqrt{7} \approx 2.646$. To check, multiply 2.646 times 2.646. Do you get 7 as the result? No, you get 7.001316, which is very close to 7. The difference is due to rounding.

1 Find each square root.

(a) $\sqrt{36}$

(b) $\sqrt{25}$

(c) $\sqrt{9}$

(d) $\sqrt{100}$

(e) $\sqrt{121}$

ANSWERS

1. **(a)** 6 **(b)** 5 **(c)** 3 **(d)** 10 **(e)** 11

2 Use a calculator with a square root key to find each square root. Round to the nearest thousandth if necessary.

(a) $\sqrt{11}$

(b) $\sqrt{40}$

(c) $\sqrt{56}$

(d) $\sqrt{196}$

(e) $\sqrt{147}$

ANSWERS

2. **(a)** $\sqrt{11} \approx 3.317$ **(b)** $\sqrt{40} \approx 6.325$
   (c) $\sqrt{56} \approx 7.483$ **(d)** $\sqrt{196} = 14$
   (e) $\sqrt{147} \approx 12.124$

EXAMPLE 1 **Finding the Square Root of Numbers**

Use a calculator to find each square root. Round answers to the nearest thousandth.

Your calculator may show more or fewer digits.

(a) $\sqrt{35}$ Calculator shows 5.916079783; round to 5.916

(b) $\sqrt{124}$ Calculator shows 11.13552873; round to 11.136

(c) $\sqrt{200}$ Calculator shows 14.14213562; round to 14.142

◀ Work Problem **2** at the Side.

OBJECTIVE 2 Find the unknown length in a right triangle. One place you will use square roots is when working with the *Pythagorean Theorem.* This theorem applies only to *right* triangles (triangles with a 90° angle). The longest side of a right triangle is called the **hypotenuse.** It is opposite the right angle. The other two sides are called *legs*. The legs form the right angle.

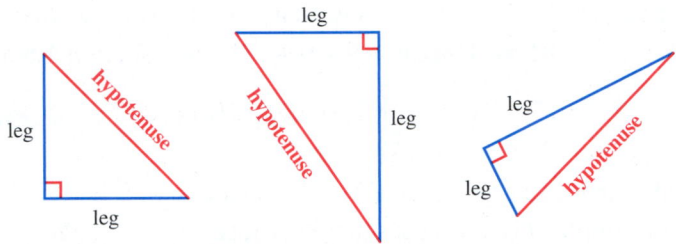

Examples of right triangles

Pythagorean Theorem

$$(\text{hypotenuse})^2 = (\text{leg})^2 + (\text{leg})^2$$

In other words, square the length of each side. After you have squared all the sides, the sum of the squares of the two legs will equal the square of the hypotenuse. An example is shown below.

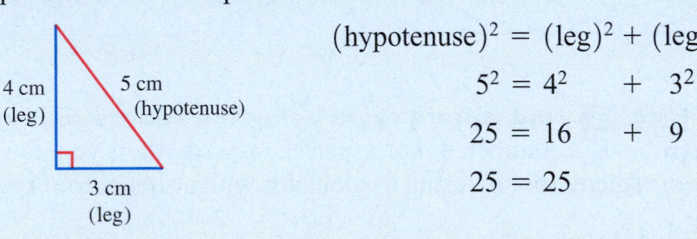

$$(\text{hypotenuse})^2 = (\text{leg})^2 + (\text{leg})^2$$
$$5^2 = 4^2 + 3^2$$
$$25 = 16 + 9$$
$$25 = 25$$

The theorem is named after Pythagoras, a Greek mathematician who lived about 2500 years ago. He and his followers may have used floor tiles to prove the theorem, as shown below.

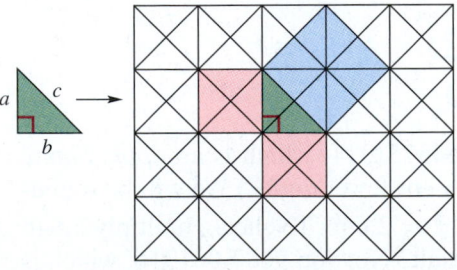

The green right triangle in the center of the floor tiles has sides a, b, and c. The pink square drawn on side a contains four triangular tiles. The pink square on side b contains four tiles. The blue square on side c contains eight tiles. The number of tiles in the square on side c equals the sum of the number of tiles in the squares on sides a and b, that is, 8 tiles = 4 tiles + 4 tiles. As a result, you often see the Pythagorean Theorem written as $c^2 = a^2 + b^2$.

If you know the lengths of any two sides in a right triangle, you can use the Pythagorean Theorem to find the length of the third side.

> ### Formulas Based on the Pythagorean Theorem
> To find the hypotenuse, use this formula:
> $$\text{hypotenuse} = \sqrt{(\text{leg})^2 + (\text{leg})^2}$$
> To find a leg, use this formula:
> $$\text{leg} = \sqrt{(\text{hypotenuse})^2 - (\text{leg})^2}$$

> **CAUTION**
> *Remember:* A small square drawn in one angle of a triangle indicates a right angle. You can use the Pythagorean Theorem *only* on triangles that have a right angle.

> **EXAMPLE 2**   **Finding the Unknown Length in Right Triangles**

Find the unknown length in each right triangle. Round answers to the nearest tenth if necessary.

(a)

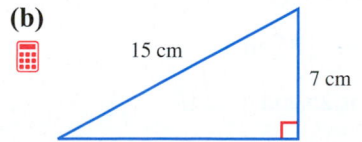

The unknown length is the side opposite the right angle, which is the hypotenuse. Use the formula for finding the hypotenuse.

$\text{hypotenuse} = \sqrt{(\text{leg})^2 + (\text{leg})^2}$ Find the hypotenuse.

$\text{hypotenuse} = \sqrt{(3)^2 + (4)^2}$ Legs are 3 and 4

$\qquad = \sqrt{9 + 16}$ 3 • 3 is 9   and   4 • 4 is 16

$\qquad = \sqrt{25}$

$\qquad = 5$ Write **ft** in the answer (**not** ft²).

The hypotenuse is 5 ft long.

(b)

We *do* know the length of the hypotenuse (15 cm), so it is the length of one of the legs that is unknown. Use the formula for finding a leg.

$\text{leg} = \sqrt{(\text{hypotenuse})^2 - (\text{leg})^2}$ Find a leg.

$\text{leg} = \sqrt{(15)^2 - (7)^2}$ Hypotenuse is 15, one leg is 7

$\qquad = \sqrt{225 - 49}$ 15 • 15 is 225   and   7 • 7 is 49

$\qquad = \sqrt{176}$ Use calculator to find $\sqrt{176}$

$\qquad \approx 13.3$ Round 13.26649916 to 13.3

The length of the leg is approximately 13.3 cm.

Work Problem **3** *at the Side.* ▶

---

3 Find the unknown length in each right triangle. Round your answers to the nearest tenth if necessary.

(a)

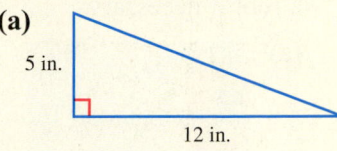

(b)

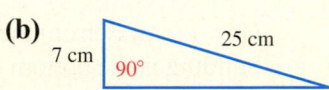

(c)

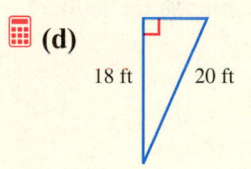

(d)

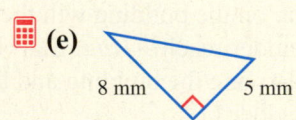

(e)

ANSWERS

3. **(a)** $\sqrt{169} = 13$ in.   **(b)** $\sqrt{576} = 24$ cm
   (c) $\sqrt{458} \approx 21.4$ m   **(d)** $\sqrt{76} \approx 8.7$ ft
   (e) $\sqrt{89} \approx 9.4$ mm

4 These problems show ladders leaning against buildings. Find the unknown lengths. Round to the nearest tenth of a foot if necessary.

(a)

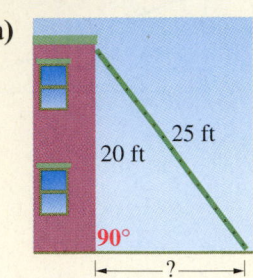

How far away from the building is the bottom of the ladder?

(b)

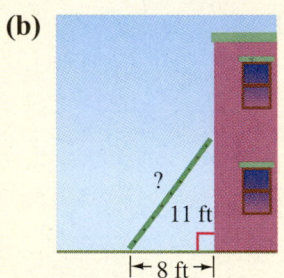

How long is the ladder?

(c) A 17 ft ladder is leaning against a building. The bottom of the ladder is 10 ft from the building. How high up on the building will the ladder reach? (*Hint:* Start by drawing the building and the ladder.)

ANSWERS

4. **(a)** leg = $\sqrt{225}$ = 15 ft
   (b) hypotenuse = $\sqrt{185}$ ≈ 13.6 ft
   (c) leg = $\sqrt{189}$ ≈ 13.7 ft

OBJECTIVE 3 **Solve application problems involving right triangles.** The next example shows an application of the Pythagorean Theorem.

EXAMPLE 3 **Using the Pythagorean Theorem**

A television antenna is on the roof of a house, as shown below. Find the length of the support wire. Round your answer to the nearest tenth of a meter if necessary.

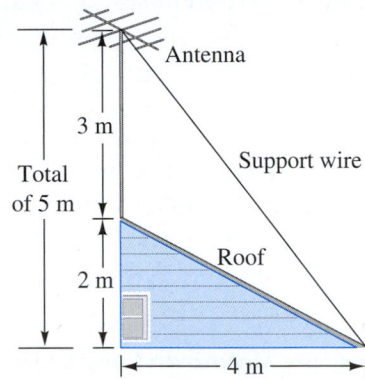

A right triangle is formed. The total length of the leg on the left is 3 m + 2 m = 5 m.

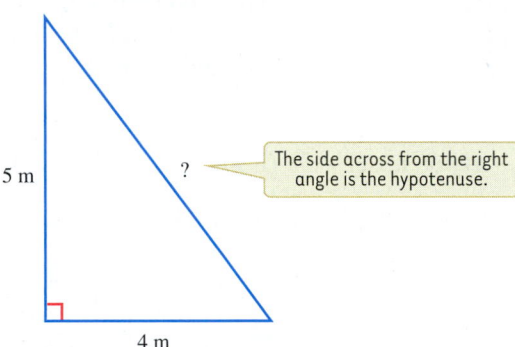

The side across from the right angle is the hypotenuse.

Notice that the support wire is opposite the right angle, so it is the hypotenuse of the right triangle.

hypotenuse = $\sqrt{(\text{leg})^2 + (\text{leg})^2}$ Find the hypotenuse.

hypotenuse = $\sqrt{(5)^2 + (4)^2}$ Legs are 5 and 4

= $\sqrt{25 + 16}$ 5^2 is 25 and 4^2 is 16

= $\sqrt{41}$ Use a calculator to find $\sqrt{41}$

≈ 6.4   Round 6.403124237 to 6.4

The length of the support wire is approximately 6.4 m. Write **m** in the answer (**not** m²).

CAUTION

You use the Pythagorean Theorem to find the *length* of one side, *not* the area of the triangle. Your answer will be in linear units, such as ft, yd, cm, m, and so on (*not* ft², yd², cm², m²).

◀ *Work Problem* **4** *at the Side.*

7.7 ▶▶▷ Exercises

Find each square root. Starting with Exercise 5, use the square root key on a calculator. Round your answers to the nearest thousandth if necessary. See Example 1.

1. $\sqrt{16}$ **2.** $\sqrt{4}$ **3.** $\sqrt{64}$ **4.** $\sqrt{81}$

5. $\sqrt{11}$ **6.** $\sqrt{23}$ **7.** $\sqrt{5}$ **8.** $\sqrt{2}$

9. $\sqrt{73}$ **10.** $\sqrt{80}$ **11.** $\sqrt{101}$ **12.** $\sqrt{125}$

13. $\sqrt{190}$ **14.** $\sqrt{160}$ **15.** $\sqrt{1000}$ **16.** $\sqrt{2000}$

17. You know that $\sqrt{25} = 5$ and $\sqrt{36} = 6$. Using just that information (no calculator), describe how you could *estimate* $\sqrt{30}$. How would you estimate $\sqrt{26}$ or $\sqrt{35}$? Now check your estimates using a calculator.

18. Explain the relationship between *squaring* a number and finding the *square root* of a number. Include two examples to illustrate your explanation.

Find the unknown length in each right triangle. Use a calculator to find square roots. Round your answers to the nearest tenth if necessary. See Example 2.

19.
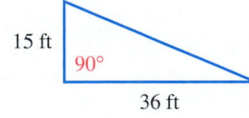
15 ft   90°   36 ft

20.
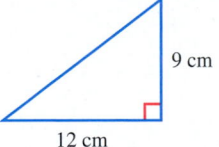
9 cm   12 cm

21.
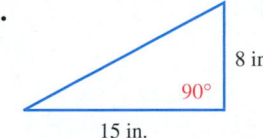
8 in.   90°   15 in.

22.

30 in.   72 in.

23.
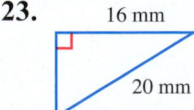
16 mm   20 mm

24.
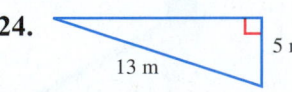
13 m   5 m

25.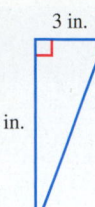

3 in.

8 in.

26.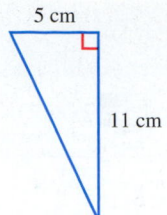

5 cm

11 cm

27.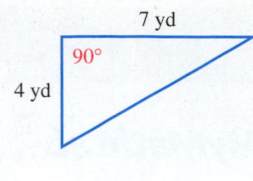

7 yd

90°

4 yd

28.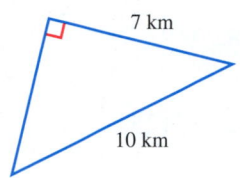

7 km

10 km

29.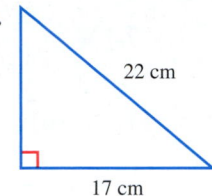

22 cm

17 cm

30.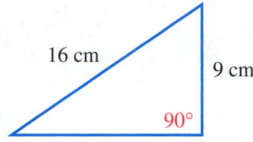

16 cm

9 cm

90°

31.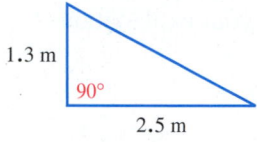

1.3 m

90°

2.5 m

32.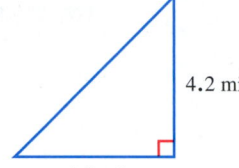

4.2 mi

4.2 mi

33.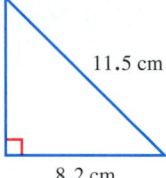

11.5 cm

8.2 cm

34.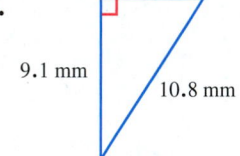

9.1 mm

10.8 mm

35.

13.2 km   90°

21.6 km

36.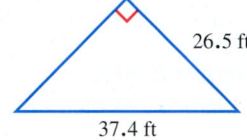

26.5 ft

37.4 ft

Solve each application problem. Round your answers to the nearest tenth if necessary. See Example 3.

37. Find the length of this loading ramp.

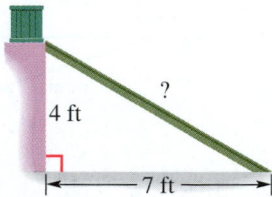

?

4 ft

7 ft

38. Find the unknown length in this roof plan.

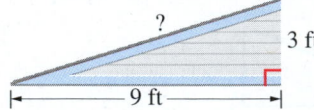

?

3 ft

9 ft

39. How high is the airplane above the ground?

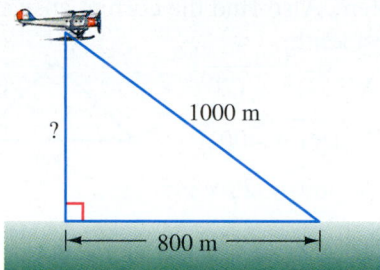

1000 m

?

800 m

40. Find the height of this farm silo.

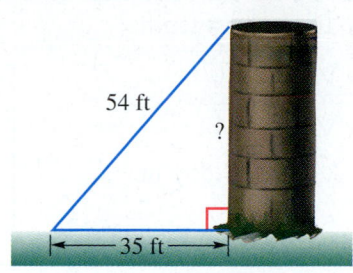

54 ft

?

35 ft

41. How long is the diagonal brace on this rectangular gate?

3.5 ft

4.5 ft

?

42. Find the height of this rectangular television screen. (*Source:* Sears.)

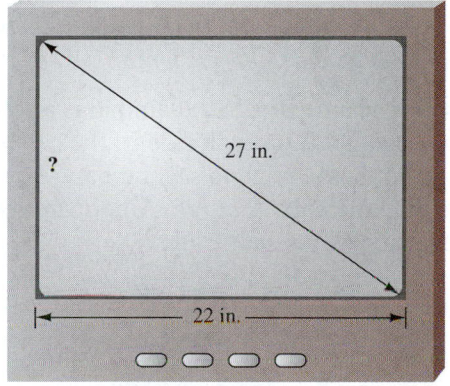

?

27 in.

22 in.

43. To reach his ladylove, a knight placed a 12 ft ladder against the castle wall. If the base of the ladder is 3 ft from the building, how high on the castle will the top of the ladder reach? Draw a sketch of the castle and ladder and solve the problem.

44. William drove his car 15 miles north, then made a right turn and drove 7 miles east. How far is he, in a straight line, from his starting point? Draw a sketch to illustrate the problem and solve it.

45. Explain the *two* errors made by a student in solving this problem. Also find the correct answer. Round to the nearest tenth.

$$? = \sqrt{(9)^2 + (7)^2}$$
$$= \sqrt{18 + 14}$$
$$= \sqrt{32} \approx 5.657 \text{ in.}$$

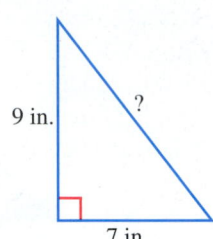

9 in.   ?   7 in.

46. Explain the *two* errors made by a student in solving this problem. Also find the correct answer. Round to the nearest tenth.

$$? = \sqrt{(13)^2 + (20)^2}$$
$$= \sqrt{169 + 400}$$
$$= \sqrt{569} \approx 23.9 \text{ m}^2$$

?   13 m   20 m

Relating Concepts (Exercises 47–50) For Individual or Group Work

*Use your knowledge of the Pythagorean Theorem to **work Exercises 47–50 in order.** Round answers to the nearest tenth.*

47. A major league baseball diamond is a square shape measuring 90 ft on each side. If the catcher throws a ball from home plate to second base, how far is he throwing the ball? (*Source:* American League of Professional Baseball Clubs.)

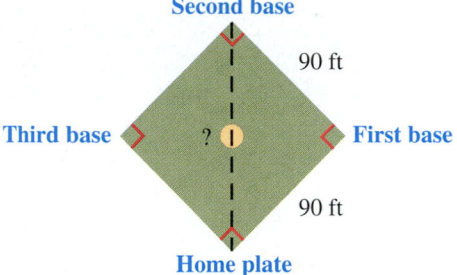

Second base

90 ft

Third base   ?   First base

90 ft

Home plate

48. A softball diamond is only 60 ft on each side. (*Source:* Amateur Softball Association.)

(a) Draw a sketch of the softball diamond and label the bases and the lengths of the sides.

(b) How far is it to throw a ball from home plate to second base?

49. Look back at your answer to Exercise 47. Explain how you can tell the distance from third base to first base without doing any further calculations.

50. Show how you could set up a proportion to answer Exercise 48 instead of using the Pythagorean Theorem. (You'll need your answer from Exercise 47.)

7.8 ▶▶▶ Congruent and Similar Triangles

Two useful concepts in geometry are *congruence* and *similarity*. If two figures are *identical,* both in *shape* and in *size,* we say the figures are **congruent.** In other words, the figures are perfect duplicates of each other, like getting two identical prints made from your digital camera. If two figures have the *same shape* but are *different sizes,* we say the figures are **similar,** like getting a print made from your digital camera and then an enlargement of the same print. We'll explore the ideas of congruence and similarity using triangles.

OBJECTIVES

1 Identify corresponding parts of congruent triangles.

2 Prove that triangles are congruent using SAS, SSS, and ASA.

3 Identify corresponding parts of similar triangles.

4 Find the unknown lengths of sides in similar triangles.

5 Solve application problems involving similar triangles.

OBJECTIVE 1   Identify corresponding parts of congruent triangles.   The two triangles shown below are *congruent* because they are the *same shape* and the *same size.*

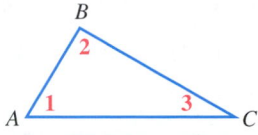
One way to name this triangle is △*ABC.*

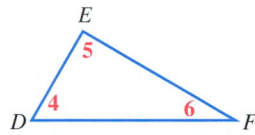
One way to name this triangle is △*DEF.*

Suppose you picked up △*ABC* and slid it over on top of △*DEF.* You would see that the two triangles are a perfect match. ∠1 would be on top of ∠4, so they are called *corresponding angles.* Similarly, ∠2 and ∠5 are corresponding angles, and ∠3 and ∠6 are corresponding angles. You would see that corresponding angles have the same measure, as indicated below.

$$m\angle 1 = m\angle 4 \qquad m\angle 2 = m\angle 5 \qquad m\angle 3 = m\angle 6$$

The abbreviation for measure is *m,* so $m\angle 1$ is read, "the measure of angle 1."

When you put △*ABC* on top of △*DEF,* you would also see that side *AB* is on top of side *DE.* We say that $\overline{AB}$ and $\overline{DE}$ are *corresponding sides.* Similarly, $\overline{BC}$ and $\overline{EF}$ are corresponding sides, and $\overline{AC}$ and $\overline{DF}$ are corresponding sides. You would see that corresponding sides have the same length.

$$AB = DE \qquad BC = EF \qquad AC = DF$$

Because corresponding angles have the same measure, and corresponding sides have the same length, we know that △*ABC* **is congruent to** △*DEF.* We can write this as △*ABC* ≅ △*DEF.*

> **Congruent Triangles**
>
> If two triangles are congruent, then
> 1. Corresponding angles have the same measure, and
> 2. Corresponding sides have the same length.

EXAMPLE 1   Identifying Corresponding Parts in Congruent Triangles

Each pair of triangles is congruent. List the corresponding angles and corresponding sides.

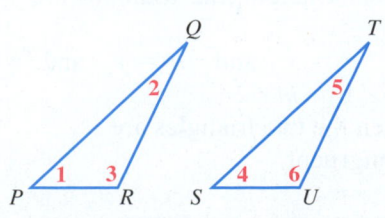

(a) If you picked up △*PQR* and slid it over on top of △*STU,* the two triangles would match.
The corresponding parts are:

∠1 and ∠4 $\overline{PQ}$ and $\overline{ST}$
∠2 and ∠5 $\overline{PR}$ and $\overline{SU}$
∠3 and ∠6 $\overline{QR}$ and $\overline{TU}$

Continued on Next Page

1 Each pair of triangles is congruent. List the corresponding angles and the corresponding sides.

(a)

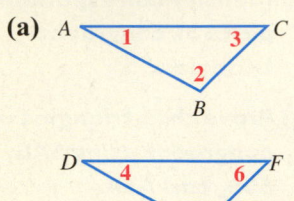

(b)

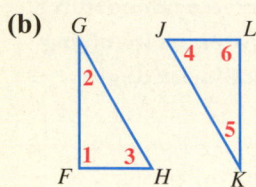

(*Hint:* Rotate △*FGH*, then slide it on top of △*JLK*.)

(c)

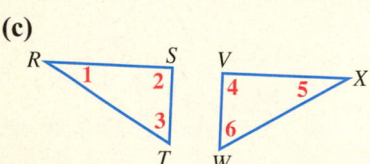

(*Hint:* Flip △*RST* over, then slide it on top of △*VWX*.)

(b) If you picked up △*ABC* and slid it over on top of △*DEF*, it wouldn't match. But if you *rotate* △*ABC* before sliding it on top of △*DEF*, it *will* match.

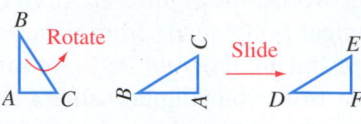

> Be careful identifying corresponding parts when one triangle is rotated or flipped.

The corresponding parts are:

∠1 and ∠6	$\overline{BC}$ and $\overline{DE}$
∠2 and ∠4	$\overline{BA}$ and $\overline{DF}$
∠3 and ∠5	$\overline{CA}$ and $\overline{EF}$

◀ *Work Problem* **1** *at the Side.*

OBJECTIVE **2** **Prove that triangles are congruent using SAS, SSS, and ASA.** One way to prove that two triangles are congruent would be to measure all the angles and all the sides. If the measures of the corresponding angles and sides are equal, then the triangles are congruent. But here are three quicker methods to prove that two triangles are congruent.

Proving That Two Triangles Are Congruent

1. Angle–Side–Angle (ASA) Method
If two angles and the side between them on one triangle measure the same as the corresponding parts on another triangle, the triangles are congruent.

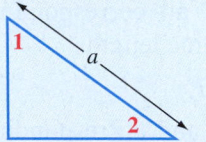 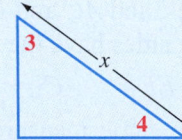

If $m\angle 1 = m\angle 3$ and $m\angle 2 = m\angle 4$ and $a = x$ then the two triangles are congruent.

2. Side–Side–Side (SSS) Method
If three sides of one triangle measure the same as the corresponding sides of another triangle, the triangles are congruent.

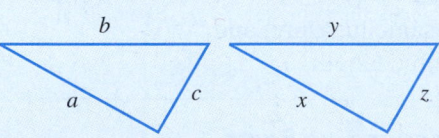

If $a = x$ and $b = y$ and $c = z$ then the two triangles are congruent.

3. Side–Angle–Side (SAS) Method
If two sides and the angle between them on one triangle measure the same as the corresponding parts on another triangle, the triangles are congruent.

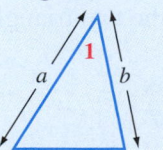 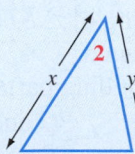

If $a = x$ and $b = y$ and $m\angle 1 = m\angle 2$ then the two triangles are congruent.

ANSWERS

1. (a) ∠1 and ∠4, ∠2 and ∠5, ∠3 and ∠6;
   $\overline{AC}$ and $\overline{DF}$, $\overline{AB}$ and $\overline{DE}$, $\overline{BC}$ and $\overline{EF}$,
(b) ∠1 and ∠6, ∠2 and ∠5, ∠3 and ∠4;
   $\overline{GF}$ and $\overline{KL}$, $\overline{FH}$ and $\overline{LJ}$, $\overline{GH}$ and $\overline{KJ}$,
(c) ∠1 and ∠5, ∠2 and ∠4, ∠3 and ∠6;
   $\overline{RS}$ and $\overline{XV}$, $\overline{RT}$ and $\overline{XW}$, $\overline{ST}$ and $\overline{VW}$

EXAMPLE 2   **Proving That Two Triangles Are Congruent**

Explain which method can be used to prove that each pair of triangles is congruent. Choose from ASA, SSS, and SAS.

(a)

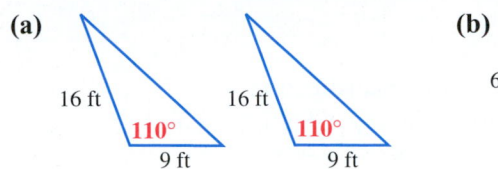

(b)

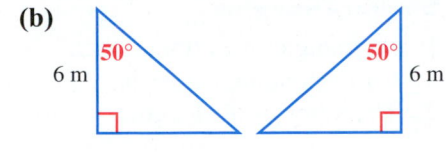

(c)

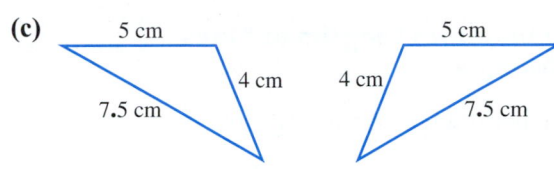

(a) On both triangles, two corresponding sides and the angle between them measure the same, so the Side–Angle–Side (SAS) method can be used to prove that the triangles are congruent.

(b) On both triangles, two corresponding angles and the side between them measure the same, so the Angle–Side–Angle (ASA) method can be used to prove that the triangles are congruent.

(c) Each pair of corresponding sides has the same length, so the Side–Side–Side (SSS) method can be used to prove that the triangles are congruent.

Work Problem **2** *at the Side.* ▶

OBJECTIVE **3** **Identify corresponding parts of similar triangles.**
Now that you've worked with *congruent* triangles, let's look at *similar* triangles. Remember that congruent triangles match exactly, both in shape and in size. Similar triangles, on the other hand, have the same shape but are *different sizes*. Three pairs of similar triangles are shown here.

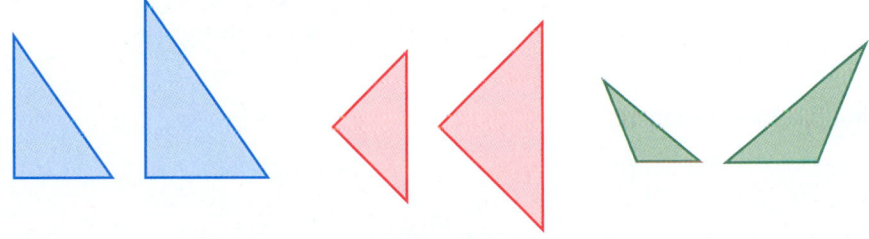

Each pair of triangles has the same shape because the corresponding angles have the same measure. But the corresponding sides are *not* the same length, so the triangles are of *different sizes*.

Two similar triangles are shown to the right. Notice that corresponding angles have the same measure, but corresponding sides have different lengths.

$\overline{CB}$ corresponds to $\overline{RQ}$. Similarly, $\overline{CA}$ corresponds to $\overline{RP}$, and $\overline{BA}$ corresponds to $\overline{QP}$. Notice that each side in the larger triangle is *twice* the length of the corresponding side in the smaller triangle.

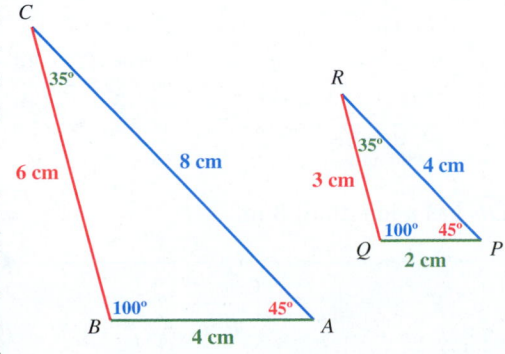

Work Problem **3** *at the Side.* ▶

2 Determine which method can be used to prove that each pair of triangles is congruent.

(a)

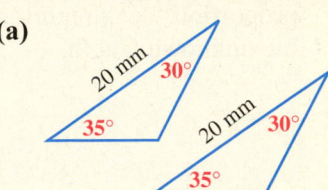

(b)

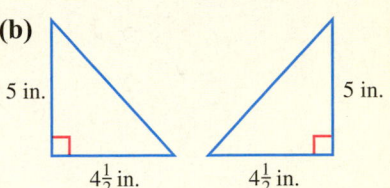

(c)

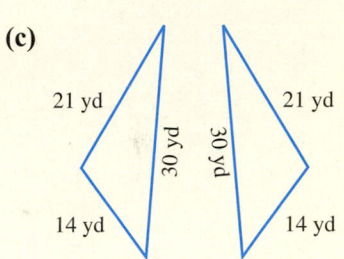

3 Identify corresponding angles and sides in these similar triangles.

(a)

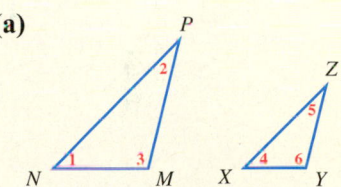

Angles: Sides:
1 and _____ $\overline{PN}$ and _____
2 and _____ $\overline{PM}$ and _____
3 and _____ $\overline{NM}$ and _____

(b)

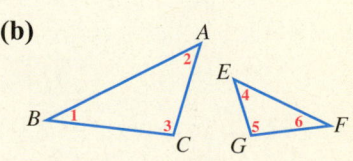

Angles: Sides:
1 and _____ $\overline{AB}$ and _____
2 and _____ $\overline{BC}$ and _____
3 and _____ $\overline{AC}$ and _____

ANSWERS

2. **(a)** ASA   **(b)** SAS   **(c)** SSS
3. **(a)** 4; 5; 6; $\overline{ZX}$; $\overline{ZY}$; $\overline{XY}$
   (b) 6; 4; 5; $\overline{EF}$; $\overline{FG}$; $\overline{EG}$

4 Find the length of $\overline{EF}$ in Example 3 at the right by setting up and solving a proportion. Let x represent the unknown length.

OBJECTIVE **4** **Find the unknown lengths of sides in similar triangles.** Similar triangles are useful because of the following definition.

> **Similar Triangles**
>
> If two triangles are similar, then
> 1. Corresponding angles have the same measure, and
> 2. The ratios of the lengths of corresponding sides are equal.

EXAMPLE 3 Finding the Unknown Lengths of Sides in Similar Triangles

Find the length of $\overline{DF}$ in the smaller triangle. Assume the triangles are similar.

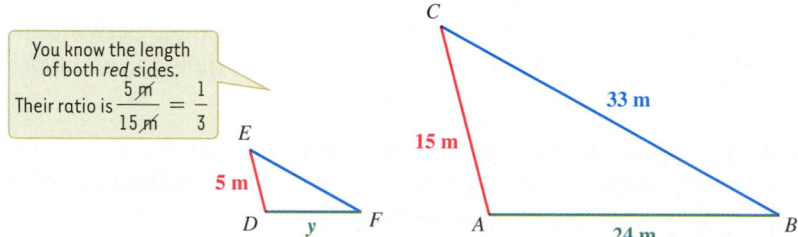

You know the length of both *red* sides. Their ratio is $\dfrac{5 \text{ m}}{15 \text{ m}} = \dfrac{1}{3}$

The length you want to find in the smaller triangle is $\overline{DF}$, and it corresponds to $\overline{AB}$ in the *larger* triangle. Then, notice that $\overline{ED}$ in the smaller triangle corresponds to $\overline{CA}$ in the larger triangle, and you know both of their lengths. Since the *ratios* of the lengths of corresponding sides are equal, you can set up a proportion. (Recall that a proportion states that two ratios are equal.)

$$\text{Corresponding sides} \begin{cases} DF \to \dfrac{y}{AB} \to \dfrac{y}{24} = \dfrac{5}{15} \leftarrow ED \\ \hphantom{AB \to} 24 \hphantom{=} 15 \leftarrow CA \end{cases} \text{Corresponding sides}$$

$$\frac{y}{24} = \frac{1}{3}$$

Write $\frac{5}{15}$ in lowest terms as $\frac{1}{3}$

Find the cross products.

$$24 \cdot 1 = 24$$

$$\frac{y}{24} \bowtie \frac{1}{3}$$

$$y \cdot 3$$

$$y \cdot 3 = 24 \qquad \text{Show that the cross products are equivalent.}$$

$$\frac{y \cdot \overset{1}{\cancel{3}}}{\cancel{3}_1} = \frac{24}{3} \qquad \text{Divide both sides by 3.}$$

Write **m** in the answer.

$$y = 8$$

$\overline{DF}$ has a length of 8 m.

◄ *Work Problem* **4** *at the Side.*

EXAMPLE 4   **Finding an Unknown Length and the Perimeter**

Find the perimeter of the smaller triangle. Assume the triangles are similar.

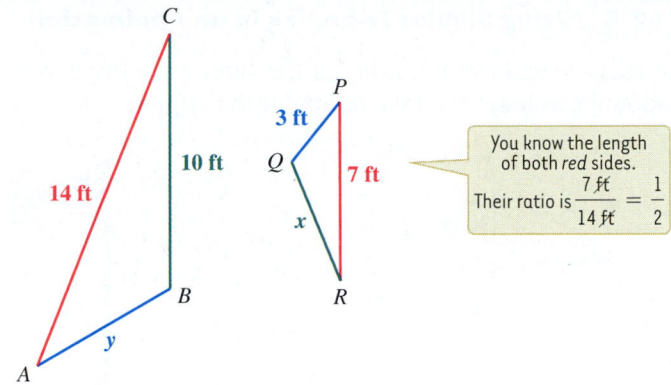

First find x, the length of $\overline{QR}$ in the smaller triangle, then add the lengths of all three sides to find the perimeter.

The smaller triangle is turned "upside down" compared to the larger triangle, so be careful when identifying corresponding sides. $\overline{PR}$ is the longest side in the smaller triangle, and $\overline{AC}$ is the longest side in the larger triangle. So $\overline{PR}$ and $\overline{AC}$ are corresponding sides and you know both of their lengths. $\overline{QR}$, the length you want to find in the smaller triangle, corresponds to $\overline{BC}$ in the larger triangle. The ratios of the lengths of corresponding sides are equal, so you can set up a proportion.

$$\begin{array}{c} QR \rightarrow \\ BC \rightarrow \end{array} \dfrac{x}{10} = \dfrac{7}{14} \begin{array}{c} \leftarrow PR \\ \leftarrow AC \end{array}$$

$$\dfrac{x}{10} = \dfrac{1}{2} \qquad \text{Write } \tfrac{7}{14} \text{ in lowest terms as } \tfrac{1}{2}$$

Find the cross products.

$$10 \cdot 1 = 10$$

$$\dfrac{x}{10} \ \diagdown\!\!\!\!\diagup \ \dfrac{1}{2}$$

$$x \cdot 2$$

$$x \cdot 2 = 10 \qquad \text{Show that the cross products are equivalent.}$$

$$\dfrac{x \cdot \overset{1}{\cancel{2}}}{\underset{1}{\cancel{2}}} = \dfrac{10}{2} \qquad \text{Divide both sides by 2.}$$

Write **ft** for the length.

$$x = 5$$

$\overline{QR}$ has a length of 5 ft.

Now add the lengths of all three sides to find the perimeter of the smaller triangle.

$$\text{Perimeter} = 5 \text{ ft} + 3 \text{ ft} + 7 \text{ ft} = 15 \text{ ft}$$

Work Problem **5** *at the Side.* ▶

5   **(a)** Find the perimeter of triangle ABC in Example 4 at the left.

(b) Find the perimeter of each triangle. Assume the triangles are similar.

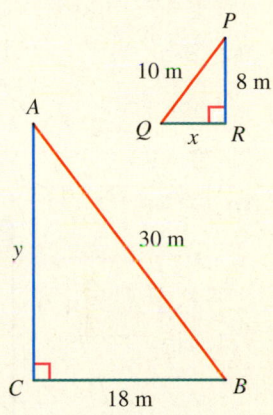

6 Find the height of each flagpole.

(a)

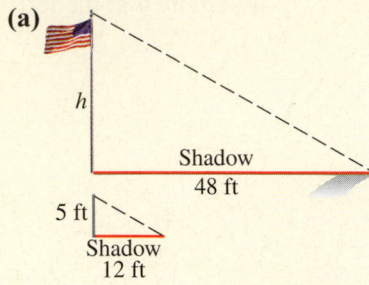

(b)

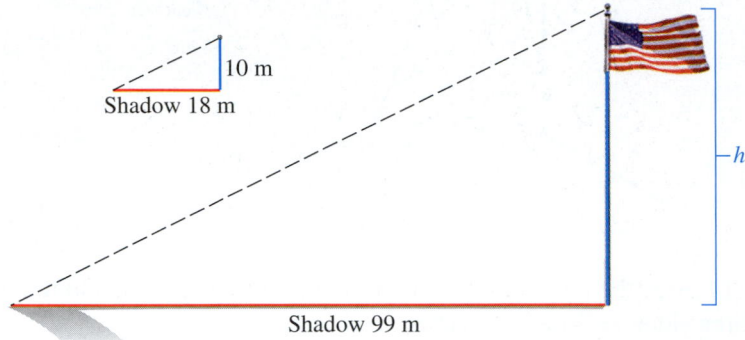

OBJECTIVE 5   **Solve application problems involving similar triangles.**   The next example shows an application of similar triangles.

EXAMPLE 5   **Using Similar Triangles in an Application**

A flagpole casts a shadow 99 m long at the same time that a pole 10 m tall casts a shadow 18 m long. Find the height of the flagpole.

The triangles shown are similar, so write a proportion to find h.

Height in larger triangle $\rightarrow$ $\dfrac{h}{10} = \dfrac{99}{18}$ $\leftarrow$ Shadow in larger triangle
Height in smaller triangle $\rightarrow$ $\leftarrow$ Shadow in smaller triangle

Find the cross products and show that they are equivalent.

$$h \cdot 18 = 10 \cdot 99$$
$$h \cdot 18 = 990$$

$$\frac{h \cdot \cancel{18}^{1}}{\cancel{18}_{1}} = \frac{990}{18} \qquad \text{Divide both sides by 18.}$$

$$h = 55$$

The flagpole is 55 m high.

> **Note**
>
> There are several other correct ways to set up the proportion in Example 5 above. One way is to simply flip the ratios on *both* sides of the equal sign.
>
> $$\frac{10}{h} = \frac{18}{99}$$
>
> But there is another option, shown below.
>
> Height in larger triangle $\rightarrow$ $\dfrac{h}{99} = \dfrac{10}{18}$ $\leftarrow$ Height in smaller triangle
> Shadow in larger triangle $\rightarrow$ $\leftarrow$ Shadow in smaller triangle
>
> Notice that both ratios compare *height* to *shadow* in the same order. The ratio on the left describes the larger triangle, and the ratio on the right describes the smaller triangle.

◀ *Work Problem* **6** *at the Side.*

ANSWERS

6. (a) $h = 20$ ft   **(b)** $h = 18$ m

7.8 ▶▶▶ Exercises

MyMathLab   Math XL PRACTICE   WATCH   DOWNLOAD   READ   REVIEW

Each pair of triangles is congruent. List the corresponding angles and the corresponding sides. See Example 1.

1.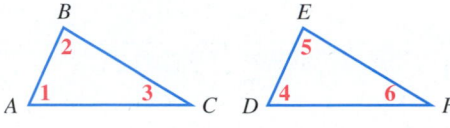

2.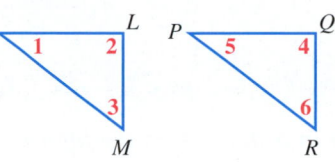

3.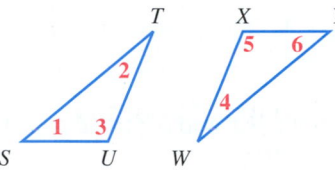

4.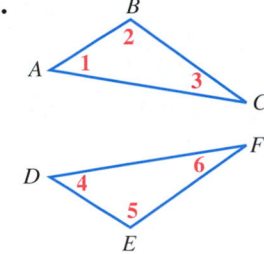

5.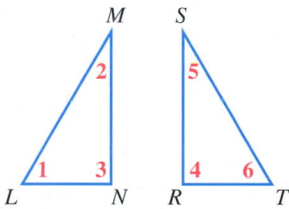

6.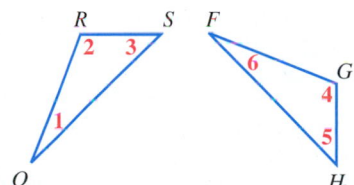

Determine which of these methods can be used to prove that each pair of triangles is congruent; Angle–Side–Angle (ASA), Side–Side–Side (SSS), or Side–Angle–Side (SAS). See Example 2.

7.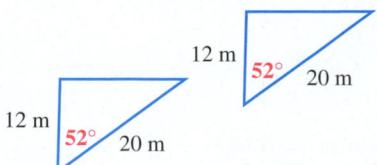

8.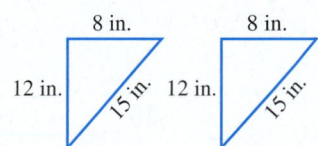

9.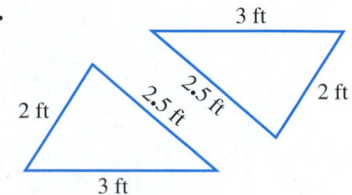

10.

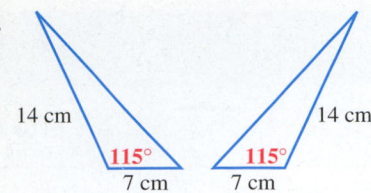

14 cm   14 cm
115°   115°
7 cm   7 cm

11.

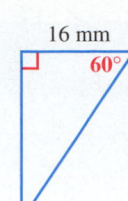

16 mm   16 mm
60°   60°

12.

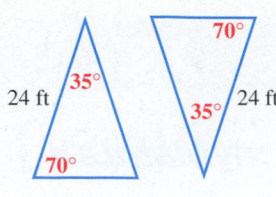

70°
35°   35°
24 ft   24 ft
70°

Relating Concepts (Exercises 13—16) For Individual or Group Work

Work Exercises 13–16 in order. Given the information in each exercise, explain how you can prove that the indicated triangles are congruent. Note: A midpoint divides a segment into two congruent parts.

13.

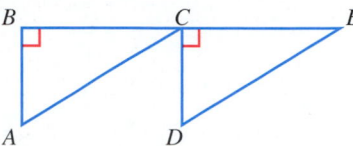

B   C   E
A   D

C is the midpoint of $\overline{BE}$ and $CD = BA$. Prove that $\triangle ABC \cong \triangle DCE$.

14.

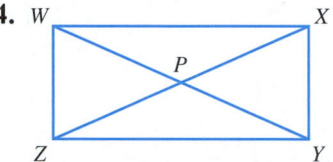

W   X
P
Z   Y

P is the midpoint of both $\overline{WY}$ and $\overline{XZ}$; $WZ = XY$. Prove that $\triangle WPZ \cong \triangle YPX$.

15.

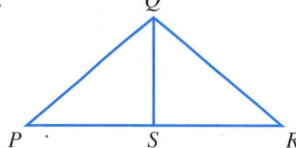

Q
P   S   R

$\overline{QS} \perp \overline{PR}$ and S is the midpoint of $\overline{PR}$. Prove that $\triangle PQS \cong \triangle RQS$.

16.

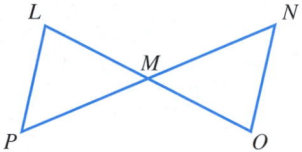

L   N
M
P   O

M is the midpoint of both $\overline{LO}$ and $\overline{PN}$. Prove that $\triangle PLM \cong \triangle NOM$.

Find the unknown lengths in each pair of similar triangles. See Example 3.

17.

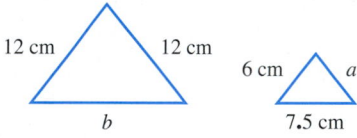

12 cm   12 cm
6 cm   a
b   7.5 cm

18.

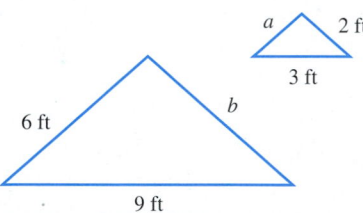

a   2 ft
3 ft
b
6 ft
9 ft

19.

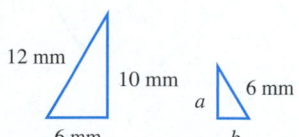

12 mm
10 mm   6 mm
a
6 mm   b

20.

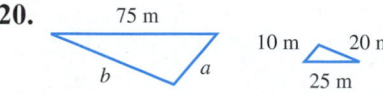

75 m
10 m   20 m
b   a
25 m

21.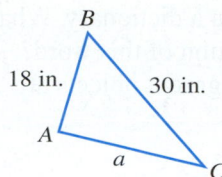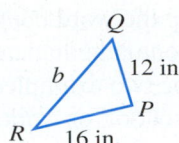

22.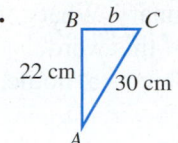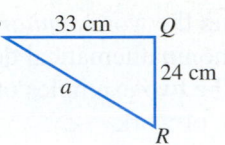

In Exercises 23 and 24, find the perimeter of each triangle. Assume the triangles are similar.
See Example 4.

23.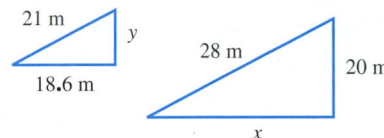

24.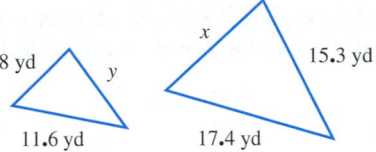

25. Triangles *CDE* and *FGH* are similar. Find the perimeter and area of triangle *FGH*. *Note:* The heights of similar triangles have the same ratio as corresponding sides. Round to the nearest tenth when necessary.

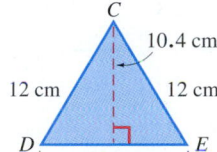 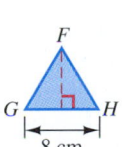

26. Triangles *JKL* and *MNO* are similar. Find the perimeter and area of triangle *MNO*.

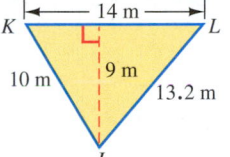 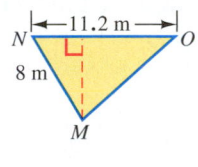

Solve each application problem. See Example 5.

27. The height of the house shown here can be found by comparing its shadow to the shadow cast by a 3-foot stick. Find the height of the house by writing a proportion and solving it.

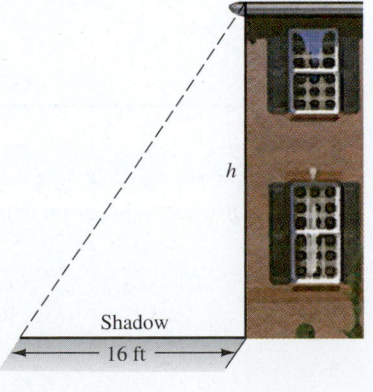

28. A fire lookout tower provides an excellent view of the surrounding countryside. The height of the tower can be found by lining up the top of the tower with the top of a 2-meter stick. Use similar triangles to find the height of the tower.

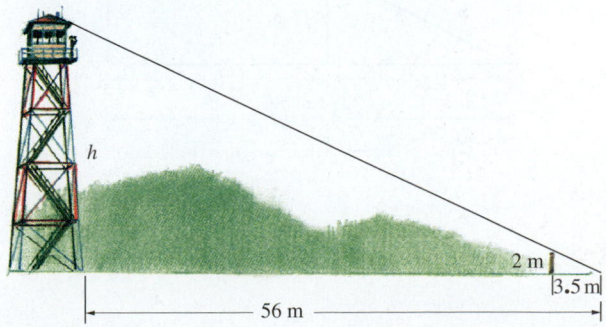

29. Look up the word *similar* in a dictionary. What is the nonmathematical definition of this word? Describe two examples of similar objects at home, school, or work.

30. Look up the word *congruent* in a dictionary. What is the nonmathematical definition of this word? Describe two examples of congruent objects at home, school, or work.

Find the unknown length in Exercises 31–34. Round your answers to the nearest tenth. Note: When a line is drawn parallel to one side of a triangle, the smaller triangle that is formed will be similar to the original triangle. In Exercises 31–32, the red segments are parallel.

31.

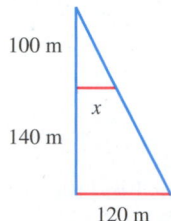

Hint: Redraw the two triangles and label the sides.

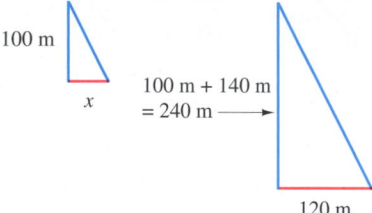

32.

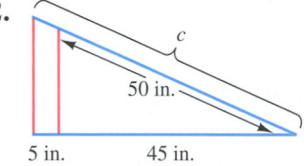

Hint: Redraw the two triangles and label the sides.

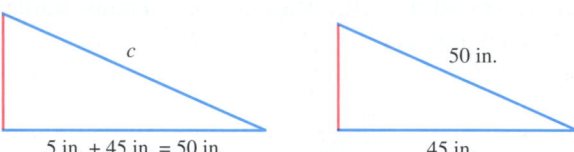

33. Use similar triangles and a proportion to find the length of the lake shown here. (*Hint:* The side 100 m long in the smaller triangle corresponds to the side of 100 m + 120 m = 220 m in the larger triangle.)

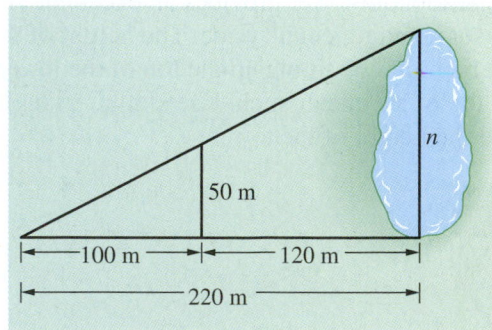

34. To find the height of the tree, find y and then add $5\frac{1}{2}$ ft for the distance from the ground to the eye level of the person.

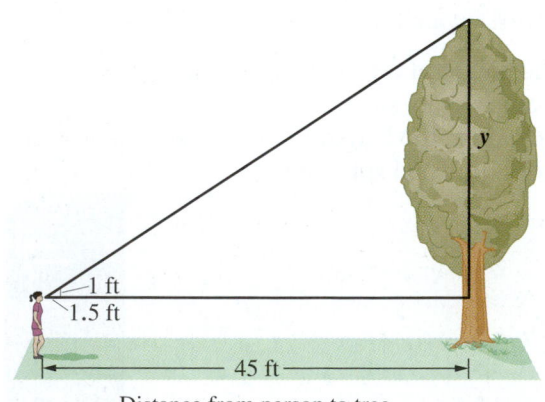

Distance from person to tree

Chapter 7 ▶▶▶ Summary

▶ Key Terms

7.1

point	A point is a location in space. *Example:* Point *P* at the right.	
line	A line is a straight row of points that goes on forever in both directions. *Example:* Line *AB*, written $\overleftrightarrow{AB}$, at the right.	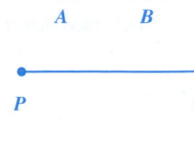
line segment	A line segment is a piece of a line with two endpoints. *Example:* Line segment *PQ*, written $\overline{PQ}$, at the right.	
ray	A ray is a part of a line that has one endpoint and extends forever in one direction. *Example:* Ray *RS*, written $\overrightarrow{RS}$, at the right.	
parallel lines	Parallel lines are two lines in the same plane that never intersect (never cross). *Example:* $\overleftrightarrow{AB}$ is parallel to $\overleftrightarrow{ST}$ at the right.	
intersecting lines	Intersecting lines cross. *Example:* $\overleftrightarrow{RQ}$ intersects $\overleftrightarrow{AB}$ at point *P* at the right.	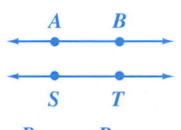
angle	An angle is made up of two rays that have a common endpoint called the vertex. *Example:* Angle 1 at the right.	
degrees	Degrees are used to measure angles; a complete circle is 360 degrees, written 360°.	
straight angle	A straight angle is an angle that measures *exactly* 180°; its sides form a straight line. *Example:* Angle *G* at the right.	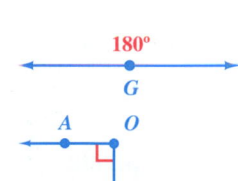
right angle	A right angle is an angle that measures *exactly* 90°. *Example:* Angle *AOB* at the right.	
acute angle	An acute angle is an angle that measures less than 90°. *Example*: Angle *E* at the right.	
obtuse angle	An obtuse angle is an angle that measures more than 90° but less than 180°. *Example:* Angle *F* at the right.	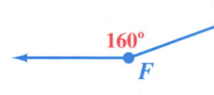
perpendicular lines	Perpendicular lines are two lines that intersect to form a right angle. *Example*: $\overleftrightarrow{PQ}$ is perpendicular to $\overleftrightarrow{RS}$ at the right.	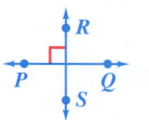
complementary angles	Complementary angles are two angles whose measures add up to 90°.	
supplementary angles	Supplementary angles are two angles whose measures add up to 180°.	
congruent angles	Congruent angles are angles that measure the same number of degrees.	
vertical angles	Vertical angles are two nonadjacent congruent angles formed by two intersecting lines. *Example*: ∠*COA* and ∠*EOF* are vertical angles at the right.	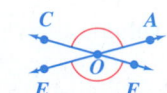

(continued)

▶ **Key Terms** (*continued*)

corresponding angles	Corresponding angles are formed when two parallel lines are crossed by a transversal; corresponding angles are congruent and are on the same side of the transversal and in the same relative position. *Example*: In the figure at the right, line *m* is parallel to line *n*. The pairs of corresponding angles are ∠1 and ∠5, ∠2 and ∠6, ∠3 and ∠7, ∠4 and ∠8.	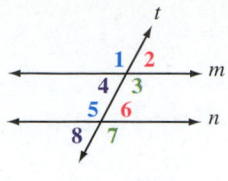
alternate interior angles	When two parallel lines are crossed by a transversal, there are two pairs of alternate interior angles and each pair is congruent. They are on opposite sides of the transversal. *Example*: In the figure at the right, line *m* is parallel to line *n*. The pairs of alternate interior angles are ∠3 and ∠5, ∠4 and ∠6.	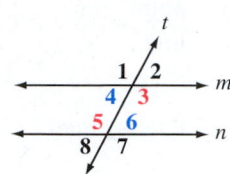

7.2–7.4	**perimeter**	Perimeter is the distance around the outside edges of a flat shape. It is measured in linear units such as ft, yd, cm, m, km, and so on.	
7.2–7.5	**area**	Area is the surface inside a two-dimensional (flat) shape. It is measured by determining the number of squares of a certain size needed to cover the surface inside the shape. Some of the commonly used units for measuring area are square inches (in.^{2}), square feet (ft^2), square yards (yd^2), square centimeters (cm^2), and square meters (m^2).	
7.2	**rectangle**	A rectangle is a four-sided figure with all sides meeting at 90° angles. The opposite sides are the same length. *Example:* The rectangle measuring 12 cm by 7 cm at the right.	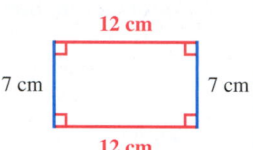
	square	A square is a rectangle with all four sides the same length. *Example:* The square with a side measurement of 20 inches at the right.	
7.3	**parallelogram**	A parallelogram is a four-sided figure with both pairs of opposite sides parallel and equal in length. *Example:* See the parallelogram at the right with sides measuring 8 m and 12 m.	
	trapezoid	A trapezoid is a four-sided figure with exactly one pair of parallel sides. *Example:* Trapezoid *PQRS* at the right; $\overline{PQ}$ is parallel to $\overline{SR}$.	
7.4	**triangle**	A triangle is a figure with exactly three sides. *Example:* Triangle *ABC* at the right.	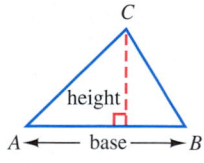
7.5	**circle**	A circle is a figure with all points the same distance from a fixed center point. *Example:* See figure at the right.	
	radius	Radius is the distance from the center of a circle to any point on the circle. *Example:* See the red radius in the circle at the right.	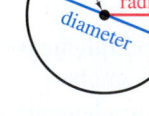
	diameter	Diameter is the distance across a circle, passing through the center. *Example:* See the blue diameter in the circle at the right.	
	circumference	Circumference is the distance around a circle.	
	π (pi)	π is the ratio of the circumference to the diameter of any circle. It is approximately equal to 3.14.	
7.6	**volume**	Volume is a measure of the space inside a three-dimensional (solid) shape. Volume is measured in cubic units such as in.^{3}, ft^3, yd^3, mm^3, cm^3, and so on.	
	surface area	Surface area is the area on the surface of a three-dimensional object (a solid). Surface area is measured in square units.	

▶ Key Terms (*continued*)

7.7	**hypotenuse**	The hypotenuse is the side of a right triangle opposite the 90° angle; it is the longest side. *Example:* See the red side in the triangle at the right.

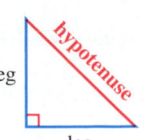

7.8	**congruent figures**	Congruent figures are identical both in shape and in size.
	similar figures	Similar figures have the same shape but are different sizes.
	congruent triangles	Congruent triangles are triangles with the same shape and the same size; corresponding angles measure the same number of degrees and corresponding sides have the same length.
	similar triangles	Similar triangles are triangles with the same shape but not necessarily the same size; corresponding angles measure the same number of degrees and the *ratios* of the lengths of corresponding sides are equal.

▶ New Symbols

$\overleftrightarrow{AB}$	line AB	**right angle:** (90° angle)	**Square units:** (for measuring area)	in.^{2} ft^2 yd^2 mi^2 mm^2 cm^2 m^2 km^2
$\overline{EF}$	line segment EF	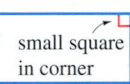 small square in corner	**Cubic units:** (for measuring volume)	in.^{3} ft^3 yd^3 mm^3 cm^3 m^3
$\overrightarrow{RS}$	ray RS	$\perp$ is perpendicular to		
$\angle MRN$	angle MRN	$\cong$ is congruent to		
1°	one degree	π Greek letter pi; ratio of circumference to diameter of any circle		

▶ New Formulas

Perimeter of a rectangle: $P = 2 \cdot l + 2 \cdot w$

Area of a rectangle: $A = l \cdot w$

Perimeter of a square: $P = 4 \cdot s$

Area of a square: $A = s^2$

Area of a parallelogram: $A = b \cdot h$

Area of a trapezoid: $A = \dfrac{1}{2} \cdot h \cdot (b + B)$

$\quad$ or $\quad A = 0.5 \cdot h \cdot (b + B)$

Area of a triangle: $A = \dfrac{1}{2} \cdot b \cdot h$

$\quad$ or $\quad A = 0.5 \cdot b \cdot h$

Diameter of a circle: $d = 2 \cdot r$

Radius of a circle: $r = \dfrac{d}{2}$

Circumference of a circle: $C = \pi \cdot d$

$\quad$ or $\quad C = 2 \cdot \pi \cdot r$

Area of a circle: $A = \pi \cdot r^2$

Area of semicircle: $A = \dfrac{\pi \cdot r^2}{2}$

Volume of rectangular solid: $V = l \cdot w \cdot h$

Volume of a sphere: $V = \dfrac{4}{3} \cdot \pi \cdot r^3$

$\quad$ or $\quad V = \dfrac{4 \cdot \pi \cdot r^3}{3}$

Volume of a hemisphere: $V = \dfrac{2}{3} \cdot \pi \cdot r^3$

$\quad$ or $\quad V = \dfrac{2 \cdot \pi \cdot r^3}{3}$

Volume of a cylinder: $V = \pi \cdot r^2 \cdot h$

Volume of a cone: $V = \dfrac{1}{3} \cdot B \cdot h \quad$ or $\quad V = \dfrac{B \cdot h}{3}$

Volume of a pyramid: $V = \dfrac{1}{3} \cdot B \cdot h \quad$ or $\quad V = \dfrac{B \cdot h}{3}$

Surface Area of a rectangular solid:
$$SA = 2lw + 2lh + 2wh$$

Surface Area of a cylinder: $\quad SA = 2\pi rh + 2\pi r^2$

Right triangle: hypotenuse $= \sqrt{(\text{leg})^2 + (\text{leg})^2}$

$\qquad\qquad$ leg $= \sqrt{(\text{hypotenuse})^2 - (\text{leg})^2}$

▶ Test Your Word Power

See how well you have learned the vocabulary in this chapter. Answers follow the Quick Review.

1. Two angles that are **complementary**
   A. have measures that add up to 180°
   B. are always congruent
   C. form a straight angle
   D. have measures that add up to 90°.

2. The **perimeter** of a flat shape is
   A. measured in square units
   B. the distance around the out-side edges
   C. the number of squares needed to cover the space inside the shape
   D. measured in cubic units.

3. An **obtuse angle**
   A. is formed by perpendicular lines
   B. is congruent to a right angle
   C. measures more than 90° but less than 180°
   D. measures less than 90°.

4. The **hypotenuse** is
   A. the long base in a trapezoid
   B. the height line in a parallelo-gram
   C. the longest side in a right triangle
   D. the distance across a circle, passing through the center.

5. π is the ratio of
   A. the diameter to the radius of a circle
   B. the circumference to the diameter of a circle
   C. the circumference to the radius of a circle
   D. the diameter to the circum-ference of a circle.

6. **Perpendicular lines**
   A. intersect to form a right angle
   B. intersect to form an acute angle
   C. never intersect
   D. have a common endpoint called the vertex.

7. In a pair of **similar triangles,**
   A. corresponding sides have the same length
   B. all the angles have the same measure
   C. the perimeters are equal
   D. the ratios of the lengths of corresponding sides are equal.

8. The **area of a rectangle** is found by
   A. using the formula $P = 2 \cdot l + 2 \cdot w$
   B. multiplying length times width
   C. adding the lengths of the sides
   D. using the formula $V = l \cdot w \cdot h$.

▶ Quick Review

| Concepts | Examples |

7.1   Lines

A *line* is a straight row of points that goes on forever in both directions. If a piece of a line has one endpoint, it is a *ray.* If it has two endpoints, it is a *line segment.*

Identify each of the following as a line, line segment, or ray and name it using the appropriate symbol.

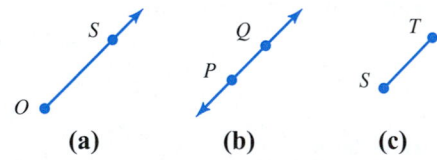

(a)   (b)   (c)

Figure **(a)** shows a ray named $\overrightarrow{OS}$.

Figure **(b)** shows a line named $\overleftrightarrow{PQ}$ or $\overleftrightarrow{QP}$.

Figure **(c)** shows a line segment named $\overline{ST}$ or $\overline{TS}$.

Concepts	Examples

7.1 Lines (continued)

If two lines intersect at right angles, they are *perpendicular*.

If two lines in the same plane never intersect, they are *parallel*.

Label each pair of lines as appearing to be parallel or as perpendicular.

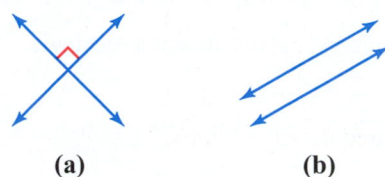

(a)   **(b)**

Figure **(a)** shows perpendicular lines (they intersect at 90°).

Figure **(b)** shows lines that appear to be parallel (they never intersect).

7.1 Angles

If the sum of the measures of two angles is 90°, they are *complementary*.

If the sum of the measures of two angles is 180°, they are *supplementary*.

If two angles measure the same number of degrees, the angles are *congruent*. The symbol for congruent is ≅.

Two nonadjacent angles formed by two intersecting lines are called *vertical angles*. Vertical angles are congruent.

Find the complement and supplement of a 35° angle.

$$90° - 35° = 55° \text{ (the complement)}$$
$$180° - 35° = 145° \text{ (the supplement)}$$

Identify the vertical angles in this figure. Which angles are congruent?

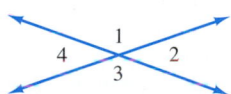

∠1 and ∠3 are vertical angles.

∠2 and ∠4 are vertical angles.

Vertical angles are congruent, so ∠1 ≅ ∠3 and ∠2 ≅ ∠4.

7.1 Parallel Lines

When two parallel lines are crossed by a transversal, corresponding angles are congruent, and alternate interior angles are congruent. Use this information to find the measures of the other angles.

Read $m\angle 1$ as "the measure of angle 1."

Line m is parallel to line n and the measure of ∠4 is 125°. Find the measures of the other angles.

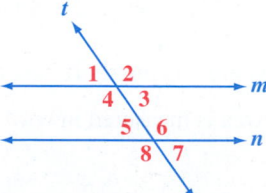

∠4 ≅ ∠8 (corresponding angles), so $m\angle 8 = 125°$.

∠4 ≅ ∠6 (alternate interior angles), so $m\angle 6 = 125°$.

∠6 ≅ ∠2 (corresponding angles), so $m\angle 2 = 125°$.

∠4 and ∠3 are supplements, so $m\angle 3 = 180° - 125° = 55°$.

∠3 ≅ ∠7 (corresponding angles), so $m\angle 7 = 55°$.

∠3 ≅ ∠5 (alternate interior angles), so $m\angle 5 = 55°$.

∠5 ≅ ∠1 (corresponding angles), so $m\angle 1 = 55°$.

(continued)

Concepts	Examples

7.2 Rectangles and Squares

Use this formula to find the perimeter of a *rectangle*.

$$P = 2 \cdot l + 2 \cdot w$$

Use this formula to find the area of a rectangle.

$$A = l \cdot w$$

Area is measured in **square units**.

Use these formulas to find the perimeter and area of a *square*.

$$P = 4 \cdot s$$
$$A = s^2$$

Area is measured in **square units**.

Find the perimeter and area of this rectangle.

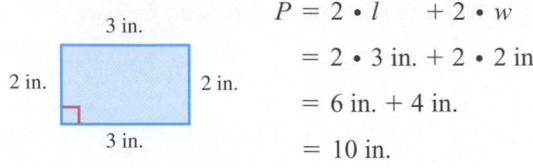

$$P = 2 \cdot l \quad + 2 \cdot w$$
$$= 2 \cdot 3 \text{ in.} + 2 \cdot 2 \text{ in.}$$
$$= 6 \text{ in.} + 4 \text{ in.}$$
$$= 10 \text{ in.}$$
$$A = l \cdot w = 3 \text{ in.} \cdot 2 \text{ in.} = 6 \text{ in.}^2$$

Find the perimeter and area of this square.

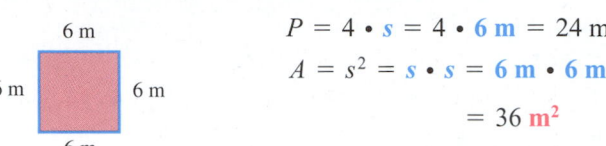

$$P = 4 \cdot s = 4 \cdot \mathbf{6\,m} = 24 \text{ m}$$
$$A = s^2 = s \cdot s = \mathbf{6\,m} \cdot \mathbf{6\,m}$$
$$= 36 \text{ m}^2$$

7.3 Parallelograms

Use these formulas to find the perimeter and area of a *parallelogram*.

$$P = \text{sum of the lengths of the sides}$$
$$A = b \cdot h$$

Area is measured in **square units**.

Find the perimeter and area of this parallelogram.

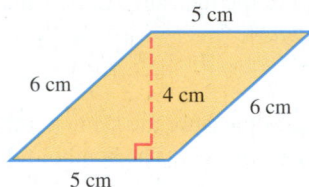

$$P = 5 \text{ cm} + 6 \text{ cm} + 5 \text{ cm} + 6 \text{ cm} = 22 \text{ cm}$$
$$A = 5 \text{ cm} \cdot 4 \text{ cm} = 20 \text{ cm}^2$$

7.3 Trapezoids

Use these formulas to find the perimeter and area of a *trapezoid*.

$$P = \text{sum of the lengths of the sides}$$
$$A = \frac{1}{2} \cdot h \cdot (b + B)$$

where *b* is the short base and *B* is the long base.

Area is measured in **square units**.

Find the perimeter and area of this trapezoid.

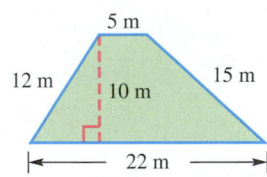

$$P = 5 \text{ m} + 15 \text{ m} + 22 \text{ m} + 12 \text{ m} = 54 \text{ m}$$
$$A = \frac{1}{2} \cdot \overset{5}{\cancel{10}} \text{ m} \cdot (5 \text{ m} + 22 \text{ m})$$
$$= 5 \text{ m} \cdot (27 \text{ m}) = 135 \text{ m}^2$$

7.4 Triangles

Use these formulas to find the perimeter and area of a *triangle*.

$$P = \text{sum of the lengths of the sides}$$
$$A = \frac{1}{2} \cdot b \cdot h$$

or $A = 0.5 \cdot b \cdot h$

Area is measured in **square units**.

Find the perimeter and area of this triangle.

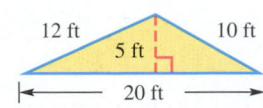

$$P = 12 \text{ ft} + 10 \text{ ft} + 20 \text{ ft} = 42 \text{ ft}$$
$$A = \frac{1}{2} \cdot b \cdot h$$
$$A = \frac{1}{2} \cdot \overset{10}{\cancel{20}} \text{ ft} \cdot 5 \text{ ft} = 50 \text{ ft}^2$$

or $A = 0.5 \cdot 20 \text{ ft} \cdot 5 \text{ ft} = 50 \text{ ft}^2$

Concepts	Examples

7.5 Circles

Use this formula to find the *diameter* of a circle when you are given the radius.

$$d = 2 \cdot r$$

Find the diameter of a circle if the radius is 7 yd.

$$d = 2 \cdot r = 2 \cdot \mathbf{7 \ yd} = 14 \ yd$$

Use this formula to find the *radius* of a circle when you are given the diameter.

$$r = \frac{d}{2}$$

Find the radius of a circle if the diameter is 5 cm.

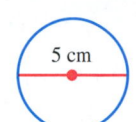

$$r = \frac{d}{2} = \frac{\mathbf{5 \ cm}}{2} = 2.5 \ cm$$

Use these formulas to find the *circumference* of a circle.

When you know the radius, use $C = 2 \cdot \pi \cdot r$

When you know the diameter, use $C = \pi \cdot d$

Use 3.14 as the approximate value for π.

Find the circumference of a circle with a radius of 3 cm.

$$C = 2 \cdot \pi \cdot r$$
$$C \approx 2 \cdot \mathbf{3.14} \cdot \mathbf{3 \ cm} \approx 18.8 \ cm \quad \leftarrow \text{Rounded}$$

Use this formula to find the *area* of a circle.

$$A = \pi \cdot r^2$$

Area is measured in **square units**.

Find the area of this circle.

$$A = \pi \cdot r^2$$
$$A \approx \mathbf{3.14} \cdot \mathbf{3 \ cm} \cdot \mathbf{3 \ cm}$$
$$A \approx 28.3 \ \mathbf{cm^2} \leftarrow \text{Rounded;} \\ \text{square units for area}$$

7.6 Volume of a Rectangular Solid

Use this formula to find the volume of *rectangular solids* (box-like solids).

$$V = l \cdot w \cdot h$$

Volume is measured in **cubic units**.

Find the volume of this box.

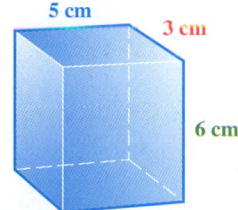

$$V = l \cdot w \cdot h$$
$$V = \mathbf{5 \ cm} \cdot \mathbf{3 \ cm} \cdot \mathbf{6 \ cm}$$
$$V = 90 \ \mathbf{cm^3} \quad \text{Cubic units} \\ \text{for volume}$$

7.6 Volume of a Sphere and Hemisphere

Use this formula to find the volume of a *sphere* (a ball-shaped solid).

$$V = \frac{4}{3} \cdot \pi \cdot r^3$$

or $$V = \frac{4 \cdot \pi \cdot r^3}{3}$$

where r is the radius of the sphere.

Volume is measured in **cubic units**.

Find the volume of a sphere with a radius of 5 m.

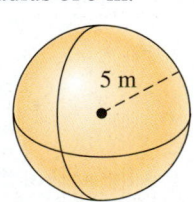

$$V = \frac{4 \cdot \pi \cdot (\mathbf{radius})^3}{3}$$

$$V \approx \frac{4 \cdot \mathbf{3.14} \cdot \mathbf{5 \ m} \cdot \mathbf{5 \ m} \cdot \mathbf{5 \ m}}{3}$$

$$V \approx 523.3 \ \mathbf{m^3} \leftarrow \text{Rounded}$$

(continued)

Concepts	Examples

7.6 Volume of a Sphere and Hemisphere *(continued)*

Use this formula to find the volume of a *hemisphere* (half of a sphere).

$$V = \frac{2}{3} \bullet \pi \bullet r^3$$

$$\text{or} \quad V = \frac{2 \bullet \pi \bullet r^3}{3}$$

where r is the radius of the hemisphere.
Volume is measured in **cubic units**.

Find the volume of a hemisphere with a radius of 20 cm.

$$V = \frac{2 \bullet \pi \bullet \textbf{(radius)}^3}{3}$$

$$V \approx \frac{2 \bullet \textbf{3.14} \bullet \textbf{20 cm} \bullet \textbf{20 cm} \bullet \textbf{20 cm}}{3}$$

$$V \approx 16{,}746.7 \textbf{ cm}^3 \leftarrow \text{Rounded}$$

7.6 Volume of a Cylinder

Use this formula to find the volume of a *cylinder*.

$$V = \pi \bullet r^2 \bullet h$$

where r is the radius of the circular base and h is the height of the cylinder.
Volume is measured in **cubic units**.

Find the volume of this cylinder.

First, find the radius. $r = \dfrac{8 \text{ m}}{2} = 4 \text{ m}$

$$V = \pi \bullet r^2 \bullet h$$
$$V \approx \textbf{3.14} \bullet \textbf{4 m} \bullet \textbf{4 m} \bullet \textbf{10 m}$$
$$V \approx 502.4 \textbf{ m}^3$$

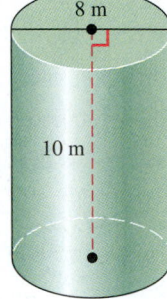

7.6 Volume of a Cone

Use this formula to find the volume of a *cone*.

$$V = \frac{1}{3} \bullet B \bullet h$$

$$\text{or} \quad V = \frac{B \bullet h}{3}$$

where B is the area of the circular base and h is the height of the cone.

Volume is measured in **cubic units**.

Find the volume of this cone.
Area of circular ***Base*** $\approx 3.14 \bullet 4$ in. $\bullet \ 4$ in.

$$\textbf{\textit{B}} \approx 50.24 \text{ in.}^2$$

$$V = \frac{\textbf{\textit{B}} \bullet \textbf{\textit{h}}}{3}$$

$$V \approx \frac{\textbf{50.24 in.}^2 \bullet \textbf{9 in.}}{3}$$

$$V \approx 150.7 \textbf{ in.}^3 \leftarrow \text{Rounded}$$

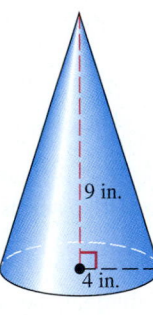

7.6 Volume of a Pyramid

Use this formula to find the volume of a *pyramid*.

$$V = \frac{1}{3} \bullet B \bullet h$$

$$\text{or} \quad V = \frac{B \bullet h}{3}$$

where B is the area of the square or rectangular base and h is the height of the pyramid.

Volume is measured in **cubic units**.

Find the volume of this pyramid.
Area of square ***Base*** $= 2$ cm $\bullet \ 2$ cm
$$\textbf{\textit{B}} = 4 \text{ cm}^2$$

$$V = \frac{\textbf{\textit{B}} \bullet \textbf{\textit{h}}}{3}$$

$$V = \frac{\textbf{4 cm}^2 \bullet \textbf{6 cm}}{3}$$

$$V = 8 \textbf{ cm}^3$$

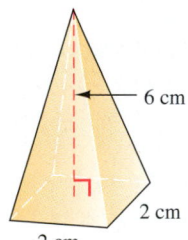

Concepts	Examples

7.6 Surface Area of a Rectangular Solid

Use this formula to find the surface area of a rectangular solid.

Surface Area $= (2 \cdot l \cdot w) + (2 \cdot l \cdot h) + (2 \cdot w \cdot h)$

or $\qquad SA = 2lw + 2lh + 2wh$

where l is the length, w is the width, and h is the height of the solid.

Surface area is measured in **square units**.

Find the surface area of this packing crate.

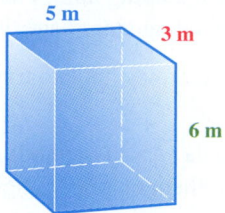

$SA = 2lw + 2lh + 2wh$

$SA = (2 \cdot \textbf{5 m} \cdot \textbf{3 m}) + (2 \cdot \textbf{5 m} \cdot \textbf{6 m}) + (2 \cdot \textbf{3 m} \cdot \textbf{6 m})$

$SA = 30 \text{ m}^2 + 60 \text{ m}^2 + 36 \text{ m}^2$

$SA = 126 \text{ m}^2 \leftarrow$ Square units for surface area

---

7.6 Surface Area of a Cylinder

Use this formula to find the surface area of a cylinder.

Surface Area $= (2 \cdot \pi \cdot r \cdot h) + (2 \cdot \pi \cdot r \cdot r)$

or $\qquad SA = 2\pi rh + 2\pi r^2$

where r is the radius of the circular base and h is the height of the cylinder. Use 3.14 as the approximate value of π.

Surface area is measured in **square units**.

Find the surface area of a hot water tank with a height of 4.5 ft and a diameter of 1.8 ft. Round your answer to the nearest tenth.

First, find the radius. $r = \dfrac{1.8 \text{ ft}}{2} = 0.9 \text{ ft}$

$SA = 2\pi rh + 2\pi r^2$

$SA \approx (2 \cdot \textbf{3.14} \cdot \textbf{0.9 ft} \cdot \textbf{4.5 ft}) + (2 \cdot \textbf{3.14} \cdot \textbf{0.9 ft} \cdot \textbf{0.9 ft})$

$SA \approx 25.434 \text{ ft}^2 + 5.0868 \text{ ft}^2$

$SA \approx 30.5208 \text{ ft}^2 \qquad$ Now round to tenths.

$SA \approx 30.5 \text{ ft}^2 \leftarrow$ Square units for surface area

---

7.7 Finding the Square Root of a Number

Use the square root key on a calculator, $\boxed{\sqrt{}}$ or $\boxed{\sqrt{x}}$. Round to the nearest thousandth if necessary.

$\sqrt{64} = 8 \qquad$ A perfect square

$\sqrt{43} \approx 6.557 \qquad$ 6.557438524 is rounded to nearest thousandth.

7.7 Finding the Unknown Length in a Right Triangle

To find the *hypotenuse,* use this formula.

$$\text{hypotenuse} = \sqrt{(\text{leg})^2 + (\text{leg})^2}$$

The hypotenuse is the side opposite the right angle; it is the longest side in a right triangle.

Find the unknown length in this right triangle. Round to the nearest tenth.

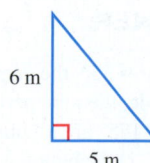

$\text{hypotenuse} = \sqrt{(6)^2 + (5)^2}$

$= \sqrt{36 + 25}$

$= \sqrt{61} \approx 7.8 \text{ m}$

To find a *leg,* use this formula.

$$\text{leg} = \sqrt{(\text{hypotenuse})^2 - (\text{leg})^2}$$

The legs are the sides that form the right angle.

Find the unknown length in this right triangle. Round to the nearest tenth.

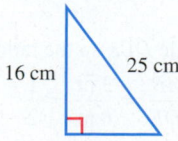

$\text{leg} = \sqrt{(25)^2 - (16)^2}$

$= \sqrt{625 - 256}$

$= \sqrt{369} \approx 19.2 \text{ cm}$

(continued)

Concepts

Examples

7.8 Proving That Two Triangles Are Congruent

Congruent triangles are identical both in shape and in size. This means that corresponding angles have the same measure and corresponding sides have the same length.

Here are three ways to prove that two triangles are congruent.

1. **Angle–Side–Angle (ASA) method:** If two angles and the side between them on one triangle measure the same as the corresponding parts on another triangle, the triangles are congruent.

2. **Side–Side–Side (SSS) method:** If three sides of one triangle measure the same as the corresponding sides of another triangle, the triangles are congruent.

3. **Side–Angle–Side (SAS) method:** If two sides and the angle between them on one triangle measure the same as the corresponding parts on another triangle, the triangles are congruent.

Determine which method can be used to prove that each pair of triangles is congruent.

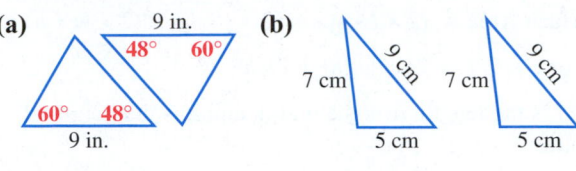

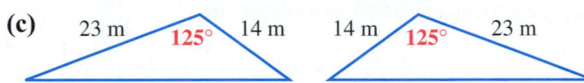

(a) On both triangles, two corresponding angles and the side between them measure the same, so use ASA.

(b) Each pair of corresponding sides has the same length, so use SSS.

(c) On both triangles, two corresponding sides and the angle between them measure the same, so use SAS.

7.8 Finding the Unknown Lengths in Similar Triangles

Use the fact that in similar triangles, the *ratios* of the lengths of corresponding sides are equal. Write a proportion. Then find the cross products and show that they are equivalent. Finish solving for the unknown length.

Find the unknown lengths in this pair of similar triangles.

$$\frac{x}{8} = \frac{5}{10}$$

$$x \cdot 10 = 8 \cdot 5$$

$$\frac{x \cdot \overset{1}{\cancel{10}}}{\underset{1}{\cancel{10}}} = \frac{40}{10}$$

$$x = 4 \text{ m}$$

$$\frac{y}{12} = \frac{5}{10}$$

$$y \cdot 10 = 12 \cdot 5$$

$$\frac{y \cdot \overset{1}{\cancel{10}}}{\underset{1}{\cancel{10}}} = \frac{60}{10}$$

$$y = 6 \text{ m}$$

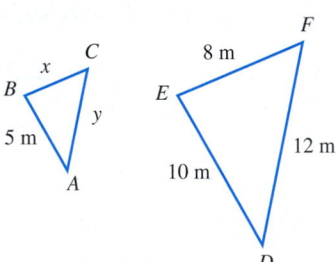

ANSWERS TO TEST YOUR WORD POWER

1. D; *Example:* If ∠1 measures 35° and ∠2 measures 55°, the angles are complementary because 35° + 55° = 90°.

2. B; *Example:* If a square measures 5 ft on each side, then the perimeter is 5 ft + 5 ft + 5 ft + 5 ft = 20 ft.

3. C; *Examples:* Angles that measure 91°, 120°, and 175° are all obtuse angles.

4. C; *Example:* In triangle *ABC* at the right, side *AC* is the hypotenuse; sides *AB* and *BC* are the legs.

5. B; *Example:* The ratio of a circumference of 12.57 cm to a diameter of 4 cm is $\frac{12.57}{4} \approx 3.14$ (rounded).

6. A; *Example:* $\overleftrightarrow{EF}$ is perpendicular to $\overleftrightarrow{GH}$, at the right.

7. D; *Example:* Triangle *ABC* is similar to triangle *DEF,* so the ratios of corresponding sides are equal.

$$\frac{AB}{DE} = \frac{3 \text{ m}}{6 \text{ m}} = \frac{1}{2} \qquad \frac{BC}{EF} = \frac{2 \text{ m}}{4 \text{ m}} = \frac{1}{2} \qquad \frac{AC}{DF} = \frac{3.5 \text{ m}}{7 \text{ m}} = \frac{1}{2}$$

8. B; *Example:* In a rectangle with a length of 8 in. and a width of 5 in., Area = 8 in. • 5 in. = 40 in.²

Chapter 7 ▷▷▷ Review Exercises

[7.1] *Identify each figure as a line, line segment, or ray, and name it using the appropriate symbol.*

1.

2.

3.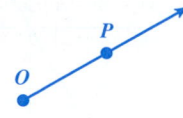

Label each pair of lines as appearing to be parallel, as perpendicular, or as intersecting.

4.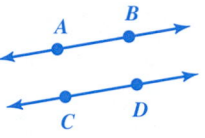

5.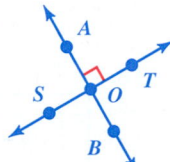

6.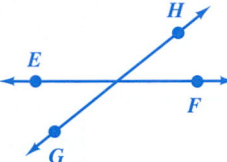

Label each angle as acute, right, obtuse, *or* straight. *For right and straight angles, indicate the number of degrees in the angle.*

7.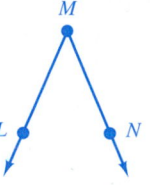

8.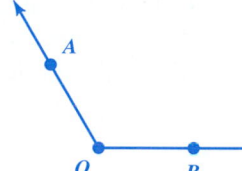

9.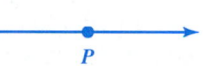

10.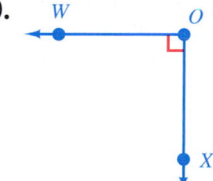

Name the pairs of supplementary angles in each figure.

11.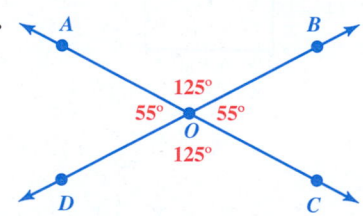

12.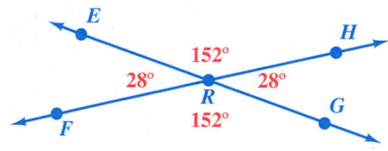

Find the complement or supplement of each angle.

13. Find the complement of:
  (a) 80°
  (b) 45°
  (c) 7°

14. Find the supplement of:
  (a) 155°
  (b) 90°
  (c) 33°

15. In the figure below, ∠2 measures 60°. Find the measure of each of the other angles.

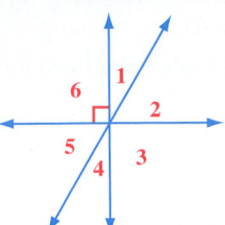

16. Line *m* is parallel to line *n* and ∠8 measures 160°. Find the measures of the other angles.

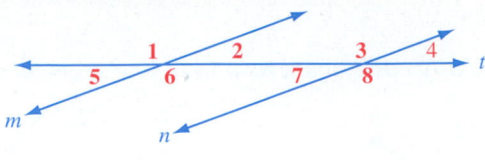

[7.2] *Find the perimeter of each rectangle or square.*

17.

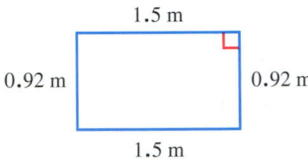

18.

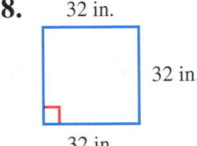

19. A square-shaped pillow measures 38 cm along each side. How much lace is needed to trim all the edges?

20. A rectangular garden plot is $8\frac{1}{2}$ ft wide and 12 ft long. How much fencing is needed to surround the garden?

Find the area of each rectangle or square. Round your answers to the nearest tenth when necessary.

21.

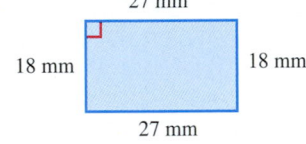

22.

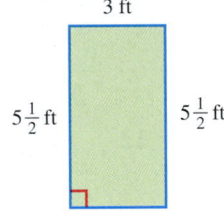

23.

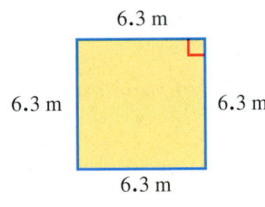

[7.3] *Find the perimeter and area of each parallelogram or trapezoid. Round your answers to the nearest tenth when necessary.*

24.

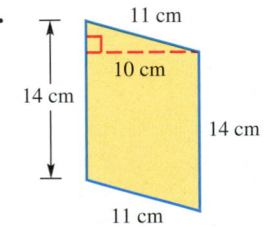

25.

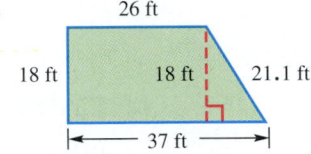

26.

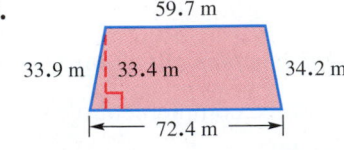

[7.4] *Find the perimeter and area of each triangle.*

27.
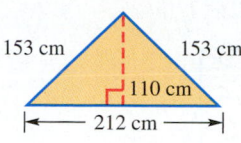

28.
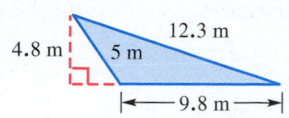

29.
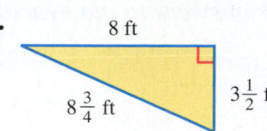

Find the number of degrees in the third angle of each triangle.

30.
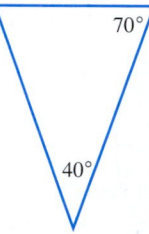

31.

[7.5] *Find the unknown length.*

32. The radius of a circular irrigation field is 68.9 m. What is the diameter of the field?

33. The diameter of a juice can is 3 in. What is the radius of the can?

Find the circumference and area of each circle. Use 3.14 as the approximate value for π. Round your answers to the nearest tenth.

34.
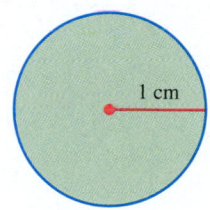

35.
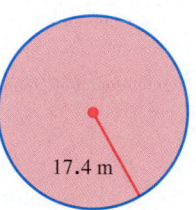

36.
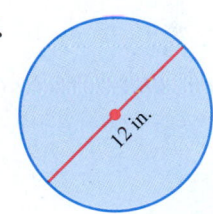

[7.2–7.5] *Find each shaded area. Use 3.14 as the approximate value for π. Round your answers to the nearest tenth when necessary.*

37.
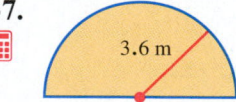

38.
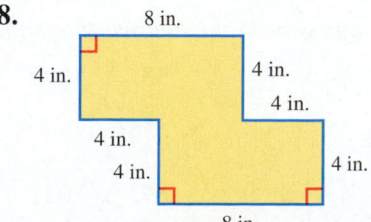

39.
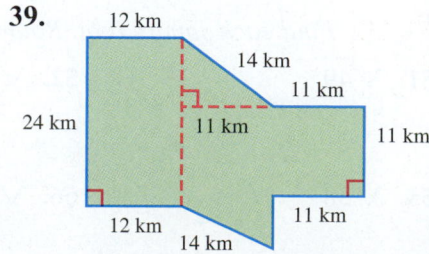

[7.6] *Name each solid and find its volume. Use 3.14 as the approximate value for π.*
Round your answers to the nearest tenth when necessary.

40.
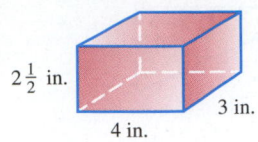
$2\frac{1}{2}$ in.
3 in.
4 in.

41.
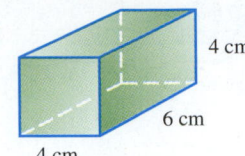
4 cm
6 cm
4 cm

42.
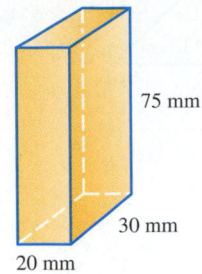
75 mm
30 mm
20 mm

43.
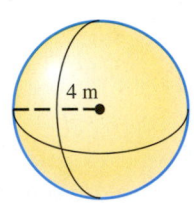
4 m

44.
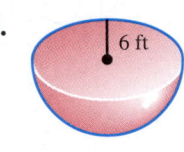
6 ft

45.
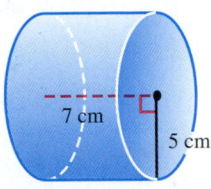
7 cm
5 cm

46.
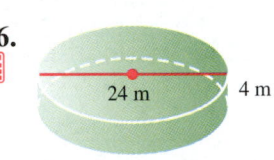
24 m
4 m

47.
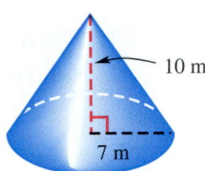
10 m
7 m

48.
4 yd
4 yd
3 yd

Find the volume and surface area of each solid. Use 3.14 as the approximate value for π.
Round your answers to the nearest tenth when necessary.

49.
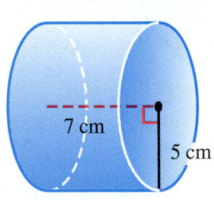
7 cm
5 cm

50. A rectangular cooler that
measures 3.5 ft by 1.5 ft and
is 1.5 ft high

[7.7] *Find each square root. Round your answers to the nearest thousandth when necessary.*

51. $\sqrt{49}$

52. $\sqrt{8}$

53. $\sqrt{3000}$

54. $\sqrt{144}$

55. $\sqrt{58}$

56. $\sqrt{625}$

57. $\sqrt{105}$

58. $\sqrt{80}$

Find the unknown length in each right triangle. Use a calculator to find square roots. *Round your answers to the nearest tenth when necessary.*

59.

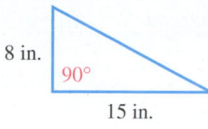

8 in. 90° 15 in.

60.

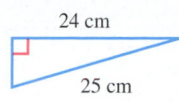

24 cm 25 cm

61.

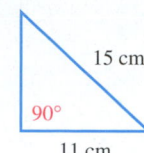

15 cm 90° 11 cm

62.

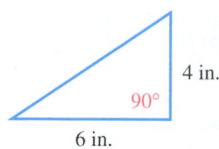

4 in. 90° 6 in.

63.

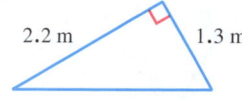

2.2 m 1.3 m

64.

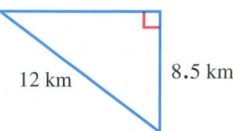

12 km 8.5 km

[7.8] *Determine which method can be used to prove that each pair of triangles is congruent.*

65.

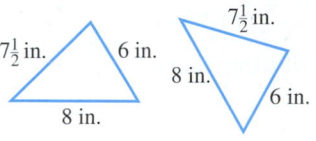

$7\frac{1}{2}$ in. 6 in. 8 in. $7\frac{1}{2}$ in. 8 in. 6 in. 6 in.

66.

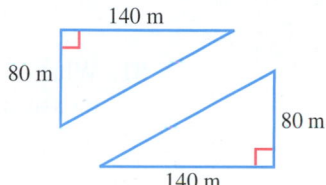

140 m 80 m 80 m 140 m

67.

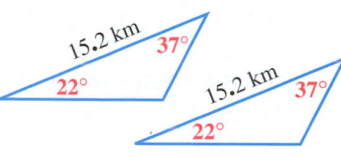

15.2 km 37° 22° 15.2 km 37° 22°

Find the unknown lengths in each pair of similar triangles. Then find the perimeter of the larger triangle in each pair.

68.

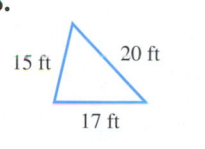

15 ft 20 ft 17 ft y 40 ft x

69.

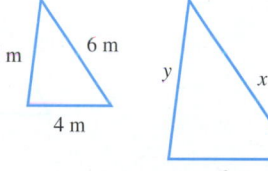

5 m 6 m 4 m y x 6 m

70.

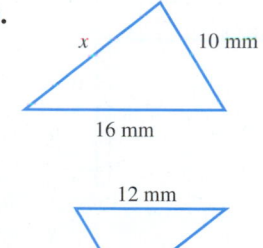

x 10 mm 16 mm 12 mm y 9 mm

> ▶▶▶ **Mixed Review Exercises**

Label each figure. Choose from these labels: line segment, ray, parallel lines, perpendicular lines, intersecting lines, acute angle, right angle, straight angle, obtuse angle. Indicate the number of degrees in the right angle and the straight angle.

71.

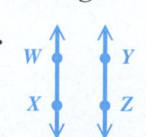

W Y X Z

72.

Q R

73.

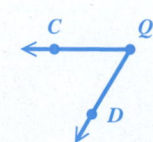

C Q D

74.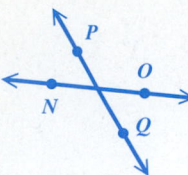

75.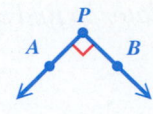

76.

77.

78.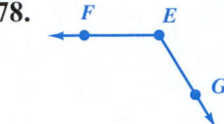

79.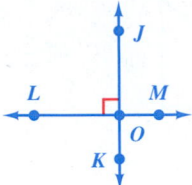

80. What is the complement of an angle measuring 9°?

81. What is the supplement of an angle measuring 42°?

Find the perimeter and area of each figure. In Exercise 82, assume that all angles are 90°.

82.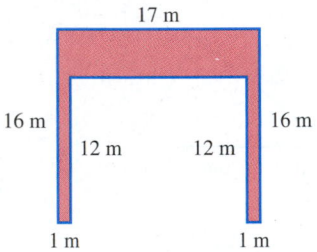

83.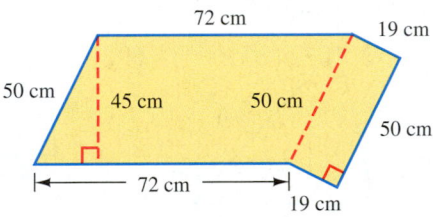

Name each figure and find its perimeter (or circumference) and area. Use 3.14 as the approximate value for π. Round your answers to the nearest tenth when necessary.

84.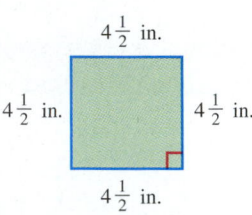

85.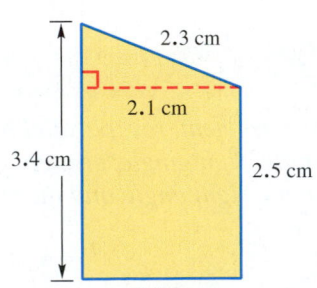

86.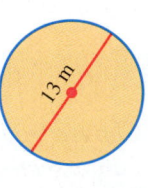

87.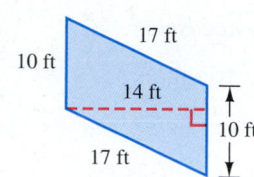

88.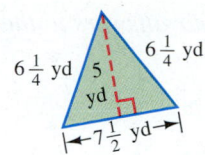

89.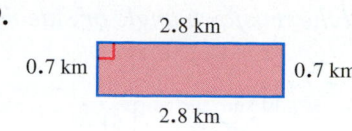

90.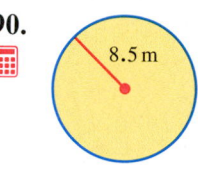

91.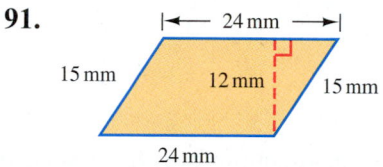

92.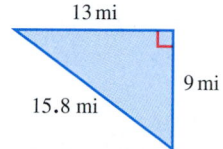

Find the volume of each solid. Use **3.14** *as the approximate value for* π*. Round your answers to the nearest tenth when necessary.*

93.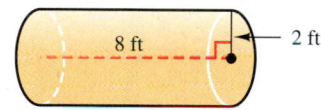

94.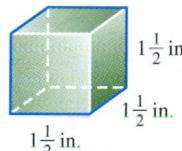

95.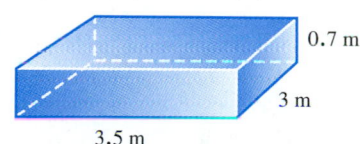

96.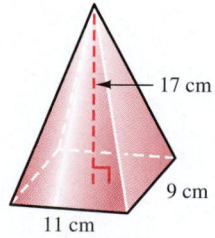

97.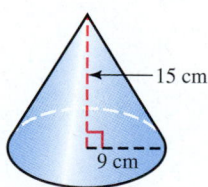

98.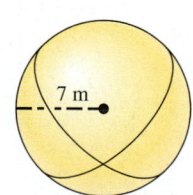

Find the unknown angle or side measurement. Round your answers to the nearest tenth when necessary.

99.

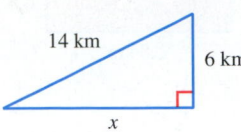

100.

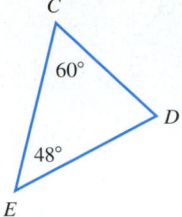

101. similar triangles

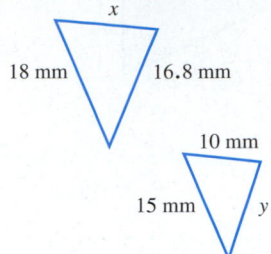

102. Explain how you could use the information about prefixes from **Section 7.5** to solve a problem that asks, "How many decades are in two centuries?"

Chapter 7 ▶▶▶ **Test** CHAPTER **Test Prep** VIDEO CD Use the Chapter Test Prep Video CD to see fully worked-out solutions to any of the exercises you want to review.

Choose the figure that matches each label. For right and straight angles, indicate the number of degrees in the angle.

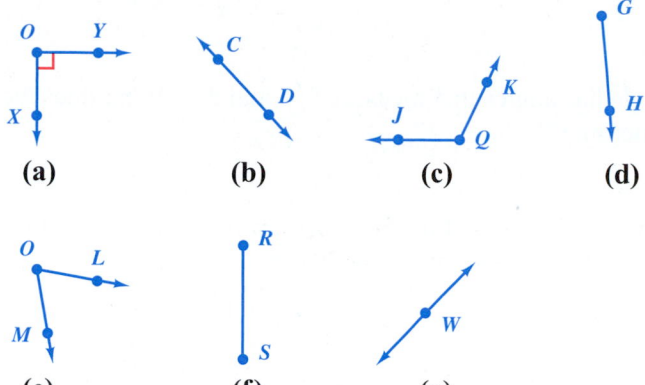

(a) (b) (c) (d)

(e) (f) (g)

1. Acute angle is figure _____.

2. Right angle is figure _____ and its measure is _____.

3. Ray is figure _____.

4. Straight angle is figure _____ and its measure is _____.

5. Write a definition of parallel lines and a definition of perpendicular lines. Make a sketch to illustrate each definition.

6. Find the complement of an 81° angle.

7. Find the supplement of a 20° angle.

8. Find the measure of each unlabeled angle in the figure at the right.

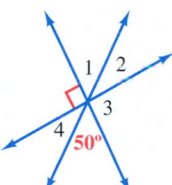

Name each figure and find its perimeter and area.

9.

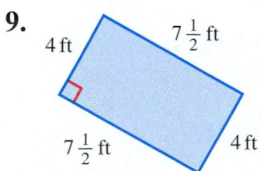

4 ft $7\frac{1}{2}$ ft

$7\frac{1}{2}$ ft   4 ft

10.

18 mm

18 mm   18 mm

18 mm

11.

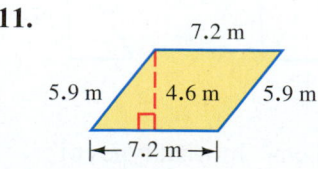

7.2 m

5.9 m   4.6 m   5.9 m

|← 7.2 m →|

12.

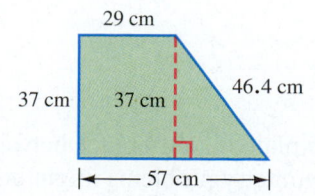

29 cm

37 cm   37 cm   46.4 cm

|← 57 cm →|

1. _____

2. _____

3. _____

4. _____

5. _____

6. _____

7. _____

8. _____

9. _____
   _____

10. _____
    _____

11. _____
    _____

12. _____
    _____

13. _____

_____

14. _____

_____

15. _____

16. _____

17. _____

18. _____

19. _____

20. _____

21. _____

22. _____

23. _____

24. _____

25. _____

Find the perimeter and area of each triangle.

13.

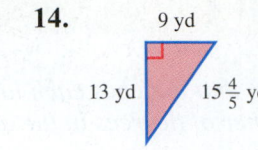

11.8 m   8 m   8.65 m

← 12 m →

14.

9 yd

13 yd   15 $\frac{4}{5}$ yd

15. A triangle has angles that measure 90° and 35°. What does the third angle measure?

In Problems 16–22, use **3.14** *as the approximate value for* π*. Round your answers to the nearest tenth when necessary.*

16. Find the radius.

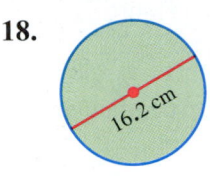

25 in.

17. Find the circumference.

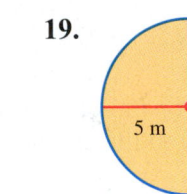

0.9 km

Find the area of each figure.

18.

16.2 cm

19.

5 m

Name each solid and find its volume.

20.

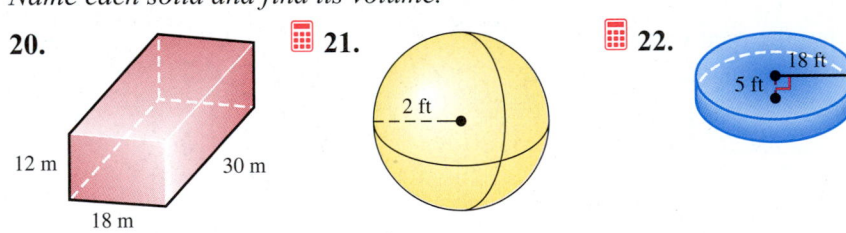

12 m   30 m

18 m

21.

2 ft

22.

18 ft

5 ft

Find the unknown lengths. Round your answers to the nearest tenth when necessary.

23.

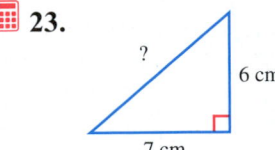

?   6 cm

7 cm

24. Similar triangles

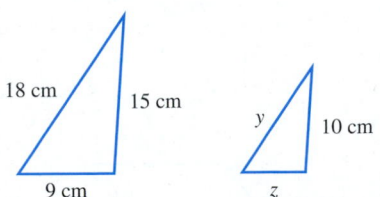

18 cm   15 cm

9 cm

y   10 cm

z

25. Explain the difference between cm, cm², and cm³. In what types of geometry problems might you use each of these units?

Statistics

The old saying "A picture is worth a thousand words" was never more true than when applied to the understanding of data. As an information society, we are constantly being bombarded with facts and numbers. The ability to understand and interpret the many types of information and the many ways it is presented has become essential. For example, if you own or manage a business, you need to understand and interpret the sales history of the business and the performance of employees to make wise managerial decisions. (See **Section 8.2,** Exercises 25–32 and 35–42, and **Section 8.3,** Exercises 29–36.)

8.1 ▶▶▶ Circle Graphs

OBJECTIVES

1 Read and understand a circle graph.

2 Use a circle graph.

3 Draw a circle graph.

The word *statistics* originally came from words that mean *state numbers.* State numbers refer to numerical information, or *data,* gathered by the government such as the number of births, deaths, or marriages in a population. Today, the word *statistics* has a much broader application; data from the fields of economics, social science, and business can all be organized and studied under the branch of mathematics called *statistics.*

OBJECTIVE 1 Read and understand a circle graph. It can be hard to understand a large collection of data. The graphs described in this section help you make sense of such data. For example, a **circle graph** shows how a total amount is divided into parts. The circle graph below shows you how 24 hours in the life of a college student are divided among different activities.

1 Use the circle graph to answer each question.

(a) The greatest number of hours is spent in which activity?

(b) How many more hours are spent working than studying?

(c) Find the total number of hours spent studying, working, and attending classes.

THE DAY OF A COLLEGE STUDENT

College classes **3 hr** · Studying **4 hr** · Other **2 hr** · Driving **2 hr** · Working **6 hr** · Sleeping **7 hr**

The circle represents *1 whole day* (24 hours).

◀ *Work Problem* **1** *at the Side.*

OBJECTIVE 2 Use a circle graph. The above circle graph uses pie-shaped pieces called *sectors* to show the amount of time spent on each activity (the total must be one day, which is 24 hours); the circle graph can therefore be used to compare the time spent on any one activity to the total number of hours in the day.

2 Use the circle graph to find each ratio. Write the ratios as fractions in lowest terms.

(a) Hours spent driving to whole day

(b) Hours spent sleeping to whole day

(c) Hours spent attending class and studying to whole day

(d) Hours spent driving and working to whole day

EXAMPLE 1 Using a Circle Graph

Find the ratio of hours spent in college classes to the total number of hours in the day. Write the ratio as a fraction in lowest terms. (See **Section 5.1.**)

The circle graph shows that 3 of the 24 hours in a day are spent in class. The ratio of class time to the hours in a day is shown below.

$$\frac{3 \text{ hours (college classes)}}{24 \text{ hours (whole day)}} = \frac{3 \text{ hours}}{24 \text{ hours}} = \frac{3 \div 3}{24 \div 3} = \frac{1}{8} \longleftarrow \text{Lowest terms}$$

◀ *Work Problem* **2** *at the Side.*

The circle graph above can also be used to find the ratio of the time spent on one activity to the time spent on any other activity.

EXAMPLE 2   **Finding a Ratio from a Circle Graph**

Use the circle graph on a student's day to find the ratio of study time to class time. Write the ratio as a fraction in lowest terms.

The circle graph shows 4 hours spent studying and 3 hours spent in class. The ratio of study time to class time is shown below.

$$\frac{4 \text{ hours (study)}}{3 \text{ hours (class)}} = \frac{4 \text{ hours}}{3 \text{ hours}} = \frac{4}{3}$$

> The common units (hours) divide out.

Work Problem **3** *at the Side.* ▶

A circle graph often shows data as percents. For example, suppose that the yearly vending machine snack food sales in the United States were $36 billion. The circle graph below shows how sales were divided among various types of snack foods. The entire circle represents the total $36 billion in sales. Each sector represents the sales of one snack item as a percent of the total sales (the total must be 100%).

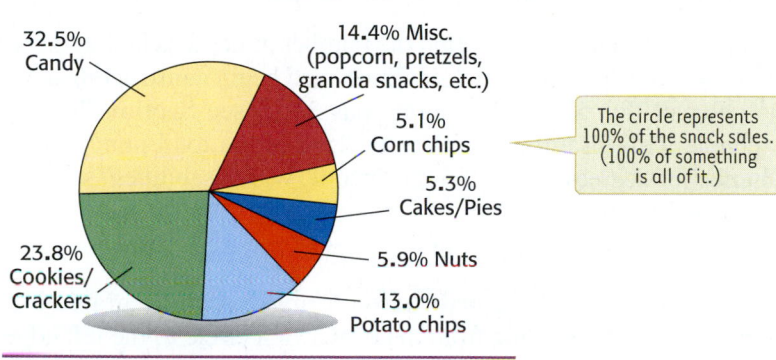

YEARLY U.S. SNACK MARKET SALES
($36 BILLION)

32.5% Candy

14.4% Misc. (popcorn, pretzels, granola snacks, etc.)

5.1% Corn chips

5.3% Cakes/Pies

5.9% Nuts

13.0% Potato chips

23.8% Cookies/Crackers

> The circle represents 100% of the snack sales. (100% of something is all of it.)

Source: Natural Choice—USA.

EXAMPLE 3   **Calculating an Amount Using a Circle Graph**

Use the circle graph above on vending machine snack sales to find the amount spent on candy for the year.

Recall the percent equation.

part = percent • whole

The total sales are $36 billion, so the whole is $36 billion. The percent is 32.5% or, as a decimal, 0.325. Find the part.

part = percent • whole

$32.5\% = 0.325$

$x = (0.325)(36 \text{ billion})$

$x = 11.7 \text{ billion}$

The amount spent on candy was $11.7 billion or $11,700,000,000.

Work Problem **4** *at the Side.* ▶

---

3 Use the circle graph on a student's day to find the following ratios. Write the ratios as fractions in lowest terms.

(a) Hours spent in class to hours spent studying

(b) Hours spent working to hours spent sleeping

(c) Hours spent driving to hours spent working

(d) Hours spent in class to hours spent for "Other"

4 Use the circle graph on vending machine snack sales to find the following.

(a) The amount spent on corn chips

(b) The amount spent on miscellaneous (popcorn, pretzels, granola snacks, etc.)

(c) The amount spent on cakes/pies

(d) The amount spent on cookies/crackers

ANSWERS

3. **(a)** $\frac{3}{4}$ **(b)** $\frac{6}{7}$ **(c)** $\frac{1}{3}$ **(d)** $\frac{3}{2}$

4. **(a)** $1.836 billion or $1,836,000,000
   (b) $5.184 billion or $5,184,000,000
   (c) $1.908 billion or $1,908,000,000
   (d) $8.568 billion or $8,568,000,000

OBJECTIVE 3 Draw a circle graph.   Last year, Goodwill Industries donors helped fund programs that let nearly 1 million people take their first financial steps toward new and better jobs and financial independence. The following table shows those who were served.

A HELPING HAND

Group Served	Percent of Total
People with disabilities	25%
Welfare recipients	15%
Working poor	10%
Ex-offenders	10%
At-risk youth	5%
Unemployed	35%
Total	100%

Source: Goodwill Industries.

You can show these percents visually by using a circle graph. The entire circle will represent all of the groups served (all 100%).

EXAMPLE 4 Drawing a Circle Graph

Using the data in the table, find the number of degrees in the sector that would represent the "people with disabilities" and begin constructing a circle graph.

Recall that a complete circle has 360° (see **Section 7.1**). Because the "people with disabilities" make up 25% of the total number of people, the number of degrees needed for the "people with disabilities" sector of the circle graph is 25% of 360°.

$$(360°)(25\%) = (360°)(0.25) = 90°$$

Use a tool called a **protractor** to make a circle graph. First, using a straightedge, draw a line from the center of a circle to the left edge. Place the hole in the protractor over the center of the circle, making sure that 0 on the protractor lines up with the line that was drawn. Find 90° and make a mark as shown in the illustration. Then remove the protractor and use the straightedge to draw a line from the center of the circle to the 90° mark at the edge of the circle. This sector is 90° and represents "people with disabilities." Label the sector with the group name and percent.

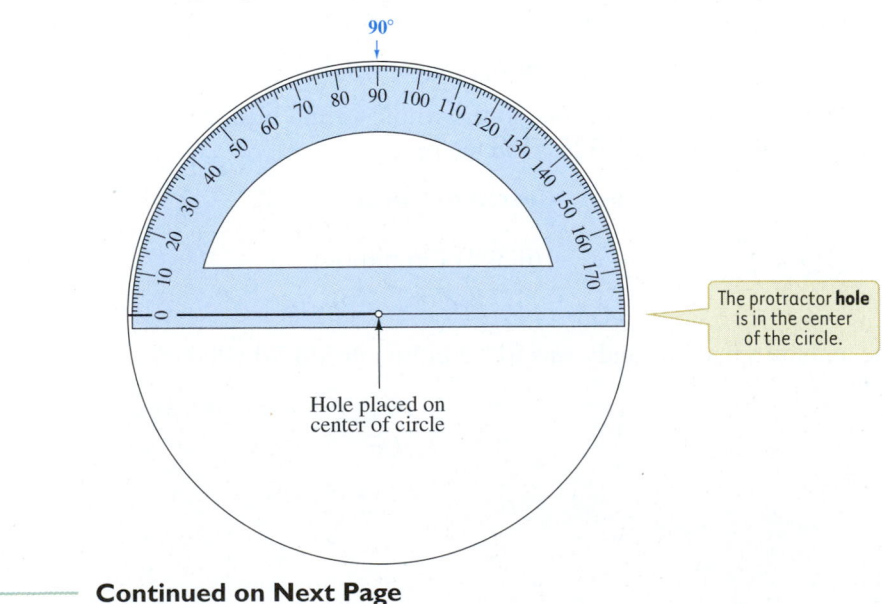

90°

The protractor **hole** is in the center of the circle.

Hole placed on center of circle

Continued on Next Page

To draw the "Welfare recipients" sector, begin by finding the number of degrees in the sector. From the table, you see that "welfare recipients" represent 15% of the total.

$$(360°)(\textbf{15\%}) = (360°)(\textbf{0.15}) = 54°$$

Again, place the hole of the protractor over the center of the circle, but this time align 0 on the second line that was drawn. Make a mark at 54° and draw a line as before. This sector is 54° and represents those who are "Welfare recipients."

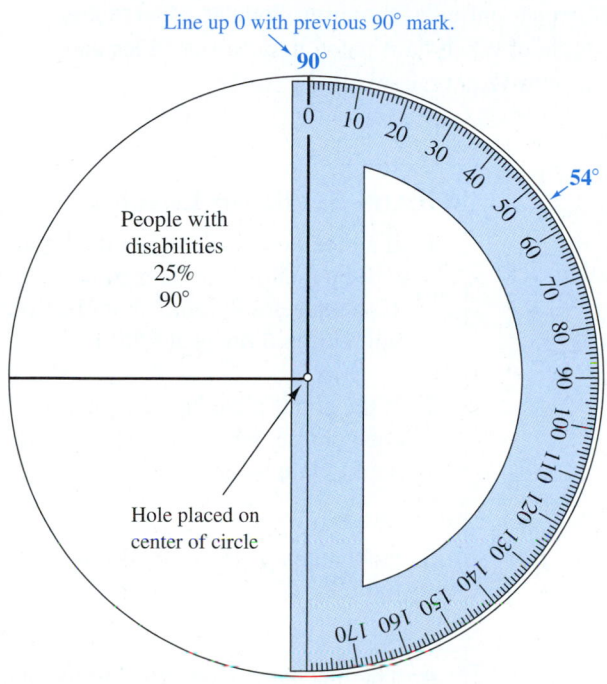

Line up 0 with previous 90° mark.

People with disabilities 25% 90°

Hole placed on center of circle

> **CAUTION**
> You must be certain that the hole in the protractor is placed over the exact center of the circle each time you measure the size of a sector.

Work Problem **5** *at the Side.* ▶

Use this circle for Problem 5 at the side.

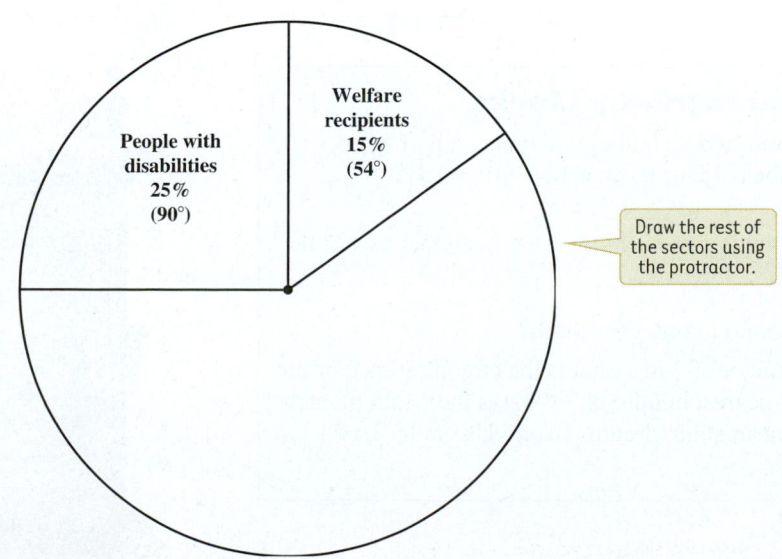

People with disabilities 25% (90°)

Welfare recipients 15% (54°)

Draw the rest of the sectors using the protractor.

5 Using the information in the table on the groups served, find the number of degrees needed for each sector. Complete the circle graph at the bottom left. Label each sector with the group name and percent.

(a) Working poor

(b) Ex-offenders

(c) At-risk youth

(d) Unemployed

ANSWERS

5. **(a)** 36° **(b)** 36° **(c)** 18° **(d)** 126°

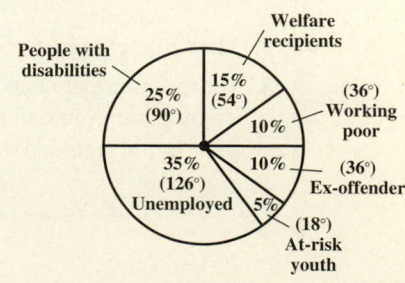

People with disabilities 25% (90°)

Welfare recipients 15% (54°)

(36°) Working poor

10%

35% (126°) Unemployed

10% (36°) Ex-offender

5% (18°) At-risk youth

Math in the Media

The *Antiques Road Show*, a popular international television program, is shown on the Public Broadcasting System (PBS) in the United States. People bring antique items and collectables to the show and receive free appraisals of value from a professional antiques appraiser. Popular collectables appraised on the show have been antique clocks, some of which have dated back to the 1800s and have ranged in value from $1000 to $100,000. (*Source:* www.antiquesroadshow.com)

Antique clocks usually chime either on the hour or on both the hour and the half hour. The mechanisms that control the chimes are a set of gears called the count plate and hammer wheel and a lever called the count hook.

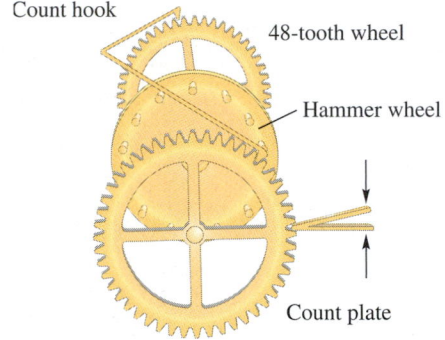

Count hook

48-tooth wheel

Hammer wheel

Count plate

Hour-Striking Clocks

1. If a clock chimes only on the hour, what will be the total number of chimes in a 12-hour period? (*Hint:* For example, it will chime 6 times at 6:00.)

2. If the count plate has one gear tooth for each chime, what fractional part of the count plate is one tooth?

3. How many degrees correspond to one gear tooth?

4. The mechanism designed to move the count plate has two wheels. The hammer wheel has 13 pins, which is combined with a wheel with 48 teeth and a pinion that fits 8 teeth. That gives a 6:1 ratio. What is significant about the combination of 13 and 6 that causes the count plate to move one gear tooth?

Hour- and Half-Hour-Striking Clocks

5. If a clock chimes on the hour and also one time on each half hour, what will be the total number of chimes in a 12-hour period?

6. If the count plate has one tooth for each chime, what fractional part of the count plate makes one tooth?

7. How many degrees correspond to one gear tooth?

8. If the count plate has a diameter of 2 in., what is the circumference of the count plate, rounded to the nearest hundredth? What is the width of each gear tooth, rounded to the nearest hundredth? (*Note:* Use $\pi \approx 3.14$.)

8.1 ▶▶▶   Exercises

This circle graph shows the number of pets owned in the United States. Use this circle graph to answer Exercises 1–6. Write ratios as fractions in lowest terms. See Examples 1 and 2.

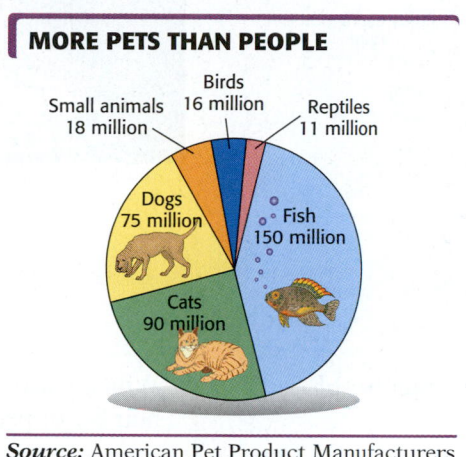

MORE PETS THAN PEOPLE

Birds 16 million
Small animals 18 million
Reptiles 11 million
Dogs 75 million
Fish 150 million
Cats 90 million

Source: American Pet Product Manufacturers Association

1. Find the number of pets owned in the United States.

2. Which type of pet is owned by the greatest number of people? How many of these pets were owned?

3. Find the ratio of the number of cats owned to the total number of pets.

4. Find the ratio of the number of small animals owned to the total number of pets.

5. Find the ratio of the number of cats owned to the number of dogs.

6. Find the ratio of the number of fish owned to the number of cats.

This circle graph, adapted from USA Today, *shows the number of people in a survey who gave various reasons for eating dinner at restaurants. Use this circle graph to answer Exercises 7–14. See Examples 1 and 2.*

ON THE TOWN

When asked in a survey why they ate dinner in restaurants, a group of people gave these reasons.

Wanted food they couldn't cook at home **1740**

Atmosphere **1200**

Enjoy eating out **1140**

Quicker **720**

Less work/ no cleanup **1020**

180 Don't know

Source: Market Facts for Tyson Foods.

7. Which reason was given by the least number of people? How many gave that reason?

8. Which reason was given by the second-highest number of people? How many gave that reason?

Answer Exercises 9–14 by writing a ratio as a fraction in lowest terms.

9. Those who said dining out is "Quicker" to total people in the survey

10. Those who said "Enjoy eating out" to the total people in the survey

11. Those who said "Less work/no cleanup" to those who said "Atmosphere"

12. Those who said "Don't know" to those who said "Quicker"

13. Those who said "Wanted food they couldn't cook at home" to those who said "Less work/no cleanup"

14. Those who said "Atmosphere" to those who said "Enjoy eating out"

This circle graph shows the favorite hot dog toppings in the United States. Each topping is expressed as a percent of the 3200 people in the survey. Use the graph to find the number of people in the survey who favored each of the toppings in Exercises 15–20. See Example 3.

15. Onions

16. Ketchup

17. Sauerkraut

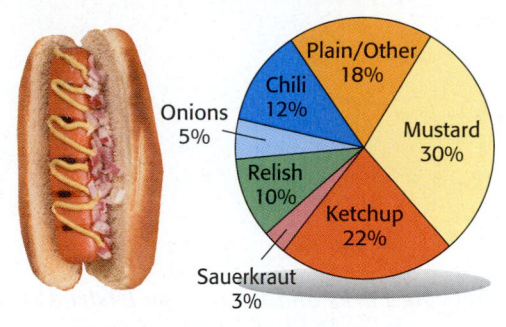

FAVORITE HOT DOG TOPPINGS

Plain/Other 18%
Chili 12%
Onions 5%
Mustard 30%
Relish 10%
Ketchup 22%
Sauerkraut 3%

Source: National Hot Dog and Sausage Council.

18. Relish   **19.** Mustard   **20.** Chili

The National Academy of Sciences recommends that adults ages 19–50 get 1000 mg of calcium daily, or about three 8-ounce glasses of milk. The circle graph, adapted from USA Today, *shows the daily consumption of milk products by adults. If 5540 adults were surveyed in this study, find the number of people giving each response in Exercises 21–26. Round to the nearest whole number. See Example 3.*

21. Consume none or very few milk products

GOT MILK

Consume two servings 27%
Consume none or very few milk products 22%
Consume recommended servings 13%
Consume one serving 33%
Don't know 5%

Source: Market Facts for Milk Mustache Mobile 100-City Cruise for Calcium.

22. Consume recommended servings

23. Consume one serving   **24.** Consume two servings

25. Don't know   **26.** Consume less than the recommended servings (Do not include those who don't know.)

27. Describe the procedure for determining how large each sector must be to represent each of the items in a circle graph.

28. A protractor is the tool used to draw a circle graph. Give a brief explanation of what the protractor does and how you would use it to measure and draw each sector in the circle graph.

During one month the Orangevale Parks and Recreation District spent $5460 for the activities shown in the following chart. Find all numbers missing from the chart.

Item	Dollar Amount	Percent of Total	Degrees of a Circle
29. Adult sports	$1365	25%	_____
30. Children's sports	$1092	_____	72°
31. Day camp	$546	_____	_____
32. Senior fitness	$546	10%	_____
33. Annual egg hunt	$819	15%	_____
34. Arts and crafts	$273	_____	_____
35. Mommy and baby exercise	$819	_____	54°

36. Draw a circle graph by using the information from Exercises 29–35. Label each sector in your graph. See Example 4.

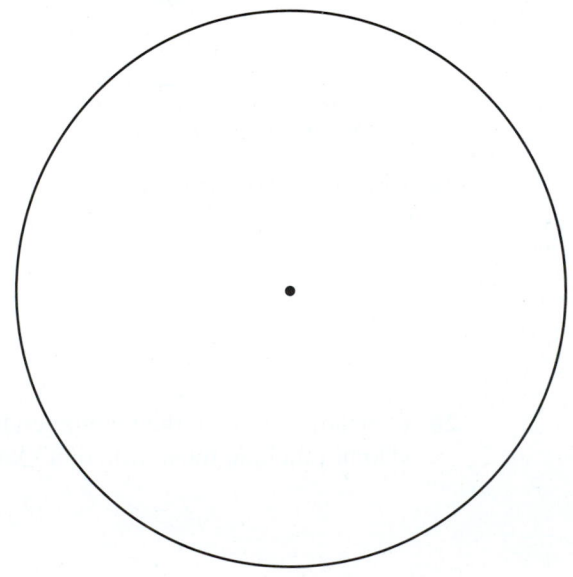

37. White Water Rafting Company divides its annual sales into five categories as follows.

Category	Annual Sales
Adventure classes	$12,500
Grocery and provision sales	$40,000
Equipment rentals	$60,000
Rafting tours	$50,000
Equipment sales	$37,500

(a) Find the total sales for the year.

(b) Find the number of degrees in a circle graph for each item.

(c) Make a circle graph showing the percent for each category. Label each sector in your graph.

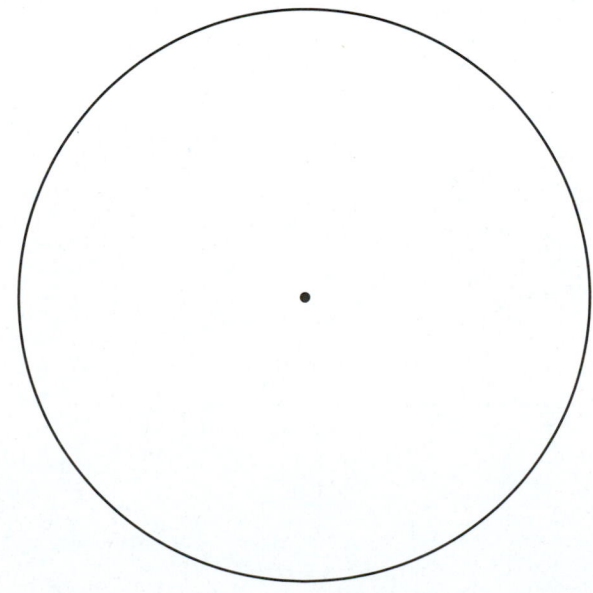

38. Online retail sales in the United States are growing rapidly. The top online retail sales categories are 50% for travel; 20% for apparel, accessories, and footwear; 15% for computer hardware and software; 10% for autos and auto parts; and 5% for home furnishings. (*Source*: Forester Research.)

(a) Find the number of degrees in a circle graph for each online retail sales category.

(b) Draw a circle graph showing the percent for each category. Label each sector in your graph.

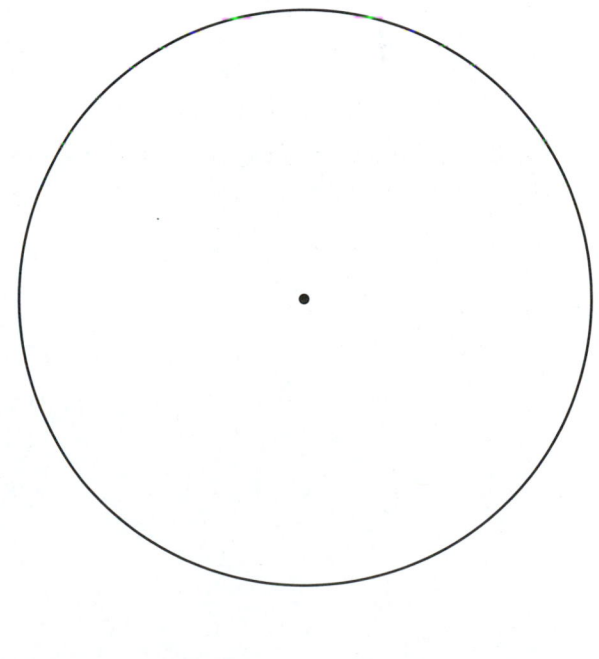

39. The Pathfinder Research Group asked 4488 Americans how they fall asleep, and the results are shown in the figure on the right.

(a) Use this information to complete the chart and draw a circle graph. Round to the nearest whole percent and to the nearest degree. Label each sector in your graph.

Sleeping Position	Number of Americans	Percent of Total	Number of Degrees
Side	_____	_____	_____
Not sure	_____	_____	_____
Stomach	_____	_____	_____
Varies	_____	_____	_____
Back	_____	_____	_____

(b) Add up the percents. Is the total 100%? Explain why or why not.

(c) Add up the degrees. Is the total 360°? Explain why or why not.

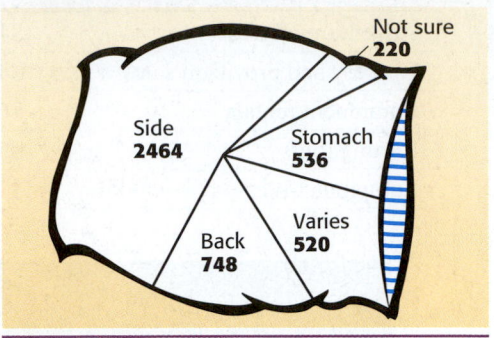

SET TO SLEEP

Number of Americans surveyed who fall asleep on their:

Not sure **220**

Side **2464**

Stomach **536**

Varies **520**

Back **748**

Source: Pathfinder Research Group.

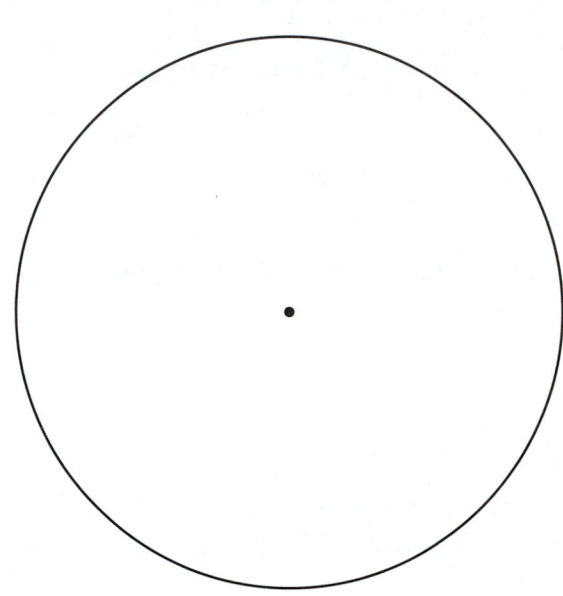

8.2 ▷▷▷ Bar Graphs and Line Graphs

OBJECTIVE **1** **Read and understand a bar graph.** Bar graphs are useful when showing comparisons. For example, the bar graph below compares the total number of members in all the Fitness Center locations during each of five years.

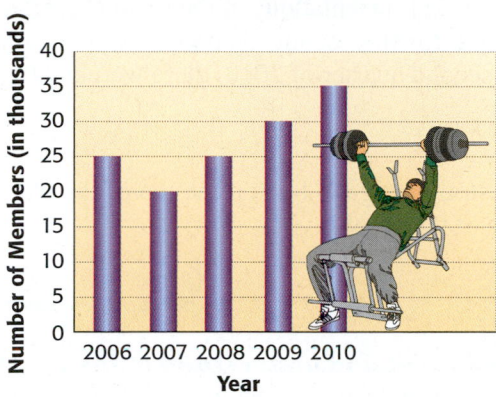

FITNESS CENTER MEMBERSHIP

Notice that the label says "in thousands."

EXAMPLE 1 **Using a Bar Graph**

How many members did the Fitness Center have in 2008?

The bar for 2008 rises to 25. Notice the label along the left side of the graph that says "Number of Members (in thousands)." The phrase *in thousands* means you have to multiply 25 by 1000 to get 25,000. So, there were 25,000 (not 25) members in the Fitness Center locations in 2008.

Work Problem **1** *at the Side.* ▶

OBJECTIVE **2** **Read and understand a double-bar graph.** A **double-bar graph** can be used to compare two sets of data. The graph below shows the number of DSL (digital subscriber line) installations each quarter for two different years.

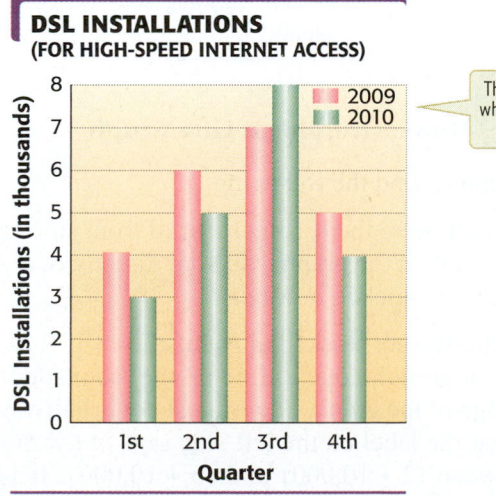

DSL INSTALLATIONS
(FOR HIGH-SPEED INTERNET ACCESS)

The color key shows which bars represent each year.

2009
2010

1 Use the bar graph at the left to find the number of members in the Fitness Centers in each of these years.

(a) 2006

(b) 2007

(c) 2009

(d) 2010

ANSWERS

1. **(a)** 25,000 members
   (b) 20,000 members
   (c) 30,000 members
   (d) 35,000 members

2 Use the double-bar graph to find the number of DSL installations in 2009 and 2010 for each quarter.

(a) 1st quarter

(b) 3rd quarter

(c) 4th quarter

(d) Find the greatest number of installations. Identify the quarter and the year in which they occurred.

3 Use the line graph at the right to find the number of trout stocked in each month.

(a) June

(b) May

(c) April

(d) July

EXAMPLE 2 **Reading a Double-Bar Graph**

Use the double-bar graph on the previous page to find the following.

(a) The number of DSL installations in the second quarter of 2009

There are two bars for the second quarter. The color code in the upper right-hand corner of the graph tells you that the **red bars** represent 2009. So the **red bar** on the *left* is for the 2nd quarter of 2009. It rises to 6. Multiply 6 by 1000 because the label on the left side of the graph says *in thousands*. So there were 6000 DSL installations for the second quarter in 2009.

(b) The number of DSL installations in the second quarter of 2010

The **green bar** for the second quarter rises to 5 and 5 times 1000 is 5000. So, in the second quarter of 2010, there were 5000 DSL installations.

> **CAUTION**
> Use a ruler or straightedge to line up the top of the bar with the number on the left side of the graph.

◀ *Work Problem* **2** *at the Side.*

OBJECTIVE 3 Read and understand a line graph. A **line graph** is often useful for showing a trend. The line graph below shows the number of trout stocked along the Feather River over a 5-month period. Each dot indicates the number of trout stocked during the month directly below that dot.

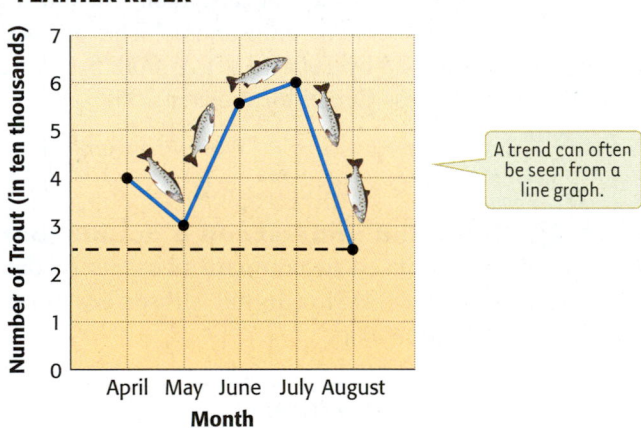

TROUT STOCKED IN THE FEATHER RIVER

A trend can often be seen from a line graph.

EXAMPLE 3 **Understanding a Line Graph**

Use the line graph to find the following.

(a) In which month were the least number of trout stocked?

The lowest point on the graph is the dot directly over August, so the least number of trout were stocked in August.

(b) How many trout were stocked in August?

Use a ruler or straightedge to line up the August dot with the numbers along the left edge of the graph. The August dot is halfway between the 2 and the 3. Notice that the label on the left side says *in ten thousands*. So August is halfway between (2 • 10,000) and (3 • 10,000). It is halfway between 20,000 and 30,000. That means 25,000 trout were stocked in August.

◀ *Work Problem* *at the Side.*

ANSWERS

2. **(a)** 4000; 3000 installations
   (b) 7000; 8000 installations
   (c) 5000; 4000 installations
   (d) 8000 installations; 3rd quarter of 2010
3. **(a)** 55,000 trout   **(b)** 30,000 trout
   (c) 40,000 trout   **(d)** 60,000 trout

OBJECTIVE **4** **Read and understand a comparison line graph.**
Two sets of data can also be compared by drawing two line graphs together as a **comparison line graph.** For example, the line graph below compares the number of plasma high-definition televisions (Plasma HDTVs) and the number of liquid crystal diode high-definition televisions (LCD HDTVs) sold during each of 5 years.

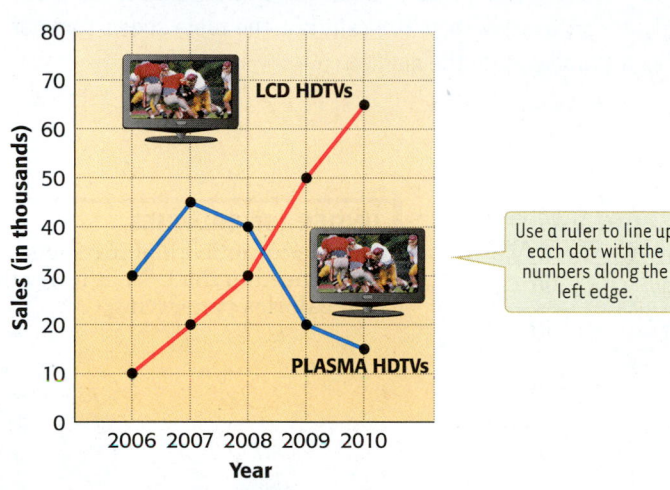

THE BIG PICTURE
SALES OF PLASMA HDTVs TO LCD HDTVs

Use a ruler to line up each dot with the numbers along the left edge.

4 Use the comparison line graph at the left to find the following.

   (a) The number of plasma HDTVs sold in 2006, 2008, 2009, and 2010

   (b) The number of LCD HDTVs sold in 2006, 2007, 2008, and 2009

EXAMPLE 4   **Interpreting a Comparison Line Graph**

Use the comparison line graph above to find the following.

(a) The number of plasma HDTVs sold in 2007
   Find the dot on the **blue line** above 2007. Use a ruler or straightedge to line up the dot with the numbers along the left edge. The dot is halfway between 40 and 50, which is 45. Then, 45 times 1000 is 45,000 plasma HDTVs sold in 2007.

(b) The number of LCD HDTVs sold in 2010
   The **red line** on the graph shows that 65,000 LCD HDTVs were sold in 2010.

(c) The first full year in which the number of LCD HDTVs sold was greater than the number of plasma HDTVs sold

> **Note**
> Both the double-bar graph and the comparison line graph are used to compare two or more sets of data.

Work Problem **4** *at the Side.* ▶

Math in the Media

SNACKS: HOW LOW CAN THEY GO?

Packaged-food giants from Quaker to Kraft had such great sales success with their 100-calorie snack packs that they are now going even lower. The bar graph shows the number of individual snack products introduced each year in the United States with fewer than 100 calories. The table shows some of the new snack-pack products that have been introduced to the market.

NEW SNACKS UNDER 100 CALORIES

Manufacturer	Product Name	Number of Calories
Quaker	Mini Delight Bar	90
Kellogg	Special K Bliss Bar	70
ConAgra	Hunts Fat Free Pudding	80
Kraft	Jet-Puffed Marshmallows	90
General Mills	Yoplait Fiber One Yogurt	80
Hershey's	Hershey Sticks in four flavors	60
Del Monte Pet Products	Pup-Peroni (for dogs)	50

Source: USA Today.

UNDER 100 CALORIES
Number of individual products* introduced in the USA at fewer than 100 calories per serving:

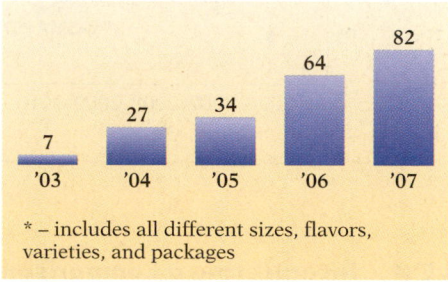

* – includes all different sizes, flavors, varieties, and packages

Source: Datamonitor's Production Online.

1. What is the title of the bar graph? Is the number of new snack products having fewer than 100 calories increasing or decreasing?

2. The greatest number of new products introduced was in which year? How many new products were introduced in that year?

3. What were the two new snack products with the highest number of calories? How many calories do they have?

4. Which product has the fewest number of calories? This product will most likely be eaten by _____ .

5. Sometimes people "jump to conclusions" without enough information. Which of these conclusions are reasonable, based on the information given in the table?

   (a) The snacks in the table are healthy snacks.

   (b) Some snacks with fewer than 100 calories could be healthy.

   (c) A person can have a snack without consuming a large number of calories

   (d) Low-calorie snacks don't taste good.

   (e) Of the snacks included in the table, dogs will eat only Pup-Peronis.

6. Conduct a survey of your class members. Make a list of their favorite snacks that are healthy while also being low in calories.

574

8.2 ▶▶▶ Exercises

The American Farm Bureau Federation reports that the average adult in the United States will work 40 days (rounded to the nearest day) to earn enough to pay the annual household food bill. This was found by multiplying the average percent of household income spent on food by 365 (the number of days in a year). This bar graph shows the percent of income spent in various countries of the world. Use this graph to answer Exercises 1–6. See Example 1.

TAKING A BITE OUT OF HOUSEHOLD INCOME

The average American adult will work 40 days each year to earn enough to pay the household food bill. Percent of household income spent on food in:

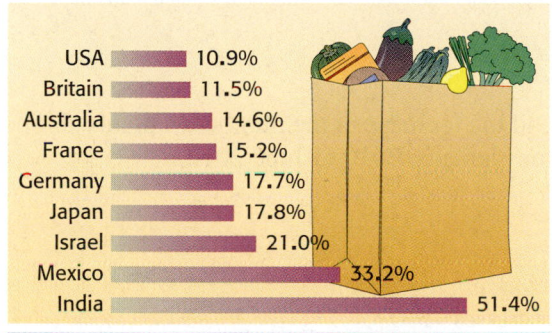

USA 10.9%
Britain 11.5%
Australia 14.6%
France 15.2%
Germany 17.7%
Japan 17.8%
Israel 21.0%
Mexico 33.2%
India 51.4%

Source: American Farm Bureau Federation.

1. In which country is the highest percent of income spent on food? What percent is this?

2. In which country is the lowest percent of income spent on food? What percent is this?

3. List all countries in the graph in which less than 15% of household income is spent, on average, for food.

4. List all countries in the graph in which more than 20% of household income is spent, on average, for food.

5. How many days each year will the average adult have to work to earn enough to pay for food in Mexico? Round to the nearest day.

6. How many days each year will the average adult have to work to earn enough to pay for food in Israel? Round to the nearest day.

This double-bar graph shows the number of outdoor plants shipped by Capital Growers during the first six months of 2009 and 2010. Use this graph to answer Exercises 7–12. See Example 2.

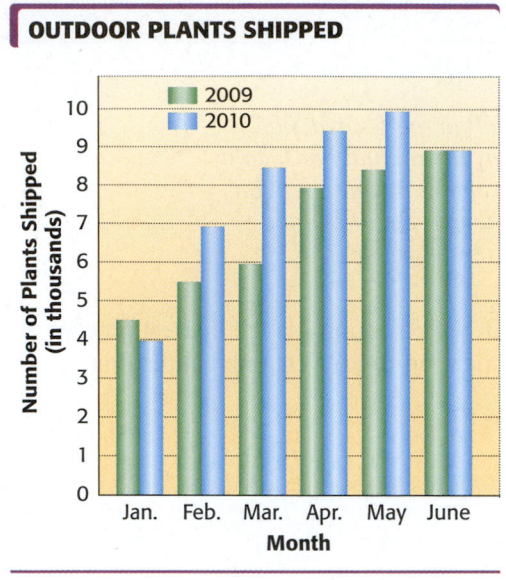

OUTDOOR PLANTS SHIPPED

7. In which month in 2010 were the greatest number of plants shipped? What was the total number of plants shipped in that month?

8. How many plants were shipped in January of 2009?

9. How many more plants were shipped in February of 2010 than in February of 2009?

10. How many fewer plants were shipped in March of 2009 than in March of 2010?

11. Find the increase in the number of plants shipped from February 2009 to April 2010.

12. Find the increase in the number of plants shipped from January 2010 to June 2010.

This double-bar graph shows sales of super unleaded and supreme unleaded gasoline at a service station for each of 5 years. Use this graph to answer Exercises 13–18. See Example 2.

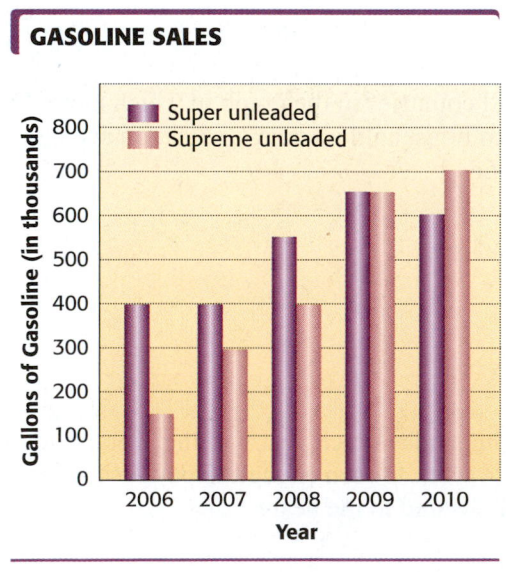

GASOLINE SALES

13. How many gallons of supreme unleaded gasoline were sold in 2006?

14. How many gallons of super unleaded gasoline were sold in 2009?

15. In which year did the greatest difference in sales between super unleaded and supreme unleaded gasoline occur? Find the difference.

16. In which year did the sales of supreme unleaded gasoline surpass the sales of super unleaded gasoline?

17. Find the increase in supreme unleaded gasoline sales from 2006 to 2010.

18. Find the increase in super unleaded gasoline sales from 2006 to 2010.

This line graph shows how the personal computer (PC) has evolved over the quarter century of its existence. What began as a technician's dream is now a common tool of business and home life. Use this line graph to answer Exercises 19–24. See Example 3.

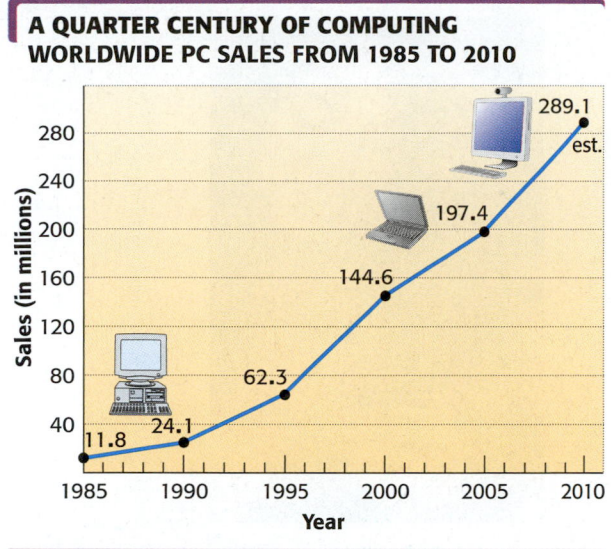

A QUARTER CENTURY OF COMPUTING
WORLDWIDE PC SALES FROM 1985 TO 2010

Source: Gartner Dataquest and ARS Technic.

19. Find the number of PCs shipped in 1990.

20. What was the number of PCs shipped in 1995?

21. Find the increase in the number of PCs shipped in 2010 from the number shipped in 2000.

22. How many more PCs were shipped in 2000 than in 1990?

23. Give two possible explanations for the increase in the number of PCs shipped.

24. Give two possible conditions that could result in a decrease in PC shipments in the future.

This comparison line graph shows the number of Apple iPods sold by two different chain stores during each of 5 years. Use this graph to find the annual number of Apple iPods sold each year in Exercises 25–30. See Example 4.

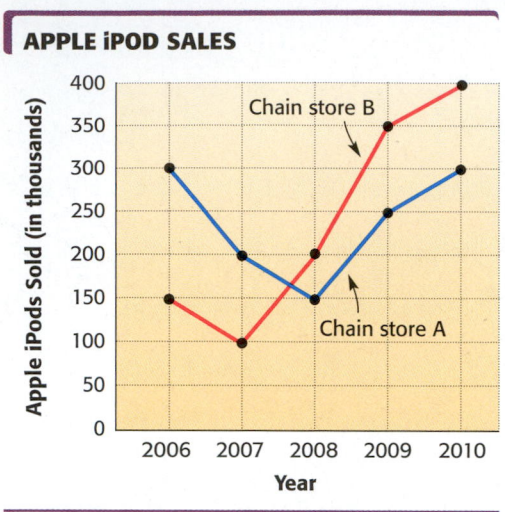

25. Store A in 2010

26. Store A in 2009

27. Store A in 2008

28. Store B in 2010

29. Store B in 2009

30. Store B in 2008

31. Looking at the comparison line graph above, which store would you like to own? Explain why. Based on the graph, what amount of sales would you predict for your store in 2011?

32. In the comparison line graph above, Store B used to have lower sales than Store A. What might have happened to cause this change? Give two possible explanations.

33. Explain in your own words why a bar graph or a line graph (not a double-bar graph or comparison line graph) can be used to show only one set of data.

34. The double-bar graph and the comparison line graph are both useful for comparing two sets of data. Explain how this works and give your own example.

This comparison line graph shows the sales and profits of Tacos-to-Go for each of 4 years. Use the graph to answer Exercises 35–42. See Example 4.

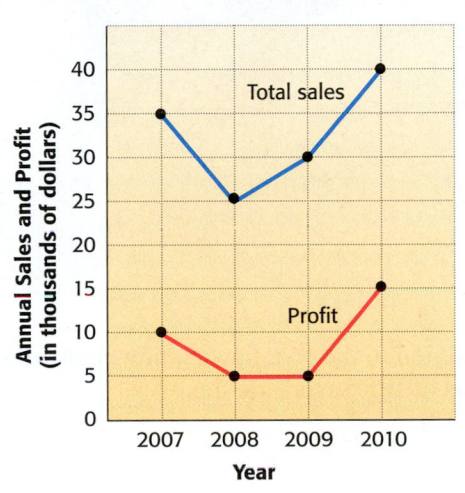

35. Total sales in 2010

36. Total sales in 2009

37. Total sales in 2008

38. Profit in 2010

39. Profit in 2009

40. Profit in 2008

41. Give two possible explanations for the decrease in sales from 2007 to 2008 and two possible explanations for the increase in sales from 2008 to 2010.

42. Based on the graph, what conclusion can you make about the relationship between sales and profits?

Relating Concepts (Exercises 43–48)   For Individual or Group Work

This newspaper clipping, adapted from USA Today, *shows some statistics regarding LifeSavers. Use this information to* **work Exercises 43–48 in order.**

ROLL WITH IT

Sweet-toothed fans recently voted to change three of the original flavors. The new five-flavor roll includes cherry, watermelon, pineapple, raspberry, and blackberry. The orange, lemon, and lime flavors have been replaced.

LifeSavers by the numbers:

Year first flavor invented: **1912 (Pep-O-Mint)**

Year five-flavor roll invented: **1935**

Total number of flavors today: **25**

Number of candies per roll: **14**

LifeSavers produced daily: **3 million**

Pounds of sugar used per day: **250,000**

Number of miniature rolls given out at Halloween: **88 million**

Source: USA Today *and Kraft Foods.*

43. The first LifeSaver flavor was Pep-O-Mint. How long after the Pep-O-Mint LifeSaver was invented was the five-flavor roll invented?

44. In addition to the five flavors, how many more flavors of LifeSavers are there?

45. Find the number of rolls of LifeSavers produced daily. Round to the nearest whole number.

46. Use your answer from Exercise 45 to find the amount of sugar in one roll of LifeSavers. Round to the nearest hundredth of a pound.

47. Check your work in Exercise 46. Is the answer reasonable? Explain why or why not.

48. Name three possible causes of errors in statistics.

8.3 ▶▶▶ Frequency Distributions and Histograms

The owner of Towne Insurance Agency has kept track of her personal phone sales call activity over the past 50 weeks. The number of sales calls made for each of the weeks is given below. Read down the columns, beginning with the left column, for successive weeks of the year.

75	65	40	50	45	30	30	35	45	25
75	70	60	55	30	25	44	30	35	30
75	70	50	30	50	20	30	30	20	25
60	62	45	45	48	40	35	25	20	25
75	45	50	40	35	40	40	30	27	40

OBJECTIVE 1 Understand a frequency distribution. A long list of numbers can be confusing. You can make the data easier to read by putting it in a special type of table called a **frequency distribution.**

EXAMPLE 1 Preparing a Frequency Distribution

Using the data above, construct a table that shows each possible number of sales calls. Then go through the original data and place a *tally* mark (I) in the tally column next to each corresponding value. Total the tally marks and place the totals in the third column. The result is a frequency distribution table.

Number of Sales Calls	Tally	Frequency	Number of Sales Calls	Tally	Frequency
20	III	3	48	I	1
25	IIII I	5	50	IIII	4
27	I	1	55	I	1
30	IIII IIII	9	60	II	2
35	IIII	4	62	I	1
40	IIII I	6	65	I	1
44	I	1	70	II	2
45	IIII	5	75	IIII	4

Work Problem ▢1 *at the Side.* ▶

OBJECTIVE 2 Arrange data in class intervals. The frequency distribution given in Example 1 above contains a great deal of information— perhaps too much to digest. It can be simplified by combining the number of sales calls into groups, forming the class intervals shown below in the left column of the table.

GROUPED DATA

Class Intervals (Number of Sales Calls)	Class Frequency (Number of Weeks)
20–29	9
30–39	13
40–49	13
50–59	5
60–69	4
70–79	6

In the table above, look at how many weeks had 20 to 29 sales calls: $3 + 5 + 1 = 9$ weeks.

OBJECTIVES

▢1 **Understand a frequency distribution.**

▢2 **Arrange data in class intervals.**

▢3 **Read and understand a histogram.**

▢1 Use the frequency distribution table at the left to find the following.

(a) The least number of sales calls made in a week

(b) The most common number of sales calls made in a week

(c) The number of weeks in which 35 calls were made

(d) The number of weeks in which 45 calls were made

ANSWERS

1. **(a)** 20 calls **(b)** 30 calls
(c) 4 weeks **(d)** 5 weeks

2 Use the grouped data for the insurance agency on the previous page to answer each question.

(a) During how many weeks were fewer than 50 calls made?

(b) During how many weeks were 50 or more calls made?

3 Use the histogram at the right to answer each question.

(a) During how many weeks were fewer than 60 calls made?

(b) During how many weeks were 60 or more calls made?

> **Note**
> The number of class intervals in the left column of the grouped data table is arbitrary. Grouped data usually has between 5 and 15 class intervals.

EXAMPLE 2 Analyzing a Frequency Distribution

Use the grouped data for the insurance agency (on the preceding page) to answer the following questions.

(a) During how many weeks were fewer than 30 calls made?

The first interval in the grouped data table (20–29) is the number of weeks during which fewer than 30 calls were made. Therefore, the owner made fewer than 30 calls during 9 weeks out of the 50 weeks shown.

(b) During how many weeks were 40 or more calls made?

The last four intervals in the grouped data table are the number of weeks during which 40 or more calls were made.

$$13 + 5 + 4 + 6 = 28 \text{ weeks}$$

◀ *Work Problem* **2** *at the Side.*

OBJECTIVE 3 Read and understand a histogram. The results in the grouped data table have been used to draw the special bar graph below, which is called a **histogram.** In a histogram, the width of each bar represents a range of numbers (*class interval*). The height of each bar in a histogram gives the *class frequency,* that is, the number of occurrences in each class interval.

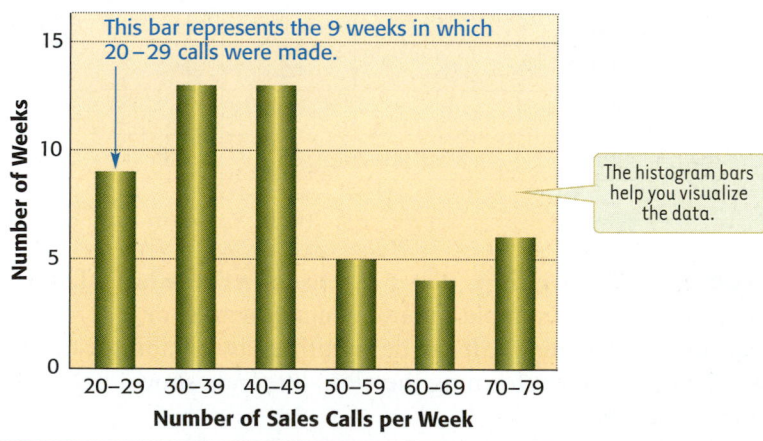

SALES CALL DATA FOR THE PAST 50 WEEKS (GROUPED DATA)

This bar represents the 9 weeks in which 20–29 calls were made.

The histogram bars help you visualize the data.

EXAMPLE 3 Using a Histogram

Use the histogram to find the number of weeks in which fewer than 40 calls were made.

Because 20–29 calls were made during 9 of the weeks and 30–39 calls were made during 13 of the weeks, the number of weeks in which fewer than 40 calls were made is $9 + 13 = 22$ weeks.

◀ *Work Problem* **3** *at the Side.*

ANSWERS
2. (a) 35 weeks **(b)** 15 weeks
3. (a) 40 weeks **(b)** 10 weeks

The Quilters Club of America recorded the ages of its members and used the results to construct this histogram. Use the histogram to answer Exercises 1–6. See Example 3.

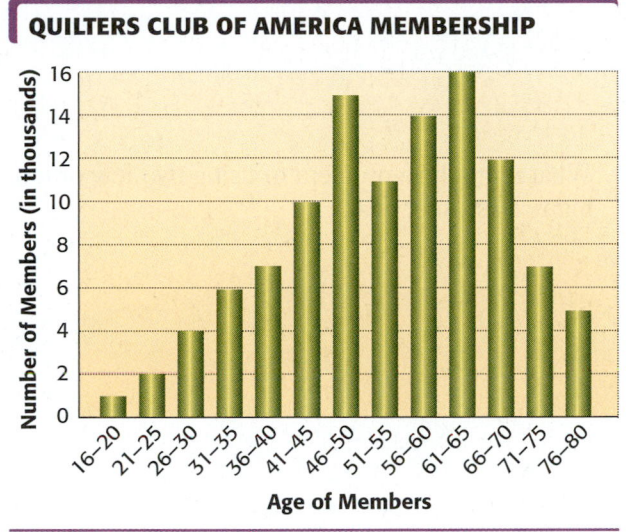

QUILTERS CLUB OF AMERICA MEMBERSHIP

Age of Members (horizontal axis): 16–20, 21–25, 26–30, 31–35, 36–40, 41–45, 46–50, 51–55, 56–60, 61–65, 66–70, 71–75, 76–80

Number of Members (in thousands) (vertical axis)

1. The greatest number of members are in which age group? How many members are in that group?

2. The least number of members are in which age group? How many members are in that group?

3. Find the number of members 35 years of age and under.

4. Find the number of members ages 61 to 80.

5. How many members are 41 to 60 years of age?

6. How many members are 46 to 55 years of age?

This histogram shows the annual earnings for the part-time employees of Wally World Amusement Park. Use this histogram to answer Exercises 7–12. See Example 3.

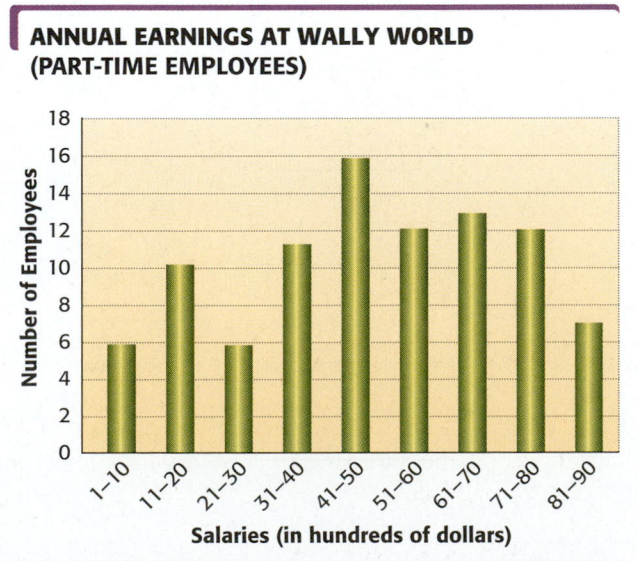

ANNUAL EARNINGS AT WALLY WORLD (PART-TIME EMPLOYEES)

Salaries (in hundreds of dollars) (horizontal axis): 1–10, 11–20, 21–30, 31–40, 41–50, 51–60, 61–70, 71–80, 81–90

Number of Employees (vertical axis)

7. The greatest number of employees are in which earnings group? How many are in that group?

8. The fewest number of employees are in which earnings groups? How many are in each group?

9. Find the number of employees who earn $3100 to $4000.

10. Find the number of employees who earn $1100 to $2000.

11. How many employees earn $5000 or less?

12. How many employees earn $6100 or more?

13. Describe class interval and class frequency. How are they used when preparing a histogram?

14. What might be a problem of using two few or too many class intervals?

This list shows the number of new accounts opened annually by the employees of the Schools Credit Union. Use it to complete the table. See Example 1.

186	191	144	198	147	158	174
193	142	155	174	162	151	178
145	151	199	182	147	195	146

Class Intervals (Number of New Accounts)	Tally	Class Frequency (Number of Employees)
15. 140–149	_____	_____
16. 150–159	_____	_____
17. 160–169	_____	_____
18. 170–179	_____	_____
19. 180–189	_____	_____
20. 190–199	_____	_____

A college professor asked her 30 students how many hours they worked each week. Use her list of student responses to complete the following table. See Examples 1–3.

14	8	12	28	33	14
6	34	17	20	13	20
25	33	32	4	7	14
0	6	10	8	35	31
25	4	32	18	0	24

	Class Intervals (Number of Hours Worked)	Tally	Class Frequency (Number of Students)
21.	0–5	_____	_____
22.	6–10	_____	_____
23.	11–15	_____	_____
24.	16–20	_____	_____
25.	21–25	_____	_____
26.	26–30	_____	_____
27.	31–35	_____	_____

28. Construct a histogram by using the data in Exercises 21–27.

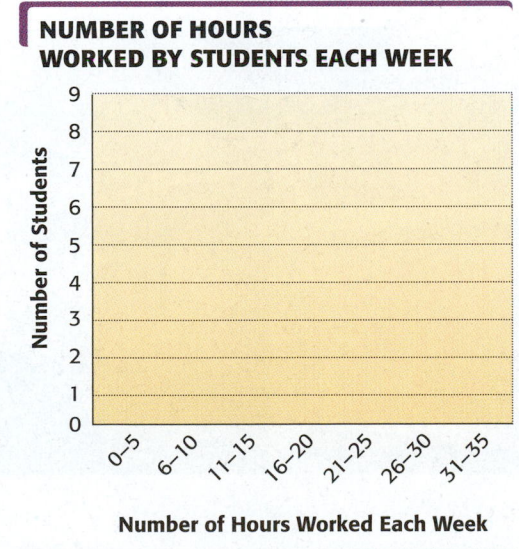

NUMBER OF HOURS WORKED BY STUDENTS EACH WEEK

Number of Students

9
8
7
6
5
4
3
2
1
0

0–5 6–10 11–15 16–20 21–25 26–30 31–35

Number of Hours Worked Each Week

Century 21 All Professional Realty has 80 salespeople spread over its five offices. The number of homes sold by each of these salespeople during the past year is shown below. Use these numbers to complete the following table. See Example 1.

9	33	14	8	17	10	25	11	4	16	3	9	5	7	14	18
15	24	19	30	16	31	21	20	30	2	6	6	27	17	3	32
3	8	5	11	15	26	7	18	29	10	7	3	12	9	25	15
11	6	10	4	2	35	10	25	5	19	34	2	4	14	11	28
8	13	25	15	23	26	12	4	22	12	21	12	22	10	18	21

Class Intervals (New Homes Sold)	Tally	Class Frequency (Number of Salespeople)
29. 1–5	_____	_____
30. 6–10	_____	_____
31. 11–15	_____	_____
32. 16–20	_____	_____
33. 21–25	_____	_____
34. 26–30	_____	_____
35. 31–35	_____	_____

36. Make a histogram showing the results from Exercises 29–35.

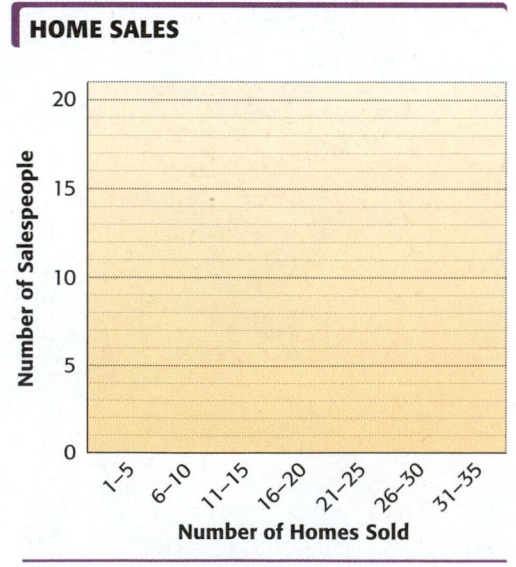

HOME SALES

Number of Salespeople

Number of Homes Sold

1–5 6–10 11–15 16–20 21–25 26–30 31–35

8.4 ▶▶▶ Mean, Median, and Mode

Businesses, governments, laboratories, colleges, and others working with lists of numbers are often faced with the problem of analyzing great amounts of raw data. The measures discussed in this section are helpful for such analyses.

OBJECTIVE 1 Find the mean of a list of numbers. When analyzing data, one of the first things to look for is a *measure of central tendency*— a single number that we can use to represent the entire list of numbers. One such measure is the *average* or **mean.** The mean can be found with the following formula.

> **Finding the Mean (Average)**
>
> $$\text{mean} = \frac{\text{sum of all values}}{\text{number of values}}$$

EXAMPLE 1 Finding the Mean

David had test scores of 84, 90, 95, 98, and 88. Find the mean (average) of his scores.

Use the formula for finding the mean. Add up all the test scores and then divide by the number of tests.

$$\text{mean} = \frac{84 + 90 + 95 + 98 + 88}{5} \quad \text{◁ Add up all the test scores.}$$

$$\text{mean} = \frac{455}{5} \quad \text{Divide.} \qquad \text{Divide by the number of tests.}$$

$$\text{mean} = 91$$

David has a mean score of 91.

Work Problem **1** *at the Side.* ▶

1 Tanya has test scores of 96, 98, 84, 88, 82, and 92. Find her mean (average) score.

2 Find the mean for each list of numbers. Round to the nearest cent if necessary.

(a) Monthly gasoline expenses of $50.28, $85.16, $110.50, $78, $120.70, $58.64, $73.80, $86.24, $67.85, $96.56, $138.65, $48.90

EXAMPLE 2 Applying the Average or Mean

Milk sales at a local 7-Eleven for each of the days last week were

$$\$86, \$103, \$118, \$117, \$126, \$158, \text{ and } \$149$$

Find the mean milk sales (rounded to the nearest cent) as shown below.

$$\text{mean} = \frac{\$86 + \$103 + \$118 + \$117 + \$126 + \$158 + \$149}{7}$$

$$\text{mean} = \frac{\$857}{7}$$

$$\text{mean} \approx \$122.43$$

The mean daily sales amount for milk was $122.43.

Work Problem **2** *at the Side.* ▶

(b) The attendance at eight major league home games this season was: 48,076; 58,595; 37,874; 46,289; 29,235; 59,311; 45,675; 39,721

ANSWERS

1. 90

2. **(a)** $\dfrac{\$1015.28}{12} \approx \$84.61$ (rounded)

   (b) $\dfrac{364{,}776}{8} = 45{,}597$ people

3 The numbers below show the amount that Pat Dunn spent for lunch and the number of days that he spent that amount. Find the weighted mean.

Value	Frequency
$ 2	4
$ 4	6
$ 6	5
$ 8	6
$10	12
$12	5
$14	8
$16	4

OBJECTIVE **2** **Find a weighted mean.** Some items in a list might appear more than once. In this case, we find a **weighted mean,** in which each value is "weighted" by multiplying it by the number of times it occurs.

EXAMPLE 3 **Understanding the Weighted Mean**

The following table shows the family size and the number of families (frequency) who were given groceries in one morning at the Twin Lakes Food Bank. Find the weighted mean.

Family Size	Frequency	
1	7	← 7 families had 1 person
2	12	
3	8	
4	8	
5	5	
6	9	← 9 families had 6 people
7	1	
8	6	

The same number of people were in more than one family: for example, there were 2 people in 12 of the families and there were 4 people in 8 of the families. There were 7 people in just 1 of the families. To find the mean, multiply the family size by its frequency. Then add the products. Next, add the numbers in the frequency column to find the total number of families.

Size	Frequency	Product
1	7	$(1 \cdot 7) = 7$
2	12	$(2 \cdot 12) = 24$
3	8	$(3 \cdot 8) = 24$
4	8	$(4 \cdot 8) = 32$
5	5	$(5 \cdot 5) = 25$
6	9	$(6 \cdot 9) = 54$
7	1	$(7 \cdot 1) = 7$
8	6	$(8 \cdot 6) = 48$
Totals	**56**	**221**

Finally, divide the totals. Round to the nearest hundredth.

$$\text{mean} = \frac{221}{56} \approx 3.95 \text{ (rounded)}$$

The mean family size of those using the food bank was 3.95 people.

◀ *Work Problem* **3** *at the Side.*

A common use of the weighted mean is to find a student's *grade point average* (GPA), as shown by the next example.

EXAMPLE 4   **Applying the Weighted Mean**

Find the grade point average for a student earning the following grades. Assume A = 4, B = 3, C = 2, D = 1, and F = 0. The number of credits determines how many times the grade is counted (the frequency).

Course	Credits	Grade	Credits · Grade
Mathematics	4	A (= 4)	4 · 4 = 16
Speech	3	C (= 2)	3 · 2 = 6
English	3	B (= 3)	3 · 3 = 9
Computer science	2	A (= 4)	2 · 4 = 8
Art history	2	D (= 1)	2 · 1 = 2
Totals	**14**		**41**

It is common to round grade point averages to the nearest hundredth. So the grade point average for this student is rounded to 2.93.

$$\text{GPA} = \frac{41}{14} \approx 2.93$$

Work Problem **4** *at the Side.* ▶

OBJECTIVE 3   Find the median.   Because it can be affected by extremely high or low numbers, the mean is often a poor indicator of central tendency for a list of numbers. In cases like this, another measure of central tendency, called the *median,* can be used. The **median** divides a group of numbers in half; half the numbers lie above the median, and half lie below the median.

Find the median by listing the numbers *in order* from *smallest* to *largest.* If the list contains an *odd* number of items, the median is the *middle number.*

EXAMPLE 5   **Finding the Median for an Odd Number of Items**

Find the median for the following list of prices for women's T-shirts.

$9, $23, $15, $8, $18, $12, $24

First arrange the numbers in numerical order from smallest to largest.

Smallest → 8, 9, 12, 15, 18, 23, 24 ← Largest

Next, find the middle number in the list.

8, 9, 12, **15**, 18, 23, 24

Three are below. ↓ Three are above.
Middle number.

> Remember to list the numbers from smallest to largest **before** finding the median.

The median price is $15.

Work Problem **5** *at the Side.* ▶

---

4 Find the grade point average (GPA) for Greg Barnes, who earned the following grades last semester. Round to the nearest hundredth.

Course	Credits	Grade
Mathematics	3	A (= 4)
P.E.	1	C (= 2)
English	3	C (= 2)
Keyboarding	2	B (= 3)
Biology	4	B (= 3)

5 Find the median for the following weights of bagged groceries.

14 lb, 18 lb, 10 lb, 17 lb, 15 lb, 19 lb, 20 lb

---

6 Find the median for the following list of measurements.

125 m, 87 m, 96 m, 108 m, 136 m, 74 m

If a list contains an *even* number of items, there is no single middle number. In this case, the median is defined as the mean (average) of the *middle two* numbers.

EXAMPLE 6 Finding the Median for an Even Number of Items

Find the median for the following list of ages.

74, 7, 15, 13, 25, 28, 47, 59, 32, 68

First arrange the numbers in numerical order from least to greatest. Then find the middle two numbers.

Smallest → 7, 13, 15, 25, **28, 32,** 47, 59, 68, 74 ← Largest

Middle two numbers

The median age is the mean of the two middle numbers.

$$\text{median} = \frac{28 + 32}{2} = \frac{60}{2} = 30 \text{ years}$$

◀ *Work Problem* **6** *at the Side.*

7 Find the mode for each list of numbers.

(a) Ages of summer work applicants (in years): 19, 18, 22, 20, 18

(b) Total points on a screening exam: 312, 219, 782, 312, 219, 426

(c) Monthly commissions of sales people: $1706, $1289, $1653, $1892, $1301, $1782

OBJECTIVE **4** **Find the mode.** The last important statistical measure is the **mode,** the number that occurs *most often* in a list of numbers. For example, if the test scores for 10 students were

↓ ↓ ↓

74, 81, 39, **74,** 82, 80, 100, 92, **74,** and 85,

then the mode is 74. Three students earned a score of 74, so 74 appears more times on the list than any other score. (It is not necessary to place the numbers in numerical order when looking for the mode.)

A list can have two modes; such a list is sometimes called **bimodal.** If no number occurs more frequently than any other number in a list, the list has *no mode.*

EXAMPLE 7 Finding the Mode

Find the mode for each list of numbers.

> The **mode** is the value that occurs most often.

(a) 51, 32, 49, 73, 49, 90

The number **49** occurs more often than any other number; therefore, 49 is the mode.

(b) 482, 485, 483, 485, 487, 487, 489

Because both **485** and **487** occur twice, each is a mode. This list is *bimodal.*

(c) $10,708; $11,519; $10,972; $12,546; $13,905; $12,182

No number occurs more than once. This list has *no mode.*

◀ *Work Problem* **7** *at the Side.*

Measures of Central Tendency

The **mean** is the sum of all the values divided by the number of values. It is the mathematical average.

The **median** is the middle number (or the average of the middle two numbers) in a group of values that are listed from least to greatest. It divides a group of numbers in half.

The **mode** is the value that occurs most often in a group of values.

ANSWERS

6. $\frac{96 + 108}{2} = \frac{204}{2} = 102$ m

7. **(a)** 18 yr
   (b) bimodal, 219 points and 312 points (this list has two modes)
   (c) no mode (no number occurs more than once)

8.4 ▶▶▶ Exercises

Find the mean for each list of numbers. Round answers to the nearest tenth if necessary.
See Example 1.

1. Shopping center ages (in years) of 6, 22, 15, 2, 8, 13

2. Minutes of cell phone use each day: 53, 77, 38, 29, 46, 48, 52

3. Inches of rain per month of 3.1, 1.5, 2.8, 0.8, 4.1

4. Algebra quiz scores of 32, 26, 30, 19, 51, 46, 38, 39

5. Annual salaries of $38,500; $39,720; $42,183; $21,982; $43,250

6. Numbers of students enrolled in Community Colleges: 27,500; 18,250; 17,357; 14,298; 33,110

Solve each application problem. See Example 2.

7. The Sunrise Pharmacy filled prescriptions that sold at the following amounts: $18.38, $168.75, $28.63, $72.85, $39.60, $183.74, $15.82, $33.18, $87.45, $98.72, and $50.70. Find the average price (mean) of the prescriptions sold.

8. In one evening, a waitress collected the following checks from her dinner customers: $30.10, $42.80, $91.60, $51.20, $88.30, $21.90, $43.70, $51.20. Find the average (mean) dinner check amount.

Find the weighted mean. Round answers to the nearest tenth. See Example 3.

9.

Customers Each Hour	Frequency
8	2
11	12
15	5
26	1

10.

Deliveries Each Week	Frequency
4	1
8	3
16	5
20	1

11.

Fish per Boat	Frequency
12	4
13	2
15	5
19	3
22	1
23	5

12.

Patients per Clinic	Frequency
25	1
26	2
29	5
30	4
32	3
33	5

Solve each application problem. See Example 4.

13. The table below shows the face value (policy amount) of life insurance policies sold and the number of policies sold for each amount by the New World Life Company during one week. Find the weighted mean amount for the policies sold.

Policy Amount	Number of Policies Sold
$ 10,000	6
$ 20,000	24
$ 25,000	12
$ 30,000	8
$ 50,000	5
$100,000	3
$250,000	2

14. A national health survey provided the information for this table. It shows how often each of the adults in the survey engages in a vigorous, leisure time activity. Find the weighted mean to determine the weekly hours of vigorous activity for the adults surveyed, to the nearest tenth.

Hours per week	Number of Adults
0	224
1	86
2	62
3	45
4	25
5	18
6	24
7	16

Find the median for each list of numbers. See Examples 5 and 6.

15. Number of books loaned: 125, 100, 150, 135, 114

16. Number of hits for a World Wide Web site (in thousands): 140, 85, 122, 114, 98

17. Calories in fast-food menu items: 501, 412, 521, 515, 298, 621, 346, 528

18. Number of cars in the parking lot each day: 520, 523, 513, 1283, 338, 509, 290, 420, 320, 980

Find the mode or modes for each list of numbers. See Example 7.

19. Porosity of soil samples: 21%, 18%, 21%, 28%, 22%, 21%, 25%

20. Low daily temperatures (in degrees Fahrenheit): 21, 32, 46, 32, 49, 32, 49

21. Ages of residents (in years) at Leisure Village: 74, 68, 68, 68, 75, 75, 74, 74, 70

22. Number of pages read: 86, 84, 79, 75, 88, 66, 72, 85, 71

23. When is the median a better measure of central tendency than the mean to describe a set of data? Make up a list of numbers to illustrate your explanation. Calculate both the mean and the median.

24. Suppose you own a hat shop and can order a certain hat in only one size. You look at last year's sales to decide on the size to order. Should you find the mean, median, or mode for these sales? Explain your answer.

Find the grade point average for students earning the following grades. Assume A = 4, B = 3, C = 2, D = 1, *and* F = 0. *Round answers to the nearest hundredth.*

25.

Credits	Grade
4	B
2	C
2	A
1	C
3	D

26.

Credits	Grade
1	C
3	A
4	B
3	C
2	A

27.

Credits	Grade
4	B
2	A
5	C
1	F
3	B

28.

Credits	Grade
3	A
3	B
4	B
3	C
3	C

29.

Credits	Grade
2	A
3	C
4	A
1	C
4	B

30.

Credits	Grade
3	A
2	A
5	B
4	A
1	A

Relating Concepts (Exercises 31–40) For Individual or Group Work

Gluco Industries manufactures and sells glucose monitors and other diabetes-related products. The number of sales calls made over an 8-week period by two sales representatives, Scott Samuels and Rob Stricker, is shown below. Use this information to **work Exercises 31–40 in order.**

Week	Number of Calls	
	Samuels	Stricker
1	39	21
2	15	22
3	40	20
4	22	23
5	13	19
6	22	24
7	17	25
8	8	22

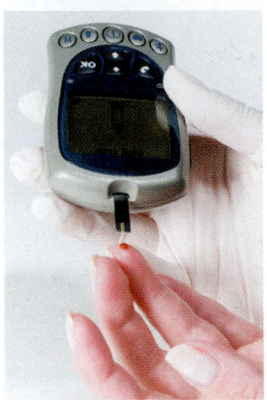

31. Find the total number of sales calls made by each of the sales representatives.

32. Find the mean number of sales calls made by Samuels and by Stricker.

33. Find the median number of sales calls made by Samuels and by Stricker.

34. Find the mode for the number of sales calls for each of the sales representatives.

35. How do the mean, median, and mode for the two sales representatives compare?

36. Describe how the pattern in the number of weekly sales calls made by Samuels differs from the pattern in Stricker's weekly sales calls.

To show the variation *or* spread *of the number of sales calls made by each of the sales representatives requires some* **measure of the dispersion,** *or spread of the numbers around the mean. A common measure of the dispersion is the* **range.** *The range is the* difference between the largest value and the smallest value in the set of numbers.

37. Find the range for the number of sales calls made by Samuels.

38. Find the range for the number of sales calls made by Stricker.

39. Is the performance data for these two sales representatives sufficient to accurately determine which is the best? What else might you want to know?

40. List three possible explanations for the wide variation (range) in the number of sales calls for Samuels.

Chapter 8 ▶▶▶ Summary

▶ Key Terms

8.1	**circle graph**	A circle graph shows how a total amount is divided into parts or sectors. It is based on percents of 360°.
	protractor	A protractor is a device (usually in the shape of a half-circle) used to measure the number of degrees in angles or parts of a circle.
8.2	**bar graph**	A bar graph uses bars of various heights or lengths to show quantity or frequency.
	double-bar graph	A double-bar graph compares two sets of data by showing two sets of bars.
	line graph	A line graph uses dots connected by lines to show trends.
	comparison line graph	A comparison line graph shows how two sets of data relate to each other by showing a line graph for each set of data.
8.3	**frequency distribution**	A frequency distribution is a table that includes a column showing each possible number in the data collected. The original data is then entered in another column using a tally mark for each corresponding value. The tally marks are counted and the totals are placed in a third column.
	histogram	A histogram is a bar graph in which the width of each bar represents a range of numbers (class interval) and the height represents the quantity or frequency of items that fall within the interval.
8.4	**mean**	The mean is the sum of all the values divided by the number of values. It is often called the *average*.
	weighted mean	The weighted mean is a mean calculated so that each value is multiplied by its frequency.
	median	The median is the middle number in a group of values that are listed from smallest to largest. It divides a group of values in half. If there are an even number of values, the median is the mean (average) of the two middle values.
	mode	The mode is the value that occurs most often in a group of values.
	bimodal	A list of numbers is bimodal when it has two modes. The two values occur the same number of times.
	dispersion	The dispersion is the variation or spread of the numbers around the mean.
	range	The range is a common measure of the dispersion of numbers. It is the difference between the largest value and the smallest value in the set of numbers.

▶ New Formula

Mean or average: $\text{mean} = \dfrac{\text{sum of all values}}{\text{number of values}}$

▶ Test Your Word Power

See how well you have learned the vocabulary in this chapter. Answers follow the Quick Review.

1. A **circle graph**
   A. uses bars of various heights to show quantity or frequency
   B. shows how a total amount is divided into parts or sectors
   C. uses dots connected by lines to show trends
   D. uses bars of various widths to represent a range of numbers.

2. A **bar graph**
   A. uses bars of various heights to show quantity or frequency
   B. shows how a total amount is divided into parts or sectors
   C. uses dots connected by lines to show trends
   D. uses bars of various widths to represent a range of numbers.

3. A **histogram** is a graph in which
   A. tally marks are used to record original data
   B. two sets of data are compared using two sets of bars
   C. dots are connected by lines to show trends
   D. the width of each bar represents a range of numbers and the height represents the frequency of items within that range.

4. A **protractor** is a device used to
   A. construct a histogram
   B. calculate measures of central tendency
   C. measure the number of degrees in angles or parts of a circle
   D. compare two sets of data.

5. The **mean** is
   A. calculated so that each value is multiplied by its frequency
   B. the sum of all values divided by the number of values
   C. the middle number in a group of values that are listed from smallest to largest
   D. the value that occurs most often in a group of values.

6. The **mode** is
   A. calculated so that each value is multiplied by its frequency
   B. the sum of all values divided by the number of values
   C. the middle number in a group of values that are listed from smallest to largest
   D. the value that occurs most often in a group of values.

▶ Quick Review

Concepts

Examples

8.1 Constructing a Circle Graph

Step 1 Determine the percent of the total for each item.
Step 2 Find the number of degrees out of 360° that each percent represents.
Step 3 Use a protractor to measure the number of degrees for each item in the circle.

Construct a circle graph from the following table, which lists the costs of options (billing) on a new luxury sports car.

Item	Amount
Leather interior	$1600
Wheels/tires	$2400
Sun roof	$1200
Sport package	$2800
Total	$8000

Item	Amount	Percent of Total	Sector Size
Leather interior	$1600	$\frac{\$1600}{\$8000} = \frac{1}{5} = $ **20%** so 360° • 20% $= 360 \cdot 0.20$	$= 72°$
Wheels/tires	$2400	$\frac{\$2400}{\$8000} = \frac{3}{10} = $ **30%** so 360° • 30% $= 360 \cdot 0.30$	$= 108°$
Sun roof	$1200	$\frac{\$1200}{\$8000} = \frac{3}{20} = $ **15%** so 360° • 15% $= 360 \cdot 0.15$	$= 54°$
Sport package	$2800	$\frac{\$2800}{\$8000} = \frac{7}{20} = $ **35%** so 360° • 35% $= 360 \cdot 0.35$	$= 126°$

See completed circle graph on the next page.

(continued)

Concepts	Examples

8.1 Constructing a Circle Graph (*continued*)

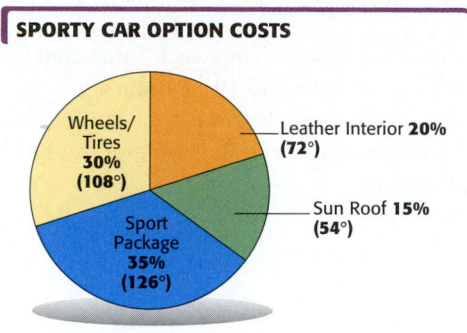

SPORTY CAR OPTION COSTS

8.2 Reading a Bar Graph

The height of the bar is used to show the quantity or frequency (number) in a specific category. Use a ruler or straightedge to line up the top of each bar with the numbers on the left side of the graph.

Use the bar graph below to determine the number of students who earned each letter grade.

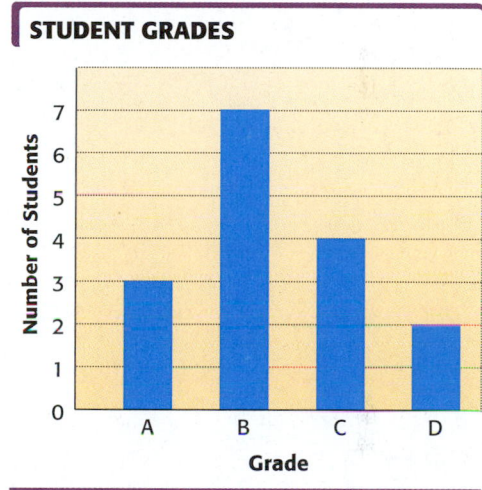

STUDENT GRADES

Grade of A: 3 students; B: 7 students; C: 4 students; D: 2 students.

8.2 Reading a Line Graph

A dot is used to show the number or quantity in a specific class. The dots are connected with lines. This kind of graph is used to show a trend.

The line graph below shows the annual sales for the Fabric Supply Center for each of 4 years.

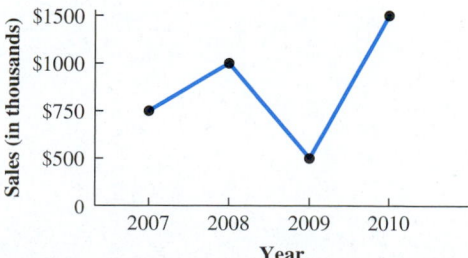

Find the sales in 2009.
The dot above 2009 lines up with $500 on the left edge.
Then $500 • 1000 = $500,000 in sales.

Concepts

8.3 **Preparing a Frequency Distribution and a Histogram from Raw Data**

Step 1 Construct a table listing each value, and the number of times this value occurs.

Step 2 Divide the data into groups, categories, or classes.

Step 3 Draw bars representing these groups to make a histogram.

Examples

Draw a histogram for these student quiz scores.

12	15	15	14
13	20	10	12
11	9	10	12
17	20	16	17
14	18	19	13

Quiz Score	Tally	Frequency	
9	I	1	1st
10	II	2	class
11	I	1	interval
12	III	3	2nd
13	II	2	class
14	II	2	interval
15	II	2	3rd
16	I	1	class
17	II	2	interval
18	I	1	4th
19	I	1	class
20	II	2	interval

Class Interval (Quiz Scores)	Frequency (Number of Students)
9–11	4
12–14	7
15–17	5
18–20	4

STUDENT QUIZ SCORES

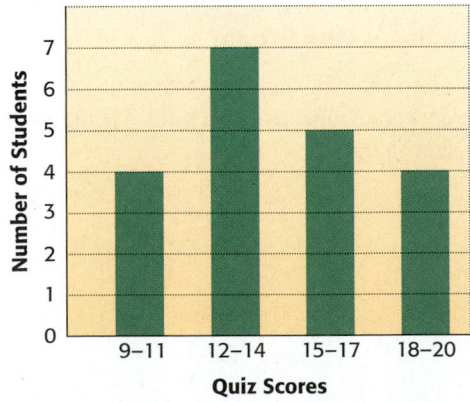

Concepts

Examples

8.4 Finding the Mean (Average) of a Set of Numbers

Step 1 Add all values to obtain a total.
Step 2 Divide the total by the number of values.

$$\text{mean (average)} = \frac{\text{sum of all values}}{\text{number of values}}$$

The test scores for Keith Zagorin in his algebra course were as follows:

80	92	92	94
76	88	84	93

Find Keith's mean (average) test score to the nearest tenth.

$$\text{mean} = \frac{80 + 92 + 92 + 94 + 76 + 88 + 84 + 93}{8}$$

$$= \frac{699}{8} \approx 87.4$$

Keith's mean test score is approximately 87.4.

8.4 Finding the Weighted Mean

Step 1 Multiply frequency by value.
Step 2 Add all the products from Step 1.
Step 3 Divide the sum in Step 2 by the total number of pieces of data.

This table shows the distribution of the number of school-age children in a survey of 30 families.

Number of School-Age Children	Frequency (Number of Families)
0	12
1	6
2	7
3	3
4	2
Total of 30 families	

Find the mean number of school-age children per family. Round to the nearest hundredth.

Value	Frequency	Product
0	12	$(0 \cdot 12) = 0$
1	6	$(1 \cdot 6) = 6$
2	7	$(2 \cdot 7) = 14$
3	3	$(3 \cdot 3) = 9$
4	2	$(4 \cdot 2) = 8$
Totals	**30**	**37**

$$\text{mean} = \frac{37}{30} \approx 1.23$$

The mean number of school-age children per family is approximately 1.23.

Concepts	Examples

(8.4) Finding the Median of a Set of Numbers

Step 1 Arrange the data from least to greatest.
Step 2 Select the middle value, or, if there is an even number of values, find the average of the two middle values.

Find the median for Keith Zagorin's test scores from the previous page.

The data arranged from smallest to largest is as follows:

76   80   84   <u>88   92</u>   92   93   94

Middle values

The middle two values are 88 and 92. The average of these two values is

$$\frac{88 + 92}{2} = 90$$

Keith's median test score is 90.

(8.4) Finding the Mode of a Set of Values

Find the value that appears most often in the list of values. If no value appears more than once, there is no mode. If two different values appear the same number of times, the list is bimodal.

Find the mode for Keith's test scores shown above.

The most frequently occurring score is 92 (it occurs twice). Therefore, the mode is 92.

ANSWERS TO TEST YOUR WORD POWER

1. (B) *Example:*

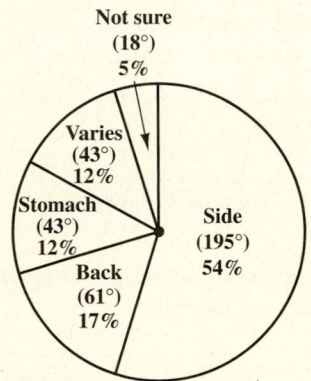

2. (A) *Example:*

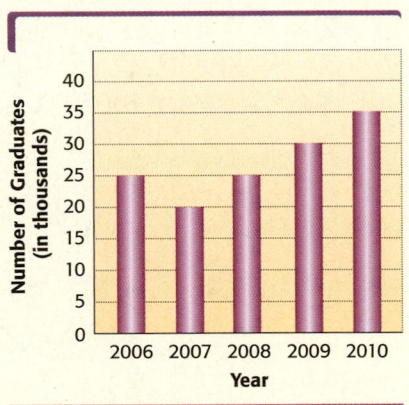

3. (D) *Example:*

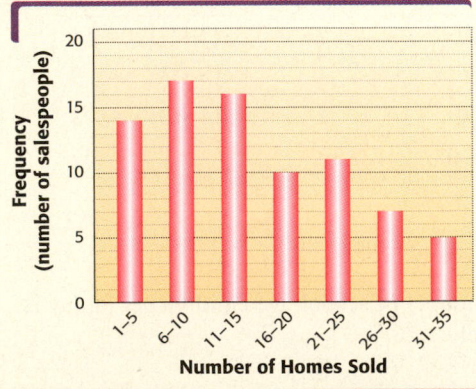

4. (C) *Example:*

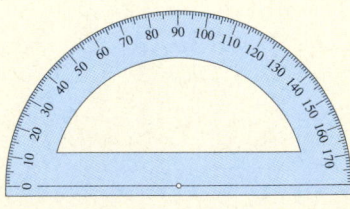

5. (B) *Example:* The mean of the values $5, $9, $7, $5, $2, and $8, is

$$\frac{\$5 + \$9 + \$7 + \$5 + \$2 + \$8}{6} = \frac{\$36}{6} = \$6.$$

6. (D) *Example:* The mode of the values $5, $9, $7, $5, $2, and $8 is $5 because $5 appears twice in the list.

Chapter 8 ▶▶▶ Review Exercises

[8.1] **1.** The number of girls participating in high school sports in the United States exceeds 3 million. The circle graph shows the number of high school girls' teams in the most popular sports. What women's sport has the greatest number of teams? How many are there?

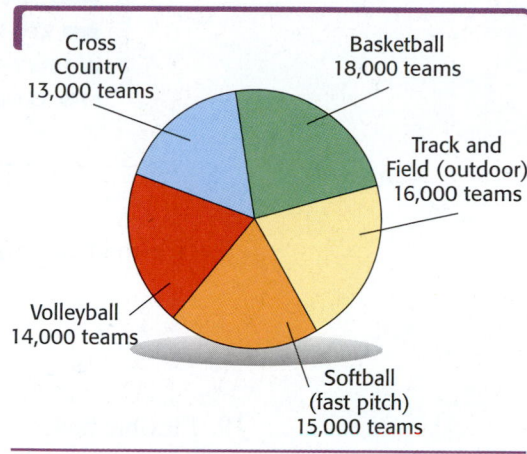

Source: National Federation of State High School Associations

Using the circle graph in Exercise 1, find each ratio. Write the ratios as fractions in lowest terms.

2. Number of track and field teams to the total number of teams.

3. Number of softball teams to the total number of teams.

4. Number of volleyball teams to the total number of teams.

5. Number of basketball teams to the number of track and field teams.

6. Number of track and field teams to the number of volleyball teams.

[8.2] *This bar graph shows the most frequently offered "work perks" and the percent of the responding companies offering them. The survey was conducted on-line and included 4800 companies ranging in size from 2 to 5000 employees. Use this graph to find the number of companies offering each work perk listed in Exercises 7–10 and to answer Exercises 11 and 12. (Source: Work Perks Survey, Ceridian Employer Services.)*

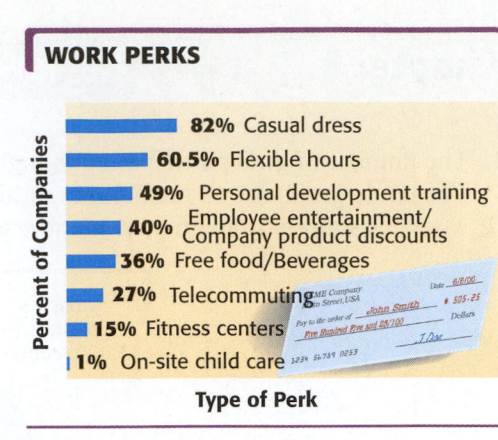

WORK PERKS

- 82% Casual dress
- 60.5% Flexible hours
- 49% Personal development training
- 40% Employee entertainment/ Company product discounts
- 36% Free food/Beverages
- 27% Telecommuting
- 15% Fitness centers
- 1% On-site child care

Percent of Companies

Type of Perk

7. Casual dress

8. Free food/Beverages

9. Fitness centers

10. Flexible hours

11. Which two work perks do companies offer least often? Give one possible explanation why these work perks are not offered.

12. Which two work perks do companies offer most often? Give one possible explanation why these work perks are so popular.

This double-bar graph shows the number of acre-feet of water in Lake Natoma for each of the first six months of 2009 and 2010. Use this graph to answer Exercises 13–18.

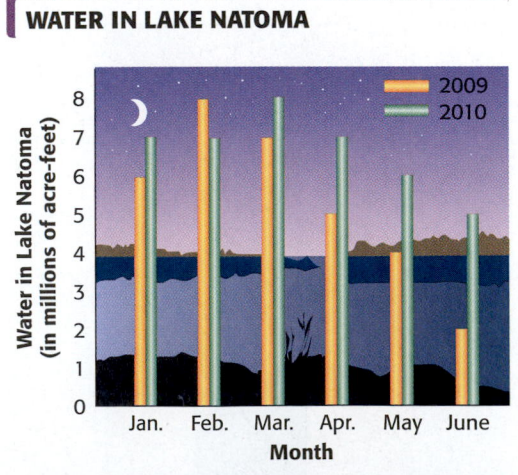

WATER IN LAKE NATOMA

Water in Lake Natoma (in millions of acre-feet)

2009
2010

Jan. Feb. Mar. Apr. May June

Month

13. During which month in 2010 was the greatest amount of water in the lake? How much was there?

14. During which month in 2009 was the least amount of water in the lake? How much was there?

15. How many acre-feet of water were in the lake in June of 2010?

16. How many acre-feet of water were in the lake in May of 2009?

17. Find the decrease in the amount of water in the lake from March 2009 to June 2009.

18. Find the decrease in the amount of water in the lake from April 2010 to June 2010.

This comparison line graph shows the annual floor-covering sales of two different home improvement centers during each of 5 years. Use this graph to find the amount of annual floor-covering sales in each year shown in Exercises 19–22 and to answer Exercises 23 and 24.

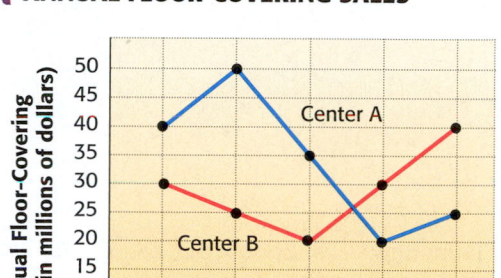

ANNUAL FLOOR-COVERING SALES

19. Center A in 2007

20. Center A in 2009

21. Center B in 2008

22. Center B in 2010

23. What trend do you see in center A's sales from 2007 to 2010? Why might this have happened?

24. What trend do you see in center B's sales starting in 2008? Why might this have happened?

[8.4] *Find the mean for each list of numbers. Round answers to the nearest tenth if necessary.*

25. Digital cameras sold: 18, 12, 15, 24, 9, 42, 54, 87, 21, 3

26. Number of harassment complaints filed: 31, 9, 8, 22, 46, 51, 48, 42, 53, 42

Find the weighted mean for each list. Round to the nearest tenth if necessary.

27.

Dollar Value	Frequency
$42	3
$47	7
$53	2
$55	3
$59	5

28.

Total Points	Frequency
243	1
247	3
251	5
255	7
263	4
271	2
279	2

Find the median for each list of numbers.

29. The number of accident forms filed: 43, 37, 13, 68, 54, 75, 28, 35, 39

30. Commissions of $576, $578, $542, $151, $559, $565, $525, $590

Find the mode or modes for each list of numbers.

31. Running shoes priced at $79, $56, $110, $79, $72, $86, $79

32. Boat launchings: 18, 25, 63, 32, 28, 37, 32, 26, 18

▶▶▶ **Mixed Review Exercises**

In her senior year of college Ally Romao had expenses of $17,920. This amount was spent as shown below. Find all the missing numbers in Exercises 33–37.

Item	Dollar Amount	Percent of Total	Degrees of Circle
33. Books and supplies	$1792	10%	_____
34. Rent	$6272	_____	126°
35. Food	$3584	_____	_____
36. Tuition/fees	$4480	_____	_____
37. Miscellaneous	$1792	_____	_____

38. Draw a circle graph using the information in Exercises 33–37.

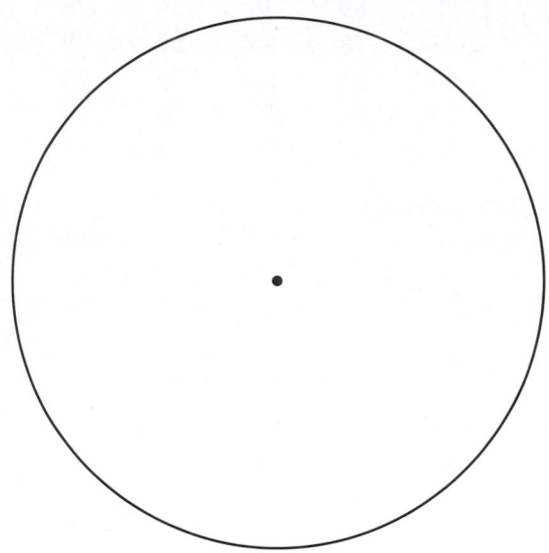

Find the mean for each list of numbers. Round answers to the nearest tenth if necessary.

39. Number of volunteers for the project: 48, 72, 52, 148, 180

40. Number of flu vaccinations in a day: 122, 135, 146, 159, 128, 147, 168, 139, 158

Find the mode or modes for each list of numbers.

41. Job applicants meeting the qualifications: 48, 43, 46, 47, 48, 48, 43

42. Number of two-bedroom apartments in each building: 26, 31, 31, 37, 43, 51, 31, 43, 43

Find the median for each list of numbers.

43. Hours worked: 4.7, 3.2, 2.9, 5.3, 7.1, 8.2, 9.4, 1.0

44. Number of e-mails each day: 35, 51, 9, 2, 17, 12, 46, 23, 3, 19, 39, 27

Here are the scores of 40 students on a computer science exam. Complete the table.

78	89	36	59	78	99	92	86
73	78	85	57	99	95	82	76
63	93	53	76	92	79	72	62
74	81	77	76	59	84	76	94
58	37	76	54	80	30	45	38

	Class Intervals (Scores)	Tally	Class Frequency (Number of Students)
45.	30–39	_____	_____
46.	40–49	_____	_____
47.	50–59	_____	_____
48.	60–69	_____	_____
49.	70–79	_____	_____
50.	80–89	_____	_____
51.	90–99	_____	_____

52. Construct a histogram by using the data in Exercises 45–51.

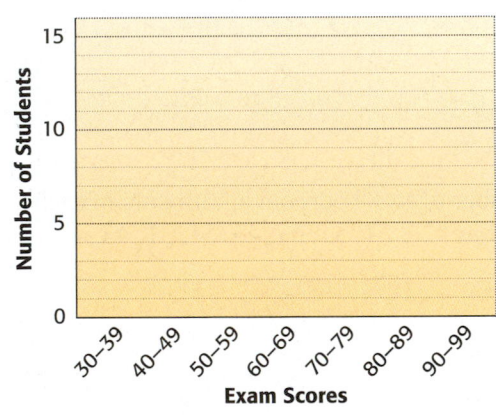

COMPUTER SCIENCE EXAM SCORES

Number of Students — *Exam Scores* (30–39, 40–49, 50–59, 60–69, 70–79, 80–89, 90–99)

Find each weighted mean. Round answers to the nearest tenth if necessary.

53.

Test Score	Frequency
46	4
54	10
62	8
70	12
78	10

54.

Units Sold	Frequency
104	6
112	14
115	21
119	13
123	22
127	6
132	9

Chapter 8 ▶▶▶ Test

 CHAPTER **Test Prep** VIDEO CD

Use the Chapter Test Prep Video CD to see fully worked-out solutions to any of the exercises you want to review.

The circle graph shows the sources of electricity generated in the United States. If the total cost of all electricity generated in one year was $298 billion, find the dollar amount spent on electricity generated by each of these sources. Round to the nearest tenth of a billion.

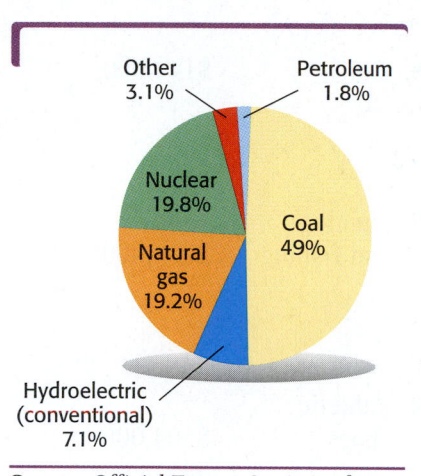

Other 3.1%
Petroleum 1.8%
Nuclear 19.8%
Coal 49%
Natural gas 19.2%
Hydroelectric (conventional) 7.1%

Source: Official Energy Statistics from the U.S. Government.

1. Nuclear

1. _____

2. Hydroelectric (conventional)

2. _____

3. Petroleum

3. _____

4. Natural gas

4. _____

5. Coal

5. _____

6. Other

6. _____

During a one-year period, Big 5 Sporting Goods had the following sales in each department. Find all numbers missing from the table.

Item	Dollar Amount	Percent of Total	Degrees of a Circle
7. Team sports	$432,000	30%	_____
8. Golf	$144,000	10%	_____
9. Hunting and fishing	$288,000	20%	_____
10. Athletic shoes	$504,000	35%	_____
11. Water sports	$72,000	_____	18°

7. _____

8. _____

9. _____

10. _____

11. _____

12.

12. Draw a circle graph using the information in Problems 7–11. Label each sector of the graph.

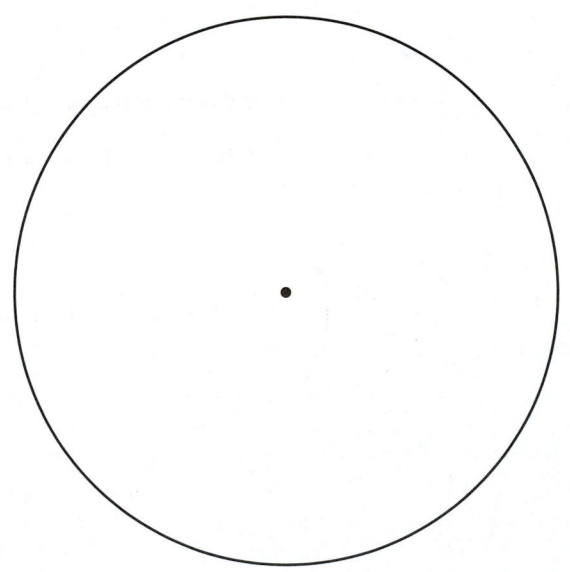

Here are the profits for each of the past 20 weeks from Alan's Snack Bar vending machines. Complete the table.

$142 $137 $125 $132 $147 $129 $151 $172 $175 $129
$159 $148 $173 $160 $152 $174 $169 $163 $149 $173

Profit	**Number of Weeks**
13. $120–129	_____
14. $130–139	_____
15. $140–149	_____
16. $150–159	_____
17. $160–169	_____
18. $170–179	_____

13. _____

14. _____

15. _____

16. _____

17. _____

18. _____

19. Use the information in Problems 13–18 to draw a histogram.

19. _____

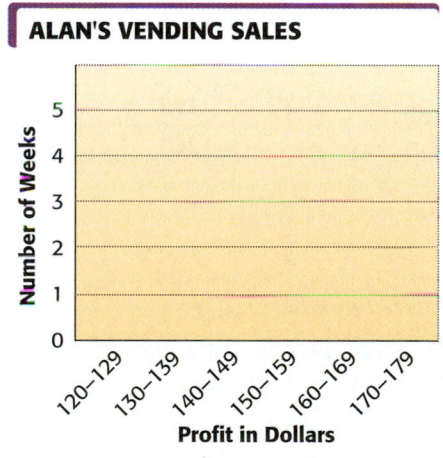

ALAN'S VENDING SALES

Find the mean for each list of numbers. Round answers to the nearest tenth if necessary.

20. Number of miles run each week while training: 52, 61, 68, 69, 73, 75, 79, 84, 91, 98

20. _____

21. Weight in pounds for the largest bass caught in the lake: 11, 14, 12, 14, 20, 16, 17, 18

21. _____

22. Airplane speeds in miles per hour: 458, 432, 496, 491, 500, 508, 512, 396, 492, 504

22. _____

23. Explain why a weighted mean is used to find a student's grade point average. Calculate your own grade point average for last semester or quarter. If you are a new student, make up a grade point average problem of your own and solve it. Round to the nearest hundredth.

23. _____

24. Explain in your own words the procedure for finding the median when there are an odd number of values in a list. Make up a problem with a list of five numbers and solve for the median.

24. _____

Find the weighted mean for the following. Round answers to the nearest whole number if necessary.

25. _____

25.

Cost	Frequency
$12	5
$20	6
$22	8
$28	4
$38	6
$48	2

26.

Points Scored	Frequency
150	15
160	17
170	21
180	28
190	19
200	7

26. _____

Find the median for each list of numbers.

27. _____

27. Lowest daily temperatures in degrees Fahrenheit:
32, 41, 28, 28, 37, 35, 16, 31

28. _____

28. The length of steel beams in meters:
7.6, 11.4, 6.2, 12.5, 31.7, 22.8, 9.1, 10.0, 9.5

Find the mode or modes for each list of numbers.

29. _____

29. Blood sample amounts in milliliters:
72, 46, 52, 37, 28, 18, 52, 61

30.

_____

30. Hot tub temperatures in degrees Fahrenheit of
96, 104, 103, 104, 103, 104, 91, 74, 103

Cumulative Review Exercises ▶▶▶ Chapters 1–8

1. Write these numbers in words.
   (a) 45.0203

   (b) 30,000,650,008

2. Write these numbers using digits.
   (a) One hundred sixty million, five hundred

   (b) Seventy-five thousandths

Round each number as indicated.

3. 46,908 to the nearest hundred

4. 6.197 to the nearest hundredth

5. 0.66148 to the nearest thousandth

6. 9951 to the nearest hundred

First use front end rounding to round each number and estimate the answer. Then find the exact answer.

7. *Estimate:* _____   *Exact:*
$$\begin{array}{r} 75{,}078 \\ -\ 46{,}090 \\ \hline \end{array}$$

8. *Estimate:* ___   *Exact:*
$$\begin{array}{r} 7.8 \\ -\ 3.5029 \\ \hline \end{array}$$

9. *Estimate:* $\times$ _____   *Exact:*
$$\begin{array}{r} 6538 \\ \times\ \ 708 \\ \hline \end{array}$$

10. *Estimate:* $\times$ _____   *Exact:*
$$\begin{array}{r} 65.3 \\ \times\ \ 8.7 \\ \hline \end{array}$$

11. *Estimate:* $\overline{)\ \ \ \ \ \ }$ *Exact:* $43\overline{)38{,}786}$

12. *Estimate:* $\overline{)\ \ \ }$ *Exact:* $0.8\overline{)6.76}$

Simplify. Write answers in lowest terms and as whole or mixed numbers when possible.

13. $4\dfrac{3}{5} + 5\dfrac{2}{3}$

14. $6\dfrac{2}{3} - 4\dfrac{3}{4}$

15. $\left(9\dfrac{3}{5}\right)\left(4\dfrac{5}{8}\right)$

16. $22\left(\dfrac{2}{5}\right)$

17. $3\dfrac{1}{3} \div 8\dfrac{3}{4}$

18. $\dfrac{2}{3}\left(\dfrac{7}{8} - \dfrac{3}{4}\right)$

19. $4 + 10 \div 2 + 7(2)$

20. $\sqrt{81} - 4(2) + 9$

21. $2^2 \cdot 3^3$

22. $\dfrac{2}{3}\left(\dfrac{7}{8} - \dfrac{1}{2}\right)$

23. $\dfrac{7}{8} \div \left(\dfrac{3}{4} + \dfrac{1}{8}\right)$

24. $\left(\dfrac{5}{6} - \dfrac{5}{12}\right) - \left(\dfrac{1}{2}\right)^2 \cdot \dfrac{2}{3}$

Write each fraction as a decimal. Round to the nearest thousandth if necessary.

25. $\dfrac{3}{4}$

26. $\dfrac{3}{8}$

27. $\dfrac{7}{12}$

28. $\dfrac{11}{20}$

Write in order, from least to greatest.

29. 0.218, 0.22, 0.199, 0.207, 0.2215

30. 0.6319, $\frac{5}{8}$, 0.608, $\frac{13}{20}$, 0.58

Write each ratio in lowest terms. Be sure to make all necessary conversions.

31. $5\frac{1}{2}$ in. to 44 in.

32. 3 hr to 45 min

Find the unknown number in each proportion.

33. $\frac{1}{5} = \frac{x}{30}$

34. $\frac{15}{x} = \frac{390}{156}$

35. $\frac{200}{135} = \frac{24}{x}$

36. $\frac{x}{208} = \frac{6.5}{26}$

Solve each percent problem.

37. Find 5.4% of 6000 homes.

38. $8\frac{1}{2}$% of what number of people is 238 people?

39. What percent of $555 is $1443?

Find the unknown value in each proportion.

40. $\frac{1}{5} = \frac{x}{30}$

41. $\frac{224}{32} = \frac{28}{x}$

42. $\frac{8}{x} = \frac{72}{144}$

43. $\frac{x}{120} = \frac{7.5}{30}$

Write each percent as a decimal. Write each decimal as a percent.

44. 3%

45. 200%

46. 0.87

47. 3.8

Write each percent as a fraction or mixed number in lowest terms. Write each fraction as a percent.

48. 8%

49. 62.5%

50. 175%

51. $\frac{7}{8}$

52. $4\frac{1}{5}$

Solve each problem.

53. Diane McKinney paid $29.90 in sales tax on a $460 purchase. What was the tax rate?

54. Find the commission on $12,538 in sales when the commission rate is 8%.

55. A box spring and mattress originally priced at $456 is discounted 45%. Find the amount of discount and the sale price.

56. A loan of $22,250 is made at 7% simple interest for 6 months. Find the total amount to be repaid.

Set up and solve a proportion for each problem.

57. Carol can test the hearing of 11 patients in 4 hours. Find the number of hearing tests that she can do in 12 hours.

58. If 12.5 ounces of Roundup® weed and grass killer is needed to make 5 gallons of spray, how much Roundup is needed for 102 gallons of spray?

Name each figure and find its perimeter (or circumference) and the area. Use 3.14 *as the approximate value of* π.

59.

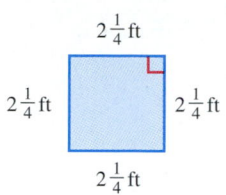

60.

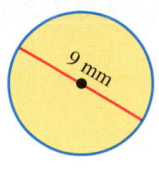

61.

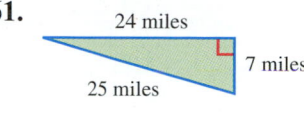

62.

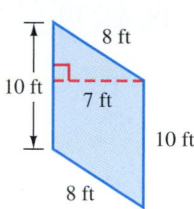

Find the unknown length, perimeter, area, or volume. Use 3.14 *as the approximate value of* π *and round answers to the nearest tenth.*

63. Find the perimeter and the area.

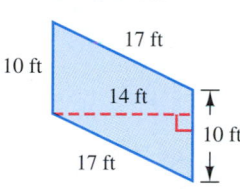

64. Find the circumference and the area.

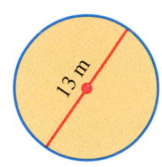

65. Find the length of the third side and the area.

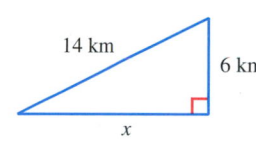

66. Find the volume and surface area.

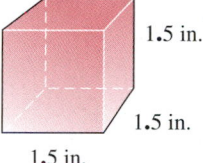

67. Find the perimeter and the area.

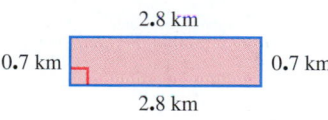

68. Find the diameter, circumference, and area.

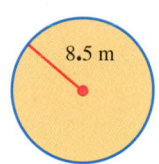

69. Find the volume and surface area.

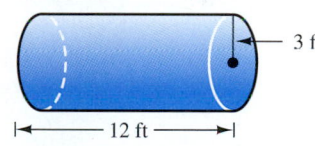

In Exercises 70 and 71, name each solid and find its volume. Use 3.14 *as the approximate value for* π. *In Exercise 72, find the unknown length. Round answers to the nearest tenth if necessary.*

70.

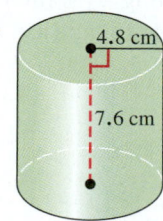

71.

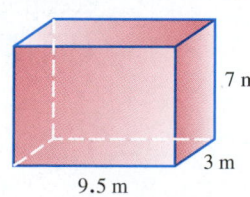

72.

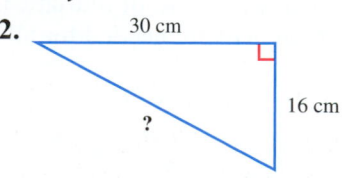

Find the mean, the median, and the mode for each list of numbers. Round to the nearest tenth if necessary.

73. Cable hookups per installer: 16, 37, 27, 31, 19, 25, 15, 38, 43, 19

74. Number of acres plowed each hour: 10.3, 4.3, 1.65, 2.85, 5.3, 5.7, 2.3, 4.35, 2.85

Solve each application problem.

75. In a study of 2082 workers it was found that only 874 of them used all of their paid time-off. What percent used all of their time-off? Round to the nearest tenth of a percent. (*Source:* Hudson Time-Off Survey.)

76. The average increase in residential winter heating bills will be 9.8%. If the increase amounts to $87.20, find (a) the average heating bill before the increase, and (b) the average heating bill after the increase. (*Source:* Energy Information Administration.)

77. Breathe Right™ nasal strips are sold in four package sizes: 12 nasal strips $6.50; 24 nasal strips $7.50; 30 nasal strips $8.95; 38 nasal strips $9.95. You have a $2-off coupon for the 12-strip size and a $1-off coupon for the 30-strip size. Which choice is the best buy?

78. The sketch below shows the plans for a lobby in a large commercial complex. What is the cost of carpeting the lobby, excluding the atrium, if the contractor charges $43.50 per square yard? Use 3.14 for π.

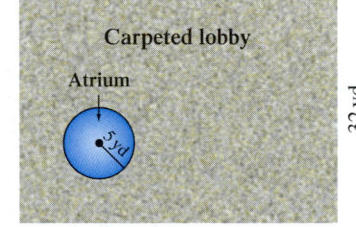

Carpeted lobby

Atrium

5 yd

32 yd

45 yd

79. The Spa Service Center services 140 spas. If each spa needs 125 mL of muriatic acid, how many liters of acid are needed for all the spas?

80. Linda Vincent has $11,400 in an 8-month CD (certificate of deposit) account earning $4\frac{3}{4}$% interest. Find the total amount in her account at maturity.

9

The Real Number System

T he personal savings rate of Americans has declined steadily since 1984, when it stood at 10.8% of after-tax income. In 2006, Americans spent everything they earned and more, causing the personal savings rate to drop to −1%, the worst personal savings rate since 1933. That year, Americans depleted their savings to cope with the job layoffs and business failures of the Great Depression. (*Source:* Commerce Department.)

In this chapter, we examine *positive* and *negative numbers* and apply them to situations such as the personal savings rate of Americans in Exercises 77 and 78 of **Section 9.5.**

9.1 ▶▶▶ Exponents, Order of Operations, and Inequality

OBJECTIVE 1 Use exponents. In algebra, we use a raised dot for multiplication as shown in **Section 2.3,** where we factored a number as the product of its prime factors. For example, 81 is written in prime factored form as

$$81 = 3 \cdot 3 \cdot 3 \cdot 3,$$

where the factor 3 appears four times. Repeated factors are written in an abbreviated form by using an *exponent*. The prime factored form of 81 is written with an exponent as

$$\underbrace{3 \cdot 3 \cdot 3 \cdot 3}_{4 \text{ factors of } 3} = 3^{\overset{\displaystyle\text{Exponent}}{4}}.$$
Base

The number 4 is the **exponent,** or **power,** and 3 is the **base** in the **exponential expression** 3^4. The exponent always follows the base and tells how many times the base is used as a factor. We read 3^4 as "3 to the fourth power," or simply "3 to the fourth." *A number raised to the first power is simply that number.* For example, $6^1 = 6$ and $(2.5)^1 = 2.5$.

EXAMPLE 1 Evaluating Exponential Expressions

Find the value of each exponential expression.

(a) $5^2 = \underbrace{5 \cdot 5}_{} = 25$

5 is used as a factor 2 times.

Read 5^2 as "5 to the second power" or, more commonly, "5 squared."

(b) $6^3 = \underbrace{6 \cdot 6 \cdot 6}_{} = 216$

6 is used as a factor 3 times.

Read 6^3 as "6 to the third power" or, more commonly, "6 cubed."

(c) $2^5 = 2 \cdot 2 \cdot 2 \cdot 2 \cdot 2 = 32$ 2 is used as a factor 5 times.

Read 2^5 as "2 to the fifth power."

(d) $7^4 = 7 \cdot 7 \cdot 7 \cdot 7 = 2401$ 7 is used as a factor 4 times.

Read 7^4 as "7 to the fourth power."

(e) $\left(\dfrac{2}{3}\right)^3 = \dfrac{2}{3} \cdot \dfrac{2}{3} \cdot \dfrac{2}{3} = \dfrac{8}{27}$ $\frac{2}{3}$ is used as a factor 3 times.

◀ *Work Problem* 1 *at the Side.*

1 Find the value of each exponential expression.

(a) 6^2

(b) 3^5

(c) $\left(\dfrac{3}{4}\right)^2$

(d) $\left(\dfrac{1}{2}\right)^4$

(e) $(0.4)^3$

ANSWERS
1. **(a)** 36 **(b)** 243 **(c)** $\dfrac{9}{16}$
   (d) $\dfrac{1}{16}$ **(e)** 0.064

CAUTION

Squaring, or raising a number to the second power, is not the same as doubling the number. For example,

$$3^2 \quad \text{means} \quad 3 \cdot 3, \quad not \quad 2 \cdot 3.$$

Thus $3^2 = 9$, not 6. Similarly, cubing, or raising a number to the third power, does *not* mean tripling the number.

OBJECTIVE **2** **Use the rules for order of operations.** Many problems involve more than one operation. To indicate the order in which the operations should be performed, we often use **grouping symbols.** If no grouping symbols are used, we apply the rules for order of operations.

Consider the expression $5 + 2 \cdot 3$. To show that the multiplication should be performed before the addition, parentheses can be used to write

$$5 + (2 \cdot 3), \quad \text{which equals} \quad 5 + 6, \quad \text{or} \quad 11.$$

If addition is to be performed first, the parentheses should group $5 + 2$ as follows:

$$(5 + 2) \cdot 3, \quad \text{which equals} \quad 7 \cdot 3, \quad \text{or} \quad 21.$$

Other grouping symbols used in more complicated expressions are brackets [ ], braces { }, and fraction bars. (For example, in $\frac{8 - 2}{3}$, the expression $8 - 2$ is considered to be grouped in the numerator.)

To work problems with more than one operation, use the following **order of operations.** This order is used by most calculators and computers.

Order of Operations

If grouping symbols are present, simplify within them, innermost first (and above and below fraction bars separately), in the following order.

Step 1 Apply all **exponents.**

Step 2 Do any **multiplications** or **divisions** in the order in which they occur, working from left to right.

Step 3 Do any **additions** or **subtractions** in the order in which they occur, working from left to right.

If no grouping symbols are present, start with Step 1.

A dot has been used to show multiplication; another way to show multiplication is with parentheses. For example, $3(7)$ means $3 \cdot 7$, or 21. Also, $3(4 + 5)$ means "3 times the sum of 4 and 5." By the order of operations, the sum in parentheses must be found first, then the product.

EXAMPLE 2   **Using the Rules for Order of Operations**

Find the value of each expression.

(a) $24 - 12 \div 3$

Use the order of operations given in the box.

$$24 - \mathbf{12 \div 3} \quad \text{←} \quad \boxed{\text{Be careful! Divide first.}}$$

$$= 24 - \mathbf{4} \qquad \text{Divide.}$$

$$= 20 \qquad \text{Subtract.}$$

(b) $9\,(\mathbf{6 + 11})$

$$= 9\,(\mathbf{17}) \qquad \text{Add inside parentheses.}$$

$$= 153 \qquad \text{Multiply.}$$

Continued on Next Page

2 Find the value of each expression.

(a) $7 + 3 \cdot 8$

(b) $2 \cdot 9 + 7 \cdot 3$

(c) $7 \cdot 6 - 3(8 + 1)$

(d) $2 + 3^2 - 5$

3 Find the value of each expression.

(a) $9[(4 + 8) - 3]$

(b) $\dfrac{2(7 + 8) + 2}{3 \cdot 5 + 1}$

(c) $6 \cdot 8 + 5 \cdot 2$

$\quad = \mathbf{48} + \mathbf{10}$ Multiply, working from left to right.

$\quad = 58$ Add.

(d) $2(\mathbf{5 + 6}) + 7 \cdot 3$

[Start here.]

$\quad = 2\,(\mathbf{11}) + \mathbf{7 \cdot 3}$ Add inside parentheses.

$\quad = 22 + \mathbf{21}$ Multiply.

$\quad = 43$ Add.

(e) $9 + \mathbf{2^3} - 5$

$\quad = 9 + \mathbf{8} - 5$ Apply the exponent.

$\quad = 12$ Add, and then subtract.

◀ *Work Problem* **2** *at the Side.*

OBJECTIVE **3** **Use more than one grouping symbol.** An expression with double parentheses, such as the expression $2(8 + 3(6 + 5))$, can be confusing. For clarity, we often use brackets, [   ], in place of the outer pair of parentheses.

EXAMPLE 3 **Using Brackets and Fraction Bars as Grouping Symbols**

Find the value of each expression.

[Start here.]

(a) $2[8 + 3\,(\mathbf{6 + 5})]$

$\quad = 2[8 + 3\,(\mathbf{11})]$ Add inside parentheses.

$\quad = 2[8 + \mathbf{33}]$ Multiply inside brackets.

$\quad = 2[\mathbf{41}]$ Add inside brackets.

$\quad = 82$ Multiply.

(b) $\dfrac{4\,(\mathbf{5 + 3}) + 3}{2\,(3) - 1}$ Simplify the numerator and denominator separately.

$\quad = \dfrac{4\,(\mathbf{8}) + 3}{2\,(3) - 1}$ Add inside parentheses.

$\quad = \dfrac{\mathbf{32} + 3}{\mathbf{6} - 1}$ Multiply.

$\quad = \dfrac{35}{5}$ Add and subtract.

$\quad = 7$ Divide.

◀ *Work Problem* **3** *at the Side.*

▦ **Calculator Tip** Calculators follow the order of operations given in this section. Try some of the examples to see that your calculator gives the same answers. Be sure to use the parentheses keys to insert parentheses where they are needed. To work Example 3(b) with a calculator, you must put parentheses around the numerator and the denominator.

ANSWERS

2. (a) 31   (b) 39   (c) 15   (d) 6
3. (a) 81   (b) 2

OBJECTIVE 4 Know the meanings of ≠, <, >, ≤, and ≥. So far, we have used only the symbols of arithmetic, such as $+$, $-$, $\cdot$, and $\div$ and the equality symbol $=$. The equality symbol with a slash through it means "is *not* equal to." For example,

$$7 \neq 8 \qquad \text{7 is not equal to 8.}$$

indicates that 7 is not equal to 8.

If two numbers are not equal, then one of the numbers must be less than the other. The symbol $<$ represents "is less than," so "7 is less than 8" is written as

$$7 < 8. \qquad \text{7 is less than 8.}$$

The symbol $>$ means "is greater than." For example,

$$8 > 2. \qquad \text{8 is greater than 2.}$$

The statement "17 is greater than 11" becomes $17 > 11$.

To keep the meanings of the symbols $<$ and $>$ clear, remember that the symbol always points to the lesser number.

$$\text{Lesser number} \rightarrow 8 < 15$$

$$15 > 8 \leftarrow \text{Lesser number}$$

> **Work Problem 4 at the Side.** ▶

Two other symbols, $\leq$ and $\geq$, also represent the idea of inequality. The symbol $\leq$ means "is less than or equal to," so

$$5 \leq 9 \qquad \text{5 is less than or equal to 9.}$$

means "5 is less than or equal to 9." *If either the $<$ part or the $=$ part is true, then the inequality $\leq$ is true.* The statement $5 \leq 9$ is true because $5 < 9$ is true. Also, $8 \leq 8$ is true because $8 = 8$ is true. But $13 \leq 9$ is not true because neither $13 < 9$ nor $13 = 9$ is true.

The symbol $\geq$ means "is greater than or equal to," so

$$9 \geq 5 \qquad \text{9 is greater than or equal to 5}$$

is true because $9 > 5$ is true.

EXAMPLE 4 Using the Symbols ≤ and ≥

Determine whether each statement is *true* or *false.*

(a) $15 \leq 20$ ⠀⠀The statement $15 \leq 20$ is true because $15 < 20$.

(b) $12 \geq 12$ ⠀⠀Since $12 = 12$, this statement is true.

(c) $\dfrac{6}{15} \geq \dfrac{2}{3}$

To compare fractions, write them with a common denominator. Here, 15 is a common denominator and $\frac{2}{3} = \frac{10}{15}$. Now decide whether $\frac{6}{15} \geq \frac{10}{15}$ is true or false. Both statements $\frac{6}{15} > \frac{10}{15}$ and $\frac{6}{15} = \frac{10}{15}$ are false; therefore, $\frac{6}{15} \geq \frac{2}{3}$ is false.

> **Work Problem 5 at the Side.** ▶

OBJECTIVE 5 Translate word statements to symbols. An important part of algebra deals with translating words into algebraic notation.

---

4 Write each statement in words. Then decide whether it is *true* or *false.*

(a) $7 < 5$

(b) $12 > 6$

(c) $4 \neq 10$

(d) $28 \neq 4 \cdot 7$

5 Tell whether each statement is *true* or *false.*

(a) $30 \leq 40$

(b) $25 \geq 10$

(c) $40 \leq 10$

(d) $21 \leq 21$

(e) $3 \geq 3$

ANSWERS

4. **(a)** Seven is less than five. False
   (b) Twelve is greater than six. True
   (c) Four is not equal to ten. True
   (d) Twenty-eight is not equal to four times seven. False
5. **(a)** true ⠀**(b)** true ⠀**(c)** false
   (d) true ⠀**(e)** true

6 Write in symbols.

(a) Nine is equal to eleven minus two.

(b) Seventeen is less than thirty.

(c) Eight is not equal to ten.

(d) Fourteen is greater than twelve.

(e) Thirty is less than or equal to fifty.

(f) Two is greater than or equal to two.

7 Write each statement with the inequality symbol reversed.

(a) $8 < 10$

(b) $3 > 1$

(c) $9 \leq 15$

(d) $6 \geq 2$

EXAMPLE 5 **Converting Words to Symbols**

Write each word statement in symbols.

(a) Twelve **is equal to** ten **plus** two. $12 = 10 + 2$

(b) Nine **is less than** ten. $9 < 10$
Compare this with "9 less than 10," which is written $10 - 9$.

(c) Fifteen **is not equal to** eighteen. $15 \neq 18$

(d) Seven **is greater than** four. $7 > 4$

(e) Thirteen **is less than or equal to** forty. $13 \leq 40$

(f) Six **is greater than or equal to** six. $6 \geq 6$

◀ *Work Problem* **6** *at the Side.*

OBJECTIVE **6** **Write statements that change the direction of inequality symbols.** Any statement with $<$ can be converted to one with $>$, and any statement with $>$ can be converted to one with $<$. *We do this by reversing both the order of the numbers and the direction of the symbol.* For example, the statement $6 < 10$ can be written as $10 > 6$.

$$6 < 10 \quad \text{becomes} \quad 10 > 6$$

Interchange numbers.
Reverse symbol.

EXAMPLE 6 **Converting between $<$ and $>$**

Parts (a)–(d) show the same statements written in two equally correct ways. In each inequality, the symbol points toward the lesser number.

(a) $9 < 16$, $16 > 9$ **(b)** $5 > 2$, $2 < 5$

(c) $3 \leq 8$, $8 \geq 3$ **(d)** $12 \geq 5$, $5 \leq 12$

◀ *Work Problem* **7** *at the Side.*

Here is a summary of the symbols of equality and inequality.

Symbol	Meaning	Example
$=$	Is equal to	$0.5 = \frac{1}{2}$ means 0.5 is equal to $\frac{1}{2}$.
$\neq$	Is not equal to	$3 \neq 7$ means 3 is not equal to 7.
$<$	Is less than	$6 < 10$ means 6 is less than 10.
$>$	Is greater than	$15 > 14$ means 15 is greater than 14.
$\leq$	Is less than or equal to	$4 \leq 8$ means 4 is less than or equal to 8.
$\geq$	Is greater than or equal to	$1 \geq 0$ means 1 is greater than or equal to 0.

CAUTION
The symbols of equality and inequality are used to write mathematical *sentences.* They differ from the symbols for operations ($+$, $-$, $\cdot$, and $\div$), discussed earlier, which are used to write mathematical *expressions* that represent a number. For example, compare the sentence $4 < 10$, which gives the relationship between 4 and 10, with the expression $4 + 10$, which tells how to operate on 4 and 10 to get the number 14.

9.1 ▶▶▶ Exercises

Decide whether each statement is true *or* false. *If it is false,* explain why.

1. An exponent tells how many times its base is used as a factor.

2. Some grouping symbols are $+, -, \cdot,$ and $\div$.

3. When evaluated, $4 + 3(8 - 2)$ is equal to 42.

4. $3^3 = 9$

5. The statement "4 is 12 less than 16" is interpreted $4 = 12 - 16$.

6. The statement "6 is 4 less than 10" is interpreted $6 < 10 - 4$.

Find the value of each exponential expression. See Example 1.

7. 7^2

8. 4^2

9. 12^2

10. 14^2

11. 4^3

12. 5^3

13. 10^3

14. 11^3

15. 3^4

16. 6^4

17. 4^5

18. 3^5

19. $\left(\dfrac{2}{3}\right)^4$

20. $\left(\dfrac{3}{4}\right)^3$

21. $(0.04)^3$

22. $(0.05)^4$

23. When evaluating $(4^2 + 3^3)^4$, what is the *last* exponent that would be applied? Explain your answer.

24. Which are not grouping symbols—parentheses, brackets, fraction bars, exponents?

Find the value of each expression. See Examples 2 and 3.

25. $13 + 9 \cdot 5$

26. $11 + 7 \cdot 6$

27. $20 - 4 \cdot 3 + 5$

28. $18 - 7 \cdot 2 + 6$

29. $9 \cdot 5 - 13$

30. $7 \cdot 6 - 11$

31. $18 - 2 + 3$

32. $22 - 8 + 9$

33. $\dfrac{1}{4} \cdot \dfrac{2}{3} + \dfrac{2}{5} \cdot \dfrac{11}{3}$

34. $\dfrac{9}{4} \cdot \dfrac{2}{3} + \dfrac{4}{5} \cdot \dfrac{5}{3}$

35. $9 \cdot 4 - 8 \cdot 3$

36. $11 \cdot 4 + 10 \cdot 3$

37. $2.5(1.9) + 4.3(7.3)$

38. $4.3(1.2) + 2.1(8.5)$

39. $10 + 40 \div 5 \cdot 2$

40. $12 + 8^2 \div 8 - 4$

41. $18 - 2(3 + 4)$

42. $30 - 3(4 + 2)$

43. $5[3 + 4(2^2)]$

44. $6\left[\dfrac{3}{4} + 8\left(\dfrac{1}{2}\right)^3\right]$

45. $\left(\dfrac{3}{2}\right)^2\left[\left(11 + \dfrac{1}{3}\right) - 6\right]$

46. $4^2[(13 + 4) - 8]$

47. $\dfrac{8 + 6(3^2 - 1)}{3 \cdot 2 - 2}$

48. $\dfrac{8 + 2(8^2 - 4)}{4 \cdot 3 - 10}$

49. $\dfrac{4(7+2)+8(8-3)}{6(4-2)-2^2}$

50. $\dfrac{6(5+1)-9(1+1)}{5(8-4)-2^3}$

Tell whether each statement is true *or* false. *In Exercises 53–62, first simplify each expression involving an operation. See Example 4.*

51. $8 \geq 17$

52. $10 \geq 41$

53. $17 \leq 18 - 1$

54. $12 \geq 10 + 2$

55. $6 \cdot 8 + 6 \cdot 6 \geq 0$

56. $4 \cdot 20 - 16 \cdot 5 \geq 0$

57. $6[5 + 3(4 + 2)] \leq 70$

58. $6[2 + 3(2 + 5)] \leq 135$

59. $\dfrac{9(7-1)-8\cdot2}{4(6-1)} > 3$

60. $\dfrac{2(5+3)+2\cdot2}{2(4-1)} > 1$

61. $8 \leq 4^2 - 2^2$

62. $10^2 - 8^2 > 6^2$

Write each word statement in symbols. See Example 5.

63. Fifteen is equal to five plus ten.

64. Twelve is equal to twenty minus eight.

65. Nine is greater than five minus four.

66. Ten is greater than six plus one.

67. Sixteen is not equal to nineteen.

68. Three is not equal to four.

69. Two is less than or equal to three.

70. Five is less than or equal to nine.

Write each statement in words and decide whether it is true *or* false. *(Hint: To compare fractions, write them with the same denominator.)*

71. $7 < 19$

72. $9 < 10$

73. $\dfrac{1}{3} \neq \dfrac{3}{10}$

74. $\dfrac{10}{7} \neq \dfrac{3}{2}$

75. $8 \geq 11$

76. $4 \leq 2$

Write each statement with the inequality symbol reversed. See Example 6.

77. $5 < 30$

78. $8 > 4$

79. $12 \geq 3$

80. $25 \leq 41$

The table shows the number of pupils per teacher in U.S. public schools in selected states in a recent year. Use this table to answer the questions in Exercises 81–84.

81. Which states had a number greater than 13.9?

82. Which states had a number that was at most 14.7?

83. Which states had a number not less than 13.9?

84. Which states had a number greater than 20.5?

State	Pupils per Teacher
Alaska	16.7
Texas	14.7
California	20.5
Wyoming	12.5
Maine	12.3
Idaho	17.8
Missouri	13.9

Source: National Center for Education Statistics.

9.2 ▷▷▷ Variables, Expressions, and Equations

A **variable** is a symbol, usually a letter such as x, y, or z, used to represent an unknown number. Different numbers can replace the variables to form specific statements. For example, in **Section 9.7** we will see that

$$a + b = b + a.$$

This statement is true for any replacements of the variables a and b, such as 2 for a and 5 for b, which gives the true statement

$$2 + 5 = 5 + 2.$$

An **algebraic expression** is a collection of numbers, variables, operation symbols, and grouping symbols, such as parentheses, square brackets, or fraction bars.

$$x + 5, \quad 2m - 9, \quad \text{and} \quad 8p^2 + 6(p - 2) \qquad \text{Algebraic expressions}$$

In $2m - 9$, the $2m$ means $2 \cdot m$, the product of 2 and m; $8p^2$ represents the product of 8 and p^2. Also, $6(p - 2)$ means the product of 6 and $p - 2$.

OBJECTIVE 1 Evaluate algebraic expressions, given values for the variables. An algebraic expression can have different numerical values for different values of the variables.

EXAMPLE 1 Evaluating Expressions

Find the value of each algebraic expression if $m = 5$ and then if $m = 9$.

(a) $8m$

$8m$	
$= 8 \cdot 5$	Let $m = 5$.
$= 40$	Multiply.

$8m$	
$= 8 \cdot 9$	Let $m = 9$.
$= 72$	Multiply.

(b) $3m^2$

$3m^2$	
$= 3 \cdot 5^2$	Let $m = 5$.
$= 3 \cdot 25$	Square 5.
$= 75$	Multiply.

$3m^2$	
$= 3 \cdot 9^2$	Let $m = 9$.
$= 3 \cdot 81$	Square 9.
$= 243$	Multiply.

CAUTION

In Example 1(b), $3m^2$ means $3 \cdot m^2$, **not** $3m \cdot 3m$. **Unless parentheses are used, the exponent refers only to the variable or number just before it.** Use parentheses to write $3m \cdot 3m$ with exponents as $(3m)^2$.

Work Problem **1** *at the Side.* ▶

EXAMPLE 2 Evaluating Expressions

Find the value of each expression if $x = 5$ and $y = 3$.

(a)

$2x + 5y$

> Follow the rules for order of operations.

$= 2 \cdot 5 + 5 \cdot 3$	Replace x with 5 and y with 3.
$= 10 + 15$	Multiply.
$= 25$	Add.

Continued on Next Page

OBJECTIVES

1 Evaluate algebraic expressions, given values for the variables.

2 Translate phrases from words to algebraic expressions.

3 Identify solutions of equations.

4 Translate sentences to equations.

5 Distinguish between *expressions* and *equations*.

1 Find the value of each expression if $p = 3$.

(a) $6p$

(b) $p + 12$

(c) $5p^2$

2 Find the value of each expression if $x = 6$ and $y = 9$.

(a) $4x + 7y$

(b) $\dfrac{4x - 2y}{x + 1}$

(c) $2x^2 + y^2$

(b) $\dfrac{9x - 8y}{2x - y}$

$= \dfrac{9 \cdot 5 - 8 \cdot 3}{2 \cdot 5 - 3}$ Replace x with 5 and y with 3.

$= \dfrac{45 - 24}{10 - 3}$ Multiply.

$= \dfrac{21}{7}$ Subtract.

$= 3$ Divide.

(c) $x^2 - 2y^2$

$\boxed{3^2 = 3 \cdot 3}$

$\boxed{5^2 = 5 \cdot 5}$

$= 5^2 - 2 \cdot 3^2$ Replace x with 5 and y with 3.

$= 25 - 2 \cdot 9$ Apply the exponents.

$= 25 - 18$ Multiply.

$= 7$ Subtract.

◀ *Work Problem* **2** *at the Side.*

OBJECTIVE **2** **Translate phrases from words to algebraic expressions.**

> **Problem-Solving Hint**
>
> Sometimes variables must be used to change word phrases into algebraic expressions. This process will be important later for solving applied problems.

EXAMPLE 3 **Using Variables to Write Word Phrases as Algebraic Expressions**

Write each word phrase as an algebraic expression, using x as the variable.

(a) The **sum** of a number and 9
"Sum" is the answer to an addition problem. This phrase translates as

$$x + 9, \quad \text{or} \quad 9 + x.$$

(b) 7 **minus** a number
"Minus" indicates subtraction, so the translation is

$$7 - x.$$

*Note that $x - 7$ would **not** be correct because we cannot subtract in either order and get the same results.*

(c) A number **subtracted from 12**
Since a number is subtracted *from* 12, write this as

$$12 - x.$$

Compare this result with "12 subtracted from a number," which is $x - 12$.

(d) The **product** of 11 and a number

$$11 \cdot x, \quad \text{or} \quad 11x$$

Continued on Next Page

(e) 5 **divided by** a number

$$5 \div x, \quad \text{or} \quad \frac{5}{x}$$

$\frac{x}{5}$ is not correct here.

(f) The **product of** 2 and the **difference** between a number and 8

$$2(x - 8)$$

We are multiplying 2 times another number. This number is the difference between some number and 8, written $x - 8$. Using parentheses around this difference, the final expression is $2(x - 8)$.

> **CAUTION**
>
> Notice that in translating the words "the difference between a number and 8" in Example 3(f), the order is kept the same: $x - 8$. "The difference between 8 and a number" would be written $8 - x$.

Work Problem **3** *at the Side.* ▶

OBJECTIVE **3** **Identify solutions of equations.** An **equation** is a statement that two expressions are equal. *Therefore, an equation always includes the equality symbol, =.*

$$x + 4 = 11, \quad 2y = 16, \quad \text{and} \quad 4p + 1 = 25 - p \qquad \text{Equations}$$

To **solve** an equation, we must find all values of the variable that make the equation true. Such values of the variable are called the **solutions** of the equation.

EXAMPLE 4 **Deciding Whether a Number Is a Solution of an Equation**

Decide whether the given number is a solution of the equation.

(a) Is 7 a solution of $5p + 1 = 36$?

$$5p + 1 = 36$$
$$5 \cdot 7 + 1 \stackrel{?}{=} 36 \qquad \text{Replace } p \text{ with 7.}$$
$$35 + 1 \stackrel{?}{=} 36 \qquad \text{Multiply.}$$
$$36 = 36 \qquad \text{True}$$

The number 7 is a solution of the equation.

(b) Is $\frac{14}{3}$ a solution of $9m - 6 = 32$?

$$9m - 6 = 32$$
$$9 \cdot \frac{14}{3} - 6 \stackrel{?}{=} 32 \qquad \text{Replace } m \text{ with } \frac{14}{3}.$$
$$42 - 6 \stackrel{?}{=} 32 \qquad \text{Multiply.}$$
$$36 = 32 \qquad \text{False}$$

The number $\frac{14}{3}$ is not a solution of the equation.

Work Problem **4** *at the Side.* ▶

---

3 Write each word phrase as an algebraic expression. Use x as the variable.

(a) The sum of 5 and a number

(b) A number minus 4

(c) A number subtracted from 48

(d) The product of 6 and a number

(e) 9 multiplied by the sum of a number and 5

4 Decide whether the given number is a solution of the equation.

(a) $p - 1 = 3; 2$

(b) $2k + 3 = 15; 7$

(c) $8p - 11 = 5; 2$

5 Write each word sentence as an equation. Use x as the variable.

(a) Three times the sum of a number and 13 is 19.

(b) Five times a number is subtracted from 21, giving 15.

6 Decide whether each is an *equation* or an *expression*.

(a) $2x + 5y - 7$

(b) $\dfrac{3x - 1}{5}$

(c) $2x + 5 = 7$

(d) $\dfrac{x}{y - 3} = 4x$

OBJECTIVE 4 Translate sentences to equations. We have seen how to translate phrases from words to expressions. Sentences given in words are translated as equations.

EXAMPLE 5 **Translating Sentences to Equations**

Write each word sentence as an equation. Use x as the variable.

(a) Twice the sum of a number and four is six.

"Twice" means two times. The word *is* suggests equals. With x representing the number, translate as follows.

$$\underset{\displaystyle 2\,\cdot}{\text{Twice}} \quad \underset{\displaystyle (x + 4)}{\text{the sum of a number and four}} \quad \underset{\displaystyle =}{\text{is}} \quad \underset{\displaystyle 6}{\text{six.}}$$

$$2(x + 4) = 6$$

(b) Nine more than five times a number is 49.

Use x to represent the unknown number. Start with $5x$ and then add 9 to it. The word *is* translates as $=$.

$$5x + 9 = 49$$

(c) Seven less than three times a number is eleven.

Here, 7 is *subtracted* from three times a number to get 11.

$$\underset{\displaystyle 3x}{\text{Three times a number}} \quad \underset{\displaystyle -}{\text{less}} \quad \underset{\displaystyle 7}{\text{seven}} \quad \underset{\displaystyle =}{\text{is}} \quad \underset{\displaystyle 11}{\text{eleven.}}$$

$$3x - 7 = 11$$

◀ *Work Problem* **5** *at the Side.*

OBJECTIVE 5 Distinguish between *expressions* and *equations*. Students often have trouble distinguishing between equations and expressions. *Remember that an equation is a sentence (with an = symbol); an expression is a phrase that represents a number.*

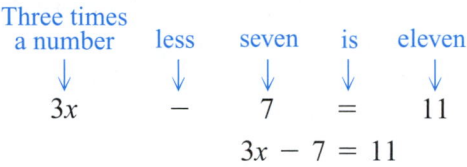

EXAMPLE 6 **Distinguishing between Equations and Expressions**

Decide whether each is an *equation* or an *expression*.

(a) $2x - 5y$

There is no equals symbol, so this is an expression.

(b) $2x = 5y$

Because there is an equals symbol with something on either side of it, this is an equation.

◀ *Work Problem* **6** *at the Side.*

9.2 ▶▶▶ Exercises

Fill in each blank with the correct response.

1. If $x = 3$, then the value of $x + 8$ is _____.

2. If $x = 1$ and $y = 2$, then the value of $5xy$ is _____.

3. "The sum of 13 and x" is represented by the expression _____. If $x = 3$, the value

of that expression is _____.

4. Will the equation $x = x + 5$ ever have a solution? _____.

5. $2x + 6$ is an _____, while $2x + 6 = 8$ is an _____.
    (equation/expression)   (equation/expression)

Exercises 6–10 cover some of the concepts introduced in this section. Give a short explanation for each.

6. Why is $2x^3$ not the same as $2x \cdot 2x \cdot 2x$? Explain, using an exponent to write $2x \cdot 2x \cdot 2x$.

7. If the words *more than* in Example 5(b) were changed to *less than,* how would the equation be changed?

8. Explain in your own words why, when evaluating the expression $4x^2$ for $x = 3$, 3 must be squared *before* multiplying by 4.

9. There are many pairs of values of x and y for which $2x + y$ will equal 6. Name two such pairs and describe how you determined them.

10. Suppose that for the equation $3x - y = 9$, the value of x is given as 4. What would be the corresponding value of y? How do you know this?

Find the numerical value of each expression if (a) x = 4 and (b) x = 6. See Example 1.

11. $4x^2$

    (a)   **(b)**

12. $5x^2$

    (a)   **(b)**

13. $\dfrac{3x - 5}{2x}$

    (a)   **(b)**

14. $\dfrac{4x - 1}{3x}$

    (a)   **(b)**

15. $\dfrac{6.459x}{2.7}$ (to the nearest thousandth)

    (a)   **(b)**

16. $\dfrac{0.74x^2}{0.85}$ (to the nearest thousandth)

    (a)   **(b)**

17. $3x^2 + x$

    (a)   **(b)**

18. $2x + x^2$

    (a)   **(b)**

Find the value of each expression if (a) x = 2 and y = 1 and (b) x = 1 and y = 5. See Example 2.

19. $3(x + 2y)$

    (a)   **(b)**

20. $2(2x + y)$

    (a)   **(b)**

21. $x + \dfrac{4}{y}$

    (a)   **(b)**

22. $y + \dfrac{8}{x}$

    (a)   **(b)**

23. $\dfrac{x}{2} + \dfrac{y}{3}$

    (a)   **(b)**

24. $\dfrac{x}{5} + \dfrac{y}{4}$

    (a)   **(b)**

25. $\dfrac{2x + 4y - 6}{5y + 2}$

    (a)   **(b)**

26. $\dfrac{4x + 3y - 1}{2x + y}$

    (a)   **(b)**

27. $2y^2 + 5x$

    (a)   **(b)**

28. $6x^2 + 4y$

    (a)   **(b)**

29. $\dfrac{3x + y^2}{2x + 3y}$

    (a)   **(b)**

30. $\dfrac{x^2 + 1}{4x + 5y}$

    (a)   **(b)**

31. $0.841x^2 + 0.32y^2$

(a)   (b)

32. $0.941x^2 + 0.2y^2$

(a)   (b)

Write each word phrase as an algebraic expression, using x as the variable.
See Example 3.

33. Twelve times a number

34. Thirteen added to a number

35. Two subtracted from a number

36. Eight subtracted from a number

37. One-third of a number, subtracted from seven

38. One-fifth of a number, subtracted from fourteen

39. The difference between twice a number and 6

40. The difference between 6 and half a number

41. 12 divided by the sum of a number and 3

42. The difference between a number and 5, divided by 12

43. The product of 6 and four less than a number

44. The product of 9 and five more than a number

45. In the phrase "four more than the product of a number and 6," does the word *and* signify the operation of addition? Explain.

46. Suppose that the directions on a test read "Solve the following expressions." How would you politely correct the person who wrote these directions?

Decide whether the given number is a solution of the equation. See Example 4.

47. Is 7 a solution of $x - 5 = 12$?

48. Is 10 a solution of $x + 6 = 15$?

49. Is 1 a solution of $5x + 2 = 7$?

50. Is 1 a solution of $3x + 5 = 8$?

51. Is $\frac{1}{5}$ a solution of $6x + 4x + 9 = 11$?

52. Is $\frac{12}{5}$ a solution of $2x + 3x + 8 = 20$?

53. Is 3 a solution of $2y + 3(y - 2) = 14$?

54. Is 2 a solution of $6a + 2(a + 3) = 14$?

55. Is $\frac{1}{3}$ a solution of $\frac{z + 4}{2 - z} = \frac{13}{5}$?

56. Is $\frac{13}{4}$ a solution of $\frac{x + 6}{x - 2} = \frac{37}{5}$?

57. Is 4.3 a solution of $3r^2 - 2 = 53.47$?

58. Is 3.7 a solution of $2x^2 + 1 = 28.38$?

Write each word sentence as an equation. Use x as the variable. See Example 5.

59. The sum of a number and 8 is 18.

60. A number minus three equals 1.

61. Five more than twice a number is 5.

62. The product of 2 and the sum of a number and 5 is 14.

63. Sixteen minus three-fourths of a number is 13.

64. The sum of six-fifths of a number and 2 is 14.

65. Three times a number is equal to 8 more than twice the number.

66. Twelve divided by a number equals $\frac{1}{3}$ times that number.

Identify each as an expression or an equation. See Example 6.

67. $3x + 2(x - 4)$

68. $5y - (3y + 6)$

69. $7t + 2(t + 1) = 4$

70. $9r + 3(r - 4) = 2$

Relating Concepts (Exercises 71–74) For Individual or Group Work

A **mathematical model** is an equation that describes the relationship between two quantities. For example, the life expectancy of Americans at birth can be approximated by the equation

$$y = 0.212x - 347,$$

where x is a year between 1943 and 2005 and y is age in years. (Source: Centers for Disease Control and Prevention.)

Use this model to approximate life expectancy (to the nearest tenth of a year) in each of the following years.

71. 1943

72. 1960

73. 1980

74. 2005

9.3 ▶▶▶ Real Numbers and the Number Line

OBJECTIVES

1 **Classify numbers and graph them on number lines.**

2 **Tell which of two real numbers is less than the other.**

3 **Find the opposite of a real number.**

4 **Find the absolute value of a real number.**

A **set** is a collection of objects. In mathematics, these objects are usually numbers. The objects that belong to the set, called **elements** of the set, are written between braces. For example, the set of numbers 1, 2, 3, 4, 5 is written

$$\{1, 2, 3, 4, 5\}.$$

OBJECTIVE **1** **Classify numbers and graph them on number lines.** The set of numbers used for counting is called the **natural numbers**.

Natural Numbers

$$\{1, 2, 3, 4, 5, \ldots\}$$

In **Chapter 1,** we introduced the set of **whole numbers.**

Whole Numbers

$$\{0, 1, 2, 3, 4, 5, \ldots\}$$

These numbers, along with many others, can be represented on a **number line** like the one in Figure 1. We draw a number line by choosing any point on the line and labeling it 0. Choose any point to the right of 0 and label it 1. The distance between 0 and 1 gives a unit of measure used to locate other points, as shown in Figure 1. The points labeled in Figure 1 correspond to the first few whole numbers.

The indicated points correspond to natural numbers.

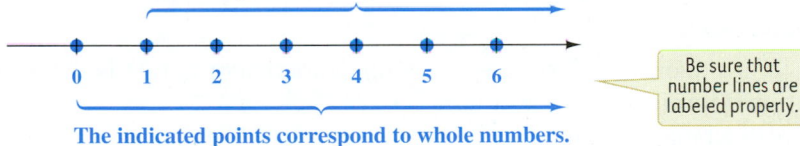

Be sure that number lines are labeled properly.

The indicated points correspond to whole numbers.

Figure 1

The natural numbers are located to the right of 0 on the number line. But numbers may also be placed to the left of 0. For each natural number we can place a corresponding number to the left of 0. These numbers, written $-1, -2, -3, -4$, and so on, are shown in Figure 2 on the next page. Each is the **opposite,** or **negative,** of a natural number. The natural numbers, their opposites, and 0 form a new set of numbers called the **integers.**

Integers

$$\{\ldots, -3, -2, -1, 0, 1, 2, 3, \ldots\}$$

1 Use an integer to express the boldface italic number(s) in each application.

(a) Erin discovers that she has spent $**53** more than she has in her checking account.

(b) The record high Fahrenheit temperature in the United States was **134**° in Death Valley, California, on July 10, 1913. (*Source: World Almanac and Book of Facts*.)

(c) A football team gained **5** yd, then lost **10** yd on the next play.

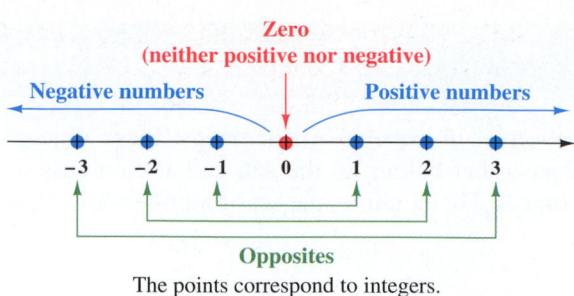

Zero
(neither positive nor negative)

Negative numbers   Positive numbers

Opposites
The points correspond to integers.

Figure 2

There are many practical applications of negative numbers. For example, a Fahrenheit temperature on a cold January day might be $-10°$, and a business that spends more than it takes in has a negative "profit" (a loss).

EXAMPLE 1   **Using Negative Numbers in Applications**

Use an integer to express the boldface italic number in each application.

(a) The lowest Fahrenheit temperature ever recorded in meteorological records was **129**° below zero at Vostok, Antarctica, on July 21, 1983. (*Source: World Almanac and Book of Facts.*)
Use -129 because "below zero" indicates a negative number.

(b) The shore surrounding the Dead Sea is **1348** ft below sea level. (*Source: World Almanac and Book of Facts.*)
Again, "below sea level" indicates a negative number, -1348.

◀ Work Problem **1** at the Side.

Not all numbers are integers. For example, $\frac{1}{2}$ is not; it is a number halfway between the integers 0 and 1. Also, $3\frac{1}{4}$ is not an integer. These numbers and others that are quotients of integers are *rational numbers*. (The name comes from the word *ratio*, which indicates a quotient.)

Rational Numbers

$\{x \,|\, x$ is a quotient of two integers, with denominator not 0$\}$ is the set of **rational numbers.**

(Read the part in the braces as "the set of all numbers x such that x is a quotient of two integers, with denominator not 0.")

Note

The set symbolism used in the definition of rational numbers,

$$\{x \,|\, x \text{ has a certain property}\},$$

is called **set-builder notation.** This notation is convenient to use when it is not possible to list all the elements of a set.

Since any integer can be written as the quotient of itself and 1, **all integers are rational numbers.** For example, $-5 = \frac{-5}{1}$. A decimal number that comes to an end (terminates), such as 0.23, is a rational number. For example, $0.23 = \frac{23}{100}$. Decimal numbers that repeat in a fixed block of digits, such as $0.3333\ldots = 0.\overline{3}$ and $0.454545\ldots = 0.\overline{45}$, are also rational numbers. For example, $0.\overline{3} = \frac{1}{3}$.

As shown in Figures 1 and 2 on the preceding pages, to **graph** a number, we place a dot on the number line at the point that corresponds to the number. The number is called the **coordinate** of the point. Think of the graph of a set of numbers as a picture of the set.

EXAMPLE 2 **Graphing Rational Numbers**

Graph each number on the number line.

$$-\frac{3}{2},\quad -\frac{2}{3},\quad \frac{1}{2},\quad 1\frac{1}{3},\quad \frac{23}{8},\quad 3\frac{1}{4}$$

To locate the improper fractions on the number line, write them as mixed numbers or decimals. The graph is shown in Figure 3.

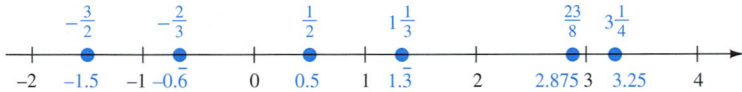

Figure 3

Work Problem **2** *at the Side.* ▶

The number system contains numbers that are not rational. For example, a square that measures one unit on a side has a diagonal whose length is the square root of 2, written $\sqrt{2}$. See Figure 4. It can be shown that $\sqrt{2}$ cannot be written as a quotient of integers. Because of this, $\sqrt{2}$ is not rational; it is *irrational*. Other examples of irrational numbers are $\sqrt{3}, \sqrt{7}, -\sqrt{10}$, and π (the ratio of the *circumference* of a circle to its *diameter*).

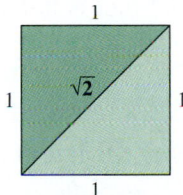

Figure 4

> **Irrational Numbers**
>
> $\{x \mid x \text{ is a nonrational number represented by a point on the number line}\}$ is the set of **irrational numbers.**

The decimal form of an irrational number neither terminates nor repeats. Irrational numbers are discussed in **Chapter 16.**

Both rational and irrational numbers can be represented by points on the number line and are called **real numbers.**

> **Real Numbers**
>
> $\{x \mid x \text{ is a rational or an irrational number}\}$ is the set of **real numbers.**

All the numbers mentioned so far are real numbers.* The relationships between the various types of numbers are shown in Figure 5 on the next page. Every real number is either a rational number or an irrational number.

---

* An example of a number that is not real is the square root of a negative number, such as $\sqrt{-4}$.

---

2 Graph each number on the number line.

$$-3,\ \frac{17}{8},\ -2.75,\ 1\frac{1}{2},\ -\frac{3}{4}$$

ANSWER

2.

3 Tell whether each statement is *true* or *false*.

(a) $-2 < 4$

(b) $6 > -3$

(c) $-9 < -12$

(d) $-4 \geq -1$

(e) $-6 \leq 0$

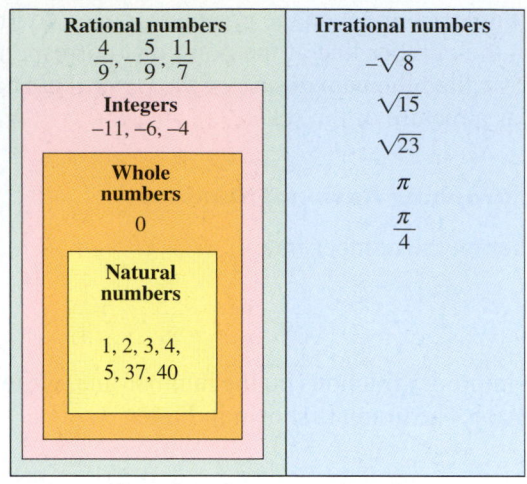

All numbers shown are real numbers.

Figure 5

OBJECTIVE 2 **Tell which of two real numbers is less than the other.** Given any two whole numbers, we can tell which number is less than the other. But what about two negative numbers, as in the set of integers? Moving from 0 to the right along a number line, the positive numbers corresponding to the points on the number line *increase*. For example, $8 < 12$, and 8 is to the left of 12 on a number line. We extend this ordering to all real numbers.

Ordering of the Real Numbers

For any two real numbers a and b, **a is less than b** if a is to the left of b on a number line.

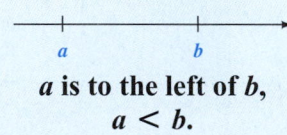

a is to the left of b,
$a < b$.

Thus, any negative number is less than 0, and any negative number is less than any positive number. Also, 0 is less than any positive number.

EXAMPLE 3 **Determining the Order of Real Numbers**

Is it true that $-3 < -1$?

To find out, locate -3 and -1 on a number line, as shown in Figure 6. Because -3 is to the left of -1 on the number line, -3 is less than -1. The statement $-3 < -1$ is true.

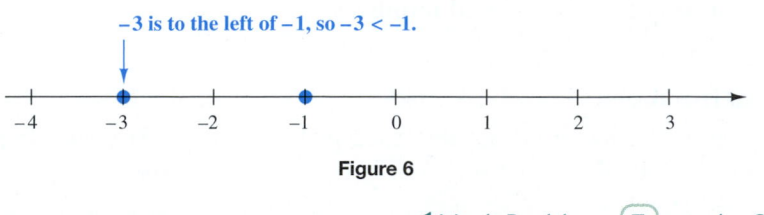

Figure 6

◀ Work Problem **3** at the Side.

We can also say that, for any real numbers a and b, **a is greater than b** if a is to the right of b on the number line.

OBJECTIVE **3** **Find the opposite of a real number.** Earlier, we saw that every positive integer has a negative integer that is its opposite, or negative. This is true for every real number except 0, which is its own opposite.* A characteristic of pairs of opposites is that they are the same distance from 0 on the number line but in opposite directions. See Figure 7.

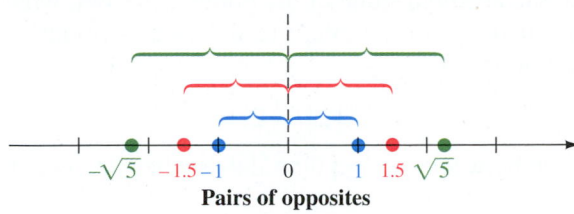

Pairs of opposites

Figure 7

We indicate the opposite of a number by writing the symbol $-$ in front of the number. For example, the opposite of 7 is -7 (read "negative 7"). We could write the opposite of -4 as $-(-4)$, but we know that 4 is the opposite of -4. Since a number can have only one opposite, $-(-4)$ and 4 must represent the same number, so

$$-(-4) = 4.$$

This idea can be generalized.

> **Double Negative Rule**
>
> For any real number a,
>
> $$-(-a) = a.$$

The following chart shows several numbers and their opposites.

Number	Opposite
-4	$-(-4)$, or 4
0	0
5	-5
$-\frac{2}{3}$	$\frac{2}{3}$
0.52	-0.52

The chart suggests the following rule.

> *Except for 0, the opposite of a number is found by changing the sign of the number.*

Work Problem **4** *at the Side.* ▶

---

*The opposite (or negative) of a number is also called the *additive inverse* of the number, as we shall see in **Section 1.7**.

4 Find the opposite of each number.

(a) 6

(b) 15

(c) -9

(d) -12

(e) 0

5 Simplify.

(a) $|-6|$

(b) $|9|$

(c) $-|15|$

(d) $-|-9|$

(e) $|9 - 4|$

(f) $-|32 - 2|$

OBJECTIVE **4** **Find the absolute value of a real number.** As previously mentioned, opposites are numbers the same distance from 0 on the number line but on opposite sides of 0. Another way to say this is to say that opposites have the same *absolute value*. The **absolute value** of a number is the undirected distance between 0 and the number on the number line. The symbol for the absolute value of the number a is $|a|$, which is read **"the absolute value of a."** For example, the distance between 2 and 0 on the number line is 2 units, so

$$|2| = 2.$$

Also, the distance between -2 and 0 on the number line is 2, so

$$|-2| = 2.$$

Because distance is a physical measurement, which is never negative, we can make the following statement.

> *The absolute value of a number can never be negative.*

For example,

$$|12| = 12 \quad \text{and} \quad |-12| = 12$$

because both 12 and -12 lie at a distance of 12 units from 0 on the number line. Since the distance of 0 from 0 is 0 units, we have

$$|0| = 0.$$

EXAMPLE 4 **Evaluating Absolute Value**

Simplify by finding the absolute value.

(a) $|5| = 5$

(b) $|-5| = 5$

(c) $-|5| = -(5) = -5$

(d) $-|-5| = -(5) = -5$ Replace $|-5|$ with 5.

(e) $|8 - 5|$

Simplify within the absolute value bars first.

$$|8 - 5| = |3| = 3$$

(f) $-|8 - 5| = -|3| = -3$

(g) $-|12 - 3| = -|9| = -9$

Parts (e)–(g) in Example 4 show that absolute value bars also act as grouping symbols. We must perform any operations within absolute value bars before finding the absolute value.

◀ *Work Problem* **5** *at the Side.*

ANSWERS

5. (a) 6   **(b)** 9   **(c)** -15 **(d)** -9
    (e) 5   **(f)** -30

9.3 ▶▶▶ Exercises

In Exercises 1–6, give an example of a number that satisfies each given condition.

1. An integer between 3.6 and 4.6

2. A rational number between 2.8 and 2.9

3. A whole number that is not positive and is less than 1

4. A whole number greater than 3.5

5. An irrational number that is between $\sqrt{12}$ and $\sqrt{14}$

6. A real number that is neither negative nor positive

*List all numbers from each set that are **(a)** natural numbers, **(b)** whole numbers,*
(c) integers, (d) rational numbers, (e) irrational numbers, (f) real numbers.

7. $\left\{ -9, -\sqrt{7}, -1\frac{1}{4}, -\frac{3}{5}, 0, \sqrt{5}, 3, 5.9, 7 \right\}$

8. $\left\{ -5.3, -5, -\sqrt{3}, -1, -\frac{1}{9}, 0, 1.2, 4, \sqrt{12} \right\}$

Use an integer to express each boldface italic number representing a change in the
following applications. See Example 1.

9. Between July 1, 2004, and July 1, 2005, the population of the United States increased by approximately ***2,845,000.*** (*Source:* U.S. Census Bureau.)

10. From 2000 to 2005, the mean SAT verbal score for Massachusetts residents increased by ***9,*** while the mathematics score increased by ***14.*** (*Source:* The College Board.)

11. From 1995 to 2005, the number of cable TV systems in the United States went from 11,218 to 8409, representing a decrease of ***2809.*** (*Source: Television and Cable Factbook.*)

12. In 1935, there were 15,295 banks in the United States. By 2004, the number was 8975, representing a decrease of ***6320*** banks. (*Source:* Federal Deposit Insurance Corporation.)

Graph each group of numbers on a number line. See Example 2.

13. $0, 3, -5, -6$

14. $2, 6, -2, -1$

15. $-2, -6, -4, 3, 4$

16. $-5, -3, -2, 0, 4$

17. $\frac{1}{4}, 2\frac{1}{2}, -3\frac{4}{5}, -4, -\frac{13}{8}$

18. $5\frac{1}{4}, \frac{41}{9}, -2\frac{1}{3}, 0, -3\frac{2}{5}$

Select the lesser number in each pair. See Example 3.

19. $-11, -4$ **20.** $-9, -16$ **21.** $-21, 1$ **22.** $-57, 3$

23. $0, -100$ **24.** $-215, 0$ **25.** $-\dfrac{2}{3}, -\dfrac{1}{4}$ **26.** $-\dfrac{3}{8}, -\dfrac{9}{16}$

Decide whether each statement is true *or* false. *See Example 3.*

27. $8 < -16$ **28.** $12 < -24$ **29.** $-3 < -2$ **30.** $-10 < -9$

*For each number, **(a)** find its opposite and **(b)** find its absolute value.*

31. -2 **32.** -8 **33.** 6

   (a)   **(b)**   **(a)**   **(b)**   **(a)**   **(b)**

34. 11 **35.** $-\dfrac{3}{4}$ **36.** $-\dfrac{1}{3}$

   (a)   **(b)**   **(a)**   **(b)**   **(a)**   **(b)**

Simplify. See Example 4.

37. $|-7|$ **38.** $|-3|$ **39.** $-|12|$ **40.** $-|23|$

41. $-\left|-\dfrac{2}{3}\right|$ **42.** $-\left|-\dfrac{4}{5}\right|$ **43.** $|13 - 4|$ **44.** $|8 - 7|$

Decide whether each statement is true *or* false.

45. $|-8| < 7$ **46.** $|-6| \geq -|6|$ **47.** $4 \leq |4|$ **48.** $-|-3| > 2$

49. Students often say "The absolute value of a number is always positive." Is this true? If not, explain.

50. If the absolute value of a number is equal to the number itself, what must be true about the number?

To answer the questions in Exercises 51–54, refer to the table, which gives the changes in producer price indexes for two recent periods.

51. What commodity for which period represents the greatest decrease?

Commodity	Change from 2002 to 2003	Change from 2003 to 2004
Farm products	12.6	10.6
Gasoline	23.3	24.7
Machinery and equipment	−0.8	0.2
Iron and steel	6.5	33.7
Electronic components and accessories	−1.7	−2.2

Source: U.S Bureau of Labor Statistics.

52. What commodity for which period represents the least change?

53. Which has lesser absolute value, the change for electronic components and accessories from 2002 to 2003 or from 2003 to 2004?

54. Which has greater absolute value, the change for machinery and equipment from 2002 to 2003 or from 2003 to 2004?

9.4 ▶▶▶ Adding Real Numbers

OBJECTIVES

1 Add two numbers with the same sign.

2 Add numbers with different signs.

3 Add mentally.

4 Use the rules for order of operations with real numbers.

5 Translate words and phrases that indicate addition.

OBJECTIVE **1** **Add two numbers with the same sign.** We can use the number line to explain addition of real numbers. Later, we will give the rules for addition. Recall that the answer to an addition problem is called the **sum.**

EXAMPLE 1 Adding Two Positive Numbers on a Number Line

Use a number line to find the sum $2 + 3$.

Add the positive numbers 2 and 3 by starting at 0 and drawing an arrow two units to the *right,* as shown in Figure 8. This arrow represents the number 2 in the sum $2 + 3$. Next, from the right end of this arrow draw another arrow three units to the right. The number below the end of this second arrow is 5, so $2 + 3 = 5$.

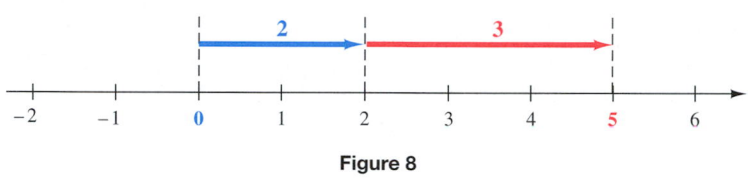

Figure 8

EXAMPLE 2 Adding Two Negative Numbers on a Number Line

Use a number line to find the sum $-2 + (-4)$. (Parentheses are placed around the -4 to avoid the confusing use of $+$ and $-$ next to each other.)

To add the negative numbers -2 and -4 on the number line, we start at 0 and draw an arrow two units to the *left,* as shown in Figure 9. From the left end of this first arrow, we draw a second arrow four units to the left. We draw the arrow to the left to represent the addition of the *negative* number, -4. The number below the end of this second arrow is -6, so $-2 + (-4) = -6$.

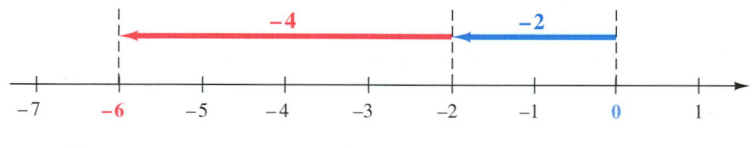

Figure 9

Work Problem **1** *at the Side.* ▶

In Example 2, we found that the sum of the two negative numbers -2 and -4 is a negative number whose distance from 0 is the sum of the distance of -2 from 0 and the distance of -4 from 0. ***That is, the sum of two negative numbers is the opposite of the sum of their absolute values.***

$$-2 + (-4) = -(|-2| + |-4|) = -(2 + 4) = -6$$

Adding Real Numbers with the Same Sign

To add two numbers with the same sign, add the absolute values of the numbers. Give the result the same sign as the numbers being added.

Example: $-4 + (-3) = -7$

---

1 Use a number line to find each sum.

(a) $1 + 4$

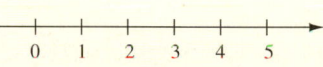

(b) $-2 + (-5)$

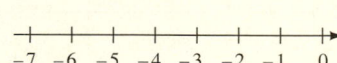

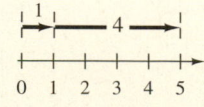
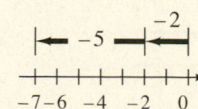

2 Find each sum.

(a) $-7 + (-3)$

(b) $-12 + (-18)$

(c) $-15 + (-4)$

3 Use a number line to find each sum.

(a) $6 + (-3)$

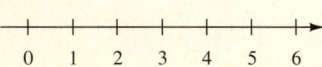

(b) $-5 + 1$

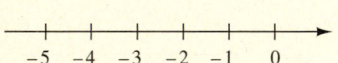

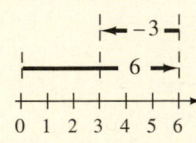
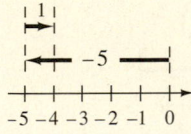

EXAMPLE 3 Adding Two Negative Numbers

Find each sum.

(a) $-2 + (-9) = -11$ The sum of two negative numbers is negative.

(b) $-8 + (-12) = -20$

(c) $-15 + (-3) = -18$

◀ Work Problem **2** at the Side.

OBJECTIVE 2 Add numbers with different signs. We use the number line again to illustrate the sum of a positive number and a negative number.

EXAMPLE 4 Adding Numbers with Different Signs

Use the number line to find the sum $-2 + 5$.

    We find the sum $-2 + 5$ on the number line by starting at 0 and drawing an arrow two units to the left. From the left end of this arrow, we draw a second arrow five units to the right, as shown in Figure 10. The number below the end of this second arrow is 3, so $-2 + 5 = 3$.

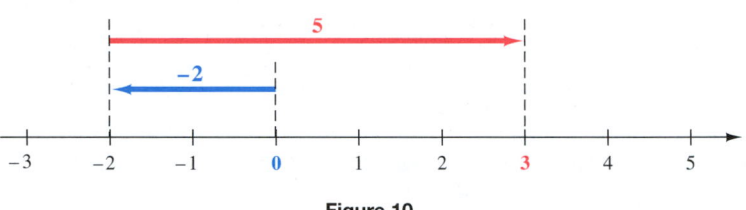

Figure 10

◀ Work Problem **3** at the Side.

    Addition of numbers with different signs also can be defined using absolute value.

> **Adding Real Numbers with Different Signs**
>
> To add numbers with different signs, find the absolute values of the numbers, and subtract the lesser absolute value from the greater. Give the answer the same sign as the number with the greater absolute value.
>
> *Example:* $-12 + 6 = -6$

    For example, to add -12 and 5, we find their absolute values:

$$|-12| = 12 \quad \text{and} \quad |5| = 5.$$

Then we find the difference between these absolute values: $12 - 5 = 7$. Since $|-12| > |5|$, the sum will be negative, so

$$-12 + 5 = -7.$$

> 🖩 **Calculator Tip**   The ⊝ or ⊕⁄⊝ key is used to input a negative number in some scientific calculators. Try using your calculator to add negative numbers.

OBJECTIVE **3** **Add mentally.** While a number line is useful in showing the rules for addition, it is important to be able to find sums mentally.

EXAMPLE 5 **Adding a Positive Number and a Negative Number**

Check each answer, trying to work the addition mentally. If you have trouble, use a number line.

(a) $7 + (-4) = 3$

(b) $-8 + 12 = 4$

(c) $-\dfrac{1}{2} + \dfrac{1}{8} = -\dfrac{4}{8} + \dfrac{1}{8} = -\dfrac{3}{8}$

> Remember to find a common denominator.

(d) $\dfrac{5}{6} + \left(-1\dfrac{1}{3}\right) = \dfrac{5}{6} + \left(-\dfrac{4}{3}\right) = \dfrac{5}{6} + \left(-\dfrac{8}{6}\right) = -\dfrac{3}{6} = -\dfrac{1}{2}$

(e) $-4.6 + 8.1 = 3.5$

(f) $-16 + 16 = 0$

Work Problem **4** *at the Side.* ▶

The rules for adding signed numbers are summarized below.

Adding Signed Numbers

Same sign Add the absolute values of the numbers. Give the sum the same sign as the numbers being added.

Different signs Find the absolute values of the numbers, and subtract the lesser absolute value from the greater. Give the answer the sign of the number having the greater absolute value.

OBJECTIVE **4** **Use the rules for order of operations with real numbers.** Sometimes a problem involves square brackets, [ ]. As we mentioned earlier, brackets are treated just like parentheses. We do the calculations inside the brackets until a single number is obtained. Remember to use the rules for order of operations given in **Section 9.1** for adding more than two numbers.

EXAMPLE 6 **Adding with Brackets**

Find each sum.

> Start here.

(a) $-3 + [4 + (-8)]$
$= -3 + (-4)$
$= -7$

(b) $8 + [(-2 + 6) + (-3)]$
$= 8 + [4 + (-3)]$
$= 8 + 1$
$= 9$

Work Problem **5** *at the Side.* ▶

OBJECTIVE **5** **Translate words and phrases that indicate addition.** We now interpret words and phrases that involve addition. Problem solving often requires translating such words and phrases into symbols. We began this process in **Section 9.1**.

---

4 Check each answer, trying to work the addition mentally. If you have trouble, use a number line.

(a) $-8 + 2 = -6$

(b) $-15 + 4 = -11$

(c) $17 + (-10) = 7$

(d) $\dfrac{3}{4} + \left(-1\dfrac{3}{8}\right) = -\dfrac{5}{8}$

(e) $-9.5 + 3.8 = -5.7$

(f) $42 + (-42) = 0$

5 Find each sum.

(a) $2 + [7 + (-3)]$

(b) $6 + [(-2 + 5) + 7]$

(c) $-9 + [-4 + (-8 + 6)]$

6 Write a numerical expression for each phrase, and simplify the expression.

(a) 4 more than -12

(b) The sum of 6 and -7

(c) -12 added to -31

(d) 7 increased by the sum of 8 and -3

The word *sum* indicates addition. The table lists other key words and phrases that also indicate addition.

Word or Phrase	Example	Numerical Expression and Simplification
Sum of	The **sum of** -3 and 4	$-3 + 4 = 1$
Added to	5 **added to** -8	$-8 + 5 = -3$
More than	12 **more than** -5	$(-5) + 12 = 7$
Increased by	-6 **increased by** 13	$-6 + 13 = 7$
Plus	3 **plus** 14	$3 + 14 = 17$

EXAMPLE 7 **Translating Words and Phrases (Addition)**

Write a numerical expression for each phrase, and simplify the expression.

(a) The **sum of** -8 and 4 and 6

$$-8 + 4 + 6$$

To simplify, add in order from left to right, to obtain

$$-4 + 6, \quad \text{or} \quad 2.$$

(b) 3 **more than** -5, **increased by** 12

$$-5 + 3 + 12 \quad \text{simplifies to} \quad -2 + 12, \quad \text{or} \quad 10.$$

◀ *Work Problem* **6** *at the Side.*

Gains (or increases) and losses (or decreases) sometimes appear in applied problems. When they do, the gains may be interpreted as positive numbers and the losses as negative numbers.

7 Solve the problem.

A football team lost 8 yd on first down, lost 5 yd on second down, and then gained 7 yd on third down. How many yards did the team gain or lose altogether on these plays?

EXAMPLE 8 **Interpreting Gains and Losses**

The Carolina Panthers football team gained 3 yd on first down, lost 12 yd on second down, and then gained 13 yd on third down. How many yards did the team gain or lose altogether on these plays?

The gains are represented by positive numbers and the loss by a negative number.

$$3 + (-12) + 13$$
$$= [3 + (-12)] + 13 \quad \text{Add from left to right.}$$
$$= (-9) + 13$$
$$= 4$$

The team gained 4 yd altogether on these plays.

◀ *Work Problem* **7** *at the Side.*

9.4 ▶▶▶ Exercises

By the order of operations, what is the first step you would use to simplify each expression?

1. $4[3(-2 + 5) - 1]$

2. $[-4 + 7(-6 + 2)]$

3. $9 + ([-1 + (-3)] + 5)$

4. $[(-8 + 4) + (-6)] + 5$

Find each sum. See Examples 1–6.

5. $6 + (-4)$

6. $8 + (-5)$

7. $12 + (-15)$

8. $4 + (-8)$

9. $-7 + (-3)$

10. $-11 + (-4)$

11. $-10 + (-3)$

12. $-16 + (-7)$

13. $-12.4 + (-3.5)$

14. $-21.3 + (-2.5)$

15. $10 + [-3 + (-2)]$

16. $13 + [-4 + (-5)]$

17. $5 + [14 + (-6)]$

18. $7 + [3 + (-14)]$

19. $-3 + [5 + (-2)]$

20. $-7 + [10 + (-3)]$

21. $-8 + [3 + (-1) + (-2)]$

22. $-7 + [5 + (-8) + 3]$

23. $\dfrac{9}{10} + \left(-\dfrac{3}{5}\right)$

24. $\dfrac{5}{8} + \left(-\dfrac{17}{12}\right)$

25. $-\dfrac{1}{6} + \dfrac{2}{3}$

26. $-\dfrac{6}{25} + \dfrac{19}{20}$

27. $2\dfrac{1}{2} + \left(-3\dfrac{1}{4}\right)$

28. $-4\dfrac{3}{8} + 6\dfrac{1}{2}$

29. $7.8 + (-9.4)$

30. $14.7 + (-10.1)$

31. $-7.1 + [3.3 + (-4.9)]$

32. $-9.5 + [-6.8 + (-1.3)]$

33. $[-8 + (-3)] + [-7 + (-7)]$

34. $[-5 + (-4)] + [9 + (-2)]$

35. $\left(-\dfrac{1}{2} + 0.25\right) - \left(-\dfrac{3}{4} + 0.75\right)$

36. $\left(-\dfrac{3}{2} - 0.75\right) - \left(2.25 - \dfrac{1}{2}\right)$

Perform each operation, and then determine whether the statement is true *or* false. *Try to do all work mentally. See Examples 5 and 6.*

37. $-11 + 13 = 13 + (-11)$

38. $16 + (-9) = -9 + 16$

39. $-10 + 6 + 7 = -3$

40. $-12 + 8 + 5 = -1$

41. $\dfrac{7}{3} + \left(-\dfrac{1}{3}\right) + \left(-\dfrac{6}{3}\right) = 0$

42. $-\dfrac{3}{2} + 1 + \dfrac{1}{2} = 0$

43. $|-8 + 10| = -8 + (-10)$

44. $|-4 + 6| = -4 + (-6)$

45. $2\dfrac{1}{5} + \left(-\dfrac{6}{11}\right) = -\dfrac{6}{11} + 2\dfrac{1}{5}$

46. $-1\dfrac{1}{2} + \dfrac{5}{8} = \dfrac{5}{8} + \left(-1\dfrac{1}{2}\right)$

47. $-7 + [-5 + (-3)] = [(-7) + (-5)] + 3$

48. $6 + [-2 + (-5)] = [(-4) + (-2)] + 5$

Relating Concepts (Exercises 49—52) For Individual or Group Work

Recall the rules for adding signed numbers introduced in this section, and **work Exercises 49–52 in order.**

49. Suppose that the sum of two numbers is negative, and you know that one of the numbers is positive. What can you conclude about the other number?

50. If you are solving the equation $x + 5 = -7$ from a set of numbers, why could you immediately eliminate any positive numbers as possible solutions? (Remember how you answered Exercise 49.)

51. Suppose that the sum of two numbers is positive, and you know that one of the numbers is negative. What can you conclude about the other number?

52. If you are solving the equation $x + (-8) = 2$ from a set of numbers, why could you immediately eliminate any negative numbers as possible solutions? (Remember how you answered Exercise 51.)

---

53. In your own words, explain how to add two negative numbers.

54. In your own words, explain how to add a positive number and a negative number. Give two cases.

Write a numerical expression for each phrase, and simplify the expression. See Example 7.

55. The sum of −5 and 12 and 6

56. The sum of −3 and 5 and −12

57. 14 added to the sum of −19 and −4

58. −2 added to the sum of −18 and 11

59. The sum of −4 and −10, increased by 12

60. The sum of −7 and −13, increased by 14

61. $\frac{2}{7}$ more than the sum of $\frac{5}{7}$ and $-\frac{9}{7}$

62. 0.85 more than the sum of −1.25 and −4.75

Solve each problem. See Example 8.

63. Nathaniel owed his older sister Jenna $24 for his share of the bill when they took their mother out to dinner for her birthday. He later borrowed $38 from his younger sister Ilana to buy two DVDs. What positive or negative number represents Nathaniel's financial situation with his siblings?

64. Bonika's checking account balance is $54.00. She then takes a gamble by writing a check for $89.00. What is her new balance? (Write the balance as a signed number.)

65. The surface, or rim, of a canyon is at altitude 0. On a hike down into the canyon, a party of hikers stops for a rest at 130 m below the surface. They then descend another 54 m. What is their new altitude? (Write the altitude as a signed number.)

66. A pilot announces to the passengers that the current altitude of their plane is 34,000 ft. Because of some unexpected turbulence, the pilot is forced to descend 2100 ft. What is the new altitude of the plane? (Write the altitude as a signed number.)

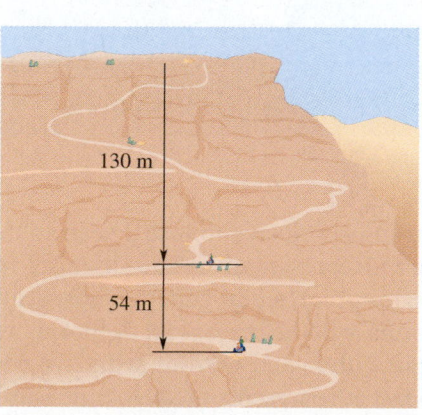

130 m

54 m

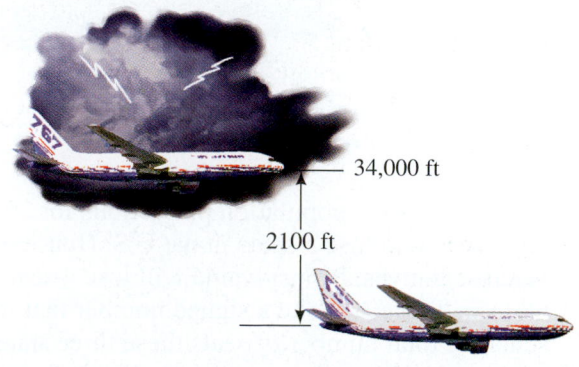

34,000 ft

2100 ft

67. On three consecutive passes, Drew Brees of the New Orleans Saints passed for a gain of 6 yd, was sacked for a loss of 12 yd, and passed for a gain of 43 yd. What positive or negative number represents the total net yardage for the plays?

68. On a series of three consecutive running plays, Peyton Manning of the Indianapolis Colts gained 4 yd, lost 3 yd, and lost 2 yd. What positive or negative number represents his total net yardage for the series of plays?

69. The lowest temperature ever recorded in Arkansas was −29°F. The highest temperature ever recorded there was 149°F more than the lowest. What was this highest temperature? (*Source: World Almanac and Book of Facts.*)

70. On January 23, 1943, the temperature rose 49°F in two minutes in Spearfish, South Dakota. If the starting temperature was −4°F, what was the temperature two minutes later?

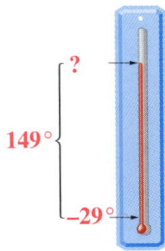

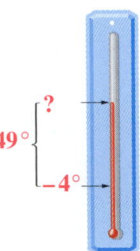

71. Dana Weightman owes $153 to a credit card company. She makes a $14 purchase with the card, and then pays $60 on the account. What is her current balance as a signed number?

72. A female polar bear weighed 660 lb when she entered her winter den. She lost 45 lb during each of the first two months of hibernation, and another 205 lb before leaving the den with her two cubs in March. How much did she weigh when she left the den?

73. Based on census population projections for 2020, New York will lose 5 seats in the U.S. House of Representatives, Pennsylvania will lose 4 seats, and Ohio will lose 3. Write a signed number that represents the total number of seats these three states are projected to lose. (*Source:* Population Reference Bureau.)

74. Michigan is projected to lose 3 seats in the U.S. House of Representatives and Illinois 2 in 2020. The states projected to gain the most seats are California with 9, Texas with 5, Florida with 3, Georgia with 2, and Arizona with 2. Write a signed number that represents the algebraic sum of these changes. (*Source:* Population Reference Bureau.)

9.5 ▶▶▶ Subtracting Real Numbers

OBJECTIVES

1 Find a difference.

2 Use the definition of subtraction.

3 Work subtraction problems that involve brackets.

4 Translate words and phrases that indicate subtraction.

OBJECTIVE **1** **Find a difference.**   In the operation $a - b$, a is called the **minuend** and b is called the **subtrahend.** As we mentioned earlier, the answer to a subtraction problem is called a **difference.**

Differences between signed numbers can be found by using a number line. Addition and subtraction are opposite operations. Thus, because *addition* of a positive number on the number line is shown by drawing an arrow to the *right,* *subtraction* of a positive number is shown by drawing an arrow to the *left.*

EXAMPLE 1   **Subtracting with the Number Line**

Use a number line to find the difference $7 - 4$.

To find the difference $7 - 4$ on the number line, begin at 0 and draw an arrow 7 units to the *right.* From the right end of this arrow, draw an arrow 4 units to the *left,* as shown in Figure 11. The number at the end of the second arrow shows that $7 - 4 = 3$.

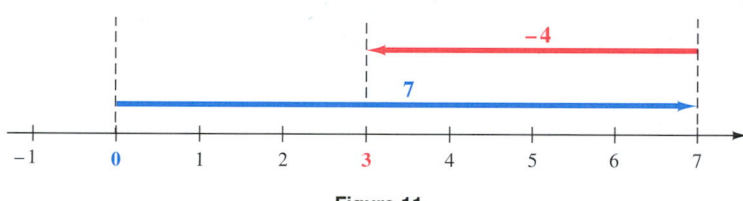

Figure 11

Work Problem **1** *at the Side.* ▶

OBJECTIVE **2** **Use the definition of subtraction.**   The procedure used in Example 1 to find $7 - 4$ is exactly the same procedure that would be used to find $7 + (-4)$, so

$$7 - 4 = 7 + (-4).$$

This shows that *subtracting* a positive number from a larger positive number is the same as *adding* the opposite of the smaller number to the larger. We use this idea to define subtraction for all real numbers.

> **Subtraction**
> For any real numbers a and b,
> $$a - b = a + (-b).$$
> *Example:* $4 - 9 = 4 + (-9) = -5$

To subtract b from a, add the opposite (or negative) of b to a. In other words, change the subtrahend to its opposite and add.

> **Subtracting Signed Numbers**
> *Step 1*   Change the subtraction symbol to addition and change the sign of the subtrahend.
>
> *Step 2*   Add, as in the previous section.

1 Use the number line to find each difference.

(a) $5 - 1$

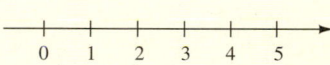

(b) $6 - 2$

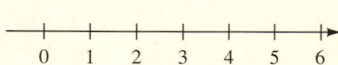

ANSWERS

1. **(a)** $5 - 1 = 4$

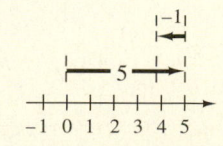

(b) $6 - 2 = 4$

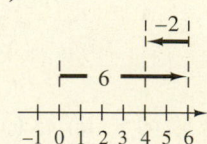

2 Subtract.

(a) $6 - 10$

(b) $-2 - 4$

(c) $3 - (-5)$

(d) $-8 - (-12)$

(e) $\dfrac{5}{4} - \left(-\dfrac{3}{7}\right)$

EXAMPLE 2 Using the Definition of Subtraction

Subtract.

> Change − to +.
> No change
> Opposite of 3

> −7 has the greater absolute value, so the sum is negative.

(a) $12 - 3 = 12 + (-3) = 9$ **(b)** $5 - 7 = 5 + (-7) = -2$

(c) $-8 - 15 = -8 + (-15) = -23$

> Change − to +.
> No change
> Opposite of −5

(d) $-3 - (-5) = -3 + (5) = 2$

(e) $\dfrac{3}{8} - \left(-\dfrac{4}{5}\right) = \dfrac{15}{40} - \left(-\dfrac{32}{40}\right) = \dfrac{15}{40} + \dfrac{32}{40} = \dfrac{47}{40}$

◀ Work Problem **2** at the Side.

Subtraction can be used to reverse the result of an addition problem. For example, if 4 is added to a number and then subtracted from the sum, the original number is the result.

$$12 + 4 = 16 \quad \text{and} \quad 16 - 4 = 12$$

Uses of the Symbol −

The symbol − has now been used for three purposes:

1. *to represent subtraction,* as in $9 - 5 = 4$;
2. *to represent negative numbers,* such as -10, -2, and -3;
3. *to represent the opposite (or negative) of a number,* as in "the opposite (or negative) of 8 is -8."

We may see more than one use in the same problem, such as $-6 - (-9)$, where -9 is subtracted from -6. The meaning of the symbol depends on its position in the algebraic expression.

OBJECTIVE **3** **Work subtraction problems that involve brackets.** As before, with problems that have both parentheses and brackets, first do any operations inside the parentheses and brackets. Work from the inside out.

EXAMPLE 3 Subtracting with Grouping Symbols

Perform each operation.

> Start here.

(a) $-6 - [2 - (8 + 3)]$

$\quad = -6 - [2 - 11]$ Add.

$\quad = -6 - [2 + (-11)]$ Change − to +.

$\quad = -6 - (-9)$ Add.

$\quad = -6 + (9)$ Change − to +.

$\quad = 3$

Continued on Next Page

ANSWERS

2. (a) -4 **(b)** -6 **(c)** 8 **(d)** 4 **(e)** $\dfrac{47}{28}$

(b) $5 - \left[\left(-\dfrac{1}{3} - \dfrac{1}{2}\right) - (4 - 1)\right]$

$= 5 - \left[\left(-\dfrac{1}{3} + \left(-\dfrac{1}{2}\right)\right) - 3\right]$ Work within the parentheses inside the brackets.

$= 5 - \left[\left(-\dfrac{5}{6}\right) - 3\right]$ Use 6 as the common denominator; $-\frac{1}{3} + \left(-\frac{1}{2}\right) = -\frac{2}{6} + \left(-\frac{3}{6}\right) = -\frac{5}{6}$

$= 5 - \left[\left(-\dfrac{5}{6}\right) + (-3)\right]$ Definition of subtraction

$= 5 - \left[\left(-\dfrac{5}{6}\right) + \left(-\dfrac{18}{6}\right)\right]$ $-\frac{3}{1} \cdot \frac{6}{6} = -\frac{18}{6}$

$= 5 - \left(-\dfrac{23}{6}\right)$ Find a common denominator.

$= 5 + \dfrac{23}{6}$ Definition of subtraction

$= \dfrac{53}{6}$ $5 = \frac{30}{6}$

Work Problem 3 at the Side. ▶

OBJECTIVE 4 Translate words and phrases that indicate subtraction. *Difference* is one word that indicates subtraction of real numbers. Some others are given in the table.

Word or Phrase	Example	Numerical Expression and Simplification
Difference between	The ***difference between*** -3 and -8	$-3 - (-8) = -3 + 8 = 5$
Subtracted from	12 ***subtracted from*** 18	$18 - 12 = 6$
Less than	6 ***less than*** 5	$5 - 6 = 5 + (-6) = -1$
Decreased by	9 ***decreased by*** -4	$9 - (-4) = 9 + 4 = 13$
Minus	-8 ***minus*** 5	$-8 - 5 = -8 + (-5) = -13$

CAUTION

When subtracting two numbers, it is important to write them in the correct order, because, in general, $a - b \neq b - a$. For example, $5 - 3 \neq 3 - 5$. *Think carefully before interpreting an expression involving subtraction!* Subtracting a larger number from a smaller number *always* produces a negative number.

EXAMPLE 4   **Translating Words and Phrases (Subtraction)**

Write a numerical expression for each phrase, and simplify the expression.

(a) The **difference between** -8 and 5

When "difference between" is used, write the numbers in the order they are given. The expression is

$-8 - 5$,   which simplifies to $-8 + (-5)$,   or -13.

Continued on Next Page

---

3 Perform each operation.

(a) $2 - [(-3) - (4 + 6)]$

(b) $[(5 - 7) + 3] - 8$

(c) $6 - [(-1 - 4) - 2]$

ANSWERS

3. (a) 15   **(b)** -7 **(c)** 13

4 Write a numerical expression for each phrase, and simplify the expression.

(a) The difference between -5 and -12

(b) -2 subtracted from the sum of 4 and -4

(c) 7 less than -2

(d) 9, decreased by 10 less than 7

(b) 4 **subtracted from** the sum of 8 and -3

Here the operation of addition is also used, as indicated by the word *sum.* First, add 8 and -3. Next, subtract 4 from this sum. The expression is

$$[8 + (-3)] - 4, \quad \text{which simplifies to} \quad 5 - 4, \quad \text{or} \quad 1.$$

(c) 4 **less than** -6

Here 4 must be taken *from* -6, so write -6 first.

> Be careful with order. $\quad -6 - 4 \quad$ simplifies to $\quad -6 + (-4), \quad$ or $\quad -10.$

Notice that "4 less than -6" differs from "4 *is less than* -6." The statement "4 is less than -6" is symbolized as $4 < -6$ (which is a false statement).

(d) 8, **decreased by** 5 **less than** 12

First, write "5 less than 12" as $12 - 5$. Next, subtract $12 - 5$ from 8.

$$8 - (12 - 5) \quad \text{simplifies to} \quad 8 - 7, \quad \text{or} \quad 1.$$

◀ *Work Problem* **4** *at the Side.*

We have seen a few applications of signed numbers in earlier sections. The next example involves subtraction of signed numbers.

EXAMPLE 5   **Solving a Problem Involving Subtraction**

The record high temperature of 134°F in the United States was recorded at Death Valley, California, in 1913. The record low was -80°F, at Prospect Creek, Alaska, in 1971. See Figure 12. What is the difference between these highest and lowest temperatures? (*Source: World Almanac and Book of Facts.*)

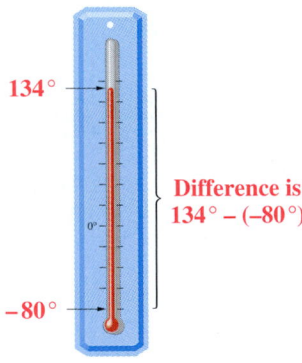

134°

Difference is
134° − (−80°)

0°

-80°

Figure 12

We must subtract the lowest temperature from the highest temperature.

> Order of numbers matters in subtraction.

$$134 - (-80)$$
$$= 134 + 80 \qquad \text{Use the definition of subtraction.}$$
$$= 214 \qquad \text{Add.}$$

The difference between the two temperatures is 214°F.

◀ *Work Problem* **5** *at the Side.*

---

5 Solve the problem.

The highest elevation in Argentina is Mt. Aconcagua, which is 6960 m above sea level. The lowest point in Argentina is the Valdes Peninsula, 40 m below sea level. Find the difference between the highest and lowest elevations.

Mt. Aconcagua

Buenos Aires

ARGENTINA

Valdes Peninsula

ANSWERS

4. **(a)** $-5 - (-12); 7$
  (b) $[4 + (-4)] - (-2); 2$
  (c) $-2 - 7; -9$
  (d) $9 - (7 - 10); 12$
5. 7000 m

9.5 ▶▶▶ **Exercises**

FOR
EXTRA
HELP

PRACTICE   WATCH   DOWNLOAD   READ   REVIEW

Fill in each blank with the correct response.

1. By the definition of subtraction, in order to perform the subtraction
$-6 - (-8)$, we must add the opposite of _____ to _____ .

2. By the rules for order of operations, to simplify $8 - [3 - (-4 - 5)]$,
the first step is to subtract _____ from _____ .

3. "The difference between 7 and 12" translates as _____ , while "the
difference between 12 and 7" translates as _____ .

4. $-9 - (-3) = -9 +$ _____

5. $-8 - 4 = -8 +$ _____

6. $-19 - 22 = -19 +$ _____

Find each difference. See Examples 1–3.

7. $-7 - 3$

8. $-12 - 5$

9. $-10 - 6$

10. $-13 - 16$

11. $7 - (-4)$

12. $9 - (-6)$

13. $6 - (-13)$

14. $13 - (-3)$

15. $-7 - (-3)$

16. $-8 - (-6)$

17. $3 - (4 - 6)$

18. $6 - (7 - 14)$

19. $-3 - (6 - 9)$

20. $-4 - (5 - 12)$

21. $\dfrac{1}{2} - \left(-\dfrac{1}{4}\right)$

22. $\dfrac{1}{3} - \left(-\dfrac{4}{3}\right)$

23. $-\dfrac{3}{4} - \dfrac{5}{8}$

24. $-\dfrac{5}{6} - \dfrac{1}{2}$

25. $\dfrac{5}{8} - \left(-\dfrac{1}{2} - \dfrac{3}{4}\right)$

26. $\dfrac{9}{10} - \left(\dfrac{1}{8} - \dfrac{3}{10}\right)$

27. $4.4 - (-9.2)$

28. $6.7 - (-12.6)$

29. $-7.4 - 4.5$

30. $-5.4 - 9.6$

31. $-5.2 - (8.4 - 10.8)$

32. $-9.6 - (3.5 - 12.6)$

33. $[(-3.1) - 4.5] - (0.8 - 2.1)$

34. $[(-7.8) - 9.3] - (0.6 - 3.5)$

35. $-12 - [(9 - 2) - (-6 - 3)]$

36. $-4 + [(-6 - 9) - (-7 + 4)]$

37. $-8 + [(-3 - 10) - (-4 + 1)]$

38. $\left(-\dfrac{3}{4} - \dfrac{5}{2}\right) - \left(-\dfrac{1}{8} - 1\right)$

39. $\left(-\dfrac{3}{8} - \dfrac{2}{3}\right) - \left(-\dfrac{9}{8} - 3\right)$

40. $[-34.99 + (6.59 - 12.25)] - 8.33$

41. $[-12.25 - (8.34 + 3.57)] - 17.88$

42. Explain in your own words how to subtract signed numbers.

43. We know that, in general, $a - b \neq b - a$. Find two pairs of values for a and b so that $a - b = b - a$.

Simplify each expression. Use the rules for order of operations.

44. $-3 - (-4) - 5$

45. $8 - (-3) - 9 + 6$

46. $-5 - 2 + 4 - 8 - (-6)$

47. Make up a subtraction problem so that the difference between two negative numbers is a negative number.

48. Make up a subtraction problem so that the difference between two negative numbers is a positive number.

Write a numerical expression for each phrase and simplify. See Example 4.

49. The difference between 4 and -8

50. The difference between 7 and -14

51. 8 less than -2

52. 9 less than -13

53. The sum of 9 and -4, decreased by 7

54. The sum of 12 and -7, decreased by 14

55. 12 less than the difference between 8 and -5

56. 19 less than the difference between 9 and -2

Solve each problem. See Example 5.

57. The coldest temperature recorded in Chicago, Illinois, was $-35°$F in 1996. The record low in South Dakota was set in 1936 and was 23°F lower than $-35°$F. What was the record low in South Dakota? (*Source: World Almanac and Book of Facts.*)

58. No one knows just why humpback whales love to heave their 45-ton bodies out of the water, but leap they do. This is called *breaching*. Chantelle, a researcher based on the island of Maui, noticed that one of her favorite whales, "Pineapple," breached 15 ft above the surface of the ocean while her mate cruised 12 ft below the surface. What is the difference between these two heights?

59. The top of Mount Whitney, visible from Death Valley, has an altitude of 14,494 ft above sea level. The bottom of Death Valley is 282 ft below sea level. Using 0 as sea level, find the difference between these two elevations. (*Source: World Almanac and Book of Facts.*)

60. A chemist is running an experiment under precise conditions. At first, she runs it at $-174.6°F$. She then lowers the temperature by $2.3°F$. What is the new temperature for the experiment?

61. Samir owed his brother $10. He later borrowed $70. What positive or negative number represents his present financial status?

62. Francesca has $15 in her purse, and Emilio has a debt of $12. Find the difference between these amounts.

63. For the year 2007, one health club showed a profit of $86,000, while another showed a loss of $19,000. Find the difference between these amounts.

64. At 2:00 A.M., a plant worker found that a dial reading was 7.904. At 3:00 A.M., she found the reading to be -3.291. Find the difference between these two readings.

65. J. D. Patin enjoys playing Triominoes every Wednesday night. Last Wednesday, on four successive turns, his scores were $-19, 28, -5,$ and 13. What was his final score for the four turns?

66. Gay Aguillard also enjoys playing Triominoes. On five successive turns, her scores were $-13, 15, -12, 24,$ and 14. What was her total score for the five turns?

67. In August, Kari Heen began with a checking account balance of $904.89. Her checks and deposits for August are given below:

Checks	Deposits
$35.84	$85.00
$26.14	$120.76
$3.12	

Assuming no other transactions, what was her account balance at the end of August?

68. In September, Derek Bowen began with a checking account balance of $904.89. His checks and deposits for September are given below:

Checks	Deposits
$41.29	$80.59
$13.66	$276.13
$84.40	

Assuming no other transactions, what was his account balance at the end of September?

69. A certain Greek mathematician was born in 426 B.C. His father was born 43 years earlier. In what year was his father born?

70. A certain Roman philosopher was born in 325 B.C. Her mother was born 35 years earlier. In what year was her mother born?

71. Kim Falgout owes $870.00 on her MasterCard account. She returns two items costing $35.90 and $150.00 and receives credits for these on the account. Next, she makes a purchase of $82.50, and then two more purchases of $10.00 each. She makes a payment of $500.00. She then incurs a finance charge of $37.23. How much does she still owe?

72. Charles Vosburg owes $679.00 on his Visa account. He returns three items costing $36.89, $29.40, and $113.55 and receives credits for these on the account. Next, he makes purchases of $135.78 and $412.88, and two purchases of $20.00 each. He makes a payment of $400. He then incurs a finance charge of $24.57. How much does he still owe?

73. José Martinez enjoys scuba diving. He dives to 34 ft below the surface of a lake. His partner, Sean O'Malley, dives to 40 ft below the surface, but then ascends 20 ft. What is the vertical distance between José and Sean?

74. Rhonda Alessi also enjoys diving. She dives to 12 ft below the surface of False River. Her sister, Sandy, dives to 20 ft below the surface, but then ascends 10 ft. What is the vertical distance between Rhonda and Sandy?

75. The height of Mt. Foraker is 17,400 ft, while the depth of the Java Trench is 23,376 ft. What is the vertical distance between the top of Mt. Foraker and the bottom of the Java Trench? (*Source: World Almanac and Book of Facts.*)

76. The height of Mt. Wilson is 14,246 ft, while the depth of the Cayman Trench is 24,721 ft. What is the vertical distance between the top of Mt. Wilson and the bottom of the Cayman Trench? (*Source: World Almanac and Book of Facts.*)

77. In 1984, Americans saved 10.8% of their after-tax incomes. In 2005, they saved −0.5%, the first negative personal savings rate since 1933. Find the difference between those two amounts.

78. Refer to Exercise 77. How is it possible that Americans had a negative personal savings rate in 2005?

79. In 2000, the federal budget had a surplus of $236 billion. In 2004, the federal budget had a deficit of $413 billion. Find the difference between these amounts. (*Source:* Treasury Department.)

80. In 1998, undergraduate college students had an average credit card balance of $1879. The average balance increased $869 by 2000 and then dropped $579 by 2004. What was the average credit card balance of undergraduate college students in 2004? (*Source:* Nellie Mae.)

Median sales prices for existing single-family homes in the United States for the years 2003 through 2007 are shown in the table. Complete the table, determining the change from one year to the next by subtraction.

	Year	Median Sales Price	Change from Previous Year
	2003	$180,200	—
81.	2004	$195,200	
82.	2005	$219,000	
83.	2006	$221,900	
84.	2007*	$217,800	

*Projected.
Source: National Association of Realtors.

In Exercises 85–88, suppose that x represents a positive number and y represents a negative number. Determine whether the given expression must represent a positive number or a negative number.

85. $x - y$ **86.** $y - x$ **87.** $x + |y|$ **88.** $y - |x|$

9.6 ▶▶▶ Multiplying and Dividing Real Numbers

The result of multiplication is called the **product.** We already know how to multiply positive numbers and that the product of two positive numbers is positive. We also know that the product of 0 and any positive number is 0, so we extend that property to all real numbers.

> **Multiplication Property of 0**
> For any real number a,
> $$a \cdot 0 = 0 \cdot a = 0.$$

OBJECTIVE 1 Find the product of a positive number and a negative number. To define the product of numbers with different signs so that the result is consistent with multiplication of positive numbers, look at the following pattern.

$$3 \cdot 5 = 15$$
$$3 \cdot 4 = 12$$
$$3 \cdot 3 = 9$$
$$3 \cdot 2 = 6$$
$$3 \cdot 1 = 3$$
$$3 \cdot 0 = 0$$
$$3 \cdot (-1) = ?$$

The products decrease by 3.

What should $3(-1)$ equal? Since multiplication can also be considered repeated addition, the product $3(-1)$ represents the sum

$$-1 + (-1) + (-1) = -3,$$

so the product should be -3, which fits the pattern. Also,

$$3(-2) = -2 + (-2) + (-2) = -6.$$

Work Problem **1** *at the Side.* ▶

These results suggest the following rule.

> **Multiplying Numbers with Different Signs**
> The product of a positive number and a negative number is negative.
> *Examples:* $6(-3) = -18$ and $-6(3) = -18$

EXAMPLE 1 Multiplying a Positive Number and a Negative Number

Find each product using the multiplication rule.

(a) $8(-5) = -(8 \cdot 5) = -40$ **(b)** $-7(2) = -(7 \cdot 2) = -14$

(c) $-9\left(\dfrac{1}{3}\right) = -3$ **(d)** $-6.2(4.1) = -25.42$

Work Problem **2** *at the Side.* ▶

OBJECTIVES

1 Find the product of a positive number and a negative number.

2 Find the product of two negative numbers.

3 Use the reciprocal of a number to apply the definition of division.

4 Use the rules for order of operations when multiplying and dividing signed numbers.

5 Evaluate expressions involving variables.

6 Translate words and phrases involving multiplication and division.

7 Translate simple sentences into equations.

---

1 Find each product by finding the sum of three numbers.

(a) $3(-3)$

(b) $3(-4)$

(c) $3(-5)$

2 Find each product.

(a) $2(-6)$

(b) $7(-8)$

(c) $-9(2)$

(d) $-16\left(\dfrac{5}{32}\right)$

(e) $4.56(-10)$

ANSWERS

1. (a) -9 (b) -12 (c) -15
2. (a) -12 (b) -56 (c) -18
   (d) $-\dfrac{5}{2}$ (e) -45.6

3 Find each product.

(a) $-5(-6)$

(b) $-7(-3)$

(c) $-8(-5)$

(d) $-11(-2)$

(e) $-17(3)(-7)$

(f) $-41(2)(-13)$

OBJECTIVE **2** **Find the product of two negative numbers.** The product of two positive numbers is positive, and the product of a positive number and a negative number is negative. What about the product of two negative numbers? Look at another pattern.

$$-5\,(4) = -20$$
$$-5\,(3) = -15$$
$$-5\,(2) = -10$$
$$-5\,(1) = -5$$
$$-5\,(0) = 0$$
$$-5\,(-1) = ?$$

The products increase by 5.

The numbers on the left of the equals signs (in color) decrease by 1 for each step down the list. The products on the right increase by 5 for each step down the list. To maintain this pattern, $-5(-1)$ should be 5 more than $-5\,(0)$, or 5 more than 0, so

$$-5\,(-1) = 5.$$

The pattern continues with

$$-5\,(-2) = 10$$
$$-5\,(-3) = 15$$
$$-5\,(-4) = 20$$
$$-5\,(-5) = 25,$$

and so on. This pattern suggests the next rule.

> **Multiplying Two Negative Numbers**
>
> The product of two negative numbers is positive.
>
> *Example:* $-5(-4) = 20$

EXAMPLE 2 **Multiplying Two Negative Numbers**

Find each product using the multiplication rule.

(a) $-9(-2) = 18$ **(b)** $-6(-12) = 72$

(c) $-2(4)(-1) = -8(-1) = 8$ **(d)** $3(-5)(-2) = -15(-2) = 30$

◀ *Work Problem* **3** *at the Side.*

> **Multiplying Signed Numbers**
>
> The product of two numbers having the *same* sign is *positive.*
>
> The product of two numbers having *different* signs is *negative.*

OBJECTIVE **3** **Use the reciprocal of a number to apply the definition of division.** Recall that the result of division is called the **quotient.** In **Section 9.5** we saw that the difference between two numbers is found by adding the opposite of the subtrahend to the minuend. Similarly, the *quotient* of two numbers involves multiplying by the *reciprocal* of the second number, which is the *divisor.*

Reciprocals

Pairs of numbers whose product is 1 are called **reciprocals** of each other.

Since $\quad 8 \cdot \dfrac{1}{8} = \dfrac{8}{8} = 1 \quad$ and $\quad \dfrac{5}{4} \cdot \dfrac{4}{5} = \dfrac{20}{20} = 1,$

the reciprocal of 8 is $\frac{1}{8}$, and the reciprocal of $\frac{5}{4}$ is $\frac{4}{5}$. The following table shows several numbers and their reciprocals.

Number	Reciprocal
4	$\frac{1}{4}$
-5	$\frac{1}{-5}$, or $-\frac{1}{5}$
0.3, or $\frac{3}{10}$	$\frac{10}{3}$
$-\frac{5}{8}$	$-\frac{8}{5}$
0	None

By definition, the product of a number and its reciprocal is 1. But the multiplication property of 0 says that the product of 0 and any number is 0. Thus, **0 has no reciprocal.**

Work Problem **4** *at the Side.* ▶

By definition, the quotient of a and b is the product of a and the reciprocal of b.

Division

The quotient $\frac{a}{b}$ of real numbers a and b, with $b \neq 0$, is

$$\frac{a}{b} = a \cdot \frac{1}{b}.$$

Example: $\quad \dfrac{8}{-4} = 8\left(-\dfrac{1}{4}\right) = -2$

This definition indicates that b, the divisor, cannot be 0. Since 0 has no reciprocal,

> $\frac{a}{0}$ **is not a number and *division by 0 is undefined.* If a division problem requires division by 0, write "undefined."**

To illustrate, $\frac{6}{2} = \mathbf{3}$ since $2 \cdot 3 = 6$, but there is no number to represent $\frac{6}{0}$, since there is no number that when multiplied by 0 gives 6 as a product.

Note

Although *division by zero is undefined,* we may divide zero by any *nonzero real number,* to obtain the quotient 0.

If $\quad b \neq 0, \quad$ then $\quad \dfrac{0}{b} = 0.$

4 Complete the table.

Number	Reciprocal
(a) 6	
(b) -2	
(c) $\frac{2}{3}$	
(d) $-\frac{1}{4}$	
(e) 0.75	
(f) 0	

5 Find each quotient.

(a) $\dfrac{42}{7}$

(b) $\dfrac{-36}{(-2)(-3)}$

(c) $\dfrac{-12.56}{-0.4}$

(d) $\dfrac{10}{7} \div \left(-\dfrac{24}{5}\right)$

(e) $\dfrac{-3}{0}$

(f) $\dfrac{0}{-53}$

6 Find each quotient.

(a) $\dfrac{-8}{-2}$

(b) $\dfrac{-16.4}{2.05}$

(c) $\dfrac{1}{4} \div \left(-\dfrac{2}{3}\right)$

(d) $\dfrac{12}{-4}$

ANSWERS

5. (a) 6 (b) −6 (c) 31.4
(d) $-\dfrac{25}{84}$ (e) undefined (f) 0

6. (a) 4 (b) −8 (c) $-\dfrac{3}{8}$ (d) −3

EXAMPLE 3 Using the Definition of Division

Find each quotient.

(a) $\dfrac{12}{3} = 12 \cdot \dfrac{1}{3} = 4$

(b) $\dfrac{5(-2)}{2} = -10 \cdot \dfrac{1}{2} = -5$

(c) $\dfrac{-1.47}{-7} = -1.47 \cdot \left(-\dfrac{1}{7}\right) = 0.21$

(d) $-\dfrac{2}{3} \div \left(-\dfrac{5}{4}\right) = -\dfrac{2}{3} \cdot \left(-\dfrac{4}{5}\right) = \dfrac{8}{15}$

(e) $\dfrac{-10}{0}$ Undefined

(f) $\dfrac{0}{13} = 0$ $\dfrac{0}{b} = 0$ $(b \neq 0)$

◀ Work Problem **5** at the Side.

When dividing fractions, multiplying by the reciprocal of the divisor works well. However, using the definition of division directly with integers is awkward. It is easier to divide in the usual way, and then determine the sign of the answer.

Dividing Signed Numbers

The quotient of two numbers having the *same* sign is *positive*.

The quotient of two numbers having *different* signs is *negative*.

Examples: $\dfrac{-15}{-5} = 3,$ $\dfrac{15}{-5} = -3,$ and $\dfrac{-15}{5} = -3$

EXAMPLE 4 Dividing Signed Numbers

Find each quotient.

(a) $\dfrac{8}{-2} = -4$

(b) $\dfrac{-10}{2} = -5$

> Remember to write in lowest terms.

(c) $\dfrac{-4.5}{-0.09} = 50$

(d) $-\dfrac{1}{8} \div \left(-\dfrac{3}{4}\right) = -\dfrac{1}{8} \cdot \left(-\dfrac{4}{3}\right) = \dfrac{1}{6}$

◀ Work Problem **6** at the Side.

From the definitions of multiplication and division of real numbers,

$$\dfrac{-40}{8} = -40 \cdot \dfrac{1}{8} = -5 \quad \text{and} \quad \dfrac{40}{-8} = 40\left(\dfrac{1}{-8}\right) = -5, \text{ so}$$

$$\dfrac{-40}{8} = \dfrac{40}{-8}.$$

Based on this example, the quotient of a positive number and a negative number can be written in any of the following three forms.

Equivalent Forms

For any positive real numbers a and b,

$$\dfrac{-a}{b} = \dfrac{a}{-b} = -\dfrac{a}{b}.$$

Similarly, the quotient of two negative numbers can be expressed as the quotient of two positive numbers.

Equivalent Forms

For any positive real numbers a and b,

$$\frac{-a}{-b} = \frac{a}{b}.$$

OBJECTIVE **4** **Use the rules for order of operations when multiplying and dividing signed numbers.**

EXAMPLE 5 **Using the Rules for Order of Operations**

Simplify.

(a) $-9(2) - (-3)(2)$

$\quad = -18 - (-6)$ Multiply.

$\quad = -18 + 6$ Definition of subtraction

$\quad = -12$ Add.

(b) $-6(-2) - 3(-4)$

$\quad = 12 - (-12)$

$\quad = 12 + 12$

$\quad = 24$

(c) $\dfrac{5(-2) - 3(4)}{2(1 - 6)}$

$\quad = \dfrac{-10 - 12}{2(-5)}$ Simplify the numerator and denominator separately.

$\quad = \dfrac{-22}{-10}$

$\quad = \dfrac{11}{5}$ Remember to write in lowest terms.

Work Problem **7** *at the Side.* ▶

The rules for operations with signed numbers are summarized here.

Operations with Signed Numbers

Addition

Same sign Add the absolute values of the numbers. The sum has the same sign as the numbers.

$$-4 + (-6) = -10$$

Different signs Find the absolute values of the numbers, and subtract the lesser absolute value from the greater. Give the sum the sign of the number having the greater absolute value.

$$4 + (-6) = -(6 - 4) = -2$$

(continued)

7 Perform the indicated operations.

(a) $-3(4) - 2(6)$

(b) $-8[-1 - (-4)(-5)]$

(c) $\dfrac{6(-4) - 2(5)}{3(2 - 7)}$

(d) $\dfrac{-6(-8) + 3(9)}{-2[4 - (-3)]}$

ANSWERS

7. **(a)** -24 **(b)** 168 **(c)** $\dfrac{34}{15}$ **(d)** $-\dfrac{75}{14}$

8 Evaluate each expression.

(a) $2x - 7(y + 1)$,
if $x = -4$ and $y = 3$

(b) $2x^2 - 4y^2$,
if $x = -2$ and $y = -3$

(c) $\dfrac{4x - 2y}{-3x}$,
if $x = 2$ and $y = -1$

Operations with Signed Numbers (continued)

Subtraction

Add the opposite of the subtrahend to the minuend.

$$8 - (-3) = 8 + 3 = 11$$

Multiplication and Division

Same sign The product or quotient of two numbers with the same sign is positive.

$$-5(-6) = 30 \quad \text{and} \quad \frac{-36}{-12} = 3$$

Different signs The product or quotient of two numbers with different signs is negative.

$$-5(6) = -30 \quad \text{and} \quad \frac{18}{-6} = -3$$

Division by 0 is undefined.

OBJECTIVE 5 Evaluate expressions involving variables. To *evaluate* an expression means to find its *value*.

EXAMPLE 6 Evaluating Expressions for Numerical Values

Evaluate each expression, given that $x = -1$, $y = -2$, and $m = -3$.

(a) $(3x + 4y)(-2m)$ — Use parentheses around substituted negative values to avoid errors.

$= [3(-1) + 4(-2)][-2(-3)]$ Substitute the given values for the variables.

$= [-3 + (-8)][6]$ Multiply.

$= [-11]6$ Add inside the brackets.

$= -66$ Multiply.

(b) $2x^2 - 3y^2$

$= 2(-1)^2 - 3(-2)^2$ Substitute.

$= 2(1) - 3(4)$ Apply the exponents.

$= 2 - 12$ Multiply.

$= -10$ Subtract.

(c) $\dfrac{4y^2 + x}{m}$

$= \dfrac{4(-2)^2 + (-1)}{-3}$ Substitute.

$= \dfrac{4(4) + (-1)}{-3}$ Apply the exponent.

$= \dfrac{16 + (-1)}{-3}$ Multiply.

$= \dfrac{15}{-3}, \text{ or } -5$ Add, and then divide.

◀ *Work Problem* **8** *at the Side.*

ANSWERS

8. (a) -36 (b) -28 (c) $-\dfrac{5}{3}$

OBJECTIVE **6** **Translate words and phrases involving multiplication and division.**   Just as there are words and phrases that indicate addition or subtraction, certain words and phrases indicate multiplication or division. The chart gives some phrases indicating multiplication.

Word or Phrase	Example	Numerical Expression and Simplification
Product of	The *product of* -5 and -2	$-5(-2) = 10$
Times	13 *times* -4	$13(-4) = -52$
Twice (meaning "2 times")	*Twice* 6	$2(6) = 12$
Of (used with fractions)	$\frac{1}{2}$ *of* 10	$\frac{1}{2}(10) = 5$
Percent of	12% *of* -16	$0.12(-16) = -1.92$

EXAMPLE 7   **Translating Words and Phrases (Multiplication)**

Write a numerical expression for each phrase, and simplify the expression.

(a) The **product of** 12 and the sum of 3 and -6
Here, 12 is multiplied by "the sum of 3 and -6." The expression is

$$12[3 + (-6)], \quad \text{which simplifies to} \quad 12[-3], \quad \text{or} \quad -36.$$

(b) **Twice** the difference between 8 and -4

$$2[8 - (-4)], \quad \text{simplifies to} \quad 2[12], \quad \text{or} \quad 24.$$

(c) Two-thirds **of** the sum of -5 and -3

$$\frac{2}{3}[-5 + (-3)] \quad \text{simplifies to} \quad \frac{2}{3}[-8], \quad \text{or} \quad -\frac{16}{3}.$$

(d) 15% **of** the difference between 14 and -2

$$0.15[14 - (-2)] \quad \text{simplifies to} \quad 0.15[16], \quad \text{or} \quad 2.4.$$

> Remember that $15\% = 0.15$.

Work Problem **9** *at the Side.* ▶

The word *quotient* refers to the answer in a division problem. In algebra, a quotient is usually represented with a fraction bar; the symbol $\div$ is seldom used. The chart gives some phrases associated with division.

Word or Phrase	Example	Numerical Expression and Simplification
Quotient of	The *quotient of* -24 and 3	$\frac{-24}{3} = -8$
Divided by	-16 *divided by* -4	$\frac{-16}{-4} = 4$
Ratio of	The *ratio of* 2 to 3	$\frac{2}{3}$

When translating a phrase involving division, we write the first number named as the numerator and the second as the denominator.

9 Write a numerical expression for each phrase, and simplify the expression.

(a) The product of 6 and the sum of -5 and -4

(b) Three times the difference between 4 and -6

(c) Three-fifths of the sum of 2 and -7

(d) 20% of the sum of 9 and -4

ANSWERS

9. **(a)** $6[(-5) + (-4)]$; -54
 (b) $3[4 - (-6)]$; 30
 (c) $\frac{3}{5}[2 + (-7)]$; -3
 (d) $0.20[9 + (-4)]$; 1

10 Write a numerical expression for each phrase, and simplify the expression.

(a) The quotient of 20 and the sum of 8 and −3

(b) The product of −9 and 2, divided by the difference between 5 and −1

11 Write each sentence in symbols, using x to represent the number.

(a) Twice a number is −6.

(b) The difference between −8 and a number is −11.

(c) The sum of 5 and a number is 8.

(d) The quotient of a number and −2 is 6.

EXAMPLE 8 **Translating Words and Phrases (Division)**

Write a numerical expression for each phrase, and simplify the expression.

(a) The **quotient of** 14 and the sum of −9 and 2

"Quotient" indicates division. The number 14 is the numerator and "the sum of −9 and 2" is the denominator. The expression is

$$\frac{14}{-9+2}, \quad \text{which simplifies to} \quad \frac{14}{-7}, \quad \text{or} \quad -2.$$

(b) The product of 5 and −6, **divided by** the difference between −7 and 8

The numerator of the fraction representing the division is found by multiplying 5 and −6. The denominator is found by subtracting −7 and 8. The expression is

$$\frac{5(-6)}{-7-8}, \quad \text{which simplifies to} \quad \frac{-30}{-15}, \quad \text{or} \quad 2.$$

◄ *Work Problem* **10** *at the Side.*

OBJECTIVE 7 Translate simple sentences into equations. In this section and the previous two sections, important words and phrases involving the four operations of arithmetic have been introduced. We can use these words and phrases to translate sentences into equations.

EXAMPLE 9 **Translating Sentences into Equations**

Write each sentence in symbols, using x to represent the number.

(a) Three **times** a number **is** −18.

The word *times* indicates multiplication, and the word *is* translates as the equals sign (=).

$$3x = -18 \qquad 3 \cdot x = 3x$$

(b) The **sum** of a number and 9 **is** 12.

$$x + 9 = 12$$

(c) The **difference between** a number and 5 **is** 0.

$$x - 5 = 0$$

(d) The **quotient of** 24 and a number **is** −2.

$$\frac{24}{x} = -2$$

◄ *Work Problem* **11** *at the Side.*

CAUTION
It is important to recognize the distinction between the types of problems found in Examples 7 and 8 and those in Example 9. In Examples 7 and 8, the phrases translate as *expressions,* while in Example 9, the sentences translate as *equations.* ***Remember that an expression is a phrase, while an equation is a sentence.***

$$\frac{5(-6)}{-7-8} \qquad\qquad 3x = -18$$

Expression   Equation

9.6 ▶▶▶ Exercises

Fill in each blank with one of the following: greater than 0, less than 0, equal to 0.

1. The product or the quotient of two numbers with the same sign is _____.

2. The product or the quotient of two numbers with different signs is _____.

3. If three negative numbers are multiplied together, the product is _____.

4. If two negative numbers are multiplied and then their product is divided by a negative number, the result is _____.

5. If a negative number is squared and the result is added to a positive number, the final answer is _____.

6. The reciprocal of a negative number is _____.

Find each product. See Examples 1 and 2.

7. $-7(4)$

8. $-8(5)$

⊙ **9.** $-5(-6)$

10. $-4(-20)$

11. $-8(0)$

12. $0(-12)$

⊙ **13.** $-\dfrac{3}{8}\left(-\dfrac{20}{9}\right)$

14. $-\dfrac{5}{4}\left(-\dfrac{6}{25}\right)$

15. $-6.8(0.35)$

16. $-4.6(0.24)$

17. $-6\left(-\dfrac{1}{4}\right)$

18. $-8\left(-\dfrac{1}{2}\right)$

Find each quotient. See Examples 3 and 4.

19. $\dfrac{-15}{5}$

20. $\dfrac{-18}{6}$

21. $\dfrac{20}{-10}$

22. $\dfrac{28}{-4}$

23. $\dfrac{-160}{-10}$

24. $\dfrac{-260}{-20}$

25. $\dfrac{0}{-3}$

26. $\dfrac{0}{-5}$

27. $\dfrac{-10.252}{0}$

28. $\dfrac{-29.584}{0}$

29. $\left(-\dfrac{3}{4}\right) \div \left(-\dfrac{1}{2}\right)$

30. $\left(-\dfrac{3}{16}\right) \div \left(-\dfrac{5}{8}\right)$

31. Which expression is undefined?

A. $\dfrac{5-5}{5+5}$ B. $\dfrac{5+5}{5+5}$ C. $\dfrac{5-5}{5-5}$ D. $\dfrac{5-5}{5}$

32. What is the reciprocal of 0.4?

Perform each indicated operation. See Example 5.

33. $\dfrac{-5(-6)}{9-(-1)}$

34. $\dfrac{-12(-5)}{7-(-5)}$

35. $\dfrac{-21(3)}{-3-6}$

36. $\dfrac{-40(3)}{-2-3}$

37. $\dfrac{-10(2)+6(2)}{-3-(-1)}$

38. $\dfrac{8(-1)+6(-2)}{-6-(-1)}$

39. $\dfrac{-27(-2)-(-12)(-2)}{-2(3)-2(2)}$

40. $\dfrac{-13(-4)-(-8)(-2)}{(-10)(2)-4(-2)}$

41. $\dfrac{3^2-4^2}{7(-8+9)}$

42. $\dfrac{5^2-7^2}{2(3+3)}$

43. $\dfrac{4(2^3-5)-5(-3^3+21)}{3[6-(-2)]}$

44. $\dfrac{-3(-2^4+10)+4(2^5-12)}{-2[8-(-7)]}$

Evaluate each expression if $x = 6$, $y = -4$, and $a = 3$. See Example 6.

45. $6x - 5y + 4a$

46. $5x - 2y + 3a$

47. $(5x - 2y)(-2a)$

48. $(2x + y)(3a)$

49. $\left(\dfrac{5}{6}x + \dfrac{3}{2}y\right)\left(-\dfrac{1}{3}a\right)$

50. $\left(\dfrac{1}{3}x - \dfrac{4}{5}y\right)\left(-\dfrac{1}{5}a\right)$

51. $(6 - x)(5 + y)(3 + a)$

52. $(-5 + x)(-3 + y)(3 - a)$

53. $5x - 4a^2$

54. $-2y^2 + 3a$

55. $\dfrac{xy + 9a}{x + y - 2}$

56. $\dfrac{2y^2 - x}{a - 3}$

Write a numerical expression for each phrase and simplify. See Examples 7 and 8.

57. The product of 4 and -7, added to -12

58. The product of -9 and 2, added to 9

59. Twice the product of -8 and 2, subtracted from -1

60. Twice the product of -1 and 6, subtracted from -4

61. The product of -3 and the difference between 3 and -7

62. The product of 12 and the difference between 9 and -8

63. Three-tenths of the sum of -2 and -28

64. Four-fifths of the sum of -8 and -2

65. The quotient of -20 and the sum of -8 and -2

66. The quotient of -12 and the sum of -5 and -1

67. The sum of -18 and -6, divided by the product of 2 and -4

68. The sum of 15 and -3, divided by the product of 4 and -3

69. The product of $-\frac{2}{3}$ and $-\frac{1}{5}$, divided by $\frac{1}{7}$

70. The product of $-\frac{1}{2}$ and $\frac{3}{4}$, divided by $-\frac{2}{3}$

Write each sentence with symbols, using x to represent the number. See Example 9.

71. Nine times a number is -36.

72. Seven times a number is -42.

73. The quotient of a number and 4 is -1.

74. The quotient of a number and 3 is -3.

75. $\frac{9}{11}$ less than a number is 5.

76. $\frac{1}{2}$ less than a number is 2.

77. When 6 is divided by a number, the result is -3.

78. When 15 is divided by a number, the result is -5.

Relating Concepts (Exercises 79–84) For Individual or Group Work

*To find the **average** of a group of numbers, we add the numbers and then divide the sum by the number of terms added. **Work Exercises 79–82 in order,** to find the average of 23, 18, 13, -4, and -8. Then find the averages in Exercises 83 and 84.*

79. Find the sum of the given group of numbers.

80. How many numbers are in the group?

81. Divide your answer for Exercise 79 by your answer for Exercise 80. Give the quotient as a mixed number.

82. What is the average of the given group of numbers?

83. What is the average of all integers between -10 and 14, including both -10 and 14?

84. What is the average of the integers between -15 and -10, including -15 and -10?

9.7 ▷▷▷ Properties of Real Numbers

If you are asked to find the sum

$$3 + 89 + 97,$$

you might mentally add $3 + 97$ to get 100, and then add $100 + 89$ to get 189. While the rules for order of operations say to add (or multiply) from left to right, the fact is we may change the order of the terms (or factors) and group them in any way we choose without affecting the sum (or product).

This is an example of a shortcut we use in everyday mathematics that is justified by the properties of real numbers introduced in this section. In the following statements, a, b, and c represent real numbers.

OBJECTIVE 1 Use the commutative properties. The word *commute* means to go back and forth. Many people commute to work or to school. If you travel from home to work and follow the same route from work to home, you travel the same distance each time. The **commutative properties** say that if two numbers are added or multiplied in any order, they give the same result.

$$a + b = b + a \qquad \text{Addition}$$
$$ab = ba \qquad \text{Multiplication}$$

EXAMPLE 1 **Using the Commutative Properties**

Use a commutative property to complete each statement.

(a) $-8 + 5 = 5 +$ _____
By the commutative property for addition, the missing number is -8 because $-8 + 5 = 5 + (-8)$.

(b) $-2(7) =$ _____ (-2)
By the commutative property for multiplication, the missing number is 7, since $-2(7) = 7(-2)$.

Work Problem ① at the Side. ▶

OBJECTIVE 2 Use the associative properties. When we *associate* one object with another, we tend to think of those objects as being grouped together. The **associative properties** say that when we add or multiply three numbers, we can group them in any manner and get the same answer.

$$(a + b) + c = a + (b + c) \qquad \text{Addition}$$
$$(ab)c = a(bc) \qquad \text{Multiplication}$$

EXAMPLE 2 **Using the Associative Properties**

Use an associative property to complete each statement.

(a) $8 + (-1 + 4) = (8 +$ _____ $) + 4$
The missing number is -1.

(b) $[2 \cdot (-7)] \cdot 6 = 2 \cdot$ _____
The missing expression on the right should be $[(-7) \cdot 6]$.

Work Problem ② at the Side. ▶

---

① Complete each statement. Use a commutative property.

(a) $x + 9 = 9 +$ _____

(b) $-12(4) =$ _____ (-12)

(c) $5x = x \cdot$ _____

② Complete each statement. Use an associative property.

(a) $(9 + 10) + (-3)$
$= 9 + [$ _____ $+ (-3)]$

(b) $-5 + (2 + 8)$
$= ($ _____ $) + 8$

(c) $10 \cdot [-8 \cdot (-3)] =$ _____

ANSWERS

1. (a) x (b) 4 (c) 5
2. (a) 10 (b) $-5 + 2$ (c) $[10 \cdot (-8)] \cdot (-3)$

By the associative property of addition, the sum of three numbers will be the same no matter how the numbers are "associated" in groups. For this reason, parentheses can be left out in many addition problems. For example, both

$$(-1 + 2) + 3 \quad \text{and} \quad -1 + (2 + 3)$$

can be written as

$$-1 + 2 + 3.$$

In the same way, parentheses also can be left out of many multiplication problems.

> **3** Decide whether each statement is an example of a commutative property, an associative property, or both.
>
> **(a)** $2(4 \cdot 6) = (2 \cdot 4)6$
>
> **(b)** $(2 \cdot 4)6 = (4 \cdot 2)6$
>
> **(c)** $(2 + 4) + 6 = 4 + (2 + 6)$

EXAMPLE 3 Distinguishing between Associative and Commutative Properties

(a) Is $(2 + 4) + 5 = 2 + (4 + 5)$ an example of an associative or a commutative property?

The order of the three numbers is the same on both sides of the equals sign. The only change is in the *grouping,* or association, of the numbers. Therefore, this is an example of an associative property.

(b) Is $6(3 \cdot 10) = 6(10 \cdot 3)$ an example of an associative or a commutative property?

The same numbers, 3 and 10, are grouped on each side. On the left, however, 3 appears first in $(3 \cdot 10)$. On the right, 10 appears first. Since the only change involves the *order* of the numbers, this statement is an example of a commutative property.

(c) Is $(8 + 1) + 7 = 8 + (7 + 1)$ an example of an associative or a commutative property, or both?

In the statement, both the order and the grouping are changed. On the left, the order of the three numbers is 8, 1, and 7. On the right, it is 8, 7, and 1. On the left, 8 and 1 are grouped, and on the right, 7 and 1 are grouped. Therefore, both associative and commutative properties are used.

◀ *Work Problem* **3** *at the Side.*

> **4** Find each sum or product.
>
> **(a)** $5 + 18 + 29 + 31 + 12$
>
> **(b)** $5(37)(20)$

EXAMPLE 4 Using Commutative and Associative Properties

Find each sum or product.

(a) $23 + 41 + 2 + 9 + 25$

$= (41 + 9) + (23 + 2) + 25$ Commutative and associative properties

$= 50 + 25 + 25$

$= 100$

(b) $25(69)(4)$

$= 25(4)(69)$

$= 100(69)$

$= 6900$

◀ *Work Problem* **4** *at the Side.*

OBJECTIVE 3 Use the identity properties. If a child wears a costume on Halloween, the child's appearance is changed, but his or her *identity* is unchanged. In the same way, the identity, or value, of a real number is left unchanged when identity properties are applied. The **identity properties** say that the sum of 0 and any number equals that number, and the product of 1 and any number equals that number.

$$a + 0 = a \quad \text{and} \quad 0 + a = a \quad \text{Addition}$$
$$a \cdot 1 = a \quad \text{and} \quad 1 \cdot a = a \quad \text{Multiplication}$$

The number 0 leaves the *identity,* or value, of any real number unchanged by addition. For this reason, 0 is called the **identity element for addition,** or the **additive identity.** Since multiplication by 1 leaves any real number unchanged, 1 is the **identity element for multiplication,** or the **multiplicative identity.**

EXAMPLE 5 **Using Identity Properties**

These statements are examples of identity properties.

(a) $-3 + 0 = -3$ Addition

(b) $1 \cdot 25 = 25$ Multiplication

Work Problem **5** *at the Side.* ▶

We use the identity property for multiplication to write fractions in lowest terms and to find common denominators.

EXAMPLE 6 **Using the Identity Element for Multiplication to Simplify Expressions**

Simplify each expression.

(a) $\dfrac{49}{35}$

$= \dfrac{7 \cdot 7}{5 \cdot 7}$ Factor.

$= \dfrac{7}{5} \cdot \dfrac{7}{7}$ Write as a product.

$= \dfrac{7}{5} \cdot 1$ Property of 1

$= \dfrac{7}{5}$ Identity property

(b) $\dfrac{3}{4} + \dfrac{5}{24} = \dfrac{3}{4} \cdot 1 + \dfrac{5}{24}$ Identity property

$= \dfrac{3}{4} \cdot \dfrac{6}{6} + \dfrac{5}{24}$ Use $1 = \frac{6}{6}$ to get a common denominator.

$= \dfrac{18}{24} + \dfrac{5}{24}$ Multiply.

$= \dfrac{23}{24}$ Add.

Work Problem **6** *at the Side.* ▶

5 Use an identity property to complete each statement.

(a) $9 + 0 =$ _____

(b) _____ $+ (-7) = -7$

(c) _____ $\cdot 1 = 5$

6 Use an identity property to simplify each expression.

(a) $\dfrac{85}{105}$

(b) $\dfrac{9}{10} - \dfrac{53}{50}$

ANSWERS

5. **(a)** 9 **(b)** 0 **(c)** 5

6. **(a)** $\dfrac{17}{21}$ **(b)** $-\dfrac{4}{25}$

7 Complete each statement so that it is an example of either an identity property or an inverse property. Tell which property is used.

(a) $-6 + \underline{\hspace{1.5cm}} = 0$

(b) $\dfrac{4}{3} \cdot \underline{\hspace{1.5cm}} = 1$

(c) $-\dfrac{1}{9} \cdot \underline{\hspace{1.5cm}} = 1$

(d) $275 + \underline{\hspace{1.5cm}} = 275$

(e) $-0.75 + \dfrac{3}{4} = \underline{\hspace{1.5cm}}$

(f) $0.2(5) = \underline{\hspace{1.5cm}}$

OBJECTIVE 4 Use the inverse properties. Each day before you go to work or school, you probably put on your shoes before you leave. When you get home or before you go to sleep at night, you probably take them off, and this leads to the same situation that existed before you put them on. These operations from everyday life are examples of inverse operations.

The **inverse properties** of addition and multiplication lead to the additive and multiplicative identities, respectively. The *opposite* of a, $-a$, is the **additive inverse** of a and the *reciprocal* of a, $\frac{1}{a}$, is the **multiplicative inverse** of the nonzero number a. The sum of the numbers a and $-a$ is 0, and the product of the nonzero numbers a and $\frac{1}{a}$ is 1.

$$a + (-a) = 0 \quad \text{and} \quad -a + a = 0 \qquad \text{Addition}$$

$$a \cdot \frac{1}{a} = 1 \quad \text{and} \quad \frac{1}{a} \cdot a = 1 \quad (a \neq 0) \qquad \text{Multiplication}$$

EXAMPLE 7 Using the Inverse Properties

The following statements are examples of the inverse properties.

(a) $(-5)\left(-\dfrac{1}{5}\right) = 1$ ⠀⠀Multiplication ⠀⠀**(b)** $4 + (-4) = 0$ ⠀⠀Addition

(c) $\dfrac{1}{2} + (-0.5) = 0$ ⠀⠀Addition ⠀⠀**(d)** $\dfrac{2}{3} \cdot \dfrac{3}{2} = 1$ ⠀⠀Multiplication

(e) $-\dfrac{1}{2} + \dfrac{1}{2} = 0$ ⠀⠀Addition ⠀⠀**(f)** $4(0.25) = 1$ ⠀⠀Multiplication

◀ *Work Problem* **7** *at the Side.*

OBJECTIVE 5 Use the distributive property. The everyday meaning of the word *distribute* is "to give out from one to several." An important property of real number operations involves this idea.

Look at the value of the following expressions:

$$2(5 + 8), \quad \text{which equals} \quad 2(13), \quad \text{or} \quad 26;$$

$$2(5) + 2(8), \quad \text{which equals} \quad 10 + 16, \quad \text{or} \quad 26.$$

Since both expressions equal 26,

$$2(5 + 8) = 2(5) + 2(8).$$

This result is an example of the *distributive property of multiplication with respect to addition,* the only property involving *both* addition and multiplication. With this property, a product can be changed to a sum or difference. This idea is illustrated by the divided rectangle in Figure 13.

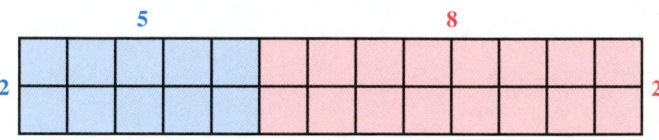

The area of the left part is 2(5) = 10.
The area of the right part is 2(8) = 16.
The total area is 2(5 + 8) = 26 or the total area is
2(5) + 2(8) = 10 + 16 = 26.
Thus, 2(5 + 8) = 2(5) + 2(8).

Figure 13

The **distributive property** says that multiplying a number a by a sum of numbers $b + c$ gives the same result as multiplying a by b and a by c and then adding the two products.

$$a(b + c) = ab + ac \quad \text{and} \quad (b + c)a = ba + ca$$

As the arrows show, the a outside the parentheses is "distributed" over the b and c inside. The distributive property is also valid for subtraction.

$$a(b - c) = ab - ac \quad \text{and} \quad (b - c)a = ba - ca$$

The distributive property also can be extended to the sum (or difference) of more than two numbers.

$$a(b + c + d) = ab + ac + ad$$

The distributive property can also be written "in reverse."

$$ab + ac = a(b + c)$$

EXAMPLE 8 **Using the Distributive Property**

Use the distributive property to rewrite each expression.

(a) $5(9 + 6)$

$\quad = 5 \cdot 9 + 5 \cdot 6$ Distributive property

Multiply first. $\quad = 45 + 30$ Multiply.

$\quad = 75$ Add.

(b) $4(x + 5 + y)$

$\quad = 4x + 4 \cdot 5 + 4y$ Distributive property

$\quad = 4x + 20 + 4y$ Multiply.

(c) $-2(x + 3)$

$\quad = -2x + (-2)(3)$ Distributive property

$\quad = -2x + (-6)$ Multiply.

$\quad = -2x - 6$

(d) $3(k - 9)$

$\quad = 3[k + (-9)]$ Definition of subtraction

$\quad = 3k + 3(-9)$ Distributive property

$\quad = 3k - 27$ Multiply.

(e) $8(3r + 11t + 5z)$

$\quad = 8(3r) + 8(11t) + 8(5z)$ Distributive property

$\quad = (8 \cdot 3)r + (8 \cdot 11)t + (8 \cdot 5)z$ Associative property

$\quad = 24r + 88t + 40z$ Multiply.

(f) $6 \cdot 8 + 6 \cdot 2$

$\quad = 6(8 + 2)$ Distributive property in reverse

$\quad = 6(10)$ Add.

$\quad = 60$ Multiply.

(g) $4x - 4m$

$\quad = 4(x - m)$ Distributive property in reverse

Work Problem **8** *at the Side.* ▶

---

8 Use the distributive property to rewrite each expression.

(a) $2(p + 5)$

(b) $-4(y + 7)$

(c) $5(m - 4)$

(d) $7(2y + 7k - 9m)$

(e) $9 \cdot k + 9 \cdot 5$

(f) $3a - 3b$

ANSWERS

8. **(a)** $2p + 10$ **(b)** $-4y - 28$
    (c) $5m - 20$ **(d)** $14y + 49k - 63m$
    (e) $9(k + 5)$ **(f)** $3(a - b)$

9 Write each expression without parentheses.

(a) $-(3k - 5)$

The symbol $-a$ may be interpreted as $-1 \cdot a$. Similarly, when a negative sign precedes an expression within parentheses, it may also be interpreted as a factor of -1. Thus, we can use the distributive property to remove (or clear) the parentheses from expressions such as $-(2y + 3)$.

$$-(2y + 3)$$
$$= \mathbf{-1} \cdot (2y + 3) \qquad -a = -1 \cdot a$$
$$= -1 \cdot (2y) + (-1) \cdot (3) \qquad \text{Distributive property}$$
$$= -2y - 3 \qquad \text{Multiply.}$$

EXAMPLE 9 **Using the Distributive Property to Remove (Clear) Parentheses**

Write each expression without parentheses.

(a) $-(7r - 8)$
$$= \mathbf{-1}(7r - 8) \qquad -a = -1 \cdot a$$
$$= \mathbf{-1}(7r) + (\mathbf{-1})(-8) \qquad \text{Distributive property}$$
$$= -7r + 8 \qquad \text{Multiply.}$$

(b) $-(2 - r)$

(b) $-(-9w + 2)$
$$= -1(-9w + 2)$$
$$= 9w - 2$$

We can interpret the $-$ sign in front of the parentheses to mean -1, yielding the opposite of each term inside the parentheses after it is distributed.

◀ **Work Problem** **9** **at the Side.**

Here is a summary of the basic properties of real numbers.

(c) $-(-5y + 8)$

Properties of Addition and Multiplication
For any real numbers a, b, and c, the following properties hold.

Commutative properties	$a + b = b + a \qquad ab = ba$
Associative properties	$(a + b) + c = a + (b + c)$
	$(ab)c = a(bc)$
Identity properties	There is a real number 0 such that
	$a + 0 = a \quad \text{and} \quad 0 + a = a.$
	There is a real number 1 such that
	$a \cdot 1 = a \quad \text{and} \quad 1 \cdot a = a.$
Inverse properties	For each real number a, there is a single real number $-a$ such that
	$a + (-a) = 0 \quad \text{and} \quad (-a) + a = 0.$
	For each nonzero real number a, there is a single real number $\frac{1}{a}$ such that
	$a \cdot \dfrac{1}{a} = 1 \quad \text{and} \quad \dfrac{1}{a} \cdot a = 1.$
Distributive property	$a(b + c) = ab + ac$
	$(b + c)a = ba + ca$

(d) $-(-z + 4)$

Match each item in Column I with the correct choice from Column II. Choices may be used once, more than once, or not at all.

I	**II**
1. Identity element for addition	**A.** $(5 \cdot 4) \cdot 3 = 5 \cdot (4 \cdot 3)$
2. Identity element for multiplication	**B.** 0
3. Additive inverse of a	**C.** $-a$
4. Multiplicative inverse, or reciprocal, of the nonzero number a	**D.** -1
5. The only number that has no multiplicative inverse	**E.** $5 \cdot 4 \cdot 3 = 60$
6. An example of an associative property	**F.** 1
7. An example of a commutative property	**G.** $(5 \cdot 4) \cdot 3 = 3 \cdot (5 \cdot 4)$
8. An example of the distributive property	**H.** $5(4 + 3) = 5 \cdot 4 + 5 \cdot 3$
	I. $\dfrac{1}{a}$

Decide whether each statement is an example of a commutative, associative, identity, or inverse property, or of the distributive property. See Examples 1, 2, 3, and 5–8.

9. $\dfrac{2}{3}(-4) = -4\left(\dfrac{2}{3}\right)$

10. $6\left(-\dfrac{5}{6}\right) = \left(-\dfrac{5}{6}\right)6$

11. $-6 + (12 + 7) = (-6 + 12) + 7$

12. $(-8 + 13) + 2 = -8 + (13 + 2)$

13. $-6 + 6 = 0$

14. $12 + (-12) = 0$

15. $\left(\dfrac{2}{3}\right)\left(\dfrac{3}{2}\right) = 1$

16. $\left(\dfrac{5}{8}\right)\left(\dfrac{8}{5}\right) = 1$

17. $2.34 \cdot 1 = 2.34$

18. $-8.456 \cdot 1 = -8.456$

19. $(4 + 17) + 3 = 3 + (4 + 17)$

20. $(-8 + 4) + (-12) = -12 + (-8 + 4)$

21. $6(x + y) = 6x + 6y$

22. $14(t + s) = 14t + 14s$

23. $-\dfrac{5}{9} = -\dfrac{5}{9} \cdot \dfrac{3}{3} = -\dfrac{15}{27}$

24. $\dfrac{13}{12} = \dfrac{13}{12} \cdot \dfrac{7}{7} = \dfrac{91}{84}$

25. $5(2x) + 5(3y) = 5(2x + 3y)$

26. $3(5t) - 3(7r) = 3(5t - 7r)$

27. What number(s) satisfy each condition? **(a)** a number that is its own additive inverse **(b)** two numbers that are their own multiplicative inverses

28. The distributive property holds for multiplication with respect to addition. Is there a distributive property for addition with respect to multiplication? That is, does $a + b \cdot c = (a + b)(a + c)$? If not, give an example to show why.

29. Evaluate $25 - (6 - 2)$ and $(25 - 6) - 2$. Use the results to explain why subtraction is or is not associative.

30. Suppose that a classmate shows you the following work.

$$-2(5 - 6)$$
$$= -2(5) - 2(6)$$
$$= -10 - 12$$
$$= -22$$

The classmate made a very common error. ***WHAT WENT WRONG?***

Write a new expression that is equal to the given expression, using the given property. Then simplify the new expression if possible. See Examples 1, 2, 5, 7, and 8.

31. $r + 7$; commutative

32. $t + 9$; commutative

33. $s + 0$; identity

34. $w + 0$; identity

35. $-6(x + 7)$; distributive

36. $-5(y + 2)$; distributive

37. $(w + 5) + (-3)$; associative

38. $(b + 8) + (-10)$; associative

39. Explain how the procedure of changing $\frac{3}{4}$ to $\frac{9}{12}$ requires the use of the multiplicative identity element, 1.

Use the properties of this section to simplify each expression. See Example 4.

40. $26 + 8 - 26 + 12$

41. $-\frac{3}{8} + \frac{2}{5} + \frac{8}{5} + \frac{3}{8}$

42. $\frac{9}{7}(-0.38)\left(\frac{7}{9}\right)$

Use the distributive property to rewrite each expression. Simplify if possible. See Example 8.

43. $4(t + 3)$

44. $5(w + 4)$

45. $-8(r + 3)$

46. $-11(x + 4)$

47. $-5(y - 4)$

48. $-9(g - 4)$

◐ 49. $-\frac{4}{3}(12y + 15z)$

50. $-\frac{2}{5}(10b + 20a)$

51. $8 \cdot z + 8 \cdot w$

52. $4 \cdot s + 4 \cdot r$

53. $5 \cdot 3 + 5 \cdot 17$

54. $15 \cdot 6 + 5 \cdot 6$

55. $7(2v) + 7(5r)$

56. $13(5w) + 13(4p)$

57. $8(3r + 4s - 5y)$

58. $2(5u - 3v + 7w)$

59. $-3(8x + 3y + 4z)$

60. $-5(2x - 5y + 6z)$

Use the distributive property to write each expression without parentheses. See Example 9.

61. $-(4t + 5m)$

62. $-(9x + 12y)$

63. $-(-5c - 4d)$

64. $-(-13x - 15y)$

65. $-(-3q + 5r - 8s)$

66. $-(-4z + 5w - 9y)$

67. "Starting a car" and "driving away in a car" are not commutative. Give an example of another pair of everyday activities that are not commutative.

68. Are "undressing" and "taking a shower" commutative?

69. *True* or *false:* "preparing a meal" and "eating a meal" are commutative.

70. The phrase "dog biting man" has two different meanings, depending on how the words are associated.

(dog biting) man   or   dog (biting man)

Give another example of a three-word phrase that has different meanings depending on how the words are associated.

71. Use parentheses to show how the associative property can be used to give two different meanings to "foreign sales clerk."

72. Use parentheses to show two different meanings for "hot pink pants."

Relating Concepts (Exercises 73–76) For Individual or Group Work

*In **Section 9.6** we used a pattern to see that the product of two negative numbers is a positive number. In the exercises that follow, we show another justification for determining the sign of the product of two negative numbers. **Work Exercises 73–76 in order.***

73. Evaluate the expression $-3[5 + (-5)]$ by using the order of operations.

74. Write the expression in Exercise 73 using the distributive property. Do not simplify the products.

75. The product $-3(5)$ should be one of the terms you wrote when answering Exercise 74. Based on the results in **Section 9.6,** what is this product?

76. In Exercise 73, you should have obtained 0 as the answer. Now, consider the following, using the results of Exercises 73 and 75.

$$-3[5 + (-5)] = -3(5) + (-3)(-5)$$
$$0 = -15 + ?$$

The question mark represents the product $-3(-5)$. When added to -15, it must give a sum of 0. Therefore, $-3(-5)$ must equal what?

9.8 ▶▶▶ Simplifying Expressions

OBJECTIVE 1 Simplify expressions. We now simplify expressions using the properties of addition and multiplication introduced in **Section 9.7.**

EXAMPLE 1 Simplifying Expressions

Simplify each expression.

(a) $4x + 8 + 9$ simplifies to $4x + 17$.

(b) $\mathbf{4}(3m - 2n)$

$\quad = \mathbf{4}(3m) - \mathbf{4}(2n) \qquad$ Distributive property

$\quad = 12m - 8n \qquad\qquad$ Associative property

(c) $\qquad\qquad 6 + 3(4k + 5)$

> Don't start by adding.

$\quad = 6 + 3(4k) + 3(5) \qquad$ Distributive property

$\quad = 6 + 12k + 15 \qquad\qquad$ Multiply.

$\quad = 21 + 12k \qquad\qquad$ Add.

(d) $\qquad\qquad 5 - (2y - 8)$

> Be careful with signs.

$\quad = 5 - \mathbf{1}(2y - 8) \qquad -a = -1 \cdot a$

$\quad = 5 - 2y + 8 \qquad\qquad$ Distributive property

$\quad = 13 - 2y \qquad\qquad$ Add.

Note

In Examples 1(c) and 1(d), we mentally used the commutative and associative properties to add in the last step. In practice, these steps are usually left out, but we should realize that they are used whenever the ordering and grouping in a sum are rearranged.

Work Problem **1** *at the Side.* ▶

OBJECTIVE 2 Identify terms and numerical coefficients. A **term** is a number, a variable, or a product or quotient of a number and one or more variables raised to powers. Examples of terms include

$$-9x^2, \quad 15y, \quad -3, \quad 8m^2n, \quad \frac{2}{p}, \quad \text{and} \quad k. \quad \text{Terms}$$

The **numerical coefficient,** or simply **coefficient,** of the term $\mathbf{9}m$ is $\mathbf{9}$; the numerical coefficient of $\mathbf{-15}x^3y^2$ is $\mathbf{-15}$; the numerical coefficient of x is 1; and the numerical coefficient of 8 is 8. In the expression $\frac{x}{3}$, the numerical coefficient of x is $\frac{1}{3}$ since $\frac{x}{3} = \frac{1x}{3} = \frac{1}{3}x$.

CAUTION

It is important to be able to distinguish between *terms* and *factors.* For example, in the expression $8x^3 + 12x^2$, there are two *terms,* $8x^3$ and $12x^2$. Terms are separated by a $+$ or $-$ sign. On the other hand, in the one-term expression $(8x^3)(12x^2)$, $8x^3$ and $12x^2$ are *factors.* Factors are multiplied.

OBJECTIVES

1 Simplify expressions.

2 Identify terms and numerical coefficients.

3 Identify like terms.

4 Combine like terms.

5 Simplify expressions from word phrases.

---

1 Simplify each expression.

(a) $9k + 12 - 5$

(b) $7(3p + 2q)$

(c) $2 + 5(3z - 1)$

(d) $-3 - (2 + 5y)$

ANSWERS

1. (a) $9k + 7$ **(b)** $21p + 14q$
    (c) $15z - 3$ **(d)** $-5 - 5y$

Here are some examples of terms and their numerical coefficients.

2 Give the numerical coefficient of each term.

Term	Numerical Coefficient
$-7y$	-7
$34r^3$	34
$-26x^5yz^4$	-26
$-k$, or $-1 \cdot k$	-1
r, or $1r$	1
$\frac{3x}{8} = \frac{3}{8}x$	$\frac{3}{8}$

(a) $15q$

(b) $-2m^3$

(c) $-18m^7q^4$

(d) $-r$

◀ *Work Problem* **2** *at the Side.*

OBJECTIVE **3** **Identify like terms.** Terms with exactly the same variables (including the same exponents) are called **like terms.** For example, $9m$ and $4m$ have the same variables and are like terms. Also, $6x^3$ and $-5x^3$ are like terms. The terms $-4y^3$ and $4y^2$ have different exponents and are **unlike terms.** Here are some additional examples:

(e) $\dfrac{5x}{4}$

$5x$ and $-12x$,	$3x^2y$ and $5x^2y$	Like terms
$4xy^2$ and $5xy$,	$8x^2y^3$ and $7x^3y^2$.	Unlike terms

3 Identify each pair of terms as *like* or *unlike*.

◀ *Work Problem* **3** *at the Side.*

(a) $9x, 4x$

OBJECTIVE **4** **Combine like terms.** Recall the distributive property:

$$x(y + z) = xy + xz.$$

As seen in **Section 9.7,** this statement can also be written as

(b) $-8y^3, 12y^2$

$$xy + xz = x(y + z) \quad \text{or} \quad yx + zx = (y + z)x.$$

Thus, the distributive property may be used to find the sum or difference of like terms. For example,

(c) $5x^2y^4, 5x^4y^2$

$$3x + 5x = (3 + 5)x = 8x.$$

(d) $7x^2y^4, -7x^2y^4$

This process is called **combining like terms.**

(e) $13kt, 4tk$

EXAMPLE 2 **Combining Like Terms**

Combine like terms in each expression.

4 Combine like terms.

(a) $9m + 5m$ **(b)** $6r + 3r + 2r$

(a) $4k + 7k$

$$= (9 + 5)m \qquad\qquad = (6 + 3 + 2)r$$
$$= 14m \qquad\qquad\quad = 11r$$

(b) $4r - r$

(c) $\dfrac{3}{4}x + x$ **(d)** $16y^2 - 9y^2$

(c) $5z + 9z - 4z$

$$\qquad\qquad\qquad = (16 - 9)y^2$$
$$= \dfrac{3}{4}x + 1x \qquad\qquad = 7y^2$$

(d) $8p + 8p^2$

$$= \left(\dfrac{3}{4} + 1\right)x$$

(e) $5x - 3y + 2x - 5y - 3$

$$= \dfrac{7}{4}x \qquad\qquad 1 = \tfrac{4}{4}$$

ANSWERS

1. **(a)** 15 **(b)** -2 **(c)** -18
   (d) -1 **(e)** $\dfrac{5}{4}$

2. **(a)** like   **(b)** unlike   **(c)** unlike
   (d) like   **(e)** like

3. **(a)** $11k$ **(b)** $3r$ **(c)** $10z$
   (d) cannot be combined   **(e)** $7x - 8y - 3$

(e) $32y + 10y^2$ cannot be combined because $32y$ and $10y^2$ are unlike terms. The distributive property cannot be used here to combine coefficients.

◀ *Work Problem* **4** *at the Side.*

CAUTION
Remember that only like terms may be combined.

EXAMPLE 3 **Simplifying Expressions Involving Like Terms**

Simplify each expression.

(a) $14y + 2(6 + 3y)$

$= 14y + 2(6) + 2(3y)$ Distributive property

$= 14y + 12 + 6y$ Multiply.

$= 20y + 12$ Combine like terms.

(b) $9k - 6 - 3(2 - 5k)$ Be careful with signs.

$= 9k - 6 - 3(2) - 3(-5k)$ Distributive property

$= 9k - 6 - 6 + 15k$ Multiply.

$= 24k - 12$ Combine like terms.

(c) $-(2 - r) + 10r$

$= -1(2 - r) + 10r$ $-(2 - r) = -1(2 - r)$

$= -1(2) - 1(-r) + 10r$ Distributive property   Be careful with signs.

$= -2 + r + 10r$ Multiply.

$= -2 + 11r$ Combine like terms.

(d) $5(2a^2 - 6a) - 3(4a^2 - 9)$

$= 10a^2 - 30a - 12a^2 + 27$ Distributive property

$= -2a^2 - 30a + 27$ Combine like terms.

Work Problem **5** *at the Side.* ▶

OBJECTIVE 5 Simplify expressions from word phrases. We now can simplify translated expressions by combining like terms.

EXAMPLE 4 **Translating Words into a Mathematical Expression**

Write the phrase as a mathematical expression and simplify.

Four times a number, subtracted from the sum of twice the number and 4

Let x represent the number. The expression is

The sum of twice the number and 4   Four times the number

$(2x + 4) - 4x,$ Write with symbols.

which simplifies to

$-2x + 4.$ Combine like terms.

Work Problem **6** *at the Side.* ▶

CAUTION
In Example 4, we are dealing with an expression to be simplified, *not* an equation to be solved.

5 Simplify.

(a) $10p + 3(5 + 2p)$

(b) $7z - 2 - (1 + z)$

(c) $-(3k^2 + 5k) + 7(k^2 - 4k)$

6 Write each phrase as a mathematical expression, and simplify by combining like terms.

(a) Three times a number, subtracted from the sum of the number and 8

(b) Twice a number added to the sum of 6 and the number

ANSWERS
5. **(a)** $16p + 15$ **(b)** $6z - 3$ **(c)** $4k^2 - 33k$
6. **(a)** $(x + 8) - 3x; -2x + 8$
(b) $2x + (6 + x); 3x + 6$

Math in the Media

THE MAGIC NUMBER IN SPORTS

The climax of any sports season is the playoffs. Baseball fans eagerly debate predictions of which team will win the pennant for their division. The *magic number* for each first-place team is often reported in media outlets. The **magic number** (sometimes called the **elimination number**) is the combined number of wins by the first-place team and losses by the second-place team that would clinch the title for the first-place team.

American League

East	W	L	PCT	GB
Boston	90	63	.588	—
New York	88	64	.579	1.5
Toronto	77	75	.507	12.5
Baltimore	65	87	.428	24.5
Tampa Bay	63	90	.412	27.0

Central	W	L	PCT	GB
Cleveland	90	62	.592	—
Detroit	83	70	.542	7.5
Minnesota	75	77	.493	15.0
Kansas City	66	86	.434	24.0
Chicago	66	87	.431	24.5

West	W	L	PCT	GB
Los Angeles	91	62	.595	—
Seattle	81	71	.533	9.5
Oakland	74	80	.481	17.5
Texas	70	83	.458	21.0

Source: mlb.com

To calculate the magic number, consider the following conditions.

The number of wins for the first-place team (W_1) plus the magic number (M) is one more than the sum of the number of wins to date (W_2) and the number of games remaining in the season (N_2) for the second-place team.

1. First, use the variable definitions to write an equation involving the magic number. Second, solve the equation for the magic number. Write the formula for the magic number.

2. The American League standings on September 10, 2007, are shown above. There were 162 regulation games in the 2007 season. Find the magic number for each team. The number of games remaining in the season for the second-place team is calculated as

$$N_2 = 162 - (W_2 + L_2),$$

where L_2 represents the number of losses for the second-place team.

(a) AL East: Boston vs New York

   Magic Number _____

(b) AL Central: Cleveland vs Detroit

   Magic Number _____

(c) AL West: Los Angeles vs Seattle

   Magic Number _____

3. Try to calculate the magic number for Toronto vs Boston. (Treat Toronto as if it were the second-place team.) How can you interpret the result?

9.8 ▶▶▶ Exercises

In Exercises 1–4, choose the letter of the correct response.

1. Which is true for all real numbers x?

   A. $6 + 2x = 8x$ **B.** $6 - 2x = 4x$

   C. $6x - 2x = 4x$ **D.** $3 + 8(4x - 6) = 11(4x - 6)$

2. Which is an example of a pair of like terms?

   A. $6t, 6w$ **B.** $-8x^2y, 9xy^2$

   C. $5ry, 6yr$ **D.** $-5x^2, 2x^3$

3. Which is an example of a term with numerical coefficient 5?

   A. $5x^3y^7$ **B.** x^5 **C.** $\dfrac{x}{5}$ **D.** 5^2xy^3

4. Which is a correct translation for "six times a number, subtracted from the product of eleven and the number" (if x represents the number)?

   A. $6x - 11x$ **B.** $11x - 6x$

   C. $(11 + x) - 6x$ **D.** $6x - (11 + x)$

Simplify each expression. See Example 1.

5. $3x + 12x$

6. $4y + 9y$

7. $8t - 5t + 2t$

8. $6s - 9s + 4s$

☻ 9. $4r + 19 - 8$

10. $7t + 18 - 4$

11. $5 + 2(x - 3y)$

12. $8 + 3(s - 6t)$

☻ 13. $-2 - (5 - 3p)$

14. $-10 - (7 - 14r)$

Give the numerical coefficient of each term.

15. $-12k$ **16.** $-23y$ **17.** $5m^2$ **18.** $-3n^6$ **19.** xw

20. pq **21.** $-x$ **22.** $-t$ **23.** 74 **24.** 98

25. Give an example of a pair of like terms with the variable x, such that one of them has a negative numerical coefficient, one has a positive numerical coefficient, and their sum has a positive numerical coefficient.

26. Give an example of a pair of unlike terms such that each term has x as the only variable.

Identify each group of terms as like *or* unlike.

27. $8r, -13r$ **28.** $-7a, 12a$ **29.** $5z^4, 9z^3$ **30.** $8x^5, -10x^3$

31. $4, 9, -24$ **32.** $7, 17, -83$ **33.** x, y **34.** t, s

35. There is an old saying "You can't add apples and oranges." Explain how this saying can be applied to Objective 3 in this section.

36. Explain how the distributive property is used in combining $6t + 5t$ to get $11t$.

Simplify each expression. See Examples 2 and 3.

37. $-5 - 2(x - 3)$

38. $-8 - 3(2x + 4)$

39. $-\dfrac{4}{3} + 2t + \dfrac{1}{3}t - 8 - \dfrac{8}{3}t$

40. $-\dfrac{5}{6} + 8x + \dfrac{1}{6}x - 7 - \dfrac{7}{6}$

41. $-5.3r + 4.9 - (2r + 0.7) + 3.2r$

42. $2.7b + 5.8 - (3b + 0.5) - 4.4b$

43. $2y^2 - 7y^3 - 4y^2 + 10y^3$

44. $9x^4 - 7x^6 + 12x^4 + 14x^6$

45. $13p + 4(4 - 8p)$

46. $5x + 3(7 - 2x)$

47. $-\dfrac{4}{3}(y - 12) - \dfrac{1}{6}y$

48. $-\dfrac{7}{5}(t - 15) - \dfrac{3}{2}$

49. $-5(5y - 9) + 3(3y + 6)$

50. $-3(2t + 4) + 8(2t - 4)$

Write each phrase as a mathematical expression. Use x to represent the number. Combine like terms when possible. See Example 4.

51. Five times a number, added to the sum of the number and three

52. Six times a number, added to the sum of the number and six

53. A number multiplied by -7, subtracted from the sum of 13 and six times the number

54. A number multiplied by 5, subtracted from the sum of 14 and eight times the number

55. Six times a number added to -4, subtracted from twice the sum of three times the number and 4

56. Nine times a number added to 6, subtracted from triple the sum of 12 and 8 times the number

57. Write the expression $9x - (x + 2)$ using words, as in Exercises 51–56.

58. Write the expression $2(3x + 5) - 2(x + 4)$ using words, as in Exercises 51–56.

Relating Concepts (Exercises 59–62) For Individual or Group Work

A manufacturer has fixed costs of $1000 to produce widgets. Each widget costs $5 to make. The fixed cost to produce gadgets is $750, and each gadget costs $3 to make.
Work Exercises 59–62 in order.

59. Write an expression for the cost to make x widgets. (*Hint:* The cost will be the sum of the fixed cost and the cost per item times the number of items.)

60. Write an expression for the cost to make y gadgets.

61. Write an expression for the total cost to make x widgets and y gadgets.

62. Simplify the expression you wrote in Exercise 61.

Chapter 9 ▶▶▶ Summary

▶ Key Terms

9.1	**exponent**	An exponent, or **power**, is a number that indicates how many times a factor is repeated.

$$3^4 \xleftarrow{} \text{Exponent}$$
$$\xleftarrow{} \text{Base}$$
$$\left.\right\} \text{Exponential expression}$$

	base	The base is the number that is a repeated factor when written with an exponent.
	exponential expression	A number written with an exponent is an exponential expression.
9.2	**variable**	A variable is a symbol, usually a letter, used to represent an unknown number.
	algebraic expression	An algebraic expression is a collection of numbers, variables, operation symbols, and grouping symbols.
	equation	An equation is a statement that says two expressions are equal.
	solution	A solution of an equation is any value of the variable that makes the equation true.
9.3	**natural numbers**	The set of natural numbers is $\{1, 2, 3, 4, \ldots\}$.
	whole numbers	The set of whole numbers is $\{0, 1, 2, 3, 4, \ldots\}$.
	number line	The number line shows the ordering of the real numbers on an infinite line.
	opposite	The opposite of a number a is the number that is the same distance from 0 on the number line as a, but on the opposite side of 0. This number is also called the **negative** of a or the **additive inverse** of a.

Number line

Negative numbers | Positive numbers

$$-3 \ -2 \ -1 \quad 0 \quad 1 \quad 2 \quad 3$$

Opposites

	integers	The set of integers is $\{\ldots, -3, -2, -1, 0, 1, 2, 3, \ldots\}$.
	negative number	A negative number is located to the *left* of 0 on the number line.
	positive number	A positive number is located to the *right* of 0 on the number line.
	rational numbers	A rational number is a number that can be written as the quotient of two integers, with denominator not 0.
	set-builder notation	Set builder notation uses a variable and a description to describe a set. It is often used to describe sets whose elements cannot easily be listed.
	coordinate	The number that corresponds to a point on the number line is the coordinate of that point.
	irrational numbers	An irrational number is a real number that is not a rational number.
	real numbers	Real numbers are numbers that can be represented by points on the number line (that is, all rational and irrational numbers).
	absolute value	The absolute value of a number is the distance between 0 and the number on the number line.
9.4	**sum**	The answer to an addition problem is called the sum.
9.5	**minuend**	In the operation $a - b$, a is called the minuend.
	subtrahend	In the operation $a - b$, b is called the subtrahend.
	difference	The answer to a subtraction problem is called the difference.
9.6	**product**	The answer to a multiplication problem is called the product.
	quotient	The answer to a division problem is called the quotient.
	reciprocal	Pairs of numbers whose product is 1 are called reciprocals, or **multiplicative inverses,** of each other.
9.7	**identity element for addition**	When the identity element for addition, which is 0, is added to a number, the number is unchanged.
	identity element for multiplication	When a number is multiplied by the identity element for multiplication, which is 1, the number is unchanged.

(continued)

9.8	**term**	A term is a number, a variable, or a product or quotient of a number and one or more variables raised to powers.
	numerical coefficient	The numerical factor in a term is its numerical coefficient, or **coefficient.**
	like terms	Terms with exactly the same variables (including the same exponents) are called like terms.

▶ New Symbols

Symbol	Meaning	Symbol	Meaning
a^n	n factors of a	$a(b)$, $(a)b$, $(a)(b)$, $a \cdot b$, or ab	a times b
$=$	is equal to	$\dfrac{a}{b}$, a/b, or $a \div b$	a divided by b
$\neq$	is not equal to	$\{\ \}$	set braces
		$\{x \mid x$ **has a certain property**$\}$	set-builder notation
$<$	is less than	$[\ ]$	square brackets
$\leq$	is less than or equal to	$\lvert x \rvert$	absolute value of x
$>$	is greater than	$-x$	additive inverse, or opposite, of x
$\geq$	is greater than or equal to	$\dfrac{1}{x}$	multiplicative inverse, or reciprocal, of x ($x \neq 0$)

▶ Test Your Word Power

See how well you have learned the vocabulary in this chapter. Answers, with examples, follow the Quick Review.

1. The **product** is
   A. the answer in an addition problem
   B. the answer in a multiplication problem
   C. one of two or more numbers that are added to get another number
   D. one of two or more numbers that are multiplied to get another number.

2. A number is **prime** if
   A. it cannot be factored
   B. it has just one factor
   C. it has exactly two different factors (itself and 1)
   D. it has at least two different factors.

3. An **exponent** is
   A. a symbol that tells how many numbers are being multiplied
   B. a number raised to a power
   C. a number that tells how many times a factor is repeated
   D. one of two or more numbers that are multiplied.

4. A **variable** is
   A. a symbol used to represent an unknown number
   B. a value that makes an equation true
   C. a solution of an equation
   D. the answer in a division problem.

5. An **integer** is
   A. a positive or negative number
   B. a natural number, its opposite, or zero
   C. any number that can be graphed on a number line
   D. the quotient of two numbers.

6. A **coordinate** is
   A. the number that corresponds to a point on a number line
   B. the graph of a number
   C. any point on a number line
   D. the distance from 0 on a number line.

7. The **absolute value** of a number is
   A. the graph of the number
   B. the reciprocal of the number

   C. the opposite of the number
   D. the distance between 0 and the number on a number line.

8. A **term** is
   A. a numerical factor
   B. a number, a variable, or a product or quotient of numbers and variables raised to powers
   C. one of several variables with the same exponents
   D. a sum of numbers and variables raised to powers.

9. A **numerical coefficient** is
   A. the numerical factor of the variables in a term
   B. the number of terms in an expression
   C. a variable raised to a power
   D. the variable factor in a term.

10. The **subtrahend** in $a - b = c$ is
    A. a
    B. b
    C. c
    D. $a - b$.

▶ Quick Review

Concepts	Examples

9.1 Exponents, Order of Operations, and Inequality

Order of Operations

Simplify within any parentheses or brackets and above and below fraction bars, using the following steps.

Step 1 Apply all exponents.

Step 2 Multiply or divide from left to right.

Step 3 Add or subtract from left to right.

$$\frac{9(2+6)}{2} - 2(2^3 + 3)$$
$$= 36 - 2(8 + 3)$$
$$= 36 - 2(11)$$
$$= 36 - 22$$
$$= 14$$

9.2 Variables, Expressions, and Equations

Evaluate an expression with a variable by substituting a given number for the variable.

Evaluate $2x + y^2$ if $x = 3$ and $y = -4$.

$$2x + y^2$$
$$= 2(3) + (-4)^2$$
$$= 6 + 16$$
$$= 22$$

Values of a variable that make an equation true are solutions of the equation.

Is 2 a solution of $5x + 3 = 18$?

$$5(2) + 3 \stackrel{?}{=} 18$$
$$13 = 18 \qquad \text{False}$$

2 is not a solution.

9.3 Real Numbers and the Number Line

Ordering Real Numbers

a is less than b if a is to the left of b on the number line.

The opposite, or additive inverse, of a is $-a$.

The absolute value of a, written $|a|$, is the distance between a and 0 on the number line.

$$-2 < 3 \qquad\qquad 3 > 0 \qquad\qquad 1 < 3$$
$$-(5) = -5 \qquad -(-7) = 7 \qquad -0 = 0$$
$$|13| = 13 \qquad\qquad |0| = 0 \qquad |-5| = 5$$

9.4 Adding Real Numbers

To add two numbers with the *same sign*, add their absolute values. The sum has that same sign.

To add two numbers with *different signs*, subtract their absolute values. The sum has the sign of the number with greater absolute value.

$$9 + 4 = 13$$
$$-8 + (-5) = -13$$
$$7 + (-12) = -5$$
$$-5 + 13 = 8$$

9.5 Subtracting Real Numbers

For any real numbers a and b,

$$a - b = a + (-b).$$

$$5 - (-2) = 5 + 2 = 7$$
$$-3 - 4 = -3 + (-4) = -7$$
$$-2 - (-6) = -2 + 6 = 4$$

Concepts	Examples

9.6 Multiplying and Dividing Real Numbers

The product (or quotient) of two numbers having the *same sign* is *positive*.

$$6 \cdot 5 = 30 \qquad -7(-8) = 56 \qquad \frac{-24}{-6} = 4$$

The product (or quotient) of two numbers having *different signs* is *negative*.

$$-6(5) = -30 \qquad 6(-5) = -30$$

$$-18 \div 9 = \frac{-18}{9} = -2 \qquad 49 \div (-7) = \frac{49}{-7} = -7$$

To divide a by b, multiply a by the reciprocal of b.

$$\frac{10}{\frac{2}{3}} = 10 \div \frac{2}{3} = 10 \cdot \frac{3}{2} = 15$$

0 divided by a nonzero number is 0.
Division by 0 is undefined.

$$\frac{0}{5} = 0 \qquad \frac{5}{0} \text{ is undefined.}$$

9.7 Properties of Real Numbers

Commutative Properties
$$a + b = b + a$$
$$ab = ba$$

$$7 + (-1) = -1 + 7$$
$$5(-3) = (-3)5$$

Associative Properties
$$(a + b) + c = a + (b + c)$$
$$(ab)c = a(bc)$$

$$(3 + 4) + 8 = 3 + (4 + 8)$$
$$[-2(6)]4 = -2[6(4)]$$

Identity Properties
$$a + 0 = a \quad 0 + a = a$$
$$a \cdot 1 = a \quad 1 \cdot a = a$$

$$-7 + 0 = -7 \qquad 0 + (-7) = -7$$
$$9 \cdot 1 = 9 \qquad\qquad 1 \cdot 9 = 9$$

Inverse Properties
$$a + (-a) = 0 \quad -a + a = 0$$
$$a \cdot \frac{1}{a} = 1 \quad \frac{1}{a} \cdot a = 1 \quad (a \neq 0)$$

$$7 + (-7) = 0 \qquad -7 + 7 = 0$$
$$-2\left(-\frac{1}{2}\right) = 1 \quad -\frac{1}{2}(-2) = 1$$

Distributive Properties
$$a(b + c) = ab + ac$$
$$(b + c)a = ba + ca$$
$$a(b - c) = ab - ac$$

$$5(4 + 2) = 5(4) + 5(2)$$
$$(4 + 2)5 = 4(5) + 2(5)$$
$$9(5 - 4) = 9(5) - 9(4)$$

9.8 Simplifying Expressions

Only like terms may be combined. We use the distributive property.

$$4(3 + 2x) - 6(5 - x)$$
$$= 12 + 8x - 30 + 6x$$
$$= 14x - 18$$

ANSWERS TO TEST YOUR WORD POWER

1. B; *Example:* The product of 2 and 5, or 2 times 5, is 10. **2.** C; *Examples:* 2, 3, 11, 41, 53

3. C; *Example:* In 2^3, the number 3 is the exponent (or power), so 2 is a factor three times; $2^3 = 2 \cdot 2 \cdot 2 = 8$.

4. A; *Examples:* a, b, c **5.** B; *Examples:* $-9, 0, 6$

6. A; *Example:* The point graphed three units to the right of 0 on a number line has coordinate 3.

7. D; *Examples:* $|2| = 2$ and $|-2| = 2$ **8.** B; *Examples:* $6, \frac{x}{2}, -4ab^2$

9. A; *Examples:* The term 3 has numerical coefficient 3, $8z$ has numerical coefficient 8, and $-10x^4y$ has numerical coefficient -10.

10. B; *Example:* In $5 - 3 = 2$, 5 is the minuend, 3 is the subtrahend, and 2 is the difference.

Chapter 9 ▶▶▶ Review Exercises

If you need help with any of these Review Exercises, look in the section indicated in brackets.

[9.1] *Find the value of each exponential expression.*

1. 5^4

2. $(0.03)^4$

3. 0.21^3

4. $\left(\dfrac{5}{2}\right)^3$

Find the value of each expression.

5. $8 \cdot 5 - 13$

6. $5[4^2 + 3(2^3)]$

7. $\dfrac{7(3^2 - 5)}{16 - 2 \cdot 6}$

8. $\dfrac{3(9 - 4) + 5(8 - 3)}{2^3 - (5 - 3)}$

Write each word sentence in symbols.

9. Thirteen is less than seventeen.

10. Five plus two is not equal to ten.

11. Write $6 < 15$ in words.

12. Construct a false statement that involves addition on the left side, the symbol $\geq$, and division on the right side.

[9.2] *Evaluate each expression if $x = 6$ and $y = 3$.*

13. $2x + 6y$

14. $4(3x - y)$

15. $\dfrac{x}{3} + 4y$

16. $\dfrac{x^2 + 3}{3y - x}$

Change each word phrase to an algebraic expression. Use x to represent the number.

17. Six added to a number

18. A number subtracted from eight

19. Nine subtracted from six times a number

20. Three-fifths of a number added to 12

Decide whether the given number is a solution of the equation.

21. $5x + 3(x + 2) = 22; 2$

22. $\dfrac{x + 5}{3x} = 1; 6$

Change each word sentence to an equation. Use x to represent the number.

23. Six less than twice a number is 10.

24. The product of a number and 4 is 8.

Identify each of the following as either an equation *or an* expression.

25. $5r - 8(r + 7) = 2$

26. $2y + (5y - 9) + 2$

[9.3] *Graph each group of numbers on a number line.*

27. $-4, -\dfrac{1}{2}, 0, 2.5, 5$

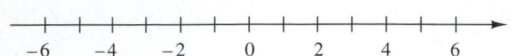

28. $-2, -3, |-3|, |-1|$

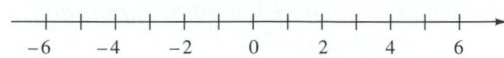

29. $-3\dfrac{1}{4}, \dfrac{14}{5}, -1\dfrac{1}{8}, \dfrac{5}{6}$

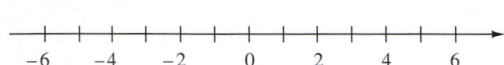

30. $|-4|, -|-3|, -|-5|, -6$

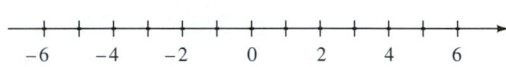

Select the lesser number in each pair.

31. $-10, 5$ **32.** $-8, -9$ **33.** $-\dfrac{2}{3}, -\dfrac{3}{4}$ **34.** $0, -|23|$

Decide whether each statement is true *or* false.

35. $12 > -13$ **36.** $0 > -5$ **37.** $-9 < -7$ **38.** $-13 > -13$

Simplify by finding the absolute value.

39. $-|3|$ **40.** $-|-19|$ **41.** $-|9 - 2|$ **42.** $|15 - 6|$

[9.4] *Find each sum.*

43. $-10 + 4$ **44.** $14 + (-18)$ **45.** $-8 + (-9)$ **46.** $\dfrac{4}{9} + \left(-\dfrac{5}{4}\right)$

47. $[-6 + (-8) + 8] + [9 + (-13)]$ **48.** $(-4 + 7) + (-11 + 3) + (-15 + 1)$

Write a numerical expression for each phrase, and simplify the expression.

49. 19 added to the sum of -31 and 12

50. 13 more than the sum of -4 and -8

Solve each problem.

51. Like many people, Otis Taylor neglects to keep up his checkbook balance. When he finally balanced his account, he found that the balance was $-\$23.75$, so he deposited $\$50.00$. What is his new balance?

52. The low temperature in Yellowknife, in the Canadian Northwest Territories, one January day was $-26°F$. It rose $16°$ that day. What was the high temperature?

[9.5] *Find each difference.*

53. $-7 - 4$

54. $-12 - (-11)$

55. $5 - (-2)$

56. $-\dfrac{3}{7} - \dfrac{4}{5}$

57. $2.56 - (-7.75)$

58. $(-10 - 4) - (-2)$

59. $(-3 + 4) - (-1)$

60. $|5 - 9| - |-3 + 6|$

Write a numerical expression for each phrase, and simplify the expression.

61. The difference between -4 and -6

62. Five less than the sum of 4 and -8

63. The difference between 18 and -23, decreased by 15

64. Nineteen, decreased by 12 less than -7

Solve each problem.

65. Peyton Manning of the Indianapolis Colts passed for a gain of 8 yd, was sacked for a loss of 12 yd, and then threw a 42 yd touchdown pass. What positive or negative number represents the total net yardage for the plays?

66. On Friday, February 22, 2008, the Dow Jones Industrial Average closed at 12,381.02, up 96.72 from the previous day. What was the closing price on Thursday, February 21, 2008? (*Source: The Washington Post.*)

The table shows the number of people naturalized in the United States (that is, made citizens of the United States) for the years 1999 through 2006. In Exercises 67–70, use a signed number to represent the change in the number of people naturalized for each time period.

Year	Number of People (in thousands)
1999	843
2000	899
2001	606
2002	574
2003	463
2004	337
2005	604
2006	702

Source: U.S. Department of Homeland Security.

67. 1999–2000

68. 2000–2001

69. 2003–2004

70. 2005–2006

[9.6] *Perform the indicated operations.*

71. $(-12)(-3)$

72. $15(-7)$

73. $-\dfrac{4}{3}\left(-\dfrac{3}{8}\right)$

74. $-4.8(-2.1)$

75. $5(8-12)$

76. $(5-7)(8-3)$

77. $2(-6)-(-4)(-3)$

78. $3(-10)-5$

79. $\dfrac{-36}{-9}$

80. $\dfrac{220}{-11}$

81. $-\dfrac{1}{2}\div\dfrac{2}{3}$

82. $-33.9\div(-3)$

83. $\dfrac{-5(3)-1}{8-4(-2)}$

84. $\dfrac{5(-2)-3(4)}{-2[3-(-2)]+10}$

85. $\dfrac{10^2-5^2}{8^2+3^2-(-2)}$

86. $\dfrac{4^2-8\cdot2}{(-1.2)^2-(-0.56)}$

Evaluate each expression if $x=-5$, $y=4$, and $z=-3$.

87. $6x-4z$

88. $5x+y-z$

89. $5x^2$

90. $z^2(3x-8y)$

Write a numerical expression for each phrase, and simplify the expression.

91. Nine less than the product of -4 and 5

92. Five-sixths of the sum of 12 and -6

93. The quotient of 12 and the sum of 8 and -4

94. The product of -20 and 12, divided by the difference between 15 and -15

Translate each sentence to an equation, using x to represent the number.

95. The quotient of a number and the sum of the number and 5 is -2.

96. 3 less than 8 times a number is -7.

[9.7] *Decide whether each statement is an example of a commutative, associative, identity, or inverse property, or of the distributive property.*

97. $6 + 0 = 6$

98. $5 \cdot 1 = 5$

99. $-\dfrac{2}{3}\left(-\dfrac{3}{2}\right) = 1$

100. $17 + (-17) = 0$

101. $5 + (-9 + 2) = [5 + (-9)] + 2$

102. $w(xy) = (wx)y$

103. $3x + 3y = 3(x + y)$

104. $(1 + 2) + 3 = 3 + (1 + 2)$

Use the distributive property to rewrite each expression. Simplify if possible.

105. $7y + y$

106. $-12(4 - t)$

107. $3(2s) + 3(4y)$

108. $-(-4r + 5s)$

[9.8] *Use the distributive property as necessary and combine like terms.*

109. $16p^2 - 8p^2 + 9p^2$

110. $4r^2 - 3r + 10r + 12r^2$

111. $-8(5k - 6) + 3(7k + 2)$

112. $2s - (-3s + 6)$

113. $-7(2t - 4) - 4(3t + 8) - 19(t + 1)$

114. $3.6t^2 + 9t - 8.1(6t^2 + 4t)$

Translate each phrase into a mathematical expression. Use x to represent the number, and combine like terms when possible.

115. Seven times a number, subtracted from the product of -2 and three times the number

116. The quotient of 9 more than a number and 6 less than the number

117. In Exercise 115, does the word *and* signify addition? Explain.

118. Write the expression $3(4x - 6)$ using words, as in Exercises 115 and 116.

>>> **Mixed Review Exercises**

Perform the indicated operations.

119. $[(-2) + 7 - (-5)] + [-4 - (-10)]$

120. $\left(-\dfrac{5}{6}\right)^2$

121. $-|(-7)(-4)| - (-2)$

122. $\dfrac{6(-4) + 2(-12)}{5(-3) + (-3)}$

123. $\dfrac{3}{8} - \dfrac{5}{12}$

124. $\dfrac{12^2 + 2^2 - 8}{10^2 - (-4)(-15)}$

125. $\dfrac{8^2 + 6^2}{7^2 + 1^2}$

126. $-16(-3.5) - 7.2(-3)$

127. $2\dfrac{5}{6} - 4\dfrac{1}{3}$

128. $-8 + [(-4 + 17) - (-3 - 3)]$

129. $-\dfrac{12}{5} \div \dfrac{9}{7}$

130. $(-8 - 3) - 5(2 - 9)$

131. $[-7 + (-2) - (-3)] + [8 + (-13)]$

132. $\dfrac{15}{2} \cdot \left(-\dfrac{4}{5}\right)$

Solve each problem.

133. The highest temperature ever recorded in Iowa was 118°F at Keokuk on July 20, 1934. The lowest temperature ever recorded in the state was at Elkader on February 3, 1996, and was 165° lower than the highest temperature. What is the record low temperature for Iowa? (*Source:* National Climatic Data Center.)

134. For a certain system of rating relief pitchers, 3 points are awarded for a save, 3 points are awarded for a win, 2 points are subtracted for a loss, and 2 points are subtracted for a blown save. If Mariano Rivera of the New York Yankees has 4 saves, 3 wins, 2 losses, and 1 blown save, how many points does he have?

Chapter 9 ▶▶▶ Test

Use the Chapter Test Prep Video CD to see fully worked-out solutions to any of the exercises you want to review.

Decide whether each statement is true *or* false.

1. $4[-20 + 7(-2)] \leq -135$

1. _____

2. $\left(\dfrac{1}{2}\right)^2 + \left(\dfrac{2}{3}\right)^2 = \left(\dfrac{1}{2} + \dfrac{2}{3}\right)^2$

2. _____

3. Graph the numbers -1, -3, $|-4|$, and $|-1|$ on the number line.

3.
```
 +--+--+--+--+--+--+--+--+-->
-3 -2 -1  0  1  2  3  4
```

Select the lesser number from each pair.

4. $6, -|-8|$

4. _____

5. $-0.742, -1.277$

5. _____

6. Write in symbols: The quotient of -6 and the sum of 2 and -8. Simplify the expression.

6. _____

7. If a and b are both negative, is $\dfrac{a + b}{a \cdot b}$ positive or negative?

7. _____

Perform the indicated operations whenever possible.

8. $-2 - (5 - 17) + (-6)$

8. _____

9. $-5\dfrac{1}{2} + 2\dfrac{2}{3}$

9. _____

10. $-6.2 - [-7.1 + (2.0 - 3.1)]$

10. _____

11. $4^2 + (-8) - (2^3 - 6)$

11. _____

12. $(-5)(-12) + 4(-4) + (-8)^2$

12. _____

13. $\dfrac{-7 - |-6 + 2|}{-5 - (-4)}$

13. _____

14. _____

14. $\dfrac{30(-1-2)}{-9[3-(-2)]-12(-2)}$

In Exercises 15 and 16, evaluate each expression if $x = -2$ and $y = 4$.

15. _____

15. $3x - 4y^2$

16. _____

16. $\dfrac{5x + 7y}{3(x + y)}$

17. _____

17. The highest Fahrenheit temperature ever recorded in Idaho was 118°F, while the lowest was -60°F. What is the difference between these highest and lowest temperatures? (_Source: World Almanac and Book of Facts._)

Match each example in Column I with a property in Column II.

I	II

18. _____

18. $3x + 0 = 3x$ **A.** Commutative

19. _____

19. $(5 + 2) + 8 = 8 + (5 + 2)$ **B.** Associative

20. _____

20. $-3(x + y) = -3x + (-3y)$ **C.** Inverse

21. _____

21. $-5 + (3 + 2) = (-5 + 3) + 2$ **D.** Identity

22. _____

22. $-\dfrac{5}{3}\left(-\dfrac{3}{5}\right) = 1$ **E.** Distributive

23. _____

23. Simplify $-2(3x^2 + 4) - 3(x^2 + 2x)$ by using the distributive property and combining like terms.

24. _____

24. Which properties are used to show that $-(3x + 1) = -3x - 1$?

25. (a) _____

(b) _____

25. Consider the expression $-6[5 + (-2)]$.
    (a) Evaluate it by first working within the brackets.
    (b) Evaluate it by using the distributive property.
    (c) Why must the answers in parts (a) and (b) be the same?

(c) _____

10

Equations, Inequalities, and Applications

In 1896, 241 competitors from 14 countries gathered in Athens, Greece, for the first modern Olympic Games. What began as a small, mainly European, sports competition has become the world's largest global sporting event. The Games of the XXIX Olympiad, hosted in 2008 by Beijing, China, attracted 10,500 athletes from over 200 countries. First introduced at the 1920 Games in Antwerp, Belgium, the five interlocking rings on the Olympic flag symbolize unity among the nations of Africa, the Americas, Asia, Australia, and Europe. (*Source:* www.olympic.org, *Microsoft Encarta Encyclopedia.*)

Throughout this chapter we use *linear equations* to solve applications about the Olympics.

10.1 ▶▶▶ The Addition Property of Equality

An *equation* is a statement that two algebraic expressions are equal.

CAUTION
Remember that an equation includes an equals sign.

Equation (to solve)   Expression (to simplify or evaluate)

$$x - 5 = 2 \qquad\qquad x - 5$$

Left   Right
side   side

OBJECTIVE 1 Identify linear equations.

Linear Equation in One Variable
A **linear equation in one variable** can be written in the form

$$Ax + B = C,$$

where A, B, and C are real numbers, with $A \neq 0$.

Some examples of linear and *non*linear equations follow.

$$4x + 9 = 0, \quad 2x - 3 = 5, \quad \text{and} \quad x = 7 \qquad \text{Linear equations}$$

$$x^2 + 2x = 5, \quad \frac{1}{x} = 6, \quad \text{and} \quad |2x + 6| = 0 \qquad \text{Nonlinear equations}$$

As we saw in **Section 9.2,** a *solution* of an equation is a number that makes the equation true when it replaces the variable. An equation is solved by finding its **solution set,** the set of all solutions. Equations that have exactly the same solution sets are **equivalent equations.** A linear equation in x is solved by using a series of steps to produce a simpler equivalent equation of the form

$$x = \textbf{a number} \quad \text{or} \quad \textbf{a number} = x.$$

OBJECTIVE 2 Use the addition property of equality. In the equation $x - 5 = 2$, both $x - 5$ and 2 represent the same number because that is the meaning of the equals sign. To solve the equation, we change the left side from $x - 5$ to just x. We do this by adding 5 to $x - 5$. We use 5 because 5 is the opposite (additive inverse) of -5, and $-5 + 5 = 0$. To keep the two sides equal, we must also add 5 to the right side.

$$x - 5 = 2$$
$$x - 5 + 5 = 2 + 5 \qquad \text{Add 5 to each side.}$$
$$x + 0 = 7 \qquad \text{Additive inverse property}$$
$$x = 7 \qquad \text{Additive identity property}$$

The solution is 7. We check by replacing x with 7 in the original equation.

Check
$$x - 5 = 2 \qquad \text{Original equation}$$
$$7 - 5 \overset{?}{=} 2 \qquad \text{Let } x = 7.$$
$$2 = 2 \qquad \text{True}$$

Since the final equation is true, 7 checks as the solution and $\{7\}$ is the solution set.

To solve the equation $x - 5 = 2$, we added the same number, 5, to each side. The **addition property of equality** justifies this step.

> **Addition Property of Equality**
>
> If A, B, and C represent real numbers, then the equations
>
> $$A = B \quad \text{and} \quad A + C = B + C$$
>
> are equivalent equations.
>
> In words, we can add the same number to each side of an equation without changing the solution.

In this property, C represents a real number. Any quantity that represents a real number can be added to each side of an equation to obtain an equivalent equation.

> **Note**
>
> Equations can be thought of in terms of a balance. Thus, adding the same quantity to each side does not affect the balance. See Figure 1.
>
>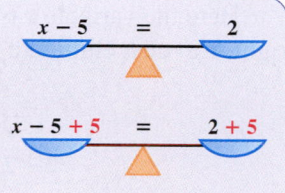
>
> **Figure 1**

EXAMPLE 1   **Using the Addition Property of Equality**

Solve $x - 16 = 7$.

Our goal is to get an equivalent equation of the form $x =$ **a number.**

$$x - 16 = 7$$
$$x - 16 + \mathbf{16} = 7 + \mathbf{16} \qquad \text{Add 16 to each side.}$$
$$\mathbf{x = 23} \qquad \text{Combine like terms.}$$

Check   Substitute 23 for x in the *original* equation.

$$x - 16 = 7 \qquad \text{Original equation}$$
$$\mathbf{23} - 16 \overset{?}{=} 7 \qquad \text{Let } x = 23.$$

> 7 is *not* the solution.

$$7 = 7 \qquad \text{True}$$

Since a true statement results, **23** is the solution and $\{23\}$ is the solution set.

──────────   *Work Problem* ▢**1** *at the Side.* ▶

EXAMPLE 2   **Using the Addition Property of Equality**

Solve $x - 2.9 = -6.4$.

> Our goal is to isolate x.

$$x - 2.9 = -6.4$$
$$x - 2.9 + \mathbf{2.9} = -6.4 + \mathbf{2.9} \qquad \text{Add 2.9 to each side.}$$
$$x = -3.5$$

Check
$$x - 2.9 = -6.4 \qquad \text{Original equation}$$
$$\mathbf{-3.5} - 2.9 \overset{?}{=} -6.4 \qquad \text{Let } x = -3.5.$$
$$-6.4 = -6.4 \qquad \text{True}$$

Since a true statement results, the solution set is $\{-3.5\}$.

──────────   *Work Problem* ▢**2** *at the Side.* ▶

▢**1** Solve.

(a) $x - 12 = 9$

(b) $x - 25 = -18$

▢**2** Solve.

(a) $x - 3.7 = -8.1$

(b) $a - 4.1 = 6.3$

ANSWERS

1. (a) $\{21\}$ (b) $\{7\}$
2. (a) $\{-4.4\}$ (b) $\{10.4\}$

3 Solve.

(a) $-3 = a + 2$

The addition property of equality says that the same number may be *added* to each side of an equation. In **Section 9.5,** subtraction was defined as addition of the opposite. Thus, we can also use the following rule when solving an equation.

> **The same number may be *subtracted* from each side of an equation without changing the solution.**

For example, to solve the equation $x + 4 = 10$, we *subtract* 4 from each side, which is the same as adding -4. The result is $x = 6$.

EXAMPLE 3 **Using the Addition Property of Equality**

Solve $-7 = x + 22$.

Here the variable x is on the right side of the equation. To isolate x on the right, we must eliminate the 22 by subtracting 22 from each side.

$$-7 = x + 22 \quad \text{◁} \boxed{\text{The variable can be isolated on } \textit{either} \text{ side.}}$$

$$-7 - \mathbf{22} = x + 22 - \mathbf{22} \qquad \text{Subtract 22 from each side.}$$

$$-29 = x, \quad \text{or} \quad x = -29 \qquad \begin{array}{l}\text{Rewrite; a number} = x,\\ \text{or } x = \text{a number.}\end{array}$$

Check

$$-7 = \mathbf{x} + 22 \qquad \text{Original equation}$$

$$-7 \overset{?}{=} \mathbf{-29} + 22 \qquad \text{Let } x = -29.$$

$$-7 = -7 \qquad \text{True}$$

The check confirms that the solution set is $\{-29\}$.

(b) $22 = -16 + r$

◁ *Work Problem* **3** *at the Side.*

> **CAUTION**
> *The final line of the check does not give the solution to the problem,* only a confirmation that the solution found is correct.

EXAMPLE 4 **Subtracting a Variable Term**

Solve $\frac{3}{5}k + 15 = \frac{8}{5}k$.

To get all terms with variables on the same side of the equation, subtract $\frac{3}{5}k$ from each side.

$$\frac{3}{5}k + 15 = \frac{8}{5}k$$

$$\frac{3}{5}k + 15 - \frac{3}{5}\mathbf{k} = \frac{8}{5}k - \frac{3}{5}\mathbf{k} \qquad \text{Subtract } \tfrac{3}{5}k \text{ from each side.}$$

$$15 = 1k \qquad \tfrac{3}{5}k - \tfrac{3}{5}k = 0;\ \tfrac{8}{5}k - \tfrac{3}{5}k = \tfrac{5}{5}k = 1k$$

$$15 = k \qquad \text{Multiplicative identity property}$$

From now on we will skip the step that changes $1k$ to k. Check the solution by replacing k with 15 in the original equation. The solution set is $\{15\}$.

What happens if we solve the equation in Example 4 by first subtracting $\frac{8}{5}k$ from each side?

$$\frac{3}{5}k + 15 = \frac{8}{5}k \qquad \text{Equation from Example 4}$$

$$\frac{3}{5}k + 15 - \frac{8}{5}k = \frac{8}{5}k - \frac{8}{5}k \qquad \text{Subtract } \frac{8}{5}k \text{ from each side.}$$

$$15 - k = 0 \qquad \frac{3}{5}k - \frac{8}{5}k = -\frac{5}{5}k = -1k = -k; \frac{8}{5}k - \frac{8}{5}k = 0$$

$$15 - k - 15 = 0 - 15 \qquad \text{Subtract 15 from each side.}$$

$$-k = -15 \qquad \text{Combine like terms; additive inverse}$$

This result gives the value of $-k$, but not of k itself. However, it does say that the additive inverse of k is -15, which means that k must be 15.

$$-k = -15$$

$$k = 15 \qquad \text{Same result as in Example 4}$$

(This result can also be justified using the multiplication property of equality, covered in **Section 10.2.**) We can make the following generalization.

If a is a number and $-x = a$, then $x = -a$.

Work Problem **4** *at the Side.* ▶

EXAMPLE 5 **Using the Addition Property of Equality Twice**

Solve $8 - 6p = -7p + 5$.

We must get all terms with variables on the same side of the equation and all terms without variables on the other side of the equation.

$$8 - 6p = -7p + 5$$

$$8 - 6p + 7p = -7p + 5 + 7p \qquad \text{Add } 7p \text{ to each side.}$$

$$8 + p = 5 \qquad \text{Combine like terms.}$$

$$8 + p - 8 = 5 - 8 \qquad \text{Subtract 8 from each side.}$$

$$p = -3 \qquad \text{Combine like terms.}$$

Check   Substitute -3 for p in the original equation.

$$8 - 6p = -7p + 5 \qquad \text{Original equation}$$

$$8 - 6(-3) \stackrel{?}{=} -7(-3) + 5 \qquad \text{Let } p = -3.$$

Use parentheses when substituting to avoid errors.

$$8 + 18 \stackrel{?}{=} 21 + 5 \qquad \text{Multiply.}$$

$$26 = 26 \qquad \text{True}$$

The check results in a true statement, so the solution set is $\{-3\}$.

Work Problem **5** *at the Side.* ▶

Note

There are often several equally correct ways to solve an equation. In Example 5, we could begin by adding $6p$, instead of $7p$, to each side. Combining like terms and subtracting 5 from each side gives $3 = -p$. (Try this.) If $3 = -p$, then $-3 = p$, and the variable has been isolated on the right side of equation. The same solution results.

---

4 **(a)** Solve $5m + 4 = 6m$.

(b) Solve $\frac{7}{2}m + 1 = \frac{9}{2}m$.

(c) What is the solution set of $-x = 6$?

(d) What is the solution set of $-x = -12$?

5 Solve.

(a) $10 - a = -2a + 9$

(b) $6x - 8 = 12 + 5x$

ANSWERS

4. (a) $\{4\}$ **(b)** $\{1\}$ **(c)** $\{-6\}$ **(d)** $\{12\}$

5. (a) $\{-1\}$ **(b)** $\{20\}$

6 Solve.

(a) $4x + 6 + 2x - 3$
$= 9 + 5x - 4$

(b) $9r + 4r + 6 - 2$
$= 9r + 4 + 3r$

OBJECTIVE 3 **Simplify, and then use the addition property of equality.** Sometimes the terms of an equation must be simplified as a first step in its solution.

EXAMPLE 6 **Combining Like Terms before Solving**

Solve $3t - 12 + t + 2 = 5 + 3t + 2$.

Begin by combining like terms on each side of the equation.

$$3t - 12 + t + 2 = 5 + 3t + 2$$

$$4t - 10 = 7 + 3t \qquad \text{Combine like terms.}$$

$$4t - 10 - 3t = 7 + 3t - 3t \qquad \text{Subtract } 3t \text{ from each side.}$$

$$t - 10 = 7 \qquad \text{Combine like terms.}$$

$$t - 10 + 10 = 7 + 10 \qquad \text{Add 10 to each side.}$$

$$t = 17 \qquad \text{Combine like terms.}$$

Check Substitute 17 for t in the original equation.

$$3t - 12 + t + 2 = 5 + 3t + 2 \qquad \text{Original equation}$$

$$3(17) - 12 + 17 + 2 \stackrel{?}{=} 5 + 3(17) + 2 \qquad \text{Let } t = 17.$$

$$51 - 12 + 17 + 2 \stackrel{?}{=} 5 + 51 + 2 \qquad \text{Multiply.}$$

$$58 = 58 \qquad \text{True}$$

The check results in a true statement, so the solution set is $\{17\}$.

◀ *Work Problem* **6** *at the Side.*

7 Solve.

(a) $4(r + 1) - (3r + 5) = 1$

(b) $-3(m - 4) + 2(5 + 2m)$
$= 29$

EXAMPLE 7 **Using the Distributive Property before Solving**

Solve $3(2 + 5x) - (1 + 14x) = 6$.

$$3(2 + 5x) - (1 + 14x) = 6$$

Be careful here!

$$3(2 + 5x) - 1(1 + 14x) = 6 \qquad -(1 + 14x) = -1(1 + 14x)$$

$$3(2) + 3(5x) - 1(1) - 1(14x) = 6 \qquad \text{Distributive property}$$

$$6 + 15x - 1 - 14x = 6 \qquad \text{Multiply.}$$

$$x + 5 = 6 \qquad \text{Combine like terms.}$$

$$x + 5 - 5 = 6 - 5 \qquad \text{Subtract 5 from each side.}$$

$$x = 1 \qquad \text{Combine like terms.}$$

Check by substituting 1 for x in the original equation. The solution set is $\{1\}$.

CAUTION

Be careful to apply the distributive property correctly in a problem like that in Example 7, or a sign error may result.

◀ *Work Problem* **7** *at the Side.*

1. Decide whether each is an *expression* or an *equation*. If it is an expression, simplify it. If it is an equation, solve it.

(a) $5x + 8 - 4x + 7$

(b) $-6m + 12 + 7m - 5$

(c) $5x + 8 - 4x = 7$

(d) $-6m + 12 + 7m = -5$

2. Which of the pairs of equations are equivalent equations?

A. $x + 2 = 6$ and $x = 4$

B. $10 - x = 5$ and $x = -5$

C. $x + 3 = 9$ and $x = 6$

D. $4 + x = 8$ and $x = -4$

3. Which of the following are not linear equations in one variable?

A. $x^2 - 5x + 6 = 0$

B. $x^3 = x$

C. $3x - 4 = 0$

D. $7x - 6x = 3 + 9x$

4. Explain how to check a solution of an equation.

Solve each equation, and check your solution. See Examples 1–5.

5. $x - 4 = 8$

6. $x - 8 = 9$

7. $x - 5 = -8$

8. $x - 7 = -9$

9. $r + 9 = 13$

10. $t + 6 = 10$

11. $x + 26 = 17$

12. $x + 45 = 24$

13. $x + \dfrac{1}{4} = -\dfrac{1}{2}$

14. $x + \dfrac{2}{3} = -\dfrac{1}{6}$

15. $x - 8.4 = -2.1$

16. $z - 15.5 = -5.1$

17. $t + 12.3 = -4.6$

18. $x + 21.5 = -13.4$

19. $7 + r = -3$

20. $8 + k = -4$

21. $2 = p + 15$

22. $3 = z + 17$

23. $-\dfrac{1}{3} = x - \dfrac{3}{5}$

24. $-\dfrac{1}{4} = x - \dfrac{2}{3}$

25. $3x = 2x + 7$

26. $5x = 4x + 9$

27. $10x + 4 = 9x$

28. $8t + 5 = 7t$

29. $\dfrac{9}{7}r - 3 = \dfrac{2}{7}r$

30. $\dfrac{8}{5}w - 6 = \dfrac{3}{5}w$

31. $5.6x + 2 = 4.6x$

32. $9.1x - 5 = 8.1x$

33. $3p + 6 = 10 + 2p$

34. $8x - 4 = -6 + 7x$

35. $5 - x = -2x - 11$

36. $3 - 8x = -9x - 1$

37. $-4z + 7 = -5z + 9$

38. $-6q + 3 = -7q + 10$

Solve each equation, and check your solution. See Examples 6 and 7.

39. $3x + 6 - 10 = 2x - 2$

40. $8k - 4 + 6 = 7k + 1$

41. $6x + 5 + 7x + 3 = 12x + 4$

42. $4x - 3 - 8x + 1 = -5x + 9$

43. $10x + 5x + 7 - 8 = 12x + 3 + 2x$

44. $7p + 4p + 13 - 7 = 7p + 9 + 3p$

45. $5.2q - 4.6 - 7.1q = -0.9q - 4.6$

46. $-4.0x + 2.7 - 1.6x = -4.6x + 2.7$

47. $\dfrac{5}{7}x + \dfrac{1}{3} = \dfrac{2}{5} - \dfrac{2}{7}x + \dfrac{2}{5}$

48. $\dfrac{6}{7}s - \dfrac{3}{4} = \dfrac{4}{5} - \dfrac{1}{7}s + \dfrac{1}{6}$

49. $(5x + 6) - (3 + 4x) = 10$

50. $(8r - 3) - (7r + 1) = -6$

51. $2(p + 5) - (9 + p) = -3$

52. $4(k - 6) - (3k + 2) = -5$

53. $-6(2x + 1) + (13x - 7) = 0$

54. $-5(3w - 3) + (1 + 16w) = 0$

55. $10(-2x + 1) = -19(x + 1)$

56. $2(2 - 3r) = -5(r - 3)$

57. $-2(8p + 2) - 3(2 - 7p) = 2(4 + 2p)$

58. $-5(1 - 2z) + 4(3 - z) = 7(3 + z)$

59. Write an equation that requires the use of the addition property of equality, in which 6 must be added to each side to solve the equation and the solution is a negative number.

60. Write an equation that requires the use of the addition property of equality, in which $\frac{1}{2}$ must be subtracted from each side and the solution is a positive number.

10.2 ▶▶▶ The Multiplication Property of Equality

OBJECTIVE 1 Use the multiplication property of equality. The addition property of equality alone is not enough to solve some equations, such as $3x + 2 = 17$.

$$3x + 2 = 17$$
$$3x + 2 - 2 = 17 - 2 \qquad \text{Subtract 2 from each side.}$$
$$3x = 15 \qquad \text{Combine like terms.}$$

Notice that the coefficient of x on the left side is 3, not 1 as desired. Another property, the **multiplication property of equality,** is needed to change $3x = 15$ to an equation of the form

x = a number.

Since $3x = 15$, both $3x$ and 15 must represent the same number. Multiplying both $3x$ and 15 by the same number will also result in an equality.

> **Multiplication Property of Equality**
>
> If A, B, and C ($C \neq 0$) represent real numbers, then the equations
>
> $$A = B \quad \text{and} \quad AC = BC$$
>
> are equivalent equations.
>
> In words, we can multiply each side of an equation by the same nonzero number without changing the solution.

In $3x = 15$, we must change $3x$ to $1x$, or x. To isolate x, we multiply each side by $\frac{1}{3}$, the reciprocal of 3, because $\frac{1}{3} \cdot 3 = \frac{3}{3} = 1$.

$$3x = 15$$

$$\frac{1}{3}(3x) = \frac{1}{3}(15) \qquad \text{Multiply each side by } \tfrac{1}{3}.$$

$$\left(\frac{1}{3} \cdot 3\right)x = \frac{1}{3}(15) \qquad \text{Associative property}$$

The product of a number and its reciprocal is 1.

$$1x = 5 \qquad \text{Multiplicative inverse property}$$

$$x = 5 \qquad \text{Multiplicative identity property}$$

The solution is 5. We can check this result in the original equation.

Work Problem **1** *at the Side.* ▶

Just as the addition property of equality permits *subtracting* the same number from each side of an equation, the multiplication property of equality permits *dividing* each side of an equation by the same nonzero number.

$$3x = 15$$

$$\frac{3x}{3} = \frac{15}{3} \qquad \text{Divide each side by 3.}$$

$$x = 5 \qquad \text{Same result as above}$$

> We can divide each side of an equation by the same nonzero number without changing the solution. ***Do not, however, divide each side by a variable, since the variable might be equal to 0.***

OBJECTIVES

1. Use the multiplication property of equality.

2. Simplify, and then use the multiplication property of equality.

1 Check that 5 is the solution of $3x = 15$, and write the solution set.

ANSWER

1. Since $3(5) = 15$, the solution of $3x = 15$ is 5, and the solution set is $\{5\}$.

2 Solve.

(a) $6p = 60$

(b) $3r = -12$

(c) $15x = 75$

3 Solve.

(a) $2m = 15$

(b) $-6x = 14$

(c) $10z = -45$

Note

In practice, it is usually easier to multiply on each side if the coefficient of the variable is a fraction, and divide on each side if the coefficient is an integer or a decimal. For example, to solve

$$-\frac{3}{4}x = 12,$$

it is easier to multiply by $-\frac{4}{3}$, the reciprocal of $-\frac{3}{4}$, than to divide by $-\frac{3}{4}$. On the other hand, to solve

$$-5x = -20,$$

it is easier to divide by -5 than to multiply by $-\frac{1}{5}$.

EXAMPLE 1 Dividing Each Side of an Equation by a Nonzero Number

Solve $5x = 60$.

Isolate x on the left by using the multiplication property of equality. Divide each side of the equation by 5, the coefficient of x.

$$5x = 60$$

$$\frac{5x}{5} = \frac{60}{5} \qquad \text{Divide each side by 5.}$$

$$x = 12 \qquad \frac{5x}{5} = \frac{5}{5}x = 1x = x$$

Check Substitute 12 for x in the original equation.

$$5x = 60 \qquad \text{Original equation}$$

$$5(12) \overset{?}{=} 60 \qquad \text{Let } x = 12.$$

$$60 = 60 \qquad \text{True}$$

Since a true statement results, the solution set is $\{12\}$.

◀ *Work Problem* **2** *at the Side.*

EXAMPLE 2 Using the Multiplication Property of Equality

Solve $-25p = 30$.

$$-25p = 30$$

$$\frac{-25p}{-25} = \frac{30}{-25} \qquad \text{Divide by } -25.$$

$$p = \frac{30}{-25} = -\frac{6}{5} \qquad \frac{a}{-b} = -\frac{a}{b}; \text{ Write in lowest terms.}$$

Check

$$-25p = 30 \qquad \text{Original equation}$$

$$\frac{-25}{1}\left(-\frac{6}{5}\right) \overset{?}{=} 30 \qquad \text{Let } p = -\frac{6}{5}.$$

$$30 = 30 \qquad \text{True}$$

The check confirms that the solution set is $\{-\frac{6}{5}\}$.

◀ *Work Problem* **3** *at the Side.*

ANSWERS

2. **(a)** $\{10\}$ **(b)** $\{-4\}$ **(c)** $\{5\}$

3. **(a)** $\left\{\frac{15}{2}\right\}$ **(b)** $\left\{-\frac{7}{3}\right\}$ **(c)** $\left\{-\frac{9}{2}\right\}$

EXAMPLE 3   **Solving an Equation with Decimals**

Solve $6.09 = 2.1x$.

$$6.09 = 2.1x \quad \boxed{\text{Isolate } x \text{ on the right.}}$$

$$\frac{6.09}{\textbf{2.1}} = \frac{2.1x}{\textbf{2.1}} \qquad \text{Divide each side by 2.1.}$$

$$2.9 = x, \quad \text{or} \quad x = 2.9 \qquad \text{Divide; you may use a calculator.}$$

Check that the solution set is $\{2.9\}$.

Work Problem **4** *at the Side.* ▶

EXAMPLE 4   **Using the Multiplication Property of Equality**

Solve $\frac{x}{4} = 3$.

Replace $\frac{x}{4}$ by $\frac{1}{4}x$, since dividing by 4 is the same as multiplying by $\frac{1}{4}$. To isolate x, multiply each side by 4, the *reciprocal* of the coefficient of x.

$$\frac{x}{4} = 3$$

$$\frac{\textbf{1}}{\textbf{4}}x = 3 \qquad \text{Change } \tfrac{x}{4} \text{ to } \tfrac{1}{4}x.$$

$$\textbf{4} \cdot \frac{1}{4}x = \textbf{4} \cdot 3 \qquad \text{Multiply each side by 4.}$$

$\boxed{4 \cdot \tfrac{1}{4}x = 1x = x}$ ▶ $\quad x = 12 \qquad$ Multiplicative inverse property; multiplicative identity property

Check

$$\frac{x}{4} = 3 \qquad \text{Original equation}$$

$$\frac{12}{4} \overset{?}{=} 3 \qquad \text{Let } x = 12.$$

$$3 = 3 \qquad \text{True}$$

Since a true statement results, the solution set is $\{12\}$.

Work Problem **5** *at the Side.* ▶

EXAMPLE 5   **Using the Multiplication Property of Equality**

Solve $\frac{3}{4}h = 6$.

To isolate h on the left, multiply each side of the equation by $\frac{4}{3}$, the reciprocal of $\frac{3}{4}$, since $\frac{4}{3} \cdot \frac{3}{4}h = 1 \cdot h = h$.

$$\frac{3}{4}h = 6$$

$$\frac{\textbf{4}}{\textbf{3}} \cdot \frac{3}{4}h = \frac{\textbf{4}}{\textbf{3}} \cdot 6 \qquad \text{Multiply each side by } \tfrac{4}{3}.$$

$$1 \cdot h = \frac{4}{3} \cdot \frac{6}{1} \qquad \text{Multiplicative inverse property}$$

$$h = 8 \qquad \text{Multiplicative identity property; multiply fractions.}$$

Check that the solution set is $\{8\}$.

Work Problem **6** *at the Side.* ▶

---

4 Solve.

(a) $-0.7m = -5.04$

(b) $-63.75 = 12.5k$

5 Solve.

(a) $\dfrac{x}{5} = 5$

(b) $\dfrac{p}{4} = -6$

6 Solve.

(a) $-\dfrac{5}{6}t = -15$

(b) $\dfrac{3}{5}k = -21$

ANSWERS

4. **(a)** $\{7.2\}$ **(b)** $\{-5.1\}$
5. **(a)** $\{25\}$ **(b)** $\{-24\}$
6. **(a)** $\{18\}$ **(b)** $\{-35\}$

7 Solve.

(a) $-m = 2$

(b) $-p = -7$

In **Section 10.1,** we obtained the equation $-k = -15$ in our alternate solution to Example 4. We reasoned that since this equation says that the additive inverse (or opposite) of k is -15, then k must equal 15. We can also use the multiplication property of equality to obtain the same result, as shown in the next example.

EXAMPLE 6   **Using the Multiplication Property of Equality When the Coefficient of the Variable Is −1**

Solve $-k = -15$.
    Work on the left side, first writing $-k$ as $-1 \cdot k$.

$$-\boldsymbol{k} = -15$$

$$-\boldsymbol{1} \cdot \boldsymbol{k} = -15 \qquad\qquad -k = -1 \cdot k$$

$$-\boldsymbol{1}(-1 \cdot k) = -\boldsymbol{1}(-15) \qquad \text{Multiply by } -1, \text{ since } -1(-1) = 1.$$

$$[-1(-1)] \cdot k = 15 \qquad\qquad \text{Associative property; multiply.}$$

$$1 \cdot k = 15 \qquad\qquad \text{Multiplicative inverse property}$$

$$k = 15 \qquad\qquad \text{Multiplicative identity property}$$

Check
$$-\boldsymbol{k} = -15 \qquad\qquad \text{Original equation}$$
$$-(\boldsymbol{15}) \overset{?}{=} -15 \qquad\qquad \text{Let } k = 15.$$
$$-15 = -15 \qquad\qquad \text{True}$$

The solution, 15, checks, so the solution set is $\{15\}$.

◀ *Work Problem* **7** *at the Side.*

---

8 Solve.

(a) $7m - 5m = -12$

(b) $4r - 9r = 20$

OBJECTIVE **2** **Simplify, and then use the multiplication property of equality.**   In the next example, it is necessary to combine like terms before using the multiplication property of equality.

EXAMPLE 7   **Combining Like Terms before Solving**

Solve $5m + 6m = 33$.

$$5m + 6m = 33$$

$$11m = 33 \qquad \text{Combine like terms.}$$

$$\frac{11m}{\boldsymbol{11}} = \frac{33}{\boldsymbol{11}} \qquad \text{Divide each side by 11.}$$

$$m = 3 \qquad \text{Multiplicative identity property}$$

Check
$$5\boldsymbol{m} + 6\boldsymbol{m} = 33 \qquad \text{Original equation}$$
$$5(\boldsymbol{3}) + 6(\boldsymbol{3}) \overset{?}{=} 33 \qquad \text{Let } m = 3.$$
$$15 + 18 \overset{?}{=} 33 \qquad \text{Multiply.}$$
$$33 = 33 \qquad \text{True}$$

The solution, 3, checks, so the solution set is $\{3\}$.

◀ *Work Problem* **8** *at the Side.*

ANSWERS

7. **(a)** $\{-2\}$ **(b)** $\{7\}$
8. **(a)** $\{-6\}$ **(b)** $\{-4\}$

10.2 ▶▶▶ Exercises

By what number is it necessary to multiply each side of each equation in order to isolate x on the left side? Do not actually solve.

1. $\frac{2}{3}x = 8$

2. $\frac{4}{5}x = 6$

3. $\frac{x}{10} = 3$

4. $\frac{x}{100} = 8$

5. $-\frac{9}{2}x = -4$

6. $-\frac{8}{3}x = -11$

7. $-x = 0.36$

8. $-x = 0.29$

By what number is it necessary to divide each side of each equation in order to isolate x on the left side? Do not actually solve.

9. $6x = 5$

10. $7x = 10$

11. $-4x = 13$

12. $-13x = 6$

13. $0.12x = 48$

14. $0.21x = 63$

15. $-x = 23$

16. $-x = 49$

17. Tell whether you would use the addition or multiplication property of equality to solve each equation. *Do not actually solve.*

    (a) $3x = 12$ **(b)** $3 + x = 12$ **(c)** $-x = 4$ **(d)** $-12 = 6 + x$

18. Which equation does *not* require the use of the multiplication property of equality?

    A. $3x - 5x = 6$ **B.** $-\frac{1}{4}x = 12$ **C.** $5x - 4x = 7$ **D.** $\frac{x}{3} = -2$

19. In the solution of a linear equation, the next-to-the-last step reads "$-x = -\frac{3}{4}$." Which of the following would be the solution to this equation?

    A. $-\frac{3}{4}$ **B.** $\frac{3}{4}$ **C.** -1 **D.** $\frac{4}{3}$

20. A student tried to solve the equation $4x = 8$ by dividing each side by 8. What is a better approach?

Solve each equation, and check your solution. See Examples 1–7.

21. $5x = 30$

22. $7x = 56$

23. $2m = 15$

24. $3m = 10$

25. $3a = -15$

26. $5k = -70$

27. $10t = -36$

28. $4s = -34$

29. $-6x = -72$

30. $-8x = -64$

31. $2r = 0$

32. $5x = 0$

33. $-x = 12$ **34.** $-t = 14$ **35.** $0.2t = 8$ **36.** $0.9x = 18$

37. $-2.1m = 25.62$ **38.** $-3.9a = 31.2$ **39.** $\dfrac{1}{4}x = -12$ **40.** $\dfrac{1}{5}p = -3$

41. $\dfrac{z}{6} = 12$ **42.** $\dfrac{x}{5} = 15$ **43.** $\dfrac{x}{7} = -5$ **44.** $\dfrac{k}{8} = -3$

45. $\dfrac{2}{7}p = 4$ **46.** $\dfrac{3}{8}x = 9$ **47.** $-\dfrac{7}{9}c = \dfrac{3}{5}$ **48.** $-\dfrac{5}{6}d = \dfrac{4}{9}$

49. $4x + 3x = 21$ **50.** $9x + 2x = 121$ **51.** $3r - 5r = 10$ **52.** $9p - 13p = 24$

53. $\dfrac{2}{5}x - \dfrac{3}{10}x = 2$ **54.** $\dfrac{2}{3}x - \dfrac{5}{9}x = 4$ **55.** $x + x - 3x = 12$ **56.** $z - 3z + z = -16$

57. $5m + 6m - 2m = 63$ **58.** $11r - 5r + 6r = 168$ **59.** $-6x + 4x - 7x = 0$

60. $-5x + 4x - 8x = 0$ **61.** $0.9w - 0.5w + 0.1w = -3$ **62.** $0.5x - 0.6x + 0.3x = -1$

63. Write an equation that requires the use of the multiplication property of equality, where each side must be multiplied by $\frac{2}{3}$ and the solution is a negative number.

64. Write an equation that requires the use of the multiplication property of equality, where each side must be divided by 100 and the solution is not an integer.

Write an equation using the information given in the problem. Use x as the variable. Then solve the equation.

65. When a number is multiplied by -4, the result is 10. Find the number.

66. When a number is divided by -5, the result is 2. Find the number.

10.3 ▶▶▶ More on Solving Linear Equations

In this section, we solve linear equations using *both* properties of equality introduced in **Sections 10.1 and 10.2**.

Work Problem **1** *at the Side.* ▶

OBJECTIVE 1 **Learn and use the four steps for solving a linear equation.** *Remember that when we solve an equation, our primary goal is to isolate the variable on one side of the equation.* We use the following four-step method.

Solving a Linear Equation

Step 1 **Simplify each side separately.** Clear (eliminate) parentheses, fractions, and decimals, using the distributive property as needed, and combine like terms.

Step 2 **Isolate the variable term on one side.** Use the addition property so that the variable term is on one side of the equation and a number is on the other.

Step 3 **Isolate the variable.** Use the multiplication property to get the equation in the form $x =$ a number, or a number $= x$. (Other letters may be used for the variable.)

Step 4 **Check.** Substitute the proposed solution into the *original* equation to see if a true statement results.

EXAMPLE 1 **Using Both Properties of Equality to Solve an Equation**

Solve $-6x + 5 = 17$.

Step 1 There are no parentheses, fractions, or decimals in this equation, so this step is not necessary.

$$-6x + 5 = 17$$

Step 2
$$-6x + 5 - 5 = 17 - 5 \qquad \text{Subtract 5 from each side.}$$
$$-6x = 12 \qquad \text{Combine like terms.}$$

Step 3
$$\frac{-6x}{-6} = \frac{12}{-6} \qquad \text{Divide each side by } -6.$$
$$x = -2$$

Step 4 Check by substituting -2 for x in the original equation.

Check
$$-6x + 5 = 17 \qquad \text{Original equation}$$
$$-6(-2) + 5 \stackrel{?}{=} 17 \qquad \text{Let } x = -2.$$
$$12 + 5 \stackrel{?}{=} 17 \qquad \text{Multiply.}$$
$$17 = 17 \qquad \text{True}$$

The solution, -2, checks, so the solution set is $\{-2\}$.

Work Problem **2** *at the Side.* ▶

OBJECTIVES

1 Learn and use the four steps for solving a linear equation.

2 Solve equations with fractions or decimals as coefficients.

3 Solve equations that have no solution or infinitely many solutions.

4 Write expressions for two related unknown quantities.

1 As a review, tell whether you would use the addition or multiplication property of equality to solve each equation. *Do not actually solve.*

(a) $7 + x = -9$

(b) $-13x = 26$

(c) $-x = \dfrac{3}{4}$

(d) $-12 = x - 4$

2 Solve.

(a) $-5p + 4 = 19$

(b) $7 + 2m = -3$

3 Solve.

(a) $2q + 3 = 4q - 9$

(b) $5 - 8k = 2k - 5$

EXAMPLE 2   **Using Both Properties of Equality to Solve an Equation**

Solve $3x + 2 = 5x - 8$.

Step 1   Again, there are no parentheses, fractions, or decimals in the equation, so we begin with Step 2.

> Our goal is to isolate x.

$$3x + 2 = 5x - 8$$

Step 2

$3x + 2 - \mathbf{5x} = 5x - 8 - \mathbf{5x}$	Subtract $5x$ from each side.
$-2x + 2 = -8$	Combine like terms.
$-2x + 2 - \mathbf{2} = -8 - \mathbf{2}$	Subtract 2 from each side.
$-2x = -10$	Combine like terms.

Step 3

$\dfrac{-2x}{\mathbf{-2}} = \dfrac{-10}{\mathbf{-2}}$	Divide each side by -2.
$x = 5$	

Step 4   Check by substituting 5 for x in the original equation.

Check

$3x + 2 = 5x - 8$	Original equation
$3(\mathbf{5}) + 2 \stackrel{?}{=} 5(\mathbf{5}) - 8$	Let $x = 5$.
$15 + 2 \stackrel{?}{=} 25 - 8$	Multiply.
$17 = 17$	True

The solution, 5, checks, so the solution set is $\{5\}$.

Note

Remember that the variable can be isolated on either side of the equation. In Example 2, x will be isolated on the right if we begin by subtracting $3x$, instead of $5x$, from each side of the equation.

$3x + 2 = 5x - 8$	Equation from Example 2
$3x + 2 - \mathbf{3x} = 5x - 8 - \mathbf{3x}$	Subtract $3x$ from each side.
$2 = 2x - 8$	Combine like terms.
$2 + \mathbf{8} = 2x - 8 + \mathbf{8}$	Add 8 to each side.
$10 = 2x$	Combine like terms.
$\dfrac{10}{\mathbf{2}} = \dfrac{2x}{\mathbf{2}}$	Divide each side by 2.
$5 = x$	The same solution results.

There are often several equally correct ways to solve an equation.

◀ *Work Problem* **3** *at the Side.*

EXAMPLE 3   **Using the Four Steps to Solve an Equation**

Solve $4(k - 3) - k = k - 6$.

Step 1   Clear the parentheses using the distributive property.

$$4(k - 3) - k = k - 6$$

$4(k) + 4(-3) - k = k - 6$	Distributive property
$4k - 12 - k = k - 6$	Multiply.
$3k - 12 = k - 6$	Combine like terms.

Step 2

$3k - 12 - k = k - 6 - k$	Subtract k.
$2k - 12 = -6$	Combine like terms.
$2k - 12 + 12 = -6 + 12$	Add 12.
$2k = 6$	Combine like terms.

Step 3

$\dfrac{2k}{2} = \dfrac{6}{2}$	Divide by 2.
$k = 3$	

Step 4   Check by substituting 3 for k in the original equation.

Check

$4(k - 3) - k = k - 6$	Original equation
$4(3 - 3) - 3 \overset{?}{=} 3 - 6$	Let $k = 3$.
$4(0) - 3 \overset{?}{=} 3 - 6$	Work inside the parentheses.
$-3 = -3$	True

The solution is 3, so $\{3\}$ is the solution set.

Work Problem 4 *at the Side.* ▶

EXAMPLE 4   **Using the Four Steps to Solve an Equation**

Solve $8a - (3 + 2a) = 3a + 1$.

Step 1

$8a - (3 + 2a) = 3a + 1$	
$8a - 1(3 + 2a) = 3a + 1$	Multiplicative identity property
$8a - 3 - 2a = 3a + 1$	Distributive property
$6a - 3 = 3a + 1$	Combine like terms.

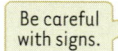 Be careful with signs.

Step 2

$6a - 3 - 3a = 3a + 1 - 3a$	Subtract $3a$.
$3a - 3 = 1$	Combine like terms.
$3a - 3 + 3 = 1 + 3$	Add 3.
$3a = 4$	Combine like terms.

Step 3

$\dfrac{3a}{3} = \dfrac{4}{3}$	Divide by 3.
$a = \dfrac{4}{3}$	

Step 4   Check that the solution set is $\{\frac{4}{3}\}$.

4 Solve.

(a) $7(p - 2) + p = 2p + 4$

(b) $11 + 3(x + 1) = 5x + 16$

5 Solve.

(a) $7m - (2m - 9) = 39$

(b) $4x - (x + 7) = 9$

CAUTION
Be very careful with signs when solving an equation like the one in Example 4. When clearing parentheses in the expression

$$8a - (3 + 2a),$$

remember that the $-$ sign acts like a factor of -1 and affects the sign of *every* term inside the parentheses. Thus,

$$8 - (3 + 2a)$$
$$= 8 - 1(3 + 2a)$$
$$= 8a + (-1)(3 + 2a)$$
$$= 8a - 3 - 2a.$$

Change to $-$ in *both* terms.

◀ *Work Problem* **5** *at the Side.*

6 Solve.

(a) $2(4 + 3r)$
   $= 3(r + 1) + 11$

(b) $2 - 3(2 + 6z)$
   $= 4(z + 1) + 18$

EXAMPLE 5 **Using the Four Steps to Solve an Equation**

Solve $4(8 - 3t) = 32 - 8(t + 2)$.

Step 1 $4(8 - 3t) = 32 - 8(t + 2)$ Be careful with signs.

$32 - 12t = 32 - 8t - 16$ Distributive property

$32 - 12t = 16 - 8t$ Combine like terms.

Step 2 $32 - 12t + 8t = 16 - 8t + 8t$ Add $8t$.

$32 - 4t = 16$ Combine like terms.

$32 - 4t - 32 = 16 - 32$ Subtract 32.

$-4t = -16$ Combine like terms.

Step 3 $\dfrac{-4t}{-4} = \dfrac{-16}{-4}$ Divide by -4.

$t = 4$

Step 4   Check this solution in the original equation.

Check $4(8 - 3t) = 32 - 8(t + 2)$ Original equation

$4[8 - 3(4)] \stackrel{?}{=} 32 - 8(4 + 2)$ Let $t = 4$.

$4[8 - 12] \stackrel{?}{=} 32 - 8(6)$ Simplify.

$4[-4] \stackrel{?}{=} 32 - 48$

$-16 = -16$ True

The solution, 4, checks. The solution set is $\{4\}$.

◀ *Work Problem* **6** *at the Side.*

OBJECTIVE **2** **Solve equations with fractions or decimals as coefficients.** We clear an equation of fractions by multiplying each side by the least common denominator (LCD) of all the fractions in the equation. It is a good idea to do this in Step 1 to avoid messy computations, as shown in Examples 6–8.

ANSWERS

5. (a) $\{6\}$ (b) $\left\{\dfrac{16}{3}\right\}$

6. (a) $\{2\}$ (b) $\left\{-\dfrac{13}{11}\right\}$

EXAMPLE 6   **Solving an Equation with Fractions as Coefficients**

Solve $\frac{2}{3}x - \frac{1}{2}x = -\frac{1}{6}x - 2$.

Step 1   The LCD of all the fractions in the equation is 6, so we multiply each side by 6 to clear the fractions.

$$\frac{2}{3}x - \frac{1}{2}x = -\frac{1}{6}x - 2$$

$$\mathbf{6}\left(\frac{2}{3}x - \frac{1}{2}x\right) = \mathbf{6}\left(-\frac{1}{6}x - 2\right) \qquad \text{Multiply by 6.}$$

> Pay particular attention here.

$$\mathbf{6}\left(\frac{2}{3}x\right) + \mathbf{6}\left(-\frac{1}{2}x\right) = \mathbf{6}\left(-\frac{1}{6}x\right) + \mathbf{6}(-2) \qquad \text{Distributive property}$$

$$4x - 3x = -x - 12 \qquad \text{Multiply.}$$

$$x = -x - 12 \qquad \text{Combine like terms.}$$

Step 2 $\quad x + \mathbf{x} = -x - 12 + \mathbf{x} \qquad \text{Add } x.$

$$2x = -12 \qquad \text{Combine like terms.}$$

Step 3 $\quad \dfrac{2x}{\mathbf{2}} = \dfrac{-12}{\mathbf{2}} \qquad \text{Divide by 2.}$

$$x = -6$$

Step 4   Check by substituting -6 for x in the original equation.

Check $\qquad \frac{2}{3}\mathbf{x} - \frac{1}{2}\mathbf{x} = -\frac{1}{6}\mathbf{x} - 2 \qquad \text{Original equation}$

$$\frac{2}{3}(\mathbf{-6}) - \frac{1}{2}(\mathbf{-6}) \overset{?}{=} -\frac{1}{6}(\mathbf{-6}) - 2 \qquad \text{Let } x = -6.$$

$$-4 + 3 \overset{?}{=} 1 - 2$$

$$-1 = -1 \qquad \text{True}$$

The solution, -6, checks. The solution set is $\{-6\}$.

Work Problem 7 *at the Side.* ▶

EXAMPLE 7   **Solving an Equation with Fractions as Coefficients**

Solve $\frac{1}{3}(x + 5) - \frac{3}{5}(x + 2) = 1$.

Step 1   We first clear the fractions by multiplying by the LCD, 15.

$$\frac{1}{3}(x + 5) - \frac{3}{5}(x + 2) = 1$$

$$\mathbf{15}\left[\frac{1}{3}(x + 5) - \frac{3}{5}(x + 2)\right] = \mathbf{15}(1) \qquad \text{Multiply by 15.}$$

$$\mathbf{15}\left[\frac{1}{3}(x + 5)\right] + \mathbf{15}\left[-\frac{3}{5}(x + 2)\right] = \mathbf{15} \qquad \text{Distributive property}$$

$$5(x + 5) - 9(x + 2) = 15 \qquad \text{Multiply.}$$

> $15\left[\frac{1}{3}(x + 5)\right]$
> $= 15 \cdot \frac{1}{3} \cdot (x + 5)$
> $= 5(x + 5)$

$$5x + 25 - 9x - 18 = 15 \qquad \text{Distributive property}$$

$$-4x + 7 = 15 \qquad \text{Combine like terms.}$$

Continued on Next Page

7 Solve

$$\frac{1}{4}x - 4 = \frac{3}{2}x + \frac{3}{4}x.$$

8 Solve
$$\frac{1}{4}(x+3) - \frac{2}{3}(x+1) = -2.$$

Step 2 $\qquad -4x + 7 \mathbf{\ -\ 7} = 15 \mathbf{\ -\ 7}$ — Subtract 7.

$\qquad\qquad\qquad\quad -4x = 8$ — Combine like terms.

Step 3 $\qquad\qquad\quad \dfrac{-4x}{\mathbf{-4}} = \dfrac{8}{\mathbf{-4}}$ — Divide by −4.

$\qquad\qquad\qquad\qquad x = -2$

Step 4 Check to confirm that $\{-2\}$ is the solution set.

◀ *Work Problem* **8** *at the Side.*

> **CAUTION**
> Be sure you understand how to multiply by the LCD to clear an equation of fractions. Study Step 1 in Examples 6 and 7 carefully.

EXAMPLE 8 **Solving an Equation with Decimals as Coefficients**

Solve $0.1t + 0.05(20 - t) = 0.09(20)$.

Step 1 The decimals here are expressed as tenths (0.1) and hundredths (0.05 and 0.09). We choose the least exponent on 10 needed to eliminate the decimals; in this case, we use $10^2 = 100$.

9 Solve $0.06(100 - x) + 0.04x$
$= 0.05(92)$.

$\qquad \mathbf{0.1}t + 0.05(20 - t) = 0.09(20)$

$\qquad \mathbf{0.10}t + 0.05(20 - t) = 0.09(20)$ — $0.1 = 0.10$

$\qquad \mathbf{100}[0.10t + 0.05(20 - t)] = \mathbf{100}[0.09(20)]$ — Multiply by 100.

$\qquad \mathbf{100}(0.10t) + \mathbf{100}[0.05(20 - t)] = \mathbf{100}[0.09(20)]$ — Distributive property

$\qquad 10t + \mathbf{5}(20 - t) = 9(20)$ — Multiply.

$\qquad 10t + \mathbf{5}(20) + \mathbf{5}(-t) = 180$ — Distributive property

$\qquad 10t + 100 - 5t = 180$ — Multiply.

$\qquad 5t + 100 = 180$ — Combine like terms.

Step 2 $\qquad 5t + 100 \mathbf{\ -\ 100} = 180 \mathbf{\ -\ 100}$ — Subtract 100.

$\qquad\qquad\qquad 5t = 80$ — Combine like terms.

Step 3 $\qquad\qquad\quad \dfrac{5t}{\mathbf{5}} = \dfrac{80}{\mathbf{5}}$ — Divide by 5.

$\qquad\qquad\qquad t = 16$

Step 4 Check to confirm that $\{16\}$ is the solution set.

◀ *Work Problem* **9** *at the Side.*

> **Note**
> In Example 8, multiplying by 100 is the same as moving the decimal point two places to the right.
>
> $\qquad 0.\mathbf{10}t + 0.\mathbf{05}(20 - t) = 0.\mathbf{09}(20)$
>
> $\qquad\qquad 10t + 5(20 - t) = 9(20)$ — Multiply by 100.

OBJECTIVE **3** **Solve equations that have no solution or infinitely many solutions.** Every equation we have solved so far has had exactly one solution. Such an equation, which is true for some values of the variable and false for others, is a **conditional equation**—it is true only under certain conditions. Some equations have no solution or infinitely many solutions.

> **EXAMPLE 9** **Solving an Equation That Has Infinitely Many Solutions**
>
> Solve $5x - 15 = 5(x - 3)$.
>
> $$5x - 15 = 5(x - 3)$$
> $$5x - 15 = 5x - 15 \qquad \text{Distributive property}$$
> $$5x - 15 - \mathbf{5x} = 5x - 15 - \mathbf{5x} \qquad \text{Subtract } 5x.$$
> $$-15 = -15$$
> $$-15 + \mathbf{15} = -15 + \mathbf{15} \qquad \text{Add 15.}$$
> $$\mathbf{0 = 0} \qquad \text{True}$$
>
> **Solution set: {all real numbers}**
>
> The variable has "disappeared." Since the last statement ($0 = 0$) is true, *any* real number is a solution. (We could have predicted this from the line in the solution that says $5x - 15 = 5x - 15$, which is true for *any* value of x.) Try several values for x in the original equation to see that they all satisfy it.
>
> An equation with both sides exactly the same, like $0 = 0$, is called an **identity.** An identity is true for all replacements of the variables. As shown above, we write the solution set as **{all real numbers}.**

> **CAUTION**
> In Example 9, do not write {0} as the solution set of the equation. While 0 is a solution, there are infinitely many other solutions.

> **EXAMPLE 10** **Solving an Equation That Has No Solution**
>
> Solve $2x + 3(x + 1) = 5x + 4$.
>
> $$2x + 3(x + 1) = 5x + 4$$
> $$2x + 3x + 3 = 5x + 4 \qquad \text{Distributive property}$$
> $$5x + 3 = 5x + 4 \qquad \text{Combine like terms.}$$
> $$5x + 3 - \mathbf{5x} = 5x + 4 - \mathbf{5x} \qquad \text{Subtract } 5x.$$
> $$\mathbf{3 = 4} \qquad \text{False}$$
>
> **There is no solution. Solution set: $\emptyset$**
>
> Again, the variable has disappeared, but this time a false statement ($3 = 4$) results. This is a signal that the equation, called a **contradiction,** has no solution. Its solution set is the **empty set,** or **null set,** symbolized $\emptyset$.

> **CAUTION**
> **Do not** write {$\emptyset$} to represent the empty set.

10 Solve each equation.

(a) $2(x - 6) = 2x - 12$

(b) $3x + 6(x + 1) = 9x - 4$

Work Problem **10** *at the Side.* ▶

11 Perform each translation.

(a) Two numbers have a sum of 36. One of the numbers is r. Write an expression for the other number.

OBJECTIVE **4** **Write expressions for two related unknown quantities.**

Problem-Solving Hint

Often we are given a problem in which the sum of two quantities is a particular number, and we are asked to find the values of the two quantities. Example 11 shows how to express the unknown quantities in terms of a single variable.

EXAMPLE 11 **Translating a Phrase into an Algebraic Expression**

Two numbers have a sum of 23. If one of the numbers is represented by k, write an expression for the other number.

First, suppose that the sum of two numbers is 23, and one of the numbers is **10**. How would you find the other number? You would subtract **10** from 23 to get 13.

$$23 - 10 = 13$$

So instead of using **10** as one of the numbers, use k as stated in the problem. The other number would be obtained in the same way. You must subtract k from 23. Therefore, an expression for the other number is

$$23 - k.$$

(b) The product of two numbers is -6. One of the numbers is q. Write an expression for the other number.

CAUTION

Since the sum of the two numbers in Example 11 is 23, the expression for the other number must be $23 - k$, *not* $k - 23$. (Subtraction is *not* commutative.) To check, find the sum of the two numbers:

$$k + (23 - k) = 23, \quad \text{as required.}$$

◀ *Work Problem* **11** *at the Side.*

ANSWERS

11. (a) $36 - r$ **(b)** $\dfrac{-6}{q}$

10.3 ▶▶▶ Exercises

Using the methods of this section, what should we do first when solving each equation?
Do not actually solve.

1. $7x + 8 = 1$

2. $7x - 5x + 15 = 8 + x$

3. $3(2t - 4) = 20 - 2t$

4. $\dfrac{3}{4}z = -15$

5. $\dfrac{2}{3}x - \dfrac{1}{6} = \dfrac{3}{2}x + 1$

6. $0.9x + 0.3(x + 12) = 6$

Solve each equation, and check your solution. See Examples 1–5, 9, and 10.

7. $3x + 2 = 14$

8. $4x + 3 = 27$

9. $-5z - 4 = 21$

10. $-7w - 4 = 10$

11. $4p - 5 = 2p$

12. $6q - 2 = 3q$

13. $5m + 8 = 7 + 3m$

14. $4r + 2 = r - 6$

15. $10p + 6 = 12p - 4$

16. $-5x + 8 = -3x + 10$

17. $7r - 5r + 2 = 5r - r$

18. $9p - 4p + 6 = 7p - 3p$

19. $x + 3 = -(2x + 2)$

20. $2x + 1 = -(x + 3)$

21. $4(2x - 1) = -6(x + 3)$

22. $6(3w + 5) = 2(10w + 10)$

23. $6(4x - 1) = 12(2x + 3)$

24. $6(2x + 8) = 4(3x - 6)$

25. $3(2x - 4) = 6(x - 2)$

26. $3(6 - 4x) = 2(-6x + 9)$

27. $3(4x - 2) + 5x = 30 - x$

28. $5(2m + 3) - 4m = 8m + 27$

29. $-2p + 7 = 3 - (5p + 1)$

30. $4x + 9 = 3 - (x - 2)$

31. Which linear equation does *not* have all real numbers as solutions?

    A. $5x = 4x + x$ **B.** $2(x + 6) = 2x + 12$ **C.** $\dfrac{1}{2}x = 0.5x$ **D.** $3x = 2x$

32. The expression $100[0.03(x - 10)]$ is equivalent to which of the following?

    A. $0.03x - 0.3$ **B.** $3x - 3$ **C.** $3x - 10$ **D.** $3x - 30$

Solve each equation, and check your solution. See Examples 6–8.

33. $-\dfrac{2}{7}r + 2r = \dfrac{1}{2}r + \dfrac{17}{2}$

34. $\dfrac{3}{5}t - \dfrac{1}{10}t = t - \dfrac{5}{2}$

35. $\dfrac{3}{4}x - \dfrac{1}{3}x + 5 = \dfrac{5}{6}x$

36. $\dfrac{1}{5}x - \dfrac{2}{3}x - 2 = -\dfrac{2}{5}x$

37. $\dfrac{1}{7}(3x + 2) - \dfrac{1}{5}(x + 4) = 2$

38. $\dfrac{1}{4}(3x - 1) + \dfrac{1}{6}(x + 3) = 3$

39. $\dfrac{1}{9}(x + 18) + \dfrac{1}{3}(2x + 3) = x + 3$

40. $-\dfrac{1}{4}(x - 12) + \dfrac{1}{2}(x + 2) = x + 4$

41. $-\dfrac{5}{6}q - \left(q - \dfrac{1}{2}\right) = \dfrac{1}{4}(q + 1)$

42. $\dfrac{2}{3}k - \left(k + \dfrac{1}{4}\right) = \dfrac{1}{12}(k + 4)$

43. $0.3(30) + 0.15x = 0.2(30 + x)$

44. $0.2(60) + 0.05x = 0.1(60 + x)$

45. $0.92x + 0.98(12 - x) = 0.96(12)$

46. $1.00x + 0.05(12 - x) = 0.10(63)$

47. $0.02(5000) + 0.03x = 0.025(5000 + x)$

48. $0.06(10{,}000) + 0.08x = 0.072(10{,}000 + x)$

Solve each equation, and check your solution. See Examples 1–10.

49. $-3(5z + 24) + 2 = 2(3 - 2z) - 4$

50. $-2(2s - 4) - 8 = -3(4s + 4) - 1$

51. $-(6k - 5) - (-5k + 8) = -3$

52. $-(4x + 2) - (-3x - 5) = 3$

53. $8(t - 3) + 4t = 6(2t + 1) - 10$

54. $9(v + 1) - 3v = 2(3v + 1) - 8$

55. $4(x + 3) = 2(2x + 8) - 4$

56. $4(x + 8) = 2(2x + 6) + 20$

57. $\dfrac{1}{3}(x + 3) + \dfrac{1}{6}(x - 6) = x + 3$

58. $\dfrac{1}{2}(x + 2) + \dfrac{3}{4}(x + 4) = x + 5$

59. $0.3(x + 15) + 0.4(x + 25) = 25$

60. $0.1(x + 80) + 0.2x = 14$

Write the answer to each problem in terms of the variable. See Example 11.

61. Two numbers have a sum of 12. One number is q. What expression represents the other number?

62. Two numbers have a sum of 26. One of the numbers is r. What expression represents the other number?

63. The product of two numbers is 9. One of the numbers is z. What expression represents the other number?

64. The product of two numbers is 13. One number is k. What expression represents the other number?

65. Monica is a years old. What expression represents her age 12 yr from now? 2 yr ago?

66. Chandler is b years old. What expression represents his age 3 yr ago? 5 yr from now?

67. Tom has r quarters. Express the value of the quarters in cents.

68. Jean has y dimes. Express the value of the dimes in cents.

10.4 ⟫⟫ An Introduction to Applications of Linear Equations

OBJECTIVE 1 Learn the six steps for solving applied problems.
We now look at how algebra is used to solve applied problems. Some of the problems may seem contrived, but the skills you develop in solving them will help you solve more realistic problems in chemistry, biology, business, and other fields.

While there is not one specific method that enables you to solve all kinds of applied problems, we suggest the following six-step method.

Solving an Applied Problem

Step 1 **Read** the problem, several times if necessary, until you *understand* what is given and what is to be found.

Step 2 **Assign a variable** to represent the unknown value, using diagrams or tables as needed. Write down what the variable represents. Express any other unknown values in terms of the variable.

Step 3 **Write an equation** using the variable expression(s).

Step 4 **Solve** the equation.

Step 5 **State the answer.** Does it seem reasonable?

Step 6 **Check** the answer in the words of the *original* problem.

OBJECTIVE 2 Solve problems involving unknown numbers.

Problem-Solving Hint

The third step in solving an applied problem is often the hardest. To translate the problem into an equation, write the given phrases as mathematical expressions. Replace any words that mean *equal* or *same* with an $=$ sign. Other forms of the verb "to be," such as *is, are, was,* and *were,* also translate as an $=$ sign. The $=$ sign leads to an equation to be solved.

EXAMPLE 1 **Finding the Value of an Unknown Number**

The product of 4, and a number decreased by 7, is 100. What is the number?

Step 1 **Read** the problem carefully. We are asked to find a number.

Step 2 **Assign a variable** to represent the unknown quantity. In this problem, we are asked to find a number, so we write

$$\text{Let } x = \text{ the number}.$$

There are no other unknown quantities to find.

Continued on Next Page

1 Use the six steps to solve the problem. Give the equation, using x as the variable, and give the answer.

If 5 is added to the product of 9 and a number, the result is 19 less than the number. Find the number.

Step 3 **Write an equation.**

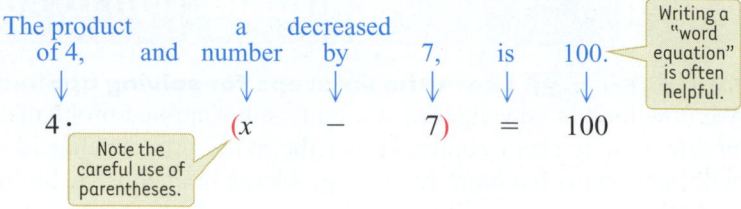

Because of the commas in the given problem, writing the equation as $4x - 7 = 100$ is *incorrect*. The equation $4x - 7 = 100$ corresponds to the statement "The product of 4 and a number, decreased by 7, is 100."

Step 4 **Solve** the equation.

$$4(x - 7) = 100$$

$$4x - 28 = 100 \qquad \text{Distributive property}$$

$$4x - 28 + 28 = 100 + 28 \qquad \text{Add 28.}$$

$$4x = 128 \qquad \text{Combine like terms.}$$

$$\frac{4x}{4} = \frac{128}{4} \qquad \text{Divide by 4.}$$

$$x = 32$$

Step 5 **State the answer.** The number is 32.

Step 6 **Check** When 32 is decreased by 7, we get $32 - 7 = 25$. If 4 is multiplied by 25, we get 100, as required. The answer, 32, is correct.

◀ *Work Problem* **1** *at the Side.*

OBJECTIVE 3 **Solve problems involving sums of quantities.**

Problem-Solving Hint

In general, to solve problems involving sums of quantities, choose a variable to represent one of the unknowns and then *represent the other quantity in terms of the same variable.* (See Example 11 in **Section 10.3.**)

EXAMPLE 2 **Finding Numbers of Olympic Medals**

In the 2006 Winter Olympics in Torino, Italy, the United States won 11 more medals than Sweden. The two countries won a total of 39 medals. How many medals did each country win? (*Source:* U.S. Olympic Committee.)

Step 1 **Read** the problem. We are given information about the total number of medals and asked to find the number each country won.

Step 2 **Assign a variable.**

Let x = the number of medals Sweden won.

Then $x + 11$ = the number of medals the U.S. won.

Continued on Next Page

ANSWER

1. $9x + 5 = x - 19; -3$

Step 3   **Write an equation.**

The total	is	the number of medals Sweden won	plus	the number of medals the U.S. won.
↓	↓	↓	↓	↓
39	=	x	+	$(x + 11)$

Step 4   **Solve** the equation.

$$39 = 2x + 11 \qquad \text{Combine like terms.}$$

$$39 - \mathbf{11} = 2x + 11 - \mathbf{11} \qquad \text{Subtract 11.}$$

$$28 = 2x \qquad \text{Combine like terms.}$$

$$\frac{28}{2} = \frac{2x}{2} \qquad \text{Divide by 2.}$$

$$14 = x, \quad \text{or} \quad x = 14$$

Step 5   **State the answer.** The variable x represents the number of medals Sweden won, so Sweden won 14 medals. Then the number of medals the United States won is $x + 11 = 14 + 11 = 25$.

Step 6   **Check.** Since the United States won 25 medals and Sweden won 14, the total number of medals was $25 + 14 = 39$. Because $25 - 14 = 11$, the United States won 11 more medals than Sweden. This information agrees with what is given in the problem, so the answer checks.

Problem-Solving Hint

The problem in Example 2 could also be solved by letting x represent the number of medals the United States won. Then $x - 11$ would represent the number of medals Sweden won. The equation would be

$$39 = x + (x - 11).$$

The solution of this equation is 25, which is the number of U.S. medals. The number of Swedish medals would be $25 - 11 = 14$. *The answers are the same,* whichever approach is used, even though the equation and its solution are different.

Work Problem **2** *at the Side.* ▶

EXAMPLE 3   **Analyzing a Gasoline/Oil Mixture**

A lawn trimmer uses a mixture of gasoline and oil. The mixture contains 16 oz of gasoline for each ounce of oil. If the tank holds 68 oz of the mixture, how many ounces of oil and how many ounces of gasoline does it require when it is full?

Step 1   **Read** the problem. We must find how many ounces of oil and gasoline are needed to fill the tank.

Step 2   **Assign a variable.**

Let $x =$ the number of ounces of oil required.

Then $16x =$ the number of ounces of gasoline required.

Continued on Next Page

---

2 Solve the problem.

In the 2006 Winter Olympics in Torino, Italy, Canada won 5 more medals than Norway. The two countries won a total of 43 medals. How many medals did each country win? (*Source:* U.S. Olympic Committee.)

3 Solve the problem.

At a meeting of the local coin club, each member brought two nonmembers. If a total of 27 people attended, how many were members and how many were nonmembers?

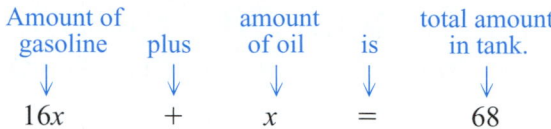
Meeting

| Members x | Nonmembers $2x$ | = 27 |

A diagram like the following is sometimes helpful.

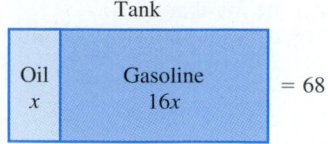
Tank

| Oil x | Gasoline $16x$ | = 68 |

Step 3 **Write an equation.**

Amount of gasoline	plus	amount of oil	is	total amount in tank.
↓	↓	↓	↓	↓
$16x$	$+$	x	$=$	68

Step 4 **Solve.**

$$17x = 68 \qquad \text{Combine like terms.}$$

$$\frac{17x}{17} = \frac{68}{17} \qquad \text{Divide by 17.}$$

$$x = 4$$

Step 5 **State the answer.** The lawn trimmer requires 4 oz of oil and $16(4) = 64$ oz of gasoline when full.

Step 6 **Check.** Since $4 + 64 = 68$, and 64 is 16 times 4, the answer checks.

◀ *Work Problem* **3** *at the Side.*

Problem-Solving Hint

Sometimes it is necessary to find three unknown quantities in an applied problem. Frequently the three unknowns are compared in *pairs*. When this happens, it is usually best to ***let the variable represent the unknown found in both pairs.*** The next example illustrates this.

EXAMPLE 4 **Dividing a Board into Pieces**

The instructions for a woodworking project call for three pieces of wood. The longest piece must be twice the length of the middle-sized piece, and the shortest piece must be 10 in. shorter than the middle-sized piece. Maria Gonzales has a board 70 in. long that she wishes to use. How long must each piece be?

Step 1 **Read** the problem. Three lengths must be found.

Step 2 **Assign a variable.** Since the middle-sized piece appears in both pairs of comparisons, let x represent the length, in inches, of the middle-sized piece. We have

$$x = \text{the length of the middle-sized piece},$$

$$2x = \text{the length of the longest piece, and}$$

$$x - 10 = \text{the length of the shortest piece}.$$

Continued on Next Page

A sketch is helpful here. See Figure 2.

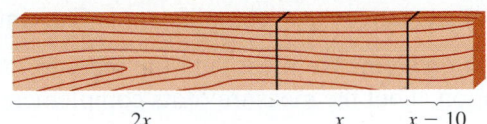

Figure 2

Step 3 **Write an equation.**

Longest	plus	middle-sized	plus	shortest	is	total length.
↓	↓	↓	↓	↓	↓	↓
$2x$	$+$	x	$+$	$(x - 10)$	$=$	70

Step 4 **Solve.**

$$4x - 10 = 70 \qquad \text{Combine like terms.}$$
$$4x - 10 + 10 = 70 + 10 \qquad \text{Add 10.}$$
$$4x = 80 \qquad \text{Combine like terms.}$$
$$\frac{4x}{4} = \frac{80}{4} \qquad \text{Divide by 4.}$$
$$x = 20$$

Step 5 **State the answer.** The middle-sized piece is 20 in. long, the longest piece is $2(20) = 40$ in. long, and the shortest piece is $20 - 10 = 10$ in. long.

Step 6 **Check.** The sum of the lengths is 70 in. All conditions of the problem are satisfied.

Work Problem **4** *at the Side.* ▶

OBJECTIVE **4** **Solve problems involving supplementary and complementary angles.** Recall from **Section 7.1** that an angle can be measured by a unit called the degree (°), which is $\frac{1}{360}$ of a complete rotation. Two angles whose sum is 90° are said to be **complementary**, or *complements* of each other. An angle that measures 90° is a **right angle.** Two angles whose sum is 180° are said to be **supplementary**, or supplements of each other. One angle *supplements* the other to form a **straight angle** of 180°. See Figure 3.

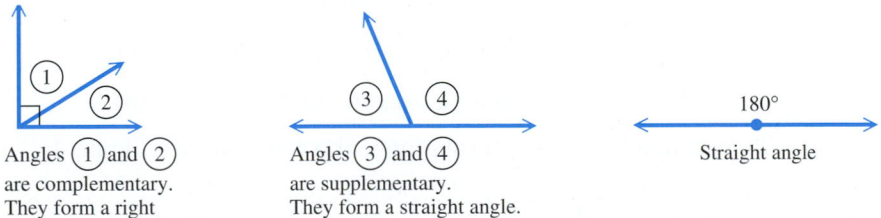

Angles ① and ② are complementary. They form a right angle, indicated by ⌐.

Angles ③ and ④ are supplementary. They form a straight angle.

180°
Straight angle

Figure 3

> **Problem-Solving Hint**
>
> If x represents the degree measure of an angle, then
>
> $\qquad$ **90 − x** represents the degree measure of its complement, and
>
> $\qquad$ **180 − x** represents the degree measure of its supplement.

Work Problem **5** *at the Side.* ▶

4 Solve the problem.
$\qquad$ A piece of pipe is 50 in. long. It is cut into three pieces. The longest piece is 10 in. longer than the middle-sized piece, and the shortest piece measures 5 in. less than the middle-sized piece. Find the lengths of the three pieces.

5 Find each angle measure.

(a) Fill in the blank below the figure. Then find the complement of an angle that measures 26°.

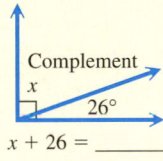

$x + 26 = $ _____

(b) Fill in the blank below the figure. Then find the supplement of an angle that measures 92°.

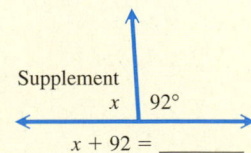

$x + 92 = $ _____

6 Solve the problem.
Find the measure of an angle whose complement is eight times its measure.

EXAMPLE 5   **Finding the Measure of an Angle**

Find the measure of an angle whose complement is five times its measure.

Step 1   **Read** the problem. We must find the measure of an angle, given information about the measure of its complement.

Step 2   **Assign a variable.**

Let $x =$ the degree measure of the angle.

Then $90 - x =$ the degree measure of its complement.

Step 3   **Write an equation.**

Measure of the complement   is   5 times the measure of the angle.

$$90 - x \quad = \quad 5x$$

Step 4   **Solve.**

$$90 - x + x = 5x + x \qquad \text{Add } x.$$

$$90 = 6x \qquad \text{Combine like terms.}$$

$$\frac{90}{6} = \frac{6x}{6} \qquad \text{Divide by 6.}$$

$$15 = x, \quad \text{or} \quad x = 15$$

Step 5   **State the answer.** The measure of the angle is 15°.

Step 6   **Check.** If the angle measures 15°, then its complement measures $90° - 15° = 75°$, which is equal to five times 15°, as required.

◀ *Work Problem* **6** *at the Side.*

EXAMPLE 6   **Finding the Measure of an Angle**

Find the measure of an angle whose supplement is 10° more than twice its complement.

Step 1   **Read** the problem. We are to find the measure of an angle, given information about its complement and its supplement.

Step 2   **Assign a variable.**

Let $x =$ the degree measure of the angle.

Then $90 - x =$ the degree measure of its complement;

$180 - x =$ the degree measure of its supplement.

We can visualize this information using a sketch. See Figure 4.

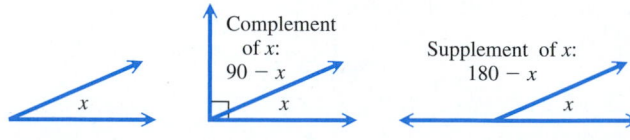

Figure 4

Continued on Next Page

Step 3 **Write an equation.**

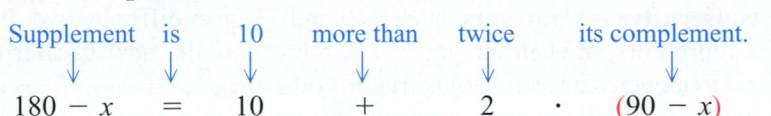

Supplement   is   10   more than   twice   its complement.

$$180 - x \;=\; 10 \;+\; 2 \;\cdot\; (90 - x)$$

Be sure to use parentheses here.

Step 4 **Solve.**

$$180 - x = 10 + 2(90 - x)$$

$$180 - x = 10 + 180 - 2x \qquad \text{Distributive property}$$

$$180 - x = 190 - 2x \qquad \text{Combine like terms.}$$

$$180 - x + 2x = 190 - 2x + 2x \qquad \text{Add } 2x.$$

$$180 + x = 190 \qquad \text{Combine like terms.}$$

$$180 + x - 180 = 190 - 180 \qquad \text{Subtract 180.}$$

$$x = 10$$

Step 5 **State the answer.** The measure of the angle is $10°$.

Step 6 **Check.** The complement of $10°$ is $80°$ and the supplement of $10°$ is $170°$. Also, $170°$ is equal to $10°$ more than twice $80°$ (that is, $170 = 10 + 2(80)$ is true). Therefore, the answer is correct.

Work Problem **7** *at the Side.* ▶

OBJECTIVE 5 Solve problems involving consecutive integers. Two integers that differ by 1 are called **consecutive integers.** For example, 3 and 4, 6 and 7, and -2 and -1 are pairs of consecutive integers. *In general, if x represents an integer, x + 1 represents the next greater consecutive integer.*

EXAMPLE 7 **Finding Consecutive Integers**

Two pages that face each other in this book have 317 as the sum of their page numbers. What are the page numbers?

Step 1 **Read** the problem. Because the two pages face each other, they must have page numbers that are consecutive integers.

Step 2 **Assign a variable.**

Let $x =$ the lesser page number.

Then $x + 1 =$ the greater page number.

Step 3 **Write an equation.** The sum of the page numbers is 317, so

$$x + (x + 1) = 317.$$

Step 4 **Solve.** $2x + 1 = 317 \qquad \text{Combine like terms.}$

$$2x = 316 \qquad \text{Subtract 1.}$$

$$x = 158 \qquad \text{Divide by 2.}$$

Step 5 **State the answer.** The lesser page number is 158, and the greater page number is $158 + 1 = 159$. (Your book is opened to these two pages!)

Step 6 **Check.** The sum of 158 and 159 is 317. The answer is correct.

Work Problem **8** *at the Side.* ▶

7 Solve the problem.
Find the measure of an angle such that twice its complement is $30°$ less than its supplement.

8 Solve the problem.
Two back-to-back page numbers in this book have a sum of 569. What are the page numbers?

9 Solve the problem.

Find two consecutive even integers such that six times the lesser added to the greater gives a sum of 86.

Consecutive *even* integers, such as 8 and 10, differ by 2. Similarly, **consecutive *odd* integers,** such as 9 and 11, also differ by two. In general, if x represents an even integer, $x + 2$ represents the next greater consecutive even integer. The same holds true for odd integers; that is, if x is an odd integer, $x + 2$ is the next greater odd integer. In this book, we list consecutive integers in increasing order when solving applications.

> **Problem-Solving Hint**
>
> When solving consecutive integer problems, if $x =$ the lesser integer, then for any
>
> two consecutive integers, use $\qquad$ $x, \quad x + 1;$
>
> two consecutive *even* integers, use $\qquad$ $x, \quad x + 2;$
>
> two consecutive *odd* integers, use $\qquad$ $x, \quad x + 2.$

In Example 8, see if you can identify the six steps.

EXAMPLE 8 **Finding Consecutive Odd Integers**

If the lesser of two consecutive odd integers is doubled, the result is 7 more than the greater of the two integers. Find the two integers.

Let x be the lesser integer. Since the two numbers are consecutive *odd* integers, then $x + 2$ is the greater. Now write an equation.

If the lesser is doubled,	the result is	7	more than	the greater.
↓	↓	↓	↓	↓
$2x$	$=$	7	$+$	$x + 2$

$$2x = 9 + x \qquad \text{Combine like terms.}$$
$$x = 9 \qquad \text{Subtract } x.$$

The lesser integer is 9 and the greater is $9 + 2 = 11$. To check, we see that when 9 is doubled, we get 18, which is 7 more than the greater odd integer, 11. The answers are correct.

◀ *Work Problem* **9** *at the Side.*

1. In your own words, write the general procedure for solving applications as outlined in this section.

2. List some of the words that translate as "=" when writing an equation to solve an applied problem.

3. Suppose that a problem requires you to find the number of cars on a dealer's lot. Which one of the following would not be a reasonable answer? Justify your answer.

   A. 0   **B.** 45   **C.** 1   **D.** $6\frac{1}{2}$

4. Suppose that a problem requires you to find the number of hours a light bulb is on during a day. Which one of the following would not be a reasonable answer? Justify your answer.

   A. 0   **B.** 4.5   **C.** 13   **D.** 25

5. Suppose that a problem requires you to find the distance traveled in miles. Which one of the following would not be a reasonable answer? Justify your answer.

   A. −10   **B.** 1.8   **C.** $10\frac{1}{2}$ **D.** 50

6. Suppose that a problem requires you to find the time in minutes. Which one of the following would not be a reasonable answer? Justify your answer.

   A. 0   **B.** 10.5   **C.** −5   **D.** 90

Solve each problem. See Example 1.

7. The product of 8, and a number increased by 6, is 104. What is the number?

8. The product of 5, and 3 more than twice a number, is 85. What is the number?

9. Two less than three times a number is equal to 14 more than five times the number. What is the number?

10. Nine more than five times a number is equal to 3 less than seven times the number. What is the number?

11. If 2 is subtracted from a number and this difference is tripled, the result is 6 more than the number. Find the number.

12. If 3 is added to a number and this sum is doubled, the result is 2 more than the number. Find the number.

13. The sum of three times a number and 7 more than the number is the same as the difference between −11 and twice the number. What is the number?

14. If 4 is added to twice a number and this sum is multiplied by 2, the result is the same as if the number is multiplied by 3 and 4 is added to the product. What is the number?

Solve each problem. See Example 2.

15. The number of drive-in movie screens has declined steadily in the United States. Pennsylvania and Ohio had the most remaining drive-in movie screens in 2007. Pennsylvania had 2 more screens than Ohio, and there were 68 screens total in the two states. How many drive-in movie screens remained in each state? (*Source:* Drive-Ins.com)

16. Two of the most watched episodes in television were the final episode of *M*A*S*H,* broadcast on February 23, 1983, and the "Who Shot J. R.?" episode of *Dallas*, broadcast on November 21, 1980. The total number of viewers for these two episodes was about 91 million, with 9 million more people watching the *M*A*S*H* episode than the *Dallas* one. How many people watched each show? (*Source:* Nielsen Media Research.)

17. During the 109th session (2005–2006), the U.S. Senate had a total of 99 Democrats and Republicans. There were 11 more Republicans than Democrats. How many Democrats and Republicans were there in the Senate? (*Source: World Almanac and Book of Facts.*)

18. The total number of Democrats and Republicans in the U.S. House of Representatives during the 109th session was 434. There were 30 more Republicans than Democrats. How many members of each party were there? (*Source: World Almanac and Book of Facts.*)

19. The Police and Kenny Chesney had the two top-grossing North American concert tours in 2007, together generating $204.3 million in ticket sales. If Kenny Chesney took in $62.1 million less than The Police, how much did each tour generate? (*Source:* Pollstar.)

20. The Toyota Camry was the top-selling passenger car in the United States in 2005, followed by the Honda Accord. Honda Accord sales were 65 thousand less than Toyota Camry sales, and 803 thousand of these two cars were sold. How many of each make of car were sold? (*Source:* www.wikipedia.org)

21. In the 2006–2007 NBA regular season, the Phoenix Suns won 19 more than twice as many games as they lost. The Suns played 82 games. How many wins and losses did the team have? (*Source:* nba.com)

22. In the 2007 regular baseball season, the Boston Red Sox won 36 less than twice as many games as they lost. They played 162 regular season games. How many wins and losses did the team have? (*Source:* www.mlb.com)

23. A one-cup serving of orange juice contains 3 mg less than four times the amount of vitamin C as a one-cup serving of pineapple juice. Servings of the two juices contain a total of 122 mg of vitamin C. How many milligrams of vitamin C are in a serving of each type of juice? (*Source:* U.S. Agriculture Department.)

24. A one-cup serving of pineapple juice has 9 more than three times as many calories as a one-cup serving of tomato juice. Servings of the two juices contain a total of 173 calories. How many calories are in a serving of each type of juice? (*Source:* U.S. Agriculture Department.)

Solve each problem. See Example 3.

25. The value of a "Mint State-63" (uncirculated) 1950 Jefferson nickel minted at Denver is $\frac{4}{3}$ the value of a similar condition 1944 nickel minted at Philadelphia. Together, the value of the two coins is $28.00. What is the value of each coin? (*Source:* Yeoman, R., *A Guide Book of United States Coins,* edited by K. Bressett, 61st edition, 2008.)

26. In one day, a store sold $\frac{8}{5}$ as many DVDs as CDs. The total number of DVDs and CDs sold that day was 273. How many DVDs were sold?

27. The world's largest taco was made in the city of Mexicali, Mexico. The taco contained approximately 1 kg of onion for every 6.6 kg of grilled steak. The total weight of these two ingredients was 617.6 kg. To the nearest tenth of a kilogram, how many kilograms of onions and how many kilograms of grilled steak were used to make the taco? (*Source: Guinness World Records.*)

28. The world's most populous countries are China and India. As of mid-2005, the combined population of these two countries was estimated at 2.4 billion. If there were about $\frac{4}{5}$ as many people living in India as China, what was the population of each country, to the nearest tenth of a billion? (*Source:* U.S. Census Bureau.)

29. U.S. five-cent coins are made from a combination of two metals: nickel and copper. For every pound of nickel, 3 lb of copper are used. How many pounds of copper would be needed to make 560 lb of five-cent coins? (*Source:* The United States Mint.)

30. A bakery makes a special whole-grain bread using two kinds of flour: whole wheat and rye. The recipe for this bread calls for 1 oz of rye flour for every 4 oz of whole-wheat flour. How many ounces of each kind of flour should be used to make a loaf of bread weighing 32 oz?

Solve each problem. See Example 4.

31. Al Moser, an office manager, books airline tickets for business trips that employees need to make. In one week, he booked 55 tickets. He booked 7 more tickets on American Airlines than United Airlines. On Southwest Airlines, he booked 4 more than twice as many tickets as on United. How many tickets did he book on each airline?

32. Lauren Morse, a mathematics textbook editor, works 7.5 hr a day. She spent a recent day making telephone calls, writing e-mails, and attending meetings. She spent twice as much time attending meetings as making telephone calls and 0.5 hr longer writing e-mails than making telephone calls. How many hours did she spend on each task?

33. The United States earned 103 medals at the 2004 Summer Olympics in Athens. The number of silver medals earned was 4 more than the number of gold medals. The number of bronze medals earned was 6 less than the number of gold medals. How many of each kind of medal did the United States earn? (*Source: The Gazette,* August 30, 2004.)

34. Nagaraj Nanjappa has a party-length submarine sandwich 59 in. long. He wants to cut it into three pieces so that the middle piece is 5 in. longer than the shortest piece and the shortest piece is 9 in. shorter than the longest piece. How long should the three pieces be?

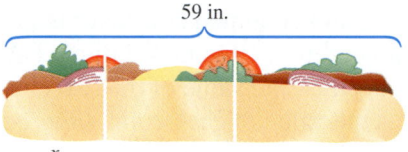

35. Venus is 31.2 million mi farther from the sun than Mercury, while Earth is 57 million mi farther from the sun than Mercury. If the total of the distances from these three planets to the sun is 196.2 million mi, how far away from the sun is Mercury? (All distances given here are mean (*average*) distances.) (*Source: The New York Times Almanac.*)

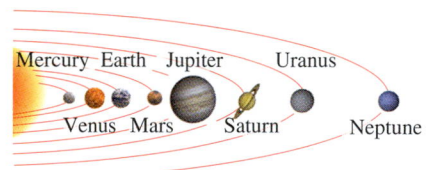

36. Saturn, Jupiter, and Uranus together have a total of 137 known satellites (moons). Jupiter has 16 more satellites than Saturn, and Uranus has 20 fewer satellites than Saturn. How many known satellites does Uranus have? (*Source: The New York Times Almanac.*)

37. The sum of the measures of the angles of any triangle is 180°. In triangle *ABC,* angles *A* and *B* have the same measure, while the measure of angle *C* is 60° greater than each of *A* and *B.* What are the measures of the three angles?

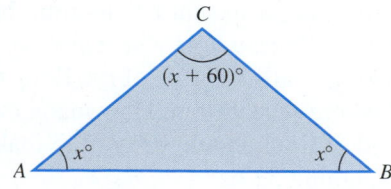

38. In triangle *ABC,* the measure of angle *A* is 141° more than the measure of angle *B.* The measure of angle *B* is the same as the measure of angle *C.* Find the measure of each angle. (*Hint:* See Exercise 37.)

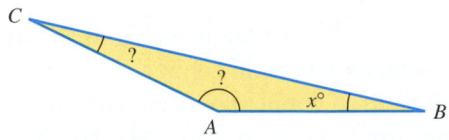

Use the concepts of this section to answer each question.

39. Is there an angle that is equal to its supplement? Is there an angle that is equal to its complement? If the answer is yes to either question, give the measure of the angle.

40. If x represents an integer, how can you express the next smaller consecutive integer in terms of x? The next smaller even integer?

Solve each problem. See Examples 5 and 6.

41. Find the measure of an angle whose complement is four times its measure.

42. Find the measure of an angle whose complement is five times its measure.

43. Find the measure of an angle whose supplement is eight times its measure.

44. Find the measure of an angle whose supplement is three times its measure.

45. Find the measure of an angle whose supplement measures 39° more than twice its complement.

46. Find the measure of an angle whose supplement measures 38° less than three times its complement.

47. Find the measure of an angle such that the difference between the measures of its supplement and three times its complement is 10°.

48. Find the measure of an angle such that the sum of the measures of its complement and its supplement is 160°.

Solve each problem. See Examples 7 and 8.

49. The numbers on two consecutively numbered gym lockers have a sum of 137. What are the locker numbers?

50. The sum of two consecutive check numbers is 357. Find the numbers.

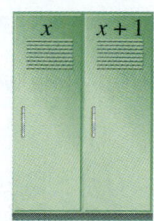

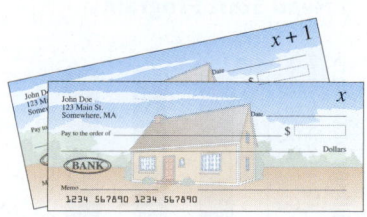

51. Two pages that are back-to-back in this book have 293 as the sum of their page numbers. What are the page numbers?

52. Two houses on the same side of the street have house numbers that are consecutive even integers. The sum of the integers is 58. What are the two house numbers?

53. Find two consecutive even integers such that the lesser added to three times the greater gives a sum of 46.

54. Find two consecutive odd integers such that twice the greater is 17 more than the lesser.

55. When the lesser of two consecutive integers is added to three times the greater, the result is 43. Find the integers.

56. If five times the lesser of two consecutive integers is added to three times the greater, the result is 59. Find the integers.

57. If the sum of three consecutive even integers is 60, what is the first of the three even integers? (*Hint:* If x and $x + 2$ represent the first two consecutive even integers, how would you represent the third consecutive even integer?)

58. If the sum of three consecutive odd integers is 69, what is the third of the three odd integers?

Apply the ideas of this section to solve Exercises 59 and 60, which are based on the graphs.

59. In 2003, federal funding for Head Start programs increased by $0.13 billion from the previous year. The increase from 2003 to 2004 was $0.10 billion. Over the three-year period 2002–2004, the total funding was $19.98 billion. What was federal Head Start funding for each of these years? (*Source:* U.S. Department of Health and Human Services.)

60. In a typical group of 1000 workers from each of the boatbuilding, iron foundry, and amusement park/arcade industries, there were 30 more injuries in iron foundries (I) than in amusement parks/arcades (A). There were 12 more injuries in amusement parks/arcades than in boatbuilding (B). Among these workers, there were 387 nonfatal occupational injuries. How many injuries took place in each industry? (*Source:* U.S. Bureau of Labor Statistics.)

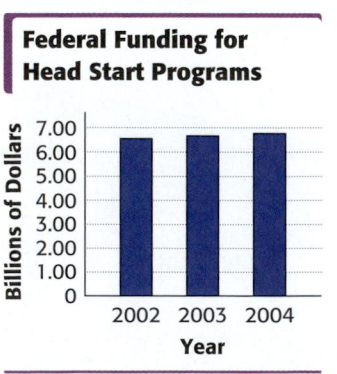

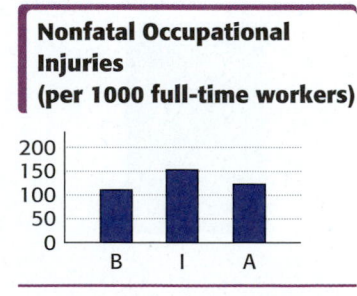

10.5 ▶▶▶ Formulas and Additional Applications from Geometry

A **formula** is an equation in which variables are used to describe a relationship. For example, formulas exist for finding perimeters and areas of geometric figures such as squares and circles, for calculating money earned on bank savings, and for converting among measurements.

$$P = 4s, \quad A = \pi r^2, \quad I = prt, \quad F = \frac{9}{5}C + 32 \qquad \text{Formulas}$$

The formulas used in this book are given on the inside covers.

OBJECTIVE 1 Solve a formula for one variable, given the values of the other variables. In Example 1, we use the idea of *area*. The **area** of a plane (two-dimensional) geometric figure is a measure of the surface covered by the figure.

EXAMPLE 1 **Using Formulas to Evaluate Variables**

Find the value of the remaining variable in each formula.

(a) $A = LW$; $A = 64, L = 10$

As shown in Figure 5, this formula gives the area of a rectangle with length L and width W. Substitute the given values into the formula.

$$A = \mathbf{LW} \qquad \text{Solve for } W.$$
$$64 = \mathbf{10}W \qquad \text{Let } A = 64 \text{ and } L = 10.$$
$$\frac{64}{10} = \frac{10W}{10} \qquad \text{Divide by 10.}$$
$$6.4 = W$$

The width is 6.4. Since $10(6.4) = 64$, the given area, the answer checks.

(b) $A = \frac{1}{2}h(b + B)$; $A = 210, B = 27, h = 10$

This formula gives the area of a trapezoid with parallel sides of lengths b and B and distance h between the parallel sides. See Figure 6.

$$A = \frac{1}{2}\mathbf{h}(b + \mathbf{B}) \qquad \text{Solve for } b.$$
$$210 = \frac{1}{2}(\mathbf{10})(b + \mathbf{27}) \qquad \text{Let } A = 210, h = 10, B = 27.$$
$$210 = 5(b + 27) \qquad \text{Multiply.}$$
$$210 = 5b + 135 \qquad \text{Distributive property}$$
$$210 - \mathbf{135} = 5b + 135 - \mathbf{135} \qquad \text{Subtract 135.}$$
$$75 = 5b \qquad \text{Combine like terms.}$$
$$\frac{75}{5} = \frac{5b}{5} \qquad \text{Divide by 5.}$$
$$15 = b$$

The length of the shorter parallel side, b, is 15. Since $\frac{1}{2}(10)(15 + 27) = 210$, the given area, the answer checks.

Work Problem **1** at the Side. ▶

OBJECTIVES

1. Solve a formula for one variable, given the values of the other variables.

2. Use a formula to solve an applied problem.

3. Solve problems involving vertical angles and straight angles.

4. Solve a formula for a specified variable.

---

1 Find the value of the remaining variable in each formula.

(a) $I = prt$; $I = \$246,$ $r = 0.06, t = 2$

(b) $P = 2L + 2W$; $P = 126,$ $W = 25$

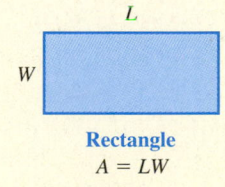

Rectangle
$A = LW$

Figure 5

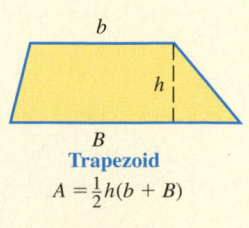

Trapezoid
$A = \frac{1}{2}h(b + B)$

Figure 6

ANSWERS
1. **(a)** $p = \$2050$ **(b)** $L = 38$

2 Solve the problem.

A farmer has 800 m of fencing material to enclose a rectangular field. The width of the field is 175 m. Find the length of the field.

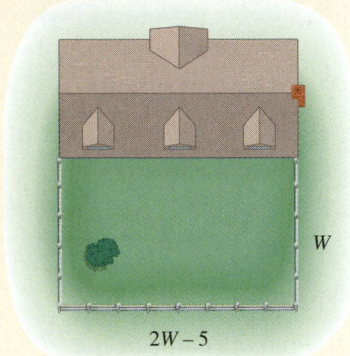

2W − 5

Figure 7

OBJECTIVE 2 Use a formula to solve an applied problem.
Formulas are often used to solve applied problems. *It is a good idea to draw a sketch when a geometric figure is involved.*

Examples 2 and 3 use the idea of *perimeter*. The **perimeter** of a plane (two-dimensional) geometric figure is the distance around the figure. For a polygon (such as a rectangle, square, or triangle), the perimeter is the sum of the lengths of the sides.

> **EXAMPLE 2** **Finding the Dimensions of a Rectangular Yard**
>
> Kari Heen's backyard is in the shape of a rectangle. The length is 5 m less than twice the width, and the perimeter is 80 m. Find the dimensions of the yard.
>
> **Step 1** **Read** the problem. We must find the dimensions of the yard.
>
> **Step 2** **Assign a variable.** Let W = the width of the lot, in meters. Since the length is 5 m less than twice the width, the length is given by $L = 2W - 5$. See Figure 7.
>
> **Step 3** **Write an equation.** The formula for the perimeter of a rectangle is
> $$P = 2L + 2W.$$
>
> Perimeter $= 2 \cdot$ Length $+ 2 \cdot$ Width
>
> $80 = 2(2W - 5) + 2W$ Substitute $2W - 5$ for length L.
>
> **Step 4** **Solve.** $80 = 4W - 10 + 2W$ Distributive property
>
> $80 = 6W - 10$ Combine like terms.
>
> $80 + 10 = 6W - 10 + 10$ Add 10.
>
> $90 = 6W$ Combine like terms.
>
> $\dfrac{90}{6} = \dfrac{6W}{6}$ Divide by 6.
>
> $15 = W$
>
> **Step 5** **State the answer.** The width is 15 m and the length is $2(15) - 5 = 25$ m.
>
> **Step 6** **Check.** If the width of the yard is 15 m and the length is 25 m, the perimeter is $2(25) + 2(15) = 50 + 30 = 80$ m, as required.

◄ *Work Problem* **2** *at the Side.*

> **EXAMPLE 3** **Finding the Dimensions of a Triangle**
>
> The longest side of a triangle is 3 ft longer than the shortest side. The medium side is 1 ft longer than the shortest side. If the perimeter of the triangle is 16 ft, what are the lengths of the three sides?
>
> **Step 1** **Read** the problem. We are given the perimeter of a triangle and want to find the lengths of the three sides.
>
> **Step 2** **Assign a variable.**
>
> Let s = the length of the shortest side, in feet,
>
> $s + 1$ = the length of the medium side, in feet, and,
>
> $s + 3$ = the length of the longest side, in feet.
>
> See Figure 8.

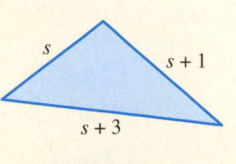

s $s + 1$

$s + 3$

Figure 8

Continued on Next Page

ANSWER

2. 225 m

Step 3   **Write an equation.** Use the formula for the perimeter of a triangle.

$$P = a + b + c$$

$$16 = s + (s + 1) + (s + 3) \qquad \text{Substitute.}$$

Step 4   **Solve.** $16 = 3s + 4$ Combine like terms.

$$12 = 3s \qquad \text{Subtract 4.}$$

$$4 = s \qquad \text{Divide by 3.}$$

Step 5   **State the answer.** Since s represents the length of the shortest side, its measure is 4 ft. The medium side measures

$$s + 1 = \mathbf{4} + 1 = 5 \text{ ft,}$$

and the longest side measures

$$s + 3 = \mathbf{4} + 3 = 7 \text{ ft.}$$

Step 6   **Check.** The medium side, 5 ft, is 1 ft longer than the shortest side, and the longest side, 7 ft, is 3 ft longer than the shortest side. Furthermore, the perimeter is $4 + 5 + 7 = 16$ ft, as required.

Work Problem ③ *at the Side.* ▶

EXAMPLE 4   **Finding the Height of a Triangular Sail**

The area of a triangular sail of a sailboat is 126 ft². (Recall that ft² means "square feet.") The base of the sail is 12 ft. Find the height of the sail.

Step 1   **Read.** We must find the height of the triangular sail.

Step 2   **Assign a variable.** Let h = the height of the sail, in feet. See Figure 9.

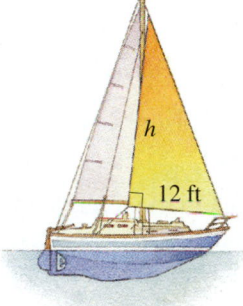

Figure 9

Step 3   **Write an equation.** The formula for the area of a triangle is $A = \frac{1}{2}bh$, where A is the area, b is the base, and h is the height.

$$A = \frac{1}{2}bh$$

$$\mathbf{126} = \frac{1}{2}(\mathbf{12})h \qquad \text{Substitute } A = 126, b = 12.$$

Step 4   **Solve.** $126 = 6h$ Multiply.

$$21 = h \qquad \text{Divide by 6.}$$

Step 5   **State the answer.** The height of the sail is 21 ft.

Step 6   **Check** to see that the values $A = 126$, $b = 12$, and $h = 21$ satisfy the formula for the area of a triangle.

Work Problem ④ *at the Side.* ▶

③ Solve the problem.
   The longest side of a triangle is 1 in. longer than the medium side. The medium side is 5 in. longer than the shortest side. If the perimeter is 32 in., what are the lengths of the three sides?

④ Solve the problem.
   The area of a triangle is 120 m². The height is 24 m. Find the length of the base of the triangle.

ANSWERS

3. 7 in.; 12 in.; 13 in.

4. 10 m

5 Find the measure of each marked angle.

(a)

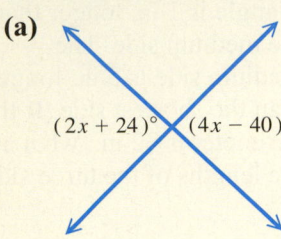

$(2x + 24)°$ $(4x - 40)°$

(b)

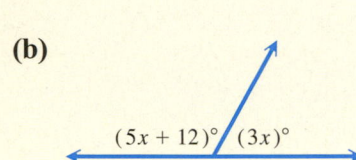

$(5x + 12)°$ $(3x)°$

(c)

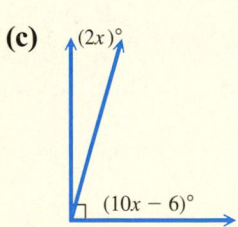

$(2x)°$

$(10x - 6)°$

OBJECTIVE 3 Solve problems involving vertical angles and straight angles. Figure 10 shows two intersecting lines forming angles that are numbered ①, ②, ③, and ④. Angles ① and ③ lie "opposite" each other. They are called **vertical angles.** Another pair of vertical angles is ② and ④. *Vertical angles have equal measures.*

Now look at angles ① and ②. When their measures are added, we get the measure of a **straight angle,** which is **180°.** There are three other such pairs of angles: ② and ③, ③ and ④, and ① and ④.

The next example uses these ideas.

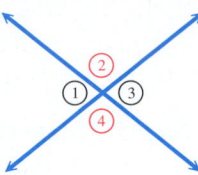

Figure 10

EXAMPLE 5 Finding Angle Measures

Refer to the appropriate figure in each part.

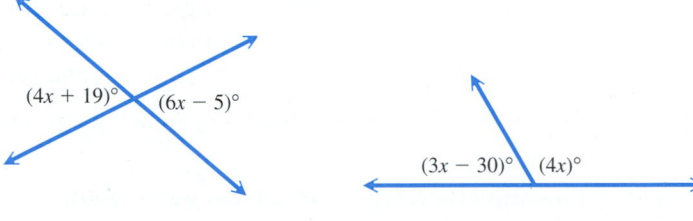

$(4x + 19)°$ $(6x - 5)°$

$(3x - 30)°$ $(4x)°$

Figure 11   **Figure 12**

(a) Find the measure of each marked angle in Figure 11.
Since the marked angles are vertical angles, they have equal measures.

$$4x + 19 = 6x - 5 \quad \text{Set } 4x + 19 \text{ equal to } 6x - 5.$$
$$19 = 2x - 5 \quad \text{Subtract } 4x.$$
$$24 = 2x \quad \text{Add 5.}$$

Don't stop here! $\quad \mathbf{12} = x \quad \text{Divide by 2.}$

Since $x = 12$, one angle has measure $4(\mathbf{12}) + 19 = \mathbf{67}$ degrees. The other has the same measure, since $6(\mathbf{12}) - 5 = \mathbf{67}$ as well. Each angle measures 67°.

(b) Find the measure of each marked angle in Figure 12.
The measures of the marked angles must add to 180° because together they form a straight angle. (They are also supplements of each other.)

$$(3x - 30) + 4x = 180$$
$$7x - 30 = 180 \quad \text{Combine like terms.}$$
$$7x = 210 \quad \text{Add 30.}$$
$$x = \mathbf{30} \quad \text{Divide by 7.}$$

To find the measures of the angles, replace x with 30 in the two expressions.

$$3x - 30 = 3(\mathbf{30}) - 30 = 90 - 30 = \mathbf{60}$$
$$4x = 4(\mathbf{30}) = \mathbf{120}$$

The two angle measures are 60° and 120°.

CAUTION
In Example 5, the answer is *not* the value of x. *Remember to substitute the value of the variable into the expression given for each angle.*

ANSWERS

5. (a) Both measure 88°. **(b)** 117° and 63°
(c) 16° and 74°

◀ *Work Problem* **5** *at the Side.*

OBJECTIVE **4** **Solve a formula for a specified variable.** Sometimes it is necessary to solve a number of problems that use the same formula. For example, a surveying class might need to solve several problems that involve the formula for the area of a rectangle, $A = LW$. Suppose that in each problem the area (A) and the length (L) of a rectangle are given, and the width (W) must be found. Rather than solving for W each time the formula is used, it would be simpler to *rewrite the formula* so that it is solved for W. This process is called **solving for a specified variable** or **solving a literal equation.**

We use the *same* steps to solve a formula for a specified variable that we used to solve an equation with just one variable. Consider the parallel reasoning to solve the following for x.

6 Solve $I = prt$ for t.

$3x + 4 = 13$	$ax + b = c$
$3x + 4 - 4 = 13 - 4$ Subtract 4.	$ax + b - b = c - b$ Subtract b.
$3x = 9$	$ax = c - b$
$\dfrac{3x}{3} = \dfrac{9}{3}$ Divide by 3.	$\dfrac{ax}{a} = \dfrac{c - b}{a}$ Divide by a.
$x = 3$	$x = \dfrac{c - b}{a}$

In solving a formula for a specified variable, we treat the specified variable as if it were the ONLY variable in the equation, and treat the other variables as if they were numbers.

EXAMPLE 6   **Solving for a Specified Variable**

Solve $A = LW$ for W.

Think of undoing what has been done to W. Since W is multiplied by L, undo the multiplication by dividing each side of $A = LW$ by L.

$$A = LW \quad \text{Our goal is to isolate } W.$$

$$\frac{A}{L} = \frac{LW}{L} \qquad \text{Divide by } L.$$

$$\frac{A}{L} = W, \quad \text{or} \quad W = \frac{A}{L} \qquad \tfrac{LW}{L} = \tfrac{L}{L} \cdot W = 1 \cdot W = W$$

Work Problem **6** *at the Side.* ▶

7 Solve $P = a + b + c$ for a.

EXAMPLE 7   **Solving for a Specified Variable**

Solve $P = 2L + 2W$ for L.

$$P = 2L + 2W \quad \text{Our goal is to isolate } L.$$

$$P - 2W = 2L + 2W - 2W \qquad \text{Subtract } 2W.$$

$$P - 2W = 2L \qquad \text{Combine like terms.}$$

$$\frac{P - 2W}{2} = \frac{2L}{2} \qquad \text{Divide by 2.}$$

$$\frac{P - 2W}{2} = L, \quad \text{or} \quad L = \frac{P - 2W}{2}$$

Work Problem **7** *at the Side.* ▶

8 **(a)** Solve $A = p + prt$ for t.

EXAMPLE 8   **Solving for a Specified Variable**

Solve $F = \frac{9}{5}C + 32$ for C. (This is the formula for converting temperatures from Celsius to Fahrenheit.)

We need to isolate C on one side of the equation. First undo the addition of 32 to $\frac{9}{5}C$ by subtracting 32 from each side.

$$F = \frac{9}{5}C + 32$$

$$F - 32 = \frac{9}{5}C + 32 - 32 \qquad \text{Subtract 32.}$$

$$F - 32 = \frac{9}{5}C$$

$$\frac{5}{9}(F - 32) = \frac{5}{9} \cdot \frac{9}{5}C \qquad \text{Multiply by } \tfrac{5}{9}.$$

> Be sure to use parentheses.

$$\frac{5}{9}(F - 32) = C, \quad \text{or} \quad C = \frac{5}{9}(F - 32)$$

This last result is the formula for converting temperatures from Fahrenheit to Celsius.

◀ *Work Problem* **8** *at the Side.*

(b) Solve $Ax + B = C$ for x.

10.5 ▶▶▶ Exercises

1. In your own words, explain what is meant by each term.

   (a) Perimeter of a plane geometric figure

   (b) Area of a plane geometric figure

2. In parts (a)–(c), choose one of the following words to make the statement true: *linear, square,* or *cubic.*

   (a) If the dimensions of a plane geometric figure are given in feet, then the **area** is given in _____ feet.

   (b) If the dimensions of a rectangle are given in yards, then the **perimeter** is given in _____ yards.

   (c) If the dimensions of a pyramid are given in meters, then the **volume** (which is defined on the next page) is given in _____ meters.

3. If a formula has exactly five variables, how many values would you need to be given in order to find the value of any one variable?

4. The formula for changing Celsius temperature to Fahrenheit is given in Example 8 as $F = \frac{9}{5}C + 32$. Sometimes it is seen as $F = \frac{9C}{5} + 32$. These are both correct. Why is it true that $\frac{9}{5}C$ is equal to $\frac{9C}{5}$?

Decide whether perimeter or area would be used to solve a problem concerning the measure of the quantity.

5. Sod for a lawn

6. Carpeting for a bedroom

7. Baseboards for a living room

8. Fencing for a yard

9. Fertilizer for a garden

10. Tile for a bathroom

11. Determining the cost of planting rye grass in a lawn for the winter

12. Determining the cost of replacing a linoleum floor with a wood floor

In the following exercises a formula is given, along with the values of all but one of the variables in the formula. Find the value of the variable that is not given. (When necessary, use 3.14 as an approximation for π (pi).) See Example 1.

13. $P = 2L + 2W$ (perimeter of a rectangle); $L = 8$, $W = 5$

14. $P = 2L + 2W$; $L = 6$, $W = 4$

15. $A = \dfrac{1}{2}bh$ (area of a triangle); $b = 8, h = 16$

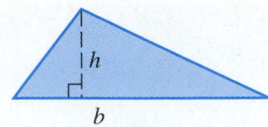

16. $A = \dfrac{1}{2}bh$; $b = 10, h = 14$

17. $P = a + b + c$ (perimeter of a triangle); $P = 12, a = 3, c = 5$

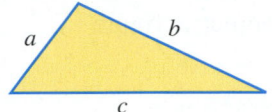

18. $P = a + b + c$; $P = 15, a = 3, b = 7$

19. $d = rt$ (distance formula); $d = 252, r = 45$ **20.** $d = rt$; $d = 100, t = 2.5$

21. $I = prt$ (simple interest); $p = 7500, r = 0.035, t = 6$ **22.** $I = prt$; $p = 5000, r = 0.025, t = 7$

23. $C = 2\pi r$ (circumference of a circle); $C = 16.328$ **24.** $C = 2\pi r$; $C = 8.164$

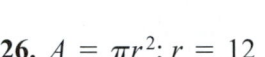
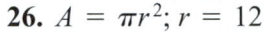

25. $A = \pi r^2$ (area of a circle); $r = 4$ **26.** $A = \pi r^2$; $r = 12$

*The **volume** of a three-dimensional object is a measure of the space occupied by the object. For example, we would need to know the volume of a gasoline tank in order to know how many gallons of gasoline it would take to completely fill the tank. In the following exercises, a formula for the volume (V) of a three-dimensional object is given, along with values for the other variables. Evaluate V. (Use 3.14 as an approximation for π.) See Example 1.*

27. $V = LWH$ (volume of a rectangular box); $L = 10, W = 5, H = 3$

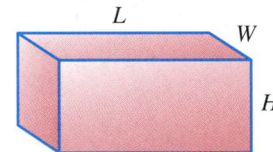

28. $V = LWH$; $L = 12, W = 8, H = 4$

29. $V = \dfrac{1}{3}Bh$ (volume of a pyramid); $B = 12, h = 13$

30. $V = \dfrac{1}{3}Bh$; $B = 36, h = 4$

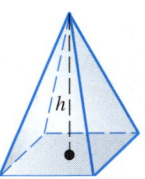

31. $V = \dfrac{4}{3}\pi r^3$ (volume of a sphere); $r = 12$

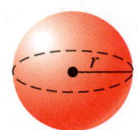

32. $V = \dfrac{4}{3}\pi r^3$; $r = 6$

Solve each perimeter problem. See Examples 2 and 3.

33. The length of a rectangle is 9 in. more than the width. The perimeter is 54 in. Find the length and the width of the rectangle.

34. The width of a rectangle is 3 ft less than the length. The perimeter is 62 ft. Find the length and the width of the rectangle.

35. The perimeter of a rectangle is 36 m. The length is 2 m more than three times the width. Find the length and the width of the rectangle.

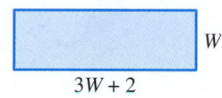

36. The perimeter of a rectangle is 36 yd. The width is 18 yd less than twice the length. Find the length and the width of the rectangle.

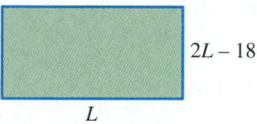

37. The longest side of a triangle is 3 in. longer than the shortest side. The medium side is 2 in. longer than the shortest side. If the perimeter of the triangle is 20 in., what are the lengths of the three sides?

38. The perimeter of a triangle is 28 ft. The medium side is 4 ft longer than the shortest side, while the longest side is twice as long as the shortest side. What are the lengths of the three sides?

39. Two sides of a triangle have the same length. The third side measures 4 m less than twice that length. The perimeter of the triangle is 24 m. Find the lengths of the three sides.

40. A triangle is such that its medium side is twice as long as its shortest side and its longest side is 7 yd less than three times its shortest side. The perimeter of the triangle is 47 yd. What are the lengths of the three sides?

Use a formula to write an equation for each application, and then use the problem-solving method of Section 10.4 to solve. (Use 3.14 as an approximation for π.) Formulas are found on the inside covers of this book. See Examples 2–4.

41. A prehistoric ceremonial site dating to about 3000 B.C. was discovered at Stanton Drew in southwestern England. The site, which is larger than Stonehenge, is a nearly perfect circle, consisting of nine concentric rings that probably held upright wooden posts. Around this timber temple is a wide, encircling ditch enclosing an area with a diameter of 443 ft. Find this enclosed area to the nearest thousand square feet. (*Source: Archaeology,* vol. 51, no. 1, Jan./Feb. 1998.)

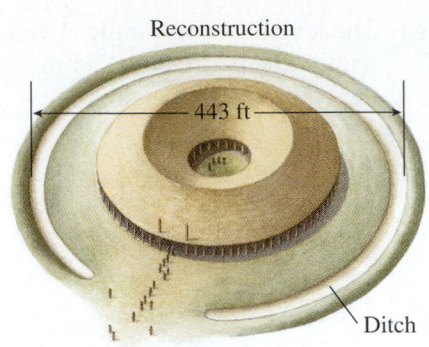

Reconstruction

443 ft

Ditch

42. The Rogers Centre in Toronto, Canada, is the first stadium with a hard-shell, retractable roof. The steel dome is 630 ft in diameter. To the nearest foot, what is the circumference of this dome? (*Source:* www.ballparks.com)

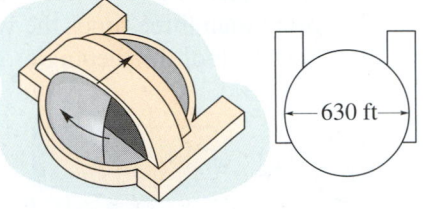

630 ft

43. The largest fashion catalogue in the world was published in Hamburg, Germany. Each of the 212 pages in the catalogue measured 1.2 m by 1.5 m. What was the perimeter of a page? What was the area? (*Source: Guinness World Records.*)

Hohe
Springen Sie
Ausgabe
Mode
NEU
Sommer
Gestaltet!

1.5 m

1.2 m

44. The world's largest sand painting was created by Buddhist monks in the Singapore Expo Hall. The painting measured 12.24 m by 12.24 m. What was the perimeter of the sand painting? To the nearest hundredth of a square meter, what was the area? (*Source: Guinness World Records.*)

45. The largest drum ever constructed was made from Japanese cedar and cowhide, with diameter 15.74 ft. What was the area of the circular face of the drum? Round your answer to the nearest hundredth of a square foot. (*Hint:* Use $A = \pi r^2$.) (*Source: Guinness World Records.*)

46. What was the circumference of the drum described in Exercise 45? Round your answer to the nearest hundredth of a foot. (*Hint:* Use $C = 2\pi r$.)

47. The area of a triangular road sign is 70 ft². If the base of the sign measures 14 ft, what is the height of the sign?

48. The area of a triangular advertising banner is 96 ft². If the height of the banner measures 12 ft, find the measure of the base.

49. The survey plat depicted here shows two lots that form a trapezoid. The measures of the parallel sides are 115.80 ft and 171.00 ft. The height of the trapezoid is 165.97 ft. Find the combined area of the two lots. Round your answer to the nearest hundredth of a square foot.

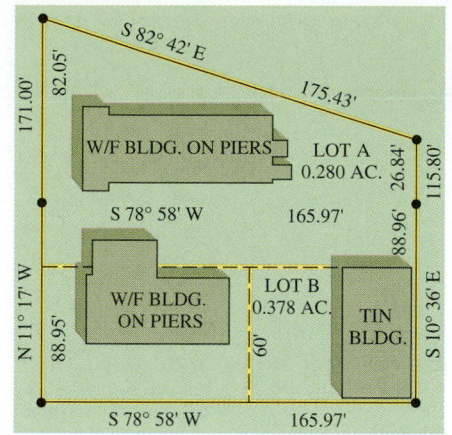

50. Lot A in the figure is in the shape of a trapezoid. The parallel sides measure 26.84 ft and 82.05 ft. The height of the trapezoid is 165.97 ft. Find the area of Lot A. Round your answer to the nearest hundredth of a square foot.

Source: Property survey in New Roads, Louisiana.

51. The U.S. Postal Service requires that any box sent by Priority Mail® have length plus girth (distance around) totaling no more than 108 in. The maximum volume that meets this condition is contained by a box with a square end 18 in. on each side. What is the length of the box? What is the maximum volume? (*Source:* United States Postal Service.)

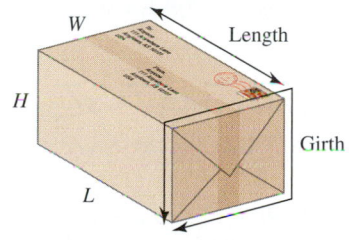

52. On March 17, 2005, a new record was set for the world's largest sandwich. The fillings of this sandwich were corned beef, cheese, lettuce, and mustard. The sandwich, made by Wild Woody's Chill and Grill in Roseville, Michigan, was 12 ft long, 12 ft wide, and $17\frac{1}{2}$ in. $(1\frac{11}{24}$ ft) thick. What was the volume of the sandwich? (*Source:* Guinness World Records.)

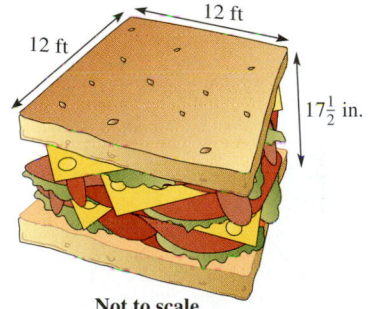

Not to scale

Find the measure of each marked angle. See Example 5.

53.

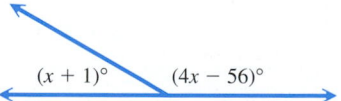

$(x + 1)°$ $(4x - 56)°$

54.

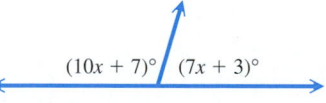

$(10x + 7)°$ $(7x + 3)°$

55.

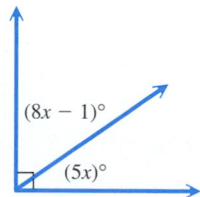

$(8x - 1)°$

$(5x)°$

56.

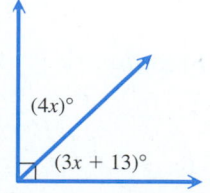

$(4x)°$

$(3x + 13)°$

57.

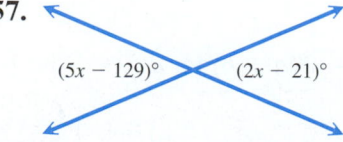

$(5x - 129)°$ $(2x - 21)°$

58.

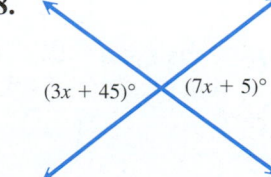

$(3x + 45)°$ $(7x + 5)°$

59.

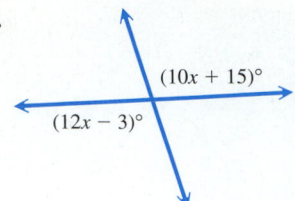

$(10x + 15)°$

$(12x - 3)°$

60.

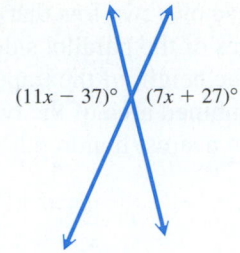

$(11x - 37)°$ $(7x + 27)°$

Solve each formula for the specified variable. See Examples 6–8.

61. $d = rt$ for t

62. $d = rt$ for r

63. $V = LWH$ for H

64. $V = LWH$ for L

65. $P = a + b + c$ for b

66. $P = a + b + c$ for c

67. $C = 2\pi r$ for r

68. $C = \pi d$ for d

69. $I = prt$ for r

70. $I = prt$ for p

71. $A = \dfrac{1}{2}bh$ for h

72. $A = \dfrac{1}{2}bh$ for b

73. $V = \dfrac{1}{3}\pi r^2 h$ for h

74. $V = \pi r^2 h$ for h

75. $P = 2L + 2W$ for W

76. $A = p + prt$ for r

77. $y = mx + b$ for m

78. $y = mx + b$ for x

79. $Ax + By = C$ for y

80. $Ax + By = C$ for x

81. $M = C(1 + r)$ for r

82. $C = \dfrac{5}{9}(F - 32)$ for F

83. $P = 2(a + b)$ for a

84. $P = 2(a + b)$ for b

10.6 ▶▶▶ Solving Linear Inequalities

Inequalities are algebraic expressions related by

< "is less than,"	≤ "is less than or equal to,"
> "is greater than,"	≥ "is greater than or equal to."

We solve an inequality by finding all real number solutions for it. For example, the solutions of $x \leq 2$ include all *real numbers* that are less than or equal to 2, not just the *integers* less than or equal to 2.

OBJECTIVE 1 Graph intervals on a number line. Graphing is a good way to show the solution set of an inequality. To graph all real numbers satisfying $x \leq 2$, we place a square bracket at 2 on a number line and draw an arrow extending from the bracket to the left (since all numbers less than 2 are also part of the graph). The graph is shown in Figure 13.

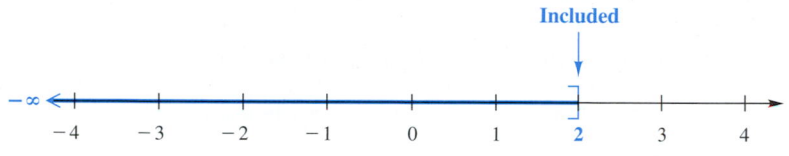

Figure 13 Graph of the interval $(-\infty, 2]$

The set of numbers less than or equal to 2 is an example of an **interval** on the number line. To write intervals, we use **interval notation**. For example, the interval of all numbers less than or equal to 2 is written $(-\infty, 2]$. The **negative infinity** symbol $-\infty$ does not indicate a number but shows that the interval includes all real numbers less than 2. As on the number line, the square bracket indicates that 2 is part of the solution.

EXAMPLE 1 Graphing an Interval on a Number Line

Graph $x > -5$.

The statement $x > -5$ says that x can represent any value greater than -5 but cannot equal -5. The interval is written $(-5, \infty)$. We show this interval on a graph by placing a parenthesis at -5 and drawing an arrow to the right, as in Figure 14. The parenthesis at -5 indicates that -5 is *not* part of the graph.

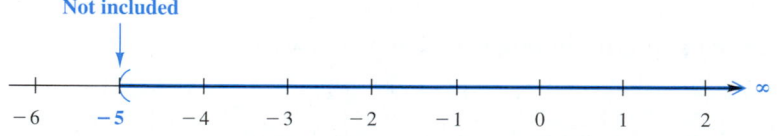

Figure 14 Graph of the interval $(-5, \infty)$

Keep the following important concepts regarding interval notation in mind:

1. A parenthesis indicates that an endpoint is *not included* in a solution set.

2. A bracket indicates that an endpoint is *included* in a solution set.

3. A parenthesis is *always* used next to an infinity symbol, $-\infty$ or ∞.

4. The set of all real numbers is written in interval notation as $(-\infty, \infty)$.

Work Problem **1** *at the Side.* ▶

OBJECTIVES

1. Graph intervals on a number line.

2. Use the addition property of inequality.

3. Use the multiplication property of inequality.

4. Solve inequalities using both properties of inequality.

5. Use inequalities to solve applied problems.

6. Solve linear inequalities with three parts.

---

1 Write each inequality in interval notation and graph the interval.

(a) $x \leq 3$

(b) $x > -4$

(c) $x \leq -\dfrac{3}{4}$

ANSWERS

1. **(a)** $(-\infty, 3]$

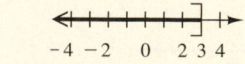

   (b) $(-4, \infty)$

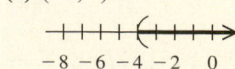

   (c) $\left(-\infty, -\dfrac{3}{4}\right]$

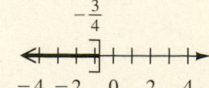

2 Write each inequality in interval notation and graph the interval.

(a) $-4 \geq x$

(b) $0 < x$

3 Write each inequality in interval notation and graph the interval.

(a) $-7 < x < -2$

(b) $-6 < x \leq -4$

2. (a) $(-\infty, -4]$

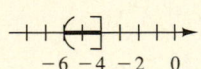
$-8 \ -6 \ -4 \ -2 \quad 0$

(b) $(0, \infty)$

$-4 \ -2 \quad 0 \quad 2 \quad 4$

3. (a) $(-7, -2)$

-7
$-8 \ -6 \ -4 \ -2 \quad 0$

(b) $(-6, -4]$

$-6 \ -4 \ -2 \quad 0$

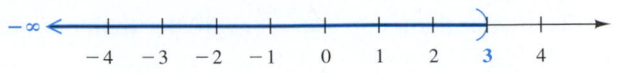

EXAMPLE 2 **Graphing an Interval on a Number Line**

Graph $3 > x$.

The statement $3 > x$ means the same as $x < 3$. The graph of $x < 3$, written in interval notation as $(-\infty, 3)$, is shown in Figure 15.

$-\infty$

$-4 \quad -3 \quad -2 \quad -1 \quad 0 \quad 1 \quad 2 \quad 3 \quad 4$

Figure 15

◄ *Work Problem* **2** *at the Side.*

EXAMPLE 3 **Graphing an Interval on a Number Line**

Graph $-3 \leq x < 2$.

The statement $-3 \leq x < 2$ is read "-3 is less than or equal to x *and x is less than 2."* We want the set of numbers that are *between* -3 and 2, with -3 included and 2 excluded. In interval notation, we write $[-3, 2)$, using a square bracket at -3 because -3 is part of the graph and a parenthesis at 2 because 2 is not part of the graph. The graph is shown in Figure 16.

Included   **Not included**

$-5 \quad -4 \quad -3 \quad -2 \quad -1 \quad 0 \quad 1 \quad 2 \quad 3 \quad 4 \quad 5$

Figure 16 Graph of the interval $[-3, 2)$

◄ *Work Problem* **3** *at the Side.*

Note

Some texts use a solid circle ● rather than a square bracket to indicate that an endpoint is included in a number line graph. An open circle ○ is used to indicate noninclusion, rather than a parenthesis.

OBJECTIVE 2 Use the addition property of inequality. Solving an inequality is similar to solving an equation.

Linear Inequality in One Variable

A **linear inequality in one variable** can be written in the form

$$Ax + B < C,$$

where A, B, and C are real numbers, with $A \neq 0$.

(All definitions and rules are also valid for $>$, $\leq$, and $\geq$.) Examples of linear inequalities in one variable include

$x + 5 < 2$, $t - 3 \geq 5$, and $2k + 5 \leq 10$.   Linear inequalities

Consider the inequality $2 < 5$. If 4 is added to each side, the result is

$$2 + 4 < 5 + 4 \qquad \text{Add 4.}$$

$$6 < 9, \qquad \text{True}$$

a true sentence. This example suggests the **addition property of inequality.**

Addition Property of Inequality

For any real numbers A, B, and C, the inequalities

$$A < B \quad \text{and} \quad A + C < B + C$$

have exactly the same solutions.

In words, the same number may be added to each side of an inequality without changing the solutions.

As with the addition property of equality, the same number may be subtracted from each side of an inequality.

EXAMPLE 4 **Using the Addition Property of Inequality**

Solve $7 + 3k \geq 2k - 5$, and graph the solution set.

$$7 + 3k \geq 2k - 5$$

$$7 + 3k \mathbf{- 2k} \geq 2k - 5 \mathbf{- 2k} \qquad \text{Subtract } 2k.$$

$$7 + k \geq -5 \qquad \text{Combine like terms.}$$

$$7 + k \mathbf{- 7} \geq -5 \mathbf{- 7} \qquad \text{Subtract } 7.$$

$$k \geq -12 \qquad \text{Combine like terms.}$$

The solution set is $[-12, \infty)$. Its graph is shown in Figure 17.

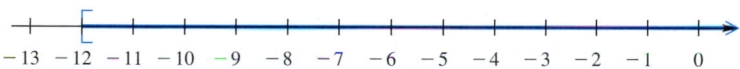

Figure 17

Work Problem **4** *at the Side.* ▶

Note

Because an inequality has many solutions, we cannot check all of them by substitution as we did with the single solution of an equation. To check the solutions in Example 4, we use a multistep process. First, we substitute -12 for k in the related *equation*.

Check $\qquad 7 + 3k = 2k - 5 \qquad$ Related equation

$$7 + 3(\mathbf{-12}) \stackrel{?}{=} 2(\mathbf{-12}) - 5 \qquad \text{Let } k = -12.$$

$$7 - 36 \stackrel{?}{=} -24 - 5 \qquad \text{Multiply.}$$

$$-29 = -29 \qquad \text{True}$$

A true statement results, so -12 is indeed the "boundary" point. Now we test a number other than -12 from the interval $[-12, \infty)$. We choose 0 since it is easy to substitute.

Check $\qquad 7 + 3k \geq 2k - 5 \qquad$ Original inequality

$$7 + 3(\mathbf{0}) \stackrel{?}{\geq} 2(\mathbf{0}) - 5 \qquad \text{Let } k = 0.$$

$$7 \geq -5 \qquad \text{True}$$

Again, a true statement results, so the checks confirm that solutions to the inequality are in the interval $[-12, \infty)$. Any number "outside" the interval $[-12, \infty)$, that is, any number in $(-\infty, -12)$, will give a false statement when tested. (Try this.)

4 Solve each inequality, and graph the solution set.

(a) $-1 + 8r < 7r + 2$

(b) $5m - \dfrac{4}{3} \leq 4m$

5 **(a)** Multiply each side of

$$-2 < 8$$

by 6 and then by -5. Reverse the direction of the inequality symbol if necessary to make a true statement.

OBJECTIVE **3** **Use the multiplication property of inequality.**
The addition property of inequality cannot be used to solve an inequality such as $4x \geq 28$. The *multiplication property of inequality* is required. To see how this property works, we look at some examples.

Multiply each side of the inequality $3 < 7$ by the positive number 2.

$$3 < 7$$
$$\mathbf{2}(3) < \mathbf{2}(7) \qquad \text{Multiply by 2.}$$
$$6 < 14 \qquad \text{True}$$

Now multiply each side of $3 < 7$ by the negative number -5.

$$3 < 7$$
$$\mathbf{-5}(3) < \mathbf{-5}(7) \qquad \text{Multiply by } -5.$$
$$-15 < -35 \qquad \text{False}$$

To get a true statement when multiplying each side by -5, we must reverse the direction of the inequality symbol.

$$3 < 7$$
$$\mathbf{-5}(3) \, \mathbf{>} \, \mathbf{-5}(7) \qquad \begin{array}{l}\text{Multiply by } -5; \text{ reverse the}\\\text{direction of the symbol.}\end{array}$$
$$-15 > -35 \qquad \text{True}$$

Now multiply each side of $-6 < 2$ by the positive number 4.

$$-6 < 2$$
$$\mathbf{4}(-6) < \mathbf{4}(2) \qquad \text{Multiply by 4.}$$
$$-24 < 8 \qquad \text{True}$$

(b) Multiply each side of

$$-4 > -9$$

by 2 and then by -8. Reverse the direction of the inequality symbol if necessary to make a true statement.

Multiplying each side of $-6 < 2$ by -5 *and at the same time reversing the direction of the inequality symbol* gives

$$-6 < 2$$
$$\mathbf{-5}(-6) \, \mathbf{>} \, \mathbf{-5}(2) \qquad \begin{array}{l}\text{Multiply by } -5; \text{ reverse the}\\\text{direction of the symbol.}\end{array}$$
$$30 > -10. \qquad \text{True}$$

◀ *Work Problem* **5** *at the Side.*

In summary, the **multiplication property of inequality** has two parts.

Multiplication Property of Inequality

For any real numbers A, B, and C ($C \neq 0$),

1. if C is *positive,* then the inequalities

$$A < B \quad \text{and} \quad AC < BC$$

have exactly the same solutions;

2. if C is *negative,* then the inequalities

$$A < B \quad \text{and} \quad AC > BC$$

have exactly the same solutions.

In words, each side of an inequality may be multiplied by the same positive number without changing the solutions. *If the multiplier is negative, we must reverse the direction of the inequality symbol.*

As with the multiplication property of equality, the same nonzero number may be divided into each side.

It is important to remember the differences in the multiplication property for positive and negative numbers.

1. When each side of an inequality is multiplied or divided by a *positive number,* the direction of the inequality symbol *does not change.* (Adding or subtracting terms on each side also does not change the symbol.)

2. When each side of an inequality is multiplied or divided by a *negative number,* the direction of the symbol *does change.* **Reverse the direction of the inequality symbol only when multiplying or dividing each side by a negative number.**

EXAMPLE 5 **Using the Multiplication Property of Inequality**

Solve $3r < -18$, and graph the solution set.

Using the multiplication property of inequality, we divide each side by 3. Since 3 is a positive number, the direction of the inequality symbol *does not change.* **It does not matter that the number on the right side of the inequality is negative.**

$$3r < -18$$

> 3 is *positive.* Do NOT reverse the direction of the symbol.

$$\frac{3r}{3} < \frac{-18}{3} \qquad \text{Divide by 3.}$$

$$r < -6$$

The graph of the solution set, $(-\infty, -6)$, is shown in Figure 18.

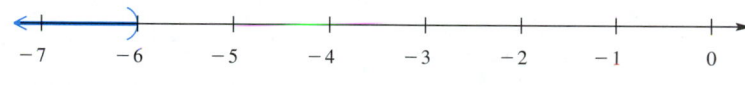

Figure 18

EXAMPLE 6 **Using the Multiplication Property of Inequality**

Solve $-4t \geq 8$, and graph the solution set.

Here each side of the inequality must be divided by -4, a negative number, which *does* require changing the direction of the inequality symbol.

$$-4t \geq 8$$

> -4 is *negative.* Change $\geq$ to $\leq$.

$$\frac{-4t}{-4} \leq \frac{8}{-4} \qquad \text{Divide by } -4.$$

$$t \leq -2$$

The solution set, $(-\infty, -2]$, is graphed in Figure 19.

Figure 19

Work Problem **6** *at the Side.* ▶

---

6 Solve each inequality. Graph the solution set.

(a) $9x < -18$

_____

(b) $-2r > -12$

_____

(c) $-5p \leq 0$

_____

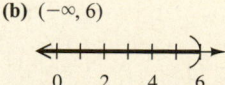
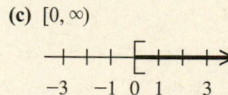

7 Solve

$$7x - 6 + 1 \geq 5x - x + 2.$$

Graph the solution set.

⟶

4   Solve inequalities using both properties of inequality.   The steps to solve a linear inequality are summarized below.

> ### Solving a Linear Inequality
>
> **Step 1   Simplify each side separately.** Use the distributive property to clear parentheses and combine like terms on each side as needed.
>
> **Step 2   Isolate the variable term on one side.** Use the addition property to get all terms with variables on one side of the inequality and all numbers on the other side.
>
> **Step 3   Isolate the variable.** Use the multiplication property to write the inequality in the form $x < c$ or $x > c$.
>
> *Remember: Reverse the direction of the inequality symbol only when multiplying or dividing each side of an inequality by a negative number.*

EXAMPLE 7   Solving a Linear Inequality

Solve $3x + 2 - 5 > -x + 7 + 2x$. Graph the solution set.

Step 1   Combine like terms and simplify.

$$3x + 2 - 5 > -x + 7 + 2x$$

$$3x - 3 > x + 7$$

Step 2   Use the addition property of inequality.

$3x - 3 \mathbf{- x} > x + 7 \mathbf{- x}$	Subtract x.
$2x - 3 > 7$	Combine like terms.
$2x - 3 \mathbf{+ 3} > 7 \mathbf{+ 3}$	Add 3.
$2x > 10$	Combine like terms.

Step 3   Use the multiplication property of inequality.

Because 2 is positive, keep the symbol $>$.

$$\frac{2x}{2} > \frac{10}{2} \qquad \text{Divide by 2.}$$

$$x > 5$$

The solution set, $(5, \infty)$, is graphed in Figure 20.

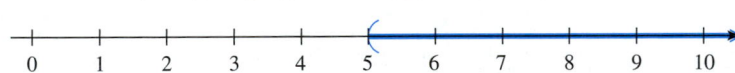

Figure 20

◀ Work Problem **7** at the Side.

EXAMPLE 8   Solving a Linear Inequality

Solve $5(k - 3) - 7k \geq 4(k - 3) + 9$. Graph the solution set.

Step 1   Clear parentheses; then combine like terms.

$5(k - 3) - 7k \geq 4(k - 3) + 9$	
$5k - 15 - 7k \geq 4k - 12 + 9$	Distributive property
$-2k - 15 \geq 4k - 3$	Combine like terms.

Continued on Next Page

ANSWER

7. $\left[\dfrac{7}{3}, \infty \right)$

$$\overset{\tfrac{7}{3}}{\underset{-2 \quad 0 \quad 2\ 3\ 4}{\longleftrightarrow}}$$

Step 2 Use the addition property.

$$-2k - 15 - \mathbf{4k} \geq 4k - 3 - \mathbf{4k}$$ Subtract $4k$.

$$-6k - 15 \geq -3$$ Combine like terms.

$$-6k - 15 + \mathbf{15} \geq -3 + \mathbf{15}$$ Add 15.

$$-6k \geq 12$$ Combine like terms.

Step 3 Divide each side by -6, a negative number. Change the direction of the inequality symbol.

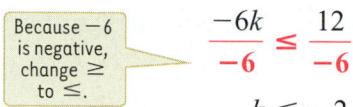 Because -6 is negative, change $\geq$ to $\leq$.

$$\frac{-6k}{-6} \leq \frac{12}{-6}$$ Divide by -6.

$$k \leq -2$$

A graph of the solution set, $(-\infty, -2]$, is shown in Figure 21.

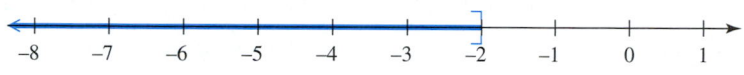

Figure 21

Work Problem **8** *at the Side.* ▶

OBJECTIVE **5** **Use inequalities to solve applied problems.** The table below gives some of the more common phrases that suggest inequality.

Phrase/Word	Example	Inequality
Is greater than	A number *is greater than* 4	$x > 4$
Is less than	A number *is less than* -12	$x < -12$
Exceeds	A number *exceeds* 3.5	$x > 3.5$
Is at least	A number *is at least* 6	$x \geq 6$
Is at most	A number *is at most* 8	$x \leq 8$

Work Problem **9** *at the Side.* ▶

The next example uses the idea of finding the average of a number of scores. ***In general, to find the average of n numbers, add the numbers and divide by n.*** We use the six problem-solving steps from **Section 2.4**, changing Step 3 to "Write an inequality."

EXAMPLE 9 **Finding an Average Test Score**

Brent has grades of 86, 88, and 78 on his first three tests in geometry. If he wants an average of at least 80 after his fourth test, what are the possible scores he can make on that test?

Step 1 **Read** the problem again.

Step 2 **Assign a variable.** Let $x =$ Brent's score on his fourth test.

Step 3 **Write an inequality.** To find his average after four tests, add the test scores and divide by 4.

$$\underbrace{\frac{86 + 88 + 78 + x}{4}}_{\text{Average}} \underbrace{\geq}_{\substack{\text{is at} \\ \text{least}}} \underbrace{80.}_{}$$

Continued on Next Page

---

8 Solve

$$-15 - (2x + 1) \geq 4(x - 1) - 3x.$$

Graph the solution set.

_____→

9 Translate each statement into an inequality, using x as the variable.

(a) The total cost is less than $10.

(b) Chicago received at most 5 in. of snow.

(c) The car's speed exceeded 60 mph.

(d) You must be at least 18 yr old to vote.

ANSWERS

8. $(-\infty, -4]$

−6 −4 −2 0

9. **(a)** $x < 10$ **(b)** $x \leq 5$
   (c) $x > 60$ **(d)** $x \geq 18$

10 Solve the problem.

Maggie has scores of 98, 86, and 88 on her first three tests in algebra. If she wants an average of at least 90 after her fourth test, what score must she make on her fourth test?

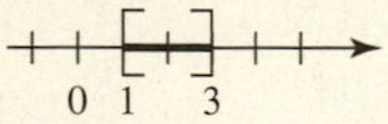

Step 4　Solve.　$\dfrac{252 + x}{4} \geq 80$　Add the known scores.

$$4\left(\dfrac{252 + x}{4}\right) \geq 4\,(80)\qquad \text{Multiply by 4.}$$

$$252 + x \geq 320$$

$$252 + x - 252 \geq 320 - 252\qquad \text{Subtract 252.}$$

$$x \geq 68\qquad \text{Combine like terms.}$$

Step 5　State the answer. He must score 68 or more on the fourth test to have an average of *at least* 80.

Step 6　Check.　$\dfrac{86 + 88 + 78 + 68}{4} = \dfrac{320}{4} = 80$

(Also show that any number in (68, 100] makes the average greater than 80.)

◀ *Work Problem* **10** *at the Side.*

OBJECTIVE **6** **Solve linear inequalities with three parts.** An inequality that says that one number is *between* two other numbers is a **three-part inequality.** For example, $-3 < 5 < 7$ says that 5 is between -3 and 7. Similarly, the three-part inequality

$$3 < x + 2 < 8$$

says that $x + 2$ is between 3 and 8. We solve this inequality as follows.

$$3 - 2 < x + 2 - 2 < 8 - 2\qquad \text{Subtract 2 } \textit{from each part.}$$

$$1 < x < 6$$

The idea is to get the inequality in the form

a number $< x <$ **another number.**

> **CAUTION**
> *Three-part inequalities are written so that the symbols point in the same direction and both point toward the lesser number.* It would be *wrong* to write an inequality as $8 < x + 2 < 3$, since this would imply that $8 < 3$, a false statement.

11 Solve

$$2 \leq 3x - 1 \leq 8.$$

Graph the solution set.

EXAMPLE 10　**Solving a Three-Part Inequality**

Solve $4 \leq 3x - 5 < 10$. Graph the solution set.

$$4 \leq 3x - 5 < 10$$

$$4 + 5 \leq 3x - 5 + 5 < 10 + 5\qquad \text{Add 5 to each part.}$$

$$9 \leq 3x < 15$$

Remember to divide all *three* parts by 3. ▶ $\dfrac{9}{3} \leq \dfrac{3x}{3} < \dfrac{15}{3}$　Divide each part by 3.

$$3 \leq x < 5$$

The solution set, [3, 5), is graphed in Figure 22.

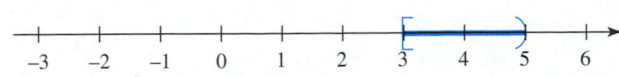

Figure 22

◀ *Work Problem* **11** *at the Side.*

10.6 ▶▶▶ Exercises

Write an inequality using the variable x that corresponds to each graph of solutions on a number line.

1.
$-4\ -3\ -2\ -1\ \ 0\ \ 1\ \ 2\ \ 3$

2.
$-4\ -3\ -2\ -1\ \ 0\ \ 1\ \ 2\ \ 3$

3.
$-2\ -1\ \ 0\ \ 1\ \ 2\ \ 3\ \ 4\ \ 5$

4.
$-2\ -1\ \ 0\ \ 1\ \ 2\ \ 3\ \ 4\ \ 5$

5.
$-1\ \ \ \ 0\ \ \ \ 1\ \ \ \ 2$

6.
$-1\ \ \ \ 0\ \ \ \ 1\ \ \ \ 2$

Write each inequality in interval notation, and graph the interval. See Examples 1–3.

7. $k \le 4$

8. $r \le -10$

9. $x > -3$

10. $x > 3$

11. $8 \le x \le 10$

12. $3 \le x \le 5$

13. $0 < x \le 10$

14. $-3 \le x < 5$

Solve each inequality. Write the solution set in interval notation, and graph it. See Example 4.

15. $z - 8 \ge -7$

16. $p - 3 \ge -11$

17. $2k + 3 \ge k + 8$

18. $3x + 7 \ge 2x + 11$

19. $3n + 5 < 2n - 1$

20. $5x - 2 < 4x - 5$

21. Under what conditions must the inequality symbol be reversed when using the multiplication property of inequality?

22. Explain the steps you would use to solve the inequality $-5x > 20$.

Solve each inequality. Write the solution set in interval notation, and graph it. See Examples 5 and 6.

23. $3x < 18$

24. $5x < 35$

25. $2x \ge -20$

26. $6m \ge -24$

27. $-8t > 24$ ←———————————→

28. $-7x > 49$ ←———————————→

29. $-x \geq 0$ ←———————————→

30. $-k < 0$ ←———————————→

31. $-\dfrac{3}{4}r < -15$ ←———————————→

32. $-\dfrac{7}{8}t < -14$ ←———————————→

33. $-0.02x \leq 0.06$ ←———————————→

34. $-0.03v \geq -0.12$ ←———————————→

Solve each inequality. Write the inequality in interval notation, and graph it. See Examples 4–8.

35. $8x + 9 \leq -15$

←———————————→

36. $6x + 7 \leq -17$

←———————————→

37. $-4x - 3 < 1$

←———————————→

38. $-5x - 4 < 6$

←———————————→

39. $5r + 1 \geq 3r - 9$

←———————————→

40. $6t + 3 < 3t + 12$

←———————————→

41. $6x + 3 + x < 2 + 4x + 4$

←———————————→

42. $-4w + 12 + 9w \geq w + 9 + w$

←———————————→

43. $-x + 4 + 7x \leq -2 + 3x + 6$

←———————————→

44. $14x - 6 + 7x > 4 + 10x - 10$

←———————————→

45. $5(t - 1) > 3(t - 2)$

←———————————→

46. $7(m - 2) < 4(m - 4)$

←———————————→

47. $5(x + 3) - 6x \leq 3(2x + 1) - 4x$

←———————————→

48. $2(x - 5) + 3x < 4(x - 6) + 1$

←———————————→

49. $\dfrac{2}{3}(p + 3) > \dfrac{5}{6}(p - 4)$

50. $\dfrac{7}{9}(n - 4) \le \dfrac{4}{3}(n + 5)$

51. $4x - (6x + 1) \le 8x + 2(x - 3)$

52. $2x - (4x + 3) < 6x + 3(x + 4)$

53. $5(2k + 3) - 2(k - 8) > 3(2k + 4) + k - 2$

54. $2(3z - 5) + 4(z + 6) \ge 2(3z + 2) + 3z - 15$

Solve each problem. See Example 9.

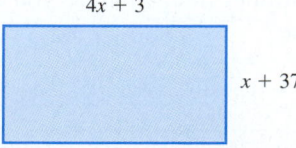 **55.** John Douglas has grades of 84 and 98 on his first two history tests. What must he score on his third test so that his average is at least 90?

56. Elizabeth Gainey has scores of 74 and 82 on her first two algebra tests. What must she score on her third test so that her average is at least 80?

57. When 2 is added to the difference between six times a number and 5, the result is greater than 13 added to 5 times the number. Find all such numbers.

58. When 8 is subtracted from the sum of three times a number and 6, the result is less than 4 more than the number. Find all such numbers.

59. The formula for converting Celsius temperature to Fahrenheit is

$$F = \dfrac{9}{5}C + 32.$$

The Fahrenheit temperature of Providence, Rhode Island, has never exceeded 104°. How would you describe this using Celsius temperature?

60. The formula for converting Fahrenheit temperature to Celsius is

$$C = \dfrac{5}{9}(F - 32).$$

If the Celsius temperature on a certain day in San Diego, California, is never more than 25°, how would you describe the corresponding Fahrenheit temperature?

61. For what values of x would the rectangle have perimeter of at least 400?

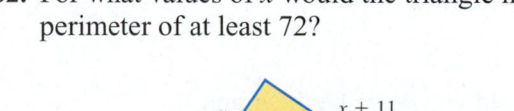

$4x + 3$

$x + 37$

62. For what values of x would the triangle have perimeter of at least 72?

x $x + 11$

$2x + 5$

63. An international phone call costs $2.00 for the first three minutes plus $0.30 per minute for each minute or fractional part of a minute after the first three minutes. If x represents the number of minutes of the length of the call after the first three minutes, then $2 + 0.30x$ represents the cost of the call. If Jorge has $5.60 to spend on a call, what is the maximum total time he can use the phone?

64. At the Speedy Gas 'n Go, a car wash costs $4.50, and gasoline is selling for $4.20 per gal. Terri Hoelker has $48.60 to spend, and her car is so dirty that she must have it washed. What is the maximum number of gallons of gasoline that she can purchase?

Solve each inequality. Write the solution set in interval notation, and graph it. See Example 10.

65. $-5 \le 2x - 3 \le 9$

66. $-7 \le 3x - 4 \le 8$

67. $10 < 7p + 3 < 24$

68. $-8 \le 3r - 1 \le -1$

69. $-12 < -1 + 6m \le -5$

70. $-14 \le 1 + 5q < 3$

Relating Concepts (Exercises 71–75) For Individual or Group Work

Work Exercises 71–75 in order, *to see the connection between the solution of an equation and the solutions of the corresponding inequalities. Graph the solutions in Exercises 71–73.*

71. $3x + 2 = 14$

72. $3x + 2 < 14$

73. $3x + 2 > 14$

74. Now graph all the solutions together on the following number line.

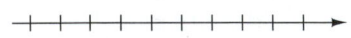

How would you describe the graph?

75. Based on your results from Exercises 71–74, if you were to graph the solutions of

$$-4x + 3 = -1, \quad -4x + 3 > -1,$$
$$\text{and} \quad -4x + 3 < -1$$

on the same number line, what do you think the graph would be?

Chapter 10 ▷▷▷ Summary

▶ Key Terms

10.1 linear equation
A linear equation in one variable is an equation that can be written in the form $Ax + B = C$, where A, B, and C are real numbers, with $A \neq 0$.

solution set
The set of all solutions of an equation is its solution set.

equivalent equations
Equations that have exactly the same solution sets are equivalent equations.

10.3 conditional equation
A conditional equation is an equation that is true for some values of the variable and false for others.

identity
An identity is an equation that is true for all values of the variable.

contradiction
A contradiction is an equation with no solution.

10.4 complementary angles
Two angles whose measures have a sum of 90° are complementary angles.

right angle
A right angle measures 90°.

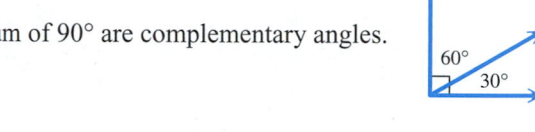

supplementary angles
Two angles whose measures have a sum of 180° are supplementary angles.

straight angle
A straight angle measures 180°.

consecutive integers
Two integers that differ by 1 are consecutive integers.

10.5 formula
A formula is an equation in which variables are used to describe a relationship.

area
The area of a plane geometric figure is a measure of the surface covered by the figure.

perimeter
The perimeter of a plane geometric figure is the distance around the figure.

vertical angles
Vertical angles are angles formed by intersecting lines. They have the same measure.

10.6 inequality
Inequalities are algebraic expressions related by $<$, $\leq$, $>$, or $\geq$.

interval
An interval is a portion of a number line.

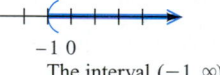
The interval $(-1, \infty)$

interval notation
Interval notation is a special notation that uses parentheses ( ) and/or brackets [ ] to describe an interval on a number line.

linear inequality
A linear inequality in one variable can be written in the form $Ax + B < C$, $Ax + B \leq C$, $Ax + B > C$, or $Ax + B \geq C$, where A, B, and C are real numbers, with $A \neq 0$.

three-part inequality
An inequality that says that one number is between two other numbers is a three-part inequality.

▶ New Symbols

$\emptyset$ empty (null) set

(a, b) interval notation for $a < x < b$

∞ infinity

$[a, b]$ interval notation for $a \leq x \leq b$

$-\infty$ negative infinity

$(-\infty, \infty)$ set of all real numbers

▶ Test Your Word Power

See how well you have learned the vocabulary in this chapter. Answers, with examples, follow the Quick Review.

1. A **solution** of an equation is a number that
   - A. makes an expression undefined
   - B. makes the equation false
   - C. makes the equation true
   - D. makes an expression equal to 0.

2. **Complementary angles** are angles
   - A. formed by two parallel lines
   - B. whose sum is 90°
   - C. whose sum is 180°
   - D. formed by perpendicular lines.

3. **Supplementary angles** are angles
   - A. formed by two parallel lines
   - B. whose sum is 90°
   - C. whose sum is 180°
   - D. formed by perpendicular lines.

4. An **inequality** is
   - A. a statement that two algebraic expressions are equal
   - B. a point on a number line
   - C. an equation with no solutions
   - D. a statement with algebraic expressions related by $<, \leq, >,$ or $\geq$.

▶ Quick Review

Concepts	Examples

10.1 The Addition Property of Equality

The same number may be added to (or subtracted from) each side of an equation without changing the solution.

Solve.
$$x - 6 = 12$$
$$x - 6 + 6 = 12 + 6 \qquad \text{Add 6.}$$
$$x = 18 \qquad \text{Combine like terms.}$$

Solution set: $\{18\}$

10.2 The Multiplication Property of Equality

Each side of an equation may be multiplied (or divided) by the same nonzero number without changing the solution.

Solve.
$$\frac{3}{4}x = -9$$
$$\frac{4}{3} \cdot \frac{3}{4}x = \frac{4}{3}(-9) \qquad \text{Multiply by } \tfrac{4}{3}.$$
$$x = -12$$

Solution set: $\{-12\}$

Concepts	Examples

10.3 More on Solving Linear Equations

Step 1   Simplify each side separately.

Step 2   Isolate the variable term on one side.

Step 3   Isolate the variable.

Step 4   Check.

Solve. $2x + 2(x + 1) = 14 + x$

$$2x + 2x + 2 = 14 + x \qquad \text{Distributive property}$$
$$4x + 2 = 14 + x \qquad \text{Combine like terms.}$$
$$4x + 2 - x - 2 = 14 + x - x - 2$$
$$\text{Subtract } x; \text{ subtract 2.}$$
$$3x = 12 \qquad \text{Combine like terms.}$$
$$\frac{3x}{3} = \frac{12}{3} \qquad \text{Divide by 3.}$$
$$x = 4$$

To check, substitute 4 for x in the original equation.

$$2(4) + 2(4 + 1) \overset{?}{=} 14 + 4 \qquad \text{Let } x = 4.$$
$$18 = 18 \qquad \text{True}$$

Solution set: $\{4\}$

10.4 An Introduction to Applications of Linear Equations

Step 1   Read.

Step 2   Assign a variable.

Step 3   Write an equation.

Step 4   Solve the equation.

Step 5   State the answer.

Step 6   Check.

One number is 5 more than another. Their sum is 21. What are the numbers?

We are looking for two numbers.

Let x represent the lesser number. Then $x + 5$ represents the greater number.

$$x + (x + 5) = 21$$
$$2x + 5 = 21 \qquad \text{Combine like terms.}$$
$$2x = 16 \qquad \text{Subtract 5.}$$
$$x = 8 \qquad \text{Divide by 2.}$$

The numbers are 8 and 13.

13 is 5 more than 8, and $8 + 13 = 21$. It checks.

10.5 Formulas and Additional Applications from Geometry

To find the value of one of the variables in a formula, given values for the others, substitute the known values into the formula.

Find L if $A = LW$, given that $A = 24$ and $W = 3$.

$$24 = L \cdot 3 \qquad A = 24, W = 3$$
$$\frac{24}{3} = \frac{L \cdot 3}{3} \qquad \text{Divide by 3.}$$
$$8 = L$$

To solve a formula for one of the variables, isolate that variable by treating the other variables as numbers and using the steps for solving equations.

Solve $P = 2a + 2b$ for b.

$$P - 2a = 2a + 2b - 2a \qquad \text{Subtract } 2a.$$
$$P - 2a = 2b \qquad \text{Combine like terms.}$$
$$\frac{P - 2a}{2} = \frac{2b}{2} \qquad \text{Divide by 2.}$$
$$\frac{P - 2a}{2} = b, \quad \text{or} \quad b = \frac{P - 2a}{2}$$

Concepts	Examples

10.6 Solving Linear Inequalities

Step 1 Simplify each side separately.

Solve and graph the solution set.

$$3(1 - x) + 5 - 2x > 9 - 6$$

$$3 - 3x + 5 - 2x > 9 - 6 \quad \text{Distributive property}$$

$$8 - 5x > 3 \quad \text{Combine like terms.}$$

Step 2 Isolate the variable term on one side.

$$8 - 5x - 8 > 3 - 8 \quad \text{Subtract 8.}$$

$$-5x > -5 \quad \text{Combine like terms.}$$

Step 3 Isolate the variable.

$$\frac{-5x}{-5} < \frac{-5}{-5} \quad \begin{array}{l}\text{Divide by } -5; \\ \text{change} > \text{to} <.\end{array}$$

Be sure to reverse the direction of the inequality symbol when multiplying or dividing by a negative number.

$$x < 1$$

Solution set: $(-\infty, 1)$

To solve a three-part inequality such as

$$4 < 2x + 6 < 8,$$

work with all three parts at the same time.

Solve.

$$4 < 2x + 6 < 8$$

$$4 - 6 < 2x + 6 - 6 < 8 - 6 \quad \text{Subtract 6.}$$

$$-2 < 2x < 2 \quad \text{Combine like terms.}$$

$$\frac{-2}{2} < \frac{2x}{2} < \frac{2}{2} \quad \text{Divide by 2.}$$

$$-1 < x < 1$$

Solution set: $(-1, 1)$

▶ Answers to Test Your Word Power

1. C; *Example:* 8 is the solution of $2x + 5 = 21$.
2. B; *Example:* Angles with measures $35°$ and $55°$ are complementary angles.
3. C; *Example:* Angles with measures $112°$ and $68°$ are supplementary angles.
4. D; *Examples:* $x < 5, 7 + 2y \geq 11$

Chapter 10 ▶▶▶ Review Exercises

[10.1–10.3] *Solve each equation. Check the solution.*

1. $x - 7 = 2$

2. $4r - 6 = 10$

3. $5x + 8 = 4x + 2$

4. $8t = 7t + \dfrac{3}{2}$

5. $(4r - 8) - (3r + 12) = 0$

6. $7(2x + 1) = 6(2x - 9)$

7. $-\dfrac{6}{5}y = -18$

8. $\dfrac{1}{2}r - \dfrac{1}{6}r + 3 = 2 + \dfrac{1}{6}r + 1$

9. $3x - (-2x + 6) = 4(x - 4) + x$

10. $0.10(x + 80) + 0.20x = 8 + 0.30x$

[10.4] *Solve each problem.*

11. If 7 is added to five times a number, the result is equal to three times the number. Find the number.

12. If 4 is subtracted from twice a number, the result is 36. Find the number.

13. The land area of Hawaii is 5213 mi² greater than that of Rhode Island. Together, the areas total 7637 mi². What is the area of each state?

14. The height of Seven Falls in Colorado is $\frac{5}{2}$ the height (in feet) of Twin Falls in Idaho. The sum of the heights is 420 ft. Find the height of each.

15. The supplement of an angle measures 10 times the measure of its complement. What is the measure of the angle (in degrees)?

16. Find two consecutive odd integers such that when the lesser is added to twice the greater, the result is 24 more than the greater integer.

[10.5] *A formula is given in each exercise, along with the values for all but one of the variables. Find the value of the variable that is not given. (For Exercises 19 and 20, use 3.14 as an approximation for π.)*

17. $A = \dfrac{1}{2}bh$; $A = 44$, $b = 8$

18. $A = \dfrac{1}{2}h(b + B)$; $b = 3$, $B = 4$, $h = 8$

19. $C = 2\pi r$; $C = 29.83$

20. $V = \dfrac{4}{3}\pi r^3$; $r = 9$

Solve each formula for the specified variable.

21. $A = bh$ for h

22. $A = \dfrac{1}{2}h(b + B)$ for h

Find the measure of each marked angle.

23.

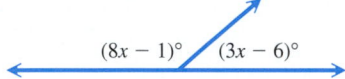

24.

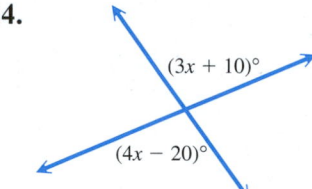

Solve each application of geometry.

25. A cinema screen in Indonesia has length 92.75 ft and width 70.5 ft. What is the perimeter? What is the area? (*Source: Guinness World Records.*)

26. There is a Montezuma cypress in Mexico that is 137 ft tall and has a circumference of about 146.9 ft. What is the diameter of the tree? What is the radius of the tree? Use 3.14 as an approximation for π. Round your answers to the nearest hundredth. (*Source: Guinness World Records.*)

[10.6] *Write each inequality in interval notation, and graph it.*

27. $p \geq -4$

28. $x < 7$

29. $-5 \leq k < 6$

30. $r \geq \dfrac{1}{2}$

Solve each inequality. Write the solution set in interval notation, and graph it.

31. $x + 6 \geq 3$

32. $5t < 4t + 2$

33. $-6x \leq -18$

34. $8(k - 5) - (2 + 7k) \geq 4$

35. $4x - 3x > 10 - 4x + 7x$

36. $3(2w + 5) + 4(8 + 3w) < 5(3w + 2) + 2w$

37. $-3 \leq 2x + 1 < 4$

38. $8 < 3x + 5 \leq 20$

39. Justin Sudak has grades of 94 and 88 on his first two calculus tests. What possible scores on a third test will give him an average of at least 90?

40. If nine times a number is added to 6, the result is at most 3. Find all such numbers.

▶▶▶ Mixed Review Exercises

Solve.

41. $\dfrac{x}{7} = \dfrac{x - 5}{2}$

42. $d = 2r$ for r

43. $-2x > -4$

44. $2k - 5 = 4k + 13$

45. $0.05x + 0.02x = 4.9$

46. $2 - 3(t - 5) = 4 + t$

47. $9x - (7x + 2) = 3x + (2 - x)$

48. $\dfrac{1}{3}s + \dfrac{1}{2}s + 7 = \dfrac{5}{6}s + 5 + 2$

49. Pizza Hut and Domino's, the top-selling pizza restaurants in 2006, together had sales of $14.45 billion. Domino's sales were $4.25 billion less than Pizza Hut's. What were sales in billions for each restaurant? (*Source:* Pizza Today.)

50. Of the 29 medals earned by Germany during the 2006 Winter Olympic games, there were two times as many silver as bronze medals and 5 more gold than bronze medals. How many of each medal did Germany earn? (*Source:* www.torin2006.org)

51. In triangle *DEF*, the measure of angle *E* is twice the measure of angle *D*. Angle *F* has measure 18 less than six times the measure of angle *D*. Find the measure of each angle.

52. The perimeter of a triangle is 96 m. One side is twice as long as another, and the third side is 30 m long. What is the length of the longest side?

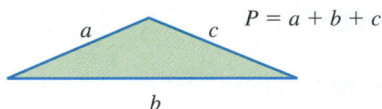

$$P = a + b + c$$

53. The perimeter of a rectangle is 288 ft. The length is 4 ft longer than the width. Find the width.

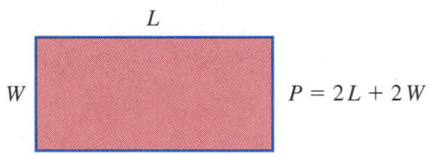

$$P = 2L + 2W$$

54. Find the measure of each marked angle.

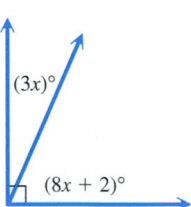

$(3x)°$

$(8x + 2)°$

55. Latarsha has grades of 82 and 96 on her first two English tests. What must she make on her third test so that her average will be at least 90?

56. If nine pairs of jeans cost $121.50, find the cost of five pairs. (Assume all are equally priced.)

Chapter 10 ▶▶▶ Test

Use the Chapter Test Prep Video CD to see fully worked-out solutions to any of the exercises you want to review.

Solve each equation, and check the solution.

1. $3x - 7 = 11$

1. _____

2. $5x + 9 = 7x + 21$

2. _____

3. $2 - 3(x - 5) = 3 + (x + 1)$

3. _____

4. $2.3x + 13.7 = 1.3x + 2.9$

4. _____

5. $7 - (m - 4) = -3m + 2(m + 1)$

5. _____

6. $-\dfrac{4}{7}x = -12$

6. _____

7. $0.06(x + 20) + 0.08(x - 10) = 4.6$

7. _____

8. $-8(2x + 4) = -4(4x + 8)$

8. _____

Solve each problem.

9. The Dallas Mavericks finished with the best record for the 2006–2007 NBA regular season. They won 7 more than four times as many games as they lost. They played 82 games. How many games did they win and lose? (*Source:* www.sports.yahoo.com)

9. _____

10. Three islands in the Hawaiian island chain are Hawaii (the Big Island), Maui, and Kauai. Together, their areas total 5300 mi². The island of Hawaii is 3293 mi² larger than the island of Maui, and Maui is 177 mi² larger than Kauai. What is the area of each island?

10. _____

Kauai
Oahu   Molokai
Lanai   Maui
The Big
Island
HAWAII

11. _____

11. Find the measure of an angle if its supplement measures 10° more than three times its complement.

12. (a) _____

    (b) _____

12. The formula for the perimeter of a rectangle is $P = 2L + 2W$.

    (a) Solve for W.

    (b) If $P = 116$ and $L = 40$, find the value of W.

Find the measure of each marked angle.

13. _____

14. _____

13.

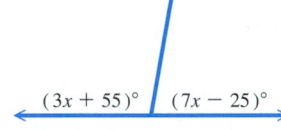

$(3x + 55)°$　$(7x - 25)°$

14.

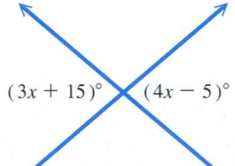

$(3x + 15)°$　$(4x - 5)°$

15. (a) _____

    (b) _____

15. Write an inequality involving x that describes the numbers graphed.

    (a)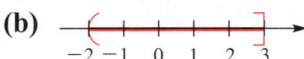

    (b)

Solve each inequality. Write the solution set in interval notation, and graph it.

16. ⟵—+—+—+—+—+—+—⟶

16. $-3x > -33$

17. ⟵—+—+—+—+—+—+—⟶

17. $-0.04x \leq 0.12$

18. ⟵—+—+—+—+—+—+—⟶

18. $-4x + 2(x - 3) \geq 4x - (3 + 5x) - 7$

19. ⟵—+—+—+—+—+—+—⟶

19. $-10 < 3x - 4 \leq 14$

20. _____

20. Shania Johnson has scores of 76 and 81 on her first two algebra tests. If she wants an average of at least 80 after her third test, what score must she make on her third test?

Graphs of Linear Equations and Inequalities in Two Variables

While U.S. consumers in general continue to pile up credit card debt, fewer undergraduate college students are carrying credit cards, and those with cards are using them less. In 2004, 76% of undergraduates carried at least one credit card, down from a peak of 83% in 2001. The average outstanding balance also dropped to $2169, from a high of $2748 in 2000. These declines are attributed to increased financial education aimed specifically at high school and college students. (*Source:* Nellie Mae.)

In Example 7 of **Section 11.2,** we examine a *linear equation in two variables* that models credit card debt in the United States.

11.1 ▸▸▸ Reading Graphs; Linear Equations in Two Variables

As we saw in **Section 8.1,** circle graphs (pie charts) provide a convenient way to organize and communicate information. Along with *bar graphs* and *line graphs,* they can be used to analyze data, make predictions, or simply to entertain us.

OBJECTIVE 1 Interpret graphs. A **bar graph** is used to show comparisons. It consists of a series of bars (or simulations of bars) arranged either vertically or horizontally. In a bar graph, values from two categories are paired with each other.

EXAMPLE 1 Interpreting a Bar Graph

The bar graph in Figure 1 shows U.S. sales of motor scooters, which have gained popularity due to their fuel efficiency. The graph compares sales in thousands.

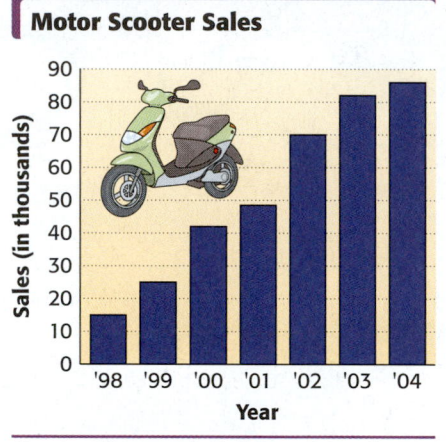

Source: Motorcycle Industry Council.

Figure 1

(a) In what years were sales greater than 50 thousand?

Locate 50 on the vertical scale and follow the line across to the right. Three years—2002, 2003, and 2004—have bars that extend above the line for 50, so sales were greater than 50 thousand in those years.

(b) Estimate sales in 2000 and 2004.

Locate the top of the bar for 2000, and move horizontally across to the vertical scale to see that it is about 40. Sales in 2000 were about 40 thousand. Follow the top of the bar for 2004 across to the vertical scale to see that it lies about halfway between 80 and 90 thousand, so sales in 2004 were about 85,000.

(c) Describe the change in sales as the years progressed.

As the years progressed, sales increased steadily, from about 15 thousand in 1998 to about 85 thousand in 2004.

1 Refer to the bar graph in Figure 1.

(a) Which years had sales less than 50 thousand?

(b) Estimate sales of motor scooters in 1999 and 2001.

(c) Describe the change in sales of motor scooters from 1999 to 2001.

◄ Work Problem 1 at the Side.

ANSWERS

1. **(a)** 1998, 1999, 2000, 2001
   (b) 1999: about 25 thousand; 2001: about 50 thousand
   (c) Sales approximately doubled from 1999 to 2001.

A **line graph** is used to show changes or trends in data over time. To form a line graph, we connect a series of points representing data with line segments.

EXAMPLE 2 **Interpreting a Line Graph**

Current projections indicate that funding for Medicare will not cover its costs unless the program changes. The line graph in Figure 2 shows Medicare funds in billions of dollars for the years 2004 through 2013.

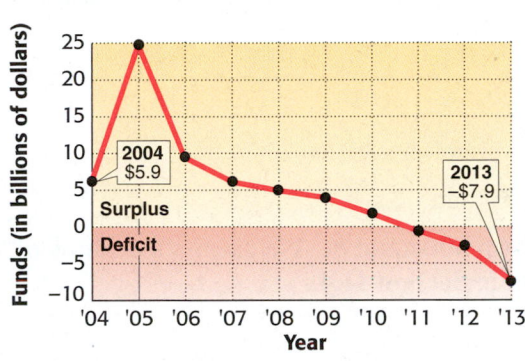

Source: Centers for Medicare and Medicaid Services.
*Projected

Figure 2

(a) Which is the only period in which Medicare funds increased?
Because the graph *rises* from 2004 to 2005 and falls in every other case, funds increased between these two years.

(b) What is the projected trend from 2005 to 2013?
Funds will decrease, since the graph *falls* during this period.

(c) In which year is it projected that funds will first show a deficit?
From 2004 to 2010, the graph is always above 0, but in 2011, it falls slightly below 0 for the first time, indicating a deficit.

(d) Based on the figures shown in the graph, what is the difference in Medicare funds from 2004 to 2013?

$$\underset{\text{2013 amount}}{-\$7.9 \text{ billion}} - \underset{\text{2004 amount}}{\$5.9 \text{ billion}} = \underset{\text{Difference}}{-\$13.8 \text{ billion}}$$

The fund amount will have *decreased* $13.8 billion (as indicated by the negative sign in −$13.8).

Work Problem **2** *at the Side.* ▶

The line graph in Figure 2 relates years to Medicare funds. We can also represent these two related quantities using a table of data, as shown at the side. Notice that in table form, we can see specific data rather than estimating it. Trends in the data are easier to see from the graph, however, which gives us a "picture" of the data.

We can extend these ideas to the subject of this chapter, *linear equations in two variables*. A linear equation in two variables, one for each of the quantities being related, can be used to represent the data in the table or graph. *The graph of a linear equation in two variables is a line.*

2 Refer to the line graph in Figure 2.

(a) Which year has the greatest amount of Medicare funds?

(b) Estimate projected Medicare funds for 2010. Is there a surplus or a deficit in 2010?

(c) About how much is it projected that funds will decrease from 2006 to 2011?

Year	Medicare Funds (in billions of dollars)*
2004	5.9
2005	25.0
2006	9.5
2007	6.0
2008	5.0
2009	4.0
2010	2.0
2011	−0.5
2012	−2.5
2013	−7.9

*Projected

ANSWERS

2. (a) 2005
   (b) $2 billion; surplus
   (c) $10 billion

3 Write each solution as an ordered pair.

(a) $x = 5$ and $y = 7$

(b) $y = 6$ and $x = -1$

(c) $y = 4$ and $x = -3$

(d) $x = \dfrac{2}{3}$ and $y = -12$

(e) $y = 1.5$ and $x = -2.4$

(f) $x = 0$ and $y = 0$

> **Linear Equation in Two Variables**
>
> A **linear equation in two variables** is an equation that can be written in the form
>
> $$Ax + By = C,$$
>
> where A, B, and C are real numbers and A and B are not both 0.

Some examples of linear equations in two variables in this form, called *standard form,* are

$$3x + 4y = 9, \quad x - y = 0, \quad \text{and} \quad x + 2y = -8. \qquad \text{Linear equations in two variables}$$

> **Note**
>
> Other linear equations in two variables, such as
>
> $$y = 4x + 5 \quad \text{and} \quad 3x = 7 - 2y,$$
>
> are not written in standard form but could be. We discuss the forms of linear equations in more detail in **Section 11.4.**

OBJECTIVE 2 Write a solution as an ordered pair. Recall from **Section 9.2** that a *solution* of an equation is a number that makes the equation true when it replaces the variable. For example, the linear equation in one variable $x - 2 = 5$ has solution **7**, since replacing x with 7 gives a true statement.

A solution of a linear equation in two variables requires two numbers, one for each variable. For example, a true statement results when we replace x with 2 and y with 13 in the equation $y = 4x + 5$ since

$$13 = 4(2) + 5. \qquad \text{Let } x = 2, y = 13.$$

The pair of numbers $x = 2$ and $y = 13$ gives one solution of the equation $y = 4x + 5$. The phrase "$x = 2$ and $y = 13$" is abbreviated

x-value ⟶   ⟵ y-value

$$(2, 13)$$

Ordered pair

with the x-value, 2, and the y-value, 13, given as a pair of numbers written inside parentheses. *The x-value is always given first.* A pair of numbers such as (2, 13) is called an **ordered pair.** As the name indicates, the order in which the numbers are written is important. The ordered pairs (**2**, **13**) and (**13**, **2**) are *not* the same. The second pair indicates that $x = 13$ and $y = 2$. *For two ordered pairs to be equal, their x-values must be equal and their y-values must be equal.*

◀ *Work Problem* **3** *at the Side.*

OBJECTIVE 3 Decide whether a given ordered pair is a solution of a given equation. We substitute the x- and y-values of an ordered pair into a linear equation in two variables to see whether the ordered pair is a solution.

ANSWERS

3. **(a)** $(5, 7)$ **(b)** $(-1, 6)$ **(c)** $(-3, 4)$
(d) $\left(\dfrac{2}{3}, -12\right)$ **(e)** $(-2.4, 1.5)$ **(f)** $(0, 0)$

EXAMPLE 3 Deciding Whether Ordered Pairs Are Solutions of an Equation

Decide whether each ordered pair is a solution of the equation $2x + 3y = 12$.

(a) $(3, 2)$

To see whether $(3, 2)$ is a solution of the given equation $2x + 3y = 12$, substitute 3 for x and 2 for y in the equation.

$$2x + 3y = 12$$
$$2(3) + 3(2) \stackrel{?}{=} 12 \qquad \text{Let } x = 3; \text{ let } y = 2.$$
$$6 + 6 \stackrel{?}{=} 12 \qquad \text{Multiply.}$$
$$12 = 12 \qquad \text{True}$$

This result is true, so $(3, 2)$ is a solution of $2x + 3y = 12$.

(b) $(-2, -7)$

$$2x + 3y = 12$$
$$2(-2) + 3(-7) \stackrel{?}{=} 12 \qquad \text{Let } x = -2; \text{ let } y = -7.$$
$$-4 + (-21) \stackrel{?}{=} 12 \qquad \text{Multiply.}$$
$$-25 = 12 \qquad \text{False}$$

> Use parentheses to avoid errors.

This result is false, so $(-2, -7)$ is *not* a solution of $2x + 3y = 12$.

——————— Work Problem **4** at the Side. ▶

OBJECTIVE 4 Complete ordered pairs for a given equation. Choosing a number for one variable in a linear equation makes it possible to find the value of the other variable.

EXAMPLE 4 Completing Ordered Pairs

Complete each ordered pair for the equation $y = 4x + 5$.

(a) $(7, \_)$

> The x-value always comes first.

In this ordered pair, $x = 7$. To find the corresponding value of y, replace x with 7 in the equation.

$$y = 4x + 5$$
$$y = 4(7) + 5 \qquad \text{Let } x = 7.$$
$$y = 28 + 5 \qquad \text{Multiply.}$$
$$y = 33 \qquad \text{Add.}$$

The ordered pair is $(7, 33)$.

(b) $(\_, -3)$

In this ordered pair, $y = -3$. Find the value of x by replacing y with -3 in the equation; then solve for x.

$$y = 4x + 5$$
$$-3 = 4x + 5 \qquad \text{Let } y = -3.$$
$$-8 = 4x \qquad \text{Subtract 5 from each side.}$$
$$-2 = x \qquad \text{Divide each side by 4.}$$

The ordered pair is $(-2, -3)$.

——————— Work Problem **5** at the Side. ▶

4 Decide whether each ordered pair is a solution of the equation $5x + 2y = 20$.

(a) $(0, 10)$

$$5x + 2y = 20$$
$$5(\_) + 2(\_) \stackrel{?}{=} 20$$
$$\_\_\_\_ + 20 \stackrel{?}{=} 20$$
$$\_\_\_\_ = 20$$

Is $(0, 10)$ a solution?

(b) $(2, -5)$

(c) $(3, 2)$

(d) $(-4, 20)$

5 Complete each ordered pair for the equation $y = 2x - 9$.

(a) $(5, \_)$

$$y = 2(\_) - 9$$
$$y = \_\_\_\_ - 9$$
$$y = \_\_\_\_$$

The ordered pair is _____.

(b) $(2, \_)$

(c) $(\_, 7)$

(d) $(\_, -13)$

6 Complete the table of values for each equation.

(a) $2x - 3y = 12$

x	y
0	
	0
3	
	-3

(b) $x = -1$

x	y
	-4
	0
	2

(c) $y = 4$

x	y
-3	
2	
5	

ANSWERS

6. **(a)**

x	y
0	-4
6	0
3	-2
$\frac{3}{2}$	-3

(b)

x	y
-1	-4
-1	0
-1	2

(c)

x	y
-3	4
2	4
5	4

OBJECTIVE 5 Complete a table of values. Ordered pairs are often displayed in a **table of values**. The table may be written either vertically or horizontally.

EXAMPLE 5 Completing Tables of Values

Complete the table of values for each equation. Then write the results as ordered pairs.

(a) $x - 2y = 8$

x	y
2	
10	
	0
	-2

To complete the first two ordered pairs of the table, let $x = 2$ and $x = 10$, respectively.

If $x = 2$,	If $x = 10$,
then $x - 2y = 8$	then $x - 2y = 8$
becomes $2 - 2y = 8$	becomes $10 - 2y = 8$
$-2y = 6$	$-2y = -2$
$y = -3$.	$y = 1$.

Now complete the last two ordered pairs by letting $y = 0$ and $y = -2$, respectively.

If $y = 0$,	If $y = -2$,
then $x - 2y = 8$	then $x - 2y = 8$
becomes $x - 2(0) = 8$	becomes $x - 2(-2) = 8$
$x - 0 = 8$	$x + 4 = 8$
$x = 8$.	$x = 4$.

The completed table of values follows.

x	y
2	-3
10	1
8	0
4	-2

Write y-values here.

Write x-values here.

The corresponding ordered pairs are $(2, -3)$, $(10, 1)$, $(8, 0)$, and $(4, -2)$. Each ordered pair is a solution of the given equation $x - 2y = 8$.

(b) $x = 5$

x	y
	-2
	6
	3

The given equation is $x = 5$. No matter which value of y is chosen, the value of x is *always* 5.

x	y
5	-2
5	6
5	3

The corresponding ordered pairs are $(5, -2)$, $(5, 6)$, and $(5, 3)$.

◀ *Work Problem* **6** *at the Side.*

Note

We can think of $x = 5$ in Example 5(b) as an equation in two variables by rewriting $x = 5$ as $x + 0y = 5$. This form of the equation shows that for any value of y, the value of x is 5. Similarly, $y = 4$ in Problem 6(c) in the margin on the preceding page is the same as $0x + y = 4$.

OBJECTIVE 6 Plot ordered pairs.   In **Section 10.3,** we saw that linear equations in *one* variable had either one, zero, or an infinite number of real number solutions. These solutions could be graphed on *one* number line. For example, the linear equation in one variable $x - 2 = 5$ has solution 7, which is graphed on the number line in Figure 3.

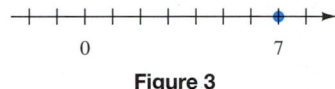

Figure 3

Every linear equation in *two* variables has an infinite number of ordered pairs as solutions. Each choice of a number for one variable leads to a particular real number for the other variable.

To graph these solutions, represented as the ordered pairs (x, y), we need *two* number lines, one for each variable, as drawn in Figure 4. The horizontal number line is called the **x-axis,** and the vertical line is called the **y-axis.** Together, the x-axis and y-axis form a **rectangular coordinate system,** also called the **Cartesian coordinate system,** in honor of René Descartes, the French mathematician who is credited with its invention.

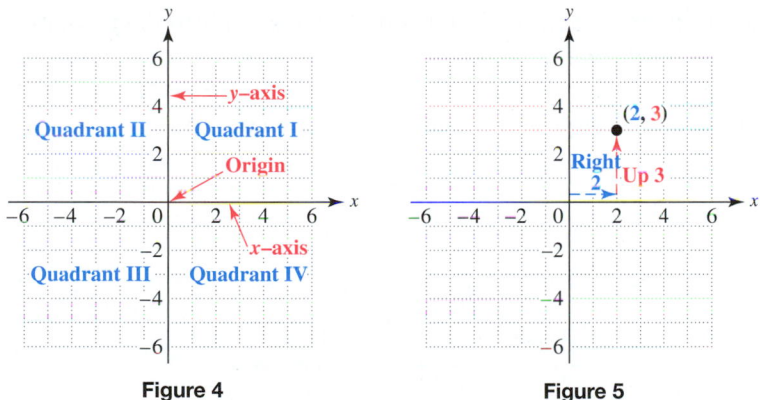

Figure 4   **Figure 5**

The coordinate system is divided into four regions, called **quadrants.** These quadrants are numbered counterclockwise, as shown in Figure 4. *Points on the axes themselves are not in any quadrant.* The point at which the x-axis and y-axis meet is called the **origin.** The origin, which is labeled 0 in Figure 4, is the point corresponding to $(0, 0)$.

Work Problem ⑦ at the Side. ▶

The x-axis and y-axis determine a **plane**—a flat surface illustrated by a sheet of paper. By referring to the two axes, every point in the plane can be associated with an ordered pair. The numbers in the ordered pair are called the **coordinates** of the point.

For example, we locate the point associated with the ordered pair $(2, 3)$ by starting at the origin. Since the x-coordinate is 2, we go 2 units to the right along the x-axis. Then, since the y-coordinate is 3, we turn and go up 3 units on a line parallel to the y-axis. The point $(2, 3)$ is **plotted** in Figure 5. From now on, we will refer to the point with x-coordinate 2 and y-coordinate 3 as the point $(2, 3)$.

⑦ Name the quadrant in which each point in the figure is located.

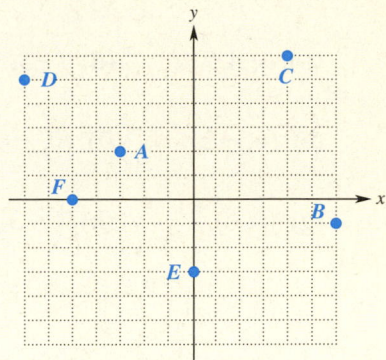

René Descartes (1596–1650)

ANSWER

7. *A*: II; *B*: IV; *C*: I; *D*: II; *E*: no quadrant; *F*: no quadrant

8 Plot each ordered pair on a coordinate system.

(a) $(3, 5)$ **(b)** $(-2, 6)$

(b) $(-4.5, 0)$ **(d)** $(-5, -2)$

(e) $(6, -2)$ **(f)** $(0, -6)$

(g) $(0, 0)$ **(h)** $\left(-3, \dfrac{5}{2}\right)$

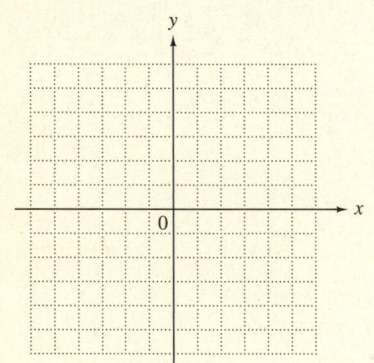

Note

When we graph on a number line (as in Figure 3), one number corresponds to each point. On a plane, however, *both* numbers in an ordered pair are needed to locate a point (as in Figure 5). The ordered pair is a name for the point.

We mentioned that René Descartes is credited with inventing the Cartesian coordinate system. Legend has it that Descartes, who was lying in bed ill, was watching a fly crawl about on the ceiling near a corner of the room. It occurred to him that the location of the fly could be described by determining its distances from the two adjacent walls. See the figure.

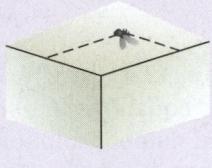

Locating a fly on a ceiling

EXAMPLE 6 **Plotting Ordered Pairs**

Plot each ordered pair on a coordinate system.

(a) $(1, 5)$ **(b)** $(-2, 3)$ **(c)** $(-1, -4)$ **(d)** $(3, -2)$

(e) $\left(\dfrac{3}{2}, 2\right)$ **(f)** $(5, 0)$ **(g)** $(0, -3)$ **(h)** $(4, -3.75)$

See Figure 6. In each case, begin at the origin. Move right or left the number of units that corresponds to the x-coordinate in the ordered pair—*right if the x-coordinate is positive or left if it is negative.* Then turn and move up or down the number of units that corresponds to the y-coordinate—*up if the y-coordinate is positive or down if it is negative.* So in part (c), locate the point $(-1, -4)$ by first going 1 unit to the *left* along the x-axis. Then turn and go 4 units *down,* parallel to the y-axis.

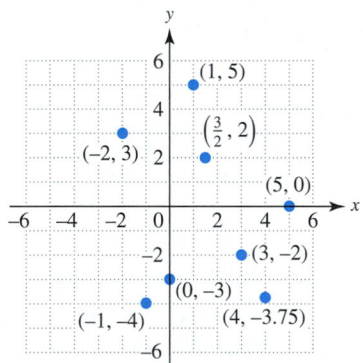

Figure 6

Notice the difference in the locations of the points $(-2, 3)$ and $(3, -2)$ in parts (b) and (d). The point $(-2, 3)$ is in quadrant II, whereas the point $(3, -2)$ is in quadrant IV. ***The order of the coordinates is important. Remember that the x-coordinate is always given first in an ordered pair.***

To plot the point $(\frac{3}{2}, 2)$ in part (e), think of the improper fraction $\frac{3}{2}$ as the mixed number $1\frac{1}{2}$ and move $\frac{3}{2}$ (or $1\frac{1}{2}$) units to the right along the x-axis. Then turn and go 2 units up, parallel to the y-axis. The point $(4, -3.75)$ in part (h) is plotted similarly, by approximating the location of the decimal y-coordinate.

In part (f), the point $(5, 0)$ lies on the x-axis since the y-coordinate is 0. In part (g), the point $(0, -3)$ lies on the y-axis since the x-coordinate is 0.

◀ *Work Problem* **8** *at the Side.*

ANSWERS

8.

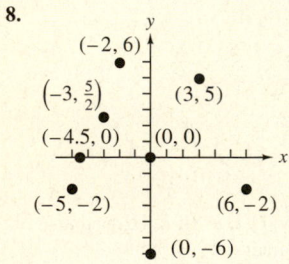

Sometimes we can use a linear equation in two variables to mathematically describe, or *model*, a real-life situation, as shown in the next example.

EXAMPLE 7   **Completing Ordered Pairs to Estimate the Number of Twin Births**

The number of twin births in the United States has increased steadily in recent years. The annual number of twin births from 2000 through 2005 can be closely approximated by the linear equation

Number of twin births ⎯⎯⎯⎯⎯   ⎯ Year

$$y = 3.074x - 6029.7,$$

which relates x, the year, and y, the number of twin births in thousands. (*Source: National Vital Statistics Reports,* Vol. 56, No. 6, December 5, 2007.)

(a) Complete the table of values for the given linear equation.

x (Year)	y (Number of Twin Births, in thousands)
2000	
2002	
2005	

To find y when $x = 2000$, substitute into the equation.

$$y = 3.074x - 6029.7$$

$\approx$ means "is approximately equal to." $y = 3.074(\mathbf{2000}) - 6029.7$ Let $x = 2000$.

$$y \approx 118$$ Use a calculator.

This means that in 2000, there were about 118 thousand (or 118,000) twin births in the United States.

Work Problem **9** *at the Side.* ▶

Including the results from Problem 9 at the side gives the completed table that follows.

x (Year)	y (Number of Twin Births, in thousands)
2000	118
2002	124
2005	134

We can write the results from the table of values as ordered pairs (x, y). Each year x is paired with its number of twin births y (in thousands):

$$(2000, 118), \quad (2002, 124), \quad \text{and} \quad (2005, 134).$$

Continued on Next Page

9 Refer to the linear equation in Example 7.

(a) Find the y-value for $x = 2002$. Round to the nearest whole number.

(b) Find the y-value for $x = 2005$. Interpret your result.

(b) Graph the ordered pairs found in part (a).

The ordered pairs (2000, 118), (2002, 124), and (2005, 134) are graphed in Figure 7. This graph of ordered pairs of data is called a **scatter diagram.** Notice how the axes are labeled: x represents the year, and y represents the number of twin births in thousands. Different scales are used on the two axes. Here, each square represents one unit in the horizontal direction and 5 units in the vertical direction. Because the numbers in the first ordered pair are large, we show a break in the axes near the origin.

x (Year)	y (Number of Twin Births, in thousands)
2000	118
2002	124
2005	134

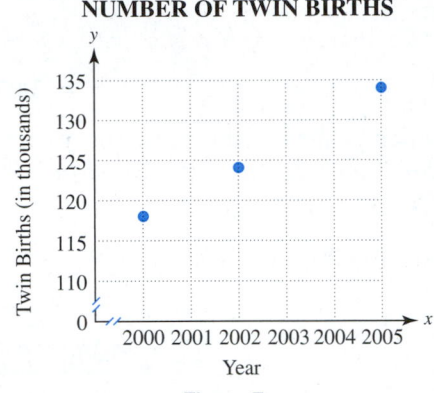

NUMBER OF TWIN BIRTHS

Figure 7

A scatter diagram enables us to tell whether two quantities are related to each other. In Figure 7, the plotted points could be connected to closely approximate a straight *line,* so the variables x (year) and y (number of twin births) have a *line*ar relationship. The increase in the number of twin births is also reflected.

> **CAUTION**
> The equation in Example 7 is valid only for the years 2000 through 2005 because it was based on data for those years. *Do not assume that this equation would provide reliable data for other years since the data for those years may not follow the same pattern.*

The bar graph shows total U.S. milk production in billions of pounds for the years 2001 through 2007. Use the bar graph to work Exercises 1–4. See Example 1.

1. In what years was U.S. milk production greater than 175 billion pounds?

2. In what years was U.S. milk production about the same?

3. Estimate U.S. milk production in 2001 and 2007.

4. Describe the change in U.S. milk production from 2001 to 2007.

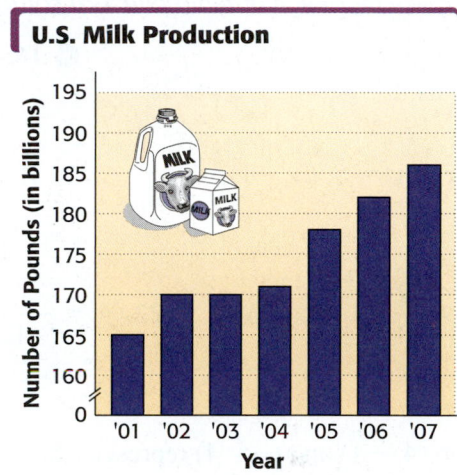

U.S. Milk Production

Source: U.S. Department of Agriculture.

The line graph shows the average price, adjusted for inflation, that Americans have paid for a gallon of gasoline for selected years since 1970. Use the line graph to work Exercises 5–8. See Example 2.

5. Over which period of years did the greatest increase in the price of a gallon of gas occur? About how much was this increase?

6. Estimate the price of a gallon of gas during 1985, 1990, 1995, and 2000.

7. Describe the trend in gas prices from 1980 to 1995.

8. During which year(s) did a gallon of gas cost approximately $1.50?

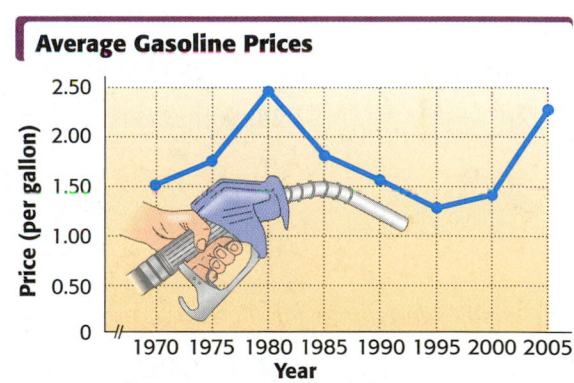

Average Gasoline Prices

Source: Energy Information Administration.

Use the concepts of this section to fill in each blank with the correct response.

9. The symbol (x, y) _____ represent an ordered pair, while the
   (does/does not)

   symbols $[x, y]$ and $\{x, y\}$ _____ represent ordered pairs.
   (do/do not)

10. The point whose graph has coordinates $(-4, 2)$ is in quadrant _____.

11. The point whose graph has coordinates $(0, 5)$ lies on the _____-axis.

12. The ordered pair $(4, \underline{\hspace{1cm}})$ is a solution of the equation $y = 3$.

13. The ordered pair $(\underline{\hspace{1cm}}, -2)$ is a solution of the equation $x = 6$.

14. The ordered pair $(3, 2)$ is a solution of the equation $2x - 5y = \underline{\hspace{0.6cm}}$.

Decide whether each ordered pair is a solution of the given equation. See Example 3.

15. $x + y = 9$; $(0, 9)$ **16.** $x + y = 8$; $(0, 8)$ **17.** $2x - y = 6$; $(4, 2)$

18. $2x + y = 5$; $(3, -1)$ **19.** $4x - 3y = 6$; $(2, 1)$ **20.** $5x - 3y = 15$; $(5, 2)$

21. $y = \dfrac{2}{3}x$; $(-6, -4)$ **22.** $y = -\dfrac{1}{4}x$; $(-8, 2)$ **23.** $x = -6$; $(5, -6)$ **24.** $y = 2$; $(2, 4)$

25. Do $(4, -1)$ and $(-1, 4)$ represent the same ordered pair? Explain.

26. Explain why it would be easier to find the corresponding y-value for $x = \frac{1}{3}$ in the equation $y = 6x + 2$ than it would be for $x = \frac{1}{7}$.

Complete each ordered pair for the equation $y = 2x + 7$. See Example 4.

27. $(2, \underline{\hspace{0.5cm}})$ **28.** $(0, \underline{\hspace{0.5cm}})$ **29.** $(\underline{\hspace{0.5cm}}, 0)$ **30.** $(\underline{\hspace{0.5cm}}, -3)$

Complete each ordered pair for the equation $y = -4x - 4$. See Example 4.

31. $(0, \underline{\hspace{0.5cm}})$ **32.** $(\underline{\hspace{0.5cm}}, 0)$ **33.** $(\underline{\hspace{0.5cm}}, 16)$ **34.** $(\underline{\hspace{0.5cm}}, 24)$

Complete each table of values. In Exercises 35–38, write the results as ordered pairs. See Example 5.

35. $2x + 3y = 12$

x	y
0	
	0
	8

36. $4x + 3y = 24$

x	y
0	
	0
	4

37. $3x - 5y = -15$

x	y
0	
	0
	-6

38. $4x - 9y = -36$

x	y
	0
0	
	8

39. $x = -9$

x	y
	6
	2
	-3

40. $x = 12$

x	y
	3
	8
	0

41. $y = -6$

x	y
8	
4	
-2	

42. $y = -10$

x	y
4	
0	
-4	

43. $x - 8 = 0$

x	y
	8
	3
	0

44. $y + 2 = 0$

x	y
9	
2	
0	

Give the ordered pairs for the points labeled A–F in the figure. Tell the quadrant in which each point is located.

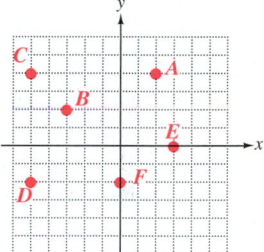

45. A

46. B

47. C

48. D

49. E

50. F

Fill in each blank with the word positive *or the word* negative.

The point with coordinates (x, y) is in

51. quadrant III if x is _____ and y is _____.

52. quadrant II if x is _____ and y is _____.

53. quadrant IV if x is _____ and y is _____.

54. quadrant I if x is _____ and y is _____.

55. A point (x, y) has the property that $xy < 0$. In which quadrant(s) must the point lie? Explain.

56. A point (x, y) has the property that $xy > 0$. In which quadrant(s) must the point lie? Explain.

Plot each ordered pair on the rectangular coordinate system provided. See Example 6.

57. $(6, 2)$

58. $(5, 3)$

59. $(-4, 2)$

60. $(-3, 5)$

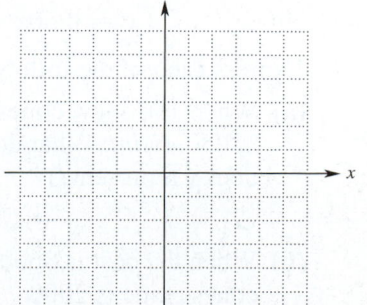

61. $\left(-\dfrac{4}{5}, -1\right)$

62. $\left(-\dfrac{3}{2}, -4\right)$

63. $(3, -1.75)$

64. $(5, -4.25)$

65. $(0, 4)$

66. $(0, -3)$

67. $(4, 0)$

68. $(-3, 0)$

Complete each table of values, and then plot the ordered pairs. See Examples 5 and 6.

69. $x - 2y = 6$

x	y
0	
	0
2	
	-1

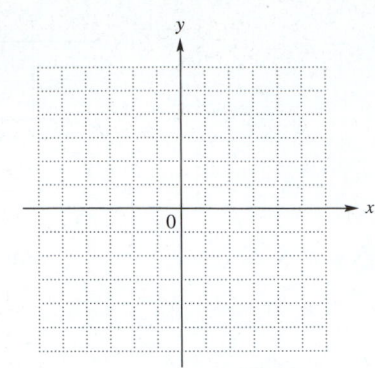

70. $2x - y = 4$

x	y
0	
	0
1	
	-6

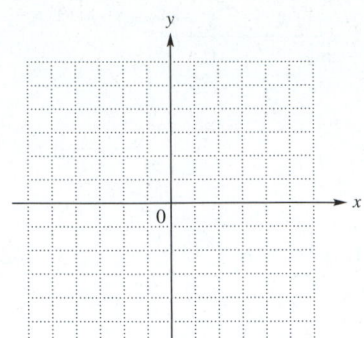

71. $3x - 4y = 12$

x	y
0	
	0
-4	
	-4

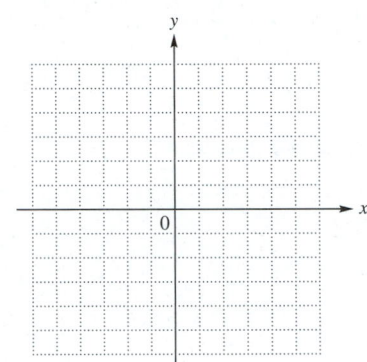

72. $2x - 5y = 10$

x	y
0	
	0
-5	
	-3

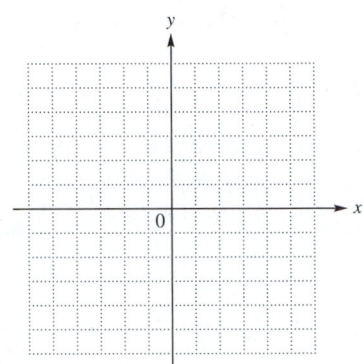

73. $y + 4 = 0$

x	y
0	
5	
-2	
-3	

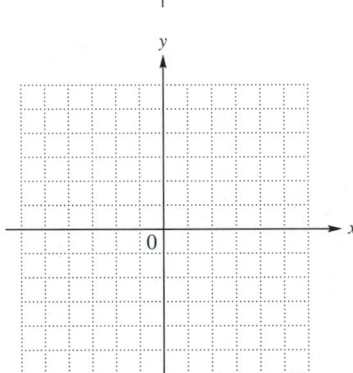

74. $x - 5 = 0$

x	y
	1
	0
	6
	-4

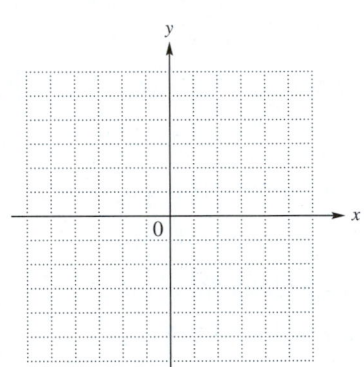

75. Look at the graphs of the ordered pairs in Exercises 69–74. Describe the pattern indicated by the plotted points.

Work each problem. See Example 7.

76. Suppose that it costs $5000 to start up a business selling snow cones. Furthermore, it costs $0.50 per cone in labor, ice, syrup, and overhead. Then the cost to make *x* snow cones is given by *y* dollars, where

$$y = 0.50x + 5000.$$

Express each of the following as an ordered pair.

(a) When 100 snow cones are made, the cost is $5050. (*Hint:* What does *x* represent? What does *y* represent?)

(b) When the cost is $6000, the number of snow cones made is 2000.

77. It costs a flat fee of $20 plus $5 per day to rent a pressure washer. Therefore, the cost to rent the pressure washer for *x* days is given by

$$y = 5x + 20,$$

where *y* is in dollars. Express each of the following as an ordered pair.

(a) When the washer is rented for 5 days, the cost is $45. (*Hint:* What does *x* represent? What does *y* represent?)

(b) I paid $50 when I returned the washer, so I must have rented it for 6 days.

78. The table shows the number of U.S. students studying abroad (in thousands) for several academic years.

Academic Year	Number of Students (in thousands)
2000	154
2001	161
2002	175
2003	191
2004	206
2005	224

Source: Institute of International Education.

(a) Write the data from the table as ordered pairs (x, y), where x represents the year and y represents the number of U.S. students studying abroad.

(b) What does the ordered pair (2004, 206) mean in the context of this problem?

(c) Make a scatter diagram of the data using the ordered pairs from part (a).

U.S. STUDENTS STUDYING ABROAD

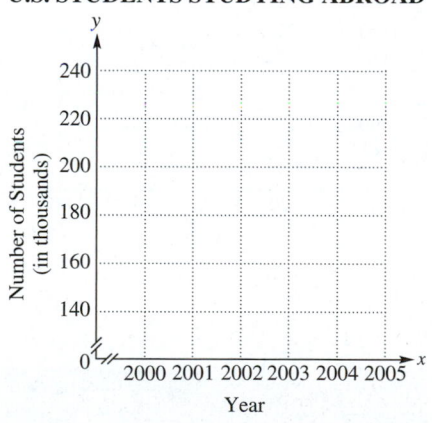

(d) Describe the pattern indicated by the points on the scatter diagram. What is the trend in the number of U.S. students studying abroad?

79. The table shows the rate (in percent) at which 2-year college students (public) complete a degree within 3 years.

Year	Percent
2000	32.4
2001	31.6
2002	31.6
2003	30.1
2004	29.0
2005	27.5

Source: ACT.

(a) Write the data from the table as ordered pairs (x, y), where x represents the year and y represents the percent.

(b) What would the ordered pair (2007, 27.1) mean in the context of this problem?

(c) Make a scatter diagram of the data using the ordered pairs from part (a).

2-YEAR COLLEGE STUDENTS COMPLETING A DEGREE WITHIN 3 YEARS

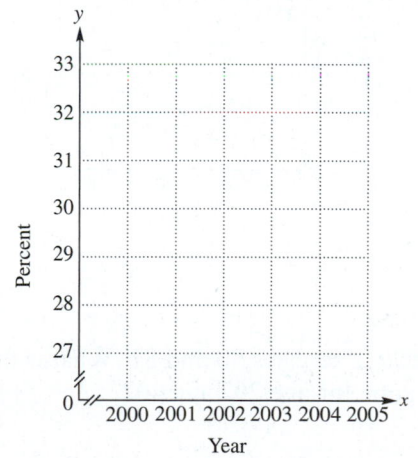

(d) Describe the pattern indicated by the points on the scatter diagram. What is happening to the rates at which 2-year college students complete a degree within 3 years?

80. The maximum benefit for the heart from exercising occurs if the heart rate is in the target heart rate zone. The lower limit of this target zone can be approximated by the linear equation

$$y = -0.5x + 108,$$

where x represents age and y represents heartbeats per minute. (*Source:* www.fitresource.com)

(a) Complete the table of values for this linear equation.

Age	Heartbeats (per minute)
20	
40	
60	
80	

(b) Write the data from the table of values as ordered pairs.

(c) Make a scatter diagram of the data. Do the points lie in a linear pattern?

TARGET HEART RATE ZONE
(Lower Limit)

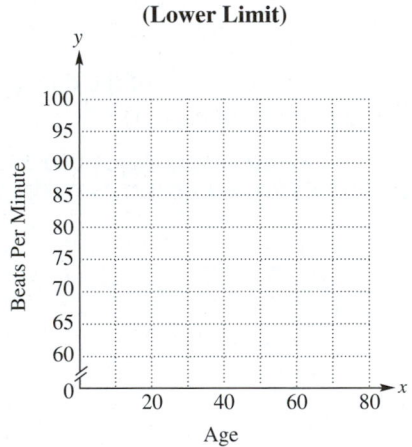

81. (See Exercise 80.) The upper limit of the target heart rate zone can be approximated by the linear equation

$$y = -0.8x + 173,$$

where x represents age and y represents heartbeats per minute. (*Source:* www.fitresource.com)

(a) Complete the table of values for this linear equation.

Age	Heartbeats (per minute)
20	
40	
60	
80	

(b) Write the data from the table of values as ordered pairs.

(c) Make a scatter diagram of the data. Describe the pattern indicated by the data.

TARGET HEART RATE ZONE
(Upper Limit)

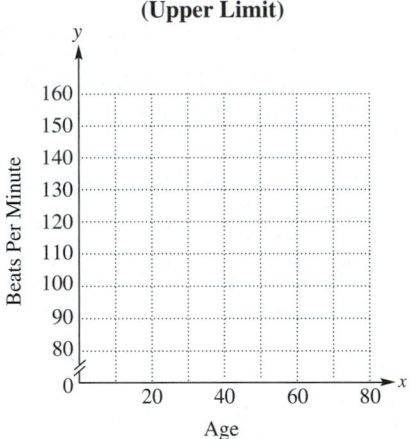

82. Refer to Exercises 80 and 81. What is the target heart rate zone for age 20? age 40?

11.2 ▶▶▶ Graphing Linear Equations in Two Variables

OBJECTIVE 1 Graph linear equations by plotting ordered pairs. There are infinitely many ordered pairs that satisfy an equation in two variables. We find these ordered-pair solutions by choosing as many values of x (or y) as we wish and then completing each ordered pair.

For example, consider the equation $x + 2y = 7$. If we choose $x = 1$, then $y = 3$, so the ordered pair $(1, 3)$ is a solution of the equation $x + 2y = 7$.

$$1 + 2(3) = 7$$

Work Problem ① *at the Side.* ▶

Figure 8 shows a graph of all the ordered-pair solutions found above and in Problem 1 at the side for $x + 2y = 7$.

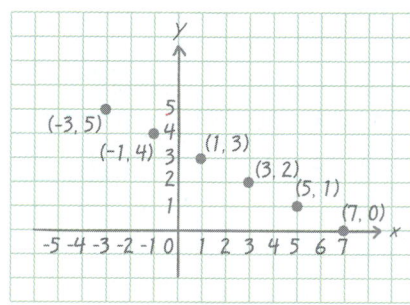

Figure 8

Notice that the points plotted in Figure 8 all appear to lie on a straight line, as shown in Figure 9. In fact, the following is true.

Every point on the line represents a solution of the equation $x + 2y = 7$, and every solution of the equation corresponds to a point on the line.

The line gives a "picture" of all the solutions of the equation $x + 2y = 7$. Only a portion of the line is shown here, but it extends indefinitely in both directions, as suggested by the arrowhead on each end.

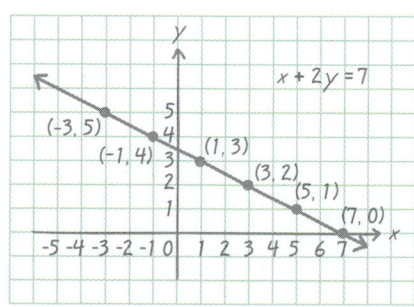

Figure 9

The line in Figure 9 is called the **graph** of the equation $x + 2y = 7$, and the process of plotting the ordered pairs and drawing the line through the corresponding points is called **graphing.**

OBJECTIVES

1. Graph linear equations by plotting ordered pairs.

2. Find intercepts.

3. Graph linear equations of the form $Ax + By = 0$.

4. Graph linear equations of the form $y = k$ or $x = k$.

5. Use a linear equation to model data.

① Complete each ordered pair for the equation $x + 2y = 7$.

(a) $(-3, \underline{\quad})$

(b) $(-1, \underline{\quad})$

(c) $(3, \underline{\quad})$

(d) $(5, \underline{\quad})$

(e) $(7, \underline{\quad})$

ANSWERS

1. (a) $(-3, 5)$ (b) $(-1, 4)$ (c) $(3, 2)$
(d) $(5, 1)$ (e) $(7, 0)$

The preceding discussion can be generalized.

2 Complete the table of values, and graph the linear equation.

$x + y = 6$

x	y
0	
	0
2	

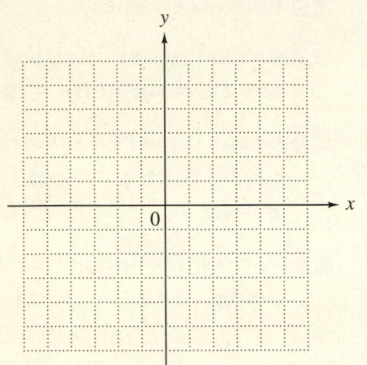

> ### Graph of a Linear Equation
> The graph of any linear equation in two variables is a straight line.

(Notice that the word **line** appears in the term "**line**ar equation.")

Because two distinct points determine a line, a straight line can be graphed by finding any two different points on the line. However, it is a good idea to plot a third point as a check.

EXAMPLE 1 Graphing a Linear Equation

Graph the linear equation $4x - 5y = 20$.

At least two different points are needed to draw the graph. First let $x = 0$ and then let $y = 0$ to complete two ordered pairs.

$$4x - 5y = 20$$
$$4(0) - 5y = 20 \quad \text{Let } x = 0.$$
$$0 - 5y = 20$$
$$-5y = 20$$
$$y = -4$$

$$4x - 5y = 20$$
$$4x - 5(0) = 20 \quad \text{Let } y = 0.$$
$$4x - 0 = 20$$
$$4x = 20$$
$$x = 5$$

Write each x-value first.

The ordered pairs are $(0, -4)$ and $(5, 0)$. Find a third ordered pair (as a check) by choosing a number other than 0 for x or y. We choose $y = 2$.

$$4x - 5y = 20$$
$$4x - 5(2) = 20 \quad \text{Let } y = 2.$$
$$4x - 10 = 20$$
$$4x = 30 \quad \text{Add 10.}$$
$$x = \frac{30}{4}, \quad \text{or} \quad \frac{15}{2} \quad \text{Divide by 4; lowest terms}$$

This gives the ordered pair $(\frac{15}{2}, 2)$, or $(7\frac{1}{2}, 2)$. Plot the three ordered pairs $(0, -4)$, $(5, 0)$, and $(7\frac{1}{2}, 2)$, and draw a line through them. This line, shown in Figure 10, is the graph of $4x - 5y = 20$.

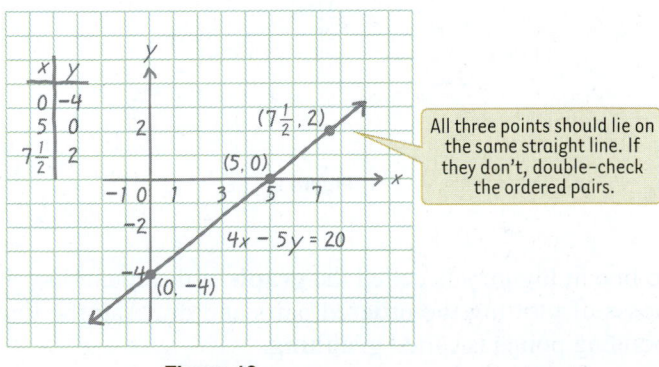

All three points should lie on the same straight line. If they don't, double-check the ordered pairs.

Figure 10

◀ *Work Problem* **2** *at the Side.*

EXAMPLE 2 **Graphing a Linear Equation**

Graph the linear equation $y = -\frac{3}{2}x + 3$.

Although this equation is not in the form $Ax + By = C$, it *could* be written in that form, so it is a linear equation. Two different points on the graph can be found by first letting $x = 0$ and then letting $y = 0$.

If $x = 0$, then

$$y = -\frac{3}{2}x + 3$$

$$y = -\frac{3}{2}(0) + 3 \qquad \text{Let } x = 0.$$

$$y = 0 + 3 \qquad \text{Multiply.}$$

$$y = 3. \qquad \text{Add.}$$

If $y = 0$, then

$$y = -\frac{3}{2}x + 3$$

$$0 = -\frac{3}{2}x + 3 \qquad \text{Let } y = 0.$$

$$\frac{3}{2}x = 3 \qquad \text{Add } \frac{3}{2}x.$$

$$x = 2. \qquad \text{Multiply by } \frac{2}{3}.$$

This gives the ordered pairs $(0, 3)$ and $(2, 0)$. We find a third point (as a check) by letting x or y equal some other number. For example, let $x = -2$.

$$y = -\frac{3}{2}x + 3$$

> Choosing a multiple of 2 makes multiplying by $-\frac{3}{2}$ easier.

$$y = -\frac{3}{2}(-2) + 3 \qquad \text{Let } x = -2.$$

$$y = 3 + 3 \qquad \text{Multiply.}$$

$$y = 6 \qquad \text{Add.}$$

This gives the ordered pair $(-2, 6)$. These three ordered pairs are shown in the table with Figure 11. Plot the corresponding points, and then draw a line through them. This line, shown in Figure 11, is the graph of $y = -\frac{3}{2}x + 3$.

x	y
0	3
2	0
-2	6

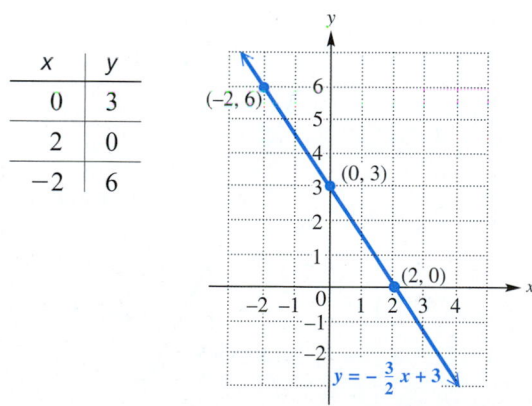

Figure 11

Work Problem **3** *at the Side.* ▶

3 Make a table of values, and graph the linear equation.

$$y = \frac{2}{3}x - 2$$

x	y

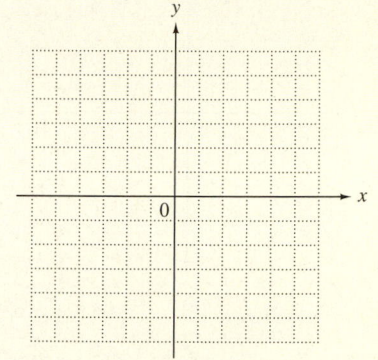

4 Find the intercepts for the graph of $5x + 2y = 10$. Then draw the graph. (Be sure to get a third point as a check.)

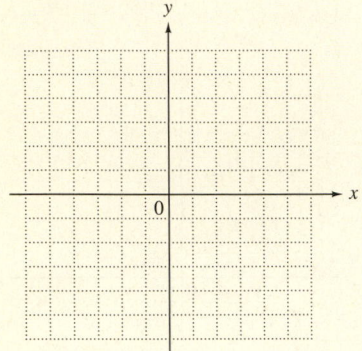

OBJECTIVE 2 Find intercepts. In Figure 11, the graph crosses, or intersects, the y-axis at $(0, 3)$ and the x-axis at $(2, 0)$. For this reason, $(0, 3)$ is called the **y-intercept,** and $(2, 0)$ is called the **x-intercept** of the graph.

The intercepts are particularly useful for graphing linear equations. The intercepts are found by replacing, in turn, each variable with 0 in the equation and solving for the value of the other variable.

Finding Intercepts

To find the x-intercept, let $y = 0$ in the given equation and solve for x. Then $(x, 0)$ is the x-intercept.

To find the y-intercept, let $x = 0$ in the given equation and solve for y. Then $(0, y)$ is the y-intercept.

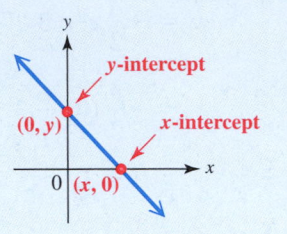

EXAMPLE 3 Finding Intercepts

Find the intercepts for the graph of $2x + y = 4$. Then draw the graph.

To find the y-intercept, let $x = 0$; to find the x-intercept, let $y = 0$.

$$2x + y = 4$$
$$2(0) + y = 4 \quad \text{Let } x = 0.$$
$$0 + y = 4$$
$$y = 4$$

$$2x + y = 4$$
$$2x + 0 = 4 \quad \text{Let } y = 0.$$
$$2x = 4$$
$$x = 2$$

The y-intercept is $(0, 4)$. The x-intercept is $(2, 0)$. Find a third point as a check. For example, choosing $x = 1$ gives $y = 2$. Plot $(0, 4)$, $(2, 0)$, and $(1, 2)$ and draw the line through them. This line, shown in Figure 12, is the graph.

x	y
0	4
2	0
1	2

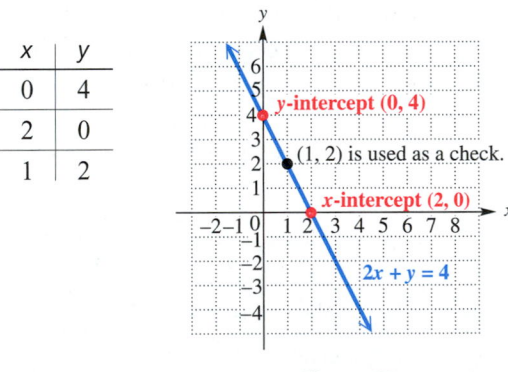

Figure 12

◀ *Work Problem* **4** *at the Side.*

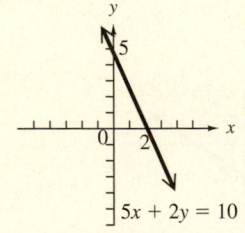

OBJECTIVE 3 Graph linear equations of the form Ax + By = 0. In the preceding examples, the x- and y-intercepts were used to help draw the graphs. This is not always possible. Example 4 shows what to do when the x- and y-intercepts are the same point (that is, coincide).

EXAMPLE 4 **Graphing an Equation of the Form Ax + By = 0**

Graph the linear equation $x - 3y = 0$.

    If we let $x = 0$, then $y = 0$, giving the ordered pair $(0, 0)$. Letting $y = 0$ also gives $(0, 0)$. This is the same ordered pair, so we choose two *other* values for x or y. Choosing 2 for y gives $x - 3 \cdot 2 = 0$, leading to $x = 6$, so another ordered pair is $(6, 2)$. Choosing -2 for y gives $x - 3(-2) = 0$, leading to $x = -6$, so a third ordered pair is $(-6, -2)$. We use the ordered pairs $(-6, -2)$, $(0, 0)$, and $(6, 2)$ to sketch the graph in Figure 13.

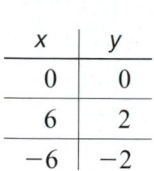

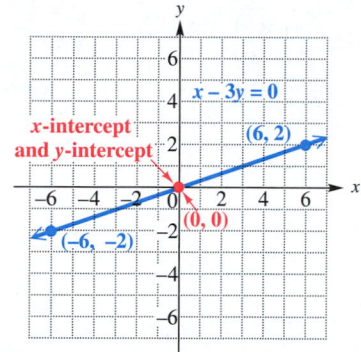

Figure 13

Work Problem **5** *at the Side.* ▶

Line through the Origin

If A and B are nonzero real numbers, the graph of a linear equation of the form

$$Ax + By = 0$$

passes through the origin $(0, 0)$.

OBJECTIVE **4** **Graph linear equations of the form $y = k$ or $x = k$.** The equation $y = -4$ is a linear equation in which the coefficient of x is 0. (To see this, write $y = -4$ as $0x + y = -4$.) Also, $x = 3$ is a linear equation in which the coefficient of y is 0. These equations lead to horizontal or vertical straight lines, as the next examples show.

EXAMPLE 5 **Graphing an Equation of the Form $y = k$**

Graph $y = -4$.

    As the equation states, for any value of x, y is always equal to -4. Three ordered pairs that satisfy the equation are shown. The graph is the horizontal line in Figure 14. The y-intercept is $(0, -4)$; there is no x-intercept.

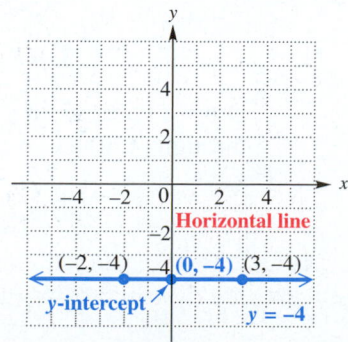

Figure 14

Work Problem **6** *at the Side.* ▶

5 Graph $2x - y = 0$.

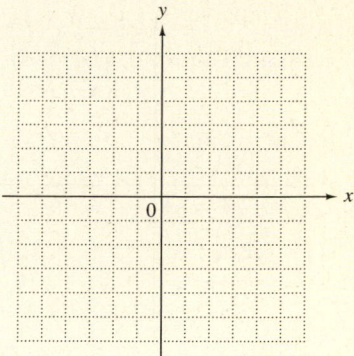

6 Graph $y = -5$.

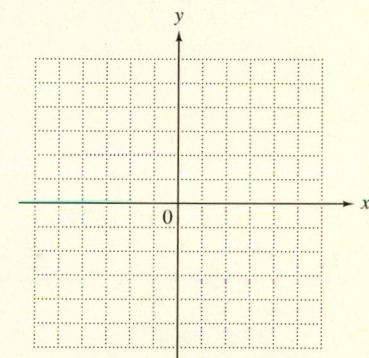

ANSWERS

5.

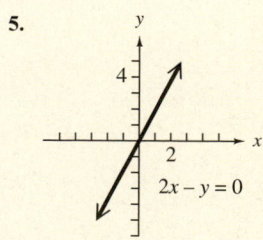

6.

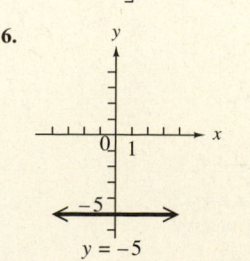

7 Graph $x + 4 = 6$.

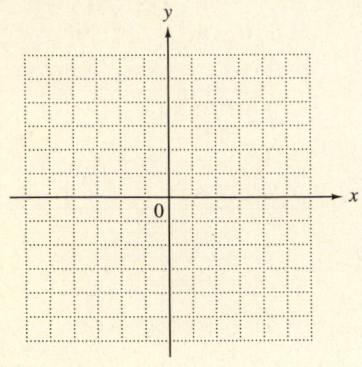

EXAMPLE 6 Graphing an Equation of the Form $x = k$

Graph $x - 3 = 0$.

First add 3 to each side of $x - 3 = 0$ to get $x = 3$. All the ordered pairs that satisfy this equation have x-coordinate 3. Any number can be used for y. See Figure 15 for the graph of this vertical line, along with a table of values. The x-intercept is $(3, 0)$; there is no y-intercept.

x	y
3	3
3	0
3	-2

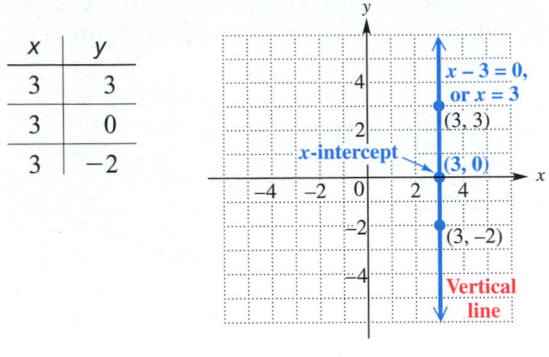

Figure 15

◀ Work Problem **7** at the Side.

From the results in Examples 5 and 6, we make the following observations.

Horizontal and Vertical Lines

The graph of the linear equation $y = k$, where k is a real number, is the horizontal line with y-intercept $(0, k)$ and no x-intercept.

The graph of the linear equation $x = k$, where k is a real number, is the vertical line with x-intercept $(k, 0)$ and no y-intercept.

In particular, notice that the horizontal line $y = 0$ is the x-axis and the vertical line $x = 0$ is the y-axis. The different forms of linear equations from this section and the methods of graphing them are summarized below.

Graphing a Linear Equation

Equation	*Graphing Method*	*Example*
$y = k$	Draw a horizontal line through $(0, k)$.	
$x = k$	Draw a vertical line through $(k, 0)$.	

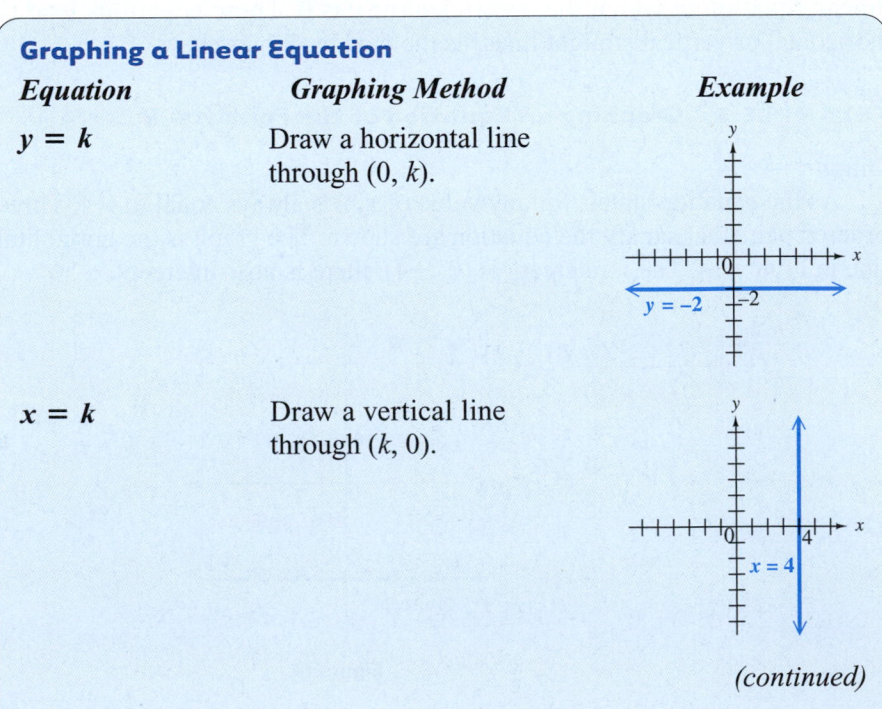

(continued)

ANSWER

7.

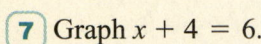

$x + 4 = 6$, or $x = 2$

Equation	Graphing Method	Example
$Ax + By = 0$	Graph passes through $(0, 0)$. To get additional points that lie on the graph, choose any values for x or y, except 0.	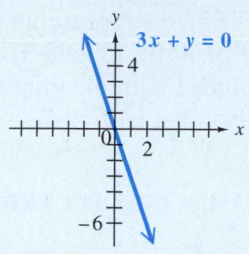
$Ax + By = C$ (but not of the types above)	Find any two points on the line. A good choice is to find the intercepts. Let $x = 0$, and find the corresponding value of y; then let $y = 0$, and find x. As a check, get a third point by choosing a value of x or y that has not yet been used.	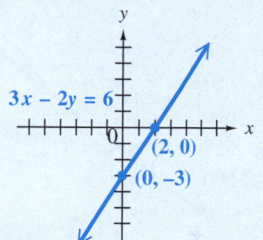

Work Problem *at the Side.* ▶

Note

Another method of graphing linear equations, using the concepts of slope and y-intercept, will be covered in Objective 2 of **Section 11.4**.

OBJECTIVE 5 Use a linear equation to model data.

EXAMPLE 7 **Using a Linear Equation to Model Credit Card Debt**

Credit card debt in the United States has increased steadily during recent years. The amount of debt y in billions of dollars can be modeled by the linear equation

$$y = 38.7x + 450,$$

where $x = 0$ represents the year 1995, $x = 1$ represents 1996, and so on. (*Source:* Board of Governors of the Federal Reserve System.)

(a) Use the equation to approximate credit card debt in the years 1995, 2000, and 2003.

Substitute the appropriate value for each year x to find credit card debt in that year.

For 1995: $y = 38.7(\mathbf{0}) + 450$ Replace x with 0.
             $y = 450$ billion dollars

For 2000: $y = 38.7(\mathbf{5}) + 450$ $2000 - 1995 = 5;$
             $y = 643.5$ billion dollars   Replace x with 5.

For 2003: $y = 38.7(\mathbf{8}) + 450$ $2003 - 1995 = 8;$
             $y = 759.6$ billion dollars   Replace x with 8.

Continued on Next Page

9 Use the graph and then the equation in Example 7 to approximate credit card debt in 1997.

(b) Write the information from part (a) as three ordered pairs, and use them to graph the given linear equation.

Since *x* represents the year and *y* represents the debt, the ordered pairs are (0, 450), (5, 643.5), and (8, 759.6). See Figure 16. (Arrowheads are not included with the graphed line, since the data are for the years 1995 to 2003 only—that is, from $x = 0$ to $x = 8$.)

U.S. CREDIT CARD DEBT

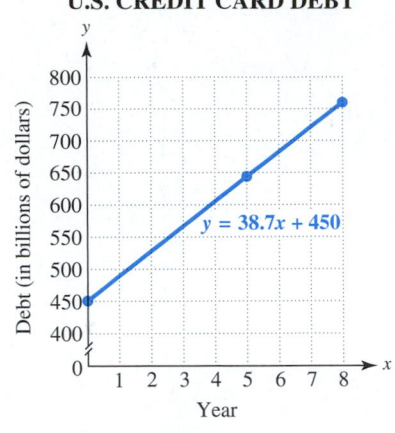

$y = 38.7x + 450$

Figure 16

(c) Use the graph and then the equation to approximate credit card debt in 2002.

For 2002, $x = 7$. On the graph, find 7 on the horizontal axis, move up to the graphed line and then across to the vertical axis. It appears that credit card debt in 2002 was about 725 billion dollars. To use the equation, substitute 7 for *x*.

$$y = 38.7x + 450$$

$$y = 38.7(7) + 450 \qquad \text{Let } x = 7.$$

$$y = 720.9 \text{ billion dollars}$$

This result for 2002 is close to our estimate of 725 billion dollars from the graph.

◀ *Work Problem* **9** *at the Side.*

11.2 ▶▶▶ **Exercises**

Complete the given ordered pairs for each equation. Then graph each equation by plotting the points and drawing the line through them. See Examples 1 and 2.

1. $x + y = 5$

$(0, \underline{\hspace{0.5cm}}), (\underline{\hspace{0.5cm}}, 0), (2, \underline{\hspace{0.5cm}})$

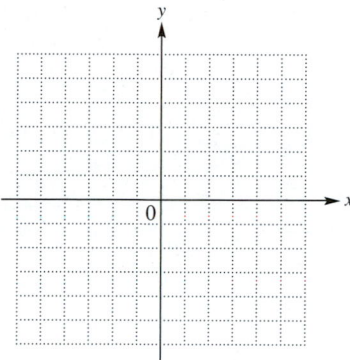

2. $x - y = 2$

$(0, \underline{\hspace{0.5cm}}), (\underline{\hspace{0.5cm}}, 0), (5, \underline{\hspace{0.5cm}})$

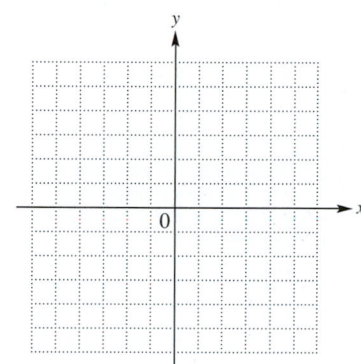

3. $y = \dfrac{2}{3}x + 1$

$(0, \underline{\hspace{0.5cm}}), (3, \underline{\hspace{0.5cm}}), (-3, \underline{\hspace{0.5cm}})$

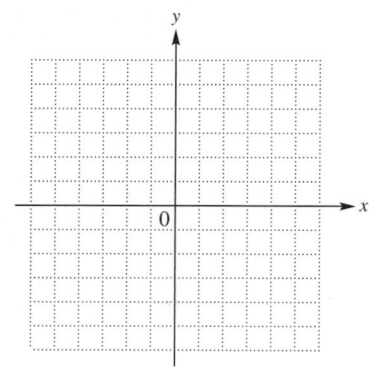

4. $y = -\dfrac{3}{4}x + 2$

$(0, \underline{\hspace{0.5cm}}), (4, \underline{\hspace{0.5cm}}), (-4, \underline{\hspace{0.5cm}})$

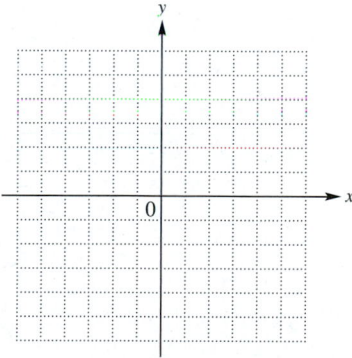

5. $3x = -y - 6$

$(0, \underline{\hspace{0.5cm}}), (\underline{\hspace{0.5cm}}, 0), \left(-\dfrac{1}{3}, \underline{\hspace{0.5cm}}\right)$

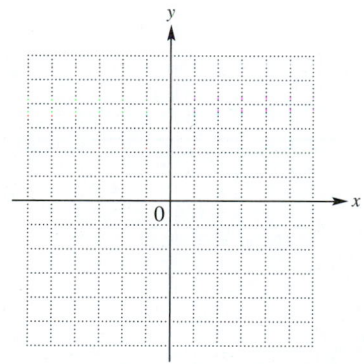

6. $x = 2y + 3$

$(\underline{\hspace{0.5cm}}, 0), (0, \underline{\hspace{0.5cm}}), \left(\underline{\hspace{0.5cm}}, \dfrac{1}{2}\right)$

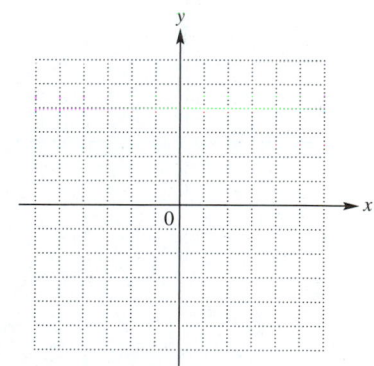

7. Match the information about each graph in Column I with the correct linear equation in Column II.

I	II
(a) The graph of the equation has y-intercept $(0, -4)$.	**A.** $3x + y = -4$
(b) The graph of the equation has $(0, 0)$ as x-intercept and y-intercept.	**B.** $x - 4 = 0$
(c) The graph of the equation does not have an x-intercept.	**C.** $y = 4x$
(d) The graph of the equation has x-intercept $(4, 0)$.	**D.** $y = 4$

8. Write a few sentences summarizing how to graph a linear equation in two variables.

Find the intercepts for the graph of each equation. See Example 3.

9. $2x - 3y = 24$

    x-intercept:

    y-intercept:

10. $-3x + 8y = 48$

    x-intercept:

    y-intercept:

11. $x + 6y = 0$

    x-intercept:

    y-intercept:

12. $3x - y = 0$

    x-intercept:

    y-intercept:

Graph each linear equation. See Examples 1–6.

13. $y = x - 2$

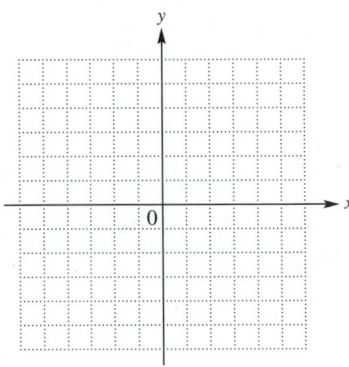

14. $y = -x + 6$

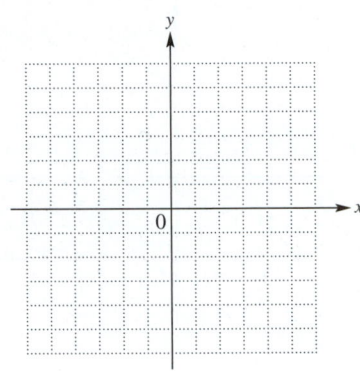

15. $x - y = 4$

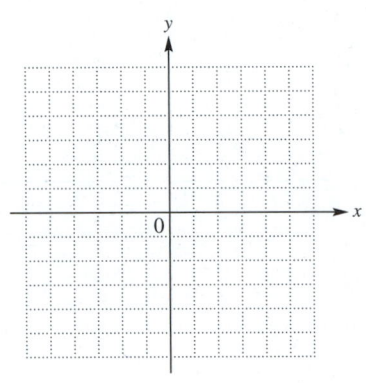

16. $x - y = 5$

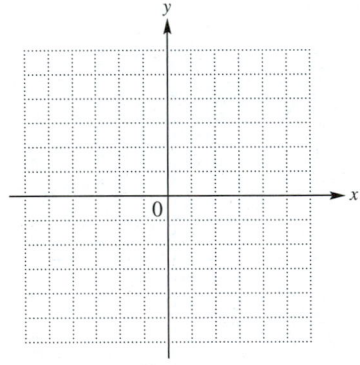

17. $2x + y = 6$

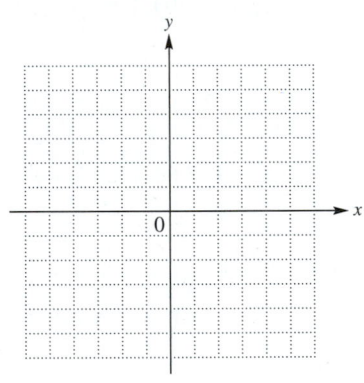

18. $-3x + y = -6$

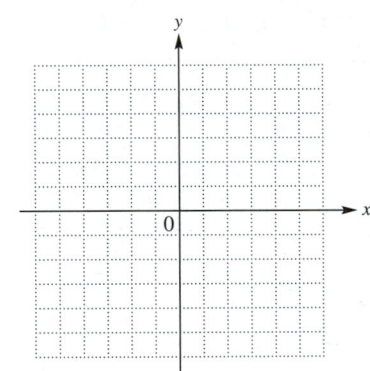

19. $3x + 7y = 14$

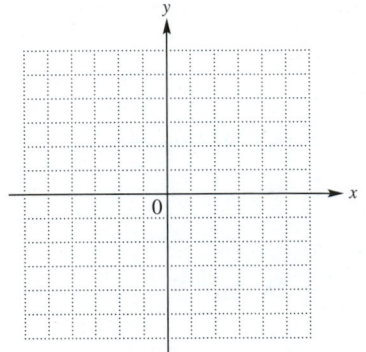

20. $6x - 5y = 18$

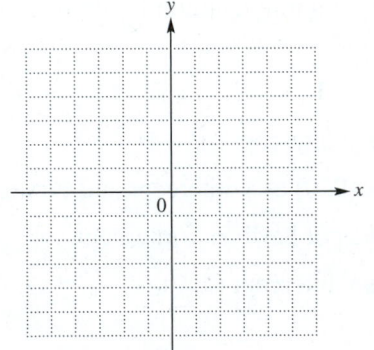

21. $y - 2x = 0$

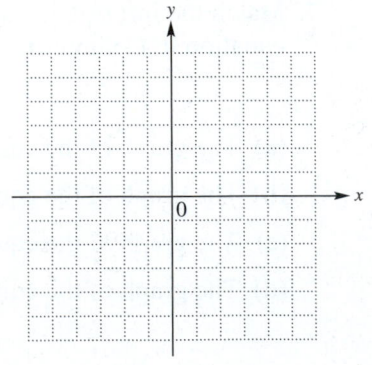

22. $y + 3x = 0$

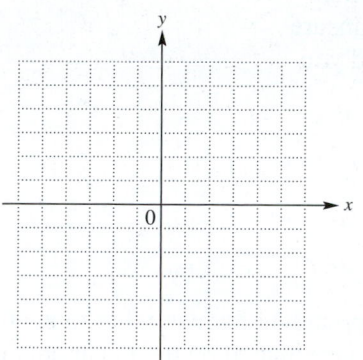

23. $y = -6x$

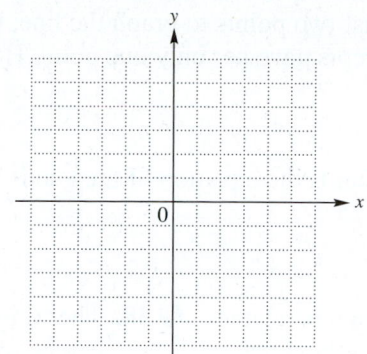

24. $y = 4x$

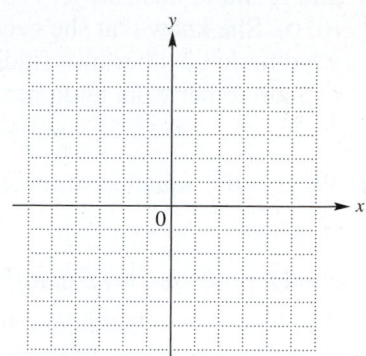

🌐 **25.** $x = -2$

26. $x = 4$

27. $y - 3 = 0$

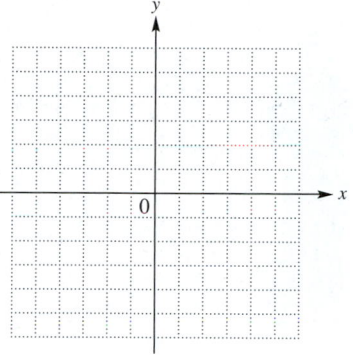

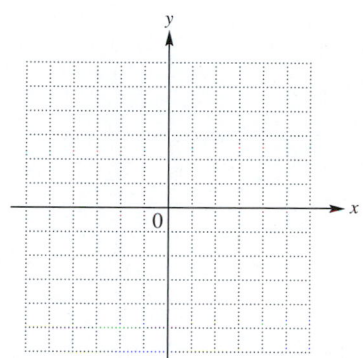

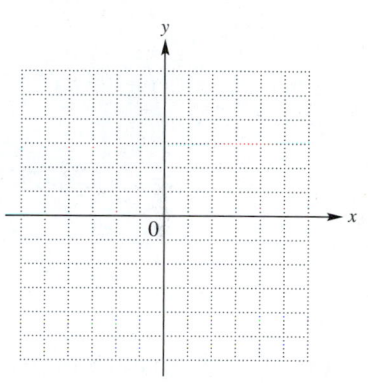

28. $y + 1 = 0$

29. $-3y = 15$

30. $-2y = 12$

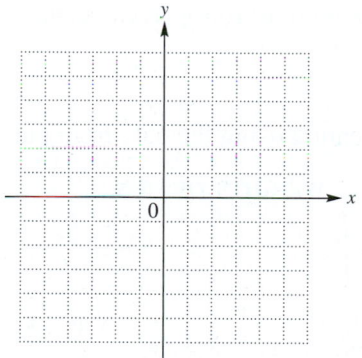

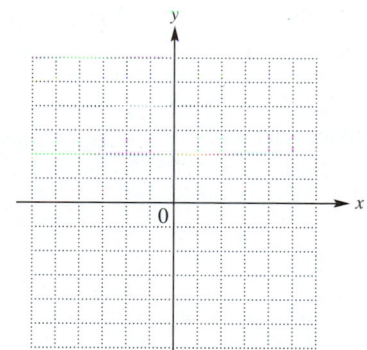

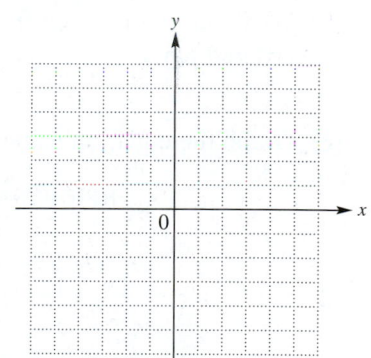

In Exercises 31–34, describe what the graph of each linear equation will look like on the coordinate plane. (Hint: Rewrite the equation if necessary so that it is in a more recognizable form.)

31. $3x = y - 9$

32. $x - 10 = 1$

33. $3y = -6$

34. $2x = 4y$

35. A student attempted to graph $4x + 5y = 0$ by finding intercepts. She first let $x = 0$ and found y; then she let $y = 0$ and found x. In both cases, the resulting point was $(0, 0)$. She knew that she needed at least two points to graph the line, but was unsure what to do next because finding intercepts gave her only one point. How would you explain to her what to do next?

36. What is the equation of the x-axis? What is the equation of the y-axis?

 *Solve each problem. See Example 7.*

37. The height y (in centimeters) of a woman is related to the length of her radius bone x (from the wrist to the elbow) and is approximated by the linear equation

$$y = 3.9x + 73.5.$$

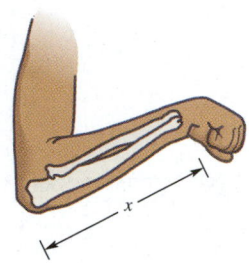

(a) Use the equation to find the approximate heights of women with radius bones of lengths 20 cm, 22 cm, and 26 cm.

(b) Write the information from part (a) as three ordered pairs.

(c) Graph the equation using the data from part (b).

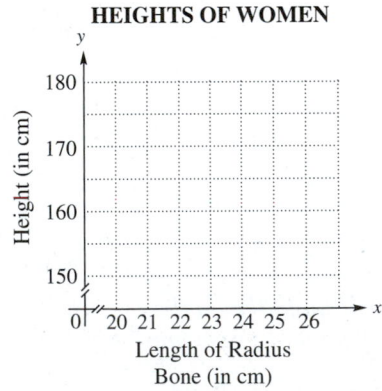

HEIGHTS OF WOMEN

(d) Use the graph to estimate the length of the radius bone in a woman who is 167 cm tall. Then use the equation to find the length of this radius bone to the nearest centimeter. (*Hint:* Substitute for y in the equation.)

38. The weight y (in pounds) of a man taller than 60 in. can be roughly approximated by the linear equation

$$y = 5.5x - 220,$$

where x is the height of the man in inches.

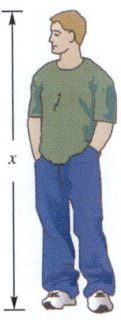

(a) Use the equation to approximate the weights of men whose heights are 62 in., 66 in., and 72 in.

(b) Write the information from part (a) as three ordered pairs.

(c) Graph the equation using the data from part (b).

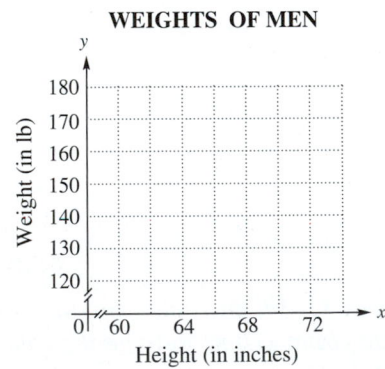

WEIGHTS OF MEN

(d) Use the graph to estimate the height of a man who weighs 155 lb. Then use the equation to find the height of this man to the nearest inch. (*Hint:* Substitute for y in the equation.)

39. As a fundraiser, a school club is selling posters. The printer charges a $25 set-up fee, plus $0.75 for each poster. Then the cost y in dollars to print x posters is given by the linear equation

$$y = 0.75x + 25.$$

(a) What is the cost y in dollars to print 50 posters? to print 100 posters?

(b) Find the number of posters x if the printer billed the club for costs of $175.

(c) Write the information from parts (a) and (b) as three ordered pairs.

(d) Use the data from part (c) to graph the equation.

40. A gas station is selling gasoline for $4.50 per gallon and charges $7 for a car wash. Then the cost y in dollars for x gallons of gasoline and a car wash is given by the linear equation

$$y = 4.50x + 7.$$

(a) What is the cost y in dollars for 9 gallons of gasoline and a car wash?   for 4 gallons of gasoline and a car wash?

(b) Find the number of gallons of gasoline x if the cost for the gasoline and a car wash is $43.00.

(c) Write the information from parts (a) and (b) as three ordered pairs.

(d) Use the data from part (c) to graph the equation.

POSTER COSTS

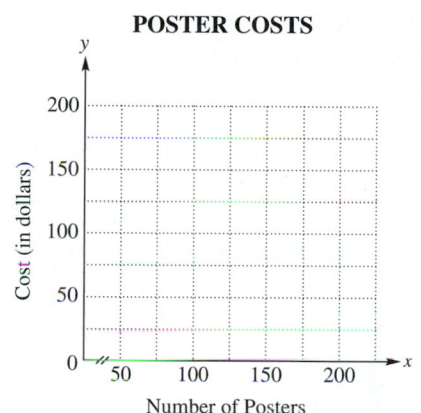

GASOLINE AND CAR WASH COSTS

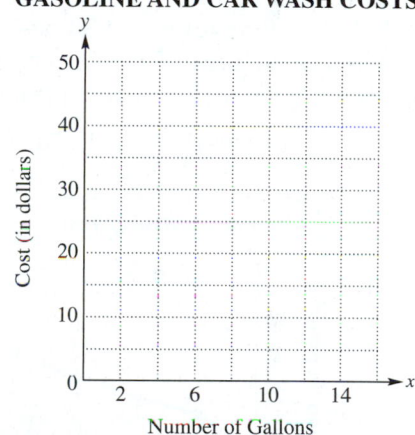

41. The graph shows the value of a certain sport-utility vehicle over the first 5 yr of ownership. Use the graph to do the following.

(a) Determine the initial value of the SUV.

(b) Find the **depreciation** (loss in value) from the original value after the first 3 yr.

(c) What is the annual or yearly depreciation in each of the first 5 yr?

SUV VALUE

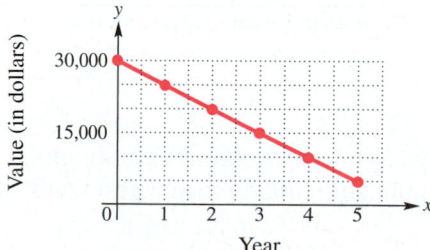

(d) What does the ordered pair (5, 5000) mean in the context of this problem?

42. Demand for an item is often closely related to its price. As price increases, demand decreases, and as price decreases, demand increases. Suppose demand for a video game is 2000 units when the price is $40, and demand is 2500 units when the price is $30.

(a) Let x be the price and y be the demand for the game. Graph the two given pairs of prices and demands.

VIDEO GAME PRICE/DEMAND

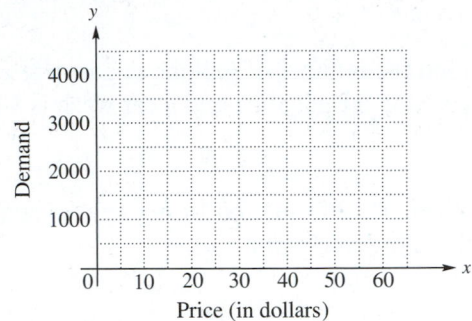

(b) Assume the relationship is linear. Draw a line through the two points from part (a). From your graph, estimate the demand if the price drops to $20.

(c) Use the graph to estimate the price if the demand is 3500 units.

(d) Write the prices and demands from parts (b) and (c) as ordered pairs.

43. U.S. per capita consumption of cheese increased for the years 1980 through 2005 as shown in the graph. If $x = 0$ represents 1980, $x = 5$ represents 1985, and so on, per capita consumption y in pounds can be modeled by the linear equation

$$y = 0.5383x + 18.74.$$

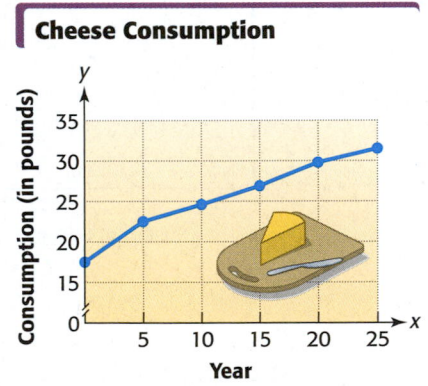

Cheese Consumption

Source: U.S. Department of Agriculture.

(a) Use the equation to approximate consumption in 1990, 2000, and 2005 to the nearest tenth.

(b) Use the graph to estimate consumption for the same years.

(c) How do the approximations using the equation compare to the estimates from the graph?

44. In the United States, sporting goods sales y (in billions of dollars) from 2000 through 2006 are shown in the graph and modeled by the linear equation

$$y = 3.018x + 72.52,$$

where $x = 0$ corresponds to 2000, $x = 1$ corresponds to 2001, and so on.

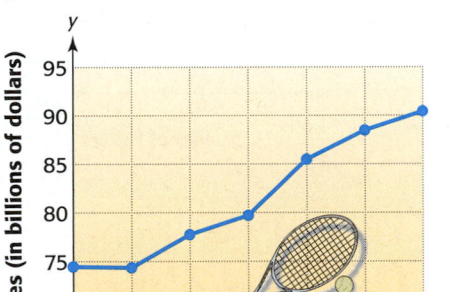

Sporting Goods Sales

Source: National Sporting Goods Association.

(a) Use the equation to approximate sporting goods sales in 2000, 2004, and 2006. Round your answers to the nearest billion dollars.

(b) Use the graph to estimate sales for the same years.

(c) How do the approximations using the equation compare to the estimates using the graph?

11.3 ▶▶▶ Slope of a Line

An important characteristic of the lines we graphed in the previous section is their slant or "steepness", as viewed from *left to right*. See Figure 17.

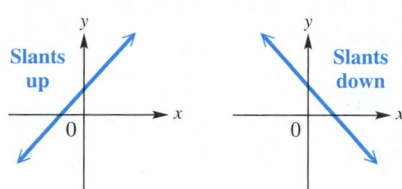

Figure 17

One way to measure the steepness of a line is to compare the vertical change in the line to the horizontal change while moving along the line from one fixed point to another. This measure of steepness is called the *slope* of the line.

OBJECTIVE 1   Find the slope of a line given two points.   To find the steepness, or slope, of the line in Figure 18, we begin at point Q and move to point P. The vertical change, or **rise,** is the change in the y-values, which is the difference $6 - 1 = 5$ units. The horizontal change, or **run,** from Q to P is the change in the x-values, which is the difference $5 - 2 = 3$ units.

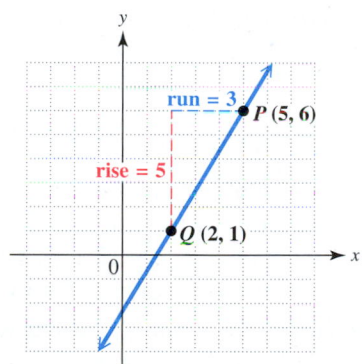

Figure 18

Remember from **Section 5.1** that one way to compare two numbers is by using a ratio. The **slope** is the ratio of the vertical change in y to the horizontal change in x. The line in Figure 18 has

$$\text{slope} = \frac{\text{vertical change in } y \text{ (rise)}}{\text{horizontal change in } x \text{ (run)}} = \frac{5}{3}.$$

To confirm this ratio, we can count grid squares. We start at point Q in Figure 18 and count *up* 5 grid squares to find the vertical change (rise). To find the horizontal change (run) and arrive at point P, we count to the *right* 3 grid squares. The slope is $\frac{5}{3}$, as found above. ***Slope is a single number that allows us to determine the direction in which a line is slanting from left to right, as well as how much slant there is to the line.***

1 Find the slope ratio of each line.

(a)

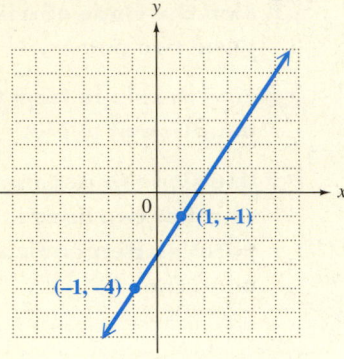

(b)

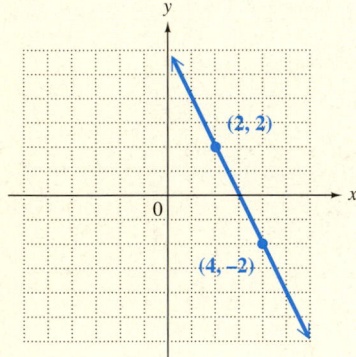

EXAMPLE 1 **Finding the Slope of a Line**

Find the slope of the line in Figure 19.

We use the two points shown on the line. The vertical change is the difference in the y-values, or $-1 - 3 = -4$, and the horizontal change is the difference in the x-values, or $6 - 2 = 4$. Thus, the line has

$$\text{slope} = \frac{\text{change in } y \text{ (rise)}}{\text{change in } x \text{ (run)}} = \frac{-4}{4}, \quad \text{or} \quad -1.$$

Counting grid squares, we begin at point P and count *down* 4 grid squares. Because we counted down, we write the vertical change as a negative number, -4 here. Then we count to the *right* 4 grid squares to reach point Q. The slope is $\frac{-4}{4}$, or -1.

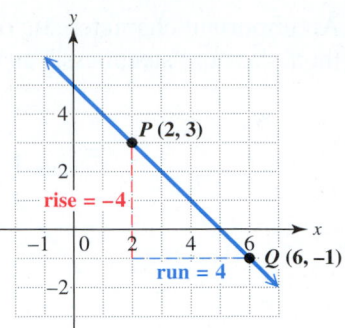

Figure 19

◀ *Work Problem* **1** *at the Side.*

Note

The slope of a line is the same for any two points on the line. To see this, refer to Figure 19. Find the points $(3, 2)$ and $(5, 0)$, which also lie on the line. If we start at $(3, 2)$ and count *down* 2 units and then to the *right* 2 units, we arrive at $(5, 0)$. The slope is $\frac{-2}{2}$, or -1, the same slope we found in Example 1.

The concept of slope is used in many everyday situations. See Figure 20. For example, a highway with a 10%, or $\frac{1}{10}$, grade (or slope) rises 1 m for every 10 m horizontally. Architects specify the pitch of a roof by using slope; a $\frac{5}{12}$ roof means that the roof rises 5 ft for every 12 ft that it runs in the horizontal direction. The slope of a stairwell also indicates the ratio of the vertical rise to the horizontal run. In the figure, the slope of the stairwell is $\frac{8}{12}$, or $\frac{2}{3}$.

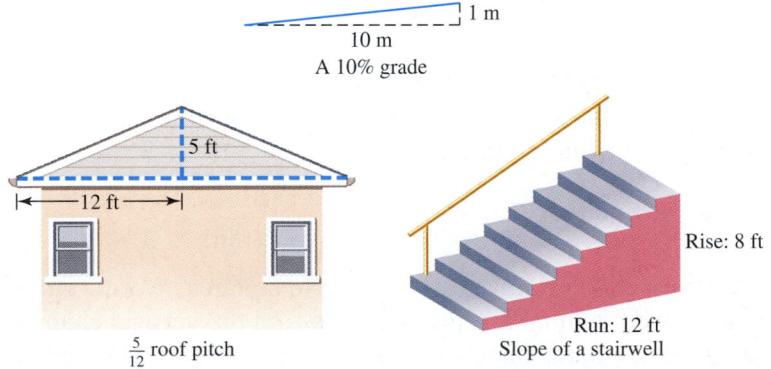

Figure 20

We can generalize the preceding discussion and find the slope of a line through two nonspecific points (x_1, y_1) and (x_2, y_2). (This notation is called **subscript notation.** Read x_1 as "*x*-sub-one" and x_2 as "*x*-sub-two.") See Figure 21.

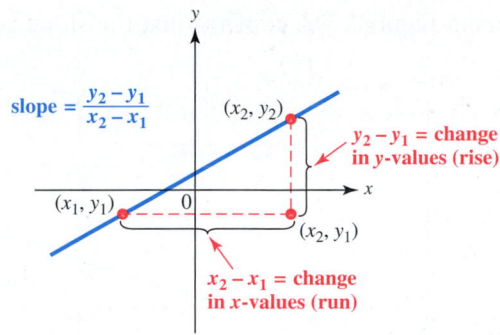

Figure 21

Moving along the line from the point (x_1, y_1) to the point (x_2, y_2), we see that y changes by $y_2 - y_1$ units. This is the vertical change (rise). Similarly, x changes by $x_2 - x_1$ units, which is the horizontal change (run). The slope of the line is the ratio of $y_2 - y_1$ to $x_2 - x_1$.

> **Note**
>
> Subscript notation is used to identify a point. It does *not* indicate any operation. Note the difference between x_2, a nonspecific value, and x^2, which means $x \cdot x$. Read x_2 as "x-sub-two," *not* "x squared."

Traditionally, the letter m represents slope. The slope m of a line is defined as follows.

> **Slope Formula**
>
> The **slope** of the line through the points (x_1, y_1) and (x_2, y_2) is
>
> $$m = \frac{\text{change in } y}{\text{change in } x} = \frac{y_2 - y_1}{x_2 - x_1} \quad (x_1 \neq x_2).$$

The slope gives the change in y for each unit of change in x.

Work Problem **2** *at the Side.* ▶

EXAMPLE 2 **Finding Slopes of Lines**

Find the slope of each line.

(a) The line through $(-4, 7)$ and $(1, -2)$

Use the slope formula. Let $(-4, 7) = (x_1, y_1)$ and $(1, -2) = (x_2, y_2)$.

$$\text{slope } m = \frac{\text{change in } y}{\text{change in } x} = \frac{y_2 - y_1}{x_2 - x_1} = \frac{-2 - 7}{1 - (-4)} = \frac{-9}{5} = -\frac{9}{5}$$

> Substitute carefully here.

Continued on Next Page

2 Find $\dfrac{y_2 - y_1}{x_2 - x_1}$ for the following values.

(a) $y_2 = 4, y_1 = -1,$
$x_2 = 3, x_1 = 4$

(b) $x_1 = 3, x_2 = -5,$
$y_1 = 7, y_2 = -9$

(c) $x_1 = 2, x_2 = 7,$
$y_1 = 4, y_2 = 9$

3 Find the slope of each line.

(a) The line through $(6, -2)$ and $(5, 4)$

(b) The line through $(-3, 5)$ and $(-4, -7)$

(c) The line through $(6, -8)$ and $(-2, 4)$

(Find this slope in two different ways as in Example 2(b).)

Count grid squares in Figure 22 to confirm that the slope is $-\frac{9}{5}$.

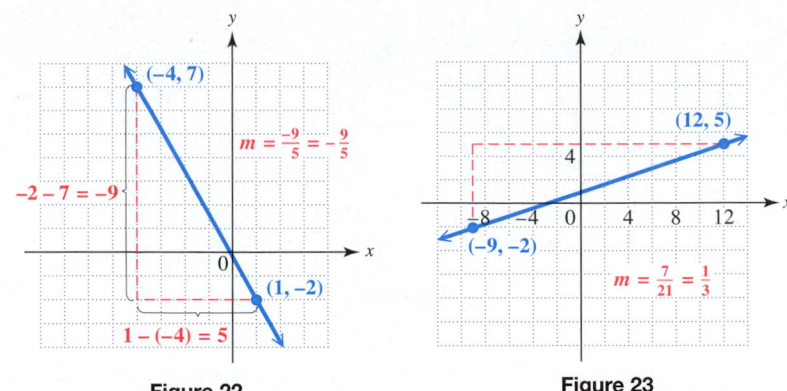

Figure 22   **Figure 23**

(b) The line through $(-9, -2)$ and $(12, 5)$

y-value

$$\text{slope } m = \frac{5 - (-2)}{12 - (-9)} = \frac{7}{21} = \frac{1}{3}$$

x-value from the *same* ordered pair

See Figure 23. Note that the same slope is obtained by subtracting in reverse order.

y-value

$$\text{slope } m = \frac{-2 - 5}{-9 - 12} = \frac{-7}{-21} = \frac{1}{3}$$

x-value from the *same* ordered pair

> **CAUTION**
> *It makes no difference which point is (x_1, y_1) or (x_2, y_2); however, be consistent.* Start with the x- and y-values of one point (either one), and subtract the corresponding values of the other point.

◀ *Work Problem* **3** *at the Side.*

The slopes we found for the lines in Figures 22 and 23 suggest the following generalization.

> **Positive and Negative Slopes**
> A line with positive slope rises (slants up) from left to right.
> A line with negative slope falls (slants down) from left to right.

ANSWERS

3. (a) -6 (b) 12 (c) $-\frac{3}{2}$; $-\frac{3}{2}$

EXAMPLE 3 **Showing that the Slope of a Horizontal Line Is Zero**

Find the slope of the line through $(-8, 4)$ and $(2, 4)$.

$$m = \frac{y_2 - y_1}{x_2 - x_1} = \frac{4 - 4}{-8 - 2} = \frac{0}{-10} = 0 \qquad \text{Zero slope}$$

As shown in Figure 24, the line through the given points is horizontal, with equation $y = 4$. **All horizontal lines have slope 0** since the difference in their y-values is always 0.

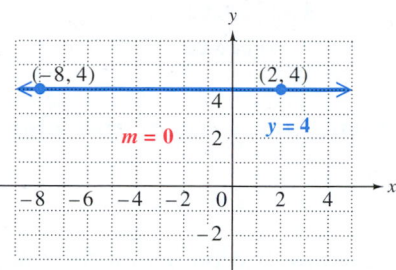

Figure 24

EXAMPLE 4 **Showing that a Vertical Line Has Undefined Slope**

Find the slope of the line through $(6, 2)$ and $(6, -4)$.

$$m = \frac{y_2 - y_1}{x_2 - x_1} = \frac{2 - (-4)}{6 - 6} = \frac{6}{0} \qquad \text{Undefined slope}$$

Because division by 0 is undefined, this line has undefined slope. (This is why the slope formula at the beginning of this section had the restriction $x_1 \neq x_2$.) The graph in Figure 25 shows that this line is vertical, with equation $x = 6$. All points on a vertical line have the same x-value, so **all vertical lines have undefined slope.**

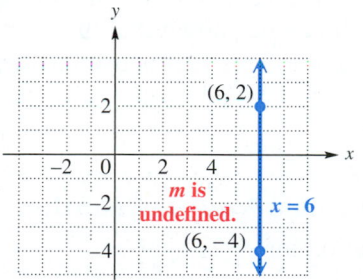

Figure 25

Slopes of Horizontal and Vertical Lines

Horizontal lines, which have equations of the form $y = k$, have **slope 0.**

Vertical lines, which have equations of the form $x = k$, have **undefined slope.**

Work Problem **4** *at the Side.* ▶

---

4 Find the slope of each line.

(a) The line through $(2, 5)$ and $(-1, 5)$

(b) The line through $(3, 1)$ and $(3, -4)$

(c) The line with equation $y = -1$

(d) The line with equation $x - 4 = 0$

ANSWERS

4. (a) 0 **(b)** undefined
(c) 0 **(d)** undefined

Figure 26 summarizes the four cases for slopes of lines.

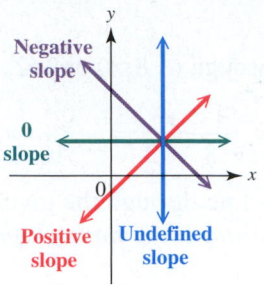

Slopes of lines

Figure 26

OBJECTIVE 2 Find the slope from the equation of a line. Consider the equation

$$y = -3x + 5.$$

We can find its slope using any two points on the line. We get these two points by first choosing two different values of x and then finding the corresponding values of y. For example, choose $x = -2$ and $x = 4$.

$y = -3x + 5$	$y = -3x + 5$
$y = -3(-2) + 5$ Let $x = -2$.	$y = -3(4) + 5$ Let $x = 4$.
$y = 6 + 5$	$y = -12 + 5$
$y = 11$	$y = -7$

The ordered pairs are $(-2, 11)$ and $(4, -7)$. Now use the slope formula.

$$m = \frac{11 - (-7)}{-2 - 4} = \frac{18}{-6} = -3$$

The slope, $m = -3$, is the same number as the coefficient of x in the equation $y = -3x + 5$. It can be shown that this always happens, *as long as the equation is solved for y.* This fact is used to find the slope of a line from its equation.

Finding the Slope of a Line from Its Equation

Step 1   Solve the equation for y.

Step 2   The slope is given by the coefficient of x.

Note

We will see in **Section 11.4** that the equation $y = -3x + 5$ is written using a special form of the equation of a line, called *slope-intercept form,*

$$y = mx + b.$$

EXAMPLE 5 **Finding Slopes from Equations**

Find the slope of each line.

(a) $2x - 5y = 4$

  Step 1 Solve the equation for y.

$$2x - 5y = 4 \quad \text{[Isolate } y \text{ on one side.]}$$

$$-5y = -2x + 4 \qquad \text{Subtract } 2x.$$

$$y = \frac{2}{5}x - \frac{4}{5} \qquad \text{Divide by } -5.$$

  Step 2 The slope is given by the coefficient of x, so the slope is $\frac{2}{5}$.

(b)
$$8x + 4y = 1$$

$$\text{[Solve for } y.\text{]} \quad 4y = -8x + 1 \qquad \text{Subtract } 8x.$$

$$y = -2x + \frac{1}{4} \qquad \text{Divide by } 4.$$

The slope of this line is given by the coefficient of x, which is -2.

Work Problem **5** *at the Side.* ▶

OBJECTIVE 3 Use slope to determine whether two lines are parallel, perpendicular, or neither. Recall that two lines in a plane that never intersect are **parallel.** We use slopes to tell whether two lines are parallel. For example, Figure 27 shows the graphs of $x + 2y = 4$ and $x + 2y = -6$. These lines appear to be parallel. Solving $x + 2y = 4$ for y gives $y = -\frac{1}{2}x + 2$. Solving $x + 2y = -6$ for y gives $y = -\frac{1}{2}x - 3$. Both lines have slope $-\frac{1}{2}$. ***Nonvertical parallel lines always have equal slopes***.

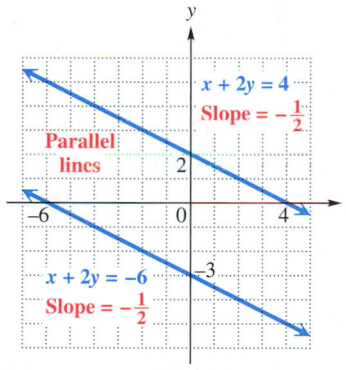

Figure 27

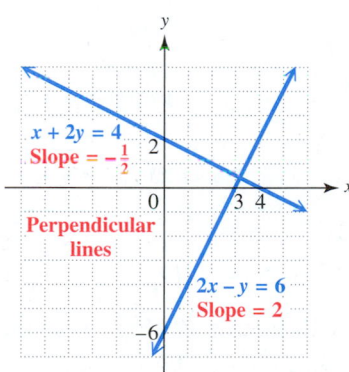

Figure 28

Figure 28 shows the graphs of $x + 2y = 4$ and $2x - y = 6$. These lines appear to be **perpendicular** (that is, they intersect at a 90° angle). Solving $x + 2y = 4$ for y gives $y = -\frac{1}{2}x + 2$, with slope $-\frac{1}{2}$. Solving $2x - y = 6$ for y gives $y = 2x - 6$, with slope **2**. The product of $-\frac{1}{2}$ and 2 is

$$-\frac{1}{2}(2) = -1.$$

This condition is true in general. ***The product of the slopes of two perpendicular lines, neither of which is vertical, is always −1.*** This means that the slopes of perpendicular lines are negative (or opposite) reciprocals—if one slope is the nonzero number a, then the other is $-\frac{1}{a}$. The table in the margin shows several examples.

5 Find the slope of each line.

(a) $y = -\frac{7}{2}x + 1$

(b) $3x + 2y = 9$

(c) $y + 4 = 0$

(d) $x + 3 = 7$

Number	Negative Reciprocal
$\frac{3}{4}$	$-\frac{4}{3}$
$\frac{1}{2}$	$-\frac{2}{1}$, or -2
-6, or $-\frac{6}{1}$	$\frac{1}{6}$
-0.4, or $-\frac{4}{10}$	$\frac{10}{4}$, or 2.5

The product of each number and its negative reciprocal is −1.

ANSWERS

5. **(a)** $-\frac{7}{2}$ **(b)** $-\frac{3}{2}$ **(c)** 0 **(d)** undefined

6 Decide whether each pair of lines is *parallel, perpendicular,* or *neither.*

(a) $x + y = 6$

$x + y = 1$

(b) $3x - y = 4$

$x + 3y = 9$

(c) $2x - y = 5$

$2x + y = 3$

(d) $3x - 7y = 35$

$7x - 3y = -6$

> **Slopes of Parallel and Perpendicular Lines**
>
> Two lines with the same slope are parallel.
>
> Two lines whose slopes have a product of -1 are perpendicular.

EXAMPLE 6 Deciding Whether Lines Are Parallel, Perpendicular, or Neither

Decide whether each pair of lines is *parallel, perpendicular,* or *neither.*

(a) $x + 2y = 7$

$-2x + y = 3$

Find the slope of each line by first solving each equation for y.

$x + 2y = 7$

$2y = -x + 7$ Subtract x.

$y = -\dfrac{1}{2}x + \dfrac{7}{2}$ Divide by 2.

Slope is $-\frac{1}{2}$.

$-2x + y = 3$

$y = 2x + 3$ Add $2x$.

Slope is 2.

Because the slopes are not equal, the lines are not parallel. Check the product of the slopes: $-\frac{1}{2}(2) = -1$. The two lines are perpendicular because the product of their slopes is -1. See Figure 29.

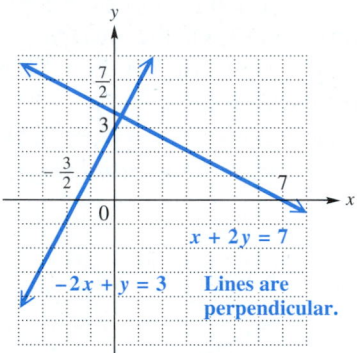

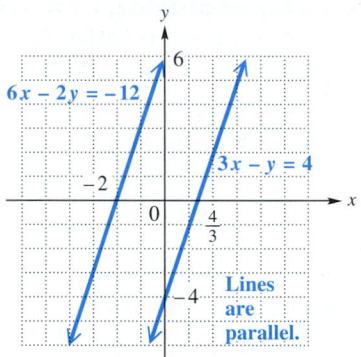

Figure 29   Figure 30

(b) $3x - y = 4$ Solve for y. $y = 3x - 4$

$6x - 2y = -12$ $y = 3x + 6$

Both lines have slope 3, so the lines are parallel. See Figure 30.

(c) $4x + 3y = 6$ $y = -\dfrac{4}{3}x + 2$

     Solve for y.

$2x - y = 5$ $y = 2x - 5$

Here the slopes are $-\frac{4}{3}$ and 2. Because $-\frac{4}{3} \neq 2$ and $-\frac{4}{3}(2) \neq -1$, these lines are neither parallel nor perpendicular.

(d) $5x - y = 1$ $y = 5x - 1$

     Solve for y.

$x - 5y = -10$ $y = \dfrac{1}{5}x + 2$

The slopes are 5 and $\frac{1}{5}$. The lines are not parallel, nor are they perpendicular. (*Be careful!* $5\left(\frac{1}{5}\right) = 1$, *not* -1.)

◀ **Work Problem** **6** at the Side.

11.3 ▶▶▶ Exercises

Use the coordinates of the indicated points to find the slope of each line. See Example 1.

1.
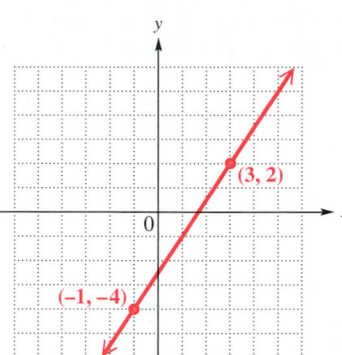

2.
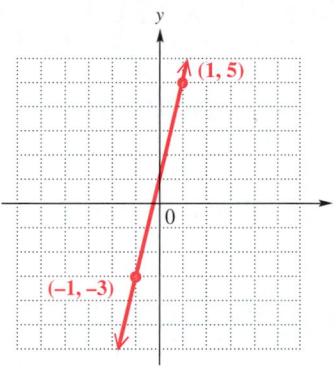

3.
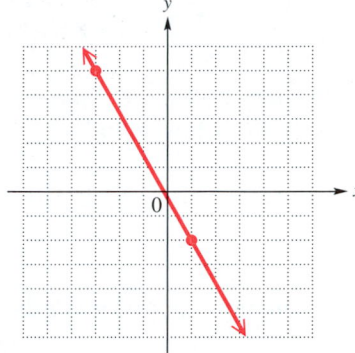

4.
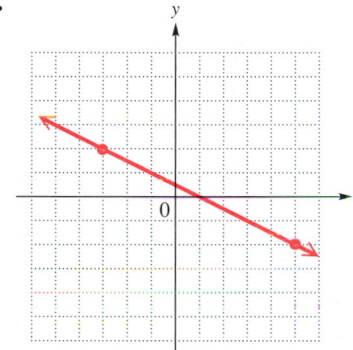

5.
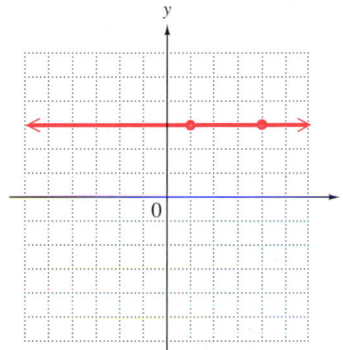

6.
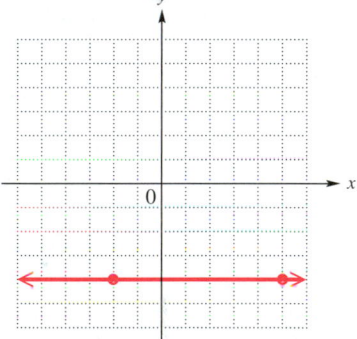

7. In the context of the graph of a straight line, what is meant by "rise"? What is meant by "run"?

8. Look at the graph in Exercise 1, and answer the following.

   (a) Start at the point $(-1, -4)$ and count vertically up to the horizontal line that goes through the other plotted point. What is this vertical change? (Remember: "up" means positive, "down" means negative.) _____

   (b) From this new position, count horizontally to the other plotted point. What is this horizontal change? (Remember: "right" means positive, "left" means negative.) _____

   (c) What is the quotient of the numbers found in parts (a) and (b)? _____ What do we call this number? _____

   (d) If we were to *start* at the point $(3, 2)$ and *end* at the point $(-1, -4)$, would the answer to part (c) be the same? Explain why or why not.

On the given coordinate system, sketch the graph of a straight line with the indicated slope.

9. Negative

10. Positive

11. Undefined

12. Zero

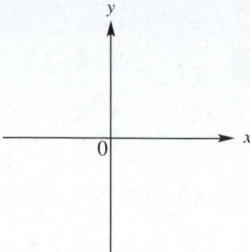

13. Decide whether the line with the given slope rises from left to right, falls from left to right, is horizontal, or is vertical.

(a) $m = -4$ **(b)** $m = 0$ **(c)** m is undefined.   **(d)** $m = \dfrac{3}{7}$

14. Explain in your own words what is meant by the *slope* of a line.

15. A student found the slope of the line through the points $(2, 5)$ and $(-1, 3)$ and got $-\frac{2}{3}$ as his answer. He showed his work as

$$\frac{3 - 5}{2 - (-1)} = \frac{-2}{3} = -\frac{2}{3}.$$

WHAT WENT WRONG? Give the correct slope.

Find the slope of the line through each pair of points. See Examples 2–4.

16. $(4, -1)$ and $(-2, -8)$  **17.** $(1, -2)$ and $(-3, -7)$ **18.** $(-8, 0)$ and $(0, -5)$

19. $(0, 3)$ and $(-2, 0)$ **20.** $(-4, -5)$ and $(-5, -8)$ **21.** $(-2, 4)$ and $(-3, 7)$

22. $(6, -5)$ and $(-12, -5)$ 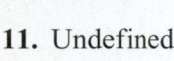 **23.** $(4, 3)$ and $(-6, 3)$ **24.** $(-8, 6)$ and $(-8, -1)$

25. $(-12, 3)$ and $(-12, -7)$ **26.** $(3.1, 2.6)$ and $(1.6, 2.1)$ **27.** $\left(-\dfrac{7}{5}, \dfrac{3}{10}\right)$ and $\left(\dfrac{1}{5}, -\dfrac{1}{2}\right)$

Find the slope of each line. See Example 5.

28. $y = 2x - 3$ **29.** $y = 5x + 12$ **30.** $2y = -x + 4$ **31.** $4y = x + 1$

32. $-6x + 4y = 4$ **33.** $3x - 2y = 3$ **34.** $y = 4$ **35.** $y = 6$

36. $x = 5$ **37.** $x = -2$ **38.** $x + y = 0$ **39.** $x - y = 0$

The figure at the right shows a line that has a positive slope (because it rises from left to right) and a positive y-value for the y-intercept (because it intersects the y-axis above the origin).

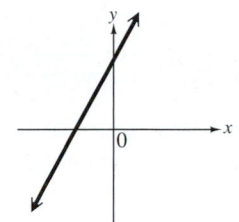

For each figure in Exercises 40–45, decide whether (a) the slope is positive, negative, *or 0 and whether (b) the y-value of the y-intercept is* positive, negative, *or 0.*

40. (a) _____

   (b) _____

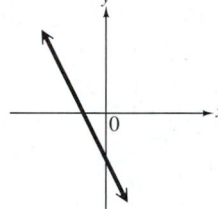

41. (a) _____

   (b) _____

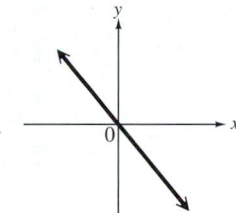

42. (a) _____

   (b) _____

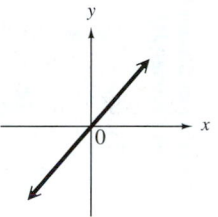

43. (a) _____

   (b) _____

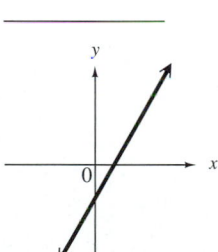

44. (a) _____

   (b) _____

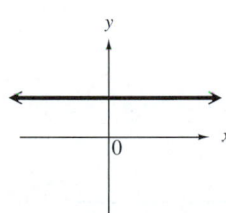

45. (a) _____

   (b) _____

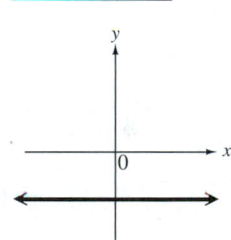

In each pair of equations, give the slope of each line, and then determine whether the two lines are parallel, perpendicular, *or* neither parallel nor perpendicular. *See Example 6.*

46. $2x + 5y = 4$
   $4x + 10y = 1$

🌐 **47.** $-4x + 3y = 4$
   $-8x + 6y = 0$

48. $8x - 9y = 6$
   $8x + 6y = -5$

49. $5x - 3y = -2$
   $3x - 5y = -8$

50. $3x - 2y = 6$
   $2x + 3y = 3$

🌐 **51.** $3x - 5y = -1$
   $5x + 3y = 2$

52. What is the slope (or pitch) of this roof?

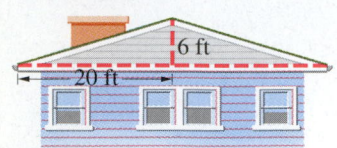

53. What is the slope (or grade) of this hill?

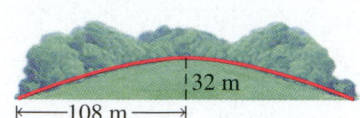

54. What is the slope (or grade) of this ski slope?

Relating Concepts (Exercises 55–60) For Individual or Group Work

Figure A gives public school enrollment (in thousands) in grades 9–12 in the United States. Figure B gives the (average) number of public school students per computer.

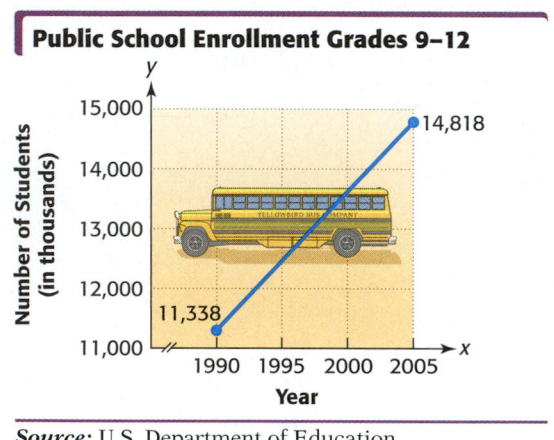

Source: U.S. Department of Education.

Figure A

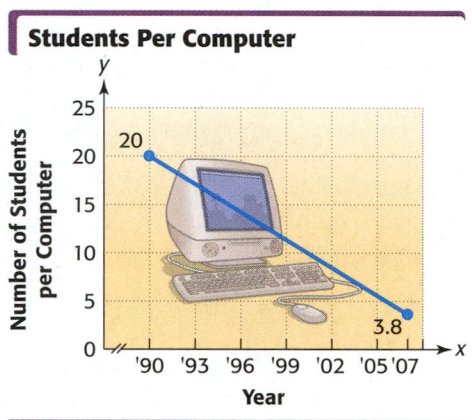

Source: Quality Education Data, Inc.

Figure B

Work Exercises 55–60 in order.

55. Use the ordered pairs (1990, 11,338) and (2005, 14,818) to find the slope of the line in Figure A.

56. The slope of the line in Figure A is _____. This means that during (positive/negative)

the period represented, enrollment _____.
(increased/decreased)

57. The slope of a line represents its *rate of change*. Based on Figure A, what was the increase in students *per year* during the period shown?

58. Use the given information to find the slope, to the nearest hundredth, of the line in Figure B.

59. The slope of the line in Figure B is _____. This means that during (positive/negative)

the period represented, the number of students per computer _____.
(increased/decreased)

60. Based on Figure B, what was the decrease in students per computer *per year* during the period shown?

11.4 ▶▶▶ Equations of Lines

In **Section 11.3,** we found the slope (steepness) of a line from the equation of the line by solving the equation for y. In that form, the slope is the coefficient of x. For example, the slope of the line with equation $y = 2x + 3$ is 2, the coefficient of x. What does the number **3** represent? If $x = 0$, the equation becomes

$$y = 2(0) + 3$$
$$y = 3.$$

Since $y = 3$ corresponds to $x = 0$, $(0, 3)$ is the y-intercept of the graph of $y = 2x + 3$. An equation like $y = 2x + 3$ that is solved for y is said to be in **slope-intercept form** because both the slope and the y-intercept of the line can be read directly from the equation.

> **Slope-Intercept Form**
>
> The slope-intercept form of the equation of a line with slope m and y-intercept $(0, b)$ is
>
> $$y = mx + b.$$
>
> Slope ⟶   �↑___ $(0, b)$ is the
> y-intercept.

REMEMBER: The intercept in slope-intercept form is the y-intercept.

> **Note**
>
> The slope-intercept form is the most useful form for a linear equation because of the information we can determine from it. It is also the form used by graphing calculators and the one that describes a *linear function,* an important concept in mathematics.

OBJECTIVE **1** **Write an equation of a line given its slope and y-intercept.**   Given the slope and y-intercept of a line, we can use the slope-intercept form to write an equation of the line.

EXAMPLE 1 **Writing an Equation of a Line**

Write an equation of the line with slope $\frac{2}{3}$ and y-intercept $(0, -1)$.
    Here $m = \frac{2}{3}$ and $b = -1$, so an equation is

Slope ⟶   ⟶ y-intercept $(0, b)$

$$y = mx + b$$

$$y = \frac{2}{3}x + (-1), \quad \text{or} \quad y = \frac{2}{3}x - 1.$$

Work Problem **1** *at the Side.* ▶

OBJECTIVES

1 Write an equation of a line given its slope and y-intercept.

2 Graph a line given its slope and a point on the line.

3 Write an equation of a line given its slope and any point on the line.

4 Write an equation of a line given two points on the line.

5 Find an equation of a line that fits a data set.

1 Write an equation of the line with the given slope and y-intercept.

(a) slope $\frac{1}{2}$; y-intercept $(0, -4)$

(b) slope -1; y-intercept $(0, 8)$

(c) slope 3; y-intercept $(0, 0)$

(d) slope 0; y-intercept $(0, 2)$

(e) slope 1; y-intercept $(0, 0.75)$

ANSWERS

1. **(a)** $y = \frac{1}{2}x - 4$ **(b)** $y = -x + 8$
   (c) $y = 3x$ **(d)** $y = 2$
   (e) $y = x + 0.75$

2 Graph $3x - 4y = 8$ by using the slope and y-intercept.

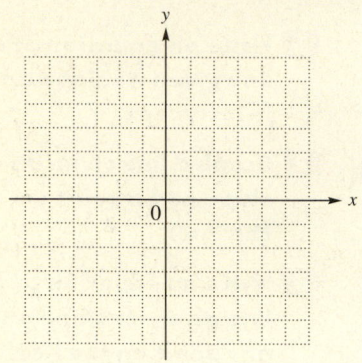

OBJECTIVE 2 Graph a line given its slope and a point on the line. We can use the slope and y-intercept to graph a line.

> **Graphing a Line by Using the Slope and y-Intercept**
>
> **Step 1** Write the equation in slope-intercept form, if necessary, by solving for y.
>
> **Step 2** Identify the y-intercept. Graph the point $(0, b)$.
>
> **Step 3** Identify slope m of the line. Use the geometric interpretation of slope ("rise over run") to find another point on the graph by counting from the y-intercept.
>
> **Step 4** Join the two points with a line to obtain the graph.

EXAMPLE 2 Graphing a Line by Using the Slope and y-Intercept

Graph $2x - 3y = 3$ by using the slope and y-intercept.

Step 1 Solve for y to write the equation in slope-intercept form.

$$2x - 3y = 3 \quad \text{[Isolate } y \text{ on one side.]} \quad \text{Given equation}$$

$$-3y = -2x + 3 \quad \text{Subtract } 2x.$$

$$y = \frac{2}{3}x - 1 \quad \text{Divide by } -3.$$

Slope m ⟶ ⟵ y-intercept $(0, b)$

Step 2 The y-intercept is $(0, -1)$. Graph this point. See Figure 31.

Step 3 The slope is $\frac{2}{3}$. By the definition of slope,

$$m = \frac{\textbf{change in } y}{\textbf{change in } x} = \frac{2}{3}.$$

Counting from the y-intercept 2 units up and 3 units to the right, we obtain another point on the graph, $(3, 1)$.

Step 4 Draw the line through the points $(0, -1)$ and $(3, 1)$ to obtain the graph of the given equation $2x - 3y = 3$. See Figure 31.

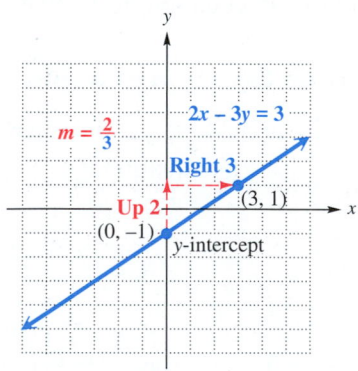

Figure 31

◀ Work Problem **2** at the Side.

ANSWER

2.

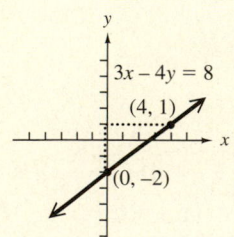

The method of Example 2 can be extended to graph a line given its slope and *any* point on the line.

3 Graph the line passing through the point $(2, -3)$, with slope $-\frac{1}{3}$.

EXAMPLE 3   **Graphing a Line by Using the Slope and a Point**

Graph the line passing through the point $(-2, 3)$, with slope -4.

First, locate the point $(-2, 3)$. See Figure 32. Then write the slope -4 as

$$\text{slope } m = \frac{\text{change in } y}{\text{change in } x} = -4 = \frac{-4}{1}.$$

Locate another point on the line by counting 4 units *down* (because of the negative sign) from $(-2, 3)$ and then 1 unit to the right. Finally, draw the line through this new point P and the given point $(-2, 3)$. See Figure 32.

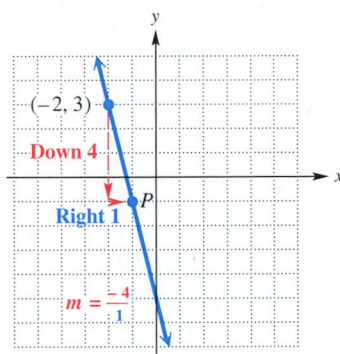

Figure 32

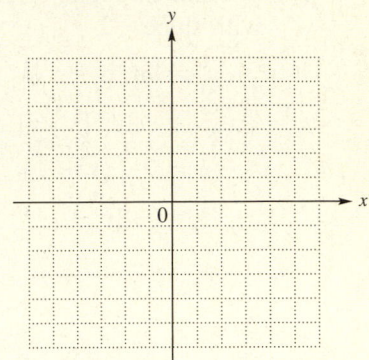

> **Note**
>
> In Example 3, we could have written the slope as $\frac{4}{-1}$ instead. In this case, we would move 4 units up from $(-2, 3)$ and then 1 unit to the *left* (because of the negative sign). Verify that this produces the same line.

Work Problem **3** *at the Side.* ▶

OBJECTIVE 3 Write an equation of a line given its slope and any point on the line.   We can use the slope-intercept form to write the equation of a line if we know the slope and any point on the line.

EXAMPLE 4   **Using the Slope-Intercept Form to Write an Equation of a Line**

Write an equation, in slope-intercept form, of the line having slope 4 passing through the point $(2, 5)$.

Since the line passes through the point $(2, 5)$, we can substitute $x = 2, y = 5$, and the given slope $m = 4$ into $y = mx + b$ and solve for b.

$$
\begin{array}{lll}
y = mx + b & \text{Slope-intercept form} \\
5 = 4(2) + b & \text{Let } x = 2, y = 5, \text{ and } m = 4. \\
5 = 8 + b & \text{Multiply.} \\
-3 = b & \text{Subtract 8.}
\end{array}
$$

Remember: $(0, b)$ is the y-intercept. Don't stop here.

The y-intercept is $(0, -3)$. Using the given slope, 4, an equation of the line is

$$y = 4x - 3. \qquad \text{Slope-intercept form}$$

Work Problem **4** *at the Side.* ▶

4 Write an equation, in slope-intercept form, of the line having slope -2 and passing through the point $(-1, 4)$.

ANSWERS

3.

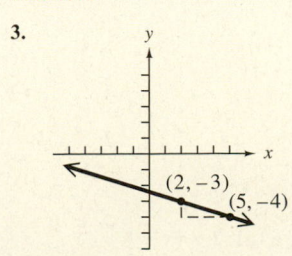

4. $y = -2x + 2$

5 Write an equation of each line. Give the final answer in slope-intercept form.

(a) The line through $(-1, 3)$, with slope -2

$$y - y_1 = m(x - x_1)$$
$$y - \underline{\quad} = \underline{\quad} [x - (\underline{\quad})]$$
$$y - 3 = -2(x + \underline{\quad})$$
$$y - 3 = -2x - \underline{\quad}$$
$$y = \underline{\qquad}$$

(b) The line through $(5, 2)$, with slope $-\frac{1}{3}$

There is another form that can be used to write the equation of a line. To develop this form, let m represent the slope of a line and let (x_1, y_1) represent a given point on the line. Let (x, y) represent any other point on the line. See Figure 33. Then,

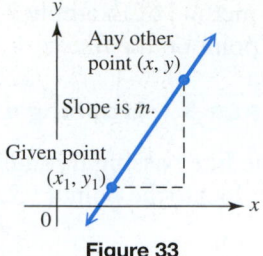

Figure 33

$$m = \frac{y - y_1}{x - x_1} \qquad \text{Definition of slope}$$
$$m(x - x_1) = y - y_1 \qquad \text{Multiply each side by } x - x_1.$$
$$y - y_1 = m(x - x_1). \qquad \text{Rewrite.}$$

This result is the **point-slope form** of the equation of a line.

Point-Slope Form

The point-slope form of the equation of a line with slope m passing through the point (x_1, y_1) is

$$\overset{\text{Slope}}{\underset{\underset{\text{point}}{\text{Given}}}{y - y_1 = m(x - x_1).}}$$

EXAMPLE 5   **Using the Point-Slope Form to Write Equations**

Write an equation of each line. Give the final answer in slope-intercept form.

(a) The line through $(-2, 4)$, with slope -3

The given point is $(-2, 4)$ so $x_1 = -2$ and $y_1 = 4$. Also, $m = -3$. Substitute these values into the point-slope form.

$$y - y_1 = m(x - x_1) \qquad \text{Point-slope form}$$
$$y - 4 = -3[x - (-2)] \qquad \text{Let } x_1 = -2, y_1 = 4, m = -3.$$
$$y - 4 = -3(x + 2) \qquad \boxed{\text{Be careful substituting.}}$$
$$y - 4 = -3x - 6 \qquad \text{Distributive property}$$
$$y = -3x - 2 \qquad \text{Add 4.}$$

(b) The line through $(4, 2)$, with slope $\frac{3}{5}$

$$y - y_1 = m(x - x_1) \qquad \text{Point-slope form}$$
$$y - 2 = \frac{3}{5}(x - 4) \qquad \text{Let } x_1 = 4, y_1 = 2, m = \frac{3}{5}.$$
$$y - 2 = \frac{3}{5}x - \frac{12}{5} \qquad \text{Distributive property}$$
$$y = \frac{3}{5}x - \frac{12}{5} + \frac{10}{5} \qquad \text{Add 2} = \frac{10}{5}.$$
$$y = \frac{3}{5}x - \frac{2}{5} \qquad \text{Combine like terms.}$$

ANSWERS

5. **(a)** $3; -2; -1; 1; 2; -2x + 1$
(b) $y = -\frac{1}{3}x + \frac{11}{3}$

We did not clear fractions after the substitution step because we want the equation in slope-intercept form—that is, solved for y.

◀ *Work Problem* **5** *at the Side.*

OBJECTIVE 4 **Write an equation of a line given two points on the line.** We can also use the point-slope form to find an equation of a line when two points on the line are known.

EXAMPLE 6 **Writing an Equation of a Line Given Two Points**

Write an equation of the line through the points $(-2, 5)$ and $(3, 4)$. Give the final answer in slope-intercept form.

First, find the slope of the line, using the slope formula.

$$\text{slope } m = \frac{y_2 - y_1}{x_2 - x_1} = \frac{5 - 4}{-2 - 3} = \frac{1}{-5} = -\frac{1}{5}$$

Now use either $(-2, 5)$ or $(3, 4)$ and the point-slope form. Using $(3, 4)$ gives

$$y - y_1 = m(x - x_1) \qquad \text{Point-slope form}$$

$$y - 4 = -\frac{1}{5}(x - 3) \qquad \text{Let } x_1 = 3, y_1 = 4, m = -\frac{1}{5}.$$

$$y - 4 = -\frac{1}{5}x + \frac{3}{5} \qquad \text{Distributive property}$$

$$y = -\frac{1}{5}x + \frac{3}{5} + \frac{20}{5} \qquad \text{Add } 4 = \frac{20}{5}.$$

$$y = -\frac{1}{5}x + \frac{23}{5}. \qquad \text{Combine like terms.}$$

The same result would be found using $(-2, 5)$ for (x_1, y_1).

Work Problem **6** *at the Side.* ▶

Note

In Example 6, the same result would also be found by substituting the slope and either given point in slope-intercept form $y = mx + b$ and then solving for b, as in Example 4. Try this.

Many of the linear equations in **Sections 11.1–11.3** were given in the form

$$Ax + By = C,$$

called **standard form,** where A, B, and C are real numbers and A and B are not both 0. In most cases, A, B, and C are rational numbers. For consistency in this book, we give answers so that A, B, and C are integers with greatest common factor 1 and $A \geq 0$.

Note

The definition of standard form is not the same in all texts. A linear equation can be written in many different, equally correct, ways. For example,

$$3x + 4y = 12, \quad 6x + 8y = 24, \quad \text{and} \quad -9x - 12y = -36$$

all represent the same set of ordered pairs. When giving answers, let us agree that $3x + 4y = 12$ is preferable to the other forms because the greatest common factor of 3, 4, and 12 is 1 and $A \geq 0$.

6 Write an equation in slope-intercept form of the line through each pair of points.

(a) $(-3, 1)$ and $(2, 4)$

(b) $(2, 5)$ and $(-1, 6)$

A summary of the forms of linear equations follows.

Forms of Linear Equations

Equation	Description	Example
$x = k$	**Vertical line** Slope is undefined; x-intercept is $(k, 0)$.	$x = 3$
$y = k$	**Horizontal line** Slope is 0; y-intercept is $(0, k)$.	$y = 3$
$y = mx + b$	**Slope-intercept form** Slope is m; y-intercept is $(0, b)$.	$y = \dfrac{3}{2}x - 6$
$y - y_1 = m(x - x_1)$	**Point-slope form** Slope is m; line passes through (x_1, y_1).	$y + 3 = \dfrac{3}{2}(x - 2)$
$Ax + By = C$	**Standard form** Slope is $-\dfrac{A}{B}$; x-intercept is $\left(\dfrac{C}{A}, 0\right)$; y-intercept is $\left(0, \dfrac{C}{B}\right)$.	$3x - 2y = 12$

OBJECTIVE 5 **Find an equation of a line that fits a data set.** Earlier in this chapter, we gave linear equations that modeled real data, such as number of twin births and amounts of credit card debt, and then used these equations to estimate or predict values. We now develop a procedure to find such an equation if the given set of data fits a linear pattern—that is, its graph consists of points lying close to a straight line.

EXAMPLE 7 **Finding an Equation of a Line That Describes Data**

The table lists the average annual cost (in dollars) of tuition and fees for in-state students at public 4-year colleges and universities for selected years. Year 1 represents 2001, year 3 represents 2003, and so on.

Year	Cost (in dollars)
1	3766
3	4645
5	5491
7	6185

Source: The College Board.

Plot the data and find an equation that approximates it.

Letting y represent the cost in year x, we plot the data as shown in Figure 34 on the next page.

Continued on Next Page

**AVERAGE ANNUAL COSTS AT
PUBLIC 4-YEAR COLLEGES**

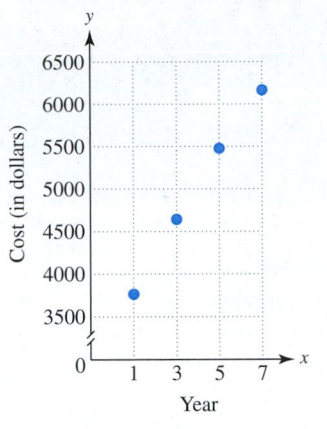

Figure 34

The points appear to lie approximately in a straight line. We can use two of the data pairs and the slope-intercept form of the equation of a line to get an equation that describes the relationship between the year and the cost. We choose the ordered pairs (5, 5491) and (7, 6185) from the table and find the slope of the line through these points.

$$m = \frac{y_2 - y_1}{x_2 - x_1} = \frac{6185 - 5491}{7 - 5} = \mathbf{347} \qquad \text{Let } (7, 6185) = (x_2, y_2)$$
$$\text{and } (5, 5491) = (x_1, y_1).$$

As we might expect, the slope, 347, is positive, indicating that tuition and fees *increased* \$347 each year. Now use this slope and the point (5, 5491) in the slope-intercept form to find an equation of the line.

$$y = mx + b \qquad \text{Slope-intercept form}$$

Solve for b, the y-value of the y-intercept.

$$5491 = \mathbf{347}(5) + b \qquad \text{Substitute for } x, y, \text{ and } m.$$
$$5491 = 1735 + b \qquad \text{Multiply.}$$
$$\mathbf{3756} = b \qquad \text{Subtract 1735.}$$

Thus, $m = 347$ and $b = 3756$, so an equation of the line is

$$y = \mathbf{347}x + \mathbf{3756}.$$

To see how well this equation approximates the ordered pairs in the data table, let $x = 3$ (for 2003) and find y.

$$y = 347x + 3756 \qquad \text{Equation of the line}$$
$$y = 347(\mathbf{3}) + 3756 \qquad \text{Substitute 3 for } x.$$
$$y = 4797 \qquad \text{Multiply; add.}$$

The corresponding value in the table for $x = 3$ is 4645, so the equation approximates the data reasonably well. With caution, the equation could be used to predict values for years that are not included in the table.

Note

In Example 7, if we had chosen two different data points, we would have gotten a slightly different equation.

Work Problem **7** *at the Side.* ▶

Math in the Media

The graph shown here is typical of many graphs that appear in magazines and newspapers. This one shows that between 1996 and 2006, the number of McDonald's restaurants worldwide rose from 20,000 to 31,000. This is depicted by the line segment joining the two points labeled *A* and *B*.

Use the graph to answer each of the following.

1. To represent point *A*, write an ordered pair in the form

   (year, number of restaurants in thousands).

   Do this for point *B* also.

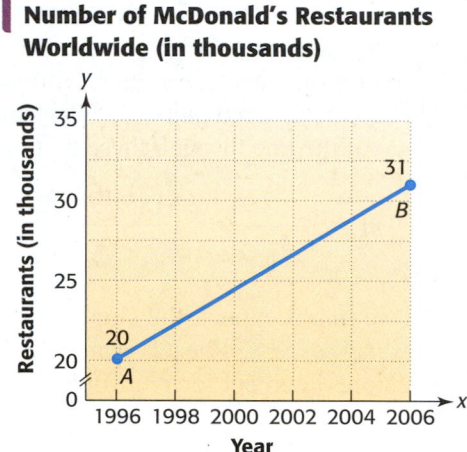

Number of McDonald's Restaurants Worldwide (in thousands)

Source: McDonald's Corp.; Hoovers.

2. The points *A* and *B* as well as the points on line *AB* between them make up line segment *AB*. We can find the coordinates of a point *M* on line segment *AB* that is exactly halfway between *A* and *B*. (This point is called the **midpoint** of the segment.)

   (a) To find the *x*-coordinate of *M*, we find the average of the *x*-coordinates of *A* and *B* by adding them and dividing by 2. What is the *x*-coordinate of *M*?

   (b) To find the *y*-coordinate of *M*, we find the average of the *y*-coordinates of *A* and *B*. What is the *y*-coordinate of *M*?

3. Fill in the blanks with the appropriate responses: The ordered pair that represents *M*, the midpoint of segment *AB*, is (_____, _____). This suggests that in the year _____, there were _____ thousand McDonald's restaurants worldwide.

4. Use the points (1996, 20) and (2006, 31) to find the $y = mx + b$ form of the equation of the line containing *A* and *B*.

5. Use the result of Exercise 4 to find the value of *y* when $x = 2001$. Does this correspond to the result you found in Exercise 2(b)?

6. The actual number of McDonald's restaurants worldwide was 30 thousand in 2001. How does this compare to your answers in Exercises 3 and 5? Explain how a line graph such as this one can be misleading. Use the concept of *slope* in your explanation.

820

11.4 ▶▶▶ Exercises

1. Match the correct equation in Column II with the description given in Column I.

I

(a) Slope -2, the line through the point $(4, 1)$

(b) Slope -2, y-intercept $(0, 1)$

(c) The line through the points $(0, 0)$ and $(4, 1)$

(d) The line through the points $(0, 0)$ and $(1, 4)$

II

A. $y = 4x$

B. $y = \dfrac{1}{4}x$

C. $y = -2x + 1$

D. $y - 1 = -2(x - 4)$

2. In the summary box on page 266, we give the equations $y = \frac{3}{2}x - 6$ and $y + 3 = \frac{3}{2}(x - 2)$ as examples of equations in slope-intercept form and point-slope form, respectively. Write each of these equations in standard form. What do you notice?

*Use the geometric interpretation of slope (rise divided by run, from **Section 11.3**) to find the slope of each line. Then, by identifying the y-intercept from the graph, write the slope-intercept form of the equation of the line.*

3.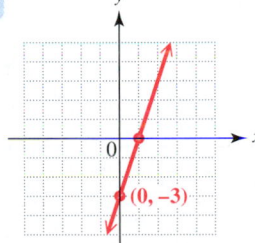

4.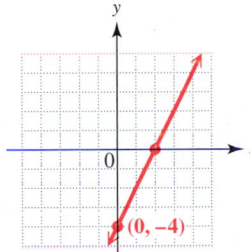

5.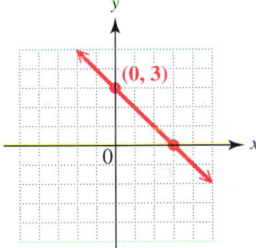

6.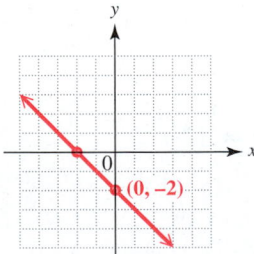

7.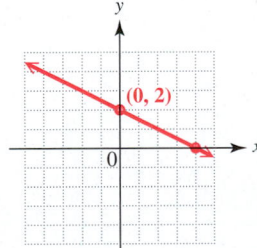

8.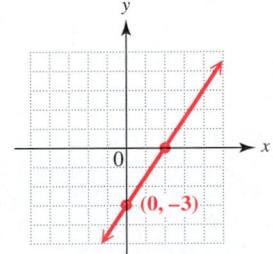

Write the equation of the line with the given slope and y-intercept. See Example 1.

9. slope 4;
   y-intercept $(0, -3)$

10. slope -5;
   y-intercept $(0, 6)$

11. slope 0;
   y-intercept $(0, 3)$

12. slope 3;
   y-intercept $(0, 0)$

13. Match each equation with the graph that would most closely resemble its graph.

(a) $y = x + 3$

(b) $y = -x + 3$

(c) $y = x - 3$

(d) $y = -x - 3$

A.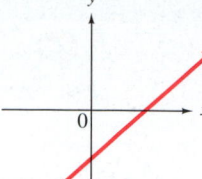

B.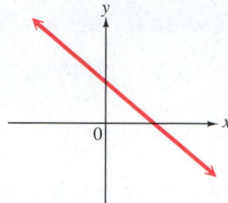

C.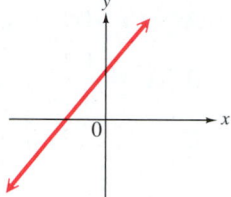

D.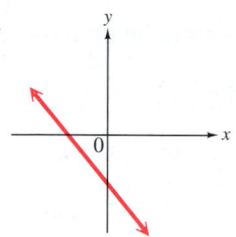

14. Explain why the equation of a vertical line cannot be written in the form $y = mx + b$.

Graph each equation by finding the slope and y-intercept, and using their definitions to find two points on the line. See Example 2.

15. $y = 3x + 2$

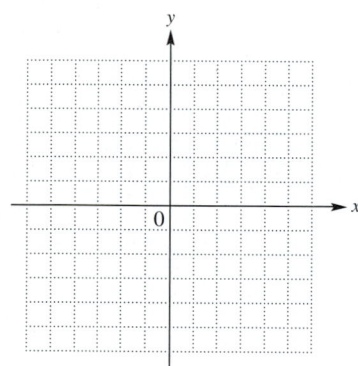

16. $y = 4x - 4$

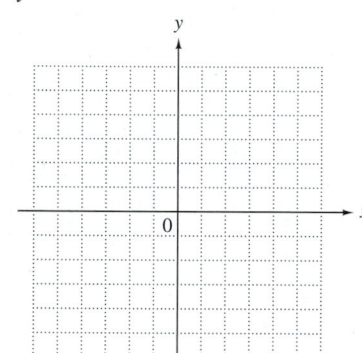

17. $2x + y = -5$

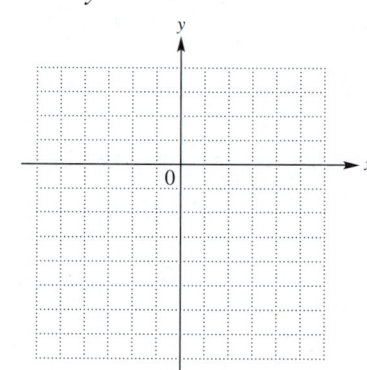

18. $3x + y = -2$

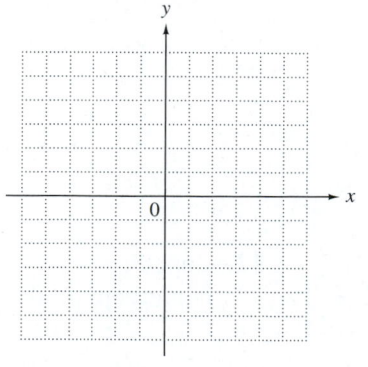

19. $x + 2y = 4$

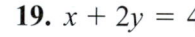

20. $x + 3y = 12$

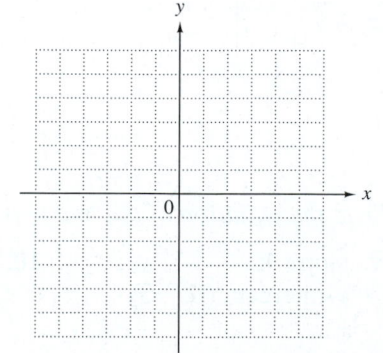

Graph each line passing through the given point and having the given slope. (In Exercises 25–28, recall the types of lines having slope 0 and undefined slope.) Give the slope-intercept form of the equation of the line if possible. See Example 3.

21. $(-2, 3)$, $m = \dfrac{1}{2}$

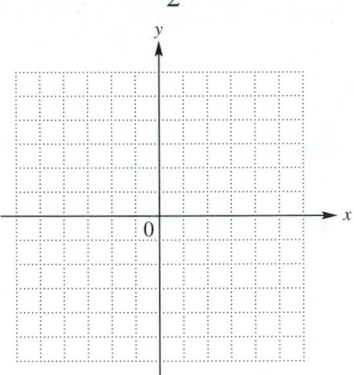

22. $(-4, -1)$, $m = \dfrac{3}{4}$

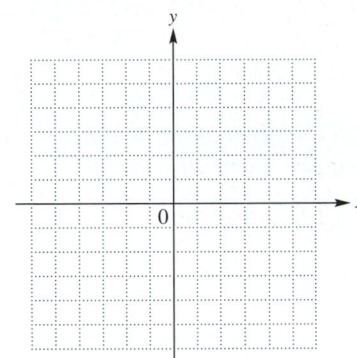

23. $(1, -5)$, $m = -\dfrac{2}{5}$

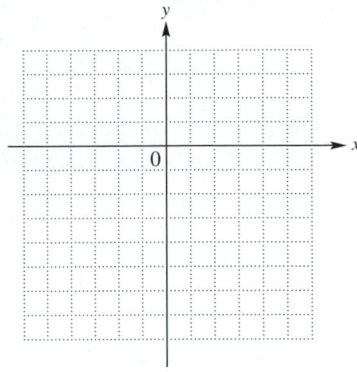

24. $(2, -1)$, $m = -\dfrac{1}{3}$

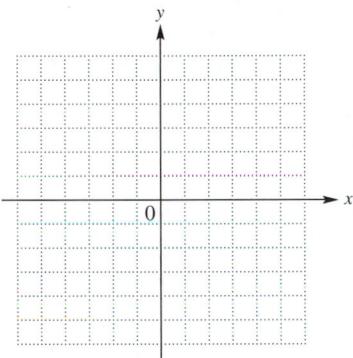

25. $(3, 2)$, $m = 0$

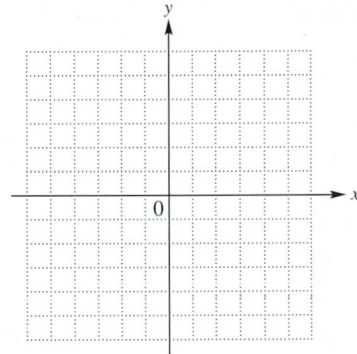

26. $(-2, 3)$, $m = 0$

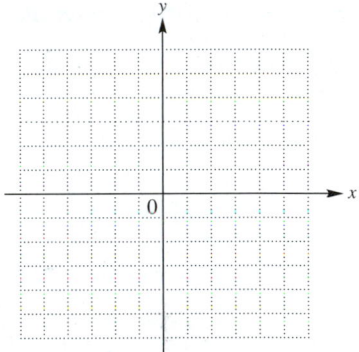

27. $(3, -2)$, undefined slope

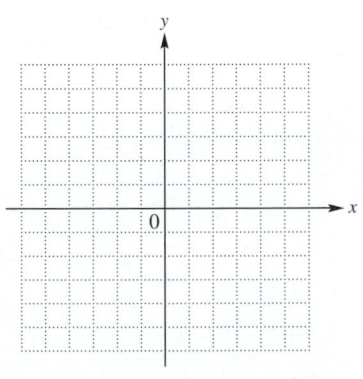

28. $(2, 4)$, undefined slope

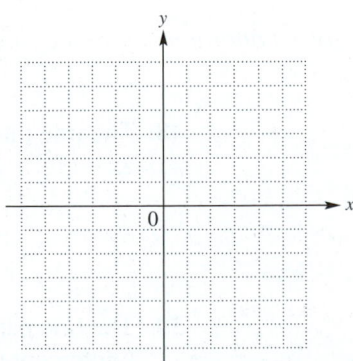

29. $(0, 0)$, $m = \dfrac{2}{3}$

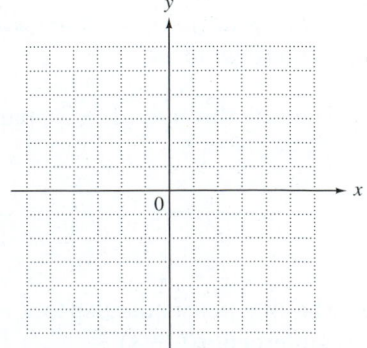

30. (a) What is the common name given to the vertical line whose x-intercept is the origin?

(b) What is the common name given to the line with slope 0 whose y-intercept is the origin?

Write an equation of the line passing through the given point and having the given slope. Give the final answer in slope-intercept form. See Examples 4 and 5.

31. $(4, 1)$, $m = 2$

32. $(2, 7)$, $m = 3$

33. $(3, -10)$, $m = -2$

34. $(2, -5)$, $m = -4$

35. $(-2, 5)$, $m = \dfrac{2}{3}$

36. $(-4, 1)$, $m = \dfrac{3}{4}$

Write an equation of the line passing through each pair of points. Give the final answer in slope-intercept form, if possible. See Example 6.

37. $(8, 5)$ and $(9, 6)$

◐ 38. $(4, 10)$ and $(6, 12)$

39. $(-1, -7)$ and $(-8, -2)$

40. $(-2, -1)$ and $(3, -4)$

41. $(0, -2)$ and $(-3, 0)$

42. $(-4, 0)$ and $(0, 2)$

43. $(3, 5)$ and $(3, -2)$

44. $(3, -5)$ and $(-1, -5)$

45. $\left(\dfrac{1}{2}, \dfrac{3}{2}\right)$ and $\left(-\dfrac{1}{4}, \dfrac{5}{4}\right)$

46. $\left(-\dfrac{2}{3}, \dfrac{8}{3}\right)$ and $\left(\dfrac{1}{3}, \dfrac{7}{3}\right)$

Write an equation of the line satisfying the given conditions. Give the final answer in slope-intercept form.

47. The line through $(2, -3)$, parallel to $3x = 4y + 5$

48. The line through $(-1, 4)$, perpendicular to $2x + 3y = 8$

49. The line perpendicular to $x - 2y = 7$, y-intercept $(0, -3)$

50. The line parallel to $5x = 2y + 10$, y-intercept $(0, 4)$

Relating Concepts (Exercises 51–58) For Individual or Group Work

If we think of ordered pairs of the form (C, F), then the two most common methods of measuring temperature, Celsius and Fahrenheit, can be related as follows: When C = 0, F = 32, and when C = 100, F = 212. **Work Exercises 51–58 in order.**

51. Write two ordered pairs relating these two temperature scales.

52. Find the slope of the line through the two points.

53. Use the point-slope form to find an equation of the line. (Your variables should be C and F rather than x and y.)

54. Write an equation for F in terms of C.

55. Use the equation from Exercise 54 to write an equation for C in terms of F.

56. Use the equation from Exercise 54 to find the Fahrenheit temperature when $C = 30$.

57. Use the equation from Exercise 55 to find the Celsius temperature when $F = 50$.

58. For what temperature is $F = C$?

---

*The cost to produce x items is, in some cases, expressed as y = mx + b. The number b gives the **fixed cost** (the cost that is the same no matter how many items are produced), and the number m is the **variable cost** (the cost to produce an additional item). Use this information to work Exercises 59 and 60.*

59. It costs $400 to start up a business selling campaign buttons. Each button costs $0.25 to produce.

  (a) What is the fixed cost?

  (b) What is the variable cost?

  (c) Write the cost equation.

  (d) What will be the cost to produce 100 campaign buttons, based on the cost equation?

  (e) How many campaign buttons will be produced if total cost is $775?

60. It costs $2000 to purchase a copier, and each copy costs $0.02 to make.

  (a) What is the fixed cost?

  (b) What is the variable cost?

  (c) Write the cost equation.

  (d) What will be the cost to produce 10,000 copies, based on the cost equation?

  (e) How many copies will be produced if total cost is $2600?

Solve each problem. See Example 7.

61. The table lists the average annual cost (in dollars) of tuition and fees at 2-year colleges for selected years, where year 1 represents 2003, year 2 represents 2004, and so on.

Year	Cost (in dollars)
1	1909
2	2079
3	2182
4	2272
5	2361

Source: The College Board.

(a) Write five ordered pairs for the data.

(b) Plot the ordered pairs. Do the points lie approximately in a straight line?

**AVERAGE ANNUAL COSTS AT
2-YEAR COLLEGES**

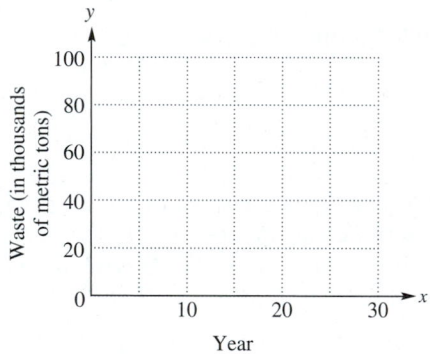

(c) Use the ordered pairs (2, 2079) and (5, 2361) to find the equation of a line that approximates the data. Write the final equation in slope-intercept form.

(d) Use the equation from part (c) to estimate the average annual cost at 2-year colleges in 2008 to the nearest dollar. (*Hint:* What is the value of x for 2008?)

62. The table gives heavy-metal nuclear waste (in thousands of metric tons) from spent reactor fuel now stored temporarily at reactor sites, awaiting permanent storage. (*Source:* "Burial of Radioactive Nuclear Waste Under the Seabed," *Scientific American*, January 1998.)

Year x	Waste y
1995	32
2000	42
2010*	61
2020*	76

*Estimates by the U.S.
Department of Energy.

Let $x = 0$ represent 1995, $x = 5$ represent 2000 (since $2000 - 1995 = 5$), and so on.

(a) For 1995, the ordered pair is (0, 32). Write ordered pairs for the data for the other years given in the table.

(b) Plot the ordered pairs (x, y). Do the points lie approximately in a straight line?

**HEAVY-METAL NUCLEAR
WASTE AWAITING STORAGE**

(c) Use the ordered pairs (0, 32) and (25, 76) to find the equation of a line that approximates the data. Write the equation in slope-intercept form.

(d) Use the equation from part (c) to estimate the amount of nuclear waste in 2015. (*Hint:* What is the value of x for 2015?)

11.5 ▶▶▶ Graphing Linear Inequalities in Two Variables

In **Section 11.2** we graphed linear equations, such as $2x + 3y = 6$. Now we extend this work to include *linear inequalities in two variables,* such as $2x + 3y \leq 6$. (Recall that $\leq$ is read "is less than or equal to.")

> **Linear Inequality in Two Variables**
>
> An inequality that can be written as
>
> $$Ax + By < C \quad \text{or} \quad Ax + By > C,$$
>
> where A, B, and C are real numbers and A and B are not both 0, is a **linear inequality in two variables.**

The symbols $\leq$ and $\geq$ may replace $<$ and $>$ in the definition.

OBJECTIVE 1 Graph linear inequalities. The linear inequality $2x + 3y \leq 6$ means that

$$2x + 3y < 6 \quad \textbf{or} \quad 2x + 3y = 6.$$

As we found earlier, the graph of $2x + 3y = 6$ is a line. This **boundary line** divides the plane into two regions. The graph of the solutions of the inequality $2x + 3y < 6$ will include only *one* of these regions. We find the required region by solving the original inequality for y.

$$2x + 3y \leq 6 \quad \text{◀ Isolate } y \text{ on one side.}$$
$$3y \leq -2x + 6 \qquad \text{Subtract } 2x.$$
$$y \leq -\frac{2}{3}x + 2 \qquad \text{Divide by 3.}$$

All ordered pairs in which y is *less than or equal to* $-\frac{2}{3}x + 2$ will be solutions of the inequality. The ordered pairs in which y is equal to $-\frac{2}{3}x + 2$ are on the boundary line, so the ordered pairs in which y *is less than* $-\frac{2}{3}x + 2$ will be *below* that line. (As we move *down* vertically, the y-values *decrease*.) To indicate the solutions, we shade the region below the line, as in Figure 35. The shaded region, along with the boundary line, is the desired graph.

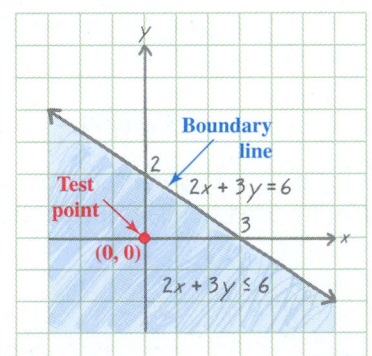

Figure 35

Work Problem **1** *at the Side.* ▶

Alternatively, a test point gives a quick way to find the correct region to shade. We choose any point *not* on the boundary line. Because $(0, 0)$ is easy to substitute, we often use it. We substitute 0 for x and 0 for y in the original inequality to see whether the resulting statement is true or false.

OBJECTIVES

1 Graph linear inequalities.

2 Graph an inequality with boundary through the origin.

1 Shade the appropriate region for each linear inequality.

(a) $x + 2y \geq 6$

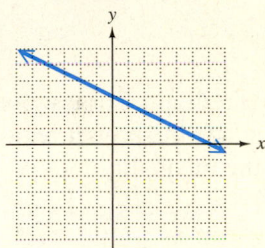

(b) $3x + 4y \leq 12$

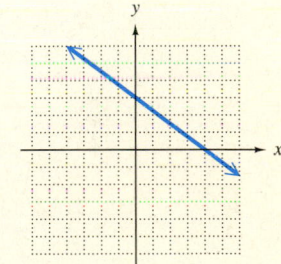

ANSWERS

1. (a)

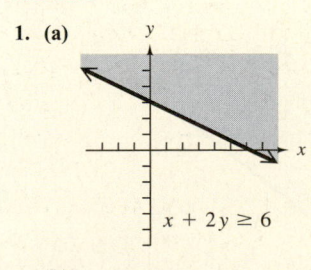

$x + 2y \geq 6$

(b)

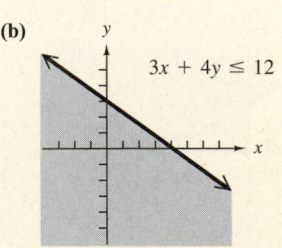

$3x + 4y \leq 12$

2 Use $(0, 0)$ as a test point to shade the proper region for the inequality

$$4x - 5y \le 20.$$

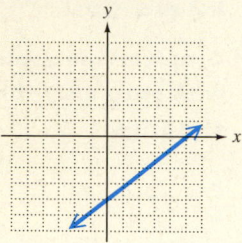

3 Use $(1, 1)$ as a test point to shade the proper region for the inequality

$$3x + 5y > 15.$$

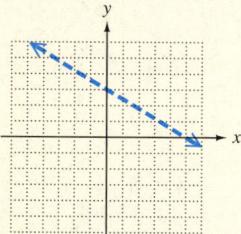

$$2x + 3y \le 6 \qquad \text{Original inequality}$$

$$2(0) + 3(0) \overset{?}{\le} 6 \qquad \text{Let } x = 0 \text{ and } y = 0.$$

> Use $(0, 0)$ as a test point.

$$0 + 0 \overset{?}{\le} 6$$

$$0 \le 6 \qquad \text{True}$$

Since the last statement is true, we shade the region that includes the test point $(0, 0)$. This agrees with the result shown in Figure 35 on the preceding page.

◀ **Work Problem** **2** **at the Side.**

EXAMPLE 1 **Graphing a Linear Inequality**

Graph the inequality $x - y > 5$.

This inequality does *not* include the equals sign. Therefore, the points on the line $x - y = 5$ do *not* belong to the graph. However, the line still serves as a boundary for two regions, one of which satisfies the inequality. To graph the inequality, first graph the equation $x - y = 5$. Use a *dashed line* to show that the points on the line are *not* solutions of the inequality $x - y > 5$. See Figure 36.

Now choose a test point to see which side of the line satisfies the inequality. Again, $(0, 0)$ is a convenient choice.

$$x - y > 5 \qquad \text{Original inequality}$$

$$0 - 0 \overset{?}{>} 5 \qquad \text{Let } x = 0 \text{ and } y = 0.$$

$$0 > 5 \qquad \text{False}$$

Since $0 > 5$ is false, the graph of the inequality is the region that *does not* contain $(0, 0)$. Shade the *other* region, as shown in Figure 36, to obtain the required graph.

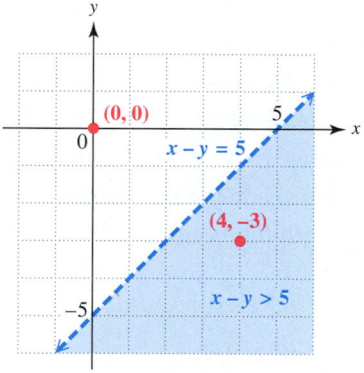

Figure 36

To check that the proper region is shaded, we test a point in the shaded region. For example, we use $(4, -3)$ from the shaded region as follows.

$$x - y > 5$$

$$4 - (-3) \overset{?}{>} 5 \qquad \text{Let } x = 4 \text{ and } y = -3.$$

> Use parentheses to avoid errors.

$$7 > 5 \qquad \text{True}$$

This verifies that the correct region is shaded in Figure 36.

◀ **Work Problem** **3** **at the Side.**

ANSWERS

2.

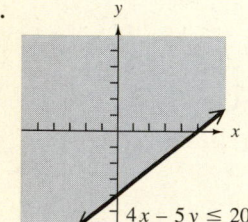

$$4x - 5y \le 20$$

3.

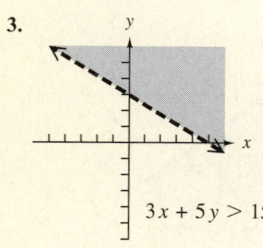

$$3x + 5y > 15$$

A summary of the steps used to graph a linear inequality in two variables follows.

> ### Graphing a Linear Inequality
>
> *Step 1* **Graph the boundary.** Graph the line that is the boundary of the region. Use the methods of **Section 11.2.** Draw a solid line if the inequality has $\leq$ or $\geq$; draw a dashed line if the inequality has $<$ or $>$.
>
> *Step 2* **Shade the appropriate side.** Use any point not on the line as a test point. Substitute for x and y in the *inequality*. If a true statement results, shade the side containing the test point. If a false statement results, shade the other side.

4 Graph $2x - y \geq -4$.

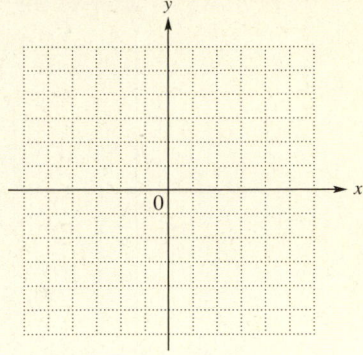

EXAMPLE 2 **Graphing a Linear Inequality**

Graph the inequality $2x - 5y \geq 10$.

Start by graphing the equation $2x - 5y = 10$. Use a solid line to show that the points on the line are solutions of the inequality $2x - 5y \geq 10$. Choose any test point not on the line. Again, we choose $(0, 0)$.

$$2x - 5y \geq 10$$
$$2(0) - 5(0) \stackrel{?}{\geq} 10 \qquad \text{Let } x = 0 \text{ and } y = 0.$$
$$0 - 0 \stackrel{?}{\geq} 10$$
$$0 \geq 10 \qquad \text{False}$$

Because $0 \geq 10$ is false, shade the region *not* containing $(0, 0)$. See Figure 37. Verify that a point in the shaded region satisfies the inequality.

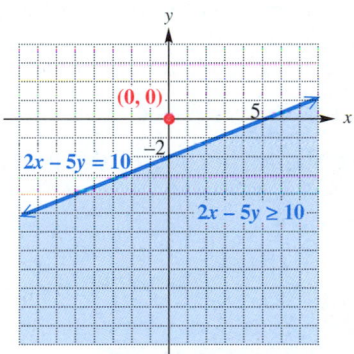

Figure 37

Work Problem **4** *at the Side.* ▶

EXAMPLE 3 **Graphing a Linear Inequality with a Vertical Boundary Line**

Graph the inequality $x < 3$.

First graph $x = 3$, a vertical line through the point $(3, 0)$. Use a dashed line (why?) and choose $(0, 0)$ as a test point.

$$x < 3 \qquad \text{Original inequality}$$
$$0 \stackrel{?}{<} 3 \qquad \text{Let } x = 0.$$
$$0 < 3 \qquad \text{True}$$

Continued on Next Page

ANSWER

4.

$2x - y \geq -4$

Because $0 < 3$ is true, we shade the region containing $(0, 0)$, as in Figure 38.

5 Graph $y < 4$.

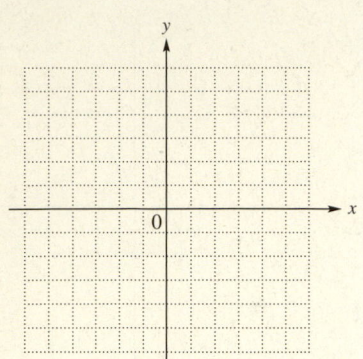

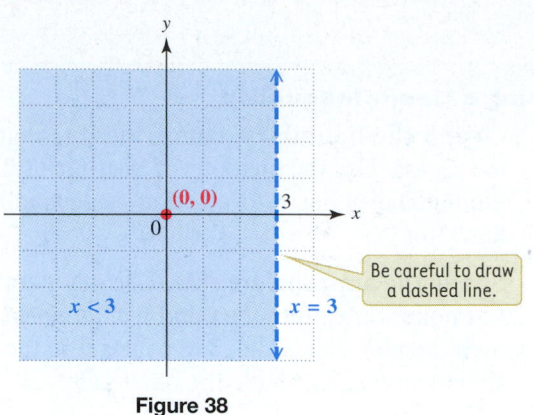

Figure 38

◀ *Work Problem* **5** *at the Side.*

OBJECTIVE 2 Graph an inequality with boundary through the origin. *If the graph of an inequality has a boundary line through the origin, $(0, 0)$ cannot be used as a test point.*

6 Graph $x \geq -3y$.

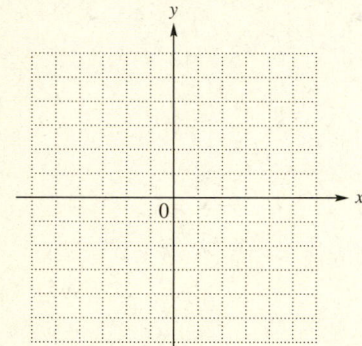

EXAMPLE 4 **Graphing a Linear Inequality with a Boundary Line through the Origin**

Graph the inequality $x \leq 2y$.

We begin by graphing $x = 2y$, using a solid line. Some ordered pairs that can be used to graph this line are $(0, 0)$, $(6, 3)$, and $(4, 2)$. We cannot use $(0, 0)$ as a test point because $(0, 0)$ is *on* the line $x = 2y$. Instead, we choose a test point *off* the line, say $(1, 3)$.

$$x \leq 2y \qquad \text{Original inequality}$$
$$1 \stackrel{?}{\leq} 2(3) \qquad \text{Let } x = 1 \text{ and } y = 3.$$
$$1 \leq 6 \qquad \text{True}$$

Because $1 \leq 6$ is true, we shade the side of the graph containing the test point $(1, 3)$. See Figure 39.

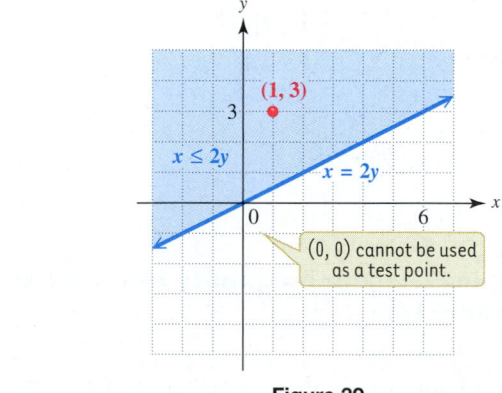

Figure 39

◀ *Work Problem* **6** *at the Side.*

ANSWERS

5.

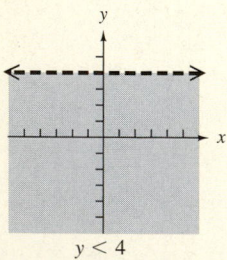

$y < 4$

6.

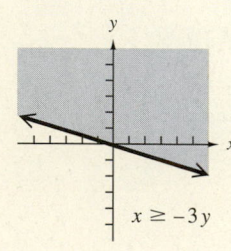

$x \geq -3y$

Decide whether each statement is true *or* false. *If false, explain why.*

1. The point $(4, 0)$ lies on the graph of $3x - 4y < 12$.

2. The point $(4, 0)$ lies on the graph of $3x - 4y \le 12$.

3. Both points $(4, 1)$ and $(0, 0)$ lie on the graph of $3x - 2y \ge 0$.

4. The graph of $y > x$ does not contain points in quadrant IV.

The following statements were taken from various media. Each includes a phrase that can be symbolized with one of the inequality symbols $<, \le, >,$ *or* $\ge$. *In Exercises 5–8, give the inequality symbol for the bold faced words.*

5. Since it was recognized in 1981, HIV/AIDS has killed **more than** 25 million people worldwide and infected **more than** 60 million, about two-thirds of whom live in Africa. (*Source:* The President's Emergency Plan for AIDS Relief, February, 2008.)

6. The average national automobile insurance premium of $1896 in 2007 was $20 **less than** the 2006 average premium. (*Source:* 2007 Mid-Year Auto Insurance Pricing Report.)

7. As of December 2007, airline passengers were allowed one carry-on bag, with dimensions totaling **at most** 45 in. (*Source: The Gazette.*)

8. As of February 2008, all major airlines except US Airways award **at least** 500 frequent flier miles per flight. (*Source: USA Today.*)

In Exercises 9–16, the straight-line boundary has been drawn. Complete each graph by shading the correct region. See Examples 1–4.

9. $x + y \geq 4$

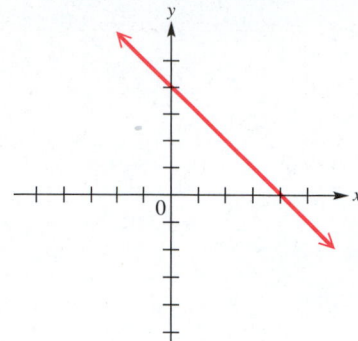

10. $x + y \leq 2$

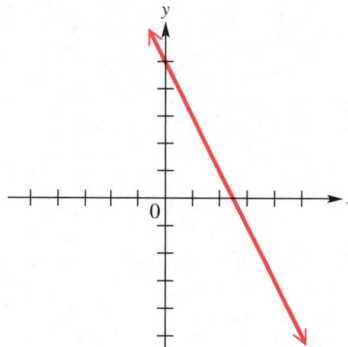

11. $x + 2y \geq 7$

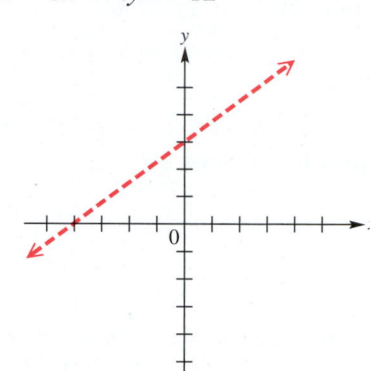

12. $2x + y \geq 5$

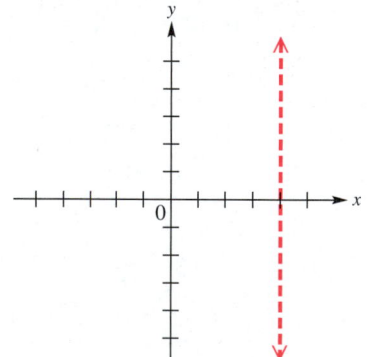

13. $-3x + 4y > 12$

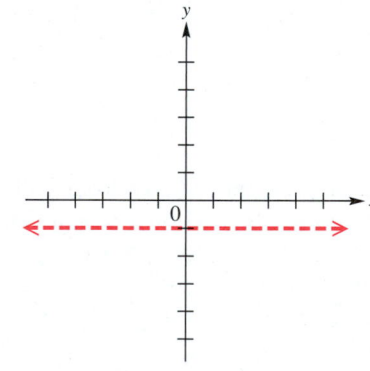

14. $4x - 5y < 20$

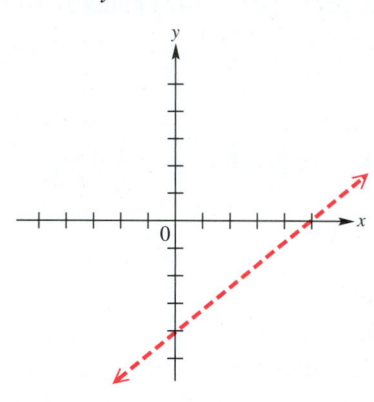
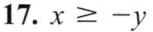

15. $x > 4$

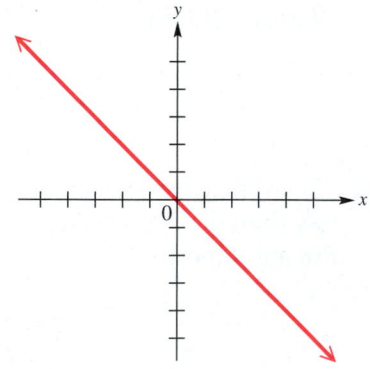

16. $y < -1$

17. $x \geq -y$

18. Explain how to determine whether to use a dashed line or a solid line when graphing a linear inequality in two variables.

19. Explain why the point $(0, 0)$ is not an appropriate choice for a test point when graphing an inequality whose boundary goes through the origin.

Graph each linear inequality. See Examples 1–4.

20. $x + y \geq 3$

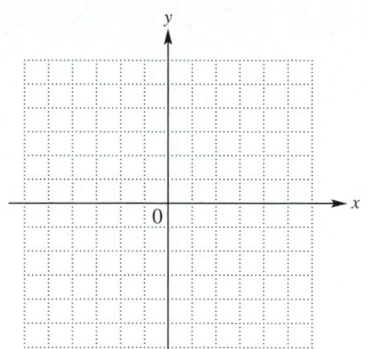

21. $x + y \leq 5$

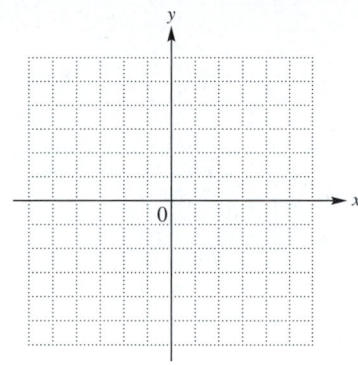

22. $x + 3y > 6$

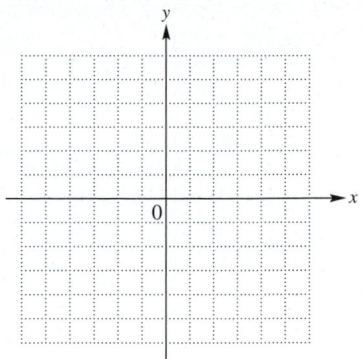

23. $x + 2y < 4$

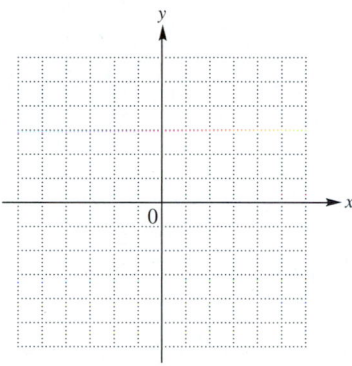

24. $-4y > 3x - 12$

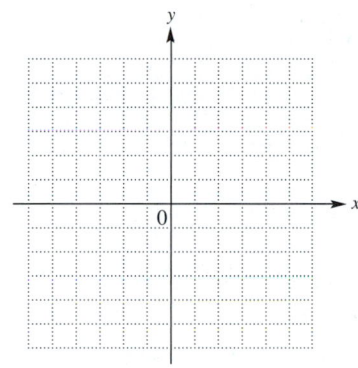

25. $2x + 6 > -3y$

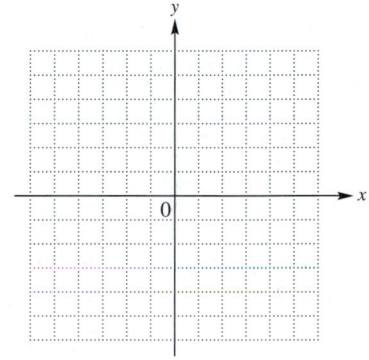

26. $y < -3x + 1$

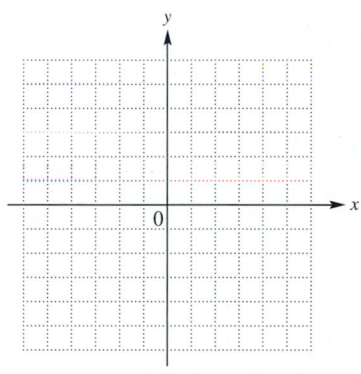

27. $y \geq 2x + 1$

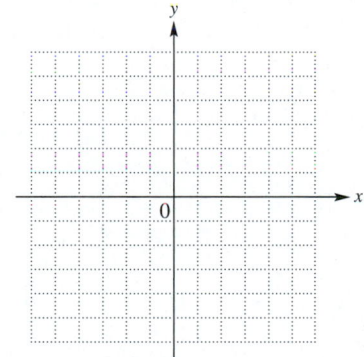

28. $x \geq 1$

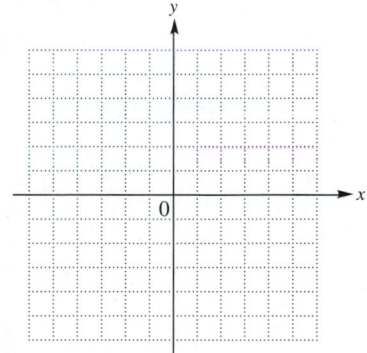

29. $x \leq -2$

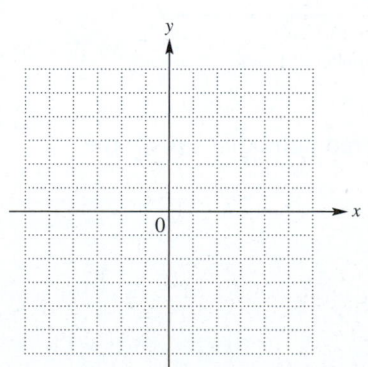

30. $y < -3$

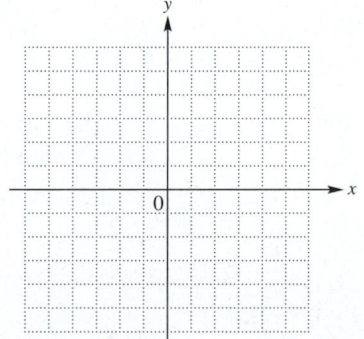

31. $y < 5$

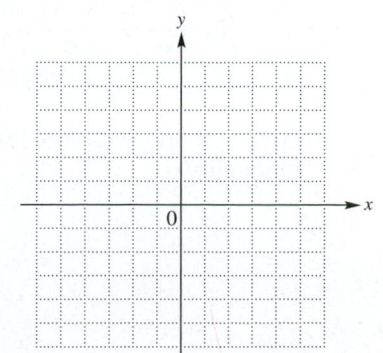

32. $y \leq 2x$

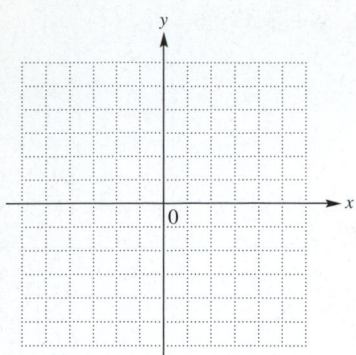

⊕ 33. $y \geq 4x$

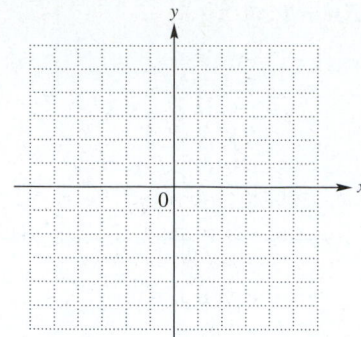

34. $x > -5y$

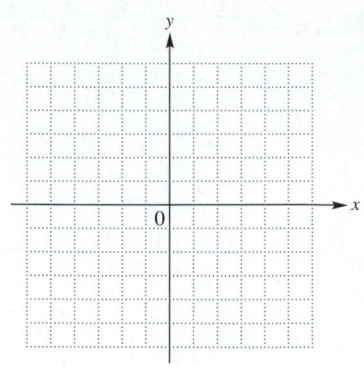

35. Explain why the graph of $y > x$ cannot lie in quadrant IV.

36. Explain why the graph of $y < x$ cannot lie in quadrant II.

Solve each problem. In part (a), $x \geq 0$ and $y \geq 0$, so graph only the part of the inequality in quadrant I.

37. A company will ship x units of merchandise to outlet I and y units of merchandise to outlet II. The company must ship a total of at least 500 units to these two outlets. This can be expressed by writing

$$x + y \geq 500.$$

(a) Graph the inequality.

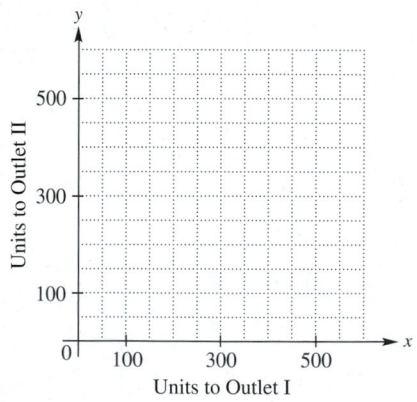

(b) Give two ordered pairs that satisfy the inequality.

38. A toy manufacturer makes stuffed bears and geese. It takes 20 min to sew a bear and 30 min to sew a goose. There is a total of 480 min of sewing time available to make x bears and y geese. These restrictions lead to the inequality

$$20x + 30y \leq 480.$$

(a) Graph the inequality.

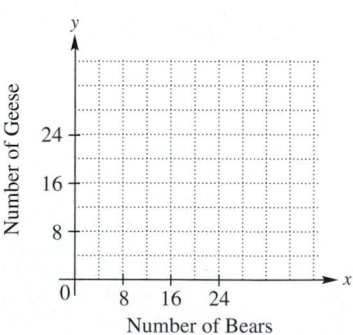

(b) Give two ordered pairs that satisfy the inequality.

Chapter 11 ▷▷▷ Summary

▶ Key Terms

11.1

bar graph	A bar graph is a series of bars used to show comparisons between two categories of data.
line graph	A line graph consists of a series of points that are connected with line segments and is used to show changes or trends in data.
linear equation in two variables	An equation that can be written in the form $Ax + By = C$ is a linear equation in two variables. (A and B are real numbers that cannot both be 0.)
ordered pair	A pair of numbers written between parentheses in which order is important is called an ordered pair.
table of values	A table showing selected ordered pairs of numbers that satisfy an equation is called a table of values.
x-axis	The horizontal axis in a coordinate system is called the x-axis.
y-axis	The vertical axis in a coordinate system is called the y-axis.
rectangular (Cartesian) coordinate system	An x-axis and y-axis at right angles form a coordinate system.
quadrants	A coordinate system divides the plane into four regions called quadrants.
origin	The point at which the x-axis and y-axis intersect is called the origin.
plane	A flat surface determined by two intersecting lines is a plane.
coordinates	The numbers in an ordered pair are called the coordinates of the corresponding point.
plot	To plot an ordered pair is to find the corresponding point on a coordinate system.
scatter diagram	A graph of ordered pairs of data is a scatter diagram.

x	y
0	4
2	0
1	2

Table of values
for $2x + y = 4$

11.2

graph	The graph of an equation is the set of all points that correspond to the ordered pairs that satisfy the equation.
graphing	The process of plotting the ordered pairs that satisfy a linear equation and drawing a line through them is called graphing.
y-intercept	If a graph intersects the y-axis at k, then the y-intercept is $(0, k)$.
x-intercept	If a graph intersects the x-axis at k, then the x-intercept is $(k, 0)$.

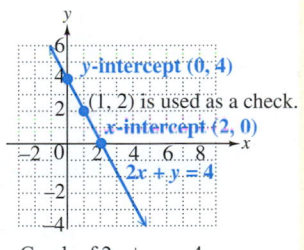

Graph of $2x + y = 4$

11.3

rise	Rise is the vertical change between two different points on a line.
run	Run is the horizontal change between two different points on a line.
slope	The slope of a line is the ratio of the change in y compared to the change in x when moving along the line from one point to another.
parallel lines	Two lines in a plane that never intersect are parallel.
perpendicular lines	Perpendicular lines intersect at a 90° angle.

11.5

linear inequality in two variables	An inequality that can be written in the form $Ax + By < C$, $Ax + By > C$, $Ax + By \leq C$, or $Ax + By \geq C$ is a linear inequality in two variables.
boundary line	In the graph of a linear inequality, the boundary line separates the region that satisfies the inequality from the region that does not satisfy the inequality.

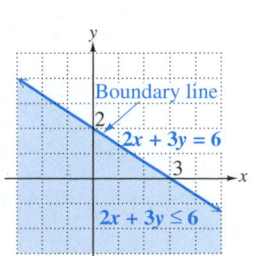

▶ New Symbols

(x, y) ordered pair

(x_1, y_1) subscript notation; x-sub-one, y-sub-one

m slope

▶ Test Your Word Power

See how well you have learned the vocabulary in this chapter. Answers, with examples, follow the Quick Review.

1. An **ordered pair** is a pair of numbers written
   A. in numerical order between brackets
   B. between parentheses or brackets
   C. between parentheses in which order is important
   D. between parentheses in which order does not matter.

2. The **coordinates** of a point are
   A. the numbers in the corresponding ordered pair
   B. the solution of an equation
   C. the values of the x- and y-intercepts
   D. the graph of the point.

3. An **intercept** is
   A. the point where the x-axis and y-axis intersect
   B. a pair of numbers written in parentheses in which order is important
   C. one of the four regions determined by a rectangular coordinate system
   D. the point where a graph intersects the x-axis or the y-axis.

4. The **slope** of a line is
   A. the measure of the run over the rise of the line
   B. the distance between two points on the line
   C. the ratio of the change in y to the change in x along the line
   D. the horizontal change compared to the vertical change of two points on the line.

5. Two lines in a plane are **parallel** if
   A. they represent the same line
   B. they never intersect
   C. they intersect at a 90° angle
   D. one has a positive slope and one has a negative slope.

6. Two lines in a plane are **perpendicular** if
   A. they represent the same line
   B. they never intersect
   C. they intersect at a 90° angle
   D. one has a positive slope and one has a negative slope.

▶ Quick Review

Concepts	Examples

11.1 Reading Graphs; Linear Equations in Two Variables

Bar graphs and line graphs are ways to "picture", or represent, the relationship between two variables.

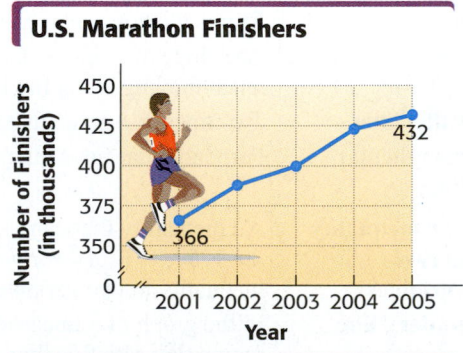

U.S. Marathon Finishers

Source: Running USA.

The line graph illustrates the number of U.S. runners in thousands who finished marathons in the years 2001–2005.

Concepts	Examples

11.1 Reading Graphs; Linear Equations in Two Variables *(continued)*

An ordered pair is a solution of an equation if it makes the equation a true statement.

Is $(2, -5)$ or $(0, -6)$ a solution of $4x - 3y = 18$?

$4(2) - 3(-5) = 23 \neq 18$	$4(0) - 3(-6) = 18$
$(2, -5)$ is not a solution.	$(0, -6)$ is a solution.

If a value of either variable in an equation is given, the value of the other variable can be found by substitution.

Complete the ordered pair $(0, \underline{})$ for $3x = y + 4$.

$$3(0) = y + 4 \qquad \text{Let } x = 0.$$
$$0 = y + 4 \qquad \text{Multiply.}$$
$$-4 = y \qquad \text{Subtract 4.}$$

The ordered pair is $(0, -4)$.

To plot the ordered pair $(-3, 4)$, start at the origin, go 3 units to the left, and from there go 4 units up.

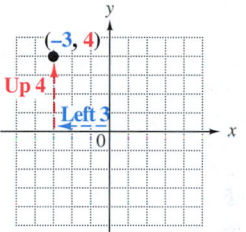

11.2 Graphing Linear Equations in Two Variables

To graph a linear equation:

Step 1 Find at least two ordered pairs that are solutions of the equation.

Step 2 Plot the corresponding points.

Step 3 Draw a straight line through the points.

The graph of $y = k$ is a horizontal line through $(0, k)$.

The graph of $x = k$ is a vertical line through $(k, 0)$.

Graph $x - 2y = 4$.

x	y
0	-2
4	0

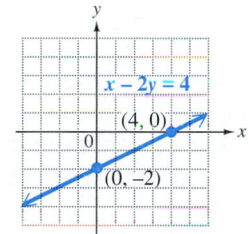

11.3 Slope of a Line

The slope of the line through (x_1, y_1) and (x_2, y_2) is

$$m = \frac{\text{change in } y}{\text{change in } x} = \frac{y_2 - y_1}{x_2 - x_1} \quad (x_1 \neq x_2).$$

The line through $(-2, 3)$ and $(4, -5)$ has slope

$$m = \frac{-5 - 3}{4 - (-2)} = \frac{-8}{6} = -\frac{4}{3}.$$

Horizontal lines have slope 0.

The line $y = -2$ has slope **0**.

Vertical lines have undefined slope.

The line $x = 4$ has undefined slope.

To find the slope of a line from its equation, solve for y. The slope is the coefficient of x.

Find the slope of the graph of $3x - 4y = 12$.

$$-4y = -3x + 12 \qquad \text{Add } -3x.$$
$$y = \frac{3}{4}x - 3 \qquad \text{Divide by } -4.$$

$\uparrow$
Slope

Concepts	Examples

11.4 Equations of Lines

Slope-Intercept Form

$y = mx + b$

m is the slope.

$(0, b)$ is the y-intercept.

Write an equation of the line with slope **2** and y-intercept $(0, -5)$.

$$y = 2x - 5$$

Point-Slope Form

$y - y_1 = m(x - x_1)$

m is the slope.

(x_1, y_1) is a point on the line.

Write an equation of the line with slope $-\frac{1}{2}$ through $(-4, 5)$.

$$y - 5 = -\frac{1}{2}[x - (-4)] \qquad \text{Substitute.}$$

$$y - 5 = -\frac{1}{2}(x + 4)$$

$$y - 5 = -\frac{1}{2}x - 2 \qquad \text{Distributive property}$$

$$y = -\frac{1}{2}x + 3 \qquad \text{Add 5.}$$

Standard Form

$Ax + By = C$

This equation is written in standard form as

$$x + 2y = 6,$$

with $A = 1$, $B = 2$, and $C = 6$.

11.5 Graphing Linear Inequalities in Two Variables

Step 1   Graph the line that is the boundary of the region. Make it solid if the inequality is $\leq$ or $\geq$; make it dashed if the inequality is $<$ or $>$.

Step 2   Use any point not on the line as a test point. Substitute for x and y in the inequality. If the result is true, shade the side of the line containing the test point; if the result is false, shade the other side.

Graph $2x + y \leq 5$.

Graph the line $2x + y = 5$. Make it solid because the symbol $\leq$ includes equality.

Use $(0, 0)$ as a test point.

$$2(0) + 0 \overset{?}{\leq} 5$$

$$0 \leq 5 \qquad \text{True}$$

Shade the side of the line containing $(0, 0)$.

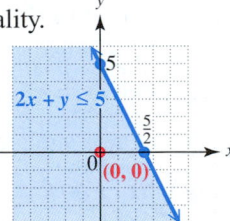

ANSWERS TO TEST YOUR WORD POWER

1. C; *Examples:* $(0, 3)$, $(3, 8)$, $(4, 0)$
2. A; *Example:* The point associated with the ordered pair $(1, 2)$ has x-coordinate 1 and y-coordinate 2.
3. D; *Example:* The graph of the equation $4x - 3y = 12$ has x-intercept at $(3, 0)$ and y-intercept at $(0, -4)$.
4. C; *Example:* The line through $(3, 6)$ and $(5, 4)$ has slope $\frac{4 - 6}{5 - 3} = \frac{-2}{2} = -1$.
5. B; *Example:* See Figure 27 in **Section 11.3**.
6. C; *Example:* See Figure 28 in **Section 11.3**.

Chapter 11 ▶▶▶ Review Exercises

[11.1] *The percent of first-year college students at two-year public institutions who returned for a second year for the years 2001 through 2007 are shown in the graph.*

1. Write ordered pairs of the form (year, percent) for the data shown in the graph.

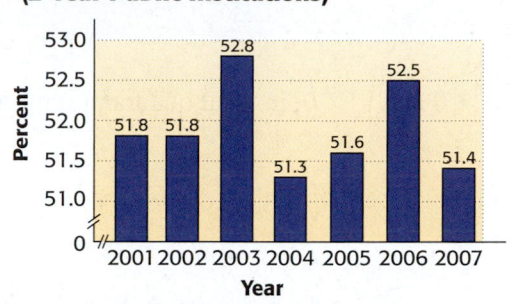

Percent of Students Who Return for Second Year (2-Year Public Institutions)

Source: ACT.

2. What does the ordered pair (2006, 52.5) mean in the context of these problems?

3. In what year did the percent show the greatest decrease from the previous year? What was this decrease?

4. In what year did the percent show the greatest increase from the previous year? What was this increase?

Complete the given ordered pairs for each equation.

5. $y = 3x + 2$ $(-1, \_\_), (0, \_\_), (\_\_, 5)$

6. $4x + 3y = 6$ $(0, \_\_), (\_\_, 0), (-2, \_\_)$

7. $x = 3y$ $(0, \_\_), (8, \_\_), (\_\_, -3)$

8. $x - 7 = 0$ $(\_\_, -3) (\_\_, 0), (\_\_, 5)$

Decide whether each ordered pair is a solution of the given equation.

9. $x + y = 7; (2, 5)$

10. $2x + y = 5; (-1, 3)$

11. $3x - y = 4; \left(\dfrac{1}{3}, -3\right)$

Plot each ordered pair on the given coordinate system.

12. $(2, 3)$

13. $(-4, 2)$

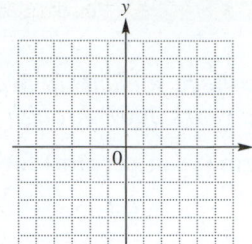

14. $(3, 0)$

15. $(0, -6)$

16. If $x > 0$ and $y < 0$, in what quadrant(s) must (x, y) lie? Explain.

17. On what axis does the point $(k, 0)$ lie for any real value of k? the point $(0, k)$? Explain.

Without plotting the given point, name the quadrant in which each point lies.

18. $(-2, 3)$

19. $(-1, -4)$

20. $\left(0, -5\frac{1}{2}\right)$

[11.2] *Find the intercepts for the graph of each equation.*

21. $y = 2x + 5$

x-intercept:

y-intercept:

22. $2x + y = -7$

x-intercept:

y-intercept:

23. $3x + 2y = 8$

x-intercept:

y-intercept:

Graph each linear equation.

24. $2x - y = 3$

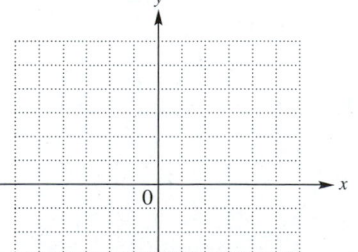

25. $x + 2y = -4$

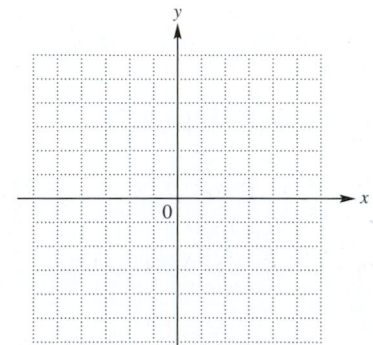

26. $x + y = 0$

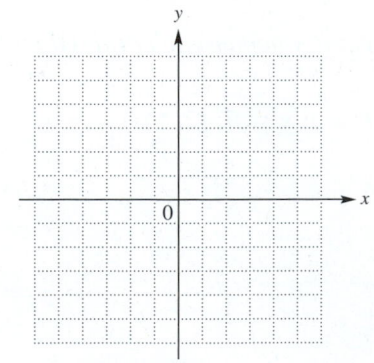

[11.3] *Find the slope of each line.*

27. The line through (2, 3) and
(−4, 6)

28. The line through (0, 0) and
(−3, 2)

29. The line through (0, 6) and
(1, 6)

30. The line through (2, 5) and
(2, 8)

31. $y = 3x - 4$

32. $y = \dfrac{2}{3}x + 1$

33.

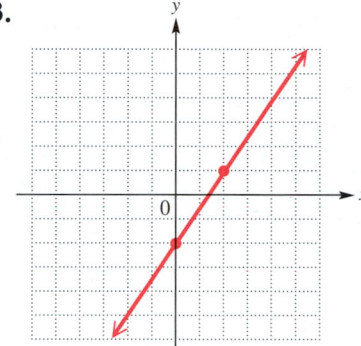

34.

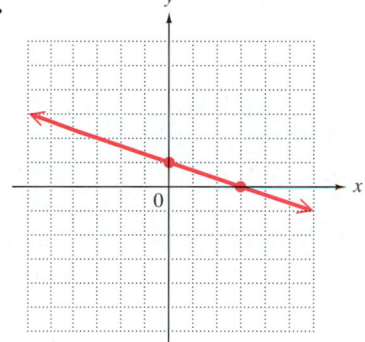

35. $x = 0$

36. $y = 4$

37. The line having these points

x	y
0	1
2	4
6	10

38. **(a)** A line parallel to the graph of $y = 2x + 3$

(b) A line perpendicular to the graph of
$y = -3x + 3$

Decide whether each pair of lines is parallel, perpendicular, *or neither.*

39. $3x + 2y = 6$
$6x + 4y = 8$

40. $x - 3y = 1$
$3x + y = 4$

41. $x - 2y = 8$
$x + 2y = 8$

42. What is the slope of a line perpendicular to a line with undefined slope?

[11.4] *Write an equation of each line. Give the final answer in slope-intercept form (if possible).*

43. $m = -1, b = \dfrac{2}{3}$

44. The line in Exercise 34

45. The line through (4, −3),
$m = 1$

46. The line through $(-1, 4)$, $m = \dfrac{2}{3}$

47. The line through $(1, -1)$, $m = -\dfrac{3}{4}$

48. The line through $(2, 1)$ and $(-2, 2)$

49. The line through $(-4, 1)$, slope 0

50. The line through $\left(\dfrac{1}{3}, -\dfrac{3}{4}\right)$, undefined slope

51. Consider the equation $x + 3y = 15$.

(a) Write it in the form $y = mx + b$.

(b) What is the slope? What is the y-intercept?

(c) Use the slope and y-intercept to graph the line. Indicate two points on the graph.

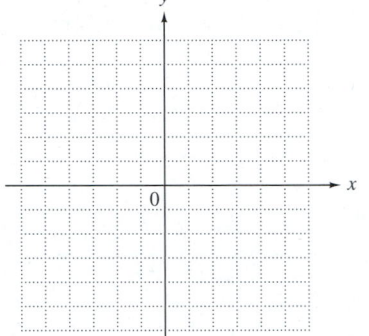

52. Match the description in Column I with the correct equation in Column II.

I	II
(a) Slope -0.5, $b = -2$	A. $y = -\dfrac{1}{2}x$
(b) x-intercept $(4, 0)$, y-intercept $(0, 2)$	B. $y = -\dfrac{1}{2}x - 2$
(c) The line through $(4, -2)$ and $(0, 0)$	C. $x - 2y = 2$
(d) $m = \dfrac{1}{2}$, passes through $(-2, -2)$	D. $x + 2y = 4$
	E. $x = 2y$

[11.5] *Graph each linear inequality.*

53. $3x + 5y > 9$

54. $2x - 3y > -6$

55. $x \geq -4$

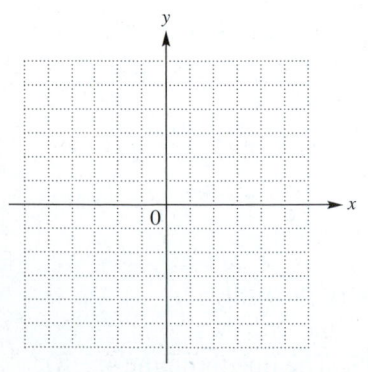

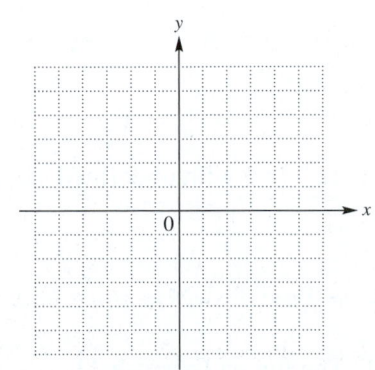

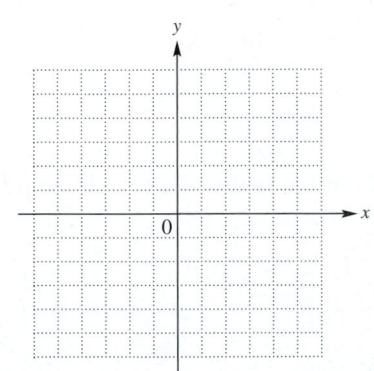

In Exercises 56–61, match each statement to the appropriate graph or graphs in A–D. Graphs may be used more than once.

A.
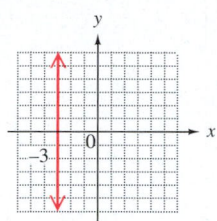

B.
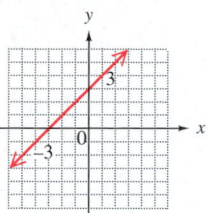

C.
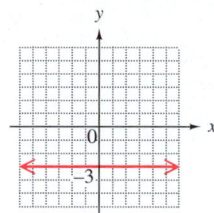

D.
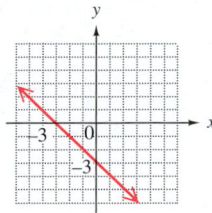

56. The line shown in the graph has undefined slope.

57. The graph of the equation has *y*-intercept $(0, -3)$.

58. The graph of the equation has *x*-intercept $(-3, 0)$.

59. The line shown in the graph has negative slope.

60. The graph is that of the equation $y = -3$.

61. The line shown in the graph has slope 1.

Find the intercepts and the slope of each line. Then graph the line.

62. $y = -2x - 5$

    x-intercept:

    y-intercept:

    slope:

63. $x + 3y = 0$

    x-intercept:

    y-intercept:

    slope:

64. $y - 5 = 0$

    x-intercept:

    y-intercept:

    slope:

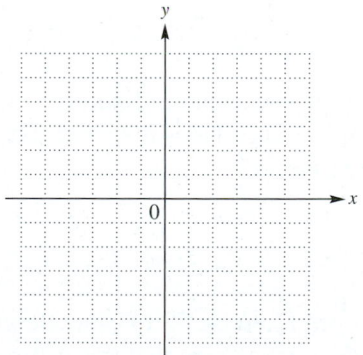

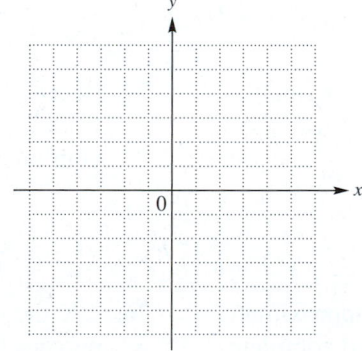

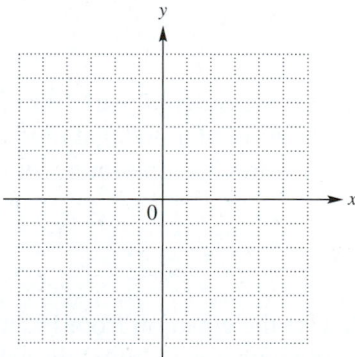

Write an equation of each line. Give the final answer in slope-intercept form.

65. $m = -\dfrac{1}{4}, b = -\dfrac{5}{4}$

66. The line through $(8, 6)$, $m = -3$

67. The line through $(3, -5)$ and $(-4, -1)$

Graph each inequality.

68. $y < -4x$

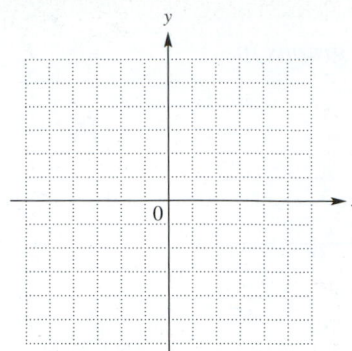

69. $x - 2y \leq 6$

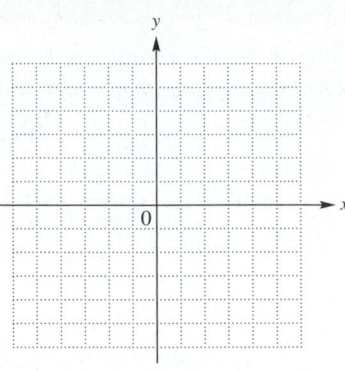

Relating Concepts (Exercises 70–76) For Individual or Group Work

*The percents of four-year college students in public schools who earned a degree within five years of entry between 2002 and 2007 are shown in the graph. Use the graph to **work Exercises 70–76 in order.***

70. What was the percent increase from 2002 to 2007?

71. Since the points of the graph lie approximately in a linear pattern, a straight line can be used to model the data. Will this line have positive or negative slope? Explain.

Percents of Students Graduating Within 5 Years (Public Institutions)

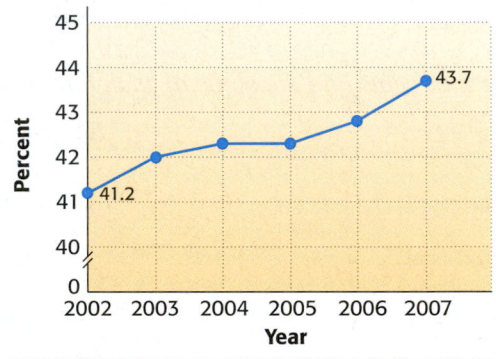

Source: ACT.

72. Write two ordered pairs for the data for 2002 and 2007.

73. Use the ordered pairs from Exercise 72 to find the equation of a line that models the data. Write the equation in slope-intercept form.

74. Based on the equation you found in Exercise 73, what is the slope of the line? Does it agree with your answer in Exercise 71?

75. Use the equation from Exercise 73 to approximate the percents for 2003 through 2006, and complete the table.

Year	Percent
2003	
2004	
2005	
2006	

76. Use the equation from Exercise 73 to estimate the percent for 2008. Can we be sure that this estimate is accurate?

Chapter 11 ▶▶▶ Test

 Use the Chapter Test Prep Video CD to see fully worked-out solutions to any of the exercises you want to review.

The line graph shows the overall unemployment rate in the U.S. civilian labor force for the years 1998 through 2005. Use the graph to work Exercises 1–3.

Unemployment Rate

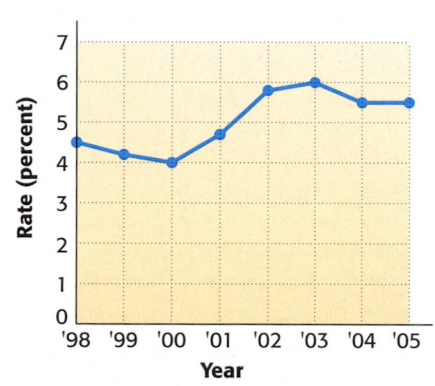

Source: U.S. Department of Labor.

1. Between which pairs of consecutive years did the unemployment rate decrease?

2. What was the general trend in the unemployment rate between 2000 and 2003?

3. Estimate the overall unemployment rate in 2003 and 2004. About how much did the unemployment rate decline between 2003 and 2004?

Graph each linear equation. Give the x- and y-intercepts.

4. $3x + y = 6$

5. $y - 2x = 0$

1. _____

2. _____

3. _____

4. *x*-intercept: _____

   y-intercept: _____

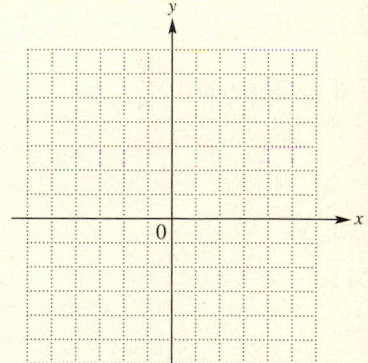

5. *x*-intercept: _____

   y-intercept: _____

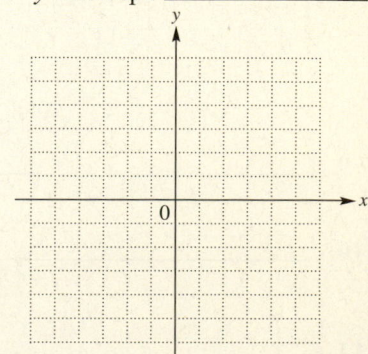

6. *x*-intercept: _____

   y-intercept: _____

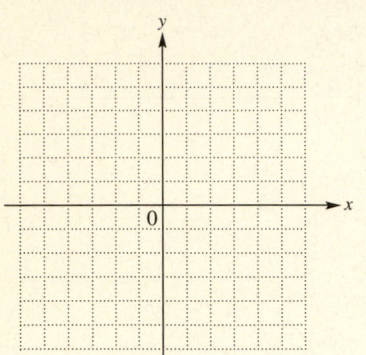

6. $x + 3 = 0$

7. *x*-intercept: _____

   y-intercept: _____

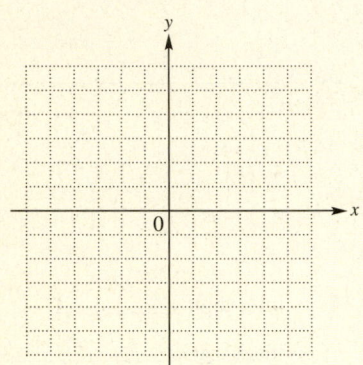

7. $y = 1$

8. *x*-intercept: _____

   y-intercept: _____

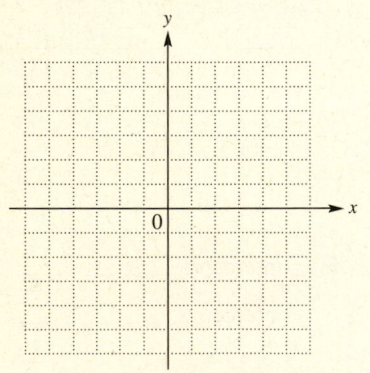

8. $x - y = 4$

Find the slope of each line.

9. _____

9. The line through $(-4, 6)$ and $(-1, -2)$

10. _____

10. $2x + y = 10$

11. _____

11. $x + 12 = 0$

12.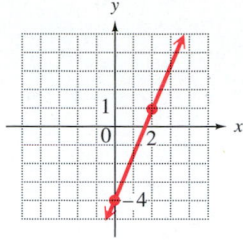

12. _____

13. A line parallel to the graph of $y - 4 = 6$

13. _____

Write an equation for each line. Give the final answer in slope-intercept form.

14. The line through $(-1, 4)$, $m = 2$

14. _____

15. The line in Exercise 12

15. _____

16. The line through $(2, -6)$ and $(1, 3)$

16. _____

17. x-intercept: $(3, 0)$; y-intercept: $\left(0, \dfrac{9}{2}\right)$

17. _____

Graph each linear inequality.

18. $x + y \leq 3$

18.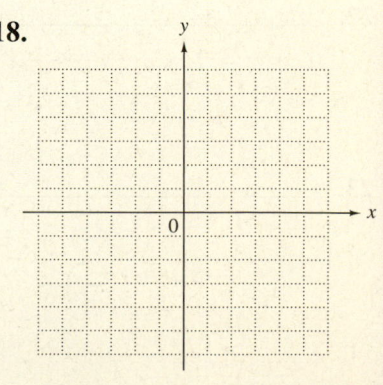

19.

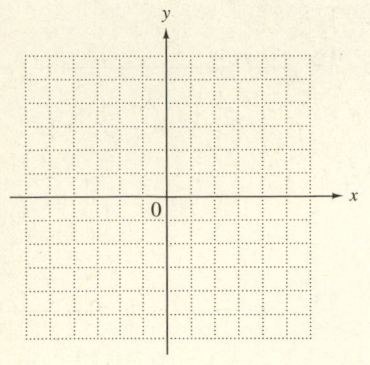

19. $3x - y > 0$

The graph shows worldwide snowmobile sales from 2000 through 2007, where 2000 corresponds to $x = 0$. Use the graph to work Exercises 20–24.

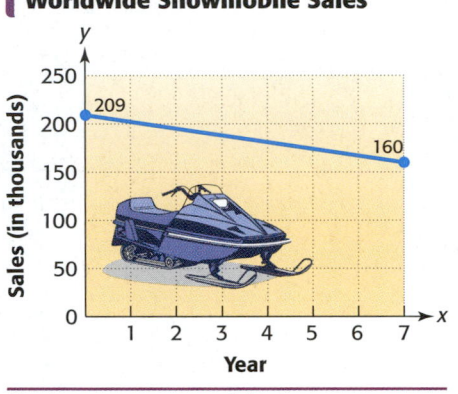

Worldwide Snowmobile Sales

Source: www.snowmobile.org

20. _____

20. Is the slope of the line in the graph positive or negative? Explain.

21. _____

21. Write two ordered pairs for the data points shown in the graph. Use them to find the slope of the line.

22. _____

22. Use the ordered pairs and slope from Exercise 21 to find an equation of a line that models the data. Write the equation in slope-intercept form.

23. _____

23. Use the equation from Exercise 22 to approximate worldwide snowmobile sales for 2005. How does your answer compare to the actual sales of 173.7 thousand?

24. _____

24. What does the ordered pair (7, 160) mean in the context of this problem?

12

Exponents and Polynomials

Just how much is a *trillion*? A trillion, written 1,000,000,000,000, is a million million, or a thousand billion. A trillion seconds would last more than 31,000 years—that is, 310 centuries. By 2017, the U.S. government projects that consumers and taxpayers will spend more than $4 trillion on health care, accounting for $1 of every $5 spent. (*Source:* Centers for Medicare and Medicaid Services.)

In **Section 12.8,** we use *exponents* and *scientific notation* to write and calculate with large numbers, such as the national debt, tax revenue, and the distances of a double-helix nebula and the star Pollux from Earth.

12.1 ▶▶▶ Adding and Subtracting Polynomials

OBJECTIVES

1. Review combining like terms.
2. Know the vocabulary for polynomials.
3. Evaluate polynomials.
4. Add polynomials.
5. Subtract polynomials.
6. Add and subtract polynomials with more than one variable.

Term	Numerical Coefficient
$-7y$	-7
$34r^3$	34
$-26x^5yz^4$	-26
$-k = -1k$	-1
$r = 1r$	1
$\frac{3x}{8} = \frac{3}{8}x$	$\frac{3}{8}$
$\frac{x}{3} = \frac{1x}{3} = \frac{1}{3}x$	$\frac{1}{3}$

1 Add like terms.

(a) $5x^4 + 7x^4$

(b) $9pq + 3pq - 2pq$

(c) $r^2 + 3r + 5r^2$

(d) $x + \frac{1}{2}x$

(e) $8t + 6w$

(f) $3x^4 - 3x^2$

ANSWERS

1. (a) $12x^4$ (b) $10pq$ (c) $6r^2 + 3r$ (d) $\frac{3}{2}x$
   (e) These are unlike terms. They cannot be added.
   (f) These are unlike terms. They cannot be added.

Recall from **Section 9.8** that in an expression such as
$$4x^3 + 6x^2 + 5x + 8,$$
the quantities that are added, $4x^3$, $6x^2$, $5x$, and 8, are called **terms.** In the term $4x^3$, the number **4** is called the **numerical coefficient,** or simply the **coefficient,** of x^3. In the same way, **6** is the coefficient of x^2 in the term $6x^2$, **5** is the coefficient of x in the term $5x$, and **8** is the **constant** term. Other examples are given in the table at the side.

OBJECTIVE 1 **Review combining like terms.** Recall from **Section 9.8** that **like terms** have exactly the same combination of variables, with the same exponents on the variables. *Only the coefficients may differ.*

$$\left.\begin{array}{ll} 19m^5 & \text{and} \quad 14m^5 \\ -37y^9 & \text{and} \quad y^9 \\ 3pq & \text{and} \quad -2pq \\ 2xy^2 & \text{and} \quad -xy^2 \end{array}\right\} \begin{array}{c} \text{Examples} \\ \text{of} \\ \text{like terms} \end{array} \qquad \left.\begin{array}{ll} 7x & \text{and} \quad 7y \\ z^4 & \text{and} \quad z \\ 2pq & \text{and} \quad 2p \\ -4xy^2 & \text{and} \quad 5x^2y \end{array}\right\} \begin{array}{c} \text{Examples} \\ \text{of} \\ \text{unlike terms} \end{array}$$

Using the distributive property, we combine, or add, like terms by adding their coefficients.

EXAMPLE 1 Adding Like Terms

Simplify each expression by adding like terms.

(a) $-4x^3 + 6x^3$
$$= (-4 + 6)x^3 \qquad \text{Distributive property}$$
$$= 2x^3$$

(b) $9x^6 - 14x^6 + x^6$
$$= (9 - 14 + 1)x^6 \qquad x^6 = 1x^6$$
$$= -4x^6$$

(c) $12m^2 + 5m + 4m^2$ (d) $3x^2y + 4x^2y - x^2y$
$$= (12 + 4)m^2 + 5m \qquad\qquad = (3 + 4 - 1)x^2y$$
$$= 16m^2 + 5m \qquad\qquad\qquad = 6x^2y$$

In Example 1(c), we cannot combine $16m^2$ and $5m$. These two terms are unlike because the exponents on the variables are different. *Unlike terms have different variables or different exponents on the same variables.*

◀ Work Problem **1** at the Side.

OBJECTIVE 2 **Know the vocabulary for polynomials.** A **polynomial in x** is a term or the sum of a finite number of terms of the form ax^n, for any real number a and any whole number n. For example,
$$16x^8 - 7x^6 + 5x^4 - 3x^2 + 4 \qquad \text{Polynomial}$$
is a polynomial in x. This polynomial is written in **descending powers,** because the exponents on x decrease from left to right.

On the other hand, $2x^3 - x^2 + \frac{4}{x}$ is not a polynomial, since a variable appears in a denominator. We can define a *polynomial* using any variable, not just x, as in Example 1(c). Polynomials may have terms with more than one variable, as in Example 1(d).

Work Problem 2 at the Side. ▶

The **degree of a term** is the sum of the exponents on the variables. A constant term has degree 0. For example, $3x^4$ has degree **4**, while $6x^{17}$ has degree **17**. The term $5x$ (or $5x^1$) has degree **1**, -7 has degree 0, and $2x^2y$ has degree $2 + 1 = 3$ (y has an exponent of 1).

The **degree of a polynomial** is the greatest degree of any nonzero term of the polynomial. For example, $3x^4 - 5x^2 + 6$ is of degree **4**, the polynomial $5x + 7$ is of degree 1, 3 is of degree 0, and $x^2y + xy - 5xy^2$ is of degree 3.

Three types of polynomials are very common and are given special names. A polynomial with only one term is called a **monomial.** (*Mono*- means "one," as in *mono*rail.) Examples are

$$9m, \quad -6y^5, \quad a^2, \quad \text{and} \quad 6. \qquad \text{\textcolor{blue}{Monomials}}$$

A polynomial with exactly two terms is called a **binomial.** (*Bi*- means "two," as in *bi*cycle.) Examples are

$$-9x^4 + 9x^3, \quad 8m^2 + 6m, \quad \text{and} \quad 3m^5 - 9m^2. \qquad \text{\textcolor{blue}{Binomials}}$$

A polynomial with exactly three terms is called a **trinomial.** (*Tri*- means "three," as in *tri*angle.) Examples are

$$9m^3 - 4m^2 + 6, \quad \frac{19}{3}y^2 + \frac{8}{3}y + 5, \quad \text{and} \quad -3m^5 - 9m^2 + 2. \qquad \text{\textcolor{blue}{Trinomials}}$$

EXAMPLE 2 **Classifying Polynomials**

Simplify each polynomial if possible. Then give the degree and tell whether the polynomial is a *monomial,* a *binomial,* a *trinomial,* or *none of these.*

(a) $2x^3 + 5$ We cannot simplify further. This is a binomial of degree 3.

(b) $4x - 5x + 2x$

Add like terms to simplify: $4x - 5x + 2x = x$. The degree is 1 (since $x = x^1$). The simplified polynomial is a monomial.

Work Problem 3 at the Side. ▶

OBJECTIVE **3** **Evaluate polynomials.**   A polynomial usually represents different numbers for different values of the variable.

EXAMPLE 3 **Evaluating a Polynomial**

Find the value of $3x^4 + 5x^3 - 4x - 4$ when $x = -2$ and when $x = 3$.

First, substitute -2 for x.

$$3x^4 + 5x^3 - 4x - 4$$

> **Use parentheses to avoid errors.**

$$= 3(-2)^4 + 5(-2)^3 - 4(-2) - 4 \qquad \text{\textcolor{blue}{Let } } x = -2.$$

$$= 3(16) + 5(-8) - 4(-2) - 4 \qquad \text{\textcolor{blue}{Apply the exponents.}}$$

$$= 48 - 40 + 8 - 4 \qquad \text{\textcolor{blue}{Multiply.}}$$

$$= 12 \qquad \text{\textcolor{blue}{Add and subtract.}}$$

Continued on Next Page

---

2 Choose all descriptions that apply for each of the expressions in parts (a)–(d).

  A. Polynomial
  B. Polynomial written in descending powers
  C. Not a polynomial

(a) $3m^5 + 5m^2 - 2m + 1$

(b) $2p^4 + p^6$

(c) $\dfrac{1}{x} + 2x^2 + 3$

(d) $x - 3$

3 Simplify each polynomial if possible. Then give the degree and tell whether the polynomial is a *monomial, binomial, trinomial,* or *none of these.*

(a) $3x^2 + 2x - 4$

(b) $x^3 + 4x^3$

(c) $x^8 - x^7 + 2x^8$

ANSWERS

2. (a) A and B   **(b)** A   **(c)** C   **(d)** A and B
3. (a) degree 2; trinomial
  (b) degree 3; monomial (simplify to $5x^3$)
  (c) degree 8; binomial (simplify to $3x^8 - x^7$)

4 Find the value of
$2x^3 + 8x - 6$ in each case.

(a) When $x = -1$

(b) When $x = 4$

Next, replace x with 3.

$$3x^4 + 5x^3 - 4x - 4$$
$$= 3\,(3)^4 + 5\,(3)^3 - 4\,(3) - 4 \qquad \text{Let } x = 3.$$
$$= 3\,(81) + 5\,(27) - 4\,(3) - 4 \qquad \text{Apply the exponents.}$$
$$= 243 + 135 - 12 - 4 \qquad \text{Multiply.}$$
$$= 362 \qquad \text{Add and subtract.}$$

> **CAUTION**
> Use parentheses around the numbers that are substituted for the variable in Example 3, particularly when substituting a negative number for a variable that is raised to a power. Otherwise, a sign error may result.

◀ *Work Problem* **4** *at the Side.*

OBJECTIVE 4 Add polynomials. Polynomials may be added, subtracted, multiplied, and divided.

> **Adding Polynomials**
> To add two polynomials, add like terms.

5 Add each pair of polynomials.

(a) $4x^3 - 3x^2 + 2x$ and
$6x^3 + 2x^2 - 3x$

(b) $x^2 - 2x + 5$ and
$4x^2 - 2$

EXAMPLE 4 **Adding Polynomials Vertically**

(a) Add $6x^3 - 4x^2 + 3$ and $-2x^3 + 7x^2 - 5$.
Write like terms in columns.

$$\begin{array}{r} 6x^3 - 4x^2 + 3 \\ -2x^3 + 7x^2 - 5 \\ \hline \end{array}$$

Now add, column by column.

$$\begin{array}{rrr} 6x^3 & -4x^2 & 3 \\ -2x^3 & 7x^2 & -5 \\ \hline \mathbf{4x^3} & \mathbf{3x^2} & \mathbf{-2} \end{array}$$

Add the three sums together.

$$4x^3 + 3x^2 + (-2) = \mathbf{4x^3 + 3x^2 - 2}$$

(b) Add $2x^2 - 4x + 3$ and $x^3 + 5x$.
Write like terms in columns and add column by column.

$$\begin{array}{r} 2x^2 - 4x + 3 \\ x^3 \qquad\ + 5x \\ \hline x^3 + 2x^2 + \ x + 3 \end{array}$$

Leave spaces for missing terms.

◀ *Work Problem* **5** *at the Side.*

The polynomials in Example 4 also could be added horizontally.

ANSWERS

4. **(a)** -16 **(b)** 154
5. **(a)** $10x^3 - x^2 - x$ **(b)** $5x^2 - 2x + 3$

EXAMPLE 5 Adding Polynomials Horizontally

(a) Add $6x^3 - 4x^2 + 3$ and $-2x^3 + 7x^2 - 5$.
Combine like terms.

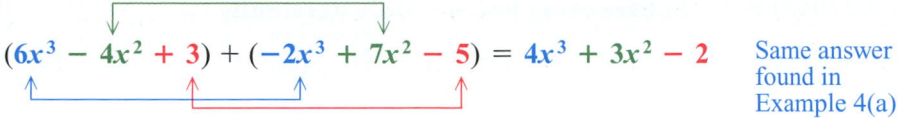

$(6x^3 - 4x^2 + 3) + (-2x^3 + 7x^2 - 5) = 4x^3 + 3x^2 - 2$ Same answer found in Example 4(a)

(b) Add $2x^2 - 4x + 3$ and $x^3 + 5x$.

$$(2x^2 - 4x + 3) + (x^3 + 5x)$$
$$= x^3 + 2x^2 - 4x + 5x + 3 \qquad \text{Commutative property}$$
$$= x^3 + 2x^2 + x + 3 \qquad \text{Combine like terms.}$$

Work Problem **6** *at the Side.* ▶

OBJECTIVE **5** **Subtract polynomials.** In **Section 9.5**, the difference $x - y$ was defined as $x + (-y)$. (We find the difference $x - y$ by adding x and the opposite of y.) For example,

$$7 - 2 = 7 + (-2) = 5 \quad \text{and} \quad -8 - (-2) = -8 + 2 = -6.$$

A similar method is used to subtract polynomials.

> **Subtracting Polynomials**
>
> To subtract two polynomials, change all the signs of the second polynomial and add the result to the first polynomial.

EXAMPLE 6 Subtracting Polynomials

(a) Perform the subtraction $(5x - 2) - (3x - 8)$.
Change the signs in the second polynomial and add.

$$(5x - 2) - (3x - 8)$$
$$= (5x - 2) + (-3x + 8)$$
$$= 2x + 6$$

(b) Subtract $6x^3 - 4x^2 + 2$ from $11x^3 + 2x^2 - 8$.

$$(11x^3 + 2x^2 - 8) - (6x^3 - 4x^2 + 2) \quad \text{◀ Write the problem in the correct order.}$$
$$= (11x^3 + 2x^2 - 8) + (-6x^3 + 4x^2 - 2)$$
$$= 5x^3 + 6x^2 - 10$$

To check a subtraction problem, use the following fact:

$$\text{If} \quad a - b = c, \quad \text{then} \quad a = b + c.$$

For example, $6 - 2 = 4$, so we check by writing $6 = 2 + 4$, which is correct. We check the polynomial subtraction above as follows:

$$(6x^3 - 4x^2 + 2) + (5x^3 + 6x^2 - 10)$$
$$= 11x^3 + 2x^2 - 8.$$

Since the sum is $11x^3 + 2x^2 - 8$, the subtraction was performed correctly.

Work Problem **7** *at the Side.* ▶

---

6 Find each sum.

(a) $(2x^4 - 6x^2 + 7)$
$\qquad + (-3x^4 + 5x^2 + 2)$

(b) $(3x^2 + 4x + 2)$
$\qquad + (6x^3 - 5x - 7)$

7 Subtract, and check your answers by addition.

(a) $(14y^3 - 6y^2 + 2y - 5)$
$\qquad - (2y^3 - 7y^2 - 4y + 6)$

(b) Subtract

$$\left(-\frac{3}{2}y^2 + \frac{4}{3}y + 6\right)$$

from $\left(\frac{7}{2}y^2 - \frac{11}{3}y + 8\right)$.

ANSWERS

6. **(a)** $-x^4 - x^2 + 9$
 (b) $6x^3 + 3x^2 - x - 5$
7. **(a)** $12y^3 + y^2 + 6y - 11$
 (b) $5y^2 - 5y + 2$

Subtraction also can be done in columns. We use vertical subtraction in **Section 12.7** when we study polynomial division.

8 Subtract by columns.

$$(4y^3 - 16y^2 + 2y)$$
$$- (12y^3 - 9y^2 + 16)$$

EXAMPLE 7 **Subtracting Polynomials Vertically**

Subtract by columns: $(14y^3 - 6y^2 + 2y - 5) - (2y^3 - 7y^2 - 4y + 6)$.

$$14y^3 - 6y^2 + 2y - 5$$
$$\underline{2y^3 - 7y^2 - 4y + 6}$$ *Arrange like terms in columns.*

Change all signs in the second row, and then add.

$$14y^3 - 6y^2 + 2y - 5$$
$$\underline{-2y^3 + 7y^2 + 4y - 6}$$ *Change signs.*
$$12y^3 + y^2 + 6y - 11$$ *Add.*

◀ *Work Problem* **8** *at the Side.*

---

9 Perform the indicated operations.

$$(6p^4 - 8p^3 + 2p - 1)$$
$$- (-7p^4 + 6p^2 - 12)$$
$$+ (p^4 - 3p + 8)$$

EXAMPLE 8 **Adding and Subtracting More Than Two Polynomials**

Perform the indicated operations to simplify the expression

$$(4 - x + 3x^2) - (2 - 3x + 5x^2) + (8 + 2x - 4x^2).$$

Rewrite, changing the subtraction to adding the opposite.

$$(4 - x + 3x^2) - (2 - 3x + 5x^2) + (8 + 2x - 4x^2)$$
$$= (4 - x + 3x^2) + (-2 + 3x - 5x^2) + (8 + 2x - 4x^2)$$
$$= (2 + 2x - 2x^2) + (8 + 2x - 4x^2)$$ *Combine like terms.*
$$= 10 + 4x - 6x^2$$ *Combine like terms.*

◀ *Work Problem* **9** *at the Side.*

---

10 Add or subtract.

(a) $(3mn + 2m - 4n)$
$+ (-mn + 4m + n)$

OBJECTIVE 6 Add and subtract polynomials with more than one variable.   Polynomials in more than one variable are added and subtracted by combining like terms, just as with single-variable polynomials.

EXAMPLE 9 **Adding and Subtracting Multivariable Polynomials**

Add or subtract as indicated.

(a) $(4a + 2ab - b) + (3a - ab + b)$

$$= 4a + 2ab - b + 3a - ab + b$$

$$= 7a + ab$$ *Combine like terms.*

(b) $(5p^2q^2 - 4p^2 + 2q)$
$- (2p^2q^2 - p^2 - 3q)$

(b) $(2x^2y + 3xy + y^2) - (3x^2y - xy - 2y^2)$

$$= 2x^2y + 3xy + y^2 - 3x^2y + xy + 2y^2$$

$$= -x^2y + 4xy + 3y^2$$ *Be careful with signs.*

◀ *Work Problem* **10** *at the Side.*

ANSWERS

8. $-8y^3 - 7y^2 + 2y - 16$
9. $14p^4 - 8p^3 - 6p^2 - p + 19$
10. (a) $2mn + 6m - 3n$
 **(b)** $3p^2q^2 - 3p^2 + 5q$

12.1 ▶▶▶ Exercises

Fill in each blank with the correct response.

1. In the term $7x^5$, the coefficient is _____ and the exponent is _____.

2. The expression $5x^3 - 4x^2$ has _____ term(s).
    (how many?)

3. The degree of the term $-4x^8$ is _____.

4. The polynomial $4x^2 - y^2$ _____ an example of a trinomial.
    (is/is not)

5. When $x^2 + 10$ is evaluated for $x = 4$, the result is _____.

6. _____ is an example of a monomial with coefficient 5, in the variable x, having degree 9.

For each polynomial, determine the number of terms, and name the coefficient of each term.

7. $6x^4$ **8.** $-9y^5$ **9.** t^4 **10.** s^7 **11.** $\dfrac{x}{5}$ **12.** $\dfrac{z}{8}$

13. $-19r^2 - r$ **14.** $2y^3 - y$ **15.** $x - 8x^2 + \dfrac{2}{3}x^3$ **16.** $v - 2v^3 + \dfrac{3}{4}v^2$

In each polynomial, combine like terms whenever possible. Write the result with descending powers. See Example 1.

17. $-3m^5 + 5m^5$ **18.** $-4y^3 + 3y^3$ **19.** $2r^5 + (-3r^5)$ **20.** $-19y^2 + 9y^2$

21. $\dfrac{1}{2}x^4 + \dfrac{1}{6}x^4$ **22.** $\dfrac{3}{10}x^6 + \dfrac{1}{5}x^6$ **23.** $0.2m^5 - 0.5m^2$ **24.** $-0.9y + 0.9y^2$

25. $-3x^5 + 2x^5 - 4x^5$ **26.** $6x^3 - 8x^3 + 9x^3$ **27.** $-4p^7 + 8p^7 + 5p^9$

28. $-3a^8 + 4a^8 - 3a^2$ **29.** $-4y^2 + 3y^2 - 2y^2 + y^2$ **30.** $3r^5 - 8r^5 + r^5 + 2r^5$

For each polynomial, first simplify, if possible, and write it with descending powers. Then give the degree of the resulting polynomial, and tell whether it is a monomial, a binomial, a trinomial, or none of these. See Example 2.

31. $6x^4 - 9x$ **32.** $7t^3 - 3t$ **33.** $5m^4 - 3m^2 + 6m^5 - 7m^3$

34. $6p^5 + 4p^3 - 8p^4 + 10p^2$

35. $\dfrac{5}{3}x^4 - \dfrac{2}{3}x^4 + \dfrac{1}{3}x^2 - 4$

36. $\dfrac{4}{5}r^6 + \dfrac{1}{5}r^6 - r^4 + \dfrac{2}{5}r$

37. $0.8x^4 - 0.3x^4 - 0.5x^4 + 7$

38. $1.2t^3 - 0.9t^3 - 0.3t^3 + 9$

39. $2.5x^2 + 0.5x + x^2 - x - 2x^2$

Find the value of each polynomial **(a)** *when* $x = 2$ *and* **(b)** *when* $x = -1$. *See Example 3.*

40. $5x - 4$

41. $-2x + 3$

42. $-3x^2 + 14x - 2$

43. $2x^2 + 5x + 1$

44. $x^4 - 6x^3 + x^2 + 1$

45. $2x^5 - 4x^4 + 5x^3 - x^2$

46. $2x^6 - 4x$

47. $-4x^5 + x^2$

Relating Concepts (Exercises 48–52) For Individual or Group Work

A polynomial can model the distance in feet that a car going approximately 68 mph will skid in t seconds. If we let D represent this distance, then

$$D = 100t - 13t^2.$$

Each time we evaluate this polynomial for a value of t, we get one and only one output value D. This idea is basic to the concept of a **function,** *an important concept in mathematics. Exercises 48–52 illustrate this idea with this polynomial and three others.* **Work them in order.**

48. Evaluate the given polynomial when $t = 5$. Use the result to fill in the blanks: In _____ seconds, the car will skid _____ feet.

49. Use the polynomial equation $D = 100t - 13t^2$ to find the distance the car will skid in 1 sec. Write an ordered pair of the form (t, D).

50. If gasoline costs $4.00 per gal, then the monomial $4.00x$ gives the cost, in dollars, of x gallons. How much would 4 gal cost?

51. If it costs $15 plus $2 per day to rent a chain saw, the binomial $2x + 15$ gives the cost in dollars to rent the chain saw for x days. How much would it cost to rent the saw for 6 days?

52. If an object is projected upward under certain conditions, its height in feet is given by the trinomial $-16t^2 + 60t + 80$, where t is in seconds. Evaluate this trinomial for $t = 2.5$, and then use the result to fill in the blanks: If _____ seconds have elapsed, the height of the object is _____ feet.

Add or subtract as indicated. See Examples 4 and 7.

53. Add.
$$3m^2 + 5m$$
$$2m^2 - 2m$$

54. Add.
$$4a^3 - 4a^2$$
$$6a^3 + 5a^2$$

55. Subtract.
$$12x^4 - x^2$$
$$8x^4 + 3x^2$$

56. Subtract.
$$13y^5 - y^3$$
$$7y^5 + 5y^3$$

57. Add.
$$\frac{2}{3}x^2 + \frac{1}{5}x + \frac{1}{6}$$
$$\frac{1}{2}x^2 - \frac{1}{3}x + \frac{2}{3}$$

58. Add.
$$\frac{4}{7}y^2 - \frac{1}{5}y + \frac{7}{9}$$
$$\frac{1}{3}y^2 - \frac{1}{3}y + \frac{2}{5}$$

59. Subtract.
$$12m^3 - 8m^2 + 6m + 7$$
$$5m^2 \qquad - 4$$

60. Subtract.
$$5a^4 - 3a^3 + 2a^2 - a + 6$$
$$-6a^4 \qquad - a^2 + a - 1$$

61. Subtract.
$$4.3x^3 - 6.1x^2 - 3.0x - 5$$
$$1.4x^3 - 2.6x^2 - 1.5x + 4$$

Perform the indicated operations. See Examples 5, 6, and 8.

62. $(3r^2 + 5r - 6) + (2r - 5r^2)$

63. $(2r^2 + 3r - 12) + (6r^2 + 2r)$

64. $(x^2 + x) - (3x^2 + 2x - 1)$

65. $(8m^2 - 7m) - (3m^2 + 7m - 6)$

66. $(-2b^6 + 3b^4 - b^2) + (b^6 + 2b^4 + 2b^2)$

67. $(16x^3 - x^2 + 3x) + (-12x^3 + 3x^2 + 2x)$

68. $(8t^5 + 3t^3 + 5t) - (19t^4 - 6t^2 + t)$

69. $(7y^4 + 3y^2 + 2y) - (18y^5 - 5y^3 + y)$

70. $[(9b^3 - 4b^2 + 3b + 2) - (-2b^3 + b)] - (8b^3 + 6b + 4)$

71. $[(8m^2 + 4m - 7) - (2m^3 - 5m + 2)] - (m^2 + m)$

72. Subtract $-5w^3 + 5w^2 - 7$ from $6w^3 + 8w + 5$.

73. Subtract $9x^2 - 3x + 7$ from $-2x^2 - 6x + 4$.

74. Find the difference when $9x^4 + 3x^2 + 5$ is subtracted from $8x^4 - 2x^3 + x - 1$.

Find a polynomial that represents the perimeter of each square, rectangle, or triangle.

75.

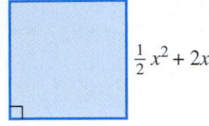

$\frac{1}{2}x^2 + 2x$

76.

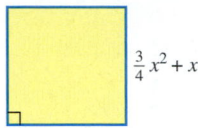

$\frac{3}{4}x^2 + x$

77.

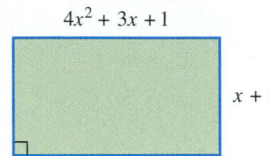

$4x^2 + 3x + 1$

$x + 2$

78.

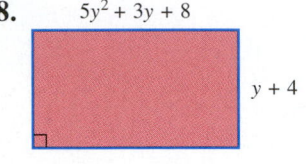

$5y^2 + 3y + 8$

$y + 4$

79.

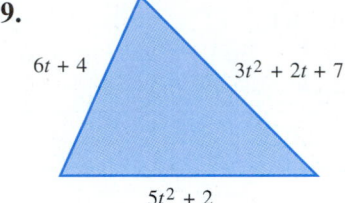

$6t + 4$

$3t^2 + 2t + 7$

$5t^2 + 2$

80.

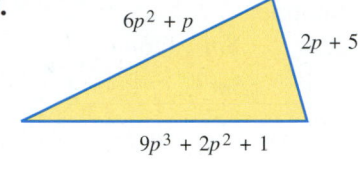

$6p^2 + p$

$2p + 5$

$9p^3 + 2p^2 + 1$

Add or subtract as indicated. See Example 9.

81. $(9a^2b - 3a^2 + 2b) + (4a^2b - 4a^2 - 3b)$

82. $(4xy^3 - 3x + y) + (5xy^3 + 13x - 4y)$

83. $(2c^4d + 3c^2d^2 - 4d^2) - (c^4d + 8c^2d^2 - 5d^2)$

84. $(3k^2h^3 + 5kh + 6k^3h^2) - (2k^2h^3 - 9kh + k^3h^2)$

85. Subtract.

$$9m^3n - 5m^2n^2 + 4mn^2$$
$$\underline{-3m^3n + 6m^2n^2 + 8mn^2}$$

86. Subtract.

$$12r^5t + 11r^4t^2 - 7r^3t^3$$
$$\underline{-8r^5t + 10r^4t^2 + 3r^3t^3}$$

*Find **(a)** a polynomial that represents the perimeter of each triangle and **(b)** the measures of the angles of the triangle. (Hint: In part (b), the sum of the measures of the angles of any triangle is 180°.)*

87.

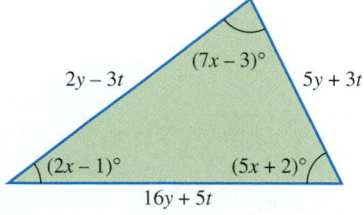

$2y - 3t$

$(7x - 3)°$

$5y + 3t$

$(2x - 1)°$

$(5x + 2)°$

$16y + 5t$

88.

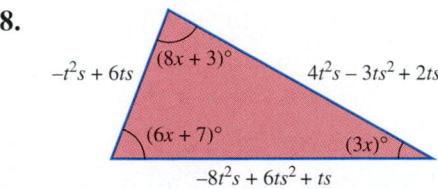

$-t^2s + 6ts$

$(8x + 3)°$

$4t^2s - 3ts^2 + 2ts$

$(6x + 7)°$

$(3x)°$

$-8t^2s + 6ts^2 + ts$

12.2 ▶▶▶ The Product Rule and Power Rules for Exponents

OBJECTIVE 1 Use exponents. In **Section 9.1,** we used exponents to write repeated products. Recall that in the expression 5^2, the number 5 is called the **base** and 2 is called the **exponent,** or **power.** The expression 5^2 is called an **exponential expression.** Although we do not usually write a quantity with an exponent of 1, in general, for any quantity a, $a = a^1$.

EXAMPLE 1 Using Exponents

Write $3 \cdot 3 \cdot 3 \cdot 3 \cdot 3$ in exponential form and evaluate.

  Since 3 occurs as a factor five times, the base is **3** and the exponent is **5.** The exponential expression is 3^5, read "3 to the fifth power," or simply "3 to the fifth."

$$\underbrace{3 \cdot 3 \cdot 3 \cdot 3 \cdot 3}_{\text{5 factors of 3}} \quad \text{means} \quad 3^5, \quad \text{or} \quad 243.$$

Work Problem ① *at the Side.* ▶

EXAMPLE 2 Evaluating Exponential Expressions

Evaluate. Name the base and the exponent.

		Base	**Exponent**
(a) $5^4 = 5 \cdot 5 \cdot 5 \cdot 5 = 625$		5	4
(b) $-5^4 = -1 \cdot 5^4 = -1 \cdot (5 \cdot 5 \cdot 5 \cdot 5) = -625$		5	4
(c) $(-5)^4 = (-5)(-5)(-5)(-5) = 625$		-5	4

CAUTION

Look at Examples 2(b) and (c). In -5^4, the absence of parentheses shows that the exponent 4 applies only to the base 5, and not -5. In $(-5)^4$, the parentheses show that the exponent 4 applies to the base -5. In summary, $-a^n$ and $(-a)^n$ are not necessarily the same.

Expression	Base	Exponent	Example
$-a^n$	a	n	$-3^2 = -(3 \cdot 3) = -9$
$(-a)^n$	$-a$	n	$(-3)^2 = (-3)(-3) = 9$

Work Problem ② *at the Side.* ▶

OBJECTIVE 2 Use the product rule for exponents. To develop the product rule, we use the definition of an exponent.

$$2^4 \cdot 2^3 = \overbrace{(2 \cdot 2 \cdot 2 \cdot 2)}^{\text{4 factors}} \overbrace{(2 \cdot 2 \cdot 2)}^{\text{3 factors}}$$

$$= \underbrace{2 \cdot 2 \cdot 2 \cdot 2 \cdot 2 \cdot 2 \cdot 2}_{4 + 3 = 7 \text{ factors}}$$

$$= 2^7$$

OBJECTIVES

1 Use exponents.

2 Use the product rule for exponents.

3 Use the rule $(a^m)^n = a^{mn}$.

4 Use the rule $(ab)^m = a^m b^m$.

5 Use the rule $\left(\dfrac{a}{b}\right)^m = \dfrac{a^m}{b^m}$.

6 Use combinations of the rules for exponents.

7 Use the rules for exponents in a geometry application.

---

① Write $2 \cdot 2 \cdot 2 \cdot 2$ in exponential form and evaluate.

② Evaluate. Name the base and the exponent.

  (a) $(-2)^5$ **(b)** -2^5

  (c) -4^2 **(d)** $(-4)^2$

ANSWERS

1. 2^4, or 16
2. **(a)** -32; -2; 5 **(b)** -32; 2; 5
   (c) -16; 4; 2 **(d)** 16; -4; 2

3 Simplify by using the product rule, if possible.

(a) $8^2 \cdot 8^5$

(b) $(-7)^5(-7)^3$

(c) $y^3 \cdot y$

(d) $z^2z^5z^6$

(e) $4^2 \cdot 3^5$

(f) $6^4 + 6^2$

Also,
$$6^2 \cdot 6^3 = (6 \cdot 6)(6 \cdot 6 \cdot 6)$$
$$= 6 \cdot 6 \cdot 6 \cdot 6 \cdot 6$$
$$= 6^5.$$

Generalizing from these examples,
$$2^4 \cdot 2^3 = 2^{4+3} = 2^7 \quad \text{and} \quad 6^2 \cdot 6^3 = 6^{2+3} = 6^5.$$

In each case, adding the exponents gives the exponent of the product, suggesting the **product rule for exponents.**

> **Product Rule for Exponents**
>
> For any positive integers m and n, $\quad a^m \cdot a^n = a^{m+n}$.
> (Keep the same base and add the exponents.)
>
> *Example:* $\quad 6^2 \cdot 6^5 = 6^{2+5} = 6^7$

> **CAUTION**
> Do not multiply the bases when using the product rule. ***Keep the same base and add the exponents.*** For example,
> $$6^2 \cdot 6^5 = 6^7, \quad \textit{not} \quad 36^7.$$

EXAMPLE 3 **Using the Product Rule**

Use the product rule for exponents to simplify, if possible.

(a) $6^3 \cdot 6^5 = 6^{3+5} = 6^8$ **(b)** $(-4)^7(-4)^2 = (-4)^{7+2} = (-4)^9$

> Keep the same base.

(c) $x^2 \cdot x = x^2 \cdot x^1 = x^{2+1} = x^3$ **(d)** $m^4m^3m^5 = m^{4+3+5} = m^{12}$

(e) $2^3 \cdot 3^2$

The product rule does not apply to the product $2^3 \cdot 3^2$ because the bases are different.

$$2^3 \cdot 3^2 = 8 \cdot 9 = 72 \qquad \text{Evaluate } 2^3 \text{ and } 3^2; \text{ then multiply.}$$

> Think: 2^3 means $2 \cdot 2 \cdot 2$.

> Think: 3^2 means $3 \cdot 3$.

(f) $2^3 + 2^4$

The product rule does not apply to $2^3 + 2^4$ because it is a *sum,* not a *product.*

$$2^3 + 2^4 = 8 + 16 = 24 \qquad \text{Evaluate } 2^3 \text{ and } 2^4; \text{ then add.}$$

> **CAUTION**
> ***The bases of the factors must be the same*** before we can apply the product rule for exponents.

ANSWERS

3. (a) 8^7 **(b)** $(-7)^8$ **(c)** y^4
   (d) z^{13} **(e)** The product rule does not apply.
(product: 3888)   **(f)** The product rule does not apply. (sum: 1332)

◀ *Work Problem* **3** *at the Side.*

EXAMPLE 4 Using the Product Rule

Multiply $2x^3$ and $3x^7$.

$2x^3 \cdot 3x^7$ $2x^3 = 2 \cdot x^3; 3x^7 = 3 \cdot x^7$

$= (2 \cdot 3) \cdot (x^3 \cdot x^7)$ Commutative and associative properties

$= 6x^{3+7}$ Multiply; product rule

$= 6x^{10}$ Add the exponents.

CAUTION

Be sure you understand the difference between *adding* and *multiplying* exponential expressions. For example,

$$8x^3 + 5x^3 \quad \text{means} \quad (8 + 5)x^3, \quad \text{or} \quad 13x^3,$$

but $\quad (8x^3)(5x^3) \quad$ means $\quad (8 \cdot 5)x^{3+3}, \quad$ or $\quad 40x^6$.

Work Problem **4** *at the Side.* ▶

OBJECTIVE **3** **Use the rule $(a^m)^n = a^{mn}$.** We can simplify an expression such as $(8^3)^2$ with the product rule for exponents, as follows.

$$(8^3)^2 = (8^3)(8^3) = 8^{3+3} = 8^6$$

The product of the exponents in $(8^3)^2$, **3 · 2**, gives the exponent in 8^6. Also,

$(5^2)^4 = 5^2 \cdot 5^2 \cdot 5^2 \cdot 5^2$ Definition of exponent

$= 5^{2+2+2+2}$ Product rule

$= 5^8,$ Add the exponents.

and $2 \cdot 4 = 8$. These examples suggest **power rule (a) for exponents.**

Power Rule (a) for Exponents

For any positive integers m and n, $(a^m)^n = a^{mn}$.
(Raise a power to a power by multiplying exponents.)

Example: $(3^2)^4 = 3^{2 \cdot 4} = 3^8$

EXAMPLE 5 Using Power Rule (a)

Use power rule (a) for exponents to simplify.

(a) $(2^5)^3 = 2^{5 \cdot 3} = 2^{15}$ **(b)** $(5^7)^2 = 5^{7 \cdot 2} = 5^{14}$ **(c)** $(x^2)^5 = x^{2 \cdot 5} = x^{10}$

Work Problem **5** *at the Side.* ▶

OBJECTIVE **4** **Use the rule $(ab)^m = a^m b^m$.** We can rewrite the expression $(4x)^3$ as shown below.

$(4x)^3 = (4x)(4x)(4x)$ Definition of exponent

$= 4 \cdot 4 \cdot 4 \cdot x \cdot x \cdot x$ Commutative and associative properties

$= 4^3x^3$ Definition of exponent

This example suggests **power rule (b) for exponents.**

---

4 Multiply.

(a) $5m^2 \cdot 2m^6$

(b) $3p^5 \cdot 9p^4$

(c) $-7p^5 \cdot (3p^8)$

5 Simplify.

(a) $(5^3)^4$

(b) $(6^2)^5$

(c) $(3^2)^4$

(d) $(a^6)^5$

ANSWERS

4. (a) $10m^8$ **(b)** $27p^9$ **(c)** $-21p^{13}$
5. (a) 5^{12} **(b)** 6^{10} **(c)** 3^8 **(d)** a^{30}

6 Simplify.

(a) $(2ab)^4$

Power Rule (b) for Exponents

For any positive integer m, $(ab)^m = a^m b^m$.

(Raise a product to a power by raising each factor to the power.)

Example: $(2p)^5 = 2^5 p^5$

EXAMPLE 6   **Using Power Rule (b)**

Use power rule (b) for exponents to simplify.

(a) $(3xy)^2$

$= 3^2 x^2 y^2$ Power rule (b)

$= 9x^2 y^2$ $3^2 = 3 \cdot 3 = 9$

(b) $9(pq)^2$

$= 9(p^2 q^2)$ Power rule (b)

$= 9p^2 q^2$ Multiply.

(b) $5(mn)^3$

(c) $5(2m^2 p^3)^4$

$= 5[2^4 (m^2)^4 (p^3)^4]$ Power rule (b)

$= 5(2^4 m^8 p^{12})$ Power rule (a)

$= 5 \cdot 2^4 m^8 p^{12}$

$= 80 m^8 p^{12}$ $5 \cdot 2^4 = 5 \cdot 16 = 80$

(d)

$(-5^6)^3$

$= (-1 \cdot 5^6)^3$ $-a = -1 \cdot a$

Raise -1 to the designated power.

$= (-1)^3 (5^6)^3$ Power rule (b)

$= -1 \cdot 5^{18}$ Power rule (a)

$= -5^{18}$

(c) $(3a^2 b^4)^5$

CAUTION

Power rule (b) does not apply to a sum:

$$(4x)^2 = 4^2 x^2, \quad \text{but} \quad (4 + x)^2 \neq 4^2 + x^2.$$

◀ *Work Problem* **6** *at the Side.*

(d) $(-5m^2)^3$

OBJECTIVE **5** **Use the rule** $\left(\frac{a}{b}\right)^m = \frac{a^m}{b^m}$.   Since the quotient $\frac{a}{b}$ can be written as $a \cdot \frac{1}{b}$, we can use power rule (b), together with some of the properties of real numbers, to get **power rule (c) for exponents.**

Power Rule (c) for Exponents

For any positive integer m, $\left(\dfrac{a}{b}\right)^m = \dfrac{a^m}{b^m}$ $(b \neq 0)$.

(Raise a quotient to a power by raising both the numerator and the denominator to the power.)

Example: $\left(\dfrac{5}{3}\right)^2 = \dfrac{5^2}{3^2}$

ANSWERS

6. **(a)** $16a^4 b^4$ **(b)** $5m^3 n^3$ **(c)** $243a^{10} b^{20}$
  (d) $-125m^6$

EXAMPLE 7 Using Power Rule (c)

Use power rule (c) for exponents to simplify.

(a) $\left(\dfrac{2}{3}\right)^5 = \dfrac{2^5}{3^5} = \dfrac{32}{243}$

(b) $\left(\dfrac{m}{n}\right)^4 = \dfrac{m^4}{n^4}, \quad n \neq 0$

(c) $\left(\dfrac{1}{5}\right)^4 = \dfrac{1^4}{5^4} = \dfrac{1}{5^4} = \dfrac{1}{625}$ $1^4 = 1 \cdot 1 \cdot 1 \cdot 1 = 1$

Note

In Example 7(c), we used the fact that $1^4 = 1$.

In general, $1^n = 1$, for any integer n.

Work Problem 7 at the Side. ▶

The rules for exponents discussed in this section are basic to the study of algebra and should be *memorized*.

Rules for Exponents

For positive integers m and n: *Examples*

Product rule $\quad a^m \cdot a^n = a^{m+n}$ $\qquad 6^2 \cdot 6^5 = 6^{2+5} = 6^7$

Power rules (a) $(a^m)^n = a^{mn}$ $\qquad (3^2)^4 = 3^{2 \cdot 4} = 3^8$

$\qquad$ **(b)** $(ab)^m = a^m b^m$ $\qquad (2p)^5 = 2^5 p^5$

$\qquad$ **(c)** $\left(\dfrac{a}{b}\right)^m = \dfrac{a^m}{b^m}$ **$(b \neq 0)$.** $\qquad \left(\dfrac{5}{3}\right)^2 = \dfrac{5^2}{3^2}$

OBJECTIVE 6 Use combinations of the rules for exponents.
More than one rule may be needed to simplify an exponential expression.

EXAMPLE 8 Using Combinations of Rules

Simplify each expression.

(a) $\left(\dfrac{2}{3}\right)^2 \cdot 2^3$

$= \dfrac{2^2}{3^2} \cdot \dfrac{2^3}{1}$ $\quad$ Power rule (c)

$= \dfrac{2^2 \cdot 2^3}{3^2 \cdot 1}$ $\quad$ Multiply fractions.

$= \dfrac{2^{2+3}}{3^2}$ $\quad$ Product rule

$= \dfrac{2^5}{3^2}$

$= \dfrac{32}{9}$

(b) $(5x)^3 (5x)^4$

$\qquad = (5x)^7 \quad$ Product rule

$\qquad = 5^7 x^7 \quad$ Power rule (b)

Continued on Next Page

---

7 Simplify. Assume that all variables represent nonzero real numbers.

(a) $\left(\dfrac{5}{2}\right)^4$

(b) $\left(\dfrac{p}{q}\right)^2$

(c) $\left(\dfrac{r}{t}\right)^3$

(d) $\left(\dfrac{1}{3}\right)^5$

(e) $\left(\dfrac{1}{x}\right)^{10}$

ANSWERS

7. **(a)** $\dfrac{625}{16}$ **(b)** $\dfrac{p^2}{q^2}$ **(c)** $\dfrac{r^3}{t^3}$ **(d)** $\dfrac{1}{243}$ **(e)** $\dfrac{1}{x^{10}}$

8 Simplify.

(a) $(2m)^3 (2m)^4$

(b) $\left(\dfrac{5k^3}{3}\right)^2$

(c) $\left(\dfrac{1}{5}\right)^4 (2x)^2$

(d) $(-3xy^2)^3 (x^2y)^4$

(c) $(2x^2y^3)^4 (3xy^2)^3$

$= 2^4 (x^2)^4 (y^3)^4 \cdot 3^3 x^3 (y^2)^3$ Power rule (b)

$= 2^4 x^8 y^{12} \cdot 3^3 x^3 y^6$ Power rule (a)

$= 2^4 \cdot 3^3 x^8 x^3 y^{12} y^6$ Commutative and associative properties

$= 16 \cdot 27 x^{11} y^{18}$ Product rule

$= 432 x^{11} y^{18}$ Multiply.

Notice that $(2x^2y^3)^4$ means $2^4 x^{2 \cdot 4} y^{3 \cdot 4}$, **not** $(2 \cdot 4) x^{2 \cdot 4} y^{3 \cdot 4}$.

> Do *not* multiply the coefficient 2 and the exponent 4.

(d) $\qquad\qquad (-x^3y)^2 (-x^5y^4)^3$

$= (-1x^3y)^2 (-1x^5y^4)^3$ $-a = -1 \cdot a$

> Think of the negative sign in each factor as -1.

$= (-1)^2 (x^3)^2 (y^2) \cdot (-1)^3 (x^5)^3 (y^4)^3$ Power rule (b)

$= (-1)^2 (x^6)(y^2) \cdot (-1)^3 (x^{15})(y^{12})$ Power rule (a)

$= (-1)^5 (x^{6+15})(y^{2+12})$ Product rule

$= -1x^{21}y^{14}$

$= -x^{21}y^{14}$

◀ *Work Problem* **8** *at the Side.*

OBJECTIVE 7 Use the rules for exponents in a geometry application.

EXAMPLE 9 Using Area Formulas

Find a polynomial that represents the area of each geometric figure.

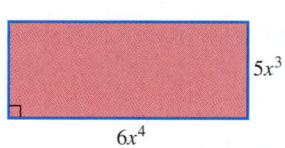

Figure 1

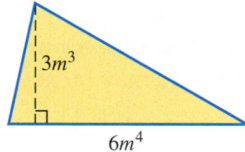

Figure 2

For Figure 1, use the formula for the area of a rectangle, $A = LW$.

$A = (6x^4)(5x^3)$ Area formula

$A = 6 \cdot 5 \cdot x^{4+3}$ Product rule

$A = 30x^7$

Figure 2 is a triangle with base $6m^4$ and height $3m^3$. Substitute into the formula for the area of a triangle and simplify.

$A = \dfrac{1}{2}bh$ Area formula

$A = \dfrac{1}{2}(6m^4)(3m^3)$ Substitute.

$A = \dfrac{1}{2}(18m^7)$,   or $9m^7$ Product rule; multiply.

◀ *Work Problem* **9** *at the Side.*

9 Find a polynomial that represents the area of the figure.

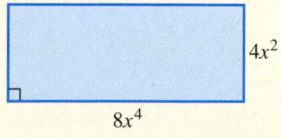

ANSWERS

8. **(a)** 2^7m^7, or $128m^7$ **(b)** $\dfrac{5^2k^6}{3^2}$, or $\dfrac{25k^6}{9}$

(c) $\dfrac{2^2x^2}{5^4}$, or $\dfrac{4x^2}{625}$

(d) $-3^3x^{11}y^{10}$, or $-27x^{11}y^{10}$

9. $32x^6$

12.2 ▶▶▶ **Exercises**

FOR EXTRA HELP

  **MyMathLab**

 Math XL
PRACTICE

 WATCH

 DOWNLOAD

READ

 REVIEW

1. What exponent is understood on the base x in the expression xy^2?

2. How are the expressions 3^2, 5^3, and 7^4 read?

Decide whether each statement is true *or* false.

3. $3^3 = 9$

4. $(-2)^4 = 2^4$

5. $(a^2)^3 = a^5$

6. $\left(\dfrac{1}{4}\right)^2 = \dfrac{1}{4^2}$

Write each expression using exponents. See Example 1.

7. $t \cdot t \cdot t \cdot t \cdot t \cdot t \cdot t \cdot t$

8. $w \cdot w \cdot w \cdot w \cdot w \cdot w$

9. $\left(\dfrac{1}{2}\right)\left(\dfrac{1}{2}\right)\left(\dfrac{1}{2}\right)\left(\dfrac{1}{2}\right)\left(\dfrac{1}{2}\right)$

10. $\left(-\dfrac{1}{4}\right)\left(-\dfrac{1}{4}\right)\left(-\dfrac{1}{4}\right)\left(-\dfrac{1}{4}\right)$

11. $(-8p)(-8p)$

12. $(-7x)(-7x)(-7x)$

13. Explain how the expressions $(-3)^4$ and -3^4 are different.

14. Explain how the expressions $(5x)^3$ and $5x^3$ are different.

Identify the base and the exponent for each exponential expression. In Exercises 15–18, also evaluate the expression. See Example 2.

15. 3^5

16. 2^7

17. $(-3)^5$

18. $(-2)^7$

19. $(-6x)^4$

20. $(-8x)^4$

21. $-6x^4$

22. $-8x^4$

23. Explain why the product rule does not apply to the expression $5^2 + 5^3$. Then evaluate the expression.

24. Explain why the product rule does not apply to the expression $3^2 \cdot 4^3$. Then evaluate the expression.

Use the product rule for exponents to simplify each expression, if possible. Write each answer in exponential form. See Examples 3 and 4.

25. $5^2 \cdot 5^6$

26. $3^6 \cdot 3^7$

27. $4^2 \cdot 4^7 \cdot 4^3$

28. $5^3 \cdot 5^8 \cdot 5^2$

29. $(-7)^3(-7)^6$

30. $(-9)^8(-9)^5$

31. $t^3 t^8 t^{13}$

32. $n^5 n^6 n^9$

33. $(-8r^4)(7r^3)$

34. $(10a^7)(-4a^3)$

35. $(-6p^5)(-7p^5)$

36. $(-5w^8)(-9w^8)$

37. $3^8 + 3^9$

38. $4^{12} + 4^5$

39. $5^8 \cdot 3^8$

40. $6^3 \cdot 8^3$

Use the power rules for exponents to simplify each expression. See Examples 5–7.

41. $(4^3)^2$

42. $(8^3)^6$

43. $(t^4)^5$

44. $(y^6)^5$

45. $(7r)^3$

46. $(11x)^4$

47. $(-5^2)^6$

48. $(-9^4)^8$

49. $(-8^3)^5$

50. $(-7^5)^7$

51. $(5xy)^5$

52. $(9pq)^6$

53. $8(qr)^3$

54. $4(vw)^5$

55. $\left(\dfrac{1}{2}\right)^3$

56. $\left(\dfrac{1}{3}\right)^5$

57. $\left(\dfrac{a}{b}\right)^3, \quad b \neq 0$

58. $\left(\dfrac{r}{t}\right)^4, \quad t \neq 0$

59. $\left(\dfrac{9}{5}\right)^8$

60. $\left(\dfrac{12}{7}\right)^6$

61. $(-2x^2y)^3$

62. $(-5m^4p^2)^3$

63. $(3a^3b^2)^2$

64. $(4x^3y^5)^4$

Simplify each expression. See Example 8.

65. $\left(\dfrac{5}{2}\right)^3 \cdot \left(\dfrac{5}{2}\right)^2$

66. $\left(\dfrac{3}{4}\right)^5 \cdot \left(\dfrac{3}{4}\right)^6$

67. $\left(\dfrac{9}{8}\right)^3 \cdot 9^2$

68. $\left(\dfrac{8}{5}\right)^4 \cdot 8^3$

69. $(2x)^9 (2x)^3$

70. $(6y)^5 (6y)^8$

71. $(-6p)^4 (-6p)$

72. $(-13q)^3 (-13q)$

73. $(6x^2y^3)^5$

74. $(5r^5t^6)^7$

75. $(x^2)^3 (x^3)^5$

76. $(y^4)^5 (y^3)^5$

77. $(2w^2x^3y)^2 (x^4y)^5$

78. $(3x^4y^2z)^3 (yz^4)^5$

79. $(-r^4s)^2 (-r^2s^3)^5$

80. $(-ts^6)^4 (-t^3s^5)^3$

81. $\left(\dfrac{5a^2b^5}{c^6}\right)^3, \quad c \neq 0$

82. $\left(\dfrac{6x^3y^9}{z^5}\right)^4, \quad z \neq 0$

83. $(-5m^3p^4q)^2 (p^2q)^3$

84. $(-a^4b^5)(-6a^3b^3)^2$

85. $(2x^2y^3z)^4 (xy^2z^3)^2$

Find a polynomial that represents the area of each figure. See Example 9.

86.

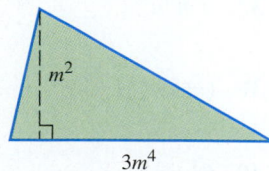

m^2

$3m^4$

87.

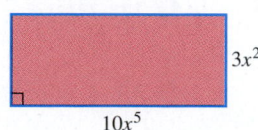

$3x^2$

$10x^5$

88.

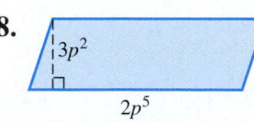

$3p^2$

$2p^5$

12.3 ▶▶▶ Multiplying Polynomials

OBJECTIVES

1 Multiply a monomial and a polynomial.

2 Multiply two polynomials.

3 Multiply binomials by the FOIL method.

OBJECTIVE 1 Multiply a monomial and a polynomial. As shown in **Section 12.2,** we find the product of two monomials by using the rules for exponents and the commutative and associative properties. For example,

$$(-8m^6)(-9n^6)$$
$$= (-8)(-9)(m^6)(n^6)$$
$$= 72m^6n^6.$$

CAUTION
Do not confuse addition of terms with multiplication of terms.
$$7q^5 + 2q^5 = 9q^5, \quad \text{but} \quad (7q^5)(2q^5) = 7 \cdot 2q^{5+5} = 14q^{10}.$$

To find the product of a monomial and a polynomial with more than one term, we use the distributive property and multiplication of monomials.

EXAMPLE 1 Multiplying Monomials and Polynomials

Find each product.

(a) $4x^2(3x + 5)$

$$4x^2(3x + 5) = 4x^2(3x) + 4x^2(5) \quad \text{Distributive property}$$
$$= 12x^3 + 20x^2 \quad \text{Multiply monomials.}$$

(b) $-8m^3(4m^3 + 3m^2 + 2m - 1)$

$$= -8m^3(4m^3) + (-8m^3)(3m^2)$$
$$+ (-8m^3)(2m) + (-8m^3)(-1) \quad \text{Distributive property}$$
$$= -32m^6 - 24m^5 - 16m^4 + 8m^3 \quad \text{Multiply monomials.}$$

Work Problem **1** *at the Side.* ▶

OBJECTIVE 2 Multiply two polynomials. We can use the distributive property repeatedly to find the product of any two polynomials. For example, to find the product of the polynomials $x^2 + 3x + 5$ and $x - 4$, think of $x - 4$ as a single quantity and use the distributive property as follows.

$$(x^2 + 3x + 5)(x - 4)$$
$$= x^2(x - 4) + 3x(x - 4) + 5(x - 4) \quad \text{Distributive property}$$
$$= x^2(x) + x^2(-4) + 3x(x) + 3x(-4) + 5(x) + 5(-4)$$
$$\quad \text{Distributive property again}$$
$$= x^3 - 4x^2 + 3x^2 - 12x + 5x - 20 \quad \text{Multiply monomials.}$$
$$= x^3 - x^2 - 7x - 20 \quad \text{Combine like terms.}$$

This example suggests the following rule.

Multiplying Polynomials
To multiply two polynomials, multiply each term of the second polynomial by each term of the first polynomial and add the products.

1 Find each product.

(a) $5m^3(2m + 7)$

(b) $2x^4(3x^2 + 2x - 5)$

(c) $-4y^2(3y^3 + 2y^2 - 4y + 8)$

ANSWERS
1. (a) $10m^4 + 35m^3$
(b) $6x^6 + 4x^5 - 10x^4$
(c) $-12y^5 - 8y^4 + 16y^3 - 32y^2$

2 Multiply.

(a) $(m + 3)(m^2 - 2m + 1)$

(b) $(6p^2 + 2p - 4)(3p^2 - 5)$

3 Find the product.

$$3x^2 + 4x - 5$$
$$\underline{ x + 4}$$

4 Use the rectangle method to find each product.

(a) $(4x + 3)(x + 2)$

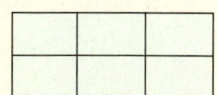

(b) $(x + 5)(x^2 + 3x + 1)$

EXAMPLE 2 **Multiplying Two Polynomials**

Multiply $(m^2 + 5)(4m^3 - 2m^2 + 4m)$.

Multiply each term of the second polynomial by each term of the first.

$(\mathbf{m^2} + \mathbf{5})(4m^3 - 2m^2 + 4m)$

$= \mathbf{m^2}(4m^3) + \mathbf{m^2}(-2m^2) + \mathbf{m^2}(4m) + \mathbf{5}(4m^3) + \mathbf{5}(-2m^2) + \mathbf{5}(4m)$

$= 4m^5 - 2m^4 + 4m^3 + 20m^3 - 10m^2 + 20m$

$= 4m^5 - 2m^4 + 24m^3 - 10m^2 + 20m$ $\quad$ Combine like terms.

◀ **Work Problem** **2** **at the Side.**

EXAMPLE 3 **Multiplying Polynomials Vertically**

Multiply $(x^3 + 2x^2 + 4x + 1)(3x + 5)$ vertically.

Write the polynomials as follows.

$$x^3 + 2x^2 + 4x + 1$$
$$\underline{ 3x + 5}$$

Begin by multiplying each of the terms in the top row by 5.

$$\mathbf{x^3 + 2x^2 + 4x + 1}$$
$$\underline{ \mathbf{3x + 5}}$$
$$5x^3 + 10x^2 + 20x + 5 \quad 5(x^3 + 2x^2 + 4x + 1)$$

Notice how this process is similar to multiplication of whole numbers. Now multiply each term in the top row by $3x$. Then add like terms.

$$\mathbf{x^3 + 2x^2 + 4x + 1}$$

Place *like* terms in columns so they can be added.

$$\underline{ \mathbf{3x + 5}}$$
$$5x^3 + 10x^2 + 20x + 5$$
$$\underline{3x^4 + 6x^3 + 12x^2 + 3x} \quad\quad 3x(x^3 + 2x^2 + 4x + 1)$$
$$3x^4 + 11x^3 + 22x^2 + 23x + 5 \quad \text{Add.}$$

The product is $3x^4 + 11x^3 + 22x^2 + 23x + 5$.

◀ **Work Problem** **3** **at the Side.**

We can use a rectangle to model polynomial multiplication. For example, to find the product

$$(\mathbf{2x + 1})(\mathbf{3x + 2}),$$

label a rectangle with each term as shown below on the left. Then put the product of each pair of monomials in the appropriate box as shown on the right.

	$3x$	2
$2x$	$6x^2$	$4x$
1	$3x$	2

The product of the binomials is the sum of these four monomial products.

$$(2x + 1)(3x + 2)$$
$$= 6x^2 + 4x + 3x + 2$$
$$= 6x^2 + 7x + 2$$

◀ **Work Problem** **4** **at the Side.**

ANSWERS

2. **(a)** $m^3 + m^2 - 5m + 3$
$\quad$ **(b)** $18p^4 + 6p^3 - 42p^2 - 10p + 20$
3. $3x^3 + 16x^2 + 11x - 20$
4. **(a)** $4x^2 + 11x + 6$
$\quad$ **(b)** $x^3 + 8x^2 + 16x + 5$

OBJECTIVE **3** **Multiply binomials by the FOIL method.** In algebra, many of the polynomials to be multiplied are both binomials (with just two terms). For these products, the **FOIL method** reduces the rectangle method to a systematic approach without the rectangle. To develop the FOIL method, we use the distributive property to find $(x + 3)(x + 5)$.

$$(x + 3)(x + 5)$$

$= (x + 3)x + (x + 3)5$	Distributive property
$= x(x) + 3(x) + x(5) + 3(5)$	Distributive property again
$= x^2 + 3x + 5x + 15$	Multiply.
$= x^2 + 8x + 15$	Combine like terms.

Here is where the letters of the word FOIL originate.

$(x + 3)(x + 5)$ Multiply the **First terms**: $x(x)$.   **F**

$(x + 3)(x + 5)$ Multiply the **Outer terms**: $x(5)$.   **O**
This is the **outer product.**

$(x + 3)(x + 5)$ Multiply the **Inner terms**: $3(x)$.   **I**
This is the **inner product.**

$(x + 3)(x + 5)$ Multiply the **Last terms**: $3(5)$.   **L**

The outer product, $5x$, and the inner product, $3x$, should be added mentally so that the three terms of the answer can be written without extra steps.

$$(x + 3)(x + 5)$$
$$= x^2 + 8x + 15$$

A summary of the steps in the FOIL method follows.

> **Multiplying Binomials by the FOIL Method**
>
> **Step 1** Multiply the two **F**irst terms of the binomials to get the first term of the answer.
>
> **Step 2** Find the **O**uter product and the **I**nner product and add them (when possible) to get the middle term of the answer.
>
> **Step 3** Multiply the two **L**ast terms of the binomials to get the last term of the answer.
>
>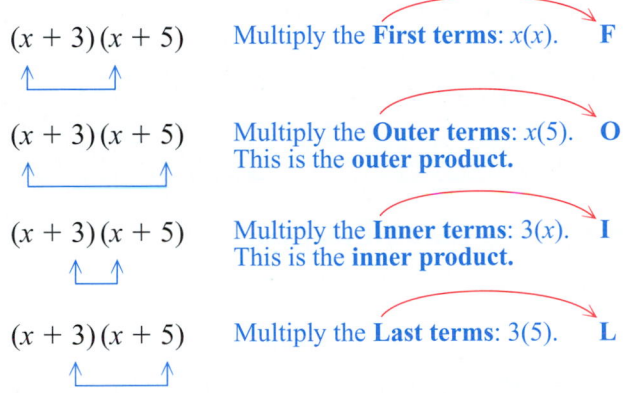
>
> $\mathbf{F} = x^2 \qquad \mathbf{L} = 15$
>
> $(x + 3)(x + 5)$
>
> $\mathbf{I} = 3x$
> $\mathbf{O} = 5x$
> $\overline{8x}$ \quad Add.

Work Problem **5** *at the Side.* ▶

5 For the product

$$(2p - 5)(3p + 7),$$

find the following.

(a) Product of first terms

(b) Outer product

(c) Inner product

(d) Product of last terms

(e) Complete product in simplified form

6 Use the FOIL method to find each product.

(a) $(m + 4)(m - 3)$

(b) $(y + 7)(y + 2)$

(c) $(r - 8)(r - 5)$

7 Find the product.

$$(4x - 3)(2y + 5)$$

8 Find each product.

(a) $(6m + 5)(m - 4)$

(b) $(3r + 2t)(3r + 4t)$

(c) $y^2(8y + 3)(2y + 1)$

EXAMPLE 4 Using the FOIL Method

Use the FOIL method to find the product $(x + 8)(x - 6)$.

Step 1 **F** Multiply the **first** terms: $x(x) = x^2$.

Step 2 **O** Find the **outer** product: $x(-6) = -6x$.

I Find the **inner** product: $8(x) = 8x$.

Add the outer and inner products mentally: $-6x + 8x = \mathbf{2x}$.

Step 3 **L** Multiply the **last** terms: $8(-6) = \mathbf{-48}$.

The product $(x + 8)(x - 6)$ is $x^2 + \mathbf{2x} - \mathbf{48}$, the sum of the terms found in Steps 1–3. As a shortcut, this product can be found as follows.

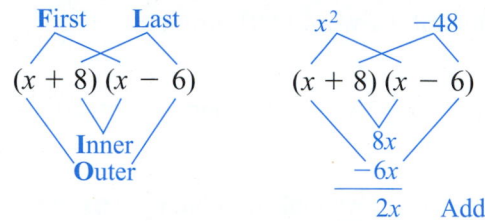

◀ Work Problem **6** at the Side.

EXAMPLE 5 Using the FOIL Method

Multiply $(9x - 2)(3y + 1)$.

First	$(\mathbf{9x} - 2)(\mathbf{3y} + 1)$	$\mathbf{27xy}$
Outer	$(\mathbf{9x} - 2)(3y + \mathbf{1})$	$\mathbf{9x}$
Inner	$(9x - \mathbf{2})(\mathbf{3y} + 1)$	$\mathbf{-6y}$
Last	$(9x - \mathbf{2})(3y + \mathbf{1})$	$\mathbf{-2}$

These unlike terms cannot be added.

F O I L

The product $(9x - 2)(3y + 1)$ is $27xy + 9x - 6y - 2$.

◀ Work Problem **7** at the Side.

EXAMPLE 6 Using the FOIL Method

Find each product.

(a) $(2k + 5y)(k + 3y)$

$$\begin{align}
&\quad\text{F}\qquad\text{O}\qquad\text{I}\qquad\text{L}\\
&= 2k(k) + 2k(3y) + 5y(k) + 5y(3y)\\
&= 2k^2 + 6ky + 5ky + 15y^2\\
&= 2k^2 + 11ky + 15y^2
\end{align}$$

(b) $(7p + 2q)(3p - q)$

$$\begin{align}
&= 21p^2 - 7pq + 6pq - 2q^2\\
&= 21p^2 - pq - 2q^2
\end{align}$$

(c) $2x^2(x - 3)(3x + 4)$

$$\begin{align}
&= 2x^2(3x^2 - 5x - 12)\\
&= 6x^4 - 10x^3 - 24x^2
\end{align}$$

◀ Work Problem **8** at the Side.

Note

In Example 6(c), we could have multiplied $2x^2$ and $x - 3$ first.

$$2x^2(x - 3)(3x + 4)$$
$$= (2x^3 - 6x^2)(3x + 4)$$
$$= 6x^4 - 10x^3 - 24x^2 \qquad \text{Same answer}$$

ANSWERS

6. (a) $m^2 + m - 12$ (b) $y^2 + 9y + 14$
   (c) $r^2 - 13r + 40$
7. $8xy + 20x - 6y - 15$
8. (a) $6m^2 - 19m - 20$
   (b) $9r^2 + 18rt + 8t^2$
   (c) $16y^4 + 14y^3 + 3y^2$

12.3 ▶▶▶ **Exercises**

  FOR EXTRA HELP    PRACTICE   WATCH   DOWNLOAD   READ REVIEW

Find each product using the rectangle method shown in the text.

1. $(x + 3)(x + 4)$ 　　　**2.** $(x + 5)(x + 2)$ 　　　**3.** $(2x + 1)(x^2 + 3x + 2)$ 　　**4.** $(x + 4)(3x^2 + 2x + 1)$

5. In multiplying a monomial by a polynomial, such as in $4x(3x^2 + 7x^3) = 4x(3x^2) + 4x(7x^3)$, the first property that is used is the _____ property.

6. Match each product in parts (a)–(d) with the correct polynomial in choices A–D.

　(a) $(x - 5)(x + 3)$ 　　**(b)** $(x + 5)(x + 3)$ 　　**(c)** $(x - 5)(x - 3)$ 　　**(d)** $(x + 5)(x - 3)$

　A. $x^2 + 8x + 15$ 　　**B.** $x^2 - 8x + 15$ 　　**C.** $x^2 - 2x - 15$ 　　**D.** $x^2 + 2x - 15$

Find each product. See Example 1.

7. $-2m(3m + 2)$ 　　　　　**8.** $-5p(6 + 3p)$ 　　　　　**9.** $\dfrac{3}{4}p(8 - 6p + 12p^3)$

10. $\dfrac{4}{3}x(3 + 2x + 5x^3)$ 　　　**11.** $2y^5(3 + 2y + 5y^4)$ 　　　**12.** $2m^4(3m^2 + 5m + 6)$

Find each product. See Examples 2 and 3.

13. $(6x + 1)(2x^2 + 4x + 1)$ 　　　　　　　　**14.** $(9y - 2)(8y^2 - 6y + 1)$

15. $(2r - 1)(3r^2 + 4r - 4)$ 　　　　　　　　**16.** $(9a + 2)(9a^2 + a + 1)$

17. $(4m + 3)(5m^3 - 4m^2 + m - 5)$ 　　　　　**18.** $(y + 4)(3y^3 - 2y^2 + y + 3)$

19. $(5x^2 + 2x + 1)(x^2 - 3x + 5)$ 　　　　　**20.** $(2m^2 + m - 3)(m^2 - 4m + 5)$

Find each product. See Examples 4–6.

21. $(m + 7)(m + 5)$ 　　**22.** $(x + 4)(x + 7)$ 　　**23.** $(n - 2)(n + 3)$ 　　**24.** $(r - 6)(r + 8)$

25. $(4r + 1)(2r - 3)$ 　　**26.** $(5x + 2)(2x - 7)$ 　　**27.** $(3x + 2)(3x - 2)$ 　　**28.** $(7x + 3)(7x - 3)$

29. $(3q + 1)(3q + 1)$

30. $(4w + 7)(4w + 7)$

31. $(5x + 7)(3y - 8)$

32. $(4x + 3)(2y - 1)$

33. $(3t + 4s)(2t + 5s)$

34. $(8v + 5w)(2v + 3w)$

35. $(-0.3t + 0.4)(t + 0.6)$

36. $(-0.5x + 0.9)(x - 0.2)$

37. $\left(x - \dfrac{2}{3}\right)\left(x + \dfrac{1}{4}\right)$

38. $\left(-\dfrac{8}{3} + 3k\right)\left(-\dfrac{2}{3} - k\right)$

39. $\left(-\dfrac{5}{4} + 2r\right)\left(-\dfrac{3}{4} - r\right)$

40. $2m^3(4m - 1)(2m + 3)$

41. $x(2x - 5)(x + 3)$

42. $5t^4(t + 3)(3t - 1)$

43. $3y^3(2y + 3)(y - 5)$

Relating Concepts (Exercises 44–48) For Individual or Group Work

Work Exercises 44–48 in order. (All units are in yards.)

44. Find a polynomial that represents the area of the rectangle.

$3x + 6$

10

45. Suppose you know that the area of the rectangle is 600 yd². Use this information and the polynomial from Exercise 44 to write an equation in x, and solve it.

46. (a) What are the dimensions of the rectangle?

(b) Use the result of part (a) to find the perimeter of the lawn.

47. Suppose the rectangle represents a lawn and it costs \$3.50 per square yard to lay sod on the lawn. How much will it cost to sod the entire lawn?

48. Again, suppose the rectangle represents a lawn and it costs \$9.00 per yard to fence the lawn. How much will it cost to fence the lawn?

---

49. Perform the following multiplications: $(x + 4)(x - 4)$; $(y + 2)(y - 2)$; $(r + 7)(r - 7)$. Observe your answers, and explain the pattern that can be found in the answers.

50. Repeat Exercise 49 for the following: $(x + 4)(x + 4)$; $(y - 2)(y - 2)$; $(r + 7)(r + 7)$.

12.4 ▶▶▶ Special Products

In this section, we develop shortcuts to find certain binomial products.

OBJECTIVE **1** **Square binomials.** The square of a binomial can be found quickly by using the method shown in Example 1.

OBJECTIVES

1 Square binomials.

2 Find the product of the sum and difference of two terms.

3 Find greater powers of binomials.

EXAMPLE 1 Squaring a Binomial

Find $(m + 3)^2$.

> $(m + 3)^2$ means $(m + 3)(m + 3)$.

$$(m + 3)(m + 3)$$

$$= m^2 + 3m + 3m + 9 \quad \text{FOIL}$$

$$= m^2 + 6m + 9 \quad \text{Combine like terms.}$$

This result has the squares of the first and the last terms of the binomial:

$$\boldsymbol{m^2 = m^2} \quad \text{and} \quad \boldsymbol{3^2 = 9.}$$

The middle term, 6m, is twice the product of the two terms of the binomial, since the outer and inner products are $m(3)$ and $3(m)$, and

$$m(3) + 3(m) = 2(\boldsymbol{m})(\boldsymbol{3}) = 6m.$$

Work Problem **1** *at the Side.* ▶

Example 1 suggests the following rules.

Square of a Binomial

The square of a binomial is a trinomial consisting of the square of the first term, plus twice the product of the two terms, plus the square of the last term of the binomial. For a and b,

$$(a + b)^2 = a^2 + 2ab + b^2.$$

Also, $\qquad\qquad (a - b)^2 = a^2 - 2ab + b^2.$

EXAMPLE 2 Squaring Binomials

Square each binomial.

$$(\boldsymbol{a} - \boldsymbol{b})^2 = \boldsymbol{a}^2 - 2 \cdot \boldsymbol{a} \cdot \boldsymbol{b} + \boldsymbol{b}^2$$

(a) $(\boldsymbol{5z} - \boldsymbol{1})^2 = (\boldsymbol{5z})^2 - 2(\boldsymbol{5z})(\boldsymbol{1}) + (\boldsymbol{1})^2$

$$= 25z^2 - 10z + 1 \qquad (5z)^2 = 5^2 z^2 = 25z^2$$

(b) $(\boldsymbol{3b} + \boldsymbol{5r})^2$

$$= (\boldsymbol{3b})^2 + 2(\boldsymbol{3b})(\boldsymbol{5r}) + (\boldsymbol{5r})^2$$

$$= 9b^2 + 30br + 25r^2$$

(c) $(2a - 9x)^2$

$$= (2a)^2 - 2(2a)(9x) + (9x)^2$$

$$= 4a^2 - 36ax + 81x^2$$

Continued on Next Page

1 Consider the binomial $x + 4$.

(a) What is the first term of the binomial? Square it.

(b) What is the last term of the binomial? Square it.

(c) Find twice the product of the two terms of the binomial.

(d) Find $(x + 4)^2$.

2 Square each binomial.

(a) $(t - 6)^2$

(b) $(2m - p)^2$

(c) $(4p + 3q)^2$

(d) $(5r - 6s)^2$

(e) $\left(3k - \dfrac{1}{2}\right)^2$

(f) $x(2x + 7)^2$

(d) $\left(4m + \dfrac{1}{2}\right)^2$

$$= (4m)^2 + 2(4m)\left(\dfrac{1}{2}\right) + \left(\dfrac{1}{2}\right)^2 \qquad (a+b)^2 = a^2 + 2ab + b^2$$

$$= 16m^2 + 4m + \dfrac{1}{4}$$

(e) $x(4x - 3)^2$ *Remember the middle term.*

$$= x(16x^2 - 24x + 9) \qquad \text{Square the binomial.}$$

$$= 16x^3 - 24x^2 + 9x \qquad \text{Distributive property}$$

Notice that in the square of a sum, all of the terms are positive, as in Examples 2(b) and (d). *In the square of a difference, the middle term is negative,* as in Examples 2(a) and (c).

> **CAUTION**
> A common error when squaring a binomial is to forget the middle term of the product. In general,
> $$(a + b)^2 = a^2 + \mathbf{2ab} + b^2, \quad \textbf{not} \quad a^2 + b^2,$$
> and
> $$(a - b)^2 = a^2 - \mathbf{2ab} + b^2, \quad \textbf{not} \quad a^2 - b^2.$$

◀ *Work Problem* **2** *at the Side.*

OBJECTIVE 2 Find the product of the sum and difference of two terms. In binomial products of the form $(a + b)(a - b)$, one binomial is the sum of two terms, and the other is the difference of the *same* two terms. For example, the product of $x + 2$ and $x - 2$ is

$$(x + 2)(x - 2)$$
$$= x^2 - 2x + 2x - 4 \qquad \text{FOIL}$$
$$= x^2 - 4. \qquad \text{Combine like terms.}$$

As the above example suggests, the product of $a + b$ and $a - b$ is the difference of two squares.

> **Product of the Sum and Difference of Two Terms**
> $$(a + b)(a - b) = a^2 - b^2$$

> **Note**
> The expressions $a + b$ and $a - b$, the sum and difference of the *same* two terms, are called **conjugates.** In the example above, $x + 2$ and $x - 2$ are conjugates.

EXAMPLE 3   **Finding the Product of the Sum and Difference of Two Terms**

Find each product.

(a) $(x + 4)(x - 4)$

Use the rule for the product of the sum and difference of two terms.

$$(x + 4)(x - 4)$$
$$= x^2 - 4^2$$
$$= x^2 - 16$$

(b) $\left(\dfrac{2}{3} - w\right)\left(\dfrac{2}{3} + w\right)$

$$= \left(\dfrac{2}{3} + w\right)\left(\dfrac{2}{3} - w\right) \qquad \text{Commutative property}$$

$$= \left(\dfrac{2}{3}\right)^2 - w^2 \qquad \text{Multiply.}$$

$$= \dfrac{4}{9} - w^2 \qquad \text{Square } \tfrac{2}{3}.$$

(c) $x(x + 2)(x - 2)$

$$= x(x^2 - 4) \qquad \begin{array}{l}\text{Find the product of the sum}\\ \text{and difference of two terms.}\end{array}$$

$$= x^3 - 4x \qquad \text{Distributive property}$$

EXAMPLE 4   **Finding the Product of the Sum and Difference of Two Terms**

Find each product.

$$(a \ + \ b) \ (a \ - \ b)$$

(a) $(5m + 3)(5m - 3)$

Use the rule for the product of the sum and difference of two terms.

$$(5m + 3)(5m - 3)$$
$$= (5m)^2 - 3^2 \qquad (a + b)(a - b) = a^2 - b^2$$
$$= 25m^2 - 9 \qquad \text{Apply the exponents.}$$

(b) $(4x + y)(4x - y)$

$$= (4x)^2 - y^2$$
$$= 16x^2 - y^2$$

(c) $\left(z - \dfrac{1}{4}\right)\left(z + \dfrac{1}{4}\right)$

$$= z^2 - \left(\dfrac{1}{4}\right)^2$$

$$= z^2 - \dfrac{1}{16}$$

(d) $2p(p^2 + 3)(p^2 - 3)$

$$= 2p(p^4 - 9) \qquad \text{Multiply the conjugates.}$$

$$= 2p^5 - 18p \qquad \text{Distributive property}$$

Work Problem **3** *at the Side.* ▶

---

3 Find each product.

(a) $(y + 3)(y - 3)$

(b) $(10m + 7)(10m - 7)$

(c) $(7p + 2q)(7p - 2q)$

(d) $\left(3r - \dfrac{1}{2}\right)\left(3r + \dfrac{1}{2}\right)$

(e) $3x(x^3 - 4)(x^3 + 4)$

ANSWERS

3. **(a)** $y^2 - 9$ **(b)** $100m^2 - 49$

   (c) $49p^2 - 4q^2$ **(d)** $9r^2 - \dfrac{1}{4}$

   (e) $3x^7 - 48x$

The product rules of this section will be important in **Chapters 13** and **14** and should be *memorized*.

4 Find each product.

(a) $(m + 1)^3$

OBJECTIVE 3 Find greater powers of binomials. The methods used in the previous section and this section can be combined to find greater powers of binomials.

EXAMPLE 5 **Finding Greater Powers of Binomials**

Find each product.

(a) $(x + 5)^3$

$$= (x + 5)^2(x + 5) \qquad a^3 = a^2 \cdot a$$
$$= (x^2 + 10x + 25)(x + 5) \qquad \text{Square the binomial.}$$
$$= x^3 + 10x^2 + 25x + 5x^2 + 50x + 125 \qquad \text{Multiply polynomials.}$$
$$= x^3 + 15x^2 + 75x + 125 \qquad \text{Combine like terms.}$$

(b) $(3k - 2)^4$

(b) $(2y - 3)^4$

$$= (2y - 3)^2(2y - 3)^2 \qquad a^4 = a^2 \cdot a^2$$
$$= (4y^2 - 12y + 9)(4y^2 - 12y + 9) \qquad \text{Square each binomial.}$$
$$= 16y^4 - 48y^3 + 36y^2 - 48y^3 + 144y^2 \qquad \text{Multiply polynomials.}$$
$$\quad - 108y + 36y^2 - 108y + 81$$
$$= 16y^4 - 96y^3 + 216y^2 - 216y + 81 \qquad \text{Combine like terms.}$$

(c) $-2r\,(r + 2)^3$

$$= -2r\,(r + 2)(r + 2)^2$$
$$= -2r\,(r + 2)(r^2 + 4r + 4)$$
$$= -2r\,(r^3 + 4r^2 + 4r + 2r^2 + 8r + 8)$$
$$= -2r\,(r^3 + 6r^2 + 12r + 8)$$
$$= -2r^4 - 12r^3 - 24r^2 - 16r$$

(c) $-3x\,(x - 4)^3$

◄ *Work Problem* **4** *at the Side.*

12.4 ▶▶▶ Exercises

1. Consider the square $(2x + 3)^2$.

   (a) What is the square of the first term, $(2x)^2$?

   (b) What is twice the product of the two terms, $2(2x)(3)$?

   (c) What is the square of the last term, 3^2?

   (d) Write the final product, which is a trinomial, using your results from parts (a)–(c).

2. Repeat Exercise 1 for the square $(3x - 2)^2$.

Find each square. See Examples 1 and 2.

3. $(p + 2)^2$ **4.** $(r + 5)^2$ **5.** $(z - 5)^2$ **6.** $(x - 3)^2$

7. $(4x - 3)^2$ **8.** $(5y + 2)^2$ **9.** $(2p + 5q)^2$ **10.** $(8a - 3b)^2$

11. $(0.8t + 0.7s)^2$ **12.** $(0.7z - 0.3w)^2$ **13.** $\left(5x + \dfrac{2}{5}y\right)^2$ **14.** $\left(6m - \dfrac{4}{5}n\right)^2$

15. $t(3t - 1)^2$ **16.** $x(2x + 5)^2$ **17.** $-(4r - 2)^2$ **18.** $-(3y - 8)^2$

19. Consider the product $(7x + 3y)(7x - 3y)$.

   (a) What is the product of the first terms, $7x(7x)$?

   (b) Multiply the outer terms, $7x(-3y)$. Then multiply the inner terms, $3y(7x)$. Add the results. What is this sum?

   (c) What is the product of the last terms, $3y(-3y)$?

   (d) Write the complete product using your answers in parts (a) and (c). Why is the sum found in part (b) omitted here?

20. Repeat Exercise 19 for the product $(5x + 7y)(5x - 7y)$.

Find each product. See Examples 3 and 4.

21. $(q + 2)(q - 2)$ **22.** $(x + 8)(x - 8)$ **23.** $(2w + 5)(2w - 5)$ **24.** $(3z + 8)(3z - 8)$

25. $(10x + 3y)(10x - 3y)$ **26.** $(13r + 2z)(13r - 2z)$ **27.** $(2x^2 - 5)(2x^2 + 5)$ **28.** $(9y^2 - 2)(9y^2 + 2)$

29. $\left(7x + \dfrac{3}{7}\right)\left(7x - \dfrac{3}{7}\right)$ **30.** $\left(9y + \dfrac{2}{3}\right)\left(9y - \dfrac{2}{3}\right)$ **31.** $p(3p + 7)(3p - 7)$ **32.** $q(5q - 1)(5q + 1)$

Relating Concepts (Exercises 33–42) For Individual or Group Work

*Special products can be illustrated by using areas of rectangles. Use the figure and **work Exercises 33–38 in order,** to justify the special product $(a + b)^2 = a^2 + 2ab + b^2$.*

33. Express the area of the large square as the square of a binomial.

34. Give the monomial that represents the area of the red square.

35. Give the monomial that represents the sum of the areas of the blue rectangles.

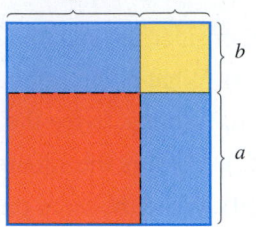

36. Give the monomial that represents the area of the yellow square.

37. What is the sum of the monomials you obtained in Exercises 34–36?

38. Explain why the binomial square you found in Exercise 33 must equal the polynomial you found in Exercise 37.

*To understand how the special product $(a + b)^2 = a^2 + 2ab + b^2$ can be applied to a purely numerical problem, **work Exercises 39–42 in order.***

39. Evaluate 35^2 using either traditional paper-and-pencil methods or a calculator.

40. The number 35 can be written as $30 + 5$. Therefore, $35^2 = (30 + 5)^2$. Use the special product for squaring a binomial with $a = 30$ and $b = 5$ to write an expression for $(30 + 5)^2$. Do not simplify at this time.

41. Use the order of operations to simplify the expression you found in Exercise 40.

42. How do the answers in Exercises 39 and 41 compare?

Find each product. See Example 5.

43. $(m - 5)^3$

44. $(p + 3)^3$

45. $(y + 2)^3$

46. $(x - 7)^3$

47. $(2a + 1)^3$

48. $(3m - 1)^3$

49. $(3r - 2t)^4$

50. $(2z + 5y)^4$

51. $3x^2(x - 3)^3$

52. $4p^3(p + 4)^3$

53. $-8x^2y(x + y)^4$

In Exercises 54 and 55, refer to the figure shown here.

54. Find a polynomial that represents the volume of the cube.

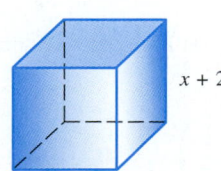

$x + 2$

55. If the value of x is 6, what is the volume of the cube?

12.5 ▷▷▷ Integer Exponents and the Quotient Rule

In all our earlier work, exponents were positive integers. Now we want to develop meaning for exponents that are *not* positive integers.

Consider the following list.

$$2^4 = 16$$
$$2^3 = 8$$
$$2^2 = 4$$

Do you see the pattern in the values? Each time we reduce the exponent by 1, the value is divided by 2 (the base). Using this pattern, we can continue the list to smaller and smaller integer exponents.

$$2^1 = 2$$
$$2^0 = 1$$
$$2^{-1} = \frac{1}{2}$$

Work Problem ▢**1** *at the Side.* ▶

From the preceding list and the answers to Problem 1 at the side, it appears that we should define 2^0 as 1 and negative exponents as reciprocals.

OBJECTIVE ▮**1**▮ **Use 0 as an exponent.**   We want the definitions of 0 and negative exponents to satisfy the rules for exponents from **Section 12.2.** For example, if $6^0 = 1$,

$$6^0 \cdot 6^2 = \mathbf{1} \cdot 6^2 = 6^2 \quad \text{and} \quad 6^0 \cdot 6^2 = 6^{0+2} = 6^2,$$

so the product rule is satisfied. Check that the power rules are also valid for a 0 exponent. Thus, we define a 0 exponent as follows.

> **Zero Exponent**
>
> For any nonzero real number a, $a^0 = 1$.
>
> *Example:* $17^0 = 1$

EXAMPLE 1   **Using Zero Exponents**

Evaluate.

(a) $60^0 = 1$ **(b)** $(-60)^0 = 1$

(c) $-60^0 = -(1) = -1$ **(d)** $y^0 = 1, \quad y \neq 0$

(e) $6y^0 = 6(1) = 6, \quad y \neq 0$ **(f)** $(6y)^0 = 1, \quad y \neq 0$

> **CAUTION**
> Look again at Examples 1(b) and (c). In $(-60)^0$, the base is -60 and the exponent is 0. Any nonzero base raised to the exponent 0 is 1. In -60^0, the base is 60. Then $60^0 = 1$, and $-60^0 = -1$.

Work Problem ▢**2** *at the Side.* ▶

OBJECTIVES

▮**1**▮ Use 0 as an exponent.

▮**2**▮ Use negative numbers as exponents.

▮**3**▮ Use the quotient rule for exponents.

▮**4**▮ Use combinations of rules.

▢**1** Continue the list of exponentials using $-2, -3,$ and -4 as exponents.

$2^{-2} =$ _____

$2^{-3} =$ _____

$2^{-4} =$ _____

▢**2** Evaluate.

(a) 28^0

(b) $(-16)^0$

(c) -7^0

(d) $m^0, \quad m \neq 0$

(e) $-p^0, \quad p \neq 0$

ANSWERS

1. $2^{-2} = \frac{1}{4}; 2^{-3} = \frac{1}{8}; 2^{-4} = \frac{1}{16}$

2. **(a)** 1 **(b)** 1 **(c)** -1 **(d)** 1 **(e)** -1

OBJECTIVE 2 Use negative numbers as exponents. From the lists at the beginning of this section and margin Problem 1, since $2^{-2} = \frac{1}{4}$ and $2^{-3} = \frac{1}{8}$, we can deduce that 2^{-n} should equal $\frac{1}{2^n}$. Is the product rule valid in such cases? For example, if we multiply 6^{-2} by 6^2, we get

$$6^{-2} \cdot 6^2 = 6^{-2+2} = 6^0 = 1.$$

The expression 6^{-2} behaves as if it were the reciprocal of 6^2, because their product is 1. The reciprocal of 6^2 may be written $\frac{1}{6^2}$, leading us to define 6^{-2} as $\frac{1}{6^2}$. This is a particular case of the definition of negative exponents.

Negative Exponents

For any nonzero real number a and any integer n, $\quad a^{-n} = \dfrac{1}{a^n}.$

Example: $\quad 3^{-2} = \dfrac{1}{3^2}$

By definition, a^{-n} and a^n are reciprocals, since

$$a^n \cdot a^{-n} = a^n \cdot \frac{1}{a^n} = 1.$$

Since $1^n = 1$, the definition of a^{-n} can also be written

$$a^{-n} = \frac{1}{a^n} = \frac{1^n}{a^n} = \left(\frac{1}{a}\right)^n.$$

For example, $\quad 6^{-3} = \left(\dfrac{1}{6}\right)^3 \quad$ and $\quad \left(\dfrac{1}{3}\right)^{-2} = 3^2.$

EXAMPLE 2 Using Negative Exponents

Simplify by writing with positive exponents. Assume that all variables represent nonzero real numbers.

(a) $3^{-2} = \dfrac{1}{3^2} = \dfrac{1}{9} \qquad a^{-n} = \frac{1}{a^n}$ **(b)** $5^{-3} = \dfrac{1}{5^3} = \dfrac{1}{125}$

(c) $\left(\dfrac{1}{2}\right)^{-3} = 2^3 = 8 \qquad \frac{1}{2}$ and 2 are reciprocals.

Notice that we can change the base to its reciprocal if we also change the sign of the exponent.

(d) $\left(\dfrac{2}{5}\right)^{-4} = \left(\dfrac{5}{2}\right)^4 = \dfrac{5^4}{2^4} = \dfrac{625}{16}$ **(e)** $\left(\dfrac{4}{3}\right)^{-5} = \left(\dfrac{3}{4}\right)^5 = \dfrac{3^5}{4^5} = \dfrac{243}{1024}$

$\frac{2}{5}$ and $\frac{5}{2}$ are reciprocals.

(f) $4^{-1} - 2^{-1}$

$$= \frac{1}{4} - \frac{1}{2} = \frac{1}{4} - \frac{2}{4} = -\frac{1}{4} \qquad \text{Apply the exponents first; then subtract.}$$

Continued on Next Page

(g) $p^{-2} = \dfrac{1}{p^2}$

(h) $\dfrac{1}{x^{-4}} = \dfrac{1^{-4}}{x^{-4}}$ $1^n = 1$, for any integer n

$= \left(\dfrac{1}{x}\right)^{-4}$ Power rule (c)

$= x^4$ $\frac{1}{x}$ and x are reciprocals.

Notice that, in general, $\dfrac{1}{a^{-n}} = a^n$.

(i) $x^3 y^{-4} = \dfrac{x^3}{1} \cdot \dfrac{1}{y^4} = \dfrac{x^3}{y^4}$

CAUTION

A negative exponent does not indicate a negative number. Negative exponents lead to reciprocals.

Expression	Example	
a^{-n}	$3^{-2} = \dfrac{1}{3^2} = \dfrac{1}{9}$	Not negative
$-a^{-n}$	$-3^{-2} = -\dfrac{1}{3^2} = -\dfrac{1}{9}$	Negative

Work Problem **3** *at the Side.* ▶

Consider the following:

$\dfrac{2^{-3}}{3^{-4}} = \dfrac{\frac{1}{2^3}}{\frac{1}{3^4}} = \dfrac{1}{2^3} \div \dfrac{1}{3^4} = \dfrac{1}{2^3} \cdot \dfrac{3^4}{1} = \dfrac{3^4}{2^3}.$ To divide by a fraction, multiply by its reciprocal.

Therefore,

$$\dfrac{2^{-3}}{3^{-4}} = \dfrac{3^4}{2^3}.$$

Changing from Negative to Positive Exponents

For any nonzero numbers a and b, and any integers m and n,

$$\dfrac{a^{-m}}{b^{-n}} = \dfrac{b^n}{a^m} \quad \text{and} \quad \left(\dfrac{a}{b}\right)^{-m} = \left(\dfrac{b}{a}\right)^m.$$

Examples: $\dfrac{3^{-5}}{2^{-4}} = \dfrac{2^4}{3^5}$ and $\left(\dfrac{4}{5}\right)^{-3} = \left(\dfrac{5}{4}\right)^3$

---

3 Simplify by writing with positive exponents. Assume that all variables represent nonzero real numbers.

(a) 4^{-3}

(b) 6^{-2}

(c) $\left(\dfrac{1}{4}\right)^{-2}$

(d) $\left(\dfrac{2}{3}\right)^{-2}$

(e) $2^{-1} + 5^{-1}$

(f) m^{-5}

(g) $\dfrac{1}{z^{-4}}$

(h) $p^2 q^{-5}$

ANSWERS

3. **(a)** $\dfrac{1}{4^3} = \dfrac{1}{64}$ **(b)** $\dfrac{1}{6^2} = \dfrac{1}{36}$ **(c)** $4^2 = 16$

(d) $\left(\dfrac{3}{2}\right)^2 = \dfrac{9}{4}$ **(e)** $\dfrac{1}{2} + \dfrac{1}{5} = \dfrac{7}{10}$

(f) $\dfrac{1}{m^5}$ **(g)** z^4 **(h)** $\dfrac{p^2}{q^5}$

4 Simplify. Assume that all variables represent nonzero real numbers.

(a) $\dfrac{7^{-1}}{5^{-4}}$

(b) $\dfrac{x^{-3}}{y^{-2}}$

(c) $\dfrac{4h^{-5}}{m^{-2}k}$

(d) $\left(\dfrac{3m}{p}\right)^{-2}$

EXAMPLE 3 **Changing from Negative to Positive Exponents**

Simplify. Assume that all variables represent nonzero real numbers.

(a) $\dfrac{4^{-2}}{5^{-3}} = \dfrac{5^3}{4^2} = \dfrac{125}{16}$

(b) $\dfrac{m^{-5}}{p^{-1}} = \dfrac{p^1}{m^5} = \dfrac{p}{m^5}$

(c) $\dfrac{a^{-2}b}{3d^{-3}} = \dfrac{bd^3}{3a^2}$ Notice that b in the numerator and the coefficient 3 in the denominator are not affected.

(d) $\left(\dfrac{x}{2y}\right)^{-4}$

$= \left(\dfrac{2y}{x}\right)^4$ Negative-to-positive rule

$= \dfrac{2^4 y^4}{x^4}$ Power rule (c)

$= \dfrac{16y^4}{x^4}$

◀ *Work Problem* **4** *at the Side.*

CAUTION

Be careful. We cannot use the rule $\dfrac{a^{-m}}{b^{-n}} = \dfrac{b^n}{a^m}$ to change negative exponents to positive exponents if the exponents occur in a *sum* or *difference* of terms. For example,

$\dfrac{5^{-2} + 3^{-1}}{7 - 2^{-3}}$ would be written with positive exponents as $\dfrac{\frac{1}{5^2} + \frac{1}{3}}{7 - \frac{1}{2^3}}$.

OBJECTIVE **3** **Use the quotient rule for exponents.**   Consider a quotient of two exponential expressions with the same base.

$$\dfrac{6^5}{6^3} = \dfrac{6 \cdot 6 \cdot 6 \cdot 6 \cdot 6}{6 \cdot 6 \cdot 6} = 6^2$$

Notice that the difference between the exponents, $5 - 3 = 2$, is the exponent in the quotient. Also,

$$\dfrac{6^2}{6^4} = \dfrac{6 \cdot 6}{6 \cdot 6 \cdot 6 \cdot 6} = \dfrac{1}{6^2} = 6^{-2}.$$

Here, $2 - 4 = -2$. These examples suggest the **quotient rule for exponents.**

Quotient Rule for Exponents

For any nonzero real number a and any integers m and n,

$$\dfrac{a^m}{a^n} = a^{m-n}.$$

(Keep the same base and subtract the exponents.)

Example: $\dfrac{5^8}{5^4} = 5^{8-4} = 5^4$

CAUTION
A common **error** is to write $\dfrac{5^8}{5^4} = 1^{8-4} = 1^4$. **_This is incorrect._** By the quotient rule, the quotient must have the *same base,* 5, so

$$\frac{5^8}{5^4} = 5^{8-4} = 5^4.$$

We can confirm this by using the definition of exponents to write out the factors:

$$\frac{5^8}{5^4} = \frac{5 \cdot 5 \cdot 5 \cdot 5 \cdot 5 \cdot 5 \cdot 5 \cdot 5}{5 \cdot 5 \cdot 5 \cdot 5} = 5^4.$$

EXAMPLE 4 **Using the Quotient Rule**

Simplify. Assume that all variables represent nonzero real numbers.

(a) $\dfrac{5^8}{5^6} = 5^{8-6} = 5^2 = 25$

> Keep the same base.

(b) $\dfrac{4^2}{4^9} = 4^{2-9} = 4^{-7} = \dfrac{1}{4^7}$

(c) $\dfrac{5^{-3}}{5^{-7}} = 5^{-3-(-7)} = 5^4 = 625$

> Be careful with signs.

(d) $\dfrac{q^5}{q^{-3}} = q^{5-(-3)} = q^8$

(e) $\dfrac{3^2 x^5}{3^4 x^3}$

$= \dfrac{3^2}{3^4} \cdot \dfrac{x^5}{x^3}$

$= 3^{2-4} \cdot x^{5-3}$

$= 3^{-2} x^2$

$= \dfrac{x^2}{3^2}$

$= \dfrac{x^2}{9}$

(f) $\dfrac{(m+n)^{-2}}{(m+n)^{-4}}$

$= (m+n)^{-2-(-4)}$

$= (m+n)^{-2+4}$

$= (m+n)^2, \quad m \neq -n$

The restriction $m \neq -n$ is necessary to prevent a denominator of 0 in the original expression. Division by 0 is undefined.

(g) $\dfrac{7x^{-3}y^2}{2^{-1}x^2 y^{-5}}$

$= \dfrac{7 \cdot 2^1 y^2 y^5}{x^2 x^3}$ Definition of negative exponent

$= \dfrac{14y^7}{x^5}$ Multiply; product rule

Work Problem **5** *at the Side.* ▶

The definitions and rules for exponents given in this section and **Section 12.2** are summarized on the next page.

5 Simplify. Assume that all variables represent nonzero real numbers.

(a) $\dfrac{5^{11}}{5^8}$

(b) $\dfrac{4^7}{4^{10}}$

(c) $\dfrac{6^{-5}}{6^{-2}}$

(d) $\dfrac{8^4 m^9}{8^5 m^{10}}$

(e) $\dfrac{3^{-1}(x+y)^{-3}}{2^{-2}(x+y)^{-4}}, \quad x \neq -y$

ANSWERS

5. **(a)** 125   **(b)** $\dfrac{1}{64}$ **(c)** $\dfrac{1}{216}$ **(d)** $\dfrac{1}{8m}$

   (e) $\dfrac{4}{3}(x+y)$

Definitions and Rules for Exponents

For any integers m and n: **Examples**

Product rule	$a^m \cdot a^n = a^{m+n}$	$7^4 \cdot 7^5 = 7^{4+5} = 7^9$
Zero exponent	$a^0 = 1 \quad (a \neq 0)$	$(-3)^0 = 1$
Negative exponent	$a^{-n} = \dfrac{1}{a^n} \quad (a \neq 0)$	$5^{-3} = \dfrac{1}{5^3}$
Quotient rule	$\dfrac{a^m}{a^n} = a^{m-n} \quad (a \neq 0)$	$\dfrac{2^2}{2^5} = 2^{2-5} = 2^{-3} = \dfrac{1}{2^3}$
Power rules (a)	$(a^m)^n = a^{mn}$	$(4^2)^3 = 4^{2 \cdot 3} = 4^6$
(b)	$(ab)^m = a^m b^m$	$(3k)^4 = 3^4 k^4$
(c)	$\left(\dfrac{a}{b}\right)^m = \dfrac{a^m}{b^m} \quad (b \neq 0)$	$\left(\dfrac{2}{3}\right)^2 = \dfrac{2^2}{3^2}$
Negative-to-positive rules	$\dfrac{a^{-m}}{b^{-n}} = \dfrac{b^n}{a^m} \quad (a, b \neq 0)$	$\dfrac{2^{-4}}{5^{-3}} = \dfrac{5^3}{2^4}$
	$\left(\dfrac{a}{b}\right)^{-m} = \left(\dfrac{b}{a}\right)^m.$	$\left(\dfrac{4}{7}\right)^{-2} = \left(\dfrac{7}{4}\right)^2$

OBJECTIVE 4 Use combinations of rules. We sometimes need to use more than one rule to simplify an expression.

EXAMPLE 5 **Using a Combination of Rules**

Simplify each expression. Assume that all variables represent nonzero real numbers.

(a) $\dfrac{(4^2)^3}{4^5}$

$= \dfrac{4^6}{4^5}$ Power rule (a)

$= 4^{6-5}$ Quotient rule

$= 4^1$

$= 4$

(b) $(2x)^3 (2x)^2$

$= (2x)^5$ Product rule

$= 2^5 x^5$ Power rule (b)

$= 32x^5$

(c) $\left(\dfrac{2x^3}{5}\right)^{-4}$

$= \left(\dfrac{5}{2x^3}\right)^4$ Negative-to-positive rule

$= \dfrac{5^4}{2^4 x^{12}}$ Power rules (a)–(c)

$= \dfrac{625}{16x^{12}}$

(d) $\left(\dfrac{3x^{-2}}{4^{-1}y^3}\right)^{-3}$

$= \dfrac{3^{-3} x^6}{4^3 y^{-9}}$ Power rules (a)–(c)

$= \dfrac{x^6 y^9}{4^3 \cdot 3^3}$ Negative-to-positive rule

$= \dfrac{x^6 y^9}{1728}$ $4^3 \cdot 3^3 = 64 \cdot 27 = 17$

Continued on Next Page

(e) $\dfrac{(4m)^{-3}}{(3m)^{-4}}$

$= \dfrac{4^{-3}m^{-3}}{3^{-4}m^{-4}}$ Power rule (b)

$= \dfrac{3^4 m^4}{4^3 m^3}$ Negative-to-positive rule

$= \dfrac{3^4 m^{4-3}}{4^3}$ Quotient rule

$= \dfrac{3^4 m}{4^3}$

$= \dfrac{81m}{64}$

Note

Since the steps can be done in several different orders, there are many equally correct ways to simplify expressions like those in Examples 5(c) through 5(e).

Work Problem ⑥ *at the Side.* ▶

⑥ Simplify each expression. Assume that all variables represent nonzero real numbers.

(a) $\dfrac{(3^4)^2}{3^3}$

(b) $(4x)^2 (4x)^4$

(c) $\dfrac{(6x)^{-1}}{(3x^2)^{-2}}$

(d) $\dfrac{3^9 \cdot (x^2 y)^{-2}}{3^3 \cdot x^{-4} y}$

Math in the Media

MORE POWER TO YOU, CAPTAIN KIRK

The original *Star Trek* series first aired during the 1966 to 1967 television season and started the phenomenon that continues today. There have been five different television series and 10 feature movies with the *Star Trek* theme.

Captain James T. Kirk, portrayed by William Shatner, led the Starship Enterprise during its first three seasons. During the first season, the February 2, 1967, episode "Court Martial" told the story of Kirk being put on trial. He was accused of negligence in the death of a crewmember, because the computer records of the ship contradicted Kirk's logs. As the trial begins, Kirk explains how the sounds on the ship can be recorded and magnified:

Kirk: *Gentlemen, this computer has an auditory sensor. It can, in effect, hear sounds. By installing a booster we can increase that capability on an order of one to the fourth power. The computer should be able to bring us every sound occurring on the ship.*

1. Read Captain Kirk's statement carefully. What error did he make?

2. What is the result if we raise the number 1 to any whole number power?

3. It is possible that Kirk meant "10 to the fourth power." Express 10^4 in expanded form.

4. The word **googol** was invented to express a very large power of 10. The search engine Google was named in honor of it. Look up the meaning of googol, and write it in exponential form.

5. Investigate the meaning of the word **googolplex.**

12.5 ▶▶▶ Exercises

Decide whether each expression is positive, negative, or 0.

1. $(-2)^{-3}$

2. $(-3)^{-2}$

3. -2^4

4. -3^6

5. $\left(\dfrac{1}{4}\right)^{-2}$

6. $\left(\dfrac{1}{5}\right)^{-2}$

7. $1 - 5^0$

8. $1 - 7^0$

Decide whether each expression is equal to either 0, 1, or −1. See Example 1.

9. 9^0

10. 5^0

11. $(-4)^0$

12. $(-10)^0$

13. -9^0

14. -5^0

15. $(-2)^0 - 2^0$

16. $(-8)^0 - 8^0$

17. $\dfrac{0^{10}}{10^0}$

18. $\dfrac{0^5}{5^0}$

Evaluate each expression. See Examples 1 and 2.

19. $7^0 + 9^0$

20. $8^0 + 6^0$

🌐 **21.** 4^{-3}

22. 5^{-4}

23. $\left(\dfrac{1}{2}\right)^{-4}$

24. $\left(\dfrac{1}{3}\right)^{-3}$

🌐 **25.** $\left(\dfrac{6}{7}\right)^{-2}$

26. $\left(\dfrac{2}{3}\right)^{-3}$

27. $(-3)^{-4}$

28. $(-4)^{-3}$

29. $5^{-1} + 3^{-1}$

30. $6^{-1} + 2^{-1}$

31. $-2^{-1} + 3^{-2}$

32. $(-3)^{-2} + (-4)^{-1}$

Relating Concepts (Exercises 33–36) For Individual or Group Work

*In Objective 1, we used the product rule to motivate the definition of a 0 exponent. We can also use the quotient rule. To see this, **work Exercises 33–36 in order.***

33. Consider the expression $\dfrac{25}{25}$. What is its simplest form?

34. Write the quotient in Exercise 33 using the fact that $25 = 5^2$.

35. Apply the quotient rule for exponents to your answer for Exercise 34. Give the answer as a power of 5.

36. Because your answers for Exercises 33 and 35 both represent $\dfrac{25}{25}$, they must be equal. Write this equality. What definition does it support?

Simplify by writing each expression with positive exponents. Assume that all variables represent nonzero real numbers. See Examples 2–4.

37. $\dfrac{9^4}{9^5}$

38. $\dfrac{7^3}{7^4}$

39. $\dfrac{6^{-3}}{6^2}$

40. $\dfrac{4^{-2}}{4^3}$

41. $\dfrac{1}{6^{-3}}$

42. $\dfrac{1}{5^{-2}}$

43. $\dfrac{2}{r^{-4}}$

44. $\dfrac{3}{s^{-8}}$

45. $\dfrac{4^{-3}}{5^{-2}}$

46. $\dfrac{6^{-2}}{5^{-4}}$

47. $p^5 q^{-8}$

48. $x^{-8} y^4$

49. $\dfrac{r^5}{r^{-4}}$

50. $\dfrac{a^6}{a^{-4}}$

51. $\dfrac{6^4 x^8}{6^5 x^3}$

52. $\dfrac{3^8 y^5}{3^{10} y^2}$

53. $\dfrac{6y^3}{2y}$

54. $\dfrac{5m^2}{m}$

55. $\dfrac{3x^5}{3x^2}$

56. $\dfrac{10p^8}{2p^4}$

57. $\dfrac{x^{-3} y}{4z^{-2}}$

58. $\dfrac{p^{-5} q^{-4}}{9r^{-3}}$

59. $\dfrac{(a+b)^{-3}}{(a+b)^{-4}}$

60. $\dfrac{(x+y)^{-8}}{(x+y)^{-9}}$

Simplify by writing each expression with positive exponents. Assume that all variables represent nonzero real numbers. See Example 5.

61. $\dfrac{(7^4)^3}{7^9}$

62. $\dfrac{(5^3)^2}{5^2}$

63. $x^{-3} \cdot x^5 \cdot x^{-4}$

64. $y^{-8} \cdot y^5 \cdot y^{-2}$

65. $\dfrac{(3x)^{-2}}{(4x)^{-3}}$

66. $\dfrac{(2y)^{-3}}{(5y)^{-4}}$

67. $\left(\dfrac{x^{-1} y}{z^2}\right)^{-2}$

68. $\left(\dfrac{p^{-4} q}{r^{-3}}\right)^{-3}$

69. $(6x)^4 (6x)^{-3}$

70. $(10y)^9 (10y)^{-8}$

71. $\dfrac{(m^7 n)^{-2}}{m^{-4} n^3}$

72. $\dfrac{(m^8 n^{-4})^2}{m^{-2} n^5}$

73. $\dfrac{5x^{-3}}{(4x)^2}$

74. $\dfrac{-3k^5}{(2k)^2}$

75. $\left(\dfrac{2p^{-1} q}{3^{-1} m^2}\right)^2$

76. $\left(\dfrac{4xy^2}{x^{-1} y}\right)^{-2}$

12.6 ▶▶▶ Dividing a Polynomial by a Monomial

OBJECTIVE 1 Divide a polynomial by a monomial. We add two fractions with a common denominator as follows.

$$\frac{a}{c} + \frac{b}{c} = \frac{a+b}{c}$$

In reverse, this statement gives a rule for dividing a polynomial by a monomial.

OBJECTIVE

1 Divide a polynomial by a monomial.

Dividing a Polynomial by a Monomial

To divide a polynomial by a monomial, divide each term of the polynomial by the monomial:

$$\frac{a+b}{c} = \frac{a}{c} + \frac{b}{c} \quad (c \neq 0).$$

Examples: $\dfrac{2+5}{3} = \dfrac{2}{3} + \dfrac{5}{3}$ and $\dfrac{x+3z}{2y} = \dfrac{x}{2y} + \dfrac{3z}{2y}$ $(y \neq 0)$

The parts of a division problem are named here.

$$\text{Dividend} \rightarrow \frac{12x^2 + 6x}{6x} \leftarrow \text{Divisor} = 2x + 1 \leftarrow \text{Quotient}$$

1 Divide.

(a) $\dfrac{6p^4 + 18p^7}{3p^2}$

EXAMPLE 1 Dividing a Polynomial by a Monomial

Divide $5m^5 - 10m^3$ by $5m^2$.

$$\frac{5m^5 - 10m^3}{5m^2}$$

$$= \frac{5m^5}{5m^2} - \frac{10m^3}{5m^2} \quad \text{Use the preceding rule, with } + \text{ replaced by } -.$$

$$= m^3 - 2m \quad \text{Quotient rule}$$

Check Multiply: $5m^2 \cdot (m^3 - 2m) = 5m^5 - 10m^3.$
Divisor · Quotient = Original polynomial (Dividend)

Because division by 0 is undefined, the quotient $\frac{5m^5 - 10m^3}{5m^2}$ is undefined if $m = 0$. From now on, we assume that no denominators are 0.

Work Problem **1** *at the Side.* ▶

(b) $\dfrac{12m^6 + 18m^5 + 30m^4}{6m^2}$

(c) $(18r^7 - 9r^2) \div (3r)$

EXAMPLE 2 Dividing a Polynomial by a Monomial

Divide $\dfrac{16a^5 - 12a^4 + 8a^2}{4a^3}$.

$$\frac{16a^5 - 12a^4 + 8a^2}{4a^3}$$

$$= \frac{16a^5}{4a^3} - \frac{12a^4}{4a^3} + \frac{8a^2}{4a^3} \quad \text{Divide each term by } 4a^3.$$

$$= 4a^2 - 3a + \frac{2}{a} \quad \text{Quotient rule}$$

Continued on Next Page

2 Divide.

(a) $\dfrac{20x^4 - 25x^3 + 5x}{5x^2}$

(b) $\dfrac{50m^4 - 30m^3 + 20m}{10m^3}$

3 Divide.

(a) $\dfrac{-9y^6 + 8y^7 - 11y - 4}{y^2}$

(b) $\dfrac{-8p^4 - 6p^3 - 12p^5}{-3p^3}$

4 Divide.

$\dfrac{45x^4y^3 + 30x^3y^2 - 60x^2y}{-15x^2y}$

The quotient $4a^2 - 3a + \frac{2}{a}$ is not a polynomial because of the expression $\frac{2}{a}$, which has a variable in the denominator. While the sum, difference, and product of two polynomials are always polynomials, the quotient of two polynomials may not be.

Check $4a^3\left(4a^2 - 3a + \dfrac{2}{a}\right)$ Divisor × Quotient should equal Dividend.

$= 4a^3(4a^2) + 4a^3(-3a) + 4a^3\left(\dfrac{2}{a}\right)$ Distributive property

$= 16a^5 - 12a^4 + 8a^2$ Dividend

◀ *Work Problem* **2** *at the Side.*

EXAMPLE 3 Dividing a Polynomial by a Monomial with a Negative Coefficient

Divide $-7x^3 + 12x^4 - 4x$ by $-4x$.

Write the polynomial in descending powers as $12x^4 - 7x^3 - 4x$ before dividing.

Write in descending powers.

$$\dfrac{12x^4 - 7x^3 - 4x}{-4x}$$

$= \dfrac{12x^4}{-4x} - \dfrac{7x^3}{-4x} - \dfrac{4x}{-4x}$ Divide each term by $-4x$.

$= -3x^3 - \dfrac{7x^2}{-4} - (-1)$ Quotient rule

$= -3x^3 + \dfrac{7}{4}x^2 + \mathbf{1}$ Be sure to include the 1 in the answer.

Check by multiplying.

◀ *Work Problem* **3** *at the Side.*

EXAMPLE 4 Dividing a Polynomial by a Monomial

Divide $180x^4y^{10} - 150x^3y^8 + 120x^2y^6 - 90xy^4 + 100y$ by $-30xy^2$.

$$\dfrac{180x^4y^{10} - 150x^3y^8 + 120x^2y^6 - 90xy^4 + 100y}{-30xy^2}$$

$= \dfrac{180x^4y^{10}}{-30xy^2} - \dfrac{150x^3y^8}{-30xy^2} + \dfrac{120x^2y^6}{-30xy^2} - \dfrac{90xy^4}{-30xy^2} + \dfrac{100y}{-30xy^2}$

$= -6x^3y^8 + 5x^2y^6 - 4xy^4 + 3y^2 - \dfrac{10}{3xy}$

◀ *Work Problem* **4** *at the Side.*

ANSWERS

2. (a) $4x^2 - 5x + \dfrac{1}{x}$ (b) $5m - 3 + \dfrac{2}{m^2}$

3. (a) $8y^5 - 9y^4 - \dfrac{11}{y} - \dfrac{4}{y^2}$

   (b) $4p^2 + \dfrac{8p}{3} + 2$

4. $-3x^2y^2 - 2xy + 4$

12.6 ▶▶▶ **Exercises**

Fill in each blank with the correct response.

1. In the statement $\dfrac{6x^2 + 8}{2} = 3x^2 + 4$, _____ is the dividend, _____ is the

   divisor, and _____ is the quotient.

2. The expression $\dfrac{3x + 12}{x}$ is undefined if $x =$ _____ .

3. To check the division shown in Exercise 1, multiply _____ by _____ and show

   that the product is _____ .

4. The expression $5x^2 - 3x + 6 + \frac{2}{x}$ _____ a polynomial.
   $\underset{\text{(is/is not)}}{}$

5. Explain why the division problem $\dfrac{16m^3 - 12m^2}{4m}$ can be performed using the method

   of this section, while the division problem $\dfrac{4m}{16m^3 - 12m^2}$ cannot.

6. Evaluate $\dfrac{5y + 6}{2}$ when $y = 2$. Evaluate $5y + 3$ when $y = 2$. Does $\dfrac{5y + 6}{2}$ equal $5y + 3$?

Perform each division. See Examples 1–4.

7. $\dfrac{60x^4 - 20x^2 + 10x}{2x}$

8. $\dfrac{120x^6 - 60x^3 + 80x^2}{2x}$

9. $\dfrac{20m^5 - 10m^4 + 5m^2}{-5m^2}$

10. $\dfrac{12t^5 - 6t^3 + 6t^2}{-6t^2}$

11. $\dfrac{8t^5 - 4t^3 + 4t^2}{2t}$

12. $\dfrac{8r^4 - 4r^3 + 6r^2}{2r}$

13. $\dfrac{4a^5 - 4a^2 + 8}{4a}$

14. $\dfrac{5t^8 + 5t^7 + 15}{5t}$

15. $\dfrac{12x^5 - 4x^4 + 6x^3}{-6x^2}$

16. $\dfrac{24x^6 - 12x^5 + 30x^4}{-6x^2}$

17. $\dfrac{4x^2 + 20x^3 - 36x^4}{4x^2}$

18. $\dfrac{5x^2 - 30x^4 + 30x^5}{5x^2}$

19. $\dfrac{-3x^3 - 4x^4 + 2x}{-3x^2}$

20. $\dfrac{-8x + 6x^3 - 5x^4}{-3x^2}$

21. $\dfrac{27r^4 - 36r^3 - 6r^2 + 3r - 2}{3r}$

22. $\dfrac{8k^4 - 12k^3 - 2k^2 - 2k - 3}{2k}$

23. $\dfrac{2m^5 - 6m^4 + 8m^2}{-2m^3}$

24. $\dfrac{6r^5 - 8r^4 + 10r^2}{-2r^4}$

25. $(120x^{11} - 60x^{10} + 140x^9 - 100x^8) \div (10x^{12})$

26. $(120x^{12} - 84x^9 + 60x^8 - 36x^7) \div (12x^9)$

27. $(20a^4b^3 - 15a^5b^2 + 25a^3b) \div (-5a^4b)$

28. $(16y^5z - 8y^2z^2 + 12yz^3) \div (-4y^2z^2)$

29. What polynomial represents the length of the rectangle?

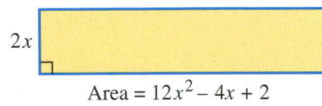

$2x$

Area = $12x^2 - 4x + 2$

30. What polynomial represents the length of the base of the triangle?

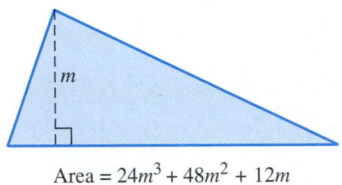

m

Area = $24m^3 + 48m^2 + 12m$

31. What polynomial, when divided by $5x^3$, yields $3x^2 - 7x + 7$ as a quotient?

32. The quotient of a certain polynomial and $-12y^3$ is $6y^3 - 5y^2 + 2y - 3 + \frac{7}{y}$. Find the polynomial.

Relating Concepts (Exercises 33–36) For Individual or Group Work

Our system of numeration is called a decimal system. It is based on powers of ten. In a whole number such as 2846, each digit is understood to represent the number of powers of ten for its place value. The 2 represents two thousands (2×10^3), the 8 represents eight hundreds (8×10^2), the 4 represents four tens (4×10^1), and the 6 represents six ones (or units) (6×10^0). In expanded form we write

$$2846 = (2 \times 10^3) + (8 \times 10^2) + (4 \times 10^1) + (6 \times 10^0).$$

Keeping this information in mind, **work Exercises 33–36 in order.**

33. Divide 2846 by 2, using paper-and-pencil methods: $2\overline{)2846}$.

34. Write your answer in Exercise 33 in expanded form.

35. Use the methods of this section to divide the polynomial $2x^3 + 8x^2 + 4x + 6$ by 2.

36. Compare your answers in Exercises 34 and 35. How are they similar? How are they different? For what value of x does the answer in Exercise 35 equal the answer in Exercise 34?

12.7 ▶▶▶ Dividing a Polynomial by a Polynomial

OBJECTIVES

1 Divide a polynomial by a polynomial.

2 Apply division to a geometry problem.

OBJECTIVE 1 Divide a polynomial by a polynomial. We use a method of "long division" to divide a polynomial by a polynomial (other than a monomial). *Both polynomials must be written in descending powers.*

Dividing Whole Numbers	Dividing Polynomials

Step 1

Divide 6696 by 27.

$$27\overline{)6696}$$

Divide $8x^3 - 4x^2 - 14x + 15$ by $2x + 3$.

$$2x + 3\overline{)8x^3 - 4x^2 - 14x + 15}$$

Step 2

66 divided by 27 = **2**;
$2 \cdot 27 = $ **54**.

$$\begin{array}{r} 2 \\ 27\overline{)6696} \\ 54 \end{array}$$

$8x^3$ divided by $2x =$ **$4x^2$**;
$4x^2(2x + 3) = $ **$8x^3 + 12x^2$**.

$$\begin{array}{r} 4x^2 \\ 2x + 3\overline{)8x^3 - 4x^2 - 14x + 15} \\ 8x^3 + 12x^2 \end{array}$$

Step 3

Subtract; then bring down the next digit.

$$\begin{array}{r} 2 \\ 27\overline{)6696} \\ 54\downarrow \\ \hline 129 \end{array}$$

Subtract; then bring down the next term.

$$\begin{array}{r} 4x^2 \\ 2x + 3\overline{)8x^3 - 4x^2 - 14x + 15} \\ 8x^3 + 12x^2 \hspace{0.3em}\downarrow \\ \hline -16x^2 - 14x \end{array}$$

(To subtract two polynomials, change the signs of the second and then add.)

Step 4

129 divided by 27 = **4**;
$4 \cdot 27 = $ **108**.

$$\begin{array}{r} 24 \\ 27\overline{)6696} \\ 54 \\ \hline 129 \\ 108 \end{array}$$

$-16x^2$ divided by $2x = $ **$-8x$**;
$-8x(2x + 3) = $ **$-16x^2 - 24x$**.

$$\begin{array}{r} 4x^2 - 8x \\ 2x + 3\overline{)8x^3 - 4x^2 - 14x + 15} \\ 8x^3 + 12x^2 \\ \hline -16x^2 - 14x \\ -16x^2 - 24x \end{array}$$

Step 5

Subtract; then bring down the next digit.

$$\begin{array}{r} 24 \\ 27\overline{)6696} \\ 54\downarrow \\ 129\downarrow \\ 108\downarrow \\ \hline 216 \end{array}$$

Subtract; then bring down the next term.

$$\begin{array}{r} 4x^2 - 8x \\ 2x + 3\overline{)8x^3 - 4x^2 - 14x + 15} \\ 8x^3 + 12x^2 \\ \hline -16x^2 - 14x \\ -16x^2 - 24x \\ \hline 10x + 15 \end{array}$$

(continued)

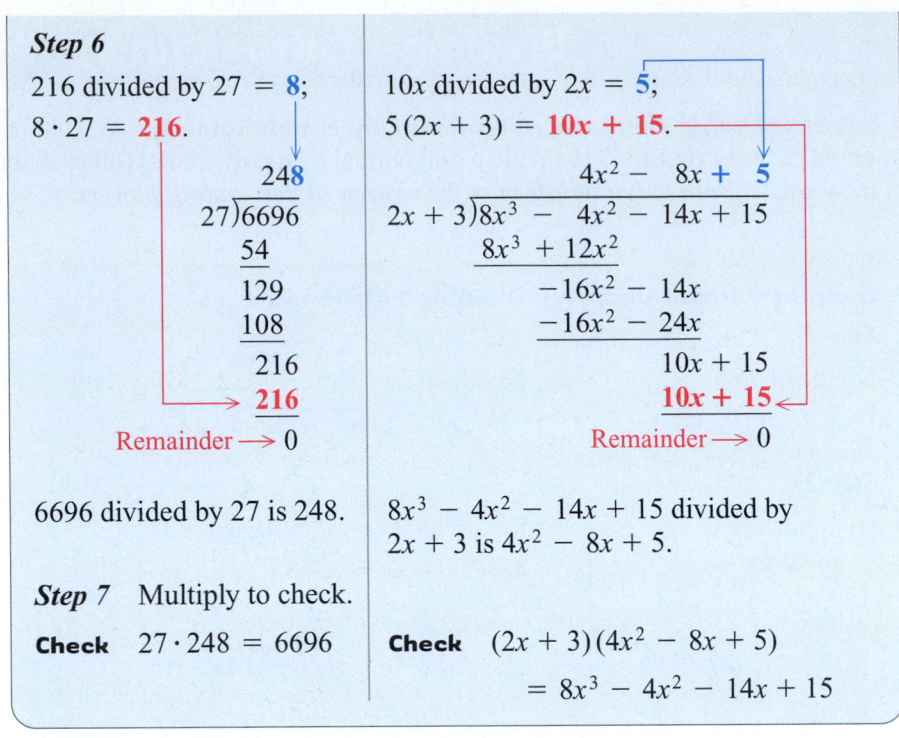

Step 6

216 divided by 27 = **8**;

$8 \cdot 27 = \textbf{216}$.

$$
\begin{array}{r}
248 \\
27\overline{)6696} \\
54 \\
\hline
129 \\
108 \\
\hline
216 \\
\textbf{216} \\
\hline
\end{array}
$$

Remainder ⟶ 0

6696 divided by 27 is 248.

$10x$ divided by $2x = \textbf{5}$;

$5(2x + 3) = \textbf{10x + 15}$.

$$
\begin{array}{r}
4x^2 - 8x + 5 \\
2x + 3\overline{)8x^3 - 4x^2 - 14x + 15} \\
8x^3 + 12x^2 \\
\hline
-16x^2 - 14x \\
-16x^2 - 24x \\
\hline
10x + 15 \\
\textbf{10x + 15} \\
\hline
\end{array}
$$

Remainder ⟶ 0

$8x^3 - 4x^2 - 14x + 15$ divided by $2x + 3$ is $4x^2 - 8x + 5$.

Step 7 Multiply to check.

Check $27 \cdot 248 = 6696$

Check $(2x + 3)(4x^2 - 8x + 5)$

$= 8x^3 - 4x^2 - 14x + 15$

EXAMPLE 1 **Dividing a Polynomial by a Polynomial**

Divide $5x + 4x^3 - 8 - 4x^2$ by $2x - 1$.

 The first polynomial must be written with the exponents in descending powers as $4x^3 - 4x^2 + 5x - 8$. Then divide by $2x - 1$.

$$
\begin{array}{r}
\textbf{2x}^2 - \textbf{x} + \textbf{2} \\
2x - 1\overline{)4x^3 - 4x^2 + 5x - 8} \\
4x^3 - 2x^2 \\
\hline
-2x^2 + 5x \\
-2x^2 + x \\
\hline
4x - 8 \\
4x - 2 \\
\hline
-6 \leftarrow \text{Remainder}
\end{array}
$$

To subtract, add the opposite.

Write in descending powers.

Step 1 $4x^3$ divided by $2x = \textbf{2x}^2$; $2x^2(2x - 1) = 4x^3 - 2x^2$.

Step 2 Subtract; bring down the next term.

Step 3 $-2x^2$ divided by $2x = \boldsymbol{-x}$; $-x(2x - 1) = -2x^2 + x$.

Step 4 Subtract; bring down the next term.

Step 5 $4x$ divided by $2x = \textbf{2}$; $2(2x - 1) = 4x - 2$.

Step 6 Subtract. The remainder is $\boldsymbol{-6}$. Write the remainder as the numerator of a fraction that has $2x - 1$ as its denominator. The answer is not a polynomial because of the nonzero remainder.

Dividend ⟶
Divisor ⟶
$$
\frac{4x^3 - 4x^2 + 5x - 8}{2x - 1} = \underbrace{2x^2 - x + 2}_{\substack{\text{Quotient} \\ \text{polynomial}}} + \frac{-6}{2x - 1}
$$

Continued on Next Page

Step 7   Multiply to check.

Check $(2x - 1)\left(2x^2 - x + 2 + \dfrac{-6}{2x - 1}\right)$

$$= (2x - 1)(2x^2) + (2x - 1)(-x) + (2x - 1)(2)$$

$$+ (2x - 1)\left(\dfrac{-6}{2x - 1}\right)$$

$$= 4x^3 - 2x^2 - 2x^2 + x + 4x - 2 - 6$$

$$= 4x^3 - 4x^2 + 5x - 8$$

Work Problem **1** *at the Side.* ▶

EXAMPLE 2   **Dividing into a Polynomial with Missing Terms**

Divide $x^3 - 1$ by $x - 1$.

   Here the polynomial $x^3 - 1$ is missing the x^2-term and the x-term. When terms are missing, use **0** as the coefficient for each missing term. (Zero acts as a placeholder here, just as it does in our numeration system.) Thus, $x^3 - 1 = x^3 + 0x^2 + 0x - 1$. Now divide.

$$
\begin{array}{r}
x^2 + x + 1 \\
x - 1 \overline{) x^3 + 0x^2 + 0x - 1} \\
\underline{x^3 - x^2} \\
x^2 + 0x \\
\underline{x^2 - x} \\
x - 1 \\
\underline{x - 1} \\
0
\end{array}
$$

> Insert placeholders for the missing terms.

The remainder is 0. The quotient is $x^2 + x + 1$.

Check

$$(x - 1)(x^2 + x + 1)$$

$$= x^3 + x^2 + x - x^2 - x - 1$$

$$= x^3 - 1$$

Work Problem **2** *at the Side.* ▶

EXAMPLE 3   **Dividing by a Polynomial with Missing Terms**

Divide $x^4 + 2x^3 + 2x^2 - x - 1$ by $x^2 + 1$.

   Since $x^2 + 1$ has a missing x-term, write it as $x^2 + 0x + 1$.

$$
\begin{array}{r}
x^2 + 2x + 1 \\
x^2 + 0x + 1 \overline{) x^4 + 2x^3 + 2x^2 - x - 1} \\
\underline{x^4 + 0x^3 + x^2} \\
2x^3 + x^2 - x \\
\underline{2x^3 + 0x^2 + 2x} \\
x^2 - 3x - 1 \\
\underline{x^2 + 0x + 1} \\
-3x - 2 \leftarrow \text{Remainder}
\end{array}
$$

> Insert a placeholder for the missing term.

Continued on Next Page

---

1 Divide.

(a) $(x^3 + x^2 + 4x - 6)$
$\div (x - 1)$

(b) $\dfrac{p^3 - 2p^2 - 5p + 9}{p + 2}$

2 Divide.

(a) $\dfrac{r^2 - 5}{r + 4}$

(b) $(x^3 - 8) \div (x - 2)$

ANSWERS

1. (a) $x^2 + 2x + 6$

   (b) $p^2 - 4p + 3 + \dfrac{3}{p + 2}$

2. (a) $r - 4 + \dfrac{11}{r + 4}$

   (b) $x^2 + 2x + 4$

3 Divide.

(a)

$(2x^4 + 3x^3 - x^2 + 6x + 5)$
  $\div (x^2 - 1)$

When the result of subtracting ($-3x - 2$, in this case) is a constant or a polynomial of degree less than the divisor ($x^2 + 0x + 1$), that constant or polynomial is the remainder. We write the answer as

$$x^2 + 2x + 1 + \frac{-3x - 2}{x^2 + 1}.$$

> Remember to include " $+ \frac{\text{remainder}}{\text{divisor}}$."

Multiply to check that this is the correct quotient.

◀ *Work Problem* **3** *at the Side.*

EXAMPLE 4 **Dividing a Polynomial when the Quotient Has Fractional Coefficients**

Divide $4x^3 + 2x^2 + 3x + 2$ by $4x - 4$.

(b)

$$\frac{2m^5 + m^4 + 6m^3 - 3m^2 - 18}{m^2 + 3}$$

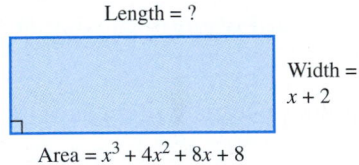

$\frac{6x^2}{4x} = \frac{3}{2}x$

$\frac{9x}{4x} = \frac{9}{4}$

$$\begin{array}{r} x^2 + \dfrac{3}{2}x + \dfrac{9}{4} \\ 4x - 4 \overline{)4x^3 + 2x^2 + 3x + 2} \\ \underline{4x^3 - 4x^2} \\ 6x^2 + 3x \\ \underline{6x^2 - 6x} \\ 9x + 2 \\ \underline{9x - 9} \\ 11 \end{array}$$

The answer is $x^2 + \dfrac{3}{2}x + \dfrac{9}{4} + \dfrac{11}{4x - 4}$.

4 Divide $3x^3 + 7x^2 + 7x + 10$ by $3x + 6$.

◀ *Work Problem* **4** *at the Side.*

OBJECTIVE **2** **Apply division to a geometry problem.**

EXAMPLE 5 **Using an Area Formula**

The area of the rectangle in Figure 3 is given by $x^3 + 4x^2 + 8x + 8$ sq. units and the width by $x + 2$ units. What is its length?

5 Divide $x^3 + 4x^2 + 8x + 8$ by $x + 2$.

Length = ?

Width = $x + 2$

Area = $x^3 + 4x^2 + 8x + 8$

Figure 3

Since $A = LW$, solving for L gives $L = \frac{A}{W}$. Divide $x^3 + 4x^2 + 8x + 8$ by the width, $x + 2$.

◀ *Work Problem* **5** *at the Side.*

The quotient from Problem 5 at the side, $x^2 + 2x + 4$, represents the length of the rectangle in units.

ANSWERS

3. **(a)** $2x^2 + 3x + 1 + \dfrac{9x + 6}{x^2 - 1}$

  (b) $2m^3 + m^2 - 6$

4. $x^2 + \dfrac{1}{3}x + \dfrac{5}{3}$

5. $x^2 + 2x + 4$

1. In the division problem $(4x^4 + 2x^3 - 14x^2 + 19x + 10) \div (2x + 5) = 2x^3 - 4x^2 + 3x + 2$, which polynomial is the divisor? Which is the quotient?

2. When dividing one polynomial by another, how do you know when to stop dividing?

3. In dividing $12m^2 - 20m + 3$ by $2m - 3$, what is the first step?

4. In the division in Exercise 3, what is the second step?

Perform each division. See Example 1.

5. $\dfrac{x^2 - x - 6}{x - 3}$

6. $\dfrac{m^2 - 2m - 24}{m - 6}$

7. $\dfrac{2y^2 + 9y - 35}{y + 7}$

8. $\dfrac{2y^2 + 9y + 7}{y + 1}$

9. $\dfrac{p^2 + 2p + 20}{p + 6}$

10. $\dfrac{x^2 + 11x + 16}{x + 8}$

11. $(r^2 - 8r + 15) \div (r - 3)$

12. $(t^2 + 2t - 35) \div (t - 5)$

13. $\dfrac{4a^2 - 22a + 32}{2a + 3}$

14. $\dfrac{9w^2 + 6w + 10}{3w - 2}$

15. $\dfrac{8x^3 - 10x^2 - x + 3}{2x + 1}$

16. $\dfrac{12t^3 - 11t^2 + 9t + 18}{4t + 3}$

Perform each division. See Examples 2–4.

17. $\dfrac{3y^3 + y^2 + 2}{y + 1}$

18. $\dfrac{2r^3 - 6r - 36}{r - 3}$

19. $\dfrac{2x^3 + x + 2}{x + 1}$

20. $\dfrac{3x^3 + x + 5}{x + 1}$

21. $\dfrac{3k^3 - 4k^2 - 6k + 10}{k^2 - 2}$

22. $\dfrac{5z^3 - z^2 + 10z + 2}{z^2 + 2}$

23. $(x^4 - x^2 - 2) \div (x^2 - 2)$

24. $(r^4 + 2r^2 - 3) \div (r^2 - 1)$

25. $\dfrac{x^4 - 1}{x^2 - 1}$

26. $\dfrac{y^3 + 1}{y + 1}$

27. $\dfrac{6p^4 - 15p^3 + 14p^2 - 5p + 10}{3p^2 + 1}$

28. $\dfrac{6r^4 - 10r^3 - r^2 + 15r - 8}{2r^2 - 3}$

29. $\dfrac{2x^5 + x^4 + 11x^3 - 8x^2 - 13x + 7}{2x^2 + x - 1}$

30. $\dfrac{4t^5 - 11t^4 - 6t^3 + 5t^2 - t + 3}{4t^2 + t - 3}$

31. $(10x^3 + 13x^2 + 4x + 1) \div (5x + 5)$

32. $(6x^3 - 19x^2 - 19x - 4) \div (2x - 8)$

Work each problem. See Example 5.

33. Give the length of the rectangle.

The area is $5x^3 + 7x^2 - 13x - 6$ sq. units.

34. Find the measure of the base of the parallelogram.

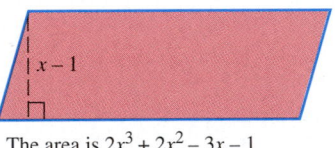

The area is $2x^3 + 2x^2 - 3x - 1$ sq. units.

Relating Concepts (Exercises 35–38) For Individual or Group Work

We can find the value of a polynomial in x for a given value of x by substituting that number for x. Surprisingly, we can accomplish the same thing by division. For example, to find the value of $2x^2 - 4x + 3$ for $x = -3$, we would divide $2x^2 - 4x + 3$ by $x - (-3)$. The remainder will give the value of the polynomial for $x = -3$. **Work Exercises 35–38 in order.**

35. Find the value of $2x^2 - 4x + 3$ for $x = -3$ by substitution.

36. Divide $2x^2 - 4x + 3$ by $x + 3$. Give the remainder.

37. Compare your answers to Exercises 35 and 36. What do you notice?

38. Choose another polynomial and evaluate it both ways for some value of the variable. Do the answers agree?

12.8 ▷▷▷ An Application of Exponents: Scientific Notation

OBJECTIVES

1 Express numbers in scientific notation.

2 Convert numbers in scientific notation to numbers without exponents.

3 Use scientific notation in calculations.

OBJECTIVE 1 Express numbers in scientific notation. Numbers occurring in science are often extremely large (such as the distance from Earth to the sun, 93,000,000 mi) or extremely small (the wavelength of yellow-green light, approximately 0.0000006 m). Because of the difficulty of working with many zeros, scientists often express such numbers with exponents, using a form called *scientific notation.*

> **Scientific Notation**
>
> A number is written in **scientific notation** when it is expressed in the form
>
> $$a \times 10^n,$$
>
> where $1 \leq |a| < 10$ and n is an integer.

In **scientific notation,** there is always one nonzero digit before the decimal point. This is shown in the following examples.

$3.19 \times 10^1 = 3.19 \times 10 = 31.9$ Decimal point moves 1 place to the right.

$3.19 \times 10^2 = 3.19 \times 100 = 319.$ Decimal point moves 2 places to the right.

$3.19 \times 10^3 = 3.19 \times 1000 = 3190.$ Decimal point moves 3 places to the right.

$3.19 \times 10^{-1} = 3.19 \times 0.1 = 0.319$ Decimal point moves 1 place to the left.

$3.19 \times 10^{-2} = 3.19 \times 0.01 = 0.0319$ Decimal point moves 2 places to the left.

$3.19 \times 10^{-3} = 3.19 \times 0.001 = 0.00319$ Decimal point moves 3 places to the left.

> **Note**
> In scientific notation, the times symbol, $\times$, is commonly used.

A number in scientific notation is always written with the decimal point after the first nonzero digit and then multiplied by the appropriate power of 10. For example, 56,200 is written 5.62×10^4, since

$$56{,}200 = 5.62 \times \mathbf{10{,}000} = 5.62 \times \mathbf{10^4}.$$

Other examples include

       42,000,000   written 4.2×10^7,

       0.000586   written 5.86×10^{-4},

and   2,000,000,000   written 2×10^9.

> It is not necessary to write 2.0.

To write a number in scientific notation, follow the steps given on the next page. (For a negative number, follow these steps using the *absolute value* of the number; then make the result negative.)

1 Write each number in scientific notation.

(a) 63,000

(b) 5,870,000

(c) 7.0065

(d) 0.0571

(e) −0.00062

Writing a Number in Scientific Notation

Step 1 Move the decimal point to the right of the first nonzero digit.

Step 2 Count the number of places you moved the decimal point.

Step 3 The number of places in Step 2 is the absolute value of the exponent on 10.

Step 4 The exponent on 10 is positive if the original number is greater than the number in Step 1; the exponent is negative if the original number is less than the number in Step 1. If the decimal point is not moved, the exponent is 0.

EXAMPLE 1 **Using Scientific Notation**

Write each number in scientific notation.

(a) 93,000,000

Move the decimal point to follow the first nonzero digit (the 9). Count the number of places the decimal point was moved.

$$93{,}000{,}000. \leftarrow \text{Decimal point}$$
$$\text{7 places}$$

The number will be written in scientific notation as 9.3×10^n. To find the value of n, first compare the original number, 93,000,000, with 9.3. Since 93,000,000 is *greater* than 9.3, we must multiply by a *positive* power of 10 so that the product 9.3×10^n will equal the larger number.

Since the decimal point was moved 7 places, and since n is positive,

$$93{,}000{,}000 = 9.3 \times 10^7.$$

(b) $63{,}200{,}000{,}000 = 6.3200000000 = 6.32 \times 10^{10}$
$$\text{10 places}$$

(c) $3.021 = 3.021 \times 10^0$

(d) 0.00462

Move the decimal point to the right of the first nonzero digit and count the number of places the decimal point was moved.

$$0.00462 \qquad \text{3 places}$$

Since 0.00462 is *less* than 4.62, the exponent must be *negative*.

$$0.00462 = 4.62 \times 10^{-3}$$

(e) $-0.0000762 = -7.62 \times 10^{-5}$
$$\text{5 places} \qquad \text{Remember the negative sign.}$$

◀ Work Problem **1** at the Side.

Note

To choose the exponent when you write a number in scientific notation, think: If the original number is "large," like 93,000,000, use a *positive* exponent on 10, since positive is greater than negative. However, if the original number is "small," like 0.00462, use a *negative* exponent on 10, since negative is less than positive.

OBJECTIVE 2 Convert numbers in scientific notation to numbers without exponents. To convert a number written in scientific notation to a number without exponents, work in reverse. *Multiplying a number by a positive power of 10 will make the number greater; multiplying by a negative power of 10 will make the number less.*

EXAMPLE 2 Writing Numbers without Exponents

Write each number without exponents.

(a) 6.2×10^3
   Since the exponent is positive, make 6.2 greater by moving the decimal point 3 places to the right. It is necessary to attach two 0s.

$$6.2 \times \mathbf{10^3} = 6.2\mathbf{00} = 6200$$

(b) $4.283 \times 10^\mathbf{5} = 4.283\mathbf{00} = 428{,}300$ $\quad$ Move 5 places to the right; attach 0s as necessary.

(c) $-9.73 \times 10^{\mathbf{-2}} = -\mathbf{0}9.73 = -0.0973$ $\quad$ Move 2 places to the left.

The exponent tells the number of places and the direction that the decimal point is moved.

————————————— Work Problem **2** at the Side. ▶

OBJECTIVE 3 Use scientific notation in calculations. The next example uses scientific notation with products and quotients.

EXAMPLE 3 Multiplying and Dividing with Scientific Notation

Perform each calculation. Write answers in scientific notation and also without exponents.

(a) $\quad (7 \times 10^3)(5 \times 10^4)$

$\quad = (7 \times 5)(10^3 \times 10^4)$ $\quad$ Commutative and associative properties

$\quad = \mathbf{35} \times 10^7$ $\quad$ Multiply; product rule

Don't stop! This number is not in scientific notation, since 35 is not between 1 and 10.

$\quad = (\mathbf{3.5 \times 10^1}) \times 10^7$ $\quad$ Write 35 in scientific notation.

$\quad = 3.5 \times (\mathbf{10^1 \times 10^7})$ $\quad$ Associative property

$\quad = 3.5 \times \mathbf{10^8}$ $\quad$ Product rule

$\quad = 350{,}000{,}000$ $\quad$ Write without exponents.

(b) $\dfrac{4 \times 10^{-5}}{2 \times 10^3} = \dfrac{4}{2} \times \dfrac{10^{-5}}{10^3} = 2 \times 10^{-8} = 0.00000002$

————————————— Work Problem **3** at the Side. ▶

Note

Multiplying or dividing numbers written in scientific notation may produce an answer in the form $a \times \mathbf{10^0}$. Since $10^0 = 1$, $a \times 10^0 = a$. For example,

$$(8 \times 10^{-4})(5 \times 10^4) = 40 \times \mathbf{10^0} = 40. \quad \textcolor{blue}{10^0 = 1}$$

Also, if $a = 1$, then $a \times 10^n = 10^n$. For example, we could write $1{,}000{,}000$ as 10^6 instead of 1×10^6.

---

2 Write without exponents.

(a) 4.2×10^3

(b) 8.7×10^5

(c) 6.42×10^{-3}

3 Perform each calculation. Write answers in scientific notation and also without exponents.

(a) $(2.6 \times 10^4)(2 \times 10^{-6})$

(b) $(3 \times 10^5)(5 \times 10^{-2})$

(c) $\dfrac{4.8 \times 10^2}{2.4 \times 10^{-3}}$

ANSWERS

2. (a) 4200 **(b)** 870,000 **(c)** 0.00642
3. (a) 5.2×10^{-2}; 0.052
   (b) 1.5×10^4; 15,000
   (c) 2×10^5; 200,000

4 The speed of light is approximately 3.0×10^5 km per sec. How far does light travel in 6.0×10^1 sec? (*Source: World Almanac and Book of Facts.*)

⊞ **Calculator Tip** Calculators usually have a key labeled EE or EXP for scientific notation. See your owner's manual for more information.

EXAMPLE 4 Using Scientific Notation to Solve an Application

A *nanometer* is a very small unit of measure that is equivalent to about 0.00000003937 in. About how much would 700,000 nanometers measure in inches? (*Source: World Almanac and Book of Facts.*)

Write each number in scientific notation, and then multiply.

$$700,000\,(0.00000003937)$$

$$= (7 \times 10^5)(3.937 \times 10^{-8}) \qquad \text{Write in scientific notation.}$$

$$= (7 \times 3.937)(10^5 \times 10^{-8}) \qquad \text{Properties of real numbers}$$

$$= \mathbf{27.559} \times 10^{-3} \qquad \text{Multiply; product rule}$$

Don't stop here.

$$= (\mathbf{2.7559 \times 10^1}) \times 10^{-3} \qquad \text{Write 27.559 in scientific notation.}$$

$$= 2.7559 \times 10^{-2} \qquad \text{Product rule}$$

$$= 0.027559 \qquad \text{Write without exponents.}$$

Thus, 700,000 nanometers would measure

$$2.7559 \times 10^{-2} \text{ in., or } 0.027559 \text{ in.}$$

◀ *Work Problem* **4** *at the Side.*

---

5 If the speed of light is approximately 3.0×10^5 km per sec, how many seconds does it take light to travel approximately 1.5×10^8 km from the sun to Earth? (*Source: World Almanac and Book of Facts.*)

EXAMPLE 5 Using Scientific Notation to Solve an Application

In 2003, the national debt was $\$3.9136 \times 10^{12}$ (which is more than \$3 trillion). The population of the United States was approximately 290 million that year. About how much would each person have had to contribute in order to pay off the national debt? (*Source:* U.S. Office of Management and Budget; U.S. Census Bureau.)

Write the population in scientific notation. Then divide to obtain the per person contribution.

$$\frac{3.9136 \times 10^{12}}{290,000,000}$$

$$= \frac{3.9136 \times 10^{12}}{\mathbf{2.9 \times 10^8}} \qquad \text{Write 290 million in scientific notation.}$$

$$= \frac{3.9136}{2.9} \times 10^4 \qquad \text{Quotient rule}$$

$$\approx 1.3495 \times 10^4 \qquad \text{Divide; round to 4 decimal places.}$$

$$\approx 13,495 \qquad \text{Write without exponents.}$$

Each person would have to pay about \$13,495.

◀ *Work Problem* **5** *at the Side.*

ANSWERS

4. 1.8×10^7 km, or 18,000,000 km
5. 5×10^2 sec, or 500 sec

12.8 ▶▶▶ Exercises

Write the numbers (other than dates) mentioned in the following statements in scientific notation.

1. NASA has budgeted $6,130,900,000 for 2003 and $5,868,900,000 for 2004 for the international space station. (*Source:* U.S. National Aeronautics and Space Administration.)

2. The mass of Pluto is 0.0021 times that of Earth; the mass of Jupiter is 317.83 times that of Earth. (*Source: World Almanac and Book of Facts.*)

Determine whether or not the given number is written in scientific notation as defined in Objective 1. If it is not, write it as such.

3. 4.56×10^3

4. 7.34×10^5

5. 5,600,000

6. 34,000

7. 0.004

8. 0.0007

9. 0.8×10^2

10. 0.9×10^3

11. Explain in your own words what it means for a number to be written in scientific notation.

12. Explain how to multiply a number by a positive power of ten. Then explain how to multiply a number by a negative power of ten.

Write each number in scientific notation. See Example 1.

🌐 **13.** 5,876,000,000

14. 9,994,000,000

15. 82,350

16. 78,330

17. 0.000007

18. 0.0000004

19. −0.00203

20. −0.0000578

Write each number without exponents. See Example 2.

🌐 **21.** 7.5×10^5

22. 8.8×10^6

23. 5.677×10^{12}

24. 8.766×10^9

25. 1×10^{12}

26. 1×10^7

27. -6.21×10^0

28. -8.56×10^0

29. 7.8×10^{-4}

30. 8.9×10^{-5}

31. 5.134×10^{-9}

32. 7.123×10^{-10}

Perform the indicated operations. Write the answers in scientific notation and then without exponents. See Example 3.

33. $(2 \times 10^8)(3 \times 10^3)$

34. $(3 \times 10^7)(3 \times 10^3)$

35. $(5 \times 10^4)(3 \times 10^2)$

36. $(8 \times 10^5)(2 \times 10^3)$

37. $(4 \times 10^{-6})(2 \times 10^3)$

38. $(3 \times 10^{-7})(2 \times 10^2)$

39. $(6 \times 10^3)(4 \times 10^{-2})$

40. $(7 \times 10^5)(3 \times 10^{-4})$

41. $(9 \times 10^4)(7 \times 10^{-7})$

42. $(6 \times 10^4)(8 \times 10^{-8})$

43. $(3.15 \times 10^{-4})(2.04 \times 10^8)$

44. $(4.92 \times 10^{-3})(2.25 \times 10^7)$

45. $\dfrac{9 \times 10^{-5}}{3 \times 10^{-1}}$

46. $\dfrac{12 \times 10^{-4}}{4 \times 10^{-3}}$

47. $\dfrac{8 \times 10^3}{2 \times 10^2}$

48. $\dfrac{15 \times 10^4}{3 \times 10^3}$

49. $\dfrac{2.6 \times 10^{-3}}{2 \times 10^2}$

50. $\dfrac{9.5 \times 10^{-1}}{5 \times 10^3}$

51. $\dfrac{4 \times 10^5}{8 \times 10^2}$

52. $\dfrac{3 \times 10^9}{6 \times 10^5}$

53. $\dfrac{2.6 \times 10^{-3} \times 7.0 \times 10^{-1}}{2 \times 10^2 \times 3.5 \times 10^{-3}}$

54. $\dfrac{9.5 \times 10^{-1} \times 2.4 \times 10^4}{5 \times 10^3 \times 1.2 \times 10^{-2}}$

55. $\dfrac{(1.65 \times 10^8)(5.24 \times 10^{-2})}{(6 \times 10^4)(2 \times 10^7)}$

Work each problem. In Exercises 58–60, give answers without exponents. See Examples 4 and 5.

56. Pollux, one of the brightest stars in the night sky, is 33.7 light-years from Earth. If one light-year is about 6,000,000,000,000 mi (that is, 6 trillion mi), about how many miles is Pollux from Earth? (*Source: World Almanac and Book of Facts.*)

57. In March 2006, astronomers using the Spitzer Space Telescope discovered a twisted double-helix nebula, a conglomeration of dust and gas stretching across the center of the Milky Way galaxy. This nebula is 25,000 light-years from Earth. If one light-year is about 6,000,000,000,000 mi, about how many miles is the twisted double-helix nebula from Earth? (*Source*: http://articles.news.aol.com)

58. In 2003, the U.S. government collected about $6730 per person in taxes. If the population at that time was 290,000,000, how much did the government collect in taxes for 2003? (*Source*: U.S. Internal Revenue Service.)

59. In 2000, the population of the United States was about 281.4 million. To the nearest dollar, calculate how much each person in the United States would have had to contribute in order to make one lucky person a trillionaire (that is, to give that person $1,000,000,000,000). (*Source*: U.S. Census Bureau.)

60. In 2006, Congress raised the government's debt limit to 9×10^{12}. When this national debt limit is reached, about how much is it for every man, woman, and child in the country? Use 300 million as the population of the United States. (*Source: The Gazette*, Cedar Rapids, Iowa, March 17, 2006.)

Chapter 12 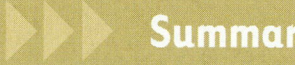 Summary

▶ Key Terms

12.1 term A term is a number, a variable, or a product or quotient of a number and one or more variables raised to powers.

like terms Terms with exactly the same variables (including the same exponents) are called like terms.

polynomial A polynomial is a term or the sum of a finite number of terms with whole number exponents.

descending powers A polynomial in x is written in descending powers if the exponents on x in its terms are in decreasing order.

degree of a term The degree of a term is the sum of the exponents on the variables.

degree of a polynomial The degree of a polynomial is the greatest degree of any term of the polynomial.

monomial A monomial is a polynomial with exactly one term.

binomial A binomial is a polynomial with exactly two terms.

trinomial A trinomial is a polynomial with exactly three terms.

12.2 exponential expression A number written with an exponent is an exponential expression.

3^4 ← Exponent $\left.\begin{array}{l}\end{array}\right\}$ Exponential expression ↑—— Base

12.3 FOIL FOIL is a shortcut method for finding the product of two binomials. The letters of the word **FOIL** originate as follows: Multiply the **F**irst terms, multiply the **O**uter terms (to get the outer product), multiply the **I**nner terms (to get the inner product), and multiply the **L**ast terms.

outer product The outer product of $(2x + 3)(x - 5)$ is $2x(-5)$.

inner product The inner product of $(2x + 3)(x - 5)$ is $3x$.

12.4 conjugate The conjugate of $a + b$ is $a - b$.

12.8 scientific notation A number written as $a \times 10^n$, where $1 \le |a| < 10$ and n is an integer, is in scientific notation.

▶ New Symbols

x^{-n} x to the negative n power

▶ Test Your Word Power

See how well you have learned the vocabulary in this chapter. Answers, with examples, follow the Quick Review.

1. A **polynomial** is an algebraic expression made up of
  A. a term or a finite product of terms with positive coefficients and exponents
  B. a term or a finite sum of terms with real coefficients and whole number exponents
  C. the product of two or more terms with positive exponents
  D. the sum of two or more terms with whole number coefficients and exponents.

2. The **degree of a term** is
  A. the number of variables in the term
  B. the product of the exponents on the variables
  C. the least exponent on the variables
  D. the sum of the exponents on the variables.

3. A **trinomial** is a polynomial with
  A. only one term
  B. exactly two terms
  C. exactly three terms
  D. more than three terms.

4. A **binomial** is a polynomial with
  A. only one term
  B. exactly two terms
  C. exactly three terms
  D. more than three terms.

5. A **monomial** is a polynomial with
  A. only one term
  B. exactly two terms
  C. exactly three terms
  D. more than three terms.

6. **FOIL** is a method for
  A. adding two binomials
  B. adding two trinomials
  C. multiplying two binomials
  D. multiplying two trinomials.

► Quick Review

Concepts	Examples

12.1 Adding and Subtracting Polynomials

Addition
Add like terms.

Add.
$$2x^2 + 5x - 3$$
$$\underline{5x^2 - 2x + 7}$$
$$7x^2 + 3x + 4$$

Subtraction
Change the signs of the terms in the second polynomial and add to the first polynomial.

Subtract. $(2x^2 + 5x - 3) - (5x^2 - 2x + 7)$
$$= (2x^2 + 5x - 3) + (-5x^2 + 2x - 7)$$
$$= -3x^2 + 7x - 10$$

12.2 The Product Rule and Power Rules for Exponents

For any integers m and n:

Product rule $\quad a^m \cdot a^n = a^{m+n}$

Power rules (a) $\quad (a^m)^n = a^{mn}$

(b) $\quad (ab)^m = a^m b^m$

(c) $\quad \left(\dfrac{a}{b}\right)^m = \dfrac{a^m}{b^m} \quad (b \neq 0)$.

Simplify.

$$2^4 \cdot 2^5 = 2^{4+5} = 2^9$$

$$(3^4)^2 = 3^{4 \cdot 2} = 3^8$$

$$(6a)^5 = 6^5 a^5$$

$$\left(\frac{2}{3}\right)^4 = \frac{2^4}{3^4}$$

12.3 Multiplying Polynomials

Multiply each term of the first polynomial by each term of the second polynomial. Then add like terms.

Multiply.
$$3x^3 - 4x^2 + 2x - 7$$
$$\underline{4x + 3}$$
$$9x^3 - 12x^2 + 6x - 21$$
$$\underline{12x^4 - 16x^3 + 8x^2 - 28x}$$
$$12x^4 - 7x^3 - 4x^2 - 22x - 21$$

FOIL Method
Step 1 Multiply the two **First** terms to get the first term of the answer.
Step 2 Find the **Outer** product and the **Inner** product and mentally add them, when possible, to get the middle term of the answer.
Step 3 Multiply the two **Last** terms to get the last term of the answer.
Add the terms found in Steps 1–3.

Multiply $(2x + 3)(5x - 4)$.
$$2x(5x) = \mathbf{10x^2}$$

$$2x(-4) + 3(5x) = \mathbf{7x}$$

$$3(-4) = \mathbf{-12}$$

The product is $\mathbf{10x^2 + 7x - 12}$.

12.4 Special Products

Square of a Binomial

$$(a + b)^2 = a^2 + 2ab + b^2$$

$$(a - b)^2 = a^2 - 2ab + b^2$$

Product of the Sum and Difference of Two Terms

$$(a + b)(a - b) = a^2 - b^2$$

Multiply.

$$(3x + 1)^2$$
$$= (3x)^2 + 2(3x)(1) + 1^2$$
$$= 9x^2 + 6x + 1$$

$$(2m - 5n)^2$$
$$= (2m)^2 - 2(2m)(5n) + (5n)^2$$
$$= 4m^2 - 20mn + 25n^2$$

$$(4a + 3)(4a - 3)$$
$$= (4a)^2 - 3^2$$
$$= 16a^2 - 9$$

Concepts	Examples

(12.5) Integer Exponents and the Quotient Rule

If $a, b \neq 0$, for integers m and n:

Zero exponent $a^0 = 1$

Negative exponent $a^{-n} = \dfrac{1}{a^n}$

Quotient rule $\dfrac{a^m}{a^n} = a^{m-n}$

Negative-to-positive rules $\dfrac{a^{-m}}{b^{-n}} = \dfrac{b^n}{a^m}$ $\left(\dfrac{a}{b}\right)^{-m} = \left(\dfrac{b}{a}\right)^m$.

Simplify.

$$15^0 = 1$$

$$5^{-2} = \frac{1}{5^2} = \frac{1}{25}$$

$$\frac{4^8}{4^3} = 4^{8-3} = 4^5$$

$$\frac{6^{-2}}{7^{-3}} = \frac{7^3}{6^2} \qquad \left(\frac{5}{3}\right)^{-4} = \left(\frac{3}{5}\right)^4$$

(12.6) Dividing a Polynomial by a Monomial

Divide each term of the polynomial by the monomial:

$$\frac{a+b}{c} = \frac{a}{c} + \frac{b}{c}.$$

Divide. $\dfrac{4x^3 - 2x^2 + 6x - 8}{2x}$

$$= \frac{4x^3}{2x} - \frac{2x^2}{2x} + \frac{6x}{2x} - \frac{8}{2x}$$

$$= 2x^2 - x + 3 - \frac{4}{x}$$

(12.7) Dividing a Polynomial by a Polynomial

Use "long division."

Divide.

$$3x + 4 \overline{)\, 6x^2 - 7x - 21} \quad \leftarrow 2x - 5 + \frac{-1}{3x+4}$$

$$\underline{6x^2 + 8x}$$
$$-15x - 21$$
$$\underline{-15x - 20}$$
$$-1 \leftarrow \text{Remainder}$$

(12.8) An Application of Exponents: Scientific Notation

To write a number in scientific notation (as $a \times 10^n$, where $1 \leq |a| < 10$), move the decimal point to the right of the first nonzero digit. If the decimal point is moved n places, and this makes the number smaller, n is positive; if it makes the number larger, n is negative. If the decimal point is not moved, n is 0.

Write in scientific notation.

$$247 = 2.47 \times 10^2$$
$$0.0051 = 5.1 \times 10^{-3}$$

Write without exponents.

$$3.25 \times 10^5 = 325{,}000$$
$$8.44 \times 10^{-6} = 0.00000844$$

ANSWERS TO TEST YOUR WORD POWER

1. B; *Example:* $5x^3 + 2x^2 - 7$

2. D; *Examples:* The term 6 has degree 0, $3x$ has degree 1, $-2x^8$ has degree 8, and $5x^2y^4$ has degree 6.

3. C; *Example:* $2a^2 - 3ab + b^2$

4. B; *Example:* $3t^3 + 5t$

5. A; *Examples:* -5 and $4xy^5$

6. C; *Example:* $(m+4)(m-3)$

$$\overset{\text{F}\quad\text{O}\quad\text{I}\quad\text{L}}{= m(m) - 3m + 4m + 4(-3)}$$

$$= m^2 + m - 12$$

Math in the Media

In recent years, the number of natural disasters seems to be on the increase. Charles F. Richter devised a scale in 1935 to compare the intensities, or relative power, of earthquakes. The **intensity** of an earthquake (often mentioned in media reports) is measured relative to the intensity of a standard **zero-level** earthquake of intensity I_0. The relationship is equivalent to $I = I_0 \times 10^R$, where R is the **Richter scale** measure. For example, if an earthquake has magnitude 5.0 on the Richter scale, then its intensity is calculated as $I = I_0 \times 10^{5.0} = I_0 \times 100,000$, which is 100,000 times as intense as a zero-level earthquake.

To compare an earthquake that measures 8.0 on the Richter scale to one that measures 5.0, find the ratio of the intensities:

$$\frac{\text{intensity } 8.0}{\text{intensity } 5.0} = \frac{I_0 \times 10^{8.0}}{I_0 \times 10^{5.0}} = \frac{10^8}{10^5} = 10^{8-5} = 10^3 = 1000.$$

Therefore, an earthquake that measures 8.0 is 1000 times as intense as one that measures 5.0.

The Gazette
TUESDAY
April 19, 2008

Quake rattles Iowans

*Chances of 'big one'
happening here remote
UNI professor says*

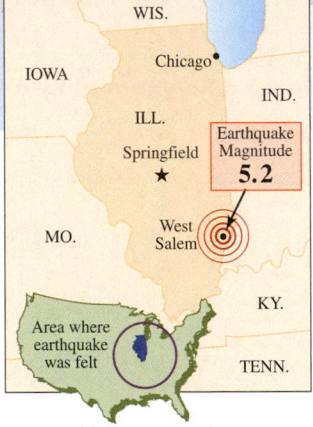

Source: ESRI; USGS.

The table gives Richter scale measurements for several earthquakes.

Year	Earthquake	Richter Scale Measurement
2008	West Salem, IL	5.2
2005	Northern Sumatra, Indonesia	8.6
2004	West coast of Northern Sumatra	9.1
2003	Southeastern Iran	6.6
1998	Balleny Islands region	8.1
1906	San Francisco, CA	7.7

Source: U.S. Geological Survey.

1. Compare the intensity of the 2004 west coast of northern Sumatra earthquake to that of the 1998 Balleny Islands region earthquake.

2. Compare the intensity of the 2005 northern Sumatra, Indonesia earthquake to that of the 2003 southeastern Iran earthquake.

3. Compare the intensity of the 1906 San Francisco earthquake, the most powerful to strike the United States, to that of the 2008 West Salem earthquake. (*Hint*: Use the exponential key of a scientific calculator to compute the required power of 10.)

4. Suppose an earthquake measures a value of x on the Richter scale. How would the intensity of a second earthquake compare if its Richter scale measure is $x + 4.0$? How would it compare if its Richter scale measure is $x - 1.0$?

Chapter 12 ▷▷▷ Review Exercises

[12.1] *Combine terms where possible in each polynomial. Write the answer in descending powers of the variable. Give the degree of the answer. Identify the polynomial as a* monomial, binomial, trinomial, *or* none of these.

1. $9m^2 + 11m^2 + 2m^2$

2. $-4p + p^3 - p^2 + 8p + 2$

3. $12a^5 - 9a^4 + 8a^3 + 2a^2 - a + 3$

4. $-7y^5 - 8y^4 - y^5 + y^4 + 9y$

Add or subtract as indicated.

5. Add.
$$\begin{array}{r} -2a^3 + 5a^2 \\ -3a^3 - a^2 \\ \hline \end{array}$$

6. Add.
$$\begin{array}{r} 4r^3 - 8r^2 + 6r \\ -2r^3 + 5r^2 + 3r \\ \hline \end{array}$$

7. Subtract.
$$\begin{array}{r} 6y^2 - 8y + 2 \\ -5y^2 + 2y - 7 \\ \hline \end{array}$$

8. Subtract.
$$\begin{array}{r} -12k^4 - 8k^2 + 7k - 5 \\ k^4 + 7k^2 + 11k + 1 \\ \hline \end{array}$$

9. $(2m^3 - 8m^2 + 4) + (8m^3 + 2m^2 - 7)$

10. $(-5y^2 + 3y + 11) + (4y^2 - 7y + 15)$

11. $(6p^2 - p - 8) - (-4p^2 + 2p + 3)$

12. $(12r^4 - 7r^3 + 2r^2) - (5r^4 - 3r^3 + 2r^2 + 1)$

[12.2] *Simplify each expression.*

13. $4^3 \cdot 4^8$

14. $(-5)^6(-5)^5$

15. $(-8x^4)(9x^3)$

16. $(2x^2)(5x^3)(x^9)$

17. $(19x)^5$

18. $(-4y)^7$

19. $5(pt)^4$

20. $\left(\dfrac{7}{5}\right)^6$

21. $(3x^2y^3)^3$

22. $(t^4)^8(t^2)^5$

23. $(6x^2z^4)^2(x^3yz^2)^4$

24. $\left(\dfrac{2m^3n}{p^2}\right)^3$

25. Find a polynomial that represents the volume of the figure. (If necessary, refer to the formulas on the inside covers.)

$5x^2$
$5x^2$
$5x^2$

26. Explain why the product rule for exponents does not apply to the expression $7^2 + 7^4$.

[12.3] *Find each product.*

27. $5x(2x + 14)$

28. $-3p^3(2p^2 - 5p)$

29. $(3r - 2)(2r^2 + 4r - 3)$

30. $(2y + 3)(4y^2 - 6y + 9)$

31. $(5p^2 + 3p)(p^3 - p^2 + 5)$

32. $(x + 6)(x - 3)$

33. $(3k - 6)(2k + 1)$

34. $(6p - 3q)(2p - 7q)$

35. $(m^2 + m - 9)(2m^2 + 3m - 1)$

[12.4] *Find each product.*

36. $(a + 4)^2$

37. $(3p - 2)^2$

38. $(2r + 5s)^2$

39. $(r + 2)^3$

40. $(2x - 1)^3$

41. $(2z + 7)(2z - 7)$

42. $(6m - 5)(6m + 5)$

43. $(5a + 6b)(5a - 6b)$

44. $(2x^2 + 5)(2x^2 - 5)$

45. The square of a binomial leads to a polynomial with how many terms? The product of the sum and difference of two terms leads to a polynomial with how many terms?

46. Explain why $(a + b)^2$ is not equal to $a^2 + b^2$.

[12.5] *Evaluate each expression.*

47. $5^0 + 8^0$

48. 2^{-5}

49. $\left(\dfrac{6}{5}\right)^{-2}$

50. $4^{-2} - 4^{-1}$

Simplify each expression. Assume that all variables represent nonzero numbers.

51. $\dfrac{6^{-3}}{6^{-5}}$

52. $\dfrac{x^{-7}}{x^{-9}}$

53. $\dfrac{p^{-8}}{p^4}$

54. $\dfrac{r^{-2}}{r^{-6}}$

55. $(2^4)^2$

56. $(9^3)^{-2}$

57. $(5^{-2})^{-4}$

58. $(8^{-3})^4$

59. $\dfrac{(m^2)^3}{(m^4)^2}$

60. $\dfrac{y^4 \cdot y^{-2}}{y^{-5}}$

61. $\dfrac{r^9 \cdot r^{-5}}{r^{-2} \cdot r^{-7}}$

62. $(-5m^3)^2$

63. $(2y^{-4})^{-3}$

64. $\dfrac{ab^{-3}}{a^4 b^2}$

65. $\dfrac{(6r^{-1})^2 \cdot (2r^{-4})}{r^{-5}(r^2)^{-3}}$

66. $\dfrac{(2m^{-5}n^2)^3 (3m^2)^{-1}}{m^{-2}n^{-4}(m^{-1})^2}$

[12.6] *Perform each division.*

67. $\dfrac{-15y^4}{-9y^2}$ **68.** $\dfrac{-12x^3y^2}{6xy}$ **69.** $\dfrac{6y^4 - 12y^2 + 18y}{-6y}$ **70.** $\dfrac{2p^3 - 6p^2 + 5p}{2p^2}$

71. $(5x^{13} - 10x^{12} + 20x^7 - 35x^5) \div (-5x^4)$ **72.** $(-10m^4n^2 + 5m^3n^3 + 6m^2n^4) \div (5m^2n)$

[12.7] *Perform each division.*

73. $(2r^2 + 3r - 14) \div (r - 2)$ **74.** $\dfrac{12m^2 - 11m - 10}{3m - 5}$

75. $\dfrac{10a^3 + 5a^2 - 14a + 9}{5a^2 - 3}$ **76.** $\dfrac{2k^4 + 4k^3 + 9k^2 - 8}{2k^2 + 1}$

[12.8] *Write each number in scientific notation.*

77. 48,000,000 **78.** 28,988,000,000 **79.** 0.000065 **80.** 0.0000000824

Write each number without exponents.

81. 2.4×10^4 **82.** 7.83×10^7 **83.** 8.97×10^{-7} **84.** 9.95×10^{-12}

Perform the indicated operations. Write the answers in scientific notation and then without exponents.

85. $(2 \times 10^{-3})(4 \times 10^5)$ **86.** $\dfrac{8 \times 10^4}{2 \times 10^{-2}}$

87. $\dfrac{12 \times 10^{-5} \times 5 \times 10^4}{4 \times 10^3 \times 6 \times 10^{-2}}$ **88.** $\dfrac{2.5 \times 10^5 \times 4.8 \times 10^{-4}}{7.5 \times 10^8 \times 1.6 \times 10^{-5}}$

89. A computer can perform 466,000,000 calculations per second. How many calculations can it perform per minute? Per hour?

90. There are 1×10^9 Social Security numbers. The population of the United States is about 3×10^8. How many Social Security numbers are available for each person? (*Source*: U.S. Census Bureau.)

▶▶▶ **Mixed Review Exercises**

Perform each indicated operation. Assume that all variables represent nonzero real numbers.

91. $19^0 - 3^0$

92. $(3p)^4 (3p^{-7})$

93. 7^{-2}

94. $(-7 + 2k)^2$

95. $\dfrac{2y^3 + 17y^2 + 37y + 7}{2y + 7}$

96. $\left(\dfrac{6r^2 s}{5}\right)^4$

97. $-m^5(8m^2 + 10m + 6)$

98. $\left(\dfrac{1}{2}\right)^{-5}$

99. $(25x^2y^3 - 8xy^2 + 15x^3y) \div (5x)$

100. $(6r^{-2})^{-1}$

101. $(2x + y)^3$

102. $2^{-1} + 4^{-1}$

103. $(a + 2)(a^2 - 4a + 1)$

104. $(5y^3 - 8y^2 + 7) - (-3y^3 + y^2 + 2)$

105. $(2r + 5)(5r - 2)$

106. $(12a + 1)(12a - 1)$

107. What polynomial represents the area of this rectangle?

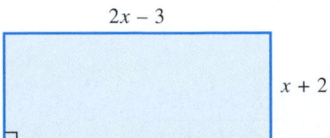

108. What polynomial represents the perimeter of this square? The area?

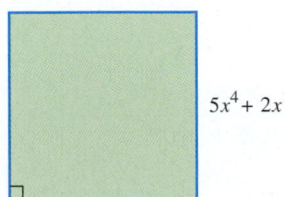

Perform the indicated operations.

1. $(5t^4 - 3t^2 + 7t + 3) - (t^4 - t^3 + 3t^2 + 8t + 3)$

1. _____

2. $(2y^2 - 8y + 8) + (-3y^2 + 2y + 3) - (y^2 + 3y - 6)$

2. _____

3. Subtract.

$$9t^3 - 4t^2 + 2t + 2$$
$$\underline{9t^3 + 8t^2 - 3t - 6}$$

3. _____

Simplify.

4. $(-2)^3 (-2)^2$

4. _____

5. $\left(\dfrac{6}{m^2}\right)^3, \quad m \neq 0$

5. _____

6. $3x^2 (-9x^3 + 6x^2 - 2x + 1)$

6. _____

7. $(2r - 3)(r^2 + 2r - 5)$

7. _____

8. $(t - 8)(t + 3)$

8. _____

9. $(4x + 3y)(2x - y)$

9. _____

10. $(5x - 2y)^2$

10. _____

11. $(10v + 3w)(10v - 3w)$

11. _____

12. $(x + 1)^3$

12. _____

13. What polynomial represents the perimeter of this square? The area?

13. _____

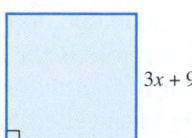

$3x + 9$

14. _____

15. _____

Evaluate each expression.

14. 5^{-4}

15. $(-3)^0 + 4^0$

16. $4^{-1} + 3^{-1}$

16. _____

Perform the indicated operations. In Exercises 17 and 18, write each answer using only positive exponents. Assume that variables represent nonzero numbers.

17. _____

17. $\dfrac{8^{-1} \cdot 8^4}{8^{-2}}$

18. _____

18. $\dfrac{(x^{-3})^{-2}(x^{-1}y)^2}{(xy^{-2})^2}$

19. _____

19. $\dfrac{8y^3 - 6y^2 + 4y + 10}{2y}$

20. _____

20. $(-9x^2y^3 + 6x^4y^3 + 12xy^3) \div (3xy)$

21. _____

21. $\dfrac{2x^2 + x - 36}{x - 4}$

22. _____

22. $(3x^3 - x + 4) \div (x - 2)$

Write each number in scientific notation.

23. (a) _____

  (b) _____

23. (a) 344,000,000,000

   (b) 0.00000557

Write each number without exponents.

24. (a) _____

  (b) _____

24. (a) 2.96×10^7

   (b) 6.07×10^{-8}

25. _____

25. A satellite galaxy of our own Milky Way, known as the Large Magellanic Cloud, is 1000 light-years across. If a light-year is equal to 5,890,000,000,000 mi, how many miles across is the Large Magellanic Cloud? (*Source:* "Images of Brightest Nebula Unveiled," *USA Today,* June 12, 2002.)

13

Factoring and Applications

W ireless communication uses radio waves to carry signals and messages across distances. Cellular phones, one of the most popular forms of wireless communication, have become an invaluable tool for people to stay connected to family, friends, and work while on the go. In 2007, there were about 243 million cell phone subscribers in the United States, with 81% of the population having cell phone service. Total revenue from this service was about $133 billion. (*Source:* CITA–The Wireless Association.)

In Exercise 31 of **Section 13.8,** we use a *quadratic equation* to model the number of cell phone subscribers in the United States in recent years.

13.1 ▶▶▶ Factors; The Greatest Common Factor

OBJECTIVES

1 Find the greatest common factor of a list of numbers.

2 Find the greatest common factor of a list of variable terms.

3 Factor out the greatest common factor.

4 Factor by grouping.

To **factor** a number means to write it as the product of two or more numbers. The product is called the **factored form** of the number. For example,

Factors

$$12 = \overbrace{6 \cdot 2}.$$

Factored form

Factoring is a process that "undoes" multiplying. We multiply $6 \cdot 2$ to get 12, but we factor 12 by writing it as $6 \cdot 2$.

OBJECTIVE 1 Find the greatest common factor of a list of numbers. An integer that is a factor of two or more integers is a **common factor** of those integers. For example, 6 is a common factor of 18 and 24 because 6 is a factor of both 18 and 24. Other common factors of 18 and 24 are 1, 2, and 3. The **greatest common factor (GCF)** of a list of integers is the largest common factor of those integers. This means 6 is the greatest common factor of 18 and 24, since it is the largest of their common factors.

> **Note**
>
> Factors of a number are also divisors of the number. The greatest common factor is the same as the greatest common divisor.

EXAMPLE 1 **Finding the Greatest Common Factor for Numbers**

Find the greatest common factor for each list of numbers.

(a) 30, 45

First write each number in prime factored form.

$$30 = 2 \cdot 3 \cdot 5$$
$$45 = 3 \cdot 3 \cdot 5$$

Use each prime the *least* number of times it appears in *all* the factored forms. There is no 2 in the prime factored form of 45, so there will be no 2 in the greatest common factor. The least number of times 3 appears in all the factored forms is 1. The least number of times 5 appears is also 1. From this, the

$$\text{GCF} = 3^1 \cdot 5^1 = 3 \cdot 5 = 15.$$

(b) 72, 120, 432

Find the prime factored form of each number.

$$72 = 2 \cdot 2 \cdot 2 \cdot 3 \cdot 3$$
$$120 = 2 \cdot 2 \cdot 2 \cdot 3 \cdot 5$$
$$432 = 2 \cdot 2 \cdot 2 \cdot 2 \cdot 3 \cdot 3 \cdot 3$$

The least number of times 2 appears in all the factored forms is 3, and the least number of times 3 appears is 1. There is no 5 in the prime factored form of either 72 or 432, so the

$$\text{GCF} = 2^3 \cdot 3^1 = 24.$$

Continued on Next Page

(c) 10, 11, 14

Write the prime factored form of each number.

$$10 = 2 \cdot 5$$
$$11 = 11$$
$$14 = 2 \cdot 7$$

There are no primes common to all three numbers, so the GCF is 1.

Work Problem **1** *at the Side.* ▶

OBJECTIVE **2** **Find the greatest common factor of a list of variable terms.** The terms x^4, x^5, x^6, and x^7 have x^4 as the greatest common factor because the least exponent on the variable x is 4.

$$x^4 = 1 \cdot \boldsymbol{x^4}, \quad x^5 = x \cdot \boldsymbol{x^4}, \quad x^6 = x^2 \cdot \boldsymbol{x^4}, \quad x^7 = x^3 \cdot \boldsymbol{x^4}$$

> **Note**
>
> The exponent on a variable in the GCF is the *least* exponent that appears on that variable in *all* the terms.

EXAMPLE 2 **Finding the Greatest Common Factor for Variable Terms**

Find the greatest common factor for each list of terms.

(a) $21m^7, -18m^6, 45m^8$

$$21m^7 = \boldsymbol{3} \cdot 7 \cdot \boldsymbol{m^7}$$
$$-18m^6 = -1 \cdot 2 \cdot \boldsymbol{3} \cdot 3 \cdot \boldsymbol{m^6}$$
$$45m^8 = \boldsymbol{3} \cdot 3 \cdot 5 \cdot \boldsymbol{m^8}$$

First, 3 is the greatest common factor of the coefficients 21, -18, and 45. The least exponent on m is 6, so the

$$\text{GCF} = 3m^6.$$

(b) $x^4y^2, x^7y^5, x^3y^7, y^{15}$

$$\boldsymbol{x^4y^2}, \quad x^7\boldsymbol{y^5}, \quad x^3\boldsymbol{y^7}, \quad \boldsymbol{y^{15}}$$

There is no x in the last term, y^{15}, so x will not appear in the greatest common factor. There is a y in each term, however, and 2 is the least exponent on y. The GCF is y^2.

(c) $-a^2b, -ab^2$

$$-a^2b = -1a^2b = \boldsymbol{-1} \cdot 1 \cdot \boldsymbol{a^2b}$$
$$-ab^2 = -1ab^2 = \boldsymbol{-1} \cdot \boldsymbol{1} \cdot ab^2$$

The factors of -1 are -1 and 1. Since $1 > -1$, the GCF is $1ab$, or ab.

> **Note**
>
> In a list of negative terms, sometimes a negative common factor is preferable (even though it is not the greatest common factor). In Example 2(c), for instance, we might prefer $-ab$ as the common factor. In factoring exercises like this, either answer will be acceptable.

1 Find the greatest common factor for each list of numbers.

(a) 30, 20, 15

$$30 = 2 \cdot 3 \cdot 5$$
$$20 = 2 \cdot \underline{\quad} \cdot \underline{\quad}$$
$$15 = 3 \cdot \underline{\quad}$$
$$\text{GCF} = \underline{\quad}$$

(b) 42, 28, 35

(c) 12, 18, 26, 32

(d) 10, 15, 21

2 Find the greatest common factor for each list of terms.

(a) $6m^4, 9m^2, 12m^5$

$6m^4 = 2 \cdot \underline{\hspace{0.8cm}} \cdot m^4$

$9m^2 = 3 \cdot \underline{\hspace{0.8cm}} \cdot \underline{\hspace{0.8cm}}$

$12m^5 = 2 \cdot 2 \cdot \underline{\hspace{0.8cm}} \cdot \underline{\hspace{0.8cm}}$

$GCF = \underline{\hspace{0.8cm}}$

Finding the Greatest Common Factor (GCF)

Step 1 **Factor.** Write each number in prime factored form.

Step 2 **List common factors.** List each prime number or each variable that is a factor of every term in the list. (If a prime does not appear in one of the prime factored forms, it cannot appear in the greatest common factor.)

Step 3 **Choose least exponents.** Use as exponents on the common prime factors the *least* exponents from the prime factored forms.

Step 4 **Multiply.** Multiply the primes from Step 3. If there are no primes left after Step 3, the greatest common factor is 1.

◀ *Work Problem* **2** *at the Side.*

OBJECTIVE 3 Factor out the greatest common factor. The polynomial

$$3m + 12$$

has two terms, $3m$ and 12. The greatest common factor of these two terms is 3. We can write $3m + 12$ so that each term is a product with 3 as one factor.

(b) $-12p^5, -18q^4$

$$3m + 12$$
$$= \mathbf{3} \cdot m + \mathbf{3} \cdot 4$$
$$= \mathbf{3}(m + 4) \qquad \text{Distributive property}$$

The factored form of $3m + 12$ is $3(m + 4)$. This process is called **factoring out the greatest common factor.**

CAUTION

The polynomial $3m + 12$ is *not* in factored form when written as the *sum*

$$3 \cdot m + 3 \cdot 4. \qquad \text{Not in factored form}$$

(c) y^4z^2, y^6z^8, z^9

The terms are factored, but the polynomial is not. The factored form of $3m + 12$ is the *product*

$$3(m + 4). \qquad \text{In factored form}$$

Writing a polynomial as a product, that is, in factored form, is called **factoring** the polynomial.

(d) $12p^{11}, 17q^5$

EXAMPLE 3 **Factoring Out the Greatest Common Factor**

Factor out the greatest common factor.

(a) $5y^2 + 10y$
$$= \mathbf{5y}(y) + \mathbf{5y}(2) \qquad \text{GCF} = 5y$$
$$= \mathbf{5y}(y + 2) \qquad \text{Distributive property}$$

Check Multiply the factored form.

$$5y(y + 2)$$
$$= 5y(y) + 5y(2) \qquad \text{Distributive property}$$
$$= 5y^2 + 10y \qquad \text{Original polynomial}$$

Continued on Next Page

ANSWERS

2. (a) $3; 3; m^2; 3; m^5; 3m^2$
(b) 6 (c) z^2 (d) 1

(b) $20m^5 + 10m^4 - 15m^3$

$= \textbf{5}\textbf{\textit{m}}^{\textbf{3}}(4m^2) + \textbf{5}\textbf{\textit{m}}^{\textbf{3}}(2m) - \textbf{5}\textbf{\textit{m}}^{\textbf{3}}(3)$ GCF $= 5m^3$

$= \textbf{5}\textbf{\textit{m}}^{\textbf{3}}(4m^2 + 2m - 3)$ Factor out $5m^3$.

Check $5m^3(4m^2 + 2m - 3)$

$= 20m^5 + 10m^4 - 15m^3$ Original polynomial

(c) $x^5 + x^3$

$= \textbf{\textit{x}}^{\textbf{3}}(x^2) + \textbf{\textit{x}}^{\textbf{3}}(\textbf{1})$

$= \textbf{\textit{x}}^{\textbf{3}}(x^2 + \textbf{1})$ — Don't forget the 1.

(d) $20m^7p^2 - 36m^3p^4$

$= 4m^3p^2(5m^4) - 4m^3p^2(9p^2)$ GCF $= 4m^3p^2$

$= 4m^3p^2(5m^4 - 9p^2)$ Factor out $4m^3p^2$.

(e) $\dfrac{1}{6}n^2 + \dfrac{5}{6}n$

$= \dfrac{\textbf{1}}{\textbf{6}}\textbf{\textit{n}}(n) + \dfrac{\textbf{1}}{\textbf{6}}\textbf{\textit{n}}(5)$ GCF $= \frac{1}{6}n$

$= \dfrac{\textbf{1}}{\textbf{6}}\textbf{\textit{n}}(n + 5)$

CAUTION
Be sure to include the **1** in a problem like Example 3(c). ***Check that the factored form can be multiplied out to give the original polynomial.***

Work Problem ③ at the Side. ▶

| **EXAMPLE 4** | **Factoring Out a Negative Common Factor** |

Factor $-8x^4 + 16x^3 - 4x^2$.

We can factor out either $4x^2$ or $-4x^2$ here. We factor out $-4x^2$ so that the coefficient of the first term in the trinomial factor will be positive.

$-8x^4 + 16x^3 - 4x^2$ — Be careful with signs.

$= -4x^2(2x^2) - 4x^2(-4x) - 4x^2(1)$ $-4x^2$ is a common factor.

$= -4x^2(2x^2 - 4x + 1)$ Factor out $-4x^2$.

Check $-4x^2(2x^2 - 4x + 1)$

$= -4x^2(2x^2) - 4x^2(-4x) - 4x^2(1)$ Distributive property

$= -8x^4 + 16x^3 - 4x^2$ Original polynomial

Work Problem ④ at the Side. ▶

Note
Whenever we factor a polynomial in which the coefficient of the first term of a polynomial is negative, we will factor out the negative common factor, even if it is just -1. However, it would also be correct to factor out $4x^2$ in Example 4 to obtain $4x^2(-2x^2 + 4x - 1)$.

③ Factor out the greatest common factor.

(a) $4x^2 + 6x$

(b) $10y^5 - 8y^4 + 6y^2$

(c) $m^7 + m^9$

(d) $8p^5q^2 + 16p^6q^3 - 12p^4q^7$

(e) $\dfrac{1}{3}b^2 - \dfrac{2}{3}b$

④ Factor

$-14a^3b^2 - 21a^2b^3 + 7ab$

by factoring out a negative common factor.

ANSWERS

3. **(a)** $2x(2x + 3)$
   (b) $2y^2(5y^3 - 4y^2 + 3)$
   (c) $m^7(1 + m^2)$
   (d) $4p^4q^2(2p + 4p^2q - 3q^5)$
   (e) $\dfrac{1}{3}b(b - 2)$

4. $-7ab(2a^2b + 3ab^2 - 1)$

5 Factor out the greatest common factor.

(a) $r(t - 4) + 5(t - 4)$

(b) $y^2(y + 2) - 3(y + 2)$

(c) $x(x - 1) - 5(x - 1)$

EXAMPLE 5 **Factoring Out a Common Binomial Factor**

Factor out the greatest common factor.

(a) $a(a + 3) + 4(a + 3)$
Sometimes the GCF has a factor with more than one term. The binomial $a + 3$ is the greatest common factor here.

Same

$$a(a + 3) + 4(a + 3)$$
$$= (a + 3)(a + 4)$$

(b) $x^2(x + 1) - 5(x + 1)$
$$= (x + 1)(x^2 - 5) \qquad \text{Factor out } x + 1.$$

◀ *Work Problem* **5** *at the Side.*

OBJECTIVE 4 Factor by grouping. *When a polynomial has four terms, common factors can sometimes be used to factor by grouping.*

EXAMPLE 6 **Factoring by Grouping**

Factor by grouping.

(a) $2x + 6 + ax + 3a$
Group the first two terms and the last two terms, since the first two terms have a common factor of 2 and the last two terms have a common factor of a.

$$2x + 6 + ax + 3a$$
$$= (2x + 6) + (ax + 3a)$$
$$= 2(x + 3) + a(x + 3)$$

The expression is still not in factored form because it is the *sum* of two terms. Now, however, $x + 3$ is a common factor and can be factored out.

$$2x + 6 + ax + 3a$$
$$= (2x + 6) + (ax + 3a) \qquad \text{Group the terms.}$$
$$= 2(x + 3) + a(x + 3) \qquad \text{Factor each group.}$$
$$= (x + 3)(2 + a) \qquad \text{Factor out } x + 3.$$

The final result is in factored form because it is a *product.* Note that the goal in factoring by grouping is to get a common factor, $x + 3$ here, so that the last step is possible. Check by multiplying the binomials using the FOIL method from **Section 12.3.**

Check $(x + 3)(2 + a)$
$$= 2x + ax + 6 + 3a \qquad \text{FOIL}$$
$$= 2x + 6 + ax + 3a, \qquad \text{Rearrange terms.}$$

which is the original polynomial.

Continued on Next Page

ANSWERS

5. (a) $(t - 4)(r + 5)$
(b) $(y + 2)(y^2 - 3)$
(c) $(x - 1)(x - 5)$

(b) $6ax + 24x + a + 4$

$$= (6ax + 24x) + (a + 4) \qquad \text{Group the terms.}$$

$$= 6x(a + 4) + 1(a + 4) \qquad \text{Factor each group.}$$

> Remember the 1.

$$= (a + 4)(6x + 1) \qquad \text{Factor out } a + 4.$$

Check $(a + 4)(6x + 1)$

$$= 6ax + a + 24x + 4 \qquad \text{FOIL}$$

$$= 6ax + 24x + a + 4, \qquad \text{Rearrange terms.}$$

which is the original polynomial.

(c) $2x^2 - 10x + 3xy - 15y$

$$= (2x^2 - 10x) + (3xy - 15y) \qquad \text{Group the terms.}$$

$$= 2x(x - 5) + 3y(x - 5) \qquad \text{Factor each group.}$$

$$= (x - 5)(2x + 3y) \qquad \text{Factor out the common factor, } x - 5.$$

Check $(x - 5)(2x + 3y)$

$$= 2x^2 + 3xy - 10x - 15y \qquad \text{FOIL}$$

$$= 2x^2 - 10x + 3xy - 15y \qquad \text{Original polynomial}$$

(d) $t^3 + 2t^2 - 3t - 6$

> Be sure to write a $+$ sign between the groups.

$$= (t^3 + 2t^2) + (-3t - 6) \qquad \text{Group the terms.}$$

$$= t^2(t + 2) - 3(t + 2) \qquad \text{Factor out } -3 \text{ so there is a common factor,}$$
$$\qquad\qquad t + 2; -3(t + 2) = -3t - 6.$$

> Be careful with signs.

$$= (t + 2)(t^2 - 3) \qquad \text{Factor out } t + 2.$$

Check by multiplying.

CAUTION

Be careful with signs when grouping in a problem like Example 6(d). It is wise to check the factoring in the second step, as shown in the example side comment, before continuing.

Work Problem ⑥ *at the Side.* ▶

Factoring a Polynomial with Four Terms by Grouping

Step 1 **Group terms.** Collect the terms into two groups so that each group has a common factor.

Step 2 **Factor within groups.** Factor out the greatest common factor from each group.

Step 3 **Factor the entire polynomial.** Factor a common binomial factor from the results of Step 2.

Step 4 **If necessary, rearrange terms.** If Step 2 does not result in a common binomial factor, try a different grouping.

---

⑥ Factor by grouping.

(a) $pq + 5q + 2p + 10$

(b) $2xy + 3y + 2x + 3$

(c) $2a^2 - 4a + 3ab - 6b$

(d) $x^3 + 3x^2 - 5x - 15$

7 Factor by grouping.

(a) $6y^2 - 20w + 15y - 8yw$

(b) $9mn - 4 + 12m - 3n$

EXAMPLE 7 Rearranging Terms Before Factoring by Grouping

Factor by grouping.

(a) $10x^2 - 12y + 15x - 8xy$

Factoring out the common factor of 2 from the first two terms and the common factor of x from the last two terms gives

$$10x^2 - 12y + 15x - 8xy$$
$$= 2(5x^2 - 6y) + x(15 - 8y).$$

This does not lead to a common factor, so we try rearranging the terms. There is usually more than one way to do this. We try the following.

$$10x^2 - 12y + 15x - \mathbf{8xy}$$

$= 10x^2 - \mathbf{8xy} - 12y + 15x$	Commutative property
$= (10x^2 - 8xy) + (-12y + 15x)$	Group the terms.
$= 2x(5x - 4y) + 3\mathbf{(-4y + 5x)}$	Factor each group.
$= 2x\mathbf{(5x - 4y)} + 3\mathbf{(5x - 4y)}$	Rewrite $-4y + 5x$.
$= \mathbf{(5x - 4y)}(2x + 3)$	Factor out $5x - 4y$.

Check $\quad (5x - 4y)(2x + 3)$

$= 10x^2 + 15x - 8xy - 12y$	FOIL
$= 10x^2 - 12y + 15x - 8xy$	Original polynomial

(b) $2xy + 12 - 3y - 8x$

We need to rearrange these terms to get two groups that each have a common factor. Trial and error suggests the following grouping.

$$2xy + 12 - 3y - 8x$$

Always write a + sign between the two groups.

$= (2xy - 3y) + (-8x + 12)$	Group the terms.
$= y(2x - 3) \mathbf{-} 4(2x - 3)$	Factor each group.

Be careful with signs.

$= (2x - 3)(y - 4)$	Factor out $2x - 3$.

Since the quantities in parentheses in the second step must be the same, we factored out -4 rather than 4.

Check $\quad (2x - 3)(y - 4)$

$= 2xy - 8x - 3y + 12$	FOIL
$= 2xy + 12 - 3y - 8x$	Original polynomial

CAUTION
Use negative signs carefully when grouping, as in Example 7(b), or a sign error will occur. *Always check by multiplying.*

◀ *Work Problem* **7** *at the Side.*

ANSWERS

7. (a) $(2y + 5)(3y - 4w)$
  (b) $(3m - 1)(3n + 4)$

13.1 ▶▶▶ Exercises

Find the greatest common factor for each list of numbers. See Example 1.

1. 12, 16

2. 18, 24

3. 40, 20, 4

4. 50, 30, 5

5. 18, 24, 36, 48

6. 15, 30, 45, 75

7. 4, 9, 12

8. 9, 16, 24

Find the greatest common factor for each list of terms. See Example 2.

9. $16y$, 24

10. $18w$, 27

11. $30x^3$, $40x^6$, $50x^7$

12. $60z^4$, $70z^8$, $90z^9$

13. $-x^4y^3$, $-xy^2$

14. $-a^4b^5$, $-a^3b$

15. $42ab^3$, $-36a$, $90b$, $-48ab$

16. $45c^3d$, $75c$, $90d$, $-105cd$

Complete each factoring.

17. $9m^4 = 3m^2 (\quad)$

18. $12p^5 = 6p^3 (\quad)$

19. $-8z^9 = -4z^5 (\quad)$

20. $-15k^{11} = -5k^8 (\quad)$

21. $6m^4n^5 = 3m^3n (\quad)$

22. $27a^3b^2 = 9a^2b (\quad)$

23. $12y + 24 = 12 (\quad)$

24. $18p + 36 = 18 (\quad)$

25. $10a^2 - 20a = 10a (\quad)$

26. $15x^2 - 30x = 15x (\quad)$

27. $8x^2y + 12x^3y^2 = 4x^2y (\quad)$

28. $18s^3t^2 + 10st = 2st (\quad)$

Factor out the greatest common factor, or a negative common factor if the coefficient of the term of greatest degree is negative. See Examples 3–5.

29. $x^2 - 4x$

30. $m^2 - 7m$

31. $6t^2 + 15t$

32. $8x^2 + 6x$

33. $\frac{1}{4}d^2 - \frac{3}{4}d$

34. $\frac{1}{5}z^2 + \frac{3}{5}z$

35. $-12x^3 - 6x^2$

36. $-21b^3 + 7b^2$

37. $65y^{10} + 35y^6$

38. $100a^5 + 16a^3$

39. $11w^3 - 100$

40. $13z^5 - 80$

41. $8m^2n^3 + 24m^2n^2$

42. $19p^2y - 38p^2y^3$

43. $-4x^3 + 10x^2 - 6x$

44. $-9z^3 + 6z^2 - 12z$

45. $13y^8 + 26y^4 - 39y^2$

46. $5x^5 + 25x^4 - 20x^3$

47. $45q^4p^5 + 36qp^6 + 81q^2p^3$

48. $125a^3z^5 + 60a^4z^4 - 85a^5z^2$

49. $c(x + 2) + d(x + 2)$

50. $r(5 - x) + t(5 - x)$

51. $a^2(2a + b) - b(2a + b)$

52. $3x(x^2 + 5) - y(x^2 + 5)$

53. $q(p + 4) - 1(p + 4)$

54. $y^2(x - 4) + 1(x - 4)$

Factor by grouping. See Examples 6 and 7.

55. $5m + mn + 20 + 4n$

56. $ts + 5t + 2s + 10$

⊕ **57.** $6xy - 21x + 8y - 28$

58. $2mn - 8n + 3m - 12$

59. $3xy + 9x + y + 3$

60. $6n + 4mn + 3 + 2m$

61. $7z^2 + 14z - az - 2a$

62. $2b^2 + 3b - 8ab - 12a$

63. $18r^2 + 12ry - 3xr - 2xy$

64. $5m^2 + 15mp - 2mr - 6pr$

65. $w^3 + w^2 + 9w + 9$

66. $y^3 + y^2 + 6y + 6$

67. $3a^3 + 6a^2 - 2a - 4$

68. $10x^3 + 15x^2 - 8x - 12$

69. $16m^3 - 4m^2p^2 - 4mp + p^3$

70. $10t^3 - 2t^2s^2 - 5ts + s^3$

71. $y^2 + 3x + 3y + xy$

72. $m^2 + 14p + 7m + 2mp$

73. $2z^2 + 6w - 4z - 3wz$

74. $2a^2 + 20b - 8a - 5ab$

Relating Concepts (Exercises 75–78) For Individual or Group Work

In many cases, the choice of which pairs of terms to group when factoring by grouping can be made in different ways. To see this for Example 7(b), **work Exercises 75–78 in order.**

75. Start with the polynomial from Example 7(b), $2xy + 12 - 3y - 8x$, and rearrange the terms as follows: $2xy - 8x - 3y + 12$. What property from **Section 1.7** allows this?

76. Group the first two terms and the last two terms of the rearranged polynomial in Exercise 75. Then factor each group.

77. Is your result from Exercise 76 in factored form? Explain your answer.

78. If your answer to Exercise 77 is *no*, factor the polynomial. Is the result the same as the one shown for Example 7(b)?

13.2 ▶▶▶ Factoring Trinomials

OBJECTIVES

1 Factor trinomials with a coefficient of 1 for the squared term.

2 Factor trinomials after factoring out the greatest common factor.

Using FOIL, the product of the binomials $k - 3$ and $k + 1$ is

$$(k - 3)(k + 1) = k^2 - 2k - 3. \quad \text{Multiplying}$$

Suppose instead that we are given the polynomial $k^2 - 2k - 3$ and want to rewrite it as the product $(k - 3)(k + 1)$. That is,

$$k^2 - 2k - 3 = (k - 3)(k + 1). \quad \text{Factoring}$$

Recall from **Section 13.1** that this process is called *factoring* the polynomial. Factoring reverses or, "undoes," multiplying.

OBJECTIVE 1 Factor trinomials with a coefficient of 1 for the squared term. When factoring polynomials with integer coefficients, we use only integers in the factors. For example, we can factor $x^2 + 5x + 6$ by finding integers m and n such that

$$x^2 + 5x + 6 \quad \text{is written as} \quad (x + \boldsymbol{m})(x + \boldsymbol{n}).$$

To find these integers m and n, we first use FOIL to multiply the two binomials on the right above:

$$(x + m)(x + n)$$
$$= x^2 + nx + mx + mn$$
$$= x^2 + (n + m)x + mn. \quad \text{Distributive property}$$

Comparing this result with $x^2 + 5x + 6$ shows that we must find integers m and n having a sum of 5 and a product of 6.

Product of *m* and *n* is 6.

$$x^2 + \boldsymbol{5}x + \boldsymbol{6} = x^2 + (\boldsymbol{n} + \boldsymbol{m})x + \boldsymbol{mn}$$

Sum of *m* and *n* is 5.

Because many pairs of integers have a sum of 5, it is best to begin by listing those pairs of integers whose product is 6. Both 5 and 6 are positive, so we consider only pairs in which both integers are positive.

Work Problem **1** *at the Side.* ▶

1 (a) List all pairs of positive integers whose product is 6.

(b) Find the pair from part (a) whose sum is 5.

From Problem 1 at the side, we see that the numbers 1 and 6 and the numbers 2 and 3 both have a product of 6, but only the pair 2 and 3 has a sum of 5. So 2 and 3 are the required integers, and

$$x^2 + 5x + 6 \quad \text{is factored as} \quad (x + \boldsymbol{2})(x + \boldsymbol{3}).$$

Check by multiplying the binomials using FOIL. ***Make sure that the sum of the outer and inner products produces the correct middle term.***

Check $(x + 2)(x + 3) = x^2 + \boldsymbol{5}x + 6$ Correct

$$\frac{\begin{array}{c} 2x \\ 3x \end{array}}{5x} \quad \text{Add.}$$

This method of factoring can be used only for trinomials that have 1 as the coefficient of the squared term.

2 Factor each trinomial.

(a) $y^2 + 12y + 20$

First complete the given list of numbers.

Factors of 20	Sums of Factors
20, 1	$20 + 1 \quad = 21$
10, ___	$10 + \text{___} = \text{___}$
5, ___	$5 + \text{___} = \text{___}$

(b) $x^2 + 9x + 18$

3 Factor each trinomial.

(a) $t^2 - 12t + 32$

First complete the given list of numbers.

Factors of 32	Sums of Factors
−32, −1	$-32 + (-1) = -33$
−16, ___	$-16 + (\text{___}) = \text{___}$
−8, ___	$-8 + (\text{___}) = \text{___}$

(b) $y^2 - 10y + 24$

2. (a) 2; 2; 12; 4; 4; 9; $(y + 10)(y + 2)$
   (b) $(x + 3)(x + 6)$
3. (a) −2; −2; −18; −4; −4; −12;
   $(t - 8)(t - 4)$
   (b) $(y - 6)(y - 4)$

EXAMPLE 1   **Factoring a Trinomial with All Positive Terms**

Factor $m^2 + 9m + 14$.

Look for two integers whose product is **14** and whose sum is **9**. List the pairs of integers whose products are 14. Then examine the sums. Only positive integers are needed since all signs in $m^2 + 9m + 14$ are positive.

Factors of 14	Sums of Factors	
14, 1	$14 + 1 = 15$	
7, 2	$7 + 2 = 9$	Sum is 9.

From the list, 7 and 2 are the required integers, since $7 \cdot 2 = 14$ and $7 + 2 = 9$. Thus,

$$m^2 + 9m + 14 \quad \text{factors as} \quad (m + 2)(m + 7).$$

Check $(m + 2)(m + 7)$

$$= m^2 + 7m + 2m + 14 \qquad \text{FOIL}$$

$$= m^2 + 9m + 14 \qquad \text{Original polynomial}$$

Note

In Example 1, the answer $(m + 2)(m + 7)$ also could have been written

$$(m + 7)(m + 2).$$

Because of the commutative property of multiplication, the order of the factors does not matter. ***Always check by multiplying.***

◀ *Work Problem* **2** *at the Side.*

EXAMPLE 2   **Factoring a Trinomial with a Negative Middle Term**

Factor $x^2 - 9x + 20$.

Find two integers whose product is **20** and whose sum is **−9**. Since the numbers we are looking for have a *positive product* and a *negative sum*, we consider only pairs of negative integers.

Factors of 20	Sums of Factors	
−20, −1	$-20 + (-1) = -21$	
−10, −2	$-10 + (-2) = -12$	
−5, −4	$-5 + (-4) = -9$	Sum is −9.

The required integers are −5 and −4, so

$$x^2 - 9x + 20 \quad \text{factors as} \quad (x - 5)(x - 4).$$

Check $(x - 5)(x - 4)$

$$= x^2 - 4x - 5x + 20 \qquad \text{FOIL}$$

$$= x^2 - 9x + 20 \qquad \text{Original polynomial}$$

◀ *Work Problem* **3** *at the Side.*

EXAMPLE 3 **Factoring a Trinomial with Two Negative Terms**

Factor $p^2 - 2p - 15$.

Find two integers whose product is -15 and whose sum is -2. If these numbers do not come to mind right away, find them (if they exist) by listing all the pairs of integers whose product is -15. Because the last term, -15, is negative, we need pairs of integers with different signs.

Factors of -15	Sums of Factors
$15, -1$	$15 + (-1) = 14$
$-15, 1$	$-15 + 1 = -14$
$5, -3$	$5 + (-3) = 2$
$\mathbf{-5, 3}$	$-5 + 3 = \mathbf{-2}$ Sum is -2.

The required integers are -5 and 3, so

$$p^2 - 2p - 15 \quad \text{factors as} \quad (p - \mathbf{5})(p + \mathbf{3}).$$

Check Multiply $(p - 5)(p + 3)$ to obtain $p^2 - 2p - 15$.

Note

In Examples 1–3, notice that we listed factors in descending order (disregarding sign) when we were looking for the required pair of integers. This helps avoid skipping the correct combination.

Work Problem **4** *at the Side.* ▶

As shown in the next example, some trinomials cannot be factored using only integers. We call such trinomials **prime polynomials.**

EXAMPLE 4 **Deciding Whether Polynomials Are Prime**

Factor each trinomial.

(a) $x^2 - 5x + 12$

As in Example 2, both factors must be negative to give a positive product and a negative sum. First, list all the pairs of negative integers whose product is 12. Then examine the sums.

Factors of 12	Sums of Factors
$-12, -1$	$-12 + (-1) = -13$
$-6, -2$	$-6 + (-2) = -8$
$-4, -3$	$-4 + (-3) = -7$

None of the pairs of integers has a sum of -5. Therefore, the trinomial $x^2 - 5x + 12$ *cannot be factored using only integers; it is a prime polynomial.*

(b) $k^2 - 8k + 11$

There is no pair of integers whose product is 11 and whose sum is -8, so $k^2 - 8k + 11$ is a prime polynomial.

Work Problem **5** *at the Side.* ▶

4 Factor each trinomial.

(a) $a^2 - 9a - 22$

(b) $r^2 - 6r - 16$

5 Factor each trinomial, if possible.

(a) $r^2 - 3r - 4$

(b) $m^2 - 2m + 5$

ANSWERS
4. **(a)** $(a - 11)(a + 2)$ **(b)** $(r - 8)(r + 2)$
5. **(a)** $(r - 4)(r + 1)$ **(b)** prime

6 Factor each trinomial.

(a) $b^2 - 3ab - 4a^2$

(b) $r^2 - 6rs + 8s^2$

Guidelines for factoring a trinomial of the form $x^2 + bx + c$ are summarized here.

> ### Factoring $x^2 + bx + c$
>
> Find two integers whose product is c and whose sum is b.
>
> **1.** Both integers must be positive if b and c are positive.
> **2.** Both integers must be negative if c is positive and b is negative.
> **3.** One integer must be positive and one must be negative if c is negative.

EXAMPLE 5 **Factoring a Trinomial with Two Variables**

Factor $z^2 - 2bz - 3b^2$.

Here, the coefficient of z in the middle term is $-2b$, so we need to find two expressions whose product is $-3b^2$ and whose sum is $-2b$. The expressions are $-3b$ and b, so

$$z^2 - 2bz - 3b^2 \quad \text{factors as} \quad (z - 3b)(z + b).$$

Check $(z - 3b)(z + b)$

$$= z^2 + zb - 3bz - 3b^2 \qquad \text{FOIL}$$

$$= z^2 + 1bz - 3bz - 3b^2 \qquad \text{Identity and commutative properties}$$

$$= z^2 - 2bz - 3b^2 \qquad \text{Combine like terms.}$$

◀ Work Problem **6** at the Side.

---

7 Factor each trinomial completely.

(a) $2p^3 + 6p^2 - 8p$

OBJECTIVE 2 Factor trinomials after factoring out the greatest common factor. The trinomial in the next example does not have a coefficient of 1 for the squared term. (In fact, there is no squared term.) However, there may be a common factor.

EXAMPLE 6 **Factoring a Trinomial with a Common Factor**

Factor $4x^5 - 28x^4 + 40x^3$.

First, factor out the greatest common factor, $4x^3$.

$$4x^5 - 28x^4 + 40x^3$$

$$= 4x^3(x^2 - 7x + 10)$$

(b) $-3x^4 + 15x^3 - 18x^2$

Now factor $x^2 - 7x + 10$. The integers -5 and -2 have a product of 10 and a sum of -7. The complete factored form is

Include $4x^3$. ⟶ $4x^3(x - 5)(x - 2)$.

Check $4x^3(x - 5)(x - 2)$

$$= 4x^3(x^2 - 7x + 10) \qquad \text{FOIL}$$

$$= 4x^5 - 28x^4 + 40x^3 \qquad \text{Distributive property}$$

◀ Work Problem **7** at the Side.

ANSWERS

6. (a) $(b - 4a)(b + a)$
(b) $(r - 4s)(r - 2s)$
7. (a) $2p(p + 4)(p - 1)$
(b) $-3x^2(x - 3)(x - 2)$

> **CAUTION**
> When factoring, *always look for a common factor first.* Remember to include the common factor as part of the answer. As a check, multiplying out the complete factored form should give the original polynomial.

13.2 ▶▶▶ Exercises

1. When factoring a trinomial in x as $(x + a)(x + b)$, what must be true of a and b, if the last term of the trinomial is negative?

2. In Exercise 1, what must be true of a and b if the last term is positive?

3. What is meant by a *prime polynomial*?

4. How can you check your work when factoring a trinomial? Does the check ensure that the trinomial is *completely* factored?

In Exercises 5–8, list all pairs of integers with the given product. Then find the pair whose sum is given. See the tables in Examples 1–4.

5. Product: 12;   Sum: 7

6. Product: 18;   Sum: 9

7. Product: -24;   Sum: -5

8. Product: -36;   Sum: -16

9. Which one of the following is the correct factored form of $x^2 - 12x + 32$?

A. $(x - 8)(x + 4)$ **B.** $(x + 8)(x - 4)$
C. $(x - 8)(x - 4)$ **D.** $(x + 8)(x + 4)$

10. What would be the first step in factoring
$$2x^3 + 8x^2 - 10x?$$

Complete each factoring.

11. $x^2 + 15x + 44 = (x + 4)(\qquad)$

12. $r^2 + 15r + 56 = (r + 7)(\qquad)$

13. $x^2 - 9x + 8 = (x - 1)(\qquad)$

14. $t^2 - 14t + 24 = (t - 2)(\qquad)$

15. $y^2 - 2y - 15 = (y + 3)(\qquad)$

16. $t^2 - t - 42 = (t + 6)(\qquad)$

17. $x^2 + 9x - 22 = (x - 2)(\qquad)$

18. $x^2 + 6x - 27 = (x - 3)(\qquad)$

19. $y^2 - 7y - 18 = (y + 2)(\qquad)$

20. $y^2 - 2y - 24 = (y + 4)(\qquad)$

Factor completely. If a polynomial cannot be factored, write prime. See Examples 1–4.

21. $y^2 + 9y + 8$

22. $a^2 + 9a + 20$

23. $b^2 + 8b + 15$

24. $x^2 + 6x + 8$

25. $m^2 + m - 20$

26. $p^2 + 4p - 5$

27. $x^2 + 3x - 40$

28. $d^2 + 4d - 45$

29. $y^2 - 8y + 15$

30. $y^2 - 6y + 8$

31. $z^2 - 15z + 56$

32. $x^2 - 13x + 36$

33. $r^2 - r - 30$

34. $q^2 - q - 42$

35. $a^2 - 8a - 48$

36. $m^2 - 10m - 24$

37. $x^2 + 4x + 5$

38. $t^2 + 11t + 12$

Factor completely. See Examples 5 and 6.

39. $r^2 + 3ra + 2a^2$

40. $x^2 + 5xa + 4a^2$

41. $x^2 + 4xy + 3y^2$

42. $p^2 + 9pq + 8q^2$

43. $t^2 - tz - 6z^2$

44. $a^2 - ab - 12b^2$

45. $v^2 - 11vw + 30w^2$

46. $v^2 - 11vx + 24x^2$

47. $4x^2 + 12x - 40$

48. $5y^2 - 5y - 30$

49. $2t^3 + 8t^2 + 6t$

50. $3t^3 + 27t^2 + 24t$

51. $-2x^6 - 8x^5 + 42x^4$

52. $-4y^5 - 12y^4 + 40y^3$

53. $a^5 + 3a^4b - 4a^3b^2$

54. $z^{10} - 4z^9y - 21z^8y^2$

55. $m^3n - 10m^2n^2 + 24mn^3$

56. $y^3z + 3y^2z^2 - 54yz^3$

57. Use the FOIL method from **Section 12.3** to show that $(2x + 4)(x - 3) = 2x^2 - 2x - 12$. Why, then, is it incorrect to completely factor $2x^2 - 2x - 12$ as $(2x + 4)(x - 3)$?

58. Why is it incorrect to completely factor $3x^2 + 9x - 12$ as the product $(x - 1)(3x + 12)$?

13.3 ▶▶▶ Factoring Trinomials by Grouping

Trinomials like $2x^2 + 7x + 6$, in which the coefficient of the squared term is *not* 1, are factored with extensions of the methods from the previous sections. One such method uses factoring by grouping from **Section 13.1.**

OBJECTIVE 1 Factor trinomials by grouping when the coefficient of the squared term is not 1. Recall that a trinomial such as $m^2 + 3m + 2$ is factored by finding two integers whose product is 2 and whose sum is 3. To factor $2x^2 + 7x + 6$, we look for two integers whose product is $2 \cdot 6 = 12$ and whose sum is 7.

OBJECTIVE

1 **Factor trinomials by grouping when the coefficient of the squared term is not 1.**

$$\begin{array}{c} \text{Sum is 7.} \\ 2x^2 + 7x + \mathbf{6} \\ \text{Product is } 2 \cdot 6 = 12. \end{array}$$

By considering pairs of positive integers whose product is 12, the necessary integers are found to be 3 and 4. We use these integers to write the middle term, $7x$, as $7x = 3x + 4x$. The trinomial $2x^2 + 7x + 6$ becomes

$$2x^2 + \mathbf{7x} + 6$$
$$= 2x^2 + \underbrace{\mathbf{3x + 4x}}_{\mathbf{7x}} + 6$$
$$= (2x^2 + 3x) + (4x + 6) \qquad \text{Group the terms.}$$
$$= x\,(\mathbf{2x + 3}) + 2\,(\mathbf{2x + 3}) \qquad \text{Factor each group.}$$
$$\qquad\qquad \text{Must be the same.}$$
$$= (\mathbf{2x + 3})(x + 2). \qquad \text{Factor out } 2x + 3.$$

Check Multiply $(2x + 3)(x + 2)$ to obtain $2x^2 + 7x + 6$.

In the preceding example, we could have written $7x$ as $4x + 3x$. Factoring by grouping this way would give the same answer.

Work Problem **1** *at the Side.* ▶

EXAMPLE 1 **Factoring Trinomials by Grouping**

Factor each trinomial.

(a) $6r^2 + r - 1$

We must find two integers with a product of $6(-1) = -6$ and a sum of 1. The integers are -2 and 3. We write the middle term, r, as $-2r + 3r$.

$$6r^2 + \mathbf{r} - 1$$
$$= 6r^2 - \mathbf{2r} + \mathbf{3r} - 1 \qquad r = -2r + 3r$$
$$= (6r^2 - 2r) + (3r - 1) \qquad \text{Group the terms.}$$
$$= 2r\,(\mathbf{3r - 1}) + \mathbf{1}\,(\mathbf{3r - 1}) \qquad \text{The binomials must be the same.}$$
$$\qquad\qquad \boxed{\text{Remember the 1.}}$$
$$= (\mathbf{3r - 1})(2r + 1) \qquad \text{Factor out } 3r - 1.$$

Check Multiply $(3r - 1)(2r + 1)$ to obtain $6r^2 + r - 1$.

Continued on Next Page

---

1 (a) Factor $2x^2 + 7x + 6$ by writing $7x$ as $4x + 3x$. Complete the following.

$2x^2 + 7x + 6$

$= 2x^2 + 4x + 3x + 6$

$= (2x^2 + \underline{}) + (3x + \underline{})$

$= 2x(x + \underline{}) + 3(x + \underline{})$

$= (\underline{})(2x + 3)$

(b) Is the answer in part (a) the same as in the example? (Remember that the order of the factors does not matter.)

2 Factor each trinomial by grouping.

(a) $2m^2 + 7m + 3$

(b) $5p^2 - 2p - 3$

(c) $15k^2 - km - 2m^2$

3 Factor each trinomial completely.

(a) $-4x^2 + 2x + 30$

(b) $18p^4 + 63p^3 + 27p^2$

(c) $6a^2 + 3ab - 18b^2$

(b) $12z^2 - 5z - 2$

Look for two integers whose product is $12(-2) = -24$ and whose sum is -5. The required integers are 3 and -8, so

$$12z^2 - 5z - 2$$
$$= 12z^2 + 3z - 8z - 2 \qquad -5z = 3z - 8z$$
$$= (12z^2 + 3z) + (-8z - 2) \qquad \text{Group the terms.}$$
$$= 3z(4z + 1) - 2(4z + 1) \qquad \text{Factor each group.}$$

> Be careful with signs.

$$= (4z + 1)(3z - 2). \qquad \text{Factor out } 4z + 1.$$

Check Multiply $(4z + 1)(3z - 2)$ to obtain $12z^2 - 5z - 2$.

(c) $10m^2 + mn - 3n^2$

Two integers whose product is $10(-3) = -30$ and whose sum is 1 are -5 and 6. Rewrite the trinomial with four terms.

$$10m^2 + mn - 3n^2$$
$$= 10m^2 - 5mn + 6mn - 3n^2 \qquad mn = -5mn + 6mn$$
$$= 5m(2m - n) + 3n(2m - n) \qquad \text{Group the terms; factor each group.}$$
$$= (2m - n)(5m + 3n) \qquad \text{Factor out } 2m - n.$$

> Check by multiplying.

◀ *Work Problem* **2** *at the Side.*

EXAMPLE 2 **Factoring a Trinomial with a Common Factor by Grouping**

Factor $28x^5 - 58x^4 - 30x^3$.

First factor out the greatest common factor, $2x^3$.

$$28x^5 - 58x^4 - 30x^3$$
$$= 2x^3(14x^2 - 29x - 15)$$

To factor $14x^2 - 29x - 15$, find two integers whose product is $14(-15) = -210$ and whose sum is -29. Factoring 210 into prime factors gives

$$210 = 2 \cdot 3 \cdot 5 \cdot 7.$$

Combine these prime factors in pairs in different ways, using one positive factor and one negative factor to get -210. The factors 6 and -35 have the correct sum, -29. Now rewrite the given trinomial and factor it.

$$28x^5 - 58x^4 - 30x^3$$

> Remember the common factor.

$$= 2x^3(14x^2 - 29x - 15)$$
$$= 2x^3(14x^2 + 6x - 35x - 15)$$
$$= 2x^3[(14x^2 + 6x) + (-35x - 15)]$$
$$= 2x^3[2x(7x + 3) - 5(7x + 3)]$$
$$= 2x^3[(7x + 3)(2x - 5)]$$
$$= 2x^3(7x + 3)(2x - 5)$$

◀ *Work Problem* **3** *at the Side.*

13.3 ▶▶▶ Exercises

The middle term of each trinomial has been rewritten. Now factor by grouping.
See Example 1.

1. $m^2 + 8m + 12$
$= m^2 + 6m + 2m + 12$

2. $x^2 + 9x + 14$
$= x^2 + 7x + 2x + 14$

3. $a^2 + 3a - 10$
$= a^2 + 5a - 2a - 10$

4. $y^2 - 2y - 24$
$= y^2 + 4y - 6y - 24$

5. $10t^2 + 9t + 2$
$= 10t^2 + 5t + 4t + 2$

6. $6x^2 + 13x + 6$
$= 6x^2 + 9x + 4x + 6$

7. $15z^2 - 19z + 6$
$= 15z^2 - 10z - 9z + 6$

8. $12p^2 - 17p + 6$
$= 12p^2 - 9p - 8p + 6$

9. $8s^2 + 2st - 3t^2$
$= 8s^2 - 4st + 6st - 3t^2$

10. $3x^2 - xy - 14y^2$
$= 3x^2 - 7xy + 6xy - 14y^2$

11. $15a^2 + 22ab + 8b^2$
$= 15a^2 + 10ab + 12ab + 8b^2$

12. $25m^2 + 25mn + 6n^2$
$= 25m^2 + 15mn + 10mn + 6n^2$

13. Which pair of integers would be used to rewrite the middle term when factoring $12y^2 + 5y - 2$ by grouping?

A. $-8, 3$ **B.** $8, -3$ **C.** $-6, 4$ **D.** $6, -4$

14. Which pair of integers would be used to rewrite the middle term when factoring $20b^2 - 13b + 2$ by grouping?

A. $10, 3$ **B.** $-10, -3$ **C.** $8, 5$ **D.** $-8, -5$

Complete the steps to factor each trinomial by grouping.

15. $2m^2 + 11m + 12$

(a) Find two integers whose product is

_____ · _____ = _____ and whose
sum is _____.

(b) The required integers are _____ and
_____.

(c) Write the middle term $11m$ as _____ +
_____.

(d) Rewrite the given trinomial using four terms.

(e) Factor the polynomial in part (d) by grouping.

(f) Check by multiplying.

16. $6y^2 - 19y + 10$

(a) Find two integers whose product is

_____ · _____ = _____ and whose
sum is _____.

(b) The required integers are _____ and
_____.

(c) Write the middle term $-19y$ as _____ +
_____.

(d) Rewrite the given trinomial using four terms.

(e) Factor the polynomial in part (d) by grouping.

(f) Check by multiplying.

Factor each trinomial by grouping. See Examples 1 and 2.

17. $2x^2 + 7x + 3$

18. $3y^2 + 13y + 4$

19. $4r^2 + r - 3$

20. $4r^2 + 3r - 10$

21. $8m^2 - 10m - 3$

22. $20x^2 - 28x - 3$

23. $21m^2 + 13m + 2$

24. $38x^2 + 23x + 2$

25. $6b^2 + 7b + 2$

26. $6w^2 + 19w + 10$

27. $12y^2 - 13y + 3$

28. $15a^2 - 16a + 4$

29. $24x^2 - 42x + 9$

30. $48b^2 - 74b - 10$

31. $2m^3 + 2m^2 - 40m$

32. $3x^3 + 12x^2 - 36x$

33. $-32z^5 + 20z^4 + 12z^3$

34. $-18x^5 - 15x^4 + 75x^3$

35. $12p^2 + 7pq - 12q^2$

36. $6m^2 - 5mn - 6n^2$

37. $6a^2 - 7ab - 5b^2$

38. $25g^2 - 5gh - 2h^2$

39. $5 - 6x + x^2$

40. $7 + 8x + x^2$

41. On a quiz, a student factored $16x^2 - 24x + 5$ by grouping as follows.

$$16x^2 - 24x + 5$$
$$= 16x^2 - 4x - 20x + 5$$
$$= 4x(4x - 1) - 5(4x - 1) \qquad \text{His answer}$$

He thought his answer was correct since it checked by multiplying. Why was his answer marked wrong? What is the correct factored form?

42. On the same quiz, another student factored $3k^3 - 12k^2 - 15k$ by first factoring out the common factor $3k$ to get $3k(k^2 - 4k - 5)$. Then she wrote

$$k^2 - 4k - 5$$
$$= k^2 - 5k + k - 5$$
$$= k(k - 5) + 1(k - 5)$$
$$= (k - 5)(k + 1). \qquad \text{Her answer}$$

Why was her answer marked wrong? What is the correct factored form?

13.4 ▶▶▶ Factoring Trinomials Using FOIL

OBJECTIVE **1** **Factor trinomials using FOIL.**   This section shows an alternative method of factoring trinomials in which the coefficient of the squared term is not 1. This method uses trial and error.

To factor $2x^2 + 7x + 6$ (the same trinomial factored at the beginning of **Section 13.3**) by trial and error, we use FOIL backwards. We want to write $2x^2 + 7x + 6$ as the product of two binomials.

$$(\qquad)(\qquad)$$

The product of the two first terms of the binomials is $2x^2$. The possible factors of $2x^2$ are $2x$ and x or $-2x$ and $-x$. Since all terms of the trinomial are positive, we consider only positive factors. Thus, we have

$$(2x \qquad)(x \qquad).$$

The product of the two last terms, 6, can be factored as $1 \cdot 6$, $6 \cdot 1$, $2 \cdot 3$, or $3 \cdot 2$. Try each pair to find the pair that gives the correct middle term, $7x$.

Work Problem **1** *at the Side.* ▶

In part (b) at the side, since $2x + 6 = 2(x + 3)$, the binomial $2x + 6$ has a common factor of 2, while $2x^2 + 7x + 6$ has no common factor other than 1. The product $(2x + 6)(x + 1)$ cannot be correct. (Part (c) also has one binomial factor with a common factor.)

> **Note**
> If the original polynomial has no common factor, then none of its binomial factors will either.

Now try the remaining numbers 3 and 2 as factors of 6.

$$(2x + 3)(x + 2) = 2x^2 + \mathbf{7x} + 6 \qquad \text{Correct}$$

$$\begin{array}{c} 3x \\ 4x \\ \hline 7x \qquad \text{Add.} \end{array}$$

Finally, we see that $2x^2 + 7x + 6$ factors as $(2x + 3)(x + 2)$.

Check   Multiply $(2x + 3)(x + 2)$ to obtain $2x^2 + 7x + 6$.

EXAMPLE 1   **Factoring a Trinomial with All Positive Terms Using FOIL**

Factor $8p^2 + 14p + 5$.

The number 8 has several possible pairs of factors, but 5 has only 1 and 5 or -1 and -5. For this reason, it is easier to begin by considering the factors of 5. Ignore the negative factors, since all coefficients in the trinomial are positive. If $8p^2 + 14p + 5$ can be factored, the factors will have the form

$$(\qquad + 5)(\qquad + 1).$$

Continued on Next Page

OBJECTIVE

1 **Factor trinomials using FOIL.**

1 Multiply to decide whether each factored form is correct or incorrect for

$$2x^2 + 7x + 6.$$

(a) $(2x + 1)(x + 6)$

(b) $(2x + 6)(x + 1)$

(c) $(2x + 2)(x + 3)$

ANSWERS

1. (a) incorrect **(b)** incorrect **(c)** incorrect

2 Factor each trinomial.

(a) $2p^2 + 9p + 9$

(b) $6p^2 + 19p + 10$

(c) $8x^2 + 14x + 3$

3 Factor each trinomial.

(a) $4y^2 - 11y + 6$

(b) $9x^2 - 21x + 10$

When factoring $8p^2 + 14p + 5$, the possible pairs of factors of $8p^2$ are $8p$ and p, or $4p$ and $2p$. Try various combinations, checking to see if the middle term is $14p$ in each case.

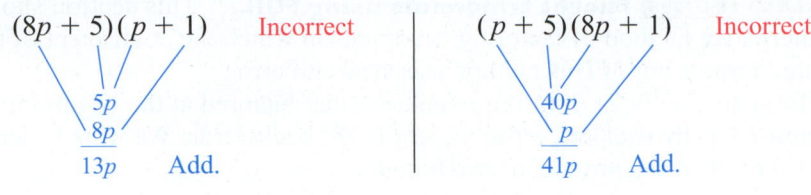

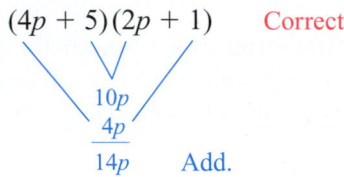

Since $14p$ is the correct middle term,

$$8p^2 + 14p + 5 \quad \text{factors as} \quad (4p + 5)(2p + 1).$$

Check Multiply $(4p + 5)(2p + 1)$ to obtain $8p^2 + 14p + 5$.

◀ *Work Problem* **2** *at the Side.*

EXAMPLE 2 **Factoring a Trinomial with a Negative Middle Term Using FOIL**

Factor $6x^2 - 11x + 3$.

Since 3 has only 1 and 3 or -1 and -3 as factors, it is better here to begin by factoring 3. The last term of the trinomial $6x^2 - 11x + 3$ is positive and the middle term has a negative coefficient, so we consider only negative factors. We need two negative factors because the *product* of two negative factors is positive and their *sum* is negative, as required.

Try -3 and -1 as factors of 3:

$$(\quad - 3)(\quad - 1).$$

The factors of $6x^2$ may be either $6x$ and x, or $2x$ and $3x$.

$(6x - 3)(x - 1)$ Incorrect $(2x - 3)(3x - 1)$ Correct
$\quad -3x$ $\quad -9x$
$\quad -6x$ $\quad -2x$
$\quad -9x$ Add. $\quad -11x$ Add.

The factors $2x$ and $3x$ produce $-11x$, the correct middle term, so

$$6x^2 - 11x + 3 \quad \text{factors as} \quad (2x - 3)(3x - 1).$$

Check Multiply $(2x - 3)(3x - 1)$ to obtain $6x^2 - 11x + 3$.

Note

In Example 2, we might also realize that our initial attempt to factor $6x^2 - 11x + 3$ as $(6x - 3)(x - 1)$ *cannot* be correct since $6x - 3$ has a common factor of 3 and the original polynomial does not.

◀ *Work Problem* **3** *at the Side.*

EXAMPLE 3   **Factoring a Trinomial with a Negative Last Term Using FOIL**

Factor $8x^2 + 6x - 9$.

The integer 8 has several possible pairs of factors, as does -9. Since the last term is negative, one positive factor and one negative factor of -9 are needed. Since the coefficient of the middle term is small, it is wise to avoid large factors such as 8 or 9. We try 4 and 2 as factors of 8, and 3 and -3 as factors of -9, and check the middle term.

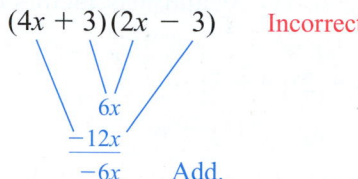

$(4x + 3)(2x - 3)$ Incorrect

Now we try interchanging 3 and -3, since only the sign of the middle term is incorrect.

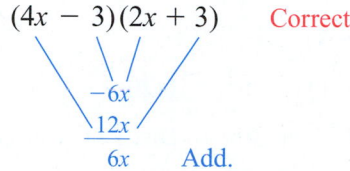

$(4x - 3)(2x + 3)$ Correct

This combination produces $6x$, the correct middle term, so

$$8x^2 + 6x - 9 \quad \text{factors as} \quad (4x - 3)(2x + 3).$$

Work Problem **4** at the Side. ▶

EXAMPLE 4   **Factoring a Trinomial with Two Variables**

Factor $12a^2 - ab - 20b^2$.

There are several pairs of factors of $12a^2$, including

$$12a \text{ and } a, \quad 6a \text{ and } 2a, \quad \text{and} \quad 4a \text{ and } 3a,$$

just as there are many possible pairs of factors of $-20b^2$, including

$$20b \text{ and } -b, \quad -20b \text{ and } b, \quad 10b \text{ and } -2b,$$

$$-10b \text{ and } 2b, \quad 4b \text{ and } -5b, \quad \text{and} \quad -4b \text{ and } 5b.$$

Once again, since the coefficient of the middle term is small, avoid the larger factors. Try the factors $6a$ and $2a$ and $4b$ and $-5b$.

$$(6a + 4b)(2a - 5b)$$

This cannot be correct, as mentioned before, since $6a + 4b$ has 2 as a common factor, while the given trinomial does not. Try $3a$ and $4a$ with $4b$ and $-5b$.

$$(3a + 4b)(4a - 5b)$$

$$= 12a^2 + ab - 20b^2 \quad \text{Incorrect}$$

Here the middle term is ab, rather than $-ab$. Interchange the signs of the last two terms in the factors.

$$(3a - 4b)(4a + 5b)$$

$$= 12a^2 - ab - 20b^2 \quad \text{Correct}$$

Work Problem **5** at the Side. ▶

4 Factor each trinomial, if possible.

(a) $6x^2 + 5x - 4$

(b) $6m^2 - 11m - 10$

(c) $4x^2 - 3x - 7$

(d) $3y^2 + 8y - 6$

5 Factor each trinomial.

(a) $2x^2 - 5xy - 3y^2$

(b) $8a^2 + 2ab - 3b^2$

ANSWERS

4. **(a)** $(3x + 4)(2x - 1)$
    (b) $(2m - 5)(3m + 2)$
    (c) $(4x - 7)(x + 1)$
    (d) prime
5. **(a)** $(2x + y)(x - 3y)$
    (b) $(4a + 3b)(2a - b)$

6 Factor each trinomial.

(a) $36z^3 - 6z^2 - 72z$

(b) $-24x^3 + 32x^2y + 6xy^2$

EXAMPLE 5   **Factoring Trinomials with Common Factors**

Factor each trinomial.

(a) $15y^3 + 55y^2 + 30y$

First factor out the greatest common factor, $5y$.

$$15y^3 + 55y^2 + 30y$$
$$= \mathbf{5y}(3y^2 + 11y + 6)$$

Now factor $3y^2 + 11y + 6$. Try $3y$ and y as factors of $3y^2$ and 2 and 3 as factors of 6.

$$(3y + 2)(y + 3)$$
$$= 3y^2 + 11y + 6 \qquad \text{Correct}$$

The complete factored form of $15y^3 + 55y^2 + 30y$ is

$$5y(3y + 2)(y + 3).$$

Remember the common factor.

Check $5y(3y + 2)(y + 3)$

$$= 5y(3y^2 + 11y + 6) \qquad \text{FOIL}$$
$$= 15y^3 + 55y^2 + 30y \qquad \text{Distributive property}$$

(b) $-24a^3 - 42a^2 + 45a$

The common factor could be $3a$ or $-3a$. If we factor out $-3a$, the first term of the trinomial will be positive, which makes it easier to factor.

$$-24a^3 - 42a^2 + 45a$$
$$= \mathbf{-3a}(8a^2 + 14a - 15) \qquad \text{Factor out } -3a.$$
$$= -3a(4a - 3)(2a + 5) \qquad \text{Use trial and error.}$$

Check $-3a(4a - 3)(2a + 5)$

$$= -3a(8a^2 + 14a - 15)$$
$$= -24a^3 - 42a^2 + 45a$$

CAUTION

This caution bears repeating: *Remember to include the common factor in the final factored form.*

◀ *Work Problem* **6** *at the Side.*

ANSWERS

6. (a) $6z(2z - 3)(3z + 4)$
   (b) $-2x(6x + y)(2x - 3y)$

13.4 ▶▶▶ Exercises

Decide which is the correct factored form of the given polynomial.

1. $2x^2 - x - 1$

   A. $(2x - 1)(x + 1)$ **B.** $(2x + 1)(x - 1)$

2. $3a^2 - 5a - 2$

   A. $(3a + 1)(a - 2)$ **B.** $(3a - 1)(a + 2)$

3. $4y^2 + 17y - 15$

   A. $(y + 5)(4y - 3)$ **B.** $(2y - 5)(2y + 3)$

4. $12c^2 - 7c - 12$

   A. $(6c - 2)(2c + 6)$ **B.** $(4c + 3)(3c - 4)$

5. $4k^2 + 13mk + 3m^2$

   A. $(4k + m)(k + 3m)$ **B.** $(4k + 3m)(k + m)$

6. $2x^2 + 11x + 12$

   A. $(2x + 3)(x + 4)$ **B.** $(2x + 4)(x + 3)$

Complete each factoring.

7. $6a^2 + 7ab - 20b^2$

   $= (3a - 4b)(\qquad)$

8. $9m^2 - 3mn - 2n^2$

   $= (3m + n)(\qquad)$

9. $2x^2 + 6x - 8$

   $= 2(\qquad)$

   $= 2(\qquad)(\qquad)$

10. $3x^2 - 9x - 30$

   $= 3(\qquad)$

   $= 3(\qquad)(\qquad)$

11. $4z^3 - 10z^2 - 6z$

   $= 2z(\qquad)$

   $= 2z(\qquad)(\qquad)$

12. $15r^3 - 39r^2 - 18r$

   $= 3r(\qquad)$

   $= 3r(\qquad)(\qquad)$

13. For the polynomial $12x^2 + 7x - 12$, 2 is not a common factor. Explain why the binomial $2x - 6$, then, cannot be a factor of the polynomial.

14. How are the signs of the last terms of the two binomial factors of a trinomial determined?

Factor each trinomial completely. See Examples 1–5.

⊕ **15.** $3a^2 + 10a + 7$

16. $7r^2 + 8r + 1$

17. $2y^2 + 7y + 6$

18. $5z^2 + 12z + 4$

19. $15m^2 + m - 2$

20. $6x^2 + x - 1$

21. $12s^2 + 11s - 5$

22. $20x^2 + 11x - 3$

⊕ **23.** $10m^2 - 23m + 12$

24. $6x^2 - 17x + 12$

25. $8w^2 - 14w + 3$

26. $9p^2 - 18p + 8$

27. $20y^2 - 39y - 11$

28. $10x^2 - 11x - 6$

29. $3x^2 - 15x + 16$

30. $2t^2 + 13t - 18$

31. $20x^2 + 22x + 6$

32. $36y^2 + 81y + 45$

33. $-40m^2q - mq + 6q$

34. $-15a^2b - 22ab - 8b$

35. $15n^4 - 39n^3 + 18n^2$

36. $24a^4 + 10a^3 - 4a^2$

37. $-15x^2y^2 + 7xy^2 + 4y^2$

38. $-14a^2b^3 - 15ab^3 + 9b^3$

39. $5a^2 - 7ab - 6b^2$

40. $6x^2 - 5xy - y^2$

41. $12s^2 + 11st - 5t^2$

42. $25a^2 + 25ab + 6b^2$

43. $6m^6n + 7m^5n^2 + 2m^4n^3$

44. $12k^3q^4 - 4k^2q^5 - kq^6$

If a trinomial has a negative coefficient for the squared term, such as $-2x^2 + 11x - 12$, *it may be easier to factor by first factoring out the common factor* -1:

$$-2x^2 + 11x - 12$$
$$= \mathbf{-1}(2x^2 - 11x + 12)$$
$$= -1(2x - 3)(x - 4).$$

Use this method to factor the trinomials in Exercises 45–50.

45. $-x^2 - 4x + 21$

46. $-x^2 + x + 72$

47. $-3x^2 - x + 4$

48. $-5x^2 + 2x + 16$

49. $-2a^2 - 5ab - 2b^2$

50. $-3p^2 + 13pq - 4q^2$

Relating Concepts (Exercises 51–56) For Individual or Group Work

One of the most common problems that beginning algebra students face is this: If an answer obtained doesn't look exactly like the one given in the back of the book, is it necessarily incorrect? Often there are several different equivalent forms of an answer that are all correct. **Work Exercises 51–56 in order,** *to see how and why this is possible for factoring problems.*

51. Factor the integer 35 as the product of two prime numbers.

52. Factor the integer 35 as the product of the negatives of two prime numbers.

53. Verify that $6x^2 - 11x + 4$ factors as $(3x - 4)(2x - 1)$.

54. Verify that $6x^2 - 11x + 4$ factors as $(4 - 3x)(1 - 2x)$.

55. Compare the two valid factored forms in Exercises 53 and 54. How do the factors in each case compare?

56. Suppose you know that the correct factored form of a particular trinomial is $(7t - 3)(2t - 5)$. Based on your observations in Exercises 51–55, what is another valid factored form?

13.5 ▶▶▶ Special Factoring Techniques

By reversing the rules for multiplication of binomials from **Section 12.4,** we obtain rules for factoring polynomials in certain forms.

OBJECTIVE 1 Factor a difference of squares. The formula for the product of the sum and difference of the same two terms is

$$(a + b)(a - b) = a^2 - b^2.$$

Reversing this rule leads to the following special factoring rule.

> **Factoring a Difference of Squares**
> $$a^2 - b^2 = (a + b)(a - b)$$

For example, $m^2 - 16$

$$= m^2 - 4^2$$

$$= (m + 4)(m - 4).$$

As the next examples show, the following conditions must be true for a binomial to be a difference of squares.

1. Both terms of the binomial must be squares, such as

$$x^2, \quad 9y^2, \quad 25, \quad 1, \quad m^4.$$

2. The terms of the binomial must have different signs (one positive and one negative).

EXAMPLE 1 **Factoring Differences of Squares**

Factor each binomial, if possible. (In part (c), use fractions.)

$$a^2 - b^2 = (a + b)(a - b)$$

(a) $x^2 - 49 = x^2 - 7^2 = (x + 7)(x - 7)$

(b) $y^2 - m^2$

$$= (y + m)(y - m)$$

(c) $z^2 - \dfrac{9}{16}$

$$= \left(z + \dfrac{3}{4}\right)\left(z - \dfrac{3}{4}\right) \qquad \tfrac{9}{16} = \left(\tfrac{3}{4}\right)^2$$

(d) $x^2 - 8$

Because 8 is not the square of an integer, this binomial is not a difference of squares. It is a prime polynomial.

(e) $p^2 + 16$

Since $p^2 + 16$ is a *sum* of squares, it is not equal to $(p + 4)(p - 4)$. Also, using FOIL,

$$(p - 4)(p - 4)$$

$$= p^2 - 8p + 16, \quad \text{not} \quad p^2 + 16,$$

and $(p + 4)(p + 4)$

$$= p^2 + 8p + 16, \quad \text{not} \quad p^2 + 16,$$

so $p^2 + 16$ is a prime polynomial.

1 Factor, if possible. (In part (b), use fractions.)

(a) $p^2 - 100$

(b) $x^2 - \dfrac{25}{36}$

(c) $x^2 + y^2$

(d) $9m^2 - 49$

(e) $64a^2 - 25$

2 Factor completely.

(a) $50r^2 - 32$

(b) $27y^2 - 75$

(c) $25a^2 - 64b^2$

(d) $k^4 - 49$

(e) $81r^4 - 16$

CAUTION

As Example 1(e) suggests, *after any common factor is removed, a sum of squares cannot be factored.*

EXAMPLE 2 **Factoring Differences of Squares**

Factor each difference of squares.

$$a^2 \; - \; b^2 \; = \; (a + b) \; (a \; - \; b)$$

(a) $25m^2 - 16 = (5m)^2 - 4^2 = (5m + 4)(5m - 4)$

(b) $49z^2 - 64$

$= (7z)^2 - 8^2$

$= (7z + 8)(7z - 8)$

◀ *Work Problem* **1** *at the Side.*

Note

Always check a factored form by multiplying.

EXAMPLE 3 **Factoring More Complex Differences of Squares**

Factor completely.

(a) $81y^2 - 36$

$= 9(9y^2 - 4)$ Factor out the GCF, 9.

$= 9[(3y)^2 - 2^2]$

$= 9(3y + 2)(3y - 2)$ Difference of squares

(b) $9x^2 - 4z^2$

$= (3x)^2 - (2z)^2$

$= (3x + 2z)(3x - 2z)$

(c) $p^4 - 36$

$= (p^2)^2 - 6^2$

$= (p^2 + 6)(p^2 - 6)$

(d) $m^4 - 16$

$= (m^2)^2 - 4^2$

$= (m^2 + 4)(m^2 - 4)$ Difference of squares

Don't stop here.

$= (m^2 + 4)(m + 2)(m - 2)$ Difference of squares again

CAUTION

Factor again when any of the factors is a difference of squares, as in Example 3(d). Check by multiplying.

◀ *Work Problem* **2** *at the Side.*

OBJECTIVE **2** **Factor a perfect square trinomial.** The expressions 144, $4x^2$, and $81m^6$ are called *perfect squares* because

$$144 = \mathbf{12^2}, \quad 4x^2 = \mathbf{(2x)^2}, \quad \text{and} \quad 81m^6 = \mathbf{(9m^3)^2}.$$

A **perfect square trinomial** is a trinomial that is the square of a binomial. For example, $x^2 + 8x + 16$ is a perfect square trinomial because it is the square of the binomial $x + 4$:

$$x^2 + 8x + 16$$
$$= (x + 4)(x + 4)$$
$$= \mathbf{(x + 4)^2}.$$

On the one hand, a necessary condition for a trinomial to be a perfect square is that *two of its terms must be perfect squares.* For this reason, $16x^2 + 4x + 15$ is not a perfect square trinomial because only the term $16x^2$ is a perfect square.

On the other hand, even if two of the terms are perfect squares, the trinomial may not be a perfect square trinomial. For example, $x^2 + 6x + 36$ has two perfect square terms, but it is not a perfect square trinomial. (Try to find a binomial that can be squared to give $x^2 + 6x + 36$.)

We can multiply to see that the square of a binomial gives one of the following perfect square trinomials.

> **Factoring Perfect Square Trinomials**
>
> $$a^2 + 2ab + b^2 = (a + b)^2$$
> $$a^2 - 2ab + b^2 = (a - b)^2$$

The middle term of a perfect square trinomial is always twice the product of the two terms in the squared binomial. (See **Section 12.4.**) Use this to check any attempt to factor a trinomial that appears to be a perfect square.

EXAMPLE 4 **Factoring a Perfect Square Trinomial**

Factor $x^2 + 10x + 25$.

The term x^2 is a perfect square, and so is 25. Try to factor the trinomial

$$x^2 + 10x + 25 \quad \text{as} \quad (x + 5)^2.$$

To check, take twice the product of the two terms in the squared binomial.

$$\mathbf{2} \cdot \mathbf{x} \cdot \mathbf{5} = 10x$$

Twice   First term
        of binomial   Last term
                         of binomial

Since $10x$ is the middle term of the trinomial, the trinomial is a perfect square and can be factored as $(x + 5)^2$. Thus,

$$x^2 + 10x + 25 \quad \text{factors as} \quad (x + 5)^2.$$

Work Problem **3** *at the Side.* ▶

---

3 Factor each trinomial.

(a) $p^2 + 14p + 49$

(b) $m^2 + 8m + 16$

(c) $x^2 + 2x + 1$

---

④ Factor each trinomial.

(a) $p^2 - 18p + 81$

(b) $16a^2 + 56a + 49$

(c) $121p^2 + 110p + 100$

(d) $64x^2 - 48x + 9$

(e) $27y^3 + 72y^2 + 48y$

EXAMPLE 5 **Factoring Perfect Square Trinomials**

Factor each trinomial.

(a) $x^2 - 22x + 121$

The first and last terms are perfect squares ($121 = 11^2$ or $(-11)^2$). Check to see whether the middle term of $x^2 - 22x + 121$ is twice the product of the first and last terms of the binomial $x - 11$.

$$2 \cdot x \cdot (-11) = -22x$$

Twice ⟶ First term   Last term

Since twice the product of the first and last terms of the binomial is the middle term, $x^2 - 22x + 121$ is a perfect square trinomial and

$$x^2 - 22x + 121 \quad \text{factors as} \quad (x - 11)^2.$$

Same sign

Notice that the sign of the second term in the squared binomial is the same as the sign of the middle term in the trinomial.

(b) $9m^2 - 24m + 16 = (3m)^2 + 2(3m)(-4) + (-4)^2 = (3m - 4)^2$

Twice ⟶ First term   Last Term

(c) $25y^2 + 20y + 16$

The first and last terms are perfect squares.

$$25y^2 = (5y)^2 \quad \text{and} \quad 16 = 4^2$$

However, twice the product of the first and last terms of the binomial $5y + 4$ is $2 \cdot 5y \cdot 4 = 40y$, which is not the middle term of $25y^2 + 20y + 16$. This trinomial is not a perfect square. In fact, the trinomial cannot be factored, even with the methods of the previous sections. It is a prime polynomial.

(d) $12z^3 + 60z^2 + 75z$

$= 3z(4z^2 + 20z + 25)$ Factor out $3z$.

$= 3z[(2z)^2 + 2(2z)(5) + 5^2]$ $4z^2 + 20z + 25$ is a perfect square trinomial.

$= 3z(2z + 5)^2$ Factor.

Note

1. The sign of the second term in the squared binomial is always the same as the sign of the middle term in the trinomial.

2. The first and last terms of a perfect square trinomial must be *positive,* because they are squares. For example, the polynomial $x^2 - 2x - 1$ cannot be a perfect square because the last term is negative.

3. Perfect square trinomials can also be factored using grouping or FOIL, although using the method of this section is often easier.

ANSWERS

4. **(a)** $(p - 9)^2$ **(b)** $(4a + 7)^2$ **(c)** prime
(d) $(8x - 3)^2$ **(e)** $3y(3y + 4)^2$

◀ *Work Problem* ④ *at the Side.*

OBJECTIVE **3** **Factor a difference of cubes.** We factored a difference of squares at the beginning of this section. We can also factor a **difference of cubes.** Use the following pattern.

> **Difference of Cubes**
>
> $$a^3 - b^3 = (a - b)(a^2 + ab + b^2)$$

$$
\begin{array}{r}
a^2 + ab + b^2 \\
a - b \\
\hline
- a^2b - ab^2 - b^3 \\
a^3 + a^2b + ab^2 \\
\hline
a^3 \qquad\qquad - b^3
\end{array}
$$

This pattern should be memorized. Multiply on the right to see that the pattern gives the correct factors, as shown in the margin.

Notice the pattern of the terms in the factored form of $a^3 - b^3$.

- $a^3 - b^3 =$ (a binomial factor) (a trinomial factor)
- The binomial factor has the difference of the cube roots of the given terms.
- The terms in the trinomial factor are all positive.
- What you write in the binomial factor determines the trinomial factor.

$$
a^3 - b^3 = (a - b)(\underbrace{a^2}_{\text{First term squared}} + \underbrace{ab}_{\substack{\text{positive} \\ \text{product of} \\ \text{the terms}}} + \underbrace{b^2}_{\substack{\text{second term} \\ \text{squared}}})
$$

EXAMPLE 6 **Factoring Differences of Cubes**

Factor each difference of cubes.

(a) $m^3 - 125$

Use the pattern for a difference of cubes.

$$a^3 - b^3 = (a - b)(a^2 + ab + b^2)$$

$$m^3 - 125 = m^3 - 5^3 = (m - 5)(m^2 + 5m + 5^2)$$

$$= (m - 5)(m^2 + 5m + 25)$$

(b) $8p^3 - 27$

$$= (2p)^3 - 3^3 \qquad\qquad 8p^3 = (2p)^3;\ 27 = 3^3$$

$$= (2p - 3)[(2p)^2 + (2p)3 + 3^2] \qquad a^3 - b^3 = (a - b)(a^2 + ab + b^2)$$

$$= (2p - 3)(4p^2 + 6p + 9) \qquad (ab)^2 = a^2b^2$$

(c) $4m^3 - 32n^3$

$$= 4(m^3 - 8n^3) \qquad \text{Factor out the common factor.}$$

$$= 4[m^3 - (2n)^3] \qquad 8n^3 = (2n)^3$$

$$= 4(m - 2n)[m^2 + m(2n) + (2n)^2]$$

$$= 4(m - 2n)(m^2 + 2mn + 4n^2)$$

Work Problem **5** *at the Side.* ▶

5 Factor each difference of cubes.

(a) $t^3 - 64$

(b) $2x^3 - 54$

(c) $8k^3 - y^3$

> **CAUTION**
> A common error in factoring $a^3 - b^3 = (a - b)(a^2 + ab + b^2)$ is to try to factor $a^2 + ab + b^2$. This particular trinomial cannot be factored.

6 Factor each sum of cubes.

(a) $x^3 + 8$

(b) $64y^3 + 1$

(c) $27m^3 + 343n^3$

OBJECTIVE **4** **Factor a sum of cubes.** A sum of squares, such as $m^2 + 25$, cannot be factored using real numbers, but a **sum of cubes** can.

> **Sum of Cubes**
> $$a^3 + b^3 = (a + b)(a^2 - ab + b^2)$$

Observe the positive and negative signs in these factoring patterns.

Positive

$$a^3 - b^3 = (a - b)(a^2 + ab + b^2) \quad \text{Difference of cubes}$$

Same sign   Opposite sign

Positive

$$a^3 + b^3 = (a + b)(a^2 - ab + b^2) \quad \text{Sum of cubes}$$

Same sign   Opposite sign

EXAMPLE 7 **Factoring Sums of Cubes**

Factor each sum of cubes.

(a) $k^3 + 27$

$\quad = k^3 + 3^3$

$\quad = (k + 3)(k^2 - 3k + 3^2)$

$\quad = (k + 3)(k^2 - 3k + 9)$

(b) $8m^3 + 125p^3$

$\quad = (2m)^3 + (5p)^3 \qquad 8m^3 = (2m)^3; 125p^3 = (5p)^3$

$\quad = (2m + 5p)[(2m)^2 - (2m)(5p) + (5p)^2]$

$\quad = (2m + 5p)(4m^2 - 10mp + 25p^2)$

◀ *Work Problem* **6** *at the Side.*

The methods of factoring discussed in this section are summarized here.

> **Special Factoring Rules**
>
> **Difference of squares** $\qquad a^2 - b^2 = (a + b)(a - b)$
>
> **Perfect square trinomials** $\quad a^2 + 2ab + b^2 = (a + b)^2$
>
> $\qquad\qquad\qquad\qquad\qquad\qquad a^2 - 2ab + b^2 = (a - b)^2$
>
> **Difference of cubes** $\qquad a^3 - b^3 = (a - b)(a^2 + ab + b^2)$
>
> **Sum of cubes** $\qquad\qquad a^3 + b^3 = (a + b)(a^2 - ab + b^2)$
>
> *Remember that a sum of squares cannot be factored using real numbers unless the terms have a common factor.*

ANSWERS

6. (a) $(x + 2)(x^2 - 2x + 4)$
 (b) $(4y + 1)(16y^2 - 4y + 1)$
 (c) $(3m + 7n)(9m^2 - 21mn + 49n^2)$

1. To help you factor a difference of squares, complete the following list of squares.

$1^2 =$ _____ $2^2 =$ _____ $3^2 =$ _____ $4^2 =$ _____ $5^2 =$ _____

$6^2 =$ _____ $7^2 =$ _____ $8^2 =$ _____ $9^2 =$ _____ $10^2 =$ _____

$11^2 =$ _____ $12^2 =$ _____ $13^2 =$ _____ $14^2 =$ _____ $15^2 =$ _____

$16^2 =$ _____ $17^2 =$ _____ $18^2 =$ _____ $19^2 =$ _____ $20^2 =$ _____

2. To use the factoring techniques described in this section, you will sometimes need to recognize fourth powers of integers. Complete the following list of fourth powers.

$1^4 =$ _____ $2^4 =$ _____ $3^4 =$ _____ $4^4 =$ _____ $5^4 =$ _____

3. The following powers of x are all perfect squares: $x^2, x^4, x^6, x^8, x^{10}$. Based on this observation, we may make a conjecture (an educated guess) that if the power of a variable is divisible by _____ (with 0 remainder), then it is a perfect square.

4. Which of the following are differences of squares?

   A. $x^2 - 4$ **B.** $y^2 + 9$ **C.** $2a^2 - 25$ **D.** $9m^2 - 1$

Factor each binomial completely. In Exercises 7, 8, 13, and 14, use fractions. See Examples 1–3.

 5. $y^2 - 25$

6. $t^2 - 16$

7. $p^2 - \dfrac{1}{9}$

8. $q^2 - \dfrac{1}{4}$

9. $m^2 - 12$

10. $k^2 - 18$

 11. $9r^2 - 4$

12. $4x^2 - 9$

13. $4m^2 - \dfrac{9}{25}$

14. $100b^2 - \dfrac{49}{81}$

 15. $36x^2 - 16$

16. $32a^2 - 8$

17. $196p^2 - 225$

18. $361q^2 - 400$

19. $16r^2 - 25a^2$

20. $49m^2 - 100p^2$

21. $100x^2 + 49$

22. $81w^2 + 16$

23. $p^4 - 49$

24. $r^4 - 25$

25. $x^4 - 1$

26. $y^4 - 10,000$

27. $p^4 - 256$

28. $16k^4 - 1$

29. When a student was directed to factor $x^4 - 81$ completely, his teacher did not give him full credit for the answer $(x^2 + 9)(x^2 - 9)$. The student argued that because his answer does indeed give $x^4 - 81$ when multiplied out, he should be given full credit. **WHAT WENT WRONG?** Give the correct factored form.

30. The binomial $4x^2 + 16$ is a sum of squares that *can* be factored. How is this binomial factored? When can a sum of squares be factored?

31. In the polynomial $9y^2 + 14y + 25$, the first and last terms are perfect squares. Can the polynomial be factored? If it can, factor it. If it cannot, explain why it is not a perfect square trinomial.

32. Which of the following are perfect square trinomials?

   A. $y^2 - 13y + 36$ **B.** $x^2 + 6x + 9$ **C.** $4z^2 - 4z + 1$ **D.** $16m^2 + 10m + 1$

Factor each trinomial completely. It may be necessary to factor out the greatest common factor first. In Exercises 37–40, use fractions or decimals, as appropriate. See Examples 4 and 5.

33. $w^2 + 2w + 1$

34. $p^2 + 4p + 4$

35. $x^2 - 8x + 16$

36. $x^2 - 10x + 25$

37. $t^2 + t + \dfrac{1}{4}$

38. $m^2 + \dfrac{2}{3}m + \dfrac{1}{9}$

39. $x^2 - 1.0x + 0.25$

40. $y^2 - 1.4y + 0.49$

41. $2x^2 + 24x + 72$

42. $3y^2 - 48y + 192$

43. $16x^2 - 40x + 25$

44. $36y^2 - 60y + 25$

45. $49x^2 - 28xy + 4y^2$

46. $4z^2 - 12zw + 9w^2$

47. $64x^2 + 48xy + 9y^2$

48. $9t^2 + 24tr + 16r^2$

49. $-50h^3 + 40h^2y - 8hy^2$

50. $-18x^3 - 48x^2y - 32xy^2$

51. To help you factor a sum or difference of cubes, complete the following list of cubes.

$1^3 = $ _____ $2^3 = $ _____ $3^3 = $ _____ $4^3 = $ _____ $5^3 = $ _____

$6^3 = $ _____ $7^3 = $ _____ $8^3 = $ _____ $9^3 = $ _____ $10^3 = $ _____

52. The following powers of x are all perfect cubes: $x^3, x^6, x^9, x^{12}, x^{15}$. Based on this observation, we may make a conjecture that if the power of a variable is divisible by _____ (with 0 remainder), then we have a perfect cube.

53. Which of the following are differences of cubes?

A. $9x^3 - 125$ **B.** $x^3 - 16$ **C.** $x^3 - 1$ **D.** $8x^3 - 27y^3$

54. Which of the following are sums of cubes?

A. $x^3 + 1$ **B.** $x^3 + 36$ **C.** $12x^3 + 27$ **D.** $64x^3 + 216y^3$

Factor. Use your answers in Exercises 51 and 52 as necessary. See Examples 6 and 7.

55. $a^3 + 1$

56. $m^3 + 8$

57. $a^3 - 1$

58. $m^3 - 8$

59. $p^3 + q^3$

60. $w^3 + z^3$

61. $y^3 - 216$

62. $x^3 - 343$

63. $k^3 + 1000$

64. $p^3 + 512$

65. $27x^3 - 1$

66. $64y^3 - 27$

67. $125a^3 + 8$

68. $216b^3 + 125$

69. $y^3 - 8x^3$

70. $w^3 - 216z^3$

71. $27a^3 - 64b^3$

72. $125m^3 - 8n^3$

73. $8p^3 + 729q^3$

74. $27x^3 + 1000y^3$

75. $16t^3 - 2$

76. $3p^3 - 81$

77. $40w^3 + 135$

78. $32z^3 + 500$

79. $x^3 + y^6$

80. $p^9 + q^3$

81. $125k^3 - 8m^9$

82. $125c^6 - 216d^3$

Relating Concepts (Exercises 83–90) For Individual or Group Work

A binomial may be both a difference of squares and a difference of cubes. One example of such a binomial is $x^6 - 1$. Using the techniques of this section, one factoring method will give the complete factored form, while the other will not. **Work Exercises 83–90 in order, to determine the method to use.**

83. Factor $x^6 - 1$ as the difference of two squares.

84. The factored form obtained in Exercise 83 consists of a difference of cubes multiplied by a sum of cubes. Factor each binomial further.

85. Now start over and factor $x^6 - 1$ as a difference of cubes.

86. The factored form obtained in Exercise 85 consists of a binomial that is a difference of squares and a trinomial. Factor the binomial further.

87. Compare your results in Exercises 84 and 86. Which one of these is the completely factored form?

88. Verify that the trinomial in the factored form in Exercise 86 is the product of the two trinomials in the factored form in Exercise 84.

89. Use the results of Exercises 83–88 to complete the following statement:

In general, if I must choose between factoring first using the method for a difference

of squares or the method for a difference of cubes, I should choose the

_____ method to eventually obtain the complete factored form.

90. Find the *complete* factored form of $x^6 - 729$ using the knowledge you have gained in Exercises 83–89.

13.6 ▶▶▶ A General Approach to Factoring

A polynomial is completely factored when **(1)** it is written as a *product* of prime polynomials with integer coefficients, and **(2)** none of the polynomial factors can be factored further.

OBJECTIVES

1 Factor out any common factor.

2 Factor binomials.

3 Factor trinomials.

4 Factor polynomials with more than three terms.

Factoring a Polynomial

Step 1 **Factor out any common factor.**

Step 2 **If the polynomial is a binomial,** check to see if it is a difference of squares, a difference of cubes, or a sum of cubes.

If the polynomial is a trinomial, check to see if it is a perfect square trinomial. If it is not, factor as in **Sections 13.2–13.4.**

If the polynomial has more than three terms, try to factor by grouping as in **Section 13.1.**

Step 3 **Check the factored form by multiplying.**

OBJECTIVE 1 Factor out any common factor.   *This step is always the same, regardless of the number of terms in the polynomial.*

1 Factor each polynomial.

(a) $8x - 80$

(b) $2x^3 + 10x^2 - 2x$

(c) $12m(p - q) - 7n(p - q)$

EXAMPLE 1   Factoring Out a Common Factor

Factor each polynomial.

(a) $9p + 45$

$= \mathbf{9}(p + 5)$ GCF = 9

(b) $8m^2p^2 + 4mp$

$= 4mp(2mp + \mathbf{1})$

(c) $5x(a + b) - y(a + b)$

$= (a + b)(5x - y)$ Factor out $a + b$.

> Check the factored form by multiplying.

Work Problem **1** *at the Side.* ▶

OBJECTIVE 2 Factor binomials.   Check for the following patterns.

2 Factor each binomial if possible.

(a) $36x^2 - y^2$

Factoring a Binomial (Two Terms)

Difference of squares	$x^2 - y^2 = (x + y)(x - y)$
Difference of cubes	$x^3 - y^3 = (x - y)(x^2 + xy + y^2)$
Sum of cubes	$x^3 + y^3 = (x + y)(x^2 - xy + y^2)$

(b) $4t^2 + 1$

(c) $125x^3 - 27y^3$

EXAMPLE 2 Factoring Binomials

Factor each binomial if possible.

(a) $64m^2 - 9n^2$

$= (\mathbf{8m})^2 - (\mathbf{3n})^2$ Difference of squares

$= (\mathbf{8m + 3n})(\mathbf{8m - 3n})$

(b) $27x^3 - 8$ $27x^3 = (3x)^3; 8 = 2^3$

$= (\mathbf{3x - 2})[(\mathbf{3x})^2 + (\mathbf{3x})(\mathbf{2}) + \mathbf{2}^2]$

$= (3x - 2)(9x^2 + 6x + 4)$

(d) $x^3 + 343y^3$

(c) $1000m^3 + 1$

$= (10m)^3 + 1^3$ Sum of cubes

$= (10m + 1)[(10m)^2 - (10m)(1) + 1^2]$

$= (10m + 1)(100m^2 - 10m + 1)$ ◀ Check by multiplying.

(d) $25m^2 + 121$ is prime.

It is a *sum* of squares.

Work Problem **2** *at the Side.* ▶

ANSWERS

1. (a) $8(x - 10)$
   (b) $2x(x^2 + 5x - 1)$
   (c) $(p - q)(12m - 7n)$
2. (a) $(6x + y)(6x - y)$
   (b) prime
   (c) $(5x - 3y)(25x^2 + 15xy + 9y^2)$
   (d) $(x + 7y)(x^2 - 7xy + 49y^2)$

3 Factor each trinomial.

(a) $16m^2 + 56m + 49$

(b) $r^2 + 18r + 72$

(c) $8t^2 - 13t + 5$

(d) $6x^2 - 3x - 63$

4 Factor each polynomial.

(a) $20 - 5m - 12n + 3mn$

(b) $p^3 - 2pq^2 + p^2q - 2q^3$

(c) $9x^2 + 24x + 16 - y^2$

> **Note**
> The binomial $25m^2 + 625$ is a sum of squares. It *can* be factored, however, as **25**$(m^2 + 25)$ because it has a common factor, 25.

OBJECTIVE **3** **Factor trinomials.** Consider the following.

> **Factoring a Trinomial (Three Terms)**
> Decide whether it is a perfect square trinomial of the form
> $$x^2 + 2xy + y^2 = (x + y)^2 \quad \text{or} \quad x^2 - 2xy + y^2 = (x - y)^2.$$
> If not, use the general factoring methods of **Sections 13.2–13.4.**

EXAMPLE 3 **Factoring Trinomials**

Factor each trinomial.

(a) $p^2 + 10p + 25$

    $= (p + 5)^2$ $2(p)(5) = 10p$

(b) $49z^2 - 42z + 9$

    $= (7z - 3)^2$ $2(7z)(3) = 42z$

(c) $y^2 - 5y - 6$

    $= (y - 6)(y + 1)$

The integers -6 and 1 have a product of -6 and a sum of -5. **(Section 6.2)**

(d) $2k^2 - k - 6$

    $= (2k + 3)(k - 2)$

    (Section 13.4)

(e) $28z^2 + 6z - 10$

    $= \mathbf{2}(14z^2 + 3z - 5)$

    $= \mathbf{2}(7z + 5)(2z - 1)$

◀ *Work Problem* **3** *at the Side.*

OBJECTIVE **4** **Factor polynomials with more than three terms.** Consider factoring by grouping from **Section 13.2.**

EXAMPLE 4 **Factoring Polynomials with More than Three Terms**

Factor each polynomial.

(a) $4 - 2q - 6p + 3pq$

    $= (4 - 2q) + (-6p + 3pq)$ Group the terms.

    $= 2\mathbf{(2 - q)} - 3p\mathbf{(2 - q)}$ Factor each group.

    $= \mathbf{(2 - q)}(2 - 3p)$ Factor out $2 - q$.

(b) $20k^3 + 4k^2 - 45k - 9$

    $= (20k^3 + 4k^2) + (-45k - 9)$ Group the terms.

    $= 4k^2\mathbf{(5k + 1)} - 9\mathbf{(5k + 1)}$ Factor each group.

    $= \mathbf{(5k + 1)}(4k^2 - 9)$ $5k + 1$ is a common factor.

    $= (5k + 1)(2k + 3)(2k - 3)$ Difference of squares

(c) $4a^2 + 4a + 1 - b^2$

    $= (4a^2 + 4a + 1) - b^2$ Group the first three terms.

    $= (2a + 1)^2 - b^2$ Perfect square trinomial

    $= (2a + 1 + b)(2a + 1 - b)$ Difference of squares

◀ *Work Problem* **4** *at the Side.*

ANSWERS

3. (a) $(4m + 7)^2$
   (b) $(r + 6)(r + 12)$
   (c) $(8t - 5)(t - 1)$
   (d) $3(2x - 7)(x + 3)$

4. (a) $(4 - m)(5 - 3n)$
   (b) $(p + q)(p^2 - 2q^2)$
   (c) $(3x + 4 + y)(3x + 4 - y)$

Match each polynomial in Column I with the method you would use to factor it in Column II. The choices in Column II may be used once, more than once, or not at all.

I	**II**
1. $12x^2 + 20x + 8$	**A.** Factor out the GCF; no further factoring is possible.
2. $x^2 - 17x + 72$	**B.** Factor a difference of squares.
3. $-16m^2n + 24mn - 40mn^2$	**C.** Factor a difference of cubes.
4. $64a^2 - 121b^2$	**D.** Factor a perfect square trinomial.
5. $36p^2 - 60pq + 25q^2$	**E.** Factor by grouping.
6. $z^2 - 4z + 6$	**F.** Factor out the GCF; then factor a trinomial by grouping or trial and error.
7. $8p^3 - 1$	**G.** Factor into two binomials by finding two integers whose product is the constant in the trinomial and whose sum is the coefficient of the middle term.
8. $x^6 + 4x^4 - 3x^2 - 12$	
9. $4w^2 + 49$	
10. $144 - 24z + z^2$	**H.** The polynomial is prime.

Factor each polynomial completely. We have randomly included all the different types of factoring exercises here to give you practice in applying factoring strategies. See Examples 1–4.

11. $100a^2 - 9b^2$

12. $10r^2 + 13r - 3$

13. $18p^5 - 24p^3 + 12p^6$

14. $15x^2 - 20x$

15. $x^2 + 2x - 35$

16. $108m^2 - 36m + 3$

17. $225p^2 + 256$

18. $x^3 + 1000$

19. $6b^2 - 17b - 3$

20. $k^2 - 6k + 16$

21. $18m^3n + 3m^2n^2 - 6mn^3$

22. $6t^2 + 19tu - 77u^2$

23. $m^3 + 4m^2 - 6m - 24$

24. $9m^2 - 45m + 18m^3$

25. $4k^2 + 28kr + 49r^2$

26. $54m^3 - 2000$

27. $mn - 2n + 5m - 10$

28. $9m^2 - 30mn + 25n^2 - p^2$

29. $x^3 + 3x^2 - 9x - 27$

30. $56k^3 - 875$

31. $9r^2 + 100$

32. $6k^2 - k - 1$

33. $8p^3 - 125$

34. $27m^2 + 144mn + 192n^2$

35. $x^4 - 625$

36. $p^3 + 64$

37. $ab + 6b + ac + 6c$

38. $12z^3 - 6z^2 + 18z$

39. $14z^2 - 3zk - 2k^2$

40. $z^2 - zp + 20p^2$

41. $256b^2 - 400c^2$

42. $64m^2 - 25n^2$

43. $1000z^3 + 512$

44. $y^4 - 81$

45. $10r^2 + 23rs - 5s^2$

46. $2a^3 + a^2 - 14a - 7$

47. $32x^2 + 16x^3 - 24x^5$

48. $y^2 + 3y - 10$

49. $14x^2 - 25xq - 25q^2$

50. $b^2 - 7ba - 18a^2$

51. $4(p + 2) + m(p + 2)$

52. $kq - 9q + kr - 9r$

53. $50p^2 - 162$

54. $25x^2 - 20xy + 4y^2$

55. $16a^2 + 8ab + b^2$

13.7 ▶▶▶ Solving Quadratic Equations by Factoring

OBJECTIVES

1 Solve quadratic equations by factoring.

2 Solve other equations by factoring.

Galileo Galilei developed theories to explain physical phenomena and set up experiments to test his ideas. According to legend, Galileo dropped objects of different weights from the Leaning Tower of Pisa to disprove the belief that heavier objects fall faster than lighter objects. He developed a formula for freely falling objects described by $d = 16t^2$, where d is the distance in feet that an object falls (disregarding air resistance) in t seconds, regardless of weight.

The equation $d = 16t^2$ is a *quadratic equation*. A quadratic equation contains a squared term and no terms of higher degree.

> **Quadratic Equation**
>
> A **quadratic equation** is an equation that can be written in the form
>
> $$ax^2 + bx + c = 0,$$
>
> where a, b, and c are real numbers, with $a \neq 0$. The given form is called **standard form**.

$x^2 + 5x + 6 = 0, \quad 2t^2 - 5t = 3, \quad y^2 = 4$ Quadratic equations

In these examples, only $x^2 + 5x + 6 = 0$ is in standard form.

Work Problems **1** *and* **2** *at the Side.* ▶

Galileo Galilei (1564–1642)

Up to now, we have factored *expressions,* including many quadratic expressions of the form $ax^2 + bx + c$. In this section, we use factored quadratic expressions to solve quadratic *equations.*

OBJECTIVE 1 Solve quadratic equations by factoring. We use the **zero-factor property** to solve a quadratic equation by factoring.

> **Zero-Factor Property**
>
> **If a and b are real numbers and $ab = 0$, then $a = 0$ or $b = 0$.**
>
> In words, if the product of two numbers is 0, then at least one of the numbers must be 0. One number *must* be 0, but both *may* be 0.

EXAMPLE 1   **Using the Zero-Factor Property**

Solve each equation.

(a) $(x + 3)(2x - 1) = 0$

The product $(x + 3)(2x - 1)$ is equal to 0. By the zero-factor property, the only way that the product of these two factors can be 0 is if at least one of the factors equals 0. Therefore, either $x + 3 = 0$ or $2x - 1 = 0$.

$$x + 3 = 0 \quad \text{or} \quad 2x - 1 = 0 \quad \text{Zero-factor property}$$
$$x = -3 \quad \text{or} \quad 2x = 1 \quad \text{Solve each equation.}$$
$$x = \frac{1}{2}$$

Continued on Next Page

1 Which of the following equations are quadratic equations?

A. $y^2 - 4y - 5 = 0$

B. $x^3 - x^2 + 16 = 0$

C. $2z^2 + 7z = -3$

D. $x + 2y = -4$

2 Write each quadratic equation in standard form.

(a) $x^2 - 3x = 4$

(b) $y^2 = 9y - 8$

ANSWERS

1. A, C
2. **(a)** $x^2 - 3x - 4 = 0$
   (b) $y^2 - 9y + 8 = 0$

3 Solve each equation. Check your solutions.

(a) $(x - 5)(x + 2) = 0$

(b) $(3x - 2)(x + 6) = 0$

(c) $z(2z + 5) = 0$

The given equation, $(x + 3)(2x - 1) = 0$, has two solutions, -3 and $\frac{1}{2}$. *Check* these solutions by substituting -3 for x in the original equation, $(x + 3)(2x - 1) = 0$. Then start over and substitute $\frac{1}{2}$ for x.

Check   If $x = -3$, then

$$(x + 3)(2x - 1) = 0$$

$$(-3 + 3)[2(-3) - 1] \stackrel{?}{=} 0$$

$$0(-7) = 0. \quad \text{True}$$

If $x = \frac{1}{2}$, then

$$(x + 3)(2x - 1) = 0$$

$$\left(\frac{1}{2} + 3\right)\left(2 \cdot \frac{1}{2} - 1\right) \stackrel{?}{=} 0$$

$$\frac{7}{2}(1 - 1) \stackrel{?}{=} 0$$

$$\frac{7}{2} \cdot 0 = 0. \quad \text{True}$$

Both -3 and $\frac{1}{2}$ result in true equations, so the solution set is $\{-3, \frac{1}{2}\}$.

(b) $$y(3y - 4) = 0$$

$$y = 0 \quad \text{or} \quad 3y - 4 = 0 \quad \text{Zero-factor property}$$

> Don't forget that 0 is a solution.

$$3y = 4$$

$$y = \frac{4}{3}$$

Check these solutions by substituting each one in the original equation. The solution set is $\{0, \frac{4}{3}\}$.

◀ *Work Problem* **3** *at the Side.*

> **Note**
> The word *or* as used in Example 1 means "one or the other or both."

In Example 1, each equation to be solved was given with the polynomial in factored form. If the polynomial in an equation is not already factored, first make sure that the equation is in standard form. Then factor and solve.

EXAMPLE 2 Solving Quadratic Equations

Solve each equation.

(a) $x^2 - 5x = -6$

First, write the equation in standard form by adding 6 to each side.

> Don't factor x out at this step.

$$x^2 - 5x = -6$$

$$x^2 - 5x + 6 = 0 \quad \text{Add 6.}$$

Now factor $x^2 - 5x + 6$. Find two numbers whose product is 6 and whose sum is -5. These two numbers are -2 and -3, so the equation becomes

$$(x - 2)(x - 3) = 0. \quad \text{Factor the trinomial.}$$

$$x - 2 = 0 \quad \text{or} \quad x - 3 = 0 \quad \text{Zero-factor property}$$

$$x = 2 \quad \text{or} \quad x = 3 \quad \text{Solve each equation.}$$

Continued on Next Page

Check If $x = 2$, then

$$2^2 - 5(2) \stackrel{?}{=} -6$$

$$4 - 10 \stackrel{?}{=} -6$$

$$-6 = -6. \quad \text{True}$$

If $x = 3$, then

$$3^2 - 5(3) \stackrel{?}{=} -6$$

$$9 - 15 \stackrel{?}{=} -6$$

$$-6 = -6. \quad \text{True}$$

Both solutions check, so the solution set is $\{2, 3\}$.

(b)

$$y^2 = y + 20$$

$$y^2 - y - 20 = 0 \qquad \text{Write in standard form.}$$

$$(y - 5)(y + 4) = 0 \qquad \text{Factor the trinomial.}$$

$$y - 5 = 0 \quad \text{or} \quad y + 4 = 0 \qquad \text{Zero-factor property}$$

$$y = 5 \quad \text{or} \qquad y = -4 \qquad \text{Solve each equation.}$$

Check by substituting in the original equation. The solution set is $\{-4, 5\}$.

Work Problem **4** *at the Side.* ▶

Solving a Quadratic Equation by Factoring

Step 1 **Write the equation in standard form,** that is, with all terms on one side of the equals sign in descending powers of the variable and 0 on the other side.

Step 2 **Factor** completely.

Step 3 **Use the zero-factor property** to set each factor with a variable equal to 0.

Step 4 **Solve** the resulting equations.

Step 5 **Check** each solution in the original equation.

EXAMPLE 3 **Solving a Quadratic Equation (Common Factor)**

Solve $4p^2 + 40 = 26p$.

$$4p^2 - 26p + 40 = 0 \qquad \text{Standard form}$$

$$\mathbf{2}(2p^2 - 13p + 20) = 0 \qquad \text{Factor out 2.}$$

$$2p^2 - 13p + 20 = 0 \qquad \text{Divide each side by 2.}$$

$$(2p - 5)(p - 4) = 0 \qquad \text{Factor.}$$

$$2p - 5 = 0 \quad \text{or} \quad p - 4 = 0 \qquad \text{Zero-factor property}$$

$$2p = 5 \quad \text{or} \qquad p = 4 \qquad \text{Solve each equation.}$$

$$p = \frac{5}{2}$$

Check that the solution set is $\{\frac{5}{2}, 4\}$ by substituting in the original equation.

CAUTION

A common error is to include the common factor **2** as a solution in Example 3. *Only factors containing variables lead to solutions.*

Work Problem **5** *at the Side.* ▶

---

4 Solve each equation. Check your solutions.

(a) $m^2 - 3m - 10 = 0$

(b) $r^2 + 2r = 8$

5 Solve each equation. Check your solutions.

(a) $10a^2 - 5a - 15 = 0$

(b) $4x^2 - 2x = 42$

ANSWERS

4. **(a)** $\{-2, 5\}$ **(b)** $\{-4, 2\}$

5. **(a)** $\left\{-1, \frac{3}{2}\right\}$ **(b)** $\left\{-3, \frac{7}{2}\right\}$

6 Solve each equation. Check your solutions.

(a) $49m^2 - 9 = 0$

(b) $p(4p + 7) = 2$

(c) $m^2 = 3m$

EXAMPLE 4 Solving Quadratic Equations

Solve each equation.

(a) $16m^2 - 25 = 0$

We can factor the left side of the equation as the difference of squares (**Section 13.5**).

$$16m^2 - 25 = 0$$
$$(4m + 5)(4m - 5) = 0 \qquad \text{Factor the difference of squares.}$$
$$4m + 5 = 0 \quad \text{or} \quad 4m - 5 = 0 \qquad \text{Zero-factor property}$$
$$4m = -5 \quad \text{or} \qquad 4m = 5 \qquad \text{Solve each equation.}$$
$$m = -\frac{5}{4} \quad \text{or} \qquad m = \frac{5}{4}$$

Check the solutions $-\frac{5}{4}$ and $\frac{5}{4}$ in the original equation. The solution set is $\{-\frac{5}{4}, \frac{5}{4}\}$.

(b) $k(2k + 5) = 3$

We need to write this equation in standard form.

$$k(2k + 5) = 3 \qquad \boxed{\text{To be in standard form, 0 must be on one side.}}$$
$$2k^2 + 5k = 3 \qquad \text{Multiply.}$$
$$2k^2 + 5k - 3 = 0 \qquad \text{Subtract 3.}$$
$$(2k - 1)(k + 3) = 0 \qquad \text{Factor.}$$
$$2k - 1 = 0 \quad \text{or} \quad k + 3 = 0 \qquad \text{Zero-factor property}$$
$$2k = 1 \quad \text{or} \qquad k = -3$$
$$k = \frac{1}{2}$$

The solution set is $\{-3, \frac{1}{2}\}$.

(c) $y^2 = 2y$

$$\boxed{\text{Don't forget to set the variable factor } y \text{ equal to 0.}} \qquad y^2 - 2y = 0 \qquad \text{Standard form}$$
$$y(y - 2) = 0 \qquad \text{Factor.}$$
$$y = 0 \quad \text{or} \quad y - 2 = 0 \qquad \text{Zero-factor property}$$
$$y = 2$$

The solution set is $\{0, 2\}$.

CAUTION

In Example 4(b), the zero-factor property could not be used to solve the equation $k(2k + 5) = 3$ in its given form because of the 3 on the right. *The zero-factor property applies only to a product that equals 0.*

In Example 4(c), it is tempting to begin by dividing each side of the equation $y^2 = 2y$ by y to get $y = 2$. Note that we do not get the other solution, 0, if we divide by a variable. (We *may* divide each side of an equation by a *nonzero* real number, however. For instance, in Example 3 we divided each side by 2.)

ANSWERS

6. **(a)** $\{-\frac{3}{7}, \frac{3}{7}\}$ **(b)** $\{-2, \frac{1}{4}\}$ **(c)** $\{0, 3\}$

◀ *Work Problem* **6** *at the Side.*

EXAMPLE 5 **Solving a Quadratic Equation with a Double Solution**

Solve each equation.

(a)
$$z^2 + 121 = 22z$$
$$z^2 - 22z + 121 = 0 \qquad \text{Standard form}$$

Because $121 = 11^2$ and $22z = 2 \cdot z \cdot 11$, the trinomial on the left is a perfect square.

$$(z - 11)^2 = 0 \qquad \text{Factor.}$$

To apply the zero-product property, write $(z - 11)^2$ as two separate factors.

$$(z - 11)(z - 11) = 0 \qquad a^2 = a \cdot a$$
$$z - 11 = 0 \quad \text{or} \quad z - 11 = 0 \qquad \text{Zero-factor property}$$

Because the two factors are identical, they both lead to the same solution. (This is called a **double solution.**) Thus,

$$z - 11 = 0$$
$$z = 11. \qquad \text{Add 11.}$$

Check
$$z^2 + 121 = 22z$$
$$11^2 + 121 \overset{?}{=} 22(11) \qquad \text{Let } z = 11.$$
$$121 + 121 \overset{?}{=} 242$$
$$242 = 242 \qquad \text{True}$$

The solution set is $\{11\}$.

(b)
$$9t^2 - 30t = -25$$
$$9t^2 - 30t + 25 = 0 \qquad \text{Standard form}$$
$$(3t - 5)^2 = 0 \qquad \text{Factor the perfect square trinomial.}$$
$$3t - 5 = 0 \quad \text{or} \quad 3t - 5 = 0 \qquad \text{Zero-factor property}$$
$$3t = 5 \qquad \text{Solve the equation.}$$
$$t = \frac{5}{3}$$

Check the double solution by substituting $\frac{5}{3}$ in the original equation. The solution set is $\left\{\frac{5}{3}\right\}$.

Work Problem **7** at the Side. ▶

CAUTION
When a trinomial has two identical factors (a perfect square trinomial), as in Examples 5(a) and (b), it is common for students to write the solution of the corresponding quadratic equation twice in the solution set. Each of these equations has only *one* distinct solution. ***There is no need to write the same number more than once in a solution set.***

Note
Not all quadratic equations can be solved by factoring. A more general method for solving such equations is given in **Chapter 17.**

7 Solve each equation. Check your solutions.

(a) $x^2 + 16x = -64$

(b) $4x^2 - 4x + 1 = 0$

ANSWERS

7. **(a)** $\{-8\}$ **(b)** $\left\{\frac{1}{2}\right\}$

8 Solve each equation. Check your solutions.

(a) $r^3 - 16r = 0$

(b) $x^3 - 3x^2 - 18x = 0$

OBJECTIVE **2** **Solve other equations by factoring.** We can extend the zero-factor property to solve equations that involve more than two factors with variables, as shown in Examples 6 and 7. (These equations are *not* quadratic equations. Why not?)

EXAMPLE 6 **Solving an Equation with More Than Two Variable Factors**

Solve $6z^3 - 6z = 0$.

$$6z^3 - 6z = 0$$
$$6z(z^2 - 1) = 0 \qquad \text{Factor out } 6z.$$
$$6z(z + 1)(z - 1) = 0 \qquad \text{Factor } z^2 - 1.$$

By an extension of the zero-factor property, this product can equal 0 only if at least one of the factors equals 0. Write and solve three equations, one for each factor with a variable.

$$6z = 0 \quad \text{or} \quad z + 1 = 0 \quad \text{or} \quad z - 1 = 0$$
$$z = 0 \quad \text{or} \qquad z = -1 \quad \text{or} \qquad z = 1$$

Check by substituting, in turn, 0, -1, and 1 in the original equation. The solution set is $\{-1, 0, 1\}$.

◀ Work Problem **8** at the Side.

9 Solve each equation. Check your solutions.

(a) $(m + 3)(m^2 - 11m + 10) = 0$

EXAMPLE 7 **Solving an Equation with a Quadratic Factor**

Solve $(2x - 1)(x^2 - 9x + 20) = 0$.

$$(2x - 1)(x^2 - 9x + 20) = 0$$
$$(2x - 1)(x - 5)(x - 4) = 0 \qquad \text{Factor } x^2 - 9x + 20.$$
$$2x - 1 = 0 \quad \text{or} \quad x - 5 = 0 \quad \text{or} \quad x - 4 = 0 \qquad \text{Zero-factor property}$$
$$x = \frac{1}{2} \quad \text{or} \qquad x = 5 \quad \text{or} \qquad x = 4$$

Check to verify that the solution set is $\{\frac{1}{2}, 4, 5\}$.

◀ Work Problem **9** at the Side.

(b) $(2x + 5)(4x^2 - 9) = 0$

CAUTION
In Example 7, it would be unproductive to begin by multiplying the two factors together. Keep in mind that the zero-factor property and its extension requires the product of two or more factors to equal 0. *Always consider first whether an equation is given in the appropriate form to apply the zero-factor property.*

ANSWERS

8. (a) $\{-4, 0, 4\}$ **(b)** $\{-3, 0, 6\}$

9. (a) $\{-3, 1, 10\}$ **(b)** $\left\{-\frac{5}{2}, -\frac{3}{2}, \frac{3}{2}\right\}$

Solve each equation, and check your solutions. See Example 1.

1. $(x + 5)(x - 2) = 0$

2. $(x - 1)(x + 8) = 0$

3. $(2m - 7)(m - 3) = 0$

4. $(6k + 5)(k + 4) = 0$

5. $t(6t + 5) = 0$

6. $w(4w + 1) = 0$

7. $2x(3x - 4) = 0$

8. $6x(4x + 9) = 0$

9. $\left(x + \dfrac{1}{2}\right)\left(2x - \dfrac{1}{3}\right) = 0$

10. $\left(a + \dfrac{2}{3}\right)\left(5a - \dfrac{1}{2}\right) = 0$

11. $(x - 9)(x - 9) = 0$

12. $(2x + 1)(2x + 1) = 0$

13. Look at this "solution." ***WHAT WENT WRONG?***

$$2x(3x - 4) = 0$$

$$x = 2 \quad \text{or} \quad x = 0 \quad \text{or} \quad 3x - 4 = 0$$

$$x = \frac{4}{3}$$

The solution set is $\{2, 0, \frac{4}{3}\}$.

14. Look at this "solution." ***WHAT WENT WRONG?***

$$x(7x - 1) = 0$$

$$7x - 1 = 0 \qquad \text{Zero-factor property}$$

$$x = \frac{1}{7}$$

The solution set is $\{\frac{1}{7}\}$.

Solve each equation, and check your solutions. See Examples 2–7.

15. $y^2 + 3y + 2 = 0$

16. $p^2 + 8p + 7 = 0$

17. $y^2 - 3y + 2 = 0$

18. $r^2 - 4r + 3 = 0$

19. $x^2 = 24 - 5x$

20. $t^2 = 2t + 15$

21. $x^2 = 3 + 2x$

22. $m^2 = 4 + 3m$

23. $z^2 + 3z = -2$

24. $p^2 - 2p = 3$

25. $m^2 + 8m + 16 = 0$

26. $b^2 - 6b + 9 = 0$

27. $3x^2 + 5x - 2 = 0$

28. $6r^2 - r - 2 = 0$

29. $6p^2 = 4 - 5p$

30. $6x^2 = 4 + 5x$

31. $9s^2 + 12s = -4$

32. $36x^2 + 60x = -25$

33. $y^2 - 9 = 0$

34. $m^2 - 100 = 0$

35. $16k^2 - 49 = 0$

36. $4w^2 - 9 = 0$

37. $n^2 = 121$

38. $x^2 = 400$

39. $x^2 = 7x$

40. $t^2 = 9t$

41. $6r^2 = 3r$

42. $10y^2 = -5y$

43. $g(g - 7) = -10$

44. $r(r - 5) = -6$

45. $z(2z + 7) = 4$

46. $b(2b + 3) = 9$

47. $2(y^2 - 66) = -13y$

48. $3(t^2 + 4) = 20t$

49. $5x^3 - 20x = 0$

50. $3x^3 - 48x = 0$

51. $9y^3 - 49y = 0$

52. $16r^3 - 9r = 0$

53. $(2r + 5)(3r^2 - 16r + 5) = 0$

54. $(3m + 4)(6m^2 + m - 2) = 0$

55. $(2x + 7)(x^2 + 2x - 3) = 0$

56. $(x + 1)(6x^2 + x - 12) = 0$

57. Galileo's formula for freely falling objects, **$d = 16t^2$**, was given at the beginning of this section. The distance d in feet an object falls depends on the time elapsed t in seconds. (This is an example of a **function**, to be introduced in **Section 17.5**.)

(a) Use Galileo's formula and complete the following table. (*Hint:* Substitute each given value into the formula and solve for the unknown value.)

t in seconds	0	1	2	3		
d in feet	0	16			256	576

(b) When $t = 0$, $d = 0$. Explain this in the context of the problem.

58. In Exercise 57, when you substituted 256 for d and solved for t, you should have found two solutions: 4 and -4. Why doesn't -4 make sense as an answer?

13.8 ▶▶▶ Applications of Quadratic Equations

We can use factoring to solve quadratic equations that arise in applications. We follow the same six problem-solving steps given in **Section 10.4**.

OBJECTIVES
1 Solve problems about geometric figures.
2 Solve problems about consecutive integers.
3 Solve problems using the Pythagorean formula.
4 Solve problems using given quadratic models.

> **Solving an Applied Problem**
>
> *Step 1* **Read** the problem, several times if necessary, until you *understand* what is given and what is to be found.
>
> *Step 2* **Assign a variable** to represent the unknown value, using diagrams or tables as needed. Write a statement that tells what the variable represents. Express any other unknown values in terms of the variable.
>
> *Step 3* **Write an equation** using the variable expression(s).
>
> *Step 4* **Solve** the equation.
>
> *Step 5* **State the answer.** Does it seem reasonable?
>
> *Step 6* **Check** the answer in the words of the original problem.

OBJECTIVE 1 Solve problems about geometric figures. Refer to the formulas given on the inside covers of the text, if necessary.

EXAMPLE 1 Solving an Area Problem

The Monroes want to plant a rectangular garden in their yard. The width of the garden will be 4 ft less than its length, and they want it to have an area of 96 ft^2. (ft^2 means square feet.) Find the length and width of the garden.

Step 1 **Read** the problem carefully. We need to find the dimensions of a garden with area 96 ft^2.

Step 2 **Assign a variable.**

Let x = the length of the garden.

Then $x - 4$ = the width. (The width is 4 ft less than the length.)

See Figure 1.

Figure 1

Step 3 **Write an equation.** The area of a rectangle is given by

$$\text{Area} = LW = \text{Length} \times \text{Width}. \qquad \textcolor{blue}{\text{Area formula}}$$

Substitute 96 for area, x for length, and $x - 4$ for width.

$$\textcolor{blue}{A = LW}$$

$$96 = x(x - 4) \qquad \textcolor{blue}{\text{Let } A = 96, L = x, W = x - 4.}$$

Continued on Next Page

1 Solve each problem.

(a) The length of a rectangular room is 2 m more than the width. The area of the floor is 48 m². Find the length and width of the room.

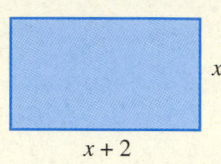

x

$x + 2$

Step 4 **Solve.**

$$96 = x(x - 4)$$
$$96 = x^2 - 4x \qquad \text{Distributive property}$$
$$x^2 - 4x - 96 = 0 \qquad \text{Standard form}$$
$$(x - 12)(x + 8) = 0 \qquad \text{Factor.}$$
$$x - 12 = 0 \quad \text{or} \quad x + 8 = 0 \qquad \text{Zero-factor property}$$
$$x = 12 \quad \text{or} \qquad x = -8$$

Step 5 **State the answer.** The solutions are 12 and −8. A rectangle cannot have a side of negative length, so discard −8. The length of the garden will be 12 ft. The width will be 12 − 4 = 8 ft.

Step 6 **Check.** The width of the garden is 4 ft less than the length; the area is $12 \cdot 8 = 96$ ft².

Problem-Solving Hint

When solving applied problems, ***always check solutions against physical facts*** and discard any answers that are not appropriate.

◄ *Work Problem* **1** *at the Side.*

(b) The length of each side of a square is increased by 4 in. The sum of the areas of the original square and the larger square is 106 in². What is the length of a side of the original square?

OBJECTIVE **2** **Solve problems about consecutive integers.** Recall from **Section 10.4** that **consecutive integers** are integers that are next to each other on a number line, such as 5 and 6, or −11 and −10. **Consecutive odd integers** are *odd* integers that are next to each other, such as 5 and 7, or −13 and −11. **Consecutive even integers** are defined similarly; 4 and 6 are consecutive even integers, as are −10 and −8. (In this book, we will list consecutive integers in increasing order from left to right.)

Problem-Solving Hint

In consecutive integer problems, if x represents the first integer, then for

two consecutive integers, use	$x, \quad x + 1$;
three consecutive integers, use	$x, \quad x + 1, \quad x + 2$;
two consecutive even or odd integers, use	$x, \quad x + 2$;
three consecutive even or odd integers, use	$x, \quad x + 2, \quad x + 4.$

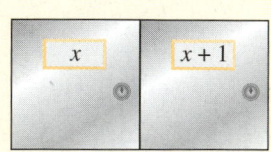

x $x + 1$

Figure 2

EXAMPLE 2 **Solving a Consecutive Integer Problem**

The product of the numbers on two consecutive post-office boxes is 210. Find the box numbers.

Step 1 **Read** the problem. Note that the boxes are numbered consecutively.

Step 2 **Assign a variable.**

Let $x =$ the first box number.

Then $x + 1 =$ the next consecutive box number.

See Figure 2.

Step 3 **Write an equation.** The product of the box numbers is 210, so

$$x(x + 1) = 210.$$

Continued on Next Page

Step 4 **Solve.**

$$x(x + 1) = 210$$
$$x^2 + x = 210 \qquad \text{Distributive property}$$
$$x^2 + x - 210 = 0 \qquad \text{Standard form}$$
$$(x + 15)(x - 14) = 0 \qquad \text{Factor.}$$
$$x + 15 = 0 \quad \text{or} \quad x - 14 = 0 \qquad \text{Zero-factor property}$$
$$x = -15 \quad \text{or} \qquad x = 14$$

Step 5 **State the answer.** The solutions are -15 and 14. Discard the solution -15 since a box number cannot be negative. When $x = 14$, then $x + 1 = 15$, so the post-office boxes have the numbers 14 and 15.

Step 6 **Check.** The numbers 14 and 15 are consecutive and their product is $14 \cdot 15 = 210$, as required.

Work Problem ⟮**2**⟯ *at the Side.* ▶

⌐ **EXAMPLE 3** **Solving a Consecutive Integer Problem**

The product of two consecutive odd integers is 1 less than five times their sum. Find the integers.

Step 1 **Read** carefully. This problem is a little more complicated.

Step 2 **Assign a variable.** We must find two consecutive *odd* integers.

Let $x =$ the lesser integer.

Then $x + 2 =$ the next greater odd integer.

Step 3 **Write an equation.** According to the problem, the product is 1 less than five times the sum.

The product	is	five times the sum	less 1.
↓	↓	↓	↓
$x(x + 2)$	$=$	$5(x + x + 2)$	$-\ 1$

Step 4 **Solve.**

$$x^2 + 2x = 5x + 5x + 10 - 1 \qquad \text{Distributive property}$$
$$x^2 + 2x = 10x + 9 \qquad \text{Combine like terms.}$$
$$x^2 - 8x - 9 = 0 \qquad \text{Standard form}$$
$$(x - 9)(x + 1) = 0 \qquad \text{Factor.}$$
$$x - 9 = 0 \quad \text{or} \quad x + 1 = 0 \qquad \text{Zero-factor property}$$
$$x = 9 \quad \text{or} \qquad x = -1$$

Step 5 **State the answer.** We need to find two consecutive odd integers.

If $x = $ **9** is the lesser, then $x + 2 = 9 + 2 = $ **11** is the greater.

If $x = $ **−1** is the lesser, then $x + 2 = -1 + 2 = $ **1** is the greater.

There are two sets of answers here since integers can be positive or negative.

Step 6 **Check.** The product of the first pair of integers is $9 \cdot 11 = 99$. One less than five times their sum is $5(9 + 11) - 1 = 99$. Thus 9 and 11 satisfy the problem. Repeat the check with -1 and 1.

Work Problem ⟮**3**⟯ *at the Side.* ▶

⟮**2**⟯ Solve the problem.
    The product of the numbers on two consecutive lockers at a health club is 132. Find the locker numbers.

⟮**3**⟯ Solve each problem.

(a) The product of two consecutive even integers is 4 more than two times their sum. Find the integers.

(b) Find three consecutive odd integers such that the product of the least and greatest is 16 more than the middle integer.

CAUTION
Do *not* use $x, x + 1, x + 3$, and so on to represent consecutive odd integers. To see why, let $x = 3$. Then $x + 1 = 3 + 1 = 4$ and $x + 3 = 3 + 3 = 6$, and 3, 4, and 6 are not consecutive odd integers.

OBJECTIVE **3** **Solve problems using the Pythagorean formula.**
The next example requires the Pythagorean formula from geometry.

Pythagorean Formula

If a right triangle (a triangle with a 90° angle) has longest side of length c and two other sides of lengths a and b, then

$$a^2 + b^2 = c^2.$$

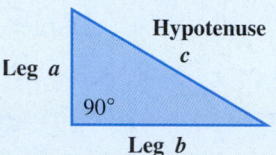

The longest side, the **hypotenuse,** is opposite the right angle. The two shorter sides are the **legs** of the triangle.

EXAMPLE 4 **Using the Pythagorean Formula**

Amy and Kevin leave their office, with Amy traveling north and Kevin traveling east. When Kevin is 1 mi farther than Amy from the office, the distance between them is 2 mi more than Amy's distance from the office. Find their distances from the office and the distance between them.

Step 1 **Read** the problem again. We must find three distances.

Step 2 **Assign a variable.** Let x represent Amy's distance from the office, $x + 1$ represent Kevin's distance from the office, and $x + 2$ represent the distance between them. Place these on a right triangle, as in Figure 3.

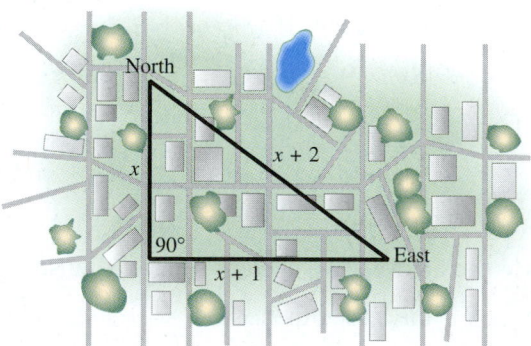

Figure 3

Step 3 **Write an equation.** Substitute into the Pythagorean formula.

$$a^2 + b^2 = c^2$$

$$x^2 + (x + 1)^2 = (x + 2)^2$$

> Be careful to substitute properly.

Continued on Next Page

Step 4 **Solve.** $x^2 + x^2 + 2x + 1 = x^2 + 4x + 4$

$\qquad x^2 - 2x - 3 = 0$ Standard form

$\qquad (x - 3)(x + 1) = 0$ Factor.

$\qquad x - 3 = 0 \quad \text{or} \quad x + 1 = 0$ Zero-factor property

$\qquad x = 3 \quad \text{or} \qquad x = -1$

Step 5 **State the answer.** Since -1 cannot represent a distance, 3 is the only possible answer. Amy's distance is 3 mi, Kevin's distance is $3 + 1 = 4$ mi, and the distance between them is $3 + 2 = 5$ mi.

Step 6 **Check.** Since $3^2 + 4^2 = 5^2$, the answer is correct.

CAUTION

When solving a problem involving the Pythagorean formula, be sure that the expressions for the sides are properly placed.

$$\textbf{leg}^2 + \textbf{leg}^2 = \textbf{hypotenuse}^2$$

Work Problem **4** *at the Side.* ▶

OBJECTIVE **4** **Solve problems using given quadratic models.**
In Examples 1–4, we wrote quadratic equations to model, or mathematically describe, various situations and then solved the equations. Now we are given the quadratic models and must use them to determine data.

EXAMPLE 5   **Finding the Height of a Ball**

A tennis player can hit a ball 180 ft per sec. If she hits a ball directly upward, the height h of the ball in feet at time t in seconds is modeled by the quadratic equation

$$h = -16t^2 + 180t + 6.$$

When will the ball be 206 ft above the ground?

A height of 206 ft means $h = 206$, so we substitute 206 for h in the equation and then solve for t.

$\qquad \textbf{206} = -16t^2 + 180t + 6$ Let $h = 206$.

$\qquad -16t^2 + 180t + 6 = 206$ Interchange sides.

$\qquad -16t^2 + 180t - 200 = 0$ Standard form

$\qquad 4t^2 - 45t + 50 = 0$ Divide by -4.

$\qquad (4t - 5)(t - 10) = 0$ Factor.

$\qquad 4t - 5 = 0 \quad \text{or} \quad t - 10 = 0$ Zero-factor property

$\qquad t = \dfrac{5}{4} \quad \text{or} \qquad t = 10$

Since we found two acceptable answers, the ball will be 206 ft above the ground twice (once on its way up and once on its way down)—at $\frac{5}{4}$ sec and at 10 sec after it is hit. See Figure 4.

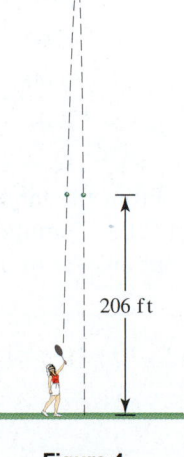

206 ft

Figure 4

Work Problem **5** *at the Side.* ▶

---

4 Solve the problem.

The hypotenuse of a right triangle is 3 in. longer than the longer leg. The shorter leg is 3 in. shorter than the longer leg. Find the lengths of the sides of the triangle.

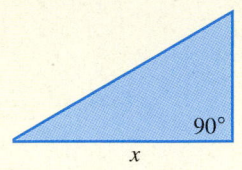

90°

x

5 Solve the problem.

The number of impulses fired after a nerve has been stimulated is modeled by

$$I = -x^2 + 2x + 60,$$

where x is in milliseconds (ms) after the stimulation. When will 45 impulses occur? Do you get two solutions? Why is only one answer given?

ANSWERS

4. 9 in., 12 in., 15 in.
5. After 5 ms; There are two solutions, -3 and 5; Only one answer makes sense here because a negative answer is not appropriate.

6 Use the model in Example 6 to find the foreign-born population of the United States in 1990. Give your answer to the nearest tenth of a million. How does it compare to the actual value from the table?

EXAMPLE 6 Modeling the Foreign-Born Population of the United States

After decreasing in the middle of the 20th century, the foreign-born population of the United States started to increase in the later part of the century and is now increasing rapidly. The foreign-born population over the years 1930–2004 can be modeled by the quadratic equation

$$y = 0.01036x^2 - 0.5316x + 15.36,$$

where $x = 0$ represents 1930, $x = 10$ represents 1940, and so on, and y is the number of people in millions. (*Source:* U.S. Census Bureau.)

(a) Use the model to find the foreign-born population in 1980 to the nearest tenth of a million.

Since $x = 0$ represents 1930, $x = 50$ represents 1980. Substitute 50 for x in the equation.

$y = 0.01036(50)^2 - 0.5316(50) + 15.36$ Let $x = 50$.

$y = 14.7$ Round to the nearest tenth.

In 1980, the foreign-born population of the United States was about 14.7 million.

(b) Repeat part (a) for 2004.

$y = 0.01036(74)^2 - 0.5316(74) + 15.36$ For 2004, let $x = 74$.

$y = 32.8$ Round to the nearest tenth.

In 2004, the foreign-born population of the United States was about 32.8 million.

(c) The model used in parts (a) and (b) was developed using the data in the table below. How do the results in parts (a) and (b) compare to the actual data from the table?

Year	Foreign-Born Population (millions)
1930	14.2
1940	11.6
1950	10.3
1960	9.7
1970	9.6
1980	14.1
1990	19.8
2000	28.4
2004	34.2

From the table, the actual value for 1980 is 14.1 million. Our answer in part (a), 14.7 million, is slightly high. For 2004, the actual value is 34.2 million, so our answer of 32.8 million in part (b) is somewhat low.

◀ *Work Problem* **6** *at the Side.*

ANSWER

6. 20.8 million; The actual value is 19.8 million, so our answer using the model is somewhat high.

13.8 ▶▶▶ Exercises

1. To review the six problem-solving steps first introduced in **Section 2.4,** complete each statement.

Step 1: _____ the problem, several times if necessary, until you understand what is given and what must be found.

Step 2: Assign a _____ to represent the unknown value.

Step 3: Write a(n) _____ using the variable expression(s).

Step 4: _____ the equation.

Step 5: State the _____ .

Step 6: _____ the answer in the words of the _____ problem.

2. A student solves an applied problem and gets 6 or -3 for the length of the side of a square. Which of these answers is reasonable? Explain.

In Exercises 3–6, a figure and a corresponding geometric formula are given. Using x as the variable, complete Steps 3–6 for each problem. (Refer to the steps in Exercise 1 as needed.)

3.

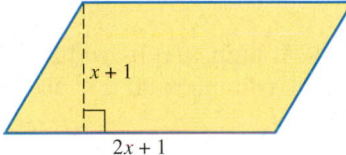

Area of a parallelogram: $A = bh$

The area of this parallelogram is 45 sq. units. Find its base and height.

4.

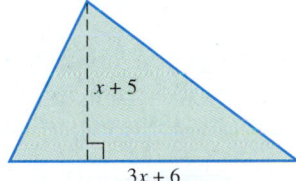

Area of a triangle: $A = \frac{1}{2}bh$

The area of this triangle is 60 sq. units. Find its base and height.

5.

Volume of a rectangular Chinese box: $V = LWH$

The volume of this box is 192 cu. units. Find its length and width.

6.

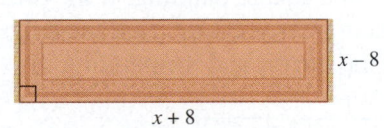

Area of a rectangular rug: $A = LW$

The area of this rug is 80 sq. units. Find its length and width.

Solve each problem. Check your answers to be sure they are reasonable. Refer to the formulas on the inside covers. See Example 1.

7. The length of a standard jewel case is 2 cm more than its width. The area of the rectangular top of the case is 168 cm². Find the length and width of the jewel case.

8. A standard DVD case is 6 cm longer than it is wide. The area of the rectangular top of the case is 247 cm². Find the length and width of the case.

9. The dimensions of an HP f1905 flat-panel monitor are such that its length is 3 in. more than its width. If the length were doubled and if the width were decreased by 1 in., the area would be increased by 150 in.². What are the length and width of the flat panel?

10. The keyboard of the computer in Exercise 9 is 11 in. longer than it is wide. If both its length and width are increased by 2 in., the area of the top of the keyboard is increased by 54 in.². Find the length and width of the keyboard. (*Source:* Author's computer.)

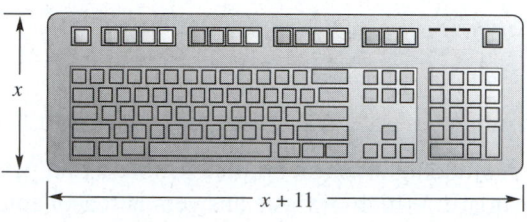

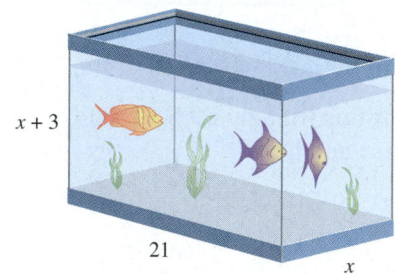

11. A 10-gal aquarium is 3 in. higher than it is wide. Its length is 21 in., and its volume is 2730 in.³. What are the height and width of the aquarium?

12. A toolbox is 2 ft high, and its width is 3 ft less than its length. If its volume is 80 ft³, find the length and width of the box.

13. A square mirror has sides measuring 2 ft less than the sides of a square painting. If the difference between their areas is 32 ft², find the lengths of the sides of the mirror and the painting.

14. The sides of one square have length 3 m more than the sides of a second square. If the area of the larger square is subtracted from 4 times the area of the smaller square, the result is 36 m². What are the lengths of the sides of each square?

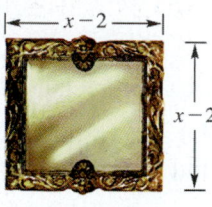

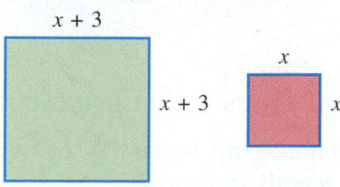

Solve each problem about consecutive integers. See Examples 2 and 3.

15. The product of the numbers on two consecutive volumes of research data is 420. Find the volume numbers.

16. The product of the page numbers on two facing pages of a book is 600. Find the page numbers.

17. The product of two consecutive integers is 11 more than their sum. Find the integers.

18. The product of two consecutive integers is 4 less than four times their sum. Find the integers.

19. Find two consecutive odd integers such that their product is 15 more than three times their sum.

20. Find two consecutive odd integers such that five times their sum is 23 less than their product.

21. Find three consecutive even integers such that the sum of the squares of the lesser two is equal to the square of the greatest.

22. Find three consecutive even integers such that the square of the sum of the lesser two is equal to twice the greatest.

Use the Pythagorean formula to solve each problem. See Example 4.

23. The hypotenuse of a right triangle is 1 cm longer than the longer leg. The shorter leg is 7 cm shorter than the longer leg. Find the length of the longer leg of the triangle.

24. The longer leg of a right triangle is 1 m longer than the shorter leg. The hypotenuse is 1 m shorter than twice the shorter leg. Find the length of the shorter leg of the triangle.

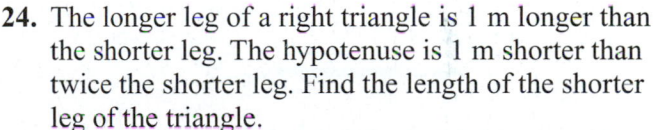

25. Terri works due north of home. Her husband Denny works due east. They leave for work at the same time. By the time Terri is 5 mi from home, the distance between them is 1 mi more than Denny's distance from home. How far from home is Denny?

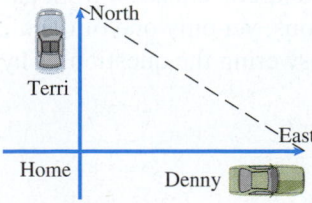

26. Two cars left an intersection at the same time. One traveled north. The other traveled 14 mi farther, but to the east. How far apart were they then, if the distance between them was 4 mi more than the distance traveled east?

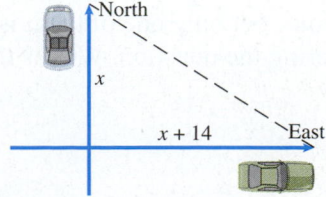

27. A ladder is leaning against a building. The distance from the bottom of the ladder to the building is 4 ft less than the length of the ladder. How high up the side of the building is the top of the ladder if that distance is 2 ft less than the length of the ladder?

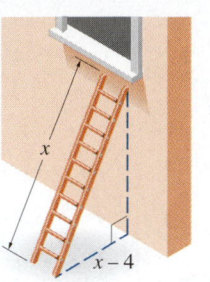

28. A lot has the shape of a right triangle with one leg 2 m longer than the other. The hypotenuse is 2 m less than twice the length of the shorter leg. Find the length of the shorter leg.

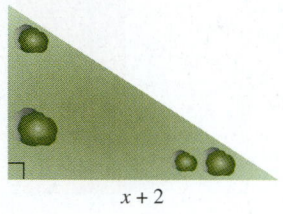

Solve each problem. See Examples 5 and 6.

29. An object projected from a height of 48 ft with an initial velocity of 32 ft per sec after t seconds has height

$$h = -16t^2 + 32t + 48.$$

48 ft

(a) After how many seconds is the height 64 ft? (*Hint:* Let $h = 64$ and solve.)

(b) After how many seconds is the height 60 ft?

(c) After how many seconds does the object hit the ground? (*Hint:* When the object hits the ground, $h = 0$.)

(d) The quadratic equation from part (c) has two solutions, yet only one of them is appropriate for answering the question. Why is this so?

30. If an object is projected upward from ground level with an initial velocity of 64 ft per sec, its height h in feet t seconds later is

$$h = -16t^2 + 64t.$$

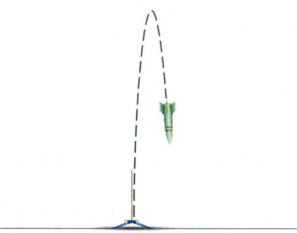

(a) After how many seconds is the height 48 ft?

(b) The object reaches its maximum height 2 sec after it is projected. What is this maximum height?

(c) After how many seconds does the object hit the ground?

(d) The quadratic equation from part (c) has two solutions, yet only one of them is appropriate for answering the question. Why is this so?

31. The table shows the number of cellular phone subscribers (in millions) in the United States.

Year	Subscribers (in millions)
1990	5
1992	11
1994	24
1996	44
1998	69
2000	109
2002	141
2004	182
2006	233

Source: CTIA-The Wireless Association.

We used the data to develop the quadratic equation

$$y = 0.734x^2 + 2.62x + 3.37,$$

which models the number of cellular phone subscribers y (in millions) in the year x, where $x = 0$ represents 1990, $x = 2$ represents 1992, and so on.

(a) Use the model to find the number of cellular phones in 1996 to the nearest million. How does the result compare to the actual data in the table?

(b) What value of x corresponds to 2004?

(c) Use the model to find the number of cellular phones in 2004 to the nearest million. How does the result compare to the actual data in the table?

(d) Assuming that the trend in the data continues, use the quadratic equation to estimate the number of cellular phones in 2009 to the nearest million.

Relating Concepts (Exercises 32–40) For Individual or Group Work

The U.S. trade deficit represents the amount by which exports are less than imports. It provides not only a sign of economic prosperity but also a warning of potential decline. The data in the table shows the U.S. trade deficit in goods and services for 2001 through 2005.

Year	Deficit (in billions of dollars)
2001	365.1
2002	423.7
2003	496.9
2004	612.1
2005	714.4

Source: U.S. Census Bureau.

Use the data to **work Exercises 32–40 in order.**

32. How much did the trade deficit in goods and services increase from 2001 to 2002? What percent increase is this (to the nearest percent)?

33. The U.S. trade deficit might be approximated by the linear equation

$$y = 88.7x + 256,$$

where y is the deficit in billions of dollars. Here $x = 1$ represents 2001, $x = 2$ represents 2002, and so on. Use this equation to approximate the trade deficits in 2003, 2004, and 2005.

34. How do your answers from Exercise 33 compare to the actual data in the table?

35. The trade deficit y (in billions of dollars) might also be approximated by the quadratic equation

$$y = 9.24x^2 + 33.24x + 321,$$

where $x = 1$ again represents 2001, $x = 2$ represents 2002, and so on. Use this equation to approximate the trade deficits in 2003, 2004, and 2005.

36. Compare your answers from Exercise 35 to the actual data in the table. Which equation, the linear one in Exercise 33 or the quadratic one in Exercise 35, models the data better?

37. Write the data from the table as a set of ordered pairs (x, y), where x represents the years starting with 2001, such that $x = 1$ for 2001, $x = 2$ for 2002, and so on, and y represents the trade deficit in billions of dollars.

Year	Deficit (in billions of dollars)
2001	365.1
2002	423.7
2003	496.9
2004	612.1
2005	714.4

Source: U.S. Census Bureau.

38. Plot the ordered pairs from Exercise 37 on the graph.

U.S. TRADE DEFICIT
(Goods and Services)

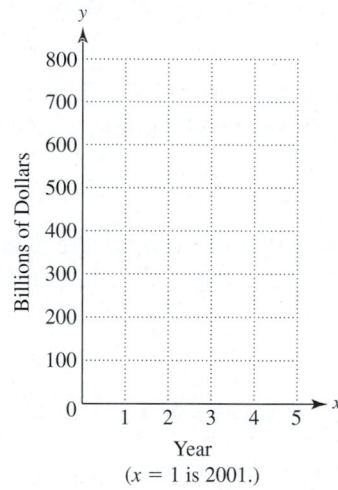

39. Assuming that the trend in the data continues, use the quadratic equation from Exercise 35 to estimate the trade deficit for the year 2006.

40. The actual trade deficit for 2006 was 758.2 billion dollars.

(a) How does the actual deficit for 2006 compare to your estimate from Exercise 39?

(b) Should the quadratic equation be used to estimate the U.S. trade deficit for years after 2005? Explain.

Chapter 13 ▷▷▷ Summary

▶ Key Terms

13.1 factor   An expression A is a factor of an expression B if B can be divided by A with 0 remainder.

factored form   An expression is in factored form when it is written as a product.

greatest common factor (GCF)   The greatest common factor is the largest quantity that is a factor of each of a group of quantities.

factoring   The process of writing a polynomial as a product is called factoring.

13.2 prime polynomial   A prime polynomial is a polynomial that cannot be factored using only integers.

13.5 perfect square trinomial   A perfect square trinomial is a trinomial that can be factored as the square of a binomial.

13.7 quadratic equation   A quadratic equation is an equation that can be written in the form $ax^2 + bx + c = 0$, with $a \neq 0$.

standard form   The form $ax^2 + bx + c = 0$ is the standard form of a quadratic equation.

13.8 hypotenuse   The longest side of a right triangle, opposite the right angle, is the hypotenuse.

legs   The two shorter sides of a right triangle are the legs.

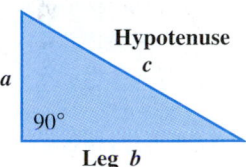

▶ Test Your Word Power

See how well you have learned the vocabulary in this chapter. Answers, with examples, follow the Quick Review.

1. **Factoring** is
   A. a method of multiplying polynomials
   B. the process of writing a polynomial as a product
   C. the answer in a multiplication problem
   D. a way to add the terms of a polynomial.

2. A polynomial is in **factored form** when
   A. it is prime
   B. it is written as a sum
   C. the squared term has a coefficient of 1
   D. it is written as a product.

3. The **greatest common factor** of a polynomial is
   A. the least integer that divides evenly into all the terms of the polynomial
   B. the least term that is a factor of all the terms in the polynomial
   C. the greatest term that is a factor of all the terms in the polynomial
   D. the variable that is common to all the terms in the polynomial.

4. A **perfect square trinomial** is a trinomial
   A. that can be factored as the square of a binomial
   B. that cannot be factored
   C. that is multiplied by a binomial
   D. where all terms are perfect squares.

5. A **quadratic equation** is an equation that can be written in the form
   A. $y = mx + b$
   B. $ax^2 + bx + c = 0 \, (a \neq 0)$
   C. $Ax + By = C$
   D. $x = k$.

6. A **hypotenuse** is
   A. either of the two shorter sides of a triangle
   B. the shortest side of a right triangle
   C. the side opposite the right angle in a right triangle
   D. the longest side in any triangle.

▶ Quick Review

Concepts | ### Examples

13.1 Factors; The Greatest Common Factor

Finding the Greatest Common Factor (GCF)

Step 1 Write each number in prime factored form.

Step 2 List each prime number or each variable that is a factor of every term in the list.

Step 3 Use as exponents on the common prime factors the least exponents from the prime factored forms.

Step 4 Multiply the primes from Step 3.

Find the greatest common factor of $4x^2y$, $-6x^2y^3$, and $2xy^2$.

$$4x^2y = \mathbf{2} \cdot 2 \cdot \mathbf{x^2} \cdot \mathbf{y}$$

$$-6x^2y^3 = -1 \cdot \mathbf{2} \cdot 3 \cdot \mathbf{x^2} \cdot \mathbf{y^3}$$

$$2xy^2 = \mathbf{2} \cdot \mathbf{x} \cdot \mathbf{y^2}$$

The greatest common factor is $2xy$.

Factoring by Grouping

Step 1 Group the terms.

Step 2 Factor out the greatest common factor from each group.

Step 3 Factor a common binomial factor from the results of Step 2.

Step 4 If necessary, rearrange terms and try a different grouping.

Factor by grouping.

$$2a^2 + 2ab + a + b$$

$$= (2a^2 + 2ab) + (a + b)$$

$$= 2a\,(\mathbf{a + b}) + 1\,(\mathbf{a + b})$$

$$= (\mathbf{a + b})(2a + 1)$$

13.2 Factoring Trinomials

To factor $x^2 + bx + c$, find m and n such that $mn = c$ and $m + n = b$.

$$x^2 + \mathbf{b}x + \mathbf{c}$$

$mn = c$

$m + n = b$

Then $x^2 + bx + c$ factors as $(x + m)(x + n)$.

Check by multiplying.

Factor $x^2 + 6x + 8$.

$$x^2 + \mathbf{6}x + \mathbf{8}$$

$mn = 8$

$m + n = 6$

$m = 2$ and $n = 4$

$x^2 + 6x + 8$ factors as $(x + 2)(x + 4)$.

Check $(x + 2)(x + 4)$

$$= x^2 + 4x + 2x + 8$$

$$= x^2 + 6x + 8$$

13.3 Factoring Trinomials by Grouping

To factor $ax^2 + bx + c$, find m and n such that $mn = ac$.

$$m + n = b$$

$$ax^2 + bx + c$$

$$mn = ac$$

Then factor $ax^2 + mx + nx + b$ by grouping.

Factor $3x^2 + 14x - 5$.

-15

Find two integers with a product of $3(-5) = -15$ and a sum of 14. The integers are -1 and 15.

$$3x^2 + \mathbf{14}x - 5$$

$$= 3x^2 - \mathbf{x} + \mathbf{15x} - 5$$

$$= (3x^2 - x) + (15x - 5)$$

$$= x(3x - 1) + 5(3x - 1)$$

$$= (3x - 1)(x + 5)$$

Concepts	Examples

13.4 Factoring Trinomials Using FOIL

To factor $ax^2 + bx + c$ by trial and error, use FOIL backwards.

By trial and error,

$$3x^2 + 14x - 5 \quad \text{factors as} \quad (3x - 1)(x + 5).$$

13.5 Special Factoring Techniques

Difference of Squares

$$a^2 - b^2 = (a + b)(a - b)$$

Perfect Square Trinomials

$$a^2 + 2ab + b^2 = (a + b)^2$$

$$a^2 - 2ab + b^2 = (a - b)^2$$

Difference of Cubes

$$x^3 - y^3 = (x - y)(x^2 + xy + y^2)$$

Sum of Cubes

$$x^3 + y^3 = (x + y)(x^2 - xy + y^2)$$

Factor.

$$4x^2 - 9$$
$$= (2x + 3)(2x - 3)$$

$$9x^2 + 6x + 1 \qquad\qquad 4x^2 - 20x + 25$$
$$= (3x + 1)^2 \qquad\qquad = (2x - 5)^2$$

$$8 - 27a^3$$
$$= (2 - 3a)(4 + 6a + 9a^2)$$

$$64z^3 + 1$$
$$= (4z + 1)(16z^2 - 4z + 1)$$

13.6 A General Approach to Factoring

See pages 951–952 for guidelines and examples.

13.7 Solving Quadratic Equations by Factoring

Zero-Factor Property

If a and b are real numbers and $ab = 0$, then $a = 0$ or $b = 0$.

Solving a Quadratic Equation by Factoring

Step 1 Write the equation in standard form.

Step 2 Factor.

Step 3 Use the zero-factor property.

Step 4 Solve the resulting equations.

Step 5 Check.

If $(x - 2)(x + 3) = 0$, then $x - 2 = 0$ or $x + 3 = 0$.

Solve $2x^2 = 7x + 15$.

$$2x^2 - 7x - 15 = 0$$
$$(2x + 3)(x - 5) = 0$$
$$2x + 3 = 0 \quad \text{or} \quad x - 5 = 0$$
$$2x = -3 \qquad\qquad x = 5$$
$$x = -\frac{3}{2}$$

The solutions $-\frac{3}{2}$ and 5 satisfy the original equation. The solution set is $\{-\frac{3}{2}, 5\}$.

Concepts	Examples

(13.8) Applications of Quadratic Equations

Pythagorean Formula
In a right triangle, the square of the hypotenuse equals the sum of the squares of the legs.

$$a^2 + b^2 = c^2$$

In a right triangle, one leg measures 2 ft longer than the other. The hypotenuse measures 4 ft longer than the shorter leg. Find the lengths of the three sides of the triangle.

Let $x =$ the length of the shorter leg. Then

$$x^2 + (x + 2)^2 = (x + 4)^2.$$

Solve this equation to get $x = 6$ or $x = -2$. Discard -2 as a solution. Check that the sides measure 6 ft, $6 + 2 = 8$ ft, and $6 + 4 = 10$ ft.

ANSWERS TO TEST YOUR WORD POWER

1. B; *Example:* $x^2 - 5x - 14$ factors as $(x - 7)(x + 2)$.
2. D; *Example:* The factored form of $x^2 - 5x - 14$ is $(x - 7)(x + 2)$.
3. C; *Example:* The greatest common factor of $8x^2$, $22xy$, and $16x^3y^2$ is $2x$.
4. A; *Example:* $a^2 + 2a + 1$ is a perfect square trinomial; its factored form is $(a + 1)^2$.
5. B; *Examples:* $y^2 - 3y + 2 = 0, x^2 - 9 = 0, 2m^2 = 6m + 8$
6. C; *Example:* See the triangle included in the Quick Review above for **Section 13.8.**

Chapter 13 ▶▶▶ Review Exercises

[13.1] *Factor out the greatest common factor or factor by grouping.*

1. $15t + 45$

2. $60z^3 + 30z$

3. $44x^3 + 55x^2$

4. $100m^2n^3 - 50m^3n^4 + 150m^2n^2$

5. $2xy - 8y + 3x - 12$

6. $6y^2 + 9y + 4xy + 6x$

[13.2] *Factor completely.*

7. $x^2 + 10x + 21$

8. $y^2 - 13y + 40$

9. $x^2 + x + 1$

10. $r^2 - r - 56$

11. $r^2 - 4rs - 96s^2$

12. $p^2 + 2pq - 120q^2$

13. $-8p^3 + 24p^2 + 80p$

14. $3x^4 + 30x^3 + 48x^2$

15. $p^7 - p^6q - 2p^5q^2$

[13.3–13.4] *Factor completely.*

16. $3r^2 + 11r - 4$

17. $6r^2 - 5r - 6$

18. $5t^2 - 11t + 12$

19. $24x^5 - 20x^4 + 4x^3$

20. $-3m^3n + 19m^2n + 40mn$

21. $14a^2 - 27ab - 20b^2$

[13.5]

22. Which one of the following is a difference of squares?

    A. $32x^2 - 1$ **B.** $4x^2y^2 - 25z^2$

    C. $x^2 + 36$ **D.** $25y^3 - 1$

23. Which one of the following is a perfect square trinomial?

    A. $x^2 + x + 1$ **B.** $y^2 - 4y + 9$

    C. $4x^2 + 10x + 25$ **D.** $x^2 - 20x + 100$

Factor completely. In Exercise 28, use fractions.

24. $n^2 - 64$

25. $25b^2 - 121$

26. $144p^2 - 36q^2$

27. $x^2 + 100$

28. $x^2 - \dfrac{49}{100}$

29. $r^2 - 12r + 36$

30. $16m^2 + 40mn + 25n^2$

31. $125x^3 - 1$

32. $1000p^3 + 27$

[13.7] *Solve each equation, and check the solutions.*

33. $(4t + 3)(t - 1) = 0$

34. $x(2x - 5) = 0$

35. $z^2 + 4z + 3 = 0$

36. $x^2 = -15 + 8x$

37. $3z^2 - 11z - 20 = 0$

38. $81t^2 - 64 = 0$

39. $y^2 = 8y$

40. $n(n - 5) = 6$

41. $t^2 = 12(t - 3)$

42. $(5z + 2)(z^2 + 3z + 2) = 0$

43. $49x^3 - 9x = 0$

44. $25w^2 - 90w + 81 = 0$

[13.8] *Solve each problem.*

45. The length of a rug is 6 ft more than the width. The area is 40 ft². Find the length and width of the rug.

46. The surface area S of a box is given by

$$S = 2WH + 2WL + 2LH.$$

A treasure chest from a sunken galleon has dimensions as shown in the figure. Its surface area is 650 ft². Find its width.

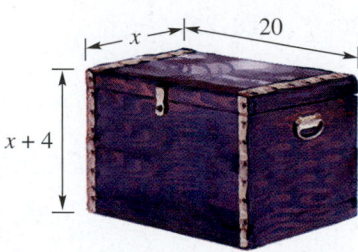

47. The product of two consecutive integers is 29 more than their sum. What are the integers?

48. Two cars left an intersection at the same time. One traveled west, and the other traveled 14 mi less, but to the south. How far apart were they then, if the distance between them was 16 mi more than the distance traveled south?

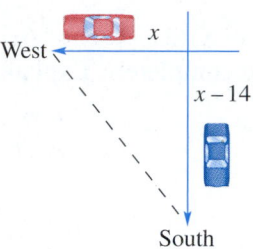

If an object is projected upward with an initial velocity of 128 *ft per sec, its height h in feet after t seconds is*

$$h = 128t - 16t^2.$$

Find the height of the object after each period of time.

49. 1 sec   **50.** 2 sec   **51.** 4 sec

52. For the object described above, when does it return to the ground?

53. Annual revenue in billions of dollars for eBay is shown in the table.

Year	Annual Revenue (in billions of dollars)
2002	1.21
2003	2.17
2004	3.27
2005	4.55
2006	5.77

Source: eBay.

Using the data, we developed the quadratic equation

$$y = 0.05x^2 + 0.95x + 1.19$$

to model eBay revenues y in year x, where $x = 0$ represents 2002, $x = 1$ represents 2003, and so on.

(a) Use the model to find eBay revenue (to the nearest hundredth) in 2005. How does your answer compare to the actual data from the table?

(b) Use the model to estimate annual revenue (to the nearest hundredth) for eBay in 2007.

▶▶▶ Mixed Review Exercises

54. Which of the following is *not* factored completely?

   A. $3(7t)$ **B.** $3x(7t + 4)$ **C.** $(3 + x)(7t + 4)$ **D.** $3(7t + 4) + x(7t + 4)$

55. Although $(2x + 8)(3x - 4) = 6x^2 + 16x - 32$ is a true statement, the polynomial is not factored completely. Explain why and give the complete factored form.

Factor completely.

56. $z^2 - 11zx + 10x^2$

57. $3k^2 + 11k + 10$

58. $15m^2 + 20mp - 12m - 16p$

59. $y^4 - 625$

60. $6m^3 - 21m^2 - 45m$

61. $24ab^3c^2 - 56a^2bc^3 + 72a^2b^2c$

62. $25a^2 + 15ab + 9b^2$

63. $12x^2yz^3 + 12xy^2z - 30x^3y^2z^4$

64. $2a^5 - 8a^4 - 24a^3$

65. $-12r^2 - 8rq + 15q^2$

66. $8z^3 + 64y^3$

67. $49t^2 + 56t + 16$

Solve.

68. $t(t - 7) = 0$

69. $x^2 + 3x = 10$

70. $25x^2 + 20x + 4 = 0$

Solve each problem.

71. The floor plan for a house is a rectangle with length 7 m more than its width. The area is 170 m². Find the width and length of the house.

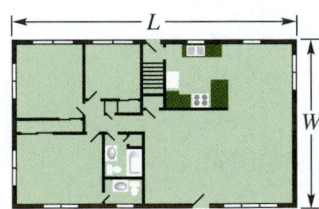

72. The triangular sail of a schooner has an area of 30 m². The height of the sail is 4 m more than the base. Find the length of the base of the sail.

1. Which one of the following is the correct, completely factored form of $2x^2 - 2x - 24$?

   A. $(2x + 6)(x - 4)$ **B.** $(x + 3)(2x - 8)$

   C. $2(x + 4)(x - 3)$ **D.** $2(x + 3)(x - 4)$

1. _____

Factor each polynomial completely.

2. $12x^2 - 30x$

2. _____

3. $2m^3n^2 + 3m^3n - 5m^2n^2$

3. _____

4. $2ax - 2bx + ay - by$

4. _____

5. $x^2 - 9x + 14$

5. _____

6. $2x^2 + x - 3$

6. _____

7. $6x^2 - 19x - 7$

7. _____

8. $3x^2 - 12x - 15$

8. _____

9. $10z^2 - 17z + 3$

9. _____

10. $t^2 + 6t + 10$

10. _____

11. $x^2 + \dfrac{1}{36}$

11. _____

12. $y^2 - 49$

12. _____

13. $81a^2 - 121b^2$

13. _____

14. $x^2 + 16x + 64$

14. _____

15. $4x^2 - 28xy + 49y^2$

15. _____

16. $-2x^2 - 4x - 2$

16. _____

17. $x^3 - 512$

17. _____

18. $8k^3 + 64$

18. _____

19. _____

20. _____

21. _____

22. _____

23. _____

24. _____

25. _____

26. _____

27. _____

28. _____

29. _____

30. _____

19. $4t^3 + 32t^2 + 64t$

20. $x^4 - 81$

Solve each equation.

21. $(x + 3)(x - 9) = 0$

22. $2r^2 - 13r + 6 = 0$

23. $25x^2 - 4 = 0$

24. $x(x - 20) = -100$

25. $t^2 = 3t$

26. $(s + 8)(6s^2 + 13s - 5) = 0$

Solve each problem.

27. The length of a rectangular flower bed is 3 ft less than twice its width. The area of the bed is 54 ft². Find the dimensions of the flower bed.

28. Find two consecutive integers such that the square of the sum of the two integers is 11 more than the lesser integer.

29. A carpenter needs to cut a brace to support a wall stud, as shown in the figure. The brace should be 7 ft less than three times the length of the stud. If the brace will be anchored on the floor 15 ft away from the stud, how long should the brace be?

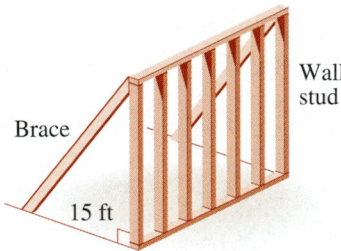

30. The number of Americans using broadband Internet access at home from 2001 through 2006 can be approximated by the quadratic equation

$$y = 2.36x^2 + 4.81x + 13.81,$$

where $x = 0$ represents 2001, $x = 1$ represents 2002, and so on, and y is number in millions. (*Source:* Pew Internet Project, Nielsen/Net Ratings.) Use the model to estimate the number of Americans using broadband Internet at home in 2004. Round your answer to the nearest million.

Rational Expressions and Applications

I n 2006, Earth's temperature was within 1.8°F of its highest level in 12,000 years. This temperature increase is causing ocean levels to rise as ice fields in Greenland and elsewhere melt. To demonstrate the effects that such global warming is having, British adventurer and endurance swimmer Lewis Gordon Pugh swam at the North Pole in July 2007 in waters that were completely frozen 10 years ago. The swim, in a water hole where polar ice had melted, was in 28.8°F waters, the coldest ever endured by a human. Pugh, who has also swum in the waters of Antarctica, hopes his efforts will inspire world leaders to take climate change seriously. (*Source:* www.breitbart.com, IPCC.)

In Exercise 11 of **Section 14.7**, we use a *rational expression* to determine the rate at which Pugh swam at the North Pole.

14

14.1 ▶▶▶ The Fundamental Property of Rational Expressions

OBJECTIVES

1 Find the values of the variable for which a rational expression is undefined.

2 Find the numerical value of a rational expression.

3 Write rational expressions in lowest terms.

4 Recognize equivalent forms of rational expressions.

The quotient of two integers (with denominator not 0), such as $\frac{2}{3}$ or $-\frac{3}{4}$, is called a *rational number*. In the same way, the quotient of two polynomials with denominator not equal to 0 is called a *rational expression*.

> **Rational Expression**
>
> A **rational expression** is an expression of the form $\frac{P}{Q}$, where P and Q are polynomials, with $Q \neq 0$.

Examples of rational expressions include

$$\frac{-6x}{x^3 + 8}, \quad \frac{9x}{y + 3}, \quad \text{and} \quad \frac{2m^3}{8}. \qquad \text{Rational expressions}$$

Our work with rational expressions will require much of what we learned in **Chapters 12 and 13** on polynomials and factoring, as well as the rules for fractions from **Chapters 2 and 3**.

OBJECTIVE 1 Find the values of the variable for which a rational expression is undefined. In the definition of a rational expression $\frac{P}{Q}$, Q cannot equal 0. *The denominator of a rational expression cannot equal 0 because division by 0 is undefined.*

For instance, in the rational expression

$$\frac{8x^2}{x - 3}, \quad \leftarrow \text{Denominator cannot equal 0.}$$

the variable x can take on any value except 3. When x is 3, the denominator becomes $3 - 3 = \mathbf{0}$, making the expression undefined. Thus, x cannot equal 3. We can indicate this restriction by writing $x \neq 3$.

To determine the values for which a rational expression is undefined, use the following procedure.

> **Determining When a Rational Expression Is Undefined**
>
> *Step 1* Set the denominator of the rational expression equal to 0.
>
> *Step 2* Solve this equation.
>
> *Step 3* The solutions of the equation are the values that make the rational expression undefined.

EXAMPLE 1 Finding Values That Make Rational Expressions Undefined

Find any values of the variable for which each rational expression is undefined.

(a) $\dfrac{p + 5}{3p + 2}$

Remember that the *numerator* may be any number; we must find any value of p that makes the *denominator* equal to 0 since division by 0 is undefined.

Continued on Next Page

Step 1 Set the denominator equal to 0.

$$3p + 2 = 0$$

Step 2 Solve this equation.

$$3p = -2$$

$$p = -\frac{2}{3}$$

Step 3 The given expression is undefined for $-\frac{2}{3}$, since substituting $-\frac{2}{3}$ for p makes the denominator 0. Thus, $p \neq -\frac{2}{3}$.

(b) $\dfrac{9m^2}{m^2 - 5m + 6}$

$$m^2 - 5m + 6 = 0 \qquad \text{Set the denominator equal to 0.}$$

$$(m - 2)(m - 3) = 0 \qquad \text{Factor.}$$

$$m - 2 = 0 \quad \text{or} \quad m - 3 = 0 \qquad \text{Zero-factor property}$$

$$m = 2 \quad \text{or} \qquad m = 3 \qquad \text{Solve.}$$

The original expression is undefined for 2 and 3, so $m \neq 2$ and $m \neq 3$.

(c) $\dfrac{2r}{r^2 + 1}$

This denominator will not equal 0 for any value of r because r^2 is always greater than or equal to 0, and adding 1 makes the sum greater than 0. Thus, there are no values for which this rational expression is undefined.

Work Problem ① *at the Side.* ▶

OBJECTIVE **2** **Find the numerical value of a rational expression.**
We use substitution to evaluate a rational expression for a given value of the variable.

EXAMPLE 2 **Evaluating Rational Expressions**

Find the numerical value of $\dfrac{3x + 6}{2x - 4}$ for each value of x.

(a) $x = 1$

$$\frac{3\mathbf{x} + 6}{2\mathbf{x} - 4}$$

$$= \frac{3(\mathbf{1}) + 6}{2(\mathbf{1}) - 4} \qquad \text{Let } x = 1.$$

$$= \frac{9}{-2}$$

$$= -\frac{9}{2} \qquad \frac{a}{-b} = -\frac{a}{b}$$

(b) $x = 0$

$$\frac{3\mathbf{x} + 6}{2\mathbf{x} - 4}$$

$$= \frac{3(\mathbf{0}) + 6}{2(\mathbf{0}) - 4} \qquad \text{Let } x = 0.$$

$$= \frac{6}{-4}$$

$$= -\frac{3}{2} \qquad \text{Lowest terms}$$

Continued on Next Page

---

1 Find any values of the variable for which each rational expression is undefined. Write answers with $\neq$.

(a) $\dfrac{x + 2}{x - 5}$

(b) $\dfrac{3r}{r^2 + 6r + 8}$

(c) $\dfrac{-5m}{m^2 + 4}$

2 Find the numerical value of each rational expression when $x = -3$, $x = 0$, and $x = 3$.

(a) $\dfrac{x}{2x + 1}$

(b) $\dfrac{2x + 6}{x - 3}$

(c) $x = 2$

$$\frac{3x + 6}{2x - 4}$$

$$= \frac{3(2) + 6}{2(2) - 4} \quad \text{Let } x = 2.$$

$$= \frac{12}{\mathbf{0}}$$

Substituting 2 for x makes the denominator 0, so the expression is undefined when $x = 2$.

(d) $x = -2$

$$\frac{3x + 6}{2x - 4}$$

$$= \frac{3(-2) + 6}{2(-2) - 4} \quad \text{Let } x = -2.$$

$$= \frac{0}{-8}$$

$$= 0 \qquad \frac{0}{b} = 0$$

> **Note**
>
> *The numerator of a rational expression may be any real number.* If the numerator equals 0 and the denominator does *not* equal 0, then the rational expression equals 0. See Example 2(d).

◀ *Work Problem* **2** *at the Side.*

OBJECTIVE 3 **Write rational expressions in lowest terms.** A fraction such as $\frac{2}{3}$ is said to be in *lowest terms*. How can "lowest terms" be defined? We use the idea of greatest common factor for this definition, which applies to all rational expressions.

> **Lowest Terms**
>
> A rational expression $\frac{P}{Q}$ ($Q \neq 0$) is in **lowest terms** if the greatest common factor of its numerator and denominator is 1.

The properties of rational numbers also apply to rational expressions. We use the **fundamental property of rational expressions** to write a rational expression in lowest terms.

> **Fundamental Property of Rational Expressions**
>
> If $\frac{P}{Q}$ ($Q \neq 0$) is a rational expression and if K represents any polynomial, where $K \neq 0$, then
>
> $$\frac{PK}{QK} = \frac{P}{Q}.$$

This property is based on the identity property of multiplication, since

$$\frac{PK}{QK} = \frac{P}{Q} \cdot \frac{K}{K} = \frac{P}{Q} \cdot 1 = \frac{P}{Q}.$$

The next example shows how to write both a rational number and a rational expression in lowest terms. Notice the similarity in the procedures. In both cases, we factor and then divide out the greatest common factor.

ANSWERS

2. **(a)** $\dfrac{3}{5}$; 0; $\dfrac{3}{7}$ **(b)** 0; -2; undefined

EXAMPLE 3 **Writing in Lowest Terms**

Write each expression in lowest terms.

(a) $\dfrac{30}{72}$

Begin by factoring.

$$= \dfrac{2 \cdot 3 \cdot 5}{2 \cdot 2 \cdot 2 \cdot 3 \cdot 3}$$

Group any factors common to the numerator and denominator.

$$= \dfrac{5 \cdot (\mathbf{2 \cdot 3})}{2 \cdot 2 \cdot 3 \cdot (\mathbf{2 \cdot 3})}$$

Use the fundamental property.

$$= \dfrac{5}{2 \cdot 2 \cdot 3}, \quad \text{or} \quad \dfrac{5}{12}$$

(b) $\dfrac{14k^2}{2k^3}$

Write k^2 as $k \cdot k$ and k^3 as $k \cdot k \cdot k$.

$$= \dfrac{2 \cdot 7 \cdot k \cdot k}{2 \cdot k \cdot k \cdot k}$$

$$= \dfrac{7 (\mathbf{2 \cdot k \cdot k})}{k (\mathbf{2 \cdot k \cdot k})}$$

$$= \dfrac{7}{k}$$

Work Problem **3** *at the Side.* ▶

Writing a Rational Expression in Lowest Terms

Step 1 **Factor** the numerator and denominator completely.

Step 2 **Use the fundamental property** to divide out any common factors.

EXAMPLE 4 **Writing in Lowest Terms**

Write each rational expression in lowest terms.

(a) $\dfrac{3x - 12}{5x - 20}$

$$= \dfrac{3 (x - 4)}{5 (x - 4)} \qquad \text{Factor. (Step 1)}$$

$$= \dfrac{3}{5} \qquad \text{Fundamental property (Step 2)}$$

The given expression is equal to $\frac{3}{5}$ for all values of x, where $x \neq 4$ (since the denominator of the original rational expression is 0 when x is 4).

(b) $\dfrac{m^2 + 2m - 8}{2m^2 - m - 6}$

$$= \dfrac{(m + 4)(m - 2)}{(2m + 3)(m - 2)} \qquad \text{Factor. (Step 1)}$$

$$= \dfrac{m + 4}{2m + 3} \qquad \text{Fundamental property (Step 2)}$$

Here $m \neq -\frac{3}{2}$ and $m \neq 2$, since the denominator of the original expression is 0 for these values of m.

3 Write each expression in lowest terms.

(a) $\dfrac{15}{40}$

(b) $\dfrac{5x^4}{15x^2}$

(c) $\dfrac{6p^3}{2p^2}$

4 Write each rational expression in lowest terms.

(a) $\dfrac{4y + 2}{6y + 3}$

(b) $\dfrac{8p + 8q}{5p + 5q}$

(c) $\dfrac{x^2 + 4x + 4}{4x + 8}$

(d) $\dfrac{a^2 - b^2}{a^2 + 2ab + b^2}$

From now on, we write statements of equality of rational expressions with the understanding that they apply only to those real numbers that make neither denominator equal to 0.

> **CAUTION**
> *Rational expressions cannot be written in lowest terms until after the numerator and denominator have been factored. Only common factors can be divided out, not common terms.* For example,
>
> $$\dfrac{6x + 9}{4x + 6} = \dfrac{3\,(2x + 3)}{2\,(2x + 3)} = \dfrac{3}{2} \qquad \bigg| \qquad \dfrac{6 + x}{4x} \leftarrow \text{Numerator cannot be factored.}$$
>
> Divide out the common factor. This expression is already in lowest terms.

◀ **Work Problem** **4** *at the Side.*

EXAMPLE 5 Writing in Lowest Terms (Factors Are Opposites)

Write $\dfrac{x - y}{y - x}$ in lowest terms.

At first glance, there might not seem to be any way in which $x - y$ and $y - x$ can be factored to get a common factor. However, $y - x$ can be factored as

$$y - x$$

Be careful with signs.

$$= -1\,(-y + x) \qquad \text{Factor out } -1.$$
$$= -1\,(x - y) \qquad \text{Commutative property}$$

With this result in mind, we simplify as follows.

$$\dfrac{x - y}{y - x}$$

$$= \dfrac{1\,(x - y)}{-1\,(x - y)} \qquad y - x = -1\,(x - y) \text{ from above.}$$

$$= \dfrac{1}{-1}, \quad \text{or} \quad -1 \qquad \text{Fundamental property}$$

◀ **Work Problem** **5** *at the Side.*

5 Write $\frac{x - y}{y - x}$ from Example 5 in lowest terms by factoring -1 from the numerator. How does the result compare to the result in Example 5?

In Example 5, notice that $y - x$ is the **opposite** (or **additive inverse**) of $x - y$. A general rule for this situation follows.

> If the numerator and the denominator of a rational expression are opposites, such as in $\frac{x - y}{y - x}$, then the rational expression is equal to -1.

> **CAUTION**
> Although x and y appear in both the numerator and denominator in Example 5, we cannot use the fundamental property right away because they are *terms,* not *factors.* ***Terms are added, while factors are multiplied.***

ANSWERS

4. **(a)** $\dfrac{2}{3}$ **(b)** $\dfrac{8}{5}$ **(c)** $\dfrac{x + 2}{4}$ **(d)** $\dfrac{a - b}{a + b}$

5. The result is -1, the same as in Example 5.

EXAMPLE 6   **Writing in Lowest Terms (Factors Are Opposites)**

Write each rational expression in lowest terms.

(a) $\dfrac{2 - m}{m - 2}$

Since $2 - m$ and $m - 2$ (or $-2 + m$) are opposites, this expression equals -1.

(b) $\dfrac{4x^2 - 9}{6 - 4x}$

$\quad = \dfrac{(2x + 3)(2x - 3)}{2(3 - 2x)}$ Factor the numerator and denominator.

$\quad = \dfrac{(2x + 3)(2x - 3)}{2(-1)(2x - 3)}$ Write $3 - 2x$ as $-1(2x - 3)$.

$\quad = \dfrac{2x + 3}{2(-1)}$ Fundamental property

$\quad = \dfrac{2x + 3}{-2},$ or $-\dfrac{2x + 3}{2}$ $\dfrac{a}{-b} = -\dfrac{a}{b}$

(c) $\dfrac{3 + r}{3 - r}$

The quantity $3 - r$ *is not* the opposite of $3 + r$. This rational expression is already in lowest terms.

Work Problem **6** *at the Side.* ▶

OBJECTIVE **4**   **Recognize equivalent forms of rational expressions.**   When working with rational expressions, it is important to be able to recognize equivalent forms of an expression. For example, the common fraction $-\frac{5}{6}$ can also be written as $\frac{-5}{6}$ and as $\frac{5}{-6}$.

Consider the final rational expression from Example 6(b),

$$-\dfrac{2x + 3}{2}.$$

The $-$ sign representing the factor -1 is in front of the expression, even with the fraction bar. As with the common fraction $-\frac{5}{6}$, the factor -1 may instead be placed in the numerator or in the denominator. Some other equivalent forms of this rational expression are

Use parentheses.

$$\dfrac{-(2x + 3)}{2} \quad \text{and} \quad \dfrac{2x + 3}{-2}.$$

> Multiply *each* term in the parentheses by -1.

By the distributive property,

$$\dfrac{-(2x + 3)}{2} \quad \text{can also be written} \quad \dfrac{-2x - 3}{2}.$$

CAUTION
$\frac{-2x + 3}{2}$ is *not* an equivalent form of $\frac{-(2x + 3)}{2}$. The sign preceding 3 in the numerator of $\frac{-2x + 3}{2}$ should be $-$ rather than $+$. ***Be careful to apply the distributive property correctly.***

---

6 Write each rational expression in lowest terms.

(a) $\dfrac{5 - y}{y - 5}$

(b) $\dfrac{m - n}{n - m}$

(c) $\dfrac{25x^2 - 16}{12 - 15x}$

(d) $\dfrac{9 - k}{9 + k}$

ANSWERS

6. **(a)** -1 **(b)** -1 **(c)** $\dfrac{5x + 4}{-3}$, or $-\dfrac{5x + 4}{3}$

   (d) already in lowest terms

7 Decide whether each rational expression is equivalent to

$$-\frac{2x-6}{x+3}.$$

(a) $\dfrac{-(2x-6)}{x+3}$

(b) $\dfrac{-2x+6}{x+3}$

(c) $\dfrac{-2x-6}{x+3}$

(d) $\dfrac{2x-6}{-(x+3)}$

(e) $\dfrac{2x-6}{-x-3}$

(f) $\dfrac{2x-6}{x-3}$

EXAMPLE 7   **Writing Equivalent Forms of a Rational Expression**

Write four equivalent forms of the rational expression

$$-\frac{3x+2}{x-6}.$$

If we apply the negative sign to the numerator, we obtain the equivalent forms

$$\frac{-(3x+2)}{x-6}, \quad \text{and, by the distributive property,} \quad \frac{-3x-2}{x-6}.$$

If we apply the negative sign to the denominator of the given rational expression, we obtain the equivalent forms

$$\frac{3x+2}{-(x-6)} \quad \text{and, by distributing again,} \quad \frac{3x+2}{-x+6}.$$

CAUTION
Recall that $-\frac{5}{6} \neq \frac{-5}{-6}$. Thus, in Example 7, it would be incorrect to distribute the negative sign in $-\frac{3x+2}{x-6}$ to *both* the numerator *and* the denominator. (Doing this would actually lead to the *opposite* of the original expression.)

◀ *Work Problem* **7** *at the Side.*

ANSWERS
7. **(a)** equivalent **(b)** equivalent
   (c) not equivalent **(d)** equivalent
   (e) equivalent **(f)** not equivalent

14.1 ▶▶▶ **Exercises**

FOR EXTRA HELP

 MyMathLab

 Math XL
PRACTICE

WATCH

DOWNLOAD

READ

REVIEW

1. Fill in each blank with the correct response.

   (a) The rational expression $\dfrac{x+5}{x-3}$ is undefined when $x =$ _____, and is equal to 0 when $x =$ _____.

   (b) The rational expression $\dfrac{p-q}{q-p}$ is undefined when $p =$ _____, and in all other cases when written in lowest terms is equal to _____.

2. Make the correct choice for each blank.

   (a) $\dfrac{4-r^2}{4+r^2}$ _____ equal to -1.
   (is/is not)

   (b) $\dfrac{5+2x}{3-x}$ and $\dfrac{-5-2x}{x-3}$ _____ equivalent rational expressions.
   (are/are not)

3. Define *rational expression* in your own words, and give an example.

4. Why can't the denominator of a rational expression equal 0?

Find any value(s) of the variable for which each rational expression is undefined. Write answers with ≠. See Example 1.

5. $\dfrac{2}{5y}$

6. $\dfrac{7}{3z}$

7. $\dfrac{x+1}{x+6}$

8. $\dfrac{m+2}{m+5}$

9. $\dfrac{4x^2}{3x-5}$

10. $\dfrac{2x^3}{3x-4}$

11. $\dfrac{m+2}{m^2+m-6}$

12. $\dfrac{r-5}{r^2-5r+4}$

13. $\dfrac{x^2-3x}{4}$

14. $\dfrac{x^2-4x}{6}$

15. $\dfrac{3x}{x^2+2}$

16. $\dfrac{4q}{q^2+9}$

*Find the numerical value of each rational expression when **(a)** $x = 2$ and **(b)** $x = -3$. See Example 2.*

17. $\dfrac{5x-2}{4x}$

18. $\dfrac{3x+1}{5x}$

19. $\dfrac{x^2-4}{2x+1}$

20. $\dfrac{2x^2-4x}{3x-1}$

21. $\dfrac{(-3x)^2}{4x+12}$

22. $\dfrac{(-2x)^3}{3x+9}$

23. $\dfrac{5x+2}{2x^2+11x+12}$

24. $\dfrac{7-3x}{3x^2-7x+2}$

Write each rational expression in lowest terms. See Examples 3 and 4.

25. $\dfrac{18r^3}{6r}$

26. $\dfrac{27p^2}{3p}$

27. $\dfrac{4(y-2)}{10(y-2)}$

28. $\dfrac{15\,(m\,-\,1)}{9\,(m\,-\,1)}$

29. $\dfrac{(x\,+\,1)\,(x\,-\,1)}{(x\,+\,1)^2}$

30. $\dfrac{(t\,+\,5)\,(t\,-\,3)}{(t\,-\,1)\,(t\,+\,5)}$

31. $\dfrac{7m\,+\,14}{5m\,+\,10}$

32. $\dfrac{8z\,-\,24}{4z\,-\,12}$

33. $\dfrac{m^2\,-\,n^2}{m\,+\,n}$

34. $\dfrac{a^2\,-\,b^2}{a\,-\,b}$

35. $\dfrac{12m^2\,-\,3}{8m\,-\,4}$

36. $\dfrac{20p^2\,-\,45}{6p\,-\,9}$

37. $\dfrac{3m^2\,-\,3m}{5m\,-\,5}$

38. $\dfrac{6t^2\,-\,6t}{2t\,-\,2}$

39. $\dfrac{9r^2\,-\,4s^2}{9r\,+\,6s}$

40. $\dfrac{16x^2\,-\,9y^2}{12x\,-\,9y}$

41. $\dfrac{2x^2\,-\,3x\,-\,5}{2x^2\,-\,7x\,+\,5}$

42. $\dfrac{3x^2\,+\,8x\,+\,4}{3x^2\,-\,4x\,-\,4}$

43. $\dfrac{zw\,+\,4z\,-\,3w\,-\,12}{zw\,+\,4z\,+\,5w\,+\,20}$

44. $\dfrac{km\,+\,4k\,+\,4m\,+\,16}{km\,+\,4k\,+\,5m\,+\,20}$

45. $\dfrac{ac\,-\,ad\,+\,bc\,-\,bd}{ac\,-\,ad\,-\,bc\,+\,bd}$

46. Which rational expression can be simplified?

A. $\dfrac{x^2\,+\,2}{x^2}$ **B.** $\dfrac{x^2\,+\,2}{2}$ **C.** $\dfrac{x^2\,+\,y^2}{y^2}$ **D.** $\dfrac{x^2\,-\,5x}{x}$

Write each rational expression in lowest terms. See Examples 5 and 6.

47. $\dfrac{6\,-\,t}{t\,-\,6}$

48. $\dfrac{2\,-\,k}{k\,-\,2}$

49. $\dfrac{m^2\,-\,1}{1\,-\,m}$

50. $\dfrac{a^2\,-\,b^2}{b\,-\,a}$

51. $\dfrac{q^2\,-\,4q}{4q\,-\,q^2}$

52. $\dfrac{z^2\,-\,5z}{5z\,-\,z^2}$

Write four equivalent forms for each rational expression. See Example 7.

53. $-\dfrac{x\,+\,4}{x\,-\,3}$

54. $-\dfrac{x\,+\,6}{x\,-\,1}$

55. $-\dfrac{2x\,-\,3}{x\,+\,3}$

56. $-\dfrac{5x\,-\,6}{x\,+\,4}$

57. $-\dfrac{3x\,-\,1}{5x\,-\,6}$

58. $-\dfrac{2x\,-\,9}{7x\,-\,1}$

59. The area of the rectangle is represented by $x^4\,+\,10x^2\,+\,21$. What is the width? $\left(\textit{Hint}:\text{ Use }W\,=\,\frac{A}{L}.\right)$

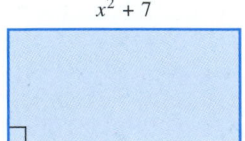

$x^2 + 7$

60. The volume of the box is represented by

$$(x^2\,+\,8x\,+\,15)\,(x\,+\,4).$$

Find the polynomial that represents the area of the bottom of the box.

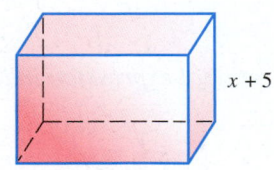

$x + 5$

14.2 ▶▶▶ Multiplying and Dividing Rational Expressions

OBJECTIVE 1 Multiply rational expressions. The product of two fractions is found by multiplying the numerators and multiplying the denominators. Rational expressions are multiplied in the same way.

> **Multiplying Rational Expressions**
>
> The product of the rational expressions $\frac{P}{Q}$ and $\frac{R}{S}$ is
>
> $$\frac{P}{Q} \cdot \frac{R}{S} = \frac{PR}{QS}.$$
>
> In words: To multiply rational expressions, multiply the numerators and multiply the denominators.

OBJECTIVES

1 Multiply rational expressions.

2 Find reciprocals.

3 Divide rational expressions.

EXAMPLE 1 Multiplying Rational Expressions

Multiply. Write each answer in lowest terms.

(a) $\dfrac{3}{10} \cdot \dfrac{5}{9}$

(b) $\dfrac{6}{x} \cdot \dfrac{x^2}{12}$

Indicate the product of the numerators and the product of the denominators.

$= \dfrac{3 \cdot 5}{10 \cdot 9}$

$= \dfrac{6 \cdot x^2}{x \cdot 12}$

Leave the products in factored form because common factors are needed to write the product in lowest terms. Factor the numerator and denominator to further identify any common factors. Then use the fundamental property to write each product in lowest terms.

$= \dfrac{3 \cdot 5}{2 \cdot 5 \cdot 3 \cdot 3}$

$= \dfrac{6 \cdot x \cdot x}{2 \cdot 6 \cdot x}$

$= \dfrac{1}{6}$

$= \dfrac{x}{2}$

Work Problem ① at the Side. ▶

EXAMPLE 2 Multiplying Rational Expressions

Multiply. Write the answer in lowest terms.

$\dfrac{x + y}{2x} \cdot \dfrac{x^2}{(x + y)^2}$

$= \dfrac{(x + y)x^2}{2x(x + y)^2}$ Multiply numerators.
Multiply denominators.

$= \dfrac{(x + y)\, x \cdot x}{2x(x + y)(x + y)}$ Factor; identify common factors.

$= \dfrac{x}{2(x + y)}$ $\dfrac{(x + y)x}{x(x + y)} = 1$; lowest terms

Work Problem ② at the Side. ▶

① Multiply. Write each answer in lowest terms.

(a) $\dfrac{2}{7} \cdot \dfrac{5}{10}$

(b) $\dfrac{3m^2}{2} \cdot \dfrac{10}{m}$

(c) $\dfrac{8p^2q}{3} \cdot \dfrac{9}{q^2p}$

② Multiply. Write each answer in lowest terms.

(a) $\dfrac{a + b}{5} \cdot \dfrac{30}{2(a + b)}$

(b) $\dfrac{3(p - q)}{q^2} \cdot \dfrac{q}{2(p - q)^2}$

ANSWERS

1. (a) $\dfrac{1}{7}$ (b) $15m$ (c) $\dfrac{24p}{q}$

2. (a) 3 (b) $\dfrac{3}{2q(p - q)}$

3 Multiply. Write each answer in lowest terms.

(a)

$$\frac{x^2 + 7x + 10}{3x + 6} \cdot \frac{6x - 6}{x^2 + 2x - 15}$$

(b)

$$\frac{m^2 + 4m - 5}{m + 5} \cdot \frac{m^2 + 8m + 15}{m - 1}$$

4 Find the reciprocal of each rational expression.

(a) $\dfrac{5}{8}$

(b) $\dfrac{6b^5}{3r^2b}$

(c) $\dfrac{t^2 - 4t}{t^2 + 2t - 3}$

EXAMPLE 3 Multiplying Rational Expressions

Multiply. Write the answer in lowest terms.

$$\frac{x^2 + 3x}{x^2 - 3x - 4} \cdot \frac{x^2 - 5x + 4}{x^2 + 2x - 3}$$

$$= \frac{(x^2 + 3x)(x^2 - 5x + 4)}{(x^2 - 3x - 4)(x^2 + 2x - 3)} \qquad \text{Definition of multiplication}$$

$$= \frac{x(x + 3)(x - 4)(x - 1)}{(x - 4)(x + 1)(x + 3)(x - 1)} \qquad \text{Factor.}$$

$$= \frac{x}{x + 1} \qquad \text{Divide out the common factors.}$$

The quotients $\frac{x + 3}{x + 3}, \frac{x - 4}{x - 4},$ and $\frac{x - 1}{x - 1}$ are all equal to 1, justifying the final product $\frac{x}{x + 1}$.

◀ Work Problem **3** at the Side.

OBJECTIVE 2 **Find reciprocals.** If the product of two rational expressions is 1, the rational expressions are called **reciprocals** (or **multiplicative inverses**) of each other. The reciprocal of a rational expression is found by interchanging the numerator and the denominator. For example,

$$\frac{2x - 1}{x - 5} \quad \text{has reciprocal} \quad \frac{x - 5}{2x - 1}.$$

EXAMPLE 4 Finding Reciprocals of Rational Expressions

Find the reciprocal of each rational expression.

(a) $\dfrac{4p^3}{9q}$ has reciprocal $\dfrac{9q}{4p^3}$.   Interchange the numerator and denominator.

(b) $\dfrac{k^2 - 9}{k^2 - k - 20}$ has reciprocal $\dfrac{k^2 - k - 20}{k^2 - 9}$.

◀ Work Problem **4** at the Side.

OBJECTIVE 3 **Divide rational expressions.** Suppose we have $\frac{7}{8}$ gal of milk and want to find how many quarts we have. Since 1 qt is $\frac{1}{4}$ gal, we ask, "How many $\frac{1}{4}$s are there in $\frac{7}{8}$?" This would be interpreted as

$$\frac{7}{8} \div \frac{1}{4}, \quad \text{or} \quad \frac{\frac{7}{8}}{\frac{1}{4}} \leftarrow \text{The fraction bar means division.}$$

The fundamental property of rational expressions discussed earlier can be applied to rational number values of P, Q, and K.

$$\frac{P}{Q} = \frac{P \cdot K}{Q \cdot K} = \frac{\frac{7}{8} \cdot 4}{\frac{1}{4} \cdot 4} = \frac{\frac{7}{8} \cdot 4}{1} = \frac{7}{8} \cdot \frac{4}{1}. \quad \begin{array}{l}\text{Let } P = \frac{7}{8}, Q = \frac{1}{4}, \\ \text{and } K = 4 \text{ (the} \\ \text{reciprocal of } Q).\end{array}$$

So, to divide $\frac{7}{8}$ by $\frac{1}{4}$, we multiply $\frac{7}{8}$ by the reciprocal of $\frac{1}{4}$, namely $\frac{4}{1}$, or 4. Since $\frac{7}{8}(4) = \frac{7}{2}$, there are $\frac{7}{2}$ qt, or $3\frac{1}{2}$ qt, in $\frac{7}{8}$ gal.

ANSWERS

3. **(a)** $\dfrac{2(x - 1)}{x - 3}$ **(b)** $(m + 5)(m + 3)$

4. **(a)** $\dfrac{8}{5}$ **(b)** $\dfrac{3r^2b}{6b^5}$ **(c)** $\dfrac{t^2 + 2t - 3}{t^2 - 4t}$

The preceding discussion illustrates dividing common fractions. Division of rational expressions is defined in the same way.

> ### Dividing Rational Expressions
>
> If $\frac{P}{Q}$ and $\frac{R}{S}$ are any two rational expressions, with $\frac{R}{S} \neq 0$, then
>
> $$\frac{P}{Q} \div \frac{R}{S} = \frac{P}{Q} \cdot \frac{S}{R} = \frac{PS}{QR}.$$
>
> In words: To divide one rational expression by another rational expression, multiply the first rational expression (dividend) by the reciprocal of the second rational expression (divisor).

The next example shows the division of two rational numbers and the division of two rational expressions.

EXAMPLE 5 **Dividing Rational Expressions**

Divide. Write each answer in lowest terms.

(a) $\dfrac{5}{8} \div \dfrac{7}{16}$ | **(b)** $\dfrac{y}{y+3} \div \dfrac{4y}{y+5}$

Multiply the first expression by the reciprocal of the second.

$= \dfrac{5}{8} \cdot \dfrac{16}{7}$ ← Reciprocal of $\frac{7}{16}$ | $= \dfrac{y}{y+3} \cdot \dfrac{y+5}{4y}$ ← Reciprocal of $\frac{4y}{y+5}$

$= \dfrac{5 \cdot 16}{8 \cdot 7}$ | $= \dfrac{y(y+5)}{(y+3)(4y)}$

$= \dfrac{5 \cdot 8 \cdot 2}{8 \cdot 7}$ | $= \dfrac{y+5}{4(y+3)}$

$= \dfrac{10}{7}$

Work Problem **5** *at the Side.* ▶

EXAMPLE 6 **Dividing Rational Expressions**

Divide. Write the answer in lowest terms.

$$\frac{(3m)^2}{(2p)^3} \div \frac{6m^3}{16p^2}$$

$(3m)^2 = 3^2 m^2;$
$(2p)^3 = 2^3 p^3$

$= \dfrac{(3m)^2}{(2p)^3} \cdot \dfrac{16p^2}{6m^3}$ Multiply by the reciprocal.

$= \dfrac{9m^2}{8p^3} \cdot \dfrac{16p^2}{6m^3}$ Power rule for exponents

$= \dfrac{9 \cdot 16 m^2 p^2}{8 \cdot 6 p^3 m^3}$ Multiply numerators. Multiply denominators.

$= \dfrac{3}{mp}$ Lowest terms

Work Problem **6** *at the Side.* ▶

---

5 Divide. Write each answer in lowest terms.

(a) $\dfrac{3}{4} \div \dfrac{5}{16}$

(b) $\dfrac{r}{r-1} \div \dfrac{3r}{r+4}$

(c) $\dfrac{6x-4}{3} \div \dfrac{15x-10}{9}$

6 Divide. Write each answer in lowest terms.

(a) $\dfrac{5a^2 b}{2} \div \dfrac{10ab^2}{8}$

(b) $\dfrac{(3t)^2}{w} \div \dfrac{3t^2}{5w^4}$

ANSWERS

5. (a) $\dfrac{12}{5}$ (b) $\dfrac{r+4}{3(r-1)}$ (c) $\dfrac{6}{5}$

6. (a) $\dfrac{2a}{b}$ (b) $15w^3$

7 Divide. Write each answer in lowest terms.

(a)
$$\frac{y^2 + 4y + 3}{y + 3} \div \frac{y^2 - 4y - 5}{y - 3}$$

(b) $\dfrac{4x(x + 3)}{2x + 1} \div \dfrac{-x^2(x + 3)}{4x^2 - 1}$

8 Divide. Write each answer in lowest terms.

(a) $\dfrac{ab - a^2}{a^2 - 1} \div \dfrac{a - b}{a - 1}$

(b) $\dfrac{x^2 - 9}{2x + 6} \div \dfrac{9 - x^2}{4x - 12}$

EXAMPLE 7   **Dividing Rational Expressions**

Divide. Write the answer in lowest terms.

$$\frac{x^2 - 4}{(x + 3)(x - 2)} \div \frac{(x + 2)(x + 3)}{-2x}$$

$$= \frac{x^2 - 4}{(x + 3)(x - 2)} \cdot \frac{-2x}{(x + 2)(x + 3)} \qquad \text{Multiply by the reciprocal.}$$

$$= \frac{-2x(x^2 - 4)}{(x + 3)(x - 2)(x + 2)(x + 3)} \qquad \begin{array}{l}\text{Multiply numerators.}\\ \text{Multiply denominators.}\end{array}$$

$$= \frac{-2x(x + 2)(x - 2)}{(x + 3)(x - 2)(x + 2)(x + 3)} \qquad \text{Factor the numerator.}$$

$$= \frac{-2x}{(x + 3)^2}, \quad \text{or} \quad -\frac{2x}{(x + 3)^2} \qquad \text{Lowest terms; } \tfrac{-a}{b} = -\tfrac{a}{b}$$

◀ *Work Problem* **7** *at the Side.*

EXAMPLE 8   **Dividing Rational Expressions (Factors Are Opposites)**

Divide. Write the answer in lowest terms.

$$\frac{m^2 - 4}{m^2 - 1} \div \frac{2m^2 + 4m}{1 - m}$$

$$= \frac{m^2 - 4}{m^2 - 1} \cdot \frac{1 - m}{2m^2 + 4m} \qquad \text{Multiply by the reciprocal.}$$

$$= \frac{(m^2 - 4)(1 - m)}{(m^2 - 1)(2m^2 + 4m)} \qquad \begin{array}{l}\text{Multiply numerators.}\\ \text{Multiply denominators.}\end{array}$$

$$= \frac{(m + 2)(m - 2)(1 - m)}{(m + 1)(m - 1)(2m)(m + 2)} \qquad \begin{array}{l}\text{Factor; } 1 - m \text{ and } m - 1 \text{ are}\\ \text{opposites.}\end{array}$$

$$= \frac{-1(m - 2)}{2m(m + 1)} \qquad \text{From \textbf{Section 7.1}, } \tfrac{1 - m}{m - 1} = -1.$$

$$= \frac{-m + 2}{2m(m + 1)}, \quad \text{or} \quad \frac{2 - m}{2m(m + 1)} \qquad \text{Distribute } -1 \text{ in the numerator.}$$

◀ *Work Problem* **8** *at the Side.*

In summary, follow these steps to multiply or divide rational expressions.

Multiplying or Dividing Rational Expressions

Step 1 **Note the operation.** If the operation is division, use the definition of division to rewrite as multiplication.

Step 2 **Multiply** numerators and multiply denominators.

Step 3 **Factor** all numerators and denominators completely.

Step 4 **Write in lowest terms** using the fundamental property.

Steps 2 and 3 may be interchanged based on personal preference.

ANSWERS

7. **(a)** $\dfrac{y - 3}{y - 5}$ **(b)** $-\dfrac{4(2x - 1)}{x}$

8. **(a)** $\dfrac{-a}{a + 1}$ **(b)** $\dfrac{-2x + 6}{3 + x}$

14.2 ▶▶▶ Exercises

1. Match each multiplication problem in Column I with the correct product in Column II.

I

(a) $\dfrac{5x^3}{10x^4} \cdot \dfrac{10x^7}{2x}$

(b) $\dfrac{10x^4}{5x^3} \cdot \dfrac{10x^7}{2x}$

(c) $\dfrac{5x^3}{10x^4} \cdot \dfrac{2x}{10x^7}$

(d) $\dfrac{10x^4}{5x^3} \cdot \dfrac{2x}{10x^7}$

II

A. $\dfrac{2}{5x^5}$

B. $\dfrac{5x^5}{2}$

C. $\dfrac{1}{10x^7}$

D. $10x^7$

2. Match each division problem in Column I with the correct quotient in Column II.

I

(a) $\dfrac{5x^3}{10x^4} \div \dfrac{10x^7}{2x}$

(b) $\dfrac{10x^4}{5x^3} \div \dfrac{10x^7}{2x}$

(c) $\dfrac{5x^3}{10x^4} \div \dfrac{2x}{10x^7}$

(d) $\dfrac{10x^4}{5x^3} \div \dfrac{2x}{10x^7}$

II

A. $\dfrac{5x^5}{2}$

B. $10x^7$

C. $\dfrac{2}{5x^5}$

D. $\dfrac{1}{10x^7}$

Multiply. Write each answer in lowest terms. See Examples 1 and 2.

3. $\dfrac{10m^2}{7} \cdot \dfrac{14}{15m}$

4. $\dfrac{36z^3}{6z} \cdot \dfrac{28}{z^2}$

5. $\dfrac{16y^4}{18y^5} \cdot \dfrac{15y^5}{y^2}$

6. $\dfrac{20x^5}{-2x^2} \cdot \dfrac{8x^4}{35x^3}$

7. $\dfrac{2(c+d)}{3} \cdot \dfrac{18}{6(c+d)^2}$

8. $\dfrac{4(y-2)}{x} \cdot \dfrac{3x}{6(y-2)^2}$

Find the reciprocal of each rational expression. See Example 4.

9. $\dfrac{3p^3}{16q}$

10. $\dfrac{6x^4}{9y^2}$

11. $\dfrac{r^2+rp}{7}$

12. $\dfrac{16}{9a^2+36a}$

13. $\dfrac{z^2+7z+12}{z^2-9}$

14. $\dfrac{p^2-4p+3}{p^2-3p}$

Divide. Write each answer in lowest terms. See Examples 5 and 6.

15. $\dfrac{9z^4}{3z^5} \div \dfrac{3z^2}{5z^3}$

16. $\dfrac{35q^8}{9q^5} \div \dfrac{25q^6}{10q^5}$

17. $\dfrac{4t^4}{2t^5} \div \dfrac{(2t)^3}{-6}$

18. $\dfrac{-12a^6}{3a^2} \div \dfrac{(2a)^3}{27a}$

19. $\dfrac{3}{2y-6} \div \dfrac{6}{y-3}$

20. $\dfrac{4m+16}{10} \div \dfrac{3m+12}{18}$

21. Explain in your own words how to multiply rational expressions.

22. Explain in your own words how to divide rational expressions.

Multiply or divide. Write each answer in lowest terms. See Examples 3, 7, and 8.

23. $\dfrac{5x - 15}{3x + 9} \cdot \dfrac{4x + 12}{6x - 18}$

24. $\dfrac{8r + 16}{24r - 24} \cdot \dfrac{6r - 6}{3r + 6}$

25. $\dfrac{2 - t}{8} \div \dfrac{t - 2}{6}$

26. $\dfrac{4}{m - 2} \div \dfrac{16}{2 - m}$

27. $\dfrac{5 - 4x}{5 + 4x} \cdot \dfrac{4x + 5}{4x - 5}$

28. $\dfrac{5 - x}{5 + x} \cdot \dfrac{x + 5}{x - 5}$

29. $\dfrac{6(m - 2)^2}{5(m + 4)^2} \cdot \dfrac{15(m + 4)}{2(2 - m)}$

30. $\dfrac{7(q - 1)}{3(q + 1)^2} \cdot \dfrac{6(q + 1)}{3(1 - q)^2}$

31. $\dfrac{p^2 + 4p - 5}{p^2 + 7p + 10} \div \dfrac{p - 1}{p + 4}$

32. $\dfrac{z^2 - 3z + 2}{z^2 + 4z + 3} \div \dfrac{z - 1}{z + 1}$

33. $\dfrac{2k^2 - k - 1}{2k^2 + 5k + 3} \div \dfrac{4k^2 - 1}{2k^2 + k - 3}$

34. $\dfrac{2m^2 - 5m - 12}{m^2 + m - 20} \div \dfrac{4m^2 - 9}{m^2 + 4m - 5}$

35. $\dfrac{2k^2 + 3k - 2}{6k^2 - 7k + 2} \cdot \dfrac{4k^2 - 5k + 1}{k^2 + k - 2}$

36. $\dfrac{2m^2 - 5m - 12}{m^2 - 10m + 24} \div \dfrac{4m^2 - 9}{m^2 - 9m + 18}$

37. $\dfrac{m^2 + 2mp - 3p^2}{m^2 - 3mp + 2p^2} \div \dfrac{m^2 + 4mp + 3p^2}{m^2 + 2mp - 8p^2}$

38. $\dfrac{r^2 + rs - 12s^2}{r^2 - rs - 20s^2} \div \dfrac{r^2 - 2rs - 3s^2}{r^2 + rs - 30s^2}$

39. $\left(\dfrac{x^2 + 10x + 25}{x^2 + 10x} \cdot \dfrac{10x}{x^2 + 15x + 50} \right) \div \dfrac{x + 5}{x + 10}$

40. $\left(\dfrac{m^2 - 12m + 32}{8m} \cdot \dfrac{m^2 - 8m}{m^2 - 8m + 16} \right) \div \dfrac{m - 8}{m - 4}$

41. If the rational expression $\dfrac{5x^2y^3}{2pq}$ represents the area of a rectangle and $\dfrac{2xy}{p}$ represents the length, what rational expression represents the width?

Width

Length $= \dfrac{2xy}{p}$

The area is $\dfrac{5x^2y^3}{2pq}$.

42. Consider the division problem $\dfrac{x - 6}{x + 4} \div \dfrac{x + 7}{x + 5}$.

We know that division by 0 is undefined, so the restrictions on x are $x \neq -4$, $x \neq -5$, and $x \neq -7$. Why is the last restriction needed?

14.3 ▸▸▸ Least Common Denominators

OBJECTIVE 1 Find the least common denominator for a list of fractions. Just as with common fractions, adding or subtracting rational expressions (to be discussed in the next section) often requires a **least common denominator (LCD)**, the simplest expression that is divisible by all denominators. For example, the least common denominator for $\frac{2}{9}$ and $\frac{5}{12}$ is 36, because 36 is the smallest positive number divisible by both 9 and 12.

We can often find least common denominators by inspection. For example, the LCD for $\frac{1}{6}$ and $\frac{2}{3m}$ is $6m$. In other cases, we find the LCD by a procedure similar to that used in **Section 13.1** for finding the greatest common factor.

> **Finding the Least Common Denominator (LCD)**
>
> *Step 1* **Factor** each denominator into prime factors.
>
> *Step 2* **List each different denominator factor** the *greatest* number of times it appears in any of the denominators.
>
> *Step 3* **Multiply** the denominator factors from Step 2 to get the LCD.

When each denominator is factored into prime factors, every prime factor must be a factor of the least common denominator.

In Example 1, we find the LCD for both numerical and algebraic denominators.

EXAMPLE 1 Finding Least Common Denominators

Find the LCD for each pair of fractions.

(a) $\frac{1}{24}, \frac{7}{15}$ **(b)** $\frac{1}{8x}, \frac{3}{10x}$

Step 1 Write each denominator in factored form with numerical coefficients in prime factored form.

$24 = 2 \cdot 2 \cdot 2 \cdot 3 = 2^3 \cdot 3$ | $8x = 2 \cdot 2 \cdot 2 \cdot x = 2^3 \cdot x$
$15 = 3 \cdot 5$ | $10x = 2 \cdot 5 \cdot x$

Step 2 We find the LCD by taking each different factor the *greatest* number of times it appears as a factor in any of the denominators.

The factor 2 appears three times in one product and not at all in the other, so the greatest number of times 2 appears is three. The greatest number of times both 3 and 5 appear is one. | Here, 2 appears three times in one product and once in the other, so the greatest number of times 2 appears is three. The greatest number of times 5 appears is one, and the greatest number of times x appears in either product is one.

Step 3 LCD = $2 \cdot 2 \cdot 2 \cdot 3 \cdot 5$ | LCD = $2 \cdot 2 \cdot 2 \cdot 5 \cdot x$
$= 2^3 \cdot 3 \cdot 5$ | $= 2^3 \cdot 5 \cdot x$
$= 120$ | $= 40x$

Work Problem **1** *at the Side.* ▸

OBJECTIVES

1 Find the least common denominator for a list of fractions.

2 Write equivalent rational expressions.

1 Find the LCD for each pair of fractions.

(a) $\frac{7}{10}, \frac{1}{25}$

(b) $\frac{7}{20p}, \frac{11}{30p}$

(c) $\frac{4}{5x}, \frac{12}{10x}$

2 Find the LCD.

(a) $\dfrac{4}{16m^3n}, \dfrac{5}{9m^5}$

(b) $\dfrac{3}{25a^2}, \dfrac{2}{10a^3b}$

EXAMPLE 2 Finding the LCD

Find the LCD for $\dfrac{5}{6r^2}$ and $\dfrac{3}{4r^3}$.

Step 1 Factor each denominator.

$$6r^2 = 2 \cdot \mathbf{3} \cdot r^2$$
$$4r^3 = \mathbf{2^2} \cdot \mathbf{r^3}$$

Step 2 The greatest number of times 2 appears is two, the greatest number of times 3 appears is one, and the greatest number of times r appears is three; therefore,

Step 3 $\qquad$ LCD $= \mathbf{2^2} \cdot \mathbf{3} \cdot \mathbf{r^3} = \mathbf{12r^3}$.

◀ Work Problem **2** at the Side.

> **CAUTION**
>
> When finding the LCD, use each factor the *greatest* number of times it appears in any *single* denominator, not the *total* number of times it appears. For instance, the greatest number of times r appears as a factor in one denominator in Example 2 is 3, *not* 5.

---

3 Find the LCD for the fractions in each list.

(a) $\dfrac{7}{3a}, \dfrac{11}{a^2 - 4a}$

(b)

$\dfrac{2m}{m^2 - 3m + 2}, \dfrac{5m - 3}{m^2 + 3m - 10},$

$\dfrac{4m + 7}{m^2 + 4m - 5}$

(c) $\dfrac{6}{x - 4}, \dfrac{3x - 1}{4 - x}$

EXAMPLE 3 Finding LCDs

Find the LCD for the fractions in each list.

(a) $\dfrac{6}{5m}, \dfrac{4}{m^2 - 3m}$

$$\left. \begin{array}{l} 5m = \mathbf{5} \cdot \mathbf{m} \\ m^2 - 3m = \mathbf{m(m - 3)} \end{array} \right\} \text{Factor each denominator.}$$

Use each different factor the greatest number of times it appears.

$$\text{LCD} = \mathbf{5} \cdot \mathbf{m} \cdot \mathbf{(m - 3)} = 5m(m - 3)$$

> Be sure to include m as a factor in the LCD.

Because m is not a *factor* of $m - 3$, both factors, m and $m - 3$, must appear in the LCD.

(b) $\dfrac{1}{r^2 - 4r - 5}, \dfrac{3}{r^2 - r - 20}, \dfrac{1}{r^2 - 10r + 25}$

$$\left. \begin{array}{l} r^2 - 4r - 5 = (r - 5)(\mathbf{r + 1}) \\ r^2 - r - 20 = (r - 5)(\mathbf{r + 4}) \\ r^2 - 10r + 25 = (\mathbf{r - 5})^2 \end{array} \right\} \text{Factor each denominator.}$$

Use each different factor the greatest number of times it appears as a factor.

$$\text{LCD} = (r - 5)^2(r + 1)(r + 4)$$

(c) $\dfrac{1}{q - 5}, \dfrac{3}{5 - q}$

The expressions $q - 5$ and $5 - q$ are opposites of each other because

$$-(q - 5) = -q + 5 = 5 - q.$$

Therefore, either $q - 5$ or $5 - q$ can be used as the LCD.

◀ Work Problem **3** at the Side.

ANSWERS

2. (a) $144m^5n$ (b) $50a^3b$
3. (a) $3a(a - 4)$
   (b) $(m - 1)(m - 2)(m + 5)$
   (c) either $x - 4$ or $4 - x$

OBJECTIVE **2** **Write equivalent rational expressions.**   Once the LCD has been found, the next step in preparing to add or subtract two rational expressions is to use the fundamental property to write equivalent rational expressions. We use the following steps.

> **Writing a Rational Expression with a Specified Denominator**
>
> **Step 1** **Factor** both denominators.
>
> **Step 2** **Decide what factor(s) the denominator must be multiplied by** in order to equal the specified denominator.
>
> **Step 3** **Multiply** the rational expression by that factor divided by itself. (That is, multiply by 1.)

EXAMPLE 4 **Writing Equivalent Rational Expressions**

Write each rational expression as an equivalent expression with the indicated denominator.

(a) $\dfrac{3}{8} = \dfrac{?}{40}$

(b) $\dfrac{9k}{25} = \dfrac{?}{50k}$

Step 1 For each example, first factor the denominator on the right. Then compare the denominator on the left with the one on the right to decide what factors are missing.

$$\dfrac{3}{8} = \dfrac{?}{\mathbf{5 \cdot 8}}$$

$$\dfrac{9k}{25} = \dfrac{?}{25 \cdot \mathbf{2k}}$$

Step 2 A factor of 5 is missing.   Factors of 2 and k are missing.

Step 3 Multiply $\frac{3}{8}$ by $\frac{5}{5}$.   Multiply $\frac{9k}{25}$ by $\frac{2k}{2k}$.

$$\dfrac{3}{8} = \dfrac{3}{8} \cdot \dfrac{\mathbf{5}}{\mathbf{5}} = \dfrac{\mathbf{15}}{40}$$

$$\dfrac{9k}{25} = \dfrac{9k}{25} \cdot \dfrac{\mathbf{2k}}{\mathbf{2k}} = \dfrac{\mathbf{18k^2}}{50k}$$

$\frac{5}{5} = 1 \longrightarrow$ $\frac{2k}{2k} = 1 \longrightarrow$

Work Problem **4** *at the Side.* ▶

EXAMPLE 5 **Writing Equivalent Rational Expressions**

Write each rational expression as an equivalent expression with the indicated denominator.

(a)

$$\dfrac{8}{3x + 1} = \dfrac{?}{12x + 4}$$

$$\dfrac{8}{3x + 1} = \dfrac{?}{\mathbf{4}(3x + 1)}$$ Factor the denominator on the right.

The missing factor is 4, so multiply the fraction on the left by $\frac{4}{4}$.

$$\dfrac{8}{3x + 1} \cdot \dfrac{\mathbf{4}}{\mathbf{4}} = \dfrac{\mathbf{32}}{12x + 4}$$ Fundamental property

Continued on Next Page

---

4 Write each rational expression as an equivalent expression with the indicated denominator.

(a) $\dfrac{3}{4} = \dfrac{?}{36}$

(b) $\dfrac{7k}{5} = \dfrac{?}{30p}$

5 Write each rational expression as an equivalent expression with the indicated denominator.

(a) $\dfrac{9}{2a+5} = \dfrac{?}{6a+15}$

(b) $\dfrac{5k+1}{k^2+2k} = \dfrac{?}{k^3+k^2-2k}$

(b) $\dfrac{12p}{p^2+8p} = \dfrac{?}{p^3+4p^2-32p}$

Factor the denominator in each rational expression.

$$\frac{12p}{p(p+8)} = \frac{?}{p(p+8)(p-4)}$$

p^3+4p^2-32p
$= p(p^2+4p-32)$
$= p(p+8)(p-4)$

The factor $p-4$ is missing, so multiply $\frac{12p}{p(p+8)}$ by $\frac{p-4}{p-4}$.

$$\frac{12p}{p^2+8p} = \frac{12p}{p(p+8)} \cdot \frac{p-4}{p-4} \quad \text{Fundamental property}$$

$$= \frac{12p(p-4)}{p(p+8)(p-4)} \quad \begin{array}{l}\text{Multiply numerators.}\\ \text{Multiply denominators.}\end{array}$$

$$= \frac{12p^2-48p}{p^3+4p^2-32p} \quad \text{Multiply the factors.}$$

Note

In the next section we add and subtract rational expressions, which sometimes requires the steps illustrated in Examples 4 and 5. While it may be beneficial to leave the equivalent expression in factored form, we multiplied out the factors in the numerator and the denominator in Example 5(b),

$$\frac{12p(p-4)}{p(p+8)(p-4)},$$

in order to give the answer,

$$\frac{12p^2-48p}{p^3+4p^2-32p},$$

in the same form as the original problem.

◀ Work Problem **5** at the Side.

14.3 ▶▶▶ Exercises

Choose the correct response in Exercises 1–4.

1. Suppose that the greatest common factor of a and b is 1. Then the least common denominator for $\frac{1}{a}$ and $\frac{1}{b}$ is

   A. a **B.** b **C.** ab **D.** 1.

2. If a is a factor of b, then the least common denominator for $\frac{1}{a}$ and $\frac{1}{b}$ is

   A. a **B.** b **C.** ab **D.** 1.

3. The least common denominator for $\frac{11}{20}$ and $\frac{1}{2}$ is

   A. 40 **B.** 2 **C.** 20 **D.** none of these.

4. Suppose that we wish to write the rational expression $\dfrac{1}{(x-4)^2(y-3)}$ with denominator $(x-4)^3(y-3)^2$. We must multiply both the numerator and the denominator by

   A. $(x-4)(y-3)$ **B.** $(x-4)^2$

   C. $x-4$ **D.** $(x-4)^2(y-3)$.

Find the least common denominator for the fractions in each list. See Examples 1 and 2.

5. $\dfrac{2}{15}, \dfrac{3}{10}, \dfrac{7}{30}$

6. $\dfrac{5}{24}, \dfrac{7}{12}, \dfrac{9}{28}$

7. $\dfrac{3}{x^4}, \dfrac{5}{x^7}$

8. $\dfrac{2}{y^5}, \dfrac{3}{y^6}$

9. $\dfrac{5}{36q}, \dfrac{17}{24q}$

10. $\dfrac{4}{30p}, \dfrac{9}{50p}$

 11. $\dfrac{6}{21r^3}, \dfrac{8}{12r^5}$

12. $\dfrac{9}{35t^2}, \dfrac{5}{49t^6}$

13. If the denominators of two fractions in prime factored form are $2^3 \cdot 3$ and $2^2 \cdot 5$, what is the factored form of their LCD?

14. Suppose two rational expressions have denominators $(t+4)^3(t-3)$ and $(t+4)^2(t+8)$. Find the factored form of their LCD. What is the similarity between the answers for this problem and for Exercise 13?

15. If two denominators have greatest common factor equal to 1, how can you easily find their least common denominator?

16. Suppose two fractions have denominators a^k and a^r, where k and r are natural numbers, with $k > r$. What is their least common denominator?

Find the least common denominator for the fractions in each list. See Examples 1–3.

17. $\dfrac{9}{28m^2}, \dfrac{3}{12m-20}$

18. $\dfrac{15}{27a^3}, \dfrac{8}{9a-45}$

19. $\dfrac{7}{5b-10}, \dfrac{11}{6b-12}$

20. $\dfrac{3}{7x^2+21x}, \dfrac{1}{5x^2+15x}$

21. $\dfrac{5}{c-d}, \dfrac{8}{d-c}$

22. $\dfrac{4}{y-x}, \dfrac{7}{x-y}$

23. $\dfrac{3}{k^2 + 5k}, \dfrac{2}{k^2 + 3k - 10}$

24. $\dfrac{1}{z^2 - 4z}, \dfrac{4}{z^2 - 3z - 4}$

25. $\dfrac{5}{p^2 + 8p + 15}, \dfrac{3}{p^2 - 3p - 18}, \dfrac{2}{p^2 - p - 30}$

26. $\dfrac{10}{y^2 - 10y + 21}, \dfrac{2}{y^2 - 2y - 3}, \dfrac{5}{y^2 - 6y - 7}$

Write each rational expression as an equivalent expression with the indicated denominator.
See Examples 4 and 5.

27. $\dfrac{4}{11} = \dfrac{?}{55}$

28. $\dfrac{6}{7} = \dfrac{?}{42}$

29. $\dfrac{-5}{k} = \dfrac{?}{9k}$

30. $\dfrac{-3}{q} = \dfrac{?}{6q}$

31. $\dfrac{13}{40y} = \dfrac{?}{80y^3}$

32. $\dfrac{5}{27p} = \dfrac{?}{108p^4}$

33. $\dfrac{5t^2}{6r} = \dfrac{?}{42r^4}$

34. $\dfrac{8y^2}{3x} = \dfrac{?}{30x^3}$

35. $\dfrac{5}{2(m + 3)} = \dfrac{?}{8(m + 3)}$

36. $\dfrac{7}{4(y - 1)} = \dfrac{?}{16(y - 1)}$

37. $\dfrac{-4t}{3t - 6} = \dfrac{?}{12 - 6t}$

38. $\dfrac{-7k}{5k - 20} = \dfrac{?}{40 - 10k}$

39. $\dfrac{14}{z^2 - 3z} = \dfrac{?}{z(z - 3)(z - 2)}$

40. $\dfrac{12}{x(x + 4)} = \dfrac{?}{x(x + 4)(x - 9)}$

41. $\dfrac{2(b - 1)}{b^2 + b} = \dfrac{?}{b^3 + 3b^2 + 2b}$

42. $\dfrac{3(c + 2)}{c(c - 1)} = \dfrac{?}{c^3 - 5c^2 + 4c}$

14.4 ▶▶▶ Adding and Subtracting Rational Expressions

To add and subtract rational expressions, we find least common denominators and write equivalent fractions with the LCD.

OBJECTIVE 1 Add rational expressions having the same denominator. We find the sum of two such rational expressions with the same procedure that we used in **Section 3.1** for adding two fractions having the same denominator.

> **Adding Rational Expressions (Same Denominator)**
>
> If $\frac{P}{Q}$ and $\frac{R}{Q}$ $(Q \neq 0)$ are rational expressions, then
>
> $$\frac{P}{Q} + \frac{R}{Q} = \frac{P+R}{Q}.$$
>
> In words: To add rational expressions with the same denominator, add the numerators and keep the same denominator.

EXAMPLE 1 Adding Rational Expressions (Same Denominator)

Add. Write each answer in lowest terms.

(a) $\dfrac{4}{9} + \dfrac{2}{9}$ **(b)** $\dfrac{3x}{x+1} + \dfrac{3}{x+1}$

The denominators are the same, so the sum is found by adding the two numerators and keeping the same (common) denominator.

$= \dfrac{4+2}{9}$ Add.

$= \dfrac{6}{9}$

$= \dfrac{2 \cdot 3}{3 \cdot 3}$ Factor.

$= \dfrac{2}{3}$ Lowest terms

$= \dfrac{3x+3}{x+1}$ Add.

$= \dfrac{3(x+1)}{x+1}$ Factor.

$= 3$ Lowest terms

Work Problem ① *at the Side.* ▶

OBJECTIVE 2 Add rational expressions having different denominators. As in **Section 3.3**, we use the following steps to add two rational expressions having different denominators.

> **Adding Rational Expressions (Different Denominators)**
>
> **Step 1** **Find the least common denominator (LCD).**
>
> **Step 2** **Write each rational expression** as an equivalent rational expression with the LCD as the denominator.
>
> **Step 3** **Add the numerators** to get the numerator of the sum. The LCD is the denominator of the sum.
>
> **Step 4** **Write in lowest terms** using the fundamental property.

OBJECTIVES

1 **Add rational expressions having the same denominator.**

2 **Add rational expressions having different denominators.**

3 **Subtract rational expressions.**

① Add. Write each answer in lowest terms.

(a) $\dfrac{7}{15} + \dfrac{3}{15}$

(b) $\dfrac{3}{y+4} + \dfrac{2}{y+4}$

(c) $\dfrac{x}{x+y} + \dfrac{1}{x+y}$

(d) $\dfrac{a}{a+b} + \dfrac{b}{a+b}$

(e) $\dfrac{x^2}{x+1} + \dfrac{x}{x+1}$

ANSWERS

1. **(a)** $\dfrac{2}{3}$ **(b)** $\dfrac{5}{y+4}$ **(c)** $\dfrac{x+1}{x+y}$
   (d) 1 **(e)** x

2 Add. Write each answer in lowest terms.

(a) $\dfrac{1}{10} + \dfrac{1}{15}$

(b) $\dfrac{6}{5x} + \dfrac{9}{2x}$

(c) $\dfrac{m}{3n} + \dfrac{2}{7n}$

EXAMPLE 2 Adding Rational Expressions (Different Denominators)

Add. Write each answer in lowest terms.

(a) $\dfrac{1}{12} + \dfrac{7}{15}$ **(b)** $\dfrac{2}{3y} + \dfrac{1}{4y}$

Step 1 First find the LCD, using the methods of the previous section.

$$12 = 2 \cdot 2 \cdot 3 = 2^2 \cdot 3 \qquad\qquad 3y = 3 \cdot y$$
$$15 = 3 \cdot 5 \qquad\qquad\qquad 4y = 2 \cdot 2 \cdot y = 2^2 \cdot y$$
$$\text{LCD} = 2^2 \cdot 3 \cdot 5 = 60 \qquad \text{LCD} = 2^2 \cdot 3 \cdot y = 12y$$

Step 2 Now write each rational expression as an equivalent expression with the LCD (either 60 or $12y$) as the denominator.

$$\dfrac{1}{12} + \dfrac{7}{15} = \dfrac{1(5)}{12(5)} + \dfrac{7(4)}{15(4)} \qquad \dfrac{2}{3y} + \dfrac{1}{4y} = \dfrac{2(4)}{3y(4)} + \dfrac{1(3)}{4y(3)}$$

$$= \dfrac{5}{60} + \dfrac{28}{60} \qquad\qquad\qquad = \dfrac{8}{12y} + \dfrac{3}{12y}$$

Step 3 Add the numerators. The LCD is the denominator.

Step 4 Write in lowest terms if necessary.

$$= \dfrac{5 + 28}{60} \qquad\qquad\qquad\qquad = \dfrac{8 + 3}{12y}$$

$$= \dfrac{33}{60}, \text{ or } \dfrac{11}{20} \qquad\qquad\qquad = \dfrac{11}{12y}$$

◀ *Work Problem* **2** *at the Side.*

EXAMPLE 3 Adding Rational Expressions

Add. Write the answer in lowest terms.

$$\dfrac{2x}{x^2 - 1} + \dfrac{-1}{x + 1}$$

Step 1 Since the denominators are different, find the LCD.

$$\left.\begin{array}{l} x^2 - 1 = (x + 1)(x - 1) \\ x + 1 \text{ is prime.} \end{array}\right\} \text{ The LCD is } (x + 1)(x - 1).$$

Step 2 Write each rational expression as an equivalent expression with the LCD as the denominator.

$$\dfrac{2x}{x^2 - 1} + \dfrac{-1}{x + 1} \qquad \text{The LCD is } (x + 1)(x - 1).$$

$$= \dfrac{2x}{(x + 1)(x - 1)} + \dfrac{-1(x - 1)}{(x + 1)(x - 1)} \qquad \text{Multiply the second fraction by } \tfrac{x-1}{x-1}.$$

$$= \dfrac{2x}{(x + 1)(x - 1)} + \dfrac{-x + 1}{(x + 1)(x - 1)} \qquad \text{Distributive property}$$

Step 3 $= \dfrac{2x - x + 1}{(x + 1)(x - 1)} \qquad \text{Add numerators; keep the same denominator.}$

Continued on Next Page

$$= \frac{x + 1}{(x + 1)(x - 1)}$$ Combine like terms in the numerator.

Step 4 $$= \frac{1(x + 1)}{(x + 1)(x - 1)}$$ Identity property for multiplication

Remember to write 1 in the numerator.

$$= \frac{1}{x - 1}$$ Divide out the common factors.

Work Problem **3** *at the Side.* ▶

EXAMPLE 4 **Adding Rational Expressions**

Add. Write the answer in lowest terms.

$$\frac{2x}{x^2 + 5x + 6} + \frac{x + 1}{x^2 + 2x - 3}$$

$$= \frac{2x}{(x + 2)(x + 3)} + \frac{x + 1}{(x + 3)(x - 1)}$$ Factor the denominators.

$$= \frac{2x(x - 1)}{(x + 2)(x + 3)(x - 1)} + \frac{(x + 1)(x + 2)}{(x + 2)(x + 3)(x - 1)}$$ The LCD is $(x + 2) \cdot (x + 3)(x - 1)$.

$$= \frac{2x(x - 1) + (x + 1)(x + 2)}{(x + 2)(x + 3)(x - 1)}$$ Add numerators; keep the same denominator.

$$= \frac{2x^2 - 2x + x^2 + 3x + 2}{(x + 2)(x + 3)(x - 1)}$$ Multiply.

$$= \frac{3x^2 + x + 2}{(x + 2)(x + 3)(x - 1)}$$ Combine like terms.

It is usually more convenient to leave the denominator in factored form. The numerator cannot be factored here, so the expression is in lowest terms.

Work Problem **4** *at the Side.* ▶

EXAMPLE 5 **Adding Rational Expressions with Denominators That Are Opposites**

Add. Write the answer in lowest terms.

$$\frac{y}{y - 2} + \frac{8}{2 - y}$$ The denominators are opposites.

$$= \frac{y}{y - 2} + \frac{8(-1)}{(2 - y)(-1)}$$ Multiply $\frac{8}{2 - y}$ by $\frac{-1}{-1}$ to get a common denominator.

$$= \frac{y}{y - 2} + \frac{-8}{-2 + y}$$ Distributive property

$$= \frac{y}{y - 2} + \frac{-8}{y - 2}$$ Rewrite $-2 + y$ as $y - 2$.

$$= \frac{y - 8}{y - 2}$$ Add numerators; keep the same denominator.

If we had chosen to use $2 - y$ as the common denominator, the final answer would be in the form $\frac{8 - y}{2 - y}$, which is equivalent to $\frac{y - 8}{y - 2}$.

Work Problem **5** *at the Side.* ▶

---

3 Add. Write each answer in lowest terms.

(a) $\dfrac{2p}{3p + 3} + \dfrac{5p}{2p + 2}$

(b) $\dfrac{4}{y^2 - 1} + \dfrac{6}{y + 1}$

(c) $\dfrac{-2}{p + 1} + \dfrac{4p}{p^2 - 1}$

4 Add. Write each answer in lowest terms.

(a) $\dfrac{2k}{k^2 - 5k + 4} + \dfrac{3}{k^2 - 1}$

(b) $\dfrac{4m}{m^2 + 3m + 2} + \dfrac{2m - 1}{m^2 + 6m + 5}$

5 Add. Write the answer in lowest terms.

$$\frac{m}{2m - 3n} + \frac{n}{3n - 2m}$$

ANSWERS

3. (a) $\dfrac{19p}{6(p + 1)}$ (b) $\dfrac{2(3y - 1)}{(y + 1)(y - 1)}$

(c) $\dfrac{2}{p - 1}$

4. (a) $\dfrac{(2k - 3)(k + 4)}{(k - 4)(k - 1)(k + 1)}$

(b) $\dfrac{6m^2 + 23m - 2}{(m + 2)(m + 1)(m + 5)}$

5. $\dfrac{m - n}{2m - 3n}$, or $\dfrac{n - m}{3n - 2m}$

6 Subtract. Write each answer in lowest terms.

(a) $\dfrac{3}{m^2} - \dfrac{2}{m^2}$

OBJECTIVE **3** **Subtract rational expressions.**   To subtract rational expressions having the same denominator, use the following rule.

> **Subtracting Rational Expressions (Same Denominator)**
>
> If $\dfrac{P}{Q}$ and $\dfrac{R}{Q}$ $(Q \neq 0)$ are rational expressions, then
>
> $$\frac{P}{Q} - \frac{R}{Q} = \frac{P - R}{Q}.$$
>
> In words: To subtract rational expressions with the same denominator, subtract the numerators and keep the same denominator.

EXAMPLE 6   **Subtracting Rational Expressions (Same Denominator)**

Subtract. Write the answer in lowest terms.

$$\frac{2m}{m-1} - \frac{m+3}{m-1}$$

Use parentheses around the quantity being subtracted.

$$= \frac{2m - (m+3)}{m-1}$$

Subtract numerators; keep the same denominator.

Be careful with signs.

$$= \frac{2m - m - 3}{m-1}$$

Distributive property

$$= \frac{m-3}{m-1}$$

Combine like terms.

(b) $\dfrac{x}{2x+3} - \dfrac{3x+4}{2x+3}$

> **CAUTION**
> Sign errors often occur in problems like the one in Example 6. The numerator of the fraction being subtracted must be treated as a single quantity. ***Be sure to use parentheses after the subtraction sign.***

◀ *Work Problem* **6** *at the Side.*

In the remaining examples, we subtract rational expressions having different denominators.

EXAMPLE 7   **Subtracting Rational Expressions (Different Denominators)**

Subtract. Write the answer in lowest terms.

$$\frac{9}{x-2} - \frac{3}{x}$$

The LCD is $x(x-2)$.

$$= \frac{9x}{x(x-2)} - \frac{3(x-2)}{x(x-2)}$$

Rewrite each expression with the LCD.

$$= \frac{9x - 3(x-2)}{x(x-2)}$$

Subtract numerators; keep the same denominator.

Continued on Next Page

$\boxed{\text{Be careful here.}}$ $= \dfrac{9x - 3x + 6}{x(x - 2)}$ Distributive property

$= \dfrac{6x + 6}{x(x - 2)}$,   or $\dfrac{6(x + 1)}{x(x - 2)}$ Combine like terms; factor the numerator.

Note

We factored the final numerator in Example 7 to get $\dfrac{6(x + 1)}{x(x - 2)}$. The fundamental property does not apply, however, since there are no common factors to divide out. The answer is in lowest terms.

Work Problem $\boxed{7}$ *at the Side.* ▶

$\boxed{\text{EXAMPLE 8}}$ **Subtracting Rational Expressions with Denominators That Are Opposites**

Subtract. Write the answer in lowest terms.

$\dfrac{3x}{x - 5} - \dfrac{2x - 25}{5 - x}$ The denominators are opposites. We choose $x - 5$ as the common denominator.

$= \dfrac{3x}{x - 5} - \dfrac{(2x - 25)(-1)}{(5 - x)(-1)}$ Multiply $\frac{2x - 25}{5 - x}$ by $\frac{-1}{-1}$ to get a common denominator.

$= \dfrac{3x}{x - 5} - \dfrac{-2x + 25}{x - 5}$ $(5 - x)(-1) = -5 + x = x - 5$

$= \dfrac{3x - (-2x + 25)}{x - 5}$ $\boxed{\text{Use parentheses.}}$ Subtract numerators.

$= \dfrac{3x + 2x - 25}{x - 5}$ Distributive property

$= \dfrac{5x - 25}{x - 5}$ Combine like terms.

$= \dfrac{5(x - 5)}{x - 5}$ Factor.

$= 5$ Divide out the common factor.

Work Problem $\boxed{8}$ *at the Side.* ▶

$\boxed{\text{EXAMPLE 9}}$ **Subtracting Rational Expressions**

Subtract. Write the answer in lowest terms.

$\dfrac{6x}{x^2 - 2x + 1} - \dfrac{1}{x^2 - 1}$

Begin by factoring the denominators.

$x^2 - 2x + 1 = (x - 1)^2$ and $x^2 - 1 = (x - 1)(x + 1)$

From the factored denominators, we identify the LCD, $(x - 1)^2(x + 1)$. *We use the factor $x - 1$ twice because it appears twice in the first denominator.*

— **Continued on Next Page**

$\boxed{7}$ Subtract. Write each answer in lowest terms.

(a) $\dfrac{1}{k + 4} - \dfrac{2}{k}$

(b) $\dfrac{6}{a + 2} - \dfrac{1}{a - 3}$

$\boxed{8}$ Subtract. Write each answer in lowest terms.

(a) $\dfrac{5}{x - 1} - \dfrac{3x}{1 - x}$

(b) $\dfrac{2y}{y - 2} - \dfrac{1 + y}{2 - y}$

ANSWERS

7. (a) $\dfrac{-k - 8}{k(k + 4)}$ (b) $\dfrac{5(a - 4)}{(a + 2)(a - 3)}$

8. (a) $\dfrac{5 + 3x}{x - 1}$,   or $\dfrac{-5 - 3x}{1 - x}$

   (b) $\dfrac{3y + 1}{y - 2}$,   or $\dfrac{-3y - 1}{2 - y}$

9 Subtract. Write each answer in lowest terms.

(a) $\dfrac{4y}{y^2 - 1} - \dfrac{5}{y^2 + 2y + 1}$

(b) $\dfrac{3r}{r^2 - 5r} - \dfrac{4}{r^2 - 10r + 25}$

10 Subtract. Write each answer in lowest terms.

(a) $\dfrac{2}{p^2 - 5p + 4} - \dfrac{3}{p^2 - 1}$

(b)

$\dfrac{q}{2q^2 + 5q - 3} - \dfrac{3q + 4}{3q^2 + 10q + 3}$

$$\dfrac{6x}{(x - 1)^2} - \dfrac{1}{(x - 1)(x + 1)} \qquad \text{The LCD is } (x-1)^2(x+1).$$

$$= \dfrac{6x(x + 1)}{(x - 1)^2(x + 1)} - \dfrac{1(x - 1)}{(x - 1)(x - 1)(x + 1)} \qquad \text{Fundamental property}$$

$$= \dfrac{6x(x + 1) - 1(x - 1)}{(x - 1)^2(x + 1)} \qquad \text{Subtract numerators.}$$

$$= \dfrac{6x^2 + 6x - x + 1}{(x - 1)^2(x + 1)} \qquad \text{Distributive property}$$

$$= \dfrac{6x^2 + 5x + 1}{(x - 1)^2(x + 1)}, \quad \text{or} \quad \dfrac{(2x + 1)(3x + 1)}{(x - 1)^2(x + 1)} \qquad \text{Combine like terms; factor the numerator.}$$

Verify that the final answer is in lowest terms.

◄ **Work Problem** **9** at the Side.

EXAMPLE 10 **Subtracting Rational Expressions**

Subtract. Write the answer in lowest terms.

$$\dfrac{q}{q^2 - 4q - 5} - \dfrac{3}{2q^2 - 13q + 15}$$

To find the LCD, factor each denominator.

$$q^2 - 4q - 5 = (q + 1)(q - 5)$$
$$2q^2 - 13q + 15 = (q - 5)(2q - 3)$$

The LCD here is $(q + 1)(q - 5)(2q - 3)$. Write each rational expression with the LCD, using the fundamental property.

$$\dfrac{q}{(q + 1)(q - 5)} - \dfrac{3}{(q - 5)(2q - 3)} \qquad \text{The LCD is } (q + 1) \cdot (q - 5)(2q - 3).$$

$$= \dfrac{q(2q - 3)}{(q + 1)(q - 5)(2q - 3)} - \dfrac{3(q + 1)}{(q + 1)(q - 5)(2q - 3)}$$

$$= \dfrac{q(2q - 3) - 3(q + 1)}{(q + 1)(q - 5)(2q - 3)} \qquad \text{Subtract numerators.}$$

$$= \dfrac{2q^2 - 3q - 3q - 3}{(q + 1)(q - 5)(2q - 3)} \qquad \text{Distributive property}$$

$$= \dfrac{2q^2 - 6q - 3}{(q + 1)(q - 5)(2q - 3)} \qquad \text{Combine like terms.}$$

Verify that the final answer is in lowest terms.

◄ **Work Problem** **10** at the Side.

ANSWERS

9. **(a)** $\dfrac{4y^2 - y + 5}{(y + 1)^2(y - 1)}$

   (b) $\dfrac{3r - 19}{(r - 5)^2}$

10. **(a)** $\dfrac{14 - p}{(p - 4)(p - 1)(p + 1)}$

   (b) $\dfrac{-3q^2 - 4q + 4}{(2q - 1)(q + 3)(3q + 1)}$

Match the expression in Column I with the correct sum or difference in Column II.

<table>
<tr><td colspan="2" align="center">**I**</td><td colspan="2" align="center">**II**</td></tr>
</table>

1. $\dfrac{x}{x+6} + \dfrac{6}{x+6}$ 　　　 **2.** $\dfrac{2x}{x-6} - \dfrac{12}{x-6}$ 　　　 **A.** 2 　　　 **B.** $\dfrac{x-6}{x+6}$

3. $\dfrac{6}{x-6} - \dfrac{x}{x-6}$ 　　　 **4.** $\dfrac{6}{x+6} - \dfrac{x}{x+6}$ 　　　 **C.** -1 　　　 **D.** $\dfrac{6+x}{6x}$

5. $\dfrac{x}{x+6} - \dfrac{6}{x+6}$ 　　　 **6.** $\dfrac{1}{x} + \dfrac{1}{6}$ 　　　 **E.** 1 　　　 **F.** 0

7. $\dfrac{1}{6} - \dfrac{1}{x}$ 　　　 **8.** $\dfrac{1}{6x} - \dfrac{1}{6x}$ 　　　 **G.** $\dfrac{x-6}{6x}$ 　　　 **H.** $\dfrac{6-x}{x+6}$

Note: When adding and subtracting rational expressions, several different equivalent forms of the answer often exist. If your answer does not look exactly like the one given in the back of the book, check to see whether you have written an equivalent form.

Add or subtract. Write each answer in lowest terms. See Examples 1 and 6.

9. $\dfrac{4}{m} + \dfrac{7}{m}$ 　　　 **10.** $\dfrac{5}{p} + \dfrac{11}{p}$ 　　　 **11.** $\dfrac{a+b}{2} - \dfrac{a-b}{2}$

12. $\dfrac{x-y}{2} - \dfrac{x+y}{2}$ 　　 **13.** $\dfrac{5}{y+4} - \dfrac{1}{y+4}$ 　　 **14.** $\dfrac{4}{y+3} - \dfrac{1}{y+3}$

15. $\dfrac{5m}{m+1} - \dfrac{1+4m}{m+1}$ 　 **16.** $\dfrac{4x}{x+2} - \dfrac{2+3x}{x+2}$ 　 **17.** $\dfrac{x^2}{x+5} + \dfrac{5x}{x+5}$

18. $\dfrac{t^2}{t-3} + \dfrac{-3t}{t-3}$ 　 **19.** $\dfrac{y^2-3y}{y+3} + \dfrac{-18}{y+3}$ 　 **20.** $\dfrac{r^2-8r}{r-5} + \dfrac{15}{r-5}$

21. Explain with an example how to add or subtract rational expressions with the same denominator.

22. Explain with an example how to add or subtract rational expressions with different denominators.

Add or subtract. Write each answer in lowest terms. See Examples 2, 3, 4, and 7.

23. $\dfrac{z}{5} + \dfrac{1}{3}$

24. $\dfrac{p}{8} + \dfrac{3}{5}$

25. $\dfrac{5}{7} - \dfrac{r}{2}$

26. $\dfrac{10}{9} - \dfrac{z}{3}$

27. $-\dfrac{3}{4} - \dfrac{1}{2x}$

28. $-\dfrac{5}{8} - \dfrac{3}{2a}$

29. $\dfrac{3}{5x} + \dfrac{9}{4x}$

30. $\dfrac{3}{2x} + \dfrac{4}{7x}$

31. $\dfrac{x+1}{6} + \dfrac{3x+3}{9}$

32. $\dfrac{2x-6}{4} + \dfrac{x+5}{6}$

33. $\dfrac{x+3}{3x} + \dfrac{2x+2}{4x}$

34. $\dfrac{x+2}{5x} + \dfrac{6x+3}{3x}$

35. $\dfrac{2}{x+3} + \dfrac{1}{x}$

36. $\dfrac{3}{x-4} + \dfrac{2}{x}$

37. $\dfrac{1}{k+5} - \dfrac{2}{k}$

38. $\dfrac{3}{m+1} - \dfrac{4}{m}$

39. $\dfrac{x}{x-2} + \dfrac{-8}{x^2-4}$

40. $\dfrac{2x}{x-1} + \dfrac{-4}{x^2-1}$

41. $\dfrac{x}{x-2} + \dfrac{4}{x+2}$

42. $\dfrac{2x}{x-1} + \dfrac{3}{x+1}$

43. $\dfrac{t}{t+2} + \dfrac{5-t}{t} - \dfrac{4}{t^2+2t}$

44. $\dfrac{2p}{p-3} + \dfrac{2+p}{p} - \dfrac{-6}{p^2-3p}$

45. What are the two possible LCDs that could be used for the sum

$$\frac{10}{m-2}+\frac{5}{2-m}?$$

46. If one form of the correct answer to a sum or difference of rational expressions is $\frac{4}{k-3}$, what would be an alternative form of the answer if the denominator is $3-k$?

Add or subtract. Write each answer in lowest terms. See Examples 5 and 8.

47. $\dfrac{4}{x-5}+\dfrac{6}{5-x}$

48. $\dfrac{10}{m-2}+\dfrac{5}{2-m}$

49. $\dfrac{-1}{1-y}+\dfrac{3-4y}{y-1}$

50. $\dfrac{-4}{p-3}-\dfrac{p+1}{3-p}$

51. $\dfrac{2}{x-y^2}+\dfrac{7}{y^2-x}$

52. $\dfrac{-8}{p-q^2}+\dfrac{3}{q^2-p}$

53. $\dfrac{x}{5x-3y}-\dfrac{y}{3y-5x}$

54. $\dfrac{t}{8t-9s}-\dfrac{s}{9s-8t}$

55. $\dfrac{3}{4p-5}+\dfrac{9}{5-4p}$

56. $\dfrac{8}{3-7y}-\dfrac{2}{7y-3}$

*In each subtraction problem, the rational expression that follows the subtraction sign has a numerator with more than one term. **Be careful with signs** and find each difference. See Examples 6–10.*

57. $\dfrac{2m}{m-n}-\dfrac{5m+n}{2m-2n}$

58. $\dfrac{5p}{p-q}-\dfrac{3p+1}{4p-4q}$

59. $\dfrac{5}{x^2 - 9} - \dfrac{x + 2}{x^2 + 4x + 3}$

60. $\dfrac{1}{a^2 - 1} - \dfrac{a - 1}{a^2 + 3a - 4}$

61. $\dfrac{2q + 1}{3q^2 + 10q - 8} - \dfrac{3q + 5}{2q^2 + 5q - 12}$

62. $\dfrac{4y - 1}{2y^2 + 5y - 3} - \dfrac{y + 3}{6y^2 + y - 2}$

Perform the indicated operations. See Examples 1–10.

63. $\dfrac{4}{r^2 - r} + \dfrac{6}{r^2 + 2r} - \dfrac{1}{r^2 + r - 2}$

64. $\dfrac{6}{k^2 + 3k} - \dfrac{1}{k^2 - k} + \dfrac{2}{k^2 + 2k - 3}$

65. $\dfrac{x + 3y}{x^2 + 2xy + y^2} + \dfrac{x - y}{x^2 + 4xy + 3y^2}$

66. $\dfrac{m}{m^2 - 1} + \dfrac{m - 1}{m^2 + 2m + 1}$

67. $\dfrac{r + y}{18r^2 + 9ry - 2y^2} + \dfrac{3r - y}{36r^2 - y^2}$

68. $\dfrac{2x - z}{2x^2 + xz - 10z^2} - \dfrac{x + z}{x^2 - 4z^2}$

69. Refer to the rectangle in the figure.

    (a) Find an expression that represents its perimeter. Give the simplified form.

    (b) Find an expression that represents its area. Give the simplified form.

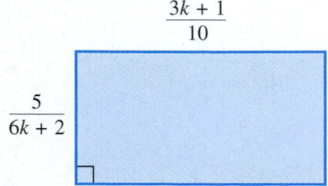

70. Refer to the triangle in the figure. Find an expression that represents its perimeter.

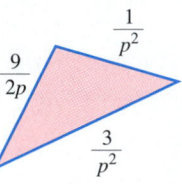

14.5 ▶▶▶ Complex Fractions

OBJECTIVES

1 **Simplify a complex fraction by writing it as a division problem (Method 1).**

2 **Simplify a complex fraction by multiplying numerator and denominator by the least common denominator (Method 2).**

The quotient of two mixed numbers in arithmetic, such as $2\frac{1}{2} \div 3\frac{1}{4}$, can be written as a fraction:

$$2\frac{1}{2} \div 3\frac{1}{4} = \frac{2\frac{1}{2}}{3\frac{1}{4}} = \frac{2+\frac{1}{2}}{3+\frac{1}{4}}.$$

The last expression is the quotient of expressions that involve fractions. In algebra, some rational expressions also have fractions in the numerator, or denominator, or both.

> **Complex Fraction**
>
> A rational expression with one or more fractions in the numerator, or denominator, or both, is called a **complex fraction.**

Examples of complex fractions include

$$\frac{2+\frac{1}{2}}{3+\frac{1}{4}}, \quad \frac{\frac{3x^2-5x}{6x^2}}{2x-\frac{1}{x}}, \quad \text{and} \quad \frac{3+x}{5-\frac{2}{x}}. \quad \text{Complex fractions}$$

The parts of a complex fraction are named as follows.

$$\begin{array}{l} \left.\dfrac{2}{p} - \dfrac{1}{q}\right\} \leftarrow \text{Numerator of complex fraction} \\ \quad\quad\quad \leftarrow \text{Main fraction bar} \\ \left.\dfrac{3}{p} + \dfrac{5}{q}\right\} \leftarrow \text{Denominator of complex fraction} \end{array}$$

OBJECTIVE 1 Simplify a complex fraction by writing it as a division problem (Method 1). Since the main fraction bar represents division in a complex fraction, one method of simplifying a complex fraction involves division.

> **Method 1 for Simplifying a Complex Fraction**
>
> *Step 1* Write both the numerator and denominator as single fractions.
>
> *Step 2* Change the complex fraction to a division problem.
>
> *Step 3* Perform the indicated division.

Once again, the first example shows complex fractions from both arithmetic and algebra.

1 Simplify each complex fraction using Method 1.

(a) $\dfrac{\dfrac{2}{5} + \dfrac{1}{4}}{\dfrac{1}{2} + \dfrac{1}{3}}$

(b) $\dfrac{6 + \dfrac{1}{x}}{5 - \dfrac{2}{x}}$

(c) $\dfrac{9 - \dfrac{4}{p}}{\dfrac{2}{p} + 1}$

2 Simplify each complex fraction using Method 1.

(a) $\dfrac{\dfrac{rs^2}{t}}{\dfrac{r^2 s}{t^2}}$

(b) $\dfrac{\dfrac{m^2 n^3}{p}}{\dfrac{m^4 n}{p^2}}$

ANSWERS

1. (a) $\dfrac{39}{50}$ (b) $\dfrac{6x + 1}{5x - 2}$ (c) $\dfrac{9p - 4}{2 + p}$

2. (a) $\dfrac{st}{r}$ (b) $\dfrac{n^2 p}{m^2}$

EXAMPLE 1 Simplifying Complex Fractions (Method 1)

Simplify each complex fraction.

(a) $\dfrac{\dfrac{2}{3} + \dfrac{5}{9}}{\dfrac{1}{4} + \dfrac{1}{12}}$

(b) $\dfrac{6 + \dfrac{3}{x}}{\dfrac{x}{4} + \dfrac{1}{8}}$

Step 1 First, write each numerator as a single fraction.

$$\frac{2}{3} + \frac{5}{9} = \frac{2(3)}{3(3)} + \frac{5}{9} \qquad 6 + \frac{3}{x} = \frac{6}{1} + \frac{3}{x}$$

$$= \frac{6}{9} + \frac{5}{9} = \frac{11}{9} \qquad = \frac{6x}{x} + \frac{3}{x} = \frac{6x + 3}{x}$$

Do the same thing with each denominator.

$$\frac{1}{4} + \frac{1}{12} = \frac{1(3)}{4(3)} + \frac{1}{12} \qquad \frac{x}{4} + \frac{1}{8} = \frac{x(2)}{4(2)} + \frac{1}{8}$$

$$= \frac{3}{12} + \frac{1}{12} = \frac{4}{12} \qquad = \frac{2x}{8} + \frac{1}{8} = \frac{2x + 1}{8}$$

Step 2 Write the equivalent complex fraction as a division problem.

$$\frac{\dfrac{11}{9}}{\dfrac{4}{12}} = \frac{11}{9} \div \frac{4}{12} \qquad \frac{\dfrac{6x + 3}{x}}{\dfrac{2x + 1}{8}} = \frac{6x + 3}{x} \div \frac{2x + 1}{8}$$

Step 3 Now use the definition of division and multiply by the reciprocal. Then write in lowest terms using the fundamental property.

$$= \frac{11}{9} \cdot \frac{12}{4} \qquad = \frac{6x + 3}{x} \cdot \frac{8}{2x + 1}$$

$$= \frac{11 \cdot 3 \cdot 4}{3 \cdot 3 \cdot 4} \qquad = \frac{3(2x + 1)}{x} \cdot \frac{8}{2x + 1}$$

$$= \frac{11}{3} \qquad = \frac{24}{x}$$

◀ Work Problem **1** at the Side.

EXAMPLE 2 Simplifying a Complex Fraction (Method 1)

Simplify the complex fraction.

$\dfrac{\dfrac{xp}{q^3}}{\dfrac{p^2}{qx^2}}$ The numerator and denominator are single fractions, so use the definition of division and then the fundamental property.

$$\frac{xp}{q^3} \div \frac{p^2}{qx^2}$$

$$= \frac{xp}{q^3} \cdot \frac{qx^2}{p^2}$$

$$= \frac{x^3}{q^2 p}$$

◀ Work Problem **2** at the Side.

EXAMPLE 3 Simplifying a Complex Fraction (Method 1)

Simplify the complex fraction.

$$\frac{\dfrac{3}{x+2} - 4}{\dfrac{2}{x+2} + 1}$$

$$= \frac{\dfrac{3}{x+2} - \dfrac{4(x+2)}{x+2}}{\dfrac{2}{x+2} + \dfrac{1(x+2)}{x+2}}$$ Write both second terms with a denominator of $x+2$.

$$= \frac{\dfrac{3 - 4(x+2)}{x+2}}{\dfrac{2 + 1(x+2)}{x+2}}$$ Subtract in the numerator.

Add in the denominator.

$$= \frac{\dfrac{3 - 4x - 8}{x+2}}{\dfrac{2 + x + 2}{x+2}}$$ Be careful with signs.

Distributive property

$$= \frac{\dfrac{-5 - 4x}{x+2}}{\dfrac{4 + x}{x+2}}$$ Combine like terms.

$$= \frac{-5 - 4x}{x+2} \cdot \frac{x+2}{4+x}$$ Multiply by the reciprocal.

$$= \frac{-5 - 4x}{4 + x}$$ Lowest terms

CAUTION

$$\frac{\dfrac{a}{b} + \dfrac{c}{d}}{\dfrac{e}{f} + \dfrac{g}{h}} \quad \textit{does not equal} \quad \left(\frac{a}{b} + \frac{c}{d}\right) \cdot \left(\frac{f}{e} + \frac{h}{g}\right).$$

Work Problem **3** *at the Side.* ▶

OBJECTIVE 2 Simplify a complex fraction by multiplying numerator and denominator by the least common denominator (Method 2). Since any expression can be multiplied by a form of 1 to get an equivalent expression, we can multiply both the numerator and the denominator of a complex fraction by the same nonzero expression to get an equivalent complex fraction. If we choose the expression to be the LCD of all the fractions within the complex fraction, the result will no longer be complex. This is Method 2.

3 Simplify using Method 1.

$$\frac{\dfrac{2}{x-1} + \dfrac{1}{x+1}}{\dfrac{3}{x-1} - \dfrac{4}{x+1}}$$

ANSWER

3. $\dfrac{3x+1}{-x+7}$

4 Simplify each complex fraction using Method 2.

(a) $\dfrac{\dfrac{2}{3} - \dfrac{1}{4}}{\dfrac{4}{9} + \dfrac{1}{2}}$

(b) $\dfrac{2 - \dfrac{6}{a}}{3 + \dfrac{4}{a}}$

(c) $\dfrac{\dfrac{p}{5 - p}}{\dfrac{4p}{2p + 1}}$

> **Method 2 for Simplifying a Complex Fraction**
>
> **Step 1** Find the LCD of all fractions within the complex fraction.
>
> **Step 2** Multiply both the numerator and the denominator of the complex fraction by this LCD using the distributive property as necessary. Write in lowest terms.

In the next example, Method 2 is used to simplify the complex fractions from Example 1.

EXAMPLE 4 Simplifying Complex Fractions (Method 2)

Simplify each complex fraction.

(a) $\dfrac{\dfrac{2}{3} + \dfrac{5}{9}}{\dfrac{1}{4} + \dfrac{1}{12}}$

(b) $\dfrac{6 + \dfrac{3}{x}}{\dfrac{x}{4} + \dfrac{1}{8}}$

Step 1 Find the LCD for all denominators in the complex fraction.

The LCD for 3, 9, 4, and 12 is 36. | The LCD for x, 4, and 8 is $8x$.

Step 2 Multiply the numerator and denominator of the complex fraction by the LCD.

$\dfrac{\dfrac{2}{3} + \dfrac{5}{9}}{\dfrac{1}{4} + \dfrac{1}{12}}$

$= \dfrac{36\left(\dfrac{2}{3} + \dfrac{5}{9}\right)}{36\left(\dfrac{1}{4} + \dfrac{1}{12}\right)}$

Multiply each term by 36.

$= \dfrac{36\left(\dfrac{2}{3}\right) + 36\left(\dfrac{5}{9}\right)}{36\left(\dfrac{1}{4}\right) + 36\left(\dfrac{1}{12}\right)}$

$= \dfrac{24 + 20}{9 + 3}$

$= \dfrac{44}{12}$

$= \dfrac{4 \cdot 11}{4 \cdot 3}$

$= \dfrac{11}{3}$

$\dfrac{6 + \dfrac{3}{x}}{\dfrac{x}{4} + \dfrac{1}{8}}$

$= \dfrac{8x\left(6 + \dfrac{3}{x}\right)}{8x\left(\dfrac{x}{4} + \dfrac{1}{8}\right)}$

Multiply each term by 8x.

$= \dfrac{8x(6) + 8x\left(\dfrac{3}{x}\right)}{8x\left(\dfrac{x}{4}\right) + 8x\left(\dfrac{1}{8}\right)}$

$= \dfrac{48x + 24}{2x^2 + x}$

$= \dfrac{24(2x + 1)}{x(2x + 1)}$

$= \dfrac{24}{x}$

◀ **Work Problem** **4** **at the Side.**

EXAMPLE 5 **Simplifying a Complex Fraction (Method 2)**

Simplify the complex fraction.

$$\dfrac{\dfrac{3}{5m} - \dfrac{2}{m^2}}{\dfrac{9}{2m} + \dfrac{3}{4m^2}}$$ The LCD for $5m$, m^2, $2m$, and $4m^2$ is $20m^2$.

$$= \dfrac{20m^2\left(\dfrac{3}{5m} - \dfrac{2}{m^2}\right)}{20m^2\left(\dfrac{9}{2m} + \dfrac{3}{4m^2}\right)}$$ Multiply numerator and denominator by $20m^2$.

$$= \dfrac{20m^2\left(\dfrac{3}{5m}\right) - 20m^2\left(\dfrac{2}{m^2}\right)}{20m^2\left(\dfrac{9}{2m}\right) + 20m^2\left(\dfrac{3}{4m^2}\right)}$$ Distributive property

$$= \dfrac{12m - 40}{90m + 15}$$ Multiply and simplify.

Work Problem **5** at the Side. ▶

Some students use Method 1 for problems like Example 2, which is the quotient of two fractions, and Method 2 for problems like Examples 1, 3, 4, and 5, which have sums or differences in the numerators or denominators.

EXAMPLE 6 **Simplifying Complex Fractions**

Simplify each complex fraction. Use either method.

(a) $\dfrac{\dfrac{1}{y} + \dfrac{2}{y+2}}{\dfrac{4}{y} - \dfrac{3}{y+2}}$ We use Method 2, since there are sums and differences in the numerator and denominator.

$$= \dfrac{\left(\dfrac{1}{y} + \dfrac{2}{y+2}\right)y(y+2)}{\left(\dfrac{4}{y} - \dfrac{3}{y+2}\right)y(y+2)}$$ Multiply numerator and denominator by the LCD, $y(y+2)$.

$$= \dfrac{\left(\dfrac{1}{y}\right)y(y+2) + \left(\dfrac{2}{y+2}\right)y(y+2)}{\left(\dfrac{4}{y}\right)y(y+2) - \left(\dfrac{3}{y+2}\right)y(y+2)}$$ Distributive property

$$= \dfrac{1(y+2) + 2y}{4(y+2) - 3y}$$ Fundamental property

$$= \dfrac{y + 2 + 2y}{4y + 8 - 3y}$$ Distributive property

$$= \dfrac{3y + 2}{y + 8}$$ Combine like terms.

Continued on Next Page

---

5 Simplify using Method 2.

$$\dfrac{\dfrac{2}{5x} - \dfrac{3}{x^2}}{\dfrac{7}{4x} + \dfrac{1}{2x^2}}$$

ANSWER

5. $\dfrac{8x - 60}{35x + 10}$

6 Simplify each complex fraction. Use either method.

(a) $\dfrac{\dfrac{1}{x} + \dfrac{2}{x-1}}{\dfrac{2}{x} - \dfrac{4}{x-1}}$

(b) $\dfrac{1 - \dfrac{2}{x} - \dfrac{15}{x^2}}{1 + \dfrac{5}{x} + \dfrac{6}{x^2}}$

(c) $\dfrac{\dfrac{2x+3}{x-4}}{\dfrac{4x^2-9}{x^2-16}}$

(b) $\dfrac{1 - \dfrac{2}{x} - \dfrac{3}{x^2}}{1 - \dfrac{5}{x} + \dfrac{6}{x^2}}$ There are sums and differences in the numerator and denominator, so we use Method 2.

$= \dfrac{\left(1 - \dfrac{2}{x} - \dfrac{3}{x^2}\right)x^2}{\left(1 - \dfrac{5}{x} + \dfrac{6}{x^2}\right)x^2}$ Multiply numerator and denominator by the LCD, x^2.

$= \dfrac{x^2 - 2x - 3}{x^2 - 5x + 6}$ Distributive property

$= \dfrac{(x-3)(x+1)}{(x-3)(x-2)}$ Factor.

$= \dfrac{x+1}{x-2}$ Divide out the common factor.

You may wish to verify that in this example, like the others, *either* method can be used to simplify the complex fractions.

(c) $\dfrac{\dfrac{x+2}{x-3}}{\dfrac{x^2-4}{x^2-9}}$ This is a quotient of two rational expressions, so we use Method 1.

$= \dfrac{x+2}{x-3} \div \dfrac{x^2-4}{x^2-9}$ Write as a division problem.

$= \dfrac{x+2}{x-3} \cdot \dfrac{x^2-9}{x^2-4}$ Multiply by the reciprocal.

$= \dfrac{(x+2)(x+3)(x-3)}{(x-3)(x+2)(x-2)}$ Multiply; factor.

$= \dfrac{x+3}{x-2}$ Divide out the common factors.

◀ *Work Problem* **6** *at the Side.*

14.5 ▶▶▶ Exercises

Note: In many problems involving complex fractions, several different equivalent forms of the answer often exist. If your answer does not look exactly like the one given in the back of the book, check to see whether you have written an equivalent form.

1. Consider the complex fraction $\dfrac{\frac{1}{2} - \frac{1}{3}}{\frac{5}{6} - \frac{1}{12}}$. Answer each part, outlining Method 1 for simplifying this complex fraction.

   (a) To combine the terms in the numerator, we must find the LCD of $\frac{1}{2}$ and $\frac{1}{3}$. What is this LCD? Determine the simplified form of the numerator of the complex fraction.

   (b) To combine the terms in the denominator, we must find the LCD of $\frac{5}{6}$ and $\frac{1}{12}$. What is this LCD? Determine the simplified form of the denominator of the complex fraction.

   (c) Now use the results from parts (a) and (b) to write the complex fraction as a division problem using the symbol ÷.

   (d) Perform the operation from part (c) to obtain the final simplification.

2. Consider the same complex fraction given in Exercise 1, $\dfrac{\frac{1}{2} - \frac{1}{3}}{\frac{5}{6} - \frac{1}{12}}$. Answer each part, outlining Method 2 for simplifying this complex fraction.

   (a) We must determine the LCD of all the fractions within the complex fraction. What is this LCD?

   (b) Multiply every term in the complex fraction by the LCD found in part (a), but at this time do not combine the terms in the numerator and the denominator.

   (c) Now combine the terms from part (b) to obtain the simplified form of the complex fraction.

Simplify each complex fraction. Use either method. See Examples 1–6.

3. $\dfrac{-\dfrac{4}{3}}{\dfrac{2}{9}}$

4. $\dfrac{-\dfrac{5}{6}}{\dfrac{5}{4}}$

5. $\dfrac{\dfrac{p}{q^2}}{\dfrac{p^2}{q}}$

6. $\dfrac{\dfrac{a}{x}}{\dfrac{a^2}{2x}}$

7. $\dfrac{\dfrac{x}{y^2}}{\dfrac{x^2}{y}}$

8. $\dfrac{\dfrac{p^4}{r}}{\dfrac{p^2}{r^2}}$

9. $\dfrac{\dfrac{4a^4b^3}{3a}}{\dfrac{2ab^4}{b^2}}$

10. $\dfrac{\dfrac{2r^4t^2}{3t}}{\dfrac{5r^2t^5}{3r}}$

11. $\dfrac{\dfrac{m+2}{3}}{\dfrac{m-4}{m}}$

12. $\dfrac{\dfrac{q-5}{q}}{\dfrac{q+5}{3}}$

13. $\dfrac{\dfrac{2}{x}-3}{\dfrac{2-3x}{2}}$

14. $\dfrac{6+\dfrac{2}{r}}{\dfrac{3r+1}{4}}$

15. $\dfrac{\dfrac{1}{x}+x}{\dfrac{x^2+1}{8}}$

16. $\dfrac{\dfrac{3}{m}-m}{\dfrac{3-m^2}{4}}$

17. $\dfrac{a-\dfrac{5}{a}}{a+\dfrac{1}{a}}$

18. $\dfrac{q+\dfrac{1}{q}}{q+\dfrac{4}{q}}$

19. $\dfrac{\dfrac{1}{2}+\dfrac{1}{p}}{\dfrac{2}{3}+\dfrac{1}{p}}$

20. $\dfrac{\dfrac{3}{4}-\dfrac{1}{r}}{\dfrac{1}{5}+\dfrac{1}{r}}$

21. $\dfrac{\dfrac{2}{p^2}-\dfrac{3}{5p}}{\dfrac{4}{p}+\dfrac{1}{4p}}$

22. $\dfrac{\dfrac{2}{m^2}-\dfrac{3}{m}}{\dfrac{2}{5m^2}+\dfrac{1}{3m}}$

23. $\dfrac{\dfrac{t}{t+2}}{\dfrac{4}{t^2-4}}$

24. $\dfrac{\dfrac{m}{m+1}}{\dfrac{3}{m^2-1}}$

25. $\dfrac{\dfrac{1}{k+1}-1}{\dfrac{1}{k+1}+1}$

26. $\dfrac{\dfrac{2}{p-1}+2}{\dfrac{3}{p-1}-2}$

27. $\dfrac{2+\dfrac{1}{x}-\dfrac{28}{x^2}}{3+\dfrac{13}{x}+\dfrac{4}{x^2}}$

28. $\dfrac{4-\dfrac{11}{x}-\dfrac{3}{x^2}}{2-\dfrac{1}{x}-\dfrac{15}{x^2}}$

29. $\dfrac{\dfrac{1}{m-1}+\dfrac{2}{m+2}}{\dfrac{2}{m+2}-\dfrac{1}{m-3}}$

30. $\dfrac{\dfrac{5}{r+3}-\dfrac{1}{r-1}}{\dfrac{2}{r+2}+\dfrac{3}{r+3}}$

31. $2-\dfrac{2}{2+\dfrac{2}{2+2}}$

32. $3-\dfrac{2}{4+\dfrac{2}{4-2}}$

14.6 ▸▸▸ Solving Equations with Rational Expressions

OBJECTIVE 1 Distinguish between operations with rational expressions and equations with terms that are rational expressions.
Before solving equations with rational expressions, you must understand the difference between sums and differences of terms with rational coefficients, or rational *expressions*, and *equations* with terms that are rational expressions. *Sums and differences lead to expressions to simplify, while equations are solved.*

EXAMPLE 1 Distinguishing between Expressions and Equations

Identify each of the following as an *expression* or an *equation*. Then simplify the expression or solve the equation.

(a) $\dfrac{3}{4}x - \dfrac{2}{3}x$ This is a difference of two terms. It represents an *expression* to simplify since there is no equals sign.

$= \dfrac{3 \cdot 3}{3 \cdot 4}x - \dfrac{4 \cdot 2}{4 \cdot 3}x$ The LCD is 12. Write each coefficient with this LCD.

$= \dfrac{9}{12}x - \dfrac{8}{12}x$ Multiply.

$= \dfrac{1}{12}x$ Combine like terms, using the distributive property: $\dfrac{9}{12}x - \dfrac{8}{12}x = \left(\dfrac{9}{12} - \dfrac{8}{12}\right)x.$

(b) $\dfrac{3}{4}x - \dfrac{2}{3}x = \dfrac{1}{2}$ Because of the equals sign, this is an *equation* to be solved.

$12\left(\dfrac{3}{4}x - \dfrac{2}{3}x\right) = 12\left(\dfrac{1}{2}\right)$ Use the multiplication property of equality to clear fractions. Multiply by 12, the LCD.

$12\left(\dfrac{3}{4}x\right) - 12\left(\dfrac{2}{3}x\right) = 12\left(\dfrac{1}{2}\right)$ Distributive property

$9x - 8x = 6$ Multiply.

$x = 6$ Combine like terms.

Check $\dfrac{3}{4}x - \dfrac{2}{3}x = \dfrac{1}{2}$ Original equation

$\dfrac{3}{4}(6) - \dfrac{2}{3}(6) \overset{?}{=} \dfrac{1}{2}$ Let $x = 6$.

$\dfrac{9}{2} - 4 \overset{?}{=} \dfrac{1}{2}$ Multiply.

$\dfrac{1}{2} = \dfrac{1}{2}$ True

Since a true statement results, {6} is the solution set of the equation.

The ideas of Example 1 can be summarized as follows.

1 Identify each as an *expression* or an *equation*. Then simplify the expression or solve the equation.

(a) $\dfrac{2x}{3} - \dfrac{4x}{9} = 2$

> ### Uses of the LCD
> When adding or subtracting rational expressions, keep the LCD throughout the simplification. (See Example 1(a).)
>
> When solving an equation with terms that are rational expressions, multiply each side by the LCD so that denominators are eliminated. (See Example 1(b).)

◄ *Work Problem* **1** *at the Side.*

> **OBJECTIVE 2 Solve equations with rational expressions.**
> When an equation involves fractions as in Example 1(b), we use the multiplication property of equality to clear it of fractions. Choose as multiplier the LCD of all denominators in the fractions of the equation.

(b) $\dfrac{2x}{3} - \dfrac{4x}{9}$

EXAMPLE 2 Solving an Equation with Rational Expressions

Solve $\dfrac{x}{3} + \dfrac{x}{4} = 10 + x$. Check the solution.

$$12\left(\dfrac{x}{3} + \dfrac{x}{4}\right) = 12(10 + x) \qquad \text{Multiply by the LCD, 12, to clear fractions.}$$

$$12\left(\dfrac{x}{3}\right) + 12\left(\dfrac{x}{4}\right) = 12(10) + 12x \qquad \text{Distributive property}$$

$$4x + 3x = 120 + 12x \qquad \text{Multiply.}$$

$$7x = 120 + 12x \qquad \text{Combine like terms.}$$

$$-5x = 120 \qquad \text{Subtract } 12x.$$

$$x = -24 \qquad \text{Divide by } -5.$$

2 Solve each equation, and check your solutions.

(a) $\dfrac{x}{5} + 3 = \dfrac{3}{5}$

Check

$$\dfrac{x}{3} + \dfrac{x}{4} = 10 + x \qquad \text{Original equation}$$

$$\dfrac{-24}{3} + \dfrac{-24}{4} \stackrel{?}{=} 10 - 24 \qquad \text{Let } x = -24.$$

$$-8 - 6 \stackrel{?}{=} -14$$

$$-14 = -14 \qquad \text{True}$$

(b) $\dfrac{x}{2} - \dfrac{x}{3} = \dfrac{5}{6}$

The solution set is $\{-24\}$.

◄ *Work Problem* **2** *at the Side.*

> ### CAUTION
> The use of the LCD here is different than in **Section 14.5.** Here, we use the multiplication property of equality to multiply each side of an *equation* by the least common multiple of the denominators, which is the LCD. Earlier, we used the fundamental property to multiply a *fraction* by another fraction that had the LCD as both its numerator and denominator. Be careful not to confuse these two methods.

ANSWERS

1. **(a)** equation; $\{9\}$ **(b)** expression; $\dfrac{2x}{9}$

2. **(a)** $\{-12\}$ **(b)** $\{5\}$

EXAMPLE 3 Solving an Equation with Rational Expressions

Solve $\dfrac{p}{2} - \dfrac{p-1}{3} = 1$.

$$6\left(\dfrac{p}{2} - \dfrac{p-1}{3}\right) = 6 \cdot 1 \qquad \text{Multiply by the LCD, 6.}$$

$$6\left(\dfrac{p}{2}\right) - 6\left(\dfrac{p-1}{3}\right) = 6 \qquad \text{Distributive property}$$

$$3p - 2(p-1) = 6 \qquad \text{Use parentheses around } p - 1 \text{ to avoid errors.}$$

$$3p - 2(p) - 2(-1) = 6 \qquad \text{Distributive property}$$

Be careful with signs.

$$3p - 2p + 2 = 6 \qquad \text{Multiply.}$$

$$p + 2 = 6 \qquad \text{Combine like terms.}$$

$$p = 4 \qquad \text{Subtract 2.}$$

Check that $\{4\}$ is the solution set by replacing p with 4 in the original equation.

Work Problem ③ *at the Side.* ▶

Recall from **Section 14.1** that the denominator of a rational expression cannot equal 0, since division by 0 is undefined. *Therefore, when solving an equation with rational expressions that have variables in the denominator, the solution cannot be a number that makes the denominator equal 0.*

EXAMPLE 4 Solving an Equation with Rational Expressions

Solve $\dfrac{x}{x-2} = \dfrac{2}{x-2} + 2$. Check the proposed solution.

$$\dfrac{x}{x-2} = \dfrac{2}{x-2} + 2 \qquad \begin{array}{l} x \neq 2, \text{ since both denominators} \\ \text{equal 0 if } x \text{ is 2.} \end{array}$$

$$(x-2)\left(\dfrac{x}{x-2}\right) = (x-2)\left(\dfrac{2}{x-2} + 2\right) \qquad \begin{array}{l}\text{Multiply each side} \\ \text{by the LCD, } x - 2.\end{array}$$

$$(x-2)\left(\dfrac{x}{x-2}\right) = (x-2)\left(\dfrac{2}{x-2}\right) + (x-2)(2) \qquad \text{Distributive property}$$

$$x = 2 + 2x - 4 \qquad \text{Simplify.}$$

$$x = -2 + 2x \qquad \text{Combine like terms.}$$

$$-x = -2 \qquad \text{Subtract } 2x.$$

$$x = 2 \qquad \text{Divide by } -1.$$

Replacing x with 2, however, causes denominators to equal 0.

Check

$$\dfrac{x}{x-2} = \dfrac{2}{x-2} + 2 \qquad \text{Original equation}$$

$$\dfrac{2}{2-2} \overset{?}{=} \dfrac{2}{2-2} + 2 \qquad \text{Let } x = 2.$$

Division by 0 is undefined.

$$\dfrac{2}{0} \overset{?}{=} \dfrac{2}{0} + 2$$

Thus, the proposed solution 2 must be rejected, and the solution set is $\varnothing$.

Work Problem ④ *at the Side.* ▶

③ Solve each equation, and check your solutions.

(a) $\dfrac{k}{6} - \dfrac{k+1}{4} = -\dfrac{1}{2}$

(b) $\dfrac{2m-3}{5} - \dfrac{m}{3} = -\dfrac{6}{5}$

④ Solve the equation, and check the proposed solution.

$$1 - \dfrac{2}{x+1} = \dfrac{2x}{x+1}$$

ANSWERS

3. (a) $\{3\}$ (b) $\{-9\}$
4. $\varnothing$ (When the equation is solved, -1 is a proposed solution. However, because $x = -1$ leads to a 0 denominator in the original equation, there is no solution.)

5 Solve each equation, and check the proposed solutions.

(a) $\dfrac{4}{x^2 - 3x} = \dfrac{1}{x^2 - 9}$

While it is always a good idea to check solutions to guard against arithmetic and algebraic errors, *it is essential to check proposed solutions when variables appear in denominators in the original equation.* Some students like to determine which numbers cannot be solutions *before* solving the equation, as we did in Example 4.

The steps used to solve an equation with rational expressions follow.

Solving an Equation with Rational Expressions

Step 1 **Multiply each side of the equation by the LCD.** (This clears the equation of fractions.)

Step 2 **Solve** the resulting equation for proposed solutions.

Step 3 **Check** each proposed solution by substituting it in the original equation. Reject any that cause a denominator to equal 0.

EXAMPLE 5 **Solving an Equation with Rational Expressions**

Solve $\dfrac{2}{x^2 - x} = \dfrac{1}{x^2 - 1}$. Check the proposed solution.

Step 1 Factor the denominators to find the LCD.

$$\frac{2}{x(x - 1)} = \frac{1}{(x + 1)(x - 1)} \qquad \text{The LCD is } x(x + 1)(x - 1).$$

Notice that 0, -1, and 1 cannot be solutions of this equation. Multiply each side of the equation by $x(x + 1)(x - 1)$.

$$x(x + 1)(x - 1)\frac{2}{x(x - 1)} = x(x + 1)(x - 1)\frac{1}{(x + 1)(x - 1)}$$

Multiply by the LCD.

(b) $\dfrac{2}{p^2 - 2p} = \dfrac{3}{p^2 - p}$

Step 2

$$2(x + 1) = x \qquad \text{Divide out the common factors.}$$
$$2x + 2 = x \qquad \text{Distributive property}$$
$$x + 2 = 0 \qquad \text{Subtract } x.$$
$$x = -2 \qquad \text{Subtract 2.}$$

Step 3 The proposed solution is -2, which does not make any denominator equal 0.

Check

$$\frac{2}{x^2 - x} = \frac{1}{x^2 - 1} \qquad \text{Original equation}$$

$$\frac{2}{(-2)^2 - (-2)} \stackrel{?}{=} \frac{1}{(-2)^2 - 1} \qquad \text{Let } x = -2.$$

$$\frac{2}{4 + 2} \stackrel{?}{=} \frac{1}{4 - 1} \qquad \text{Apply the exponents.}$$

$$\frac{1}{3} = \frac{1}{3} \qquad \text{True}$$

The solution set is $\{-2\}$.

◀ *Work Problem* **5** *at the Side.*

ANSWERS

5. (a) $\{-4\}$ (b) $\{4\}$

EXAMPLE 6 **Solving an Equation with Rational Expressions**

Solve $\dfrac{2m}{m^2 - 4} + \dfrac{1}{m - 2} = \dfrac{2}{m + 2}$.

Factor the first denominator on the left.

$$\dfrac{2m}{(m + 2)(m - 2)} + \dfrac{1}{m - 2} = \dfrac{2}{m + 2} \qquad \text{The LCD is } (m + 2)(m - 2).$$

Notice that -2 and 2 cannot be solutions of the equation.

$$(m + 2)(m - 2)\left(\dfrac{2m}{(m + 2)(m - 2)} + \dfrac{1}{m - 2}\right)$$

$$= (m + 2)(m - 2)\dfrac{2}{m + 2} \qquad \text{Multiply by the LCD.}$$

$$(m + 2)(m - 2)\dfrac{2m}{(m + 2)(m - 2)} + (m + 2)(m - 2)\dfrac{1}{m - 2}$$

$$= (m + 2)(m - 2)\dfrac{2}{m + 2} \qquad \text{Distributive property}$$

$$2m + m + 2 = 2(m - 2) \qquad \text{Divide out the common factors.}$$

$$3m + 2 = 2m - 4 \qquad \text{Combine like terms; distributive property}$$

$$m + 2 = -4 \qquad \text{Subtract } 2m.$$

$$m = -6 \qquad \text{Subtract } 2.$$

Check to see that $\{-6\}$ is the solution set of the given equation.

Work Problem **6** *at the Side.* ▶

EXAMPLE 7 **Solving an Equation with Rational Expressions**

Solve $\dfrac{1}{x - 1} + \dfrac{1}{2} = \dfrac{2}{x^2 - 1}$.

The denominator $x^2 - 1$ factors as $(x + 1)(x - 1)$. Multiply each side by the LCD, $2(x + 1)(x - 1)$. Notice that -1 and 1 cannot be solutions.

$$2(x + 1)(x - 1)\left(\dfrac{1}{x - 1} + \dfrac{1}{2}\right) = 2(x + 1)(x - 1)\dfrac{2}{(x + 1)(x - 1)}$$

$$2(x + 1)(x - 1)\dfrac{1}{x - 1} + 2(x + 1)(x - 1)\dfrac{1}{2}$$

$$= 2(x + 1)(x - 1)\dfrac{2}{(x + 1)(x - 1)}$$

$$2(x + 1) + (x + 1)(x - 1) = 2(2) \qquad \text{Divide out the common factors.}$$

$$2x + 2 + x^2 - 1 = 4 \qquad \text{Distributive property; multiply.}$$

$$x^2 + 2x + 1 = 4 \qquad \text{Combine like terms.}$$

> Write in standard form.

$$x^2 + 2x - 3 = 0 \qquad \text{Subtract } 4.$$

$$(x + 3)(x - 1) = 0 \qquad \text{Factor.}$$

$$x + 3 = 0 \quad \text{or} \quad x - 1 = 0 \qquad \text{Zero-factor property}$$

$$x = -3 \quad \text{or} \qquad x = 1 \qquad \text{Solve for } x.$$

Continued on Next Page

6 Solve each equation, and check the proposed solutions.

(a)

$$\dfrac{2p}{p^2 - 1} = \dfrac{2}{p + 1} - \dfrac{1}{p - 1}$$

(b)

$$\dfrac{8r}{4r^2 - 1} = \dfrac{3}{2r + 1} + \dfrac{3}{2r - 1}$$

ANSWERS

6. **(a)** $\{-3\}$ **(b)** $\{0\}$

7 Solve the equation, and check the proposed solutions.

$$\frac{2}{3x + 1} - \frac{1}{x} = \frac{-6x}{3x + 1}$$

8 Solve each equation, and check the proposed solutions.

(a) $\dfrac{1}{x - 2} + \dfrac{1}{5} = \dfrac{2}{5(x^2 - 4)}$

(b) $\dfrac{6}{5a + 10} - \dfrac{1}{a - 5}$

$= \dfrac{4}{a^2 - 3a - 10}$

7. $\left\{\dfrac{1}{2}\right\}$

8. **(a)** $\{-4, -1\}$ **(b)** $\{60\}$

Proposed solutions are -3 and 1. However, 1 makes an original denominator equal 0, so 1 is *not* a solution. Check that -3 is a solution.

Check

$$\frac{1}{x - 1} + \frac{1}{2} = \frac{2}{x^2 - 1} \qquad \text{Original equation}$$

$$\frac{1}{-3 - 1} + \frac{1}{2} \stackrel{?}{=} \frac{2}{(-3)^2 - 1} \qquad \text{Let } x = -3.$$

$$\frac{1}{-4} + \frac{1}{2} \stackrel{?}{=} \frac{2}{9 - 1} \qquad \text{Simplify.}$$

$$\frac{1}{4} \stackrel{?}{=} \frac{2}{8}$$

$$\frac{1}{4} = \frac{1}{4} \qquad \text{True}$$

The check shows that $\{-3\}$ is the solution set.

◀ *Work Problem* **7** *at the Side.*

EXAMPLE 8 **Solving an Equation with Rational Expressions**

Solve $\dfrac{1}{k^2 + 4k + 3} + \dfrac{1}{2k + 2} = \dfrac{3}{4k + 12}.$

$$\frac{1}{k^2 + 4k + 3} + \frac{1}{2k + 2} = \frac{3}{4k + 12}$$

$$\frac{1}{(k + 1)(k + 3)} + \frac{1}{2(k + 1)} = \frac{3}{4(k + 3)} \qquad \begin{array}{l}\text{Factor each denominator.}\\ \text{The LCD is } 4(k+1)(k+3).\end{array}$$

Notice that -1 and -3 cannot be solutions of this equation, since replacing k with -1 or -3 causes denominators to equal 0.

$$4(k + 1)(k + 3)\left(\frac{1}{(k + 1)(k + 3)} + \frac{1}{2(k + 1)}\right)$$

$$= 4(k + 1)(k + 3)\frac{3}{4(k + 3)} \qquad \text{Multiply by the LCD.}$$

$$4(k + 1)(k + 3)\frac{1}{(k + 1)(k + 3)} + 2 \cdot 2(k + 1)(k + 3)\frac{1}{2(k + 1)}$$

$$= 4(k + 1)(k + 3)\frac{3}{4(k + 3)} \qquad \text{Distributive property}$$

$$4 + 2(k + 3) = 3(k + 1) \qquad \text{Simplify.}$$

$$4 + 2k + 6 = 3k + 3 \qquad \text{Distributive property}$$

$$2k + 10 = 3k + 3 \qquad \text{Combine like terms.}$$

$$10 = k + 3 \qquad \text{Subtract } 2k.$$

$$7 = k \qquad \text{Subtract } 3.$$

The proposed solution, 7, does not make an original denominator equal 0. A check shows that the algebra is correct, so $\{7\}$ is the solution set.

◀ *Work Problem* **8** *at the Side.*

OBJECTIVE **3** **Solve a formula for a specified variable.** Solving a formula for a specified variable was first discussed in **Section 10.5.** *Remember to treat the variable for which you are solving as if it were the only variable, and all others as if they were constants.*

EXAMPLE 9 **Solving for Specified Variables**

Solve each formula for the specified variable.

(a) $a = \dfrac{v - w}{t}$ for v

> Our goal is to isolate v.

$$a = \frac{v - w}{t}$$

$$at = v - w \qquad \text{Multiply by } t.$$

$$at + w = v, \quad \text{or} \quad v = at + w \qquad \text{Add } w.$$

Check Substitute $at + w$ for v in the original equation. The final result will be the identity $a = a$, indicating that the result obtained is correct.

(b) $F = \dfrac{k}{d - D}$ for d

$$F = \frac{k}{d - D}$$

> Our goal is to isolate d.

$$F(d - D) = \frac{k}{d - D}(d - D) \qquad \begin{array}{l}\text{Multiply by } d - D \\ \text{to clear the fraction.}\end{array}$$

$$F(d - D) = k \qquad \text{Simplify.}$$

$$Fd - FD = k \qquad \text{Distributive property}$$

$$Fd = k + FD \qquad \text{Add } FD.$$

$$d = \frac{k + FD}{F} \qquad \text{Divide by } F.$$

If desired, we can write an equivalent form of this answer by expressing the right side as the sum of two fractions.

$$d = \frac{k + FD}{F} \qquad \text{Answer from above}$$

$$d = \frac{k}{F} + \frac{FD}{F} \qquad \begin{array}{l}\text{Definition of addition of} \\ \text{fractions: } \frac{a + b}{c} = \frac{a}{c} + \frac{b}{c}\end{array}$$

$$d = \frac{k}{F} + D \qquad \begin{array}{l}\text{Divide out the common} \\ \text{factor from } \frac{FD}{F}.\end{array}$$

Either answer is correct:

$$d = \frac{k + FD}{F}, \quad \text{or} \quad d = \frac{k}{F} + D.$$

Work Problem **9** *at the Side.* ▶

---

9 Solve each formula for the specified variable.

(a) $r = \dfrac{A - p}{pt}$ for A

(b) $z = \dfrac{x}{x + y}$ for y

---

10 Solve $\dfrac{2}{x} = \dfrac{1}{y} + \dfrac{1}{z}$ for z.

EXAMPLE 10 Solving for a Specified Variable

Solve the formula $\dfrac{1}{a} = \dfrac{1}{b} + \dfrac{1}{c}$ for c.

$$\frac{1}{a} = \frac{1}{b} + \frac{1}{c}$$

Goal: Isolate c, the specified variable.

$$abc\left(\frac{1}{a}\right) = abc\left(\frac{1}{b} + \frac{1}{c}\right) \qquad \text{Multiply by the LCD, } abc.$$

$$abc\left(\frac{1}{a}\right) = abc\left(\frac{1}{b}\right) + abc\left(\frac{1}{c}\right) \qquad \text{Distributive property}$$

$$bc = ac + ab \qquad \text{Simplify.}$$

$$bc - ac = ab \qquad \begin{array}{l}\text{Subtract } ac \text{ to get both terms} \\ \text{with } c \text{ on the same side.}\end{array}$$

Pay careful attention here.

$$c(b - a) = ab \qquad \text{Factor out } c.$$

$$c = \frac{ab}{b - a} \qquad \text{Divide by } b - a.$$

CAUTION

Students often have trouble in the step that involves factoring out the variable for which they are solving. In Example 10, we needed to get both terms with c on the same side of the equation. This allowed us to factor out c on the left, and then isolate it by dividing each side by $b - a$.

$$bc = ac + ab \qquad \text{From Example 10}$$

$$bc - ac = ab \qquad \text{Get } c\text{-terms on the same side.}$$

$$c(b - a) = ab \qquad \text{Factor out } c.$$

$$c = \frac{ab}{b - a} \qquad \text{Divide by } b - a.$$

When solving an equation for a specified variable, be sure that the specified variable appears alone on only one side of the equals sign in the final equation.

◀ *Work Problem* **10** *at the Side.*

14.6 ▶▶▶ **Exercises**

Identify each as an expression *or an* equation. *Then simplify the expression or solve the equation. See Example 1.*

1. $\dfrac{7}{8}x + \dfrac{1}{5}x$

2. $\dfrac{4}{7}x + \dfrac{3}{5}x$

3. $\dfrac{7}{8}x + \dfrac{1}{5}x = 1$

4. $\dfrac{4}{7}x + \dfrac{3}{5}x = 1$

5. $\dfrac{3}{5}y - \dfrac{7}{10}y$

6. $\dfrac{3}{5}y - \dfrac{7}{10}y = 1$

7. Explain how the LCD is used in a different way when adding and subtracting rational expressions compared to solving equations with rational expressions.

8. If we multiply each side of the equation $\dfrac{6}{x+5} = \dfrac{6}{x+5}$ by $x + 5$, we get $6 = 6$. Are all real numbers solutions of this equation? Explain.

Solve each equation, and check your solutions. See Examples 2 and 3.

9. $\dfrac{2}{3}x + \dfrac{1}{2}x = -7$

10. $\dfrac{1}{4}x - \dfrac{1}{3}x = 1$

11. $\dfrac{3x}{5} - 6 = x$

12. $\dfrac{5t}{4} + t = 9$

13. $\dfrac{4m}{7} + m = 11$

14. $a - \dfrac{3a}{2} = 1$

15. $\dfrac{z-1}{4} = \dfrac{z+3}{3}$

16. $\dfrac{r-5}{2} = \dfrac{r+2}{3}$

17. $\dfrac{3p+6}{8} = \dfrac{3p-3}{16}$

18. $\dfrac{2z+1}{5} = \dfrac{7z+5}{15}$

19. $\dfrac{2x+3}{-6} = \dfrac{3}{2}$

20. $\dfrac{4x+3}{6} = \dfrac{5}{2}$

21. $\dfrac{q+2}{3} + \dfrac{q-5}{5} = \dfrac{7}{3}$

22. $\dfrac{b+7}{8} - \dfrac{b-2}{3} = \dfrac{4}{3}$

23. $\dfrac{t}{6} + \dfrac{4}{3} = \dfrac{t-2}{3}$

24. $\dfrac{x}{2} = \dfrac{5}{4} + \dfrac{x-1}{4}$

25. $\dfrac{3m}{5} - \dfrac{3m-2}{4} = \dfrac{1}{5}$

26. $\dfrac{8p}{5} = \dfrac{3p-4}{2} + \dfrac{5}{2}$

27. What values of x would have to be rejected as possible solutions of the equation $\dfrac{1}{x-4} = \dfrac{3}{2x}$?

28. What is wrong with the following problem? "Solve $\dfrac{2}{3x} + \dfrac{1}{5x}$."

Solve each equation, and check your solutions. See Examples 4–8.

29. $\dfrac{5-2x}{x} = \dfrac{1}{4}$

30. $\dfrac{2x+3}{x} = \dfrac{3}{2}$

31. $\dfrac{k}{k-4} - 5 = \dfrac{4}{k-4}$

32. $\dfrac{-5}{a+5} = \dfrac{a}{a+5} + 2$

33. $\dfrac{3}{x-1} + \dfrac{2}{4x-4} = \dfrac{7}{4}$

34. $\dfrac{2}{p+3} + \dfrac{3}{8} = \dfrac{5}{4p+12}$

35. $\dfrac{x}{3x+3} = \dfrac{2x-3}{x+1} - \dfrac{2x}{3x+3}$

36. $\dfrac{2k+3}{k+1} - \dfrac{3k}{2k+2} = \dfrac{-2k}{2k+2}$

37. $\dfrac{2}{m} = \dfrac{m}{5m+12}$

38. $\dfrac{x}{4-x} = \dfrac{2}{x}$

39. $\dfrac{5x}{14x+3} = \dfrac{1}{x}$

40. $\dfrac{m}{8m+3} = \dfrac{1}{3m}$

41. $\dfrac{2}{z-1} - \dfrac{5}{4} = \dfrac{-1}{z+1}$

42. $\dfrac{5}{p-2} = 7 - \dfrac{10}{p+2}$

43. $\dfrac{4}{x^2-3x} = \dfrac{1}{x^2-9}$

44. $\dfrac{2}{t^2-4} = \dfrac{3}{t^2-2t}$

45. $\dfrac{-2}{z+5} + \dfrac{3}{z-5} = \dfrac{20}{z^2-25}$

46. $\dfrac{3}{r+3} - \dfrac{2}{r-3} = \dfrac{-12}{r^2-9}$

47. $\dfrac{1}{x+4} + \dfrac{x}{x-4} = \dfrac{-8}{x^2-16}$

48. $\dfrac{x}{x-3} + \dfrac{4}{x+3} = \dfrac{18}{x^2-9}$

49. $\dfrac{4}{3x+6} - \dfrac{3}{x+3} = \dfrac{8}{x^2+5x+6}$

50. $\dfrac{-13}{t^2+6t+8} + \dfrac{4}{t+2} = \dfrac{3}{2t+8}$

51. $\dfrac{3x}{x^2+5x+6} = \dfrac{5x}{x^2+2x-3} - \dfrac{2}{x^2+x-2}$

52. $\dfrac{x+4}{x^2-3x+2} - \dfrac{5}{x^2-4x+3} = \dfrac{x-4}{x^2-5x+6}$

53. If you are solving a formula for the letter k, and your steps lead to the equation $kr - mr = km$, what would be your next step?

54. If you are solving a formula for the letter k, and your steps lead to the equation $kr - km = mr$, what would be your next step?

Solve each formula for the specified variable. See Example 9.

55. $m = \dfrac{kF}{a}$ for F

56. $I = \dfrac{kE}{R}$ for E

57. $m = \dfrac{kF}{a}$ for a

58. $I = \dfrac{kE}{R}$ for R

59. $m = \dfrac{y - b}{x}$ for y

60. $y = \dfrac{C - Ax}{B}$ for C

61. $I = \dfrac{E}{R + r}$ for R

62. $I = \dfrac{E}{R + r}$ for r

63. $h = \dfrac{2A}{B + b}$ for b

64. $h = \dfrac{2A}{B + b}$ for B

65. $d = \dfrac{2S}{n(a + L)}$ for a

66. $d = \dfrac{2S}{n(a + L)}$ for L

Solve each equation for the specified variable. See Example 10.

67. $\dfrac{2}{r} + \dfrac{3}{s} + \dfrac{1}{t} = 1$ for t

68. $\dfrac{5}{p} + \dfrac{2}{q} + \dfrac{3}{r} = 1$ for r

69. $\dfrac{1}{a} - \dfrac{1}{b} - \dfrac{1}{c} = 2$ for c

70. $\dfrac{-1}{x} + \dfrac{1}{y} + \dfrac{1}{z} = 4$ for y

71. $9x + \dfrac{3}{z} = \dfrac{5}{y}$ for z

72. $-3t - \dfrac{4}{p} = \dfrac{6}{s}$ for p

14.7 ▶▶▶ Applications of Rational Expressions

In **Section 14.6** we solved equations with rational expressions; now we can solve applications that involve this type of equation. The six-step problem-solving method of **Section 10.4** still applies.

OBJECTIVE 1 Solve problems about numbers.

EXAMPLE 1 Solving a Problem about an Unknown Number

If the same number is added to both the numerator and the denominator of the fraction $\frac{2}{5}$, the result is equivalent to $\frac{2}{3}$. Find the number.

Step 1 **Read** the problem carefully. We are trying to find a number.

Step 2 **Assign a variable.**

Let x = the number added to the numerator and the denominator.

Step 3 **Write an equation.** The fraction

$$\frac{2 + x}{5 + x}$$

represents the result of adding the same number x to both the numerator and the denominator. Since this result is equivalent to $\frac{2}{3}$, the equation is

$$\frac{2 + x}{5 + x} = \frac{2}{3}.$$

Step 4 **Solve** this equation.

$$\mathbf{3\,(5 + x)}\frac{2 + x}{5 + x} = \mathbf{3\,(5 + x)}\frac{2}{3} \qquad \text{Multiply by the LCD, } 3\,(5 + x).$$

$$3\,(2 + x) = 2\,(5 + x) \qquad \text{Divide out the common factors.}$$

$$6 + 3x = 10 + 2x \qquad \text{Distributive property}$$

$$x = 4 \qquad \text{Subtract } 2x; \text{ subtract } 6.$$

Step 5 **State the answer.** The number is 4.

Step 6 **Check** the solution in the words of the original problem. If 4 is added to both the numerator and the denominator of $\frac{2}{5}$, the result is $\frac{6}{9}$, or $\frac{2}{3}$, as required.

Work Problem ① *at the Side.* ▶

OBJECTIVE 2 Solve problems about distance, rate, and time.
If an automobile travels at an average rate of 65 mph for 2 hr, then it travels $65 \times 2 = 130$ mi. This is an example of the basic relationship between distance, rate, and time given by the formula $d = rt$. By solving, in turn, for r and t in the formula, we obtain two other equivalent forms of the formula. The three forms are given below.

Distance, Rate, and Time Relationship

$$d = rt \qquad r = \frac{d}{t} \qquad t = \frac{d}{r}$$

OBJECTIVES

① Solve problems about numbers.

② Solve problems about distance, rate, and time.

③ Solve problems about work.

---

① Solve each problem.

(a) A certain number is added to the numerator and subtracted from the denominator of $\frac{5}{8}$. The new fraction equals the reciprocal of $\frac{5}{8}$. Find the number.

(b) The denominator of a fraction is 1 more than the numerator. If 6 is added to the numerator and subtracted from the denominator, the result is $\frac{15}{4}$. Find the original fraction.

ANSWERS

1. (a) 3 **(b)** $\frac{9}{10}$

The next example illustrates the uses of these formulas.

2 Solve each problem.

(a) The world record in the men's 100-m dash was set by Justin Gatlin of the United States in 2006. He ran it in 9.77 sec. What was his speed in meters per second, to the nearest hundredth? (*Source: Guinness World Records.*)

(b) The world record for the women's 3000-m run was set by Julnara Samitova of Russia in 2005. Her speed was 5.539 m per sec. To the nearest second, what was her time? (*Source: Guinness World Records.*)

(c) A small plane flew from Chicago to St. Louis averaging 145 mph. The trip took 2 hr. What is the distance between Chicago and St. Louis?

EXAMPLE 2 Finding Distance, Rate, or Time

(a) The speed of sound is 1088 ft per sec at sea level at 32°F. In 5 sec, under these conditions, sound travels

$$1088 \; \times \; 5 \; = 5440 \text{ ft.}$$
$$\uparrow \qquad \uparrow \qquad \uparrow$$
$$\text{Rate} \; \times \; \text{Time} \; = \; \text{Distance}$$

Here, we found distance, given rate and time, using $d = rt$.

(b) The winner of the first Indianapolis 500 race (in 1911) was Ray Harroun, driving a Marmon Wasp at an average speed of 74.59 mph. (*Source: Universal Almanac.*)
To complete the 500 mi, it took him

$$\begin{array}{c}\text{Distance} \rightarrow \\ \text{Rate} \quad \rightarrow \end{array} \frac{500}{74.59} \approx 6.70 \text{ hr (rounded).} \; \leftarrow \text{Time}$$

Here, we found time, given distance and rate, using $t = \frac{d}{r}$. To convert the decimal 0.70 hr to minutes, multiply by 60 to get $0.70(60) = 42$. It took Harroun about 6 hr, 42 min, to complete the race.

(c) At the 2004 Olympic Games in Athens, Greece, Chinese swimmer Luo Xuejuan set an Olympic record of 66.64 sec in the women's 100-m breast-stroke swimming event. (*Source: World Almanac and Book of Facts.*) Her rate was

$$\text{Rate} \; = \; \begin{array}{c}\text{Distance} \rightarrow \\ \text{Time} \quad \rightarrow\end{array} \frac{100}{66.64} \approx 1.50 \text{ m per sec (rounded).}$$

Here, we found rate, given distance and time, using $r = \frac{d}{t}$.

◀ *Work Problem* **2** *at the Side.*

> **Problem-Solving Hint**
> Many applied problems use the formulas just discussed. The next two ex-amples show how to solve typical applications of the formula $d = rt$.
> A helpful strategy for solving such problems is to **first make a sketch** showing what is happening in the problem. **Then make a table** using the information given, along with the unknown quantities. The table will help organize the information, and the sketch will help set up the equation.

EXAMPLE 3 Solving a Problem about Distance, Rate, and Time

Two cars leave Baton Rouge, Louisiana, at the same time and travel east on Interstate 10. One travels at a constant speed of 55 mph and the other travels at a constant speed of 63 mph. In how many hours will the distance between them be 24 mi?

Step 1 **Read** the problem. We are trying to find the time when the dis-tance between the cars will be 24 mi.

Continued on Next Page

Step 2 **Assign a variable.** Since we are looking for time,

> let t = the number of hours until the distance between them is 24 mi.

The sketch in Figure 1 shows what is happening in the problem.

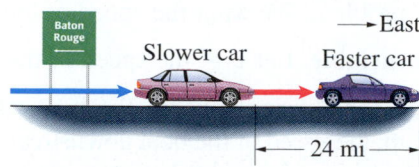

Figure 1

Now, construct a table like the one that follows. Fill in the information given in the problem, and use t for the time traveled by each car. Multiply rate by time to get the expressions for distances traveled.

	Rate	× Time	= Distance
Faster Car	63	t	**$63t$**
Slower Car	55	t	**$55t$**

Difference is 24 mi.

The quantities $63t$ and $55t$ represent the two distances. Refer to Figure 1, and notice that the *difference* between the larger distance and the smaller distance is 24 mi.

Step 3 **Write an equation.** Subtract the smaller distance from the larger distance.

$$63t - 55t = 24$$

Step 4 **Solve** the equation.

$$63t - 55t = 24$$
$$8t = 24 \qquad \text{Combine like terms.}$$
$$t = 3 \qquad \text{Divide by 8.}$$

Step 5 **State the answer.** It will take the cars 3 hr to be 24 mi apart.

Step 6 **Check.** After 3 hr the faster car will have traveled

$$63 \times 3 = 189 \text{ mi,}$$

and the slower car will have traveled

$$55 \times 3 = 165 \text{ mi.}$$

Since $189 - 165 = 24$, the conditions of the problem are satisfied.

Problem-Solving Hint

In motion problems like the one in Example 3, once we have filled in two pieces of information in each row of the table, we can automatically fill in the third piece of information, using the appropriate form of the formula relating distance, rate, and time. Then we set up the equation based on our sketch and the information in the table.

Work Problem **3** *at the Side.* ▶

---

3 Solve each problem.

(a) From a point on a straight road, Lupe and Maria ride bicycles in opposite directions. Lupe rides 10 mph and Maria rides 12 mph. In how many hours will they be 55 mi apart?

(b) At a given hour, two steamboats leave a city in the same direction on a straight canal. One travels at 18 mph, and the other travels at 25 mph. In how many hours will the boats be 35 mi apart?

ANSWERS

3. (a) $2\frac{1}{2}$ hr   (b) 5 hr

EXAMPLE 4 **Solving a Problem about Distance, Rate, and Time**

The Tickfaw River has a current of 3 mph. A motorboat takes as long to go 12 mi downstream as to go 8 mi upstream. What is the speed of the boat in still water?

Step 1 **Read** the problem. We want the speed of the boat in still water.

Step 2 **Assign a variable.** Let x = the speed of the boat in still water.

Because the current pushes the boat when the boat is going downstream, the speed of the boat downstream will be the *sum* of the speed of the boat and the speed of the current, $(x + 3)$ mph. Because the current slows down the boat when the boat is going upstream, the boat's speed upstream is given by the *difference* between the speed of the boat in still water and the speed of the current, $(x - 3)$ mph. See Figure 2.

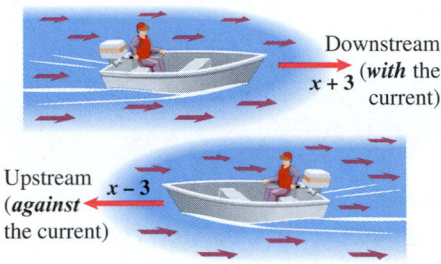

Downstream
$x + 3$ (*with* the current)

Upstream $x - 3$
(*against* the current)

Figure 2

This information is summarized in the following table.

	d	r	t
Downstream	12	$x + 3$	
Upstream	8	$x - 3$	

Fill in the column representing time by using the formula $t = \frac{d}{r}$. Then the time downstream is the distance divided by the rate, or

$$t = \frac{d}{r} = \frac{12}{x + 3}, \qquad \text{Time downstream}$$

and the time upstream is also the distance divided by the rate, or

$$t = \frac{d}{r} = \frac{8}{x - 3}. \qquad \text{Time upstream}$$

The completed table follows.

	d	r	t
Downstream	12	$x + 3$	$\dfrac{12}{x + 3}$
Upstream	8	$x - 3$	$\dfrac{8}{x - 3}$

Times are equal.

Step 3 **Write an equation.** According to the original problem, the time downstream equals the time upstream. The two times from the table must therefore be equal, giving the equation

$$\frac{12}{x + 3} = \frac{8}{x - 3}.$$

Continued on Next Page

Step 4 **Solve.** Begin by multiplying each side by the LCD, $(x + 3)(x - 3)$.

$$\frac{12}{x + 3} = \frac{8}{x - 3}$$

$$(x + 3)(x - 3)\frac{12}{x + 3} = (x + 3)(x - 3)\frac{8}{x - 3}$$

$$12(x - 3) = 8(x + 3) \qquad \text{Divide out the common factors.}$$

$$12x - 36 = 8x + 24 \qquad \text{Distributive property}$$

$$4x = 60 \qquad \text{Subtract } 8x; \text{ add } 36.$$

$$x = 15 \qquad \text{Divide by 4.}$$

Step 5 **State the answer.** The speed of the boat in still water is 15 mph.

Step 6 **Check.** First find the speed of the boat downstream, which is $15 + 3 = 18$ mph. Traveling 12 mi would take

$$t = \frac{d}{r} = \frac{12}{18} = \frac{2}{3} \textbf{ hr.}$$

The speed of the boat upstream is $15 - 3 = 12$ mph, and traveling 8 mi would take

$$t = \frac{d}{r} = \frac{8}{12} = \frac{2}{3} \textbf{ hr.}$$

The time upstream equals the time downstream, as required.

Work Problem **4** *at the Side.* ▶

OBJECTIVE **3** **Solve problems about work.**   Suppose that you can mow your lawn in 4 hr. Then after 1 hr, you will have mowed $\frac{1}{4}$ of the lawn. After 2 hr, you will have mowed $\frac{2}{4}$, or $\frac{1}{2}$, of the lawn, and so on. This idea is generalized as follows.

> **Rate of Work**
> If a job can be completed in t units of time, then the rate of work is
> $$\frac{1}{t} \textbf{ job per unit of time.}$$

> **Problem-Solving Hint**
> The relationship between problems involving work and problems involving distance is a very close one. Recall that the formula $d = rt$ says that distance traveled is equal to rate of travel multiplied by time traveled. Similarly, the fractional part of a job accomplished is equal to the rate of work multiplied by the time worked. In the lawn mowing example, after 3 hr, the fractional part of the job done is
> $$\underbrace{\frac{1}{4}}_{\substack{\text{Rate of} \\ \text{work}}} \cdot \underbrace{3}_{\substack{\text{Time} \\ \text{worked}}} = \underbrace{\frac{3}{4}}_{\substack{\text{Fractional part} \\ \text{of job done}}}.$$
> After 4 hr, $\frac{1}{4}(4) = 1$ whole job has been done.

4 Solve each problem.

(a) A boat can go 10 mi against the current in the same time it can go 30 mi with the current. The current is flowing at 4 mph. Find the speed of the boat with no current.

(b) An airplane, maintaining a constant airspeed, takes as long to go 450 mi with the wind as it does to go 375 mi against the wind. If the wind is blowing at 15 mph, what is the speed of the plane?

> **EXAMPLE 5** Solving a Problem about Work Rates

With spraying equipment, Mateo can paint the woodwork in a small house in 8 hr. His assistant, Chet, needs 14 hr to complete the same job painting by hand. If both Mateo and Chet work together, how long will it take them to paint the woodwork?

Step 1 **Read** the problem again. We are looking for time working together.

Step 2 **Assign a variable.** Let $x =$ the number of hours it will take for Mateo and Chet to paint the woodwork, working together.

Begin by making a table. Based on the previous discussion, Mateo's rate alone is $\frac{1}{8}$ job per hour, and Chet's rate is $\frac{1}{14}$ job per hour.

	Rate	Time Working Together	Fractional Part of the Job Done When Working Together	
Mateo	$\frac{1}{8}$	x	$\frac{1}{8}x$	⟵
Chet	$\frac{1}{14}$	x	$\frac{1}{14}x$	⟵

Sum is 1 whole job.

Step 3 **Write an equation.** Together Mateo and Chet complete 1 whole job. We must add the fractional parts and set the sum equal to 1.

$$\underbrace{\text{Fractional part}}_{\text{done by Mateo}} + \underbrace{\text{Fractional part}}_{\text{done by Chet}} = \underbrace{1 \text{ whole job}}$$

$$\frac{1}{8}x \quad + \quad \frac{1}{14}x \quad = \quad 1$$

Step 4 **Solve.**

$$56\left(\frac{1}{8}x + \frac{1}{14}x\right) = 56\,(1) \qquad \text{Multiply by the LCD, 56.}$$

$$56\left(\frac{1}{8}x\right) + 56\left(\frac{1}{14}x\right) = 56\,(1) \qquad \text{Distributive property}$$

$$7x + 4x = 56$$

$$11x = 56 \qquad \text{Combine like terms.}$$

$$x = \frac{56}{11} \qquad \text{Divide by 11.}$$

Step 5 **State the answer.** Working together, Mateo and Chet can paint the woodwork in $\frac{56}{11}$ hr, or $5\frac{1}{11}$ hr.

Step 6 **Check.** Substitute $\frac{56}{11}$ for x in the equation from Step 3.

$$\frac{1}{8}x + \frac{1}{14}x = 1 \qquad \text{Equation from Step 3}$$

$$\frac{1}{8}\left(\frac{56}{11}\right) + \frac{1}{14}\left(\frac{56}{11}\right) \stackrel{?}{=} 1 \qquad \text{Let } x = \frac{56}{11}.$$

$$\frac{7}{11} + \frac{4}{11} = 1 \qquad \text{True}$$

Our answer, $\frac{56}{11}$ hr, or $5\frac{1}{11}$ hr, seems correct. See the Problem-Solving Hint on the next page for additional strategies for checking.

Problem-Solving Hint

A common error students make when solving a work problem like that in Example 5 is to add the two times, 8 hr and 14 hr, to get an answer of 22 hr. We reason, however, that x, the time it will take Mateo and Chet working together, must be *less than* 8 hr, since Mateo can complete the job by himself in 8 hr.

Another common error students make is to try to split the job in half between the two workers so that Mateo would work $\frac{1}{2}(8)$, or 4 hr, and Chet would work $\frac{1}{2}(14)$, or 7 hr. In this case, Mateo finishes 3 hr before Chet and they have not worked together to get the entire job done as quickly as possible. If Mateo, when he finishes, helps Chet, the job should actually be completed in a time between 4 hr and 7 hr.

Based on this reasoning, does our answer of $5\frac{1}{11}$ hr in Example 5 hold up?

Note

An alternative approach in work problems is to consider the part of the job that can be done in 1 hr. For instance, in Example 5 Mateo can do the entire job in 8 hr, and Chet can do it in 14 hr. Thus, their work rates, as we saw in Example 5, are $\frac{1}{8}$ and $\frac{1}{14}$, respectively. Since it takes them x hours to complete the job when working together, in 1 hr they can paint $\frac{1}{x}$ of the woodwork. The amount painted by Mateo in 1 hr plus the amount painted by Chet in 1 hr must equal the amount they can do together. This leads to the equation

Amount by Chet

Amount by Mateo $\rightarrow \dfrac{1}{8} + \dfrac{1}{14} = \dfrac{1}{x}.$ $\leftarrow$ Amount together

Compare this with the equation in Step 3 of Example 5. Multiplying each side by $56x$ leads to

$$7x + 4x = 56,$$

the same equation found in the third line of Step 4 in the example. The same solution results.

Work Problem **5** at the Side. ▶

5 Solve each problem.

(a) Michael can paint a room, working alone, in 8 hr. Lindsay can paint the same room, working alone, in 6 hr. How long will it take them if they work together?

(b) Roberto can detail his Camaro in 2 hr working alone. His brother Marco can do the job in 3 hr working alone. How long would it take them if they worked together?

Math in the Media

TO PLAY BASEBALL, YOU MUST *WORK* AT IT!

In the 1994 movie *Little Big League,* the young Billy Heywood inherits the Minnesota Twins baseball team and becomes its manager. He leads the team to the Division Championship and then to the playoffs. But before the final playoff game, the biggest game of the year, he can't keep his mind on his job because a homework problem is giving him trouble.

If Joe can paint a house in 3 hours, and Sam can paint the same house in 5 hours, how long does it take for them to do it together?

With the help of one of his players, he is able to solve the problem, and the team goes on to victory.

1. Use the method described in Example 5 of **Section 14.7** to solve this problem.

2. Before the player was able to solve the problem correctly, Billy got "help" from some of the other players. The incorrect answers they gave him were

    (a) 15 hr   **(b)** 8 hr   **(c)** 4 hr.

    Explain the faulty reasoning behind each of these incorrect answers.

3. The player who gave Billy the correct answer solved the problem as follows:

    Using the simple formula a times b over a plus b, we get our answer of one and seven-eighths.

    Show that if it takes one person *a* hours to complete one job and another *b* hours to complete the same job, then the expression stated by the player,

    $$\frac{a \cdot b}{a + b}$$

    actually does give the number of hours it would take them to do the job together. (*Hint:* Refer to Example 5 and use *a* and *b* rather than 8 and 14. Then solve the resulting formula for *x*.)

14.7 ▶▶▶ **Exercises**

FOR EXTRA HELP

MyMathLab

Math XL
PRACTICE

WATCH

DOWNLOAD

READ

REVIEW

Use Steps 2 and 3 of the six-step problem solving method to set up the equation you would use to solve each problem. (Remember that Step 1 is to read the problem carefully.) Do not actually solve the equation. See Example 1.

1. The numerator of the fraction $\frac{5}{6}$ is increased by an amount so that the value of the resulting fraction is equivalent to $\frac{13}{3}$. By what amount was the numerator increased?

   (a) Let $x =$ _____ . (*Step 2*)

   (b) Write an expression for "the numerator of the fraction $\frac{5}{6}$ is increased by an amount."

   (c) Set up an equation to solve the problem.

   (*Step 3*)

2. If the same number is added to the numerator and subtracted from the denominator of $\frac{23}{12}$, the resulting fraction is equivalent to $\frac{3}{2}$. What is the number?

   (a) Let $x =$ _____ . (*Step 2*)

   (b) Write an expression for "a number is added to the numerator of $\frac{23}{12}$." Then write an expression for "the same number is subtracted from the denominator of $\frac{23}{12}$."

   (c) Set up an equation to solve the problem.

   (*Step 3*)

Solve each problem. See Example 1.

3. In a certain fraction, the denominator is 4 less than the numerator. If 3 is added to both the numerator and the denominator, the resulting fraction is equivalent to $\frac{3}{2}$. What was the original fraction?

4. In a certain fraction, the denominator is 6 more than the numerator. If 3 is added to both the numerator and the denominator, the resulting fraction is equivalent to $\frac{5}{7}$. What was the original fraction (*not* written in lowest terms)?

5. The denominator of a certain fraction is three times the numerator. If 2 is added to the numerator and subtracted from the denominator, the resulting fraction is equivalent to 1. What was the original fraction (*not* written in lowest terms)?

6. The numerator of a certain fraction is four times the denominator. If 6 is added to both the numerator and the denominator, the resulting fraction is equivalent to 2. What was the original fraction (*not* written in lowest terms)?

7. One-sixth of a number is 5 more than the same number. What is the number?

8. One-third of a number is 2 more than one-sixth of the same number. What is the number?

9. A quantity, its $\frac{3}{4}$, its $\frac{1}{2}$, and its $\frac{1}{3}$, added together, become 93. What is the quantity? (*Source: Rhind Mathematical Papyrus.*)

10. A quantity, its $\frac{2}{3}$, its $\frac{1}{2}$, and its $\frac{1}{7}$, added together, become 33. What is the quantity? (*Source: Rhind Mathematical Papyrus.*)

Solve each problem. See Example 2.

11. In July 2007, British explorer and endurance swimmer Lewis Gordon Pugh became the first person to swim at the North Pole. To highlight climate change, he swam 0.6 mi in 18.833 min in waters created by melted sea ice. What was his rate (to three decimal places)? (*Source: The Gazette,* July 16, 2007.)

12. In the 2004 Summer Olympics in Athens, Greece, Jody Henry of Australia won the women's 100-m freestyle swimming event. Her rate was 1.854 m per sec. What was her time (to the nearest hundredth of a second)? (*Source: World Almanac and Book of Facts.*)

13. The winner of the 2007 Daytona 500 (mile) race was Kevin Harvick who drove his Chevrolet to victory with a rate of 149.335 mph. What was his time (to the nearest thousandth of an hour)? (*Source:* www.daytona500.com)

14. In 2008, Scott Dixon drove his Target Chip Ganassi Racing car to victory in the Indianapolis 500 (mile) race. His rate was 143.567 mph. What was his time (to the nearest thousandth of an hour)? (*Source:* www.indy500.com)

15. In the 2006 Winter Olympics in Torino, Italy, Svetlana Zhurova of Russia won the 500-m speed skating event for women. Her time was 76.57 sec. What was her rate (to three decimal places)? (*Source:* www.espn.com)

16. The winner of the women's 1500-m race in the 2004 Olympics was Kelly Holmes of Great Britain with a time of 3.965 min. What was her rate (to three decimal places)? (*Source:* www.olympics.com)

Set up the equation you would use to solve each problem. Do not actually solve the equation. See Examples 3 and 4.

17. Luvenia can row 4 mph in still water. She takes as long to row 8 mi upstream as 24 mi downstream. How fast is the current? (Let x = speed of the current.)

	d	r	t
Upstream	8	$4 - x$	
Downstream	24	$4 + x$	

18. Julio flew his airplane 500 mi against the wind in the same time it took him to fly it 600 mi with the wind. If the speed of the wind was 10 mph, what was the average speed of his plane in still air? (Let x = speed of the plane in still air.)

	d	r	t
Against the Wind	500	$x - 10$	
With the Wind	600	$x + 10$	

Solve each problem. See Examples 3 and 4.

19. If a migrating hawk travels m mph in still air, what is its rate when it flies into a steady headwind of 5 mph? What is its rate with a tailwind of 5 mph?

20. Suppose Stephanie walks D miles at R mph in the same time that Wally walks d miles at r mph. Give an equation relating D, R, d, and r.

21. A boat can go 20 mi against a current in the same time that it can go 60 mi with the current. The current is 4 mph. Find the speed of the boat in still water.

22. A plane flies 350 mi with the wind in the same time that it can fly 310 mi against the wind. The plane has a still-air speed of 165 mph. Find the speed of the wind.

23. The sanderling is a small shorebird about 6.5 in. long, with a thin, dark bill and a wide, white wing stripe. If a sanderling can fly 30 mi with the wind in the same time it can fly 18 mi against the wind when the wind speed is 8 mph, what is the speed of the bird in still air? (*Source:* U.S. Geological Survey.)

24. Airplanes usually fly faster from west to east than from east to west because the prevailing winds go from west to east. The air distance between Chicago and London is about 4000 mi, while the air distance between New York and London is about 3500 mi. If a jet can fly eastbound from Chicago to London in the same time it can fly westbound from London to New York in a 35-mph wind, what is the speed of the plane in still air? (*Source: Encyclopaedia Britannica.*)

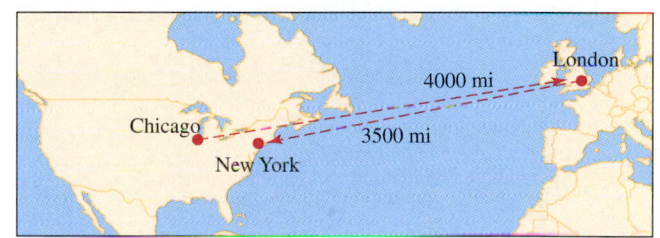

25. Perian Herring's boat goes 12 mph. Find the rate of the current of the river if she can go 6 mi upstream in the same amount of time she can go 10 mi downstream.

26. Sarah Sponholz can travel 8 mi upstream in the same time it takes her to go 12 mi downstream. Her boat goes 15 mph in still water. What is the rate of the current?

Set up the equation you would use to solve each problem. Do not actually solve the equation. See Example 5.

27. Edwin Bedford can tune up his Chevy in 2 hr working alone. His son, Beau, can do the job in 3 hr working alone. How long would it take them if they worked together? (Let t represent the time working together.)

	r	t	w
Edwin		t	
Beau		t	

28. Working alone, Helio can paint a room in 8 hr. Julianne can paint the same room working alone in 6 hr. How long will it take them if they work together? (Let t represent the time working together.)

	r	t	w
Helio		t	
Julianne		t	

Solve each problem. See Example 5.

29. Ms. Tseng, a high school mathematics teacher, gave a test on perimeter, area, and volume to her geometry classes. Working alone, it would take her 4 hr to grade the tests. Her student teacher, Jonah Schmidt, would take 6 hr to grade the same tests. How long would it take them to grade these tests if they work together?

30. Geraldo and Luisa Hernandez operate a small laundry. Luisa, working alone, can clean a day's laundry in 9 hr. Geraldo can clean a day's laundry in 8 hr. How long would it take them if they work together?

31. Todd's copier can do a printing job in 7 hr. Scott's copier can do the same job in 12 hr. How long would it take to do the job using both copiers?

32. A pump can pump the water out of a flooded basement in 10 hr. A smaller pump takes 12 hr. How long would it take to pump the water from the basement using both pumps?

33. Hilda can paint a room in 6 hr. Working together with Brenda, they can paint the room in $3\frac{3}{4}$ hr. How long would it take Brenda to paint the room by herself?

34. Grant can completely mess up his room in 15 min. If his cousin Wade helps him, they can completely mess up the room in $8\frac{4}{7}$ min. How long would it take Wade to mess up the room by himself?

35. An inlet pipe can fill a swimming pool in 9 hr, and an outlet pipe can empty the pool in 12 hr. Through an error, both pipes are left open. How long will it take to fill the pool?

36. One pipe can fill a swimming pool in 6 hr, and another pipe can do it in 9 hr. How long will it take the two pipes working together to fill the pool $\frac{3}{4}$ full?

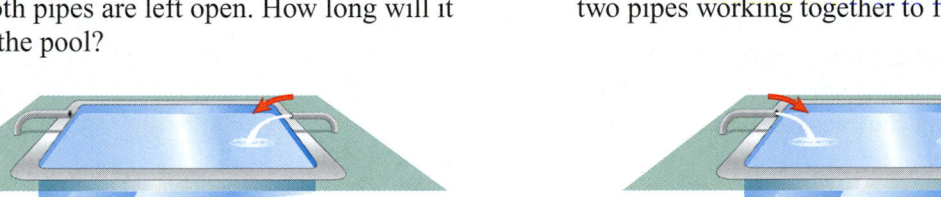

37. Refer to Exercise 35. Assume the error was discovered after both pipes had been running for 3 hr, and the outlet pipe was then closed. How much more time would then be required to fill the pool? (*Hint:* Consider how much of the job had been done when the error was discovered.)

38. A cold water faucet can fill a sink in 12 min, and a hot water faucet can fill it in 15 min. The drain can empty the sink in 25 min. If both faucets are on and the drain is open, how long will it take to fill the sink?

14.8 ▷▷▷ Variation

OBJECTIVE 1 Solve direct variation problems. Suppose that gasoline costs $2.25 per gal. Then 1 gal costs $2.25, 2 gal cost $2($2.25) = $4.50, 3 gal cost $3($2.25) = $6.75, and so on. Each time, the total cost is obtained by multiplying the number of gallons by the price per gallon. In general, if k equals the price per gallon and x equals the number of gallons, then the total cost y is equal to kx. Notice that

As the *number of gallons* **increases,** the *total cost* **increases.**

The reverse is also true:

As the *number of gallons* **decreases,** the *total cost* **decreases.**

The preceding discussion is an example of *variation.* ***Two variables vary directly if one is a constant multiple of the other.***

Direct Variation

y **varies directly as** *x* if there exists a constant k such that

$$y = kx.$$

Also, y is said to be *proportional to x.* The constant k in the equation for direct variation is a numerical value, such as 4.50 in the gasoline price discussion. This value is called the **constant of variation.**

EXAMPLE 1 Using Direct Variation

Suppose y varies directly as x, and $y = 20$ when $x = 4$. Find y when $x = 9$.

Since y varies directly as x, there is a constant k such that $y = kx$. We know that $y = 20$ when $x = 4$. Substituting these values into $y = kx$ and solving for k gives

$$y = k\textbf{\textit{x}} \qquad \text{Equation for direct variation}$$

$$\textbf{20} = k \cdot \textbf{4} \qquad \text{Substitute the given values.}$$

$$k = \textbf{5}. \longleftarrow \text{Constant of variation}$$

Since $y = kx$ and $k = 5$,

$$y = \textbf{5}x. \qquad \text{Let } k = 5.$$

Therefore, when $x = 9$,

$$y = 5x = 5 \cdot \textbf{9} = 45. \qquad \text{Let } x = 9.$$

Thus, $y = 45$ when $x = 9$.

Work Problem **1** *at the Side.* ▶

OBJECTIVE 2 Solve inverse variation problems. In direct variation, where $k > 0$, as x increases, y increases. Similarly, as x decreases, y decreases. Another type of variation is *inverse variation.* With inverse variation, where $k > 0$,

As one variable ***increases,*** the other variable ***decreases.***

1 Solve each problem.

(a) If z varies directly as t, and $z = 11$ when $t = 4$, find z when $t = 32$.

(b) The circumference of a circle varies directly as the radius. A circle with a radius of 7 cm has a circumference of 43.96 cm. Find the circumference if the radius is 11 cm.

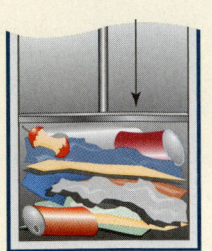

As pressure increases, volume decreases.

Figure 3

2 Solve the problem.

Suppose z varies inversely as t, and $z = 8$ when $t = 2$. Find z when $t = 32$.

For example, in a closed space, volume decreases as pressure increases, as illustrated by a trash compactor. See Figure 3. As the compactor presses down, the pressure on the trash increases; in turn, the trash occupies a smaller space.

> **Inverse Variation**
>
> y **varies inversely as** x if there exists a constant k such that
> $$y = \frac{k}{x}.$$

EXAMPLE 2 **Using Inverse Variation**

Suppose y varies inversely as x, and $y = 3$ when $x = 8$. Find y when $x = 6$.

Since y varies inversely as x, there is a constant k such that $y = \frac{k}{x}$. We know that $y = 3$ when $x = 8$, so we can find k.

$$y = \frac{k}{x} \qquad \text{Equation for inverse variation}$$

$$3 = \frac{k}{8} \qquad \text{Substitute the given values.}$$

$$k = \mathbf{24} \qquad \text{Solve for } k.$$

Since $y = \frac{24}{x}$, we let $x = 6$ and solve for y.

$$y = \frac{24}{x} = \frac{24}{6} = 4$$

Therefore, when $x = 6$, $y = 4$.

◀ **Work Problem** **2** **at the Side.**

3 Solve the problem.

The current in a simple electrical circuit varies inversely as the resistance. If the current is 80 amps when the resistance is 10 ohms, find the current if the resistance is 16 ohms.

EXAMPLE 3 **Using Inverse Variation**

In the manufacturing of a certain medical syringe, the cost of producing the syringe varies inversely as the number produced. If 10,000 syringes are produced, the cost is \$2 per unit. Find the cost per unit to produce 25,000 syringes.

Let $x = $ the number of syringes produced

and $c = $ the cost per unit.

Since c varies inversely as x, there is a constant k such that

$$c = \frac{k}{x} \qquad \text{Equation for inverse variation}$$

$$2 = \frac{k}{10,000} \qquad \text{Substitute the given values.}$$

$$20,000 = k. \qquad \text{Multiply by 10,000.}$$

Since $c = \frac{k}{x}$,

$$c = \frac{20,000}{25,000} = 0.80. \qquad \text{Let } k = 20,000 \text{ and } x = 25,000.$$

The cost per unit to make 25,000 syringes is \$0.80.

◀ **Work Problem** **3** **at the Side.**

ANSWERS

2. $\frac{1}{2}$

3. 50 amps

14.8 ▷▷▷ Exercises

1. **(a)** If the constant of variation is positive and y varies directly as x, then as

    x increases, y _____ .
    (increases/decreases)

    (b) If the constant of variation is positive and y varies inversely as x, then as

    x increases, y _____ .
    (increases/decreases)

2. Bill Veeck was the owner of several major league baseball teams in the 1950s and 1960s. He was known to often sit in the stands and enjoy games with his paying customers. Here is a quote attributed to him:

    "I have discovered in 20 years of moving around a ballpark, that the knowledge of the game is usually in inverse proportion to the price of the seats."

    Explain in your own words the meaning of this statement. (To prove his point, Veeck once allowed the fans to vote on managerial decisions.)

Solve each problem involving direct variation. See Example 1.

🌐 3. If z varies directly as x, and $z = 30$ when $x = 8$, find z when $x = 4$.

4. If y varies directly as x, and $x = 27$ when $y = 6$, find x when $y = 2$.

5. If d varies directly as r, and $d = 200$ when $r = 40$, find d when $r = 60$.

6. If d varies directly as t, and $d = 150$ when $t = 3$, find d when $t = 5$.

Solve each problem involving inverse variation. See Example 2.

🌐 7. If z varies inversely as x, and $z = 50$ when $x = 2$, find z when $x = 25$.

8. If x varies inversely as y, and $x = 3$ when $y = 8$, find y when $x = 4$.

9. If m varies inversely as r, and $m = 12$ when $r = 8$, find m when $r = 16$.

10. If p varies inversely as q, and $p = 7$ when $q = 6$, find p when $q = 2$.

Solve each variation problem. See Examples 1–3.

11. For a given base, the area of a triangle varies directly as its height. Find the area of a triangle with a height of 6 in., if the area is 10 in.^{2} when the height is 4 in.

12. The interest on an investment varies directly as the rate of interest. If the interest is $48 when the interest rate is 5%, find the interest when the rate is 4.2%.

13. Hooke's law for an elastic spring states that the distance a spring stretches varies directly with the force applied. If a force of 75 lb stretches a certain spring 16 in., how much will a force of 200 lb stretch the spring?

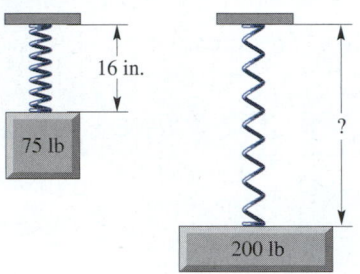

14. The pressure exerted by water at a given point varies directly with the depth of the point beneath the surface of the water. Water exerts 4.34 lb per in.^{2} for every 10 ft traveled below the water's surface. What is the pressure exerted on a scuba diver at 20 ft?

15. For a constant area, the length of a rectangle varies inversely as the width. The length of a rectangle is 27 ft when the width is 10 ft. Find the width of a rectangle with the same area if the length is 18 ft.

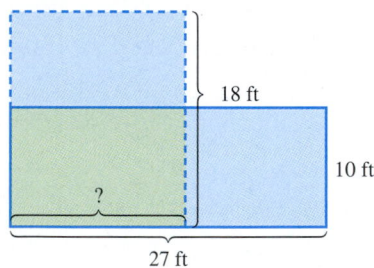

16. Over a specified distance, speed varies inversely with time. If a Dodge Viper on a test track goes a certain distance in one-half minute at 160 mph, what speed is needed to go the same distance in three-fourths minute?

17. If the temperature is constant, the pressure of a gas in a container varies inversely as the volume of the container. If the pressure is 10 lb per ft^2 in a container with volume 3 ft^3, what is the pressure in a container with volume 1.5 ft^3?

18. The current in a simple electrical circuit varies inversely as the resistance. If the current is 20 amps when the resistance is 5 ohms, find the current when the resistance is 8 ohms.

19. In the inversion of raw sugar, the rate of change of the amount of raw sugar varies directly as the amount of raw sugar remaining. The rate is 200 kg per hr when there are 800 kg left. What is the rate of change per hour when only 100 kg are left?

20. The force required to compress a spring varies directly as the change in the length of the spring. If a force of 12 lb is required to compress a certain spring 3 in., how much force is required to compress the spring 5 in.?

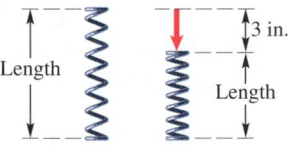

*Use personal experience or intuition to determine whether the situation suggests direct or inverse variation.**

21. The rate and the distance traveled by a pickup truck in 3 hr

22. The number of different lottery tickets you buy and your probability of winning that lottery

23. The number of days from now until December 25 and the magnitude of the frenzy of Christmas shopping

24. Your age and the probability that you believe in Santa Claus

25. The amount of gasoline that you pump and the amount of empty space left in your tank

26. The surface area of a balloon and its diameter

27. The amount of pressure put on the accelerator of a car and the speed of the car

28. The number of days until the end of the baseball season and the number of home runs that Alex Rodriguez has

*The authors thank Linda Kodama of Kapi'olani Community College for suggesting the inclusion of exercises of this type.

*Two triangles are **similar** if they have the same shape (but not necessarily the same size).*
Similar triangles have side lengths that are proportional—that is, vary directly.
The figure shows two similar triangles. Notice that the ratios of the corresponding
sides are all equal to $\frac{3}{2}$:

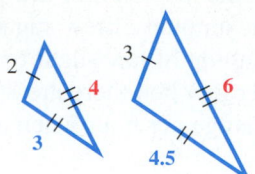

$$\frac{3}{2} = \frac{3}{2} \qquad \frac{4.5}{3} = \frac{3}{2} \qquad \frac{6}{4} = \frac{3}{2}.$$

If we know that two triangles are similar, we can set up a direct variation equation to solve
for the length of an unknown side.

Find the length x, given that the pair of triangles are similar.

29.

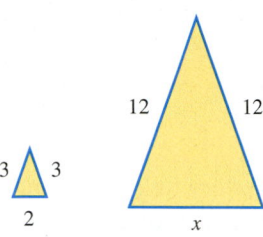

30.

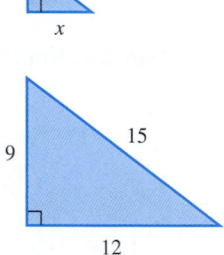

31.

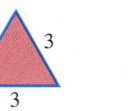

32.

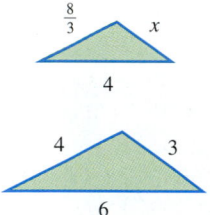

33.

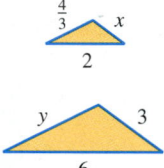

34.

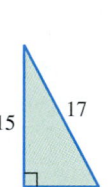

Use similar triangles and direct variation to solve each problem. (Source: Guinness
World Records.)

35. One of the tallest candles ever constructed was
exhibited at the 1897 Stockholm Exhibition. If it
cast a shadow 5 ft long at the same time a vertical
pole 32 ft high cast a shadow 2 ft long, how tall
was the candle?

36. An enlarged version of the chair used by George
Washington at the Constitutional Convention casts
a shadow 18 ft long at the same time a vertical pole
12 ft high casts a shadow 4 ft long. How tall is the
chair?

▶ Key Terms

14.1	**rational expression**	The quotient of two polynomials with denominator not 0 is called a rational expression.
	lowest terms	A rational expression is written in lowest terms if the greatest common factor of its numerator and denominator is 1.
14.3	**least common denominator (LCD)**	The simplest expression that is divisible by all denominators is called the least common denominator.
14.5	**complex fraction**	A rational expression with one or more fractions in the numerator, denominator, or both, is called a complex fraction.
14.8	**direct variation**	y varies directly as x if there is a constant k such that $y = kx$.
	constant of variation	In the equation $y = kx$ or $y = \frac{k}{x}$, the number k is called the constant of variation.
	inverse variation	y varies inversely as x if there is a constant k such that $y = \frac{k}{x}$.

▶ Test Your Word Power

See how well you have learned the vocabulary in this chapter. Answers, with examples, follow the Quick Review.

1. A **rational expression** is
   A. an algebraic expression made up of a term or the sum of a finite number of terms with real coefficients and whole number exponents
   B. a polynomial equation of degree 2
   C. an expression with one or more fractions in the numerator, denominator, or both
   D. the quotient of two polynomials with denominator not 0.

2. A **complex fraction** is
   A. an algebraic expression made up of a term or the sum of a finite number of terms with real coefficients and whole number exponents
   B. a polynomial equation of degree 2

C. a rational expression with one or more fractions in the numerator, denominator, or both
   D. the quotient of two polynomials with denominator not 0.

3. In a given set of fractions, the **least common denominator** is
   A. the simplest denominator of all the denominators
   B. the simplest expression that is divisible by all the denominators
   C. the greatest integer that evenly divides the numerator and denominator of all the fractions
   D. the greatest denominator of all the denominators.

4. If two positive quantities x and y are in **direct variation,** and the constant of variation is positive, then
   A. as x increases, y decreases
   B. as x increases, y increases
   C. as x increases, y remains constant
   D. as x decreases, y remains constant.

5. If two positive quantities x and y are in **inverse variation,** and the constant of variation is positive, then
   A. as x increases, y decreases
   B. as x increases, y increases
   C. as x increases, y remains constant
   D. as x decreases, y remains constant.

▶ **Quick Review**

Concepts	Examples

14.1 The Fundamental Property of Rational Expressions

To find the value(s) for which a rational expression is undefined, set the denominator equal to 0 and solve the equation.

Find the values for which $\dfrac{x-4}{x^2-16}$ is undefined.

$$x^2 - 16 = 0$$
$$(x-4)(x+4) = 0 \qquad \text{Factor.}$$
$$x - 4 = 0 \quad \text{or} \quad x + 4 = 0 \qquad \text{Zero-factor property}$$
$$x = 4 \quad \text{or} \qquad x = -4 \qquad \text{Solve for } x.$$

The rational expression is undefined for 4 and −4, so $x \neq 4$ and $x \neq -4$.

Writing a Rational Expression in Lowest Terms

Step 1 Factor the numerator and denominator.

Step 2 Use the fundamental property to divide out common factors from the numerator and denominator.

Write the expression in lowest terms.

$$\frac{x^2 - 1}{(x-1)^2}$$
$$= \frac{(x-1)(x+1)}{(x-1)(x-1)}$$
$$= \frac{x+1}{x-1}$$

There are often several different equivalent forms of a rational expression.

To find four equivalent forms of $-\dfrac{x-1}{x+2}$, distribute the − sign in the numerator to get $\dfrac{-(x-1)}{x+2}$, or $\dfrac{-x+1}{x+2}$; do so in the denominator to get $\dfrac{x-1}{-(x+2)}$, or $\dfrac{x-1}{-x-2}$.

14.2 Multiplying and Dividing Rational Expressions

Multiplying or Dividing Rational Expressions

Step 1 Note the operation. If the operation is division, use the definition of division to rewrite as multiplication.

Step 2 Multiply numerators and multiply denominators.

Step 3 Factor numerators and denominators completely.

Step 4 Write in lowest terms, using the fundamental property.

Steps 2 and 3 may be interchanged based on personal preference.

Multiply. $\dfrac{3x+9}{x-5} \cdot \dfrac{x^2-3x-10}{x^2-9}$

$$= \frac{(3x+9)(x^2-3x-10)}{(x-5)(x^2-9)}$$
$$= \frac{3(x+3)(x-5)(x+2)}{(x-5)(x+3)(x-3)}$$
$$= \frac{3(x+2)}{x-3}$$

Divide. $\dfrac{2x+1}{x+5} \div \dfrac{6x^2-x-2}{x^2-25}$

$$= \frac{(2x+1)(x^2-25)}{(x+5)(6x^2-x-2)}$$
$$= \frac{(2x+1)(x+5)(x-5)}{(x+5)(2x+1)(3x-2)}$$
$$= \frac{x-5}{3x-2}$$

Concepts	Examples

14.3 Least Common Denominators

Finding the LCD

Step 1 Factor each denominator into prime factors.

Step 2 List each different factor the greatest number of times it appears.

Step 3 Multiply the factors from Step 2 to get the LCD.

Find the LCD for $\dfrac{3}{k^2 - 8k + 16}$ and $\dfrac{1}{4k^2 - 16k}$.

$$k^2 - 8k + 16 = (k - 4)^2$$
$$4k^2 - 16k = 4k(k - 4)$$
Factor each denominator.

$$\text{LCD} = (k - 4)^2 \cdot 4 \cdot k$$
$$= 4k(k - 4)^2$$

Writing Equivalent Rational Expressions

Step 1 Factor both denominators.

Step 2 Decide what factors the denominator must be multiplied by to equal the specified denominator.

Step 3 Multiply the rational expression by that factor divided by itself. (That is, multiply by 1.)

Find the numerator: $\dfrac{5}{2z^2 - 6z} = \dfrac{?}{4z^3 - 12z^2}$.

$$\frac{5}{2z(z - 3)} = \frac{?}{4z^2(z - 3)}$$

$2z(z - 3)$ must be multiplied by $2z$.

$$\frac{5}{2z(z - 3)} \cdot \frac{2z}{2z} = \frac{10z}{4z^2(z - 3)} = \frac{10z}{4z^3 - 12z^2}$$

14.4 Adding and Subtracting Rational Expressions

Adding Rational Expressions

Step 1 Find the LCD.

Step 2 Rewrite each rational expression with the LCD as denominator.

Step 3 Add the numerators to get the numerator of the sum. The LCD is the denominator of the sum.

Step 4 Write in lowest terms.

Add. $\dfrac{2}{3m + 6} + \dfrac{m}{m^2 - 4}$

$$3m + 6 = 3(m + 2)$$
$$m^2 - 4 = (m + 2)(m - 2)$$
The LCD is $3(m + 2)(m - 2)$.

$$= \frac{2(m - 2)}{3(m + 2)(m - 2)} + \frac{3m}{3(m + 2)(m - 2)}$$

$$= \frac{2m - 4 + 3m}{3(m + 2)(m - 2)}$$

$$= \frac{5m - 4}{3(m + 2)(m - 2)}$$

Subtracting Rational Expressions

Follow the same steps as for addition, but subtract in Step 3.

Subtract. $\dfrac{6}{k + 4} - \dfrac{2}{k}$ The LCD is $k(k + 4)$.

$$= \frac{6k}{(k + 4)k} - \frac{2(k + 4)}{k(k + 4)}$$

$$= \frac{6k - 2(k + 4)}{k(k + 4)}$$

$$= \frac{6k - 2k - 8}{k(k + 4)}$$

$$= \frac{4k - 8}{k(k + 4)}, \quad \text{or} \quad \frac{4(k - 2)}{k(k + 4)}$$

Concepts	Examples

14.5 Complex Fractions

Simplifying Complex Fractions

Method 1 Simplify the numerator and denominator separately. Then divide the simplified numerator by the simplified denominator.

Simplify.

Method 1

$$\dfrac{\dfrac{1}{a} - a}{1 - a} = \dfrac{\dfrac{1}{a} - \dfrac{a^2}{a}}{1 - a} = \dfrac{\dfrac{1 - a^2}{a}}{1 - a}$$

$$= \dfrac{1 - a^2}{a} \div (1 - a)$$

$$= \dfrac{1 - a^2}{a} \cdot \dfrac{1}{1 - a}$$

$$= \dfrac{(1 - a)(1 + a)}{a(1 - a)}$$

$$= \dfrac{1 + a}{a}$$

Method 2 Multiply the numerator and denominator of the complex fraction by the LCD of all the denominators in the complex fraction. Write in lowest terms.

Method 2

$$\dfrac{\dfrac{1}{a} - a}{1 - a} = \dfrac{\left(\dfrac{1}{a} - a\right)a}{(1 - a)a} = \dfrac{\dfrac{a}{a} - a^2}{(1 - a)a}$$

$$= \dfrac{1 - a^2}{(1 - a)a} = \dfrac{(1 + a)(1 - a)}{(1 - a)a}$$

$$= \dfrac{1 + a}{a}$$

14.6 Solving Equations with Rational Expressions

Solving Equations with Rational Expressions

Step 1 Multiply each side of the equation by the LCD. (This clears the equation of fractions.)

Solve $\dfrac{x}{x - 3} + \dfrac{4}{x + 3} = \dfrac{18}{x^2 - 9}$.

$$\dfrac{x}{x - 3} + \dfrac{4}{x + 3} = \dfrac{18}{(x - 3)(x + 3)} \qquad \text{Factor.}$$

The LCD is $(x - 3)(x + 3)$. Note that 3 and -3 cannot be solutions, as they cause a denominator to equal 0.

$$(x - 3)(x + 3)\left(\dfrac{x}{x - 3} + \dfrac{4}{x + 3}\right)$$

$$= (x - 3)(x + 3)\dfrac{18}{(x - 3)(x + 3)} \qquad \begin{array}{l}\text{Multiply by}\\ \text{the LCD.}\end{array}$$

Step 2 Solve the resulting equation.

$$x(x + 3) + 4(x - 3) = 18 \qquad \text{Distributive property}$$

$$x^2 + 3x + 4x - 12 = 18 \qquad \text{Distributive property}$$

$$x^2 + 7x - 30 = 0 \qquad \text{Standard form}$$

$$(x - 3)(x + 10) = 0 \qquad \text{Factor.}$$

$$x - 3 = 0 \quad \text{or} \quad x + 10 = 0$$

$$\text{Zero-factor property}$$

Step 3 Check each proposed solution by substituting it in the original equation. Reject any value that causes an original denominator to equal 0.

$$\text{Reject} \longrightarrow x = 3 \quad \text{or} \qquad x = -10 \qquad \text{Solve for } x.$$

Since 3 causes denominators to equal 0, the only solution is -10. Thus, $\{-10\}$ is the solution set.

Concepts

| | Examples |

14.7 Applications of Rational Expressions

Solving Problems about Distance, Rate, and Time
Use the six-step method.

Step 1 **Read** the problem carefully.

Step 2 **Assign a variable.** Use a table to identify distance, rate, and time. Solve $d = rt$ for the unknown quantity in the table.

On a trip from Sacramento to Monterey, Marge traveled at an average speed of 60 mph. The return trip, at an average speed of 64 mph, took $\frac{1}{4}$ hr less. How far did she travel between the two cities?

Let x = the unknown distance.

	d	r	$t = \dfrac{d}{r}$
Going	x	60	$\dfrac{x}{60}$
Returning	x	64	$\dfrac{x}{64}$

Step 3 **Write an equation.** From the wording in the problem, decide the relationship between the quantities. Use those expressions to write an equation.

Since the time for the return trip was $\frac{1}{4}$ hr less, the time going equals the time returning plus $\frac{1}{4}$.

$$\frac{x}{60} = \frac{x}{64} + \frac{1}{4}$$

Step 4 **Solve** the equation.

$$16x = 15x + 240 \qquad \text{Multiply by the LCD, 960.}$$

$$x = 240 \qquad \text{Subtract } 15x.$$

Step 5 **State the answer.**

She traveled 240 mi.

Step 6 **Check** the solution.

The trip there took $\frac{240}{60} = 4$ hr, while the return trip took $\frac{240}{64} = 3\frac{3}{4}$ hr, which is $\frac{1}{4}$ hr less time. The solution checks.

Solving Problems about Work

Step 1 **Read** the problem carefully.

It takes the regular mail carrier 6 hr to cover her route. A substitute takes 8 hr to cover the same route. How long would it take them to cover the route together?

Step 2 **Assign a variable.** State what the variable represents. Put the information from the problem in a table. If a job is done in t units of time, then the rate is $\frac{1}{t}$.

Let x = the number of hours to cover the route together.

The rate of the regular carrier is $\frac{1}{6}$ job per hour; the rate of the substitute is $\frac{1}{8}$ job per hour. Multiply rate by time to get the fractional part of the job done.

	Rate	Time	Part of the Job Done
Regular	$\dfrac{1}{6}$	x	$\dfrac{1}{6}x$
Substitute	$\dfrac{1}{8}$	x	$\dfrac{1}{8}x$

Step 3 **Write an equation.** The sum of the fractional parts should equal 1 (whole job).

The equation is $\dfrac{1}{6}x + \dfrac{1}{8}x = 1$.

Step 4 **Solve** the equation.

The solution of the equation is $\frac{24}{7}$, or $3\frac{3}{7}$. The solution checks, because $\frac{1}{6}\left(\frac{24}{7}\right) + \frac{1}{8}\left(\frac{24}{7}\right) = 1$ is true.

Steps 5 and 6 **State the answer** and **check** the solution.

It would take them $3\frac{3}{7}$ hr to cover the route together.

Concepts	Examples

14.8 Variation

Solving Variation Problems

Step 1   Write the variation equation. Use

$$y = kx \qquad \text{Direct variation}$$

$$\text{or} \quad y = \frac{k}{x}. \qquad \text{Inverse variation}$$

Step 2   Find k by substituting the given values of x and y into the equation.

Step 3   Write the equation with the value of k from Step 2 and the given value of x or y. Solve for the remaining variable.

If y varies inversely as x, and $y = 4$ when $x = 9$, find y when $x = 6$.

$$y = \frac{k}{x} \qquad \text{Equation for inverse variation}$$

$$4 = \frac{k}{9} \qquad \text{Substitute given values.}$$

$$k = 36 \qquad \text{Solve for } k.$$

$$y = \frac{36}{x} \qquad k = 36$$

$$y = \frac{36}{6} \qquad \text{Let } x = 6.$$

$$y = 6$$

ANSWERS TO TEST YOUR WORD POWER

1. D; *Examples:* $-\dfrac{3}{4y}, \; \dfrac{5x^3}{x+2}, \; \dfrac{a+3}{a^2-4a-5}$

2. C; *Examples:* $\dfrac{\frac{2}{3}}{\frac{4}{7}}, \; \dfrac{x - \frac{1}{y}}{x + \frac{1}{y}}, \; \dfrac{\frac{2}{a+1}}{a^2-1}$

3. B; *Examples:* The least common denominator of $\dfrac{1}{2}, \dfrac{1}{3},$ and $\dfrac{1}{4}$ is 12. The least common denominator of $\dfrac{1}{x}$ and $\dfrac{1}{x+1}$ is $x(x+1)$.

4. B; *Example:* The equation $y = 3x$ represents direct variation. When $x = 2$, $y = 6$. If x increases to 3, then y increases to $3(3) = 9$.

5. A; *Example:* The equation $y = \dfrac{3}{x}$ represents inverse variation. When $x = 1$, $y = 3$. If x increases to 2, then y decreases to $\dfrac{3}{2}$, or $1\dfrac{1}{2}$.

Chapter 14 ▶▶▶ Review Exercises

[14.1] *Find the value(s) of the variable for which each rational expression is undefined. Write answers with ≠.*

1. $\dfrac{4}{x - 3}$

2. $\dfrac{x + 3}{2x}$

3. $\dfrac{m - 2}{m^2 - 2m - 3}$

4. $\dfrac{2k + 1}{3k^2 + 17k + 10}$

*Find the numerical value of each rational expression when **(a)** $x = -2$ and **(b)** $x = 4$.*

5. $\dfrac{x^2}{x - 5}$

6. $\dfrac{4x - 3}{5x + 2}$

7. $\dfrac{3x}{x^2 - 4}$

8. $\dfrac{x - 1}{x + 2}$

Write each rational expression in lowest terms.

9. $\dfrac{5a^3 b^3}{15a^4 b^2}$

10. $\dfrac{m - 4}{4 - m}$

11. $\dfrac{4x^2 - 9}{6 - 4x}$

12. $\dfrac{4p^2 + 8pq - 5q^2}{10p^2 - 3pq - q^2}$

Write four equivalent expressions for each fraction.

13. $-\dfrac{4x - 9}{2x + 3}$

14. $-\dfrac{8 - 3x}{3 - 6x}$

[14.2] *Multiply or divide. Write each answer in lowest terms.*

15. $\dfrac{8x^2}{12x^5} \cdot \dfrac{6x^4}{2x}$

16. $\dfrac{9m^2}{(3m)^4} \div \dfrac{6m^5}{36m}$

17. $\dfrac{x - 3}{4} \cdot \dfrac{5}{2x - 6}$

18. $\dfrac{2r + 3}{r - 4} \cdot \dfrac{r^2 - 16}{6r + 9}$

19. $\dfrac{3q + 3}{5 - 6q} \div \dfrac{4q + 4}{2(5 - 6q)}$

20. $\dfrac{y^2 - 6y + 8}{y^2 + 3y - 18} \div \dfrac{y - 4}{y + 6}$

21. $\dfrac{2p^2 + 13p + 20}{p^2 + p - 12} \cdot \dfrac{p^2 + 2p - 15}{2p^2 + 7p + 5}$

22. $\dfrac{3z^2 + 5z - 2}{9z^2 - 1} \cdot \dfrac{9z^2 + 6z + 1}{z^2 + 5z + 6}$

[14.3] *Find the least common denominator for the fractions in each list.*

23. $\dfrac{1}{8}, \dfrac{5}{12}, \dfrac{7}{32}$

24. $\dfrac{4}{9y}, \dfrac{7}{12y^2}, \dfrac{5}{27y^4}$

25. $\dfrac{1}{m^2 + 2m}, \dfrac{4}{m^2 + 7m + 10}$

26. $\dfrac{3}{x^2 + 4x + 3}, \dfrac{5}{x^2 + 5x + 4}, \dfrac{2}{x^2 + 7x + 12}$

Write each rational expression as an equivalent expression with the indicated denominator.

27. $\dfrac{5}{8} = \dfrac{?}{56}$

28. $\dfrac{10}{k} = \dfrac{?}{4k}$

29. $\dfrac{3}{2a^3} = \dfrac{?}{10a^4}$

30. $\dfrac{9}{x - 3} = \dfrac{?}{18 - 6x}$

31. $\dfrac{-3y}{2y - 10} = \dfrac{?}{50 - 10y}$

32. $\dfrac{4b}{b^2 + 2b - 3} = \dfrac{?}{(b + 3)(b - 1)(b + 2)}$

[14.4] *Add or subtract. Write each answer in lowest terms.*

33. $\dfrac{10}{x} + \dfrac{5}{x}$

34. $\dfrac{6}{3p} - \dfrac{12}{3p}$

35. $\dfrac{9}{k} - \dfrac{5}{k - 5}$

36. $\dfrac{4}{y} + \dfrac{7}{7 + y}$

37. $\dfrac{m}{3} - \dfrac{2 + 5m}{6}$

38. $\dfrac{12}{x^2} - \dfrac{3}{4x}$

39. $\dfrac{5}{a - 2b} + \dfrac{2}{a + 2b}$

40. $\dfrac{4}{k^2 - 9} - \dfrac{k + 3}{3k - 9}$

41. $\dfrac{8}{z^2 + 6z} - \dfrac{3}{z^2 + 4z - 12}$

42. $\dfrac{11}{2p - p^2} - \dfrac{2}{p^2 - 5p + 6}$

[14.5] *Simplify each complex fraction.*

43. $\dfrac{\dfrac{a^4}{b^2}}{\dfrac{a^3}{b}}$

44. $\dfrac{\dfrac{y - 3}{y}}{\dfrac{y + 3}{4y}}$

45. $\dfrac{\dfrac{3m + 2}{m}}{\dfrac{2m - 5}{6m}}$

46. $\dfrac{\dfrac{1}{p} - \dfrac{1}{q}}{\dfrac{1}{q - p}}$

47. $\dfrac{x + \dfrac{1}{w}}{x - \dfrac{1}{w}}$

48. $\dfrac{\dfrac{1}{r + t} - 1}{\dfrac{1}{r + t} + 1}$

[14.6] *Solve each equation. Check your solutions.*

49. $\dfrac{k}{5} - \dfrac{2}{3} = \dfrac{1}{2}$

50. $\dfrac{4 - z}{z} + \dfrac{3}{2} = \dfrac{-4}{z}$

51. $\dfrac{x}{2} - \dfrac{x - 3}{7} = -1$

52. $\dfrac{3y - 1}{y - 2} = \dfrac{5}{y - 2} + 1$

53. $\dfrac{3}{m - 2} + \dfrac{1}{m - 1} = \dfrac{7}{m^2 - 3m + 2}$

Solve for the specified variable.

54. $m = \dfrac{Ry}{t}$ for t

55. $x = \dfrac{3y - 5}{4}$ for y

56. $\dfrac{1}{r} - \dfrac{1}{s} = \dfrac{1}{t}$ for t

[14.7] *Solve each problem.*

57. In a certain fraction, the denominator is 5 less than the numerator. If 5 is added to both the numerator and the denominator, the resulting fraction is equivalent to $\frac{5}{4}$. Find the original fraction (*not* written in lowest terms).

58. The denominator of a certain fraction is six times the numerator. If 3 is added to the numerator and subtracted from the denominator, the resulting fraction is equivalent to $\frac{2}{5}$. Find the original fraction (*not* written in lowest terms).

59. On June 24, 2007, Dario Franchitti won the first Iowa Corn Indy 250. He drove a Dallara-Honda the 218.75 mi distance, with an average speed of 123.896 mph. What was his time (to the nearest thousandth of an hour)? (*Source:* www.iowaspeedway.com)

60. In the 2006 Winter Olympics in Torino, Italy, Chad Hedrick of the United States won the men's 5000-m speed skating event in 6.245 min. What was his rate (to three decimal places)? (*Source: World Almanac and Book of Facts.*)

61. Zachary and Samuel are brothers who share a bedroom. By himself, Zachary can completely mess up their room in 20 min, while it would take Samuel only 12 min to do the same thing. How long would it take them to mess up the room together?

62. A man can plant his garden in 5 hr, working alone. His daughter can do the same job in 8 hr. How long would it take them if they worked together?

[14.8] *Solve each problem.*

63. If y varies directly as x, and $x = 12$ when $y = 5$, find x when $y = 3$.

64. If a parallelogram has a fixed area, the height varies inversely as the base. A parallelogram has a height of 8 cm and a base of 12 cm. Find the height if the base is changed to 24 cm.

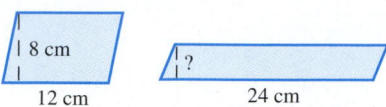

Mixed Review Exercises

Perform the indicated operations.

65. $\dfrac{4}{m-1} - \dfrac{3}{m+1}$

66. $\dfrac{8p^5}{5} \div \dfrac{2p^3}{10}$

67. $\dfrac{r-3}{8} \div \dfrac{3r-9}{4}$

68. $\dfrac{\dfrac{5}{x} - 1}{\dfrac{5-x}{3x}}$

69. $\dfrac{4}{z^2 - 2z + 1} - \dfrac{3}{z^2 - 1}$

Solve.

70. $a = \dfrac{v - w}{t}$ for v

71. $\dfrac{2}{z} - \dfrac{z}{z+3} = \dfrac{1}{z+3}$

72. $\dfrac{2x}{x^2 - 16} - \dfrac{2}{x-4} = \dfrac{4}{x+4}$

73. Anne Kelly flew her plane 400 km with the wind in the same time it took her to go 200 km against the wind. The speed of the wind is 50 km per hr. Find the speed of the plane in still air.

74. At a given hour, two steamboats leave a city in the same direction on a straight canal. One travels at 18 mph, and the other travels at 25 mph. In how many hours will the boats be 70 mi apart?

75. In rectangles of constant area, length and width vary inversely. When the length is 24, the width is 2. What is the width when the length is 12?

76. The longer the term of your subscription to *ESPN: The Magazine*, the less you will have to pay per year. Is this an example of direct or inverse variation?

Chapter 14 ▶▶▶ Test

🎓 CHAPTER Test Prep VIDEO CD — Use the Chapter Test Prep Video CD to see fully worked-out solutions to any of the exercises you want to review

1. Find any values for which $\dfrac{3x - 1}{x^2 - 2x - 8}$ is undefined. Write your answer with $\neq$.

1. _____

2. Find the numerical value of $\dfrac{6r + 1}{2r^2 - 3r - 20}$ when

   (a) $r = -2$ and **(b)** $r = 4$.

2. **(a)** _____

   (b) _____

3. Write four rational expressions equivalent to $-\dfrac{6x - 5}{2x + 3}$.

3. _____

Write each rational expression in lowest terms.

4. $\dfrac{-15x^6 y^4}{5x^4 y}$

5. $\dfrac{6a^2 + a - 2}{2a^2 - 3a + 1}$

4. _____

5. _____

Multiply or divide. Write each answer in lowest terms.

6. $\dfrac{5(d - 2)}{9} \div \dfrac{3(d - 2)}{5}$

7. $\dfrac{6k^2 - k - 2}{8k^2 + 10k + 3} \cdot \dfrac{4k^2 + 7k + 3}{3k^2 + 5k + 2}$

6. _____

7. _____

8. $\dfrac{4a^2 + 9a + 2}{3a^2 + 11a + 10} \div \dfrac{4a^2 + 17a + 4}{3a^2 + 2a - 5}$

8. _____

Find the least common denominator for each list of fractions.

9. $\dfrac{-3}{10p^2}, \dfrac{21}{25p^3}, \dfrac{-7}{30p^5}$

10. $\dfrac{r + 1}{2r^2 + 7r + 6}, \dfrac{-2r + 1}{2r^2 - 7r - 15}$

9. _____

10. _____

Write each rational expression as an equivalent expression with the indicated denominator.

11. $\dfrac{15}{4p} = \dfrac{?}{64p^3}$

12. $\dfrac{3}{6m - 12} = \dfrac{?}{42m - 84}$

11. _____

12. _____

Add or subtract. Write each answer in lowest terms.

13. $\dfrac{4x + 2}{x + 5} + \dfrac{-2x + 8}{x + 5}$

14. $\dfrac{-4}{y + 2} + \dfrac{6}{5y + 10}$

13. _____

14. _____

15. $\dfrac{x + 1}{3 - x} - \dfrac{x^2}{x - 3}$

16. $\dfrac{3}{2m^2 - 9m - 5} - \dfrac{m + 1}{2m^2 - m - 1}$

15. _____

16. _____

17. _____

18. _____

19. _____

20. _____

21. _____

22. _____

23. _____

24. _____

25. _____

26. _____

Simplify each complex fraction.

17. $\dfrac{\dfrac{2p}{k^2}}{\dfrac{3p^2}{k^3}}$

18. $\dfrac{\dfrac{1}{x+3}-1}{1+\dfrac{1}{x+3}}$

Solve each equation.

19. $\dfrac{2}{x-1}-\dfrac{2}{3}=\dfrac{-1}{x+1}$

20. $\dfrac{2x}{x-3}+\dfrac{1}{x+3}=\dfrac{-6}{x^2-9}$

21. Solve the formula $F=\dfrac{k}{d-D}$ for D.

Solve each problem.

22. If the same number is added to the numerator and subtracted from the denominator of $\frac{5}{6}$, the resulting fraction is equivalent to $\frac{1}{10}$. What is the number?

23. A boat goes 7 mph in still water. It takes as long to go 20 mi upstream as 50 mi downstream. Find the speed of the current.

24. A man can paint a room in his house, working alone, in 5 hr. His wife can do the job in 4 hr. How long will it take them to paint the room if they work together?

25. If x varies directly as y, and $x=12$ when $y=4$, find x when $y=9$.

26. Under certain conditions, the length of time that it takes for fruit to ripen during the growing season varies inversely as the average maximum temperature during the season. If it takes 25 days for fruit to ripen with an average maximum temperature of 80°F, find the number of days it would take at 75°F. Round your answer to the nearest whole number.

15

Systems of Linear Equations and Inequalities

Over the years, Americans have continued their fascination with Hollywood and the movies. Although many people now prefer to watch movies at home on DVD, the number of tickets sold at domestic movie theaters increased in 2007 to about 1.4 billion, with a total gross of over $9.6 billion. The top box office draws of that year—*Spider-Man 3* and *Shrek the Third*—attracted millions of adults and children wishing to get away from it all for a few hours. (*Source:* www.boxofficemojo.com)

In Exercises 13 and 14 of **Section 15.4,** we use *systems of equations* to find out how much money these top films earned in total and on their opening weekends.

15.1 ▶▶▶ Solving Systems of Linear Equations by Graphing

OBJECTIVES

1 Decide whether a given ordered pair is a solution of a system.

2 Solve linear systems by graphing.

3 Solve special systems by graphing.

4 Identify special systems without graphing.

A **system of linear equations,** often called a **linear system,** consists of two or more linear equations with the same variables.

$$2x + 3y = 4 \qquad x + 3y = 1 \qquad x - y = 1$$
$$3x - y = -5 \qquad -y = 4 - 2x \qquad y = 3$$

Linear systems

In the system on the right, think of $y = 3$ as an equation in two variables by writing it as $0x + y = 3$.

OBJECTIVE 1 Decide whether a given ordered pair is a solution of a system. A **solution of a system** of linear equations is an ordered pair that makes both equations true at the same time. A solution of an equation is said to *satisfy* the equation.

EXAMPLE 1 Determining Whether an Ordered Pair Is a Solution

Is $(4, -3)$ a solution of each system?

(a) $x + 4y = -8$
$3x + 2y = 6$

To decide whether $(4, -3)$ is a solution of the system, substitute 4 for x and -3 for y in each equation.

$$x + 4y = -8 \qquad\qquad 3x + 2y = 6$$
$$4 + 4(-3) \stackrel{?}{=} -8 \qquad\qquad 3(4) + 2(-3) \stackrel{?}{=} 6$$
$$4 + (-12) \stackrel{?}{=} -8 \quad \text{Multiply.} \qquad 12 + (-6) \stackrel{?}{=} 6 \quad \text{Multiply.}$$
$$-8 = -8 \quad \text{True} \qquad\qquad 6 = 6 \quad \text{True}$$

Because $(4, -3)$ satisfies both equations, it is a solution of the system.

(b) $2x + 5y = -7$
$3x + 4y = 2$

Again, substitute 4 for x and -3 for y in both equations.

$$2x + 5y = -7 \qquad\qquad 3x + 4y = 2$$
$$2(4) + 5(-3) \stackrel{?}{=} -7 \qquad\qquad 3(4) + 4(-3) \stackrel{?}{=} 2$$
$$8 + (-15) \stackrel{?}{=} -7 \quad \text{Multiply.} \qquad 12 + (-12) \stackrel{?}{=} 2 \quad \text{Multiply.}$$
$$-7 = -7 \quad \text{True} \qquad\qquad 0 = 2 \quad \text{False}$$

The ordered pair $(4, -3)$ is not a solution of this system because it does not satisfy the second equation.

◀ *Work Problem* **1** *at the Side.*

1 Fill in the blanks, and decide whether the given ordered pair is a solution of the system.

(a) $(2, 5)$

$3x - 2y = -4$
$5x + y = 15$

$$3x - 2y = -4$$
$$3(\underline{\quad}) - 2(\underline{\quad}) \stackrel{?}{=} -4$$

$$5x + y = 15$$
$$5(2) + \underline{\quad} \stackrel{?}{=} \underline{\quad}$$

$(2, 5)$ _____ a solution.
(is/is not)

(b) $(1, -2)$

$x - 3y = 7$
$4x + y = 5$

$(1, -2)$ _____ a solution.
(is/is not)

OBJECTIVE 2 Solve linear systems by graphing. The set of all ordered pairs that are solutions of a system is its **solution set.** One way to find the solution set of a system of two linear equations is to graph both equations on the same axes. The graph of each line shows points whose coordinates satisfy the equation of that line. Any intersection point would be on both lines and would therefore be a solution of *both* equations. ***Thus, the coordinates of any point where the lines intersect give a solution of the system.***

ANSWERS

1. (a) 2; 5; 5; 15; is **(b)** is not

The graph in Figure 1 shows that the solution of the system in Example 1(a) is the intersection point $(4, -3)$. Because *two different* straight lines can intersect at no more than one point, there can never be more than one solution for such a system.

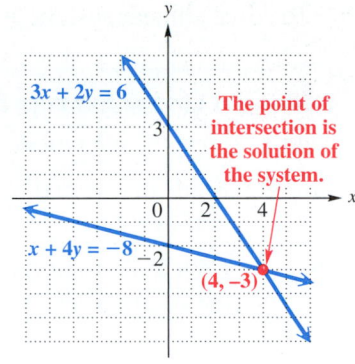

Figure 1

---

2 Solve each system of equations by graphing both equations on the same axes. Check your solutions.

(a) $5x - 3y = 9$
$\quad\;\; x + 2y = 7$

(One of the lines is already graphed.)

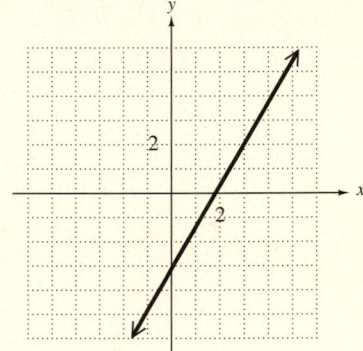

EXAMPLE 2   **Solving a System by Graphing**

Solve the system of equations by graphing both equations on the same axes.

$$2x + 3y = 4$$
$$3x - y = -5$$

We graph these two equations by plotting several points for each line. Recall from **Section 3.2** that the intercepts are often convenient choices.

$2x + 3y = 4$

x	y
0	$\frac{4}{3}$
2	0
-2	$\frac{8}{3}$

Find a third ordered pair as a check.

$3x - y = -5$

x	y
0	5
$-\frac{5}{3}$	0
-2	-1

The lines in Figure 2 suggest that the graphs intersect at the point $(-1, 2)$. We check this by substituting -1 for x and 2 for y in both equations.

Check
$$2x + 3y = 4$$
$$2(-1) + 3(2) \overset{?}{=} 4$$
$$4 = 4 \qquad \text{True}$$

$$3x - y = -5$$
$$3(-1) - 2 \overset{?}{=} -5$$
$$-5 = -5 \qquad \text{True}$$

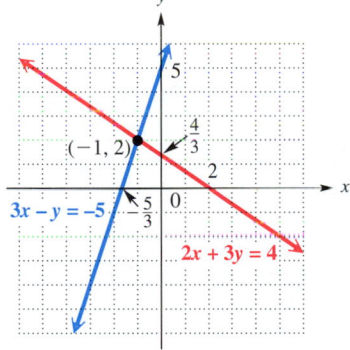

Figure 2

Because $(-1, 2)$ satisfies both equations, the solution set of this system is $\{(-1, 2)\}$.

(b) $x + y = 4$
$\quad\; 2x - y = -1$

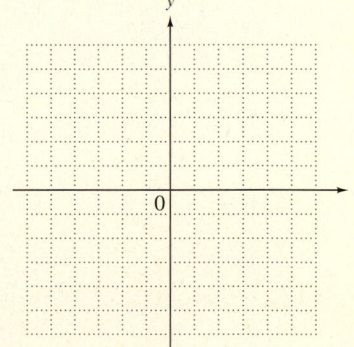

Work Problem **2** *at the Side.* ▶

Note

We can also graph a linear system by writing each equation in the system in slope-intercept form and using the slope and y-intercept to graph each line. For Example 2,

$2x + 3y = 4$ becomes $y = -\frac{2}{3}x + \frac{4}{3}$ y-intercept $(0, \frac{4}{3})$; slope $-\frac{2}{3}$

$3x - y = -5$ becomes $y = 3x + 5$. y-intercept $(0, 5)$; slope 3, or $\frac{3}{1}$

Confirm that graphing these equations results in the same lines and the same solution shown in Figure 2.

To solve a linear system by graphing, follow these steps.

Solving a Linear System by Graphing

Step 1   **Graph each equation** of the system on the same coordinate axes.

Step 2   **Find the coordinates of the point of intersection** of the graphs if possible. This is the solution of the system.

Step 3   **Check** the solution in *both* of the original equations. Then write the solution set.

CAUTION

A difficulty with the graphing method of solution is that it may not be possible to determine from the graph the exact coordinates of the point that represents the solution, particularly if these coordinates are not integers. For this reason, algebraic methods of solution are explained later in this chapter. The graphing method does, however, show geometrically how solutions are found and is useful when approximate answers will do.

OBJECTIVE 3 **Solve special systems by graphing.**   Sometimes the graphs of the two equations in a system either do not intersect at all or are the same line, as in the systems in Example 3.

EXAMPLE 3   **Solving Special Systems**

Solve each system by graphing.

(a) $2x + y = 2$

$2x + y = 8$

The graphs of these lines are shown in Figure 3. The two lines are parallel and have no points in common. For such a system, there is no solution. We write the solution set as $\emptyset$.

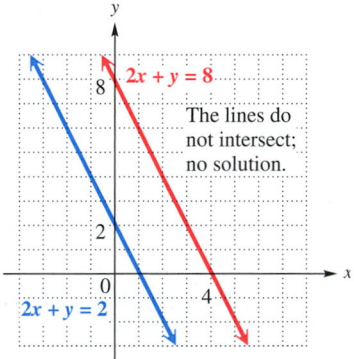

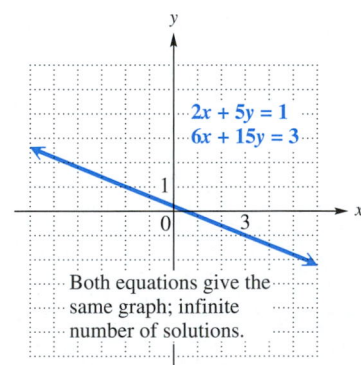

Figure 3   Figure 4

(b) $2x + 5y = 1$

$6x + 15y = 3$

The graphs of these two equations are the same line. See Figure 4. The second equation can be obtained by multiplying each side of the first equation by 3. In this case, every point on the line is a solution of the system, and the solution set contains an infinite number of ordered pairs that satisfy the equations.

Continued on Next Page

We write the solution set as

$$\{(x, y) \mid 2x + 5y = 1\},$$

> This is the first equation in the system. See the Note below.

read "the set of ordered pairs (x, y) such that $2x + 5y = 1$." Recall from **Section 9.3** that this notation is called **set-builder notation.**

Note

When a system has an infinite number of solutions, as in Example 3(b), either equation of the system could be used to write the solution set. *We prefer to use the equation in standard form with coefficients that are integers having greatest common factor 1.* If neither of the given equations of the system is in this form, use an *equivalent* equation that is in standard form with coefficients that are integers having greatest common factor 1 to write the solution set with set-builder notation.

Work Problem **3** *at the Side.* ▶

The system in Example 2 has exactly one solution. A system with at least one solution is called a **consistent system.** A system of equations with no solution, such as the one in Example 3(a), is called an **inconsistent system.** The equations in Example 2 are **independent equations** with different graphs. The equations of the system in Example 3(b) have the same graph and are equivalent. Because they are different forms of the same equation, these equations are called **dependent equations.**

Examples 2 and 3 show the three cases that may occur when solving a system of two equations with two variables.

Three Cases for Solutions of Systems

1. The graphs intersect at exactly one point, which gives the (single) ordered-pair solution of the system. The **system is consistent** and the **equations are independent.** See Figure 5(a).

2. The graphs are parallel lines, so there is no solution and the solution set is ∅. The **system is inconsistent** and the **equations are independent.** See Figure 5(b).

3. The graphs are the same line. There is an infinite number of solutions, and the solution set is written in set-builder notation as $\{(x, y) \mid \underline{\hspace{2cm}}\}$, where one of the equations is written after the | symbol. The **system is consistent** and the **equations are dependent.** See Figure 5(c).

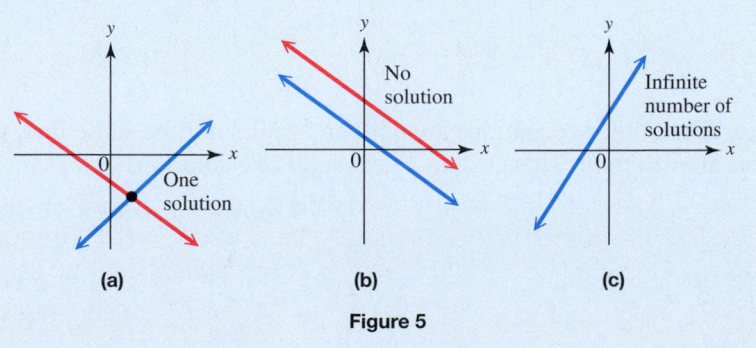

(a)　　　(b)　　　(c)

Figure 5

---

3 Solve each system of equations by graphing both equations on the same axes.

(a) $3x - y = 4$

$6x - 2y = 12$

(One of the lines is already graphed.)

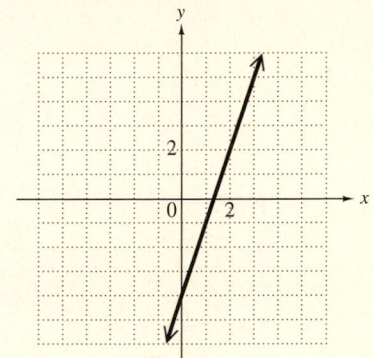

(b) $x - 3y = -2$

$2x - 6y = -4$

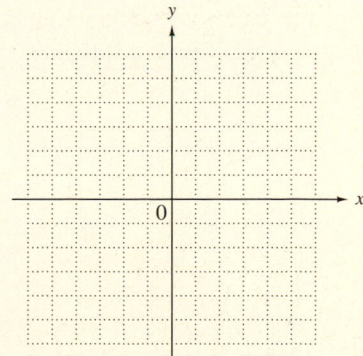

4 Describe each system without graphing. State the number of solutions.

(a) $2x - 3y = 5$
$3y = 2x - 7$

(b) $-x + 3y = 2$
$2x - 6y = -4$

(c) $6x + y = 3$
$2x - y = -11$

OBJECTIVE 4 **Identify special systems without graphing.** Example 3 showed that the graphs of an inconsistent system are parallel lines and the graphs of a system of dependent equations are the same line. We can recognize these special kinds of systems without graphing by using slopes.

EXAMPLE 4 **Identifying the Three Cases by Using Slopes**

Describe each system without graphing. State the number of solutions.

(a) $3x + 2y = 6$
$-2y = 3x - 5$
    Write each equation in slope-intercept form, $y = mx + b$, by solving for y.

$3x + 2y = 6$	$-2y = 3x - 5$
$2y = -3x + 6$ Subtract $3x$.	
$y = -\dfrac{3}{2}x + 3$ Divide by 2.	$y = -\dfrac{3}{2}x + \dfrac{5}{2}$ Divide by -2.

Both equations have slope $-\frac{3}{2}$ but they have different y-intercepts, 3 and $\frac{5}{2}$. In **Section 11.3,** we found that lines with the same slope are parallel, so these equations have graphs that are parallel lines. Thus, the system has no solution.

(b) $2x - y = 4$
$x = \dfrac{y}{2} + 2$

Again, write the equations in slope-intercept form.

$2x - y = 4$	$x = \dfrac{y}{2} + 2$
$-y = -2x + 4$	$\dfrac{y}{2} + 2 = x$
$y = 2x - 4$	$\dfrac{y}{2} = x - 2$
	$y = 2x - 4$

The equations are exactly the same—their graphs are the same line. Thus, the system has an infinite number of solutions.

(c) $x - 3y = 5$
$2x + y = 8$
In slope-intercept form, the equations are as follows.

$x - 3y = 5$	$2x + y = 8$
$-3y = -x + 5$	$y = -2x + 8$
$y = \dfrac{1}{3}x - \dfrac{5}{3}$	

The graphs of these equations are neither parallel nor the same line, since the slopes are different. This system has exactly one solution.

◀ *Work Problem* **4** *at the Side.*

15.1 ▷▷▷ Exercises

1. Which ordered pair could be a solution of the system graphed? Why is it the only valid choice?

A. $(2, 2)$

B. $(-2, 2)$

C. $(-2, -2)$

D. $(2, -2)$

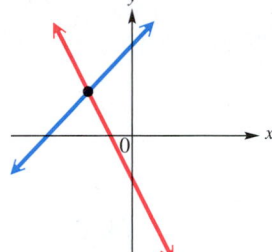

2. Which ordered pair could be a solution of the system graphed? Why is it the only valid choice?

A. $(2, 0)$

B. $(0, 2)$

C. $(-2, 0)$

D. $(0, -2)$

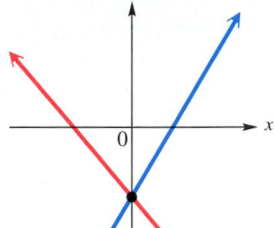

3. How can you tell without graphing that this system has no solution?

$$x + y = 2$$
$$x + y = 4$$

4. Explain why a system of two linear equations cannot have exactly two solutions.

Decide whether the given ordered pair is a solution of the given system. See Example 1.

5. $(2, -3)$
$$x + y = -1$$
$$2x + 5y = 19$$

6. $(4, 3)$
$$x + 2y = 10$$
$$3x + 5y = 3$$

7. $(-1, -3)$
$$3x + 5y = -18$$
$$4x + 2y = -10$$

8. $(-9, -2)$
$$2x - 5y = -8$$
$$3x + 6y = -39$$

9. $(7, -2)$
$$4x = 26 - y$$
$$3x = 29 + 4y$$

10. $(9, 1)$
$$2x = 23 - 5y$$
$$3x = 24 + 3y$$

11. $(6, -8)$
$$-2y = x + 10$$
$$3y = 2x + 30$$

12. $(-5, 2)$
$$5y = 3x + 20$$
$$3y = -2x - 4$$

Solve each system of equations by graphing. If the system is inconsistent or the equations are dependent, say so. See Examples 2 and 3.

13. $x - y = 2$
$x + y = 6$

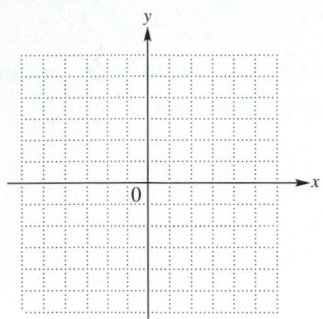

14. $x - y = 3$
$x + y = -1$

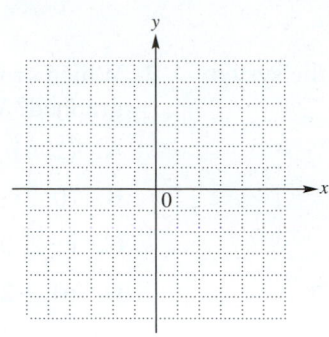

15. $x + y = 4$
$y - x = 4$

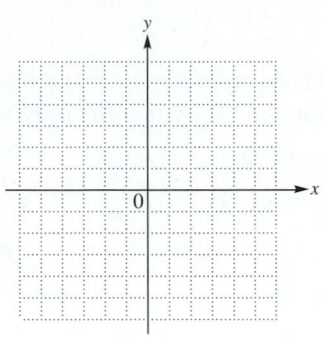

16. $x + y = -5$
$x - y = 5$

17. $x - 2y = 6$
$x + 2y = 2$

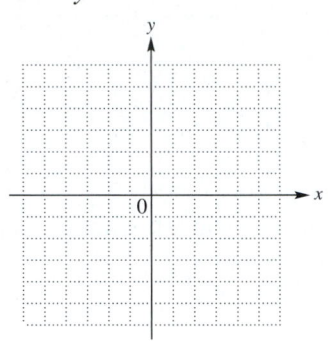

18. $2x - y = 4$
$4x + y = 2$

19. $3x - 2y = -3$
$-3x - y = -6$

20. $2x - y = 4$
$2x + 3y = 12$

21. $2x - 3y = -6$
$y = -3x + 2$

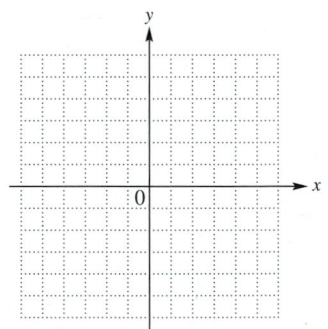

22. $-3x + y = -3$
$y = x - 3$

23. $x + 2y = 6$
$2x + 4y = 8$

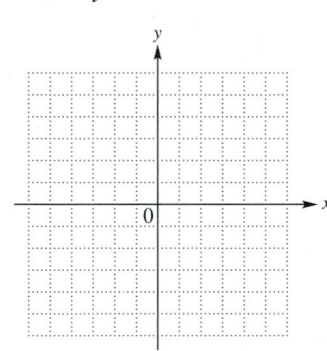

24. $2x - y = 6$
$6x - 3y = 12$

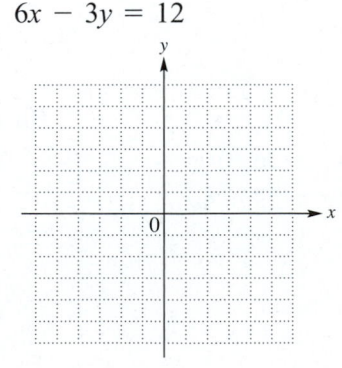

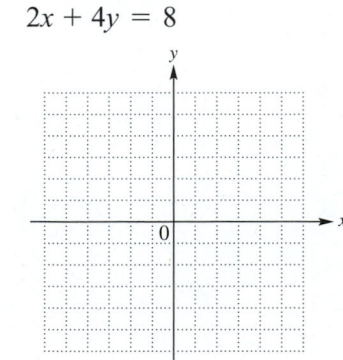

25. $4x - 2y = 8$
$2x = y + 4$

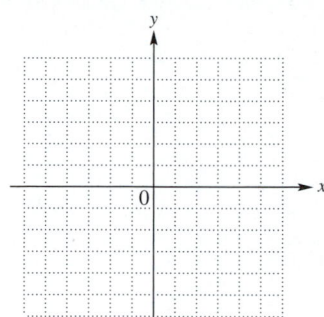

26. $3x = 5 - y$
$6x + 2y = 10$

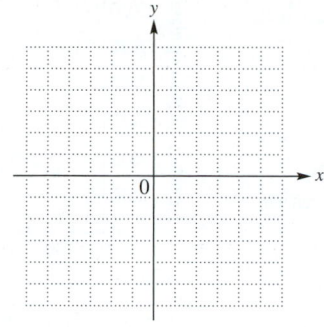

27. $3x - 4y = 24$
$$y = -\frac{3}{2}x + 3$$

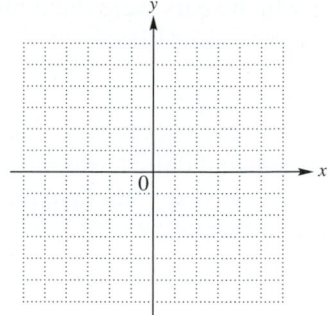

28. $3x - 2y = 12$
$y = -4x + 5$

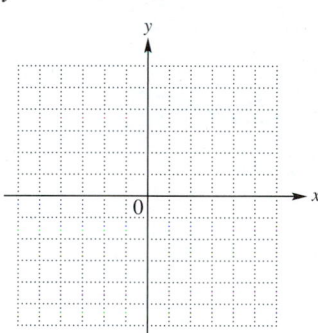

29. $3x = y + 5$
$6x - 5 = 2y$

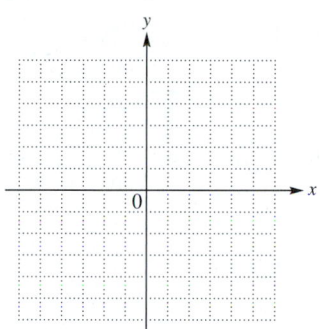

30. $2x = y - 4$
$4x - 2y = -4$

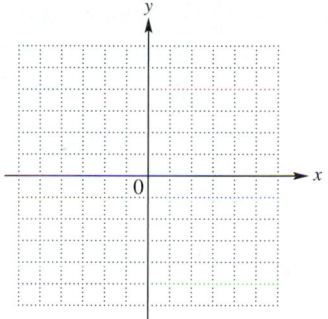

Without graphing, answer the following questions for each linear system. See Example 4.

(a) *Is the system inconsistent, are the equations dependent, or neither?*
(b) *Is the graph a pair of intersecting lines, a pair of parallel lines, or one line?*
(c) *Does the system have one solution, no solution, or an infinite number of solutions?*

31. $y - x = -5$
$x + y = 1$

32. $2x + y = 6$
$x - 3y = -4$

33. $x + 2y = 0$
$4y = -2x$

34. $y = 3x$
$y + 3 = 3x$

35. $5x + 4y = 7$
$10x + 8y = 4$

36. $4x - 6y = 10$
$-6x + 9y = -15$

The numbers of daily morning and evening newspapers in the United States in selected years are shown in the graph. Use the graph to work Exercises 37 and 38.

37. For which years were there more evening dailies than morning dailies?

38. Estimate the year in which the number of evening and morning dailies was closest to the same. About how many newspapers of each type were there in that year?

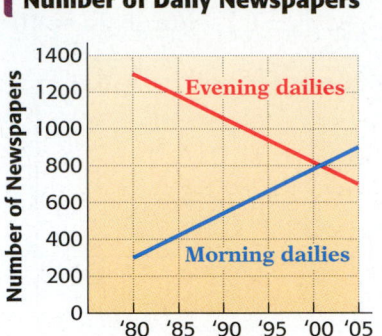

Number of Daily Newspapers

Source: Editor & Publisher International Year Book.

Work Exercises 39 and 40 using the graphs provided.

39. The graph shows how college students managed their money during the period 1997–2004.

(a) During what period did ATM use dominate both credit card *and* debit card use?

(b) In what year did debit card use overtake credit card use?

(c) In what year did debit card use overtake ATM use?

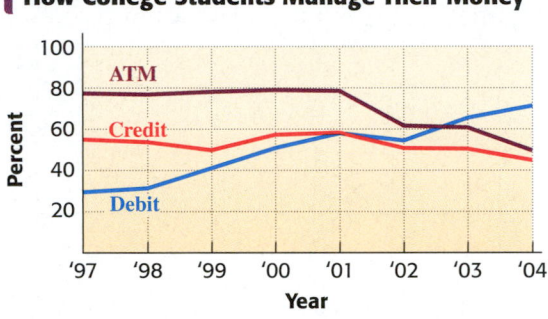

How College Students Manage Their Money

Source: Georgetown University Credit Research Center.

(d) Write an ordered pair for the debit card use data in the year 2004.

(e) Describe the trend in debit card use over this period.

40. The graph shows how the average viewing hours for broadcast TV and cable/satellite TV in the United States has changed during the period 1998–2004.

(a) In approximately what year did Americans spend almost the same number of hours watching broadcast and cable/satellite TV? How many hours per year was this?

(b) Express the point of intersection of the two graphs as an ordered pair of the form (year, hours).

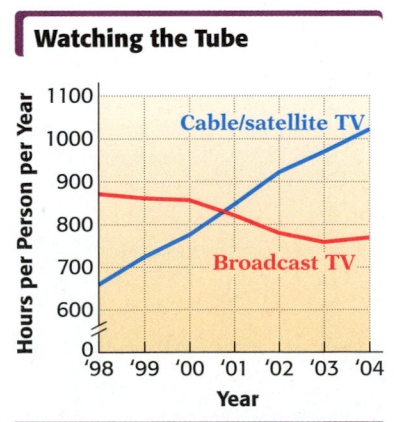

Watching the Tube

Source: Veronis Suhler Stevenson.

(c) During what period was the time spent watching broadcast TV almost constant?

15.2 ▶▶▶ Solving Systems of Linear Equations by Substitution

OBJECTIVE 1 Solve linear systems by substitution.

Work Problem ① *at the Side.* ▶

As we saw in Problem 1 at the side, graphing to solve a system of equations has a serious drawback: It is difficult to accurately find a solution such as $\left(\frac{11}{3}, -\frac{4}{9}\right)$ from a graph. One algebraic method for solving a system of equations is the **substitution method.**

EXAMPLE 1 **Using the Substitution Method**

Solve the system by the substitution method.

$$3x + 5y = 26$$
$$y = 2x$$

The second equation is already solved for y. This equation says that $y = 2x$. Substituting $2x$ for y in the first equation gives

$$3x + 5y = 26$$
$$3x + 5(\mathbf{2x}) = 26 \qquad \text{Let } y = 2x.$$
$$3x + 10x = 26 \qquad \text{Multiply.}$$
$$13x = 26 \qquad \text{Combine like terms.}$$
$$x = 2. \qquad \text{Divide by 13.}$$

Because $x = 2$, we find y from the equation $y = 2\mathbf{x}$ by substituting 2 for x.

$$y = 2(\mathbf{2}) = \mathbf{4} \qquad \text{Let } x = 2.$$

Check that the solution set of the given system is $\{(\mathbf{2}, \mathbf{4})\}$ by substituting 2 for x and 4 for y in *both* equations.

Work Problem ② *at the Side.* ▶

EXAMPLE 2 **Using the Substitution Method**

Solve the system by the substitution method.

$$2x + 5y = 7$$
$$x = -1 - y$$

The second equation gives x in terms of y. Substitute $-1 - y$ for x in the first equation.

$$2x + 5y = 7$$
$$2(\mathbf{-1 - y}) + 5y = 7 \qquad \text{Let } x = -1 - y.$$

> Distribute 2 to *both* -1 and $-y$.

$$-2 - 2y + 5y = 7 \qquad \text{Distributive property}$$
$$-2 + 3y = 7 \qquad \text{Combine like terms.}$$
$$3y = 9 \qquad \text{Add 2.}$$
$$y = 3 \qquad \text{Divide by 3.}$$

To find x, substitute 3 for y in the equation $x = -1 - y$ to get

$$x = -1 - \mathbf{3} = \mathbf{-4}.$$

> Write the x-coordinate first.

Check that the solution set of the given system is $\{(\mathbf{-4}, \mathbf{3})\}$.

Work Problem ③ *at the Side.* ▶

OBJECTIVES

1 Solve linear systems by substitution.

2 Solve special systems by substitution.

3 Solve linear systems with fractions and decimals by substitution.

① Solve the system by graphing.

$$2x + 3y = 6$$
$$x - 3y = 5$$

Can you determine the answer? Why or why not?

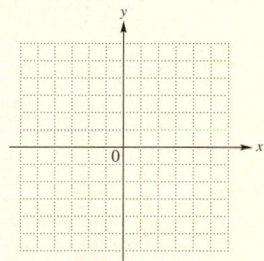

② Fill in the blanks to solve by the substitution method. Check your solution.

$$3x + 5y = 69$$
$$y = 4x$$
$$3x + 5(\underline{\quad}) = 69$$
$$\underline{\quad} = 69$$
$$x = \underline{\quad}$$
$$y = 4(\underline{\quad}) = \underline{\quad}$$

The solution set is $\underline{\quad}$.

③ Solve by the substitution method. Check your solution.

$$2x + 7y = -12$$
$$x = 3 - 2y$$

ANSWERS

1. The answer cannot be determined from the graph because it is too difficult to read the exact coordinates.
2. $4x$; $23x$; 3; 3; 12; $\{(3, 12)\}$
3. $\{(15, -6)\}$

4 Solve each system by substitution. Check each solution.

(a) Fill in the blanks to solve

$$x + 4y = -1$$
$$2x - 5y = 11.$$

Solve the first equation for x.

$$x = -1 - \underline{\quad}$$

Substitute into the second equation to find y.

$$2(\underline{\quad}) - 5y = 11$$
$$-2 - 8y - 5y = 11$$
$$-2 - \underline{\quad} y = 11$$
$$\underline{\quad} y = 13$$
$$y = \underline{\quad}$$

Find x.

$$x = -1 - \underline{\quad}$$
$$x = \underline{\quad}$$

The solution set is $\underline{\quad}$.

(b) $2x + 5y = 4$
$x + \ y = -1$

CAUTION

Even though we found y first in Example 2, *the x-coordinate is always written first in the ordered-pair solution of a system.*

To solve a system by substitution, follow these steps.

Solving a Linear System by Substitution

Step 1   **Solve one equation for either variable.** If one of the variables has coefficient 1 or -1, choose it, since it usually makes the substitution easier.

Step 2   **Substitute** for that variable in the other equation. The result should be an equation with just one variable.

Step 3   **Solve** the equation from Step 2.

Step 4   **Substitute** the result from Step 3 into the equation from Step 1 to find the value of the other variable.

Step 5   **Check** the solution in *both* of the original equations. Then write the solution set.

EXAMPLE 3   **Using the Substitution Method**

Use substitution to solve the system.

$$2x = 4 - y \qquad (1)$$
$$5x + 3y = 10 \qquad (2)$$

Step 1   For the substitution method, we must solve one of the equations for either x or y. Because the coefficient of y in equation (1) is -1, we choose equation (1) and solve for y.

$$2x = 4 - y \qquad (1)$$
$$y + 2x = 4 \qquad \text{Add } y.$$
$$y = -2x + 4 \qquad \text{Subtract } 2x.$$

Step 2   Now substitute $-2x + 4$ for y in equation (2).

$$5x + 3y = 10 \qquad (2)$$
$$5x + 3(\mathbf{-2x + 4}) = 10 \qquad \text{Let } y = -2x + 4.$$

Step 3   Now solve the equation from Step 2.

> Distribute 3 to both $-2x$ and 4.

$$5x - 6x + 12 = 10 \qquad \text{Distributive property}$$
$$-x + 12 = 10 \qquad \text{Combine like terms.}$$
$$-x = -2 \qquad \text{Subtract 12.}$$
$$x = 2 \qquad \text{Multiply by } -1.$$

Step 4   Since $y = -2x + 4$ and $x = 2$, $y = -2(\mathbf{2}) + 4 = \mathbf{0}$.

Step 5   Check that $(\mathbf{2}, \mathbf{0})$ is the solution.

Check

$$2x = 4 - y \qquad (1) \qquad\qquad 5x + 3y = 10 \qquad (2)$$
$$2(\mathbf{2}) \stackrel{?}{=} 4 - \mathbf{0} \qquad\qquad\qquad 5(\mathbf{2}) + 3(\mathbf{0}) \stackrel{?}{=} 10$$
$$4 = 4 \qquad \text{True} \qquad\qquad\qquad 10 = 10 \qquad \text{True}$$

Since both results are true, the solution set of the system is $\{(2, 0)\}$.

◀ **Work Problem** **4** at the Side.

EXAMPLE 4   **Using the Substitution Method**

Use substitution to solve the system.

$$2x + 3y = 10 \quad (1)$$
$$-3x - 2y = 0 \quad (2)$$

Step 1   To use the substitution method, we must solve one of the equations for one of the variables. We choose equation (1) and solve for x.

$$2x + 3y = 10 \qquad (1)$$
$$2x = 10 - 3y \qquad \text{Subtract } 3y.$$
$$x = 5 - \frac{3}{2}y \qquad \text{Divide by 2.}$$

Step 2   Substitute this expression for x in equation (2).

$$-3x - 2y = 0 \qquad (2)$$
$$-3\left(5 - \frac{3}{2}y\right) - 2y = 0 \qquad \text{Let } x = 5 - \tfrac{3}{2}y.$$

Step 3 $$-15 + \frac{9}{2}y - 2y = 0 \qquad \text{Distributive property}$$

$\boxed{\begin{aligned}-3(-\tfrac{3}{2}) &= (-\tfrac{3}{1})(-\tfrac{3}{2})\\ &= \tfrac{9}{2}\end{aligned}}$

$$-15 + \frac{5}{2}y = 0 \qquad \text{Combine like terms.}$$
$$\frac{5}{2}y = 15 \qquad \text{Add 15.}$$

$\boxed{\begin{aligned}\tfrac{2}{5}(\tfrac{5}{2}y) &= (\tfrac{2}{5}\cdot\tfrac{5}{2})y\\ &= 1\cdot y = y\end{aligned}}$

$$y = \frac{30}{5} = 6 \qquad \text{Multiply by } \tfrac{2}{5}.$$

Step 4   Find x by substituting **6** for y in $x = 5 - \tfrac{3}{2}y$.

$$x = 5 - \frac{3}{2}(6) = -4$$

Step 5   Check that $(-4, 6)$ is the solution.

Check

$$2x + 3y = 10 \quad (1) \qquad\qquad -3x - 2y = 0 \quad (2)$$
$$2(-4) + 3(6) \stackrel{?}{=} 10 \qquad\qquad -3(-4) - 2(6) \stackrel{?}{=} 0$$
$$-8 + 18 \stackrel{?}{=} 10 \qquad\qquad 12 - 12 \stackrel{?}{=} 0$$
$$10 = 10 \quad \text{True} \qquad\qquad 0 = 0 \quad \text{True}$$

Both results are true, so the solution set of the system is $\{(-4, 6)\}$.

Note

In Example 4, we could have started the solution by solving the second equation for either x or y and then substituting the result into the first equation. The solution would be the same.

Work Problem **5** *at the Side.* ▶

---

5 Solve the system by substitution. Check your solution.

$$3x + 2y = 1$$
$$3x - 4y = -11$$

OBJECTIVE 2 Solve special systems by substitution. We can solve inconsistent systems with graphs that are parallel lines and systems of dependent equations with graphs that are the same line using the substitution method.

EXAMPLE 5 Solving an Inconsistent System by Substitution

Use substitution to solve the system.

$$x = 5 - 2y \qquad (1)$$
$$2x + 4y = 6 \qquad (2)$$

Substitute $5 - 2y$ for x in equation (2).

$$2x + 4y = 6 \qquad (2)$$
$$2(5 - 2y) + 4y = 6 \qquad \text{Let } x = 5 - 2y.$$
$$10 - 4y + 4y = 6 \qquad \text{Distributive property}$$
$$10 = 6 \qquad \text{False}$$

This false result means that the equations in the system have graphs that are parallel lines. The system is inconsistent, and the solution set is $\emptyset$. See Figure 6.

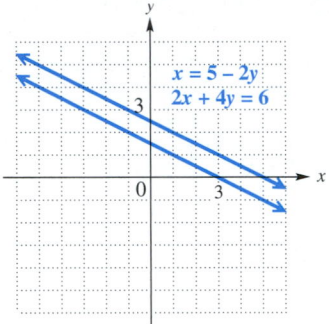

Figure 6

CAUTION
It is a common error to give "false" as the answer to an inconsistent system. The correct response is $\emptyset$.

EXAMPLE 6 Solving a System with Dependent Equations by Substitution

Solve the system by the substitution method.

$$3x - y = 4 \qquad (1)$$
$$-9x + 3y = -12 \qquad (2)$$

Begin by solving equation (1) for y to get $y = 3x - 4$. Substitute $3x - 4$ for y in equation (2) and solve the resulting equation.

$$-9x + 3y = -12 \qquad (2)$$
$$-9x + 3(3x - 4) = -12 \qquad \text{Let } y = 3x - 4.$$
$$-9x + 9x - 12 = -12 \qquad \text{Distributive property}$$
$$0 = 0 \qquad \text{Add 12; combine like terms.}$$

Continued on Next Page

This true result means that every solution of one equation is also a solution of the other, so the system has an infinite number of solutions—all the ordered pairs corresponding to points that lie on the common graph. The solution set is $\{(x, y) \mid 3x - y = 4\}$. A graph of the equations of this system is shown in Figure 7.

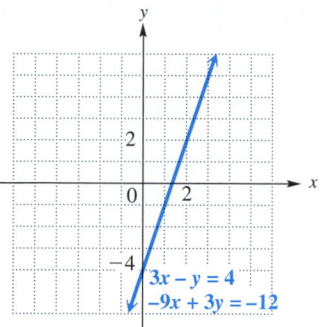

Figure 7

CAUTION

It is a common error to give "true" as the solution of a system of dependent equations. Remember that we give the solution set in set-builder notation using an equation that is in standard form, with integer coefficients having greatest common factor 1.

Work Problem ⑥ *at the Side.* ▶

OBJECTIVE 3 Solve linear systems with fractions and decimals by substitution.   When a system includes an equation with fractions as coefficients, eliminate the fractions by multiplying each side of the equation by a common denominator. Then solve the resulting system.

EXAMPLE 7 Using the Substitution Method with Fractions as Coefficients

Solve the system by the substitution method.

$$3x + \frac{1}{4}y = 2 \qquad (1)$$

$$\frac{1}{2}x + \frac{3}{4}y = -\frac{5}{2} \qquad (2)$$

Clear equation (1) of fractions by multiplying each side by 4.

$$4\left(3x + \frac{1}{4}y\right) = 4\,(2) \qquad \text{Multiply by 4.}$$

$$4\,(3x) + 4\left(\frac{1}{4}y\right) = 4\,(2) \qquad \text{Distributive property}$$

$$12x + y = 8 \qquad (3)$$

Continued on Next Page

---

⑥ Solve each system by substitution.

(a) $8x - y = 4$
$\qquad\quad y = 8x + 4$

(b) $\qquad 7x - 6y = 10$
$\qquad -14x + 20 = -12y$

7 Solve the system by substitution. First clear all fractions.

$$\frac{2}{3}x + \frac{1}{2}y = 6$$

$$\frac{1}{2}x - \frac{3}{4}y = 0$$

Now clear equation (2) of fractions by multiplying each side by the common denominator 4.

$$\frac{1}{2}x + \frac{3}{4}y = -\frac{5}{2} \qquad (2)$$

$$4\left(\frac{1}{2}x + \frac{3}{4}y\right) = 4\left(-\frac{5}{2}\right) \qquad \text{Multiply by 4.}$$

$$4\left(\frac{1}{2}x\right) + 4\left(\frac{3}{4}y\right) = 4\left(-\frac{5}{2}\right) \qquad \text{Distributive property}$$

$$2x + 3y = -10 \qquad (4)$$

The given system of equations has been simplified to the equivalent system

$$12x + y = 8 \qquad (3)$$

$$2x + 3y = -10. \qquad (4)$$

To solve this system by substitution, equation (3) can be solved for y.

$$12x + y = 8 \qquad (3)$$

$$y = -12x + 8 \qquad \text{Subtract } 12x.$$

Now substitute this result for y in equation (4).

$$2x + 3y = -10 \qquad (4)$$

$$2x + 3(-12x + 8) = -10 \qquad \text{Let } y = -12x + 8.$$

$$2x - 36x + 24 = -10 \qquad \text{Distributive property}$$

Distribute 3 to both $-12x$ and 8.

$$-34x = -34 \qquad \text{Combine like terms; subtract 24.}$$

$$x = 1 \qquad \text{Divide by } -34.$$

Substitute 1 for x in $y = -12x + 8$ to get

$$y = -12(1) + 8 = -4.$$

8 Complete the process of solving the system given in Example 8 by substitution. (*Hint:* Solve equation (4) for x. Then substitute this result for x in equation (3) to find y.)

Check by substituting 1 for x and -4 for y in both of the original equations. The solution set is $\{(1, -4)\}$.

◀ *Work Problem* **7** *at the Side.*

If any of the coefficients in the equations of a system are decimals, we can eliminate the decimals by multiplying by a power of 10, as we did when solving linear equations with decimal coefficients in **Section 10.3.**

EXAMPLE 8 **Using the Substitution Method with Decimals as Coefficients**

Solve the system by the substitution method.

$$0.5x + 2.4y = 4.2 \qquad (1)$$

$$-0.1x + 1.5y = 5.1 \qquad (2)$$

To eliminate (or "clear") decimals, multiply each equation by 10.

$$5x + 24y = 42 \qquad (3)$$

$$-x + 15y = 51 \qquad (4) \quad (-0.1x) \cdot 10 = -1x = -x$$

Now we can solve this equivalent system by substitution.

◀ *Work problem* **8** *at the Side.*

1. A student solves the system

$$5x - y = 15$$
$$7x + y = 21$$

and finds that $x = 3$, which is the correct value for x. The student gives the solution set as $\{3\}$. **WHAT WENT WRONG?**

2. When you use the substitution method, how can you tell that a system has

(a) no solution?   **(b)** an infinite number of solutions?

Solve each system by the substitution method. Check each solution. See Examples 1–8.

3. $x + y = 12$
   $y = 3x$

4. $x + 3y = -28$
   $y = -5x$

5. $3x + 2y = 27$
   $x = y + 4$

6. $4x + 3y = -5$
   $x = y - 3$

7. $3x + 5y = 14$
   $x - 2y = -10$

8. $5x + 2y = -1$
   $2x - y = -13$

9. $3x + 4 = -y$
   $2x + y = 0$

10. $2x - 5 = -y$
   $x + 3y = 0$

11. $7x + 4y = 13$
   $x + y = 1$

12. $3x - 2y = 19$
   $x + y = 8$

13. $3x - y = 5$
   $y = 3x - 5$

14. $4x - y = -3$
   $y = 4x + 3$

15. $6x - 8y = 6$
   $2y = -2 + 3x$

16. $3x + 2y = 6$
   $6x = 8 + 4y$

17. $2x + 8y = 3$
   $x = 8 - 4y$

18. $2x + 10y = 3$
   $x = 1 - 5y$

19. $12x - 16y = 8$
   $3x = 4y + 2$

20. $6x + 9y = 6$
   $2x = 2 - 3y$

21. $\dfrac{1}{5}x + \dfrac{2}{3}y = -\dfrac{8}{5}$
   $3x - y = 9$

22. $\dfrac{1}{3}x - \dfrac{1}{2}y = \dfrac{1}{6}$
   $3x - 2y = 9$

23. $\dfrac{x}{2} - \dfrac{y}{3} = 9$
   $\dfrac{x}{5} - \dfrac{y}{4} = 5$

24. $\dfrac{1}{6}x + \dfrac{1}{6}y = 2$
   $-\dfrac{1}{2}x - \dfrac{1}{3}y = -8$

25. $\dfrac{x}{5} + 2y = \dfrac{16}{5}$
   $\dfrac{3x}{5} + \dfrac{y}{2} = -\dfrac{7}{5}$

26. $\dfrac{x}{3} - \dfrac{3y}{4} = -\dfrac{1}{2}$
   $\dfrac{x}{6} + \dfrac{y}{8} = \dfrac{3}{4}$

27. $0.1x + 0.9y = -2$
   $0.5x - 0.2y = 4.1$

28. $0.2x - 1.3y = -3.2$
   $-0.1x + 2.7y = 9.8$

29. $0.08x - 0.01y = 1.3$
   $0.22x + 0.15y = 8.9$

Relating Concepts (Exercises 30–33)   For Individual or Group Work

A system of linear equations can be used to model the cost and the revenue of a business. **Work Exercises 30–33 in order.**

30. Suppose that you start a business manufacturing and selling bicycles, and it costs you $5000 to get started. You determine that each bicycle will cost $400 to manufacture. Explain why the linear equation $y_1 = 400x + 5000$ gives your *total* cost to manufacture x bicycles (y_1 in dollars).

31. You decide to sell each bike for $600. What expression in x represents the revenue you will take in if you sell x bikes? Write an equation using y_2 to express your revenue when you sell x bikes (y_2 in dollars).

32. Form a system from the two equations in Exercises 30 and 31, and then solve the system, assuming $y_1 = y_2$, that is, cost = revenue.

33. The value of x from Exercise 32 is the number of bikes it takes to *break even*. Fill in the blanks: When _____ bikes are sold, the break-even point is reached. At that point, you have spent _____ dollars and taken in _____ dollars.

15.3 ▶▶▶ Solving Systems of Linear Equations by Elimination

OBJECTIVE 1 Solve linear systems by elimination. An algebraic method that depends on the addition property of equality can be used to solve systems. As mentioned earlier, adding the same quantity to each side of an equation results in equal sums.

$$\text{If } A = B, \quad \text{then} \quad A + C = B + C.$$

This addition can be taken a step further. Adding *equal* quantities, rather than the *same* quantity, to both sides of an equation also results in equal sums.

$$\text{If } A = B \quad \text{and} \quad C = D, \quad \text{then} \quad A + C = B + D.$$

Using the addition property to solve systems is called the **elimination method.** When using this method, the idea is to *eliminate* one of the variables. *To do this, one of the variables in the two equations must have coefficients that are opposites*.

OBJECTIVES

1 Solve linear systems by elimination.

2 Multiply when using the elimination method.

3 Use an alternative method to find the second value in a solution.

4 Use the elimination method to solve special systems.

EXAMPLE 1 Using the Elimination Method

Use the elimination method to solve the system.

$$x + y = 5$$
$$x - y = 3$$

Each equation in this system is a statement of equality, so the sum of the left sides equals the sum of the right sides. Adding in this way gives

$$(x + y) + (x - y) = 5 + 3.$$

Combine like terms and simplify to get

$$2x = 8$$
$$x = 4. \quad \text{Divide by 2.}$$

Notice that y has been eliminated. The result, $x = 4$, gives the x-value of the solution of the given system. To find the y-value of the solution, substitute 4 for x in either of the two equations of the system.

Work Problem **1** *at the Side.* ▶

Check the solution set found at the side, $\{(4, 1)\}$, by substituting 4 for x and 1 for y in both equations of the given system.

Check

$x + y = 5$	$x - y = 3$
$4 + 1 \overset{?}{=} 5$	$4 - 1 \overset{?}{=} 3$
$5 = 5$ True	$3 = 3$ True

Since both results are true, the solution set of the system is $\{(4, 1)\}$.

1 (a) Substitute 4 for x in the equation $x + y = 5$ to find the value of y.

(b) Give the solution set of the system.

CAUTION

A system is not completely solved until values for both x and y are found. Do not stop after finding the value of only one variable. Remember to write the solution set as a set containing an ordered pair.

ANSWERS

1. (a) $y = 1$ **(b)** $\{(4, 1)\}$

◀ *Work Problem* ② *at the Side.*

② Solve each system by the elimination method. Check each solution.

(a) Fill in the blanks to solve the following system.

$$x + y = 8$$
$$x - y = 2$$

Add.

$$(x + y) + (x - y) = 8 + \underline{}$$
$$2\underline{} = \underline{}$$
$$x = \underline{}$$

Find y.

$$x - y = 2$$
$$\underline{} - y = 2$$
$$-y = \underline{}$$
$$y = \underline{}$$

The solution set is ___.

(b) $3x - y = 7$
$2x + y = 3$

In general, to solve a system by elimination, follow these steps.

> **Solving a Linear System by Elimination**
>
> *Step 1* **Write both equations in standard form** $Ax + By = C$.
>
> *Step 2* **Transform so that the coefficients of one pair of variable terms are opposites.** Multiply one or both equations by appropriate numbers so that the sum of the coefficients of either the x- or y-terms is 0.
>
> *Step 3* **Add** the new equations to eliminate a variable. The sum should be an equation with just one variable.
>
> *Step 4* **Solve** the equation from Step 3 for the remaining variable.
>
> *Step 5* **Substitute** the result from Step 4 into *either* of the original equations and solve for the other variable.
>
> *Step 6* **Check** the solution in *both* of the original equations. Then write the solution set.

It does not matter which variable is eliminated first. Usually we choose the one that is more convenient to work with.

EXAMPLE 2 **Using the Elimination Method**

Solve the system.

$$y + 11 = 2x$$
$$5x = y + 26$$

Step 1 Rewrite both equations in the form $Ax + By = C$ to get the system

$$-2x + y = -11 \qquad \text{Subtract } 2x \text{ and } 11.$$
$$5x - y = 26. \qquad \text{Subtract } y.$$

Step 2 Because the coefficients of y are 1 and -1, adding will eliminate y. It is not necessary to multiply either equation by a number.

Step 3 Add the two equations. This time we use vertical addition.

$$\begin{array}{rcl} -2x + y &=& -11 \\ 5x - y &=& 26 \\ \hline 3x &=& 15 \end{array} \qquad \text{Add in columns.}$$

Step 4 Solve the equation.

$$3x = 15$$

Don't stop here. → $x = 5$ Divide by 3.

Step 5 Find the value of y by substituting 5 for x in either of the original equations. Choosing the first equation gives

$$y + 11 = 2x$$
$$y + 11 = 2(5) \qquad \text{Let } x = 5.$$
$$y + 11 = 10$$
$$y = -1. \qquad \text{Subtract 11.}$$

Continued on Next Page

Step 6 Check the solution by substituting $x = 5$ and $y = -1$ into both of the original equations.

Check

$$y + 11 = 2x \qquad\qquad 5x = y + 26$$
$$-1 + 11 \overset{?}{=} 2(5) \qquad\qquad 5(5) \overset{?}{=} -1 + 26$$
$$10 = 10 \quad \text{True} \qquad\qquad 25 = 25 \quad \text{True}$$

Since $(5, -1)$ is a solution of *both* equations, the solution set is $\{(5, -1)\}$.

Work Problem **3** *at the Side.* ▶

OBJECTIVE 2 Multiply when using the elimination method. Sometimes we need to multiply each side of one or both equations in a system by some number before adding the equations will eliminate a variable.

EXAMPLE 3 **Multiplying Both Equations When Using the Elimination Method**

Solve the system.

$$2x + 3y = -15 \qquad (1)$$
$$5x + 2y = 1 \qquad (2)$$

Adding the two equations gives $7x + 5y = -14$, which does not eliminate either variable. However, we can multiply each equation by a suitable number so that the coefficients of one of the two variables are opposites. For example, to eliminate x, multiply each side of equation (1) by 5, and each side of equation (2) by -2.

$$\begin{array}{ll} 10x + 15y = -75 & \text{Multiply equation (1) by 5.} \\ -10x - 4y = -2 & \text{Multiply equation (2) by } -2. \\ \hline 11y = -77 & \text{Add.} \\ y = -7 & \text{Divide by 11.} \end{array}$$

Substituting -7 for y in either equation (1) or (2) gives $x = 3$. Check that the solution set of the system is $\{(3, -7)\}$.

Work Problem **4** *at the Side.* ▶

OBJECTIVE 3 Use an alternative method to find the second value in a solution. Sometimes it is easier to find the value of the second variable in a solution by using the elimination method twice.

EXAMPLE 4 **Finding the Second Value Using an Alternative Method**

Solve the system.

$$4x = 9 - 3y \qquad (1)$$
$$5x - 2y = 8 \qquad (2)$$

Rearrange the terms in equation (1) so that like terms are aligned in columns. To do this, add $3y$ to each side to get the following system.

$$4x + 3y = 9 \qquad (3)$$
$$5x - 2y = 8 \qquad (2)$$

One way to proceed is to eliminate y by multiplying each side of equation (3) by 2 and each side of equation (2) by 3, and then adding.

Continued on Next Page

---

3 Solve each system by the elimination method. Check each solution.

(a) $2x - y = 2$
$4x + y = 10$

(b) $8x - 5y = 32$
$4x + 5y = 4$

4 (a) Solve the system in Example 3 by first eliminating the variable y. Check your solution.

(b) Solve

$$6x + 7y = 4$$
$$5x + 8y = -1,$$

and check your solution.

ANSWERS

3. (a) $\{(2, 2)\}$ (b) $\left\{\left(3, -\dfrac{8}{5}\right)\right\}$

4. (a) $\{(3, -7)\}$ (b) $\{(3, -2)\}$

5 Solve each system of equations.

(a) $5x = 7 + 2y$
$5y = 5 - 3x$

(b) $3y = 8 + 4x$
$6x = 9 - 2y$

$8x + 6y = 18$ Multiply equation (3) by 2.

$\dfrac{15x - 6y = 24}{23x \qquad = 42}$ Multiply equation (2) by 3.

               Add.

$x = \dfrac{42}{23}$ Divide by 23.

Substituting $\frac{42}{23}$ for x in one of the given equations would give y, but the arithmetic involved would be messy. Instead, solve for y by starting again with the original equations and eliminating x. Multiply each side of equation (3) by 5 and each side of equation (2) by -4, and then add.

$\mathbf{20x} + 15y = \quad 45$ Multiply equation (3) by 5.

$\dfrac{\mathbf{-20x} + \quad 8y = -32}{23y = \quad 13}$ Multiply equation (2) by -4.

              Add.

$y = \dfrac{13}{23}$ Divide by 23.

Check that the solution set is $\{(\frac{42}{23}, \frac{13}{23})\}$.

◀ **Work Problem 5 at the Side.**

When the value of the first variable is a fraction, the method used in Example 4 helps avoid arithmetic errors. Of course, this method could be used to solve any system of equations.

OBJECTIVE 4 **Use the elimination method to solve special systems.**

6 Solve each system by the elimination method.

(a) $4x + 3y = 10$

$2x + \dfrac{3}{2}y = 12$

(b) $\quad 4x - 6y = 10$
$-10x + 15y = -25$

EXAMPLE 5 **Using the Elimination Method for an Inconsistent System or Dependent Equations**

Solve each system by the elimination method.

(a) $2x + 4y = 5$

$4x + 8y = -9$

Multiply each side of $2x + 4y = 5$ by -2; then add to $4x + 8y = -9$.

$-4x - 8y = -10$

$\dfrac{4x + 8y = \quad -9}{\mathbf{0 = -19}}$ **False**

The false statement $0 = -19$ indicates that the solution set is $\emptyset$.

(b) $\quad 3x - y = 4$
$-9x + 3y = -12$

Multiply each side of the first equation by 3; then add the two equations.

$9x - 3y = \quad 12$

$\dfrac{-9x + 3y = -12}{\mathbf{0 = \quad 0}}$ **True**

A true statement occurs when the equations are equivalent. As before, this indicates that every solution of one equation is also a solution of the other. The solution set is $\{(x, y) \mid 3x - y = 4\}$. (See **Section 15.2,** Example 6, where the same system was solved using substitution.)

◀ **Work Problem 6 at the Side.**

In Exercises 1–4, answer true *or* false *for each statement. If* false, *tell why.*

1. The ordered pair $(0, 0)$ *must* be a solution of a system of the form

$$Ax + By = 0$$
$$Cx + Dy = 0.$$

2. To eliminate the y-terms in the system

$$2x + 12y = 7$$
$$3x + 4y = 1,$$

we should multiply the bottom equation by 3 and then add.

3. The system

$$x + y = 1$$
$$x + y = 2$$

has $\emptyset$ as its solution set.

4. The ordered pair $(4, -5)$ cannot be a solution of a system that contains the equation $5x - 4y = 0$.

Solve each system by the elimination method. Check each solution. See Examples 1 and 2.

5. $x + y = 2$
$2x - y = -5$

6. $3x - y = -12$
$x + y = 4$

7. $2x + y = -5$
$x - y = 2$

8. $2x + y = -15$
$-x - y = 10$

9. $3x + 2y = 0$
$-3x - y = 3$

10. $5x - y = 5$
$-5x + 2y = 0$

11. $6x - y = -1$
$5y = 17 + 6x$

12. $y = 9 - 6x$
$-6x + 3y = 15$

Solve each system by the elimination method. Check each solution. See Examples 3–5.

13. $2x - y = 12$
$3x + 2y = -3$

14. $x + y = 3$
$-3x + 2y = -19$

15. $x + 3y = 19$
$2x - y = 10$

16. $4x - 3y = -19$
$2x + y = 13$

17. $x + 4y = 16$
$3x + 5y = 20$

18. $2x + y = 8$
$5x - 2y = -16$

19. $5x - 3y = -20$
$-3x + 6y = 12$

20. $4x + 3y = -28$
$5x - 6y = -35$

21. $2x - 8y = 0$
$4x + 5y = 0$

22. $3x - 15y = 0$
$6x + 10y = 0$

23. $x + y = 7$
$x + y = -3$

24. $x - y = 4$
$x - y = -3$

25. $-x + 3y = 4$
$-2x + 6y = 8$

26. $6x - 2y = 24$
$-3x + y = -12$

27. $4x - 3y = -19$
$3x + 2y = 24$

28. $5x + 4y = 12$
    $3x + 5y = 15$

29. $3x - 7 = -5y$
    $5x + 4y = -10$

30. $2x + 3y = 13$
    $6 + 2y = -5x$

31. $2x + 3y = 0$
    $4x + 12 = 9y$

32. $-4x + 3y = 2$
    $5x + 3 = -2y$

33. $24x + 12y = -7$
    $16x - 17 = 18y$

34. $9x + 4y = -3$
    $6x + 7 = -6y$

35. $3x = 3 + 2y$
    $-\dfrac{4}{3}x + y = \dfrac{1}{3}$

36. $3x = 27 + 2y$
    $x - \dfrac{7}{2}y = -25$

37. $5x - 2y = 3$
    $10x - 4y = 5$

38. $3x - 5y = 1$
    $6x - 10y = 4$

39. $6x + 3y = 0$
    $-18x - 9y = 0$

40. $3x - 5y = 0$
    $9x - 15y = 0$

Attending the movies is one of America's favorite forms of entertainment. The graph shows U.S. movie attendance from 1996 through 2004. In 1996, attendance was 1339 million, as represented by the point P(1996, 1339). In 2004, attendance was 1536 million, as represented by the point Q(2004, 1536). We can find an equation of line segment PQ by using a system of equations. Then we use the equation we found to approximate the attendance in any of the years between 1996 and 2004. **Work Exercises 41–46 in order.**

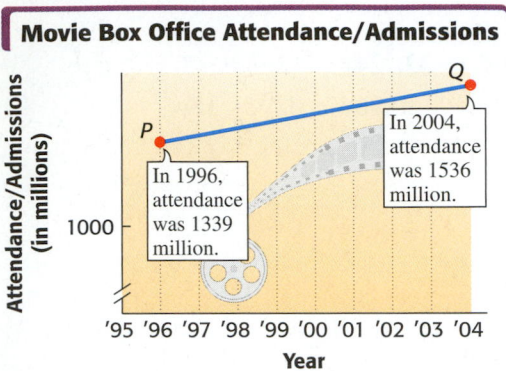

Movie Box Office Attendance/Admissions

Source: Motion Picture Association of America.

41. The line segment has an equation that can be written in the form $y = ax + b$. Using the coordinates of point P with $x = 1996$ and $y = 1339$, write an equation in the variables a and b.

42. Using the coordinates of point Q with $x = 2004$ and $y = 1536$, write a second equation in the variables a and b.

43. Write the system of equations formed from the two equations in Exercises 41 and 42, and solve the system using the elimination method.

44. What is the equation of the segment PQ?

45. Let $x = 2002$ in the equation of Exercise 44, and solve for y to the nearest tenth. How does the result compare with the actual figure of 1639 million?

46. The actual data points for the years 1996 through 2004 do not lie in a perfectly straight line. Explain the pitfalls of relying too heavily on using the equation in Exercise 44 to predict attendance.

15.4 ▶▶▶ Applications of Linear Systems

Recall from **Section 10.4** the six-step method for solving applied problems. We modify those steps slightly to allow for two variables and two equations.

> **Solving an Applied Problem with Two Variables**
>
> *Step 1*   **Read** the problem, several times if necessary, until you understand what is given and what is to be found.
>
> *Step 2*   **Assign variables** to represent the unknown values, using diagrams or tables as needed. Write down what each variable represents.
>
> *Step 3*   **Write two equations** using both variables.
>
> *Step 4*   **Solve** the system of two equations.
>
> *Step 5*   **State the answer** to the problem. Is the answer reasonable?
>
> *Step 6*   **Check** the answer in the words of the original problem.

OBJECTIVE 1 Solve problems about unknown numbers.

EXAMPLE 1 Solving a Problem about Two Unknown Numbers

In 2004, sales of athletic/sports footwear were \$3551 million more than sales of sports clothing. Together, total sales for these items were \$25,953 million. (*Source:* National Sporting Goods Association.) What were the sales for each?

Step 1   **Read** the problem carefully. We must find the 2004 sales (in millions of dollars) for athletic/sports footwear and clothing. We know how much more footwear sales were than clothing sales. Also, we know the total sales.

Step 2   **Assign variables.**

Let x = sales of footwear in millions of dollars,

and y = sales of clothing in millions of dollars.

Step 3   **Write two equations.**

$x = 3551 + y$ Sales of footwear were \$3551 million more than sales of clothing.

$x + y = 25{,}953$ Total sales were \$25,953 million.

Step 4   **Solve** the system for x and y from Step 3. The substitution method works well here since the first equation is already solved for x.

Work Problem **1** *at the Side.* ▶

Step 5   **State the answer.** Footwear sales were \$14,752 million, and clothing sales were \$11,201 million.

Step 6   **Check** the answer in the original problem. Since

$14{,}752 - 11{,}201 = 3551$ and $14{,}752 + 11{,}201 = 25{,}953,$

the answer satisfies the information in the problem.

1 Solve the system.

$x = 3551 + y$
$x + y = 25{,}953$

2 Set up a system of equations for the following problem. Do not solve the system.

Two of the most popular movies of 2007 were *Ratatouille* and *The Simpsons Movie.* Together, their domestic gross was $389.5 million. *The Simpsons Movie* grossed $23.3 million less than *Ratatouille.* How much did each movie gross? (*Source:* www.boxofficemojo.com)

Let x = the amount (in millions) that *Ratatouille* grossed, and y = the amount (in millions) that _____ grossed.

CAUTION

If an applied problem asks for *two* values as in Example 1, be sure to give both of them in your answer.

◀ *Work Problem* **2** *at the Side.*

OBJECTIVE 2 Solve problems about quantities and their costs. We can also use a linear system to solve an applied problem involving two quantities and their costs.

EXAMPLE 2 Solving a Problem about Quantities and Costs

Musicals have long been the most popular shows on Broadway. The musical *Wicked,* based on a "re-imagining" of *The Wizard of Oz,* has played to sold-out houses in cities around the world. (*Source:* www.broadway.com)

For the production playing at the Ford Center in Chicago, orchestra (main floor) seats cost $148, while the best balcony tickets cost $65. (*Source:* www.ticketmaster.com) Suppose that the members of a club spent a total of $2614 for 30 tickets to *Wicked.* How many tickets of each kind did they buy?

Step 1 **Read** the problem several times.

Step 2 **Assign variables.**

Let x = the number of orchestra seats,

and y = the number of balcony seats.

Summarize the information given in the problem in a table. The entries in the first two rows of the Total Value column were found by multiplying the number of tickets sold by the price per ticket.

	Number of Tickets	Price per Ticket (in dollars)	Total Value
Orchestra	x	148	$148x$
Balcony	y	65	$65y$
Total	30	✕✕✕✕✕	2614

Step 3 **Write two equations.** The total number of tickets was 30, so

$$x + y = 30. \quad \text{Total number of tickets}$$

Since the total value was $2614, the final column leads to

$$148x + 65y = 2614. \quad \text{Total value of tickets}$$

These two equations form the system

$$x + \quad y = 30 \qquad (1)$$
$$148x + 65y = 2614. \qquad (2)$$

Step 4 **Solve** the system using the elimination method. To eliminate the y-terms, multiply each side of equation (1) by -65 to get

$$-65x - 65y = -1950.$$

Continued on Next Page

Then add this result to equation (2).

$$-65x - 65y = -1950$$
$$\underline{148x + 65y = \quad 2614} \quad \text{(2)}$$
$$83x \qquad = \quad 664 \qquad \text{Add.}$$
$$x = \mathbf{8} \qquad \text{Divide by 83.}$$

Substitute 8 for x in equation (1) to get

$$\mathbf{x} + y = 30 \qquad \text{(1)}$$
$$\mathbf{8} + y = 30 \qquad \text{Let } x = 8.$$
$$y = 22$$

Step 5 **State the answer.** The club members bought 8 orchestra tickets and 22 balcony tickets.

Step 6 **Check.** The sum of 8 and 22 is 30, so the total number of tickets is correct. Since 8 tickets were purchased at $148 each and 22 at $65 each, the total of all the ticket prices is

$$\$148(8) + \$65(22) = \$2614,$$

which agrees with the total amount stated in the problem.

Work Problem **3** *at the Side.* ▶

OBJECTIVE **3** **Solve problems about mixtures.** In **Section 2.6** we solved percent problems using one variable. Many problems about mixtures that involve percent can be solved using a system of two equations in two variables.

EXAMPLE 3 **Solving a Mixture Problem Involving Percent**

A pharmacist needs 100 L of 50% alcohol solution. She has on hand 30% alcohol solution and 80% alcohol solution, which she can mix. How many liters of each will be required to make the 100 L of 50% alcohol solution?

Step 1 **Read** the problem. Note the percent of each solution and of the mixture.

Step 2 **Assign variables.**

Let x = the number of liters of 30% alcohol needed,

and y = the number of liters of 80% alcohol needed.

Summarize the information given in the problem in a table. Percents are written as decimals.

Liters of Mixture	Percent	Liters of Pure Alcohol
x	0.30	$0.30x$
y	0.80	$0.80y$
100	0.50	$0.50(100)$

Continued on Next Page

---

3 For the production of *Wicked* playing at the Pantages Theatre in Los Angeles, orchestra seats cost $96 and mid-priced mezzanine tickets cost $58. (*Source:* www.ticketmaster.com) If a group of 18 people attended the show and spent a total of $1234 for their tickets, how many of each kind of ticket did they buy?

(a) Complete the table.

	Number of Tickets Sold	Price (in dollars)	Total Value
Orchestra	x		
Mezzanine	y		
Total		XXXXX	

(b) Write a system of equations.

(c) Solve the system and check your answer in the words of the original problem.

4 How many liters of 25% alcohol solution must be mixed with 12% solution to get 13 L of 15% solution?

(a) Complete the table.

Liters	Percent	Liters of Pure Alcohol
x	0.25	$0.25x$
y	0.12	
13	0.15	

(b) Write a system of equations, and solve it.

5 Solve the problem.

Joe needs 60 milliliters (mL) of 20% acid solution for a chemistry experiment. The lab has on hand only 10% and 25% solutions. How much of each should he mix to get the desired amount of 20% solution?

6 Solve using the formula $d = rt$.

A small plane traveled from Stockholm, Sweden, to Oslo, Norway, averaging 244 km per hr. The trip took 1.7 hr. To the nearest kilometer, what is the distance between the two cities?

4. (a)

Liters	Percent	Liters of Pure Alcohol
x	0.25	$0.25x$
y	0.12	$0.12y$
13	0.15	$0.15\,(13)$

(b) $\quad x + \quad y = 13$
$\quad 0.25x + 0.12y = 0.15\,(13)$
$\quad$ 3 L of 25%, 10 L of 12%

5. 20 mL of 10%, 40 mL of 25%

6. 415 km

Figure 8 gives an idea of what is actually happening in this problem.

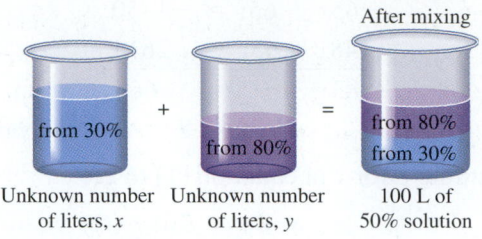

After mixing

Figure 8

Step 3 **Write two equations.** Since the total number of liters in the final mixture will be 100, the first equation is

$$x + y = 100.$$

To find the amount of pure alcohol in each mixture, multiply the number of liters by the concentration. The amount of pure alcohol in the 30% solution added to the amount of pure alcohol in the 80% solution will equal the amount of pure alcohol in the final 50% solution. This gives the second equation,

$$0.30x + 0.80y = 0.50\,(100).$$

These two equations form the system

> Be sure to write two equations.

$$x + \quad y = 100$$
$$0.30x + 0.80y = 50. \quad \textcolor{blue}{0.50\,(100) = 50}$$

Step 4 **Solve** this system by the substitution method. Solving the first equation of the system for x gives $x = 100 - y$. Substitute $100 - y$ for x in the second equation.

$$0.30\textcolor{blue}{x} + 0.80y = 50$$
$$0.30\,(\textcolor{blue}{100 - y}) + 0.80y = 50 \qquad \textcolor{blue}{\text{Let } x = 100 - y.}$$
$$30 - 0.30y + 0.80y = 50 \qquad \textcolor{blue}{\text{Distributive property}}$$
$$30 + 0.50y = 50 \qquad \textcolor{blue}{\text{Combine like terms.}}$$
$$0.50y = 20 \qquad \textcolor{blue}{\text{Subtract 30.}}$$
$$y = 40 \qquad \textcolor{blue}{\text{Divide by 0.50.}}$$

Then $x = 100 - \textcolor{blue}{y} = 100 - \textcolor{blue}{40} = 60.$

Step 5 **State the answer.** The pharmacist should use 60 L of the 30% solution and 40 L of the 80% solution.

Step 6 Since $60 + 40 = 100$ and $0.30\,(60) + 0.80\,(40) = 50$, this mixture will give the 100 L of 50% solution, as required in the original problem.

◀ *Work Problems* **4** *and* **5** *at the Side.*

OBJECTIVE **4** **Solve problems about distance, rate (or speed), and time.** If an automobile travels at an average rate of 50 mph for 2 hr, then it travels $50 \times 2 = 100$ mi. This is an example of the basic relationship between distance, rate, and time:

$$\textbf{distance} = \textbf{rate} \times \textbf{time,} \quad \text{given by the formula} \quad \textbf{\textit{d} = \textit{rt}.}$$

◀ *Work Problem* **6** *at the Side.*

EXAMPLE 4 **Solving a Problem about Distance, Rate, and Time**

Two executives in cities 400 mi apart drive to a business meeting at a location on the line between their cities. They meet after 4 hr. Find the speed of each car if one car travels 20 mph faster than the other.

Step 1 **Read** the problem carefully.

Step 2 **Assign variables.** Let x = the speed of the faster car,

and y = the speed of the slower car.

We use the formula $d = rt$. Since each car travels for 4 hr, the time, t, for each car is 4. The distance is found by using the formula $d = rt$ and the expressions already entered in the table.

	r	t	d
Faster Car	x	4	**4x**
Slower Car	y	4	**4y**

Find d from $d = rt$.

Figure 9 shows what is happening in the problem.

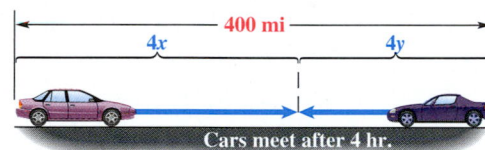

Figure 9

Step 3 **Write two equations.** As shown in the figure, since the total distance traveled by both cars is 400 mi, one equation is

4x + 4y = 400.

Because the faster car goes 20 mph faster than the slower car, the second equation is

$$x = 20 + y.$$

Step 4 **Solve** the system of equations,

$$4x + 4y = 400 \quad (1)$$
$$x = 20 + y, \quad (2)$$

by substitution. Replace x with $20 + y$ in equation (1) and then solve for y.

$4(\mathbf{20 + y}) + 4y = 400$	Let $x = 20 + y$.
$80 + 4y + 4y = 400$	Distributive property
$80 + 8y = 400$	Combine like terms.
$8y = 320$	Subtract 80.
$y = 40$	Divide by 8.

Since $x = 20 + y$ and $y = \mathbf{40}$,

$$x = 20 + \mathbf{40} = 60.$$

Step 5 **State the answer.** The speeds of the cars are 40 mph and 60 mph.

Step 6 **Check** the answer. Since each car travels for 4 hr, total distance is

$$4(60) + 4(40) = 240 + 160 = 400 \text{ mi}, \quad \text{as required.}$$

Work Problem **7** *at the Side.* ▶

7 Two cars that were 450 mi apart traveled toward each other. They met after 5 hr. If one car traveled twice as fast as the other, what were their speeds?

(a) Complete this table.

	r	t	d
Faster Car	x	5	
Slower Car	y	5	

(b) Write a system, and solve it.

ANSWERS

7. **(a)**

	r	t	d
Faster Car	x	5	$5x$
Slower Car	y	5	$5y$

(b) $5x + 5y = 450$
$x = 2y$
faster car: $x = 60$ mph;
slower car: $y = 30$ mph

8 Solve the system.

$$x + y = 320$$
$$x - y = 280$$

EXAMPLE 5 **Solving a Problem about Distance, Rate, and Time**

A plane flies 560 mi in 1.75 hr traveling with the wind. The return trip against the same wind takes the plane 2 hr. Find the speed of the plane and the speed of the wind.

Step 1 **Read** the problem several times.

Step 2 **Assign variables.**

Let x = the speed of the plane,

and y = the speed of the wind.

The speed (rate) of the plane *with* the wind is $(x + y)$ mph, and the speed (rate) of the plane *against* the wind is $(x - y)$ mph. See Figure 10.

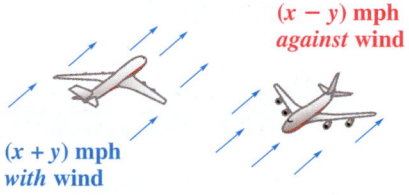

$(x - y)$ **mph**
against **wind**

$(x + y)$ **mph**
with **wind**

Figure 10

We use this information and the formula $d = rt$ (or $rt = d$) to complete a table.

	r	t	d
With Wind	$x + y$	1.75	560
Against Wind	$x - y$	2	560

Step 3 **Write two equations.** From the table,

$$1.75(x + y) = 560 \xrightarrow{\text{Divide by 1.75.}} x + y = 320 \quad (1)$$
$$2(x - y) = 560 \xrightarrow{\text{Divide by 2.}} x - y = 280. \quad (2)$$

Step 4 **Solve** the system of equations (1) and (2).

◀ Work Problem **8** at the Side.

Step 5 **State the answer.** From Problem 8 at the side, the speed of the plane is 300 mph and the speed of the wind is 20 mph.

Step 6 **Check.** The answer seems reasonable, and true statements result when the values are substituted into the equations of the system.

◀ Work Problem **9** at the Side.

---

9 Solve the problem.

In 1 hr, Gigi can row 2 mi against the current or 10 mi with the current. Find the speed of the current and Gigi's speed in still water. (*Hint:* Let x = the speed of the current and y = Gigi's speed in still water. Then her rate against the current is $(y - x)$ mph, and her rate with the current is $(y + x)$ mph.)

15.4 ▷▷▷ Exercises

Choose the correct response in Exercises 1–7.

1. Which expression represents the monetary value of x 20-dollar bills?

   A. $\dfrac{x}{20}$ dollars $\qquad$ **B.** $\dfrac{20}{x}$ dollars $\qquad$ **C.** $(20 + x)$ dollars $\qquad$ **D.** $20x$ dollars

2. Which expression represents the cost of t pounds of candy that sells for \$1.95 per lb?

   A. \$1.95t $\qquad$ **B.** $\dfrac{\$1.95}{t}$ $\qquad$ **C.** $\dfrac{t}{\$1.95}$ $\qquad$ **D.** \$1.95 + t

3. Which expression represents the amount of interest earned on d dollars at an interest rate of 2%?

   A. $2d$ dollars $\qquad$ **B.** $0.02d$ dollars $\qquad$ **C.** $0.2d$ dollars $\qquad$ **D.** $200d$ dollars

4. Suppose that x liters of a 40% acid solution are mixed with y liters of a 35% solution to obtain 100 L of a 38% solution. One equation in a system for solving this problem is $x + y = 100$. Which one of the following is the other equation?

   A. $0.35x + 0.40y = 0.38(100)$ $\qquad$ **B.** $0.40x + 0.35y = 0.38(100)$

   C. $35x + 40y = 38$ $\qquad$ **D.** $40x + 35y = 0.38(100)$

5. According to *Natural History* magazine, the speed of a cheetah is 70 mph. If a cheetah runs for x hours, how many miles does the cheetah cover?

   A. $(70 + x)$ miles $\quad$ **B.** $(70 - x)$ miles $\quad$ **C.** $\dfrac{70}{x}$ miles $\qquad$ **D.** $70x$ miles

6. What is the speed of a plane that travels at a rate of 560 mph *against* a wind of r mph?

   A. $(560 + r)$ mph $\quad$ **B.** $\dfrac{560}{r}$ mph $\qquad$ **C.** $(560 - r)$ mph $\qquad$ **D.** $(r - 560)$ mph

7. What is the speed of a plane that travels at a rate of 560 mph *with* a wind of r mph?

   A. $\dfrac{r}{560}$ mph $\qquad$ **B.** $(560 - r)$ mph $\quad$ **C.** $(560 + r)$ mph $\qquad$ **D.** $(r - 560)$ mph

8. Using the list of steps for solving an applied problem with two variables, describe the general procedure you will use to solve the problems that follow in this exercise set.

Exercises 9 and 10 are good warm-up problems. In each case, refer to the six-step problem-solving method, fill in the blanks for Steps 2 and 3, and then complete the solution by applying Steps 4–6.

9. The sum of two numbers is 98 and the difference between them is 48. Find the two numbers.

   Step 1 **Read** the problem carefully.

   Step 2 **Assign variables.**

   Let $x =$ the first number and let

   $y = $ _____ .

   Step 3 **Write two equations.**

   First equation: $x + y = 98$

   Second equation: _____

10. The sum of two numbers is 201 and the difference between them is 11. Find the two numbers.

    Step 1 **Read** the problem carefully.

    Step 2 **Assign variables.**

    Let $x =$ the first number and let

    $y = $ _____ .

    Step 3 **Write two equations.**

    First equation: $x + y = 201$

    Second equation: _____

Write a system of equations for each problem, and then solve the problem. See Example 1.

11. As of 2008, the two longest-running shows in Broadway history were *The Phantom of the Opera* and *Cats.* As of September 26, 2007, there had been a total of 15,682 Broadway performances of the two shows, with 712 more performances of *The Phantom of the Opera* than *Cats.* How many performances were there of each show? (*Source:* The Broadway League.)

12. Two other musicals that had very long Broadway runs were *A Chorus Line* and *Beauty and the Beast.* During their runs, there were 676 fewer performances of *Beauty and the Beast* than of *A Chorus Line,* and a total of 11,598 performances of the two shows. How many performances were there of each show? (*Source:* The Broadway League.)

13. The two domestic top-grossing movies of 2007 were *Spider-Man 3* and *Shrek the Third. Shrek the Third* grossed $13.8 million less than *Spider-Man 3*, and together the two films took in $659.2 million. How much did each of these movies earn? (*Source:* www.boxofficemojo.com)

14. During their opening weekends, *Spider-Man 3* and *Shrek the Third* grossed a total of $272.7 million, with *Spider-Man 3* grossing $29.5 million more than *Shrek the Third*. How much did each of these movies earn during their opening weekends? (*Source:* www.boxofficemojo.com)

15. The Terminal Tower in Cleveland, Ohio, is 242 ft shorter than the Key Tower, also in Cleveland. The total of the heights of the two buildings is 1658 ft. Find the heights of the buildings. (*Source: World Almanac and Book of Facts.*)

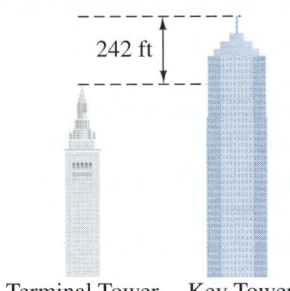

Terminal Tower   Key Tower

16. In 2006, a total of 646.3 thousand people lived in the Twin Cities of Minneapolis and St. Paul, Minnesota. Minneapolis had 99.3 thousand more residents than St. Paul. What was the population of each city? (*Source: U.S. Census Bureau.*)

If x units of a product cost C dollars to manufacture and earn revenue of R dollars, the value of x where the expressions for C and R are equal is called the **break-even quantity,** *the number of units that produce 0 profit. In Exercises 17 and 18,* **(a)** *find the break-even quantity, and* **(b)** *decide whether the product should be produced based on whether it will earn a profit. (Profit equals revenue minus cost.)*

17. $C = 85x + 900$; $R = 105x$; no more than 38 units can be sold.

18. $C = 105x + 6000$; $R = 255x$; no more than 400 units can be sold.

Write a system of equations for each problem, and then solve the system. See Example 2.

19. A motel clerk counts his $1 and $10 bills at the end of a day. He finds that he has a total of 74 bills having a combined monetary value of $326. Find the number of bills of each denomination that he has.

Number of Bills	Denomination of Bill	Total Value
x	$1	
y	$10	
74	✕✕✕✕✕	$326

20. Carly is a bank teller. At the end of a day, she has a total of 69 $5 and $10 bills. The total value of the money is $590. How many of each denomination does she have?

Number of Bills	Denomination of Bill	Total Value
x	$5	$5x$
y	$10	
	✕✕✕✕✕	

21. A newspaper advertised DVDs and CDs. Tracy Sudak went shopping and bought each of her seven nephews a gift, either a DVD of the movie *Night at the Museum* or the latest Linkin Park CD. The DVD cost $14.95 and the CD cost $16.88, and she spent a total of $114.30. How many DVDs and how many CDs did she buy?

22. Terry Wong saw the ad (see Exercise 21) and he, too, went shopping. He bought each of his five nieces a gift, either a DVD of *Hairspray* or the CD soundtrack to *High School Musical 2*. The DVD cost $14.99 and the soundtrack cost $13.88, and he spent a total of $70.51. How many DVDs and CDs did he buy?

23. Maria Lopez has twice as much money invested at 5% simple annual interest as she does at 4%. If her yearly income from these two investments is $350, how much does she have invested at each rate?

24. Charles Miller invested his textbook royalty income in two accounts, one paying 3% annual simple interest and the other paying 2% interest. He earned a total of $11 interest. If he invested three times as much in the 3% account as he did in the 2% account, how much did he invest at each rate?

25. The two top-grossing North American concert tours in 2007 were The Police and Van Halen. Based on the average ticket prices for these tours, it cost a total of $1217 to buy six tickets for The Police and five tickets to a Van Halen concert. Three tickets for The Police and four tickets for Van Halen cost a total of $781. How much did an average ticket cost for each tour? (*Source:* Pollstar.)

26. Two other popular North American concert tours in 2007 were Billy Joel and Neil Young. Based on the average ticket prices for these tours, it cost a total of $986 to buy eight tickets for Billy Joel and three tickets to a Neil Young concert. Four tickets for Billy Joel and five tickets for Neil Young cost a total of $878. How much did an average ticket cost for each tour? (*Source:* Pollstar.)

Write a system of equations for each problem, and then solve the system. See Example 3.

27. A 40% dye solution is to be mixed with a 70% dye solution to get 120 L of a 50% solution. How many liters of the 40% and 70% solutions will be needed?

Liters of Solution	Percent (as a Decimal)	Liters of Pure Dye
x	0.40	
y	0.70	
120	0.50	

28. A 90% antifreeze solution is to be mixed with a 75% solution to make 120 L of a 78% solution. How many liters of the 90% and 75% solutions will be used?

Liters of Solution	Percent (as a Decimal)	Liters of Pure Antifreeze
x	0.90	
y	0.75	
120	0.78	

29. Ahmad Hashemi wishes to mix coffee worth $6 per lb with coffee worth $3 per lb to get 90 lb of a mixture worth $4 per lb. How many pounds of the $6 and the $3 coffees will be needed?

Pounds	Dollars per Pound	Cost
x	6	
y		
90		

30. Mariana Coanda wishes to blend candy selling for $1.20 per lb with candy selling for $1.80 per lb to get a mixture that will be sold for $1.40 per lb. How many pounds of the $1.20 and the $1.80 candies should be used to get 45 lb of the mixture?

Pounds	Dollars per Pound	Cost
x		
y	1.80	
45		

31. How many pounds of nuts selling for $6 per lb and raisins selling for $3 per lb should Kelli Hammer combine to obtain 60 lb of a trail mix selling for $5 per lb?

32. Avis Proctor works at a gourmet delicatessen. She is preparing cheese trays for a large reception. She is using some cheeses that sell for $8 per lb and others that sell for $12 per lb. How many pounds of cheese at each price should she use in order for the mixed cheeses on the trays to weigh a total of 56 lb and sell for $10.50 per lb?

Write a system of equations for each problem, and then solve the system.
See Examples 4 and 5.

33. RAGBRAI®, the Des Moines **R**egister's **A**nnual **G**reat **B**icycle **R**ide **A**cross **I**owa, is the longest and oldest touring bicycle ride in the world. Suppose a cyclist began the 471 mi ride on July 20, 2008, in western Iowa at the same time that a car traveling toward it left eastern Iowa. If the bicycle and the car met after 7.5 hr and the car traveled 35.8 mph faster than the bicycle, find the average speed of each. (*Source:* www.ragbrai.org)

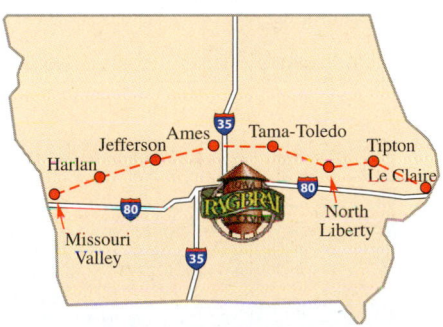

34. In 2006, Atlanta's Hartsfield Airport was the nation's busiest. Suppose two planes leave the airport at the same time, one traveling east and the other traveling west. If the planes are 2100 mi apart after 2 hr and one plane travels 50 mph faster than the other, find the speed of each plane. (*Source:* Airports Council International.)

35. Toledo and Cincinnati are 200 mi apart. A car leaves Toledo traveling toward Cincinnati, and another car leaves Cincinnati at the same time, traveling toward Toledo. The car leaving Toledo averages 15 mph faster than the other, and they meet after 1 hr and 36 min. What are the rates of the cars?

36. Kansas City and Denver are 600 mi apart. Two cars start from these cities, traveling toward each other. They meet after 6 hr. Find the rate of each car if one travels 30 mph slower than the other.

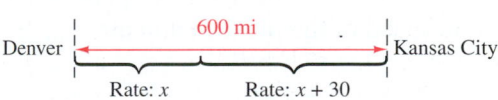

37. At the beginning of a bicycle ride for charity, Roberto and Juana are 30 mi apart. If they leave at the same time and ride in the same direction, Roberto overtakes Juana in 6 hr. If they ride toward each other, they meet in 1 hr. What are their speeds?

38. Mr. Abbot left Farmersville in a plane at noon to travel to Exeter. Mr. Baker left Exeter in his automobile at 2 P.M. to travel to Farmersville. It is 400 mi from Exeter to Farmersville. If the sum of their speeds was 120 mph, and if they crossed paths at 4 P.M., find the speed of each.

39. A boat takes 3 hr to go 24 mi upstream. It can go 36 mi downstream in the same time. Find the speed of the current and the speed of the boat in still water if $x =$ the speed of the boat in still water and $y =$ the speed of the current.

	r	t	d
Downstream	$x + y$		36
Upstream	$x - y$		24

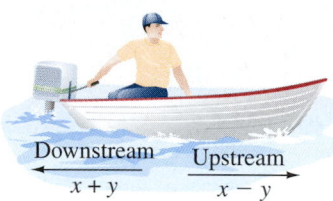

Downstream $x + y$ Upstream $x - y$

40. It takes a boat $1\frac{1}{2}$ hr to go 12 mi downstream, and 6 hr to return. Find the speed of the boat in still water and the speed of the current. Let $x =$ the speed of the boat in still water and $y =$ the speed of the current.

	r	t	d
Downstream	$x + y$	$\frac{3}{2}$	12
Upstream		6	

41. If a plane can travel 440 mph against the wind and 500 mph with the wind, find the speed of the wind and the speed of the plane in still air.

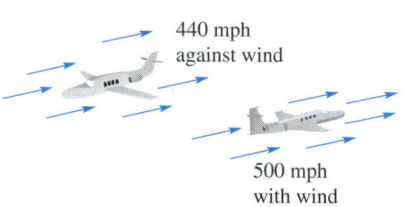

440 mph against wind

500 mph with wind

42. A small plane travels 200 mph with the wind and 120 mph against it. Find the speed of the wind and the speed of the plane in still air.

15.5 ▶▶▶ Solving Systems of Linear Inequalities

We graphed the solutions of a linear inequality in **Section 11.5.** Recall that to graph the solutions of $x + 3y > 12$, for example, we first graph $x + 3y = 12$ by finding and plotting a few ordered pairs that satisfy the equation. Because the points on the line do *not* satisfy the inequality, we use a dashed line. To decide which side of the line includes the points that are solutions, we choose a test point not on the line, such as $(0, 0)$. Substituting these values for x and y in the inequality gives

$$x + 3y > 12$$
$$0 + 3(0) \overset{?}{>} 12$$
$$0 > 12. \qquad \text{False}$$

This false result indicates that the solutions are those points on the side of the line that does not include $(0, 0)$, as shown in Figure 11.

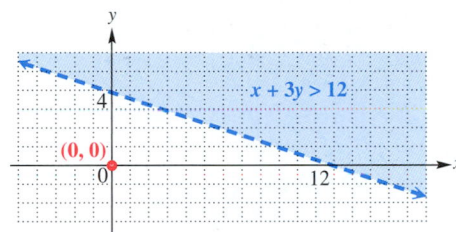

Figure 11

Now we use the same techniques to solve systems of linear inequalities.

OBJECTIVE 1 Solve systems of linear inequalities by graphing.
A **system of linear inequalities** consists of two or more linear inequalities. The **solution set of a system of linear inequalities** includes all points that make all inequalities of the system true at the same time. To solve a system of linear inequalities, use the following steps.

Solving a System of Linear Inequalities

Step 1 **Graph the inequalities.** Graph each inequality using the method of **Section 11.5.**

Step 2 **Choose the intersection.** Indicate the solution set of the system by shading the intersection of the graphs (the region where the graphs overlap).

EXAMPLE 1 **Solving a System of Two Linear Inequalities**

Graph the solution set of the system.

$$3x + 2y \leq 6$$
$$2x - 5y \geq 10$$

To graph $3x + 2y \leq 6$, graph the solid boundary line $3x + 2y = 6$ and shade the region containing $(0, 0)$, as shown in Figure 12(a) on the next page. Then graph $2x - 5y \geq 10$ with the solid boundary line $2x - 5y = 10$. The test point $(0, 0)$ makes this inequality false, so shade the region on the other side of the boundary line. See Figure 12(b).

Continued on Next Page

1 Graph the solution set of the system.

$$x - 2y \leq 8$$
$$3x + y \geq 6$$

To get you started, the graphs of $x - 2y = 8$ and $3x + y = 6$ are shown.

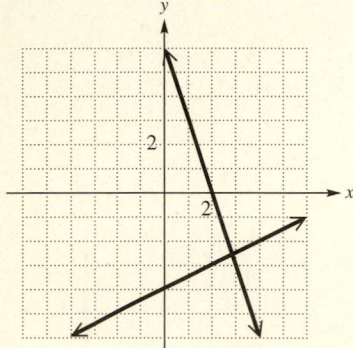

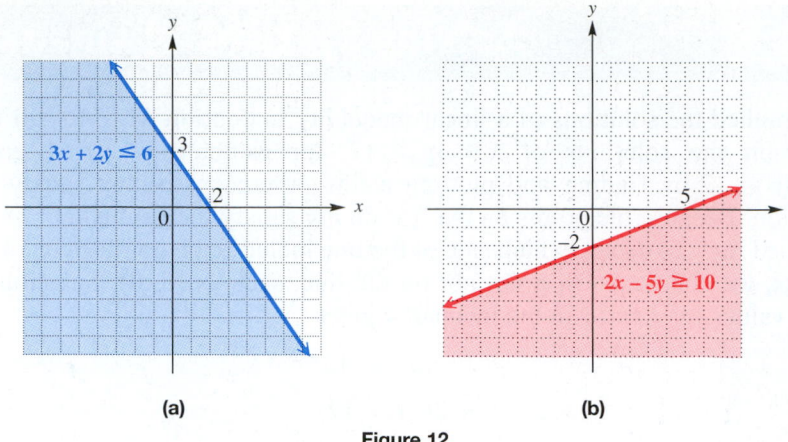

(a)   (b)

Figure 12

The solution set of this system includes all points in the intersection (overlap) of the graphs of the two inequalities. It includes the shaded region and portions of the two boundary lines shown in Figure 13.

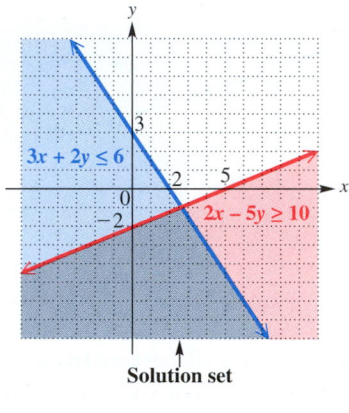

Solution set

Figure 13

◀ *Work Problem* **1** *at the Side.*

Note

We usually do all the work on one set of axes. In the following examples, only one graph is shown. Be sure that the region of the final solution set is clearly indicated.

EXAMPLE 2 **Solving a System of Two Linear Inequalities**

Graph the solution set of the system.

$$x - y > 5$$
$$2x + y < 2$$

Figure 14 shows the graphs of both $x - y > 5$ and $2x + y < 2$. Dashed lines show that the graphs of the inequalities do not include their boundary lines. The solution set of the system is the region with the darkest shading. The solution set does not include either boundary line.

Continued on Next Page

ANSWER

1.

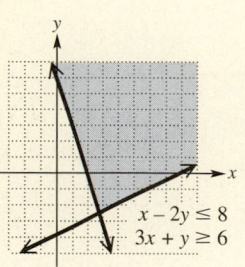

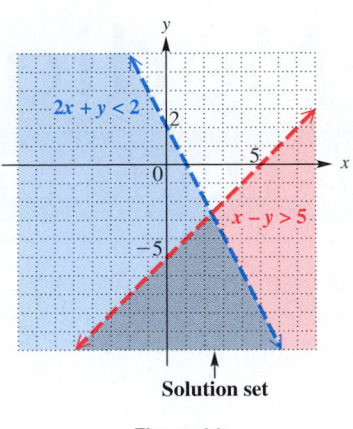

Figure 14

EXAMPLE 3 **Solving a System of Three Linear Inequalities**

Graph the solution set of the system.

$$4x - 3y \leq 8$$
$$x \geq 2$$
$$y \leq 4$$

Recall that $x = 2$ is a vertical line through the point $(2, 0)$, and $y = 4$ is a horizontal line through $(0, 4)$. The graph of the solution set is the shaded region in Figure 15, including all boundary lines.

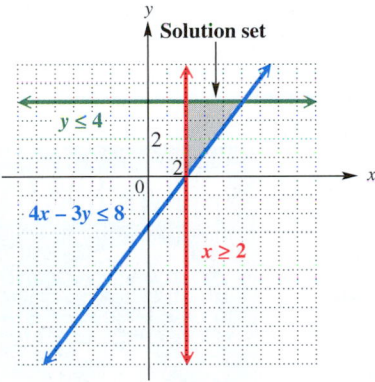

Figure 15

Work Problem **2** *at the Side.* ▶

---

2 Graph the solution set of each system.

(a) $x + 2y < 0$
$3x - 4y < 12$

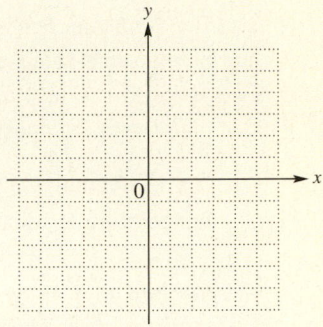

(b) $3x + 2y \leq 12$
$x \leq 2$
$y \leq 4$

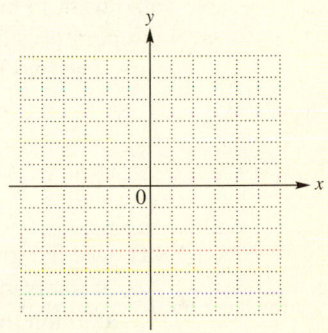

ANSWERS

2. (a)

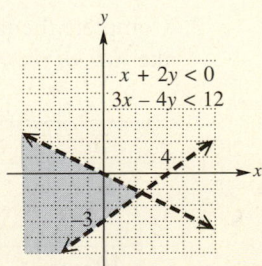

(b)

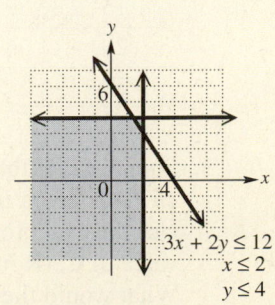

Math in the Media

AT WHAT TEMPERATURE DOES PAPER BURN?

The 1966 film *Fahrenheit 451*, directed by François Truffaut and based on the Ray Bradbury book of the same name, is a science fiction classic. In a future dominated by oppression, a fireman is assigned the task of burning books to discourage independent thinking. He eventually begins to read them and questions the motive of the government.

The Fahrenheit scale for temperature is used in the United States, but in many other countries temperature is routinely reported in Celsius. The formula for converting Celsius to Fahrenheit is $F = \frac{9}{5}C + 32$, but a quick "rule of thumb" given by travel books is *"Double the Celsius temperature and add 30,"* which is mathematically stated by the approximation formula

$$F = 2C + 30.$$

1. Suppose you are interested in knowing for what temperature the rule of thumb and the actual formulas give the same result. You also want to know if the rule of thumb formula is predicting temperatures that are lower or higher than the actual temperature. The two formulas can be written as the system of equations

$$F = \frac{9}{5}C + 32$$

$$F = 2C + 30.$$

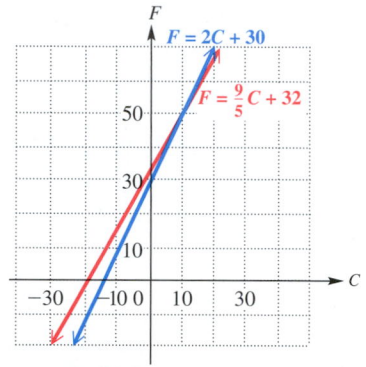

   (a) Use the graph of the system of equations to find the point of intersection. (*Hint:* To check your answer, use substitution to see if it satisfies both formulas.)

   (b) For what temperature in degrees Celsius do the two formulas agree?

   (c) For what temperature in degrees Fahrenheit do the two formulas agree?

2. Complete the table of values to compare the *actual* and the *rule of thumb* formulas for temperature conversion.

°C	°F (Actual)	°F (Rule of Thumb)
0		
5		
10		
15		
20		
30		

3. Suppose that the movie and book title *Fahrenheit 451* was given in Celsius rather than Fahrenheit. Use the exact formula $F = \frac{9}{5}C + 32$ to solve for this value of C. (Round to the nearest whole number.) What would the title then be?

15.5 ▶▶▶ Exercises

Match each system of inequalities with the correct graph from choices A–D.

1. $x \geq 5$
$y \leq -3$

2. $x \leq 5$
$y \geq -3$

3. $x > 5$
$y < -3$

4. $x < 5$
$y > -3$

A.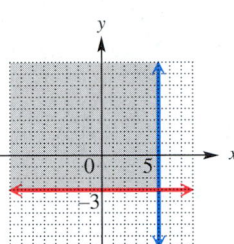

B.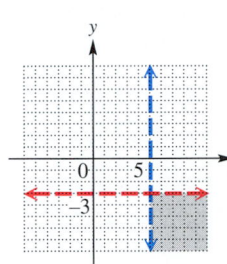

C.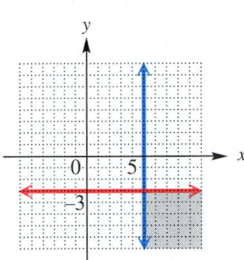

D.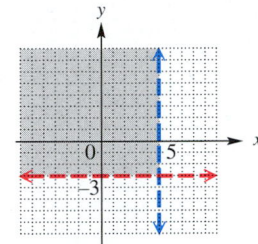

Graph the solution set of each system of linear inequalities. See Examples 1–3.

◉ 5. $x + y \leq 6$
$x - y \geq 1$

6. $x + y \leq 2$
$x - y \geq 3$

7. $4x + 5y \geq 20$
$x - 2y \leq 5$

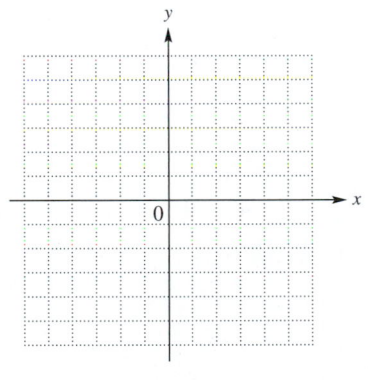

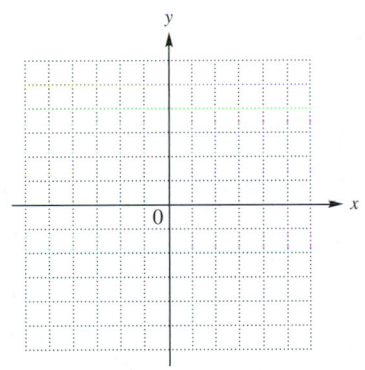

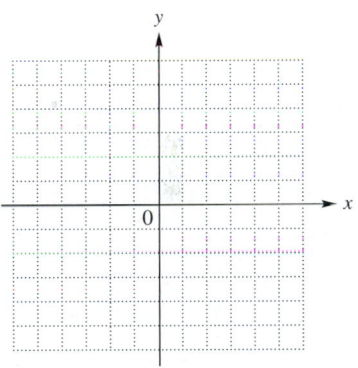

8. $x + 4y \leq 8$
$2x - y \geq 4$

◉ 9. $2x + 3y < 6$
$x - y < 5$

10. $x + 2y < 4$
$x - y < -1$

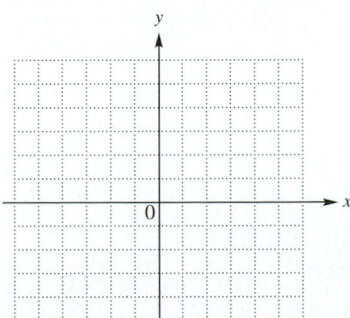

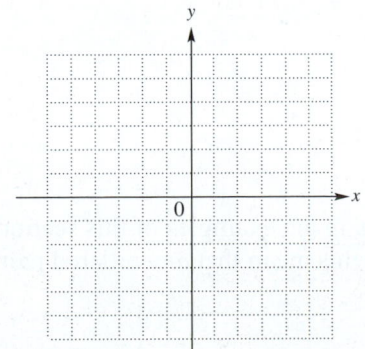

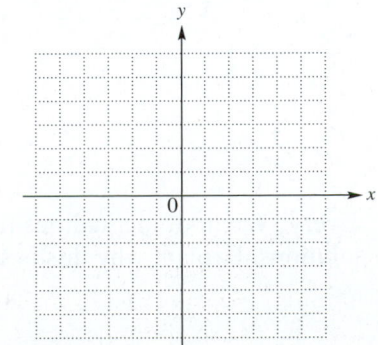

11. $y \leq 2x - 5$
$x < 3y + 2$

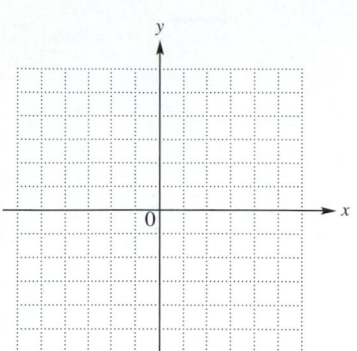

12. $x \geq 2y + 6$
$y > -2x + 4$

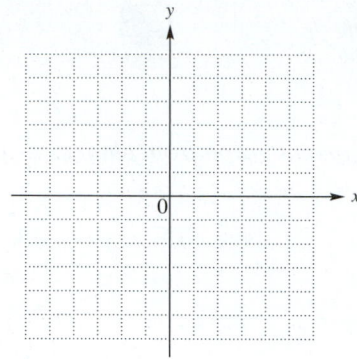

13. $4x + 3y < 6$
$x - 2y > 4$

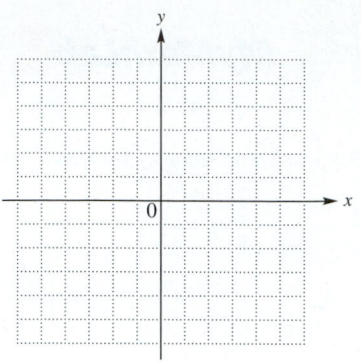

14. $3x + y > 4$
$x + 2y < 2$

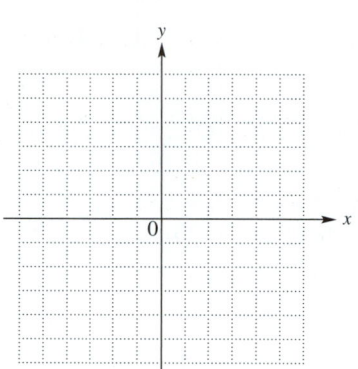

15. $x \leq 2y + 3$
$x + y < 0$

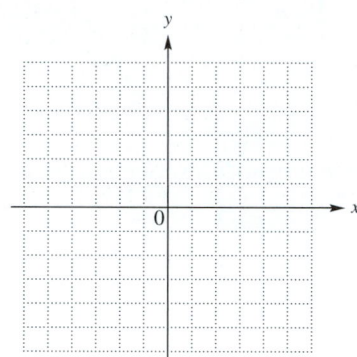

16. $x \leq 4y + 3$
$x + y > 0$

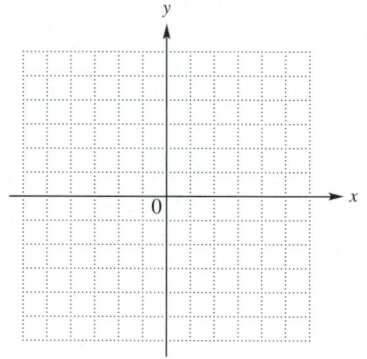

17. $4x + 5y < 8$
$y > -2$
$x > -4$

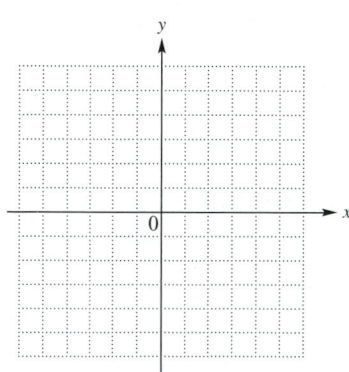

18. $x + y \geq -3$
$x - y \leq 3$
$y \leq 3$

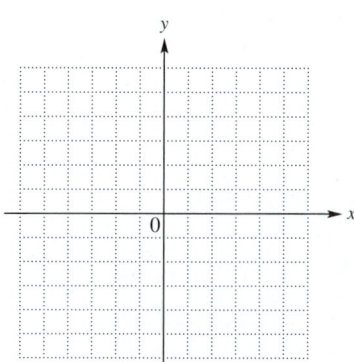

19. $3x - 2y \geq 6$
$x + y \leq 4$
$x \geq 0$
$y \geq -4$

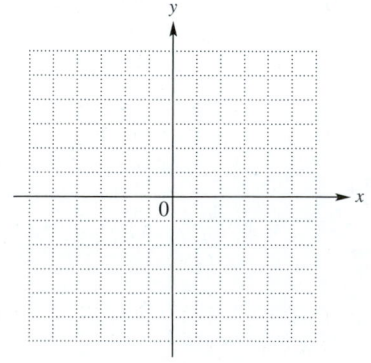

20. Every system of inequalities illustrated in the examples of this section has infinitely many solutions. Explain why this is so. Does this mean that *any* ordered pair is a solution?

Chapter 15 ▶▶▶ Summary

▶ **Key Terms**

15.1 **system of linear equations**
A system of linear equations (or **linear system**) consists of two or more linear equations with the same variables.

solution of a system
The solution of a system of linear equations includes all the ordered pairs that make all the equations of the system true at the same time.

solution set of a system
The set of all ordered pairs that are solutions of a system is its solution set.

consistent system
A system of equations with at least one solution is a consistent system.

inconsistent system
An inconsistent system of equations is a system with no solution.

independent equations
Equations of a system that have different graphs are called independent equations.

dependent equations
Equations of a system that have the same graph (because they are different forms of the same equation) are called dependent equations.

15.5 **system of linear inequalities**
A system of linear inequalities contains two or more linear inequalities (and no other kinds of inequalities).

solution set of a system of linear inequalities
The solution set of a system of linear inequalities includes all points that make all inequalities of the system true at the same time.

▶ **Test Your Word Power**

See how well you have learned the vocabulary in this chapter. Answers, with examples, follow the Quick Review.

1. A **system of linear equations** consists of
   A. at least two linear equations with different variables
   B. two or more linear equations that have an infinite number of solutions
   C. two or more linear equations with the same variables
   D. two or more linear inequalities.

2. A **solution of a system** of linear equations is
   A. an ordered pair that makes one equation of the system true

   B. an ordered pair that makes all the equations of the system true at the same time
   C. any ordered pair that makes one or the other or both equations of the system true
   D. the set of values that make all the equations of the system false.

3. A **consistent system** is a system of equations
   A. with at least one solution
   B. with no solution
   C. with an infinite number of solutions
   D. that have the same graph.

4. An **inconsistent system** is a system of equations
   A. with one solution
   B. with no solution
   C. with an infinite number of solutions
   D. that have the same graph.

5. **Dependent equations**
   A. have different graphs
   B. have no solution
   C. have one solution
   D. are different forms of the same equation.

▶ Quick Review

Concepts	Examples

15.1 Solving Systems of Linear Equations by Graphing

An ordered pair is a solution of a system if it makes all equations of the system true at the same time.

Is $(4, -1)$ a solution of the system $\begin{array}{l} x + y = 3 \\ 2x - y = 9 \end{array}$?

Because $4 + (-1) = 3$ and $2(4) - (-1) = 9$ are both true, $(4, -1)$ is a solution.

If the graphs of the equations of a system are both sketched on the same axes, then the points of intersection, if any, are solutions of the system.

Solve by graphing.
$$x + y = 5$$
$$2x - y = 4$$

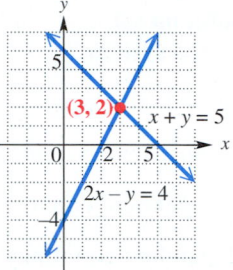

If the graphs of the equations do not intersect (that is, the lines are parallel), then the system has no solution and the solution set is $\emptyset$.

If the graphs of the equations are the same line, then the system has an infinite number of solutions. Use set-builder notation to write the solution set: $\{(x, y) \mid \underline{\qquad}\}$.

The ordered pair $(3, 2)$ satisfies both equations, so $\{(3, 2)\}$ is the solution set.

15.2 Solving Systems of Linear Equations by Substitution

Step 1 Solve one equation for either variable.

Solve by substitution.
$$x + 2y = -5 \qquad (1)$$
$$y = -2x - 1 \qquad (2)$$

Equation (2) is already solved for y.

Step 2 Substitute for that variable in the other equation to get an equation in one variable.

Substitute $-2x - 1$ for y in equation (1).
$$x + 2(-2x - 1) = -5$$

Step 3 Solve the equation from Step 2.

$$x - 4x - 2 = -5$$
$$-3x - 2 = -5$$
$$-3x = -3$$
$$x = 1$$

Step 4 Substitute the result into the equation from Step 1 to get the value of the other variable.

To find y, let $x = 1$ in equation (2):
$$y = -2(1) - 1 = -3.$$

Step 5 Check. Write the solution set.

The solution $(1, -3)$ checks, so $\{(1, -3)\}$ is the solution set.

15.3 Solving Systems of Linear Equations by Elimination

Step 1 Write both equations in standard form $Ax + By = C$.

Solve by elimination.
$$x + 3y = 7 \qquad (1)$$
$$3x - y = 1 \qquad (2)$$

Step 2 If necessary, multiply one or both equations by appropriate numbers so that the sum of the coefficients of either the x- or y-terms is 0.

Multiply equation (1) by -3 to eliminate the x-terms.

Concepts	Examples

15.3 Solving Systems of Linear Equations by Elimination (*continued*)

Step 3 Add the equations to get an equation with only one variable (or no variable).

$$\begin{aligned} -3x - 9y &= -21 \\ \underline{3x - \ y} &= \underline{\ \ 1} \\ -10y &= -20 \qquad \text{Add.} \\ y &= 2 \qquad \ \text{Divide by } -10. \end{aligned}$$

Step 4 Solve the equation from Step 3.

Step 5 Substitute the solution from Step 4 into either of the original equations to find the value of the remaining variable.

Substitute to get the value of x.

$$\begin{aligned} x + 3(2) &= 7 \qquad (1) \\ x + 6 &= 7 \\ x &= 1 \end{aligned}$$

Step 6 Check. Write the solution set.

Since $1 + 3(2) = 7$ and $3(1) - 2 = 1$, the solution $(1, 2)$ checks, so the solution set is $\{(1, 2)\}$.

15.4 Applications of Linear Systems

Use the modified six-step method.

Step 1 **Read** the problem carefully.

Step 2 **Assign variables** for each unknown value. Use diagrams or tables as needed.

Step 3 **Write two equations** using both variables.

Step 4 **Solve** the system.

The sum of two numbers is 30. Their difference is 6. Find the numbers.

Let x represent one number.

Let y represent the other number.

$$\begin{aligned} x + y &= 30 \\ \underline{x - y} &= \underline{\ 6} \\ 2x \quad &= 36 \qquad \text{Add.} \\ x &= 18 \qquad \text{Divide by 2.} \end{aligned}$$

Step 5 **State the answer.**

Let $x = 18$ in the first equation: $18 + y = 30$. Solve to get $y = 12$. The numbers are 18 and 12.

Step 6 **Check** the answer in the words of the original problem.

The sum of 18 and 12 is 30, and the difference between 18 and 12 is 6, so the answer checks.

15.5 Solving Systems of Linear Inequalities

To solve a system of two or more linear inequalities, graph the inequalities on the same axes. (This was explained in **Section 11.5.**) The solution of the system is the intersection (overlap) of the regions of the graphs. The portions of the boundary lines that bound the region of solutions are included for a $\leq$ or $\geq$ inequality and excluded for a $<$ or $>$ inequality.

The shaded region is the solution of the system

$$\begin{aligned} 2x + 4y &\geq 5 \\ x &\geq 1. \end{aligned}$$

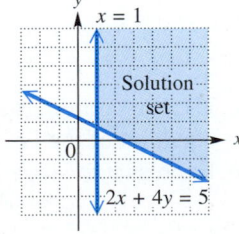

ANSWERS TO TEST YOUR WORD POWER

1. C; *Example:* $2x + y = 7$, $3x - y = 3$

2. B; *Example:* The ordered pair $(2, 3)$ satisfies both equations of the system in the Answer 1 example, so it is a solution of the system.

3. A; *Example:* The system in the Answer 1 example is consistent. The graphs of the equations intersect at exactly one point, in this case the solution $(2, 3)$.

4. B; *Example:* The equations of two parallel lines make up an inconsistent system; their graphs never intersect, so there is no solution to the system.

5. D; *Example:* The equations $4x - y = 8$ and $8x - 2y = 16$ are dependent because their graphs are the same line.

Math in the Media

The March 16, 2008, headline and accompanying graph shown here indicate the Port of New Orleans cargo volume declined since 1998, after a period of growth between 1991 and 1998. The graph shows year-to-year fluctuations, as it consists of line segments describing the data from each year to the next.

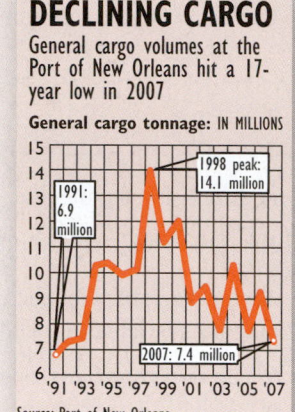

N.O. port fights to grow as cargo lags

DECLINING CARGO

General cargo volumes at the Port of New Orleans hit a 17-year low in 2007

General cargo tonnage: IN MILLIONS

1998 peak: 14.1 million

1991: 6.9 million

2007: 7.4 million

'91 '93 '95 '97 '99 '01 '03 '05 '07

Source: Port of New Orleans

1. Suppose that we wish to depict the basic idea but provide a less detailed graph of the data consisting of only three data points: the point for the year 1991, represented by $A(1991, 6.9)$; the point for the year 1998, represented by $B(1998, 14.1)$; and the point for the year 2007, represented by $C(2007, 7.4)$. We could simply graph the segment AB and then graph the segment BC. Do this on the axes provided.

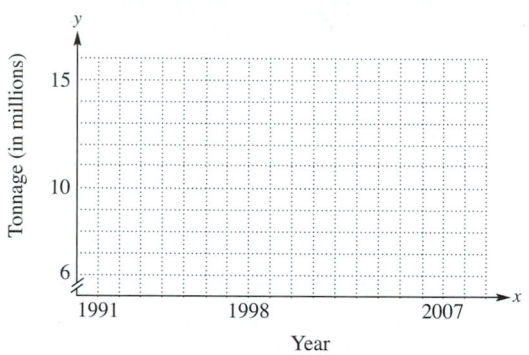

2. Use the coordinates for points A and B to find the equation of the line on which segment AB lies. Express it in slope-intercept form.

3. Use the coordinates for points B and C to find the equation of the line on which segment BC lies. Express it in slope-intercept form.

4. Consider the two equations you found in Exercises 3 and 4 as a system of linear equations. Solve the system, and confirm that the solution of the system is (1998, 14.1).

Chapter 15  Review Exercises

[15.1] *Decide whether the given ordered pair is a solution of the given system.*

1. $(3, 4)$

$$4x - 2y = 4$$
$$5x + y = 19$$

2. $(-5, 2)$

$$x - 4y = -13$$
$$2x + 3y = 4$$

Solve each system by graphing.

3. $x + y = 4$
$2x - y = 5$

4. $x - 2y = 4$
$2x + y = -2$

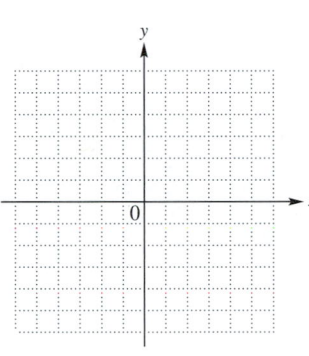

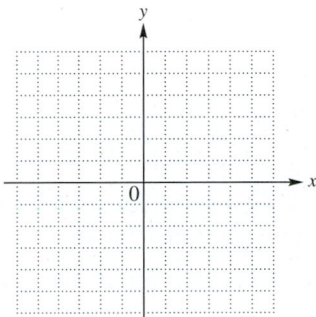

5. $x - 2 = 2y$
$2x - 4y = 4$

6. $2x + 4 = 2y$
$y - x = -3$

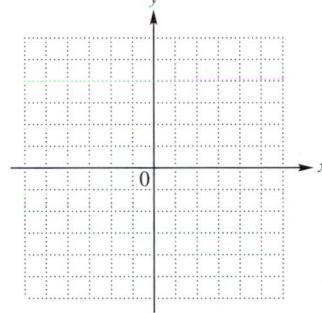

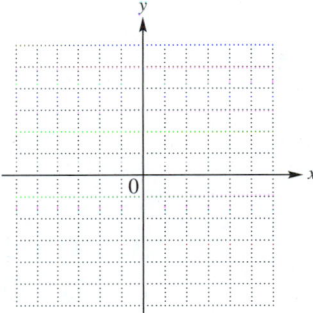

7. When a student was asked to determine whether the ordered pair $(1, -2)$ is a solution of the system

$$x + y = -1$$
$$2x + y = 4,$$

he answered "yes." His reasoning was that the ordered pair satisfies the equation $x + y = -1$; that is, $1 + (-2) = -1$ is true. Why is his answer wrong?

[15.2] *Solve each system by the substitution method.*

8. $3x + y = 7$
$x = 2y$

9. $2x - 5y = -19$
$y = x + 2$

10. $4x + 5y = 44$
$x + 2 = 2y$

11. $5x + 15y = 3$
$x + 3y = 2$

[15.3] *Solve each system by the elimination method.*

12. $2x - y = 13$
$x + y = 8$

13. $3x - y = -13$
$x - 2y = -1$

14. $-4x + 3y = 25$
$6x - 5y = -39$

15. $3x - 4y = 9$
$6x - 8y = 18$

16. For the system

$$2x + 12y = 7$$
$$3x + 4y = 1,$$

if we were to multiply the first (top) equation by -3, by what number would we have to multiply the second (bottom) equation in order to

(a) eliminate the *x*-terms when solving by the elimination method?

(b) eliminate the *y*-terms when solving by the elimination method?

Solve each system by any method.

17. $x - 2y = 5$
$y = x - 7$

18. $5x - 3y = 11$
$2y = x - 4$

19. $\dfrac{x}{2} + \dfrac{y}{3} = 7$
$\dfrac{x}{4} + \dfrac{2y}{3} = 8$

20. $\dfrac{3x}{4} - \dfrac{y}{3} = \dfrac{7}{6}$
$\dfrac{x}{2} + \dfrac{2y}{3} = \dfrac{5}{3}$

21. $2.4x + 1.7y = 7.6$
$1.2x - 0.5y = 9.2$

22. $0.5x + 3.4y = 13$
$1.5x - 2.6y = -25$

[15.4] *Solve each problem by using a system of equations.*

23. At the end of 2006, Subway topped McDonald's as the largest restaurant chain in the United States. Subway operated 6981 more restaurants than McDonald's, and together the two chains had 34,529 restaurants. How many restaurants did each company operate? (*Source:* Technomic.)

24. In 2006, the two magazines with the largest circulations in the United States were *AARP The Magazine* and *Reader's Digest*. Together, the average total circulation of these two magazines was 33.5 million copies. The circulation of *Reader's Digest* was 13.3 million less than that of *AARP The Magazine*. What were the circulation figures for each magazine? (*Source:* www.myjobsource.com)

25. The perimeter of a rectangle is 90 m. Its length is $1\frac{1}{2}$ times its width. Find the length and width of the rectangle.

26. A cashier has 20 bills, all of which are $10 or $20 bills. The total value of the money is $330. How many of each type does the cashier have?

Number of Bills	Denomination of Bills	Total Value
x	$10	$10x$
	$20	
	XXXXXX	$330

27. Candy that sells for $1.30 per lb is to be mixed with candy selling for $0.90 per lb to get 100 lb of a mix that will sell for $1 per lb. How much of each type should be used?

28. A certain plane flying with the wind travels 540 mi in 2 hr. Later, flying against the same wind, the plane travels 690 mi in 3 hr. Find the speed of the plane in still air and the speed of the wind.

29. After taxes, Ms. Cesar's game show winnings were $18,000. She invested part of it at 3% annual simple interest and the rest at 4%. Her interest income for the first year was $650. How much did she invest at each rate?

Amount of Principal	Percent (as a Decimal)	Interest
x	0.03	
y	0.04	
$18,000	XXXXXX	

30. A 40% antifreeze solution is to be mixed with a 70% solution to get 90 L of a 50% solution. How many liters of the 40% and 70% solutions will be needed?

Number of Liters	Percent (as a Decimal)	Amount of Pure Antifreeze
x	0.40	
y	0.70	
90	0.50	XXXXXX

[15.5] *Graph the solution set for each system of linear inequalities.*

31. $x + y \geq 2$
  $x - y \leq 4$

32. $y \geq 2x$
  $2x + 3y \leq 6$

33. $x + y < 3$
  $2x > y$

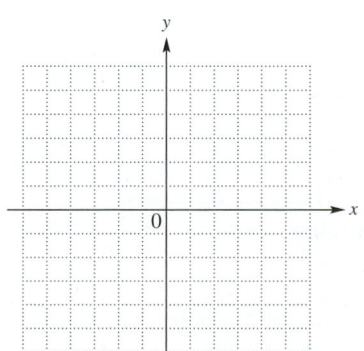

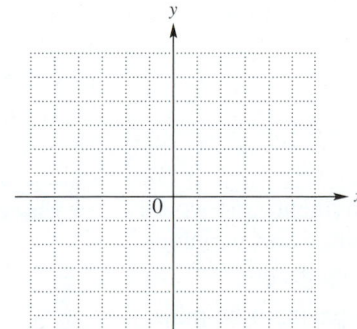

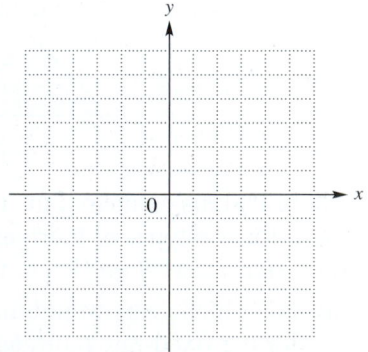

34. Which system of linear inequalities is graphed in the figure?

   A. $x \leq 3$
      $y \leq 1$

   B. $x \leq 3$
      $y \geq 1$

   C. $x \geq 3$
      $y \leq 1$

   D. $x \geq 3$
      $y \geq 1$

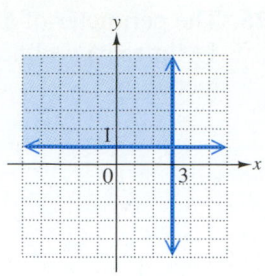

35. Without actually graphing, determine which system of inequalities has no solution.

   A. $x \geq 4$
      $y \leq 3$

   B. $x + y > 4$
      $x + y < 3$

   C. $x > 2$
      $y < 1$

   D. $x + y < 4$
      $x - y < 3$

▶▶▶ Mixed Review Exercises

Solve each system.

36. $3x + 4y = 6$
    $4x - 5y = 8$

37. $\dfrac{3x}{2} + \dfrac{y}{5} = -3$

    $4x + \dfrac{y}{3} = -11$

38. $x + 6y = 3$
    $2x + 12y = 2$

39. $x + y < 5$
    $x - y \geq 2$

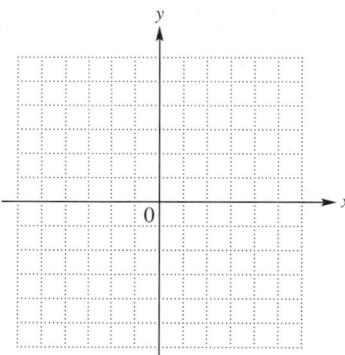

40. $y \leq 2x$
    $x + 2y > 4$

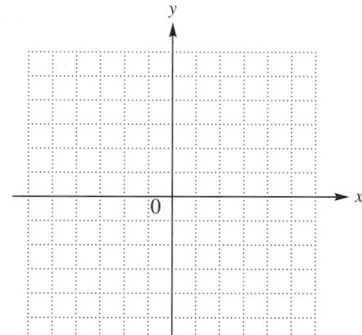

41. $y < -4x$
    $y < -2$

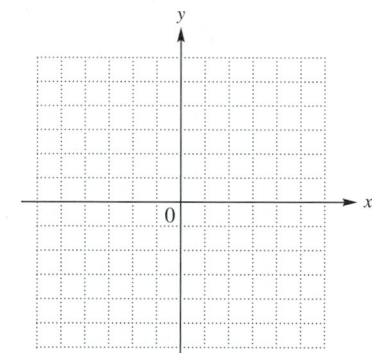

42. The perimeter of an isosceles triangle is 29 in. One side of the triangle is 5 in. longer than each of the two equal sides. Find the lengths of the sides of the triangle.

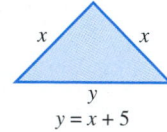

43. Super Bowl XLII was played in Glendale, Arizona, on February 3, 2008. The New York Giants beat the New England Patriots by 3 points, and the winning score was 11 points less than twice the losing score. What was the final score of the game? (*Source:* NFL.)

44. Eboni Perkins compared the monthly payments she would incur for two types of mortgages: fixed-rate and variable-rate. Her observations led to the following graph.

   (a) For which years would the monthly payment be more for the fixed-rate mortgage than for the variable-rate mortgage?

   (b) In what year would the payments be the same, and what would those payments be?

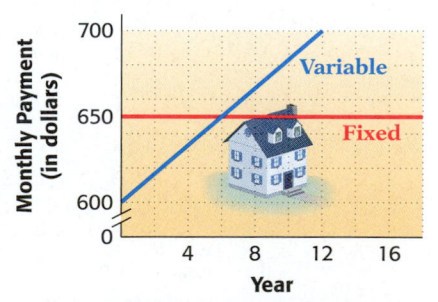

Chapter 15 ▶▶▶ Test

Use the Chapter Test Prep Video CD to see fully worked-out solutions to any of the exercises you want to review.

1. Solve the system by graphing.

$$2x + y = 1$$
$$3x - y = 9$$

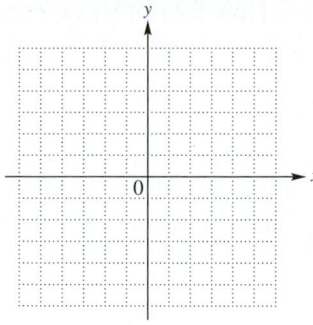

1. _____

2. _____

3. _____

2. Suppose that the graph of a system of two linear equations consists of lines that have the same slope but different y-intercepts. How many solutions does the system have?

4. _____

Solve each system by the substitution method.

3. $2x + y = -4$
   $x = y + 7$

4. $4x + 3y = -35$
   $x + y = 0$

5. _____

6. _____

Solve each system by the elimination method.

5. $2x - y = 4$
   $3x + y = 21$

6. $4x + 2y = 2$
   $5x + 4y = 7$

7. _____

7. $6x - 5y = 0$
   $-2x + 3y = 0$

8. $4x + 5y = 2$
   $-8x - 10y = 6$

8. _____

Solve each system by any method.

9. _____

9. $3x = 6 + y$
   $6x - 2y = 12$

10. $\dfrac{x}{2} - \dfrac{y}{4} = 7$

    $\dfrac{2x}{3} + \dfrac{5y}{4} = 3$

10. _____

Solve each problem.

11. _____

11. The distance between Memphis and Atlanta is 782 mi less than the distance between Minneapolis and Houston. Together, the two distances total 1570 mi. How far is it between Memphis and Atlanta? How far is it between Minneapolis and Houston? (*Source: Rand McNally Road Atlas.*)

12. _____

12. In 2004, a total of 9.0 million people visited the Statue of Liberty and the National World War II Memorial, two popular tourist attractions. The Statue of Liberty had 1.8 million fewer visitors than the National World War II Memorial. How many visitors did each of these attractions have? (*Source:* National Park Service, Department of the Interior.)

13. _____

13. A 15% solution of alcohol is to be mixed with a 40% solution to get 50 L of a final mixture that is 30% alcohol. How much of each of the original solutions should be used?

14. _____

14. Two cars leave from Perham, Minnesota, and travel in the same direction. One car travels $1\frac{1}{3}$ times as fast as the other. After 3 hr they are 45 mi apart. What are the speeds of the cars?

15.

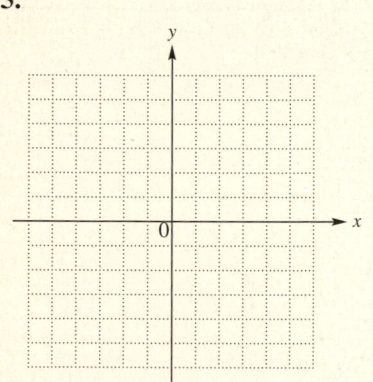

Graph the solution set of each system of inequalities.

15. $2x + 7y \leq 14$
     $x - y \geq 1$

16. $2x - y > 6$
     $4y + 12 \geq -3x$

16.

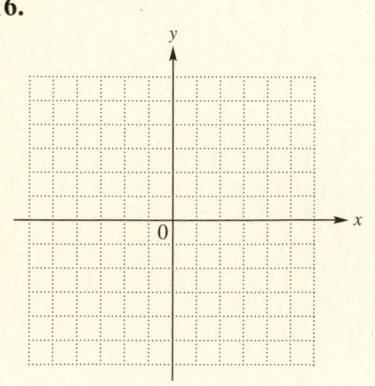

Roots and Radicals

The London Eye opened on New Year's Eve in 1999. This unique structure features 32 observation capsules and has a diameter of 135 meters. Located on the bank of the Thames River, it faces the Houses of Parliament and is the fourth-tallest structure in London. (*Source:* www.londoneye.com)

The formula

$$\text{sight distance} = 111.7\sqrt{\text{height of structure in kilometers}}$$

involves *radicals*, the subject of this chapter, and can be used to determine how far one can see (in kilometers) from the top of a structure on a clear day. In Exercise 55 of **Section 16.6,** we use this formula to determine the truth of the claim that passengers on the London Eye can see Windsor Castle, 25 miles away.

16.1 ▶▶▶ Evaluating Roots

OBJECTIVES

1 Find square roots.

2 Decide whether a given root is rational, irrational, or not a real number.

3 Find decimal approximations for irrational square roots.

4 Use the Pythagorean formula.

5 Find cube, fourth, and other roots.

In **Section 9.1,** we discussed the idea of the *square* of a number. Recall that squaring a number means multiplying the number by itself.

$$\text{If } a = 8, \quad \text{then} \quad a^2 = \mathbf{8 \cdot 8} = 64.$$
$$\text{If } a = -4, \quad \text{then} \quad a^2 = (-4)(-4) = 16.$$
$$\text{If } a = -\frac{1}{2}, \quad \text{then} \quad a^2 = \left(-\frac{1}{2}\right)\left(-\frac{1}{2}\right) = \frac{1}{4}.$$

In this chapter, we consider the opposite process.

$$\text{If } a^2 = 64, \quad \text{then} \quad a = \mathbf{?}.$$
$$\text{If } a^2 = 16, \quad \text{then} \quad a = \mathbf{?}.$$
$$\text{If } a^2 = \frac{1}{4}, \quad \text{then} \quad a = \mathbf{?}.$$

OBJECTIVE 1 Find square roots. To find a in the three preceding statements, we must find a number that when multiplied by itself results in the given number. The number a is called a **square root** of the number a^2.

EXAMPLE 1 Finding All Square Roots of a Number

Find all square roots of 49.

To find a square root of 49, think of a number that when multiplied by itself gives 49. One square root is 7 because $7 \cdot 7 = 49$. Another square root of 49 is -7 because $(-7)(-7) = 49$. The number 49 has two square roots, 7 and -7; one is positive, and one is negative.

◀ *Work Problem* **1** *at the Side.*

---

1 Find all square roots.

(a) 100

(b) 25

(c) 36

(d) $\dfrac{25}{36}$

Early radical symbol

The **positive** or **principal square root** of a number is written with the symbol $\sqrt{}$. For example, the positive square root of 121 is 11, written

$$\sqrt{121} = \mathbf{11}.$$

The symbol $-\sqrt{}$ is used for the **negative square root** of a number. For example, the negative square root of 121 is -11, written

$$-\sqrt{121} = \mathbf{-11}.$$

The symbol $\sqrt{}$, called a **radical sign,** always represents the positive square root (except that $\sqrt{0} = 0$). The number inside the radical sign is called the **radicand,** and the entire expression, radical sign and radicand, is called a **radical.**

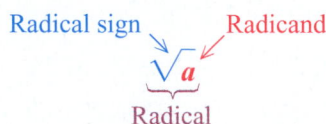

An algebraic expression containing a radical is called a **radical expression.**

Radicals have a long mathematical history. The radical sign $\sqrt{}$ has been used since sixteenth-century Germany and was probably derived from the letter R. The radical symbol in the margin comes from the Latin word for root, *radix*. It was first used by Leonardo da Pisa (Fibonnaci) in 1220.

We summarize our discussion of square roots as follows.

Square Roots of a

If a is a positive real number, then

$\sqrt{a}$ is the positive or principal square root of a,

and $\quad -\sqrt{a}$ is the negative square root of a.

For nonnegative a,

$$\sqrt{a} \cdot \sqrt{a} = \left(\sqrt{a}\right)^2 = a \quad \text{and} \quad -\sqrt{a} \cdot \left(-\sqrt{a}\right) = \left(-\sqrt{a}\right)^2 = a.$$

Also, $\sqrt{0} = 0$.

▦ **Calculator Tip** Most calculators have a square root key, usually labeled ⟨√x⟩, that allows us to find the square root of a number. On some models, the square root key must be used in conjunction with the key marked ⟨INV⟩ or ⟨2nd⟩.

EXAMPLE 2 **Finding Square Roots**

Find each square root.

(a) $\sqrt{144}$

The radical $\sqrt{144}$ represents the positive or principal square root of 144. Think of a positive number whose square is 144.

$$12^2 = 144, \quad \text{so} \quad \sqrt{144} = 12.$$

(b) $-\sqrt{1024}$

This symbol represents the negative square root of 1024. A calculator with a square root key can be used to find $\sqrt{1024} = 32$. Then, $-\sqrt{1024} = -32$.

(c) $\sqrt{\dfrac{4}{9}} = \dfrac{2}{3}$ **(d)** $-\sqrt{\dfrac{16}{49}} = -\dfrac{4}{7}$ **(e)** $\sqrt{0.81} = 0.9$

Work Problem ▣ **2** *at the Side.* ▶

As noted above, when the square root of a positive real number is squared, the result is that positive real number. $\left(\text{Also}, \left(\sqrt{0}\right)^2 = 0.\right)$

EXAMPLE 3 **Squaring Radical Expressions**

Find the *square* of each radical expression.

(a) $\sqrt{13}$

$\left(\sqrt{13}\right)^2 = 13$ Definition of square root

(b) $-\sqrt{29}$

$\left(-\sqrt{29}\right)^2 = 29$ The square of a *negative* number is positive.

(c) $\sqrt{p^2 + 1}$

$\left(\sqrt{p^2 + 1}\right)^2 = p^2 + 1$

Work Problem ▣ **3** *at the Side.* ▶

▣ 2 Find each square root.

(a) $\sqrt{16}$

(b) $-\sqrt{169}$

(c) $-\sqrt{225}$

(d) $\sqrt{729}$

(e) $\sqrt{\dfrac{36}{25}}$

(f) $\sqrt{0.49}$

▣ 3 Find the *square* of each radical expression.

(a) $\sqrt{41}$

(b) $-\sqrt{39}$

(c) $\sqrt{2x^2 + 3}$

ANSWERS

2. (a) 4   **(b)** -13 **(c)** -15
    (d) 27   **(e)** $\dfrac{6}{5}$ **(f)** 0.7

3. (a) 41   **(b)** 39   **(c)** $2x^2 + 3$

4 Tell whether each square root is *rational, irrational,* or *not a real number.*

(a) $\sqrt{9}$

(b) $\sqrt{7}$

(c) $\sqrt{\dfrac{4}{9}}$

(d) $\sqrt{72}$

(e) $\sqrt{-43}$

OBJECTIVE 2 Decide whether a given root is rational, irrational, or not a real number. All numbers with square roots that are rational are called **perfect squares.**

Perfect squares		Rational square roots
25		$\sqrt{25} = 5$
144	are perfect squares since	$\sqrt{144} = 12$
$\dfrac{4}{9}$		$\sqrt{\dfrac{4}{9}} = \dfrac{2}{3}$

A number that is not a perfect square has a square root that is not a rational number. For example, $\sqrt{5}$ is not a rational number because it cannot be written as the ratio of two integers. Its decimal equivalent (or approximation) neither terminates nor repeats. However, $\sqrt{5}$ is a real number and corresponds to a point on the number line. As mentioned in **Section 9.3,** a real number that is not rational is called an **irrational number.** The number $\sqrt{5}$ is irrational. Many square roots of integers are irrational.

> If a is a *positive* real number that is *not* a perfect square, then
> $$\sqrt{a} \text{ is irrational.}$$

Not every number has a real number square root. For example, there is no real number that can be squared to obtain -36. (The square of a real number can never be negative.) Because of this, $\sqrt{-36}$ *is not a real number.*

> If a is a *negative* real number, then $\sqrt{a}$ is *not* a real number.

CAUTION
Be careful not to confuse $\sqrt{-36}$ and $-\sqrt{36}$. $\sqrt{-36}$ is not a real number since there is no real number that can be squared to obtain -36. However, $-\sqrt{36}$ is the negative square root of 36, which is -6.

EXAMPLE 4 Identifying Types of Square Roots

Tell whether each square root is *rational, irrational,* or *not a real number.*

(a) $\sqrt{17}$
Because 17 is not a perfect square, $\sqrt{17}$ is irrational.

(b) $\sqrt{64}$
The number 64 is a perfect square, 8^2, so $\sqrt{64} = 8$ is a rational number.

(c) $\sqrt{-25}$
There is no real number whose square is -25. Therefore, $\sqrt{-25}$ is not a real number.

◀ *Work Problem* **4** *at the Side.*

ANSWERS
4. (a) rational (b) irrational (c) rational (d) irrational (e) not a real number

Note

Not all irrational numbers are square roots of integers. For example, π (approximately 3.14159) is an irrational number that is not a square root of any integer.

OBJECTIVE **3** **Find decimal approximations for irrational square roots.**   Even if a number is irrational, a decimal that approximates the number can be found using a calculator. For example, if we use a calculator to find $\sqrt{10}$, the display might show 3.16227766, which is only an *approximation* of $\sqrt{10}$, not an exact rational value.

EXAMPLE 5   **Approximating Irrational Square Roots**

Find a decimal approximation for each square root. Round answers to the nearest thousandth.

(a) $\sqrt{11}$
   Using the square root key of a calculator gives 3.31662479 $\approx$ 3.317, where $\approx$ means "is approximately equal to."

(b) $\sqrt{39} \approx 6.245$ Use a calculator.   **(c)** $-\sqrt{740} \approx -27.203$

Work Problem **5** *at the Side.* ▶

OBJECTIVE **4** **Use the Pythagorean formula.**   Many applications of square roots use the Pythagorean formula. Recall from **Section 13.8** that by this formula if c is the length of the hypotenuse of a right triangle, and a and b are the lengths of the two legs, as shown in Figure 1, then

$$a^2 + b^2 = c^2.$$

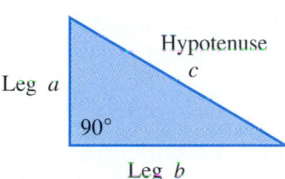

Figure 1

EXAMPLE 6   **Using the Pythagorean Formula**

Find the length of the unknown side of each right triangle with sides a, b, and c, where c is the hypotenuse.

(a) $a = 3, b = 4$
   Use the Pythagorean formula to find c^2 first.

$$c^2 = a^2 + b^2$$
$$c^2 = 3^2 + 4^2 \quad \text{Let } a = 3 \text{ and } b = 4.$$
$$c^2 = 9 + 16 \quad \text{Square.}$$
$$c^2 = 25 \quad \text{Add.}$$

Since the length of a side of a triangle must be a positive number, find the positive square root of 25 to get c.

$$c = \sqrt{25} = 5$$

Continued on Next Page

5 Find a decimal approximation for each square root. Round answers to the nearest thousandth.

(a) $\sqrt{28}$

(b) $\sqrt{63}$

(c) $-\sqrt{190}$

(d) $\sqrt{1000}$

ANSWERS
5. **(a)** 5.292   **(b)** 7.937   **(c)** -13.784
   (d) 31.623

6 Find the length of the unknown side in each right triangle. Give any decimal approximations to the nearest thousandth.

(a) $a = 7, b = 24$

(b) $c = 15, b = 13$

(b) $c = 9, b = 5$

Substitute the given values in the Pythagorean formula. Then solve for a^2.

$$c^2 = a^2 + b^2$$
$$9^2 = a^2 + 5^2 \qquad \text{Let } c = 9 \text{ and } b = 5.$$
$$81 = a^2 + 25 \qquad \text{Square.}$$
$$56 = a^2 \qquad \text{Subtract 25.}$$

Use a calculator to find the positive square root of 56 to get a.

$$a = \sqrt{56} \approx 7.483$$

CAUTION

Be careful not to make the common mistake of thinking that $\sqrt{a^2 + b^2}$ equals $a + b$. As Example 6(a) shows, $\sqrt{9 + 16} = \sqrt{25} = 5$. However, $\sqrt{9} + \sqrt{16} = 3 + 4 = 7$. Since $5 \neq 7$, in general,

$$\sqrt{a^2 + b^2} \neq a + b.$$

◀ Work Problem **6** at the Side.

The Pythagorean formula can be used to solve applied problems that involve right triangles. Use the same six problem-solving steps that we have been using throughout the text.

EXAMPLE 7 Using the Pythagorean Formula to Solve an Application

A ladder 10 ft long leans against a wall. The foot of the ladder is 6 ft from the base of the wall. How high up the wall does the top of the ladder rest?

Step 1 **Read** the problem again.

Step 2 **Assign a variable.** As shown in Figure 2, a right triangle is formed with the ladder as the hypotenuse. Let a represent the height of the top of the ladder when measured straight down to the ground.

(c)

A right triangle with side 8 on the left (vertical), hypotenuse 11, and unknown side labeled ? on the bottom.

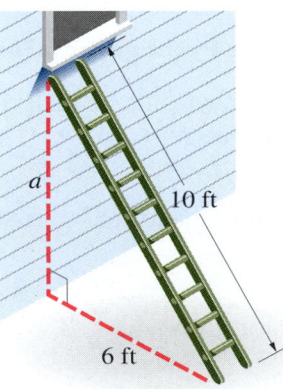

Figure 2

Continued on Next Page

Step 3 **Write an equation** using the Pythagorean formula.

$$c^2 = a^2 + b^2$$

> Substitute carefully.

$$\mathbf{10}^2 = a^2 + \mathbf{6}^2 \qquad \text{Let } c = 10 \text{ and } b = 6.$$

Step 4 **Solve.**

$$100 = a^2 + 36 \qquad \text{Square.}$$
$$64 = a^2 \qquad \text{Subtract 36.}$$
$$\sqrt{64} = a$$
$$a = 8 \qquad \sqrt{64} = 8$$

Choose the positive square root of 64 since a represents a length.

Step 5 **State the answer.** The top of the ladder rests 8 ft up the wall.

Step 6 **Check.** From Figure 2, we see that we must have

$$8^2 + 6^2 \overset{?}{=} 10^2$$
$$64 + 36 = 100. \qquad \text{True}$$

The check confirms that the top of the ladder rests 8 ft up the wall.

Work Problem ⑦ *at the Side.* ▶

OBJECTIVE 5 Find cube, fourth, and other roots. Finding the square root of a number is the inverse (reverse) of squaring a number. In a similar way, there are inverses to finding the cube of a number, or finding the fourth or higher power of a number. These inverses are the **cube root,** written $\sqrt[3]{a}$, and the **fourth root,** written $\sqrt[4]{a}$. Similar symbols are used for higher roots. In general, we have the following.

$\sqrt[n]{a}$

The *n*th root of a is written $\sqrt[n]{a}.$

In $\sqrt[n]{a}$, the number n is the **index,** or **order,** of the radical.

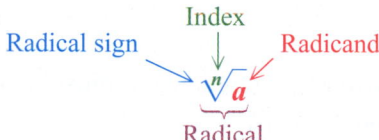

Index

Radical sign   Radicand

$\sqrt[n]{a}$

Radical

We could write $\sqrt[2]{a}$ instead of $\sqrt{a}$, but the simpler symbol $\sqrt{a}$ is customary since the square root is the most commonly used root.

🖩 **Calculator Tip** A calculator that has a key marked $\boxed{\sqrt[x]{y}}$, $\boxed{x^y}$, or $\boxed{y^x}$ (again perhaps in conjunction with the $\boxed{\text{INV}}$ or $\boxed{\text{2nd}}$ key) can be used to find other roots.

When working with cube roots or fourth roots, it is helpful to memorize the first few *perfect cubes* ($1^3 = 1, 2^3 = 8, 3^3 = 27$, and so on) and the first few *perfect fourth powers* ($1^4 = 1, 2^4 = 16, 3^4 = 81$, and so on).

Work Problem ⑧ *at the Side.* ▶

⑦ A rectangle has dimensions 5 ft by 12 ft. Find the length of its diagonal.

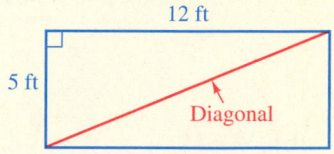

12 ft

5 ft

Diagonal

(Note that the diagonal divides the rectangle into two right triangles with itself as the hypotenuse.)

⑧ Complete the following list of perfect cubes and perfect fourth powers.

Perfect Cubes	Perfect Fourth Powers
$1^3 = 1$	$1^4 = 1$
$2^3 = 8$	$2^4 = 16$
$3^3 = 27$	$3^4 = 81$
$4^3 = \underline{\quad}$	$4^4 = \underline{\quad}$
$5^3 = \underline{\quad}$	$5^4 = \underline{\quad}$
$6^3 = \underline{\quad}$	$6^4 = \underline{\quad}$
$7^3 = \underline{\quad}$	$7^4 = \underline{\quad}$
$8^3 = \underline{\quad}$	$8^4 = \underline{\quad}$
$9^3 = \underline{\quad}$	$9^4 = \underline{\quad}$
$10^3 = \underline{\quad}$	$10^4 = \underline{\quad}$

ANSWERS

7. 13 ft
8. Perfect cubes: 64; 125; 216; 343; 512; 729; 1000
   Perfect fourth powers: 256; 625; 1296; 2401; 4096; 6561; 10,000

9 Find each cube root.

(a) $\sqrt[3]{27}$

(b) $\sqrt[3]{64}$

(c) $\sqrt[3]{-125}$

10 Find each root.

(a) $\sqrt[4]{81}$

(b) $\sqrt[4]{-81}$

(c) $-\sqrt[4]{81}$

(d) $\sqrt[5]{243}$

(e) $\sqrt[5]{-243}$

EXAMPLE 8 Finding Cube Roots

Find each cube root.

(a) $\sqrt[3]{8}$

Look for a number that can be cubed to give 8. Because $2^3 = 8, \sqrt[3]{8} = 2$.

(b) $\sqrt[3]{-8} = -2$ because $(-2)^3 = -8$.

(c) $\sqrt[3]{216} = 6$ because $6^3 = 216$.

Notice in Example 8(b) that we can find the cube root of a negative number. (Contrast this with the square root of a negative number, which is not real.) In fact, the cube root of a positive number is positive, and the cube root of a negative number is negative. ***There is only one real number cube root for each real number.***

◀ *Work Problem* **9** *at the Side.*

When a radical has an ***even index*** (square root, fourth root, and so on), ***the radicand must be nonnegative*** to yield a real number root. Also,

$$\sqrt{a}, \sqrt[4]{a}, \sqrt[6]{a}, \text{ and so on are positive (principal) roots;}$$

$$-\sqrt{a}, -\sqrt[4]{a}, -\sqrt[6]{a}, \text{ and so on are negative roots.}$$

EXAMPLE 9 Finding Other Roots

Find each root.

(a) $\sqrt[4]{16} = 2$ because 2 is positive and $2^4 = 16$.

(b) $-\sqrt[4]{16}$

From part (a), $\sqrt[4]{16} = 2$, so the negative root is $-\sqrt[4]{16} = -2$.

(c) $\sqrt[4]{-16}$

For a real number fourth root, the radicand must be nonnegative. There is no real number that equals $\sqrt[4]{-16}$.

(d) $-\sqrt[5]{32}$

First find $\sqrt[5]{32}$. Because 2 is the number whose fifth power is 32, $\sqrt[5]{32} = 2$. Since $\sqrt[5]{32} = 2$, it follows that

$$-\sqrt[5]{32} = -2.$$

(e) $\sqrt[5]{-32}$

Because $(-2)^5 = -32$, $\sqrt[5]{-32} = -2$.

◀ *Work Problem* **10** *at the Side.*

ANSWERS

9. (a) 3 (b) 4 (c) −5
10. (a) 3 (b) not a real number
    (c) −3 (d) 3 (e) −3

16.1 ▶▶▶   **Exercises**

Decide whether each statement is true *or* false. *If false,* tell why.

1. Every positive number has two real square roots.

2. A negative number has negative square roots.

3. Every nonnegative number has two real square roots.

4. The positive square root of a positive number is its principal square root.

5. The cube root of every real number has the same sign as the number itself.

6. Every positive number has three real cube roots.

Find all square roots of each number. See Example 1.

7. 9

8. 16

9. 64

10. 100

11. 169

12. 225

13. $\dfrac{25}{196}$

14. $\dfrac{81}{400}$

15. 900

16. 1600

Find each square root. See Examples 2 and 4(c).

17. $\sqrt{1}$

18. $\sqrt{4}$

19. $\sqrt{49}$

20. $\sqrt{81}$

21. $-\sqrt{256}$

22. $-\sqrt{196}$

23. $-\sqrt{\dfrac{144}{121}}$

24. $-\sqrt{\dfrac{49}{36}}$

25. $\sqrt{0.64}$

26. $\sqrt{0.16}$

27. $\sqrt{-121}$

28. $\sqrt{-64}$

29. $-\sqrt{-49}$

30. $-\sqrt{-100}$

Find the square of each radical expression. See Example 3.

31. $\sqrt{100}$

32. $\sqrt{36}$

33. $-\sqrt{19}$

34. $-\sqrt{99}$

35. $\sqrt{\dfrac{2}{3}}$

36. $\sqrt{\dfrac{5}{7}}$

37. $\sqrt{3x^2 + 4}$

38. $\sqrt{9y^2 + 3}$

What must be true about the value of a for each statement in Exercises 39–42 to be true?

39. $\sqrt{a}$ represents a positive number.

40. $-\sqrt{a}$ represents a negative number.

41. $\sqrt{a}$ is not a real number.

42. $-\sqrt{a}$ is not a real number.

Write rational, irrational, *or* not a real number *for each number. If a number is rational, give its exact value. If a number is irrational, give a decimal approximation to the nearest thousandth. Use a calculator as necessary. See Examples 4 and 5.*

43. $\sqrt{25}$

44. $\sqrt{169}$

45. $\sqrt{29}$

46. $\sqrt{33}$

47. $-\sqrt{64}$

48. $-\sqrt{81}$

49. $-\sqrt{300}$

50. $-\sqrt{500}$

51. $\sqrt{-29}$

52. $\sqrt{-47}$

53. $\sqrt{1200}$

54. $\sqrt{1500}$

Work Exercises 55 and 56 without using a calculator.

55. Choose the best estimate for the length and width (in meters) of this rectangle.

    A. 11 by 6   **B.** 11 by 7   **C.** 10 by 7   **D.** 10 by 6

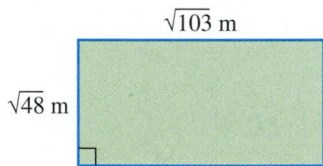

56. Choose the best estimate for the base and height (in feet) of this triangle.

    A. $b = 8, h = 5$ **B.** $b = 8, h = 4$

    C. $b = 9, h = 5$ **D.** $b = 9, h = 4$

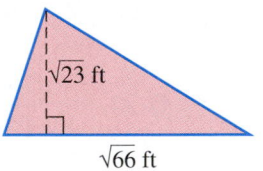

Find the length of the unknown side of each right triangle with sides a, b, and c, where c is the hypotenuse. See Figure 1 and Example 6. Give any decimal approximations to the nearest thousandth.

57. $a = 8, b = 15$

58. $a = 24, b = 10$

59. $a = 6, c = 10$

60. $b = 12, c = 13$

61. $a = 11, b = 4$

62. $a = 13, b = 9$

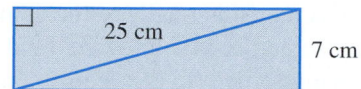 *Solve each problem. See Example 7.*

63. The diagonal of a rectangle measures 25 cm. The width of the rectangle is 7 cm. Find the length of the rectangle.

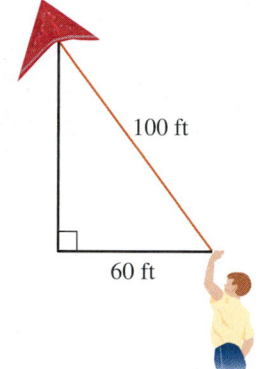

25 cm

7 cm

64. The length of a rectangle is 40 m, and the width is 9 m. Find the measure of the diagonal of the rectangle.

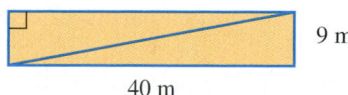

9 m

40 m

65. Tyler is flying a kite on 100 ft of string. How high is it above his hand (vertically) if the horizontal distance between Tyler and the kite is 60 ft?

100 ft

60 ft

66. A guy wire is attached to the mast of a short-wave transmitting antenna. It is attached 96 ft above ground level. If the wire is staked to the ground 72 ft from the base of the mast, how long is the wire?

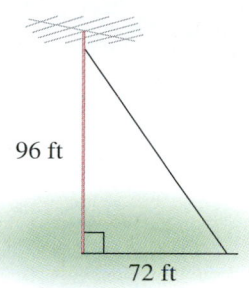

96 ft

72 ft

67. A surveyor measured the distances shown in the figure. Find the distance across the lake between points R and S.

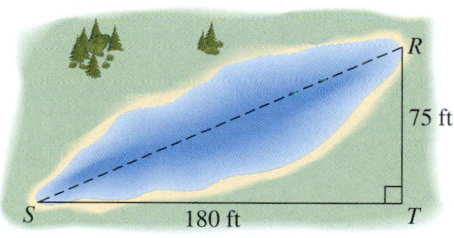

R

75 ft

S 180 ft T

68. A boat is being pulled toward a dock with a rope attached at water level. When the boat is 24 ft from the dock, 30 ft of rope is extended. What is the height of the dock above the water?

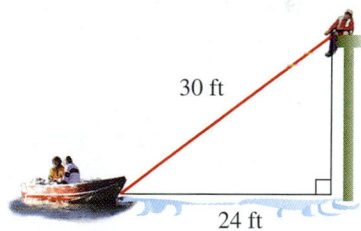

30 ft

24 ft

69. A surveyor wants to find the height of a building. At a point 110.0 ft from the base of the building he sights to the top of the building and finds the distance to be 193.0 ft. How high is the building (to the nearest tenth)?

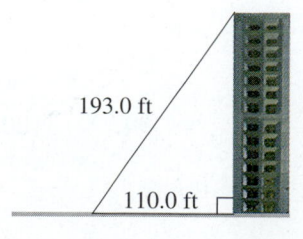

193.0 ft

110.0 ft

70. Two towns are separated by a dense forest. To go from Town B to Town A, it is necessary to travel due west for 19.0 mi, then turn due north and travel for 14.0 mi. How far apart are the towns (to the nearest tenth)?

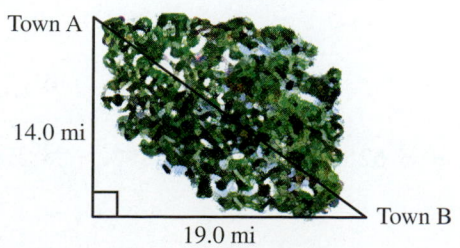

Town A

14.0 mi

Town B

19.0 mi

71. Following Hurricane Katrina, thousands of pine trees in southeastern Louisiana formed right triangles as shown in the photo. Suppose that, for a small such tree, the vertical distance from the base of the broken tree to the point of the break is 4.5 ft. The length of the broken part is 12.0 ft. How far along the ground (to the nearest tenth) is it from the base of the tree to the point where the broken part touches the ground?

72. One of the authors of this text purchased a new rear-projection Toshiba 51H84 television. A television set is "sized" according to the diagonal measurement of the viewing screen. The author purchased a 51-in. TV, so the TV measures 51 in. from one corner of the viewing screen diagonally to the other corner. The viewing screen is 44.5 in. wide. Find the height of the viewing screen (to the nearest tenth).

73. What is the value of x (to the nearest thousandth) in the figure?

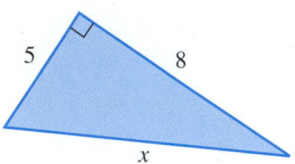

74. What is the value of y (to the nearest thousandth) in the figure?

Find each root. See Examples 8 and 9.

75. $\sqrt[3]{1}$

76. $\sqrt[3]{729}$

77. $\sqrt[3]{125}$

78. $\sqrt[3]{1000}$

79. $\sqrt[3]{-27}$

80. $\sqrt[3]{-64}$

81. $\sqrt[3]{-216}$

82. $\sqrt[3]{-343}$

83. $-\sqrt[3]{-8}$

84. $-\sqrt[3]{-216}$

85. $\sqrt[4]{256}$

86. $\sqrt[4]{625}$

87. $\sqrt[4]{1296}$

88. $\sqrt[4]{10,000}$

89. $\sqrt[4]{-1}$

90. $\sqrt[4]{-625}$

91. $-\sqrt[4]{625}$

92. $-\sqrt[4]{256}$

93. $\sqrt[5]{-1024}$

94. $\sqrt[5]{-100,000}$

16.2 ▶▶▶ Multiplying, Dividing, and Simplifying Radicals

OBJECTIVE 1 Multiply square root radicals. We now develop several rules for finding products and quotients of radicals. Notice that

$$\sqrt{4} \cdot \sqrt{9} = 2 \cdot 3 = \mathbf{6} \quad \text{and} \quad \sqrt{4 \cdot 9} = \sqrt{36} = \mathbf{6},$$

showing that

$$\sqrt{4} \cdot \sqrt{9} = \sqrt{4 \cdot 9}.$$

This result is a particular case of the **product rule for radicals.**

Product Rule for Radicals

For nonnegative real numbers a and b,

$$\sqrt{a} \cdot \sqrt{b} = \sqrt{a \cdot b} \quad \text{and} \quad \sqrt{a \cdot b} = \sqrt{a} \cdot \sqrt{b}.$$

In words, the product of two square roots is the square root of the product, and the square root of a product is the product of the two square roots.

EXAMPLE 1 Using the Product Rule to Multiply Radicals

Use the product rule for radicals to find each product.

(a) $\sqrt{2} \cdot \sqrt{3}$
$= \sqrt{2 \cdot 3}$
$= \sqrt{6}$

(b) $\sqrt{7} \cdot \sqrt{5}$
$= \sqrt{35}$

(c) $\sqrt{11} \cdot \sqrt{a}$
$= \sqrt{11a}$ Assume $a \geq 0$.

Work Problem **1** *at the Side.* ▶

OBJECTIVE 2 Simplify radicals using the product rule. *A square root radical is simplified when no perfect square factor other than 1 remains under the radical sign.* This is accomplished by using the product rule.

EXAMPLE 2 Using the Product Rule to Simplify Radicals

Simplify each radical.

(a) $\sqrt{20}$

Because 20 has a perfect square factor of 4, we can write

$$\sqrt{20}$$
$$= \sqrt{4 \cdot 5} \qquad \text{4 is a perfect square.}$$
$$= \sqrt{4} \cdot \sqrt{5} \qquad \text{Product rule}$$
$$= 2\sqrt{5}. \qquad \sqrt{4} = 2$$

Thus, $\sqrt{20} = 2\sqrt{5}$. Because 5 has no perfect square factor (other than 1), $2\sqrt{5}$ is called the **simplified form** of $\sqrt{20}$. Note that $2\sqrt{5}$ represents a product, where the factors are 2 and $\sqrt{5}$.

We could also factor 20 into prime factors and look for pairs of like factors. Each pair of like factors produces one factor outside the radical. Thus,

$$\sqrt{20} = \sqrt{2 \cdot 2 \cdot 5} = 2\sqrt{5}.$$

Continued on Next Page

OBJECTIVES

1 Multiply square root radicals.

2 Simplify radicals using the product rule.

3 Simplify radicals using the quotient rule.

4 Simplify radicals involving variables.

5 Simplify other roots.

---

1 Use the product rule for radicals to find each product.

(a) $\sqrt{6} \cdot \sqrt{11}$

(b) $\sqrt{2} \cdot \sqrt{5}$

(c) $\sqrt{10} \cdot \sqrt{r}, \quad r \geq 0$

ANSWERS

1. (a) $\sqrt{66}$ **(b)** $\sqrt{10}$ **(c)** $\sqrt{10r}$

2 Simplify each radical.

(a) $\sqrt{8}$

(b) $\sqrt{27}$

(c) $\sqrt{50}$

(d) $\sqrt{60}$

(e) $\sqrt{30}$

(b) $\sqrt{72}$

Look for the *largest* perfect square factor of 72. This number is 36, so

$$\sqrt{72}$$
$$= \sqrt{36 \cdot 2} \qquad \text{36 is a perfect square.}$$
$$= \sqrt{36} \cdot \sqrt{2} \qquad \text{Product rule}$$
$$= 6\sqrt{2}. \qquad \sqrt{36} = 6$$

We could also factor 72 into its prime factors and look for pairs of like factors.

$$\sqrt{72} = \sqrt{2 \cdot 2 \cdot 2 \cdot 3 \cdot 3} = 2 \cdot 3 \cdot \sqrt{2} = 6\sqrt{2}$$

In either case, we obtain $6\sqrt{2}$ as the simplified form of $\sqrt{72}$. However, our work is simpler if we begin with the largest perfect square factor.

(c) $\sqrt{300}$

$$= \sqrt{100 \cdot 3} \qquad \text{100 is a perfect square.}$$
$$= \sqrt{100} \cdot \sqrt{3} \qquad \text{Product rule}$$
$$= 10\sqrt{3} \qquad \sqrt{100} = 10$$

(d) $\sqrt{15}$

The number 15 has no perfect square factors (except 1), so $\sqrt{15}$ cannot be simplified further.

◀ *Work Problem* **2** *at the Side.*

EXAMPLE 3   **Multiplying and Simplifying Radicals**

Find each product and simplify.

(a) $\sqrt{9} \cdot \sqrt{75}$

$$= 3\sqrt{75} \qquad \sqrt{9} = 3$$
$$= 3\sqrt{25 \cdot 3} \qquad \text{Factor; 25 is a perfect square.}$$
$$= 3\sqrt{25} \cdot \sqrt{3} \qquad \text{Product rule}$$
$$= 3 \cdot 5 \cdot \sqrt{3} \qquad \sqrt{25} = 5$$
$$= 15\sqrt{3} \qquad \text{Multiply.}$$

Notice that we could have used the product rule to get $\sqrt{9} \cdot \sqrt{75} = \sqrt{675}$, and then simplified. However, the product rule as used here allows us to obtain the final answer without using a large number like 675.

(b) $\sqrt{8} \cdot \sqrt{12}$

$$= \sqrt{8 \cdot 12} \qquad \text{Product rule}$$
$$= \sqrt{4 \cdot 2 \cdot 4 \cdot 3} \qquad \text{Factor; 4 is a perfect square.}$$
$$= \sqrt{4} \cdot \sqrt{4} \cdot \sqrt{2 \cdot 3} \qquad \text{Commutative property; product rule}$$
$$= 2 \cdot 2 \cdot \sqrt{6} \qquad \sqrt{4} = 2$$
$$= 4\sqrt{6} \qquad \text{Multiply.}$$

Continued on Next Page

(c) $2\sqrt{3} \cdot 3\sqrt{6}$

$\quad = 2 \cdot 3 \cdot \sqrt{3 \cdot 6}$ Commutative property; product rule

$\quad = 6\sqrt{18}$ Multiply.

$\quad = 6\sqrt{9 \cdot 2}$ Factor; 9 is a perfect square.

$\quad = 6\sqrt{9} \cdot \sqrt{2}$ Product rule

$\quad = 6 \cdot 3 \cdot \sqrt{2}$ $\sqrt{9} = 3$

$\quad = 18\sqrt{2}$ Multiply.

Note

We could also simplify Example 3(b) as follows.

$\qquad \sqrt{8} \cdot \sqrt{12}$

$\qquad = \sqrt{4 \cdot 2} \cdot \sqrt{4 \cdot 3}$ Factor.

$\qquad = 2\sqrt{2} \cdot 2\sqrt{3}$ $\sqrt{4} = 2$

$\qquad = 2 \cdot 2 \cdot \sqrt{2} \cdot \sqrt{3}$ Commutative property

$\qquad = 4\sqrt{6}$ Same result

There is often more than one way to find such a product.

Work Problem ③ *at the Side.* ▶

OBJECTIVE 3 Simplify radicals using the quotient rule. The **quotient rule for radicals** is very similar to the product rule.

Quotient Rule for Radicals

If a and b are nonnegative real numbers and $b \neq 0$, then

$$\sqrt{\frac{a}{b}} = \frac{\sqrt{a}}{\sqrt{b}} \quad \text{and} \quad \frac{\sqrt{a}}{\sqrt{b}} = \sqrt{\frac{a}{b}}.$$

In words, the square root of a quotient is the quotient of the two square roots, and the quotient of two square roots is the square root of the quotient.

EXAMPLE 4 Using the Quotient Rule to Simplify Radicals

Use the quotient rule to simplify each radical.

(a) $\sqrt{\dfrac{25}{9}}$

$\quad = \dfrac{\sqrt{25}}{\sqrt{9}}$

$\quad = \dfrac{5}{3}$

(b) $\dfrac{\sqrt{288}}{\sqrt{2}}$

$\quad = \sqrt{\dfrac{288}{2}}$

$\quad = \sqrt{144}$

$\quad = 12$

(c) $\sqrt{\dfrac{3}{4}}$

$\quad = \dfrac{\sqrt{3}}{\sqrt{4}}$

$\quad = \dfrac{\sqrt{3}}{2}$

Work Problem ④ *at the Side.* ▶

---

③ Find each product and simplify.

(a) $\sqrt{3} \cdot \sqrt{15}$

(b) $\sqrt{10} \cdot \sqrt{50}$

(c) $\sqrt{12} \cdot \sqrt{2}$

(d) $\sqrt{7} \cdot \sqrt{14}$

(e) $3\sqrt{5} \cdot 4\sqrt{10}$

④ Use the quotient rule to simplify each radical.

(a) $\sqrt{\dfrac{81}{16}}$

(b) $\dfrac{\sqrt{192}}{\sqrt{3}}$

(c) $\sqrt{\dfrac{10}{49}}$

ANSWERS

3. (a) $3\sqrt{5}$ **(b)** $10\sqrt{5}$ **(c)** $2\sqrt{6}$
(d) $7\sqrt{2}$ **(e)** $60\sqrt{2}$

4. (a) $\dfrac{9}{4}$ **(b)** 8 **(c)** $\dfrac{\sqrt{10}}{7}$

5 Simplify $\dfrac{8\sqrt{50}}{4\sqrt{5}}$.

EXAMPLE 5 **Using the Quotient Rule to Divide Radicals**

Simplify.

$$\frac{27\sqrt{15}}{9\sqrt{3}}$$

$$= \frac{27}{9} \cdot \frac{\sqrt{15}}{\sqrt{3}} \qquad \text{Multiplication of fractions}$$

$$= \frac{27}{9} \cdot \sqrt{\frac{15}{3}} \qquad \text{Quotient rule}$$

$$= 3\sqrt{5} \qquad \text{Divide.}$$

◀ *Work Problem* **5** *at the Side.*

EXAMPLE 6 **Using Both the Product and Quotient Rules**

Simplify.

$$\sqrt{\frac{3}{5}} \cdot \sqrt{\frac{1}{5}}$$

6 Simplify.

(a) $\sqrt{\dfrac{5}{6}} \cdot \sqrt{120}$

$$= \sqrt{\frac{3}{5} \cdot \frac{1}{5}} \qquad \text{Product rule}$$

$$= \sqrt{\frac{3}{25}} \qquad \text{Multiply fractions.}$$

$$= \frac{\sqrt{3}}{\sqrt{25}} \qquad \text{Quotient rule}$$

$$= \frac{\sqrt{3}}{5} \qquad \sqrt{25} = 5$$

◀ *Work Problem* **6** *at the Side.*

(b) $\sqrt{\dfrac{3}{8}} \cdot \sqrt{\dfrac{7}{2}}$

OBJECTIVE **4** **Simplify radicals involving variables.** Simplifying radicals with variable radicands, such as $\sqrt{x^2}$, requires careful analysis. If x represents a nonnegative number, then $\sqrt{x^2} = x$. If x represents a negative number, then $\sqrt{x^2} = -x$, the *opposite* of x (which is positive). For example,

$$\sqrt{5^2} = 5, \qquad \text{but} \qquad \sqrt{(-5)^2} = \sqrt{25} = 5, \quad \text{the } opposite \text{ of } -5.$$

This means that the square root of a squared number is always nonnegative. We can use absolute value to express this.

> $\sqrt{a^2}$
>
> For any real number a, $\qquad \sqrt{a^2} = |a|.$

The product and quotient rules apply when variables appear under the radical sign, as long as the variables represent only *nonnegative* real numbers. ***To avoid negative radicands, variables under radical signs are assumed to be nonnegative in this text.*** Therefore, absolute value bars are not necessary, since for $x \geq 0$, $|x| = x$.

EXAMPLE 7 **Simplifying Radicals Involving Variables**

Simplify each radical. Assume that all variables represent nonnegative real numbers.

(a) $\sqrt{x^4} = x^2$ since $(x^2)^2 = x^4$.

(b) $\sqrt{25m^6}$

$\quad = \sqrt{25} \cdot \sqrt{m^6}$ Product rule

$\quad = 5m^3$ $(m^3)^2 = m^6$

(c) $\sqrt{8p^{10}}$

$\quad = \sqrt{4 \cdot 2 \cdot p^{10}}$ Factor; 4 is a perfect square.

$\quad = \sqrt{4} \cdot \sqrt{2} \cdot \sqrt{p^{10}}$ Product rule

$\quad = 2 \cdot \sqrt{2} \cdot p^5$ $(p^5)^2 = p^{10}$

$\quad = 2p^5\sqrt{2}$

(d) $\sqrt{r^9}$

$\quad = \sqrt{r^8 \cdot r}$

$\quad = \sqrt{r^8} \cdot \sqrt{r}$ Product rule

$\quad = r^4\sqrt{r}$ $(r^4)^2 = r^8$

(e) $\sqrt{\dfrac{5}{x^2}}$

$\quad = \dfrac{\sqrt{5}}{\sqrt{x^2}}$ Quotient rule

$\quad = \dfrac{\sqrt{5}}{x}$ $x \neq 0$

> **7** Simplify each radical. Assume that all variables represent nonnegative real numbers.
>
> **(a)** $\sqrt{x^8}$
>
> **(b)** $\sqrt{36y^6}$
>
> **(c)** $\sqrt{100p^{12}}$
>
> **(d)** $\sqrt{12z^2}$
>
> **(e)** $\sqrt{a^5}$
>
> **(f)** $\sqrt{\dfrac{10}{n^4}}$, $n \neq 0$

Note

A quick way to find the square root of a variable raised to an even power is to divide the exponent by the index, 2. For example,

$$\sqrt{x^6} = x^3 \quad \text{and} \quad \sqrt{x^{10}} = x^5.$$

$$6 \div 2 = 3 \qquad 10 \div 2 = 5$$

Work Problem **7** *at the Side.* ▶

OBJECTIVE **5** **Simplify other roots.** The product and quotient rules for radicals also work for other roots. To simplify cube roots, look for factors that are *perfect cubes*. A **perfect cube** is a number with a rational cube root. For example, $\sqrt[3]{64} = 4$, and because 4 is a rational number, 64 is a perfect cube. Other roots are handled in a similar manner.

Properties of Radicals

For all real numbers where the indicated roots exist,

$$\sqrt[n]{a} \cdot \sqrt[n]{b} = \sqrt[n]{ab} \quad \text{and} \quad \frac{\sqrt[n]{a}}{\sqrt[n]{b}} = \sqrt[n]{\frac{a}{b}} \quad (b \neq 0).$$

ANSWERS

7. **(a)** x^4 **(b)** $6y^3$ **(c)** $10p^6$ **(d)** $2z\sqrt{3}$

    (e) $a^2\sqrt{a}$ **(f)** $\dfrac{\sqrt{10}}{n^2}$

8 Simplify each radical.

(a) $\sqrt[3]{108}$

(b) $\sqrt[4]{160}$

(c) $\sqrt[4]{\dfrac{16}{625}}$

9 Simplify each radical.

(a) $\sqrt[3]{z^9}$

(b) $\sqrt[3]{8x^6}$

(c) $\sqrt[3]{54t^5}$

(d) $\sqrt[3]{\dfrac{a^{15}}{64}}$

EXAMPLE 8   Simplifying Other Roots

Simplify each radical.

(a) $\sqrt[3]{32}$

> Remember to write the root index 3 in each radical.

$= \sqrt[3]{8 \cdot 4}$ Factor; 8 is a perfect cube.

$= \sqrt[3]{8} \cdot \sqrt[3]{4}$ Product rule

$= 2\sqrt[3]{4}$

(b) $\sqrt[4]{32}$

> Remember to write the root index 4 in each radical.

$= \sqrt[4]{16 \cdot 2}$ Factor; 16 is a perfect fourth power.

$= \sqrt[4]{16} \cdot \sqrt[4]{2}$ Product rule

$= 2\sqrt[4]{2}$

(c) $\sqrt[3]{\dfrac{27}{125}}$

$= \dfrac{\sqrt[3]{27}}{\sqrt[3]{125}}$ Quotient rule

$= \dfrac{3}{5}$

◀ *Work Problem* **8** *at the Side.*

Other roots of radicals involving variables can also be simplified. To simplify cube roots with variables, use the fact that for any real number a,

$$\sqrt[3]{a^3} = a.$$

This is true whether a is positive or negative.

EXAMPLE 9   Simplifying Cube Roots Involving Variables

Simplify each radical.

(a) $\sqrt[3]{m^6}$

$= m^2$ $(m^2)^3 = m^6$

(b) $\sqrt[3]{27x^{12}}$

$= \sqrt[3]{27} \cdot \sqrt[3]{x^{12}}$ Product rule

$= 3x^4$ $3^3 = 27;$ $(x^4)^3 = x^{12}$

(c) $\sqrt[3]{32a^4}$

$= \sqrt[3]{8a^3 \cdot 4a}$ Factor; 8 is a perfect cube.

$= \sqrt[3]{8a^3} \cdot \sqrt[3]{4a}$ Product rule

$= 2a\sqrt[3]{4a}$ $(2a)^3 = 8a^3$

(d) $\sqrt[3]{\dfrac{y^3}{125}}$

$= \dfrac{\sqrt[3]{y^3}}{\sqrt[3]{125}}$ Quotient rule

$= \dfrac{y}{5}$

◀ *Work Problem* **9** *at the Side.*

ANSWERS

8. (a) $3\sqrt[3]{4}$ (b) $2\sqrt[4]{10}$ (c) $\dfrac{2}{5}$

9. (a) z^3 (b) $2x^2$ (c) $3t\sqrt[3]{2t^2}$ (d) $\dfrac{a^5}{4}$

16.2 ▶▶▶ Exercises

Decide whether each statement is true *or* false. *If* false, *show why.*

1. $\sqrt{(-6)^2} = -6$

2. $\sqrt[3]{(-6)^3} = -6$

Use the product rule for radicals to find each product. See Example 1.

3. $\sqrt{3} \cdot \sqrt{5}$

4. $\sqrt{3} \cdot \sqrt{7}$

5. $\sqrt{2} \cdot \sqrt{11}$

6. $\sqrt{2} \cdot \sqrt{15}$

7. $\sqrt{6} \cdot \sqrt{7}$

8. $\sqrt{5} \cdot \sqrt{6}$

9. $\sqrt{13} \cdot \sqrt{r}, r \geq 0$

10. $\sqrt{19} \cdot \sqrt{k}, k \geq 0$

11. Which one of the following radicals is simplified? See Example 2.

  A. $\sqrt{47}$ **B.** $\sqrt{45}$ **C.** $\sqrt{48}$ **D.** $\sqrt{44}$

12. If p is a prime number, is $\sqrt{p}$ in simplified form? Explain your answer.

Simplify each radical. See Example 2.

13. $\sqrt{45}$

14. $\sqrt{200}$

15. $\sqrt{24}$

16. $\sqrt{44}$

17. $\sqrt{90}$

18. $\sqrt{56}$

19. $\sqrt{75}$

20. $\sqrt{18}$

21. $\sqrt{125}$

22. $\sqrt{80}$

23. $\sqrt{145}$

24. $\sqrt{110}$

25. $\sqrt{160}$

26. $\sqrt{128}$

27. $-\sqrt{700}$

28. $-\sqrt{600}$

Find each product and simplify. See Example 3.

29. $\sqrt{3} \cdot \sqrt{18}$ **30.** $\sqrt{3} \cdot \sqrt{21}$ **31.** $\sqrt{12} \cdot \sqrt{48}$ **32.** $\sqrt{50} \cdot \sqrt{72}$

33. $\sqrt{12} \cdot \sqrt{30}$ **34.** $\sqrt{30} \cdot \sqrt{24}$ **35.** $2\sqrt{10} \cdot 3\sqrt{2}$

36. $5\sqrt{6} \cdot 2\sqrt{10}$ **37.** $5\sqrt{3} \cdot 2\sqrt{15}$ **38.** $4\sqrt{6} \cdot 3\sqrt{2}$

39. Simplify the product $\sqrt{8} \cdot \sqrt{32}$ in two ways. First, multiply 8 by 32 and simplify the square root of this product. Second, simplify $\sqrt{8}$, simplify $\sqrt{32}$, and then multiply. How do the answers compare? Make a conjecture (an educated guess) about whether the correct answer can always be obtained using either method when simplifying a product such as this.

40. Simplify the radical $\sqrt{288}$ in two ways. First, factor 288 as $144 \cdot 2$ and then simplify. Second, factor 288 as $48 \cdot 6$ and then simplify. How do the answers compare? Make a conjecture concerning the quickest way to simplify such a radical.

Simplify each radical expression. See Examples 4–6.

41. $\sqrt{\dfrac{16}{225}}$ **42.** $\sqrt{\dfrac{9}{100}}$ **43.** $\sqrt{\dfrac{7}{16}}$ **44.** $\sqrt{\dfrac{13}{25}}$

45. $\dfrac{\sqrt{75}}{\sqrt{3}}$ **46.** $\dfrac{\sqrt{200}}{\sqrt{2}}$ **47.** $\sqrt{\dfrac{5}{2}} \cdot \sqrt{\dfrac{125}{8}}$

48. $\sqrt{\dfrac{8}{3}} \cdot \sqrt{\dfrac{512}{27}}$ **49.** $\dfrac{30\sqrt{10}}{5\sqrt{2}}$ **50.** $\dfrac{50\sqrt{20}}{2\sqrt{10}}$

Simplify each radical. Assume that all variables represent nonnegative real numbers.
See Example 7.

51. $\sqrt{m^2}$

52. $\sqrt{k^2}$

53. $\sqrt{y^4}$

54. $\sqrt{s^4}$

55. $\sqrt{36z^2}$

56. $\sqrt{49n^2}$

57. $\sqrt{400x^6}$

58. $\sqrt{900y^8}$

59. $\sqrt{18x^8}$

60. $\sqrt{20r^{10}}$

61. $\sqrt{45c^{14}}$

62. $\sqrt{50d^{20}}$

63. $\sqrt{z^5}$

64. $\sqrt{y^3}$

65. $\sqrt{a^{13}}$

66. $\sqrt{p^{17}}$

67. $\sqrt{64x^7}$

68. $\sqrt{25t^{11}}$

69. $\sqrt{x^6y^{12}}$

70. $\sqrt{a^8b^{10}}$

71. $\sqrt{81m^4n^2}$

72. $\sqrt{100c^4d^6}$

73. $\sqrt{\dfrac{7}{x^{10}}}, \quad x \neq 0$

74. $\sqrt{\dfrac{14}{z^{12}}}, \quad z \neq 0$

75. $\sqrt{\dfrac{y^4}{100}}$

76. $\sqrt{\dfrac{w^8}{144}}$

77. $\sqrt{\dfrac{x^6}{y^8}}, \quad y \neq 0$

78. $\sqrt{\dfrac{a^4}{b^6}}, \quad b \neq 0$

Simplify each radical. See Example 8.

79. $\sqrt[3]{40}$

80. $\sqrt[3]{48}$

81. $\sqrt[3]{54}$

82. $\sqrt[3]{135}$

83. $\sqrt[3]{128}$

84. $\sqrt[3]{192}$

85. $\sqrt[4]{80}$

86. $\sqrt[4]{243}$

87. $\sqrt[3]{\dfrac{8}{27}}$

88. $\sqrt[3]{\dfrac{64}{125}}$

89. $\sqrt[3]{-\dfrac{216}{125}}$

90. $\sqrt[3]{-\dfrac{1}{64}}$

Simplify each radical. See Example 9.

91. $\sqrt[3]{p^3}$

92. $\sqrt[3]{w^3}$

93. $\sqrt[3]{x^9}$

94. $\sqrt[3]{y^{18}}$

95. $\sqrt[3]{64z^6}$

96. $\sqrt[3]{125a^{15}}$

97. $\sqrt[3]{343a^9b^3}$

98. $\sqrt[3]{216m^3n^6}$

99. $\sqrt[3]{16t^5}$

100. $\sqrt[3]{24x^4}$

101. $\sqrt[3]{\dfrac{m^{12}}{8}}$

102. $\sqrt[3]{\dfrac{n^9}{27}}$

The volume of a cube is found with the formula $V = s^3$, where s is the length of an edge of the cube. Use this information in Exercises 103 and 104.

103. A container in the shape of a cube has a volume of 216 cm³. What is the depth of the container?

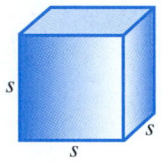

104. A cube-shaped box must be constructed to contain 128 ft³. What should the dimensions (height, width, and length) of the box be?

The volume of a sphere is found with the formula $V = \frac{4}{3}\pi r^3$, where r is the length of the radius of the sphere. Use this information in Exercises 105 and 106.

105. A ball in the shape of a sphere has a volume of 288π in.³. What is the radius of the ball?

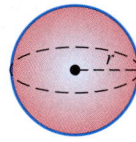

106. Suppose that the volume of the ball described in Exercise 105 is multiplied by 8. How is the radius affected?

Work Exercises 107 and 108 without using a calculator.

107. Choose the best estimate for the area (in square inches) of this rectangle.

A. 45   **B.** 72   **C.** 80   **D.** 90

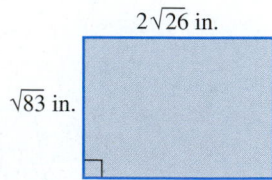

108. Choose the best estimate for the area (in square feet) of the triangle.

A. 20   **B.** 40   **C.** 60   **D.** 80

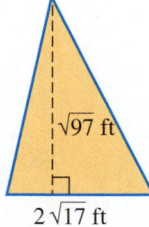

16.3 ▶▶▶ Adding and Subtracting Radicals

OBJECTIVES

1 Add and subtract radicals.

2 Simplify radical sums and differences.

3 Simplify more complicated radical expressions.

OBJECTIVE 1 Add and subtract radicals. We add or subtract radicals by using the distributive property. For example,

$$8\sqrt{3} + 6\sqrt{3}$$
$$= (8+6)\sqrt{3}$$
$$= 14\sqrt{3}$$

$$2\sqrt{11} - 7\sqrt{11}$$
$$= (2-7)\sqrt{11}$$
$$= -5\sqrt{11}.$$

Only **like radicals**—those that are *multiples of the same root of the same number*—can be combined. In the examples above, $8\sqrt{3}$ and $6\sqrt{3}$ are like radicals, as are $2\sqrt{11}$ and $-7\sqrt{11}$. Examples of **unlike radicals** are

$2\sqrt{5}$ and $2\sqrt{3}$,   Radicands are different.

as well as $2\sqrt{3}$ and $2\sqrt[3]{3}$.   Indexes are different.

Work Problem **1** *at the Side.* ▶

1 Indicate whether the radicals in each pair are *like* or *unlike*.

(a) $5\sqrt{6}$ and $4\sqrt{6}$

(b) $2\sqrt{3}$ and $3\sqrt{2}$

(c) $\sqrt{10}$ and $\sqrt[3]{10}$

(d) $7\sqrt{2x}$ and $8\sqrt{2x}$

(e) $\sqrt{3y}$ and $\sqrt{6y}$

EXAMPLE 1 Adding and Subtracting Like Radicals

Add or subtract, as indicated.

(a) $3\sqrt{6} + 5\sqrt{6}$
$$= (3+5)\sqrt{6}$$
$$= 8\sqrt{6}$$

(b) $5\sqrt{10} - 7\sqrt{10}$
$$= (5-7)\sqrt{10}$$
$$= -2\sqrt{10}$$

(c) $\sqrt{7} + 2\sqrt{7}$
$$= 1\sqrt{7} + 2\sqrt{7}$$
$$= (1+2)\sqrt{7}$$
$$= 3\sqrt{7}$$

(d) $\sqrt{5} + \sqrt{5}$
$$= 1\sqrt{5} + 1\sqrt{5}$$
$$= (1+1)\sqrt{5}$$
$$= 2\sqrt{5}$$

(e) $\sqrt{3} + \sqrt{7}$ cannot be added using the distributive property.

Work Problem **2** *at the Side.* ▶

OBJECTIVE 2 Simplify radical sums and differences.

EXAMPLE 2 Simplifying Radicals to Add or Subtract

Add or subtract, as indicated.

(a) $3\sqrt{2} + \sqrt{8}$
$$= 3\sqrt{2} + \sqrt{4 \cdot 2}$$ Factor.
$$= 3\sqrt{2} + \sqrt{4} \cdot \sqrt{2}$$ Product rule
$$= 3\sqrt{2} + 2\sqrt{2}$$ $\sqrt{4} = 2$
$$= 5\sqrt{2}$$ Add like radicals.

(b)
$$\sqrt{18} - \sqrt{27}$$
$$= \sqrt{9 \cdot 2} - \sqrt{9 \cdot 3}$$ Factor.
$$= \sqrt{9} \cdot \sqrt{2} - \sqrt{9} \cdot \sqrt{3}$$ Product rule
$$= 3\sqrt{2} - 3\sqrt{3}$$ $\sqrt{9} = 3$

These terms cannot be combined.

2 Add or subtract, as indicated.

(a) $8\sqrt{5} + 2\sqrt{5}$

(b) $-4\sqrt{3} + 9\sqrt{3}$

(c) $12\sqrt{11} - 3\sqrt{11}$

(d) $\sqrt{15} + \sqrt{15}$

(e) $2\sqrt{7} + 2\sqrt{10}$

ANSWERS

1. (a) like (b) unlike (c) unlike
   (d) like (e) unlike
2. (a) $10\sqrt{5}$ (b) $5\sqrt{3}$ (c) $9\sqrt{11}$
   (d) $2\sqrt{15}$ (e) cannot be added

Continued on Next Page

3 Add or subtract, as indicated.

(a) $\sqrt{8} + 4\sqrt{2}$

(b) $\sqrt{27} + \sqrt{12}$

(c) $5\sqrt{200} - 6\sqrt{18}$

4 Simplify each radical expression. Assume that all variables represent non-negative real numbers.

(a) $\sqrt{7} \cdot \sqrt{21} + 2\sqrt{27}$

(b) $\sqrt{3r} \cdot \sqrt{6} + \sqrt{8r}$

(c) $y\sqrt{72} - \sqrt{18y^2}$

(d) $\sqrt[3]{81x^4} + 5\sqrt[3]{24x^4}$

(c) $2\sqrt{12} + 3\sqrt{75}$

$$= 2\left(\sqrt{4} \cdot \sqrt{3}\right) + 3\left(\sqrt{25} \cdot \sqrt{3}\right) \qquad \text{Product rule}$$

$$= 2\left(2\sqrt{3}\right) + 3\left(5\sqrt{3}\right) \qquad \sqrt{4} = 2;\ \sqrt{25} = 5$$

$$= 4\sqrt{3} + 15\sqrt{3} \qquad \text{Multiply.}$$

$$= 19\sqrt{3} \qquad \text{Add like radicals.}$$

◀ *Work Problem* **3** *at the Side.*

OBJECTIVE **3** **Simplify more complicated radical expressions.**

EXAMPLE 3 **Simplifying Radical Expressions**

Simplify each radical expression. Assume that all variables represent non-negative real numbers.

(a) $\sqrt{5} \cdot \sqrt{15} + 4\sqrt{3}$

$$= \sqrt{5 \cdot 15} + 4\sqrt{3} \qquad \text{Product rule}$$

$$= \sqrt{75} + 4\sqrt{3} \qquad \text{Multiply.}$$

$$= \sqrt{25 \cdot 3} + 4\sqrt{3} \qquad \text{Factor; 25 is a perfect square.}$$

$$= \sqrt{25} \cdot \sqrt{3} + 4\sqrt{3} \qquad \text{Product rule}$$

$$= 5\sqrt{3} + 4\sqrt{3} \qquad \sqrt{25} = 5$$

$$= 9\sqrt{3} \qquad \text{Add like radicals.}$$

(b) $\sqrt{2} \cdot \sqrt{6k} + \sqrt{27k}$

$$= \sqrt{12k} + \sqrt{27k} \qquad \text{Product rule}$$

$$= \sqrt{4} \cdot \sqrt{3k} + \sqrt{9} \cdot \sqrt{3k} \qquad \text{Factor; product rule}$$

$$= 2\sqrt{3k} + 3\sqrt{3k} \qquad \sqrt{4} = 2;\ \sqrt{9} = 3$$

$$= 5\sqrt{3k} \qquad \text{Add like radicals.}$$

(c) $3x\sqrt{50} + \sqrt{2x^2}$

$$= 3x\sqrt{25 \cdot 2} + \sqrt{x^2 \cdot 2} \qquad \text{Factor.}$$

$$= 3x\sqrt{25} \cdot \sqrt{2} + \sqrt{x^2} \cdot \sqrt{2} \qquad \text{Product rule}$$

$$= 3x \cdot 5\sqrt{2} + x\sqrt{2} \qquad \sqrt{25} = 5;\ \sqrt{x^2} = x$$

$$= 15x\sqrt{2} + x\sqrt{2} \qquad \text{Multiply.}$$

$$= 16x\sqrt{2} \qquad \text{Add like radicals.}$$

(d) $2\sqrt[3]{32m^3} - \sqrt[3]{108m^3}$

$$= 2\sqrt[3]{(8m^3)4} - \sqrt[3]{(27m^3)4} \qquad \text{Factor.}$$

$$= 2(2m)\sqrt[3]{4} - 3m\sqrt[3]{4} \qquad \sqrt[3]{8m^3} = 2m;\ \sqrt[3]{27m^3} = 3m$$

$$= 4m\sqrt[3]{4} - 3m\sqrt[3]{4} \qquad \text{Multiply.}$$

$$= m\sqrt[3]{4} \qquad \text{Subtract like radicals.}$$

◀ *Work Problem* **4** *at the Side.*

16.3 ▶▶▶ Exercises

Fill in each blank with the correct response.

1. $5\sqrt{2} + 6\sqrt{2} = (5 + 6)\sqrt{2} = 11\sqrt{2}$ is an example of the _____ property.

2. The radicals $\sqrt[4]{3xy^3}$ and $-6\sqrt[4]{3xy^3}$ are examples of like radicals because both radicals have the same root index, _____, and the same radicand, _____ .

3. $\sqrt{5} + 5\sqrt{3}$ cannot be simplified because the _____ are different.

4. $4\sqrt[3]{2} + 3\sqrt{2}$ cannot be simplified because the _____ are different.

Simplify and add or subtract wherever possible. See Examples 1 and 2.

5. $14\sqrt{7} - 19\sqrt{7}$

6. $16\sqrt{2} - 18\sqrt{2}$

⊙ 7. $\sqrt{17} + 4\sqrt{17}$

8. $5\sqrt{19} + \sqrt{19}$

9. $6\sqrt{7} - \sqrt{7}$

10. $11\sqrt{14} - \sqrt{14}$

⊙ 11. $\sqrt{45} + 4\sqrt{20}$

12. $\sqrt{24} + 6\sqrt{54}$

13. $5\sqrt{72} - 3\sqrt{50}$

14. $6\sqrt{18} - 5\sqrt{32}$

15. $-5\sqrt{32} + 2\sqrt{98}$

16. $-4\sqrt{75} + 3\sqrt{12}$

17. $5\sqrt{7} - 3\sqrt{28} + 6\sqrt{63}$

18. $3\sqrt{11} + 5\sqrt{44} - 8\sqrt{99}$

19. $2\sqrt{8} - 5\sqrt{32} - 2\sqrt{48}$

20. $5\sqrt{72} - 3\sqrt{48} + 4\sqrt{128}$

21. $4\sqrt{50} + 3\sqrt{12} - 5\sqrt{45}$

22. $6\sqrt{18} + 2\sqrt{48} + 6\sqrt{28}$

23. $\frac{1}{4}\sqrt{288} + \frac{1}{6}\sqrt{72}$

24. $\frac{2}{3}\sqrt{27} + \frac{3}{4}\sqrt{48}$

Find the perimeter of each figure.

25.
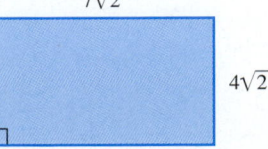
$7\sqrt{2}$
$4\sqrt{2}$

26.
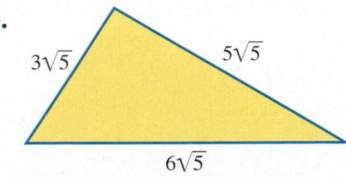
$3\sqrt{5}$ $5\sqrt{5}$
$6\sqrt{5}$

Perform the indicated operations. Assume that all variables represent nonnegative real numbers. See Example 3.

27. $\sqrt{6} \cdot \sqrt{2} + 9\sqrt{3}$

28. $4\sqrt{15} \cdot \sqrt{3} + 4\sqrt{5}$

29. $\sqrt{9x} + \sqrt{49x} - \sqrt{25x}$

30. $\sqrt{4a} - \sqrt{16a} + \sqrt{100a}$

31. $\sqrt{6x^2} + x\sqrt{24}$

32. $\sqrt{75x^2} + x\sqrt{108}$

33. $3\sqrt{8x^2} - 4x\sqrt{2} - x\sqrt{8}$

34. $\sqrt{2b^2} + 3b\sqrt{18} - b\sqrt{200}$

35. $-8\sqrt{32k} + 6\sqrt{8k}$

36. $4\sqrt{12x} + 2\sqrt{27x}$

37. $2\sqrt{125x^2z} + 8x\sqrt{80z}$

38. $\sqrt{48x^2y} + 5x\sqrt{27y}$

39. $4\sqrt[3]{16} - 3\sqrt[3]{54}$

40. $5\sqrt[3]{128} + 3\sqrt[3]{250}$

41. $6\sqrt[3]{8p^2} - 2\sqrt[3]{27p^2}$

42. $8k\sqrt[3]{54k} + 6\sqrt[3]{16k^4}$

43. $5\sqrt[4]{m^3} + 8\sqrt[4]{16m^3}$

44. $5\sqrt[4]{m^5} + 3\sqrt[4]{81m^5}$

Relating Concepts (Exercises 45–48)   For Individual or Group Work

Adding and subtracting like radicals is no different than adding and subtracting like terms. **Work Exercises 45–48 in order.**

45. Combine like terms: $5x^2y + 3x^2y - 14x^2y$.

46. Combine like terms: $5(p - 2q)^2(a + b) + 3(p - 2q)^2(a + b) - 14(p - 2q)^2(a + b)$.

47. Combine like radicals: $5a^2\sqrt{xy} + 3a^2\sqrt{xy} - 14a^2\sqrt{xy}$.

48. Compare your answers in Exercises 45–47. How are they alike? How are they different?

16.4 ▶▶▶ Rationalizing the Denominator

OBJECTIVE 1 Rationalize denominators with square roots. Although calculators now make it fairly easy to divide by a radical in an expression such as $\frac{1}{\sqrt{2}}$, it is sometimes easier to work with radical expressions if the denominators do not contain any radicals. For example, the radical in the denominator of $\frac{1}{\sqrt{2}}$ can be eliminated by multiplying the numerator and denominator by $\sqrt{2}$, since $\sqrt{2} \cdot \sqrt{2} = \sqrt{4} = 2$.

$$\frac{1}{\sqrt{2}} = \frac{1 \cdot \sqrt{2}}{\sqrt{2} \cdot \sqrt{2}} = \frac{\sqrt{2}}{2} \qquad \text{Multiply by } \frac{\sqrt{2}}{\sqrt{2}} = 1.$$

This process of changing the denominator from a radical (irrational number) to a rational number is called **rationalizing the denominator.** *The value of the radical expression is not changed; only the form is changed, because the expression has been multiplied by 1 in the form of $\frac{\sqrt{2}}{\sqrt{2}}$.*

EXAMPLE 1 Rationalizing Denominators

Rationalize each denominator.

(a) $\dfrac{9}{\sqrt{6}}$

$$= \frac{9 \cdot \sqrt{6}}{\sqrt{6} \cdot \sqrt{6}} \qquad \text{Multiply by } \frac{\sqrt{6}}{\sqrt{6}} = 1.$$

$$= \frac{9\sqrt{6}}{6} \qquad \sqrt{6} \cdot \sqrt{6} = \sqrt{36} = 6$$

$$= \frac{3\sqrt{6}}{2} \qquad \text{Lowest terms}$$

(b) $\dfrac{12}{\sqrt{8}}$

The denominator could be rationalized by multiplying by $\sqrt{8}$. However, first simplifying the denominator is more direct.

$$\frac{12}{\sqrt{8}}$$

$$= \frac{12}{2\sqrt{2}} \qquad \sqrt{8} = \sqrt{4} \cdot \sqrt{2} = 2\sqrt{2}$$

$$= \frac{12 \cdot \sqrt{2}}{2\sqrt{2} \cdot \sqrt{2}} \qquad \text{Multiply by } \frac{\sqrt{2}}{\sqrt{2}} = 1.$$

$$= \frac{12 \cdot \sqrt{2}}{2 \cdot 2} \qquad \sqrt{2} \cdot \sqrt{2} = \sqrt{4} = 2$$

$$= \frac{12\sqrt{2}}{4} \qquad \text{Multiply.}$$

$$= 3\sqrt{2} \qquad \text{Lowest terms}$$

1 Rationalize each denominator.

(a) $\dfrac{3}{\sqrt{5}}$

(b) $\dfrac{-6}{\sqrt{11}}$

(c) $-\dfrac{\sqrt{7}}{\sqrt{2}}$

(d) $\dfrac{20}{\sqrt{18}}$

Note

In Example 1(b), we could also have rationalized the original denominator $\sqrt{8}$ by multiplying by $\sqrt{2}$, since $\sqrt{8} \cdot \sqrt{2} = \sqrt{16} = 4$.

$$\frac{12}{\sqrt{8}} = \frac{12 \cdot \sqrt{2}}{\sqrt{8} \cdot \sqrt{2}} = \frac{12\sqrt{2}}{\sqrt{16}} = \frac{12\sqrt{2}}{4} = 3\sqrt{2}$$

Both approaches are correct.

◀ *Work Problem* **1** *at the Side.*

OBJECTIVE 2 Write radicals in simplified form. A radical is considered to be in simplified form if the following three conditions are met.

Conditions for Simplified Form of a Radical

1. The radicand contains no factor (except 1) that is a perfect square (when dealing with square roots), a perfect cube (when dealing with cube roots), and so on.
2. The radicand has no fractions.
3. No denominator contains a radical.

---

2 Simplify.

(a) $\sqrt{\dfrac{16}{11}}$

(b) $\sqrt{\dfrac{5}{18}}$

(c) $\sqrt{\dfrac{8}{32}}$

EXAMPLE 2 Simplifying a Radical

Simplify.

$$\sqrt{\frac{27}{5}}$$

$$= \frac{\sqrt{27}}{\sqrt{5}} \qquad \text{Quotient rule}$$

$$= \frac{\sqrt{27} \cdot \sqrt{5}}{\sqrt{5} \cdot \sqrt{5}} \qquad \text{Rationalize the denominator.}$$

$$= \frac{\sqrt{27} \cdot \sqrt{5}}{5} \qquad \sqrt{5} \cdot \sqrt{5} = \sqrt{25} = 5$$

$$= \frac{\sqrt{9 \cdot 3} \cdot \sqrt{5}}{5} \qquad \text{Factor.}$$

$$= \frac{\sqrt{9} \cdot \sqrt{3} \cdot \sqrt{5}}{5} \qquad \text{Product rule}$$

$$= \frac{3 \cdot \sqrt{3} \cdot \sqrt{5}}{5} \qquad \sqrt{9} = 3$$

$$= \frac{3\sqrt{15}}{5} \qquad \text{Product rule}$$

◀ *Work Problem* **2** *at the Side.*

ANSWERS

1. (a) $\dfrac{3\sqrt{5}}{5}$ (b) $\dfrac{-6\sqrt{11}}{11}$

(c) $-\dfrac{\sqrt{14}}{2}$ (d) $\dfrac{10\sqrt{2}}{3}$

2. (a) $\dfrac{4\sqrt{11}}{11}$ (b) $\dfrac{\sqrt{10}}{6}$ (c) $\dfrac{1}{2}$

EXAMPLE 3 **Simplifying a Product of Radicals**

Simplify.

$$\sqrt{\frac{5}{8}} \cdot \sqrt{\frac{1}{6}}$$

$$= \sqrt{\frac{5}{8} \cdot \frac{1}{6}}$$ Product rule

$$= \sqrt{\frac{5}{48}}$$ Multiply fractions.

$$= \frac{\sqrt{5}}{\sqrt{48}}$$ Quotient rule

$$= \frac{\sqrt{5}}{\sqrt{16} \cdot \sqrt{3}}$$ Product rule

$$= \frac{\sqrt{5}}{4\sqrt{3}}$$ $\sqrt{16} = 4$

$$= \frac{\sqrt{5} \cdot \sqrt{3}}{4\sqrt{3} \cdot \sqrt{3}}$$ Rationalize the denominator.

$$= \frac{\sqrt{15}}{4 \cdot 3}$$ Product rule; $\sqrt{3} \cdot \sqrt{3} = 3$

$$= \frac{\sqrt{15}}{12}$$ Multiply.

Work Problem **3** at the Side. ▶

EXAMPLE 4 **Simplifying Quotients Involving Radicals**

Simplify. Assume that x and y represent positive real numbers.

(a) $\dfrac{\sqrt{4x}}{\sqrt{y}}$

$$= \frac{\sqrt{4x} \cdot \sqrt{y}}{\sqrt{y} \cdot \sqrt{y}}$$ Rationalize the denominator.

$$= \frac{\sqrt{4xy}}{y}$$ Product rule; $\sqrt{y} \cdot \sqrt{y} = y$

$$= \frac{2\sqrt{xy}}{y}$$ $\sqrt{4} = 2$

(b) $\sqrt{\dfrac{2x^2y}{3}}$

$$= \frac{\sqrt{2x^2y}}{\sqrt{3}}$$ Quotient rule

$$= \frac{\sqrt{2x^2y} \cdot \sqrt{3}}{\sqrt{3} \cdot \sqrt{3}}$$ Rationalize the denominator.

$$= \frac{\sqrt{6x^2y}}{3}$$ Product rule; $\sqrt{3} \cdot \sqrt{3} = 3$

$$= \frac{\sqrt{x^2}\sqrt{6y}}{3}$$ Product rule

$$= \frac{x\sqrt{6y}}{3}$$ $\sqrt{x^2} = x$, since $x > 0$.

Work Problem **4** at the Side. ▶

3 Simplify.

(a) $\sqrt{\dfrac{1}{2}} \cdot \sqrt{\dfrac{5}{6}}$

(b) $\sqrt{\dfrac{1}{10}} \cdot \sqrt{20}$

(c) $\sqrt{\dfrac{5}{8}} \cdot \sqrt{\dfrac{24}{10}}$

4 Simplify. Assume that all variables represent positive real numbers.

(a) $\dfrac{\sqrt{5p}}{\sqrt{q}}$

(b) $\sqrt{\dfrac{5r^2t^2}{7}}$

ANSWERS

3. **(a)** $\dfrac{\sqrt{15}}{6}$ **(b)** $\sqrt{2}$ **(c)** $\dfrac{\sqrt{6}}{2}$

4. **(a)** $\dfrac{\sqrt{5pq}}{q}$ **(b)** $\dfrac{rt\sqrt{35}}{7}$

5 Rationalize each denominator.

(a) $\sqrt[3]{\dfrac{5}{7}}$

OBJECTIVE **3** **Rationalize denominators with cube roots.** A denominator with a cube root is rationalized by changing the radicand in the denominator to a perfect cube, as shown in the next example.

EXAMPLE 5 **Rationalizing Denominators with Cube Roots**

Rationalize each denominator.

(a) $\sqrt[3]{\dfrac{3}{2}}$

First write the expression as a quotient of radicals. Then multiply the numerator and denominator by the appropriate number of factors of 2 to make the radicand in the denominator a perfect cube. This will eliminate the radical in the denominator. Here, multiply by $\sqrt[3]{2^2}$.

$$\sqrt[3]{\frac{3}{2}} = \frac{\sqrt[3]{3}}{\sqrt[3]{2}} = \frac{\sqrt[3]{3} \cdot \sqrt[3]{2^2}}{\sqrt[3]{2} \cdot \sqrt[3]{2^2}} = \frac{\sqrt[3]{3 \cdot 2^2}}{\sqrt[3]{2^3}} = \frac{\sqrt[3]{12}}{2} \qquad \sqrt[3]{2^3} = \sqrt[3]{8} = 2$$

Be careful not to multiply by $\sqrt[3]{2}$ here.

⬆ Denominator is a perfect cube.

(b) $\dfrac{\sqrt[3]{5}}{\sqrt[3]{9}}$

(b) $\dfrac{\sqrt[3]{3}}{\sqrt[3]{4}}$

Since $\sqrt[3]{4} \cdot \sqrt[3]{2} = \sqrt[3]{2^2} \cdot \sqrt[3]{2} = \sqrt[3]{2^3} = 2$, multiply the numerator and denominator by $\sqrt[3]{2}$.

$$\frac{\sqrt[3]{3}}{\sqrt[3]{4}} = \frac{\sqrt[3]{3} \cdot \sqrt[3]{2}}{\sqrt[3]{2^2} \cdot \sqrt[3]{2}} = \frac{\sqrt[3]{6}}{\sqrt[3]{2^3}} = \frac{\sqrt[3]{6}}{2}$$

(c) $\dfrac{\sqrt[3]{2}}{\sqrt[3]{3x^2}}, \quad x \neq 0$

Multiply the numerator and denominator by the appropriate number of factors of 3 and of x to get a perfect cube in the radicand of the denominator. Here, multiply by $\sqrt[3]{3^2 x}$ (that is, $\sqrt[3]{9x}$) since $\sqrt[3]{3x^2} \cdot \sqrt[3]{3^2 x} = \sqrt[3]{(3x)^3} = 3x$.

(c) $\dfrac{\sqrt[3]{4}}{\sqrt[3]{25y}}, \quad y \neq 0$

$$\frac{\sqrt[3]{2}}{\sqrt[3]{3x^2}} = \frac{\sqrt[3]{2} \cdot \sqrt[3]{3^2 x}}{\sqrt[3]{3x^2} \cdot \sqrt[3]{3^2 x}} = \frac{\sqrt[3]{18x}}{\sqrt[3]{(3x)^3}} = \frac{\sqrt[3]{18x}}{3x}$$

Be careful not to multiply by $\sqrt[3]{3x^2}$ here.

⬆ Denominator is a perfect cube.

CAUTION

A common error in a problem like the one in Example 5(a) is to multiply by $\sqrt[3]{2}$ instead of $\sqrt[3]{2^2}$. Doing this would give a denominator of $\sqrt[3]{2} \cdot \sqrt[3]{2} = \sqrt[3]{4}$. Because 4 is not a perfect cube, the denominator is still not rationalized.

◀ *Work Problem* **5** *at the Side.*

ANSWERS

5. (a) $\dfrac{\sqrt[3]{245}}{7}$ (b) $\dfrac{\sqrt[3]{15}}{3}$ (c) $\dfrac{\sqrt[3]{20y^2}}{5y}$

16.4 ▶▶▶ Exercises

Rationalize each denominator. See Examples 1 and 2.

1. $\dfrac{8}{\sqrt{2}}$

2. $\dfrac{12}{\sqrt{3}}$

3. $\dfrac{-\sqrt{11}}{\sqrt{3}}$

4. $\dfrac{-\sqrt{13}}{\sqrt{5}}$

5. $\dfrac{7\sqrt{3}}{\sqrt{5}}$

6. $\dfrac{4\sqrt{6}}{\sqrt{5}}$

7. $\dfrac{24\sqrt{10}}{16\sqrt{3}}$

8. $\dfrac{18\sqrt{15}}{12\sqrt{2}}$

9. $\dfrac{16}{\sqrt{27}}$

10. $\dfrac{24}{\sqrt{18}}$

11. $\dfrac{-3}{\sqrt{50}}$

12. $\dfrac{-5}{\sqrt{75}}$

13. $\dfrac{63}{\sqrt{45}}$

14. $\dfrac{27}{\sqrt{32}}$

15. $\dfrac{\sqrt{24}}{\sqrt{8}}$

16. $\dfrac{\sqrt{36}}{\sqrt{18}}$

17. $\sqrt{\dfrac{1}{2}}$

18. $\sqrt{\dfrac{1}{3}}$

19. $\sqrt{\dfrac{13}{5}}$

20. $\sqrt{\dfrac{17}{11}}$

21. When we rationalize the denominator in an expression such as $\dfrac{4}{\sqrt{3}}$, we multiply both the numerator and denominator by $\sqrt{3}$. By what number are we actually multiplying the given expression, and what property of real numbers justifies the fact that our result is equal to the given expression?

22. In Example 1(a), we show algebraically that $\dfrac{9}{\sqrt{6}}$ is equal to $\dfrac{3\sqrt{6}}{2}$. Support this result numerically by finding the decimal approximation of $\dfrac{9}{\sqrt{6}}$ on your calculator, and then finding the decimal approximation of $\dfrac{3\sqrt{6}}{2}$. What do you notice?

Simplify each product of radicals. See Example 3.

23. $\sqrt{\dfrac{7}{13}} \cdot \sqrt{\dfrac{13}{3}}$

24. $\sqrt{\dfrac{19}{20}} \cdot \sqrt{\dfrac{20}{3}}$

25. $\sqrt{\dfrac{21}{7}} \cdot \sqrt{\dfrac{21}{8}}$

26. $\sqrt{\dfrac{5}{8}} \cdot \sqrt{\dfrac{5}{6}}$

27. $\sqrt{\dfrac{1}{12}} \cdot \sqrt{\dfrac{1}{3}}$

28. $\sqrt{\dfrac{1}{8}} \cdot \sqrt{\dfrac{1}{2}}$

29. $\sqrt{\dfrac{2}{9}} \cdot \sqrt{\dfrac{9}{2}}$

30. $\sqrt{\dfrac{4}{3}} \cdot \sqrt{\dfrac{3}{4}}$

Simplify each radical. Assume that all variables represent positive real numbers. See Example 4.

31. $\dfrac{\sqrt{7}}{\sqrt{x}}$

32. $\dfrac{\sqrt{19}}{\sqrt{y}}$

33. $\dfrac{\sqrt{4x^3}}{\sqrt{y}}$

34. $\dfrac{\sqrt{9t^3}}{\sqrt{s}}$

35. $\sqrt{\dfrac{5x^3z}{6}}$

36. $\sqrt{\dfrac{3st^3}{5}}$

37. $\sqrt{\dfrac{9a^2r^5}{7t}}$

38. $\sqrt{\dfrac{16x^3y^2}{13z}}$

39. Which one of the following would be an appropriate choice for multiplying the numerator and denominator of $\dfrac{\sqrt[3]{2}}{\sqrt[3]{5}}$ by in order to rationalize the denominator?

A. $\sqrt[3]{5}$ **B.** $\sqrt[3]{25}$ **C.** $\sqrt[3]{2}$ **D.** $\sqrt[3]{4}$

40. In Example 5(b), we multiplied the numerator and denominator of $\dfrac{\sqrt[3]{3}}{\sqrt[3]{4}}$ by $\sqrt[3]{2}$ to rationalize the denominator. Suppose we had chosen to multiply by $\sqrt[3]{16}$ instead. Would we have obtained the correct answer after all simplifications were done?

Rationalize each denominator. Assume that variables in the denominator represent nonzero real numbers. See Example 5.

41. $\sqrt[3]{\dfrac{5}{9}}$

42. $\sqrt[3]{\dfrac{2}{5}}$

43. $\dfrac{\sqrt[3]{4}}{\sqrt[3]{7}}$

44. $\dfrac{\sqrt[3]{5}}{\sqrt[3]{10}}$

45. $\sqrt[3]{\dfrac{3}{4y^2}}$

46. $\sqrt[3]{\dfrac{3}{25x^2}}$

47. $\dfrac{\sqrt[3]{7m}}{\sqrt[3]{36n}}$

48. $\dfrac{\sqrt[3]{11p}}{\sqrt[3]{49q}}$

*In Exercises 49 and 50, **(a)** give the answer as a simplified radical and **(b)** use a calculator to give the answer correct to the nearest thousandth.*

49. The period p of a pendulum is the time it takes for it to swing from one extreme to the other and back again. The value of p in seconds is given by

$$p = k \cdot \sqrt{\dfrac{L}{g}},$$

where L is the length of the pendulum, g is the acceleration due to gravity, and k is a constant. Find the period when $k = 6$, $L = 9$ ft, and $g = 32$ ft per sec per sec.

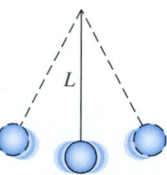

L

50. The velocity v of a meteorite approaching Earth is given by

$$v = \dfrac{k}{\sqrt{d}}$$

km per sec, where d is its distance from the center of Earth and k is a constant. What is the velocity of a meteorite that is 6000 km away from the center of Earth, if $k = 450$?

16.5 ▶▶▶ More Simplifying and Operations with Radicals

We now present a set of guidelines to simplify radical expressions.

Simplifying Radical Expressions

1. If a radical represents a rational number, use that rational number in place of the radical.

   Examples: $\sqrt{49} = 7$; $\sqrt{\dfrac{169}{9}} = \dfrac{13}{3}$

2. If a radical expression contains products of radicals, use the product rule for radicals, $\sqrt[n]{x} \cdot \sqrt[n]{y} = \sqrt[n]{xy}$, to get a single radical.

   Examples: $\sqrt{3} \cdot \sqrt{2} = \sqrt{6}$; $\sqrt[3]{5} \cdot \sqrt[3]{x} = \sqrt[3]{5x}$

3. If a radicand has a factor that is a perfect square, express the radical as the product of the positive square root of the perfect square and the remaining radical factor. A similar statement applies to higher roots.

   Examples: $\sqrt{20} = \sqrt{4 \cdot 5} = \sqrt{4} \cdot \sqrt{5} = 2\sqrt{5}$;
   $\sqrt[3]{16} = \sqrt[3]{8 \cdot 2} = \sqrt[3]{8} \cdot \sqrt[3]{2} = 2\sqrt[3]{2}$

4. If a radical expression contains sums or differences of radicals, use the distributive property to combine like radicals.

   Examples: $3\sqrt{2} + 4\sqrt{2} = 7\sqrt{2}$;
   $3\sqrt{2} + 4\sqrt{3}$ cannot be simplified further.

5. Rationalize any denominator containing a radical.

   Examples: $\dfrac{5}{\sqrt{3}} = \dfrac{5 \cdot \sqrt{3}}{\sqrt{3} \cdot \sqrt{3}} = \dfrac{5\sqrt{3}}{3}$;

   $\sqrt[3]{\dfrac{1}{4}} = \dfrac{\sqrt[3]{1}}{\sqrt[3]{4}} = \dfrac{\sqrt[3]{1} \cdot \sqrt[3]{2}}{\sqrt[3]{4} \cdot \sqrt[3]{2}} = \dfrac{\sqrt[3]{2}}{\sqrt[3]{8}} = \dfrac{\sqrt[3]{2}}{2}$

OBJECTIVE 1 Simplify products of radical expressions.

EXAMPLE 1 Multiplying Radical Expressions

Find each product, and simplify.

(a) $\sqrt{5}\left(\sqrt{8} - \sqrt{32}\right)$ — Simplify inside the parentheses.

$= \sqrt{5}\left(2\sqrt{2} - 4\sqrt{2}\right)$ $\sqrt{8} = 2\sqrt{2}$; $\sqrt{32} = 4\sqrt{2}$

$= \sqrt{5}\left(-2\sqrt{2}\right)$ Subtract like radicals.

$= -2\sqrt{5 \cdot 2}$ Product rule

$= -2\sqrt{10}$ Multiply.

Continued on Next Page

1 Find each product, and simplify.

(a) $\sqrt{7}\left(\sqrt{2} + \sqrt{5}\right)$

(b) $\sqrt{2}\left(\sqrt{8} + \sqrt{20}\right)$

(c)
$\left(\sqrt{2} + 5\sqrt{3}\right)\left(\sqrt{3} - 2\sqrt{2}\right)$

(d)
$\left(\sqrt{2} - \sqrt{5}\right)\left(\sqrt{10} + \sqrt{2}\right)$

(b) $\left(\sqrt{3} + 2\sqrt{5}\right)\left(\sqrt{3} - 4\sqrt{5}\right)$ ⟵ Use the FOIL method to multiply.

$$= \underbrace{\sqrt{3}\left(\sqrt{3}\right)}_{\text{First}} + \underbrace{\sqrt{3}\left(-4\sqrt{5}\right)}_{\text{Outer}} + \underbrace{2\sqrt{5}\left(\sqrt{3}\right)}_{\text{Inner}} + \underbrace{2\sqrt{5}\left(-4\sqrt{5}\right)}_{\text{Last}}$$

$= 3 - 4\sqrt{15} + 2\sqrt{15} - 8 \cdot 5$ Product rule

This does **not** equal $-39\sqrt{15}$.

$= 3 - 2\sqrt{15} - 40$ Add like radicals; multiply.

$= -37 - 2\sqrt{15}$ Combine like terms.

(c) $\left(\sqrt{3} + \sqrt{21}\right)\left(\sqrt{3} - \sqrt{7}\right)$

$$= \sqrt{3}\left(\sqrt{3}\right) + \sqrt{3}\left(-\sqrt{7}\right) + \sqrt{21}\left(\sqrt{3}\right) + \sqrt{21}\left(-\sqrt{7}\right)$$
FOIL

$= 3 - \sqrt{21} + \sqrt{63} - \sqrt{147}$ Product rule

$= 3 - \sqrt{21} + \sqrt{9} \cdot \sqrt{7} - \sqrt{49} \cdot \sqrt{3}$ Factor; 9 and 49 are perfect squares.

$= 3 - \sqrt{21} + 3\sqrt{7} - 7\sqrt{3}$ $\sqrt{9} = 3$; $\sqrt{49} = 7$

Since there are no like radicals, no terms can be combined.

◀ Work Problem **1** at the Side.

Example 2 uses the rules for the square of a binomial from **Section 12.4,** $(a + b)^2 = a^2 + 2ab + b^2$ and $(a - b)^2 = a^2 - 2ab + b^2$.

EXAMPLE 2 Using Special Products with Radicals

Find each product.

(a) $\left(\sqrt{10} - 7\right)^2$ $(a - b)^2 = a^2 - 2ab + b^2$

$= \left(\sqrt{10}\right)^2 - 2\left(\sqrt{10}\right)(7) + 7^2$ Let $a = \sqrt{10}$ and $b = 7$.

Do **not** try to combine further here.

$= 10 - 14\sqrt{10} + 49$ $\left(\sqrt{10}\right)^2 = 10$; $7^2 = 49$

$= 59 - 14\sqrt{10}$ Combine like terms.

(b) $\left(2\sqrt{3} + 4\right)^2$ $(a + b)^2 = a^2 + 2ab + b^2$

$= \left(2\sqrt{3}\right)^2 + 2\left(2\sqrt{3}\right)(4) + 4^2$ $a = 2\sqrt{3}$; $b = 4$

$= 12 + 16\sqrt{3} + 16$ $\left(2\sqrt{3}\right)^2 = 4 \cdot 3 = 12$

$= 28 + 16\sqrt{3}$ ⟵ Do **not** try to combine further here.

(c) $\left(5 - \sqrt{x}\right)^2$

$= 5^2 - 2(5)\left(\sqrt{x}\right) + \left(\sqrt{x}\right)^2$

$= 25 - 10\sqrt{x} + x, \quad x \geq 0$

◀ Work Problem **2** at the Side.

2 Find each product. Simplify the answers.

(a) $\left(\sqrt{5} - 3\right)^2$

(b) $\left(4\sqrt{2} + 5\right)^2$

(c) $\left(6 + \sqrt{m}\right)^2, \quad m \geq 0$

ANSWERS

1. (a) $\sqrt{14} + \sqrt{35}$
  (b) $4 + 2\sqrt{10}$
  (c) $11 - 9\sqrt{6}$
  (d) $2\sqrt{5} + 2 - 5\sqrt{2} - \sqrt{10}$
2. (a) $14 - 6\sqrt{5}$ **(b)** $57 + 40\sqrt{2}$
  (c) $36 + 12\sqrt{m} + m$

CAUTION

Only like radicals can be combined. In Examples 2(a) and (b),

$$59 - 14\sqrt{10} \neq 45\sqrt{10} \quad \text{and} \quad 28 + 16\sqrt{3} \neq 44\sqrt{3}.$$

Example 3 uses the rule for the product of the sum and difference of two terms, $(a + b)(a - b) = a^2 - b^2$.

EXAMPLE 3 Using a Special Product with Radicals

Find each product.

(a) $\left(4 + \sqrt{3}\right)\left(4 - \sqrt{3}\right)$ $(a + b)(a - b) = a^2 - b^2$

$\quad = 4^2 - \left(\sqrt{3}\right)^2$ Let $a = 4$ and $b = \sqrt{3}$.

$\quad = 16 - 3$ $4^2 = 16; \left(\sqrt{3}\right)^2 = 3$

$\quad = 13$

(b) $\left(\sqrt{x} - \sqrt{6}\right)\left(\sqrt{x} + \sqrt{6}\right)$

$\quad = \left(\sqrt{x}\right)^2 - \left(\sqrt{6}\right)^2$

$\quad = x - 6, \quad x \geq 0$ $\left(\sqrt{x}\right)^2 = x; \left(\sqrt{6}\right)^2 = 6$

Work Problem ③ at the Side. ▶

The pairs of expressions being multiplied in Example 3, $4 + \sqrt{3}$ and $4 - \sqrt{3}$, and $\sqrt{x} - \sqrt{6}$ and $\sqrt{x} + \sqrt{6}$, are **conjugates** of each other. Recall from **Section 12.4** that the expressions $a + b$ and $a - b$ are conjugates.

OBJECTIVE 2 Use conjugates to rationalize denominators of radical expressions. To rationalize the denominator in a quotient such as

$$\frac{2}{4 - \sqrt{3}},$$

we multiply the numerator and denominator by $4 + \sqrt{3}$ to obtain

$$\frac{2\left(4 + \sqrt{3}\right)}{\left(4 - \sqrt{3}\right)\left(4 + \sqrt{3}\right)}, \quad \text{or} \quad \frac{2\left(4 + \sqrt{3}\right)}{13}.$$

EXAMPLE 4 Using Conjugates to Rationalize Denominators

Simplify by rationalizing each denominator.

(a) $\dfrac{5}{3 + \sqrt{5}}$

$\quad = \dfrac{5\left(3 - \sqrt{5}\right)}{\left(3 + \sqrt{5}\right)\left(3 - \sqrt{5}\right)}$ Multiply the numerator and denominator by the conjugate of the denominator.

$\quad = \dfrac{5\left(3 - \sqrt{5}\right)}{3^2 - \left(\sqrt{5}\right)^2}$ $(a + b)(a - b) = a^2 - b^2$

$\quad = \dfrac{5\left(3 - \sqrt{5}\right)}{9 - 5}$ $3^2 = 9; \left(\sqrt{5}\right)^2 = 5$

$\quad = \dfrac{5\left(3 - \sqrt{5}\right)}{4}$ Subtract.

Continued on Next Page

---

③ Find each product. Simplify the answers.

(a) $\left(3 + \sqrt{5}\right)\left(3 - \sqrt{5}\right)$

(b) $\left(\sqrt{3} - 2\right)\left(\sqrt{3} + 2\right)$

(c) $\left(\sqrt{5} + \sqrt{3}\right)\left(\sqrt{5} - \sqrt{3}\right)$

(d) $\left(\sqrt{10} - \sqrt{y}\right)\left(\sqrt{10} + \sqrt{y}\right),$ $y \geq 0$

ANSWERS

3. (a) 4   (b) −1   (c) 2   (d) $10 - y$

4 Rationalize each denominator.

(a) $\dfrac{5}{4 + \sqrt{2}}$

(b) $\dfrac{\sqrt{5} + 3}{2 - \sqrt{5}}$

(c) $\dfrac{7}{5 - \sqrt{x}}$

(b) $\dfrac{6 + \sqrt{2}}{\sqrt{2} - 5}$

$= \dfrac{\left(6 + \sqrt{2}\right)\left(\sqrt{2} + 5\right)}{\left(\sqrt{2} - 5\right)\left(\sqrt{2} + 5\right)}$ — Multiply the numerator and denominator by the conjugate of the denominator.

$= \dfrac{6\sqrt{2} + 30 + 2 + 5\sqrt{2}}{2 - 25}$ — FOIL; $(a + b)(a - b) = a^2 - b^2$

$= \dfrac{11\sqrt{2} + 32}{-23}$ — Combine like terms.

$= \dfrac{-11\sqrt{2} - 32}{23}$ — $\dfrac{a}{-b} = \dfrac{-a}{b}$

(c) $\dfrac{4}{3 - \sqrt{x}}$

$= \dfrac{4\left(3 + \sqrt{x}\right)}{\left(3 - \sqrt{x}\right)\left(3 + \sqrt{x}\right)}$ — Multiply by $\dfrac{3 + \sqrt{x}}{3 + \sqrt{x}} = 1$.

> We assume here that $x \geq 0$ and $x \neq 9$.

$= \dfrac{4\left(3 + \sqrt{x}\right)}{9 - x}$ — $3^2 = 9; \left(\sqrt{x}\right)^2 = x$

◀ *Work Problem* **4** *at the Side.*

OBJECTIVE **3** **Write radical expressions with quotients in lowest terms.**

5 Write each quotient in lowest terms.

(a) $\dfrac{5\sqrt{3} - 15}{10}$

(b) $\dfrac{12 + 8\sqrt{5}}{16}$

EXAMPLE 5 **Writing a Radical Quotient in Lowest Terms**

Write $\dfrac{3\sqrt{3} + 9}{12}$ in lowest terms.

$\dfrac{3\sqrt{3} + 9}{12}$ — Don't simplify yet!

$= \dfrac{3\left(\sqrt{3} + 3\right)}{3(4)}$ — Factor first.

$= 1 \cdot \dfrac{\sqrt{3} + 3}{4}$ — Now divide out the common factor; $\frac{3}{3} = 1$

$= \dfrac{\sqrt{3} + 3}{4}$ — Lowest terms

◀ *Work Problem* **5** *at the side.*

CAUTION

An expression like the one in Example 5 can only be simplified by factoring a common factor from the denominator and *each* term of the numerator. For example, *first factor*

$$\dfrac{4 + 8\sqrt{5}}{4} \quad \text{as} \quad \dfrac{4\left(1 + 2\sqrt{5}\right)}{4} \quad \text{to get} \quad 1 + 2\sqrt{5}.$$

ANSWERS

4. (a) $\dfrac{5\left(4 - \sqrt{2}\right)}{14}$ **(b)** $-11 - 5\sqrt{5}$

(c) $\dfrac{7\left(5 + \sqrt{x}\right)}{25 - x}$

5. (a) $\dfrac{\sqrt{3} - 3}{2}$ **(b)** $\dfrac{3 + 2\sqrt{5}}{4}$

In Exercises 1–4, perform the operations mentally, and write the answers without doing intermediate steps.

1. $\sqrt{49} + \sqrt{36}$

2. $\sqrt{100} - \sqrt{81}$

3. $\sqrt{2} \cdot \sqrt{8}$

4. $\sqrt{8} \cdot \sqrt{8}$

Simplify each expression. Use the five guidelines given in this section. Assume that all variables represent nonnegative real numbers. See Examples 1–3.

5. $\sqrt{5}\left(\sqrt{3} - \sqrt{7}\right)$

6. $\sqrt{7}\left(\sqrt{10} + \sqrt{3}\right)$

7. $2\sqrt{5}\left(\sqrt{2} + 3\sqrt{5}\right)$

8. $3\sqrt{7}\left(2\sqrt{7} + 4\sqrt{5}\right)$

9. $3\sqrt{14} \cdot \sqrt{2} - \sqrt{28}$

10. $7\sqrt{6} \cdot \sqrt{3} - 2\sqrt{18}$

11. $\left(2\sqrt{6} + 3\right)\left(3\sqrt{6} + 7\right)$

12. $\left(4\sqrt{5} - 2\right)\left(2\sqrt{5} - 4\right)$

13. $\left(5\sqrt{7} - 2\sqrt{3}\right)\left(3\sqrt{7} + 4\sqrt{3}\right)$

14. $\left(2\sqrt{10} + 5\sqrt{2}\right)\left(3\sqrt{10} - 3\sqrt{2}\right)$

15. $\left(8 - \sqrt{7}\right)^2$

16. $\left(6 - \sqrt{11}\right)^2$

17. $\left(2\sqrt{7} + 3\right)^2$

18. $\left(4\sqrt{5} + 5\right)^2$

19. $\left(\sqrt{a} + 1\right)^2$

20. $\left(\sqrt{y} + 4\right)^2$

21. $\left(5 - \sqrt{2}\right)\left(5 + \sqrt{2}\right)$

22. $\left(3 - \sqrt{5}\right)\left(3 + \sqrt{5}\right)$

23. $\left(\sqrt{8} - \sqrt{7}\right)\left(\sqrt{8} + \sqrt{7}\right)$ **24.** $\left(\sqrt{12} - \sqrt{11}\right)\left(\sqrt{12} + \sqrt{11}\right)$ **25.** $\left(\sqrt{y} - \sqrt{10}\right)\left(\sqrt{y} + \sqrt{10}\right)$

26. $\left(\sqrt{t} - \sqrt{13}\right)\left(\sqrt{t} + \sqrt{13}\right)$ **27.** $\left(\sqrt{2} + \sqrt{3}\right)\left(\sqrt{6} - \sqrt{2}\right)$ **28.** $\left(\sqrt{3} + \sqrt{5}\right)\left(\sqrt{15} - \sqrt{5}\right)$

29. $\left(\sqrt{10} - \sqrt{5}\right)\left(\sqrt{5} + \sqrt{20}\right)$ **30.** $\left(\sqrt{6} - \sqrt{3}\right)\left(\sqrt{3} + \sqrt{18}\right)$ **31.** $\left(\sqrt{5} + \sqrt{30}\right)\left(\sqrt{6} + \sqrt{3}\right)$

32. $\left(\sqrt{10} - \sqrt{20}\right)\left(\sqrt{2} - \sqrt{5}\right)$ **33.** $\left(\sqrt{5} - \sqrt{10}\right)\left(\sqrt{x} - \sqrt{2}\right)$ **34.** $\left(\sqrt{x} + \sqrt{6}\right)\left(\sqrt{10} + \sqrt{3}\right)$

35. In Example 1(b), the original expression simplifies to $-37 - 2\sqrt{15}$. Students often try to simplify such expressions by combining -37 and -2 to get $-39\sqrt{15}$, which is incorrect. Explain why.

36. If you try to rationalize the denominator of $\dfrac{2}{4 + \sqrt{3}}$ by multiplying the numerator and denominator by $4 + \sqrt{3}$, what problem arises? What should you multiply by?

Rationalize each denominator. Write quotients in lowest terms. Assume that all variables represent nonnegative real numbers. See Examples 4 and 5.

37. $\dfrac{1}{3 + \sqrt{2}}$ **38.** $\dfrac{1}{4 - \sqrt{3}}$ **39.** $\dfrac{14}{2 - \sqrt{11}}$ **40.** $\dfrac{19}{5 - \sqrt{6}}$

41. $\dfrac{\sqrt{2}}{2 - \sqrt{2}}$

42. $\dfrac{\sqrt{7}}{7 - \sqrt{7}}$

43. $\dfrac{\sqrt{5}}{\sqrt{2} + \sqrt{3}}$

44. $\dfrac{\sqrt{3}}{\sqrt{2} + \sqrt{3}}$

45. $\dfrac{\sqrt{5} + 2}{2 - \sqrt{3}}$

46. $\dfrac{\sqrt{7} + 3}{4 - \sqrt{5}}$

47. $\dfrac{12}{\sqrt{x} + 1}$

48. $\dfrac{10}{\sqrt{x} - 4}$

49. $\dfrac{3}{7 - \sqrt{x}}$

50. $\dfrac{1}{6 + \sqrt{z}}$

Write each quotient in lowest terms. See Example 5.

51. $\dfrac{6\sqrt{11} - 12}{6}$

52. $\dfrac{12\sqrt{5} - 24}{12}$

53. $\dfrac{2\sqrt{3} + 10}{16}$

54. $\dfrac{4\sqrt{6} + 24}{20}$

55. $\dfrac{12 - \sqrt{40}}{4}$

56. $\dfrac{9 - \sqrt{72}}{12}$

Relating Concepts (Exercises 57–62) For Individual or Group Work

Work Exercises 57–62 in order, to see why a common student error is indeed an error.

57. Use the distributive property to write $6(5 + 3x)$ as a sum.

58. Your answer in Exercise 57 should be $30 + 18x$. Why can't we combine these two terms to get $48x$?

59. Repeat Exercise 14 from earlier in this exercise set.

60. Your answer in Exercise 59 should be $30 + 18\sqrt{5}$. Many students will, in error, try to combine these terms to get $48\sqrt{5}$. Why is this wrong?

61. Write the expression similar to $30 + 18x$ that simplifies to $48x$. Then write the expression similar to $30 + 18\sqrt{5}$ that simplifies to $48\sqrt{5}$.

62. Write a short paragraph explaining the similarities between combining like terms and combining like radicals.

Solve each problem.

63. The radius of the circular top or bottom of a tin can with a surface area S and a height h is given by

$$r = \frac{-h + \sqrt{h^2 + 0.64S}}{2}.$$

What radius should be used to make a can with a height of 12 in. and a surface area of 400 in.²?

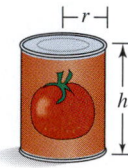

64. If an investment of P dollars grows to A dollars in 2 yr, the annual rate of return on the investment is given by

$$r = \frac{\sqrt{A} - \sqrt{P}}{\sqrt{P}}.$$

Rationalize the denominator. Then find the annual rate of return r (as a percent) if \$50,000 increases to \$58,320.

16.6 ▶▶▶ Solving Equations with Radicals

A **radical equation** is an equation with a variable in the radicand, such as

$$\sqrt{x + 1} = 3 \quad \text{or} \quad 3\sqrt{x} = \sqrt{8x + 9}. \qquad \text{Radical equations}$$

OBJECTIVE **1** **Solve radical equations having square root radicals.** The addition and multiplication properties of equality are not enough to solve radical equations. We need a new property, called the *squaring property*.

> ### Squaring Property of Equality
> If each side of a given equation is squared, all solutions of the original equation are *among* the solutions of the squared equation.

> **CAUTION**
> Be very careful with the squaring property: Using this property can give a new equation with *more* solutions than the original equation. For example, starting with the equation $x = 4$ and squaring each side gives
>
> $$x^2 = 4^2, \quad \text{or} \quad x^2 = 16.$$
>
> This last equation, $x^2 = 16$, has *two* solutions, 4 or -4, while the original equation, $x = 4$, has only *one* solution, 4. Because of this possibility, checking is more than just a guard against algebraic errors when solving an equation with radicals. It is an essential part of the solution process. *All proposed solutions from the squared equation must be checked in the original equation.*

EXAMPLE 1 **Using the Squaring Property of Equality**

Solve $\sqrt{p + 1} = 3$.

Use the squaring property of equality to square each side of the equation.

$$\sqrt{p + 1} = 3$$
$$\left(\sqrt{p + 1}\right)^2 = 3^2$$
$$p + 1 = 9 \qquad \left(\sqrt{p + 1}\right)^2 = p + 1$$
$$\boldsymbol{p = 8} \qquad \text{Subtract 1.}$$

Now check this proposed solution in the original equation.

Check $\qquad \sqrt{p + 1} = 3 \qquad$ Original equation

A check is essential. $\quad \sqrt{8 + 1} \overset{?}{=} 3 \qquad$ Let $p = 8$.

$$\sqrt{9} \overset{?}{=} 3$$
$$3 = 3 \qquad \text{True}$$

Because this statement is true, $\{8\}$ is the solution set of $\sqrt{p + 1} = 3$. In this case the equation obtained by squaring had just one solution, which also satisfied the original equation.

Work Problem **1** at the Side. ▶

OBJECTIVES

1 Solve radical equations having square root radicals.

2 Identify equations with no solutions.

3 Solve equations by squaring a binomial.

4 Solve problems using formulas that involve radicals.

1 Solve each equation. Be sure to check your solutions.

(a) $\sqrt{k} = 3$

(b) $\sqrt{x - 2} = 4$

(c) $\sqrt{9 - t} = 4$

ANSWERS

1. **(a)** $\{9\}$ **(b)** $\{18\}$ **(c)** $\{-7\}$

2 Solve each equation.

(a) $\sqrt{3x + 9} = 2\sqrt{x}$

EXAMPLE 2 Using the Squaring Property with a Radical on Each Side

Solve $3\sqrt{x} = \sqrt{x + 8}$.

$$3\sqrt{x} = \sqrt{x + 8}$$

$$\left(3\sqrt{x}\right)^2 = \left(\sqrt{x + 8}\right)^2 \qquad \text{Squaring property}$$

$$3^2\left(\sqrt{x}\right)^2 = \left(\sqrt{x + 8}\right)^2 \qquad (ab)^2 = a^2b^2$$

Be careful here.

$$9x = x + 8 \qquad \left(\sqrt{x}\right)^2 = x; \left(\sqrt{x + 8}\right)^2 = x + 8$$

$$8x = 8 \qquad \text{Subtract } x.$$

$$x = 1 \qquad \text{Divide by 8.}$$

Check $\qquad 3\sqrt{x} = \sqrt{x + 8} \qquad \text{Original equation}$

$$3\sqrt{1} \overset{?}{=} \sqrt{1 + 8} \qquad \text{Let } x = 1.$$

$$3(1) \overset{?}{=} \sqrt{9}$$

This is not the solution

$$3 = 3 \qquad \text{True}$$

The solution set of $3\sqrt{x} = \sqrt{x + 8}$ is $\{1\}$.

◀ *Work Problem* **2** *at the Side.*

(b) $5\sqrt{x} = \sqrt{20x + 5}$

OBJECTIVE **2** **Identify equations with no solutions.** Not all radical equations have solutions, as shown in Examples 3 and 4.

EXAMPLE 3 Using the Squaring Property When One Side Is Negative

Solve $\sqrt{x} = -3$.

$$\sqrt{x} = -3$$

$$\left(\sqrt{x}\right)^2 = (-3)^2 \qquad \text{Squaring property}$$

$$x = 9 \longleftarrow \text{Proposed solution}$$

3 Solve $\sqrt{x} = -4$.

Check $\qquad \sqrt{x} = -3 \qquad \text{Original equation}$

$$\sqrt{9} \overset{?}{=} -3 \qquad \text{Let } x = 9.$$

$$3 = -3 \qquad \text{False}$$

Because the statement $3 = -3$ is false, the number 9 is *not* a solution of the given equation and is said to be an **extraneous solution**; it must be discarded. In fact, $\sqrt{x} = -3$ has no solution. The solution set is $\emptyset$.

Note

Because $\sqrt{x}$ represents the *principal* or *nonnegative* square root of x in Example 3, we might have seen immediately that there is no solution.

ANSWERS

2. (a) $\{9\}$ (b) $\{1\}$
3. $\emptyset$

◀ *Work Problem* **3** *at the Side.*

We use the following steps when solving an equation with radicals.

> **Solving a Radical Equation**
>
> **Step 1** **Isolate a radical.** Arrange the terms so that a radical is alone on one side of the equation.
>
> **Step 2** **Square each side.**
>
> **Step 3** **Combine like terms.**
>
> **Step 4** **Repeat Steps 1–3,** if there is still a term with a radical.
>
> **Step 5** **Solve the equation.** Find all proposed solutions.
>
> **Step 6** **Check all proposed solutions** in the original equation.

EXAMPLE 4 **Using the Squaring Property with a Quadratic Expression**

Solve $x = \sqrt{x^2 + 5x + 10}$.

Step 1 The radical is already isolated on the right side of the equation.

Step 2 Square each side.

$$x^2 = \left(\sqrt{x^2 + 5x + 10}\right)^2 \quad \text{Squaring property}$$

$$x^2 = x^2 + 5x + 10 \qquad \left(\sqrt{x^2 + 5x + 10}\right)^2 = x^2 + 5x + 10$$

Step 3 $\quad 0 = 5x + 10 \qquad$ Subtract x^2.

Step 4 This step is not needed.

Step 5 $\quad -10 = 5x \qquad$ Subtract 10.

$\qquad -2 = x \qquad$ Divide by 5.

Step 6 **Check** $\qquad x = \sqrt{x^2 + 5x + 10}$

> The principal square root of a quantity *cannot* be negative.

$$-2 \overset{?}{=} \sqrt{(-2)^2 + 5(-2) + 10} \quad \text{Let } x = -2.$$

$$-2 \overset{?}{=} \sqrt{4 - 10 + 10} \qquad \text{Multiply.}$$

$$-2 = 2 \qquad\qquad\qquad \text{False}$$

Since substituting -2 for x leads to a false result, the equation has no solution, and the solution set is $\emptyset$.

Work Problem **4** *at the Side.* ▶

OBJECTIVE **3** **Solve equations by squaring a binomial.** The next examples use the following rules from **Section 12.4.**

$$(a + b)^2 = a^2 + 2ab + b^2 \quad \text{and} \quad (a - b)^2 = a^2 - 2ab + b^2$$

By the second rule, for example,

$$(x - 3)^2$$

$$= x^2 - 2x(3) + 3^2$$

$$= x^2 - 6x + 9.$$

Work Problem **5** *at the Side.* ▶

---

4 Solve $x = \sqrt{x^2 - 4x - 16}$.

5 Square each expression.

(a) $w - 5$

(b) $2k - 5$

(c) $3m - 2p$

6 Solve each equation.

(a) $\sqrt{6w + 6} = w + 1$

EXAMPLE 5 Using the Squaring Property When One Side Has Two Terms

Solve $\sqrt{2x - 3} = x - 3$.

Square each side, using the rule $(a - b)^2 = a^2 - 2ab + b^2$ to square the binomial on the right side of the equation.

$$\left(\sqrt{2x - 3}\right)^2 = (x - 3)^2$$

> Remember the middle term when squaring.

$$2x - 3 = x^2 - 6x + 9$$

This equation is quadratic because of the x^2-term. To solve it, as shown in **Section 13.7,** we must write the equation in standard form.

$$x^2 - 8x + 12 = 0 \qquad \text{Subtract } 2x; \text{ add } 3; \text{ standard form}$$
$$(x - 6)(x - 2) = 0 \qquad \text{Factor.}$$
$$x - 6 = 0 \quad \text{or} \quad x - 2 = 0 \qquad \text{Zero-factor property}$$
$$x = 6 \quad \text{or} \qquad x = 2 \qquad \text{Solve.}$$

Check *both* of these proposed solutions in the original equation.

Check If $x = 6$, then

$$\sqrt{2x - 3} = x - 3$$
$$\sqrt{2(6) - 3} \overset{?}{=} 6 - 3$$
$$\sqrt{12 - 3} \overset{?}{=} 3$$
$$\sqrt{9} \overset{?}{=} 3$$
$$3 = 3. \qquad \text{True}$$

If $x = 2$, then

$$\sqrt{2x - 3} = x - 3$$
$$\sqrt{2(2) - 3} \overset{?}{=} 2 - 3$$
$$\sqrt{4 - 3} \overset{?}{=} -1$$
$$\sqrt{1} \overset{?}{=} -1$$
$$1 = -1. \qquad \text{False}$$

(b) $2u - 1 = \sqrt{10u + 9}$

Only 6 is a valid solution. (2 is extraneous.) The solution set is $\{6\}$.

◀ *Work Problem* **6** *at the Side.*

In Example 6, we must isolate the radical on one side of the equation *before* squaring each side. This is Step 1 given in the steps for solving a radical equation. If we forget to do this and square each side, we actually get a more complicated equation that still contains a radical.

EXAMPLE 6 Rewriting before Using the Squaring Property

Solve $3\sqrt{x} - 1 = 2x$.

$$3\sqrt{x} - 1 = 2x$$

> This is a key step.

$$3\sqrt{x} = 2x + 1 \qquad \text{Add 1 to isolate the radical.}$$
$$\left(3\sqrt{x}\right)^2 = (2x + 1)^2 \qquad \text{Square each side.}$$
$$9x = 4x^2 + 4x + 1 \qquad (a + b)^2 = a^2 + 2ab + b^2$$
$$4x^2 - 5x + 1 = 0 \qquad \text{Subtract } 9x; \text{ standard form}$$
$$(4x - 1)(x - 1) = 0 \qquad \text{Factor.}$$
$$4x - 1 = 0 \quad \text{or} \quad x - 1 = 0 \qquad \text{Zero-factor property}$$
$$x = \frac{1}{4} \quad \text{or} \qquad x = 1 \qquad \text{Solve.}$$

Continued on Next Page

Check   If $x = \dfrac{1}{4}$, then

$$3\sqrt{x} - 1 = 2x$$

$$3\sqrt{\dfrac{1}{4}} - 1 \overset{?}{=} 2\left(\dfrac{1}{4}\right)$$

$$3\left(\dfrac{1}{2}\right) - 1 \overset{?}{=} \dfrac{1}{2}$$

$$\dfrac{1}{2} = \dfrac{1}{2}. \qquad \text{True}$$

If $x = 1$, then

$$3\sqrt{x} - 1 = 2x$$

$$3\sqrt{1} - 1 \overset{?}{=} 2(1)$$

$$3 - 1 \overset{?}{=} 2$$

$$2 = 2. \qquad \text{True}$$

Both proposed solutions check, so the solution set is $\left\{\frac{1}{4}, 1\right\}$.

> **CAUTION**
> Errors often occur when each side of an equation is squared. For instance, in Example 6 when each side of
> $$3\sqrt{x} = 2x + 1$$
> is squared, *the entire binomial on the right must be squared* to get $4x^2 + 4x + 1$. It would be *incorrect* to square the $2x$ and the 1 separately to get $4x^2 + 1$.

Work Problem **7** *at the Side.* ▶

Some radical equations require squaring twice, as in the next example.

EXAMPLE 7   **Using the Squaring Property Twice**

Solve $\sqrt{21 + x} = 3 + \sqrt{x}$.

$$\sqrt{21 + x} = 3 + \sqrt{x}$$

$$\left(\sqrt{21 + x}\right)^2 = \left(3 + \sqrt{x}\right)^2 \qquad \text{Square each side.}$$

$$21 + x = 9 + 6\sqrt{x} + x \quad \overleftarrow{\text{Be careful here.}}$$

$$12 = 6\sqrt{x} \qquad \text{Subtract 9; subtract } x.$$

$$2 = \sqrt{x} \qquad \text{Divide by 6.}$$

$$2^2 = \left(\sqrt{x}\right)^2 \qquad \text{Square each side again.}$$

$$4 = x$$

Check   If $x = 4$, then

$$\sqrt{21 + x} = 3 + \sqrt{x} \qquad \text{Original equation}$$

$$\sqrt{21 + 4} \overset{?}{=} 3 + \sqrt{4}$$

$$5 = 5. \qquad \text{True}$$

The solution set is $\{4\}$.

Work Problem **8** *at the Side.* ▶

---

7 Solve each equation.

(a) $\sqrt{x} - 3 = x - 15$

(b) $\sqrt{z + 5} + 2 = z + 5$

8 Solve each equation.

(a) $\sqrt{p + 1} - \sqrt{p - 4} = 1$

(b) $\sqrt{2x + 1} + \sqrt{x + 4} = 3$

ANSWERS

7. (a) $\{16\}$ (b) $\{-1\}$
8. (a) $\{8\}$ (b) $\{0\}$

9 Find the area (to the nearest tenth) of a triangle with sides of lengths 12 cm, 15 cm, and 21 cm.

OBJECTIVE 4 Solve problems using formulas that involve radicals. In **Section 10.5**, we worked with a variety of formulas, including some area formulas from geometry. Many useful formulas involve radicals.

The most common formula for the area of a triangle is $A = \frac{1}{2}bh$, where b is the length of the base and h is the height. What if the height is not known? What if we know only the lengths of the sides? Another formula, known as **Heron's formula,** allows us to calculate the area of a triangle if we know the lengths of the sides a, b, and c. Using the lengths of the sides, we must first find the **semiperimeter s,** which is one-half the perimeter.

$$s = \frac{1}{2}(a + b + c) \qquad \text{Semiperimeter}$$

The area $\mathcal{A}$ is given by the formula

$$\mathcal{A} = \sqrt{s(s-a)(s-b)(s-c)}. \qquad \text{Heron's formula}$$

For example, the familiar 3-4-5 right triangle has area $\mathcal{A} = \frac{1}{2}(3)(4) = 6$ square units, calculated using the familiar formula. Using Heron's formula, $s = \frac{1}{2}(3 + 4 + 5) = 6$, and

$$\mathcal{A} = \sqrt{6(6-3)(6-4)(6-5)}$$

$$\mathcal{A} = \sqrt{6 \cdot 3 \cdot 2 \cdot 1}$$

$$\mathcal{A} = \sqrt{36}$$

$$\mathcal{A} = 6.$$

The area is again 6 square units, as expected.

EXAMPLE 8 Using Heron's Formula to Find the Area of a Triangle

The sides of a triangular garden plot have lengths 20 ft, 34 ft, and 42 ft. See Figure 3. Find its area.

First find the semiperimeter of the triangle.

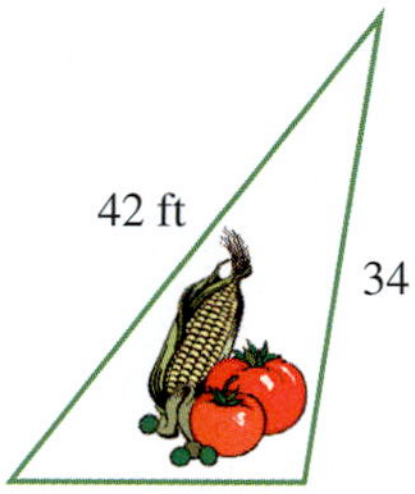

$$s = \frac{1}{2}(a + b + c)$$

$$s = \frac{1}{2}(20 + 34 + 42) \qquad \text{Let } a = 20, b = 34, \text{ and } c = 42.$$

$$s = \frac{1}{2}(96) \qquad \text{Add.}$$

$$s = 48 \qquad \text{Multiply.}$$

42 ft

34 ft

20 ft

Figure 3

Now use Heron's formula to find the area.

$$\mathcal{A} = \sqrt{s(s-a)(s-b)(s-c)}$$

$$\mathcal{A} = \sqrt{48(48-20)(48-34)(48-42)} \qquad \text{Substitute.}$$

$$\mathcal{A} = \sqrt{48(28)(14)(6)} \qquad \text{Subtract.}$$

$$\mathcal{A} = \sqrt{112{,}896} \qquad \text{Multiply.}$$

$$\mathcal{A} = 336 \qquad \text{Use a calculator.}$$

The area of the garden plot is 336 ft².

ANSWER

9. 88.2 cm²

◄ *Work Problem* **9** *at the Side.*

16.6 ▶▶▶ Exercises

Solve each equation. See Examples 1–4.

1. $\sqrt{x} = 7$

2. $\sqrt{k} = 10$

3. $\sqrt{t + 2} = 3$

4. $\sqrt{x + 7} = 5$

5. $\sqrt{r - 4} = 9$

6. $\sqrt{k - 12} = 3$

7. $\sqrt{4 - t} = 7$

8. $\sqrt{9 - s} = 5$

9. $\sqrt{2t + 3} = 0$

10. $\sqrt{5x - 4} = 0$

11. $\sqrt{3x - 8} = -2$

12. $\sqrt{6x + 4} = -3$

13. $\sqrt{w} - 4 = 7$

14. $\sqrt{t} + 3 = 10$

15. $\sqrt{10x - 8} = 3\sqrt{x}$

16. $\sqrt{17t - 4} = 4\sqrt{t}$

17. $5\sqrt{x} = \sqrt{10x + 15}$

18. $4\sqrt{z} = \sqrt{20z - 16}$

19. $\sqrt{3x - 5} = \sqrt{2x + 1}$

20. $\sqrt{5x + 2} = \sqrt{3x + 8}$

21. $k = \sqrt{k^2 - 5k - 15}$

22. $s = \sqrt{s^2 - 2s - 6}$

23. $7x = \sqrt{49x^2 + 2x - 10}$

24. $6x = \sqrt{36x^2 + 5x - 5}$

25. Consider the following "solution"? *WHAT WENT WRONG?* Give the correct solution set.

$$-\sqrt{x-1} = -4$$
$$-(x-1) = 16 \qquad \text{Square each side.}$$
$$-x+1 = 16 \qquad \text{Distributive property}$$
$$-x = 15 \qquad \text{Subtract 1.}$$
$$x = -15 \qquad \text{Multiply by } -1.$$

26. The first step in solving the equation

$$\sqrt{2x+1} = x - 7$$

is to square each side of the equation. Errors often occur in solving equations such as this one when the right side of the equation is squared incorrectly. What is the square of the right side?

Solve each equation. See Examples 5 and 6.

27. $\sqrt{2x+1} = x - 7$

28. $\sqrt{3x+3} = x - 5$

29. $\sqrt{3k+10} + 5 = 2k$

30. $\sqrt{4t+13} + 1 = 2t$

31. $\sqrt{5x+1} - 1 = x$

32. $\sqrt{x+1} - x = 1$

33. $\sqrt{6t+7} + 3 = t + 5$

34. $\sqrt{10x+24} = x + 4$

35. $x - 4 - \sqrt{2x} = 0$

36. $x - 3 - \sqrt{4x} = 0$

37. $\sqrt{x+6} = 2x$

38. $\sqrt{k+12} = k$

Solve each equation. See Example 7.

39. $\sqrt{x+1} - \sqrt{x-4} = 1$

40. $\sqrt{2x+3} + \sqrt{x+1} = 1$

41. $\sqrt{x} = \sqrt{x-5} + 1$

42. $\sqrt{2x} = \sqrt{x+7} - 1$

43. $\sqrt{3x+4} - \sqrt{2x-4} = 2$

44. $\sqrt{1-x} + \sqrt{x+9} = 4$

45. $\sqrt{2x+11} + \sqrt{x+6} = 2$

46. $\sqrt{x+9} + \sqrt{x+16} = 7$

Solve each problem.

47. The square root of the sum of a number and 4 is 5. Find the number.

48. A certain number is the same as the square root of the product of 8 and the number. Find the number.

49. Three times the square root of 2 equals the square root of the sum of some number and 10. Find the number.

50. The negative square root of a number equals that number decreased by 2. Find the number.

Use Heron's formula to solve each problem. See Example 8.

51. A surveyor has measured the lengths of the sides of a triangular plot of land as 180 ft, 200 ft, and 240 ft. Find the area of the plot. (Round to the nearest whole number.)

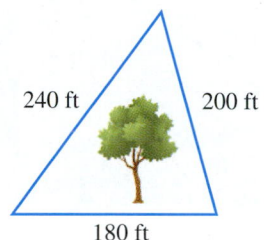

240 ft   200 ft

180 ft

52. A triangular indoor play space in a day care center has sides of length 30 ft, 45 ft, and 65 ft. How much carpeting would be needed to cover this space? (Round to the nearest whole number.)

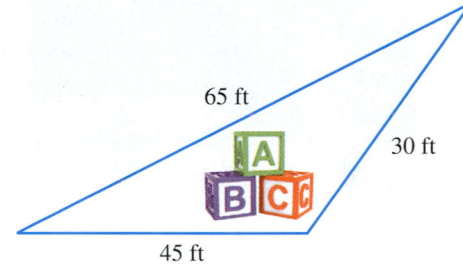

65 ft   30 ft

45 ft

Solve each problem. Give answers to the nearest tenth.

53. To estimate the speed at which a car was traveling at the time of an accident, a police officer drives the car involved in the accident under conditions similar to those during which the accident took place and then skids to a stop. If the car is driven at 30 mph, then the speed at the time of the accident is given by

$$s = 30\sqrt{\dfrac{a}{p}},$$

where a is the length of the skid marks left at the time of the accident and p is the length of the skid marks in the police test. Find s for the following values of a and p.

(a) $a = 862$ ft; $p = 156$ ft

(b) $a = 382$ ft; $p = 96$ ft

(c) $a = 84$ ft; $p = 26$ ft

54. A formula for calculating the distance, d, one can see from an airplane to the horizon on a clear day is

$$d = 1.22\sqrt{x},$$

where x is the altitude of the plane in feet and d is given in miles.

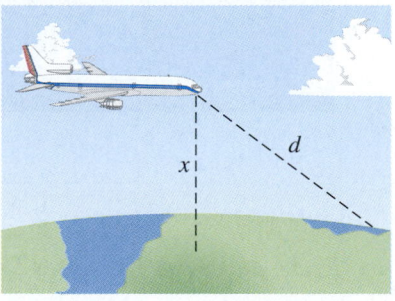

How far can one see to the horizon in a plane flying at the following altitudes?

(a) 15,000 ft

(b) 18,000 ft

(c) 24,000 ft

On a clear day, the maximum distance in kilometers that you can see from a structure is given by the formula

$$\text{sight distance} = 111.7\sqrt{\text{height of structure in kilometers}}.$$

Use this formula and the conversion equations 1 ft ≈ 0.3048 m *and* 1 km ≈ 0.621371 mi *as necessary to solve each problem. Round your answers to the nearest mile. (Source: A Sourcebook of Applications of School Mathematics, NCTM, 1980.)*

55. The London Eye is a unique structure that features 32 observation capsules. It is the fourth-tallest structure in London, with a diameter of 135 m. (*Source:* www.londoneye.com) Does the formula justify the claim that on a clear day passengers on the London Eye can see Windsor Castle, 25 mi away?

56. The Empire State Building in New York City is 1250 ft high. (The antenna reaches to 1454 ft.) The observation deck, located on the 102nd floor, is at a height of 1050 ft. (*Source:* www.esbnyc.com) How far could you see on a clear day from the observation deck?

57. The twin Petronas Towers in Kuala Lumpur, Malaysia, were built in 1998. Both towers are 1483 ft high (including the spires). How far would one of the builders have been able to see on a clear day from the top of a spire?

58. The Khufu Pyramid in Giza was built in about 2566 B.C. to a height, at that time, of 482 ft. It is now only about 450 ft high. How far would one of the original builders of the pyramid have been able to see from the top of the pyramid?

Relating Concepts (Exercises 59–64) For Individual or Group Work

*Consider the figure, and **work Exercises 59–64 in order**.*

59. The lengths of the sides of the entire triangle are 7, 7, and 12. Find the semiperimeter *s*.

60. Now use Heron's formula to find the area of the entire triangle. Write it as a simplified radical.

61. Find the value of *h* by using the Pythagorean formula.

62. Find the area of each of the congruent right triangles forming the entire triangle by using the formula $A = \frac{1}{2}bh$.

63. Double your result from Exercise 62 to determine the area of the entire triangle.

64. How do your answers in Exercises 60 and 63 compare?

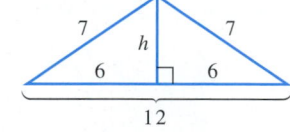

Chapter 16 ▷▷▷ Summary

▶ Key Terms

16.1	**square root**	A number a is a square root of b if $a^2 = b$.
	principal square root	The positive square root of a number is its principal square root.
	radicand	The number or expression inside a radical sign is called the radicand.
	radical	A radical sign with a radicand is called a radical.
	radical expression	An algebraic expression containing a radical is called a radical expression.
	perfect square	A number with a rational square root is called a perfect square.
	irrational number	A real number that is not rational is called an irrational number.
	cube root	A number a is a cube root of b if $a^3 = b$.
	index (order)	In a radical of the form $\sqrt[n]{a}$, the number n is the index or order.
16.2	**perfect cube**	A number with a rational cube root is called a perfect cube.
16.3	**like radicals**	Like radicals are multiples of the same root of the same number.
16.4	**rationalizing the denominator**	The process of changing the denominator of a fraction from a radical (irrational number) to an expression not involving a radical is called rationalizing the denominator.
16.5	**conjugate**	The conjugate of $a + b$ is $a - b$.
16.6	**radical equation**	An equation with a variable in the radicand is a radical equation.
	extraneous solution	A proposed solution that does not satisfy the given equation is an extraneous solution.

Radical sign — Index — $\sqrt[n]{a}$ ← Radicand — Radical

▶ New Symbols

 radical sign $\approx$ is approximately equal to   cube root of a nth root of a

▶ Test Your Word Power

See how well you have learned the vocabulary in this chapter. Answers, with examples, follow the Quick Review.

1. A **square root** of a number is
 A. the number raised to the second power
 B. the number under a radical sign
 C. a number that when multiplied by itself gives the original number
 D. the inverse of the number.

2. A **radicand** is
 A. the index of a radical
 B. the number or expression inside the radical sign
 C. the positive root of a number
 D. the radical sign.

3. A **radical** is
 A. a symbol that indicates the nth root
 B. an algebraic expression containing a square root
 C. the positive nth root of a number
 D. a radical sign and the number or expression inside it.

4. The **principal root** of a positive number with even index n is
 A. the positive nth root of the number
 B. the negative nth root of the number

 C. the square root of the number
 D. the cube root of the number.

5. An **irrational number** is
 A. the quotient of two integers, with denominator not 0
 B. a decimal number that neither terminates nor repeats
 C. the principal square root of a number
 D. a nonreal number.

(continued)

▶ Test Your Word Power (*continued*)

6. The **Pythagorean formula** states that, in a right triangle,
 A. the sum of the measures of the angles is 180°
 B. the sum of the lengths of the two shorter sides equals the length of the longest side
 C. the longest side is opposite the right angle
 D. the square of the length of the longest side equals the sum of the squares of the lengths of the two shorter sides.

7. **Like radicals** are
 A. radicals in simplest form
 B. algebraic expressions containing radicals

 C. multiples of the same root of the same number
 D. radicals with the same index.

8. The **conjugate** of $a + b$ is
 A. $a - b$
 B. $a \cdot b$
 C. $a \div b$
 D. $(a + b)^2$.

9. **Rationalizing the denominator** is the process of
 A. eliminating fractions from a radical expression
 B. changing the denominator of a fraction from a radical expression to an expression not involving a radical

 C. clearing a radical expression of radicals
 D. multiplying radical expressions.

10. An **extraneous solution** is a value
 A. that makes an equation false and must be discarded
 B. that makes an equation true
 C. that makes an expression equal 0
 D. that checks in the original equation.

▶ Quick Review

Concepts

Examples

16.1 Evaluating Roots

If a is a positive real number, then

 $\sqrt{a}$ is the positive or principal square root of a;

 $-\sqrt{a}$ is the negative square root of a; $\sqrt{0} = 0$.

If a is a negative real number, then $\sqrt{a}$ is not a real number.

If a is a positive rational number, then $\sqrt{a}$ is rational if a is a perfect square. $\sqrt{a}$ is irrational if a is not a perfect square.

Every real number has exactly one real cube root.

$$\sqrt{49} = 7$$

$$-\sqrt{81} = -9$$

$$\sqrt{-25} \text{ is not a real number.}$$

$\sqrt{\dfrac{4}{9}}$ and $\sqrt{16}$ are rational. $\sqrt{\dfrac{2}{3}}$ and $\sqrt{21}$ are irrational.

$$\sqrt[3]{27} = 3; \quad \sqrt[3]{-8} = -2$$

Pythagorean Formula

If c is the length of the longest side (hypotenuse) of a right triangle and a and b are the lengths of the shorter sides (legs), then

$$a^2 + b^2 = c^2.$$

Find b for the triangle in the figure.

$$10^2 + b^2 = \left(2\sqrt{61}\right)^2$$

$$100 + b^2 = 4(61)$$

$$100 + b^2 = 244$$

$$b^2 = 144$$

$$b = 12$$

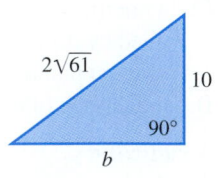

Concepts	Examples

16.2 Multiplying, Dividing, and Simplifying Radicals

Product Rule for Radicals

For nonnegative real numbers a and b,

$$\sqrt{a} \cdot \sqrt{b} = \sqrt{ab} \quad \text{and} \quad \sqrt{ab} = \sqrt{a} \cdot \sqrt{b}.$$

$$\sqrt{5} \cdot \sqrt{7} = \sqrt{5 \cdot 7} = \sqrt{35}$$

$$\sqrt{48} = \sqrt{16 \cdot 3} = \sqrt{16} \cdot \sqrt{3} = 4\sqrt{3}$$

Quotient Rule for Radicals

If a and b are nonnegative real numbers and $b \neq 0$, then

$$\sqrt{\frac{a}{b}} = \frac{\sqrt{a}}{\sqrt{b}} \quad \text{and} \quad \frac{\sqrt{a}}{\sqrt{b}} = \sqrt{\frac{a}{b}}.$$

If all indicated roots are real, then

$$\sqrt[n]{a} \cdot \sqrt[n]{b} = \sqrt[n]{ab} \quad \text{and} \quad \frac{\sqrt[n]{a}}{\sqrt[n]{b}} = \sqrt[n]{\frac{a}{b}} \quad (b \neq 0).$$

$$\sqrt{\frac{25}{64}} = \frac{\sqrt{25}}{\sqrt{64}} = \frac{5}{8}; \quad \frac{\sqrt{8}}{\sqrt{2}} = \sqrt{\frac{8}{2}} = \sqrt{4} = 2$$

$$\sqrt[3]{5} \cdot \sqrt[3]{3} = \sqrt[3]{15}; \quad \frac{\sqrt[4]{12}}{\sqrt[4]{4}} = \sqrt[4]{\frac{12}{4}} = \sqrt[4]{3}$$

16.3 Adding and Subtracting Radicals

Add and subtract like radicals by using the distributive property. *Only like radicals can be combined in this way.*

$$2\sqrt{5} + 4\sqrt{5}$$
$$= (2 + 4)\sqrt{5}$$
$$= 6\sqrt{5}$$

$$\sqrt{8} - \sqrt{32}$$
$$= 2\sqrt{2} - 4\sqrt{2}$$
$$= -2\sqrt{2}$$

16.4 Rationalizing the Denominator

The denominator of a radical expression can be rationalized by multiplying both the numerator and denominator by a number that will eliminate the radical from the denominator. In some cases, the number will be the conjugate of the denominator.

$$\frac{2}{\sqrt{3}} = \frac{2 \cdot \sqrt{3}}{\sqrt{3} \cdot \sqrt{3}} = \frac{2\sqrt{3}}{3}$$

$$\sqrt[3]{\frac{5}{121}} = \frac{\sqrt[3]{5} \cdot \sqrt[3]{11}}{\sqrt[3]{11^2} \cdot \sqrt[3]{11}} = \frac{\sqrt[3]{55}}{11}$$

16.5 More Simplifying and Operations with Radicals

When appropriate, use the rules for adding and multiplying polynomials to simplify radical expressions.

$$\sqrt{6}(\sqrt{5} - \sqrt{7}) = \sqrt{30} - \sqrt{42}$$

$$(\sqrt{3} + 1)(\sqrt{3} - 2)$$
$$= 3 - 2\sqrt{3} + \sqrt{3} - 2 \qquad \text{FOIL}$$
$$= 1 - \sqrt{3} \qquad\qquad \text{Combine like terms.}$$

The formulas

$$(a + b)^2 = a^2 + 2ab + b^2,$$
$$(a - b)^2 = a^2 - 2ab + b^2,$$

and

$$(a + b)(a - b) = a^2 - b^2$$

are useful when simplifying radical expressions.

$$(\sqrt{13} - \sqrt{2})^2$$
$$= (\sqrt{13})^2 - 2(\sqrt{13})(\sqrt{2}) + (\sqrt{2})^2$$
$$= 13 - 2\sqrt{26} + 2$$
$$= 15 - 2\sqrt{26}$$

$$(\sqrt{5} + \sqrt{3})(\sqrt{5} - \sqrt{3})$$
$$= 5 - 3 = 2$$

(continued)

Concepts	Examples

16.5 More Simplifying and Operations with Radicals (continued)

Any denominators with radicals should be rationalized.

$$\frac{3}{\sqrt{6}} = \frac{3 \cdot \sqrt{6}}{\sqrt{6} \cdot \sqrt{6}} = \frac{3\sqrt{6}}{6} = \frac{\sqrt{6}}{2}$$

If a radical expression contains two terms in the denominator and at least one of those terms is a square root radical, multiply both the numerator and denominator by the conjugate of the denominator.

$$\frac{6}{\sqrt{7} - \sqrt{2}}$$

$$= \frac{6\left(\sqrt{7} + \sqrt{2}\right)}{\left(\sqrt{7} - \sqrt{2}\right)\left(\sqrt{7} + \sqrt{2}\right)}$$

$$= \frac{6\left(\sqrt{7} + \sqrt{2}\right)}{7 - 2} \qquad \text{Multiply.}$$

$$= \frac{6\left(\sqrt{7} + \sqrt{2}\right)}{5} \qquad \text{Subtract.}$$

16.6 Solving Equations with Radicals

Solving a Radical Equation

Step 1 Isolate a radical.

Step 2 Square each side. (By the squaring property of equality, all solutions of the original equation are *among* the solutions of the squared equation.)

Step 3 Combine like terms.

Step 4 If there is still a term with a radical, repeat Steps 1–3.

Step 5 Solve the equation for proposed solutions.

Step 6 Check all proposed solutions from Step 5 in the original equation.

Solve $\sqrt{2x - 3} + x = 3$

$\sqrt{2x - 3} = 3 - x$ Isolate the radical.

$\left(\sqrt{2x - 3}\right)^2 = (3 - x)^2$ Square each side.

$2x - 3 = 9 - 6x + x^2$

$0 = x^2 - 8x + 12$ Standard form

$0 = (x - 2)(x - 6)$ Factor.

$x - 2 = 0 \quad \text{or} \quad x - 6 = 0$ Zero-factor property

$x = 2 \quad \text{or} \quad x = 6$ Solve.

A check is essential here. Verify that 2 is the only solution. (6 is extraneous.) The solution set is $\{2\}$.

ANSWERS TO TEST YOUR WORD POWER

1. C; *Examples:* 6 is a square root of 36 since $6^2 = 6 \cdot 6 = 36$; -6 is also a square root of 36.

2. B; *Example:* In $\sqrt{3xy}$, $3xy$ is the radicand.   **3.** D; *Examples:* $\sqrt{144}$, $\sqrt{4xy^2}$, and $\sqrt{4 + t^2}$

4. A; *Examples:* $\sqrt{36} = 6$, $\sqrt[4]{81} = 3$, and $\sqrt[6]{64} = 2$ **5.** B; *Examples:* π, $\sqrt{2}$, $-\sqrt{5}$

6. D; *Example:* In a right triangle where $a = 6$, $b = 8$, and $c = 10$, $6^2 + 8^2 = 10^2$.

7. C; *Examples:* $\sqrt{7}$ and $3\sqrt{7}$ are like radicals; so are $2\sqrt[3]{6k}$ and $5\sqrt[3]{6k}$.

8. A; *Example:* The conjugate of $\sqrt{3} + 1$ is $\sqrt{3} - 1$.

9. B; *Example:* To rationalize the denominator of $\dfrac{5}{\sqrt{3} + 1}$, multiply the numerator and denominator by $\sqrt{3} - 1$

(the conjugate of the denominator) to get $\dfrac{5\left(\sqrt{3} - 1\right)}{2}$.

10. A; *Example:* The proposed solution 2 is extraneous when $\sqrt{5q - 1} + 3 = 0$ is solved using the method of **Section 16.6.**

Chapter 16 ▶▶▶ Review Exercises

[16.1] *Find all square roots of each number.*

1. 49
2. 81
3. 196
4. 121
5. 256
6. 729

Find each root.

7. $\sqrt{16}$
8. $-\sqrt{0.36}$
9. $\sqrt[3]{-512}$
10. $\sqrt[4]{81}$

11. $\sqrt{-8100}$
12. $-\sqrt{4225}$
13. $\sqrt{\dfrac{144}{169}}$
14. $-\sqrt{\dfrac{100}{81}}$

15. Find the value of x.

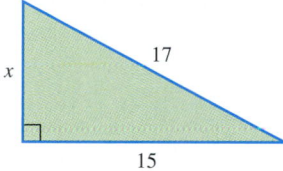

16. An HP f1905 computer monitor has viewing screen dimensions as shown in the figure. Find the diagonal measure of the viewing screen to the nearest tenth. (*Source:* Author's computer.)

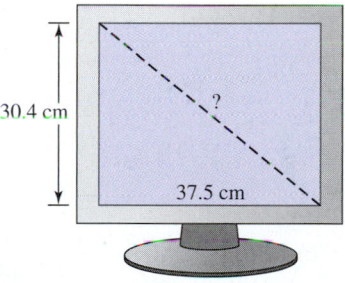

Write rational, irrational, *or* not a real number *for each number. If a number is rational, give its exact value. If a number is irrational, give a decimal approximation for the number. Round approximations to the nearest thousandth.*

17. $\sqrt{73}$
18. $\sqrt{169}$
19. $-\sqrt{625}$
20. $\sqrt{-19}$

[16.2] *Simplify each expression.*

21. $\sqrt{48}$
22. $-\sqrt{288}$
23. $\sqrt[3]{16}$
24. $\sqrt[3]{375}$

25. $\sqrt{12} \cdot \sqrt{27}$
26. $\sqrt{32} \cdot \sqrt{48}$
27. $-\sqrt{\dfrac{121}{400}}$
28. $\sqrt{\dfrac{7}{169}}$

29. $\sqrt{\dfrac{1}{6}} \cdot \sqrt{\dfrac{5}{6}}$ **30.** $\sqrt{\dfrac{2}{5}} \cdot \sqrt{\dfrac{2}{45}}$ **31.** $\dfrac{3\sqrt{10}}{\sqrt{5}}$ **32.** $\dfrac{8\sqrt{150}}{4\sqrt{75}}$

Simplify each expression. Assume that all variables represent nonnegative real numbers.

33. $\sqrt{r^{18}}$ **34.** $\sqrt{x^{10}y^{16}}$ **35.** $\sqrt{162x^9}$ **36.** $\sqrt{\dfrac{36}{p^2}}, \quad p \neq 0$

37. $\sqrt{a^{15}b^{21}}$ **38.** $\sqrt{121x^6y^{10}}$ **39.** $\sqrt[3]{y^6}$ **40.** $\sqrt[3]{216x^{15}}$

[16.3] *Simplify and combine terms where possible.*

41. $7\sqrt{11} + \sqrt{11}$ **42.** $3\sqrt{2} + 6\sqrt{2}$ **43.** $3\sqrt{75} + 2\sqrt{27}$

44. $4\sqrt{12} + \sqrt{48}$ **45.** $4\sqrt{24} - 3\sqrt{54} + \sqrt{6}$ **46.** $2\sqrt{7} - 4\sqrt{28} + 3\sqrt{63}$

47. $\dfrac{2}{5}\sqrt{75} + \dfrac{3}{4}\sqrt{160}$ **48.** $\dfrac{1}{3}\sqrt{18} + \dfrac{1}{4}\sqrt{32}$ **49.** $\sqrt{15} \cdot \sqrt{2} + 5\sqrt{30}$

Simplify each expression. Assume that all variables represent nonnegative real numbers.

50. $\sqrt{4x} + \sqrt{36x} - \sqrt{9x}$ **51.** $\sqrt{16p} + 3\sqrt{p} - \sqrt{49p}$ **52.** $3k\sqrt{8k^2n} + 5k^2\sqrt{2n}$

[16.4] *Perform the indicated operations, and write all answers in simplest form. Rationalize all denominators. Assume that all variables represent nonnegative real numbers.*

53. $\dfrac{10}{\sqrt{3}}$ **54.** $\dfrac{8\sqrt{2}}{\sqrt{5}}$ **55.** $\dfrac{12}{\sqrt{24}}$ **56.** $\sqrt{\dfrac{2}{5}}$

57. $\sqrt{\dfrac{5}{14}} \cdot \sqrt{28}$ **58.** $\sqrt{\dfrac{2}{7}} \cdot \sqrt{\dfrac{1}{3}}$ **59.** $\sqrt{\dfrac{r^2}{16x}}, \quad x \neq 0$ **60.** $\sqrt[3]{\dfrac{1}{3}}$

Solve each problem.

61. The radius r of a cone in terms of its volume V is given by the formula

$$r = \sqrt{\dfrac{3V}{\pi h}}.$$

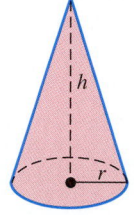

Rationalize the denominator of the radical expression.

62. The radius r of a sphere in terms of its surface area S is given by the formula

$$r = \sqrt{\dfrac{S}{4\pi}}.$$

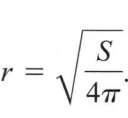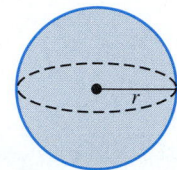

Rationalize the denominator of the radical expression.

[16.5] *Simplify each expression.*

63. $-\sqrt{3}\left(\sqrt{5} + \sqrt{27}\right)$ **64.** $3\sqrt{2}\left(\sqrt{3} + 2\sqrt{2}\right)$ **65.** $\left(2\sqrt{3} - 4\right)\left(5\sqrt{3} + 2\right)$

66. $\left(\sqrt{7} + 2\sqrt{6}\right)\left(\sqrt{12} - \sqrt{2}\right)$ **67.** $\left(2\sqrt{3} + 5\right)\left(2\sqrt{3} - 5\right)$ **68.** $\left(\sqrt{x} + 2\right)^2, \quad x \geq 0$

Rationalize each denominator.

69. $\dfrac{1}{2 + \sqrt{5}}$ **70.** $\dfrac{3}{1 + \sqrt{x}}, \quad x \geq 0, x \neq 1$ **71.** $\dfrac{\sqrt{5} - 1}{\sqrt{2} + 3}$

Write each quotient in lowest terms.

72. $\dfrac{15 + 10\sqrt{6}}{15}$ **73.** $\dfrac{3 + 9\sqrt{7}}{12}$ **74.** $\dfrac{6 + \sqrt{192}}{2}$

[16.6] *Solve each equation.*

75. $\sqrt{x} + 5 = 0$

76. $\sqrt{k+1} = 7$

77. $\sqrt{5t+4} = 3\sqrt{t}$

78. $\sqrt{2p+3} = \sqrt{5p-3}$

79. $\sqrt{4x+1} = x - 1$

80. $\sqrt{13+4t} = t + 4$

81. $\sqrt{2-x} + 3 = x + 7$

82. $\sqrt{x} - x + 2 = 0$

83. $\sqrt{x+2} - \sqrt{x-3} = 1$

▶▶▶ Mixed Review Exercises

Simplify each expression if possible. Assume that all variables represent nonnegative real numbers.

84. $2\sqrt{27} + 3\sqrt{75} - \sqrt{300}$

85. $\dfrac{1}{5+\sqrt{2}}$

86. $\sqrt{\dfrac{1}{3}} \cdot \sqrt{\dfrac{24}{5}}$

87. $\sqrt[3]{54a^7b^{10}}$

88. $-\sqrt{5}\left(\sqrt{2} + \sqrt{75}\right)$

89. $\sqrt{\dfrac{16r^3}{3r}}, \quad s \neq 0$

90. $\dfrac{12 + 6\sqrt{13}}{12}$

91. $\left(\sqrt{5} - \sqrt{2}\right)^2$

92. $\left(6\sqrt{7} + 2\right)\left(4\sqrt{7} - 1\right)$

Solve each equation.

93. $\sqrt{x+2} = x - 4$

94. $\sqrt{k} + 3 = 0$

95. $\sqrt{1+3t} - t = -3$

96. Match each radical in Column I with the equivalent choice in Column II. Choices may be used once, more than once, or not at all.

I		II	
(a) $\sqrt{64}$	**(b)** $-\sqrt{64}$	**A.** 4	**B.** 8
(c) $\sqrt{-64}$	**(d)** $\sqrt[3]{64}$	**C.** -4	**D.** Not a real number
(e) $\sqrt[3]{-64}$	**(f)** $-\sqrt[3]{-64}$	**E.** 16	**F.** -8

Chapter 16 ▷▷▷ Test

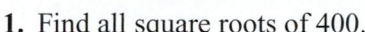

Use the Chapter Test Prep Video CD to see fully worked-out solutions to any of the exercises you want to review.

1. Find all square roots of 400.

1. _____

2. Consider $\sqrt{142}$.

   (a) Determine whether it is rational or irrational.

   (b) Find a decimal approximation to the nearest thousandth.

2. (a) _____

   (b) _____

3. If $\sqrt{a}$ is not a real number, then what kind of number must a be?

3. _____

Simplify where possible. Assume that all variables represent nonnegative real numbers.

4. $\sqrt[3]{216}$

4. _____

5. $-\sqrt{54}$

5. _____

6. $\sqrt{\dfrac{128}{25}}$

6. _____

7. $\sqrt[3]{32}$

7. _____

8. $\dfrac{20\sqrt{18}}{5\sqrt{3}}$

8. _____

9. $3\sqrt{28} + \sqrt{63}$

9. _____

10. $3\sqrt{27x} - 4\sqrt{48x} + 2\sqrt{3x}$

10. _____

11. $\sqrt{32x^2y^3}$

11. _____

12. $\left(6 - \sqrt{5}\right)\left(6 + \sqrt{5}\right)$

12. _____

13. $\left(2 - \sqrt{7}\right)\left(3\sqrt{2} + 1\right)$

13. _____

14. $\left(\sqrt{5} + \sqrt{6}\right)^2$

14. _____

Solve each problem.

15. (a) _____

   (b) _____

15. Find the measure of the unknown leg of this right triangle.

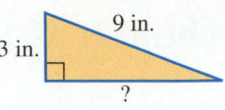

  (a) Give its length in simplified radical form.

  (b) Round the answer to the nearest thousandth.

16. _____

16. In electronics, the impedance Z of an alternating current circuit is given by the formula

$$Z = \sqrt{R^2 + X^2},$$

where R is the resistance and X is the reactance, both in ohms. Find the value of the impedance Z if $R = 40$ ohms and $X = 30$ ohms. (*Source:* Cooke, N., and J. Orleans, *Mathematics Essential to Electricity and Radio*, McGraw-Hill, 1943.)

Rationalize each denominator.

17. _____

17. $\dfrac{5\sqrt{2}}{\sqrt{7}}$

18. $\sqrt{\dfrac{2}{3x}}, \quad x \neq 0$

18. _____

19. _____

19. $\dfrac{-2}{\sqrt[3]{4}}$

20. $\dfrac{-3}{4 - \sqrt{3}}$

20. _____

21. _____

21. Write in lowest terms: $\dfrac{\sqrt{12} + 3\sqrt{128}}{6}$.

Solve each equation.

22. _____

22. $\sqrt{p} + 4 = 0$

23. $\sqrt{x + 1} = 5 - x$

23. _____

24. _____

24. $3\sqrt{x} - 2 = x$

25. $\sqrt{x + 7} - \sqrt{x} = 1$

25. _____

26. _____

26. Consider the following "solution." ***WHAT WENT WRONG?*** Give the correct solution set.

$$\sqrt{2x + 1} + 5 = 0$$

$\sqrt{2x + 1} = -5$ Subtract 5.

$2x + 1 = 25$ Square each side.

$2x = 24$ Subtract 1.

$x = 12$ Divide by 2.

The solution set is $\{12\}$.

17

Quadratic Equations

Recreational fishing is big business in the United States. In 2006, nearly 40 million anglers generated over $45.3 billion in retail sales and had a $125.0 billion impact on the U.S. economy. The sportfishing industry created employment for over one million people. If sportfishing were a corporation, it would rank 47th on the 2007 Fortune 500 list of largest American companies based on total sales, ahead of such global giants as Time Warner, IBM, and Microsoft. More Americans fish than play golf and tennis combined. (*Source:* American Sportfishing Association.)

In Example 5 of **Section 17.1**, we apply the *square root property*, a topic of this chapter, to a formula for calculating the length of a bass.

17.1 ▶▶▶ Solving Quadratic Equations by the Square Root Property

OBJECTIVES

1 Solve equations of the form $x^2 = k$, where $k > 0$.

2 Solve equations of the form $(ax + b)^2 = k$, where $k > 0$.

3 Use formulas involving squared variables.

A **quadratic equation** is an equation that can be written in the form

$$ax^2 + bx + c = 0 \qquad \text{Standard form}$$

for real numbers a, b, and c, with $a \neq 0$. As we saw in **Section 13.7**, we can solve a quadratic equation such as $x^2 + 4x + 3 = 0$ by first factoring and then applying the zero-factor property.

$$x^2 + 4x + 3 = 0$$

$$(x + 3)(x + 1) = 0 \qquad \text{Factor.}$$

$$x + 3 = 0 \quad \text{or} \quad x + 1 = 0 \qquad \text{Zero-factor property}$$

$$x = -3 \quad \text{or} \qquad x = -1 \qquad \text{Solve each equation.}$$

The solution set is $\{-3, -1\}$.

◀ Work Problem **1** at the Side.

1 Solve each equation by factoring.

(a) $x^2 - 2x - 15 = 0$

OBJECTIVE 1 Solve equations of the form $x^2 = k$, where $k > 0$.
We can solve an equation such as $x^2 = 9$ by factoring as follows.

$$x^2 = 9$$

$$x^2 - 9 = 0 \qquad \text{Subtract 9.}$$

$$(x + 3)(x - 3) = 0 \qquad \text{Factor.}$$

$$x + 3 = 0 \quad \text{or} \quad x - 3 = 0 \qquad \text{Zero-factor property}$$

$$x = -3 \quad \text{or} \qquad x = 3 \qquad \text{Solve each equation.}$$

We might also solve $x^2 = 9$ by noticing that x must be a number whose square is 9. Thus, $x = \sqrt{9} = 3$ or $x = -\sqrt{9} = -3$. This approach is generalized as the **square root property.**

> **Square Root Property**
> If k is a positive number and if $x^2 = k$, then
> $$x = \sqrt{k} \quad \text{or} \quad x = -\sqrt{k}.$$

(b) $2x^2 - 3x + 1 = 0$

EXAMPLE 1 Solving Quadratic Equations of the Form $x^2 = k$

Solve each equation. Write radicals in simplified form.

(a) $x^2 = 16$
By the square root property, if $x^2 = 16$, then

$$x = \sqrt{16} = 4 \quad \text{or} \quad x = -\sqrt{16} = -4.$$

An abbreviation for $x = 4$ or $x = -4$ is $x = \pm 4$ (read "positive or negative 4," or "plus or minus 4"). Check each solution by substituting it for x in the original equation. The solution set is $\{-4, 4\}$, or $\{\pm 4\}$.

ANSWERS

1. (a) $\{-3, 5\}$

   (b) $\left\{\frac{1}{2}, 1\right\}$

(b) $z^2 = 5$
The solutions are $z = \sqrt{5}$ or $z = -\sqrt{5}$, so the solution set is $\{-\sqrt{5}, \sqrt{5}\}$, or $\{\pm\sqrt{5}\}$.

Continued on Next Page

(c)

$$5m^2 - 32 = 8$$

$$5m^2 = 40 \qquad \text{Add 32.}$$

$$m^2 = 8 \qquad \text{Divide by 5.}$$

> Don't stop here. Simplify the radicals.

$$m = \sqrt{8} \quad \text{or} \quad m = -\sqrt{8} \qquad \text{Square root property}$$

$$m = 2\sqrt{2} \quad \text{or} \quad m = -2\sqrt{2} \qquad \sqrt{8} = \sqrt{4} \cdot \sqrt{2} = 2\sqrt{2}$$

The solution set is $\left\{ -2\sqrt{2}, 2\sqrt{2} \right\}$, or $\left\{ \pm 2\sqrt{2} \right\}$.

(d) $x^2 = -4$

Because -4 is a negative number and because the square of a real number cannot be negative, ***there is no real number solution*** of this equation. (In this book, we are concerned with finding only *real number* solutions. To use the square root property to find both real number solutions, k must be positive.) The solution set is $\emptyset$.

— **Work Problem 2 at the Side.** ▶

OBJECTIVE 2 **Solve equations of the form $(ax + b)^2 = k$, where $k > 0$.** In each equation in Example 1, the exponent 2 had a single variable as its base. We can extend the square root property to solve equations in which the base is a binomial, as shown in the next example.

EXAMPLE 2 **Solving Quadratic Equations of the Form $(x + b)^2 = k$**

Solve each equation.

(a) ⟨ Use $x - 3$ as the base. ⟩ $(x - 3)^2 = 16$

$$x - 3 = \sqrt{16} \quad \text{or} \quad x - 3 = -\sqrt{16} \qquad \text{Square root property}$$

$$x - 3 = 4 \quad \text{or} \quad x - 3 = -4 \qquad \sqrt{16} = 4$$

$$x = 7 \quad \text{or} \qquad x = -1 \qquad \text{Add 3.}$$

Check Substitute each solution in the original equation.

$$(x - 3)^2 = 16 \qquad\qquad (x - 3)^2 = 16$$

$$(7 - 3)^2 \stackrel{?}{=} 16 \quad \text{Let } x = 7. \qquad (-1 - 3)^2 \stackrel{?}{=} 16 \quad \text{Let } x = -1.$$

$$4^2 \stackrel{?}{=} 16 \qquad\qquad\qquad (-4)^2 \stackrel{?}{=} 16$$

$$16 = 16 \quad \text{True} \qquad\qquad 16 = 16 \quad \text{True}$$

The solutions are 7 and -1, and the solution set is $\{-1, 7\}$.

(b)

$$(x + 1)^2 = 6$$

$$x + 1 = \sqrt{6} \qquad \text{or} \quad x + 1 = -\sqrt{6} \qquad \text{Square root property}$$

$$x = -1 + \sqrt{6} \quad \text{or} \qquad x = -1 - \sqrt{6} \qquad \text{Add } -1.$$

Check

$$\left(-1 + \sqrt{6} + 1 \right)^2 = \left(\sqrt{6} \right)^2 = 6;$$

$$\left(-1 - \sqrt{6} + 1 \right)^2 = \left(-\sqrt{6} \right)^2 = 6.$$

The solution set is $\left\{ -1 + \sqrt{6}, -1 - \sqrt{6} \right\}$, or $\left\{ -1 \pm \sqrt{6} \right\}$.

— **Work Problem 3 at the Side.** ▶

---

2 Solve each equation. Write radicals in simplified form.

(a) $x^2 = 49$

(b) $x^2 = 11$

(c) $2x^2 + 8 = 32$

(d) $x^2 = -9$

3 Solve each equation.

(a) $(x + 2)^2 = 36$

(b) $(x - 4)^2 = 3$

ANSWERS

2. **(a)** $\{-7, 7\}$ **(b)** $\left\{ -\sqrt{11}, \sqrt{11} \right\}$
   (c) $\left\{ -2\sqrt{3}, 2\sqrt{3} \right\}$ **(d)** $\emptyset$

3. **(a)** $\{-8, 4\}$ **(b)** $\left\{ 4 + \sqrt{3}, 4 - \sqrt{3} \right\}$

4 Solve $(2x - 5)^2 = 18$.

5 Solve each equation.

(a) $(5x + 1)^2 = 7$

(b) $(7x - 1)^2 = -1$

6 Use the formula in Example 5 to approximate the length of a bass weighing 2.80 lb and having girth 11 in.

EXAMPLE 3 Solving a Quadratic Equation of the Form $(ax + b)^2 = k$

Solve $(3r - 2)^2 = 27$.

$3r - 2 = \sqrt{27}$	or	$3r - 2 = -\sqrt{27}$

Square root property
$$\sqrt{27} = \sqrt{9} \cdot \sqrt{3}$$
$$= 3\sqrt{3}$$

$3r - 2 = 3\sqrt{3}$	or	$3r - 2 = -3\sqrt{3}$

$3r = 2 + 3\sqrt{3}$	or	$3r = 2 - 3\sqrt{3}$

Add 2.

$r = \dfrac{2 + 3\sqrt{3}}{3}$	or	$r = \dfrac{2 - 3\sqrt{3}}{3}$

Divide by 3.

The solution set is $\left\{ \dfrac{2 + 3\sqrt{3}}{3}, \dfrac{2 - 3\sqrt{3}}{3} \right\}$.

◀ Work Problem **4** at the Side.

> **CAUTION**
> The solutions in Example 3 are fractions that cannot be simplified. Note that 3 is *not* a common factor in the numerator.

EXAMPLE 4 Recognizing When There Is No Real Solution

Solve $(x + 3)^2 = \mathbf{-9}$.

Because the square root of -9 is not a real number, there is no real number solution for this equation. The solution set is $\emptyset$.

◀ Work Problem **5** at the Side.

OBJECTIVE **3** Use formulas involving squared variables.

EXAMPLE 5 Finding the Length of a Bass

We can approximate the weight of a bass, in pounds, given its length L and its girth (distance around) g, where both are measured in inches, using the formula

$$w = \frac{L^2 g}{1200}.$$

Approximate the length of a bass weighing 2.20 lb and having girth 10 in. (*Source: Sacramento Bee*, November 29, 2000.)

$w = \dfrac{L^2 g}{1200}$	Given formula
$\mathbf{2.20} = \dfrac{L^2 \cdot \mathbf{10}}{1200}$	$w = 2.20, g = 10$
$2640 = 10L^2$	Multiply by 1200.
$L^2 = 264$	Divide by 10; interchange the sides.
$L = \sqrt{264}$ or $L = -\sqrt{264}$	Square root property

A calculator shows that $\sqrt{264} \approx 16.25$, so the length of the bass is a little more than 16 in. (We discard the negative solution $-\sqrt{264} \approx -16.25$, since L represents length.)

◀ Work Problem **6** at the Side.

17.1 ▶▶▶ Exercises

Decide whether each statement is true *or* false. *If* false, *tell why.*

1. If k is a prime number, then $x^2 = k$ has two irrational solutions.

2. If k is a positive perfect square, then $x^2 = k$ has two rational solutions.

3. If k is a positive integer, then $x^2 = k$ must have two rational solutions.

4. If $-10 < k < 0$, then $x^2 = k$ has no real solution.

5. If $-10 < k < 10$, then $x^2 = k$ has no real solution.

6. If k is an integer greater than 24 and less than 26, then $x^2 = k$ has two solutions, -5 and 5.

Solve each equation by using the square root property. Write all radicals in simplest form. See Example 1.

7. $x^2 = 81$

8. $x^2 = 121$

9. $k^2 = 14$

10. $m^2 = 22$

11. $t^2 = 48$

12. $x^2 = 54$

13. $x^2 = \dfrac{25}{4}$

14. $m^2 = \dfrac{36}{121}$

15. $x^2 = -100$

16. $x^2 = -64$

17. $z^2 = 2.25$

18. $w^2 = 56.25$

19. $r^2 - 3 = 0$

20. $x^2 - 13 = 0$

21. $7x^2 = 4$

22. $3x^2 = 10$

23. $4x^2 - 72 = 0$

24. $5z^2 - 200 = 0$

25. $3x^2 - 8 = 64$

26. $2x^2 + 7 = 61$

27. $5x^2 + 4 = 8$

28. $4x^2 - 3 = 7$

Solve each equation by using the square root property. Express all radicals in simplest form. See Examples 2–4.

29. $(x - 3)^2 = 25$

30. $(x - 7)^2 = 16$

31. $(x + 5)^2 = -13$

32. $(x + 2)^2 = -17$

33. $(x - 8)^2 = 27$

34. $(x - 5)^2 = 40$

35. $(3x + 2)^2 = 49$

36. $(5x + 3)^2 = 36$

37. $(4x - 3)^2 = 9$

38. $(7x - 5)^2 = 25$

39. $(5 - 2x)^2 = 30$

40. $(3 - 2x)^2 = 70$

41. $(3x + 1)^2 = 18$

42. $(5x + 6)^2 = 75$

43. $\left(\dfrac{1}{2}x + 5\right)^2 = 12$

44. $\left(\dfrac{1}{3}x + 4\right)^2 = 27$

45. $(4x - 1)^2 - 48 = 0$

46. $(2x - 5)^2 - 180 = 0$

47. Johnny solved the equation in Exercise 39 and wrote his answer as $\left\{ \dfrac{5 + \sqrt{30}}{2}, \dfrac{5 - \sqrt{30}}{2} \right\}$.

Terry solved the same equation and wrote her answer as $\left\{ \dfrac{-5 + \sqrt{30}}{-2}, \dfrac{-5 - \sqrt{30}}{-2} \right\}$.

The teacher gave them both full credit. Explain why both students were correct, although their answers seem to differ.

48. In the solutions found in Example 3 of this section, why is it not valid to simplify by dividing out the 3s in the numerators and denominators?

Solve each problem. See Example 5.

49. One expert at marksmanship can hold a silver dollar at forehead level, drop it, draw his gun, and shoot the coin as it passes waist level. The distance traveled by a falling object is given by

$$d = 16t^2,$$

where d is the distance (in feet) the object falls in t seconds. If the coin falls about 4 ft, use the formula to estimate the time that elapses between the dropping of the coin and the shot.

50. The illumination produced by a light source depends on the distance from the source. For a particular light source, this relationship can be expressed as

$$d^2 = \frac{4050}{I},$$

where d is the distance from the source (in feet) and I is the amount of illumination in foot-candles. How far from the source is the illumination equal to 50 foot-candles?

51. The area A of a circle with radius r is given by the formula

$$A = \pi r^2.$$

If a circle has area 81π in.², what is its radius?

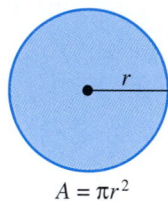

$A = \pi r^2$

52. The surface area S of a sphere with radius r is given by the formula

$$S = 4\pi r^2.$$

If a sphere has surface area 36π ft², what is its radius?

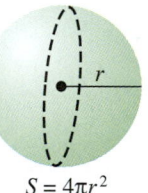

$S = 4\pi r^2$

The amount A that P dollars invested at an annual rate of interest r will grow to in 2 yr is $A = P(1 + r)^2$.

53. At what interest rate will $100 grow to $110.25 in 2 yr?

54. At what interest rate will $500 grow to $572.45 in 2 yr?

55. At what interest rate will $200 grow to $208.08 in 2 yr?

17.2 ▶▶▶ Solving Quadratic Equations by Completing the Square

OBJECTIVE **1** **Solve quadratic equations by completing the square when the coefficient of the second-degree term is 1.** The methods we have studied so far are not enough to solve the equation

$$x^2 + 6x + 7 = 0.$$

If we could write the equation in the form $(x + 3)^2$ equals a constant, we could solve it with the square root property discussed in **Section 17.1.** To do that, we need to have a perfect square trinomial on one side of the equation.

Recall from **Section 13.5** that a perfect square trinomial has the form

$$x^2 \mathbf{\color{red}{+ 2kx}} + k^2 \quad \text{or} \quad x^2 \mathbf{\color{red}{- 2kx}} + k^2,$$

where k represents a positive number.

EXAMPLE 1 **Creating Perfect Square Trinomials**

Complete each trinomial so that it is a perfect square. Then factor the trinomial.

(a) $x^2 \mathbf{\color{red}{+ 8x}} +$ _____

The perfect square trinomial will have the form $x^2 \mathbf{\color{red}{+ 2kx}} + k^2$. Thus, the middle term, $8x$, must equal $2kx$.

$$8x = 2kx \quad \leftarrow \text{Solve this equation for } k.$$
$$4 = k \qquad \text{Divide each side by } 2x.$$

Therefore, $k = 4$ and $k^2 = 4^2 = \mathbf{\color{blue}{16}}$. The required perfect square trinomial is

$$x^2 + 8x + \mathbf{\color{blue}{16}}, \quad \text{which factors as} \quad (x + 4)^2.$$

(b) $x^2 \mathbf{\color{red}{- 18x}} +$ _____

Here the perfect square trinomial will have the form $x^2 \mathbf{\color{red}{- 2kx}} + k^2$. The middle term, $-18x$, must equal $-2kx$.

$$-18x = -2kx \quad \leftarrow \text{Solve this equation for } k.$$
$$9 = k \qquad \text{Divide each side by } -2x.$$

Thus, $k = 9$ and $k^2 = 9^2 = \mathbf{\color{blue}{81}}$. The required perfect square trinomial is

$$x^2 - 18x + \mathbf{\color{blue}{81}}, \quad \text{which factors as} \quad (x - 9)^2.$$

Work Problem **1** *at the Side.* ▶

EXAMPLE 2 **Rewriting an Equation to Use the Square Root Property**

Solve $x^2 + 6x + 7 = 0$.

$$x^2 + 6x = -7 \qquad \text{Subtract 7 from each side.}$$

To solve this equation with the square root property, the quantity on the left side, $x^2 + 6x$, must be written as a perfect square trinomial in the form $x^2 \mathbf{\color{red}{+ 2kx}} + k^2$.

$$x^2 \mathbf{\color{red}{+ 6x}} + \text{_____}$$

Here, $2kx = 6x$, so $k = 3$ and $k^2 = \mathbf{\color{blue}{9}}$. The required perfect square trinomial is

$$x^2 + 6x + \mathbf{\color{blue}{9}}, \quad \text{which factors as} \quad (x + 3)^2.$$

Continued on Next Page

OBJECTIVES

1 Solve quadratic equations by completing the square when the coefficient of the second-degree term is 1.

2 Solve quadratic equations by completing the square when the coefficient of the second-degree term is not 1.

3 Simplify the terms of an equation before solving.

4 Solve applied problems that require quadratic equations.

1 Complete each trinomial so that it is a perfect square. Then factor the trinomial.

(a) $x^2 + 12x +$ _____

(b) $x^2 - 14x +$ _____

(c) $x^2 - 2x +$ _____

ANSWERS

1. **(a)** $36; (x + 6)^2$
(b) $49; (x - 7)^2$
(c) $1; (x - 1)^2$

2 Solve $x^2 - 4x - 1 = 0$.

3 Solve each equation by completing the square.

(a) $x^2 + 4x = 1$

(b) $z^2 + 6z - 3 = 0$

2. $\{2 + \sqrt{5}, 2 - \sqrt{5}\}$

3. (a) $\{-2 + \sqrt{5}, -2 - \sqrt{5}\}$

(b) $\{-3 + 2\sqrt{3}, -3 - 2\sqrt{3}\}$

Therefore, if we add 9 to *each* side of $x^2 + 6x = -7$, the equation will have a perfect square trinomial on the left side, as needed.

> This is a key step.

$$x^2 + 6x + 9 = -7 + 9 \qquad \text{Add 9.}$$
$$(x + 3)^2 = 2 \qquad \text{Factor; add.}$$

Now use the square root property to complete the solution.

$$x + 3 = \sqrt{2} \qquad \text{or} \quad x + 3 = -\sqrt{2}$$
$$x = -3 + \sqrt{2} \quad \text{or} \qquad x = -3 - \sqrt{2}$$

Check by substituting $-3 + \sqrt{2}$ and $-3 - \sqrt{2}$ for x in the original equation. The solution set is $\{-3 + \sqrt{2}, -3 - \sqrt{2}\}$.

◄ Work Problem **2** at the Side.

The process of changing the form of the equation in Example 2 from

$$x^2 + 6x + 7 = 0 \quad \text{to} \quad (x + 3)^2 = 2$$

is called **completing the square.** Completing the square changes only the form of the equation. To see this, multiply out the left side of $(x + 3)^2 = 2$. Then write the equation in standard form to get $x^2 + 6x + 7 = 0$.

Look again at the original equation in Example 2,

$$x^2 + 6x + 7 = 0.$$

If we take half of 6, the coefficient of x, and square it, we get 9.

$$\frac{1}{2} \cdot 6 = 3 \quad \text{and} \quad 3^2 = 9$$

Coefficient of x \qquad Quantity added to each side

To complete the square in Example 2, we added **9** to each side.

EXAMPLE 3 **Completing the Square to Solve a Quadratic Equation**

Complete the square to solve $x^2 - 8x = 5$.

To complete the square on $x^2 - 8x$, take half of -8, the coefficient of x, and square it.

$$\frac{1}{2}(-8) = -4 \quad \text{and} \quad (-4)^2 = 16$$

Coefficient of x

Add the result, **16**, to each side of the equation.

$$x^2 - 8x = 5 \qquad\qquad \text{Given equation}$$
$$x^2 - 8x + 16 = 5 + 16 \qquad \text{Add 16.}$$
$$(x - 4)^2 = 21 \qquad\qquad \text{Factor; add.}$$
$$x - 4 = \sqrt{21} \qquad \text{or} \quad x - 4 = -\sqrt{21} \qquad \text{Square root property}$$
$$x = 4 + \sqrt{21} \quad \text{or} \qquad x = 4 - \sqrt{21} \qquad \text{Add 4.}$$

A check indicates that the solution set is $\{4 + \sqrt{21}, 4 - \sqrt{21}\}$.

◄ Work Problem **3** at the Side.

OBJECTIVE 2 Solve quadratic equations by completing the square when the coefficient of the second-degree term is not 1.

If a quadratic equation has the form

$$ax^2 + bx + c = 0, \quad \text{where } a \neq 1,$$

then to obtain 1 as the coefficient of x^2, we first divide each side of the equation by a.

EXAMPLE 4 Solving a Quadratic Equation by Completing the Square

Solve $4x^2 + 16x = 9$.

Before completing the square, the coefficient of x^2 must be 1, not 4. We get 1 as the coefficient of x^2 here by dividing each side by 4.

$$4x^2 + 16x = 9$$

> The coefficient of x^2 must be 1. $\longrightarrow$

$$x^2 + 4x = \frac{9}{4} \qquad \text{Divide by 4.}$$

Next, we begin to complete the square by taking half the coefficient of x, and squaring it:

$$\frac{1}{2}(4) = 2 \quad \text{and} \quad 2^2 = 4.$$

We add the result, **4**, to each side of the equation.

$$x^2 + 4x + 4 = \frac{9}{4} + 4 \qquad \text{Add 4.}$$

$$(x + 2)^2 = \frac{25}{4} \qquad \text{Factor; } \frac{9}{4} + 4 = \frac{9}{4} + \frac{16}{4} = \frac{25}{4}.$$

$$x + 2 = \sqrt{\frac{25}{4}} \quad \text{or} \quad x + 2 = -\sqrt{\frac{25}{4}} \qquad \text{Square root property}$$

$$x + 2 = \frac{5}{2} \quad \text{or} \quad x + 2 = -\frac{5}{2} \qquad \text{Take square roots.}$$

$$x = -2 + \frac{5}{2} \quad \text{or} \quad x = -2 - \frac{5}{2} \qquad \text{Subtract 2.}$$

$$x = \frac{1}{2} \quad \text{or} \quad x = -\frac{9}{2} \qquad -2 = -\frac{4}{2}$$

Check

$$4x^2 + 16x = 9 \qquad\qquad\qquad 4x^2 + 16x = 9$$

$$4\left(\frac{1}{2}\right)^2 + 16\left(\frac{1}{2}\right) \overset{?}{=} 9 \qquad 4\left(-\frac{9}{2}\right)^2 + 16\left(-\frac{9}{2}\right) \overset{?}{=} 9$$

$$4\left(\frac{1}{4}\right) + 8 \overset{?}{=} 9 \qquad\qquad 4\left(\frac{81}{4}\right) - 72 \overset{?}{=} 9$$

$$1 + 8 \overset{?}{=} 9 \qquad\qquad\qquad 81 - 72 \overset{?}{=} 9$$

$$9 = 9 \quad \text{True} \qquad\qquad\quad 9 = 9 \quad \text{True}$$

The two solutions, $\frac{1}{2}$ and $-\frac{9}{2}$, check, so the solution set is $\left\{-\frac{9}{2}, \frac{1}{2}\right\}$.

Work Problem **4** *at the Side.* ▶

---

4 Solve each equation by completing the square.

(a) $9x^2 + 18x = -5$

(b) $4t^2 - 24t + 11 = 0$

ANSWERS

4. (a) $\left\{-\frac{1}{3}, -\frac{5}{3}\right\}$ (b) $\left\{\frac{11}{2}, \frac{1}{2}\right\}$

The steps used to solve a quadratic equation $ax^2 + bx + c = 0$ by completing the square are summarized here.

5 Solve each equation by completing the square.

(a) $3x^2 + 5x - 2 = 0$

> ### Solving a Quadratic Equation by Completing the Square
>
> **Step 1** **Be sure the second-degree term has coefficient 1.** If the coefficient of the second-degree term is 1, proceed to Step 2. If it is not 1, but some other nonzero number a, divide each side of the equation by a.
>
> **Step 2** **Write in correct form.** Make sure that all terms with variables are on one side of the equals sign and that all constant terms are on the other side.
>
> **Step 3** **Complete the square.** Take half the coefficient of the first-degree term, and square it. Add the square to each side of the equation. Factor the variable side, and simplify on the other side.
>
> **Step 4** **Solve** the equation by using the square root property.

EXAMPLE 5 **Solving a Quadratic Equation by Completing the Square**

Solve $2x^2 - 7x - 9 = 0$.

Step 1 Get 1 as the coefficient of the x^2-term.

$$x^2 - \frac{7}{2}x - \frac{9}{2} = 0 \qquad \text{Divide by 2.}$$

(b) $2x^2 - 4x - 1 = 0$

Step 2 Add $\frac{9}{2}$ to each side to get the variable terms on the left and the constant on the right.

$$x^2 - \frac{7}{2}x = \frac{9}{2} \qquad \text{Add } \tfrac{9}{2}.$$

Step 3 To complete the square, take half the coefficient of x and square it: $\left[\frac{1}{2}\left(-\frac{7}{2}\right)\right]^2 = \left(-\frac{7}{4}\right)^2 = \frac{49}{16}$.

$$x^2 - \frac{7}{2}x + \frac{49}{16} = \frac{9}{2} + \frac{49}{16} \qquad \boxed{\text{Be sure to add } \tfrac{49}{16} \text{ to } \textit{each} \text{ side.}}$$

$$\left(x - \frac{7}{4}\right)^2 = \frac{121}{16} \qquad \text{Factor; } \tfrac{9}{2} + \tfrac{49}{16} = \tfrac{72}{16} + \tfrac{49}{16} = \tfrac{121}{16}.$$

Step 4 Solve by using the square root property.

$$x - \frac{7}{4} = \sqrt{\frac{121}{16}} \quad \text{or} \quad x - \frac{7}{4} = -\sqrt{\frac{121}{16}} \qquad \text{Square root property}$$

$$x = \frac{7}{4} + \frac{11}{4} \quad \text{or} \quad x = \frac{7}{4} - \frac{11}{4} \qquad \text{Add } \tfrac{7}{4}; \sqrt{\tfrac{121}{16}} = \tfrac{11}{4}.$$

$$x = \frac{18}{4} = \frac{9}{2} \quad \text{or} \quad x = -\frac{4}{4} = -1 \qquad \text{Simplify.}$$

A check confirms that the solution set is $\left\{-1, \frac{9}{2}\right\}$.

◀ *Work Problem* **5** *at the Side.*

ANSWERS

5. (a) $\left\{-2, \frac{1}{3}\right\}$ **(b)** $\left\{\dfrac{2 + \sqrt{6}}{2}, \dfrac{2 - \sqrt{6}}{2}\right\}$

EXAMPLE 6   **Solving a Quadratic Equation by Completing the Square**

Solve $4p^2 + 8p + 5 = 0$.

$$4p^2 + 8p + 5 = 0$$

> The coefficient of the second-degree term must be 1.

$$p^2 + 2p + \frac{5}{4} = 0 \qquad \text{Divide by 4.}$$

$$p^2 + \mathbf{2}p = -\frac{5}{4} \qquad \text{Subtract } \tfrac{5}{4}.$$

The coefficient of p is 2. Take half of 2; square the result: $\left[\frac{1}{2}(\mathbf{2})\right]^2 = 1^2 = \mathbf{1}$. Add this result to each side. Then write the left side as a perfect square.

$$p^2 + 2p + \mathbf{1} = -\frac{5}{4} + \mathbf{1} \qquad \text{Add 1.}$$

$$(p+1)^2 = -\frac{\mathbf{1}}{\mathbf{4}} \qquad \text{Factor; add.}$$

We cannot use the square root property to solve this equation, because the square root of $-\frac{1}{4}$ is not a real number. This equation has no real number solution.* The solution set is $\emptyset$.

Work Problem **6** *at the Side.* ▶

OBJECTIVE **3** **Simplify the terms of an equation before solving.**

EXAMPLE 7   **Simplifying before Completing the Square**

Solve $(x+3)(x-1) = 2$.

$$(x+3)(x-1) = 2 \qquad \text{Given equation}$$
$$x^2 + 2x - 3 = 2 \qquad \text{Multiply using the FOIL method.}$$
$$x^2 + 2x = 5 \qquad \text{Add 3.}$$
$$x^2 + 2x + \mathbf{1} = 5 + \mathbf{1} \qquad \text{Complete the square—add } [\tfrac{1}{2}(2)]^2 = 1.$$
$$(x+1)^2 = 6 \qquad \text{Factor on the left; add on the right.}$$

$$x + 1 = \sqrt{6} \qquad \text{or} \qquad x + 1 = -\sqrt{6} \qquad \text{Square root property}$$
$$x = -1 + \sqrt{6} \quad \text{or} \qquad x = -1 - \sqrt{6} \qquad \text{Add } -1.$$

The solution set is $\{-1 + \sqrt{6}, \ -1 - \sqrt{6}\}$.

Work Problem **7** *at the Side.* ▶

> **Note**
> The solutions $-1 \pm \sqrt{6}$ given in Example 7 are *exact*. In applications, decimal solutions are more appropriate. Using the square root key of a calculator, $\sqrt{6} \approx 2.449$. Approximating the two solutions gives
> $$x \approx 1.449 \quad \text{and} \quad x \approx -3.449.$$

---

*The equation in Example 6 has no solution over the *real number system*. In the **complex number system,** however, this equation does have solutions. The complex numbers include numbers whose squares are negative. These numbers are discussed in intermediate and college algebra courses.

6 Solve $5x^2 + 3x + 1 = 0$ by completing the square.

7 Solve each equation.

   (a) $r(r-3) = -1$

   (b) $(x+2)(x+1) = 5$

ANSWERS

6. $\emptyset$

7. **(a)** $\left\{\dfrac{3 + \sqrt{5}}{2}, \dfrac{3 - \sqrt{5}}{2}\right\}$

    (b) $\left\{\dfrac{-3 + \sqrt{21}}{2}, \dfrac{-3 - \sqrt{21}}{2}\right\}$

8 Suppose a ball is projected upward from ground level with an initial velocity of 128 ft per sec. Its height at time t (in seconds) is given by

$$s = -16t^2 + 128t,$$

where s is in feet. At what times will it be 48 ft above the ground? Give answers to the nearest tenth.

OBJECTIVE 4 Solve applied problems that require quadratic equations. There are many practical applications of quadratic equations. The next example illustrates an application from physics.

EXAMPLE 8 Solving a Velocity Problem

If a ball is projected into the air from ground level with an initial velocity of 64 ft per sec, its altitude (height) s in feet in t seconds is given by the formula

$$s = -16t^2 + 64t.$$

How long will it take the ball to be 48 ft above the ground?

Since s represents the height, we substitute **48** for s in the formula and then solve this equation for time t by completing the square.

$48 = -16t^2 + 64t$	Let $s = 48$.
$-3 = t^2 - 4t$	Divide by -16.
$t^2 - 4t = -3$	Interchange the sides.
$t^2 - 4t + 4 = -3 + 4$	Add $\left[\frac{1}{2}(-4)\right]^2 = 4$.
$(t - 2)^2 = 1$	Factor; add.
$t - 2 = 1$ or $t - 2 = -1$	Square root property
$t = 3$ or $t = 1$	Add 2.

The ball reaches a height of 48 ft twice, once on the way up and again on the way down. It takes 1 sec to reach 48 ft on the way up, and then after 3 sec, the ball reaches 48 ft again on the way down.

◀ *Work Problem* **8** *at the Side.*

Complete each trinomial so that it is a perfect square. Then factor the trinomial.
See Example 1.

1. $x^2 + 10x +$ _____

2. $x^2 + 16x +$ _____

3. $x^2 + 2x +$ _____

4. $m^2 - 2m +$ _____

5. $p^2 - 5p +$ _____

6. $x^2 + 3x +$ _____

7. Which step is an appropriate way to begin solving the quadratic equation

$$2x^2 - 4x = 9$$

by completing the square?

A. Add 4 to each side of the equation.

B. Factor the left side as $2x(x - 2)$.

C. Factor the left side as $x(2x - 4)$.

D. Divide each side by 2.

8. In Example 3 of **Section 13.7,** we solved the quadratic equation

$$4p^2 - 26p + 40 = 0$$

by factoring. If we were to solve by completing the square, would we get the same solutions, $\frac{5}{2}$ and 4?

Solve each equation by completing the square. See Examples 2 and 3.

9. $x^2 - 4x = -3$

10. $x^2 - 2x = 8$

11. $x^2 + 5x + 6 = 0$

12. $x^2 + 6x + 5 = 0$

13. $x^2 + 2x - 5 = 0$

14. $x^2 + 4x + 1 = 0$

15. $x^2 - 8x = -4$

16. $m^2 - 4m = 14$

17. $t^2 + 6t + 9 = 0$

18. $k^2 - 8k + 16 = 0$

19. $x^2 + x - 1 = 0$

20. $x^2 + x - 3 = 0$

Solve each equation by completing the square. See Examples 4–7.

21. $4x^2 + 4x - 3 = 0$

22. $9x^2 + 3x - 2 = 0$

23. $2x^2 - 4x = 5$

24. $2x^2 - 6x = 3$

25. $2p^2 - 2p + 3 = 0$

26. $3q^2 - 3q + 4 = 0$

27. $3k^2 + 7k = 4$

28. $2k^2 + 5k = 1$

29. $(x + 3)(x - 1) = 5$

30. $(y - 8)(y + 2) = 24$

31. $(r - 3)(r - 5) = 2$

32. $(k - 1)(k - 7) = 1$

33. $-x^2 + 2x = -5$

34. $-r^2 + 3r = -2$

Solve each problem. See Example 8.

35. If an object is projected upward from ground level on Earth with an initial velocity of 96 ft per sec, its altitude (height) s in feet in t seconds is given by the formula $s = -16t^2 + 96t$. At what times will the object be 80 ft above the ground?

36. At what times will the object described in Exercise 35 be 100 ft above the ground? Round your answers to the nearest tenth.

37. If an object is projected upward on the surface of Mars from ground level with an initial velocity of 104 ft per sec, its altitude (height) s in feet in t seconds is given by the formula $s = -13t^2 + 104t$. At what times will the object be 195 ft above the surface?

38. After how many seconds will the object in Exercise 37 return to the surface? (*Hint:* When it returns to the surface, $s = 0$.)

39. A farmer has a rectangular cattle pen with perimeter 350 ft and area 7500 ft². What are the dimensions of the pen? (*Hint:* Use the figure to set up the equation.)

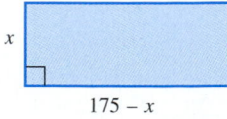

x

$175 - x$

40. The base of a triangle measures 1 m more than three times the height of the triangle. Its area is 15 m². Find the lengths of the base and the height.

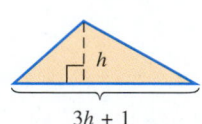

h

$3h + 1$

17.3 ▶▶▶ Solving Quadratic Equations by the Quadratic Formula

We can solve any quadratic equation by completing the square, but the method can be tedious. In this section we complete the square on the general quadratic equation

$$ax^2 + bx + c = 0, \qquad a \neq 0, \qquad \text{Standard form}$$

to obtain the *quadratic formula,* which gives the solution(s) of any quadratic equation.

> **Note**
>
> In $ax^2 + bx + c = 0$, there is a restriction that a is not zero. If it were, the equation would be linear, not quadratic.

OBJECTIVE 1 Identify the values of a, b, and c in a quadratic equation. To solve a quadratic equation by this new method, we must first identify the values of a, b, and c in the standard form.

EXAMPLE 1 Identifying Values of a, b, and c in Quadratic Equations

Identify the values of the variables a, b, and c in each quadratic equation $ax^2 + bx + c = 0$.

(a) $\underset{\downarrow}{2}x^2 + \underset{\downarrow}{3}x - \underset{\downarrow}{5} = 0$

> This equation is in standard form $ax^2 + bx + c = 0$.

In this example, $a = 2$, $b = 3$, and $c = -5$.

(b) $-x^2 + 2 = 6x$

First write the equation in standard form $ax^2 + bx + c = 0$.

$$-x^2 + 2 = 6x$$

> $-x^2$ means $-1x^2$.

$$-x^2 - 6x + 2 = 0 \qquad \text{Subtract } 6x.$$

Here, $a = -1$, $b = -6$, and $c = 2$.

(c) $5x^2 - 12 = 0$

The x-term is missing, so write the equation as

$$5x^2 + 0x - 12 = 0.$$

Then $a = 5$, $b = 0$, and $c = -12$.

(d) $-4x^2 = -x$

In $ax^2 + bx + c = 0$ form, this equation is written $-4x^2 + x = 0$. The constant term c is missing, so $a = -4$, $b = 1$, and $c = 0$.

(e) $\qquad (2x - 7)(x + 4) = -23$

$$2x^2 + x - 28 = -23 \qquad \text{Multiply using the FOIL method.}$$

$$2x^2 + x - 5 = 0 \qquad \text{Add 23; standard form}$$

Now, identify the required values: $a = 2$, $b = 1$, and $c = -5$.

Work Problem **1** at the Side. ▶

OBJECTIVES

1 Identify the values of a, b, and c in a quadratic equation.

2 Use the quadratic formula to solve quadratic equations.

3 Solve quadratic equations with only one solution.

4 Solve quadratic equations with fractions.

---

1 Identify the values of the variables a, b, and c in each quadratic equation $ax^2 + bx + c = 0$.

(a) $5x^2 + 2x - 1 = 0$

(b) $3x^2 = x - 2$

(c) $9x^2 - 13 = 0$

(d) $-x^2 + x = 0$

(e) $(3x + 2)(x - 1) = 8$

ANSWERS

1. (a) $a = 5, b = 2, c = -1$
   (b) $a = 3, b = -1, c = 2$
   (c) $a = 9, b = 0, c = -13$
   (d) $a = -1, b = 1, c = 0$
   (e) $a = 3, b = -1, c = -10$

OBJECTIVE **2** **Use the quadratic formula to solve quadratic equations.** To develop the quadratic formula, we complete the square on $ax^2 + bx + c = 0$ $(a > 0)$. For comparison, we also show the corresponding steps for solving $2x^2 + x - 5 = 0$ (from Example 1(e)).

Step 1 Transform so that the coefficient of x^2 is equal to 1.

$2x^2 + x - 5 = 0$	$ax^2 + bx + c = 0$ Standard form
$x^2 + \dfrac{1}{2}x - \dfrac{5}{2} = 0$ Divide by 2.	$x^2 + \dfrac{b}{a}x + \dfrac{c}{a} = 0$ Divide by a.

Step 2 Write so that the variable terms with x are alone on the left side.

$$x^2 + \frac{1}{2}x = \frac{5}{2} \qquad \text{Add } \tfrac{5}{2}. \qquad\qquad x^2 + \frac{b}{a}x = -\frac{c}{a} \qquad \text{Subtract } \tfrac{c}{a}.$$

Step 3 Add the square of half the coefficient of x to each side, factor the left side, and combine terms on the right.

$$x^2 + \frac{1}{2}x + \frac{1}{16} = \frac{5}{2} + \frac{1}{16} \quad \text{Add } \tfrac{1}{16}. \qquad x^2 + \frac{b}{a}x + \frac{b^2}{4a^2} = -\frac{c}{a} + \frac{b^2}{4a^2} \quad \text{Add } \tfrac{b^2}{4a^2}.$$

$$\left(x + \frac{1}{4}\right)^2 = \frac{41}{16} \quad \begin{array}{l}\text{Factor;}\\ \text{add on}\\ \text{right.}\end{array} \qquad \left(x + \frac{b}{2a}\right)^2 = \frac{b^2 - 4ac}{4a^2} \quad \begin{array}{l}\text{Factor;}\\ \text{add on}\\ \text{right.}\end{array}$$

Step 4 Use the square root property to complete the solution.

$$x + \frac{1}{4} = \pm\sqrt{\frac{41}{16}} \qquad\qquad x + \frac{b}{2a} = \pm\sqrt{\frac{b^2 - 4ac}{4a^2}}$$

$$x + \frac{1}{4} = \pm\frac{\sqrt{41}}{4} \qquad\qquad x + \frac{b}{2a} = \pm\frac{\sqrt{b^2 - 4ac}}{2a}$$

$$x = -\frac{1}{4} \pm \frac{\sqrt{41}}{4} \qquad\qquad x = -\frac{b}{2a} \pm \frac{\sqrt{b^2 - 4ac}}{2a}$$

$$x = \frac{-1 \pm \sqrt{41}}{4} \qquad\qquad \boldsymbol{x = \frac{-b \pm \sqrt{b^2 - 4ac}}{2a}}$$

The final result on the right is called the **quadratic formula.** (It is also valid for $a < 0$.) *It is a key result that should be memorized.* Notice that there are two values: one for the $+$ sign and one for the $-$ sign.

Quadratic Formula

The solutions of the quadratic equation $ax^2 + bx + c = 0$, $a \neq 0$, are

$$x = \frac{-b + \sqrt{b^2 - 4ac}}{2a} \quad \text{and} \quad x = \frac{-b - \sqrt{b^2 - 4ac}}{2a},$$

or in compact form, $\quad x = \dfrac{-b \pm \sqrt{b^2 - 4ac}}{2a}.$

CAUTION

Notice in the quadratic formula that the fraction bar is under $-b$ as well as the radical. *Be sure to find the values of $-b \pm \sqrt{b^2 - 4ac}$ first, and then divide those results by the value of $2a$.*

EXAMPLE 2 Solving a Quadratic Equation by the Quadratic Formula

Solve $2x^2 - 7x - 9 = 0$. (This equation was solved by completing the square in **Section 17.2**, Example 5.)

In this equation, $a = 2$, $b = -7$, and $c = -9$.

$$x = \frac{-b \pm \sqrt{b^2 - 4ac}}{2a} \qquad \text{Quadratic formula}$$

$$x = \frac{-(-7) \pm \sqrt{(-7)^2 - 4(2)(-9)}}{2(2)} \qquad \begin{array}{l} \text{Substitute } a = 2, b = -7, \\ \text{and } c = -9. \end{array}$$

$$x = \frac{7 \pm \sqrt{49 + 72}}{4} \qquad \text{Simplify.}$$

$$x = \frac{7 \pm \sqrt{121}}{4}$$

$$x = \frac{7 \pm 11}{4} \qquad \sqrt{121} = 11$$

$$x = \frac{7 + 11}{4} = \frac{18}{4} = \frac{9}{2} \quad \text{or} \quad x = \frac{7 - 11}{4} = \frac{-4}{4} = -1 \qquad \begin{array}{l} \text{Find } both \\ \text{solutions.} \end{array}$$

Check each solution in the original equation. The solution set is $\left\{-1, \frac{9}{2}\right\}$.

Work Problem **2** *at the Side.* ▶

2 Solve each equation by using the quadratic formula.

(a) $2x^2 + 3x - 5 = 0$

(b) $6x^2 + x - 1 = 0$

EXAMPLE 3 Rewriting a Quadratic Equation before Solving

Solve $x^2 = 2x + 1$.

Write the equation in standard form as $x^2 - 2x - 1 = 0$.

$$x = \frac{-b \pm \sqrt{b^2 - 4ac}}{2a} \qquad \text{Quadratic formula}$$

$$x = \frac{-(-2) \pm \sqrt{(-2)^2 - 4(1)(-1)}}{2(1)} \qquad \begin{array}{l} \text{Substitute } a = 1, b = -2, \\ \text{and } c = -1. \end{array}$$

$$x = \frac{2 \pm \sqrt{8}}{2} \qquad \text{Simplify.}$$

$$x = \frac{2 \pm 2\sqrt{2}}{2} \qquad \sqrt{8} = \sqrt{4} \cdot \sqrt{2} = 2\sqrt{2}$$

$$x = \frac{2\left(1 \pm \sqrt{2}\right)}{2} \qquad \boxed{\text{Factor first. Then divide out the common factor.}}$$

$$x = 1 \pm \sqrt{2}$$

The solution set is $\left\{1 + \sqrt{2}, \ 1 - \sqrt{2}\right\}$.

Work Problem **3** *at the Side.* ▶

3 Solve $x^2 + 1 = -8x$.

ANSWERS

2. (a) $\left\{1, -\frac{5}{2}\right\}$ **(b)** $\left\{-\frac{1}{2}, \frac{1}{3}\right\}$

3. $\left\{-4 + \sqrt{15}, -4 - \sqrt{15}\right\}$

④ Solve $9x^2 - 12x + 4 = 0$.

EXAMPLE 4 **Solving a Quadratic Equation with One Solution**

Solve $4x^2 + 25 = 20x$.

Write the equation in standard form as

$$4x^2 - 20x + 25 = 0. \quad \text{Subtract } 20x.$$

Here, $a = 4$, $b = -20$, and $c = 25$. By the quadratic formula,

$$x = \frac{-(-20) \pm \sqrt{(-20)^2 - 4(4)(25)}}{2(4)} = \frac{20 \pm 0}{8} = \frac{5}{2}.$$

In this case, $b^2 - 4ac = 0$, and the trinomial $4x^2 - 20x + 25$ is a perfect square. There is just one solution in the solution set $\left\{\frac{5}{2}\right\}$.

◀ *Work Problem* **4** *at the Side.*

⑤ Solve.

(a) $x^2 - \dfrac{4}{3}x + \dfrac{2}{3} = 0$

Note

The single solution of the equation in Example 4 is a rational number. If all solutions of a quadratic equation are rational, the equation can be solved by factoring as well.

EXAMPLE 5 **Solving a Quadratic Equation with Fractions**

Solve $\dfrac{1}{10}t^2 = \dfrac{2}{5}t - \dfrac{1}{2}$.

$$\frac{1}{10}t^2 = \frac{2}{5}t - \frac{1}{2}$$

$$10\left(\frac{1}{10}t^2\right) = 10\left(\frac{2}{5}t - \frac{1}{2}\right) \quad \begin{array}{l}\text{Clear fractions; multiply}\\ \text{by the LCD, 10.}\end{array}$$

(b) $x^2 - \dfrac{9}{5}x = \dfrac{2}{5}$

$$10\left(\frac{1}{10}t^2\right) = 10\left(\frac{2}{5}t\right) - 10\left(\frac{1}{2}\right) \quad \text{Distributive property}$$

$$t^2 = 4t - 5 \quad \text{Multiply.}$$

$$t^2 - 4t + 5 = 0 \quad \text{Standard form}$$

Identify $a = 1$, $b = -4$, and $c = 5$. By the quadratic formula,

$$t = \frac{-(-4) \pm \sqrt{(-4)^2 - 4(1)(5)}}{2(1)} \quad \text{Substitute in the quadratic formula.}$$

$$t = \frac{4 \pm \sqrt{16 - 20}}{2} \quad \text{Simplify.}$$

$$t = \frac{4 \pm \sqrt{-4}}{2} \quad \boxed{\text{Stop here.}}$$

Because $\sqrt{-4}$ does not represent a real number, the solution set is $\emptyset$.

◀ *Work Problem* **5** *at the Side.*

17.3 ▶▶▶ Exercises

Write each equation in standard form $ax^2 + bx + c = 0$, if necessary. Then identify the values of a, b, and c. Do not actually solve the equation. See Example 1.

1. $4x^2 + 5x - 9 = 0$

$a = $ _____, $b = $ _____, $c = $ _____

2. $8x^2 + 3x - 4 = 0$

$a = $ _____, $b = $ _____, $c = $ _____

3. $3x^2 = 4x + 2$

$a = $ _____, $b = $ _____, $c = $ _____

4. $5x^2 = 3x - 6$

$a = $ _____, $b = $ _____, $c = $ _____

5. $3x^2 = -7x$

$a = $ _____, $b = $ _____, $c = $ _____

6. $9x^2 = 8x$

$a = $ _____, $b = $ _____, $c = $ _____

Use the quadratic formula to solve each equation. Write all radicals in simplified form, and write all answers in lowest terms. See Examples 2–4.

7. $k^2 + 12k - 13 = 0$

8. $r^2 - 8r - 9 = 0$

9. $p^2 - 4p + 4 = 0$

10. $9x^2 + 6x + 1 = 0$

11. $2x^2 = 5 + 3x$

12. $2z^2 = 30 + 7z$

13. $2x^2 + 12x = -5$

14. $5m^2 + m = 1$

15. $6x^2 + 6x = 0$

16. $4n^2 - 12n = 0$

17. $-2x^2 = -3x + 2$

18. $-x^2 = -5x + 20$

19. $3x^2 + 5x + 1 = 0$

20. $6x^2 - 6x + 1 = 0$

21. $7x^2 = 12x$

22. $9r^2 = 11r$

23. $x^2 - 24 = 0$

24. $z^2 - 96 = 0$

25. $25x^2 - 4 = 0$

26. $16x^2 - 9 = 0$

27. $3x^2 - 2x + 5 = 10x + 1$

28. $4x^2 - x + 4 = x + 7$

29. $2x^2 + x + 5 = 0$

30. $3x^2 + 2x + 8 = 0$

31. If we apply the quadratic formula and find that the value of $b^2 - 4ac$ is negative, what can we conclude?

32. A student writes the quadratic formula as $x = -b \pm \dfrac{\sqrt{b^2 - 4ac}}{2a}$. Is this correct?
If not, explain the error, and give the correct formula.

Use the quadratic formula to solve each equation. See Example 5.

33. $\dfrac{3}{2}k^2 - k - \dfrac{4}{3} = 0$

34. $\dfrac{2}{5}x^2 - \dfrac{3}{5}x - 1 = 0$

35. $\dfrac{1}{2}x^2 + \dfrac{1}{6}x = 1$

36. $\dfrac{2}{3}t^2 - \dfrac{4}{9}t = \dfrac{1}{3}$

⊙ 37. $\dfrac{3}{8}x^2 - x + \dfrac{17}{24} = 0$

38. $\dfrac{1}{3}x^2 + \dfrac{8}{9}x + \dfrac{7}{9} = 0$

39. $0.6x - 0.4x^2 = -1$

40. $0.5x^2 = x + 0.5$

41. $0.25x^2 = -1.5x - 1$

Solve each problem.

42. An astronaut on the moon throws a baseball upward. The altitude (height) h of the ball, in feet, x seconds after he throws it, is given by the equation
$$h = -2.7x^2 + 30x + 6.5.$$
At what times is the ball 12 ft above the moon's surface? Give answer(s) to the nearest tenth.

43. A frog is sitting on a stump 3 ft above the ground. He hops off the stump and lands on the ground 4 ft away. During his leap, his height h is given by the equation
$$h = -0.5x^2 + 1.25x + 3,$$
where x is the distance in feet from the base of the stump, and h is in feet. How far was the frog from the base of the stump when he was 1.25 ft above the ground?

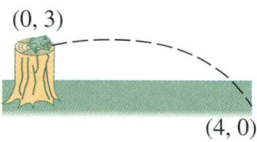

(0, 3)

(4, 0)

44. An old Babylonian problem asks for the length of the side of a square, given that the area of the square minus the length of a side is 870. Find the length of the side. (*Source:* Eves, H., *An Introduction to the History of Mathematics*, Sixth Edition, Saunders College Publishing, 1990.)

45. A rule for estimating the number of board feet of lumber that can be cut from a log depends on the diameter of the log. To find the diameter d required to get 9 board feet of lumber, we use the equation
$$\left(\dfrac{d - 4}{4}\right)^2 = 9.$$
Solve this equation for d. Are both answers reasonable?

17.4 ▶▶▶ Graphing Quadratic Equations

OBJECTIVE **1** **Graph quadratic equations.** In **Chapter 11,** we saw that the graph of a linear equation in two variables is a straight line that represents all the solutions of the equation. In this section, we graph quadratic equations in two variables, of the form

$$y = ax^2 + bx + c.$$

The simplest quadratic equation, $y = x^2$ (or $y = 1x^2 + 0x + 0$), can be graphed in much the same way that straight lines were graphed, by finding ordered pairs that satisfy the equation.

EXAMPLE 1 **Graphing a Quadratic Equation**

Graph $y = x^2$.

Select several values for x; then find the corresponding y-values. For example, selecting $x = 2$ and substituting in $y = x^2$ gives

$$y = 2^2 = 4,$$

and so the point $(2, 4)$ is on the graph of $y = x^2$. (***Recall that in an ordered pair such as (2, 4), the x-value comes first and the y-value second.***)

Work Problem **1** *at the Side.* ▶

If we plot the points from Problem 1 at the side on a coordinate system and draw a smooth curve through them, we obtain the graph in Figure 1. A table of values is shown with the graph.

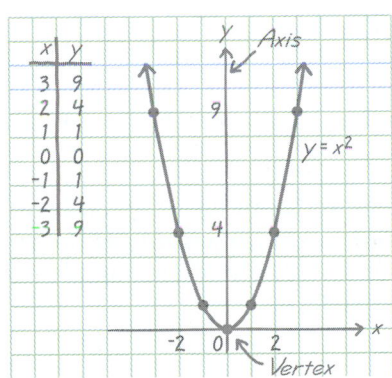

Figure 1

Work Problem **2** *at the Side.* ▶

The curve in Figure 1 is called a **parabola.** The point $(0, 0)$, the *lowest* point on this graph, is called the **vertex** of the parabola. The vertical line through the vertex (the y-axis here) is called the **axis** of the parabola. The axis of a parabola is a **line of symmetry** for the graph, because if the graph is folded on this line, the two halves will coincide.

Every equation of the form $y = ax^2 + bx + c$, with $a \neq 0$, has a graph that is a parabola. Because of its many useful properties, the parabola occurs frequently in real-life applications. For example, the cross sections of radar, spotlight, and telescope reflectors form parabolas.

OBJECTIVES

1 Graph quadratic equations.

2 Find the vertex of a parabola.

1 Complete the table of values for $y = x^2$.

x	y
3	
2	4
1	
0	
−1	
−2	
−3	

2 Graph $y = \frac{1}{2}x^2$ by first completing the table of values.

x	y
−2	
−1	
0	
1	
2	

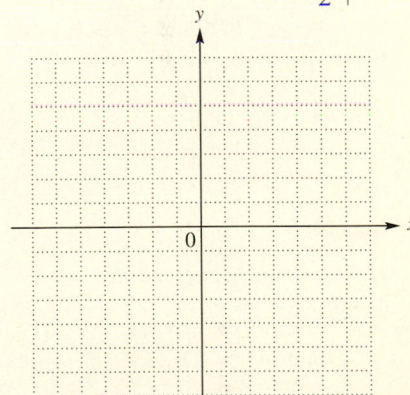

ANSWERS

1. See the table beside Figure 1.

2.
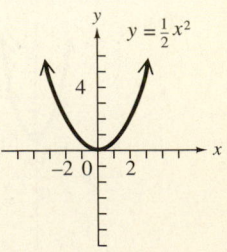

x	y
−2	2
−1	$\frac{1}{2}$
0	0
1	$\frac{1}{2}$
2	2

3 Complete each ordered pair for $y = -x^2 + 3$.

$(-2, \_\_)$, $(-1, \_\_)$,

$(1, \_\_)$, $(2, \_\_)$

4 Graph each equation, and identify each vertex.

(a) $y = -x^2 - 3$

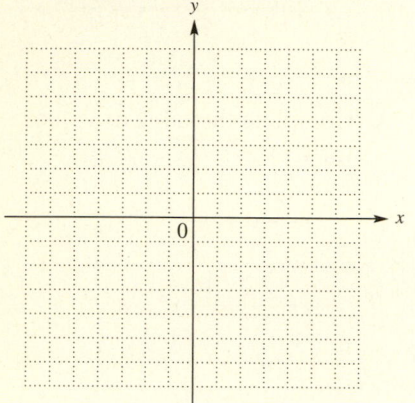

(b) $y = x^2 + 3$

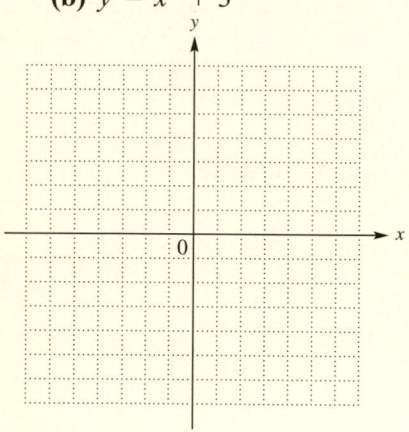

ANSWERS

3. $(-2, -1), (-1, 2), (1, 2), (2, -1)$

4. **(a)**

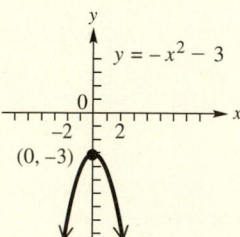

(b)

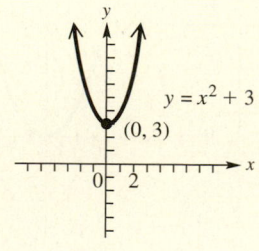

EXAMPLE 2 **Graphing a Parabola by Plotting Points**

Graph $y = -x^2 + 3$.

Find several ordered pairs. Let $x = 0$ to find the y-intercept:

$$y = -x^2 + 3 = -0^2 + 3 = 3.$$

This gives the ordered pair $(0, 3)$.

◀ Work Problem **3** at the Side.

The ordered pair $(0, 3)$ and the ordered pairs from Problem 3 at the side are listed in the table shown with Figure 2. Plot these points and connect them with a smooth curve as shown in Figure 2. The vertex of this parabola is $(0, 3)$. The graph opens downward because x^2 has a negative coefficient, so the vertex is the *highest* point of the graph.

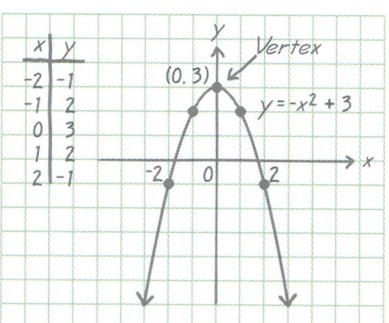

Figure 2

◀ Work Problem **4** at the Side.

OBJECTIVE **2** **Find the vertex of a parabola.** The vertex is the most important point to locate when graphing a quadratic equation.

EXAMPLE 3 **Finding the Vertex to Graph a Parabola**

Graph $y = x^2 - 2x - 3$.

We want to find the vertex of the graph. *Because of its symmetry, if a parabola has two x-intercepts, the x-value of the vertex is exactly halfway between them.* Therefore, we begin by finding the x-intercepts. We let $y = 0$ in the equation, and solve for x.

$$0 = x^2 - 2x - 3$$

$$0 = (x + 1)(x - 3) \qquad \text{Factor.}$$

$$x + 1 = 0 \quad \text{or} \quad x - 3 = 0 \qquad \text{Zero-factor property}$$

$$x = -1 \quad \text{or} \qquad x = 3$$

There are two x-intercepts, $(-1, 0)$ and $(3, 0)$. Since the x-value of the vertex is halfway between the x-values of the two x-intercepts, it is half their sum.

$$x = \frac{1}{2}(-1 + 3) = \frac{1}{2}(2) = 1 \longleftarrow x\text{-value of the vertex}$$

We find the corresponding y-value by substituting 1 for x in the equation.

$$y = 1^2 - 2(1) - 3 = -4 \longleftarrow y\text{-value of the vertex}$$

The vertex is $(1, -4)$. The axis is the line $x = 1$.

Continued on Next Page

To find the *y*-intercept, substitute 0 for *x* in the equation.

$$y = 0^2 - 2(0) - 3 = -3 \qquad \text{Let } x = 0.$$

The *y*-intercept is $(0, -3)$.

We plot the three intercepts and the vertex, and find additional ordered pairs as needed. For example, if $x = 2$, then

$$y = 2^2 - 2(2) - 3 = -3,$$

leading to the ordered pair $(2, -3)$. A table that includes the ordered pairs we found is shown with the graph in Figure 3.

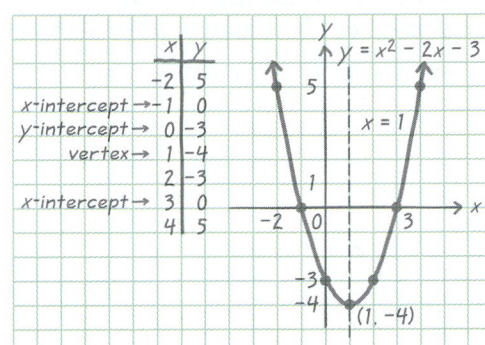

Figure 3

Work Problem **5** *at the Side.* ▶

We can generalize from Example 3. The *x*-coordinates of the *x*-intercepts for $y = ax^2 + bx + c$, by the quadratic formula, are

$$x = \frac{-b + \sqrt{b^2 - 4ac}}{2a} \quad \text{and} \quad x = \frac{-b - \sqrt{b^2 - 4ac}}{2a}.$$

Thus, the *x*-value of the vertex is

$$x = \frac{1}{2}\left(\frac{-b + \sqrt{b^2 - 4ac}}{2a} + \frac{-b - \sqrt{b^2 - 4ac}}{2a} \right)$$

$$x = \frac{1}{2}\left(\frac{-b + \sqrt{b^2 - 4ac} - b - \sqrt{b^2 - 4ac}}{2a} \right)$$

$$x = \frac{1}{2}\left(\frac{-2b}{2a} \right) \qquad \text{Combine like terms.}$$

$$x = -\frac{b}{2a}. \qquad \text{Multiply; lowest terms}$$

For the equation in Example 3, $y = x^2 - 2x - 3$, $a = 1$, and $b = -2$. Thus, the *x*-value of the vertex is

$$x = -\frac{b}{2a} = -\frac{-2}{2(1)} = 1,$$

which is the same *x*-value for the vertex we found in Example 3. (The *x*-value of the vertex is $x = -\frac{b}{2a}$, even if the graph has no *x*-intercepts.)

5 Graph $y = x^2 + 2x - 8$.

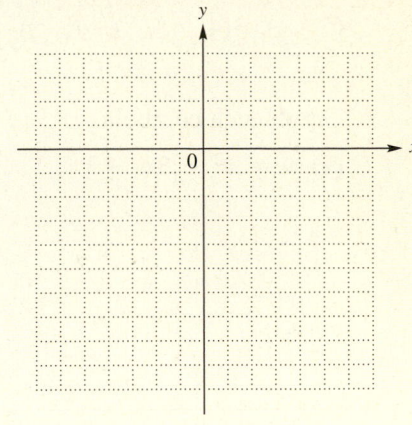

ANSWER

5.

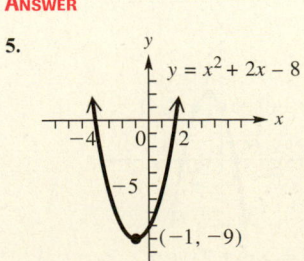

6 Complete each ordered pair for $y = x^2 - 4x + 1$.

$(5,\_\_)$, $(4,\_\_)$, $(-1,\_\_)$

7 Graph each parabola.

(a) $y = x^2 - 3x - 3$

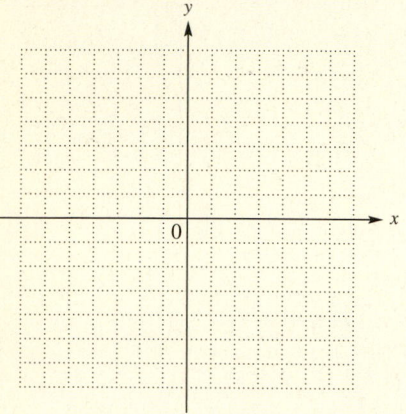

(b) $y = -x^2 + 2x + 4$

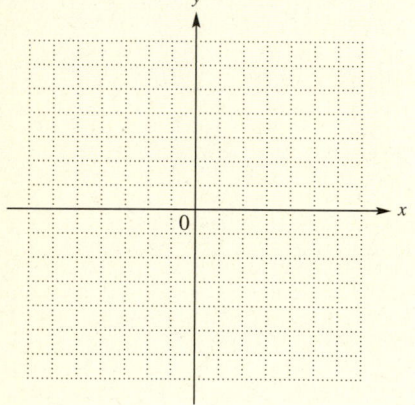

ANSWERS

6. $(5, 6), (4, 1), (-1, 6)$

7. **(a)**

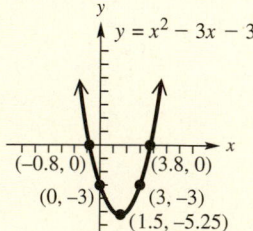

(b)

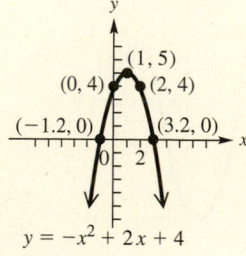

EXAMPLE 4 **Graphing a Parabola**

Graph $y = x^2 - 4x + 1$.

Find the vertex. Here, $a = 1$ and $b = -4$, so the x-value of the vertex is

$$x = -\frac{b}{2a} = -\frac{-4}{2(1)} = \mathbf{2}.$$

The y-value is

$$y = \mathbf{2}^2 - 4(\mathbf{2}) + 1 = \mathbf{-3},$$

so the vertex is $(\mathbf{2}, \mathbf{-3})$. The axis is the line $x = \mathbf{2}$.

Now find the intercepts. Let $x = 0$ in $y = x^2 - 4x + 1$.

$$y = \mathbf{0}^2 - 4(\mathbf{0}) + 1 = \mathbf{1}$$

The y-intercept is $(0, \mathbf{1})$. Let $y = 0$ to get the x-intercepts. If $y = 0$, the equation becomes $0 = x^2 - 4x + 1$, which cannot be solved by factoring. Use the quadratic formula to solve for x.

$$x = \frac{-(-4) \pm \sqrt{(-4)^2 - 4(1)(1)}}{2(1)} \quad \text{Let } a = 1, b = -4, c = 1 \text{ in the quadratic formula.}$$

$$x = \frac{4 \pm \sqrt{12}}{2} \quad \text{Simplify.}$$

$$x = \frac{4 \pm 2\sqrt{3}}{2} \quad \sqrt{12} = \sqrt{4} \cdot \sqrt{3} = 2\sqrt{3}$$

$$x = \frac{\mathbf{2}(2 \pm \sqrt{3})}{\mathbf{2}} \quad \text{Factor.}$$

$$x = \mathbf{2} \pm \sqrt{3} \quad \text{Divide out 2.}$$

Use a calculator to find that the x-intercepts are $(\mathbf{3.7}, 0)$ and $(\mathbf{0.3}, 0)$, to the nearest tenth.

◀ *Work Problem* **6** *at the Side.*

Plot the intercepts, vertex, and the points found in Problem 6. Connect these points with a smooth curve. The graph is shown in Figure 4.

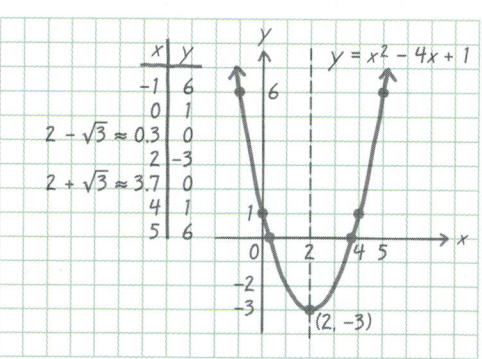

Figure 4

◀ *Work Problem* **7** *at the Side.*

17.4 ▶▶▶ Exercises

1. In your own words, explain what is meant by the vertex of a parabola.

2. In your own words, explain what is meant by the line of symmetry of a parabola that opens upward or downward.

Graph each equation. Give the coordinates of the vertex in each case. See Examples 1–4.

3. $y = 2x^2$

4. $y = 3x^2$

5. $y = x^2 - 4$

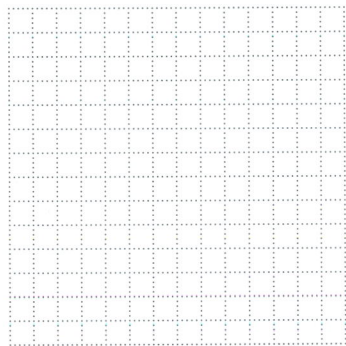

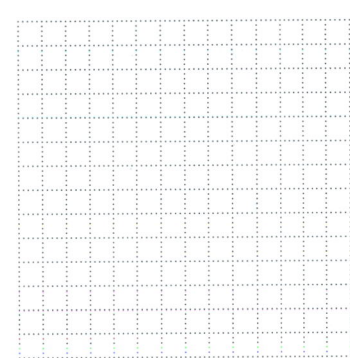

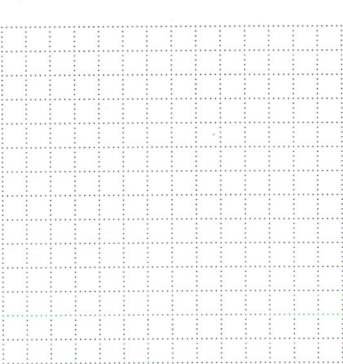

6. $y = x^2 - 6$

⊙ **7.** $y = -x^2 + 2$

8. $y = -x^2 + 4$

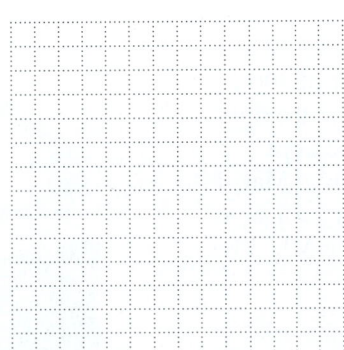

9. $y = (x + 3)^2$

10. $y = (x - 4)^2$

◉ **11.** $y = x^2 + 2x + 3$

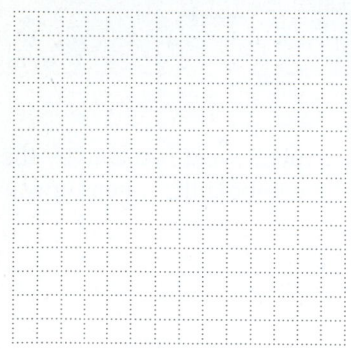  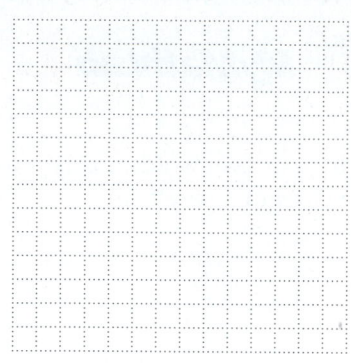

12. $y = x^2 - 4x + 3$

13. $y = -x^2 + 6x - 5$

14. $y = -x^2 - 4x - 3$

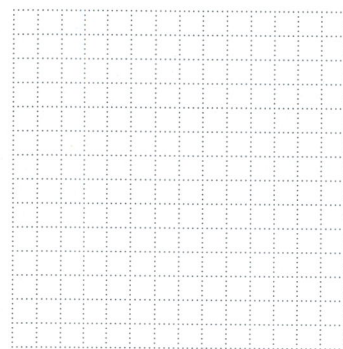  

15. Based on your work in Exercises 3–14, what seems to be the direction in which the parabola $y = ax^2 + bx + c$ opens if $a > 0$?   if $a < 0$?

16. See Exercises 10–12. How many real solutions does a quadratic equation have if its corresponding graph has

(a) no x-intercepts

(b) one x-intercept

(c) two x-intercepts?

Solve each problem.

17. The U.S. Naval Research Laboratory designed a
◉ giant radio telescope that had a diameter of 300 ft and maximum depth of 44 ft. The graph on the right below describes a cross section of this telescope. Find the equation of this parabola. (*Source:* Mar, J. and Liebowitz, H., *Structure Technology for Large Radio and Radar Telescope Systems*, The MIT Press, 1969.)

18. Suppose the telescope in Exercise 17 had a diameter of 400 ft and maximum depth of 50 ft. Find the equation of this parabola.

17.5 ⟩⟩⟩   Introduction to Functions

If gasoline costs $3.00 per gal and you buy **1** gal, then you must pay $3.00(**1**) = $3.00. If you buy **2** gal, your cost is $3.00(**2**) = $6.00; for **3** gal, your cost is $3.00(**3**) = $9.00, and so on. Generalizing, if *x* represents the number of gallons, then the cost is $3.00*x*. If we let *y* represent the cost, then the equation

$$y = 3.00x$$

relates the number of gallons, *x*, to the cost in dollars, *y*. The set of ordered pairs (*x*, *y*) that satisfy this equation forms a *relation*.

OBJECTIVE 1 Understand the definition of a relation.   In an ordered pair (*x*, *y*), *x* and *y* are called the **components** of the ordered pair. Any set of ordered pairs is called a **relation.** The set of all first components in the ordered pairs of a relation is the **domain** of the relation, and the set of all second components in the ordered pairs is the **range** of the relation.

EXAMPLE 1   Using Ordered Pairs to Define Relations

(a) {(0, 1), (2, 5), (3, 8), (4, 2)}
This relation has domain {0, 2, 3, 4} and range {1, 2, 5, 8}. The correspondence between the elements of the domain and the elements of the range is shown in Figure 5.

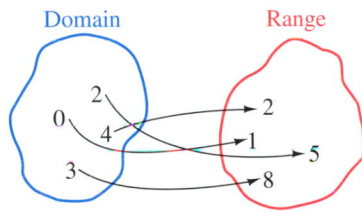

Figure 5

(b) {(**3**, **5**), (**3**, **6**), (**3**, **7**), (**3**, **8**)}
The relation has domain {**3**} and range {**5**, **6**, **7**, **8**}.

⸻⸻⸻⸻⸻⸻   *Work Problem* **1** *at the Side.* ▶

OBJECTIVE 2 Understand the definition of a function.   We now investigate an important type of relation, called a *function.*

> **Function**
>
> A **function** is a set of ordered pairs in which each distinct first component corresponds to exactly one second component.

By this definition, the relation in Example 1(a) is a function. The relation in Example 1(b) is *not* a function, because the first component, 3, corresponds to more than one second component. Notice, however, that if the components of the ordered pairs in Example 1(b) were interchanged, giving the relation

$$\{(5, 3), (6, 3), (7, 3), (8, 3)\},$$

then the relation *would* be a function. In that case, each domain element (first component) corresponds to *exactly one* range element (second component).

OBJECTIVES

1 Understand the definition of a relation.

2 Understand the definition of a function.

3 Decide whether an equation defines a function.

4 Use function notation.

5 Apply the function concept in an application.

1 Give the domain and the range of each relation.

(a) {(5, 10), (15, 20), (25, 30), (35, 40)}

(b) {(1, 4), (2, 4), (3, 4)}

ANSWERS

1. (a) domain: {5, 15, 25, 35};
      range: {10, 20, 30, 40}
   (b) domain: {1, 2, 3};
      range: {4}

2 Decide whether each relation is a function.

(a) $\{(-2, 8), (-1, 1), (0, 0), (1, 1), (2, 8)\}$

(b) $\{(5, 2), (5, 1), (5, 0)\}$

EXAMPLE 2   **Determining Whether Relations Are Functions**

Determine whether each relation is a function.

(a) $\{(-2, 4), (-1, 1), (0, 0), (1, 1), (2, 4)\}$

Notice that each first component appears once and only once. Because of this, the relation is a function.

(b) $\{(\mathbf{9}, 3), (\mathbf{9}, -3), (4, 2)\}$

The first component 9 appears in two ordered pairs, and corresponds to two different second components. Therefore, this relation is not a function.

◀ *Work Problem* **2** *at the Side.*

The simple relations given in Examples 1 and 2 were defined by listing the ordered pairs or by showing the correspondence with a figure. Most useful functions have an infinite number of ordered pairs and are usually defined with equations that tell how to get the second components, given the first. We have been using equations with x and y as the variables, where x represents the first component (input) and y the second component (output) in the ordered pairs.

Here are some everyday examples of functions.

1. The **cost y** in dollars charged by an express mail company is a function of the **weight x** in pounds determined by the equation $y = 1.5(x - 1) + 9$.

2. In one state, the sales tax is 6% of the price of an item. The **tax y** on a particular item is a function of the **price x**, because $y = 0.06x$.

3. The **distance d** traveled by a car moving at a constant speed of 45 mph is a function of the **time t**. Thus, $d = 45t$.

The function concept can be illustrated by an input-output "machine," as seen in Figure 6. It shows how the express mail company equation $y = 1.5(x - 1) + 9$ provides an output (the cost, represented by y) for a given input (the weight in pounds, given by x).

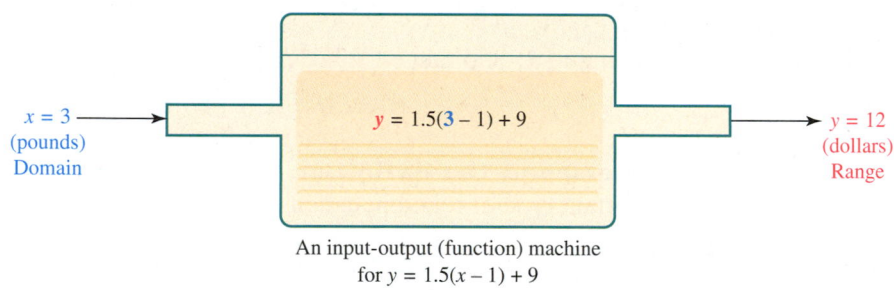

$x = 3$ (pounds) Domain → $y = 1.5(3 - 1) + 9$ → $y = 12$ (dollars) Range

An input-output (function) machine
for $y = 1.5(x - 1) + 9$

Figure 6

OBJECTIVE **3** **Decide whether an equation defines a function.** Given the graph of an equation, the definition of a function can be used to decide whether or not the graph represents a function. By the definition of a function, each x-value must lead to exactly one y-value. In Figure 7(a) on the next page, the indicated x-value leads to two y-values, so this graph is not the graph of a function. A vertical line can be drawn that intersects this graph in more than one point.

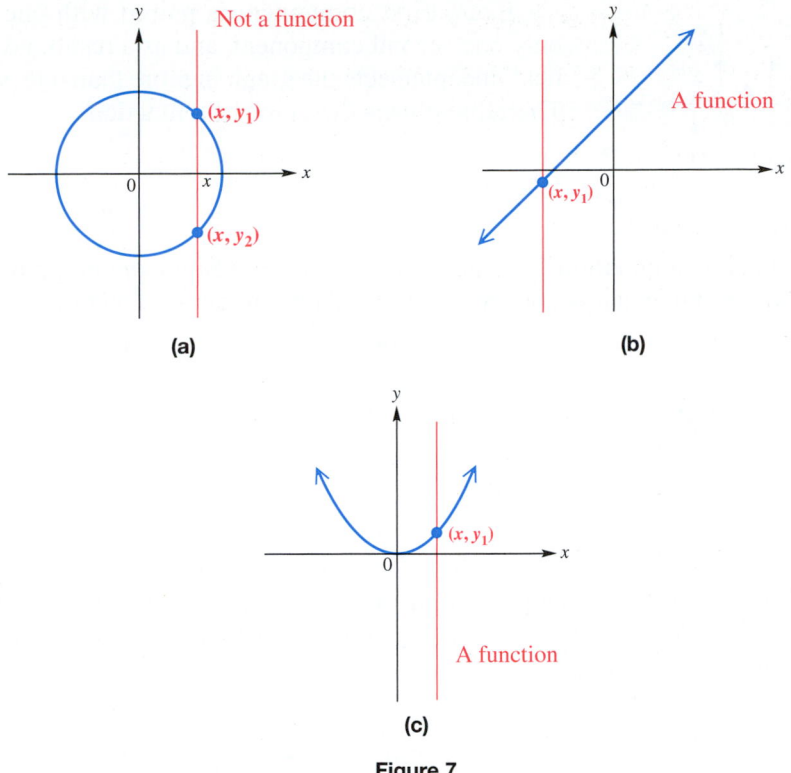

Figure 7

On the other hand, in Figures 7(b) and 7(c) any vertical line will intersect each graph in no more than one point. Because of this, the graphs in Figures 7(b) and 7(c) are graphs of functions. This idea leads to the **vertical line test** for a function.

> **Vertical Line Test**
>
> If a vertical line intersects a graph in more than one point, the graph is not the graph of a function.

As Figure 7(b) suggests, any nonvertical line is the graph of a function. For this reason, *any linear equation of the form $y = mx + b$ defines a function.* (Recall that a vertical line has undefined slope.) Also, any vertical parabola, as in Figure 7(c), is the graph of a function, so *any quadratic equation of the form $y = ax^2 + bx + c \ (a \neq 0)$ defines a function.*

EXAMPLE 3 **Deciding Whether Relations Define Functions**

Decide whether each relation graphed or defined is a function.

(a)

Because there are two ordered pairs with first component -4, as shown in red, this is not the graph of a function.

Continued on Next Page

3 Decide whether each relation graphed or defined is a function.

(a)

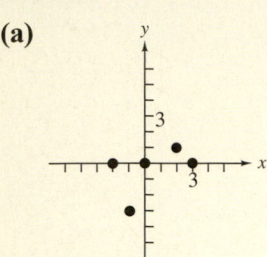

(b)

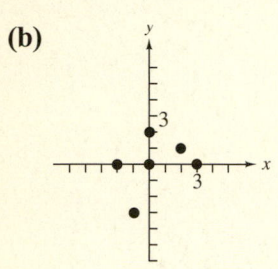

(c)

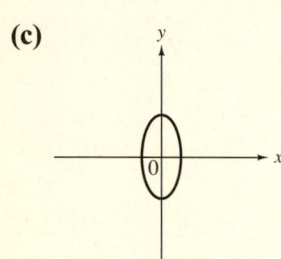

(d)

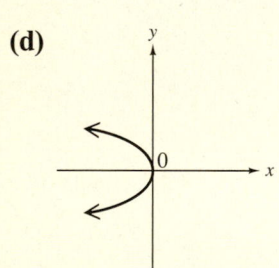

(e) $y = 3$

(b)

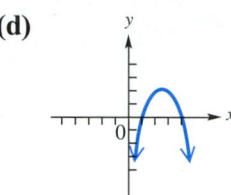

Every first component is paired with one and only one second component, and as a result, no vertical line intersects the graph in more than one point. Therefore, this is the graph of a function.

(c) $y = 2x - 9$

This linear equation is in the form $y = mx + b$. Since the graph of this equation is a line that is not vertical, the equation defines a function.

(d)

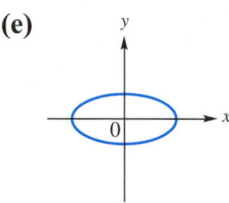

(e)

Use the vertical line test. Any vertical line intersects the graph of a vertical parabola just once, so this is the graph of a function.

The vertical line test shows that this graph is not the graph of a function; a vertical line could intersect the graph twice.

(f) $x = 4$

The graph of $x = 4$ is a vertical line, so the equation does *not* define a function. (Every ordered pair has x-value 4.)

◀ Work Problem **3** at the Side.

OBJECTIVE **4** **Use function notation.**   The letters f, g, and h are commonly used to name functions. For example, the function with defining equation $y = 3x + 5$ may be written

$$f(x) = 3x + 5,$$

where $f(x)$ is read "f of x." The notation $f(x)$ is another way of writing y in a function. For the function defined by $f(x) = 3x + 5$, if $x = 7$ then

$$f(7) = 3 \cdot 7 + 5 \quad \text{Let } x = 7.$$
$$f(7) = 21 + 5 \quad \text{Multiply.}$$
$$f(7) = 26. \quad \text{Add.}$$

Read this result, $f(7) = 26$, as "f of 7 equals 26." The notation $f(7)$ means the value of y when x is 7. The statement $f(7) = 26$ says that the value of y is 26 when x is 7. It also indicates that the point $(7, 26)$ lies on the graph of f.

To find $f(-3)$ for $f(x) = 3x + 5$, substitute -3 for x.

Use parentheses to avoid errors.

$$f(-3) = 3(-3) + 5 \quad \text{Let } x = -3.$$
$$f(-3) = -9 + 5$$
$$f(-3) = -4$$

CAUTION
The notation $f(x)$ does *not* mean f times x. *The symbol f(x) means the value of the function f when evaluated for x. It represents the y-value that corresponds to x.*

Function Notation

In the notation $f(x)$,

f is the name of the function,

x is the domain value,

and $f(x)$ is the range value y for the domain value x.

EXAMPLE 4   **Using Function Notation**

For the function defined by $f(x) = x^2 - 3$, find the following.

(a) $f(4)$

Substitute 4 for x.

$$f(x) = x^2 - 3$$

$$f(4) = 4^2 - 3 \qquad \text{Let } x = 4.$$

Think:
$4^2 = 4 \cdot 4$

$$= 16 - 3 \qquad \text{Apply the exponent.}$$

$$f(4) = 13 \qquad \text{Subtract.}$$

(b) $f(0) = 0^2 - 3$ **(c)** $f(-3) = (-3)^2 - 3$

    $f(0) = 0 - 3$ $f(-3) = 9 - 3$

    $f(0) = -3$ $f(-3) = 6$

Work Problem (**4**) *at the Side.* ▶

OBJECTIVE 5 **Apply the function concept in an application.**
Because a function assigns to each element in its domain exactly one element in its range, the function concept is used in real-data applications where two quantities are related. Our final example discusses such an application using a table of values.

EXAMPLE 5   **Applying the Function Concept to Population**

Asian-American populations (in millions) are shown in the table.

ASIAN-AMERICAN POPULATION

Year	Population (in millions)
1996	9.7
2000	11.2
2004	13.1
2006	14.9

Source: U.S. Census Bureau.

(a) Use the table to write a set of ordered pairs that defines a function f.

If we choose the years as the domain elements and the populations as the range elements, the information in the table can be written as a set of four ordered pairs. In set notation, the function f is

$$f = \{(1996, 9.7), (2000, 11.2), (2004, 13.1), (2006, 14.9)\}.$$

Continued on Next Page

4 For $f(x) = 6x - 2$, find each function value.

(a) $f(-1)$

(b) $f(0)$

(c) $f(1)$

5 The numbers of U.S. children (in millions) educated at home for selected years are given in the table.

School Year	Number of Children
1997	1.1
1999	1.3
2001	1.7
2003	2.2

Source: National Home Education Research Institute, Salem, OR.

(a) Write a set of ordered pairs that defines a function *f* for the data.

(b) Give the domain and range of *f*.

(c) Find $f(1999)$.

(d) For what *x*-value does $f(x)$ equal 1.7 million?

(b) What is the domain of *f*? What is the range?

We repeat the table and the set of ordered pairs from part (a),

ASIAN-AMERICAN POPULATION

Year	Population (in millions)
1996	9.7
2000	11.2
2004	13.1
2006	14.9

Source: U.S. Census Bureau.

$$f = \{(1996, 9.7), (2000, 11.2), (2004, 13.1), (2006, 14.9)\}$$

The domain is the set of years, or *x*-values:

$$\{1996, 2000, 2004, 2006\}.$$

The range is the set of populations, in millions, or *y*-values:

$$\{9.7, 11.2, 13.1, 14.9\}.$$

(c) Find $f(1996)$ and $f(2004)$.

We refer to the table or the ordered pairs repeated in part (b) to find that

$$f(\mathbf{1996}) = 9.7 \text{ million} \quad \text{and} \quad f(\mathbf{2004}) = 13.1 \text{ million}.$$

(d) For what *x*-value does $f(x)$ equal 14.9 million? 11.2 million?

We use the table or the ordered pairs found in part (b) to determine

$$f(\mathbf{2006}) = 14.9 \text{ million} \quad \text{and} \quad f(\mathbf{2000}) = 11.2 \text{ million}.$$

◀ *Work Problem* **5** *at the Side.*

ANSWERS

5. **(a)** $f = \{(1997, 1.1), (1999, 1.3),$ $(2001, 1.7), (2003, 2.2)\}$
(b) domain: $\{1997, 1999, 2001, 2003\}$; range: $\{1.1, 1.3, 1.7, 2.2\}$
(c) 1.3 million
(d) 2001

Complete the following table for the function defined by $f(x) = x + 2$.

x	x + 2	f(x)	(x, y)
0	2	2	(0, 2)
1. 1			
2. 2			
3. 3			
4. 4			

5. Describe the graph of function f in Exercises 1–4 if the domain is $\{0, 1, 2, 3, 4\}$.

6. Describe the graph of function f in Exercises 1–4 if the domain is the set of all real numbers.

Determine whether each relation is a function. Give the domain and the range in Exercises 7–12. See Examples 1–3.

7. $\{(-4, 3), (-2, 1), (0, 5), (-2, -8)\}$

8. $\{(3, 7), (1, 4), (0, -2), (-1, -1), (-2, 5)\}$

9.

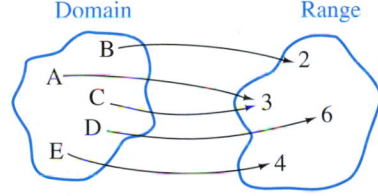

 10.

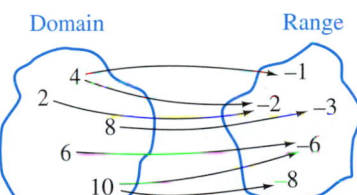

11.

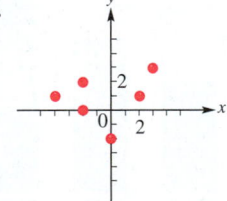

12.

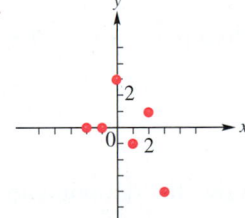

 13.

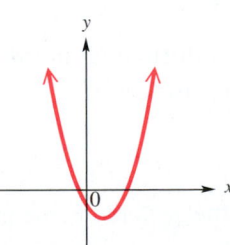

14.

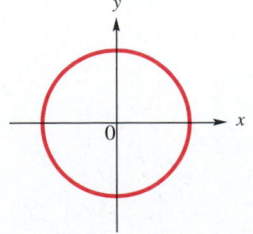

 15.

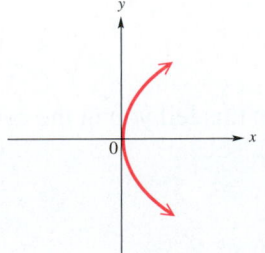

16.

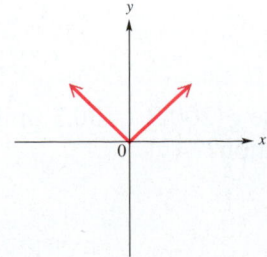

Decide whether each equation defines y as a function of x. (Remember that to be a function, every value of x must give one and only one value of y.) See Example 3.

17. $y = 5x + 3$ **18.** $y = -7x + 12$ **19.** $x = |y|$ **20.** $x = y^2$

Relating Concepts (Exercises 21–24) For Individual or Group Work

*A function defined by $f(x) = 3x - 4$, called a **linear function** because its graph is a straight line, can be graphed by replacing $f(x)$ with y and then using the methods described earlier. Let us assume that some function is written in the form $f(x) = mx + b$, for particular values of m and b.* **Work Exercises 21–24 in order.**

21. Since $f(2) = 4$, name the coordinates of one point on the line.

22. Since $f(-1) = -4$, name the coordinates of another point on the line.

23. Use the results of Exercises 21 and 22 to find the slope of the line.

24. Use the slope-intercept form of the equation of a line to write the function in the form $f(x) = mx + b$.

For each function f, find **(a)** $f(2)$, **(b)** $f(0)$, *and* **(c)** $f(-3)$. *See Example 4.*

25. $f(x) = 4x + 3$ **26.** $f(x) = -3x + 5$ **27.** $f(x) = x^2 - x + 2$

28. $f(x) = x^3 + x$ **29.** $f(x) = |x|$ **30.** $f(x) = |x + 7|$

The number of U.S. foreign-born residents has grown by more than 43% since 1990. The graph shows the number of such residents (in millions) for selected years. Use the information in the graph for Exercises 31–35. See Example 5.

31. Write the information in the graph as a set of ordered pairs. Does this set define a function?

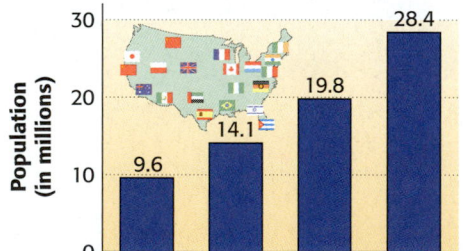

Growth of U.S. Foreign-Born Population

Source: U.S. Census Bureau.

32. Suppose that g is the name given to this relation. Give the domain and range of g.

33. Find $g(1980)$ and $g(1990)$.

34. For what value of x does $g(x) = 28.4$ (million)?

35. Suppose $g(2002) = 30.3$ (million). What does this tell you in the context of the application?

Chapter 17 ▶▶▶ Summary

▶ Key Terms

17.4	**parabola**	The graph of the quadratic equation $y = ax^2 + bx + c$ is called a parabola.
	vertex	The vertex of a parabola that opens upward or downward is the lowest or highest point on the graph.
	axis	The axis of a parabola that opens upward or downward is a vertical line through the vertex.
	line of symmetry	If a graph is folded on its line of symmetry, the two sides coincide.
17.5	**components**	In an ordered pair (x, y), x and y are the components.
	relation	Any set of ordered pairs is called a relation.
	domain	The set of all first components in the ordered pairs of a relation is the domain of the relation.
	range	The set of all second components in the ordered pairs of a relation is the range of the relation.
	function	A function is a set of ordered pairs in which each first component corresponds to exactly one second component.

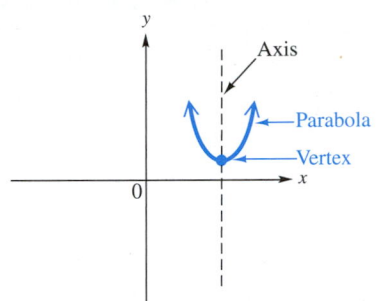

▶ New Symbols

$\pm$	positive or negative (plus or minus)
$f(x)$	function f of x

▶ Test Your Word Power

See how well you have learned the vocabulary in this chapter. Answers, with examples, follow the Quick Review.

1. A **parabola** is the graph of
   A. any equation in two variables
   B. a linear equation
   C. an equation of degree three
   D. a quadratic equation in two variables.

2. The **vertex** of a parabola is
   A. the point where the graph intersects the y-axis
   B. the point where the graph intersects the x-axis
   C. the lowest point on a parabola that opens up or the highest point on a parabola that opens down
   D. the origin.

3. A **relation** is
   A. any set of ordered pairs
   B. a set of ordered pairs in which each distinct first component corresponds to exactly one second component

   C. two sets of ordered pairs that are related
   D. a graph of ordered pairs.

4. The **domain** of a relation is
   A. the set of all x- and y-values in the ordered pairs of the relation
   B. the difference between the components in an ordered pair of the relation
   C. the set of all first components in the ordered pairs of the relation
   D. the set of all second components in the ordered pairs of the relation.

5. The **range** of a relation is
   A. the set of all x- and y-values in the ordered pairs of the relation
   B. the difference between the components in an ordered pair of the relation

   C. the set of all first components in the ordered pairs of the relation
   D. the set of all second components in the ordered pairs of the relation.

6. A **function** is
   A. any set of ordered pairs
   B. a set of ordered pairs in which each distinct first component corresponds to exactly one second component
   C. two sets of ordered pairs that are related
   D. a graph of ordered pairs.

▶ Quick Review

Concepts	Examples

[17.1] Solving Quadratic Equations by the Square Root Property

Square Root Property

If k is positive, and if $x^2 = k$, then

$$x = \sqrt{k} \quad \text{or} \quad x = -\sqrt{k}.$$

The solution set $\{-\sqrt{k}, \sqrt{k}\}$ can also be written $\{\pm\sqrt{k}\}$.

Solve $(2x + 1)^2 = 5$.

$$2x + 1 = \sqrt{5} \qquad \text{or} \quad 2x + 1 = -\sqrt{5}$$
$$2x = -1 + \sqrt{5} \quad \text{or} \qquad 2x = -1 - \sqrt{5}$$
$$x = \frac{-1 + \sqrt{5}}{2} \quad \text{or} \qquad x = \frac{-1 - \sqrt{5}}{2}$$

Solution set: $\left\{\dfrac{-1 + \sqrt{5}}{2}, \dfrac{-1 - \sqrt{5}}{2}\right\}$

[17.2] Solving Quadratic Equations by Completing the Square

Solving a Quadratic Equation by Completing the Square

Step 1 If the coefficient of the second-degree term is 1, go to Step 2. If it is not 1, divide each side of the equation by this coefficient.

Step 2 Make sure that all variable terms are on one side of the equation and all constant terms are on the other.

Step 3 Take half the coefficient of x, square it, and add the square to each side of the equation. Factor the variable side, and combine terms on the other side.

Step 4 Use the square root property to solve the equation.

Solve $2x^2 + 4x - 1 = 0$.

$$x^2 + 2x - \frac{1}{2} = 0 \qquad \text{Divide by 2.}$$

$$x^2 + 2x = \frac{1}{2} \qquad \text{Add } \tfrac{1}{2}.$$

$$x^2 + 2x + 1 = \frac{1}{2} + 1 \qquad \left[\tfrac{1}{2}(2)\right]^2 = 1^2 = 1$$

$$(x + 1)^2 = \frac{3}{2} \qquad \text{Factor; add.}$$

$$x + 1 = \sqrt{\frac{3}{2}} \qquad \text{or} \quad x + 1 = -\sqrt{\frac{3}{2}}$$

$$x + 1 = \frac{\sqrt{3}\cdot\sqrt{2}}{\sqrt{2}\cdot\sqrt{2}} \quad \text{or} \quad x + 1 = -\frac{\sqrt{3}\cdot\sqrt{2}}{\sqrt{2}\cdot\sqrt{2}}$$

$$x + 1 = \frac{\sqrt{6}}{2} \qquad \text{or} \quad x + 1 = -\frac{\sqrt{6}}{2}$$

$$x = -1 + \frac{\sqrt{6}}{2} \quad \text{or} \qquad x = -1 - \frac{\sqrt{6}}{2}$$

$$x = \frac{-2}{2} + \frac{\sqrt{6}}{2} \quad \text{or} \qquad x = \frac{-2}{2} - \frac{\sqrt{6}}{2}$$

$$x = \frac{-2 + \sqrt{6}}{2} \quad \text{or} \qquad x = \frac{-2 - \sqrt{6}}{2}$$

Solution set: $\left\{\dfrac{-2 + \sqrt{6}}{2}, \dfrac{-2 - \sqrt{6}}{2}\right\}$

Concepts

Examples

17.3 **Solving Quadratic Equations by the Quadratic Formula**

Quadratic Formula

The solutions of $ax^2 + bx + c = 0\,(a \neq 0)$ are

$$x = \frac{-b \pm \sqrt{b^2 - 4ac}}{2a}.$$

Solve $3x^2 - 4x - 2 = 0$.

$$x = \frac{-(-4) \pm \sqrt{(-4)^2 - 4(3)(-2)}}{2(3)}$$

$$x = \frac{4 \pm \sqrt{40}}{6} \qquad \text{Simplify.}$$

$$x = \frac{4 \pm 2\sqrt{10}}{6} \qquad \sqrt{40} = \sqrt{4} \cdot \sqrt{10} = 2\sqrt{10}$$

$$x = \frac{2(2 \pm \sqrt{10})}{2(3)} \qquad \text{Factor.}$$

$$x = \frac{2 \pm \sqrt{10}}{3} \qquad \text{Divide out 2.}$$

Solution set: $\left\{ \dfrac{2 + \sqrt{10}}{3}, \dfrac{2 - \sqrt{10}}{3} \right\}$

17.4 **Graphing Quadratic Equations**

To graph $y = ax^2 + bx + c$:

Step 1 Find the vertex. Use $x = -\frac{b}{2a}$ and find y by substituting this value for x in the equation.

Graph $y = 2x^2 - 5x - 3$.

$$x = -\frac{b}{2a} = -\frac{-5}{2(2)} = \frac{5}{4} \qquad a = 2, b = -5$$

$$y = 2\left(\frac{5}{4}\right)^2 - 5\left(\frac{5}{4}\right) - 3$$

$$y = 2\left(\frac{25}{16}\right) - \frac{25}{4} - 3$$

$$y = \frac{25}{8} - \frac{50}{8} - \frac{24}{8}$$

$$y = -\frac{49}{8}$$

The vertex is $\left(\frac{5}{4}, -\frac{49}{8}\right)$.

Step 2 Let $x = 0$ to find the y-intercept.

$$y = 2(0)^2 - 5(0) - 3 = -3$$

The y-intercept is $(0, -3)$.

Step 3 Let $y = 0$ to find the x-intercepts (if they exist).

$$0 = 2x^2 - 5x - 3 \qquad \text{Let } y = 0.$$

$$0 = (2x + 1)(x - 3)$$

$$2x + 1 = 0 \qquad \text{or} \quad x - 3 = 0$$

$$2x = -1 \qquad \text{or} \qquad x = 3$$

$$x = -\frac{1}{2}$$

The x-intercepts are $\left(-\frac{1}{2}, 0\right)$ and $(3, 0)$.

Concepts

17.4 Graphing Quadratic Equations (continued)

Step 4   Plot the intercepts and the vertex.

Step 5   Find and plot additional ordered pairs near the vertex and intercepts as needed.

Examples

x	y
$-\frac{1}{2}$	0
0	-3
1	-6
$\frac{5}{4}$	$-\frac{49}{8}$
2	-5
3	0

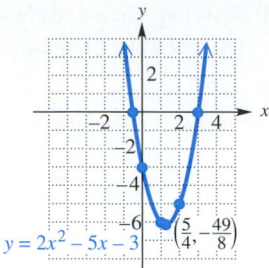

$y = 2x^2 - 5x - 3$ $\left(\frac{5}{4}, -\frac{49}{8}\right)$

17.5 Introduction to Functions

Vertical Line Test
If a vertical line intersects a graph in more than one point, the graph is not the graph of a function.

By the vertical line test, the graph shown is not the graph of a function.

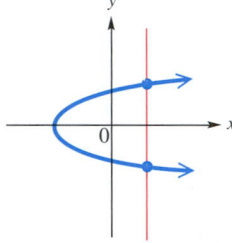

The set of all first components in the ordered pairs of a relation is the **domain** of the relation, and the set of all second components in the ordered pairs is the **range** of the relation.

The function

$$\{(10, 5), (20, 15), (30, 25)\}$$

has domain $\{10, 20, 30\}$ and range $\{5, 15, 25\}$.

To find $f(x)$ for a specific value of x, replace x by that value in the expression for the function f.

If $f(x) = 2x + 7$, then

$$f(\mathbf{3}) = 2(\mathbf{3}) + 7$$
$$f(3) = 13.$$

ANSWERS TO TEST YOUR WORD POWER

1. D; *Examples*: See Figures 1–4 in **Section 17.4**.
2. C; *Example*: The graph of $y = (x + 3)^2$ has vertex $(-3, 0)$, which is the lowest point on the graph.
3. A; *Example*: $\{(0, 2), (2, 4), (3, 6), (-1, 3)\}$
4. C; *Example*: The domain in the relation given in Answer 3 is the set of x-values $\{0, 2, 3, -1\}$.
5. D; *Example*: The range of the relation given in Answer 3 is the set of y-values $\{2, 4, 6, 3\}$.
6. B; *Example*: The relation given in Answer 3 is a function since each x-value corresponds to exactly one y-value.

Chapter 17 ▶▶▶ Review Exercises

[17.1] *Solve each equation by using the square root property. Express all radicals in simplest form.*

1. $y^2 = 144$

2. $x^2 = 37$

3. $m^2 = 128$

4. $(k + 2)^2 = 25$

5. $(r - 3)^2 = 10$

6. $(2p + 1)^2 = 14$

7. $(3k + 2)^2 = -3$

8. $(3x + 5)^2 = 0$

[17.2] *Solve each equation by completing the square.*

9. $m^2 + 6m + 5 = 0$

10. $p^2 + 4p = 7$

11. $-x^2 + 5 = 2x$

12. $2x^2 - 3 = -8x$

13. $4(x^2 + 7x) + 29 = -20$

14. $(4x + 1)(x - 1) = -7$

Solve each problem.

15. If an object is projected upward on Earth from a height of 50 ft, with an initial velocity of 32 ft per sec, then its height h after t seconds is given by $h = -16t^2 + 32t + 50$, where h is in feet. At what time(s) will it be at a height of 30 ft?

16. Find the lengths of the three sides of the right triangle shown.

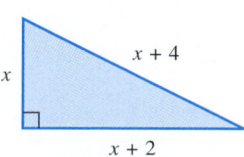

17. What must be added to $x^2 + 3x$ to make it a perfect square?

[17.3]

18. Consider the equation $x^2 - 9 = 0$.

    (a) Solve the equation by factoring.

    (b) Solve the equation by the square root property.

    (c) Solve the equation by the quadratic formula.

    (d) Compare your answers. If a quadratic equation can be solved by both factoring and the quadratic formula, should we always get the same results? Explain.

Solve each equation by using the quadratic formula.

19. $-4x^2 - 2x + 7 = 0$ **20.** $2x^2 + 8 = 4x + 11$ **21.** $x(5x - 1) = 1$

22. $\dfrac{1}{4}x^2 = 2 - \dfrac{3}{4}x$ **23.** $\dfrac{1}{2}x^2 + 3x = 5$

24. A student writes the quadratic formula as $x = -b \pm \sqrt{\dfrac{b^2 - 4ac}{2a}}$. Is this correct?

If not, explain the error, and give the correct formula.

[17.4] *Sketch the graph of each equation. Identify each vertex.*

25. $y = -3x^2$ **26.** $y = -x^2 + 5$ **27.** $y = x^2 - 2x + 1$

  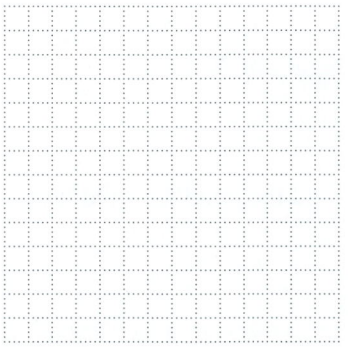

28. $y = -x^2 + 2x + 3$

29. $y = x^2 + 4x + 2$

30. $y = (x + 4)^2$

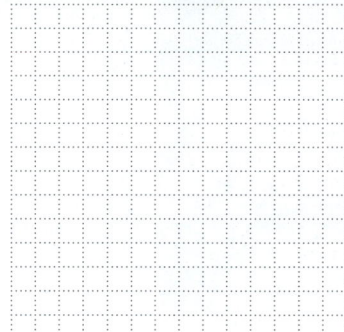

31. Refer to Exercise 17 in **Section 17.4.** Suppose that a telescope has a diameter of 200 ft and a maximum depth of 30 ft. Find the equation for a cross section of the parabolic dish.

[17.5] *Decide whether each relation is or is not a function. In Exercises 32 and 33, give the domain and the range.*

32. $\{(-2, 4), (0, 8), (2, 5), (2, 3)\}$

33. $\{(8, 3), (7, 4), (6, 5), (5, 6), (4, 7)\}$

34.

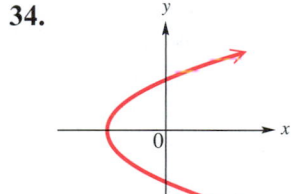

35.

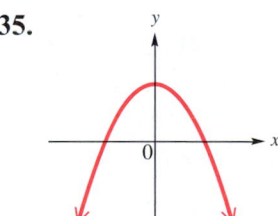

36. $2x + 3y = 12$

37. $y = x^2$

38. $x = 2|y|$

*Find **(a)** $f(2)$ and **(b)** $f(-1)$.*

39. $f(x) = 3x + 2$

40. $f(x) = 2x^2 - 1$

41. $f(x) = |x + 3|$

Becky and Brad are the owners of Cole's Baseball Cards. They have found that the price y, in dollars, of a particular Cal Ripken baseball card depends on the demand x, in hundreds, for the card, according to the function defined by

$$y = -x^2 + 12x - 26.$$

42. What demand produces a price of $6 for the card?

43. Find the vertex of the parabola $y = -x^2 + 12x - 26$.

44. Give the demand and price that correspond to the vertex.

▶▶▶ **Mixed Review Exercises**

Solve by any method.

45. $(2t - 1)(t + 1) = 54$

46. $(2p + 1)^2 = 100$

47. $(k + 2)(k - 1) = 3$

48. $6t^2 + 7t - 3 = 0$

49. $2x^2 + 3x + 2 = x^2 - 2x$

50. $x^2 + 2x + 5 = 7$

51. $m^2 - 4m + 10 = 0$

52. $k^2 - 9k + 10 = 0$

53. $(5x + 6)^2 = 0$

54. $\frac{1}{2}r^2 = \frac{7}{2} - r$

55. $x^2 + 4x = 1$

56. $7x^2 - 8 = 5x^2 + 8$

Chapter 17 ▶▶▶ Test

 Use the Chapter Test Prep Video CD to see fully worked-out solutions to any of the exercises you want to review.

Solve by using the square root property.

1. $x^2 = 39$

1. _____

2. $(x + 3)^2 = 64$

2. _____

3. $(4x + 3)^2 = 24$

3. _____

Solve by completing the square.

4. $x^2 - 4x = 6$

4. _____

5. $2x^2 + 12x - 3 = 0$

5. _____

Solve by the quadratic formula.

6. $2x^2 + 5x - 3 = 0$

6. _____

7. $3w^2 + 2 = 6w$

7. _____

8. $4x^2 + 8x + 11 = 0$

8. _____

9. $t^2 - \dfrac{5}{3}t + \dfrac{1}{3} = 0$

9. _____

Solve by the method of your choice.

10. $p^2 - 2p - 1 = 0$

10. _____

11. $(2x + 1)^2 = 18$

11. _____

12. $(x - 5)(2x - 1) = 1$

12. _____

13. $t^2 + 25 = 10t$

13. _____

Solve each problem.

14. _____

14. If an object is projected into the air from ground level with an initial velocity of 64 ft per sec, its altitude (height) s in feet after t seconds is given by the formula $s = -16t^2 + 64t$. At what time(s) will the object be at a height of 64 ft?

15. _____

15. Find the lengths of the three sides of the right triangle.

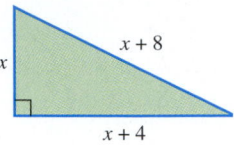

Sketch the graph of each equation. Identify each vertex.

16. _____

16. $y = (x - 3)^2$

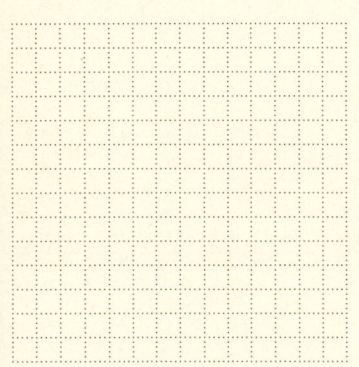

17. _____

17. $y = -x^2 - 2x - 4$

18. (a) _____

18. Decide whether each relation represents a function. If it does, give the domain and the range.

(a) $\{(2, 3), (2, 4), (2, 5)\}$ **(b)** $\{(0, 2), (1, 2), (2, 2)\}$

(b) _____

19. _____

19. Use the vertical line test to determine whether the graph at the right is that of a function.

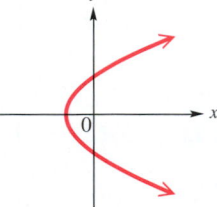

20. _____

20. If $f(x) = 3x + 7$, find $f(-2)$.

Cumulative Review Exercises  Chapters 9–17

Find the value of each expression if $x = -2$ and $y = 4$.

1. $\dfrac{4x - 2y}{x + y}$

2. $x^3 - 4xy$

Perform the indicated operations.

3. $\dfrac{(-13 + 15) - (3 + 2)}{6 - 12}$

4. $-7 - 3[2 + (5 - 8)]$

Decide what property justifies each statement.

5. $(9 + 2) + 3 = 9 + (2 + 3)$

6. $-7 + 7 = 0$

7. $6(4 + 2) = 6(4) + 6(2)$

Solve each equation.

8. $2x - 7x + 8x = 30$

9. $2 - 3(t - 5) = 4 + t$

10. $2(5h + 1) = 10h + 4$

11. $d = rt$ for r

12. $\dfrac{x}{5} = \dfrac{x - 2}{7}$

13. $\dfrac{1}{3}p - \dfrac{1}{6}p = -2$

14. $0.05x + 0.15(50 - x) = 5.50$

15. $4 - (3x + 12) = (2x - 9) - (5x - 1)$

Solve each inequality. Write the solution set in interval notation and graph it.

16. $-2.5x < 6.5$

17. $4(x + 3) - 5x < 12$

18. $\dfrac{2}{3}t - \dfrac{1}{6}t \le -2$

Graph the following.

19. $2x + 3y = 6$

20. $y = 3$

21. $2x - 5y < 10$

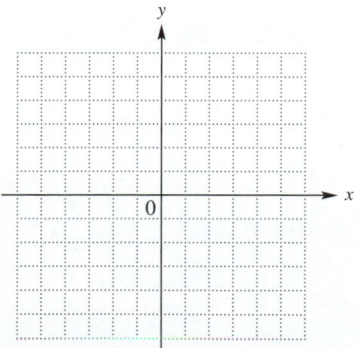

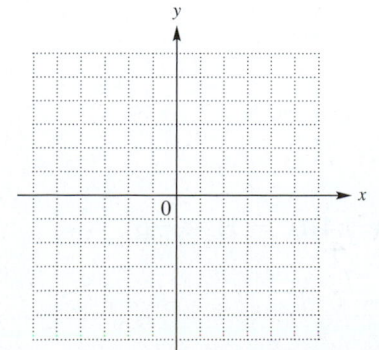

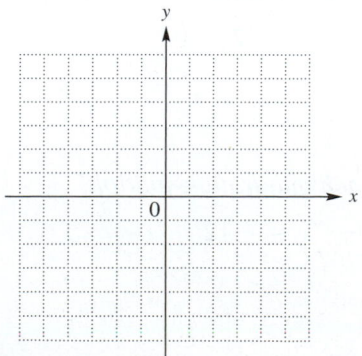

Find the slope of each line.

22. Through $(-5, 6)$ and $(1, -2)$

23. Perpendicular to the line $y = 4x - 3$

Write an equation for each line in slope-intercept form.

24. Through $(-4, 1)$ with slope $\frac{1}{2}$

25. Through the points $(1, 3)$ and $(-2, -3)$

26. (a) Write an equation of the vertical line through $(9, -2)$.

  (b) Write an equation of the horizontal line through $(4, -1)$.

Evaluate each expression.

27. $2^{-3} \cdot 2^5$

28. $\left(\dfrac{3}{4}\right)^{-2}$

29. $\dfrac{6^5 \cdot 6^{-2}}{6^3}$

30. $\left(\dfrac{4^{-3} \cdot 4^4}{4^5}\right)^{-1}$

Simplify each expression and write the answer using only positive exponents. Assume no denominators are 0.

31. $\dfrac{(p^2)^3 p^{-4}}{(p^{-3})^{-1} p}$

32. $\dfrac{(m^{-2})^3 m}{m^5 m^{-4}}$

Perform the indicated operations.

33. $(2k^2 + 4k) - (5k^2 - 2) - (k^2 + 8k - 6)$

34. $(9x + 6)(5x - 3)$

35. $(3p + 2)^2$

36. $\dfrac{8x^4 + 12x^3 - 6x^2 + 20x}{2x}$

Factor completely.

37. $2a^2 + 7a - 4$

38. $10m^2 + 19m + 6$

39. $8p^3 + 125$

40. $4p^2 - 12p + 9$

41. $25r^2 - 81t^2$

42. $2pq + 6p^3q + 8p^2q$

Solve each equation.

43. $6m^2 + m - 2 = 0$

44. $8x^2 = 64x$

45. $49x^2 - 56x + 16 = 0$

Solve each quadratic equation.

46. $x^2 - 7x = -12$

47. $(x + 4)(x - 1) = -6$

48. $z^2 + 144 = 24z$

Perform each operation, and write the answer in lowest terms.

49. $\dfrac{5}{q} - \dfrac{1}{q}$

50. $\dfrac{3}{7} + \dfrac{4}{r}$

51. $\dfrac{4}{5q - 20} - \dfrac{1}{3q - 12}$

52. $\dfrac{2}{k^2 + k} - \dfrac{3}{k^2 - k}$

53. $\dfrac{7z^2 + 49z + 70}{16z^2 + 72z - 40} \div \dfrac{3z + 6}{4z^2 - 1}$

54. Simplify the complex fraction $\dfrac{\dfrac{4}{a} + \dfrac{5}{2a}}{\dfrac{7}{6a} - \dfrac{1}{5a}}$.

Solve each equation. Check your solutions.

55. $\dfrac{r + 2}{5} = \dfrac{r - 3}{3}$

56. $\dfrac{1}{x} = \dfrac{1}{x + 1} + \dfrac{1}{2}$

Solve each system by any method.

57. $2x - y = -8$
$\quad\ x + 2y = 11$

58. $4x + 5y = -8$
$\quad\ 3x + 4y = -7$

59. $3x + 4y = 2$
$\quad\ 6x + 8y = 1$

60. Graph the solution set of the system.

$$x + 2y \le 12$$
$$2x - y \le 8$$

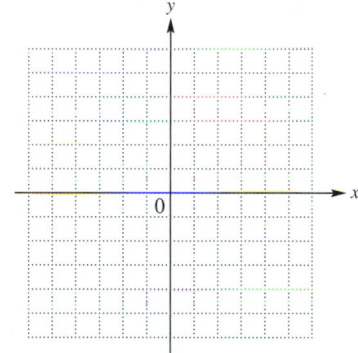

Simplify each expression as completely as possible.

61. $\dfrac{6\sqrt{6}}{\sqrt{5}}$

62. $3\sqrt{5} - 2\sqrt{20} + \sqrt{125}$

63. $\sqrt[3]{16a^3b^4} - \sqrt[3]{54a^3b^4}$

64. $\sqrt{27} - 2\sqrt{12} + 6\sqrt{75}$

65. $\dfrac{2}{\sqrt{3} + \sqrt{5}}$

66. $\left(3\sqrt{2} + 1\right)\left(4\sqrt{2} - 3\right)$

Solve each equation.

67. $\sqrt{x + 2} = -8$

68. $\sqrt{x + 2} = x - 10$

69. $2a^2 - 2a = 1$

70. Graph the parabola $y = x^2 - 4$. Identify the vertex. What are the x-intercepts?

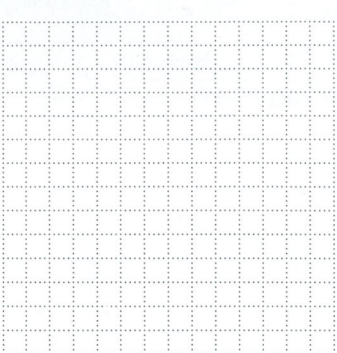

71. (a) Is $\{(0, 4), (1, 2), (3, 5)\}$ a function?

(b) Give the domain and the range of the relation in part (a).

72. If $f(x) = -2x + 7$, find $f(-2)$.

Solve each problem.

73. A 1-oz mouse takes about 16 times as many breaths as does a 3-ton elephant. (*Source: Dinosaurs, Spitfires, and Sea Dragons*, McGowan, C., Harvard University Press, 1991.) If the two animals take a combined total of 170 breaths per minute, how many breaths does each take during that time period?

74. If a number is subtracted from 8 and this difference is tripled, the result is three times the number. Find this number, and you will learn how many times a dolphin rests during a 24-hr period.

The graph shows a line that models the number of cell phone subscribers in millions in the United States from 2000 to 2006.

75. Use the ordered pairs shown on the graph to find the slope of the line to the nearest tenth. Interpret the slope.

76. Use the slope from Exercise 75 and the ordered pair (0, 109.5) to find the equation of the line that models the data, where $x = 0$ represents 2000.

77. Use the equation from Exercise 76 to estimate the number of cell phone subscribers in 2008. Round to the nearest tenth.

Cell Phone Subscribers

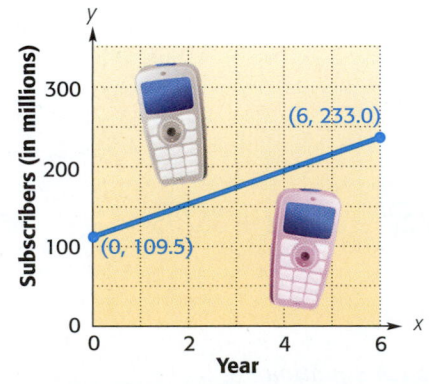

Source: CITA–The Wireless Association.

Use a system of equations to solve each problem.

78. Admission prices at a football game were $6 for adults and $2 for children. The total value of the tickets sold was $2528, and 454 tickets were sold. How many adults and how many children attended the game?

Kind of Ticket	Number Sold	Cost of Each (in dollars)	Total Value (in dollars)
Adult	x	6	$6x$
Child	y		
Total	454		

79. The perimeter of a triangle is 53 in. If two sides are of equal length, and the third side measures 4 in. less than each of the equal sides, what are the lengths of the three sides?

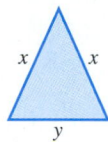

80. Des Moines and Chicago are 345 mi apart. Two cars start from these cities traveling toward each other. They meet after 3 hr. The car from Chicago has an average speed 7 mph faster than the other car. Find the average speed of each car. (*Source: State Farm Road Atlas.*)

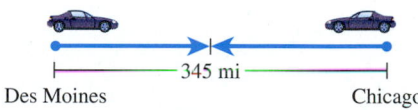

Appendices

◄ The world's best-preserved crater made by a meteor crashing into Earth is located in northern Arizona. Notice the buildings at the upper left edge of the crater for size comparison.

W e are constantly measuring things, from very large to very small.

For example, a meteor crashed into Earth in northern Arizona nearly 50,000 years ago. The bowl-shaped crater from that crash is sixty stories deep and 4150 feet across. (See **Appendix A.1**, Exercises 63–66.)

On the other hand, computer microchips often measure a tiny 5 mm long by 1 mm wide. (Learn about millimeters in **Appendix A.2** and then try Exercise 39.)

A greatly enlarged photo of a computer microchip.

A.1 ▶▶▶ Problem Solving with U.S. Customary Measurements

OBJECTIVES

1 Learn the basic U.S. customary measurement units.

2 Convert among measurement units using multiplication or division.

3 Convert among measurement units using unit fractions.

4 Solve application problems using U.S. customary measurement units.

We measure things all the time: the distance traveled on vacation, the floor area we want to cover with carpet, the amount of milk in a recipe, the weight of the bananas we buy at the store, the number of hours we work, and many more.

In the United States, we still use **U.S. customary measurement units** for many everyday activities. Examples are inches, feet, quarts, ounces, and pounds. However, science, medicine, sports, and manufacturing use the **metric system** (meters, liters, and grams). And, because the rest of the world uses *only* the metric system, U.S. businesses have been changing to the metric system in order to compete internationally.

OBJECTIVE 1 Learn the basic U.S. customary measurement units. Until the switch to the metric system is complete, we still need to know how to use U.S. customary measurement units. The table below lists the relationships you should memorize. The time relationships are used in both the U.S. customary and metric systems.

1 After memorizing the measurement conversions, answer these questions.

(a) 1 c = _____ fl oz

(b) _____ qt = 1 gal

(c) 1 wk = _____ days

(d) _____ ft = 1 yd

(e) 1 ft = _____ in.

(f) _____ oz = 1 lb

(g) 1 ton = _____ lb

(h) _____ min = 1 hr

(i) 1 pt = _____ c

(j) _____ hr = 1 day

(k) 1 min = _____ sec

(l) 1 qt = _____ pt

(m) _____ ft = 1 mi

U.S. Customary Measurement Relationships	
Length	**Weight**
12 inches (in.) = 1 foot (ft)	16 ounces (oz) = 1 pound (lb)
3 feet (ft) = 1 yard (yd)	2000 pounds (lb) = 1 ton
5280 feet (ft) = 1 mile (mi)	
Capacity	**Time**
8 fluid ounces (fl oz) = 1 cup (c)	60 seconds (sec) = 1 minute (min)
2 cups (c) = 1 pint (pt)	60 minutes (min) = 1 hour (hr)
2 pints (pt) = 1 quart (qt)	24 hours (hr) = 1 day
4 quarts (qt) = 1 gallon (gal)	7 days = 1 week (wk)

As you can see, there is no simple way to convert among these various measures. The units evolved over hundreds of years and were based on a variety of "standards." For example, one yard was the distance from the tip of a king's nose to his thumb when his arm was outstretched. An inch was three dried barleycorns laid end to end.

EXAMPLE 1 Knowing U.S. Customary Measurement Units

Memorize the U.S. customary measurement conversions shown above. Then answer these questions.

(a) 24 hr = _____ day   Answer: 1 day

(b) 1 yd = _____ ft   Answer: 3 ft

◀ Work Problem **1** at the Side.

OBJECTIVE 2 Convert among measurement units using multiplication or division. You often need to convert from one unit of measure to another. Two methods of converting measurements are shown here. Study each way and use the method you prefer. The first method involves deciding whether to multiply or divide.

ANSWERS

1. (a) 8 **(b)** 4 **(c)** 7 **(d)** 3 **(e)** 12
(f) 16 **(g)** 2000 **(h)** 60 **(i)** 2 **(j)** 24
(k) 60 **(l)** 2 **(m)** 5280

> **Converting among Measurement Units**
> 1. *Multiply* when converting from a larger unit to a smaller unit.
> 2. *Divide* when converting from a smaller unit to a larger unit.

EXAMPLE 2 **Converting from One Unit of Measure to Another**

Convert each measurement.

(a) 7 ft to inches

You are converting from a *larger* unit to a *smaller* unit (a *foot* is longer than an *inch*), so multiply.

Because *1 ft =* **12** *in.*, multiply by 12.

$$7 \text{ ft} = 7 \cdot 12 = 84 \text{ in.}$$

(b) $3\frac{1}{2}$ lb to ounces

You are converting from a *larger* unit to a *smaller* unit (a *pound* is heavier than an *ounce*), so multiply.

Because *1 lb =* **16** *oz,* multiply by 16.

> Divide 16 and 2 by their common factor of 2.
> 16 ÷ 2 is 8.
> 2 ÷ 2 is 1.

$$3\frac{1}{2} \text{ lb} = 3\frac{1}{2} \cdot 16 = \frac{7}{\underset{1}{\cancel{2}}} \cdot \frac{\overset{8}{\cancel{16}}}{1} = \frac{56}{1} = 56 \text{ oz}$$

(c) 20 qt to gallons

You are converting from a *smaller* unit to a *larger* unit (a *quart* is smaller than a *gallon*) so divide.

Because **4** *qt = 1 gal,* divide by 4.

$$20 \text{ qt} = \frac{20}{4} = 5 \text{ gal}$$

Divide by 4. ⬆

(d) 45 min to hours

You are converting from a *smaller* unit to a *larger* unit (a *minute* is less than an *hour*), so divide.

Because **60** *min = 1 hr,* divide by 60 and write the fraction in lowest terms.

$$45 \text{ min} = \frac{45}{60} = \frac{45 \div 15}{60 \div 15} = \frac{3}{4} \text{ hr} \leftarrow \text{Lowest terms}$$

Divide by 60. ⬆

Work Problem **2** *at the Side.* ▶

OBJECTIVE **3** **Convert among measurement units using unit fractions.** If you have trouble deciding whether to multiply or divide when converting measurements, use *unit fractions* to solve the problem. You'll also use this method in science courses. A **unit fraction** is equivalent to 1. Here is an example.

$$\frac{12 \text{ in.}}{12 \text{ in.}} = \frac{\overset{1}{\cancel{12} \text{ in.}}}{\underset{1}{\cancel{12} \text{ in.}}} = 1$$

Use the table of measurement relationships on the previous page to find that 12 in. is the same as 1 ft. So you can substitute 1 ft for 12 in. in the numerator, or you can substitute 1 ft for 12 in. in the denominator. This makes two useful unit fractions.

$$\frac{\textbf{1 ft}}{12 \text{ in.}} = 1 \qquad \text{or} \qquad \frac{12 \text{ in.}}{\textbf{1 ft}} = 1$$

2 Convert each measurement using multiplication or division.

(a) $5\frac{1}{2}$ ft to inches

(b) 64 oz to pounds

(c) 6 yd to feet

(d) 2 tons to pounds

(e) 35 pt to quarts

(f) 20 min to hours

(g) 4 wk to days

3 First write the unit fraction needed to make each conversion. Then complete the conversion.

(a) 36 in. to feet

unit $\Big\}$ $\dfrac{1 \text{ ft}}{12 \text{ in.}}$
fraction

(b) 14 ft to inches

unit $\Big\}$ $\dfrac{\text{in.}}{\text{ft}}$
fraction

(c) 60 in. to feet

unit $\Big\}$ _____
fraction

(d) 4 yd to feet

unit $\Big\}$ _____
fraction

(e) 39 ft to yards

unit $\Big\}$ _____
fraction

(f) 2 mi to feet

unit $\Big\}$ _____
fraction

ANSWERS

3. **(a)** 3 ft   **(b)** $\dfrac{12 \text{ in.}}{1 \text{ ft}}$; 168 in.

(c) $\dfrac{1 \text{ ft}}{12 \text{ in.}}$; 5 ft   **(d)** $\dfrac{3 \text{ ft}}{1 \text{ yd}}$; 12 ft

(e) $\dfrac{1 \text{ yd}}{3 \text{ ft}}$; 13 yd   **(f)** $\dfrac{5280 \text{ ft}}{1 \text{ mi}}$; 10,560 ft

---

To convert from one measurement unit to another, just multiply by the appropriate unit fraction. Remember, a unit fraction is equivalent to 1. Multiplying something by 1 does *not* change its value.

Use these guidelines to choose the correct unit fraction.

> **Choosing a Unit Fraction**
>
> The **numerator** should use the measurement unit you want in the *answer*.
> The **denominator** should use the measurement unit you want to *change*.

EXAMPLE 3   **Using Unit Fractions with Length Measurements**

(a) Convert 60 in. to feet.

Use a unit fraction with feet (the unit for your answer) in the numerator, and inches (the unit being changed) in the denominator. Because *1 ft = 12 in.*, the necessary unit fraction is

$$\dfrac{1 \text{ ft}}{12 \text{ in.}} \quad \begin{array}{l}\leftarrow \text{ Unit for your answer is feet.}\\ \leftarrow \text{ Unit being changed is inches.}\end{array}$$

Next, multiply 60 in. times this unit fraction. Write 60 in. as the fraction $\dfrac{60 \text{ in.}}{1}$ and divide out common units and factors wherever possible.

$$60 \text{ in.} \cdot \dfrac{1 \text{ ft}}{12 \text{ in.}} = \dfrac{\overset{5}{\cancel{60} \text{ in.}}}{1} \cdot \dfrac{1 \text{ ft}}{\underset{1}{\cancel{12} \text{ in.}}} = \dfrac{5 \cdot 1\text{ ft}}{1} = 5 \text{ ft}$$

These units should match. — Divide out inches. — Divide 60 and 12 by 12.

(b) Convert 9 ft to inches.

Select the correct unit fraction to change 9 ft to inches.

$$\dfrac{12 \text{ in.}}{1 \text{ ft}} \quad \begin{array}{l}\leftarrow \text{ Unit for your answer is inches.}\\ \leftarrow \text{ Unit being changed is feet.}\end{array}$$

Multiply 9 ft times the unit fraction.

$$9 \text{ ft} \cdot \dfrac{12 \text{ in.}}{1 \text{ ft}} = \dfrac{9 \cancel{\text{ ft}}}{1} \cdot \dfrac{12 \text{ in.}}{1 \cancel{\text{ ft}}} = \dfrac{9 \cdot 12 \text{ in.}}{1} = 108 \text{ in.}$$

These units should match. — Divide out feet.

> **CAUTION**
> If no units will divide out, you made a mistake in choosing the unit fraction.

◀ *Work Problem* **3** *at the Side.*

EXAMPLE 4   **Using Unit Fractions with Capacity and Weight Measurements**

(a) Convert 9 pt to quarts.

First select the correct unit fraction.

$$\dfrac{1 \text{ qt}}{2 \text{ pt}} \quad \begin{array}{l}\leftarrow \text{ Unit for your answer is quarts.}\\ \leftarrow \text{ Unit being changed is pints.}\end{array}$$

Continued on Next Page

Next multiply.

Write as a mixed number.

$$9 \text{ pt} \cdot \frac{1 \text{ qt}}{2 \text{ pt}} = \frac{9 \text{ pt}}{1} \cdot \frac{1 \text{ qt}}{2 \text{ pt}} = \frac{9}{2} \text{ qt} = 4\frac{1}{2} \text{ qt}$$

These units should match. Divide out pints.

(b) Convert $7\frac{1}{2}$ gal to quarts.

Write as an improper fraction.

$$\frac{7\frac{1}{2} \text{ gal}}{1} \cdot \frac{4 \text{ qt}}{1 \text{ gal}} = \frac{15}{2} \cdot \frac{4}{1} \text{ qt}$$

Divide out gallons.

> Divide 4 and 2 by their common factor of 2.
> $4 \div 2$ is 2.
> $2 \div 2$ is 1.

$$= \frac{15}{\underset{1}{\cancel{2}}} \cdot \frac{\overset{2}{\cancel{4}}}{1} \text{ qt}$$

$$= 30 \text{ qt}$$

(c) Convert 36 oz to pounds.

> Notice that **oz** divides out, leaving **lb**, the unit you want for the answer.

$$\frac{\overset{9}{\cancel{36}} \text{ oz}}{1} \cdot \frac{1 \text{ lb}}{\underset{4}{\cancel{16}} \text{ oz}} = \frac{9}{4} \text{ lb} = 2\frac{1}{4} \text{ lb}$$

> **Note**
> In Example 4(c) above you get $\frac{9}{4}$ lb. Recall that $\frac{9}{4}$ means $9 \div 4$. If you do $9 \div 4$ on your calculator, you get 2.25 lb. U.S. customary measurements usually use fractions or mixed numbers, like $2\frac{1}{4}$ lb. However, 2.25 lb is also correct and is the way grocery stores often show weights of produce, meat, and cheese.

Work Problem **4** *at the Side.* ▶

> **EXAMPLE 5** **Using Several Unit Fractions**

Sometimes you may need to use two or three unit fractions to complete a conversion.

(a) Convert 63 in. to yards.

Use the unit fraction $\frac{1 \text{ ft}}{12 \text{ in.}}$ to change inches to feet and the unit fraction $\frac{1 \text{ yd}}{3 \text{ ft}}$ to change feet to yards. Notice how all the units divide out except yards, which is the unit you want in the answer.

$$\frac{63 \text{ in.}}{1} \cdot \frac{1 \text{ ft}}{12 \text{ in.}} \cdot \frac{1 \text{ yd}}{3 \text{ ft}} = \frac{63}{36} \text{ yd} = \frac{63 \div 9}{36 \div 9} \text{ yd} = \frac{7}{4} \text{ yd} = 1\frac{3}{4} \text{ yd}$$

Continued on Next Page

---

4 Convert using unit fractions.

(a) 16 qt to gallons

(b) 3 c to pints

(c) $3\frac{1}{2}$ tons to pounds

(d) $1\frac{3}{4}$ lb to ounces

(e) 4 oz to pounds

ANSWERS

4. (a) 4 gal **(b)** $1\frac{1}{2}$ pt or 1.5 pt **(c)** 7000 lb

(d) 28 oz **(e)** $\frac{1}{4}$ lb or 0.25 lb

5 Convert using two or three unit fractions.

(a) 4 tons to ounces

You can also divide out common factors in the numbers.

$$\frac{\overset{7}{\cancel{\overset{2}{\cancel{63}}}}}{1} \cdot \frac{1}{\underset{4}{\cancel{12}}} \cdot \frac{1}{\underset{1}{\cancel{3}}} = \frac{7}{4} = 1\frac{3}{4} \text{ yd}$$

Instead of changing $\frac{7}{4}$ to $1\frac{3}{4}$, you can enter $7 \div 4$ on your calculator to get 1.75 yd. Both answers are correct because 1.75 is equivalent to $1\frac{3}{4}$.

(b) Convert 2 days to seconds.

Use three unit fractions. The first one changes days to hours, the next one changes hours to minutes, and the last one changes minutes to seconds. All the units divide out except seconds, which is the unit you want in your answer.

(b) 3 mi to inches

$$\frac{2 \text{ days}}{1} \cdot \frac{24 \text{ hr}}{1 \text{ day}} \cdot \frac{60 \text{ min}}{1 \text{ hr}} \cdot \frac{60 \text{ sec}}{1 \text{ min}} = 172{,}800 \text{ sec}$$

Divide out **days**.
Divide out **hr**.
Divide out **min**.

◀ *Work Problem* **5** *at the Side.*

(c) 36 pt to gallons

OBJECTIVE **4** **Solve application problems using U.S. customary measurement units.** To solve application problems, we will use the steps you learned in **Section 1.10.** Those steps are summarized here.

Step 1 **Read** the problem carefully.

Step 2 **Work out a plan.**

Step 3 **Estimate** a reasonable answer.

Step 4 **Solve** the problem.

Step 5 **State the answer.**

Step 6 **Check** your work.

(d) 2 wk to minutes

EXAMPLE 6 **Solving U.S. Customary Measurement Applications**

(a) A 36 oz can of coffee is on sale at Jerry's Foods for $7.89. What is the cost per pound, to the nearest cent? (*Source:* Jerry's Foods.)

Step 1 **Read** the problem. The problem asks for the cost per *pound* of coffee.

Step 2 **Work out a plan.** The weight of the coffee is given in *ounces*, but the answer must be cost *per pound*. Convert ounces to pounds. The word *per* indicates division. You need to divide the cost by the number of pounds.

Step 3 **Estimate** a reasonable answer. To estimate, round $7.89 to $8. Then, there are 16 oz in a pound, so 36 oz are a little more than 2 pounds. So, $8 \div 2 = $4 per pound as our estimate.

ANSWERS

5. (a) 128,000 oz **(b)** 190,080 in.

(c) $4\frac{1}{2}$ gal or 4.5 gal **(d)** 20,160 min

Continued on Next Page

Step 4 **Solve** the problem. Use a unit fraction to convert 36 oz to pounds.

Notice that **oz** divides out, leaving **lb** for the answer.

On your calculator, $9 \div 4 = 2.25$

$$\frac{\overset{9}{\cancel{36}} \cancel{oz}}{1} \cdot \frac{1 \text{ lb}}{\underset{4}{\cancel{16}} \cancel{oz}} = \frac{9}{4} \text{ lb} = 2.25 \text{ lb}$$

Then divide to find the *cost* per *pound*.

$$\begin{array}{c} \text{Cost} \rightarrow \\ \text{per} \rightarrow \\ \text{pound} \rightarrow \end{array} \frac{\$7.89}{2.25 \text{ lb}} = 3.50\overline{6} \approx 3.51 \quad \text{(rounded)}$$

Step 5 **State the answer.** The coffee costs $3.51 per pound (to the nearest cent).

Step 6 **Check** your work. The exact answer of $3.51 is close to our estimate of $4.

(b) Bilal's favorite cake recipe uses $1\frac{2}{3}$ cups of milk. If he makes six cakes for a bake sale at his son's school, how many quarts of milk will he need?

Step 1 **Read** the problem. The problem asks for the number of *quarts* of milk needed for six cakes.

Step 2 **Work out a plan.** Multiply to find the number of *cups* of milk for six cakes. Then convert *cups* to *quarts* (the unit required in the answer).

Step 3 **Estimate** a reasonable answer. To estimate, round $1\frac{2}{3}$ cups to 2 cups. Then, 2 cups times $6 = 12$ cups. There are 4 cups in a quart, so 12 cups $\div 4 = 3$ quarts as our estimate.

Step 4 **Solve** the problem. First multiply. Then use unit fractions to convert.

$$1\frac{2}{3} \cdot 6 = \frac{5}{\cancel{3}} \cdot \frac{\overset{2}{\cancel{6}}}{1} = \frac{10}{1} = 10 \text{ cups} \quad \left\{ \begin{array}{l} \text{Milk needed for} \\ \text{six cakes} \end{array} \right.$$

$$\frac{\overset{5}{\cancel{10} \cancel{\text{cups}}}}{1} \cdot \frac{1 \cancel{pt}}{\underset{1}{\cancel{2} \cancel{\text{cups}}}} \cdot \frac{1 \text{ qt}}{2 \cancel{pt}} = \frac{5}{2} \text{ qt} = 2\frac{1}{2} \text{ qt}$$

Both **cups** and **pt** divide out, leaving **qt**, the unit you want for the answer.

Step 5 **State the answer.** Bilal needs $2\frac{1}{2}$ qt (or 2.5 qt) of milk.

Step 6 **Check** your work. The exact answer of $2\frac{1}{2}$ qt is close to our estimate of 3 qt.

> **Note**
>
> In Step 2 above, we *first multiplied* $1\frac{2}{3}$ cups by 6 to find the number of cups needed, then *converted* 10 cups to $2\frac{1}{2}$ quarts. It would also work to *first convert* $1\frac{2}{3}$ cups to $\frac{5}{12}$ qt, then *multiply* $\frac{5}{12}$ qt by 6 to get $2\frac{1}{2}$ qt.

Work Problem **6** *at the Side.* ▶

---

6 Solve each application problem using the six problem-solving steps.

(a) Kristin paid $3.29 for 12 oz of extra sharp cheddar cheese. What is the price per pound, to the nearest cent?

(b) A moving company estimates 11,000 lb of furnishings for an average 3-bedroom house. If the company made five such moves last week, how many tons of furnishings did they move? (*Source:* North American Van Lines.)

ANSWERS

6. **(a)** $4.39 per pound (rounded)

   (b) 27.5 or $27\frac{1}{2}$ tons

Math in the Media

GROWING SUNFLOWERS

The front and back of a seed packet for sunflowers are shown at the right. Look at the top of the packet first. (Ignore sales tax in Questions 1 and 2.)

1. There were 42 seeds in the packet. If 40 of the seeds sprouted, what was the cost per sprout, to the nearest cent?

2. If vegetable and flower seeds were on sale at 30% off, what was the cost per sprout, to the nearest cent?

3. What percent of the seeds sprouted, to the nearest whole percent?

4. How many seeds would add up to a weight of 1 gram?

5. The table at the bottom of the packet uses the symbol (′) for feet, and the symbol (″) for inches.

   (a) How tall will the plants grow, in feet?

   (b) How tall will they grow in inches?

   (c) How tall will they grow in yards?

6. If you plant all 42 seeds in one long row, using the spacing given on the package, how long will your row be in feet?

7. How many inches tall should the plants be when you thin them (remove less vigorous plants to give others room to grow)? How tall is that in feet?

8. What is the range in the diameter of the flowers, in inches, and in feet? Diameter is the distance across the circular flower.

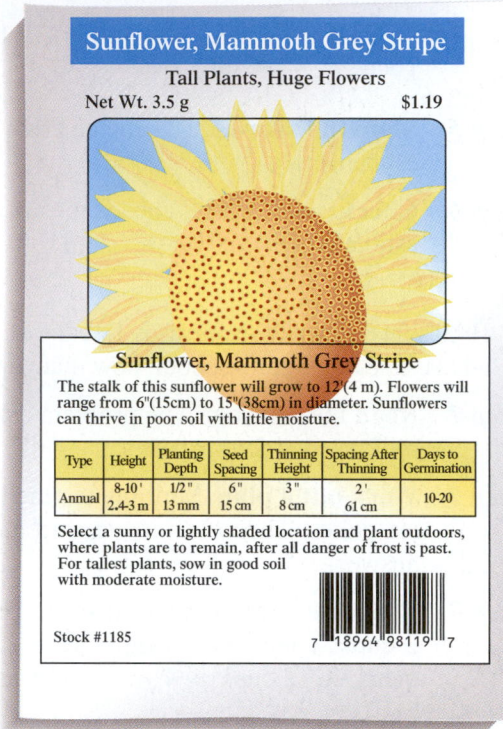

Sunflower, Mammoth Grey Stripe

Tall Plants, Huge Flowers

Net Wt. 3.5 g $1.19

Sunflower, Mammoth Grey Stripe

The stalk of this sunflower will grow to 12′(4 m). Flowers will range from 6″(15cm) to 15″(38cm) in diameter. Sunflowers can thrive in poor soil with little moisture.

Type	Height	Planting Depth	Seed Spacing	Thinning Height	Spacing After Thinning	Days to Germination
Annual	8-10′ 2.4-3 m	1/2″ 13 mm	6″ 15 cm	3″ 8 cm	2′ 61 cm	10-20

Select a sunny or lightly shaded location and plant outdoors, where plants are to remain, after all danger of frost is past. For tallest plants, sow in good soil with moderate moisture.

Stock #1185   7 18964 98119 7

Source: Olds Seed Solutions.

A.1 ▶▶▶ Exercises

 PRACTICE WATCH DOWNLOAD READ REVIEW

Fill in the blanks with the measurement relationships you have memorized. See Example 1.

1. 1 yd = _____ ft; _____ in. = 1 ft

2. 1 ft = _____ in.; _____ ft = 1 mi

3. _____ fl oz = 1 c; 1 qt = _____ pt

4. _____ qt = 1 gal; 1 pt = _____ c

5. 1 mi = _____ ft; _____ ft = 1 yd

6. 1 wk = _____ days; _____ sec = 1 min

7. _____ lb = 1 ton; 1 lb = _____ oz

8. _____ oz = 1 lb; 1 ton = _____ lb

9. 1 min = _____ sec; _____ min = 1 hr

10. 1 day = _____ hr; _____ sec = 1 min

Convert each measurement by multiplying or dividing. See Example 2.

11. **(a)** 120 sec = _____ min

(b) 4 hr = _____ min

12. **(a)** 180 min = _____ hr

(b) 5 min = _____ sec

13. **(a)** 2 qt = _____ gal

(b) $6\frac{1}{2}$ ft = _____ in.

14. **(a)** $4\frac{1}{2}$ gal = _____ qt

(b) 12 oz = _____ lb

15. An adult African elephant could weigh 7 to 8 tons. How many pounds could it weigh? (*Source: The Top 10 of Everything.*)

16. A reticulated python snake is the world's longest snake. It grows to a length of 18 to 33 feet. How many yards long can the snake be? (*Source: The Top 10 of Everything.*)

Convert each measurement in Exercises 17–38 using unit fractions. See Examples 3 and 4.

17. 9 yd = _____ ft

18. 20,000 lb = _____ tons

19. 7 lb = _____ oz

20. 96 oz = _____ lb

21. 5 qt = _____ pt

22. 26 pt = _____ qt

23. 90 min = _____ hr

24. 45 sec = _____ min

25. 3 in. = _____ ft

26. 30 in. = _____ ft

27. 24 oz = _____ lb

28. 36 oz = _____ lb

29. 5 c = _____ pt

30. 15 qt = _____ gal

Use the information in the bar graph below to answer Exercises 31–32.

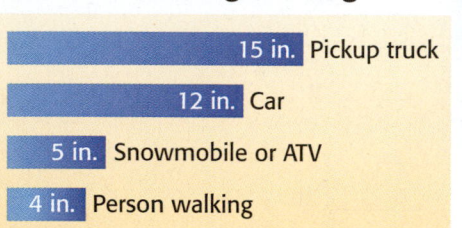

Thickness of Lake Ice Needed for Safe Walking/Driving

15 in. Pickup truck

12 in. Car

5 in. Snowmobile or ATV

4 in. Person walking

Source: Wisconsin DNR.

31. If the ice on a lake is $\frac{1}{2}$ ft thick, what will it safely support?

32. How many feet of ice are needed to safely drive a pickup truck on a lake?

33. $2\frac{1}{2}$ tons = _____ lb

34. $4\frac{1}{2}$ pt = _____ c

35. $4\frac{1}{4}$ gal = _____ qt

36. $2\frac{1}{4}$ hr = _____ min

37. After 15 years, a saguaro cactus is still only one-third to two-thirds of a foot tall, depending upon rainfall. How tall could the cactus be in inches? (*Source: Ecology of the Saguaro III.*)

38. Yao Ming, an NBA basketball player from China, is $7\frac{1}{2}$ ft tall. What is his height in inches? (*Source:* www.NBA.com)

Use two or three unit fractions to make each conversion. See Example 5.

39. 6 yd = _____ in.

40. 2 tons = _____ oz

41. 112 c = _____ qt

42. 336 hr = _____ wk

43. 6 days = _____ sec

44. 5 gal = _____ c

45. $1\frac{1}{2}$ tons = _____ oz

46. $3\frac{1}{3}$ yd = _____ in.

47. The statement 8 = 2 is *not* true. But with appropriate measurement units, it *is* true.

$$8 \text{ quarts} = 2 \text{ gallons}$$

Attach measurement units to these numbers to make the statement true.

(a) 1 _____ = 16 _____

(b) 10 _____ = 20 _____

(c) 120 _____ = 2 _____

(d) 2 _____ = 24 _____

(e) 6000 _____ = 3 _____

(f) 35 _____ = 5 _____

48. Explain in your own words why you can add 2 feet + 12 inches to get 3 feet, but you cannot add 2 feet + 12 pounds.

Convert each measurement. See Example 5.

49. $2\frac{3}{4}$ mi = _____ in.

50. $5\frac{3}{4}$ tons = _____ oz

51. $6\frac{1}{4}$ gal = _____ fl oz

52. $3\frac{1}{2}$ days = _____ sec

53. 24,000 oz = _____ ton

54. 57,024 in. = _____ mi

Solve each application problem. Show your work. See Example 6.

55. Geralyn bought 20 oz of strawberries for $2.29. What was the price per pound for the strawberries, to the nearest cent?

56. Zach paid $0.90 for a 0.8 oz candy bar. What was the cost per pound?

57. Dan orders supplies for the science labs. Each of the 24 stations in the chemistry lab needs 2 ft of rubber tubing. If rubber tubing sells for $8.75 per yard, how much will it cost to equip all the stations?

58. In 2006, Marquette, Michigan, had 170 inches of snowfall, while Detroit, Michigan, had 15 inches. What was the difference in snowfall between the two cities, in feet? Round to the nearest tenth. (*Source: World Almanac.*)

59. Tropical cockroaches are the fastest land insects. They can run about 5 feet per second. At this rate, how long would it take the cockroach to travel one mile? (*Source: Guinness World Records.*)

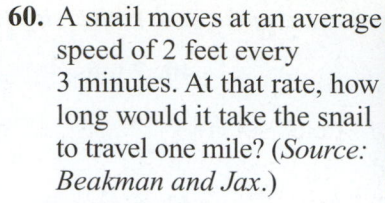

Give your answer

(a) in seconds;

(b) in minutes.

60. A snail moves at an average speed of 2 feet every 3 minutes. At that rate, how long would it take the snail to travel one mile? (*Source: Beakman and Jax.*)

Give your answer

(a) in hours;

(b) in days.

61. At the day care center, each of the 15 toddlers drinks about $\frac{2}{3}$ cup of milk with lunch. The center is open 5 days a week.

(a) How many quarts of milk will the center need for one week of lunches?

(b) If the center buys milk in gallon containers, how many containers should be ordered for one week?

62. Bob's Candies in Albany, Georgia, makes 135,000 pounds of candy canes each day. (*Source:* Bob's Candies, Inc.)

(a) How many tons of candy canes are produced during a 5-day workweek?

(b) The plant operates 24 hours per day. How many tons of candy canes are produced each hour, to the nearest tenth?

Relating Concepts (Exercises 63–66)   For Individual or Group Work

*On the first page of this chapter, we said that the bowl-shaped crater in northern Arizona made by a meteor crash was sixty stories deep and 4150 ft across. Use this information as you **work Exercises 63–66 in order.** (Source: The Meteor Crater Story.)*

63. (a) The distance across the crater is what part of a mile, to the nearest tenth?

(b) The crater is nearly circular. In a circle, the distance around the outside edge is about 3.14 times the distance across the circle. How far is it to walk around the edge of the crater in feet?

(c) How far is it to walk around the edge of the crater in miles, to the nearest tenth?

64. (a) The crater is 550 ft deep. The depth is how many yards, to the nearest whole number?

(b) How many inches deep is the crater?

(c) The depth of the crater is what part of a mile, to the nearest tenth?

(d) When we say that the crater is as deep as a sixty-story building, we are assuming that each story is how many feet tall, to the nearest foot?

65. On one side of the crater there are a few small juniper trees. The trees are 700 years old but only 18 inches to 30 inches tall because of the strong winds and lack of rain.

(a) How tall are the trees in feet?

(b) How many months old are the trees?

(c) At this rate of growth, how long would it take a 30-inch tree to reach a height of three feet?

66. Evidence of two huge meteor crashes has been found on the floor of the Caribbean Sea. One giant circular crater is 90 miles across and the other is 120 miles across.

(a) Using the information about circles in Exercise 63, what is the approximate distance around the edge of the smaller crater, to the nearest mile?

(b) Around the larger crater?

A.2 ▶▶▶ The Metric System—Length

Around 1790, a group of French scientists developed the metric system of measurement. It is an organized system based on multiples of 10, like our number system and our money. After you are familiar with metric units, you will see that they are easier to use than the hodgepodge of U.S. customary measurement relationships you used in **Appendix A.1.**

> **Note**
>
> The metric system information in this text is consistent with usage guidelines from the National Institute of Standards and Technology, www.nist.gov/metric.

OBJECTIVE 1 Learn the basic metric units of length. The basic unit of length in the metric system is the **meter** (also spelled *metre*). Use the symbol **m** for meter; do not put a period after it. If you put five of the pages from this textbook side by side, they would measure about 1 meter. Or, look at a yardstick—a meter is just a little longer. A yard is 36 inches long; a meter is about 39 inches long.

book page	book page	book page	book page	book page

|← 1 meter (about 39 in.) →|
|← 1 yard (36 in.) →|

In the metric system, you use meters for things like measuring the length of your living room, talking about heights of buildings, or describing track and field athletic events.

6 m
(about 20 ft)

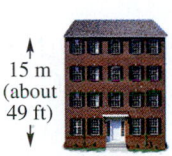
15 m
(about
49 ft)

Run the
100 m dash.

Work Problem **1** *at the Side.* ▶

To make longer or shorter length units in the metric system, **prefixes** are written in front of the word *meter*. For example, the prefix *kilo* means **1000**, so a *kilo*meter is **1000** meters. The table below shows how to use the prefixes for length measurements. It is helpful to memorize the prefixes because they are also used with weight and capacity measurements. The blue boxes are the units you will use most often in daily life.

Prefix	kilo-meter	hecto-meter	deka-meter	meter	deci-meter	centi-meter	milli-meter
Meaning	1000 meters	100 meters	10 meters	1 meter	$\frac{1}{10}$ of a meter	$\frac{1}{100}$ of a meter	$\frac{1}{1000}$ of a meter
Symbol	**k**m	**h**m	**da**m	m	**d**m	**c**m	**m**m

Length units that are used most often

OBJECTIVES

1 Learn the basic metric units of length.

2 Use unit fractions to convert among units.

3 Move the decimal point to convert among units.

1 Circle the items that measure about 1 meter.

Length of a pencil

Length of a baseball bat

Height of doorknob from the floor

Height of a house

Basketball player's arm length

Length of a paper clip

ANSWER

1. baseball bat, height of doorknob, basketball player's arm length

2 Write the most reasonable metric unit in each blank. Choose from km, m, cm, and mm.

(a) The woman's height is 168 _____ .

(b) The man's waist is 90 _____ around.

(c) Louise ran the 100 _____ dash in the track meet.

(d) A postage stamp is 22 _____ wide.

(e) Michael paddled his canoe 2 _____ down the river.

(f) The pencil lead is 1 _____ thick.

(g) A stick of gum is 7 _____ long.

(h) The highway speed limit is 90 _____ per hour.

(i) The classroom was 12 _____ long.

(j) A penny is about 18 _____ across.

Here are some comparisons to help you get acquainted with the commonly used length units: km, m, cm, mm.

Kilometers are used instead of miles. A kilometer is **1000** meters. It is about 0.6 mile (a little more than half a mile) or about 5 to 6 city blocks. If you participate in a 10 km run, you'll run about 6 miles.

A meter is divided into 100 smaller pieces called **centi**meters. Each centimeter is $\frac{1}{100}$ of a meter. Centimeters are used instead of inches. A centimeter is a little shorter than $\frac{1}{2}$ inch. The cover of this textbook is about 21 cm wide. A nickel is about 2 cm across. Measure the width and length of your little finger on this centimeter ruler. The width of your little finger is probably about 1 cm, or a little more.

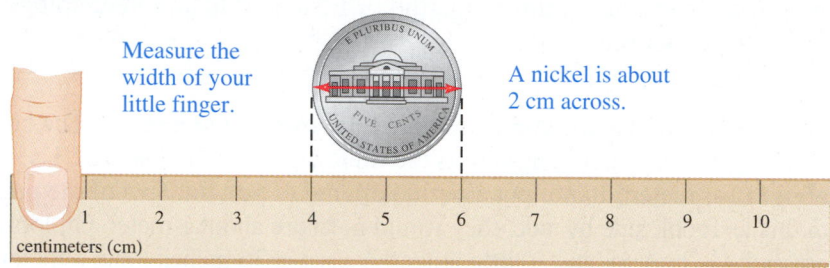

A meter is divided into 1000 smaller pieces called **milli**meters. Each millimeter is $\frac{1}{1000}$ of a meter. It takes 10 mm to equal 1 cm, so it is a very small length. The thickness of a dime is about 1 mm. Measure the width of your pen or pencil and the width of your little finger on this millimeter ruler.

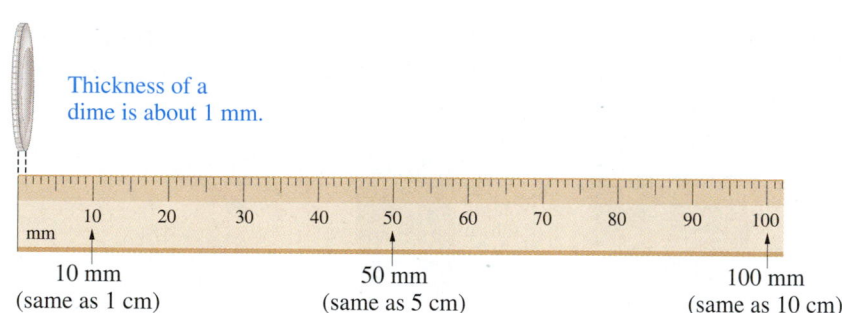

EXAMPLE 1   **Using Metric Length Units**

Write the most reasonable metric unit in each blank. Choose from km, m, cm, and mm.

(a) The distance from home to work is 20 _____ .

   20 **km** because kilometers are used instead of miles. 20 km is about 12 miles.

(b) My wedding ring is 4 _____ wide.

   4 **mm** because the width of a ring is very small.

(c) The newborn baby is 50 _____ long.

   50 **cm**; which is half of a meter; a meter is about 39 inches so half a meter is around 20 inches.

◀ *Work Problem* **2** *at the Side.*

OBJECTIVE **2** **Use unit fractions to convert among units.** You can convert among metric length units using unit fractions. Keep these relationships in mind when setting up the unit fractions.

Metric Length Relationships

1 km = 1000 m so the unit fractions are:	**1 m = 1000 mm** so the unit fractions are:
$\dfrac{1 \text{ km}}{1000 \text{ m}}$ or $\dfrac{1000 \text{ m}}{1 \text{ km}}$	$\dfrac{1 \text{ m}}{1000 \text{ mm}}$ or $\dfrac{1000 \text{ mm}}{1 \text{ m}}$
1 m = 100 cm so the unit fractions are:	**1 cm = 10 mm** so the unit fractions are:
$\dfrac{1 \text{ m}}{100 \text{ cm}}$ or $\dfrac{100 \text{ cm}}{1 \text{ m}}$	$\dfrac{1 \text{ cm}}{10 \text{ mm}}$ or $\dfrac{10 \text{ mm}}{1 \text{ cm}}$

EXAMPLE 2 **Using Unit Fractions to Convert Length Measurements**

Convert each measurement using unit fractions.

(a) 5 km to m

Put the unit for the answer (meters) in the numerator of the unit fraction; put the unit you want to change (km) in the denominator.

Unit fraction equivalent to 1 $\begin{cases} \dfrac{1000 \text{ m}}{1 \text{ km}} \end{cases}$ ← Unit for answer
← Unit being changed

Multiply. Divide out common units where possible.

$$5 \text{ km} \cdot \frac{1000 \text{ m}}{1 \text{ km}} = \frac{5 \text{ km}}{1} \cdot \frac{1000 \text{ m}}{1 \text{ km}} = \frac{5 \cdot 1000 \text{ m}}{1} = 5000 \text{ m}$$

These units should match.

Here, **km** divides out leaving **m**, the unit you want for your answer.

5 km = 5000 m

The answer makes sense because a kilometer is much longer than a meter, so 5 km will contain many meters.

(b) 18.6 cm to m

Multiply by a unit fraction that allows you to divide out centimeters.

Unit fraction

$$\frac{18.6 \text{ cm}}{1} \cdot \frac{1 \text{ m}}{100 \text{ cm}} = \frac{18.6}{100} \text{ m} = 0.186 \text{ m}$$

Do **not** write a period here.

18.6 cm = 0.186 m

There are 100 cm in a meter, so 18.6 cm will be a small part of a meter. The answer makes sense.

Work Problem **3** *at the Side.* ▶

3 First write the unit fraction needed to make each conversion. Then complete the conversion.

(a) 3.67 m to cm

unit fraction $\Big\}$ $\dfrac{100 \text{ cm}}{1 \text{ m}}$

(b) 92 cm to m

unit fraction $\Big\}$ $\dfrac{\text{m}}{\text{cm}}$

(c) 432.7 cm to m

unit fraction $\Big\}$ _____

(d) 65 mm to cm

unit fraction $\Big\}$ _____

(e) 0.9 m to mm

unit fraction $\Big\}$ _____

(f) 2.5 cm to mm

unit fraction $\Big\}$ _____

ANSWERS

3. **(a)** 367 cm **(b)** $\dfrac{1 \text{ m}}{100 \text{ cm}}$; 0.92 m

(c) $\dfrac{1 \text{ m}}{100 \text{ cm}}$; 4.327 m

(d) $\dfrac{1 \text{ cm}}{10 \text{ mm}}$; 6.5 cm

(e) $\dfrac{1000 \text{ mm}}{1 \text{ m}}$; 900 mm

(f) $\dfrac{10 \text{ mm}}{1 \text{ cm}}$; 25 mm

④ Do each multiplication or division by hand or on a calculator. Compare your answer to the one you get by moving the decimal point.

(a) $(43.5)(10) =$ _____

43.5 gives 435

(b) $43.5 \div 10 =$ _____

43.5 gives _____

(c) $(28)(100) =$ _____

28.00 gives _____

(d) $28 \div 100 =$ _____

28. gives _____

(e) $(0.7)(1000) =$ _____

0.700 gives _____

(f) $0.7 \div 1000 =$ _____

000.7 gives _____

OBJECTIVE **3** **Move the decimal point to convert among units.**
By now you have probably noticed that conversions among metric units are made by multiplying or dividing by 10, by 100, or by 1000. A quick way to *multiply* by 10 is to move the decimal point one place to the *right*. Move it two places to the right to multiply by 100, three places to multiply by 1000. *Dividing* is done by moving the decimal point to the *left* in the same manner.

◀ *Work Problem* ④ *at the Side.*

An alternate conversion method to unit fractions is moving the decimal point using this **metric conversion line.**

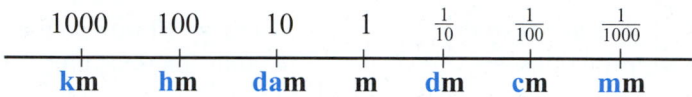

Here are the steps for using the conversion line.

Using the Metric Conversion Line

Step 1   Find the unit you are given on the metric conversion line.

Step 2   Count the number of places to get from the unit you are given to the unit you want in the answer.

Step 3   Move the decimal point the **same number of places** and in the **same direction** as you did on the conversion line.

EXAMPLE 3   **Using the Metric Conversion Line**

Use the metric conversion line to make the following conversions.

(a) 5.702 km to m

Find **km** on the metric conversion line. To get to **m**, you move *three places* to the *right*. So move the decimal point in 5.702 *three places* to the *right*.

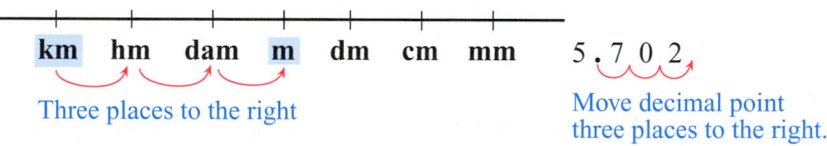

Three places to the right

Move decimal point three places to the right.

5.702 km = 5702 m

(b) 69.5 cm to m

Find **cm** on the conversion line. To get to **m**, move *two places* to the *left*.

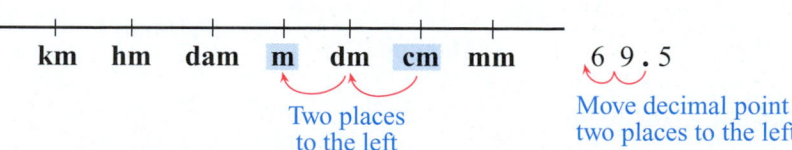

Two places to the left

Move decimal point two places to the left.

69.5 cm = 0.695 m

Continued on Next Page

(c) 8.1 cm to mm

From **cm** to **mm** is *one place* to the *right*.

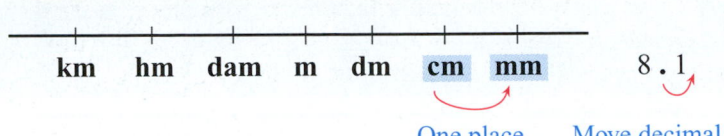

km   hm   dam   m   dm   **cm**   **mm**   8 . 1

One place   Move decimal point
to the right   one place to the right.

8.1 cm = 81 mm

────────────────────── Work Problem **5** at the Side. ▶

EXAMPLE 4 | **Practicing Length Conversions**

Convert using the metric conversion line.

(a) 1.28 m to mm

Moving from **m** to **mm** is going *three places to the right*. In order to move the decimal point in 1.28 three places to the right, you must write a 0 as a placeholder.

1.28**0**   Zero is written in as a placeholder.

Move decimal point
three places to the right.   The answer is a *whole number*,
                                so you do not need to write
1.28 m = 1280 mm   the decimal point.

(b) 60 cm to m

From **cm** to **m** is two places to the left. The decimal point in 60 starts at the *far right side* because 60 is a whole number. Then move it two places to the left.

60.   60.

Decimal point starts here.   Move decimal point
                              two places to the left.

60 cm = 0.60 m, and 0.60 m is equivalent to 0.6 m.

(c) 8 m to km

From **m** to **km** is three places to the left. The decimal point in 8 starts at the far right side. In order to move it three places to the left, you must write two zeros as placeholders.

                              ┌───── Two zeros are written
                              ↓↓   in as placeholders.
8.   **008.**

Decimal point starts here.   Move decimal point
                              three places to the left.

8 m = 0.008 km

────────────────────── Work Problem **6** at the Side. ▶

---

5 Convert using the metric conversion line.

(a) 12.008 km to m

(b) 561.4 m to km

(c) 20.7 cm to m

(d) 20.7 cm to mm

(e) 4.66 m to cm

(f) 85.6 mm to cm

6 Convert using the metric conversion line.

(a) 9 m to mm

(b) 3 cm to m

(c) 14.6 km to m

(d) 5 mm to cm

(e) 70 m to km

(f) 0.8 m to cm

ANSWERS

5. (a) 12,008 m **(b)** 0.5614 km
  (c) 0.207 m **(d)** 207 mm
  (e) 466 cm **(f)** 8.56 cm
6. (a) 9000 mm **(b)** 0.03 m
  (c) 14,600 m **(d)** 0.5 cm
  (e) 0.07 km **(f)** 80 cm

Math in the Media

MEASURING UP

Hair and Nail Growth

Q: How fast do hair and nails grow? Do they grow faster in the summer?

A: Fingernails grow, on average, about one-tenth of a millimeter per day, although there is considerable variation among individuals. Fingernails grow faster than toenails, and nails on the longest fingers appear to grow the fastest.

Fingernails, as well as hair and skin, grow faster in the summer, presumably under the influence of sunlight, which expands blood vessels, bringing more oxygen and nutrients to the area and allowing for faster growth.

The rate the scalp hair grows is 0.3 to 0.4 millimeter per day, or about 6 inches a year.

Source: Minneapolis Star Tribune.

1. **(a)** How much do fingernails grow in one week?

   (b) In one month?

   (c) In one year?

2. **(a)** Using metric units, how much does hair grow in one week?

   (b) In one month?

   (c) In one year?

3. When you have finished **Appendix A.4,** come back to this article. Is the statement about hair growing 6 inches a year accurate? Explain your answer.

4. **(a)** According to *Guinness World Records,* the longest toenails measure 15.2 cm. How many millimeters is that length? How many meters?

   (b) The longest eyelash is listed as 5.08 cm. How long is the eyelash in millimeters? In meters?

A.2 ▶▶▶ Exercises

Use your knowledge of the meaning of metric prefixes to fill in the blanks.

1. *kilo* means _____ so

1 km = _____ m

2. *deka* means _____ so

1 dam = _____ m

3. *milli* means _____ so

1 mm = _____ m

4. *deci* means _____ so

1 dm = _____ m

5. *centi* means _____ so

1 cm = _____ m

6. *hecto* means _____ so

1 hm = _____ m

Use this ruler to measure the width of your thumb and hand for Exercises 7–10.

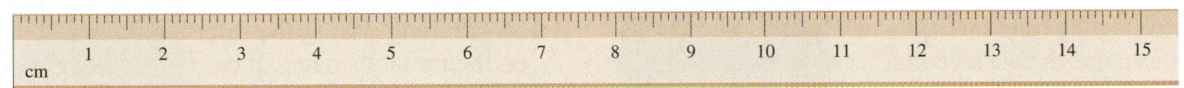

7. The width of your hand in centimeters

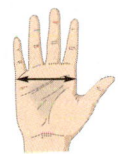

8. The width of your hand in millimeters

9. The width of your thumb in millimeters

10. The width of your thumb in centimeters

Write the most reasonable metric length unit in each blank. Choose from km, m, cm, and mm.
See Example 1.

11. The child was 91 _____ tall.

12. The cardboard was 3 _____ thick.

13. Ming-Na swam in the 200 _____ backstroke race.

14. The bookcase is 75 _____ wide.

15. Adriana drove 400 _____ on her vacation.

16. The door is 2 _____ high.

17. An aspirin tablet is 10 _____ across.

18. Lamard jogs 4 _____ every morning.

19. A paper clip is about 3 _____ long.

20. My pen is 145 _____ long.

21. Dave's truck is 5 _____ long.

22. Wheelchairs need doorways that are at least 80 _____ wide.

23. Describe at least three examples of metric length units that you have come across in your daily life.

24. Explain one reason the metric system would be easier for a child to learn than the U.S. customary system.

Convert each measurement. Use unit fractions or the metric conversion line. See Examples 2–4.

25. 7 m to cm

26. 18 m to cm

27. 40 mm to m

28. 6 mm to m

⊙ **29.** 9.4 km to m

30. 0.7 km to m

31. 509 cm to m

32. 30 cm to m

33. 400 mm to cm

34. 25 mm to cm

35. 0.91 m to mm

36. 4 m to mm

37. Is 82 cm greater than or less than 1 m? What is the difference in the lengths?

38. Is 1022 m greater than or less than 1 km? What is the difference in the lengths?

39. On the first page of this appendix, we said that computer microchips may be only 5 mm long and 1 mm wide. Using the ruler on the previous page, draw a rectangle that measures 5 mm by 1 mm. Then convert each measurement to centimeters.

A greatly enlarged photo of a computer microchip.

40. The world's smallest butterfly has a wingspan of 15 mm. The smallest mouse is 50 mm long. Using the ruler on the previous page, draw a line that is 15 mm long and a line 50 mm long. Then convert each measurement to centimeters. (*Source*: *Top 10 of Everything.*)

41. The Roe River near Great Falls, Montana, is the shortest river in the world, with a north fork that is just under 18 m long. How many kilometers long is the north fork of the river? (*Source: Guinness Book of Amazing Nature.*)

42. There are 60,000 km of blood vessels in the human body. How many meters of blood vessels are in the body? (*Source: Big Book of Knowledge.*)

43. The median height for U.S. females who are 20 to 29 years old is about 1.64 m. Convert this height to centimeters and to millimeters. (*Source*: U.S. National Center for Health Statistics.)

44. The median height for 20- to 29-year-old males in the United States is about 177 cm. Convert this height to meters and to millimeters. (*Source*: U.S. National Center for Health Statistics.)

45. Use two unit fractions to convert 5.6 mm to km.

46. Use two unit fractions to convert 16.5 km to mm.

A.3 ▶▶▶ The Metric System—Capacity and Weight (Mass)

We use capacity units to measure liquids, such as the amount of milk in a recipe, the gasoline in our car tank, and the water in an aquarium. (The capacity units in the U.S. customary system are cups, pints, quarts, and gallons.) The basic metric unit for capacity is the **liter** (also spelled *litre*). The capital letter L is the symbol for liter, to avoid confusion with the numeral 1.

OBJECTIVE 1 **Learn the basic metric units of capacity.** The liter is related to metric length in this way: a box that measures 10 cm on every side holds exactly one liter. (The volume of the box is 10 cm • 10 cm • 10 cm = 1000 cubic centimeters. Volume is discussed in **Section 7.6**.) A liter is just a little more than 1 quart.

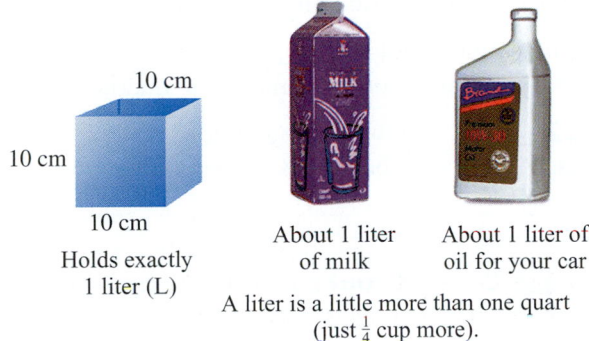

10 cm
10 cm
10 cm
Holds exactly 1 liter (L)

About 1 liter of milk

About 1 liter of oil for your car

A liter is a little more than one quart (just $\frac{1}{4}$ cup more).

In the metric system you use liters for things like buying milk and soda at the store, filling a pail with water, and describing the size of your home aquarium.

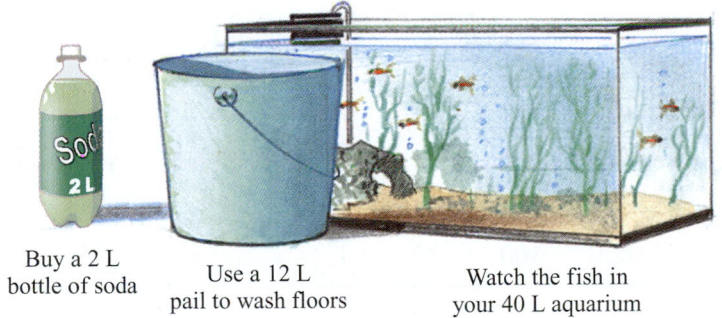

Buy a 2 L bottle of soda

Use a 12 L pail to wash floors

Watch the fish in your 40 L aquarium

Work Problem 1 *at the Side.* ▶

To make larger or smaller capacity units, we use the same **prefixes** as we did with length units. For example, *kilo* means 1000 so a *kilo*meter is 1000 meters. In the same way, a *kilo*liter is 1000 liters.

Prefix	kilo-liter	hecto-liter	deka-liter	liter	deci-liter	centi-liter	milli-liter
Meaning	1000 liters	100 liters	10 liters	1 liter	$\frac{1}{10}$ of a liter	$\frac{1}{100}$ of a liter	$\frac{1}{1000}$ of a liter
Symbol	**k**L	**h**L	**da**L	L	**d**L	**c**L	**m**L

↑ ↑
Capacity units used most often

1 Which things can be measured in liters?

Amount of water in the bathtub

Length of the bathtub

Width of your car

Amount of gasoline you buy for your car

Weight of your car

Height of a pail

Amount of water in a pail

ANSWER

1. water in bathtub, gasoline, water in a pail

2 Write the most reasonable metric unit in each blank. Choose from L and mL.

(a) I bought 8 _____ of soda at the store.

(b) The nurse gave me 10 _____ of cough syrup.

(c) This is a 100 _____ garbage can.

(d) It took 10 _____ of paint to cover the bedroom walls.

(e) My car's gas tank holds 50 _____.

(f) I added 15 _____ of oil to the pancake mix.

(g) The can of orange soda holds 350 _____.

(h) My friend gave me a 30 _____ bottle of expensive perfume.

The capacity units you will use most often in daily life are liters (L) and *milli*liters (mL). A tiny box that measures 1 cm on every side holds exactly one milliliter. (In medicine, this small amount is also called 1 cubic centimeter, or 1 cc for short.) It takes 1000 mL to make 1 L. Here are some useful comparisons.

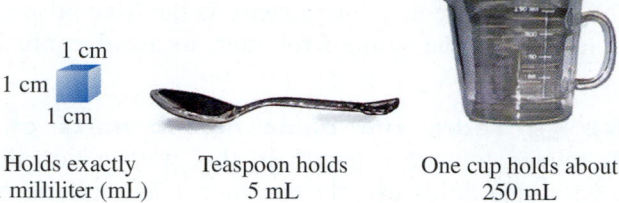

| Holds exactly 1 milliliter (mL) | Teaspoon holds 5 mL | One cup holds about 250 mL |

EXAMPLE 1 **Using Metric Capacity Units**

Write the most reasonable metric unit in each blank. Choose from L and mL.

(a) The bottle of shampoo held 500 _____.
    500 **mL** because 500 L would be about 500 quarts, which is too much.

(b) I bought a 2 _____ carton of orange juice.
    2 **L** because 2 mL would be less than a teaspoon.

◀ *Work Problem* **2** *at the Side.*

OBJECTIVE **2** **Convert among metric capacity units.**   Just as with length units, you can convert between milliliters and liters using unit fractions.

Metric Capacity Relationships

1 L = 1000 mL, so the unit fractions are:

$$\frac{1\ L}{1000\ mL} \quad \text{or} \quad \frac{1000\ mL}{1\ L}$$

Or you can use a metric conversion line to decide how to move the decimal point.

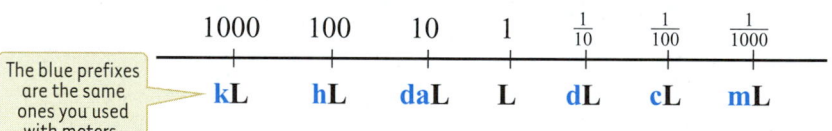

The blue prefixes are the same ones you used with meters.

EXAMPLE 2 **Converting among Metric Capacity Units**

Convert using the metric conversion line or unit fractions.

(a) 2.5 L to mL
    Using the metric conversion line:
    From **L** to **mL** is *three places* to the *right*.

    2.5̲0̲0̲,   Write two zeros as placeholders.

    2.5 L = 2500 mL

    Using unit fractions:

    Multiply by a unit fraction that allows you to divide out liters.

    $$\frac{2.5\ \cancel{L}}{1} \cdot \frac{1000\ mL}{1\ \cancel{L}} = 2500\ mL$$

    L divides out, leaving **mL** for your answer.

Continued on Next Page

(b) 80 mL to L

Using the metric conversion line:

From **mL** to **L** is *three places* to the *left*.

80.   080.

Decimal point starts here.   Move decimal point three places to the left.

80 mL = 0.080 L or 0.08 L

Using unit fractions:

Multiply by a unit fraction that allows you to divide out mL.

$$\frac{80 \text{ mL}}{1} \cdot \frac{1 \text{ L}}{1000 \text{ mL}}$$

$$= \frac{80}{1000} \text{ L} = 0.08 \text{ L}$$

Do **not** write a period here.

Work Problem **3** *at the Side.* ▶

OBJECTIVE **3** **Learn the basic metric units of weight (mass).**
The **gram** is the basic metric unit for *mass*. Although we often call it "weight," there is a difference. Weight is a measure of the pull of gravity; the farther you are from the center of Earth, the less you weigh. In outer space you become weightless, but your mass, the amount of matter in your body, stays the same regardless of where you are. In science courses, it will be important to distinguish between the weight of an object and its mass. But for everyday purposes, we will use the word *weight*.

The gram is related to metric length in this way: The weight of the water in a box measuring 1 cm on every side is 1 gram. This is a very tiny amount of water (1 mL) and a very small weight. One gram is also the weight of a dollar bill or a single raisin. A nickel weighs 5 grams. A plain, regular-sized hamburger and bun weighs from 175 to 200 grams.

The 1 mL of water in this tiny box weighs 1 gram.

A nickel weighs 5 grams.

A dollar bill weighs 1 gram.

A plain hamburger weighs 175 to 200 grams.

Work Problem **4** *at the Side.* ▶

---

3 Convert.

(a) 9 L to mL

(b) 0.75 L to mL

(c) 500 mL to L

(d) 5 mL to L

(e) 2.07 L to mL

(f) 3275 mL to L

4 Which things would weigh about 1 gram?

A small paper clip

A pair of scissors

One playing card from a deck of cards

A calculator

An average-sized apple

The check you wrote to the cable company

ANSWERS

3. **(a)** 9000 mL   **(b)** 750 mL   **(c)** 0.5 L
   (d) 0.005 L   **(e)** 2070 mL   **(f)** 3.275 L
4. paper clip, playing card, check

5 Write the most reasonable metric unit in each blank. Choose from kg, g, and mg.

(a) A thumbtack weighs

800 _____.

(b) A teenager weighs

50 _____.

(c) This large cast-iron frying pan weighs

1 _____.

(d) Jerry's basketball weighed 600 _____.

(e) Tamlyn takes a

500 _____ calcium tablet every morning.

(f) On his diet, Greg can eat

90 _____ of meat for lunch.

(g) One strand of hair weighs

2 _____.

(h) One banana might weigh

150 _____.

ANSWERS

5. **(a)** mg   **(b)** kg   **(c)** kg
  (d) g   **(e)** mg   **(f)** g
  (g) mg   **(h)** g

To make larger or smaller weight units, we use the same **prefixes** as we did with length and capacity units. For example, *kilo* means 1000 so a *kilo*meter is 1000 meters, a *kilo*liter is 1000 liters, and a *kilo*gram is 1000 grams.

Prefix	kilo-gram	hecto-gram	deka-gram	gram	deci-gram	centi-gram	milli-gram
Meaning	1000 grams	100 grams	10 grams	1 gram	$\frac{1}{10}$ of a gram	$\frac{1}{100}$ of a gram	$\frac{1}{1000}$ of a gram
Symbol	**kg**	**hg**	**da**g	g	**dg**	**cg**	**mg**

Weight (mass) units that are used most often

The units you will use most often in daily life are kilograms (kg), grams (g), and milligrams (mg). *Kilo*grams are used instead of pounds. A kilogram is 1000 grams. It is about 2.2 pounds. Two packages of butter plus one stick of butter weigh about 1 kg. An average newborn baby weighs 3 to 4 kg; a college football player might weigh 100 to 130 kg.

1 kilogram is about 2.2 pounds   100 to 130 kg   3 to 4 kg

Extremely small weights are measured in *milli*grams. It takes 1000 mg to make 1 g. Recall that a dollar bill weighs about 1 g. Imagine cutting it into 1000 pieces; the weight of one tiny piece would be 1 mg. Dosages of medicine and vitamins are given in milligrams. You will also use milligrams in science classes.

Cut a dollar bill into 1000 pieces.
One tiny piece weighs 1 milligram.

EXAMPLE 3 **Using Metric Weight Units**

Write the most reasonable metric unit in each blank. Choose from kg, g, and mg.

(a) Ramon's suitcase weighed 20 _____.
20 **kg** because kilograms are used instead of pounds.
20 kg is about 44 pounds.

(b) LeTia took a 350 _____ aspirin tablet.
350 **mg** because 350 g would be more than the weight of a hamburger, which is too much.

(c) Jenny mailed a letter that weighed 30 _____.
30 **g** because 30 kg would be much too heavy and 30 mg is less than the weight of a dollar bill.

◀ *Work Problem* **5** *at the Side.*

OBJECTIVE 4 Convert among metric weight (mass) units. As with length and capacity, you can convert among metric weight units by using unit fractions. The unit fractions you need are shown here.

Metric Weight (Mass) Relationships

1 kg = 1000 g so the unit fractions are:	**1 g = 1000 mg** so the unit fractions are:
$\dfrac{1 \text{ kg}}{1000 \text{ g}}$ or $\dfrac{1000 \text{ g}}{1 \text{ kg}}$	$\dfrac{1 \text{ g}}{1000 \text{ mg}}$ or $\dfrac{1000 \text{ mg}}{1 \text{ g}}$

Or you can use a metric conversion line to decide how to move the decimal point.

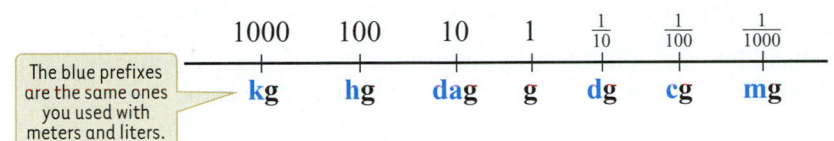

1000	100	10	1	$\frac{1}{10}$	$\frac{1}{100}$	$\frac{1}{1000}$
kg	**hg**	**dag**	**g**	**dg**	**cg**	**mg**

The blue prefixes are the same ones you used with meters and liters.

EXAMPLE 4   Converting among Metric Weight Units

Convert using the metric conversion line or unit fractions.

(a) 7 mg to g

Using the metric conversion line:
From **mg** to **g** is *three places* to the *left.*

7.   007.
↑
Decimal point   Move decimal
starts here.   point three places
                   to the left.

7 mg = 0.007 g

Using unit fractions:

Multiply by a unit fraction that allows you to divide out mg.

$$\frac{7 \text{ mg}}{1} \cdot \frac{1 \text{ g}}{1000 \text{ mg}} = \frac{7}{1000} \text{ g}$$

$$= 0.007 \text{ g}$$

Three decimal places for thousandths.

(b) 13.72 kg to g

Using the metric conversion line:
From **kg** to **g** is *three places* to the *right.*

13.720   Decimal point moves
          three places to
          the right.

13.72 kg = 13,720 g

A comma (not a decimal point)

Using unit fractions:

Multiply by a unit fraction that allows you to divide out kg.

$$\frac{13.72 \text{ kg}}{1} \cdot \frac{1000 \text{ g}}{1 \text{ kg}} = 13,720 \text{ g}$$

A comma (not a decimal point)

Work Problem **6** *at the Side.* ▶

---

6 Convert.

(a) 10 kg to g

(b) 45 mg to g

(c) 6.3 kg to g

(d) 0.077 g to mg

(e) 5630 g to kg

(f) 90 g to kg

7 First decide which type of units are needed: length, capacity, or weight. Then write the most appropriate unit in the blank. Choose from km, m, cm, mm, L, mL, kg, g, and mg.

(a) Gail bought a 4 _____ can of paint.

Use _____ units.

(b) The bag of chips weighed 450 _____.

Use _____ units.

(c) Give the child 5 _____ of cough syrup.

Use _____ units.

(d) The width of the window is 55 _____.

Use _____ units.

(e) Akbar drives 18 _____ to work.

Use _____ units.

(f) The laptop computer weighs 2 _____.

Use _____ units.

(g) A credit card is 55 _____ wide.

Use _____ units.

OBJECTIVE 5 Distinguish among basic metric units of length, capacity, and weight (mass). As you encounter things to be measured at home, on the job, or in your classes at school, be careful to use the correct type of measurement unit.

Use *length units* (kilometers, meters, centimeters, millimeters) to measure:

how long	how high	how far away
how wide	how tall	how far around (perimeter)
how deep	distance	

Use *capacity units* (liters, milliliters) to measure liquids (things that can be poured) such as:

water	shampoo	gasoline
milk	perfume	oil
soft drinks	cough syrup	paint

Also use liters and milliliters to describe how much liquid something can hold, such as an eyedropper, measuring cup, pail, or bathtub.

Use *weight units* (kilograms, grams, milligrams) to measure:

the weight of something   how heavy something is

In **Chapter 7** you used square units (such as square meters) to measure area, and cubic units (such as cubic centimeters) to measure volume.

EXAMPLE 5   Using a Variety of Metric Units

First decide which type of units are needed: length, capacity, or weight. Then write the most appropriate metric unit in the blank. Choose from km, m, cm, mm, L, mL, kg, g, and mg.

(a) The letter needs another stamp because it weighs 40 _____.

Use _____ units.

The letter weighs 40 **g** because 40 mg is less than the weight of a dollar bill and 40 kg would be about 88 pounds.

Use **weight** units because of the word "weighs."

(b) The swimming pool is 3 _____ deep at the deep end.

Use _____ units.

The pool is 3 **m** deep because 3 cm is only about an inch and 3 km is more than a mile.

Use **length** units because of the word "deep."

(c) This is a 340 _____ can of juice.

Use _____ units.

It is a 340 **mL** can because 340 liters would be more than 340 quarts.

Use **capacity** units because juice is a liquid.

◀ *Work Problem* **7** *at the Side.*

A.3 ▶▶▶ Exercises

Write the most reasonable metric unit in each blank. Choose from L, mL, kg, g, and mg. See Examples 1 and 3.

1. The glass held 250 _____ of water.

2. Hiromi used 12 _____ of water to wash the kitchen floor.

3. Dolores can make 10 _____ of soup in that pot.

4. Jay gave 2 _____ of vitamin drops to the baby.

5. Our yellow Labrador dog grew up to weigh 40 _____.

6. A small safety pin weighs 750 _____.

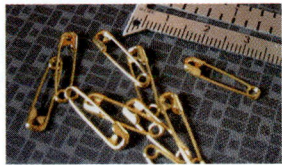

7. Lori caught a small sunfish weighing 150 _____.

8. One dime weighs 2 _____.

9. Andre donated 500 _____ of blood today.

10. Barbara bought the 2 _____ bottle of cola.

11. The patient received a 250 _____ tablet of medication each hour.

12. The 8 people on the elevator weighed a total of 500 _____.

13. The gas can for the lawn mower holds 4 _____.

14. Kevin poured 10 _____ of vanilla into the mixing bowl.

15. Pam's backpack weighs 5 _____ when it is full of books.

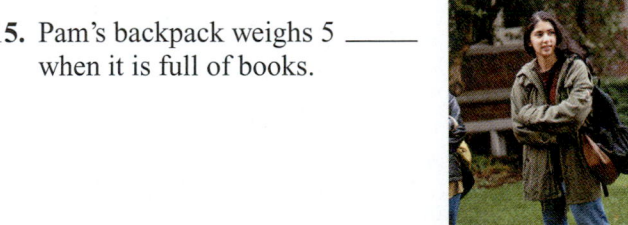

16. One grain of salt weighs 2 _____.

Today, medical measurements are usually given in the metric system. Since we convert among metric units of measure by moving the decimal point, it is possible that mistakes can be made. Examine the following dosages and indicate whether they are reasonable or unreasonable. If a dose is unreasonable, indicate whether it is too much or too little.

17. Drink 4.1 L of Kaopectate after each meal.

18. Drop 1 mL of solution into the eye twice a day.

19. Soak your feet in 5 kg of Epsom salts per liter of water.

20. Inject 0.5 L of insulin each morning.

21. Take 15 mL of cough syrup every four hours.

22. Take 200 mg of vitamin C each day.

23. Take 350 mg of aspirin three times a day.

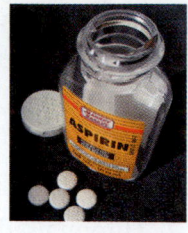

24. Buy a tube of ointment weighing 0.002 g.

25. Describe at least two examples of metric capacity units and two examples of metric weight units that you have come across in your daily life.

26. Explain in your own words how the meter, liter, and gram are related.

27. Describe how you decide which unit fraction to use when converting 6.5 kg to grams.

28. Write an explanation of each step you would use to convert 20 mg to grams using the metric conversion line.

Convert each measurement. Use unit fractions or the metric conversion line. See Examples 2 and 4.

29. 15 L to mL

30. 6 L to mL

31. 3000 mL to L

32. 18,000 mL to L

33. 925 mL to L

34. 200 mL to L

35. 8 mL to L

36. 25 mL to L

37. 4.15 L to mL

38. 11.7 L to mL

39. 8000 g to kg

40. 25,000 g to kg

41. 5.2 kg to g

42. 12.42 kg to g

43. 0.85 g to mg

44. 0.2 g to mg

45. 30,000 mg to g

46. 7500 mg to g

47. 598 mg to g

48. 900 mg to g

49. 60 mL to L

50. 6.007 kg to g

51. 3 g to kg

52. 12 mg to g

53. 0.99 L to mL

54. 13,700 mL to L

Write the most appropriate metric unit in each blank. Choose from km, m, cm, mm, L, mL, kg, g, and mg. See Example 5.

55. The masking tape is 19 _____ wide.

56. The roll has 55 _____ of tape on it.

57. Buy a 60 _____ jar of acrylic paint for art class.

58. One onion weighs 200 _____.

59. My waist measurement is 65 _____.

60. Add 2 _____ of windshield washer fluid to your car.

61. A single postage stamp weighs 90 _____.

62. The hallway is 10 _____ long.

Solve each application problem. Show your work. (Source for Exercises 63–68: Top 10 of Everything.)

63. Human skin has about 3 million sweat glands, which release an average of 300 mL of sweat per day. How many liters of sweat are released each day?

64. In hot climates, the sweat glands in a person's skin may release up to 3.5 L of sweat in one day. How many milliliters is that?

65. The average weight of a human brain is 1.34 kg. How many grams is that?

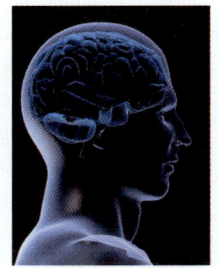

66. A healthy human heart pumps about 70 mL of blood per beat. How many liters of blood does it pump per beat?

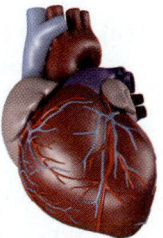

67. On average, we breathe in and out roughly 900 mL of air every 10 seconds. How many liters of air is that?

68. In the Victorian era, people believed that heavier brains meant greater intelligence. They were impressed that Otto von Bismarck's brain weighed 1907 g, which is how many kilograms?

69. A small adult cat weighs from 3000 g to 4000 g. How many kilograms is that? (*Source: Lyndale Animal Hospital.*)

70. If the letter you are mailing weighs 29 g, you must put additional postage on it. How many kilograms does the letter weigh? (*Source:* U.S. Postal Service.)

71. Is 1005 mg greater than or less than 1 g? What is the difference in the weights?

72. Is 990 mL greater than or less than 1 L? What is the difference in the amounts?

73. One nickel weighs 5 g. How many nickels are in 1 kg of nickels?

74. The ratio of the total length of all the fish to the amount of water in an aquarium can be 3 cm of fish for every 4 L of water. What is the total length of all fish you can put in a 40 L aquarium? (*Source: Tropical Aquarium Fish.*)

Relating Concepts (Exercises 75–78) For Individual or Group Work

*Recall that the prefix **kilo** means 1000, so a **kilo**meter is 1000 meters. You'll learn about other prefixes for numbers greater than 1000 as you **work Exercises 75–78 in order.***

75. (a) The prefix *mega* means one million. Use the symbol M (capitalized) for *mega*. So a megameter (Mm) is how many meters?

   1 Mm = _____ m

   (b) Figure out a unit fraction that you can use to convert megameters to meters. Then use it to convert 3.5 Mm to meters.

76. (a) The prefix *giga* means one billion. Use the symbol G (capitalized) for *giga*. So a gigameter (Gm) is how many meters?

   1 Gm = _____ m

   (b) Figure out a unit fraction you can use to convert meters to gigameters. Then use it to convert 2500 m to gigameters.

77. (a) The prefix *tera* means one trillion. Use the symbol T (capitalized) for *tera*. So a *tera*meter (Tm) is how many meters?

   1 Tm = _____ m.

   (b) Think carefully before you fill in the blanks:

   1 Tm = _____ Gm

   1 Tm = _____ Mm

78. A computer's memory is measured in *bytes*. A byte can represent a single letter, a digit, or a punctuation mark. The memory for a desktop computer may be measured in megabytes (abbreviated MB) or gigabytes (abbreviated GB). Using the meanings of *mega* and *giga*, it would seem that

   1 MB = _____ bytes and

   1 GB = _____ bytes.

   However, because computers use a base 2 or binary system, 1 MB is actually 2^{20} and 1 GB is 2^{30}. Use your calculator to find the actual values.

   2^{20} = _____ 2^{30} = _____

A.4 ▶▶▶ Metric-U.S. Customary Conversions and Temperature

OBJECTIVE 1 **Use unit fractions to convert between metric and U.S. customary units.** Until the United States has switched completely from customary units to the metric system, it will be necessary to make conversions from one system to the other. *Approximate* conversions can be made with the help of the table below, in which the values have been rounded to the nearest hundredth or thousandth. (The only value that is exact, not rounded, is 1 inch = 2.54 cm.)

Metric to U.S. Customary		U.S. Customary to Metric	
1 kilometer	≈ 0.62 mile	1 mile	≈ 1.61 kilometers
1 meter	≈ 1.09 yards	1 yard	≈ 0.91 meter
1 meter	≈ 3.28 feet	1 foot	≈ 0.30 meter
1 centimeter	≈ 0.39 inch	1 inch	= 2.54 centimeters
1 liter	≈ 0.26 gallon	1 gallon	≈ 3.79 liters
1 liter	≈ 1.06 quarts	1 quart	≈ 0.95 liter
1 kilogram	≈ 2.20 pounds	1 pound	≈ 0.45 kilogram
1 gram	≈ 0.035 ounce	1 ounce	≈ 28.35 grams

EXAMPLE 1 **Converting between Metric and U.S. Customary Length Units**

Convert 10 m to yards using unit fractions. Round your answer to the nearest tenth if necessary.

We're changing from a *metric* unit to a *U.S. customary* unit. In the "Metric to U.S. Customary" side of the table above, you see that 1 meter ≈ 1.09 yards. Two unit fractions can be written using that information.

$$\frac{1 \text{ m}}{1.09 \text{ yd}} \quad \text{or} \quad \frac{1.09 \text{ yd}}{1 \text{ m}}$$

Multiply by the unit fraction that allows you to divide out meters (that is, meters is in the denominator).

$$10 \text{ m} \cdot \frac{1.09 \text{ yd}}{1 \text{ m}} = \frac{10 \text{ m}}{1} \cdot \frac{1.09 \text{ yd}}{1 \text{ m}} = \frac{(10)(1.09 \text{ yd})}{1} = 10.9 \text{ yd}$$

These units should match.

Meters (**m**) divide out leaving **yd**, the unit you want for the answer.

10 m ≈ 10.9 yd

Note

In Example 1 above, you could also use the numbers from the "U.S. Customary to Metric" side of the table that involve meters and yards: 1 yard ≈ 0.91 meter.

$$\frac{10 \text{ m}}{1} \cdot \frac{1 \text{ yd}}{0.91 \text{ m}} = \frac{10}{0.91} \text{ yd} ≈ 10.99 \text{ yd}$$

The answer is slightly different because the values in the table are rounded. Also, you have to divide instead of multiply, which is usually more difficult to do without a calculator. We will use the first method in this chapter.

Work Problem ① at the Side. ▶

OBJECTIVES

1 Use unit fractions to convert between metric and U.S. customary units.

2 Learn common temperatures on the Celsius scale.

3 Use formulas to convert between Celsius and Fahrenheit temperatures.

1 Convert using unit fractions. Round your answers to the nearest tenth.

(a) 23 m to yards

(b) 40 cm to inches

(c) 5 mi to kilometers (Look at the "U.S. Customary to Metric" side of the table.)

(d) 12 in. to centimeters

ANSWERS

1. (a) 23 m ≈ 25.1 yd **(b)** 40 cm ≈ 15.6 in.
(c) 5 mi ≈ 8.1 km **(d)** 12 in. ≈ 30.5 cm

2 Convert. Use the values from the table on the previous page to make unit fractions. Round answers to the nearest tenth.

(a) 17 kg to pounds

(b) 5 L to quarts

(c) 90 g to ounces

(d) 3.5 gal to liters

(e) 145 lb to kilograms

(f) 8 oz to grams

EXAMPLE 2   **Converting between Metric and U.S. Customary Weight and Capacity Units**

Convert using unit fractions. Round your answers to the nearest tenth.

(a) 3.5 kg to pounds

Look in the "Metric to U.S. Customary" side of the table on the previous page to see that 1 kilogram ≈ 2.20 pounds. Use this information to write a unit fraction that allows you to divide out kilograms.

$$\frac{3.5 \ \cancel{kg}}{1} \cdot \frac{2.20 \ lb}{1 \ \cancel{kg}} = \frac{(3.5)(2.20 \ lb)}{1} = 7.7 \ lb$$

3.5 kg ≈ 7.7 lb   *The conversion value is approximate, so use the ≈ symbol in your answer.*

(b) 18 gal to liters

Look in the "U.S. Customary to Metric" side of the table to see that 1 gallon ≈ 3.79 liters. Write a unit fraction that allows you to divide out gallons.

gal divides out, leaving L for your answer.

$$\frac{18 \ \cancel{gal}}{1} \cdot \frac{3.79 \ L}{1 \ \cancel{gal}} = \frac{(18)(3.79 \ L)}{1} = 68.22 \ L$$

68.22 rounded to the nearest tenth is 68.2

18 gal ≈ 68.2 L

(c) 300 g to ounces

In the "Metric to U.S. Customary" side of the table, 1 gram ≈ 0.035 ounce.

$$\frac{300 \ \cancel{g}}{1} \cdot \frac{0.035 \ oz}{1 \ \cancel{g}} = \frac{(300)(0.035 \ oz)}{1} = 10.5 \ oz$$

300 g ≈ 10.5 oz

> **CAUTION**
> Because the metric and U.S. customary systems were developed independently, almost all comparisons are approximate. Your answers should be written with the "≈" symbol to show they are approximate.

◀ *Work Problem* **2** *at the Side.*

OBJECTIVE **2** **Learn common temperatures on the Celsius scale.** In the metric system, temperature is measured on the **Celsius** scale. On the Celsius scale, water freezes at 0 °C and boils at 100 °C. The small raised circle stands for "degrees" and the capital **C** is for Celsius. Read the temperatures like this:

Water freezes at 0 degrees Celsius (0 °C).

Water boils at 100 degrees Celsius (100 °C).

The U.S. customary temperature system, used only in the United States, is measured on the **Fahrenheit** scale. On this scale:

Water freezes at 32 degrees Fahrenheit (32 °F).

Water boils at 212 degrees Fahrenheit (212 °F).

ANSWERS

2. (a) 17 kg ≈ 37.4 lb   **(b)** 5 L ≈ 5.3 qt
   (c) 90 g ≈ 3.2 oz   **(d)** 3.5 gal ≈ 13.3 L
   (e) 145 lb ≈ 65.3 kg   **(f)** 8 oz ≈ 226.8 g

The thermometer below shows some typical temperatures in both Celsius and Fahrenheit. For example, comfortable room temperature is about 20 °C or 68 °F, and normal body temperature is about 37 °C or 98.6 °F.

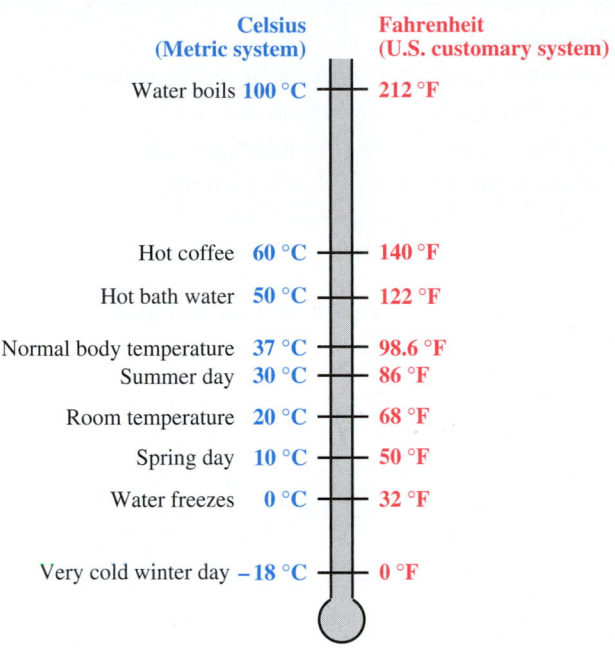

	Celsius (Metric system)	Fahrenheit (U.S. customary system)
Water boils	100 °C	212 °F
Hot coffee	60 °C	140 °F
Hot bath water	50 °C	122 °F
Normal body temperature	37 °C	98.6 °F
Summer day	30 °C	86 °F
Room temperature	20 °C	68 °F
Spring day	10 °C	50 °F
Water freezes	0 °C	32 °F
Very cold winter day	−18 °C	0 °F

Note

The freezing and boiling temperatures are exact. The other temperatures are approximate. Even normal body temperature varies slightly from person to person.

EXAMPLE 3 Using Celsius Temperatures

Circle the Celsius temperature that is most reasonable for each situation.

(a) Warm summer day   29 °C   64 °C   90 °C

29 °C is reasonable. 64 °C and 90 °C are too hot; they're both above the temperature of hot bath water (above 122 °F).

(b) Inside a freezer   −10 °C   3 °C   25 °C

−10 °C is the reasonable temperature because it is the only one below the freezing point of water (0 °C). Your frozen foods would start thawing at 3 °C or 25 °C.

Work Problem **3** *at the Side.* ▶

OBJECTIVE 3 Use formulas to convert between Celsius and Fahrenheit temperatures. You can use these formulas to convert between Celsius and Fahrenheit temperatures.

Celsius–Fahrenheit Conversion Formulas

Converting from Fahrenheit (F) to Celsius (C)

$$C = \frac{5(F - 32)}{9}$$

Converting from Celsius (C) to Fahrenheit (F)

$$F = \frac{9 \cdot C}{5} + 32$$

3 Circle the Celsius temperature that is *most* reasonable for each situation.

(a) Set the living room thermostat at:
11 °C   21 °C   71 °C

(b) The baby has a fever of:
29 °C   39 °C   49 °C

(c) Wear a sweater outside because it's:
15 °C   25 °C   50 °C

(d) My iced tea is:
−5 °C   5 °C   30 °C

(e) Time to go swimming! It's:
95 °C   65 °C   35 °C

(f) Inside a refrigerator (not the freezer) it's:
−15 °C   0 °C   3 °C

(g) There's a blizzard outside. It's:
12 °C   4 °C   −20 °C

(h) I need hot water to get these clothes clean. It should be:
55 °C   105 °C   200 °C

ANSWERS

3. **(a)** 21 °C   **(b)** 39 °C   **(c)** 15 °C
   (d) 5 °C   **(e)** 35 °C   **(f)** 3 °C
   (g) −20 °C   **(h)** 55 °C

As you use these formulas, be sure to follow the order of operations from **Section 1.8**.

4 Convert to Celsius.

(a) 59 °F

(b) 41 °F

(c) 212 °F

(d) 98.6 °F

> **Order of Operations**
> 1. Do all operations inside **parentheses** or **other grouping symbols**.
> 2. Simplify any expressions with **exponents** and find any **square roots**.
> 3. **Multiply** or **divide**, proceeding from left to right.
> 4. **Add** or **subtract**, proceeding from left to right.

EXAMPLE 4   **Converting Fahrenheit to Celsius**

Convert 68 °F to Celsius.

Use the formula and follow the order of operations.

$$C = \frac{5(\mathbf{F} - 32)}{9}$$ Fahrenheit to Celsius formula.

$$= \frac{5(\mathbf{68} - 32)}{9}$$ Work inside parentheses first; $68 - 32$ is 36.

$$= \frac{5(\mathbf{36})}{9}$$

$$= \frac{5(3\overset{4}{\cancel{6}})}{\underset{1}{\cancel{9}}}$$ Divide out the common factor. Multiply in the numerator.

$$= 20$$

Thus, 68 °F = 20 °C.

◀ *Work Problem* **4** *at the Side.*

5 Convert to Fahrenheit.

(a) 100 °C

(b) 25 °C

(c) 80 °C

(d) 5 °C

EXAMPLE 5   **Converting Celsius to Fahrenheit**

Convert 15 °C to Fahrenheit.

Use the formula and follow the order of operations.

$$F = \frac{9 \cdot \mathbf{C}}{5} + 32$$ Celsius to Fahrenheit formula.

$$= \frac{9 \cdot \mathbf{15}}{5} + 32$$

$$= \frac{9 \cdot \overset{3}{\cancel{15}}}{\underset{1}{\cancel{5}}} + 32$$ Divide out the common factor. Multiply in the numerator.

$$= 27 + 32$$ Add.

$$= 59$$

Thus, 15 °C = 59 °F.

◀ *Work Problem* **5** *at the Side.*

Use the table on the first page of this section and unit fractions to make approximate conversions from metric to U.S. customary or U.S. customary to metric. Round your answers to the nearest tenth. See Examples 1 and 2.

1. 20 m to yards

2. 8 km to miles

3. 80 m to feet

4. 85 cm to inches

5. 16 ft to meters

6. 3.2 yd to meters

7. 150 g to ounces

8. 2.5 oz to grams

9. 248 lb to kilograms

10. 7.68 kg to pounds

11. 28.6 L to quarts

12. 15.75 L to gallons

13. For the 2000 Olympics, the 3M Company used 5 g of pure gold to coat Michael Johnson's track shoes. (*Source:* 3M Company.)

(a) How many ounces of gold were used, to the nearest tenth?

(b) Was this enough extra weight to slow him down?

14. The label on a Van Ness auto feeder for cats and dogs says it holds 1.4 kg of dry food. How many pounds of food does it hold, to the nearest tenth? (*Source:* Van Ness Plastics.)

15. The heavy-duty wash cycle in a dishwater uses 8.4 gal of water. How many liters does it use, to the nearest tenth? (*Source:* Frigidaire.)

16. The rinse-and-hold cycle in a dishwasher uses only 4.5 L of water. How many gallons does it use, to the nearest tenth? (*Source:* Frigidaire.)

17. The smallest pet fish are dwarf gobies, which are half an inch long. How many centimeters long is a dwarf gobie, to the nearest tenth? (*Hint*: Write half an inch in decimal form.)

18. The fastest nerve signals in the human body travel 120 meters per second. How many feet per second do the signals travel? (*Source: Big Book of Knowledge.*)

Circle the most reasonable Celsius temperature for each situation. See Example 3.

19. A snowy day

    12 °C   28 °C   −8 °C

20. Brewing coffee

    80 °C   180 °C   15 °C

21. A high fever

    21 °C   40 °C   103 °C

22. Swimming pool water

    90 °C   78 °C   25 °C

23. Oven temperature

    150 °C   50 °C   30 °C

24. Light jacket weather

    0 °C   10 °C   −10 °C

25. Would a drop in temperature of 20 Celsius degrees be more or less than a drop of 20 Fahrenheit degrees? Explain your answer.

26. Describe one advantage of switching from the Fahrenheit temperature scale to the Celsius scale. Describe one disadvantage.

Use the conversion formulas from this section and the order of operations to convert Fahrenheit temperatures to Celsius and Celsius temperatures to Fahrenheit. Round your answers to the nearest degree if necessary. See Examples 4 and 5.

27. 60 °F

28. 80 °F

29. 104 °F

30. 36 °F

31. 8 °C

32. 18 °C

33. 35 °C

34. 0 °C

Solve each application problem. Round your answers to the nearest degree if necessary.

35. The highest temperature ever recorded on Earth was 136 °F at El Azizia, Libya, in 1922. Convert this temperature to Celsius. (*Source: The World Almanac.*)

36. Hummingbirds have a normal body temperature of 107 °F. But on cold nights they go into a state of torpor where their body temperature drops to 39 °F. What are these temperatures in Celsius? (*Source: Wildbird.*)

37. **(a)** Here is the tag on a pair of Sorel boots. In what kind of weather would you wear these boots?

Comfort range 24 °C to 4 °C

Source: Sorel.

(b) For what Fahrenheit temperatures are the boots designed?

(c) What range of metric temperatures do you have in January where you live? Would you be comfortable in these boots?

38. Sleeping bags made by Eddie Bauer are sold around the world. Each type of sleeping bag is designed for outdoor camping in certain temperatures.

Junior bag	5 °C or warmer
Removable liner bag	0 °C to 15 °C
Conversion bag	−7 °C to 0 °C

Source: Eddie Bauer.

(a) At what Fahrenheit temperatures should you use the Junior bag?

(b) What Fahrenheit temperatures is the removable liner bag designed for?

The article below appeared in American newspapers. However, both Newfoundland (part of Canada) and Ireland use the metric system. Their newspapers would have reported all the measurements in metric units. Complete the conversions to metric, rounding answers to the nearest tenth.

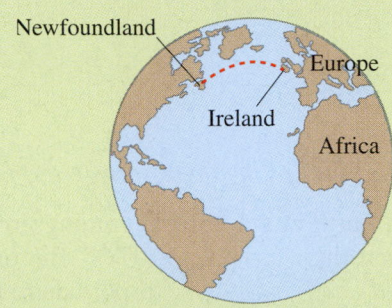

Q. A recent news brief reported on some men who flew a model airplane from Newfoundland to Ireland. Can you provide some details of the flight?

A. The model plane is 6 feet long and weighs 11 pounds. Made of balsa wood and mylar, it crossed the Atlantic—the flight path took it 1,888.3 miles—in 38 hours, 23 minutes. It soared at a cruising altitude of 1,000 feet. The plane used a souped-up piston engine and carried less than a gallon of fuel, as mandated by rules of the Federation Aeronautique Internationale, the governing body of model airplane building. When it landed in County Galway, Ireland, it had less than 2 fluid ounces of fuel left. The plane was built by Maynard Hill of Silver Spring, Maryland.

(*Source: New York Times.*)

39. Length of model plane

40. Weight of plane

41. Length of flight path

42. Time of flight

43. Cruising altitude

44. Fuel at the start, in milliliters

45. Fuel left after landing, in milliliters
(*Hint:* First convert 2 fl oz to quarts.)

46. What *percent* of the fuel was left at the end of the flight?

Appendix B: Inductive and Deductive Reasoning

Inductive and Deductive Reasoning

◀◀◀ **Appendix B**

OBJECTIVE 1 Use inductive reasoning to analyze patterns. In many scientific experiments, conclusions are drawn from specific outcomes. After many repetitions and similar outcomes, the findings are generalized into statements that appear to be true. When general conclusions are drawn from specific observations, we are using a type of reasoning called **inductive reasoning.** The next several examples illustrate this type of reasoning.

EXAMPLE 1 Using Inductive Reasoning

Find the next number in the sequence 3, 7, 11, 15, ...

To discover a pattern, calculate the difference between each pair of successive numbers.

$$7 - 3 = 4$$
$$11 - 7 = 4$$
$$15 - 11 = 4$$

Notice that the difference is always 4. Each number is 4 greater than the previous one. Thus, the next number in the pattern is $15 + 4$, or 19.

Work Problem **1** *at the Side.* ▶

EXAMPLE 2 Using Inductive Reasoning

Find the next number in this sequence.

$$7, 11, 8, 12, 9, 13, . . .$$

The pattern in this example involves addition and subtraction.

$$7 + 4 = 11$$
$$11 - 3 = 8$$
$$8 + 4 = 12$$
$$12 - 3 = 9$$
$$9 + 4 = 13$$

To get the second number, we add 4 to the first number. To get the third number, we subtract 3 from the second number. To obtain subsequent numbers, we continue the pattern. The next number is $13 - 3 = 10$.

Work Problem **2** *at the Side.* ▶

OBJECTIVES

1 Use inductive reasoning to analyze patterns.

2 Use deductive reasoning to analyze arguments.

3 Use deductive reasoning to solve problems.

1 Find the next number in the sequence 2, 8, 14, 20, . . . Describe the pattern.

2 Find the next number in the sequence 6, 11, 7, 12, 8, 13, . . . Describe the pattern.

ANSWERS

1. 26; add 6 each time.
2. 9; add 5, subtract 4.

AP-39

3 Find the next number in the sequence 2, 6, 18, 54, ... Describe the pattern.

EXAMPLE 3 **Using Inductive Reasoning**

Find the next number in the sequence 1, 2, 4, 8, 16, ...

Each number after the first is obtained by multiplying the previous number by 2. So the next number would be $16 \cdot 2 = 32$.

◀ *Work Problem* **3** *at the Side.*

EXAMPLE 4 **Using Inductive Reasoning**

(a) Find the next geometric shape in this sequence.

The figures alternate between a blue circle and a red triangle. Also, the number of dots increases by 1 in each subsequent figure. Thus, the next figure should be a blue circle with five dots inside it.

(b) Find the next geometric shape in this sequence.

The first two shapes consist of vertical lines with horizontal lines at the bottom extending first *left* and then *right*. The third shape is a vertical line with a horizontal line at the top extending to the *left*. Therefore, the next shape should be a vertical line with a horizontal line at the top extending to the *right*.

◀ *Work Problem* **4** *at the Side.*

---

4 Find the next shape in this sequence.

⌐ ⌐ ⌐

OBJECTIVE **2** **Use deductive reasoning to analyze arguments.**
In the previous discussion, specific cases were used to find patterns and predict the next event. There is another type of reasoning called **deductive reasoning,** which moves from general cases to specific conclusions.

EXAMPLE 5 **Using Deductive Reasoning**

Does the conclusion follow from the premises in this argument?

All Buicks are automobiles.	← Premise
All automobiles have horns.	← Premise
∴ All Buicks have horns.	← Conclusion

In this example, the first two statements are called *premises* and the third statement (below the line) is called a *conclusion*. The symbol ∴ is a mathematical symbol meaning "**therefore**." The entire set of statements is called an *argument*.

ANSWERS

3. 162; multiply by 3.

4.

⌐

— **Continued on Next Page**

The focus of deductive reasoning is to determine whether the conclusion follows (is valid) from the premises. A set of circles called **Euler circles** is used to analyze the argument.

In Example 5, the statement "All Buicks are automobiles" can be represented by two circles, one for Buicks and one for automobiles. Note that the circle representing Buicks is totally inside the circle representing automobiles because the first premise states that *all* Buicks are automobiles.

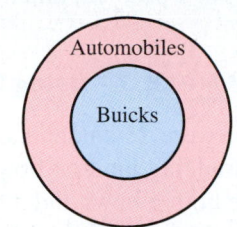

Now, a circle is added to represent the second statement, vehicles with horns. This circle must completely surround the circle representing automobiles because the second premise states that *all* automobiles have horns.

To analyze the conclusion, notice that the circle representing Buicks is *completely* inside the circle representing vehicles with horns. Therefore, it must follow that all Buicks have horns. *The conclusion is valid.*

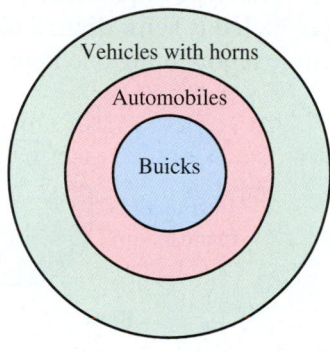

Work Problem **5** *at the Side.* ▶

⑤ Does the conclusion follow from the premises in the following argument?

All cars have four wheels.
All Fords are cars.
∴ All Fords have four wheels.

EXAMPLE 6 **Using Deductive Reasoning**

Does the conclusion follow from the premises in this argument?

All tables are round.
All glasses are round.
∴   All glasses are tables.

Use Euler circles. Draw a circle representing tables *inside* a circle representing round objects, because the first premise states that *all* tables are round.

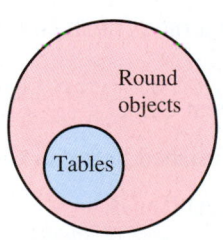

The second statement requires that a circle representing glasses must now be drawn inside the circle representing round objects, but not necessarily inside the circle representing tables. Therefore, the conclusion does *not* follow from the premises. This means that *the conclusion is invalid.*

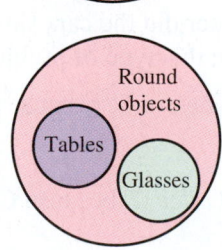

Work Problem **6** *at the Side.* ▶

⑥ Does each conclusion follow from the premises?

(a) All animals are wild.
All cats are animals.
∴ All cats are wild.

(b) All students use math.
All adults use math.
∴ All adults are students.

7 In a college class of 100 students, 35 take both math and history, 50 take history, and 40 take math. How many take neither math nor history? Draw a Venn diagram.

8 A Chevy, BMW, Cadillac, and Ford are parked side by side. The known facts are:

(a) The Ford is on the right end.

(b) The BMW is next to the Cadillac.

(c) The Chevy is between the Ford and the Cadillac.

Which car is parked on the left end?

OBJECTIVE 3 Use deductive reasoning to solve problems.
Another type of deductive reasoning problem occurs when a set of facts is given in a problem and a conclusion must be drawn using these facts.

EXAMPLE 7 Using Deductive Reasoning

There were 25 students enrolled in a ceramics class. During the class, 10 of the students made a bowl and 8 students made a birdbath. Three students made both a bowl and a birdbath. How many students did not make either a bowl or a birdbath?

This type of problem is best solved by organizing the data using a drawing called a **Venn diagram.** Two overlapping circles are drawn, with each circle representing one item made by the students, as shown below.

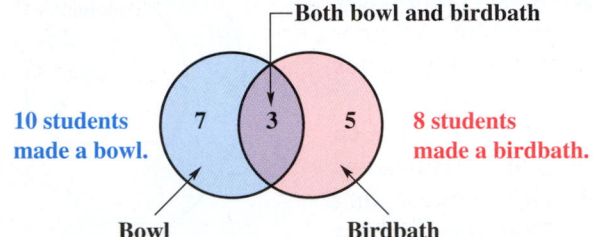

In the region where the circles overlap, write the number of students who made *both* items, namely, 3. In the remaining portion of the birdbath circle, write the number 5, which when added to 3 will give the total number of students who made a birdbath, namely, 8. In a similar manner, write 7 in the remaining portion of the bowl circle, since 7 + 3 = 10, the total number of students who made a bowl. The total of all three numbers written in the circles is 15. Since there were 25 students in the class, this means 25 − 15 or 10 students did not make either a birdbath or a bowl.

◀ *Work Problem* **7** *at the Side.*

EXAMPLE 8 Using Deductive Reasoning

Four cars in a race finish first, second, third, and fourth. The following facts are known.

(a) Car A beat Car C.

(b) Car D finished between Cars C and B.

(c) Car C beat Car B.

In which order did the cars finish?
To solve this type of problem, it is helpful to use a line diagram.

1. *Write A before C,* because Car A beat Car C (fact **a**).

A   C

2. *Write B after C,* because Car C beat Car B (fact **c**).

A   C   B

3. *Write D between C and B,* because Car D finished between Car C and Car B (fact **b**).

The correct order of finish is shown below.

A   C   D   B

◀ *Work Problem* **8** *at the Side.*

Appendix B ▶▶▶ Exercises

Find the next number in each sequence. Describe the pattern in each sequence. See Examples 1–3.

1. 2, 9, 16, 23, 30, . . .

2. 5, 8, 11, 14, 17, . . .

3. 0, 10, 8, 18, 16, . . .

4. 3, 9, 7, 13, 11, . . .

5. 1, 2, 4, 8, . . .

6. 1, 4, 16, 64, . . .

7. 1, 3, 9, 27, 81, . . .

8. 3, 6, 12, 24, 48, . . .

9. 1, 4, 9, 16, 25, . . .

10. 6, 7, 9, 12, 16, . . .

Find the next shape in each sequence. See Example 4.

11.

12.
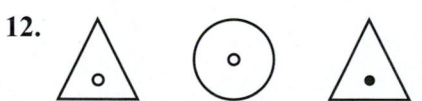

13.

14.

For each argument, draw Euler circles and then state whether or not the conclusion follows from the premises. See Examples 5 and 6.

15. All animals are wild.
     All lions are animals.
     ∴ All lions are wild.

16. All students are hard workers.
     All business majors are students.
     ∴ All business majors are hard workers.

17. All teachers are serious.
     All mathematicians are serious.
     ∴ All mathematicians are teachers.

18. All boys ride bikes.
     All Americans ride bikes.
     ∴ All Americans are boys.

Solve each application problem. See Examples 7 and 8.

19. In a given 30-day period, a husband watched television 20 days and his wife watched television 25 days. If they watched television together 18 days, how many days did neither watch television? Draw a Venn diagram.

20. In a class of 40 students, 21 students take both calculus and physics. If 30 students take calculus and 25 students take physics, how many do not take either calculus or physics? Draw a Venn diagram.

21. Tom, Dick, Mary, and Joan all work for the same company. One is a secretary, one is a computer operator, one is a receptionist, and one is a mail clerk.

   (a) Tom and Joan eat dinner with the computer operator.

   (b) Dick and Mary carpool with the secretary.

   (c) Mary works on the same floor as the computer operator and the mail clerk.

   Who is the computer operator?

22. Four cars—a Ford, a Buick, a Mercedes, and an Audi—are parked in a garage in four spaces.

   (a) The Ford is in the last space.

   (b) The Buick and Mercedes are next to each other.

   (c) The Audi is next to the Ford but not next to the Buick.

   Which car is in the first space?

ANSWERS TO SELECTED EXERCISES

In this section we provide the answers that we think most students will obtain when they work the exercises using the methods explained in the text. If your answer does not look exactly like the one given here, it is not necessarily wrong. In many cases there are equivalent forms of the answer that are correct. For example, if the answer section shows $\frac{3}{4}$ and your answer is 0.75 you have obtained the right answer but written it in a different (yet equivalent) form. Unless the directions specify otherwise, 0.75 is just as valid an answer as $\frac{3}{4}$.

In general, if your answer does not agree with the one given in the text, see whether it can be transformed into the other form. If it can, then it is the correct answer. If you still have doubts, talk with your instructor.

Chapter 1 Whole Numbers

Section 1.1 (pages 7–8)

1. 3; 6 **3.** 1; 0 **5.** 8; 2 **7.** 3; 561; 435 **9.** 60; 000; 502; 109
11. Evidence suggests that this is true. It is common to count using fingers. **13.** twenty-three thousand, one hundred fifteen
15. three hundred forty-six thousand, nine **17.** twenty-five million, seven hundred fifty-six thousand, six hundred sixty-five
19. 63,163 **21.** 10,000,223 **23.** 3,200,000 **25.** 50,051,507
27. 35,079,448 **29.** 800,000,621,020,215 **31.** public transportation; six million, sixty-nine thousand, five hundred eighty-nine **33.** seven million, eight hundred ninety-four thousand, nine hundred eleven

Section 1.2 (pages 17–20)

1. 97 **3.** 89 **5.** 889 **7.** 889 **9.** 7785 **11.** 1589 **13.** 7676
15. 78,446 **17.** 8928 **19.** 59,224 **21.** 150 **23.** 155
25. 121 **27.** 145 **29.** 102 **31.** 1651 **33.** 1154 **35.** 413
37. 1771 **39.** 1410 **41.** 6391 **43.** 11,624 **45.** 17,611
47. 15,954 **49.** 10,648 **51.** 15,594 **53.** 11,557 **55.** 12,078
57. 4250 **59.** 12,268 **61.** correct **63.** incorrect; should be 769
65. correct **67.** incorrect; should be 11,577 **69.** correct
71. Changing the order in which numbers are added does not change the sum. You can add from bottom to top when checking addition. **73.** 33 miles **75.** 38 miles **77.** $16,342
79. 699 people **81.** 20,157 students **83.** 970 ft **85.** 72 ft
87. 9421 **88.** 1249 **89.** 77,762 **90.** 22,267 **91.** 9,994,433
92. 3,334,499 **93.** Write the largest digits on the left, using the smaller digits as you move right. **94.** Write the smallest digits on the left, using the larger digits as you move right.

Section 1.3 (pages 27–30)

1. 16 **3.** 33 **5.** 17 **7.** 213 **9.** 101 **11.** 7111 **13.** 3412
15. 2111 **17.** 13,160 **19.** 41,110 **21.** correct **23.** incorrect; should be 62 **25.** incorrect; should be 121 **27.** correct
29. incorrect; should be 7222 **31.** 38 **33.** 45 **35.** 19
37. 281 **39.** 519 **41.** 7059 **43.** 7589 **45.** 8859 **47.** 3
49. 23 **51.** 19 **53.** 2833 **55.** 7775 **57.** 503 **59.** 156
61. 2184 **63.** 5687 **65.** 19,038 **67.** 31,556 **69.** 6584
71. correct **73.** correct **75.** correct **77.** correct **79.** Possible answers are 1. $3 + 2 = 5$ could be changed to $5 - 2 = 3$ or $5 - 3 = 2$ 2. $6 - 4 = 2$ could be changed to $2 + 4 = 6$ or $4 + 2 = 6$.
81. 47 calories **83.** 367 ft **85.** 467 passengers **87.** 3270 jobs eliminated **89.** 9539 flags **91.** $263 **93.** 758 more people visited on Tuesday **95.** $57,500 **97.** 330 fewer calories; 27 fewer fat grams **99.** 380 calories; 10 fat grams

Section 1.4 (pages 37–40)

1. 24 **3.** 48 **5.** 0 **7.** 24 **9.** 40 **11.** 0 **13.** Factors may be multiplied in any order to get the same answer. They are the same; you may add or multiply numbers in any order. **15.** 210 **17.** 238
19. 3210 **21.** 1872 **23.** 8612 **25.** 10,084 **27.** 20,488
29. 258,447 **31.** 280 **33.** 480 **35.** 2220 **37.** 3600 **39.** 3750
41. 65,400 **43.** 270,000 **45.** 86,000,000 **47.** 48,500 **49.** 720,000
51. 1,940,000 **53.** 476 **55.** 2400 **57.** 3735 **59.** 2378 **61.** 6164
63. 15,792 **65.** 21,665 **67.** 15,730 **69.** 82,320 **71.** 183,996
73. 2,468,928 **75.** 66,005 **77.** 86,028 **79.** 19,422,180
81. 2,278,410 **83.** To multiply by 10, 100, or 1000, just add one, two, or three zeros, respectively, to the number you are multiplying and that's your answer. **85.** 3000 balls **87.** 540 eggs **89.** 24,090 gallons
91. $600 **93.** $1560 **95.** $112,888 **97.** 50,568 **99.** 38,250 trees
101. 7,623,663 people **103.** $7390 **105.** (a) 452 (b) 452
106. commutative **107.** (a) 281 (b) 281 **108.** associative
109. (a) 15,840 (b) 15,840 **110.** commutative **111.** (a) 6552
(b) 6552 **112.** associative **113.** No. Some examples are 1. $7 - 5 = 2$, but $5 - 7$ does not equal 2 2. $12 - 6 = 6$, but $6 - 12$ does not equal 6
3. $(8 - 2) - 5 = 1$, but $8 - (2 - 5)$ does not equal 1. **114.** No.
Some examples are 1. $10 \div 2 = 5$, but $2 \div 10$ does not equal 5
2. $(16 \div 8) \div 2 = 1$, but $16 \div (8 \div 2)$ does not equal 1.

Section 1.5 (pages 51–54)

1. $4\overline{)24}^{\,6}$ $\dfrac{24}{4} = 6$ **3.** $9\overline{)45}^{\,5}$ $45 \div 9 = 5$ **5.** $16 \div 2 = 8$ $\dfrac{16}{2} = 8$
7. 1 **9.** 7 **11.** undefined **13.** 24 **15.** 0 **17.** undefined **19.** 15
21. 8 **23.** 25 **25.** 18 **27.** 304 **29.** 627 R1 **31.** 1522 R5 **33.** 309
35. 3005 **37.** 5006 **39.** 811 R1 **41.** 2589 R2 **43.** 7324 R2
45. 3157 R2 **47.** 2630 **49.** 12,458 R3 **51.** 10,253 R5 **53.** 18,377 R6
55. correct **57.** incorrect; should be 1908 R1 **59.** incorrect; should be 670 R2 **61.** incorrect; should be 3568 R1 **63.** correct
65. correct **67.** incorrect; should be 9628 R3 **69.** correct
71. Multiply the quotient by the divisor and add any remainder. The result should be the dividend. **73.** 328 tables **75.** 9600 each hour **77.** $48,500 **79.** 135 acres **81.** $1,137,500 **83.** $5429
85. ✓ ✓ ✓ ✓ **87.** ✓ X X X **89.** X X ✓ X
91. X ✓ X X **93.** ✓ ✓ X X **95.** X X X X

Section 1.6 (pages 61–63)

1. 53 **3.** 250 **5.** 120 R7 **7.** 1105 R5 **9.** 7134 R12 **11.** 900 R100
13. 73 R5 **15.** 476 R15 **17.** 2407 R1 **19.** 1146 R15 **21.** 3331 R82
23. 850 **25.** incorrect; should be 101 R14 **27.** incorrect; should be 658
29. incorrect; should be 62 **31.** When dividing by 10, 100, or 1000, drop the same number of zeros from the dividend as there are in the divisor to get the quotient. One example is $2500 \div 100 = 25$. **33.** 50 miles
35. 56 floor clocks **37.** $355 **39.** 43,200 rings **41.** $39 per week
43. $0 **44.** 0 **45.** undefined **46.** impossible; if you have 6 cookies, it is not possible to divide them among 0 people. **47.** (a) 14 (b) 17
(c) 38 **48.** Yes. Some examples are $18 \cdot 1 = 18$; $26 \cdot 1 = 26$; $43 \cdot 1 = 43$. **49.** (a) 3200 (b) 320 (c) 32 **50.** Drop the same number of zeros that appear in the divisor. The result is the quotient. With the divisor 10, drop one 0; with 100, drop two zeros; with 1000, drop three zeros.

Section 1.7 (pages 71–74)

1. 620 **3.** 860 **5.** 6800 **7.** 86,800 **9.** 28,500 **11.** 6000
13. 16,000 **15.** 78,000 **17.** 8,000,000,000 **19.** 10,000 **21.** 600,000
23. 5,000,000 **25.** 4480; 4500; 4000 **27.** 3370; 3400; 3000
29. 6050; 6000; 6000 **31.** 5340; 5300; 5000 **33.** 19,540; 19,500; 20,000 **35.** 26,290; 26,300; 26,000 **37.** 93,710; 93,700; 94,000

39. 1. Locate the place to be rounded and underline it. 2. Look only at the next digit to the right. If this digit is 5 or more, increase the underlined digit by 1. 3. Change all digits to the right of the underlined place to zeros. **41.** 30 60 50 80 220; 219 **43.** 80 40 40; 35 **45.** 70 30 2100; 2278 **47.** 900 700 400 800 2800; 2828 **49.** 900 400 500; 435 **51.** 800 400 320,000; 282,000 **53.** 8000 60 700 4000 12,760; 12,605 **55.** 700 500 200; 158 **57.** 900 30 27,000; 27,231 **59.** Perhaps the best explanation is that 3492 is closer to 3500 than 3400, but 3492 is closer to 3000 than 4000. **61.** 80 million people; 300 million people **63.** 349,000; 350,000 **65.** 1,670,000 pounds; 1,700,000 pounds; 2,000,000 pounds **67.** $25,765,500,000; $25,800,000,000; $26,000,000,000 **69.** 71,500 **70.** 72,499 **71.** 7500 **72.** 8499 **73.** 3930; 11,240; 15,970; 17,920; 534,880; 2,788,000 **74.** 4000; 10,000; 20,000; 20,000; 500,000; 3,000,000 **75. (a)** When using front end rounding, all digits are 0 except the first digit. These numbers are easier to work with when estimating answers. **(b)** When using front end rounding to estimate an answer, the estimated answer can vary greatly from the exact answer.

Section 1.8 (pages 77–80)

1. 2; 3; 9 **3.** 2; 5; 25 **5.** 2; 8; 64 **7.** 2; 15; 225 **9.** 4 **11.** 8 **13.** 10 **15.** 12 **17.** 36; 36 **19.** 400; 400 **21.** 1225; 1225 **23.** 625; 625 **25.** 10,000; 10,000 **27.** A perfect square is the square of a whole number. The number 25 is the square of 5 because 5 • 5 = 25. The number 50 is not a perfect square. There is no whole number that can be squared to get 50. **29.** 12 **31.** 15 **33.** 4 **35.** 20 **37.** 45 **39.** 63 **41.** 118 **43.** 22 **45.** 30 **47.** 102 **49.** 9 **51.** 63 **53.** 33 **55.** 70 **57.** 7 **59.** 17 **61.** 55 **63.** 108 **65.** 26 **67.** 26 **69.** 27 **71.** 16 **73.** 16 **75.** 21 **77.** 7 **79.** 20 **81.** 14 **83.** 25 **85.** 16 **87.** 23 **89.** 233

Section 1.9 (pages 85–88)

1. 4500 stores **3.** Dollar General; about 5750 stores **5.** 1250 fewer stores **7.** 9 people **9. (a)** Saw ad **(b)** 25 people **11.** 9 people **13.** 2011; 7000 trees **15.** 4500 trees **17.** Possible answers are 1. shortage of trees to plant 2. lack of qualified workers 3. poor economy 4. less demand for planting. **19.** $(7 - 2) \cdot 3 - 6$ **20.** $(4 + 2) \cdot (5 + 1)$

21. $36 \div (3 \cdot 3) \cdot 4$ **22.** $56 \div (2 \cdot 2 \cdot 2) + \dfrac{0}{6}$

23. (a) 7920 + 1320 + 2640 + (5280 − 1320 − 1320) + 2640 + 1320 + 7920 + 5280 **(b)** 31,680 × 3 = 95,040 ft **(c)** 18 miles

Section 1.10 (pages 93–96)

1. *Estimate:* 600 + 900 + 1000 + 800 + 2000 = 5300 sandwiches; *Exact:* 5208 sandwiches **3.** *Estimate:* 70 − 60 = 10 more recalls; *Exact:* 13 more recalls **5.** *Estimate:* 200 × 20 = 4000 kits; *Exact:* 5664 kits **7.** *Estimate:* 3000 ÷ 700 ≈ 4 toys; *Exact:* 4 toys **9.** *Estimate:* 8000 − 4000 = 4000 people; *Exact:* 4174 people **11.** *Estimate:* $30 × 5 = $150; *Exact:* $170 **13.** *Estimate:* $50,000 − $40,000 = $10,000; *Exact:* $14,100 **15. (a)** *Estimate:* $50,000 + $40,000 = $90,000 White; $40,000 + $50,000 = $90,000 Easterly; *Exact:* $86,950 White; $93,370, Easterly; Mr. and Mrs. Easterly **(b)** *Estimate:* $90,000 − $90,000 = $0; *Exact:* $6420 **17.** *Estimate:* $2000 − $700 − $300 − $400 − $200 − $200 = $200; *Exact:* $350 **19.** *Estimate:* 40,000 × 100 = 4,000,000 square feet; *Exact:* 6,011,280 square feet **21.** *Estimate:* $400 + $1000 + $200 + $400 + $200 + $200 = $2400; *Exact:* $2680 **23.** *Estimate:* $400 + $200 + $200 + $400 = $1200; $1200 − $1000 = $200; *Exact:* $140 **25.** *Estimate:* ($1000 × 6) + ($900 × 20) = $24,000; *Exact:* $20,961 **27.** Possible answers are Addition: more; total; gain of Subtraction: less; loss of; decreased by Multiplication: twice; of; product Division: divided by; goes into; per Equals: is; are **29.** Estimating the answer can help you avoid careless mistakes like decimal or calculation errors. Examples of reasonable answers in daily life might be a $35 bag of groceries, $50 to fill the gas tank, or $45 for a phone bill. **31.** $20,009 **33.** 2477 pounds **35.** $378 **37.** $375 **39.** 20 seats

Chapter 1 Review Exercises (pages 101–108)

1. 6; 573 **2.** 36; 215 **3.** 105; 724 **4.** 1; 768; 710; 618 **5.** seven hundred twenty-eight **6.** fifteen thousand, three hundred ten **7.** three hundred nineteen thousand, two hundred fifteen **8.** sixty-two million, five hundred thousand, five **9.** 10,008 **10.** 200,000,455 **11.** 110 **12.** 121 **13.** 5464 **14.** 15,657 **15.** 10,986 **16.** 9845 **17.** 40,602 **18.** 49,855 **19.** 36 **20.** 27 **21.** 189 **22.** 184 **23.** 6849 **24.** 4327 **25.** 224 **26.** 25,866 **27.** 49 **28.** 0 **29.** 32 **30.** 64 **31.** 45 **32.** 42 **33.** 56 **34.** 81 **35.** 40 **36.** 45 **37.** 48 **38.** 8 **39.** 0 **40.** 42 **41.** 48 **42.** 0 **43.** 84 **44.** 368 **45.** 522 **46.** 98 **47.** 5000 **48.** 2992 **49.** 5396 **50.** 45,815 **51.** 14,912 **52.** 20,160 **53.** 465,525 **54.** 174,984 **55.** 875 **56.** 2368 **57.** 1176 **58.** 5100 **59.** 15,576 **60.** 30,184 **61.** 887,169 **62.** 500,856 **63.** $360 **64.** $1064 **65.** $20,352 **66.** $684 **67.** 14,000 **68.** 23,800 **69.** 206,800 **70.** 318,500 **71.** 128,000,000 **72.** 90,300,000 **73.** 5 **74.** 7 **75.** 6 **76.** 2 **77.** 6 **78.** 4 **79.** 7 **80.** 0 **81.** undefined **82.** 0 **83.** 8 **84.** 9 **85.** 82 **86.** 98 **87.** 4422 **88.** 352 **89.** 150 R4 **90.** 124 R25 **91.** 820 **92.** 15,200 **93.** 21,000 **94.** 70,000 **95.** 3490; 3500; 3000 **96.** 20,070; 20,100; 20,000 **97.** 98,200; 98,200; 98,000 **98.** 352,120; 352,100; 352,000 **99.** 4 **100.** 7 **101.** 12 **102.** 14 **103.** 3; 7; 343 **104.** 6; 3; 729 **105.** 3; 5; 125 **106.** 5; 4; 1024 **107.** 34 **108.** 26 **109.** 9 **110.** 4 **111.** 9 **112.** 6 **113.** 8 parents **114.** 5 parents **115.** Keeping bedroom clean; 25 parents **116.** Hanging up wet bath towels; 3 parents **117.** *Estimate:* 40 million × 400 = 16,000 million or 16,000,000,000 checks; *Exact:* 14,600 million or 14,600,000,000 checks **118.** *Estimate:* 1000 × 60 = 60,000 revolutions; *Exact:* 84,000 revolutions **119.** *Estimate:* 100 × 20 = 2000 forks; *Exact:* 2160 forks **120.** *Estimate:* 6000 × 30 = 180,000 brackets; *Exact:* 180,000 brackets **121.** *Estimate:* 2000 × 10 = 20,000 hr; *Exact:* 24,000 hr **122.** *Estimate:* 80 × 5 = 400 mi; *Exact:* 400 mi **123.** *Estimate:* (30 × $2000) + (30 × $900) = $87,000; *Exact:* $74,052 **124.** *Estimate:* (60 × $20) + (20 × $7) = $1340; *Exact:* $1139 **125.** *Estimate:* $600 − $100 = $500; *Exact:* = $513 **126.** *Estimate:* 30,000 ÷ 1000 = 30 hr; *Exact:* 33 hr **127.** *Estimate:* 9000 ÷ 200 = 45 pounds; *Exact:* 50 pounds **128.** *Estimate:* $2000 − $500 − $400 = $1100; *Exact:* $1019 **129.** *Estimate:* 30,000 ÷ 600 = 50 acres; *Exact:* 52 acres **130.** *Estimate:* 6000 ÷ 200 = 30 homes; *Exact:* 32 homes **131.** 332 **132.** 448 **133.** 253 **134.** 588 **135.** 1041 **136.** 1661 **137.** 32,062 **138.** 24,947 **139.** 3 **140.** 7 **141.** 93,635 **142.** 83,178 **143.** undefined **144.** 7 **145.** 6900 **146.** 2310 **147.** 1,079,040 **148.** 130,212 **149.** 108 **150.** 207 **151.** three hundred seventy-six thousand, eight hundred fifty-three **152.** four hundred eight thousand, six hundred ten **153.** 8700 **154.** 401,000 **155.** 8 **156.** 9 **157.** $5544 **158.** $31,080 **159.** $2288 **160.** $15,782 **161.** 468 cards **162.** 3600 textbooks **163.** $280 **164.** $114,635 **165.** $1905 **166.** $12,420 **167.** 869 ft **168.** 162 ft **169. (a)** 8393 ft **(b)** greater than a mile by 3113 ft **170.** 4 football fields

Chapter 1 Test (pages 109–110)

1. nine thousand, two hundred five **2.** twenty-five thousand, sixty-five **3.** 426,005 **4.** 8530 **5.** 112,630 **6.** 1045 **7.** 6206 **8.** 168 **9.** 171,000 **10.** 1615 **11.** 4,450,743 **12.** 7047 **13.** undefined **14.** 458 R5 **15.** 160 **16.** 6350 **17.** 77,000 **18.** 41 **19.** 28 **20.** *Estimate:* $500 + $500 + $500 + $400 − $800 = $1100; *Exact:* $1140 **21.** *Estimate:* 90,000 ÷ 400 = 225 acres; *Exact:* 231 acres **22.** *Estimate:* $2000 − $500 − $200 − $200 = $1100; *Exact:* $948 **23.** *Estimate:* (50 × 60 × 4) + (40 × 60 × 3) = 19,200 chicks; *Exact:* 18,000 chicks **24.** 1. Locate the place to which you are rounding and underline it. 2. Look only at the next digit to the right. If this digit is a 4 or less, do not change the underlined digit. If the digit is 5 or more, increase the underlined digit by 1. 3. Change all digits to the right of the underlined place to zeros. Each person's rounding example will vary. **25.** 1. Read the problem carefully. 2. Work out a plan. 3. Estimate a reasonable answer. 4. Solve the problem. 5. State the answer. 6. Check your work.

Chapter 2 Multiplying and Dividing Fractions

Section 2.1 (pages 115–116)

1. $\frac{3}{4}; \frac{1}{4}$　**3.** $\frac{1}{3}; \frac{2}{3}$　**5.** $\frac{7}{5}; \frac{3}{5}$　**7.** $\frac{5}{6}; \frac{4}{6}; \frac{2}{6}$　**9.** $\frac{8}{25}$　**11.** $\frac{303}{520}$　**13.** 4; 5

15. 9; 8　**17.** Proper $\frac{1}{3}, \frac{5}{8}, \frac{7}{16}$　Improper $\frac{8}{5}, \frac{6}{6}, \frac{12}{2}$

19. Proper $\frac{3}{4}, \frac{9}{11}, \frac{7}{15}$　Improper $\frac{3}{2}, \frac{5}{5}, \frac{19}{18}$

21. One possibility is

$\frac{3}{4}$ ← Numerator
　← Denominator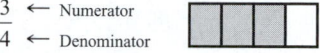

The denominator shows the number of equal parts in the whole and the numerator shows how many of the parts are being considered.

23. 3; 8　**25.** 5; 24

Section 2.2 (pages 121–124)

1. $\frac{5}{4}$　**3.** $\frac{23}{5}$　**5.** $\frac{13}{2}$　**7.** $\frac{33}{4}$　**9.** $\frac{18}{11}$　**11.** $\frac{19}{3}$　**13.** $\frac{81}{8}$　**15.** $\frac{43}{4}$

17. $\frac{27}{8}$　**19.** $\frac{43}{5}$　**21.** $\frac{54}{11}$　**23.** $\frac{131}{4}$　**25.** $\frac{221}{12}$　**27.** $\frac{269}{15}$　**29.** $\frac{187}{24}$

31. $1\frac{1}{3}$　**33.** $2\frac{1}{4}$　**35.** 8　**37.** $7\frac{3}{5}$　**39.** $4\frac{7}{8}$　**41.** 9　**43.** $15\frac{3}{4}$

45. $5\frac{2}{9}$　**47.** $8\frac{1}{8}$　**49.** $16\frac{4}{5}$　**51.** 28　**53.** $26\frac{1}{7}$

55. Multiply the denominator by the whole number and add the numerator. The result becomes the new numerator, which is placed over the original denominator.

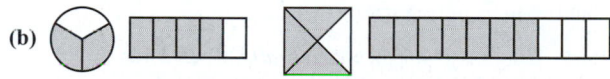

$$2\frac{1}{2} \quad 2 \times 2 + 1 = 5 \quad \frac{5}{2}$$

57. $\frac{501}{2}$　**59.** $\frac{1000}{3}$　**61.** $\frac{4179}{8}$　**63.** $154\frac{1}{4}$　**65.** 171

67. $122\frac{13}{32}$　**69.** $\frac{2}{3}, \frac{4}{5}, \frac{3}{4}, \frac{7}{10}$　**70. (a)** numerator; denominator

(b)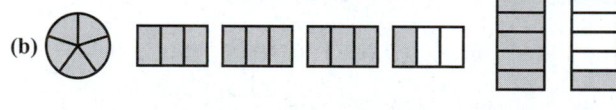

(c) less　**71.** $\frac{5}{5}, \frac{10}{3}, \frac{6}{5}$　**72. (a)** numerator; denominator

(b)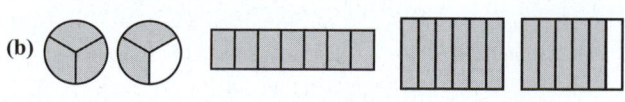

(c) greater　**73.** $\frac{5}{3} = 1\frac{2}{3}; \frac{7}{7} = 1; \frac{11}{6} = 1\frac{5}{6}$

74. (a) improper; greater than or equal to

(b)

(c) Divide the numerator by the denominator. Use the quotient as the whole number and place the remainder over the denominator.

Section 2.3 (pages 129–130)

1. 1, 2, 4, 8　**3.** 1, 3, 5, 15　**5.** 1, 2, 3, 4, 6, 8, 12, 16, 24, 48
7. 1, 2, 3, 4, 6, 9, 12, 18, 36　**9.** 1, 2, 4, 5, 8, 10, 20, 40
11. 1, 2, 4, 8, 16, 32, 64　**13.** composite　**15.** prime　**17.** composite
19. prime　**21.** prime　**23.** composite　**25.** composite　**27.** composite
29. 2^3　**31.** $2^2 \cdot 5$　**33.** $2^2 \cdot 3^2$　**35.** 5^2　**37.** $2^2 \cdot 17$　**39.** $2^3 \cdot 3^2$

41. $2^2 \cdot 11$　**43.** $2^2 \cdot 5^2$　**45.** 5^3　**47.** $2^2 \cdot 3^2 \cdot 5$　**49.** $2^6 \cdot 5$
51. $2^3 \cdot 3^2 \cdot 5$　**53.** A composite number has a factor(s) other than itself or 1. Examples include 4, 6, 8, 9, 10. A prime number is a whole number that has exactly two *different* factors, itself and 1. Examples include 2, 3, 5, 7, 11. The numbers 0 and 1 are neither prime nor composite.
55. All the possible factors of 24 are 1, 2, 3, 4, 6, 8, 12, and 24. This list includes both prime numbers and composite numbers. The prime factors of 24 include only prime numbers. The prime factorization of 24 is $2 \cdot 2 \cdot 2 \cdot 3 = 2^3 \cdot 3$　**57.** $2 \cdot 5^2 \cdot 7$　**59.** $2^6 \cdot 3 \cdot 5$
61. $2^3 \cdot 3 \cdot 5 \cdot 13$　**63.** $2^2 \cdot 3^2 \cdot 5 \cdot 7$
65. 2, 3, 5, 7, 11, 13, 17, 19, 23, 29, 31, 37, 41, 43, 47
66. A prime number is a whole number that is evenly divisible by itself and 1 only.　**67.** No. Every other even number is divisible by 2 in addition to being divisible by itself and 1.　**68.** No. A multiple of a prime number can never be prime because it will always be divisible by the prime number.　**69.** $2 \cdot 2 \cdot 3 \cdot 5 \cdot 5 \cdot 7$　**70.** $2^2 \cdot 3 \cdot 5^2 \cdot 7$

Section 2.4 (pages 135–136)

1. ✓ ✓ ✓ ✓　**3.** ✓ ✓ ✗ ✗　**5.** ✓ ✗ ✓ ✓　**7.** ✓ ✓ ✗ ✗　**9.** $\frac{3}{4}$

11. $\frac{1}{4}$　**13.** $\frac{3}{5}$　**15.** $\frac{6}{7}$　**17.** $\frac{7}{8}$　**19.** $\frac{6}{7}$　**21.** $\frac{4}{7}$　**23.** $\frac{1}{50}$　**25.** $\frac{8}{11}$

27. $\frac{5}{9}$　**29.** $\frac{\overset{1}{\cancel{2}} \cdot \overset{1}{\cancel{3}} \cdot 3}{\cancel{2} \cdot 2 \cdot 2 \cdot \cancel{3}} = \frac{3}{4}$　**31.** $\frac{\overset{1}{\cancel{8}} \cdot 7}{2 \cdot 2 \cdot 2 \cdot \cancel{8}} = \frac{7}{8}$

33. $\frac{\overset{1}{\cancel{2}} \cdot \overset{1}{\cancel{3}} \cdot \overset{1}{\cancel{3}} \cdot \overset{1}{\cancel{3}}}{\cancel{2} \cdot 2 \cdot \cancel{3} \cdot \cancel{3} \cdot \cancel{3}} = \frac{1}{2}$　**35.** $\frac{\overset{1}{\cancel{2}} \cdot \overset{1}{\cancel{2}} \cdot \overset{1}{\cancel{3}} \cdot 3}{\cancel{2} \cdot \cancel{2} \cdot \cancel{3}} = 3$

37. $\frac{2 \cdot 2 \cdot 2 \cdot \overset{1}{\cancel{3}} \cdot \overset{1}{\cancel{3}}}{\cancel{3} \cdot \cancel{3} \cdot 5 \cdot 5} = \frac{8}{25}$　**39.** $\frac{1}{2} = \frac{1}{2}$; equivalent

41. $\frac{5}{12} \neq \frac{2}{5}$; not equivalent　**43.** $\frac{5}{8} \neq \frac{35}{52}$; not equivalent

45. $\frac{7}{8} = \frac{7}{8}$; equivalent　**47.** $8 \neq 9$; not equivalent

49. $\frac{5}{6} = \frac{5}{6}$; equivalent　**51.** A fraction is in lowest terms when the numerator and the denominator have no common factors other than 1. Some examples are $\frac{1}{2}, \frac{3}{8}$, and $\frac{2}{3}$.　**53.** $\frac{5}{8}$　**55.** $\frac{2}{1} = 2$

Section 2.5 (pages 145–148)

1. $\frac{1}{4}$　**3.** $\frac{2}{35}$　**5.** $\frac{3}{4}$　**7.** $\frac{1}{4}$　**9.** $\frac{5}{12}$　**11.** $\frac{9}{32}$　**13.** $\frac{2}{5}$　**15.** $\frac{13}{32}$　**17.** $\frac{21}{128}$

19. 4　**21.** 40　**23.** 24　**25.** $13\frac{1}{2}$　**27.** $31\frac{1}{2}$　**29.** 80　**31.** $94\frac{2}{3}$

33. 400　**35.** 810　**37.** $\frac{1}{4}$ m²　**39.** 9 m²　**41.** $\frac{3}{10}$ in.²

43. Multiply the numerators and multiply the denominators. An example is

$\frac{3}{4} \cdot \frac{1}{2} = \frac{3 \cdot 1}{4 \cdot 2} = \frac{3}{8}$　**45.** $1\frac{1}{2}$ yd²　**47.** $3\frac{1}{2}$ mi²　**49.** They are both the

same size: $\frac{3}{64}$ mi²　**51.** 60,000 stores　**52.** 64,615 stores

53. *Estimate:* $\frac{2}{3} \cdot 9000 = 6000$; *Exact:* 6267 stores (rounded)

54. *Estimate:* $\frac{2}{5} \cdot 8000 = 3200$; *Exact:* 3013 stores (rounded)

55. $\frac{2}{3} \cdot 9300 = 6200$ stores

56. $\frac{2}{5} \cdot 7500 = 3000$ stores

ANSWERS (side tab)

Section 2.6 (pages 153–156)

1. $\frac{1}{2}$ yd² 3. $\frac{8}{9}$ ft² 5. $\frac{3}{10}$ yd² 7. $2568 9. $36

11. 910 women 13. Don't know; 50 people

15. 100 much less + 250 somewhat less = 350 people

17. Because everyone is included and fractions are given for *all* groups, the sum of the fractions must be *1* or *all* of the people.

19. $58,000 21. $11,600 23. $3625

25. The correct solution is $\frac{9}{10} \times \frac{20}{21} = \frac{\overset{3}{\cancel{9}}}{\underset{1}{\cancel{10}}} \times \frac{\overset{2}{\cancel{20}}}{\underset{7}{\cancel{21}}} = \frac{6}{7}$

27. $750 29. 2 ft 31. 9000 votes 33. $\frac{1}{32}$ of the estate

Section 2.7 (pages 163–166)

1. $\frac{8}{3}$ 3. $\frac{6}{5}$ 5. $\frac{5}{8}$ 7. $\frac{1}{4}$ 9. $\frac{2}{3}$ 11. $2\frac{5}{8}$ 13. $\frac{9}{20}$ 15. 4 17. 6

19. $\frac{13}{16}$ 21. $\frac{4}{5}$ 23. 18 25. 24 27. $\frac{1}{14}$ 29. $\frac{2}{9}$ quart 31. 15 times

33. 88 dispensers 35. 60 trips 37. You can divide two fractions by using the reciprocal of the second fraction (divisor) and then multiplying

39. 12 pounds 41. 208 homes 43. 124 visits 45. 2432 towels

47. double, twice, times, product, twice as much 48. goes into, divide, per, quotient, divided by 49. reciprocal 50. $\frac{4}{3}; \frac{8}{7}; \frac{1}{5}; \frac{19}{12}$

51. (a) Multiply the length of one side by 3, 4, 5, or 6.

(b) $\frac{15}{16} \times 4 = \frac{15}{\underset{4}{\cancel{16}}} \times \frac{\overset{1}{\cancel{4}}}{1} = \frac{15}{4} = 3\frac{3}{4}$ in.

52. $\frac{225}{256}$ in.²; Multiply the length by the width.

Section 2.8 (pages 173–176)

1. *Exact:* $7\frac{7}{8}$; *Estimate:* 5 • 2 = 10 3. *Exact:* $4\frac{1}{2}$;

Estimate: 2 • 3 = 6 5. *Exact:* 4; *Estimate:* 3 • 1 = 3

7. *Exact:* 50; *Estimate:* 8 • 6 = 48 9. *Exact:* $49\frac{1}{2}$;

Estimate: 5 • 2 • 5 = 50 11. *Exact:* 12;

Estimate: 3 • 2 • 3 = 18 13. *Exact:* $\frac{1}{3}$; *Estimate:* 1 ÷ 4 = $\frac{1}{4}$

15. *Exact:* $\frac{5}{6}$; *Estimate:* 3 ÷ 3 = 1 17. *Exact:* $3\frac{3}{5}$;

Estimate: 9 ÷ 3 = 3 19. *Exact:* $\frac{5}{12}$; *Estimate:* 1 ÷ 2 = $\frac{1}{2}$

21. *Exact:* $\frac{3}{10}$; *Estimate:* 2 ÷ 6 = $\frac{1}{3}$ 23. *Exact:* $\frac{17}{18}$; *Estimate:*

6 ÷ 6 = 1 25. (a) *Estimate:* 1 • 3 = 3 cups; *Exact:* $1\frac{7}{8}$ cups of applesauce (b) *Estimate:* 1 • 3 = 3 teaspoons;

Exact: $1\frac{1}{4}$ teaspoons of salt (c) *Estimate:* 2 • 3 = 6 cups;

Exact: $4\frac{3}{8}$ cups of flour 27. (a) *Estimate:* 1 ÷ 2 = $\frac{1}{2}$ teaspoon;

Exact: $\frac{1}{4}$ teaspoon vanilla extract (b) *Estimate:* 1 ÷ 2 = $\frac{1}{2}$ cup;

Exact: $\frac{3}{8}$ cup applesauce (c) *Estimate:* 2 ÷ 2 = 1 cup;

Exact: $\frac{7}{8}$ cup of flour 29. *Estimate:* 1316 ÷ 12 ≈ 110 units;

Exact: 112 units 31. *Estimate:* 135 • 20 = 2700 in.;

Exact: $2632\frac{1}{2}$ in. 33. The answer should include *Step 1* Change mixed

numbers to improper fractions. *Step 2* Multiply the fractions. *Step 3* Write the answer in lowest terms, changing to mixed or whole numbers where possible. 35. *Estimate:* 51,460 ÷ 20 = 2573 tires; *Exact:* 2480 tires

37. *Estimate:* 10 ÷ 1 = 10 spacers; *Exact:* 13 spacers 39. *Estimate:*

(a) 18 • 4 = 72 in.; *Exact:* 71 in. (b) No, because 6 ft = 72 in.

41. *Estimate:* 7 • 26 = 182 gallons; *Exact:* $172\frac{1}{8}$ gallons

Chapter 2 Review Exercises (pages 183–186)

1. $\frac{1}{3}$ 2. $\frac{5}{8}$ 3. $\frac{2}{4}$ 4. Proper $\frac{1}{8}, \frac{3}{4}, \frac{2}{3}$; Improper $\frac{4}{3}, \frac{5}{5}$

5. Proper $\frac{15}{16}, \frac{1}{8}$; Improper $\frac{6}{5}, \frac{16}{13}, \frac{5}{3}$ 6. $\frac{19}{4}$ 7. $\frac{59}{6}$ 8. $3\frac{3}{8}$

9. $12\frac{3}{5}$ 10. 1, 2, 3, 6 11. 1, 2, 3, 4, 6, 8, 12, 24 12. 1, 5, 11, 55

13. 1, 2, 3, 5, 6, 9, 10, 15, 18, 30, 45, 90 14. 3^3 15. $2 \cdot 3 \cdot 5^2$

16. $2^2 \cdot 3 \cdot 5 \cdot 7$ 17. 25 18. 288 19. 1728 20. 2048

21. $\frac{24}{24} = 1$ 22. $\frac{18}{24} = \frac{3}{4}$ 23. $\frac{14}{24} = \frac{7}{12}$ 24. $\frac{10}{24} = \frac{5}{12}$

25. $\frac{\overset{1}{\cancel{5}} \cdot 5}{2 \cdot 2 \cdot 3 \cdot \underset{1}{\cancel{5}}}; \frac{5}{12}$ 26. $\frac{\cancel{2} \cdot \cancel{2} \cdot \cancel{2} \cdot \cancel{2} \cdot \cancel{2} \cdot 2 \cdot 2 \cdot \cancel{3}}{\cancel{2} \cdot \cancel{2} \cdot \cancel{2} \cdot \cancel{2} \cdot \cancel{2} \cdot \cancel{3}}; 4$

27. equivalent 28. not equivalent 29. equivalent

30. $\frac{3}{5}$ 31. $\frac{3}{16}$ 32. $\frac{1}{7}$ 33. $\frac{4}{21}$ 34. 15 35. 625 36. $1\frac{1}{3}$

37. $\frac{5}{3} = 1\frac{2}{3}$ 38. $\frac{5}{2} = 2\frac{1}{2}$ 39. 2 40. 8 41. 24 42. $\frac{5}{24}$

43. $\frac{2}{15}$ 44. $\frac{4}{13}$ 45. $\frac{9}{40}$ ft² 46. $\frac{7}{12}$ in.² 47. 7857 ft²

48. $5\frac{1}{2}$ ft² 49. *Exact:* $6\frac{7}{8}$; *Estimate:* 6 • 1 = 6

50. *Exact:* $21\frac{3}{8}$; *Estimate:* 2 • 7 • 1 = 14

51. *Exact:* $5\frac{1}{6}$; *Estimate:* 16 ÷ 3 = $5\frac{1}{3}$

52. *Exact:* $\frac{3}{4}$; *Estimate:* 5 ÷ 6 = $\frac{5}{6}$

53. 512 bins 54. $\frac{2}{15}$ of the estate

55. *Estimate:* 158 ÷ 4 ≈ 40 pull cords; *Exact:* 36 pull cords

56. *Estimate:* 8 • 2 • 50 = 800 pounds; *Exact:* $833\frac{1}{3}$ pounds

57. 25 pounds 58. $186 59. $\frac{3}{32}$ of the budget

60. $\frac{4}{25}$ ton 61. $\frac{3}{8}$ 62. $\frac{2}{5}$ 63. $31\frac{1}{4}$ 64. $28\frac{1}{8}$ 65. $\frac{1}{10}$ 66. $\frac{5}{32}$

67. 30 68. $2\frac{3}{5}$ 69. $1\frac{3}{5}$ 70. $38\frac{1}{4}$ 71. $\frac{17}{3}$ 72. $\frac{307}{8}$

73. $\frac{\cancel{2} \cdot \cancel{2} \cdot 2}{\cancel{2} \cdot \cancel{2} \cdot 3} = \frac{2}{3}$ 74. $\frac{\cancel{2} \cdot 2 \cdot \cancel{3} \cdot 3 \cdot 3}{\cancel{2} \cdot \cancel{3} \cdot 5 \cdot 7} = \frac{18}{35}$ 75. $\frac{5}{6}$ 76. $\frac{2}{3}$ 77. $\frac{2}{5}$

78. $\frac{1}{3}$ 79. *Estimate:* 4 • 44 = 176 ounces; *Exact:* $157\frac{1}{2}$ ounces

80. *Estimate:* 7 • 26 = 182 quarts; *Exact:* $184\frac{7}{8}$ quarts

81. $1\frac{17}{32}$ in.² 82. $\frac{5}{16}$ yd²

Chapter 2 Test (pages 189–190)

1. $\frac{5}{6}$ 2. $\frac{3}{8}$ 3. $\frac{2}{3}, \frac{6}{7}, \frac{1}{4}, \frac{5}{8}$ 4. $\frac{27}{8}$ 5. $30\frac{3}{4}$ 6. 1, 2, 3, 6, 9, 18

7. $3^2 \cdot 5$ 8. $2^4 \cdot 3^2$ 9. $2^2 \cdot 5^3$ 10. $\frac{3}{4}$ 11. $\frac{5}{6}$

12. $\dfrac{56}{84} = \dfrac{\overset{1}{\cancel{2}} \cdot \overset{1}{\cancel{2}} \cdot 2 \cdot \overset{1}{\cancel{7}}}{\underset{1}{\cancel{2}} \cdot \underset{1}{\cancel{2}} \cdot 3 \cdot \underset{1}{\cancel{7}}} = \dfrac{2}{3}$

13. Answers will vary. One possibility is: Multiply fractions by multiplying the numerators and multiplying the denominators. Divide two fractions by using the reciprocal of the second fraction (divisor) and then multiplying.
14. $\dfrac{1}{3}$ **15.** 36 **16.** $\dfrac{5}{12}$ yd² **17.** 5475 seedlings **18.** $\dfrac{9}{10}$ **19.** $15\dfrac{3}{4}$

20. 24 pieces **21.** *Estimate:* $4 \cdot 4 = 16$; *Exact:* $14\dfrac{7}{16}$

22. *Estimate:* $2 \cdot 4 = 8$; *Exact:* $7\dfrac{17}{18}$ **23.** *Estimate:* $10 \div 2 = 5$;
Exact: $4\dfrac{4}{15}$ **24.** *Estimate:* $9 \div 2 = 4\dfrac{1}{2}$; *Exact:* $5\dfrac{1}{10}$

25. *Estimate:* $3 \cdot 12 = 36$; *Exact:* $30\dfrac{5}{8}$ grams

Chapter 3 Adding and Subtracting Fractions

Section 3.1 (pages 197–198)
1. $\dfrac{5}{8}$ **3.** $\dfrac{5}{6}$ **5.** $\dfrac{1}{2}$ **7.** $1\dfrac{1}{5}$ **9.** $\dfrac{1}{3}$ **11.** $\dfrac{13}{20}$ **13.** $\dfrac{11}{15}$ **15.** $1\dfrac{1}{2}$
17. $\dfrac{11}{27}$ **19.** $\dfrac{3}{8}$ **21.** $\dfrac{6}{11}$ **23.** $\dfrac{3}{5}$ **25.** $1\dfrac{1}{7}$ **27.** $\dfrac{1}{5}$ **29.** $1\dfrac{1}{6}$ **31.** $1\dfrac{1}{10}$
33. Three steps to add like fractions are: 1. Add the numerators of the fractions to find the numerator of the sum (the answer). 2. Use the denominator of the fractions as the denominator of the sum. 3. Write the answer in lowest terms. **35.** $\dfrac{7}{9}$ **37.** $\dfrac{5}{7}$ **39.** $\dfrac{1}{2}$ acre

Section 3.2 (pages 207–210)
1. 6 **3.** 15 **5.** 36 **7.** 14 **9.** 30 **11.** 100 **13.** 20 **15.** 60
17. 36 **19.** 120 **21.** 180 **23.** 720 **25.** $\dfrac{16}{24}$ **27.** $\dfrac{18}{24}$ **29.** $\dfrac{20}{24}$
31. 3 **33.** 12 **35.** 28 **37.** 12 **39.** 32 **41.** 72 **43.** 84 **45.** 96
47. 27 **49.** It probably depends on how large the numbers are. If the numbers are small, the method using multiples of the largest number seems best. If the numbers are larger, or there are more than two numbers, then the factorization method will be better. **51.** 3600 **53.** 10,584
55. like; unlike **56.** numerators; denominator; lowest **57.** least; smallest
58. 40 is the least common multiple. **59.** 70 **60.** 450 **61.** 240 is a common multiple but twice as large as the least common multiple; 120 is the LCM. **62.** The least common multiple can be no smaller than the largest number in a group and the number 1760 is a multiple of 55.

Section 3.3 (pages 215–218)
1. $\dfrac{7}{8}$ **3.** $\dfrac{8}{9}$ **5.** $\dfrac{3}{4}$ **7.** $\dfrac{39}{40}$ **9.** $\dfrac{23}{36}$ **11.** $\dfrac{14}{15}$ **13.** $\dfrac{29}{36}$ **15.** $\dfrac{17}{20}$
17. $\dfrac{23}{30}$ **19.** $\dfrac{3}{8}$ **21.** $\dfrac{23}{48}$ **23.** $\dfrac{1}{2}$ **25.** $\dfrac{1}{2}$ **27.** $\dfrac{7}{15}$ **29.** $\dfrac{1}{6}$ **31.** $\dfrac{19}{45}$
33. $\dfrac{3}{40}$ **35.** $\dfrac{17}{48}$ **37.** $\dfrac{1}{4}$ in. **39.** $\dfrac{17}{40}$ for reserved seating **41.** $\dfrac{31}{40}$ in.
43. $\dfrac{1}{24}$ of the tank **45.** You cannot add or subtract until all the fractional pieces are the same size. For example, halves are larger than fourths, so you cannot add $\dfrac{1}{2} + \dfrac{1}{4}$ until you rewrite $\dfrac{1}{2}$ as $\dfrac{2}{4}$. **47.** $\dfrac{7}{24}$
49. work and travel; 8 hr; $\dfrac{1}{2}$ **51.** $\dfrac{3}{16}$ in.

Section 3.4 (pages 225–232)
1. *Estimate:* $6 + 3 = 9$; *Exact:* $8\dfrac{5}{6}$

3. *Estimate:* $7 + 4 = 11$; *Exact:* $11\dfrac{1}{2}$

5. *Estimate:* $1 + 4 = 5$; *Exact:* $4\dfrac{5}{24}$

7. *Estimate:* $25 + 19 = 44$; *Exact:* $43\dfrac{2}{3}$

9. *Estimate:* $34 + 19 = 53$; *Exact:* $52\dfrac{1}{10}$

11. *Estimate:* $23 + 15 = 38$; *Exact:* $38\dfrac{5}{28}$

13. *Estimate:* $13 + 19 + 15 = 47$; *Exact:* $45\dfrac{5}{6}$

15. *Estimate:* $15 - 12 = 3$; *Exact:* $2\dfrac{5}{8}$

17. *Estimate:* $13 - 1 = 12$; *Exact:* $11\dfrac{7}{15}$

19. *Estimate:* $28 - 6 = 22$; *Exact:* $22\dfrac{7}{30}$

21. *Estimate:* $17 - 7 = 10$; *Exact:* $10\dfrac{3}{8}$

23. *Estimate:* $19 - 6 = 13$; *Exact:* $12\dfrac{19}{20}$

25. *Estimate:* $20 - 12 = 8$; *Exact:* $7\dfrac{11}{12}$

27. $9\dfrac{1}{8}$ **29.** $11\dfrac{1}{2}$ **31.** $3\dfrac{5}{6}$ **33.** $6\dfrac{11}{12}$ **35.** $8\dfrac{1}{8}$ **37.** $\dfrac{5}{6}$ **39.** $2\dfrac{7}{8}$
41. $2\dfrac{7}{12}$ **43.** $5\dfrac{9}{20}$ **45.** $3\dfrac{16}{21}$ **47.** Find the least common denominator. Change the fraction parts so that they have the same denominator. Add the fraction parts. Add the whole number parts. Write the answer as a mixed number.

49. *Estimate:* $26 - 15 = 11$ ft; *Exact:* $11\dfrac{1}{4}$ ft

51. *Estimate:* $23 - 19 = 4$ in.; *Exact:* $4\dfrac{1}{8}$ in.

53. *Estimate:* $23 + 22 + 21 = 66$ in.; *Exact:* 66 in.

55. *Estimate:* $3 - 1 = 2$ in.; *Exact:* $2\dfrac{3}{16}$ in.

57. *Estimate:* $16 + 19 + 24 + 31 = 90$ ft; *Exact:* $87\dfrac{8}{4} = 89$ ft

59. *Estimate:* $24 + 35 + 24 + 35 = 118$ in.; *Exact:* $116\dfrac{1}{2}$ in.

61. *Estimate:* $100 - 10 - 14 - 9 - 19 - 12 - 10 - 14 = 12$ gallons; *Exact:* $12\dfrac{5}{8}$ gallons

63. *Estimate:* $527 - 108 - 151 - 139 = 129$ ft; *Exact:* 130 ft

65. *Estimate:* $59 + 24 + 17 + 29 + 58 = 187$ tons; *Exact:* $186\dfrac{13}{24}$ tons

67. $4\dfrac{11}{16}$ in. **69.** $21\dfrac{3}{8}$ in. **71. (a)** 30 **(b)** 28 **(c)** 25
(d) 264

72. least common denominator **73. (a)** $\dfrac{23}{24}$ **(b)** $\dfrac{8}{15}$ **(c)** $\dfrac{43}{48}$ **(d)** $\dfrac{4}{21}$

74. fraction parts **75.** improper; large
76. (a) $4\dfrac{5}{8} + 3\dfrac{6}{8} = 7\dfrac{11}{8} = 8\dfrac{3}{8}; \dfrac{37}{8} + \dfrac{30}{8} = \dfrac{67}{8} = 8\dfrac{3}{8}$
(b) $11\dfrac{56}{40} - 8\dfrac{35}{40} = 3\dfrac{21}{40}; \dfrac{496}{40} - \dfrac{355}{40} = \dfrac{141}{40} = 3\dfrac{21}{40}$

Section 3.5 (pages 237–240)

1.–12. $\frac{1}{4}$ $\frac{1}{2}$ $\frac{7}{8}$ $\frac{5}{4}$ $\frac{3}{2}$ $1\frac{7}{8}$ $2\frac{1}{6}$ $\frac{7}{3}$ $\frac{11}{4}$ $3\frac{1}{4}$ $\frac{7}{2}$ $3\frac{4}{5}$

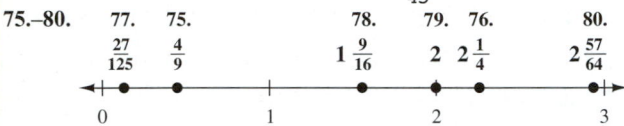

2. 1. 10. 4. 3. 12. 7. 5. 6. 11. 9. 8.

13. > **15.** < **17.** > **19.** < **21.** > **23.** > **25.** $\frac{1}{9}$ **27.** $\frac{25}{64}$

29. $\frac{9}{16}$ **31.** $\frac{64}{125}$ **33.** $\frac{81}{16} = 5\frac{1}{16}$ **35.** $\frac{81}{256}$

37. A number line is a horizontal line with a range of equally spaced whole numbers placed on it. The lowest number is on the left and the greatest number is on the right. It can be used to compare the size or value of numbers.

$$0 \quad \frac{1}{2} \quad 1 \quad 1\frac{1}{2} \quad 2 \quad 2\frac{1}{2} \quad 3$$

39. 4 **41.** 10 **43.** 1 **45.** $\frac{3}{16}$ **47.** $\frac{4}{9}$ **49.** $\frac{1}{3}$ **51.** $\frac{1}{2}$ **53.** $\frac{3}{8}$

55. $\frac{1}{4}$ **57.** $1\frac{1}{2}$ **59.** $\frac{1}{12}$ **61.** 3 **63.** $\frac{5}{16}$ **65.** $\frac{1}{4}$ **67.** $\frac{1}{32}$

69. $\frac{5}{30}$ in Atlanta is greater. **71.** <; > **72. (a)** like; numerators; numerator **(b)** Answers will vary. **73.** parentheses; exponents; square; multiply; divide; add; subtract **74.** $\frac{2}{45}$

75.–80. **77. 75.** **78. 79. 76. 80.**
$\frac{27}{125}$ $\frac{4}{9}$ $1\frac{9}{16}$ 2 $2\frac{1}{4}$ $2\frac{57}{64}$

$$0 \qquad\qquad 1 \qquad\qquad 2 \qquad\qquad 3$$

Chapter 3 Review Exercises (pages 245–250)

1. $\frac{6}{7}$ **2.** $\frac{7}{9}$ **3.** $\frac{3}{4}$ **4.** $\frac{1}{8}$ **5.** $\frac{4}{5}$ **6.** $\frac{1}{6}$ **7.** $\frac{13}{31}$ **8.** $\frac{1}{3}$

9. $\frac{11}{12}$ of his total income **10.** $\frac{1}{4}$ Web page less **11.** 10 **12.** 12

13. 60 **14.** 24 **15.** 120 **16.** 180 **17.** 8 **18.** 21 **19.** 10

20. 45 **21.** 32 **22.** 20 **23.** $\frac{5}{6}$ **24.** $\frac{7}{8}$ **25.** $\frac{5}{8}$ **26.** $\frac{5}{12}$ **27.** $\frac{13}{24}$

28. $\frac{17}{36}$ **29.** $\frac{9}{10}$ of the students **30.** $\frac{23}{24}$ of her business

31. *Estimate:* $19 + 14 = 33$; *Exact:* $32\frac{3}{8}$

32. *Estimate:* $23 + 15 = 38$; *Exact:* $38\frac{1}{9}$

33. *Estimate:* $13 + 9 + 10 = 32$; *Exact:* $31\frac{43}{80}$

34. *Estimate:* $32 - 15 = 17$; *Exact:* $17\frac{1}{12}$

35. *Estimate:* $34 - 16 = 18$; *Exact:* $18\frac{1}{3}$

36. *Estimate:* $215 - 136 = 79$; *Exact:* $79\frac{7}{16}$

37. $9\frac{1}{10}$ **38.** $10\frac{5}{12}$ **39.** $3\frac{1}{4}$ **40.** $1\frac{2}{3}$ **41.** $5\frac{1}{2}$ **42.** $2\frac{19}{24}$

43. *Estimate:* $19 - 6 - 7 = 6$ miles; *Exact:* $5\frac{19}{24}$ miles

44. *Estimate:* $29 + 25 = 54$ tons; *Exact:* $53\frac{5}{12}$ tons

45. *Estimate:* $9 + 10 + 7 = 26$ pounds; *Exact:* $25\frac{1}{8}$ pounds

46. *Estimate:* $9 - 2 - 3 = 4$ acres; *Exact:* $4\frac{1}{16}$ acres

47.–50.

$$\frac{3}{8} \quad \frac{7}{4} \quad \frac{8}{3} \; 3\frac{1}{5}$$
$$0 \quad 1 \quad 2 \quad 3 \quad 4$$
47. 48. 49. 50.

51. < **52.** < **53.** > **54.** > **55.** < **56.** > **57.** <

58. > **59.** $\frac{1}{4}$ **60.** $\frac{4}{9}$ **61.** $\frac{27}{1000}$ **62.** $\frac{81}{4096}$ **63.** $\frac{1}{2}$ **64.** $6\frac{3}{4}$

65. $\frac{1}{16}$ **66.** 1 **67.** $\frac{3}{16}$ **68.** $1\frac{25}{64}$ **69.** $\frac{3}{4}$ **70.** $\frac{2}{5}$ **71.** $\frac{19}{32}$

72. $\frac{11}{16}$ **73.** $2\frac{1}{6}$ **74.** $26\frac{1}{4}$ **75.** $5\frac{3}{8}$ **76.** $11\frac{43}{80}$ **77.** $15\frac{5}{12}$

78. $\frac{8}{11}$ **79.** $\frac{1}{250}$ **80.** $\frac{1}{2}$ **81.** $\frac{2}{9}$ **82.** $\frac{11}{27}$ **83.** > **84.** < **85.** <

86. > **87.** 36 **88.** 120 **89.** 126 **90.** 18 **91.** 108 **92.** 60

93. *Estimate:* $93 - 14 - 22 = 57$ ft; *Exact:* $56\frac{7}{8}$ ft

94. *Estimate:* $4 \cdot 50 = 200$ pounds of sugar; $200 - 69 - 77 - 33 = 21$ pounds; *Exact:* $21\frac{5}{8}$ pounds

Chapter 3 Test (pages 251–252)

1. $\frac{3}{4}$ **2.** $\frac{1}{2}$ **3.** $\frac{2}{5}$ **4.** $\frac{1}{6}$ **5.** 12 **6.** 30 **7.** 108 **8.** $\frac{5}{8}$

9. $\frac{23}{36}$ **10.** $\frac{5}{24}$ **11.** $\frac{1}{40}$

12. *Estimate:* $8 + 5 = 13$; *Exact:* $12\frac{1}{2}$

13. *Estimate:* $16 - 12 = 4$; *Exact:* $4\frac{11}{15}$

14. *Estimate:* $19 + 9 + 12 = 40$; *Exact:* $40\frac{29}{60}$

15. *Estimate:* $24 - 18 = 6$; *Exact:* $5\frac{5}{8}$

16. Answers will vary. Probably addition and subtraction of fractions is more difficult because you have to find the least common denominator and then change the fractions to the same denominator. **17.** Answers will vary. Round mixed numbers to the nearest whole number. Then add or subtract to estimate the answer. The estimate may vary from the exact answer but it lets you know if your answer is reasonable.

18. *Estimate:* $10 + 85 + 37 + 8 = 140$ pounds; *Exact:* $140\frac{1}{24}$ pounds

19. *Estimate:* $148 - 69 - 37 - 6 = 36$ gal; *Exact:* $35\frac{7}{8}$ gal

20. > **21.** > **22.** 2 **23.** $\frac{13}{48}$ **24.** $1\frac{3}{4}$ **25.** $1\frac{1}{3}$

Chapter 4 Decimals

Section 4.1 (pages 261–264)

1. 7; 0; 4 **3.** 5; 1; 8 **5.** 4; 7; 0 **7.** 1; 6; 3 **9.** 1; 8; 9 **11.** 6; 2; 1

13. 410.25 **15.** 6.5432 **17.** 5406.045 **19.** $\frac{7}{10}$ **21.** $13\frac{2}{5}$ **23.** $\frac{1}{4}$

25. $\frac{33}{50}$ **27.** $10\frac{17}{100}$ **29.** $\frac{3}{50}$ **31.** $\frac{41}{200}$ **33.** $5\frac{1}{500}$ **35.** $\frac{343}{500}$

37. five tenths **39.** seventy-eight hundredths **41.** one hundred five thousandths **43.** twelve and four hundredths **45.** one and seventy-five thousandths **47.** 6.7 **49.** 0.32 **51.** 420.008 **53.** 0.0703 **55.** 75.030

57. Anne should not say "and" because that denotes a decimal point.

59. ten thousandths inch; $\frac{10}{1000} = \frac{1}{100}$ inch **61.** 12 pounds **63.** 3-C
65. 4-A **67.** One and six hundred two thousandths centimeters
69. millionths, ten-millionths, hundred-millionths, billionths; these match the words on the left side of the chart with "ths" attached. **70.** The first place to the left of the decimal point is ones, so the first place to the right could be one*ths*, like tens and ten*ths*. But anything that is 1 or more is to the *left* of the decimal point. **71.** Seventy-two million four hundred thirty-six thousand nine hundred fifty-five hundred-millionths **72.** six hundred seventy-eight thousand five hundred fifty-four billionths **73.** eight thousand six and five hundred thousand one millionths **74.** twenty thousand sixty and five hundred five millionths **75.** 0.0302040 **76.** 9,876,543,210.100200300

Section 4.2 (pages 271–272)

1. 16.9 **3.** 0.956 **5.** 0.80 **7.** 3.661 **9.** 794.0 **11.** 0.0980
13. 49 **15.** 9.09 **17.** 82.0002 **19.** $0.82 **21.** $1.22 **23.** $0.50
25. $48,650 **27.** $310 **29.** $849 **31.** $500 **33.** $1.00 **35.** $1000
37. (a) 322 miles per hour **(b)** 107 miles per hour **39. (a)** 186.0 miles per hour **(b)** 763.0 miles per hour **41.** Rounds to $0 (zero dollars) because $0.499 is closer to $0 than to $1. **42.** Round amounts less than $1.00 to the nearest cent instead of the nearest dollar. **43.** Rounds to $0.00 (zero cents) because $0.0015 is closer to $0.00 than to $0.01. **44.** Both round to $0.60. Rounding to nearest thousandth (tenth of a cent) would allow you to identify $0.597 as less than $0.601.

Section 4.3 (pages 277–280)

1. 17.48 **3.** 23.013 **5.** 7.763 **7.** 77.006 **9.** 20.104
11. 0.109 **13.** 330.86895 **15. (a)** 24.75 in. **(b)** 3.95 in.
17. (a) 62.27 in. **(b)** 0.39 in. **19.** 6 should be written 6.00; sum is 46.22. **21.** *Estimate:* $20 − 7 = $13; *Exact:* $13.16 **23.** *Estimate:* 400 + 1 + 20 = 421; *Exact:* 414.645 **25.** *Estimate:* 9 − 4 = 5; *Exact:* 4.849 **27.** *Estimate:* 60 + 500 + 6 = 566; *Exact:* 608.4363 **29.** 0.275 **31.** 6.507 **33.** 1.81 **35.** 6056.7202 **37.** *Estimate:* 30 − 20 = 10 million people; *Exact:* 12.1 million people **39.** *Estimate:* 200 + 100 + 90 + 50 + 30 + 20 + 20 = 510 million people; *Exact:* 527.0 million people **41.** *Estimate:* 2 + 2 + 2 = 6 meters; *Exact:* 6.1 meters, which is less than the rhino by 0.3 meter **43.** *Estimate:* 11 − 10 = 1 ounce; *Exact:* 0.65 ounce **45.** *Estimate:* 20 + 6 + 20 + 6 = 52 in.; *Exact:* 52.1 in. **47.** *Estimate:* $5 − $5 = $0; *Exact:* $0.30
49. *Estimate:* $19 + 2 + 2 + 10 + 2 = $35; *Exact:* $35.25
51. $1939.36 **53.** $3.97 **55.** $598.22 **57.** $b = 1.39$ centimeters
59. $q = 23.843$ ft

Section 4.4 (pages 283–286)

1. 0.1344 **3.** 159.10 **5.** 15.5844 **7.** $34,500.20 **9.** 43.2
11. 0.432 **13.** 0.0432 **15.** 0.00432 **17.** 0.0000312 **19.** 0.000025
21. 59.6; 32; 4.76; 803.5; 7226; 9. Multiplying by 10, decimal point moves one place to the right; by 100, two places to the right; by 1000, three places to the right. **22.** 5.96; 0.32; 0.0476; 8.035; 6.5; 52.3. Multiplying by 0.1, decimal point moves one place to the left; by 0.01, two places to the left; by 0.001, three places to the left. **23.** *Estimate:* 40 × 5 = 200; *Exact:* 190.08 **25.** *Estimate:* 40 × 40 = 1600; *Exact:* 1558.2 **27.** *Estimate:* 7 × 5 = 35; *Exact:* 30.038 **29.** *Estimate:* 3 × 7 = 21; *Exact:* 19.24165 **31.** unreasonable; $289.00
33. reasonable **35.** unreasonable; $4.19 **37.** unreasonable; 9.5 pounds
39. $945.87 (rounded) **41.** $2.45 (rounded) **43.** $90.02 (rounded)
45. $20,265 **47. (a)** Area before 1929 ≈ 23.2 in.²; Area today ≈ 16.0 in.²
(b) 7.2 in.² **49. (a)** 0.43 in. **(b)** 4.3 in. **51.** $984.04; $2207.80
53. $76.50 **55.** $4.09 (rounded) **57.** $129.25 **59. (a)** $70.05
(b) $25.80

Section 4.5 (pages 293–296)

1. 3.9 **3.** 0.47 **5.** 400.2 **7.** 36 **9.** 0v06 **11.** 6000 **13.** 25.3
15. 516.67 (rounded) **17.** 26.756 (rounded) **19.** 10,082.647
(rounded) **21.** 0.377; 0.91; 0.0886; 3.019; 40.65; 662.57. **(a)** Dividing by 10, decimal point moves one place to the left; by 100, two places to the left; by 1000, three places to the left. **(b)** The decimal point moved to the *right* when *multiplying* by 10 or 100 or 1000. Here it moves to the *left* when *dividing* by those numbers. **22.** 402; 71; 3.39; 157.7; 460; 8730. **(a)** Dividing by 0.1, decimal point moves one place to the right; by 0.01, two places to the right; by 0.001, three places to the right. **(b)** The decimal point moved to the *left* when *multiplying* by 0.1 or 0.01 or 0.001. Here the decimal point moves to the *right* when *dividing* by those numbers.
23. unreasonable; 40 ÷ 8 = 5; Correct answer is 4.725. **25.** reasonable; 50 ÷ 50 = 1 **27.** unreasonable; 300 ÷ 5 = 60; Correct answer is 60.2.
29. unreasonable; 9 ÷ 1 = 9; Correct answer is 7.44. **31.** $4.00 (rounded) **33.** $67.08 **35.** $0.30 **37.** $11.92 per hour
39. 21.2 miles per gallon (rounded) **41.** 7.37 meters (rounded)
43. 0.08 meter **45.** 22.49 meters **47.** 14.25 **49.** 73.4 **51.** 1.205
53. 0.334 **55.** $0.03 per can (rounded) **57. (a)** 1,583,333 pieces (rounded) **(b)** 26,389 pieces (rounded) **(c)** 440 pieces (rounded).
59. 100,000 box tops **61.** 2632 box tops (rounded)

Section 4.6 (pages 301–304)

1. 0.5 **3.** 0.75 **5.** 0.3 **7.** 0.9 **9.** 0.6 **11.** 0.875 **13.** 2.25
15. 14.7 **17.** 3.625 **19.** 6.333 (rounded) **21.** 0.833 (rounded)
23. 1.889 (rounded) **25. (a)** A proper fraction like $\frac{5}{9}$ is less than 1, so it cannot be equivalent to a decimal number that is greater than 1.
(b) $\frac{5}{9}$ means 5 ÷ 9 or $9\overline{)5}$ so correct answer is 0.556 (rounded). This makes sense because both the fraction and decimal are less than 1.
26. (a) $2.035 = 2\frac{35}{1000} = 2\frac{7}{200}$, not $2\frac{7}{20}$. **(b)** Adding the whole number part gives 2 + 0.35, which is 2.35, not 2.035. To check, $2.35 = 2\frac{35}{100} = 2\frac{7}{20}$. **27.** Just add the whole number part to 0.375. So $1\frac{3}{8} = 1.375$; $3\frac{3}{8} = 3.375$; $295\frac{3}{8} = 295.375$. **28.** It works only when the fraction part has a one-digit numerator and a denominator of 10, a two-digit numerator and a denominator of 100, and so on.
29. $\frac{2}{5}$ **31.** $\frac{5}{8}$ **33.** $\frac{7}{20}$ **35.** 0.35 **37.** $\frac{1}{25}$ **39.** $\frac{3}{20}$ **41.** 0.2
43. $\frac{9}{100}$ **45.** shorter; 0.72 inch **47.** too much; 0.005 gram
49. 0.9991 cm, 1.0007 cm **51.** more; 0.05 inch
53. 0.5399, 0.54, 0.5455 **55.** 5.0079, 5.79, 5.8, 5.804
57. 0.6009, 0.609, 0.628, 0.62812 **59.** 2.8902, 3.88, 4.876, 5.8751
61. 0.006, 0.043, $\frac{1}{20}$, 0.051 **63.** 0.37, $\frac{3}{8}$, $\frac{2}{5}$, 0.4001 **65.** red box
67. green box **69.** 1.4 in. (rounded) **71.** 0.3 in. (rounded)
73. 0.4 in. (rounded)

Chapter 4 Review Exercises (pages 309–312)

1. 0; 5 **2.** 0; 6 **3.** 8; 9 **4.** 5; 9 **5.** 7; 6 **6.** $\frac{1}{2}$ **7.** $\frac{3}{4}$ **8.** $4\frac{1}{20}$
9. $\frac{7}{8}$ **10.** $\frac{27}{1000}$ **11.** $27\frac{4}{5}$ **12.** eight tenths **13.** four hundred and twenty-nine hundredths **14.** twelve and seven thousandths **15.** three hundred six ten-thousandths **16.** 8.3 **17.** 0.205 **18.** 70.0066
19. 0.30 **20.** 275.6 **21.** 72.79 **22.** 0.160 **23.** 0.091 **24.** 1.0
25. $15.83 **26.** $0.70 **27.** $17,625.79 **28.** $350 **29.** $130
30. $100 **31.** $29 **32.** *Estimate:* 6 + 400 + 20 = 426; *Exact:* 444.86

33. *Estimate:* $80 + 1 + 100 + 1 + 30 = 212$; *Exact:* 233.515
34. Estimate: $300 - 20 = 280$; Exact: 290.7
35. Estimate: $9 - 8 = 1$; *Exact:* 1.2684
36. *Estimate:* 90 million $-$ 20 million $=$ 70 million; *Exact:* 73.9 million more cats **37.** *Estimate:* $\$300 - \$200 - \$40 = \$60$; Exact: $\$45.80$
38. *Estimate:* $\$2 + \$5 + \$20 = \$27$; $\$30 - \$27 = \$3$;
Exact: $\$4.14$ **39.** *Estimate:* $2 + 4 + 5 = 11$ kilometers;
Exact: 11.55 kilometers **40.** *Estimate:* $6 \times 4 = 24$; *Exact:* 22.7106
41. *Estimate:* $40 \times 3 = 120$; *Exact:* 141.57 **42.** 0.0112 **43.** 0.000355
44. reasonable; $700 \div 10 = 70$ **45.** unreasonable; $30 \div 3 = 10$;
Correct answer is 9.5. **46.** 14.467 (rounded) **47.** 1200 **48.** 0.4
49. $\$708$ (rounded) **50.** $\$2.99$ (rounded) **51.** 133 shares (rounded)
52. $\$3.47$ (rounded) **53.** 29.215 **54.** 10.15 **55.** 3.8 **56.** 0.64
57. 1.875 **58.** 0.111 (rounded) **59.** $3.6008, 3.68, 3.806$
60. $0.209, 0.2102, 0.215, 0.22$ **61.** $\dfrac{1}{8}, \dfrac{3}{20}, 0.159, 0.17$ **62.** 404.865
63. 254.8 **64.** 3583.261 (rounded) **65.** 29.0898 **66.** 0.03066
67. 9.4 **68.** 175.675 **69.** 9.04 **70.** 19.50 **71.** 8.19
72. 0.928 **73.** 35 **74.** 0.259 **75.** 0.3 **76.** $\$3.00$ (rounded)
77. $\$2.17$ (rounded) **78.** $\$35.96$ **79.** $\$199.71$ **80.** $\$78.50$
81. (a) baked potato with skin **(b)** cup of orange juice
(c) 0.59 milligram **82. (a)** 2.06 milligrams
(b) more, by 0.06 milligram

Chapter 4 Test (pages 313–314)

1. $18\dfrac{2}{5}$ **2.** $\dfrac{3}{40}$ **3.** sixty and seven thousandths **4.** two hundred eight
ten-thousandths **5.** 725.6 **6.** 0.630 **7.** $\$1.49$ **8.** $\$7860$ **9.** *Estimate:*
$8 + 80 + 40 + 128$; *Exact:* 129.2028 **10.** *Estimate:* $80 - 4 = 76$;
Exact: 75.498 **11.** *Estimate:* $6(1) = 6$; *Exact:* 6.948 **12.** *Estimate:*
$20 \div 5 = 4$; *Exact:* 4.175 **13.** 839.762 **14.** 669.004 **15.** 0.0000483
16. 480 **17.** 2.625 **18.** $0.44, \dfrac{9}{20}, 0.4506, 0.451$ **19.** 35.49
20. $\$446.87$ **21.** pintails, wigeons, gadwalls **22.** $\$5.35$ (rounded)
23. 2.8 degrees **24.** $\$1.79$ per foot (rounded) **25.** Answer varies.

Chapter 5 Ratio and Proportion

Section 5.1 (pages 323–326)

1. $\dfrac{8}{9}$ **3.** $\dfrac{2}{1}$ **5.** $\dfrac{1}{3}$ **7.** $\dfrac{8}{5}$ **9.** $\dfrac{3}{8}$ **11.** $\dfrac{9}{7}$ **13.** $\dfrac{6}{1}$ **15.** $\dfrac{5}{6}$ **17.** $\dfrac{8}{5}$
19. $\dfrac{1}{12}$ **21.** $\dfrac{5}{16}$ **23.** $\dfrac{4}{1}$ **25.** $\dfrac{1}{2}$ **27.** $\dfrac{36}{1}$ **29.** Answers will vary. One
possibility is stocking cards of various types in the same ratios as those in
the table. **31.** $\dfrac{6}{5}; \dfrac{36}{17}$ **33.** *White Christmas* to *It's Now or Never*; *White*
Christmas to *I Will Always Love You*; *Candle in the Wind* to *I Want to Hold*
Your Hand. **35. (a)** $\dfrac{6}{1}$ **(b)** $\dfrac{5}{2}$ **(c)** $\dfrac{7}{16}$ **37.** $\dfrac{7}{5}$ **39.** $\dfrac{6}{1}$ **41.** $\dfrac{38}{17}$ **43.** $\dfrac{1}{4}$
45. $\dfrac{34}{35}$ **47.** $\dfrac{1}{1}$; as long as the sides all have the same length, any measure-
ment you choose will maintain the ratio. **48.** Answers will vary. Some
possibilities are $\dfrac{4}{5} = \dfrac{8}{10} = \dfrac{12}{15} = \dfrac{16}{20} = \dfrac{20}{25} = \dfrac{24}{30} = \dfrac{28}{35}$. **49.** It is not
possible. Amelia would have to be older than her mother to have a ratio of
5 to 3. **50.** Answers will vary, but a ratio of 3 to 1 means your income is
3 times your friend's income.

Section 5.2 (pages 331–334)

1. $\dfrac{5 \text{ cups}}{3 \text{ people}}$ **3.** $\dfrac{3 \text{ feet}}{7 \text{ seconds}}$ **5.** $\dfrac{18 \text{ miles}}{1 \text{ gallon}}$
7. $\$12$ per hour or $\$12$/hour **9.** 1.25 pounds/person **11.** $\$103.30$/day
13. $325.9; 21.0$ (rounded) **15.** $338.6; 20.9$ (rounded) **17.** 4 ounces for
$\$3.65$, about $\$0.913$/ounce **19.** 14 ounces for $\$2.89$, about $\$0.206$/ounce
21. 18 ounces for $\$1.79$, about $\$0.099$/ounce **23.** Answers will vary. For
example, you might choose Brand B because you like more chicken, so the
cost per chicken chunk may actually be the same or less than Brand A.
25. 1.75 pounds/week **27.** $\$12.26$/hour **29. (a)** Radiant, $\$0.44$;
IDT, $\$0.25$; Access; $\$0.235$ **(b)** Radiant, $\$0.088$/min; IDT, $\$0.05$/min;
Access, $\$0.047$/min; Access America is the best buy. **31.** For a 15-minute
call: Radiant, $\$0.036$/min; IDT, $\$0.031$/min; Access, $\$0.047$/min; IDT is the
best buy. For a 25-minute call: Radiant, $\$0.026$/min; IDT, $\$0.028$/min;
Access, $\$0.047$/min; Radiant is the best buy. **33.** one battery for $\$1.79$;
like getting 3 batteries so $\$1.79 \div 3 \approx \$0.597$ per battery **35.** Brand P
with the 50¢ coupon is the best buy. ($\$3.39 - \$0.50 = \$2.89$;
$\$2.89 \div 16.5$ ounces $\approx \$0.175$ per ounce) **37.** $\$2.92$ (rounded) per
month for Verizon, T-Mobile, and Nextel; $\$3$ per month for Sprint.
38. Average number of weekdays per month is about 21.7;
Verizon $\approx$ 18 min/weekday; T-Mobile $\approx$ 28 min/weekday; Nextel
and Sprint $\approx$ 23 min/weekday. **39.** Round to hundredths (nearest
cent) to see that T-Mobile is best buy at $\$0.07$ per "anytime minute";
Verizon $\approx \$0.16$; Sprint $\approx \$0.12$; Nextel $\approx \$0.10$.
40. Verizon $\approx \$1.65$ per "anytime minute"; T-Mobile $\approx \$1.57$;
Nextel $\approx \$1.63$; Sprint $= \$1.48$.

Section 5.3 (pages 339–340)

1. $\dfrac{\$9}{12 \text{ cans}} = \dfrac{\$18}{24 \text{ cans}}$ **3.** $\dfrac{200 \text{ adults}}{450 \text{ children}} = \dfrac{4 \text{ adults}}{9 \text{ children}}$ **5.** $\dfrac{120}{150} = \dfrac{8}{10}$
7. $\dfrac{3}{5} = \dfrac{3}{5}$; true **9.** $\dfrac{5}{8} = \dfrac{5}{8}$; true **11.** $\dfrac{3}{4} \neq \dfrac{2}{3}$; false **13.** $\dfrac{14}{5} = \dfrac{14}{5}$; true
15. $\dfrac{16}{9} = \dfrac{16}{9}$; true **17.** $\dfrac{7}{6} \neq \dfrac{9}{8}$; false **19.** $54 = 54$; True
21. $336 \neq 320$; False **23.** $2880 \neq 2970$; False **25.** $28 = 28$; True
27. $44.8 \neq 45$; False **29.** $66 = 66$; True **31.** $68\dfrac{1}{4} = 68\dfrac{1}{4}$; True
33. $5\dfrac{2}{5} \neq 5\dfrac{1}{3}$; False **35.** $2\dfrac{1}{80} \neq 2\dfrac{7}{100}$ or $2.0125 \neq 2.07$; False
37. $\dfrac{17 \text{ hits}}{50 \text{ at bats}} = \dfrac{153 \text{ hits}}{450 \text{ at bats}}$ $\begin{array}{l} 50 \cdot 153 = 7650 \\ 17 \cdot 450 = 7650 \end{array}$ Cross products
are *equal* so the proportion is *true*; they hit equally well.

Section 5.4 (pages 345–346)

1. $x = 4$ **3.** $x = 2$ **5.** $x = 88$ **7.** $x = 91$ **9.** $x = 5$ **11.** $x = 10$
13. $x \approx 24.44$ (rounded) **15.** $x = 50.4$ **17.** $x \approx 17.64$ (rounded)
19. $x = 1$ **21.** $x = 3\dfrac{1}{2}$ **23.** $x = 0.2$ or $x = \dfrac{1}{5}$
25. $x = 0.005$ or $x = \dfrac{1}{200}$
27. Find cross products: $20 \neq 30$, so the proportion is false.
$\dfrac{6\frac{2}{3}}{4} = \dfrac{5}{3}$ or $\dfrac{10}{6} = \dfrac{5}{3}$ or $\dfrac{10}{4} = \dfrac{7.5}{3}$ or $\dfrac{10}{4} = \dfrac{5}{2}$
28. Find cross products: $192 \neq 180$, so the proportion is false.
$\dfrac{6.4}{8} = \dfrac{24}{30}$ or $\dfrac{6}{7.5} = \dfrac{24}{30}$ or $\dfrac{6}{8} = \dfrac{22.5}{30}$ or $\dfrac{6}{8} = \dfrac{24}{32}$

Section 5.5 (pages 351–354)

1. 22.5 hours **3.** $\$7.20$ **5.** 42 pounds **7.** $\$403.68$ **9.** 10 ounces
(rounded) **11.** 5 quarts **13.** 14 ft, 10 ft **15.** 14 ft, 8 ft

17. 96 pieces chicken; 33.6 pounds lasagna; 10.8 pounds deli meats; $5\frac{3}{5}$ pounds cheese; 7.2 dozen (about 86) buns; 14.4 pounds of salad
19. 2065 students (reasonable); about 4214 students with incorrect setup (only 2950 students in the group)   **21.** about 83 people (reasonable); about 750 people with incorrect setup (only 250 people attended)
23. 110,838,000 households (reasonable); about 115,408,163 households with incorrect setup (only 113,100,000 U.S. households).   **25.** 625 stocks
27. 4.06 meters (rounded)   **29.** 311 calories (rounded)   **31.** 10.53 meters (rounded)   **33.** You cannot solve this problem using a proportion because the ratio of age to weight is not constant. As Jim's age increases, his weight may decrease, stay the same, or increase.   **35.** 5050 students use cream
37. 120 calories and 12 grams of fiber   **39.** $1\frac{3}{4}$ cups water, 3 Tbsp margarine, $\frac{3}{4}$ cup milk, 2 cups flakes   **40.** $5\frac{1}{4}$ cups water, 9 Tbsp margarine, $2\frac{1}{4}$ cups milk, 6 cups flakes   **41.** $\frac{7}{8}$ cup water, $1\frac{1}{2}$ Tbsp margarine, $\frac{3}{8}$ cup milk, 1 cup flakes   **42.** $2\frac{5}{8}$ cups water, $4\frac{1}{2}$ Tbsp margarine, $1\frac{1}{8}$ cups milk, 3 cups flakes

Chapter 5 Review Exercises (pages 361–364)

1. $\frac{3}{4}$ **2.** $\frac{4}{1}$ **3.** great white shark to whale shark; whale shark to blue whale   **4.** $\frac{2}{1}$ **5.** $\frac{2}{3}$ **6.** $\frac{5}{2}$ **7.** $\frac{1}{6}$ **8.** $\frac{3}{1}$ **9.** $\frac{3}{8}$ **10.** $\frac{4}{3}$
11. $\frac{1}{9}$ **12.** $\frac{10}{7}$ **13.** $\frac{7}{5}$ **14.** $\frac{5}{6}$ **15.** $\frac{\$11}{1 \text{ dozen}}$ **16.** $\frac{12 \text{ children}}{5 \text{ families}}$
17. 0.2 page/minute or $\frac{1}{5}$ page/minute; 5 minutes/page
18. \$8/hour; 0.125 hour/dollar or $\frac{1}{8}$ hour/dollar
19. 8 ounces for \$4.98, about \$0.623/ounce   **20.** 17.6 pounds for \$18.69 − \$1 coupon, about \$1.005/pound
21. $\frac{3}{5} = \frac{3}{5}$ or 90 = 90; true   **22.** $\frac{1}{8} \neq \frac{1}{4}$ or 432 ≠ 216; false
23. $\frac{47}{10} \neq \frac{49}{10}$ or 980 ≠ 940; false   **24.** $\frac{16}{9} = \frac{16}{9}$ or 3456 = 3456; true   **25.** 4.8 = 4.8; true   **26.** 14 = 14; true   **27.** $x = 1575$
28. $x = 20$ **29.** $x = 400$ **30.** $x = 12.5$ **31.** $x \approx 14.67$ (rounded)
32. $x \approx 8.17$ (rounded)   **33.** $x = 50.4$ **34.** $x \approx 0.57$ (rounded)
35. $x \approx 2.47$ (rounded)   **36.** 27 cats   **37.** 46 hits   **38.** \$15.63 (rounded)
39. 3299 students (rounded)   **40.** 68 feet   **41.** $27\frac{1}{2}$ hours or 27.5 hours
42. 511 calories (rounded)   **43.** 14.7 milligrams   **44.** $x = 105$
45. $x = 0$ **46.** $x = 128$ **47.** $x \approx 23.08$ (rounded)   **48.** $x = 6.5$
49. $x \approx 117.36$ (rounded)   **50.** 1440 ≠ 1485; False
51. 10.8 ≠ 10.864; False   **52.** $2\frac{1}{3} = 2\frac{1}{3}$; True   **53.** $\frac{8}{5}$
54. $\frac{33}{80}$ **55.** $\frac{15}{4}$ **56.** $\frac{4}{1}$ **57.** $\frac{4}{5}$ **58.** $\frac{37}{7}$ **59.** $\frac{3}{8}$ **60.** $\frac{1}{12}$
61. $\frac{45}{13}$ **62.** 24,900 fans (rounded)   **63.** $\frac{8}{3}$ **64.** 75 ft for \$1.99 − \$0.50 coupon, about \$0.020/ft   **65.** 21 ft long; 15 ft wide   **66.** 7.5 hours or $7\frac{1}{2}$ hours   **67.** $\frac{1}{2}$ teaspoon or 0.5 teaspoon   **68.** 21 points (rounded)
69. Set up the proportion to compare teaspoons to pounds on both sides.
$$\frac{1.5 \text{ teaspoons}}{24 \text{ pounds}} = \frac{x \text{ teaspoons}}{8 \text{ pounds}}$$
Show that cross products are equal. $(24)(x) = (1.5)(8)$

Divide both sides by 24. $\dfrac{\overset{1}{\cancel{24}}(x)}{\underset{1}{\cancel{24}}} = \dfrac{12}{24}$
so $x = \frac{1}{2}$ teaspoon or 0.5 teaspoon
70. (a) 1400 milligrams   **(b)** 100 milligrams

Chapter 5 Test (pages 365–366)

1. $\frac{4}{5}$ **2.** $\frac{20 \text{ miles}}{1 \text{ gallon}}$ **3.** $\frac{\$1}{5 \text{ minutes}}$ **4.** $\frac{9}{2}$ **5.** $\frac{15}{4}$
6. Best buy is 8 in. sub, about \$0.724/inch
7. 16 ounces for \$1.89 − \$0.50 coupon, about \$0.087/ounce
8. You earned less this year. An example is: $\begin{array}{c}\text{Last year}\\\text{This year}\end{array} \quad \dfrac{\$30,000}{\$20,000} = \dfrac{3}{2}$
9. $\frac{3}{7} \neq \frac{2}{5}$ or   252 ≠ 270; false   **10.** 5.88 = 5.88; true   **11.** $x = 25$
12. $x \approx 2.67$ (rounded)   **13.** $x = 325$ **14.** $x = 10\frac{1}{2}$ **15.** 24 orders
16. 3.6 ounces   **17.** 87 students (rounded)   **18.** No, 4875 cannot be correct because there are only 650 students in the whole school.
19. 23.8 grams (rounded)   **20.** 60 ft

Chapter 6 Percent

Section 6.1 (pages 373–378)

1. 0.12   **3.** 0.70 or 0.7   **5.** 0.25   **7.** 1.40 or 1.4   **9.** 0.055
11. 1.00 or 1   **13.** 0.005   **15.** 0.0035   **17.** 60%   **19.** 58%
21. 1%   **23.** 12.5%   **25.** 37.5%   **27.** 200%   **29.** 370%
31. 3.12%   **33.** 416.2%   **35.** 0.28%   **37.** Answers will vary. Some possibilities are: No common denominators are needed with percents. The denominator is always 100 with percent, which makes comparisons easier to understand.   **39.** 0.38   **41.** 0.47   **43.** 3.5%   **45.** 200%
47. 0.5%   **49.** 1.536   **51.** 20 children   **53.** 420 employees
55. 270 chairs   **57.** \$377.50   **59.** 820 commuters   **61.** 26 plants
63. (a) Since 100% means 100 parts out of 100 parts, 100% is all of the number.   **(b)** Answers will vary. For example, 100% of \$72 is \$72.
65. (a) Since 200% is two times a number, find 200% of the number by multiplying the number by 2 (double it).   **(b)** Answers will vary. For example, 200% of \$20 is 2 • \$20 = \$40.   **67. (a)** Since 10% means 10 parts out of 100 parts or $\frac{1}{10}$, the shortcut for finding 10% of a number is to move the decimal point in the number one place to the left.
(b) Answers will vary. For example, 10% of \$90 is \$9.   **69.** 12%; 0.12   **71. (a)** witch costume   **(b)** 4.5%; 0.045   **73.** 26%; 0.26
75. (a) Marine Corps   **(b)** 5%; 0.05   **77.** 16.8%; 0.168   **(a)** Germany; Bulgaria   **(b)** 16.5%; 0.165   **81.** 95% shaded; 5% unshaded
83. 30% shaded; 70% unshaded   **85.** 55% shaded; 45% unshaded
87. 64% shaded; 36% unshaded

Section 6.2 (pages 385–390)

1. $\frac{1}{4}$ **3.** $\frac{3}{4}$ **5.** $\frac{17}{20}$ **7.** $\frac{5}{8}$ **9.** $\frac{1}{16}$ **11.** $\frac{1}{6}$ **13.** $\frac{1}{15}$ **15.** $\frac{1}{200}$
17. $1\frac{4}{5}$ **19.** $3\frac{3}{4}$ **21.** 50%   **23.** 80%   **25.** 70%   **27.** 37%
29. 62.5%   **31.** 87.5%   **33.** 48%   **35.** 46%   **37.** 35%   **39.** 83.3% (rounded)   **41.** 55.6% (rounded)   **43.** 14.3% (rounded)   **45.** $\frac{1}{2}$; 50%
47. $\frac{7}{8}$; 0.875   **49.** $\frac{4}{5}$; 80%   **51.** 0.167 (rounded); 16.7% (rounded)
53. $\frac{7}{10}$; 70%   **55.** $\frac{1}{8}$; 0.125   **57.** 0.667 (rounded); 66.7% (rounded)
59. 0.06; 6%   **61.** 0.08; 8%   **63.** 0.005; 0.5%   **65.** $2\frac{1}{2}$; 250%

67. 3.25; 325%   **69.** There are many possible answers. Examples 2 and 3 show the steps that students should include in their answers.
71. $\frac{9}{50}$; 0.18; 18%   **73.** $\frac{13}{100}$; 0.13; 13%   **75.** $\frac{1}{5}$; 0.2; 20%   **77.** $\frac{1}{5}$; 0.20; 20%   **79.** $\frac{1}{10}$; 0.1; 10%   **81.** $\frac{1}{4}$; 0.25; 25%   **83.** $\frac{1}{20}$; 0.05; 5%
85. 100; 100   **86. (a)** 765 workers   **(b)** 96 letters   **(c)** 21 videos
87. 50; 100; half or $\frac{1}{2}$ **88.** 10; 100; 1; left   **89.** 1; 100; 2; left
90. (a) 525 homes   **(b)** 37 printers   **(c)** \$0.08
91. Find 10% of \$160, then add $\frac{1}{2}$ of 10%.

$$10\% + 5\% = 15\%$$
$$\downarrow \qquad \downarrow \qquad \downarrow$$
$$\$16 + \$8 = \$24$$

92. Find 100% of \$160, then add 50% of 160.
$$100\% + 50\% = 150\%$$
$$\downarrow \qquad \downarrow \qquad \downarrow$$
$$\$160 + \$80 = \$240$$
93. From 100% of \$450, subtract 10% of \$450.
$$100\% - 10\% = 90\%$$
$$\downarrow \qquad \downarrow \qquad \downarrow$$
$$\$450 - \$45 = \$405$$
94. To 100% of \$800, add 100% of \$800, and then add 10% of \$800.
$$100\% + 100\% + 10\% = 210\%$$
$$\downarrow \qquad \downarrow \qquad \downarrow \qquad \downarrow$$
$$\$800 + \$800 + \$80 = \$1680$$

Section 6.3 (pages 395–398)

1. whole = 50   **3.** whole = 150   **5.** whole = 70   **7.** percent = 25%
9. percent = 150%   **11.** percent = 33.3% (rounded)   **13.** part = 26
15. part = 21.6   **17.** percent = 26.5% (rounded)   **19.** 115
21. 0.4%   **23.** 2.5%   **25.** 0.3%
27. Part $\rightarrow$ $\dfrac{60}{\text{unknown}}$ = $\dfrac{10}{100}$ $\leftarrow$ Percent
Whole $\rightarrow$ unknown $\quad$ 100 $\leftarrow$ Always 100
29. $\dfrac{600}{800} = \dfrac{75}{100}$ **31.** $\dfrac{\text{unknown}}{970} = \dfrac{25}{100}$ **33.** $\dfrac{12}{\text{unknown}} = \dfrac{20}{100}$
35. $\dfrac{34}{68} = \dfrac{50}{100}$ **37.** $\dfrac{177}{296} = \dfrac{\text{unknown}}{100}$ **39.** $\dfrac{54.34}{\text{unknown}} = \dfrac{3.25}{100}$
41. $\dfrac{\text{unknown}}{487} = \dfrac{0.68}{100}$
43. Percent—the ratio of the part to the whole. It appears with the word *percent* or "%" after it. Whole—the entire quantity. Often appears after the word *of*. Part—the part being compared with the whole.
45. $\dfrac{535}{1082} = \dfrac{\text{unknown}}{100}$ **47.** $\dfrac{86}{142} = \dfrac{\text{unknown}}{100}$ **49.** $\dfrac{\text{unknown}}{610} = \dfrac{23}{100}$
51. $\dfrac{680}{2000} = \dfrac{\text{unknown}}{100}$ **53.** $\dfrac{\text{unknown}}{480} = \dfrac{55}{100}$ **55.** $\dfrac{\text{unknown}}{822} = \dfrac{49.5}{100}$
57. $\dfrac{504}{\text{unknown}} = \dfrac{16.8}{100}$ **59.** $\dfrac{\text{unknown}}{680} = \dfrac{45}{100}$

Section 6.4 (pages 407–412)

1. 42 test tubes   **3.** 1836 military personnel   **5.** 4.8 ft   **7.** 315 files
9. 819 trucks   **11.** \$3.28   **13.** 1530 tables   **15.** 182 cell phones
17. \$21.60   **19.** 320 e-mails   **21.** 160 hay bales   **23.** 550 students
25. 330 mountain bikes   **27.** 2800   **29.** 50%   **31.** 52%   **33.** 8%
35. 1.5%   **37.** 18.6% (rounded)   **39.** 9.2%   **41.** 150% of \$30 cannot be less than \$30 because 150% is greater than 1 (100%). The answer must be greater than \$30. 25% of \$16 cannot be greater than \$16 because 25% is less than 1 (100%). The answer must be less than \$16.   **43.** \$52.80

45. 44 females (rounded)   **47.** 220 students   **49. (a)** 28.6% female lawyers   **(b)** 313,099 female lawyers (rounded)   **51.** 33%   **53.** 2074 children   **55.** 2%   **57.** \$3200 monthly earnings; \$38,400 yearly earnings   **59.** Breyers   **61.** 76 people (rounded)   **63.** Hardee's
65. \$18.06 billion   **67.** \$645   **69.** 2156 products   **71.** whole; 100
72. whole   **73.** 108 calories   **74.** 300 grams   **75.** 67 grams   **76.** Yes, since they would eat 28% $\times$ 4 servings = 112% of the daily value.
77. 17 packages (rounded). It may be possible but would result in a diet that is high in total fat, saturated fat, sodium, and total carbohydrates.

Section 6.5 (pages 417–420)

1. 270 donors   **3.** 1350 bath towels   **5.** 83.2 quarts   **7.** 3500 airbags
9. 1029.2 meters   **11.** \$4.16   **13.** 160 patients   **15.** 325 salads
17. 680 circuits   **19.** 1080 people   **21.** 300 gallons   **23.** 50%
25. 76%   **27.** 125%   **29.** 1.5%   **31.** 250%
33. You must first change the fraction in the percent to a decimal, then divide the percent by 100 to change it to a decimal.

$$\underbrace{2\tfrac{1}{2}\% = 2.5\%}_{\text{Write } 2\frac{1}{2} \text{ as } 2.5.} = 0.025 \leftarrow 2.5\% \text{ as a decimal}$$

35. 3.78 million or 3,780,000 office workers   **37.** 91.5 million homes (rounded)   **39. (a)** 17,640 have a refrigerator   **(b)** 360 do not have a refrigerator   **41. (a)** 4%   **(b)** 96%   **43.** 36.9% (rounded)
45. 2106 people   **47.** 2916 people   **49.** \$1817.2 million (rounded)
51. 478,175 Mustangs (rounded)   **53.** \$510,390   **55.** \$439.61

Section 6.6 (pages 427–430)

1. \$0.24; \$6.24   **3.** 3%; \$437.75   **5.** 5%; \$298.20   **7.** \$672.60 (rounded); \$12,901.60 (rounded)   **9.** \$22.40   **11.** 20%   **13.** \$185.51 (rounded)   **15.** \$6615   **17.** \$20 (rounded); \$179.99 (rounded)   **19.** 30%; \$126   **21.** \$4.38 (rounded); \$13.12   **23.** \$8.76; \$49.64   **25.** On the basis of commission alone you would choose Company A. Other considerations might be: reputation of the company; expense allowances; other fringe benefits; travel; promotion and training, to name a few.
27. \$203.93 (rounded)   **29.** \$90.62 (rounded)   **31.** 5%   **33.** 35.9% (rounded)   **35.** 22.5% (rounded)   **37.** \$134   **39.** \$568.80   **41.** 4%
43. \$85.80; \$304.20   **45.** 5.6% (rounded)   **47.** \$18.43 (rounded)
49. \$4276.97 (rounded)   **51.** \$14,032.12 (rounded)   **53.** percent; whole
54. rate (percent) of tax; cost of item   **55.** \$146.25 (rounded)   **56.** Yes, the same; (12.4% • \$123) + (6.5% • \$123) = \$23.25 (rounded) and (12.4% + 6.5%) • (\$123) = \$23.25   **57.** \$1358.12   **58.** No. In Exercise 57 the excise tax of \$13.40 cannot be added to the sales tax rate of $7\tfrac{3}{4}\%$ because the excise tax is an amount, not a percent.

Section 6.7 (pages 433–436)

1. \$6   **3.** \$105   **5.** \$28.80   **7.** \$488.75   **9.** \$2227.50   **11.** \$10
13. \$49.20   **15.** \$42.30   **17.** \$16.84 (rounded)   **19.** \$647.06 (rounded)
21. \$210   **23.** \$773.30   **25.** \$2043   **27.** \$3412.50   **29.** \$17,797.81 (rounded)   **31.** The answer should include: amount of principal—This is the amount of money borrowed or loaned. Interest rate—This is the percent used to calculate the interest. Time of loan—The length of time that money is loaned or borrowed is an important factor in determining interest.
33. \$41.25   **35.** \$26,250   **37.** \$1240   **39.** \$277.50   **41.** \$159.50
43. \$12,254.69 (rounded)   **45.** \$2165.63 (rounded)

Section 6.8 (pages 441–444)

1. \$540.80   **3.** \$1966.91 (rounded)   **5.** \$4587.79 (rounded)
7. \$1102.50   **9.** \$1873.52 (rounded)   **11.** \$2027.46 (rounded)
13. \$17,360.41 (rounded)   **15.** \$1216.70; \$216.70   **17.** \$14,326.40; \$6326.40   **19.** \$10,976.01 (rounded); \$2547.84 (rounded)   **21.** Interest paid on past interest as well as on the principal. Many people describe compound interest as "interest on interest."   **23.** \$13,467.61

25. (a) $128,402 **(b)** $52,402 **27. (a)** $87,786 **(b)** $17,786
29. principal; rate; time p; r; t **30.** principal; interest **31.** principal;
interest **32.** principal; interest **33. (a)** $254.48 **(b)** No; it has
more than doubled (about five times). **(c)** Examples will vary.
34. (a) the compound interest account; $1091.20 more **(b)** Greater
amounts of interest are earned with compound interest than simple
interest because interest is earned on both principal and past interest.

Chapter 6 Review Exercises (pages 451–456)

1. 0.35 **2.** 1.5 **3.** 0.9944 **4.** 0.00085 **5.** 315% **6.** 2%
7. 87.5% **8.** 0.2% **9.** $\dfrac{3}{20}$ **10.** $\dfrac{3}{8}$ **11.** $1\dfrac{3}{4}$ **12.** $\dfrac{1}{400}$ **13.** 75%
14. 62.5% or $62\dfrac{1}{2}$% **15.** 325% **16.** 0.5% **17.** 0.125 **18.** 12.5%
19. $\dfrac{1}{4}$ **20.** 25% **21.** $1\dfrac{4}{5}$ **22.** 1.8 **23.** whole = 250 **24.** part = 24
25. $\dfrac{\text{Part} \rightarrow 287}{\text{Whole} \rightarrow 820} = \dfrac{35 \leftarrow \text{Percent}}{100 \leftarrow \text{Always } 100}$
26. $\dfrac{73}{90} = \dfrac{\text{unknown}}{100}$ **27.** $\dfrac{\text{unknown}}{160} = \dfrac{14}{100}$ **28.** $\dfrac{418}{\text{unknown}} = \dfrac{16}{100}$
29. $\dfrac{3}{8} = \dfrac{\text{unknown}}{100}$ **30.** $\dfrac{\text{unknown}}{1280} = \dfrac{88}{100}$ **31.** 171 programs
32. 870 reference books **33.** 31.2 acres **34.** 2.8 kilograms
35. 750 crates **36.** 2320 test tubes **37.** 484 miles **38.** 17,000 cases
39. 55% **40.** 5.2% (rounded) **41.** 9.5% (rounded) **42.** 30.8%
(rounded) **43.** $2.7 million (rounded) **44.** 10.4% (rounded)
45. $145.28 **46.** 186 trucks **47.** 0.4% **48.** 175% **49.** 120 miles
50. $575 **51.** $31.50; $661.50 **52.** $7\dfrac{1}{2}$%; $838.50 **53.** $276 **54.** 5%
55. $33.75; $78.75 **56.** 25%; $189 **57.** $8 **58.** $67.50 **59.** $7
60. $152.10 **61.** $832.50 **62.** $1598.85 **63.** $5375.60; $1375.60
64. $2187.71 (rounded); $317.71 (rounded) **65.** $4534.92; $934.92
66. $16,337.50; $3837.50 **67.** part = 12 **68.** whole = 1640
69. 23.28 meters **70.** 150% **71.** $0.51 **72.** 1980 employees
73. 40% **74.** 498.8 liters **75.** 0.55 **76.** 3 **77.** 500% **78.** 471%
79. 0.086 **80.** 62.1% **81.** 0.00375 **82.** 0.06% **83.** 75% **84.** $\dfrac{21}{50}$
85. $\dfrac{7}{8}$ **86.** 37.5% or $37\dfrac{1}{2}$% **87.** $\dfrac{13}{40}$ **88.** 60% **89.** $\dfrac{1}{400}$ **90.** 375%
91. $3331.25 **92.** $16,520 **93.** 32.4% (rounded)
94. (a) $15,780.96 (rounded) **(b)** $3280.96 (rounded) **95.** $6225
96. 18.2% (rounded) **97.** $848.40 (rounded) **98.** 1100 boaters
99. $13,005 **100.** 33.4% (rounded)

Chapter 6 Test (pages 457–458)

1. 0.65 **2.** 80% **3.** 175% **4.** 87.5% or $87\dfrac{1}{2}$% **5.** 3.00 or 3
6. 0.02 **7.** $\dfrac{1}{8}$ **8.** $\dfrac{1}{400}$ **9.** 60% **10.** 62.5% or $62\dfrac{1}{2}$%
11. 250% **12.** 800 sacks **13.** 20% **14.** 125,000,000 households
15. $3450.60 **16.** 8% **17.** 25%
18. $\dfrac{\text{amount of increase}}{\text{last year's salary}} = \dfrac{p}{100}$
A possible answer is: Part is the increase in salary. Whole is last year's
salary. Percent of increase is unknown.
19. $I = p \cdot r \cdot t$
Some possibilities are
$$I = (1000)(0.05)\left(\dfrac{9}{12}\right) = \$37.50$$
$$I = (1000)(0.05)(2.5) = \$125$$

The interest formula is $I = p \cdot r \cdot t$. If time is in months, it is expressed
as a fraction with 12 as the denominator. If time is expressed in years, it is
placed over 1 or shown as a decimal number. **20.** $11.52; $84.48
21. $91; $189 **22.** $378 **23.** $192 **24.** $5657.75
25. (a) $10,668 (rounded) **(b)** $1668

Chapter 7 Geometry

Section 7.1 (pages 472–475)

1. line named $\overleftrightarrow{CD}$ or $\overleftrightarrow{DC}$ **3.** line segment named $\overline{GF}$ or $\overline{FG}$
5. ray named $\overrightarrow{PQ}$ **7.** perpendicular **9.** parallel **11.** intersecting
13. $\angle AOS$ or $\angle SOA$ **15.** $\angle CRT$ or $\angle TRC$ **17.** $\angle AQC$ or $\angle CQA$
19. right (90°) **21.** acute **23.** straight (180°) **25.** $\angle EOD$ and $\angle COD$;
$\angle AOB$ and $\angle BOC$ **27.** $\angle HNE$ and $\angle ENF$; $\angle ACB$ and $\angle KOL$
29. 50° **31.** 4° **33.** 50° **35.** 90° **37.** $\angle SON \cong \angle TOM$;
$\angle TOS \cong \angle MON$ **39.** $\angle GOH$ measures 63°; $\angle EOF$ measures 37°;
$\angle AOC$ and $\angle GOF$ both measure 80°. **41.** True, because $\overleftrightarrow{UQ}$ is
perpendicular to $\overleftrightarrow{ST}$. **42.** True, because they form a 90° angle, as indicated by the small red square. **43.** False; the angles have the same mea-
sure (both are 180°). **44.** False; $\overleftrightarrow{ST}$ and $\overleftrightarrow{PR}$ are parallel. **45.** False; $\overleftrightarrow{QU}$
and $\overleftrightarrow{TS}$ are perpendicular. **46.** True, because both angles are formed by
perpendicular lines, so they both measure 90°. **47.** corresponding an-
gles; $\angle 1$ and $\angle 8$, $\angle 2$ and $\angle 5$, $\angle 3$ and $\angle 6$, $\angle 4$ and $\angle 7$; alternate
interior angles; $\angle 4$ and $\angle 5$, $\angle 3$ and $\angle 8$. **49.** $\angle 2$, $\angle 4$, $\angle 6$, $\angle 8$ all
measure 130°; $\angle 1$, $\angle 3$, $\angle 5$, $\angle 7$ all measure 50°. **51.** $\angle 6$, $\angle 1$, $\angle 3$,
$\angle 8$ all measure 47°; $\angle 5$, $\angle 2$, $\angle 7$, $\angle 4$ all measure 133°. **53.** $\angle 6$, $\angle 8$,
$\angle 4$, $\angle 2$ all measure 114°; $\angle 7$, $\angle 5$, $\angle 3$, $\angle 1$ all measure 66°.
55. $\angle 1 \cong \angle 3$, both are 138°; $\angle 2 \cong \angle ABC$, both are 42°.

Section 7.2 (pages 482–485)

1. $P = 28$ yd; $A = 48$ yd^2 **3.** $P = 3.6$ km; $A = 0.81$ km^2
5. $P = 40$ ft; $A = 100$ ft^2 **7.** $P = 29$ m; $A = 7$ m^2 **9.** $P = 196.2$ ft;
$A = 1674.2$ ft^2 **11.** $P = 12$ mi; $A = 9$ mi^2 **13.** $P = 38$ m; $A = 39$ m^2
15. $P = 98$ m; $A = 492$ m^2 **17.** unlabeled side = 12 in.; $P = 78$ in.;
$A = 234$ in.^{2} **19.** $P = 48$ m; $A = 144$ m^2 **21.** $460 **23.** $94.81
25. Area of largest field is 13,000 yd^2; Area of smallest field is 5000 yd^2;
difference is 13,000 yd^2 − 5000 yd^2 = 8000 yd^2. **27.** 53 yd
29. $A = 6528$ ft^2

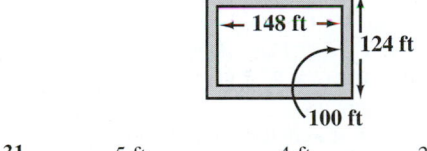

31.
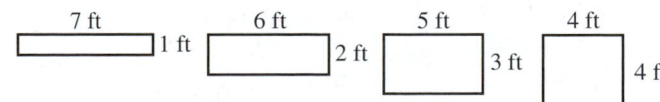

32. (a) 5 ft by 1 ft has area of 5 ft^2; 4 ft by 2 ft has area of 8 ft^2; 3 ft by
3 ft has area of 9 ft^2 **(b)** The square plot 3 ft by 3 ft has greatest area.
33.
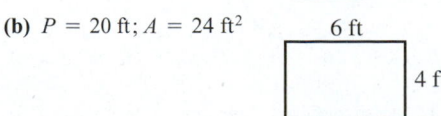

34. (a) 7 ft^2; 12 ft^2; 15 ft^2; 16 ft^2 **(b)** Square plots have the greatest area.
35. (a) $P = 10$ ft; $A = 6$ ft^2

(b) $P = 20$ ft; $A = 24$ ft^2

(c) Perimeter is twice the original; area is four times the original.
36. (a) $P = 30$ ft; $A = 54$ ft^2

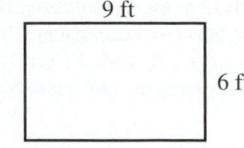

(b) Perimeter is three times the original; area is nine times the original.
(c) Perimeter will be four times the original; area will be 16 times the original.

Section 7.3 (pages 490–491)

1. $P = 208$ m   **3.** $P = 207.2$ m   **5.** $P = 6.06$ km   **7.** $A = 775$ mm^2
9. $A = 19.25$ ft^2 **11.** $A = 3099.6$ cm^2
13. $1410.75

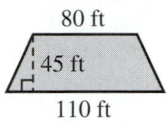

15. $A = 437.5$ in.^{2}

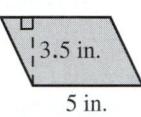

17. Height is not part of perimeter; square units are used for area, not perimeter. $P = 2.5$ cm $+ 2.5$ cm $+ 2.5$ cm $+ 2.5$ cm $= 10$ cm
19. $A = 3.02$ m^2 **21.** $A = 25,344$ ft^2

Section 7.4 (pages 496–498)

1. $P = 202$ m; $A = 1914$ m^2 **3.** $P = 58.9$ cm; $A = 139.15$ cm^2
5. $P = 26\frac{1}{4}$ yd or 26.25 yd; $A = 30\frac{3}{4}$ yd^2 or 30.75 yd^2
7. $P = 85.2$ cm; $A = 302.46$ cm^2 **9.** $A = 198$ m^2 **11.** $A = 1664$ m^2
13. $32°$ **15.** $48°$ **17.** No. Right angles are $90°$, so two right angles are $180°$, and the sum of all *three* angles in a triangle equals $180°$.
19. $7\frac{7}{8}$ ft^2 or 7.875 ft^2 **21. (a)** 126.8 m of curb   **(b)** 672 m^2 of sod
23. (a) $A = 32$ m^2 **(b)** $A = 13.41$ m^2 **25. (a)** 175 yd of frontage
(b) $A = 8750$ yd^2

Section 7.5 (pages 506–509)

1. $d = 18$ mm   **3.** $r = 0.35$ km   **5.** $C \approx 69.1$ ft; $A \approx 379.9$ ft^2
7. $C \approx 8.2$ m; $A \approx 5.3$ m^2 **9.** $C \approx 47.1$ cm; $A \approx 176.6$ cm^2
11. $C \approx 23.6$ ft; $A \approx 44.2$ ft^2 **13.** $C \approx 27.2$ km; $A \approx 58.7$ km^2
15. $A \approx 76.9$ in.^{2} **17.** $A \approx 57$ cm^2
19. π is the ratio of circumference of a circle to its diameter. If you divide the circumference of any circle by its diameter, the answer is always a little more than 3. The approximate value is 3.14, which we call π (pi). Your test question could involve finding the circumference or the area of a circle.   **21.** $A \approx 7850$ yd^2 **23.** $C \approx 91.4$ in.;
Bonus: 693 revolutions/mile (rounded)
25. $A \approx 70,650$ mi^2

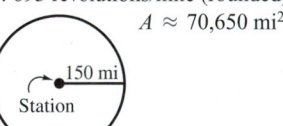

27. Watch $C \approx 3.1$ in.   Wall Clock $C \approx 18.8$ in.
$A \approx 0.8$ in.^{2} $A \approx 28.3$ in.^{2}

29.

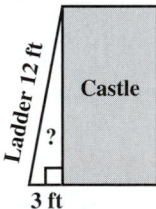

$A \approx 78.5$ mi^2 $- 12.6$ mi^2
Difference ≈ 65.9 mi^2

31. (a) $d \approx 45.9$ cm   **(b)** Divide the circumference by π (3.14).
33. $325.83 (rounded).   **35.** The prefix *rad* tells you that radius is a ray from the center of the circle. The prefix *dia* means the diameter goes through the circle, and the prefix *circum* means the circumference is the distance around.   **37.** $A \approx 44.2$ in.^{2} **38.** $A \approx 132.7$ in.^{2}
39. $A \approx 201.0$ in.^{2} **40.** small: $0.063 (rounded); medium: $0.049 (rounded); large: $0.046 (rounded) Best Buy   **41.** small: $0.084 (rounded); medium: $0.067 (rounded) Best Buy; large: $0.071 (rounded)
42. small: $0.077 (rounded) Best Buy; medium: $0.083 (rounded); large: $0.078 (rounded)

Section 7.6 (pages 518–520)

1. Rectangular solid; $V = 550$ cm^3 **3.** Sphere; $V \approx 44,579.6$ m^3
5. Hemisphere; $V \approx 3617.3$ in.^{3} **7.** Cylinder; $V \approx 471$ ft^3 **9.** Cone; $V \approx 418.7$ m^3 **11.** Pyramid; $V = 800$ cm^3 **13.** $V \approx 513$ cm^3
15. $V \approx 471$ ft^3; $SA \approx 345.4$ ft^2 **17.** $V = 1617$ m^3; $SA = 849.4$ m^2
19. $V \approx 763.0$ in.^{3}; $SA \approx 678.2$ in.^{2} **21.** $V = 5550$ mm^3;
$SA = 2150$ mm^2 **23.** $V = 18$ in.^{3} **25.** $V \approx 2481.5$ cm^3
27. $V = 651,775$ m^3 **29.** $V \approx 3925$ ft^3 **31.** $SA = 163.6$ in.^{2}
33. Student used diameter of 7 cm; should use radius of 3.5 cm in formula. Units for volume are cm^3, not cm^2. Correct answer is $V \approx 192.3$ cm^3.

Section 7.7 (pages 525–528)

1. 4   **3.** 8   **5.** 3.317   **7.** 2.236   **9.** 8.544   **11.** 10.050   **13.** 13.784
15. 31.623   **17.** 30 is about halfway between 25 and 36, so $\sqrt{30}$ should be about halfway between 5 and 6, or about 5.5. Using a calculator, $\sqrt{30} \approx 5.477$. Similarly, $\sqrt{26}$ should be a little more than $\sqrt{25}$; by calculator $\sqrt{26} \approx 5.099$. And $\sqrt{35}$ should be a little less than $\sqrt{36}$; by calculator $\sqrt{35} \approx 5.916$.   **19.** $\sqrt{1521} = 39$ ft   **21.** $\sqrt{289} = 17$ in.
23. $\sqrt{144} = 12$ mm   **25.** $\sqrt{73} \approx 8.5$ in.   **27.** $\sqrt{65} \approx 8.1$ yd
29. $\sqrt{195} \approx 14.0$ cm   **31.** $\sqrt{7.94} \approx 2.8$ m   **33.** $\sqrt{65.01} \approx 8.1$ cm
35. $\sqrt{292.32} \approx 17.1$ km   **37.** hypotenuse $= \sqrt{65} \approx 8.1$ ft
39. leg $= \sqrt{360,000} = 600$ m   **41.** hypotenuse $= \sqrt{32.5} \approx 5.7$ ft
43. leg $= \sqrt{135} \approx 11.6$ ft

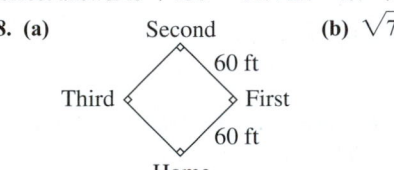

45. The student did not square the numbers correctly: 9^2 is 81 and 7^2 is 49. Also, the final answer is rounded to thousandths instead of tenths. Correct answer is $\sqrt{130} \approx 11.4$ in.   **47.** $\sqrt{16200} \approx 127.3$ ft
48. (a)   Second   **(b)** $\sqrt{7200} \approx 84.9$ ft

Third   First
60 ft   60 ft

Home

49. The distance from third to first is the same as the distance from home to second because the baseball diamond is a square.
50. One possibility is:

$$\text{major league} \left\{ \frac{90 \text{ ft}}{127.3 \text{ ft}} = \frac{60 \text{ ft}}{x} \right\} \text{softball}$$

$$x \approx 84.9 \text{ ft}$$

Section 7.8 (pages 535–538)

1. ∠1 and ∠4; ∠2 and ∠5; ∠3 and ∠6; $\overline{AB}$ and $\overline{DE}$; $\overline{BC}$ and $\overline{EF}$; $\overline{AC}$ and $\overline{DF}$.   **3.** ∠1 and ∠6; ∠2 and ∠4; ∠3 and ∠5; $\overline{ST}$ and $\overline{YW}$; $\overline{TU}$ and $\overline{WX}$; $\overline{SU}$ and $\overline{YX}$ **5.** ∠1 and ∠6; ∠2 and ∠5; ∠3 and ∠4; $\overline{LM}$ and $\overline{TS}$; $\overline{LN}$ and $\overline{TR}$; $\overline{MN}$ and $\overline{SR}$ **7.** SAS   **9.** SSS   **11.** ASA
13. use SAS: $BC = CE$, ∠$ABC \cong$ ∠DCE, $BA = CD$ **14.** use SSS:
$WP = YP, ZP = XP, WZ = YX$ **15.** use SAS: $PS = RS, m\angle QSP = m\angle QSR = 90°, QS = QS$ (common side)   **16.** use SAS: $LM = OM$,
$PM = NM$, ∠$LMP \cong$ ∠OMN (vertical angles)   **17.** $a = 6$ cm;
$b = 15$ cm   **19.** $a = 5$ mm; $b = 3$ mm   **21.** $a = 24$ in.; $b = 20$ in.
23. $x = 24.8$ m; Perimeter $= 72.8$ m; $y = 15$ m; Perimeter $= 54.6$ m
25. Perimeter $= 8$ cm $+ 8$ cm $+ 8$ cm $= 24$ cm; Area $\approx$
$(0.5)(8 \text{ cm})(6.9 \text{ cm}) \approx 27.6$ cm²; The area is approximate because the height was rounded to the nearest tenth.   **27.** $h = 24$ ft   **29.** One dictionary definition is "resembling, but not identical." Examples of similar objects are sets of different size pots or measuring cups; small and large size cans of beans; child's tennis shoe and adult's tennis shoe.
31. $x = 50$ m   **33.** $n = 110$ m

Chapter 7 Review Exercises (pages 549–556)

1. line segment named $\overline{AB}$ or $\overline{BA}$ **2.** line named $\overleftrightarrow{CD}$ or $\overleftrightarrow{DC}$
3. ray named $\overrightarrow{OP}$ **4.** parallel   **5.** perpendicular   **6.** intersecting
7. acute   **8.** obtuse   **9.** straight; 180°   **10.** right; 90°   **11.** ∠AOB and ∠BOC; ∠BOC and ∠COD; ∠COD and ∠DOA; ∠DOA and ∠AOB **12.** ∠ERH and ∠HRG; ∠HRG and ∠GRF; ∠FRG and ∠FRE; ∠FRE and ∠ERH **13. (a)** 10° **(b)** 45° **(c)** 83°
14. (a) 25° **(b)** 90° **(c)** 147°   **15.** ∠1 and ∠4 measure 30°; ∠3 and ∠6 measure 90°; ∠5 measures 60°.   **16.** ∠8, ∠3, ∠6, ∠1 all measure 160°; ∠4, ∠7, ∠2, ∠5 all measure 20°.   **17.** $P = 4.84$ m
18. $P = 128$ in.   **19.** 152 cm   **20.** 41 ft   **21.** $A = 486$ mm²
22. $A = 16.5$ ft² or $16\frac{1}{2}$ ft²   **23.** $A \approx 39.7$ m²   **24.** $P = 50$ cm;
$A = 140$ cm²   **25.** $P = 102.1$ ft; $A = 567$ ft²   **26.** $P = 200.2$ m;
$A \approx 2206.1$ m²   **27.** $P = 518$ cm; $A = 11{,}660$ cm²
28. $P = 27.1$ m; $A = 23.52$ m²   **29.** $P = 20\frac{1}{4}$ ft or 20.25 ft;
$A = 14$ ft²   **30.** 70°   **31.** 24°   **32.** $d = 137.8$ m   **33.** $r = 1\frac{1}{2}$ in.
or 1.5 in.   **34.** $C \approx 6.3$ cm; $A \approx 3.1$ cm²   **35.** $C \approx 109.3$ m;
$A \approx 950.7$ m²   **36.** $C \approx 37.7$ in.; $A \approx 113.0$ in.²
37. $A \approx 20.3$ m²   **38.** $A = 64$ in.²   **39.** $A \approx 673$ km²
40. Rectangular solid; $V = 30$ in.³   **41.** Rectangular solid; $V = 96$ cm³
42. Rectangular solid; $V = 45{,}000$ mm³   **43.** Sphere; $V \approx 267.9$ m³
44. Hemisphere; $V \approx 452.3$ ft³   **45.** Cylinder; $V \approx 549.5$ cm³
46. Cylinder; $V \approx 1808.6$ m³   **47.** Cone; $V \approx 512.9$ m³
48. Pyramid; $V = 16$ yd³   **49.** $V \approx 549.5$ cm³; $SA \approx 376.8$ cm²
50. $V \approx 7.9$ ft³; $SA = 25.5$ ft²   **51.** 7   **52.** 2.828 (rounded)
53. 54.772 (rounded)   **54.** 12   **55.** 7.616 (rounded)   **56.** 25
57. 10.247 (rounded)   **58.** 8.944 (rounded)   **59.** $\sqrt{289} = 17$ in.
60. $\sqrt{49} = 7$ cm   **61.** $\sqrt{104} \approx 10.2$ cm   **62.** $\sqrt{52} \approx 7.2$ in.
63. $\sqrt{6.53} \approx 2.6$ m   **64.** $\sqrt{71.75} \approx 8.5$ km   **65.** SSS
66. SAS   **67.** ASA   **68.** $y = 30$ ft; $x = 34$ ft; $P = 104$ ft

69. $y = 7.5$ m; $x = 9$ m; $P = 22.5$ m   **70.** $x = 12$ mm;
$y = 7.5$ mm; $P = 38$ mm   **71.** parallel lines   **72.** line segment
73. acute angle   **74.** intersecting lines   **75.** right angle; 90°
76. ray   **77.** straight angle; 180°   **78.** obtuse angle
79. perpendicular lines   **80.** 81°   **81.** 138°   **82.** $P = 90$ m;
$A = 92$ m²   **83.** $P = 282$ cm; $A = 4190$ cm²
84. Square; $P = 18$ in.; $A \approx 20.3$ in.² or $A = 20\frac{1}{4}$ in.²
85. Trapezoid; $P = 10.3$ cm; $A \approx 6.2$ cm²
86. Circle; $C \approx 40.8$ m; $A \approx 132.7$ m²   **87.** Parallelogram;
$P = 54$ ft; $A = 140$ ft²   **88.** Triangle; $P = 20$ yd; $A = 18\frac{3}{4}$ yd² or
18.8 yd² (rounded)   **89.** Rectangle; $P = 7$ km; $A \approx 2.0$ km²
90. Circle; $C \approx 53.4$ m; $A \approx 226.9$ m²   **91.** Parallelogram;
$P = 78$ mm; $A = 288$ mm²   **92.** Triangle; $P = 37.8$ mi;
$A = 58.5$ mi²   **93.** $V \approx 100.5$ ft³   **94.** $V \approx 3.4$ in.³ or $V = 3\frac{3}{8}$ in.³
95. $V \approx 7.4$ m³   **96.** $V = 561$ cm³   **97.** $V \approx 1271.7$ cm³
98. $V \approx 1436.0$ m³   **99.** $x \approx 12.6$ km   **100.** ∠$D = 72°$
101. $x = 12$ mm; $y = 14$ mm   **102.** The prefix *dec* in *dec*ade means 10 and the prefix *cent* in *cent*ury means 100, so divide 200 (two centuries) by 10. The answer is 20 decades.

Chapter 7 Test (pages 557–558)

1. (e)   **2. (a)**; 90°   **3. (d)**   **4. (g)**; 180°   **5.** Parallel lines are lines in the same plane that never intersect. Perpendicular lines intersect to form a right angle.

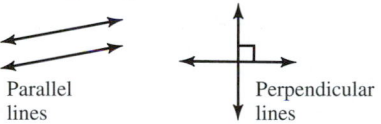

Parallel lines   Perpendicular lines

6. 9°   **7.** 160°   **8.** ∠1 measures 50°; ∠3 measures 90°; ∠2 and ∠4 measure 40°.   **9.** Rectangle; $P = 23$ ft; $A = 30$ ft²
10. Square; $P = 72$ mm; $A = 324$ mm²   **11.** Parallelogram;
$P = 26.2$ m; $A = 33.12$ m²   **12.** Trapezoid; $P = 169.4$ cm;
$A = 1591$ cm²   **13.** $P = 32.45$ m; $A = 48$ m²   **14.** $P = 37.8$ yd or
$37\frac{4}{5}$ yd; $A = 58.5$ yd²   **15.** 55°   **16.** $r = 12.5$ in. or $12\frac{1}{2}$ in.
17. $C \approx 5.7$ km (rounded)   **18.** $A \approx 206.0$ cm² (rounded)
19. $A \approx 39.3$ m² (rounded)   **20.** Rectangular solid; $V = 6480$ m³
21. Sphere; $V \approx 33.5$ ft³ (rounded)   **22.** Cylinder; $V \approx 5086.8$ ft³
23. $\sqrt{85} \approx 9.2$ cm (rounded)   **24.** $y = 12$ cm; $z = 6$ cm
25. Linear units like cm are used to measure perimeter, radius, diameter, and circumference. Area is measured in square units like cm² (squares that measure 1 cm on each side). Volume is measured in cubic units like cm³.

Chapter 8   Statistics

Section 8.1 (pages 565–570)

1. 360 million or 360,000,000 pets   **3.** $\frac{90}{360} = \frac{1}{4}$ **5.** $\frac{90}{75} = \frac{6}{5}$ **7.** Don't know; 180 people   **9.** $\frac{720}{6000} = \frac{3}{25}$ **11.** $\frac{1020}{1200} = \frac{17}{20}$ **13.** $\frac{1740}{1020} = \frac{29}{17}$
15. 160 people   **17.** 96 people   **19.** 960 people   **21.** 1219 people (rounded)   **23.** 1828 people (rounded)   **25.** 277 people   **27.** First find the percent of the total that is to be represented by each item. Next, multiply the percent by 360° to find the size of each sector. Finally, use a protractor to draw each sector.   **29.** 90°   **31.** 10%; 36°   **33.** 54°   **35.** 15%

37. (a) $200,000 **(b)** 22.5°; 72°; 108°; 90°; 67.5°
(c)

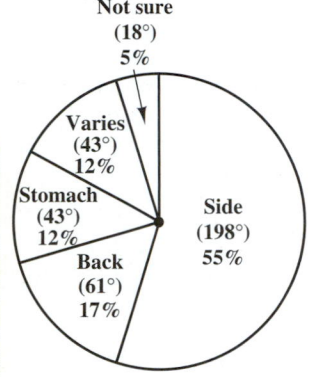

39. (a) 2464; 55% (rounded); 198°; 220; 5% (rounded); 18°; 536; 12% (rounded); 43° (rounded); 520; 12% (rounded); 43° (rounded); 748; 17% (rounded); 61° (rounded)

(b) No. The total is 101% due to rounding.
(c) No. The total is 363° due to rounding.

Section 8.2 (pages 575–580)

1. India; 51.4% **3.** USA, Britain, and Australia **5.** 121 days (rounded) **7.** May; 10,000 plants **9.** 1500 plants **11.** 4000 plants **13.** 150,000 gal **15.** 2006; 250,000 gal **17.** 550,000 gal **19.** 24.1 million or 24,100,000 PCs **21.** 144.5 million or 144,500,000 PCs **23.** Answers will vary. Some possibilities are: greater demand as a result of lower price; more uses and applications for the owner; improved technology; multiple computers in each location; more laptop computers sold. **25.** 300,000 iPods **27.** 150,000 iPods **29.** 350,000 iPods **31.** Probably Store B with greater sales. Predicted sales might be 450,000 iPods to 500,000 iPods in 2011. **33.** A single bar or a single line must be used for each set of data. To show multiple sets of data, multiple sets of bars or lines must be used. **35.** $40,000 **37.** $25,000 **39.** $5000 **41.** Answers will vary. Some possibilities are that the decrease in sales may have resulted from poor service or greater competition. The increase in sales may have been a result of more advertising or better service. **43.** 23 years **44.** 20 additional flavors **45.** 214,286 rolls each day (rounded) **46.** 1.17 lb (rounded) **47.** The weight of one roll of LifeSavers is much less than 1.17 lb. The answer is "correct" using the information given, but some of the data given must not be accurate. Perhaps 3 million **rolls** of Life-Savers are produced daily. **48.** Answers will vary. Possible answers are: Misprints or typographical errors; careless reporting of data; math errors.

Section 8.3 (pages 583–586)

1. 61–65 years; 16,000 members **3.** 13,000 members **5.** 50,000 members **7.** $4100 to $5000; 16 employees **9.** 11 employees **11.** 49 employees **13.** Class intervals are the result of combining data into groupings. Class frequency is the number of data items that fit in each class interval. These are used to group data and to have multiple responses (frequency) in a class interval—this makes the data easier to interpret. **15.** ⅢⅠ; 6 **17.** Ⅰ; 1 **19.** ⅠⅠ; 2 **21.** ⅢⅠⅠ; 4 **23.** ⅢⅠ; 5 **25.** ⅢⅠ; 3 **27.** ⅢⅠ; 7 **29.** ⅢⅡ ⅢⅠ ⅢⅠ; 14 **31.** ⅢⅠ ⅢⅠ ⅢⅠ Ⅰ; 16 **33.** ⅢⅠ ⅢⅠ Ⅰ; 11 **35.** ⅢⅠ; 5

Section 8.4 (pages 591–594)

1. 11 yr **3.** 2.5 in. of rain (rounded) **5.** $37,127 **7.** $72.53 **9.** 12.5 customers (rounded) **11.** 17.2 fish (rounded) **13.** $35,500 **15.** 125 books **17.** 508 calories **19.** 21% **21.** 68 and 74 yr (bimodal) **23.** The median is a better measure of central tendency when the list contains one or more extreme values. Find the mean and the median of the following home values: $182,000; $164,000; $191,000; $115,000; $982,000: mean home value = $326,800; median home value = $182,000. **25.** 2.42 (rounded) **27.** 2.60 **29.** 3.14 (rounded) **31.** 176 sales calls **32.** mean for Samuels: 22; mean for Stricker: 22 **33.** median for Samuels: 19.5; median for Stricker: 22 **34.** mode for Samuels: 22; mode for Stricker: 22 **35.** The mean and mode are identical for both sales representatives and the medians are close. **36.** The number of weekly sales calls made by Samuels varies greatly from week to week, while the number of weekly sales calls made by Stricker remains fairly constant. **37.** range = 40 − 8 = 32 **38.** range = 25 − 19 = 6 **39.** No, not with any certainty. There probably are additional questions that need to be answered before a determination could be made, such as the dollar amount of sales, number of repeat customers, and so on. **40.** Answers will vary. Some possible answers are: He works hard one week, then takes it easy the next week; the characteristics of the sales territories vary greatly; illness or personal problems may be affecting performance.

Chapter 8 Review Exercises (pages 601–606)

1. Basketball; 18,000 teams **2.** $\dfrac{16,000}{76,000} = \dfrac{4}{19}$ **3.** $\dfrac{15,000}{76,000} = \dfrac{15}{76}$

4. $\dfrac{14,000}{76,000} = \dfrac{7}{38}$ **5.** $\dfrac{18,000}{16,000} = \dfrac{9}{8}$ **6.** $\dfrac{16,000}{14,000} = \dfrac{8}{7}$ **7.** 3936 companies

8. 1728 companies **9.** 720 companies **10.** 2904 companies **11.** On-site child care and Fitness centers Answers will vary. Perhaps employers feel that they are not needed or would not be used. Or, it may be that they would be too expensive for the benefit derived. **12.** Flexible hours and casual dress Answers will vary. Perhaps employees request them and appreciate them. Or, it may be that neither of them cost the employer anything to offer. **13.** March; 8,000,000 acre-feet **14.** June; 2,000,000 acre-feet **15.** 5,000,000 acre-feet **16.** 4,000,000 acre-feet **17.** 5,000,000 acre-feet **18.** 2,000,000 acre-feet **19.** $50,000,000 **20.** $20,000,000 **21.** $20,000,000 **22.** $40,000,000 **23.** The floor-covering sales decreased for 2 years and then moved up slightly. Answers will vary. Perhaps there is less new construction, remodeling, and home improvement in the area near Center A, or a better product selection and service have reversed the decline in sales. **24.** The floor-covering sales are increasing. Answers will vary. Perhaps new construction and home remodeling have increased in the area near Center B, or greater advertising has attracted more customers. **25.** 28.5 digital cameras **26.** 35.2 complaints **27.** $51.05 **28.** 257.3 points (rounded) **29.** 39 forms **30.** $562 **31.** $79 **32.** 18 and 32 launchings (bimodal) **33.** 36° **34.** 35% **35.** 20%; 72° **36.** 25%; 90° **37.** 10%; 36° **38.**

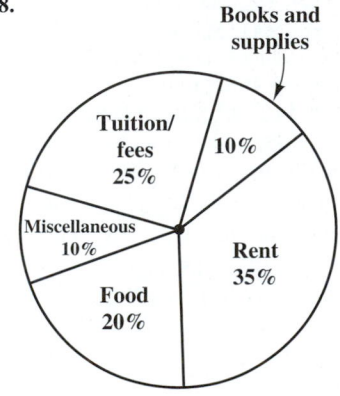

39. 100 volunteers   **40.** 144.7 vaccinations (rounded)
41. 48 applicants   **42.** 31 and 43 two-bedroom apartments
(bimodal)   **43.** 5.0 hr   **44.** 21 e-mails   **45.** ||||; 4   **46.** |; 1
47. ||||| |; 6   **48.** ||; 2   **49.** ||||| ||||| |||; 13   **50.** ||||| ||; 7   **51.** ||||| ||; 7
52.

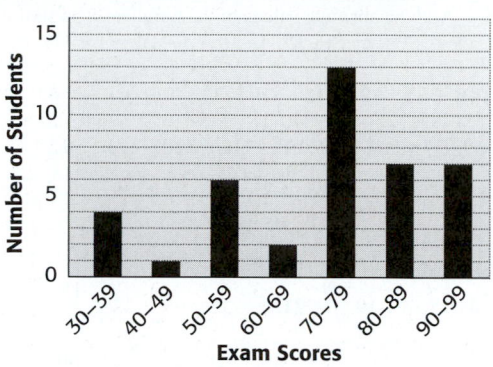

COMPUTER SCIENCE EXAM SCORES

53. 64.5 (rounded)   **54.** 118.8 units (rounded)

Chapter 8 Test (pages 607–610)

1. $59.0 billion (rounded)   **2.** $21.2 billion (rounded)   **3.** $5.4 billion
(rounded)   **4.** $57.2 billion (rounded)   **5.** $146.0 billion (rounded)
6. $9.2 billion (rounded)   **7.** 108°   **8.** 36°   **9.** 72°   **10.** 126°   **11.** 5%
12.

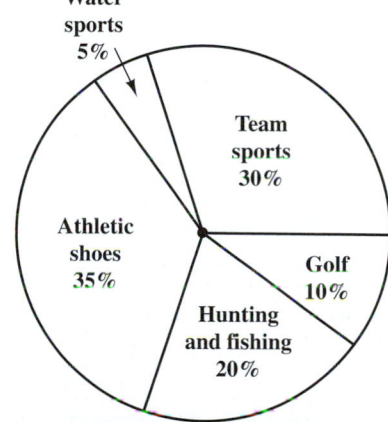

Water
sports
5%

Team
sports
30%

Golf
10%

Hunting
and fishing
20%

Athletic
shoes
35%

13. 3   **14.** 2   **15.** 4   **16.** 3   **17.** 3   **18.** 5
19.

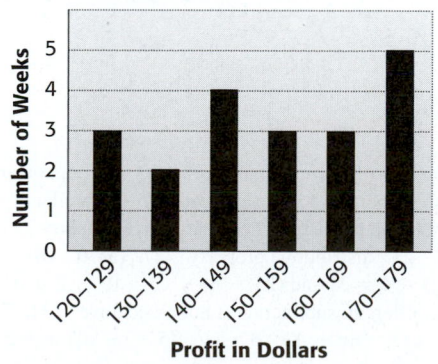

ALAN'S VENDING SALES

20. 75 mi   **21.** 15.3 lb (rounded)   **22.** 478.9 mph

23. The weighted mean must be used because different classes are worth
different numbers of credits. GPA problems will vary; one possibility is
shown.

Credits	Grade	
3	A	$3 \times 4 = 12$
2	C	$2 \times 2 = 4$
4	B	$4 \times 3 = 12$
9		28

$$28 \div 9 \approx 3.11$$

24. Arrange the values in order, from smallest to largest. When there is
an odd number of values in a list, the median is the middle value.
8, 17, 23, 32, 64
        └─ Median
25. $26 (rounded)   **26.** 174 points (rounded)   **27.** 31.5 degrees
28. 10.0 meters   **29.** 52 milliliters   **30.** 103 and 104 degrees (bimodal)

Cumulative Review Exercises: Chapters 1–8 (pages 611–614)

1. (a) forty-five and two hundred three ten-thousandths   **(b)** thirty billion,
six hundred fifty thousand, eight   **2. (a)** 160,000,500   **(b)** 0.075
3. 46,900   **4.** 6.20   **5.** 0.661   **6.** 10,000
7. *Estimate:* $80,000 - 50,000 = 30,000;$ *Exact:* 28,988
8. *Estimate:* $8 - 4 = 4;$ *Exact:* 4.2971
9. *Estimate:* $7000 \times 700 = 4,900,000;$ *Exact:* 4,628,904
10. *Estimate:* $70 \times 9 = 630;$ *Exact:* 568.11
11. *Estimate:* $40,000 \div 40 = 1000;$ *Exact:* 902
12. *Estimate:* $7 \div 1 = 7;$ *Exact:* 8.45
13. $10\frac{4}{15}$ **14.** $1\frac{11}{12}$ **15.** $44\frac{2}{5}$ **16.** $8\frac{4}{5}$ **17.** $\frac{8}{21}$ **18.** $\frac{1}{12}$
19. 23   **20.** 10   **21.** 108   **22.** $\frac{1}{4}$ **23.** 1   **24.** $\frac{1}{4}$ **25.** 0.75
26. 0.375   **27.** 0.583 (rounded)   **28.** 0.55
29. 0.199, 0.207, 0.218, 0.22, 0.2215   **30.** 0.58, 0.608, $\frac{5}{8}$, 0.6319, $\frac{13}{20}$
31. $\frac{1}{8}$ **32.** $\frac{4}{1}$ **33.** $x = 6$ **34.** $x = 6$ **35.** $x = 16.2$
36. $x = 52$ **37.** 324 homes   **38.** 2800 people   **39.** 260%
40. $x = 6$ **41.** $x = 4$ **42.** $x = 16$ **43.** $x = 30$
44. 0.03   **45.** 2.00 or 2   **46.** 87%   **47.** 380%   **48.** $\frac{2}{25}$
49. $\frac{5}{8}$ **50.** $1\frac{3}{4}$ **51.** 87.5% or $87\frac{1}{2}$%   **52.** 420%
53. 6.5% or $6\frac{1}{2}$%   **54.** $1003.04   **55.** $205.20; $250.80
56. $23,028.75   **57.** 33 hearing tests   **58.** 255 ounces
59. Square; $P = 9$ ft; $A = 5\frac{1}{16}$ or 5.0625 ft^2
60. Circle; $C \approx 28.26$ mm; $A \approx 63.585$ mm^2 **61.** Right triangle;
$P = 56$ mi; $A = 84$ mi^2 **62.** Parallelogram; $P = 36$ ft; $A = 70$ ft^2
63. $P = 54$ ft; $A = 140$ ft^2 **64.** $C \approx 40.8$ m; $A \approx 132.7$ m^2
65. $x \approx 12.6$ km; $A \approx 37.8$ km^2 **66.** $V \approx 3.4$ in.^{3}; $SA = 13.5$ in.^{2}
67. $P = 7$ km; $A \approx 2.0$ km^2 **68.** $d = 17$ m; $C \approx 53.4$ m;
$A \approx 226.9$ m^2 **69.** $V \approx 339.1$ ft^3; $SA \approx 282.6$ ft^2 **70.** Cylinder:
$V \approx 549.8$ cm^3 (rounded)   **71.** Rectangular solid: $V = 199.5$ m^3
72. 34 cm   **73.** mean = 27 hookups; median = 26 hookups;
mode = 19 hookups   **74.** mean = 4.4 acres; median = 4.3
acres; mode = 2.85 acres   **75.** 42.0%   **76. (a)** $889.80   **(b)** $977
77. 38 nasal strips for $9.95   **78.** $59,225.25   **79.** 17.5 L
80. $11,761

Chapter 9 The Real Number System

Section 9.1 (pages 621–622)

1. true   **3.** false; Using the guidelines for order of operations gives $4 + 3(8 - 2) = 4 + 3(6) = 4 + 18 = 22.$ **5.** false; The correct translation is $4 = 16 - 12.$ **7.** 49   **9.** 144   **11.** 64   **13.** 1000   **15.** 81
17. 1024   **19.** $\dfrac{16}{81}$ **21.** 0.000064   **23.** The 4 would be applied last because we work first inside the parentheses.   **25.** 58   **27.** 13   **29.** 32
31. 19   **33.** $\dfrac{49}{30}$ **35.** 12   **37.** 36.14   **39.** 26   **41.** 4   **43.** 95   **45.** 12
47. 14   **49.** $\dfrac{19}{2}$ **51.** false   **53.** true   **55.** true   **57.** false   **59.** false
61. true   **63.** $15 = 5 + 10$ **65.** $9 > 5 - 4$ **67.** $16 \neq 19$ **69.** $2 \leq 3$
71. Seven is less than nineteen; true   **73.** One-third is not equal to three-tenths; true   **75.** Eight is greater than or equal to eleven; false
77. $30 > 5$ **79.** $3 \leq 12$ **81.** Alaska, Texas, California, Idaho
83. Alaska, Texas, California, Idaho, Missouri

Section 9.2 (pages 627–630)

1. 11   **3.** $13 + x$; 16   **5.** expression; equation   **7.** The equation would be $5x - 9 = 49.$ **9.** Answers will vary. Two such pairs are $x = 0, y = 6$ and $x = 1, y = 4.$ To find a pair, choose one number, substitute it for a variable, then calculate the value for the other variable.   **11. (a)** 64   **(b)** 144
13. (a) $\dfrac{7}{8}$ **(b)** $\dfrac{13}{12}$ **15. (a)** 9.569   **(b)** 14.353   **17. (a)** 52   **(b)** 114
19. (a) 12   **(b)** 33   **21. (a)** 6   **(b)** $\dfrac{9}{5}$ **23. (a)** $\dfrac{4}{3}$ **(b)** $\dfrac{13}{6}$ **25. (a)** $\dfrac{2}{7}$
(b) $\dfrac{16}{27}$ **27. (a)** 12   **(b)** 55   **29. (a)** 1   **(b)** $\dfrac{28}{17}$ **31. (a)** 3.684
(b) 8.841   **33.** $12x$ **35.** $x - 2$ **37.** $7 - \dfrac{1}{3}x$ **39.** $2x - 6$ **41.** $\dfrac{12}{x + 3}$
43. $6(x - 4)$ **45.** The word *and* does not signify addition here. In the phrase "the product of a number and 6," *and* connects two quantities to be multiplied.   **47.** no   **49.** yes   **51.** yes   **53.** no   **55.** yes   **57.** yes
59. $x + 8 = 18$ **61.** $2x + 5 = 5$ **63.** $16 - \dfrac{3}{4}x = 13$
65. $3x = 2x + 8$ **67.** expression   **69.** equation   **71.** 64.9 yr
72. 68.5 yr   **73.** 72.8 yr   **74.** 78.1 yr

Section 9.3 (pages 637–638)

1. 4   **3.** 0   **5.** One example is $\sqrt{13}$. There are others.   **7. (a)** 3, 7
(b) 0, 3, 7   **(c)** $-9, 0, 3, 7$ **(d)** $-9, -1\dfrac{1}{4}, -\dfrac{3}{5}, 0, 3, 5.9, 7$
(e) $-\sqrt{7}, \sqrt{5}$ **(f)** All are real numbers.   **9.** 2,845,000   **11.** -2809
13. **15.**
$-3\dfrac{4}{5}$ $-\dfrac{13}{8}$ $\dfrac{1}{4}$ $2\dfrac{1}{2}$
17.   **19.** -11 **21.** -21 **23.** -100
25. $-\dfrac{2}{3}$ **27.** false   **29.** true   **31. (a)** 2   **(b)** 2   **33. (a)** -6 **(b)** 6
35. (a) $\dfrac{3}{4}$ **(b)** $\dfrac{3}{4}$ **37.** 7   **39.** -12 **41.** $-\dfrac{2}{3}$ **43.** 9   **45.** false
47. true   **49.** No; the statement is false for one number, 0.   **51.** electronic components and accessories, 2003 to 2004   **53.** 2002 to 2003

Section 9.4 (pages 643–646)

1. Add -2 and 5.   **3.** Add -1 and -3.   **5.** 2   **7.** -3 **9.** -10
11. -13 **13.** -15.9 **15.** 5   **17.** 13   **19.** 0   **21.** -8 **23.** $\dfrac{3}{10}$

25. $\dfrac{1}{2}$ **27.** $-\dfrac{3}{4}$ **29.** -1.6 **31.** -8.7 **33.** -25 **35.** $-\dfrac{1}{4}$, or -0.25
37. true   **39.** false   **41.** true   **43.** false   **45.** true   **47.** false
49. It must be negative and have the greater absolute value.   **50.** The sum of a positive number and 5 cannot be -7.   **51.** It must be positive and have the greater absolute value.   **52.** The sum of a negative number and -8 cannot be 2.   **53.** Add the absolute values of the numbers. The sum will be negative.   **55.** $-5 + 12 + 6$; 13   **57.** $[-19 + (-4)] + 14$; -9
59. $[-4 + (-10)] + 12$; -2 **61.** $\left[\dfrac{5}{7} + \left(-\dfrac{9}{7}\right)\right] + \dfrac{2}{7}$; $-\dfrac{2}{7}$ **63.** $-\$62$
65. -184 m   **67.** 37 yd   **69.** 120°F   **71.** $-\$107$ **73.** -12

Section 9.5 (pages 651–656)

1. -8; -6 **3.** $7 - 12$; $12 - 7$ **5.** -4 **7.** -10 **9.** -16 **11.** 11
13. 19   **15.** -4 **17.** 5   **19.** 0   **21.** $\dfrac{3}{4}$ **23.** $-\dfrac{11}{8}$ **25.** $\dfrac{15}{8}$ **27.** 13.6
29. -11.9 **31.** -2.8 **33.** -6.3 **35.** -28 **37.** -18 **39.** $\dfrac{37}{12}$
41. -42.04 **43.** For example, let $a = 1, b = 1$ or let $a = 2, b = 2$. In general, choose $a = b$.   **45.** 8   **47.** For example, $-8 - (-2) = -6$.
49. $4 - (-8)$; 12   **51.** $-2 - 8$; -10 **53.** $[9 + (-4)] - 7$; -2
55. $[8 - (-5)] - 12$; 1   **57.** -58°F   **59.** 14,776 ft   **61.** $-\$80$
63. \$105,000   **65.** 17   **67.** \$1045.55   **69.** 469 B.C.   **71.** \$323.83
73. 14 ft   **75.** 40,776 ft   **77.** 11.3%   **79.** \$649 billion   **81.** \$15,000
83. \$2900   **85.** positive   **87.** positive

Section 9.6 (pages 665–668)

1. greater than 0   **3.** less than 0   **5.** greater than 0   **7.** -28 **9.** 30
11. 0   **13.** $\dfrac{5}{6}$ **15.** -2.38 **17.** $\dfrac{3}{2}$ **19.** -3 **21.** -2 **23.** 16   **25.** 0
27. undefined   **29.** $\dfrac{3}{2}$ **31.** C   **33.** 3   **35.** 7   **37.** 4   **39.** -3 **41.** -1
43. $\dfrac{7}{4}$ **45.** 68   **47.** -228 **49.** 1   **51.** 0   **53.** -6 **55.** undefined
57. $-12 + 4(-7)$; -40 **59.** $-1 - 2(-8)(2)$; 31
61. $-3[3 - (-7)]$; -30 **63.** $\dfrac{3}{10}[-2 + (-28)]$; -9

65. $\dfrac{-20}{-8 + (-2)}$; 2   **67.** $\dfrac{-18 + (-6)}{2(-4)}$; 3   **69.** $\dfrac{-\dfrac{2}{3}\left(-\dfrac{1}{5}\right)}{\dfrac{1}{7}}$; $\dfrac{14}{15}$
71. $9x = -36$ **73.** $\dfrac{x}{4} = -1$ **75.** $x - \dfrac{9}{11} = 5$ **77.** $\dfrac{6}{x} = -3$
79. 42   **80.** 5   **81.** $8\dfrac{2}{5}$ **82.** $8\dfrac{2}{5}$ **83.** 2   **84.** $-12\dfrac{1}{2}$

Section 9.7 (pages 675–678)

1. B   **3.** C   **5.** B   **7.** G   **9.** commutative property   **11.** associative property   **13.** inverse property   **15.** inverse property   **17.** identity property   **19.** commutative property   **21.** distributive property
23. identity property   **25.** distributive property   **27. (a)** 0   **(b)** 1, -1
29. $25 - (6 - 2) = 25 - 4 = 21$ and $(25 - 6) - 2 = 19 - 2 = 17$. Since these results are different, subtraction is not associative.   **31.** $7 + r$
33. s **35.** $-6x + (-6)7$; $-6x - 42$ **37.** $w + [5 + (-3)]$; $w + 2$
39. We must multiply $\dfrac{3}{4}$ by 1 in the form $\dfrac{3}{3}$: $\dfrac{3}{4} \cdot \dfrac{3}{3} = \dfrac{9}{12}$.   **41.** 2
43. $4t + 12$ **45.** $-8r - 24$ **47.** $-5y + 20$ **49.** $-16y - 20z$
51. $8(z + w)$ **53.** $5(3 + 17)$; 100   **55.** $7(2v + 5r)$ **57.** $24r + 32s - 40y$
59. $-24x - 9y - 12z$ **61.** $-4t - 5m$ **63.** $5c + 4d$ **65.** $3q - 5r + 8s$

67. Answers will vary. For example, "putting on your socks" and "putting on your shoes" **69.** false **71.** (foreign sales) clerk; foreign (sales clerk) **73.** 0 **74.** $-3(5) + (-3)(-5)$ **75.** -15 **76.** The product $-3(-5)$ must equal 15, since it is the additive inverse of -15.

Section 9.8 (pages 683–684)

1. C **3.** A **5.** $15x$ **7.** $5t$ **9.** $4r + 11$ **11.** $5 + 2x - 6y$ **13.** $-7 + 3p$ **15.** -12 **17.** 5 **19.** 1 **21.** -1 **23.** 74 **25.** Answers will vary. For example, $-3x$ and $4x$ **27.** like **29.** unlike **31.** like **33.** unlike **35.** We cannot "add" two unlike terms to obtain a single term, so we must be able to identify like terms in order to combine them. **37.** $1 - 2x$ **39.** $-\frac{1}{3}t - \frac{28}{3}$ **41.** $-4.1r + 4.2$ **43.** $-2y^2 + 3y^3$ **45.** $-19p + 16$ **47.** $-\frac{3}{2}y + 16$ **49.** $-16y + 63$

51. $(x + 3) + 5x$; $6x + 3$ **53.** $(13 + 6x) - (-7x)$; $13 + 13x$ **55.** $2(3x + 4) - (-4 + 6x)$; 12 **57.** Wording may vary. One example is "the difference between 9 times a number and the sum of the number and 2." **59.** $1000 + 5x$ (dollars) **60.** $750 + 3y$ (dollars) **61.** $1000 + 5x + 750 + 3y$ (dollars) **62.** $1750 + 5x + 3y$ (dollars)

Chapter 9 Review Exercises (pages 689–694)

1. 625 **2.** 0.00000081 **3.** 0.009261 **4.** $\frac{125}{8}$ **5.** 27 **6.** 200 **7.** 7 **8.** $\frac{20}{3}$ **9.** $13 < 17$ **10.** $5 + 2 \neq 10$ **11.** Six is less than fifteen. **12.** Answers will vary. One example is $2 + 5 \geq \frac{16}{2}$. **13.** 30 **14.** 60 **15.** 14 **16.** 13 **17.** $x + 6$ **18.** $8 - x$ **19.** $6x - 9$ **20.** $12 + \frac{3}{5}x$ **21.** yes **22.** no **23.** $2x - 6 = 10$ **24.** $4x = 8$ **25.** equation

26. expression

27. [number line: $-\frac{1}{2}$ and 2.5 marked, -6 -4 -2 0 2 4 5 6]

28. [number line: -6 -4 -3 -2 0 1 2 3 4 6]

29. [number line: $-3\frac{1}{4}$ $-1\frac{1}{8}$ $\frac{5}{6}$ $\frac{14}{5}$ marked, -6 -4 -2 0 2 4 6]

30. [number line: -6 -5 -4 -3 -2 0 2 4 6]

31. -10 **32.** -9 **33.** $-\frac{3}{4}$ **34.** $-|23|$ **35.** true **36.** true **37.** true **38.** false **39.** -3 **40.** -19 **41.** -7 **42.** 9 **43.** -6 **44.** -4 **45.** -17 **46.** $-\frac{29}{36}$ **47.** -10 **48.** -19 **49.** $(-31 + 12) + 19$; 0 **50.** $[-4 + (-8)] + 13$; 1 **51.** \$26.25 **52.** $-10°$F **53.** -11 **54.** -1 **55.** 7 **56.** $-\frac{43}{35}$ **57.** 10.31 **58.** -12 **59.** 2 **60.** 1 **61.** $-4 - (-6)$; 2 **62.** $[4 + (-8)] - 5$; -9 **63.** $[18 - (-23)] - 15$; 26 **64.** $19 - (-7 - 12)$; 38 **65.** 38 **66.** 12,284.30 **67.** 56 thousand **68.** -293 thousand **69.** -126 thousand **70.** 98 thousand **71.** 36 **72.** -105 **73.** $\frac{1}{2}$ **74.** 10.08 **75.** -20 **76.** -10 **77.** -24 **78.** -35 **79.** 4 **80.** -20 **81.** $-\frac{3}{4}$ **82.** 11.3 **83.** -1 **84.** undefined **85.** 1 **86.** 0 **87.** -18 **88.** -18 **89.** 125

90. -423 **91.** $-4(5) - 9$; -29 **92.** $\frac{5}{6}[12 + (-6)]$; 5 **93.** $\frac{12}{8 + (-4)}$; 3 **94.** $\frac{-20(12)}{15 - (-15)}$; -8 **95.** $\frac{x}{x + 5} = -2$ **96.** $8x - 3 = -7$ **97.** identity property **98.** identity property **99.** inverse property **100.** inverse property **101.** associative property **102.** associative property **103.** distributive property **104.** commutative property **105.** $(7 + 1)y$; $8y$ **106.** $-12 \cdot 4 - 12(-t)$; $-48 + 12t$ **107.** $3(2s + 4y)$; $6s + 12y$ **108.** $-1(-4r) + (-1)(5s)$; $4r - 5s$ **109.** $17p^2$ **110.** $16r^2 + 7r$ **111.** $-19k + 54$ **112.** $5s - 6$ **113.** $-45t - 23$ **114.** $-45t^2 - 23.4t$ **115.** $-2(3x) - 7x$; $-13x$ **116.** $\frac{x + 9}{x - 6}$ **117.** No. The use of *and* there indicates the two quantities that are to be multiplied. **118.** Answers may vary. For example, "3 times the difference between 4 times a number and 6"

119. 16 **120.** $\frac{25}{36}$ **121.** -26 **122.** $\frac{8}{3}$ **123.** $-\frac{1}{24}$ **124.** $\frac{7}{2}$ **125.** 2 **126.** 77.6 **127.** $-1\frac{1}{2}$ **128.** 11 **129.** $-\frac{28}{15}$ **130.** 24 **131.** -11 **132.** -6 **133.** $-47°$F **134.** 15 points

Chapter 9 Test (pages 695–696)

1. true **2.** false **3.** [number line: -3 -1 0 1 4] **4.** $-|-8|$ (or -8) **5.** -1.277 **6.** $\frac{-6}{2 + (-8)}$; 1 **7.** negative **8.** 4 **9.** $-2\frac{5}{6}$ **10.** 2 **11.** 6 **12.** 108 **13.** 11 **14.** $\frac{30}{7}$ **15.** -70 **16.** 3 **17.** $178°$F **18.** D **19.** A **20.** E **21.** B **22.** C **23.** $-9x^2 - 6x - 8$ **24.** identity and distributive properties **25.** (a) -18 (b) -18 (c) The distributive property tells us that the two methods produce equal results.

Chapter 10 Equations, Inequalities, and Applications

Section 10.1 (pages 703–704)

1. (a) expression; $x + 15$ (b) expression; $m + 7$ (c) equation; $\{-1\}$ (d) equation; $\{-17\}$ **3.** A and B **5.** $\{12\}$ **7.** $\{-3\}$ **9.** $\{4\}$ **11.** $\{-9\}$ **13.** $\left\{-\frac{3}{4}\right\}$ **15.** $\{6.3\}$ **17.** $\{-16.9\}$ **19.** $\{-10\}$ **21.** $\{-13\}$ **23.** $\left\{\frac{4}{15}\right\}$ **25.** $\{7\}$ **27.** $\{-4\}$ **29.** $\{3\}$ **31.** $\{-2\}$ **33.** $\{4\}$ **35.** $\{-16\}$ **37.** $\{2\}$ **39.** $\{2\}$ **41.** $\{-4\}$ **43.** $\{4\}$ **45.** $\{0\}$ **47.** $\left\{\frac{7}{15}\right\}$ **49.** $\{7\}$ **51.** $\{-4\}$ **53.** $\{13\}$ **55.** $\{29\}$ **57.** $\{18\}$ **59.** Answers will vary. One example is $x - 6 = -8$.

Section 10.2 (pages 709–710)

1. $\frac{3}{2}$ **3.** 10 **5.** $-\frac{2}{9}$ **7.** -1 **9.** 6 **11.** -4 **13.** 0.12 **15.** -1 **17.** (a) and (c): multiplication property of equality; (b) and (d): addition property of equality **19.** B **21.** $\{6\}$ **23.** $\left\{\frac{15}{2}\right\}$ **25.** $\{-5\}$

27. $\left\{-\dfrac{18}{5}\right\}$ **29.** $\{12\}$ **31.** $\{0\}$ **33.** $\{-12\}$ **35.** $\{40\}$ **37.** $\{-12.2\}$

39. $\{-48\}$ **41.** $\{72\}$ **43.** $\{-35\}$ **45.** $\{14\}$ **47.** $\left\{-\dfrac{27}{35}\right\}$ **49.** $\{3\}$

51. $\{-5\}$ **53.** $\{20\}$ **55.** $\{-12\}$ **57.** $\{7\}$ **59.** $\{0\}$ **61.** $\{-6\}$

63. Answers will vary. One example is $\dfrac{3}{2}x = -6$.   **65.** $-4x = 10; -\dfrac{5}{2}$

Section 10.3 (pages 719–722)

1. Use the addition property of equality to subtract 8 from each side.
3. Clear parentheses by using the distributive property.
5. Clear fractions by multiplying by the LCD, 6.

7. $\{4\}$ **9.** $\{-5\}$ **11.** $\left\{\dfrac{5}{2}\right\}$ **13.** $\left\{-\dfrac{1}{2}\right\}$ **15.** $\{5\}$ **17.** $\{1\}$ **19.** $\left\{-\dfrac{5}{3}\right\}$

21. $\{-1\}$ **23.** $\varnothing$ **25.** $\{$all real numbers$\}$ **27.** $\{2\}$ **29.** $\left\{-\dfrac{5}{3}\right\}$ **31.** D

33. $\{7\}$ **35.** $\{12\}$ **37.** $\{11\}$ **39.** $\{0\}$ **41.** $\left\{\dfrac{3}{25}\right\}$ **43.** $\{60\}$ **45.** $\{4\}$

47. $\{5000\}$ **49.** $\left\{-\dfrac{72}{11}\right\}$ **51.** $\{0\}$ **53.** $\varnothing$ **55.** $\{$all real numbers$\}$

57. $\{-6\}$ **59.** $\{15\}$ **61.** $12 - q$ **63.** $\dfrac{9}{z}$ **65.** $a + 12; a - 2$ **67.** $25r$

Section 10.4 (pages 731–736)

1. The procedure should include the following steps: read the problem carefully; assign a variable to represent the unknown to be found; write down variable expressions for any other unknown quantities; translate into an equation; solve the equation; state the answer; check your solution.
3. D; There cannot be a fractional number of cars.   **5.** A; Distance cannot be negative.   **7.** 7   **9.** -8 **11.** 6   **13.** -3 **15.** Pennsylvania: 35 screens; Ohio: 33 screens   **17.** Democrats: 44; Republicans: 55   **19.** The Police: $133.2 million; Kenny Chesney: $71.1 million   **21.** wins: 61; losses: 21
23. orange: 97 mg; pineapple: 25 mg   **25.** 1950 Denver nickel: $16.00; 1944 Philadelphia nickel: $12.00   **27.** onions: 81.3 kg; grilled steak: 536.3 kg   **29.** 420 lb   **31.** American: 18; United: 11; Southwest: 26
33. gold: 35; silver: 39; bronze: 29   **35.** 36 million mi   **37.** A and B: 40°; C: 100°   **39.** yes, 90°; yes, 45°   **41.** 18°   **43.** 20°   **45.** 39°
47. 50°   **49.** 68, 69   **51.** 146, 147   **53.** 10, 12   **55.** 10, 11   **57.** 18
59. 2002: $6.54 billion; 2003: $6.67 billion; 2004: $6.77 billion

Section 10.5 (pages 743–748)

1. (a) The perimeter of a plane geometric figure is the distance around the figure.   **(b)** The area of a plane geometric figure is the measure of the surface covered or enclosed by the figure.   **3.** four   **5.** area   **7.** perimeter
9. area   **11.** area   **13.** $P = 26$ **15.** $A = 64$ **17.** $b = 4$ **19.** $t = 5.6$
21. $I = 1575$ **23.** $r = 2.6$ **25.** $A = 50.24$ **27.** $V = 150$
29. $V = 52$ **31.** $V = 7234.56$ **33.** length: 18 in.; width: 9 in.
35. length: 14 m; width: 4 m   **37.** shortest: 5 in.; medium: 7 in., longest: 8 in.   **39.** two equal sides: 7 m; third side: 10 m   **41.** about 154,000 ft^2
43. perimeter: 5.4 m; area: 1.8 m^2 **45.** 194.48 ft^2 **47.** 10 ft
49. 23,800.10 ft^2 **51.** length: 36 in.; maximum volume: 11,664 in.^{3}
53. 48°, 132°   **55.** 55°, 35°   **57.** 51°, 51°   **59.** 105°, 105°   **61.** $t = \dfrac{d}{r}$

63. $H = \dfrac{V}{LW}$ **65.** $b = P - a - c$ **67.** $r = \dfrac{C}{2\pi}$ **69.** $r = \dfrac{I}{pt}$

71. $h = \dfrac{2A}{b}$ **73.** $h = \dfrac{3V}{\pi r^2}$ **75.** $W = \dfrac{P - 2L}{2}$ **77.** $m = \dfrac{y - b}{x}$

79. $y = \dfrac{C - Ax}{B}$ **81.** $r = \dfrac{M - C}{C}$ **83.** $a = \dfrac{P - 2b}{2}$

Section 10.6 (pages 757–760)

1. $x > -4$ **3.** $x \le 4$ **5.** $-1 < x \le 2$

7. $(-\infty, 4]$

9. $(-3, \infty)$

11. $[8, 10]$

13. $(0, 10]$

15. $[1, \infty)$

17. $[5, \infty)$

19. $(-\infty, -6)$

21. It must be reversed when multiplying or dividing by a negative number.

23. $(-\infty, 6)$

25. $[-10, \infty)$

27. $(-\infty, -3)$

29. $(-\infty, 0]$

31. $(20, \infty)$

33. $[-3, \infty)$

35. $(-\infty, -3]$

37. $(-1, \infty)$

39. $[-5, \infty)$

41. $(-\infty, 1)$

43. $(-\infty, 0]$

45. $\left(-\dfrac{1}{2}, \infty\right)$

47. $[4, \infty)$

49. $(-\infty, 32)$

51. $\left[\dfrac{5}{12}, \infty\right)$

53. $(-21, \infty)$

55. 88 or more   **57.** all numbers greater than 16   **59.** It has never exceeded 40°C.   **61.** 32 or greater   **63.** 15 min

65. $[-1, 6]$

67. $(1, 3)$

69. $\left(-\dfrac{11}{6}, -\dfrac{2}{3}\right]$

71.

72.

73.

74. It is the set of all real numbers.

75. The graph would be the set of all real numbers.

Chapter 10 Review Exercises (pages 765–768)

1. $\{9\}$ **2.** $\{4\}$ **3.** $\{-6\}$ **4.** $\left\{\dfrac{3}{2}\right\}$ **5.** $\{20\}$ **6.** $\left\{-\dfrac{61}{2}\right\}$ **7.** $\{15\}$

8. $\{0\}$ **9.** $\varnothing$ **10.** {all real numbers}   **11.** $-\dfrac{7}{2}$ **12.** 20

13. Hawaii: 6425 mi²; Rhode Island: 1212 mi²   **14.** Seven Falls: 300 ft; Twin Falls: 120 ft   **15.** 80°   **16.** 11, 13   **17.** $h = 11$ **18.** $A = 28$

19. $r = 4.75$ **20.** $V = 3052.08$ **21.** $h = \dfrac{A}{b}$ **22.** $h = \dfrac{2A}{b+B}$

23. 135°, 45°   **24.** 100°, 100°   **25.** perimeter: 326.5 ft; area: 6538.875 ft²

26. diameter: 46.78 ft; radius: 23.39 ft

27. $[-4, \infty)$

28. $(-\infty, 7)$

29. $[-5, 6)$

30. $\left[\dfrac{1}{2}, \infty\right)$

31. $[-3, \infty)$

32. $(-\infty, 2)$

33. $[3, \infty)$

34. $[46, \infty)$

35. $(-\infty, -5)$

36. $(-\infty, -37)$

37. $\left[-2, \dfrac{3}{2}\right)$

38. $(1, 5]$

39. 88 or more   **40.** all numbers less than or equal to $-\dfrac{1}{3}$ **41.** $\{7\}$

42. $r = \dfrac{d}{2}$ **43.** $(-\infty, 2)$ **44.** $\{-9\}$ **45.** $\{70\}$ **46.** $\left\{\dfrac{13}{4}\right\}$

47. $\varnothing$ **48.** {all real numbers}   **49.** Pizza Hut: \$9.35 billion; Domino's: \$5.1 billion   **50.** gold: 11; silver: 12; bronze: 6   **51.** D: 22°; E: 44°; F: 114°   **52.** 44 m   **53.** 70 ft   **54.** 24°, 66°   **55.** 92 or more   **56.** \$67.50

Chapter 10 Test (pages 769–770)

1. $\{6\}$ **2.** $\{-6\}$ **3.** $\left\{\dfrac{13}{4}\right\}$ **4.** $\{-10.8\}$ **5.** $\varnothing$ **6.** $\{21\}$ **7.** $\{30\}$

8. {all real numbers}   **9.** 67 wins, 15 losses   **10.** Hawaii: 4021 mi²; Maui: 728 mi²; Kauai: 551 mi²   **11.** 50°   **12.** (a) $W = \dfrac{P - 2L}{2}$ (b) 18

13. 100°, 80°   **14.** 75°, 75°   **15.** (a) $x < 0$ (b) $-2 < x \le 3$

16. $(-\infty, 11)$

17. $[-3, \infty)$

18. $(-\infty, 4]$

19. $(-2, 6]$

20. 83 or more

Chapter 11 Graphs of Linear Equations and Inequalities in Two Variables

Section 11.1 (pages 781–786)

1. 2005, 2006, 2007   **3.** 2001: about 165 billion lb; 2007: about 185 billion lb   **5.** from 2000 to 2005; about \$0.85   **7.** The price of a gallon of gas was decreasing.   **9.** does; do not   **11.** y
13. 6   **15.** yes   **17.** yes   **19.** no   **21.** yes   **23.** no
25. No. For two ordered pairs (x, y) to be equal, the x-values must be equal and the y-values must be equal. Here we have $4 \ne -1$ and $-1 \ne 4$.

27. 11   **29.** $-\dfrac{7}{2}$ **31.** -4 **33.** -5 **35.** 4; 6; -6; (0, 4); (6, 0); $(-6, 8)$
37. 3; -5; -15; (0, 3); $(-5, 0)$; $(-15, -6)$ **39.** -9; -9; -9
41. -6; -6; -6 **43.** 8; 8; 8   **45.** (2, 4); I   **47.** $(-5, 4)$; II
49. (3, 0); no quadrant   **51.** negative; negative   **53.** positive; negative
55. If $xy < 0$, then either $x < 0$ and $y > 0$ or $x > 0$ and $y < 0$. If $x < 0$ and $y > 0$, then the point lies in quadrant II. If $x > 0$ and $y < 0$, then the point lies in quadrant IV.

57.–68.

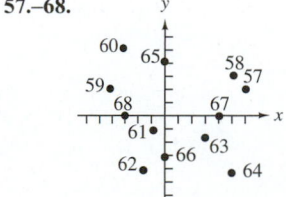

69. $-3; 6; -2; 4$

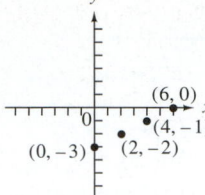

71. $-3; 4; -6; -\dfrac{4}{3}$

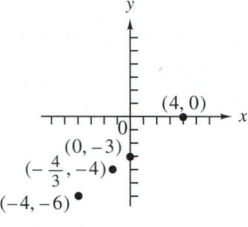

73. $-4; -4; -4; -4$

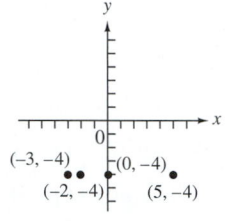

75. The points in each graph appear to lie on a straight line.   **77. (a)** $(5, 45)$
(b) $(6, 50)$ **79. (a)** $(2000, 32.4), (2001, 31.6), (2002, 31.6), (2003, 30.1),$
$(2004, 29.0), (2005, 27.5)$ **(b)** $(2007, 27.1)$ means that 27.1 percent of
2-year college students in 2007 received a degree within 3 years.

(c)   **2-YEAR COLLEGE STUDENTS**
COMPLETING A DEGREE
WITHIN 3 YEARS

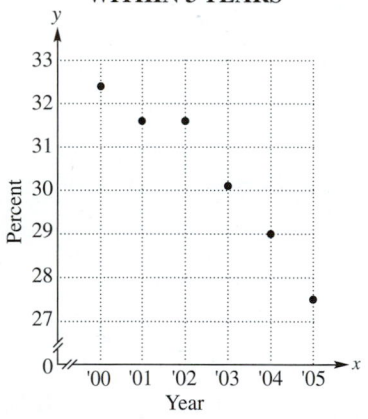

(d) With the exception of the point for 2002, the points lie approximately
on a straight line. Rates at which 2-year college students complete a degree
within 3 years are generally decreasing.
81. (a) $157, 141, 125, 109$ **(b)** $(20, 157), (40, 141), (60, 125), (80, 109)$

(c)   **TARGET HEART RATE ZONE**
(Upper Limit)

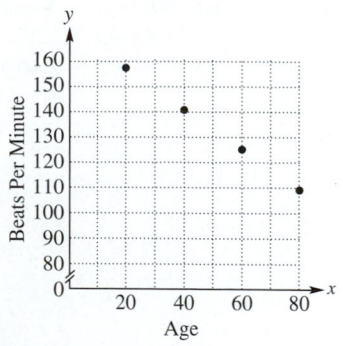

The points lie in a linear pattern.

Section 11.2 (pages 795–800)

1. $5; 5; 3$

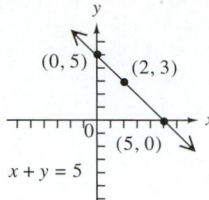

3. $1; 3; -1$

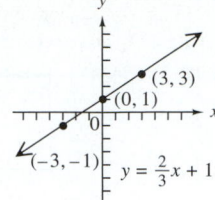

5. $-6; -2; -5$

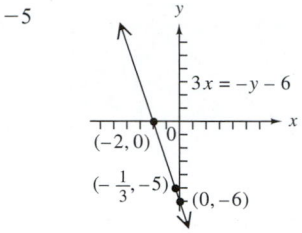

7. (a) A   **(b)** C   **(c)** D   **(d)** B   **9.** $(12, 0); (0, -8)$ **11.** $(0, 0); (0, 0)$

13.

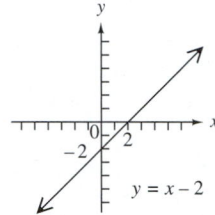

15.

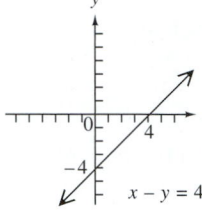

17.

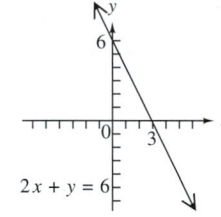

19.

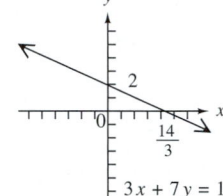

21.

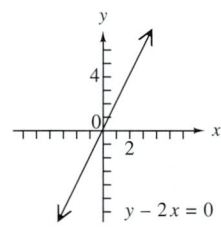

23.

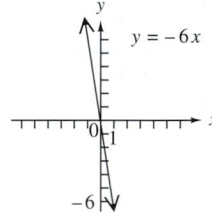

25.

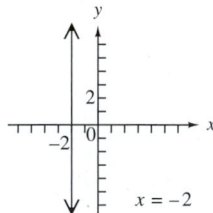

27.

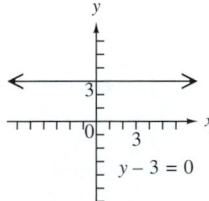

29.

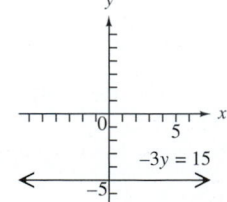

In Exercises 31 and 33, descriptions may vary.
31. The graph is a line with *x*-intercept $(-3, 0)$ and *y*-intercept $(0, 9)$.
33. The graph is a horizontal line with *y*-intercept $(0, -2)$.
35. Choose a value *other than* 0 for either *x* or *y*. For example,
if $x = -5, y = 4$.
37. **(a)** 151.5 cm, 159.3 cm, 174.9 cm   **(b)** (20, 151.5), (22, 159.3),
(26, 174.9)
(c)

HEIGHTS OF WOMEN

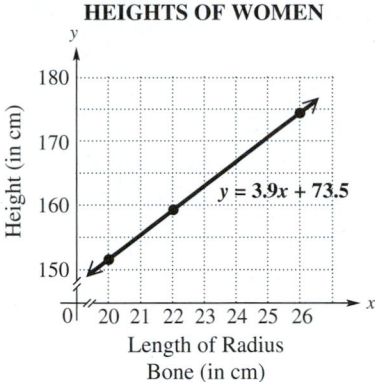

(d) 24 cm; 24 cm
39. **(a)** \$62.50; \$100 **(b)** 200 **(c)** (50, 62.50), (100, 100), (200, 175)
(d)

POSTER COSTS

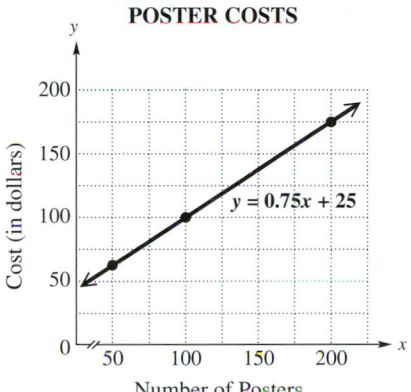

41. **(a)** \$30,000 **(b)** \$15,000 **(c)** \$5000 **(d)** After 5 yr, the SUV
has a value of \$5000. **43.** **(a)** 1990: 24.1 lb; 2000: 29.5 lb; 2005: 32.2 lb
(b) 1990: 25 lb; 2000: 30 lb; 2005: 32 lb **(c)** The values are quite close.

Section 11.3 (pages 809–812)

1. $\dfrac{3}{2}$ **3.** $-\dfrac{7}{4}$ **5.** 0 **7.** Rise is the vertical change between two different
points on a line. Run is the horizontal change between two different points
on a line.

9.–12. Answers will vary.

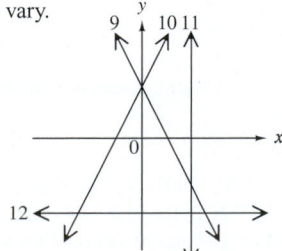

13. **(a)** falls from left to right **(b)** horizontal **(c)** vertical **(d)** rises
from left to right **15.** Because he found the difference $3 - 5 = -2$ in
the numerator, he should have subtracted in the same order in the denominator to get $-1 - 2 = -3$. The correct slope is $\dfrac{-2}{-3} = \dfrac{2}{3}$.

17. $\dfrac{5}{4}$ **19.** $\dfrac{3}{2}$ **21.** -3 **23.** 0 **25.** undefined **27.** $-\dfrac{1}{2}$ **29.** 5
31. $\dfrac{1}{4}$ **33.** $\dfrac{3}{2}$ **35.** 0 **37.** undefined **39.** 1 **41.** **(a)** negative
(b) 0 **43.** **(a)** positive **(b)** negative **45.** **(a)** 0 **(b)** negative
47. $\dfrac{4}{3}$; $\dfrac{4}{3}$; parallel **49.** $\dfrac{5}{3}$; $\dfrac{3}{5}$; neither **51.** $\dfrac{3}{5}$; $-\dfrac{5}{3}$; perpendicular
53. $\dfrac{8}{27}$ **55.** 232 thousand, or 232,000 **56.** positive; increased
57. 232,000 students **58.** -0.95 **59.** negative; decreased
60. 0.95 student per computer

Section 11.4 (pages 821–826)

1. **(a)** D **(b)** C **(c)** B **(d)** A **3.** $y = 3x - 3$
5. $y = -x + 3$ **7.** $y = -\dfrac{1}{2}x + 2$ **9.** $y = 4x - 3$ **11.** $y = 3$
13. **(a)** C **(b)** B **(c)** A **(d)** D

15.

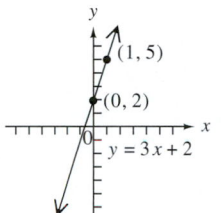

17.

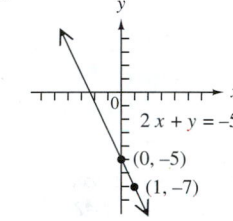

19.

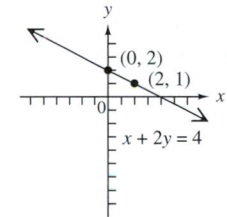

21. $y = \dfrac{1}{2}x + 4$

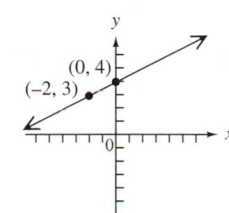

23. $y = -\dfrac{2}{5}x - \dfrac{23}{5}$

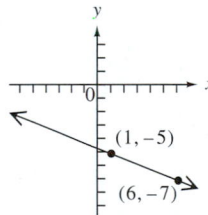

25. $y = 2$

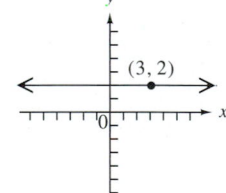

27. $x = 3$ (no slope-intercept form) **29.** $y = \dfrac{2}{3}x$

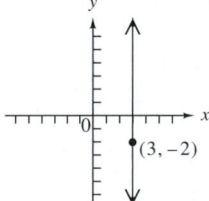

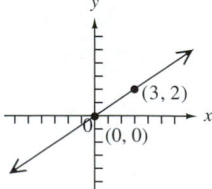

31. $y = 2x - 7$ **33.** $y = -2x - 4$ **35.** $y = \dfrac{2}{3}x + \dfrac{19}{3}$
37. $y = x - 3$ **39.** $y = -\dfrac{5}{7}x - \dfrac{54}{7}$ **41.** $y = -\dfrac{2}{3}x - 2$

43. $x = 3$ (no slope-intercept form) **45.** $y = \frac{1}{3}x + \frac{4}{3}$

47. $y = \frac{3}{4}x - \frac{9}{2}$ **49.** $y = -2x - 3$ **51.** $(0, 32); (100, 212)$ **52.** $\frac{9}{5}$

53. $F - 32 = \frac{9}{5}(C - 0)$ **54.** $F = \frac{9}{5}C + 32$ **55.** $C = \frac{5}{9}(F - 32)$

56. $86°$ **57.** $10°$ **58.** $-40°$ **59.** (a) $400 (b) $0.25
(c) $y = 0.25x + 400$ (d) $425 (e) 1500 **61.** (a) $(1, 1909), (2, 2079),$
$(3, 2182), (4, 2272), (5, 2361)$
(b) yes

**AVERAGE ANNUAL COSTS AT
2-YEAR COLLEGES**

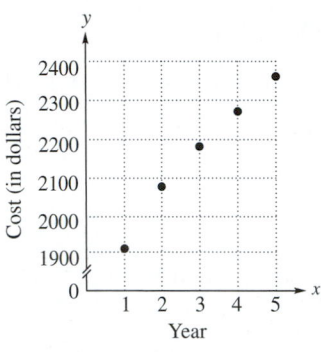

(c) $y = 94x + 1891$ (d) $2455

Section 11.5 (pages 831–834)

1. false; The point $(4, 0)$ lies on the boundary line $3x - 4y = 12$, which
is *not* part of the graph because the symbol $<$ does not involve equality.
3. true **5.** $>, >$ **7.** $\leq$

9. 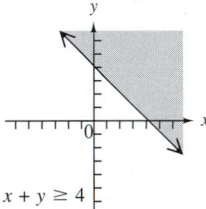 **11.**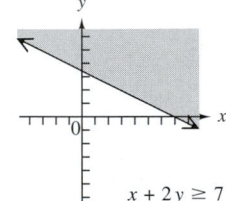

13. 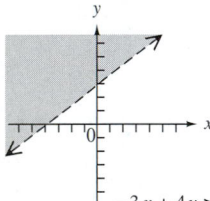 **15.**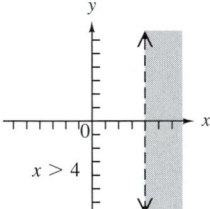

17.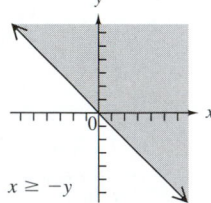

19. A test point cannot lie on the boundary line. It must lie on one side of
the boundary.

21. 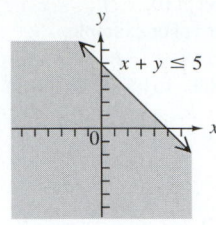 **23.**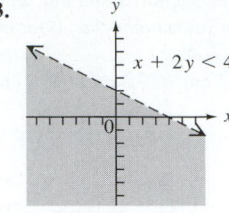

25. 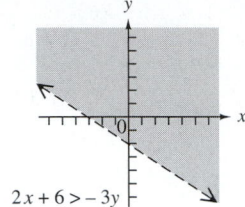 **27.**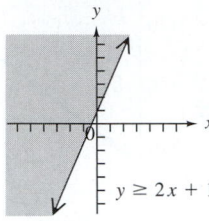

29. 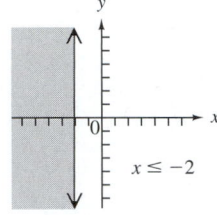 **31.**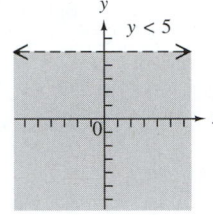

33.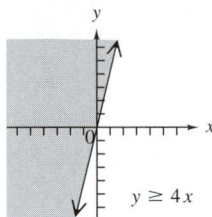

35. Every point in quadrant IV has a positive x-value and a negative
y-value. Substituting into $y > x$ would imply that a negative number is
greater than a positive number, which is always false. Thus, the graph of
$y > x$ cannot lie in quadrant IV.

37. (a)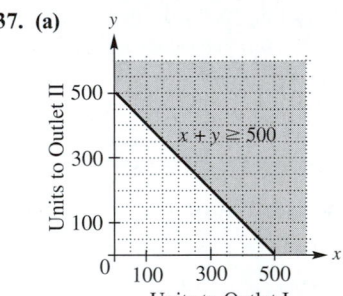

(b) $(500, 0)$ and $(200, 400)$; Other answers are possible.

Chapter 11 Review Exercises (pages 839–844)

1. $(2001, 51.8), (2002, 51.8), (2003, 52.8), (2004, 51.3), (2005, 51.6),$
$(2006, 52.5), (2007, 51.4)$ **2.** In the year 2006, 52.5% of first-year college
students at two-year public institutions returned for a second year.

3. 2004; 1.5% **4.** 2003; 1.0% **5.** $-1; 2; 1$ **6.** $2; \frac{3}{2}; \frac{14}{3}$

7. $0; \frac{8}{3}; -9$ **8.** $7; 7; 7$ **9.** yes **10.** no **11.** yes

12.–15.

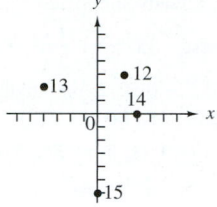

16. x is positive in quadrants I and IV; y is negative in quadrants III and IV. Thus, if x is positive and y is negative, (x, y) must lie in quadrant IV. **17.** In the ordered pair $(k, 0)$, the y-value is 0, so the point lies on the x-axis. In the ordered pair $(0, k)$, the x-value is 0, so the point lies on the y-axis. **18.** II **19.** III **20.** no quadrant

21. $\left(-\dfrac{5}{2}, 0\right); (0, 5)$ **22.** $\left(-\dfrac{7}{2}, 0\right); (0, -7)$ **23.** $\left(\dfrac{8}{3}, 0\right); (0, 4)$

24. **25.**

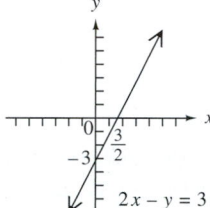

26.

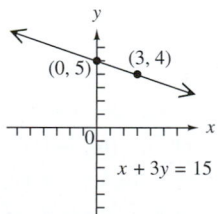

27. $-\dfrac{1}{2}$ **28.** $-\dfrac{2}{3}$ **29.** 0 **30.** undefined **31.** 3 **32.** $\dfrac{2}{3}$ **33.** $\dfrac{3}{2}$

34. $-\dfrac{1}{3}$ **35.** undefined **36.** 0 **37.** $\dfrac{3}{2}$ **38.** (a) 2 (b) $\dfrac{1}{3}$ **39.** parallel

40. perpendicular **41.** neither **42.** 0 **43.** $y = -x + \dfrac{2}{3}$

44. $y = -\dfrac{1}{3}x + 1$ **45.** $y = x - 7$ **46.** $y = \dfrac{2}{3}x + \dfrac{14}{3}$

47. $y = -\dfrac{3}{4}x - \dfrac{1}{4}$ **48.** $y = -\dfrac{1}{4}x + \dfrac{3}{2}$ **49.** $y = 1$ **50.** $x = \dfrac{1}{3}$

51. (a) $y = -\dfrac{1}{3}x + 5$ (b) slope: $-\dfrac{1}{3}$; y-intercept: $(0, 5)$

(c)

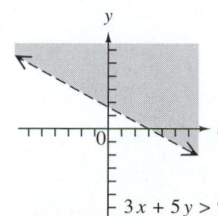

52. (a) B (b) D (c) A (d) C

53. **54.**

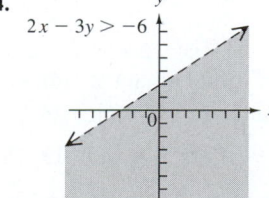

55.

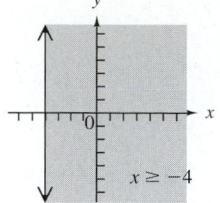

56. A **57.** C, D **58.** A, B, D **59.** D **60.** C **61.** B

62. $\left(-\dfrac{5}{2}, 0\right); (0, -5); -2$

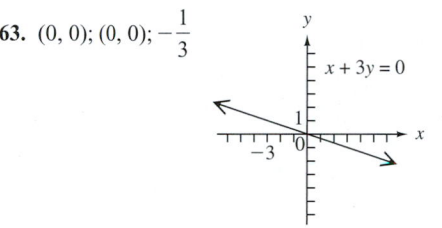

63. $(0, 0); (0, 0); -\dfrac{1}{3}$

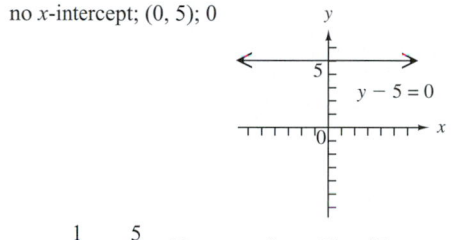

64. no x-intercept; $(0, 5)$; 0

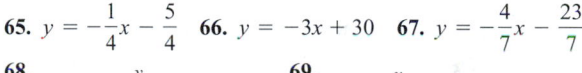

65. $y = -\dfrac{1}{4}x - \dfrac{5}{4}$ **66.** $y = -3x + 30$ **67.** $y = -\dfrac{4}{7}x - \dfrac{23}{7}$

68. **69.**

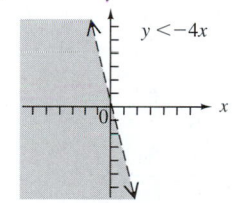 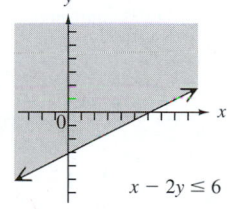

70. 2.5% **71.** Since the graph rises from left to right, the slope is positive. **72.** (2002, 41.2), (2007, 43.7) **73.** $y = 0.5x - 959.8$ **74.** 0.5; yes **75.** 41.7, 42.2, 42.7, 43.2 **76.** 44.2%; No. The equation is based on data only for 2002 through 2007.

Chapter 11 Test (pages 845–848)

1. between 1998 and 1999, 1999 and 2000, and 2003 and 2004
2. The unemployment rate was increasing.
3. 2003: 6.0%; 2004: 5.5%; decline: 0.5%
4. x-intercept: $(2, 0)$; y-intercept: $(0, 6)$

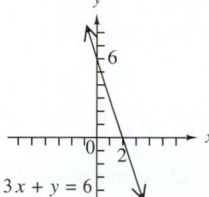

5. x-intercept: $(0, 0)$; y-intercept: $(0, 0)$

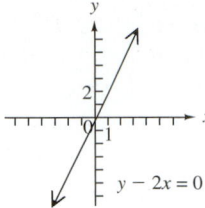

6. x-intercept: $(-3, 0)$; y-intercept: none

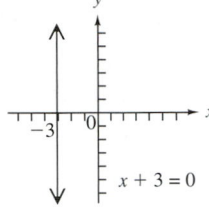

7. x-intercept: none; y-intercept: $(0, 1)$

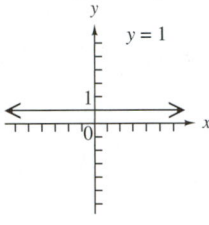

8. x-intercept: $(4, 0)$; y-intercept: $(0, -4)$

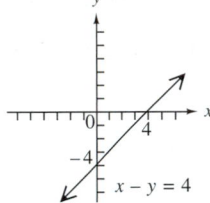

9. $-\dfrac{8}{3}$ **10.** -2 **11.** undefined **12.** $\dfrac{5}{2}$ **13.** 0 **14.** $y = 2x + 6$

15. $y = \dfrac{5}{2}x - 4$ **16.** $y = -9x + 12$ **17.** $y = -\dfrac{3}{2}x + \dfrac{9}{2}$

18.
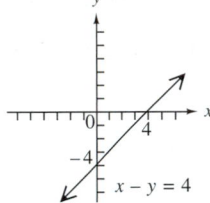

19.

20. The slope is negative since sales are decreasing. **21.** $(0, 209)$, $(7, 160)$; -7 **22.** $y = -7x + 209$ **23.** 174 thousand; The equation gives a good approximation of the actual sales. **24.** In 2007, worldwide snowmobile sales were 160 thousand.

Chapter 12 Exponents and Polynomials

Section 12.1 (pages 855–858)

1. $7; 5$ **3.** 8 **5.** 26 **7.** $1; 6$ **9.** $1; 1$ **11.** $1; \dfrac{1}{5}$ **13.** $2; -19, -1$

15. $3; 1, -8, \dfrac{2}{3}$ **17.** $2m^5$ **19.** $-r^5$ **21.** $\dfrac{2}{3}x^4$ **23.** cannot be simplified; $0.2m^5 - 0.5m^2$ **25.** $-5x^5$ **27.** $5p^9 + 4p^7$ **29.** $-2y^2$

31. already simplified; 4; binomial **33.** already simplified; $6m^5 + 5m^4 - 7m^3 - 3m^2$; 5; none of these **35.** $x^4 + \dfrac{1}{3}x^2 - 4$; 4; trinomial **37.** 7; 0; monomial **39.** $1.5x^2 - 0.5x$; 2; binomial **41.** **(a)** -1 **(b)** 5 **43.** **(a)** 19 **(b)** -2 **45.** **(a)** 36 **(b)** -12 **47.** **(a)** -124 **(b)** 5 **48.** 5; 175 **49.** 87 ft; $(1, 87)$ **50.** \$16.00 **51.** \$27 **52.** 2.5; 130 **53.** $5m^2 + 3m$ **55.** $4x^4 - 4x^2$ **57.** $\dfrac{7}{6}x^2 - \dfrac{2}{15}x + \dfrac{5}{6}$ **59.** $12m^3 - 13m^2 + 6m + 11$ **61.** $2.9x^3 - 3.5x^2 - 1.5x - 9$ **63.** $8r^2 + 5r - 12$ **65.** $5m^2 - 14m + 6$ **67.** $4x^3 + 2x^2 + 5x$ **69.** $-18y^5 + 7y^4 + 5y^3 + 3y^2 + y$ **71.** $-2m^3 + 7m^2 + 8m - 9$ **73.** $-11x^2 - 3x - 3$ **75.** $2x^2 + 8x$ **77.** $8x^2 + 8x + 6$ **79.** $8t^2 + 8t + 13$ **81.** $13a^2b - 7a^2 - b$ **83.** $c^4d - 5c^2d^2 + d^2$ **85.** $12m^3n - 11m^2n^2 - 4mn^2$ **87.** **(a)** $23y + 5t$ **(b)** $25°, 67°, 88°$

Section 12.2 (pages 865–866)

1. 1 **3.** false **5.** false **7.** t^7 **9.** $\left(\dfrac{1}{2}\right)^5$ **11.** $(-8p)^2$ **13.** The expression $(-3)^4$ means $(-3)(-3)(-3)(-3) = 81$, while -3^4 means $-(3 \cdot 3 \cdot 3 \cdot 3) = -81$. **15.** base: 3; exponent: 5; 243 **17.** base: -3; exponent: 5; -243 **19.** base: $-6x$; exponent: 4 **21.** base: x; exponent: 4 **23.** The product rule does not apply to $5^2 + 5^3$ because it is a *sum*, not a product. $5^2 + 5^3 = 25 + 125 = 150$ **25.** 5^8 **27.** 4^{12} **29.** $(-7)^9$ **31.** t^{24} **33.** $-56r^7$ **35.** $42p^{10}$ **37.** The product rule does not apply. **39.** The product rule does not apply. **41.** 4^6 **43.** t^{20} **45.** $343r^3$ **47.** 5^{12} **49.** -8^{15} **51.** $5^5x^5y^5$ **53.** $8q^3r^3$ **55.** $\dfrac{1}{8}$ **57.** $\dfrac{a^3}{b^3}$ **59.** $\dfrac{9^8}{5^8}$ **61.** $-8x^6y^3$ **63.** $9a^6b^4$ **65.** $\dfrac{5^5}{2^5}$ **67.** $\dfrac{9^5}{8^3}$ **69.** $2^{12}x^{12}$ **71.** -6^5p^5 **73.** $6^5x^{10}y^{15}$ **75.** x^{21} **77.** $4w^4x^{26}y^7$ **79.** $-r^{18}s^{17}$ **81.** $\dfrac{125a^6b^{15}}{c^{18}}$ **83.** $25m^6p^{14}q^5$ **85.** $16x^{10}y^{16}z^{10}$ **87.** $30x^7$

Section 12.3 (pages 871–872)

1. $x^2 + 7x + 12$ **3.** $2x^3 + 7x^2 + 7x + 2$ **5.** distributive **7.** $-6m^2 - 4m$ **9.** $6p - \dfrac{9}{2}p^2 + 9p^4$ **11.** $6y^5 + 4y^6 + 10y^9$ **13.** $12x^3 + 26x^2 + 10x + 1$ **15.** $6r^3 + 5r^2 - 12r + 4$ **17.** $20m^4 - m^3 - 8m^2 - 17m - 15$ **19.** $5x^4 - 13x^3 + 20x^2 + 7x + 5$ **21.** $m^2 + 12m + 35$ **23.** $n^2 + n - 6$ **25.** $8r^2 - 10r - 3$ **27.** $9x^2 - 4$ **29.** $9q^2 + 6q + 1$ **31.** $15xy - 40x + 21y - 56$ **33.** $6t^2 + 23st + 20s^2$ **35.** $-0.3t^2 + 0.22t + 0.24$ **37.** $x^2 - \dfrac{5}{12}x - \dfrac{1}{6}$ **39.** $\dfrac{15}{16} - \dfrac{1}{4}r - 2r^2$ **41.** $2x^3 + x^2 - 15x$ **43.** $6y^5 - 21y^4 - 45y^3$ **44.** $(30x + 60)$ yd^2 **45.** $30x + 60 = 600$; $\{18\}$ **46.** **(a)** 10 yd by 60 yd **(b)** 140 yd **47.** \$2100 **48.** \$1260 **49.** The answers are $x^2 - 16$, $y^2 - 4$, and $r^2 - 49$. Each product is the difference of the square of the first term and the square of the last term of the binomials.

Section 12.4 (pages 877–878)

1. **(a)** $4x^2$ **(b)** $12x$ **(c)** 9 **(d)** $4x^2 + 12x + 9$ **3.** $p^2 + 4p + 4$ **5.** $z^2 - 10z + 25$ **7.** $16x^2 - 24x + 9$ **9.** $4p^2 + 20pq + 25q^2$ **11.** $0.64t^2 + 1.12ts + 0.49s^2$ **13.** $25x^2 + 4xy + \dfrac{4}{25}y^2$ **15.** $9t^3 - 6t^2 + t$ **17.** $-16r^2 + 16r - 4$ **19.** **(a)** $49x^2$ **(b)** 0 **(c)** $-9y^2$ **(d)** $49x^2 - 9y^2$; Because 0 is the identity element for addition, it is not necessary to write "+ 0." **21.** $q^2 - 4$ **23.** $4w^2 - 25$

25. $100x^2 - 9y^2$ **27.** $4x^4 - 25$ **29.** $49x^2 - \dfrac{9}{49}$ **31.** $9p^3 - 49p$

33. $(a+b)^2$ **34.** a^2 **35.** $2ab$ **36.** b^2 **37.** $a^2 + 2ab + b^2$

38. They both represent the area of the entire large square.

39. 1225 **40.** $30^2 + 2(30)(5) + 5^2$ **41.** 1225 **42.** They are equal. **43.** $m^3 - 15m^2 + 75m - 125$ **45.** $y^3 + 6y^2 + 12y + 8$

47. $8a^3 + 12a^2 + 6a + 1$

49. $81r^4 - 216r^3t + 216r^2t^2 - 96rt^3 + 16t^4$

51. $3x^5 - 27x^4 + 81x^3 - 81x^2$

53. $-8x^6y - 32x^5y^2 - 48x^4y^3 - 32x^3y^4 - 8x^2y^5$ **55.** 512 cu. units

Section 12.5 (pages 887–888)

1. negative **3.** negative **5.** positive **7.** 0 **9.** 1 **11.** 1 **13.** -1

15. 0 **17.** 0 **19.** 2 **21.** $\dfrac{1}{64}$ **23.** 16 **25.** $\dfrac{49}{36}$ **27.** $\dfrac{1}{81}$ **29.** $\dfrac{8}{15}$

31. $-\dfrac{7}{18}$ **33.** 1 **34.** $\dfrac{5^2}{5^2}$ **35.** $5^{2-2} = 5^0$ **36.** $5^0 = 1$; This supports the

definition of a 0 exponent. **37.** $\dfrac{1}{9}$ **39.** $\dfrac{1}{6^5}$, or $\dfrac{1}{7776}$ **41.** 216 **43.** $2r^4$

45. $\dfrac{25}{64}$ **47.** $\dfrac{p^5}{q^8}$ **49.** r^9 **51.** $\dfrac{x^5}{6}$ **53.** $3y^2$ **55.** x^3 **57.** $\dfrac{yz^2}{4x^3}$

59. $a+b$ **61.** 343 **63.** $\dfrac{1}{x^2}$ **65.** $\dfrac{64x}{9}$ **67.** $\dfrac{x^2z^4}{y^2}$ **69.** $6x$

71. $\dfrac{1}{m^{10}n^5}$ **73.** $\dfrac{5}{16x^5}$ **75.** $\dfrac{36q^2}{m^4p^2}$

Section 12.6 (pages 891–892)

1. $6x^2 + 8$; 2; $3x^2 + 4$ **3.** $3x^2 + 4$; 2 (These may be reversed.); $6x^2 + 8$

5. To use the method of this section, the divisor must be just one term. This is true of the first problem, but not the second. **7.** $30x^3 - 10x + 5$

9. $-4m^3 + 2m^2 - 1$ **11.** $4t^4 - 2t^2 + 2t$ **13.** $a^4 - a + \dfrac{2}{a}$

15. $-2x^3 + \dfrac{2x^2}{3} - x$ **17.** $-9x^2 + 5x + 1$ **19.** $\dfrac{4x^2}{3} + x - \dfrac{2}{3x}$

21. $9r^3 - 12r^2 - 2r + 1 - \dfrac{2}{3r}$ **23.** $-m^2 + 3m - \dfrac{4}{m}$

25. $\dfrac{12}{x} - \dfrac{6}{x^2} + \dfrac{14}{x^3} - \dfrac{10}{x^4}$ **27.** $-4b^2 + 3ab - \dfrac{5}{a}$ **29.** $6x - 2 + \dfrac{1}{x}$

31. $15x^5 - 35x^4 + 35x^3$ **33.** 1423

34. $(1 \times 10^3) + (4 \times 10^2) + (2 \times 10^1) + (3 \times 10^0)$

35. $x^3 + 4x^2 + 2x + 3$ **36.** They are similar in that the coefficients of the powers of ten are equal to the coefficients of the powers of x. They are different in that one is a number while the other is a polynomial. They are equal if $x = 10$.

Section 12.7 (pages 897–898)

1. The divisor is $2x + 5$; the quotient is $2x^3 - 4x^2 + 3x + 2$. **3.** Divide $12m^2$ by $2m$ to get $6m$. **5.** $x + 2$ **7.** $2y - 5$ **9.** $p - 4 + \dfrac{44}{p+6}$

11. $r - 5$ **13.** $2a - 14 + \dfrac{74}{2a+3}$ **15.** $4x^2 - 7x + 3$

17. $3y^2 - 2y + 2$ **19.** $2x^2 - 2x + 3 + \dfrac{-1}{x+1}$ **21.** $3k - 4 + \dfrac{2}{k^2-2}$

23. $x^2 + 1$ **25.** $x^2 + 1$ **27.** $2p^2 - 5p + 4 + \dfrac{6}{3p^2+1}$

29. $x^3 + 6x - 7$ **31.** $2x^2 + \dfrac{3}{5}x + \dfrac{1}{5}$ **33.** $(x^2 + x - 3)$ units **35.** 33

36. 33 **37.** They are the same. **38.** The answers should agree.

Section 12.8 (pages 903–904)

1. 6.1309×10^9; 5.8689×10^9 **3.** in scientific notation **5.** not in scientific notation; 5.6×10^6 **7.** not in scientific notation; 4×10^{-3} **9.** not in scientific notation; 8×10^1 **11.** A number is written in scientific notation if it is the product of a number whose absolute value is between 1 and 10 (inclusive of 1) and a power of 10. **13.** 5.876×10^9

15. 8.235×10^4 **17.** 7×10^{-6} **19.** -2.03×10^{-3} **21.** 750,000

23. 5,677,000,000,000 **25.** 1,000,000,000,000 **27.** -6.21

29. 0.00078 **31.** 0.000000005134 **33.** 6×10^{11}; 600,000,000,000

35. 1.5×10^7; 15,000,000 **37.** 8×10^{-3}; 0.008 **39.** 2.4×10^2; 240

41. 6.3×10^{-2}; 0.063 **43.** 6.426×10^4; 64,260 **45.** 3×10^{-4}; 0.0003 **47.** 4×10^1; 40 **49.** 1.3×10^{-5}; 0.000013 **51.** 5×10^2; 500 **53.** 2.6×10^{-3}; 0.0026 **55.** 7.205×10^{-6}; 0.000007205

57. 1.5×10^{17} mi **59.** \$3554

Chapter 12 Review Exercises (pages 909–912)

1. $22m^2$; degree 2; monomial **2.** $p^3 - p^2 + 4p + 2$; degree 3; none of these **3.** already in descending powers; degree 5; none of these **4.** $-8y^5 - 7y^4 + 9y$; degree 5; trinomial **5.** $-5a^3 + 4a^2$

6. $2r^3 - 3r^2 + 9r$ **7.** $11y^2 - 10y + 9$ **8.** $-13k^4 - 15k^2 - 4k - 6$

9. $10m^3 - 6m^2 - 3$ **10.** $-y^2 - 4y + 26$ **11.** $10p^2 - 3p - 11$

12. $7r^4 - 4r^3 - 1$ **13.** 4^{11} **14.** -5^{11} **15.** $-72x^7$ **16.** $10x^{14}$

17. 19^5x^5 **18.** -4^7y^7 **19.** $5p^4t^4$ **20.** $\dfrac{7^6}{5^6}$ **21.** $27x^6y^9$ **22.** t^{42}

23. $36x^{16}y^4z^{16}$ **24.** $\dfrac{8m^9n^3}{p^6}$ **25.** $125x^6$ **26.** The product rule for exponents does not apply here because we want the sum of 7^2 and 7^4, not their product. **27.** $10x^2 + 70x$ **28.** $-6p^5 + 15p^4$

29. $6r^3 + 8r^2 - 17r + 6$ **30.** $8y^3 + 27$ **31.** $5p^5 - 2p^4 - 3p^3 + 25p^2 + 15p$ **32.** $x^2 + 3x - 18$ **33.** $6k^2 - 9k - 6$

34. $12p^2 - 48pq + 21q^2$ **35.** $2m^4 + 5m^3 - 16m^2 - 28m + 9$

36. $a^2 + 8a + 16$ **37.** $9p^2 - 12p + 4$ **38.** $4r^2 + 20rs + 25s^2$

39. $r^3 + 6r^2 + 12r + 8$ **40.** $8x^3 - 12x^2 + 6x - 1$ **41.** $4z^2 - 49$

42. $36m^2 - 25$ **43.** $25a^2 - 36b^2$ **44.** $4x^4 - 25$ **45.** three; two

46. $(a+b)^2 = (a+b)(a+b) = a^2 + 2ab + b^2$. The term $2ab$ is not in

$a^2 + b^2$. **47.** 2 **48.** $\dfrac{1}{32}$ **49.** $\dfrac{25}{36}$ **50.** $-\dfrac{3}{16}$ **51.** 36 **52.** x^2 **53.** $\dfrac{1}{p^{12}}$

54. r^4 **55.** 2^8 **56.** $\dfrac{1}{9^6}$ **57.** 5^8 **58.** $\dfrac{1}{8^{12}}$ **59.** $\dfrac{1}{m^2}$ **60.** y^7 **61.** r^{13}

62. $25m^6$ **63.** $\dfrac{y^{12}}{8}$ **64.** $\dfrac{1}{a^3b^5}$ **65.** $72r^5$ **66.** $\dfrac{8n^{10}}{3m^{13}}$ **67.** $\dfrac{5y^2}{3}$

68. $-2x^2y$ **69.** $-y^3 + 2y - 3$ **70.** $p - 3 + \dfrac{5}{2p}$

71. $-x^9 + 2x^8 - 4x^3 + 7x$ **72.** $-2m^2n + mn^2 + \dfrac{6n^3}{5}$ **73.** $2r + 7$

74. $4m + 3 + \dfrac{5}{3m-5}$ **75.** $2a + 1 + \dfrac{-8a+12}{5a^2-3}$

76. $k^2 + 2k + 4 + \dfrac{-2k-12}{2k^2+1}$ **77.** 4.8×10^7 **78.** 2.8988×10^{10}

79. 6.5×10^{-5} **80.** 8.24×10^{-8} **81.** 24,000 **82.** 78,300,000

83. 0.000000897 **84.** 0.00000000000995 **85.** 8×10^2; 800

86. 4×10^6; 4,000,000 **87.** 2.5×10^{-2}; 0.025 **88.** 1×10^{-2}; 0.01

89. 2.796×10^{10} calculations; 1.6776×10^{12} calculations

90. about 3.3 **91.** 0 **92.** $\dfrac{243}{p^3}$ **93.** $\dfrac{1}{49}$ **94.** $49 - 28k + 4k^2$

95. $y^2 + 5y + 1$ **96.** $\dfrac{1296r^8s^4}{625}$ **97.** $-8m^7 - 10m^6 - 6m^5$ **98.** 32

99. $5xy^3 - \dfrac{8y^2}{5} + 3x^2y$ **100.** $\dfrac{r^2}{6}$ **101.** $8x^3 + 12x^2y + 6xy^2 + y^3$

102. $\dfrac{3}{4}$ **103.** $a^3 - 2a^2 - 7a + 2$ **104.** $8y^3 - 9y^2 + 5$

105. $10r^2 + 21r - 10$ **106.** $144a^2 - 1$ **107.** $2x^2 + x - 6$

108. $20x^4 + 8x^2;\ 25x^8 + 20x^6 + 4x^4$

Chapter 12 Test (pages 913–914)

1. $4t^4 + t^3 - 6t^2 - t$ **2.** $-2y^2 - 9y + 17$ **3.** $-12t^2 + 5t + 8$

4. -32 **5.** $\dfrac{216}{m^6}$ **6.** $-27x^5 + 18x^4 - 6x^3 + 3x^2$

7. $2r^3 + r^2 - 16r + 15$ **8.** $t^2 - 5t - 24$ **9.** $8x^2 + 2xy - 3y^2$

10. $25x^2 - 20xy + 4y^2$ **11.** $100v^2 - 9w^2$ **12.** $x^3 + 3x^2 + 3x + 1$

13. $12x + 36;\ 9x^2 + 54x + 81$ **14.** $\dfrac{1}{625}$ **15.** 2 **16.** $\dfrac{7}{12}$ **17.** 8^5

18. x^2y^6 **19.** $4y^2 - 3y + 2 + \dfrac{5}{y}$ **20.** $-3xy^2 + 2x^3y^2 + 4y^2$

21. $2x + 9$ **22.** $3x^2 + 6x + 11 + \dfrac{26}{x-2}$ **23. (a)** 3.44×10^{11}

(b) 5.57×10^{-6} **24. (a)** $29{,}600{,}000$ **(b)** 0.0000000607

25. 5.89×10^{15} mi

Chapter 13 Factoring and Applications

Section 13.1 (pages 923–924)

1. 4 **3.** 4 **5.** 6 **7.** 1 **9.** 8 **11.** $10x^3$ **13.** xy^2 **15.** 6 **17.** $3m^2$

19. $2z^4$ **21.** $2mn^4$ **23.** $y + 2$ **25.** $a - 2$ **27.** $2 + 3xy$

29. $x(x - 4)$ **31.** $3t(2t + 5)$ **33.** $\dfrac{1}{4}d(d - 3)$ **35.** $-6x^2(2x + 1)$

37. $5y^6(13y^4 + 7)$ **39.** no common factor (except 1) **41.** $8m^2n^2(n + 3)$

43. $-2x(2x^2 - 5x + 3)$ **45.** $13y^2(y^6 + 2y^2 - 3)$

47. $9qp^3(5q^3p^2 + 4p^3 + 9q)$ **49.** $(x + 2)(c + d)$

51. $(2a + b)(a^2 - b)$ **53.** $(p + 4)(q - 1)$ **55.** $(5 + n)(m + 4)$

57. $(2y - 7)(3x + 4)$ **59.** $(y + 3)(3x + 1)$ **61.** $(z + 2)(7z - a)$

63. $(3r + 2y)(6r - x)$ **65.** $(w + 1)(w^2 + 9)$ **67.** $(a + 2)(3a^2 - 2)$

69. $(4m - p^2)(4m^2 - p)$ **71.** $(y + 3)(y + x)$ **73.** $(z - 2)(2z - 3w)$

75. commutative property **76.** $2x(y - 4) - 3(y - 4)$

77. No, because it is not a product. It is the difference between
$2x(y - 4)$ and $3(y - 4)$. **78.** $(2x - 3)(y - 4)$; yes

Section 13.2 (pages 929–930)

1. a and b must have different signs. **3.** A prime polynomial is one that
cannot be factored using only integers in the factors. **5.** 1 and 12, -1 and
-12, 2 and 6, -2 and -6, 3 and 4, -3 and -4; The pair with a sum of 7 is
3 and 4. **7.** 1 and -24, -1 and 24, 2 and -12, -2 and 12, 3 and -8, -3
and 8, 4 and -6, -4 and 6; The pair with a sum of -5 is 3 and -8. **9.** C

11. $x + 11$ **13.** $x - 8$ **15.** $y - 5$ **17.** $x + 11$ **19.** $y - 9$

21. $(y + 8)(y + 1)$ **23.** $(b + 3)(b + 5)$ **25.** $(m + 5)(m - 4)$

27. $(x + 8)(x - 5)$ **29.** $(y - 5)(y - 3)$ **31.** $(z - 8)(z - 7)$

33. $(r - 6)(r + 5)$ **35.** $(a - 12)(a + 4)$ **37.** prime

39. $(r + 2a)(r + a)$ **41.** $(x + y)(x + 3y)$ **43.** $(t + 2z)(t - 3z)$

45. $(v - 5w)(v - 6w)$ **47.** $4(x + 5)(x - 2)$ **49.** $2t(t + 1)(t + 3)$

51. $-2x^4(x - 3)(x + 7)$ **53.** $a^3(a + 4b)(a - b)$

55. $mn(m - 6n)(m - 4n)$ **57.** The factored form $(2x + 4)(x - 3)$
is incorrect because $2x + 4$ has a common factor of 2, which must be
factored out for the trinomial to be *completely* factored.

Section 13.3 (pages 933–934)

1. $(m + 6)(m + 2)$ **3.** $(a + 5)(a - 2)$ **5.** $(2t + 1)(5t + 2)$

7. $(3z - 2)(5z - 3)$ **9.** $(2s - t)(4s + 3t)$ **11.** $(3a + 2b)(5a + 4b)$

13. B **15. (a)** 2; 12; 24; 11 **(b)** 3; 8 (Order is irrelevant.)

(c) $3m$; $8m$ **(d)** $2m^2 + 3m + 8m + 12$ **(e)** $(2m + 3)(m + 4)$

(f) $(2m + 3)(m + 4) = 2m^2 + 11m + 12$ **17.** $(2x + 1)(x + 3)$

19. $(4r - 3)(r + 1)$ **21.** $(4m + 1)(2m - 3)$ **23.** $(3m + 1)(7m + 2)$

25. $(2b + 1)(3b + 2)$ **27.** $(4y - 3)(3y - 1)$ **29.** $3(4x - 1)(2x - 3)$

31. $2m(m - 4)(m + 5)$ **33.** $-4z^3(z - 1)(8z + 3)$

35. $(3p + 4q)(4p - 3q)$ **37.** $(3a - 5b)(2a + b)$ **39.** $(5 - x)(1 - x)$

41. The student stopped too soon. He needs to factor out the common
factor $4x - 1$ to get $(4x - 1)(4x - 5)$ as the correct answer.

Section 13.4 (pages 939–940)

1. B **3.** A **5.** A **7.** $2a + 5b$ **9.** $x^2 + 3x - 4$; $x + 4$, $x - 1$, or
$x - 1$, $x + 4$ **11.** $2z^2 - 5z - 3$; $3z + 1$, $z - 3$, or $z - 3$, $3z + 1$

13. The binomial $2x - 6$ cannot be a factor because it has a common
factor of 2, but the polynomial does not. **15.** $(3a + 7)(a + 1)$

17. $(2y + 3)(y + 2)$ **19.** $(3m - 1)(5m + 2)$ **21.** $(3s - 1)(4s + 5)$

23. $(5m - 4)(2m - 3)$ **25.** $(4w - 1)(2w - 3)$

27. $(4y + 1)(5y - 11)$ **29.** prime **31.** $2(5x + 3)(2x + 1)$

33. $-q(5m + 2)(8m - 3)$ **35.** $3n^2(5n - 3)(n - 2)$

37. $-y^2(5x - 4)(3x + 1)$ **39.** $(5a + 3b)(a - 2b)$

41. $(4s + 5t)(3s - t)$ **43.** $m^4n(3m + 2n)(2m + n)$

45. $-1(x + 7)(x - 3)$ **47.** $-1(3x + 4)(x - 1)$

49. $-1(a + 2b)(2a + b)$ **51.** $5 \cdot 7$ **52.** $(-5)(-7)$

53. The product of $3x - 4$ and $2x - 1$ is $6x^2 - 11x + 4$.

54. The product of $4 - 3x$ and $1 - 2x$ is $6x^2 - 11x + 4$.

55. The factors in Exercise 53 are the opposites of the factors
in Exercise 54. **56.** $(3 - 7t)(5 - 2t)$

Section 13.5 (pages 947–950)

1. 1; 4; 9; 16; 25; 36; 49; 64; 81; 100; 121; 144; 169; 196; 225; 256;

289; 324; 361; 400 **3.** 2 **5.** $(y + 5)(y - 5)$ **7.** $\left(p + \dfrac{1}{3}\right)\left(p - \dfrac{1}{3}\right)$

9. prime **11.** $(3r + 2)(3r - 2)$ **13.** $\left(2m + \dfrac{3}{5}\right)\left(2m - \dfrac{3}{5}\right)$

15. $4(3x + 2)(3x - 2)$ **17.** $(14p + 15)(14p - 15)$

19. $(4r + 5a)(4r - 5a)$ **21.** prime **23.** $(p^2 + 7)(p^2 - 7)$

25. $(x^2 + 1)(x + 1)(x - 1)$ **27.** $(p^2 + 16)(p + 4)(p - 4)$

29. The teacher was justified, because it was not factored *completely*;
$x^2 - 9$ can be factored as $(x + 3)(x - 3)$. The complete factored form
is $(x^2 + 9)(x + 3)(x - 3)$. **31.** No, it is not a perfect square since the
middle term would have to be $30y$. **33.** $(w + 1)^2$ **35.** $(x - 4)^2$

37. $\left(t + \dfrac{1}{2}\right)^2$ **39.** $(x - 0.5)^2$ **41.** $2(x + 6)^2$ **43.** $(4x - 5)^2$

45. $(7x - 2y)^2$ **47.** $(8x + 3y)^2$ **49.** $-2h(5h - 2y)^2$

51. 1; 8; 27; 64; 125; 216; 343; 512; 729; 1000 **53.** C, D

55. $(a + 1)(a^2 - a + 1)$ **57.** $(a - 1)(a^2 + a + 1)$

59. $(p + q)(p^2 - pq + q^2)$ **61.** $(y - 6)(y^2 + 6y + 36)$

63. $(k + 10)(k^2 - 10k + 100)$ **65.** $(3x - 1)(9x^2 + 3x + 1)$

67. $(5a + 2)(25a^2 - 10a + 4)$ **69.** $(y - 2x)(y^2 + 2xy + 4x^2)$

71. $(3a - 4b)(9a^2 + 12ab + 16b^2)$

73. $(2p + 9q)(4p^2 - 18pq + 81q^2)$ **75.** $2(2t - 1)(4t^2 + 2t + 1)$
77. $5(2w + 3)(4w^2 - 6w + 9)$ **79.** $(x + y^2)(x^2 - xy^2 + y^4)$
81. $(5k - 2m^3)(25k^2 + 10km^3 + 4m^6)$ **83.** $(x^3 - 1)(x^3 + 1)$
84. $(x - 1)(x^2 + x + 1)(x + 1)(x^2 - x + 1)$
85. $(x^2 - 1)(x^4 + x^2 + 1)$ **86.** $(x - 1)(x + 1)(x^4 + x^2 + 1)$
87. The result in Exercise 84 is completely factored.   **88.** Show that
$x^4 + x^2 + 1 = (x^2 + x + 1)(x^2 - x + 1)$.   **89.** difference of squares
90. $(x - 3)(x^2 + 3x + 9)(x + 3)(x^2 - 3x + 9)$

Section 13.6 (pages 953–954)

1. F   **3.** A   **5.** D   **7.** C   **9.** H   **11.** $(10a + 3b)(10a - 3b)$
13. $6p^3(3p^2 - 4 + 2p^3)$ **15.** $(x + 7)(x - 5)$ **17.** prime
19. $(6b + 1)(b - 3)$ **21.** $3mn(3m + 2n)(2m - n)$
23. $(m + 4)(m^2 - 6)$ **25.** $(2k + 7r)^2$ **27.** $(m - 2)(n + 5)$
29. $(x + 3)^2(x - 3)$ **31.** prime   **33.** $(2p - 5)(4p^2 + 10p + 25)$
35. $(x^2 + 25)(x + 5)(x - 5)$ **37.** $(a + 6)(b + c)$
39. $(7z + 2k)(2z - k)$ **41.** $16(4b + 5c)(4b - 5c)$
43. $8(5z + 4)(25z^2 - 20z + 16)$ **45.** $(5r - s)(2r + 5s)$
47. $8x^2(4 + 2x - 3x^3)$ **49.** $(2x - 5q)(7x + 5q)$
51. $(p + 2)(4 + m)$ **53.** $2(5p + 9)(5p - 9)$ **55.** $(4a + b)^2$

Section 13.7 (pages 961–962)

1. $\{-5, 2\}$ **3.** $\left\{3, \frac{7}{2}\right\}$ **5.** $\left\{-\frac{5}{6}, 0\right\}$ **7.** $\left\{0, \frac{4}{3}\right\}$ **9.** $\left\{-\frac{1}{2}, \frac{1}{6}\right\}$
11. $\{9\}$ **13.** Set each *variable* factor equal to 0, to get $2x = 0$ or
$3x - 4 = 0$. The solution set is $\left\{0, \frac{4}{3}\right\}$.   **15.** $\{-2, -1\}$ **17.** $\{1, 2\}$
19. $\{-8, 3\}$ **21.** $\{-1, 3\}$ **23.** $\{-2, -1\}$ **25.** $\{-4\}$ **27.** $\left\{-2, \frac{1}{3}\right\}$
29. $\left\{-\frac{4}{3}, \frac{1}{2}\right\}$ **31.** $\left\{-\frac{2}{3}\right\}$ **33.** $\{-3, 3\}$ **35.** $\left\{-\frac{7}{4}, \frac{7}{4}\right\}$ **37.** $\{-11, 11\}$
39. $\{0, 7\}$ **41.** $\left\{0, \frac{1}{2}\right\}$ **43.** $\{2, 5\}$ **45.** $\left\{-4, \frac{1}{2}\right\}$ **47.** $\left\{-12, \frac{11}{2}\right\}$
49. $\{-2, 0, 2\}$ **51.** $\left\{-\frac{7}{3}, 0, \frac{7}{3}\right\}$ **53.** $\left\{-\frac{5}{2}, \frac{1}{3}, 5\right\}$ **55.** $\left\{-\frac{7}{2}, -3, 1\right\}$
57. (a) 64; 144; 4; 6   **(b)** No time has elapsed, so the object hasn't fallen
(been released) yet.

Section 13.8 (pages 969–974)

1. Read; variable; equation; Solve; answer; Check; original
3. *Step 3:* $45 = (2x + 1)(x + 1)$; *Step 4:* $x = 4$ or $x = -\frac{11}{2}$;
Step 5: base: 9 units; height: 5 units; *Step 6:* $9 \cdot 5 = 45$
5. *Step 3:* $192 = 4x(x + 2)$; *Step 4:* $x = 6$ or $x = -8$; *Step 5:* length:
8 units; width: 6 units; *Step 6:* $8 \cdot 6 \cdot 4 = 192$ **7.** length: 14 cm; width:
12 cm   **9.** length: 15 in.; width: 12 in.   **11.** height: 13 in.; width: 10 in.
13. mirror: 7 ft; painting: 9 ft   **15.** 20, 21   **17.** $-3, -2$ or 4, 5
19. $-3, -1$ or 7, 9   **21.** $-2, 0, 2$ or 6, 8, 10   **23.** 12 cm   **25.** 12 mi
27. 8 ft   **29. (a)** 1 sec   **(b)** $\frac{1}{2}$ sec and $1\frac{1}{2}$ sec   **(c)** 3 sec
(d) The negative solution, -1, does not make sense since t represents time,
which cannot be negative.   **31. (a)** 46 million; The result using the
model is a little more than 44 million, the actual number for 1996.
(b) 14   **(c)** 184 million; The result is a little more than 182 million, the
actual number for 2004.   **(d)** 318 million   **32.** $58.6 billion; 16%
33. 2003: $522.1 billion; 2004: $610.8 billion; 2005: $699.5 billion.

34. The answer using the linear equation is close to the actual data for 2004,
but not for the other years.   **35.** 2003: $503.9 billion; 2004: $601.8 billion;
2005: $718.2 billion   **36.** The answers in Exercise 35 are fairly close
to the actual data. The quadratic equation models the data better.
37. (1, 365.1), (2, 423.7), (3, 496.9), (4, 612.1), (5, 714.4)
38.   **U.S. TRADE DEFICIT**
(Goods and Services)

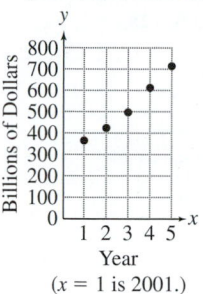

Year
($x = 1$ is 2001.)
39. $853.1 billion   **40. (a)** The actual deficit is quite a bit less than the
estimate.   **(b)** No, data for later years might not follow the same pattern.

Chapter 13 Review Exercises (pages 979–982)

1. $15(t + 3)$ **2.** $30z(2z^2 + 1)$ **3.** $11x^2(4x + 5)$
4. $50m^2n^2(2n - mn^2 + 3)$ **5.** $(x - 4)(2y + 3)$ **6.** $(2y + 3)(3y + 2x)$
7. $(x + 3)(x + 7)$ **8.** $(y - 5)(y - 8)$ **9.** prime
10. $(r - 8)(r + 7)$ **11.** $(r + 8s)(r - 12s)$ **12.** $(p + 12q)(p - 10q)$
13. $-8p(p + 2)(p - 5)$ **14.** $3x^2(x + 2)(x + 8)$
15. $p^5(p - 2q)(p + q)$ **16.** $(3r - 1)(r + 4)$ **17.** $(3r + 2)(2r - 3)$
18. prime   **19.** $4x^3(3x - 1)(2x - 1)$ **20.** $-mn(3m + 5)(m - 8)$
21. $(2a - 5b)(7a + 4b)$ **22.** B   **23.** D   **24.** $(n + 8)(n - 8)$
25. $(5b + 11)(5b - 11)$ **26.** $36(2p + q)(2p - q)$ **27.** prime
28. $\left(x + \frac{7}{10}\right)\left(x - \frac{7}{10}\right)$ **29.** $(r - 6)^2$ **30.** $(4m + 5n)^2$
31. $(5x - 1)(25x^2 + 5x + 1)$ **32.** $(10p + 3)(100p^2 - 30p + 9)$
33. $\left\{-\frac{3}{4}, 1\right\}$ **34.** $\left\{0, \frac{5}{2}\right\}$ **35.** $\{-3, -1\}$ **36.** $\{3, 5\}$ **37.** $\left\{-\frac{4}{3}, 5\right\}$
38. $\left\{-\frac{8}{9}, \frac{8}{9}\right\}$ **39.** $\{0, 8\}$ **40.** $\{-1, 6\}$ **41.** $\{6\}$ **42.** $\left\{-\frac{2}{5}, -2, -1\right\}$
43. $\left\{-\frac{3}{7}, 0, \frac{3}{7}\right\}$ **44.** $\left\{\frac{9}{5}\right\}$ **45.** length: 10 ft; width: 4 ft   **46.** 5 ft
47. 6, 7 or $-5, -4$ **48.** 26 mi   **49.** 112 ft   **50.** 192 ft   **51.** 256 ft
52. after 8 sec   **53. (a)** $4.49 billion; The answer using the model is
very close.   **(b)** $7.19 billion   **54.** D   **55.** The factor $2x + 8$ has a
common factor of 2. The complete factored form is $2(x + 4)(3x - 4)$.
56. $(z - x)(z - 10x)$ **57.** $(3k + 5)(k + 2)$ **58.** $(3m + 4p)(5m - 4)$
59. $(y^2 + 25)(y + 5)(y - 5)$ **60.** $3m(2m + 3)(m - 5)$
61. $8abc(3b^2c - 7ac^2 + 9ab)$ **62.** prime
63. $6xyz(2xz^2 + 2y - 5x^2yz^3)$ **64.** $2a^3(a + 2)(a - 6)$
65. $-(2r + 3q)(6r - 5q)$ **66.** $8(z + 2y)(z^2 - 2zy + 4y^2)$
67. $(7t + 4)^2$ **68.** $\{0, 7\}$ **69.** $\{-5, 2\}$ **70.** $\left\{-\frac{2}{5}\right\}$
71. width: 10 m; length: 17 m   **72.** 6 m

Chapter 13 Test (pages 983–984)

1. D   **2.** $6x(2x - 5)$ **3.** $m^2n(2mn + 3m - 5n)$ **4.** $(2x + y)(a - b)$
5. $(x - 7)(x - 2)$ **6.** $(2x + 3)(x - 1)$ **7.** $(3x + 1)(2x - 7)$

8. $3(x+1)(x-5)$ **9.** $(5z-1)(2z-3)$ **10.** prime **11.** prime
12. $(y+7)(y-7)$ **13.** $(9a+11b)(9a-11b)$ **14.** $(x+8)^2$
15. $(2x-7y)^2$ **16.** $-2(x+1)^2$ **17.** $(x-8)(x^2+8x+64)$
18. $8(k+2)(k^2-2k+4)$ **19.** $4t(t+4)^2$ **20.** $(x^2+9)(x+3)(x-3)$

21. $\{-3,9\}$ **22.** $\left\{\dfrac{1}{2},6\right\}$ **23.** $\left\{-\dfrac{2}{5},\dfrac{2}{5}\right\}$ **24.** $\{10\}$

25. $\{0,3\}$ **26.** $\left\{-8,-\dfrac{5}{2},\dfrac{1}{3}\right\}$ **27.** 6 ft by 9 ft **28.** $-2,-1$

29. 17 ft **30.** 49 million

Chapter 14 Rational Expressions and Applications

Section 14.1 (pages 993–994)

1. (a) $3;-5$ **(b)** $q;-1$ **3.** A rational expression is a quotient of
polynomials, such as $\dfrac{x+3}{x^2-4}$. **5.** $y\neq 0$ **7.** $x\neq -6$ **9.** $x\neq \dfrac{5}{3}$
11. $m\neq -3, m\neq 2$ **13.** never undefined **15.** never undefined
17. (a) 1 **(b)** $\dfrac{17}{12}$ **19. (a)** 0 **(b)** -1 **21. (a)** $\dfrac{9}{5}$ **(b)** undefined
23. (a) $\dfrac{2}{7}$ **(b)** $\dfrac{13}{3}$ **25.** $3r^2$ **27.** $\dfrac{2}{5}$ **29.** $\dfrac{x-1}{x+1}$ **31.** $\dfrac{7}{5}$ **33.** $m-n$
35. $\dfrac{3(2m+1)}{4}$ **37.** $\dfrac{3m}{5}$ **39.** $\dfrac{3r-2s}{3}$ **41.** $\dfrac{x+1}{x-1}$ **43.** $\dfrac{z-3}{z+5}$
45. $\dfrac{a+b}{a-b}$ **47.** -1 **49.** $-(m+1)$ **51.** -1

Answers may vary in Exercises 53–57.

53. $\dfrac{-(x+4)}{x-3},\dfrac{-x-4}{x-3},\dfrac{x+4}{-(x-3)},\dfrac{x+4}{-x+3}$

55. $\dfrac{-(2x-3)}{x+3},\dfrac{-2x+3}{x+3},\dfrac{2x-3}{-(x+3)},\dfrac{2x-3}{-x-3}$

57. $\dfrac{-(3x-1)}{5x-6},\dfrac{-3x+1}{5x-6},\dfrac{3x-1}{-(5x-6)},\dfrac{3x-1}{-5x+6}$ **59.** x^2+3

Section 14.2 (pages 999–1000)

1. (a) B **(b)** D **(c)** C **(d)** A **3.** $\dfrac{4m}{3}$ **5.** $\dfrac{40y^2}{3}$ **7.** $\dfrac{2}{c+d}$
9. $\dfrac{16q}{3p^3}$ **11.** $\dfrac{7}{r^2+rp}$ **13.** $\dfrac{z^2-9}{z^2+7z+12}$ **15.** 5 **17.** $-\dfrac{3}{2t^4}$ **19.** $\dfrac{1}{4}$
21. To multiply two rational expressions, multiply the numerators
and multiply the denominators. Write the answer in lowest terms.
23. $\dfrac{10}{9}$ **25.** $-\dfrac{3}{4}$ **27.** -1 **29.** $\dfrac{9(m-2)}{-(m+4)}$, or $\dfrac{-9(m-2)}{m+4}$ **31.** $\dfrac{p+4}{p+2}$
33. $\dfrac{(k-1)^2}{(k+1)(2k-1)}$ **35.** $\dfrac{4k-1}{3k-2}$ **37.** $\dfrac{m+4p}{m+p}$ **39.** $\dfrac{10}{x+10}$ **41.** $\dfrac{5xy^2}{4q}$

Section 14.3 (pages 1005–1006)

1. C **3.** C **5.** 30 **7.** x^7 **9.** $72q$ **11.** $84r^5$ **13.** $2^3\cdot 3\cdot 5$
15. The least common denominator is their product. **17.** $28m^2(3m-5)$
19. $30(b-2)$ **21.** $c-d$ or $d-c$ **23.** $k(k+5)(k-2)$
25. $(p+3)(p+5)(p-6)$ **27.** $\dfrac{20}{55}$ **29.** $\dfrac{-45}{9k}$ **31.** $\dfrac{26y^2}{80y^3}$
33. $\dfrac{35t^2r^3}{42r^4}$ **35.** $\dfrac{20}{8(m+3)}$ **37.** $\dfrac{8t}{12-6t}$ **39.** $\dfrac{14(z-2)}{z(z-3)(z-2)}$
41. $\dfrac{2(b-1)(b+2)}{b^3+3b^2+2b}$

Section 14.4 (pages 1013–1016)

1. E **3.** C **5.** B **7.** G **9.** $\dfrac{11}{m}$ **11.** b **13.** $\dfrac{4}{y+4}$ **15.** $\dfrac{m-1}{m+1}$
17. x **19.** $y-6$ **21.** Combine the numerators and keep the same
denominator. For example, $\dfrac{3x+2}{x-6}+\dfrac{-2x-8}{x-6}=\dfrac{x-6}{x-6}$. Then write in
lowest terms: $\dfrac{x-6}{x-6}=1$. **23.** $\dfrac{3z+5}{15}$ **25.** $\dfrac{10-7r}{14}$ **27.** $\dfrac{-3x-2}{4x}$
29. $\dfrac{57}{20x}$ **31.** $\dfrac{x+1}{2}$ **33.** $\dfrac{5x+9}{6x}$ **35.** $\dfrac{3x+3}{x(x+3)}$ **37.** $\dfrac{-k-10}{k(k+5)}$
39. $\dfrac{x+4}{x+2}$ **41.** $\dfrac{x^2+6x-8}{(x-2)(x+2)}$ **43.** $\dfrac{3}{t}$ **45.** $m-2$ or $2-m$
47. $\dfrac{-2}{x-5}$, or $\dfrac{2}{5-x}$ **49.** -4 **51.** $\dfrac{-5}{x-y^2}$, or $\dfrac{5}{y^2-x}$
53. $\dfrac{x+y}{5x-3y}$, or $\dfrac{-x-y}{3y-5x}$ **55.** $\dfrac{-6}{4p-5}$, or $\dfrac{6}{5-4p}$ **57.** $\dfrac{-(m+n)}{2(m-n)}$
59. $\dfrac{-x^2+6x+11}{(x+3)(x-3)(x+1)}$ **61.** $\dfrac{-5q^2-13q+7}{(3q-2)(q+4)(2q-3)}$
63. $\dfrac{9r+2}{r(r+2)(r-1)}$ **65.** $\dfrac{2x^2+6xy+8y^2}{(x+y)(x+y)(x+3y)}$, or $\dfrac{2x^2+6xy+8y^2}{(x+y)^2(x+3y)}$
67. $\dfrac{15r^2+10ry-y^2}{(3r+2y)(6r-y)(6r+y)}$ **69. (a)** $\dfrac{9k^2+6k+26}{5(3k+1)}$ **(b)** $\dfrac{1}{4}$

Section 14.5 (pages 1023–1024)

1. (a) $6;\dfrac{1}{6}$ **(b)** $12;\dfrac{3}{4}$ **(c)** $\dfrac{1}{6}\div\dfrac{3}{4}$ **(d)** $\dfrac{2}{9}$ **3.** -6 **5.** $\dfrac{1}{pq}$ **7.** $\dfrac{1}{xy}$
9. $\dfrac{2a^2b}{3}$ **11.** $\dfrac{m(m+2)}{3(m-4)}$ **13.** $\dfrac{2}{x}$ **15.** $\dfrac{8}{x}$ **17.** $\dfrac{a^2-5}{a^2+1}$ **19.** $\dfrac{3(p+2)}{2(2p+3)}$
21. $\dfrac{40-12p}{85p}$ **23.** $\dfrac{t(t-2)}{4}$ **25.** $\dfrac{-k}{2+k}$ **27.** $\dfrac{2x-7}{3x+1}$
29. $\dfrac{3m(m-3)}{(m-1)(m-8)}$ **31.** $\dfrac{6}{5}$

Section 14.6 (pages 1033–1036)

1. expression; $\dfrac{43}{40}x$ **3.** equation; $\left\{\dfrac{40}{43}\right\}$ **5.** expression; $-\dfrac{1}{10}y$
7. When solving an equation, we multiply each side by the LCD, which
eliminates all denominators. When adding or subtracting fractions, we
multiply by a form of 1 so that the result has the LCD in the denominator.
The denominators are not eliminated. **9.** $\{-6\}$ **11.** $\{-15\}$ **13.** $\{7\}$
15. $\{-15\}$ **17.** $\{-5\}$ **19.** $\{-6\}$ **21.** $\{5\}$ **23.** $\{12\}$ **25.** $\{2\}$
27. 0 and 4 **29.** $\left\{\dfrac{20}{9}\right\}$ **31.** $\emptyset$ **33.** $\{3\}$ **35.** $\{3\}$ **37.** $\{-2,12\}$
39. $\left\{-\dfrac{1}{5},3\right\}$ **41.** $\left\{-\dfrac{3}{5},3\right\}$ **43.** $\{-4\}$ **45.** $\emptyset$ **47.** $\{-1\}$
49. $\{-6\}$ **51.** $\left\{-6,\dfrac{1}{2}\right\}$ **53.** Transform the equation so that the terms
with k are on one side and the remaining term is on the other.
55. $F=\dfrac{ma}{k}$ **57.** $a=\dfrac{kF}{m}$ **59.** $y=mx+b$ **61.** $R=\dfrac{E-Ir}{I}$, or
$R=\dfrac{E}{I}-r$ **63.** $b=\dfrac{2A-hB}{h}$, or $b=\dfrac{2A}{h}-B$ **65.** $a=\dfrac{2S-dnL}{dn}$,
or $a=\dfrac{2S}{dn}-L$ **67.** $t=\dfrac{rs}{rs-2s-3r}$, or $t=\dfrac{-rs}{-rs+2s+3r}$
69. $c=\dfrac{ab}{b-a-2ab}$, or $c=\dfrac{-ab}{-b+a+2ab}$

71. $z = \dfrac{3y}{5 - 9xy}$, or $z = \dfrac{-3y}{9xy - 5}$

Section 14.7 (pages 1045–1048)

1. (a) the amount **(b)** $5 + x$ **(c)** $\dfrac{5 + x}{6} = \dfrac{13}{3}$ **3.** $\dfrac{9}{5}$ **5.** $\dfrac{2}{6}$ **7.** -6

9. 36 **11.** 0.032 mi per min **13.** 3.348 hr **15.** 6.530 m per sec

17. $\dfrac{8}{4 - x} = \dfrac{24}{4 + x}$ **19.** into a headwind: $(m - 5)$ mph; with a tailwind:

$(m + 5)$ mph **21.** 8 mph **23.** 32 mph **25.** 3 mph **27.** $\dfrac{1}{2}t + \dfrac{1}{3}t = 1$, or

$\dfrac{1}{2} + \dfrac{1}{3} = \dfrac{1}{t}$ **29.** $2\dfrac{2}{5}$ hr **31.** $4\dfrac{8}{19}$ hr **33.** 10 hr **35.** 36 hr **37.** $8\dfrac{1}{4}$ hr

Section 14.8 (pages 1051–1054)

1. (a) increases **(b)** decreases **3.** 15 **5.** 300 **7.** 4 **9.** 6

11. 15 in.² **13.** $42\dfrac{2}{3}$ in. **15.** 15 ft **17.** 20 lb per ft² **19.** 25 kg per hr

21. direct **23.** inverse **25.** inverse **27.** direct **29.** 8 **31.** 2

33. $x = 1; y = 4$ **35.** 80 ft

Chapter 14 Review Exercises (pages 1061–1064)

1. $x \neq 3$ **2.** $x \neq 0$ **3.** $m \neq -1, m \neq 3$ **4.** $k \neq -5, k \neq -\dfrac{2}{3}$

5. (a) $-\dfrac{4}{7}$ **(b)** -16 **6. (a)** $\dfrac{11}{8}$ **(b)** $\dfrac{13}{22}$ **7. (a)** undefined **(b)** 1

8. (a) undefined **(b)** $\dfrac{1}{2}$ **9.** $\dfrac{b}{3a}$ **10.** -1 **11.** $\dfrac{-(2x + 3)}{2}$

12. $\dfrac{2p + 5q}{5p + q}$

Answers may vary in Exercises 13 and 14.

13. $\dfrac{-(4x - 9)}{2x + 3}, \dfrac{-4x + 9}{2x + 3}, \dfrac{4x - 9}{-(2x + 3)}, \dfrac{4x - 9}{-2x - 3}$

14. $\dfrac{-(8 - 3x)}{3 - 6x}, \dfrac{-8 + 3x}{3 - 6x}, \dfrac{8 - 3x}{-(3 - 6x)}, \dfrac{8 - 3x}{-3 + 6x}$ **15.** 2 **16.** $\dfrac{2}{3m^6}$

17. $\dfrac{5}{8}$ **18.** $\dfrac{r + 4}{3}$ **19.** $\dfrac{3}{2}$ **20.** $\dfrac{y - 2}{y - 3}$ **21.** $\dfrac{p + 5}{p + 1}$ **22.** $\dfrac{3z + 1}{z + 3}$

23. 96 **24.** $108y^4$ **25.** $m(m + 2)(m + 5)$ **26.** $(x + 3)(x + 1)(x + 4)$

27. $\dfrac{35}{56}$ **28.** $\dfrac{40}{4k}$ **29.** $\dfrac{15a}{10a^4}$ **30.** $\dfrac{-54}{18 - 6x}$ **31.** $\dfrac{15y}{50 - 10y}$

32. $\dfrac{4b(b + 2)}{(b + 3)(b - 1)(b + 2)}$ **33.** $\dfrac{15}{x}$ **34.** $-\dfrac{2}{p}$ **35.** $\dfrac{4k - 45}{k(k - 5)}$

36. $\dfrac{28 + 11y}{y(7 + y)}$ **37.** $\dfrac{-2 - 3m}{6}$ **38.** $\dfrac{3(16 - x)}{4x^2}$ **39.** $\dfrac{7a + 6b}{(a - 2b)(a + 2b)}$

40. $\dfrac{-k^2 - 6k + 3}{3(k + 3)(k - 3)}$ **41.** $\dfrac{5z - 16}{z(z + 6)(z - 2)}$ **42.** $\dfrac{-13p + 33}{p(p - 2)(p - 3)}$

43. $\dfrac{a}{b}$ **44.** $\dfrac{4(y - 3)}{y + 3}$ **45.** $\dfrac{6(3m + 2)}{2m - 5}$ **46.** $\dfrac{(q - p)^2}{pq}$ **47.** $\dfrac{xw + 1}{xw - 1}$

48. $\dfrac{1 - r - t}{1 + r + t}$ **49.** $\left\{\dfrac{35}{6}\right\}$ **50.** $\{-16\}$ **51.** $\{-4\}$ **52.** $\varnothing$ **53.** $\{3\}$

54. $t = \dfrac{Ry}{m}$ **55.** $y = \dfrac{4x + 5}{3}$ **56.** $t = \dfrac{rs}{s - r}$ **57.** $\dfrac{20}{15}$ **58.** $\dfrac{3}{18}$

59. 1.766 hr **60.** 800.641 m per min **61.** $7\dfrac{1}{2}$ min **62.** $3\dfrac{1}{13}$ hr **63.** $\dfrac{36}{5}$

64. 4 cm **65.** $\dfrac{m + 7}{(m - 1)(m + 1)}$ **66.** $8p^2$ **67.** $\dfrac{1}{6}$ **68.** 3

69. $\dfrac{z + 7}{(z + 1)(z - 1)^2}$ **70.** $v = at + w$ **71.** $\{-2, 3\}$ **72.** $\{2\}$

73. 150 km per hr **74.** 10 hr **75.** 4 **76.** inverse

Chapter 14 Test (pages 1065–1066)

1. $x \neq -2, x \neq 4$ **2. (a)** $\dfrac{11}{6}$ **(b)** undefined **3.** (Answers may vary.)

$\dfrac{-(6x - 5)}{2x + 3}, \dfrac{-6x + 5}{2x + 3}, \dfrac{6x - 5}{-(2x + 3)}, \dfrac{6x - 5}{-2x - 3}$ **4.** $-3x^2y^3$ **5.** $\dfrac{3a + 2}{a - 1}$

6. $\dfrac{25}{27}$ **7.** $\dfrac{3k - 2}{3k + 2}$ **8.** $\dfrac{a - 1}{a + 4}$ **9.** $150p^5$ **10.** $(2r + 3)(r + 2)(r - 5)$

11. $\dfrac{240p^2}{64p^3}$ **12.** $\dfrac{21}{42m - 84}$ **13.** 2 **14.** $\dfrac{-14}{5(y + 2)}$

15. $\dfrac{x^2 + x + 1}{3 - x}$, or $\dfrac{-x^2 - x - 1}{x - 3}$ **16.** $\dfrac{-m^2 + 7m + 2}{(2m + 1)(m - 5)(m - 1)}$

17. $\dfrac{2k}{3p}$ **18.** $\dfrac{-2 - x}{4 + x}$ **19.** $\left\{-\dfrac{1}{2}, 5\right\}$ **20.** $\left\{-\dfrac{1}{2}\right\}$

21. $D = \dfrac{dF - k}{F}$, or $D = d - \dfrac{k}{F}$ **22.** -4 **23.** 3 mph **24.** $2\dfrac{2}{9}$ hr

25. 27 **26.** 27 days

Chapter 15 Systems of Linear Equations and Inequalities

Section 15.1 (pages 1073–1076)

1. B, because the ordered pair must be in quadrant II. **3.** There is no way that the sum of two numbers can be both 2 and 4 at the same time.
5. no **7.** yes **9.** yes **11.** no
We show the graphs here only for Exercises 13–17.
13. $\{(4, 2)\}$ **15.** $\{(0, 4)\}$

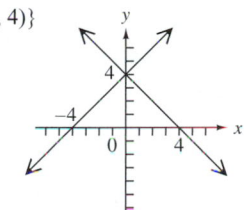

17. $\{(4, -1)\}$

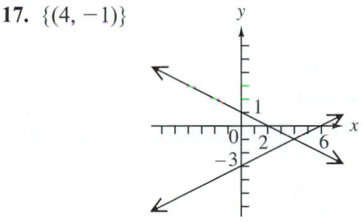

19. $\{(1, 3)\}$ **21.** $\{(0, 2)\}$ **23.** $\varnothing$ (inconsistent system)
25. $\{(x, y) \mid 2x - y = 4\}$ (dependent equations) **27.** $\{(4, -3)\}$
29. $\varnothing$ (inconsistent system) **31. (a)** neither **(b)** intersecting lines
(c) one solution **33. (a)** dependent **(b)** one line **(c)** infinite number of
solutions **35. (a)** inconsistent **(b)** parallel lines **(c)** no solution
37. 1980–2000 **39. (a)** 1997–2002 **(b)** 2001 **(c)** 2002 **(d)** (2004, 70)
(The y-value is approximate.) **(e)** During the period 1997–2004, debit card
use went from least popular to most popular of the three methods depicted.

Section 15.2 (pages 1083–1084)

1. The y-value must also be determined. The solution set is $\{(3, 0)\}$.
3. $\{(3, 9)\}$ **5.** $\{(7, 3)\}$ **7.** $\{(-2, 4)\}$ **9.** $\{(-4, 8)\}$ **11.** $\{(3, -2)\}$

13. $\{(x, y) \mid 3x - y = 5\}$ **15.** $\left\{\left(\dfrac{1}{3}, -\dfrac{1}{2}\right)\right\}$ **17.** $\varnothing$

19. $\{(x, y) \mid 3x - 4y = 2\}$ **21.** $\{(2, -3)\}$ **23.** $\{(10, -12)\}$

25. $\{(-4, 2)\}$ **27.** $\{(7, -3)\}$ **29.** $\{(20, 30)\}$ **30.** To find the total cost, multiply the number of bicycles (x) by the cost per bicycle (400 dollars) and add the fixed cost (5000 dollars). Thus, $y_1 = 400x + 5000$ gives this total cost (in dollars). **31.** $y_2 = 600x$ **32.** $y_1 = 400x + 5000, y_2 = 600x$; solution set: $\{(25, 15,000)\}$ **33.** 25; 15,000; 15,000

Section 15.3 (pages 1089–1092)

1. true **3.** true **5.** $\{(-1, 3)\}$ **7.** $\{(-1, -3)\}$ **9.** $\{(-2, 3)\}$
11. $\left\{\left(\frac{1}{2}, 4\right)\right\}$ **13.** $\{(3, -6)\}$ **15.** $\{(7, 4)\}$ **17.** $\{(0, 4)\}$
19. $\{(-4, 0)\}$ **21.** $\{(0, 0)\}$ **23.** $\emptyset$ **25.** $\{(x, y) \mid x - 3y = -4\}$
27. $\{(2, 9)\}$ **29.** $\{(-6, 5)\}$ **31.** $\left\{\left(-\frac{6}{5}, \frac{4}{5}\right)\right\}$ **33.** $\left\{\left(\frac{1}{8}, -\frac{5}{6}\right)\right\}$
35. $\{(11, 15)\}$ **37.** $\emptyset$ **39.** $\{(x, y) \mid 2x + y = 0\}$ **41.** $1339 = 1996a + b$
42. $1536 = 2004a + b$ **43.** $1996a + b = 1339, 2004a + b = 1536$;
solution set: $\{(24.625, -47,812.5)\}$ **44.** $y = 24.625x - 47,812.5$
45. 1486.8 (million); This is quite a bit less than the actual figure.
46. Since the data do not lie in a perfectly straight line, the quantity obtained from an equation determined in this way will probably be "off" a bit. We cannot put too much faith in models such as this one, because not all sets of data points are linear in nature.

Section 15.4 (pages 1099–1104)

1. D **3.** B **5.** D **7.** C **9.** the second number; $x - y = 48$; The two numbers are 73 and 25. **11.** *The Phantom of the Opera:* 8197; *Cats:* 7485
13. *Spider-Man 3:* $336.5 million; *Shrek the Third:* $322.7 million
15. Terminal Tower: 708 ft; Key Tower: 950 ft **17. (a)** 45 units
(b) Do not produce; the product will lead to a loss. **19.** 46 ones; 28 tens **21.** 2 DVDs of *Night at the Museum* and 5 Linkin Park CDs
23. $2500 at 4%; $5000 at 5% **25.** The Police: $107; Van Halen: $115
27. 80 L of 40% solution; 40 L of 70% solution **29.** 30 lb at $6 per lb; 60 lb at $3 per lb **31.** nuts: 40 lb; raisins: 20 lb **33.** bicycle: 13.5 mph; car: 49.3 mph **35.** car leaving Cincinnati: 55 mph; car leaving Toledo: 70 mph **37.** Roberto: 17.5 mph; Juana: 12.5 mph
39. boat: 10 mph; current: 2 mph **41.** plane: 470 mph; wind: 30 mph

Section 15.5 (pages 1109–1110)

1. C **3.** B
5.

7.
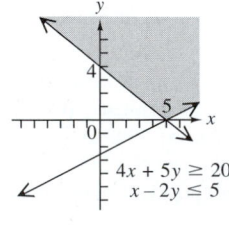
9.
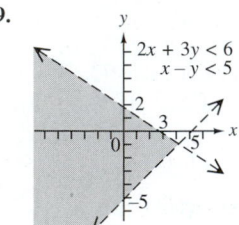
11.
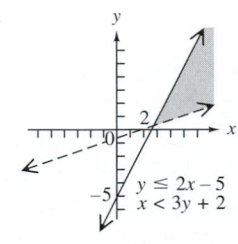

13.
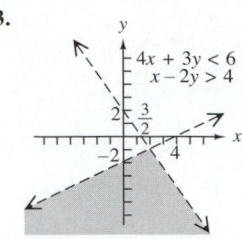
15.
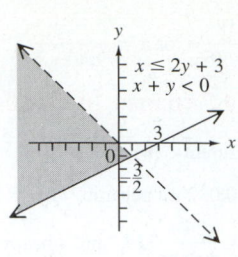
17.
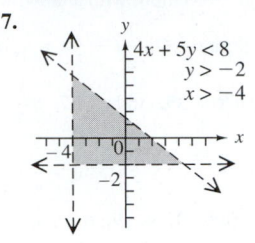
19.
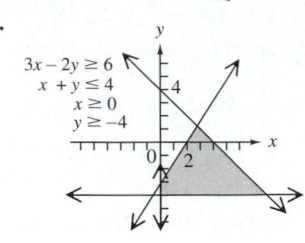

Chapter 15 Review Exercises (pages 1115–1118)

1. yes **2.** no **3.** $\{(3, 1)\}$ **4.** $\{(0, -2)\}$ **5.** $\{(x, y) \mid x - 2y = 2\}$
6. $\emptyset$ **7.** It is not a solution of the system because it is not also a solution of the second equation, $2x + y = 4$. **8.** $\{(2, 1)\}$ **9.** $\{(3, 5)\}$
10. $\{(6, 4)\}$ **11.** $\emptyset$ **12.** $\{(7, 1)\}$ **13.** $\{(-5, -2)\}$ **14.** $\{(-4, 3)\}$
15. $\{(x, y) \mid 3x - 4y = 9\}$ **16. (a)** 2 **(b)** 9 **17.** $\{(9, 2)\}$
18. $\left\{\left(\frac{10}{7}, -\frac{9}{7}\right)\right\}$ **19.** $\{(8, 9)\}$ **20.** $\{(2, 1)\}$ **21.** $\{(6, -4)\}$
22. $\{(-8, 5)\}$ **23.** Subway: 20,755 restaurants; McDonald's: 13,774 restaurants **24.** *AARP The Magazine:* 23.4 million; *Reader's Digest:* 10.1 million **25.** length: 27 m; width: 18 m **26.** 13 twenties; 7 tens
27. 25 lb of $1.30 candy; 75 lb of $0.90 candy **28.** plane: 250 mph; wind: 20 mph **29.** $7000 at 3%; $11,000 at 4% **30.** 60 L of 40% solution; 30 L of 70% solution

31.
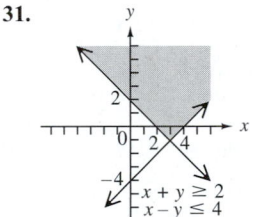
32.
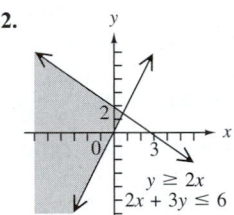
33.
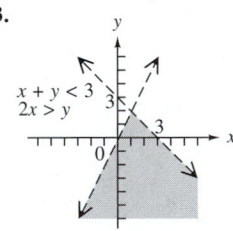

34. B **35.** B **36.** $\{(2, 0)\}$ **37.** $\{(-4, 15)\}$ **38.** $\emptyset$
39.
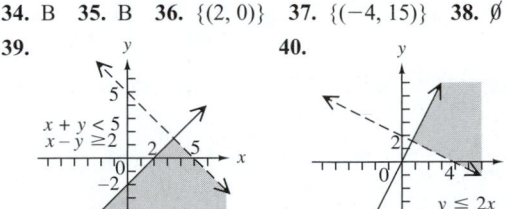
40.

41.

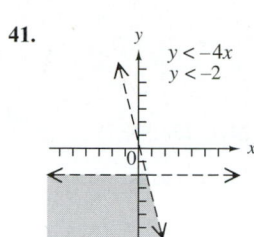

$y < -4x$
$y < -2$

42. 8 in., 8 in., and 13 in.   **43.** Giants: 17; Patriots: 14
44. (a) years 0–6   **(b)** year 6; about $650

Chapter 15 Test (pages 1119–1120)

1. $\{(2, -3)\}$ **2.** It has no solution.   **3.** $\{(1, -6)\}$ **4.** $\{(-35, 35)\}$
5. $\{(5, 6)\}$ **6.** $\{(-1, 3)\}$ **7.** $\{(0, 0)\}$ **8.** $\emptyset$ **9.** $\{(x, y) \mid 3x - y = 6\}$
10. $\{(12, -4)\}$ **11.** Memphis and Atlanta: 394 mi; Minneapolis and
Houston: 1176 mi   **12.** Statue of Liberty: 3.6 million; National World
War II Memorial: 5.4 million   **13.** 20 L of 15% solution; 30 L of 40%
solution   **14.** slower car: 45 mph; faster car: 60 mph
15.

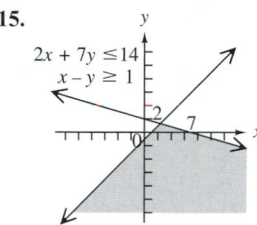

$2x + 7y \le 14$
$x - y \ge 1$

16.

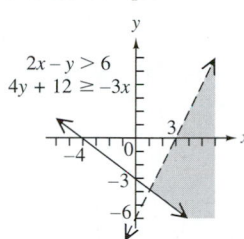

$2x - y > 6$
$4y + 12 \ge -3x$

Chapter 16   Roots and Radicals

Section 16.1 (pages 1129–1132)

1. true   **3.** false; Zero has only one square root.   **5.** true
7. $-3, 3$ **9.** $-8, 8$ **11.** $-13, 13$ **13.** $-\dfrac{5}{14}, \dfrac{5}{14}$ **15.** $-30, 30$
17. 1   **19.** 7   **21.** -16 **23.** $-\dfrac{12}{11}$ **25.** 0.8   **27.** not a real number
29. not a real number   **31.** 100   **33.** 19   **35.** $\dfrac{2}{3}$ **37.** $3x^2 + 4$
39. a must be positive.   **41.** a must be negative.   **43.** rational; 5
45. irrational; 5.385   **47.** rational; -8 **49.** irrational; -17.321
51. not a real number   **53.** irrational; 34.641   **55.** C   **57.** $c = 17$
59. $b = 8$ **61.** $c = 11.705$ **63.** 24 cm   **65.** 80 ft   **67.** 195 ft
69. 158.6 ft   **71.** 11.1 ft   **73.** 9.434   **75.** 1   **77.** 5   **79.** -3
81. -6 **83.** 2   **85.** 4   **87.** 6   **89.** not a real number
91. -5 **93.** -4

Section 16.2 (pages 1139–1142)

1. false; $\sqrt{(-6)^2} = \sqrt{36} = 6$ **3.** $\sqrt{15}$ **5.** $\sqrt{22}$ **7.** $\sqrt{42}$
9. $\sqrt{13r}$ **11.** A   **13.** $3\sqrt{5}$ **15.** $2\sqrt{6}$ **17.** $3\sqrt{10}$ **19.** $5\sqrt{3}$
21. $5\sqrt{5}$ **23.** cannot be simplified   **25.** $4\sqrt{10}$ **27.** $-10\sqrt{7}$
29. $3\sqrt{6}$ **31.** 24   **33.** $6\sqrt{10}$ **35.** $12\sqrt{5}$ **37.** $30\sqrt{5}$
39. $\sqrt{8} \cdot \sqrt{32} = \sqrt{8 \cdot 32} = \sqrt{256} = 16$. Also, $\sqrt{8} = 2\sqrt{2}$ and
$\sqrt{32} = 4\sqrt{2}$, so $\sqrt{8} \cdot \sqrt{32} = 2\sqrt{2} \cdot 4\sqrt{2} = 8 \cdot 2 = 16$.
Both methods give the same answer, and the correct answer can
always be obtained using either method.   **41.** $\dfrac{4}{15}$ **43.** $\dfrac{\sqrt{7}}{4}$
45. 5   **47.** $\dfrac{25}{4}$ **49.** $6\sqrt{5}$ **51.** m **53.** y^2 **55.** $6z$ **57.** $20x^3$

59. $3x^4\sqrt{2}$ **61.** $3c^7\sqrt{5}$ **63.** $z^2\sqrt{z}$ **65.** $a^6\sqrt{a}$ **67.** $8x^3\sqrt{x}$
69. x^3y^6 **71.** $9m^2n$ **73.** $\dfrac{\sqrt{7}}{x^5}$ **75.** $\dfrac{y^2}{10}$ **77.** $\dfrac{x^3}{y^4}$ **79.** $2\sqrt[3]{5}$
81. $3\sqrt[3]{2}$ **83.** $4\sqrt[3]{2}$ **85.** $2\sqrt[4]{5}$ **87.** $\dfrac{2}{3}$ **89.** $-\dfrac{6}{5}$ **91.** p
93. x^3 **95.** $4z^2$ **97.** $7a^3b$ **99.** $2t\sqrt[3]{2t^2}$ **101.** $\dfrac{m^4}{2}$ **103.** 6 cm
105. 6 in.   **107.** D

Section 16.3 (pages 1145–1146)

1. distributive   **3.** radicands   **5.** $-5\sqrt{7}$ **7.** $5\sqrt{17}$ **9.** $5\sqrt{7}$
11. $11\sqrt{5}$ **13.** $15\sqrt{2}$ **15.** $-6\sqrt{2}$ **17.** $17\sqrt{7}$
19. $-16\sqrt{2} - 8\sqrt{3}$ **21.** $20\sqrt{2} + 6\sqrt{3} - 15\sqrt{5}$ **23.** $4\sqrt{2}$
25. $22\sqrt{2}$ **27.** $11\sqrt{3}$ **29.** $5\sqrt{x}$ **31.** $3x\sqrt{6}$ **33.** 0
35. $-20\sqrt{2k}$ **37.** $42x\sqrt{5z}$ **39.** $-\sqrt[3]{2}$ **41.** $6\sqrt[3]{p^2}$ **43.** $21\sqrt[4]{m^3}$
45. $-6x^2y$ **46.** $-6(p - 2q)^2(a + b)$ **47.** $-6a^2\sqrt{xy}$
48. The answers are alike because the numerical coefficient of the three
answers is the same: -6. Also, the first variable factor is raised to the
second power, and the second variable factor is raised to the first power.
The answers are different because the variables are different: x and y,
then $p - 2q$ and $a + b$, and then a and $\sqrt{xy}$.

Section 16.4 (pages 1151–1152)

1. $4\sqrt{2}$ **3.** $\dfrac{-\sqrt{33}}{3}$ **5.** $\dfrac{7\sqrt{15}}{5}$ **7.** $\dfrac{\sqrt{30}}{2}$ **9.** $\dfrac{16\sqrt{3}}{9}$
11. $\dfrac{-3\sqrt{2}}{10}$ **13.** $\dfrac{21\sqrt{5}}{5}$ **15.** $\sqrt{3}$ **17.** $\dfrac{\sqrt{2}}{2}$ **19.** $\dfrac{\sqrt{65}}{5}$
21. We are actually multiplying by 1. The identity property of
multiplication justifies our result.   **23.** $\dfrac{\sqrt{21}}{3}$ **25.** $\dfrac{3\sqrt{14}}{4}$ **27.** $\dfrac{1}{6}$
29. 1   **31.** $\dfrac{\sqrt{7x}}{x}$ **33.** $\dfrac{2x\sqrt{xy}}{y}$ **35.** $\dfrac{x\sqrt{30xz}}{6}$ **37.** $\dfrac{3ar^2\sqrt{7rt}}{7t}$
39. B   **41.** $\dfrac{\sqrt[3]{15}}{3}$ **43.** $\dfrac{\sqrt[3]{196}}{7}$ **45.** $\dfrac{\sqrt[3]{6y}}{2y}$ **47.** $\dfrac{\sqrt[3]{42mn^2}}{6n}$
49. (a) $\dfrac{9\sqrt{2}}{4}$ sec   **(b)** 3.182 sec

Section 16.5 (pages 1157–1160)

1. 13   **3.** 4   **5.** $\sqrt{15} - \sqrt{35}$ **7.** $2\sqrt{10} + 30$ **9.** $4\sqrt{7}$
11. $57 + 23\sqrt{6}$ **13.** $81 + 14\sqrt{21}$ **15.** $71 - 16\sqrt{7}$
17. $37 + 12\sqrt{7}$ **19.** $a + 2\sqrt{a} + 1$ **21.** 23   **23.** 1
25. $y - 10$ **27.** $2\sqrt{3} - 2 + 3\sqrt{2} - \sqrt{6}$ **29.** $15\sqrt{2} - 15$
31. $\sqrt{30} + \sqrt{15} + 6\sqrt{5} + 3\sqrt{10}$
33. $\sqrt{5x} - \sqrt{10} - \sqrt{10x} + 2\sqrt{5}$ **35.** Because multiplication must
be performed before addition, it is incorrect to add -37 and -2. Only
like radicals can be combined.   **37.** $\dfrac{3 - \sqrt{2}}{7}$ **39.** $-4 - 2\sqrt{11}$
41. $1 + \sqrt{2}$ **43.** $-\sqrt{10} + \sqrt{15}$ **45.** $2\sqrt{5} + \sqrt{15} + 4 + 2\sqrt{3}$
47. $\dfrac{12(\sqrt{x} - 1)}{x - 1}$ **49.** $\dfrac{3(7 + \sqrt{x})}{49 - x}$ **51.** $\sqrt{11} - 2$ **53.** $\dfrac{\sqrt{3} + 5}{8}$
55. $\dfrac{6 - \sqrt{10}}{2}$ **57.** $30 + 18x$ **58.** They are not like terms.
59. $30 + 18\sqrt{5}$ **60.** They are not like radicals.   **61.** Make the first
term $30x$, so that $30x + 18x = 48x$; make the first term $30\sqrt{5}$, so that
$30\sqrt{5} + 18\sqrt{5} = 48\sqrt{5}$.   **62.** Both like terms and like radicals are
combined by adding their numerical coefficients. The variables in like
terms are replaced by radicals in like radicals.   **63.** 4 in.

Section 16.6 (pages 1167–1170)

1. {49} **3.** {7} **5.** {85} **7.** {−45} **9.** $\left\{-\frac{3}{2}\right\}$ **11.** ∅

13. {121} **15.** {8} **17.** {1} **19.** {6} **21.** ∅ **23.** {5}

25. When the left side is squared, the result should be $x - 1$, not $-(x - 1)$. The correct solution set is {17}.

27. {12} **29.** {5} **31.** {0, 3} **33.** {−1, 3} **35.** {8} **37.** {4}

39. {8} **41.** {9} **43.** {4, 20} **45.** {−5} **47.** 21 **49.** 8

51. 17,616 ft^2 **53.** (a) 70.5 mph (b) 59.8 mph (c) 53.9 mph

55. yes; 26 mi **57.** 47 mi **59.** $s = 13$ units **60.** $6\sqrt{13}$ sq. units

61. $h = \sqrt{13}$ units **62.** $3\sqrt{13}$ sq. units **63.** $6\sqrt{13}$ sq. units

64. They are both $6\sqrt{13}$.

Chapter 16 Review Exercises (pages 1175–1178)

1. −7, 7 **2.** −9, 9 **3.** −14, 14 **4.** −11, 11 **5.** −16, 16

6. −27, 27 **7.** 4 **8.** −0.6 **9.** −8 **10.** 3 **11.** not a real

number **12.** −65 **13.** $\frac{12}{13}$ **14.** $-\frac{10}{9}$ **15.** 8 **16.** 48.3 cm

17. irrational; 8.544 **18.** rational; 13 **19.** rational; −25

20. not a real number **21.** $4\sqrt{3}$ **22.** $-12\sqrt{2}$ **23.** $2\sqrt[3]{2}$

24. $5\sqrt[3]{3}$ **25.** 18 **26.** $16\sqrt{6}$ **27.** $-\frac{11}{20}$ **28.** $\frac{\sqrt{7}}{13}$ **29.** $\frac{\sqrt{5}}{6}$

30. $\frac{2}{15}$ **31.** $3\sqrt{2}$ **32.** $2\sqrt{2}$ **33.** r^9 **34.** $x^5 y^8$ **35.** $9x^4\sqrt{2x}$

36. $\frac{6}{p}$ **37.** $a^7 b^{10}\sqrt{ab}$ **38.** $11x^3 y^5$ **39.** y^2 **40.** $6x^5$ **41.** $8\sqrt{11}$

42. $9\sqrt{2}$ **43.** $21\sqrt{3}$ **44.** $12\sqrt{3}$ **45.** 0 **46.** $3\sqrt{7}$

47. $2\sqrt{3} + 3\sqrt{10}$ **48.** $2\sqrt{2}$ **49.** $6\sqrt{30}$ **50.** $5\sqrt{x}$ **51.** 0

52. $11k^2\sqrt{2n}$ **53.** $\frac{10\sqrt{3}}{3}$ **54.** $\frac{8\sqrt{10}}{5}$ **55.** $\sqrt{6}$ **56.** $\frac{\sqrt{10}}{5}$

57. $\sqrt{10}$ **58.** $\frac{\sqrt{42}}{21}$ **59.** $\frac{r\sqrt{x}}{4x}$ **60.** $\frac{\sqrt[3]{9}}{3}$ **61.** $r = \frac{\sqrt{3V\pi h}}{\pi h}$

62. $r = \frac{\sqrt{S\pi}}{2\pi}$ **63.** $-\sqrt{15} - 9$ **64.** $3\sqrt{6} + 12$ **65.** $22 - 16\sqrt{3}$

66. $2\sqrt{21} - \sqrt{14} + 12\sqrt{2} - 4\sqrt{3}$ **67.** −13 **68.** $x + 4\sqrt{x} + 4$

69. $-2 + \sqrt{5}$ **70.** $\frac{3(1 - \sqrt{x})}{1 - x}$ **71.** $\frac{-\sqrt{10} + 3\sqrt{5} + \sqrt{2} - 3}{7}$

72. $\frac{3 + 2\sqrt{6}}{3}$ **73.** $\frac{1 + 3\sqrt{7}}{4}$ **74.** $3 + 4\sqrt{3}$ **75.** ∅ **76.** {48}

77. {1} **78.** {2} **79.** {6} **80.** {−3, −1} **81.** {−2} **82.** {4}

83. {7} **84.** $11\sqrt{3}$ **85.** $\frac{5 - \sqrt{2}}{23}$ **86.** $\frac{2\sqrt{10}}{5}$ **87.** $3a^2 b^3\sqrt[3]{2ab}$

88. $-\sqrt{10} - 5\sqrt{15}$ **89.** $\frac{4r\sqrt{3rs}}{3s}$ **90.** $\frac{2 + \sqrt{13}}{2}$ **91.** $7 - 2\sqrt{10}$

92. $166 + 2\sqrt{7}$ **93.** {7} **94.** ∅ **95.** {8} **96.** (a) B (b) F
(c) D (d) A (e) C (f) A

Chapter 16 Test (pages 1179–1180)

1. −20, 20 **2.** (a) irrational (b) 11.916 **3.** a must be negative.

4. 6 **5.** $-3\sqrt{6}$ **6.** $\frac{8\sqrt{2}}{5}$ **7.** $2\sqrt[3]{4}$ **8.** $4\sqrt{6}$ **9.** $9\sqrt{7}$

10. $-5\sqrt{3x}$ **11.** $4xy\sqrt{2y}$ **12.** 31

13. $6\sqrt{2} + 2 - 3\sqrt{14} - \sqrt{7}$ **14.** $11 + 2\sqrt{30}$

15. (a) $6\sqrt{2}$ in. (b) 8.485 in. **16.** 50 ohms **17.** $\frac{5\sqrt{14}}{7}$

18. $\frac{\sqrt{6x}}{3x}$ **19.** $-\sqrt[3]{2}$ **20.** $\frac{-3(4 + \sqrt{3})}{13}$

21. $\frac{\sqrt{3} + 12\sqrt{2}}{3}$ **22.** ∅ **23.** {3} **24.** {1, 4} **25.** {9}

26. 12 is not a solution. A check shows that it does not satisfy the original equation. The solution set is ∅.

Chapter 17 Quadratic Equations

Section 17.1 (pages 1185–1188)

1. true **3.** false; If k is a positive integer that is not a perfect square, then the solutions will be irrational. **5.** false; For values of k that satisfy $0 \le k < 10$, there are real solutions. **7.** {−9, 9}

9. $\left\{-\sqrt{14}, \sqrt{14}\right\}$ **11.** $\left\{-4\sqrt{3}, 4\sqrt{3}\right\}$ **13.** $\left\{-\frac{5}{2}, \frac{5}{2}\right\}$

15. ∅ **17.** {−1.5, 1.5} **19.** $\left\{-\sqrt{3}, \sqrt{3}\right\}$ **21.** $\left\{-\frac{2\sqrt{7}}{7}, \frac{2\sqrt{7}}{7}\right\}$

23. $\left\{-3\sqrt{2}, 3\sqrt{2}\right\}$ **25.** $\left\{-2\sqrt{6}, 2\sqrt{6}\right\}$ **27.** $\left\{-\frac{2\sqrt{5}}{5}, \frac{2\sqrt{5}}{5}\right\}$

29. {−2, 8} **31.** ∅ **33.** $\left\{8 + 3\sqrt{3}, 8 - 3\sqrt{3}\right\}$ **35.** $\left\{-3, \frac{5}{3}\right\}$

37. $\left\{0, \frac{3}{2}\right\}$ **39.** $\left\{\frac{5 + \sqrt{30}}{2}, \frac{5 - \sqrt{30}}{2}\right\}$

41. $\left\{\frac{-1 + 3\sqrt{2}}{3}, \frac{-1 - 3\sqrt{2}}{3}\right\}$ **43.** $\left\{-10 + 4\sqrt{3}, -10 - 4\sqrt{3}\right\}$

45. $\left\{\frac{1 + 4\sqrt{3}}{4}, \frac{1 - 4\sqrt{3}}{4}\right\}$ **47.** The answers are equivalent. If the

answer of either student is multiplied by $\frac{-1}{-1}$, it will look like the answer

of the other student. **49.** about $\frac{1}{2}$ sec **51.** 9 in. **53.** 5% **55.** 2%

Section 17.2 (pages 1195–1196)

1. 25; $(x + 5)^2$ **3.** 1; $(x + 1)^2$ **5.** $\frac{25}{4}$; $\left(p - \frac{5}{2}\right)^2$ **7.** D

9. {1, 3} **11.** {−3, −2} **13.** $\left\{-1 + \sqrt{6}, -1 - \sqrt{6}\right\}$

15. $\left\{4 + 2\sqrt{3}, 4 - 2\sqrt{3}\right\}$ **17.** {−3}

19. $\left\{\frac{-1 + \sqrt{5}}{2}, \frac{-1 - \sqrt{5}}{2}\right\}$ **21.** $\left\{-\frac{3}{2}, \frac{1}{2}\right\}$

23. $\left\{\frac{2 + \sqrt{14}}{2}, \frac{2 - \sqrt{14}}{2}\right\}$ **25.** ∅

27. $\left\{\frac{-7 + \sqrt{97}}{6}, \frac{-7 - \sqrt{97}}{6}\right\}$ **29.** {−4, 2}

31. $\left\{4 + \sqrt{3}, 4 - \sqrt{3}\right\}$ **33.** $\left\{1 + \sqrt{6}, 1 - \sqrt{6}\right\}$

35. 1 sec and 5 sec **37.** 3 sec and 5 sec **39.** 75 ft by 100 ft

Section 17.3 (pages 1201–1202)

1. 4; 5; −9 **3.** 3; −4; −2 **5.** 3; 7; 0 **7.** {−13, 1} **9.** {2}

11. $\left\{-1, \frac{5}{2}\right\}$ **13.** $\left\{\frac{-6 + \sqrt{26}}{2}, \frac{-6 - \sqrt{26}}{2}\right\}$ **15.** {−1, 0}

17. $\emptyset$ **19.** $\left\{\dfrac{-5+\sqrt{13}}{6},\dfrac{-5-\sqrt{13}}{6}\right\}$ **21.** $\left\{0,\dfrac{12}{7}\right\}$

23. $\left\{-2\sqrt{6},2\sqrt{6}\right\}$ **25.** $\left\{-\dfrac{2}{5},\dfrac{2}{5}\right\}$ **27.** $\left\{\dfrac{6+2\sqrt{6}}{3},\dfrac{6-2\sqrt{6}}{3}\right\}$

29. $\emptyset$ **31.** There is no real number solution, so the solution set is $\emptyset$.

33. $\left\{-\dfrac{2}{3},\dfrac{4}{3}\right\}$ **35.** $\left\{\dfrac{-1+\sqrt{73}}{6},\dfrac{-1-\sqrt{73}}{6}\right\}$ **37.** $\emptyset$

39. $\left\{-1,\dfrac{5}{2}\right\}$ **41.** $\left\{-3+\sqrt{5},-3-\sqrt{5}\right\}$ **43.** 3.5 ft

45. $-8, 16$; Only 16 board feet is a reasonable answer.

Section 17.4 (pages 1207–1208)

1. The vertex of a parabola is the lowest or highest point on the graph.

3. $(0, 0)$ **5.** $(0, -4)$

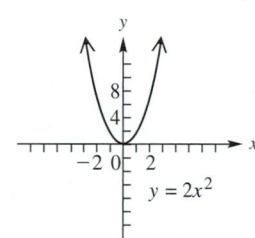

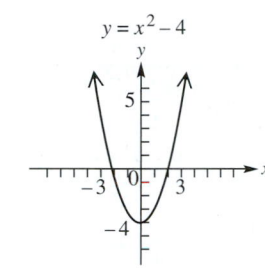

7. $(0, 2)$ **9.** $(-3, 0)$

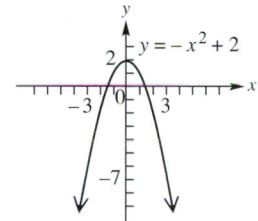

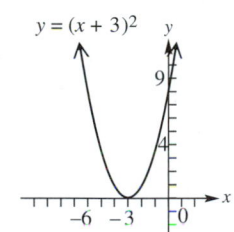

11. $(-1, 2)$ **13.** $(3, 4)$

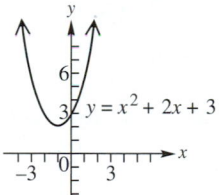

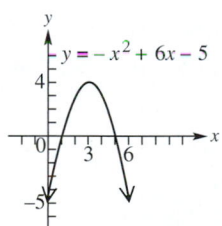

15. If $a > 0$, it opens upward, and if $a < 0$, it opens downward.

17. $y = \dfrac{11}{5625}x^2$

Section 17.5 (pages 1215–1216)

1. $3; 3; (1, 3)$ **3.** $5; 5; (3, 5)$ **5.** The graph consists of the five points $(0, 2), (1, 3), (2, 4), (3, 5)$, and $(4, 6)$. **7.** not a function; domain: $\{-4, -2, 0\}$; range: $\{3, 1, 5, -8\}$ **9.** function; domain: $\{A, B, C, D, E\}$; range: $\{2, 3, 6, 4\}$ **11.** not a function; domain: $\{-4, -2, 0, 2, 3\}$; range: $\{-2, 0, 1, 2, 3\}$ **13.** function **15.** not a function **17.** function **19.** not a function **21.** $(2, 4)$

22. $(-1, -4)$ **23.** $\dfrac{8}{3}$ **24.** $f(x) = \dfrac{8}{3}x - \dfrac{4}{3}$ **25. (a)** 11 **(b)** 3

(c) -9 **27. (a)** 4 **(b)** 2 **(c)** 14 **29. (a)** 2 **(b)** 0 **(c)** 3

31. $\{(1970, 9.6), (1980, 14.1), (1990, 19.8), (2000, 28.4)\}$; yes

33. $g(1980) = 14.1$ (million); $g(1990) = 19.8$ (million)

35. For the year 2002, the function gives 30.3 million foreign-born residents in the United States.

Chapter 17 Review Exercises (pages 1221–1224)

1. $\{-12, 12\}$ **2.** $\left\{-\sqrt{37}, \sqrt{37}\right\}$ **3.** $\left\{-8\sqrt{2}, 8\sqrt{2}\right\}$

4. $\{-7, 3\}$ **5.** $\left\{3 + \sqrt{10}, 3 - \sqrt{10}\right\}$

6. $\left\{\dfrac{-1+\sqrt{14}}{2},\dfrac{-1-\sqrt{14}}{2}\right\}$ **7.** $\emptyset$ **8.** $\left\{-\dfrac{5}{3}\right\}$ **9.** $\{-5, -1\}$

10. $\left\{-2+\sqrt{11}, -2-\sqrt{11}\right\}$ **11.** $\left\{-1+\sqrt{6}, -1-\sqrt{6}\right\}$

12. $\left\{\dfrac{-4+\sqrt{22}}{2},\dfrac{-4-\sqrt{22}}{2}\right\}$ **13.** $\left\{-\dfrac{7}{2}\right\}$ **14.** $\emptyset$ **15.** 2.5 sec

16. $6, 8, 10$ **17.** $\left(\dfrac{3}{2}\right)^2$, or $\dfrac{9}{4}$ **18. (a)** $\{-3, 3\}$ **(b)** $\{-3, 3\}$

(c) $\{-3, 3\}$ **(d)** We will always get the same results, no matter which method of solution is used.

19. $\left\{\dfrac{-1+\sqrt{29}}{4},\dfrac{-1-\sqrt{29}}{4}\right\}$ **20.** $\left\{\dfrac{2+\sqrt{10}}{2},\dfrac{2-\sqrt{10}}{2}\right\}$

21. $\left\{\dfrac{1+\sqrt{21}}{10},\dfrac{1-\sqrt{21}}{10}\right\}$ **22.** $\left\{\dfrac{-3+\sqrt{41}}{2},\dfrac{-3-\sqrt{41}}{2}\right\}$

23. $\left\{-3+\sqrt{19}, -3-\sqrt{19}\right\}$ **24.** No, because the fraction bar should be under both $-b$ and $\sqrt{b^2 - 4ac}$. The term $2a$ should not be in the radicand. The correct formula is $x = \dfrac{-b \pm \sqrt{b^2 - 4ac}}{2a}$.

25. $(0, 0)$ **26.** $(0, 5)$

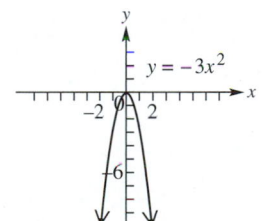

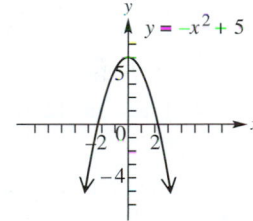

27. $(1, 0)$ **28.** $(1, 4)$

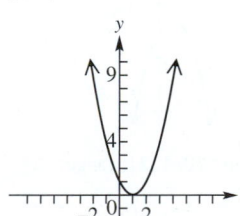

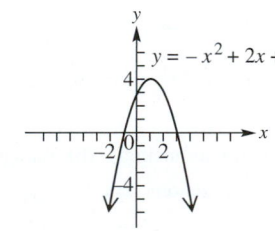

29. $(-2, -2)$ **30.** $(-4, 0)$

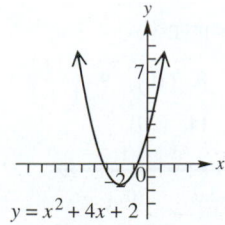

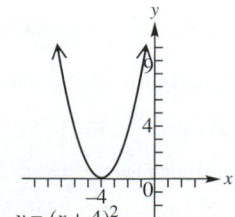

31. $y = \dfrac{3}{1000}x^2$ **32.** not a function; domain: $\{-2, 0, 2\}$;

range: $\{4, 8, 5, 3\}$ **33.** function; domain: $\{8, 7, 6, 5, 4\}$;

range: $\{3, 4, 5, 6, 7\}$ **34.** not a function **35.** function

36. function **37.** function **38.** not a function **39.** (a) 8

(b) -1 **40.** (a) 7 (b) 1 **41.** (a) 5 (b) 2 **42.** 400 or 800

43. $(6, 10)$ **44.** demand: 600; price: \$10 **45.** $\left\{-\dfrac{11}{2}, 5\right\}$

46. $\left\{-\dfrac{11}{2}, \dfrac{9}{2}\right\}$ **47.** $\left\{\dfrac{-1 + \sqrt{21}}{2}, \dfrac{-1 - \sqrt{21}}{2}\right\}$ **48.** $\left\{-\dfrac{3}{2}, \dfrac{1}{3}\right\}$

49. $\left\{\dfrac{-5 + \sqrt{17}}{2}, \dfrac{-5 - \sqrt{17}}{2}\right\}$ **50.** $\left\{-1 + \sqrt{3}, -1 - \sqrt{3}\right\}$

51. $\varnothing$ **52.** $\left\{\dfrac{9 + \sqrt{41}}{2}, \dfrac{9 - \sqrt{41}}{2}\right\}$ **53.** $\left\{-\dfrac{6}{5}\right\}$

54. $\left\{-1 + 2\sqrt{2}, -1 - 2\sqrt{2}\right\}$ **55.** $\left\{-2 + \sqrt{5}, -2 - \sqrt{5}\right\}$

56. $\left\{-2\sqrt{2}, 2\sqrt{2}\right\}$

Chapter 17 Test (pages 1225–1226)

1. $\left\{-\sqrt{39}, \sqrt{39}\right\}$ **2.** $\{-11, 5\}$ **3.** $\left\{\dfrac{-3 + 2\sqrt{6}}{4}, \dfrac{-3 - 2\sqrt{6}}{4}\right\}$

4. $\left\{2 + \sqrt{10}, 2 - \sqrt{10}\right\}$ **5.** $\left\{\dfrac{-6 + \sqrt{42}}{2}, \dfrac{-6 - \sqrt{42}}{2}\right\}$

6. $\left\{-3, \dfrac{1}{2}\right\}$ **7.** $\left\{\dfrac{3 + \sqrt{3}}{3}, \dfrac{3 - \sqrt{3}}{3}\right\}$ **8.** $\varnothing$

9. $\left\{\dfrac{5 + \sqrt{13}}{6}, \dfrac{5 - \sqrt{13}}{6}\right\}$ **10.** $\left\{1 + \sqrt{2}, 1 - \sqrt{2}\right\}$

11. $\left\{\dfrac{-1 + 3\sqrt{2}}{2}, \dfrac{-1 - 3\sqrt{2}}{2}\right\}$ **12.** $\left\{\dfrac{11 + \sqrt{89}}{4}, \dfrac{11 - \sqrt{89}}{4}\right\}$

13. $\{5\}$ **14.** 2 sec **15.** 12, 16, 20

16. vertex: $(3, 0)$ **17.** vertex: $(-1, -3)$

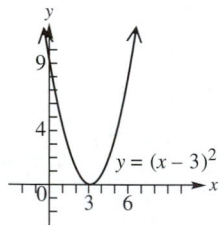

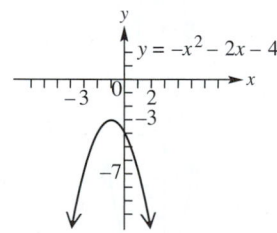

18. (a) not a function (b) function; domain: $\{0, 1, 2\}$; range: $\{2\}$

19. not a function **20.** 1

Cumulative Review Exercises: Chapters 9–17 (pages 1227–1231)

1. -8 **2.** 24 **3.** $\dfrac{1}{2}$ **4.** -4 **5.** associative property

6. inverse property **7.** distributive property **8.** $\{10\}$ **9.** $\left\{\dfrac{13}{4}\right\}$

10. $\varnothing$ **11.** $r = \dfrac{d}{t}$ **12.** $\{-5\}$ **13.** $\{-12\}$ **14.** $\{20\}$

15. {all real numbers} **16.** $(-2.6, \infty)$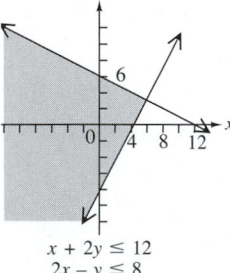

17. $(0, \infty)$

18. $(-\infty, -4]$

19.

20.

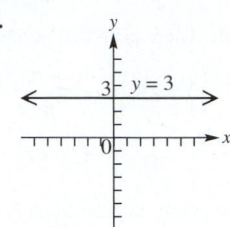

21.

22. $-\dfrac{4}{3}$ **23.** $-\dfrac{1}{4}$ **24.** $y = \dfrac{1}{2}x + 3$ **25.** $y = 2x + 1$ **26.** (a) $x = 9$

(b) $y = -1$ **27.** 4 **28.** $\dfrac{16}{9}$ **29.** 1 **30.** 256 **31.** $\dfrac{1}{p^2}$ **32.** $\dfrac{1}{m^6}$

33. $-4k^2 - 4k + 8$ **34.** $45x^2 + 3x - 18$ **35.** $9p^2 + 12p + 4$

36. $4x^3 + 6x^2 - 3x + 10$ **37.** $(2a - 1)(a + 4)$

38. $(2m + 3)(5m + 2)$ **39.** $(2p + 5)(4p^2 - 10p + 25)$

40. $(2p - 3)^2$ **41.** $(5r + 9t)(5r - 9t)$ **42.** $2pq(3p + 1)(p + 1)$

43. $\left\{-\dfrac{2}{3}, \dfrac{1}{2}\right\}$ **44.** $\{0, 8\}$ **45.** $\left\{\dfrac{4}{7}\right\}$ **46.** $\{3, 4\}$

47. $\{-2, -1\}$ **48.** $\{12\}$ **49.** $\dfrac{4}{q}$ **50.** $\dfrac{3r + 28}{7r}$ **51.** $\dfrac{7}{15(q - 4)}$

52. $\dfrac{-k - 5}{k(k + 1)(k - 1)}$ **53.** $\dfrac{7(2z + 1)}{24}$ **54.** $\dfrac{195}{29}$ **55.** $\left\{\dfrac{21}{2}\right\}$

56. $\{-2, 1\}$ **57.** $\{(-1, 6)\}$ **58.** $\{(3, -4)\}$ **59.** $\varnothing$

60.

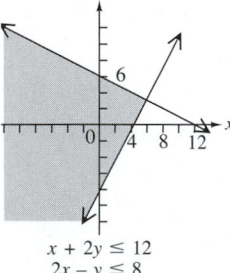

61. $\dfrac{6\sqrt{30}}{5}$ **62.** $4\sqrt{5}$ **63.** $-ab\sqrt[3]{2b}$

64. $29\sqrt{3}$ **65.** $-\sqrt{3} + \sqrt{5}$ **66.** $21 - 5\sqrt{2}$ **67.** $\varnothing$ **68.** $\{16\}$

69. $\left\{\dfrac{1 + \sqrt{3}}{2}, \dfrac{1 - \sqrt{3}}{2}\right\}$

70. $(0, -4)$; $(2, 0)$ and $(-2, 0)$

$y = x^2 - 4$

71. (a) yes **(b)** domain: $\{0, 1, 3\}$; range: $\{4, 2, 5\}$ **72.** 11
73. mouse: 160; elephant: 10 **74.** 4 **75.** 20.6; The number of subscribers increased by an average of 20.6 million per yr.
76. $y = 20.6x + 109.5$ **77.** 274.3 million **78.** 405 adults and 49 children **79.** 19 in., 19 in., 15 in. **80.** from Chicago: 61 mph; from Des Moines; 54 mph

Appendix A Measurement

Appendix A.1 (pages AP-9–AP-12)

1. 3; 12 **3.** 8; 2 **5.** 5280; 3 **7.** 2000; 16 **9.** 60; 60 **11. (a)** 2
(b) 240 **13. (a)** $\frac{1}{2}$ or 0.5 **(b)** 78 **15.** 14,000 to 16,000 lb
17. 27 **19.** 112 **21.** 10 **23.** $1\frac{1}{2}$ or 1.5 **25.** $\frac{1}{4}$ or 0.25
27. $1\frac{1}{2}$ or 1.5 **29.** $2\frac{1}{2}$ or 2.5 **31.** snowmobile/ATV; person
walking **33.** 5000 **35.** 17 **37.** 4 to 8 in. **39.** 216 **41.** 28
43. 518,400 **45.** 48,000 **47. (a)** pound; ounces **(b)** quarts/pints;
pints/cups **(c)** minutes/seconds; hours/minutes **(d)** feet; inches
(e) pounds; tons **(f)** days; weeks **49.** 174,240 **51.** 800
53. 0.75 or $\frac{3}{4}$ **55.** \$1.83 (rounded) **57.** \$140 **59. (a)** 1056 sec;
(b) 17.6 min **61. (a)** $12\frac{1}{2}$ qt **(b)** 4 containers, because you can't buy
part of a container **63. (a)** 0.8 mi (rounded) **(b)** 13,031 ft **(c)** 2.5 mi
(rounded) **64. (a)** 183 yd (rounded) **(b)** 6600 in. **(c)** 0.1 mi (rounded)
(d) 9 ft (rounded) **65. (a)** $1\frac{1}{2}$ to $2\frac{1}{2}$ ft, or 1.5 to 2.5 ft **(b)** 8400 months
(c) 140 more years, for a total of 840 years in all **66. (a)** 283 mi (rounded)
(b) 377 mi (rounded)

Appendix A.2 (pages AP-19–AP-20)

1. 1000; 1000 **3.** $\frac{1}{1000}$ or 0.001; $\frac{1}{1000}$ or 0.001 **5.** $\frac{1}{100}$ or 0.01;
$\frac{1}{100}$ or 0.01 **7.** Answers will vary; about 8 to 10 cm. **9.** Answers will
vary; about 20 to 25 mm. **11.** cm **13.** m **15.** km **17.** mm **19.** cm
21. m **23.** Some possible answers are: track and field events, metric
auto parts, and lead refills for mechanical pencils. **25.** 700 cm
27. 0.040 m or 0.04 m **29.** 9400 m **31.** 5.09 m **33.** 40 cm
35. 910 mm **37.** less; 18 cm or 0.18 m
39. ⌒ 5 mm = 0.5 cm
▭ ← 1 mm = 0.1 cm
41. 0.018 km **43.** 164 cm; 1640 mm **45.** 0.0000056 km

Appendix A.3 (pages AP-27–AP-30)

1. mL **3.** L **5.** kg **7.** g **9.** mL **11.** mg **13.** L **15.** kg
17. unreasonable; too much **19.** unreasonable; too much **21.** reasonable
23. reasonable **25.** Some capacity examples are 2 L bottles of soda and
shampoo bottles marked in mL; weight examples are grams of fat listed
on food packages and vitamin doses in milligrams. **27.** Unit for your
answer (g) is in numerator; unit being changed (kg) is in denominator so

it will divide out. The unit fraction is $\frac{1000 \text{ g}}{1 \text{ kg}}$. **29.** 15,000 mL **31.** 3 L
33. 0.925 L **35.** 0.008 L **37.** 4150 mL **39.** 8 kg **41.** 5200 g
43. 850 mg **45.** 30 g **47.** 0.598 g **49.** 0.06 L **51.** 0.003 kg
53. 990 mL **55.** mm **57.** mL **59.** cm **61.** mg **63.** 0.3 L
65. 1340 g **67.** 0.9 L **69.** 3 kg to 4 kg **71.** greater; 5 mg or 0.005 g
73. 200 nickels **75. (a)** 1,000,000
(b) $\frac{3.5 \text{ M\!m}}{1} \cdot \frac{1,000,000 \text{ m}}{1 \text{ M\!m}} = 3,500,000 \text{ m}$
76. (a) 1,000,000,000
(b) $\frac{2500 \text{ m\!}}{1} \cdot \frac{1 \text{ Gm}}{1,000,000,000 \text{ m\!}} = 0.0000025 \text{ Gm}$
77. (a) 1,000,000,000,000 **(b)** 1000; 1,000,000
78. 1,000,000; 1,000,000,000; $2^{20} = 1,048,576$; $2^{30} = 1,073,741,824$

Appendix A.4 (pages AP-35–AP-38)

1. 21.8 yd **3.** 262.4 ft **5.** 4.8 m **7.** 5.3 oz **9.** 111.6 kg **11.** 30.3 qt
13. (a) about 0.2 oz **(b)** probably not **15.** about 31.8 L **17.** about
1.3 cm **19.** $-8\,°C$ **21.** $40\,°C$ **23.** $150\,°C$ **25.** More. There are
180 degrees between freezing and boiling on the Fahrenheit scale, but
only 100 degrees on the Celsius scale, so each Celsius degree is a greater
change in temperature. **27.** $16\,°C$ (rounded) **29.** $40\,°C$ **31.** $46\,°C$
(rounded) **33.** $95\,°F$ **35.** $58\,°C$ (rounded) **37. (a)** pleasant weather,
above freezing but not hot **(b)** $75\,°F$ to $39\,°F$ (rounded) **(c)** Answers
will vary. In Minnesota, it's $0\,°C$ to $-40\,°C$; in California, $24\,°C$ to $0\,°C$.
39. about 1.8 m **40.** about 5.0 kg **41.** about 3040.2 km
42. 38 hr 23 min **43.** about 300 m **44.** less than 3790 mL
45. about 59.4 mL **46.** about 1.6%

Appendix B Inductive and Deductive Reasoning

Appendix B (pages AP-43–AP-44)

1. 37; Add 7. **3.** 26; Add 10, subtract 2. **5.** 16; Multiply by 2.
7. 243; Multiply by 3. **9.** 36; add 3, add 5, add 7, etc.;
or $1^2, 2^2, 3^2$, etc.
11. ⌐‾‾ **13.** ⌐⌐
15. Conclusion follows. **17.** Conclusion does not follow.

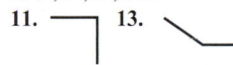

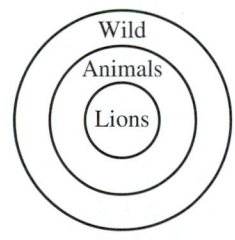

Neither watched TV on 3 days.
21. Dick is the computer operator.

$2 + 18 + 7 = 27$
$30 - 27 = 3$

Chapter 1 Whole Numbers

Section 1.1 (pages 7–8)

9. $\underline{60},\underline{000},\underline{502},\underline{109}$; billions: 60; millions: 000; thousands: 502; ones: 109

15. 346,009 is three hundred forty-six thousand, nine.

31. The least-used method of transportation is public transportation. 6,069,589 in words is six million, sixty-nine thousand, five hundred eighty-nine.

Section 1.2 (pages 17–20)

11. Line up the numbers in columns. Then start at the right and add the ones digits. Add the ten digits next, and, finally, the hundreds digits.

$$\begin{array}{r} 932 \\ 44 \\ +\ 613 \\ \hline 1589 \end{array}$$

49. *Step 1*
Add the digits in the ones column.

Step 2
Add the digits in the tens column, including the regrouped 2.

Step 3
Add the hundreds column, including the regrouped 1.

Step 4
Add the thousands column, including the regrouped 1.

$$\begin{array}{r} ^{112} \\ 18 \\ 708 \\ 9\,286 \\ +\ 636 \\ \hline 10{,}648 \end{array}$$

63. Add up to check addition.

$$\begin{array}{r} 769 \\ \hline 179 \\ 214 \\ +\ 376 \\ \hline 759 \end{array}$$ incorrect; should be 769

77. To find the total amount raised, add the amounts raised at each event.

$3482	flea market
+ 12,860	annual auction
$16,342	total amount raised

Section 1.3 (pages 27–30)

31. In the ones column, 5 is less than 7, so in order to subtract, regroup 1 ten as 10 ones. Then subtract 7 ones from 15 ones in the ones column. Finally, subtract 3 tens from 6 tens in the tens column.

$$\begin{array}{r} ^{6}\cancel{7}^{15} \\ \cancel{7}\cancel{5} \\ -3\,7 \\ \hline 3\,8 \end{array}$$

67.

$$\begin{array}{r} ^{9}\ ^{9} \\ ^{5}\ ^{10}\ ^{10}\ ^{10} \\ 6\cancel{6},\ \cancel{0}\ \cancel{0}\ \cancel{0} \\ -34,\ 4\ 4\ 4 \\ \hline 31,\ 5\ 5\ 6 \end{array}$$

81. To find how many fewer calories a woman burns, subtract the number of calories a woman burns from the number of calories a man burns.

187	calories man burns
−140	calories woman burns
47	fewer calories for a woman

The woman burned 47 fewer calories.

Section 1.4 (pages 37–40)

9. $(4)(5)(2) = [(4)(5)](2) = 20(2) = 40$
or $(4)[(5)(2)] = (4)(10) = 40$.

37.

$$\begin{array}{ccc} 600 & 6 & 600 \\ \underline{\times\ 6} & \underline{\times\ 6} & \underline{\times\ 6} \\ 36 & 3600 & 3600 \end{array}$$ Attach 00.

73. First multiply 9352 by 4. Then multiply 9352 by 6, making sure to line up the tens. Then multiply 9352 by 2, making sure to line up the hundreds. Then add the partial products.

$$\begin{array}{r} 9\,3\,5\,2 \\ \times\ \ 2\,6\,4 \\ \hline 3\,7\,4\,0\,8 \quad \leftarrow 4\times 9352 \\ 5\,6\,1\,1\,2 \quad\ \ \leftarrow 6\times 9352 \\ 1\,8\,7\,0\,4 \qquad\ \leftarrow 2\times 9352 \\ \hline 2{,}4\,6\,8{,}9\,2\,8 \end{array}$$

85. *Approach* To find the number of balls purchased, multiply the number of cartons (300) by the number of balls per carton (10).

Solution

300	cartons
× 10	balls per carton
3000	balls

3000 balls were purchased.

103. Multiply the number of laptop computers purchased (6) by the price per laptop ($880). Then multiply the number of printers purchased (6) by the price per printer ($235). Then multiply the number of fax machines purchased (5) by the price per fax machine ($140).

$$\begin{array}{ccc} 880 & 235 & 140 \\ \underline{\times\ 6} & \underline{\times\ 6} & \underline{\times\ 5} \\ 5280 & 1410 & 700 \end{array}$$

The total cost is the sum of these values:
$5280 + $1410 + $700 = $7390

Section 1.5 (pages 51–54)

1. $24 \div 4 = 6$: $4\overline{)24}^{\ 6}$ or $\dfrac{24}{4} = 6$

31.

$$6\overline{)9\,^{3}1\,^{1}3\,^{1}7}\ \ ^{1\ 5\ 2\ 2\ R5}$$

Check $(6 \times 1522) + 5 = 9132 + 5$
$= 9137$

51.

$$7\overline{)7\,1,\,^{1}7\,^{3}7\,^{2}6}\ \ ^{1\,0,\ 2\ 5\ 3\ R5}$$

Check $(7 \times 10{,}253) + 5 = 71{,}771 + 5$
$= 71{,}776$

67.

$$9\overline{)8\,6,6\,5\,5}\ \ ^{9\ 6\,2\,8\ R7}$$

Check $(9 \times 9628) + 7 = 86{,}652 + 7$
$= 86{,}659$

Does not match dividend, incorrect
Rework:

$$9\overline{)8\,6,\,^{5}6\,^{2}5\,^{7}5}\ \ ^{9\ 6\,2\,8\ R3}$$

Check $(9 \times 9628) + 3 = 86{,}652 + 3$
$= 86{,}655$

Matches dividend, correct

73. To find the number of tables that can be set, divide the number of napkins (2624) by the number of napkins it takes to set each table (8).

$$8\overline{)2\,6\,^{2}2\,^{6}4}\ \ ^{3\ 2\ 8}$$

328 tables can be set.

Section 1.6 (pages 61–63)

13. Use 2 as a trial divisor since 18 is closer to 20 than to 10.

$$\dfrac{13}{2} = 6 \text{ with 1 left over}$$

Because $6 \times 18 = 108$ and $131 - 108 = 23$, which is greater than the divisor, use 7 instead. To find the next digit in the quotient, use 2 as a trial divisor again.

$$\dfrac{5}{2} = 2 \text{ with 1 left over}$$

Because $2 \times 18 = 36$ and $59 - 36 = 23$, which is greater than the divisor, use 3 instead.

$$\begin{array}{r} 7\ 3\ R5 \\ 18\overline{)1\,3\,1\,9} \\ \underline{1\,2\,6} \\ 5\,9 \\ \underline{5\,4} \\ 5 \end{array} \qquad \textbf{Check} \qquad \begin{array}{r} 7\,3 \\ \times\ \ 1\,8 \\ \hline 5\,8\,4 \\ 7\,3 \\ \hline 1\,3\,1\,4 \\ +\ \ \ \ 5 \\ \hline 1\,3\,1\,9 \end{array}$$

S-1

SOLUTIONS

25.

$$35\overline{\smash{)}3549} \quad \underline{101}\,\mathbf{R}4$$

Check 101

$$\begin{array}{r} 101 \\ \times \ 35 \\ \hline 505 \\ 303 \\ \hline 3535 \\ + \quad 4 \\ \hline 3539 \end{array}$$ incorrect

Rework:

$$35\overline{\smash{)}3549} \quad \underline{101}\,\mathbf{R}14$$

$$\begin{array}{r} 35 \\ \hline 049 \\ 35 \\ \hline 14 \end{array}$$

Check 101

$$\begin{array}{r} 101 \\ \times \ 35 \\ \hline 505 \\ 303 \\ \hline 3535 \\ + \quad 14 \\ \hline 3549 \end{array}$$ Matches

The correct answer is 101 **R**14.

41. Divide the yearly amount spent for eating away from home by 52 (the number of weeks in one year).

$$52\overline{\smash{)}2028} \quad \underline{39}$$

$$\begin{array}{r} 156 \\ \hline 468 \\ 468 \\ \hline 0 \end{array}$$

The average weekly amount spent for eating away from home is $39 per household.

Section 1.7 (pages 71–74)

1. 624 rounded to the nearest ten: 620

6<u>2</u>4   Underline the 2 in the tens place. Next digit is 4 or less, so tens place does not change. All digits to the right of the underlined place change to zero.

25. To the nearest ten: 44<u>7</u>6

Underline the 7 in the tens place. Next digit is 5 or more, so add 1 to the 7. All digits to the right of the underlined place are changed to zero.

4480

To the nearest hundred: 4<u>4</u>76

Underline the 4 in the hundreds place. Next digit is 5 or more, so add 1 to the 4. All digits to the right of the underlined place are changed to zero.

4500

To the nearest thousand: <u>4</u>476

Underline the 4 in the thousands place. Next digit is 4 or less, so leave 4 as 4. All digits to the right of the underlined place are changed to zero.

4000

41.

Estimate:	Exact:
30	25
60	63
50	47
+ 80	+ 84
220	219

47.

Estimate:	Exact:
900	863
700	735
400	438
+ 800	+ 792
2800	2828

53.

Estimate:	Exact:
8000	8 215
60	56
700	729
+ 4000	+ 3 605
12,760	12,605

61. <u>7</u>6,000,000   Next digit is 5 or more, so add 1 to 7. Change digit to the right of the underlined place to zero. Add 1 to 7.
80 million people

3<u>0</u>3,000,000   Next digit is 4 or less, so leave 0 as 0. Change digit to the right of the underlined place to zero.
300 million people

Section 1.8 (pages 77–80)

1. 3^2: exponent is 2, base is 3.
$3^2 = 3 \cdot 3 = 9$

21. $35^2 = \underline{1225}$, so $\sqrt{1225} = 35$.

37.

$5 \cdot 3^2 + \dfrac{0}{8}$	Exponent
$5 \cdot 9 + \dfrac{0}{8}$	Multiply
$45 + \dfrac{0}{8}$	Divide
$45 + 0 = 45$	Add

67.

$8 \cdot \sqrt{49} - 6(9 - 4)$	Parentheses
$8 \cdot \sqrt{49} - 6(5)$	Square root
$8 \cdot 7 - 6(5)$	Multiply
$56 - 6(5)$	Multiply.
$56 - 30 = 26$	Subtract

85. $8 \cdot 9 \div \sqrt{36} - 4 \div 2 + (14 - 8)$

	Parentheses
$8 \cdot 9 \div \sqrt{36} - 4 \div 2 + 6$	Square root
$8 \cdot 9 \div 6 - 4 \div 2 + 6$	Multiply
$72 \div 6 - 4 \div 2 + 6$	Divide
$12 - 4 \div 2 + 6$	Divide
$12 - 2 + 6$	Subtract
$10 + 6 = 16$	Add

Section 1.9 (pages 85–88)

5. According the pictograph, Family Dollar has 4500 stores. Subtract 4500 from 5750 to find the difference.

$5750 - 4500 = 1250$ fewer stores

11. According to the bar graph, $18 - 9 = 9$ more people out of 100 found their career as a result of "Studied in school" than "Luck or chance."

15. According to the line graph the increase in the number of trees planted from 2010 to 2011 is $7000 - 2500 = 4500$.

19. We want to insert parentheses in $7 - 2 \cdot 3 - 6$ so that it simplifies to 9.

There isn't a method to use, so just use trial and error.

$(7 - 2) \cdot 3 - 6 = 5 \cdot 3 - 6$
$\quad = 15 - 6 = 9$

Section 1.10 (pages 93–96)

7. *Step 1*

Find the number of toys each child will receive.

Step 2

"Same number of toys to each" indicates division should be used.

Step 3

3000 divided by 700 gives an estimate of about 4 toys.

Step 4

$$657\overline{\smash{)}2628} \quad \underline{4}$$
$$\begin{array}{r} 2628 \\ \hline 0 \end{array}$$

Step 5

Each child will receive 4 toys.

Step 6

Exact answer is close to estimate. Also, $4 \times 657 = 2628$ which matches the number given in the problem.

17. *Step 1*

Find her monthly savings.

Step 2

Her monthly take home pay and expenses are given. "Remainder" indicates subtraction may be used.

Step 3

Estimate:

$2000 - $700 - $300 - $400 - $200 - $200 = $200

Step 4

$695	$2240
340	−1890
435	$350
240	
+ 180	
1890	

Step 5

Her monthly savings are $350.

Step 6

The exact answer seems reasonable compared to the estimate.

Check: $350 + $1890 = $2240. Re-add the expenses to check the sum of $1890.

25. *Step 1*

The number and cost of wheelchairs and recorder-players are given, and the total cost of all items must be found.

Step 2

Find the cost of all wheel chairs and the cost of all recorder-players. Then add these costs to get the total cost.

Step 3

The cost of the wheelchairs is about $1000 \times 6 = \$6000$.

The cost of the recorder-players is about $900 \times 20 = \$18,000$.

Estimate: $\$6000 + \$18,000 = \$24,000$

Step 4

cost of wheelchairs: $1256

$$\begin{array}{r} 1256 \\ \times\ \ 6 \\ \hline \$7536 \end{array}$$

cost of recorder-players:

$$\begin{array}{r} \$8\ 9\ 5 \\ \times\ \ 1\ 5 \\ \hline 4\ 4\ 7\ 5 \\ 8\ 9\ 5\ \ \ \\ \hline \$1\ 3{,}4\ 2\ 5 \end{array}$$

Step 5

The total cost is $\$13,425 + \$7536 = \$20,961$.

Step 6

The exact answer is reasonably close to the estimate. Check by repeating Step 4.

39. *Step 1*

The total seating is given along with the information to find the number of seats on the main floor. The number of rows of seats in the balcony is given, and the number of seats in each row of the balcony must be found.

Step 2

First, the number of seats on the main floor must be found. Next, the number of seats on the main floor must be subtracted from the total number of seats to find the number of seats in the balcony. Finally, the number of seats in the balcony must be divided by the number of rows of seats in the balcony.

Step 3

Seats on main floor: $30 \times 25 = 750$

Seats in balcony: $1250 - 750 = 500$

Seats in each row in balcony:
$500 \div 25 = 20$

Since we didn't round, our estimate matches our exact answer.

Step 4

See the calculations in Step 3.

Step 5

The number of seats in each row of the balcony is 20.

Step 6

The answer matches the estimate, as expected.

Check

Seats in balcony: $20 \times 25 = 500$
Seats on main floor: $30 \times 25 = 750$
Total seats: $500 + 750 = 1250$, which matches the number given in the problem.

Chapter 2 Multiplying and Dividing Fractions

Section 2.1 (pages 115–116)

5. Each of the two figures is divided into 5 parts and 7 are shaded: $\dfrac{7}{5}$

Three are unshaded: $\dfrac{3}{5}$

17. Proper fractions: numerator *smaller* than denominator.

$$\frac{1}{3},\ \frac{5}{8},\ \frac{7}{16}$$

Improper fractions: numerator *greater than or equal to* denominator.

$$\frac{8}{5},\ \frac{6}{6},\ \frac{12}{2}$$

Section 2.2 (pages 121–124)

16. Write $5\dfrac{4}{5}$ as an improper fraction.

$5 \cdot 5 = 25$ Multiply 5 and 5.
$25 + 4 = 29$ Add 4. The numerator is 29.

$5\dfrac{4}{5} = \dfrac{29}{5}$ Use the same denominator.

43. $\dfrac{63}{4}$

Divide 63 by 4.

$$\begin{array}{r} 1\ 5\ \ \leftarrow \text{Whole number part} \\ 4\overline{)6\ 3} \\ 4\ \ \ \\ \hline 2\ 3 \\ 2\ 0 \\ \hline 3\ \ \leftarrow \text{Remainder} \end{array}$$

The quotient 15 is the whole number part of the mixed number. The remainder 3 is the numerator of the fraction, and the denominator remains as 4.

$$\frac{63}{4} = 15\frac{3}{4}$$

61. Write $522\dfrac{3}{8}$ as an improper fraction.

$522 \cdot 8 = 4176$

$4176 + 3 = 4179$ Add 3. The numerator is 4179.

$522\dfrac{3}{8} = \dfrac{4179}{8}$ Use the same denominator.

Section 2.3 (pages 129–130)

5. Factorizations of 48:

$1 \cdot 48 = 48$ $2 \cdot 24 = 48$ $3 \cdot 16 = 48$
$4 \cdot 12 = 48$ $6 \cdot 8 = 48$

The factors of 48 are 1, 2, 3, 4, 6, 8, 12, 16, 24, and 48.

41. 44

$$\begin{array}{r} 22 \\ 2\overline{)44} \end{array}$$ Divide 44 by 2.

$$\begin{array}{r} 11 \\ 2\overline{)22} \end{array}$$ Divide 22 by 2.

$$\begin{array}{r} 1 \\ 11\overline{)11} \end{array}$$ Divide 11 by 11.

Quotient is 1.

Because all factors (divisors) are prime, the prime factorization of 44 is

$2 \cdot 2 \cdot 11 = 2^2 \cdot 11$

Section 2.4 (pages 135–136)

17. Because the greatest common factor of 56 and 64 is 8, divide both numerator and denominator by 8.

$$\frac{56}{64} = \frac{56 \div 8}{64 \div 8} = \frac{7}{8}$$

29. Write the prime factorization of both numerator and denominator. Then divide both numerator and denominator by any common factors, and write a 1 by each factor that has been divided. Finally, multiply the remaining factors in both numerator and denominator.

$$\frac{18}{24} = \frac{\overset{1}{\cancel{2}} \cdot \overset{1}{\cancel{3}} \cdot 3}{\underset{1}{\cancel{2}} \cdot 2 \cdot 2 \cdot \underset{1}{\cancel{3}}} = \frac{1 \cdot 1 \cdot 3}{1 \cdot 2 \cdot 2 \cdot 1} = \frac{3}{4}$$

39. $\dfrac{3}{6}$ and $\dfrac{18}{36}$

Write each fraction in lowest terms.

$$\frac{3}{6} = \frac{1 \cdot \overset{1}{\cancel{3}}}{2 \cdot \underset{1}{\cancel{3}}} = \frac{1}{2}$$

$$\frac{18}{36} = \frac{\overset{1}{\cancel{2}} \cdot \overset{1}{\cancel{3}} \cdot \overset{1}{\cancel{3}}}{\underset{1}{\cancel{2}} \cdot 2 \cdot \underset{1}{\cancel{3}} \cdot \underset{1}{\cancel{3}}} = \frac{1}{2}$$

The fractions are *equivalent* $\left(\dfrac{1}{2} = \dfrac{1}{2}\right)$.

53. Write the prime factorization of both numerator and denominator. Then divide both numerator and denominator by any common factors, and write a 1 by each factor that has been divided. Finally, multiply the remaining factors in both numerator and denominator.

$$\frac{160}{256} = \frac{\overset{1}{\cancel{2}} \cdot \overset{1}{\cancel{2}} \cdot \overset{1}{\cancel{2}} \cdot \overset{1}{\cancel{2}} \cdot \overset{1}{\cancel{2}} \cdot 5}{\underset{1}{\cancel{2}} \cdot \underset{1}{\cancel{2}} \cdot \underset{1}{\cancel{2}} \cdot \underset{1}{\cancel{2}} \cdot \underset{1}{\cancel{2}} \cdot 2 \cdot 2 \cdot 2} = \frac{5}{8}$$

Section 2.5 (pages 145–148)

7. Divide a numerator and denominator by the same number to use the multiplication shortcut.

$$\frac{2}{3} \cdot \frac{7}{12} \cdot \frac{9}{14} = \frac{\overset{1}{\cancel{2}}}{\underset{1}{\cancel{3}}} \cdot \frac{\overset{1}{\cancel{7}}}{\underset{6}{\cancel{12}}} \cdot \frac{\overset{\overset{1}{\cancel{3}}}{\cancel{9}}}{\underset{2}{\cancel{14}}}$$

$$= \frac{1 \cdot 1 \cdot 1}{1 \cdot 2 \cdot 2} = \frac{1}{4}$$

25. Write 36 as $\dfrac{36}{1}$ and multiply.

$$36 \cdot \frac{5}{8} \cdot \frac{9}{15} = \frac{\overset{9}{\cancel{36}}}{1} \cdot \frac{\overset{1}{\cancel{5}}}{\underset{2}{\cancel{8}}} \cdot \frac{\overset{3}{\cancel{9}}}{\underset{\underset{1}{\cancel{3}}}{\cancel{15}}} = \frac{27}{2} = 13\frac{1}{2}$$

41. Area = length • width

Area $= \dfrac{7}{5} \cdot \dfrac{3}{14}$

$= \dfrac{\overset{1}{7}}{5} \cdot \dfrac{3}{\underset{2}{14}}$ Divide numerator and denominator by 7.

$= \dfrac{3}{10}$ in.^{2}

45. Area = length • width

Area $= 2 \cdot \dfrac{3}{4}$

$= \dfrac{\overset{1}{2}}{1} \cdot \dfrac{3}{\underset{2}{4}}$ Divide numerator and denominator by 2.

$= \dfrac{3}{2}$ Write as a mixed number.

$= 1\dfrac{1}{2}$ yd^2

Section 2.6 (pages 153–156)

1. *Step 1*
The problem asks for the area of the top of a file cabinet.

Step 2
To find area, multiply the length (depth) $\dfrac{3}{4}$ yd by the width $\dfrac{2}{3}$ yd.

Step 3
An estimate is $\dfrac{6}{12}$ yd^2.

$\dfrac{3}{4} \cdot \dfrac{2}{3} = \dfrac{\overset{1}{3} \cdot \overset{1}{2}}{\underset{2}{4} \cdot \underset{1}{3}} = \dfrac{1}{2}$

Step 4
The exact value is $\dfrac{1}{2}$, which is the same as the estimate since we didn't round.

Step 5
The area of the file cabinet top is $\dfrac{1}{2}$ yd^2.

Step 6
The answer, $\dfrac{1}{2}$ yd^2, matches our estimate.

11. *Step 1*
The problem asks for the number of runners who are women.

Step 2
$\dfrac{7}{12}$ of the 1560 runners are women, so multiply $\dfrac{7}{12}$ by 1560 to find the number of runners who are women.

Step 3
Round the number of runners to 1600 then, $\dfrac{1}{2}$ of 1600 is 800. Since $\dfrac{7}{12}$ is more than $\dfrac{1}{2}$, our estimate is that "more than 800 runners" are women.

Step 4

$\dfrac{7}{12} \cdot 1560 = \dfrac{7 \cdot \overset{130}{1560}}{\underset{1}{12} \cdot 1} = 910$

Step 5 910 runners are women.

Step 6
The exact answer, 910, fits our estimate of "more than 800."

21. From Exercise 19 the total income is $58,000. The circle graph shows that $\dfrac{1}{5}$ of the income is for rent.

$\dfrac{1}{5} \cdot 58{,}000 = \dfrac{1}{\underset{1}{5}} \cdot \dfrac{\overset{11{,}600}{58{,}000}}{1} = 11{,}600$

The amount of their rent is $11,600.

33. To find the remaining amount of the estate subtract $\dfrac{7}{8}$ from 1.

$1 - \dfrac{7}{8} = \dfrac{8}{8} - \dfrac{7}{8} = \dfrac{1}{8}$

Multiply the remaining $\dfrac{1}{8}$ of the estate by the fraction going to the American Cancer Society.

$\dfrac{1}{4} \cdot \dfrac{1}{8} = \dfrac{1}{32}$

$\dfrac{1}{32}$ of the estate goes to the American Cancer Society.

Section 2.7 (pages 163–166)

1. The reciprocal of $\dfrac{3}{8}$ is $\dfrac{8}{3}$ because

$\dfrac{3}{8} \cdot \dfrac{8}{3} = \dfrac{24}{24} = 1$.

25. $\dfrac{18}{\frac{3}{4}} = 18 \div \dfrac{3}{4}$ Rewrite by using the ÷ symbol for division.

$= \dfrac{18}{1} \div \dfrac{3}{4}$ Write 18 as $\dfrac{18}{1}$.

$= \dfrac{\overset{6}{18}}{1} \cdot \dfrac{4}{\underset{1}{3}}$ The reciprocal of $\dfrac{3}{4}$ is $\dfrac{4}{3}$. Change "÷" to " • ". Divide the numerator and denominator by 3.

$= \dfrac{6 \cdot 4}{1 \cdot 1}$ Multiply.

$= 24$

31. *Step 1*
The problem asks for the number of times a measuring cup needs to be filled.

Step 2
Solve the problem by dividing the total number of cups (5) by the size of the measuring cup $\left(\dfrac{1}{3}\right)$.

Step 3
Because there are three $\dfrac{1}{3}$-cups in one cup, multiply 3 by 5 to get 15. So, our estimate is 15.

Step 4

$5 \div \dfrac{1}{3} = \dfrac{5}{1} \div \dfrac{1}{3} = \dfrac{5}{1} \cdot \dfrac{3}{1} = 15$

Step 5
They need to fill the measuring cup 15 times.

Step 6
The exact answer, 15, is the same as our estimate.

45. *Step 1*
The problem asks for the number of towels that can be made.

Step 2
Solve the problem by dividing the 912 yards of fabric by the fraction of a yard $\left(\dfrac{3}{8}\right)$ needed for each dish towel.

Step 3
Round 912 yards to 900 yards. Round $\dfrac{3}{8}$ to $\dfrac{1}{2}$.

Multiply 900 by 2, the reciprocal of $\dfrac{1}{2}$, to get the estimate of 1800 towels.

Step 4

$912 \div \dfrac{3}{8} = \dfrac{912}{1} \cdot \dfrac{8}{3} = \dfrac{\overset{304}{912}}{1} \cdot \dfrac{8}{\underset{1}{3}} = 2432$

Step 5 2432 towels can be made.

Step 6
The exact answer, 2432, is close to our estimate, 1800.

Section 2.8 Exercises (pages 173–176)

1. $4\dfrac{1}{2} \cdot 1\dfrac{3}{4}$

Estimate: $5 \cdot 2 = 10$

To find the exact answer, change each mixed number to an improper fraction and then multiply.

Exact: $4\dfrac{1}{2} \cdot 1\dfrac{3}{4} = \dfrac{9}{2} \cdot \dfrac{7}{4} = \dfrac{63}{8} = 7\dfrac{7}{8}$

13. $1\dfrac{1}{4} \div 3\dfrac{3}{4}$

Estimate: $1 \div 4 = \dfrac{1}{4}$

To find the exact answer, first change each mixed number to an improper fraction. Then, use the reciprocal of the second fraction and multiply.

Exact:

$1\dfrac{1}{4} \div 3\dfrac{3}{4} = \dfrac{5}{4} \div \dfrac{15}{4} = \dfrac{\overset{1}{5}}{\underset{1}{4}} \cdot \dfrac{\overset{1}{4}}{\underset{3}{15}} = \dfrac{1}{3}$

Chapter 3 Adding and Subtracting Fractions

Section 3.1 (pages 197–198)

13. Add the numerators. Keep the denominator.

$$\frac{4}{15} + \frac{2}{15} + \frac{5}{15} = \frac{4+2+5}{15} = \frac{11}{15}$$

29. First, subtract the numerators and keep the denominator. Then write the fraction in lowest terms.

$$\frac{47}{36} - \frac{5}{36} = \frac{47-5}{36} = \frac{42}{36}$$

$$= \frac{42 \div 6}{36 \div 6} = \frac{7}{6} = 1\frac{1}{6}$$

39. First, add the two fractions of land that were purchased.

$$\frac{9}{10} + \frac{3}{10} = \frac{9+3}{10} = \frac{12}{10}$$

Next, subtract the fraction of land that was planted in carrots.

$$\frac{12}{10} - \frac{7}{10} = \frac{12-7}{10} = \frac{5}{10} = \frac{1}{2}$$

$\frac{1}{2}$ acre is planted in squash.

Section 3.2 (pages 207–210)

11. 20 and 50

Multiples of 50:

50, 100, 150, 200, 250, . . .

100 is the first multiple of 50 that is divisible by 20. $(100 \div 20 = 5)$

The least common multiple of 20 and 50 is 100.

19. Find the prime factorization for each number.

4, 6, 8, 10

$$4 = 2 \cdot 2$$
$$6 = 2 \cdot 3$$
$$8 = 2 \cdot 2 \cdot 2$$
$$10 = 2 \cdot 5$$

$$\text{LCM} = 2 \cdot 2 \cdot 2 \cdot 3 \cdot 5 = 120$$

The LCM of 4, 6, 8, and 10 is 120.

37. $\frac{3}{16} = \frac{?}{64}$ Divide 64 by 16, getting 4. Now multiply both numerator and denominator of $\frac{3}{16}$ by 4.

$$\frac{3}{16} = \frac{3 \cdot 4}{16 \cdot 4} = \frac{12}{64}$$

53. $\frac{109}{1512}, \frac{23}{392}$

```
2 | 1512  392
2 |  756  196
2 |  378   98
3 |  189   49
3 |   63   49
3 |   21   49
7 |    7   49
7 |    1    7
        1    1
```

The LCM of the denominators is

$2 \cdot 2 \cdot 2 \cdot 3 \cdot 3 \cdot 3 \cdot 7 \cdot 7 = 10{,}584$.

59. Find the prime factorization for each number.

5, 7, 14, 10

$$5 = 1 \cdot 5$$
$$7 = 1 \cdot 7$$
$$14 = 2 \cdot 7$$
$$10 = 2 \cdot 5$$

$$\text{LCM} = 2 \cdot 5 \cdot 7 = 70$$

The LCM of 5, 7, 14, and 10 is 70.

Section 3.3 (pages 215–218)

1. The least common multiple of 4 and 8 is 8. Rewrite both fractions as fractions with a least common denominator of 8. Then add the numerators.

$$\frac{3}{4} + \frac{1}{8} = \frac{6}{8} + \frac{1}{8}$$

$$= \frac{6+1}{8}$$

$$= \frac{7}{8}$$

29. The least common multiple of 12 and 4 is 12. Rewrite both fractions as fractions with a least common denominator of 12. Then subtract the numerators and write the resulting fraction in lowest terms.

$$\frac{5}{12} - \frac{1}{4} = \frac{5}{12} - \frac{3}{12}$$

$$= \frac{5-3}{12}$$

$$= \frac{2}{12} = \frac{1}{6}$$

41. Add the fractions to find the total length of the screw. The least common denominator of the fractions is 40. Rewrite all three fractions with a denominator of 40. Then add the numerators.

$$\frac{1}{8} + \frac{1}{4} + \frac{2}{5} = \frac{5}{40} + \frac{10}{40} + \frac{16}{40}$$

$$= \frac{5+10+16}{40}$$

$$= \frac{31}{40}$$

The total length of the screw is $\frac{31}{40}$ inch.

49. One way to compare fractions accurately is to rewrite each fraction with a common denominator.

$$\frac{1}{3} = \frac{8}{24}, \frac{1}{6} = \frac{4}{24}, \frac{1}{8} = \frac{3}{24}, \frac{1}{12} = \frac{2}{24}, \frac{7}{24}$$

The fraction with the largest numerator is the largest fraction. This is $\frac{8}{24}$ or $\frac{1}{3}$, which is "Work and Travel."

To find the number of hours, multiply.

$$\frac{1}{3} \cdot 24 = \frac{1}{\overset{1}{\cancel{3}}} \cdot \frac{\overset{8}{\cancel{24}}}{1} = \frac{8}{1} = 8$$

8 hours were spent on work and travel.

Note that this is just the numerator in the equivalent fraction $\frac{8}{24}$.

To find what fraction of the day was spent on "Work and Travel" and "Class," add the fractions $\frac{1}{3}$ and $\frac{1}{6}$. The least common denominator is 6. Rewrite $\frac{1}{3}$ with a denominator of 6. Then add the numerators and write in lowest terms.

$$\frac{1}{3} + \frac{1}{6} = \frac{2}{6} + \frac{1}{6} = \frac{2+1}{6} = \frac{3}{6} = \frac{1}{2}$$

Section 3.4 (pages 225–232)

11. *Estimate:*   *Exact:*

$$23 \xleftarrow{\text{Rounds to}} \begin{cases} 22\frac{3}{4} = 22\frac{21}{28} \end{cases}$$

$$+15 \xleftarrow{\text{Rounds to}} \begin{cases} + 15\frac{3}{7} = 15\frac{12}{28} \end{cases}$$

$$\overline{}$$

$$38 \qquad\qquad 37\frac{33}{28}$$

$$37\frac{33}{28} = 37 + 1\frac{5}{28} = 38\frac{5}{28}$$

21. *Estimate:*   *Exact:*

$$17 \xleftarrow{\text{Rounds to}} \{ \ 17$$

$$-7 \xleftarrow{\text{Rounds to}} \{ - 6\frac{5}{8}$$

$$\overline{10}$$

To subtract $\frac{5}{8}$, first regroup the whole number 17 into $16 + 1$.

$$17 = 16 + 1 = 16 + \frac{8}{8}$$

Now you can subtract.

$$\begin{array}{r} 16\frac{8}{8} \\ - \ 6\frac{5}{8} \\ \hline 10\frac{3}{8} \end{array}$$

39.

$$\begin{array}{r} 8\frac{3}{4} = \frac{35}{4} = \frac{70}{8} \\ - \ 5\frac{7}{8} = \frac{47}{8} = \frac{47}{8} \\ \hline \frac{23}{8} = 2\frac{7}{8} \end{array}$$

55. Subtract the size of the smallest hose clamp from the size of the largest hose clamp.

Estimate: $3 - 1 = 2$ in.

$$\textit{Exact:} \quad 2\frac{3}{4} = \frac{11}{4} = \frac{44}{16}$$

$$- \quad \frac{9}{16} = \frac{9}{16} = \frac{9}{16}$$

$$\overline{} \quad \frac{35}{16} = 2\frac{3}{16}$$

The difference in size is $2\frac{3}{16}$ inches.

65. Add the weights.

Estimate:

$59 + 24 + 17 + 29 + 58 = 187$ tons

Exact:

$$58\frac{1}{2} = 58\frac{12}{24}$$

$$23\frac{5}{8} = 23\frac{15}{24}$$

$$16\frac{5}{6} = 16\frac{20}{24}$$

$$29\frac{1}{4} = 29\frac{6}{24}$$

$$+ \; 58\frac{1}{3} = 58\frac{8}{24}$$

$$184\frac{61}{24} = 184 + 2\frac{13}{24} = 186\frac{13}{24}$$

The total weight is $186\frac{13}{24}$ tons.

Section 3.5 (pages 237–240)

7.

33. $\left(\dfrac{3}{2}\right)^4 = \dfrac{3}{2} \cdot \dfrac{3}{2} \cdot \dfrac{3}{2} \cdot \dfrac{3}{2} = \dfrac{81}{16} = 5\dfrac{1}{16}$

49. $6\left(\dfrac{2}{3}\right)^2\left(\dfrac{1}{2}\right)^3 = 6\left(\dfrac{2}{3} \cdot \dfrac{2}{3}\right)\left(\dfrac{1}{2} \cdot \dfrac{1}{2} \cdot \dfrac{1}{2}\right)$

Simplify the expressions with exponents.

$$= \frac{\overset{2}{\cancel{6}} \cdot \overset{1}{\cancel{2}} \cdot \overset{1}{\cancel{2}} \cdot 1 \cdot 1 \cdot 1}{\underset{1}{\cancel{3}} \cdot 3 \cdot \underset{1}{\cancel{2}} \cdot \underset{1}{\cancel{2}} \cdot \underset{1}{\cancel{2}}}$$

$$= \frac{1}{3}$$

63. $\left(\dfrac{3}{4}\right)^2 - \left(\dfrac{1}{2} - \dfrac{1}{6}\right) \div \dfrac{4}{3}$

$$= \left(\frac{3}{4}\right)^2 - \left(\frac{3}{6} - \frac{1}{6}\right) \div \frac{4}{3}$$

Work inside parentheses first.

$$= \left(\frac{3}{4}\right)^2 - \left(\frac{2}{6}\right) \cdot \frac{3}{4}$$

Simplify the expression with the exponent.

$$= \frac{3}{4} \cdot \frac{3}{4} - \left(\frac{2}{6}\right) \cdot \frac{3}{4} \quad \text{Multiply.}$$

$$= \frac{9}{16} - \frac{\overset{1}{\cancel{2}}}{\underset{2}{\cancel{6}}} \cdot \frac{\overset{1}{\cancel{3}}}{\underset{2}{\cancel{4}}}$$

$$= \frac{9}{16} - \frac{1}{4} \quad \begin{array}{l}\text{The LCD is 16. Rewrite } \frac{1}{4} \\ \text{as a fraction with a} \\ \text{denominator of 16.}\end{array}$$

$$= \frac{9}{16} - \frac{4}{16} \quad \text{Subtract.}$$

$$= \frac{5}{16}$$

Chapter 4 Decimals

Section 4.1 (pages 261–264)

1.

tens	ones		tenths	
7	0	.	4	8	9

---

17. Reorder as: 5 thousands, 4 hundreds,
0 tens, 0 tenths, 4 hundredths,
5 thousandths: 5406.045

31. $0.205 = \dfrac{205}{1000} = \dfrac{205 \div 5}{1000 \div 5} = \dfrac{41}{200}$

45. Three decimal places is *thousandths,* so
1.075 is one and seventy-five *thousandths.*

53. seven hundred three ten-thousandths:

$$\frac{703}{10,000} = 0.0703$$

59. 8-pound test line has a diameter of
0.010 inch. 0.010 inch is read
ten thousandths of an inch.

$$0.010 = \frac{10}{1000} = \frac{10 \div 10}{1000 \div 10} = \frac{1}{100} \text{ inch}$$

Section 4.2 (pages 271–272)

9. Draw a cut-off line after the tenths
place: 793.9|88

The first digit cut is 8, which is 5 or more,
so round up the tenths place.

$$\begin{array}{r} 793.9 \\ + \quad 0.1 \\ \hline 794.0 \end{array}$$

Answer: $793.988 \approx 794.0$

27. Round $310.08 to the nearest dollar.
Draw a cut-off line: $310.|08

The first digit cut is 0, which is 4 or less.
The part you keep stays the same.
Union dues: $310.08 \approx$ \$310

35. Round $999.73 to the nearest dollar.
Draw a cut-off line: $999.|73

The first digit cut is 7, which is 5 or more,
so round up.

$$\begin{array}{r} \$999 \\ + \quad 1 \\ \hline \$1000 \end{array}$$

Answer: $\$999.73 \approx \$1000$

Section 4.3 (pages 277–280)

11. Subtract 0.291 from 0.4

Line up decimal points.

$$\begin{array}{r} 0.400 \quad \text{Write two zeros.} \\ - \; 0.291 \\ \hline 0.109 \end{array}$$

23.

Estimate:	*Exact:*
400	392.700
1	0.865
+ 20	+ 21.080
421	414.645

41. First add the three players' heights.

Estimate:	*Exact:*
2	1.83
2	2.16
+ 2	+ 2.11
6 meters	6.10 meters

Now subtract the players' combined height
of 6.1 meters from the rhino's height of
6.4 meters.

$6.4 - 6.1 = 0.3$

So, the NBA stars' combined height is
0.3 meter less than the rhino's height.

47. Subtract the price of the regular fishing
line, $4.84, from the price of the fluores-
cent fishing line, $5.14.

Estimate:	*Exact:*
$5	$5.14
− 5	− 4.84
$0	$0.30

The fluorescent fishing line costs $0.30
more than the regular fishing line.

Section 4.4 (pages 283–286)

11. (7.2) (0.06)

72	7.2	← 1 decimal place
× 6	× 0.06	← 2 decimal places
432	0.432	← 3 decimal places

17. (0.006) (0.0052)

0.0052	← 4 decimal places
× 0.006	← 3 decimal places
0.0000312	← 7 decimal places

43. Multiply the number of gallons that she
pumped into her pickup truck by the price
per gallon. Use a calculator.

$20.510 \times 4.389 = 90.01839$

Round 90.01839 to the nearest cent.
Michelle paid $90.02 for the gas.

59. (a) Find the cost for three short-sleeved,
solid-color shirts

$$\begin{array}{r} \$14.75 \\ \times \quad 3 \\ \hline \$44.25 \end{array}$$

Based on this subtotal, shipping is $5.95.

Next, find the cost of the three monograms.

$$\begin{array}{r} \$4.95 \\ \times \quad 3 \\ \hline \$14.85 \end{array}$$

Add these amounts, plus $5.00 for a
gift box.

$44.25	← Shirts
5.95	← Shipping
14.85	← Monograms
+ 5.00	← Gift box
$70.05	

The total cost is $70.05.

(b) Subtract the cost of the shirts to find
the difference.

$$\begin{array}{r} \$70.05 \\ - \; 44.25 \\ \hline \$25.80 \end{array}$$

The monograms, gift box, and shipping
added $25.80 to the cost of the gift.

Section 4.5 (pages 293–296)

7.

$$\begin{array}{r} 3\,6. \\ 1.5 \,)\overline{54.0} \\ \underline{4\,5} \\ 9\,0 \\ \underline{9\,0} \\ 0 \end{array}$$

Move decimal point in
divisor and dividend
1 place; write 0 in the
dividend.

9. Given: $108 \div 18 = 6$

Find: $0.108 \div 1.8$ by moving the decimal points.

$$1.8\overline{)0.1\,08}\quad \frac{.06}{}$$

So, $0.108 \div 1.8 = 0.06$

15. $\dfrac{3.1}{0.006}$

$$0.006\overline{)3.100000}\quad\frac{516.666}{}$$

Line up decimal points.

Move decimal point in divisor and dividend three places. Write 00 in dividend.

$\begin{array}{r}30\\10\\6\\40\\36\end{array}$

$\begin{array}{r}40\quad\text{Write 0 in dividend.}\\36\end{array}$

$\begin{array}{r}40\quad\text{Write 0 in dividend.}\\36\end{array}$

$\begin{array}{r}40\quad\text{Write 0 in dividend.}\\36\end{array}$

4   Stop and round answer to the nearest hundredth.

The quotient is 516.67 (rounded).

27. $307.02 \div 5.1 = 6.2$

Estimate: $300 \div 5 = 60$

The answer 6.2 is *unreasonable* because it is so much less than 60.

$$5.1\overline{)307.02}\quad\frac{60.2}{}$$

$\begin{array}{r}306\\10\\0\\102\\102\\0\end{array}$

The correct answer is 60.2, which is close to the estimate of 60.

53. $33 - 3.2(0.68 + 9) - 1.3^2$ Parentheses; Exponent

$33 - 3.2(9.68) - 1.69$ Multiply

$33 - 30.976 - 1.69$ Subtract

$2.024 - 1.69 = 0.334$ Subtract

61. Divide 100,000 box tops (the answer from Exercise 59) by 38 weeks.

$$38\overline{)100,000.0}\quad\frac{2631.5}{}$$

$\begin{array}{r}76\\240\\228\\120\\114\\60\\38\\220\\190\\30\end{array}$

2631.5 rounds to 2632.

The school needs to collect 2632 box tops (rounded) during each of the 38 weeks.

Section 4.6 (pages 301–304)

17. $3\dfrac{5}{8} = \dfrac{29}{8} = 3.625$

$$8\overline{)29.000}\quad\frac{3.625}{}$$

$\begin{array}{r}24\\50\\48\\20\\16\\40\\40\\0\end{array}$

33. $0.35 = \dfrac{35}{100} = \dfrac{35 \div 5}{100 \div 5} = \dfrac{7}{20}$

49. Write zeros so that all the numbers have four decimal places. The acceptable lengths must be greater than 0.9980 cm and less than 1.0020 cm.

$1.0100 > 1.0020$ *unacceptable*

$0.9991 > 0.9980$ and
$0.9991 < 1.0020$ *acceptable*

$1.0007 > 0.9980$ and
$1.0007 < 1.0020$ *acceptable*

$0.9900 < 0.9980$ *unacceptable*

The lengths of 0.9991 cm and 1.0007 cm are acceptable.

63. $\dfrac{3}{8}, \dfrac{2}{5}, 0.37, 0.4001$

$\dfrac{3}{8} = 0.3750$

$\dfrac{2}{5} = 0.4000$

$0.37 = 0.3700 \leftarrow$ least

$0.4001 = 0.4001 \leftarrow$ greatest

From least to greatest: $0.37, \dfrac{3}{8}, \dfrac{2}{5}, 0.4001$

Chapter 5 Ratio and Proportion

Section 5.1 (pages 323–326)

3. $100 to $50

$\dfrac{\$100}{\$50} = \dfrac{100}{50} = \dfrac{100 \div 50}{50 \div 50} = \dfrac{2}{1}$

15. $1\dfrac{1}{4}$ to $1\dfrac{1}{2}$

$\dfrac{1\frac{1}{4}}{1\frac{1}{2}} = \dfrac{\frac{5}{4}}{\frac{3}{2}} = \dfrac{5}{4} \div \dfrac{3}{2} = \dfrac{5}{\cancel{4}} \cdot \dfrac{\cancel{2}}{3} = \dfrac{5}{6}$

35. (a) rent to utilities

$\dfrac{\$750}{\$125} = \dfrac{750}{125} = \dfrac{750 \div 125}{125 \div 125} = \dfrac{6}{1}$

(b) rent to food

$\dfrac{\$750}{\$300} = \dfrac{750}{300} = \dfrac{750 \div 150}{300 \div 150} = \dfrac{5}{2}$

(c) rent and utilities to total budget

rent and utilities: $750 + $125 = $875

total budget

$225 + $125 + $200 + $400

$+ \$300 + \$750 = \$2000$

$\dfrac{\$875}{\$2000} = \dfrac{875}{2000} = \dfrac{875 \div 125}{2000 \div 125} = \dfrac{7}{16}$

43. The increase in price is
$12.50 - $10 = $2.50

$\dfrac{\$2.50}{\$10} = \dfrac{2.50}{10} = \dfrac{2.50 \cdot 10}{10 \cdot 10} = \dfrac{25}{100}$

$= \dfrac{25 \div 25}{100 \div 25} = \dfrac{1}{4}$

The ratio of the increase in price to the original price is $\dfrac{1}{4}$.

Section 5.2 (pages 331–334)

15. Miles traveled:
$28,396.7 - 28,058.1 = 338.6$

Miles per gallon:

$\dfrac{338.6 \text{ miles} \div 16.2}{16.2 \text{ gallons} \div 16.2} \approx 20.90 \approx 20.9$

19.

Size	Cost per Unit
12 ounces	$\dfrac{\$2.49}{12 \text{ ounces}} \approx \$0.208$
14 ounces	$\dfrac{\$2.89}{14 \text{ ounces}} \approx \$0.206$
18 ounces	$\dfrac{\$3.96}{18 \text{ ounces}} = \$0.22$

Because $0.206 is the lowest cost per ounce, the best buy is 14 ounces for $2.89.

33. One battery for $1.79 is like getting 3 batteries for $1.79 \div 3 \approx $0.597 per battery.

An eight-pack of AAA batteries for $4.99 is $4.99 \div 8 \approx $0.624 per battery.

Because $0.597 is the lower cost per battery, the better buy is the one battery package.

Section 5.3 (pages 339–340)

11. $\dfrac{150}{200} = \dfrac{200}{300}$

$\dfrac{150 \div 50}{200 \div 50} = \dfrac{3}{4}$ and $\dfrac{200 \div 100}{300 \div 100} = \dfrac{2}{3}$

Because $\dfrac{3}{4}$ is *not* equivalent to $\dfrac{2}{3}$, the proportion is *false*.

31. $\dfrac{2\frac{5}{8}}{3\frac{1}{4}} = \dfrac{21}{26}$

Cross products:

$2\dfrac{5}{8} \cdot 26 = \dfrac{21}{\cancel{8}} \cdot \dfrac{\cancel{26}^{13}}{1} = \dfrac{273}{4} = 68\dfrac{1}{4}$

$3\dfrac{1}{4} \cdot 21 = \dfrac{13}{4} \cdot \dfrac{21}{1} = \dfrac{273}{4} = 68\dfrac{1}{4}$

The cross products are *equal*, so the proportion is *true*.

35. $\dfrac{2\frac{3}{10}}{8.05} = \dfrac{\frac{1}{4}}{0.9}$

(continued)

Cross products:

$$2\frac{3}{10}(0.9) = \frac{23}{10} \cdot \frac{9}{10} = \frac{207}{100} = 2\frac{7}{100}$$

or $(2.3)(0.9) = 2.07$

$$(8.05)\frac{1}{4} = (8.05)(0.25) = 2.0125$$

or $8\frac{5}{100} \cdot \frac{1}{4} = \frac{805}{100} \cdot \frac{1}{4} = \frac{805}{400} = 2\frac{1}{80}$

The cross products are *unequal,*
so the proportion is *false.*

$$\left(2\frac{7}{100} \neq 2\frac{1}{80} \quad \text{or} \quad 2.07 \neq 2.0125\right)$$

Section 5.4   (pages 345–346)

7. $\dfrac{42}{x} = \dfrac{18}{39}$

$x \cdot 18 = 42 \cdot 39$ Cross products are equivalent

$\dfrac{x \cdot \overset{1}{\cancel{18}}}{\cancel{18}} = \dfrac{1638}{18}$ Divide both sides by 18.

$x = 91$

Check

$42 \cdot 39 = 1638$

$91 \cdot 18 = 1638$

13. $\dfrac{99}{55} = \dfrac{44}{x}$ **OR** $\dfrac{9}{5} = \dfrac{44}{x}$

$9 \cdot x = 5 \cdot 44$ Cross products are equivalent

$\dfrac{\overset{1}{\cancel{9}} \cdot x}{\underset{1}{\cancel{9}}} = \dfrac{220}{9}$ Divide both sides by 9.

$x = \dfrac{220}{9}$

$x \approx 24.44$

Check

$55 \cdot 44 = 2420$

$99 \cdot 24.44 = 2419.56$

Slightly different because of rounding.

21. $\dfrac{2\frac{1}{3}}{1\frac{1}{2}} = \dfrac{x}{2\frac{1}{4}}$

$1\frac{1}{2} \cdot x = 2\frac{1}{3} \cdot 2\frac{1}{4}$

$\dfrac{3}{2} \cdot x = \dfrac{7}{\underset{1}{\cancel{3}}} \cdot \dfrac{\overset{3}{\cancel{9}}}{4}$

$\dfrac{3}{2} \cdot x = \dfrac{21}{4}$

$\dfrac{\frac{3}{2} \cdot x}{\frac{3}{2}} = \dfrac{\frac{21}{4}}{\frac{3}{2}}$

$x = \dfrac{21}{4} \div \dfrac{3}{2} = \dfrac{\overset{7}{\cancel{21}}}{\underset{2}{\cancel{4}}} \cdot \dfrac{\overset{1}{\cancel{2}}}{\underset{1}{\cancel{3}}} = \dfrac{7}{2} = 3\frac{1}{2}$

25. $\dfrac{x}{\frac{3}{50}} = \dfrac{0.15}{1\frac{4}{5}}$

Change to decimals:

$$\frac{3}{50} = 3 \div 50 = 0.06$$

$1\frac{4}{5} = \dfrac{9}{5}$ and $9 \div 5 = 1.8$

$\dfrac{x}{0.06} = \dfrac{0.15}{1.8}$

$x \cdot 1.8 = (0.06)(0.15)$

$\dfrac{x \cdot \cancel{1.8}}{\cancel{1.8}} = \dfrac{0.009}{1.8}$

$x = 0.005$

Change to fractions:

$$0.15 = \frac{15 \div 5}{100 \div 5} = \frac{3}{20}$$

$\dfrac{x}{\frac{3}{50}} = \dfrac{\frac{3}{20}}{1\frac{4}{5}}$

$1\frac{4}{5} \cdot x = \dfrac{3}{50} \cdot \dfrac{3}{20}$

$\dfrac{1\frac{4}{5} \cdot x}{1\frac{4}{5}} = \dfrac{\frac{9}{1000}}{1\frac{4}{5}}$

$x = \dfrac{9}{1000} \div 1\frac{4}{5} = \dfrac{\overset{1}{\cancel{9}}}{\underset{200}{\cancel{1000}}} \cdot \dfrac{\overset{1}{\cancel{5}}}{\underset{1}{\cancel{9}}}$

$= \dfrac{1}{200}$

Compare answers.

$0.005 = \dfrac{5 \div 5}{1000 \div 5} = \dfrac{1}{200}$ ← Matches

Section 5.5   (pages 351–354)

9. $\dfrac{6 \text{ ounces}}{7 \text{ servings}} = \dfrac{x \text{ ounces}}{12 \text{ servings}}$

$7 \cdot x = 6 \cdot 12$

$\dfrac{\cancel{7} \cdot x}{\cancel{7}} = \dfrac{72}{7}$

$x \approx 10.3 \approx 10$

You need about 10 ounces for 12 servings.

15.   The length of the dining area is the same as the length of the kitchen, which is 14 feet (from Exercise 13).

Find the width of the dining area.

4.5 inches − 2.5 inches = 2 inches
on the floor plan

$\dfrac{1 \text{ inch}}{4 \text{ feet}} = \dfrac{2 \text{ inches}}{x \text{ feet}}$

$1 \cdot x = 4 \cdot 2$

$x = 8$

The dining area is 8 feet wide.

19. $\dfrac{7 \text{ refresher}}{10 \text{ entering}} = \dfrac{x \text{ refresher}}{2950 \text{ entering}}$

$10 \cdot x = 7 \cdot 2950$

$\dfrac{\cancel{10} \cdot x}{\cancel{10}} = \dfrac{20,650}{10}$

$x = 2065$

2065 students will probably need a refresher course. This is a reasonable answer because it's more than half the students, but not all the students.

Incorrect setup

$\dfrac{10 \text{ entering}}{7 \text{ refresher}} = \dfrac{x \text{ refresher}}{2950 \text{ entering}}$

$7 \cdot x = 10 \cdot 2950$

$\dfrac{\cancel{7} \cdot x}{\cancel{7}} = \dfrac{29,500}{7}$

$x \approx 4214$

The *incorrect* setup gives an *unreasonable* estimate of 4214 entering students since there are only 2950 entering students.

31.   Coretta $\left\{ \dfrac{1.05 \text{ meters}}{1.68 \text{ meters}} = \dfrac{6.58 \text{ meters}}{x \text{ meters}} \right\}$ tree

$1.05 \cdot x = 1.68(6.58)$

$\dfrac{\cancel{1.05} \cdot x}{\cancel{1.05}} = \dfrac{11.0544}{1.05}$

$x = 10.528 \approx 10.53$

The height of the tree is about 10.53 meters.

37.   First find the number of calories in a $\frac{1}{2}$-cup serving of bran cereal.

$\dfrac{\frac{1}{3} \text{ cup}}{80 \text{ calories}} = \dfrac{\frac{1}{2} \text{ cup}}{x \text{ calories}}$

$\dfrac{1}{3} \cdot x = 80 \cdot \dfrac{1}{2}$

$\dfrac{\frac{1}{3} \cdot x}{\frac{1}{3}} = \dfrac{40}{\frac{1}{3}}$

$x = \dfrac{40}{1} \cdot \dfrac{3}{1}$

$x = 120$ calories

Then find the number of grams of fiber in a $\frac{1}{2}$-cup serving of bran cereal.

$\dfrac{\frac{1}{3} \text{ cup}}{8 \text{ grams}} = \dfrac{\frac{1}{2} \text{ cup}}{x \text{ grams}}$

$\dfrac{1}{3} \cdot x = 8 \cdot \dfrac{1}{2}$

$\dfrac{\frac{1}{3} \cdot x}{\frac{1}{3}} = \dfrac{4}{\frac{1}{3}}$

$x = \dfrac{4}{1} \cdot \dfrac{3}{1}$

$x = 12$ grams

A $\frac{1}{2}$-cup serving of bran cereal provides 120 calories and 12 grams of fiber.

Chapter 6 Percent

Section 6.1 (pages 373–378)

13. $0.5\% = 0.005$

Drop the percent sign. Attach two zeros so the decimal point can be moved two places to the left.

27. $2 = 200\%$

Two zeros are attached so the decimal point can be moved two places to the right. Attach a percent sign.

51. 100% is all of the children.

So, 100% of 20 children is 20 children.

20 children are served both meals.

77. 16.8% of the population of Spain is 65 or older. To write 16.8% as a decimal, drop the percent sign and move the decimal point two places to the left, resulting in 0.168. So, 0.168 of the population of Spain is 65 or older.

Section 6.2 (pages 385–390)

9. First write 6.25 over 100. Then to get a whole number in the numerator, multiply the numerator and denominator by 100. Finally, write the fraction in lowest terms.

$$6.25\% = \frac{6.25}{100} = \frac{6.25(100)}{100(100)}$$
$$= \frac{625 \div 625}{10,000 \div 625} = \frac{1}{16}$$

41. $\dfrac{5}{9} = \dfrac{p}{100}$

$9 \cdot p = 5 \cdot 100$ Find cross products.

$\dfrac{\cancel{9} \cdot p}{\cancel{9}} = \dfrac{500}{9}$ Divide both sides by 9.

$p = \dfrac{500}{9}$

$p \approx 55.5\overline{5}$

Thus, $\dfrac{5}{9} = 55.6\%$ (rounded).

47. $87.5\% = 0.875$ decimal

$= \dfrac{875 \div 125}{1000 \div 125} = \dfrac{7}{8}$ fraction

63. $\dfrac{1}{200} = \dfrac{1 \cdot 5}{200 \cdot 5} = \dfrac{5}{1000} = 0.005$ decimal

$= 0.5\%$ percent

67. $3\dfrac{1}{4} = 3\dfrac{1 \cdot 25}{4 \cdot 25} = 3\dfrac{25}{100} = 3.25$ decimal

$= 325\%$ percent

77. 64 out of 80 employees have cell phones. Therefore, $80 - 64 = 16$ employees do *not* have cell phones.

$\dfrac{16}{80} = \dfrac{16 \div 16}{80 \div 16} = \dfrac{1}{5}$ fraction

$\dfrac{1}{5} = \dfrac{1 \cdot 2}{5 \cdot 2} = \dfrac{2}{10} = 0.2$ decimal

$0.2 = 0.20 = 20\%$ percent

Section 6.3 (pages 395–398)

7. part $= 15$, whole $= 60$

$\dfrac{15}{60} = \dfrac{x}{100}$

$\dfrac{1}{4} = \dfrac{x}{100}$ $\dfrac{15}{60}$ is $\dfrac{1}{4}$ in lowest terms.

$4 \cdot x = 1 \cdot 100$ Find cross products.

$\dfrac{\cancel{4} \cdot x}{\cancel{4}} = \dfrac{100}{4}$ Divide both sides by 4.

$x = 25$

The percent is 25, written as 25%.

21. whole $= 5000$, part $= 20$

$\dfrac{20}{5000} = \dfrac{x}{100}$ Percent proportion.

$5000 \cdot x = 20 \cdot 100$ Find cross products.

$\dfrac{5000 \cdot x}{5000} = \dfrac{2000}{5000}$ Divide both sides by 5000.

$x = 0.4$

The percent is 0.4, written as 0.4%.

33. $\underbrace{\text{12 injections}}_{\text{part}}$ is $\underbrace{\text{20\%}}_{\text{percent}}$ of what number of injections?

$\underbrace{}_{\substack{\text{whole} \\ \text{(unknown)}}}$

$\dfrac{12}{\text{unknown}} = \dfrac{20}{100}$

47. $\underbrace{86}_{\text{part}}$ of $\underbrace{\text{142 people}}_{\text{whole}}$ is $\underbrace{\text{what percent?}}_{\substack{\text{percent} \\ \text{(unknown)}}}$

$\dfrac{86}{142} = \dfrac{\text{unknown}}{100}$

Section 6.4 (pages 407–412)

13. Write the percent 225% as a decimal, 2.25. 225% of 680 tables

$(2.25)(680) = 1530$

part $= 1530$ tables

27. part is 350; percent is 12.5

$\dfrac{\text{part}}{\text{whole}} = \dfrac{\text{percent}}{100}$

so $\dfrac{350}{x} = \dfrac{12.5}{100}$

$12.5 \cdot x = 35,000$ Cross products

$\dfrac{12.5 \cdot x}{12.5} = \dfrac{35,000}{12.5}$ Divide both sides by 12.5.

$x = 2800$

$12\dfrac{1}{2}\%$ of 2800 is 350.

37. part is 64; whole is 344

$\dfrac{\text{part}}{\text{whole}} = \dfrac{\text{percent}}{100}$

$\dfrac{64}{344} = \dfrac{x}{100}$ OR $\dfrac{8}{43} = \dfrac{x}{100}$

Cross products $43 \cdot x = 8 \cdot 100$

Divide both sides by 43. $\dfrac{\cancel{43} \cdot x}{\cancel{43}} = \dfrac{800}{43}$

$x \approx 18.6$

$64 is 18.6\%$ (rounded) of \$344.

43. part is unknown; whole is 240; percent is 22

$\dfrac{\text{part}}{\text{whole}} = \dfrac{\text{percent}}{100}$

$\dfrac{x}{240} = \dfrac{22}{100}$ OR $\dfrac{x}{240} = \dfrac{11}{50}$

Cross products $x \cdot 50 = 240 \cdot 11$

Divide both sides by 50. $\dfrac{x \cdot \cancel{50}}{\cancel{50}} = \dfrac{2640}{50}$

$x = 52.8$

The amount withheld is \$52.80.

55. part is 960; whole is 48,000; percent is unknown

$\dfrac{\text{part}}{\text{whole}} = \dfrac{\text{percent}}{100}$

$\dfrac{960}{48,000} = \dfrac{x}{100}$ OR $\dfrac{1}{50} = \dfrac{x}{100}$

Cross products $50 \cdot x = 1 \cdot 100$

Divide both sides by 50. $\dfrac{50 \cdot x}{50} = \dfrac{100}{50}$

$x = 2$

There are 2% of these jobs filled by women.

67. percent is 20; whole is 3100; part is unknown

$\dfrac{\text{part}}{\text{whole}} = \dfrac{\text{percent}}{100}$

$\dfrac{x}{3100} = \dfrac{20}{100}$ OR $\dfrac{x}{3100} = \dfrac{1}{5}$

Find cross products. $\dfrac{\cancel{5} \cdot x}{\cancel{5}} = \dfrac{3100}{5}$

Divide both sides by 5.

$x = 620$

Add the application fee.

$\$620 + \$25 = \$645$

The total charge is \$645.

Section 6.5 (pages 417–420)

1. Write 25% as the decimal 0.25. The whole is 1080. Let x represent the unknown part.

part $=$ percent $\cdot$ whole

$x = (0.25)(1080)$

$x = 270$

25% of 1080 blood donors is 270 blood donors.

21. The part is 3.75 and the percent is $1\dfrac{1}{4}\% = 1.25\%$ or 0.0125 as a decimal. The whole is unknown.

part $=$ percent $\cdot$ whole

$3.75 = (0.0125)(x)$

$\dfrac{3.75}{0.0125} = \dfrac{(0.0125)(x)}{0.0125}$

$300 = x$

$1\dfrac{1}{4}\%$ of 300 gallons in 3.75 gallons.

29. Because 160 follows *of*, the whole is 160. The part is 2.4, and the percent is unknown.

part = percent • whole

$2.4 = x \cdot 160$

$\dfrac{2.4}{160} = \dfrac{x \cdot \cancel{160}}{\cancel{160}}$

$0.015 = x$

0.015 is 1.5%

1.5% of 160 liters is 2.4 liters.

43. Because 1250 follows *of*, the whole is 1250. The part is 461, and the percent is unknown.

part = percent • whole

$461 = x \cdot 1250$

$\dfrac{461}{1250} = \dfrac{x \cdot \cancel{1250}}{\cancel{1250}}$

$0.369 \approx x$

0.369 is 36.9%

36.9% (rounded) of these Americans rate their health as excellent.

55. Find the sales tax on the new Polaris unit. The whole is 524. The percent is $7\frac{3}{4}\% = 7.75\%$, which is 0.0775 as a decimal

part = percent • whole

$x = (0.0775)(524)$

$x = 40.61$

Add the sales tax to the purchase price.

$\$524 + \$40.61 = \$564.61$

Subtract the trade-in.

$\$564.61 - \$125 = \$439.61$

The total cost to the customer is $439.61.

Section 6.6 (pages 427–430)

3. Find the sales tax rate using the sales tax formula. The sales tax is $12.75, and the cost of the item is $425.

sales tax = rate of tax • cost of item

$\$12.75 = r \cdot \$425$

$\dfrac{12.75}{425} = \dfrac{r \cdot \overset{1}{\cancel{425}}}{\underset{1}{\cancel{425}}}$ Divide both sides by 425.

$0.03 = r$

0.03 is 3%   Write the decimal as a percent.

The tax rate is 3% and the total cost is $425 + \$12.75 = \$437.75$.

9. The problem asks for the amount of commission. Use the commission formula. The rate of commission is 8%, and the sales are $280.

commission = rate of commission • sales

$= (8\%)(\$280)$

$= (0.08)(\$280)$

$= \$22.40$

The amount of commission is $22.40.

23. The problem asks for the amount of the discount and the sale price. First, find the amount of discount using the discount formula. The rate of discount is 15%, and the original price is $58.40.

amount of discount = rate of discount • original price

$= (15\%)(\$58.40)$

$= (0.15)(\$58.40)$ Write 15% as a decimal.

$= \$8.76$ Amount of discount

Now find the sale price by subtracting the amount of the discount ($8.76) from the original price.

The amount of discount is $8.76 and the sale price is $49.64.

31. The problem asks for the rate of sales tax. Use the sales tax formula. The sales tax is $99. The cost of the door is $1980. Use *r* to represent the unknown rate of tax.

sales tax = rate of tax • cost of item

$\$99 = r \cdot \$1980$

$\dfrac{99}{1980} = \dfrac{r \cdot \overset{1}{\cancel{1980}}}{\underset{1}{\cancel{1980}}}$ Divide both sides by 1980.

$0.05 = r$

0.05 is 5%   Write the decimal as a percent.

The rate of sales tax is 5%.

47. The problem asks for the cost of the dictionary. First find the amount of discount using the discount formula. The rate of discount is 6%, and the original price is $18.50.

discount = rate of discount • original price

$= (6\%)(\$18.50)$

$= (0.06)(\$18.50)$ Write 6% as a decimal.

$= \$1.11$ Amount of discount

To find the sale price, subtract the amount of discount ($1.11) from the original price.

Sale price

$=$ original price $-$ amount of discount

$= \$18.50 - \$1.11 = \$17.39$ sale price

To find the sales tax, use the sales tax formula. The rate of tax is 6% and the cost of the dictionary is $17.39.

sales tax = rate of tax • cost of item

$= (6\%)(\$17.39)$

$= (0.06)(\$17.39)$ Write 6% as a decimal.

$\approx \$1.04$ Sales tax

The total cost of the dictionary is $17.39 + \$1.04 = \$18.43$.

57. To find the sales tax, use the sales tax formula.

sales tax = rate of tax • cost of item

$= \left(7\frac{3}{4}\%\right)(\$1248)$

$= (0.0775)(\$1248)$ Write $7\frac{3}{4}\%$ as a decimal.

$= \$96.72$ Sales tax

The total cost is the sum of the ticket, the sales tax, and the excise tax.

$\$1248 + \$96.72 + \$13.40 = \$1358.12$

Section 6.7 (pages 433–436)

7. $2300 at $8\frac{1}{2}\%$ for $2\frac{1}{2}$ years

The principal (*p*) is $2300. The rate (*r*) is $8\frac{1}{2}\%$, or 0.085 as a decimal, and the time (*t*) is $2\frac{1}{2}$ or 2.5 years.

$I = p \cdot r \cdot t$

$= (2300)(0.085)(2.5)$

$= 488.75$

The interest is $488.75.

15. $940 at 3% for 18 months

The principal is $940. The rate is 3% or 0.03, and the time is $\frac{18}{12}$ of a year.

$I = p \cdot r \cdot t$

$= \underbrace{(940)(0.03)}\left(\dfrac{18}{12}\right)$

18 months $= \dfrac{18}{12}$ of a year.

$= (28.2)(1.5)$ $\dfrac{18}{12} = (1.5)$.

$= 42.3$

The interest is $42.30.

29. $16,850 at $7\frac{1}{2}\%$ for 9 months

First find the interest. The principal is $16,850. The rate is $7\frac{1}{2}\%$ or 0.075, and the time is $\frac{9}{12}$ of a year.

$I = p \cdot r \cdot t$

$= (16,850)(0.075)\left(\dfrac{9}{12}\right)$

$= (1263.75)\left(\dfrac{3}{4}\right)$ Write $\dfrac{9}{12}$ in lowest terms as $\dfrac{3}{4}$.

$= 947.81$ (rounded)

The interest is $947.81.

To find the total amount due, add the principal and the interest.

amount due = principal + interest

$= \$16,850 + \$947.81$

$= \$17,797.81$

The total amount due is $17,797.81.

39. $14,800 at $2\frac{1}{4}$% for 10 months

The principal is $14,800. The rate is $2\frac{1}{4}$%

or 0.0225, and the time is $\frac{10}{12}$ of a year.

$I = p \cdot r \cdot t$

$= \underbrace{(14,800)(0.0225)}_{}\left(\frac{10}{12}\right)$

$= (333)\left(\frac{5}{6}\right)$ Write $\frac{10}{12}$ in

lowest terms as $\frac{5}{6}$.

$= 277.50$

She will earn $277.50 in interest.

Section 6.8 (pages 441–444)

1. $500 at 4% for 2 years

Year	Interest	Compound Amount
1	($500)(0.04)(1) = $20	
		$500 + $20 = $520
2	($520)(0.04)(1) = $20.80	
		$520 + $20.80 = $540.80

The compound amount is $540.80.

9. $1400 at 6% for 5 years

Year 1 Year 2 Year 3 Year 4 Year 5

($1400) $\underbrace{(1.06)(1.06)(1.06)(1.06)(1.06)}_{} \approx 1873.52

$\uparrow$ $\uparrow$

Original   100% + 6% = 106% = 1.06   Compound
deposit   amount

The compound amount is $1873.52.

15. $1000 at 4% for 5 years

Look down the column headed 4%, and across to row 5 (because 5 years = 5 time periods). At the intersection of the column and the row, read the compound amount, 1.2167.

compound amount = ($1000)(1.2167)

$= 1216.70

Find the interest by subtracting the principal ($1000) from the compound amount.

interest = $1216.70 − $1000

$= 216.70

25. (a) 6% column, row 9 from the table is 1.6895.

compound amount = ($76,000)(1.6895)

$= $128,402$

The total amount that should be repaid is $128,402.

(b) To find the amount of interest earned, subtract the principal ($76,000) from the compound amount.

The amount of interest earned is

$128,402 − $76,000 = $52,402

Chapter 7 Geometry

Section 7.1 (pages 472–475)

5. The figure starts at point P and goes on forever in one direction, so it is a ray named $\overrightarrow{PQ}$.

7. The lines are *perpendicular* because they intersect at right angles. The small red square indicates the right angle (90°).

23. Two rays in a straight line pointing in opposite directions measure 180°. An angle that measures 180° is called a *straight angle*.

31. Find the complement of 86° by subtracting.

90° − 86° = 4° ← Complement

35. Find the supplement of 90° by subtracting.

180° − 90° = 90° ← Supplement

37. $\angle SON \cong \angle TOM$ because they are vertical angles.

$\angle TOS \cong \angle MON$ because they are vertical angles.

39. Because $\angle COE$ and $\angle GOH$ are vertical angles, they are also congruent. This means they have the same measure. $\angle COE$ measures 63°, so $\angle GOH$ also measures 63°.

The sum of the measures of $\angle COE$, $\angle AOC$, and $\angle AOH$ equals 180°. Therefore, $\angle AOC$ measures 180° − (63° + 37°) = 180° − 100° = 80°.

Since $\angle AOC$ and $\angle GOF$ are vertical angles, they are congruent, so $\angle GOF$ also measures 80°.

Since $\angle AOH$ and $\angle EOF$ are vertical angles, they are congruent, so $\angle EOF$ measures 37°.

51. $\angle 6 \cong \angle 1$ (vertical angles), so the measure of $\angle 1$ is also 47°.

$\angle 6 \cong \angle 8$ (corresponding angles), so the measure of $\angle 8$ is also 47°.

$\angle 6 \cong \angle 3$ (alternate interior angles), so the measure of $\angle 3$ is also 47°.

Notice that the exterior sides of $\angle 6$ and $\angle 5$ form a straight angle of 180°. Therefore, $\angle 6$ and $\angle 5$ are supplementary angles and the sum of their measures is 180°. If $\angle 6$ is 47° then $\angle 5$ must be 133° because 180° − 47° = 133°. So the measure of $\angle 5$ is 133°.

$\angle 5 \cong \angle 2$ (vertical angles), so the measure of $\angle 2$ is also 133°.

$\angle 5 \cong \angle 7$ (corresponding angles), so the measure of $\angle 7$ is also 133°.

$\angle 7 \cong \angle 4$ (vertical angles), so the measure of $\angle 4$ is also 133°.

Section 7.2 (pages 482–485)

3. $P = 4 \cdot s$

$= 4 \cdot 0.9$ km

$= 3.6$ km

$A = s \cdot s$

$= 0.9$ km $\cdot 0.9$ km

$= 0.81$ km^2 ← Write km^2 for area (**not** km).

---

9. A storage building that is 76.1 ft by 22 ft

76.1 ft

| | 22 ft |

The length of this rectangle is 76.1 ft and the width is 22 ft. Use the formula $P = 2 \cdot l + 2 \cdot w$ to find the perimeter.

$P = 2 \cdot 76.1$ ft $+ 2 \cdot 22$ ft

$= 152.2$ ft $+ 44$ ft

$= 196.2$ ft

Use the formula $A = l \cdot w$ to find the area.

$A = 76.1$ ft $\cdot 22$ ft

$= 1674.2$ ft^2

17. The unlabeled side is 6 in. + 6 in. or 12 in. Add the lengths of the sides to find the perimeter.

$P = 15$ in. $+ 6$ in. $+ 6$ in. $+ 6$ in. $+ 6$ in.
    $+ 6$ in. $+ 9$ in. $+ 12$ in. $+ 6$ in. $+ 6$ in.

$= 78$ in.

Draw two horizontal lines to break up the figure into two smaller squares and one larger rectangle. Add the areas of these squares and rectangle to find the total area.

$A = (6$ in. $\cdot 6$ in.$) + (18$ in. $\cdot 9$ in.$)$
   $+ (6$ in. $\cdot 6$ in.$)$

$= 36$ in.$^2 + 162$ in.$^2 + 36$ in.^{2}

$= 234$ in.^{2}

You could also draw two vertical lines to break up the figure into three rectangles.

$A = (15$ in. $\cdot 6$ in.$) + (15$ in. $\cdot 6$ in.$)$
   $+ (9$ in. $\cdot 6$ in.$)$

$= 90$ in.$^2 + 90$ in.$^2 + 54$ in.^{2}

$= 234$ in.^{2} ← Same result

23. Find the perimeter of the room.

$P = 2 \cdot 4.4$ m $+ 2 \cdot 5.1$ m

$= 8.8$ m $+ 10.2$ m $= 19$ m

She will need 19 m of the strip. Multiply by the cost per meter to find the total cost.

$\text{Cost} = \frac{19 \text{ m}}{1} \cdot \frac{\$4.99}{1 \text{ m}} = \$94.81$

Tyra will have to spend $94.81 for the strip.

27. $A = \text{length} \cdot \text{width}$

5300 yd^2 = 100 yd $\cdot$ width

$\frac{5300 \text{ yd} \cdot \text{yd}}{100 \text{ yd}} = \frac{100 \text{ yd} \cdot \text{width}}{100 \text{ yd}}$

53 yd = width

The width is 53 yd.

Section 7.3 (pages 490–491)

11. The figure is a trapezoid. The height (h) is 42 cm, the short base (b) is 61.4 cm, and the long base (B) is 86.2 cm.

$A = \frac{1}{2} \cdot h \cdot (b + B)$

$= 0.5 \cdot 42$ cm $\cdot (61.4$ cm $+ 86.2$ cm$)$

$= 0.5 \cdot 42$ cm $\cdot (147.6$ cm$)$

$= 3099.6$ cm^2

15. Area of one parallelogram:

$A = b \cdot h$

$\quad = 5 \text{ in.} \cdot 3.5 \text{ in.}$

$\quad = 17.5 \text{ in.}^2$

To find the total area of the 25 parallelogram pieces, multiply the area of one parallelogram (17.5 in.²) by the number of pieces (25).

Total area $= 25 \cdot 17.5 \text{ in.}^2 = 437.5 \text{ in.}^2$

19. Break the figure into two pieces, a rectangle on the left side and a parallelogram.

Find the area of each piece, and then add the areas.

Area of rectangle:

$A = \text{length} \cdot \text{width}$

$\quad = 1.3 \text{ m} \cdot 0.8 \text{ m}$

$\quad = 1.04 \text{ m}^2$

Area of parallelogram:

$A = \text{base} \cdot \text{height}$

$\quad = 1.8 \text{ m} \cdot 1.1 \text{ m}$

$\quad = 1.98 \text{ m}^2$

Total area $= 1.04 \text{ m}^2 + 1.98 \text{ m}^2$

$\qquad\qquad = 3.02 \text{ m}^2$

Section 7.4 (pages 496–498)

7. To find the perimeter, add the lengths of the three sides.

$P = 35.5 \text{ cm} + 21.3 \text{ cm} + 28.4 \text{ cm}$

$\quad = 85.2 \text{ cm}$

When finding the area of this right triangle, the base and height are the perpendicular sides. So the base is 28.4 cm and the height is 21.3 cm.

$A = \dfrac{1}{2} \cdot b \cdot h$

$\quad = \dfrac{1}{2} \cdot 28.4 \text{ cm} \cdot 21.3 \text{ cm}$

$\quad = 302.46 \text{ cm}^2$

11. First, find the area of the entire figure, which is a rectangle.

$A = l \cdot w$

$\quad = 52 \text{ m} \cdot 37 \text{ m}$

$\quad = 1924 \text{ m}^2$

Find the area of the unshaded triangle.

$A = \dfrac{1}{2} \cdot b \cdot h$

$\quad = \dfrac{1}{2} \cdot 52 \text{ m} \cdot 10 \text{ m}$

$\quad = 260 \text{ m}^2$

Subtract the unshaded area from the area of the entire figure to find the shaded area.

Shaded area $= 1924 \text{ m}^2 - 260 \text{ m}^2$

$\qquad\qquad\quad = 1664 \text{ m}^2$

13. *Step 1* Add the two angles given.

$\quad 90° + 58° = 148°$

Step 2 Subtract the sum from 180°.

$\quad 180° - 148° = 32°$

The third angle measures 32°.

21. (a) To find the amount of curbing needed to go around the triangular space, find the perimeter.

$P = 42 \text{ m} + 32 \text{ m} + 52.8 \text{ m}$

$\quad = 126.8 \text{ m}$

126.8 m of curbing will be needed.

(b) To find the amount of sod needed to cover the space, find the area. It is a right triangle, so the perpendicular sides are the base and height.

$A = \dfrac{1}{2} \cdot 32 \text{ m} \cdot 42 \text{ m} = 672 \text{ m}^2$

672 m² of sod will be needed.

Section 7.5 (pages 506–509)

11. $d = 7\dfrac{1}{2} \text{ ft or } 7.5 \text{ ft}$

$C = \pi \cdot d$

$\quad \approx 3.14 \cdot 7.5 \text{ ft}$

$\quad \approx 23.6 \text{ ft}$

To find the area, first find the radius.

$r = \dfrac{7.5 \text{ ft}}{2} = 3.75 \text{ ft}$

$A = \pi \cdot r \cdot r$

$\quad \approx 3.14 \cdot 3.75 \text{ ft} \cdot 3.75 \text{ ft}$

$\quad \approx 44.2 \text{ ft}^2$

17. Find the area of a whole circle with a radius of 10 cm:

$A = \pi \cdot r \cdot r$

$\quad \approx 3.14 \cdot 10 \text{ cm} \cdot 10 \text{ cm}$

$\quad = 314 \text{ cm}^2$

Divide the area of the whole circle by 2 to find the area of the semicircle:

$\dfrac{314 \text{ cm}^2}{2} = 157 \text{ cm}^2$

Area of the triangle:

$A = \dfrac{1}{2} \cdot b \cdot h$

$\quad = \dfrac{1}{2} \cdot 20 \text{ cm} \cdot 10 \text{ cm}$

$\quad = 100 \text{ cm}^2$

Subtract to find the shaded area.

$157 \text{ cm}^2 - 100 \text{ cm}^2 = 57 \text{ cm}^2$

23. A point on the tire tread moves the length of the circumference in one complete turn.

$C = \pi \cdot d$

$\quad \approx 3.14 \cdot 29.10 \text{ in.}$

$\quad \approx 91.4 \text{ in.}$

Bonus question:

$\dfrac{1 \text{ revolution}}{91.4 \cancel{\text{ inches}}} \cdot \dfrac{12 \cancel{\text{ inches}}}{1 \cancel{\text{ foot}}} \cdot \dfrac{5280 \cancel{\text{ feet}}}{1 \text{ mile}}$

$\approx 693 \text{ revolutions/mile}$

25.

Find the area of a circle with a radius of 150 miles.

$A = \pi \cdot r \cdot r$

$\quad \approx 3.14 \cdot 150 \text{ mi} \cdot 150 \text{ mi}$

$\quad = 70,650 \text{ mi}^2$

There are about 70,650 mi² in the broadcast area.

31. (a) $C = 144 \text{ cm}$

$\qquad C = \pi \cdot d$

$144 \text{ cm} = \pi \cdot d$

$144 \text{ cm} \approx 3.14 \cdot d$

$\dfrac{144 \text{ cm}}{3.14} \approx \dfrac{3.14 \cdot d}{3.14}$

$45.9 \text{ cm} \approx d$

The diameter is about 45.9 cm.

(b) Divide the circumference by π (3.14).

Section 7.6 (pages 518–520)

5. The figure is a hemisphere.

$V = \dfrac{2}{3} \cdot \pi \cdot r^3 \quad \text{or} \quad \dfrac{2 \cdot \pi \cdot r^3}{3}$

$\quad \approx \dfrac{2 \cdot 3.14 \cdot 12 \text{ in.} \cdot 12 \text{ in.} \cdot 12 \text{ in.}}{3}$

$\quad \approx 3617.3 \text{ in.}^3$

9. The figure is a cone.

First find B, the area of the circular base.

$B = \pi \cdot r \cdot r$

$\quad \approx 3.14 \cdot 5 \text{ m} \cdot 5 \text{ m}$

$\quad \approx 78.5 \text{ m}^2$

Now find the volume of the cone.

$V = \dfrac{B \cdot h}{3}$

$\quad \approx \dfrac{78.5 \text{ m}^2 \cdot 16 \text{ m}}{3}$

$\quad \approx 418.7 \text{ m}^3$

19. First, find the radius of the cylinder.

$r = \dfrac{d}{2} = \dfrac{18 \text{ in.}}{2} = 9 \text{ in.}$

Then find the volume.

$V = \pi \cdot r \cdot r \cdot h$

$\quad \approx 3.14 \cdot 9 \text{ in.} \cdot 9 \text{ in.} \cdot 3 \text{ in.}$

$\quad \approx 763.0 \text{ in.}^3$

Now find the surface area

$SA = (2 \cdot \pi \cdot r \cdot h)$

$\qquad + (2 \cdot \pi \cdot r \cdot r)$

$\quad \approx (2 \cdot 3.14 \cdot 9 \text{ in.} \cdot 3 \text{ in.})$

$\qquad + (2 \cdot 3.14 \cdot 9 \text{ in.} \cdot 9 \text{ in.})$

$\quad \approx 169.56 + 508.68$

$\quad \approx 678.2 \text{ in.}^2$

27. Use the formula for the volume of a pyramid. First find B, the area of the square base.

$B = s \cdot s$

$\quad = 145 \text{ m} \cdot 145 \text{ m}$

$\quad = 21,025 \text{ m}^2$

Now find the volume of the pyramid.

$V = \dfrac{B \cdot h}{3}$

$\quad = \dfrac{21,025 \text{ m}^2 \cdot 93 \text{ m}}{3}$

$\quad = 651,775 \text{ m}^3$

The volume of the ancient stone pyramid is 651,775 m³.

29. Use the formula for the volume of a cylinder.

First find the radius of the pipe.

$$\text{radius} = \frac{5 \text{ ft}}{2} = 2.5 \text{ ft}$$

Now find the volume.

$$V = \pi \cdot r^2 \cdot h$$
$$\approx 3.14 \cdot 2.5 \text{ ft} \cdot 2.5 \text{ ft} \cdot 200 \text{ ft}$$
$$\approx 3925 \text{ ft}^3$$

The volume of the city sewer pipe is about 3925 ft^3.

31. Use the formula for the surface area of a rectangular solid.

$$SA = 2lw + 2lh + 2wh$$
$$= (2 \cdot 5.5 \text{ in.} \cdot 2.8 \text{ in.})$$
$$+ (2 \cdot 5.5 \text{ in.} \cdot 8 \text{ in.})$$
$$+ (2 \cdot 2.8 \text{ in.} \cdot 8 \text{ in.})$$
$$= 30.8 + 88 + 44.8$$
$$= 163.6 \text{ in.}^2$$

The amount of cardboard needed is 163.6 in.^2

Section 7.7 (pages 525–528)

25. The unknown length is the side opposite the right angle, which is the hypotenuse. The lengths of the legs are 8 in. and 3 in.

$$\text{hypotenuse} = \sqrt{(\text{leg})^2 + (\text{leg})^2}$$
$$= \sqrt{(8)^2 + (3)^2}$$
$$= \sqrt{64 + 9}$$
$$= \sqrt{73}$$
$$\approx 8.5 \text{ in.}$$

35. The length of the hypotenuse is 21.6 km. The length of one of the legs is 13.2 km.

$$\text{leg} = \sqrt{(\text{hypotenuse})^2 - (\text{leg})^2}$$
$$= \sqrt{(21.6)^2 - (13.2)^2}$$
$$= \sqrt{466.56 - 174.24}$$
$$= \sqrt{292.32}$$
$$\approx 17.1 \text{ km}$$

41. The diagonal brace is the hypotenuse. The lengths of the legs are 4.5 ft and 3.5 ft.

$$\text{hypotenuse} = \sqrt{(\text{leg})^2 + (\text{leg})^2}$$
$$= \sqrt{(4.5)^2 + (3.5)^2}$$
$$= \sqrt{20.25 + 12.25}$$
$$= \sqrt{32.5}$$
$$\approx 5.7 \text{ ft}$$

The diagonal brace is about 5.7 ft long.

43.

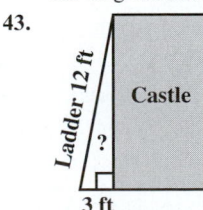

The ladder is opposite the right angle so it is the hypotenuse. Thus, the length of the hypotenuse is 12 ft and the length of one of the legs is 3 ft.

$$\text{leg} = \sqrt{(\text{hypotenuse})^2 - (\text{leg})^2}$$
$$= \sqrt{(12)^2 - (3)^2}$$
$$= \sqrt{144 - 9}$$
$$= \sqrt{135}$$
$$\approx 11.6 \text{ ft}$$

The ladder will reach about 11.6 ft high on the building.

Section 7.8 (pages 535–538)

3. Notice that *rotating* $\triangle STU$ makes it possible to slide it on top of $\triangle WXY$ so the triangles match.

The corresponding angles are:

$\angle 1$ and $\angle 6$; $\angle 2$ and $\angle 4$; $\angle 3$ and $\angle 5$

The corresponding sides are:

$\overline{ST}$ and $\overline{YW}$; $\overline{TU}$ and $\overline{WX}$; $\overline{SU}$ and $\overline{YX}$

11. On both triangles, two corresponding angles and the side that connects them measure the same, so the Angle-Side-Angle (ASA) method can be used to prove that the triangles are congruent.

19. You want to find the values of a and b in the smaller triangle. Notice that the side that is 6 mm long in the smaller triangle corresponds to the side that is 12 mm long in the larger triangle. Set up a ratio of corresponding sides.

$$\frac{6 \text{ mm}}{12 \text{ mm}} = \frac{6}{12} = \frac{1}{2} \text{ in lowest terms}$$

Write a proportion to find a.

$$\frac{a}{10} = \frac{1}{2}$$

$a \cdot 2 = 10 \cdot 1$ Cross products are equivalent.

$$\frac{a \cdot \overset{1}{\cancel{2}}}{\underset{1}{\cancel{2}}} = \frac{10}{2}$$ Divide both sides by 2.

$$a = 5 \text{ mm}$$

Write a proportion to find b.

$$\frac{b}{2} = \frac{1}{2}$$

$b \cdot 2 = 6 \cdot 1$ Cross products are equivalent.

$$\frac{b \cdot \overset{1}{\cancel{2}}}{\underset{1}{\cancel{2}}} = \frac{6}{2}$$ Divide both sides by 2.

$$b = 3 \text{ mm}$$

25. Since triangles CDE and FGH are similar and all of the sides of CDE are the same length, all the sides of FGH are also the same length. Therefore, each missing side of triangle FGH is 8 cm.

Now add the lengths of all three sides to find the perimeter of triangle FGH.

$$P = 8 \text{ cm} + 8 \text{ cm} + 8 \text{ cm}$$

$$P = 24 \text{ cm}$$

Set up a ratio of corresponding sides to find the height (h) of triangle FGH.

$$\frac{10.4}{12} = \frac{h}{8}$$

$12 \cdot h = 8 \cdot 10.4$ Cross products are equivalent.

$$\frac{\overset{1}{\cancel{12}} \cdot h}{\underset{1}{\cancel{12}}} = \frac{83.2}{12}$$ Divide both sides by 12.

$$h \approx 6.9 \text{ cm} \quad \text{Rounded}$$

Area of triangle FGH
$$= 0.5 \cdot b \cdot h$$
$$\approx (0.5)(8 \text{ cm})(6.9 \text{ cm})$$
$$\approx 27.6 \text{ cm}^2$$

Chapter 8 Statistics

Section 8.1 (pages 565–570)

1. The number of pets owned in the United States is 11 million + 150 million + 90 million + 75 million + 18 million + 16 million = 360 million.

13. "Wanted food they couldn't cook at home" to "Less work/No clean up":

$$\frac{1740}{1020} = \frac{1740 \div 60}{1020 \div 60} = \frac{29}{17}$$

25. Don't know:

$$x = 5\% \text{ of } 5540$$
$$= (0.05)(5540)$$
$$= 277 \text{ people}$$

31. Percent for day camp $= \dfrac{\$546}{\$5460}$

$$= 0.10 = 10\%$$

Degrees of a circle $= 10\%$ of $360°$
$$= (0.10)(360°)$$
$$= 36°$$

37. (a) Total sales $= \$12{,}500 + \$40{,}000$
$+ \$60{,}000 + \$50{,}000 + \$37{,}500$
$= \$200{,}000$

(b) Adventure classes $= \$12{,}500$

percent of total $= \dfrac{12{,}500}{200{,}000} = 0.0625$

$$= 6.25\%$$

number of degrees $= (0.0625)(360°)$
$$= 22.5°$$

Grocery and provision sales $= \$40{,}000$

percent of total $= \dfrac{40{,}000}{200{,}000} = 0.2 = 20\%$

number of degrees $= (0.2)(360°) = 72°$

Equipment rentals $= \$60{,}000$

percent of total $= \dfrac{60{,}000}{200{,}000} = 0.3 = 30\%$

number of degrees $= (0.3)(360°) = 108°$

Rafting tours $= \$50{,}000$

percent of total $= \dfrac{50{,}000}{200{,}000} = 0.25 = 25\%$

number of degrees $= (0.25)(360°) = 90°$

Equipment sales $= \$37{,}500$

percent of total $= \dfrac{37{,}500}{200{,}000} = 0.1875$

$$= 18.75\%$$

number of degrees $= (0.1875)(360°)$
$$= 67.5°$$

(c)

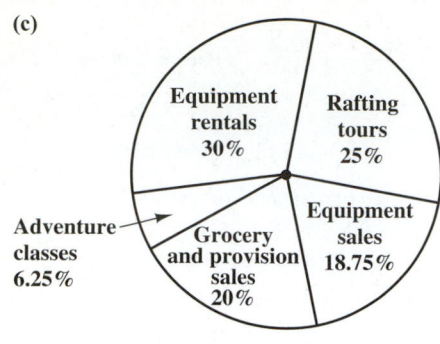

Equipment rentals 30%
Rafting tours 25%
Adventure classes 6.25%
Grocery and provision sales 20%
Equipment sales 18.75%

Section 8.2 (pages 575–580)

3. The countries in which less than 15% of household income is spent, on average, for food are USA, Britain, and Australia.

9. There are two bars for February. One is for 2009 and one is for 2010. The bar for 2010 rises to 7. Multiply 7 by 1000 because the label on the left side of the graph says *in thousands*. The bar for 2009 rises to halfway between 5 and 6. So multiply 5.5 by 1000.

Plants shipped in February of 2010 = 7000

Plants shipped in February of 2009 = 5500

$$7000 - 5500 = 1500$$

There were 1500 more plants shipped in February of 2010.

19. Above the dot for 1990 is 24.1. Looking at the left edge of the graph, notice that this number is in millions.

The number of PCs shipped in 1990 was 24.1 million or 24,100,000.

29. Find the dot above 2009 on the line for Store B. Use a ruler or straightedge to line up the dot with the numbers along the left edge. The dot is aligned with 350, so chain Store B sold 350 · 1000 = 350,000 iPods in 2009.

35. Find the dot above 2010 on the line for total sales. Use a ruler or straightedge to line up the dot with the numbers along the left edge. The dot is aligned with 40. Multiply 40 by 1000 because the label on the left side of the graph says *in thousands*. The total sales in 2010 were

$$40 \cdot \$1000 = \$40,000.$$

Section 8.3 (pages 583–586)

1. Because the bar associated with 61–65 is the highest, the age group of 61–65 years has the greatest number of members, which is 16,000.

9. The bar for 31–40 rises to 11, so there are 11 employees who earn $3100 to $4000.

21. 0–5; 4 tally marks and a class frequency of 4

Section 8.4 (pages 591–594)

1. $\text{Mean} = \dfrac{\text{sum of all values}}{\text{number of values}}$

$$= \frac{6 + 22 + 15 + 2 + 8 + 13}{6}$$

$$= \frac{66}{6}$$

$$= 11 \text{ years}$$

The mean (average) age of the shopping centers was 11 years.

9.

Customers Each Hour	Frequency	Product
8	2	(8 · 2) = 16
11	12	(11 · 12) = 132
15	5	(15 · 5) = 75
26	1	(26 · 1) = 26
Totals	**20**	**249**

weighted mean

$$= \frac{\text{sum of products}}{\text{total number of customers}}$$

$$= \frac{249}{20} \approx 12.5 \text{ customers (rounded)}$$

17. Arrange the numbers in numerical order from least to greatest.

298, 346, 412, 501, 515, 521, 528, 621

The list has 8 numbers. The middle numbers are the 4th and 5th numbers, so the median is

$$\frac{501 + 515}{2} = \frac{1016}{2} = 508 \text{ calories.}$$

21. 74, 68, 68, 68, 75, 75, 74, 74, 70

Because both 68 and 74 occur three times, each is a mode. This list is *bimodal*.

29.

Credits	Grade	Credits · Grade
2	A (= 4)	2 · 4 = 8
3	C (= 2)	3 · 2 = 6
4	A (= 4)	4 · 4 = 16
1	C (= 2)	1 · 2 = 2
4	B (= 3)	4 · 3 = 12
14		**44**

$$\text{GPA} = \frac{\text{sum of (Credits · Grade)}}{\text{total number of credits}}$$

$$= \frac{44}{14} \approx 3.14 \text{ (rounded)}$$

Chapter 9 The Real Number System

Section 9.1 (pages 621–622)

33. $\dfrac{1}{4} \cdot \dfrac{2}{3} + \dfrac{2}{5} \cdot \dfrac{11}{3}$

$$= \frac{1}{6} + \frac{22}{15} \qquad \text{Multiply; } \frac{2}{12} = \frac{1}{6}$$

$$= \frac{5}{30} + \frac{44}{30} \qquad \text{LCD} = 30$$

$$= \frac{49}{30}, \text{ or } 1\frac{19}{30} \qquad \text{Add.}$$

45. $\left(\dfrac{3}{2}\right)^2 \left[\left(11 + \dfrac{1}{3}\right) - 6\right]$

$$= \frac{9}{4}\left[\left(11 + \frac{1}{3}\right) - 6\right] \qquad \begin{array}{l}\text{Apply the}\\\text{exponent.}\end{array}$$

$$= \frac{9}{4}\left[\left(\frac{33}{3} + \frac{1}{3}\right) - 6\right] \qquad \text{LCD} = 3$$

$$= \frac{9}{4}\left(\frac{34}{3} - 6\right)$$

$$= \frac{9}{4}\left(\frac{34}{3} - \frac{18}{3}\right) \qquad \text{LCD} = 3$$

$$= \frac{9}{4}\left(\frac{16}{3}\right) \qquad \begin{array}{l}\text{Subtract inside}\\\text{parentheses.}\end{array}$$

$$= \frac{144}{12} \qquad \text{Multiply.}$$

$$= 12 \qquad \text{Divide.}$$

59. $\dfrac{9(7 - 1) - 8 \cdot 2}{4(6 - 1)} > 3$

$$\frac{9(6) - 8 \cdot 2}{4(5)} > 3 \qquad \begin{array}{l}\text{Work inside}\\\text{parentheses.}\end{array}$$

$$\frac{54 - 16}{20} > 3 \qquad \text{Multiply.}$$

$$\frac{38}{20} > 3 \qquad \text{Subtract.}$$

$$1\frac{9}{10} > 3 \qquad \begin{array}{l}\text{Write as a}\\\text{mixed number}\end{array}$$

Since $1\dfrac{9}{10} < 3$, the statement is *false*.

Section 9.2 (pages 627–630)

29. (a) $\dfrac{3x + y^2}{2x + 3y}$

$$= \frac{3(2) + 1^2}{2(2) + 3(1)} \qquad \text{Let } x = 2 \text{ and } y = 1.$$

$$= \frac{3(2) + 1}{4 + 3} \qquad \begin{array}{l}\text{Work separately}\\\text{above and below the}\\\text{fraction bar.}\end{array}$$

$$= \frac{6 + 1}{7}$$

$$= \frac{7}{7}, \text{ or } 1$$

(b) $\dfrac{3x + y^2}{2x + 3y}$

$$= \frac{3(1) + 5^2}{2(1) + 3(5)} \qquad \begin{array}{l}\text{Let } x = 1 \text{ and}\\y = 5.\end{array}$$

$$= \frac{3 + 25}{2 + 15}$$

$$= \frac{28}{17}$$

31. (a) $0.841x^2 + 0.32y^2$

$$= 0.841 \cdot 2^2 + 0.32 \cdot 1^2$$

$$\qquad \text{Let } x = 2, y = 1.$$

$$= 0.841 \cdot 4 + 0.32 \cdot 1$$

$$\qquad \text{Apply the exponents.}$$

$$= 3.364 + 0.32 \qquad \text{Multiply.}$$

$$= 3.684 \qquad \text{Add.}$$

(b) $0.841x^2 + 0.32y^2$

$$= 0.841 \cdot 1^2 + 0.32 \cdot 5^2$$

$$\qquad \text{Let } x = 1, y = 5.$$

$$= 0.841 \cdot 1 + 0.32 \cdot 25$$

$$= 0.841 + 8$$

$$= 8.841$$

55. $\dfrac{z+4}{2-z} = \dfrac{13}{5}$

$\dfrac{\frac{1}{3}+4}{2-\frac{1}{3}} \overset{?}{=} \dfrac{13}{5}$ Let $z = \dfrac{1}{3}$.

$\dfrac{\frac{1}{3}+\frac{12}{3}}{\frac{6}{3}-\frac{1}{3}} \overset{?}{=} \dfrac{13}{5}$

$\dfrac{\frac{13}{3}}{\frac{5}{3}} \overset{?}{=} \dfrac{13}{5}$

$\dfrac{13}{3} \cdot \dfrac{3}{5} \overset{?}{=} \dfrac{13}{5}$

$\dfrac{13}{5} = \dfrac{13}{5}$ True

The true result shows that $\dfrac{1}{3}$ is a solution of the equation.

Section 9.3 (pages 637–638)

25. In order to compare these two numbers, write them with a common denominator.

$$-\dfrac{2}{3} = -\dfrac{8}{12} \quad \text{and} \quad -\dfrac{1}{4} = -\dfrac{3}{12}$$

Since

$$\dfrac{8}{12} > \dfrac{3}{12},$$

$-\dfrac{2}{3}$ is farther to the left of 0 on the number line than $-\dfrac{1}{4}$, so $-\dfrac{2}{3}$ is the lesser number.

51. A decrease is represented by a negative number, and the amount of decrease is represented by its absolute value. There are three negative numbers in the table, -0.8, -1.7, and -2.2. Of these, -2.2, has the greatest absolute value, 2.2, so electronic components and accessories for 2003 to 2004 represents the greatest decrease.

Section 9.4 (pages 643–646)

35. $\left(-\dfrac{1}{2} + 0.25\right) - \left(-\dfrac{3}{4} + 0.75\right)$

$= \left(-\dfrac{1}{2} + \dfrac{1}{4}\right) - \left(-\dfrac{3}{4} + \dfrac{3}{4}\right)$

$= \left(-\dfrac{2}{4} + \dfrac{1}{4}\right) - 0$

$= -\dfrac{1}{4}, \quad \text{or} \quad -0.25$

61. "$\dfrac{2}{7}$ more than the sum of $\dfrac{5}{7}$ and $-\dfrac{9}{7}$" is

$$\left[\dfrac{5}{7} + \left(-\dfrac{9}{7}\right)\right] + \dfrac{2}{7}$$

$= -\dfrac{4}{7} + \dfrac{2}{7}$

$= -\dfrac{2}{7}.$

Section 9.5 (pages 651–656)

39. $\left(-\dfrac{3}{8} - \dfrac{2}{3}\right) - \left(-\dfrac{9}{8} - 3\right)$

$= \left[-\dfrac{3}{8} + \left(-\dfrac{2}{3}\right)\right] - \left[-\dfrac{9}{8} + (-3)\right]$

$= \left[-\dfrac{9}{24} + \left(-\dfrac{16}{24}\right)\right]$

$\quad - \left[-\dfrac{9}{8} + \left(-\dfrac{24}{8}\right)\right]$ Get a common demominator for each pair of fractions.

$= -\dfrac{25}{24} - \left(-\dfrac{33}{8}\right)$ Add inside the brackets.

$= -\dfrac{25}{24} - \left(-\dfrac{99}{24}\right)$ Get a common denominator.

$= -\dfrac{25}{24} + \dfrac{99}{24}$

$= \dfrac{74}{24}$

$= \dfrac{37}{12}, \quad \text{or} \quad 3\dfrac{1}{12}$

41. $[-12.25 - (8.34 + 3.57)] - 17.88$

$= [-12.25 - 11.91] - 17.88$

$= [-12.25 + (-11.91)] - 17.88$

$= -24.16 - 17.88$

$= -24.16 + (-17.88)$

$= -42.04$

67. Sum of checks:

$\$35.84 + \$26.14 + \$3.12$

$= \$61.98 + \$3.12$

$= \$65.10$

Sum of deposits:

$\$85.00 + \$120.76 = \$205.76$

Final balance

$= \text{Beginning balance}$
$\quad - \text{checks} + \text{deposits}$

$= \$904.89 - \$65.10 + \$205.76$

$= \$839.79 + \$205.76$

$= \$1045.55$

Her account balance at the end of August was $\$1045.55$.

Section 9.6 (pages 665–668)

43. $\dfrac{4(2^3 - 5) - 5(-3^3 + 21)}{3[6 - (-2)]}$

$= \dfrac{4(8 - 5) - 5(-27 + 21)}{3[6 - (-2)]}$

$= \dfrac{4(3) - 5(-6)}{3[8]}$

$= \dfrac{12 + 30}{24}$

$= \dfrac{42}{24}$

$= \dfrac{7}{4}, \quad \text{or} \quad 1\dfrac{3}{4}$

49. $\left(\dfrac{5}{6}x + \dfrac{3}{2}y\right)\left(-\dfrac{1}{3}a\right)$

$= \left[\dfrac{5}{6}(6) + \dfrac{3}{2}(-4)\right]\left[-\dfrac{1}{3}(3)\right]$ Let $x = 6$, $y = -4$, and $a = 3$.

$= [5 + (-6)](-1)$

$= (-1)(-1)$

$= 1$

69. "The product of $-\dfrac{2}{3}$ and $-\dfrac{1}{5}$, divided by $\dfrac{1}{7}$" is

$$\dfrac{-\dfrac{2}{3}\left(-\dfrac{1}{5}\right)}{\dfrac{1}{7}}$$

$= \dfrac{\dfrac{2}{15}}{\dfrac{1}{7}}$

$= \dfrac{2}{15} \cdot \dfrac{7}{1}$

$= \dfrac{14}{15}.$

Section 9.7 (pages 675–678)

59. $-3(8x + 3y + 4z)$

$= -3(8x) + (-3)(3y) + (-3)(4z)$

   Distributive property

$= (-3 \cdot 8)x + (-3 \cdot 3)y + (-3 \cdot 4)z$

   Associative property

$= -24x - 9y - 12z$ Multiply.

65. $-(-3q + 5r - 8s)$

$= -1(-3q + 5r - 8s)$

$= 3q - 5r + 8s$

Section 9.8 (pages 683–684)

39. $-\dfrac{4}{3} + 2t + \dfrac{1}{3}t - 8 - \dfrac{8}{3}t$

$= \left(2t + \dfrac{1}{3}t - \dfrac{8}{3}t\right) + \left(-\dfrac{4}{3} - 8\right)$

$= \left(2 + \dfrac{1}{3} - \dfrac{8}{3}\right)t + \left(-\dfrac{4}{3} - 8\right)$

$= \left(\dfrac{6}{3} + \dfrac{1}{3} - \dfrac{8}{3}\right)t + \left(-\dfrac{4}{3} - \dfrac{24}{3}\right)$

$= -\dfrac{1}{3}t - \dfrac{28}{3}$

41. $-5.3r + 4.9 - (2r + 0.7) + 3.2r$

$= -5.3r + 4.9 - 2r - 0.7 + 3.2r$

$= (-5.3r - 2r + 3.2r) + (4.9 - 0.7)$

$= (-5.3 - 2 + 3.2)r + (4.9 - 0.7)$

$= -4.1r + 4.2$

Chapter 10 Equations, Inequalities, and Applications

Section 10.1 (pages 703–704)

27.
$$10x + 4 = 9x$$
$$10x + 4 - 9x = 9x - 9x \quad \text{Subtract } 9x.$$
$$1x + 4 = 0$$
$$x + 4 - 4 = 0 - 4 \quad \text{Subtract } 4.$$
$$x = -4$$

Check
$$10(-4) + 4 \stackrel{?}{=} 9(-4) \quad \text{Let } x = -4.$$
$$-36 = -36 \quad \text{True}$$
Solution set: $\{-4\}$

47.
$$\frac{5}{7}x + \frac{1}{3} = \frac{2}{5} - \frac{2}{7}x + \frac{2}{5}$$
$$\frac{5}{7}x + \frac{1}{3} = \frac{4}{5} - \frac{2}{7}x$$
$$\frac{5}{7}x + \frac{1}{3} + \frac{2}{7}x = \frac{4}{5} - \frac{2}{7}x + \frac{2}{7}x \quad \text{Add } \frac{2}{7}x.$$
$$\frac{7}{7}x + \frac{1}{3} = \frac{4}{5}$$
$$1x + \frac{1}{3} - \frac{1}{3} = \frac{4}{5} - \frac{1}{3} \quad \text{Subtract } \frac{1}{3}.$$
$$x = \frac{12}{15} - \frac{5}{15} \quad \text{LCD} = 15$$
$$x = \frac{7}{15}$$

Check
$$\frac{5}{7}\left(\frac{7}{15}\right) + \frac{1}{3} \stackrel{?}{=} \frac{2}{5} - \frac{2}{7}\left(\frac{7}{15}\right) + \frac{2}{5}$$
$$\text{Let } x = \frac{7}{15}.$$
$$\frac{1}{3} + \frac{1}{3} \stackrel{?}{=} \frac{4}{5} - \frac{2}{15}$$
$$\frac{2}{3} \stackrel{?}{=} \frac{12}{15} - \frac{2}{15}$$
$$\frac{2}{3} = \frac{10}{15}, \quad \text{or} \quad \frac{2}{3} \quad \text{True}$$
Solution set: $\left\{\frac{7}{15}\right\}$

55.
$$10(-2x + 1) = -19(x + 1)$$
$$-20x + 10 = -19x - 19$$
$$\text{Distributive property}$$
$$-20x + 10 + 19x = -19x - 19 + 19x$$
$$\text{Add } 19x.$$
$$-x + 10 = -19$$
$$\text{Combine like terms.}$$
$$-x + 10 - 10 = -19 - 10$$
$$\text{Subtract } 10.$$
$$-x = -29$$
$$x = 29$$

A check confirms this solution.
Solution set: $\{29\}$

Section 10.2 (pages 709–710)

53.
$$\frac{2}{5}x - \frac{3}{10}x = 2$$
$$\frac{4}{10}x - \frac{3}{10}x = 2 \quad \text{LCD} = 10$$
$$\frac{1}{10}x = 2 \quad \text{Combine like terms.}$$
$$10 \cdot \frac{1}{10}x = 10 \cdot 2 \quad \text{Multiply by } 10.$$
$$x = 20$$

Check
$$\frac{2}{5}(20) - \frac{3}{10}(20) \stackrel{?}{=} 2 \quad \text{Let } x = 20.$$
$$8 - 6 \stackrel{?}{=} 2$$
$$2 = 2 \quad \text{True}$$
Solution set: $\{20\}$

61.
$$0.9w - 0.5w + 0.1w = -3$$
$$0.5w = -3 \quad \text{Combine like terms.}$$
$$\frac{0.5w}{0.5} = \frac{-3}{0.5} \quad \text{Divide by } 0.5.$$
$$w = -6$$

A check confirms this solution.
Solution set: $\{-6\}$

Section 10.3 (pages 719–722)

47. $0.02(5000) + 0.03x = 0.025(5000 + x)$

Multiply both sides by 1000, *not* 100.

$$1000\,[0.02(5000) + 0.03x]$$
$$= 1000\,[0.025(5000 + x)]$$
$$20(5000) + 30x = 25(5000 + x)$$
$$100{,}000 + 30x = 125{,}000 + 25x$$
$$\text{Distributive property}$$
$$5x + 100{,}000 = 125{,}000$$
$$\text{Subtract } 25x.$$
$$5x = 25{,}000$$
$$\text{Subtract } 100{,}000.$$
$$x = 5000 \quad \text{Divide by } 5.$$

Check   Substituting 5000 for x in the original equation results in a true statement, $250 = 250$.

Solution set: $\{5000\}$

51.
$$-(6k - 5) - (-5k + 8) = -3$$
$$-1(6k - 5) - 1(-5k + 8) = -3$$
$$-6k + 5 + 5k - 8 = -3$$
$$-k - 3 = -3$$
$$-k = 0$$
$$k = 0$$

Check   Substituting 0 for k in the original equation results in a true statement, $-3 = -3$.

Solution set: $\{0\}$

Section 10.4 (pages 731–736)

13. *Step 1*
Read the problem again.

Step 2
Let $x =$ the unknown number. Then $3x$ is three times the number, $x + 7$ is 7 more than the number, and the sum is
$$3x + (x + 7).$$
$2x$ is twice the number, and
$$-11 - 2x$$
is the difference between -11 and twice the number.

Step 3
$$3x + (x + 7) = -11 - 2x$$

Step 4
$$4x + 7 = -11 - 2x$$
$$6x + 7 = -11 \quad \text{Add } 2x.$$
$$6x = -18 \quad \text{Subtract } 7.$$
$$x = -3 \quad \text{Divide by } 6.$$

Step 5
The number is -3.

Step 6
Check that -3 is the correct answer by substituting this result into the words of the original problem. The sum of three times this number and 7 more than the number is
$$3(-3) + (-3 + 7) = -5,$$
and the difference between -11 and twice this number is
$$-11 - 2(-3) = -5.$$
The values are equal, so the number -3 is the correct answer.

37. *Step 1*
Read the problem again.

Step 2
Let $x =$ the measures of angles A and B.
$x + 60 =$ the measure of angle C.

Step 3
The sum of the measures of the angles of any triangle is 180°, so
$$x + x + (x + 60) = 180.$$

Step 4
$$3x + 60 = 180$$
$$3x = 120 \quad \text{Subtract } 60.$$
$$x = 40 \quad \text{Divide by } 3.$$

Step 5
Angles A and B have measures of 40°, and angle C has a measure of $40 + 60 = 100°$.

Step 6

The answer checks since

$$40 + 40 + 100 = 180.$$

Section 10.5 (pages 743–748)

51. Let L represent the length of the box. The girth is $4 \cdot 18 = 72$ in. Since the length plus the girth is 108, we have

$$L + 72 = 108$$

$$L = 36 \text{ in.} \quad \text{Subtract 72.}$$

The maximum volume of the box is

$$V = LWH$$

$$V = 36(18)(18) \quad \text{Substitute.}$$

$$V = 11{,}664 \text{ in.}^3.$$

55. The two angles are complementary, so the sum of their measures is $90°$.

$$(8x - 1) + 5x = 90$$

$$13x - 1 = 90 \quad \text{Combine like terms.}$$

$$13x = 91 \quad \text{Add 1.}$$

$$x = 7 \quad \text{Divide by 13.}$$

Since $x = 7$, we have

$$8x - 1 = 8(7) - 1 = 56 - 1 = 55$$

and

$$5x = 5(7) = 35.$$

The angles measure $55°$ and $35°$.

81.
$$M = C(1 + r) \text{ for } r$$

$$M = C + Cr \quad \text{Distributive property}$$

$$M - C = Cr \quad \text{Subtract } C.$$

$$\frac{M - C}{C} = \frac{Cr}{C} \quad \text{Divide by } C.$$

$$\frac{M - C}{C} = r, \quad \text{or} \quad r = \frac{M - C}{C}$$

Section 10.6 (pages 757–760)

49.
$$\frac{2}{3}(p + 3) > \frac{5}{6}(p - 4)$$

$$6\left(\frac{2}{3}\right)(p + 3) > 6\left(\frac{5}{6}\right)(p - 4)$$

$$\text{Multiply by 6, the LCD.}$$

$$4(p + 3) > 5(p - 4)$$

$$4p + 12 > 5p - 20$$

$$\text{Distributive property}$$

$$-p + 12 > -20$$

$$\text{Subtract } 5p.$$

$$-p > -32$$

$$\text{Subtract 12.}$$

$$\frac{-p}{-1} < \frac{-32}{-1}$$

$$\text{Divide by } -1; \text{ reverse the direction of the symbol.}$$

$$p < 32$$

Solution set: $(-\infty, 32)$

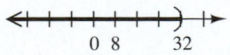

---

59. The Celsius temperature C must give a Fahrenheit temperature F that has *never exceeded* $104°$, which translates as *less than or equal to* $104°$.

$$F \le 104$$

$$\frac{9}{5}C + 32 \le 104$$

$$\frac{9}{5}C \le 72 \quad \text{Subtract 32.}$$

$$\frac{5}{9}\left(\frac{9}{5}C\right) \le \frac{5}{9}(72) \quad \text{Multiply by } \frac{5}{9}.$$

$$C \le 40$$

The temperature of Providence, Rhode Island, has never exceeded $40°$ Celsius.

Chapter 11 Graphs of Linear Equations and Inequalities in Two Variables

Section 11.1 (pages 781–786)

23. Is $(5, -6)$ a solution of the equation $x = -6$?

Since y does not appear in the equation, we just substitute 5 for x.

$$x = -6$$

$$5 \stackrel{?}{=} -6 \quad \text{Let } x = 5.$$

The result is false, so $(5, -6)$ is not a solution of the equation $x = -6$.

43. The given equation $x - 8 = 0$ may be written as $x = 8$. For any value of y, the value of x will always be 8. The completed table of values follows.

x	y
8	8
8	3
8	0

Section 11.2 (pages 795–800)

29. $-3y = 15$

$$y = \frac{15}{-3}, \text{ or } -5 \quad \text{Divide by } -3.$$

For any value of x, the value of y is -5. Three ordered pairs are $(-2, -5)$, $(0, -5)$, and $(1, -5)$. Plot these points and draw a line through them. The graph is a horizontal line.

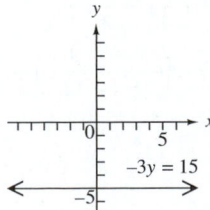

Section 11.3 (pages 809–812)

27. Use the slope formula with

$$\left(-\frac{7}{5}, \frac{3}{10}\right) = (x_1, y_1)$$

and

$$\left(\frac{1}{5}, -\frac{1}{2}\right) = (x_2, y_2).$$

$$\text{slope } m = \frac{\text{change in } y}{\text{change in } x}$$

$$= \frac{y_2 - y_1}{x_2 - x_1}$$

$$= \frac{-\frac{1}{2} - \frac{3}{10}}{\frac{1}{5} - \left(-\frac{7}{5}\right)} \quad \text{Substitute.}$$

$$= \frac{-\frac{5}{10} - \frac{3}{10}}{\frac{1}{5} + \frac{7}{5}} \quad \begin{array}{l}\text{Get a common denominator in the numerator;} \\ \text{simplify the denominator.}\end{array}$$

$$= \frac{-\frac{8}{10}}{\frac{8}{5}} \quad \begin{array}{l}\text{Subtract in the numerator;} \\ \text{add in the denominator.}\end{array}$$

$$= -\frac{8}{10} \div \frac{8}{5} \quad \frac{a}{b} = a \div b$$

$$= -\frac{8}{10} \cdot \frac{5}{8} \quad \begin{array}{l}\text{Multiply by the reciprocal.}\end{array}$$

$$= -\frac{1}{2} \quad \begin{array}{l}\text{Multiply; lowest terms}\end{array}$$

53. The slope (or grade) of the hill is the ratio of the rise to the run, or the ratio of the vertical change to the horizontal change. Since the rise is 32 and the run is 108, the slope is

$$\frac{32}{108} = \frac{8 \cdot 4}{27 \cdot 4} = \frac{8}{27}. \quad \text{Lowest terms}$$

Section 11.4 (pages 821–826)

3. The rise is 3 and the run is 1, so the slope m is given by

$$m = \frac{\text{rise}}{\text{run}} = \frac{3}{1} = 3.$$

The y-intercept is $(0, -3)$, so $b = -3$. The equation of the line, written in slope-intercept form $y = mx + b$, is

$$y = 3x - 3. \quad m = 3, b = -3$$

45. $\left(\frac{1}{2}, \frac{3}{2}\right), \left(-\frac{1}{4}, \frac{5}{4}\right)$

First, find the slope of the line.

$$m = \frac{\frac{5}{4} - \frac{3}{2}}{-\frac{1}{4} - \frac{1}{2}} = \frac{\frac{5}{4} - \frac{6}{4}}{-\frac{1}{4} - \frac{2}{4}} = \frac{-\frac{1}{4}}{-\frac{3}{4}}$$

$$= -\frac{1}{4} \div \left(-\frac{3}{4}\right) = -\frac{1}{4}\left(-\frac{4}{3}\right) = \frac{1}{3}$$

Now use the point $\left(\frac{1}{2}, \frac{3}{2}\right)$ for (x_1, y_1) and $m = \frac{1}{3}$ in the point-slope form.

$$y - y_1 = m(x - x_1)$$

$$y - \frac{3}{2} = \frac{1}{3}\left(x - \frac{1}{2}\right) \quad \text{Substitute.}$$

$$y - \frac{3}{2} = \frac{1}{3}x - \frac{1}{6} \quad \begin{array}{l}\text{Distributive property}\end{array}$$

$$y = \frac{1}{3}x - \frac{1}{6} + \frac{9}{6} \quad \text{Add } \frac{3}{2} = \frac{9}{6}.$$

$$y = \frac{1}{3}x + \frac{4}{3} \quad \begin{array}{l}\text{Combine like terms; } \frac{8}{6} = \frac{4}{3}\end{array}$$

47. Solve the given equation for y.

$$3x = 4y + 5$$
$$3x - 4y = 5 \qquad \text{Subtract } 4y.$$
$$-4y = -3x + 5 \qquad \text{Subtract } 3x.$$
$$y = \frac{3}{4}x - \frac{5}{4} \qquad \text{Divide by } -4.$$

The slope is $\frac{3}{4}$. A line parallel to this line has the same slope. Use the point-slope form with $m = \frac{3}{4}$ and $(x_1, y_1) = (2, -3)$.

$$y - y_1 = m(x - x_1)$$
$$y - (-3) = \frac{3}{4}(x - 2) \qquad \text{Substitute.}$$
$$y + 3 = \frac{3}{4}x - \frac{3}{2} \qquad \begin{array}{l}\text{Distributive}\\\text{property}\end{array}$$
$$y = \frac{3}{4}x - \frac{3}{2} - \frac{6}{2} \qquad \begin{array}{l}\text{Subtract}\\3 = \frac{6}{2}.\end{array}$$
$$y = \frac{3}{4}x - \frac{9}{2} \qquad \begin{array}{l}\text{Combine like}\\\text{terms.}\end{array}$$

Chapter 12   Exponents and Polynomials

Section 12.1 (pages 855–858)

21. $\dfrac{1}{2}x^4 + \dfrac{1}{6}x^4$

$$= \left(\frac{1}{2} + \frac{1}{6}\right)x^4 \qquad \text{Distributive property}$$
$$= \left(\frac{3}{6} + \frac{1}{6}\right)x^4 \qquad \begin{array}{l}\text{Write fractions with a}\\\text{common denominator.}\end{array}$$
$$= \frac{4}{6}x^4 \qquad \text{Add fractions.}$$
$$= \frac{2}{3}x^4 \qquad \text{Lowest terms}$$

37. $0.8x^4 - 0.3x^4 - 0.5x^4 + 7$

$$= (0.8 - 0.3 - 0.5)x^4 + 7$$
$$= 0x^4 + 7$$
$$= 7$$

Since 7 can be written as $7x^0$, the degree of the polynomial is 0. The simplified polynomial has one term, so it is a monomial.

57. Add.

$$\frac{2}{3}x^2 + \frac{1}{5}x + \frac{1}{6}$$
$$\underline{\frac{1}{2}x^2 - \frac{1}{3}x + \frac{2}{3}}$$

Rewrite so that the fractions in each column have a common denominator; then add column by column.

$$\frac{4}{6}x^2 + \frac{3}{15}x + \frac{1}{6}$$
$$\underline{\frac{3}{6}x^2 - \frac{5}{15}x + \frac{4}{6}}$$
$$\frac{7}{6}x^2 - \frac{2}{15}x + \frac{5}{6}$$

75. Use the formula for the perimeter of a square, $P = 4s$, with $s = \dfrac{1}{2}x^2 + 2x$.

$$P = 4s$$
$$P = 4\left(\frac{1}{2}x^2 + 2x\right) \qquad \text{Substitute.}$$
$$P = 4\left(\frac{1}{2}x^2\right) + 4(2x) \qquad \begin{array}{l}\text{Distributive}\\\text{property}\end{array}$$
$$P = 2x^2 + 8x \qquad \text{Multiply.}$$

Section 12.2 (pages 865–866)

47. $(-5^2)^6$

$$= (-1 \cdot 5^2)^6 \qquad \text{Write the factor } -1.$$
$$= (-1)^6 \cdot (5^2)^6 \qquad \text{Power rule (b)}$$
$$= 1 \cdot 5^{2 \cdot 6} \qquad \text{Power rule (a)}$$
$$= 5^{12}$$

81. $\left(\dfrac{5a^2b^5}{c^6}\right)^3, \quad c \neq 0$

$$= \frac{(5a^2b^5)^3}{(c^6)^3} \qquad \text{Power rule (c)}$$
$$= \frac{5^3(a^2)^3(b^5)^3}{(c^6)^3} \qquad \text{Power rule (b)}$$
$$= \frac{125a^6b^{15}}{c^{18}} \qquad \text{Power rule (a)}$$

83. $(-5m^3p^4q)^2\,(p^2q)^3$

$$= (-1 \cdot 5m^3p^4q)^2\,(p^2q)^3 \qquad \begin{array}{l}\text{Write the}\\\text{factor } -1.\end{array}$$
$$= (-1)^2 \cdot 5^2 \cdot (m^3)^2 \cdot (p^4)^2$$
$$\quad \cdot q^2 \cdot (p^2)^3 \cdot q^3 \qquad \text{Power rule (b)}$$
$$= 1 \cdot 25 \cdot m^6p^8q^2p^6q^3 \qquad \text{Power rule (a)}$$
$$= 25m^6p^{8+6}q^{2+3} \qquad \text{Product rule}$$
$$= 25m^6p^{14}q^5$$

Section 12.3 (pages 871–872)

37. $\left(x - \dfrac{2}{3}\right)\left(x + \dfrac{1}{4}\right)$

$$\quad\; \mathbf{F} \qquad \mathbf{O} \qquad \mathbf{I} \qquad \mathbf{L}$$
$$= x(x) + x\left(\frac{1}{4}\right) + \left(-\frac{2}{3}\right)x + \left(-\frac{2}{3}\right)\frac{1}{4}$$
$$= x^2 + \frac{1}{4}x - \frac{2}{3}x - \frac{1}{6}$$
$$= x^2 + \left(\frac{3}{12}x - \frac{8}{12}x\right) - \frac{1}{6}$$
$$= x^2 - \frac{5}{12}x - \frac{1}{6}$$

Section 12.4 (pages 877–878)

11. $(0.8t + 0.7s)^2$

$$= (0.8t)^2 + 2(0.8t)(0.7s) + (0.7s)^2$$
$$\qquad\qquad (a + b)^2 = a^2 + 2ab + b^2$$
$$= 0.64t^2 + 1.12ts + 0.49s^2$$

17. $-(4r - 2)^2$

First square the binomial.

$$(4r - 2)^2$$
$$= (4r)^2 - 2(4r)(2) + 2^2$$
$$= 16r^2 - 16r + 4$$

Now multiply by -1.

$$-1(16r^2 - 16r + 4)$$
$$= -16r^2 + 16r - 4$$

27. $(2x^2 - 5)(2x^2 + 5)$

$$= (2x^2)^2 - 5^2 \quad (a+b)(a-b) = a^2 - b^2$$
$$= 4x^4 - 25 \qquad (2x^2)^2 = 2^2x^4 = 4x^4$$

55. Use the formula for the volume of a cube, $V = e^3$. Here, $V = (x + 2)^3$. If $x = 6$, then

$$V = (6 + 2)^3 = 8^3 = 512.$$

The volume of the cube is 512 cu. units.

Section 12.5 (pages 887–888)

15. $(-2)^0 - 2^0$

$$= 1 - 1 \qquad a^0 = 1$$
$$= 0$$

31. $-2^{-1} + 3^{-2}$

$$= -(2^{-1}) + 3^{-2} \qquad \begin{array}{l}\text{In } -2^{-1}, \text{ the base}\\\text{is 2.}\end{array}$$
$$= -\frac{1}{2^1} + \frac{1}{3^2} \qquad a^{-n} = \frac{1}{a^n}$$
$$= -\frac{1}{2} + \frac{1}{9} \qquad \text{Apply the exponents.}$$
$$= -\frac{9}{18} + \frac{2}{18} \qquad \begin{array}{l}\text{Write fractions}\\\text{with a common}\\\text{denominator.}\end{array}$$
$$= -\frac{7}{18} \qquad \text{Add.}$$

71. $\dfrac{(m^7n)^{-2}}{m^{-4}n^3}$

$$= \frac{(m^7)^{-2}n^{-2}}{m^{-4}n^3} \qquad \text{Power rule (b)}$$
$$= \frac{m^{7(-2)}n^{-2}}{m^{-4}n^3} \qquad \text{Power rule (a)}$$
$$= \frac{m^{-14}n^{-2}}{m^{-4}n^3}$$
$$= m^{-14-(-4)}n^{-2-3} \qquad \text{Quotient rule}$$
$$= m^{-10}n^{-5}$$
$$= \frac{1}{m^{10}n^5} \qquad a^{-n} = \frac{1}{a^n}$$

Section 12.6 (pages 891–892)

19. $\dfrac{-3x^3 - 4x^4 + 2x}{-3x^2}$

$= \dfrac{-4x^4 - 3x^3 + 2x}{-3x^2}$ Write in descending powers.

$= \dfrac{-4x^4}{-3x^2} - \dfrac{3x^3}{-3x^2} + \dfrac{2x}{-3x^2}$ Divide each term by $-3x^2$.

$= \dfrac{4x^2}{3} + x - \dfrac{2}{3x}$ Quotient rule; be careful with signs.

Notice how the third term is written with x in the denominator, which is **not** the same as $-\dfrac{2}{3}x$ or $-\dfrac{2x}{3}$. In $-\dfrac{2}{3x}$, we are *dividing* by x; in $-\dfrac{2}{3}x$ and $-\dfrac{2x}{3}$, we are *multiplying* by x. Applying the quotient rule to the term $\dfrac{2x}{-3x^2}$ gives

$\dfrac{2x}{-3x^2} = \dfrac{2x^1}{-3x^2} = -\dfrac{2}{3}x^{1-2} = -\dfrac{2}{3}x^{-1}$

$= -\dfrac{2}{3}\left(\dfrac{1}{x}\right) = -\dfrac{2}{3x}.$

$\left(\dfrac{4}{3}x^2 \text{ is an acceptable form for the first term,} \right.$

$\left. \dfrac{4x^2}{3}. \text{ Why?}\right)$

29. Use the formula for the area of a rectangle, $A = LW$, with $A = 12x^2 - 4x + 2$ and $W = 2x$.

$A = LW$

$12x^2 - 4x + 2 = L(2x)$ Substitute for A and W.

$\dfrac{12x^2 - 4x + 2}{2x} = L$ Divide each side by $2x$.

$\dfrac{12x^2}{2x} - \dfrac{4x}{2x} + \dfrac{2}{2x} = L$ Divide each term by $2x$.

$6x - 2 + \dfrac{1}{x} = L$ $\dfrac{2}{2x} = \dfrac{2}{2}\cdot\dfrac{1}{x}$

$= 1\cdot\dfrac{1}{x} = \dfrac{1}{x}$

$6x - 2 + \dfrac{1}{x}$ represents the length of the rectangle.

Section 12.7 (pages 897–898)

23. $(x^4 - x^2 - 2) \div (x^2 - 2)$

Use 0 as the coefficient of the missing x^3- and x-terms in the dividend and the missing x-term in the divisor.

$$
\begin{array}{r}
x^2 \qquad\qquad + 1 \\
x^2 + 0x - 2 \overline{)x^4 + 0x^3 - x^2 + 0x - 2} \\
\underline{x^4 + 0x^3 - 2x^2} \\
x^2 + 0x - 2 \\
\underline{x^2 + 0x - 2} \\
0
\end{array}
$$

The remainder is 0. The answer is the quotient, $x^2 + 1$.

---

29. $\dfrac{2x^5 + x^4 + 11x^3 - 8x^2 - 13x + 7}{2x^2 + x - 1}$

$$
\begin{array}{r}
x^3 \qquad\qquad\quad + 6x - 7 \\
2x^2 + x - 1 \overline{)2x^5 + x^4 + 11x^3 - 8x^2 - 13x + 7} \\
\underline{2x^5 + x^4 - x^3} \\
12x^3 - 8x^2 - 13x \\
\underline{12x^3 + 6x^2 - 6x} \\
-14x^2 - 7x + 7 \\
\underline{-14x^2 - 7x + 7} \\
0
\end{array}
$$

The remainder is 0. The answer is the quotient, $x^3 + 6x - 7$.

Section 12.8 (pages 903–904)

51. $\dfrac{4 \times 10^5}{8 \times 10^2}$

$= \dfrac{4}{8} \times \dfrac{10^5}{10^2}$

$= 0.5 \times 10^3$ Divide; quotient rule

$= (5 \times 10^{-1}) \times 10^3$ Write 0.5 in scientific notation.

$= 5 \times (10^{-1} \times 10^3)$ Associative property

$= 5 \times 10^2$ Product rule

$= 500$ Write without exponents.

55. $\dfrac{(1.65 \times 10^8)(5.24 \times 10^{-2})}{(6 \times 10^4)(2 \times 10^7)}$

$= \dfrac{1.65 \times 5.24}{6 \times 2} \times \dfrac{10^8 \times 10^{-2}}{10^4 \times 10^7}$

   Associative and commutative properties

$= 0.7205 \times \dfrac{10^6}{10^{11}}$ Product rule

$= 0.7205 \times 10^{-5}$ Quotient rule

$= (7.205 \times 10^{-1}) \times 10^{-5}$

   Write 0.7205 in scientific notation.

$= 7.205 \times (10^{-1} \times 10^{-5})$

   Associative property

$= 7.205 \times 10^{-6}$ Product rule

$= 0.000007205$

   Write without exponents.

Chapter 13 Factoring and Applications

Section 13.1 (pages 923–924)

63. $18r^2 + 12ry - 3xr - 2xy$

$= (18r^2 + 12ry) + (-3xr - 2xy)$

   Group the terms.

$= 6r(3r + 2y) - x(3r + 2y)$

   Factor each group.

$= (3r + 2y)(6r - x)$

   Factor out the common factor, $3r + 2y$.

---

71. $y^2 + 3x + 3y + xy$

$= y^2 + 3y + xy + 3x$

   Rearrange terms.

$= (y^2 + 3y) + (xy + 3x)$

   Group the terms.

$= y(y + 3) + x(y + 3)$

   Factor each group.

$= (y + 3)(y + x)$

   Factor out the common factor, $y + 3$.

Section 13.2 (pages 929–930)

35. $a^2 - 8a - 48$

Find two integers whose product is -48 and whose sum is -8. Since c is negative, one integer must be positive and one must be negative.

Factors of -48	Sums of Factors
$-1, 48$	47
$1, -48$	-47
$-2, 24$	22
$2, -24$	-22
$-3, 16$	13
$3, -16$	-13
$-4, 12$	8
$4, -12$	-8 ←
$-6, 8$	2
$6, -8$	-2

Thus,

$a^2 - 8a - 48$ factors as $(a + 4)(a - 12)$.

51. $-2x^6 - 8x^5 + 42x^4$

First, factor out the negative common factor, $-2x^4$.

$-2x^6 - 8x^5 + 42x^4$

$= -2x^4(x^2 + 4x - 21)$

Now factor $x^2 + 4x - 21$.

Factors of -21	Sums of Factors
$1, -21$	-20
$-1, 21$	20
$3, -7$	-4
$-3, 7$	4 ←

Thus,

$x^2 + 4x - 21$ factors as $(x - 3)(x + 7)$.

The complete factored form is

$-2x^6 - 8x^5 + 42x^4$

$= -2x^4(x - 3)(x + 7).$

55. $m^3n - 10m^2n^2 + 24mn^3$

First, factor out the GCF, mn.

$m^3n - 10m^2n^2 + 24mn^3$

$= mn(m^2 - 10mn + 24n^2)$

The expressions $-6n$ and $-4n$ have a product of $24n^2$ and a sum of $-10n$. The complete factored form is

$m^3n - 10m^2n^2 + 24mn^3$

$= mn(m - 6n)(m - 4n).$

Section 13.3 (pages 933–934)

33. $-32z^5 + 20z^4 + 12z^3$

First factor out the negative common factor, $-4z^3$.

$$-32z^5 + 20z^4 + 12z^3$$
$$= -4z^3(8z^2 - 5z - 3)$$

To factor $8z^2 - 5z - 3$, find two integers whose product is $8(-3) = -24$ and whose sum is -5. These integers are -8 and 3. Now rewrite the given trinomial and factor it.

$$-32z^5 + 20z^4 + 12z^3$$
$$= -4z^3(8z^2 - 5z - 3)$$
$$= -4z^3(8z^2 - 8z + 3z - 3)$$
$$\quad -5z = -8z + 3z$$
$$= -4z^3[(8z^2 - 8z) + (3z - 3)]$$
Group the terms.
$$= -4z^3[8z(z - 1) + 3(z - 1)]$$
Factor each group.
$$= -4z^3[(z - 1)(8z + 3)]$$
Factor out common factor, $z - 1$.
$$= -4z^3(z - 1)(8z + 3)$$

39. $5 - 6x + x^2$

Find two integers whose product is $5(1) = 5$ and whose sum is -6. The integers are -1 and -5.

$$5 - 6x + x^2$$
$$= 5 - x - 5x + x^2$$
$$= (5 - x) + (-5x + x^2)$$
Group the terms.
$$= 1(5 - x) - x(5 - x)$$
Factor each group.
$$= (5 - x)(1 - x)$$
Factor out the common factor, $5 - x$.

Section 13.4 (pages 939–940)

43. $6m^6n + 7m^5n^2 + 2m^4n^3$

Factor out the GCF, m^4n.

$$6m^6n + 7m^5n^2 + 2m^4n^3$$
$$= m^4n(6m^2 + 7mn + 2n^2)$$

Now factor $6m^2 + 7mn + 2n^2$ by trial and error. Possible factors of $6m^2$ are $6m$ and m or $3m$ and $2m$. Possible factors of $2n^2$ are $2n$ and n.

$$(3m + 2n)(2m + n)$$
$$= 6m^2 + 7mn + 2n^2 \qquad \text{Correct}$$

The complete factored form is

$$6m^6n + 7m^5n^2 + 2m^4n^3$$
$$= m^4n(3m + 2n)(2m + n).$$

Section 13.5 (pages 947–950)

13. $4m^2 - \dfrac{9}{25}$

Because $4m^2 = (2m)^2$ and $\dfrac{9}{25} = \left(\dfrac{3}{5}\right)^2$,

$4m^2 - \dfrac{9}{25}$ is a difference of squares.

$$4m^2 - \dfrac{9}{25}$$
$$= (2m)^2 - \left(\dfrac{3}{5}\right)^2$$
$$= \left(2m + \dfrac{3}{5}\right)\left(2m - \dfrac{3}{5}\right)$$

39. $x^2 - 1.0x + 0.25$

The first and last terms are perfect squares, x^2 and $(-0.5)^2$. The trinomial is a perfect square, since the middle term is

$$2 \cdot x \cdot (-0.5) = -1.0x.$$

Therefore,

$$x^2 - 1.0x + 0.25$$
$$= (x)^2 - 2(x)(0.5) + (0.5)^2$$
$$= (x - 0.5)^2.$$

65. $27x^3 - 1$

Because $27x^3 = (3x)^3$ and $1 = 1^3$, the binomial is a difference of cubes. Use the pattern

$$a^3 - b^3 = (a - b)(a^2 + ab + b^2).$$

$$27x^3 - 1$$
$$= (3x)^3 - 1^3$$
$$= (3x - 1)[(3x)^2 + 3x(1) + 1^2]$$
$$= (3x - 1)(9x^2 + 3x + 1)$$

79. $x^3 + y^6$

Because $y^6 = (y^2)^3$, the binomial is a sum of cubes. Use the pattern

$$a^3 + b^3 = (a + b)(a^2 - ab + b^2).$$

$$x^3 + y^6$$
$$= x^3 + (y^2)^3$$
$$= (x + y^2)[x^2 - x(y^2) + (y^2)^2]$$
$$= (x + y^2)(x^2 - xy^2 + y^4)$$

Section 13.6 (pages 953–954)

21. $18m^3n + 3m^2n^2 - 6mn^3$
$$= 3mn(6m^2 + mn - 2n^2)$$
Factor out the GCF, $3mn$.
$$= 3mn(3m + 2n)(2m - n)$$
Factor the trinomial.

29. $x^3 + 3x^2 - 9x - 27$

Factor by grouping.

$$= (x^3 + 3x^2) + (-9x - 27)$$
$$= x^2(x + 3) - 9(x + 3)$$
$$= (x + 3)(x^2 - 9)$$

Since $x^2 - 9$ is the difference of two squares, $x^2 - 3^2$, factor it as $(x + 3)(x - 3)$.

$$= (x + 3)(x + 3)(x - 3)$$
$$= (x + 3)^2(x - 3)$$

35. $x^4 - 625$

Because $x^4 = (x^2)^2$ and $625 = 25^2$, this binomial is a difference of squares.

$$x^4 - 625$$
$$= (x^2)^2 - 25^2$$
$$= (x^2 + 25)(x^2 - 25)$$

Because $x^2 - 25$ is also the difference of two squares, $x^2 - 5^2$, factor it as $(x + 5)(x - 5)$.

$$= (x^2 + 25)(x + 5)(x - 5)$$

43. $1000z^3 + 512$
$$= 8(125z^3 + 64) \qquad \text{GCF} = 8$$
$$= 8[(5z)^3 + 4^3] \qquad \text{Sum of cubes}$$
$$= 8[5z + 4][(5z)^2 - (5z)(4) + 4^2]$$
$$= 8(5z + 4)(25z^2 - 20z + 16)$$

49. $14x^2 - 25xq - 25q^2$

Two integer factors whose product is $14(-25) = -350$ and whose sum is -25 are -35 and 10.

$$14x^2 - 25xq - 25q^2$$
$$= 14x^2 - 35xq + 10xq - 25q^2$$
$$= (14x^2 - 35xq) + (10xq - 25q^2)$$
$$= 7x(2x - 5q) + 5q(2x - 5q)$$
$$= (2x - 5q)(7x + 5q)$$

Section 13.7 (pages 961–962)

51. $9y^3 - 49y = 0$

To factor the polynomial, begin by factoring out the greatest common factor.

$$y(9y^2 - 49) = 0$$

Now factor $9y^2 - 49$ as the difference of two squares.

$$y(3y + 7)(3y - 7) = 0$$

Set each of the three factors equal to 0 and solve.

$$y = 0 \quad \text{or} \quad 3y + 7 = 0 \quad \text{or} \quad 3y - 7 = 0$$
$$3y = -7 \quad \text{or} \qquad 3y = 7$$
$$y = -\dfrac{7}{3} \quad \text{or} \qquad y = \dfrac{7}{3}$$

Solution set: $\left\{-\dfrac{7}{3}, 0, \dfrac{7}{3}\right\}$

53. $(2r + 5)(3r^2 - 16r + 5) = 0$

$(2r + 5)(3r - 1)(r - 5) = 0$

  Factor $3r^2 - 16r + 5$.

Set each of the three factors equal to 0, and solve the resulting equations.

$2r + 5 = 0$ or $3r - 1 = 0$ or $r - 5 = 0$

$2r = -5$ or $3r = 1$ or $r = 5$

$r = -\dfrac{5}{2}$ or $r = \dfrac{1}{3}$

Solution set: $\left\{-\dfrac{5}{2}, \dfrac{1}{3}, 5\right\}$

Section 13.8 (pages 969–974)

13. Let $x =$ the length of a side of the square painting.

Then $x - 2 =$ the length of a side of the square mirror.

Since the formula for the area of a square is $A = s^2$, the area of the painting is x^2, and the area of the mirror is $(x - 2)^2$. The difference between their areas is 32, so

$x^2 - (x - 2)^2 = 32$

$x^2 - (x^2 - 4x + 4) = 32$

$x^2 - x^2 + 4x - 4 = 32$

$4x - 4 = 32$

$4x = 36$

$x = 9.$

The length of a side of the painting is 9 ft. The length of a side of the mirror is

$9 - 2 = 7\,\text{ft}.$

As a check,

$9^2 - 7^2 = 81 - 49 = 32,$ as required.

21. Let $x =$ the least even integer.

Then $x + 2 =$ the next even integer and

$x + 4 =$ the third even integer.

$x^2 + (x + 2)^2 = (x + 4)^2$

$x^2 + x^2 + 4x + 4 = x^2 + 8x + 16$

  Square the binomials.

$2x^2 + 4x + 4 = x^2 + 8x + 16$

  Combine like terms.

$x^2 - 4x - 12 = 0$

  Standard form

$(x + 2)(x - 6) = 0$

  Factor.

$x + 2 = 0$ or $x - 6 = 0$

  Zero-factor property

$x = -2$ or $x = 6$

If $x = -2$, then $x + 2 = 0$ and $x + 4 = 2$.

If $x = 6$, then $x + 2 = 8$ and $x + 4 = 10$.

The integers are -2, 0, and 2, or 6, 8, and 10.

Check

$(-2)^2 + 0^2 \overset{?}{=} 2^2$	$6^2 + 8^2 \overset{?}{=} 10^2$
$4 + 0 \overset{?}{=} 4$	$36 + 64 \overset{?}{=} 100$
$4 = 4$	$100 = 100$
True	True

Chapter 14 Rational Expressions and Applications

Section 14.1 (pages 993–994)

21. (a) $\dfrac{(-3x)^2}{4x + 12}$

$= \dfrac{(-3 \cdot 2)^2}{4(2) + 12}$ Let $x = 2$.

$= \dfrac{(-6)^2}{8 + 12}$

$= \dfrac{36}{20}$

$= \dfrac{9}{5}$

(b) $\dfrac{(-3x)^2}{4x + 12}$

$= \dfrac{[-3(-3)]^2}{4(-3) + 12}$ Let $x = -3$.

$= \dfrac{9^2}{-12 + 12},$ or $\dfrac{81}{0}$

Since substituting -3 for x makes the denominator 0, the given rational expression is *undefined* when $x = -3$.

35. $\dfrac{12m^2 - 3}{8m - 4}$

$= \dfrac{3(4m^2 - 1)}{4(2m - 1)}$ Factor out the common factors.

$= \dfrac{3(2m + 1)(2m - 1)}{4(2m - 1)}$ Factor again in the numerator.

$= \dfrac{3(2m + 1)}{4}$ Divide out the common factors.

43. We factor the numerator and denominator by grouping in this rational expression.

$\dfrac{zw + 4z - 3w - 12}{zw + 4z + 5w + 20}$

$= \dfrac{(zw + 4z) - (3w + 12)}{(zw + 4z) + (5w + 20)}$ Group the terms.

$= \dfrac{z(w + 4) - 3(w + 4)}{z(w + 4) + 5(w + 4)}$ Factor each group.

$= \dfrac{(w + 4)(z - 3)}{(w + 4)(z + 5)}$ Factor by grouping.

$= \dfrac{z - 3}{z + 5}$ Divide out the common factor.

59. $LW = A$ Formula for the area of a rectangle

$W = \dfrac{A}{L}$ Divide by L.

$W = \dfrac{x^4 + 10x^2 + 21}{x^2 + 7}$ Substitute for A and L.

$W = \dfrac{(x^2 + 7)(x^2 + 3)}{x^2 + 7}$ Factor.

$W = x^2 + 3$ Divide out the common factor.

Note: If it is not apparent that A,

$$x^4 + 10x^2 + 21,$$

factors as $(x^2 + 7)(x^2 + 3)$, we can use "long division" to find the quotient $\dfrac{A}{L}$.

Remember to insert zeros for the coefficients of the missing terms.

$$
\begin{array}{r}
x^2 + 3 \\
x^2 + 0x + 7\overline{)x^4 + 0x^3 + 10x^2 + 0x + 21} \\
\underline{x^4 + 0x^3 + 7x^2 } \\
3x^2 + 0x + 21 \\
\underline{3x^2 + 0x + 21} \\
0
\end{array}
$$

The width of the rectangle is $x^2 + 3$.

Section 14.2 (pages 999–1000)

29. $\dfrac{6(m - 2)^2}{5(m + 4)^2} \cdot \dfrac{15(m + 4)}{2(2 - m)}$

$= \dfrac{6 \cdot 15(m - 2)^2(m + 4)}{5 \cdot 2(m + 4)^2(2 - m)}$ Multiply.

$= \dfrac{2 \cdot 3 \cdot 3 \cdot 5(m - 2)(m - 2)(m + 4)}{5 \cdot 2(m + 4)(m + 4)(-1)(m - 2)}$

  Factor.

$= \dfrac{3 \cdot 3(m - 2)}{(m + 4)(-1)}$ Divide out the common factors.

$= \dfrac{9(m - 2)}{-(m + 4)},$ or $\dfrac{-9(m - 2)}{m + 4}$

37. $\dfrac{m^2 + 2mp - 3p^2}{m^2 - 3mp + 2p^2} \div \dfrac{m^2 + 4mp + 3p^2}{m^2 + 2mp - 8p^2}$

$= \dfrac{m^2 + 2mp - 3p^2}{m^2 - 3mp + 2p^2} \cdot \dfrac{m^2 + 2mp - 8p^2}{m^2 + 4mp + 3p^2}$

  Multiply by the reciprocal.

$= \dfrac{(m + 3p)(m - p)(m + 4p)(m - 2p)}{(m - 2p)(m - p)(m + 3p)(m + p)}$

  Multiply numerators and multiply denominators; factor.

$= \dfrac{m + 4p}{m + p}$

  Divide out the common factors.

41. The formula for the area of a rectangle is $A = LW$. Solving for W gives $W = \dfrac{A}{L}$, which can be written as $W = A \div L$, since the fraction bar indicates division. Substituting the given expressions for A and L gives

$$
\begin{array}{cc}
A & L \\
\downarrow & \downarrow
\end{array}
$$

$\dfrac{5x^2y^3}{2pq} \div \dfrac{2xy}{p}$

$= \dfrac{5x^2y^3}{2pq} \cdot \dfrac{p}{2xy}$ Multiply by the reciprocal.

$= \dfrac{5xy^2}{4q}$ Multiply; divide out the common factors.

as the width.

SOLUTIONS

Section 14.3 (pages 1005–1006)

37. $\dfrac{-4t}{3t-6} = \dfrac{?}{12-6t}$

Factor each denominator. (On the right, rewrite $12-6t$ as $-6t+12$, and factor.)

$$\frac{-4t}{3(t-2)} = \frac{?}{-6(t-2)}$$

The missing factor is -2, so multiply the fraction on the left by $\dfrac{-2}{-2}$.

$$\frac{-4t}{3(t-2)} \cdot \frac{-2}{-2} = \frac{8t}{-6(t-2)}, \text{ or } \frac{8t}{12-6t}$$

Section 14.4 (pages 1013–1016)

31. $\dfrac{x+1}{6} + \dfrac{3x+3}{9}$

First, write the second fraction in lowest terms.

$$\frac{3x+3}{9} \text{ is } \frac{3(x+1)}{9}, \text{ or } \frac{x+1}{3}.$$

The problem becomes

$$\frac{x+1}{6} + \frac{x+1}{3}.$$

The LCD is 6. Thus,

$$\frac{x+1}{6} + \frac{x+1}{3}$$

$= \dfrac{x+1}{6} + \dfrac{x+1}{3} \cdot \dfrac{2}{2}$ Get a common denominator.

$= \dfrac{x+1}{6} + \dfrac{2(x+1)}{6}$

$= \dfrac{x+1+2x+2}{6}$ Add numerators; distributive property

$= \dfrac{3x+3}{6}$ Combine like terms.

$= \dfrac{3(x+1)}{3 \cdot 2}$ Factor.

$= \dfrac{x+1}{2}.$ Lowest terms

43. $\dfrac{t}{t+2} + \dfrac{5-t}{t} - \dfrac{4}{t^2+2t}$

$= \dfrac{t}{t+2} + \dfrac{5-t}{t} - \dfrac{4}{t(t+2)}$ Factor.

$= \dfrac{t}{t+2} \cdot \dfrac{t}{t} + \dfrac{5-t}{t} \cdot \dfrac{t+2}{t+2}$

$\quad - \dfrac{4}{t(t+2)}$ LCD $= t(t+2)$

$= \dfrac{t \cdot t + (5-t)(t+2) - 4}{t(t+2)}$

Write using the LCD.

$= \dfrac{t^2 + 5t + 10 - t^2 - 2t - 4}{t(t+2)}$

Multiply in the numerator.

$= \dfrac{3t+6}{t(t+2)}$ Combine like terms.

$= \dfrac{3(t+2)}{t(t+2)}$ Factor.

$= \dfrac{3}{t}$ Lowest terms

61. $\dfrac{2q+1}{3q^2+10q-8} - \dfrac{3q+5}{2q^2+5q-12}$

$= \dfrac{2q+1}{(3q-2)(q+4)} - \dfrac{3q+5}{(2q-3)(q+4)}$

Factor.

$= \dfrac{(2q+1)(2q-3)}{(3q-2)(q+4)(2q-3)}$

$\quad - \dfrac{(3q+5)(3q-2)}{(2q-3)(q+4)(3q-2)}$

LCD $= (3q-2)(q+4)(2q-3)$

$= \dfrac{(2q+1)(2q-3) - (3q+5)(3q-2)}{(3q-2)(q+4)(2q-3)}$

Subtract numerators.

$= \dfrac{(4q^2-4q-3) - (9q^2+9q-10)}{(3q-2)(q+4)(2q-3)}$

Multiply in the numerator.

$= \dfrac{4q^2-4q-3-9q^2-9q+10}{(3q-2)(q+4)(2q-3)}$

Distributive property

$= \dfrac{-5q^2-13q+7}{(3q-2)(q+4)(2q-3)}$

Combine like terms.

65. $\dfrac{x+3y}{x^2+2xy+y^2} + \dfrac{x-y}{x^2+4xy+3y^2}$

$= \dfrac{x+3y}{(x+y)(x+y)} + \dfrac{x-y}{(x+3y)(x+y)}$

$= \dfrac{(x+3y)(x+3y)}{(x+y)(x+y)(x+3y)}$

$\quad + \dfrac{(x-y)(x+y)}{(x+3y)(x+y)(x+y)}$

LCD $= (x+y)(x+y)(x+3y)$

$= \dfrac{(x^2+6xy+9y^2) + (x^2-y^2)}{(x+y)(x+y)(x+3y)}$

$= \dfrac{2x^2+6xy+8y^2}{(x+y)(x+y)(x+3y)},$

or $\dfrac{2x^2+6xy+8y^2}{(x+y)^2(x+3y)}$

There is no need to factor out 2 in the numerator, since the denominator does not have 2 as a factor. The answer is in lowest terms.

Section 14.5 (pages 1023–1024)

15. $\dfrac{\dfrac{1}{x}+x}{\dfrac{x^2+1}{8}}$ Use Method 2.

$= \dfrac{8x\left(\dfrac{1}{x}+x\right)}{8x\left(\dfrac{x^2+1}{8}\right)}$ LCD $= 8x$

$= \dfrac{8+8x^2}{x(x^2+1)}$ Distributive property

$= \dfrac{8(1+x^2)}{x(x^2+1)}$ Factor.

$= \dfrac{8}{x}$ Lowest terms

29. $\dfrac{\dfrac{1}{m-1}+\dfrac{2}{m+2}}{\dfrac{2}{m+2}-\dfrac{1}{m-3}}$ Use Method 2.

$= \dfrac{(m-1)(m+2)(m-3)\left(\dfrac{1}{m-1}+\dfrac{2}{m+2}\right)}{(m-1)(m+2)(m-3)\left(\dfrac{2}{m+2}-\dfrac{1}{m-3}\right)}$

LCD $= (m-1)(m+2)(m-3)$

$= \dfrac{(m+2)(m-3) + 2(m-1)(m-3)}{2(m-1)(m-3) - (m-1)(m+2)}$

Distributive property

$= \dfrac{(m-3)[(m+2)+2(m-1)]}{(m-1)[2(m-3)-(m+2)]}$

Factor out $m-3$ in the numerator and $m-1$ in the denominator.

$= \dfrac{(m-3)[m+2+2m-2]}{(m-1)[2m-6-m-2]}$

Distributive property

$= \dfrac{3m(m-3)}{(m-1)(m-8)}$

Combine like terms.

31. $2 - \dfrac{2}{2+\dfrac{2}{2+2}}$

$= 2 - \dfrac{2}{2+\dfrac{2}{4}}$ Simplify the denominator.

$= 2 - \dfrac{2}{\dfrac{5}{2}}$ $2+\dfrac{2}{4} = \dfrac{4}{2}+\dfrac{1}{2} = \dfrac{5}{2}$

$= 2 - \left(2 \cdot \dfrac{2}{5}\right)$ $\dfrac{2}{\dfrac{5}{2}} = 2 \div \dfrac{5}{2} = 2 \cdot \dfrac{2}{5}$

$= 2 - \dfrac{4}{5}$ Multiply.

$= \dfrac{10}{5} - \dfrac{4}{5}$ $2 = \dfrac{10}{5}$

$= \dfrac{6}{5}$ Subtract.

Section 14.6 (pages 1033–1036)

11.
$$\frac{3x}{5} - 6 = x$$

$$5\left(\frac{3x}{5} - 6\right) = 5(x) \quad \text{Multiply by the LCD, 5.}$$

$$5\left(\frac{3x}{5}\right) - 5(6) = 5x \quad \text{Distributive property}$$

$$3x - 30 = 5x$$

$$-30 = 2x \quad \text{Subtract } 3x.$$

$$-15 = x \quad \text{Divide by 2.}$$

Check by substituting -15 for x in the original equation.
Solution set: $\{-15\}$

21.
$$\frac{q + 2}{3} + \frac{q - 5}{5} = \frac{7}{3}$$

$$15\left(\frac{q + 2}{3} + \frac{q - 5}{5}\right) = 15\left(\frac{7}{3}\right)$$
$$\text{Multiply by the LCD, 15.}$$

$$15\left(\frac{q + 2}{3}\right) + 15\left(\frac{q - 5}{5}\right) = 5 \cdot 7$$

$$5(q + 2) + 3(q - 5) = 35$$

$$5q + 10 + 3q - 15 = 35$$

$$8q - 5 = 35$$

$$8q = 40$$

$$q = 5$$

A check confirms that 5 is the solution.
Solution set: $\{5\}$

37.
$$\frac{2}{m} = \frac{m}{5m + 12}$$

$$m(5m + 12)\left(\frac{2}{m}\right) = m(5m + 12)\left(\frac{m}{5m + 12}\right)$$
$$\text{Multiply by the LCD, } m(5m + 12).$$

$$(5m + 12)(2) = m(m) \quad \text{Divide out the common factors.}$$

$$10m + 24 = m^2 \quad \text{Multiply.}$$

$$m^2 - 10m - 24 = 0 \quad \text{Standard form}$$

$$(m + 2)(m - 12) = 0 \quad \text{Factor.}$$

$$m + 2 = 0 \quad \text{or} \quad m - 12 = 0$$
$$\text{Zero-factor property}$$

$$m = -2 \quad \text{or} \quad m = 12$$

The proposed solutions are -2 and 12.

Check $\quad \dfrac{2}{-2} \overset{?}{=} \dfrac{-2}{5(-2) + 12} \quad$ Let $m = -2$.

$$-1 = \frac{-2}{2} \quad \text{True}$$

Check $\quad \dfrac{2}{12} \overset{?}{=} \dfrac{12}{5(12) + 12} \quad$ Let $m = 12$.

$$\frac{1}{6} = \frac{12}{72} \quad \text{True}$$

Solution set: $\{-2, 12\}$

---

65. $d = \dfrac{2S}{n(a + L)}$ for a

We need to isolate a on one side of the equation.

$$d = \frac{2S}{n(a + L)}$$

$$d \cdot n(a + L) = \frac{2S}{n(a + L)} \cdot n(a + L)$$
$$\text{Multiply by } n(a + L).$$

$$dn(a + L) = 2S$$

$$dna + dnL = 2S \quad \text{Distributive property}$$

$$dna = 2S - dnL \quad \text{Subtract } dnL.$$

$$a = \frac{2S - dnL}{dn}, \quad \text{Divide by } dn.$$

or $a = \dfrac{2S}{dn} - L$

71.
$$9x + \frac{3}{z} = \frac{5}{y} \quad \text{for } z$$

$$yz\left(9x + \frac{3}{z}\right) = yz\left(\frac{5}{y}\right) \quad \text{Multiply by the LCD, } yz.$$

$$yz(9x) + yz\left(\frac{3}{z}\right) = yz\left(\frac{5}{y}\right) \quad \text{Distributive property}$$

$$9xyz + 3y = 5z$$

$$9xyz - 5z = -3y \quad \text{Get the } z\text{-terms on one side.}$$

$$z(9xy - 5) = -3y \quad \text{Factor out } z.$$

$$z = \frac{-3y}{9xy - 5}, \quad \text{Divide by } 9xy - 5.$$

or $z = \dfrac{3y}{5 - 9xy}$

Section 14.7 (pages 1045–1048)

25. Let x represent the rate of the current of the river. Then $(12 - x)$ is the rate upstream (against the current) and $(12 + x)$ is the rate downstream (with the current).

Use $t = \dfrac{d}{r}$ to complete the table.

	d	r	t
Upstream	6	$12 - x$	$\dfrac{6}{12 - x}$
Downstream	10	$12 + x$	$\dfrac{10}{12 + x}$

Since the times are equal, we get the following equation.

$$\frac{6}{12 - x} = \frac{10}{12 + x}$$

$$(12 + x)(12 - x)\frac{6}{12 - x}$$

$$= (12 + x)(12 - x)\frac{10}{12 + x}$$

Multiply by the LCD, $(12 + x)(12 - x)$.

$$6(12 + x) = 10(12 - x)$$

$$72 + 6x = 120 - 10x$$

$$16x = 48$$

$$x = 3$$

The rate of the current of the river is 3 mph.

33. Let $x =$ the number of hours it would take Brenda to paint the room by herself.

	Rate	Time Working Together	Fractional Part of the Job Done When Working Together
Hilda	$\dfrac{1}{6}$	$3\dfrac{3}{4} = \dfrac{15}{4}$	$\dfrac{1}{6} \cdot \dfrac{15}{4} = \dfrac{5}{8}$
Brenda	$\dfrac{1}{x}$	$3\dfrac{3}{4} = \dfrac{15}{4}$	$\dfrac{1}{x} \cdot \dfrac{15}{4} = \dfrac{15}{4x}$

Since together Hilda and Brenda complete 1 whole job, we must add their individual fractional parts and set the sum equal to 1.

$$\frac{5}{8} + \frac{15}{4x} = 1$$

$$8x\left(\frac{5}{8} + \frac{15}{4x}\right) = 8x(1) \quad \text{Multiply by the LCD, } 8x.$$

$$8x\left(\frac{5}{8}\right) + 8x\left(\frac{15}{4x}\right) = 8x$$

$$5x + 30 = 8x$$

$$30 = 3x$$

$$10 = x$$

It will take Brenda 10 hr to paint the room by herself.

35. Let $x =$ the number of hours to fill the pool with both pipes left open.

	Rate	Time to Fill the Pool	Part Done by Each Pipe
Inlet pipe	$\dfrac{1}{9}$	x	$\dfrac{1}{9}x$
Outlet pipe	$\dfrac{1}{12}$	x	$\dfrac{1}{12}x$

Part done by inlet pipe	$-$	Part done by outlet pipe	$=$	Full pool
$\downarrow$		$\downarrow$		$\downarrow$
$\dfrac{1}{9}x$	$-$	$\dfrac{1}{12}x$	$=$	1

$$36\left(\frac{1}{9}x - \frac{1}{12}x\right) = 36(1) \quad \text{Multiply by the LCD, 36.}$$

$$36\left(\frac{1}{9}x\right) - 36\left(\frac{1}{12}x\right) = 36$$

$$4x - 3x = 36$$

$$x = 36$$

It will take 36 hr to fill the pool.

Section 14.8 (pages 1051–1054)

11. For a given base, the area A of a triangle varies directly as its height h, so there is a constant k such that $A = kh$. Find the value of k.

$$10 = k(4) \qquad \text{Let } A = 10, h = 4.$$

$$k = \frac{10}{4} = 2.5 \qquad \text{Solve for } k.$$

When $k = 2.5$, $A = kh$ becomes

$$A = 2.5h.$$

Now find A when $h = 6$.

$$A = 2.5(6) = 15$$

When the height of the triangle is 6 in., the area of the triangle is 15 in.^{2}

21. For a constant time of 3 hr, if the rate of the pickup truck *increases*, then the distance traveled *increases*. Thus, the variation between the quantities is *direct*.

29. Ratios of corresponding sides are equal, so since $\frac{12}{3} = 4$, we must have

$$\frac{x}{2} = 4, \text{ and thus, } x = 8.$$

Chapter 15 Systems of Linear Equations and Inequalities

Section 15.1 (pages 1073–1076)

21. $2x - 3y = -6$

$y = -3x + 2$

To graph $2x - 3y = -6$, find the intercepts.

$$2x - 3(0) = -6 \quad \text{Let } y = 0.$$
$$2x = -6$$
$$x = -3$$

The x-intercept is $(-3, 0)$.

$$2(0) - 3y = -6 \quad \text{Let } x = 0.$$
$$-3y = -6$$
$$y = 2$$

The y-intercept is $(0, 2)$.

Plot the intercepts, $(-3, 0)$ and $(0, 2)$, and draw the line through them.

To graph the second line, start by plotting the y-intercept, $(0, 2)$. From this point, go 3 units down and 1 unit to the right (because the slope is -3) to reach the point $(1, -1)$. Draw the line through $(0, 2)$ and $(1, -1)$.

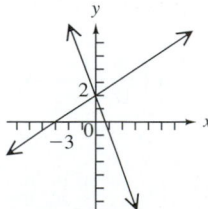

The lines intersect at their common y-intercept, $(0, 2)$.

Solution set: $\{(0, 2)\}$

25. $4x - 2y = 8$ (1)

$2x = y + 4$ (2)

Graph the line $4x - 2y = 8$ using its intercepts, $(2, 0)$ and $(0, -4)$.

Graph the equation $2x = y + 4$ using its intercepts, $(2, 0)$ and $(0, -4)$. Since both equations have the same intercepts, they are equations of the same line.

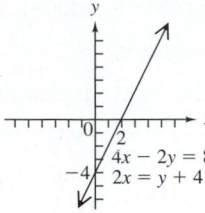

There are an infinite number of solutions, so we use set-builder notation to write the solution set. Rewrite equation (2) in standard form.

$$2x - y = 4$$

The solution set is

$$\{(x, y) \mid 2x - y = 4\}.$$

The given system consists of dependent equations.

Section 15.2 (pages 1083–1084)

15. $6x - 8y = 6$ (1)

$2y = -2 + 3x$ (2)

Solve equation (2) for y.

$$2y = -2 + 3x$$
$$y = \frac{3x - 2}{2} \qquad (3)$$

Substitute $\frac{3x - 2}{2}$ for y in equation (1).

$$6x - 8y = 6 \quad (1)$$
$$6x - 8\left(\frac{3x - 2}{2}\right) = 6$$
$$6x - 4(3x - 2) = 6$$
$$6x - 12x + 8 = 6$$
$$-6x + 8 = 6$$
$$-6x = -2$$
$$x = \frac{-2}{-6}, \text{ or } \frac{1}{3}$$

To find y, let $x = \frac{1}{3}$ in equation (3).

$$y = \frac{3x - 2}{2} = \frac{3(\frac{1}{3}) - 2}{2} = \frac{1 - 2}{2} = -\frac{1}{2}$$

Solution set: $\left\{\left(\frac{1}{3}, -\frac{1}{2}\right)\right\}$

19. $12x - 16y = 8$ (1)

$3x = 4y + 2$ (2)

Solve equation (2) for x.

$$3x = 4y + 2$$
$$x = \frac{4y + 2}{3} \qquad (3)$$

Substitute $\frac{4y + 2}{3}$ for x in equation (1).

$$12x - 16y = 8 \quad (1)$$

$$12\left(\frac{4y + 2}{3}\right) - 16y = 8$$
$$4(4y + 2) - 16y = 8$$
$$16y + 8 - 16y = 8$$
$$8 = 8 \qquad \text{True}$$

This true result means that every solution of one equation is also a solution of the other, so the system has an infinite number of solutions. To write the solution set, rewrite equation (2) in standard form.

$$3x - 4y = 2$$

Solution set: $\{(x, y) \mid 3x - 4y = 2\}$

25. $\dfrac{x}{5} + 2y = \dfrac{16}{5}$ (1)

$\dfrac{3x}{5} + \dfrac{y}{2} = -\dfrac{7}{5}$ (2)

Multiply each side of equation (1) by 5.

$$5\left(\frac{x}{5} + 2y\right) = 5\left(\frac{16}{5}\right)$$
$$x + 10y = 16 \qquad (3)$$

Multiply each side of equation (2) by 10.

$$10\left(\frac{3x}{5} + \frac{y}{2}\right) = 10\left(-\frac{7}{5}\right)$$
$$6x + 5y = -14 \qquad (4)$$

We now have the simplified system

$$x + 10y = 16 \qquad (3)$$
$$6x + 5y = -14. \qquad (4)$$

Solve equation (3) for x.

$$x = 16 - 10y \qquad (5)$$

Substitute $16 - 10y$ for x in equation (4).

$$6x + 5y = -14 \qquad (4)$$
$$6(16 - 10y) + 5y = -14$$
$$96 - 60y + 5y = -14$$
$$-55y = -110$$
$$y = 2$$

To find x, let $y = 2$ in equation (5).

$$x = 16 - 10y \qquad (5)$$
$$x = 16 - 10(2)$$
$$x = -4$$

Solution set: $\{(-4, 2)\}$

Section 15.3 (pages 1089–1092)

25. $-x + 3y = 4$ (1)

$-2x + 6y = 8$ (2)

Multiply equation (1) by -2 and add the result to equation (2).

$$\begin{array}{rl} 2x - 6y = -8 & (3) \\ -2x + 6y = 8 & (2) \\ \hline 0 = 0 & \text{True} \end{array}$$

The true statement indicates that the equations of the original system are dependent. To obtain an equation in standard form, with $A > 0$ in $Ax + By = C$, multiply equation (1) by -1.

$$x - 3y = -4$$

Solution set: $\{(x, y) \mid x - 3y = -4\}$

39.
$$6x + 3y = 0 \quad (1)$$
$$-18x - 9y = 0 \quad (2)$$

Multiply equation (1) by 3 and add the result to equation (2).

$$18x + 9y = 0 \quad (3)$$
$$\underline{-18x - 9y = 0 \quad (2)}$$
$$0 = 0 \quad \text{True}$$

This true result, $0 = 0$, means that the system has an infinite number of solutions. To obtain an equation for the solution set in which the coefficients have greatest common factor 1, divide both sides of equation (1) by 3 to obtain the equivalent equation

$$2x + y = 0.$$

Solution set: $\{(x, y) \mid 2x + y = 0\}$

Section 15.4 (pages 1099–1104)

33. *Step 1*

Read the problem again.

Step 2

Assign variables.

Let x = the average speed of the bicycle;
y = the average speed of the car.

	r	t	d
Bicycle	x	7.5	$7.5x$
Car	y	7.5	$7.5y$

↑
Use the formula
$d = rt$.

Step 3

The total distance traveled by the bicycle and the car is 471 mi, so

$$7.5x + 7.5y = 471. \quad (1)$$

The car traveled 35.8 mi faster than the bicycle, so

$$y = x + 35.8. \quad (2)$$

We now have the system

$$7.5x + 7.5y = 471 \quad (1)$$
$$y = x + 35.8. \quad (2)$$

Step 4

Because equation (2) is already solved for y, the substitution method is a good choice. Substitute $x + 35.8$ for y in equation (1).

$$7.5x + 7.5y = 471 \quad (1)$$
$$7.5x + 7.5(x + 35.8) = 471$$
Let $y = x + 35.8$.
$$7.5x + 7.5x + 268.5 = 471$$
Distributive property
$$15x + 268.5 = 471$$
Combine like terms.
$$15x = 202.5$$
Subtract 268.5.
$$x = 13.5$$
Divide by 15.

To find the value of y substitute 13.5 for x in equation (2).

$$y = x + 35.8 \quad (2)$$
$$y = 13.5 + 35.8 \quad \text{Let } x = 13.5.$$
$$y = 49.3$$

Step 5

The average speed of the bicycle is 13.5 mph, and the average speed of the car is 49.3 mph.

Step 6

In 7.5 hr, the total distance traveled by the bicycle and the car is

$$7.5(13.5) + 7.5(49.3) = 471 \text{ mi.}$$

Since $49.3 - 13.5 = 35.8$, the car traveled 35.8 mph faster than the bicycle, as stated.

Section 15.5 (pages 1109–1110)

15. $x \leq 2y + 3$
$x + y < 0$

Graph $x = 2y + 3$ as a solid line through $(3, 0)$ and $(5, 1)$. Using $(0, 0)$ as a test point will result in the true statement $0 \leq 3$, so shade the region containing the origin.

Graph $x + y = 0$ as a dashed line through $(0, 0)$ and $(1, -1)$. Using $(1, 0)$ as a test point will result in the false statement $1 < 0$, so shade the region *not* containing $(1, 0)$.

The solution set of the system is the intersection of the two shaded regions. It includes the portion of the line $x = 2y + 3$ that bounds the region but not the portion of the line $x + y = 0$.

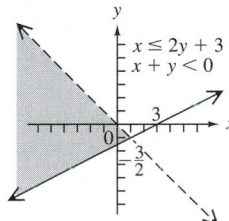

19. $3x - 2y \geq 6$
$x + y \leq 4$
$x \geq 0$
$y \geq -4$

Graph $3x - 2y = 6$, $x + y = 4$, $x = 0$, and $y = -4$ as solid lines. All four inequalities are true for $(2, -2)$. Shade the region bounded by the four lines, which contains the test point $(2, -2)$. The solution set is the shaded region.

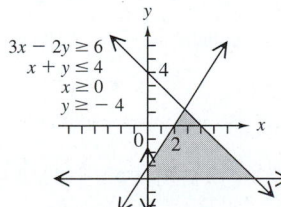

Chapter 16 Roots and Radicals

Section 16.1 (pages 1129–1132)

55. $\sqrt{103} \approx \sqrt{100} = 10$ Length
$\sqrt{48} \approx \sqrt{49} = 7$ Width

The best estimate for the length and width, in meters, of the rectangle is 10 by 7, choice C.

71. Let $a = 4.5$ and $c = 12.0$. Use the Pythagorean formula.

$$c^2 = a^2 + b^2$$
$$(12.0)^2 = (4.5)^2 + b^2 \quad \text{Substitute carefully.}$$
$$144 = 20.25 + b^2$$
$$123.75 = b^2$$
$$b = \sqrt{123.75}$$
$$b \approx 11.1 \quad \text{Use a calculator.}$$

The distance from the base of the tree to the point where the broken part touches the ground is 11.1 ft (to the nearest tenth).

83. $-\sqrt[3]{-8} = -(-2) = 2$

Section 16.2 (pages 1139–1142)

37. $5\sqrt{3} \cdot 2\sqrt{15}$ Commutative property;
$= 5 \cdot 2 \cdot \sqrt{3 \cdot 15}$ product rule
$= 10\sqrt{45}$ Multiply.
$= 10\sqrt{9 \cdot 5}$ Factor; 9 is a perfect square.
$= 10\sqrt{9} \cdot \sqrt{5}$ Product rule
$= 10 \cdot 3 \cdot \sqrt{5}$ $\sqrt{9} = 3$
$= 30\sqrt{5}$ Multiply.

105. Use the formula for the volume of a sphere,

$$V = \frac{4}{3}\pi r^3.$$

Let $V = 288\pi$ and solve for r.

$$288\pi = \frac{4}{3}\pi r^3 \quad V = 288\pi$$
$$\frac{3}{4}(288\pi) = \frac{3}{4}\left(\frac{4}{3}\pi r^3\right) \quad \text{Multiply by } \frac{3}{4}.$$
$$216\pi = \pi r^3 \quad \text{Simplify.}$$
$$216 = r^3 \quad \text{Divide by } \pi.$$
$$\sqrt[3]{216} = r \quad \text{Take cube roots.}$$
$$6 = r$$

The radius is 6 in.

107. $2\sqrt{26} \approx 2\sqrt{25} = 2 \cdot 5 = 10$ Length
$\sqrt{83} \approx \sqrt{81} = 9$ Width

Using 10 and 9 as estimates for the length and the width of the rectangle gives us

$$10 \cdot 9 = 90$$

as an estimate for the area, in square inches. Choice D is the best estimate.

Section 16.3 (pages 1145–1146)

19. $2\sqrt{8} - 5\sqrt{32} - 2\sqrt{48}$

$= 2(\sqrt{4} \cdot \sqrt{2}) - 5(\sqrt{16} \cdot \sqrt{2})$

$\quad - 2(\sqrt{16} \cdot \sqrt{3})$ Product rule

$= 2(2\sqrt{2}) - 5(4\sqrt{2}) - 2(4\sqrt{3})$

$\qquad\qquad \sqrt{4} = 2; \sqrt{16} = 4$

$= 4\sqrt{2} - 20\sqrt{2} - 8\sqrt{3}$

$\qquad\qquad$ Multiply.

$= -16\sqrt{2} - 8\sqrt{3}$ Subtract like
$\qquad\qquad\qquad$ radicals.

23. $\dfrac{1}{4}\sqrt{288} + \dfrac{1}{6}\sqrt{72}$

$= \dfrac{1}{4}\sqrt{144 \cdot 2} + \dfrac{1}{6}\sqrt{36 \cdot 2}$

$= \dfrac{1}{4}(\sqrt{144} \cdot \sqrt{2}) + \dfrac{1}{6}(\sqrt{36} \cdot \sqrt{2})$

$\qquad\qquad\qquad$ Product rule

$= \dfrac{1}{4}(12\sqrt{2}) + \dfrac{1}{6}(6\sqrt{2})$

$\qquad\qquad \sqrt{144} = 12; \sqrt{36} = 6$

$= 3\sqrt{2} + 1\sqrt{2}$ Multiply.

$= 4\sqrt{2}$ Add like radicals.

25. Use the formula for the perimeter
of a rectangle.

$P = 2L + 2W$

$P = 2(7\sqrt{2}) + 2(4\sqrt{2})$

$P = 14\sqrt{2} + 8\sqrt{2}$

$P = 22\sqrt{2}$

43. $5\sqrt[4]{m^3} + 8\sqrt[4]{16m^3}$

$= 5\sqrt[4]{m^3} + 8\sqrt[4]{16 \cdot m^3}$

$= 5\sqrt[4]{m^3} + 8\sqrt[4]{16}\sqrt[4]{m^3}$ Product rule

$= 5\sqrt[4]{m^3} + 8 \cdot 2 \cdot \sqrt[4]{m^3}$ $\sqrt[4]{16} = 2$

$= 5\sqrt[4]{m^3} + 16\sqrt[4]{m^3}$ Multiply.

$= 21\sqrt[4]{m^3}$ Add like
$\qquad\qquad\qquad$ radicals.

Section 16.4 (pages 1151–1152)

37. $\sqrt{\dfrac{9a^2r^5}{7t}}$

$= \dfrac{\sqrt{9a^2r^5}}{\sqrt{7t}}$ Quotient rule

$= \dfrac{\sqrt{9a^2r^4 \cdot r}}{\sqrt{7t}}$ Factor.

$= \dfrac{\sqrt{9a^2r^4} \cdot \sqrt{r}}{\sqrt{7t}}$ Product rule

$= \dfrac{3ar^2\sqrt{r}}{\sqrt{7t}}$ $\sqrt{9} = 3; \sqrt{a^2} = a,$
$\qquad\qquad$ since $a > 0; \sqrt{r^4} = r^2$

$= \dfrac{3ar^2\sqrt{r} \cdot \sqrt{7t}}{\sqrt{7t} \cdot \sqrt{7t}}$ Rationalize the
$\qquad\qquad\qquad$ denominator.

$= \dfrac{3ar^2\sqrt{7rt}}{7t}$ Product rule;
$\qquad\qquad \sqrt{7t} \cdot \sqrt{7t} = 7t$

41. $\sqrt[3]{\dfrac{5}{9}}$

$= \dfrac{\sqrt[3]{5}}{\sqrt[3]{9}}$ Quotient rule

$= \dfrac{\sqrt[3]{5} \cdot \sqrt[3]{3}}{\sqrt[3]{9} \cdot \sqrt[3]{3}}$ Multiply numerator and
$\qquad\qquad$ denominator by $\sqrt[3]{3}$ to
$\qquad\qquad$ make radical in denom-
$\qquad\qquad$ inator a perfect cube.

$= \dfrac{\sqrt[3]{15}}{\sqrt[3]{27}}$ Product rule

$= \dfrac{\sqrt[3]{15}}{3}$ $\sqrt[3]{27} = 3$

49. (a) $p = k \cdot \sqrt{\dfrac{L}{g}}$

$p = 6 \cdot \sqrt{\dfrac{9}{32}}$ Let $k = 6, L = 9,$
$\qquad\qquad$ $g = 32.$

$p = \dfrac{6\sqrt{9}}{\sqrt{32}}$

$p = \dfrac{6 \cdot 3}{\sqrt{16 \cdot 2}}$

$p = \dfrac{18}{4\sqrt{2}}$

$p = \dfrac{9}{2\sqrt{2}}$

$p = \dfrac{9 \cdot \sqrt{2}}{2\sqrt{2} \cdot \sqrt{2}}$ Rationalize the
$\qquad\qquad$ denominator.

$p = \dfrac{9\sqrt{2}}{4}$

The period of the pendulum is $\dfrac{9\sqrt{2}}{4}$ sec.

(b) Using a calculator, $\dfrac{9\sqrt{2}}{4} \approx 3.182$ sec.

Section 16.5 (pages 1157–1160)

13. $(5\sqrt{7} - 2\sqrt{3})(3\sqrt{7} + 4\sqrt{3})$

$= 5\sqrt{7}(3\sqrt{7}) + 5\sqrt{7}(4\sqrt{3})$

$\quad - 2\sqrt{3}(3\sqrt{7}) - 2\sqrt{3}(4\sqrt{3})$

$\qquad\qquad\qquad$ FOIL

$= 15 \cdot 7 + 20\sqrt{21} - 6\sqrt{21} - 8 \cdot 3$

$\qquad\qquad\qquad$ Multiply.

$= 105 + 14\sqrt{21} - 24$ Multiply
$\qquad\qquad$ numbers; add
$\qquad\qquad$ like radicals.

$= 81 + 14\sqrt{21}$ Subtract.

43. $\dfrac{\sqrt{5}}{\sqrt{2} + \sqrt{3}}$

$= \dfrac{\sqrt{5}(\sqrt{2} - \sqrt{3})}{(\sqrt{2} + \sqrt{3})(\sqrt{2} - \sqrt{3})}$

Multiply numerator and denominator
by the conjugate of the denominator.

$= \dfrac{\sqrt{5} \cdot \sqrt{2} - \sqrt{5} \cdot \sqrt{3}}{(\sqrt{2})^2 - (\sqrt{3})^2}$

Distributive property;
$(a + b)(a - b) = a^2 - b^2$

$= \dfrac{\sqrt{10} - \sqrt{15}}{2 - 3}$ Product rule;
$\qquad\qquad (\sqrt{a})^2 = a$

$= \dfrac{\sqrt{10} - \sqrt{15}}{-1}$

$= -\sqrt{10} + \sqrt{15}$

55. $\dfrac{12 - \sqrt{40}}{4}$

$= \dfrac{12 - \sqrt{4} \cdot \sqrt{10}}{4}$ Product rule

$= \dfrac{12 - 2\sqrt{10}}{4}$ $\sqrt{4} = 2$

$= \dfrac{2(6 - \sqrt{10})}{2(2)}$ Factor numerator
$\qquad\qquad$ and denominator.

$= \dfrac{6 - \sqrt{10}}{2}$ Lowest terms

Section 16.6 (pages 1167–1170)

19. $\sqrt{3x - 5} = \sqrt{2x + 1}$

$(\sqrt{3x - 5})^2 = (\sqrt{2x + 1})^2$

$\qquad\qquad\qquad$ Squaring property

$3x - 5 = 2x + 1$ $(\sqrt{a})^2 = a$

$x = 6$ Subtract $2x$; add 5.

Check $\sqrt{3x - 5} = \sqrt{2x + 1}$

$\sqrt{3(6) - 5} \overset{?}{=} \sqrt{2(6) + 1}$

$\qquad\qquad$ Let $x = 6.$

$\sqrt{13} = \sqrt{13}$ True

Solution set: $\{6\}$

29. $\sqrt{3k + 10} + 5 = 2k$

$\sqrt{3k + 10} = 2k - 5$ Subtract 5.

$(\sqrt{3k + 10})^2 = (2k - 5)^2$ Squaring
$\qquad\qquad\qquad$ property

$3k + 10 = 4k^2 - 20k + 25$

$\qquad\qquad (\sqrt{a})^2 = a$; square the binomial.

$4k^2 - 23k + 15 = 0$ Standard form

$(4k - 3)(k - 5) = 0$ Factor.

$k = \dfrac{3}{4}$ or $k = 5$ Zero-factor property

Checking both proposed solutions in the original equation will show that 5 is a solution, but $\dfrac{3}{4}$ is not.

Solution set: $\{5\}$

45.

$\sqrt{2x + 11} + \sqrt{x + 6} = 2$

$\sqrt{2x + 11} = 2 - \sqrt{x + 6}$ Isolate a radical.

$\left(\sqrt{2x + 11}\right)^2 = \left(2 - \sqrt{x + 6}\right)^2$ Square both sides.

$2x + 11 = 4 - 4\sqrt{x + 6} + x + 6$

$\left(\sqrt{a}\right)^2 = a^2$; square the binomial.

$x + 1 = -4\sqrt{x + 6}$ Isolate a radical.

$(x + 1)^2 = \left(-4\sqrt{x + 6}\right)^2$ Square both sides again.

$x^2 + 2x + 1 = 16(x + 6)$

Square the binomial; $\left(a\sqrt{b}\right)^2 = a^2 b$

$x^2 + 2x + 1 = 16x + 96$ Distributive property

$x^2 - 14x - 95 = 0$ Standard form

$(x + 5)(x - 19) = 0$ Factor.

$x = -5$ or $x = 19$ Zero-factor property

Checking both proposed solutions in the original equation will show that -5 is a solution, but 19 is not.

Solution set: $\{-5\}$

49. Let $x =$ the unknown number.

Write an equation.

$3\sqrt{2} = \sqrt{x + 10}$

To solve this equation, square both sides and then solve the resulting equation.

$\left(3\sqrt{2}\right)^2 = \left(\sqrt{x + 10}\right)^2$ Square both sides.

$18 = x + 10$

$\left(3\sqrt{2}\right)^2 = 9 \cdot 2 = 18$

$8 = x$ Subtract 10.

Check Since $\sqrt{8 + 10} = \sqrt{18}$, and $\sqrt{18} = 3\sqrt{2}$, the number is 8.

Chapter 17 Quadratic Equations

Section 17.1 (pages 1185–1188)

21. $7x^2 = 4$

$x^2 = \dfrac{4}{7}$ Divide by 7.

$x = \sqrt{\dfrac{4}{7}}$ or $x = -\sqrt{\dfrac{4}{7}}$

Square root property

$x = \dfrac{\sqrt{4}}{\sqrt{7}} \cdot \dfrac{\sqrt{7}}{\sqrt{7}}$ or $x = -\dfrac{\sqrt{4}}{\sqrt{7}} \cdot \dfrac{\sqrt{7}}{\sqrt{7}}$

Rationalize denominators.

$x = \dfrac{2\sqrt{7}}{7}$ or $x = -\dfrac{2\sqrt{7}}{7}$

Solution set: $\left\{-\dfrac{2\sqrt{7}}{7}, \dfrac{2\sqrt{7}}{7}\right\}$

43. $\left(\dfrac{1}{2}x + 5\right)^2 = 12$

$\dfrac{1}{2}x + 5 = \sqrt{12}$ or $\dfrac{1}{2}x + 5 = -\sqrt{12}$

$\dfrac{1}{2}x = -5 + \sqrt{12}$ or $\dfrac{1}{2}x = -5 - \sqrt{12}$

Subtract 5.

$\dfrac{1}{2}x = -5 + 2\sqrt{3}$ or $\dfrac{1}{2}x = -5 - 2\sqrt{3}$

$\sqrt{12} = \sqrt{4} \cdot \sqrt{3} = 2\sqrt{3}$

$x = -10 + 4\sqrt{3}$ or $x = -10 - 4\sqrt{3}$

Multiply each term by 2.

Solution set: $\left\{-10 + 4\sqrt{3}, -10 - 4\sqrt{3}\right\}$

53. $A = P(1 + r)^2$

$110.25 = 100(1 + r)^2$

Let $A = 110.25, P = 100$.

$1.1025 = (1 + r)^2$ Divide by 100.

$(1 + r)^2 = 1.1025$ Interchange sides.

$1 + r = \sqrt{1.1025}$ or $1 + r = -\sqrt{1.1025}$

Square root property

$r = -1 + 1.05$ or $r = -1 - 1.05$

Subtract 1; use a calculator.

$r = 0.05$ or $r = -2.05$

Reject the solution -2.05 since the rate of interest cannot be negative. The rate is 0.05, or 5%.

Section 17.2 (pages 1195–1196)

17. $t^2 + 6t + 9 = 0$

The left side of this equation is already a perfect square.

$(t + 3)^2 = 0$ Factor.

$t + 3 = 0$ Square root property

$t = -3$ Subtract 3.

A check verifies that the solution is -3.

Solution set: $\{-3\}$

19. $x^2 + x - 1 = 0$

$x^2 + x = 1$ Add 1.

Take half of 1, the coefficient of x, square it, and add the result to each side.

$x^2 + x + \dfrac{1}{4} = 1 + \dfrac{1}{4}$

Add $\left[\dfrac{1}{2}(1)\right]^2 = \dfrac{1}{4}$.

$\left(x + \dfrac{1}{2}\right)^2 = \dfrac{5}{4}$

Factor; add.

$x + \dfrac{1}{2} = \sqrt{\dfrac{5}{4}}$ or $x + \dfrac{1}{2} = -\sqrt{\dfrac{5}{4}}$

$x + \dfrac{1}{2} = \dfrac{\sqrt{5}}{2}$ or $x + \dfrac{1}{2} = -\dfrac{\sqrt{5}}{2}$

$x = -\dfrac{1}{2} + \dfrac{\sqrt{5}}{2}$ or $x = -\dfrac{1}{2} - \dfrac{\sqrt{5}}{2}$

$x = \dfrac{-1 + \sqrt{5}}{2}$ or $x = \dfrac{-1 - \sqrt{5}}{2}$

Solution set: $\left\{\dfrac{-1 + \sqrt{5}}{2}, \dfrac{-1 - \sqrt{5}}{2}\right\}$

23. $2x^2 - 4x = 5$

$x^2 - 2x = \dfrac{5}{2}$ Divide by 2.

$x^2 - 2x + 1 = \dfrac{5}{2} + 1$

Add $\left[\dfrac{1}{2}(-2)\right]^2 = 1$.

$(x - 1)^2 = \dfrac{7}{2}$ Factor; add.

$x - 1 = \sqrt{\dfrac{7}{2}}$ or $x - 1 = -\sqrt{\dfrac{7}{2}}$

$x - 1 = \dfrac{\sqrt{7}}{\sqrt{2}}$ or $x - 1 = -\dfrac{\sqrt{7}}{\sqrt{2}}$

$x - 1 = \dfrac{\sqrt{14}}{2}$ or $x - 1 = -\dfrac{\sqrt{14}}{2}$

Rationalize denominators by multiplying by $\dfrac{\sqrt{2}}{\sqrt{2}}$.

$x = 1 + \dfrac{\sqrt{14}}{2}$ or $x = 1 - \dfrac{\sqrt{14}}{2}$

$x = \dfrac{2}{2} + \dfrac{\sqrt{14}}{2}$ or $x = \dfrac{2}{2} - \dfrac{\sqrt{14}}{2}$

$x = \dfrac{2 + \sqrt{14}}{2}$ or $x = \dfrac{2 - \sqrt{14}}{2}$

Solution set: $\left\{\dfrac{2 + \sqrt{14}}{2}, \dfrac{2 - \sqrt{14}}{2}\right\}$

Section 17.3 (pages 1201–1202)

21. $7x^2 = 12x$

Write the equation in standard form.

$$7x^2 - 12x = 0$$

Substitute $a = 7$, $b = -12$, and $c = 0$ into the quadratic formula.

$$x = \frac{-b \pm \sqrt{b^2 - 4ac}}{2a}$$

$$x = \frac{-(-12) \pm \sqrt{(-12)^2 - 4(7)(0)}}{2(7)}$$

$$x = \frac{12 \pm \sqrt{144 - 0}}{14}$$

$$x = \frac{12 \pm 12}{14}$$

$$x = \frac{12 + 12}{14} = \frac{24}{14} = \frac{12}{7} \quad \text{or}$$

$$x = \frac{12 - 12}{14} = \frac{0}{14} = 0$$

Solution set: $\left\{0, \dfrac{12}{7}\right\}$

27. $3x^2 - 2x + 5 = 10x + 1$

Write the equation in standard form.

$$3x^2 - 12x + 4 = 0$$

Substitute $a = 3$, $b = -12$, and $c = 4$ into the quadratic formula.

$$x = \frac{-b \pm \sqrt{b^2 - 4ac}}{2a}$$

$$x = \frac{-(-12) \pm \sqrt{(-12)^2 - 4(3)(4)}}{2(3)}$$

$$x = \frac{12 \pm \sqrt{144 - 48}}{6}$$

$$x = \frac{12 \pm \sqrt{96}}{6}$$

$$x = \frac{12 \pm 4\sqrt{6}}{6}$$

$$\sqrt{96} = \sqrt{16} \cdot \sqrt{6} = 4\sqrt{6}$$

$$x = \frac{2\left(6 \pm 2\sqrt{6}\right)}{2 \cdot 3} \qquad \text{Factor.}$$

$$x = \frac{6 \pm 2\sqrt{6}}{3} \qquad \text{Divide out 2.}$$

Solution set: $\left\{\dfrac{6 + 2\sqrt{6}}{3}, \dfrac{6 - 2\sqrt{6}}{3}\right\}$

39. $0.6x - 0.4x^2 = -1$

To eliminate the decimals, multiply each side by 10.

$$6x - 4x^2 = -10$$

Write this equation in standard form.

$$4x^2 - 6x - 10 = 0$$

Divide each side by 2 so that we can work with smaller coefficients in the quadratic formula.

$$2x^2 - 3x - 5 = 0$$

Use the quadratic formula with $a = 2$, $b = -3$, and $c = -5$.

$$x = \frac{-(-3) \pm \sqrt{(-3)^2 - 4(2)(-5)}}{2(2)}$$

$$x = \frac{3 \pm \sqrt{9 + 40}}{4}$$

$$x = \frac{3 \pm \sqrt{49}}{4}$$

$$x = \frac{3 \pm 7}{4}$$

$$x = \frac{3 + 7}{4} = \frac{10}{4} = \frac{5}{2} \quad \text{or}$$

$$x = \frac{3 - 7}{4} = \frac{-4}{4} = -1$$

Solution set: $\left\{-1, \dfrac{5}{2}\right\}$

Section 17.4 (pages 1207–1208)

17. Because the vertex is at the origin, an equation of the parabola is of the form

$$y = ax^2.$$

As shown in the figure, one point on the graph has coordinates (150, 44).

$$y = ax^2 \qquad \text{General equation}$$

$$44 = a(150)^2 \qquad \text{Let } x = 150, y = 44.$$

$$44 = 22{,}500a \qquad \text{Apply the exponent.}$$

$$a = \frac{44}{22{,}500} \qquad \text{Solve for } a.$$

$$a = \frac{4 \cdot 11}{4 \cdot 5625}, \quad \text{or} \quad \frac{11}{5625} \qquad \text{Lowest terms}$$

Thus, an equation of the parabola is

$$y = \frac{11}{5625}x^2.$$

INDEX

Antony Dickson/AFP/Getty Images, p. 1038; **AP,** pp. 985, 1046 left; **AP Photo/Doug Mills,** p. 849; **AP Photo/Koji Sasahara,** p. 1063 right; **AP Wideworld,** pp. 116 right, 484, 642, 682, 724, 746; **Beth Anderson,** pp. 2, 8 top right, 53 right, 96 left, 230 left, 271, 310 right, 312, 388 right, 397 top right and bottom left and right, 419 left, 507 left, 562, 580, 603, 820, 825, 1116, AP-27 top right, middle, and bottom right, AP-28, AP-29 top left, AP-30 bottom left; **Bettmann/Corbis,** p. 1051; **Blend Images/Getty RF,** pp. 317, 334; **Bohemian Nomad Picturemakers/Corbis,** p. 483 right; **Brand X Pictures,** pp. AP-10 bottom left, AP-37 right; **Bruce Anderson Photos,** p. 398 left; **Bureau L.A. Collection/ Corbis,** p. 1044; **Car Cultures/Getty Images,** p. 732 left; **Comstock,** p. 296 left; **Corbis,** pp. 14, 60, 272, 303 bottom left, 483 left, 509, 583, 606, 614, 904, 1053, AP-27 top left, AP-30 bottom right; **David Gray/Reuters/Corbis,** p. 1046; **Digital Vision,** pp. 147 right, 228 right, 302 right, 303 top left and bottom right, 374 right, 375 left, 389 right, 410 right, 566, 755, 768 left, 786, 968, 973 bottom, 1049, AP-12 top right; **Digital Vision/Getty RF,** p. 1132 right; **Earl Orf/Heart and Mind, Inc.,** p. 350; **Gallo Images ROOTS/Getty RF,** p. 733; **Getty,** pp. 245 left, 246 right, 248 left, 429 bottom, 456 left, 569 right, 578, 615, 656 top, 671, 984, 1084, 1092, 1121, 1160, 1170 top left; **Getty Editorial,** pp. 156 right, 172, 452 left, 499, 691, 694, 697, 1102; **Getty Images News,** p. 1114; **Getty Images Sports,** pp. 340, 1063 left; **Golden Horseshoe Mustang Association,** p. 420 left; **Guy Motil/Corbis,** p. 1224; **Image Source/Getty RF,** p. 308; **iStockphoto,** pp. 183, 411, 429 top left, 962, AP-29 bottom right; **Courtesy of Jelly Belly Candy Company,** pp. 111, 174; **John Hornsby,** p. 1132 left; **Johner Images/Getty RF,** p. 8 bottom left, 915, 973 top; **Johns Hopkins University Applied Physics Laboratory,** p. 903; **Kara Salzman,** p. 419; **kcarlson@cardstacker.com,** pp. 193, 228 left, 231 right; **NASA,** pp. 303 top right, 507 right, 734, 914, 1152, 1202; **NationBill/Corbis Sygma,** p. 296 right; **NBC/Photofest,** p. 886; **Paul Kuroda/Superstock,** p. 819; **PhotoDisc,** pp. 8 top left, 39 bottom, 108, 248 right, 352, 354, 363 left, 364 bottom, 398 right, 409 right, 420 right, 436 left, 456 right, 520, 569 left, 630, 1170 top right and bottom left, AP-27 bottom left, AP-29 top right, AP-30 top left and right; **PhotoDisc Blue,** pp. 302 left, 310 left, 388 left, 452 right, 508 right, AP-1 bottom, AP-12 top left, AP-20 left; **PhotoDisc Red,** pp. 62 right, 363 right, 375 right, AP-37 left; **PhotoDisc/Getty RF,** p. 360, 981, 1093; **Photofest,** pp. 1067, 1094, 1100 top right and bottom; **Photographer's Choice/Getty RF,** p. 772; **Photonica/ Getty Images,** p. 436 right; **Photo Researchers,** pp. AP-1 top, AP-12 bottom; **Picture Quest/Brand X Pictures,** p. 324; **Picture Quest/PhotoDisc,** p. 326; **Purestock/Getty RF,** pp. 771, 794; **Reuters/Corbis,** p. 295; **Richard Carson/Reuters/Corbis,** p. AP-10 bottom right; **SAU St. Andrew's University MacTutor Archives,** p. 955; **Shutterstock,** pp. 1, 6, 8 bottom right, 19, 29 left, 39 top, 50, 53 left, 62 left, 63, 95, 96 right, 106, 107, 115, 116 left, 147 left, 153, 154, 162, 164, 165, 217, 218, 230 right, 231 left, 239, 245 right, 246 left, 250, 255, 263, 270, 333, 367, 376, 389 left, 397 top left, 401, 406, 429 top right, 443, 559, 564, 574, 585, 586, 593, 594, 597, 601, 605, 607, 646, 656 bottom, 848, 1041, 1047, 1052, 1103, 1104, 1170 bottom right, 1181, 1184, 1188, 1210; **Somos/Getty RF,** p. 779; **Stan Salzman,** pp. 29 right, 91, 155, 198, 374 left, 410 left, 588, AP-8, AP-9, AP-10 top, AP-18, AP-20 right, AP-29 bottom left; **Steve Starr/Corbis,** p. 1120; **Stockbyte/Getty RF,** p. 152, 768 right, 773; **Thierry Orban/Corbis Sygma,** p. AP-35; **ThinkStock,** pp. 264, 364 top; **Twentieth Century Fox/The Kobal Collection,** p. 732 right; **Twentieth Century Fox/Photofest,** p. 338; **Universal Pictures/Photofest,** p. 1108; **U.S. Department of Agriculture,** pp. 461, 508 left; **U.S. Navy,** pp. 156 left, 409 left; **U.S. Postal Service,** pp. 64, 83, 166, 186; **Warner Bros./Photofest,** p. 1100 top left